Your Learning Style

To discover what kind of learner you are, complete the Learning Styles Inventory on page xiv (or in MyMathLab). Then in the textbook, watch for the Learning Strategy boxes and the accompanying icons that provide ideas for maximizing your own learning style.

Your Learning Strategies

Learning Strategy

Developing a good study system and understanding how you best learn is essential to academic success. Make sure you familiarize your-self with the study system outlined in the To the Student sec-tion at the beginning of the text. Also take a moment to complete the Learning Styles Inventory found at the end of that section to discover your personal learning style. In these Learning Strategy boxes, we offer tips and suggestions on how to connect the study system and your learning style to help you be successful in the course.

_____ 1. I remember information better if I write it down or draw a picture of it.

_____ 2. I remember things better when I hear them instead of just reading or seeing them.

_____ 3. When I receive something that has to be assembled, I just start doing it. I don't read the directions.

_____ 4. If I am taking a test, I can visualize the page of text or lecture notes where the answer is located.

_____ 5. I would rather have the professor explain a graph, chart, or diagram to me instead of just showing it to me.

_____ 6. When learning new things, I want to do it rather than hear about it.

_____ 7. I would rather have the instructor write the information on the board or overhead instead of just lecturing.

_____ 8. I would rather listen to an audiobook than read the book.

_____ 9. I enjoy making things, putting things together, and working with my hands.

_____ 10. I am able to conceptualize quickly and visualize information.

_____ 11. I learn best by hearing words.

_____ 12. I have been called hyperactive by my parents, spouse, partner, or professor.

_____ 13. I have no trouble reading maps, charts, or diagrams.

_____ 14. I can usually pick up on small such as like bells, crickets, or frogs or on distant sounds such as train whistles.

_____ 15. I use my hands and gesture a lot when I speak to others.

Learning Strategy

In the To the Student section, we suggest that when taking notes, you use a red pen for def-initions and a blue pen for rules and procedures. Notice that we have used those colors in the design of the text to connect with your notes.

Definition **Additive inverses:** Two numbers whose sum in zero.

Examples of additive inverses: 15 and -15 because $15 + (-15) = 0$

-9 and 9 because $-9 + 9 = 0$

0 and 0 because $0 + 0 = 0$

Notice that 0 is its own additive inverse and that for numbers other than 0, the additive inverses have the same absolute value but *opposite* signs.

Rules

The sum of two additive inverse is zero.

The additive inverse of 0 is 0.

The additive inverse of a nonzero number has the same absolute value but opposite sign.

The Carson Math Study System

The Carson Math Study System is designed to help you succeed in your math course. You will discover your own learning style, and you will use study strategies that match the way you learn best. You will also learn how to organize your course materials, manage your time efficiently, and study and review effectively.

Your Math Notebook

Notes
(see pages xvii–xviii)

Section #	9/20

We can simplify an expression by combining like terms.

p. 37 def. Like terms: constant terms or variable terms that have the same variable(s) raised to the same powers.

Ex 1) 2x and 3x are like terms.

Ex 2) 5x and 7y are not like terms.

Homework
(see pages xviii–xxi)

Section # Homework	9/21

#1 – 15 odd

1. $5^2 + 3 \cdot 4 - 7$
    $= 25 + 3 \cdot 4 - 7$
    $= 25 + 12 - 7$
    $= 37 - 7$
    $= 30$

Quizzes/Tests
(see page xxv)

Chapter # Quiz	10/10

For 1-4, simplify.

1. $|-8|$ $= 8$ ✓

2. $|9|$ $= 9$ ✓

3. $-15 + 5$ $= -10$ ✓

4. $-8 + -6$ $= -14$ ✓

4/4 = 100%
Nice work!

Study Materials
(see pages xxii–xxiv)

Chapter # Study Sheets	9/10

To graph a number on a number line, draw a dot on the mark for the number.

ex) Graph -2,
$-4\ -3\ -2\ -1\ \ 0\ \ 1$

The absolute value of a positive number is positive.

ex) $|7| = 7$

The absolute value of a negative number is positive.

ex) $|-12| = 12$

Annotated Instructor's Edition

Fourth Edition

Elementary and Intermediate Algebra

Tom Carson

Bill Jordan
Seminole State College of Florida

PEARSON

Boston Columbus Indianapolis New York San Francisco Upper Saddle River
Amsterdam Cape Town Dubai London Madrid Milan Munich Paris Montréal Toronto
Delhi Mexico City São Paulo Sydney Hong Kong Seoul Singapore Taipei Tokyo

Editorial Director: Chris Hoag
Editor in Chief: Maureen O'Connor
Executive Editor: Cathy Cantin
Content Editor: Christine Whitlock
Editorial Assistant: Kerianne Okie
Senior Managing Editor: Karen Wernholm
Associate Managing Editor: Tamela Ambush
Digital Assets Manager: Marianne Groth
Media Producer: Aimee Thorne
QA Manager, Assessment Content: Marty Wright
Executive Content Manager: Rebecca Williams
Senior Content Developer: John Flanagan
Senior Marketing Manager: Rachel Ross
Marketing Assistant: Kelly Cross
Liaison Manager, Text Permissions Group: Joseph Croscup
Rights & Permissions, Associate Project Manager, Aptara, Inc.: Pamela Foley
Image Manager: Rachel Youdelman
Procurement Specialist: Debbie Rossi
Associate Director of Design, USHE North and West: Andrea Nix
Program Design Lead: Heather Scott
Text Design: Tamara Newnam
Production Coordination: PreMediaGlobal
Composition: PreMediaGlobal
Illustrations: PreMediaGlobal
Cover Design: Tamara Newnam
Cover Image: Quintanilla/Shutterstock

For permission to use copyrighted material, grateful acknowledgment is made to the copyright holders on page P-1, which is hereby made part of this copyright page.

Many of the designations used by manufacturers and sellers to distinguish their products are claimed as trademarks. Where those designations appear in this book, and Pearson Education was aware of a trademark claim, the designations have been printed in initial caps or all caps.

Library of Congress Cataloging-in-Publication Data

Carson, Tom, 1967–
    Elementary & intermediate algebra / Tom Carson, Bill Jordan—Fourth edition.
        pages cm.
    title: Elementary and intermediate algebra
    Includes index.
    ISBN-13: 978-0-321-92514-5   ISBN-10: 032192536X (Annotated Instructor's Edition)
    ISBN-10: 0-321-92514-9   ISBN-13: 9780321925367 (Annotated Instructor's Edition)
    1. Algebra—Textbooks. I. Jordan, Bill. II. Title. III. Title: Elementary and intermediate algebra.

    QA152.3.C374 2015
    512.9—dc23   2013009305

1 2 3 4 5 6 7 8 9 10—CRK— 17 16 15 14 13

www.pearsonhighered.com

ISBN 13: 978-0-321-92514-5
ISBN 10: 0-321-92514-9

Contents

Preface vi

To the Student xiii

Learning Styles Inventory xv

The Carson Math Study System xvii

1 Foundations of Algebra 1

1.1 Number Sets and the Structure of Algebra 2
1.2 Fractions 10
1.3 Adding and Subtracting Real Numbers; Properties of Real Numbers 20
1.4 Multiplying and Dividing Real Numbers; Properties of Real Numbers 34
1.5 Exponents, Roots, and Order of Operations 47
1.6 Translating Word Phrases to Expressions 58
1.7 Evaluating and Rewriting Expressions 65
Summary and Review Exercises 75
Practice Test 85

2 Solving Linear Equations and Inequalities 87

2.1 Equations, Formulas, and the Problem-Solving Process 88
2.2 The Addition Principle of Equality 101
2.3 The Multiplication Principle of Equality 112
2.4 Applying the Principles to Formulas 125
2.5 Translating Word Sentences to Equations 130
2.6 Ratios and Proportions 138
2.7 Percents 150
2.8 Solving Linear Inequalities 162
Summary and Review Exercises 173
Practice Test 184
Chapters 1–2 Cumulative Review Exercises 187

3 Graphing Linear Equations and Inequalities 189

3.1 The Rectangular Coordinate System 190
3.2 Graphing Linear Equations 197
3.3 Graphing Using Intercepts 208
3.4 Slope–Intercept Form 217
3.5 Point–Slope Form 229
3.6 Graphing Linear Inequalities 239
3.7 Introduction to Functions and Function Notation 248
Summary and Review Exercises 263
Practice Test 275
Chapters 1–3 Cumulative Review Exercises 277

4 Systems of Linear Equations and Inequalities 279

4.1 Solving Systems of Linear Equations Graphically 280
4.2 Solving Systems of Linear Equations by Substitution; Applications 289
4.3 Solving Systems of Linear Equations by Elimination; Applications 300
4.4 Solving Systems of Linear Equations in Three Variables; Applications 312
4.5 Solving Systems of Linear Equations Using Matrices 321
4.6 Solving Systems of Linear Inequalities 330
Summary and Review Exercises 337
Practice Test 345
Chapters 1–4 Cumulative Review Exercises 347

5 Polynomials 349

5.1 Exponents and Scientific Notation 350
5.2 Introduction to Polynomials 360
5.3 Adding and Subtracting Polynomials 371
5.4 Exponent Rules and Multiplying Monomials 380

5.5 Multiplying Polynomials; Special Products 390
5.6 Exponent Rules and Dividing Polynomials 402
Summary and Review Exercises 416
Practice Test 423
Chapters 1–5 Cumulative Review Exercises 425

6 Factoring 427

6.1 Greatest Common Factor and Factoring by Grouping 428
6.2 Factoring Trinomials of the Form $x^2 + bx + c$ 438
6.3 Factoring Trinomials of the Form $ax^2 + bx + c$, where $a \neq 1$ 444
6.4 Factoring Special Products 452
6.5 Strategies for Factoring 460
6.6 Solving Quadratic Equations by Factoring 464
6.7 Graphs of Quadratic Equations and Functions 476
Summary and Review Exercises 486
Practice Test 493
Chapters 1–6 Cumulative Review Exercises 495

7 Rational Expressions and Equations 497

7.1 Simplifying Rational Expressions 498
7.2 Multiplying and Dividing Rational Expressions 509
7.3 Adding and Subtracting Rational Expressions with the Same Denominator 519
7.4 Adding and Subtracting Rational Expressions with Different Denominators 525
7.5 Complex Rational Expressions 534
7.6 Solving Equations Containing Rational Expressions 542
7.7 Applications with Rational Expressions, Including Variation 551
Summary and Review Exercises 564
Practice Test 573
Chapters 1–7 Cumulative Review Exercises 575

8 More on Inequalities, Absolute Value, and Functions 577

8.1 Compound Inequalities 578
8.2 Equations Involving Absolute Value; Absolute Value Functions 589
8.3 Inequalities Involving Absolute Value 595
Summary and Review Exercises 603
Practice Test 608
Chapters 1–8 Cumulative Review Exercises 610

9 Rational Exponents, Radicals, and Complex Numbers 612

9.1 Radical Expressions and Functions 613
9.2 Rational Exponents 623
9.3 Multiplying, Dividing, and Simplifying Radicals 633
9.4 Adding, Subtracting, and Multiplying Radical Expressions 641
9.5 Rationalizing Numerators and Denominators of Radical Expressions 648
9.6 Radical Equations and Problem Solving 658
9.7 Complex Numbers 667
Summary and Review Exercises 675
Practice Test 685
Chapters 1–9 Cumulative Review Exercises 687

10 Quadratic Equations and Functions 689

10.1 The Square Root Principle and Completing the Square 690
10.2 Solving Quadratic Equations Using the Quadratic Formula 700
10.3 Solving Equations That Are Quadratic in Form 711
10.4 Graphing Quadratic Functions 720
10.5 Solving Nonlinear Inequalities 732
10.6 Function Operations 743
Summary and Review Exercises 750
Practice Test 758
Chapters 1–10 Cumulative Review Exercises 761

11 Exponential and Logarithmic Functions 763

11.1 Composite and Inverse Functions 764

11.2 Exponential Functions 776

11.3 Logarithmic Functions 785

11.4 Properties of Logarithms 793

11.5 Common and Natural Logarithms 801

11.6 Exponential and Logarithmic Equations with Applications 806

Summary and Review Exercises 817

Practice Test 829

Chapters 1–11 Cumulative Review Exercises 831

12 Conic Sections 833

12.1 Parabolas and Circles 834

12.2 Ellipses and Hyperbolas 847

12.3 Nonlinear Systems of Equations 858

12.4 Nonlinear Inequalities and Systems of Inequalities 864

Summary and Review Exercises 871

Practice Test 880

Chapters 1–12 Cumulative Review Exercises 882

R Elementary Algebra Review R-1

Appendixes

A. Mean, Median, and Mode APP-1

B. Arithmetic Sequences and Series APP-5

C. Geometric Sequences and Series APP-12

D. The Binomial Theorem APP-19

E. Synthetic Division APP-24

F. Solving Systems of Linear Equations Using Cramer's Rule APP-28

Photo Credits P-1

Collaborative Exercises C-1

Answers A-1

Glossary G-1

Index of Applications IA-1

Index I-1

Preface

The *Why* behind Understanding Algebra
+
The *Carson Math Study System* with *Learning Styles Inventory*

From the Authors

Welcome to the fourth edition of *Elementary and Intermediate Algebra*! Revising our program has been both exciting and rewarding, and it has given us the opportunity to respond to valuable instructor and student feedback. With great pride, we share with you the improvements to this edition, as well as the key features and proven style of our approach.

Elementary and Intermediate Algebra, Fourth Edition, is one title in a series of four that also includes *Prealgebra*, Fourth Edition; *Elementary Algebra*, Fourth Edition; and *Intermediate Algebra*, Fourth Edition. We have designed our program to be versatile enough for use in a variety of teaching and learning formats, including standard lecture, self-paced lab, hybrid, online, and even independent study.

We write in a relaxed, nonthreatening style, taking great care to ensure that students who have struggled with math in the past will be comfortable with our subject matter. Throughout the text, we explain *why* an algebraic process works the way it does, instead of just showing how to follow steps to solve problems. In addition, through problems from science, engineering, accounting, health, the arts, and everyday life, we link algebra to the real world.

Finally, to help students succeed in these courses, we offer the complete Carson Study System that includes a note To the Student, a Learning Styles Inventory, a Math Study System plan for developing a notebook as a personalized organizational and study tool, and integrated Learning Strategies from both the authors and students.

Tom Carson
Bill Jordan

The Carson Math Study System

The Carson Math Study System is designed to help students develop the skills (for example, time management, test prep, and note-taking) that students need to succeed in math, their college careers, and life.

1. **To the Student** (page xiii) focuses on why math is important and what students need to do to succeed in their math courses.
2. Taking the **Learning Styles Inventory** (page xv) will help students assess their particular style of learning and use that knowledge to identify helpful study skills.
3. **The Math Study System** (page xvii) offers students abundant suggestions for developing a class notebook to reflect their personal learning styles, taking effective notes, doing homework, and reviewing for quizzes and tests.
4. Throughout the text, **Learning Strategy** boxes offer advice on implementing the study system effectively based on a student's learning style (pages xv–xvi).

New to this Edition

In response to feedback from instructors and students, the presentation was refined; examples, exercises, Your Turns, and Instructor Notes were added; and real-data applications were updated or replaced with current data and topics. In addition, the order and content of some topics was improved for student comprehension and retention.

Content Changes

Chapter 1 The objectives in Sections 1.3 and 1.4 were condensed. Instead of separate objectives for integers and rational numbers, each section has a single objective for rational numbers.

Chapter 2 The former Chapter 3 on problem solving was streamlined and the topics integrated elsewhere. Former Sections 3.1 and 3.2 on ratios, proportions, and percents were moved to Chapter 2 as Sections 2.6 and 2.7. Former Sections 3.3–3.5 have been integrated into Chapter 4. Those sections involved solving problems with two unknowns using single-variable equations. In the authors' experience, however, students find it easier and more intuitive to solve those problems using systems of equations; so the topic is now covered with systems.

Chapter 3 Graphing linear equations is now covered earlier, in Chapter 3.

Chapter 4 Formerly Chapter 9, the chapter on systems of equations now follows the chapter on graphing linear equations. This keeps all of the discussion of linear equations together in a natural flow. This also introduces problems with two unknowns earlier in the text, using the more intuitive approach. Cramer's Rule has been moved to Appendix F.

Chapter 7 The discussion of LCD and the examples of variation were streamlined and are now covered in a more visual presentation.

Chapter 8 Function operations (formerly in Chapter 8) are now in Chapter 10, and the review section on graphing functions was deleted. The only functions that had not been introduced to this point were absolute value functions. Graphing absolute value functions is now an objective in Section 8.2 rather than in a separate section.

Chapter 9 The development of complex numbers is now more precise.

Chapter 10 Operations on functions are now covered in Section 10.6. The authors did this for two reasons. First, because all of the functions involved are polynomial functions, the topic fits better with the chapter on quadratic equations. Second, by placing this topic at the end of the chapter, it acts as a good transition into composite functions at the beginning of Chapter 11.

New and Revised Features

Student Learning Strategies and helpful tips written by successful students and recent college graduates were added throughout the text. These can be identified by the student's name at the end of the strategy.

The **Chapter Openers** now provide a brief topical overview of the chapter. In the instructor's edition, each opener now includes teaching suggestions for the chapter.

Each section now begins with **Warm-up Exercises** that review material presented previously and are helpful to understanding the concepts in the section.

Section Exercises are now grouped by objective to make it easier for students to connect examples with related exercises. In addition, new **Prep Exercises** help students focus on the terminology, rules, and processes corresponding to the exercises that immediately follow.

The **Chapter Summary and Review** is now interactive, with Review Exercises integrated within the summary to encourage active learning and review. For each key topic, students are prompted to complete the corresponding definitions, rules, and procedures.

Finally, we now offer two versions of MyMathLab: **Standard MyMathLab courses** allow instructors to build the course their way, offering maximum flexibility and control over all aspects of assignment creation. **New Ready-to-Go courses** provide students with all the same great MyMathLab features, but make it easier for instructors to get started.

Key Features

In addition to the new and revised features just described, the following key components round out the comprehensive guided learning approach.

An Algebra Pyramid is used throughout the text to help students see how the topic they are learning relates to the big picture of algebra—focusing particularly on the relationship between constants, variables, expressions, and equations and inequalities (page 3). In the end-of-section Review Exercises, Chapter Review Exercises, and Cumulative Review Exercises, an Algebra Pyramid icon indicates the level of the pyramid that correlates to a particular group of exercises. This helps students determine what actions are appropriate with these exercises (for example, whether to "simplify" or "solve" (pages 74, 75–76, 187–188).

Connection Boxes help students understand how math concepts are related and build on each other (pages 190, 690).

Your Turn Practice Exercises, found after most examples, give students an opportunity to work problems similar to the examples they just saw. This practice step engages students and provides immediate feedback to help them develop confidence in their problem-solving skills (pages 2, 596).

Real, Relevant, and Interesting Applications in the examples and exercises reflect real-world situations in science, engineering, health, finance, the arts, and other areas. These applications illustrate the everyday use of basic algebraic concepts and encourage students to apply mathematical concepts to solve problems (pages 305–306, 559).

Thorough Explanations are key to student understanding. The authors take great care to explain not only *how* to do the math but also *why* the math works the way it does, how concepts are related, and how the math is relevant to students' everyday lives. Knowing all of this gives students a context in which to learn and remember math concepts.

MyMathLab

Ties the Complete Learning Program Together

MyMathLab® Online Course (access code required)

MyMathLab from Pearson is the world's leading online resource in mathematics, integrating interactive homework, assessment, and media in a flexible, easy-to-use format. MyMathLab delivers **proven results** in helping individual students succeed. It provides **engaging experiences** that personalize, stimulate, and measure learning for each student. And it comes from an **experienced partner** with educational expertise and an eye on the future.

MyMathLab® for Developmental Mathematics

Prepared to go wherever you want to take your students.

Personalized Support for Students

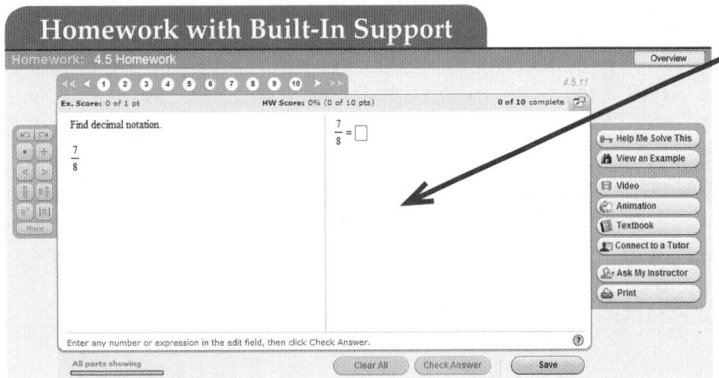

Exercises: The homework and practice exercises in MyMathLab are correlated to the exercises in the textbook, and they regenerate algorithmically to give students unlimited opportunity for practice and mastery. The software offers immediate, helpful feedback when students enter incorrect answers.

Multimedia Learning Aids: Exercises include guided solutions, sample problems, animations, videos, and eText access for extra help at point of use.

Expert Tutoring: Although many students describe the whole of MyMathLab as "like having your own personal tutor," students using MyMathLab do have access to live tutoring from Pearson, from qualified math instructors.

To help students achieve mastery, MyMathLab can generate **personalized homework** based on individual performance on tests or quizzes. Personalized homework allows students to focus on topics they have not yet mastered.

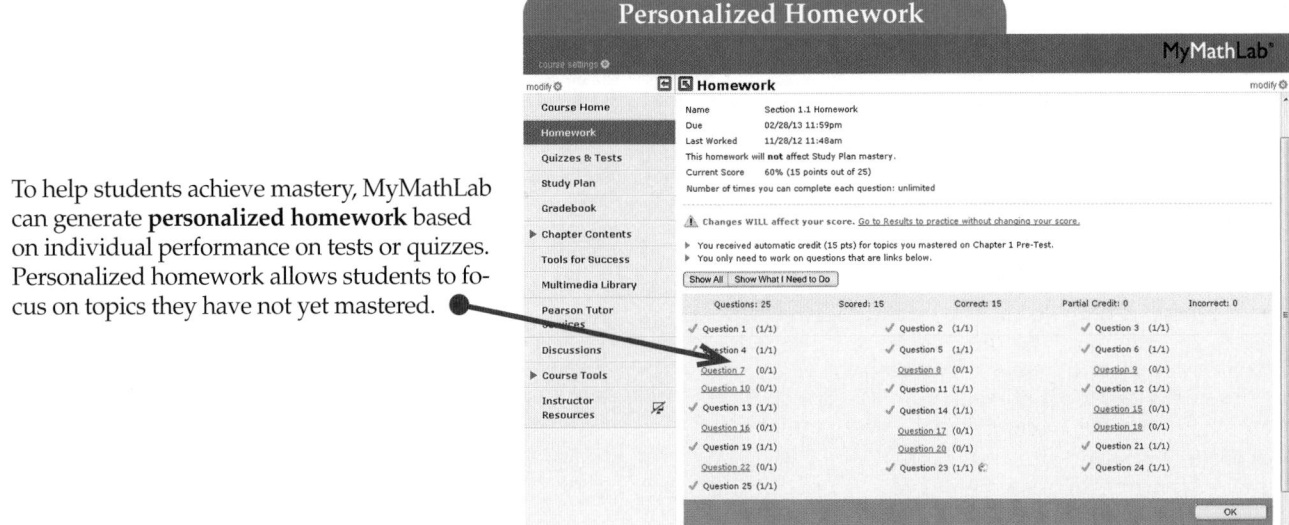

The **Adaptive Study Plan** makes studying more efficient and effective for every student. Performance and activity are assessed continually in real time. The data and analytics are used to provide personalized content—reinforcing concepts that target each student's strengths and weaknesses.

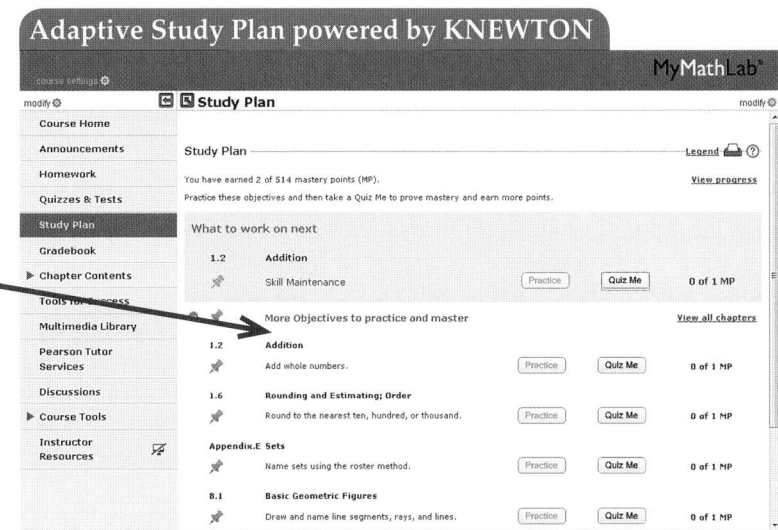

Flexible Design, Easy Start-up, and Results for Instructors

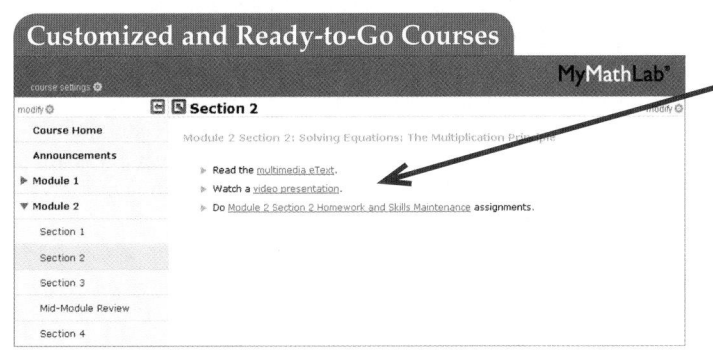

Instructors can modify the left-hand nav and insert their own directions onto course-level landing pages, and a **custom MyMathLab** course can be built to reorganize material and structure the course by chapters, modules, units—whatever the needs may be.

Ready-to-Go courses include pre-assigned homework, quizzes, and tests to make it even easier for instructors to get started.

The **comprehensive online gradebook** automatically tracks students' results on tests, quizzes, homework, and in the study plan. You can use the gradebook to intervene quickly if students have trouble or to provide positive feedback on a job well done. The data within MyMathLab is easily exported to a variety of spreadsheet programs such as Microsoft Excel. Instructors can determine which points of data to export and then analyze the results to determine success.

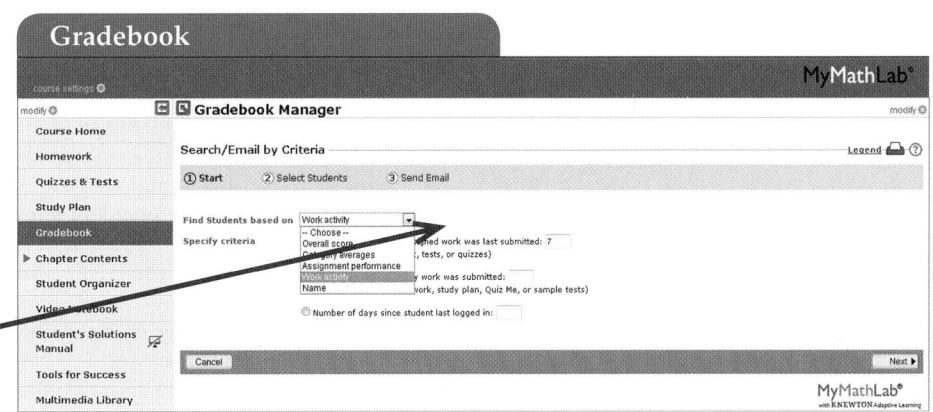

New features such as **Search/Email by criteria** make the gradebook a powerful tool for instructors. With this feature, instructors can easily communicate with both at-risk and successful students. Instructors can search by score on specific assignments, noncompletion of assignments within a given time frame, last log-in date, or overall score.

Additional Resources in MyMathLab

*In addition to the robust course delivery, the course also include the **full Carson eText, additional Carson Program Features,** and the **entire set of instructor and student resources** in one easy place to access online.*

For Students	For Instructors

Student's Solutions Manual*

ISBN-10: 0-321-92529-7
ISBN-13: 978-03-21-92529-9

- Contains complete solutions to the odd-numbered section exercises and solutions to all of the section-level Review Exercises, Chapter Review Exercises, Practice Tests, and Cumulative Review Exercises.

MyWorkBook for Intermediate Algebra*

ISBN-10: 0-321-92526-2
ISBN-13: 978-0-321-92526-8

MyWorkbook can be packaged with the textbook or with the MyMathLab access kit and includes the following resources for each section of the text:

- Key vocabulary terms and vocabulary practice problems.
- Guided Examples with stepped-out solutions and similar Practice Exercises, keyed to the text by Learning Objective.
- References to textbook Examples and Section Lecture Videos for additional help.
- Additional Exercises with ample space for students to show their work, keyed to the text by Learning Objective.

Video Resources

Within MyMathLab, students can access short lectures for each section of the text, plus Chapter Test Prep Videos, which allow students to watch an instructor work through step-by-step solutions for all of the Chapter Test exercises from the textbook. All videos include optional English and Spanish captions.

*Printed supplements are also available for separate purchase through MyMathLab, MyPearsonStore.com, or other retail outlets. They can also be value-packed with a textbook or MyMathLab code at a discount.

Annotated Instructor's Edition**

ISBN-10: 0-321-92536-X
ISBN-13: 978-0-321-92536-7

- Includes answers to all exercises, including Puzzle Problems and Collaborative Exercises, printed in bright blue near the corresponding problems.
- Useful teaching tips are printed in the margin.
- A ★ icon, found in the AIE only, indicates especially challenging exercises in the exercise sets.

Instructor's Resource Manual with Tests and Mini Lectures** (download only)

ISBN-10: 0-321-92524-6
ISBN-13: 978-0321-92524-4

- A mini-lecture for each section of the text, organized by objective, includes key examples and teaching tips.
- Designed to help both new and adjunct faculty with course preparation and classroom management.
- Contains one diagnostic test per chapter; four free-response test forms per chapter, one of which contains higher-level questions; one multiple-choice test per chapter; a midchapter check-up for each chapter; one midterm exam; and two final exams.

Instructor's Solutions Manual** (Download only)

ISBN-10: 0-321-92525-4
ISBN-13: 978-0-321-92525-1

- Contains complete solutions to all even-numbered section exercises, Puzzle Problems, and Collaborative Exercises.

PowerPoint® Lecture Slides** (download only)

Present key concepts and definitions from the text.

TestGen®

TestGen®(www.pearsoned.com/testgen) enables instructors to build, edit, print, and administer tests using a computerized bank of questions developed to cover all the objectives of the text. TestGen is algorithmically based, allowing instructors to create multiple but equivalent versions of the same question or test with the click of a button. Instructors can also modify test bank questions or add new questions. The software and test bank are available for download from Pearson Education's online catalog.

**Also available in print or for download from the Instructor Resource Center (IRC) on www.pearsonhighered.com.

To learn more about how MyMathLab combines proven learning applications with powerful assessment, visit **www.mymathlab.com** or contact your Pearson representative.

Acknowledgments

Many people gave of themselves in so many ways during the development of this text. Mere words cannot contain the fullness of our gratitude. Although the words of thanks that follow may be few, please know that our gratitude is great.

We would like to thank the following people who have given of their time in reviewing this and previous editions of this textbook. Their thoughtful input was vital to the development of the text.

Khadija Ahmed, *Monroe County Community College*

Bruce Armbrust, *Lake Tahoe Community College*

Frank Attanucci, *Scottsdale Community College*

Daniel Bacon, *Massasoit Community College*

Kerry Bailey, *Laramie Community College*

Madelaine Bates, *Bronx Community College*

William Berech, *Arapahoe Community College*

Marta Brito-Villani, *Miami-Dade College*

Denise Brown, *Collin County Community College*

Ronnie Brown, *University of Baltimore*

Debra Bryant, *Tennessee Technological University*

Baruch Cahlon, *Oakland University*

Nancy Carpenter, *Johnson County Community College*

Laurence Chernoff, *Miami-Dade College*

Delaine E. Cochran, *Indiana University Southeast*

Pat Cook, *Weatherford College*

Greg Cripe, *Spokane Falls Community College*

Patrick S. Cross, *University of Oklahoma*

Cheryl B. Davids, *Central Carolina Technical College*

Elias Deeba, *University of Houston–Downtown Campus*

Nelson de la Rosa, *Miami-Dade College*

Stephan DeLong, *Tidewater Community College–Virginia Beach Campus*

Cynthia Ellis, *Indiana University–Purdue University Fort Wayne*

Laura Ferguson, *Weatherford College*

John Garlow, *Tarrant County Junior College*

Kristin Good, *Washtenaw Community College*

Margret Hathaway, *Kansas City Kansas Community College*

Fat Lam, *Gallaudet University*

Ruby Martinez, *San Antonio College*

Diane Martling, *WM Rainey Harper College*

Theresa McChesney, *Johnson County Community College*

Allen Miller, *South Plains College*

Carol Murphy, *Miramar College*

Andrea Payne, *Montana State University Billings*

Joanne Peeples, *El Paso Community College*

Larry Pontaski, *Pueblo Community College*

Jack Sharp, *Floyd College*

Linda Shoesmith, *Scott Community College*

James Smith, *Columbia State Community College*

Chairsty Stewart, *Montana State University Billings*

Mary Lou Townsend, *Wor-Wic Community College*

Imre Tuba, *San Diego State University*

Cindy Vanderlaan, *Indiana University–Purdue University Fort Wayne*

Alexis Venter, *Arapahoe Community College*

James Vicich, *Scottsdale Community College*

Beverly Vredevelt, *Spokane Falls Community College*

xii   Preface

Linda J. Wagner, *Indiana University–Purdue University Fort Wayne*
Walter Wang, *Baruch College*
Floyd Wouters, *University of Wisconsin Oshkosh*
Patricia Youngblood, *New Mexico Junior College*
Vivian Zabrocki, *Montana State University Billings*

In addition, we would like to thank the students who provided their successful learning strategies for inclusion in the book. We hope their tips will help other students succeed as well.

Ruben C.
Matt D., *University of West Florida*
Ellyn G., *Stanly Community College*
Jason J., *University of Alabama*
Joel G., *Palo Alto College*
Judah G., *Columbus State Community College*
Lauren H., *University of Central Florida*
Biridiana N.
Stella P., *Virginia Commonwealth University*
Paul S., *Columbus State Community College*
Zheng Alick Z., *University of Southern California*

We would like to extend a heartfelt thank-you to Cathy Cantin, Kerianne Okie, Christine Whitlock, Kari Heen, Tamela Ambush, Michelle Renda, Rachel Ross, Alicia Frankel, Kelly Cross, Aimee Throne, Rebecca Williams, Greg Tobin, and all of the wonderful people at Pearson Arts & Sciences who helped so much in the revision of the text. Special thanks to Janis Cimperman and Paul Lorczak for checking the manuscript and pages for accuracy and to Becky Troutman for researching application problems and preparing the Glossary.

To Erin Donahue and all of the fabulous people at PreMediaGlobal, thank you for your attention to detail and for working so hard to put together the finished pages.

To Melinda McLaughlin and her team at GEX for their work on the solutions manuals, the Instructor's Resource Manual, and the MyWorkBook.

Finally, we'd like to thank our families for their support and encouragement during the process of developing and revising this text.

Tom Carson
Bill Jordan

To the Student

Why Do I Have to Take This Course?

Often, this is one of the first questions students ask when they find out they must take an algebra course. What a great question! But why focus on math alone? What about English, history, psychology, or science? Does anyone really use *every* topic of *every* course in the curriculum? Most jobs do not require that we write essays on Shakespeare, discuss the difference between various psychological theories, or analyze the cell structure of a frog's liver. So what's the point? The issue comes down to recognizing that general education courses are not job training. The purpose of those courses is to stretch and exercise the mind so that the educated person can better communicate, analyze situations, and solve problems, which are all valuable skills in life and any job.

Professional athletes offer a good analogy. They usually have an exercise routine apart from the sport designed to build and improve their body. They may seek a trainer to design exercises intended to improve strength, stamina, or balance and then push them in ways they would not normally push themselves. That trainer may have absolutely no experience with his or her client's sport, but can still be quite effective in designing an exercise program because the trainer is focused on building basic skills useful for any athlete in any sport. Education is similar: it is exercise for the mind. A teacher's job is like that of a physical trainer. A teacher develops exercises intended to improve communication skills, critical- thinking skills, and problem-solving skills. Different courses are like different types of fitness equipment. Some courses may focus more on communication through writing papers and through discussion and debate. Other courses, such as mathematics, focus more on critical thinking and problem solving.

Another similarity is that physical exercise must be challenging for your body to improve. Similarly, mental exercise must be challenging for the mind to improve. Expect course assignments to challenge you and push you mentally in ways you wouldn't push yourself. That's the best way to grow. So as you think about the courses you are taking and the assignments in those courses, remember the bigger picture of what you are developing: your mind. When you are writing papers, responding to questions, analyzing data, and solving problems, you are developing skills important to life and any career out there.

What Do I Need to Do to Succeed?

☑ **Adequate Time** To succeed, you must have adequate time and be willing to use that time to perform whatever is necessary. To determine if you have adequate time, use the following guide.

Step 1. Calculate your work hours per week.

Step 2. Calculate the number of hours in class each week.

Step 3. Calculate the number of hours required for study by doubling the number of hours you spend in class.

Step 4. Add the number of hours from steps 1–3.

Adequate time: If the total number of hours is below 60, then you have adequate time.

Inadequate time: If the total number of hours is 60 or more, then you do not have adequate time. You may be able to hang in there for a while, but eventually, you will find yourself overwhelmed and unable to fulfill all of your obligations. Remember, the above calculations do not consider other likely elements of life such as a commute, family, and recreation. The wise thing to do is cut back on work hours or drop some courses.

Assuming you have adequate time available, choosing to use that time to perform whatever is necessary depends on your attitude, commitment, and self-discipline.

☑ **Positive Attitude** We do not always get to choose our circumstances, but we do get to choose our reaction and behavior. A positive attitude is choosing to be cheerful, hopeful, and encouraging no matter the situation. A benefit of a positive attitude is that it tends to encourage people around you. As a result, they are more likely to want to help you achieve your goal. A negative attitude, on the other hand, tends to discourage people around you. As a result, they are less likely to want to help you. The best way to maintain a positive attitude is to keep life in perspective, recognizing that difficulties and setbacks are temporary.

☑ **Commitment** Commitment means binding yourself to a course of action. Remember, expect difficulties and setbacks, but don't give up. That's why a positive attitude is important; it helps you stay committed in the face of difficulty.

☑ **Self-Discipline** Self-discipline is choosing to do what needs to be done—even when you don't feel like it. In pursuing a goal, it is normal to get distracted or tired. It is at those times that your positive attitude and commitment to the goal help you discipline yourself to stay on task.

Thomas Edison, inventor of the lightbulb, provides an excellent example of all of these principles. Edison tried over 2000 different combinations of materials for the filament before he found a successful combination. When asked about all his failed attempts, Edison replied, "I didn't fail once, I invented the lightbulb. It was just a 2000-step process." He also said, "Our greatest weakness lies in giving up. The most certain way to succeed is always to try just one more time." In those two quotes, we see a person who obviously had time to try 2000 experiments, had a positive attitude about the setbacks, never gave up, and had the self-discipline to keep working.

Behaviors of Strong Students and Weak Students

The four requirements for success can be translated into behaviors. The following table compares the typical behaviors of strong students with the typical behaviors of weak students.

Strong Students ...	Weak Students ...
• are relaxed, patient, and work carefully.	• are rushed, impatient, and hurry through work.
• almost always arrive on time and leave the classroom only in an emergency.	• often arrive late and often leave class to "take a break."
• sit as close to the front as possible.	• sit as far away from the front as possible.
• pay attention to instruction.	• ignore instruction, chit-chat, draw, fidget, etc.

• use courteous and respectful language, encourage others, make positive comments, are cheerful and friendly.	• use disrespectful language, discourage others, make negative comments, are grumpy and unfriendly.
	
	Examples of unacceptable language include:
	    "I hate this stuff!" (or even worse!)
	    "Are we doing anything important today?"
	    "Can we leave early?"

• ask appropriate questions and answer instructor's questions during class.	• avoid asking questions and rarely answer instructor's questions in class.
• take a lot of notes, have organized notebooks, seek out and use study strategies	• take few notes, have disorganized notebooks, do not use study strategies
• begin assignments promptly and manage time wisely and almost always complete assignments on time.	• procrastinate, manage time poorly, and often complete assignments late.
• label assignments properly and show all work neatly.	• show little or no work and write sloppily.
• read and work ahead.	• rarely read or work ahead.
• contact instructors outside of class for help, and use additional resources such as study guides, solutions manuals, computer aids, videos, and tutorial services.	• avoid contacting instructors outside of class and rarely use additional resources available.

Assuming you have the prerequisites for success and understand the behaviors of a good student, our next step is to develop two major tools for success:

1. **Learning Style:** Complete the Learning Styles Inventory to determine how you tend to learn.
2. **The Study System:** This system describes a way to organize your notebook, take notes, and create study tools to complement your learning style. We've seen students transform their mathematics grades from D's and F's to A's and B's by using the Study System that follows.

Learning Styles Inventory

What Is Your Personal Learning Style?

A learning style is the way in which a person processes new information. Knowing your learning style can help you make choices in the way you focus on and study new material. Below are 15 statements that will help you assess your learning style. After reading each statement, rate your response to the statement using the scale below. There are no right or wrong answers.

3 = Often applies   **2** = Sometimes applies   **1** = Never or almost never applies

_____ 1. I remember information better if I write it down or draw a picture of it.

_____ 2. I remember things better when I hear them instead of just reading or seeing them.

_____ 3. When I receive something that has to be assembled, I just start doing it. I don't read the directions.

_____ 4. If I am taking a test, I can visualize the page of text or lecture notes where the answer is located.

_____ 5. I would rather have the professor explain a graph, chart, or diagram to me instead of just showing it to me.

_____ 6. When learning new things, I want to do it rather than hear about it.

_____ 7. I would rather have the instructor write the information on the board or overhead instead of just lecturing.

_____ 8. I would rather listen to an audiobook than read the book.

_____ 9. I enjoy making things, putting things together, and working with my hands.

_____ 10. I am able to conceptualize quickly and visualize information.

_____ 11. I learn best by hearing words.

_____ 12. I have been called hyperactive by my parents, spouse, partner, or professor.

_____ 13. I have no trouble reading maps, charts, or diagrams.

_____ 14. I can usually pick up on small sounds such as bells, crickets, frogs or on distant sounds such as train whistles.

_____ 15. I use my hands and gesture a lot when I speak to others.

Write your score for each statement beside the appropriate statement number below. Then add the scores in each column to get a total score for that column.

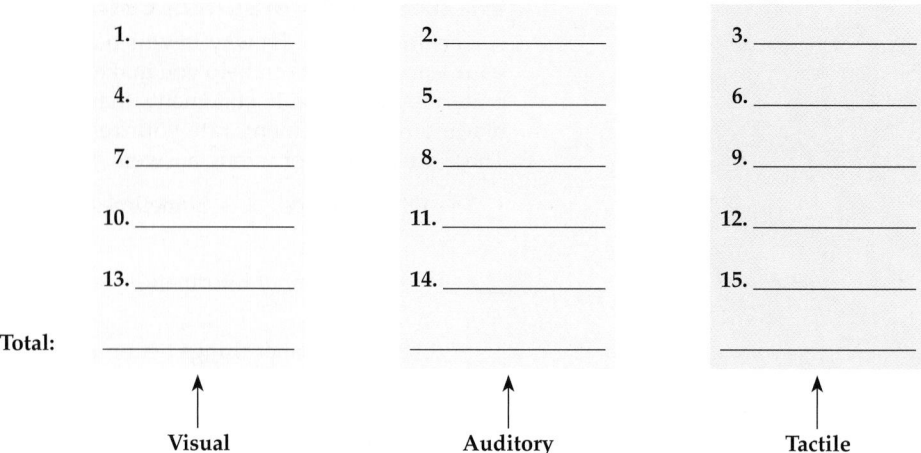

1. _____	2. _____	3. _____
4. _____	5. _____	6. _____
7. _____	8. _____	9. _____
10. _____	11. _____	12. _____
13. _____	14. _____	15. _____

Total: _____   _____   _____

Visual   **Auditory**   **Tactile**

The largest total of the three columns indicates your dominant learning style.

Visual learners learn best by seeing. If this is your dominant learning style, then you should focus on learning strategies that involve seeing. The color coding in the study system (see page xvii–xviii) will be especially important. The same color coding is used in the text. Draw diagrams, arrows, and pictures in your notes to help you see what is happening. Reading your notes, study sheets, and text repeatedly will be an important strategy.

Auditory learners learn best by hearing. If this is your dominant learning style, then you should use learning strategies that involve hearing. After getting permission from your instructor, bring a recorder to class to record the discussion. When you study your notes, play back the recording. Also, when you learn rules, say the rule over and over. As you work problems, say the rule before you do the problem. You may also find the videotapes to be beneficial because you can hear explanations of problems taken from the text.

Tactile (also known as kinesthetic) learners learn best by touching or doing. If this is your dominant learning style, you should use learning strategies that involve doing. Doing a lot of practice problems will be important. Make use of the Your Turn exercises in the text. These are designed to give you an opportunity to do problems that are similar to the examples as soon as a topic is discussed. Writing out your study sheets and doing your practice tests repeatedly will be important strategies for you.

Note that the study system developed in this text is for all learners. Your learning style will help you decide what aspects and strategies in the study system to focus on, but being predominantly an auditory learner does not mean that you shouldn't read the textbook, do a lot of practice problems, or use the color-coding system in your notes. Auditory learners can benefit from seeing and doing, and tactile learners can benefit from seeing and hearing. In other words, do not use your dominant learning style as a reason for not doing things that are beneficial to the learning process. Also remember that the Learning Strategy boxes presented throughout the text provide tips to help you use your personal learning style to your advantage.

The Carson Math Study System

Organize the notebook into four parts using dividers shown:

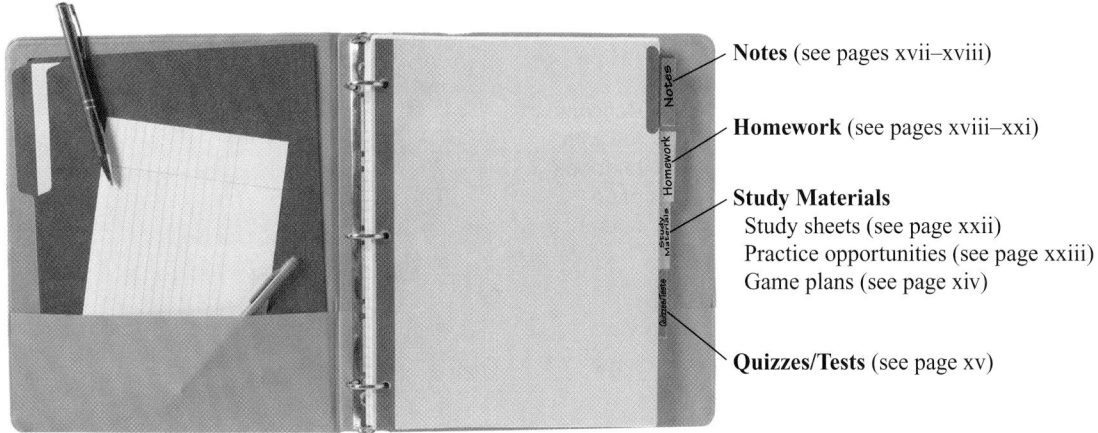

Notes (see pages xvii–xviii)

Homework (see pages xviii–xxi)

Study Materials
Study sheets (see page xxii)
Practice opportunities (see page xxiii)
Game plans (see page xiv)

Quizzes/Tests (see page xv)

Notes

- Use a color code: **red** for definitions, blue for rules or procedures, and **pencil** for all examples and other notes.

- Begin notes for each class on a new page (front and back for that day is okay). Include a topic title or section number and the date on each page.

- Try to write your instructor's spoken explanations along with the things he or she writes on the board.

- Mark examples your instructor emphasizes in some way to give them a higher priority. These problems often appear on quizzes and tests.

- Write warnings your instructor discusses about a particular situation.

- Include common errors that your instructor illustrates, but mark them clearly as errors so that you do not mistake them for correct.

- To speed note taking, eliminate unnecessary words such as *the* and use codes for common words such as + for *and* and ∴ for *therefore*. Also, instead of writing complete definitions, rules, or procedures, write the first few words and place the page reference from the text so that you can copy from the text later.

Sample Notes with Color Code

Include title.

Include date.

Section #   *9/20*

We can simplify an expression by combining like terms.

Definition in red with textbook page reference.

p. # def. Like terms: constant terms or variable terms that have the same variable(s) raised to the same powers.

Ex 1) 2x and 3x are like terms.

Ex 2) 5x and 7y are not like terms.

Consider 2x + 3x

2x means two x's are added together.   *3x means three x's are added together.*

$$2x + 3x$$

$$= x + x + x + x + x$$ *We have a total of five x's added together.*

$$= 5x$$

We can just add the coefficients.

Procedure in blue with textbook page reference.

p. # Procedure: To combine like terms, add or subtract the coefficients and keep the variables and their exponents the same.

Ex 1) 7x + 5x = 12x

Ex 2) $4y^2 - 10y^2 = -6y^2$

Homework

This section of the notebook contains all homework. Use the following guidelines whether your assignments are from the textbook, a handout, or a computer program such as MyMathLab or MathXL.

- Use pencil so that mistakes can be erased (scratching through mistakes is messy and should be avoided).

- Label according to your instructor's requirements. Usually, at least include your name, the date, and the assignment title. It is also wise to write the assigned problems at the top as they were given. For example, if your instructor writes "Section 1.5 #1–15 odd," write it that way at the top. Labeling each assignment with this much detail shows that you take the assignment seriously and leaves no doubt about what you interpreted the assignment to be.

- For each problem you solve, write the problem number and show all solution steps neatly.

Why do I need to show work and write all the steps? Isn't the right answer all that's needed?

- Mathematics is not just about getting correct answers. You really learn mathematics when you organize your thoughts and present those thoughts clearly using mathematical language.
- You can arrive at correct answers with incorrect thinking. Showing your work allows your instructor to verify that you are using correct procedures to arrive at your answers.
- Having a labeled, well-organized, and neat hard copy is a good study tool for exams.

What if I submit my answers in MyMathLab or MathXL?
Do I still need to show work?

Think of MyMathlab or MathXL as a personal tutor who provides the exercises, offers guided assistance, and checks your answers before you submit the assignment. For the same reasons as those listed above, you should still create a neatly written hard copy of your solutions, even if your instructor does not check the work. Following are some additional reasons to show your work when submitting answers in MyMathlab or MathXL.

- If you have difficulties that are unresolved by the program, you can show your instructor. Without the written work, your instructor cannot see your thinking.
- If you have a correct answer but have difficulty entering that correct answer, you have a record of it and can show your instructor. If correct, your instructor can override the score.

Sample Homework: Simplifying Expressions or Solving Equations

Suppose you are given the following exercise:

For Exercises 1–30 simplify.

1. $5^2 + 3 \cdot 4 - 7$

Your homework should look something like the following:

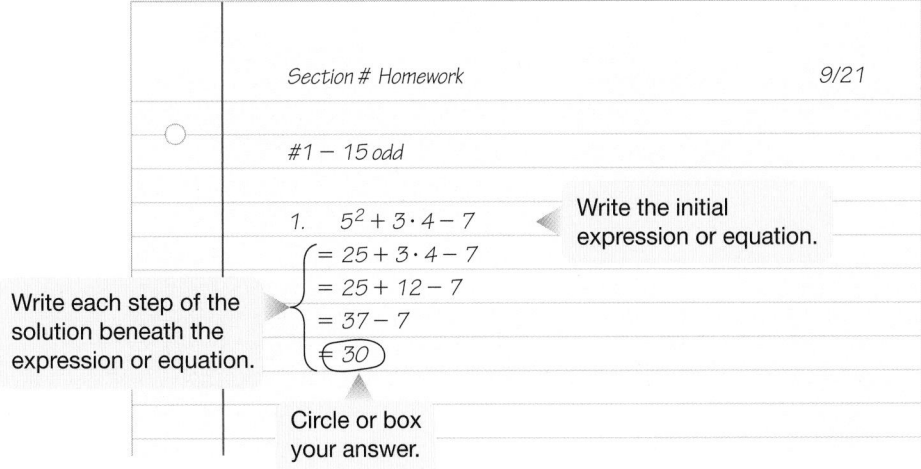

Sample Homework: Solving Application Problems

Example: Suppose you are given the following two problems.

For Exercises 1 and 2, solve.

1. Find the area of a circle with a diameter of 10 feet.
2. Two cars are traveling toward each other on the same highway. One car is traveling 65 miles per hour, and the other is traveling at 60 miles per hour. If the two cars are 20 miles apart, how long will it be until they meet?

Your homework should look something like the following:

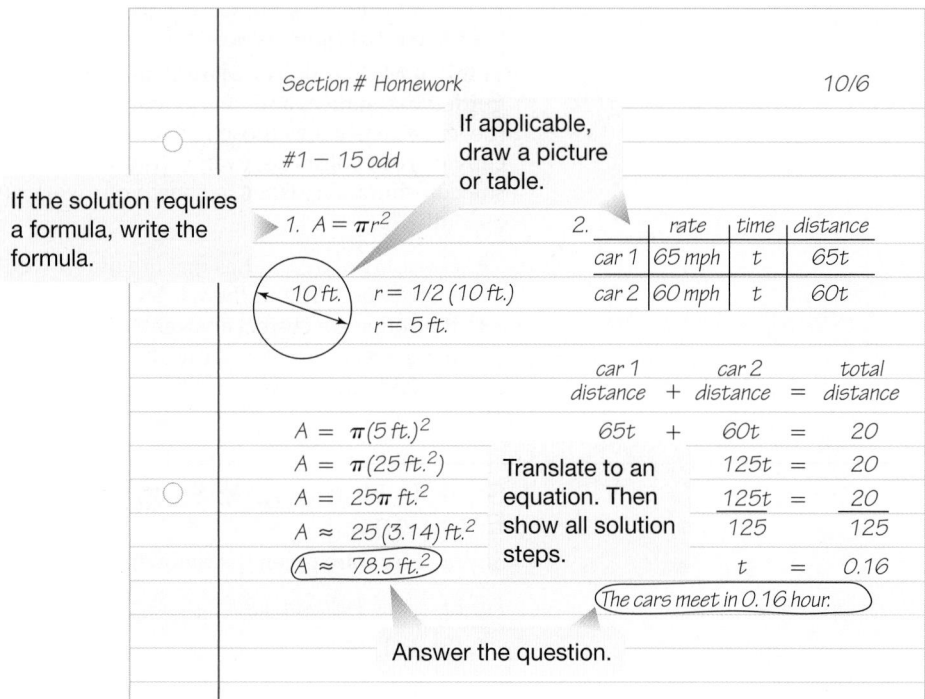

Sample Homework: Graphing

Example: Suppose you are given the following two problems.

For Exercises 1 and 2, graph the equation.

1. $y = 2x - 3$ **2.** $y = -2x + 1$

Your homework should look something like the following:

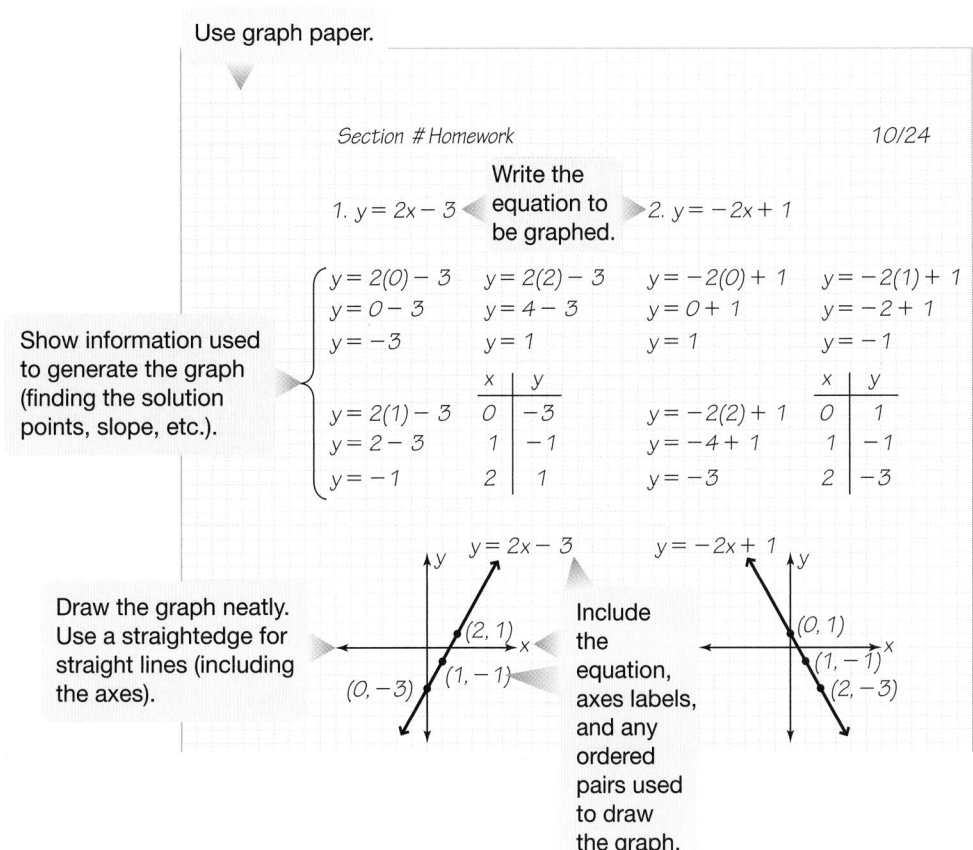

Use graph paper.

Write the equation to be graphed.

Show information used to generate the graph (finding the solution points, slope, etc.).

Draw the graph neatly. Use a straightedge for straight lines (including the axes).

Include the equation, axes labels, and any ordered pairs used to draw the graph.

Section # Homework   10/24

1. $y = 2x - 3$ 2. $y = -2x + 1$

$y = 2(0) - 3$ $y = 2(2) - 3$ $y = -2(0) + 1$ $y = -2(1) + 1$
$y = 0 - 3$ $y = 4 - 3$ $y = 0 + 1$ $y = -2 + 1$
$y = -3$ $y = 1$ $y = 1$ $y = -1$

x	y
0	-3
1	-1
2	1

$y = 2(1) - 3$
$y = 2 - 3$
$y = -1$

$y = -2(2) + 1$
$y = -4 + 1$
$y = -3$

x	y
0	1
1	-1
2	-3

$y = 2x - 3$ $y = -2x + 1$

(2, 1)
(0, -3) (1, -1)

(0, 1)
(1, -1)
(2, -3)

Study Materials

This section of the notebook contains three types of study materials for each chapter.

Study Material 1: The Study Sheet A study sheet contains *every* rule or procedure in the current chapter.

Use the chapter summary at the end of each chapter as a guide.

Write each rule or procedure studied. They are in blue in your notes and in the text.

If you are a visual or tactile learner, include a key example to illustrate what the rule or procedure says.

Include anything that helps you remember the procedures and rules. For example, auditory learners might write poems, rhymes, or jingles, as shown here.

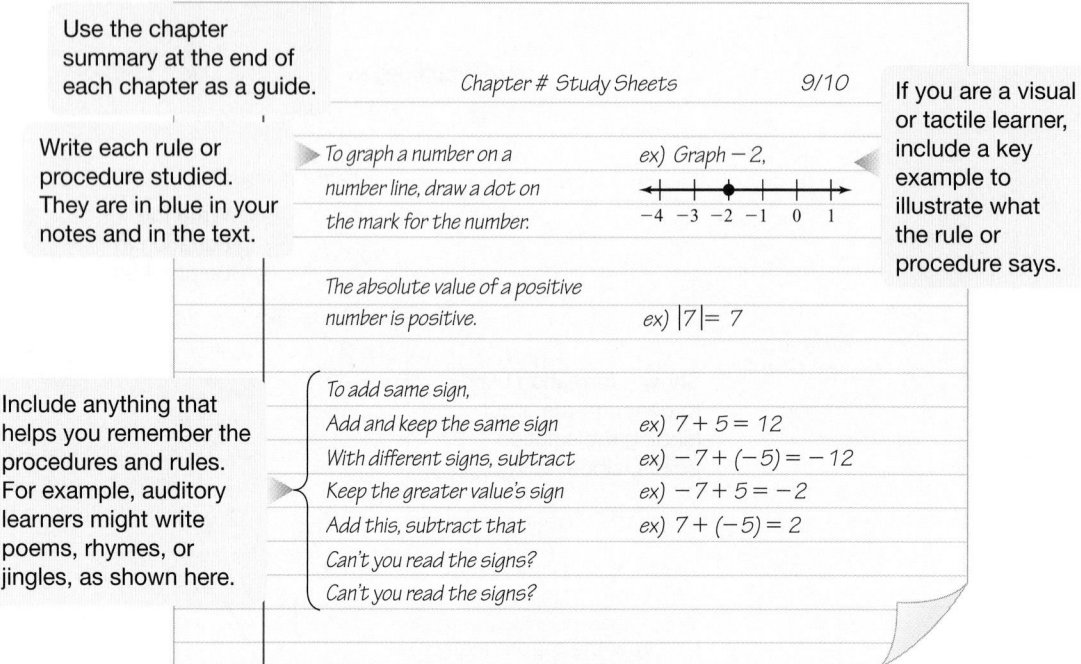

Chapter # Study Sheets   9/10

To graph a number on a number line, draw a dot on the mark for the number.

ex) Graph −2,

The absolute value of a positive number is positive.

ex) |7| = 7

To add same sign,
Add and keep the same sign   ex) 7 + 5 = 12
With different signs, subtract   ex) −7 + (−5) = −12
Keep the greater value's sign   ex) −7 + 5 = −2
Add this, subtract that   ex) 7 + (−5) = 2
Can't you read the signs?
Can't you read the signs?

Study Material 2: The Practice Test If your instructor gives you a practice test, proceed to the discussion of creating a game plan.

If your instructor does not give you a practice test, use your notes to create your own practice test from the examples given in class. Include only the instructions and the problem, not the solutions. The following sample practice test was created from examples in the notes for Chapter 2 in a prealgebra course.

Chapter # Practice Test   *10/10*

For # 1 and 2, graph on a number line.

1. 4

2. −3

For each example in your notes, write the directions and the problem but not the solution.

For #3 and 4, simplify.

3. $|-8|$

4. $|9|$

After working through the practice test, use your notes to check your solutions.

For #5 and 6, find the additive inverse of the number.

5. 13

6. −15

For #7 – 10, add.

7. $13 + (-9)$ 8. $-20 + (-6)$

9. $-15 + 8$ 10. $3 + (-24)$

Study Material 3: The Game Plan The game plan refines the study process further. It is your plan for the test based on the practice test. For each problem on your practice test, write the definition, rule, or procedure used to solve the problem.

The sample shown gives the rule or procedure used to solve each problem on the preceding sample practice test. The rules and procedures came from the sample study sheet.

Chapter # Game Plan   *9/10*

Multiple problems that use the same rule or procedure can be grouped together.

Write the rule or procedure used to solve the problems on the practice test.

#1 and 2: Draw a dot on the mark for the number.

#3 and 4: The absolute value of a positive number
is a positive number.
The absolute value of a negative number
is a positive number.
The absolute value of 0 is 0.

#5 and 6: Additive inverses are numbers whose
sum is 0.

#7 – 10: To add same sign,
Add and keep the same sign
With different signs
Subtract and keep the greater value's sign
Add this, subtract that
Can't you read the signs?
Can't you read the signs?

Quizzes/Tests

Archive all returned quizzes and tests in this section of the notebook.

- Midterm and final exam questions are often taken from the quizzes and tests, so they make excellent study tools for those cumulative exams.

- Keeping all graded quizzes and tests offers a backup system in the unlikely event your instructor should lose any of your scores.

- If a dispute arises about a particular score, you have the graded test to show your instructor.

Chapter # Quiz   10/10

For 1-4, simplify.

4/4 = 100%
Nice work!

1. $|-8|$ $= 8$ ✓

2. $|9|$ $= 9$ ✓

3. $-15 + 5$ $= -10$ ✓

4. $-8 + -6$ $= -14$ ✓

Chapter Overview

This chapter is a review of arithmetic, which is the foundation of algebra. More specifically, we review the following topics:

- ▶ Number sets.
- ▶ Operations of arithmetic.
- ▶ Properties of real numbers.
- ▶ Evaluating expressions.
- ▶ Simplifying expressions.

This review is by no means a complete instruction of arithmetic. Because a solid foundation is important, if you encounter a topic in this chapter that is a particular weakness for you, consult with your instructor or other more complete sources for extra practice.

Instructor Note

This chapter is intended to be a review. To maximize pace, consider giving a diagnostic quiz to help you determine the classes review needs. Then cover only those topics that are needed by the majority of the class. Another approach is to cover Section 1.1 and the properties of real numbers in Sections 1.3 and 1.4. Then begin working in Section 1.5 and review the arithmetic as you work through problems requiring the order of operations.

1.1 Number Sets and the Structure of Algebra

1.2 Fractions

1.3 Adding and Subtracting Real Numbers; Properties of Real Numbers

1.4 Multiplying and Dividing Real Numbers; Properties of Real Numbers

1.5 Exponents, Roots, and Order of Operations

1.6 Translating Word Phrases to Expressions

1.7 Evaluating and Rewriting Expressions

1.1 Number Sets and the Structure of Algebra

Objectives

1 Understand the structure of algebra.
2 Classify number sets.
3 Graph rational numbers on a number line.
4 Determine the absolute value of a number.
5 Compare numbers.

Warm-up *Refer to the* To the Student *Section on p. xiii.*
   1. What are the four sections of the notebook?
   2. What is the color code for notes?
   3. Complete the Learning Style Inventory. What is your learning style?

Objective 1 Understand the structure of algebra.

Learning mathematics is like learning a language. When we learn a language, we must learn the alphabet, vocabulary, and sentence structure. Similarly, mathematics has its own alphabet, vocabulary, and sentence structure. In this section, we begin the development of the foundation of algebra with an overview of its components and structure. The basic components are **variables** and **constants**.

Definitions Variable: A symbol that can vary in value.
Constant: A symbol that does not vary in value.

Variables are usually letters of the alphabet such as x or y. Usually, constants are symbols for numbers such as $1, 2, \frac{3}{4}$, and 6.74. However, constants can sometimes be symbols such as e or the Greek letter π, each having special numeric values. Variables and constants are used to make **expressions**, **equations**, and **inequalities**.

Definition Expression: A constant, a variable, or any combination of constants, variables, and arithmetic operations that describes a calculation.

Examples of expressions:

$$2 + 6 \quad 4x - 5 \quad \frac{1}{3}\pi r^2 h$$

Definition Equation: A mathematical relationship that contains an equal sign.

Examples of equations:

$$2 + 6 = 8 \quad 4x - 5 = 12 \quad V = \frac{1}{3}\pi r^2 h$$

Connection Think of expressions as phrases and equations as complete sentences. The expression $2 + 6$ is read "two plus six," which is not a complete sentence. The equation $2 + 6 = 8$ is read "two plus six is eight." Notice that the equal sign translates to the verb *is*, which makes the sentence a complete sentence.

Definition Inequality: A mathematical relationship that contains an inequality symbol

$$(\neq, <, >, \leq, \text{or} \geq).$$

Learning Strategy

It is important to sit up front in class because it helps you avoid distractions and makes it easier to hear any hints the professor might be giving.
                                    —Jason J.

Answers to Warm-up
   1. notes, homework, study materials, graded work
   2. red = definitions; blue = rules/procedures; pencil = all other notes
   3. Answers may vary.

Inequality Symbols and Their Translations

Symbolic Form	Translation
$8 \neq 3$	Eight is not equal to three.
$5 < 7$	Five is less than seven.
$7 > 5$	Seven is greater than five.
$x \leq 3$	x is less than or equal to three.
$y \geq 2$	y is greater than or equal to two.

 Learning Strategy

Developing a good study system and understanding how you best learn is essential to academic success. Make sure you familiarize yourself with the study system outlined in the To the Student section at the beginning of the text. Also take a moment to complete the Learning Styles Inventory found at the end of that section to discover your personal learning style. In these Learning Strategy boxes, we offer tips and suggestions on how to connect the study system and your learning style to help you be successful in the course.

The algebra pyramid shown illustrates how variables, constants, expressions, equations, and inequalities relate. At the foundation of algebra and our pyramid are constants and variables, which are used to build expressions, which in turn are used to build equations and inequalities.

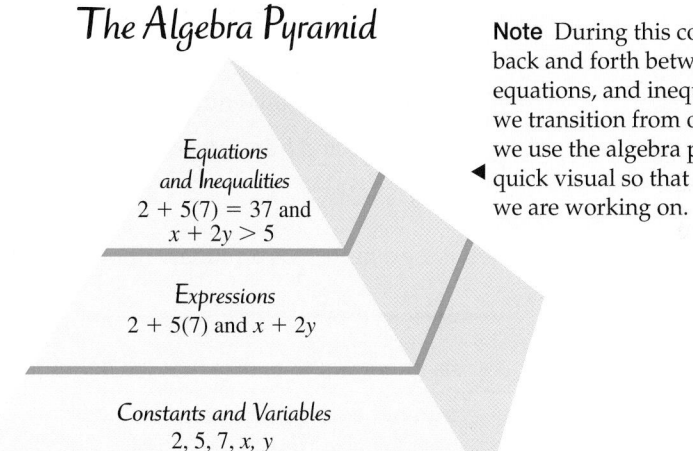

The Algebra Pyramid

Equations and Inequalities
$2 + 5(7) = 37$ and $x + 2y > 5$

Expressions
$2 + 5(7)$ and $x + 2y$

Constants and Variables
$2, 5, 7, x, y$

Note During this course, we move back and forth between expressions, equations, and inequalities. When we transition from one to the other, we use the algebra pyramid as a quick visual so that it's clear what we are working on.

Objective 2 Classify number sets.

Numbers can be placed into different categories using **sets**.

Definition **Set:** A collection of objects.

Braces are used to indicate a set. For example, the set containing the numbers 1, 2, 3, and 4 is written $\{1, 2, 3, 4\}$. The numbers 1, 2, 3, and 4 are called the *members* or *elements* of this set.

> **Procedure** Writing Sets
>
> To write a set, write the members or elements of the set separated by commas within braces, { }.

Example 1 Write the set containing the first five letters of the alphabet.

Answer: $\{A, B, C, D, E\}$

Your Turn 1 Write the set containing the first four months of the year.

Numbers are classified using number sets. The set of *natural numbers* contains the counting numbers 1, 2, 3, 4, . . . and is written $\{1, 2, 3, . . . \}$. The three dots are an *ellipsis* and indicate that the numbers continue forever in the same pattern. The set of *whole numbers* contains all of the natural numbers and 0 and is written $\{0, 1, 2, 3, . . . \}$.

Answer to Your Turn 1
$\{$January, February, March, April$\}$

The set of *integers* contains all of the whole numbers and the opposite (or negative) of every natural number and is written $\{\ldots, -3, -2, -1, 0, 1, 2, 3, \ldots\}$. Notice that the integers continue forever in both directions.

A number line is often useful in mathematics. The following number line is marked with the integers. Note that the positive numbers are to the right of 0 and the negative numbers are to the left.

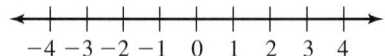

Although we can mark and view only a portion of the number line, the arrows at the ends indicate that the line and the numbers on it continue forever in both directions. There are other types of numbers that include the integers and all numbers between them. If we traveled along the number line forever in both directions, we would encounter every number in the set of *real numbers*.

Some of the real numbers that are not integers are the **rational numbers** that contain every real number that can be expressed as a ratio of integers.

Definition Rational number: Any real number that can be expressed in the form $\frac{a}{b}$, where a and b are integers and $b \neq 0$.

> **Note** In the definition for a rational number, the notation $b \neq 0$ is important because if the denominator b were to equal 0, the fraction would be undefined. The reason it is undefined will be explained later.

For example, $\frac{3}{4}$ is a rational number because 3 and 4 are integers. Note that numbers such as 0.75 and 75% are also rational numbers because they can be written as $\frac{3}{4}$. It is important to realize that the definition of a rational number does not state that the number *must* be expressed in the form $\frac{a}{b}$. Rather, if a number *can* be expressed in the form $\frac{a}{b}$, it is a rational number. For example, the number 5 is a rational number because it can be expressed as $\frac{5}{1}$.

> **Instructor Note** Consider pointing out that the method for writing these nonterminating decimal numbers (like in part c of Example 2) as fractions involves algebra. So students will not be asked to find the equivalent fractions, just to recognize that the numbers are rational numbers.

Example 2   Determine whether the given number is a rational number.

a. $\frac{2}{3}$

Answer: Yes, because 2 and 3 are integers.

b. 0.4

Answer: Yes, 0.4 is a rational number because it can be expressed as $\frac{4}{10}$ and 4 and 10 are integers.

c. $0.\overline{6}$

Answer: A bar written over a decimal digit indicates that the digit repeats without end. So $0.\overline{6} = 0.66666 \ldots$, and we say that these decimal numbers are nonterminating decimal numbers. Usually, we encounter these numbers as quotients in certain division problems, such as when we write certain fractions as decimals. In this case, $0.\overline{6}$ is the decimal equivalent of $\frac{2}{3}$. Because $0.\overline{6}$ can be expressed as the fraction $\frac{2}{3}$, it is a rational number.

> **Note** All nonterminating decimal numbers with repeating digits are rational numbers because they can be expressed as fractions with integers in both the numerator and denominator.

d. 3

Answer: Yes, because 3 can be thought of as $\frac{3}{1}, \frac{6}{2}, \frac{9}{3}$, etc. All integers are rational numbers.

Your Turn 2 Determine whether the given number is a rational number.

a. $\dfrac{4}{7}$ **b.** 0.56 **c.** $0.\overline{2}$ **d.** -7

Not all numbers can be expressed as a ratio of integers. One such number is π (pronounced "pie"). Because the exact value of π cannot be expressed as a ratio of integers, it is categorized as an **irrational number**.

Definition Irrational number: Any real number that is not rational.

Some other irrational numbers are $\sqrt{2}$ and $\sqrt{3}$. (Square roots are explained in more detail in Section 1.5.) Because an irrational number cannot be written as a ratio of integers, if a calculation involves an irrational number, we must leave it in symbolic form or use a rational number approximation. We can approximate π with rational numbers such as 3.14 and $\dfrac{22}{7}$. Any decimal representation of an irrational number is a nonrepeating, nonterminating decimal number. For example, 0.1010010001 . . . is irrational.

Every **real number** is either rational or irrational.

Definition Real number: Any rational or irrational number.

The following figure shows how the number sets relate in the real number system.

Real Numbers

Rational Numbers: Real numbers that can be expressed in the form $\dfrac{a}{b}$, where a and b are integers and $b \neq 0$, such as $-4\dfrac{3}{4}, -\dfrac{2}{3},$ 0.018, $0.\overline{3},$ and $\dfrac{5}{8}$.

Integers: . . . , $-3, -2, -1, 0, 1, 2, 3,$ . . .

Whole Numbers: 0, 1, 2, 3, . . .

Natural Numbers: 1, 2, 3, . . .

Irrational Numbers: Any real number that is not rational, such as $-\sqrt{2},$ $-\sqrt{3}, \sqrt{0.8},$ and π.

Objective 3 Graph rational numbers on a number line.

Number lines can be useful tools when comparing numbers or solving certain arithmetic problems. Let's review how to graph a number on a number line.

Example 3 Graph on a number line.

a. $1\dfrac{3}{5}$

Solution: The number $1\dfrac{3}{5}$ is located $\dfrac{3}{5}$ of the way between 1 and 2, so we divide the space between 1 and 2 into 5 equal divisions and place a dot on the 3rd mark.

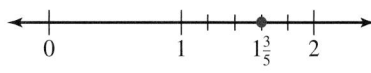

b. -2.56

Solution: Because -2.56 means $-2\dfrac{56}{100}$, we could divide the space between -2 and -3 into 100 divisions and count to the 56th mark. Because this is tedious to do, we gradually use smaller and smaller sections of the number line to graph the number.

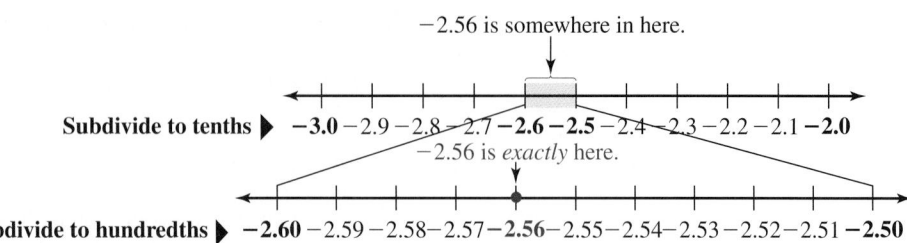

Your Turn 3   Graph on a number line.

a. $2\dfrac{3}{4}$ **b.** -1.78

Objective 4   Determine the absolute value of a number.

The word *value* indicates how much something is worth, or its magnitude. In mathematics, **absolute value** indicates the magnitude of a number.

Definition   Absolute value: A given number's distance from 0 on a number line.

For example, the absolute value of 5 is 5 because the number 5 is 5 units from 0 on a number line. Likewise, the absolute value of -5 is 5 because -5 is also 5 units from 0 on a number line.

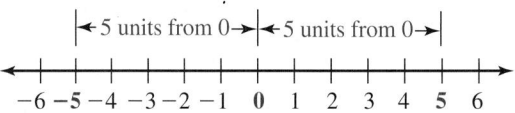

The absolute value of a number n is written $|n|$. The examples just mentioned translate this way:

The absolute value of 5 is 5.   The absolute value of -5 is 5.

Symbolic form: $|5| = 5$ $|-5| = 5$

What about the absolute value of 0? Because there are no units between 0 and itself, the absolute value of 0 is 0.

> **Rule Absolute Value**
> The absolute value of every real number is either positive or 0.

Answers to Your Turn 3

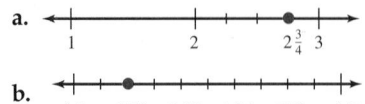

Example 4   Simplify.

a. $|-4.5|$

Answer: $|-4.5| = 4.5$ because -4.5 is 4.5 units from 0 on a number line.

b. $\left|\dfrac{3}{8}\right|$

Answer: $\left|\dfrac{3}{8}\right| = \dfrac{3}{8}$ because $\dfrac{3}{8}$ is $\dfrac{3}{8}$ of a unit from 0 on a number line.

Your Turn 4 Simplify.

a. $\left|-2\dfrac{4}{5}\right|$

b. $|12.8|$

Objective 5 Compare numbers.

Number lines can also be used to determine which of two numbers is greater. Because numbers increase from left to right on a number line, the number farther to the right will be the greater of two numbers.

> **Rule Comparing Numbers**
>
> For any two real numbers a and b, a is greater than b if a is to the right of b on a number line. Equivalently, b is less than a if b is to the left of a on a number line.

Consider the following number line where we compare 2 and 8.

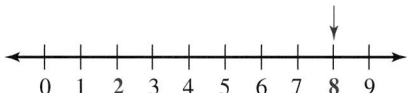

8 is farther right than 2.

Because the number 8 is farther to the right on a number line than the number 2 is, we say that 8 is greater than 2, or in symbols, $8 > 2$. Or we could say that because 2 is to the left of 8, $2 < 8$.

Example 5 Use $=$, $<$, or $>$ to write a true statement.

a. 4 ____ -4

Answer: $4 > -4$ because 4 is farther to the right on a number line than -4 is.

b. -2.7 ____ -2.5

Answer: $-2.7 < -2.5$ because -2.7 is farther to the left on a number line than -2.5 is.

c. $\left|3\dfrac{5}{6}\right|$ ____ $3\dfrac{5}{6}$

Answer: $\left|3\dfrac{5}{6}\right| = 3\dfrac{5}{6}$ because the absolute value of $3\dfrac{5}{6}$ is equal to $3\dfrac{5}{6}$.

d. $|-0.7|$ ____ -1.5

Answer: $|-0.7| > -1.5$ because the absolute value of -0.7 is equal to 0.7, which is farther to the right on a number line than -1.5 is.

Answers to Your Turn 4

a. $2\dfrac{4}{5}$ **b.** 12.8

Answers to Your Turn 5

a. $-15 > -21$ **b.** $-4.1 < 0$

c. $2\dfrac{5}{6} > 2\dfrac{1}{4}$ **d.** $|-9.5| = 9.5$

Your Turn 5 Use $=$, $<$, or $>$ to write a true statement.

a. -15 ____ -21 **b.** -4.1 ____ 0   **c.** $2\dfrac{5}{6}$ ____ $2\dfrac{1}{4}$ **d.** $|-9.5|$ ____ 9.5

1.1 Exercises   For Extra Help   MyMathLab®

Note: Exercises marked with a ★ represent challenging exercises.

Objective 1

Prep Exercise 1 Explain the difference between a constant and a variable. A constant is a symbol that does not vary in value, whereas a variable is a symbol that varies in value.

Prep Exercise 2 An expression is a constant, a variable, or any combination of constants, variables, and arithmetic operations that describes a(n) ___calculation___.

Prep Exercise 3 An equation is a mathematical relationship that contains a(n) ___equal sign___.

Prep Exercise 4 List the inequality symbols. $<, >, \le, \ge, \ne$

Prep Exercise 5 Complete the name of each level of the algebra pyramid.

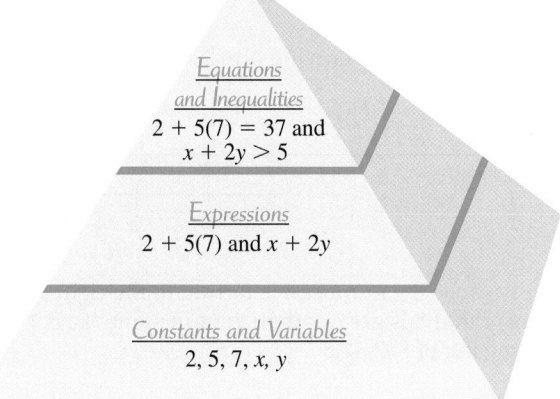

The Algebra Pyramid

Equations and Inequalities
$2 + 5(7) = 37$ and
$x + 2y > 5$

Expressions
$2 + 5(7)$ and $x + 2y$

Constants and Variables
$2, 5, 7, x, y$

Objective 2

Prep Exercise 6 To write a set, write the members or elements of the set separated by commas within ___braces___, ___{ }___.

For Exercises 1–10, write a set representing each description. See Example 1.

1. The days of the week
   {Sunday, Monday, Tuesday, Wednesday, Thursday, Friday, Saturday}

2. The last ten letters of the English alphabet
   {q, r, s, t, u, v, w, x, y, z}

3. The vowels of the English alphabet
   {a, e, i, o, u}

4. The states in the United States that do not share a border with any other state
   {Alaska, Hawaii}

5. The natural numbers that are multiples of 5
   {5, 10, 15, 20, . . . }

6. The even natural numbers
   {2, 4, 6, 8, . . . }

7. The odd natural numbers greater than 7
   {9, 11, 13, 15, . . . }

8. The even integers greater than or equal to 16
   {16, 18, 20, 22, . . . }

9. The integers greater than or equal to −6 but less than −2
   {−6, −5, −4, −3}

10. The integers greater than −2.1 and less than $\frac{3}{4}$
    {−2, −1, 0}

Prep Exercise 7 Explain the difference between a rational number and an irrational number.
A rational number can be expressed as a ratio of integers, but an irrational number cannot.

For Exercises 11–20, determine whether each number is a rational number or an irrational number. See Example 2.

11. $-\dfrac{4}{5}$ rational   **12.** $\dfrac{1}{4}$ rational   **13.** 9 rational   **14.** -12 rational   **15.** π irrational

16. $\dfrac{\pi}{4}$ irrational   **17.** -0.21 rational   **18.** -0.8 rational   **19.** $0.\overline{6}$ rational   **20.** $0.\overline{13}$ rational

Prep Exercise 8 Every real number is either ___rational___ or ___irrational___.

For Exercises 21–26, answer true or false.

21. Every rational number is a real number.
true

22. Every real number is a rational number.
False. There are real numbers that are not rational (irrational numbers).

23. Every whole number is an integer.
true

24. Every real number is a natural number.
False. There are real numbers that are not natural numbers, such as $0, -2, \dfrac{3}{4}, 0.\overline{6}$, and π.

25. A number exists that is both rational and irrational.
False. All real numbers are either rational or irrational.

26. Zero is a rational number.
true

Objective 3

For Exercises 27–34, graph each number on a number line. See Example 3.

27. $-\dfrac{5}{6}$

28. $5\dfrac{1}{2}$

29. $2\dfrac{3}{8}$
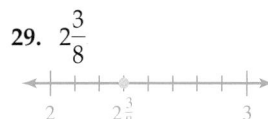

30. $-\dfrac{2}{5}$

31. -3.5

32. 7.4

33. 2.45
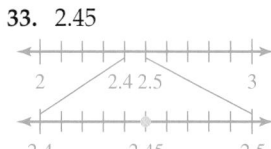

34. -7.62
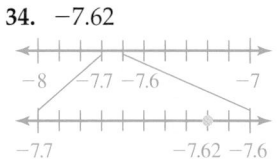

Objective 4

Prep Exercise 9 The absolute value of every real number is either ___positive___ or ___0___.

For Exercises 35–44, simplify. See Example 4.

35. $|23|$ 23   **36.** $|6|$ 6   **37.** $|-2|$ 2   **38.** $|-8|$ 8   **39.** $|-5.7|$ 5.7

40. $|-4.5|$ 4.5   **41.** $\left|-3\dfrac{1}{8}\right|$ $3\dfrac{1}{8}$ **42.** $\left|2\dfrac{3}{5}\right|$ $2\dfrac{3}{5}$ **43.** $|0|$ 0   **44.** $|-67.8|$ 67.8

Objective 5

Prep Exercise 10 When comparing two numbers, the number farther to the ___right___ on a number line is the greater number.

For Exercises 45–72, use $=$, $<$, or $>$ to write a true statement. See Example 5.

45. 9 3 >   **46.** 2 7 <   **47.** 4 -3 >   **48.** -6 5 <   **49.** -7 -8 >

50. -19 ⬚ -7 $<$ **51.** -8 ⬚ 0 $<$ **52.** 0 ⬚ -5 $>$ **53.** 6.2 ⬚ 4.5 $>$ **54.** 2.63 ⬚ 3.75 $<$

55. -4.3 ⬚ -7.6 $>$ **56.** -3.5 ⬚ -3.1 $<$ **57.** $2\frac{2}{3}$ ⬚ $2\frac{3}{4}$ $<$ **58.** $3\frac{5}{6}$ ⬚ $3\frac{1}{4}$ $>$ **59.** 5.8 ⬚ $|-5.8|$ $=$

60. $|-4.1|$ ⬚ 4.1
$=$
 61. 7.3 ⬚ $|-8.7|$
$<$
 62. $|-10.4|$ ⬚ 3.2
$>$
 63. $|4.31|$ ⬚ $|-4.31|$
$=$
 64. $|-0.59|$ ⬚ $|0.59|$
$=$

65. $\left|-6\frac{5}{8}\right|$ ⬚ $5\frac{3}{8}$ $>$ **66.** $4\frac{2}{9}$ ⬚ $\left|4\frac{5}{9}\right|$ $<$ **67.** $|-4|$ ⬚ $|-2|$ $>$ **68.** $|-10|$ ⬚ $|-8|$ $>$ **69.** $|29.5|$ ⬚ $|-29.7|$
$<$

70. $|-5.36|$ ⬚ $|5.76|$
$<$
 71. $\left|-\frac{2}{3}\right|$ ⬚ $\left|-\frac{4}{3}\right|$
$<$
 72. $\left|-\frac{9}{11}\right|$ ⬚ $\left|-\frac{7}{11}\right|$
$>$

For Exercises 73–76, list the given numbers in order from least to greatest.

73. $-2.56, 5.4, |8.3|, \left|-7\frac{1}{2}\right|, -4.7$

$-4.7, -2.56, 5.4, \left|-7\frac{1}{2}\right|, |8.3|$

74. $2.9, 1, -12.6, |-1.3|, -9.6, \left|-2\frac{3}{4}\right|$

$-12.6, -9.6, 1, |-1.3|, \left|-2\frac{3}{4}\right|, 2.9$

75. $0.4, -0.6, 0, 3\frac{1}{4}, |-0.02|, -0.44, \left|1\frac{2}{3}\right|$

$-0.6, -0.44, 0, |-0.02|, 0.4, \left|1\frac{2}{3}\right|, 3\frac{1}{4}$

76. $1.02, -0.13, -4\frac{1}{8}, -2\frac{1}{4}, |-1.06|, -2, |0.1|$

$-4\frac{1}{8}, -2\frac{1}{4}, -2, -0.13, |0.1|, 1.02, |-1.06|$

1.2 Fractions

Objectives

1 Write equivalent fractions.
2 Write equivalent fractions with the LCD.
3 Write the prime factorization of a number.
4 Simplify a fraction to lowest terms.

Connection We learned in Section 1.1 that if a number can be expressed in the form $\frac{a}{b}$, where a and b are integers, it is a rational number. So fractions that have integer numerators and denominators are rational numbers.

Warm-up
[1.1] **1.** Is $A = \pi r^2$ an expression or an equation? Why?
[1.1] **2.** Use $=$, $<$, or $>$ to write a true statement. -3.6 ⬚ -2.5
[1.1] **3.** Find $|-4|$.

In this section, we review some basic principles of **fractions**.

Definition Fraction: A quotient of two numbers or expressions a and b having the form $\frac{a}{b}$, where $b \neq 0$.

For example, $\frac{3}{4}$ is a fraction. The top number in a fraction is called the *numerator*, and the bottom number is called the *denominator*.

$$\frac{3}{4} \begin{array}{l}\leftarrow \text{Numerator} \\ \leftarrow \text{Denominator}\end{array}$$

We can use fractions to indicate a part of a whole. For example, the rectangle to the right has been divided into four equal pieces, three of which are shaded. So the shaded region represents $\frac{3}{4}$ of the rectangle.

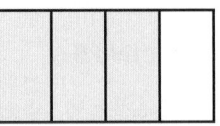

Answers to Warm-up
1. Equation. It is a mathematical expression containing an equal sign.
2. $<$
3. 4

Objective 1 Write equivalent fractions.

Now let's review how to write *equivalent fractions*, which are fractions that represent the same amount. For example, $\frac{1}{2}$ and $\frac{2}{4}$ are equivalent. We can see that they represent the same amount on a ruler.

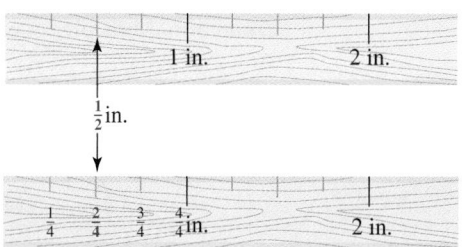

We can rewrite $\frac{1}{2}$ as $\frac{2}{4}$ by multiplying its numerator and denominator by 2.

$$\frac{1}{2} = \frac{1 \cdot 2}{2 \cdot 2} = \frac{2}{4}$$

We also can begin with $\frac{2}{4}$ and get back to $\frac{1}{2}$ by dividing both the numerator and denominator by 2.

$$\frac{2}{4} = \frac{2 \div 2}{4 \div 2} = \frac{1}{2}$$

> **Rule Writing Equivalent Fractions**
>
> For any fraction, we can write an equivalent fraction by multiplying or dividing both its numerator and denominator by the same nonzero number.

This rule makes use of the fact that multiplying or dividing both the numerator and denominator by the same nonzero number is equivalent to multiplying or dividing by 1, which does not change a number.

Instructor Note To speed the review, we have forgone a formal discussion of the multiplicative identity here.

Example 1   Find the missing number that makes the fractions equivalent.

a. $\dfrac{7}{12} = \dfrac{?}{48}$

Solution: Because $48 \div 12 = 4, 12 \cdot 4 = 48$. So we multiply the numerator and denominator by 4.

$$\frac{7 \cdot 4}{12 \cdot 4} = \frac{28}{48} \qquad \text{Multiply the numerator and denominator by 4.}$$

b. $-\dfrac{30}{42} = \dfrac{?}{7}$

Solution: Because $42 \div 7 = 6, 42 \div 6 = 7$. So divide the numerator and denominator by 6.

$$-\frac{30 \div 6}{42 \div 6} = -\frac{5}{7} \qquad \text{Divide the numerator and denominator by 6.}$$

Note When a fraction is negative, the sign can be placed to the left of the numerator, to the left of the fraction line, or to the left of the denominator.

$$-\frac{30}{42} = -\frac{5}{7} = \frac{-5}{7} = \frac{5}{-7}$$

▶

Your Turn 1   Find the missing number that makes the fractions equivalent.

a. $\dfrac{9}{16} = \dfrac{?}{32}$

b. $-\dfrac{54}{72} = -\dfrac{3}{?}$

Answers to Your Turn 1
a. $\dfrac{9}{16} = \dfrac{18}{32}$ **b.** $-\dfrac{54}{72} = -\dfrac{3}{4}$

Objective 2 Write equivalent fractions with the LCD.

It is sometimes necessary to write fractions that have different denominators, such as $\frac{1}{2}$ and $\frac{1}{3}$, as equivalent fractions with a common denominator. A common denominator can be any number that is a **multiple** of the denominators.

Definition Multiple: A multiple of a given integer n is the product of n and an integer.

We can generate multiples of a given number by multiplying the given number by the integers. For example, to generate the positive **multiples** of 2 and 3, we multiply 2 and 3 by 1, 2, 3, 4, and so on.

Note Every multiple of 2 is evenly divisible by 2. Similarly, every multiple of 3 is evenly divisible by 3. In general, every multiple of a number n is evenly divisible by n. ▶

Multiples of 2	Multiples of 3
↓	↓
$2 \cdot 1 = 2$	$3 \cdot 1 = 3$
$2 \cdot 2 = 4$	$3 \cdot 2 = 6$
$2 \cdot 3 = 6$	$3 \cdot 3 = 9$
$2 \cdot 4 = 8$	$3 \cdot 4 = 12$
$2 \cdot 5 = 10$	$3 \cdot 5 = 15$
$2 \cdot 6 = 12$	$3 \cdot 6 = 18$

Notice that some common multiples of 2 and 3 in our lists are 6 and 12. When we rewrite fractions with a common denominator, it is usually preferable to use the **least common multiple** of the denominators. Such a denominator is called the **least common denominator**. From our lists, we see that the least common multiple of 2 and 3 is 6. So 6 is the least common denominator of $\frac{1}{2}$ and $\frac{1}{3}$.

Definitions Least common multiple (LCM): The smallest number that is a multiple of each number in a given set of numbers.

Least common denominator (LCD): The least common multiple of the denominators of a given set of fractions.

We can now write equivalent fractions for $\frac{1}{2}$ and $\frac{1}{3}$ with their LCD, 6.

$$\frac{1}{2} = \frac{1 \cdot 3}{2 \cdot 3} = \frac{3}{6} \quad \text{and} \quad \frac{1}{3} = \frac{1 \cdot 2}{3 \cdot 2} = \frac{2}{6}$$

Example 2   Write each pair as equivalent fractions with the LCD.

a. $\frac{5}{8}$ and $\frac{1}{6}$

Solution: The LCD of 8 and 6 is 24.

$$\frac{5}{8} = \frac{5 \cdot 3}{8 \cdot 3} = \frac{15}{24} \quad \text{and} \quad \frac{1}{6} = \frac{1 \cdot 4}{6 \cdot 4} = \frac{4}{24}$$

b. $\frac{2}{3}$ and $\frac{5}{9}$

Solution: The LCD of 3 and 9 is 9.

$\frac{2}{3} = \frac{2 \cdot 3}{3 \cdot 3} = \frac{6}{9}$, and $\frac{5}{9}$ already has the LCD as its denominator.

Your Turn 2   Write each pair as equivalent fractions with the LCD.

a. $\frac{4}{9}$ and $\frac{5}{6}$

b. $-\frac{3}{4}$ and $-\frac{5}{8}$

Answers to Your Turn 2
a. $\frac{8}{18}$ and $\frac{15}{18}$ **b.** $-\frac{6}{8}$ and $-\frac{5}{8}$

Objective 3 Write the prime factorization of a number.

We now consider simplifying fractions. When simplifying fractions, we work with **factors**, which are numbers or variables that are multiplied together.

Definition Factors: If $a \cdot b = c$, then a and b are factors of c.

In the multiplication statement $20 = 2 \cdot 10$, the 2 and 10 are factors of 20. Notice that a factor of a given number divides the given number evenly (leaves a remainder of 0). The numbers 1, 2, 4, 5, 10, and 20 are factors of 20 because they all divide 20 evenly. When a number is expressed as a product of its factors, it is said to be in *factored form*. Factored form is sometimes referred to as a *factorization*. Here are some factorizations for 20:

$$20 = 1 \cdot 20 \quad \text{or} \quad 20 = 2 \cdot 10 \quad \text{or} \quad 20 = 4 \cdot 5$$

Some natural numbers have only two distinct factors, 1 and the number itself. These numbers are called **prime numbers**.

Definition Prime number: A natural number that has exactly two different factors, 1 and the number itself.

Note A natural number that has factors other than 1 and itself is called a **composite number**. Some examples are 4, 6, 8, and 9. Notice that 1 is not composite, so it is neither prime nor composite.

Note that 0 is not a prime number because it is not a natural number. (0 is a whole number.) Also, 1 is not a prime number because the only factor for 1 is itself and, by definition, a prime number must have *two different* factors.

The first prime number is 2 because it is the first natural number greater than 1 that has exactly two factors, 1 and 2. The number 4 is not a prime number because it is divisible by a number other than 1 and 4, namely 2. Here is a list of prime numbers:

$$2, 3, 5, 7, 11, 13, 17, 19, 23, 29, 31, 37, 41, 43, \ldots$$

When simplifying a fraction, it is useful to write the **prime factorization** of the numerator and denominator. For example, the prime factorization for 20 is $2 \cdot 2 \cdot 5$.

Definition Prime factorization: A factorization that contains only prime factors.

Of Interest

Eratosthenes (276–196 B.C.) developed a method for finding prime numbers that he called a *sieve* because it sifted out the composite numbers, leaving only prime numbers.

The method involves writing the positive integers starting with 2, the first prime number. Keep 2 and cross out all multiples of 2. The smallest remaining number is 3, the next prime number—keep it and cross out all multiples of 3. The smallest remaining number is now 5, the next prime number—keep it and cross out all multiples of 5. You can continue in this fashion indefinitely.

2, 3, 4̶, 5, 6̶, 7, 8̶, 9̶, 1̶0̶, 11, 1̶2̶, 13, 1̶4̶, 1̶5̶, 1̶6̶, 17, 1̶8̶, 19, 2̶0̶, 2̶1̶, 2̶2̶, 23, 2̶4̶, 2̶5̶, 2̶6̶, 2̶7̶, 2̶8̶, 29, 3̶0̶, 31, 3̶2̶, 3̶3̶, 3̶4̶, 3̶5̶, 3̶6̶, 37, 3̶8̶, 3̶9̶, . . .

We can find the prime factorization of a number using any factorization of the number as a starting point. Consider 20 again and suppose we factored it as $4 \cdot 5$. Because 4 is not a prime number, we can factor 4 as $2 \cdot 2$. It looks like this:

$$20 = 4 \cdot 5$$
$$= 2 \cdot 2 \cdot 5$$

We will use factor trees to find prime factorizations. The idea is to draw two branches beneath any composite number and write two factors of that number at the end of the branches. Following are two ways we can use a factor tree to find the prime factorization of 20. Notice that when a branch ends in a prime factor, we circle the prime factor. This is optional, but many people find it helpful in making the prime factors stand out.

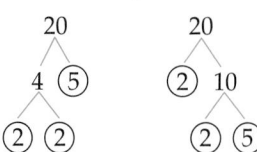

We usually write the prime numbers in a prime factorization in ascending order. So we would write $20 = 2 \cdot 2 \cdot 5$ (instead of $20 = 5 \cdot 2 \cdot 2$ or $20 = 2 \cdot 5 \cdot 2$). Note that no matter how we write the prime factorization, the product of two 2s and one 5 is always 20. No two numbers have the same prime factorization. Consequently, we say that a number's prime factorization is unique.

Example 3   Find the prime factorization of 360.

Solution: Use a factor tree.

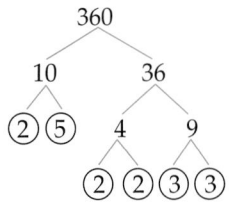

Factor 360 to 10 and 36. (Any two factors will work.)

Factor 10 to 2 and 5, which are primes. Then factor 36 to 4 and 9.

4 is then factored to 2 and 2, which are primes, and 9 is factored to 3 and 3, which are primes.

Answer: $360 = 2 \cdot 2 \cdot 2 \cdot 3 \cdot 3 \cdot 5$

Your Turn 3   Find the prime factorization.

a. 84   b. 280

Objective 4 Simplify a fraction to lowest terms.

We can use prime factorizations to simplify fractions to **lowest terms**.

Definition Lowest terms: Given a fraction $\frac{a}{b}$ and $b \neq 0$, if the only factor common to both a and b is 1, then the fraction is in lowest terms.

For example, $\frac{3}{4}$ is in lowest terms because the only factor common to both 3 and 4 is the number 1, whereas $\frac{6}{8}$ is not in lowest terms because 6 and 8 have a common factor of 2.

To develop the method of simplifying fractions to lowest terms using primes, we must understand that if the numerator and denominator are identical and not both 0, then the fraction is equivalent to 1.

$$1 = \frac{1}{1} = \frac{2}{2} = \frac{3}{3} = \frac{4}{4} = \frac{5}{5} = \cdots$$

This suggests the following rule:

> **Rule**
> Simplifying a Fraction with the Same Nonzero Numerator and Denominator
> $$\frac{n}{n} = \frac{1}{1} = 1, \text{ when } n \neq 0.$$

This rule can be extended to common factors within a fraction. It allows us to eliminate factors that are identical in the numerator and denominator of a fraction.

> **Rule**
> Eliminating a Common Factor in a Fraction
> $$\frac{an}{bn} = \frac{a \cdot 1}{b \cdot 1} = \frac{a}{b}, \text{ when } b \neq 0 \text{ and } n \neq 0.$$

Answers to Your Turn 3
a. $84 = 2 \cdot 2 \cdot 3 \cdot 7$
b. $280 = 2 \cdot 2 \cdot 2 \cdot 5 \cdot 7$

These rules allow us to write fractions in lowest terms using prime factorizations. The idea is to replace the numerator and denominator with their prime factorizations and then eliminate the prime factors that are common to both the numerator and denominator. Consider the fraction $\frac{10}{15}$. Replacing the 10 and 15 with their prime factorizations, we have the following:

$$\frac{10}{15} = \frac{2 \cdot 5}{3 \cdot 5}$$

$$= \frac{2 \cdot 1}{3 \cdot 1}$$

Note The common factor of 5 is replaced with 1. In Example 4, we leave this step out.

$$= \frac{2}{3}$$

This process of eliminating common factors is called *dividing out* the common factors and is sometimes written this way:

$$\frac{10}{15} = \frac{2 \cdot \overset{1}{\cancel{5}}}{3 \cdot \underset{1}{\cancel{5}}} = \frac{2}{3}$$

Although there are different styles of showing which common factors are eliminated, we simply highlight the common factors that are eliminated.

Procedure **Simplifying a Fraction to Lowest Terms**

To simplify a fraction to lowest terms:
1. Replace the numerator and denominator with their prime factorizations.
2. Eliminate (divide out) all prime factors common to the numerator and denominator.
3. Multiply the remaining factors.

Connection You may remember a method for simplifying fractions in which you divide out the greatest common factor. For example, to simplify $\frac{28}{70}$, we recognize that 14 is the greatest common factor of 28 and 70 and divide both of those numbers by 14 to get lowest terms:
$\frac{28}{70} = \frac{28 \div 14}{70 \div 14} = \frac{2}{5}$. This method is a perfectly good way of simplifying fractions. However, we show the prime method to prepare for simplifying rational expressions in Chapter 7.

Example 4 Simplify to lowest terms.

a. $\frac{28}{70}$

Solution: $\frac{28}{70} = \frac{2 \cdot 2 \cdot 7}{2 \cdot 5 \cdot 7} = \frac{2}{5}$ Replace the numerator and denominator with their prime factorizations; then eliminate the common prime factors.

b. $\frac{156}{210}$

Solution: $\frac{156}{210} = \frac{2 \cdot 2 \cdot 3 \cdot 13}{2 \cdot 3 \cdot 5 \cdot 7} = \frac{26}{35}$

Your Turn 4 Simplify to lowest terms.

a. $\frac{36}{48}$ b. $\frac{126}{315}$ c. $\frac{312}{378}$

Example 5 A newspaper reports that over the last year, 160 households per 1000 households were victims of a crime. Write a fraction in simplest form representing the fraction of households that were victims of crime.

Solution: If 160 out of 1000 households were victims of crime, the fraction of households that were victims of crime is $\frac{160}{1000}$. Now we can simplify the fraction.

$$\frac{160}{1000} = \frac{2 \cdot 2 \cdot 2 \cdot 2 \cdot 2 \cdot 5}{2 \cdot 2 \cdot 2 \cdot 5 \cdot 5 \cdot 5} = \frac{4}{25}$$

Answers to Your Turn 4

a. $\frac{3}{4}$ b. $\frac{2}{5}$ c. $\frac{52}{63}$

Answer: The fraction, in simplest form, of households that were victims of crime is $\frac{4}{25}$, which means that 4 out of 25 households were victims of crime.

Your Turn 5 Researchers gave 560 volunteers a new medication and found that 48 of them developed a headache within 30 minutes of taking the medication. Write a fraction in simplest form representing the portion of the volunteers that developed a headache.

1.2 Exercises For Extra Help MyMathLab®

Objective 1

Prep Exercise 1 A fraction is a quotient of two numbers or expressions a and b having the form _____, where $b \neq 0$. $\frac{a}{b}$

For Exercises 1–4, identify the fraction represented by each shaded region.

1.
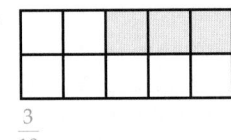
$\frac{3}{10}$

2.
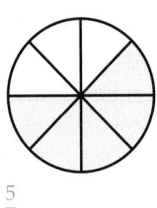
$\frac{5}{8}$

3.
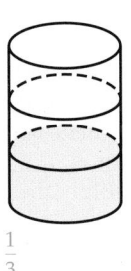
$\frac{1}{3}$

4.
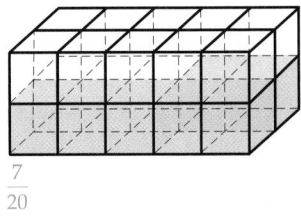
$\frac{7}{20}$

For Exercises 5–10, write the length of each line segment in lowest terms.

5.
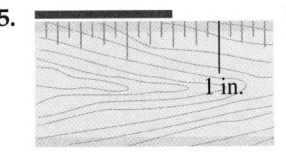
$\frac{3}{4}$ in.

6.
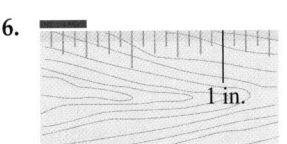
$\frac{1}{4}$ in.

7.
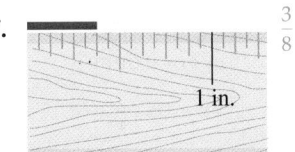
$\frac{3}{8}$ in.

8.
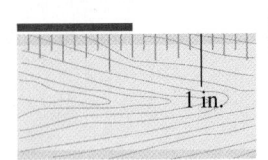
$\frac{5}{8}$ in.

9.
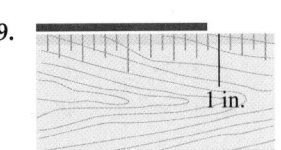
$\frac{15}{16}$ in.

10.
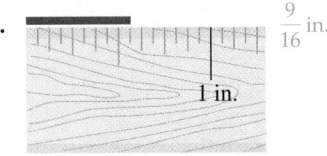
$\frac{9}{16}$ in.

Prep Exercise 2 What are two ways to generate fractions equivalent to a given fraction?
Multiply or divide the numerator and denominator by the same nonzero number.

For Exercises 11–18, find the missing number that makes the fractions equivalent. See Example 1.

11. $\frac{3}{10} = \frac{?}{40}$ 12

12. $\frac{5}{8} = \frac{?}{16}$ 10

13. $-\frac{7}{12} = \frac{?}{60}$ -35

14. $\frac{2}{5} = \frac{6}{?}$ 15

15. $\frac{4}{7} = \frac{?}{28}$ 16

16. $\frac{6}{8} = \frac{?}{4}$ 3

17. $\frac{8}{60} = \frac{2}{?}$ 15

18. $\frac{27}{30} = \frac{9}{?}$ 10

Objective 2

Prep Exercise 3 Is 4 a multiple of 12? Explain. No, a multiple of 12 is found by multiplying 12 by an integer, and we cannot multiply 12 by an integer to get 4.

Prep Exercise 4 The LCD of two or more fractions is the _____ of the denominators. least common multiple

For Exercises 19–26, write the fractions as equivalent fractions with the LCD. See Example 2.

19. $\dfrac{4}{5}$ and $\dfrac{2}{3}$
$\dfrac{12}{15}$ and $\dfrac{10}{15}$

20. $\dfrac{5}{7}$ and $\dfrac{3}{11}$
$\dfrac{55}{77}$ and $\dfrac{21}{77}$

21. $\dfrac{4}{9}$ and $\dfrac{7}{6}$
$\dfrac{8}{18}$ and $\dfrac{21}{18}$

22. $\dfrac{5}{8}$ and $\dfrac{7}{12}$
$\dfrac{15}{24}$ and $\dfrac{14}{24}$

23. $-\dfrac{11}{18}$ and $-\dfrac{17}{24}$
$-\dfrac{44}{72}$ and $-\dfrac{51}{72}$

24. $-\dfrac{9}{20}$ and $-\dfrac{7}{15}$
$-\dfrac{27}{60}$ and $-\dfrac{28}{60}$

25. $-\dfrac{1}{16}$ and $-\dfrac{15}{36}$
$-\dfrac{9}{144}$ and $-\dfrac{60}{144}$

26. $-\dfrac{13}{21}$ and $-\dfrac{9}{14}$
$-\dfrac{26}{42}$ and $-\dfrac{27}{42}$

Objective 3

Prep Exercise 5 A prime number is a natural number that has exactly two factors, _____1_____ and ____itself____.

Prep Exercise 6 Is $2 \cdot 2 \cdot 4 \cdot 7$ a prime factorization? Explain. No, 4 is not a prime number. The correct prime factorization is $2 \cdot 2 \cdot 2 \cdot 2 \cdot 7$.

For Exercises 27–34, write the prime factorization for each number. See Example 3.

27. 44 $2 \cdot 2 \cdot 11$

28. 33 $3 \cdot 11$

29. 36 $2 \cdot 2 \cdot 3 \cdot 3$

30. 42 $2 \cdot 3 \cdot 7$

31. 64 $2 \cdot 2 \cdot 2 \cdot 2 \cdot 2 \cdot 2$

32. 48 $2 \cdot 2 \cdot 2 \cdot 2 \cdot 3$

33. 250 $2 \cdot 5 \cdot 5 \cdot 5$

34. 810 $2 \cdot 3 \cdot 3 \cdot 3 \cdot 3 \cdot 5$

Objective 4

Prep Exercise 7 How do you know if a fraction is in lowest terms? The numerator and the denominator have no common factors other than 1.

Prep Exercise 8 To simplify a fraction to lowest terms:
1. Replace the numerator and denominator with their prime factorizations.
2. Eliminate (divide out) all prime factors common to the numerator and denominator.
3. Multiply the remaining factors.

For Exercises 35–42, simplify to lowest terms. See Example 4.

35. $\dfrac{72}{90}$ $\dfrac{4}{5}$

36. $\dfrac{48}{84}$ $\dfrac{4}{7}$

37. $\dfrac{63}{99}$ $\dfrac{7}{11}$

38. $\dfrac{42}{91}$ $\dfrac{6}{13}$

39. $-\dfrac{78}{104}$ $-\dfrac{3}{4}$

40. $-\dfrac{30}{54}$ $-\dfrac{5}{9}$

41. $-\dfrac{48}{90}$ $-\dfrac{8}{15}$

42. $-\dfrac{24}{162}$ $-\dfrac{4}{27}$

For Exercises 43–46, determine whether each simplification is correct. If it isn't, explain why. See Example 4.

43. $\dfrac{\overset{1}{2}+3}{\underset{1}{2}} = \dfrac{1+3}{1} = 4$

Incorrect. You may divide out only factors, not addends.

44. $\dfrac{3-4\cdot\overset{1}{2}}{\underset{1}{2}}$

Incorrect. 2 is not a factor of the numerator.

45. $\dfrac{84}{240} = \dfrac{\overset{1}{2}\cdot\overset{1}{2}\cdot\overset{1}{3}\cdot 7}{2\cdot 2\cdot\underset{1}{2}\cdot\underset{1}{3}\cdot 3\cdot 5} = \dfrac{7}{30}$

Incorrect. The prime factorization of 240 should be $2 \cdot 2 \cdot 2 \cdot 2 \cdot 3 \cdot 5$.

46. $\dfrac{300}{108} = \dfrac{\overset{1}{2}\cdot\overset{1}{2}\cdot\overset{1}{3}\cdot 5\cdot 5}{2\cdot 2\cdot\underset{1}{3}\cdot\underset{1}{3}} = \dfrac{25}{3}$

Incorrect. The prime factorization of 108 should be $2 \cdot 2 \cdot 3 \cdot 3 \cdot 3$.

For Exercises 47–50, write each fraction in lowest terms. See Example 5.

47. A student scores 294 points out of a total of 336 points in a college course. What fraction of the total points is the student's score?
$\dfrac{7}{8}$

48. The nutrition label on a package of frozen fish sticks indicates that a serving of fish sticks has 250 calories, with 130 of those calories coming from fat. What fraction of the calories in a serving of fish sticks comes from fat? $\dfrac{13}{25}$

49. At a company, 300 of the 575 employees have optional life insurance coverage as part of their benefits package. What fraction of the employees have optional life insurance coverage?
$\dfrac{12}{23}$

50. Laura uses 120 square feet of a room in her home as an office for her business. The total living space of her home is 1830 square feet. For tax purposes, she needs to compute the fraction of her home that is used as an office. What is that fraction? $\dfrac{4}{61}$

For Exercises 51–54, use the following bar graph, which shows the number of hours that Carla spends each week in each activity. Write all fractions in lowest terms. See Example 5.

51. What fraction of a week does Carla spend working?
$\dfrac{5}{21}$

52. What fraction of a week does Carla spend sleeping?
$\dfrac{25}{84}$

53. What fraction of a week does Carla spend in all of the listed activities combined?
$\dfrac{2}{3}$

54. What fraction of the week does Carla have for free time, which is time away from all of the listed activities?
$\dfrac{1}{3}$

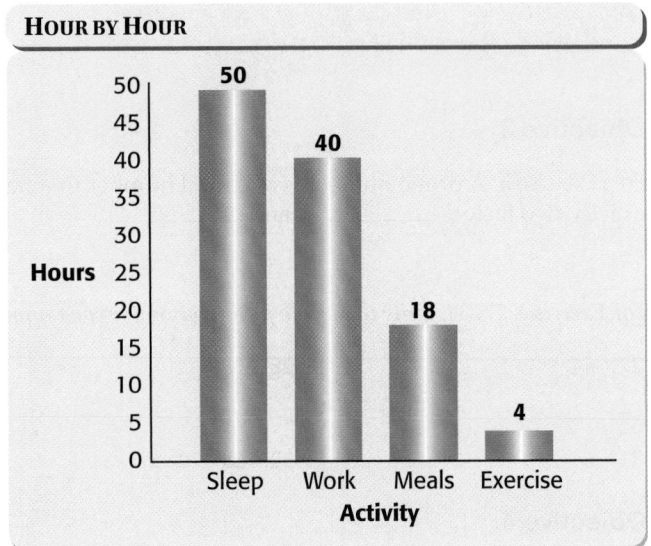

For Exercises 55–62, use the following table, which shows the number of households per 1000 that were victims of property crime in each year. Write all fractions in lowest terms. See Example 5.

National Crime Victimization Survey, Property Crime Trends, 1990–2008

Year	Total Property Crime*	Burglary	Thefts	Motor Vehicle Thefts
1990	349	65	264	21
1991	354	65	267	22
1992	325	59	248	19
1993	319	58	242	19
1994	310	56	235	19
1995	291	49	224	17
1996	266	47	206	14
1997	248	45	190	14
1998	217	39	168	11
1999	198	34	154	10
2000	178	32	138	9
2001	167	29	129	9
2002	159	28	122	9
2003	163	30	124	9
2004	161	30	123	9
2005	154	30	116	8
2006	161	30	122	8
2007	147	27	111	8
2008	135	26	102	7

*Victimizations per 1000 households

Source: U.S. Department of Justice, Bureau of Crime Statistics.

55. What fraction of households were victims of property crime in 1992?
$\frac{13}{40}$

56. What fraction of households were victims of property crime in 1994?
$\frac{31}{100}$

57. In 1992, what fraction of households were not victims of property crime?
$\frac{27}{40}$

58. In 1994, what fraction of households were not victims of property crime?
$\frac{69}{100}$

59. a. What year had the highest number of thefts? 1991
b. What fraction of households were victims in that year?
$\frac{177}{500}$

60. a. What year had the lowest number of burglaries? 2008
b. What fraction of households were victims of burglary in that year?
$\frac{13}{500}$

61. In 1999, what fraction of the total property crimes were thefts (not including motor vehicle thefts)?
$\frac{7}{9}$

62. In 2002, what fraction of total property crimes were motor vehicle theft?
$\frac{3}{53}$

For Exercises 63–70, write each fraction in lowest terms. See Example 5.

63. What fraction of a year is 30 days? Use 365 days for the number of days in a year.
$\frac{6}{73}$

64. What fraction of an hour is 8 minutes?
$\frac{2}{15}$

65. What fraction of a minute is 40 seconds?
$\frac{2}{3}$

66. What fraction of a foot is 4 inches?
$\frac{1}{3}$

67. In the first session of 2011, the U.S. House of Representatives had 193 Democrats and 242 Republicans. What fraction of the House of Representatives was Democrat? (*Source:* Official list of United States House of Representatives.)
$\frac{193}{435}$

68. In 2011, the U.S. Senate had 51 Democrats, 47 Republicans, and 2 Independents. What fraction of the Senate was not Democrat? (*Source:* Official list of United States Senators.)
$\frac{49}{100}$

69. One molecule of lactose contains 12 carbon atoms, 22 hydrogen atoms, and 11 oxygen atoms. What fraction of the atoms that make up the molecule are carbon?
$\frac{4}{15}$

70. One molecule of glucose contains 6 carbon atoms, 12 hydrogen atoms, and 6 oxygen atoms. What fraction of the atoms that make up the molecule are not carbon?
$\frac{3}{4}$

Of Interest

It takes Earth one year to complete one revolution around the Sun, which is actually $365\frac{1}{4}$ days. Because it is impractical to have $\frac{1}{4}$ days on the calendar, every four years those $\frac{1}{4}$ days add up to a full day, which is why we have a leap year with 366 days.

Of Interest

The chemical formula for a molecule of lactose is $C_{12}H_{22}O_{11}$. Notice that the subscripts in a chemical formula indicate the number of atoms of each element that are present in the molecule.

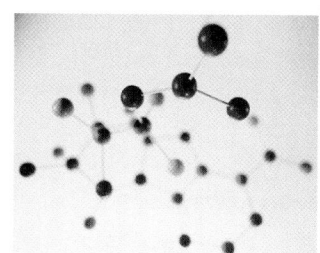

Review Exercises

Exercises 1–3  **Constants and Variables**

[1.1] 1. Write a set containing the names of the four planets closest to the Sun.

{Mercury, Venus, Earth, Mars}

[1.1] 2. Is 0.8 a rational or irrational number? Explain.

It is a rational number because it can be written as a ratio of the integers 8 and 10.

[1.1] 3. Graph 3.7 on a number line.

<-+++++++++++++->
    3   3.7   4

Exercises 4–6  **Expressions**

[1.1] 4. Is $5x + 2$ an expression or an equation? Explain.

It is an expression because it has no $=$ sign.

[1.1] 5. Simplify: $|27|$ 27

[1.1] 6. Use $<$, $>$, or $=$ to make a true statement: $|-6|$ 6

$=$

1.3 Adding and Subtracting Real Numbers; Properties of Real Numbers

Objectives

1 Add rational numbers.
2 Find the additive inverse of a number.
3 Subtract rational numbers.

Warm-up

[1.1] 1. Find $|22|$ and $|-14|$.

[1.2] 2. Write $\dfrac{1}{6}$ and $\dfrac{5}{9}$ as equivalent fractions with the LCD.

Objective 1 Add rational numbers.

We now turn our attention to adding and subtracting numbers. We begin with addition. First consider the parts of an addition statement. The numbers added are called *addends*, and the answer is called a *sum*.

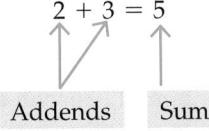

There are three important properties of addition that are true for all real numbers. First, the sum of a number and 0 is that number. Because adding 0 to a number does not change the identity of the number, 0 is called the *additive identity*.

$$2 + 0 = 2$$

Second, the order of addends can be changed without affecting the sum. This is known as the *commutative property of addition*.

$$2 + 3 = 5 \quad \text{or} \quad 3 + 2 = 5$$

Answers to Warm-up
1. 22, 14
2. $\dfrac{3}{18}, \dfrac{10}{18}$

Third, when adding three or more numbers, we can group the addends differently without affecting the sum. This is known as the *associative property of addition*.

$$2 + (3 + 4) \qquad \text{or} \qquad (2 + 3) + 4$$
$$= 2 + 7 \qquad\qquad\qquad = 5 + 4$$
$$= 9 \qquad\qquad\qquad\quad = 9$$

Following is a summary of these properties.

Properties of Addition	Symbolic Form	Word Form
Additive Identity	$a + 0 = a$	The sum of a number and 0 is that number.
Commutative Property of Addition	$a + b = b + a$	Changing the order of addends does not affect the sum.
Associative Property of Addition	$a + (b + c) = (a + b) + c$	Changing the grouping of three or more addends does not affect the sum.

Example 1   Indicate whether each equation illustrates the additive identity, commutative property of addition, or associative property of addition.

a. $(4 + 7) + 2 = 4 + (7 + 2)$

Answer: Associative property of addition because the grouping is changed.

b. $0 + (-7) = -7$

Answer: Additive identity because the sum of a number and 0 is that number.

c. $(-3 + 5) + 2 = 2 + (-3 + 5)$

Answer: Commutative property of addition because the order of the addends is changed.

Your Turn 1   Indicate whether each equation illustrates the additive identity, commutative property of addition, or associative property of addition.

a. $-6 + (3 + 5) = (-6 + 3) + 5$ **b.** $3 + 0 = 3$ **c.** $-7 + 3 = 3 + (-7)$

Adding Numbers with the Same Sign

Keeping these properties in mind, let's consider how to add signed numbers. First, we learn how to add numbers that have the same sign. It is helpful to relate adding signed numbers to money. For example, adding two positive numbers is like adding deposits, whereas adding two negative numbers is like adding debts.

$$15 + 10 = 25 \qquad\qquad\qquad -20 + (-8) = -28$$

Note This addition is like having $15 in an account and depositing another $10 into the same account, bringing the total to $25.

Note This addition is like having a debt balance of $20 on a credit card and charging another $8 on the same card, bringing the balance to a total debt of $28.

Answers to Your Turn 1
a. associative property of addition
b. additive identity
c. commutative property of addition

We can also illustrate each of these cases using a number line. The first addend is the starting point, and the second addend indicates the direction and distance to travel on the number line so that we finish at the sum.

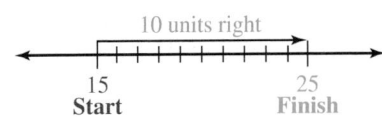

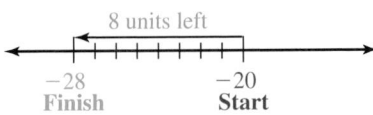

These two cases are similar and suggest the following procedure:

> **Procedure** **Adding Numbers with the Same Sign**
>
> To add two numbers that have the same sign, add their absolute values and keep the same sign.

Example 2 Add.

a. $15 + 13$

Solution: $15 + 13 = 28$

b. $-14 + (-23)$

Solution: $-14 + (-23) = -37$

Note In terms of money, part a illustrates two deposits, whereas part b illustrates two debts.

Your Turn 2 Add.

a. $32 + 19$

b. $-62 + (-13)$

Adding Numbers with Different Signs

Now consider adding two numbers with different signs, which is like making a payment toward a debt. Consider the following cases:

$$-15 + 3 = -12 \qquad\qquad -9 + 20 = 11$$

Note This addition is like having a debt of \$15 and a payment of \$3 is made toward that debt. This decreases the debt to \$12.

Note This addition is like having a debt of \$9 and a payment of \$20 is made toward that debt. Because the payment is more than the debt, we now have a credit of \$11.

Using number lines:

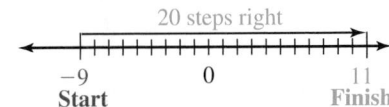

Note that the commutative property of addition indicates that in the examples, the addends can be rearranged without affecting the sum.

$$3 + (-15) = -12 \qquad\qquad 20 + (-9) = 11$$

These examples suggest the following procedure:

> **Procedure** **Adding Numbers with Different Signs**
>
> To add two numbers that have different signs, subtract the smaller absolute value from the greater absolute value and keep the sign of the number with the greater absolute value.

Answers to Your Turn 2
a. 51   **b.** −75

Example 3   Add.

a. $22 + (-14)$

Solution: $22 + (-14) = 8$

b. $-13 + 20$

Solution: $-13 + 20 = 7$

Note In terms of money, parts a and b illustrate situations in which the deposit is greater than the debt. Therefore, the result is a credit, or positive.

c. $12 + (-18)$

Solution: $12 + (-18) = -6$

d. $-24 + 5$

Solution: $-24 + 5 = -19$

Note In terms of money, parts c and d illustrate situations in which a payment toward a debt is not enough to pay off the debt. Therefore, the result is still debt, or negative.

Your Turn 3   Add.

a. $15 + (-6)$ **b.** $27 + (-35)$ **c.** $-13 + 19$ **d.** $-29 + 14$

Adding Fractions with the Same Denominator

Recall that the set of rational numbers contains any number that can be expressed as a ratio of integers. This set includes the integers themselves because every integer can be expressed as a ratio with 1 as the denominator. Now let's turn our attention to the rest of the rational number set, namely, fractions and decimals. Consider the intuitive fact that adding two quarters equals a half. (In money, this would be a half-dollar.) The following circle shows this sum.

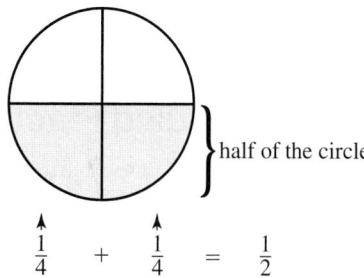

}half of the circle

$$\frac{1}{4} \quad + \quad \frac{1}{4} \quad = \quad \frac{1}{2}$$

Notice that if the numerators are added and the denominator remains the same, the result simplifies to $\frac{1}{2}$.

$$\frac{1}{4} + \frac{1}{4} = \frac{1+1}{4} = \frac{2}{4} = \frac{1}{2}$$

Procedure

To add fractions with the same denominator, add the numerators and keep the same denominator; then simplify.

Example 4   Add.

a. $\dfrac{5}{8} + \dfrac{1}{8}$

Answers to Your Turn 3
a. 9   **b.** -8 **c.** 6   **d.** -15

Solution: $\dfrac{5}{8} + \dfrac{1}{8} = \dfrac{5+1}{8} = \dfrac{6}{8}$ Add the numerators and keep the same denominator. Because $\dfrac{6}{8}$ is not in lowest terms, we must simplify.

$\qquad\qquad = \dfrac{2 \cdot 3}{2 \cdot 2 \cdot 2} = \dfrac{3}{4}$ Replace 6 and 8 with their prime factorizations; divide out the common factor, 2; and then multiply the remaining factors.

b. $-\dfrac{3}{10} + \left(-\dfrac{1}{10}\right)$

Solution: $-\dfrac{3}{10} + \left(-\dfrac{1}{10}\right) = \dfrac{-3 + (-1)}{10} = -\dfrac{4}{10}$ ◄ **Note** Because the fractions have the same sign, we add and keep the same sign.

$\qquad\qquad = -\dfrac{2 \cdot 2}{2 \cdot 5} = -\dfrac{2}{5}$ Simplify to lowest terms by dividing out the common factor, 2.

c. $\dfrac{7}{9} + \left(-\dfrac{4}{9}\right)$

Note Because the two addends have different signs, we subtract the smaller absolute value from the greater absolute value and keep the sign of the number with the greater absolute value.

Solution: $\dfrac{7}{9} + \left(-\dfrac{4}{9}\right) = \dfrac{7 + (-4)}{9} = \dfrac{3}{9}$ ◄

$\qquad\qquad = \dfrac{3}{3 \cdot 3} = \dfrac{1}{3}$ Simplify to lowest terms by dividing out the common factor, 3.

Your Turn 4   Add.

a. $\dfrac{3}{8} + \dfrac{1}{8}$ **b.** $-\dfrac{5}{9} + \left(-\dfrac{2}{9}\right)$ **c.** $-\dfrac{7}{12} + \dfrac{5}{12}$

Adding Fractions with Different Denominators

If the denominators are different, we must first find a common denominator. Recall that in Section 1.2, we wrote equivalent fractions with the LCD. Consider the addition $\dfrac{3}{4} + \dfrac{1}{6}$. The LCD of 4 and 6 is 12. We now write each fraction as an equivalent fraction with the LCD.

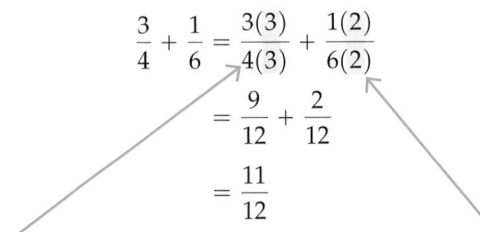

$$\dfrac{3}{4} + \dfrac{1}{6} = \dfrac{3(3)}{4(3)} + \dfrac{1(2)}{6(2)}$$

$$= \dfrac{9}{12} + \dfrac{2}{12}$$

$$= \dfrac{11}{12}$$

Note To write $\dfrac{3}{4}$ as an equivalent fraction with 12 as the denominator, we multiply the numerator and denominator by 3.

Note To write $\dfrac{1}{6}$ as an equivalent fraction with 12 as the denominator, we multiply the numerator and denominator by 2.

Procedure **Adding Fractions**

To add fractions with different denominators:
1. Write each fraction as an equivalent fraction with the LCD.
2. Add the numerators and keep the LCD.
3. Simplify.

Answers to Your Turn 4
a. $\dfrac{1}{2}$ **b.** $-\dfrac{7}{9}$ **c.** $-\dfrac{1}{6}$

Example 5 Add.

a. $\dfrac{1}{6} + \dfrac{5}{9}$

Solution: The LCD of 6 and 9 is 18.

$$\dfrac{1}{6} + \dfrac{5}{9} = \dfrac{1(3)}{6(3)} + \dfrac{5(2)}{9(2)}$$ Write equivalent fractions with 18 in the denominator.

$$= \dfrac{3}{18} + \dfrac{10}{18}$$ Add numerators and keep the common denominator.

$$= \dfrac{3 + 10}{18}$$ Because the addends have the same sign, we add and keep the same sign.

$$= \dfrac{13}{18}$$

b. $-\dfrac{4}{5} + \dfrac{2}{3}$

Solution: The LCD of 5 and 3 is 15.

$$-\dfrac{4}{5} + \dfrac{2}{3} = -\dfrac{4(3)}{5(3)} + \dfrac{2(5)}{3(5)}$$ Write equivalent fractions with 15 in the denominator.

$$= -\dfrac{12}{15} + \dfrac{10}{15}$$ Add numerators and keep the common denominator.

$$= \dfrac{-12 + 10}{15}$$ Because the addends have different signs, we subtract and keep the sign of the number with the greater absolute value.

$$= -\dfrac{2}{15}$$

c. $-\dfrac{7}{12} + \dfrac{7}{30}$

Solution: The LCD of 12 and 30 is 60.

$$-\dfrac{7}{12} + \dfrac{7}{30} = -\dfrac{7(5)}{12(5)} + \dfrac{7(2)}{30(2)}$$ Write equivalent fractions with 60 as the denominator.

$$= -\dfrac{35}{60} + \dfrac{14}{60}$$ Add numerators and keep the common denominator.

$$= \dfrac{-35 + 14}{60}$$

$$= -\dfrac{21}{60}$$ Simplify to lowest terms.

$$= -\dfrac{3 \cdot 7}{2 \cdot 2 \cdot 3 \cdot 5}$$ Write the numerator and denominator in terms of prime factors.

$$= -\dfrac{3 \cdot 7}{2 \cdot 2 \cdot 3 \cdot 5}$$ Divide out the common factor of 3.

$$= -\dfrac{7}{20}$$

Answers to Your Turn 5

a. $\dfrac{29}{60}$ **b.** $\dfrac{27}{40}$ **c.** $-\dfrac{2}{9}$

Your Turn 5 Add.

a. $\dfrac{1}{15} + \dfrac{5}{12}$

b. $\dfrac{7}{8} + \left(-\dfrac{1}{5}\right)$

c. $-\dfrac{7}{18} + \dfrac{5}{30}$

Adding Decimal Numbers

Now let's review adding decimal numbers, which are also a part of the set of rational numbers. Recall that to add decimal numbers, we add like place values.

Example 6   Angela has a balance of $578.26 and incurs a debt of $92.45. What is Angela's new balance?

Solution: A debt of $92.45 is $-\$92.45$. Her balance is

$$578.26 + (-92.45) = \$485.81$$

Your Turn 6   Daryl has a balance of $-\$452.75$ on a credit card, and he makes a payment of $120. What is his new balance?

Objective 2 Find the additive inverse of a number.

What happens if we add two numbers that have the same absolute value but different signs, such as $5 + (-5)$? In money terms, this is like making a $5 payment toward a debt of $5. Notice that the payment pays off the debt so that the balance is 0.

$$5 + (-5) = 0$$

Because their sum is 0, we say that 5 and -5 are **additive inverses**, or opposites.

Definition   Additive inverses: Two numbers whose sum is 0.

Example 7   Find the additive inverse of the given numbers.

a. 9

Answer: -9 because $9 + (-9) = 0$

b. -7

Answer: 7 because $-7 + 7 = 0$

c. 0

Answer: 0 because $0 + 0 = 0$

Your Turn 7   Find the additive inverse of the given numbers.

a. -12 **b.** 0.67   **c.** $-\dfrac{3}{8}$

A minus sign can be used to indicate additive inverse.

Example 8   Simplify.

a. $-(-8)$

Answer: $-(-8)$ indicates that we need to find the additive inverse of -8. So $-(-8) = 8$.

b. $-|6|$

Answer: $-|6|$ indicates that we need to find the additive inverse of the absolute value of 6. So $-|6| = -6$.

c. $-|-4|$

Answer: $-|-4|$ indicates that we need to find the additive inverse of the absolute value of -4. So $-|-4| = -4$.

Answer to Your Turn 6
$-\$332.75$

Answers to Your Turn 7
a. 12   **b.** -0.67 **c.** $\dfrac{3}{8}$

| Your Turn 8 | Simplify.

a. $-(-5)$ **b.** $-|8|$ **c.** $-|-3|$

Objective 3 Subtract rational numbers.

Now let's turn our attention to subtraction. The parts of subtraction are

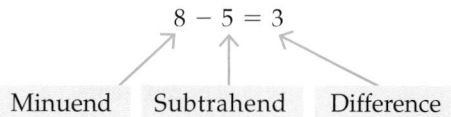

$$8 - 5 = 3$$

Minuend   Subtrahend   Difference

Recall that when we add two numbers that have different signs, we actually subtract. This implies that subtraction and addition are related. In fact, every subtraction statement can be written as an equivalent addition statement. Notice that the subtraction statement $8 - 5 = 3$ and the addition statement $8 + (-5) = 3$ are equivalent: They are both ways of finding the balance given an \$8 deposit with a \$5 debt. Note that in writing $8 - 5$ as an equivalent addition $8 + (-5)$, we change the operation sign from a minus sign to a plus sign and change the 5 to its additive inverse, -5.

Any subtraction statement can be written as an equivalent addition statement using this procedure. Consider $8 - (-5)$, which represents an \$8 deposit subtracting a debt of \$5. What does it mean to subtract a debt? If a bank records a debt and then realizes it made a mistake, it must reverse that debt. Subtracting, or reversing, a debt means that the bank must record a deposit to correct the mistake. In math terms, this means that $8 - (-5)$ is equivalent to $8 + 5$, which equals 13. Notice that this still follows the process we just discussed.

$$8 - (-5)$$

Change the operation
from minus to plus.

Change the subtrahend
to its additive inverse.

$$= 8 + 5$$
$$= 13$$

> **Procedure** **Rewriting Subtraction**
>
> To write a subtraction statement as an equivalent addition statement, change the operation symbol from a minus sign to a plus sign and change the subtrahend to its additive inverse.

| Example 9 | Subtract.

a. $-15 - (-6)$

Solution: Write the subtraction as an equivalent addition.

$$-15 - (-6)$$

Change the operation
from minus to plus.

Change the subtrahend
to its additive inverse.

$$= -15 + 6$$
$$= -9$$

Note We can think of this as subtracting a debt of \$6 from a debt of \$15. Subtracting a debt from a debt is equivalent to making a deposit or payment against the existing debt balance. So this decreases the amount of debt.

b. $-\dfrac{2}{3} - \dfrac{3}{4}$

Solution: Write an equivalent addition; then find a common denominator and continue the process of adding fractions.

$$-\frac{2}{3} - \frac{3}{4}$$

Change the operation from minus to plus. ⟶ ⟵ Change the subtrahend to its additive inverse.

$$= -\frac{2}{3} + \left(-\frac{3}{4}\right)$$

$$= -\frac{2(4)}{3(4)} + \left(-\frac{3(3)}{4(3)}\right)$$ Write equivalent fractions with the common denominator, 12.

Note If we begin with a debt and subtract a positive amount, we go deeper in debt.

$$= -\frac{8}{12} + \left(-\frac{9}{12}\right)$$

▶ $$= -\frac{17}{12}$$

c. $5.04 - 8.01$

Solution: Write an equivalent addition statement.

Note We begin with a positive balance of $5.04 and subtract $8.01. Because we are subtracting more than we have, the result is negative.

▶
$$5.04 - 8.01$$
$$= 5.04 + (-8.01)$$
$$= -2.97$$ ◀

Note To calculate the sum, the number with the smaller absolute value must be subtracted from the number with the greater absolute value.

$$\begin{array}{r} 8.01 \\ -5.04 \\ \hline 2.97 \end{array}$$

Because the number with the greater absolute value is negative, the result is negative.

Your Turn 9 Subtract.

a. $25 - (-17)$ **b.** $-\frac{3}{8} - \left(-\frac{1}{3}\right)$ **c.** $0.06 - 4.02$

Example 10 On Mars, the daytime high temperature can reach $-15°C$, while at night, the temperature can plummet to $-140°C$. What is the difference between the high and low temperatures?

Solution: *Difference* indicates that we subtract the temperatures. Because the wording in the problem is to find the "difference between the high and low temperatures," we arrange the subtraction with the low temperature subtracted from the high temperature.

$$-15 - (-140) = -15 + 140$$ Write as an equivalent addition.
$$= 125$$ Add.

Answer: The difference between the high and low temperatures is 125°C.

Answers to Your Turn 9

a. 42   **b.** $-\dfrac{1}{24}$ **c.** −3.96

Answer to Your Turn 10
34.4°C

Your Turn 10 In an experiment, a mixture begins at a temperature of 5.8°C. The mixture is then cooled to a temperature of −28.6°C. Find the difference between the initial and final temperatures.

1.3 Exercises
For Extra Help   MyMathLab®

Note: Exercises marked with a ★ represent challenging exercises.

Objective 1

Prep Exercise 1 Explain the difference between the commutative property of addition and the associative property of addition.

The commutative property of addition changes the order, while the associative property of addition changes the grouping.

Prep Exercise 2 Why is 0 called the additive identity?

Adding 0 to a number or an expression does not change the identity of the number or expression.

For Exercises 1–12, indicate whether each equation illustrates the additive identity, commutative property of addition, associative property of addition, or additive inverse. See Examples 1 and 7.

1. $5 + 6 = 6 + 5$
Commutative property of addition

2. $-3 + 7 = 7 + (-3)$
Commutative property of addition

3. $0 + 8 = 8$
Additive identity

4. $-5 + 0 = -5$
Additive identity

5. $0.8 + (-0.8) = 0$
Additive inverse

6. $-\dfrac{4}{9} + \dfrac{4}{9} = 0$
Additive inverse

7. $(3 + 5) + 2 = 3 + (5 + 2)$
Associative property of addition

8. $6 + (2 + 3) = (6 + 2) + 3$
Associative property of addition

9. $-8 + (7 + 3) = (7 + 3) + (-8)$
Commutative property of addition

10. $(5 + 3) - 4 = -4 + (5 + 3)$
Commutative property of addition

11. $6.3 + (2.1 - 2.1) = 6.3 + 0$
Additive inverse

12. $(-4.6 + 4.6) + 9.5 = 0 + 9.5$
Additive inverse

Prep Exercise 3 In your own words, explain how to add two numbers that have the same sign.

When adding two numbers that have the same sign, add their absolute values and keep the same sign.

Prep Exercise 4 In your own words, explain how to add two numbers that have different signs.

When adding two numbers that have different signs, subtract the smaller absolute value from the greater absolute value and keep the sign of the number with the greater absolute value.

Prep Exercise 5 In your own words, explain the process of adding or subtracting two fractions with different denominators.

To add (or subtract) fractions, (1) write each fraction as an equivalent fraction with the LCD, (2) add (or subtract) the numerators and keep the LCD, and (3) simplify.

For Exercises 13–48, add. See Examples 2–6.

13. $8 + 13$
21

14. $15 + 7$
22

15. $-6 + (-12)$
-18

16. $-5 + (-7)$
-12

17. $-4 + 13$
9

18. $-5 + 16$
11

19. $-14 + 5$
-9

20. $-17 + 8$
-9

21. $27 + (-13)$
14

22. $29 + (-7)$
22

23. $-28 + 12$
-16

24. $-16 + 13$
-3

25. $\dfrac{5}{8} + \dfrac{1}{8}$
$\dfrac{3}{4}$

26. $\dfrac{9}{16} + \dfrac{5}{16}$
$\dfrac{7}{8}$

27. $-\dfrac{4}{9} + \left(-\dfrac{2}{9}\right)$
$\dfrac{2}{3}$

28. $-\dfrac{3}{5} + \left(-\dfrac{1}{5}\right)$
$\dfrac{4}{5}$

29. $\dfrac{1}{6} + \left(-\dfrac{5}{6}\right)$
$\dfrac{2}{3}$

30. $-\dfrac{9}{14} + \dfrac{3}{14}$
$\dfrac{3}{7}$

31. $\dfrac{3}{4} + \dfrac{1}{6}$
$\dfrac{11}{12}$

32. $\dfrac{1}{4} + \dfrac{7}{8}$
$\dfrac{9}{8}$

33. $-\dfrac{1}{12} + \left(-\dfrac{2}{3}\right)$
$-\dfrac{3}{4}$

34. $-\dfrac{2}{5} + \left(-\dfrac{3}{20}\right)$
$-\dfrac{11}{20}$

35. $-\dfrac{5}{6} + \dfrac{4}{21}$
$-\dfrac{9}{14}$

36. $-\dfrac{5}{16} + \dfrac{3}{12}$
$-\dfrac{1}{16}$

37. $0.21 + 0.05$
0.26

38. $0.06 + 0.17$
0.23

39. $-0.18 + 6.7$
6.52

40. $-15.81 + 4.28$
-11.53

41. $-0.28 + (-4.1)$
-4.38

42. $-7.8 + (-9.16)$
-16.96

★**43.** $-42 + |-14|$
-28

★**44.** $-31 + |-54|$
23

★**45.** $|-2.4| + |-0.78|$
3.18

★**46.** $|-0.6| + |-9.1|$
9.7

★**47.** $\left|-\dfrac{3}{8}\right| + \left|\dfrac{5}{6}\right|$
$\dfrac{29}{24}$

★**48.** $\left|-\dfrac{4}{5}\right| + \left|\dfrac{3}{4}\right|$
$\dfrac{31}{20}$

Objective 2

Prep Exercise 6 Explain why 4 and -4 are additive inverses. The sum of 4 and -4 is 0.

For Exercises 49–62, find the additive inverse. See Example 7.

49. 5
-5

50. 7
-7

51. -12
12

52. -6
6

53. 0
0

54. -9
9

55. $\dfrac{5}{6}$
$-\dfrac{5}{6}$

56. $-\dfrac{6}{17}$
$\dfrac{6}{17}$

57. -0.29
0.29

58. 2.8
-2.8

59. $-x$
x

60. b
$-b$

61. $\dfrac{m}{n}$
$-\dfrac{m}{n}$

62. $-\dfrac{a}{b}$
$\dfrac{a}{b}$

For Exercises 63–70, simplify. See Example 8.

63. $-(-2)$
2

64. $-(-15)$
15

★**65.** $-(-(-4))$
-4

★**66.** $-(-(-1))$
-1

67. $-|4|$
-4

68. $-|10|$
-10

69. $-|-12|$
-12

70. $-|-5|$
-5

Objective 3

Prep Exercise 7 Explain how to write a subtraction statement as an equivalent addition statement. To write a subtraction statement as an equivalent addition statement, change the operation symbol from a minus sign to a plus sign and change the subtrahend to its additive inverse.

Prep Exercise 8 Explain why $6 - 5$ and $5 - 6$ have different answers, whereas $-6 - 5$ and $-5 - 6$ have the same answer. (*Hint:* Think about the properties of addition.) Subtraction is not commutative, so $6 - 5$ and $5 - 6$ are not equivalent. However, $-6 - 5$ and $-5 - 6$ are equivalent because $-6 - 5 = -6 + (-5)$ and by the commutative property of addition, $-6 + (-5) = -5 + (-6)$, which is equivalent to $-5 - 6$.

For Exercises 71–94, subtract. See Example 9.

71. $6 - 15$
-9

72. $8 - 20$
-12

73. $-4 - 9$
-13

74. $-7 - 15$
-22

75. $4 - (-3)$
7

76. $6 - (-7)$
13

77. $-8 - (-2)$
-6

78. $-13 - (-6)$
-7

79. $-\dfrac{1}{5} - \left(-\dfrac{1}{5}\right)$
0

80. $-\dfrac{3}{4} - \left(-\dfrac{3}{4}\right)$
0

81. $\dfrac{7}{10} - \left(-\dfrac{3}{5}\right)$
$\dfrac{13}{10}$

82. $\dfrac{3}{8} - \left(-\dfrac{5}{6}\right)$
$\dfrac{29}{24}$

83. $-\dfrac{4}{5} - \left(-\dfrac{2}{7}\right)$
$-\dfrac{18}{35}$

84. $-\dfrac{1}{2} - \left(-\dfrac{1}{3}\right)$
$-\dfrac{1}{6}$

85. $4.01 - 3.65$
0.36

86. 8.1 − 4.76
3.34

87. 0.07 − 5.82
−5.75

88. 0.107 − 5.802
−5.695

89. −6.1 − (−4.5)
−1.6

90. −7.1 − (−2.3)
−4.8

★ **91.** −|−4| − |−6|
−10

★ **92.** −|−9| − |−12|
−21

★ **93.** |6.2| − |−7.1|
−0.9

★ **94.** |4.6| − |−7.3|
−2.7

For Exercises 95–110, solve. See Example 10.

95. The Walt Disney Company's third-quarter financial report for 2012 contains the following information. Find the net income. $2,036,000,000

Third-quarter report for the quarter ending June 30, 2012

Income (in millions)	Expenditures (in millions)
Revenues = $11,088	Costs and expenses = $8128
Equity in the income of investees = $169	Net interest expense = $93
	Income taxes = $993
	Restructuring and impairment charges = $7

96. Following is a balance sheet of income and expenditures for a small business. Find the net profit or loss. −$834.26, which indicates a loss

June 2012 balance sheet

Income	Expenditures
Revenue = $24,572.88	Materials = $1545.75
Dividends = $1284.56	Lease = $2700
	Utilities = $865.45
	Employee wages = $21,580.50

97. In engineering, the resultant force on an object is the sum of all forces acting on the object. The diagram shows the forces acting on a steel beam. Find the resultant force. What does the sign of the resultant force indicate? −268.2 N; the negative indicates that the beam is moving downward.

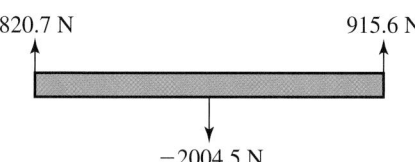

820.7 N   915.6 N

−2004.5 N

Of Interest

N is an abbreviation for the metric unit of force called the newton. The unit is named after Sir Isaac Newton (1642–1727) in honor of his development of a theory of forces.

98. A family's taxable income is its income less deductions. Following is a table that shows the Brendel family's income and deductions for the 2012 tax year. What is the Brendel family's taxable income? $47,770.99

Income	Deductions
Mr. Brendel = $31,672.88	Mortgage interest = $6545.75
Mrs. Brendel = $32,284.56	Charitable donations = $1200
Dividends = $124.75	Medical expenses = $165.45
Miscellaneous income = $2400	Exemptions = $10,800

99. On July 24, 2012, the closing price of Apple's stock was $600.92. On July 25, 2012, the closing price was $574.97. Find the difference in closing price from July 24 to July 25. $25.95

100. On July 24, 2012, the closing price of Microsoft's stock was $29.15. On July 25, 2012, the closing price was $28.83. Find the difference in closing price from July 24 to July 25. $0.32

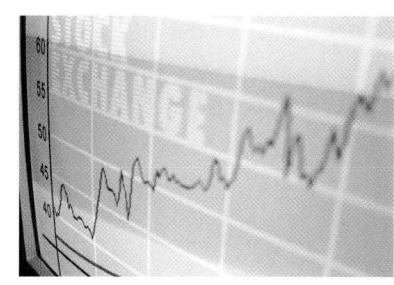

101. On July 11, 2012, the Dow Jones Industrial Average (DJIA) closed at 12,604.53. If the value had a change of −48.59 from the previous trading date's closing value, what was the closing value on July 10, 2012? 12,653.12

102. On July 11, 2012, an investor found that the NASDAQ closed at 2887.98, which was a change of −14.35 from the previous day's closing value. What was the closing value on July 10, 2012? 2902.33

103. The temperature of liquid nitrogen is −208°C. When an apple is placed in the liquid, the apple raises the temperature of the liquid to its boiling point of −196°C. Write an expression that describes the difference between the boiling point and the initial temperature; then calculate the difference. −196 − (−208); 12

104. Absolute zero is the temperature at which molecular motion is at a minimum. This temperature is −273.15°C. A piece of metal is cooled to −256.5°C. Write an expression that describes the difference between the piece of metal's current temperature and absolute zero; then calculate the difference. −256.5 − (−273.15); 16.65

Of Interest

Absolute zero refers to 0 on the Kelvin scale, which is the scale most scientists use to measure temperatures. Increments on the Kelvin scale are equal to increments on the Celsius scale, so a change of 1 K corresponds to a change of 1°C.

Use the following line graph to answer Exercises 105 and 106. The graph shows the mean composite ACT score of entering college freshmen from 1985 to 2010.

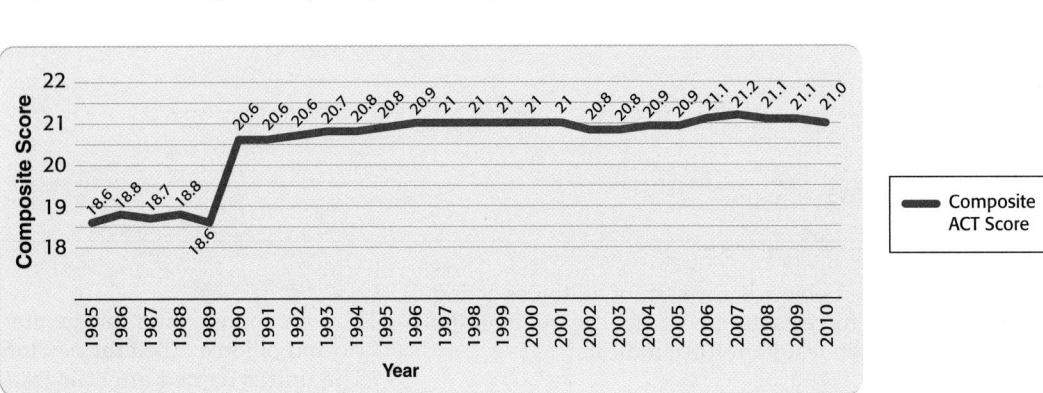

(*Source:* U.S. Department of Education.)

105. a. Write an expression that describes the difference between the mean composite scores in 1989 and 1986.
   18.6 − 18.8

   b. Calculate the difference.
   −0.2

   c. What does the sign of the difference indicate about how the scores changed?
   The negative difference indicates that the mean composite score in 1989 was less than the score in 1986.

106. a. Write an expression that describes the difference between the mean composite scores in 2010 and 1986.
   21.0 − 18.8

   b. Calculate the difference.
   2.2

   c. What does the sign of the difference indicate about how the scores changed?
   The positive difference indicates that the mean composite score in 2010 was greater than the score in 1986.

Use the following bar graph to answer Exercises 107–110. The graph shows the mean annual income by level of educational attainment in 2010.

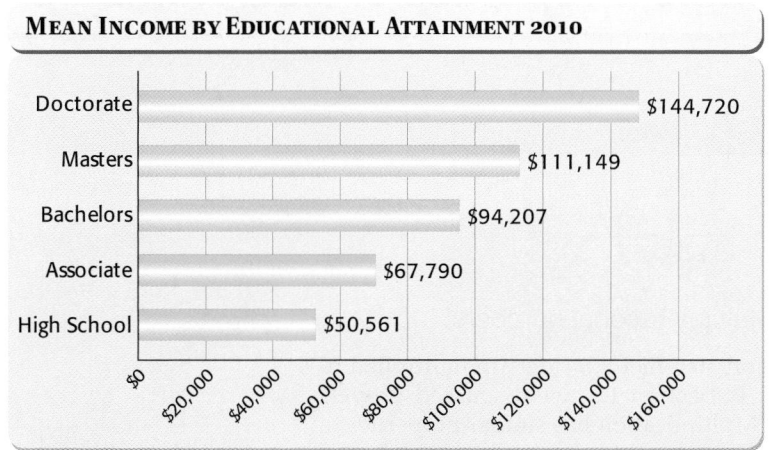

MEAN INCOME BY EDUCATIONAL ATTAINMENT 2010

Doctorate $144,720
Masters $111,149
Bachelors $94,207
Associate $67,790
High School $50,561

(*Source:* U.S. Census Bureau, Current Population Survey, Annual Social and Economic Supplements.)

107. What was the difference in mean salary of a person with an associate degree and a person with a high school diploma? $17,229

108. What was the difference in mean salary between a person with a bachelors and a person with an associate degree? $26,417

109. Which degree earned the greatest increase in mean salary compared with the next lower degree? How much was that increase in mean salary? Doctorate; $33,571

110. Which degree earned the least increase in mean salary compared with the next lower degree? How much was that increase in mean salary? Masters; $16,942

Puzzle Problem Fill in each square with a number 1 through 9 so that the sum of the numbers in each row, each column, and the two diagonals is the same.

```
2 9 4
7 5 3
6 1 8
```

Review Exercises

Exercises 1–4      **Expressions**

[1.1] 1. Write a set containing the last names of the first four presidents of the United States.
{Washington, Adams, Jefferson, Madison}

[1. 1] 2. Simplify: $|-7|$ 7

[1. 2] 3. Find the prime factorization of 100. $2 \cdot 2 \cdot 5 \cdot 5$

[1. 2] 4. A company report indicates that 78 employees out of 91 have medical flexible spending accounts. What fraction of the employees have medical flexible spending accounts? Express the fraction in lowest terms.
$\frac{6}{7}$

Exercises 5–6   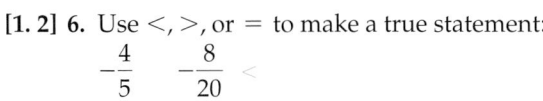   **Equations and Inequalities**

[1. 2] 5. Find the missing number that makes $\frac{40}{48} = \frac{?}{6}$ true.   5

[1. 2] 6. Use $<, >,$ or $=$ to make a true statement:
$-\frac{4}{5}$ _____ $-\frac{8}{20}$ $<$

1.4 Multiplying and Dividing Real Numbers; Properties of Real Numbers

Objectives

1 Multiply rational numbers.
2 Find the multiplicative inverse of a number.
3 Divide rational numbers.

Warm-up

[1.2] **1.** Simplify $\dfrac{15}{24}$ to lowest terms.

[1.3] **2.** Add: $-5 + (-12)$

[1.3] **3.** Subtract: $-\dfrac{2}{3} - \left(-\dfrac{5}{6}\right)$

Objective 1 Multiply rational numbers.

In a multiplication statement, *factors* are multiplied to equal a *product*. In Section 1.3, we discussed properties of addition. Multiplication has similar properties to addition. The following table shows the properties of multiplication.

$$2 \cdot 3 = 6$$

Factors   Product

Properties of Multiplication	Symbolic Form	Word Form
Multiplicative property of 0	$0 \cdot a = 0$	The product of a number multiplied by 0 is 0.
Multiplicative identity	$1 \cdot a = a$	The product of a number multiplied by 1 is the number.
Commutative property of multiplication	$ab = ba$	Changing the order of factors does not affect the product.
Associative property of multiplication	$a(bc) = (ab)c$	Changing the grouping of three or more factors does not affect the product.
Distributive property of multiplication over addition	$a(b + c) = ab + ac$	A sum multiplied by a factor is equal to the sum of that factor multiplied by each addend.

Note Because $b - c = b + (-c)$, we can say $a(b - c) = a(b + (-c)) = ab + (-ac) = ab - ac$. Although the parentheses contain subtraction, the property can still be called the distributive property ▶ of multiplication over addition.

Example 1 Give the name of the property of multiplication that is illustrated by each equation.

a. $4(-3) = -3 \cdot 4$

Answer: Commutative property of multiplication because the order of the factors is different.

b. $6(-2 \cdot 7) = [6(-2)] \cdot 7$

Answer: Associative property of multiplication because the grouping of factors is different.

c. $5(8 - 3) = 5 \cdot 8 - 5 \cdot 3$

Answer: Distributive property of multiplication over addition.

Your Turn 1 Give the name of the property of multiplication that is illustrated by each equation.

a. $-8 \cdot 0 = 0$ **b.** $9 \cdot 1 = 9$ **c.** $-2(7 + 4) = -2 \cdot 7 - 2 \cdot 4$

Answers to Your Turn 1
a. multiplicative property of 0
b. multiplicative identity
c. distributive property of multiplication over addition

Answers to Warm-up
1. $\dfrac{5}{8}$
2. -17
3. $\dfrac{1}{6}$

Multiplying Numbers with Different Signs

To determine the rules for multiplying signed numbers, consider the following pattern. Pay attention to the product as we decrease the second factor.

$2 \cdot 4 = 8$ As we decrease the second factor by 1, the product
$2 \cdot 3 = 6$ ← decreases by 2, from 8 to 6.

$2 \cdot 2 = 4$ Notice that the same pattern continues with the product
$2 \cdot 1 = 2$ decreasing by 2 each time we decrease the factor by 1.

$2 \cdot 0 = 0$ Notice that when we decrease the 0 factor to -1, we continue
                     the pattern and decrease the product by 2 so that it is -2.
$2 \cdot (-1) = -2$ ← Decreasing the factor by 1 again decreases the product by
$2 \cdot (-2) = -4$ 2, resulting in -4.

This pattern suggests that multiplying a positive number by a negative number equals a negative product. Note that by the commutative property of multiplication, we can exchange the factors and conclude that multiplying a negative number times a positive number also equals a negative product.

$$2 \cdot (-3) = (-3) \cdot 2 = -6$$

> **Rule** **Multiplying Two Numbers with Different Signs**
> When multiplying two numbers that have different signs, the product is negative.

| **Example 2** | Multiply. |

a. $9(-8)$

Solution: $9(-8) = -72$

b. $(-12)3$

Solution: $(-12)3 = -36$

Warning Make sure you see the difference between $9(-8)$, which indicates multiplication, and $9 - 8$, which indicates subtraction.

| **Your Turn 2** | Multiply. |

a. $13(-7)$ **b.** $-5 \cdot 9$

Multiplying Numbers with the Same Sign

Now let's examine multiplying numbers with the same sign. Consider the following pattern to determine the rule.

$(-2) \cdot 4 = -8$ We have already established that the product of a
                         negative number and a positive number is negative.

$(-2) \cdot 3 = -6$ ← As we decrease the positive factor by 1, the product
                       increases by 2, from -8 to -6.

$(-2) \cdot 2 = -4$ Notice that the same pattern continues with the
$(-2) \cdot 1 = -2$ product increasing by 2 each time we decrease the
                         factor by 1.

$(-2) \cdot 0 = 0$ To continue the same pattern, when we decrease
$(-2) \cdot (-1) = 2$ ← the 0 factor to -1, we must continue to increase
$(-2) \cdot (-2) = 4$ the product by 2. This means that the product must
                         become positive.

Answers to Your Turn 2
a. -91 **b.** -45

This pattern indicates that the product of two negative numbers is a positive number. We have already seen that the product of two positive numbers is a positive number. So we can draw the following conclusion:

> **Rule** **Multiplying Two Numbers with the Same Sign**
> When multiplying two numbers that have the same sign, the product is positive.

Example 3 Multiply.

a. $-6(-8)$

Solution: $-6(-8) = 48$

b. $(-9)(-7)$

Solution: $(-9)(-7) = 63$

Your Turn 3 Multiply.

a. $-6(-9)$ **b.** $(-3)(-13)$

Multiply More than Two Numbers

Suppose we have to multiply more than two numbers, such as $(-2)(-3)(5)$. The associative property of multiplication allows us to group the factors any way we like.

$$(-2)(-3)(5) = 6(5) \qquad \text{Multiply the first two factors.}$$
$$= 30$$

or

$$(-2)(-3)(5) = (-2)(-15) \qquad \text{Multiply the second and third factors.}$$
$$= 30$$

Notice that no matter which way we choose to multiply the factors, the result is the same. Also notice that the result is positive because there are two negative factors involved in the multiplication. What if we have three negative factors, as in $(-2)(-3)(-5)$?

$$(-2)(-3)(-5) = 6(-5) \qquad \text{Multiply the first two factors.}$$
$$= -30$$

or

$$(-2)(-3)(-5) = (-2)(15) \qquad \text{Multiply the second and third factors.}$$
$$= -30$$

With three negative factors, the result is negative. This suggests the following rule:

> **Rule** **Multiplying with Negative Factors**
> The product of an even number of negative factors is positive, whereas the product of an odd number of negative factors is negative.

Example 4 Multiply.

a. $(-1)(-1)(-2)(6)$

Solution: Because there are three negative factors (an odd number of negative factors), the result is negative.

$$(-1)(-1)(-2)(6) = -12$$

Answers to Your Turn 3
a. 54 **b.** 39

b. $(-1)(-3)(-7)(2)(-4)$

Solution: Because there are four negative factors (an even number of negative factors), the result is positive.

$$(-1)(-3)(-7)(2)(-4) = 168$$

Your Turn 4 Multiply.

a. $(-3)(4)(-5)(2)$ **b.** $(6)(-5)(-1)(2)(-1)$

Multiplying Fractions

On a ruler, note that half of a fourth is an eighth. When preceded by a fraction, the word *of* means multiply. So the sentence would translate as follows:

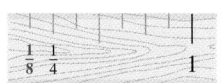

Half of a fourth is an eighth.

$$\frac{1}{2} \quad \cdot \quad \frac{1}{4} \quad = \quad \frac{1}{8}$$

Notice that the numerators multiply to equal the numerator and the denominators multiply to equal the denominator. This suggests the following rule for multiplying fractions:

> **Rule Multiplying Fractions**
>
> $$\frac{a}{b} \cdot \frac{c}{d} = \frac{ac}{bd}, \text{ where } b \neq 0 \text{ and } d \neq 0.$$

Multiplying and Simplifying

Instructor Note We use primes in an effort to be clear about what factors divide out and to prepare for rational expressions in Chapter 7, where we use the same approach. We show the other styles to help students make connections to methods with which they may be familiar.

In mathematics, we are expected to simplify results to lowest terms. Consider $\frac{3}{4} \cdot \frac{5}{6}$. We can multiply first, then simplify.

$$\frac{3}{4} \cdot \frac{5}{6} = \frac{15}{24} = \frac{3 \cdot 5}{2 \cdot 2 \cdot 2 \cdot 3} = \frac{5}{8}$$

◄ **Note** After multiplying, we divide out the common factor, 3.

Or as an alternative, we can divide out the common factors before multiplying the fractions. In our example, we can divide out the common factor of 3 and then multiply.

$$\frac{3}{4} \cdot \frac{5}{6} = \frac{3}{2 \cdot 2} \cdot \frac{5}{2 \cdot 3} = \frac{5}{8}$$

◄ **Note** Some people prefer to divide out the common factors without breaking down to primes, like this

$$\frac{\overset{1}{3}}{4} \cdot \frac{5}{\underset{2}{6}} = \frac{5}{8}$$

We can divide out common factors before or after multiplying because every factor in the original numerators becomes a factor in the product's numerator and every factor in the original denominators becomes a factor in the product's denominator. Now let's put this together with the sign rules for multiplication.

Connection Using the slash marks style, Example 5(a) looks like this:

$$-\frac{\overset{1}{2}}{3} \cdot \frac{7}{\underset{4}{8}} = -\frac{7}{12}$$

Notice that we are still dividing out a factor of 2 in the 2 and 8.

Answers to Your Turn 4
a. 120 **b.** −60

Example 5 Multiply.

a. $-\frac{2}{3} \cdot \frac{7}{8}$

Solution: $-\frac{2}{3} \cdot \frac{7}{8} = -\frac{2}{3} \cdot \frac{7}{2 \cdot 2 \cdot 2}$ *Divide out the common factor, 2.*

$$= -\frac{7}{12}$$ *Because we are multiplying two numbers that have different signs, the product is negative.*

b. $-\dfrac{4}{15} \cdot \left(-\dfrac{9}{16}\right) \cdot \dfrac{14}{15}$

Solution: $-\dfrac{4}{15} \cdot \left(-\dfrac{9}{16}\right) \cdot \dfrac{14}{15} = -\dfrac{2 \cdot 2}{3 \cdot 5} \cdot \left(-\dfrac{3 \cdot 3}{2 \cdot 2 \cdot 2 \cdot 2}\right) \cdot \dfrac{2 \cdot 7}{3 \cdot 5}$ Divide out the common factors.

$= \dfrac{7}{50}$ Because there is an even number of negative factors, the product is positive.

Your Turn 5   Multiply.

a. $\dfrac{5}{8} \cdot \left(-\dfrac{3}{10}\right)$ b. $-\dfrac{4}{5} \cdot \left(\dfrac{1}{6}\right) \cdot \left(-\dfrac{3}{9}\right)$

Multiplying Decimal Numbers

Recall that decimal numbers name fractions with denominators that are powers of 10. In the problem $0.3 \cdot 0.7$, we could multiply by replacing the decimal factors with their fraction equivalents.

$$0.3 \cdot 0.7 = \dfrac{3}{10} \cdot \dfrac{7}{10} = \dfrac{21}{100} = 0.21$$

Because we multiply the denominators, if we multiply tenths times tenths, the product will be hundredths. Notice that if we count the number of decimal places in the factors, this corresponds to the total places in the product.

$$
\begin{array}{rll}
0.7 & \longrightarrow & 1 \text{ place} \\
\times\, 0.3 & \longrightarrow & +1 \text{ place} \\
\hline
0.21 & \longrightarrow & 2 \text{ places}
\end{array}
$$

Likewise, hundredths (2 places) times tenths (1 place) equals thousandths (3 places).

$$0.03 \cdot 0.7 = \dfrac{3}{100} \cdot \dfrac{7}{10} = \dfrac{21}{1000} = 0.021 \quad \text{or}$$

$$
\begin{array}{rll}
0.7 & \longrightarrow & 1 \text{ place} \\
\times\, 0.03 & \longrightarrow & +\, 2 \text{ places} \\
\hline
0.021 & \longrightarrow & 3 \text{ places}
\end{array}
$$

Procedure   **Multiplying Decimal Numbers**

To multiply decimal numbers:
1. Multiply as if they were whole numbers.
2. Place the decimal in the product so that it has the same number of decimal places as the total number of decimal places in the factors.

Example 6   Multiply.

a. $(-5.6)(0.03)$

Solution: First, we calculate the value and disregard the signs for now.

$$
\begin{array}{rll}
0.03 & & 2 \text{ places} \\
\times\ \ 5.6 & & +1 \text{ place} \\
\hline
0\,1\,8 & & \\
+\,0\,1\,5\ \ & & \\
\hline
0.1\,6\,8 & \longleftarrow & 3 \text{ places}
\end{array}
$$

Answer: -0.168 ◀ **Note:** When we multiply two numbers with different signs, the product is negative.

Answers to Your Turn 5
a. $-\dfrac{3}{16}$ b. $\dfrac{2}{45}$

b. $(-2)(4.3)(1.8)(-7.1)$

Solution: First, we calculate the value and disregard the signs for now. Multiply from left to right.

$$\underset{\underset{(2)(4.3)}{\downarrow}}{} \quad \underset{\underset{(8.6)(1.8)}{\downarrow}}{}$$

$$(2)(4.3)(1.8)(7.1) = (8.6)(1.8)(7.1) = (15.48)(7.1) = 109.908$$

Answer: 109.908   ◄ **Note:** The product of an even number of negative factors is positive. The factors have a total of three decimal places, so the product has three decimal places.

| **Your Turn 6** | Multiply.

  a. $(-0.07)(-2.65)$ **b.** $(-1)(-0.9)(-24)(0.2)$

Objective 2 Find the multiplicative inverse of a number.

Two numbers that multiply to equal 1 are called **multiplicative inverses**.

Definition Multiplicative inverses: Two numbers whose product is 1.

For example, $\dfrac{2}{3}$ and $\dfrac{3}{2}$ are multiplicative inverses because their product is 1.

$$\frac{2}{3} \cdot \frac{3}{2} = \frac{6}{6} = 1$$

Notice that to write a number's multiplicative inverse, we simply invert the numerator and denominator. Multiplicative inverses are also known as *reciprocals*.

| **Example 7** | Find the multiplicative inverse.

a. $\dfrac{5}{8}$

Answer: The multiplicative inverse of $\dfrac{5}{8}$ is $\dfrac{8}{5}$ because $\dfrac{5}{8} \cdot \dfrac{8}{5} = 1$.

b. $-\dfrac{1}{4}$

Answer: The multiplicative inverse of $-\dfrac{1}{4}$ is -4 because $-\dfrac{1}{4} \cdot (-4) = -\dfrac{1}{4} \cdot \left(-\dfrac{4}{1}\right) = 1$.

c. -6

Answer: The multiplicative inverse of -6 is $-\dfrac{1}{6}$ because $-6 \cdot \left(-\dfrac{1}{6}\right) = -\dfrac{6}{1} \cdot \left(-\dfrac{1}{6}\right) = 1$.

| **Your Turn 7** | Find the multiplicative inverse.

  a. $\dfrac{4}{5}$ **b.** $\dfrac{1}{7}$ **c.** -9

Objective 3 Divide rational numbers.

Now let's turn our attention to division. The parts of a division statement are shown next.

$$\underset{\underset{\text{Dividend}}{\uparrow}}{8} \div \underset{\underset{\text{Divisor}}{\uparrow}}{2} = \underset{\underset{\text{Quotient}}{\uparrow}}{4}$$

Connection Division can also be expressed in fraction form. For example, the division statement $8 \div 2 = 4$ can also be written as $\dfrac{8}{2} = 4$.

Answers to Your Turn 6
  a. 0.1855   **b.** -4.32

Answers to Your Turn 7
  a. $\dfrac{5}{4}$ **b.** 7   **c.** $-\dfrac{1}{9}$

Sign Rules for Division

The sign rules for division are the same as for multiplication. We can see this by looking at the relationship between division and multiplication. We can use one operation to check the other. For example, if $8 \cdot 2 = 16$, then $16 \div 2 = 8$ or $16 \div 8 = 2$. In checking a multiplication, we divide the product by one of the factors to equal the other factor. If we place signs in the problem, we can say that if $8 \cdot (-2) = -16$, then $-16 \div (-2) = 8$ or $-16 \div 8 = -2$.

> **Rule** **Dividing Signed Numbers**
> When dividing two numbers that have the *same* sign, the quotient is positive.
> When dividing two numbers that have *different* signs, the quotient is negative.

Example 8   Divide.

a. $48 \div (-6)$

Solution: $48 \div (-6) = -8$

b. $-35 \div (-7)$

Solution: $-35 \div (-7) = 5$

Your Turn 8   Divide.

a. $(-60) \div 12$ **b.** $-42 \div (-6)$

Division Involving 0

What if 0 is involved in division? First consider when the *dividend* is 0 and the *divisor* is not 0, as in $0 \div 9$. The result should be $0 \div 9 = 0$. Remember that we can check division by multiplying the quotient and the divisor to equal the dividend. So we can verify that $0 \div 9 = 0$ by writing the following multiplication:

$$\text{Quotient} \cdot \text{Divisor} = \text{Dividend}$$
$$0 \quad \cdot \quad 9 \quad = \quad 0$$

Conclusion: If the dividend is 0 and the divisor is not zero, the quotient is 0.

What if the divisor is 0, as in $8 \div 0$? Again, think about the check statement.

$$\text{Quotient} \cdot \text{Divisor} = \text{Dividend}$$
$$? \quad \cdot \quad 0 \quad = \quad 8$$

There is no number that can make $? \cdot 0 = 8$ true because any number multiplied by 0 equals 0, not 8. So there also is no number that can make $8 \div 0 = ?$ true, and we say that dividing a nonzero number by 0 is *undefined*.

Conclusion: If the divisor is 0 with a nonzero dividend, the statement is undefined.

What if both the dividend and divisor are 0, as in $0 \div 0$? Again, think about the check statement.

$$\text{Quotient} \cdot \text{Divisor} = \text{Dividend}$$
$$? \quad \cdot \quad 0 \quad = \quad 0$$

Notice that any number will work for this quotient because any number times 0 equals 0. Because we cannot determine a unique quotient here, we say that $0 \div 0$ is *indeterminate*.

Conclusion: If both the dividend and divisor are 0, the statement is indeterminate.

Following is a summary of these rules.

Instructor Note Ask students what $\frac{0}{0}$ equals. Check their answers by multiplication. Make up some ridiculous answer like 343.79 and check it as well. They soon get the idea that $\frac{0}{0}$ is indeterminate.

Answers to Your Turn 8
a. -5 **b.** 7

> **Rule Division Involving 0**
>
> $0 \div n = 0$ when $n \neq 0$.
> $n \div 0$ is undefined when $n \neq 0$.
> $0 \div 0$ is indeterminate.

> **Connection** This rule says that the numerator of a fraction can equal 0 (unless the denominator equals 0) but that the denominator can never equal 0.

Dividing Fractions

To determine how to divide when the divisor is a fraction, consider this problem: How many quarters are in $5? A quarter is $\frac{1}{4}$ of a dollar, so this question translates to the following division problem:

$$\text{Number of quarters} = 5 \div \frac{1}{4}$$

Because there are 4 quarters in each of those 5 dollars, there must be 20 quarters in $5. Notice in analyzing the situation that we translated the division problem into a multiplication problem. We can say the following:

$$\text{Number of quarters} = 5 \div \frac{1}{4} = 5 \cdot 4 = 20$$

Notice that $\frac{1}{4}$ and 4 are multiplicative inverses. We can write a division problem as an equivalent multiplication problem by changing the divisor to its multiplicative inverse.

> **Rule Dividing Fractions**
>
> $$\frac{a}{b} \div \frac{c}{d} = \frac{a}{b} \cdot \frac{d}{c}, \text{ where } b \neq 0, c \neq 0, \text{ and } d \neq 0.$$

> **Connection** Using slash marks, Example 9 looks like this:
>
> $$-\frac{7}{8} \div \frac{5}{6} = -\frac{7}{8_4} \cdot \frac{6^3}{5} = -\frac{21}{20}$$

Example 9 Divide. $-\dfrac{7}{8} \div \dfrac{5}{6}$

Solution: $-\dfrac{7}{8} \div \dfrac{5}{6} = -\dfrac{7}{8} \cdot \dfrac{6}{5}$ Write an equivalent multiplication.

$$= -\frac{7}{2 \cdot 2 \cdot 2} \cdot \frac{2 \cdot 3}{5}$$ Divide out the common factor, 2.

$$= -\frac{21}{20}$$ Because we are dividing two numbers that have different signs, the result is negative.

Your Turn 9 Divide.

  a. $\dfrac{4}{9} \div \left(-\dfrac{8}{15}\right)$ **b.** $-\dfrac{8}{21} \div \left(-\dfrac{6}{7}\right)$

Dividing Decimal Numbers

Division with decimal numbers can be organized into two categories: (1) if the divisor is an integer and (2) if the divisor contains a decimal point. The following procedure summarizes how to divide decimal numbers.

Answers to Your Turn 9
a. $-\dfrac{5}{6}$ **b.** $\dfrac{4}{9}$

> **Procedure** **Dividing Decimal Numbers**
>
> To divide decimal numbers, set up a long division and consider the divisor.
>
> **Case 1:** If the divisor is an integer, divide as if the dividend were a whole number and place the decimal point in the quotient directly above its position in the dividend.
>
> **Case 2:** If the divisor is a decimal number:
> 1. Move the decimal point in the divisor to the right enough places to make the divisor an integer.
> 2. Move the decimal point in the dividend the same number of places.
> 3. Divide the divisor into the dividend as if both numbers were whole numbers. Make sure you align the digits in the quotient properly.
> 4. Write the decimal point in the quotient directly above its new position in the dividend.
>
> In either case, continue the division process until you get a remainder of 0 or a repeating digit (or block of digits) in the quotient.

Example 10 Divide. $-12.8 \div (-0.09)$

Solution: Because the divisor is a decimal number, we move the decimal point enough places to the right to create an integer—in this case, two places. Then we move the decimal point two places to the right in the dividend. Because we are dividing two numbers with the same sign, the result is positive.

$$
\begin{array}{r}
142.22 \\
9\overline{)1280.00} \\
-9 \\
\overline{38 } \\
-36 \\
\overline{20 } \\
-18 \\
\overline{20}
\end{array}
$$

◄ **Note** Because the 2 digit repeats without end, we write a repeat bar over the 2 just after the decimal point.

Answer: $142.\overline{2}$

Your Turn 10 Divide.

a. $32.04 \div (-12)$ b. $-13.12 \div 6.4$ c. $-0.88 \div (-0.6)$

Applications

It is often necessary to use signed numbers when solving applications problems.

Example 11 Solve.

a. For a chemistry experiment, Fred used $\frac{3}{5}$ of the contents of a bottle of hydrochloric acid that was $\frac{3}{4}$ full. What fractional part of a full bottle did he use?

Solution: To find $\frac{3}{5}$ of $\frac{3}{4}$, multiply $\frac{3}{5}$ and $\frac{3}{4}$.

$$\frac{3}{5} \cdot \frac{3}{4} = \frac{3 \cdot 3}{5 \cdot 4} = \frac{9}{20}$$

Answer: Fred used $\frac{9}{20}$ of a full bottle of hydrochloric acid.

Answers to Your Turn 10
a. -2.67 b. -2.05 c. $1.4\overline{6}$

Learning Strategy

Repetition is key! Sometimes homework is done quickly for the grade and because of time constraints, but it should be taken very seriously because it's the closest thing to the test. Do the practice test and all the materials Pearson has to offer.

—Lauren H.

b. An object's weight is a downward force due to gravity, and this force is calculated by multiplying the object's mass by the acceleration due to gravity (Force = Mass × Acceleration.) Find the force due to gravity of a truck with a mass of 1800 kilograms if the acceleration due to gravity is -9.8 m/sec^2. The force will be in newtons.

Solution: Force = Mass × Acceleration   *Substitute for mass and acceleration.*

$$\text{Force} = (1800)(-9.8)$$
$$= -17{,}640$$

Answer: The force due to gravity of the truck is $-17{,}640$ newtons.

Your Turn 11 Solve.

a. At a company picnic, $\dfrac{2}{5}$ of the people attending were men and of those men, $\dfrac{3}{4}$ were over the age of 50. What fraction of those attending were men over the age of 50?

Answers to Your Turn 11

a. $\dfrac{3}{10}$ **b.** 500 kg

b. A truck is carrying a load that has a weight of 4900 newtons, which means that the downward force due to gravity is -4900 newtons. Find the mass of the load in kilograms.

1.4 Exercises   For Extra Help   MyMathLab®

Note: Exercises marked with a ★ represent challenging exercises.

Objective 1

Prep Exercise 1 Explain the difference between $6(-7)$ and $6 - 7$.

$6(-7) = -42$ shows multiplication, whereas $6 - 7 = -1$ shows subtraction.

Prep Exercise 2 In $(-5)(-8) = 40$, -5 and -8 are called ___factors___.

For Exercises 1–12, indicate whether each equation illustrates the multiplicative property of 0, the multiplicative identity, the commutative property of multiplication, the associative property of multiplication, or the distributive property. See Example 1.

1. $6(3 + 4) = 6 \cdot 3 + 6 \cdot 4$
Distributive property

2. $7(2 - 5) = 7 \cdot 2 - 7 \cdot 5$
Distributive property

3. $1 \cdot \left(-\dfrac{5}{6}\right) = -\dfrac{5}{6}$
Multiplicative identity

4. $1 \cdot (-7) = -7$
Multiplicative identity

5. $0 \cdot (-6.3) = 0$
Multiplicative property of 0

6. $\dfrac{2}{3} \cdot 0 = 0$
Multiplicative property of 0

7. $\dfrac{3}{5} \cdot \left(-\dfrac{1}{4}\right) = -\dfrac{1}{4} \cdot \dfrac{3}{5}$
Commutative property of multiplication

8. $-\dfrac{2}{3} \cdot \dfrac{4}{5} = \dfrac{4}{5} \cdot \left(-\dfrac{2}{3}\right)$
Commutative property of multiplication

9. $[4(-2)] \cdot 3 = 4(-2 \cdot 3)$
Associative property of multiplication

10. $-6 \cdot (5 \cdot 4) = (-6 \cdot 5) \cdot 4$
Associative property of multiplication

11. $(-4.5)(2 + 0.6) = (2 + 0.6)(-4.5)$
Commutative property of multiplication

12. $-6.2(-4.3 + 7.1) = (-4.3 + 7.1)(-6.2)$
Commutative property of multiplication

Prep Exercise 3 When multiplying or dividing two numbers that have the same sign, the sign of the result is ___positive___.

Prep Exercise 4 When multiplying or dividing two numbers that have different signs, the sign of the result is ___negative___.

Prep Exercise 5 The sign of the product of an odd number of negative factors is ___negative___.

Prep Exercise 6 The sign of the product of an even number of negative factors is ___positive___.

For Exercises 13–46, multiply. See Examples 2–6.

13. $6(-3)$
-18

14. $4(-7)$
-28

15. $(-10)5$
-50

16. $(-8)(5)$
-40

17. $(9)(-6)$
-54

18. $(12)(-4)$
-48

19. $-4(-5)$
20

20. $(-4)(-3)$
12

21. $(-7)(-6)$
42

22. $(-8)(-12)$
96

23. $\dfrac{1}{2} \cdot \left(-\dfrac{1}{2}\right)$
$-\dfrac{1}{4}$

24. $-\dfrac{4}{5} \cdot \left(\dfrac{20}{3}\right)$
$-\dfrac{16}{3}$

25. $\left(-\dfrac{2}{3}\right)\left(-\dfrac{3}{4}\right)$
$\dfrac{1}{2}$

26. $\left(-\dfrac{5}{6}\right)\left(-\dfrac{6}{5}\right)$
1

27. $\left(-\dfrac{5}{6}\right)\left(\dfrac{8}{15}\right)$
$-\dfrac{4}{9}$

28. $\left(\dfrac{2}{9}\right)\left(-\dfrac{21}{26}\right)$
$-\dfrac{7}{39}$

29. $3.8(-10)$
-38

30. $8(-2.5)$
-20

31. $-1.6(-4.2)$
6.72

32. $-7.1(-0.5)$
3.55

33. $-4.3(1.52)$
-6.536

34. $8.1(-2.75)$
-22.275

35. $6(-9)(-1)$
54

36. $-4(5)(-3)$
60

37. $6(-2)(4)$
-48

38. $3(7)(-8)$
-168

39. $(-4)(-6)(-7)$
-168

40. $(-5)(-3)(-2)$
-30

41. $12(-6)(-2)(-1)$
-144

42. $-5(3)(-4)(-2)$
-120

43. $(-1)(-6)(-40)(-3)$
720

44. $(-2)(-4)(-30)(-1)$
240

45. $(-2)(3)(-4)(-1)(2)$
-48

46. $(-1)(-1)(4)(-5)(-3)$
60

Objective 2

Prep Exercise 7 Why are 5 and $\dfrac{1}{5}$ multiplicative inverses?
Their product is 1.

For Exercises 47–54, find the multiplicative inverse. See Example 7.

47. $\dfrac{2}{3}$
$\dfrac{3}{2}$

48. $\dfrac{20}{3}$
$\dfrac{3}{20}$

49. $-\dfrac{5}{2}$
$-\dfrac{2}{5}$

50. $-\dfrac{6}{7}$
$-\dfrac{7}{6}$

51. -5
$-\dfrac{1}{5}$

52. 17
$\dfrac{1}{17}$

53. 0
There is no multiplicative inverse of 0.

54. -1
-1

Objective 3

Prep Exercise 8 Why is $5 \div 0$ undefined?
There is no quotient that we can multiply by 0 to get 5.

Prep Exercise 9 Why is $0 \div 0$ indeterminate?
There is no unique quotient.

Prep Exercise 10 Explain how to divide fractions.
Write the division problem as an equivalent multiplication problem that has the dividend multiplied by the multiplicative inverse of the divisor.

For Exercises 55–82, divide. See Examples 8–10.

55. $-6 \div 2$
-3

56. $42 \div (-7)$
-6

57. $-56 \div (-4)$
14

58. $-12 \div (-4)$
3

59. $\dfrac{-18}{9}$
-2

60. $\dfrac{75}{-3}$
-25

61. $\dfrac{-63}{-7}$
9

62. $\dfrac{-48}{-6}$
8

63. $\dfrac{0}{-6}$
0

64. $\dfrac{0}{5}$
0

65. $\dfrac{-8}{0}$
undefined

66. $-21 \div 0$
undefined

67. $\dfrac{0}{0}$
indeterminate

68. $0 \div 0$
indeterminate

69. $6 \div \dfrac{2}{3}$
9

70. $-8 \div \dfrac{3}{4}$
$-\dfrac{32}{3}$

71. $\dfrac{3}{5} \div \left(-\dfrac{9}{10}\right)$
$-\dfrac{2}{3}$

72. $-\dfrac{4}{5} \div \dfrac{4}{5}$
-1

73. $-\dfrac{2}{7} \div \left(-\dfrac{8}{21}\right)$
$\dfrac{3}{4}$

74. $-\dfrac{1}{3} \div \left(-\dfrac{3}{2}\right)$
$\dfrac{2}{9}$

75. $-\dfrac{4}{9} \div \dfrac{10}{21}$
$-\dfrac{14}{15}$

76. $\dfrac{7}{15} \div \left(-\dfrac{35}{24}\right)$
$-\dfrac{8}{25}$

77. $9.03 \div 4.3$
2.1

78. $8.1 \div 0.6$
13.5

79. $-36.72 \div (-0.4)$
91.8

80. $-10.65 \div (-7.1)$
1.5

81. $-14 \div 0.3$
$-46.\overline{6}$

82. $19 \div (-0.06)$
$-316.\overline{6}$

For Exercises 83–88, solve. See Example 11a.

83. In a poll where respondents can agree, disagree, or have no opinion, $\dfrac{3}{4}$ of the respondents said that they agreed and $\dfrac{2}{3}$ of those that agreed were women. What fraction of all respondents were women who agreed with the statement in the poll?
$\dfrac{1}{2}$

84. On a standard-size guitar, the length of the strings between the saddle and nut is $25\dfrac{1}{2}$ inches. Guitar makers must place the 12th fret at exactly half of the length of the string. How far from the saddle or nut should a guitar maker measure to place the 12th fret?
$12\dfrac{3}{4}$ in.

85. A financial planner estimates that at the rate one of his clients is increasing debt, she will have $3\dfrac{1}{2}$ times her current debt in five years. If her current debt is represented by $-\$2480$, how can her debt in five years be represented?
$-\$8680$

86. Mario finds that $\dfrac{2}{3}$ of his credit card debt is from dining out. If his total credit card debt is represented by $-\$858$, how much of the debt is from dining out?
$-\$572$

87. In 2000, the Garret family's only debt was credit card balance represented by $-\$258.75$. In 2013, they have a mortgage, two car loans, and three credit cards with a total debt represented by $-\$158{,}572.85$. How many times greater is their debt in 2013 than it was in 2000?
≈ 612.8

88. A company's stock loses value by an amount represented by $-\$\dfrac{3}{8}$ each day for four days. Write a representation of the total loss in value at the close of the 4th day.
$-\$1\dfrac{1}{2}$

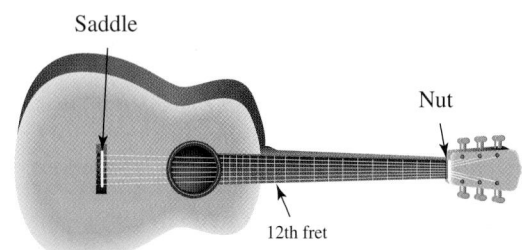

Saddle

Nut

12th fret

Of Interest

Three components affect the pitch of a string on a stringed instrument: diameter, tension, and length. Given two strings of the same diameter and same tension, if one string is half the length of the other, it will have a pitch that is one octave higher than the longer string. On a guitar, placing a finger at the 12th fret cuts the string length in half, thereby creating a tone that is an octave higher than the string's open tone.

For Exercises 89–92, use the fact that an object's weight is a downward force due to gravity and is calculated by multiplying the object's mass by the acceleration due to gravity, which is a constant (Force = Mass × Acceleration). The following table indicates the units. See Example 11b.

	Force	=	Mass	×	Acceleration
American measurement	pounds (lb)		slugs (s)		−32.2 ft./sec.^{2}
Metric measurement	newtons (N)		kilograms (kg)		−9.8 m/sec.^{2}

89. Find the force due to gravity on an object with a mass of 12.5 slugs. What does the sign of the weight indicate?   −402.5 lb.; the force is downward.

90. Find the force due to gravity on a person with a mass of 70.4 kilograms.   −689.92 N

91. The blue whale is the largest animal on Earth. The largest blue whale ever caught weighed 1,658,181.8 newtons, which means that the downward force due to gravity was −1,658,181.8 newtons. What was the whale's mass in kilograms? (*Source: The Whale Watcher's Guide: Whale Watching Trips in North America*, Corrigan, 1999.)
169,202.2245 Kg

92. The Liberty Bell weighs 2080 pounds, which means that the downward force due to gravity is −2080 pounds. What is its mass in slugs? (*Source: Ring in the Jubilee: The Epic of America's Liberty Bell*, Boland, 1973.)   ≈64.6 slugs

For Exercises 93 and 94, use the fact that in an electrical circuit, voltage (V) is equal to the product of the current, measured in amperes (A), and the resistance of the circuit, measured in ohms (Ω) (Voltage = Current × Resistance).

93. Suppose the current in a circuit is −6.4 amperes and the resistance is 8 ohms. Find the voltage.   −51.2 V

94. An electrical technician measures the voltage in a circuit to be −15 volts and the current to be −8 amperes. What is the resistance of the circuit?   1.875 Ω

For Exercises 95 and 96, use the fact that in an electrical circuit, power, which is measured in watts (W), is the product of the voltage and the current (Power = Voltage × Current).

95. An engineer measures the voltage in a circuit to be −120 volts and the current to be −30 amperes. What is the power?   3600 W

★ **96.** An electrical technician determines that the power in a circuit is 400 watts. If the current is measured to be −6.5 amperes, what is the resistance in the circuit?
9.47 Ω

Of Interest

The blue whale can reach lengths of over 100 feet and weigh from 130 to 150 tons. They were hunted to near extinction until 1967, when hunting was banned. It is estimated that about 4000 to 6000 blue whales remain.

Of Interest

Cast in 1752, the Liberty Bell cracked while being tested in September 1752 at the Pennsylvania State House in Philadelphia. It was recast and used to announce such special events as the first public reading of the *Declaration of Independence*.

Review Exercises

Exercises 1–3  Constants and Variables

[1.1] 1. Is π a rational or irrational number?
irrational

[1.1] 2. Graph −7.2 on a number line.

<!-- number line with point at −7.2 between −8 and −7 -->

[1.3] 3. Find the additive inverse of $\frac{2}{3}$. $-\frac{2}{3}$

Exercises 4–6 Expressions

[1.1] 4. Simplify: $|-6.8|$ 6.8

[1.3] 5. Add: $-7 + (-15)$ −22

[1.3] 6. Subtract: $-\frac{5}{8} - \left(-\frac{1}{6}\right)$ $-\frac{11}{24}$

1.5 Exponents, Roots, and Order of Operations

Objectives

1 Evaluate numbers in exponential form.
2 Evaluate square roots.
3 Use the order-of-operations agreement to simplify numerical expressions.
4 Find the mean of a set of data.

Warm-up

[1.4] 1. Multiply: $\left(-\dfrac{2}{3}\right)\left(-\dfrac{2}{3}\right)$

[1.4] 2. Multiply: $(-3)(-3)(-3)(-3)$

[1.4] 3. Divide: $-36 \div 3$

Objective 1 Evaluate numbers in exponential form.

Sometimes problems involve repeatedly multiplying the same number. In such problems, we can use an **exponent** to indicate that a **base** number is repeatedly multiplied.

Definitions Exponent: A symbol written to the upper right of a base number that indicates how many times to use the base as a factor.
Base: The number that is repeatedly multiplied.

When we write a number with an exponent, we say that the expression is in *exponential form*. The expression 2^4 is in exponential form, where the base is 2 and the exponent is 4. It is read "two to the fourth power," or simply "two to the fourth." To *evaluate* 2^4, write 2 as a factor 4 times; then multiply.

$$2^4 = \overbrace{2 \cdot 2 \cdot 2 \cdot 2}^{\text{Four 2s}} = 16$$

Base   Exponent

Procedure Evaluating an Exponential Form

To evaluate an exponential form raised to a natural number exponent, write the base as a factor the number of times indicated by the exponent; then multiply.

Example 1 Evaluate.

a. $(-7)^2$

Solution: The exponent 2 indicates that we have two factors of -7. Because we multiply two negative numbers, the result is positive.

$$(-7)^2 = (-7)(-7) = 49$$

◄ **Note** When a base is raised to the second power, we say it is **squared**. The expression $(-7)^2$ can be read as "negative seven **squared**."

b. $\left(-\dfrac{2}{3}\right)^3$

Solution: The exponent 3 means that we write the base as a factor three times.

$$\left(-\frac{2}{3}\right)^3 = \left(-\frac{2}{3}\right)\left(-\frac{2}{3}\right)\left(-\frac{2}{3}\right)$$
$$= -\frac{8}{27}$$

◄ **Note** When a base is raised to the third power, we can say it is **cubed**. The expression $\left(-\dfrac{2}{3}\right)^3$ can be read as "negative two-thirds **cubed**."

Answers to Warm-up

1. $\dfrac{4}{9}$

2. 81

3. -12

Example 1 suggests the following sign rules:

> **Rule Evaluating Exponential Forms with Negative Bases**
> - If the base of an exponential form is a negative number and the exponent is even, the product is positive.
> - If the base is a negative number and the exponent is odd, the product is negative.

Your Turn 1 Evaluate.

a. $\left(-\dfrac{1}{2}\right)^5$　　　**b.** $(-0.3)^4$

There is a great deal of difference between expressions of the form $(-a)^n$ and $-a^n$. The base of the exponent in $(-a)^n$ is $-a$. So $(-a)^n$ means multiply n factors of $-a$. The base of the exponent in $-a^n$ is a. So $-a^n$ means find the additive inverse of a^n.

Note This is read as "negative 2 to the sixth power." ▶

Note This is read as "the additive inverse of two to the sixth power." ▶

Instructor Note Point out that expressions of the form $(-a)^n$ with $a > 0$ are positive or negative depending on whether n is odd or even, but expressions of the form $-a^n$ with $a > 0$ are always negative.

Example 2 Evaluate.

a. $(-2)^6$

Solution: $(-2)^6 = (-2)(-2)(-2)(-2)(-2)(-2) = 64$

b. -2^6

Solution: $-2^6 = -(2 \cdot 2 \cdot 2 \cdot 2 \cdot 2 \cdot 2) = -64$

c. $(-3)^3$

Solution: $(-3)^3 = (-3)(-3)(-3) = -27$

d. -3^3

Solution: $-3^3 = -(3 \cdot 3 \cdot 3) = -27$

Your Turn 2 Evaluate.

a. $(-3)^4$　　　**b.** -3^4　　　**c.** $(-2)^3$　　　**d.** -2^3

Objective 2 Evaluate square roots.

Roots are inverses of exponents. More specifically, a square root is the inverse of a square. So a square root of a given number is a number that, when squared, equals the given number. For example, 5 is a square root of 25 because $5^2 = 25$. Notice that -5 is also a square root of 25 because $(-5)^2 = 25$.

Conclusion: For every positive number, there are *two* square roots, a positive root and a negative root.

For convenience, if asked to find all square roots of 25, we can write both the positive and negative answers in a compact expression like this: ± 5.

Example 3 Find all square roots of the given number.

a. 36

Answer: ± 6

b. -25

Answer: No real-number square roots exist.

Note We do not say that this situation is undefined or indeterminate because we eventually will define the square root of a negative number. Such roots will be defined ◀ using imaginary numbers.

Explanation Because of the sign rules for multiplication, there is no way to square a real number and get anything other than a positive result.

Answers to Your Turn 1

a. $-\dfrac{1}{32}$　　　**b.** 0.0081

Answers to Your Turn 2

a. 81　**b.** -81　**c.** -8　**d.** -8

Our work so far suggests the following rules:

> **Rule Square Roots**
> - Every positive number has two square roots, a positive root and a negative root.
> - Negative numbers have no real-number square roots.

Your Turn 3   Find all square roots of the given number.

  a. 144   b. -81

The Principal Square Root

A symbol, $\sqrt{}$, called the *radical*, is used to indicate finding the *principal* (nonnegative) square root of a given number. The given number or expression inside the radical is called the *radicand*.

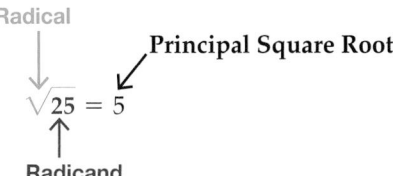

What about the square root of a fraction such as $\sqrt{\dfrac{4}{9}}$? Because $\left(\dfrac{2}{3}\right)^2 = \dfrac{2}{3} \cdot \dfrac{2}{3} = \dfrac{4}{9}$, we can say that $\sqrt{\dfrac{4}{9}} = \dfrac{2}{3}$. Notice that we can evaluate $\sqrt{\dfrac{4}{9}}$ by finding the square roots of the numerator and denominator separately, or $\sqrt{\dfrac{4}{9}} = \dfrac{\sqrt{4}}{\sqrt{9}} = \dfrac{2}{3}$. We now have the following rules for square roots involving the radical sign.

> **Rule Square Roots Involving the Radical Sign**
> - The radical symbol $\sqrt{}$ denotes the principal (nonnegative) square root.
> - $\sqrt{\dfrac{a}{b}} = \dfrac{\sqrt{a}}{\sqrt{b}}$, where $a \geq 0$ and $b > 0$.

Answers to Your Turn 3
  a. ± 12
  b. No real-number square roots exist.

Answers to Your Turn 4
  a. 11   b. $\dfrac{9}{13}$ c. 0.7
  d. not a real number

Example 4   Evaluate the square roots.

a. $\sqrt{144}$

Solution: $\sqrt{144} = 12$

c. $\sqrt{0.81}$

Solution: $\sqrt{0.81} = 0.9$

b. $\sqrt{\dfrac{49}{81}}$

Solution: $\sqrt{\dfrac{49}{81}} = \dfrac{\sqrt{49}}{\sqrt{81}} = \dfrac{7}{9}$

d. $\sqrt{-81}$

Solution: not a real number

Your Turn 4   Evaluate the square roots.

  a. $\sqrt{121}$ b. $\sqrt{\dfrac{81}{169}}$ c. $\sqrt{0.49}$ d. $\sqrt{-0.09}$

Objective 3 Use the order-of-operations agreement to simplify numerical expressions.

Some expressions are too complicated to evaluate using only the associative and commutative properties. If an expression contains a mixture of operations, we must be careful about the order in which we perform them. Performing the operations in a different order can change the results. Consider the expression $2 + 5 \cdot 3$.

Adding first: $\quad 2 + 5 \cdot 3$ $\qquad$ Multiplying first: $\quad 2 + 5 \cdot 3$

$\qquad\qquad = 7 \cdot 3$ $\qquad\qquad\qquad\qquad\qquad = 2 + 15$

$\qquad\qquad = 21$ $\qquad\qquad\qquad\qquad\qquad\quad = 17$

Changing the order of the operations changes the outcome. To ensure that everyone arrives at the same answer, mathematicians developed an order of operations that everyone agrees to follow.

Procedure Order-of-Operations Agreement

Perform operations in the following order:

1. Within grouping symbols: parentheses ( ), brackets [ ], braces { }, above/below fraction bars, absolute value | |, and radicals $\sqrt{}$.
2. Exponents/Roots from left to right, in order as they occur.
3. Multiplication/Division from left to right, in order as they occur.
4. Addition/Subtraction from left to right, in order as they occur.

If we follow this agreement to simplify $2 + 5 \cdot 3$, the correct order is to multiply first and then add; so the answer is 17.

Embedded Parentheses

When one set of grouping symbols contains another set of grouping symbols, we say that they are *nested* or *embedded*. Work from the innermost set of grouping symbols outward.

Note Recall that the expressions $(-a)^n$ and $-a^n$ are different. The expression $(-a)^n$ means "multiply n factors of $-a$" and $-a^n$ means "find the additive inverse of a^n."

Note The division and multiplication must be calculated before the addition. As we read from left to right, we see the ◄ division before the multiplication; so we divide first.

Example 5 Simplify.

a. $-34 + 12 \div (-3) \cdot 5$

Solution: $-34 + 12 \div (-3) \cdot 5$

$= -34 + (-4) \cdot 5 \qquad$ Divide $12 \div (-3) = -4.$

$= -34 + (-20) \qquad$ Multiply $(-4) \cdot 5 = -20.$

$= -54 \qquad$ Add $-34 + (-20) = -54.$

b. $-2^4 + 4|18 - 24|$

Solution: $-2^4 + 4|18 - 24|$

$= -2^4 + 4|-6| \qquad$ Subtract inside the absolute value: $18 - 24 = -6.$

$= -2^4 + 4 \cdot 6 \qquad$ Simplify the absolute value $|-6| = 6.$

$= -16 + 4 \cdot 6 \qquad$ Evaluate the exponential form $-2^4 = -16.$

$= -16 + 24 \qquad$ Multiply $4 \cdot 6 = 24.$

$= 8 \qquad$ Add $-16 + 24 = 8.$

Note When grouping symbols are imbedded, parentheses are usually placed within brackets and when the parentheses are removed the brackets are changed to parentheses.

c. $(-5)^2 - 8[4 - (6 + 3)] - \sqrt{36}$

Solution: $(-5)^2 - 8[4 - (6 + 3)] - \sqrt{36}$

$= (-5)^2 - 8(4 - 9) - \sqrt{36}$ Calculate within the innermost grouping symbol. $6 + 3 = 9$.

$= (-5)^2 - 8(-5) - \sqrt{36}$ Calculate within the parentheses. $4 - 9 = -5$.

$= 25 - 8(-5) - 6$ Evaluate the exponential form and square root.

$= 25 - (-40) - 6$ Multiply $8(-5) = -40$.

$= 25 + 40 - 6$ Write $25 - (-40)$ as an equivalent addition.

$= 65 - 6$ Add $25 + 40 = 65$.

$= 59$ Subtract $65 - 6 = 59$.

Your Turn 5   Simplify.

a. $28 - 36 \div 3(-4)$ **b.** $6|-5 - 4| + 2(-3)^3$

c. $15.8 - 0.2[(-7)^2 - (12 + 8) \div 4]$

Radical Symbols

Radical symbols can be grouping symbols. Consider $\sqrt{36 \cdot 4}$. Following the order of operations, we multiply first, then find the root of the product. However, if a radical contains multiplication, we can find the roots of the factors and multiply those roots to get the same answer.

Multiplying first: Multiplying the roots:

$\sqrt{36 \cdot 4} = \sqrt{144}$ or $\sqrt{36 \cdot 4} = \sqrt{36} \cdot \sqrt{4}$

$= 12$ $= 6 \cdot 2$

$= 12$

Both approaches yield the same result. The same holds true for division under a radical.

$\sqrt{36 \div 4} = \sqrt{9}$ or $\sqrt{36 \div 4} = \sqrt{36} \div \sqrt{4}$

$= 3$ $= 6 \div 2$

$= 3$

> **Rule   Square Root of a Product or Quotient**
>
> If a square root contains multiplication or division, we can multiply or divide first, then find the square root of the result, or we can find the square roots of the individual numbers, then multiply or divide the square roots. In symbols,
>
> $\sqrt{\dfrac{a}{b}} = \dfrac{\sqrt{a}}{\sqrt{b}}$ for $a \geq 0$ and $b > 0$ and $\sqrt{ab} = \sqrt{a} \cdot \sqrt{b}$ for $a \geq 0$ and $b \geq 0$.

Note that this rule does *not* hold for addition or subtraction under a radical, such as in $\sqrt{9 + 16}$.

Adding first: Adding the roots:

$\sqrt{9 + 16} = \sqrt{25}$ $\sqrt{9 + 16} = 3 + 4$

$= 5$ $= 7$

Because the results are different, we do not have a choice of how to approach the situation when the radical contains addition. The correct approach is to treat the radical as a grouping symbol and add, then find the root of the sum. Thus, the correct answer for $\sqrt{9 + 16}$ is 5. The same procedure applies when a radical contains subtraction.

Answers to Your Turn 5
a. 76   **b.** 0   **c.** 7

> **Rule** **Square Root of a Sum or Difference**
> When a radical contains addition or subtraction, we must add or subtract first, then find the root of the sum or difference.

Example 6 Simplify.

a. $12.4 \div 5(-3)^2 + 4\sqrt{169 - 25}$

Solution: $12.4 \div 5(-3)^2 + 4\sqrt{169 - 25}$

$= 12.4 \div 5(-3)^2 + 4\sqrt{144}$ Subtract within the radical: $169 - 25 = 144$.

$= 12.4 \div 5(9) + 4(12)$ Evaluate the exponential form and root.

$= 2.48(9) + 4(12)$ Divide $12.4 \div 5 = 2.48$. ◄ **Note** We divide before

$= 22.32 + 48$ Multiply $2.48(9) = 22.32$ we multiply here
and $4(12) = 48$.   because the order is

$= 70.32$ Add $22.32 + 48 = 70.32$.   to multiply or divide
from left to right in the
order in which those
operations occur.

b. $\left(-\dfrac{1}{4}\right)^3 - \left(\dfrac{1}{2} + \dfrac{3}{8}\right) \div \sqrt{\dfrac{48}{3}}$

Solution: $\left(-\dfrac{1}{4}\right)^3 - \left(\dfrac{1(4)}{2(4)} + \dfrac{3}{8}\right) \div \sqrt{\dfrac{48}{3}}$

$= \left(-\dfrac{1}{4}\right)^3 - \left(\dfrac{4}{8} + \dfrac{3}{8}\right) \div \sqrt{\dfrac{48}{3}}$ Write equivalent fractions with a common denominator to add within the parentheses.

$= \left(-\dfrac{1}{4}\right)^3 - \dfrac{7}{8} \div \sqrt{16}$ Add within the parentheses and divide within the radical.

$= -\dfrac{1}{64} - \dfrac{7}{8} \div 4$ Evaluate the exponential form and square root.

$= -\dfrac{1}{64} - \dfrac{7}{8} \cdot \dfrac{1}{4}$ Write an equivalent multiplication using the reciprocal of the divisor.

$= -\dfrac{1}{64} - \dfrac{7(2)}{32(2)}$ Multiply $\dfrac{7}{8} \cdot \dfrac{1}{4}$ to get $\dfrac{7}{32}$; then write it as an equivalent fraction with the common denominator 64 to subtract.

$= -\dfrac{1}{64} - \dfrac{14}{64}$

$= -\dfrac{15}{64}$ Subtract.

Your Turn 6 Simplify.

a. $-\dfrac{3}{5} \div \dfrac{1}{10} \cdot 4 + \sqrt{64 + 36}$ **b.** $5 + (0.2)^3 - 3(8 - 14)$

Fraction Lines

Sometimes fraction lines are used as grouping symbols. When they are, we simplify the numerator and denominator separately, then divide the results.

Answers to Your Turn 6
a. -14 **b.** 23.008

Connection The expression in Example 7 is equivalent to the expression
$$[5(-6) - 2^3] \div [3(7) - 2].$$

Example 7 Simplify.

a. $\dfrac{5(-6) - 2^3}{3(7) - 2}$

Solution: $\dfrac{5(-6) - 2^3}{3(7) - 2}$

$= \dfrac{5(-6) - 8}{21 - 2}$ Evaluate the exponential form in the numerator and multiply in the denominator.

$= \dfrac{-30 - 8}{19}$ Multiply in the numerator and subtract in the denominator.

$= \dfrac{-38}{19}$ Subtract in the numerator.

$= -2$ Divide.

b. $\dfrac{9(3) + 15}{4^2 + 2(-8)}$

Solution: $\dfrac{9(3) + 15}{4^2 + 2(-8)}$

$= \dfrac{27 + 15}{16 + 2(-8)}$ Multiply in the numerator and evaluate the exponential form in the denominator.

$= \dfrac{42}{16 + (-16)}$ Add in the numerator and multiply in the denominator.

$= \dfrac{42}{0}$ Add in the denominator.

Because the denominator or divisor is 0, the answer is undefined.

Your Turn 7 Simplify.

a. $\dfrac{7(3 - 7) + 1}{-1 - 2^3}$ **b.** $\dfrac{2[9 - 4(3 + 5)]}{25 - (6 - 1)^2}$

Objective 4 Find the mean of a set of data.

Fraction line notation can be used to indicate the proper sequence of operations in calculating an arithmetic mean, or average, of a set of data.

> **Procedure** **Finding the Arithmetic Mean**
>
> To find the arithmetic mean, or average, of n numbers, divide the sum of the numbers by n.
>
> $$\text{Arithmetic mean} = \frac{x_1 + x_2 + \cdots + x_n}{n}$$

Note The subscripts indicate that each x represents a different given number. So x_1 represents the first given number, x_2 represents the second, and so on, until x_n, which represents the last given number.

Example 8 Jacky has the following test scores: 86, 95, 78, 82, and 84. Find the average of her test scores.

Solution: $\dfrac{86 + 95 + 78 + 82 + 84}{5} = \dfrac{425}{5}$ Divide the sum of the 5 scores by 5.

$= 85$

Answers to Your Turn 7
a. 3   **b.** undefined

Your Turn 8 This table shows the daily rainfall accumulations for the month of April in a certain city. Find the average daily accumulation for April. (Remember, April has 30 days.)

Date	April 2	April 7	April 9	April 15	April 21	April 26
Accumulation (in inches)	0.5	1.25	1.0	1.25	0.25	1.0

Answer to Your Turn 8
0.175 in./day

1.5 Exercises For Extra Help MyMathLab®

Note: Exercises marked with a ★ represent challenging exercises.

Objective 1

Prep Exercise 1 What is another way to say "two to the third power"? Two cubed

Prep Exercise 2 What is another way to say "three to the second power"? Three squared

For Exercises 1–6, identify the base and the exponent; then translate the expression to words. See Objective 1.

1. 7^2 Base: 7; exponent: 2; "seven squared"

2. 9^4 Base: 9; exponent: 4; "nine to the fourth power"

3. $(-5)^3$ Base: −5; exponent: 3; "negative five cubed"

4. $(-8)^2$ Base: −8; exponent: 2; "negative eight squared"

5. -2^7
Base: 2; exponent: 7; "additive inverse of two to the seventh power"

6. -3^8
Base: 3; exponent: 8; "additive inverse of three to the eighth power"

Prep Exercise 3 To evaluate 4^5, write ___4___ as a factor ___5___ times, then multiply.

Prep Exercise 4 If the base of an exponential form is a negative number and the exponent is even, will the sign of the result be positive or negative? positive

For Exercises 7–28, evaluate. See Examples 1 and 2.

7. 3^4
81

8. 2^5
32

9. $(-8)^2$
64

10. $(-2)^4$
16

11. -8^2
−64

12. -2^4
−16

13. $(-5)^3$
−125

14. $(-3)^5$
−243

15. -5^3
−125

16. -3^5
−243

17. $-(-2)^3$
8

18. $-(-3)^3$
27

19. $-(-1)^6$
−1

20. $-(-1)^4$
−1

21. $\left(-\dfrac{1}{5}\right)^2$
$\dfrac{1}{25}$

22. $\left(-\dfrac{2}{7}\right)^2$
$\dfrac{4}{49}$

23. $\left(-\dfrac{3}{4}\right)^3$
$\dfrac{27}{64}$

24. $\left(-\dfrac{1}{3}\right)^5$
$\dfrac{1}{243}$

25. $(0.2)^3$
0.008

26. $(0.3)^4$
0.0081

27. $(-4.1)^2$
16.81

28. $(-0.2)^4$
0.0016

Objective 2

Prep Exercise 5 Explain the difference between squaring a number and finding its square root. Squaring a number means to multiply the number by itself; finding its square root means to find a number whose square is the given number.

Prep Exercise 6 Every positive real number has two square roots, a ___positive___ root and a ___negative___ root.

For Exercises 29–36, find all square roots of each number. See Example 3.

29. 121
± 11

30. 49
± 7

31. -81
No real-number square
roots exist.

32. -36
No real-number square
roots exist.

33. 196
± 14

34. 169
± 13

35. 256
± 16

36. 225
± 15

Prep Exercise 7 The radical symbol denotes the ___principal (nonnegative)___ square root.

For Exercises 37–48, evaluate the square roots. See Example 4.

37. $\sqrt{16}$
4

38. $\sqrt{36}$
6

39. $\sqrt{144}$
12

40. $\sqrt{289}$
17

41. $\sqrt{0.49}$
0.7

42. $\sqrt{0.01}$
0.1

43. $\sqrt{-64}$
Not a real number

44. $\sqrt{-25}$
Not a real number

45. $\sqrt{\dfrac{64}{81}}$
$\dfrac{8}{9}$

46. $\sqrt{\dfrac{9}{100}}$
$\dfrac{3}{10}$

47. $\sqrt{\dfrac{50}{2}}$
5

48. $\sqrt{\dfrac{48}{3}}$
4

Objective 3

Prep Exercise 8 What is the first step in simplifying
$9 + 40 \div 2 \cdot 5$? Divide 40 by 2.

Prep Exercise 9 When simplifying a square root of a sum of
two numbers, what is the proper order of operations?
We must add first, then find the square root of the sum.

For Exercises 49–96, simplify. See Examples 5–7.

49. $5 \cdot 3 + 4$
19

50. $4 \cdot 6 - 5$
19

51. $8 \div 2 - 4$
0

52. $18 \div 2 + 3$
12

53. $8 + 4 \div 2$
10

54. $9 + 6 \div 3$
11

55. $4 \cdot 6 - 7 \cdot 5$
-11

56. $-3 \cdot 4 - 2 \cdot 7$
-26

57. $12 - 2^4$
-4

58. $8 - 3^2$
-1

59. $8 - 3(-4)^2$
-40

60. $16 - 5(-2)^2$
-4

61. $4^2 - 36 \div 4(12 - 8)$
-20

62. $3^2 - 18 \div 3(6 - 3)$
-9

63. $-4 + 3(-1)^4 + 18 \div 3 \cdot 3$
17

64. $12 - 2(-2)^3 - 64 \div 4 \cdot 2$
-4

65. $-2^3 + 8 - 7(4 - 3)$
-7

66. $-3^3 - 16 - 5(7 - 2)$
-68

67. $24 \div (-9 + 3)(3 + 4)$
-28

68. $18 \div (-6 + 3)(4 + 1)$
-30

69. $13.02 \div (-3.1) + 6^2 - \sqrt{25}$
26.8

70. $-15.54 \div 3.7 + (-2)^4 + \sqrt{49}$
18.8

71. $18.2 + 3.4[(5 + 9) \div 7 - 6^2]$
-97.4

72. $16.3 + 2.8[(8 + 7) \div 5 - 4^2]$
-20.1

73. $-4^3 - 3^2 + 4|7 - 12|$
-53

74. $-2|9 - 15| + 5^2 - 3^2$
4

75. $-\dfrac{3}{4} \div \left(\dfrac{1}{8}\right) + \left(-\dfrac{2}{5}\right)(-3)(-4)$
$-10\dfrac{4}{5}$

76. $\dfrac{5}{6} \div \left(-\dfrac{2}{3}\right) + \left(-\dfrac{2}{7}\right)(5)(-14)$
$18\dfrac{3}{4}$

77. $-36 \div (-2)(3) + \sqrt{169 - 25} + 12$
78

78. $\sqrt{100 - 64} + 18 \div (-3)(-2)$
18

79. $8 - 3[5 - (7 + 4)] - \sqrt{49}$
19

80. $4 - 8[3 - (9 + 3)] + \sqrt{64}$
84

81. $\sqrt{71 - 35} - 3^2[6 - (4 - 9)] + 4^3$
-29

82. $\sqrt{83 - 58} - 2^2[9 - (3 - 8)] + 3^4$
30

83. $\dfrac{9}{8} \cdot \left(-\dfrac{2}{3}\right) + \left(\dfrac{1}{5} - \dfrac{2}{3}\right) \div \sqrt{\dfrac{125}{5}}$
$-\dfrac{253}{300}$

84. $\left(\dfrac{3}{4} - \dfrac{2}{3}\right) \div \sqrt{\dfrac{9}{81}} - \left(\dfrac{16}{27}\right) \div \left(\dfrac{4}{9}\right)$
$-\dfrac{13}{12}$

85. $\dfrac{2}{5} \div \left(-\dfrac{1}{10}\right) \cdot (-3) + \sqrt{64 + 36}$
22

86. $\dfrac{5}{6}(-18) \div \left(\dfrac{3}{2}\right) - \sqrt{9 + 16}$
-15

87. $-16 \cdot \left(\dfrac{3}{4}\right) \div (-2) + |9 - 3(4 + 1)|$
12

88. $18 \cdot \left(-\dfrac{5}{6}\right) \div (-3) + 2|4 + 2(7 - 3)|$
29

89. $\dfrac{7 - |4(5) - 13|}{6^2 - 3(2 - 14)}$
0

90. $\dfrac{|6(-3) + 7| - 11}{5^3 - 2(6 - 12)}$
0

91. $\dfrac{4[5 - 8(2 + 1)]}{3 - 6 - (-4)^2}$
4

92. $\dfrac{3[24 - 4(6 - 2)]}{-3^3 + 4^2 + 3}$
-3

93. $\dfrac{5^3 - 3(4^3 - 41)}{19 - (3 - 10)^2 + 38}$
7

94. $\dfrac{6^2 - 3(4 + 2^5)}{4 + 20 - (2 + 4)^2}$
6

95. $\dfrac{4(5^2 - 10) + 3}{\sqrt{25 - 16} - 3}$
undefined

96. $\dfrac{5(4 - 9) + 1}{2^3 - \sqrt{100 - 36}}$
undefined

In Exercises 97–100, a property of arithmetic was correctly used as an alternative to the order of operations. Determine what property of arithmetic was applied and explain how it is different from the order-of-operations agreement.

97. $14 - 2 \cdot 6 \cdot 3 + 8^2$
$= 14 - 2 \cdot 18 + 64$
$= 14 - 36 + 64$
$= 42$
Associative property of multiplication
The multiplication was not performed from left to right.

98. $2(3 + 8) - \sqrt{81}$
$= 6 + 16 - 9$
$= 13$
Distributive property
The parentheses were not simplified first.

99. $-6(5 + 3^2) - \sqrt{14 + 11}$
$= -6(5 + 9) - \sqrt{25}$
$= -30 + (-54) - 5$
$= -89$
Distributive property
The parentheses were not simplified first.

100. $(-4)^3 + 2[-10 + 8 + (-3)]$
$= -64 + 2(-13 + 8)$
$= -64 + 2(-5)$
$= -64 + (-10)$
$= -74$
Commutative property of addition
The addition was not performed from left to right.

Find ⊗ the Mistake   *For Exercises 101–104, explain the mistakes; then simplify.*

101. $24 \div 4 \cdot 2 - 11$

$= 24 \div 8 - 11$

$= 3 - 11$

$= -8$

Mistake: Multiplied before dividing

Correct: 1

102. $19 - 6(10 - 8)$

$= 19 - 6(2)$

$= 13(2)$

$= 26$

Mistake: Subtracted before multiplying

Correct: 7

103. $40 \div 2 + \sqrt{25 - 9}$

$= 20 + \sqrt{25 - 9}$

$= 20 + 5 - 3$

$= 22$

Mistake: Found the square roots of the subtrahend and minuend in square root of a difference

Correct: 24

104. $-3^4 + 20 \div 5 - (16 - 24)$

$= 81 + 4 - (-8)$

$= 85 + 8$

$= 93$

Mistake: Treated -3^4 as $(-3)^4$

Correct: −69

Objective 4

Prep Exercise 10 To find the arithmetic mean of n numbers, divide the ____sum____ of the numbers by ____n____.

For Exercises 105–112, solve. See Example 8.

105. Tomeka has the following test scores in a history course: 82, 76, 64, 90, and 74. What is the average of her test scores?

77.2

106. Will's math instructor gives quizzes worth 10 points each. Will has the following quiz scores: 9, 8, 4, 8, 7, 7, 6, 9, and 8. If his instructor drops the lowest quiz score, what is Will's quiz average?

7.75

★ **107.** Michael will not have to take his chemistry final if he has a test average greater than or equal to 90 on the five tests in the course. His current test scores are 96, 88, 86, and 84. Using trial and error, determine the minimum score on the last test that will give him a test average of 90.

96

★ **108.** To get an A in her psychology course, Lisa must have a test average greater than or equal to 90 on four out of five tests. (The lowest test score is dropped.) Her scores on the first four tests are 98, 68, 84, and 86. What is the minimum score on the last test that will give her a test average of 90?

92

109. The following chart shows the average weekly earnings for workers in the United States in a recent year. Find the average weekly earnings for the year.

AVERAGE WEEKLY EARNINGS			
Month	Average Earnings	Month	Average Earnings
January	$577.98	July	$589.81
February	$578.29	August	$591.50
March	$583.42	September	$592.85
April	$583.05	October	$593.87
May	$585.42	November	$596.23
June	$589.86	December	$598.60

(*Source:* Bureau of Labor Statistics.)

$588.41

110. The following graphic shows the number of people 20 years and older who were unemployed in the United States for each month during a recent year. Find the average number of people unemployed per month during that year.

ALL IN A MONTH'S WORK		YEAR 2011	
Month	Unemployed (in thousands)	Month	Unemployed (in thousands)
January	14,937	July	14,428
February	14,542	August	14,008
March	14,060	September	13,520
April	13,237	October	13,102
May	13,421	November	12,613
June	14,409	December	12,692

(*Source:* Bureau of Labor Statistics, *Labor Force Statistics from Current Population Survey*.)

13,747,000

111. The following bar graph shows the daily closing price of Walmart stock over a one-week period from July 30 to August 3, 2012. Find the average of the daily prices for that week.

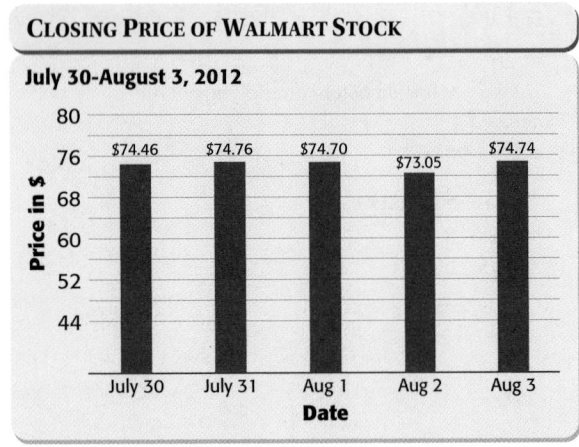

CLOSING PRICE OF WALMART STOCK

(*Source:* www.google.com.)
$74.34

112. The following bar graph shows the Dow Jones Industrial Average (DJIA) at the close of each of the days listed. What is the average of the DJIA for that week?

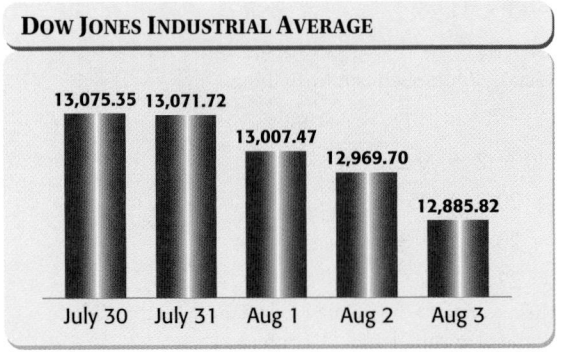

DOW JONES INDUSTRIAL AVERAGE

(*Source:* www.google.com.)
13,002.01

Review Exercises

Exercises 1–3   Constants and Variables

[1.1] 1. Write a set containing the whole numbers through 10.
{0, 1, 2, 3, 4, 5, 6, 7, 8, 9, 10}

[1.4] 3. Find the multiplicative inverse of -9. $-\dfrac{1}{9}$

[1.1] 2. Graph $-\dfrac{4}{5}$ on a number line.

Exercises 4–6   Expressions

[1.1] 4. Is $5x - 4y$ an equation or an expression? Explain.
It is an expression because it has no = sign.

[1.4] 6. Divide: $\dfrac{-15}{-5}$ 3

[1.4] 5. Multiply: $-4(3)(-2)$ 24

1.6 Translating Word Phrases to Expressions

Objective

1 Translate word phrases to expressions.

Warm-up

[1.3] 1. Subtract: $14 - (-5)$
[1.4] 2. Multiply: $(-3)(4)(-5)$
[1.5] 3. Evaluate: $-20 \div 5 \cdot 2 - 3^4$

Objective 1 Translate word phrases to expressions.

An important technique in solving problems is to translate words in the problem to math symbols. In this section, we focus on how to translate word phrases to expressions containing variables, but we begin by translating expressions containing variables to word phrases for addition, subtraction, multiplication, and division.

Answers to Warm-up
1. 19
2. 60
3. -89

Operation	Variable Expression	Word Phrases
Addition	$x + 3$	Some number plus three Three added to some number The sum of some number and three Three more than some number Some number increased by three
Subtraction	$y - 6$	Some number minus six Six subtracted from some number The difference of some number and six Six less than some number Some number decreased by six Some number less six
	$6 - y$	Six minus some number Some number subtracted from six The difference of six and some number Some number less than six Six decreased by some number Six less some number
Multiplication	$2a$	Twice some number Two times some number The product of two and some number Some number multiplied by two
Division	$\dfrac{x}{5}$	Some number divided by five The quotient of some number and five The ratio of some number and five
	$\dfrac{5}{x}$	Five divided by some number The quotient of five and some number The ratio of five and some number

The first step in translating a word phrase to math symbols is to identify the unknown amount. If a variable is not already given, select a variable to represent that unknown amount. The above translations from variable expressions to word phrases can be used as a guide.

Translating Basic Phrases

The following table contains some basic phrases and their translations.

Addition	Translation
The sum of x and three	$x + 3$
h plus k	$h + k$
t added to seven	$7 + t$
Three more than a number	$n + 3$
y increased by two	$y + 2$

Subtraction	Translation
The difference of three and x	$3 - x$
h minus k	$h - k$
Seven subtracted from t	$t - 7$
Three less than a number	$n - 3$
Two decreased by y	$2 - y$

Note Because addition is a commutative operation, it doesn't matter in what order we write the translation.

For "the sum of x and three," we can write $x + 3$ or $3 + x$.

Note Subtraction is not a commutative operation; therefore, the way we write the translation matters. We must translate each key phrase exactly as it was presented above. Notice that when we translate "less than" or "subtracted from," the translation is in reverse order from what we read.

Multiplication	Translation
The product of x and three	$3x$
h times k	hk
Twice a number	$2n$
Triple the number	$3n$
Two-thirds of a number	$\frac{2}{3}n$

Division	Translation
The quotient of x and three	$x \div 3$ or $\frac{x}{3}$
h divided by k	$h \div k$ or $\frac{h}{k}$
h divided into k	$k \div h$ or $\frac{k}{h}$
The ratio of a to b	$a \div b$ or $\frac{a}{b}$

Note Like addition, multiplication is a commutative operation. This means that we can write the translation order any way we like.

h times k can be hk or kh.

Note Division is like subtraction in that it is not a commutative operation; therefore, we must translate division phrases exactly as presented above. Notice how "divided into" is translated in reverse order of what we read.

Exponents	Translation
c squared	c^2
The square of b	b^2
k cubed	k^3
The cube of b	b^3
n to the fourth power	n^4
y raised to the fifth power	y^5

Roots	Translation
The square root of x	$\sqrt{x}$

Consider the translated phrases that follow. The key words *sum, difference, products,* and *quotient* indicate the answer for their respective operations. Also notice that all of the phrases involve the word *and*. In the translation, the word *and* separates the parts and can therefore be translated to the operation symbol indicated by the key word *sum, difference, product,* or *quotient.*

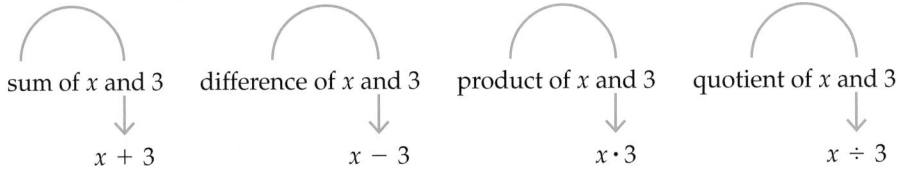

sum of x and 3	difference of x and 3	product of x and 3	quotient of x and 3
$x + 3$	$x - 3$	$x \cdot 3$	$x \div 3$

Combinations of Basic Phrases

Now that we have seen the basic phrases, let's translate phrases that involve combinations of the basic phrases.

Example 1 Translate to an algebraic expression.

a. four more than three times a number

Note The commutative property of addition allows us to write the expression in either order.

Translation: $4 + 3n$ or $3n + 4$

b. six less than the square of a number

Translation: $n^2 - 6$

Note Because subtraction is not commutative, we must translate the subtraction in a specific order. *Less than* indicates that 6 is the subtrahend, and the *square of a number* is the minuend.

c. the sum of h raised to the fifth power and fifteen

Translation: $h^5 + 15$

Note When coupled with the word *ratio*, the word *to* translates to the fraction line. The amount to the left of the word *to* goes in the numerator, and the amount to the right of the word *to* goes in the denominator.

d. the ratio of m times n to r cubed

Translation: $\dfrac{mn}{r^3}$

e. one-half of v divided by the square root of t

Translation: $\frac{1}{2}v \div \sqrt{t}$ **Note** When the word *of* is preceded ◄ by a fraction, it means multiply.

Your Turn 1   Translate to an algebraic expression.

a. two-thirds subtracted from the product of nine and a number
b. -6.2 increased by 9.8 times a number

Translating Phrases Involving Parentheses

Sometimes the word phrases imply an order of operations that would require us to use parentheses in the translation. These situations arise when the phrase indicates that a sum or difference is to be calculated before a higher-order operation such as multiplication, division, exponent, or root is performed.

Note Without the parentheses, the expression is $5x + y$, which indicates that we are to multiply 5 times x first and then add y to the result. The parentheses indicate that we add x and y first, then multiply the resulting sum by 5.

Example 2   Translate to an algebraic expression.

a. five times the sum of x and y

► **Translation:** $5(x + y)$

b. the sum of five times x and y

Translation: $5x + y$

Note The *square root of the difference* indicates that we are to calculate the difference before we calculate the square root. We indicate this symbolically by placing the entire subtraction under the radical.

c. the square root of the difference of the square of x and the square of y

► **Translation:** $\sqrt{x^2 - y^2}$

d. the product of x and y divided by the sum of x^2 and 5

Translation: $xy \div (x^2 + 5)$ or $\dfrac{xy}{x^2 + 5}$

Your Turn 2   Translate to an algebraic expression.

a. negative two times the sum of n and seven
b. the sum of eight times a and b
c. the difference of m and n, all raised to the fourth power
d. the product of 5 and m divided by the difference of m and 3

Example 3   Isaac Newton developed an expression that describes the gravitational effect between two objects. The relationship is the product of the masses of the two objects divided by the square of the distance between the objects. Translate the relationship to an expression.

Translation: $Mm \div d^2$ or $\dfrac{Mm}{d^2}$ ◄ **Note** If the same letter is to describe two different quantities, uppercase and lowercase can be used to distinguish the quantities. Subscripts can also be used, as in $m_1 m_2 \div d^2$.

Answers to Your Turn 1
a. $9n - \dfrac{2}{3}$ **b.** $-6.2 + 9.8n$

Answers to Your Turn 2
a. $-2(n + 7)$ **b.** $8a + b$
c. $(m - n)^4$ **d.** $\dfrac{5m}{m - 3}$

Answer to Your Turn 3
$d \div (v_1 + v_2)$ or $\dfrac{d}{v_1 + v_2}$

Your Turn 3   Two asteroids are on a collision course with each other. We can calculate the time to impact by dividing the distance that separates the asteroids by the sum of their velocities. Translate the relationship to an expression.

1.6 Exercises (For Extra Help) MyMathLab®

Note: Exercises marked with a ★ represent challenging exercises.

Objective 1

Prep Exercise 1 List three key words that indicate addition.
Sum, plus, added

Prep Exercise 2 List three key words that indicate subtraction.
Difference, minus, less

Prep Exercise 3 List three key words that indicate multiplication.
Product, times, twice

Prep Exercise 4 List three key words that indicate division.
Quotient, divided, ratio

Prep Exercise 5 The phrase *nine subtracted from n* translates to $n - 9$, which is in reverse order of what we read. What other key words for subtraction translate in reverse order?
Less than

Prep Exercise 6 What key words for division translate in reverse order?
Divided into

Prep Exercise 7 Translate the algebraic expression $x + 7$ into five different word phrases.
Some number plus seven, seven added to some number, the sum of some number and seven, seven more than some number, some number increased by seven

Prep Exercise 8 Translate the algebraic expression $x - 5$ into six different word phrases.
Some number minus five, five subtracted from some number, the difference of some number and five, five less than some number, some number decreased by five, some number less five

Prep Exercise 9 Translate the algebraic expression $\frac{7}{b}$ into three different word phrases.
Seven divided by some number, the quotient of seven and some number, the ratio of seven and some number

Prep Exercise 10 Translate the algebraic expression $3n$ into three different word phrases.
Three times some number, the product of three and some number, some number multiplied by three

For Exercises 1–34, translate each phrase to an algebraic expression. See Examples 1 and 2.

1. four times a number
   $4x$

2. the product of a number and four
   $4n$

3. the sum of four times x and sixteen
   $4x + 16$

4. five more than y
   $5 + y$

5. the difference of seven times x and eight
   $7x - 8$

6. six less than T
   $T - 6$

7. the quotient of negative four and the cube of y
   $-4 \div y^3$ or $\frac{-4}{y^3}$

8. the ratio of seven to the square of m
   $\frac{7}{m^2}$

9. the product of eight and p decreased by four
   $8p - 4$

10. thirteen subtracted from twice a number
    $2y - 13$

11. fourteen divided into m
    $\frac{m}{14}$

12. r divided by six
    $r \div 6$ or $\frac{r}{6}$

13. x to the fourth power increased by five
    $x^4 + 5$

14. the sum of b cubed and seven
    $b^3 + 7$

15. one-fifth subtracted from the product of seven and a number
    $7w - \frac{1}{5}$

16. two-thirds added to the product of four and a number
    $4x + \frac{2}{3}$

17. the product of negative three and the difference of a number and two
    $-3(n - 2)$

18. the product of three and the sum of the number and four
    $3(n + 4)$

19. the sum of four and n, all raised to the fifth power

$(4 + n)^5$

20. the difference of two and l, all raised to the third power

$(2 - l)^3$

21. five less than the product of m and n

$mn - 5$

22. the product of three and a number, increased by five

$3a + 5$

23. the quotient of four and a number, decreased by two

$4 \div n - 2$ or $\dfrac{4}{n} - 2$

24. seven added to the quotient of x and y

$x \div y + 7$ or $\dfrac{x}{y} + 7$

25. negative twenty-seven decreased by the sum of a and b

$-27 - (a + b)$

26. the difference of m and n subtracted from negative eight

$-8 - (m - n)$

27. six-tenths decreased by the product of four and the difference of y and two

$0.6 - 4(y - 2)$

28. eighty-one hundredths increased by the product of eight and the sum of x and three tenths

$0.81 + 8(x + 0.3)$

29. the difference of p and q decreased by the sum of m and n

$(p - q) - (m + n)$

30. the sum of a and b subtracted from the difference of c and d

$(c - d) - (a + b)$

31. the product of m and n subtracted from the square root of y

$\sqrt{y} - mn$

32. the square root of x subtracted from the product of a and b

$ab - \sqrt{x}$

33. a number minus the product of three and the difference of the number and six

$n - 3(n - 6)$

34. the product of five and a number minus the sum of the number and two

$5n - (n + 2)$

Find ⊗ the Mistake   *For Exercises 35–38, explain each mistake. Then translate the phrase correctly.*

35. seventeen less than three times t

Translation: $17 - 3t$

Mistake: Order is incorrect.

Correct: $3t - 17$

36. four subtracted from the square of m

Translation: $4 - m^2$

Mistake: Order is incorrect.

Correct: $m^2 - 4$

37. nine times the sum of x and y

Translation: $9x + y$

Mistake: Multiplied x by 9 instead of the sum of x and y.

Correct: $9(x + y)$

38. Nineteen divided into the product of h and k

Translation: $19 \div hk$

Mistake: Wrote 19 as a dividend instead of a divisor.

Correct: $\dfrac{hk}{19}$ or $hk \div 19$

For Exercises 39–48, translate each phrase. See Example 3.

39. The length of a rectangle is five more than the width. If the width is represented by w, write an expression that describes the length.

$w + 5$

40. The width of a rectangle is four less than the length. If the length is represented by l, write an expression that describes the width.

$l - 4$

41. The length of a rectangle is three times the width. If the width is represented by the variable w, write an expression that describes the length.

$3w$

42. The width of a rectangle is one-fourth of the length. If the length is represented by l, write an expression that describes the width.

$\dfrac{1}{4}l$

43. The radius of a circle is one-half of the diameter. If d represents the diameter, write an expression for the radius.

$\dfrac{1}{2}d$

44. The diameter of a circle is twice the radius. If r represents the radius, write an expression for the diameter.

$2r$

45. Lindsey has 42 coins in her change purse that are either dimes or quarters. If n represents the number of quarters she has, write an expression in terms of n that describes the number of dimes.

$42 - n$

46. Sherice owns a total of 60 shares of stock in two companies. If n represents the number of the higher-priced stock, write an expression in terms of n for the number of shares of the lower-priced stock.

$60 - n$

★ **47.** Don passes mile marker 51 on the highway. One-fourth of an hour later, a state trooper traveling in the same direction passes the same mile marker. If t represents the amount of time it takes the trooper to catch up to Don, write an expression in terms of t that describes the amount of time Don has traveled since passing marker 51.

$t + \dfrac{1}{4}$

★ **48.** Barry is jogging along a trail and passes a sign. One-third of an hour later, Dedra, who is traveling in the same direction, passes the same sign. If t represents the amount of time it takes Dedra to catch up to Barry, write an expression in terms of t that describes the amount of time Barry has traveled since passing the sign.

$t + \dfrac{1}{3}$

Exercises 49–56 contain word descriptions of expressions from mathematics and physics. Translate the descriptions to symbolic form. See Examples 1–3.

49. The perimeter of a rectangle is twice the width plus twice the length. $2w + 2l$

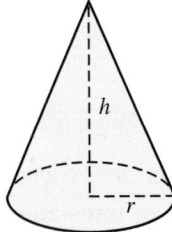

50. The area of a circle can be found by multiplying the square of the radius by π. πr^2

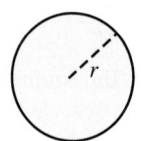

51. The volume of a cone is one-third of the product of π, the square of the radius, and the height of the cone.

$\dfrac{1}{3}\pi r^2 h$

52. The volume of a sphere is four-thirds the product of π and the cube of the radius.

$\dfrac{4}{3}\pi r^3$

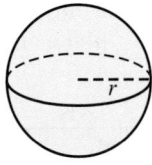

53. Albert Einstein developed an expression that describes the energy of a particle at rest, which is the product of the mass, m, of the particle and the square of the speed of light, which is represented by c.

mc^2

54. The centripetal acceleration of an object traveling in a circular path is found by an expression that is the ratio of the square of the velocity, v, to the radius, r, of the orbital path.

$\dfrac{v^2}{r}$

★ **55.** René Descartes developed an expression for the distance between two points in the coordinate plane. The distance between two points is the square root of the sum of the square of the difference of x_2 and x_1 and the square of the difference of y_2 and y_1.

$\sqrt{(x_2 - x_1)^2 + (y_2 - y_1)^2}$

★ **56.** Albert Einstein's theory of relativity includes a mathematical expression that is the square root of the difference of one and the ratio of the square of the velocity, v, of an object in motion to the square of the speed of light, c.

$\sqrt{1 - \dfrac{v^2}{c^2}}$

For Exercises 57–62, the algebraic expressions have been translated to a word phrase that is ambiguous (unclear). Indicate why the translation is unclear and give a translation that is not ambiguous. There is more than one correct answer.

57. $3x + 4$

Translation: three times x plus four

Mistake: Could be translated as $3(x + 4)$

Correct: Four more than three times a number

58. $2a - 7$

Translation: two times a minus seven

Mistake: Could be translated as $2(a - 7)$

Correct: Seven less than two times a

59. $5(x - 1)$

Translation: five times x minus one

Mistake: Could be translated as $5x - 1$

Correct: Five times the difference of x and one

60. $4(y + 6)$

Translation: four times y plus six

Mistake: Could be translated as $4y + 6$

Correct: Four times the sum of y and six

61. $n + 5(n - 6)$

Translation: n plus five times the difference of n and six

Mistake: Could be translated as $(n + 5)(n - 6)$

Correct: n plus the product of five and the difference of n and six

62. $m - 3(m + 2)$

Translation: m minus three times the sum of m and two

Mistake: Could be translated as $(m - 3)(m + 2)$

Correct: m minus the product of three and the sum of m and two

For Exercises 63–70, translate the expression to a word phrase.

63. The area of a triangle: $\frac{1}{2}bh$

One-half of the product of the base and height

64. The area of a trapezoid: $\frac{1}{2}h(a+b)$

The product of one-half of the height and the sum of a and b

65. The volume of a box: lwh

The product of the length, width, and height

66. The volume of a cylinder: $\pi r^2 h$

The product of π, the radius squared, and the height

67. The surface area of a box: $2(lw + lh + wh)$

The product of the length and width added to the product of the length and height added to the product of the width and height, all doubled

68. The surface area of a cylinder: $2\pi r(r+h)$

Twice the product of π, the radius, and the sum of the radius and the height

69. The slope of a line: $\frac{y_2 - y_1}{x_2 - x_1}$

The ratio of the difference of y_2 and y_1 to the difference of x_2 and x_1

70. A quadratic expression: $ax^2 + bx + c$

The product of a and x squared added to the product of b and x added to c

Puzzle Problem Consecutive integers follow one another in a pattern such as 1, 2, 3, . . . Let n represent the first integer in a set of consecutive integers.

a. Write expressions for the next two consecutive integers. $n+1, n+2$

b. Suppose n is odd. Write expressions for the next two consecutive odd integers.

$n+2, n+4$

c. Suppose n is even. Write expressions for the next two consecutive even integers.

$n+2, n+4$

Review Exercises

Exercises 1–6 Expressions

[1.4] 1. Rewrite and simplify $2(5+2)$ using the distributive property.

$2\cdot5 + 2\cdot2 = 14$

[1.3] For Exercises 2 and 3, simplify.

2. $4 - 8 + 6$

2

3. $\frac{2}{3} - \frac{1}{4}$

$\frac{5}{12}$

[1.5] For Exercises 4–6, simplify.

4. $-6^2 \div 3 + 4(5-9)$

-28

5. $-4|3-8| + 2^5$

12

6. -3^2

-9

1.7 Evaluating and Rewriting Expressions

Objectives

1 Evaluate an expression.
2 Determine all values that cause an expression to be undefined.
3 Rewrite an expression using the distributive property.
4 Rewrite an expression by combining like terms.

Warm-up

[1.3] 1. $\frac{1}{3} + \frac{1}{4}$

[1.5] 2. Simplify: $(-4)^2 - 0.2(6)(-4)$

[1.6] 3. *Translate to an algebraic expression.*
six less than nine times a number

To be able to solve equations in Chapter 2, we must learn how to manipulate algebraic expressions. We can perform two actions with algebraic expressions. We can *evaluate* them and *rewrite* them.

Answers to Warm-up

1. $\frac{7}{12}$ **2.** 20.8   **3.** $9n - 6$

Objective 1 Evaluate an expression.

First, we learn how to evaluate an expression.

> **Procedure** **Evaluating an Algebraic Expression**
>
> To evaluate an algebraic expression:
> 1. Replace the variables with their corresponding given values.
> 2. Calculate the numerical expression using the order of operations.

Example 1 Evaluate each expression using the given values.

a. $4m - 3(n - 4)$ when $m = 3$ and $n = 2$

Solution: $4m - 3(n - 4)$

$4(3) - 3(2 - 4)$ Replace m with 3 and n with 2.

$= 4(3) - 3(-2)$ Simplify inside the parentheses first.

$= 12 + 6$ Multiply.

$= 18$ Add.

b. $y^2 - 0.2xy + 5$ when $x = 6$ and $y = -4$

Solution: $y^2 - 0.2xy + 5$

$(-4)^2 - 0.2(6)(-4) + 5$ Replace x with 6 and y with -4.

$= 16 - 0.2(6)(-4) + 5$ Begin calculating by simplifying the exponential form.

$= 16 - (-4.8) + 5$ Multiply.

$= 16 + 4.8 + 5$ Write the subtraction as an equivalent addition.

$= 25.8$ Add from left to right.

c. $\dfrac{5r^2}{d - 3}$ when $r = 2$ and $d = -5$

Solution: $\dfrac{5r^2}{d - 3}$

$\dfrac{5(2)^2}{-5 - 3}$ Replace r with 2 and d with -5.

$= \dfrac{5(4)}{-8}$ Calculate the top and bottom expressions separately.

$= \dfrac{20}{-8}$

$= -\dfrac{5}{2}$ or -2.5

Your Turn 1 Evaluate the expression using the given values.

a. $-5a + 3(b - 6)$ when $a = -2$ and $b = 4$

b. $m^3 - 6n^2$ when $m = -2$ and $n = 3$

c. $\dfrac{3a^2 - b}{a + 6}$ when $a = -4$ and $b = 2$

Answers to Your Turn 1
a. 4   **b.** -62 **c.** 23

Objective 2 Determine all values that cause an expression to be undefined.

In Example 1c, what if $d = 3$? Notice that the denominator would become $0 \left(\dfrac{5r^2}{3 - 3} = \dfrac{5r^2}{0} \right)$ which is undefined if $r \neq 0$ and indeterminate if $r = 0$. When asked to evaluate a division expression in which the divisor or denominator contains a variable or variables, we must be careful about what values replace the variable(s). Often we need to know what values replace the variable(s) and cause the expression to be undefined or indeterminate.

Example 2 Determine all values that cause each expression to be undefined.

a. $\dfrac{5}{x + 7}$

Answer: If $x = -7$, we have $\dfrac{5}{-7 + 7} = \dfrac{5}{0}$, which is undefined because the denominator is 0.

b. $\dfrac{-y}{(y + 2)(y - 6)}$

Solution: Note that because $(y + 2)(y - 6)$ is a product, if either factor is 0, the entire denominator is 0.

Notice that if $y = -2$, we have $\dfrac{-(-2)}{(-2 + 2)(-2 - 6)}$

$$= \dfrac{2}{(0)(-8)}$$

$$= \dfrac{2}{0}, \text{ which is undefined}$$

Also, if $y = 6$, we have $\dfrac{-6}{(6 + 2)(6 - 6)}$

$$= \dfrac{-6}{(8)(0)}$$

$$= \dfrac{-6}{0}, \text{ which is undefined}$$

Answer: Both $y = -2$ and $y = 6$ cause the expression to be undefined.

> **Connection** This principle—that if either factor is 0, then the entire product is 0—is known as the *zero-factor theorem.* We use this theorem extensively in Chapter 6 when solving equations such as $(y + 2)(y - 6) = 0$.

Instructor Note Point out that an equivalent way of stating this is that $x \neq -7$ in Example 2a and $y \neq -2$ or $y \neq 6$ in Example 2b.

Your Turn 2 Determine all values that cause each expression to be undefined.

a. $\dfrac{9}{n + 6}$

b. $\dfrac{2m}{(m - 3)(m + 4)}$

Objective 3 Rewrite an expression using the distributive property.

Recall from Section 1.4 the distributive property of multiplication over addition.

> **Rule Distributive Property of Multiplication over Addition**
>
> $$a(b + c) = ab + ac$$

Answers to Your Turn 2
a. $n = -6$
b. both $m = 3$ and $m = -4$

This property gives us an alternative to the order of operations. Look at the numerical expression $2(5 + 6)$ and compare using the order of operations versus using the distributive property.

Following order of operations:

$$2(5 + 6) = 2(11)$$
$$= 22$$

or

Using the distributive property:

$$2(5 + 6) = 2 \cdot 5 + 2 \cdot 6$$
$$= 10 + 12$$
$$= 22$$

▲

Note Using the order of operations, we add within the parentheses first, then multiply.

▲

Note Using the distributive property, we multiply each addend by the factor outside the parentheses, then add the products.

The result is the same either way, so the distributive property allows us to say that $2(5 + 6)$ and $2 \cdot 5 + 2 \cdot 6$ are equivalent expressions.

Conclusion We can use the distributive property to rewrite an expression in another form that is equivalent to the original form.

Example 3 Use the distributive property to write an equivalent expression and simplify.

a. $2(x + 5)$

Solution: $2(x + 5) = 2 \cdot x + 2 \cdot 5$
$$= 2x + 10$$

b. $-6(n - 9)$

Solution: $-6(n - 9) = -6 \cdot n - (-6) \cdot 9$
$$= -6n - (-54)$$
$$= -6n + 54$$

c. $\dfrac{3}{8}\left(2m + \dfrac{4}{5}\right)$

Solution: $\dfrac{3}{8}\left(2m + \dfrac{4}{5}\right) = \dfrac{3}{8} \cdot 2m + \dfrac{3}{8} \cdot \dfrac{4}{5}$ Use the distributive property.

$$= \dfrac{3}{\overset{}{\underset{4}{8}}} \cdot \dfrac{\overset{1}{2}}{1}m + \dfrac{3}{\overset{}{\underset{2}{8}}} \cdot \dfrac{\overset{1}{4}}{5}$$ Divide out common factors.

$$= \dfrac{3}{4}m + \dfrac{3}{10}$$

> **Connection** We can apply the distributive property to $-6(n - 9)$ even though the parentheses contain subtraction instead of addition because subtraction can be expressed as addition like this:
>
> $$-6(n - 9) = -6[n + (-9)]$$
>
> In fact, it is helpful to think of the expression this way because when we multiply the -6 by the -9, we get positive 54.

Your Turn 3 Use the distributive property to write an equivalent expression and simplify.

a. $3(n + 2)$

b. $-9(7 - 3x)$

c. $\dfrac{4}{5}\left(\dfrac{1}{2}y - 10\right)$

Objective 4 Rewrite an expression by combining like terms.

Many expressions are sums of expressions called **terms**. For example, the expression $5x + 3$ is a sum of the terms $5x$ and 3.

Definition Terms: Expressions that are the addends in an expression that is a sum.

If an expression contains subtraction, as in $4x^2 - 6x + 8$, we can identify its terms by writing the subtraction as addition. Because $4x^2 - 6x + 8 = 4x^2 + (-6x) + 8$, its terms are $4x^2, -6x$, and 8. Notice we can see that the second term in $4x^2 - 6x + 8$ is

Answers to Your Turn 3
a. $3n + 6$
b. $-63 + 27x$
c. $\dfrac{2}{5}y - 8$

$-6x$ without writing the equivalent addition by recognizing that the sign to the left of the term is its sign.

The numerical factor in a term is called the numerical **coefficient** or, simply, the coefficient of the term.

Definition Coefficient: The numerical factor in a term.

The coefficient of $4x^2$ is 4.
The coefficient of $-6x$ is -6.
The coefficient of 8 is 8.

The coefficient of x is 1 because $1 \cdot x = x$.
The coefficient of $-x$ is -1 because $-1 \cdot x = -x$.
The coefficient of $\dfrac{2x}{3}$ is $\dfrac{2}{3}$ because $\dfrac{2}{3}x = \dfrac{2x}{3}$.

Another way to rewrite an expression is to combine **like terms**.

Definition Like terms: Constant terms or variable terms that have the same variable(s) raised to the same exponents.

Examples of like terms:	Examples of unlike terms:	
$3x$ and $5x$	$2x$ and $8y$	(different variables)
$4y^2$ and $9y^2$	$4t^2$ and $4t^3$	(different exponents)
$7xy$ and $3xy$	x^2y and xy^2	(different exponents)
6 and 15	12 and $12x$	(different variables)

Combining Like Terms

Now consider an expression that is a sum of like terms, such as $3x + 5x$. We can rewrite this expression in a more compact form. Multiplication of 3 times x means that x is repeatedly added three times. Likewise, 5 times x means that x is repeatedly added five times. We can expand the expression like this:

$$\overbrace{3x}^{} \quad + \quad \overbrace{5x}^{}$$
$$= \underbrace{x + x + x + x + x + x + x + x}_{}$$
$$= \underbrace{\hspace{2cm}8x\hspace{2cm}}$$

Note After expanding, we see that there are a total of eight x's repeatedly added. We can write those eight x's as a single term, $8x$.

Notice that when we combine the like terms, we get an expression that is more compact. When we rewrite an expression in a more compact form, we say that we are *simplifying* the expression. Also notice that when combining like terms, we simply add the coefficients and keep the variable the same.

> **Procedure Combining Like Terms**
> To combine like terms, add or subtract the coefficients and keep the variables and their exponents the same.

Example 4 Combine like terms.

a. $9y + 7y$

Solution: $9y + 7y = 16y$ ◀ **Note** We think: Nine y's plus seven y's equals sixteen y's.

b. $7x - 2x$

Solution: $7x - 2x = 5x$ ◀ **Note** We think: Seven x's minus two x's leaves five x's.

Connection We can use the distributive property in reverse to confirm our procedure for combining like terms. Consider the expression $3x + 5x$ again. Notice that there is a common factor of x in both terms. When applying the distributive property in reverse, we write this common factor outside the parentheses and write the remaining factors as addends within the parentheses. It looks like this:

$$3x + 5x = (3 + 5)x$$

Because the parentheses contain only the addition of two numbers, we simplify by adding the numbers.

$$3x + 5x = (3 + 5)x$$
$$= 8x$$

c. $14n^2 - n^2$

Solution: $14n^2 - n^2 = 13n^2$ ◀ **Note** The coefficient of n^2 is 1, so we think fourteen n^2's minus one n^2 leaves thirteen n^2's.

Your Turn 4 Combine like terms.

a. $6r + 8r$

b. $6.5x + 2.3x$

c. $\dfrac{1}{3}n^3 - \dfrac{3}{4}n^3$

Collecting Like Terms

Sometimes expressions are more complex and contain different sets of like terms. In such cases, we combine the like terms and copy any unlike terms in the final expression. Many people like to use the commutative property of addition and rearrange the expression first so that the like terms are together, then combine them. This type of manipulation is called *collecting* the like terms.

Example 5 Combine like terms in $4y^2 + 5 + 3y^2 - 9$.

Solution: $4y^2 + 5 + 3y^2 - 9$

$= 4y^2 + 3y^2 + 5 - 9$ Collect the like terms.

$= 7y^2 - 4$ Combine like terms.

Collecting like terms is optional. Alternatively, many people mark through the terms as they combine them as a way of keeping track of what has been combined. Using this technique with the expression from Example 5 looks like this:

$$4y^2 + 5 + 3y^2 - 9$$
$$= 7y^2 - 4$$

Example 6 Combine like terms in $15y + 8x - y - 8x$.

Solution: $15y + 8x - y - 8x$ ◀ **Note** The term $-y$ is equivalent to $-1y$.

$= 14y + 0$

$= 14y$

Sometimes collecting the like terms is beneficial, as in Example 7, where there are like terms with fraction coefficients. Some people find it easier to "see" the steps in writing the equivalent fractions when the fractions are closer together.

Example 7 Combine like terms in $\dfrac{1}{3}n - 9m + 2 - m + \dfrac{1}{4}n$.

Solution: $\dfrac{1}{3}n - 9m + 2 - m + \dfrac{1}{4}n$

$= \dfrac{1}{3}n + \dfrac{1}{4}n - 9m - m + 2$ Collect the like terms.

$= \dfrac{1(4)}{3(4)}n + \dfrac{1(3)}{4(3)}n - 9m - m + 2$ Write fraction coefficients as equivalent fractions with their LCD, 12.

$= \dfrac{4}{12}n + \dfrac{3}{12}n - 9m - m + 2$

$= \dfrac{7}{12}n - 10m + 2$ Combine like terms.

Answers to Your Turn 4

a. $14r$ **b.** $8.8x$ **c.** $-\dfrac{5}{12}n^3$

Answers to Your Turn 7
a. $10x^3 - 4y$ **b.** $4.2x^2 - 6.7x - 3$
c. $-\dfrac{3}{8}y + 4x + 2$

| **Your Turn 7** | Combine like terms. |

a. $6x^3 + 2y + 4x^3 - 6y$

b. $4.2x^2 - 6x + 9 - 0.7x - 12$

c. $\dfrac{3}{8}y - x + 2 - \dfrac{3}{4}y + 5x$

Learning Strategy

Math tends to build on prior knowledge, so it's important to review and be familiar with things you learned up to the previous class. Look at the materials in the next section in advance. Sleep well the night before, stay healthy, and be on time.

—Zheng Alick Z.

Connection One way to verify that two expressions are equivalent is to evaluate both using the same value(s) for the variable(s). Consider the expression in Example 5. If we evaluate the original expression $4y^2 + 5 + 3y^2 - 9$ and the simplified expression $7y^2 - 4$ using the same chosen value for y, we should get the same answer. Let's choose the number 2 for y.

Original Expression	Simplified Expression
$4y^2 + 5 + 3y^2 - 9$	$7y^2 - 4$
$4(2)^2 + 5 + 3(2)^2 - 9$	$7(2)^2 - 4$
$= 4(4) + 5 + 3(4) - 9$	$= 7(4) - 4$
$= 16 + 5 + 12 - 9$	$= 28 - 4$
$= 24$	$= 24$

Warning Because some expressions that are not equivalent may give the same answer using this method, this "check" does not guarantee that the expressions are equivalent. It is, however, a good sign.

Notice that the simplified expression gave the same answer, yet was easier to work with. This is one of the reasons we simplify expressions. If we can simplify an expression, we have an easier time evaluating it.

1.7 Exercises   For Extra Help   MyMathLab®

Note: Exercises marked with a ★ represent challenging exercises.

Objective 1

Prep Exercise 1 In your own words, explain how to evaluate an expression. To evaluate an expression, (1) replace the variables with their corresponding given values and (2) calculate the numerical expression using the order of operations.

For Exercises 1–22, evaluate the expressions using the given values. See Example 1.

1. $3(a + 5) - 4b; a = 2, b = 3$
9

2. $8n - 2(m + 1); m = 5, n = 3$
12

3. $4 - 0.2(x + 3); x = 1$
3.2

4. $6 - 0.4(y - 2); y = 5$
4.8

5. $2p^2 - 3p - 4; p = -3$
23

6. $n^2 - 8n + 1; n = -1$
10

7. $2y^2 - 4y + 3; y = -\dfrac{1}{2}$
$\dfrac{11}{2}$

8. $3r^2 - 9r + 6; r = -\dfrac{1}{3}$
$\dfrac{28}{3}$

9. $8 - 2(y + 4); y = -2.3$
4.6

10. $-6 - 2(l - 5); l = -0.4$
4.8

11. $-|3x| + |4y^3|; x = 5, y = -1$
-11

12. $-|2m^2| - |4n|; m = 3, n = -2$
-26

13. $|2r^2 - 3q|; r = -3, q = 5$
3

14. $|2m^2 + 2n|; m = -4, n = -5$
22

15. $\sqrt{c} + 4ab^2; a = -1, b = -2, c = 16$
-12

16. $-2x^3y + \sqrt{z}; x = -2, y = -3, z = 4$
-46

17. $-5\sqrt{a} + 2\sqrt{b}; a = 25, b = 4$
-21

18. $-3\sqrt{h} + 3\sqrt{k}; h = 16, k = 9$
-3

19. $\dfrac{6x^3}{y-8}$; $x = 2, y = 7$

-48

20. $\dfrac{4m^2}{n+4}$; $m = 2, n = 4$

2

21. $\dfrac{15-b^2}{2\sqrt{c+d}}$; $b = 3, c = 25, d = 144$

$\dfrac{3}{13}$

22. $\dfrac{5-a^2}{3\sqrt{x+y}}$; $a = 1, x = 64, y = 36$

$\dfrac{2}{15}$

23. The expression $b^2 - 4ac$ is called the *discriminant* and is used to determine the types of solutions for quadratic equations. Find the value of the discriminant given the following values.

a. $a = 2, b = 1, c = -3$

25

b. $a = 1, b = -2, c = 4$

-12

> **Connection** You will learn all of the details about each expression presented in Exercises 23–26 in future chapters (and future courses). Notice that even without a full understanding of the context in which the expressions are used, you can still evaluate them.

24. The expression $ad - bc$ is used to calculate the determinant of a matrix. Find the determinant given the following values.

a. $a = 1, b = 0.5, c = -4, d = 6$

8

b. $a = -3, b = \dfrac{4}{5}, c = 2, d = \dfrac{1}{2}$

$-\dfrac{31}{10}$

25. The expression $\dfrac{y_2 - y_1}{x_2 - x_1}$ is used to calculate the slope of a line. Find the slope given each set of the following values.

a. $x_1 = 2, y_1 = 1, x_2 = 5, y_2 = 7$

2

b. $x_1 = -1, y_1 = 2, x_2 = -7, y_2 = -2$

$\dfrac{2}{3}$

26. The expression $\sqrt{(x_2 - x_1)^2 + (y_2 - y_1)^2}$ is used to calculate the distance between two points in the coordinate plane. Evaluate the expression using the following values.

a. $x_1 = 2, y_1 = 1, x_2 = 5, y_2 = 7$

$\sqrt{45} \approx 6.7$

b. $x_1 = -1, y_1 = 2, x_2 = -7, y_2 = -2$

$\sqrt{52} \approx 7.2$

Objective 2

Prep Exercise 2 What causes an expression to be undefined?

An expression is undefined when the denominator is equal to 0.

For Exercises 27–34, determine all values that cause each expression to be undefined. See Example 2.

27. $\dfrac{-7}{5+y}$

-5

28. $\dfrac{8}{x+3}$

-3

29. $\dfrac{6m}{(m+1)(m-3)}$

$-1, 3$

30. $\dfrac{-5a}{(a-4)(a-2)}$

$4, 2$

31. $\dfrac{6+x^2}{x}$

0

32. $\dfrac{7-y}{y}$

0

★ **33.** $\dfrac{x+1}{3x-2}$

$\dfrac{2}{3}$

★ **34.** $\dfrac{3y}{2y+1}$

$\dfrac{1}{2}$

Objective 3

Prep Exercise 3 Explain the difference between evaluating and rewriting an expression.

Evaluating an expression involves replacing variables with numbers, then calculating. Rewriting an expression involves using mathematical properties to write the expression in an altered but equivalent form.

For Exercises 35–42, use the distributive property to write an equivalent expression and simplify. See Example 3.

35. $6(a + 2)$
$6a + 12$

36. $4(b - 5)$
$4b - 20$

37. $-8(4 - 3y)$
$-32 + 24y$

38. $-7(3 - 2m)$
$-21 + 14m$

39. $\dfrac{7}{8}\left(\dfrac{1}{2}c - 16\right)$
$\dfrac{7}{16}c - 14$

40. $\dfrac{4}{5}\left(-10h + \dfrac{2}{9}\right)$
$-8h + \dfrac{8}{45}$

41. $0.2(3n - 8)$
$0.6n - 1.6$

42. $-1.5(6x + 7)$
$-9x - 10.5$

Objective 4

Prep Exercise 4 What is the coefficient in a term?   The coefficient is the numerical factor in a term.

For Exercises 43–52, identify the coefficient of each term. See Objective 4.

43. $-6x^3$
-6

44. $-14y$
-14

45. m^9
1

46. y^5
1

47. $-b$
-1

48. $-n$
-1

49. $-\dfrac{2}{3}m$
$-\dfrac{2}{3}$

50. $\dfrac{5}{8}a$
$\dfrac{5}{8}$

51. $\dfrac{y}{5}$
$\dfrac{1}{5}$

52. $-\dfrac{u}{3}$
$-\dfrac{1}{3}$

Prep Exercise 5 What are like terms?   Like terms are constant terms or variable terms that have the same variable(s) raised to the same exponents.

Prep Exercise 6 Explain how to combine like terms.   To combine like terms, add or subtract the coefficients and keep the variables and their exponents the same.

For Exercises 53–78, simplify by combining like terms. See Examples 4–7.

53. $3y + 5y$
$8y$

54. $6m + 7m$
$13m$

55. $6a - 11a$
$-5a$

56. $5b - 13b$
$-8b$

57. $14x - 3x$
$11x$

58. $-5y + 12y$
$7y$

59. $-3r - 8r$
$-11r$

60. $-7m - 6m$
$-13m$

61. $6.3n - 8.2n$
$-1.9n$

62. $-5.1x^4 + 3.4x^4$
$-1.7x^4$

63. $\dfrac{1}{2}b^2 - \dfrac{5}{6}b^2$
$-\dfrac{1}{3}b^2$

64. $\dfrac{3}{4}z - \dfrac{7}{5}z$
$-\dfrac{13}{20}z$

65. $-11c - 10c - 5c$
$-26c$

66. $-15w - 6w - 11w$
$-32w$

67. $6x + 7 - x - 8$
$5x - 1$

68. $5y^2 + 6 + 3y^2 - 8$
$8y^2 - 2$

69. $-8x + 3y - 7 - 2x + y + 6$
$-10x + 4y - 1$

70. $-4a + 9b - a + 5 + 2b - 8$
$-5a + 11b - 3$

71. $-10m + 7n + 1 + m - 2n - 12 + y$
$-9m + 5n + y - 11$

72. $-3h + 7k - 5 - 8h - 7k + 19 + x$
$-11h + x + 14$

73. $1.5x + y - 2.8x + 0.3 - y - 0.7$
$-1.3x - 0.4$

74. $0.4t^2 + t - 2.8 - t^2 + 0.9t - 4$
$-0.6t^2 + 1.9t - 6.8$

75. $\dfrac{1}{6}c + 3d + \dfrac{2}{3}c + \dfrac{1}{7} - \dfrac{1}{2}d$
$\dfrac{5}{6}c + \dfrac{5}{2}d + \dfrac{1}{7}$

76. $\dfrac{5}{8}y + 4 - \dfrac{3}{4}x + \dfrac{2}{3} - \dfrac{1}{4}y$
$-\dfrac{3}{4}x + \dfrac{3}{8}y + \dfrac{14}{3}$

77. $7a + \dfrac{4}{5}b^2 - 5 - \dfrac{3}{4}a - \dfrac{2}{7}b^2 + 9$

$\dfrac{25}{4}a + \dfrac{18}{35}b^2 + 4$

78. $\dfrac{1}{2}m - 3n + 14 - \dfrac{3}{8}m - \dfrac{9}{10}n - 5$

$\dfrac{1}{8}m - \dfrac{39}{10}n + 9$

★ **79.** **a.** Translate to an algebraic expression: fourteen plus the difference of six times a number and eight times the same number.

$14 + (6n - 8n)$

 b. Simplify the expression.

$14 - 2n$

 c. Evaluate the expression when the number is -3.

20

★ **80.** **a.** Translate to an algebraic expression: the sum of negative five times a number and eight minus two times the same number.

$-5n + (8 - 2n)$

 b. Simplify the expression.

$8 - 7n$

 c. Evaluate the expression when the number is 0.2.

6.6

> Puzzle Problem Each letter in the addition problem to the right represents a different whole number, 0–9. What number does each letter represent?
>
> F = 2, O = 9, R = 7, T = 8, Y = 6, E = 5, N = 0, S = 3, I = 1, X = 4

```
  FORTY
    TEN
+   TEN
  -----
  SIXTY
```

Review Exercises

Exercise 1   Constants and Variables

[1.1] 1. Arrange in order from least to greatest. $4.2, |-6|, 4\dfrac{5}{8}, -2.5, -3$

$-3, -2.5, 4.2, 4\dfrac{5}{8}, |-6|$

Exercises 2–6   Expressions

[1.2] 2. Simplify: $\dfrac{72}{420}$

$\dfrac{6}{35}$

For Exercises 3–6, perform the indicated operation(s).

[1.3] 3. $-8 + (-14)$

-22

[1.4] 4. $\dfrac{1}{2}(12.5)(8)$

50

[1.5] 5. $4(-2) + 9$

1

[1.4] 6. $\dfrac{3}{4}\left(5\dfrac{1}{3}\right) - 1$

3

Chapter 1 Summary and Review Exercises

Complete each incomplete definition, rule, or procedure; study the key examples; then work the related exercises.

1.1 Number Sets and the Structure of Algebra

Definitions/Rules/Procedures	Key Example(s)
A **variable** is a symbol that _can vary_ in value.	Examples of variables: x, y, and z
A **constant** is a symbol that _does not vary_ in value.	Examples of constants: -14, 2.8, and π
An **expression** is a constant, a variable, or any combination of constants, variables, and arithmetic operations that _describes a calculation_ .	Examples of expressions: $2x$ and $y^2 + 7$
An **equation** is a mathematical relationship that contains a(n) _equal sign_ .	Examples of equations: $3x + 4 = 19$ and $y^2 = 16$
An **inequality** is a mathematical relationship that contains a(n) _inequality symbol_ .	Examples of inequalities: $12 < 17$ and $x - 9 \geq 14$

[1.1] 1. Complete the name of each level of the algebra pyramid.

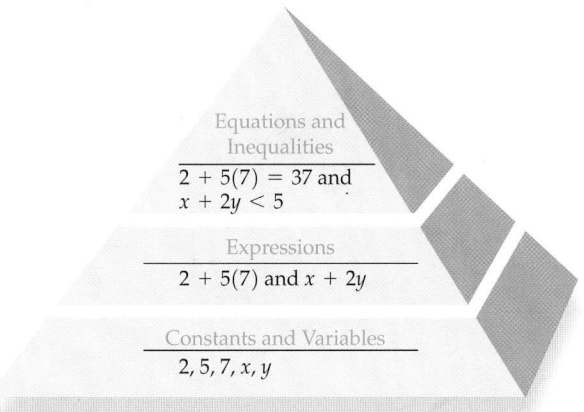

Equations and Inequalities
$2 + 5(7) = 37$ and
$x + 2y < 5$

Expressions
$2 + 5(7)$ and $x + 2y$

Constants and Variables
$2, 5, 7, x, y$

Definitions/Rules/Procedures	Key Example(s)
A **set** is a(n) _collection_ of objects. **To write a set**, write the elements or members of the set within _braces_ , _{ }_ .	Write the set containing the natural numbers divisible by 3. **Answer:** $\{3, 6, 9, 12, \ldots\}$

[1.1] For Exercises 2 and 3, write a set representing each description.

2. The months beginning with the letter M
   {March, May}

3. The even natural numbers
   $\{2, 4, 6, \ldots\}$

Definitions/Rules/Procedures	Key Example(s)
A **rational number** is a number that can be written in the form ____ $\frac{a}{b}$ ____, where a and b are ____ integers ____ and $b \neq 0$. An **irrational number** is any real number that is not ____ rational ____. A **real number** is any number that is ____ rational ____ or ____ irrational ____.	Graph -2.8 on a number line. **Solution:** The number -2.8 is graphed between -3 and -2.5 on a number line marked -3, -2.8, -2.5, -2.

Exercises 4–6 Constants and Variables

[1.1] 4. Complete the Real Numbers diagram below by writing the appropriate set name in the blanks provided.

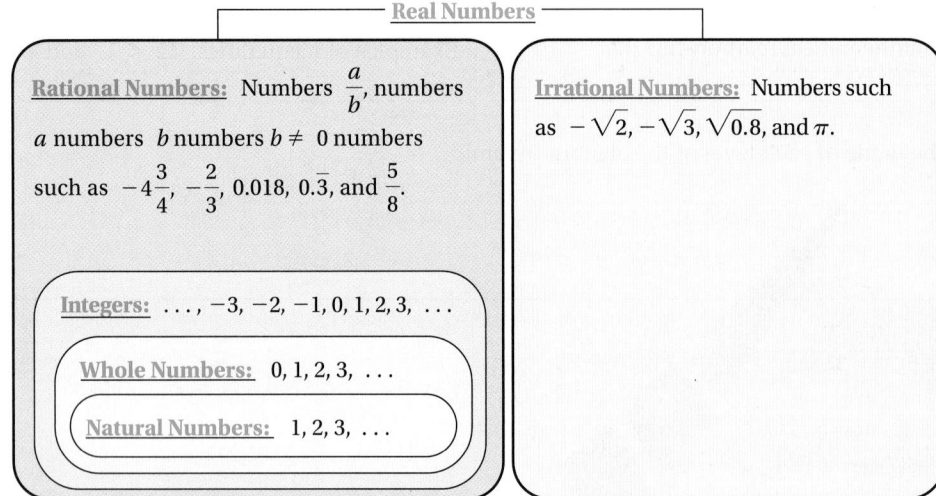

[1.1] *For Exercises 5 and 6, graph each number on a number line.*

5. $-3\frac{2}{5}$

The number $-3\frac{2}{5}$ graphed on a number line marked -4, $-3\frac{2}{5}$, -3.

6. 8.2

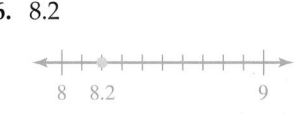

Definitions/Rules/Procedures	Key Example(s)
The **absolute value** of a number is ____ positive ____ or 0.	Find the absolute value. **a.** $|8| = 8$ **b.** $|-15| = 15$ **c.** $|0| = 0$

Exercises 7 and 8 Expressions

[1.1] *For Exercises 7 and 8, simplify.*

7. $|-6.3|$

6.3

8. $\left|2\frac{1}{6}\right|$

$2\frac{1}{6}$

Definitions/Rules/Procedures	Key Example(s)
For any two real numbers a and b, a is greater than b if a is to the ___right___ of b on the number line.	Use $<$ or $>$ to make a true statement. **a.** $-8 \quad -10$ **Answers: a.** $-8 > -10$ **b.** $-6 \quad 9$ **b.** $-6 < 9$

Exercises 9 and 10 ▲ Equations and Inequalities

[1.1] For Exercises 9 and 10, use $=, +, <,$ *or* $>$ *to write a true statement.*

9. $|-6.4| \quad 6.4$

    $=$

10. $-14 \quad -9$

    $<$

1.2 Fractions

Definitions/Rules/Procedures	Key Example(s)
A **fraction** is a quotient of two numbers or expressions a and b having the form $\dfrac{a}{b}$ where $b \neq 0$ ___. For any fraction, we can write an equivalent fraction by ___multiplying___ or ___dividing___ the numerator and denominator by the same nonzero number. A **multiple** of a given integer n is the ___product___ of n and ___an integer___. The **least common multiple (LCM)** is the ___smallest number___ that is a(n) ___multiple___ of each number in a given set of numbers ___. The **least common denominator (LCD)** is the ___least common multiple of denominators of a given set of fractions___	Find the missing number in $\dfrac{5}{6} = \dfrac{?}{24}$ so that the fractions are equivalent. **Answer:** $\dfrac{5}{6} = \dfrac{5 \cdot 4}{6 \cdot 4} = \dfrac{20}{24}$ Write $\dfrac{5}{6}$ and $\dfrac{3}{4}$ with the LCD. **Solution:** The LCD is 12. $\dfrac{5}{6} = \dfrac{5 \cdot 2}{6 \cdot 2} = \dfrac{10}{12}$ and $\dfrac{3}{4} = \dfrac{3 \cdot 3}{4 \cdot 3} = \dfrac{9}{12}$

Exercises 11-14 ◢ Constants and Variables

[1.2] For Exercises 11 and 12, find the missing number that makes the fractions equivalent.

11. $-\dfrac{6}{7} = -\dfrac{?}{14}$

    12

12. $\dfrac{15}{30} = \dfrac{?}{10}$

    5

[1.2] For Exercises 13 and 14, write each pair as equivalent fractions with the LCD.

13. $\dfrac{5}{6}$ and $\dfrac{3}{8}$ $\dfrac{20}{24}$ and $\dfrac{9}{24}$

14. $-\dfrac{7}{15}$ and $\dfrac{9}{10}$ $-\dfrac{14}{30}$ and $\dfrac{27}{30}$

Definitions/Rules/Procedures	Key Example(s)
If ___$a \cdot b = c$___, then a and b are **factors** of c.	Since $3 \cdot 4 = 12$, 3 and 4 are factors of 12.
A **prime number** is a natural number with exactly two different factors ___1___, and ___the number itself___.	Examples of prime numbers are $2, 3, 5, 7, 11, 13 \ldots$
A **prime factorization** is a factorization that contains only ___prime factors___.	Find the prime factorization of 60. **Solution:** $2 \cdot 2 \cdot 3 \cdot 5$

Exercises 15–16 ▲, Expressions

[1.2] For Exercises 15 and 16, write the prime factorization for each number.

15. 100 $2 \cdot 2 \cdot 5 \cdot 5$　　　　　　　　　　**16.** 84 $2 \cdot 2 \cdot 3 \cdot 7$

Definitions/Rules/Procedures	Key Example(s)
Given a fraction $\dfrac{a}{b}$ and $b \neq 0$, if the only factor common to a and b is 1, then the fraction is in __lowest terms__. **To simplify a fraction to lowest terms:** 1. Replace the numerator and denominator with __their prime factorizations__. 2. Eliminate (divide out) all prime factors __common__ to the numerator and denominator. 3. __Multiply__ the remaining factors	Simplify $\dfrac{54}{60}$ to lowest terms. $$\dfrac{54}{60} = \dfrac{2 \cdot 3 \cdot 3 \cdot 3}{2 \cdot 2 \cdot 3 \cdot 5} = \dfrac{9}{10}$$

Exercises 17 and 18 ▲, Expressions

[1.2] For Exercises 17 and 18, simplify to lowest terms.

17. $\dfrac{152}{200}$ $\dfrac{19}{25}$　　　　　　　　　**18.** $\dfrac{26}{39}$ $\dfrac{2}{3}$

1.3 Adding and Subtracting Real Numbers; Properties of Real Numbers

Definitions/Rules/Procedures	Key Example(s)
Two numbers are **additive inverses** if their sum is __0__.	5 and -5 are additive inverses because $5 + (-5) = 0$.
When adding two numbers that have the same sign, __add__ their absolute values and keep __the same sign__.	Add. **a.** $4 + 9 = 13$ **b.** $-4 + (-9) = -13$
When adding two numbers that have different signs, __subtract__ the smaller absolute value from the greater absolute value and keep the sign of the number with the __greater absolute value__.	**c.** $-4 + 9 = 5$ **d.** $4 + (-9) = -5$
To add fractions with the same denominator, __add__ the numerators and __keep the same denominator__; then simplify.	Add. **a.** $\dfrac{1}{10} + \dfrac{3}{10} = \dfrac{4}{10} = \dfrac{2}{5}$
To add fractions with different denominators: 1. Write each fraction as an equivalent fraction with the __LCD__. 2. Add the __numerators__ and keep the LCD. 3. Simplify.	**b.** $-\dfrac{3}{4} + \dfrac{5}{6} = -\dfrac{3(3)}{4(3)} + \dfrac{5(2)}{6(2)}$ $= -\dfrac{9}{12} + \dfrac{10}{12}$ $= \dfrac{1}{12}$
To write a subtraction statement as an equivalent addition statement, change the operation sign from a(n) __minus sign__ to a(n) __plus sign__ and change __the subtrahend to its additive inverse__.	Subtract. **a.** $4 - 9 = 4 + (-9) = -5$ **b.** $-4 - 9 = -4 + (-9) = -13$ **c.** $4 - (-9) = 4 + 9 = 13$ **d.** $-4 - (-9) = -4 + 9 = 5$

Exercises 19 and 20 Equations and Inequalities

[1.3] For Exercises 19 and 20, indicate whether the expression illustrates the commutative property of addition, associative property of addition, or additive identity.

19. $4 + (7 + 3) = (4 + 7) + 3$
Associative property of addition

20. $5.6 + (11.2 + 4.3) = 5.6 + (4.3 + 11.2)$
Commutative property of addition

Exercises 21–24 Expressions

For Exercises 21–24, add or subtract.

21. $-15 + (-12)$ -27

22. $-\dfrac{1}{4} + \dfrac{2}{3}$ $\dfrac{5}{12}$

23. $-8 - (-4.2)$ -3.8

24. $6 - (-9.1)$ 15.1

25. The credit card statement to the right shows when each transaction was posted to the account. Find the new balance.
$-\$503.59$

4-01-13	Beginning balance	$-\$685.92$
4-03-13	Dillard's	$-\$45.80$
4-05-13	Payment	$\$250.00$
4-08-13	Applebee's	$-\$36.45$
4-16-13	CVS Pharmacy	$-\$12.92$
4-24-13	Target credit	$\$32.68$
4-30-13	Finance charge	$-\$5.18$

26. The bar graph to the right shows the closing price for Target stock each day for the week beginning June 11, 2012.

   a. Find the difference between the closing price at the end of the week and the closing price at the beginning of the week.
   -0.44

   b. If a person bought shares at the close of 6/12 and sold at the close of 6/14, what would that person's net profit or loss be on each share?
   loss of $\$0.48$

CLOSING PRICE OF TARGET STOCK

$59.36	**$58.70**	**$58.21**	**$58.22**	**$58.92**
6/11/12	6/12/12	6/13/12	6/14/12	6/15/12

1.4 Multiplying and Dividing Real Numbers; Properties of Real Numbers

Definitions/Rules/Procedures	Key Example(s)
When multiplying two numbers that have the same sign, the product is ___positive___.	**Multiply.** **a.** $(5)(7) = 35$
When multiplying two numbers that have different signs, the product is ___negative___.	**b.** $(-5)(-7) = 35$ **c.** $(-1)(-2)(-5)(-7) = 70$
The product of an even number of negative factors is ___positive___, whereas the product of an odd number of negative factors is ___negative___.	**d.** $(-5)(7) = -35$ **e.** $(5)(-7) = -35$ **f.** $(-2)(-5)(-7) = -70$
Rule for multiplying fractions: $\dfrac{a}{b} \cdot \dfrac{c}{d} = \dfrac{ac}{bd}$, where $b \neq 0$ and $d \neq 0$.	**Multiply.** **a.** $\dfrac{3}{4} \cdot \dfrac{5}{7} = \dfrac{15}{28}$ **b.** $\dfrac{5}{6} \cdot \dfrac{4}{9} = \dfrac{5}{2 \cdot 3} \cdot \dfrac{2 \cdot 2}{9} = \dfrac{10}{27}$
To multiply decimal numbers: 1. Multiply as if they were ___whole numbers___. 2. Place the decimal in the product so that it has the same number of decimal places as ___the total number of decimal places in the factors___.	**Multiply.** $(3.6)(2.4)$ $\begin{array}{r} 2.4 \longrightarrow \text{ 1 place} \\ \times\ 3.6 \longrightarrow +\text{ 1 place} \\ \hline 144 \\ +\ 72 \\ \hline 8.64 \longleftarrow \text{ 2 places} \end{array}$

Exercises 27–30 ▲, Equations and Inequalities

[1.4] For Exercises 27–30, indicate whether the expression illustrates the commutative property of multiplication, associative property of multiplication, multiplicative identity, or distributive property.

27. $\dfrac{1}{3}\cdot\dfrac{5}{6}=\dfrac{5}{6}\cdot\dfrac{1}{3}$
Commutative property of multiplication

28. $3(9+1)=27+3$
Distributive property

29. $(n-8)(-3)=-3(n-8)$
Commutative property of multiplication

30. $(2\cdot3)\cdot7=2\cdot(3\cdot7)$
Associative property of multiplication

Exercises 31–34 ▲, Expressions

[1.4] For Exercises 31–34, multiply.

31. $-7(-13)$ 91

32. $6(-8)$ −48

33. $-\dfrac{3}{8}\left(\dfrac{5}{9}\right)$ $-\dfrac{5}{24}$

34. $(-6.5)(-0.4)$ 2.6

Definitions/Rules/Procedures	Key Example(s)
When dividing two numbers that have the same sign, the quotient is ___positive___.	**Divide.**
When dividing two numbers that have different signs, the quotient is ___negative___.	**a.** $24\div8=3$
Rules for division by 0: $\quad 0\div n=\underline{\ 0\ }$ when $n\neq0$. $\qquad\qquad\qquad n\div0$ is ___undefined___ when $n\neq0$. $\qquad\qquad\qquad 0\div0$ is ___indeterminate___.	**b.** $-24\div(-8)=3$ **c.** $-24\div8=-3$ **d.** $24\div(-8)=-3$ **e.** $0\div5=0$ **f.** $14\div0$ is undefined.
Multiplicative inverses are two numbers whose product is ___1___.	5 and $\dfrac{1}{5}$ are multiplicative inverses because $5\cdot\dfrac{1}{5}=1$.
Rule for dividing fractions: $\dfrac{a}{b}\div\dfrac{c}{d}=\underline{\dfrac{a\,\cdot\,d}{b\,\cdot\,c}}$, where $b\neq0, c\neq0$, and $d\neq0$.	**Divide** $\dfrac{5}{8}\div\dfrac{3}{4}$. $\dfrac{5}{8}\div\dfrac{3}{4}=\dfrac{5}{8}\cdot\dfrac{4}{3}=\dfrac{5}{2\cdot2\cdot2}\cdot\dfrac{2\cdot2}{3}=\dfrac{5}{6}$
When dividing decimal numbers, set up a long division and consider the divisor.	**Divide** $12.88\div0.06$.
Case 1: If the divisor is an integer, divide as if the dividend were a(n) ___whole number___ and place the decimal point in the quotient ___directly above its position in the dividend___.	$\begin{array}{r} 214.66 \\ 6\overline{)1288.00} \\ -12 \\ \hline 08 \\ -6 \\ \hline 28 \\ -24 \\ \hline 40 \\ -36 \\ \hline 40 \end{array}$
Case 2: If the divisor is a decimal number:	
1. Move the decimal point in the divisor to the ___right___ enough places to make the divisor a(n) ___integer___.	
2. Move the decimal place in the dividend ___the same___ number of places.	
3. Divide the ___divisor___ into the ___dividend___ as if both numbers were ___whole numbers___. Make sure you align the digits in the quotient properly.	
4. Write the decimal point in the quotient ___directly above its new position in the dividend___. In either case, continue the division process until you get a remainder of 0 or a repeating digit (or block of digits) in the quotient	**Answers:** $214.\overline{6}$

Exercises 35–40 ◢◣◢ **Expressions**

[1.4] For Exercises 35–38, divide.

35. $-25 \div 5$ -5

36. $-30 \div (-3)$ 10

37. $-\dfrac{7}{12} \div \left(-\dfrac{3}{8}\right)$ $\dfrac{14}{9}$ or $1\dfrac{5}{9}$

38. $30.66 \div (-7.3)$ -4.2

39. Refer to the graph in Exercise 26.
   a. If Kaye bought 200 shares at the close of 6/11, how much did she spend?
   $11872
   b. If she sold all 200 shares on 6/15, what was her loss?
   $88 loss, or -$88

40. In 2000, Larry has a loan balance of $-\$2405.80$. If he makes no payment for ten years, his balance will become $-\$14,896.08$. How many times greater is his debt in 2010 than it was in 2000?
   ≈ 6.2

1.5 Exponents, Roots, and Order of Operations

Definitions/Rules/Procedures	Key Example(s)
An **exponent** is a symbol written to the upper right of a base number that indicates <u>how many times to use the base as a factor</u>. The **base** of an exponent is the number that is <u>repeatedly multiplied</u>. **To evaluate an exponential form raised to a natural number exponent**, write the <u>base</u> as a factor the number of times indicated by the <u>exponent</u>; then multiply. **If the base of an exponential form is a negative number and the exponent is even**, the product is <u>positive</u>. **If the base is a negative number and the exponent is odd**, the product is <u>negative</u>.	Evaluate. **a.** $3^4 = 3 \cdot 3 \cdot 3 \cdot 3 = 81$ **b.** $-3^4 = -(3 \cdot 3 \cdot 3 \cdot 3) = -81$ **c.** $(-2)^4 = (-2)(-2)(-2)(-2) = 16$ **d.** $(-2)^5 = (-2)(-2)(-2)(-2)(-2)$ $= -32$

Exercises 41 and 42 ◢◣◢ **Expressions**

[1.5] For Exercises 41 and 42, evaluate.

41. $(-3)^2$
   9

42. -6^2
   -36

Definitions/Rules/Procedures	Key Example(s)
Every positive number has two square roots, a <u>positive</u> root and a <u>negative</u> root. Negative numbers have <u>no</u> real-number square roots. The **radical symbol** $\sqrt{}$ denotes the <u>principal (nonnegative)</u> square root.	Find all square roots. **a.** 49 **Answer:** ± 7 **b.** -36 **Answer:** No real-number square roots exist. **Simplify.** **a.** $\sqrt{81} = 9$ **b.** $\sqrt{-49}$ is not a real number. **c.** $\sqrt{\dfrac{25}{64}} = \dfrac{\sqrt{25}}{\sqrt{64}} = \dfrac{5}{8}$

Exercises 43–46 ▲ **Expressions**

[1.5] For Exercises 43 and 44, find all square roots of each number.

43. 49
± 7

44. −36
No real-number square roots exist

[1.5] For Exercises 45 and 46, evaluate the square roots.

45. $\sqrt{196}$
14

46. $\sqrt{\dfrac{25}{81}}$
$\dfrac{5}{9}$

Definitions/Rules/Procedures	Key Example(s)
Order-of-Operations Agreement Perform operations in the following order: **1.** Within _____grouping_____ symbols **2.** ___Exponents/Roots___ from left to right, in order as they occur. **3.** ___Multiplication/Division___ from left to right, in order as they occur. **4.** ___Addition/Subtraction___ from left to right, in order as they occur.	Simplify. $35 - [29 - 2(12 + 4)] + 4^3$ $= 35 - [29 - 2(16)] + 4^3$ $= 35 - (29 - 32) + 4^3$ $= 35 - (-3) + 4^3$ $= 35 - (-3) + 64$ $= 35 + 3 + 64$ $= 102$

Exercises 47–55 ▲ **Expressions**

[1.5] For Exercises 43–55, simplify.

47. $(-2)^3(4)(-5)$
160

48. $3^5 \div 3^2 \div 3^2 \div 3$
1

49. $-\dfrac{1}{3} - \dfrac{3}{2} \div \left(\dfrac{1}{6}\right)$
$-\dfrac{28}{3}$

50. $4|-2| \div (-8)$
-1

51. $10 \div 5 + (-5)(-5)$
27

52. $6 + \{3(4 - 5) + 2[6 + (-4)]\}$
7

53. $-1 - 3[4^2 - 3(-6)]$
-103

54. $\dfrac{-6 + 3^2(8 - 10)}{2^3 - 16}$
3

55. $\dfrac{3(4 - 5) - 4 \cdot 3 + (11 - 20)}{(3 - 4)^3}$
24

Definitions/Rules/Procedures	Key Example(s)
To find the arithmetic mean, or average, of n numbers, divide <u>the sum of the numbers by n</u>. **Arithmetic** mean $= \dfrac{x_1 + x_2 + \cdots + x_n}{n}$	**Example 5:** Carl has the following test scores: 68, 76, 82, and 78. Find the average of his test scores. $\dfrac{68 + 76 + 82 + 78}{4} = \dfrac{304}{4} = 76$

Exercise 56 ➤ Expressions

[1.5] **56.** A company reports the following revenues for the first six months of 2013. Find the average revenue.

$54,158\frac{1}{3}$, or $54,158.33

Month	Revenue
January	$45,320
February	$38,250
March	$61,400
April	$42,500
May	$74,680
June	$62,800

1.6 Translating Word Phrases to Expressions

Definitions/Rules/Procedures	Key Example(s)
To translate a word phrase to an expression, identify the <u>unknowns</u>, <u>constants</u>, and <u>key words</u>; then write the corresponding symbolic form.	**Translate to an algebraic expression:** **a.** Five more than seven times a number **Answer:** $7n + 5$ **b.** Six times the difference of a number and nine **Answer:** $6(n - 9)$ **c.** Four divided by a number cubed **Answer:** $4 \div n^3$

Exercises 57–60 ➤ Expressions

[1.6] *For Exercises 57–60, translate to an algebraic expression.*

57. twice a number subtracted from fourteen

$14 - 2n$

58. the ratio of y to seven

$\dfrac{y}{7}$

59. a number minus twice the sum of the number and four

$y - 2(y + 4)$

60. seven times the difference of m and n

$7(m - n)$

1.7 Evaluating and Rewriting Expressions

Definitions/Rules/Procedures	Key Example(s)
To evaluate an algebraic expression: **1.** Replace the <u>variables</u> with their corresponding values. **2.** Calculate the numerical expression using the <u>order of operations agreement</u>.	**Evaluate** $4x^2 - 5x$ when $x = -3$. $4(-3)^2 - 5(-3)$ $= 4(9) - 5(-3)$ $= 36 - (-15)$ $= 36 + 15$ $= 51$

Exercises 61–66 ▲, Expressions

[1.7] For Exercises 61–66, evaluate the expressions using the given values.

61. $b^2 - 4ac; a = -3, b = -1, c = 5$
61

62. $b^2 - 4ac; a = 2, b = 6, c = -2$
52

63. $-|4x| + |y^2|; x = -3, y = -2$
−8

64. $\sqrt{m} + \sqrt{n}; m = 16, n = 9$
7

65. $\sqrt{m + n}; m = 16, n = 9$
5

66. $\dfrac{3l^2}{4 - n}; l = -4, n = 16$
−4

Definitions/Rules/Procedures	Key Example(s)
An expression of the form $\dfrac{a}{b}$ with $a \neq 0$ is undefined if $\underline{b = 0}$.	For what value of x is $\dfrac{x}{x + 7}$ undefined? $x = -7$

Exercises 67 and 68 ▲, Expressions

[1.7] For Exercises 67 and 68, determine all value(s) that cause each expression to be undefined.

67. $\dfrac{n}{n + 6}$
−6

68. $\dfrac{y}{(y - 4)(y + 3)}$
−3, 4

Definitions/Rules/Procedures	Key Example(s)
Distributive Property of Multiplication over Addition $a(b + c) = \underline{ab + ac}$	Use the distributive property to write an equivalent expression. $$3(x + 7) = 3 \cdot x + 3 \cdot 7$$ $$= 3x + 21$$

Exercises 69–72 ▲, Equations and Inequalities

[1.7] For Exercises 69–72, use the distributive property to write an equivalent expression and simplify.

69. $5(x + 6)$
5x + 30

70. $-3(5n - 8)$
−15n + 24

71. $\dfrac{1}{4}(8y + 3)$
$2y + \dfrac{3}{4}$

72. $0.6(4.5m - 2.1)$
2.7m − 1.26

Definitions/Rules/Procedures	Key Example(s)
Terms are expressions that are the <u>addends in an expression that is a sum</u>. A **coefficient** is the <u>numerical factor in a product</u>. **Like terms** are constants or variable terms that <u>have the same variables raised to the same exponents</u>. **To combine like terms**, add or subtract the <u>coefficients</u> and keep the <u>variables and their exponents</u> the same.	Combine like terms. $$15x^2 + 8x + 3x^2 + 7 - 9x$$ $$= 15x^2 + 3x^2 + 8x - 9x + 7$$ $$= 18x^2 - x + 7$$

Exercises 73–78 ➤ Expressions

[1.7] For Exercises 73–78, simplify by combining like terms.

73. $6x + 3y - 9x - 6y - 15$
$-3x - 3y - 15$

74. $5y^2 - 6y + 4y - y^2 + 9y$
$4y^2 + 7y$

75. $-6xy + 9xy - xy$
$2xy$

76. $8x^3 - 4x - 6x^2 - 10x^3 + 4x$
$-2x^3 - 6x^2$

77. $-4m - 4m - 4n + 4n$
$-8m$

78. $14 - 6x + 4y - 8 - 10y - 8x$
$-14x - 6y + 6$

Chapter 1 **Practice Test**

For Extra Help

Step-by-step test solutions are found on the Chapter Test Prep Videos available in MyMathLab® *or on* YouTube.

1. Graph $-3\frac{2}{7}$ on a number line.

$-4 \qquad -3\frac{2}{7} \quad -3$

[1.1]

2. Simplify: $|-3.67|$
3.67 [1.1]

3. Write the prime factorization of 100.
$2 \cdot 2 \cdot 5 \cdot 5$ [1.2]

4. Simplify: $\dfrac{17}{51}$
$\frac{1}{3}$ [1.2]

5. Find the missing number: $-\dfrac{6}{5} = -\dfrac{?}{10}$
12 [1.2]

For Exercises 6 and 7, indicate whether the expression illustrates the commutative property of addition, the associative property of addition, the commutative property of multiplication, the associative property of multiplication, or the distributive property of multiplication over addition.

6. $3(2 + 5) = 6 + 15$
Distributive property [1.4]

7. $4(7 + 1) = 4(1 + 7)$
Commutative property of addition [1.3]

For Exercises 8–13, calculate.

8. $8 + (-4)$
4 [1.3]

9. $\dfrac{7}{8} - \left(-\dfrac{5}{6}\right)$
$1\frac{17}{24}$ or $\frac{41}{24}$ [1.3]

10. $(-1.5)(-0.4)$
0.6 [1.4]

11. $-\dfrac{5}{6} \div \dfrac{2}{3}$
$-\frac{5}{4}$ [1.4]

12. $(-4)^3$
-64 [1.5]

13. -3^4
-81 [1.5]

For Exercises 14–17, simplify.

14. $-12 \div 3 (6 - 2^2)$
-8 [1.5]

15. $\dfrac{6^2 + 14}{(2 - 7)^2}$
2 [1.5]

16. $-8 \div |4 - 2| + 3^2$

5 [1.5]

17. $\sqrt{25 - 16} + [(14 + 2) - 3^2]$

10 [1.5]

For Exercises 18 and 19, translate the indicated phrase.

18. twice the sum of m and n

$2(m + n)$ [1.6]

19. five less than three times w

$3w - 5$ [1.6]

20. After making a payment, Karen's balance on her credit card is $-\$854.80$. If her balance prior to making the payment was $-\$1104.80$, how much did she pay toward the balance?

$250 [1.3]

21. On six days over a three-month period, a city receives 6 inches, 10 inches, 4 inches, 3 inches, 2.5 inches, and 8 inches of snow. Calculate the average daily amount of snowfall over those days.

5.583 [1.5]

22. Evaluate $-|3x^2 + 2y|$, when $x = -1$ and $y = 4$.

−11 [1.7]

23. Evaluate $\sqrt{a - b}$, when $a = 64$ and $b = -36$.

10 [1.7]

24. Use the distributive property to write an equivalent expression: $-5(4y + 9)$

$-20y - 45$ [1.7]

25. Simplify: $3.5x - 8 + 2.1x + 9.3$

$5.6x + 1.3$ [1.7]

Solving Linear Equations and Inequalities

Chapter Overview

In Chapter 1, we saw that we can perform two actions with expressions: (1) evaluate and (2) rewrite. In this chapter, we build on that foundation by creating equations and inequalities out of those expressions; then we *solve* those equations and inequalities.

Specifically, we will learn the following objectives:

▶ Use the addition and multiplication principles of equality to solve equations.

▶ Use those two principles to isolate a variable in a formula.

▶ Solve applications such as proportions and percent problems.

▶ Modify the principles of equality to solve inequalities.

2.1 Equations, Formulas, and the Problem-Solving Process

2.2 The Addition Principle of Equality

2.3 The Multiplication Principle of Equality

2.4 Applying the Principles to Formulas

2.5 Translating Word Sentences to Equations

2.6 Ratios and Proportions

2.7 Percents

2.8 Solving Linear Inequalities

Instructor Note

Consider breaking this chapter into two units: 2.1–2.4 and 2.5–2.8. This tests the basics of solving equations before getting into applications and inequalities.

2.1 Equations, Formulas, and the Problem-Solving Process

Objectives

1 Verify solutions to equations.

2 Use formulas to solve problems.

Warm-up

[1.5] Simplify using the order of operations.

    1. $(15)(18) + (2)(18 + 9)$

    2. $(12.5)(8) - (2.8)(4.8)$

[1.7] **3.** Simplify by combining like terms. $2y + 3(y + 0.2)$

We now move to the top tier of the Algebra Pyramid and focus on equations. We repeat the definition, but state it a little differently.

Definition Equation: Two expressions set equal.

For example, $4x + 5 = 9$ is an equation made from the expressions $4x + 5$ and 9.

The Algebra Pyramid

Equations
and Inequalities
$4x + 5 = 9$

Expressions
$4x + 5, 9$

Constants and Variables
$4, 5, 9, x$

> **Connection** We can think of an equation as a complete sentence and the equal sign as the verb *is*. We can read $4x + 5 = 9$ as "Four x plus five is equal to nine" or "Four x plus five is nine."
>
> $$4x + 5 = 9$$
> $$\updownarrow$$
> "Four x plus five is 9."

Objective 1 Verify solutions to equations.

An equation can be true or false. For example, the equation $3 + 1 = 4$ is true, whereas the equation $3 + 1 = 5$ is false. When given an equation that contains a variable for a unknown number, such as $x + 2 = 7$, our goal is to solve it, which means to find every number that can replace the variable and make the equation true. Such a number is called a **solution** to the equation.

Definition Solution: A number that makes an equation true when it replaces the variable in the equation.

For example, the number 5 is a solution to the equation $x + 2 = 7$ because when 5 replaces x, it makes the equation true.

$$x + 2 = 7$$
$$\downarrow$$
$$5 + 2 = 7 \quad \text{True}$$

Answers to Warm-up
1. 324   **2.** 86.56
3. $5y + 0.6$

By showing that the equation $x + 2 = 7$ is a true statement when x is replaced with 5, we verify, or *check*, that 5 is the solution to the equation.

> **Procedure** **Checking a Possible Solution**
>
> To determine whether a value is a solution to a given equation, replace the variable in the equation with the value. If the resulting equation is true, the value is a solution.

Example 1 Check to see if the given value is a solution to the equation.

a. $4x + 9 = 1; x = -2$

Solution: $4x + 9 = 1$

$$4(-2) + 9 \stackrel{?}{=} 1 \quad \text{\textcolor{gray}{Replace } x \text{ with } -2 \text{ and see if the equation is true.}}$$
$$-8 + 9 \stackrel{?}{=} 1$$
$$1 = 1$$

Because -2 makes the equation true, it is a solution to $4x + 9 = 1$.

b. $n^2 + 0.8 = 3n - 0.1; n = 0.4$

Solution: $n^2 + 0.8 = 3n - 0.1$

$$(0.4)^2 + 0.8 \stackrel{?}{=} 3(0.4) - 0.1 \quad \text{\textcolor{gray}{Replace } n \text{ with 0.4 and see if the equation is true.}}$$
$$0.16 + 0.8 \stackrel{?}{=} 1.2 - 0.1$$
$$0.96 \neq 1.1$$

Because 0.4 does not make the equation true, it is not a solution to $n^2 + 0.8 = 3n - 0.1$.

c. $\dfrac{3}{4}y - 1 = \dfrac{y}{2} + \dfrac{1}{3}; y = 5\dfrac{1}{3}$

Note It is convenient to write $\dfrac{y}{2}$ as $\dfrac{1}{2}y$, which

Solution: $\dfrac{3}{4}y - 1 = \dfrac{1}{2}y + \dfrac{1}{3}$ ◄ we can do because $\dfrac{y}{2} = y \div 2 = y \cdot \dfrac{1}{2} = \dfrac{1}{2}y.$

$$\dfrac{3}{4}\left(5\dfrac{1}{3}\right) - 1 \stackrel{?}{=} \dfrac{1}{2}\left(5\dfrac{1}{3}\right) + \dfrac{1}{3} \quad \text{\textcolor{gray}{Replace } y \text{ with } 5\dfrac{1}{3} \text{ and see if the equation is true.}}$$

$$\dfrac{\overset{1}{3}}{\underset{1}{4}}\left(\dfrac{\overset{4}{16}}{\underset{1}{3}}\right) - 1 \stackrel{?}{=} \dfrac{1}{2}\left(\dfrac{\overset{8}{16}}{3}\right) + \dfrac{1}{3} \quad \text{\textcolor{gray}{Rewrite } 5\dfrac{1}{3} \text{ as an improper fraction, divide out common factors, and then multiply.}}$$

$$4 - 1 \stackrel{?}{=} \dfrac{8}{3} + \dfrac{1}{3}$$

$$3 = 3 \quad \text{\textcolor{gray}{Therefore, } 5\dfrac{1}{3} \text{ is a solution to } \dfrac{3}{4}y - 1 = \dfrac{y}{2} + \dfrac{1}{3}.}$$

Your Turn 1 Check to see if the given value is a solution to the equation.

a. $5n - 8 = 3n + 4; n = 6$ **b.** $4y - 3.5 = 2y + 3(y + 0.2); y = -1.5$

c. $\dfrac{2}{3}x + \dfrac{1}{4} = \dfrac{1}{5} - x^2; x = \dfrac{1}{6}$

Objective 2 Use formulas to solve problems.

A primary purpose of studying mathematics is to develop and improve problem-solving skills. George Polya proposed the idea that all problem solving follows a four-step outline: (1) Understand the problem, (2) devise a plan for solving the problem, (3) execute the plan, and (4) check the results. Following is an outline for problem solving based on Polya's four stages. We'll see Polya's four stages illustrated throughout the rest of the text in application problems.

Note The symbol $\stackrel{?}{=}$ indicates ▶ that we are asking, "How does the left side compare with the right side? Are they equal or not equal?"

Learning Strategy

As suggested in the To the Student section, set up your notebook with four sections: Notes, Homework, Study sheets, and Practice tests. Remember to use color in your notes: red for definitions and blue for rules and procedures. Organization and color coding will help you locate important concepts faster.

Answers to Your Turn 1
a. yes **b.** no **c.** no

Instructor Note In future sections, we will illustrate the outline in the more complex application examples.

Procedure **Problem-Solving Outline**

1. **Understand** the problem.
   a. Read the question(s) (not the whole problem, just the question at the end) and write a note to yourself about what you are to find.
   b. Read the whole problem, underlining the key words.
   c. If possible or useful, draw a picture, make a list or table to organize what is known and unknown, simulate the situation, or search for a related example problem.
2. **Plan** your solution by searching for a formula or using the key words to translate to an equation.
3. **Execute** the plan by solving the equation/formula.
4. **Answer** the question. Look at the note about what you were to find and make sure you answer that question. Include appropriate units.
5. **Check** results.
   a. Try finding the solution in a different way, reversing the process, or estimating the answer and making sure the estimate and actual answer are reasonably close.
   b. Make sure the answer is reasonable.

As the course develops, we will explore the various strategies listed in the outline. In this section, we focus on using **formulas** to solve problems. Later in this chapter, we solve problems by using key words to translate to an equation.

Definition **Formula:** An equation that describes a mathematical relationship.

First, let's consider using formulas from geometry. The table on page 91 lists some common geometric formulas. Before we examine the formulas, let's review a few terms.

Definitions **Perimeter:** The distance around a figure.
Area: The total number of square units that fill a figure.
Volume: The total number of cubic units that fill a space.
Circumference: The distance around a circle.
Radius: The distance from the center of a circle to any point on the circle.
Diameter: The distance across a circle through its center.

Procedure **Using a Formula**

To use a formula:
1. Replace the variables with the corresponding given values.
2. Solve for the missing value.

Geometric Formulas

Plane Figures			

Square

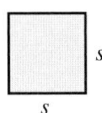

$P = 4s$
$A = s^2$

Rectangle

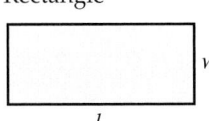

$P = 2l + 2w$
$A = lw$

Parallelogram

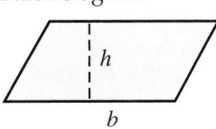

$A = bh$

▲

Note A parallelogram is a four-sided figure with two pairs of parallel sides. A trapezoid has only one pair of parallel sides.

Trapezoid

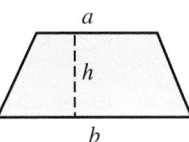

$A = \dfrac{1}{2}h(a + b)$

▲

Triangle

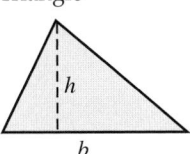

$A = \dfrac{1}{2}bh$

Circle

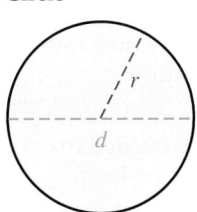

$C = \pi d$ or $C = 2\pi r$
$A = \pi r^2$

◀ **Note** A circle's radius is half its diameter. (Or its diameter is twice the radius.)

◀ **Note** Recall that π represents an irrational number. The π key on most calculators approximates the value as 3.141592654. Some people prefer to round the value to 3.14, which is simpler but less accurate.

Solids				

Box

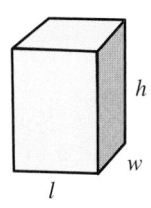
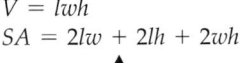

$V = lwh$
$SA = 2lw + 2lh + 2wh$
▲

Pyramid

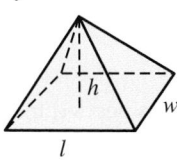

$V = \dfrac{1}{3}lwh$
▲

Cylinder

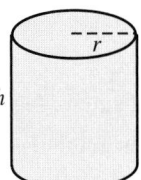

$V = \pi r^2 h$

Cone

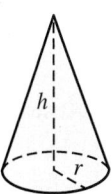

$V = \dfrac{1}{3}\pi r^2 h$

Sphere

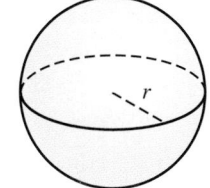

$V = \dfrac{4}{3}\pi r^3$

Connection Notice that the volume formulas for a box and a pyramid are similar. This is because a pyramid has $\dfrac{1}{3}$ the volume of a box with the same-size base and same height. The same relationship is true for cylinders and cones with the same-size base and same height.

Learning Strategy

When it comes to memorizing formulas, take it one at a time. Don't focus on memorizing them all in a day; it's too overwhelming. Take one or two and study them each day. Once you have them memorized, add a couple more.

—Ellyn G.

Problems Involving a Single Formula

Example 2

a. A mason is building a rectangular foundation wall that is to be 75 feet by 40 feet. What is the total distance around the wall?

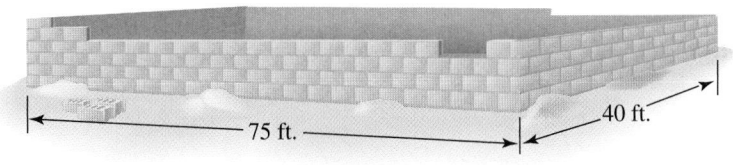

75 ft.   40 ft.

Instructor Note When finding perimeter, area, or volume, encourage students to substitute both the numerical value and the units into the formulas. This will help prevent the error of using unlike units and will help in expressing area and volume in the correct units.

Understand The "total distance around the wall" is the perimeter.

Plan Because the shape is a rectangle, we can use the formula $P = 2l + 2w$.

Execute Replace l with 75 feet and w with 40 feet; then calculate.

$$P = 2l + 2w$$
$$P = 2(75 \text{ ft.}) + 2(40 \text{ ft.})$$
$$P = 150 \text{ ft.} + 80 \text{ ft.}$$
$$P = 230 \text{ ft.}$$

Answer The mason must have enough supplies to build a wall that is 230 feet long.

Check In this case, the solution can be verified using an alternative method. One could add all four side lengths.

$$P = 75 \text{ ft.} + 75 \text{ ft.} + 40 \text{ ft.} + 40 \text{ ft.} = 230 \text{ ft.}$$

b. A tabletop is in the shape of a circle whose radius is 4 feet. If the tabletop is to be covered with Formica, how much Formica is needed?

Understand The amount of Formica needed to cover the tabletop is the same as the area of the tabletop.

Plan Because the tabletop is in the shape of a circle, we use the formula $A = \pi r^2$.

4 ft.

Execute Replace π with 3.14 and r with 4 ft. and simplify.

$$A = \pi r^2$$
$$A = (3.14)(4 \text{ ft.})^2$$
$$A = (3.14)(16 \text{ ft.}^2)$$
$$A = 50.24 \text{ ft.}^2$$

Answer About 50.24 square feet of Formica is needed to cover the tabletop.

Check Verify the reasonableness of the answer using estimation. If we round π to 3, the answer is $(3)(4 \text{ ft.})(4 \text{ ft.}) = 48 \text{ ft.}^2$. Because π is a little more than 3 and our answer is a little more than 48 ft.^{2}, our answer is reasonable.

Your Turn 2 A wallpaper hanger is to paper a bedroom wall that is 18 feet long and 12 feet high. How much wallpaper is needed to cover the wall?

Problems Involving Combinations of Formulas

A *composite figure* contains more than one geometric figure and requires a combination of formulas.

Procedure **Calculating the Area of Composite Figures**

For composite figures:
1. To calculate the area of a figure composed of two or more figures that are next to each other, add the areas of the individual figures.
2. To calculate the area of a region defined by a smaller figure within a larger figure, subtract the area of the smaller figure from the area of the larger figure.

Answer to Your Turn 2
216 square feet

Example 3   Following is a drawing of a room that is to be carpeted. Calculate the area of the room.

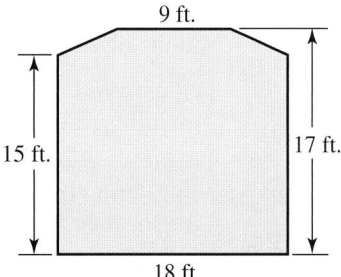

Understand   This figure can be viewed as a trapezoid and a rectangle. The height of the trapezoid is found by subtracting 15 feet from 17 feet.

Plan   Because the figure consists of a trapezoid and rectangle combined, we can add the areas of these shapes to find the total area of the figure. The area of a rectangle is found using the formula $A = lw$, and the area of a trapezoid is found using the formula $A = \frac{1}{2}h(a + b)$. The total area is

$$A = \text{Area of the rectangle} + \text{Area of the trapezoid}$$

$$A = \qquad lw \qquad + \qquad \frac{1}{2}h(a + b)$$

Execute   Replace the variables with the corresponding values and calculate.

$$A = lw + \frac{1}{2}h(a + b)$$

$$A = (18\,\text{ft.})(15\,\text{ft.}) + \frac{1}{2}(2\,\text{ft.})(18\,\text{ft.} + 9\,\text{ft.}) \quad \begin{array}{l} l = 18\,\text{ft.}, w = 15\,\text{ft.}, h = 2\,\text{ft.}, \\ a = 18\,\text{ft.}, \text{and } b = 9\,\text{ft.} \end{array}$$

$$A = (18\,\text{ft.})(15\,\text{ft.}) + \frac{1}{2}(2\,\text{ft.})(27\,\text{ft.})$$

$$A = 270\,\text{ft.}^2 + 27\,\text{ft.}^2$$

$$A = 297\,\text{ft.}^2$$

Answer   The total area is 297 square feet.

Check   We can verify the reasonableness of the answer using estimation. Suppose the figure had been a rectangle measuring 18 feet by 17 feet. We would expect the area of this rectangle to be slightly greater than the area of the actual figure. The area of an 18-foot by 17-foot rectangle is $A = (18\,\text{ft.})(17\,\text{ft.}) = 306\,\text{ft.}^2$, which indicates that 297 square feet is reasonable.

Your Turn 3   Calculate the area of the following sign. (Use $\pi \approx 3.14$.)

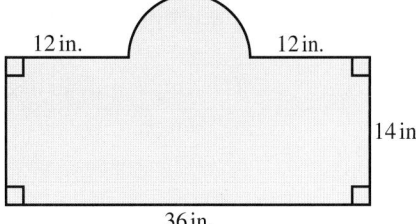

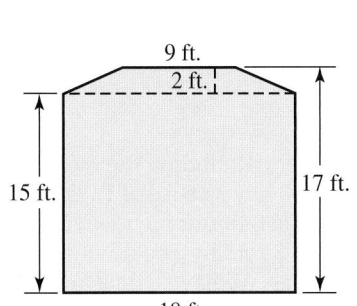

Answer to Your Turn 3
560.52 in.^{2}

Sometimes a figure may be placed within another figure and we must calculate the area of the region *between* the outside figure and inside figure.

Example 4   A portion of the front of a house is to be covered with siding (shaded in the figure). The window is 2.8 feet by 4.8 feet. Calculate the area to be covered.

Understand The shaded area is the entire triangle excluding the rectangular window.

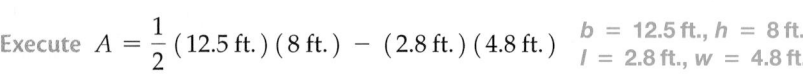

Plan We can exclude the area of the window by calculating the area of the triangle and subtracting the area of the window.

A = Area of the triangle − Area of the rectangle

$$A = \frac{1}{2}bh - lw$$

Execute $A = \dfrac{1}{2}(12.5 \text{ ft.})(8 \text{ ft.}) - (2.8 \text{ ft.})(4.8 \text{ ft.})$ $b = 12.5 \text{ ft.}, h = 8 \text{ ft.},$
$l = 2.8 \text{ ft.}, w = 4.8 \text{ ft.}$

$A = 50 \text{ ft.}^2 - 13.44 \text{ ft.}^2$

$A = 36.56 \text{ ft.}^2$

Answer The area to be covered is 36.56 square feet.

Check Estimate the area by rounding the decimal numbers. Rounding the measurements of the triangle to the nearest whole number, we have a base of 13 feet and a height of 8 feet, which means an area of 52 square feet. Rounding the window dimensions to the nearest whole number, we have a 3-foot by 5-foot window, which has an area of 15 square feet. Subtracting these areas, we have a final area of 37 square feet, which indicates that our answer of 36.56 square feet is reasonable. Because all of our estimates are greater than or equal to the actual amount, the estimated answer is larger than the actual answer.

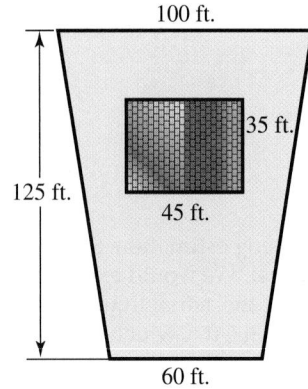

Your Turn 4   A house is to be situated on a lot as shown at the left. Once the house is built, the remaining area will be landscaped. Calculate the area to be landscaped.

Problems Involving Nongeometric Formulas

Following is a list of other formulas you may encounter.

The distance, d, an object travels given its rate, r, and the time of travel, t:	$d = rt$
The average rate of travel, r, given the total distance, d, and total time, t:	$r = \dfrac{d}{t}$
The voltage, V, in a circuit with a current, i, in amperes (A) and a resistance, R, in ohms (Ω):	$V = iR$
The temperature in degrees Celsius given degrees Fahrenheit:	$C = \dfrac{5}{9}(F - 32)$
The temperature in degrees Fahrenheit given degrees Celsius:	$F = \dfrac{9}{5}C + 32$

Example 5   A truck driver begins a delivery at 9 A.M. and travels 150 miles before taking a 30-minute break. He then travels another 128 miles, arriving at his destination at 2 P.M. What was his average driving rate?

Understand We are given travel distances and times, and we are to find the average driving rate.

Answer to Your Turn 4
8425 ft.^{2}

Plan To find the average rate, we first need the total distance traveled and the total time spent driving. We can then use the formula $r = \dfrac{d}{t}$.

Execute Total distance $= 150 + 128 = 278$ miles $= d$
Total time driving $= 5 - 0.5 = 4.5$ hours $= t$

◄ **Note** From 9 A.M. to 2 P.M. is 5 hours, but we must deduct the 0.5-hour (30-minute) break because he was not traveling during that time.

Now we can calculate the average rate.

$$r = \frac{278 \text{ miles}}{4.5 \text{ hours}}$$

$$r = 61.\overline{7} \text{ mph (miles per hour)}$$

Answer His average driving rate was $61.\overline{7}$ miles per hour.

Check We can use $d = rt$ to verify that if he traveled an average rate of $61.\overline{7}$ miles per hour for 4.5 hours, he would travel 278 miles.

$$d \approx 61.8(4.5)$$

$$d \approx 278.1 \text{ miles}$$

◄ **Note** We rounded $61.\overline{7}$, so the distance is an approximation.

Answer to Your Turn 5

$3\dfrac{3}{17}$ or ≈ 3.2 mph

Your Turn 5 Hilda hikes 10 miles then rests for 20 minutes, after which she hikes another 8 miles. If she began her hike at 10:30 A.M. and stopped at 4:30 P.M., what was her average rate?

2.1 Exercises For Extra Help MyMathLab®

Note: Exercises marked with a ★ represent challenging exercises.

Objective 1

Prep Exercise 1 What symbol is present in an equation but not in an expression?
An equation has an equal sign.

Prep Exercise 2 What is a solution to an equation?
A solution to an equation is a number that makes the equation true when it replaces the variable in the equation.

Prep Exercise 3 Explain how to check a value to see if it is a solution to a given equation.
(1) Replace the variable in the equation with the value, and (2) if the resulting equation is true, the value is a solution.

For Exercises 1–16, check to see if the given number is a solution to the given equation.
See Example 1.

1. $3x - 5 = 41; x = 2$
no

2. $4a + 7 = 51; a = 11$
yes

3. $7y - 1 = y + 3; y = \dfrac{2}{3}$
yes

4. $-8t - 3 = 2t - 15; t = -\dfrac{6}{5}$
no

5. $-4(x - 5) + 2 = 5(x - 1) + 3; x = -2$
no

6. $2(3m + 2) - 2 = 5m - 1; m = -3$
yes

7. $-\dfrac{1}{2} + \dfrac{1}{3}y = \dfrac{3}{2}; y = 6$
yes

8. $\dfrac{1}{2}p - \dfrac{1}{2} = \dfrac{2}{5}p + \dfrac{3}{2}; p = 20$
yes

9. $4.3z - 5.71 = 2.1z + 0.07; z = 2.2$
no

10. $12.7a + 12.6 = a + 5.4a; a = -2$
yes

⋆ **11.** $b^2 - 4b = b^3 + 6; b = -1$
yes

⋆ **12.** $-x^3 + 9 = 2x^2 - 6x; x = -3$
yes

⋆ **13.** $|x^2 - 9| = -x + 3; x = 3$
yes

⋆ **14.** $-|2u - 3| = -3u + 8; u = 5$
yes

⋆ **15.** $\dfrac{4x - 3}{x - 4} = \sqrt{x + 8}; x = 17$
yes

⋆ **16.** $\dfrac{-y}{10 + y} = \dfrac{\sqrt{4 - y}}{3}; y = -5$
yes

Objective 2

Prep Exercise 4 If the dimensions of a figure are given in inches, in what unit will the area be expressed?
Square inches

For Exercises 17–40, solve using geometric formulas. (Answers to exercises involving π were calculated using the π key on a calculator, which approximates the value as 3.141592654.) See Examples 2–4.

17. Janet wants to put a wallpaper border around her rectangular kitchen. The room is 13 feet by 20 feet.
    a. What is the total length of wallpaper border she needs?
       66 ft.
    b. If the wallpaper border comes in packages of 12 feet, how many packages must she buy?
       6
    c. If the packages are priced at $8.99 each, what will be the total cost of the wallpaper?
       $53.94

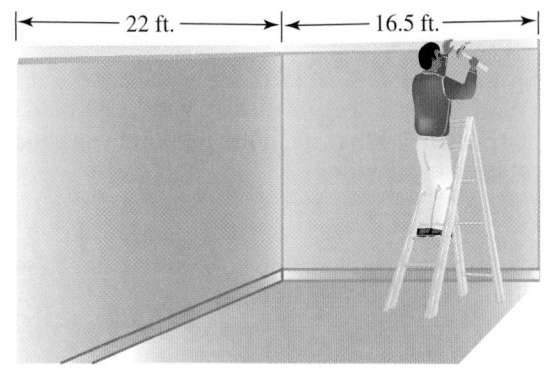
|← 22 ft. →|← 16.5 ft. →|

18. Jamal is planning to install crown molding where the walls and ceiling of his rectangular living room meet. The room is 16.5 feet by 22 feet.
    a. What is the total length of crown molding he needs?
       77 ft.
    b. If the crown molding comes in strips of 8 feet, how many strips must he purchase?
       10
    c. If the strips cost $9.99 each, what will be the cost of the crown molding?
       $99.90

19. Fermi National Accelerator Laboratory is a circular tunnel that is used to accelerate elementary particles. The radius of the tunnel is about 2 kilometers. If a particle travels one complete revolution around the circumference of the tunnel, what distance does it travel to the nearest hundredth of a kilometer? (*Source:* Cesare Emiliani, *The Scientific Companion*, 2nd ed., Wiley, 1995.)
    ≈ 12.57 km

20. The Chicxulub crater, which is buried partly beneath the Yucatan peninsula and partly beneath the Gulf of Mexico, is circular and is about 180 kilometers in diameter. What is the circumference of this crater to the nearest hundredth of a kilometer?
    ≈ 565.49 km

21. At one time, the Venetian Hotel and Casino in Las Vegas boasted having the largest hotel rooms in the world. If one of the rectangular rooms is 70 feet by 100 feet, what is the total square footage in the room?
    7000 ft.²

22. The world's largest oil production barge was built by Kvaernel Oil and Gas, Inc., of Norway and Single Bouy Moorings, Inc., of Switzerland. The rectangular deck of the barge is 273 meters long and 50 meters wide. What is the area of the deck?
    13,650 m²

Of Interest

The Chicxulub crater was created when Earth was hit by an asteroid about 65 million years ago. The devastation of the impact is thought to have caused the extinction of about 50 percent of the species on Earth, including the dinosaurs.

23. A large wall in a home is to be painted with a custom accent color. The rectangular wall measures 32.5 feet by 12 feet. A painter charges $2.50 per square foot to paint a wall with a custom color. How much will it cost to have the painter paint the wall?
$975

24. Hardwood floors are to be installed in a rectangular bedroom that measures 14 feet by 15 feet. If the contractor charges $34.50 per square foot to install the floors, what will be the cost of this improvement?
$7245

25. Tina needs to buy enough pine straw to cover a rectangular flower bed that measures 6 feet by 13 feet.
  a. What is the total area that she must cover?
  78 ft.²
  b. A landscape consultant tells Tina that she should use one bale of pine straw for every 10 square feet to be covered. If Tina follows this advice, how many bales must she purchase?
  8 bales
  c. If Tina's local garden center charges $4.50 per bale, what will be the cost of covering her flower bed?
  $36

26. Juan is considering installing carpet in his new office space. The office is rectangular and measures 42 feet by 36 feet. Juan has a budget of $3000.
  a. What is the total area that will be carpeted?
  1512 ft.²
  b. How many square yards is this area? (There are 9 square feet in 1 square yard.)
  168 yd.²
  c. If the contractor quotes Juan a price of $22.50 per square yard to install the carpet, how much will it cost to carpet the space? Is this feasible with Juan's budget?
  $3780; no

27. A city planning committee decides to install 20 historic markers around the city. The signs are triangular, measuring 26 inches wide at the base and 20 inches high.
  a. What is the area of each sign?
  260 in.²
  b. If a contractor charges $0.75 per square inch to build and install the signs, how much will it cost to have them installed?
  $3900

28. A parks department is going to install a flower bed in a triangular space created by three trails, as shown. The flower bed measures 32 feet at the base and is 24 feet high.
  a. What is the area of the flower bed?
  384 ft.²
  b. If a landscaper charges $6.50 per square foot to design and install the bed, what will be the total cost?
  $2496

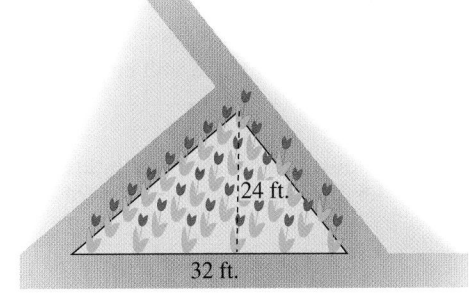

29. Conner and Ellyn purchase a new home. The area around an 8-foot by 9-foot storage building in the rectangular backyard is to be landscaped, as shown in the figure.
  a. What is the total area that is to be landscaped?
  3208 ft.²
  b. If $\frac{2}{3}$ of this area is to be grass, what amount will be covered with grass?
  $2138\frac{2}{3}$ ft.²
  c. If each pallet of grass sod covers 504 square feet, how many pallets will be needed?
  5
  d. Each pallet cost $106. What is the total cost of the sod?
  $530

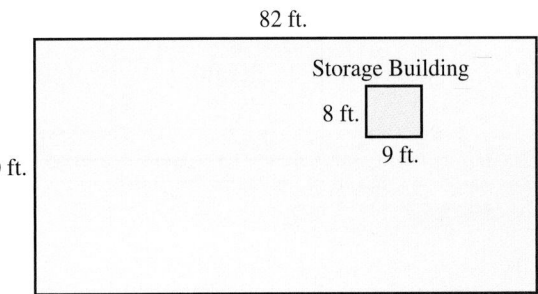

30. A kitchen floor is to be recovered with granite tiles. There is a fixed island in the center of the floor, as shown in the figure.

    a. What is the total area to be tiled?

    240.5 ft.²

    b. If each piece of tile covers $\frac{1}{4}$ of a square foot, how many pieces of tile will be needed?

    962

    c. If each tile costs $3.95, what is the cost of the tile?

    $3799.90

    d. If the contractor charges $8 per square foot to install the new floor, what will be the installation cost?

    $1924

31. A company cuts two circular pieces out of a rectangular sheet of metal, and the rest of the sheet is recycled. The sheet is 4 feet by 8 feet, and the diameter of each circle is 4 feet. Find the area of the sheet that is recycled. Round your answer to the nearest tenth.

    ≈ 6.9 ft.²

32. A compact disc has a diameter of $5\frac{3}{4}$ inches. On a CD, no information is placed within a circle in the center of the CD with a diameter of $1\frac{3}{4}$ inches. Find the area on a CD that can contain information. Round your answer to the nearest tenth.

    ≈ 23.6 in.²

Instructor Note For simplicity in Exercise 32, we have ignored the fact that on a CD, data also is placed $\frac{1}{8}$ of an inch from the outer edge. As a challenge problem, you might have students find the area that contains information if that $\frac{1}{8}$-inch band also is excluded.

★ 33. The wall shown is to be painted. The door, which will not be painted, is 2.5 feet by 7.5 feet. What area will be painted?

    82.25 ft.²

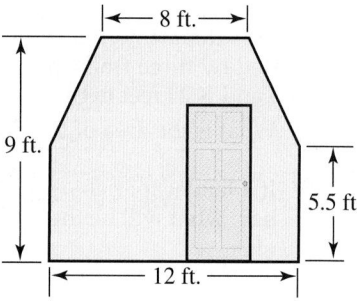

★ 34. The side of the house in the figure needs new siding. The window measures 3 feet by 4.5 feet. Find the area to be covered in siding.

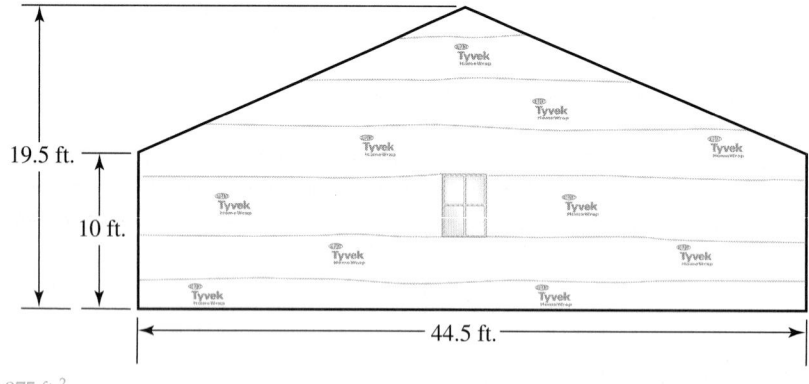

642.875 ft.²

35. A small storage building is shaped like a box that measures 2 meters by 5 meters by 4.75 meters. What is the storage building's volume?

    47.5 m³

36. A dorm refrigerator measures 2 feet by $1\frac{1}{2}$ feet by 4 feet. What is its volume?

    12 ft.³

37. A cylindrical bottle has a diameter of 8 centimeters and a height of 20 centimeters. Find the volume of the bottle. Round your answer to the nearest tenth.
$\approx$ 1005.3 cm³

Of Interest

The Great Pyramid was built by King Khufu (2589–2566 B.C.). Originally, the pyramid had a smooth outer case of stone, which caused its base to be 754 feet by 754 feet and its height to be 481 feet.

38. Earth is roughly a sphere with a radius of approximately 6370 kilometers. What is the approximate volume of Earth?
$\approx 1.1 \times 10^{12}$ km³ or 1,100,000,000,000 km³

39. The base of the Great Pyramid at Giza, Egypt, is 745 feet by 745 feet, and its height is 449 feet. Find the volume of the pyramid.
83,068,741.$\overline{6}$ ft.³ or 83,068,741$\frac{2}{3}$ ft.³

40. If a funnel is approximately a cone, approximate the volume of a funnel with a diameter of $8\frac{1}{2}$ inches and a height of 6 inches. Round your answer to the nearest tenth.
$\approx$ 113.5 in.³

For Exercises 41–46, use the formulas relating distance, rate, and time $\left(d = rt, r = \dfrac{d}{t} \right)$.

See Example 5.

41. A family began a trip of 516 miles at 8 A.M. They arrived at their final destination at 5:30 P.M. If they took two 15-minute breaks and an hour for lunch, what was their average driving rate?
64.5 mph

42. A family began a trip at 7:30 A.M. At the beginning of the trip, the car's odometer read 45,362.6. When they arrived at their final destination at 6 P.M., the odometer read 45,785.2. If they took three 15-minute breaks and an hour for lunch, what was their average driving rate?
$\approx$ 48.3 mph

43. A truck driver is to make a delivery that requires him to drive 285 miles by 4 P.M. If he starts at 10 A.M. and makes three 15-minute stops at weigh stations along the route, what must his average driving rate be to make the delivery on time?
$\approx$ 54.3 mph

★ **44.** In 2012, Bradley Wiggins of Great Britain won the Tour de France, which covered a distance of 3606 kilometers, with a time of 87 hours, 34 minutes, and 47 seconds. What was his average rate in kilometers per hour for the race? (*Source: http://www.bikeraceinfo.com/tdf/tdfindex.html.*)
$\approx$ 41.2 kph

★ **45.** A flight departs Atlanta at 8:30 A.M. EST to arrive in Las Vegas at 10 A.M. PST. If the plane flies at an average rate of $388\frac{1}{3}$ mph, what distance does it travel? (*Hint:* There is a three-hour time difference between EST and PST.)
1747.5 mi.

★ **46.** A flight departs at 7:30 A.M. EST from Philadelphia and travels to Dallas–Fort Worth, arriving at 10:10 A.M. CST. If the plane is averaging a speed of 368.2 miles per hour, what distance does it travel? (*Hint:* There is a one-hour time difference between EST and CST. Also express the time using fractions.)
1350.0$\overline{6}$ mi.

For Exercises 47 and 48, use the formula relating voltage, current, and resistance $(V = iR)$.
See Example 5.

47. A technician measures the current in a circuit to be −6.5 amperes, and the resistance is 8 ohms. Find the voltage.
−52 V

48. The current in a circuit is measured to be 4.2 amperes. If the resistance is 16 ohms, find the voltage.
67.2 V

For Exercises 49–54, use the formulas for converting degrees Fahrenheit to degrees Celsius or degrees Celsius to degrees Fahrenheit $\left(C = \dfrac{5}{9}(F - 32), F = \dfrac{9}{5}C + 32 \right).$
See Example 5.

49. On a drive to work, Tim notices that the temperature reading on a bank's digital thermometer is 34°C. What is this temperature in degrees Fahrenheit?
93.2°F

50. Iron melts at a temperature of 1535°C. What is this temperature in degrees Fahrenheit?
2795°F

51. Liquid oxygen boils at a temperature of −183°C. What is this temperature in degrees Fahrenheit?
−297.4°F

52. The average temperature on Neptune is approximately −360°F. What is this temperature in degrees Celsius?
−217.7°C

53. At the South Pole, temperatures can drop to −76°F. What is this temperature in degrees Celsius?
−60°C

54. The average temperature on Venus is approximately 890°F. What is this temperature in degrees Celsius?
476.6°C

Of Interest

Because of the tilt of Earth, the South Pole experiences daylight for six months, then night for six months. The Sun rises in mid-September and sets in mid-March.

Puzzle Problem A contest has people guess how many marbles are inside a jar. The jar is a cylinder with a diameter of 9 inches and a height of 12 inches. Suppose each marble has a diameter of 0.5 inch. Calculate the number of marbles that might fit inside the jar. Discuss the accuracy of your calculation.
≈ 11,664 marbles (≈ 15 per cubic inch)

Review Exercises

Exercises 1–6   **Expressions**

[1.5] *For Exercises 1 and 2, simplify.*

1. $6(-3 - 3) + 5(-3)$
−51

2. $6(2 + 3^2) - 4(2 + 3)^2$
−34

[1.7] *For Exercises 3 and 4, multiply using the distributive property.*

3. $\dfrac{1}{3}(x + 6)$
$\dfrac{1}{3}x + 2$

4. $-(4w - 6)$
−4w + 6

[1.7] *For Exercises 5 and 6, combine like terms.*

5. $5x - 14 + 3x + 9$
8x − 5

6. $6.2y + 7 - 1.5 - 0.8y$
5.4y + 5.5

2.2 The Addition Principle of Equality

Objectives

1. Determine whether a given equation is linear.
2. Solve linear equations in one variable using the addition principle of equality.
3. Solve equations with variables on both sides of the equal sign.
4. Solve identities and contradictions.
5. Solve application problems.

Connection Every equation has a corresponding graph. The names *linear* and *nonlinear* refer to the graphs of the equations that they describe. The graph of a linear equation is a line; the graph of a nonlinear equation is not.

Warm-up

[1.3] **1.** Simplify: $\dfrac{3}{5} - \dfrac{1}{3}$

[1.7] **2.** Simplify by combining like terms. $2(2x - 3) - 3(x + 5)$

Objective 1 Determine whether a given equation is linear.

There are many types of equations. In this section, we introduce and begin solving **linear equations**.

Definitions Linear equation: An equation in which each variable term contains a single variable raised to an exponent of 1.

Equations that are not linear are *nonlinear equations.*

Example 1 Determine whether the equation is linear or nonlinear.

a. $3x + 5 = 12$

Answer: This equation is linear because the variable x has an exponent of 1.

b. $7x^3 + x = 1$

Answer: This equation is nonlinear because the variable in the term $7x^3$ has an exponent of 3.

c. $3x + 2y = -10$

Answer: This equation is linear because the variables x and y have exponents of 1.

Your Turn 1 Determine whether the given equation is linear.

a. $3x - 4 = 11$ **b.** $-6x + y = 30$ **c.** $n^2 = 5m + 4$ **d.** $y = 3x^2 + 1$

If a linear equation has the same variable throughout, we say it is a **linear equation in one variable**.

Definition Linear equation in one variable: An equation that can be written in the form $ax + b = c$, where a, b, and c are real numbers and $a \neq 0$.

$5x + 9 = -6$ is a linear equation in one variable.
$x^2 - 2x + 5 = 8$ is a nonlinear equation in one variable.

In chapter 4, we consider linear equations in two variables, such as $x + y = 3$.

Objective 2 Solve linear equations in one variable using the addition principle of equality.

In Section 2.1, we learned how to check an equation's solution. Now we will develop a method for solving an equation (which means to find its solution(s)) called the *balance method*. Imagine an equation as scales with the equal sign as the fulcrum.

Answers to Your Turn 1
a. yes   **b.** yes   **c.** no   **d.** no

Answers to Warm-up
1. $\dfrac{4}{15}$
2. $x - 21$

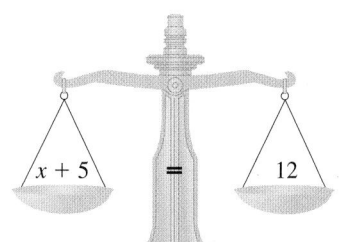

Note The solution to ◀ $x + 5 = 12$ is 7 because $7 + 5 = 12$.

Like a scale, adding an amount to one side unbalances the equation. For example, adding 4 to the left side of the equation tips the scale out of balance.

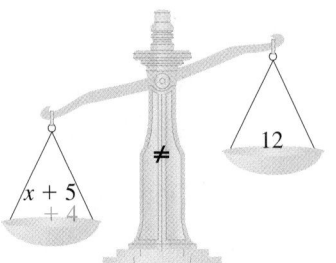

Note The equation is now unbalanced because its solution is no longer 7. Simplifying the left side gives $x + 9 = 12$, and 7 is no longer the solution because $7 + 9 \neq 12$.

However, if the same amount is added to *both* sides, like a scale, the equation stays balanced.

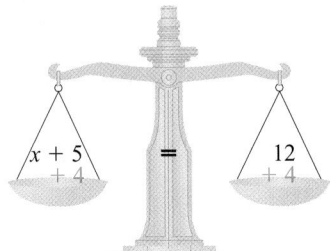

Note The equation is now balanced because its solution is still 7. Simplifying both sides gives $x + 9 = 16$, and 7 is the solution because $7 + 9 = 16$.

We say that the equation is balanced because adding 4 to both sides of the equation did not change the solution. This rule is called the *addition principle of equality*.

> **Rule The Addition Principle of Equality**
>
> If $a = b$, then $a + c = b + c$ is true for all real numbers a, b, and c.

Note We say that $a = b$ and $a + c = b + c$ are **equivalent equations** because they have the same solution.

In solving an equation, the goal is to write an equivalent equation in a simpler form, $x = c$, where the variable is alone, or isolated, on one side of the equal sign and the solution appears on the other side. For example, to isolate x in $x + 5 = 12$, we need to eliminate $+5$, so let's add -5 to both sides of the equation.

$$
\begin{array}{rcl}
x + 5 &=& 12 \\
\underline{-5} & & \underline{-5} \\
x + 0 &=& 7 \\
x &=& 7
\end{array}
$$

Note Adding -5 to both sides eliminates 5 from the left side, isolating the variable x. On the right side, we see the solution, 7.

The statement $x = 7$ is the solution statement.

Note Adding an additive inverse is equivalent to subtraction, so we can also say that we subtract the term we want to eliminate from both sides.

> **Procedure Using the Addition Principle of Equality**
>
> To use the addition principle of equality to eliminate a term from one side of an equation, add the additive inverse of that term to both sides of the equation.

Example 2 Solve and check.

a. $x - 13 = -22$

Solution: To isolate x, eliminate -13 by adding 13 to both sides.

$$
\begin{array}{rcl}
x - 13 &=& -22 \\
\underline{+13} & & \underline{+13} \qquad \text{Add 13 to both sides.} \\
x + 0 &=& -9 \\
x &=& -9
\end{array}
$$

Check: Recall from Section 2.1 that to check, we replace x in the original equation with -9 and verify that the equation is true.

$$x - 13 = -22$$
$$-9 - 13 \stackrel{?}{=} -22 \qquad \text{Replace } x \text{ with } -9.$$
$$-22 = -22$$

True; so -9 is the solution.

b. $\dfrac{3}{5} = y + \dfrac{1}{3}$

> **Note** Here, we illustrate another popular style of writing the addition principle, in which $-\dfrac{1}{3}$ is written beside the expressions on both sides of the equation instead of underneath the corresponding terms.

Solution:
$$\dfrac{3}{5} - \dfrac{1}{3} = y + \dfrac{1}{3} - \dfrac{1}{3} \qquad \text{Subtract } \dfrac{1}{3} \text{ from both sides to isolate } y.$$

$$\dfrac{9}{15} - \dfrac{5}{15} = y + 0 \qquad \text{Rewrite fractions with their LCD.}$$

$$\dfrac{4}{15} = y \qquad \text{Subtract } \dfrac{5}{15} \text{ from } \dfrac{9}{15}.$$

Check: $\dfrac{3}{5} = y + \dfrac{1}{3}$

$$\dfrac{3}{5} \stackrel{?}{=} \dfrac{4}{15} + \dfrac{1}{3} \qquad \text{Replace } y \text{ in the original equation with } \dfrac{4}{15} \text{ and verify that the equation is true.}$$

$$\dfrac{3}{5} \stackrel{?}{=} \dfrac{4}{15} + \dfrac{5}{15} \qquad \text{Rewrite fractions with their LCD.}$$

$$\dfrac{3}{5} \stackrel{?}{=} \dfrac{9}{15} \qquad \text{Add } \dfrac{4}{15} \text{ and } \dfrac{5}{15}.$$

$$\dfrac{3}{5} = \dfrac{3}{5} \qquad \text{Simplify to lowest terms.}$$

True; so $\dfrac{4}{15}$ is the solution.

> **Your Turn 2** Solve and check.
>
> **a.** $4.7 = x - 9.8$ **b.** $n + \dfrac{3}{4} = \dfrac{1}{6}$

Learning Strategy

If you are a visual learner, to become comfortable with the addition principle, consider writing the amount added or subtracted on both sides of an equation in a different color than you use in the rest of the equation. Once you are comfortable seeing the amount added or subtracted, you may not need to use the color any longer.

Simplifying before Isolating the Variable

Some equations have expressions that can be simplified. If like terms are on the same side of the equation, we combine the like terms before isolating the variable. If the equation to be solved contains parentheses, we use the distributive property to eliminate the parentheses before isolating the variable.

> **Example 3** Solve and check.
>
> **a.** $8y - 2.1 - 7y = -3.7 + 9$
>
> **Solution:** Simplify the expressions; then isolate y.
>
> $$8y - 2.1 - 7y = -3.7 + 9$$
>
> Combine $8y$ and $-7y$ to equal y. $y - 2.1 = 5.3$ Combine -3.7 and 9 to equal 5.3.
>
> $$\underline{+2.1 \quad +2.1} \qquad \text{Add 2.1 to both sides to isolate } y.$$
>
> $$y + 0 = 7.4$$
>
> $$y = 7.4$$

Answers to Your Turn 2

a. 14.5 **b.** $-\dfrac{7}{12}$

Check: Replace y in the original equation with 7.4 and verify that the equation is true. We will leave this to the reader.

b. $6(n - 3) - 5n = -25 + 4$

Solution: Use the distributive property, simplify, and then isolate n.

$$6(n - 3) - 5n = -25 + 4$$

Distribute 6 to eliminate parentheses. $6n - 18 - 5n = -21$ Combine -25 and 4 to equal -21.

Combine $6n$ and $-5n$ to equal n. $n - 18 = -21$

$$\begin{array}{r} +18 \quad +18 \\ \hline n + 0 = \quad -3 \end{array}$$ Add 18 to both sides to isolate n.

$$n = \quad -3$$

Check: Replace n with -3 in the original equation and verify that the equation is true. We will leave this to the reader.

Your Turn 3   Solve and check.

   a. $9x + 4 - 8x = 1 - 6$ **b.** $6.2 - 0.4 = 1.5(m - 6) - 0.5m$

Objective 3   Solve equations with variables on both sides of the equal sign.

Some equations have variable terms on both sides of the equal sign. To isolate the variable, we must first get the variable terms together on the same side of the equal sign. To do this, we use the addition principle.

Example 4   Solve and check. $5x - 8 = 4x - 13$

Solution: Use the addition principle to get the variable terms together on the same side of the equal sign. Then isolate the variable.

Note Remember that x means $1 \cdot x$. It does *not* mean that $x = 1$.

$$5x - 8 = 4x - 13$$

$$\begin{array}{r} -4x \qquad -4x \\ \hline x - 8 = 0 - 13 \end{array}$$ Subtract $4x$ from both sides.

$$x - 8 = -13$$

$$\begin{array}{r} +8 \qquad +8 \\ \hline x + 0 = -5 \end{array}$$ Add 8 to both sides to isolate x.

$$x = -5$$

Note By choosing to subtract $4x$ from both sides, we eliminate the $4x$ term from the right side and combine it with the $5x$ on the left side. It doesn't matter which term you move first. (See the explanation following the example.)

Check: $5x - 8 = 4x - 13$

$$5(-5) - 8 \overset{?}{=} 4(-5) - 13$$

$$-25 - 8 \overset{?}{=} -20 - 13$$

$$-33 = -33$$

Replace x in the original equation with -5 and verify that the equation is true.

True; so -5 is the solution.

When variable terms appear on both sides of the equal sign, it doesn't matter which term you choose to eliminate first, although some choices make the process easier. Suppose in Example 4 that we eliminate the $5x$ term first.

$$5x - 8 = 4x - 13$$

$$\begin{array}{r} -5x \qquad -5x \\ \hline 0 - 8 = -x - 13 \end{array}$$

$$-8 = -x - 13$$

$$\begin{array}{r} +13 \qquad +13 \\ \hline 5 = -x + 0 \end{array}$$

$$5 = -x$$

$$-5 = x$$

Note Eliminating the $5x$ term first gives $-x$, whereas when we eliminated the $4x$ term, we got x. Most people prefer working with positive coefficients.

Note $5 = -x$ means that we must find a number whose additive inverse is 5; so x must be equal to -5. In effect, we simply changed the signs of both sides. In Section 2.3, we will see another way to interpret $-x$.

Answers to Your Turn 3
   a. -9 **b.** 14.8

Notice that the term we chose to eliminate in the first step did not affect the solution, but it did affect our approach to the rest of the problem.

Conclusion: In solving an equation that has variable terms on both sides of the equal sign, when we select a variable term to eliminate, we can avoid negative coefficients by eliminating the term with the lesser coefficient.

> **Your Turn 4** | Solve and check.

  a. $7y - 3 = 6y + 10$ **b.** $2.6 + 3n = 9.5 + 4n$

Some equations that have variables on both sides require using the distributive property and combining like terms to solve.

> **Example 5** | Solve and check. $y - (3y + 11) = 5(y - 2) - 8y$

Note When we distribute a minus sign in an expression such as $-(3y + 11)$, we can think of the minus sign as -1; so we have

$$-(3y + 11)$$
$$= -1(3y + 11)$$
$$= -1(3y) + (-1)(11)$$
$$= -3y - 11$$

Solution: Simplify both sides of the equation. Then isolate y.

$$y - (3y + 11) = 5(y - 2) - 8y$$

Distribute the minus sign. $y - 3y - 11 = 5y - 10 - 8y$ Distribute 5.

Note The coefficient of $-3y$ is less than the coefficient of $-2y$, so we chose to eliminate $-3y$ by adding $3y$.

$$-2y - 11 = -3y - 10 \quad \text{Combine like terms.}$$
$$\underline{+3y \qquad\quad +3y} \quad \text{Add } 3y \text{ to both sides.}$$
$$y - 11 = \quad 0 - 10$$
$$y - 11 = -10$$
$$\underline{+11 \quad +11} \quad \text{Add 11 to both sides}$$
$$y + 0 = \quad 1 \quad \text{to isolate } y.$$
$$y = 1$$

Check: Replace y in the original equation with 1 and verify that the equation is true. We will leave this to the reader.

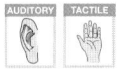 **Learning Strategy**

If you are an auditory learner, try saying the step before doing it. If you are a tactile learner, focus on writing each step in a deliberate and methodical manner. Using these suggestions will help solidify the process as you practice.

Following is an outline for solving equations based on everything you've learned so far.

> **Procedure** **Solving Linear Equations**
>
> To solve linear equations requiring the addition principle only:
> 1. Simplify both sides of the equation as needed.
>   a. Distribute to eliminate parentheses.
>   b. Combine like terms.
> 2. Use the addition principle so that all variable terms are on one side of the equation and all constants are on the other side. Then combine like terms.
>
> **Tip:** Eliminate the variable term that has the lesser coefficient to avoid negative coefficients.

> **Your Turn 5** | Solve and check.

  a. $7n - (n - 3) = 4(n + 4) + n$ **b.** $11 + 3(5m - 2) = 7(2m - 1) + 9$

Objective 4 Solve identities and contradictions.

In general, a linear equation in one variable has only one real-number solution. However, there are two special cases that we need to consider. First, if a linear equation is an identity, as in Example 6, every real number is a solution. Second, as we'll see in Example 7, some linear equations have no solution.

Answers to Your Turn 4
  a. 13   **b.** -6.9

Answers to Your Turn 5
  a. 13   **b.** -3

Equations with an Infinite Number of Solutions

Consider $3x + 9 = 3x + 9$. Because the expressions on each side of the equation are identical, each produces the same result no matter what number replaces the variable x. This means that every real number is a solution. Such an equation is called an **identity**.

Definition Identity: An equation that has every real number as a solution (excluding any numbers that cause an expression in the equation to be undefined).

The equation $5y - 4 = 7y - 3$ is not an identity because not every real number is a solution. Consider the number 2, for example. If we replace y with 2 in the equation, we see that it does not check.

$$5(2) - 4 \stackrel{?}{=} 7(2) - 3$$
$$10 - 4 \stackrel{?}{=} 14 - 3 \qquad \text{The real number we picked, 2, is not a solution for}$$
$$6 \neq 11 \qquad\qquad 5y - 4 = 7y - 3; \text{ so the equation is not an identity.}$$

Sometimes it isn't obvious that an equation is an identity. Simplifying the expressions on each side of the equation can make the identity apparent.

> ### Procedure Recognizing an Identity
>
> When solving a linear equation, if after simplifying each side of the equation the expressions are identical, the equation is an identity and every real number for which the equation is defined is a solution.

Instructor Note Suggest to students that they each pick a different number to check while you choose a number and show your check. Explain that because linear equations generally have only one solution, when two different numbers check for a linear equation, it most likely is an identity. Discuss the importance of checking **more than one number** because if it is not an identity, you might just happen to select the one solution.

| **Example 6** | Solve and check. $5(2t + 1) - 8 = 4t - 3(1 - 2t)$ |

Solution: Simplify both sides of the equation. Then isolate the variable.

$$5(2t + 1) - 8 = 4t - 3(1 - 2t)$$

Distribute 5. $\qquad 10t + 5 - 8 = 4t - 3 + 6t \qquad$ Distribute -3.

$$10t - 3 = 10t - 3 \qquad \text{Combine like terms.}$$

▲

Note We can stop here because the equation is obviously an identity. However, if we continue and apply the addition principle, the equation is still an identity.

$$10t - 3 = 10t - 3$$
$$\underline{-10t \qquad -10t}$$
$$0 - 3 = \ \ 0 - 3$$
$$-3 = -3$$

Because the linear equation is an identity, every real number is a solution.

Check: Every real number is a solution for an identity, so any number we choose should check in the original equation. We will choose to test the number 1.

$$5(2t + 1) - 8 = 4t - 3(1 - 2t)$$
$$5(2(1) + 1) - 8 \stackrel{?}{=} 4(1) - 3(1 - 2(1)) \qquad \text{Replace } t \text{ with 1.}$$
$$5(3) - 8 \stackrel{?}{=} 4 - 3(-1)$$
$$15 - 8 \stackrel{?}{=} 4 + 3$$
$$7 = 7$$

Because the equation is true, 1 is a solution, which supports but does not prove our conclusion that the equation is an identity. To be more certain, we could choose another number to check in the equation. (Remember, every number should work.) We will leave this to the reader.

Conclusion: If after simplifying, the linear equation is an identity, then every real number for which the equation is defined is a solution to the equation.

Equations with No Solution

Some equations, called **contradictions**, have no solution.

Definition Contradiction: An equation that has no real-number solution.

We recognize linear equations of this type by the fact that after the expressions on each side of the equation have been simplified, the variable terms will match but the constant terms will not. For example, $2x - 5 = 2x - 3$ is a contradiction. In Example 7, watch what happens when we try to solve the equation.

Example 7   Solve and check. $2x - 5 = 2x - 3$

Note The variable terms are eliminated, and the resulting numeric equation is false, which indicates that the given equation is a contradiction.

▶

Solution:
$$2x - 5 = 2x - 3$$
$$\underline{-2x \qquad\quad -2x} \qquad \text{Subtract 2x from both sides.}$$
$$0 - 5 = 0 - 3$$
$$-5 = -3$$

Because the equation is a contradiction, it has no solution.

Check: Because the variable terms, $2x$, are identical on both sides of the equal sign, replacing x with any number will yield an identical product. Because we subtract 5 from that product on the left side and we subtract 3 from that product on the right side, the equation cannot be true. Therefore, it has no solution.

Note The symbol $\varnothing$, which indicates the empty set, is often used to represent the solution set of an equation that has no solutions.

Conclusion: When solving a linear equation in one variable, if applying the addition principle of equality causes the variable terms to be eliminated from the equation and the resulting equation is false, then the equation is a contradiction and has no solution.

Procedure Recognizing a Contradiction

When solving a linear equation, if after simplifying the expressions on each side of the equation the expressions have the same variable term but different constant terms, the equation is a contradiction and has no solution. Equivalently, if the variable terms are eliminated and the resulting equation is false, the equation is a contradiction.

Your Turn 7   Solve and check.

a. $4x - 3(x + 5) = 2(2x - 3) - 3(x + 5)$ **b.** $2(2x + 4) - 11 = 4(x - 2) + 5$

Objective 5 Solve application problems.

Now consider some application problems that can be solved using the addition principle.

Example 8   Laura wants to buy a car stereo that costs $275. She currently has $142. How much more is needed?

Understand We are given the total required and the amount she currently has, and we must find how much she needs.

Plan Let x represent the amount Laura needs. We will write an equation, then solve.

Answers to Your Turn 7
a. contradiction—no solution
b. identity—all real numbers

Execute Current amount + Needed amount = 275

$$142 \quad + \quad x \quad = 275$$

$$142 + x = 275$$

$$\underline{-142 \qquad -142}$$ Subtract 142 from both sides

$$0 + x = 133$$ to isolate x.

$$x = 133$$

Answer Laura needs $133 to buy the stereo.

Check Does $142 plus the additional $133 equal $275?

$$142 + 133 \stackrel{?}{=} 275$$

$$275 = 275 \quad \text{It checks.}$$

Your Turn 8   Daryl has a balance of $-\$568$ on a credit card. How much must he pay to bring his balance to $-\$480$?

Example 9   The figure to the left shows the front-view design of a bookcase. Find the unknown length.

Understand In the drawing, the sum of the dimensions on the left side must be equal to 32 inches.

Plan Let d represent the unknown distance. We will write an equation and solve for d.

Execute $12\frac{1}{2} + 10\frac{3}{4} + d = 32$

$$23\frac{1}{4} + d = 32 \qquad \text{Combine like terms.}$$

$$23\frac{1}{4} - 23\frac{1}{4} + d = 32 - 23\frac{1}{4} \quad \text{Subtract } 23\frac{1}{4} \text{ from both sides to isolate } d.$$

$$d = 8\frac{3}{4}$$

Answer The unknown length is $8\frac{3}{4}$ inches.

Check Verify that the sum of the three lengths on the left side is equal to 32 inches. We will leave this to the reader.

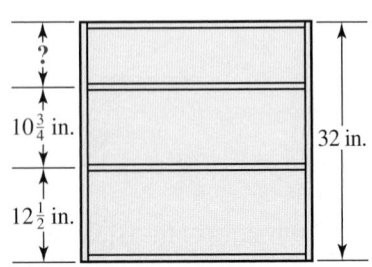

Note As in Example 2(b), because we are subtracting a fraction from both sides, we are writing the subtraction horizontally instead of vertically. Use whichever approach, horizontal or vertical, you prefer.

Answer to Your Turn 8
$88

Answer to Your Turn 9
157 students

Your Turn 9   In researching the grades given in a particular course during one semester, a department chair finds that 89 students received an A, 154 received a B, and 245 received a C. If the department does not give Ds and 645 students enrolled in the course initially, how many students did not receive a passing grade in the course?

2.2 Exercises (For Extra Help) MyMathLab®

Note: Exercises marked with a ★ represent challenging exercises.

Objective 1

Prep Exercise 1 A linear equation is an equation in which each variable term has a single variable raised to an exponent of ____1____.

For Exercises 1–16, determine whether each equation is linear. See Example 1.

1. $4y + 7 = 2y - 8$
yes

2. $6x - 5 = 3y + 64$
yes

3. $m^2 + 4 = 20$
no

4. $-1 = w^2 - 5w$
no

5. $7u^4 + u^2 = 16$
no

6. $6n^3 - 9n^2 = 4n + 6$
no

7. $4x - 8 = 2(x + 3)$
yes

8. $5(t - 2) + 7 = 3t - 1$
yes

9. $6x - y = 11$
yes

10. $3x + 2y = 12$
yes

11. $x^2 + y^2 = 1$
no

12. $3x^2 + 2y^2 = 24$
no

13. $y = 8$
yes

14. $x = -3$
yes

15. $y = \dfrac{1}{4}x - 3$
yes

16. $y = -0.5x + 2$
yes

Objectives 2–4

Prep Exercise 2 In your own words, explain what the addition principle of equality says. The addition principle of equality says that we can add (or subtract) the same amount on both sides of an equation without affecting its solution(s).

Prep Exercise 3 In the equation $5x - 9 = 4x + 1$, how do you use the addition principle to eliminate the $4x$ term? Add $-4x$ to (or subtract $4x$ from) both sides of the equation.

Prep Exercise 4 In the equation $5x - 9 = 4x + 1$, how do you use the addition principle to eliminate the -9 term? Add 9 to both sides of the equation.

Prep Exercise 5 If an equation has variable terms on both sides of the equation, what should you do first? Use the addition principle of equality to get the variable terms together on the same side of the equal sign.

Prep Exercise 6 How do you handle parentheses in an equation? Distribute to eliminate the parentheses.

For Exercises 17–68, solve and check. See Examples 2–7.

17. $x - 7 = 2$
9

18. $a - 8 = 30$
38

19. $m - 8 = -3$
5

20. $n - 6 = -2$
4

21. $r - 2 = -9$
-7

22. $a - 3 = -11$
-8

23. $x + 7 = 12$
5

24. $x + 2 = 8$
6

25. $n + 12 = 2$
-10

26. $y + 15 = 8$
-7

27. $-16 = y + 9$
-25

28. $-24 = n + 11$
-35

29. $m + \dfrac{7}{8} = \dfrac{4}{5}$
$-\dfrac{3}{40}$

30. $k + \dfrac{5}{9} = -\dfrac{1}{3}$
$-\dfrac{8}{9}$

31. $-\dfrac{5}{8} = y - \dfrac{1}{6}$
$-\dfrac{11}{24}$

32. $\dfrac{3}{4} = c - \dfrac{2}{3}$
$\dfrac{17}{12}$

33. $15.8 + y = 7.6$
-8.2

34. $b + 8.8 = 5.4$
-3.4

35. $-2.1 = n - 7.5 + 0.8$
4.6

36. $x + 0.4 - 1.6 = -12.5$
-11.3

37. $6 - 18 = 7x - 3 - 6x$
-9

38. $2z + 6 - z = 5 - 9$
-10

39. $5m = 4m + 7$
7

40. $7y = 6y - 8$
-8

41. $7x - 9 = 6x + 4$
13

42. $12y + 22 = 11y - 3$
-25

43. $-2y - 11 = -3y - 5$
6

44. $-4t + 9 = -5t + 1$
-8

45. $7x - 2 + x = 10x - 1 - x$
-1

46. $3t + 6 + 4t = 9t - 2 - t$
8

47. $8n + 7 - 12n = 4n + 5 - 9n$
-2

48. $-10x - 9 + 8x = -4x - 5 + 3x$
-4

49. $2.6 + 7a + 5 = 8a - 5.6$
13.2

50. $9c + 4.8 = 7.5 + 4.8 + 8c$
7.5

51. $12 - 7(h + 6) + 8h = 14 - 8$
36

52. $19 - 3(m + 4) + 4m = 42 - 18$
17

53. $5 - \dfrac{1}{2}(4x + 6) = -3x - 8$
-10

54. $6 - \dfrac{2}{3}(6b - 9) = -5b - 21$
−33

55. $5y - (3y + 7) = y + 9$
16

56. $-15 - 2x = 16 - (3x - 9)$
40

57. $3(3x - 5) - 2(4x - 3) = 5 - 10$
4

58. $5(5x - 3) - 6(4x - 2) = 12 - 15$
0

59. $1.4(2.5x - 4.5) - (6.2 + 2.5x) = -7.3 + 4.5$
9.7

60. $0.5(3.8x - 6.2) - (0.9x - 4) = 2.9 - 4.7$
−2.7

61. $4 + 3y + y - 12 = 5y + 9 - y - 17$
All real numbers

62. $-9 - 4v - 1 + v = -2v + 5 - v - 15$
All real numbers

63. $7.3 - 0.2x + 1.3 - 0.6x = 12 - 0.8x - 3.6$
No solution

64. $2.5y - 3.4 - 1.2y = 6.7 - 9.1 + 1.3y$
No solution

★ **65.** $20z + \dfrac{2}{3}(6z - 48) - 9z = 14z + 0.2(50z - 250) - 9z$
No solution

★ **66.** $6b - 1.5(8 + 2b) + 4 = 6b - \dfrac{1}{8}(24b + 48)$
No solution

★ **67.** $8(2m - 4) + 20(m - 5) = 4m + 5(5m - 20) - 32 + 7m$
All real numbers

★ **68.** $-3(2x + 5) + 8(x + 2) - 7 = 6(x - 5) - 4(x - 6)$
All real numbers

Objective 5

For Exercises 69–80, translate to an equation and then solve. See Examples 8 and 9.

69. Latonia is planning to buy a new car. The down payment is $2373. She has $1947 saved. How much more does she need?
$1947 + x = 2373$; $426

70. Kent owes $12,412 on his Visa card. What payment should he make so he will owe $10,500?
$10{,}500 + x = 12{,}412$; $1912

71. Robert knows that the distance from his home to work is 42 miles. Unfortunately, he gets a flat tire 16 miles from work. How far did Robert drive before his tire went flat?
$16 + x = 42$; 26 mi.

72. Susan is playing Yahtzee. The "chance" score is found by adding the value shown on five dice. If she has a total of 23 on four of the dice, what does the fifth die need to be so that her score will be 28?
$23 + x = 28$; 5

73. On May 16, the balance in Nikki's checking account was $1741.62. Afterwards, she writes four checks and doesn't record the amount of the last check. The first check is for $16.82, the second is for $150.88, and the third is for $192.71. On May 21, she finds her balance to be $1286.65. If the checks were the only transactions that could have cleared the bank during that time, what was the amount of the fourth check?
$1741.62 - 1286.65 = 16.82 + 150.88 + 192.71 + x$; $94.56

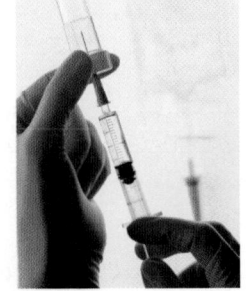

74. A patient must receive 350 cc of a medication in three injections. He has received two injections at 110 cc each. How much should the third injection be?
$110 + 110 + x = 350$; 130 cc

75. The perimeter of the trapezoid shown is 67.2 centimeters. Find the length of the unknown side.

$x + 12.4 + 16.3 + 27.2 = 67.2$; 11.3 cm

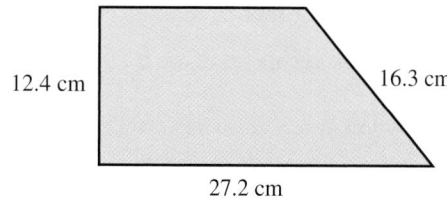

12.4 cm   16.3 cm

27.2 cm

76. The perimeter of the triangle shown is $84\frac{1}{2}$ inches. Find the length of the unknown side.

$x + 23 + 35\frac{1}{4} = 84\frac{1}{2}$; $26\frac{1}{4}$ in.

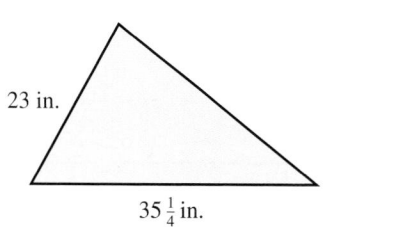

23 in.

$35\frac{1}{4}$ in.

77. In the blueprint shown, what is the distance x?

$x + 19 + 10 = 54$; 25 ft.

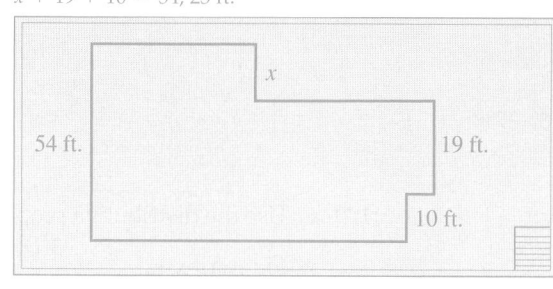

x

54 ft.   19 ft.

10 ft.

78. What is the distance x in the following figure?

$6 + x + 4 = 16$; 6 cm.

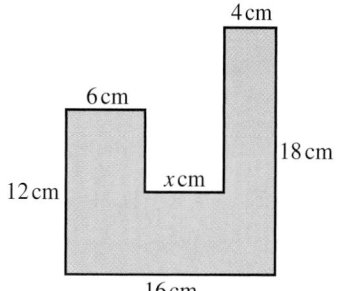

4 cm

6 cm

12 cm x cm   18 cm

16 cm

79. John sells scanning devices. The company quota is set at $8500 each month. The spreadsheet shows John's sales as of the end of the second week of April. How much more does John need to sell to make the quota? Do you think John will make the quota? Why or why not?

$x + 900 + 3 \cdot 1500 + 2 \cdot 1245 = 8500$: $610; yes, because $610 is less than all of his prior weeks.

Date	Item No.	Quantity	Price per Unit
4/2	32072	1	$900
4/9	17032	3	$1500
4/13	48013	2	$1245

80. Tamika works from 8 A.M. until 5 P.M. in the women's department of a major clothing store. She has set a goal for herself of $600 per day in sales. Thus far, she has sold 3 blouses for $25 each, 5 pairs of slacks for $30 each, and 2 pairs of shoes for $85 each. How much more in sales does she need to reach her goal? If it is now 4 P.M., is it likely that she will make her goal? Why or why not?

$x + 3 \cdot 25 + 5 \cdot 30 + 2 \cdot 85 = 600$: $205; No, because she still has over one-third of her goal to go but only one hour left to work.

★ **81.** In a survey, respondents were given a statement, and they could agree, disagree, or have no opinion regarding the statement. The results indicate that $\frac{1}{3}$ of the respondents agree with the statement while $\frac{2}{5}$ disagree. What fraction had no opinion?

$x + \frac{1}{3} + \frac{2}{5} = 1$; $\frac{4}{15}$

★ **82.** In a *Time*/CNN poll, $\frac{19}{50}$ of the respondents said that they believed that the increase in the number of divorces was due more to changes in women's attitude toward marriage, while $\frac{9}{50}$ said that the increase was due more to changes in men's attitude toward marriage. The rest thought that the increase was a result of both men and women equally. What fraction of the respondents believed that the increase was due to men and women equally?

$x + \frac{19}{50} + \frac{9}{50} = 1$; $\frac{11}{25}$

Puzzle Problem In a study of an experimental medication, half of the participants are given the medication and half are given a placebo. The results indicate that $\frac{1}{3}$ of the participants had improvement in their condition, $\frac{1}{8}$ had no improvement in their condition but experienced known side effects of the medication, and the rest showed no effects at all. What fraction of the group that received the medication showed no discernible effects from it? (Hint: Assume that all of the participants who saw improvement in their condition or experienced known side effects took the medication.)

$\frac{1}{24}$

Review Exercises

Exercises 1–6   Expressions

[1.4] 1. What is the multiplicative inverse of $-\frac{4}{5}$?

$-\frac{5}{4}$

For Exercises 2 and 3, simplify.

[1.4] 2. $4(-8.3)$

-33.2

[1.4] 3. $27 \div (-3)$

-9

[1.7] 4. Simplify: $13 - 6(x + 2)$

$-6x + 1$

[1.7] 5. Combine like terms:

$3x^2 - 6x + 5x^2 + x - 9$

$8x^2 - 5x - 9$

[1.7] 6. Use the distributive property to rewrite: $12\left(\frac{1}{3}x - \frac{3}{4}\right)$

$4x - 9$

2.3 The Multiplication Principle of Equality

Objectives

1 Solve linear equations using the multiplication principle of equality.

2 Solve linear equations using both the addition and multiplication principles.

3 Use the multiplication principle to eliminate fractions and decimals from equations.

4 Solve application problems.

Warm-up

[1.4] 1. Find the product: $-\frac{4}{5} \cdot \left(-\frac{5}{4}\right)$

[1.7] 2. Use the distributive property to write an equivalent expression and simplify.

$$12\left(\frac{5}{6}x + 2\right)$$

[2.2] 3. Solve: $17 + 4n + 5 = 8n - 3n + 21$

Objective 1 Solve linear equations using the multiplication principle of equality.

In the previous section, we learned that the addition principle of equality states that adding the same amount to or subtracting the same amount from both sides of an equation does not change the equation's solution. Likewise, we can multiply or divide both sides of an equation by the same nonzero number without affecting the equation's solution. This rule is called the *multiplication principle of equality*.

Answers to Warm-up
1. 1
2. $10x + 24$
3. 1

> **Rule The Multiplication Principle of Equality**
>
> If $a = b$, then $ac = bc$ is true for all real numbers a, b, and c, where $c \neq 0$.

While the addition principle of equality is used to eliminate terms from one side of an equation, the multiplication principle of equality is used to eliminate a variable's coefficient. For example, to isolate the variable n in the equation $2n = 14$, we need to eliminate the coefficient 2. Multiplying both sides by $\frac{1}{2}$ or dividing both sides by 2 will eliminate the coefficient of 2, thereby isolating n.

$$\frac{1}{2} \cdot 2n = 14 \cdot \frac{1}{2}$$

Note Dividing by a number is equivalent to multiplying by its multiplicative inverse (reciprocal). If the coefficient is an integer, most people prefer the look of division.

$$\frac{1}{\cancel{2}} \cdot \frac{2n}{1} = \frac{\overset{7}{\cancel{14}}}{1} \cdot \frac{1}{\cancel{2}}$$

$$n = 7$$

$$\frac{2n}{\cancel{2}} = \frac{14}{2}$$

$$1n = 7$$

$$n = 7$$

We can summarize how to use the multiplication principle of equality to eliminate a coefficient as follows:

> **Procedure** **Using the Multiplication Principle of Equality**
>
> To use the multiplication principle of equality to eliminate a coefficient in an equation, multiply both sides of the equation by the multiplicative inverse of that coefficient or divide both sides by the coefficient.

Example 1 Solve and check.

a. $-\frac{4}{5}m = \frac{6}{7}$

Solution: $-\frac{4}{5}m = \frac{6}{7}$

$$-\frac{\overset{1}{\cancel{5}}}{\cancel{4}} \cdot -\frac{\overset{1}{\cancel{4}}}{\cancel{5}}m = \frac{\overset{3}{\cancel{6}}}{7} \cdot -\frac{5}{\cancel{4}}$$

Eliminate the coefficient $-\frac{4}{5}$ by multiplying both sides by its multiplicative inverse, $-\frac{5}{4}$.

$$m = -\frac{15}{14}$$

Check: $-\frac{4}{5}m = \frac{6}{7}$

$$-\frac{\overset{2}{\cancel{4}}}{\cancel{5}}\left(-\frac{\overset{3}{\cancel{15}}}{\cancel{14}}\right) \overset{?}{=} \frac{6}{7}$$

Replace m in the original equation with $-\frac{15}{14}$ and verify that the equation is true.

$$\frac{6}{7} = \frac{6}{7}$$

True; therefore, $-\frac{15}{14}$ is correct.

b. $-1.9x = -4.56$

Solution: $-1.9x = -4.56$

$$\frac{\cancel{-1.9}x}{\cancel{-1.9}} = \frac{-4.56}{-1.9}$$

Eliminate the coefficient -1.9 by dividing both sides by -1.9.

$$x = 2.4$$

Note If the equation contains fractions, the answer will be left as a fraction or an integer. If the equation contains decimals, the answer will be left as a decimal or an integer. If the equation contains integers only, the answer will be expressed as an integer, a fraction, or a decimal, as appropriate.

Check: $-1.9x = -4.56$

$-1.9(2.4) = -4.56$ Replace *x* with 2.4 and verify that the equation is true.

$-4.56 = -4.56$

True; therefore, 2.4 is correct.

| **Your Turn 1** | Solve and check. |

a. $-56 = -7t$ 　　　　　 **b.** $-\dfrac{3}{8}y = 9$ 　　　　　 **c.** $0.4x = -0.92$

Objective 2 Solve linear equations using both the addition and multiplication principles.

Now let's put the multiplication principle together with the addition principle. We follow the same outline as in Section 2.2, with one new step: using the multiplication principle to eliminate any remaining coefficient at the end.

> **Procedure** **Solving Linear Equations**
>
> To solve linear equations in one variable:
> 1. Simplify both sides of the equation as needed.
>   a. Distribute to eliminate parentheses.
>   b. Combine like terms.
> 2. Use the addition principle so that all variable terms are on one side of the equation and all constants are on the other side. (Eliminate the variable term with the lesser coefficient to avoid negative coefficients.) Then combine like terms.
> 3. Use the multiplication principle to eliminate any remaining coefficient.

| **Example 2** | Solve and check. $-3x - 11 = 7$ |

Solution: Use the addition principle to separate the variable term and constant terms; then use the multiplication principle to eliminate any remaining coefficient.

Note Multiplying both sides by $-\dfrac{1}{3}$ would also work:

$$\left(-\frac{1}{3}\right)\left(-\frac{\overset{1}{3x}}{1}\right) = \left(\frac{\overset{6}{18}}{1}\right)\left(-\frac{1}{\underset{1}{3}}\right)$$

$$x = -6$$

$$
\begin{aligned}
-3x - 11 &= 7 \quad &&\text{Add 11 to both sides to isolate the $-3x$ term.}\\
\underline{+11} & \underline{+11}\\
-3x + 0 &= 18\\
-3x &= 18\\
\frac{-3x}{-3} &= \frac{18}{-3} \quad &&\text{Divide both sides by -3 to eliminate the -3 coefficient.}\\
x &= -6
\end{aligned}
$$

Check: $-3x - 11 = 7$

$-3(-6) - 11 \overset{?}{=} 7$ Replace *x* in the original equation with -6 and verify that the equation is true.

$18 - 11 \overset{?}{=} 7$

$7 = 7$

True; therefore, -6 is correct.

| **Your Turn 2** | Solve and check. |

a. $6y - 19 = -22$ 　　　　　 **b.** $-12 = 20 - 8t$

Answers to Your Turn 1
a. 8 　　 **b.** -24 　　 **c.** -2.3

Answers to Your Turn 2
a. $-\dfrac{1}{2}$ 　 **b.** 4

Solving Equations with Variable Terms on Both Sides

Recall that when variable terms appear on both sides of the equal sign, we use the addition principle to get the variable terms on one side of the equal sign and the constant terms on the other side.

Example 3 Solve and check. $8y - 5 = 2y - 29$

Solution: Use the addition principle to get the variable terms on one side of the equation and the constant terms on the other side; then use the multiplication principle to eliminate the remaining coefficient.

$$8y - 5 = 2y - 29 \qquad \text{Subtract } 2y \text{ from both sides. (} 2y \text{ has the lesser coefficient.)}$$
$$\underline{-2y \quad -2y}$$
$$6y - 5 = 0 - 29$$
$$6y - 5 = -29$$
$$\underline{+5 \qquad +5} \qquad \text{Add 5 to both sides to isolate the } 6y \text{ term.}$$
$$6y + 0 = -24$$
$$\frac{6y}{6} = \frac{-24}{6} \qquad \text{Divide both sides by 6 to eliminate the 6 coefficient.}$$
$$y = -4$$

Check:
$$8y - 5 = 2y - 29$$
$$8(-4) - 5 \stackrel{?}{=} 2(-4) - 29 \qquad \text{Replace } y \text{ in the original equation with } -4 \text{ and verify that the equation is true.}$$
$$-32 - 5 \stackrel{?}{=} -8 - 29$$
$$-37 = -37$$

True; therefore, -4 is correct.

Instructor Note To emphasize that it doesn't matter which of the four terms you eliminate first, redo Example 3 by first eliminating $8y$, -5, or -29.

Simplifying First

Recall that when an equation contains parentheses or like terms that appear on the same side of the equal sign, we first eliminate the parentheses and combine like terms. Then we use the addition principle to separate the variable terms and constant terms.

Example 4 Solve and check. $17 - (4n - 5) = 9n - 3(n + 7)$

Solution:
$$17 - (4n - 5) = 9n - 3(n + 7)$$
$$17 - 4n + 5 = 9n - 3n - 21 \qquad \text{Distribute to eliminate parentheses.}$$
$$22 - 4n = 6n - 21 \qquad \text{Combine like terms.}$$
$$22 - 4n = 6n - 21$$
$$\underline{+4n \quad +4n} \qquad \text{Add } 4n \text{ to both sides.}$$
$$\qquad\qquad\qquad (-4n \text{ has the lesser coefficient.})$$
$$22 + 0 = 10n - 21$$
$$22 = 10n - 21$$
$$\underline{+21 = \qquad +21} \qquad \text{Add 21 to both sides to isolate } 10n.$$
$$43 = 10n + 0$$
$$\frac{43}{10} = \frac{10n}{10} \qquad \text{Divide both sides by 10 to eliminate the 10 coefficient.}$$
$$\frac{43}{10} \text{ or } 4.3 = n$$

Check:
$$17 - (4n - 5) = 9n - 3(n + 7)$$
$$17 - (4(4.3) - 5) \stackrel{?}{=} 9(4.3) - 3(4.3 + 7) \qquad \text{Replace } n \text{ in the original equation with 4.3 and verify that the equation is true.}$$
$$17 - (17.2 - 5) \stackrel{?}{=} 38.7 - 3(11.3)$$
$$17 - 12.2 \stackrel{?}{=} 38.7 - 33.9$$
$$4.8 = 4.8$$

True; therefore, 4.3 is correct.

Your Turn 4 Solve and check.

a. $3t - 10 + 9t = 17 + 4t - 3$

b. $13 - 6(x + 2) = 3x - (11x + 5)$

Objective 3 Use the multiplication principle to eliminate fractions and decimals from equations.

The multiplication principle of equality can be used to eliminate fractions or decimals from an equation. Although equations can be solved without eliminating the fractions or decimals, most people find equations that contain only integers easier to solve.

Eliminating Fractions in an Equation

If the equation contains fractions, we multiply both sides by a number that will eliminate all of the denominators. We could multiply both sides by any multiple of the denominators; however, using the LCD (least common denominator) results in the simplest equations.

Example 5 Solve and check. $\frac{1}{3}x - \frac{3}{4} = \frac{5}{6}x + 2$

Solution: Eliminate the fractions by multiplying through by the LCD. The LCD of 3, 4, and 6 is 12.

Note We have chosen to eliminate the fractions in our first step. Remember, you can eliminate the fractions at any step in solving the equation or not eliminate them at all.

$12\left(\dfrac{1}{3}x - \dfrac{3}{4}\right) = \left(\dfrac{5}{6}x + 2\right)12$
Eliminate the fractions by multiplying both sides by the LCD, 12.

$\dfrac{\overset{4}{\cancel{12}}}{1} \cdot \dfrac{1}{\underset{1}{3}}x - \dfrac{\overset{3}{\cancel{12}}}{1} \cdot \dfrac{3}{\underset{1}{4}} = \dfrac{\overset{2}{\cancel{12}}}{1} \cdot \dfrac{5}{\underset{1}{6}}x + 12 \cdot 2$
Distribute 12; then divide out the denominators.

$4x - 9 = 10x + 24$

$4x - 9 = 10x + 24$

$\underline{-4x \qquad\qquad -4x}$
Subtract 4x from both sides.

$0 - 9 = 6x + 24$

$-9 = 6x + 24$

$\underline{\quad -24 \qquad\quad -24}$
Subtract 24 from both sides.

$-33 = 6x + 0$

$\dfrac{-33}{6} = \dfrac{6x}{6}$
Divide both sides by 6 to eliminate the coefficient.

$-\dfrac{11}{2}$ or $-5.5 = x$
Simplify.

Check: $\dfrac{1}{3}x - \dfrac{3}{4} = \dfrac{5}{6}x + 2$

$\dfrac{1}{3}\left(-\dfrac{11}{2}\right) - \dfrac{3}{4} \overset{?}{=} \dfrac{5}{6}\left(-\dfrac{11}{2}\right) + 2$
Replace x in the original equation with $-\dfrac{11}{2}$ and verify that the equation is true.

$-\dfrac{11}{6} - \dfrac{3}{4} \overset{?}{=} -\dfrac{55}{12} + 2$

$-\dfrac{22}{12} - \dfrac{9}{12} \overset{?}{=} -\dfrac{55}{12} + \dfrac{24}{12}$
Write equivalent fractions with their LCD.

$-\dfrac{31}{12} = -\dfrac{31}{12}$

True; therefore, $-\dfrac{11}{2}$ is correct.

Answers to Your Turn 4
a. 3   **b.** −3

Eliminating Decimals in an Equation

The multiplication principle can also be used to eliminate decimals in an equation. Because multiplying by a power of 10 will cause the decimal point to move to the right, we can eliminate decimals by multiplying both sides of the equation by the appropriate power of 10. The power of 10 we use depends on the decimal number with the most decimal places.

Example 6 | Solve and check. $0.2(y - 6) = 0.48y + 3$

Solution: Decimals can be eliminated at any step in the process of solving an equation. In this case, we will distribute to eliminate the parentheses first, then the decimals.

$$0.2(y - 6) = 0.48y + 3$$
$$0.2y - 1.2 = 0.48y + 3 \quad \text{Distribute to eliminate parentheses.}$$
$$100(0.2y - 1.2) = (0.48y + 3)100$$
$$20y - 120 = 48y + 300$$
$$20y - 120 = 48y + 300$$
$$\underline{-20y \qquad\qquad -20y}$$
$$0 - 120 = 28y + 300$$
$$-120 = 28y + 300$$
$$\underline{-300 \qquad\qquad -300}$$
$$-420 = 28y + 0$$
$$\frac{-420}{28} = \frac{28y}{28}$$
$$-15 = y$$

◄ **Note** The number 0.48 has two decimal places, which is more decimal places than any of the other decimal numbers; so we multiply both sides by 100 to eliminate the decimals.

Check:
$$0.2(y - 6) = 0.48y + 3$$
$$0.2(-15 - 6) \stackrel{?}{=} 0.48(-15) + 3 \quad \text{Replace } y \text{ in the original equation with}$$
$$0.2(-21) \stackrel{?}{=} -7.2 + 3 \quad\quad -15 \text{ and verify that the equation is true.}$$
$$-4.2 = -4.2$$

True; therefore, -15 is correct.

Connection Eliminating decimals and fractions uses the same process. Decimal numbers represent fractions with denominators that are powers of 10. For example, $0.48 = \frac{48}{100}$ and $0.2 = \frac{2}{10}$. The LCD for these fractions is 100.

We can amend the outline for solving equations to include eliminating fractions and decimals. Remember that this process is optional.

Procedure **Solving Linear Equations**

To solve linear equations in one variable:
1. Simplify both sides of the equation as needed.
   a. Distribute to eliminate parentheses.
   b. Eliminate fractions or decimals by multiplying through by the LCD. In the case of decimals, the LCD is the power of 10 with the same number of zero digits as decimal places in the number with the most decimal places. (Eliminating fractions and decimals is optional.)
   c. Combine like terms.
2. Use the addition principle so that all variable terms are on one side of the equation and all constants are on the other side. (Eliminate the variable term with the lesser coefficient to avoid negative coefficients.) Then combine like terms.
3. Use the multiplication principle to eliminate any remaining coefficient.

Answers to Your Turn 6
a. $-\dfrac{25}{8}$ b. 0.4

Your Turn 6 | Solve and check.

a. $\dfrac{1}{3}(y - 2) = \dfrac{3}{5}y + \dfrac{1}{6}$

b. $2.5n - 1.04 = 0.15n - 0.1$

Objective 4 Solve application problems.

In Section 2.1, we began solving application problems using formulas. Recall that to use a formula, we replace the variables with the corresponding given numbers, then solve for the unknown amount.

| **Example 7** | Solve.

a. The perimeter of the figure shown is 188 feet. Find the width and length.

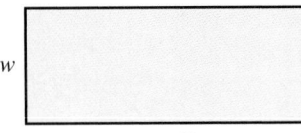

Understand The width is represented by w, and the length is represented by $w + 7$. We are given the perimeter, so we can use the formula $P = 2l + 2w$.

Plan In the perimeter formula, replace P with 188 and l with $w + 7$ and solve for w.

Execute
$$P = 2l + 2w$$
$$188 = 2(w + 7) + 2w$$
$$188 = 2w + 14 + 2w \qquad \text{Distribute.}$$
$$188 = 4w + 14 \qquad \text{Combine like terms.}$$
$$188 = 4w + 14$$
$$\underline{-14 \qquad\qquad -14} \qquad \text{Subtract 14 from both sides.}$$
$$174 = 4w + 0$$
$$\frac{174}{4} = \frac{4w}{4} \qquad \text{Divide both sides by 4.}$$
$$\frac{87}{2} \text{ or } 43.5 = w$$

Answer The width is 43.5 feet. To find the length, we evaluate the expression that represents the length, $w + 7$, with $w = 43.5$.

$$\text{Length} = 43.5 + 7 = 50.5 \text{ ft.}$$

Check First verify that the length is 7 more than the width: $50.5 = 7 + 43.5$. Then verify that a rectangle with a length of 50.5 feet and a width of 43.5 feet has a perimeter of 188 feet.

$$P = 2l + 2w$$
$$P = 2(50.5) + 2(43.5)$$
$$P = 101 + 87$$
$$P = 188$$

It checks.

b. The total material allotted for the construction of a closed (top not open) metal box is 4754 square inches. The length is to be 30 inches, and the width is to be 28 inches. Find the height.

Understand The material will be used to create the outer shell that is the box, which means that 4754 square inches is the surface area of the box. The formula for the surface area of a box is $SA = 2lw + 2lh + 2wh$.

Plan Replace SA with 4754, l with 30, and w with 28 and solve for h.

Execute
$$SA = 2lw + 2lh + 2wh$$
$$4754 = 2(30)(28) + 2(30)h + 2(28)h$$
$$4754 = 1680 + 60h + 56h \qquad \text{Simplify.}$$
$$4754 = 1680 + 116h \qquad \text{Combine like terms.}$$
$$4754 = 1680 + 116h$$
$$\underline{-1680 \quad -1680} \qquad\qquad\qquad \text{Subtract 1680 from both sides.}$$
$$3074 = \quad 0 + 116h$$

$$\frac{3074}{116} = \frac{116h}{116}$$ Divide both sides by 116.

$$26.5 = h$$

Answer The height is 26.5 inches.

Check Does a 30-inch by 28-inch by 26.5-inch box have a surface area of 4754 square inches?

$$4754 \stackrel{?}{=} 2(30)(28) + 2(30)(26.5) + 2(28)(26.5)$$
$$4754 \stackrel{?}{=} \quad 1680 \quad + \quad 1590 \quad + \quad 1484$$
$$4754 = 4754$$

It checks.

Your Turn 7 Solve.

a. The desired surface area for a box is 98 square feet. If the length of the box is to be 4 feet and the height is to be 6 feet, what will be the width?

b. The circumference of Venus at the equator is 24,170 miles. What is the diameter of Venus at the equator? (Use $C = \pi d$ with $\pi \approx 3.14$.)

Answers to Your Turn 7
a. 2.5 ft.   **b.** ≈ 7697 mi.

2.3 Exercises   For Extra Help   MyMathLab®

Note: Exercises marked with a ★ represent challenging exercises.

Objective 1

Prep Exercise 1 In your own words, explain what the multiplication principle of equality says.

We can multiply (or divide) both sides of an equation by the same amount without affecting its solution(s).

Prep Exercise 2 How do you use the multiplication principle to eliminate a coefficient that is an integer?

Divide both sides of the equation by the coefficient.

Prep Exercise 3 How do you use the multiplication principle to eliminate a coefficient that is a fraction?

Multiply both sides of the equation by the multiplicative inverse of the coefficient.

Prep Exercise 4 To eliminate a negative coefficient, do you multiply/divide by a positive or negative number? Why?

Negative because the goal is to get a positive 1 coefficient for the variable.

For Exercises 1–12, solve and check. See Example 1.

1. $3x = 12$ 4

2. $8x = -24$
-3

3. $-6x = -18$
3

4. $-5y = 20$
-4

5. $\dfrac{n}{2} = 8$
16

6. $\dfrac{t}{3} = -4$
-12

7. $\dfrac{3}{4}y = -15$
-20

8. $\dfrac{5}{6}x = 20$
24

9. $-\dfrac{4}{7}t = -\dfrac{2}{3}$
$\dfrac{7}{6}$

10. $-\dfrac{3}{8}a = \dfrac{5}{6}$
$-\dfrac{20}{9}$

11. $-\dfrac{2}{5}t = \dfrac{8}{15}$
$-\dfrac{4}{3}$

12. $-\dfrac{7}{9}t = -\dfrac{5}{12}$
$\dfrac{15}{28}$

Objective 2

For Exercises 13–34, solve and check. See Example 2.

13. $5n - 2n = 21$
7

14. $7t - 3t = 20$
5

15. $4x + 1 = 21$
5

16. $3x + 5 = 11$
2

17. $4a - 3 = 17$
5

18. $3x - 8 = 10$
6

19. $5x + 7 = -8$
−3

20. $3x + 9 = -6$
−5

21. $9 - 2n = 12$
$-\dfrac{3}{2}$

22. $1 - 7y = -8$
$\dfrac{9}{7}$

23. $\dfrac{5}{8}x - 2 = 8$
16

24. $7 = \dfrac{3}{4}x + 13$
−8

25. $2(x - 3) = -6$
0

26. $4(5x + 7) = 28$
0

27. $3(n + 7) = -6$
−9

28. $4(n - 3) = -8$
1

29. $3a - 2a + 7 + 6a = -28$
−5

30. $c + 2c + 3 + 4c = 24$
3

31. $12b - 8(2b - 3) = 16$
2

32. $2x - 6(x + 8) = -12$
9

33. $2(y - 33) + 3(y - 2) = 8$
16

34. $4(r - 8) + 2(r + 3) = -8$
3

For Exercises 35–56, solve and check. See Examples 3 and 4.

35. $8x + 5 = 2x + 17$
2

36. $10t + 1 = 6t + 13$
3

37. $5y - 7 = 2y + 13$
$\dfrac{20}{3}$

38. $9m + 1 = 3m - 14$
$-\dfrac{5}{2}$

39. $9 - 4k = 15 - k$
−2

40. $6 - 12m = -20m + 22$
2

41. $4k + 5 = 7k - 7 + 9k$
1

42. $-11b - 5 = -5b + 23 - 10$
−3

43. $3a - 4a + 9 = -12a + 43 - 6a$
2

44. $17b - 11b - 17 = -4b + 13 - 5b$
2

45. $9x + 12 - 3x - 2 = x + 8 + 5x$
No solution

46. $12 - 6r - 14 = 9 - 4r - 7 - 2r$
No solution

47. $14(w - 2) + 13 = 4w + 5$
2

48. $2x + 2(3x - 4) = -23 + 5x$
−5

49. $2n - (6n + 5) = 9 - (2n - 1)$
$-\dfrac{15}{2}$

50. $2 - (17 - 5m) = 9m - (m + 7)$
$-\dfrac{8}{3}$

51. $4 - (6x + 5) = 7 - 2(3x + 4)$
All real numbers

52. $-6 - (3z - 2) = 5z - 4(2z + 1)$
All real numbers

53. $5(a - 1) - 9(a - 2) = -3(2a + 1) - 2$
−9

54. $-4(k + 4) + 13(k - 1) = -3(k - 2) + 13$
4

55. $3(2x - 5) - 4x = 2(x - 3) - 12$
No solution

56. $4(2x - 1) - 3(x + 5) = 5(x - 2) + 7$
No solution

Objective 3

Prep Exercise 5 If an equation contains fractions, how do you transform the equation so that it contains only integers?
Multiply both sides of the equation by a common multiple of the denominators. Using the LCD results in an equation with the smallest integers possible.

Prep Exercise 6 If an equation contains decimal numbers, how do you transform the equation so that it contains only integers?
Multiply both sides by the appropriate power of 10 to eliminate decimals.

For Exercises 57–76, use the multiplication principle of equality to eliminate the fractions or decimals; then solve and check. See Examples 5 and 6.

57. $\dfrac{2}{3}n - 1 = \dfrac{1}{4}$
$\dfrac{15}{8}$

58. $\dfrac{2}{5}n - 1 = \dfrac{3}{2}$
$\dfrac{25}{4}$

59. $\dfrac{3}{4}n - 2 = -\dfrac{7}{8}n + 4$
$\dfrac{48}{13}$

60. $-\dfrac{2}{5}t + 1 = \dfrac{3}{10}t - 3$
$\dfrac{40}{7}$

61. $\dfrac{3}{4}x + \dfrac{5}{6} = \dfrac{1}{4} + \dfrac{4}{3}x$
1

62. $\dfrac{7}{9}w - \dfrac{13}{6} = \dfrac{7}{2}w + \dfrac{5}{9}$
-1

63. $\dfrac{1}{2}(y + 5) = \dfrac{3}{2} - y$
$-\dfrac{2}{3}$

64. $\dfrac{2}{3}(x - 4) = \dfrac{4}{3} + 2x$
-3

65. $\dfrac{1}{6}(m - 4) = \dfrac{2}{3}(m + 1) - \dfrac{1}{3}m$
-8

66. $\dfrac{1}{5}(y - 3) = \dfrac{3}{10}(y + 5) - \dfrac{2}{5}y$
7

67. $-2.9u + 3.6u = 6.3$
9

68. $-4.6z + 2.2z = 4.8$
-2

69. $0.2 - 1.3t - 0.8t = 1 - 2.1t - 0.8$
All real numbers

70. $4.2y - 8.2 + 2.3y = 0.9 + 6.5y - 9.1$
All real numbers

71. $0.3(a - 18) = 0.9a$
-9

72. $0.6(w - 12) = 0.2w$
18

73. $0.08(15) + 0.4x = 0.05(36 + 7x)$
12

74. $0.06(25) + 0.27x = 0.3(4 + x)$
10

75. $9 - 0.2(v - 8) = 0.3(6v - 8)$
6.5

76. $0.5 - (t - 6) = 10.4 - 0.4(1 - t)$
-2.5

Find ⊗ **the Mistake** *For Exercises 77–80, the check indicates that a mistake was made. Find and correct the mistake. See Examples 2–4.*

77.
$$6x - 11 = 8x - 15$$
$$\underline{-6x \qquad\qquad -6x}$$
$$11 = 2x - 15$$
$$\underline{+15 \qquad\qquad +15}$$
$$26 = 2x$$
$$\dfrac{26}{2} = \dfrac{2x}{2}$$
$$13 = x$$

Check $6x - 11 = 8x - 15$
$$6(13) - 11 \overset{?}{=} 8(13) - 15$$
$$78 - 11 \overset{?}{=} 104 - 15$$
$$67 \neq 89$$

Mistake: The minus sign was dropped in front of the 11.
Correct: 2

78.
$$6n + 1 = -2n + 21$$
$$\underline{-1 \qquad\qquad -1}$$
$$6n + 0 = -2n + 20$$
$$6n = -2n + 20$$
$$\underline{+2n \qquad\quad +2n}$$
$$8n = 0 + 20$$
$$\dfrac{8n}{8} = \dfrac{20}{8}$$
$$n = \dfrac{5}{2}$$

Check $6n + 1 = -2n + 21$
$$\dfrac{\overset{3}{\cancel{6}}}{1}\left(\dfrac{5}{\cancel{2}}\right) + 1 \overset{?}{=} -\dfrac{\overset{1}{\cancel{2}}}{1}\left(\dfrac{5}{\cancel{2}}\right) + 21$$
$$3 + 1 \overset{?}{=} -5 + 21$$
$$4 \neq 16$$

Mistake: In the check, neglected to multiply 5 by 3 after dividing out 2.

Correct: $\dfrac{5}{2}$ is correct; the second to the last line of the check should be $15 + 1 = -5 + 21$.

79.

$$4 - (x + 2) = 2x + 3(x - 8)$$
$$4 - x + 2 = 2x + 3x - 24$$
$$6 - x = 5x - 24$$
$$\underline{+x \quad +x}$$
$$6 + 0 = 6x - 24$$
$$6 = 6x - 24$$
$$\underline{+24 \qquad\quad +24}$$
$$30 = 6x + 0$$
$$\frac{30}{6} = \frac{6x}{6}$$
$$5 = x$$

Check $4 - (x + 2) = 2x + 3(x - 8)$
$$4 - (5 + 2) \stackrel{?}{=} 2(5) + 3(5 - 8)$$
$$4 - 7 \stackrel{?}{=} 10 + 3(-3)$$
$$-3 \stackrel{?}{=} 10 + (-9)$$
$$-3 \neq 1$$

Mistake: In the second line, the minus sign was not distributed to the 2.

Correct: $\dfrac{13}{3}$

80.

$$\frac{1}{2}n - 3 = \frac{2}{3}n + \frac{1}{4}$$

$$\frac{\overset{6}{\cancel{12}}}{1} \cdot \frac{1}{\underset{1}{\cancel{2}}}n - 3 = \frac{\overset{4}{\cancel{12}}}{1} \cdot \frac{2}{\underset{1}{\cancel{3}}}n + \frac{\overset{3}{\cancel{12}}}{1} \cdot \frac{1}{\underset{1}{\cancel{4}}}$$

$$6n - 3 = 8n + 3$$
$$\underline{-6n \qquad\quad -6n}$$
$$0 - 3 = 2n + 3$$
$$\underline{-3 \qquad\quad -3}$$
$$-6 = 2n + 0$$
$$\frac{-6}{2} = \frac{2n}{2}$$
$$-3 = n$$

Check $\dfrac{1}{2}n - 3 = \dfrac{2}{3}n + \dfrac{1}{4}$

$$\frac{1}{2}\left(-\frac{\overset{1}{\cancel{3}}}{1}\right) - 3 \stackrel{?}{=} \frac{2}{\underset{1}{\cancel{3}}}\left(-\frac{\overset{1}{\cancel{3}}}{1}\right) + \frac{1}{4}$$

$$-\frac{3}{2} - 3 \stackrel{?}{=} -2 + \frac{1}{4}$$

$$-\frac{3}{2} - \frac{6}{2} \stackrel{?}{=} -\frac{8}{4} + \frac{1}{4}$$

$$-\frac{9}{2} \neq -\frac{7}{4}$$

Mistake: Did not multiply 3 by 12.

Correct: $-\dfrac{39}{2}$

Objective 4

For Exercises 81–96, solve for the unknown amount. See Example 7.

81. The area of a rectangular deck is known to be 140 square feet. Find the length if the width must be 10 feet. (Use $A = lw$.)

14 ft.

82. A crate manufacturer receives an order for a crate with a volume of 128 cubic feet. The crate must be 5 feet 4 inches by 6 feet. What must the crate's height be? (Use $V = lwh$.)

4 ft.

83. The area of the metal plate shown in the figure is 45.6 square centimeters. What is the length of the side labeled h?

(Use $A = \dfrac{1}{2}bh$.)

19 cm

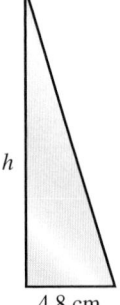

h

4.8 cm

84. The volume of a rectangular pyramid is 76.8 cubic inches, and the area of the base is 6.4 square inches.

Find the height. (Use $V = \dfrac{1}{3}Ah$.)

36 in.

85. A fence is to be installed around a rectangular field. The field's perimeter is 212 feet. Find the dimensions of the field if the length of the field is 2 feet more than the width, as shown in the figure. (Use $P = 2l + 2w$.)

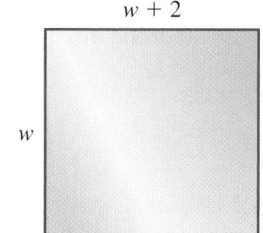

w + 2

w

$l = 54$ ft., $w = 52$ ft.

86. The base of a building has a perimeter of 241 feet. The length is 16 feet more than the width, as shown in the figure. Find the dimensions of the base of the building. (Use $P = 2l + 2w$.)

w

w + 16

$w = 52.25$ ft., $l = 68.25$ ft.

87. The area of the trapezoidal plot of land shown in the figure is 7500 square feet. What is the length of the side labeled *b*?

(Use $A = \dfrac{1}{2}h(a + b)$.)

110 ft.

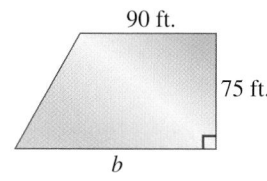

90 ft.

75 ft.

b

88. A tabletop is in the shape of a trapezoid and has an area of 1008 square inches. What is the length of the side labeled *a*?

(Use $A = \dfrac{1}{2}h(a + b)$.)

36 in.

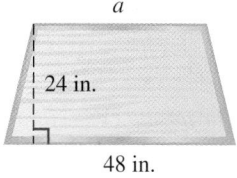

a

24 in.

48 in.

89. A small storage box is designed so that its surface area is 106 square inches. If the length is 7.5 inches and the width is 4 inches, find the height. (Use $SA = 2lw + 2lh + 2wh$.)

2 in.

90. Jerry has 1992 square inches of wood to build a chest for storing musical equipment. He wants the length to be 22 inches and the width to be 18 inches. What will the height be? (Use $SA = 2lw + 2lh + 2wh$.)

15 in.

91. The circumference of Earth along the equator is approximately 40,053.84 kilometers. What is the equatorial radius? (Use $C = 2\pi r$ and round your answer to the nearest tenth.)

≈ 6374.8 km

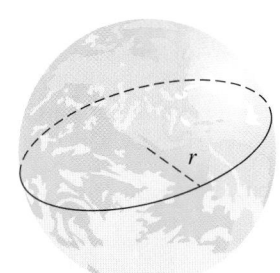

r

92. A giant sequoia is measured to have a circumference of 26.4 meters. What is the diameter? (Use $C = \pi d$ and round your answer to the nearest tenth.)

≈ 8.4 m

Of Interest

The giant sequoia tree is found along the western slopes of the Sierra Nevada range. These trees reach heights of nearly 100 meters, and some are over 4000 years old.

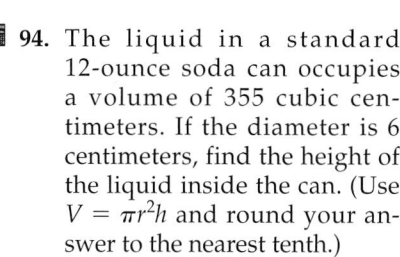

C

d

93. An engineer is designing a chemical storage tank in the shape of a cylinder. The volume of the tank is to be 60 cubic meters, and the radius is to be 2 meters. Find the height. (Use $V = \pi r^2 h$ and round your answer to the nearest tenth.)

≈ 4.8 m

2 m

94. The liquid in a standard 12-ounce soda can occupies a volume of 355 cubic centimeters. If the diameter is 6 centimeters, find the height of the liquid inside the can. (Use $V = \pi r^2 h$ and round your answer to the nearest tenth.)

≈ 12.6 cm

6 cm

Lime-Up

12 ounces (355 mL)

★ **95.** The area of the L-shaped room shown is 321 square feet. Find all of the missing dimensions.
  $w = 11$ ft., $x = 2$ ft., $w + 2 = 13$ ft., $y = 23$ ft.

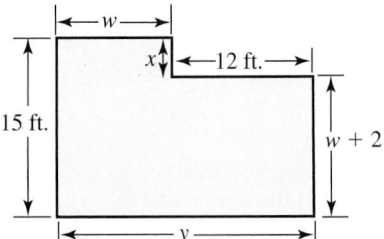

★ **96.** The area of the figure below is 192 square inches. Find all of the missing dimensions.
  $a = 5$ in., $b = 3$ in., $c = 18$ in., $a + 7 = 12$ in.

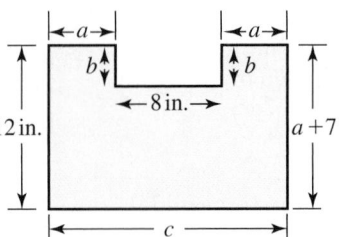

The formula $V = iR$ describes the voltage, V, in a circuit when the current, i, passes through a resistance, R. Voltage is measured in volts (V), current is measured in amps (A), and resistance is measured in ohms (Ω). Use the formula for voltage to solve Exercises 97 and 98. See Example 7.

97. The voltage in a circuit is measured to be -76 V. If the resistance is 8 Ω, find the current.
  -9.5 A

98. A technician applies 15 V to a circuit. The current is measured to be 2.5 A. What is the resistance?
  6 Ω

The formula $F = ma$ describes the force that an object with mass m experiences when accelerated an amount a. If acceleration is measured in meters per second squared ($m/sec.^2$) and the mass is measured in kilograms (kg), the force is in units of newtons (N). Use the force formula to solve Exercises 99–102. See Example 7.

99. A car with a mass of 1200 kilograms exerts a force of 4800 newtons. Find the acceleration.
  4 m/sec.^{2}

100. An object weighs 44.1 newtons; that is, it exerts a downward force of -44.1 newtons. The object is dropped so that it accelerates at -9.8 meters per second squared. Find the mass of the object.
  4.5 kg

101. The Liberty Bell weighs 9265.9 newtons, so its downward force is -9265.9 newtons. If the acceleration due to gravity is about -9.8 meters per second squared, what is the mass of the bell?
  945.5 kg

102. Jacob weighs 160 pounds (a downward force of -160 pounds). If the acceleration due to gravity is about -32.2 feet per second squared, what is his mass? (The mass will be in terms of the unit *slugs*.)
  ≈ 4.97 slugs

Of Interest

Earth's gravity accelerates all objects at the same rate, -9.8 m/sec.^{2} (or -32.2 ft./sec.^{2}). The accelerations are negative because the direction is downward.

Of Interest

The Liberty Bell was cast in 1752 and arrived in Philadelphia in August of the same year. In September 1752, the bell cracked during a test while hanging in the yard of the Pennsylvania state house.

103. The formula $w = \dfrac{s}{f}$ describes the wavelength w of a musical note if f is the frequency and s is the speed of sound. A note is sounded with a wavelength of 0.94 meters/cycle and a frequency of 440 cycles/second. Find the speed of sound.
  413.6 m/sec.

104. The formula $t = \dfrac{V}{g}$ gives the time that a free-falling object falls, where V is the velocity and g is the acceleration due to gravity. Find the velocity if the object has been falling for four seconds and the acceleration due to gravity is 9.8 meters per second squared.
  39.2 m/sec.

105. The formula $C = 0.3m + 25$ describes the total cost in dollars of using a digital phone for 100 minutes, where m represents the number of minutes beyond 100 minutes. Pam's bill shows that she used the phone for 187 minutes. What is the total cost of her use of the phone?
  $51.10

106. A rental car agency rents a van for a weekend using the formula $C = 0.27m + 200$, where m represents the number of miles and C represents the cost in dollars. Laura rents a van from the company for a weekend when the company is offering 100 free miles. If she returns the van and has driven 421 miles, what is the total cost of the rental?
  $286.67

107. The formula $d = v_i t + \dfrac{1}{2} a t^2$ describes the distance, d, an object travels if the object has an initial velocity (speed) v_i and an acceleration of a for a time t. An object travels 135 feet after being accelerated for three seconds with an acceleration of 20 feet per second squared. Find the object's initial velocity.

15 ft./sec.

108. The formula $F = \dfrac{9}{5} C + 32$ gives a relationship between the temperature in Celsius and Fahrenheit. If the temperature is 82.4°F, find the temperature in Celsius.

28°C

Review Exercises

Exercises 1–3   Expressions

[1.5] **1.** Simplify: $-6 + 3(5^2 + 2) - (6 - 1)$

70

[1.7] **2.** Combine like terms:
$$-3.2x^2 + 0.4x + 7x^2 - 8.03x + 1.5$$
$3.8x^2 - 7.63x + 1.5$

[1.7] **3.** Use the distributive property to rewrite:
$$-\frac{3}{4}(6x + 8) \quad -\frac{9}{2}x - 6$$

Exercises 4–6   Equations and Inequalities

For Exercises 4–6, solve and check.

[2.2] **4.** $6 + x = -14 + 2x$

20

[2.3] **5.** $-\dfrac{2}{5}x = 10$

−25

[2.3] **6.** $2x + 3(x - 3) = 10x - (3x - 11)$

−10

2.4 Applying the Principles to Formulas

Objective

1 Isolate a variable in a formula using the addition and multiplication principles.

Warm-up

[2.3] Solve the equations.

1. $22 = 2(3 + 4x)$

2. $10 = 4 + \dfrac{3}{2}x$

Objective 1   Isolate a variable in a formula using the addition and multiplication principles.

So far when using formulas, we have substituted given numbers for all of the variables except one. We then used the addition and multiplication principles of equality to isolate that variable. However, we can actually isolate a particular variable in a formula without being given numbers for the other variables. We follow the same steps as if we had been given numbers.

Instructor Note Emphasize that the procedure is the same whether you are working with numbers or variables. Try doing a similar example with one variable. For example, before solving $A = lw$ for w, solve $7 = 3w$ for w.

Consider the area formula, $A = lw$. If we were given values for A and l and asked to isolate w, or solve for w, we would divide both sides by the value of l. We can perform this manipulation symbolically.

$$\frac{A}{l} = \frac{lw}{l} \qquad \text{To isolate } w \text{, we eliminate } l \text{ by dividing both sides by } l.$$

$$\frac{A}{l} = w$$

Answers to Warm-up

1. $x = 2$

2. $x = 4$

This development suggests the following procedure:

> **Procedure** **Isolating a Variable in a Formula**
>
> To isolate a particular variable in a formula, treat all other variables like constants and isolate the desired variable using the outline for solving equations.

Isolate a Variable Using the Addition Principle

Recall that the addition principle is used to eliminate terms that are added or subtracted.

Example 1   Isolate R in the formula $P = R - C$.

Solution: $P = R - C$ To isolate R, we must eliminate C. Because C is subtracted from R, we add C to both sides.

$$\underline{\quad + C \qquad + C\quad}$$
$$P + C = R + 0$$
$$P + C = R$$

◀ **Note** Because P and C are not like terms, they cannot be combined.

Of Interest

The formula $P = R - C$ is used to calculate profit, given revenue, R, and cost, C.

Your Turn 1   Isolate v_i in the formula that describes the velocity of an object after being accelerated from an initial velocity (v_i), $v = v_i + at$.

Isolate a Variable Using the Multiplication Principle

Now let's use the multiplication principle to isolate a variable in a formula.

Example 2

a. Isolate w in the formula for the volume of a box, $V = lwh$.

Solution:

$$\frac{V}{lh} = \frac{lwh}{lh}$$ To isolate w, we must eliminate l and h. Because l and h are multiplying w, we divide both sides by l and h.

$$\frac{V}{lh} = w$$

Of Interest

The formula $B = P(1 + rt)$ is used in finance to calculate the balance for a simple-interest investment or loan. The variable B represents the final balance, P represents the principal, r represents the interest rate, and t represents time.

b. Isolate P in the formula $B = P(1 + rt)$.

Solution:

$$\frac{B}{1 + rt} = \frac{P(1 + rt)}{1 + rt}$$ To isolate P, we must eliminate $(1 + rt)$. Because $(1 + rt)$ is multiplying P, we divide both sides by $1 + rt$.

$$\frac{B}{1 + rt} = P$$

Your Turn 2

a. Isolate b in the formula for the area of a parallelogram, $A = bh$.

b. Isolate i in the voltage formula for a series circuit with two resistors, $V = i(R + r)$.

Answer to Your Turn 1
$v - at = v_i$

Answers to Your Turn 2

a. $\dfrac{A}{h} = b$ **b.** $\dfrac{V}{R + r} = i$

Isolate a Variable Using Both Principles

Now consider formulas that require the use of both principles to isolate the desired variable.

Example 3

a. Isolate l in the perimeter formula for rectangles, $P = 2l + 2w$.

Solution: First use the addition principle to isolate the term that contains l. Then use the multiplication principle to eliminate the coefficient.

$$P = 2l + 2w$$

$$\underline{\quad - 2w \qquad - 2w\quad} \quad \text{Isolate } 2l \text{ by subtracting } 2w \text{ from both sides.}$$

$$P - 2w = 2l + \quad 0$$

$$\frac{P - 2w}{2} = \frac{2l}{2} \qquad \text{Isolate } l \text{ by dividing both sides by 2.}$$

$$\frac{P - 2w}{2} = l$$

b. Isolate d in the formula from physics, $v^2 = v_0^2 + \dfrac{2Fd}{m}$.

Solution: First use the addition principle to isolate the term that contains d. Then use the multiplication principle to eliminate the coefficient.

$$v^2 = v_0^2 + \frac{2Fd}{m} \qquad \text{Isolate } \frac{2Fd}{m} \text{ by subtracting } v_0^2 \text{ on both sides.}$$

$$\underline{\quad - v_0^2 \qquad - v_0^2\quad}$$

$$v^2 - v_0^2 = 0 + \frac{2Fd}{m}$$

$$v^2 - v_0^2 = \frac{2Fd}{m}$$

$$\frac{m}{2F} \cdot \frac{v^2 - v_0^2}{1} = \frac{\overset{1}{\cancel{m}}}{\underset{1}{\cancel{2F}}} \cdot \frac{\overset{1}{\cancel{2F}}d}{\underset{1}{\cancel{m}}} \qquad \text{Eliminate the coefficient } \frac{2F}{m} \text{ by multiplying both sides by its reciprocal.}$$

$$\frac{m(v^2 - v_0^2)}{2F} = d$$

Of Interest

In this example, the 0 subscript is used to indicate that v_0 is an initial velocity. We have also seen v_i used to indicate initial velocity. Both forms are used in physics and engineering.

Answers to Your Turn 3

a. $\dfrac{v - v_i}{a} = t$

b. $\dfrac{r}{t}(T + R) = n$ or $\dfrac{r(T + R)}{t} = n$

Your Turn 3

a. Isolate t in the formula that describes the velocity of an object after being accelerated from an initial velocity, $v = v_i + at$.

b. Isolate n in the formula $T = \dfrac{nt}{r} - R$.

2.4 Exercises (For Extra Help) MyMathLab®

Note: Exercises marked with a ★ represent challenging exercises.

Objective 1

Prep Exercise 1 What would you do to isolate w in the formula $l = w - 2$? Why?

Add 2 to both sides. Because 2 is subtracted from w, adding 2 eliminates the 2, leaving w isolated.

Prep Exercise 2 What would you do to isolate p in the formula $C = np$? Why?

Divide both sides by n. Because p is multiplied by n, dividing by n divides out the n, thereby isolating p.

Prep Exercise 3 To isolate y in $2x + 3y = 8$, what would you do first? Why?

Subtract $2x$ from both sides. Subtracting $2x$ isolates the $3y$ term, which contains the variable we want to isolate.

Prep Exercise 4 How can you verify that your answer is correct after isolating a variable in a formula?

Substitute the same values for the variables in the original formula and in your answer; then evaluate to see if you got the same ending value.

For Exercises 1–20, solve for the indicated variable. See Examples 1–3.

1. $t - 4u = v; t$
$t = 4u + v$

2. $x = a + 3y; a$
$a = x - 3y$

3. $-5y = x; y$
$y = -\dfrac{x}{5}$

4. $2n = a; n$
$n = \dfrac{a}{2}$

5. $2x - 3 = b; x$
$x = \dfrac{b + 3}{2}$

6. $3m + b = y; m$
$m = \dfrac{y - b}{3}$

7. $y = mx + b; m$
$m = \dfrac{y - b}{x}$

8. $ab + c = d; b$
$b = \dfrac{d - c}{a}$

9. $3x + 4y = 8; y$
$y = \dfrac{8 - 3x}{4}$

10. $19 = 2l + 2w; w$
$w = \dfrac{19 - 2l}{2}$

11. $\dfrac{mn}{4} - Y = f; Y$
$Y = \dfrac{mn}{4} - f$

12. $q = \dfrac{rs}{2} - p; p$
$p = \dfrac{rs}{2} - q$

13. $6(c + 2d) = m - np; c$
$c = \dfrac{m - np - 12d}{6}$

14. $5(2n + a) = bn - c; a$
$a = \dfrac{bn - c - 10n}{5}$

15. $\dfrac{x}{3} + \dfrac{y}{5} = 1; y$
$y = \dfrac{15 - 5x}{3}$

16. $\dfrac{a}{4} + \dfrac{b}{6} = 3; a$
$a = \dfrac{36 - 2b}{3}$

17. $\dfrac{3}{4} + 5a = \dfrac{n}{c}; n$
$n = c\left(\dfrac{3}{4} + 5a\right)$

18. $\dfrac{x}{6} + 5y = \dfrac{m}{n}; m$
$m = n\left(\dfrac{x}{6} + 5y\right)$

19. $t = \dfrac{A - P}{pr}; p$
$p = \dfrac{A - P}{tr}$

20. $S = \dfrac{C}{1 - M}; M$
$M = \dfrac{S - C}{S}$

For Exercises 21–36, solve the geometric formula for the indicated variable. See Examples 1–3.

21. $P = a + b + c; a$
$a = P - b - c$

22. $a + b + c = 180; b$
$b = 180 - a - c$

23. $A = lw; l$
$l = \dfrac{A}{w}$

24. $C = 2\pi r; r$
$r = \dfrac{C}{2\pi}$

25. $\dfrac{C}{d} = \pi; d$
$d = \dfrac{C}{\pi}$

26. $\dfrac{A}{b} = h; b$
$b = \dfrac{A}{h}$

27. $V = \dfrac{1}{3}\pi r^2 h; r^2$
$r^2 = \dfrac{3V}{\pi h}$

28. $V = \dfrac{4}{3}\pi r^3; r^3$
$r^3 = \dfrac{3V}{4\pi}$

29. $A = \dfrac{1}{2}bh; b$
$b = \dfrac{2A}{h}$

30. $S = \dfrac{1}{2}gt^2; t^2$
$t^2 = \dfrac{2S}{g}$

31. $P = 2l + 2w; w$
$w = \dfrac{P - 2l}{2}$

32. $S = 2\pi r^2 - 2\pi rh; h$
$h = \dfrac{S - 2\pi r^2}{-2\pi r}$ or $h = \dfrac{2\pi r^2 - S}{2\pi r}$

33. $S = \dfrac{n}{2}(a + l); l$
$l = \dfrac{2S - na}{n}$

34. $A = \dfrac{h}{2}(a + b); b$
$b = \dfrac{2A - ah}{h}$

★ 35. $V = h(\pi r^2 - lw); w$
$w = \dfrac{\pi r^2 h - V}{lh}$

★ 36. $V = h(\pi r^2 + lw); r^2$
$r^2 = \dfrac{V - lwh}{\pi h}$

For Exercises 37–44, solve the financial formula for the indicated variable. See Examples 1–3.

37. $P = R - C; C$
$C = R - P$

38. $P = R - C; R$
$R = P + C$

39. $I = Prt; r$
$r = \dfrac{I}{Pt}$

40. $I = Prt; t$
$t = \dfrac{I}{Pr}$

41. $P = \dfrac{C}{n}; C$
$C = nP$

42. $P = \dfrac{C}{n}; n$
$n = \dfrac{C}{P}$

43. $A = P + Prt; r$
$r = \dfrac{A - P}{Pt}$

44. $A = P + Prt; t$
$t = \dfrac{A - P}{Pr}$

For Exercises 45–56, solve the physics formula for the indicated variable. See Examples 1–3.

45. $d = rt; t$
$t = \dfrac{d}{r}$

46. $F = ma; m$
$m = \dfrac{F}{a}$

47. $W = Fd; d$
$d = \dfrac{W}{F}$

48. $V = kT; k$
$k = \dfrac{V}{T}$

49. $P = \dfrac{W}{t}; t$
$t = \dfrac{W}{P}$

50. $D = \dfrac{M}{V}; V$
$V = \dfrac{M}{D}$

51. $v = -32t + v_0; t$
$t = \dfrac{v - v_0}{-32}$ or $t = \dfrac{v_0 - v}{32}$

52. $x = x_0 + vt; v$
$v = \dfrac{x - x_0}{t}$

53. $F = \dfrac{9}{5}C + 32; C$

$C = \dfrac{5}{9}(F - 32)$

54. $C = \dfrac{5}{9}(F - 32); F$

$F = \dfrac{9}{5}C + 32$

55. $F = G\dfrac{Mm}{R^2}; m$

$m = \dfrac{FR^2}{GM}$

56. $F = \dfrac{kMn}{d^2}; k$

$k = \dfrac{Fd^2}{Mn}$

Find ⊗ the Mistake   *For Exercises 57–60, find and explain the mistake in each solution and then solve correctly. See Examples 1–3.*

57. $3n + 7t = 54$; isolate t

$3n + 7t = 54$

$\underline{-3n \qquad\qquad -3n}$

$7t = 54 - 3n$

$7t = 54 - 3n$

$\underline{-7 \qquad -7}$

$t = 47 - 3n$

Mistake: Subtracted the coefficient 7 instead of dividing.

Correct: $t = \dfrac{54 - 3n}{7}$

58. $\dfrac{1}{4}kt = r$; isolate t

$\dfrac{1}{4}kt = r$

$\dfrac{4}{1} \cdot \dfrac{1}{4}kt = 4r$

$\dfrac{1}{k} \cdot kt = 4r\dfrac{k}{1}$

$t = 4kr$

Mistake: Applied the multiplication principle incorrectly, multiplying the left side by $\dfrac{1}{k}$ (which is correct) but the right side by $\dfrac{k}{1}$ (which is a different amount).

Correct: $t = \dfrac{4r}{k}$

59. $\dfrac{3m - 2}{5} = nk$; isolate m

$\dfrac{3m - 2}{5} = nk$

$\dfrac{5}{1} \cdot \dfrac{3m - 2}{5} = nk \cdot 5$

$3m - 10 = 5nk$

$3m - 10 = 5nk$

$\underline{+10 \quad +10}$

$3m = 5nk + 10$

$\dfrac{3m}{3} = \dfrac{5nk + 10}{3}$

$m = \dfrac{5nk + 10}{3}$

Mistake: Multiplied 5 by -2.

Correct: $m = \dfrac{5nk + 2}{3}$

60. $7(y - 5) = xv$; isolate y

$7(y - 5) = xv$

$7y - 5 = xv$

$7y - 5 = xv$

$\underline{+5 \qquad +5}$

$\dfrac{7y}{7} = \dfrac{xv + 5}{7}$

$y = \dfrac{xv + 5}{7}$

Mistake: Did not distribute 7 to multiply -5.

Correct: $y = \dfrac{xv + 35}{7}$

Review Exercises

Exercises 1 and 2 📶 Expressions

[1.6] *For Exercises 1 and 2, translate each phrase to an expression.*

1. four more than seven times a number

$7n + 4$

2. nine less than three times the sum of a number and two

$3(x + 2) - 9$

Exercises 3–6 📶 Equations and Inequalities

[2.3] *For Exercises 3–6, solve.*

3. $4x - 7 = -5$

0.5

4. $\dfrac{3}{4}x = -\dfrac{5}{8}$

$-\dfrac{5}{6}$

5. $3(n - 5) = 7n - 12$

$-\dfrac{3}{4}$

6. $\dfrac{1}{4}y + 6 = \dfrac{1}{5}(y + 10)$

-80

2.5 Translating Word Sentences to Equations

Objective

1 Translate sentences to equation using key words; then solve.

Warm-up
[2.3] Solve.
1. $0.6n = -25.08$

2. $\frac{2}{3}(6 - n) = \frac{n}{4}$

[1.5] 3. Simplify using the order of operations. $\frac{2}{3}\left(6 - \frac{48}{11}\right)$

Objective 1 Translate sentences to equations using key words; then solve.

In previous sections, we translated phrases to expressions (Section 1.6) and solved problems using formulas (Section 2.1). In this section, we further develop problem-solving strategies by translating sentences to equations. Because we have not learned how to solve equations containing exponents and roots, we will explore translations that involve only addition, subtraction, multiplication, and division. As in Section 1.6, we begin by translating algebraic equations to word sentences. Several key words or phrases indicate an equal sign.

Key words for an equal sign:	is equal to	is	yields
	is the same as	produces	results in

Translating Algebraic Equations to English Sentences

Equation	Translation
	Equations Involving Addition
$x + 3 = 6$	The sum of x and three is six.
	Three more than x is equal to six.
	x plus three results in six.
	Three added to x produces six.
	x increased by three is the same as six.
	Equations Involving Subtraction
$y - 2 = 7$	The difference of y and two is seven.
	y minus two equals seven.
	Two subtracted from y yields seven.
	Two less than a number results in seven.
	y minus two produces seven.
	A number decreased by two equals seven.
	Equations Involving Multiplication
$\frac{3}{4}m = 9$	The product of three-fourths and m is nine.
	Three-fourths times m is the same as nine.
	Three-fourths of a number equals nine.
	Equations Involving Division
$\frac{x}{4} = -2$	The quotient of x and four results in negative two.
	x divided by four produces negative two.
	Four divided into a number is equal to negative two.
	The ratio of x and four is the same as negative two.

Answers to Warm-up
1. $n = -41.8$
2. $n = \frac{48}{11}$
3. $\frac{12}{11}$

Equation	Translation
	Equations That Involve More Than One Operation
$4n + 2 = 8$	Two more than the product of four and n is eight.
	Four times n increased by two is equal to eight.
	The sum of four times n and two is the same as eight.
	Two more than the product of four and a number equals eight.
	Two added to four times a number results in eight.
	Equations Involving Parentheses
$4(m - 3) = 6$	Four times the difference of m and three results in six.
	The product of four and the difference of m and three is six.

You also may want to review or reference the key words and their translations in Section 1.6 on pages 59 and 60.

> **Procedure** **Translating Word Sentences**
>
> To translate a word sentence to an equation, identify the unknown(s), constants, and key words; then write the corresponding symbolic form.

Problems Involving Addition or Subtraction

Learning Strategy

If you are a visual learner, underline key words and phrases or write key words and phrases in different colors as is done in the text.

Instructor Note Point out that due to the commutative property of addition, $n + 25 = 14$ is also correct.

Instructor Note Point out that because subtraction is not commutative, $19 - n = 7$ is not correct.

Example 1 Translate to an equation and then solve.

a. The sum of twenty-five and a number is equal to fourteen.

Solution: The key word *sum* indicates addition, *is equal to* indicates an equal sign, and *a number* indicates a variable.

Translate: The sum of twenty-five and a number is equal to fourteen.

$$25 \quad + \quad n \quad = \quad 14$$

Solve:
$$25 + n = 14$$
$$\underline{-25 \qquad -25} \quad \text{Subtract 25 from both sides to isolate } n.$$
$$0 + n = -11 \qquad \text{**Note** This is the solution to the}$$
$$n = -11 \qquad \text{problem because the sum of 25 and } -11 \text{ is indeed 14.}$$

b. Nineteen less than a number is seven.

Solution: The key words *less than* indicate subtraction. The key word *is* means an equal sign.

Translate: Nineteen less than a number is seven.

$$n - 19 \qquad = \quad 7$$

Note The key words *less than* require careful translation. In the sentence, the word *nineteen* comes before the words *less than* and the words *a number* come after. In the translation, this order is reversed.

Solve:
$$n - 19 = 7$$
$$\underline{+19 \quad +19} \quad \text{Add 19 to both sides.}$$
$$n + 0 = 26$$
$$n = 26$$

Your Turn 1 Translate to an equation and then solve.

a. Fifteen more than a number is negative seven.
b. The difference of x and thirty-five is negative nine.

Problems Involving Multiplication

Example 2 Three-fourths of a number is negative five-eighths. Translate to an equation and then solve.

Solution: When *of* is preceded by a fraction, it means multiply. The word *is* means an equal sign.

Translate: Three-fourths of a number is negative five-eighths.

$$\frac{3}{4} \cdot n = -\frac{5}{8}$$

Instructor Note Some students find it easier to first multiply by 4 and then divide by 3.

Solve: $\frac{\overset{1}{4}}{\underset{1}{3}} \cdot \frac{\overset{1}{3}}{\underset{1}{4}} n = -\frac{5}{\underset{2}{8}} \cdot \frac{\overset{1}{4}}{3}$ Eliminate the coefficient $\frac{3}{4}$ by multiplying both sides by its reciprocal, $\frac{4}{3}$.

$$n = -\frac{5}{6}$$

Your Turn 2 The product of 0.6 and a number is -25.08. Translate to an equation and then solve.

Problems Involving More Than One Operation

Example 3 Seven less than the product of four and a number is equal to negative five. Translate to an equation and then solve.

Solution: *Less than* indicates subtraction in reverse order, *product* indicates multiplication, and *is equal to* indicates an equal sign.

Translate:

Seven less than the product of four and a number is equal to negative five.

$4n - 7$ $=$ -5

Solve: $4n - 7 = -5$

$\underline{+7 \quad +7}$ Add 7 to both sides to isolate $4n$.

$4n + 0 = 2$

$\frac{4n}{4} = \frac{2}{4}$ Divide both sides by 4 to eliminate the coefficient.

$n = \frac{1}{2}$ or 0.5

Answers to Your Turn 1
a. $n + 15 = -7; n = -22$
b. $x - 35 = -9; x = 26$

Answers to Your Turn 2
$0.6n = -25.08; n = -41.8$

Answers to Your Turn 3
$4n + 13 = -7; n = -5$

Your Turn 3 Thirteen more than the product of four and a number yields negative seven. Translate to an equation and then solve.

Translations Requiring Parentheses

Sometimes a translation requires parentheses. This occurs when a sum or difference is multiplied or divided. When we multiply or divide a sum or difference, the addition or subtraction is calculated first. Because addition and subtraction follow multiplication and division in the order of operations, parentheses must be inserted around the addition or subtraction so that these operations are performed first.

Example 4 Two-thirds of the difference of six and a number is the same as the number divided by four. Translate to an equation and then solve.

Solution: *Of* means to multiply, and the word *difference* indicates subtraction. Because the difference is being multiplied, we write the difference expression in parentheses. *Divided by* indicates division.

Translate:

Two-thirds of the difference of six and a number is the same as the number divided by four.

$$\frac{2}{3} \cdot (6 - n) = \frac{n}{4}$$

Solve: $\dfrac{2}{3}(6 - n) = \dfrac{n}{4}$

$\dfrac{2}{\cancel{3}} \cdot \dfrac{\overset{2}{\cancel{6}}}{1} - \dfrac{2}{3}n = \dfrac{n}{4}$ Distribute to eliminate the parentheses.

$4 - \dfrac{2}{3}n = \dfrac{n}{4}$ Simplify.

$12\left(4 - \dfrac{2}{3}n\right) = \left(\dfrac{n}{4}\right)12$ Multiply both sides by the LCD, 12, to eliminate the fractions.

$12 \cdot 4 - \dfrac{\overset{4}{\cancel{12}}}{1} \cdot \dfrac{2}{\underset{1}{\cancel{3}}}n = \dfrac{n}{\underset{1}{\cancel{4}}} \cdot \dfrac{\overset{3}{\cancel{12}}}{1}$ Distribute 12 and simplify.

$48 - 8n = 3n$ Add 8n to both sides to separate the constant term and variable terms.

$\underline{+8n \quad +8n}$

$48 + 0 = 11n$

$\dfrac{48}{11} = \dfrac{11n}{11}$ Eliminate the coefficient by dividing both sides by 11.

$\dfrac{48}{11} = n$

Answers to Your Turn 4

a. $3(n - 5) = 7n - 12$;

$n = -\dfrac{3}{4}$

b. $\dfrac{1}{4}n + 6 = \dfrac{1}{5}(n + 10)$;

$n = -80$

Your Turn 4 Translate to an equation and then solve.

a. Three times the difference of a number and five is equal to seven times the number minus twelve.

b. The sum of one-fourth of a number and six is equal to one-fifth of the sum of the number and ten.

2.5 Exercises   For Extra Help   MyMathLab®

Objective 1

Prep Exercise 1 List three key words that indicate addition.
Sum, plus, added

Prep Exercise 2 List three key words that indicate subtraction.   Difference, minus, less

Prep Exercise 3 List three key words that indicate multiplication.   Of, times, twice

Prep Exercise 4 List three key words that indicate division.   Quotient, divided, ratio

Prep Exercise 5 Why are "four minus n" and "four less than n" different?   The order of the subtraction is different.

Prep Exercise 6 Explain why "five times the sum of x and 3" requires parentheses.   The word *sum* indicates the answer to addition. The parentheses ensure that the sum is found before multiplication is done.

For Exercises 1–38, translate each sentence to an equation and then solve. See Examples 1–4.

1. Three more than y is negative 8.
   $y + 3 = -8; -11$

2. Six added to p is negative two.
   $p + 6 = -2, p = -8$

3. Six decreased by the number x is equal to negative three.
   $6 - x = -3; 9$

4. Eighteen subtracted from a number is equal to negative three.
   $w - 18 = -3; 15$

5. Eleven times m is negative ninety-nine.
   $11m = -99; -9$

6. The product of negative three and z is -18.
   $-3z = -18; 6$

7. Four divided into m is 1.6.
   $m \div 4 = 1.6; 6.4$

8. The quotient of a number and -6.5 is 4.2.
   $k \div (-6.5) = 4.2; -27.3$

9. Four-fifths of a number is five-eighths.
   $\frac{4}{5}x = \frac{5}{8}; \frac{25}{32}$

10. Three-sevenths of a number is negative nine-eighths.
    $\frac{3}{7}x = -\frac{9}{8}; -\frac{21}{8}$

11. Seven more than three times a number is equal to thirty-four.
    $7 + 3w = 34; 9$

12. Nineteen more than triple a number is one hundred.
    $19 + 3p = 100; 27$

13. Nine less than five times a number is seventy-six.
    $5a - 9 = 76; 17$

14. Twenty-five less than four times a number is eleven.
    $4r - 25 = 11; 9$

15. Eight times the sum of eight and a number is equal to one hundred sixty.
    $8(8 + t) = 160; 12$

16. Four times the sum of a number and three is equal to sixteen.
    $4(x + 3) = 16; 1$

17. Negative three times the difference of x and two is twelve.
    $-3(x - 2) = 12; -2$

18. Tripling the difference of x and five produces negative fifteen.
    $3(x - 5) = -15; 0$

19. One-third of the sum of a number and two is one.
    $\frac{1}{3}(g + 2) = 1; 1$

20. One-half of the difference of a number and two results in four.
    $\frac{1}{2}(h - 2) = 4; 10$

21. Eleven less than three times a number is equal to that number added to five.
    $3m - 11 = 5 + m; 8$

22. Five more than four times a number is equal to seven subtracted from that number.
    $4x + 5 = x - 7; -4$

23. Ten is the result when one is subtracted from the ratio of a number to four.

$10 = \frac{r}{4} - 1$; 44

24. Two less than the quotient of a number and five is six.

$\frac{x}{5} - 2 = 6$; 40

25. The product of three and the sum of a number and four subtracted from twice the number yields negative three.

$2a - 3(a + 4) = -3$; −9

26. Seven times the sum of a number and six subtracted from four times the number results in negative twenty-one.

$4x - 7(x + 6) = -21$; −7

27. The difference of a number and twelve is added to the difference of twice the number and eight so that the result is thirteen.

$(2d - 8) + (d - 12) = 13$; 11

28. The sum of a number and six is added to the difference of the number tripled and four so that the result is fourteen.

$(x + 6) + (3x - 4) = 14$; 3

29. Five times the sum of a number and two-thirds is equal to three times the number decreased by two-thirds.

$5\left(x + \frac{2}{3}\right) = 3x - \frac{2}{3}$; −2

30. The product of three and the sum of a number and two-fifths is four-fifths less than negative two times the number.

$3\left(x + \frac{2}{5}\right) = -2x - \frac{4}{5}$; $-\frac{2}{5}$

31. Two times a number decreased by four is equal to triple the number added to sixteen.

$2x - 4 = 16 + 3x$; −20

32. Six less than three times a number is equal to the difference of fourteen and twice the number.

$3x - 6 = 14 - 2x$; 4

33. Two-fifths of a number is equal to two less than one-half the number.

$\frac{2}{5}x = \frac{1}{2}x - 2$; 20

34. One-third of a number is the same as half of the number decreased by one.

$\frac{1}{3}p = \frac{1}{2}p - 1$; 6

35. The quotient of three less than a number and three is the same as the quotient of one less than the number and four.

$\frac{n - 3}{3} = \frac{n - 1}{4}$; 9

36. The quotient of one more than a number and two is the same as the quotient of six more than twice the number and eight.

$\frac{x + 1}{2} = \frac{2x + 6}{8}$; 1

37. The product of negative four and the difference of two and a number is six less than the product of two and the difference of four and three times the number.

$-4(2 - x) = 2(4 - 3x) - 6$; 1

38. Negative two times the difference of one and three times a number is equal to four added to five times the difference of one and the number.

$-2(1 - 3t) = 5(1 - t) + 4$; 1

For Exercises 39–48, translate the equation to a word sentence. There is more than one correct translation.

39. $4x + 3 = 7$

Three added to four times a number is seven.

40. $-3y + 8 = 10$

Negative three times a number plus eight is the same as ten.

41. $6(y + 4) = -10y$

Six times the sum of a number and four is equal to the product of negative ten and the number.

42. $8(w - 2) = 3w$

Eight times the difference of a number and two will yield three times the number.

43. $\frac{1}{2}(x - 3) = \frac{2}{3}(x - 8)$

One-half of the difference of a number and three will result in two-thirds of the difference of the number and eight.

44. $\frac{1}{2}(t + 1) = \frac{1}{3}(t - 5)$

Half of the sum of a number and one is one-third of the difference of the number and five.

45. $0.05m + 0.06(m - 11) = 22$

Five-hundredths of a number added to six-hundredths of the difference of the number and eleven is twenty-two.

46. $0.05v + 0.03(v - 4.5) = 0.465$

Five-hundredths of a number added to three-hundredths of the difference of the number and four and five-tenths is equal to four hundred sixty-five thousandths.

47. $\frac{2}{3}x + \frac{3}{4}x + \frac{1}{2}x = 10$

The sum of two-thirds, three-fourths, and one-half of the same number will equal ten.

48. $\frac{1}{2}a + \frac{1}{3}a + \frac{1}{6}a = 5$

The sum of one-half, one-third, and one-sixth of the same number will equal five.

Find the Mistake *For Exercises 49–54, explain the mistake in each translation; then translate correctly. See Examples 1–4.*

49. Ten less than a number is forty.

Translation: $10 - n = 40$

Mistake: Subtraction translated in reverse order.

Correct: $n - 10 = 40$

50. Twelve divided into a number is negative eight.

Translation: $12 \div n = -8$

Mistake: Division translated in reverse order.

Correct: $n \div 12 = -8$

51. Five times the difference of a number and six is equal to negative two.

Translation: $5x - 6 = -2$

Mistake: Multiplied 5 times the unknown number instead of the difference, which requires parentheses.

Correct: $5(x - 6) = -2$

52. Nine times the sum of a number and eight is the same as three subtracted from the number.

Translation: $9(y + 8) = 3 - y$

Mistake: "three subtracted from" indicates that the 3 should be after the minus sign.

Correct: $9(y + 8) = y - 3$

53. Twice a number minus the sum of the number and three is equal to negative six.

Translation: $2t - t + 3 = -6$

Mistake: Subtracted the unknown number instead of the sum, which requires parentheses.

Correct: $2t - (t + 3) = -6$

54. Four times the sum of a number and two is equal to the number minus the difference of the number and seven.

Translation: $4(y + 2) = y - (7 - y)$

Mistake: "difference" was translated in reverse order.

Correct: $4(y + 2) = y - (y - 7)$

For Exercises 55–62, translate to a formula; then use the formula to solve the problem.

55. The perimeter of a rectangle is equal to twice the sum of its length and width. Find the perimeter with the following length and width.

Length	Width
a. 24 ft.	18 ft.
b. 12.5 cm	18 cm
c. $8\frac{1}{4}$ in.	$10\frac{3}{8}$ in.

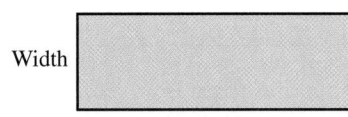

Translation: $P = 2(l + w)$

a. 84 ft. b. 61 cm c. $37\frac{1}{4}$ in.

56. The surface area of a box is equal to twice the sum of its length times its width, its length times its height, and its width times its height. Find the surface area of a box with the following length, width, and height.

Length	Width	Height
a. 15 in.	6 in.	4 in.
b. 9.2 cm	12 cm	6.5 cm
c. 8 ft.	$2\frac{1}{2}$ ft.	3 ft.

Translation: $SA = 2(lw + lh + wh)$

a. 348 in.² b. 496.4 cm² c. 103 ft.²

57. An isosceles triangle has two sides that are equal in length. The perimeter of an isosceles triangle is the sum of its base and twice the length of one of the other sides. Find the perimeter of each listed isosceles triangle.

Base	Sides
a. 5 in.	11 in.
b. 1.5 m	0.6 m
c. $15\frac{1}{4}$ in.	$9\frac{1}{2}$ in.

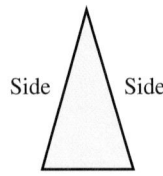

Translation: $P = b + 2s$

a. 27 in. b. 2.7 m c. $34\frac{1}{4}$ in.

58. The perimeter of a semicircle is the sum of the diameter and π times the radius. Find the perimeter of each semicircle listed. Where appropriate, round your answer to the nearest tenth.

a. Diameter = 18 in. (Use $\pi \approx 3.14$.)

b. Radius = 4.5 m (Use $\pi \approx 3.14$.)

c. Diameter = $2\frac{3}{4}$ ft. $\left(\text{Use } \pi \approx \frac{22}{7}.\right)$

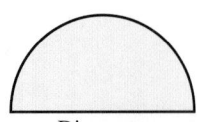

Translation: $P = d + \pi r$

a. ≈46.3 in. b. ≈23.1 m c. $\approx \frac{99}{14}$ ft.

59. The simple interest earned after investing an amount of money, called principal, is equal to the product of the principal, the interest rate, and the time in years that the money remains invested. Use the formula to calculate the interest of each of the following investments.

	Principal	Rate	Time
a.	$4000	0.05	2 years
b.	$500	0.03	$\frac{1}{2}$ year
c.	$2000	0.06	$\frac{1}{4}$ year

Translation: $I = Prt$

a. $400 b. $7.50 c. $30

60. Investors often consider a stock's price-to-earnings ratio to help them decide whether a stock is worth investing in. The price-to-earnings ratio is equal to the stock's price divided by its company's earnings per share over the last 12 months. Use your formula to calculate each company's price-to-earnings ratio. Round your answers to the nearest tenth. (*Source:* David and Tom Gardner, *The Motley Fool Investment Guide*, Simon & Schuster, 1996.)

	Company	Price	Earnings per Share
a.	Microsoft	$47.98	$1.41
b.	Intel	$18.79	$0.29
c.	General Motors	$46.55	$4.01

Translation: $R = \dfrac{P}{E}$

a. ≈34 b. ≈64.8 c. ≈11.6

61. Skydivers often use a formula to estimate free-fall time. The amount of free-fall time in seconds is approximately equal to the difference of the exit altitude and the parachute deployment altitude divided by 153.8. Use this formula to calculate free-fall times if a skydiver deploys at each of the following altitudes after falling from an exit altitude of 12,500 feet. Round your answers to the nearest second.

a. 5000 feet
b. 4500 feet
c. 3000 feet

Translation: $t = \dfrac{e - d}{153.8}$

a. ≈49 sec. b. ≈52 sec. c. ≈62 sec.

62. A person's body mass index (BMI) is a measure of the amount of body fat based on height, in inches, and weight, in pounds. A person's BMI is equal to the ratio of the product of his or her weight and 704.5 to his or her height in inches multiplied by itself. Use your formula to find the BMI of each person listed below. Round your answers to the nearest tenth. (*Source:* National Institutes of Health.)

	Height	Weight
a.	5′4″	138 lb.
b.	5′9″	155 lb.
c.	6′2″	220 lb.

Translation: $BMI = \dfrac{704.5w}{h \cdot h}$

a. ≈23.7 b. ≈22.9 c. ≈28.3

Of Interest

After exiting the plane, a skydiver accelerates from 0 to approximately 110 miles per hour (terminal velocity) in about 10 seconds and falls about 1000 feet. Terminal velocity means that air resistance balances out gravitational acceleration so that the skydiver falls at a constant 110 miles per hour until he or she deploys the parachute. At terminal velocity, the skydiver falls at a rate of about 1000 feet every 6 seconds. Of course, the speed of the fall can be changed by varying body position to increase or decrease air resistance.

Of Interest

The National Institutes of Health use the following categories when analyzing BMI:

Underweight:	BMI < 18.5
Normal Weight:	BMI = 18.5–24.9
Overweight:	BMI = 25–29.9
Obesity:	BMI ≥ 30

Note that this way of calculating BMI is a rough estimate and is not accurate for extremely fit people such as athletes and bodybuilders because their "extra" weight is muscle rather than fat.

Puzzle Problem Fill in each blank with a whole number so that the resulting rational number has the following properties:

The rational number is between 300 and 500.

The tenths placeholder is less than the tens placeholder.

The sum of the tens place and the tenths place is 8.

The product of all four placeholders is 42.

___ ____ ___ . ____

372.1

Review Exercises

Exercises 1-6 ◢◣◢ Equations and Inequalities

[1.1] 1. Is $19 \geq 19$ true or false? Explain.

True; 19 is equal to 19.

[1.1, 1.5] *For Exercises 2 and 3, use* $<$*,* $>$*, or* $=$ *to make a true sentence.*

2. $-|-6|$ $-2(3)$

$=$

3. $5(4) - 23$ $2(3 - 8)$

$>$

[2.3] *For Exercises 4–6, solve and check.*

4. $7x + 9 = 4x - 6$

-5

5. $\dfrac{5}{6} = -\dfrac{3}{4}x$

$-\dfrac{10}{9}$

6. $\dfrac{1}{3}(y - 2) = \dfrac{3}{5}y + 6$

-25

2.6 Ratios and Proportions

Objectives

1 Solve problems involving ratios.
2 Solve for an unknown number in a proportion.
3 Solve proportion problems.
4 Use proportions to solve for unknown lengths in figures that are similar.

Warm-up

[1.4] 1. Simplify: $\dfrac{2}{3} \cdot 3\dfrac{1}{2}$

[2.3] 2. Solve: $9.6a = 33.6$

Objective 1 Solve problems involving ratios.

In this section, we solve problems involving **ratios**.

Definition Ratio: A comparison of two quantities using a quotient.

Ratios are usually expressed in fraction form or with a colon. When a ratio is written in English, the word *to* separates the numerator and denominator quantities.

The ratio of 12 to 17 translates to $\dfrac{12}{17}$.

↑   ↑

Numerator   Denominator

Note The quantity preceding the word *to* is written in the numerator, and the quantity following *to* is written in the denominator.

Example 1 A small college has 2450 students and 140 faculty. Write the ratio of students to faculty in simplest form. Interpret the answer.

Solution: The ratio of students to faculty $= \dfrac{2450}{140} = \dfrac{35}{2}$.

This means that there are 35 students for every 2 faculty members.

Answer to Your Turn 1

$\dfrac{6400}{4800} = \dfrac{4}{3}$; for every four people in favor of higher taxes, three are against them.

Answers to Warm-up

1. $\dfrac{7}{3}$

2. $a = 3.5$

Your Turn 1 In a local referendum, 6400 people voted for increased school taxes and 4800 people voted against the increase. In simplest form, write the ratio of those voting for higher taxes to those voting against. Interpret the answer.

As noted, the ratio in Example 1 indicates that there are 35 students for every 2 faculty members. Expressing a ratio as a **unit ratio** often makes the ratio easier to interpret.

Definition Unit ratio: A ratio with a denominator of 1.

Example 2 Express the ratio from Example 1 as a unit ratio. Interpret the answer.

Solution: $\dfrac{2450}{140} = \dfrac{35}{2} = \dfrac{17.5}{1}$ Divide and write the denominator as 1.

The unit ratio $\dfrac{17.5}{1}$ indicates that there are 17.5 students for every faculty member at the college.

Your Turn 2 The price of a 12.5-ounce box of cereal is $2.85. Write the unit ratio of price to weight. Interpret the answer.

Objective 2 Solve for an unknown number in a proportion.

Some problems involve two equal ratios in a **proportion**.

Definition Proportion: An equation in the form $\dfrac{a}{b} = \dfrac{c}{d}$, where $b \neq 0$ and $d \neq 0$.

In Section 1.2, we discussed equivalent fractions and saw that fractions such as $\dfrac{3}{8}$ and $\dfrac{6}{16}$ are equivalent because $\dfrac{3}{8}$ can be rewritten as $\dfrac{6}{16}$ and $\dfrac{6}{16}$ can be simplified to $\dfrac{3}{8}$. Note that when fractions are equivalent, the cross products are equal.

$$16 \cdot 3 = 48 \qquad 8 \cdot 6 = 48$$

$$\frac{3}{8} \times \frac{6}{16}$$

Let's see why all proportions have equal cross products. We begin with the general form of a proportion from the definition $\dfrac{a}{b} = \dfrac{c}{d}$, where $b \neq 0$ and $d \neq 0$. We can use the multiplication principle of equality to eliminate the fractions by multiplying both sides by the LCD, which is bd.

$$\frac{\overset{1}{bd}}{1} \cdot \frac{a}{\underset{1}{b}} = \frac{\overset{1}{bd}}{1} \cdot \frac{c}{\underset{1}{d}}$$

$$ad = bc$$

Note that ad and bc are the cross products.

$$ad \qquad bc$$

$$\frac{a}{b} \times \frac{c}{d}$$

Conclusion: If two ratios are equal, their cross products are equal.

Answer to Your Turn 2
$\dfrac{0.228}{1}$; the cereal costs $0.228 per ounce.

Rule Proportions and Their Cross Products

If $\dfrac{a}{b} = \dfrac{c}{d}$, where $b \neq 0$ and $d \neq 0$, then $ad = bc$.

We can use cross products to solve for an unknown number in a proportion.

> **Procedure** **Solving a Proportion**
>
> To solve a proportion using cross products:
> 1. Find the cross products.
> 2. Set the cross products equal to each other.
> 3. Use the multiplication principle of equality to isolate the variable.

VISUAL

Learning Strategy

If you are a visual learner, draw the arrows as is done here to help visualize the pattern of cross products.

Example 3 Solve.

a. $\dfrac{7}{16} = \dfrac{x}{10}$

Solution: $10 \cdot 7 = 70$ $\qquad$ $16 \cdot x = 16x$ $\qquad$ Find the cross products.

$$\dfrac{7}{16} = \dfrac{x}{10}$$

$70 = 16x$ $\qquad$ Set the cross products equal to each other.

$\dfrac{70}{16} = \dfrac{16x}{16}$ $\qquad$ Divide both sides by 16 to isolate x.

Note We could have expressed the answer as 4.375. ▶ $\dfrac{35}{8} = x$

b. $\dfrac{3\frac{1}{5}}{12} = \dfrac{-9}{n}$

Solution:

$3\dfrac{1}{5} \cdot n = \dfrac{16}{5}n$ $\qquad$ $12 \cdot (-9) = -108$ $\qquad$ Find the cross products.

$$\dfrac{3\frac{1}{5}}{12} = \dfrac{-9}{n}$$

$\dfrac{16}{5}n = -108$ $\qquad$ Set the cross products equal to each other.

$\dfrac{\overset{1}{\cancel{5}}}{\cancel{16}} \cdot \dfrac{\overset{1}{\cancel{16}}}{\cancel{5}}n = \dfrac{\overset{-27}{\cancel{-108}}}{1} \cdot \dfrac{5}{\underset{4}{\cancel{16}}}$ $\qquad$ Multiply both sides by $\dfrac{5}{16}$ to isolate n.

$n = -\dfrac{135}{4}$ or $-33\dfrac{3}{4}$

Your Turn 3 Solve.

a. $\dfrac{9}{y} = -\dfrac{5}{6}$ $\qquad$ **b.** $\dfrac{\frac{2}{3}}{n} = \dfrac{-5}{4\frac{1}{2}}$ $\qquad$ **c.** $-\dfrac{2.4}{8.1} = -\dfrac{t}{12}$

Objective 3 Solve proportion problems.

Answers to Your Turn 3

a. $-\dfrac{54}{5}$ or -10.8 **b.** $-\dfrac{3}{5}$ **c.** $3.\overline{5}$

In a typical proportion problem, you are given a ratio and then asked to find an equivalent ratio. The key to translating problems to a proportion is to write the proportion so that the numerators and denominators correspond to each other in a logical way.

Procedure To translate an application problem to a proportion, write two ratios set equal to each other so that the information from the problem pairs both horizontally (straight across) and vertically (up and down).

Example 4 A company determines that it spends $254.68 every 5 business days on equipment failure. At this rate, how much would the company spend over 28 business days?

Solution: We are given a rate of $254.68 every 5 days spent on equipment failure, which can be written as $\dfrac{\$254.68}{5 \text{ days}}$. *At this rate* indicates that $\dfrac{\$254.68}{5 \text{ days}}$ stays the same for the 28 days, so we can use a proportion. If we let n represent the unknown amount spent over the 28 days, we have a ratio of $\dfrac{\$n}{28 \text{ days}}$. We create the proportion by setting those ratios equal to each other.

Note With proportions, it is helpful to perform a quick mental estimate to check the reasonableness of your answer. Because 28 days is a little less than 6 times the 5 days we were given, the amount of money spent will be a little less than 6 times the $254.68 spent on those 5 days. Rounding $254.68 to $250 so that we can calculate mentally, we see that $6 \cdot 250 = \$1500$. So we expect the amount spent on 28 days to be a little less than $1500. Our solution of $1426.21 fits that expectation.

$$\frac{\$254.68}{5 \text{ days}} = \frac{\$n}{28 \text{ days}}$$

$$28 \cdot 254.68 = 7131.04 \qquad 5 \cdot n = 5n$$

$$\frac{254.68}{5} \diagup\!\!\!\!\diagdown \frac{n}{28}$$

◄ **Note** This proportion is correct because the information from the problem pairs both horizontally and vertically— horizontally because the units match straight across and vertically because both have dollars over days.

$$7131.04 = 5n \qquad \text{Set the cross products equal.}$$

$$\frac{7131.04}{5} = \frac{5n}{5} \qquad \text{Divide both sides by 5 to isolate n.}$$

$$1426.208 = n$$

Answer: Because the answer is in dollars, we round to the nearest hundredth (or cent). The company spends $1426.21 over 28 business days.

Checking Proportion Problems

One of the nice things about proportion problems is that if you set up the proportion incorrectly, you usually get an unreasonable answer. A quick estimate check can be a fast way to recognize when you have set up a proportion incorrectly. For example, suppose we had incorrectly set up Example 4 this way:

$$\frac{\$254.68}{5 \text{ days}} = \frac{28 \text{ days}}{\$n}$$

Warning This setup is incorrect because the numerators and denominators do not correspond.

$$254.68 \cdot n = 254.68n \qquad 5 \cdot 28 = 140 \qquad \text{Cross multiply.}$$

$$\frac{254.68}{5} \diagup\!\!\!\!\diagdown \frac{28}{n}$$

$$254.68n = 140 \qquad\qquad \text{Set the cross products equal.}$$

$$\frac{254.68n}{254.68} = \frac{140}{254.68} \qquad\qquad \text{Divide both sides by 254.68 to isolate } n.$$

$$n \approx 0.55 \qquad \text{Warning This answer indicates that the company spends only \$0.55 over 28 business days based on spending \$254.68 every 5 business days, which is clearly too low.}$$

There are many correct ways to set up a proportion for a particular problem. For example, following are some other correct ways we could have set up the proportion for Example 4. Note that the proportions involve rates that correspond with each other.

$$\frac{5 \text{ days}}{\$254.68} = \frac{28 \text{ days}}{\$n}$$

▲
Note Think:
In 5 days, $254.68 is spent.
In 28 days, n is spent.

$$\frac{5 \text{ days}}{28 \text{ days}} = \frac{\$254.68}{\$n}$$

▲
Note Think of this
version as an analogy:
If 5 days means $254.68,
then 28 days means n.

$$\frac{28 \text{ days}}{5 \text{ days}} = \frac{\$n}{\$254.68}$$

▲
Note This version is
a rearrangement of
the previous version.

Procedure **Solving Proportion Application Problems**

To solve proportion problems:
1. Set up a proportion in which the numerators and denominators of the ratios correspond in a logical manner.
2. Solve using cross products.

Your Turn 4   Pam drove 351.6 miles using 16.4 gallons of gasoline. At this rate, how much gasoline would she need to drive 1000 miles?

Objective 4 Use proportions to solve for unknown lengths in figures that are similar.

Consider triangles *ABC* and *DEF*.

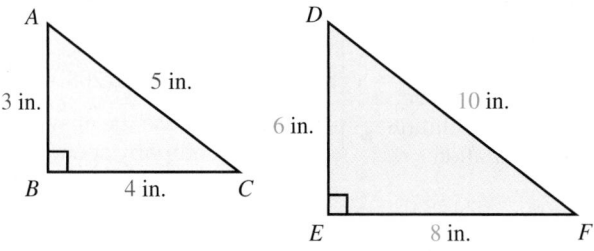

Although the triangles are different in size, their corresponding angle measurements are **congruent**.

Definition Congruent angles: Angles that have the same measure. The symbol for congruent is $\cong$.

In symbols, we write $\angle B \cong \angle E$, $\angle A \cong \angle D$, and $\angle C \cong \angle F$.

Besides having all angles congruent, notice that the sides of triangle DEF are twice as long as the corresponding sides of triangle *ABC*. Or we could say that all of the sides of triangle *ABC* are half as long as the corresponding sides of triangle *DEF*. In other words, the corresponding side lengths are proportional.

Ratio of Triangle *DEF*
to $\longrightarrow$ $\frac{6}{3} = \frac{8}{4} = \frac{10}{5} = \frac{2}{1}$
Triangle *ABC*

or

Ratio of Triangle *ABC*
to $\longrightarrow$ $\frac{3}{6} = \frac{4}{8} = \frac{5}{10} = \frac{1}{2}$
Triangle *DEF*

Answer to Your Turn 4
≈ 46.6 gal.

Because the corresponding angles in the two triangles are congruent and the side lengths are proportional, we say that these triangles are **similar figures**.

Definition **Similar figures:** Figures with congruent angles and proportional side lengths.

Example 5 The two trapezoids shown are similar. Find the unknown lengths.

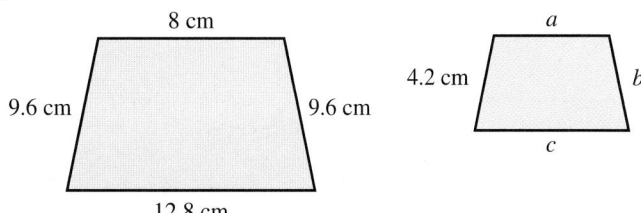

Solution: Because the figures are similar, the lengths of the corresponding sides are proportional. The corresponding sides can be shown as follows:

Ratio of larger trapezoid $\longrightarrow$ to smaller trapezoid $\longrightarrow$

$$\frac{9.6}{4.2} = \frac{8}{a} = \frac{9.6}{b} = \frac{12.8}{c}$$

It is not necessary to write a proportion to find b. Because two sides in the larger trapezoid measure 9.6 cm, the corresponding sides in the smaller trapezoid must be equal in length; so we can conclude that $b = 4.2$ cm.

To find a:

$$\frac{9.6}{4.2} = \frac{8}{a}$$

$$a \cdot 9.6 = 9.6a \qquad 4.2 \cdot 8 = 33.6$$

$$\frac{9.6}{4.2} \diagup\!\!\!\diagdown \frac{8}{a}$$

$$9.6a = 33.6$$

$$\frac{9.6a}{9.6} = \frac{33.6}{9.6}$$

$$a = 3.5$$

To find c:

$$\frac{9.6}{4.2} = \frac{12.8}{c}$$

$$c \cdot 9.6 = 9.6c \qquad 4.2 \cdot 12.8 = 53.76$$

$$\frac{9.6}{4.2} \diagup\!\!\!\diagdown \frac{12.8}{c}$$

$$9.6c = 53.76$$

$$\frac{9.6c}{9.6} = \frac{53.76}{9.6}$$

$$c = 5.6$$

Answer: The unknown lengths are $a = 3.5$ cm, $b = 4.2$ cm, and $c = 5.6$ cm.

Your Turn 5 Find the unknown lengths in the similar figures.

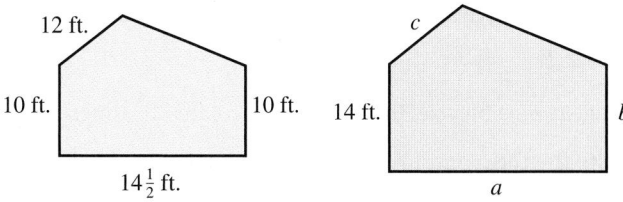

Answers to Your Turn 5

$a = 20\dfrac{3}{10}$ ft. or 20.3 ft.

$b = 14$ ft.

$c = 16\dfrac{4}{5}$ ft. or 16.8 ft.

2.6 Exercises For Extra Help MyMathLab®

Note: Exercises marked with a ★ represent challenging exercises.

Objective 1

Prep Exercise 1 In the phrase *the ratio of the longest side to shortest side*, how do you determine the numerator and denominator of the ratio?
The quantity preceding *to* (longest side) is written in the numerator, and the quantity following the word *to* (shortest side) is written in the denominator.

For Exercises 1–6, write the ratio in simplest form. See Example 1.

1. One molecule of lactose contains 12 carbon atoms, 22 hydrogen atoms, and 11 oxygen atoms.
   a. Write the ratio of carbon atoms to total atoms in the molecule.
   $\dfrac{4}{15}$
   b. Write the ratio of carbon atoms to hydrogen atoms.
   $\dfrac{6}{11}$
   c. Write the ratio of hydrogen atoms to oxygen atoms.
   $\dfrac{2}{1}$

   Of Interest
   Lactose is a molecule found in milk. Its chemical formula is $C_{12}H_{22}O_{11}$. Notice that the subscripts indicate the number of atoms of each element in the molecule.

2. One molecule of hematcin contains 16 carbon atoms, 12 hydrogen atoms, and 6 oxygen atoms.
   a. What is the ratio of hydrogen atoms to the total number of atoms in the molecule?
   $\dfrac{6}{17}$
   b. What is the ratio of carbon atoms to hydrogen atoms?
   $\dfrac{4}{3}$
   c. What is the ratio of carbon atoms to oxygen atoms?
   $\dfrac{8}{3}$

3. The roof of a house rises 0.75 foot for every 1.5 feet of horizontal length. Write the ratio of rise to horizontal length as a fraction in simplest form.
   $\dfrac{1}{2}$

4. A steep mountain road has a grade of 8%, which means that the road rises 8 feet vertically for every 100 feet it travels horizontally. Write the grade as a ratio in simplest form.
   $\dfrac{2}{25}$

5. In a certain gear on a bicycle, the back wheel rotates $1\frac{1}{3}$ times for every 2 rotations of the pedals. Write the ratio of rotations of the back wheel to pedal rotations as a fraction in simplest form.
   $\dfrac{2}{3}$

6. A recipe calls for $2\frac{1}{2}$ cups of flour and $1\frac{1}{3}$ cups of milk. Write the ratio of milk to flour as a fraction in simplest form.
   $\dfrac{8}{15}$

For Exercises 7–14, write each as a unit ratio and interpret the answer. See Example 2.

7. Jamial drove 254 miles in 4 hours. What is the unit ratio of miles to hours? Interpret the answer.
   $\dfrac{63.5}{1}$; he drives an average of 63.5 miles per hour.

8. Janine drove 199.2 miles and used 6 gallons of gas. What is the unit ratio of miles to gallons? Interpret the answer.
   $\dfrac{33.2}{1}$; for every 33.2 miles, she uses 1 gallon of gas.

9. Six bananas cost $0.75. What is the unit ratio of cost to bananas? Interpret the answer.

$\frac{0.125}{1}$; each banana cost 12.5 cents.

10. An 8-pound bag of ice costs $1.96. What is the unit ratio of cost to pounds of ice. Interpret the answer.

$\frac{0.245}{1}$; the ice costs 24.5 cents per pound.

11. In 2012, Wal-Mart stock earned $4.73 per share. The current price of the stock is $73.98. Write a unit ratio that expresses the price to earnings. What does this ratio indicate?

$\approx \frac{15.64}{1}$; the price of the stock is $15.64 for every $1 earned in 2012.

12. The price of a 10.5-ounce can of soup is $1.68. Write the unit ratio that expresses the price to weight. What does this ratio indicate?

$\frac{0.16}{1}$; the soup costs 16 cents per ounce.

★ 13. An 8-ounce container of sour cream costs $1.42. A 24-ounce container of the same brand of sour cream costs $3.59. Which is the better buy? Why?

The 24-oz. container is better because it costs less per ounce. (The unit ratio of price to quantity is less.)

★ 14. The price of a 12-ounce box of corn flakes is $3.59, and an 18-ounce box of the same cereal is $4.59. Which is the better buy? Why?

The 18-oz. box is better because it costs less per ounce. (The unit ratio of price to quantity is less.)

For Exercises 15 and 16, use the graph to the right. See Example 2.

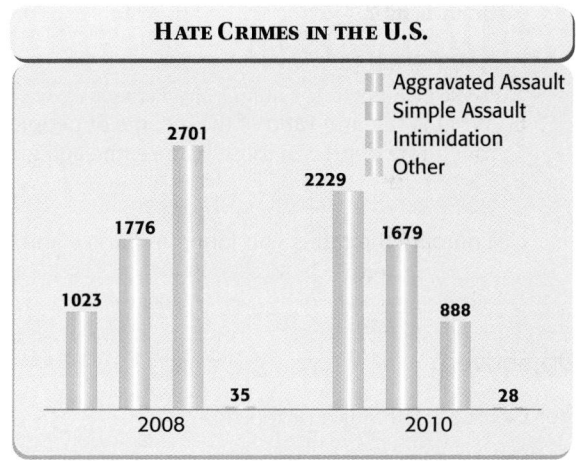

15. a. Write the unit ratio of aggravated assault to total hate crimes against persons in 2008.

$\approx \frac{0.18}{1}$

b. Write the unit ratio of aggravated assault to total hate crimes against persons in 2010.

$\approx \frac{0.46}{1}$

c. Compare the ratios for 2008 to 2010. What do these ratios indicate?

The ratio of aggravated assault hate crimes to total hate crimes was greater in 2010 than in 2008.

16. a. Write the unit ratio of intimidation crimes to total hate crimes against persons in 2008.

$\approx \frac{0.49}{1}$

b. Write the unit ratio of intimidation crimes to total hate crimes against persons in 2010.

$\approx \frac{0.18}{1}$

c. Compare the ratios for 2008 to 2010. What do these ratios indicate?

The ratio of intimidation crimes to total hate crimes against persons was greater in 2008 than in 2010.

(*Source:* Bureau of Federal Justice Statistics.)

For Exercises 17 and 18, use the following table. See Example 2.

Mean Earnings in the United States, 2008 Year-Round, Full-Time Workers by Educational Attainment and Age

	Not High School Graduate	High School Graduate or GED	Some College, No Degree	Associate Degree	Bachelor's Degree	Master's Degree	Professional Degree	Doctorate Degree
25 to 34 Years	$20,471	$28,224	$31,956	$35,541	$48,445	$55,636	$79,785	$83,219
35 to 44 Years	$23,793	$35,233	$41,416	$42,611	$66,332	$76,480	$130,730	$102,837
45 to 54 Years	$25,598	$36,916	$42,559	$44,996	$70,053	$84,495	$147,878	$111,843
55 to 64 Years	$27,393	$35,338	$41,741	$43,062	$64,807	$72,604	$141,584	$105,255
65 Years and Over	$18,550	$27,532	$32,218	$28,255	$43,378	$45,802	$112,449	$74,518

(*Source:* Census 2005a.)

17. a. What is the unit ratio of the income for people in the 25–34 year age group with an associate degree to high school graduates in the same age group? What does this ratio indicate?

$\approx \dfrac{1.26}{1}$; people in the 25–34 age group with an associate degree earn $1.26 for every $1.00 earned by high school graduates.

b. Calculate the unit ratio of income for people in the 25–34 age group with a bachelor's degree to people of the same age group with an associate degree.

$\approx \dfrac{1.36}{1}$

c. Compare the ratios you found in parts a and b. What do these ratios indicate?

In general, people 25–34 years of age who have bachelor's degree make more money than do high school graduates or people with associate degrees.

18. a. What is the unit ratio of the income of people in the 35–44 age group with a doctorate degree to those in the same age group who have a master's degree? What does this ratio indicate?

$\approx \dfrac{1.34}{1}$; people with a doctorate degree in the 35–44 age group earn $1.34 for each $1.00 earned by people in the same age group with a master's degree.

b. What is the unit ratio of the income of people in the 35–44 age group with a doctorate degree to those in the same age group with a bachelor's degree?

$\approx \dfrac{1.55}{1}$

c. Compare the ratios you found in parts a and b. What do these ratios indicate?

In general, people 35–44 years of age who have doctorate degrees make more money than people with bachelor's or master's degrees.

Objective 2

Prep Exercise 2 What is a proportion?

A proportion is an equation in the form $\dfrac{a}{b} = \dfrac{c}{d}$, where $b \neq 0$ and $d \neq 0$.

Prep Exercise 3 Why are $\dfrac{5}{8} \cdot \dfrac{x}{10}$ and $\dfrac{5}{8} = \dfrac{x}{10}$ different?

$\dfrac{5}{8} \cdot \dfrac{x}{10}$ is a product, whereas $\dfrac{5}{8} = \dfrac{x}{10}$ is a proportion.

Prep Exercise 4 If $\dfrac{a}{b} = \dfrac{c}{d}$, where $b \neq 0$ and $d \neq 0$, is true, what can we say about the cross products ad and bc?

$ad = bc$

For Exercises 19–28, determine whether the ratios are equal. See Objective 2.

19. $\dfrac{15}{25} \stackrel{?}{=} \dfrac{3}{5}$

yes

20. $\dfrac{6}{8} \stackrel{?}{=} \dfrac{15}{20}$

yes

21. $\dfrac{20}{25} \stackrel{?}{=} \dfrac{25}{30}$

no

22. $\dfrac{12}{20} \stackrel{?}{=} \dfrac{18}{30}$

yes

23. $\dfrac{7}{4} \stackrel{?}{=} \dfrac{15}{6.2}$

no

24. $\dfrac{15}{21} \stackrel{?}{=} \dfrac{130.2}{93}$

no

25. $\dfrac{10}{14} \stackrel{?}{=} \dfrac{2.5}{3.5}$

yes

26. $\dfrac{6.5}{14} \stackrel{?}{=} \dfrac{10.5}{42}$

no

27. $\dfrac{\frac{2}{5}}{\frac{3}{10}} \stackrel{?}{=} \dfrac{15}{10\frac{1}{2}}$

no

28. $\dfrac{3\frac{1}{4}}{5\frac{1}{2}} \stackrel{?}{=} \dfrac{4\frac{2}{3}}{6\frac{1}{2}}$

no

Prep Exercise 5 Explain how to solve for an unknown number in a proportion.

To solve a proportion using cross products, (a) calculate the cross products, (b) set the cross products equal to each other, and (c) use the multiplication principle of equality to isolate the variable.

For Exercises 29–44, solve for the unknown number. See Example 3.

29. $\dfrac{2}{3} = \dfrac{7}{x}$

10.5

30. $\dfrac{7}{8} = \dfrac{y}{40}$

35

31. $\dfrac{-9}{2} = \dfrac{63}{n}$

−14

32. $\dfrac{11}{m} = \dfrac{132}{-24}$

−2

33. $\dfrac{21}{5} = \dfrac{h}{2.5}$
10.5

34. $\dfrac{29}{7} = \dfrac{k}{1.75}$
7.25

35. $\dfrac{-17}{b} = \dfrac{8.5}{4}$
−8

36. $\dfrac{4.6}{3} = \dfrac{-23}{c}$
−15

37. $\dfrac{-1.5}{60} = \dfrac{-2\frac{7}{8}}{t}$
115

38. $\dfrac{u}{-4\frac{7}{16}} = \dfrac{-60}{1.5}$
$177\frac{1}{2}$

39. $\dfrac{2\frac{1}{4}}{6\frac{2}{3}} = \dfrac{d}{10\frac{2}{3}}$
$3\frac{3}{5}$

40. $\dfrac{\frac{2}{5}}{4\frac{1}{2}} = \dfrac{7\frac{2}{3}}{j}$
$86\frac{1}{4}$

★ **41.** $\dfrac{2}{3} = \dfrac{8}{x + 8}$
4

★ **42.** $\dfrac{x + 5}{6} = \dfrac{7}{3}$
9

★ **43.** $\dfrac{4x - 4}{8} = \dfrac{2x + 1}{5}$
7

★ **44.** $\dfrac{3}{4} = \dfrac{x + 1}{x + 3}$
5

Objective 3

Prep Exercise 6 How do you set up and solve a word problem that translates to a proportion?
To solve proportion problems, (a) set up a proportion in which the numerators and denominators of the ratios correspond in a logical manner and (b) solve using cross products.

For Exercises 45–68, solve. See Example 4.

45. LaTonia drives 110 miles in three hours. At this rate, how far will she go in five hours?
$183\frac{1}{3}$mi.

46. If 90 feet of wire weighs 18 pounds, what will 110 feet of the same type of wire weigh?
22 lb.

47. Jodi pays taxes of $1040 on her house, which is valued at $154,000. She is planning to buy a house valued at $200,000. At the same rate, how much will her taxes be on the new house?
$1350.65

48. Gary notices that his water bill was $24.80 for 600 cubic feet of water. At that rate, what would be the charges for 940 cubic feet of water?
$38.85

49. A company determines that it spends $1575 every 20 business days (one month) for equipment maintenance. How much does the company spend in 120 business days (six months)?
$9450

50. A quality-control manager samples 1200 units of her company's product and finds seven defects. If the company produces about 50,000 units each year, how many of those units would she expect to have a defect?
≈ 292

51. Carson is planning a trip to London. He knows that it takes 1.8946 U.S. dollars to equal a British pound. If he has $800 to take with him, how much will he have in British pounds?
≈ 422.25 pounds

52. Sydney is planning a trip to Mexico. It takes 15.3827 Mexican pesos to equal 1 U.S. dollar. If she has $400 to take with her, how many pesos will she have?
6153.08 pesos

53. Fletcher is building a model of an F-16 jet. If the scale is $1\frac{1}{2}$ inches to 1 foot, what length on the model represents 3.25 feet?
4.875 in.

54. The legend on a map indicates that 1 inch = 500 feet. If a road measures $3\frac{3}{4}$ inches on the map, how long is the road in feet?
1875 ft.

55. A recipe for pancakes calls for $1\frac{1}{2}$ teaspoons of sugar to make 12 pancakes. How many teaspoons of sugar should be used to make 20 pancakes?
$2\frac{1}{2}$ tsp.

56. A recipe for biscuits calls for $3\frac{1}{2}$ cups of flour to make 8 biscuits. How many cups of flour would be needed to make 12 biscuits?
$5\frac{1}{4}$ cups

57. Ford estimates that its 2012 Fusion hybrid will travel 630 highway miles on one tank of gas. If the gas tank of the Fusion hybrid holds 17.5 gallons, how far can a driver expect to travel on 12 gallons?
432 highway miles

58. Toyota estimates that its 2012 Prius hybrid will travel 576 highway miles on one tank of gas. If the gas tank of the Prius holds 12 gallons, how far can a driver expect to travel on 8 gallons?
384 highway miles

59. Grace Pak of the United States holds the world record for the fastest typing on a smart phone. In 2011, she typed 264 characters in 56.6 seconds. If she could maintain that rate, how long would it take her to type 300 characters? (*Source: Guinness Book of World Records.*)
≈ 64.3 sec.

60. Sean Shannon of Canada holds the world record for the fastest recital of Hamlet's soliloquy "To be or not to be." On August 30, 1995, he recited the 260-word soliloquy in 23.8 seconds. If he could maintain that rate, how long would it take him to recite Shakespeare's entire script of Hamlet, which contains 32,241 words? (*Sources: Guinness Book of World Records*; http://william-shakespeare.info.)
≈ 2951.3 sec. (or ≈ 49.2 min.)

61. A consultant recommends that about 9 pounds of rye grass seed be used for every 1000 square feet. If the client's lawn is about 25,000 square feet, how many pounds of seed should he buy?
225 lb.

62. Mark used 3 gallons of paint to cover 1200 square feet of wall space in his home. Mark's neighbor wants to paint 2000 square feet. Based on Mark's rate, how many gallons should the neighbor buy?
5 gal.

★ **63.** A carpet-cleaning company charges $49.95 to clean a rectangular living room that is 20 feet by 15 feet. The client asks the company also to clean the rectangular dining room, which measures 18 feet by 12 feet. If the company charges the same rate per square foot, how much should it cost to clean both the living room and dining room?
$85.91

★ **64.** A painter charged $320 to paint two walls that measured 12 feet by 9 feet and two walls that measured 10 feet by 9 feet. The client asks him to return to paint two walls that measure 15 feet by 12 feet and two walls that measure 18.5 feet by 12 feet. If the painter charges the same rate per square foot, what should be the price?
$649.70

★ **65.** A state wildlife department wants to estimate the number of deer in a state forest. In one month, it captures 25 deer, tags them, and releases them. Later, it captures 45 deer and finds that 17 are tagged. Assuming that the ratio of tagged deer to total deer in the forest remains constant, estimate the number of deer in the preserve.
≈ 66 deer

★ **66.** A marine biologist is studying the population of trout in a mountain lake. During one month, she catches 225 trout, tags them, and releases them. Later, she captures 300 trout and finds that 48 are tagged. Assuming that the ratio of tagged trout to total trout in the lake remains constant, estimate the population of trout in the lake.
≈ 1406

67. The Hoover Dam has 17 generators and can supply all of the electricity needed by a city of 750,000 people. If only 14 generators are running, how many people will be supplied with electricity?
617,647

68. Until the early 1950s, the erosion rate for Niagara Falls was an average of 3 feet per year. The suggested rate of erosion may now be as low as 1 foot per 10 years for Niagara Falls. Given this rate, how much erosion may be expected over the next 50 years?
5 ft.

Objective 4

For Exercises 69–72, find the unknown lengths in the similar figures. See Example 5.

69.

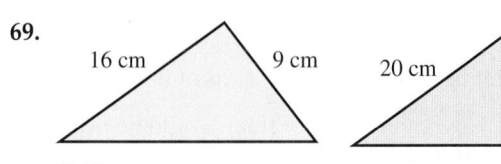

11.25 cm

70.

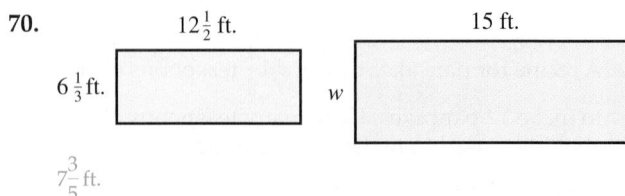

$7\frac{3}{5}$ ft.

71.

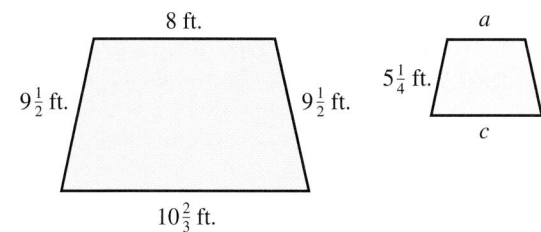

$a = 4\dfrac{8}{19}$ ft.; $b = 5\dfrac{1}{4}$ ft.; $c = 5\dfrac{17}{19}$ ft.

72.

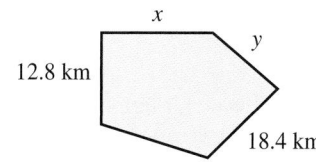

$x = 16.8$ km; $y = 10.4$ km; $z = 11.5$ km

73. To estimate the height of the Eiffel Tower, a tourist uses the concept of similar triangles. The Eiffel Tower is found to have a shadow measuring 200 meters in length. The tourist has his own shadow measured at the same time. His shadow measures 1.2 meters. If he is 1.8 meters tall, how tall is the Eiffel Tower?

300 m

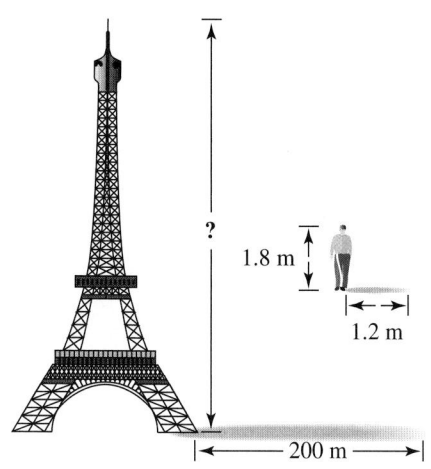

Of Interest

The Eiffel Tower was intended for demolition in 1909. It had been built 20 years before for the 1889 Universal Exhibition. However, the broadcasting antennae at the top helped ensure the Eiffel Tower's survival. (*Source: AAA France Travel Book.*)

74. To estimate the height of the Great Pyramid in Egypt, the Greek mathematician Thales performed a procedure similar to that in Exercise 73, except that he supposedly used a staff instead of his body. Suppose the shadow from the staff was measured to be 6 feet and at the same time the shadow of the pyramid was measured to be $524\dfrac{2}{3}$ feet. If the staff was $5\dfrac{1}{2}$ feet tall, what was the height of the pyramid?

480.9$\overline{4}$ft.

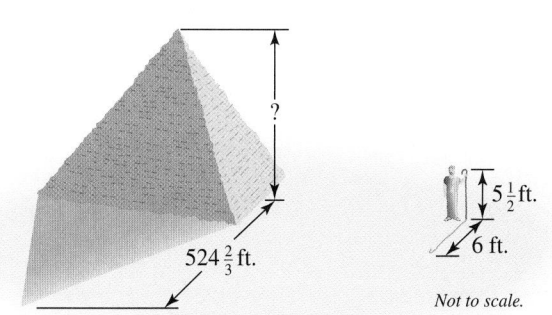

Not to scale.

Of Interest

The Greek mathematician Thales lived from 640 B.C. to 546 B.C. He is the first person to have proposed the importance of proof and is considered to be the father of science and mathematics in the Greek culture. (*Source*: D. E. Smith, *History of Mathematics*, Dover, 1951.)

75. To estimate the distance across a river, an engineer creates similar triangles. The idea is to create two similar right triangles as shown. Calculate the width of the river.

26.4 m

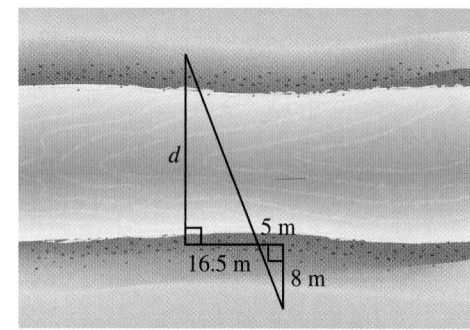

76. An engineer wants to estimate the distance across the Amazon River at his current location. The engineer creates two similar right triangles as shown. Calculate the distance across the river.

1632 m

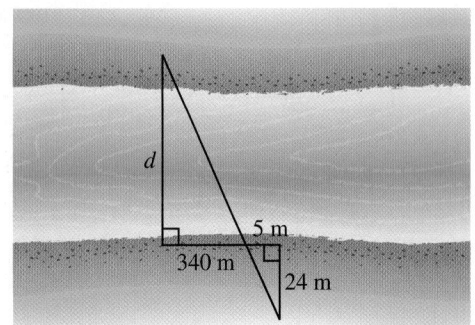

Of Interest

At 6275 km long, the Amazon River is the second-longest river in the world. Although the Nile River is slightly longer, the Amazon is the largest river in terms of watershed area, number of tributaries, and volume of water discharged.

Review Exercises

Exercise 1   **Constants and Variables**

[1.1] **1.**   Is 0.58 a rational number? Why or why not?

Yes, because it can be written as a ratio of integers, $\frac{58}{100}$.

Exercises 2–4 **Expressions**

For Exercises 2–4, simplify.

[1.2] **2.** $\dfrac{42}{350}$

$\dfrac{3}{25}$

[1.4] **3.** $(0.15)(20.4)$

3.06

[1.4] **4.** $45.2 \div 100$

0.452

Exercises 5 and 6 **Equations and Inequalities**

For Exercises 5 and 6, solve.

[2.3] **5.** $0.75n = 240$

320

[2.3] **6.** $60p = 24$

0.4 or $\dfrac{2}{5}$

2.7 Percents

Objectives

1 Write a percent as a fraction or decimal number.
2 Write a fraction or decimal number as a percent.
3 Translate and solve percent sentences.
4 Solve problems involving percents.

Warm-up

[1.4] **1.** Simplify: $30\dfrac{1}{4} \div 100$

[2.3] **2.** Solve: $0.76n = 63.84$

[2.6] **3.** Solve: $\dfrac{P}{100} = \dfrac{12}{28}$

Objective 1 Write a percent as a fraction or decimal number.

The word *percent* is a compound word made from the prefix *per* and the suffix *cent*. The prefix *per* means "for each." The suffix *cent* comes from the Latin word *centum*, which means "100." Therefore, **percent** means "for each 100." In other words, percent is a ratio with the denominator 100.

Definition Percent: A ratio representing some part out of 100.

Answers to Warm-up

1. $\dfrac{121}{400}$
2. 84
3. 42.9

The symbol for percent is %. For example, 20 percent, which means 20 out of 100, is written 20%. Note that the definition can be used to write percents as fractions or decimal numbers.

$$20\% = \mathbf{20} \text{ out of } 100$$
$$= \frac{20}{100}$$
$$= \frac{1}{5} \text{ or } 0.2$$

Procedure Rewriting a Percent

To write $n\%$ as a fraction or decimal:

1. Write $n\%$ as $\dfrac{n}{100}$.
2. Simplify to the desired form.

Example 1   Write each percent as a decimal and as a fraction in simplest form.

a. 20.5%

Solution:
$$20.5\% = \frac{20.5}{100} \qquad \text{Write 20.5 over 100.}$$
$$= 0.205 \qquad \text{Divide to write the decimal form.}$$
$$= \frac{205}{1000} \qquad \text{Write the fraction form.}$$
$$= \frac{41}{200} \qquad \text{Simplify to lowest terms.}$$

◀ **Note** Dividing a decimal number by 100 moves the decimal point two places to the left.

b. $30\frac{1}{4}\%$

Solution:
$$30\frac{1}{4}\% = \frac{30\frac{1}{4}}{100} \qquad \text{Write } 30\tfrac{1}{4} \text{ over 100.}$$
$$= 30\frac{1}{4} \div 100 \qquad \text{Rewrite the division.}$$
$$= \frac{121}{4} \cdot \frac{1}{100} \qquad \text{Write an equivalent multiplication.}$$
$$= \frac{121}{400} \qquad \text{Multiply to write the fraction form.}$$
$$= 0.3025 \qquad \text{Divide 121 by 400 to write the decimal form.}$$

Your Turn 1   Write each percent as a fraction in simplest form and as a decimal.

a. 26%   **b.** 9.25%   **c.** $30\frac{2}{3}\%$

Objective 2 Write a fraction or decimal number as a percent.

We have seen that $20\% = \dfrac{20}{100} = \dfrac{1}{5}$ or 0.2. To write a percent as a fraction or a decimal, we replace the % sign with division by 100. To write a fraction or decimal as a percent, we should multiply by 100% and simplify.

Answers to Your Turn 1
a. $\dfrac{13}{50}$ or 0.26   **b.** $\dfrac{37}{400}$ or 0.0925
c. $\dfrac{23}{75}$ or $0.30\overline{6}$

Procedure Writing a Fraction or Decimal as a Percent

To write a fraction or decimal number as a percent:
1. Multiply by 100%.
2. Simplify.

Connection Notice that $100\% = \dfrac{100}{100} = 1$. Therefore, when we multiply a number by 100%, we are multiplying by 1, which is why the resulting percent is equal to the given number.

Example 2 Write as a percent. If necessary, round your answer to the nearest tenth of a percent.

a. $\dfrac{5}{8}$

Solution: $\dfrac{5}{8} = \dfrac{5}{\underset{2}{8}} \cdot \dfrac{\overset{25}{100}}{1}\%$ Multiply by 100%.

$$= \dfrac{125}{2}\% \quad\quad\quad \text{Simplify.}$$

$$= 62\dfrac{1}{2}\% \text{ or } 62.5\%$$

b. 0.267

Solution: $0.267 = 0.267 \cdot 100\%$ Multiply by 100%.

$$= 26.7\%$$

Note Multiplying a decimal number by 100 moves the decimal point two places to the right.

Your Turn 2 Write as a percent. If necessary, round your answer to the nearest tenth of a percent.

a. $\dfrac{5}{9}$ **b.** 2.4

Objective 3 Translate and solve percent sentences.

To solve problems involving percents, it is often helpful to reduce the problem to the following simple percent sentence:

A percent of a whole is a part of the whole.

Note that there are three pieces in the simple sentence: the *percent*, the *whole*, and the *part*. Any one of those pieces could be unknown in a problem.

Instructor Note Make sure you point out the variations of wording so that students see that although there are only three cases, there are many ways to word those three cases. Also, the whole is often called the *base* and the part is called the *amount*.

	A percent	of	a whole	is	a part.
Unknown part:	42%	of	68	is	what amount?
Unknown whole:	76%	of	what number	is	63.84?
Unknown percent:	What percent	of	72	is	63?

To solve for an unknown piece of a simple percent sentence, we will translate the sentence word for word to an equation or to a proportion. The key to translating is to identify each of the three pieces. The percent always has a percent sign, or the words *what percent* are used. The whole always follows the word *of*. Once those two pieces are identified, the third piece must be the part.

Procedure Translating Simple Percent Sentences

Method 1. Translate the sentence word for word.
1. Select a variable for the unknown.
2. Translate the word *is* to an equal sign.
3. If *of* is preceded by the percent, translate it to multiplication.
   If *of* is preceded by a whole number, translate it to division.

Method 2. Translate to a proportion by writing the following form:

$$\text{Percent} = \dfrac{\text{Part}}{\text{Whole}},$$

where the percent is expressed as a fraction with a denominator of 100.

Answers to Your Turn 2
a. $55\dfrac{5}{9}\%$ or 55.6%   **b.** 240%

First, let's consider the case where the part is unknown.

Simple Percent Sentence with an Unknown Part

Example 3   42% of 68 is what number?   ◄ **Note** This question can also be worded this way: "What is 42% of 68?"

Solution: Notice that the part is the unknown.

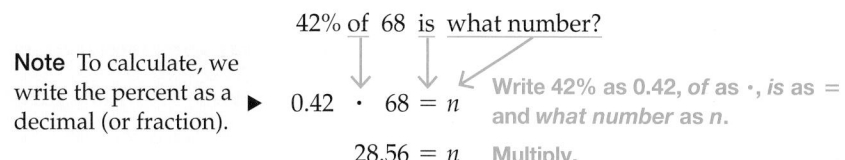

42%	of	68	is	what number?
↑	↑		↑	
Percent	of	the whole	is	the part.

Method 1: Using word-for-word translation, because *of* is preceded by the percent, it means multiply and *is* means equals.

42% of 68 is what number?

Note To calculate, we write the percent as a decimal (or fraction). ▶

$0.42 \cdot 68 = n$ Write 42% as 0.42, *of* as ·, *is* as =, and *what number* as *n*.

$28.56 = n$ Multiply.

Method 2: Using a proportion, we equate the percent written as a ratio to the ratio of the part to the whole.

$$\text{Percent} = \frac{\text{Part}}{\text{Whole}}$$

42% written as a ratio

$\dfrac{42}{100} = \dfrac{n}{68}$ ← Part ← Whole   Use percent $= \dfrac{\text{part}}{\text{whole}}$.

$2856 = 100n$ Equate the cross products.

$\dfrac{2856}{100} = \dfrac{100n}{100}$ Divide both sides by 100 to isolate *n*.

$28.56 = n$

Your Turn 3   What number is 15% of 36?

Simple Percent Sentence with an Unknown Whole

Now consider the situation in which the whole amount is unknown.

Example 4   76% of what number is 63.84?   ◄ **Note** This question can be worded "63.84 is 76% of what number?"

Solution: Note the three pieces:

76%	of	what number	is	63.84?
↑		↑		↑
Percent	of	the whole	is	the part.

Method 1:   76% of what number is 63.84?

$0.76 \cdot n = 63.84$ Write 76% as a decimal number, *of* as ·, *what number* as *n*, and *is* as =.

$0.76n = 63.84$

$\dfrac{0.76n}{0.76} = \dfrac{63.84}{0.76}$ Divide both sides by 0.76.

$n = 84$

Learning Strategy

AUDITORY

If you are an auditory learner, when translating a percent sentence to a proportion, imagine or say to yourself "*of* is underneath" to drive home the fact that the number following *of* is the denominator.

Answer to Your Turn 3
5.4

Method 2:

76% expressed as a ratio

$$\frac{76}{100} = \frac{63.84}{n}$$ ← Part ← Whole   Use percent $= \frac{\text{part}}{\text{whole}}$.

$$76n = 6384$$ Equate the cross products.

$$\frac{76n}{76} = \frac{6384}{76}$$ Divide both sides by 76 to isolate n.

$$n = 84$$

Your Turn 4 | 48 is 120% of what number?

Simple Percent Sentence with an Unknown Percent

Finally, consider the case in which the percent is the unknown.

Example 5 | What percent of 72 is 63?

Note This question can be worded "63 is what percent of 72?"
or
"What percent is 63 out of 72?"
or
"What percent is 63 of 72?"
Note how 72, the whole amount, always follows *of*.

Solution: Note the pieces.

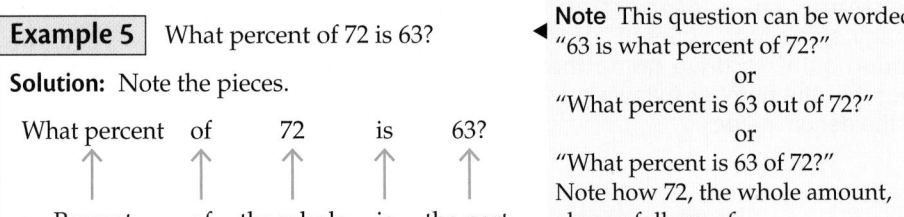

What percent   of   72   is   63?
Percent   of   the whole   is   the part.

Method 1:

What percent of 72 is 63?

$$p \quad \cdot 72 = 63$$ Write *what percent* as p, of as $\cdot$, and *is* as $=$.

$$72p = 63$$

$$\frac{72p}{72} = \frac{63}{72}$$ Divide both sides by 72.

$$p = 0.875$$ **Note** To write this decimal result as a percent, we must multiply by 100%.

Answer: $0.875 \cdot 100\% = 87.5\%$ Multiply by 100% to write the result as a percent.

Method 2:

P% written as a ratio

$$\frac{P}{100} = \frac{63}{72}$$ ← Part ← Whole   Use percent $= \frac{\text{part}}{\text{whole}}$.

$$72P = 6300$$ Equate the cross products.

$$\frac{72P}{72} = \frac{6300}{72}$$ Divide both sides by 72 to isolate p.

$$P = 87.5$$ **Note** In the original proportion, $P/100$ means $P\%$. Therefore, if the value of P is 87.5, all we need to do is attach the % sign.

Answer: 87.5% Attach the % sign.

Answer to Your Turn 4
40

Answer to Your Turn 5
24%

As Example 5 shows, when the percent is unknown, the proportion method gives the answer as a percent; so you may want to consider using proportions in this case.

Your Turn 5 | 21.6 is what percent of 90?

Objective 4   Solve problems involving percents.

If a problem involves a percent, it can be helpful to try to write the information in the form of a simple percent sentence.

> **Procedure** **Solving Percent Application Problems**
> To solve problems involving percent:
> 1. Identify the percent, whole, and part, noting which is unknown.
> 2. Write the problem as a simple percent sentence (if needed).
> 3. Translate to an equation (word for word or proportion).
> 4. Solve for the unknown.

Example 6   In an English class, 12 of the 28 students received an A. What percent of the class received an A?

Solution: The percent is unknown. The 12 students who received an A is the part out of the whole 28 students in the class. We can write the following simple percent sentence.

> What percent of 28 is 12?

We'll let P represent the unknown percent and translate to a proportion because the proportion's solution will be the percent value.

P% written as a ratio
$$\frac{P}{100} = \frac{12}{28} \begin{matrix} \leftarrow \text{ number of A students (part)} \\ \leftarrow \text{ total students (whole)} \end{matrix}$$
Use percent $= \dfrac{\text{part}}{\text{whole}}$.

$$28P = 1200 \qquad \text{Equate the cross products.}$$

$$\frac{28P}{28} = \frac{1200}{28} \qquad \text{Divide both sides by 28 to isolate } P.$$

$$P \approx 42.9 \quad \blacktriangleleft \begin{matrix}\textbf{Note} \text{ We rounded to the nearest} \\ \text{tenth. We need to add the \% sign.}\end{matrix}$$

Answer: About 42.9% of the students in the class received an A.

Your Turn 6   On Monday, January 18, 2010, the *Morning Sentinel* reported that it planned to cut 100 full-time jobs in a bid to return to profitability. This is 6% of the newspaper's workforce. How many employees does the *Sentinel* have prior to this layoff?

Percent Problems Involving Increase or Decrease

Sometimes a percent problem involves an increase or a decrease of some initial amount. Sales tax, interest, and discount are situations in which an initial amount of money is increased or decreased by an amount of money that is a percent of the initial amount.

Example 7   If the sales tax rate in a certain city is 6%, what is the total cost of the following items?

Raisin Bran	$3.79
Milk	$1.29
Orange Juice	$1.49
Yogurt	$2.59

Solution: The total cost is the purchase price plus the sales tax, which is a percent of the purchase price. To find the sale tax, we need the total purchase price; so we must calculate the total purchase price first.

$$\text{Purchase price } = 3.79 + 1.29 + 1.49 + 2.59 = \$9.16$$

Answer to Your Turn 6
1667

To find the sales tax, we can translate a simple percent sentence: The sales tax is 6% of $9.16.

$$x = 0.06 \cdot 9.16$$ Let x represent the unknown sales tax; write *is* as $=$, 6% as 0.06, and *of* as $\cdot$.

$$x = 0.5496$$ Multiply.

Sales tax is always rounded to the nearest cent, so the sales tax is $0.55.

Answer: The total cost is $9.16 + $0.55 = $9.71.

Your Turn 7 An advertisement indicates that a coat is on sale at 20% off the marked price. If the marked price is $79.95, what will be the price after the discount?

Connection A clever shortcut for Example 7 is to recognize that we pay 100% of the initial price *plus* 6% in sales tax, so the total is 106% of the initial price.

Total cost $= 1.06 \cdot 9.16 = 9.7096$, which rounds to $9.71.

A similar shortcut exists for the Your Turn problem involving a discount. If we take 20% off the marked price, we pay 80% of the marked price.

Price after discount $= 0.8 \cdot 79.95 = $63.96.

Now suppose the percent of increase or decrease is unknown.

Example 8 On May 1, 2006, the median price of a house sold in Jacksonville, Florida, reached an all-time high of $322,990. On April 7, 2008, the median price was $253,990. What was the percent of decrease? (*Source:* www.deptofnumbers.com/asking-prices.)

Solution: We are to determine the percent of decrease given the initial price and the price after the decrease. To determine the percent of decrease, we need the amount of the decrease, which is found by subtracting the final amount from the initial amount.

$$\text{Amount of decrease} = 322{,}990 - 253{,}990 = 69{,}000$$

Simple percent sentence: What percent of 322,990 is 69,000?

Note The whole is the original amount (before the decrease).

Because the percent is unknown, we translate to a proportion, letting P represent the number of the unknown percent.

P% written as a ratio.

$$\frac{P}{100} = \frac{69{,}000}{322{,}990} \begin{matrix}\leftarrow \text{amount of decrease (part)} \\ \leftarrow \text{initial amount (whole)}\end{matrix}$$

Use percent $= \dfrac{\text{part}}{\text{whole}}$.

$$322{,}990P = 6{,}900{,}000$$ Equate the cross products.

$$\frac{322{,}990P}{322{,}990} = \frac{6{,}900{,}000}{322{,}990}$$ Divide both sides by 322,990 to isolate P.

$$P \approx 21.36$$

Answer: The percent of decrease in the median price of a house in Jacksonville was about 21.36%.

In general, when solving for a percent of increase or decrease, the *part* is the amount of increase or decrease and the *whole* is the initial amount.

$$\frac{P}{100} = \frac{\text{Amount of increase/decrease}}{\text{Initial amount}}$$

Answer to Your Turn 7
$63.96

Answer to Your Turn 8
35.7%

Your Turn 8 The Air Transportation Association reports that the per-gallon price for jet fuel jumped from $2.10 in 2007 to $2.85 in 2008. What is the percent of increase?

2.7 Exercises For Extra Help MyMathLab®

Note: Exercises marked with a ★ represent challenging exercises.

Objective 1

Prep Exercise 1 Explain how to write a percent as a decimal or fraction.
To write a percent as a decimal or fraction, (a) write the percent as a ratio with 100 in the denominator and (b) simplify to the desired form.

For Exercises 1–12, write each percent as a decimal and as a fraction in simplest form. See Example 1.

1. 20%
$0.2, \dfrac{1}{5}$

2. 30%
$0.3, \dfrac{3}{10}$

3. 15%
$0.15, \dfrac{3}{20}$

4. 85%
$0.85, \dfrac{17}{20}$

5. 14.8%
$0.148, \dfrac{37}{250}$

6. 18.6%
$0.186, \dfrac{93}{500}$

7. 3.75%
$0.0375, \dfrac{3}{80}$

8. 6.25%
$0.0625, \dfrac{1}{16}$

9. $45\dfrac{1}{2}\%$
$0.455, \dfrac{91}{200}$

10. $65\dfrac{1}{4}\%$
$0.6525, \dfrac{261}{400}$

11. $33\dfrac{1}{3}\%$
$0.3\overline{3}, \dfrac{1}{3}$

12. $18\dfrac{1}{6}\%$
$0.181\overline{6}, \dfrac{109}{600}$

Objective 2

Prep Exercise 2 Explain how to write a decimal or fraction as a percent.
To write a decimal or fraction as a percent, (a) multiply by 100% and (b) simplify.

For Exercises 13–32, write as a percent rounded to the nearest tenth if necessary. See Example 2.

13. $\dfrac{3}{5}$
60%

14. $\dfrac{1}{5}$
20%

15. $\dfrac{3}{8}$
37.5%

16. $\dfrac{5}{8}$
62.5%

17. $\dfrac{5}{6}$
83.3%

18. $\dfrac{4}{9}$
44.4%

19. $\dfrac{2}{3}$
66.7%

20. $\dfrac{5}{11}$
45.5%

21. 0.96
96%

22. 0.42
42%

23. 0.8
80%

24. 0.7
70%

25. 0.09
9%

26. 0.01
1%

27. 1.2
120%

28. 3.58
358%

29. 0.028
2.8%

30. 0.065
6.5%

31. 4.051
405.1%

32. 0.007
0.7%

Objective 3

Prep Exercise 3 What is the simple percent sentence?
A percent of a whole is a part of the whole.

Prep Exercise 4 When preceded by a percent, the word *of* indicates what operation?
multiplication

Prep Exercise 5 Explain how to translate a simple percent sentence to a proportion.
Because the percent is a ratio with 100 as the denominator, we can write the percent as one of the ratios in a proportion. The equivalent ratio is also some part to a whole amount.

Prep Exercise 6 When using word-for-word translation to find a percent, the calculation will not be a percent, whereas if you use the proportion method, the result will be a percent. Why?
In a word-for-word translation, the division yields a decimal number that must be written as a percent. When using the proportion method, the decimal number is multiplied by 100, which gives the percent.

For Exercises 33–54, translate word for word or translate to a proportion; then solve.
See Examples 3–5.

33. 80% of 35 is what number?
28

34. What number is 40% of 90?
36

35. 2.5% of 124 is what number?
3.1

36. 13.5% of 940 is what number?
126.9

37. What number is $9\frac{1}{4}$% of 64?
5.92

38. $3\frac{3}{4}$% of 24 is what number?
0.9

39. 120% of 86 is what number?
103.2

40. What number is 250% of 62.8?
157

41. 30% of what number is 15?
50

42. 45 is 5% of what number?
900

43. 16.4 is 20.5% of what number?
80

44. 30.2% of what number is 18.12?
60

45. 7.8 is $12\frac{1}{2}$% of what number?
62.4

46. $5\frac{1}{4}$% of what number is 1.26?
24

47. What percent of 45 is 9?
20%

48. 15 is what percent of 40?
37.5%

49. What percent of 38 is 39.9?
105%

50. 73.44 is what percent of 68?
108%

★ **51.** What percent is 18 of 27?
66.$\overline{6}$%

★ **52.** 50 of 60 is what percent?
83.$\overline{3}$%

★ **53.** What percent is 12 of 15?
80%

★ **54.** 18 of 24 is what percent?
75%

Objective 4

Prep Exercise 7 What pieces need to be identified when solving a percent application problem?
The percent, whole, and part

Prep Exercise 8 When given an initial amount and a final amount after an increase, how do you find the percent of the increase?
First find the amount of the increase by subtracting the initial amount from the final amount. Then write a proportion in the form $\frac{P}{100} = \frac{\text{amount of increase}}{\text{initial amount}}$ and solve for P.

For Exercises 55–60, solve. See Example 6.

55. Angela answers 86% of the questions on a psychology multiple choice test correctly. If there were 50 questions on the test, how many did she answer correctly?
43

56. Terra is a server at a restaurant. She serves a table of seven people, for which the restaurant automatically adds a service charge of 15% of the cost of the meal. If the meal costs $127.40, how much is the service charge?
$19.11

57. A county's annual property taxes are 4% of the market value of the property. How much is the tax on a property valued at $120,000? If the tax is paid in monthly installments, what amount must be paid each month?
$4800; $400

58. How much is the tax on a house valued at $230,000 in the same county as that in Exercise 57? If the tax is paid in monthly installments, what amount must be paid each month?
$9200; $766.67

59. In general, lenders do not want home buyers to spend more than 28% of their gross monthly income on a house payment. If a buyer has a monthly gross income of $3210, what is the most a lender will allow for a mortgage payment?
$898.80

60. In general, lenders do not want home buyers to spend more than 36% of gross income a month on debts. Using the buyer in Exercise 59, what is the most a lender will accept in total debt payments per month? If she makes the highest mortgage payment allowed each month, what does she have left to spend on other debt and still be within the lender's guidelines?
$1155.60; $256.80

Of Interest

In finance, the ratio of the house payment to the gross monthly income is known as the *front-end ratio*. The ratio of the monthly debts to the gross monthly income is known as the *back-end ratio*. The front-end and back-end ratios, along with a person's credit history, are primary factors in determining loan qualification.

For Exercises 61 and 62, solve using the circle graph. See Example 6.

61. How many teragrams of carbon dioxide were emitted by the United States?
    ≈ 5783.5 tg

62. How many teragrams of methane were emitted by the United States?
    ≈ 596.4 tg

U.S. GREENHOUSE GAS EMISSIONS

TOTAL EMISSIONS: 6,934.6 TERAGRAMS OF CO_2 EQUIVALENT

Carbon dioxide
83.4%

Nitrous oxide
6.0%

Methane
8.6%

HFCs, PFCs, &
Sulfur hexafluoride
2.0%

(*Source:* U.S. Environmental Protection Agency.)

For Exercises 63–98, solve. See Examples 6–8.

63. Terra, the server from Exercise 56, receives a tip of $8.50. If this was 20% of the cost of the meal, what was the cost of the meal?
    $42.50

64. When a camera is sold on eBay for $50.00 or less, the seller must pay eBay 6% of the selling price. If eBay charges $2.70 for the sale of a camera, what was the selling price of the camera?
    $45.00

65. A theater charges 2.5% of the ticket price as a handling fee to hold tickets at the will-call booth. If the handling fee for four tickets is $7.50, what is the total cost of the tickets?
    $300

66. Carol's checking account earns 0.025% of her average daily balance as a dividend every month. If she receives a dividend of 15 cents one month, what was her average daily balance?
    $600

67. During his career, Detroit Pistons player Isiah Thomas made 7194 field goals out of 15,904 attempts. What percent of his field goal attempts did he score?
    ≈ 45.2%

68. Ty Cobb has the record in professional baseball for the best career batting average. During his career, he had 4189 hits out of 11,434 at bats. What percent of his at bats were hits?
    ≈ 36.6%

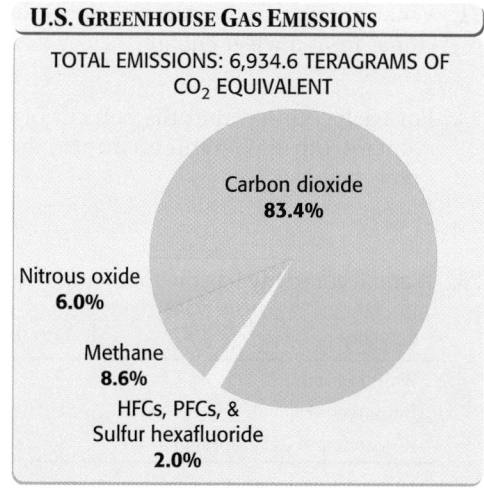

69. According to the 2010 census, the total population of the United States was 308,745,538. If there were 220,958,853 people 21 years of age and older, what percent of the total population was 21 years of age and older? (*Source:* Bureau of the Census, Department of Commerce, 2010 Census.)
    ≈ 71.6%

Of Interest

Batting averages are expressed as decimal numbers rounded to the nearest thousandth. For example, a player's batting average might be 0.285, which means that he has had a hit 28.5% of the number of times he's been at bat.

70. In 2010, 59,308,000 households out of 114,825,000 households in the United States were headed by married couples. What percent of U.S. households was headed by married couples in 2010? (*Source:* Bureau of the Census, Department of Commerce.)
    ≈ 51.7%

For Exercises 71–74, use the following table, which shows the energy use in some of the countries that use the most energy. The unit used is the amount of energy produced by one metric ton of oil. (Source: Firefly's World of Facts, 2007.)

Country	Oil	Gas	Coal	Nuclear	Hydroelectric	Total
United States	944.6	570.1	575.4	185.9	60.6	2336.6
China	327.3	42.3	1081.9	11.8	90.8	1554.0
Russia	130.0	364.6	111.6	33.9	39.6	679.6

71. What percent of the energy use in the United States comes from nuclear energy?

≈ 8.0%

72. What percent of the energy use in China comes from coal?

≈ 69.6%

73. For each country, find the percent of energy use that comes from gas. Which country has the greatest percentage?

United States ≈ 24.4%, China ≈ 2.7%, Russia ≈ 53.6%; Russia

74. For each country, find the percent of energy use that comes from the least polluting form of energy, hydro-electric. Which country has the smallest percentage?

United States ≈ 2.6%, China ≈ 5.8%, Russia ≈ 5.8%; United States

75. A small company has the following costs in one quarter.

Category	Cost (in thousands)
Plant operations	124.2
Employee wages	248.6
Research and development	115.7

What percent of the company's total cost is spent on research and development?

≈ 23.7%

76. A pharmaceutical company conducts a study on the side effects of a new drug. Following is a report of the data from the study.

Side Effect	Number of Subjects
No side effects	1462
Headache	35
Rash	24
Fever	65

What percent of the subjects in the study experienced fever?

4.1%

★ **77.** Margaret mixes 50 milliliters of a solution that is 10% HCl with 100 milliliters of a solution that is 25% HCl. What percent of the resulting solution is HCl?

20%

★ **78.** Leon invests $3400 in stock that returns 12% the first year and $2200 in stock that returns 15% during the same year. What is his total return as a percent of his total investment?

≈ 13.2%

79. Carl purchases $124.45 in groceries. If the sales tax rate is 5%, what is the amount of the tax and the cost after tax is added?

$6.22; $130.67

80. Juanita purchases several books over the Internet that have a total price of $85.79. The online store charges a handling fee of 2.5% of the total cost. What is the handling fee and the price after the fee is added?

$2.14; $87.93

81. An appliance store has all refrigerators discounted 10% off the regular price. If the regular price of a refrigerator is $949, what is the amount of the discount and the price after the discount?

$94.90; $854.10

82. A furniture store has all sofas discounted 40% off. If the original price of a particular sofa is $785.99, what is the amount of the discount and the price after the discount?

$314.40; $471.59

★ **83.** Johanna paid a total of $2675 for a sofa, which included 7% sales tax. What was the price of the sofa before the sales tax was added?

$2500

★ **84.** Paul paid a total of $1908 for a refrigerator, which included 6% sales tax. What was the price of the refrigerator before the sales tax was added?

$1800

85. Video game sales in a department store increased from $2600 in November to $4550 in December. What was the amount of increase? What was the percent of increase?

$1950; 75%

86. Fifteen years ago James inherited a marble-top table. It was appraised at $420. The insurance company asked that he have it reappraised. It is now worth $1200. What was the amount of the increase? What was the percent of increase?

$780; ≈ 185.7%

87. From 2005–2006, about 845,000 bachelor's degrees were awarded to women in the United States. From 2009–2010, about 943,381 bachelor's degrees were awarded to women. What was the percent of increase in degrees awarded to women? (*Source:* National Center of Education Statistics, U.S. Department of Education.)

≈ 11.6%

88. On May 7, 2012, the Dow Jones Industrial Average (DJIA) closed at 13,008.53, and on May 11, it closed at 12,820.60. What was the percent of decrease?

≈ 1.44%

89. The price of a TI-84 calculator drops from $120 to $79. What is the percent of decrease?

34.16%

90. After the Vallina family purchased a smaller car, their monthly gasoline bill dropped from $225 per month to $170 per month. What was the percent of decrease in the monthly gasoline bill?

24.4%

91. In 2000, the carbon monoxide concentration in Syracuse was 2.4 parts per million, and in 2010, the concentration was 1.4 parts per million. What was the percent of decrease in the carbon monoxide concentration? (*Source:* www.epa.gov/airquality.)
≈ 41.7%

92. In 2000, the population of New Orleans, Louisiana, was 484,674. In 2010, the population was 343,829. What was the percent of decrease in population? (*Source:* U.S. Bureau of the Census, Department of Commerce.)
≈ 29.1%

93. In 1938, the federal hourly minimum wage was $0.25. In 2011, the hourly minimum wage was $7.25. What was the percent of increase in the minimum wage from 1938 to 2011? (*Source:* U.S. Department of Labor.)
2800%

94. A small business owner has an audit performed and finds that the net worth of the company is $58,000. After two years of poor business, an audit shows the net worth is −$20,000. What is the percent of the decrease in the net worth of the business?
≈ 134.5%

95. The U.S. Bureau of the Census reported a 597.5% increase in the number of single fathers from 1970 to 2010. If the number of single fathers in 2010 was 2,790,000, how many single fathers were there in 1970? (*Source:* U.S. Census Report, June 18, 2006.)
400,000

96. The projected population of Florida for 2030 is 30.1 million, which is a 986.25% increase from the population in 1950. What was the population in 1950? (*Source:* Floridians for a Sustainable Population, 2008.)
≈ 2.771 million

97. On "Black Monday," October 19, 1987, the stock market closed at 1739, which was a 22.6% decrease from the previous day's closing price. What was the closing price on Friday, October 16?
≈ 2246.77

98. On August 31, 2012, the closing price of Disney stock was at $3.96, which was a 49.6% decrease from August 31, 2010. What was the closing price on August 31, 2010?
≈ $7.86

Review Exercises

Exercises 1–3 ▲ Expressions

[1.5] 1. Simplify: -3°
−1

[1.7] 2. Multiply: $-3(x + 7)$
−3x − 21

[1.7] 3. Combine like terms: $7xy + 4x - 8xy + 3y$
−xy + 4x + 3y

Exercises 4–6 ▲ Equations and Inequalities

For Exercise 4, translate to an equation and then solve.

[2.5] 4. Three times the sum of a number and nine is equal to eleven less than the number.
3(n + 9) = n − 11; −19

[2.6] 5. Solve: $\dfrac{1}{5} = \dfrac{n}{10.5}$
2.1

[2.6] 6. The figures shown are similar. Find the unknown side length.
10.5 cm

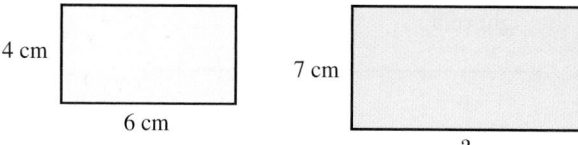

4 cm

6 cm

7 cm

?

2.8 Solving Linear Inequalities

Objectives

1 Represent solutions to inequalities graphically and using set notation.
2 Solve linear inequalities.
3 Solve problems involving linear inequalities.

Instructor Note Emphasize the difference between the symbols < and ≤ as well as > and ≥. Many students do not understand the difference.

Warm-up

[2.3] Solve the equations.

1. $\dfrac{-2}{5}t = \dfrac{9}{10}$

2. $7(x - 3) = 4x + 3$

Objective 1 Represent solutions to inequalities graphically and using set notation.

Not all problems translate to equations. Sometimes a problem can have a range of values as solutions. In mathematics, we can write inequalities to describe situations where a range of solutions is possible. Following are the inequality symbols and their meanings.

Symbol	Meaning
$<$	is less than
$>$	is greater than
$\leq$	is less than or equal to
$\geq$	is greater than or equal to

Throughout the chapter, we have explored techniques for solving linear equations. In this section, we explore how to solve **linear inequalities**.

Definition   Linear inequality: An inequality containing expressions in which each variable term contains a single variable with an exponent of 1.

Following are some examples of linear inequalities:

$$x > 5 \qquad n + 2 < 6 \qquad 2(y - 3) \leq 5y - 9 \qquad 2x + 3y \geq 6$$

The solutions of an inequality are any numbers that can replace the variable(s) in the inequality and make it true. Because inequalities have a range of solutions, we often write those solutions in set-builder notation. For example, the solution set for the inequality $x \geq 5$ contains 5 and every real number greater than 5, which we write as $\{x \mid x \geq 5\}$. We read this set-builder notation as shown here.

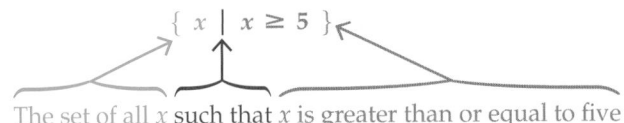

The set of all x such that x is greater than or equal to five

We can graph solution sets for inequalities on a number line. Because the solution set for $x \geq 5$ contains 5 and every real number to the right of 5, we draw a dot (or solid circle) at 5 and shade to the right of 5.

$x \geq 5$ is represented by

$$\longleftarrow \;\; \overset{\textstyle -1 \;\; 0 \;\; 1 \;\; 2 \;\; 3 \;\; 4 \;\; 5 \;\; 6 \;\; 7}{\rule{0pt}{0pt}} \longrightarrow$$

The solution set for $x < 2$ contains every real number to the left of 2, but not 2 itself; so it is written $\{x \mid x < 2\}$. To graph this solution set, we draw an open circle at 2, indicating that 2 is not included, and shade to the left of 2.

$x < 2$ is represented by

$$\longleftarrow \;\; \overset{\textstyle -1 \;\; 0 \;\; 1 \;\; 2 \;\; 3 \;\; 4 \;\; 5 \;\; 6 \;\; 7}{\rule{0pt}{0pt}} \longrightarrow$$

Another popular notation used to indicate ranges of values is *interval notation*, which uses parentheses and brackets to indicate whether end values are included. Parentheses are used for end values that are not included in the interval, and brackets are used for end values that are included. To write interval notation, we imagine traveling from left to right on the number line.

For example, as we travel from left to right on the graph for $x \geq 5$, we encounter 5 and then every real number to the right of 5; so the interval notation is $[5, \infty)$. The symbol ∞ means infinity and is used to indicate the positive extreme. Because ∞ can never be reached, it will always have a parenthesis indicating that it is not included as an end value. As we travel from left to right on the graph for $x < 2$, we go from the negative extreme up to 2, but not including 2; so the interval notation is $(-\infty, 2)$. Notice that $-\infty$ represents the leftmost extreme.

The parentheses and brackets from interval notation, instead of open and solid circles, can be used on the graphs.

$x \geq 5$ is represented by

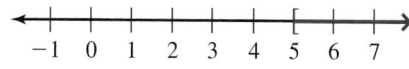

$x < 2$ is represented by

Your instructor may have a style preference. Because interval notation is a preferred notation in college algebra courses, we draw our graphs with the parentheses and brackets. Notice that the parentheses or brackets "open" in the direction that is shaded.

Procedure **Graphing Inequalities**

To graph an inequality on a number line:
1. If the symbol is $\leq$ or $\geq$, at the indicated number draw a bracket (or solid circle) on the number line that opens to the left for $\leq$ and to the right for $\geq$. If the symbol is $<$ or $>$, at the indicated number draw a parenthesis (or open circle) on the number line that opens to the left for $<$ and to the right for $>$.
2. If the variable is greater than the indicated number, shade to the right of the indicated number. If the variable is less than the indicated number, shade to the left of the indicated number.

Example 1 Write the solution set in set-builder notation and interval notation; then graph the solution set.

a. $x \leq -6$

Solution: Set-builder notation: $\{x | x \leq -6\}$
Interval notation: $(-\infty, -6]$

Note When writing interval notation, imagine traveling from left to right; so this range of values is ◀ from $-\infty$ to -6, including -6.

Graph:

b. $n > 0$

Solution: Set-builder notation: $\{n | n > 0\}$ ◀
Interval notation: $(0, \infty)$

Connection $\{n | n > 0\}$ is a way of expressing the set of all positive real numbers.

Graph:

Your Turn 1 Write the solution set in set-builder notation and interval notation; then graph the solution set.

a. $x < 2$ **b.** $t \geq -1$

Answers to Your Turn 1
a. Set-builder notation: $\{x | x < 2\}$
Interval notation: $(-\infty, 2)$

Graph: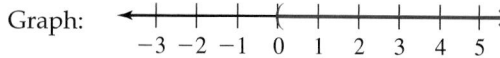
-4-3-2-1 0 1 2 3 4

b. Set-builder notation: $\{t | t \geq -1\}$
Interval notation: $[-1, \infty)$

Graph:
-4-3-2-1 0 1 2 3 4

Compound Inequalities

Inequalities containing two inequality symbols are called *compound inequalities*. Compound inequalities are useful in writing a range of values between two numbers. For example, $1 < x < 6$ is a compound inequality meaning that x can be any number that is greater than 1 and less than 6. Its solution set contains every real number between 1 and 6, but not 1 and 6.

Set-builder notation: $\{x \mid 1 < x < 6\}$

Interval notation: $(1, 6)$

Graph:

Note Because the end values 1 and 6 are not included in the solution set, we use parentheses in the interval notation and on the graph.

Example 2   Write the solution set for $-5 \le x < 1$ in set-builder notation and interval notation; then graph the solution set.

Solution: Set-builder notation: $\{x \mid -5 \le x < 1\}$

Interval notation: $[-5, 1)$

Graph:

Note Because the end value -5 is included in the solution set, we use a bracket in the interval notation and on the graph. Because 1 is not included, we use a parenthesis.

Your Turn 2   Write the solution set for $-4 < y \le -1$ in set-builder notation and interval notation; then graph the solution set.

Objective 2 Solve linear inequalities.

To solve inequalities such as $n + 2 < 6$ and $2(x - 3) \le 5x - 9$, we follow essentially the same process as for solving equations. The addition and multiplication principles of inequalities are similar to the principles of equality. First, let's examine the addition principle of inequality.

Consider the inequality $3 < 7$. According to the addition principle, if we add or subtract the same amount on both sides, the inequality should still be true. Let's test this by choosing some numbers to add and subtract on both sides.

Add 2 to both sides:	**Subtract 9 from both sides:**
$3 < 7$	$3 < 7$
$\underline{+2 \quad +2}$	$\underline{-9 \quad -9}$
$5 < 9$ Still true	$-6 < -2$ Still true

Note This principle indicates that we can add (or subtract) the same amount on both sides of an inequality without affecting its solution(s). Although the principle is written in terms of the $<$ symbol, the principle is true for ▶ any inequality symbol.

> **Rule The Addition Principle of Inequality**
>
> If $a < b$, then $a + c < b + c$ is true for all real numbers a, b, and c.
> The principle also holds true when $<$ is replaced with $>$, $\le$, or $\ge$.

Example 3   Solve $x + 2 \le -3$ and write the solution set in set-builder notation and interval notation; then graph the solution set.

Solution: **Note** Subtracting 2 from both sides does not affect the inequality.

$$x + 2 \le -3$$
$$\underline{\quad -2 \quad -2} \qquad \text{Subtract 2 from both sides.}$$
$$x \le -5$$

Answer to Your Turn 2

Set-builder notation: $\{y \mid -4 < y \le -1\}$

Interval notation: $(-4, -1]$

Graph:

Set-builder notation: $\{x \mid x \leq -5\}$

Interval notation: $(-\infty, -5]$

Graph:

Your Turn 3 Solve $n - 6 > -5$ and write the solution set in set-builder notation and interval notation; then graph the solution set.

The multiplication principle, on the other hand, does not work as neatly as it did for equations. As we shall see, sometimes multiplying or dividing both sides of an inequality by the same number can turn a true inequality into a false inequality. Let's multiply both sides of $3 < 7$ by a positive number and then by a negative number.

Multiply both sides by 4:

$4(3) < 4(7)$

$12 < 28$ Still true

Multiply both sides by -2:

$-2(3) < -2(7)$

$-6 < -14$ Not true!

Conclusion: An inequality remains true when we multiply both sides by a positive number, but not when we multiply both sides by a negative number.

Note We can multiply (or divide) both sides of an inequality by the same positive amount without affecting its solution(s). However, if we multiply (or divide) both sides of an inequality by the same negative amount, we must reverse the direction of the inequality symbol to maintain the truth of the inequality.

▶

Rule   The Multiplication Principle of Inequality

If a and b are real numbers, where $a < b$, then $ac < bc$ is true if c is a positive real number.

If a and b are real numbers, where $a < b$, then $ac > bc$ is true if c is a negative real number.

The principle also holds true for $>$, $\leq$, and $\geq$.

Example 4 Solve and write the solution set in set-builder notation and interval notation; then graph the solution set.

a. $3x > -7$

Solution:

Note Dividing both sides by a positive number does not affect the inequality.

▶ $\dfrac{3x}{3} > \dfrac{-7}{3}$ Divide both sides by 3.

$x > -\dfrac{7}{3}$

Set-builder notation: $\left\{ x \mid x > -\dfrac{7}{3} \right\}$

Interval notation: $\left(-\dfrac{7}{3}, \infty \right)$

Graph:

b. $-8x \geq 24$

Solution: $\dfrac{-8x}{-8} \leq \dfrac{24}{-8}$ Divide both sides by -8. Because we are dividing by a negative number, we also

$x \leq -3$ reverse the direction of the inequality.

Set-builder notation: $\{x \mid x \leq -3\}$

Interval notation: $(-\infty, -3]$

Graph:

Answer to Your Turn 3

$n > 1$

Set-builder notation: $\{n \mid n > 1\}$

Interval notation: $(1, \infty)$

Graph:

c. $\dfrac{5}{6} \le -\dfrac{3}{4}n$

Solution:

Note You may find it helpful to rewrite the inequality so that the variable is on the left side. The inequality $-\dfrac{10}{9} \ge n$ is the same as $n \le -\dfrac{10}{9}$.

$$-\dfrac{\overset{2}{\cancel{4}}}{3}\left(\dfrac{5}{\underset{3}{\cancel{6}}}\right) \ge -\dfrac{\overset{1}{\cancel{4}}}{3}\left(-\dfrac{\overset{1}{\cancel{3}}}{\underset{1}{\cancel{4}}}n\right)$$

$$-\dfrac{10}{9} \ge n$$

Multiply both sides by $-\dfrac{4}{3}$. **Because we are multiplying both sides by a negative number, we reverse the direction of the inequality symbol.**

Set-builder notation: $\left\{ n \,\middle|\, n \le -\dfrac{10}{9} \right\}$

Interval notation: $\left(-\infty, -\dfrac{10}{9} \right]$

Graph:

Your Turn 4   Solve $-\dfrac{2}{5}t < \dfrac{9}{10}$ and write the solution set in set-builder notation and in interval notation; then graph the solution set.

We can solve more complex linear inequalities using an outline similar to that for solving linear equations.

Procedure **Solving Linear Inequalities**

To solve linear inequalities:
1. Simplify both sides of the inequality as needed.
   a. Distribute to eliminate parentheses.
   b. Eliminate fractions or decimals by multiplying through by the LCD just as we did for equations. (Eliminating fractions and decimals is optional.)
   c. Combine like terms.
2. Use the addition principle so that all variable terms are on one side of the inequality and all constants are on the other side. Then combine like terms.
3. Use the multiplication principle to eliminate any remaining coefficient. If you multiply (or divide) both sides by a negative number, reverse the direction of the inequality symbol.

Notice that eliminating the term with the lesser coefficient in step 2 results in a positive coefficient, which means that you won't have to reverse the direction of the inequality in step 3.

Answer to Your Turn 4

$t > -\dfrac{9}{4}$ or $-2\dfrac{1}{4}$

Set-builder notation: $\left\{ t \,\middle|\, t > -\dfrac{9}{4} \right\}$

Interval notation: $\left(-\dfrac{9}{4}, \infty \right)$

Graph:

Example 5   Solve $7x + 9 > 4x - 6$ and write the solution set in set-builder notation and interval notation; then graph the solution set.

Solution:

Note Eliminating the $4x$ term results in a positive coefficient after combining like terms; so we won't have to reverse the inequality when we eliminate the 3 coefficient.

$$7x + 9 > 4x - 6$$
$$\underline{ -4x -4x }$$
$$3x + 9 > 0 - 6$$
$$3x + 9 > -6$$
$$\underline{ -9 -9 }$$
$$3x + 0 > -15$$

Subtract $4x$ from both sides.

Subtract 9 from both sides.

$$\frac{3x}{3} > \frac{-15}{3}$$ Divide both sides by 3 to isolate x.

$$x > -5$$

Set-builder notation: $\{x \mid x > -5\}$

Interval notation: $(-5, \infty)$

Graph:

$$\begin{array}{c} \leftarrow\!\!+\!\!+\!\!+\!\!\circ\!\!+\!\!+\!\!+\!\!+\!\!+\!\!+\!\!+\!\!\rightarrow \\ {\scriptstyle -7\ -6\ -5\ -4\ -3\ -2\ -1\ \ 0\ \ 1} \end{array}$$

Your Turn 5 Solve $7(x - 3) \le 4x + 3$ and write the solution set in set-builder notation and interval notation; then graph the solution set.

Objective 3 Solve problems involving linear inequalities.

Problems requiring inequalities can be translated using key words much like those we used to translate sentences to equations. The following table lists common key words that indicate inequalities.

Less Than:		Greater Than:	
A number is less than seven.	$n < 7$	A number is greater than two.	$n > 2$
A number must be smaller than five.	$n < 5$	A number must be greater than three.	$n > 3$
		A number must be more than negative six.	$n > -6$

Less Than or Equal To:		Greater Than or Equal To:	
A number is at most nine.	$n \le 9$	A number is at least two.	$n \ge 2$
The maximum is fourteen.	$n \le 14$	The minimum is eighteen.	$n \ge 18$

Instructor Note You might find it helpful to use examples with money when discussing *at least* and *at most*.

Example 6 Three-fourths of a number is at least eighteen. Translate to an inequality; then solve.

Solution: Translate the key words; then solve. We'll use n for the variable. Because the word *of* is preceded by a fraction, it means multiplication. The key words *at least* indicate a greater-than or equal-to symbol.

Translate:

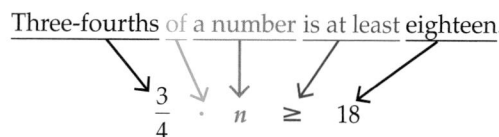

Three-fourths of a number is at least eighteen.
$$\frac{3}{4} \cdot n \ge 18$$

Solve: $\dfrac{3}{4} \cdot n \ge 18$

$$\frac{\overset{1}{4}}{\underset{1}{3}} \cdot \frac{\overset{1}{3}}{\underset{1}{4}} n \ge \frac{4}{3} \cdot \frac{\overset{6}{18}}{\underset{1}{1}}$$ Multiply both sides by $\frac{4}{3}$ to isolate n.

Answer: $n \ge 24$

Your Turn 6 0.2 times the sum of a number and eight is less than the difference of the number and five. Translate to an inequality; then solve.

Answer to Your Turn 5
$x \le 8$
Set-builder notation: $\{x \mid x \le 8\}$
Interval notation: $(-\infty, 8]$
Graph:

$$\begin{array}{c} \leftarrow\!\!+\!\!+\!\!+\!\!+\!\!+\!\!+\!\!+\!\!+\!\!\bullet\!\!+\!\!+\!\!\rightarrow \\ {\scriptstyle 0\ 1\ 2\ 3\ 4\ 5\ 6\ 7\ 8\ 9\ 10} \end{array}$$

Answer to Your Turn 6
$0.2(n + 8) < n - 5; n > 8.25$

Example 7   Darwin is planning a garden area, which he will enclose with a fence. Cost restricts him to a total of 240 feet of fencing materials. He wants the garden to span the entire width of his lot, which is 80 feet. What is the maximum length of the garden?

Understand The fence surrounds the garden; so 240 feet is the maximum perimeter for the garden. The width of the garden is to be 80 feet, and we must find the length.

Plan The formula for the perimeter of a rectangle is $P = 2l + 2w$. Because 240 feet is a *maximum* perimeter, we write the perimeter formula as an inequality so that the expression used to calculate perimeter ($2l + 2w$) is less than or equal to 240. Because the width is to be 80 feet, we replace w in the formula with 80.

Execute Translate:

The calculation for perimeter must be less than or equal to 240.

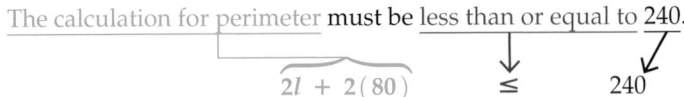

$$2l + 2(80) \leq 240$$

Solve: $2l + 160 \leq 240$ Subtract 160 from both sides.

$$\underline{-160 \quad -160}$$

$$2l + 0 \leq 80$$

$$\frac{2l}{2} \leq \frac{80}{2}$$ Divide both sides by 2 to isolate l.

$$l \leq 40$$

Answer The length must be less than or equal to 40 feet, which means that the maximum length is 40 feet.

Check Verify that a garden with a length of 40 feet or less and a width of 80 feet has a perimeter that is 240 feet or less. This will be left to the reader.

Your Turn 7   The surface area of a box is to be at least 678 square centimeters. If the length is to be 12 centimeters and the width is to be 9 centimeters, find the range of values for the height that satisfies the minimum surface area of 678 square centimeters. $(SA = 2lw + 2lh + 2wh)$

Answer to Your Turn 7

$h \geq 11$ cm

2.8 Exercises   For Extra Help   MyMathLab®

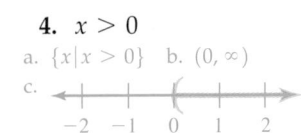

Note: Exercises marked with a ★ represent challenging exercises.

Objective 1

Prep Exercise 1 What is a solution for an inequality?

Any number that can replace the variable(s) in the inequality and make it true.

Prep Exercise 2 Explain the difference between $x < 8$ and $x \leq 8$.

$x < 8$ means any number less than 8; $x \leq 8$ means any number less than 8 *or* equal to 8.

Prep Exercise 3 In your own words, explain the meaning of $x \geq -5$.

Any number greater than -5 and -5 itself.

For Exercises 1–12: **a.** *Write the solution set in set-builder notation.*
                      b. *Write the solution set in interval notation.*   *See Examples 1 and 2.*
                      c. *Graph the solution set.*

1. $x \geq -3$

a. $\{x | x \geq -3\}$ b. $[-3, \infty)$

c. ⟵|—|—|—[—|—|—|—|⟶
   $-5\ -4\ -3\ -2\ -1\ \ 0\ \ 1$

2. $n \leq -5$

a. $\{n | n \leq -5\}$ b. $(-\infty, -5]$

c. ⟵|—|—|—]—|—|—|—|⟶
   $-7\ -6\ -5\ -4\ -3\ -2\ -1$

3. $h < 6$

a. $\{h | h < 6\}$ b. $(-\infty, 6)$

c. ⟵|—|—|—|—|—|—)—|—|⟶
   $0\ 1\ 2\ 3\ 4\ 5\ 6\ 7\ 8$

4. $x > 0$

a. $\{x | x > 0\}$ b. $(0, \infty)$

c. ⟵|—|—(—|—|—|⟶
   $-2\ \ -1\ \ 0\ \ 1\ \ 2$

5. $n < -\dfrac{2}{3}$

a. $\left\{n \mid n < -\dfrac{2}{3}\right\}$ b. $\left(-\infty, -\dfrac{2}{3}\right)$

c.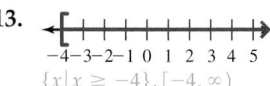

6. $a \geq \dfrac{4}{5}$

a. $\left\{a \mid a \geq \dfrac{4}{5}\right\}$ b. $\left[\dfrac{4}{5}, \infty\right)$

c.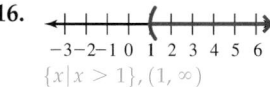

7. $t \geq 2.4$

a. $\{t \mid t \geq 2.4\}$ b. $[2.4, \infty)$
c.

8. $p \leq -0.6$

a. $\{p \mid p \leq -0.6\}$ b. $(-\infty, -0.6]$
c.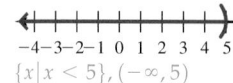

9. $-3 < x < 6$

a. $\{x \mid -3 < x < 6\}$
b. $(-3, 6)$
c.

10. $4 < c \leq 10$

a. $\{c \mid 4 < c \leq 10\}$
b. $(4, 10]$
c.

11. $0 \leq n \leq 5$

a. $\{n \mid 0 \leq n \leq 5\}$
b. $[0, 5]$
c.

12. $-1 < a < 5$

a. $\{a \mid -1 < a < 5\}$
b. $(-1, 5)$
c.

For Exercises 13–18, for each graph, write the inequality in set-builder notation and interval notation. See Objective 1.

13.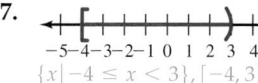

$\{x \mid x \geq -4\}, [-4, \infty)$

14.

$\{x \mid x \leq 2\}, (-\infty, 2]$

15.

$\{x \mid x < 5\}, (-\infty, 5)$

16.

$\{x \mid x > 1\}, (1, \infty)$

17.

$\{x \mid -4 \leq x < 3\}, [-4, 3)$

18.

$\{x \mid -2 < x \leq 2\}, (-2, 2]$

Objective 2

Prep Exercise 4 What action causes the direction of an inequality symbol to change?
Multiplying or dividing by a negative number.

For Exercises 19–52:   **a.** *Solve.*
b. *Write the solution set in set-builder notation.*
c. *Write the solution set in interval notation.*   *See Examples 3–5.*
d. *Graph the solution set.*

19. $n - 3 > 2$
a. $n > 5$ b. $\{n \mid n > 5\}$
c. $(5, \infty)$
d.

20. $m - 5 < 20$
a. $m < 25$ b. $\{m \mid m < 25\}$
c. $(-\infty, 25)$
d.

21. $z + 2 \leq -4$
a. $z \leq -6$ b. $\{z \mid z \leq -6\}$
c. $(-\infty, -6]$
d.

22. $x + 8 \geq -12$
a. $x \geq -20$ b. $\{x \mid x \geq -20\}$
c. $[-20, \infty)$
d.

23. $16y \geq 32$
a. $y \geq 2$ b. $\{y \mid y \geq 2\}$
c. $[2, \infty)$
d.

24. $4z \leq -20$
a. $z \leq -5$ b. $\{z \mid z \leq -5\}$
c. $(-\infty, -5]$
d.

25. $-3x \geq -12$
a. $x \leq 4$ b. $\{x \mid x \leq 4\}$
c. $(-\infty, 4]$
d.

26. $-4x < -16$
a. $x > 4$ b. $\{x \mid x > 4\}$
c. $(4, \infty)$
d.

27. $\dfrac{2}{3}x \geq 4$

a. $x \geq 6$ b. $\{x \mid x \geq 6\}$
c. $[6, \infty)$
d.

28. $\dfrac{3}{5}a < -6$

a. $a < -10$ b. $\{a \mid a < -10\}$
c. $(-\infty, -10)$
d.

29. $-\dfrac{3}{4}m < -6$

a. $m > 8$ b. $\{m \mid m > 8\}$
c. $(8, \infty)$
d.

30. $-\dfrac{5}{3}p > -10$

a. $p < 6$ b. $\{p \mid p < 6\}$
c. $(-\infty, 6)$
d.

31. $5y + 1 > 16$
a. $y > 3$ b. $\{y \mid y > 3\}$
c. $(3, \infty)$
d.

32. $3y - 2 > 10$
a. $y > 4$ b. $\{y \mid y > 4\}$
c. $(4, \infty)$
d.

33. $1 - 6x < 25$
a. $x > -4$ b. $\{x \mid x > -4\}$
c. $(-4, \infty)$
d.

34. $3 - 2c \geq 17$
a. $c \leq -7$ b. $\{c \mid c \leq -7\}$
c. $(-\infty, -7]$
d.

35. $\dfrac{a}{2} + 1 < \dfrac{3}{2}$

a. $a < 1$ b. $\{a \mid a < 1\}$
c. $(-\infty, 1)$
d.

36. $\dfrac{k}{3} - \dfrac{2}{3} > -2$

a. $k > -4$ b. $\{k \mid k > -4\}$
c. $(-4, \infty)$
d.

37. $10 + 2f \le 2 - 2f$

a. $f \le -2$ b. $\{f \mid f \le -2\}$
c. $(-\infty, -2]$
d.

38. $9 + 2n < 3 - 4n$

a. $n < -1$ b. $\{n \mid n < -1\}$
c. $(-\infty, -1)$
d.

39. $3 - 6u \ge -5 - 2u$

a. $u \le 2$ b. $\{u \mid u \le 2\}$
c. $(-\infty, 2]$
d.

40. $4 - 9k < -4k + 19$

a. $k > -3$ b. $\{k \mid k > -3\}$
c. $(-3, \infty)$
d.

41. $2(c - 3) < 3(c + 2)$

a. $c > -12$ b. $\{c \mid c > -12\}$
c. $(-12, \infty)$
d.

42. $4(d + 4) \ge 5(d + 2)$

a. $d \le 6$ b. $\{d \mid d \le 6\}$
c. $(-\infty, 6]$
d.

43. $10(9w + 13) - 5(3w - 1) \le 8(11w + 12)$

a. $w \ge 3$ b. $\{w \mid w \ge 3\}$ c. $[3, \infty)$
d.

44. $6(n + 13) - 2(4n - 2) > 3(5n - 1)$

a. $n < 5$ b. $\{n \mid n < 5\}$ c. $(-\infty, 5)$
d.

45. $\dfrac{1}{2}(3x + 1) \le \dfrac{1}{3}(4x - 5)$

a. $x \le -13$ b. $\{x \mid x \le -13\}$ c. $(-\infty, -13]$
d.

46. $\dfrac{1}{5}(6m - 7) > \dfrac{1}{2}(3m - 1)$

a. $m < -3$ b. $\{m \mid m < -3\}$ c. $(-\infty, -3)$
d.

47. $\dfrac{1}{6}(2n + 4) - \dfrac{1}{3}(3n - 15) > 1$

a. $n < 7$ b. $\{n \mid n < 7\}$ c. $(-\infty, 7)$
d.

48. $-\dfrac{1}{3}(2y - 5) + \dfrac{1}{6}(5y - 2) \ge 1$

a. $y \ge -2$ b. $\{y \mid y \ge -2\}$ c. $[-2, \infty)$
d.

49. $0.4l - 0.37 < 0.3(l + 2.1)$

a. $l < 10$ b. $\{l \mid l < 10\}$ c. $(-\infty, 10)$
d.

50. $0.05 + 0.03(v - 4.5) \ge 0.465$

a. $v \ge 18.\overline{3}$ b. $\{v \mid v \ge 18.\overline{3}\}$ c. $[18.\overline{3}, \infty)$
d.

51. $0.6t + 0.2(t - 8) \le 0.5t - 0.6$

a. $t \le 3.\overline{3}$ b. $\{t \mid t \le 3.\overline{3}\}$ c. $(-\infty, 3.\overline{3}]$
d.

52. $2.1s + 3(1.4s - 1.5) \ge 0.09s + 20.34$

a. $s \ge 4$ b. $\{s \mid s \ge 4\}$ c. $[4, \infty)$
d.

Objective 3

For Exercises 53–64, translate to an inequality and then solve. See Example 6.

53. Four subtracted from a number is greater than twenty-four.

$n - 4 > 24$
$n > 28$

54. Six less than a number is less than or equal to negative four.

$x - 6 \le -4$
$x \le 2$

55. Four-ninths of a number is less than or equal to negative eight.

$\dfrac{4}{9}x \le -8$
$x \le -18$

56. Three-fifths of a number is greater than negative twenty-one.

$\dfrac{3}{5}x > -21$
$x > -35$

57. Eight times a number less thirty-six is at least sixty.

$8y - 36 \ge 60$
$y \ge 12$

58. Three subtracted from a number times eleven is at least forty-one.

$11w - 3 \ge 41$
$w \ge 4$

59. Five added to half of a number is at most two.

$$\frac{1}{2}a + 5 \leq 2$$

$$a \leq -6$$

60. The sum of two-fifths of a number and two is at most eight.

$$\frac{2}{5}x + 2 \leq 8$$

$$x \leq 15$$

61. The difference of four times a number and eight is less than two times the number.

$$4x - 8 < 2x$$

$$x < 4$$

62. Triple a number is less than the number decreased by eight.

$$3m < m - 8$$

$$m < -4$$

63. Twenty-five is greater than or equal to seven more than six times a number.

$$25 \geq 6x + 7$$

$$x \leq 3$$

64. Negative five is greater than fifteen less than five times a number.

$$-5 > 5x - 15$$

$$x < 2$$

For Exercises 65–78, solve. See Example 7. (Formulas are on pages 91 and 94.)

65. A school is planning a playground. The area of a rectangular playground may not exceed 170 square feet. If the width is 17 feet, what range of values can the length have?

$l \leq 10$ ft.

66. Scotty needs to make a rectangular pool cover that is at least 128 square feet. If the pool length is 16 feet, what range of values must the width be?

$w \geq 8$ ft.

67. The design of a storage box calls for a width of 27 inches and a length of 41 inches. If the surface area must be at least 4254 square inches, find the range of values for the height.

$h \geq 15$ in.

68. A company builds truck containers in the shape of rectangular solids. One such container has a length of 24 feet and width of 8 feet. What is the range of values for the height if the volume is to be at least 576 cubic feet?

$h \geq 3$ ft.

69. Andre builds tables. One particular circular table requires him to glue a strip of veneer around the circumference of the top. He is designing a new circular table and has a strip of veneer that is 150 inches long that he wants to use. What range of values can the radius have? (Use $\pi \approx 3.14$.) Round your answer to the nearest hundredth.

$r \leq 23.89$ in.

70. The city parks department is putting in a circular flower garden that is to be enclosed using an edging material. If 125 feet of the edging material is available, what is the range of values for the diameter of the flower garden? (Use $\pi \approx 3.14$.) Round your answer to the nearest hundredth.

$d \leq 39.79$ ft.

71. Jon never exceeds 65 miles per hour when driving on the highway. If he is traveling 410 miles, what range of values can his time spent driving have?

$t \geq 6\frac{4}{13}$ hr.

72. A truck driver must make a delivery 298 miles away in five hours or less. What range of values can his average rate be so that he meets the schedule?

$r \geq 59.6$ mph

73. The fourth-quarter profit goal for a software company is at least $850,000. If it is known that the cost for the quarter will be $625,000, find the revenue that must be generated to achieve the goal. (Use $P = R - C$.)

$R \geq \$1,475,000$

74. Silver is a solid up to a temperature of 960.8°C. Write an inequality to describe this range of temperatures in degrees Fahrenheit. (Use $F = \frac{9}{5}C + 32$.)

$F \leq 1761.44°$

75. To get an A in her math course, Tina needs the average of five tests to be at least 90. Her current test scores are 82, 91, 95, and 84. What range of scores on the fifth test would get her an A in the course?

$x \geq 98$

76. In Aaron's English course, the final grade is determined by the average of five papers. The department requires any student whose average falls below 75 to repeat the course. Aaron's scores on the first four papers are 68, 78, 80, and 72. What range of scores on the fifth paper would result in him having to repeat the course?

$x < 77$

77. The design of a circuit specifies that the voltage cannot exceed 12 volts. If the resistance of the circuit is 8 ohms, find the range of values that the current can have. (Use $V = iR$.)

$i \leq 1.5 \, \text{A}$

★ 78. A label in an elevator indicates that the maximum load is 9000 newtons; that is, the downward force should not exceed -9000 newtons. If the acceleration due to gravity is -9.8 meters per second squared, find the range of values of mass that can be loaded onto the elevator. Round your answer to the nearest hundredth. (Use $F = ma$.)

$m \leq 918.37 \, \text{kg}$

| Puzzle Problem Write the set of all values such that
$|x| = -x$. $\{x \mid x \leq 0\}$

Review Exercises

Exercise 1   Constants and Variables

[1.1] 1. Write a set representing the natural numbers up to 10.

$\{1, 2, 3, 4, 5, 6, 7, 8, 9, 10\}$

Exercises 2–4   Expressions

[1.4] 2. Simplify: $\dfrac{2450}{140}$

$\dfrac{35}{2}$

[1.6] 3. Translate to an expression: six more than twice the difference of a number and five.

$2(x - 5) + 6$

[1.7] 4. Evaluate $b^2 - 4ac$ when $a = 1, b = -2,$ and $c = -3$.

16

Exercises 5 and 6   Equations and Inequalities

[2.3] *For Exercises 5 and 6, solve.*

5. $3\dfrac{1}{5}n = -108$

$-\dfrac{135}{4}$ or $-33\dfrac{3}{4}$

6. $5n = 7131.04$

1426.208

Chapter 2 Summary and Review Exercises

Complete each incomplete definition, rule, or procedure; study the key examples; and then work the related exercises.

Formulas

2.1

Perimeter of a rectangle: $P = \underline{\quad 2l + 2w \quad}$

Circumference of a circle: $C = \underline{\quad \pi d \quad}$ or $C = \underline{\quad 2\pi r \quad}$

Area of a parallelogram: $A = \underline{\quad bh \quad}$

Area of a rectangle: $A = \underline{\quad lw \quad}$

Area of a triangle: $A = \underline{\quad \frac{1}{2}bh \quad}$

Area of a trapezoid: $A = \underline{\quad \frac{1}{2}h(a + b) \quad}$

Area of a circle: $A = \underline{\quad \pi r^2 \quad}$

Surface area of a box: $SA = \underline{\quad 2lw + 2lh + 2wh \quad}$

Volume of a box: $V = \underline{\quad lwh \quad}$

Volume of a pyramid: $V = \underline{\quad \frac{1}{3}lwh \quad}$

Volume of a cylinder: $V = \underline{\quad \pi r^2 h \quad}$

Volume of a cone: $V = \underline{\quad \frac{1}{3}\pi r^2 h \quad}$

Volume of a sphere: $V = \underline{\quad \frac{4}{3}\pi r^3 \quad}$

Distance, d, an object travels given its rate, r, and the time of travel, t: $d = \underline{\quad rt \quad}$

The average rate of travel, r, given the total time, t: $r = \underline{\quad \frac{d}{t} \quad}$

Voltage, V, in a circuit with current, i, in amperes (A), and a resistance, R, in ohms (Ω): $V = \underline{\quad iR \quad}$

The temperature in degrees Celsius given degrees Fahrenheit: $C = \underline{\quad \frac{5}{9}(F - 32) \quad}$

The temperature in degrees Fahrenheit given degrees Celsius: $F = \underline{\quad \frac{9}{5}C + 32 \quad}$

[2.4]

The profit, P, after cost, C, is deducted from revenue, R:
$P = \underline{\quad R - C \quad}$

   Learning Strategy

The summaries in the textbook are like the study sheet suggested in the To the Student section that precedes Chapter 1. Remember that your study sheet is a list of the rules, procedures, and formulas that you need to know. If you are a tactile or visual learner, spend a lot of time reviewing and writing the rules or procedures. Try to get to the point where you can write the essence of each rule and procedure from memory. If you are an audio learner, record yourself saying each rule and procedure; then listen to the recording over and over. Consider developing a clever rhyme or song for each rule or procedure to help you remember it.

Procedure Problem-Solving Outline

1. $\underline{\text{Understand}}$ the problem.
    a. Read the question(s) (not the whole problem, just the question at the end) and write a note to yourself about what you are to find.
    b. Read the whole problem, underlining the key words.
    c. If possible or useful, draw a picture, make a list or table to organize what is known and unknown, simulate the situation, or search for a related example problem.
2. $\underline{\text{Plan}}$ your solution by searching for a formula or using the key words to translate to an equation.
3. $\underline{\text{Execute}}$ the plan by solving the equation/formula.
4. $\underline{\text{Answer}}$ the question. Look at the note about what you were to find and make sure you answer that question. Include appropriate units.
5. $\underline{\text{Check}}$ results.
    a. Try finding the solution in a different way, reversing the process, or estimating the answer and making sure the estimate and actual answer are reasonably close.
    b. Make sure the answer is reasonable.

2.1 Equations, Formulas, and the Problem-Solving Process

Definitions/Procedures/Rules	Key Example(s)
An **equation** is ___two expressions set equal___ . A **solution** of an equation is ___a number that makes an equation true when it replaces the variable in the equation___ . To **determine whether a value is a solution** to a given equation, replace the ___variable___ in the equation with the ___value___ . If the resulting equation is ___true___ , the value is a solution.	Check to see if the given value is a solution to the given equation. **a.** $7n - 13 = 3n + 11; n = 6$ $7(6) - 13 \overset{?}{=} 3(6) + 11$ $42 - 13 \overset{?}{=} 18 + 11$ $29 = 29$ Yes, 6 is a solution. **b.** $2(x - 3) = 5x + 4; x = 1$ $2(1 - 3) \overset{?}{=} 5(1) + 4$ $2(-2) \overset{?}{=} 5 + 4$ $-4 \neq 9$ No, 1 is not a solution.

Exercises 1–4 ▲   Equations and Inequalities

[2.1] *For Exercises 1–4, check to see if the given number is a solution to the equation.*

1. $y + 6 + 3y - 4 = 14; y = 3$
yes

2. $\frac{1}{4}(n - 1) = \frac{1}{2}(n + 1); n = 5$
no

3. $x^2 + 6.1 = 3x - 0.1; x = -0.1$
no

4. $\frac{5}{6}m - 1 = \frac{m}{3} + \frac{1}{2}; m = 3$
yes

Definitions/Procedures/Rules	Key Example(s)
A **formula** is an equation that describes a(n) ___mathematical relationship___ . The **perimeter** of a geometric figure is ___the distance around the figure___ . The **area** of a geometric figure is ___the total number of square units that fill the figure___ . The **volume** of a geometric figure is ___the total number of cubic units that fill a space___ . The **circumference** of a circle is ___the distance around the circle___ . The **radius** of a circle is ___the distance from the center of a circle to any point on the circle___ . The **diameter** of a circle is ___the distance across a circle through its center___ . To use a **formula**: **1.** Replace the ___variables___ with the corresponding given values. **2.** Solve for the ___unknown value___ .	 Find the volume of a box with a length of 4 centimeters, width of 5 centimeters, and height of 6 centimeters. $V = lwh$ $V = (4\,\text{cm})(5\,\text{cm})(6\,\text{cm})$ $V = 120\,\text{cm}^3$

Exercises 5–8 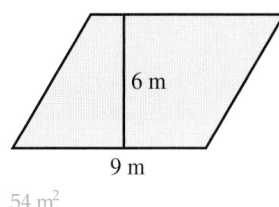 Equations and Inequalities

[2.1] *For Exercises 5 and 6, find the area of the figure. If necessary, round to the nearest hundredth. Use* $\pi \approx 3.14$.

5.

6 m

9 m

54 m^2

6.

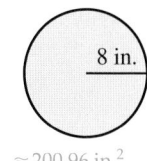

8 in.

≈ 200.96 in.^{2}

[2.1] *For Exercises 7 and 8, solve.*

7. Find the volume of a pyramid with length of 9 cm, width of 5 centimeters, and height of 10 centimeters.

150 cm^3

8. Find the volume of a jar with a radius of 2 inches and a height of 8 inches. Approximate the volume rounded to the nearest hundredth. Use $\pi \approx 3.14$.

≈ 100.48 in.^{2}

Definitions/Procedures/Rules	Key Example(s)
For **composite figures:** 1. To calculate the area of a figure composed of two or more figures that are next to each other, __add the areas__ of the individual figures. 2. To calculate the area of a region defined by a smaller figure within a larger figure, __subtract the area of the smaller figure from the area of the larger figure__.	Find the shaded area in the figure shown. 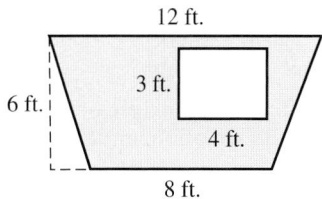 $A = $ Area of trapezoid $-$ Area of rectangle $A = \frac{1}{2}h(a + b) - lw$ $A = \frac{1}{2}(6)(12 + 8) - (3)(4)$ $A = \frac{1}{2}(6)(20) - 12$ $A = 60 - 12$ $A = 48$ ft.^{2}

Exercises 9–10 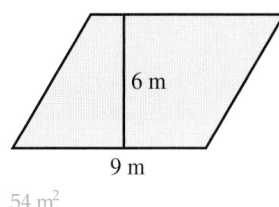 Equations and Inequalities

[2.1] 9. Best Buy is selling an oak entertainment center. Elijah is concerned it will take up too much room in his den. Using the figure shown, determine the total area.

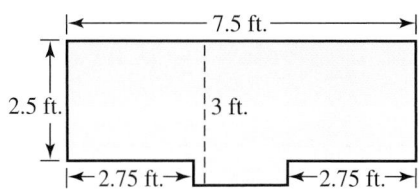

7.5 ft.

2.5 ft.   3 ft.

2.75 ft.   2.75 ft.

19.75 ft.^{2}

[2.1] 10. Find the area of the figure shown.

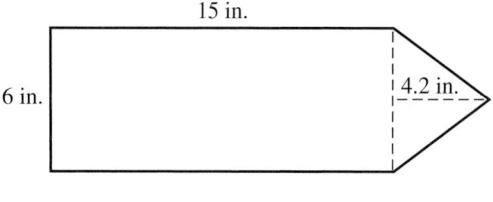

15 in.

6 in.   4.2 in.

102.6 in.^{2}

2.2 and 2.3 **Solving Linear Equations**

Definitions/Procedures/Rules	Key Example(s)
A **linear equation** is an equation in which each variable term contains a single variable raised to an exponent of $\underline{1}$.	$5x + 6 = 2$ is a linear equation, and $y^4 - 3y = 7$ is not linear.

Solve.

a. $x + 9 = 15$ **b.** $y - 5 = 7$

A **linear equation in one variable** is an equation that can be written in the form $\underline{ax + b = c}$, where a, b, and c are real numbers and $a \neq 0$.

$$\begin{array}{rr} x + 9 = 15 & y - 5 = 7 \\ \underline{-9 \quad -9} & \underline{+5 \quad +5} \\ x + 0 = 6 & y + 0 = 12 \\ x = 6 & y = 12 \end{array}$$

The **addition principle of equality** states that if $a = b$, then $a + c = \underline{b + c}$ for all real numbers a, b, and c.

To use the addition principle of equality to eliminate a term from one side of an equation, add the $\underline{additive\ inverse}$ of that term to both sides of the equation.

c. $5(n - 2) + 3n = 7n - 4$

$$5n - 10 + 3n = 7n - 4$$
$$8n - 10 = 7n - 4$$
$$\underline{-7n \qquad \quad -7n}$$
$$n - 10 = 0 - 4$$
$$n - 10 = -4$$
$$\underline{+10 \quad +10}$$
$$n + 0 = \quad 6$$
$$n = 6$$

The **multiplication principle of equality** states that if $a = b$, then $ac = \underline{bc}$ is true for all real numbers a, b, and c, where $c \neq 0$.

Solve.

To use the multiplication principle of equality to eliminate a coefficient in an equation, multiply both sides of the equation by the $\underline{multiplicative\ inverse}$ of that coefficient or divide both sides by $\underline{the\ coefficient}$.

a. $-5x = 15$ **b.** $\dfrac{3}{4}n = \dfrac{5}{6}$

$$\dfrac{-5x}{-5} = \dfrac{15}{-5} \qquad \overset{1}{\underset{3}{\cancel{\dfrac{4}{3}}}} \cdot \overset{1}{\underset{1}{\cancel{\dfrac{3}{4}}}} n = \overset{}{\underset{3}{\cancel{\dfrac{5}{6}}}} \cdot \overset{2}{\underset{}{\cancel{\dfrac{4}{3}}}}$$
$$x = -3 \qquad\qquad\qquad n = \dfrac{10}{9}$$

To solve linear equations in one variable:

c. $7n - (n - 3) = 2n - 5$

1. Simplify both sides of the equation as needed.
   a. Distribute to eliminate $\underline{parentheses}$.
   b. Eliminate fractions or decimals by multiplying through by the $\underline{LCD}$. In the case of decimals, the $\underline{LCD}$ is the power of 10 with the same number of zero digits as decimal places in the number with the $\underline{most\ decimal\ places}$.
   c. Combine like terms.

$$7n - n + 3 = 2n - 5$$
$$6n + 3 = 2n - 5$$
$$\underline{-2n \qquad\quad -2n}$$
$$4n + 3 = 0 - 5$$
$$4n + 3 = -5$$
$$\underline{-3 \quad -3}$$
$$4n + 0 = -8$$
$$\dfrac{4n}{4} = \dfrac{-8}{4}$$
$$n = -2$$

2. Use the $\underline{addition\ principle}$ so that all variable terms are on one side of the equation and all constants are on the other side. (Eliminate the variable term with the lesser coefficient to avoid negative coefficients.) Then combine like terms.

3. Use the $\underline{multiplication\ principle}$ to eliminate any remaining coefficient.

Definitions/Procedures/Rules	Key Example(s)
When **solving a linear equation**, if after simplifying each side of the equation the expressions are identical, the equation is a(n) <u>identity</u> and <u>every real number</u> for which the equation is defined is a solution.	**d.** $5x - 9 - x = 4(x - 2) - 1$ 　　　$4x - 9 = 4x - 8 - 1$ 　　　$4x - 9 = 4x - 9$　　The equation is an identity, so every real number is a solution.
When **solving a linear equation**, if after simplifying the expressions on each side of the equation the expressions have the same variable term but different constants, the equation is a(n) <u>contradiction</u> and has <u>no solution</u>.	**e.** $4(2n - 3) - 5n + 2 = 9n - 2(3n - 6)$　Distribute. 　　$8n - 12 - 5n + 2 = 9n - 6n + 12$ 　　　　　　　　　　　　　　Combine like terms. 　　　　$3n - 10 = 3n + 12$ 　　　　　　The equation is a contradiction, so it has no solution.

Exercises 11–32 ▲ Equations and Inequalities

[2.2–2.3] *For Exercises 11–28, solve and check.*

11. $x + 7 = -5$
　-12

12. $\dfrac{2}{3} = x + \dfrac{3}{4}$
　$-\dfrac{1}{12}$

13. $-6x = 30$
　-5

14. $-\dfrac{2}{3}y = -12$
　18

15. $4n - 9 = 15$
　6

16. $5(2a - 3) - 6(a + 2) = -7$
　5

17. $12t + 9 = 4t - 7$
　-2

18. $2(c + 4) - 5c = 17 - 8c$
　$\dfrac{9}{5}$

19. $1 - 2(3c - 5) = 10 - 6c$
　No solution

20. $5(b - 1) - 3b = 7b - 5(1 + b)$
　All real numbers

21. $5(u - 1) - (u - 2) = 10 - (2u + 1)$
　2

22. $\dfrac{3}{4}m - \dfrac{1}{2} = \dfrac{2}{3}m + 1$
　18

23. $\dfrac{1}{12}(4x + 9) = \dfrac{2}{3}x$
　$\dfrac{9}{4}$

24. $\dfrac{5}{3} + 7x - 2x = \dfrac{1}{3}(9x + 2)$
　$-\dfrac{1}{2}$

25. $2 - \dfrac{2}{3}m = 6\left(m - \dfrac{1}{3}\right) - 4m + 2$
　$\dfrac{3}{4}$

26. $1.4x - 0.5(9 - 6x) = 6 + 2.4x$
　5.25

27. $3(x + 2) - 4 = x + 2x + 2$
　All real numbers

28. $4x - 3(x + 5) = 2(2x - 3) - 3(x + 5)$
　No solution

[2.2–2.3] *For Exercises 29–32, solve.*

29. Bonnie is making a shower curtain. The pattern says that the curtain will use 5760 square inches. If the length is to be 72 inches, how wide will the curtain be?
80 in.

30. The area of the trapezoid shown is 58 square inches. Find the unknown length.
7.1 in.

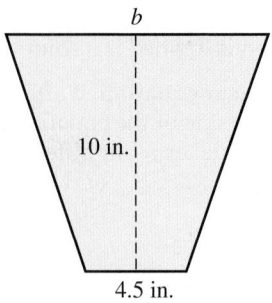

31. Randy is snowskiing at Lake Tahoe. He knows that the trail is 1.8 miles long, and he is one-third of the way down. How far does he have left to ski?
1.2 mi.

32. The surface area of a storage trunk is 2610 square inches. If the height of the trunk is 15 inches and the width is 36 inches, what is the length?
15 in.

2.4 Applying the Principles to Formulas

Definitions/Procedures/Rules	Key Example(s)
To **isolate a particular variable** in a formula, treat all other variables like ___constants___ and isolate the desired variable using the outline for solving equations.	Isolate w in the formula $P = 2l + 2w$. $$P = 2l + 2w$$ $$\underline{-2l \quad -2l}$$ $$P - 2l = 0 + 2w$$ $$\frac{P - 2l}{2} = \frac{2w}{2}$$ $$\frac{P - 2l}{2} = w$$

Exercises 33–40   Equations and Inequalities

[2.4] *For Exercises 33–40, solve for the indicated variable.*

33. $x + y = 1$; for x
$x = 1 - y$

34. $C = \pi d$; for d
$d = \dfrac{C}{\pi}$

35. $A = \dfrac{1}{2}bh$; for h
$h = \dfrac{2A}{b}$

36. $F = \dfrac{mv^2}{r}$; for m
$m = \dfrac{Fr}{v^2}$

37. $V = \dfrac{1}{3}\pi r^2 h$; for h
$h = \dfrac{3V}{\pi r^2}$

38. $y = mx + b$; for m
$m = \dfrac{y - b}{x}$

39. $P = 2l + 2w$; for w
$w = \dfrac{P - 2l}{2}$

40. $A = \dfrac{1}{2}h(a + b)$; for h
$h = \dfrac{2A}{a + b}$

2.5 Translating Word Sentences to Equations

Definitions/Procedures/Rules	Key Example(s)
To translate a word sentence to an equation, identify the <u>unknown(s)</u>, <u>constants</u>, and <u>key words</u>; then write the corresponding symbolic form.	Translate to an equation. **a.** Five more than seven times a number is twelve. **Answer:** $7n + 5 = 12$ **b.** Six times the difference of a number and nine is equal to four times the sum of a number and five. **Answer:** $6(n - 9) = 4(n + 5)$

Exercises 41–44 ▲♦ Equations and Inequalities

[2.5] *For Exercises 41–44, translate to an equation and solve.*

41. Six times a number is the same as negative eighteen.
$$6x = -18$$
$$x = -3$$

42. Three subtracted from half of a number is equal to nine less than one-fourth of the number.
$$\frac{1}{2}m - 3 = \frac{1}{4}m - 9$$
$$m = -24$$

43. One less than two times the sum of a number and four results in one.
$$2(v + 4) - 1 = 1$$
$$v = -3$$

44. Twice the difference of a number and one added to triple the number yields twenty less than six times the number.
$$2(y - 1) + 3y = 6y - 20$$
$$y = 18$$

2.6 Ratios and Proportions

Definitions/Procedures/Rules	Key Example(s)
A **ratio** is a comparison of two quantities using a(n) <u>quotient</u>. A **unit ratio** is a ratio with a denominator of <u>1</u>. A **proportion** is an equation in the form <u>$\frac{a}{b} = \frac{c}{d}$, where $b \neq 0$ and $d \neq 0$</u>. If $\frac{a}{b} = \frac{c}{d}$, where $b \neq 0$ and $d \neq 0$, then <u>$ad = bc$</u>.	Determine whether the ratios are equal. $$\frac{5}{6} \overset{?}{=} \frac{15}{18}$$ $$18 \cdot 5 = 90 \qquad 6 \cdot 15 = 90$$ $$\frac{5}{6} \overset{=}{\bcancel{\times}} \frac{15}{18}$$ Because the cross products are equal, the ratios are equal.

Exercises 45–48 ▲♦ Equations and Inequalities

[2.6] **45.** A recipe calls for $\frac{3}{4}$ of a cup of sugar and 2 cups of flour. Write the ratio of sugar to flour as a fraction in simplest form.
$$\frac{3}{8}$$

[2.6] **46.** The price of a 10.5-ounce can of vegetables is $0.89. Write the unit ratio of price to weight rounded to the nearest thousandth. Interpret the answer.
$$\approx \frac{\$0.085}{1 \text{ oz}}; \text{ each ounce costs 8.5 cents.}$$

[2.6] *For Exercises 47 and 48, determine whether the ratios are equal.*

47. $\dfrac{2}{5} \overset{?}{=} \dfrac{10}{20}$
no

48. $-\dfrac{2}{8} \overset{?}{=} -\dfrac{1}{4}$
yes

Definitions/Procedures/Rules	Key Example(s)
To solve a proportion using cross products: 1. __Find__ the cross products. 2. Set the __cross products__ equal to each other. 3. Use the __multiplication__ principle of equality to isolate the variable.	Solve for the missing number in the proportion. $$\frac{5}{12} = \frac{x}{18}$$ $90 = 12x$ Cross products $\dfrac{90}{12} = \dfrac{12x}{12}$ Divide by 12. $7.5 = x$

Exercises 49–52 Equations and Inequalities

[2.6] *For Exercises 49–52, solve for the missing number.*

49. $\dfrac{3}{5} = \dfrac{15}{x}$

25

50. $\dfrac{-11}{12} = \dfrac{n}{-24}$

22

51. $\dfrac{7}{p} = \dfrac{\frac{2}{5}}{4\frac{1}{2}}$

$78\frac{3}{4}$

52. $\dfrac{c}{12.7} = \dfrac{-5}{3}$

$-21.1\overline{6}$

Definitions/Procedures/Rules	Key Example(s)
To solve proportion problems: 1. Set up a proportion in which the __numerators__ and __denominators__ correspond in a logical manner. 2. Solve using __cross products__ .	A car can travel about 350 miles on 16 gallons of gasoline. How many gallons would be needed to travel 600 miles? $$\frac{350}{16} = \frac{600}{x}$$ $350x = 9600$ Cross products $\dfrac{350x}{350} = \dfrac{9600}{350}$ Divide by 350. $x \approx 27.4$ gal.

Exercises 53–56 Equations and Inequalities

[2.6] *For Exercises 53 and 54, translate to a proportion and solve.*

53. It costs $5.85 to mail a 4.5-pound package. At this same rate, how much will it cost to mail a 7-pound package?

$9.10

54. Ferrari estimates that the 2007 612 Scaglietti two-door coupe will travel 209 city miles on one tank of gas. If the gas tank of the automobile holds 23.7 gallons, how far can a driver expect to travel on 25 gallons?

≈ 220.5 mi.

[2.6] *For Exercises 55 and 56, find the missing lengths in the similar shapes.*

55.

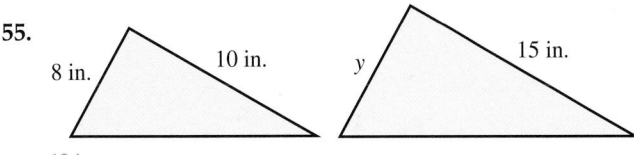

8 in.   10 in. y 15 in.

12 in.

56.

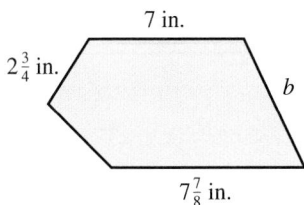 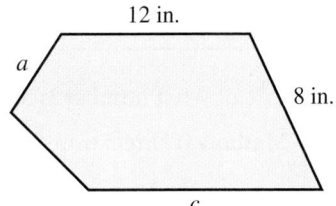

$a = 4\frac{5}{7}$ in.; $b = 4\frac{2}{3}$ in.; $c = 13\frac{1}{2}$ in.

2.7 Percents

Definitions/Procedures/Rules	Key Example(s)
To write $n\%$ as a fraction or decimal: **1.** Write $n\%$ as $\dfrac{n}{100}$. **2.** Simplify to the desired form.	Write each percent as a fraction in simplest form and as a decimal. **a.** $42\% = \dfrac{42}{100} = \dfrac{21}{50} = 0.42$ **b.** $8.5\% = \dfrac{8.5}{100} = \dfrac{85}{1000} = \dfrac{17}{200} = 0.085$ **c.** $20\frac{1}{2}\% = \dfrac{20\frac{1}{2}}{100} = \dfrac{41}{200} = 0.205$

Exercises 57–60   Expressions

[2.7] *For Exercises 57–60, write each percent as a decimal and as a fraction in simplest form.*

57. 15%

$0.15, \dfrac{3}{20}$

58. 82.5%

$0.825, \dfrac{33}{40}$

59. $12\frac{1}{2}\%$

$0.125, \dfrac{1}{8}$

60. $33\frac{1}{3}\%$

$0.\overline{3}, \dfrac{1}{3}$

Definitions/Procedures/Rules	Key Example(s)
To write a fraction or decimal as a percent: **1.** Multiply by 100%. **2.** Simplify.	Write as a percent. **a.** $0.453 = 0.453 \cdot 100\% = 45.3\%$ **b.** $\dfrac{5}{8} = \dfrac{5}{8} \cdot \dfrac{\overset{25}{100}}{1}\% = \dfrac{125}{2}\% = 62\frac{1}{2}\%$

Exercises 61–64   Expressions

[2.7] *For Exercises 61–64, write as a percent.*

61. $\dfrac{2}{5}$

40%

62. $\dfrac{4}{11}$

$36\frac{4}{11}\%$ or $36.\overline{36}\%$

63. 0.35

35%

64. 2.016

201.6%

Definitions/Procedures/Rules	Key Example(s)
To translate simple percent sentences:	15% of what number is 33?
Method 1. Translate the sentence word for word.	**Method 1:** Direct translation
1. Select a(n) <u>variable</u> for the unknown.	$0.15n = 33$
2. Translate the word *is* to a(n) <u>equal sign</u> .	$\dfrac{0.15n}{0.15} = \dfrac{33}{0.15}$
3. If *of* is preceded by the percent, translate it to <u>multiplication</u> . If *of* is preceded by a whole number, translate it to <u>division</u> .	$n = 220$
Method 2. Translate to a proportion by writing the following form.	**Method 2:** Proportion
Percent $= \dfrac{\text{Part}}{\text{Whole}},$	$\dfrac{15}{100} = \dfrac{33}{n}$
where the percent is expressed as a fraction with a denominator of 100.	$15n = 3300$
	$\dfrac{15n}{15} = \dfrac{3300}{15}$
	$n = 220$

Exercises 65–72   Equations and Inequalities

[2.7] *For Exercises 65–68, translate the percent sentence to an equation; then solve.*

65. 16% of 91 is what number?
14.56

66. 14.5% of what number is 42?
≈ 289.7

67. 6.5 is what percent of 20?
32.5%

68. What percent is 18 out of 32?
56.25%

[2.7] *For Exercises 69–72, solve.*

69. In a Gallup poll, 47% of those polled knew someone who lost a job during the past year. If 1016 Americans were polled, how many knew someone who lost a job?
478

70. Karen buys a new kitchen table for $759.99. If the tax rate in her area is 6%, what is the cost of the table after tax is added?
$805.59

71. In 2007, the Boston Red Sox won 96 out of 162 games. What percent of its games did the Red Sox win?
≈ 59.3%

72. In 1995, cotton farmers received an average price for upland cotton of 75.4 cents per pound. In 2006, the average price of upland cotton was 48.2. What was the percent of the decrease in average price? (*Source:* National Agricultural Statistics Services, U.S. Department of Agriculture.)
≈ 36.1%

2.8 Solving Linear Inequalities

Definitions/Procedures/Rules	Key Example(s)

A **linear inequality** is an inequality containing expressions in which each variable term contains a single variable with an exponent of ___1___ .

To **graph an inequality** on a number line:

1. If the symbol is ≤ or ≥, at the indicated number draw a(n) __bracket (or solid circle)__ on the number line that opens to the __left__ for ≤ and to the __right__ for ≥. If the symbol is < or >, at the indicated number draw a(n) __parenthesis (or open circle)__ on the number line that opens to the __left__ for < and to the __right__ for >.

2. If the variable is greater than the indicated number, shade to the __right of the indicated number__ . If the variable is less than the indicated number, shade to the __left of the indicated number__ .

To solve linear inequalities:

1. Simplify both sides of the inequality as needed.
   a. Distribute to eliminate __parentheses__ .
   b. Eliminate fractions or decimals by multiplying through by the __LCD__ just as we did for equations. (Eliminating fractions and decimals is optional.)
   c. __Combine__ like terms.

2. Use the addition principle so that all variable terms are on __one side of the inequality__ and all constants are on __the other side__ . Then combine like terms. (Remember, eliminating the term with the lesser coefficient results in a positive coefficient.)

3. Use the multiplication principle to eliminate any remaining coefficients. If you multiply (or divide) both sides by a negative number, __reverse the direction__ of the inequality symbol.

Write the solution set in set-builder notation and interval notation; then graph the solution set.

a. $x \leq 6$

Set-builder notation: $\{x \mid x \leq 6\}$

Interval notation: $(-\infty, 6]$

Graph:

$$0 \quad 1 \quad 2 \quad 3 \quad 4 \quad 5 \quad 6 \quad 7$$

b. $x > -2$

Set-builder notation: $\{x \mid x > -2\}$

Interval notation: $(-2, \infty)$

Graph:

$$-3 \quad -2 \quad -1 \quad 0 \quad 1 \quad 2 \quad 3 \quad 4$$

c. $-3 \leq x < 1$

Set-builder notation: $\{x \mid -3 \leq x < 1\}$

Interval notation: $[-3, 1)$

Graph:

$$-5 \quad -4 \quad -3 \quad -2 \quad -1 \quad 0 \quad 1 \quad 2$$

Solve and write the solution set in set-builder notation and interval notation. Then graph the solution set.

a. $-6x > 24$

$$\frac{-6x}{-6} < \frac{24}{-6} \qquad \text{Reverse the direction of the inequality symbol.}$$

$$x < -4$$

Set-builder notation: $\{x \mid x < -4\}$

Interval notation: $(-\infty, -4)$

Graph:

$$-7 \quad -6 \quad -5 \quad -4 \quad -3 \quad -2 \quad -1 \quad 0$$

b. $8y - 2 \geq 5y + 7$

$$8y - 2 \geq 5y + 7$$
$$\underline{-5y \qquad\quad -5y}$$
$$3y - 2 \geq 0 + 7$$
$$3y - 2 \geq 7$$
$$\underline{+2 \quad +2}$$
$$3y + 0 \geq 9$$
$$\frac{3y}{3} \geq \frac{9}{3}$$
$$y \geq 3$$

Set-builder notation: $\{y \mid y \geq 3\}$

Interval notation: $[3, \infty)$

Graph:

$$-1 \quad 0 \quad 1 \quad 2 \quad 3 \quad 4 \quad 5 \quad 6$$

Exercises 73–84   ▲   **Equations and Inequalities**

[2.8] *For Exercises 73–78:*   a.   *Solve.*
b.   *Write the solution set in set-builder notation.*
c.   *Write the solution set in interval notation.*
d.   *Graph the solution set.*

73. $-6x > 18$
a. $x < -3$ b. $\{x|x < -3\}$ c. $(-\infty, -3)$
d.

74. $-3x + 2 \leq 5$
a. $x \geq -1$ b. $\{x|x \geq -1\}$ c. $[-1, \infty)$
d.

75. $1 - 2z + 4 \leq -3(z + 1) + 7$
a. $z \leq -1$ b. $\{z|z \leq -1\}$ c. $(-\infty, -1]$
d.

76. $\frac{1}{4}v < 3 + v$
a. $v > -4$ b. $\{v|v > -4\}$ c. $(-4, \infty)$
d.

77. $\frac{1}{2}(m - 1) > 5 - \frac{7}{2}$
a. $m > 4$ b. $\{m|m > 4\}$ c. $(4, \infty)$
d.

78. $2c - (5c - 7) + 4c \geq 8 - 10$
a. $c \geq -9$ b. $\{c|c \geq -9\}$ c. $[-9, \infty)$
d.

[2.8] *For Exercises 79–82, translate to an inequality and solve.*

79. Thirteen is greater than three minus ten times a number.
$13 > 3 - 10p$
$p > -1$

80. Three increased by twice a number is less than three times the difference of a number and five.
$3 + 2z < 3(z - 5)$
$z > 18$

81. Negative one-half of a number is greater than or equal to four.
$-\frac{1}{2}x \geq 4$
$x \leq -8$

82. Negative two is less than or equal to one minus one-fourth of a number.
$-2 \leq 1 - \frac{1}{4}k$
$k \leq 12$

83. The cost of border material limits the circumference of a circular flower garden to 10 feet. What is the range of values for the diameter of the garden? (Use $\pi \approx 3.14$.)
$d \leq 3.18$ ft.

84. A company has a first-quarter profit goal of at least $350,000. If the projected cost is $475,000, what range of values can revenue have to meet the profit goal?
$R \geq \$825,000$

Learning Strategy

Practice, practice, practice makes perfect. The best way is to consistently understand the material well over the course of the semester. If we put in our effort over the long term, it benefits us over the long term with a much greater depth of understanding. If we put in effort over only a short time before an exam, it benefits us for the exam, but we can't usually remember it much longer than that.

—Zheng Alick Z.

Chapter 2 Practice Test

For Extra Help

Step-by-step test solutions are found on the Chapter Test Prep Videos available in **MyMathLab®** *or on* **YouTube**.

1. Is $5m - 14 = 21$ an expression or an equation? Why?
Equation because there is an equal sign. [2.1]

2. Is $5y + 14 = y^2 - 9$ linear or nonlinear? Why?
Nonlinear because there is a variable raised to an exponent other than 1. [2.2]

3. Check to see if 3 is a solution for $3x^2 + x = 10x$.
yes [2.1]

Learning Strategy

Because your actual test most likely will not have the same problems as the Practice Test, it is important that you understand how to solve each type of problem. Before taking the Practice Test, write the definition, rule, procedure, or formula that applies to each problem. This will help you focus on how to solve each type. Once you've done this, take a break and then take the test. Repeat this process until every problem on the Practice Test seems easy to solve.

For Exercises 4–7, solve and check.

4. $-5.6 + 8a = 7a + 7.6$

13.2 [2.2]

5. $8 + 5(x - 3) = 9x - (6x + 1)$

3 [2.3]

6. $1 - 3(c + 4) + 8c = 1 - 42 + 5(c + 7)$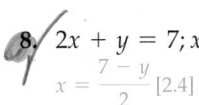

No solution [2.3]

7. $-\dfrac{1}{8}(6m - 32) = \dfrac{3}{4}m - 2$

4 [2.3]

For Exercises 8 and 9, solve for the indicated variable.

8. $2x + y = 7; x$

$x = \dfrac{7 - y}{2}$ [2.4]

9. $A = \dfrac{1}{2}bh; h$

$h = \dfrac{2A}{b}$ [2.4]

10. Solve for the missing number. $\dfrac{m}{8} = \dfrac{5}{12}$

$3\dfrac{1}{3}$ [2.6]

For Exercises 11 and 12, write each percent as a decimal and as a fraction in simplest form.

11. 22%

$0.22, \dfrac{11}{50}$ [2.7]

12. $3\dfrac{1}{3}\%$

$0.0\overline{3}, \dfrac{1}{30}$ [2.7]

For Exercises 13 and 14, write as a percent.

13. 3.2

320% [2.7]

14. $\dfrac{2}{5}$

40% [2.7]

For Exercises 15 and 16, write the solution set in set-builder notation and interval notation; then graph the solution set.

15. $x \geq 3$

Set-builder notation: $\{x \mid x \geq 3\}$
Interval notation: $[3, \infty)$
Graph: 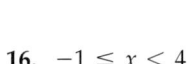 [2.8]

16. $-1 \leq x < 4$

Set-builder notation: $\{x \mid -1 \leq x < 4\}$
Interval notation: $[-1, 4)$
Graph: 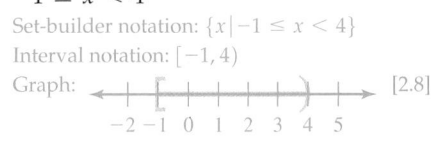 [2.8]

For Exercises 17–20:   a. *Solve.*
       b. *Write the solution set in set-builder notation.*
       c. *Write the solution set in interval notation.*
       d. *Graph the solution set.*

17. $-5m + 3 > -12$

a. $m < 3$ b. $\{m \mid m < 3\}$ c. $(-\infty, 3)$
d. 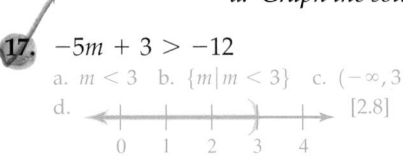 [2.8]

18. $-2(2 - x) \geq -20$

a. $x \geq -8$ b. $\{x \mid x \geq -8\}$ c. $[-8, \infty)$
d. 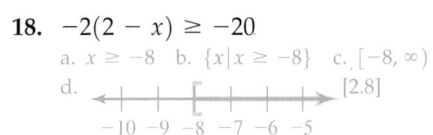 [2.8]

19. $6(p - 2) - 5 > 3 - 4(p + 6)$

a. $p > -\dfrac{2}{5}$ b. $\left\{p \mid p > -\dfrac{2}{5}\right\}$ c. $\left(-\dfrac{2}{5}, \infty\right)$
d. 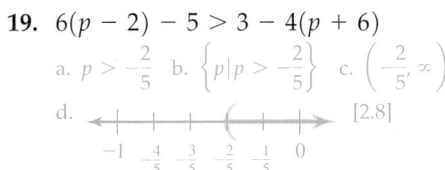 [2.8]

20. $\dfrac{1}{5}(l + 10) \geq \dfrac{1}{2}(l + 5)$

a. $l \leq -\dfrac{5}{3}$ b. $\left\{l \mid l \leq -\dfrac{5}{3}\right\}$ c. $\left(-\infty, -\dfrac{5}{3}\right]$
d. 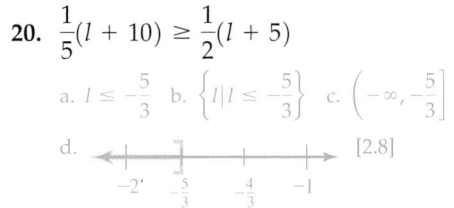 [2.8]

For Exercises 21–25, translate and then solve.

21. Two-thirds of a number minus one-sixth is the same as twice the number.

$$\frac{2}{3}n - \frac{1}{6} = 2n$$

$$n = -\frac{1}{8} \ [2.5]$$

22. Three subtracted from five times the difference of a number and two is equal to ten minus four times the difference of the number and one.

$$5(n - 2) - 3 = 10 - 4(n - 1)$$

$$n = 3 \ [2.5]$$

23. 12 is what percent of 60?
20% [2.7]

24. 70 is 40% of what number?
175 [2.7]

25. One minus a number is greater than twice the number.

$$1 - n > 2n$$

$$n < \frac{1}{3} \ [2.8]$$

For Exercises 26–34, solve.

26. A carpenter needs to determine the amount of wood required to trim a window. If the window is 4 feet by 6 feet, what is the perimeter of the window?
20 ft. [2.1]

27. Find the area of the figure shown.
68.25 in.² [2.1]

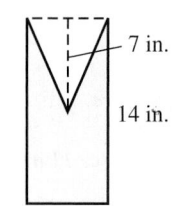

7 in.

14 in.

6.5 in.

28. A rent-to-own store advertises a television for $10 down with $10 payments each week. The formula $C = 10 + 10w$ represents the cost, C, after making payments for w weeks. If the customer signs a contract for one year, what is the total cost of the television?
$530 [2.1]

29. The roof of a building rises 8 inches for every 20 inches of horizontal length. Write the ratio of rise to horizontal length as a fraction in simplest form.
$\frac{2}{5}$ [2.6]

30. A 14.5-ounce box of cereal costs $3.77. Write the unit ratio of price to weight. Interpret the answer.
$\frac{0.26}{1}$; 1 ounce of cereal costs 26 cents. [2.6]

31. The following figures are similar. Find the missing side length.
12.96 cm [2.6]

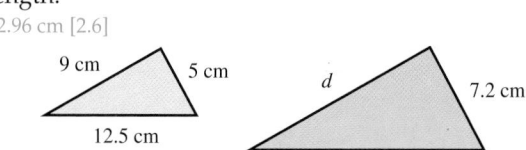

9 cm   5 cm d 7.2 cm

12.5 cm   18 cm

32. According to a survey conducted by the National Center for Health, 37.1% of Americans say that they are in excellent health. Based on these findings, in a group of 500 Americans, how many would say that they are in excellent health? (*Source:* National Center for Health Statistics.)
186 [2.7]

33. In March, Andrea notes that her electric bill is $148.40. In April, her bill is $166.95. What is the percent of the increase in her bill?
12.5% [2.7]

34. In Dan's psychology class, if his average on five tests is at least 92, he can be exempt from the final exam. His current scores are 90, 84, 89, and 96. What range of scores on the last test will allow him to be exempt from the final exam?
$x \geq 101$ [2.8]

Chapters 1–2 Cumulative Review Exercises

For Exercises 1–4, answer true or false.

[1.1] 1. $\sqrt{2}$ is an irrational number.

true

[1.5] 2. When evaluating an expression using the order of operations, you must always do multiplication before division.

false

[1.6] 3. "Four less than a number" can be translated to $4 - n$.

false

[2.2] 4. The equation $2(x - 1) - 3x = 5 - x - 7$ is an identity.

true

For Exercises 5 and 6, fill in the blank.

[1.2] 5. A prime number is a natural number that has exactly two different factors, 1 and __itself__.

[2.3] 6. To eliminate fractions from an equation, we can use the multiplication principle of equality and multiply both sides by the __LCD__.

[1.2] 7. Find the prime factorization of 120.

$2^3 \cdot 3 \cdot 5$

[1.7] 8. Determine the values that make $\dfrac{4x}{x - 3}$ undefined.

3

[1.7] 9. Which property of arithmetic is illustrated by $4(x + 3) = 4x + 12$?

Distributive property

[2.6] 10. Explain in your own words how to solve a proportion.

Set the cross products equal to each other and solve for the variable.

Exercises 11–17 **Expressions**

[1.5] *For Exercises 11–16, simplify.*

11. $\left(-\dfrac{2}{3}\right)^2$

$\dfrac{4}{9}$

12. $\sqrt{16} + \sqrt{9}$

7

13. $\dfrac{5 - 2(-1)^2}{3\sqrt{64} + 36}$

$\dfrac{1}{10}$

14. $-|3 \cdot (-5)| + |(-1)^3|$

-14

15. $2\left(-\dfrac{1}{2}\right)^2 - 4\left(\dfrac{3}{5}\right) + 3$

$\dfrac{11}{10}$

16. $-6 - 2(3 + 0.2)^2$

-26.48

[1.7] 17. Use the distributive property to simplify $-3\left(\dfrac{1}{6}x - 9\right)$.

$-\dfrac{1}{2}x + 27$

Exercises 18–30 **Equations and Inequalities**

[2.2 and 2.3] *For Exercises 18–21, solve and check.*

18. $3.23c - 8.75 = 1.41c + 7.63$

9

19. $2m + \dfrac{1}{2} = 4m - 3\dfrac{1}{2} + 8m$

$\dfrac{2}{5}$

20. $6x - 2 = 3\left(2x - \dfrac{2}{3}\right)$

All real numbers.

21. $-7 = -6(p + 2) + 5(2p - 3)$

5

[2.4] *For Exercises 22 and 23, solve for the indicated variable.*

22. $A = \dfrac{1}{2}bh$ for b

$b = \dfrac{2A}{h}$

23. $P = a + b + c$ for c

$c = P - a - b$

[2.8] *For Exercises 24 and 25:   a. Solve.*
b. Write the solution set in set-builder notation
c. Write the solution set in interval notation.
d. Graph the solution set.

24. $-3h - 8 \leq 19$
a. $h \geq -9$ b. $\{h | h \geq -9\}$ c. $[-9, \infty)$
d.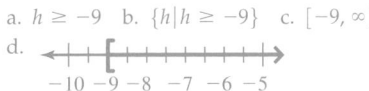

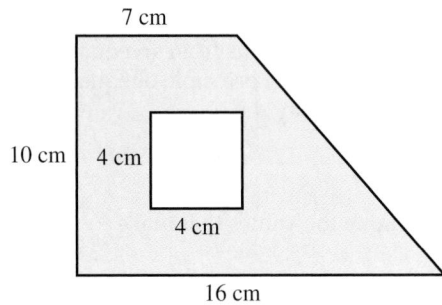. $9m - (2m + 3) < 4(m - 5) + 2$
a. $m < -5$ b. $\{m | m < -5\}$ c. $(-\infty, -5)$
d.

For Exercises 26–27, solve.

[2.1] 26. Find the area of the shaded region in the figure.
99 cm²

7 cm

10 cm   4 cm

4 cm

16 cm

[1.7] 27. The formula $h = -16t^2 + s_0$ describes the height in feet of a falling object, where s_0 represents the initial height from the ground and t represents the number of seconds after the object is dropped. If an object is dropped from 2400 feet, what will be its height after falling 8 seconds?
1376 ft.

For Exercises 28–30, solve.

[2.7] What is 15% of 42.3?
6.345

[2.8] 29. Jon scored an 82 on his first test and a 70 on his second test. What range of scores does he need on his third test to have an average of at least 80?
$t \geq 88$

[2.6] . If a car travels 120 miles on 4 gallons of gasoline, how far can it travel on 10 gallons?
300 mi.

Chapter Overview

In this chapter, we learn about the graphs that correspond to linear equations or inequalities. Graphs can be helpful in solving problems because they provide a visual representation of all solutions to an equation or inequality. The main goals are as follows:

▶ Graph linear equations or inequalities.

▶ Write the equation of a line given information about that line.

▶ Learn about special relations, called functions, that also relate to graphing.

3.1 The Rectangular Coordinate System

3.2 Graphing Linear Equations

3.3 Graphing Using Intercepts

3.4 Slope–Intercept Form

3.5 Point–Slope Form

3.6 Graphing Linear Inequalities

3.7 Introduction to Functions and Function Notation

Instructor Note

This chapter is intended to flow from general techniques to more specific techniques for graphing linear equations. Section 3.4 explains how to graph using the slope, as well as how to find the slope given two points on a line. Section 3.5 is about writing equations of lines given information about the line. The introduction to functions in the last section sets the stage for discussion of functions in future chapters. Although set relations, table relations, and general graphs are explored, the student is required to graph only linear functions. In future chapters, as various types of equations are explored, the corresponding functions will also be explored.

3.1 The Rectangular Coordinate System

Objectives

1 Determine the coordinates of a given point.

2 Plot points in the coordinate plane.

3 Determine the quadrant for a given ordered pair.

4 Determine whether the graph of a set of data points is linear.

Warm-up

[1.1] *Graph each of the following on the number line.*

1. $\{-5, -4, -1, 0\}$

2. $3\dfrac{3}{4}$

Objective 1 Determine the coordinates of a given point.

In 1619, René Descartes, the French philosopher and mathematician, recognized that positions of points in a plane could be described using two number lines that intersect at a right angle. Each number line is called an *axis*.

Two perpendicular axes form the *rectangular*, or Cartesian, *coordinate system*, named in honor of René Descartes. Usually, we call the horizontal axis the *x-axis* and the vertical axis the *y-axis*.

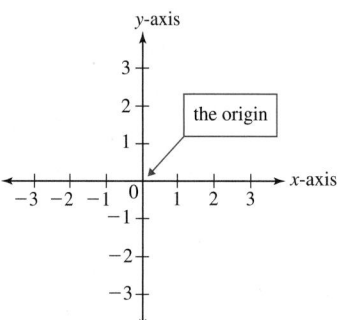

The point where the axes intersect is 0 for both the *x*-axis and *y*-axis. This position is called the *origin*. The positive numbers are to the right and up from the origin, whereas negative numbers are to the left and down from the origin.

Learning Strategy

To prepare for class, read the section being covered before class so that the information doesn't look foreign. Make sure you understand past information because math is like building blocks.

—Judah G.

Any point in the plane can be described using two numbers, one number from each axis. To avoid confusion, these two numbers are written in a specific order. The number representing a point's *horizontal distance* from the origin is given *first*, and the number representing a point's *vertical distance* from the origin is given *second*. Because the order in which we say or write these two numbers matters, we say that they form an *ordered pair*. Each number in an ordered pair is called a *coordinate* of the ordered pair. The notation for writing ordered pairs is (horizontal coordinate, vertical coordinate).

Consider the point labeled *A* in the coordinate plane shown. The point is drawn at the intersection of the 3rd line to the right of the origin and the 4th line up from the origin. The ordered pair that describes point *A* is $(3, 4)$.

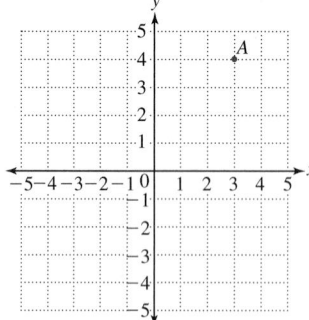

Connection Think of the intersecting lines in the grid as avenues and streets in a city such as New York, where avenues are north–south and streets are east–west. To describe an intersection, a person would say the avenue first, then the street. Point *A* is at the intersection of 3rd Avenue and 4th Street, or, as a New Yorker would say, "3rd and 4th."

Procedure Identifying the Coordinates of a Point

To determine the coordinates of a given point in the rectangular system:

1. Follow a vertical line from the point to the *x*-axis (horizontal axis). The number at this position on the *x*-axis is the first coordinate.
2. Follow a horizontal line from the point to the *y*-axis (vertical axis). The number at this position on the *y*-axis is the second coordinate.

Answers to Warm-up

1.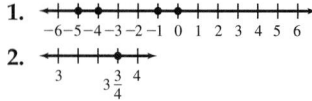

2.

Example 1 Write the coordinates for each point shown.

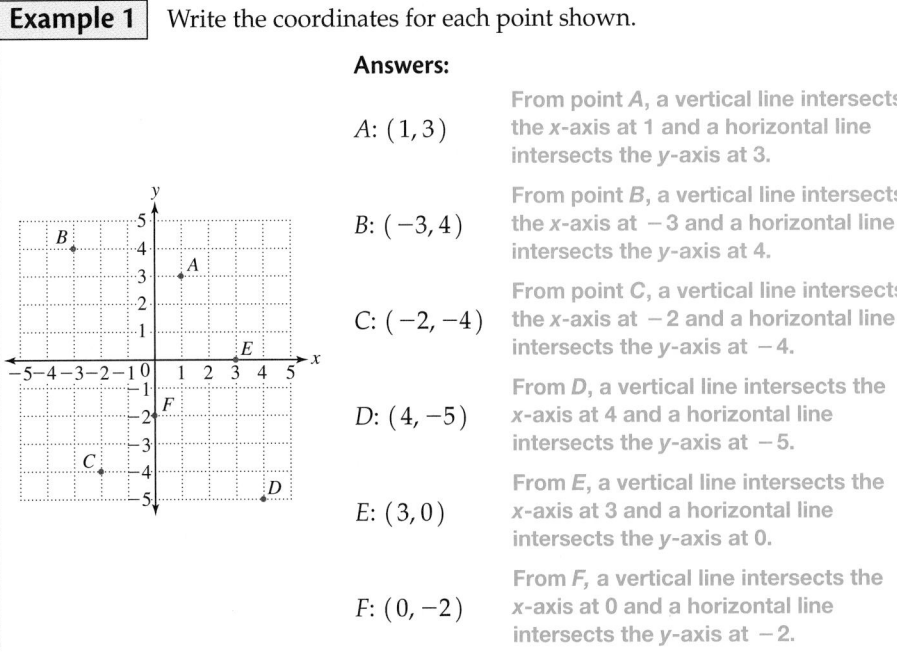

Answers:

$A: (1, 3)$

From point *A*, a vertical line intersects the *x*-axis at 1 and a horizontal line intersects the *y*-axis at 3.

$B: (-3, 4)$

From point *B*, a vertical line intersects the *x*-axis at -3 and a horizontal line intersects the *y*-axis at 4.

$C: (-2, -4)$

From point *C*, a vertical line intersects the *x*-axis at -2 and a horizontal line intersects the *y*-axis at -4.

$D: (4, -5)$

From *D*, a vertical line intersects the *x*-axis at 4 and a horizontal line intersects the *y*-axis at -5.

$E: (3, 0)$

From *E*, a vertical line intersects the *x*-axis at 3 and a horizontal line intersects the *y*-axis at 0.

$F: (0, -2)$

From *F*, a vertical line intersects the *x*-axis at 0 and a horizontal line intersects the *y*-axis at -2.

Your Turn 1 Write the coordinates for each point shown.

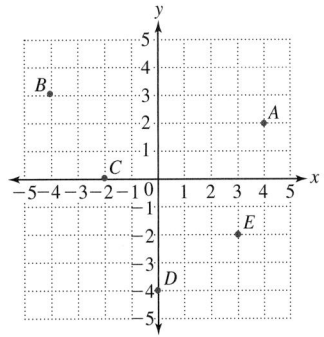

Objective 2 Plot points in the coordinate plane.

Any point can be plotted in the coordinate plane using its coordinates. Remember that the first coordinate in an ordered pair indicates the distance to move right or left and that the second coordinate indicates the distance to move up or down. To plot the ordered pair $(3, 4)$, we begin at the origin, move to the right 3, move up 4, and draw a dot to indicate the point.

> ### Procedure Plotting a Point
>
> To graph or plot a point given its coordinates:
> 1. Beginning at the origin, $(0, 0)$, move to the right or left along the *x*-axis the amount indicated by the first coordinate.
> 2. From that position on the *x*-axis, move up or down the amount indicated by the second coordinate.
> 3. Draw a dot to represent the point described by the coordinates.

Answers to Your Turn 1

$A: (4, 2)$ $B: (-4, 3)$
$C: (-2, 0)$ $D: (0, -4)$
$E: (3, -2)$

Example 2 Plot the point described by the coordinates.

a. $(3, -4)$ **b.** $(-4, 2)$ **c.** $(0, 5)$

Learning Strategy

If you are a tactile learner, when plotting a point, move your pencil along the x-axis first; then move up or down to the point location, just as we've done with the arrows.

Solution:

Note For $(-4, 2)$, we begin at the origin and move to the left 4, then up 2. ▶

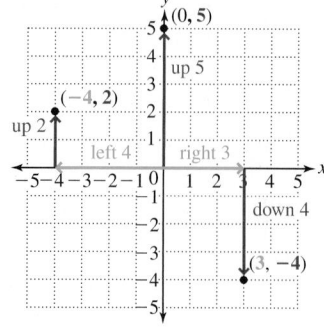

◀ **Note** For $(0, 5)$, because the first coordinate is 0, we do not move right or left. Because the second coordinate is 5, we move straight up 5, drawing the point on the y-axis.

◀ **Note** For $(3, -4)$, we begin at the origin and move to the right 3, then down 4.

Your Turn 2 Plot the point described by the coordinates.

$A: (-3, 2)$ $\qquad$ $B: (1, 4)$ $\qquad$ $C: (-2, -4)$ $\qquad$ $D: (3, 0)$

Objective 3 Determine the quadrant for a given ordered pair.

The two perpendicular axes divide the coordinate plane into four regions called *quadrants*.

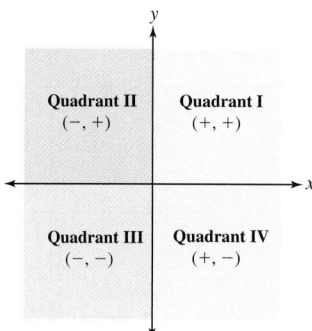

The quadrants are numbered using Roman numerals, as shown to the left. Note that the signs of the coordinates determine the quadrant in which a point lies. For points in quadrant I, both coordinates are positive. Every point in quadrant II has a negative first coordinate (horizontal) and a positive second coordinate (vertical). Points in quadrant III have both coordinates negative. Every point in quadrant IV has a positive first coordinate and a negative second coordinate. Points on axes are not in any quadrant.

Procedure **Identifying Quadrants**

To determine the quadrant for a given ordered pair, consider the signs of the coordinates.

$(+, +)$ means the point is in quadrant I.
$(-, +)$ means the point is in quadrant II.
$(-, -)$ means the point is in quadrant III.
$(+, -)$ means the point is in quadrant IV.

Note: If either coordinate is 0, the point is on an axis and not in a quadrant.

Answer to Your Turn 2

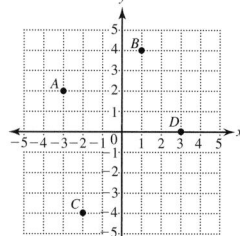

Example 3 State the quadrant in which each point is located. If the point is on an axis, state which axis.

a. $(-61, 23)$

Answer: Quadrant II because the first coordinate is negative and the second coordinate is positive.

b. $\left(14, -37\frac{2}{3}\right)$

Answer: Quadrant IV because the first coordinate is positive and the second coordinate is negative.

c. $(0, -12)$

Answer: Because the x-coordinate is 0, this point is on the y-axis and is not in a quadrant.

Your Turn 3 State the quadrant in which each point is located. If the point is on an axis, state which axis.

a. $(-42, -109)$ **b.** $(37, -15.9)$ **c.** $(4.7, 0)$ **d.** $\left(-16\frac{3}{4}, 124\right)$ **e.** $\left(4.3, 1\frac{7}{8}\right)$

Objective 4 Determine whether the graph of a set of data points is linear.

In many problems, data are listed as ordered pairs. For example, in a report on stock prices, we might have the listing in the margin of the closing price of a particular stock for each day over four days.

Day	Closing Price
1	$24.50
2	$25.75
3	$27.00
4	$28.25

Note Because the vertical axis has dollar values that involve fractions, there are two marks for every whole dollar. This allows halves to be plotted accurately and quarters to be estimated well.

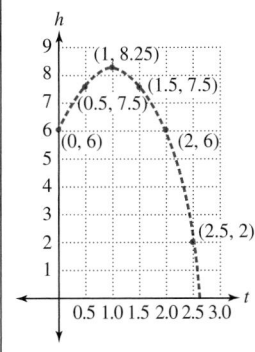

If we plot each pair of data as an ordered pair in the form (day, closing price), we see that the points lie on a straight line.

Because the points can be connected to form a straight line, they are said to be *linear*. Points that do not form a straight line are *nonlinear*.

Note The broken line in the vertical axis indicates that values between 0 and 24 were skipped to conserve space.

Time (in seconds)	Height (in feet)
0	6.00
0.5	7.50
1.0	8.25
1.5	7.50
2.0	6.00
2.5	2.00

Note The graph for Example 4 does not have the origin at the center of the graph. It is often necessary to adjust the location of the origin, and consequently the axes, to fit the data. This frequently happens when the values plotted on the vertical and/or horizontal axes are nonnegative.

Example 4 The data points in the margin track the trajectory, or path, of an object over a period of time. Plot the points with the time along the horizontal axis and the height along the vertical axis. Then state whether the trajectory is linear or nonlinear.

Solution: Because the data points do not form a straight line when connected, the trajectory is non linear.

Note In general, when a graph involves time, the time values are plotted along the horizontal axis.

Answers to Your Turn 3
a. III **b.** IV **c.** on the x-axis
d. II **e.** I

Answer to Your Turn 4
The points are linear.

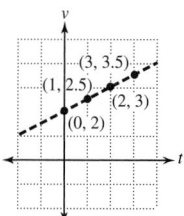

Your Turn 4 The following data points show the velocity of an object as time passes. Plot the points with the time along the horizontal axis and the velocity along the vertical axis. Then state whether the points are linear or nonlinear.

Time (in seconds)	Height (in feet)
0	2.0
1	2.5
2	3.0
3	3.5

3.1 Exercises   For Extra Help   MyMathLab®

Note: Exercises marked with a ★ represent challenging exercises.

Objective 1

Prep Exercise 1 When writing an ordered pair, which is written first, the horizontal-axis coordinate or the vertical-axis coordinate?   Horizontal-axis coordinate

Prep Exercise 2 Describe how to use the coordinates of an ordered pair to plot a point in the rectangular coordinate system.   Beginning at the origin, move to the left or right along the x-axis by the amount indicated by the first coordinate. From that position on the x-axis, move up or down by the amount indicated by the second coordinate.

For Exercises 1–4, write the coordinates for each point. See Example 1.

1.
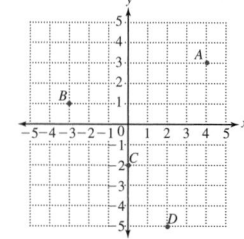
$A\,(4,3),\,B\,(-3,1),\,C\,(0,-2),$
$D\,(2,-5)$

2.
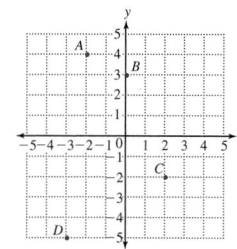
$A\,(-2,4),\,B\,(0,3),\,C\,(2,-2),$
$D\,(-3,-5)$

3.
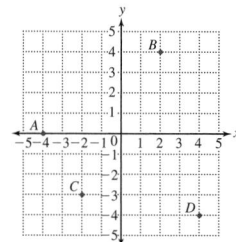
$A\,(-4,0),\,B\,(2,4),\,C\,(-2,-3),$
$D\,(4,-4)$

4.
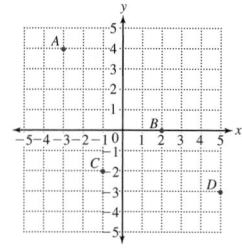
$A\,(-3,4),\,B\,(2,0),\,C\,(-1,-2),$
$D\,(5,-3)$

Objective 2

For Exercises 5–8, on a piece of graph paper, draw and label x- and y-axes. Then plot and label the points indicated by the coordinate pairs. See Example 2.

5. $(5,4),\,(2,-3),\,(-1,-3),\,(2,0)$

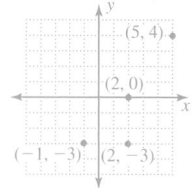

6. $(-3,0),\,(0,1),\,(-2,-2),\,(4,-1)$

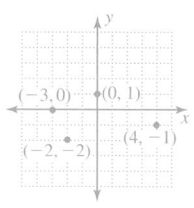

7. $(-2, 0), (3, -3), (-1, -5), (2, 2)$

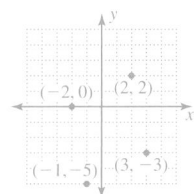

8. $(-1, -4), (0, -2), (4, -2), (-4, 2)$

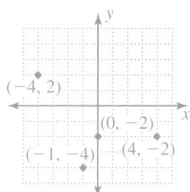

Objective 3

Prep Exercise 3 Draw the axes in the rectangular coordinate system and label the quadrants. Include the signs for coordinates in each quadrant.

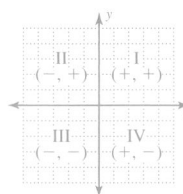

For Exercises 9–20, state the quadrant in which the point is located. If the point lies on an axis, state which axis. See Example 3.

9. $(-6.5, 1050)$
II

10. $\left(-42\frac{1}{3}, -500\right)$
III

11. $(620, 50)$
I

12. $(-42, 50)$
II

13. $(57, -82.71)$
IV

14. $(16, 27)$
I

15. $\left(-41\frac{1}{2}, -82\right)$
III

16. $\left(47, -51\frac{5}{8}\right)$
IV

17. $(0, -9)$
y-axis

18. $(0, 5.3)$
y-axis

19. $(-0.6, 0)$
x-axis

20. $(8.2, 0)$
x-axis

Objective 4

Prep Exercise 4 Describe how to determine whether a set of data points is linear.
Points that can be connected to form a straight line are said to have a *linear* relationship.

For Exercises 21–30, determine whether the set of points is linear or nonlinear. See Example 4.

21. $(-2, -9), (0, -5), (2, -1), (4, 3), (5, 5)$
linear

22. $(-5, 7), (-1, 3), (2, 0), (4, -2), (6, -4)$
linear

23. $(-4, 8), (-2, 2), (0, 0), (1, 0.5), (3, 4.5)$
nonlinear

24. $(-1, -3), (0, 0), (2, 0), (3, -3)$
nonlinear

25. (pounds of chicken, cost): $(1, 2.5), (2, 5), (3, 7.5), (4, 10)$
linear

26. (time, distance): $(0, 2), (1, 5), (2, 8), (3, 11), (4, 14)$
linear

27. (time, height): $(0, 8.0), (0.5, 9.5), (1.0, 11.0), (1.5, 9.5),$
$(2.0, 8.0), (2.5, 6.5)$
nonlinear

28. (days, stock closing price): $(1, 23), (2, 21), (3, 19), (4, 17)$
linear

★ **29.** (copy costs, number of copies): $(15, 300), (18, 400),$
$(21, 500), (24, 600)$
linear

★ **30.** (number of hours, plumbing cost): $(1, 120), (2, 160),$
$(3, 200), (4, 240)$
linear

31. List the coordinates of three points on the line shown.

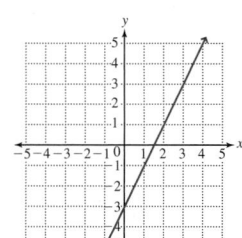

Answers may vary. Some possible answers are $(-1, -5)$, $(0, -3)$, and $(2, 1)$.

32. List the coordinates of three points on the line shown.

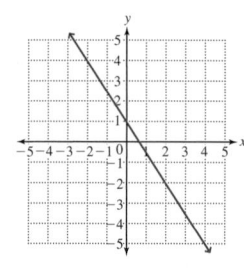

Answers may vary. Some possible answers are $(0, 1)$, $(2, -2)$, and $(4, -5)$.

A vertex on a figure is a point where two line segments form a corner or joint on the figure. For Exercises 33 and 34, find the coordinates of each labeled vertex. See Example 1.

33.

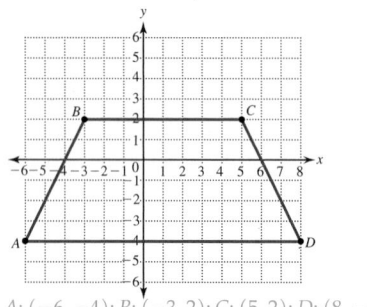

$A: (-6, -4); B: (-3, 2); C: (5, 2); D: (8, -4)$

34.

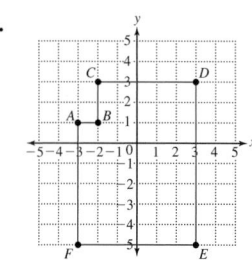

$A: (-3, 1); B: (-2, 1); C: (-2, 3); D: (3, 3); E: (3, -5); F: (-3, -5)$

★ **35.** A parallelogram has been moved from its original position, shown with the dashed lines, to a new position, shown with solid lines.

  a. List the coordinates of each vertex of the parallelogram in its original position.
  $(-3, 1), (2, 1), (1, -3), (-4, -3)$

  b. List the coordinates of each vertex of the parallelogram in its new position.
  $(0, 2), (5, 2), (4, -2), (-1, -2)$

  c. Write a rule that describes in mathematical terms how to move each point on the parallelogram in its original position to each point on the parallelogram in its new position.

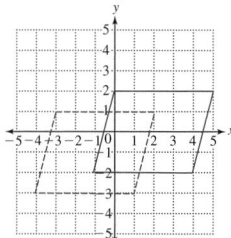

$(x + 3, y + 1)$

★ **36.** A trapezoid has been moved from its original position, shown with the dashed lines, to a new position, shown with solid lines.

  a. List the coordinates of each vertex of the trapezoid in its original position.
  $(-4, 4), (-1, 4), (0, 1), (-5, 1)$

  b. List the coordinates of each vertex of the trapezoid in its new position.
  $(1, 2), (4, 2), (5, -1), (0, -1)$.

  c. Write a rule that describes in mathematical terms how to move each point on the trapezoid in its original position to each point on the trapezoid in its new position.

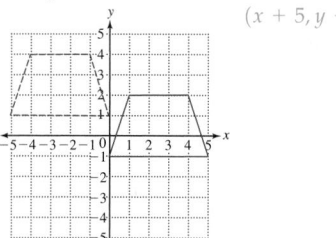

$(x + 5, y - 2)$

37. The ordered pairs $(-3, 1), (-1, 1), (-1, 3), (2, 3), (2, -2)$, and $(-3, -2)$ form the vertices of a figure.

  a. Plot the points and connect them to form the figure.

  b. Find the perimeter of the figure.
  20 units

  c. Find the area of the figure.
  21 square units

a.
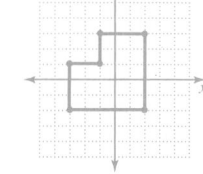

38. The ordered pairs $(-4, 2), (-2, 2), (-2, 4), (1, 4), (1, 3), (3, 3), (3, -4)$, and $(-4, -4)$ form the vertices of a figure.

  a. Plot the points and connect them to form the figure.

  b. Find the perimeter of the figure.
  30 units

  c. Find the area of the figure.
  50 square units

a.
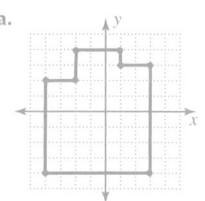

Review Exercises

Exercise 1 ➤ Constants and Variables

[1.1] 1. Graph -3 on a number line.

$$\xleftarrow{\hspace{0.5em}}\underset{-4\ -3\ -2\ -1\ \ 0\ \ \ 1\ \ \ 2}{|\ \ |\ \ |\ \ |\ \ |\ \ |\ \ |}\xrightarrow{\hspace{0.5em}}$$

Exercises 2–4 ➤ Expressions

[1.7] 2. Evaluate $2x - 3y$ when $x = 4$ and $y = 2$.
2

[1.7] 3. Evaluate $-4x - 5y$ when $x = -3$ and $y = 0$.
12

[1.7] 4. Evaluate $-\dfrac{1}{3}x + 2$ when $x = 6$.
0

Exercises 5 and 6 ➤ Equations and Inequalities

[2.3] *For Exercises 5 and 6, solve.*

5. $2x - 5 = 10$
$\dfrac{15}{2}$

6. $-\dfrac{5}{6}x = \dfrac{3}{4}$
$-\dfrac{9}{10}$

3.2 Graphing Linear Equations

Objectives

1 Determine whether a given pair of coordinates is a solution to a given equation with two unknowns.

2 Find solutions for an equation with two unknowns.

3 Graph linear equations.

Warm-up

[1.7] 1. Find the value of $\dfrac{2}{3}x + 1$ if $x = -2$.

[2.3] 2. Solve: $3x - 4 = 2$

Objective 1 Determine whether a given pair of coordinates is a solution to a given equation with two unknowns.

In Chapter 2, we considered linear equations with one variable. We now consider linear equations that have two variables, such as $x + y = 4$ and $y = 2x - 3$.

A solution for an equation with two variables is a pair of numbers, one number for each variable, that can replace the corresponding variables and make the equation true. These solutions are ordered pairs, so we can write them using coordinates. The ordered pair $(1, 3)$ is a solution for $x + y = 4$ because replacing x with 1 and y with 3 makes the equation true.

Note Unless otherwise specified, the coordinates are written in alphabetical order; so $(1, 3)$ means that $x = 1$ and $y = 3$.

$$x + y = 4$$
$$\downarrow \quad \downarrow$$
▶ $1 + 3 = 4$ The equation is true, so the ordered pair is a solution.

> **Procedure** **Checking a Potential Solution for an Equation with Two Variables**
>
> To determine whether a given ordered pair is a solution for an equation with two variables:
> 1. Replace the variables in the equation with the corresponding coordinates.
> 2. Verify that the equation is true.

Answers to Warm-up

1. $-\dfrac{1}{3}$

2. 2

Example 1 Determine whether the ordered pair is a solution for the equation.

a. $(-4, -10); y = 2x - 3$

Solution: $-10 \overset{?}{=} 2(-4) - 3$ Replace x with -4 and y with -10 and see if the equation is true.

$\qquad\qquad -10 \overset{?}{=} -8 - 3$ Simplify.

$\qquad\qquad -10 \neq -11$ Because the equation is not true, $(-4, -10)$ is not a solution for $y = 2x - 3$.

b. $(4, -1); 2m - n = 9$

Solution: $2(4) - (-1) \overset{?}{=} 9$ Unless otherwise specified, ordered pairs are expressed in alphabetical order; therefore, replace m with 4 and n with -1 and see if the equation is true.

$\qquad\qquad 8 + 1 \overset{?}{=} 9$

$\qquad\qquad\qquad 9 = 9$ The equation is true, so $(4, -1)$ is a solution for $2m - n = 9$.

c. $\left(-2, -\dfrac{1}{3}\right); y = \dfrac{2}{3}x + 1$

Solution: $-\dfrac{1}{3} \overset{?}{=} \dfrac{2}{3}(-2) + 1$ Replace x with -2 and y with $-\dfrac{1}{3}$ and see if the equation is true.

$\qquad\qquad -\dfrac{1}{3} \overset{?}{=} -\dfrac{4}{3} + \dfrac{3}{3}$

$\qquad\qquad -\dfrac{1}{3} = -\dfrac{1}{3}$ The equation is true, so $\left(-2, -\dfrac{1}{3}\right)$ is a solution for $y = \dfrac{2}{3}x + 1$.

Instructor Note Ask students to find additional solutions of $2m - n = 9$. After several correct solutions, ask students *how* they found the solutions.

Your Turn 1 Determine whether the ordered pair is a solution for the equation.

a. $(-8, -2); x - 3y = -2$ **b.** $(1, -6); q = -5p + 1$ **c.** $\left(\dfrac{3}{4}, \dfrac{9}{4}\right); y = \dfrac{x}{3} + 2$

Objective 2 Find solutions for an equation with two unknowns.

Consider the equation $x + y = 4$. We have already seen that $(1, 3)$ is a solution. But there are other solutions. We list some of those solutions in the following table.

x	y	Ordered Pair
0	4	$(0, 4)$
1	3	$(1, 3)$
2	2	$(2, 2)$
3	1	$(3, 1)$
4	0	$(4, 0)$
5	−1	$(5, -1)$

◄ **Note** Every solution for $x + y = 4$ is an ordered pair of numbers whose sum is 4.

In fact, $x + y = 4$, and every other linear equation, has an infinite number of solutions. For every x-value, there is a corresponding y-value that will add to the x-value to equal 4 and vice versa, which gives a clue about how to find solutions. We can simply choose a value for x or y and solve for the corresponding value of the other variable. (In this case, our equation was easy enough that we could solve for y mentally.)

> **Procedure** Finding Solutions to Linear Equations with Two Variables
>
> To find a solution to a linear equation with two variables:
> 1. Choose a value for one of the variables (any value).
> 2. Replace the corresponding variable with your chosen value.
> 3. Solve the equation for the value of the other variable.

Answers to Your Turn 1
a. yes   **b.** no   **c.** yes

Instructor Note Point out that three times the x-value plus the y-value must be equal to two for the ordered pair to be a solution of $3x + y = 2$. An equation is nothing more than a restriction placed on the variable(s).

Instructor Note Often, we get fractional values that can be difficult to plot. There is a way of avoiding fractional values if we can find one ordered pair that has integer values for both x and y. If you have an x-value that results in an integer y-value, then that (x-value) + (any multiple of the y-coefficient) = (another x-value that gives an integer y-value). Similarly, if you have a y-value that results in an integer x-value, then that (y-value) + (any multiple of the x-coefficient) = (another y-value that gives an integer x-value).

Connection When an equation is written with y isolated, as in the equation $y = -\dfrac{1}{3}x + 2$ from Example 2(b), we can think of x-values as inputs and the corresponding y-values as outputs. In Example 2(b), the x-input values were 0, 3, and 6 and the corresponding y-output values were 2, 1, and 0. This is the fundamental way of thinking about a mathematical concept called a function, which will be introduced in Section 3.7.

Answers to Your Turn 2
Remember, your solutions may be different.
a. $(0, 2), (4, 1), (8, 0)$
b. $(0, -3), (2, 0), \left(1, -\dfrac{3}{2}\right)$
c. $(0, 5), (2, 6), (4, 7)$

Example 2 | Find three solutions for the equation.

a. $3x + y = 2$

Solution: To find a solution, we replace one of the variables with a chosen value and then solve for the value of the other variable.

For the first solution, we will choose x to be 0.	For the second solution, we will choose x to be 1.	For the third solution, we will choose y to be -4.
$3x + y = 2$	$3x + y = 2$	$3x + y = 2$
$3(0) + y = 2$	$3(1) + y = 2$	$3x + (-4) = 2$
$y = 2$	$3 + y = 2$	$3x = 6$
Solution $(0, 2)$	$y = -1$	$x = 2$

Note Choosing x (or y) to equal 0 usually makes the equation very easy to solve.

Subtract 3 from both sides.

Solution $(1, -1)$

Divide both sides by 3.

Solution $(2, -4)$

We can summarize the solutions in a table.

If we choose x to be 0, then y is equal to 2.
If we choose x to be 1, then y is equal to -1.
If we choose y to be -4, then x is equal to 2.

x	y	Ordered Pair
0	2	$(0, 2)$
1	-1	$(1, -1)$
2	-4	$(2, -4)$

Keep in mind that there are an infinite number of correct solutions for a given equation in two variables; so your solutions may be different from the solutions someone else finds.

b. $y = -\dfrac{1}{3}x + 2$

Solution: Notice that this equation has y isolated. If we select values for x, we won't have to isolate y as we did in part a. We simply calculate the y value. Also notice that the coefficient for x is a fraction. Because we can choose any value for x, let's choose values such as 3 and 6 that will divide out nicely with the denominator of 3.

For the first solution, we will choose x to be 0.	For the second solution, we will choose x to be 3.	For the third solution, we will choose x to be 6.
$y = -\dfrac{1}{3}x + 2$	$y = -\dfrac{1}{3}x + 2$	$y = -\dfrac{1}{3}x + 2$
$y = -\dfrac{1}{3}(0) + 2$	$y = -\dfrac{1}{3}(3) + 2$	$y = -\dfrac{1}{3}(6) + 2$
$y = 2$	$y = -\dfrac{1}{3}\left(\dfrac{\overset{1}{3}}{\underset{1}{1}}\right) + 2$	$y = -\dfrac{1}{3}\left(\dfrac{\overset{2}{6}}{\underset{1}{1}}\right) + 2$
Solution $(0, 2)$	$y = -1 + 2$	$y = -2 + 2$
	$y = 1$	$y = 0$
	Solution $(3, 1)$	**Solution** $(6, 0)$

Solution: In summary:

If we choose x to be 0, then y is equal to 2.
If we choose x to be 3, then y is equal to 1.
If we choose x to be 6, then y is equal to 0.

x	y	Ordered Pair
0	2	$(0, 2)$
3	1	$(3, 1)$
6	0	$(6, 0)$

Your Turn 2 | Find three solutions for each equation. (Answers may vary.)

a. $x + 4y = 8$ **b.** $3x - 2y = 6$ **c.** $y = \dfrac{1}{2}x + 5$

Objective 3 Graph linear equations.

We have learned that equations in two variables have an infinite number of solutions. Because of this fact, there is no way that all solutions to an equation can be found, much less listed. However, all of the solutions can be represented using a graph. Consider some of the solutions for $x + y = 4$, which we listed in the table, on page 198: $(0, 4), (1, 3), (2, 2), (3, 1), (4, 0)$, and $(5, -1)$.

Note The arrows on either end of the line indicate that the solutions continue beyond our "window" in both directions. ▶

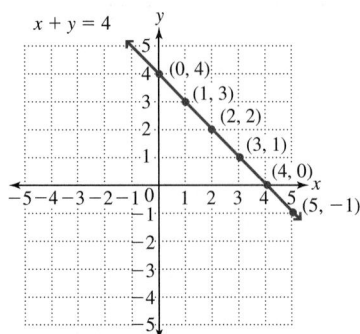

Notice that when we plot our ordered pair solutions for $x + y = 4$, the corresponding points lie in a straight line. In fact, all solutions for $x + y = 4$ are on that same line, which is why we say it is a *linear* equation. By connecting the points, we are graphically representing every possible solution for $x + y = 4$.

The graph of the solutions of every linear equation will be a straight line. Because two points determine a line in a plane, we need a minimum of two ordered pairs. This means that we must find at least two solutions in order to graph the line. However, it is wise to find three solutions, using the third solution as a check. If we plot all three points and they cannot be connected with a straight line, then we know something is wrong.

Procedure **Graphing Linear Equations**

To graph a linear equation:
1. Find at least two solutions to the equation.
2. Plot the solutions as points in the rectangular coordinate system.
3. Connect the points to form a straight line.

Example 3 Graph each equation.

a. $3x + y = 2$

Solution: We found three solutions to this equation in Example 2(a). Those solutions are listed in the following table. We plot each solution as a point in the rectangular coordinate system and then connect the points to form a straight line.

x	y	Ordered Pair
0	2	$(0, 2)$
1	-1	$(1, -1)$
2	-4	$(2, -4)$

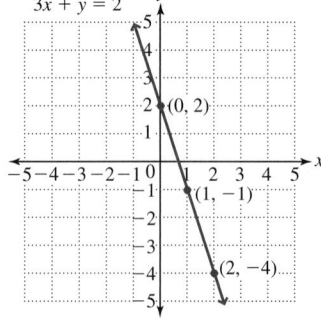

b. $y = -\dfrac{1}{3}x + 2$

Solution: We found three solutions to this equation in Example 2(b). Those solutions are listed in the following table. We plot each solution as a point in the rectangular coordinate system and then connect the points to form a straight line.

x	y	Ordered Pair
0	2	$(0, 2)$
3	1	$(3, 1)$
6	0	$(6, 0)$

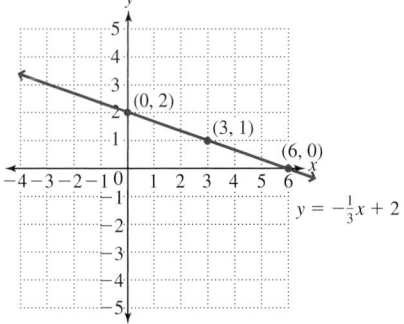

Your Turn 3 Graph each equation.

a. $x + 2y = 4$ **b.** $y = \dfrac{1}{2}x - 1$

The graph of an equation is a visual representation of the entire solution set of the equation. The graph of every ordered pair that solves the equation lies on the line, and the coordinates of every point on the line is a solution of the equation. Looking at the graph of Example 3(b), we see that $(-3, 3)$ are the coordinates of a point on the line; so $(-3, 3)$ is a solution of the equation $y = -\dfrac{1}{3}x + 2$. We verify by substituting and simplifying.

$$y = -\frac{1}{3}x + 2$$

$$3 = -\frac{1}{3}(-3) + 2$$

$$3 = 1 + 2$$

$$3 = 3$$

We have graphed linear equations in two variables. Now let's consider the graphs of equations in one variable, such as $y = 3$ or $x = 4$, in which the variable is equal to a constant.

Example 4 Graph $y = 3$.

Solution: The equation $y = 3$ indicates that y is equal to a constant, 3. To establish ordered pairs, we can rewrite the equation $y = 3$ as $0x + y = 3$. Because the coefficient of x is 0, y is always 3 no matter what we choose for x. We could complete a table of solutions like this:

If we choose x to be 0, then y equals 3.

If we choose x to be 2, then y is 3.

If we choose x to be 4, then y is still 3.

x	y	Ordered Pair
0	3	$(0, 3)$
2	3	$(2, 3)$
4	3	$(4, 3)$

Answers to Your Turn 3

a.

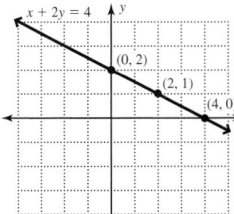

b.

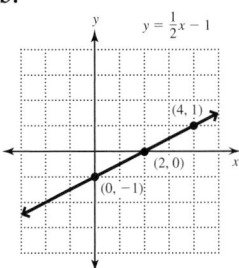

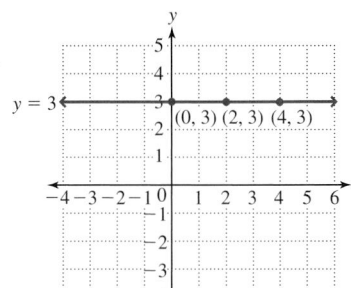

◀ **Note** The graph of $y = 3$ is a horizontal line parallel to the x-axis that passes through the y-axis at a point with coordinates $(0, 3)$.

> **Rule Horizontal Lines**
>
> The graph of $y = c$, where c is a real-number constant, is a horizontal line parallel to the x-axis that passes through the y-axis at a point with coordinates $(0, c)$.

Learning Strategy

To best do your homework, work a practice problem similar to the homework before attempting to do it and keep working on a problem even if it is difficult. Make yourself think about what you know and how to apply it to the problem.

—Stella P.

Answers to Your Turn 5

a.

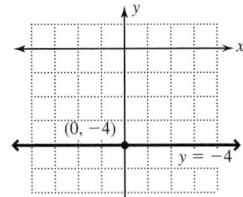

b.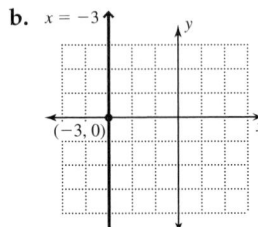

| **Example 5** | Graph $x = 4$. |

Solution: The equation $x = 4$ indicates that x is equal to a constant, 4. If the y-variable is missing from an equation, it means that its coefficient is 0; so we can write $x = 4$ as $x + 0y = 4$. Notice that x is always 4 no matter what we choose for y. We could complete a table of solutions like this:

If we choose y to be 0, then x equals 4.

If we choose y to be 1, then x is 4.

If we choose y to be 2, then x is still 4.

x	y	Ordered Pair
4	0	$(4, 0)$
4	1	$(4, 1)$
4	2	$(4, 2)$

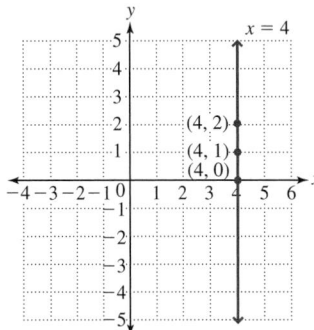

◀ **Note** The graph of $x = 4$ is a vertical line parallel to the y-axis that passes through the x-axis at a point with coordinates $(4, 0)$.

> **Rule Vertical Lines**
>
> The graph of $x = c$, where c is a real-number constant, is a vertical line parallel to the y-axis that passes through the x-axis at a point with coordinates $(c, 0)$.

| **Your Turn 5** | Graph. |

a. $y = -4$ **b.** $x = -3$

| **3.2 Exercises** | For Extra Help | MyMathLab® |

Note: Exercises marked with a ★ represent challenging exercises.

Objective 1

Prep Exercise 1 How do you determine whether an ordered pair is a solution for a given equation?

a. Replace the variables in the equation with the corresponding coordinates.   b. Verify that the equation is true.

Prep Exercise 2 If a given ordered pair is not a solution for a linear equation with two variables, does that mean the equation has no solutions? Explain.

No, because linear equations with two variables have an infinite number of solutions.

For Exercises 1–16, determine whether the given ordered pair is a solution for the equation. See Example 1.

1. $(2, 3); x + 2y = 8$

yes

2. $(3, 1); 2x - y = 5$

yes

3. $(4, 9); 3x - 2y = -6$

yes

4. $(-5, 10); 5x + 3y = 5$

yes

5. $(-5, 2); y - 4x = 3$

no

6. $(4, -6); y - 3x = -1$

no

7. $(9, 0); y = -2x + 18$
yes

8. $(0, -1); y = 2x + 1$
no

9. $(6, -2); y = -\frac{2}{3}x$
no

10. $(-10, -4); y = \frac{2}{5}x$
yes

11. $(-8, -8); y = \frac{3}{4}x - 2$
yes

12. $(-5, 0); y = -\frac{3}{5}x + 3$
no

13. $\left(-1\frac{2}{5}, 0\right); y - 3x = 5$
no

14. $\left(0, -2\frac{1}{3}\right); 4x - 3y = 7$
yes

15. $(2.2, -11.2); y + 6x = -2$
no

16. $(-1.5, -1.3); y = 0.2x - 1$
yes

Objectives 2 and 3

Prep Exercise 3 How do you find a solution for a given equation with two unknowns?
a. Choose a value for one of the variables.   b. Replace the corresponding variable with your chosen value.   c. Solve the equation for the value of the other variable.

Prep Exercise 4 What does the graph of an equation represent?
The graph of an equation represents all of the solutions to the equation.

Prep Exercise 5 What is the minimum number of points needed to draw a straight line? Explain.
A minimum of two ordered pairs are needed because two points determine a line.

Prep Exercise 6 In your own words, explain the process of graphing a linear equation.
a. Find at least two solutions to the equation.   b. Plot the solutions as points in a rectangular coordinate system.   c. Connect the points to form a straight line.

For Exercises 17–52, find three solutions for the given equation. Then graph.
(Answers may vary for the three solutions.) See Examples 2–5.

17. $y = x$
$(-1, -1)$
$(0, 0)$
$(1, 1)$
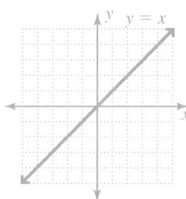

18. $y = -x$
$(-1, 1)$
$(0, 0)$
$(1, -1)$
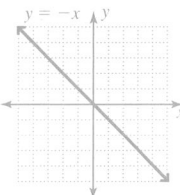

19. $y = 2x$
$(-1, -2)$
$(0, 0)$
$(1, 2)$
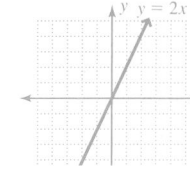

20. $y = 3x$
$(0, 0)$
$(1, 3)$
$(2, 6)$
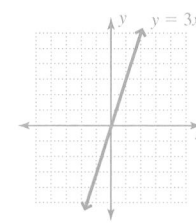

21. $y = -5x$
$(-1, 5)$
$(0, 0)$
$(1, -5)$
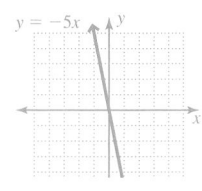

22. $y = -2x$
$(-1, 2)$
$(0, 0)$
$(1, -2)$
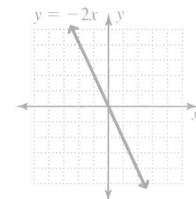

23. $y = x - 3$
$(0, -3)$
$(2, -1)$
$(4, 1)$
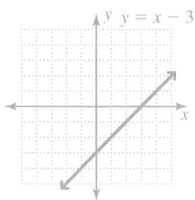

24. $y = x + 4$
$(0, 4)$
$(-2, 2)$
$(2, 6)$
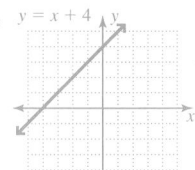

25. $y = -x - 2$
$(0, -2)$
$(4, -6)$
$(-4, 2)$
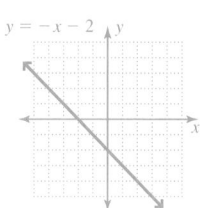

26. $y = -x + 5$
$(-1, 6)$
$(0, 5)$
$(1, 4)$
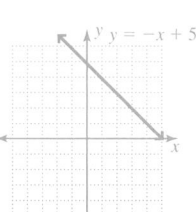

27. $y = 2x - 5$
$(0, -5)$
$(1, -3)$
$(2, -1)$
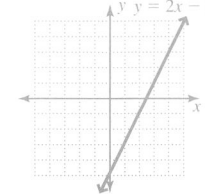

28. $y = 3x + 2$
$(-1, -1)$
$(0, 2)$
$(1, 5)$
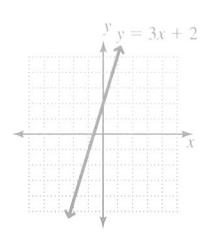

29. $y = -2x + 4$
$(-1, 6)$
$(0, 4)$
$(1, 2)$
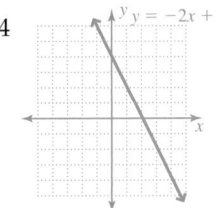

30. $y = -5x - 1$
$(-1, 4)$
$(0, -1)$
$(1, -6)$
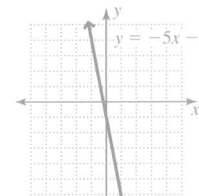

31. $y = \frac{1}{2}x$
$(-2, -1)$
$(0, 0)$
$(2, 1)$
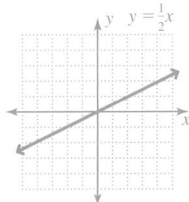

32. $y = \frac{1}{3}x$
$(0, 0)$
$(3, 1)$
$(-3, -1)$
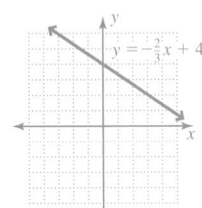

33. $y = -\frac{2}{3}x$
$(0, 0)$
$(-3, 2)$
$(3, -2)$
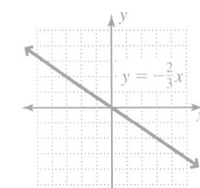

34. $y = -\frac{3}{4}x$
$(-4, 3)$
$(0, 0)$
$(4, -3)$
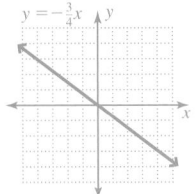

35. $y = -\frac{2}{3}x + 4$
$(-3, 6)$
$(0, 4)$
$(3, 2)$

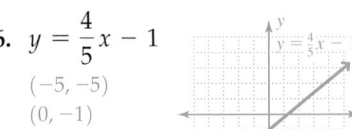

36. $y = \frac{4}{5}x - 1$
$(-5, -5)$
$(0, -1)$
$(5, 3)$
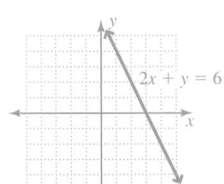

37. $x - y = 8$
$(3, -5)$
$(5, -3)$
$(6, -2)$
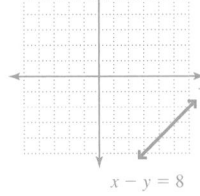

38. $x + y = -5$
$(0, -5)$
$(-2, -3)$
$(-3, -2)$
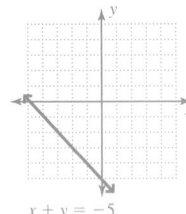

39. $2x + y = 6$
$(1, 4)$
$(3, 0)$
$(5, -4)$
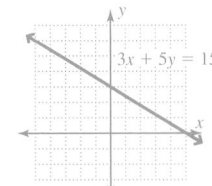

40. $3x - y = 9$
$(1, -6)$
$(2, -3)$
$(3, 0)$
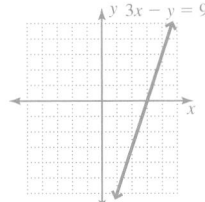

41. $2x + 3y = 12$
$(0, 4)$
$(6, 0)$
$(3, 2)$
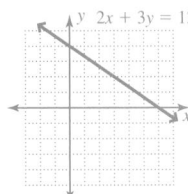

42. $3x + 5y = 15$
$(0, 3)$
$(5, 0)$
$(-5, 6)$
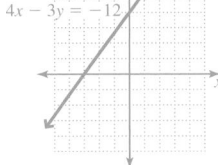

43. $4x - 3y = -12$
$(0, 4)$
$(-3, 0)$
$(3, 8)$
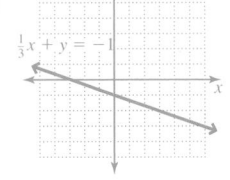

44. $3x - 2y = -6$
$(0, 3)$
$(-2, 0)$
$(-4, -3)$
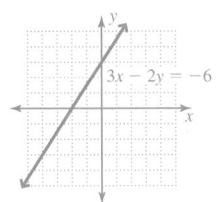

45. $x - \frac{1}{4}y = 2$
$(1, -4)$
$(2, 0)$
$(3, 4)$
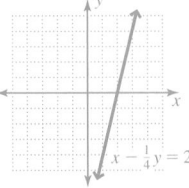

46. $\frac{1}{3}x + y = -1$
$(0, -1)$
$(3, -2)$
$(6, -3)$

47. $y = -5$
$(-1, -5), (0, -5), (1, -5)$
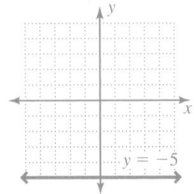

48. $y = 4$
$(-1, 4), (0, 4), (1, 4)$
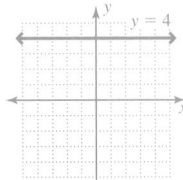

49. $x = 7$
$(7, -1), (7, 0), (7, 1)$
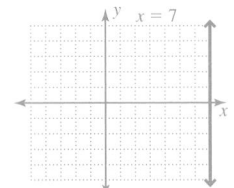

50. $x = -6$

$(-6, -1), (-6, 0), (-6, 1)$

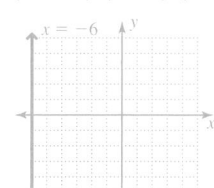

★ **51.** $y = 0.4x - 2.5$

$(-1, -2.9), (0, -2.5), (1, -2.1)$

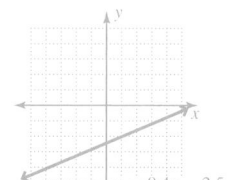

★ **52.** $1.2x + 0.5y = 6$

$(0, 12), (1, 9.6), (5, 0)$

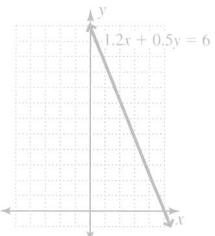

53. Compare the graphs of $y = x$, $y = 2x$, and $y = 3x$ from Exercises 17, 19, and 20. For an equation in the form $y = ax$, where $a > 0$, what effect does increasing a seem to have on the graph?

As a gets larger, the graph gets steeper.

54. Compare the graphs of $y = x$ and $y = -x$ from Exercises 17 and 18; then compare the graphs of $y = 2x$ and $y = -2x$ from Exercises 19 and 22. For an equation in the form $y = ax$, what can you conclude about the graph when a is positive versus when a is negative?

If $a > 0$, the graph increases from left to right. If $a < 0$, the graph decreases from left to right.

55. Compare the graphs of $y = 3x$ and $y = 3x + 2$ from Exercises 20 and 28; then compare the graphs of $y = -2x$ and $y = -2x + 4$ from Exercises 22 and 29. For an equation in the form $y = ax + b$, where $b > 0$, what effect does adding b to $y = ax$ seem to have on the graph?

The graph will shift up b units on the y-axis.

56. Compare the graphs of $y = x$ and $y = x - 3$ from Exercises 17 and 23; then compare the graphs of $y = -5x$ and $y = -5x - 1$ from Exercises 21 and 30. For an equation in the form $y = ax - b$, where $b > 0$, what effect does subtracting b from $y = ax$ seem to have on the graph?

The graph will shift down b units on the y-axis.

For Exercises 57–64, solve.

57. A plumber charges $80 plus $40 per hour of labor. The equation $c = 40n + 80$ describes the total cost that she would charge for a service visit, where n represents the number of hours of labor.

a. Find the total cost if labor is 2 hours.

$160

b. If a client's total charges are $240, for how many hours of labor was the client charged?

4 hr.

c. Graph the equation with n along the horizontal axis and c along the vertical axis. (*Hint:* Let each grid mark on the c-axis represent $20.) Because n and c are nonnegative, the graph is restricted to the first quadrant only.

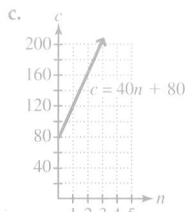

d. What does the point where the graph intersects the c-axis represent?

It represents the initial charge, which is $80.

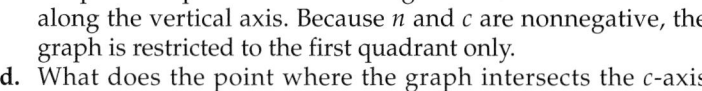

58. An academic tutor charges $20 for supplies and $25 per hour of tutoring. The equation $c = 25n + 20$ describes the total cost that she would charge for tutoring, where n represents the number of hours of tutoring.

a. Find the total cost if the tutor works 3 hours. $95

b. If a client's total charges are $145, for how many hours of tutoring was the client charged? 5 hr.

c. Graph the equation with n along the horizontal axis and c along the vertical axis. Because n and c are nonnegative, the graph is restricted to the first quadrant only.

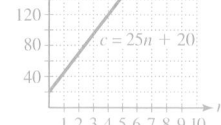

d. What does the point where the graph intersects the c-axis represent?

It represents the initial supply charge, which is $20.

59. A copy center charges $15.00 for the first 300 copies plus $0.03 per copy after that. The equation $c = 0.03n + 15$ describes the total cost that would be charged for a copy service, where n represents the number of copies above 300.
    a. Find the total cost for 400 copies.
       $18
    b. If a client's total charges are $21, for how many copies was the client charged?
       500 copies
    c. Graph the equation with n along the horizontal axis and c along the vertical axis. Because n and c are nonnegative, the graph is restricted to the first quadrant only. (*Hint:* Let each grid mark on the n-axis represent 50 copies and each grid mark on the c-axis represent $3.)

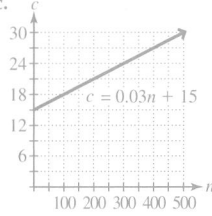

60. A wireless company charges $20.00 per month for 300 anytime minutes plus $0.10 per minute above 300. The equation $c = 0.10n + 20$ describes the total cost that the company would charge for a monthly bill, where n represents the number of minutes used above 300.
    a. Find the total cost if a customer uses 400 anytime minutes.
       $30
    b. If a client's total charges are $22, for how many additional minutes was the client charged?
       20 min.
    c. Graph the equation with n along the horizontal axis and c along the vertical axis. Because n and c are nonnegative, the graph is restricted to the first quadrant only.

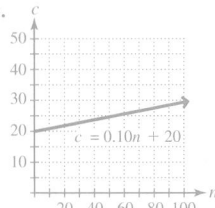

61. It is recommended that a hot tub be emptied and cleaned regularly. The hot tub Benjamin owns holds approximately 500 gallons of water. He can drain 3.5 gallons of water per minute using a garden hose. The equation $g = -3.5m + 500$ describes the amount of water in the hot tub in gallons, where g represents the number of gallons remaining after pumping out water for m minutes.
    a. After 25 minutes, what is the amount of water in the hot tub?
       412.5 gal.
    b. How long does it take to pump half of the water out of the hot tub?
       ≈ 71.4 min.
    c. How long does it take to pump all of the water out of the hot tub?
       ≈ 143 min.
    d. Graph the equation. Because m and g are nonnegative, the graph is restricted to the first quadrant only.
    e. What does the point where the graph intersects the g-axis represent?
       It represents the original amount of water in the hot tub, which is 500 gallons.
    f. What does the point where the graph intersects the m-axis represent?
       It represents the amount of time it takes the hot tub to empty (0 gallons of water).

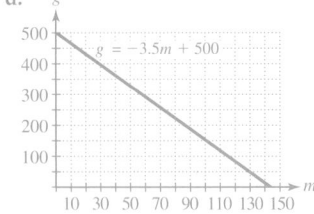

62. The battery in Jenny's laptop is fully charged (100%). While using the laptop continuously on a flight from Kansas to California, she depletes the battery at a rate of 0.75% per minute. The equation $P = -0.75m + 100$ describes the battery's power, where P represents the percent of the battery's power that remains after using the laptop for m minutes.
    a. After 1 hour, what percent of the battery's power remains?
       55%
    b. How long does it take the battery to be at one-fourth of its full power?
       100 min. (1 hr., 40 min.)
    c. How long does it take the battery to reach 0%?
       133.3̄ min. (2.2̄ hr.)
    d. Graph the equation. Because m and P are nonnegative, the graph is restricted to the first quadrant only.
    e. What does the point where the graph intersects the P-axis represent?
       It represents the initial charge in the battery, which is 100%.
    f. What does the point where the graph intersects the m-axis represent?
       It represents the number of minutes for the battery to be completely depleted of power (0%).

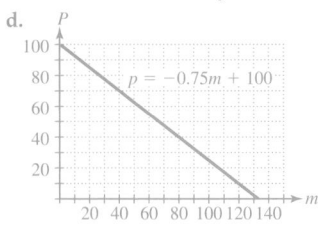

63. A small business owner buys a new computer for $2400. Each year the computer is in use, she can deduct its depreciated value when calculating her taxable income. The equation $c = -300n + 2400$ describes the value of the computer, where c is the value after n years of use.

   a. Find the value of the computer after 4 years. $1200

   b. In how many years will the computer be worth half of its initial value? 4 yr.

   c. After how many years will the computer have a value of $0? 8 yr.

   d. Graph the equation with n along the horizontal axis and c along the vertical axis. Because n and c are nonnegative, the graph is restricted to the first quadrant only.

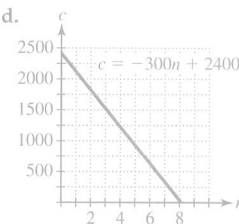

64. As of 2010, it is estimated that Earth is being deforested at a rate of about 50,000 square miles per year. The equation $A = -50,000n + 2,000,000$ describes the area of forest remaining, where A represents the area remaining n years after 2010. (*Source:* www.wikipedia.org.)

   a. Find the area of forest remaining in 2015.
      1,750,000 mi.²

   b. If the rate does not change, in what year will Earth have half of the area of the forest that was present in 2010?
      2030

   c. If the rate does not change, in what year will forest be gone from Earth?
      2050

   d. Graph the equation with n along the horizontal axis and A along the vertical axis. Because n and A are nonnegative, the graph is restricted to the first quadrant only.

Of Interest

Estimates for the rate of deforestation vary. As of 2010, the rates range from 20,000 square miles per year to 50,000 square miles per year.

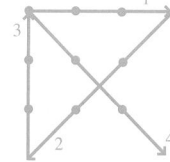

Puzzle Problem Without lifting your pencil from the paper, connect the dots using only four straight lines:

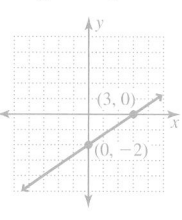

Review Exercises

Exercises 1–3 Expressions

[1.7] 1. Evaluate $\frac{2}{3}x + 2$ when $x = 3$. 4

[2.1] 2. Calculate the perimeter of a rectangle with a length of 8.5 inches and a width of 7.25 inches.
31.5 in.

[2.1] 3. Calculate the area of a triangle with a base of 4.2 inches and a height of 2.7 inches.
5.67 sq. in.

Exercises 4–6 Equations and Inequalities

[2.3] 4. Solve $0 = 3x - 5$.
$\frac{5}{3}$

[2.4] 5. Isolate x in the formula $v = mx + k$.
$x = \frac{v - k}{m}$

[3.2] 6. Graph the line that passes through the points $(3, 0)$ and $(0, -2)$.

3.3 Graphing Using Intercepts

Objectives

1 Given an equation, find the coordinates of the *x*- and *y*-intercepts.
2 Graph linear equations using intercepts.

[3.2] Given $3x - 4y = 8$, find the following and write the solution as an ordered pair in the form (x, y).
 1. y if $x = 0$
 2. x if $y = 0$
 3. y if $x = 4$

Objective 1 Given an equation, find the coordinates of the *x*- and *y*-intercepts.

Look at the following graph.

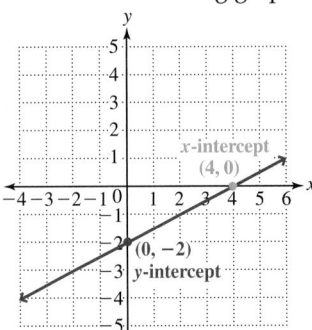

Notice that the line intersects the *x*-axis at $(4, 0)$ and the *y*-axis at $(0, -2)$. These points are called *intercepts*. A point where a graph intersects the *y*-axis is called a **y-intercept**.

Definitions **x-intercept:** A point where a graph intersects the *x*-axis.
y-intercept: A point where a graph intersects the *y*-axis.

Note that the *y*-coordinate of any *x*-intercept is always 0. Similarly, the *x*-coordinate of any *y*-intercept is always 0. These facts about intercepts suggest the following procedures.

> **Procedure** **Finding the x- and y-intercepts**
>
> To find an *x*-intercept:
> 1. Replace *y* with 0 in the given equation.
> 2. Solve for *x*.
>
> To find a *y*-intercept:
> 1. Replace *x* with 0 in the given equation.
> 2. Solve for *y*.

Note We no longer show the steps containing the addition or multiplication principles as we did in Chapters 2 and 3.

Answers to Your Turn 1
x-intercept: $\left(\dfrac{8}{3}, 0\right)$
y-intercept: $(0, -2)$

Answers to Warm-up
1. $(0, -2)$ 2. $\left(\dfrac{8}{3}, 0\right)$ 3. $(4, 1)$

▶ **Example 1** For the equation $5x - 2y = 10$, find the *x*- and *y*-intercepts.

Solution:

For the *x*-intercept, replace *y* with 0 and solve for *x*.
$$5x - 2(0) = 10$$
$$5x = 10$$
$$x = 2 \quad \text{Divide both sides by 5.}$$
x-intercept: $(2, 0)$

For the *y*-intercept, replace *x* with 0 and solve for *y*.
$$5(0) - 2y = 10$$
$$-2y = 10$$
$$y = -5 \quad \text{Divide both sides by } -2.$$
y-intercept: $(0, -5)$

Your Turn 1 For the equation $3x - 4y = 8$, find the *x*- and *y*-intercepts.

The equation in Example 1 is in the form $Ax + By = C$, where A, B, and C are real-number constants with both A and B not equal to 0. A linear equation written in this form is said to be in *standard form*. One of the advantages of having a linear equation in standard form is that it is easy to find the intercepts. Throughout this section, we introduce other forms for linear equations and explore some of the information that can be discerned from equations written in various forms.

Example 2 For the equation $y = \dfrac{2}{3}x$, find the x- and y-intercepts.

Solution: Replace x with 0 and solve for y. Then replace y with 0 and solve for x.

y-intercept: $y = \dfrac{2}{3}(0)$ x-intercept: $0 = \dfrac{2}{3}x$

$\qquad\qquad\quad y = 0$ $0 = x$ Multiply both sides by $\dfrac{3}{2}$.

x- and y-intercept: $(0, 0)$

The equation in Example 2 is a linear equation in the form $y = mx$, where m is any real number. The graph of any equation in the form $y = mx$ is always a line that passes through the origin, which means that the x- and y-intercepts are both at $(0, 0)$. So to graph an equation of the form $y = mx$, at least one other point must be found.

> **Rule**   **Intercepts for $y = mx$**
>
> If an equation can be written in the form $y = mx$, where m is a real number other than 0, then the x- and y-intercepts are at the origin, $(0, 0)$.

Example 3 For the equation $y = 2x + 1$, find the x- and y-intercepts.

Solution: Replace y with 0 and solve for x. Then replace x with 0 and solve for y.

x-intercept: $0 = 2x + 1$ y-intercept: $y = 2(0) + 1$

$\qquad\quad -1 = 2x$ Subtract 1 from both sides. $y = 1$

$\qquad\quad -\dfrac{1}{2} = x$ Divide both sides by 2. y-intercept: $(0, 1)$

x-intercept: $\left(-\dfrac{1}{2}, 0\right)$

▲ **Note** When we replace x with 0, we are left with the constant 1 for the y-intercept.

The equation in Example 3 is in the form $y = mx + b$, where m and b can be any real number. In this form, when we replace x with 0 to find the y-intercept, we are left with the constant b.

> **Rule**   **The y-intercept for $y = mx + b$**
>
> If an equation is in the form $y = mx + b$, where m and b are nonzero real numbers, then the y-intercept will be $(0, b)$.

Answers to Your Turn 3
a. x- and y-intercept: $(0, 0)$
b. x-intercept: $\left(\dfrac{5}{3}, 0\right)$
$\qquad$ y-intercept: $(0, -5)$

Your Turn 3 Find the x- and y-intercepts for the following equations.

a. $y = -\dfrac{3}{4}x$ **b.** $y = 3x - 5$

Example 4 For the equation $y = 3$, find the x- and y-intercepts.

Solution: Remember from Section 3.2, Example 4, that the graph of $y = 3$ is a horizontal line parallel to the x-axis that passes through the y-axis at the point $(0, 3)$. Notice that this point is the y-intercept. Because the line is parallel to the x-axis, it will never intersect the x-axis. Therefore, there is no x-intercept.

Example 4 suggests the following rule.

> **Rule** Intercepts for $y = c$
>
> The graph of an equation in the form $y = c$, where c is a nonzero real number, has no x-intercept and the y-intercept is $(0, c)$. The graph of $y = 0$ is the x-axis.

We can draw a similar conclusion about the graph of an equation in the form $x = c$.

> **Rule** Intercepts for $x = c$
>
> The graph of an equation in the form $x = c$, where c is a nonzero real number, has no y-intercept and the x-intercept is $(c, 0)$. The graph of $x = 0$ is the y-axis.

Your Turn 4 Find the x- and y-intercepts for the following equations.

a. $y = -5$

b. $x = -2$

Objective 2 Graph linear equations using intercepts.

Because two points determine a line in a plane, we can use the x- and y-intercepts to draw the graph of a linear equation.

Example 5 Graph using the x- and y-intercepts: $5x - 2y = 10$

Solution: In Example 1, the intercepts were found to be $(2, 0)$ and $(0, -5)$. We plot these intercepts and connect them to graph the line. It is helpful to find a third solution and verify that all three points can be connected to form a straight line. For this third point, we will choose $x = 3$ and solve for y.

Third solution: (check) If $x = 3$, then

$$5(3) - 2y = 10$$
$$15 - 2y = 10$$
$$-2y = -5 \qquad \text{Subtract 15 from both sides.}$$
$$y = \frac{5}{2} \qquad \text{Divide both sides by } -2. \quad \textbf{Third solution: } \left(3, \frac{5}{2}\right)$$

Instructor Note Emphasize that the reason we graph using intercepts is that they are often the easiest ordered pairs to find and are easy to graph because they lie on the axes.

Answers to Your Turn 4
a. no x-intercept
   y-intercept: $(0, -5)$
b. x-intercept: $(-2, 0)$
   no y-intercept

Answer to Your Turn 5

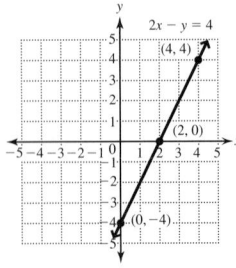

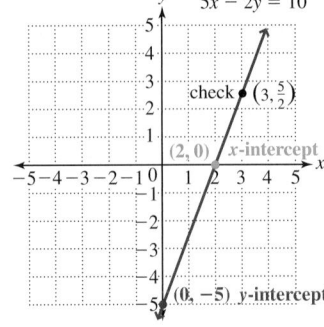

Table of Solutions

	x	y
x-intercept:	2	0
y-intercept:	0	-5
Third solution: (check)	3	$\frac{5}{2}$

Your Turn 5 Graph using the x- and y-intercepts: $2x - y = 4$

As previously mentioned, if the x- and y-intercepts are both at the origin, we must find an additional point in order to graph the linear equation.

Example 6 Graph using the x- and y-intercepts.

a. $y = \dfrac{2}{3}x$

Solution: In Example 2 we found $(0, 0)$ to be both the x- and y-intercept. Because this single point is not enough to determine the line in the plane, we must find at least one more solution. We will also find a third solution as a check.

Second solution:
Choose x to be 3.

$$y = \frac{2}{3}(3)$$

$$y = \frac{2}{3}\left(\frac{\overset{1}{\cancel{3}}}{1}\right)$$

$$y = \overset{1}{2}$$

Second solution: $(3, 2)$

Third solution (check):
Choose x to be -3.

$$y = \frac{2}{3}(-3)$$

$$y = \frac{2}{3}\left(-\frac{\overset{1}{\cancel{3}}}{1}\right)$$

$$y = \overset{1}{-2}$$

Third solution: $(-3, -2)$

Now plot the solutions and connect the points to form the line.

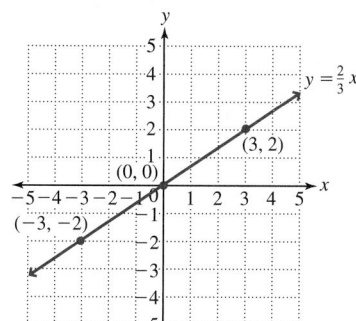

Table of Solutions

	x	y
x- and y-intercept:	0	0
Second solution:	3	2
Third solution: (check)	-3	-2

b. $y = \dfrac{2}{3}x + 2$

Solution: Notice that this equation is in the form $y = mx + b$. Because $b = 2$ in this case, we can conclude that the y-intercept is $(0, 2)$. For the x-intercept, let $y = 0$ and solve for x. Also, as a check, we will choose $x = 3$ to generate a third point.

x-intercept: $0 = \dfrac{2}{3}x + 2$

$\qquad -2 = \dfrac{2}{3}x$ Subtract 2 from both sides.

$\qquad \dfrac{3}{\underset{1}{2}} \cdot \dfrac{\overset{1}{2}}{1} = \dfrac{\overset{1}{3}}{\underset{1}{2}} \cdot \dfrac{\overset{1}{2}}{\underset{1}{3}}x$ Multiply both sides by $\dfrac{3}{2}$.

$\qquad -3 = x$

x-intercept: $(-3, 0)$

Third solution: (check) $y = \dfrac{2}{3}(3) + 2$

$$y = \frac{2}{3} \cdot \frac{\overset{1}{\cancel{3}}}{1} + 2$$

$$y = \overset{1}{2} + 2$$

$$y = 4$$

Third solution: $(3, 4)$

Now plot the solutions and connect the points to form the line.

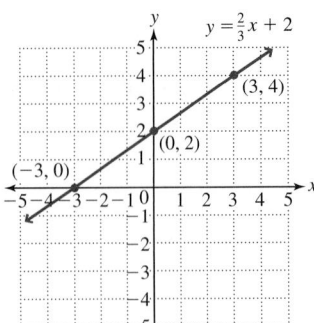

Table of Solutions

	x	y
x-intercept:	−3	0
y-intercept:	0	2
Third solution: (check)	3	4

Your Turn 6   Graph using the x- and y-intercepts.

a. $y = \dfrac{1}{2}x$

b. $y = \dfrac{2}{3}x - 2$

VISUAL

Learning Strategy

If you are a visual learner, imagine the line $y = \dfrac{2}{3}x$ sliding up the y-axis 2 units to become $y = \dfrac{2}{3}x + 2$.

Compare the graphs in Examples 6(a) and 6(b). If we place both lines on the same grid (shown here), we see that they are parallel, which means that they do not intersect at any point. Further, each point on the graph of $y = \dfrac{2}{3}x + 2$ is 2 units higher than each point with the same x-coordinate on the graph of $y = \dfrac{2}{3}x$.

Answers to Your Turn 6

a.

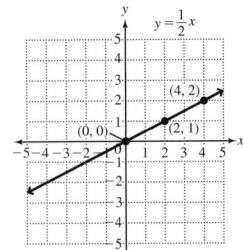

b.

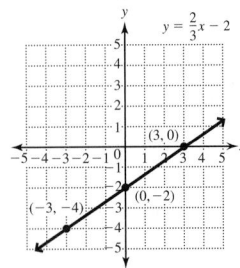

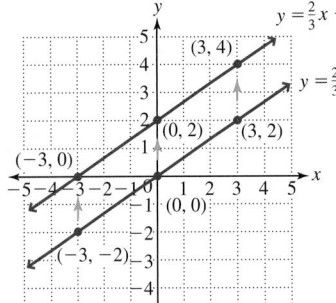

Table of Solutions

x	$y = \dfrac{2}{3}x$	$y = \dfrac{2}{3}x + 2$
−3	−2	0
0	0	2
3	2	4

From the table, we see that the difference in an equation in the form $y = mx$ and an equation in the form $y = mx + b$ is the value of the y-intercept. In other words, the line $y = mx$ shifts upward or downward by b units. The graph of $y = \dfrac{2}{3}x - 2$ is a line parallel to $y = \dfrac{2}{3}x$, only shifted down 2 units so that its y-intercept is $(0, -2)$. We will explore this further in Section 3.5.

3.3 Exercises (For Extra Help) MyMathLab®

Note: Exercises marked with a ★ represent challenging exercises.

Objective 1

Prep Exercise 1 What is an x-intercept?   A point where a graph intersects the x-axis

Prep Exercise 2 What is a y-intercept?   A point where a graph intersects the y-axis

For Exercises 1–8, find the x- and y-intercepts of each graph. See Objective 1.

1.

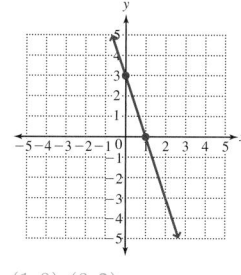

$(1, 0), (0, 3)$

2.

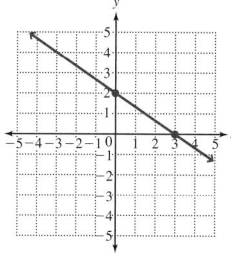

$(3, 0), (0, 2)$

3.

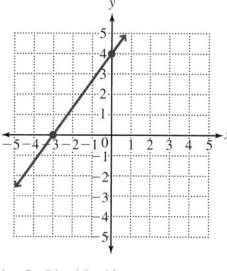

$(-3, 0), (0, 4)$

4.

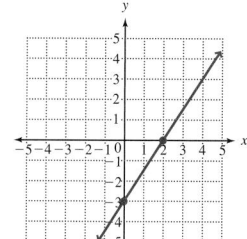

$(2, 0), (0, -3)$

5.

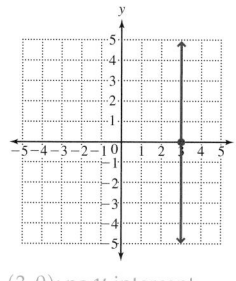

$(3, 0);$ no y-intercept

6.

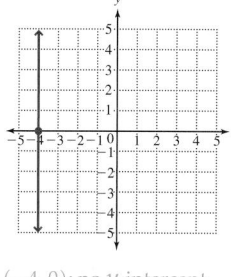

$(-4, 0);$ no y-intercept

7.

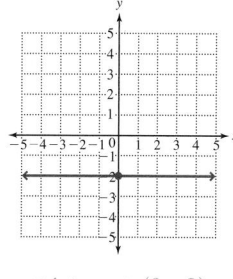

no x-intercept; $(0, -2)$

8.

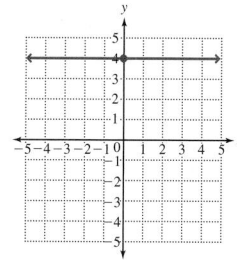

no x-intercept; $(0, 4)$

Prep Exercise 3 If given an equation, how do you find an x-intercept? (a) Replace y with 0 in the given equation. (b) Solve for x.

Prep Exercise 4 If given an equation, how do you find a y-intercept? (a) Replace x with 0 in the given equation. (b) Solve for y.

Prep Exercise 5 What are the coordinates of the y-intercept in an equation of the form $y = mx + b$? $(0, b)$

Prep Exercise 6 What are the coordinates of the x- and y-intercepts for an equation of the form $y = c$, where c is a real number? no x-intercept; $(0, c)$

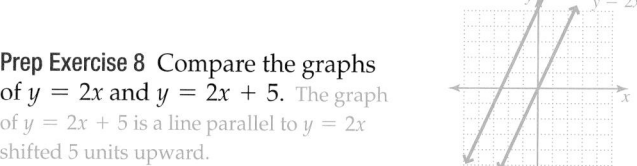

Prep Exercise 7 What are the coordinates of the x- and y-intercepts for an equation of the form $x = c$, where c is a real number? $(c, 0);$ no y-intercept

Prep Exercise 8 Compare the graphs of $y = 2x$ and $y = 2x + 5$. The graph of $y = 2x + 5$ is a line parallel to $y = 2x$ shifted 5 units upward.

For Exercises 9–26, find the x- and y-intercepts. See Examples 1–4.

9. $x - y = 4$
$(4, 0), (0, -4)$

10. $x + y = 7$
$(7, 0), (0, 7)$

11. $2x + 3y = 6$
$(3, 0), (0, 2)$

12. $2x + 3y = 12$
$(6, 0), (0, 4)$

13. $3x - 4y = -10$
$\left(-\dfrac{10}{3}, 0\right), \left(0, \dfrac{5}{2}\right)$

14. $3x - 4y = 2$
$\left(\dfrac{2}{3}, 0\right), \left(0, -\dfrac{1}{2}\right)$

15. $y = 3x - 6$
$(2, 0), (0, -6)$

16. $y = 2x + 8$
$(-4, 0), (0, 8)$

17. $y = 2x + 5$
$\left(-\dfrac{5}{2}, 0\right), (0, 5)$

18. $y = 3x - 1$
$\left(\dfrac{1}{3}, 0\right), (0, -1)$

19. $y = 2x$
$(0, 0)$ for both

20. $y = \dfrac{2}{5}x$
$(0, 0)$ for both

21. $2x + 3y = 0$
$(0, 0)$ for both

22. $x + 3y = 0$
$(0, 0)$ for both

23. $x = 5$
$(5, 0);$ no y-intercept

24. $x = -2$
$(-2, 0);$ no y-intercept

25. $y = -2$
no x-intercept; $(0, -2)$

26. $y = 7$
no x-intercept; $(0, 7)$

Objective 2

For Exercises 27–58, graph using the x- and y-intercepts. See Examples 5 and 6.

27. $x + y = 7$

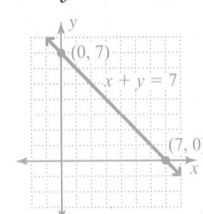

28. $x + y = 10$

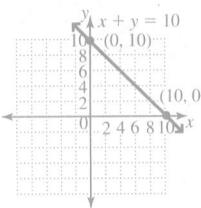

29. $3x - y = 6$

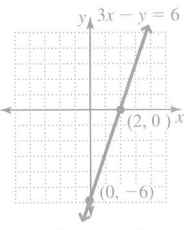

30. $x - 5y = 15$

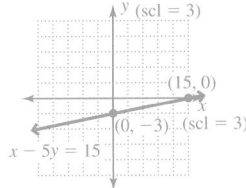

31. $2x + 3y = 6$

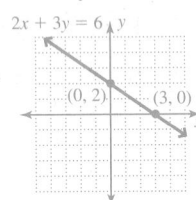

32. $3x + 4y = 12$

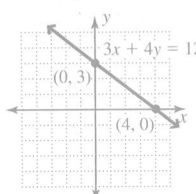

33. $5x + 2y = 10$

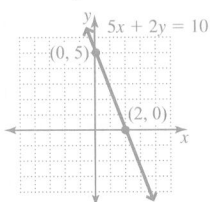

34. $3x - 2y = 18$

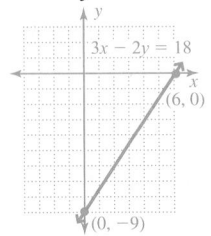

35. $4x + 3y = -12$

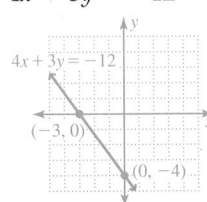

36. $3x + 2y = -6$

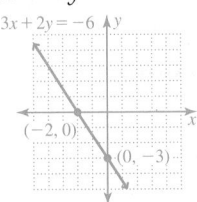

37. $5x - 3y = -15$

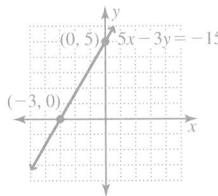

38. $2x - 5y = -10$

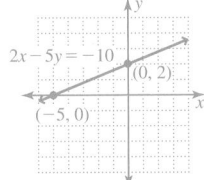

39. $y = 2x$

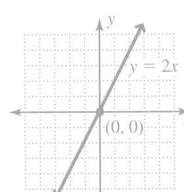

40. $y = -3x$

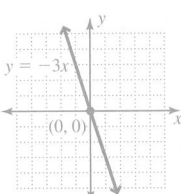

41. $y = \dfrac{2}{5}x$

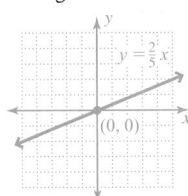

42. $-\dfrac{1}{4}x = y$

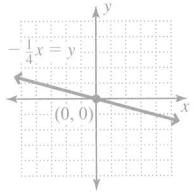

43. $y = x - 4$

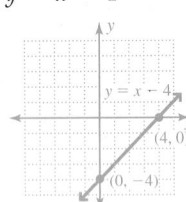

44. $y = x + 2$

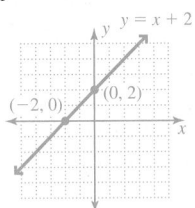

45. $y = 2x - 6$

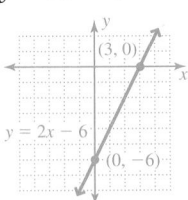

46. $y = 3x + 9$

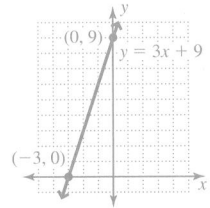

47. $6x + 3y = 9$

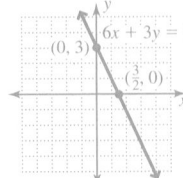

48. $2x - y = -1$

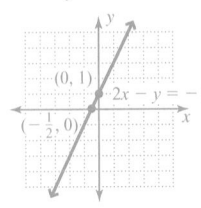

49. $4x + 2y = 5$

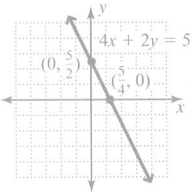

50. $3x + 2y = 7$

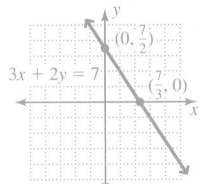

51. $x = -4$

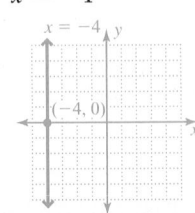

52. $x = 5$

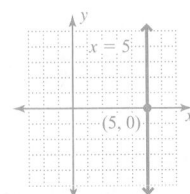

53. $y = -2$

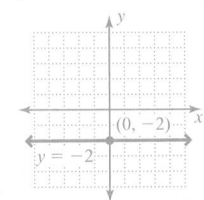

54. $y = 5$

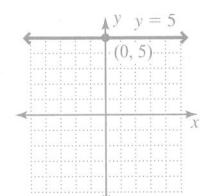

55. $y - 2 = 0$

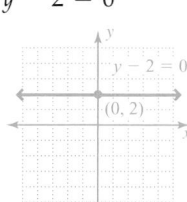

56. $y + 4 = 0$

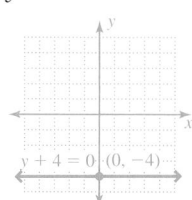

57. $x - 2 = 0$

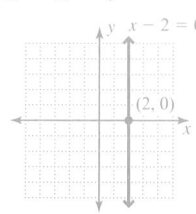

58. $x + 3 = 0$

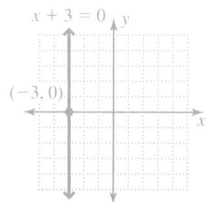

★ **59.** Which of the following could be the graph of $1.4x - 0.9y = 2.7$? Explain.

a.
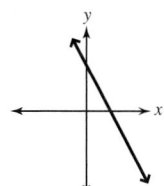

b.
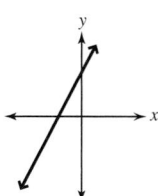

c.
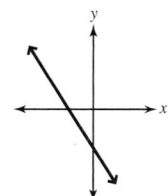

d.
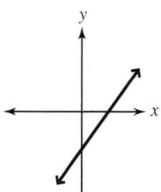

d because the x-coordinate of the x-intercept is positive and the y-coordinate of the y-intercept is negative

★ **60.** Which of the following could be the graph of $0.65x - 2.9y = -3.5$? Explain.

a.
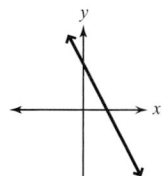

b.
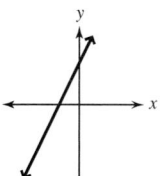

c.
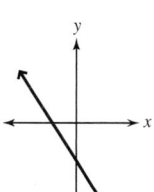

d.
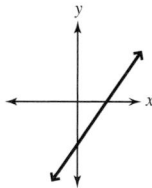

b because the x-coordinate of the x-intercept is negative and the y-coordinate of the y-intercept is positive

★ **61.** Which of the following could be the graph of $\frac{2}{3}x - \frac{3}{4}y = \frac{10}{3}$? Explain.

a.
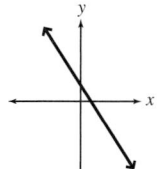

b.
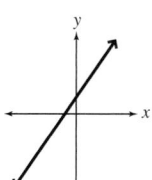

c.
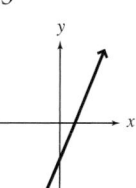

d.
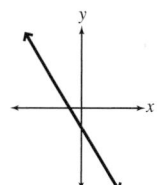

c because the x-coordinate of the x-intercept is positive and the y-coordinate of the y-intercept is negative

★ **62.** Which of the following could be the graph of $\frac{3}{2}x + \frac{6}{5}y = -\frac{8}{5}$? Explain.

a.
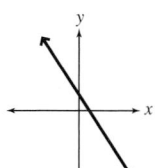

b.
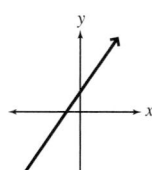

c.
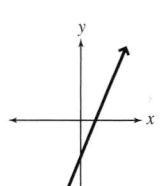

d.
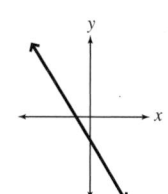

d because the x-coordinate of the x-intercept is negative and the y-coordinate of the y-intercept is negative

63. If a circle with a radius of 2.7 units was to be drawn in the coordinate plane with its center at the origin, what would be the coordinates of its x- and y-intercepts?

$(-2.7, 0), (2.7, 0), (0, -2.7), (0, 2.7)$

64. If a circle with a diameter of 6.8 units was to be drawn in the coordinate plane with its center at the origin, what would be the coordinates of its x- and y-intercepts?

$(-3.4, 0), (3.4, 0), (0, -3.4), (0, 3.4)$

★ **65.** A salesperson sells two different sizes of facial cleansing lotion. She sells the small size for $15 and the large size for $20.

   a. If x represents the number of units of the small size sold and y represents the number of units of the large size sold, write an expression that describes her total sales.

   $15x + 20y$

   b. If her supervisor sets a goal of selling $2000 worth of the cleansing lotion, use the expression from part a to write an equation describing the sales goal.

   $15x + 20y = 2000$

   c. Find the x- and y-intercepts.

   $\left(133\frac{1}{3}, 0\right), (0, 100)$

   d. What do the x- and y-intercepts mean in terms of sales?

   The number of units required to meet the goal if only one size is sold

   e. Find a third combination of sales numbers that would equal the sales goal.

   Answers may vary. Some possibilities are $(20, 85)$, $(40, 70)$, and $(60, 55)$.

   f. Graph the equation. Because x and y are nonnegative, the graph is restricted to the first quadrant only.

f.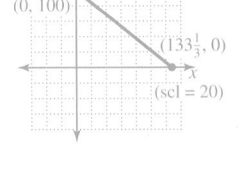

★ **66.** A comic book collector will pay $50 for any Superman comic published prior to 1970 if it is in mint condition and $30 if it is not in mint condition.

   a. If x represents the number of comic books in mint condition and y represents the number of comic books not in mint condition, write an expression that describes the amount the collector would pay for a combination of comic books in mint and nonmint condition.

   $50x + 30y$

   b. Suppose the collector visits a convention for collectors and has a budget of $1800 to spend. Use the expression from part a to write an equation describing his expenditure on a combination of the comics if he spends all of his budget.

   $50x + 30y = 1800$

   c. Find the x- and y-intercepts.

   $(36, 0), (0, 60)$

   d. What do the x- and y-intercepts mean in terms of purchases?

   The number he can buy if only one type is purchased

   e. Find a third combination of comic books he could purchase to equal his budget amount.

   Answers may vary. Some possibilities are $(3, 55)$, $(6, 50)$, and $(9, 45)$.

   f. Graph the equation. Because x and y are nonnegative, the graph is restricted to the first quadrant only.

f.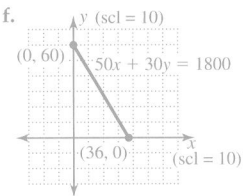

Review Exercises

Exercises 1 and 2 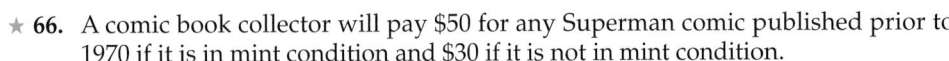 **Expressions**

[1.5] *For Exercises 1 and 2, evaluate.*

   1. $\dfrac{4 - (-2)}{-2 - 1}$

   -2

   2. $\dfrac{-3 - 4}{-2 - (-2)}$

   undefined

Exercises 3–6 **Equations and Inequalities**

[2.4] **3.** Solve for w in the equation $P = 2l + 2w$.

   $w = \dfrac{P - 2l}{2}$

[2.5] **4.** Translate to an equation and then solve. Four more than three times a number is equal to five times the sum of the number and six.

   $4 + 3x = 5(x + 6); -13$

[3.1] **5.** Where is the point located whose coordinates are $(0, 4)$?

   On the y-axis

[3.2] **6.** Determine whether $(-8, -2)$ is a solution for $x + 3y = -5$.

   no

3.4 Slope–Intercept Form

Objectives

1 Compare lines with different slopes.
2 Graph equations in slope–intercept form.
3 Find the slope of a line given two points on the line.
4 Use slope–intercept form to write the equation of a line.

Warm-up

[1.7] Evaluate $\dfrac{d-c}{b-a}$ for the following values of the variables.

1. $a = 3, b = -1, c = 5$, and $d = 7$
2. $a = 2, b = 2, c = 4$, and $d = -3$
[2.4] 3. Solve $3x + 4y = 8$ for y.

In Section 3.3, we discussed various forms of linear equations, one of which was $y = mx + b$, where m and b are real-number constants. Further, we discovered that the graph of an equation in this form is a line with its y-intercept at $(0, b)$. In this section, we explore how the coefficient m affects the graph.

Objective 1 Compare lines with different slopes.

First, we consider graphs of equations in which $b = 0$ so that the equation is $y = mx$ and the graphs all have the origin $(0, 0)$ as the x- and y-intercepts.

Example 1 Graph each of the following equations on the same grid.

$$y = x \qquad y = 2x \qquad y = 3x$$

Solution: In the following table, the same choice for x has been substituted into each equation and the corresponding y-coordinate has been found. We plot the ordered pairs and connect the points to graph the lines.

If x is	and $y = x$, then y is	and $y = 2x$, then y is	and $y = 3x$, then y is
0	0	0	0
1	1	2	3
2	2	4	6

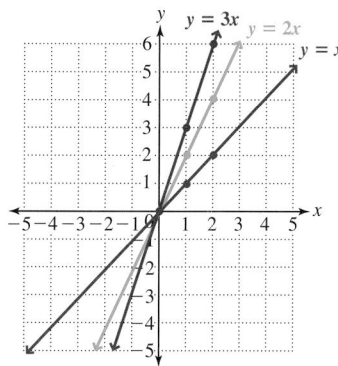

From Example 1, we see that for equations of the form $y = mx$, as the coefficient m increases, the graphs get steeper. Because the coefficient m affects how steep a line is, m is called the *slope* of the line.

> For $y = x$, the slope is 1 ($m = 1$).
> For $y = 2x$, the slope is 2 ($m = 2$).
> For $y = 3x$, the slope is 3 ($m = 3$).

We have seen that if the slope is greater than 1, the corresponding line has a steeper incline than the graph of $y = x$. What values for m would cause a line to be less inclined than $y = x$? The pattern in the graphs suggests that we should explore values less than 1.

First, consider values less than 1 but greater than 0, such as $\dfrac{1}{2}$ and $\dfrac{1}{3}$.

Answers to Warm-up

1. $-\dfrac{1}{2}$
2. Undefined
3. $y = -\dfrac{3}{4}x + 2$ or $\dfrac{8 - 3x}{4}$

Example 2 Graph each of the following equations on the same grid.

$$y = x \qquad y = \frac{1}{2}x \qquad y = \frac{1}{3}x$$

Solution: For consistency, we use the same three choices for x that we used in Example 1. After finding the corresponding y-coordinates for each equation, we plot the ordered pairs and connect the points to graph the lines.

If x is	and $y = x$, then y is	and $y = \frac{1}{2}x$, then y is	and $y = \frac{1}{3}x$, then y is
0	0	0	0
1	1	$\frac{1}{2}$	$\frac{1}{3}$
2	2	1	$\frac{2}{3}$

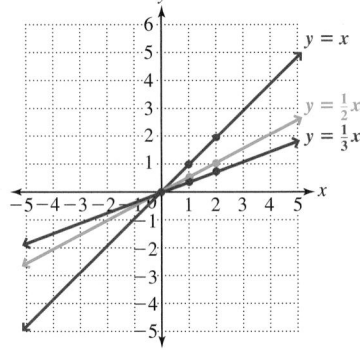

Notice that if the slope is a fraction between 0 and 1, the smaller the fraction, the less inclined or flatter the lines. So far, we have considered only positive slope values. If we "travel" from left to right along lines with positive slopes, we travel uphill. What would a line with a negative slope look like?

Example 3 Graph each of the following equations on the same grid.

$$y = -x \qquad y = -2x \qquad y = -\frac{1}{2}x$$

Solution: Again, we make a table of ordered pairs, then plot those ordered pairs and connect the points to graph the lines.

If x is	and $y = -x$, then y is	and $y = -2x$, then y is	and $y = -\frac{1}{2}x$, then y is
0	0	0	0
1	−1	−2	$-\frac{1}{2}$
2	−2	−4	−1

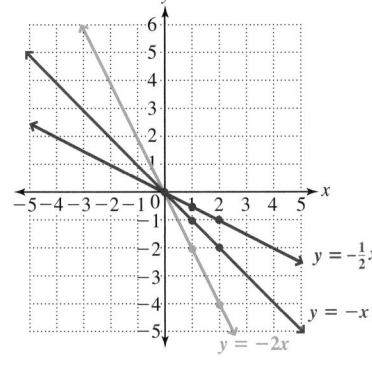

Note When we speak of traveling along a graph, the convention is to travel from left to right, just as we do when we read English.

Notice that if we "travel" from left to right along lines with negative slopes, we travel downhill. As the slope becomes more negative, the downhill incline becomes steeper.

Rule Graphs of $y = mx$

Given an equation of the form $y = mx$, the graph of the equation is a line passing through the origin and having the following characteristics:

If $m > 0$ (slope is positive), the graph is a line that slants uphill from left to right.

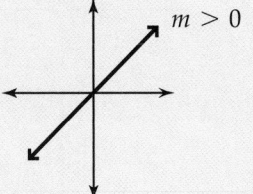

If $m < 0$ (slope is negative), the graph is a line that slants downhill from left to right.

The greater the absolute value of m, the steeper the line.

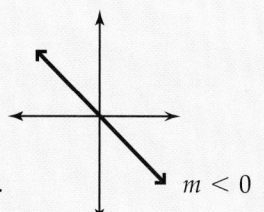

$m < 0$

Your Turn 3

a. Is $y = 5x$ steeper or less inclined than $y = x$?

b. Is the graph of $y = -\dfrac{2}{3}x$ uphill or downhill from left to right?

Objective 2 Graph equations in slope–intercept form.

We have seen that slope determines the incline of a line. What more can we discover about slope? Notice that if we isolate m in $y = mx$, we get $m = \dfrac{y}{x}$, which suggests that **slope** is a ratio of the amount of vertical change to the amount of horizontal change. A change up or to the right is a positive change, and a change down or to the left is a negative change.

Definition Slope: The ratio of the vertical change to the horizontal change between any two points on a line.

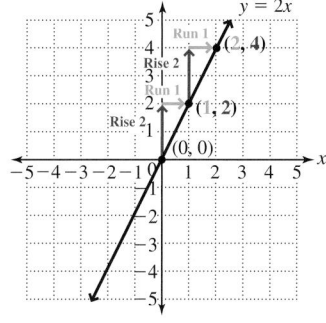

For example, the slope of $y = 2x$ is 2, which can be written as $\dfrac{2}{1}$. This slope value means that rising vertically 2 units and then running horizontally 1 unit from any point on the line locates a second point on the line. For example, if we rise 2 units and run 1 unit from $(0, 0)$, we end up at $(1, 2)$. Similarly, rising 2 units and running 1 unit from $(1, 2)$ puts us at $(2, 4)$.

We have learned that in an equation of the form $y = mx + b$, the slope is m and the y-intercept is $(0, b)$. For this reason, we say that equations of the form $y = mx + b$ are in *slope–intercept* form. We can graph equations in slope–intercept form using the y-intercept as a starting point and then using the slope to locate other points on the line.

Procedure Graphing Equations in Slope–Intercept Form

To graph an equation in slope–intercept form, $y = mx + b$:
1. Plot the y-intercept, $(0, b)$.
2. Plot a second point by rising the number of units indicated by the numerator of the slope, m, then running the number of units indicated by the denominator of the slope, m.
3. Draw a straight line through the two points.

Note: You can check by locating additional points using the slope. Every point you locate using the slope should be on the line.

Example 4 For the equation $y = -\dfrac{1}{3}x + 2$, determine the slope and the y-intercept. Then graph the equation.

Solution: Because the equation is in the form $y = mx + b$, where $m = -\dfrac{1}{3}$ and $b = 2$, the slope of the line is $-\dfrac{1}{3}$ and the y-intercept is $(0, 2)$. To graph the line, we can plot the y-intercept and then use the slope to find other points. To interpret $-\dfrac{1}{3}$ in terms of "rise" and "run," we place the negative sign in either the numerator or denominator. We can "rise" -1 and "run" 3, which gives the point $(3, 1)$. Or we can "rise" 1 and "run" -3, which gives the point $(-3, 3)$. Either interpretation gives another point on the same line.

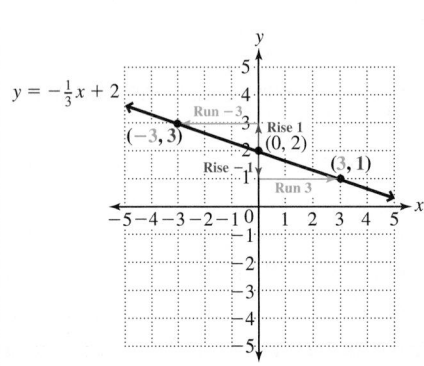

The points we found using the slope can be verified in the equation. We can check the ordered pair $(-3, 3)$ in the equation.

$$y = -\frac{1}{3}x + 2$$

$$3 \overset{?}{=} -\frac{1}{3}(-3) + 2$$

$$3 \overset{?}{=} -\frac{1}{3} \cdot -\frac{\overset{1}{\cancel{3}}}{1} + 2$$

$$3 \overset{?}{=} 1 + 2$$

$$3 = 3 \quad \text{This checks.}$$

We will leave the check of $(3, 1)$ to the reader.

Your Turn 4 For the equation $y = \dfrac{2}{5}x - 2$, determine the slope and the y-intercept. Then graph the equation.

If an equation is not in slope–intercept form, as in $3x + 4y = 8$, and we need to determine the slope and y-intercept, we can write the equation in slope–intercept form by isolating y.

Example 5 For the equation $3x + 4y = 8$, determine the slope and the y-intercept. Then graph the equation.

Solution: Write the equation in slope–intercept form by isolating y.

$$3x + 4y = 8$$

$$4y = -3x + 8 \qquad \text{Subtract 3x from both sides to isolate 4y.}$$

$$\frac{4y}{4} = \frac{-3x + 8}{4} \qquad \text{Divide both sides by 4 to isolate y.}$$

$$y = -\frac{3}{4}x + \frac{8}{4} \qquad \text{Simplify.}$$

$$y = -\frac{3}{4}x + 2$$

The slope is $-\dfrac{3}{4}$, and the y-intercept is $(0, 2)$. To graph the line, we begin at $(0, 2)$ and rise -3 and run 4, which gives the point $(4, -1)$. Or we can begin at $(0, 2)$ and rise 3 and run -4, which gives the point $(-4, 5)$.

Answers to Your Turn 4

$m = \dfrac{2}{5}$; y-intercept: $(0, -2)$

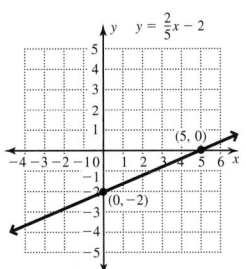

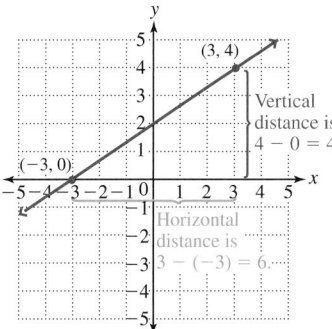

We can check the ordered pairs in the equation.

For $(-4, 5)$:

$$3x + 4y = 8$$
$$3(-4) + 4(5) \stackrel{?}{=} 8$$
$$-12 + 20 \stackrel{?}{=} 8$$
$$8 = 8$$

For $(4, -1)$:

$$3x + 4y = 8$$
$$3(4) + 4(-1) \stackrel{?}{=} 8$$
$$12 + (-4) \stackrel{?}{=} 8$$
$$8 = 8$$

Both ordered pairs check.

Your Turn 5 For the equation $3x - 2y = 6$, determine the slope and the y-intercept. Then graph the equation.

Objective 3 Find the slope of a line given two points on the line.

Given two points on a line, we can determine the slope of the line. Consider the points $(-3, 0)$ and $(3, 4)$ on the graph of $y = \frac{2}{3}x + 2$, which we graphed in Example 6(b) of Section 3.3. In the equation, we see that the slope is $\frac{2}{3}$. Recall that slope is the ratio of the vertical change between any two points on a line to the horizontal change between those points. Therefore, the ratio of the vertical distance between $(-3, 0)$ and $(3, 4)$ to the horizontal distance between those points should be $\frac{2}{3}$. Let's verify.

To find the vertical distance between the two points, we can subtract their y-coordinates: $4 - 0 = 4$. Similarly, the horizontal distance between the points is the difference of the x-coordinates: $3 - (-3) = 6$. Now we can write the ratio.

$$\frac{\text{Vertical distance}}{\text{Horizontal distance}} = \frac{4 - 0}{3 - (-3)} = \frac{4}{6} = \frac{2}{3}$$

Our example suggests the following formula:

Answers to Your Turn 5

$m = \dfrac{3}{2}$; y-intercept: $(0, -3)$

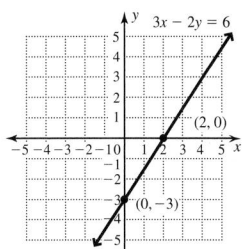

> **Procedure** **The Slope Formula**
>
> The slope of a line through two points (x_1, y_1) and (x_2, y_2), where $x_2 \neq x_1$, is given by the formula
>
> $$m = \frac{y_2 - y_1}{x_2 - x_1}.$$
>
> **Warning** When using the slope formula, make sure your x_1 is from the same ordered pair as your y_1.

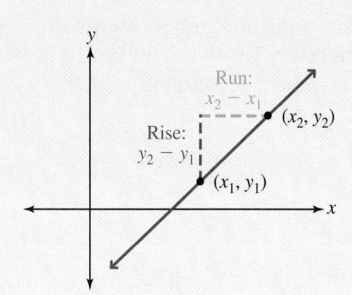

Instructor Note You could point out that

$$\frac{y_2 - y_1}{x_2 - x_1} = \frac{-1(y_2 - y_1)}{-1(x_2 - x_1)} = \frac{-y_2 + y_1}{-x_2 + x_1} = \frac{y_1 - y_2}{x_1 - x_2}.$$

Consequently, it doesn't matter which ordered pair is (x_1, y_1) or (x_2, y_2).

Learning Strategy

Many people, especially visual learners, find it helpful to write the coordinate labels beneath their corresponding value.

$$(3, 5) \qquad (-1, 7)$$
$$(x_1, y_1) \qquad (x_2, y_2)$$

Also, some people prefer to stack the ordered pairs.

$$(-1, 7)$$
$$(3, 5)$$

Example 6 Find the slope of the line through $(3, 5)$ and $(-1, 7)$.

Solution: $m = \dfrac{7 - 5}{-1 - 3} = \dfrac{2}{-4} = -\dfrac{1}{2}$ Use $m = \dfrac{y_2 - y_1}{x_2 - x_1}$ with $(3, 5)$ as (x_1, y_1) and $(-1, 7)$ as (x_2, y_2).

Note It doesn't matter which ordered pair is (x_1, y_1) and which is (x_2, y_2). If we let $(-1, 7)$ be (x_1, y_1) and $(3, 5)$ be (x_2, y_2), we get the same slope.

$$m = \frac{5 - 7}{3 - (-1)} = \frac{-2}{4} = -\frac{1}{2}$$

Your Turn 6 Find the slope of the line through the given points.

a. $(2, 5)$ and $(6, 1)$ **b.** $(-5, -2)$ and $(7, -5)$

Example 7 Find the slope of the line through the given points.

a. $(-3, -4)$ and $(1, -4)$

Solution: $m = \dfrac{-4 - (-4)}{1 - (-3)} = \dfrac{0}{4} = 0$ Use $m = \dfrac{y_2 - y_1}{x_2 - x_1}$ with $(-3, -4)$ as (x_1, y_1) and $(1, -4)$ as (x_2, y_2).

b. $(2, 4)$ and $(2, -3)$

Solution: $m = \dfrac{-3 - 4}{2 - 2}$ Use $m = \dfrac{y_2 - y_1}{x_2 - x_1}$ with $(2, 4)$ as (x_1, y_1) and $(2, -3)$ as (x_2, y_2).

$$m = \frac{-7}{0}$$ Simplify.

Because we never divide by 0, the slope is *undefined*.

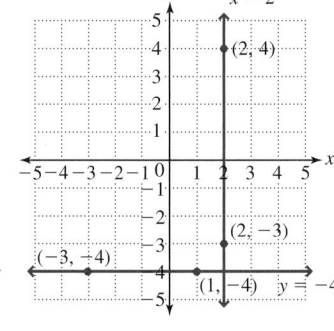

Note The line through $(-3, -4)$ and $(1, -4)$ from Example 7a is a horizontal line with equation $y = -4$. We found the slope of this line to be 0, which is the case for all horizontal lines.

Note The line through $(2, 4)$ and $(2, -3)$ from Example 7b is a vertical line with equation $x = 2$. We found its slope to be undefined, which is the case for all vertical lines.

Note All points on a horizontal line have the same y-coordinate. All points on a vertical line have the same x-coordinate.

Rules Zero Slope, Undefined Slope

The slope of a horizontal line is 0. The slope of a vertical line is undefined.

Your Turn 7 Find the slope of the line through the given points.

a. $(4, -1)$ and $(-3, -1)$ **b.** $(3, 2)$ and $(3, -4)$

Objective 4 Use slope–intercept form to write the equation of a line.

If you are given the slope, m, and the y-intercept, $(0, b)$, of a line, use the slope–intercept form of the equation to write the equation of the line.

Answers to Your Turn 6

a. -1 **b.** $-\dfrac{1}{4}$

Answers to Your Turn 7

a. 0   **b.** undefined

> ### Procedure Equation of a Line Given Its Slope and *y*-Intercept
>
> To write the equation of a line given its slope, m, and its y-intercept, $(0, b)$, use the slope–intercept form of the equation, $y = mx + b$.

Example 8   Write the equation of the lines in slope–intercept form.

a. $m = 5$; y-intercept: $(0, 3)$

Solution: $y = 5x + 3$ In $y = mx + b$, replace m with the slope, 5, and b with the y-coordinate of the y-intercept, 3.

b.

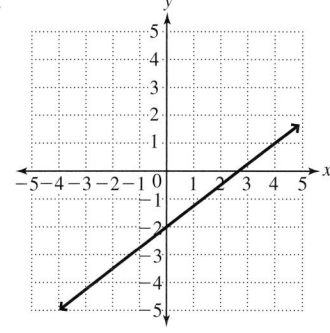

Solution: To find the slope of the line, choose two points, preferably with integer coordinates, such as $(0, -2)$ and $(4, 1)$, and determine the "rise" and "run." We see that the "rise" is 3 units and the "run" is 4 units, so the slope is $\dfrac{3}{4}$. Next, identify the y-intercept, which is the point $(0, -2)$.

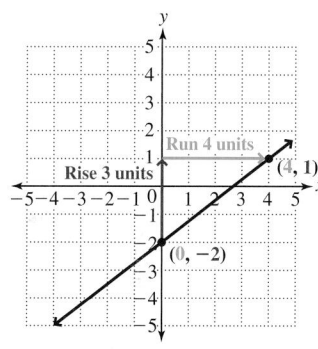

Answer: $y = \dfrac{3}{4}x - 2$ In $y = mx + b$, replace m with $\dfrac{3}{4}$ and b with -2.

Your Turn 8   Write the equation of the line in slope–intercept form.

a. $m = -3$; y-intercept: $(0, -4)$

b.

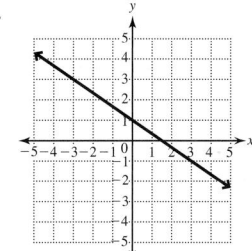

Answers to Your Turn 8

a. $y = -3x - 4$

b. $y = -\dfrac{2}{3}x + 1$

3.4 Exercises   For Extra Help   MyMathLab®

Objective 1

Prep Exercise 1 In your own words, what is the slope of a line?

How steep a line is

Prep Exercise 2 Does the graph of $y = 3x + 5$ go uphill from left to right or downhill? Why?

Uphill because the slope, 3, is positive.

Prep Exercise 3 Does the graph of $y = -2x + 1$ go uphill from left to right or downhill? Why?

Downhill because the slope, -2, is negative.

For Exercises 1–8, graph each set of equations on the same grid. For each set of equations, compare the slopes, the y-intercepts, and their effects on the graphs. See Examples 1–3.

1. $y = \dfrac{1}{2}x$

$y = x$

$y = 2x$

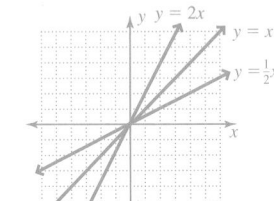

2. $y = \dfrac{1}{3}x$

$y = x$

$y = 3x$

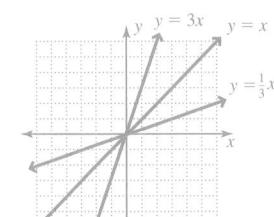

3. $y = x$

$y = \dfrac{1}{2}x$

$y = \dfrac{1}{5}x$

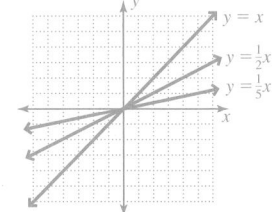

4. $y = x$

$y = \dfrac{1}{3}x$

$y = \dfrac{1}{6}x$

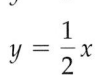

5. $y = -\dfrac{1}{2}x$

$y = -x$

$y = -2x$

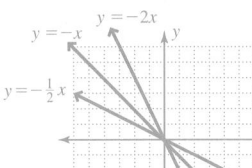

6. $y = -\dfrac{1}{3}x$

$y = -x$

$y = -3x$

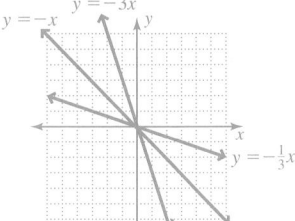

7. $y = x$

$y = x + 2$

$y = x - 2$

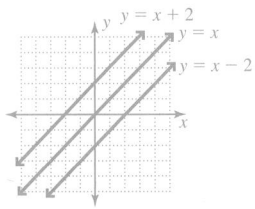

8. $y = 2x$

$y = 2x - 1$

$y = 2x - 3$

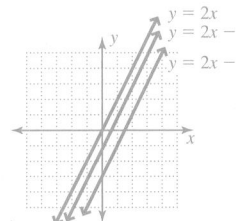

Objective 2

Prep Exercise 4 For an equation of the form $y = mx + b$, which variable represents the slope?

m

Prep Exercise 5 For an equation of the form $y = mx + b$, what are the coordinates of the y-intercept?

$(0, b)$

Prep Exercise 6 Suppose the slope of a line is $\dfrac{3}{4}$ and the y-intercept is $(0, 1)$. Explain how to use the slope to get to a second point.

"Rise" up 3 units from $(0, 1)$ and then "run" to the right 4 units.

For Exercises 9–32, determine the slope and the y-intercept. Then graph the equation. See Examples 4 and 5.

9. $y = 2x + 3$ $m = 2$
$(0, 3)$

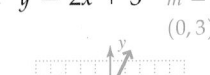
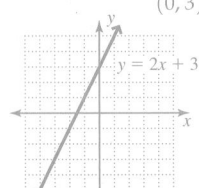

10. $y = 4x - 3$ $m = 4$
$(0, -3)$

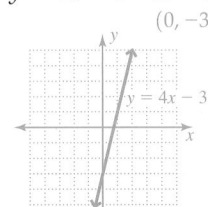

11. $y = -2x - 1$ $m = -2$
$(0, -1)$

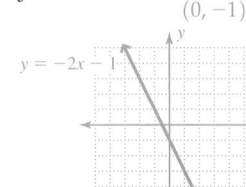

12. $y = -3x + 5$ $m = -3$
$(0, 5)$

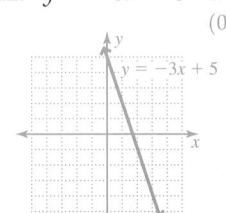

13. $y = \dfrac{1}{3}x + 2$ $m = \dfrac{1}{3}$
$(0, 2)$
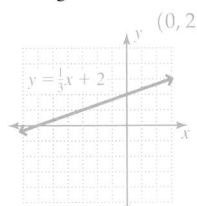

14. $y = \dfrac{1}{2}x - 4$ $m = \dfrac{1}{2}$
$(0, -4)$
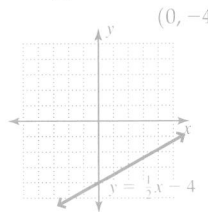

15. $y = \dfrac{3}{4}x - 2$ $m = \dfrac{3}{4}$
$(0, -2)$
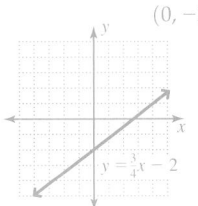

16. $y = \dfrac{2}{5}x - 5$ $m = \dfrac{2}{5}$
$(0, -5)$
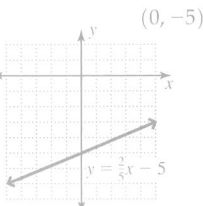

17. $y = -\dfrac{2}{3}x + 8$ $m = -\dfrac{2}{3}$
$(0, 8)$
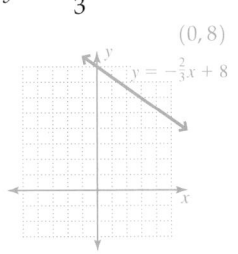

18. $y = -\dfrac{4}{3}x + 7$ $m = -\dfrac{4}{3}$
$(0, 7)$
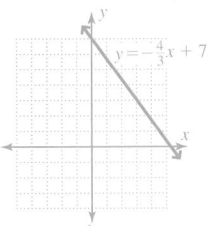

19. $2x + y = 4$ $m = -2$
$(0, 4)$
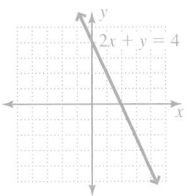

20. $3x + y = 6$ $m = -3$
$(0, 6)$
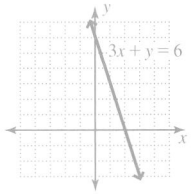

21. $3x - y = -1$ $m = 3$
$(0, 1)$
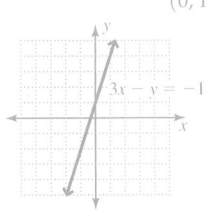

22. $2x - y = 5$ $m = 2$
$(0, -5)$
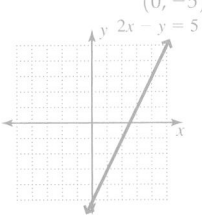

23. $2x + 3y = 6$ $m = -\dfrac{2}{3}$
$(0, 2)$
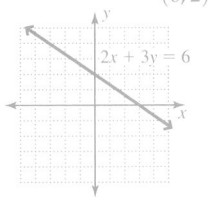

24. $4x - 3y = 12$ $m = \dfrac{4}{3}$
$(0, -4)$
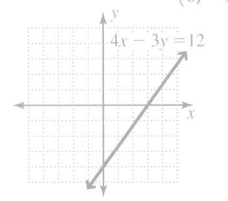

25. $x + 2y = -4$ $m = -\dfrac{1}{2}$
$(0, -2)$
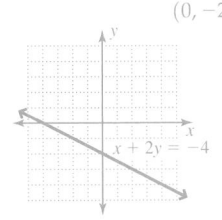

26. $2x + 3y = -6$ $m = -\dfrac{2}{3}$
$(0, -2)$
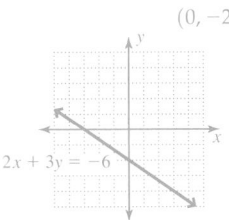

27. $3x - 2y + 7 = 0$ $m = \dfrac{3}{2}$
$\left(0, \dfrac{7}{2}\right)$
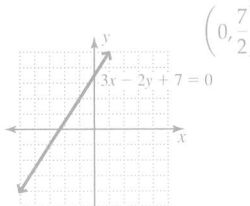

28. $2x - 3y + 8 = 0$ $m = \dfrac{2}{3}$
$\left(0, \dfrac{8}{3}\right)$
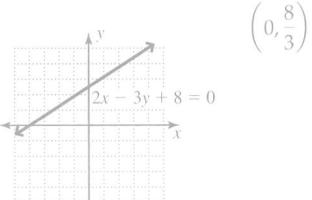

29. $2y = -3x + 5$ $m = -\dfrac{3}{2}$
$\left(0, \dfrac{5}{2}\right)$
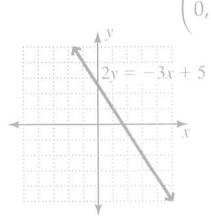

30. $3y = 2x - 4$ $m = \dfrac{2}{3}$
$\left(0, -\dfrac{4}{3}\right)$
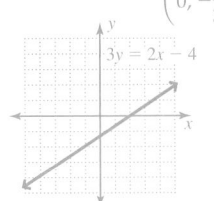

31. $0.6x - 0.2y = 1$ $m = 3$
$(0, -5)$
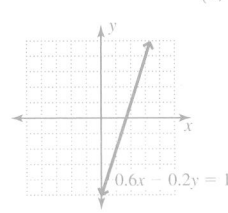

32. $0.3x + 1.5y = -6$
$m = -0.2$
$(0, -4)$
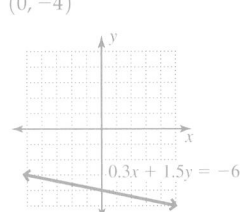

For Exercises 33–42, match the equation with the appropriate graph. See Examples 4 and 5.

33. $y = 2x + 4$
d

34. $3x + 2y = 6$
e

35. $x = y$
a

36. $y = 3$
c

37. $x - 2 = 0$
f

38. $x - 2y = 8$
b

39. $y = \dfrac{2}{3}x - 2$
g

40. $y = -\dfrac{3}{4}x$
i

41. $x = -3$
h

42. $y + 4 = 0$
j

a.
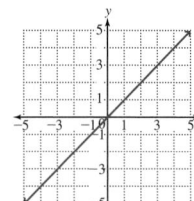

b.
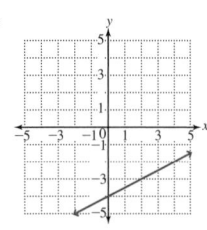

c.
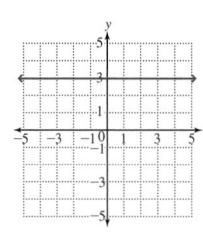

d.
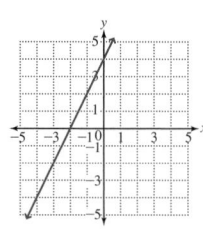

e.
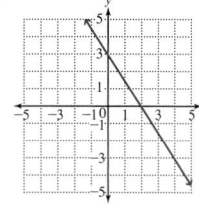

f.
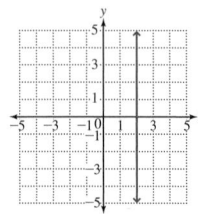

g.
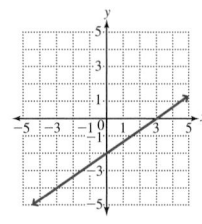

h.
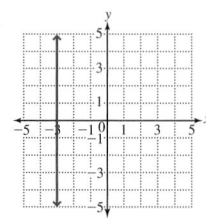

i.
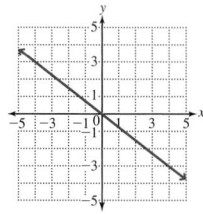

j.
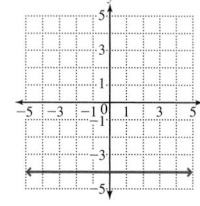

Objective 3

Prep Exercise 7 What is the formula for finding the slope of a line through two points (x_1, y_1) and (x_2, y_2), where $x_1 \neq x_2$?

$m = \dfrac{y_2 - y_1}{x_2 - x_1}$

Prep Exercise 8 What type of line has 0 slope? Explain why the slope is 0 for this type of line.

Horizontal because for any two points on the line, the y-coordinates are the same. This causes the numerator of the slope formula to be 0, which simplifies to 0.

Prep Exercise 9 What type of line has undefined slope? Explain why the slope is undefined for this type of line.

Vertical because for any two points on the line, the x-coordinates are the same. This causes the denominator of the slope formula to equal 0, which is undefined.

For Exercises 43–58, find the slope of the line through the given points. See Examples 6 and 7.

43. $(4, 1), (8, 11)$
$\dfrac{5}{2}$

44. $(1, 5), (3, 8)$
$\dfrac{3}{2}$

45. $(2, 4), (5, 2)$
$-\dfrac{2}{3}$

46. $(6, 2), (3, 7)$
$\dfrac{5}{3}$

47. $(-3, 5), (4, 7)$
$\dfrac{2}{7}$

48. $(3, -2), (5, 5)$
$\dfrac{7}{2}$

49. $(10, -12), (4, -4)$
$-\dfrac{4}{3}$

50. $(-8, 14), (4, 6)$
$-\dfrac{2}{3}$

51. $(3, -3), (-15, -15)$
$\dfrac{2}{3}$

52. $(8, -12), (-4, -18)$
$\dfrac{1}{2}$

53. $(0, 5), (4, 0)$
$-\dfrac{5}{4}$

54. $(-3, 0), (0, 5)$
$\dfrac{5}{3}$

55. $(-3, 5), (4, 5)$
0

56. $(-2, -8), (4, -8)$
0

57. $(6, 1), (6, -8)$
undefined

58. $(-3, 2), (-3, 5)$
undefined

Objective 4

Prep Exercise 10 What is slope–intercept form? $y = mx + b$

For Exercises 59–70, write the equation of the line in slope–intercept form given the slope and the coordinates of the y-intercept. See Example 8.

59. $m = 2; (0, 3)$
$y = 2x + 3$

60. $m = 3; (0, -2)$
$y = 3x - 2$

61. $m = -3; (0, -9)$
$y = -3x - 9$

62. $m = -4; (0, -5)$
$y = -4x - 5$

63. $m = \dfrac{3}{4}; (0, 5)$

$y = \dfrac{3}{4}x + 5$

64. $m = \dfrac{3}{7}; (0, -1)$

$y = \dfrac{3}{7}x - 1$

65. $m = -\dfrac{2}{5}; \left(0, \dfrac{7}{8}\right)$

$y = -\dfrac{2}{5}x + \dfrac{7}{8}$

66. $m = \dfrac{5}{2}; \left(0, \dfrac{3}{8}\right)$

$y = \dfrac{5}{2}x + \dfrac{3}{8}$

67. $m = 0.8; (0, -5.1)$
$y = 0.8x - 5.1$

68. $m = -0.75; (0, 2.5)$
$y = -0.75x + 2.5$

69. $m = -3; (0, 0)$
$y = -3x$

70. $m = 1; (0, 0)$
$y = x$

For Exercises 71–82, find the slope of the line and the y-intercept and write the equation of the line in slope–intercept form. See Example 8.

71.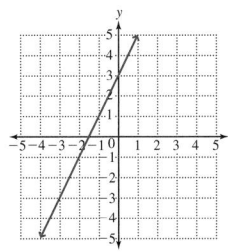

$m = 2; (0, 3); y = 2x + 3$

72.

$m = \dfrac{3}{2}; (0, 3); y = \dfrac{3}{2}x + 3$

73.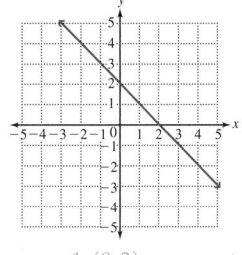

$m = -1; (0, 2); y = -x + 2$

74.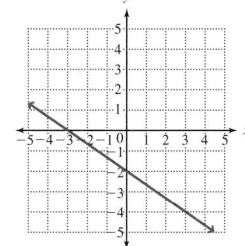

$m = -\dfrac{2}{3}; (0, -2);$

$y = -\dfrac{2}{3}x - 2$

75.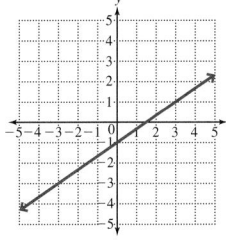

$m = \dfrac{2}{3}; (0, -1); y = \dfrac{2}{3}x - 1$

76.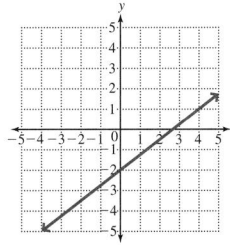

$m = \dfrac{3}{4}; (0, -2);$

$y = \dfrac{3}{4}x - 2$

77.

$m = -\dfrac{1}{4}; (0, 3);$

$y = -\dfrac{1}{4}x + 3$

78.

$m = -\dfrac{3}{2}; (0, 4);$

$y = -\dfrac{3}{2}x + 4$

79.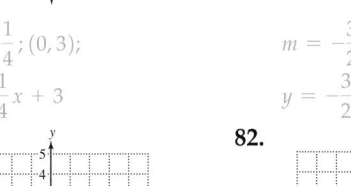

m is undefined;
no y-intercept; $x = 2$

80.

m is undefined;
no y-intercept; $x = -4$

81.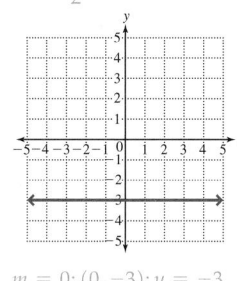

$m = 0; (0, 3); y = 3$

82.

$m = 0; (0, -3); y = -3$

83. What is the equation of a horizontal line through $(0, -2)$? A vertical line through $(3, 0)$?
$y = -2, x = 3$

84. What is the equation of a vertical line through $(-3.5, 0)$? A horizontal line through $(0, 2.6)$?
$x = -3.5, y = 2.6$

85. What equation describes the x-axis?

$y = 0$

86. What equation describes the y-axis?

$x = 0$

87. A parallelogram has vertices at $(1, -2)$, $(3, 4)$, $(-3, -2)$, and $(-1, 4)$.

a.

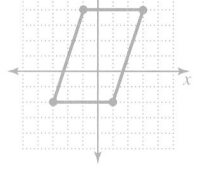

   a. Plot the vertices in a coordinate plane; then connect them to form the parallelogram.

   b. Find the slope of each side of the parallelogram.

   3 and 0

   c. What do you notice about the slopes of the parallel sides?

   They are the same.

88. A right triangle has vertices at $(-2, 3)$, $(2, 1)$, and $(3, 3)$.

a.

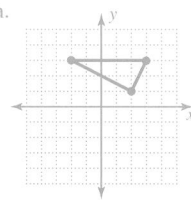

   a. Plot the vertices in a coordinate plane; then connect them to form the right triangle with the right angle at $(2, 1)$.

   b. Find the slope of each side of the triangle.

   $0, 2, -\dfrac{1}{2}$

   c. What do you notice about the slopes of the perpendicular sides?

   They are negative reciprocals.

89. In architecture, the slope of a roof is called its *pitch*. The roof of a house rises 3.5 feet for every 6 feet of horizontal distance. Find the pitch of the roof. Write the pitch as a decimal number and a fraction in simplest form.

$0.58\overline{3}, \dfrac{7}{12}$

90. The slope of a hill is referred to as the *grade* of the hill. Heartbreak Hill, which is part of the Boston Marathon, rises from an elevation of 150 feet to an elevation of 225 feet over a horizontal distance of about 5000 feet. Find the grade of Heartbreak Hill. (*Source:* Boston Athletic Association.)

$\dfrac{3}{200} = 0.015$

91. The Great Pyramid at Giza rises 29.3 meters for every 23 meters of horizontal distance. Find the slope of the pyramid.

$\dfrac{293}{230} \approx 1.27$

92. The steps of the Pyramid of the Sun at Teotihuacan, Mexico, rise 65 meters over a horizontal distance of 112.5 meters. Find the slope of the steps of the Pyramid of the Sun.

$\dfrac{26}{45} = 0.5\overline{7}$

Review Exercises

Exercise 1 ◢◣ Constants and Variables

[1.4] 1. Find the multiplicative inverse of $-\dfrac{2}{3}$.

$-\dfrac{3}{2}$

Exercises 2–3 ◢◣ Expressions

[1.7] 2. Simplify $12x - 9y + 7 - x + 9y$.

$11x + 7$

[1.7] 3. Multiply $-2(x + 7)$.

$-2x - 14$

Exercises 4–6 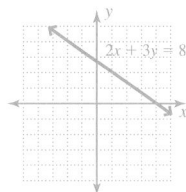 Equations and Inequalities

[2.4] 4. Isolate x in the equation $Ax + By = C$.

$$x = \frac{C - By}{A}$$

[3.3] 6. Find the x- and y-intercepts for $3x - 5y = 7$.

$$\left(\frac{7}{3}, 0\right), \left(0, -\frac{7}{5}\right)$$

[3.2] 5. Graph $2x + 3y = 8$.

3.5   Point–Slope Form

Objectives

1 Use point–slope form to write the equation of a line.

2 Write the equation of a line parallel to a given line.

3 Write the equation of a line perpendicular to a given line.

Warm-up

[3.5] Simplify. Leave the answer solved for y ($y = mx + b$).

1. $y - 5 = 2(x - 3)$

2. $y - (-1) = \frac{4}{3}(x - 3)$

[3.4] 3. Solve $5x - 2y = 6$ for y.

Objective 1 Use point–slope form to write the equation of a line.

In the previous section, we used the slope–intercept form to write the equation of a line because we were given the slope and y-intercept. However, we need a more general approach so that we can write the equation of a line given *any* two points on the line. For this more general approach, we can use the slope formula to derive a new form of the equation of a line called the *point–slope* form. Recall the slope formula.

$$m = \frac{y_2 - y_1}{x_2 - x_1}$$

$$(x_2 - x_1) \cdot m = \left(\frac{y_2 - y_1}{x_2 - x_1}\right) \cdot \left(\frac{x_2 - x_1}{1}\right) \qquad \text{Multiply both sides by } x_2 - x_1 \text{ to isolate the } y\text{'s.}$$

$$(x_2 - x_1)m = y_2 - y_1 \qquad \text{Simplify.}$$

$$y_2 - y_1 = m(x_2 - x_1) \qquad \text{Rewrite with } y\text{'s on the left side to resemble slope–intercept form.}$$

Connection We also can view the formula for slope as a proportion and cross multiply to get the point–slope form.

$$\frac{m}{1} = \frac{y_2 - y_1}{x_2 - x_1}$$

$$(x_2 - x_1)m = y_2 - y_1$$

To write the equation of a line with this formula, we replace m with the slope and substitute one of the given ordered pairs for x_1 and y_1, leaving x_2 and y_2 as variables. To indicate that x_2 and y_2 remain variables, we remove their subscripts so that we have $y - y_1 = m(x - x_1)$, which is called the point–slope form of the equation of a line.

Instructor Note Emphasize that the point–slope form is just the slope formula written in a different form.

Answers to Warm-up

1. $y = 2x - 1$

2. $y = \frac{4}{3}x - 5$

3. $y = \frac{5}{2}x - 3$

Procedure **Using the Point–Slope Form of the Equation of a Line**

To write the equation of a line given its slope and any point, (x_1, y_1), on the line, use the point–slope form of the equation of a line, $y - y_1 = m(x - x_1)$. If given a second point, (x_2, y_2), and not the slope, we first calculate the slope using

$$m = \frac{y_2 - y_1}{x_2 - x_1}; \text{ then we use } y - y_1 = m(x - x_1).$$

Example 1 Write the equation of a line with a slope of 2 passing through the point $(3, 5)$. Write the equation in slope–intercept form.

Solution: Because we are given the coordinates of a point and the slope of a line passing through the point, we begin with the point–slope form. After replacing m, x_1, and y_1 with their corresponding values, we isolate y to get slope–intercept form.

$$y - y_1 = m(x - x_1)$$
$$y - 5 = 2(x - 3) \qquad \text{Replace } m \text{ with 2, } x_1 \text{ with 3, and } y_1 \text{ with 5.}$$
$$y - 5 = 2x - 6 \qquad \text{Simplify.}$$
$$y = 2x - 1 \qquad \text{Add 5 to both sides to isolate } y.$$
$$\text{This is now slope–intercept form.}$$

Your Turn 1 Write the equation of the line in slope–intercept form with the given slope passing through the given point.

$$m = 0.6; (-4, 2)$$

Example 2 Write the equation of a line passing through the points $(3, 2)$ and $(-3, 6)$. Write the equation in slope–intercept form.

Solution: Because we do not have the slope, we calculate it.

$$m = \frac{6 - 2}{-3 - 3} = \frac{4}{-6} = -\frac{2}{3} \qquad \text{Use } m = \frac{y_2 - y_1}{x_2 - x_1}.$$

Note If one of the given points is the y-intercept, after finding the slope, you can use the slope–intercept form to write the equation of the line.

Now we can use the point–slope form, then isolate y to write the slope–intercept form. Because we are given two points, we can use either point for (x_1, y_1) in the point–slope equation. We select $(3, 2)$ to be (x_1, y_1).

$$y - y_1 = m(x - x_1)$$
$$y - 2 = -\frac{2}{3}(x - 3) \qquad \text{Replace } m \text{ with } -\frac{2}{3}, x_1 \text{ with 3, and } y_1 \text{ with 2.}$$
$$y - 2 = -\frac{2}{3}x + 2 \qquad \text{Simplify.}$$
$$y = -\frac{2}{3}x + 4 \qquad \text{Add 2 to both sides to isolate } y.$$

Your Turn 2 Write the equation of the line passing through the given points. Write the equation in slope–intercept form.

$$(-2, -4), (4, -1)$$

Writing Linear Equations in Standard Form

Equations can also be written in standard form. Recall from Section 3.3 that standard form is $Ax + By = C$, where A, B, and C are real numbers. We will manipulate the point–slope form of the equation so that the x and y terms are on the same side of the equation and a constant appears on the other side of the equation.

Answer to Your Turn 1
$y = 0.6x + 4.4$

Answer to Your Turn 2
$y = \frac{1}{2}x - 3$

Example 3 The following data points relate the velocity of an object as time passes. (You may recall this data from Your Turn 4 in Section 3.1.) The graph shows that the points are in a line. Write the equation of the line in standard form.

Time (x) (in seconds)	Velocity (y) (in ft./sec.)
0	2.0
1	2.5
2	3.0
3	3.5

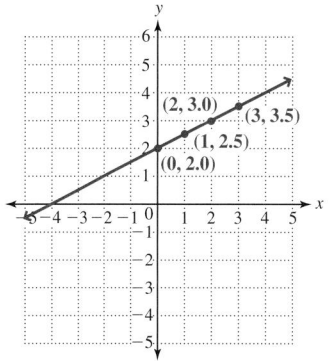

Solution: First, we find the slope of the line using any two ordered pairs in the slope formula. We will use $(0, 2)$ and $(1, 2.5)$.

$$m = \frac{2.5 - 2}{1 - 0} = 0.5 \text{ or } \frac{1}{2} \quad \text{Use } m = \frac{y_2 - y_1}{x_2 - x_1}.$$

Because $(0, 2)$ is the y-intercept, we can write the equation in slope–intercept form, then manipulate the equation so that x and y appear on the same side of the equal sign to get standard form.

$$y = \frac{1}{2}x + 2 \quad \text{Slope–intercept form}$$

$$y - \frac{1}{2}x = 2 \quad \text{Subtract } \frac{1}{2}x \text{ from both sides to get } x \text{ and } y \text{ together.}$$

When we write an equation in standard form $(Ax + By = C)$, it is customary to write the equation so that the x term is first with a positive coefficient and, if possible, so that A, B, and C are all integers. Let's manipulate $y - \frac{1}{2}x = 2$ to this more polished form.

$$-\frac{1}{2}x + y = 2 \quad \text{Use the commutative property of addition to write the } x \text{ term first.}$$

$$(-2)\left(-\frac{1}{2}x + y\right) = (-2)(2) \quad \text{Multiply both sides by } -2 \text{ so that the coefficient of } x \text{ is a positive integer.}$$

$$x - 2y = -4 \quad \text{Standard form with } A, B, \text{ and } C \text{ integers and } A > 0$$

Example 4   A line connects the points $(1, 5)$ and $(-3, 2)$. Write the equation of the line in the form $Ax + By = C$, where A, B, and C are integers and $A > 0$.

Solution: Find the slope; then write the equation of the line using point–slope form. To finish, rewrite the equation in standard form.

$$\text{Find the slope:} \quad m = \frac{2 - 5}{-3 - 1} = \frac{-3}{-4} = \frac{3}{4} \quad \text{Use } m = \frac{y_2 - y_1}{x_2 - x_1}.$$

Because we were not given the y-intercept, we use the point–slope form of the linear equation.

$$y - 5 = \frac{3}{4}(x - 1) \quad \text{Use } y - y_1 = m(x - x_1) \text{ and } (1, 5) \text{ for } (x_1, y_1).$$

Now manipulate the equation to get the form $Ax + By = C$, where A, B, and C are integers and $A > 0$.

$$y - 5 = \frac{3}{4}x - \frac{3}{4} \quad \text{Distribute } \frac{3}{4} \text{ to clear parentheses.}$$

$$4(y - 5) = 4\left(\frac{3}{4}x - \frac{3}{4}\right) \quad \text{Multiply both sides by the LCD, 4.}$$

Instructor Note Point out that you would get the same equation if you used $(-3, 2)$ instead of $(1, 5)$ for (x_1, y_1).

$$4y - 20 = 3x - 3$$

$$4y - 3x - 20 = -3 \qquad \text{Subtract } 3x \text{ from both sides to get } x \text{ and } y \text{ together.}$$

$$4y - 3x = 17 \qquad \text{Add 20 to both sides to get the constant terms on the right-hand side of the equation.}$$

$$-3x + 4y = 17 \qquad \text{Use the commutative property of addition to write the } x \text{ term first.}$$

$$(-1)(-3x + 4y) = (-1)(17) \qquad \text{Multiply both sides by } -1 \text{ so that the coefficient of } x \text{ is positive.}$$

$$3x - 4y = -17 \qquad \text{Standard form with } A, B, \text{ and } C \text{ integers and } A > 0$$

Your Turn 4   Write the equation of the line through the given points in the form $Ax + By = C$, where $A, B,$ and C are integers and $A > 0$.

$$(-3, 6), (1, -4)$$

Objective 2 Write the equation of a line parallel to a given line.

Consider the graphs of $y = 2x - 3$ and $y = 2x + 1$.

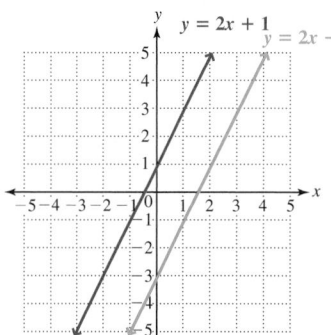

These two lines are parallel, which means that they will never intersect. Notice in the equations that their slopes are equal, which is why they are parallel.

Rule Parallel Lines
Nonvertical parallel lines have equal slopes and different y-intercepts. Vertical lines are parallel.

Example 5   Write the equation of a line that passes through $(1, -4)$ and is parallel to $y = -3x + 5$. Write the equation in slope–intercept form.

Solution: In the given equation, we see that the slope of the line is -3; so the slope of the parallel line will also be -3. We can now use the point–slope form to write the equation of the parallel line.

$$y - y_1 = m(x - x_1)$$

$$y - (-4) = -3(x - 1) \qquad \text{Replace } m \text{ with } -3, x_1 \text{ with 1, and } y_1 \text{ with } -4.$$

$$y + 4 = -3x + 3 \qquad \text{Simplify.}$$

$$y = -3x - 1 \qquad \text{Subtract 4 from both sides to isolate } y.$$

Instructor Note You might ask students to graph the given equation (in this case, $y = -3x + 5$) and the equation they found (in this case, $y = -3x - 1$) and see if it confirms their answer.

Answer to Your Turn 4
$5x + 2y = -3$

Answer to Your Turn 5
$y = \dfrac{3}{4}x + 1$

Your Turn 5   Write the equation of the line that passes through the given point and is parallel to the given line. Write the equation in slope–intercept form.

$$(-4, -2); y = \frac{3}{4}x - 7$$

Objective 3 Write the equation of a line perpendicular to a given line.

Consider the graphs of $y = \frac{2}{3}x - 4$ and $y = -\frac{3}{2}x + 1$.

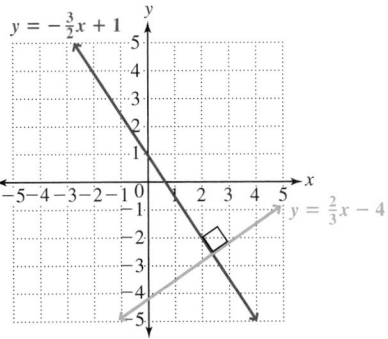

These two lines are perpendicular, which means that they intersect at a 90° angle. Notice in the equations that their slopes are negative reciprocals. In other words, $-\frac{3}{2}$ is the negative reciprocal of $\frac{2}{3}$.

Rule Perpendicular Lines

The slope of a line perpendicular to a line with a slope of $\frac{a}{b}$ will be $-\frac{b}{a}$. Horizontal and vertical lines are perpendicular.

Another way of stating the rule is that the slopes of perpendicular lines are opposites *and* reciprocals. Also note that $\left(\frac{a}{b}\right)\left(-\frac{b}{a}\right) = -1$; so two lines are perpendicular if and only if the product of their slopes is -1.

Example 6 Write the equation of a line that passes through $(3, -1)$ and is perpendicular to $3x + 4y = 8$. Write the equation in slope–intercept form.

Solution: To write the equation of the perpendicular line, we need its slope. To determine its slope, we need the slope of the given line. Because the equation of the given line is in standard form, to determine its slope, we rewrite it in slope–intercept form.

$$3x + 4y = 8$$
$$4y = -3x + 8 \qquad \text{Subtract } 3x \text{ from both sides.}$$
$$y = -\frac{3}{4}x + 2 \qquad \text{Divide both sides by 4.}$$

We see that the slope of the given line is $-\frac{3}{4}$; so the slope of any perpendicular line will be $\frac{4}{3}$. We can now use point–slope form to write the equation of the perpendicular line through $(3, -1)$.

$$y - y_1 = m(x - x_1)$$
$$y - (-1) = \frac{4}{3}(x - 3) \qquad \text{Replace } m \text{ with } \frac{4}{3}, x_1 \text{ with 3, and } y_1 \text{ with } -1.$$
$$y + 1 = \frac{4}{3}x - 4 \qquad \text{Simplify.}$$
$$y = \frac{4}{3}x - 5 \qquad \text{Subtract 1 from both sides to isolate } y.$$

Answer to Your Turn 6
$y = -\frac{2}{5}x - 1$

Your Turn 6 Write the equation of a line that passes through $(-10, 3)$ and is perpendicular to $5x - 2y = 6$. Write the equation in slope–intercept form.

By using slopes, it is possible to determine whether the graphs of two lines are parallel, perpendicular, or neither parallel nor perpendicular.

Example 7   Determine whether the given lines are parallel, perpendicular, or neither.

a. $y = \dfrac{2}{3}x - 4$ and $y = -\dfrac{3}{2}x + 2$

Solution: The slopes are $\dfrac{2}{3}$ and $-\dfrac{3}{2}$. Because the slopes are reciprocals with opposite signs, the lines are perpendicular.

b. $2x + 4y = 7$ and $3x + 6y = 8$

Solution: To determine the slopes of the lines, write the equations in slope–intercept form.

$$2x + 4y = 7 \qquad\qquad 3x + 6y = 8$$
$$4y = -2x + 7 \qquad\qquad 6y = -3x + 8$$
$$y = -\frac{1}{2}x + \frac{7}{4} \qquad\qquad y = -\frac{1}{2}x + \frac{4}{3}$$

Because the slope of both lines is $-\dfrac{1}{2}$ and the y-intercepts differ, the lines are parallel.

Learning Strategy

The best way to prepare for class is to complete all the assigned homework problems and read the material in the text that will be discussed in the next class.
—Paul S.

Note If the y-intercepts also were the same, these would be the same line, not two parallel lines. ▶

Your Turn 7   Determine whether the given lines are parallel, perpendicular, or neither.

Answer to Your Turn 7
a. neither   **b.** parallel

a. $y = 2x + 4$ and $y = -2x - 1$ **b.** $x - 2y = 6$ and $3x - 6y = 12$

3.5 Exercises (For Extra Help) MyMathLab®

Note: Exercises marked with a ★ represent challenging exercises.

Objective 1

Prep Exercise 1   What is the point–slope equation? $y - y_1 = m(x - x_1)$

For Exercises 1–12, write the equation of the line in slope–intercept form with the given slope passing through the given point. See Example 1.

1. $m = 3;\ (5, 2)$
$y = 3x - 13$

2. $m = 2;\ (3, -5)$
$y = 2x - 11$

3. $m = -1;\ (-1, -1)$
$y = -x - 2$

4. $m = -5;\ (2, 0)$
$y = -5x + 10$

5. $m = \dfrac{2}{3};\ (6, 3)$
$y = \dfrac{2}{3}x - 1$

6. $m = \dfrac{3}{2};\ (-2, 3)$
$y = \dfrac{3}{2}x + 6$

7. $m = -\dfrac{3}{4};\ (-1, -5)$
$y = -\dfrac{3}{4}x - \dfrac{23}{4}$

8. $m = -\dfrac{2}{5};\ (-2, -1)$
$y = -\dfrac{2}{5}x - \dfrac{9}{5}$

9. $m = -\dfrac{4}{3};\ (-2, 4)$
$y = -\dfrac{4}{3}x + \dfrac{4}{3}$

10. $m = \dfrac{2}{9};\ (-1, -7)$
$y = \dfrac{2}{9}x - \dfrac{61}{9}$

11. $m = 2;\ (0, 0)$
$y = 2x$

12. $m = -3;\ (0, 0)$
$y = -3x$

Prep Exercise 2 How do you rewrite an equation that is in point–slope form in standard form?

Manipulate the point–slope form so that the variables are on one side of the equation and a constant is on the other side.

For Exercises 13–16, write the equation of a line. Leave your answer in standard form. (Hint:
Find the slope and a point or two points on the line.) See Examples 1 and 2.

13.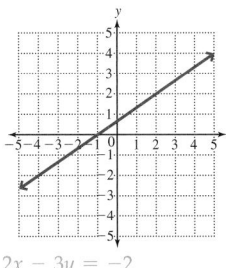

$2x - 3y = -2$

14.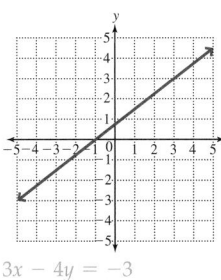

$3x - 4y = -3$

15.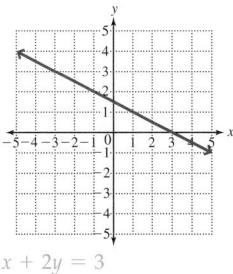

$x + 2y = 3$

16.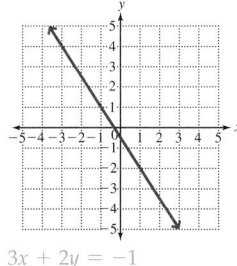

$3x + 2y = -1$

Prep Exercise 3 What is the first step when you are given two ordered pairs and asked to write the equation of the line through the two points?

Find the slope.

Prep Exercise 4 Given the ordered pairs for two points on a line, what is the formula for finding the slope of the line?

$m = \dfrac{y_2 - y_1}{x_2 - x_1}$

Prep Exercise 5 When given two ordered pairs and asked to write an equation of a line, does it matter which ordered pair is used in the point–slope equation? Explain.

No, both points produce the same equation.

For Exercises 17–28, write the equation of a line passing through the given points. Write the
equation in slope–intercept form. See Example 2.

17. $(4, -3), (-1, 7)$
$y = -2x + 5$

18. $(-1, -5), (-4, 1)$
$y = -2x - 7$

19. $(3, 1), (5, 6)$
$y = \dfrac{5}{2}x - \dfrac{13}{2}$

20. $(-5, -2), (7, 5)$
$y = \dfrac{7}{12}x + \dfrac{11}{12}$

21. $(0, 0), (-1, -7)$
$y = 7x$

22. $(0, 0), (-5, 1)$
$y = -\dfrac{1}{5}x$

23. $(8, -1), (8, -10)$
$x = 8$

24. $(-3, 2), (-3, 5)$
$x = -3$

25. $(-9, -1), (2, -1)$
$y = -1$

26. $(-2, 5), (6, 5)$
$y = 5$

★ 27. $(1.4, 5), (3, 9.8)$
$y = 3x + 0.8$

★ 28. $(3.9, -2.1), (4.1, -1)$
$y = 5.5x - 23.55$

For Exercises 29–32, write the equation of the line in slope–intercept form given the y-intercept
and one other point. See Objective 1.

29. $(3, 4), (0, -2)$
$y = 2x - 2$

30. $(0, 2), (4, 10)$
$y = 2x + 2$

31. $(0, 3), (5, -3)$
$y = -\dfrac{6}{5}x + 3$

32. $(-3, -1), (0, -5)$
$y = -\dfrac{4}{3}x - 5$

For Exercises 33–44, write the equation of a line through the given points in the form
$Ax + By = C$, *where A, B, and C are integers and* $A > 0$. *See Example 4.*

33. $(2, 3), (4, 8)$
$5x - 2y = 4$

34. $(4, 1), (8, 4)$
$3x - 4y = 8$

35. $(2, 0), (0, -3)$
$3x - 2y = 6$

36. $(-6, 0), (0, 6)$
$x - y = -6$

37. $(-4, -1), (7, -5)$
$4x + 11y = -27$

38. $(5, -2), (-3, -3)$
$x - 8y = 21$

39. $(-3, 2), (-6, 4)$
$2x + 3y = 0$

40. $(4, 8), (-4, -8)$
$2x - y = 0$

41. $(-1, 0), (-1, 8)$
$x = -1$

42. $(-2, 5), (-2, -6)$
$x = -2$

43. $(3, -1), (4, -1)$
$y = -1$

44. $(2, -4), (3, -4)$
$y = -4$

Objective 2

Prep Exercise 6 The slopes of parallel lines are ___equal___ .

For Exercises 45–56, write the equation of a line that passes through the given point and is parallel to the given line. See Example 5.
a. *Write the equation in slope–intercept form.*
b. *Write the equation in the form $Ax + By = C$, where A, B, and C are integers and $A > 0$.*

45. $(4, 2); y = 4x + 1$
a. $y = 4x - 14$
b. $4x - y = 14$

46. $(0, -1); y = 3x + 2$
a. $y = 3x - 1$
b. $3x - y = 1$

47. $(-2, 1); y = -3x + 1$
a. $y = -3x - 5$
b. $3x + y = -5$

48. $(0, 3); y = -2x + 4$
a. $y = -2x + 3$
b. $2x + y = 3$

49. $(5, 2); y = \dfrac{1}{3}x + 4$
a. $y = \dfrac{1}{3}x + \dfrac{1}{3}$
b. $x - 3y = -1$

50. $(-2, 4); y = \dfrac{3}{4}x - 3$
a. $y = \dfrac{3}{4}x + \dfrac{11}{2}$
b. $3x - 4y = -22$

51. $(-1, -3); y = -\dfrac{3}{4}x + 1$
a. $y = -\dfrac{3}{4}x - \dfrac{15}{4}$
b. $3x + 4y = -15$

52. $(-1, -1); y = -\dfrac{2}{3}x + 6$
a. $y = -\dfrac{2}{3}x - \dfrac{5}{3}$
b. $2x + 3y = -5$

53. $(2, -3); 2x + y = 5$
a. $y = -2x + 1$
b. $2x + y = 1$

54. $(5, 1); 6x + 3y = 8$
a. $y = -2x + 11$
b. $2x + y = 11$

55. $(-4, 2); 3x - 4y = 7$
a. $y = \dfrac{3}{4}x + 5$
b. $3x - 4y = -20$

56. $(2, -6); 3x - 2y = 5$
a. $y = \dfrac{3}{2}x - 9$
b. $3x - 2y = 18$

Objective 3

Prep Exercise 7 How are the slopes of perpendicular lines related? They are reciprocals with opposite signs.

For Exercises 57–68, write the equation of a line that passes through the given point and is perpendicular to the given line. See Example 6.
a. *Write the equation in slope–intercept form.*
b. *Write the equation in the form $Ax + By = C$, where A, B, and C are integers and $A > 0$.*

57. $(-4, -1); y = 2x + 3$
a. $y = -\dfrac{1}{2}x - 3$
b. $x + 2y = -6$

58. $(3, -3); y = 3x - 1$
a. $y = -\dfrac{1}{3}x - 2$
b. $x + 3y = -6$

59. $(3, -4); y = \dfrac{1}{4}x + 2$
a. $y = -4x + 8$
b. $4x + y = 8$

60. $(2, 3); y = \dfrac{1}{5}x - 4$
a. $y = -5x + 13$
b. $5x + y = 13$

61. $(2, -4); y = -2x + 7$
a. $y = \dfrac{1}{2}x - 5$
b. $x - 2y = 10$

62. $(-1, -1); y = -3x + 4$
a. $y = \dfrac{1}{3}x - \dfrac{2}{3}$
b. $x - 3y = 2$

63. $(-2, 3); y = \dfrac{3}{2}x - \dfrac{5}{2}$
a. $y = -\dfrac{2}{3}x + \dfrac{5}{3}$
b. $2x + 3y = 5$

64. $(4, 2); y = \dfrac{3}{4}x - 11$
a. $y = -\dfrac{4}{3}x + \dfrac{22}{3}$
b. $4x + 3y = 22$

65. $(-3, -1); 2x - 5y = 10$
a. $y = -\dfrac{5}{2}x - \dfrac{17}{2}$
b. $5x + 2y = -17$

66. $(-2, -4); 3x - 4y = 8$
a. $y = -\dfrac{4}{3}x - \dfrac{20}{3}$
b. $4x + 3y = -20$

67. $(3, -4); 2x + 3y = 12$
a. $y = \dfrac{3}{2}x - \dfrac{17}{2}$
b. $3x - 2y = 17$

68. $(3, 7); 5x + 2y = 3$
a. $y = \dfrac{2}{5}x + \dfrac{29}{5}$
b. $2x - 5y = -29$

For Exercises 69–84, determine whether the given lines are parallel, perpendicular, or neither. See Example 7.

69. $y = 4x - 7$
$y = 4x + 2$
parallel

70. $y = 3x + 2$
$y = 3x - 2$
parallel

71. $y = \dfrac{1}{2}x$
$y = -2x + 4$
perpendicular

72. $y = -x + 3$
$y = x - 2$
perpendicular

73. $y = \dfrac{1}{2}x + 3$
$y = 2x - 1$
neither

74. $y = 3x + 4$
$y = \dfrac{1}{3}x - 3$
neither

75. $4x + 6y = 5$
$8x + 12y = 9$
parallel

76. $3x + 6y = 10$
$2x + 4y = 9$
parallel

77. $5x + 10y = 12$
$4x - 2y = 5$
perpendicular

78. $5x - 3y = 11$
$3x + 5y = 8$
perpendicular

79. $2x + 3y = -6$
$3x + 2y = 4$
neither

80. $4x - 2y = 5$
$2x - 4y = 3$
neither

81. $y = 1$
    $x = -3$
    perpendicular

82. $x = 2$
    $y = -4$
    perpendicular

83. $x = 4$
    $x = -2$
    parallel

84. $y = 4$
    $y = -3$
    parallel

85. A taxi charges a \$7.75 initial fee plus \$0.30 for each one-eighth of a mile.
   a. Let n represent the number of eighths of a mile and c represent the total cost of the taxi. Write an equation in slope–intercept form that describes how much it costs to hire a taxi.
     $c = 0.3n + 7.75$
   b. What will be the total cost of traveling 5 miles from the airport?
     \$19.75
   c. Graph the equation with n plotted along the horizontal axis and c along the vertical axis. Because n and c are nonnegative, restrict the graph to the first quadrant only.

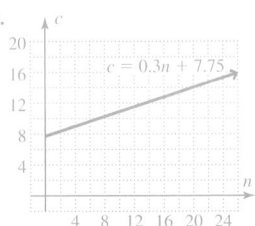

86. A company manager is examining profits. He sees that the profit in May was \$52,000. He then discovers that his company's profits have been declining \$1000 each month thereafter.
   a. Let n represent the number of months and p represent the profit. Write an equation in slope–intercept form that describes the profit n months after May.
     $p = -1000n + 52{,}000$
   b. If the decline continues at the same rate, what will be the profit in nine months?
     \$43,000
   c. Graph the equation with n plotted along the horizontal axis and p along the vertical axis. Because n and p are nonnegative, restrict the graph to the first quadrant only.

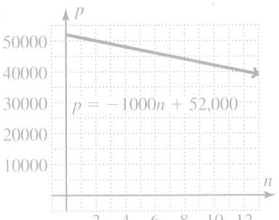

87. In the 2000–2001 school year, about 2.78 million females were participating in high school athletics. Over the next 10 years, participation increased in an approximately linear pattern so that in 2010–2011, 3.17 million females were participating in high school athletics. (*Source: 2010–2011 High School Athletics Participation Survey.*)
   a. Consider 2000–2001 to be year 0 so that 2010–2011 is year 10. Plot the two ordered pairs with the number of years along the horizontal axis and the number of females participating in athletics along the vertical axis. Then draw a line through the two points. Because the number of athletes and years are both nonnegative, restrict the graph to the first quadrant.
   b. Find the slope of the line.
     0.039
   c. Let n represent the number of years after the 2000–2001 school year and p represent the number of female participants in high school athletics. Write an equation of the line in slope–intercept form.
     $p = 0.039n + 2.78$
   d. Using your equation, how many females participated in high school athletics in the 2005–2006 school year?
     2.975 million
   e. If the linear trend continues, how many females will participate in high school athletics in the 2018–2019 school year?
     3.482 million

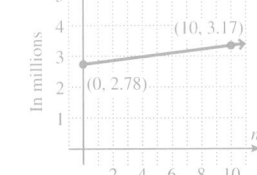

88. The population of the United States was 281 million in 2000 and continued to increase in a linear pattern until 2010, when the population was 309 million.
   a. Consider 2000 to be year 0 so that 2010 is year 10. Plot the two ordered pairs with the year along the horizontal axis and the population (in millions) along the vertical axis. Then draw a line through the two points. Because the years and the population are both nonnegative, restrict the graph to the first quadrant.
   b. Find the slope of the line.
     2.8
   c. Let n represent the number of years after 2000 and P represent the population in millions. Write an equation of the line in slope–intercept form.
     $P = 2.8n + 281$

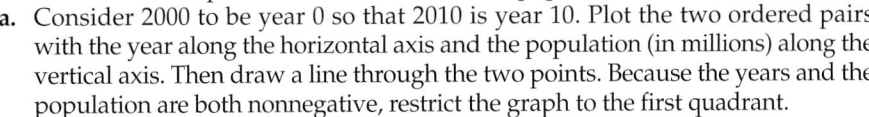
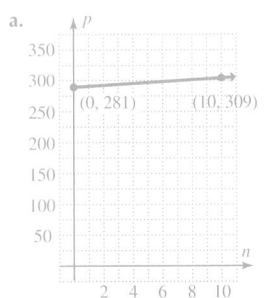

   d. Using your equation, what was the population of the United States in 2007?

     300.6 million

   e. If the linear trend continues, what will be the population in 2018?

     331.4 million

89. A stock begins a decline in price following a linear pattern of depreciation. The stock's initial price before the decline began was $36. On the 5th day, the closing price was $28.50.

   a. Graph the line with the number of days after the decline begins plotted along the horizontal axis and the price along the vertical axis. Because the number of days and price are nonnegative, the graph is restricted to the first quadrant only.

   b. Find the slope of the line.

     −1.5

   c. Write an equation of the line in slope–intercept form with n representing the number of days of decline in price and p representing the price.

     $p = -1.5n + 36$

   d. Find the p-intercept. Explain what the p-intercept represents in the problem.

     $(0, 36)$; the p-intercept indicates the initial price.

   e. Find the n-intercept. Explain what the n-intercept represents in the problem.

     $(24, 0)$; the n-intercept indicates that on the 24th day, the price will be $0.

   f. If the decline in price were to continue at the same rate, what would be the price of the stock after eight days of decline?

     $24

a.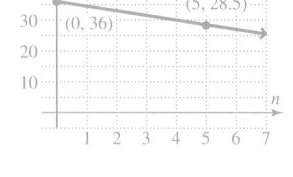

90. Keith deducts the value of his computer on his taxes. The amount of the deduction is based on the value of the computer, which depreciates each year. The initial value of the computer was $2100. After two years in service, the computer's value is $1500.

   a. Graph the line with the number of years plotted along the horizontal axis and the value along the vertical axis. Because the number of years and the value are nonnegative, the graph is restricted to the first quadrant only.

   b. Find the slope of the line.

     −300

   c. Write an equation of the line in slope–intercept form with n representing the number of years the computer is in service and v representing the value.

     $v = -300n + 2100$

   d. Find the v-intercept. Explain what the v-intercept represents in the problem.

     $(0, 2100)$; the v-intercept indicates the initial value of the computer.

   e. Find the n-intercept. Explain what the n-intercept represents in the problem.

     $(7, 0)$; the n-intercept indicates the number of years it takes for the computer's value to be $0.

   f. If the decline in value were to continue at the same rate, what would be the computer's value after five years in service? $600

a.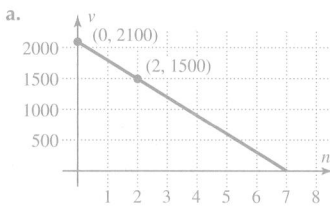

Review Exercises

Exercises 1–6 Equations and Inequalities

[1.1] **1.** Use $<, >$, or $=$ to write a true statement: $0.6 \quad \dfrac{3}{4}$

     $<$

[2.3] **2.** Solve: $\dfrac{3}{4}x - \dfrac{4}{5} = \dfrac{1}{2}x + 1$

     $\dfrac{36}{5}$

[2.8] *For Exercises 3 and 4, solve and graph the solution set on the number line.*

   3. $4x - 9 \geq 6x + 5$

     $x \leq -7$

     $-10\ -9\ -8\ -7\ -6\ -5\ -4$

   4. $2(x + 1) - 5x \geq 3x - 6$

     $x \leq \dfrac{4}{3}$

     $0 \quad 1\frac{4}{3}\ 2 \quad 3$

[3.4] **5.** What are the slope and y-intercept for $y = -\dfrac{3}{4}x + 2$?

     $m = -\dfrac{3}{4}; (0, 2)$

[3.3] **6.** Find the x- and y-intercepts for $2x - 7y = 14$.

     $(7, 0)(0, -2)$

3.6 Graphing Linear Inequalities

Objectives

1 Determine whether an ordered pair is a solution for a linear inequality with two variables.

2 Graph linear inequalities.

Warm-up

[3.2] **1.** Determine whether $(-4, 7)$ is a solution of $y = -2x + 3$.

[3.2] **2.** Graph $x - 2y = 4$.

[2.8] **3.** Graph $x \geq 3$ on a number line.

Objective 1 Determine whether an ordered pair is a solution for a linear inequality with two variables.

Now that we have learned how to graph linear equations, let's turn our attention to linear inequalities. Linear inequalities have the same form as linear equations except that they contain an inequality symbol instead of an equal sign. Examples of linear inequalities are as follows:

$$2x + 3y > 6$$
$$y \leq x - 3$$

Remember that a solution for a linear equation in two variables is an ordered pair that makes the equation true. Similarly, a solution to a linear inequality in two variables is an ordered pair that makes the inequality true. For example, the ordered pair $(2, 1)$ is a solution for $2x + 3y > 6$.

$$2x + 3y > 6$$
$$2(2) + 3(1) \overset{?}{>} 6 \qquad \text{Replace } x \text{ with 2 and } y \text{ with 1.}$$
$$4 + 3 \overset{?}{>} 6$$
$$7 > 6 \qquad \text{This is true, so } (2, 1) \text{ is a solution.}$$

Procedure Checking an Ordered Pair

To determine whether an ordered pair is a solution for an inequality, replace the variables with the corresponding coordinates. If the resulting inequality is true, the ordered pair is a solution.

Example 1 Determine whether $(-4, 7)$ is a solution for $y \geq -2x + 3$.

Solution: $y \geq -2x + 3$

$7 \overset{?}{\geq} -2(-4) + 3 \qquad$ Replace x with -4 and y with 7; then determine whether the inequality is true.

$7 \overset{?}{\geq} 8 + 3$

$7 \geq 11$

This statement is false, so $(-4, 7)$ is not a solution.

Your Turn 1 Determine whether the given ordered pair is a solution for the given inequality.

  a. $3x - 5y < 6; (-1, -2)$ **b.** $y \leq -4x + 1; (2, -7)$

Objective 2 Graph linear inequalities.

Graphing linear inequalities is very much like graphing linear equations. Consider $2x + 3y \geq 6$. The greater than or equal to sign indicates that solutions to the inequality $2x + 3y \geq 6$ will be ordered pairs that satisfy $2x + 3y > 6$ and $2x + 3y = 6$. The graph of $2x + 3y = 6$ is a line. Let's graph this line using intercepts.

Answers to Your Turn 1
a. No **b.** Yes

Answers to Warm-up
1. No
2.

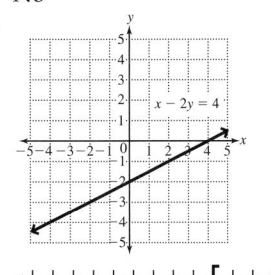

3.
$-6\ -5\ -4\ -3\ -2\ -1\ \ 0\ \ 1\ \ 2\ \ 3\ \ 4\ \ 5\ \ 6$

Find the x- and y-intercepts.

x-intercept:

$$2x + 3y = 6$$
$$2x + 3(0) = 6$$
$$2x = 6$$
$$x = 3$$

y-intercept:

$$2x + 3y = 6$$
$$2(0) + 3y = 6$$
$$3y = 6$$
$$y = 2$$

x-intercept: $(3, 0)$
y-intercept: $(0, 2)$

The "greater than" portion of the inequality consists of ordered pairs that satisfy $2x + 3y > 6$. To find one of these pairs, we select a point on one side of the line and see if its coordinates satisfy $2x + 3y > 6$. Let's select the point with coordinates $(1, 4)$.

$$2x + 3y > 6$$
$$2(1) + 3(4) \overset{?}{>} 6 \quad \text{Replace } x \text{ with 1 and } y \text{ with 4.}$$
$$2 + 12 \overset{?}{>} 6$$
$$14 > 6 \quad \text{This statement is true; therefore, } (1, 4) \text{ is a solution.}$$

It can be shown that every ordered pair in the region containing the point $(1, 4)$, which is above the line, will satisfy $2x + 3y > 6$. It also can be shown that no ordered pair on the other side of the line will satisfy $2x + 3y > 6$. For example, consider the origin $(0, 0)$.

$$2x + 3y > 6 \quad \text{Replace } x \text{ with 0 and } y \text{ with 0.}$$
$$2(0) + 3(0) > 6$$
$$0 > 6 \quad \text{This statement is false, so } (0, 0) \text{ is not a solution.}$$

The line that is the graph of $2x + 3y = 6$ is called a *boundary*. The region on one side of this boundary line contains all of the ordered pair solutions satisfying $2x + 3y > 6$. The region on the other side contains all ordered pairs that satisfy $2x + 3y < 6$. To indicate that every ordered pair above the boundary line is a solution to the inequality, we shade the region above the line.

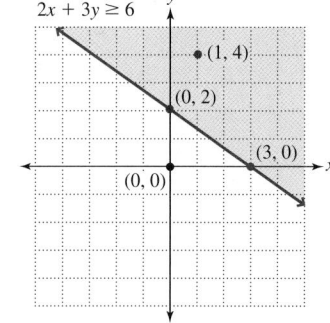

If the inequality had been $2x + 3y > 6$, the ordered pairs that satisfy $2x + 3y = 6$ would not be solutions. Because these ordered pairs form the boundary line, we draw a dashed line instead of a solid line to indicate that those ordered pairs are not solutions. The graph of $2x + 3y > 6$ is shown here.

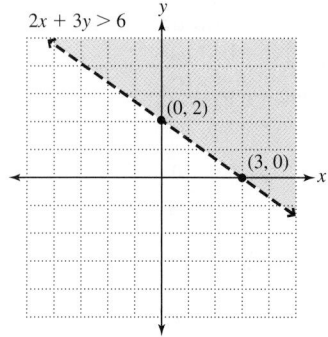

For the inequalities $2x + 3y \leq 6$ and $2x + 3y < 6$, we shade the region on the other side of the boundary line.

Connection When graphing inequalities with the symbols $\leq$ or $\geq$ on a number line, we used a bracket (or closed circle) to indicate that equality is included. When the inequalities have $<$ or $>$, we used a parenthesis (or open circle) to indicate that equality is excluded. When linear inequalities are graphed in the rectangular coordinate system, a solid line is like a bracket and a dashed line is like a parenthesis.

The graph of $2x + 3y \leq 6$:

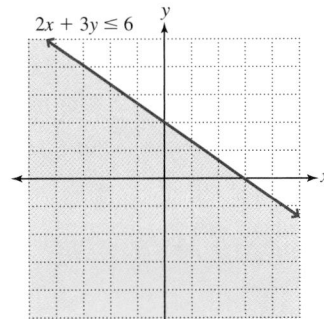

The graph of $2x + 3y < 6$:

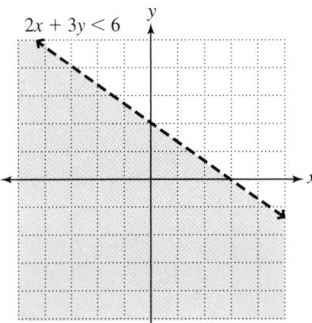

Procedure   Graphing Linear Inequalities

To graph a linear inequality in two variables:
1. Graph the related equation (the boundary line). The related equation has an equal sign in place of the inequality symbol. If the inequality symbol is $\leq$ or $\geq$, draw a solid line. If the inequality symbol is $<$ or $>$, draw a dashed line.
2. Choose an ordered pair on one side of the boundary line and test this ordered pair in the inequality. If the ordered pair satisfies the inequality, shade the region that contains it. If the ordered pair does not satisfy the inequality, shade the region on the other side of the boundary line.

Example 2   Graph the linear inequalities.

a. $y > 3x + 1$

Solution: First graph the related equation $y = 3x + 1$. Two ordered pairs that satisfy $y = 3x + 1$ are $(0, 1)$ and $(1, 4)$.

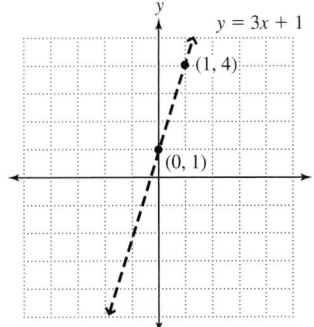

Note The equation $y = 3x + 1$ could also be graphed using the slope and y-intercept.

Because the inequality is strictly greater than, we draw a dashed line to indicate that ordered pairs on this boundary line are not solutions.

Now we choose an ordered pair on one side of the line and test this ordered pair in the inequality. We can choose any point, but for easier calculations, let's choose $(0, 0)$.

Note You cannot choose the origin if it lies on the boundary line.

$$y > 3x + 1$$
$$0 \overset{?}{>} 3(0) + 1 \qquad \text{Replace } x \text{ with } 0 \text{ and } y \text{ with } 0.$$
$$0 > 1 \qquad \text{This statement is false, so } (0, 0) \text{ is not a solution for the inequality.}$$

Because $(0, 0)$ did not satisfy the inequality, we shade the region on the other side of the boundary line.

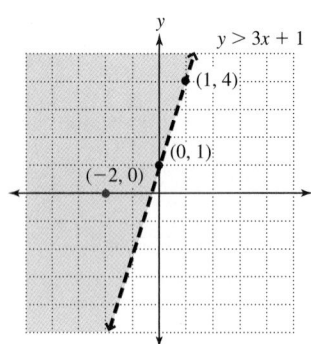

We can confirm that our shading is correct by choosing an ordered pair in this region. We will choose the ordered pair $(-2, 0)$.

$$y > 3x + 1$$
$$0 \overset{?}{>} 3(-2) + 1 \qquad \text{Replace } x \text{ with } -2 \text{ and } y \text{ with } 0.$$
$$0 > -5 \qquad \text{This statement is true, which confirms that we have shaded the correct side of the boundary line.}$$

b. $x - 2y \le 4$

Solution: First graph the related equation $x - 2y = 4$. Two ordered pairs that satisfy $x - 2y = 4$ are $(0, -2)$ and $(4, 0)$.

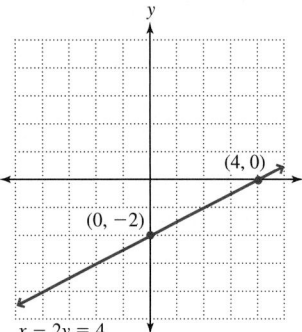

Because the inequality symbol is less than or equal to, we draw a solid line to indicate that ordered pairs on the boundary line are solutions.

Now we choose an ordered pair on one side of the line and test this ordered pair in the inequality. We will choose the origin $(0, 0)$.

$$x - 2y \le 4$$
$$0 - 2(0) \overset{?}{\le} 4 \qquad \text{Replace } x \text{ with } 0 \text{ and } y \text{ with } 0.$$
$$0 \le 4 \qquad \text{This statement is true; therefore, } (0, 0) \text{ is a solution for the inequality.}$$

Because $(0, 0)$ satisfies the inequality, we shade the region that contains it.

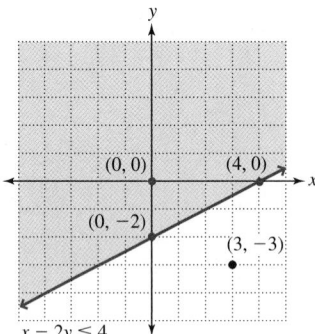

We can confirm that the region on the other side of the line should not be shaded by choosing a point in that region, such as $(3, -3)$.

$$x - 2y \le 4$$
$$3 - 2(-3) \overset{?}{\le} 4 \qquad \text{Replace } x \text{ with } 3 \text{ and } y \text{ with } -3.$$
$$3 + 6 \overset{?}{\le} 4$$
$$9 \le 4 \qquad \text{This statement is false, which confirms that the region containing the point with coordinates } (3, -3) \text{ should not be shaded.}$$

c. $x > 2$

Solution: First graph the related line $x = 2$, which is a vertical line intersecting the x-axis at $(2, 0)$.

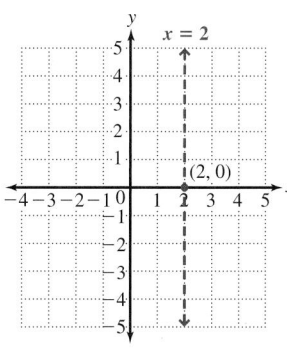

Because the inequality symbol is greater than, we draw a dashed line to indicate that the ordered pairs on the boundary line are not solutions.

Now we choose an ordered pair on one side of the line and test this ordered pair in the inequality. We will choose the origin $(0, 0)$.

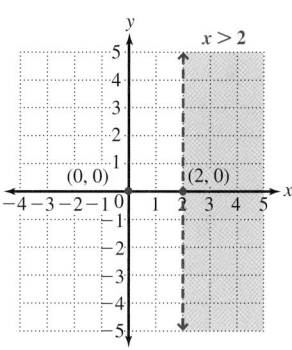

$x > 0$

$0 > 0$ Replace x with 0, which results in a false statement. So shade the side of the line opposite $(0, 0)$.

Note It is possible to determine the solutions without using a test point. We are looking for all ordered pairs whose x-values are greater than 2. The ordered pairs whose x-values are greater than 2 are to the right of 2, much like graphing on a number line.

d. $y \leq -3$

Solution: First graph the related line $y = -3$, which is a horizontal line intersecting the y-axis at $(0, -3)$.

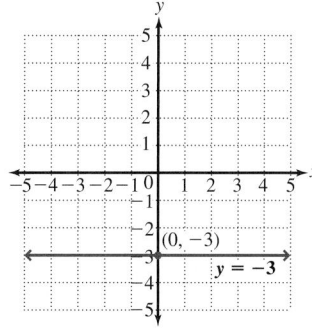

Because the inequality symbol is less than or equal to, we draw a solid line to indicate that the ordered pairs on the boundary line are solutions.

Now we choose an ordered pair on one side of the line and test this ordered pair in the inequality. We will choose the origin $(0, 0)$.

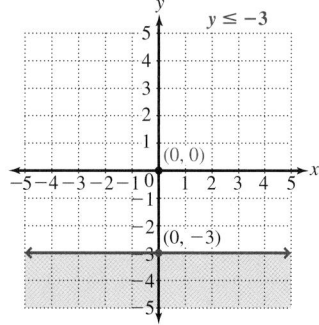

$y \leq -3$

$0 \leq -3$ Replace y with 0, which results in a false statement. So shade the side of the line on the opposite side from $(0, 0)$.

Note Again, it is possible to determine the solutions without using a test point. We are looking for all ordered pairs whose y-values are less than or equal to -3. The ordered pairs whose y-values are less than or equal to -3 are either on or below the line.

Your Turn 2   Graph the linear inequalities.

a. $y < -\dfrac{1}{3}x + 2$ **b.** $2x + 3y \geq 6$ **c.** $x \leq -3$ **d.** $y > 2$

Answers to Your Turn 2

a.

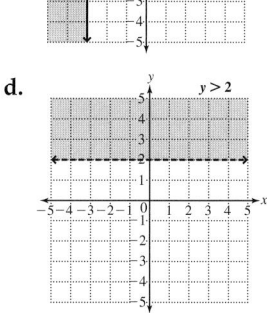

b.

c.

d.

3.6 Exercises For Extra Help MyMathLab®

Objective 1

Prep Exercise 1 Explain how to determine whether an ordered pair is a solution for an inequality. Replace the variables with the corresponding coordinates. If the resulting inequality is true, the ordered pair is a solution.

For Exercises 1–8, determine whether the ordered pair is a solution for the linear inequality. See Example 1.

1. $(5, 5)$; $y > -x + 2$
yes

2. $(0, 1)$; $y > -x + 2$
no

3. $(2, 0)$; $y \geq \dfrac{1}{2}x + 1$
no

4. $(0, 0)$; $y < -\dfrac{2}{3}x + 2$
yes

5. $(-1, 2)$; $3x - y \leq -8$
no

6. $(3, -1)$; $x - y < -2$
no

7. $(-1, -2)$; $x - 2y > -7$
yes

8. $(0, -2)$; $x + 4y \geq 8$
no

Objective 2

Prep Exercise 2 Explain how to determine the related equation for a given inequality. Replace the inequality symbol with an equal sign.

Prep Exercise 3 In graphing a linear inequality, when do you draw a dashed line and when do you draw a solid line? Draw a dashed line when the inequality is $>$ or $<$. Draw a solid line when the inequality is $\geq$ or $\leq$.

Prep Exercise 4 When graphing a linear inequality, how do you determine which side of the boundary line to shade? Choose an ordered pair on one side of the boundary line and test this ordered pair in the inequality. If the ordered pair satisfies the inequality, shade the region that contains it. If the ordered pair does not satisfy the inequality, shade the region on the other side of the boundary line.

For Exercises 9–36, graph the linear inequality. See Example 2.

9. $y \leq -x + 1$

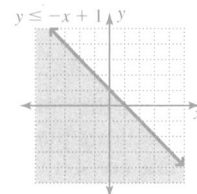

10. $y \leq x - 8$

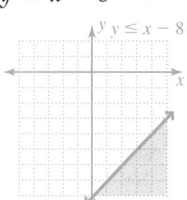

11. $y > -2x + 5$

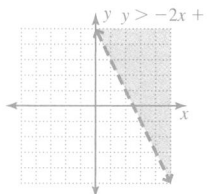

12. $y < -3x + 2$

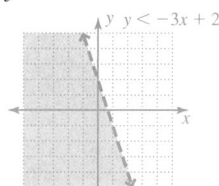

13. $y > 2x$

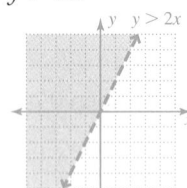

14. $y < x$

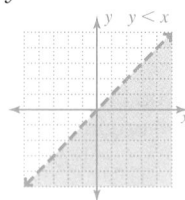

15. $y > \dfrac{1}{3}x$

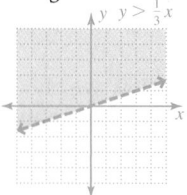

16. $y < \dfrac{2}{5}x$

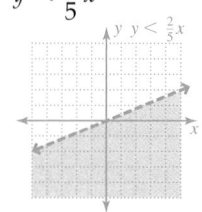

17. $y \geq \dfrac{3}{4}x + 2$

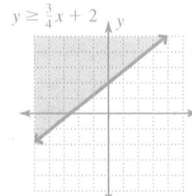

18. $y \geq -\dfrac{2}{3}x - 3$

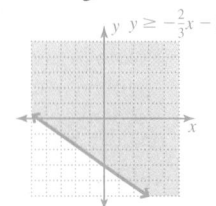

19. $3x + y \leq 9$

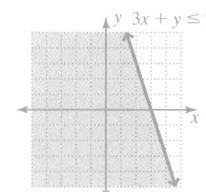

20. $2x + y > 6$

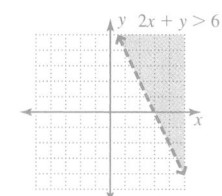

21. $5x - 2y < 10$

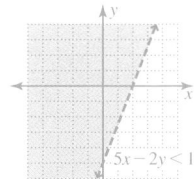

22. $3x - 2y \geq 6$

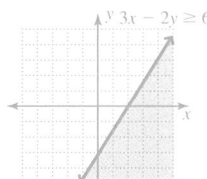

23. $3x + y \geq 8$

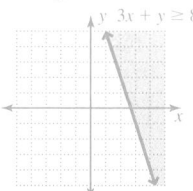

24. $5x + y \geq -10$

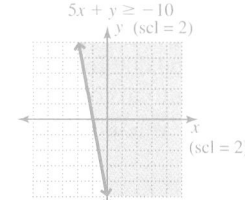

25. $2x + 3y < 9$

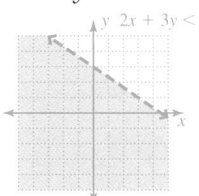

26. $4x + 2y \leq 3$

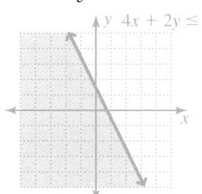

27. $3x + y \geq 0$

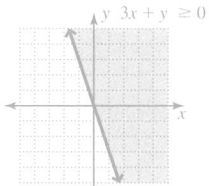

28. $-3x - 2y < 0$

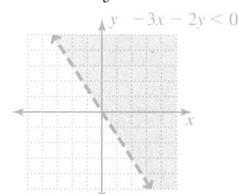

29. $x > -4$

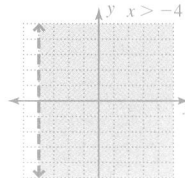

30. $x \leq 6$

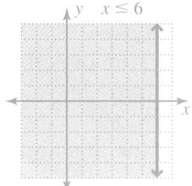

31. $y \leq 2$

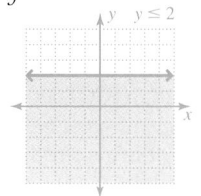

32. $y > -1$

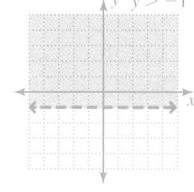

33. $x + 2 \geq 0$

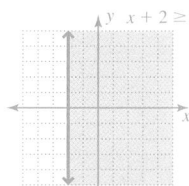

34. $x - 4 < 0$

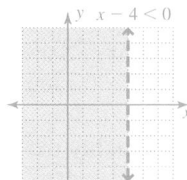

35. $y - 3 < 0$

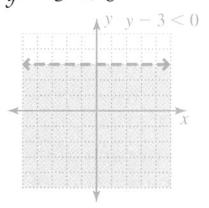

36. $y + 2 \geq 0$

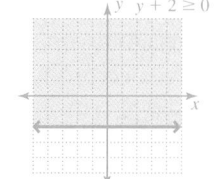

37. A company sells two sizes of bottles of hand lotion. The revenue for each unit of the large bottle is $2.00. The revenue for each unit of the small bottle is $1.50. For the company to break even during the first quarter, the company must generate $280,000 in revenue. The inequality $2x + 1.5y \geq 280,000$ describes the amount of revenue that must be generated from the sale of both size bottles for the company to break even or turn a profit.

a. What do x and y represent?

x represents the number of large bottles sold; y represents the number of small bottles sold.

b. Graph the inequality. Because x and y are both nonnegative, restrict the graph to the first quadrant.

c. What does the boundary line represent? What does the shaded region represent?

The boundary line represents combinations of bottle sizes sold to produce a revenue of exactly $280,000 (breakeven). The shaded region represents combinations that produce a revenue greater than $280,000 (profit).

b.

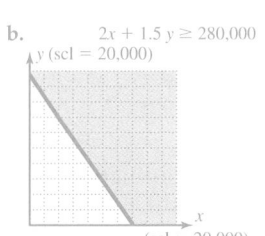

d. List three combinations of sales numbers of each size that allow the company to break even.

Answers will vary; three possible combinations are (80,000, 80,000), (20,000, 160,000) and (50,000, 120,000).

e. List three combinations of sales numbers of each size that allow the company to turn a profit.

Answers will vary; three possible combinations are (150,000, 0), (90,000, 100,000) and (80,000, 120,000).

f. In reality, is every combination of units sold represented by the line and shaded region a possibility? Explain.

No, only combinations of whole numbers are possible because the company cannot sell fractions of a bottle of hand lotion.

38. A company produces two versions of tax preparation software. The regular version costs $12.50 per unit to produce, and the deluxe version costs $15.00 per unit to produce. Management plans for the cost of production to be a maximum of $300,000 for the first quarter. The inequality $12.50x + 15y \leq 300,000$ describes the total cost as prescribed by management.

   a. What do x and y represent?

      x represents the number of regular versions produced; y represents the number of deluxe versions produced.

   b. Graph the inequality. Because x and y are both nonnegative, restrict the graph to the first quadrant.

   c. What does the line represent?

      The line represents the combinations of both versions produced that yield a total cost of exactly $300,000.

   d. What does the shaded region represent?

      The shaded region represents the cost of producing different combinations of software that yield a total cost less than $300,000.

   e. List three combinations of numbers of units produced of each version of software that yield a total cost of $300,000.

      Answers will vary; some possible combinations are $(0, 20,000)$, $(24,000, 0)$, and $(4800, 16,000)$.

   f. List three combinations of numbers of units produced of each version of software that yield a total cost less than $300,000.

      Answers will vary; some combinations are $(4000, 4000)$, $(4000, 8000)$, and $(8000, 8000)$.

   g. In reality, is every combination of units produced represented by the line and shaded region a possibility? Explain.

      No, only combinations of whole numbers are possible because the company cannot sell fractions of a piece of software.

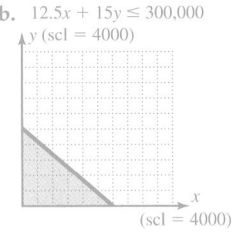
b. $12.5x + 15y \leq 300,000$
y (scl = 4000)
(scl = 4000)

39. Andrea visits a garden center to purchase new plants for her garden. She has a $100 gift certificate. The plants that she is considering come in two sizes and are priced according to size. Plants in 1-gallon containers are priced at $5.50 each, and plants in 5-gallon containers are priced at $12.50 each.

   a. Select a variable for the number of each plant size and write an inequality that has Andrea's total cost within the amount of the gift certificate.

      $5.5x + 12.5y \leq 100$

   b. Graph the inequality. Because both quantities are nonnegative, restrict the graph to the first quadrant.

   c. What does the line represent?

      The combinations of plant purchases that would cost exactly $100

   d. What does the shaded region represent?

      The combinations of plants purchased that would cost less than $100

   e. What are some possible combinations of containers she could purchase?

      Answers will vary; some combinations are $(2, 4)$, $(6, 2)$, and $(8, 4)$.

   f. In reality, could she purchase any combinations that would yield a total in the exact amount of the gift certificate? Explain.

      Yes, $(0, 8)$. Because she cannot purchase fractional numbers of plants, the combination must be whole numbers and $(0, 8)$ is the only combination of whole numbers that costs exactly $100.

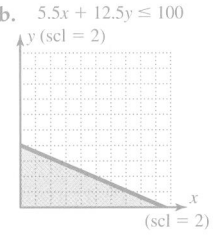
b. $5.5x + 12.5y \leq 100$
y (scl = 2)
(scl = 2)

40. Michelle is coordinating a fund-raiser that will sell two cookbooks. One cookbook sells for $10; the other, for $12. The goal is to raise at least $30,000.

   a. Select variables for each type of cookbook and write an inequality in which total sales is at least $30,000.

      $10x + 12y \geq 30,000$

   b. Graph the inequality. Because both quantities are nonnegative, restrict the graph to the first quadrant.

   c. What does the line represent?

      The combinations of cookbook purchases that would raise exactly $30,000

   d. What does the shaded region represent?

      The combinations of cookbooks purchased that would raise more than $30,000

   e. Find two combinations of book sales that yield exactly $30,000 in total sales.

      Answers will vary; some combinations are $(0, 2500)$, $(600, 2000)$, and $(1800, 1000)$.

   f. Find two combinations of units sold that yield more than $30,000.

      Answers will vary; some combinations are $(500, 2500)$, $(1000, 3000)$, and $(3500, 500)$.

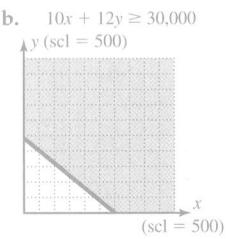
b. $10x + 12y \geq 30,000$
y (scl = 500)
(scl = 500)

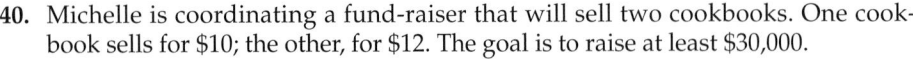

41. A building with a rectangular base is to be designed so that the maximum perimeter is 180 feet.

   a. Select variables for the length and width and write an inequality in which the maximum perimeter is 180 feet.

   $2l + 2w \leq 180$

   b. Graph the inequality. Because the length and width are both positive, restrict the graph to the first quadrant.

   c. What does the line represent?

   The combinations of lengths and widths that yield a perimeter of exactly 180 ft.

   d. What does the shaded region represent?

   The combinations of length and width that yield a perimeter that is less than 180 ft.

   e. Find three combinations of length and width that yield a perimeter of exactly 180 feet.

   Answers will vary; some combinations are $(10, 80)$, $(20, 70)$, and $(40, 50)$.

   f. Find three combinations of length and width that yield a perimeter of less than 180 feet.

   Answers will vary; some combinations are $(10, 10)$, $(20, 20)$, and $(30, 30)$.

42. Steel beams are to be welded together to form a frame that is an isosceles triangle (two sides of equal length). The perimeter of the frame cannot exceed 60 feet.

   a. Select a variable for the length of the base and a second variable for the sides of equal length. Then write an inequality in which the maximum perimeter is 60 feet.

   $b + 2l \leq 60$

   b. Graph the inequality. Because the lengths are positive, restrict the graph to the first quadrant.

   c. What does the line represent?

   The combinations of base and side lengths that yield a perimeter that is exactly 60 ft.

   d. What does the shaded region represent?

   The combinations of base and side lengths that yield a perimeter that is less than 60 ft.

   e. Find sets of dimensions that yield a perimeter of exactly 60 feet.

   Answers will vary; a couple are $(20, 20)$ and $(40, 10)$.

   f. Find sets of dimensions that yield a perimeter of less than 60 feet.

   Answers will vary; a couple are $(10, 20)$ and $(20, 10)$.

41. b. $2l + 2w \leq 180$

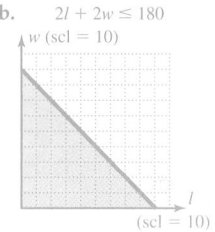

42. b. $b + 2l \leq 60$

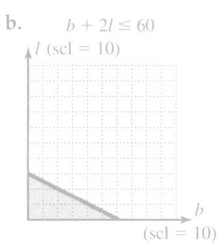

Puzzle Problem The sum of two different whole numbers is equal to 8, and the difference of the two numbers is greater than or equal to 1. Find all possible ordered pairs that satisfy both constraints.

$(8, 0)$ $(7, 1)$ $(6, 2)$ $(5, 3)$

Review Exercises

Exercises 1 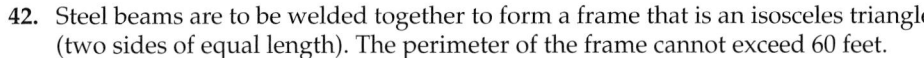 **Constants and Variables**

[1.1] 1. Write a set containing all vowels in the English alphabet.

   $\{a, e, i, o, u\}$

Exercises 2–6 **Equations and Inequalities**

[1.7] 2. Evaluate $\frac{3}{4}x - 5$ when $x = 8$.

   1

[2.3] 3. Solve and check: $3x - 4(x + 7) = -15$

   -13

[2.4] 4. The length of a rectangle is four more than the width. If the perimeter is 88 feet, find the dimensions.

   20 ft. by 24 ft.

[3.3] 5. Find the x- and y-intercepts for $3x - 2y = 8$.

   $\left(\frac{8}{3}, 0\right)$, $(0, -4)$

[3.4] 6. Graph $y = x + 7$.

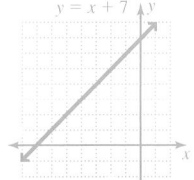

3.7 Introduction to Functions and Function Notation

Objectives

1 Identify the domain and range of a relation.
2 Identify functions and their domains and ranges.
3 Find the value of a function.
4 Graph linear functions.

Warm-up

[1.7] Evaluate $\dfrac{4}{x - 3}$ for the following values of x.

1. 2
2. -5
3. 3

In this section, we study relations and functions, which develop naturally from our work with graphing because they involve ordered pairs.

Objective 1 Identify the domain and range of a relation.

To formally define a function, we first discuss **relations**.

Definition Relation: A set of ordered pairs.

A linear equation such as $y = 2x$ is a relation because it pairs a given x-value with a corresponding y-value. You might recall that with an equation in this form, we thought of the x-values as input values and the y-values as output values. In a relation, the set of all of the input values (x-values) is called the **domain** and the set of all of the output values (y-values) is called the **range**.

Definitions Domain: The set of all input values (x-values) for a relation.
Range: The set of all output values (y-values) for a relation.

The following table lists some values in the domain and range for $y = 2x$.

Domain (x-values)	Range (y-values)
0	0
1	2
2	4
3	6

Example 1 Determine the domain and range of the relation $\{(2, -1), (3, 0), (4, 5), (5, 8)\}$.

Solution: The domain is the set containing all of the x-values $\{2, 3, 4, 5\}$, and the range is the set containing all of the y-values $\{-1, 0, 5, 8\}$.

Domain and Range with a Graph

We can determine the domain and range of a relation from its graph. Look at the graph of $y = 2x$.

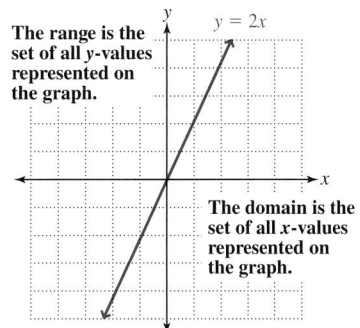

The range is the set of all *y*-values represented on the graph.

The domain is the set of all *x*-values represented on the graph.

It is convenient to imagine traveling along a graph from left to right. As we travel along $y = 2x$ from left to right, every *x*-value along the *x*-axis has a corresponding *y*-value. Thus, the domain is a set containing all real numbers. Similarly, every *y*-value is paired with an *x*-value; so the range also contains all real numbers.

Procedure   **Determining Domain and Range**

To determine the domain of a relation given its graph, answer this question: What are all of the *x*-values that have a corresponding *y*-value? To determine the range, answer this question: What are all of the *y*-values that have a corresponding *x*-value?

Example 2   Determine the domain and range of the relation.

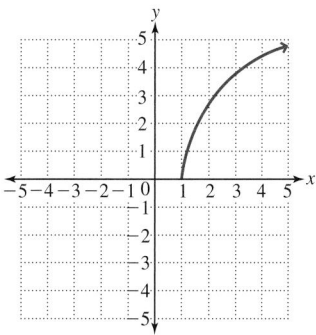

Solution: From left to right, this graph "begins" at the point $(1, 0)$. Values along the *x*-axis begin at 1 and continue infinitely; so the domain is $\{x \mid x \geq 1\}$ in set-builder notation and $[1, \infty)$ in interval notation. For the range, the *y*-values begin at 0 and continue infinitely; so the range is $\{y \mid y \geq 0\}$ in set-builder notation and $[0, \infty)$ in interval notation.

Your Turn 2   Determine the domain and range of the relation.

a. $\{(-2, -4), (0, -1), (3, 6), (7, 5)\}$ **b.**

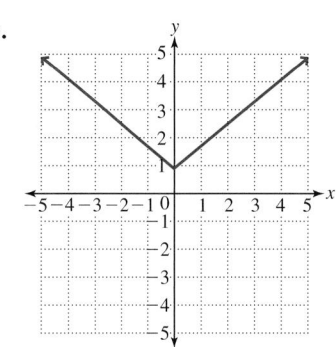

Answers to Your Turn 2
a. Domain: $\{-2, 0, 3, 7\}$
  Range: $\{-4, -1, 6, 5\}$
b. Domain: All real numbers or
  $(-\infty, \infty)$
  Range: $\{y \mid y \geq 1\}$ or $[1, \infty)$

Objective 2 Identify functions and their domains and ranges.

Now we turn to a special type of relation called a *function*. Let's reconsider the trajectory data from Example 4 of Section 3.1 (p. 193), which we have listed here as a relation in the following table.

Time in Seconds (domain)	Height in Feet (range)
0	6.00
0.5	7.50
1.0	8.25
1.5	7.50
2.0	6.00
2.5	2.00

Note that each value for time is paired with one and only one height value. This is the key feature that makes this relation a **function**.

Definition Function: A relation in which every value in the domain is paired with (or assigned to) exactly one value in the range.

From the data in the preceding table, we see that a function can have different values in the domain assigned to the same value in the range. For example, domain values 0.5 seconds and 1.5 seconds each correspond to the same range value 7.50 feet. Also, the domain values 0 seconds and 2.0 seconds each correspond to 6.00 feet in the range. Using arrows to match the values in the domain with their corresponding values in the range can be a helpful technique in determining whether a relation is a function. Such a map might look like this:

Domain: $\{0,\ \ 0.5,\ \ 1.0,\ \ 1.5,\ \ 2.0,\ \ 2.5\}$

Note Each element in the domain ◀ has a single arrow pointing to an element in the range.

Range: $\{2.00,\ \ 6.00,\ \ 7.50,\ \ 8.25\}$

Note that every function is a relation but not every relation is a function. For example, consider the following relation that assigns baseball outfield positions (the domain) to the 2013 Atlanta Braves outfielders (the range).

Note Elements in the domain have more than one arrow pointing to an ▶ element in the range.

Domain: $\{$catcher, first base, third base$\}$

Range: $\{$Laird, McCann, Freeman, Terdoslavich, Francisco, Johnson$\}$

Notice that all three positions are assigned to more than one player. Because an element in the domain is assigned to more than one element in the range, this relation is not a function.

Conclusion: If any value in the domain is assigned to more than one value in the range, the relation is not a function.

Example 3 Determine whether the relation is a function.

a. The table to the right indicates the national health expenditures, in billions of dollars, for selected years. (*Source:* www.healthcare.org.)

Year (domain)	Expenditure (range)
2000	1310.0
2001	1424.5
2002	1547.6
2003	1660.5
2008	2354.6
2010	3079.8

Solution: This relation is a function because each year value in the domain has only one expenditure value in the range.

b. The following relation assigns a birth date to the corresponding person.

Domain: {May 1, June 4, July 8, August 2}

Range: {Danielle, Gerard, Juan, René, Candice}

Solution: This relation is not a function because an element in the domain, July 8, is assigned to two people in the range.

c. {(1, 4), (3, −2), (−5, 0) (3, 4)}

Solution: This relation is not a function because an element in the domain, 3, is paired with two elements, −2 and 4, in the range.

Note The domain of Example 3(c) is {1, 3, −5}. Even though 3 occurs twice, it is listed only once.

| Your Turn 3 | Determine whether the relation is a function.

a.

Index Topic (domain)	Page Number(s) (range)
Celestial sphere	56
Celsius	36
Centigrade	36
Centimeter	6
Centripetal force	239, 305

b. The table shows the top five finishers in the 2013 Daytona 500 race.

Final Position (domain)	Car Number (range)
1st	9 (Jimmie Johnson)
2nd	19 (Dale Earnhardt Jr.)
3rd	14 (Mark Martin)
4th	15 (Brad KeselowskI)
5th	34 (Ryan Newman)

c. {(−2, −1), (5, −2), (0, 0), (3, −2)}

We can determine whether a relation is a function by its graph. Remember that in Example 4 of Section 3.1 (p. 193), we created a graph of the trajectory data (that data also appears in this section on p. 193) by plotting the ordered pairs as points in the coordinate plane.

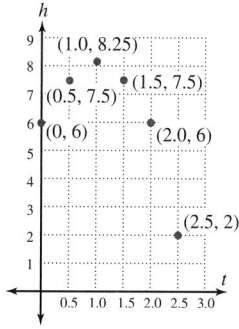

Notice that the domain (time in this case) is plotted along the horizontal axis and the range (height) is plotted along the vertical axis.

To determine whether a relation is a function from its graph, we can perform a test called the *vertical line test*. The nature of the test is to draw or imagine a vertical line through every point in the domain.

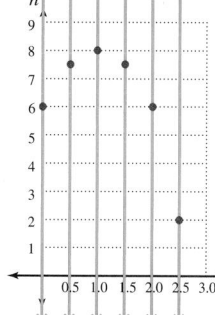

Here, we've drawn lines in blue through each value in the domain. Note that each vertical line intersects the graph at one and only one point, which means that each value in the domain corresponds to exactly one value in the range. This relation is a function.

Answers to Your Turn 3
a. not a function
b. function
c. function

Learning Strategy

If you are a tactile learner, perform the vertical line test by using a pencil as your vertical line and sweep it from left to right across the graph. If your pencil (held vertically) ever touches the graph in two or more places, the graph is *not* a function.

Procedure The Vertical Line Test

To determine whether a relation is a function from its graph, perform a vertical line test.

1. Draw or imagine vertical lines through each point in the domain.
2. If each vertical line intersects the graph in at most one point, the graph is the graph of a function.
3. If any vertical line intersects the graph at two or more different points, the graph is not the graph of a function.

Example 4 For each graph, determine the domain and range. Then state whether the relation is a function.

a.

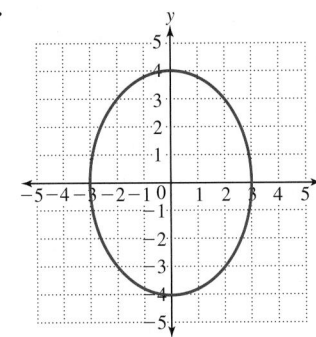 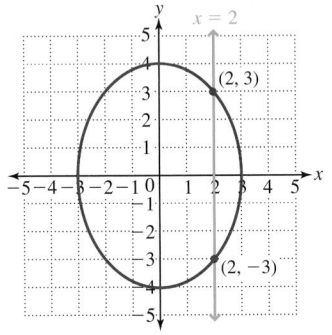

Solution: Domain: $\{ x \mid -3 \le x \le 3 \}$ or $[-3, 3]$
Range: $\{ y \mid -4 \le y \le 4 \}$ or $[-4, 4]$

This relation is not a function because a vertical line can be drawn that intersects the graph at two different points.

For example, the vertical line $x = 2$ (shown in blue) passes through two different points on the graph, $(2, 3)$ and $(2, -3)$.

b.

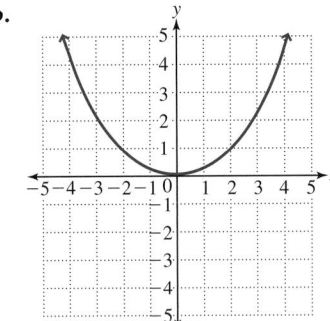 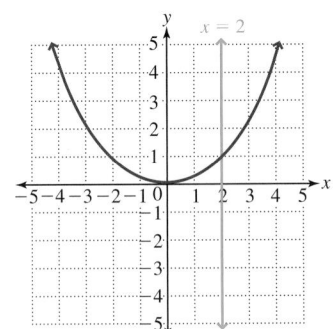

Solution: Domain: all real numbers
Range: $\{ y \mid y \ge 0 \}$ or $[0, \infty)$

This relation is a function. A vertical line through any value along the horizontal axis will intersect the graph at only one point. We have shown $x = 2$ to illustrate one such line.

c.

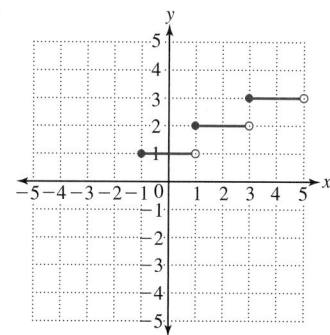 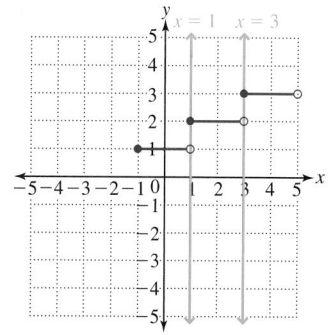

Solution: Domain: $\{x \mid -1 \le x < 5\}$ or $[-1, 5)$
Range: $\{y \mid y = 1, 2, 3\}$

This relation is a function. It may seem that the vertical lines $x = 1$ and $x = 3$ (shown in blue) intersect the graph at two different points. However, the open circles at $(1, 1)$, $(3, 2)$, and $(5, 3)$ indicate that those points are not part of the graph. The closed circles at $(-1, 1)$, $(1, 2)$, and $(3, 3)$ are part of the graph. The vertical line $x = 1$ "passes through" the open circle at $(1, 1)$ and intersects the graph at only one point, the closed circle at $(1, 2)$. Similarly, $x = 3$ passes through the open circle at $(3, 2)$ and intersects the graph at only one point, $(3, 3)$.

d.

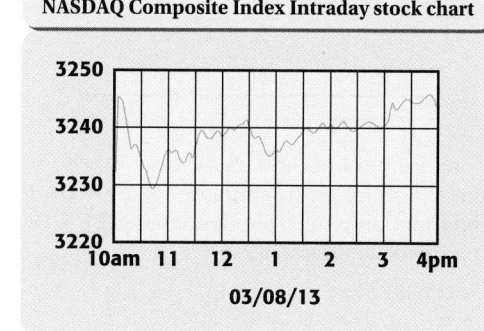

Solution:

Domain:

$\{t \mid$ 10:00 A.M. $\le t \le$ 4:00 P.M. $\}$ or
$[$ 10:00 A.M., 4:00 P.M. $]$

Range:

$\{N \mid 3228 \le N \le 3246\}$ or
$[3228, 3249]$

This relation is a function. A vertical line through any value along the horizontal axis will intersect the graph at only one point. In the terminology from the graph, at each time during this business day, the Nasdaq had only one value.

Your Turn 4   For each graph, determine the domain and range. Then state whether the relation is a function.

a.

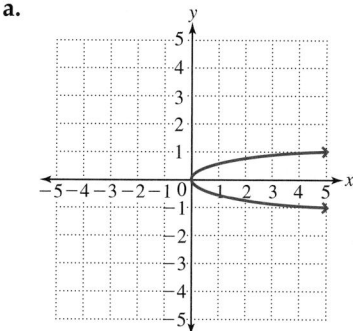

b.

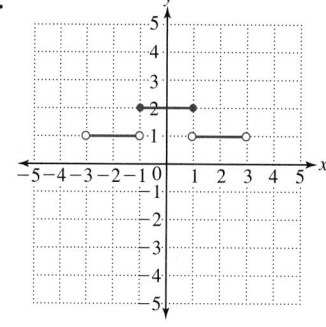

Answers to Your Turn 4
a. Domain: $\{x \mid x \ge 0\}$ or $[0, \infty)$
Range: all real numbers not a function
b. Domain: $\{x \mid -3 < x < 3\}$ or $(-3, 3)$
Range: $\{y \mid y = 1, 2\}$ function

Objective 3 Find the value of a function.

Function Notation

When written as an equation, the notation for a function is a modification of an equation in two variables. Instead of $y = 2x + 5$, in function notation, we would write $f(x) = 2x + 5$ to mean the same equation. Notice that the notation $f(x)$, which is read "a function in terms of x," or "f of x," replaces the variable y. The following information shows how equations in two variables can be written using function notation.

Warning The notation $f(x)$ does *not* mean multiply f and x.

Equation in two variables:	Function notation:
$y = 3x$	$f(x) = 3x$
$y = -4x + 1$	$f(x) = -4x + 1$
$y = \dfrac{2}{3}x - 5$	$f(x) = \dfrac{2}{3}x - 5$

Finding the Value of a Function

Remember that when we found solutions to equations with y isolated, we often spoke of the x-value as an input value and the corresponding y-value as the output value. Finding the value of a function is the same as finding a solution to an equation in two variables: We input a value for x and calculate the output value, $f(x)$. The following tables list the same ordered pairs. The table on the left is for $y = 2x + 5$, and the table on the right is for $f(x) = 2x + 5$. Notice that the only difference is in labeling the output values.

(input) x	(output) y
0	$2(0) + 5 = 5$
1	$2(1) + 5 = 7$
2	$2(2) + 5 = 9$

(input) x	(output) $f(x)$
0	$2(0) + 5 = 5$
1	$2(1) + 5 = 7$
2	$2(2) + 5 = 9$

Function notation offers a clever way to indicate that a specific x-value is to be used. For example, given a function $f(x) = 3x$, the notation $f(2)$ means to find the value of the function where $x = 2$ so that $f(2) = 3(2) = 6$, which is another way of representing the ordered pair $(2, 6)$.

Procedure **Finding the Value of a Function**

Given a function $f(x)$, to find $f(a)$, where a is a real number in the domain of f, replace x in the function with a and calculate the value.

Connection We use similar procedures to find the value of a function and to evaluate an expression.

Function language:
For $f(x) = \dfrac{4}{x - 3}$, find $f(-5)$.

Expression language:
Evaluate the expression
$\dfrac{4}{x - 3}$, where $x = -5$.

Example 5 For the function $f(x) = \dfrac{4}{x - 3}$, find the following.

a. $f(2)$

Solution: $f(2) = \dfrac{4}{2 - 3}$ Replace x in $\dfrac{4}{x - 3}$ with 2; then calculate.

$= \dfrac{4}{-1}$

$= -4$

b. $f(-5)$

Solution: $f(-5) = \dfrac{4}{-5 - 3}$ Replace x in $\dfrac{4}{x - 3}$ with -5; then calculate.

$= \dfrac{4}{-8}$

$= -\dfrac{1}{2}$

c. $f(3)$

Solution: $f(3) = \dfrac{4}{3 - 3}$ Replace x in $\dfrac{4}{x - 3}$ with 3; then calculate.

$= \dfrac{4}{0}$ Therefore, $f(3)$ is undefined.

| **Your Turn 5** | For the function $\sqrt{x + 4}$, find the following.

a. $f(5)$ **b.** $f(3)$ **c.** $f(-6)$

Finding the Value of a Function Given Its Graph

We can also find the value of a function given its graph. For a given value in the domain (x-value), we look on the graph for the corresponding value in the range (y-value).

| **Example 6** | Using the graph, find the value of the function.

a. $f(0)$

Solution: The notation $f(0)$ means to find the value of the function (y-value), where $x = 0$. On the graph, we see that when $x = 0$, the corresponding y-value is 1; so we say that $f(0) = 1$.

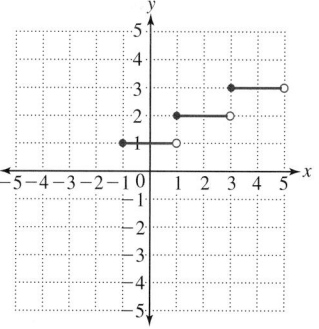

b. $f(1)$

Solution: When $x = 1$, we see that $y = 2$; so $f(1) = 2$.

c. $f(-4)$

Solution: Remember from Example 4(c) that the domain is $\{x \mid -1 \le x < 5\}$ or $[-1, 5)$. Because the value -4 is not in the domain, we say that $f(-4)$ is undefined.

| **Your Turn 6** | Use the graph in the margin to find the value of the function.

a. $f(-1)$ **b.** $f(0)$ **c.** $f(5)$ **d.** $f(-3)$

Objective 4 Graph linear functions.

We create the graph of a function the same way we create the graph of an equation in two variables. In this chapter, we graph only linear functions. Linear functions have a form similar to linear equations in slope–intercept form.

$$\text{Slope-intercept form:} \quad y = mx + b$$
$$\text{Linear function:} \quad f(x) = mx + b$$

Answers to Your Turn 5
a. 3   **b.** $\sqrt{7}$
c. not a real number

Answers to Your Turn 6
a. 0   **b.** 1   **c.** 2   **d.** undefined

| **Example 7** | Graph: $f(x) = -2x + 3$

Solution: Think of the function as the equation $y = -2x + 3$. We can make a table of ordered pairs or use the fact that the slope is -2 and the y-intercept is 3.

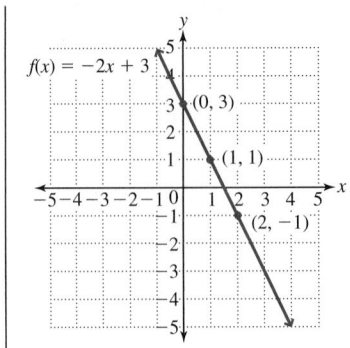

Table of ordered pairs

x	f(x)
0	3
1	1
2	−1

Answer to Your Turn 7

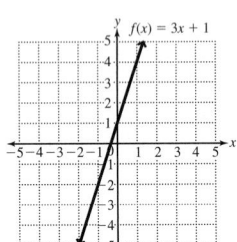

Your Turn 7 Graph $f(x) = 3x + 1$.

3.7 Exercises For Extra Help MyMathLab®

Note: Exercises marked with a ★ represent challenging exercises.

Objective 1

Prep Exercise 1 Explain domain of a relation in your own words.

The domain is a set containing all x-values.

Prep Exercise 2 Explain range of a relation in your own words.

The range is a set containing all y-values.

Prep Exercise 3 True or false: All relations are functions. Explain.

False; a function is a special type of relation.

Prep Exercise 4 When using the vertical line test on a graph, how do you determine whether the graph is the graph of a function?

Draw or imagine a vertical line through every point in the domain. If each vertical line intersects the graph in at most one point, the relation is a function.

For Exercises 1–6, determine the domain and range of the relation. See Example 1.

1. $\{(2,1),(3,-2),(1,4),(-1,-1)\}$
   Domain: $\{2,3,1,-1\}$; range: $\{1,-2,4,-1\}$

2. $\{(-2,1),(4,3),(2,5),(-3,-2)\}$
   Domain: $\{-2,4,2,-3\}$; range: $\{1,3,5,-2\}$

3. $\{(4,1),(2,1),(3,-5),(-4,0)\}$
   Domain: $\{4,2,3,-4\}$; range: $\{1,-5,0\}$

4. $\{(-2,1),(8,11),(-3,5),(4,-4)\}$
   Domain: $\{-2,8,-3,4\}$; range: $\{1,11,5,-4\}$

5. $\{(0,0),(-3,-3),(2,9),(8,-16)\}$
   Domain: $\{0,-3,2,8\}$; range: $\{0,-3,9,-16\}$

6. $\{(5,4),(-3,8),(10,-2),(0,-5)\}$
   Domain: $\{5,-3,10,0\}$; range: $\{4,8,-2,-5\}$

Objective 2

For Exercises 7–14, determine whether the relation is a function. In each table, the domain is in the left column and the range is in the right column. See Example 3.

7. Average college tuition in the United States for 2010–2011:

   yes

Sector	Cost ($)
Two-year public	2794
Two-year private	15,373
Four-year public	7249
Four-year private	14,236

   (*Source:* National Center for Education Statistics, U.S. Department of Education.)

8. Spending on selected food products for 2009–2010:

   yes

Food Product	Sales (in billions of dollars)
Beer	9.5
Soft drinks	8.2
Cereal	6.5
Cookies	4.1
Ice cream	4.0

   (*Source: The World Almanac and Book of Facts,* 2011.)

9. Results of the 2012 Masters golf tournament:

Position	Name
1st	Bubba Watson
2nd	Louis Oosthuizen
3rd	Peter Hanson, Matt Kuchar, Phil Mickelson, Lee Westwood
7th	Ian Poulter

no

(*Source:* www.masters.org.)

10. Top-five cable TV networks in 2012:

Rank	Channel
1	TBS
2	Discovery
3	USA Network
4	TNT, The Weather Channel
5	Nickelodeon

no

(*Source: The World Almanac and Book of Facts,* 2012.)

11. All-time top-grossing American movies:

Title	Units Sold (millions of dollars)
Avatar	749.8
Titanic	600.8
The Dark Knight	533.3
Star Wars	461.0
Shrek 2	436.7

yes

(*Source:* Rentrak Corporation.)

12. Top-selling video games in 2009:

Title	Units Sold (in millions)
Call of Duty: Modern Warfare 2	5.84
Wii Sports Resort w/ Wii Motion Plus	4.54
New Super Mario Bros. Wii	4.22
Wii Fit w/ balance board	3.60

yes

(*Source:* The NPD Group/Retail Tracking Service.)

13. Average high temperature for Orlando, Florida:

Average High (°F)	Month
92	July, August
91	June
90	September
88	May
85	October
83	April
79	November
78	March
73	February, December
71	January

no

14. Average rainfall for Seattle:

Avg. Rainfall (inches)	Month
6	November
5	December, January
4	March
3	February, April, October
2	May, June, September
1	July, August

no

For Exercises 15–30, determine whether the relation is a function and explain your answer. See Examples 3 and 4.

15. $\{ (2,7), (3,4), (-1,4), (0,5) \}$

Yes, every element in the domain is assigned to one element in the range.

16. $\{ (1,2), (4,-7), (9,5), (6,14) \}$

Yes, every element in the domain is assigned to one element in the range.

17. $\{ (8,2), (4,-5), (3,0), (8,-1) \}$

No, an element in the domain is assigned to more than one element in the range. (8 is assigned to both 2 and −1.)

18. $\{ (-1,5), (2,6), (2,9), (-3,-2) \}$

No, an element in the domain is assigned to more than one element in the range. (2 is assigned to both 6 and 9.)

19. The grades on a recent calculus test:

Student	Score
Jonathon	54
Nathan	79
Rebecca	84
Terri	97
Allen	97

Yes, every element in the domain is assigned to one element in the range.

20. The sources of distraction among inattentive drivers:

Distraction	Percentage of Drivers
Outside person, object, or event	29.4
Adjusting radio/CD	11.4
Other occupant	10.9
Using/dialing cell phone	1.5
Smoking related	0.9

(*Source:* AAA Foundation for Traffic Safety/UNC Safety Research Center.)

Yes, every element in the domain is assigned to one element in the range.

21. Ten most home runs hit in a single season:

Player	Home Runs
Barry Bonds	73
Mark McGwire	70
Sammy Sosa	66
Mark McGwire	65
Sammy Sosa	64
Sammy Sosa	63
Roger Maris	61
Babe Ruth	60
Babe Ruth	59
Mark McGwire	58

No, an element in the domain is assigned to more than one element in the range.

22. Number of NASCAR championships (3 or more):

Number	Driver
7	Richard Petty
	Dale Earnhardt
5	Jimmie Johnson
4	Jeff Gordon
3	David Pearson
	Darrell Waltrip
	Cale Yarborough
	Lee Petty
	Tony Stewart

No, an element in the domain is assigned to more than one element in the range.

23.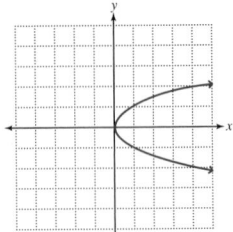

No, it fails the vertical line test.

24.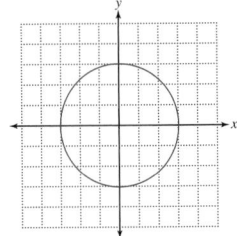

No, it fails the vertical line test.

25.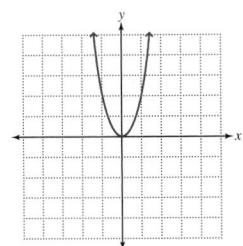

Yes, it passes the vertical line test.

26.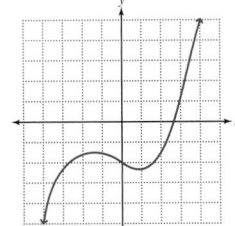

Yes, it passes the vertical line test.

27.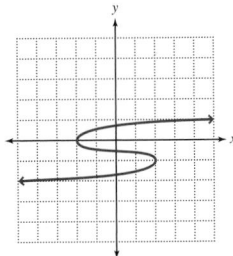

No, it fails the vertical line test.

28.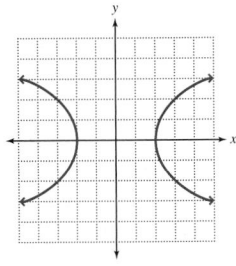

No, it fails the vertical line test.

29.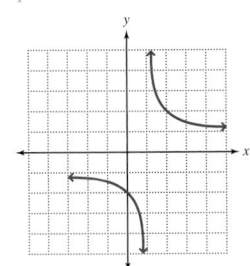

Yes, it passes the vertical line test.

30.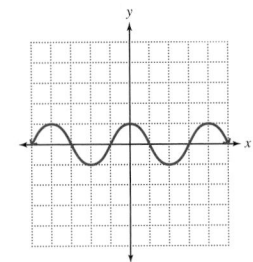

Yes, it passes the vertical line test.

Prep Exercise 5 Given a graph of a relation, how do you determine the domain?

Trail along the x-axis and determine what values of x have a corresponding y-value.

Prep Exercise 6 Given a graph of a relation, how do you determine the range?

Trail along the y-axis and determine what values of y have a corresponding x-value.

For Exercises 31–40, determine the domain and range. Then state whether the relation is a function. See Examples 2 and 4.

31. The following graph shows the number of households in the United States with cable television by year. (*Source:* Nielsen Media Research.)

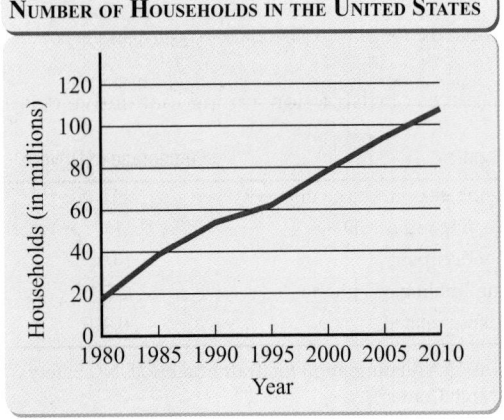

NUMBER OF HOUSEHOLDS IN THE UNITED STATES

Domain: $\{x \mid 1980 \leq x \leq 2010\}$ or $[1980, 2010]$; range: $\{y \mid 17.7 \text{ million} \leq y \leq 104.1 \text{ million}\}$ or $[17.7 \text{ million}, 104.1 \text{ million}]$; function

32. The line graph represents postage costs for first-class letters from 1991 to 2012.

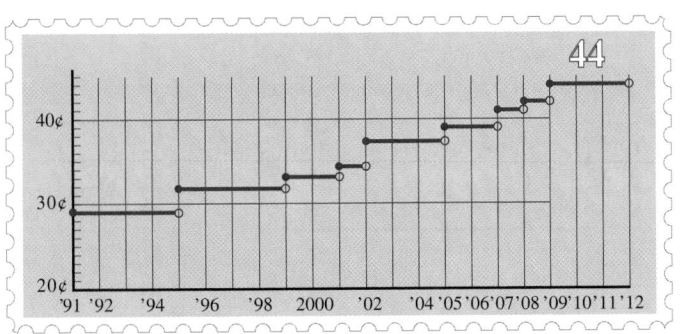

(*Source:* Postal Rate Commission.)

Domain: $\{x \mid 1991 \leq x \leq 2012\}$ or $[1991, 2012]$; range: $\{29, 32, 33, 34, 37, 39, 41, 42, 44\}$; function

33. This graph shows the percentage of the population that was foreign-born at each census.

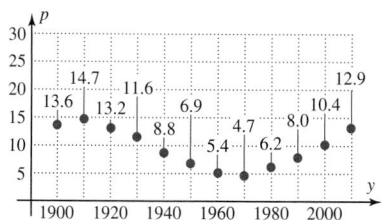

(*Source:* Bureau of the Census, U.S. Department of Commerce.)
Domain {1900, 1910, 1920, 1930, 1940, 1950, 1960, 1970, 1980, 1990, 2000, 2010};
range: {4.7, 5.4, 6.2, 6.9, 8.0, 8.8, 10.4, 11.6, 12.9, 13.2, 13.6, 14.7};
function

34. This graph shows the total number of people enrolled in a program each month after the program began.

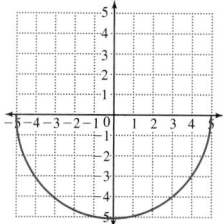

Domain: {1, 2, 3, 4, 5, 6};
range: {12, 15, 16, 22, 25};
function

35.

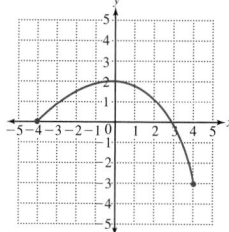

Domain: $\{x \mid -4 \le x \le 4\}$ or $[-4, 4]$;
range: $\{y \mid -3 \le y \le 2\}$ or $[-3, 2]$;
function

36.

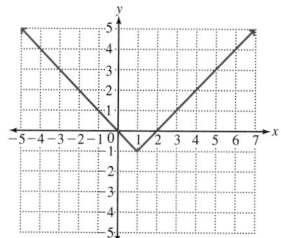

Domain: $\{x \mid -5 \le x \le 5\}$ or $[-5, 5]$;
range: $\{y \mid -5 \le y \le 0\}$ or $[-5, 0]$;
function

37.

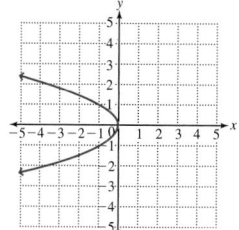

Domain: $\{x \mid x \le 0\}$ or $(-\infty, 0]$;
range: all real numbers or $(-\infty, \infty)$;
not a function

38.

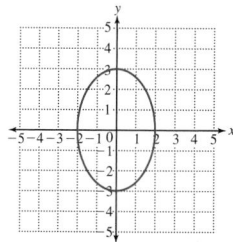

Domain: $\{x \mid -2 \le x \le 2\}$ or $[-2, 2]$;
range: $\{y \mid -3 \le y \le 3\}$ or $[-3, 3]$;
not a function

39.

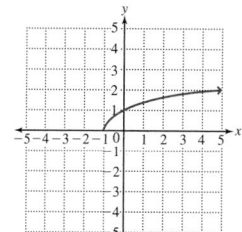

Domain: all real numbers or $(-\infty, \infty)$;
range: $\{y \mid y \ge -1\}$ or $[-1, \infty)$;
function

40.

Domain: $\{x \mid x \ge -1\}$ or $[-1, \infty)$;
range: $\{y \mid y \ge 0\}$ or $[0, \infty)$;
function

Objective 3

For Exercises 41–64, find the value of the function. See Example 5.

41. $f(x) = 2x - 5$
    a. $f(0)$ −5
    b. $f(1)$ −3
    c. $f(-2)$ −9
    d. $f\left(\dfrac{1}{2}\right)$ −4

42. $f(x) = 5x + 3$
    a. $f(1)$ 8
    b. $f(0)$ 3
    c. $f(-1)$ −2
    d. $f(t)$ $5t + 3$

43. $f(x) = x^2 - 3x + 7$
    a. $f(0)$ 7
    b. $f(1)$ 5
    c. $f(-1)$ 11
    d. $f(-2)$ 17

44. $f(x) = 3x^2 + 2x - 1$
    a. $f(-2)$ 7
    b. $f(0.2)$ −0.48
    c. $f(a)$ $3a^2 + 2a - 1$
    d. $f(3)$ 32

45. $f(x) = \sqrt{x + 3}$
    a. $f(0)$ $\sqrt{3}$
    b. $f(-4)$ Not a real number
    c. $f(-1)$ $\sqrt{2}$
    d. $f(a)$ $\sqrt{a + 3}$

46. $f(x) = \sqrt{x + 5}$
    a. $f(-2)$ $\sqrt{3}$
    b. $f(6)$ $\sqrt{11}$
    c. $f(-5)$ 0
    d. $f(a)$ $\sqrt{a + 5}$

47. $f(x) = \sqrt{x^2 - 4}$

  a. $f(2)$
   0

  b. $f(1)$
   Not a real number

  c. $f(-2)$
   0

  d. $f(3)$
   $\sqrt{5}$

48. $f(x) = \sqrt{x^2 + 2x}$

  a. $f(-2)$
   0

  b. $f(-1)$
   Not a real number

  c. $f(0)$
   0

  d. $f(1)$
   $\sqrt{3}$

49. $f(x) = \dfrac{1}{x - 1}$

  a. $f(0)$
   -1

  b. $f(1)$
   undefined

  c. $f(-1)$
   $-\dfrac{1}{2}$

  d. $f(a)$
   $\dfrac{1}{a - 1}$

50. $f(x) = \dfrac{2}{x + 2}$

  a. $f(-2)$
   undefined

  b. $f(-1)$
   2

  c. $f(3)$
   $\dfrac{2}{5}$

  d. $f(a)$
   $\dfrac{2}{a + 2}$

51. $f(x) = x^2 - 2x - 3$

  a. $f(0)$
   -3

  b. $f(0.1)$
   -3.19

  c. $f(a)$
   $a^2 - 2a - 3$

  d. $f(-a)$
   $a^2 + 2a - 3$

52. $f(x) = 2x^2 + x - 3$

  a. $f\left(\dfrac{3}{4}\right)$
   $-1\dfrac{1}{8}$

  b. $f(-1)$
   -2

  c. $f(a)$
   $2a^2 + a - 3$

  d. $f(-a)$
   $2a^2 - a - 3$

53. $f(x) = \dfrac{1}{3}x - 2$

  a. $f(0)$
   -2

  b. $f(1)$
   $-\dfrac{5}{3}$

  c. $f(-1)$
   $-\dfrac{7}{3}$

  d. $f(a)$
   $\dfrac{1}{3}a - 2$

54. $f(x) = \dfrac{2}{3}x + 1$

  a. $f(-3)$
   -1

  b. $f(-1)$
   $\dfrac{1}{3}$

  c. $f(3)$
   3

  d. $f(t)$
   $\dfrac{2}{3}t + 1$

55. $f(x) = \dfrac{x + 3}{x - 4}$

  a. $f(0)$
   $-\dfrac{3}{4}$

  b. $f(4)$
   undefined

  c. $f(-2)$
   $-\dfrac{1}{6}$

  d. $f(3)$
   -6

56. $f(x) = \dfrac{9 - x}{x + 2}$

  a. $f(0)$
   $\dfrac{9}{2}$

  b. $f(-2)$
   undefined

  c. $f(3)$
   $\dfrac{6}{5}$

  d. $f(-1)$
   10

57. $f(x) = \dfrac{1}{\sqrt{x + 2}}$

  a. $f(-2)$
   undefined

  b. $f(1)$
   $\dfrac{1}{\sqrt{3}}$

  c. $f(-3)$
   Not a real number

  d. $f(a)$
   $\dfrac{1}{\sqrt{a + 2}}$

58. $f(x) = \dfrac{2}{\sqrt{2 - x}}$

  a. $f(-2)$
   1

  b. $f(-1)$
   $\dfrac{2}{\sqrt{3}}$

  c. $f(3)$
   Not a real number

  d. $f(a)$
   $\dfrac{2}{\sqrt{2 - a}}$

59. $f(x) = |4x - 2|$

  a. $f(0)$
   2

  b. $f(1.5)$
   4

  c. $f(-1)$
   6

  d. $f(-2)$
   10

60. $f(x) = |3x - 5|$

  a. $f(-2)$
   11

  b. $f(-1)$
   8

  c. $f(2)$
   1

  d. $f(3)$
   4

61. $f(x) = x^2 - 3x + 2, g(x) = \dfrac{x - 4}{x + 2}$

  a. $f(2)$
   0

  b. $g(2)$
   $-\dfrac{1}{2}$

  c. $f(-3)$
   20

  d. $g(-3)$
   7

62. $f(x) = \sqrt{x^2 - 3x}, g(x) = \dfrac{2}{x + 4}$

  a. $g(-3)$
   2

  b. $f(-1)$
   2

  c. $g(6)$
   $\dfrac{1}{5}$

  d. $f(4)$
   2

★63. $f(x) = 2x^2 + 3x + 1$

  a. $f(2a)$
   $8a^2 + 6a + 1$

  b. $f(-3a)$
   $18a^2 - 9a + 1$

  c. $f\left(\dfrac{1}{2}a\right)$
   $\dfrac{1}{2}a^2 + \dfrac{3}{2}a + 1$

  d. $f(0.1a)$
   $0.02a^2 + 0.3a + 1$

★64. $f(x) = 3x^2 - 2x - 4$

  a. $f(3a)$
   $27a^2 - 6a - 4$

  b. $f(-2a)$
   $12a^2 + 4a - 4$

  c. $f\left(\dfrac{2}{3}a\right)$
   $\dfrac{4}{3}a^2 - \dfrac{4}{3}a - 4$

  d. $f(0.2a)$
   $0.12a^2 - 0.4a - 4$

For Exercises 65–68, use the given graph to find the value of the function. See Example 6.

65.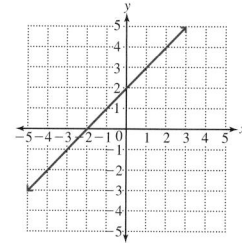

a. $f(1)$
  3
b. $f(2)$
  4
c. $f(0)$
  2

66.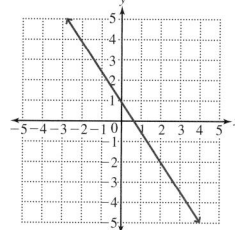

a. $f(4)$
  −5
b. $f(2)$
  −2
c. $f(0)$
  1

67.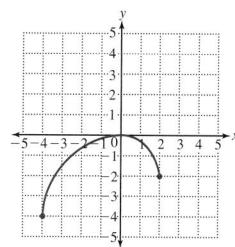

a. $f(0)$
  0
b. $f(2)$
  −2
c. $f(3)$
  undefined

68.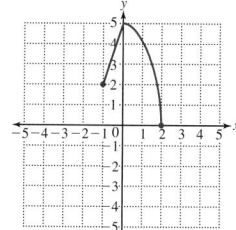

a. $f(0)$
  5
b. $f(2)$
  0
c. $f(-2)$
  undefined

Objective 4

For Exercises 69–78, graph. See Example 7.

69. $f(x) = 2x + 1$

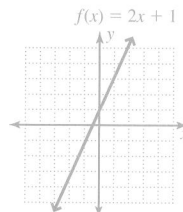

70. $f(x) = 3x - 5$

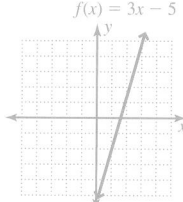

71. $f(x) = \dfrac{1}{3}x + 3$

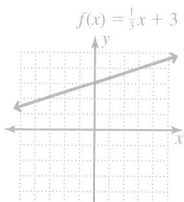

72. $f(x) = \dfrac{1}{2}x - 3$

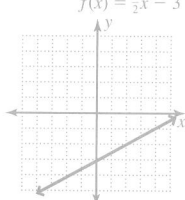

73. $f(x) = -4x + 1$

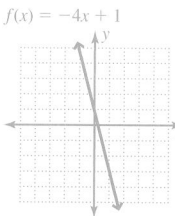

74. $f(x) = -x - 1$

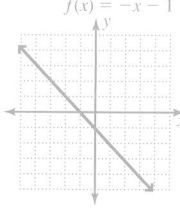

75. $f(x) = -\dfrac{2}{3}x$

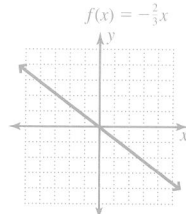

76. $f(x) = \dfrac{2}{5}x$

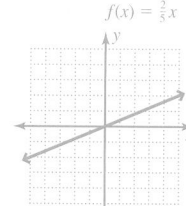

77. $f(x) = -\dfrac{1}{4}x - 2$

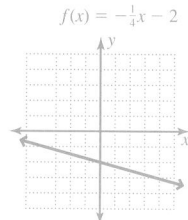

78. $f(x) = -\dfrac{1}{3}x + 4$

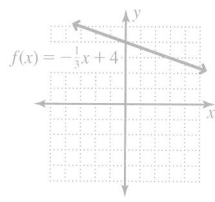

79. The cost, $C(x)$, as a function of the number of items produced, x, is given by $C(x) = 20x + 150$.
  a. Find the cost of producing 10 items.
    $350
  b. Find the cost of producing 15 items.
    $450
  c. What does $C(30) = 750$ mean?
    The cost of producing 30 items is $750.

80. For a rental car, the cost, $C(x)$, as a function of the number of miles driven, x, is given by $C(x) = 0.12x + 22$.
  a. Find the cost of driving 100 miles.
    $34
  b. Find the cost of driving 350 miles.
    $64
  c. What does $C(250) = 52$ mean?
    The cost of driving 250 miles is $52.

81. The value, $V(x)$, of a car as a function of the number of years after it was purchased, x, is given by $V(x) = -1500x + 18,000$.

   a. Find the value of the car after three years.

     $13,500

   b. Find the value of the car after six years.

     $9000

   c. What does $V(10) = 3000$ mean?

     The value of the car after ten years is $3000.

82. The monthly salary, $S(x)$, of a salesperson as a function of her sales, x, is given by $S(x) = 0.10x + 175$.

   a. Find her salary for sales of $15,000.

     $1675

   b. Find her salary for sales of $10,000.

     $1175

   c. What does $S(20,000) = 2175$ mean?

     Her salary for $20,000 of sales is $2175.

Review Exercises

Exercises 1 and 2   **Expressions**

For Exercises 1 and 2, simplify.

[1.5] **1.** $\dfrac{2}{5} - \dfrac{5}{8} \div \left(\dfrac{3}{4}\right)$

     $-\dfrac{13}{30}$

[1.7] **2.** $x + 3y - 5x + 7 - 3y - 9$

     $-4x - 2$

Exercises 3–6   **Equations and Inequalities**

[1.3] **3.** Which property is illustrated in the equation $4(x + 9) = 4(9 + x)$?

     Commutative property of addition

[3.3] **4.** Graph $2x - y = 4$ using the intercepts.

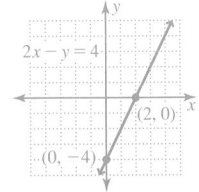

[3.4] **5.** Solve $3x - 2y = 6$ for y.

     $y = \dfrac{3}{2}x - 3$

[3.5] **6.** Are the graphs of $2x + 3y = 9$ and $4x + 6y = 12$ parallel, perpendicular, or neither?

     parallel

Chapter 3 Summary and Review Exercises

Complete each incomplete definition, rule, or procedure; study the key examples; and then work the related exercises.

Forms of Linear Equations	Formula(s)
Slope–intercept form: $y = \underline{\quad mx + b \quad}$. **Point–slope form:** $y - y_1 = \underline{\quad m(x_1 - x_2) \quad}$. **Standard form:** $Ax + \underline{\quad By \quad} = C$, where A, B, and C are integers and $A > 0$.	**Slope:** $m = \underline{\quad \dfrac{y_2 - y_1}{x_2 - x_1} \quad}$ $x_2 \neq x_1$.

3.1 The Rectangular Coordinate System

Definitions/Rules/Procedures	Key Example(s)
To determine the coordinates of a given point in the rectangular coordinate system: 1. Follow a vertical line from the point to the $\underline{\text{x-axis}}$ (horizontal axis). The number at this position on the $\underline{\text{x-axis}}$ is the first coordinate. 2. Follow a(n) $\underline{\text{horizontal}}$ line from the point to the y-axis (vertical axis). The number at this position on the y-axis is the $\underline{\text{second coordinate}}$. **To graph or plot a point given its coordinates:** 1. Beginning at the origin, $(0, 0)$, move to the $\underline{\text{right}}$ or $\underline{\text{left}}$ along the x-axis the amount indicated by the first coordinate. 2. From that position on the x-axis, move $\underline{\text{up}}$ or $\underline{\text{down}}$ the amount indicated by the second coordinate. 3. Draw a(n) $\underline{\text{dot}}$ to represent the point described by the coordinates.	Plot each of the following: $(-2, 4), (5, 0), (0, -2), (-4, -3), (3, -4), (2, 3)$ 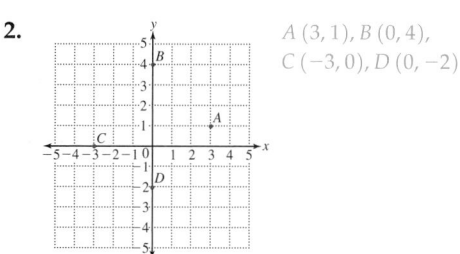

[3.1] *For Exercises 1 and 2, write the coordinates for each point.*

1. 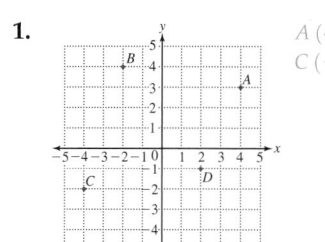 $A\,(4, 3), B\,(-2, 4),$ $C\,(-4, -2), D\,(2, -1)$

2. $A\,(3, 1), B\,(0, 4),$ $C\,(-3, 0), D\,(0, -2)$

[3.1] *For Exercises 3 and 4, plot and label the points indicated by the coordinate pairs.*

3. $(3, 4), (-2, 0), (0, 1), (-1, -5)$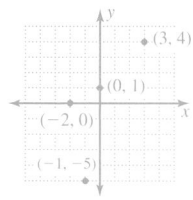

4. $(-2, 2), (0, -3), (4, 1), (5, 0)$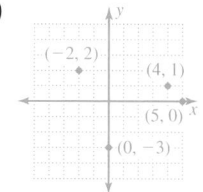

Definitions/Rules/Procedures	Key Example(s)
To determine the quadrant for a given ordered pair, consider the signs of the coordinates.	State the quadrant in which each point is located. If the point lies on an axis, state which axis.
_(+, +)_ means the point is in quadrant I.	(19, 78): quadrant I
_(−, +)_ means the point is in quadrant II.	(−67, 45): quadrant II
_(−, −)_ means the point is in quadrant III.	(−107, −36): quadrant III
_(+, −)_ means the point is in quadrant IV.	(58, −92): quadrant IV
Note: If either coordinate is 0, the point is on an axis and not in a quadrant.	(0, −16): This point is on the y-axis.

[3.1] *For Exercises 5–8, state the quadrant in which the point is located. If the point lies on an axis, state which axis.*

5. $(-2.8, 1203)$
II

6. $\left(-\dfrac{1}{8}, -16\right)$
III

7. $(421, 5300)$
I

8. $\left(52\dfrac{2}{3}, -0.36\right)$
IV

3.2 Graphing Linear Equations

Definitions/Rules/Procedures	Key Example(s)
To determine whether a given ordered pair is a solution for an equation with two variables:	Determine whether $(3, 5)$ is a solution for $4x - y = 7$.
1. Replace the variables in the equation with the corresponding _coordinates_.	**Solution:** $4(3) - 5 \stackrel{?}{=} 7$ Replace x with 3 and y with 5.
2. Verify that the equation is _true_.	$12 - 5 \stackrel{?}{=} 7$
	$7 = 7$
	$(3, 5)$ is a solution.
	Determine whether $(4, -9)$ is a solution for $4x - y = 7$.
	Solution: $4(4) - (-9) \stackrel{?}{=} 7$ Replace x with 4 and y with −9.
	$16 + 9 \stackrel{?}{=} 7$
	$25 \neq 7$
	$(4, -9)$ is not a solution.

Exercises 9–12 Equations and Inequalities

[3.2] *For Exercises 9–12, determine whether the given ordered pair is a solution for the equation.*

9. $(-1, 3); x + 3y = 8$
yes

10. $(2, 1); 2x - y = 3$
yes

11. $(0, 0); 6x = 2y + 1$
no

12. $(-2, -2); y = \dfrac{2}{3}x$
no

Definitions/Rules/Procedures	Key Example(s)
To find a solution to a linear equation with two variables:	Find two solutions for the equation $y = 2x - 3$.
1. Choose a(n) _value_ for one of the variables.	First solution: Let $x = 0$
2. Replace the corresponding _variable_ with your chosen value.	Second solution: Let $x = 1$
	$y = 2(0) - 3$ $y = 2(1) - 3$
3. Solve the equation for the value of the other _variable_.	$y = -3$ $y = 2 - 3$
	$y = -1$
	Solution: $(0, -3)$ Solution: $(1, -1)$

Definitions/Rules/Procedures	Key Example(s)
To graph a linear equation: 1. Find at least __two__ solutions to the equation. 2. Plot the solutions as __points__ in the rectangular coordinate system. 3. Connect the points to form a(n) __straight line__ .	Graph $y = 2x - 3$. We found two solutions in the previous Key Example: $(0, -3)$ and $(1, -1)$. 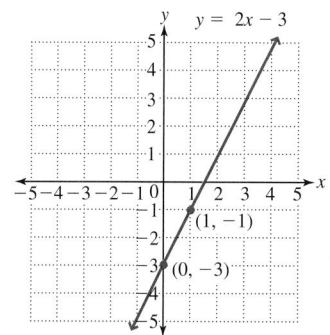

Exercises 13–16 <img_pyramid> **Equations and Inequalities**

[3.2] *For Exercises 13–16, find three solutions for the given equation. Then graph.*

13. $y = -2x$

(0, 0)
(1, −2)
(−1, 2)

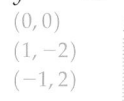

14. $y = -x + 7$

(0, 7)
(1, 6)
(3, 4)

15. $2x + 3y = 6$

(0, 2)
(3, 0)
(−3, 4)

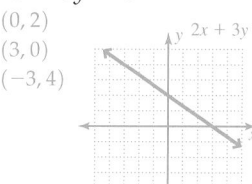

16. $y + \dfrac{1}{3}x = 3$

(0, 3)
(3, 2)
(−3, 4)

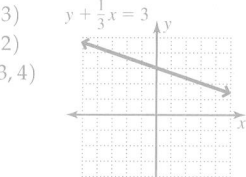

17. A salesperson receives \$1000 per month plus a commission of 5% of his total sales. The equation $p = 0.05s + 1000$ describes the salesperson's gross pay each month, where p represents the gross pay and s represents the total sales in dollars.

   a. Find the gross pay if the person sells \$24,000 worth of merchandise.

   \$2200

   b. Find the point where the graph intersects the p-axis. Explain what it indicates.

   (0, 1000); this indicates the person's gross pay if he has \$0 in sales during a month.

   c. Graph the equation.

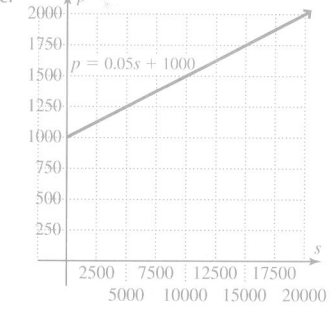

Definitions/Rules/Procedures	Key Example(s)
The graph of $y = c$, where c is a real-number constant, is a(n) __horizontal__ line parallel to the x-axis that passes through the y-axis at a point with coordinates ____(0, c)____ . **The graph of $x = c$, where c is a real-number constant,** is a(n) __vertical__ line parallel to the y-axis that passes through the x-axis at a point with coordinates ____(c, 0)____ .	Graph **a.** $y = 4$ **b.** $x = -3$ 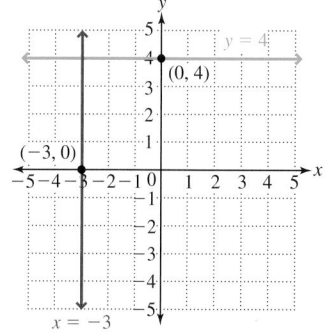

Exercises 18–21 Equations and Inequalities

[3.2] *For Exercises 18–21, graph the horizontal or vertical lines.*

18. $x = 3$

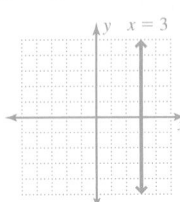

19. $x = -1$

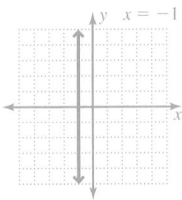

20. $y = -4$

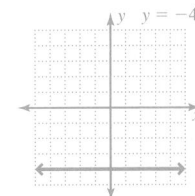

21. $y = 1$

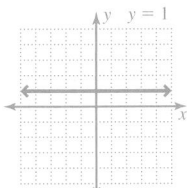

3.3 Graphing Using Intercepts

Definitions/Rules/Procedures	Key Example(s)
The **x-intercept** is a point where a graph ___intersects___ the x-axis.	
The **y-intercept** is a point where a graph ___intersects___ the y-axis.	
To find an x-intercept:	Find the x- and y-intercepts for $2x - 3y = 18$.
1. Replace ___y___ with 0 in the given equation.	Solution:
2. Solve for ___x___.	x-intercept: y-intercept:
To find a y-intercept:	$2x - 3(0) = 18$ $2(0) - 3y = 18$
1. Replace ___x___ with 0 in the given equation.	$2x = 18$ $-3y = 18$
2. Solve for ___y___.	$x = 9$ $y = -6$
	x-intercept: $(9, 0)$ y-intercept: $(0, -6)$
If an equation can be written in the form $y = mx$, where m is a real number other than 0, then the x- and y-intercepts are at ___the origin, (0, 0)___.	The graph of the equation $y = 3x$ has $(0, 0)$ as both the x- and y-intercept.
If an equation is in the form $y = mx + b$, where m and b are nonzero real numbers, then the y-intercept will be ___(0, b)___.	The y-intercept of the graph of $y = 3x - 2$ is $(0, -2)$.
The graph of an equation in the form $y = c$, where c is a nonzero real number, has no x-intercept and the y-intercept is ___(0, c)___. **The graph of $y = 0$ is** the ___x-axis___.	The graph of the equation $y = 4$ has no x-intercept, and its y-intercept is $(0, 4)$.
The graph of an equation in the form $x = c$, where c is a nonzero real number, has no ___y-intercept___ and the x-intercept is $(c, 0)$. **The graph of $x = 0$ is the** ___y-axis.___	The graph of the equation $x = -3$ has no y-intercept, and its x-intercept is $(-3, 0)$.

Exercises 22–27 Equations and Inequalities

[3.3] *For Exercises 22–27, find the x- and y-intercepts and graph.*

22. $3x + 2y = 6$

   $(2, 0), (0, 3)$

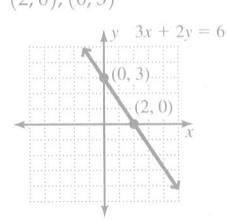

23. $y = -2x$

   $(0, 0)$ for both

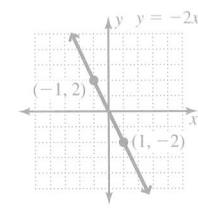

24. $y = -\dfrac{1}{5}x + 3$

   $(15, 0)\ (0, 3)$

25. $2x = 5y + 10$
$(0, -2), (5, 0)$

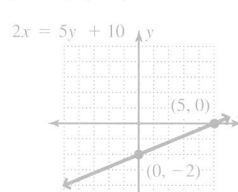

26. $7 = y$
no x-intercept, $(0, 7)$

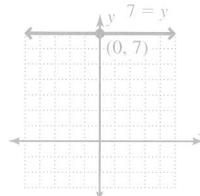

27. $x = -1$
$(-1, 0)$, no y-intercept

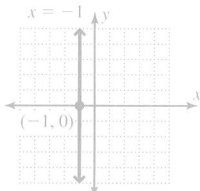

3.4 Slope–Intercept Form

Definitions/Rules/Procedures	Key Example(s)
The **slope** of a line is the ratio of the ____vertical____ change to the ____horizontal____ change between any two points on a line. **Given an equation of the form $y = mx + b$:** If $m > 0$ (slope is positive), the graph is a line that slants ____uphill____ from left to right. If $m < 0$ (slope is negative), the graph is a line that slants ____downhill____ from left to right. **To graph an equation in slope–intercept form, $y = mx + b$:** 1. Plot the ____y-intercept____, $(0, b)$. 2. Plot a second point by ____rising____ the number of units indicated by the numerator of the slope, m, then ____running____ the number of units indicated by the denominator of the slope, m. 3. Draw a(n) ____straight line____ through the two points. *Note:* You can check by locating additional points using the slope. Every point you locate using the slope should be on the line.	Graph. **a.** $y = 2x + 1$ **b.** $y = -\dfrac{3}{4}x - 2$ **Solution:** For $y = 2x + 1$, the slope is 2 and the y-intercept is $(0, 1)$. The positive slope indicates an uphill line from left to right. To graph, we can rise 2, then run 1 from the y-intercept, $(0, 1)$, to get a second point $(1, 3)$. For $y = -\dfrac{3}{4}x - 2$, the slope is $-\dfrac{3}{4}$ and the y-intercept is $(0, -2)$. The negative slope indicates a downhill line from left to right. To graph, we rise -3, then run 4 from the y-intercept, $(0, -2)$, to get a second point $(4, -5)$. 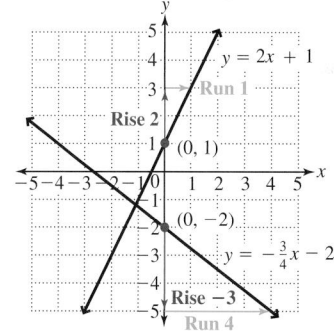

Exercises 28–33 ◢◣ Equations and Inequalities

[3.4] *For Exercises 28–33, determine the slope and the y-intercept and then graph.*

28. $y = -2x + 3$
$m = -2; (0, 3)$ $y = -2x + 3$

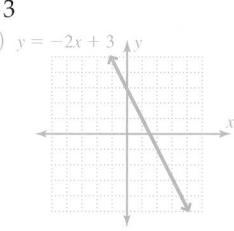

29. $y = -\dfrac{1}{4}x + 2$
$m = -\dfrac{1}{4}; (0, 2)$ $y = -\dfrac{1}{4}x + 2$

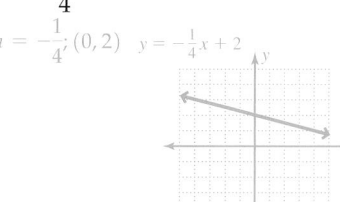

30. $y = 3x$
$m = 3; (0, 0)$

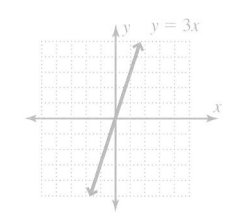

31. $x + y = 5$

$m = -1$; $x + y = 5$
$(0, 5)$

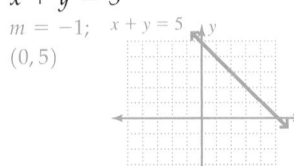

32. $x - 3y = 7$

$m = \dfrac{1}{3}$;

$\left(0, -\dfrac{7}{3}\right)$

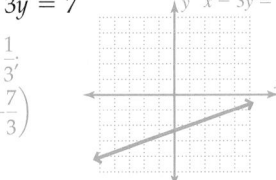

33. $2x - 3y = 8$

$m = \dfrac{2}{3}$;

$\left(0, -\dfrac{8}{3}\right)$

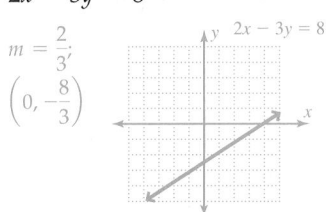

Definitions/Rules/Procedures	Key Example(s)
The slope of a line through two points (x_1, y_1) and (x_2, y_2), where $x_1 \neq x_2$, is given by the formula $m = \underline{\dfrac{y_2 - y_1}{x_2 - x_1}}$ **The slope of a horizontal line is** $\underline{\quad 0 \quad}$. **The slope of a vertical line is** $\underline{\quad\text{undefined}\quad}$.	Find the slope of a line connecting the points $(2, 7)$ and $(-8, 3)$. **Solution:** $m = \dfrac{3 - 7}{-8 - 2} = \dfrac{-4}{-10} = \dfrac{2}{5}$ The slope of a line connecting the points $(2, 4)$ and $(1, 4)$ is 0, and the equation of the line is $y = 4$. The slope of a line connecting the points $(-3, 7)$ and $(-3, -2)$ is undefined, and the equation of the line is $x = -3$.

Exercises 34–37 ▲, Equations and Inequalities

[3.4] *For Exercises 34–37, find the slope of the line connecting the given points.*

34. $(2, 7), (-1, -2)$

3

35. $(-6, 2), (-1, -1)$

$-\dfrac{3}{5}$

36. $(2, 8), (-1, 8)$

0

37. $(-7, 3), (-7, -5)$

undefined

Definitions/Rules/Procedures	Key Example(s)
To **write the equation of a line given its slope,** m, **and its** y-**intercept,** $(0, b)$, use the $\underline{\quad\text{slope–intercept}\quad}$ form of the equation, $y = mx + b$.	Write the slope–intercept form of the equation of a line with slope $\dfrac{3}{4}$ and y-intercept $(0, -1)$. **Solution:** $y = \dfrac{3}{4}x - 1$

Exercises 38–41 ▲, Equations and Inequalities

[3.4] *For Exercises 38–41, write the equation of the line in slope–intercept form given the slope and the y-intercept.*

38. $m = -1; (0, 7)$

$y = -x + 7$

39. $m = -\dfrac{1}{5}; (0, -8)$

$y = -\dfrac{1}{5}x - 8$

40. $m = 0.2; (0, 6)$

$y = 0.2x + 6$

41. $m = 1; (0, 0)$

$y = x$

3.5 Point–Slope Form

Definitions/Rules/Procedures	Key Example(s)
To write the equation of a line given its **slope** and any **point**, (x_1, y_1), on the line, use the point–slope form of the equation of a line, $\underline{y - y_1 = m(x - x_1)}$. **If given a second point**, (x_2, y_2), and **not the slope**, we first calculate the slope using $m = \dfrac{y_2 - y_1}{x_2 - x_1}$; then we use $\underline{y - y_1 = m(x - x_1)}$.	Write the slope–intercept form of the equation of a line with slope -3 that passes through $(2, 5)$.

Solution: $y - y_1 = m(x - x_1)$

$\quad\quad\quad y - 5 = -3(x - 2) \quad x_1 = 2, y_1 = 5$

$\quad\quad\quad y - 5 = -3x + 6$

$\quad\quad\quad\quad\quad y = -3x + 11$

Write the equation of a line connecting $(2, 3)$ and $(-3, -1)$ in the form $Ax + By = C$, where A, B, and C are integers and $A > 0$.

Solution: Find the slope:

$$m = \frac{-1 - 3}{-3 - 2} = \frac{-4}{-5} = \frac{4}{5}$$

Write the equation: $y - 3 = \dfrac{4}{5}(x - 2)$

$\quad\quad 5(y - 3) = 5 \cdot \dfrac{4}{5}(x - 2)$ $\quad\begin{cases}\text{Multiply by 5 to} \\ \text{clear the fraction.}\end{cases}$

$\quad\quad 5y - 15 = 4(x - 2)$

$\quad\quad 5y - 15 = 4x - 8$ $\quad\quad\quad$ Distribute 4.

$\quad 5y - 4x - 15 = -8$ $\quad\quad\quad$ Subtract $4x$ from both sides.

$\quad\quad 5y - 4x = 7$ $\quad\quad\quad\quad\quad$ Add 15 to both sides.

$\quad\quad -4x + 5y = 7$ $\quad\quad\quad\quad$ Rearrange so that the $-4x$ term is first.

$\quad -1(-4x + 5y) = -1 \cdot 7$ $\quad\quad\begin{cases}\text{Multiply by } -1 \text{ so that} \\ \text{the } x \text{ term is positive.}\end{cases}$

$\quad\quad 4x - 5y = -7$ $\quad\quad\quad\quad$ Simplify.

Exercises 42–46 Equations and Inequalities

[3.5] *For Exercises 42 and 43, write the equation of a line in slope–intercept form with the given slope that passes through the given point.*

42. $m = -2; (1, 7)$

$y = -2x + 9$

43. $m = -\dfrac{2}{5}; (3, 3)$

$y = -\dfrac{2}{5}x + \dfrac{21}{5}$

[3.5] *For Exercises 44 and 45, write the equation of a line connecting the given points in slope–intercept form and in standard form.*

44. $(7, -3), (2, 2)$

$y = -x + 4$

$x + y = 4$

45. $(5, -3), (9, 2)$

$y = \dfrac{5}{4}x - \dfrac{37}{4}$

$5x - 4y = 37$

[3.5] 46. Andrew purchased a photocopier for his business in 2009 for $18,000. In 2013, he considers selling the copier on eBay and notes that others like it are selling for $12,000.

a. Assuming that the depreciation is linear, plot the two given data points with 2009 being year 0 so that 2013 is year 4. Draw a line connecting the two points.

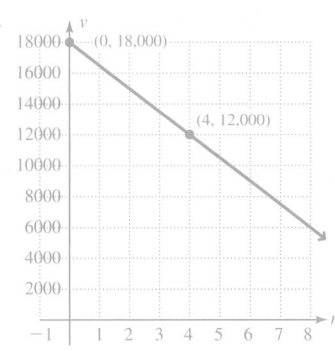

b. What is the slope of the line?

-1500

c. Let n represent the number of years the copier is in service and v represent the value of the copier. Write the equation of the line in slope–intercept form.

$v = -1500n + 18{,}000$

d. If the depreciation continues at the same rate, in what year was the copier worth half of its original value?

2015

e. In what year will the copier be worth $0?

2021

Definitions/Rules/Procedures	Key Example(s)
Nonvertical parallel lines have equal ____slopes____ and different ____y-intercepts____ . Vertical lines are parallel. The **slope of a line perpendicular** to a line with a slope of $\dfrac{a}{b}$ will be ____$-\dfrac{b}{a}$____. Horizontal and vertical lines are perpendicular.	Find the slope of a line parallel and perpendicular to $3x + 2y = 4$. **Solution:** Write $3x + 2y = 4$ in slope–intercept form. $2y = -3x + 4$ Subtract 3x from both sides. $y = -\dfrac{3}{2}x + 2$ Divide both sides by 2. The slope of $3x + 2y = 4$ is $-\dfrac{3}{2}$. The slope of a line parallel to $3x + 2y = 4$ is $-\dfrac{3}{2}$. The slope of a line perpendicular to $3x + 2y = 4$ is $\dfrac{2}{3}$.

Exercises 47–50 Equations and Inequalities

[3.5] *For Exercises 47–50, write the equations in slope–intercept form.*

47. Find the equation of a line that passes through $(0, -2)$ and is parallel to the line $y = 2x + 7$.

$y = 2x - 2$

48. Find the equation of a line that passes through $(1, 5)$ and is parallel to the line $2x + 3y = 6$.

$y = -\dfrac{2}{3}x + \dfrac{17}{3}$

49. Find the equation of a line that passes through $(-1, -2)$ and is perpendicular to the line $y = \dfrac{3}{5}x - 1$.

$y = -\dfrac{5}{3}x - \dfrac{11}{3}$

50. Find the equation of a line that passes through $(-2, 4)$ and is perpendicular to the line $y = -\dfrac{2}{5}x - 6$.

$y = \dfrac{5}{2}x + 9$

3.6 Graphing Linear Inequalities

Definitions/Rules/Procedures	Key Example(s)
To determine whether an ordered pair is a solution for an inequality, replace the variables with the corresponding coordinates. If the resulting inequality is _____true_____, the ordered pair is a solution.	Determine whether $(4, 3)$ is a solution for $2x - 4y \le 8$. $2x - 4y \le 8$ $2(4) - 4(3) \overset{?}{\le} 8 \quad x = 4, y = 3$ $8 - 12 \overset{?}{\le} 8$ $-4 \le 8 \quad$ This is true, so $(4, 3)$ is a solution. Determine whether $(3, -1)$ is a solution for $2x - 4y \le 8$. $2x - 4y \le 8$ $2(3) - 4(-1) \overset{?}{\le} 8 \quad x = 3, y = -1$ $6 + 4 \overset{?}{\le} 8$ $10 \le 8 \quad$ This is false, so $(3, -1)$ is not a solution.

Exercises 51–54   Equations and Inequalities

[3.6] *For Exercises 51–54, determine whether the ordered pair is a solution for the linear inequality.*

51. $(3, 2); x + 2y > 5$
yes

52. $(-1, 0); y < x + 2$
yes

53. $(-5, 8); 2x - 6y \le 17$
yes

54. $(0, 0); x + y \ge 5$
no

Definitions/Rules/Procedures	Key Example(s)
To graph a linear inequality in two variables: 1. Graph the related equation (the boundary line). The related equation has an equal sign in place of the inequality symbol. If the inequality symbol is $\le$ or $\ge$, draw a(n) _____solid_____ line. If the inequality symbol is $<$ or $>$, draw a(n) _____dashed_____ line. 2. Choose an ordered pair on one side of the boundary line and test this ordered pair in the inequality. **If the ordered pair satisfies the inequality**, shade the region that _____contains it_____. **If the ordered pair does not satisfy the inequality**, shade the region on the _____other side_____ of the boundary line.	Graph $2x - 4y \le 8$. **Solution:** Graph $2x - 4y = 8$ as a solid line. We found $(0, 0)$ to be a solution; therefore, shade the region containing $(0, 0)$. 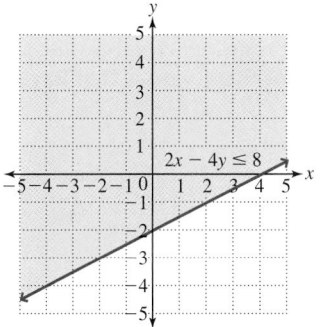

Exercises 55–60   Equations and Inequalities

[3.6] *For Exercises 55–59, graph the linear inequality.*

55. $y < 3x - 5$

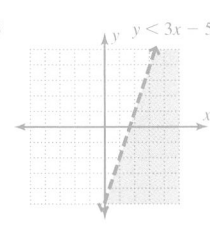

56. $y \ge -\dfrac{2}{3}x$

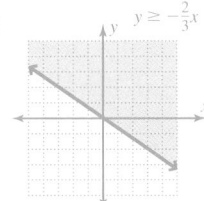

57. $x - y \ge 3$

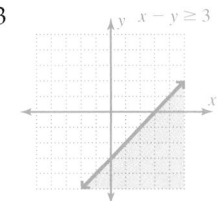

58. $-3x - 5y < -15$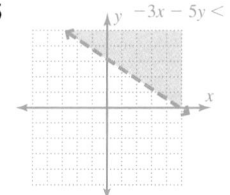

59. $x \geq -1$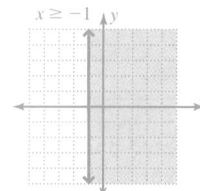

[3.6] **60.** A company produces two versions of its product. The lower-priced package costs \$6 to make, whereas the higher-priced package costs \$8 to make.

    a. Let a represent the number of the lower-priced packages produced and b represent the number of the higher-priced packages produced. Write an inequality in which the total cost is at most \$12,000.

       $6a + 8b \leq 12,000$

    b. Graph the inequality.

    c. Give a combination of the number of each package that the company could produce that has a cost equal to \$12,000.

       Answers may vary; some combinations are $(400, 1200)$, $(800, 900)$, and $(1200, 600)$.

    d. Give a combination of the number of each package that the company could produce that has a cost less than \$12,000.

       Answers may vary; some combinations are $(500, 100)$, $(1000, 200)$, and $(200, 1100)$.

b.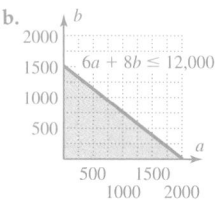

3.7 Introduction to Functions and Function Notation

Definitions/Rules/Procedures	Key Example(s)
A **relation** is a set of ____ordered pairs____.	For the set of ordered pairs $\{(7, -3), (4, 7), (-5, 2), (3, -3)\}$,
The **domain** is the set of all ____input____ values (x-values) for a relation.	**a.** Is the relation a function? **b.** Find the domain. **c.** Find the range.
The **range** is the set of all ____output____ values (y-values) for a relation.	**Solution:**
	a. yes **b.** $\{-5, 3, 4, 7\}$ **c.** $\{-3, 2, 7\}$
To determine the domain of a relation given its graph, answer this question: **What are all of the x-values that have a corresponding ____y-value____?**	Give the domain and range of the relation shown.
To determine the range, answer this question: **What are all of the y-values that have a corresponding ____x-value____?**	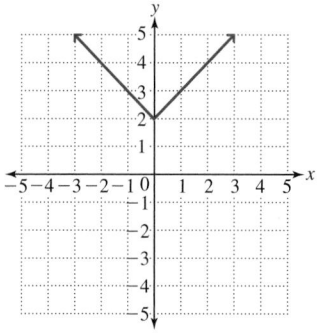
	Domain: all real numbers
	Range: $\{ y \mid y \geq 2 \}$ or $[2, \infty)$

Exercises 61–64 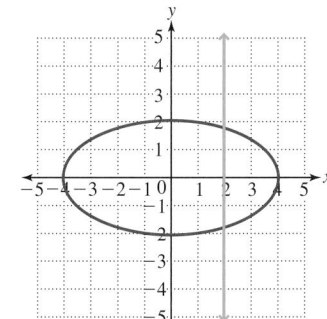 Equations and Inequalities

[3.7] *For Exercises 61 and 62, find the domain and range of the relation.*

61. $\{(2,3), (-1,5), (3,6), (-3,-4)\}$
Domain: $\{2, -1, 3, -3\}$; range: $\{3, 5, 6, -4\}$

62. Cars holding their value for resale:

Automobile	Percentage of Original Value
Volkswagen	52.2
Honda	49.7
Toyota	49.0
Subaru	47.8
Nissan	45.8

Source: Automotive Lease Guide

Domain: {Volkswagen, Honda, Toyota, Subaru, Nissan};
range: {52.2, 49.7, 49.0, 47.8, 45.8}

[3.7] *For Exercises 63 and 64, determine whether the relation is a function.*

63. The following relation shows the courses taught by each instructor during one semester. no

Domain (instructor name)	Range (courses taught)
Hames	Math 100
Carson	Math 100, Math 102
Pritchard	Math 035
Webb	Math 100, Math 110

64. The following relation shows the price of a particular brand of dog food based on the size of the bag. yes

Domain (size of bag)	Range (price)
5 lb.	$3.95
10 lb.	$7.90
20 lb.	$15.00
40 lb.	$29.95

Definitions/Rules/Procedures	Key Example(s)

To determine whether a relation is a function from its graph, perform the vertical line test.

1. Draw or imagine vertical lines through each point in the ____domain____.
2. If each vertical line intersects the graph in at most ____one____ point, the graph is the graph of a function.
3. If any vertical line intersects the graph at ____two____ or ____more____ different points, the graph is not the graph of a function.

Determine whether the graph of the relation is the graph of a function.

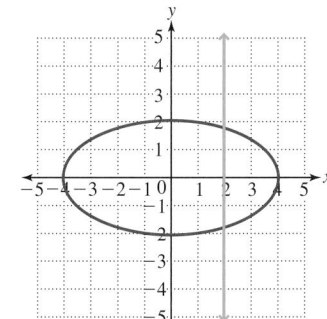

Solution: The relation in the graph is not a function because a vertical line can be drawn that intersects the graph at two points. This means that there is a domain value that corresponds to two values in the range.

Exercises 65–67 Equations and Inequalities

[3.7] *For Exercises 65–67, give the domain and range. Then determine whether the graph is the graph of a function.*

65.

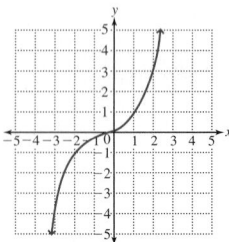

Domain: all real numbers or $(-\infty, \infty)$;
range: all real numbers or $(-\infty, \infty)$; it is a function.

66.

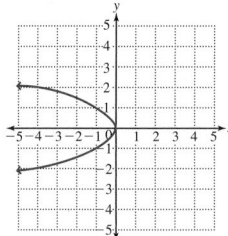

Domain: $\{x \mid x \leq 0\}$ or $(-\infty, 0]$;
range: all real numbers or $(-\infty, \infty)$; it is not a function.

67.

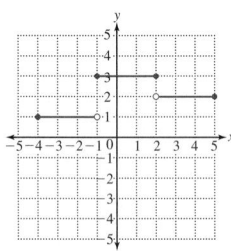

Domain: $\{x \mid -4 \leq x \leq 5\}$ or $[-4, 5]$;
range: $\{1, 2, 3\}$; it is a function.

Definitions/Rules/Procedures	Key Example(s)
Given a function $f(x)$, to find $f(a)$, where a is a real number in the domain of f, replace x in the function with _____a_____ and calculate the value.	For the function $f(x) = x^2 - 3x + 1$, find $f(-2)$. $$f(-2) = (-2)^2 - 3(-2) + 1$$ $$= 4 + 6 + 1$$ $$= 11$$

Exercises 68–70 Equations and Inequalities

[3.7] *For Exercises 68–70, find the value of the function.*

68. Use the graph in Exercise 67 to find the value of the function at the indicated values.

   a. $f(-3)$
      1
   b. $f(-1)$
      3
   c. $f(0)$
      3
   d. $f(3)$
      2

69. $f(x) = x^3 - 5x + 2$

   a. $f(2)$
      0
   b. $f(0)$
      2
   c. $f(-3)$
      -10

70. $f(x) = \dfrac{x}{x - 3}$

   a. $f(6)$
      2
   b. $f(3)$
      undefined
   c. $f(-5)$
      $\dfrac{5}{8}$

Learning Strategy

To prepare for an exam, work all the problems you cover in class, review your notes, and work some more problems! In each problem that you work, try to understand what you are doing and why.

—Stella P.

Chapter 3 Practice Test

For Extra Help

Step-by-step test solutions are found on the Chapter Test Prep Videos available in MyMathLab® *or on* You Tube.

1. Determine the coordinates for each point.

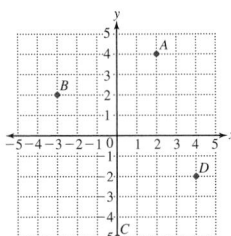

$A\,(2,4)\ B\,(-3,2)$
$C\,(0,-5)\ D\,(4,-2)$
[3.1]

2. Plot and label the points indicated by the coordinate pairs.
$(0,-3),(4,-1),(-3,2),(2,0),(-1,-5)$

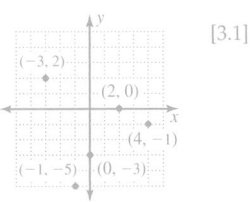

[3.1]

3. State the quadrant in which $\left(4\frac{2}{3},5115\right)$ is located.

If the point lies on an axis, state which axis.

I [3.1]

4. Determine whether $(1,7)$ is a solution for
$$y = -\frac{2}{5}x - 8.$$

no [3.2]

For Exercises 5–8, find the x- and y-intercepts and then graph.

5. $x - 2y = 4$
$(4,0),(0,-2)$ [3.3]

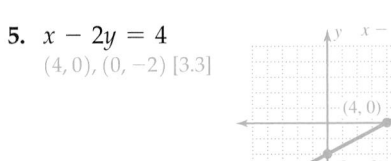

6. $2x + 5y = 10$
$(5,0),(0,2)$ [3.3]

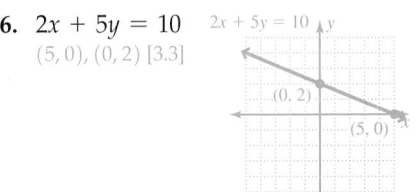

7. $y = -2x$
$(0,0)$ [3.3]

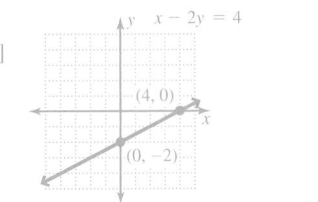

8. $y = \frac{3}{4}x - 1$

$\left(\frac{4}{3},0\right),(0,-1)$ [3.3]

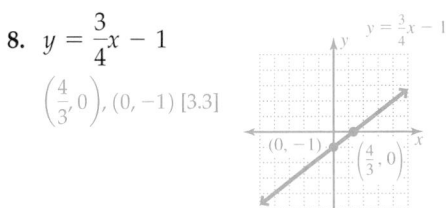

For Exercises 9 and 10, determine the slope and the y-intercept.

9. $y = \frac{3}{4}x + 11$

$m = \frac{3}{4};(0,11)$ [3.4]

10. $5x - 3y = 8$

$m = \frac{5}{3};\left(0,-\frac{8}{3}\right)$ [3.4]

For Exercises 11 and 12, determine the slope of the line connecting the given points.

11. $(-5,6),(-2,-4)$

$-\frac{10}{3}$ [3.4]

12. $(6,2),(3,2)$

0 [3.4]

For Exercises 13 and 14, write the equation of the line in slope–intercept form.

13. $m = \frac{3}{5}$; y-intercept $(0,4)$

$y = \frac{3}{5}x + 4$ [3.4]

14. Passing through the points $(8,-1),(-7,5)$

$y = -\frac{2}{5}x + \frac{11}{5}$ [3.5]

15. Write the equation of a line in the form $Ax + By = C$ through the points $(2, 1)$ and $(5, 3)$.

$2x - 3y = 1$ [3.5]

16. Write the equation of a line in the form $Ax + By = C$ through the point $(4, -2)$ and perpendicular to the line $y = -3x + 4$.

$x - 3y = 10$ [3.5]

17. From 2004 to 2011, the number of visitors to Florida increased at an approximately linear rate. In 2004, the number of visitors was 79.9 million, and in 2011, the number was 86.5 million. (*Source*: media.visitflorida.org.)

a.

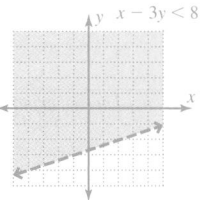

  a. Let 2004 be year 0 so that 2011 is year 7. Let x represent the number of years after 2004 and y represent the number of people visiting Florida. Plot the two data points in the coordinate plane; then draw a line connecting them.

  b. Find the slope of the line rounded to the nearest thousandths.

   0.943 [3.5]

  c. Write the equation of the line in slope–intercept form.

   $y = 0.943x + 79.9$ [3.5]

  d. If the trend continues, predict the number of visitors to Florida in 2015 to the nearest tenth of a million.

   90.3 million [3.6]

18. Determine whether $(2, 7)$ is a solution for the linear inequality $y \geq 2x - 9$.

yes [3.6]

For Exercises 19 and 20, graph the linear inequality.

19. $y \geq \dfrac{2}{5}x - 1$

[3.6]

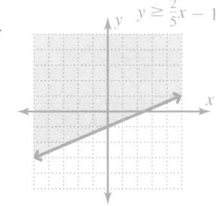

20. $x - 3y < 8$

[3.6]

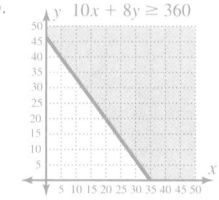

21. Alex works two part-time jobs. He receives \$10 per hour when working as a cook in a restaurant and \$8 per hour when working at a music store. To pay all of his monthly expenses, he needs to make at least \$360 per week.

  a. Let x represent the number of hours he works at the restaurant and y represent the number of hours he works at the music store. Write an inequality in which his total weekly income is at least \$360.

   $10x + 8y \geq 360$ [3.6]

  b. Graph the inequality. Because he can work only a positive number of hours, restrict the graph to the first quadrant.

   [3.6]

b.

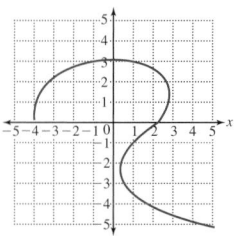

  c. Give a combination of hours that provides him with an income of exactly \$360.

   Answers may vary. One possible answer is $(0, 45)$. [3.6]

  d. Give a combination of hours that provides him with an income of more than \$360.

   Answers may vary. One possible answer is $(10, 35)$. [3.6]

22. The following relation shows U.S. per capita income for each year based on 2010 dollars. Is the relation a function? yes [3.7]

Year	Income ($)
2004	27,961
2005	28,497
2006	28,186
2007	28,305
2008	27,305
2009	26,968
2010	26,487

(*Source*: U.S. Department of Commerce.)

23. Determine whether the graph is the graph of a function. no [3.7]

24. Find the indicated value of the function $f(x) = \dfrac{x^2}{x - 4}$.

   a. $f(2)$
      -2 [3.7]

   b. $f(4)$
      undefined [3.7]

   c. $f(-3)$
      $-\dfrac{9}{7}$ [3.7]

25. **a.** Give the domain and range of the function graphed below.
      Domain: $\{x \mid -3 \le x \le 3\}$ or $[-3, 3]$; range: $\{-2, 1, 2\}$ [3.7]

   b. Find $f(1)$. 2 [3.7]

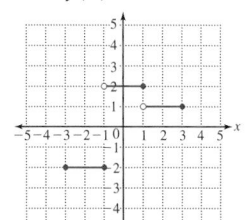

Chapters 1–3 Cumulative Review Exercises

For Exercises 1–4, answer true or false.

[1.1] **1.** π is a rational number.
   false

[3.4] **2.** $y = mx + b$ is the equation of a line in slope–intercept form.
   true

[1.5] **3.** $6° = 6$
   false

[1.5] **4.** $\sqrt{-36} = -6$
   false

For Exercises 5 and 6, fill in the blank.

[2.2] **5.** An equation that has every real number for a solution is a(n) ____identity____.

[2.7] **6.** To write a percent as a fraction or decimal:
   a. Write the percent as a ratio with ____100____ in the denominator.
   b. Simplify to the desired form.

[2.7] **7.** Write 16.5% as a fraction.
   $\dfrac{33}{200}$

Exercises 8–14 **Expressions**

[2.3] **8.** Explain in your own words how to clear decimals from an equation.
   Multiply all of the terms by the power of 10 that will clear the decimal from the number with the most decimal places.

[1.5]For Exercises 9–14, simplify.

9. $(-2)^3 - 3\sqrt{9 + 16}$
   -23

10. $\{4 - 2[6 + (-10)]\} + 3$
   15

11. $-|-12 - (-2)(-3)|$
   -18

12. -3^4
   -81

13. $-5\sqrt{25} + 20 \div (-4) \cdot 2 - 6$
   -41

14. $4^2 - 7 \cdot 3 + \sqrt{144} - 25 \div (-5)$
   12

Exercises 15–30 **Equations and Inequalities**

[3.3] **15.** Find the x- and y-intercepts for $2x - 3y = 6$.
   $(3, 0), (0, -2)$

[3.2] **16.** Find three solutions for $y = -3x + 4$.
   Answers may vary. Three solutions are $(1, 1), (0, 4)$, and $(-1, 7)$.

[3.3] **17.** Graph $2y - 3x = 9$.

   $2y - 3x = 9$

[3.3] **18.** Determine the slope of the line $x + 3y = 8$.
   $m = -\dfrac{1}{3}$

[3.5] 19. Write the equation of a line through the points $(5, 3)$ and $(-3, -1)$.

$y = \dfrac{1}{2}x + \dfrac{1}{2}$

[3.6] 20. Graph $2x + 5y > 10$.

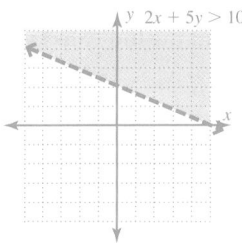

For Exercises 21–23, solve and check:

[2.3] 21. $\dfrac{3}{5}a - 8 = \dfrac{1}{2}$

$\dfrac{85}{6}$

[2.6] 22. $\dfrac{u}{14} = -\dfrac{6}{7}$

-12

[2.3] 23. $1.6x - 14 = 8 + 2.4(x - 5)$

-12.5

[2.4] 24. Solve $C = 2\pi r$ for r.

$r = \dfrac{C}{2\pi}$

For Exercises 25–30, solve.

[2.7] 25. Using the pie chart, if 1000 Internet users were polled, how many would check their e-mail weekly?

230

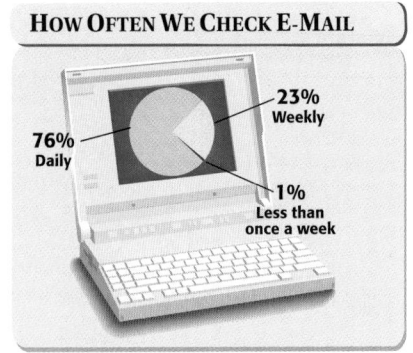

(*Source:* UCLA Center for Communication Policy.)

[2.7] 26. If a married couple makes a combined income of $48,241 per year, what percent does the couple spend on holiday gifts?

$\approx 1.9\%$

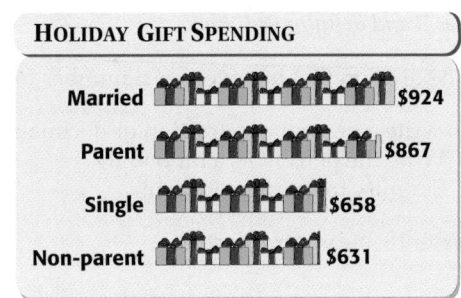

(*Source:* The Gallup Organization.)

[2.6] 27. If an automobile travels 224 miles on 8 gallons of gas, how far can it travel on 15 gallons?

420 miles

[2.7] 28. The ticket prices for the 2013 Super Bowl in New Orleans, Louisiana, ranged from $2187 for end zone tickets to $6917 for lower-level 30–50 yard line tickets. What is the percent of increase from an end zone ticket to a lower-level 30–50 yard line ticket?

$\approx 216\%$

[2.6] 29. A manufacturer tested 400 randomly chosen cell phones as they came off the production line. If she found 15 defective phones, how many defectives should she expect to find in 2000 randomly selected phones?

75 phones

[2.6] 30. On a map, $2\dfrac{1}{4}$ inches represents 450 miles. What is the distance between two points that are 4 inches apart on the map?

800 miles

Chapter Overview

In this chapter, we solve problems involving two or more unknowns. We assign a different variable to each unknown and translate the problem to several equations, called a *system of equations*. We learn several techniques for solving a system of equations:

- ▶ Graphing (Section 4.1).
- ▶ Substitution (Section 4.2).
- ▶ Elimination (Section 4.3).
- ▶ Matrices (Section 4.5).

Finally, in Section 4.6, we solve systems of linear inequalities.

Instructor Note

Except for Section 4.4, each section of this chapter is independent of the other sections and is not prerequisite for material later in the text; so you can pick and choose which sections you would like to cover. Systems involving three unknowns have been separated into Section 4.4 and a separate objective within the matrices section so that you can easily teach only two unknowns if you wish. Cramer's Rule is now located in Appendix D.

4.1 Solving Systems of Linear Equations Graphically

4.2 Solving Systems of Linear Equations by Substitution; Applications

4.3 Solving Systems of Linear Equations by Elimination; Applications

4.4 Solving Systems of Linear Equations in Three Variables; Applications

4.5 Solving Systems of Linear Equations Using Matrices

4.6 Solving Systems of Linear Inequalities

4.1 Solving Systems of Linear Equations Graphically

Objectives

1 Determine whether an ordered pair is a solution for a system of equations.

2 Solve a system of linear equations graphically.

3 Classify systems of linear equations in two unknowns.

Learning Strategy

Don't worry about writing everything the teacher says word for word. If you're obsessed with taking meticulous notes, you often miss what the teacher is saying. Write down the stuff that isn't in the text and go back and take notes later on the stuff that is in the text.

—Ellyn G.

Instructor Note Remind students that in general, or unless otherwise specified, we write ordered pairs with x first then y or in alphabetical order when variables other than x and y are used.

Answers to Warm-up

1. yes

2.

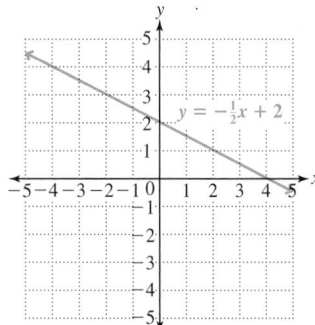

3.

Warm-up

[3.2] **1.** Determine whether $(-1, -6)$ is a solution of $y = 2x - 4$.

[3.2] For Exercises 2 and 3, graph.

2. $y = -\dfrac{1}{2}x + 2$

3. $2x - 3y = 6$

Objective 1 Determine whether an ordered pair is a solution for a system of equations.

Problems with more than one unknown can be solved using a group of equations called a **system of equations**.

Definition **System of equations:** A group of two or more equations.

For example, look at the following problem: The sum of two numbers is 3. Twice the first number plus three times the second number is 8. What are the two numbers? If x represents the first number and y represents the second number, we can translate each sentence to an equation.

Sentence in the problem:	Translation:	
The sum of two numbers is 3.	$x + y = 3$	(Equation 1)
Twice the first number plus three times the second number is 8.	$2x + 3y = 8$	(Equation 2)

The two equations together form a system of equations that describes the problem.

$$\text{System of equations} \quad \begin{cases} x + y = 3 & \text{(Equation 1)} \\ 2x + 3y = 8 & \text{(Equation 2)} \end{cases}$$

A **solution for a system of equations** is an ordered set of numbers that satisfies *all* equations in the system.

Definition **Solution for a system of equations:** An ordered set of numbers that makes all equations in the system true.

For example, in the preceding system of equations, $x = 1$ and $y = 2$ make both equations true so they form a solution to the system of equations. We can check by substituting the values in place of the corresponding variables.

Equation 1:

$x + y = 3$

$1 + 2 \overset{?}{=} 3$

$3 = 3$ True

Equation 2:

$2x + 3y = 8$

$2(1) + 3(2) \overset{?}{=} 8$

$2 + 6 = 8$ True

We can write a solution to a system of equations in two variables as an ordered pair, in this case, $(1, 2)$. This suggests the following procedure for checking solutions.

Procedure **Checking a Solution to a System of Equations**

To verify or check a solution to a system of equations:

1. Replace each variable in each equation with its corresponding value.
2. Verify that each equation is true.

Example 1 Determine whether each ordered pair is a solution to the system of equations.

$$\begin{cases} x + y = 5 & (\text{Equation 1}) \\ y = 2x - 4 & (\text{Equation 2}) \end{cases}$$

a. $(-1, 6)$

Solution:

$x + y = 5$ (Equation 1)	$y = 2x - 4$ (Equation 2)	
$-1 + 6 \stackrel{?}{=} 5$	$6 \stackrel{?}{=} 2(-1) - 4$	In both equations,
$5 = 5$ True	$6 \stackrel{?}{=} -2 - 4$	replace x with -1
	$6 = -6$ False	and y with 6.

Because $(-1, 6)$ does not satisfy both equations, it is not a solution for the system.

b. $(3, 2)$

Solution:

$x + y = 5$ (Equation 1)	$y = 2x - 4$ (Equation 2)	
$3 + 2 \stackrel{?}{=} 5$	$2 \stackrel{?}{=} 2(3) - 4$	In both equations,
$5 = 5$ True	$2 \stackrel{?}{=} 6 - 4$	replace x with 3 and y
	$2 = 2$ True	with 2.

Because $(3, 2)$ satisfies both equations, it is a solution for the system.

Your Turn 1 Determine whether each ordered pair is a solution to the system of equations.

$$\begin{cases} 2x + y = 5 & (\text{Equation 1}) \\ y = 1 - x & (\text{Equation 2}) \end{cases}$$

a. $(3, -1)$ **b.** $(4, -3)$

Objective 2 Solve a system of linear equations graphically.

A system of two linear equations in two variables can have one solution, no solution, or an infinite number of solutions. To see why, let's look at the graphs of the equations in three different systems. First, let's graph the equations in the system from Example 1.

$$\begin{cases} x + y = 5 \\ y = 2x - 4 \end{cases}$$

Note In Chapter 3, we learned that the graph of an equation represents every ordered pair solution for the equation; so when two graphs intersect, the point of intersection is a solution for both equations.

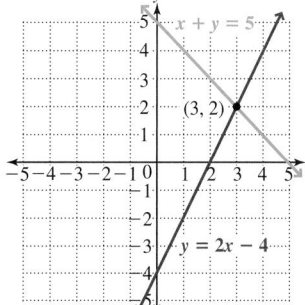

Notice that the graphs of $x + y = 5$ and $y = 2x - 4$ intersect at the point (3, 2), which is the solution for the system. If two linear graphs intersect at a single point, the system has a *single solution* at the point of intersection.

Note These lines also have different slopes, which is always the case when a system of two linear equations has a single solution.

Now let's look at a system with no solution. Look at the graphs of the equations in the system $\begin{cases} y = 3x + 1 \\ y = 3x - 2 \end{cases}$.

Answers to Your Turn 1
a. $(3, -1)$ is not a solution.
b. $(4, -3)$ is a solution.

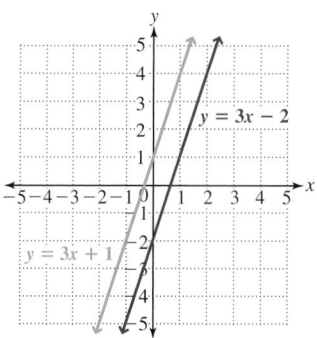

Notice that the equations $y = 3x + 1$ and $y = 3x - 2$ have the same slope, 3, and different y-intercepts; so their graphs are parallel lines. Because the graphs of the equations in this system have no point of intersection, the system has no solution.

Finally, let's examine a system with an infinite number of solutions, such as

$$\begin{cases} x + 2y = 4 & \text{(Equation 1)} \\ 2x + 4y = 8 & \text{(Equation 2)} \end{cases}$$

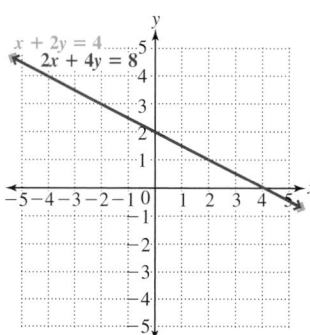

Notice that the graphs of the two equations are identical. Consequently, an infinite number of solutions to this system of equations lie on that line.

Note Having an infinite number of solutions is not the same as having all real numbers as solutions, as in the solutions of identities.

By writing the equations in slope–intercept form, we can see without graphing that the two equations describe the same line.

$x + 2y = 4$ (Equation 1)

$2y = -x + 4$	Subtract x from both sides.
$y = -\dfrac{1}{2}x + \dfrac{4}{2}$	Divide both sides by 2 to isolate y.
$y = -\dfrac{1}{2}x + 2$	Simplify the fraction.

$2x + 4y = 8$ (Equation 2)

$4y = -2x + 8$	Subtract $2x$ from both sides.
$y = -\dfrac{2}{4}x + \dfrac{8}{4}$	Divide both sides by 4 to isolate y.
$y = -\dfrac{1}{2}x + 2$	Simplify the fraction.

Our work suggests the following graphical method for solving systems of equations.

Procedure **Solving Systems of Linear Equations Graphically**

To solve a system of linear equations graphically:

1. Graph each equation.
   a. If the lines intersect at a single point, the coordinates of that point form the solution.
   b. If the lines are parallel, there is no solution.
   c. If the lines are identical, there are an infinite number of solutions, which are the coordinates of all points on that line.
2. Check your solution.

Example 2 Solve the system of equations graphically.

a. $\begin{cases} y = -3x + 1 & (\text{Equation 1}) \\ 2x - y = 4 & (\text{Equation 2}) \end{cases}$

Solution: Graph each equation.

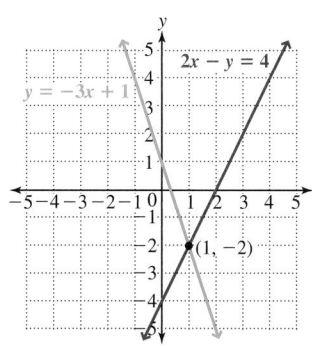

The lines intersect at a single point, which appears to be $(1, -2)$. We can verify that it is the solution by substituting into the equations.

$$y = -3x + 1 \qquad\qquad 2x - y = 4$$
$$-2 \overset{?}{=} -3(1) + 1 \qquad 2(1) - (-2) \overset{?}{=} 4$$
$$-2 \overset{?}{=} -3 + 1 \qquad\qquad 2 + 2 \overset{?}{=} 4$$
$$-2 = -2 \quad \text{True} \qquad\qquad 4 = 4 \quad \text{True}$$

Warning Graphing by hand can be imprecise, and not all solutions have integer coordinates. So you should always check your solutions by substituting them into the original equations.

Answer Because $(1, -2)$ makes both equations true, it is the solution.

b. $\begin{cases} 2x - 3y = 6 & (\text{Equation 1}) \\ y = \dfrac{2}{3}x + 1 & (\text{Equation 2}) \end{cases}$

Solution: Graph each equation.

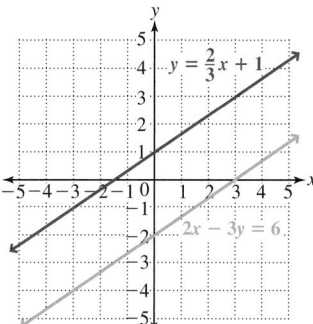

The lines appear to be parallel, which we can verify by comparing the slopes and y-intercepts. The slope of equation 2 is $\dfrac{2}{3}$, and the y-intercept is $(0, 1)$. To determine the slope and y-intercept of equation 1, we rewrite it in slope–intercept form.

$$2x - 3y = 6$$
$$-3y = -2x + 6 \qquad \text{Subtract } 2x \text{ from both sides.}$$
$$y = \frac{2}{3}x - 2 \qquad \begin{array}{l}\text{Divide both sides by } -3 \text{ to} \\ \text{isolate } y.\end{array}$$

The slope of equation 1 also is $\dfrac{2}{3}$, and the y-intercept is $(0, -2)$. Because the slopes are the same and the y-intercepts are different, the lines are indeed parallel.

Answer The system has no solution.

c. $\begin{cases} 6x - 2y = 2 & (\text{Equation 1}) \\ y = 3x - 1 & (\text{Equation 2}) \end{cases}$

Solution: Graph each equation.

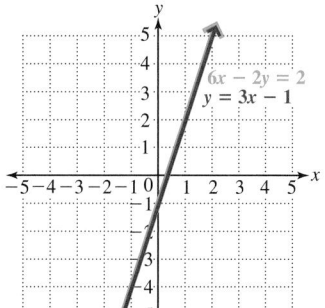

The lines appear to be identical, which we can verify by rewriting $6x - 2y = 2$ in slope–intercept form.

$$6x - 2y = 2$$
$$-2y = -6x + 2 \qquad \text{Subtract } 6x \text{ from both sides.}$$
$$y = 3x - 1 \qquad \begin{array}{l}\text{Divide both sides by } -2 \text{ to} \\ \text{isolate } y.\end{array}$$

The equations are identical.

Answer All ordered pairs along the line $6x - 2y = 2$ (or $y = 3x - 1$).

| **Your Turn 2** | Solve the system of equations graphically. |

a. $\begin{cases} y = \dfrac{3}{4}x + 1 \\ 2x - y = -6 \end{cases}$
        b. $\begin{cases} x + y = -3 \\ y = -x + 2 \end{cases}$

Objective 3 Classify systems of linear equations in two unknowns.

As we have seen, a system of two equations with two variables can have one solution, an infinite number of solutions along a common line, or no solution. Special terms are used to indicate each of these three classifications. The first terms describe whether a system has a solution.

Definitions   Consistent system of equations: A system of equations that has at least one solution.
Inconsistent system of equations: A system of equations that has no solution.

We have seen two types of consistent systems of equations. One type has a single solution because the two equations produce different lines that intersect at a single point. The other type has an infinite number of solutions because the two equations are identical and therefore produce the same line. To distinguish the two consistent cases, we use additional terms. If the equations in a consistent system produce different lines so that the system has one solution, we say that the equations are *independent*. If the equations in a consistent system produce the same line so that this system has an infinite number of solutions along that line, we say that the equations are *dependent*.

We have learned that linear equations whose graphs are different have different slopes or, in the case of parallel lines, have the same slopes and different y-intercepts. Linear equations whose graphs are identical have the same slope and same y-intercept. Therefore, by observing the slope and y-intercept of the equations in a system of equations, we can determine which of the three cases we are dealing with, which suggests the following procedure.

Answers to Your Turn 2
a. The solution is $(-4, -2)$.

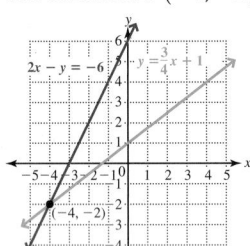

b. no solution

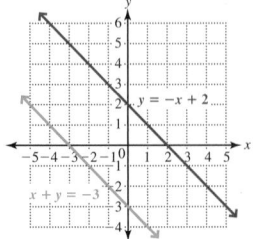

Procedure Classifying Systems of Equations

To classify a system of two linear equations in two unknowns, write the equations in slope–intercept form and compare the slopes and y-intercepts.

Consistent system with independent equations:	**Consistent system with dependent equations:**	**Inconsistent system:**
The system has a single solution at the point of intersection. The graphs are different. They have different slopes.	The system has an infinite number of solutions. The graphs are identical. They have the same slope and same y-intercept.	The system has no solution. The graphs are parallel lines. They have the same slope, but different y-intercepts.
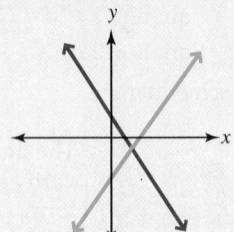	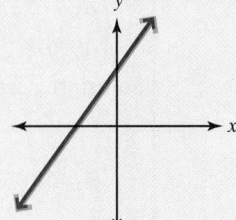	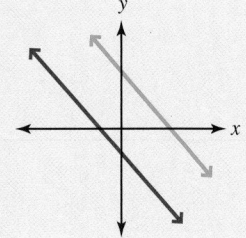

Example 3 For each of the systems of equations in Example 2, determine whether the system is consistent with independent equations, consistent with dependent equations, or inconsistent. How many solutions does the system have?

Solution: When we solved the system $\begin{cases} y = -3x + 1 \\ 2x - y = 4 \end{cases}$ in Example 2(a), we found that the graphs intersected at a single point. Therefore, the system is consistent (it has a solution) with independent equations (different graphs) and has one solution: $(1, -2)$.

When we solved the system $\begin{cases} 2x - 3y = 6 \\ y = \dfrac{2}{3}x + 1 \end{cases}$ in Example 2(b), we found the graphs to be parallel lines. Therefore, the system is inconsistent and has no solution:

When we solved the system $\begin{cases} 6x - 2y = 2 \\ y = 3x - 1 \end{cases}$ in Example 2(c), we found the graphs to coincide. Therefore, the system is consistent (it has a solution) with dependent equations (same graph) and has an infinite number of solutions.

Your Turn 3

a. Determine whether the following system is consistent with independent equations, consistent with dependent equations, or inconsistent.

b. How many solutions does the system have?

$$\begin{cases} 5x + y = 1 \\ y = -5x - 2 \end{cases}$$

Answers to Your Turn 3
a. inconsistent   b. no solution

4.1 Exercises For Extra Help MyMathLab®

Objective 1

Prep Exercise 1 How do you check a solution for a system of equations?
a. Replace each variable in each equation with its corresponding value.
b. Verify that each equation is true.

For Exercises 1–10, determine whether the given ordered pair is a solution to the given system of equations. See Example 1.

1. $(1, 1)$; $\begin{cases} x = y \\ 9y - 13 = -4x \end{cases}$
   yes

2. $(3, -2)$; $\begin{cases} 2x = -3y \\ x - 2y = 7 \end{cases}$
   yes

3. $(2, -3)$; $\begin{cases} 4x + 3y = -1 \\ 2x - 5y = -11 \end{cases}$
   no

4. $(4, 3)$; $\begin{cases} 3x + 4y = 24 \\ 3x + 2y = 18 \end{cases}$
   yes

5. $\left(\dfrac{2}{3}, \dfrac{4}{3}\right)$; $\begin{cases} x + y = 2 \\ 2x - y = 0 \end{cases}$
   yes

6. $\left(\dfrac{3}{2}, -\dfrac{3}{2}\right)$; $\begin{cases} x - y = 3 \\ x + y = 0 \end{cases}$
   yes

7. $(2, -4)$; $\begin{cases} 0.5x + 1.25y = 4 \\ x + y = -2 \end{cases}$
   no

8. $(1, 5)$; $\begin{cases} 0.25x + 0.75y = 4 \\ x + y = -7 \end{cases}$
   no

9. $(5, -2)$; $\begin{cases} x + 0.5y = 4 \\ \dfrac{1}{5}x - \dfrac{1}{2}y = 2 \end{cases}$
   yes

10. $(-1, -2)$; $\begin{cases} \dfrac{2}{3}x - y = 7 \\ 0.25x + y = -1 \end{cases}$
    no

Objective 2

Prep Exercise 2 When solving a system of linear equations by graphing, if the system has a single solution, how do you identify that solution? The solution is the ordered pair of the point of intersection of the graphs.

Prep Exercise 3 When solving a system of equations by graphing, how do you know if the system has no solution? The graphs are parallel.

Prep Exercise 4 When solving a system of linear equations by graphing, how do you know if the system has an infinite number of solutions? Their graphs are identical.

Prep Exercise 5 What are some weaknesses in using graphing to find a solution to a system of equations? Graphing by hand can be imprecise, and not all solutions are integers.

For Exercises 11–36, solve the system of linear equations graphically. See Example 2.

11. $\begin{cases} x + y = 5 \\ 2x - y = 7 \end{cases}$

$(4, 1)$

12. $\begin{cases} x - y = 4 \\ x + y = 8 \end{cases}$

$(6, 2)$

13. $\begin{cases} x + y = 1 \\ y = 7 + x \end{cases}$

$(-3, 4)$

14. $\begin{cases} x + y = 4 \\ y - x = 6 \end{cases}$

$(-1, 5)$

15. $\begin{cases} y = x - 5 \\ 2x - y = 8 \end{cases}$

$(3, -2)$

16. $\begin{cases} y = x - 1 \\ 3x + y = 11 \end{cases}$

$(3, 2)$

17. $\begin{cases} x = y - 4 \\ x + y = 2 \end{cases}$

$(-1, 3)$

18. $\begin{cases} x = y + 4 \\ x + y = 2 \end{cases}$

$(3, -1)$

19. $\begin{cases} 4x + y = 8 \\ 2x - 3y = 18 \end{cases}$

$(3, -4)$

20. $\begin{cases} 2x + 3y = 6 \\ x + 4 = 2y \end{cases}$

$(0, 2)$

21. $\begin{cases} 2x + y = 5 \\ x + 2y = -2 \end{cases}$

$(4, -3)$

22. $\begin{cases} 2x + y = 1 \\ 2x + 3y = 3 \end{cases}$

$(0, 1)$

23. $\begin{cases} 2x + 2y = -2 \\ 3x - 2y = 12 \end{cases}$

$(2, -3)$

24. $\begin{cases} 2x - 4y = 2 \\ 6x - 2y = -4 \end{cases}$

$(-1, -1)$

25. $\begin{cases} 3y = -2x + 6 \\ 4x + 6y = 18 \end{cases}$

No solution

26. $\begin{cases} 2x + y = 0 \\ 2y = 3 - 4x \end{cases}$

No solution

27. $\begin{cases} 2x + 3y = 2 \\ 6x + 9y = 6 \end{cases}$

All ordered pairs along $2x + 3y = 2$

28. $\begin{cases} x + y = 4 \\ -2x - 2y = -8 \end{cases}$

All ordered pairs along $x + y = 4$

29. $\begin{cases} 3x - 2y = -6 \\ x = 2 \end{cases}$

$(2, 6)$

30. $\begin{cases} 5x + 2y = -10 \\ x = -4 \end{cases}$

$(-4, 5)$

31. $\begin{cases} x - 2y = -6 \\ y = 2 \end{cases}$

$(-2, 2)$

32. $\begin{cases} 3x - y = 6 \\ y = -3 \end{cases}$

$(1, -3)$

33. $\begin{cases} y = 3 \\ x = -2 \end{cases}$

$(-2, 3)$

34. $\begin{cases} x = 1 \\ y = -4 \end{cases}$

$(1, -4)$

35. $\begin{cases} y = -\dfrac{2}{5}x \\ x - y = 7 \end{cases}$

$(5, -2)$

36. $\begin{cases} y = \dfrac{1}{4}x \\ x - y = 3 \end{cases}$

$(4, 1)$

Objective 3

Prep Exercise 6 What do the graphs of a consistent system with independent equations look like? The graphs intersect at a single point.

Prep Exercise 7 What do the graphs of a system of dependant equations look like? The graphs are identical.

Prep Exercise 8 What do the graphs of an inconsistent system of linear equations look like? The graphs are parallel lines.

For Exercises 37–42, (a) determine whether the graph shows a consistent system with independent equations, a consistent system with dependent equations, or an inconsistent system and (b) determine how many solutions the system has. See Example 3.

37.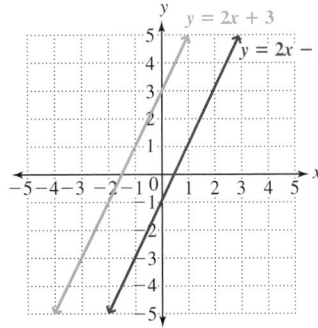

a. inconsistent
b. no solution

38.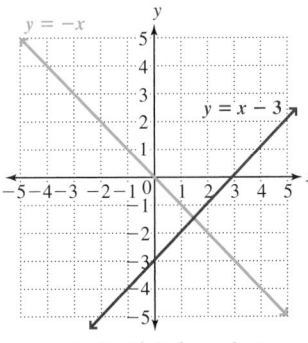

a. consistent with independent equations
b. one solution

39.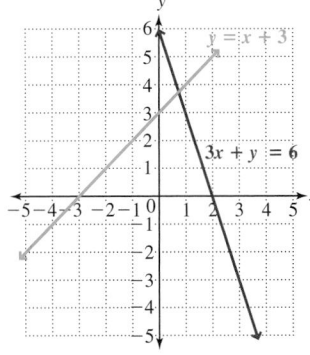

a. consistent with independent equations
b. one solution

40.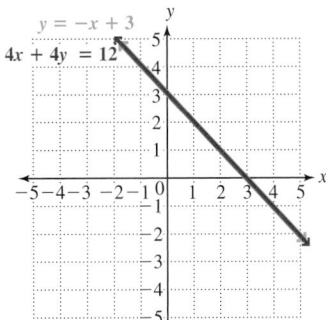

a. consistent with dependent equations
b. infinite number of solutions

41.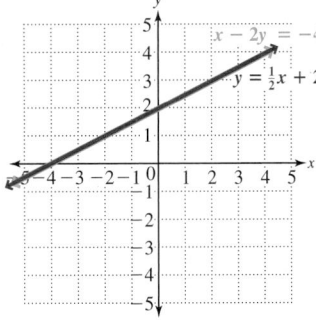

a. consistent with dependent equations
b. infinite number of solutions

42.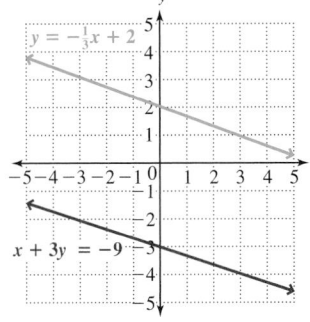

a. inconsistent
b. no solution

For Exercises 43–50, (a) determine whether the system of equations is consistent with independent equations, consistent with dependent equations, or inconsistent and (b) determine how many solutions the system has. See Example 3.

43. $\begin{cases} x - 4y = 9 \\ 2x - 3y = 8 \end{cases}$
a. consistent with independent equations
b. one solution

44. $\begin{cases} -4y = 2x - 12 \\ x - 2y = -4 \end{cases}$
a. consistent with independent equations
b. one solution

45. $\begin{cases} 3x + 2y = 5 \\ -6x - 4y = 1 \end{cases}$
a. inconsistent
b. no solution

46. $\begin{cases} 3x + 4y = 15 \\ 9x + 12y = 8 \end{cases}$
a. inconsistent
b. no solution

47. $\begin{cases} x - y = 1 \\ 2x - 2y = 2 \end{cases}$
a. consistent with dependent equations
b. infinite number of solutions

48. $\begin{cases} 2x - y = 1 \\ -4x + 2y = -2 \end{cases}$
a. consistent with dependent equations
b. infinite number of solutions

49. $\begin{cases} 2x - y = 1 \\ x + y = 4 \end{cases}$
a. consistent with independent equations
b. one solution

50. $\begin{cases} x - y = 5 \\ x + 5y = 11 \end{cases}$
a. consistent with independent equations
b. one solution

51. A business breaks even when its costs and revenue are equal. To the right is a graph of the cost of a product based on the number of units produced and the revenue based on the number of units sold.

   a. How many units must be produced and sold if the business is to break even?
5000 units

   b. What amount of revenue is needed for the business to break even?
$4000

   c. Write an inequality that describes the number of units that must be sold for the business to make a profit.
$u > 5000$

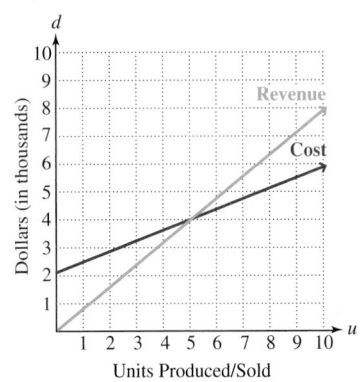

Units Produced/Sold
(in thousands)

52. John's job requires that he drive between two cities using a toll road. He has two choices for toll plans. The first plan is to stop and pay cash at each booth, which amounts to $15 per trip. The second plan is to purchase a toll pass for $12 per trip, which allows him to drive through the toll booths without stopping. However, to use the pass, he also must purchase a transponder for a one-time fee of $24. To the right, the cost of using each plan has been graphed.

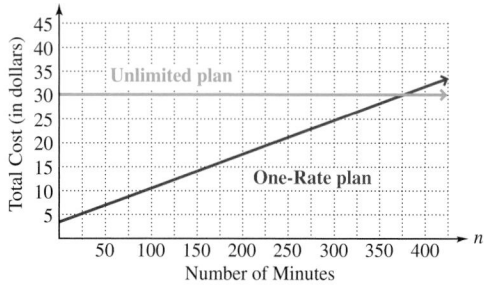

a. Which plan costs less if John makes 6 trips?

The cash plan

b. How many trips must he make for the costs to be equal?

8

c. What other factors should he consider when choosing a plan?

After 8 trips, the toll pass will cost less because the $24 fee is paid only once. Time and gas expenses are also considerations because the cash plan requires stopping at each toll booth.

53. The graph shows two long-distance plans. The Unlimited plan allows you to call anywhere in the United States at any time for $30 per month. The One-Rate plan costs $3.75 per month plus $0.07 per minute for each long-distance call.

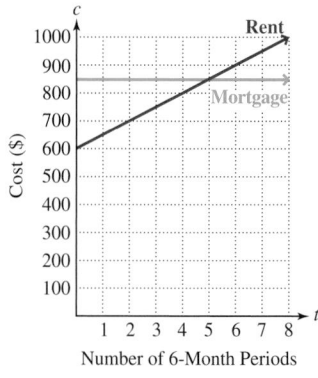

a. Which plan costs less if a person does not use more than 350 minutes of long-distance service each month?

The One-Rate plan

b. How many minutes would a person have to talk in a month for the plans to cost the same amount?

375 min.

c. How many minutes would a person need to talk per month for the Unlimited plan to cost less?

More than 375 min.

54. Deciding whether to purchase or rent a home is a difficult decision based on many factors. The graph shows the monthly mortgage payment ($850 per month) compared with the monthly rent payment (initially $600 per month and raised $50 every six months) for a 1200-square-foot house.

a. If a customer plans to stay only two years, ignoring any tax benefits, which costs less per month, renting or buying?

Renting

b. After how many months will the monthly rent be the same as the monthly mortgage payment?

After 30 months

c. In the fourth year, which costs less per month?

Buying

d. What other factors should a person consider when deciding whether to purchase or rent a home?

Down payment, closing costs, tax benefits, and equity

Review Exercises

Exercises 1 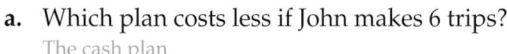 **Expressions**

[1.7] **1.** Simplify: $2y - 5(y + 7)$

$-3y - 35$

Exercises 2–6 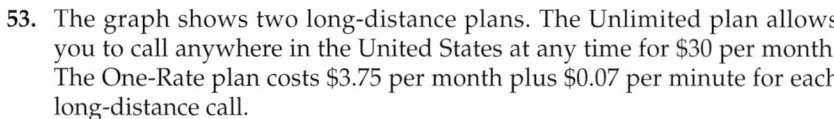 **Equations and Inequalities**

[2.4] **2.** Solve for y: $x + 3y = 10$

$y = \dfrac{10 - x}{3}$

[2.4] **3.** Solve for x: $\dfrac{1}{3}x - 2y = 10$

$x = 30 + 6y$

[2.1]   4. A cone made of marble is used as a decorative accent piece. It is 10 inches tall, and the radius of its base is 2 inches. Find its volume. $\left(\text{Use } V = \frac{1}{3}\pi r^2 h.\right)$

$\frac{40\pi}{3}$ in.^{3} or ≈ 41.89 in.^{3}

[2.5]   5. Five more than four times a number is equal to seven subtracted from that number. What is the number?

-4

[3.7]   6. Given $f(x) = 2x + 9$, find $f(x + 4)$.

$2x + 17$

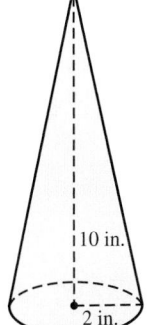

10 in.

2 in.

4.2 Solving Systems of Linear Equations by Substitution; Applications

Objectives

1 Solve systems of linear equations using substitution.
2 Solve applications involving two unknowns using a system of equations.

Warm-up

[2.4]   1. Solve: $3x + 2y = 6$ for x

[2.3]   2. Solve: $2\left(2 - \frac{2}{3}y\right) + 6y = -10$

In Section 4.1, we solved systems of equations by graphing. However, if the solution to a system of equations contains fractions or decimal numbers, it could be difficult to determine those values using the graphing method. For example, look at the following system of equations:

$$\begin{cases} x + 3y = 10 & (\text{Equation 1}) \\ y = x + 4 & (\text{Equation 2}) \end{cases}$$

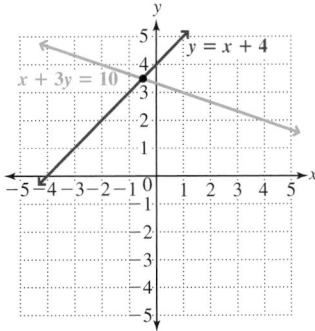

In the graph, notice that the point of intersection is between grid lines, which means the solution contains fractions. We could make a guess at the solution and check our guess in the equations, but it would be better to use a method that does not require guessing. One such method is the *substitution* method.

Objective 1 Solve systems of linear equations using substitution.

Remember that in a system's solution, the same x and y values satisfy both equations. In the preceding system, notice that Equation 2 indicates that y is equal to $x + 4$; so y must be equal to $x + 4$ in Equation 1 as well. Therefore, we can substitute $x + 4$ for y in Equation 1.

$$x + 3y = 10$$
$$x + 3(\overbrace{x + 4}) = 10$$

Answers to Warm-up

1. $x = 2 - \frac{2}{3}y$

2. $y = -3$

Now we have an equation in terms of a single variable, x, which allows us to solve for x.

$$x + 3(x + 4) = 10$$
$$x + 3x + 12 = 10 \qquad \text{Distribute 3.}$$
$$4x + 12 = 10 \qquad \text{Combine like terms.}$$
$$4x = -2 \qquad \text{Subtract 12 from both sides.}$$
$$x = -\frac{1}{2} \qquad \text{Divide both sides by 4 and simplify.}$$

We can now find the y-value by substituting $-\dfrac{1}{2}$ for x in either of the original equations. Equation 2 is easier because y is isolated.

$$y = x + 4$$
$$y = -\frac{1}{2} + 4 \qquad \text{Substitute } -\frac{1}{2} \text{ for } x.$$
$$y = 3\frac{1}{2}$$

The solution to the system of equations is $\left(-\dfrac{1}{2}, 3\dfrac{1}{2}\right)$. Notice in our graph that the lines do appear to intersect at $\left(-\dfrac{1}{2}, 3\dfrac{1}{2}\right)$, which supports our solution graphically.

Procedure **Solving Systems of Two Linear Equations Using Substitution**

To find the solution of a system of two linear equations using the substitution method:

1. Isolate one of the variables in one of the equations.
2. In the other equation, substitute the expression you found in step 1 for that variable.
3. Solve this new equation. (It will now have only one variable.)
4. Using one of the equations containing both variables, substitute the value you found in step 3 for that variable and solve for the value of the other variable.
5. Check the solution in the original equations.

Example 1 Solve the system of equations using substitution.

$$\begin{cases} x = 3 - y \\ 5x + 3y = 5 \end{cases}$$

Solution:

Step 1: We isolate a variable in one of the equations. Because x is already isolated in the first equation, we proceed to step 2.

Step 2: Substitute the expression $3 - y$ for x in the second equation.

$$5x + 3y = 5$$
$$5(3 - y) + 3y = 5 \qquad \text{Substitute } 3 - y \text{ for } x.$$

Step 3: We now have a linear equation in terms of a single variable, y; so we solve for y.

$$15 - 5y + 3y = 5 \qquad \text{Distribute 5.}$$
$$15 - 2y = 5 \qquad \text{Combine like terms.}$$
$$-2y = -10 \qquad \text{Subtract 15 from both sides.}$$
$$y = 5 \qquad \text{Divide both sides by } -2 \text{ to isolate } y.$$

Step 4: Now we solve for x by substituting 5 for y in one of the equations containing both variables. We will use $x = 3 - y$.

> **Connection** The graphs of $x = 3 - y$ and $5x + 3y = 5$ are lines that intersect at the point whose coordinates are $(-2, 5)$.

$$x = 3 - y$$
$$x = 3 - 5 \qquad \text{Substitute 5 for } y.$$
$$x = -2$$

◀ **Note** We chose $x = 3 - y$ because x is isolated.

The solution is $(-2, 5)$. We will let the reader check that $(-2, 5)$ makes both equations true.

Your Turn 1 Solve the system of equations using substitution.

$$\begin{cases} x - 3y = 11 \\ y = 2x + 3 \end{cases}$$

In step 1, if no variable is isolated in any of the given equations, we select an equation and isolate one of its variables. It is easiest to isolate a variable that has a coefficient of 1, as this will avoid fractions.

Example 2 Solve the system of equations using substitution.

$$\begin{cases} x + y = 6 \\ 5x + 2y = 8 \end{cases}$$

Solution:

Step 1: Isolate a variable in one of the equations. Because both x and y have a coefficient of 1 in $x + y = 6$, isolating either variable is easy. We will isolate y.

> **Instructor Note** Solve this system again by solving $x + y = 6$ for x to show students that it doesn't matter which variable is isolated.

$$x + y = 6$$
$$y = 6 - x \qquad \text{Subtract } x \text{ from both sides to isolate } y.$$

Step 2: Substitute $6 - x$ for y in the second equation.

$$5x + 2y = 8$$
$$5x + 2\overbrace{(6 - x)} = 8 \qquad \text{Substitute } 6 - x \text{ for } y.$$

Step 3: Solve for x using the equation we found in step 2.

$$5x + 12 - 2x = 8 \qquad \text{Distribute 2.}$$
$$3x + 12 = 8 \qquad \text{Combine like terms.}$$
$$3x = -4 \qquad \text{Subtract 12 from both sides.}$$
$$x = -\frac{4}{3} \qquad \text{Divide both sides by 3 to isolate } x.$$

Step 4: Find the value of y by substituting $-\dfrac{4}{3}$ for x in one of the equations containing both variables. Because we isolated y in $y = 6 - x$, we will use this equation.

$$y = 6 - x$$
$$y = 6 - \left(-\frac{4}{3}\right) \qquad \text{Substitute } -\frac{4}{3} \text{ for } x.$$

Answer to Your Turn 1
$(-4, -5)$

$$y = \frac{18}{3} + \frac{4}{3}$$

Rewrite as addition of equivalent fractions with a common denominator.

$$y = \frac{22}{3}$$

The solution is $\left(-\frac{4}{3}, \frac{22}{3} \right)$.

Step 5: We will leave the check to the reader.

Your Turn 2 Solve the system of equations by substitution.

$$\begin{cases} x - y = 1 \\ 3x + y = 11 \end{cases}$$

If none of the variables in the system of equations has a coefficient of 1 or −1, for step 1, select the equation that seems easiest to work with and the variable in that equation that seems easiest to isolate. When selecting that variable, recognize that you'll eventually divide out its coefficient; so choose t he variable whose coefficient will divide evenly into most, if not all, of the other numbers in its equation. In Section 5.3, we will discuss another method of solving systems that reduces the amount of work with fractions.

Example 3 Solve the system of equations using substitution.

$$\begin{cases} 3x + 4y = 6 \\ 2x + 6y = -1 \end{cases}$$

Instructor Note Point out that given a choice, substitution is not the preferred method for solving this system. The elimination method (Section 4.3) is easier.

Solution:

Step 1: We will isolate x in $3x + 4y = 6$.

$$3x + 4y = 6$$
$$3x = 6 - 4y \quad \text{Subtract } 4y \text{ from both sides.}$$
$$x = 2 - \frac{4}{3}y \quad \text{Divide both sides by 3 to isolate } x.$$

Note We chose to isolate x because its coefficient, 3, divides evenly into one of the other numbers in the equation, 6, whereas y's coefficient, 4, does not divide evenly into any of the other numbers.

Step 2: Substitute $2 - \frac{4}{3}y$ for x in $2x + 6y = -1$.

$$2x + 6y = -1$$
$$2\overbrace{\left(2 - \frac{4}{3}y \right)} + 6y = -1 \quad \text{Substitute } 2 - \frac{4}{3}y \text{ for } x.$$

Step 3: Solve for y using the equation we found in step 2.

$$4 - \frac{8}{3}y + 6y = -1 \qquad \text{Distribute to eliminate the parentheses.}$$

$$3 \cdot 4 - 3 \cdot \frac{8}{3}y + 3 \cdot 6y = 3 \cdot (-1) \qquad \text{Multiply both sides by 3 to clear the fraction.}$$

$$12 - 8y + 18y = -3$$
$$12 + 10y = -3 \qquad \text{Combine like terms.}$$
$$10y = -15 \qquad \text{Subtract 12 from both sides.}$$
$$y = -\frac{15}{10} \qquad \text{Divide both sides by 10 to isolate } y.$$
$$y = -\frac{3}{2} \qquad \text{Simplify.}$$

Step 4: Find the value of x by substituting $-\frac{3}{2}$ for y in one of the equations containing both variables. Because x is isolated in $x = 2 - \frac{4}{3}y$, we will use that equation.

Answer to Your Turn 2
(3, 2)

$$x = 2 - \frac{4}{3}y$$

$$x = 2 - \frac{\overset{2}{\cancel{4}}}{3} \cdot \left(-\frac{\overset{1}{\cancel{3}}}{\underset{1}{\cancel{2}}}\right)$$ Substitute $-\frac{3}{2}$ for y and simplify.

$$x = 2 + 2$$

$$x = 4$$

The solution is $\left(4, -\frac{3}{2}\right)$.

Step 5: We will leave the check to the reader.

Your Turn 3   Solve the system of equations using substitution.

$$\begin{cases} 5x - 2y = 10 \\ 3x - 6y = 2 \end{cases}$$

Inconsistent Systems of Equations

In Section 4.1, we learned that inconsistent systems of linear equations have no solution because the graphs of the equations are parallel lines. Let's see what happens when we solve an inconsistent system of equations using the substitution method.

Example 4   Solve the system of equations using substitution.

$$\begin{cases} x + 2y = 4 \\ y = -\frac{1}{2}x + 1 \end{cases}$$

Solution: Because y is isolated in the second equation, we substitute $-\frac{1}{2}x + 1$ in place of y in the first equation.

$$x + 2y = 4$$

$$x + 2\left(-\frac{1}{2}x + 1\right) = 4 \quad \text{Substitute } -\frac{1}{2}x + 1 \text{ for } y.$$

Now solve for x.

$$x - x + 2 = 4 \quad \text{Distribute to clear the parentheses.}$$

$$2 = 4 \quad \text{Combine like terms.}$$

Notice that $2 = 4$ no longer has a variable and is false. This false equation with no variable indicates that the system is inconsistent and has no solution.

Connection If we were to solve the system in Example 4 using graphing, we would see that the lines are parallel.

Consistent Systems with Dependent Equations

Recall from Section 4.1 that consistent systems with dependent equations have an infinite number of solutions. Let's see what happens when we use the substitution method to solve a system of dependent equations.

Example 5   Solve the system of equations using substitution.

$$\begin{cases} x = 3y + 4 \\ 2x - 6y = 8 \end{cases}$$

Answer to Your Turn 3
$$\left(\frac{7}{3}, \frac{5}{6}\right)$$

Solution: Because x is isolated in the first equation, we substitute $3y + 4$ in place of x in the second equation.

$$2x - 6y = 8$$

$$2(\overbrace{3y + 4}) - 6y = 8 \quad \text{Substitute } 3y + 4 \text{ for } x.$$

Now solve for y.

$$6y + 8 - 6y = 8 \quad \text{Distribute to clear the parentheses.}$$

$$8 = 8 \quad \text{Combine like terms.}$$

Connection If we were to solve the system in Example 5 using graphing, we would see that the lines are identical.

Notice that $8 = 8$ no longer has a variable and is true. This true equation with no variable indicates that the equations in the system are dependent; so there are an infinite number of solutions that are all of the ordered pairs along $x = 3y + 4$ (or $2x - 6y = 8$).

Your Turn 5 Solve the systems of equations using substitution.

a. $\begin{cases} y = 2x - 7 \\ 4x - 2y = -3 \end{cases}$

b. $\begin{cases} 3x - y = 4 \\ 2y - 6x = -8 \end{cases}$

Objective 2 Solve applications involving two unknowns using a system of equations.

Instructor Note Mention that eventually we will solve applications problems involving more than two unknowns. So that we can solve such problems, *the number of variables* and *number of equations in the system* must be the same to have a unique solution.

To solve a problem involving two unknowns, we must be given two relationships about those unknowns. We translate each of the two relationships to an equation and then solve that system of equations.

Procedure Solving Applications Using a System of Equations

To solve a problem with two unknowns using a system of equations:
1. Select a variable for each of the unknowns.
2. Translate each relationship to an equation.
3. Solve the system.

Example 6 A rectangular frame is to be built so that the length is 4 inches more than the width. The perimeter of the frame is to be 72 inches. Find the dimensions of the frame.

Understand We are given two relationships about the frame, and we are to find the length and width.

Plan Select a variable for the length and another variable for the width, translate the relationships to a system of equations, and then solve the system.

Execute Let l represent the length and w represent the width.

Relationship 1: The length is 4 more than the width.

Translation: $l = w + 4$

Relationship 2: The perimeter is 72 inches.

Translation: $2l + 2w = 72$

Our system: $\begin{cases} l = w + 4 \\ 2l + 2w = 72 \end{cases}$

Because the first equation has l isolated, we substitute $w + 4$ for l in the second equation.

$$2l + 2w = 72$$

$$2(w + 4) + 2w = 72 \quad \text{Substitute } w + 4 \text{ for } l.$$

$$2w + 8 + 2w = 72$$

$$4w + 8 = 72 \quad \text{Combine like terms.}$$

$$4w = 64 \quad \text{Subtract 8 from both sides of the equation.}$$

$$w = 16 \quad \text{Divide both sides by 4.}$$

Answers to Your Turn 5
a. no solution
b. all ordered pairs along
$3x - y = 4$ (or $2y - 6x = -8$)

Now we can find the value of l using $l = w + 4$.

$$l = 16 + 4 \quad \text{Substitute 16 for } w.$$
$$l = 20$$

Answer   The length should be 20 inches, and the width should be 16 inches.

Check   Verify that the length is 4 more than the width: $20 = 16 + 4$. Also verify that the perimeter is 72 inches: $2(20) + 2(16) = 40 + 32 = 72$.

| Your Turn 6 | The width of a tennis court (along the doubles lines) is 42 feet less than the length. If the perimeter is 228 feet, what are the dimensions?

Instructor Note Remind students that they need the same number of relationships as unknowns to solve problems involving two or more unknowns.

For some problems involving two unknowns, it helps to organize the given information in a table.

Example 7   A home improvement store sells two sizes of cans of wood stain. A small can sells for $8.95, and a large can sells for $15.95. One day the store sold twice as many of the large cans as the small cans. If the total revenue that day for these cans of stain was $694.45, how many cans of each size were sold?

Understand   The two unknowns are the number of each size of cans of stain sold. There are also two relationships, one involving the number sold and the other involving the total revenue.

Plan and Execute   Let x represent the number of small cans of stain sold and let y represent the number of large cans. We can use a table to organize the information.

Category	Price	Number	Revenue
Small can	8.95	x	$8.95x$
Large can	15.95	y	$15.95y$

◄ **Note** Sales revenue is found by multiplying an item's price by the number sold.

Relationship 1:   The store sold twice as many of the large cans as the small.

Translation: $y = 2x$

Relationship 2:   The total revenue was $694.45.

Translation: $8.95x + 15.95y = 694.45$

$$\text{Our system: } \begin{cases} y = 2x \\ 8.95x + 15.95y = 694.45 \end{cases}$$

To solve this system of equations, substitution is the best method because y is isolated in the first equation. We replace y in the second equation with $2x$ from the first equation.

$$8.95x + 15.95y = 694.45$$
$$8.95x + 15.95(2x) = 694.45 \quad \text{Substitute 2x for y.}$$
$$8.95x + 31.9x = 694.45$$
$$40.85x = 694.45 \quad \text{Combine like terms.}$$
$$x = 17 \quad \text{Divide both sides by 40.85 to isolate x.}$$

Now we can find the value of y using $y = 2x$ by substituting 17 for x.

$$y = 2(17) = 34$$

Answer   The home improvement store sold 17 small cans of stain and 34 large cans.

Check   Verify both given relationships. The number of large cans of stain sold, 34, is twice the number of small cans sold, 17. The total revenue, $8.95(17) + 15.95(34) = 694.45$, is also correct.

Answer to Your Turn 6
78 ft. by 36 ft.

Your Turn 7 A cookware consultant sells two sizes of a rectangular baking dish. The small size sells for $24, and the large size sells for $35. In one month, she sold 6 more of the small-size dish than the large-size dish. If she made a total of $439 from the sale of both dishes, how many of each size baking dish did she sell?

Example 8 Jasmine and Darius are traveling east in separate cars on the same highway. Jasmine is traveling at 60 miles per hour; Darius, at 70 miles per hour. Jasmine passes Exit 82 at 3:15 P.M. Darius passes the same exit at 3:30 P.M. At what time will Darius catch up with Jasmine?

Understand To determine the time at which Darius catches up with Jasmine, we must calculate the amount of time it will take him to catch up with her. We can then add that amount of time to 3:30 P.M.

Plan and Execute We will let x represent Jasmine's travel time after passing Exit 82 and y represent Darius's travel time after passing Exit 82.

Category	Rate	Time	Distance
Jasmine	60	x	$60x$
Darius	70	y	$70y$

◄ **Note** Distance is the product of rate and time.

Relationship 1: Darius passes the exit 15 minutes after Jasmine; so when he catches up with her, she will have traveled 15 minutes longer than he has.

Translation: $x = y + \dfrac{1}{4}$ ◄ **Note** We use $\dfrac{1}{4}$ of an hour instead of 15 minutes because the rates are in miles per hour and the time units must be consistent.

Relationship 2: When Darius catches up, they will have traveled the same distance from Exit 82; so we set the expressions for their individual distances equal to each other.

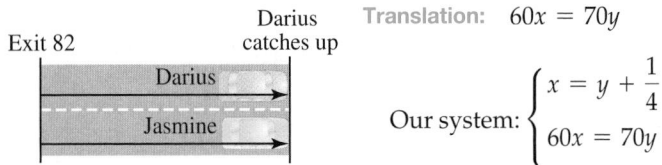

Translation: $60x = 70y$

Our system: $\begin{cases} x = y + \dfrac{1}{4} \\ 60x = 70y \end{cases}$

Because we have an isolated variable, x, we will solve the system using the substitution method.

$$60x = 70y$$

$$60\left(y + \frac{1}{4}\right) = 70y \quad \text{Substitute } y + \frac{1}{4} \text{ for x.}$$

$$60y + 15 = 70y \quad \text{Distribute 60.}$$

$$15 = 10y \quad \text{Subtract 60y from both sides.}$$

$$1.5 = y \quad \text{Divide both sides by 10 to isolate y.}$$

Answer It will take Darius 1.5 hours to catch up with Jasmine. So he catches up at 5 P.M.

Check At 60 miles per hour, in $\dfrac{1}{4}$ of an hour, Jasmine travels 15 miles. Because Darius is traveling 70 miles per hour, he is gaining on her at a rate of 10 miles per hour (the difference of their rates). So he must make up the 15 miles that separate them at a rate of 10 miles per hour, which will indeed take 1.5 hours.

Answer to Your Turn 7
11 small, 5 large

Answer to Your Turn 8
10:46 A.M.

| Your Turn 8 | At 10:30 A.M., Celeste begins walking along a trail at an average rate of 3 miles per hour. At 10:40 A.M., Chris begins bicycling along the same trail at 8 miles per hour. At what time will Chris catch up with Celeste?

4.2 Exercises For Extra Help MyMathLab®

Objective 1

Prep Exercise 1 What advantages does the method of substitution have over graphing for solving systems of equations?
When using the graphing method, if either coordinate in the solution is a fraction, we may have to guess the value. Substitution requires no guessing.

Prep Exercise 2 Suppose you are given the system of equations $\begin{cases} y = x - 6 \\ 4x + 3y = 1 \end{cases}$.
a. Which variable would you replace in the substitution process?
y
b. What expression would replace that variable?
$x - 6$

Prep Exercise 3 Suppose you are given the system of equations $\begin{cases} 5x + y = 4 \\ 3x - 2y = 9 \end{cases}$. Which variable in which equation would you isolate first? Why?
y in the first equation because it can be isolated without using division

Prep Exercise 4 Suppose you are given the system of equations $\begin{cases} 3x + 4y = 1 \\ x - 5y = 12 \end{cases}$. Which variable would you isolate first? Why?
x in the second equation because it can be isolated without using division

Prep Exercise 5 How do you know that a system has no solution (inconsistent system) when using substitution?
The resulting equation will no longer contain variables and becomes a false equation.

Prep Exercise 6 How do you know that a system has an infinite number of solutions (consistent with dependent equations) when using substitution?
The resulting equation will no longer contain variables and becomes a true equation.

For Exercises 1–28, solve the system of equations using substitution. Note that some systems may be inconsistent or consistent with dependent equations. See Examples 1–5.

1. $\begin{cases} x + y = 6 \\ x = 2y \end{cases}$
$(4, 2)$

2. $\begin{cases} x - y = 8 \\ y = 2x \end{cases}$
$(-8, -16)$

3. $\begin{cases} x = -3y \\ 2x - 5y = 44 \end{cases}$
$(12, -4)$

4. $\begin{cases} x + 2y = 9 \\ y = -2x \end{cases}$
$(-3, 6)$

5. $\begin{cases} 2x - y = 4 \\ y = 3x + 6 \end{cases}$
$(-10, -24)$

6. $\begin{cases} x = 3y - 2 \\ 2x - 3y = 2 \end{cases}$
$(4, 2)$

7. $\begin{cases} y = -2x \\ 4x + 2y = 0 \end{cases}$
All ordered pairs along $y = -2x$ (dependent)

8. $\begin{cases} x = 2y - 6 \\ -2x + 4y = 12 \end{cases}$
All ordered pairs along $x = 2y - 6$ (dependent)

9. $\begin{cases} 2x + y = -4 \\ 3x + 2y = -5 \end{cases}$
$(-3, 2)$

10. $\begin{cases} 5x + y = -4 \\ 3x + 2y = -1 \end{cases}$
$(-1, 1)$

11. $\begin{cases} 5x - y = 1 \\ 3x + 2y = 24 \end{cases}$
$(2, 9)$

12. $\begin{cases} x + 3y = 5 \\ 2y - x = 10 \end{cases}$
$(-4, 3)$

13. $\begin{cases} 5x - 3y = 4 \\ y - x = -2 \end{cases}$
$(-1, -3)$

14. $\begin{cases} 2x - y = 39 \\ 14 + 3x = 74y \end{cases}$
$(20, 1)$

15. $\begin{cases} 2x = 10 - y \\ 3x - y = 5 \end{cases}$
$(3, 4)$

16. $\begin{cases} 3y = 7 - x \\ 2x - y = 0 \end{cases}$
$(1, 2)$

17. $\begin{cases} -x + 2y = 1 \\ 3x - 2y = 1 \end{cases}$
$(1, 1)$

18. $\begin{cases} 2x + 3y = -1 \\ 5x - y = 6 \end{cases}$
$(1, -1)$

19. $\begin{cases} 3x + 2y = 8 \\ x + 4y = 1 \end{cases}$
$\left(3, -\dfrac{1}{2}\right)$

20. $\begin{cases} 4x + y = -2 \\ 8x - y = 11 \end{cases}$
$\left(\dfrac{3}{4}, -5\right)$

21. $\begin{cases} x - 3y = -4 \\ -5x + 15y = 6 \end{cases}$
No solution
(inconsistent)

22. $\begin{cases} 2x - y = -1 \\ 6x - 3y = 3 \end{cases}$
No solution
(inconsistent)

23. $\begin{cases} 4x + 3y = -2 \\ 8x - 2y = 12 \end{cases}$
$(1, -2)$

24. $\begin{cases} 2x + 3y = -1 \\ 6x + 3y = -9 \end{cases}$
$(-2, 1)$

25. $\begin{cases} 5x + 4y = 12 \\ 7x - 6y = 40 \end{cases}$
$(4, -2)$

26. $\begin{cases} 4x - 3y = 2 \\ 3x + 5y = 16 \end{cases}$
$(2, 2)$

27. $\begin{cases} 5x + 6y = 2 \\ 10x + 3y = -2 \end{cases}$
$\left(-\dfrac{2}{5}, \dfrac{2}{3}\right)$

28. $\begin{cases} 7x - 6y = 1 \\ 4x - 6y = 3 \end{cases}$
$\left(-\dfrac{2}{3}, -\dfrac{17}{18}\right)$

■ *For Exercises 29–32, solve using substitution; then verify the solution by graphing the equations on a graphing utility. See Examples 1–5.*

29. $\begin{cases} y = 2x + 1 \\ 3x + y = 11 \end{cases}$
$(2, 5)$

30. $\begin{cases} y = -4x + 3 \\ y - 2x = -3 \end{cases}$
$(1, -1)$

31. $\begin{cases} 2x + y = 4 \\ x + 2y = 5 \end{cases}$
$(1, 2)$

32. $\begin{cases} 4x - y = 12 \\ x - 3y = 14 \end{cases}$
$(2, -4)$

Find ⊗ the Mistake   *For Exercises 33 and 34, find the mistake.*

33. $\begin{cases} 3x + 2y = 5 \\ x - 3y = 6 \end{cases}$

Rewrite: $x = 6 + 3y$

Substitute: $3(6 + 3y) + 2y = 5$
$$18 + 3y + 2y = 5$$
$$18 + 5y = 5$$
$$5y = -13$$
$$y = -\dfrac{13}{5}$$

$x = 6 + 3y$

$x = 6 + 3\left(-\dfrac{13}{5}\right)$

$x = 6 - \dfrac{39}{5}$

$x = -\dfrac{9}{5}$

Solution: $\left(-\dfrac{9}{5}, -\dfrac{13}{5}\right)$

Mistake: 3 was not distributed to 3y.

Correct: $\left(\dfrac{27}{11}, -\dfrac{13}{11}\right)$

34. $\begin{cases} 2x + 3y = 8 \\ x + 2y = 6 \end{cases}$

Rewrite: $x = 6 - 2y$

Substitute: $x + 2y = 6$
$$6 - 2y + 2y = 6$$
$$6 - 6 - 2y + 2y = 6 - 6$$
$$0 = 0$$

Solution: Infinite solutions along
$x + 2y = 6$

Mistake: After x was isolated, $6 - 2y$
was substituted into the equation it
came from. Correct: $(-2, 4)$

Objective 2

Prep Exercise 7 How many relationships must be given to solve a problem involving two unknowns?
Two

For Exercises 35–54, translate the problem to a system of equations and then solve. See Examples 6–8.

35. The greater of two integers is 10 more than twice the smaller integer. The sum of the two integers is −8. Find the integers.
−6, −2

36. One integer is six less than three times another integer. The sum of the integers is −42. Find the integers.
−33, −9

37. Jon is 13 years older than Tony. The sum of their ages is 27. How old are they?
Jon is 20; Tony is 7.

38. Patrick is twice as old as his sister Sherry. The sum of their ages is 12. How old are they?
Patrick is 8; Sherry is 4.

39. The perimeter of a college basketball court is 288 feet. If the length of the court is 44 feet longer than it is wide, what are the dimensions?
Length: 94 ft.; width: 50 ft.

40. The length of a volleyball court is twice the width. If the perimeter of the court along the doubles lines is 54 meters, what are the dimensions? (*Source: USA Volleyball Rule Book.*)
Length: 18 m; width: 9 m

41. The length of the Reflecting Pool at the National Mall in Washington, D.C., is 800 feet more than ten times the width. What are the dimensions if the perimeter is 4900 feet? (*Source: www.answers.com.*)
Length: 2300 ft.; width: 150 ft.

42. The perimeter of the base of the Castillo at Chichen Itza (a pyramid in Mexico) is 300 feet. If the length is equal to the width, what are the dimensions?
Length: 75 ft.; width: 75 ft.

43. A support beam is attached to a wall and to the bottom of a ceiling truss. The angle made with the truss on one side of the support beam is three times the angle on the other side of the support beam. Find the two angles.
135°, 45°

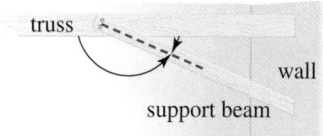

44. A laser beam is aimed at an angle at a flat photocell. The angle on one side of the beam is 25 degrees more than the angle on the other side of the beam. Find the two angles.
102.5°, 77.5°

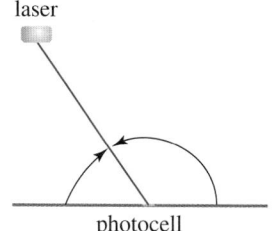

45. Troy has a checking account and a savings account with a combined balance of $4200. The balance in his savings account is four times the balance in his checking account. What are the balances of both accounts?
Checking: $840; savings: $3360

46. Susan has two credit cards with a combined balance of $2700. If she owes three times more on one of the cards than the other, how much does she owe on each card?
$2025, $675

47. U2 and Bon Jovi were the two top-grossing tours of 2011. U2 earned $101 million more than Bon Jovi, and together they earned $483 million. What were the earnings of each of these bands? (*Source: Billboard.*)
U2; $292 million; Bon Jovi; $191 million

48. The highest grossing film of all time is *Avatar*, which grossed $597 million more than the second place *Titanic*. Together the two movies grossed $4967 million. How much revenue was brought in by each movie? (*Source: Box Office Mojo.*)
Avatar: $2782 million; *Titanic*: $2185 million

49. The cashier's office at a college has collected $7020 from part-time students for a total of 42 courses. If each credit hour costs $45, how many 3-hour courses and how many 5-hour courses were paid for?
27 3-hour courses, 15 5-hour courses

50. One day a college bookstore collected $6260 for the sale of 44 beginning algebra and calculus books. If beginning algebra books sold for $130 each and calculus books sold for $160, how many of each type of book did the bookstore sell?
26 beginning algebra, 18 calculus

51. John and Karen are traveling south in separate cars on the same interstate. Karen is traveling at 65 miles per hour; John, at 70 miles per hour. Karen passes Exit 38 at 1:15 P.M. John passes the same exit at 1:30 P.M. At what time will John catch up with Karen?
4:45 P.M.

52. Diane and Curtis are running in the Cooper River Bridge Run. Diane passes the first checkpoint at 10:30 A.M.; Curtis, at 11:00 A.M. While Curtis is pushing himself to run at 5 miles per hour, Diane is pacing herself at 4 miles per hour. When will Curtis catch up with Diane?
1 P.M.

53. Sharon and Rae are riding bicycles on the same trail in the same direction. Sharon crosses a bridge at 12:18 P.M. Rae crosses the same bridge at 12:33 P.M. If Sharon is going 8 miles per hour and Rae is going 10 miles per hour, when will Rae catch up with Sharon?
1:33 P.M.

54. Holly and Rebecca are jogging on the same road in the same direction. Holly passes a McDonald's at 8:10 A.M. Rebecca passes the same McDonald's at 8:20 A.M. If Holly is going 4 miles per hour and Rebecca is going 6 miles per hour, when will Rebecca catch up with Holly?
8:40 A.M.

Review Exercises

Exercises 1–4 ▱➤ Expressions

[1.7] *For Exercises 1 and 2, multiply.*

1. $\frac{1}{4}(20x - 16y)$

$5x - 4y$

2. $-3(x - 7y)$

$-3x + 21y$

[1.7] *For Exercises 3 and 4, add or subtract.*

3. $6x + 2y + 3x - 2y$

$9x$

4. $5x + 3y - 5x + 10y$

$13y$

Exercises 5 and 6 ▱➤ Equations and Inequalities

[2.3] *For Exercises 5 and 6, solve.*

5. $\frac{1}{4}x = 20$

80

6. $-8y = 12$

$-\frac{3}{2}$

4.3 Solving Systems of Linear Equations by Elimination; Applications

Objectives

1 Solve systems of linear equations using elimination.

2 Solve applications using elimination.

Warm-up

[1.7] For Exercises 1 and 2, find the products.
  1. $-4(x - 2y)$
  2. $10(0.4x - 0.3y)$
[1.7] **3.** Combine like terms.
    $-4x + 8y + 4x - 3y$
[2.3] **4.** Solve: $5y = 4$

We have seen two methods for solving a system of equations: graphing and substitution. Each method has advantages and disadvantages. The substitution method is advantageous over the graphing method when solutions involve fractions. Also, substitution is easy when a variable's coefficient is 1. However, substitution can be tedious when no coefficients are 1. For this reason, we turn to yet a third method, the *elimination* method.

Objective 1 Solve systems of linear equations using elimination.

The elimination method uses the addition principle of equality to add equations so that a new equation emerges with one of the variables eliminated. We will work through some examples to get a sense of the method before stating it formally.

Answers to Warm-up
  1. $-4x + 8y$
  2. $4x - 3y$
  3. $5y$
  4. $y = \frac{4}{5}$

Example 1 | Solve the system of equations using elimination.

$$\begin{cases} x + y = 5 & \text{(Equation 1)} \\ 2x - y = 7 & \text{(Equation 2)} \end{cases}$$

Solution: In chapter 2, we learned that the addition principle of equality says that adding the same amount to both sides of an equation will not affect its solution(s). Because Equation 1 indicates that $x + y$ and 5 are the same amount, if we add $x + y$ to the left side of Equation 2 and 5 to the right side of Equation 2, we are applying the addition principle of equality. Most people prefer to stack the equations and combine like terms vertically like this:

$$\begin{array}{ll} x + y = 5 & \text{(Equation 1)} \\ \underline{2x - y = 7} & \text{(Equation 2)} \\ 3x + 0 = 12 & \text{Add Equation 1 to Equation 2.} \end{array}$$

Connection The terms y and $-y$ are additive inverses, or opposites, which means that their sum is 0.

Notice that y is eliminated in this new equation; so we can easily solve for the value of x.

$$3x = 12$$
$$x = 4 \qquad \text{Divide both sides by 3 to isolate } x.$$

Now that we have the value of x, we can find y by substituting 4 for x in one of the original equations. We will use $x + y = 5$.

$$x + y = 5$$
$$4 + y = 5 \qquad \text{Substitute 4 for } x.$$
$$y = 1 \qquad \text{Subtract 4 from both sides to isolate } y.$$

The solution is $(4, 1)$. We can check by verifying that $(4, 1)$ makes both of the original equations true. We will leave this to the reader.

Your Turn 1 | Solve the system of equations using elimination.

$$\begin{cases} 4x + 3y = 8 \\ x - 3y = 7 \end{cases}$$

Multiplying One Equation by a Number to Create Additive Inverses

You may have noted that the expressions $x + y$ and $2x - y$ conveniently contained the additive inverses y and $-y$, which eliminated the y's when the expressions were added. If no such pairs of additive inverses appear in a system of equations, we will use the multiplication principle of equality to multiply both sides of one equation by a number to create additive inverse pairs.

Example 2 | Solve the system of equations using elimination.

$$\begin{cases} x + y = 6 & \text{(Equation 1)} \\ 2x - 5y = -16 & \text{(Equation 2)} \end{cases}$$

Solution: Because no variables are eliminated when $x + y = 6$ and $2x - 5y = -16$ are added, we will rewrite one of the equations so that it has a term that is the additive inverse of one of the terms in the other equation. We will multiply both sides of Equation 1 by 5 so that its y term becomes $5y$, which is the opposite of the $-5y$ term in Equation 2.

$$\begin{array}{ll} x + y = 6 & \\ 5 \cdot x + 5 \cdot y = 5 \cdot 6 & \text{Multiply both sides by 5.} \\ 5x + 5y = 30 & \text{(Equation 1 rewritten)} \end{array}$$

Note We could have multiplied both sides of Equation 1 by -2. We would get a $-2x$ term in Equation 1 that is the opposite of the $2x$ term in Equation 2.

Answer to Your Turn 1
$$\left(3, -\frac{4}{3}\right)$$

Now we add the rewritten Equation 1 to Equation 2 to eliminate a variable, just as we did in Example 1.

Note Multiplying Equation 1 by 5 made the elimination of the y terms possible.

$$\begin{array}{ll} 5x + 5y = 30 & \text{(Equation 1 rewritten)} \\ 2x - 5y = -16 & \text{(Equation 2)} \\ \hline 7x + 0 = 14 & \text{Add rewritten Equation 1 to} \\ & \text{Equation 2 to eliminate } y. \end{array}$$

We can now solve for x.

$$7x = 14$$
$$x = 2 \qquad \text{Divide both sides by 7 to isolate } x.$$

To finish, we substitute 2 for x in one of the equations containing both variables. We will use $x + y = 6$.

$$x + y = 6$$
$$2 + y = 6 \qquad \text{Substitute 2 for } x.$$
$$y = 4 \qquad \text{Subtract 2 from both sides to isolate } y.$$

The solution is $(2, 4)$. We will leave the check to the reader.

Your Turn 2 Solve the system of equations using elimination.

$$\begin{cases} -3x - 5y = 6 \\ x + 2y = -1 \end{cases}$$

Multiplying Each Equation by a Number to Create Additive Inverses

If every coefficient in a system of equations is other than 1, we may have to multiply each equation by a number to generate a pair of additive inverses.

Example 3 Solve the system of equations using elimination.

$$\begin{cases} 4x - 3y = -2 & \text{(Equation 1)} \\ 6x - 7y = 7 & \text{(Equation 2)} \end{cases}$$

Solution: We will choose to eliminate x; so we must multiply both equations by numbers that make the x terms additive inverses. This is like finding the least common multiple (or denominator) of the coefficients of x, but with opposite signs. We will multiply Equation 1 by 3 and Equation 2 by -2.

Instructor Note Emphasize that students can multiply by *any* numbers that create additive inverses. Remind them that they must multiply both sides of the equation, not just the side with variables. Do this example again by eliminating y to show that you get the same answers.

$$\begin{array}{lcl} 4x - 3y = -2 & \xrightarrow{\text{Multiply by 3.}} & 12x - 9y = -6 \\ 6x - 7y = 7 & \xrightarrow{\text{Multiply by } -2.} & -12x + 14y = -14 \end{array}$$

◀ **Note** The x terms are now additive inverses, $12x$ and $-12x$.

Now we can add the rewritten equations to eliminate the x term.

$$\begin{array}{ll} 12x - 9y = -6 \\ -12x + 14y = -14 \\ \hline 0 + 5y = -20 & \text{Add the rewritten equations to eliminate } x. \\ y = -4 & \text{Divide both sides by 5 to isolate } y. \end{array}$$

Note It doesn't matter which equation we use because x should ▶ be the same value in any equation in the system.

To finish, we substitute -4 for y in one of the original equations and solve for x. We will use $4x - 3y = -2$.

$$4x - 3y = -2$$
$$4x - 3(-4) = -2 \qquad \text{Substitute } -4 \text{ for y.}$$
$$4x + 12 = -2$$
$$4x = -14 \qquad \text{Subtract 12 from both sides.}$$

Answer to Your Turn 2
$(-7, 3)$

$$x = -\frac{14}{4} \quad \text{Divide both sides by 4 to isolate } y.$$

$$x = -\frac{7}{2} \quad \text{Simplify.}$$

The solution is $\left(-\frac{7}{2}, -4\right)$. We will leave the check to the reader.

Your Turn 3   Solve the system of equations using elimination.

$$\begin{cases} 3x - 2y = 7 \\ 4x - 3y = 10 \end{cases}$$

Fractions or Decimals in a System

If any of the equations in a system of equations contains fractions or decimals, it is helpful to use the multiplication principle of equality to clear those fractions or decimals so that the equations contain only integers.

Note Because this system contains both fractions and decimals, the answers can be left as either a fraction or a decimal.

▶

Example 4   Solve the system of equations using elimination.

$$\begin{cases} \frac{1}{2}x - y = \frac{3}{4} & \text{(Equation 1)} \\ 0.4x - 0.3y = 1 & \text{(Equation 2)} \end{cases}$$

Solution: To clear the fractions in Equation 1, we can multiply both sides by the LCD, which is 4. To clear the decimals in Equation 2, we can multiply both sides by 10.

$$\frac{1}{2}x - y = \frac{3}{4} \quad \xrightarrow{\text{Multiply by 4.}} \quad 2x - 4y = 3$$

$$0.4x - 0.3y = 1 \quad \xrightarrow{\text{Multiply by 10.}} \quad 4x - 3y = 10$$

Now that both equations contain only integers, it will be easier to solve the system. We will choose to eliminate x. So we multiply the first equation by -2, then combine the equations.

$$\begin{array}{l} 2x - 4y = 3 \quad \xrightarrow{\text{Multiply by } -2.} \\ 4x - 3y = 10 \end{array} \quad \begin{array}{l} -4x + 8y = -6 \\ \underline{4x - 3y = 10} \\ 0 + 5y = 4 \end{array}$$

Add the rewritten equations to eliminate x.

$$y = \frac{4}{5} \text{ or } 0.8 \quad \text{Divide both sides by 5 to isolate } y.$$

To finish, we substitute $\frac{4}{5}$ for y in one of the original equations and solve for x. We will use $\frac{1}{2}x - y = \frac{3}{4}$.

Note We could have multiplied both sides of the equation by 20 to clear the fraction.

▶

$$\frac{1}{2}x - \frac{4}{5} = \frac{3}{4} \quad \text{Substitute } \frac{4}{5} \text{ for } y.$$

$$\frac{1}{2}x = \frac{3}{4} + \frac{4}{5} \quad \text{Add } \frac{4}{5} \text{ to both sides.}$$

$$\frac{1}{2}x = \frac{15}{20} + \frac{16}{20} \quad \text{Write the fractions with their LCD, 20.}$$

$$\frac{1}{2}x = \frac{31}{20} \quad \text{Add the fractions.}$$

Answer to Your Turn 3
$(1, -2)$

$$x = \frac{\overset{1}{\cancel{31}}}{\underset{10}{\cancel{20}}} \cdot \frac{2}{1} \qquad \text{Multiply both sides by 2 to isolate } x.$$

$$x = \frac{31}{10} \text{ or } 3.1$$

The solution is $\left(\dfrac{31}{10}, \dfrac{4}{5}\right)$ or $(3.1, 0.8)$. We will leave the check to the reader.

Your Turn 4 Solve the system of equations using elimination.

$$\begin{cases} 0.6x + y = 3 \\ \dfrac{2}{5}x + \dfrac{1}{4}y = 1 \end{cases}$$

Rewriting the Equations in the Form $Ax + By = C$

Notice in Examples 1 through 4 that all of the equations are in standard form, which we learned in Chapter 3 to be $Ax + By = C$. When the elimination method is used, the equations need to be written in standard form. For example, in the system that we solved in Example 1, the equations could have been given in a nonstandard form, as follows:

$$\begin{cases} y = 5 - x & \text{(Equation 1)} \\ 2x - y = 7 & \text{(Equation 2)} \end{cases}$$

Adding x to both sides of Equation 1 puts it in standard form so that we can use elimination as we did in Example 1.

$$\begin{cases} x + y = 5 & \text{(Equation 1)} \\ 2x - y = 7 & \text{(Equation 2)} \end{cases}$$

We can now summarize the elimination method with the following procedure.

Procedure **Solving Systems of Two Linear Equations Using Elimination**

To solve a system of two linear equations using the elimination method:
1. Write the equations in standard form ($Ax + By = C$).
2. Use the multiplication principle to clear fractions or decimals (optional).
3. If necessary, multiply one or both equations by a number (or numbers) so that they have a pair of terms that are additive inverses.
4. Add the equations. The result should be an equation in terms of one variable.
5. Solve the equation from step 4 for the value of that variable.
6. Using an equation containing both variables, substitute the value you found in step 5 for the corresponding variable and solve for the value of the other variable.
7. Check your solution in the original equations.

Inconsistent Systems and Dependent Equations

How would we recognize an inconsistent system or a consistent system with dependent equations using the elimination method?

Example 5 Solve the system of equations.

a. $\begin{cases} 2x - y = 1 \\ 2x - y = -3 \end{cases}$

Solution: Notice that the left sides of the equations match. Multiplying one of the equations by -1 and then adding the equations will eliminate both variables.

$$\begin{array}{ll} 2x - y = 1 & \xrightarrow{\text{Multiply by } -1.} & -2x + y = -1 \\ 2x - y = -3 & & \underline{2x - y = -3} \\ & & 0 = -4 \quad \text{Add the rewritten equations.} \end{array}$$

Answer to Your Turn 4
$\left(1, \dfrac{12}{5}\right)$ or $(1, 2.4)$

Connection If we graph the equations in Example 5(a), the lines will be parallel, indicating there is no solution.

Both variables have been eliminated, and the resulting equation, $0 = -4$, is false. Therefore, there is no solution. This system of equations is inconsistent.

b. $\begin{cases} 3x + 4y = 5 \\ 9x + 12y = 15 \end{cases}$

Solution: To eliminate x, we could multiply the first equation by -3, then combine the equations.

$$\begin{array}{l} 3x + 4y = 5 \\ 9x + 12y = 15 \end{array} \xrightarrow{\text{Multiply by } -3.} \begin{array}{l} -9x - 12y = -15 \\ \underline{9x + 12y = 15} \\ 0 = 0 \end{array}$$ Add the rewritten equations.

Connection If graphed, both equations in Example 5(b) generate the same line, indicating that the equations are dependent.

Both variables have been eliminated, and the resulting equation, $0 = 0$, is true. This means that the equations are dependent. So there are an infinite number of solutions, which are all of the ordered pairs along the line $3x + 4y = 5$ (or $9x + 12y = 15$).

Your Turn 5 Solve the system of equations.

a. $\begin{cases} x - 4y = 2 \\ 5x - 20y = 10 \end{cases}$ **b.** $\begin{cases} x + 2y = 3 \\ x + 2y = 1 \end{cases}$

Objective 2 Solve applications using elimination.

In the previous section, we solved problems in which the substitution method was advantageous because after translating, one of the equations had an isolated variable. In this section, we focus on problems in which elimination is advantageous. These problems translate to a system of linear equations in which every equation is in standard form $(Ax + By = C)$.

Example 6 Airplanes traveling west fly against the jet stream, which decreases their average speed. When flying east, the jet stream increases average speed. Suppose a large jet flies from Atlanta to Los Angeles, a distance of approximately 2000 miles, in 5 hours. On the return trip, the plane takes only 4 hours. Find the plane's speed in still air. How much does the jet stream change the speed of the plane?

Understand We assume that without the jet stream, the plane would average the same speed going west as going east. We also assume that the amount the jet stream adds to the plane's speed is the same as the amount subtracted depending on which way the plane is traveling.

Plan and Execute Let x represent the plane's speed in still air. Let y represent the amount the jet stream increases or decreases the plane's speed depending on which way it is traveling. We can use a table to organize the information.

Los Angeles
Atlanta
2000 miles

Category	Rate	Time	Distance
West	$x - y$	5	$5(x - y)$
East	$x + y$	4	$4(x + y)$

Relationship 1: The distance from Atlanta to Los Angeles is 2000 miles.

Translation: $5(x - y) = 2000$

$\dfrac{5(x - y)}{5} = \dfrac{2000}{5}$ Divide both sides by 5 to streamline the equation.

$x - y = 400$

Answers to Your Turn 5
a. all ordered pairs along $x - 4y = 2$ or $5x - 20y = 10$ (dependent)
b. no solution (inconsistent)

Relationship 2: The distance from Los Angeles to Atlanta is 2000 miles.

$$\text{Translation:} \quad 4(x + y) = 2000$$

$$\frac{4(x + y)}{4} = \frac{2000}{4} \quad \text{Divide both sides by 4 to streamline the equation.}$$

$$x + y = 500$$

$$\text{Our system:} \begin{cases} x - y = 400 \\ x + y = 500 \end{cases}$$

Because the equations are in standard form, we will use the elimination method.

$$\begin{aligned} x - y &= 400 \\ \underline{x + y} &= \underline{500} \\ 2x + 0 &= 900 \quad \text{Add the equations to eliminate } y. \\ 2x &= 900 \\ x &= 450 \quad \text{Divide both sides by 2 to isolate } x. \end{aligned}$$

Now we can find the value of y by substituting 450 for x in one of the equations. We will use $x + y = 500$.

$$x + y = 500$$

$$450 + y = 500 \quad \text{Substitute 450 for } x.$$

$$y = 50 \quad \text{Subtract 450 from both sides to isolate } y.$$

Answer The plane travels 450 miles per hour in still air. The jet stream changes the plane's speed by 50 miles per hour depending whether it is traveling east or west.

Check Verify the time of travel for each part of the round trip.

Going west: $450 - 50 = 400$ miles per hour

Going east: $450 + 50 = 500$ miles per hour

Time: $t = \dfrac{d}{r} = \dfrac{2000}{400} = 5$ hours

Time: $t = \dfrac{d}{r} = \dfrac{2000}{500} = 4$ hours

Your Turn 6 To train for the swimming portion of a triathlon, an athlete swims in a river to gauge how the current will affect him. When swimming against the current, he can swim a mile in 1.25 hours. When swimming with the current, he can swim a mile in 0.5 hour. What is his rate in still water? How much did the current increase or decrease his rate?

Example 7 Jeff invests $5000 in two different accounts. The first account has an annual percentage rate (APR) of 4%. The second account has an APR of 6%. If the total interest earned after one year is $252, what principal was invested in each account?

Understand The two unknowns are the principals invested in each account. One relationship involves the total principal ($5000), and the other relationship involves the total interest earned ($252).

Plan and Execute Let x and y represent the two amounts invested. We can use a table to organize the information.

Category	APR	Principal	Interest
Account 1	0.04	x	$0.04x$
Account 2	0.06	y	$0.06y$

Answer to Your Turn 6
His rate in still water is 1.4 miles per hour. The current changes his rate by 0.6 miles per hour.

Note Recall from Chapter 2 that we write percentage rates as decimal numbers. Also, interest is the product of the APR and the principal.

Relationship 1:   The total principal is $5000.

> Translation: $x + y = 5000$

Relationship 2:   The total interest is $252.

> Translation: $0.04x + 0.06y = 252$

Our system: $\begin{cases} x + y = 5000 \\ 0.04x + 0.06y = 252 \end{cases}$

Because the equations are in standard form, elimination is the better method. We will eliminate x by multiplying both sides of the first equation by -0.04.

Note We could have first eliminated the decimal numbers, but we chose not to because it is easy to eliminate x or y by the elimination method because both coefficients in $x + y = 5000$ are 1.

$\begin{aligned} x + y &= 5000 \\ 0.04x + 0.06y &= 252 \end{aligned}$ Multiply by -0.04.

$\begin{aligned} -0.04x - 0.04y &= -200 \\ 0.04x + 0.06y &= 252 \\ \hline 0.02y &= 52 \\ y &= 2600 \end{aligned}$

Add the equations to eliminate x.

Divide both sides by 0.02 to isolate y.

Now we can find x by substituting 2600 for y in one of the equations. We will use $x + y = 5000$.

$\begin{aligned} x + y &= 5000 \\ x + 2600 &= 5000 \quad \text{Substitute 2600 for } y. \\ x &= 2400 \quad \text{Subtract 2600 from both sides to isolate } x. \end{aligned}$

Answer   Jeff invested $2400 in the 4% account and $2600 in the 6% account.

Check   The total investment is $2400 + $2600 = $5000, and the total interest is $0.04(2400) + 0.06(2600) = \$96 + \$156 = \$252$.

Your Turn 7   Monique invests a total of $8000 in two different accounts. The first account returns 5%, while the second account returns 8%. If the total interest earned after one year is $532, what principal was invested in each account?

Example 8   How much 10% HCl solution and 30% HCl solution must be mixed together to make 500 milliliters of 15% HCl solution?

Understand   The two unknowns are the volumes of 10% solution and 30% solution that are mixed. One relationship involves concentrations of each solution in the mixture, and the other relationship involves the total volume of the final mixture (500 ml).

Plan and Execute   Let x and y represent the two amounts to be mixed.

Solution	Concentration	Volume	Amount of HCl
10% solution	0.10	x	$0.10x$
30% solution	0.30	y	$0.30y$
15% solution	0.15	500	$0.15(500)$

Relationship 1:   The total volume is 500 ml.

> Translation: $x + y = 500$

Note The amount of HCl is the product of the concentration and the volume. Notice that we were given the volume of the 15% solution.

Relationship 2:   The combined amount of HCl in the two mixed solutions is to be 15% of the total amount of the mixture.

> Translation: $0.10x + 0.30y = 0.15(500)$

Our system: $\begin{cases} x + y = 500 \\ 0.10x + 0.30y = 0.15(500) \end{cases}$

Because the equations are in standard form, elimination is the better method. We will eliminate x by multiplying both sides of the first equation by -0.10.

$$x + y = 500 \xrightarrow{\text{Multiply by } -0.10.} -0.10x - 0.10y = -50$$

$$0.10x + 0.30y = 75 \qquad\qquad \underline{0.10x + 0.30y = 75}$$

Add the equations to eliminate x.

$$0.20y = 25$$

$$y = 125$$

Divide both sides by 0.20 to isolate y.

Now we can find x by substituting 125 for y in one of the equations. We will use $x + y = 500$.

$$x + y = 500$$
$$x + 125 = 500 \qquad \text{Substitute 125 for } y.$$
$$x = 375 \qquad \text{Subtract 125 on both sides to isolate } x.$$

Answer Mixing 375 ml of 10% solution with 125 ml of 30% solution gives 500 ml of 15% solution.

Check The mixture volume is $125 + 375 = 500$. The amount of HCl is $0.10(375) + 0.30(125) = 37.5 + 37.5 = 75$, which means that 75 ml is HCl out of the 500 ml mixture. We can verify that 75 ml is 15% of 500 ml by multiplying: $0.15(500) = 75$.

Answer to Your Turn 8
450 ml of 40%, 150 ml of 20%

Your Turn 8 How much 20% HCl solution and 40% HCl solution must be mixed together to make 600 milliliters of 35% HCl solution?

4.3 Exercises For Extra Help MyMathLab®

Objective 1

Prep Exercise 1 What advantages does the method of elimination have over graphing and substitution? The elimination method is advantageous over graphing when the solution involves fractions. The elimination method is advantageous over substitution when no coefficients are 1.

Prep Exercise 2 Suppose you are given the system of equations
$$\begin{cases} x - y = 2 \\ 3x + y = 5 \end{cases}$$
Which variable would you eliminate? Why? y because $-y$ and y are additive inverses

Prep Exercise 3 Suppose you are given the system of equations $\begin{cases} 6x - 2y = 1 \\ 3x + 5y = 7 \end{cases}$
a. Which variable is easier to eliminate? Why?
x because multiplying only the second equation by -2 allows the x to be eliminated; whereas to eliminate y, we need to multiply each equation by a number.
b. How would you eliminate the variable you chose?
Multiply $3x + 5y = 7$ by -2 and then add the resulting equation to $6x - 2y = 1$.

Prep Exercise 4 Suppose you are given the system of equations $\begin{cases} 5x + 4y = 8 \\ 3x - 6y = 2 \end{cases}$
a. Which variable is easier to eliminate? Why?
y because $4y$ and $-6y$ have opposite signs, which means we can multiply each equation by a positive number; whereas to eliminate x, we need to multiply one equation by a positive number and the other by a negative number.
b. How would you eliminate the variable you chose?
Multiply the first equation by 3 and the second by 2 and then add the resulting equations.

Prep Exercise 5 When using elimination, how do you know if a system of equations has no solution (inconsistent system)? Both variables have been eliminated, and the resulting equation is false.

Prep Exercise 6 When using elimination, how do you know if a system of equations has an infinite number of solutions (dependent equations)? Both variables have been eliminated, and the resulting equation is true.

For Exercises 1–36, solve the system of equations using the elimination method. Note that some systems may be inconsistent or consistent with dependent equations. See Examples 1–5.

1. $\begin{cases} x - y = 14 \\ x + y = -2 \end{cases}$
$(6, -8)$

2. $\begin{cases} x + y = 9 \\ 5x - y = 3 \end{cases}$
$(2, 7)$

3. $\begin{cases} 3x + y = 9 \\ 2x - y = 1 \end{cases}$
$(2, 3)$

4. $\begin{cases} 3x + y = 10 \\ -2x - y = -7 \end{cases}$
$(3, 1)$

5. $\begin{cases} 3x + 2y = -7 \\ 5x - 2y = -1 \end{cases}$
$(-1, -2)$

6. $\begin{cases} 2x - 3y = 8 \\ 4x + 3y = 16 \end{cases}$
$(4, 0)$

7. $\begin{cases} 4x + 3y = 17 \\ 2x + 3y = 13 \end{cases}$
$(2, 3)$

8. $\begin{cases} x - 3y = 7 \\ x - 5y = 13 \end{cases}$
$(-2, -3)$

9. $\begin{cases} 5x + y = 14 \\ 2x + y = 5 \end{cases}$
$(3, -1)$

10. $\begin{cases} 3x - y = 7 \\ 5x - y = 15 \end{cases}$
$(4, 5)$

11. $\begin{cases} 3x - 5y = -17 \\ 4x + y = -15 \end{cases}$
$(-4, 1)$

12. $\begin{cases} 7x - 4y = 4 \\ 5x + y = 26 \end{cases}$
$(4, 6)$

13. $\begin{cases} 12x - 2y = -54 \\ 13x + 4y = -40 \end{cases}$
$(-4, 3)$

14. $\begin{cases} 5x + 2y = 3 \\ 7x - 6y = 13 \end{cases}$
$(1, -1)$

15. $\begin{cases} 2x + y = -2 \\ 8x + 4y = -8 \end{cases}$
All ordered pairs along $2x + y = -2$ (dependent)

16. $\begin{cases} 4x - 2y = 3 \\ -8x + 4y = -6 \end{cases}$
All ordered pairs along $4x - 2y = 3$ (dependent)

17. $\begin{cases} 2x - 5y = 4 \\ -8x + 20y = -20 \end{cases}$
No solution (inconsistent)

18. $\begin{cases} 5x + 5y = 50 \\ x + y = 2.5 \end{cases}$
No solution (inconsistent)

19. $\begin{cases} 5x + 6y = 11 \\ 2x - 4y = -2 \end{cases}$
$(1, 1)$

20. $\begin{cases} -6x + 7y = -2 \\ 9x - 5y = -8 \end{cases}$
$(-2, -2)$

21. $\begin{cases} 4x + 5y = -4 \\ 3x + 8y = -20 \end{cases}$
$(4, -4)$

22. $\begin{cases} 3x - 4y = -14 \\ 5x + 7y = 45 \end{cases}$
$(2, 5)$

23. $\begin{cases} 5x + 6y = 2 \\ 10x + 3y = -2 \end{cases}$
$\left(-\dfrac{2}{5}, \dfrac{2}{3}\right)$

24. $\begin{cases} 8x + 6y = 5 \\ 2x + y = 1 \end{cases}$
$\left(\dfrac{1}{4}, \dfrac{1}{2}\right)$

25. $\begin{cases} 10x + 5y = 3.5 \\ 3x - 4y = -0.6 \end{cases}$
$(0.2, 0.3)$

26. $\begin{cases} 2x - y = -0.8 \\ 3x - 2y = -1 \end{cases}$
$(-0.6, -0.4)$

27. $\begin{cases} \dfrac{3}{2}x - \dfrac{4}{3}y = \dfrac{11}{3} \\ \dfrac{1}{4}x - \dfrac{2}{3}y = -\dfrac{7}{6} \end{cases}$
$(6, 4)$

28. $\begin{cases} \dfrac{3}{4}x - \dfrac{2}{7}y = -5 \\ \dfrac{5}{8}x + \dfrac{1}{4}y = -\dfrac{3}{4} \end{cases}$
$(-4, 7)$

29. $\begin{cases} 0.7x - \dfrac{1}{2}y = 9.4 \\ 0.9x + \dfrac{7}{10}y = 0 \end{cases}$
$(7, -9)$

30. $\begin{cases} \dfrac{1}{3}x + 0.2y = -0.2 \\ \dfrac{2}{3}x - 0.75y = -5 \end{cases}$
$(-3, 4)$

31. $\begin{cases} y = 2x - 5 \\ x - y = 9 \end{cases}$
$(-4, -13)$

32. $\begin{cases} x + 3y = 8 \\ 2y = x + 6 \end{cases}$
$(-0.4, 2.8)$ or $-\dfrac{2}{5}, \dfrac{14}{5}$

33. $\begin{cases} y = \dfrac{3}{4}x + 1 \\ 4x = 2y + 3 \end{cases}$
$\left(2, \dfrac{5}{2}\right)$

34. $\begin{cases} y = -\dfrac{2}{3}x - 5 \\ 3y = 2x - 27 \end{cases}$
$(3, -7)$

35. $\begin{cases} 0.2x - y = -3.2 \\ y = \dfrac{3}{4}x + 1 \end{cases}$
$(4, 4)$

36. $\begin{cases} y = \dfrac{1}{5}x - 2 \\ 6y = 0.3x - 1.2 \end{cases}$
$(12, 0.4)$ or $\left(12, \dfrac{2}{5}\right)$

Find ⊗ the Mistake   *For Exercises 37 and 38, find the mistake.*

37. $\begin{cases} x - y = 1 \\ 9x + 8y = 77 \end{cases}$

$$\begin{array}{l} -9(x - y) = 1 \\ 9x + 8y = 77 \end{array} \longrightarrow \begin{array}{l} -9x + 9y = 1 \\ \underline{9x + 8y = 77} \\ 17y = 78 \\ y = \dfrac{78}{17} \end{array}$$

$$\begin{array}{l} x - y = 1 \\ x - \left(\dfrac{78}{17}\right) = 1 \\ x = \dfrac{95}{17} \end{array}$$

Solution: $\left(\dfrac{95}{17}, \dfrac{78}{17}\right)$

Mistake: Did not multiply the right side of $x - y = 1$ by -9

Correct: $(5, 4)$

38. $\begin{cases} x + y = 1 \\ x + 2y = 2 \end{cases}$

$$\begin{array}{l} x + y = 1 \\ -1(x + 2y = 2) \end{array} \longrightarrow \begin{array}{l} x + y = 1 \\ \underline{-x - 2y = 2} \\ -y = 3 \\ y = -3 \end{array}$$

$$\begin{array}{l} x + y = 1 \\ x + (-3) = 1 \\ x = 4 \end{array}$$

Solution: $(4, -3)$

Mistake: Did not multiply the right side of $x + 2y = 2$ by -1

Correct: $(0, 1)$

Objective 2

For Exercises 39–56, translate the problem to a system of equations; then solve using the elimination method. See Examples 6–8.

39. What are the length and width of a rectangle if the length exceeds the width by 14 inches and the perimeter is 60 inches?

22 in., 8 in.

40. What are the length and width of a rectangle if the width is 27 feet less than its length and the perimeter is 398 feet?

113 ft., 86 ft.

41. A boat travels 24 miles upstream in 3 hours. Going downstream, it can travel 36 miles in the same amount of time. Find the speed of the current and the speed of the boat in still water.

10 mph in still water; current is 2 mph

42. A boat travels 12 miles downstream in $1\frac{1}{2}$ hours. Going upstream, the boat takes 6 hours to return. Find the speed of the current and the speed of the boat in still water.

5 mph in still water; current is 3 mph

43. If a plane can travel 500 miles per hour with the wind and only 440 miles per hour against the wind, find the speed of the wind and the speed of the plane in still air.

470 mph in still air, 30 mph winds

44. If a plane can travel 200 miles per hour with the wind and only 120 miles per hour against the wind, find the speed of the wind and the speed of the plane in still air.

160 mph in still air, 40 mph winds

45. Adam inherited $15,500. He invested part of the money at an APR of 4% and put the rest into a savings account at an APR of 3%. If his annual income from these investments is $585, how much did he invest at each rate?

$12,000 at 4%, $3500 at 3%

46. Petunia invests some money in a plan that pays a 5% APR and invests twice as much money in a plan that pays an 8% APR. If the total interest from the two plans is $1029, how much was invested in each?

$4900 at 5%, $9800 at 8%

47. Ellyn owes $13,250 on two student loans. One loan has an APR of 5%; the other, an APR of 8%. If the total interest after one year is $727, what is the amount of each loan?

$11,100 at 5%, $2150 at 8%

48. Enrique has two credit cards. One is a Visa at an APR of 14%, and the other is a department store card with an APR of 18%. If the total amount he owes is $1536 and the interest for one year is $245.36, what is the amount owed on each card? $778 at 14%, $758 at 18%

49. Rhonda and Mike have two car loans totaling $16,250. One loan has an APR of 2.9%, and the other has an APR of 6%. If Rhonda and Mike pay total interest of $657.25 after one year, what is the amount owed on each car?

$10,250 at 2.9%, $6000 at 6%

50. Corey borrowed a total of $250,000 in two loans. One loan is at an APR of 8%, and the other is at an APR of 18%. If he paid $23,000 in interest during the first year, how much was loaned at each rate?

$220,000 at 8%, $30,000 at 18%

51. How much of a 20% acid solution is to be mixed with 80 milliliters of a 15% acid solution to get an 18% solution?

120 ml

52. How many gallons of milk containing 1% butterfat is to be mixed with 50 gallons of milk that is 6% butterfat to get milk that is 2% butterfat?

200 gallons

53. A pharmacist is preparing a 300-milliliter solution of 25% alcohol. If she has 15% and 45% alcohol solutions in stock, how much of each must she use?

200 ml of 15%, 100 ml of 45%

54. During chemistry class, Dr. Ellis needs a solution of 20% HCl, but all he has is 10% and 25% HCl solutions. How many milliliters of each does he need to mix to get 150 milliliters of the 20% HCl solution?

50 ml of 10%, 100 ml of 25%

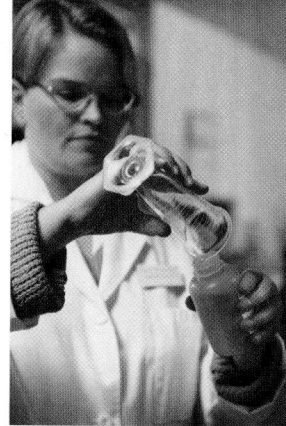

55. A coffee shop is considering a new mixture of coffee beans. It will be created from Italian Roast beans costing $9.80 per pound and the Gold Coast Blend (GCB) beans costing $11.20 per pound. The two types of beans will be mixed to create 20 pounds of coffee to be sold for $10.43 per pound. How much of each type of bean should be used?

11 lb. of Italian Roast, 9 lb. of GCB

56. A store is selling two types of jelly beans. One kind is $1.20 per pound, and the other is $0.90 per pound. If the merchant wants to sell a mixture at $1.11 per pound, how many pounds of each must be used to make 30 pounds?

21 lb. of $1.20, 9 lb. of $0.90

Puzzle Problem   Use algebraic methods to prove that 0.999 . . . is equal to 1. (Hint: Let $x = 0.\overline{9}$. Then write a second equation using the multiplication principle so that when the two equations are added, the repeated 9 is eliminated.)

Let $x = 0.\overline{9}$. Multiply both sides by -10 so that we have a second equation:

$-10x = -9.\overline{9}$. We now have the system $\begin{cases} x = 0.\overline{9} \\ -10x = -9.\overline{9} \end{cases}$.

Solve the system using elimination by adding the two equations.

$$x = 0.\overline{9}$$
$$\underline{-10x = -9.\overline{9}}$$
$$-9x = -9 \qquad \text{Note that the repeated decimal digits are eliminated.}$$
$$x = 1 \qquad \text{Divide both sides by } -9.$$

Review Exercises

Exercises 1 and 2   **Expressions**

[1.7] 1. Use the distributive property: $4(5x + 7y - z)$

20x + 28y − 4z

[1.7] 2. Simplify: $3x - 4y + z - 2x + y - z$

x − 3y

Exercises 3–6   **Equations and Inequalities**

[2.4] 3. Solve for y: $x + 3y = 6$

$y = -\dfrac{1}{3}x + 2$

[2.4] 4. Solve for x: $\dfrac{1}{3}x - 2y = 10$

x = 6y + 30

[2.2] 5. Given $x + y + 2z = 7$, find x if $y = -1$ and $z = 3$.

x = 2

[2.2] 6. Given $x - 3y + 2z = 6$, find y if $x = -2$ and $z = 4$.

y = 0

4.4 Solving Systems of Linear Equations in Three Variables; Applications

Objectives

1 Determine whether an ordered triple is a solution for a system of equations.

2 Understand the types of solution sets for systems of three equations.

3 Solve a system of three linear equations using the elimination method.

4 Solve application problems that translate to a system of three linear equations.

Learning Strategy

The best way to prepare for class is to stay up to date on homework. Seek help immediately if you have a problem that you cannot answer. In math classes, teachers explain problems that will be used again later in more complicated problems.

—Matt D.

Warm-up

[1.7] **1.** Find the value of $2x + 3y + 2z$ if $x = 3$, $y = 1$, and $z = -1$.

[1.7] **2.** Find the product: $-2(2x + y + 2z)$

[1.7] **3.** Combine like terms: $-4x - 2y - 4z + 4x + 8y + 5z$

Objective 1 Determine whether an ordered triple is a solution for a system of equations.

In this section, we solve systems of three linear equations with three unknowns. Solutions of these systems are *ordered triples* with the form (x, y, z). We check a solution for a system of three equations the same way we check a system of two equations, by replacing each variable with its corresponding value and verifying that each equation is true.

Example 1 Determine whether each ordered triple is a solution to the system of equations.

$$\begin{cases} x + y + z = 3 & \text{(Equation 1)} \\ 2x + 3y + 2z = 7 & \text{(Equation 2)} \\ 3x - 4y + z = 4 & \text{(Equation 3)} \end{cases}$$

a. $(2, 1, 0)$

Solution: In all three equations, replace x with 2, y with 1, and z with 0.

Equation 1:

$x + y + z = 3$

$2 + 1 + 0 \stackrel{?}{=} 3$

$3 = 3$ True

Equation 2:

$2x + 3y + 2z = 7$

$2(2) + 3(1) + 2(0) \stackrel{?}{=} 7$

$4 + 3 + 0 \stackrel{?}{=} 7$

$7 = 7$ True

Equation 3:

$3x - 4y + z = 4$

$3(2) - 4(1) + 0 \stackrel{?}{=} 4$

$6 - 4 + 0 \stackrel{?}{=} 4$

$2 = 4$ False

Because $(2, 1, 0)$ does not satisfy all three equations in the system, it is not a solution for the system.

b. $(3, 1, -1)$

Solution:

Equation 1:

$x + y + z = 3$

$3 + 1 + (-1) \stackrel{?}{=} 3$

$3 = 3$ True

Equation 2:

$2x + 3y + 2z = 7$

$2(3) + 3(1) + 2(-1) \stackrel{?}{=} 7$

$6 + 3 - 2 \stackrel{?}{=} 7$

$7 = 7$ True

Equation 3:

$3x - 4y + z = 4$

$3(3) - 4(1) + (-1) \stackrel{?}{=} 4$

$9 - 4 - 1 \stackrel{?}{=} 4$

$4 = 4$ True

Because $(3, 1, -1)$ satisfies all three equations in the system, it is a solution for the system.

Your Turn 1 Determine whether the following ordered triples are solutions to the system of equations.

a. $(4, 1, 1)$ **b.** $(-1, 5, 2)$

$$\begin{cases} x + y + z = 6 & \text{(Equation 1)} \\ 3x - 2y + 3z = -7 & \text{(Equation 2)} \\ 4x - 2y + z = -12 & \text{(Equation 3)} \end{cases}$$

Answers to Your Turn 1

a. not a solution

b. solution

Answers to Warm-up

1. 7

2. $-4x - 2y - 4z$

3. $6y + z$

Objective 2 Understand the types of solution sets for systems of three equations.

Recall that we plot an ordered pair using two axes: x and y. Similarly, we plot an ordered triple such as $(3, 4, 5)$ using three axes—x, y, and z— each of which is perpendicular to the other two, as shown.

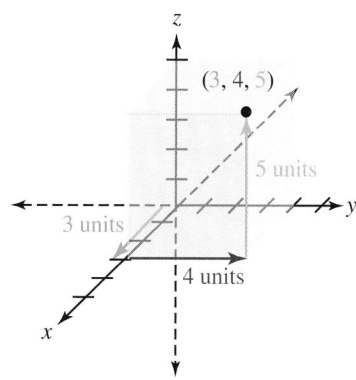

The graph of a linear equation in three variables is a *plane*, which is a flat surface much like a sheet of paper with infinite length and width. We draw a plane as a parallelogram (shown in the margin), but remember that the length and width are infinite.

In Section 4.1, we found that two lines could intersect in one point (consistent system), no points (inconsistent system), or an infinite number of points (consistent system with dependent equations). Similarly, the intersection of the planes of a three-variable system can tell us about the number of solutions for the system.

A Single Solution

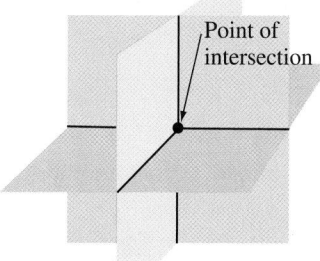

Connection Recall from Section 4.1 that a system of equations that has a solution (single or infinite) is consistent and a system that has no solution is inconsistent. This is true for systems of three equations as well. We also discussed dependent and independent equations. However, for systems of three equations, defining dependent and independent equations and discussing their graphical representations is too complicated for this course. Here, we simply say that systems with an infinite number of solutions have dependent equations. As we will see, we can identify these systems when solving them using the elimination method.

If the planes intersect at a single point, that ordered triple is the solution to the system.

Infinite Number of Solutions

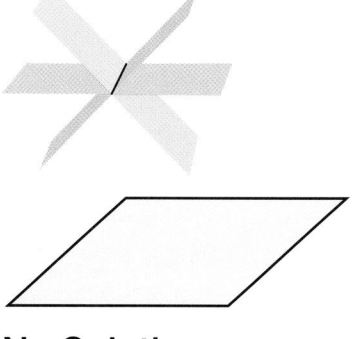

If the three planes intersect along a line, the system has an infinite number of solutions, which are the coordinates of any point along that line.

If all three graphs are the same plane, the system has an infinite number of solutions, which are the coordinates of any point in the plane.

No Solution

If all of the planes are parallel, the system has no solution.

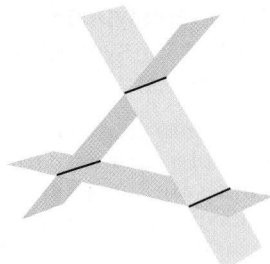

Pairs of planes also can intersect, as shown. However, because all three planes do not have a common intersection, the system has no solution.

Note Planes can be configured in still more ways so that they do not have a common intersection.

Objective 3 Solve a system of three linear equations using the elimination method.

Because graphing systems of three linear equations is usually impractical, we solve them using elimination. The process is much like what we learned in Section 4.3.

Example 2 Solve the system of equations using elimination.

a. $\begin{cases} x + y + z = 2 & \text{(Equation 1)} \\ 2x + y + 2z = 1 & \text{(Equation 2)} \\ 3x + 2y + z = 1 & \text{(Equation 3)} \end{cases}$

Solution: First, we choose a variable and eliminate that variable from two pairs of equations. Let's eliminate y using Equations 1 and 2. We multiply Equation 1 by -1, then add the equations.

Note Our initial goal is to generate two equations with the same two variables.

$$\text{(Eq. 1)} \quad x + y + z = 2 \quad \xrightarrow{\text{Multiply by } -1.} \quad -x - y - z = -2$$
$$\text{(Eq. 2)} \quad 2x + y + 2z = 1 \qquad\qquad\qquad \underline{2x + y + 2z = 1} \quad \text{Add the equations.}$$
$$x + z = -1 \quad \text{(Equation 4)}$$

Now we need to eliminate y from another pair of equations. We will choose Equations 1 and 3.

$$\text{(Eq. 1)} \quad x + y + z = 2 \quad \xrightarrow{\text{Multiply by } -2.} \quad -2x - 2y - 2z = -4$$
$$\text{(Eq. 3)} \quad 3x + 2y + z = 1 \qquad\qquad\qquad \underline{3x + 2y + z = 1} \quad \text{Add the equations.}$$
$$-x - z = -3 \quad \text{(Equation 5)}$$

Equations 4 and 5 form a system of two equations with two variables x and z. We can eliminate z by adding the equations.

$$\text{(Eq. 4)} \quad x + z = -1$$
$$\text{(Eq. 5)} \quad \underline{x - z = -3} \quad \text{Add the equations.}$$
$$2x = -4$$
$$x = -2 \quad \text{Divide both sides by 2 to isolate } x.$$

To find z, we substitute -2 for x in Equation 4 or 5. We will use Equation 4.

$$x + z = -1$$
$$-2 + z = -1 \quad \text{Substitute } -2 \text{ for } x.$$
$$z = 1 \quad \text{Add 2 to both sides.}$$

To find y, substitute -2 for x and 1 for z in any of the original equations. We will use Equation 1.

$$x + y + z = 2$$
$$-2 + y + 1 = 2 \quad \text{Substitute } -2 \text{ for } x \text{ and 1 for } z.$$
$$y - 1 = 2 \quad \text{Simplify the left side.}$$
$$y = 3 \quad \text{Add 1 to both sides to isolate } y.$$

Note Recall that ordered triples are written in the form (x, y, z). ▶

The solution is $(-2, 3, 1)$. We can check the solution by verifying that the ordered triple satisfies each of the three original equations. We will leave the check to the reader.

b. $\begin{cases} 2x + 3y \quad\;\; = 3 & \text{(Equation 1)} \\ \quad\;\; 2y - 3z = -8 & \text{(Equation 2)} \\ 4x \quad\;\; - z = 10 & \text{(Equation 3)} \end{cases}$

Solution: Notice that a variable is missing in each of the equations, which can simplify solving the system. We will eliminate z from Equations 2 and 3.

Note We chose Equation 3 because z has a coefficient of -1. ▶

(Eq. 2) $2y - 3z = -8$ $\qquad\qquad\qquad\qquad\qquad 2y - 3z = -8$

(Eq. 3) $4x - z = 10$ $\;\;\xrightarrow{\text{Multiply by } -3.}\;\; \underline{-12x \qquad\;\; + 3z = -30}$ Add the equations.

$\qquad\qquad\qquad\qquad\qquad\qquad\qquad\qquad\qquad -12x + 2y \quad\;\; = -38$ (Equation 4)

Because Equations 1 and 4 do not have z terms, we do not need to eliminate z from another pair of equations. We will use Equations 1 and 4 to eliminate x.

(Eq. 1) $\quad\;\; 2x + 3y = 3$ $\;\;\xrightarrow{\text{Multiply by } -6.}\;\; 12x + 18y = 18$

(Eq. 4) $-12x + 2y = -38$ $\qquad\qquad\qquad \underline{-12x + 2y = -38}$ Add the equations.

$\qquad\qquad\qquad\qquad\qquad\qquad\qquad\qquad\;\; 20y = -20$

$\qquad\qquad\qquad\qquad\qquad\qquad\qquad\qquad\quad\; y = -1$ **Divide both sides by 20 to isolate y.**

To find x, substitute -1 for y in Equation 1 or 4. We will use Equation 1.

$2x + 3(-1) = 3$ Substitute -1 for y in Equation 1.

$2x - 3 = 3$ Simplify.

$2x = 6$ Add 3 to both sides.

$x = 3$ Divide both sides by 2 to isolate x.

To find z, substitute in Equation 2 or 3. (Equation 1 has no z term.) We will use Equation 2.

$2(-1) - 3z = -8$ Substitute -1 for y in Equation 2.

$-2 - 3z = -8$ Simplify.

$-3z = -6$ Add 2 to both sides.

$z = 2$ Divide both sides by -3 to isolate z.

The solution is $(3, -1, 2)$. We will leave the check to the reader.

Following is a summary of the process of solving a system of three linear equations.

Procedure **Solving Systems of Three Linear Equations Using Elimination**

To solve a system of three linear equations with three unknowns using elimination:

1. Write each equation in the form $Ax + By + Cz = D$.
2. Eliminate one variable from one pair of equations using the elimination method.
3. If necessary, eliminate the same variable from another pair of equations.
4. Steps 2 and 3 result in two equations with the same two variables. Solve these equations using the elimination method.
5. To find the third variable, substitute the values of the variables found in step 4 into any of the three original equations that contain the third variable.
6. Check the ordered triple in all three original equations.

Your Turn 2 Solve the system using elimination.

a. $\begin{cases} x + 3y + 2z = 6 \\ 2x - 3y + z = -18 \\ -3x + 2y + z = 12 \end{cases}$
b. $\begin{cases} 2x - 3z = -6 \\ x + 3y = -3 \\ 2y - 3z = -16 \end{cases}$

Let's see what happens when a system of three linear equations has no solution (inconsistent) or has an infinite number of solutions (dependent).

Example 3 Solve the system using elimination.

$$\begin{cases} 2x + y + 2z = -1 & \text{(Equation 1)} \\ -3x + 2y + 3z = -13 & \text{(Equation 2)} \\ 4x + 2y + 4z = 5 & \text{(Equation 3)} \end{cases}$$

Note We mentioned in Example 5b of Section 4.3 that we can identify systems of dependent equations when using the elimination method. A system of three equations is dependent if during the solution process we add two equations and get $0 = 0$. (We saw this with two equations in Example 5(b) of Section 4.3.) For example, if Equation 4 of Example 3 were $-7x - z = -18$, the sum of Equations 4 and 5 would be $0 = 0$.

$\begin{array}{ll} \text{(Eq. 4)} & -7x - z = -18 \\ \text{(Eq. 5)} & \underline{7x + z = 18} \\ & 0 = 0 \end{array}$

In such a case, we say that the equations in the system are dependent and that the system has an infinite number of solutions.

Solution: We need to eliminate a variable from two pairs of equations. Let's eliminate y using Equations 1 and 2.

(Eq. 1) $\quad 2x + y + 2z = -1 \quad$ Multiply by -2. $\quad -4x - 2y - 4z = 2$

(Eq. 2) $-3x + 2y + 3z = -13 \quad\longrightarrow\quad \underline{-3x + 2y + 3z = -13}$ Add the equations.

$\qquad\qquad\qquad\qquad\qquad\qquad -7x - z = -11 \quad$ (Equation 4)

Now we eliminate y from Equations 2 and 3.

(Eq. 2) $-3x + 2y + 3z = -13 \quad$ Multiply by -1. $\quad 3x - 2y - 3z = 13$

(Eq. 3) $\quad 4x + 2y + 4z = 5 \quad\longrightarrow\quad \underline{4x + 2y + 4z = 5}$ Add the equations.

$\qquad\qquad\qquad\qquad\qquad\qquad 7x + z = 18 \quad$ (Equation 5)

Equations 4 and 5 form a system in x and z.

$\begin{array}{ll} \text{(Eq. 4)} & -7x - z = -11 \\ \text{(Eq. 5)} & \underline{7x + z = 18} \quad \text{Add the equations.} \\ & 0 = 7 \end{array}$

All variables are eliminated and the resulting equation is false, which means that this system has no solution; it is inconsistent.

Your Turn 3 Solve the following systems of equations using elimination.

a. $\begin{cases} 2x + y + 2z = 1 \\ x + y + z = 2 \\ 4x + 2y + 4z = 6 \end{cases}$
b. $\begin{cases} x + 3y - 6z = 9 \\ -7y + 6z = -3 \\ -2x + y + 6z = -15 \end{cases}$

Objective 4 Solve application problems that translate to a system of three linear equations.

Instructor Note For systems of three equations in three unknowns, we have chosen to avoid having students determine whether the solution set for a system of dependent equations contains all points along a line of intersection or within a plane described by the three equations. We just say that the equations are dependent and the system has an infinite number of solutions.

We follow the same procedure to solve applications involving three unknowns that we used when solving applications with two unknowns.

Example 4 At a movie theater, John buys one popcorn, one soft drink, and one candy bar, all for $7. Fred buys two popcorns, three soft drinks, and two candy bars for $16. Carla buys one popcorn, two soft drinks, and three candy bars for $12. Find the price of one popcorn, one soft drink, and one candy bar.

Understand We have three unknowns and three relationships, and we are to find the cost of each.

Plan Select a variable for each unknown, translate the relationships to a system of three equations, and solve the system.

Answers to Your Turn 2
a. $(-2, 4, -2)$ b. $(3, -2, 4)$

Answers to Your Turn 3
a. no solution (inconsistent)
b. infinite number of solutions (dependent equations)

Execute Let x represent the cost of one popcorn, y the cost of one soft drink, and z the cost of one candy bar.

Relationship 1: One popcorn, one soft drink, and one candy bar cost $7.

Translation: $x + y + z = 7$

Relationship 2: Two popcorns, three soft drinks, and two candy bars cost $16.

Translation: $2x + 3y + 2z = 16$

Relationship 3: One popcorn, two soft drinks, and three candy bars cost $12.

Translation: $x + 2y + 3z = 12$

$$\text{Our system:} \begin{cases} x + y + z = 7 & (\text{Equation 1}) \\ 2x + 3y + 2z = 16 & (\text{Equation 2}) \\ x + 2y + 3z = 12 & (\text{Equation 3}) \end{cases}$$

We will choose to eliminate z from two pairs of equations. We will start with Equations 1 and 2.

(Eq. 1) $x + y + z = 7$ $\xrightarrow{\text{Multiply by } -2.}$ $-2x - 2y - 2z = -14$
(Eq. 2) $2x + 3y + 2z = 16$ $\underline{2x + 3y + 2z = 16}$ Add the equations.
$y = 2$ (Equation 4)

Note Although we were trying to eliminate z, we ended up eliminating both x and z.

Equation 4 gives us the value of y, indicating that a soft drink costs $2. Now we choose another pair of equations and eliminate z again. We will use Equations 1 and 3.

(Eq. 1) $x + y + z = 7$ $\xrightarrow{\text{Multiply by } -3.}$ $-3x - 3y - 3z = -21$
(Eq. 3) $x + 2y + 3z = 12$ $\underline{x + 2y + 3z = 12}$ Add the equations.
$-2x - y = -9$ (Equation 5)

Because we already know that $y = 2$, we can substitute for y in $-2x - y = -9$.

$-2x - 2 = -9$ Substitute **2** for **y**.

$-2x = -7$ Add 2 to both sides.

$x = 3.5$ Divide both sides by -2 to isolate x.

Because x represents the cost of one popcorn, a popcorn costs $3.50. To find z, substitute for x and y into one of the original equations. We will use Equation 1.

$3.5 + 2 + z = 7$ Substitute 3.5 for x and **2** for **y**.

$5.5 + z = 7$ Simplify.

$z = 1.5$ Subtract 5.5 from both sides to isolate z.

Because z represents the cost of one candy bar, a candy bar costs $1.50.

Answer Popcorn costs $3.50, a soft drink costs $2.00, and a candy bar costs $1.50.

Check Verify that at these prices—the amount of money spent by John, Fred, and Carla—are correct. We will leave the check to the reader.

Answer to Your Turn 4
length: 20 in.
width: 12 in.
height: 14 in.

Your Turn 4 A small aquarium is in the shape of a rectangular solid. The sum of the length, width, and height is 46 inches. The sum of twice the length, three times the width, and the height is 90 inches. The sum of the length, twice the width, and three times the height is 86 inches. Find the dimensions of the aquarium.

4.4 Exercises For Extra Help MyMathLab®

Objectives 1 and 2

Prep Exercise 1 When solving a system of three equations with three variables, you eliminate a variable from one pair of equations, then eliminate the same variable from a second pair of equations. Does it matter which equations are chosen?

It doesn't matter which equations are chosen as long as the second pair is different from the first pair.

Prep Exercise 2 When solving a system of three equations with three variables, we eliminate a variable from one pair of equations, then eliminate the same variable from a second pair of equations. Why do we eliminate the same variable from two pairs of equations?

We eliminate the same variable from two pairs of equations to get two equations with the same two variables.

Prep Exercise 3 Suppose you are given this system of equations:

$$\begin{cases} x + y + z = 0 & \text{(Equation 1)} \\ 2x + 4y + 3z = 5 & \text{(Equation 2)} \\ 4x - 2y + 3z = -13 & \text{(Equation 3)} \end{cases}$$

Which variable would you choose to eliminate? Which pairs of equations would you use? Why?

Answers may vary. A good choice is to eliminate z using Equations 1 and 2 and 1 and 3. Multiplying Equation 1 by -3 and then adding this result to Equations 2 and 3 takes very few steps.

Prep Exercise 4 When using elimination to solve a system of three equations with three variables, how do you know whether it has no solution?

After adding two equations, if the resulting equation has no variables and is false, the system has no solution.

Prep Exercise 5 Two drawings were given in the text of inconsistent systems of equations with three unknowns. Make another drawing of an inconsistent system other than the ones given.

Two parallel planes intersecting a third plane

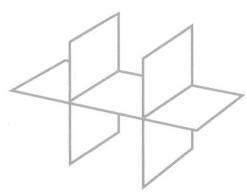

Prep Exercise 6 Where will a point be located if it solves two equations of a system of three equations in three variables but does not solve the third equation?

It will be on the line of intersection of two planes but not on the third plane.

For Exercises 1–6, determine whether the ordered triple is a solution of the system.
See Example 1.

1. $(3, -1, 1)$
$$\begin{cases} x + y + z = 3 \\ 2x - 2y - z = 7 \\ 2x + y - 2z = 3 \end{cases}$$
yes

2. $(2, -2, 1)$
$$\begin{cases} 3x + 2y + z = 3 \\ 2x - 3y - 2z = 8 \\ -2x + 4y + 3z = -9 \end{cases}$$
yes

3. $(1, 0, 2)$
$$\begin{cases} 2x + 3y - 3z = -4 \\ -2x + 4y - z = -4 \\ 3x - 4y + 2z = 5 \end{cases}$$
no

4. $(3, 4, 0)$
$$\begin{cases} x + 2y + 5z = 11 \\ 3x - 2y - 4z = 1 \\ 2x + 2y + 3z = 12 \end{cases}$$
no

5. $(2, -2, 4)$
$$\begin{cases} x + 2y - z = -6 \\ 2x - 3y + 4z = 26 \\ -x + 2y - 3z = -18 \end{cases}$$
yes

6. $(0, -2, 4)$
$$\begin{cases} 3x + 2y - 3z = -16 \\ 2x - 4y - z = 4 \\ -3x + 4y - 2z = -16 \end{cases}$$
yes

Objective 3

For Exercises 7–28, solve the systems of equations. See Examples 2 and 3.

7. $\begin{cases} x + y + z = 5 \\ 2x + y - 2z = -5 \\ x - 2y + z = 8 \end{cases}$
$(2, -1, 4)$

8. $\begin{cases} x + y + z = 2 \\ 3x + y - z = -2 \\ 2x - 2y + 3z = 15 \end{cases}$
$(1, -2, 3)$

9. $\begin{cases} x + y + z = 2 \\ 4x - 3y + 2z = 2 \\ 2x + 3y - 2z = -8 \end{cases}$
$(-1, 0, 3)$

10. $\begin{cases} x + y - z = 7 \\ -2x + 2y - z = -5 \\ 3x - 3y + 2z = 6 \end{cases}$
$(4, 0, -3)$

11. $\begin{cases} 2x + y + 2z = 5 \\ 3x - 2y + 3z = 4 \\ -2x + 3y + z = 8 \end{cases}$
$(-1, 1, 3)$

12. $\begin{cases} 2x + 3y - 2z = -4 \\ 4x - 3y + z = 25 \\ x + 2y - 4z = -12 \end{cases}$
$(4, -2, 3)$

13. $\begin{cases} x + 2y - z = 1 \\ 2x + 4y - 2z = -8 \\ 3x + y - 4z = 6 \end{cases}$
No solution (inconsistent)

14. $\begin{cases} 3x - 2y + z = 5 \\ 4x - 5y + 2z = 7 \\ 9x - 6y + 3z = 7 \end{cases}$
No solution (inconsistent)

15. $\begin{cases} 4x - 2y + 3z = 6 \\ 6x - 3y + 4.5z = 9 \\ 12x - 6y + 9z = 18 \end{cases}$
Infinite number of solutions (dependent equations)

16. $\begin{cases} -8x + 4y + 6z = -18 \\ 2x - y - 1.5z = 4.5 \\ 4x - 2y - 3z = 9 \end{cases}$
Infinite number of solutions (dependent equations)

17. $\begin{cases} x = 2y + z + 7 \\ y = -3x + 2z + 1 \\ 2x + y - z = 0 \end{cases}$
$(2, -3, 1)$

18. $\begin{cases} z = -3x + y - 10 \\ 2x + 3y - 2z = 5 \\ x = 3y - 3z - 14 \end{cases}$
$(-2, 1, -3)$

19. $\begin{cases} x = 4y - z + 1 \\ 3x + 2y - z = -8 \\ x + 6y + 2z = -3 \end{cases}$
$\left(-2, -\frac{1}{2}, 1\right)$

20. $\begin{cases} 2x - 3y - 4z = 3 \\ y = -4x + 8z - 1 \\ x = -5y - 2z - 4 \end{cases}$
$\left(\frac{1}{2}, -1, \frac{1}{4}\right)$

21. $\begin{cases} 4x + 2y + 3z = 9 \\ 2x - 4y - z = 7 \\ 3x - 2z = 4 \end{cases}$
$(2, -1, 1)$

22. $\begin{cases} 4x + 3y - 2z = 19 \\ 2x + 5z = -4 \\ 3x + 2y + 3z = 5 \end{cases}$
$(3, 1, -2)$

23. $\begin{cases} 3x - 2z = -1 \\ 4x + 5y = 23 \\ y + 2z = -1 \end{cases}$
$(-3, 7, -4)$

24. $\begin{cases} 4y + 3z = 2 \\ 3x + 4y = 2 \\ 2x - 5z = -6 \end{cases}$
$(2, -1, 2)$

25. $\begin{cases} 3x + 2y = -2 \\ 2x - 3z = 1 \\ 0.4y - 0.5z = -2.1 \end{cases}$
$(2, -4, 1)$

26. $\begin{cases} 0.2x - 0.3z = -1.8 \\ 3x + 2y = 5 \\ 3x + 2z = -1 \end{cases}$
$(-3, 7, 4)$

27. $\begin{cases} \frac{3}{2}x + y - z = 0 \\ 4y - 3z = -22 \\ -0.2x + 0.3y = -2 \end{cases}$
$(4, -4, 2)$

28. $\begin{cases} -0.2x - 0.3y + 0.1z = -0.3 \\ -3x + 2y = 13 \\ \frac{1}{4}y - \frac{1}{2}z = 2 \end{cases}$
$(-3, 2, -3)$

Objective 4

For Exercises 29–40, translate to a system of three equations; then solve. See Example 4.

29. The sum of the measures of the angles in every triangle is 180°. The measure of one angle of a triangle is three times that of a second angle, and the measure of the second angle is 5° less than the measure of the third. Find the measure of each angle of the triangle.
105°, 35°, 40°

30. The perimeter of a triangle is 33 inches. The sum of the length of the longest side and twice the length of the shortest side is 31 inches. Twice the length of the longest side minus both of the other side lengths is 12 inches. Find the side lengths.
8 in., 10 in., 15 in.

31. At a fast-food restaurant, one burger, one order of fries, and one drink cost $5.00; three burgers, two orders of fries, and two drinks cost $12.50; and two burgers, four orders of fries, and three drinks cost $14.00. Find the individual cost of one burger, one order of fries, and one drink.
Burger: $2.50; fries: $1.50; drink: $1.00

32. James went to the college bookstore and purchased two pens, two erasers, and one pack of paper for $6. Tamika purchased four pens, three erasers, and two packs of paper for $11.50. Jermaine purchased three pens, one eraser, and three packs of paper for $9.50. Find the individual costs of one pen, one eraser, and one pack of paper.
Pen: $2.00; eraser: $0.50; paper: $1.00

33. In a basketball game, John scored 14 points with a combination of 3-point field goals, 2-point field goals, and free throws (1 point each). If he scored a total of 7 times and made two more free throws than 2-point field goals, find the number of each that he made.

Three 3-point field goals, one 2-point field goal, three free throws

34. At a track meet, 10 points are awarded for each first-place finish, 5 points for each second place, and 1 point for each third place. Suppose a track team scored a total of 71 points and had two more first-place finishes than seconds and one less first-place finish than third. Find the number of first-, second-, and third-place finishes for the team.

5 first place, 3 second place, 6 third place

35. A delicatessen sells Black Forest ham for $11.96 per pound, turkey breast for $8.76 per pound, and roast beef for $9.16 per pound. Suppose the deli makes a 10-pound party tray of these three meats such that the average cost is $9.80 per pound. Find the number of pounds of each meat if the number of pounds of turkey breast is equal to the sum of the number of pounds of Black Forest ham and roast beef.

Ham: 3 lb.; turkey: 5 lb.; beef: 2 lb.

36. A coffee shop sells Jamaican Blue Mountain coffee for $45.99 per pound, Hawaiian Kona for $36.99 per pound, and Sulawesi Kalossi for $12.99 per pound. A 25-pound mixture of these three coffees sells for $30.27 per pound. Find the number of pounds of each coffee if the sum of the number of pounds of Jamaica Blue Mountain and Hawaiian Kona is 5 pounds more than the number of pounds of Sulawesi Kalossi.

Jamaican: 8 lb.; Hawaiian: 7 lb.; Sulawesi: 10 lb.

37. A total of $8000 is invested in three stocks, which paid 4%, 6%, and 7% dividends in one year. The amount invested in the stock that returned 7% dividends is $1500 more than the amount that returned 4% dividends. If the total dividends in one year were $475, find the amount invested in each stock.

$2000 at 4%, $2500 at 6%, $3500 at 7%

38. A total of $5000 is invested in three funds. The money market fund pays 5% annually, the income fund pays 6% annually, and the growth fund pays 8% annually. The total earnings for one year from the three funds are $340. If the amount invested in the growth fund is $500 less than twice the amount invested in the money market fund, find the amount in each fund.

MM: $1666.67; IF: $500; GF: $2833.33

39. The number of calories burned per hour bicycling is 120 more than the number burned from brisk walking. The number of calories burned per hour in climbing stairs is 180 more than from bicycling and is twice that from brisk walking. Find the number of calories burned per hour for each of the three activities. (*Source: Numbers: How Many, How Long, How Far, How Much.*)

Bicycling: 420; walking: 300; climbing stairs: 600

40. The intensity of sound is measured in decibels. The decibel reading of a rock concert is 60 less than three times the reading of normal conversation and 10 less than that of a jet at takeoff. The decibel reading of a jet takeoff is 10 more than twice that of normal conversation. Find the decibel reading of each. (*Source: Numbers: How Many, How Long, How Far, How Much.*)

Concert: 120 dB; conversation: 60 dB; jet: 130 dB

Review Exercises

Exercises 1 ▲▲▲, **Expressions**

[1.7] 1. Use the distributive property to rewrite
$-3(2x - 4y - z + 9)$.
$-6x + 12y + 3z - 27$

Exercises 2–6 ▲▲▲, **Equations and Inequalities**

[2.3] 2. Given the equation $x - 0.5y = 8$, let $x = 6$; then solve for y.
$y = -4$

[3.7] *For Exercises 3–6, use the function* $f(x) = -2x + 1$.

3. Find $f(1)$.
-1

4. Find the y-intercept.
$(0, 1)$

5. Find the slope.
$m = -2$

6. Graph the function.

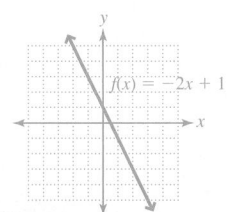

4.5 Solving Systems of Linear Equations Using Matrices

Objectives

1 Write a system of equations as an augmented matrix.

2 Solve a system of linear equations by transforming its augmented matrix to row echelon form.

3 Solve applications problems using matrices.

Note One way to remember that columns are vertical is to imagine the columns on a building, which are vertical. ▶

Warm-up

[2.2] 1. Given $x + 3y = \dfrac{1}{2}$ and $y = \dfrac{1}{2}$, find x.

[2.2] 2. Given $x + y + 2z = 7$ and $y = -1$ and $z = 3$, find x.

Objective 1 Write a system of equations as an augmented matrix.

Although the elimination method that we learned in Section 4.3 is effective for solving systems of linear equations, we can streamline the method by manipulating just the coefficients and constants in a **matrix**.

Definition **Matrix:** A rectangular array of numbers.

Following are some examples of matrices (plural of *matrix*).

$$\begin{bmatrix} 1 & -2 \\ -3 & 4 \end{bmatrix} \begin{bmatrix} 1 & -5 & 6 \\ -2 & 4 & 0 \end{bmatrix} \begin{bmatrix} -3 & 0 & 9 \\ -2 & 4 & 7 \\ 9 & -2 & 0 \end{bmatrix}$$

Matrices are made up of horizontal rows and vertical columns.

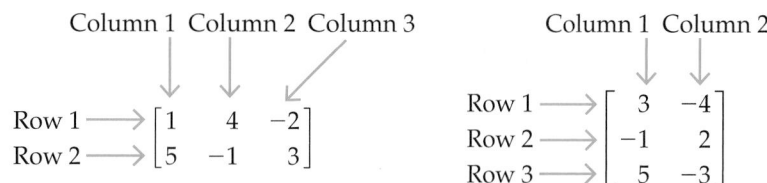

The number of rows followed by the number of columns gives the *dimensions* of the matrix. So $\begin{bmatrix} 2 & 5 & -1 \\ 5 & 4 & 3 \end{bmatrix}$ is a 2 × 3 (read "2 by 3") matrix because it has two rows and three columns; $\begin{bmatrix} 2 & 5 \\ -4 & 6 \\ 3 & -2 \end{bmatrix}$ is a 3 × 2 matrix because it has three rows and two columns.

Each number in a matrix is called an *element*. To solve a system of equations using matrices, we first rewrite the system as an **augmented matrix**.

Definition **Augmented matrix:** A matrix made up of the coefficients and the constant terms of a system. The constant terms are separated from the coefficients by a dashed vertical line.

For example, to write the system $\begin{cases} 3x - 2y = 7 \\ 4x - 3y = 10 \end{cases}$ as an augmented matrix, we omit the variables and write $\begin{bmatrix} 3 & -2 & 7 \\ 4 & -3 & 10 \end{bmatrix}$.

Example 1 Write $\begin{cases} 2x + 3y - 4z = 1 \\ 3x - 5y + z = -4 \\ 2x - 6y + 3z = 2 \end{cases}$ as an augmented matrix.

Solution: $\begin{bmatrix} 2 & 3 & -4 & 1 \\ 3 & -5 & 1 & -4 \\ 2 & -6 & 3 & 2 \end{bmatrix}$ ◀ **Note** It is helpful to think of the dashed line as the equal sign of each equation.

Your Turn 1 Write the augmented matrix for $\begin{cases} x + 3y = -2 \\ -4x - 5y = 7 \end{cases}$.

Objective 2 Solve a system of linear equations by transforming its augmented matrix to row echelon form.

In using the elimination method, we can interchange equations, add equations, or multiply equations by a number and then add the rewritten equations. Because each row of an augmented matrix contains the constants and coefficients of each equation in the system, we can interchange rows, add rows, or multiply rows by a number and then add the rewritten rows.

Rule Row Operations

The solution of a system is not affected by the following row operations in its augmented matrix.

1. Any two rows may be interchanged.
2. The elements of any row may be multiplied (or divided) by any nonzero real number.
3. Any row may be replaced by a row resulting from adding the elements of that row (or multiples of that row) to a multiple of the elements of any other row.

We use these row operations to solve a system of linear equations by transforming the augmented matrix of the system into an equivalent matrix that is in **row echelon form**.

Definition   Row echelon form: An augmented matrix whose coefficient portion has 1's on the diagonal from upper left to lower right and 0s below the 1's.

For example, $\begin{bmatrix} 1 & 3 & | & 3 \\ 0 & 1 & | & 4 \end{bmatrix}$ and $\begin{bmatrix} 1 & 2 & -4 & | & 5 \\ 0 & 1 & -5 & | & 7 \\ 0 & 0 & 1 & | & 5 \end{bmatrix}$ are in row echelon form.

In the following examples, we let R_1 mean row 1, R_2 mean row 2, and so on.

Example 2 Solve the following linear systems by transforming their augmented matrices into row echelon form.

a. $\begin{cases} x - 2y = -4 & \text{(Equation 1)} \\ -2x + 5y = 9 & \text{(Equation 2)} \end{cases}$

Solution: First, we write the augmented matrix: $\begin{bmatrix} 1 & -2 & | & -4 \\ -2 & 5 & | & 9 \end{bmatrix}$.

Now we perform row operations to transform the matrix into row echelon form. The element in the first row, first column is already 1, which is what we want. Therefore, we need to rewrite the matrix so that -2 in the second row, first column becomes 0. To do this, we multiply the first row by 2 and add it to the second row.

Note The result of the row operations replaces the affected row in the matrix. Other rows are rewritten unchanged.

$2R_1 + R_2 \longrightarrow \begin{bmatrix} 1 & -2 & | & -4 \\ 0 & 1 & | & 1 \end{bmatrix}$

$2(1 \; -2 \; | \; -4) = (2 \; -4 \; | \; -8)$ and
$(2 \; -4 \; | \; -8) + (-2 \; 5 \; | \; 9) =$
$(0 \; 1 \; | \; 1) \text{ New } R_2 = 2R_1 + R_2$

The resulting matrix represents the system $\begin{cases} x - 2y = -4 \\ y = 1 \end{cases}$.

Answers to Your Turn 1

$\begin{bmatrix} 1 & 3 & | & -2 \\ -4 & -5 & | & 7 \end{bmatrix}$

Instructor Note In Example 2(a), you can continue by replacing R_1 of $\begin{bmatrix} 1 & -2 & | & -4 \\ 0 & 1 & | & 1 \end{bmatrix}$ with $2R_2 + R_1$ and get $\begin{bmatrix} 1 & 0 & | & -2 \\ 0 & 1 & | & 1 \end{bmatrix}$, from which you can read the solutions $x = -2, y = 1$. This form is called reduced row echelon form.

Because $y = 1$, we can solve for x using substitution.

$$x - 2(1) = -4 \quad \text{Substitute 1 for } y \text{ in } x - 2y = -4.$$
$$x - 2 = -4 \quad \text{Simplify.}$$
$$x = -2 \quad \text{Add 2 to both sides.}$$

The solution is $(-2, 1)$. The check is left to the reader.

b. $\begin{cases} 2x + 6y = 1 & \text{(Equation 1)} \\ x + 8y = 3 & \text{(Equation 2)} \end{cases}$

Solution: First, we write the augmented matrix: $\begin{bmatrix} 2 & 6 & | & 1 \\ 1 & 8 & | & 3 \end{bmatrix}$.

Now we perform row operations to get row echelon form.

$$\frac{1}{2}R_1 \longrightarrow \begin{bmatrix} 1 & 3 & | & \frac{1}{2} \\ 1 & 8 & | & 3 \end{bmatrix}$$

We need the 2 in row 1, column 1 to be a 1; so multiply each number in row 1 by $\frac{1}{2}$.

Connection Row operations correspond to equation manipulations in the elimination method. For example, $-R_1 + R_2$ corresponds to multiplying $x + 3y = \frac{1}{2}$ by -1 and then adding the resulting equation to $x + 8y = 3$ so that we get $5y = \frac{5}{2}$.

$$-R_1 + R_2 \longrightarrow \begin{bmatrix} 1 & 3 & | & \frac{1}{2} \\ 0 & 5 & | & \frac{5}{2} \end{bmatrix}$$

We need the 1 in row 2, column 1 to be 0; so multiply row 1 by -1 and add it to row 2.

$$\frac{1}{5}R_2 \longrightarrow \begin{bmatrix} 1 & 3 & | & \frac{1}{2} \\ 0 & 1 & | & \frac{1}{2} \end{bmatrix}$$

We need the 5 in row 2, column 2 to be 1; so multiply row 2 by $\frac{1}{5}$.

This matrix represents the system $\begin{cases} x + 3y = \dfrac{1}{2} \\ \quad\quad y = \dfrac{1}{2} \end{cases}$.

Instructor Note Illustrate the connection between the row operations and equation manipulations by showing the corresponding equations side by side with the matrices.

Because $y = \frac{1}{2}$, we can solve for x using substitution.

$$x + 3\left(\frac{1}{2}\right) = \frac{1}{2} \quad \text{Substitute } \frac{1}{2} \text{ for } y \text{ in } x + 3y = \frac{1}{2}.$$
$$x + \frac{3}{2} = \frac{1}{2} \quad \text{Simplify.}$$
$$x = -1 \quad \text{Subtract } \frac{3}{2} \text{ from both sides.}$$

The solution is $\left(-1, \dfrac{1}{2}\right)$. The check is left to the reader.

Instructor Note Again, you can continue by replacing R_1 with $-3R_2 + R_1$, which gives $\begin{bmatrix} 1 & 0 & | & -1 \\ 0 & 1 & | & \frac{1}{2} \end{bmatrix}$, from which we get $x = -1, y = \frac{1}{2}$.

Your Turn 2 Solve the following system by transforming its augmented matrix into row echelon form.

$$\begin{cases} 3x + 4y = 6 \\ x - 3y = -11 \end{cases}$$

Now let's use row operations to solve systems of three equations.

Example 3 Use the row echelon method to solve the following system.

$$\begin{cases} 2x + y - 2z = -3 & \text{(Equation 1)} \\ x + y + 2z = 7 & \text{(Equation 2)} \\ 4y + 3z = 5 & \text{(Equation 3)} \end{cases}$$

Answer to Your Turn 2
$(-2, 3)$

Note When writing a matrix in row echelon form, if you get a row that is all zeros to the left of the dashed line and nonzero to the right, the system is inconsistent (no solution). If you get a row that is all zeros, the system is dependent (infinite solutions).

Solution: Write the augmented matrix: $\begin{bmatrix} 2 & 1 & -2 & \vdots & -3 \\ 1 & 1 & 2 & \vdots & 7 \\ 0 & 4 & 3 & \vdots & 5 \end{bmatrix}$.

Now we perform row operations to get row echelon form.

$$\begin{matrix} R_2 \longrightarrow \\ R_1 \longrightarrow \\ {} \end{matrix} \begin{bmatrix} 1 & 1 & 2 & \vdots & 7 \\ 2 & 1 & -2 & \vdots & -3 \\ 0 & 4 & 3 & \vdots & 5 \end{bmatrix}$$

Because we need a 1 in row 1, column 1, we interchange rows 1 and 2.

$$-2R_1 + R_2 \longrightarrow \begin{bmatrix} 1 & 1 & 2 & \vdots & 7 \\ 0 & -1 & -6 & \vdots & -17 \\ 0 & 4 & 3 & \vdots & 5 \end{bmatrix}$$

To get a 0 in row 2, column 1, we multiply the new row 1 by -2 and add it to row 2.

$$4R_2 + R_3 \longrightarrow \begin{bmatrix} 1 & 1 & 2 & \vdots & 7 \\ 0 & -1 & -6 & \vdots & -17 \\ 0 & 0 & -21 & \vdots & -63 \end{bmatrix}$$

Row 3 needs only a 0 in the second column, so we multiply row 2 by 4 and add it to row 3.

$$\begin{matrix} -R_2 \longrightarrow \\ -\dfrac{1}{21}R_3 \longrightarrow \end{matrix} \begin{bmatrix} 1 & 1 & 2 & \vdots & 7 \\ 0 & 1 & 6 & \vdots & 17 \\ 0 & 0 & 1 & \vdots & 3 \end{bmatrix}$$

To get a 1 in row 2, column 2, we multiply row 2 by -1.
To get a 1 in row 3, column 3, we multiply row 3 by $-\dfrac{1}{21}$.

Instructor Note Again, instead of stopping with $\begin{bmatrix} 1 & 1 & 2 & \vdots & 7 \\ 0 & 1 & 6 & \vdots & 17 \\ 0 & 0 & 1 & \vdots & 3 \end{bmatrix}$, you can replace R_2 with $R_2 - 6R_3$,

giving $\begin{bmatrix} 1 & 1 & 2 & \vdots & 7 \\ 0 & 1 & 0 & \vdots & -1 \\ 0 & 0 & 1 & \vdots & 3 \end{bmatrix}$.

Then you replace R_1 with $R_1 - R_2 - 2R_3$, which gives

$\begin{bmatrix} 1 & 0 & 0 & \vdots & 2 \\ 0 & 1 & 0 & \vdots & -1 \\ 0 & 0 & 1 & \vdots & 3 \end{bmatrix}$, from which $x = 2, y = -1$, and $z = 3$ can be read.

The resulting matrix represents the system $\begin{cases} x + y + 2z = 7 \\ y + 6z = 17. \\ z = 3 \end{cases}$

To find y, substitute 3 for z in $y + 6z = 17$.

$$y + 6(3) = 17 \quad \text{Substitute 3 for } z.$$
$$y + 18 = 17 \quad \text{Multiply 6 and 3.}$$
$$y = -1 \quad \text{Subtract 18 from both sides.}$$

To find x, substitute -1 for y and 3 for z in $x + y + 2z = 7$.

$$x + (-1) + 2(3) = 7 \quad \text{Substitute } -1 \text{ for } y \text{ and 3 for } z.$$
$$x + 5 = 7 \quad \text{Simplify the left side of the equation.}$$
$$x = 2 \quad \text{Subtract 5 from both sides of the equation.}$$

The solution is $(2, -1, 3)$. We can check by verifying that the ordered triple satisfies all three of the original equations. This is left to the reader.

Your Turn 3 Use the row echelon method to solve the following system.

$$\begin{cases} 3x + y + z = 2 \\ x + 2y - z = -4 \\ 2x - 2y + 3z = 9 \end{cases}$$

Objective 3 Solve applications problems using matrices.

Answer to Your Turn 3
$(1, -2, 1)$

We now solve an applications problem using matrices we previously solved using the elimination method.

Example 4 Jeff invests $5000 in two accounts. The first account has an annual percentage rate (APR) of 4%. The second account has an APR of 6%. If the total interest earned after one year is $252, what principal was invested in each account?

Solution: This resulted in the system of equations $\begin{cases} x + y = 5000 \\ 0.04x + 0.06y = 252 \end{cases}$, where x represents the amount invested at 4% and y represents the amount invested at 6%. The augmented matrix for this system is as follows:

$$\left[\begin{array}{cc|c} 1 & 1 & 5000 \\ 0.04 & 0.06 & 252 \end{array}\right]$$

To put the matrix in row-echelon form, we need a 0 in the second row, first column where 0.04 is presently located. So multiply the first row by -0.04, add it to the second row, and replace the second row with the result.

$$-0.04R_1 + R_2 \longrightarrow \left[\begin{array}{cc|c} 1 & 1 & 5000 \\ 0 & 0.02 & 52 \end{array}\right]$$

$-0.04(1\ \ 1\ \vdots\ 5000) = (-0.04\ -0.04\ \vdots\ -200)$ and $(-0.04\ -0.04\ \vdots\ 200) + (0.04\ 0.06\ \vdots\ 252) = (0\ \ 0.02\ \vdots\ 52)$

Dividing R_2 by 0.02 gives

$$R_2 \div 0.02 \longrightarrow \left[\begin{array}{cc|c} 1 & 1 & 5000 \\ 0 & 1 & 2600 \end{array}\right]$$, which gives us the system $\begin{cases} x + y = 5000 \\ y = 2600 \end{cases}$.

Because $y = 2600$, we can solve for x using substitution.

$$x + y = 5000$$
$$x + 2600 = 5000 \quad \text{Substitute 2600 for } y.$$
$$x = 2400 \quad \text{Subtract 2600 from both sides.}$$

So Jeff invested $2400 at 4% and $2600 at 6%.

Your Turn 4 In winning the 2013 NCAA basketball championship, Louisville scored a total of 46 times in its 82 to 76 victory over Michigan. The sum of the number of 3-point field goals and 2-point field goals was ten more than the number of free throws (1 point each). How many of each did Louisville score?

Answer to Your Turn 4
3-point $= 8$, 2-point $= 20$,
free throws $= 18$

4.5 Exercises (For Extra Help) MyMathLab®

Note: Exercises marked with a ★ represent challenging exercises.

Objective 1

Prep Exercise 1 How many rows and columns are in a 4×2 matrix?

4 rows, 2 columns

Prep Exercise 2 How do you write a system of equations as an augmented matrix?

Arrange each equation so that all variables are in the same order on the left side of the equal sign and the constant term is on the right side of the equal sign. Omit all of the variables and the equal signs. Place a vertical dashed line where the equal signs were so that it is between the coefficients of the variables and the constants.

Prep Exercise 3 To what does the dashed line in an augmented matrix correspond in a system of equations?

The dashed line corresponds to the equal signs in the equations.

Prep Exercise 4 Explain how solving a system of equations using row operations is similar to solving the system using elimination.

The rules are the same for both processes. With matrices, we use only the coefficients and the constants. With the elimination method, we also use the variables and the equal sign.

Prep Exercise 5 What is row echelon form?

A matrix is in row echelon form when the coefficient portion of the augmented matrix has 1s on the diagonal from upper left to lower right and 0's below the 1's.

Prep Exercise 6 Once an augmented matrix is in row echelon form, how do you find the values of the variables?

The bottom equation represents the value of the last variable. Substitute the known value into the equation above to determine the value of the next variable. Continue until all values for the variables are found.

For Exercises 1–8, write the augmented matrix for the system of equations. See Example 1.

1. $\begin{cases} 14x + 7y = 6 \\ 7x + 6y = 8 \end{cases}$

$$\begin{bmatrix} 14 & 7 & | & 6 \\ 7 & 6 & | & 8 \end{bmatrix}$$

2. $\begin{cases} 5x + 6y = 2 \\ 10x + 3y = -2 \end{cases}$

$$\begin{bmatrix} 5 & 6 & | & 2 \\ 10 & 3 & | & -2 \end{bmatrix}$$

3. $\begin{cases} 7x - 6y = 1 \\ -2y = 5 \end{cases}$

$$\begin{bmatrix} 7 & -6 & | & 1 \\ 0 & -2 & | & 5 \end{bmatrix}$$

4. $\begin{cases} 3x = 6 \\ 9x + 2y = -2 \end{cases}$

$$\begin{bmatrix} 3 & 0 & | & 6 \\ 9 & 2 & | & -2 \end{bmatrix}$$

5. $\begin{cases} x - 3y + z = 4 \\ 2x - 4y + 2z = -4 \\ 6x - 2y + 5z = -4 \end{cases}$

$$\begin{bmatrix} 1 & -3 & 1 & | & 4 \\ 2 & -4 & 2 & | & -4 \\ 6 & -2 & 5 & | & -4 \end{bmatrix}$$

6. $\begin{cases} 3x + 2y - 3z = 2 \\ 2x - 4y + 5z = -10 \\ 5x - 4y + z = 0 \end{cases}$

$$\begin{bmatrix} 3 & 2 & -3 & | & 2 \\ 2 & -4 & 5 & | & -10 \\ 5 & -4 & 1 & | & 0 \end{bmatrix}$$

7. $\begin{cases} 4x + 6y - 2z = -1 \\ 8x + 3y = -12 \\ -y + 2z = 4 \end{cases}$

$$\begin{bmatrix} 4 & 6 & -2 & | & -1 \\ 8 & 3 & 0 & | & -12 \\ 0 & -1 & 2 & | & 4 \end{bmatrix}$$

8. $\begin{cases} x - 3y = 3 \\ -3x + 4y + 9z = 3 \\ 2x + 7z = -9 \end{cases}$

$$\begin{bmatrix} 1 & -3 & 0 & | & 3 \\ -3 & 4 & 9 & | & 3 \\ 2 & 0 & 7 & | & -9 \end{bmatrix}$$

Objective 2

For Exercises 9–12, given the matrices in row echelon form, find the solution for the system. See Objective 2 and Example 2.

9. $\begin{bmatrix} 1 & -3 & | & -2 \\ 0 & 1 & | & 2 \end{bmatrix}$

$(4, 2)$

10. $\begin{bmatrix} 1 & -3 & | & 7 \\ 0 & 1 & | & -5 \end{bmatrix}$

$(-8, -5)$

11. $\begin{bmatrix} 1 & -4 & -8 & | & 6 \\ 0 & 1 & -2 & | & -7 \\ 0 & 0 & 1 & | & 1 \end{bmatrix}$

$(-6, -5, 1)$

12. $\begin{bmatrix} 1 & -2 & 6 & | & 16 \\ 0 & 1 & -4 & | & -13 \\ 0 & 0 & 1 & | & 3 \end{bmatrix}$

$(-4, -1, 3)$

For Exercises 13–18, complete the indicated row operation. See Objective 2 and Example 2.

13. Replace R_2 in $\begin{bmatrix} 1 & 3 & | & -1 \\ -2 & 5 & | & 6 \end{bmatrix}$ with $2R_1 + R_2$.

$$\begin{bmatrix} 1 & 3 & | & -1 \\ 0 & 11 & | & 4 \end{bmatrix}$$

14. Replace R_2 in $\begin{bmatrix} 1 & 2 & | & -2 \\ 3 & 8 & | & -4 \end{bmatrix}$ with $-3R_1 + R_2$.

$$\begin{bmatrix} 1 & 2 & | & -2 \\ 0 & 2 & | & 2 \end{bmatrix}$$

15. Replace R_3 in $\begin{bmatrix} 1 & -2 & 4 & | & 6 \\ 0 & 2 & -1 & | & -5 \\ 0 & 8 & -6 & | & -3 \end{bmatrix}$ with $-4R_2 + R_3$.

$$\begin{bmatrix} 1 & -2 & 4 & | & 6 \\ 0 & 2 & -1 & | & -5 \\ 0 & 0 & -2 & | & 17 \end{bmatrix}$$

16. Replace R_2 in $\begin{bmatrix} 1 & 5 & -3 & | & 8 \\ -3 & 2 & -4 & | & -6 \\ 0 & 1 & -2 & | & 9 \end{bmatrix}$ with $3R_1 + R_2$.

$$\begin{bmatrix} 1 & 5 & -3 & | & 8 \\ 0 & 17 & -13 & | & 18 \\ 0 & 1 & -2 & | & 9 \end{bmatrix}$$

17. Replace R_1 in $\begin{bmatrix} 4 & 8 & | & -10 \\ -1 & 3 & | & 2 \end{bmatrix}$ with $\frac{1}{4}R_1$.

$$\begin{bmatrix} 1 & 2 & | & -2.5 \\ -1 & 3 & | & 2 \end{bmatrix}$$

18. Replace R_3 in $\begin{bmatrix} 1 & 3 & 4 & | & 8 \\ 0 & 1 & -2 & | & -6 \\ 0 & 0 & -6 & | & 24 \end{bmatrix}$ with $-\frac{1}{6}R_3$.

$$\begin{bmatrix} 1 & 3 & 4 & | & 8 \\ 0 & 1 & -2 & | & -6 \\ 0 & 0 & 1 & | & -4 \end{bmatrix}$$

For Exercises 19–22, describe the next row operation that should be performed to make the matrix closer to row echelon form.

19. $\begin{bmatrix} 1 & 2 & \vdots & -1 \\ -3 & 4 & \vdots & 5 \end{bmatrix}$

Replace R_2 with $3R_1 + R_2$.

20. $\begin{bmatrix} 1 & -3 & \vdots & 2 \\ 5 & 6 & \vdots & -4 \end{bmatrix}$

Replace R_2 with $-5R_1 + R_2$.

21. $\begin{bmatrix} 1 & -4 & 3 & \vdots & 6 \\ 0 & 1 & -2 & \vdots & 4 \\ 0 & 2 & -5 & \vdots & 1 \end{bmatrix}$

Replace R_3 with $-2R_2 + R_3$.

22. $\begin{bmatrix} 1 & -2 & 4 & \vdots & 7 \\ 0 & 1 & -2 & \vdots & 3 \\ 0 & -5 & 4 & \vdots & -9 \end{bmatrix}$

Replace R_3 with $5R_2 + R_3$.

For Exercises 23–46, solve by transforming the augmented matrix into row echelon form. See Examples 2 and 3.

23. $\begin{cases} x - y = 5 \\ x + y = -1 \end{cases}$

$(2, -3)$

24. $\begin{cases} x + y = 2 \\ x - y = -8 \end{cases}$

$(-3, 5)$

25. $\begin{cases} x + y = 3 \\ 3x - y = 1 \end{cases}$

$(1, 2)$

26. $\begin{cases} x + 3y = -9 \\ -x + 2y = -11 \end{cases}$

$(3, -4)$

27. $\begin{cases} x + 2y = -7 \\ 2x - 4y = 2 \end{cases}$

$(-3, -2)$

28. $\begin{cases} x - 3y = -13 \\ 3x + 2y = 5 \end{cases}$

$(-1, 4)$

29. $\begin{cases} -2x + 5y = 4 \\ x - 2y = -2 \end{cases}$

$(-2, 0)$

30. $\begin{cases} 2x + y = 12 \\ x - 3y = 6 \end{cases}$

$(6, 0)$

31. $\begin{cases} 4x - 3y = -2 \\ 2x - 3y = -10 \end{cases}$

$(4, 6)$

32. $\begin{cases} 6x - 3y = 0 \\ 2x + 4y = 0 \end{cases}$

$(0, 0)$

33. $\begin{cases} 5x + 2y = 12 \\ 2x + 3y = -4 \end{cases}$

$(4, -4)$

34. $\begin{cases} 4x + 3y = 14 \\ 3x - 2y = 2 \end{cases}$

$(2, 2)$

35. $\begin{cases} x + y + z = 3 \\ 2x + y - 3z = -10 \\ 2x + 2y + z = 3 \end{cases}$

$(-1, 1, 3)$

36. $\begin{cases} x + 2y + z = 2 \\ 3x + y - z = 3 \\ 2x + y + 2z = 7 \end{cases}$

$(2, -1, 2)$

37. $\begin{cases} x - y + z = -1 \\ 2x - 2y + z = 0 \\ x + 3y + 2z = 1 \end{cases}$

$(2, 1, -2)$

38. $\begin{cases} x + 3y - 2z = 4 \\ 2x - 3y + 2z = -7 \\ 3x + 2y - 2z = -1 \end{cases}$

$(-1, 3, 2)$

39. $\begin{cases} 2x - y + z = 8 \\ x - 2y + 3z = 11 \\ 2x + 3y - z = -6 \end{cases}$

$(2, -3, 1)$

40. $\begin{cases} 2x + 3y - 2z = -21 \\ 2x - 4y + 3z = 15 \\ 3x + 2y - 3z = -24 \end{cases}$

$(-3, -3, 3)$

41. $\begin{cases} 3x + 2y - 3z = 1 \\ -2x + 3y - 4z = 7 \\ 5x - 2y + z = -5 \end{cases}$

$(-1, -1, -2)$

42. $\begin{cases} 2x - 2y + z = 6 \\ -2x + 4y + 3z = 4 \\ 4x - 3y - 2z = -7 \end{cases}$

$(-2, -3, 4)$

43. $\begin{cases} 3x - 6y + z = -10 \\ 7y - z = 2 \\ 2x + 4z = 14 \end{cases}$

$(-3, 1, 5)$

44. $\begin{cases} 6x - y + 5z = -28 \\ 4x + 2y = 10 \\ 5x + 6z = -23 \end{cases}$

$(-1, 7, -3)$

45. $\begin{cases} 2x + 5y - z = 10 \\ x - y - z = 14 \\ x - 6y = 20 \end{cases}$

$(8, -2, -4)$

46. $\begin{cases} 2x + 4y - 5z = 16 \\ x - 2y - 11z = 1 \\ 4x + 5y = -6 \end{cases}$

$(-9, 6, -2)$

For Exercises 47–54, solve using a matrix on a graphing calculator.

47. $\begin{cases} 2x + 5y = -14 \\ 6x + 7y = -10 \end{cases}$

$(3, -4)$

48. $\begin{cases} 4x - 3y = 11 \\ 5x + 4y = 68 \end{cases}$

$(8, 7)$

49. $\begin{cases} 4x - 7y = -80 \\ 9x + 5y = -14 \end{cases}$

$(-6, 8)$

50. $\begin{cases} 7x - 3y = 23 \\ 4x - 9y = -67 \end{cases}$

$(8, 11)$

51. $\begin{cases} 3x + 2y - 5z = -19 \\ 4x - 7y + 6z = 67 \\ 5x - 6y - 4z = 18 \end{cases}$

$(4, -3, 5)$

52. $\begin{cases} 6x - 7y + 2z = -52 \\ 8x + 5y - 6z = 132 \\ 12x + 4y - 7z = 150 \end{cases}$

$(4, 8, -10)$

53. $\begin{cases} 8x - 7y + 2z = 108 \\ 6x + 11y - 10z = -74 \\ -2y + 13z = 77 \end{cases}$

$(7, -6, 5)$

54. $\begin{cases} 8x - 11y + 14z = 40 \\ 5x + 15y - 5z = 35 \\ 17x - 9y = -157 \end{cases}$

$(-5, 8, 12)$

Find ⊗ the Mistake   *For Exercises 55 and 56, explain the mistake; then find the correct solution.*

55. $\begin{cases} x + 3y = 13 \\ -4x - y = -26 \end{cases}$

$$\begin{bmatrix} 1 & 3 & | & 13 \\ -4 & -1 & | & -26 \end{bmatrix}$$

$4R_1 + R_2 \longrightarrow \begin{bmatrix} 1 & 3 & | & 13 \\ 0 & 11 & | & -13 \end{bmatrix}$

$\dfrac{1}{11}R_2 \longrightarrow \begin{bmatrix} 1 & 3 & | & 13 \\ 0 & 1 & | & -\dfrac{13}{11} \end{bmatrix}$

Solution: $\left(\dfrac{182}{11}, -\dfrac{13}{11} \right)$

Mistake: In the second step, $4R_1 + R_2$ is calculated incorrectly.

Correct: The correct calculation is $\begin{bmatrix} 1 & 3 & | & 13 \\ 0 & 11 & | & 26 \end{bmatrix}$.

The solution is $(65/11, 26/11)$.

56. $\begin{cases} x - y + z = 8 \\ 3x - z = -9 \\ 4y + z = -6 \end{cases}$

$$\begin{bmatrix} 1 & -1 & 1 & | & 8 \\ 3 & 0 & -1 & | & -9 \\ 4 & 0 & 1 & | & -6 \end{bmatrix}$$

$-3R_1 + R_2 \longrightarrow \begin{bmatrix} 1 & -1 & 1 & | & 8 \\ 0 & 3 & -4 & | & -33 \\ 4 & 0 & 1 & | & -6 \end{bmatrix}$

$\begin{matrix} \dfrac{1}{3}R_2 \longrightarrow \\ -4R_1 + R_3 \longrightarrow \end{matrix} \begin{bmatrix} 1 & -1 & 1 & | & 8 \\ 0 & 1 & -\dfrac{4}{3} & | & -11 \\ 0 & 4 & -3 & | & -38 \end{bmatrix}$

$-4R_2 + R_3 \longrightarrow \begin{bmatrix} 1 & -1 & 1 & | & 8 \\ 0 & 1 & -\dfrac{4}{3} & | & -11 \\ 0 & 0 & \dfrac{7}{3} & | & 6 \end{bmatrix}$

$\dfrac{3}{7}R_3 \longrightarrow \begin{bmatrix} 1 & -1 & 1 & | & 8 \\ 0 & 1 & -\dfrac{4}{3} & | & -11 \\ 0 & 0 & 1 & | & \dfrac{18}{7} \end{bmatrix}$

Solution: $\left(-\dfrac{15}{7}, -\dfrac{53}{7}, \dfrac{18}{7} \right)$

Mistake: In the matrix being created, the coefficient 4 was placed in the x-position instead of the y-position.

Correct: The original matrix should be $\begin{bmatrix} 1 & -1 & 1 & | & 8 \\ 3 & 0 & -1 & | & -9 \\ 0 & 4 & 1 & | & -6 \end{bmatrix}$.

The solution is $(-1, -3, 6)$.

Objective 3

For Exercises 57–68, translate the problem to a system of equations; then solve using matrices.

57. Brad purchased three grilled chicken sandwiches and two drinks for $12.90, and Angel purchased seven grilled chicken sandwiches and four drinks for $29.30. Find the price of one grilled chicken sandwich and one drink.

Chicken sandwich: $3.50; drink: $1.20

58. Sharika purchased three general admission tickets and two student tickets to a college play for $55, and Yo Chen purchased two general admission tickets and four student tickets for $50. Find the cost of one general admission and one student ticket.

General admission: $15; student: $5

59. The two longest rivers in the world, the Nile and the Amazon, have a combined length of 8050 miles. The Nile is 250 miles longer than the Amazon. Find the length of both rivers.

Nile: 4150 mi.; Amazon: 3900 mi.

Of Interest

During the new and full moon, a small tidal wave, called a tidal bore, sweeps up the Amazon. The tidal bore can travel 450 miles upstream at speeds in excess of 40 miles per hour, causing waves of 16 feet or more along the riverbank.

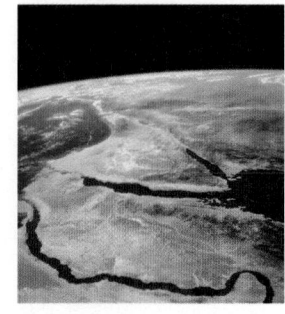

60. The longest vehicular tunnel in the world is the Saint Gotthard Tunnel in Switzerland, and the second longest is the Arlberg Tunnel in Austria. The total length of the two tunnels is 18.8 miles. The Saint Gotthard Tunnel is 1.4 miles longer than the Arlberg. Find the length of each tunnel. (*Source: Webster's New World Book of Facts.*)
Saint Gotthard: 10.1 mi.; Arlborg: 8.7 mi.

61. Nikita invested a total of $10,000 in certificates of deposit that pay 5% annually and in a money market account that pays 6% annually. If the total interest earned in one year from the two investments is $536, find the principal that was invested in each.
$6400 in CD, $3600 in money market

62. An athlete received a $60,000 signing bonus, which he invested in two funds—a money market fund paying 5% annually and a growth fund paying 7% annually. The total annual interest received from the two investments is $3440. Find the principal invested in each fund.
$38,000 in money market, $22,000 in growth fund

63. John bought two CDs, four books, and three DVDs for $164, and Tanelle bought five CDs, two books, and two DVDs for $160. If the sum of the cost of one CD and one book equals the cost of one DVD, find the cost of one of each.
CD: $16; book: $12, DVD: $28

64. Angel bought 3 pounds of salmon, 2 pounds of tuna, and 5 pounds of cod for $63. Sara bought 6 pounds of salmon, 3 pounds of tuna, and 5 pounds of cod for $94. The cost of 1 pound of salmon and 1 pound of tuna is the same as the cost of 3 pounds of cod. Find the cost of 1 pound of each.
Salmon: $8; tuna: $7; cod: $5

65. In Green Bay's 31 to 25 victory over Pittsburgh in Super Bowl XLV, Green Bay scored touchdowns (6 points), extra points (1 point), and field goals (3 points). The number of touchdowns equaled the number of extra points. Also, the number of touchdowns was three more than the number of field goals. Find how many of each type of score Green Bay had. (*Source:* www.nfl.com.)
4 touchdowns, 4 extra points, 1 field goal

66. In the 2012 NCAA women's basketball championship game, Baylor scored forty-seven times for a total of 80 points using a combination of 3-point field goals, 2-point field goals, and free throws (1 point). If the sum of the number of 3-point and 2-point field goals was eleven more than the number of free throws, find the number of each type.
3-point field goals: 4; 2-point field goals: 25; free throws: 18

67. An electrical circuit has three points of connection with different voltage measurements at each of the three connections. An engineer has written the following equations to describe the voltages. Find each voltage.

$$4v_1 - v_2 = 30$$
$$-2v_1 + 5v_2 - v_3 = 10$$
$$-v_2 + 5v_3 = 4$$

$v_1 = 9, v_2 = 6, v_3 = 2$

68. An engineer has written the following system of equations to describe the forces in pounds acting on a steel structure. Find the forces.

$$6F_1 - F_2 = 350$$
$$9F_1 - 2F_2 - F_1 = -100$$
$$3F_2 - F_3 = 250$$

$F_1 = 100, F_2 = 250, F_3 = 500$

Review Exercises

Exercises 1–6 ◢ Expressions

[1.5] *For Exercises 1–6, simplify.*

1. $(-3)(4) - (-4)(5)$
8

2. $2(2 - 8) + 3[-1 - (-6)] - 4(4 - 6)$
11

3. $\dfrac{(-5)(-1) - (9)(3)}{(2)(-1) - (3)(3)}$
2

4. $\dfrac{2\left(\dfrac{5}{4}\right) - 8\left(\dfrac{1}{4}\right)}{\left(\dfrac{1}{2}\right)\left(\dfrac{5}{4}\right) - \left(\dfrac{3}{2}\right)\left(\dfrac{1}{4}\right)}$
2

5. $\dfrac{(1.1)(-0.2) - (1.7)(-0.5)}{(0.2)(-0.2) - (0.5)(-0.5)}$
3

6. $\dfrac{1[-45 - (-9)] + 7(-9 - 2) - 2(-27 - 30)}{1(6 - 3) - 2(-9 - 2) - 2[9 - (-4)]}$
−1

4.6 Solving Systems of Linear Inequalities

Objectives

1 Graph the solution set of a system of linear inequalities.

2 Solve applications involving a system of linear inequalities.

Warm-up

For Exercises 1 and 2, graph.

[3.2] **1.** $x + 2y = 6$

[3.6] **2.** $y < 2x + 1$

Objective 1 Graph the solution set of a system of linear inequalities.

In Section 2.8, we learned to solve linear inequalities in one variable. In Section 3.6, we learned how to graph linear inequalities in two variables. In this section, we develop a graphical approach to solving *systems of linear inequalities* in two variables.

Consider the system of linear inequalities $\begin{cases} x + 2y < 6 \\ 2x - y \geq 2 \end{cases}$. First, let's graph each inequality separately.

Note Recall from Section 3.6 that we use a dashed line with $<$ or $>$ and that points on a dashed line are not in the solution set. ▶

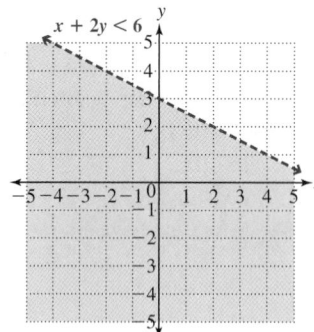

In the graph of $x + 2y < 6$, the shaded region contains all ordered pairs that make $x + 2y < 6$ true.

Note Recall from Section 3.6 that we determine which side of the line to shade by choosing an ordered pair not on the line and checking to see if it makes the inequality true. If it does, we shade on the side of the line containing that ordered pair. If it does not, we shade the other side of the line.

Answers to Warm-up

1.

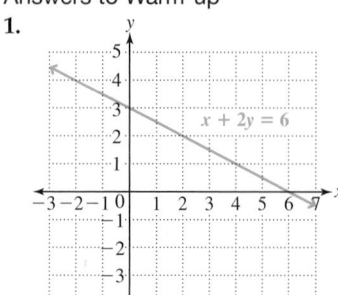

2.

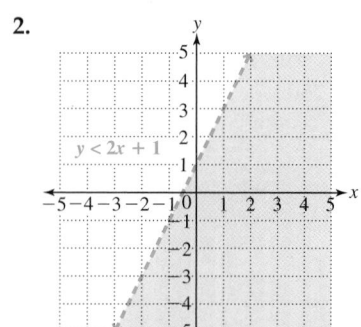

Note Recall from Section 3.6 that we use a solid line with $\leq$ or $\geq$ and that points on a solid line are part of the solution set. ▶

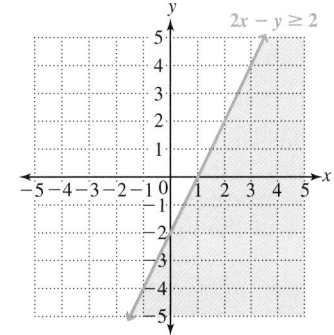

In the graph of $2x - y \geq 2$, ordered pairs in the shaded region and on the line itself make $2x - y \geq 2$ true.

Now we put the two graphs together on the same grid to determine the solution set for the system. A solution for a system of inequalities is an ordered pair that makes every inequality in the system true.

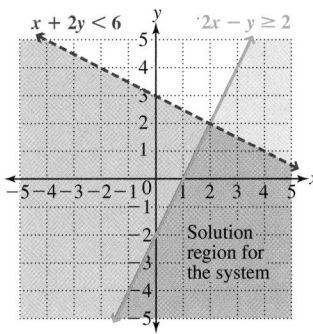

On our graph containing both $x + 2y < 6$ and $2x - y \geq 2$, the region where the shading overlaps contains ordered pairs that make both inequalities true. So all ordered pairs in this region are in the solution set for the system. Also, ordered pairs on the solid line for $2x - y \geq 2$ where it touches the region of overlap are in the solution set for the system, whereas ordered pairs on the dashed line for $x + 2y < 6$ are not. This implies the following procedure.

Procedure **Solving a System of Linear Inequalities in Two Variables**

To solve a system of linear inequalities in two variables, graph all of the inequalities on the same grid. The solution set for the system contains all ordered pairs in the region where the inequalities' solution sets overlap along with ordered pairs on the portion of any solid line that touches the region of overlap.

To check, we can select a point in the solution region such as $(3, 0)$ and verify that it makes both inequalities true.

First inequality:	**Second inequality:**	
$x + 2y < 6$	$2x - y \geq 2$	
$3 + 2(0) \overset{?}{<} 6$	$2(3) - 0 \overset{?}{\geq} 2$	Replace x with 3 and y with 0 in both inequalities.
$3 < 6$ True.	$6 \geq 2$ True.	

Because $(3, 0)$ makes both inequalities true, it is a solution to the system. Although we selected only one ordered pair in the solution region, remember that *every* ordered pair in that region is a solution.

Connection Remember that for a system of two linear equations, a solution is a point of intersection of the graphs of the two equations. Similarly, for a system of linear inequalities, the solution set is a region of intersection of the graphs of the two inequalities.

Instructor Note Technically, when talking about the lines in Example 2, we should be looking at the corresponding equations

$$y = \frac{1}{4}x + 3 \text{ and } y = \frac{1}{4}x - 2.$$

We left them as inequalities to avoid bogging down students with extra discussion. You might mention this technicality in class.

Connection Notice that the lines in Example 2 are parallel. Recall that in a system of linear equations, if the lines are parallel, the system is inconsistent because there is no point of intersection. With linear inequalities, the lines must be parallel *and* the shaded regions must not overlap for the system to be inconsistent.

Answers to Your Turn 1

a.

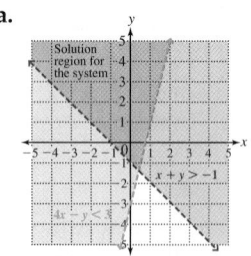

b. Solution region for the system (including the portion of the red line touching this region)

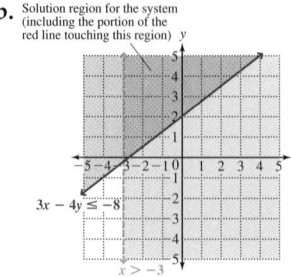

Example 1 Graph the solution set for the system of inequalities.

a. $\begin{cases} x + 3y > 6 \\ y < 2x - 1 \end{cases}$

Solution: Graph the inequalities on the same grid. Because both lines are dashed, the solution set for the system contains only those ordered pairs in the region of overlap (the purple shaded region).

Note Ordered pairs on dashed lines are not part of the solution region.

▶

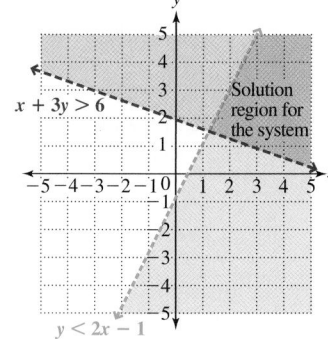

b. $\begin{cases} y < 2 \\ 5x - y \le 4 \end{cases}$

Solution: Graph the inequalities on the same grid. The solution set for this system contains all ordered pairs in the region of overlap (purple shaded region) together with all ordered pairs on the portion of the solid blue line that touches the purple shaded region.

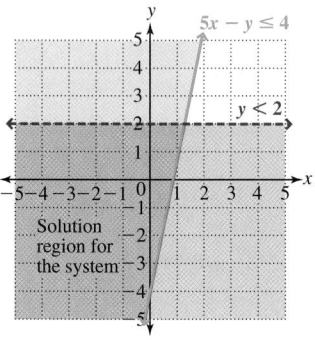

Your Turn 1 Graph the solution set for the system of inequalities.

a. $\begin{cases} x + y > -1 \\ 4x - y < 3 \end{cases}$ **b.** $\begin{cases} x > -3 \\ 3x - 4y \le -8 \end{cases}$

Inconsistent Systems

Some systems of linear inequalities have no solution. We say that these systems are inconsistent.

Example 2 Graph the solution set for the system of inequalities.

$$\begin{cases} x - 4y \ge 8 \\ y \ge \dfrac{1}{4}x + 3 \end{cases}$$

Solution: Graph the inequalities on the same grid.

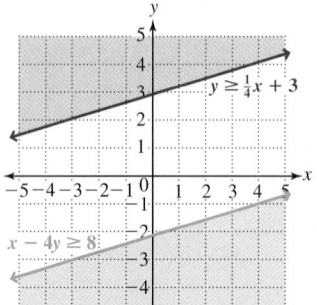

The lines appear to be parallel. We can verify by writing $x - 4y \ge 8$ in slope–intercept form and comparing the slopes and y-intercepts.

$$x - 4y \ge 8$$
$$-4y \ge -x + 8 \quad \text{Subtract } x \text{ from both sides.}$$
$$y \le \frac{1}{4}x - 2 \quad \begin{array}{l}\text{Divide both sides by } -4, \\ \text{which changes the direction} \\ \text{of the inequality symbol.}\end{array}$$

The slopes are equal, and the y-intercepts differ; so the lines are in fact parallel. Because the lines are parallel and the shaded regions do not overlap, there is no solution region for this system. The system is inconsistent.

Objective 2 Solve applications involving a system of linear inequalities.

Example 3 A home interiors store stocks two different-size prints from an artist. The manager wants to purchase at least 15 prints for the store. The artist sells the smaller print for $20 and the larger print for $30. The manager cannot spend more than $800 for the prints. Write a system of inequalities that describes the manager's order; then solve the system by graphing.

Understand We must translate to a system of inequalities, then solve the system.

Plan and Execute Let x represent the number of small prints and y represent the number of large prints ordered.

Relationship 1:   The manager wants to purchase at least 15 prints.

The words *at least* indicate that the combined number of prints is to be greater than or equal to 15; so $x + y \geq 15$.

Relationship 2:   The manager cannot spend more than $800.

Because the small prints cost $20 each, $20x$ describes the amount spent on small prints. Similarly, $30y$ describes the amount spent on large prints. The total cannot exceed $800; so $20x + 30y \leq 800$.

$$\text{Our system: } \begin{cases} x + y \geq 15 \\ 20x + 30y \leq 800 \end{cases}$$

Answer (See the graph.) Because the manager cannot order a negative number of prints, the solution set is confined to quadrant I. Any ordered pair in the solution region or on a portion of either line touching the solution region is a solution for the system. However, assuming that only whole prints can be purchased, only ordered pairs of whole numbers in the solution set, such as $(10, 15)$ and $(25, 10)$, are realistic.

Check We will check one ordered pair in the solution region. The ordered pair $(10, 15)$ indicates an order of 10 small prints and 15 large prints so that 25 prints are ordered (which is more than 15) that cost a total of $20(10) + 30(15) = \$650$ (which is less than $800).

Your Turn 3 In designing a building with a rectangular base, an architect decides that the perimeter cannot exceed 200 feet. Also, the length will be at least 10 feet more than the width. Write a system of inequalities that describes the design requirements; then solve the system graphically.

Answer to Your Turn 3

System: $\begin{cases} 2l + 2w \leq 200 \\ l \geq w + 10 \end{cases}$

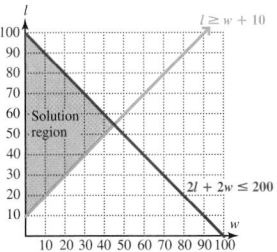

Note Because length and width must be positive numbers, the solution region is confined to quadrant 1.

4.6 Exercises For Extra Help **MyMathLab**®

Note: Exercises marked with a ★ represent challenging exercises.

Objective 1

Prep Exercise 1 Explain how to check a possible solution to a system of inequalities.
Select a point in the solution region and verify that it makes both inequalities true.

Prep Exercise 2 Explain how to determine the region that is the solution set for a system of linear inequalities.
After graphing each inequality, determine the region of overlap.

Prep Exercise 3 What circumstances would cause a system of inequalities to have no solution?
Parallel lines with shaded regions that do not overlap

Prep Exercise 4 Write a system of linear inequalities whose solution set is the entire first quadrant.
$\begin{cases} x > 0 \\ y > 0 \end{cases}$

Prep Exercise 5 Write a system of linear inequalities whose solution set is the entire third quadrant.
$\begin{cases} x < 0 \\ y < 0 \end{cases}$

Prep Exercise 6 Write a system of linear inequalities whose solution set is the entire fourth quadrant.
$\begin{cases} x > 0 \\ y < 0 \end{cases}$

For Exercises 1–24, graph the solution set for the system of inequalities. See Examples 1 and 2.

1. $\begin{cases} x + y > 4 \\ x - y < 6 \end{cases}$

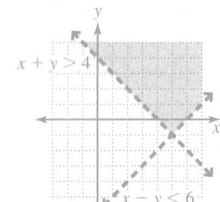

2. $\begin{cases} x - y > 2 \\ x + y > 4 \end{cases}$

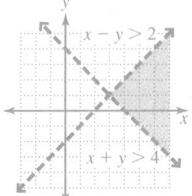

3. $\begin{cases} x + y < -2 \\ x - y > 6 \end{cases}$

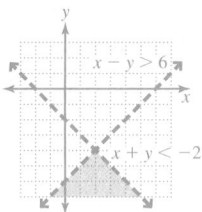

4. $\begin{cases} x - y < -5 \\ x + y < 3 \end{cases}$

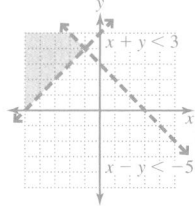

5. $\begin{cases} 2x + y \geq -1 \\ x - y \leq 5 \end{cases}$

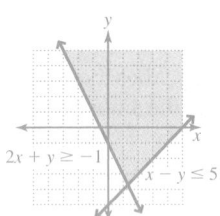

6. $\begin{cases} x - y < -5 \\ 2x - y < -7 \end{cases}$

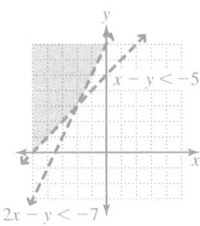

7. $\begin{cases} x + y < 3 \\ x - 2y \geq 2 \end{cases}$

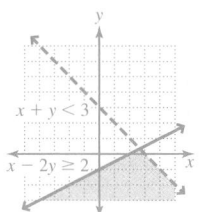

8. $\begin{cases} x + 3y \leq 6 \\ x - y > 5 \end{cases}$

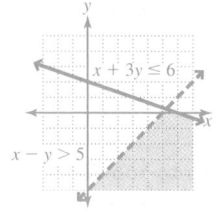

9. $\begin{cases} 2x + y > 9 \\ y > \dfrac{1}{4}x \end{cases}$

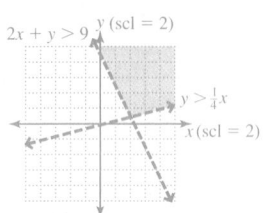

10. $\begin{cases} 3x + 4y \leq -9 \\ y < 3x \end{cases}$

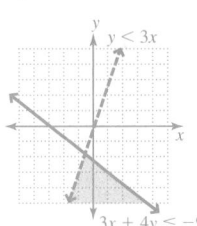

11. $\begin{cases} y < x \\ y > -x + 1 \end{cases}$

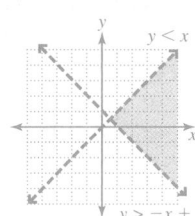

12. $\begin{cases} y < x \\ y < 2x - 3 \end{cases}$

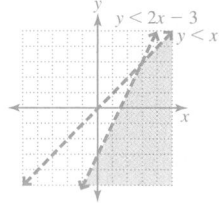

13. $\begin{cases} 3x > -4y \\ 3x + 4y \leq -8 \end{cases}$

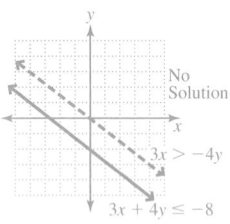

14. $\begin{cases} 2y - x \geq 6 \\ 2x - 4y \geq 5 \end{cases}$

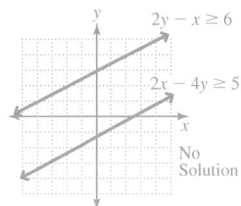

15. $\begin{cases} y < 3x + 1 \\ 3x - y \leq 4 \end{cases}$

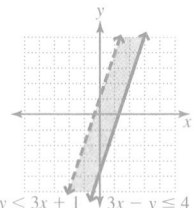

16. $\begin{cases} x - 2y > 3 \\ 2x - 4y \leq 20 \end{cases}$

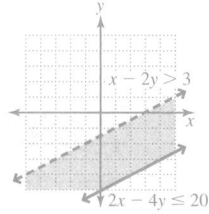

17. $\begin{cases} x > 2 \\ 3x - 2y > 6 \end{cases}$

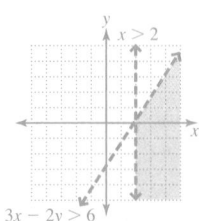

18. $\begin{cases} y \leq -1 \\ x + 2y > 3 \end{cases}$

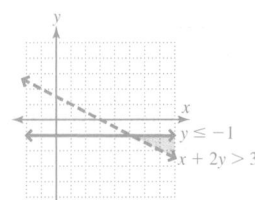

19. $\begin{cases} x \geq -3y \\ y \geq 2x \end{cases}$

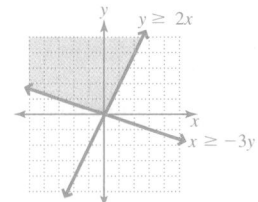

20. $\begin{cases} x > 2y \\ x + y > 6 \end{cases}$

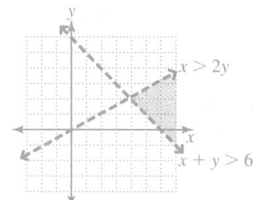

21. $\begin{cases} y > 2 \\ x < -1 \end{cases}$

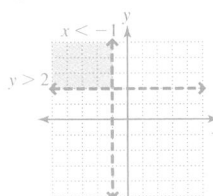

22. $\begin{cases} y \geq -1 \\ x < 2 \end{cases}$

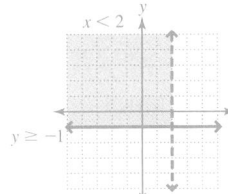

23. $\begin{cases} 2x + y \geq 1 \\ x - y > -1 \\ x > 2 \end{cases}$

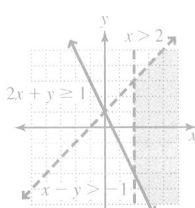

24. $\begin{cases} x + y \geq 1 \\ x - y \geq -5 \\ y > -2 \end{cases}$

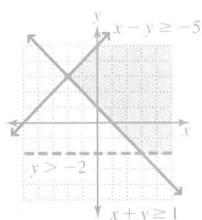

Find 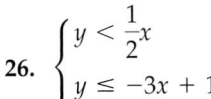 the Mistake   *For Exercises 25–28, explain the mistake.*

25. $\begin{cases} x + y > 3 \\ x - y \leq 2 \end{cases}$

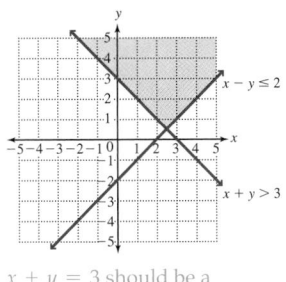

$x + y = 3$ should be a dashed line.

26. $\begin{cases} y < \dfrac{1}{2}x \\ y \leq -3x + 1 \end{cases}$

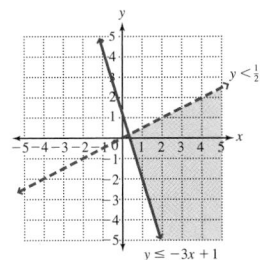

The wrong area is shaded. It should be the region containing the point $(-2, -3)$.

27. $\begin{cases} x < 3 \\ y > 4 \end{cases}$

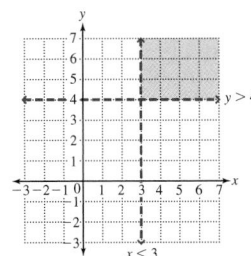

The wrong area is shaded. It should be the region containing the point $(1, 5)$.

28. $\begin{cases} x > 3y \\ x + y \geq 2 \end{cases}$

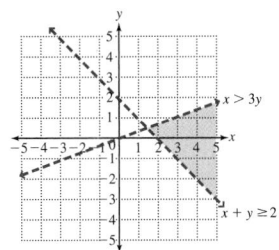

$x + y = 2$ should be a solid line.

Objective 2

Prep Exercise 7 When translating "at least," which inequality symbol is used?

$\geq$

Prep Exercise 8 When translating "at most," which inequality symbol is used?

$\leq$

For Exercises 29–32, solve. see Example 3.

29. To be admitted to a certain college, prospective students must have a combined verbal and math score of at least 1100 on the SAT. To be admitted to the engineering program at this college, students must also have a math score of at least 500.

   a. Write two inequalities describing the requirements to be admitted to the college of engineering.

   $V + M \geq 1100$

   $M \geq 500$

   b. The maximum score on each part of the SAT is 800. Write two inequalities describing these maximum scores.

   $V \leq 800$

   $M \leq 800$

   c. Solve the system of inequalities described by the four inequalities in parts a and b by graphing. Let the horizontal axis be verbal and the vertical axis be math.

   d. Give two different combinations of verbal and math scores that satisfy the admission requirements to the college of engineering.

   Answers will vary. Some possible (V, M) pairs are $(500, 700)$, $(600, 500)$, and $(700, 600)$.

c.

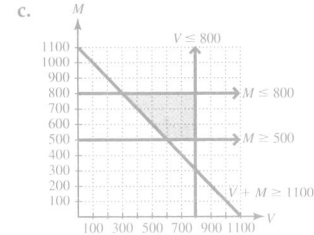

30. An architect is designing a rectangular platform for an auditorium. The client wants the length of the platform to be at least twice the width, and the perimeter must not exceed 200 feet.

   a. Write a system of inequalities that describes the specifications for the platform.
   $$\begin{cases} l \geq 2w \\ 2l + 2w \leq 200 \end{cases}$$

   b. Solve the system by graphing. Let the horizontal axis be the width and the vertical axis be the length.

   c. Give two different combinations of length and width that satisfy the requirements of the client.

   Answers will vary. Some possible (w, l) pairs are $(10, 30)$, $(20, 50)$, and $(30, 70)$.

b.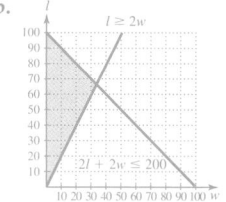

★ 31. A company sells two versions of software: the regular version and the deluxe version. The regular version sells for $15.95, and the deluxe version sells for $20.95. The company gives a bonus to any salesperson who sells at least 100 units and has at least $1700 in total sales.

   a. Write a system of inequalities that describes the requirements a salesperson must meet to receive the bonus.
   $$\begin{cases} R + D \geq 100 \\ 15.95R + 20.95D \geq 1700 \end{cases}$$

   b. Solve the system by graphing. Let the horizontal axis be the regular version and the vertical axis be the deluxe version.

   c. Give two different combinations of number of units sold that meet the requirements for receiving the bonus.

   Answers will vary. Some possible (R, D) pairs are $(30, 90)$, $(40, 80)$, and $(50, 60)$.

b.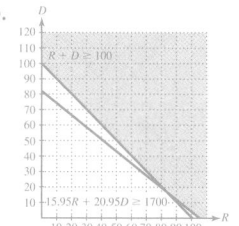

★ 32. An investor is trying to decide how much to invest in two different stocks. She has decided that she will invest at most $5000. The "safe" stock is very stable and is projected to return about 5% over the next year. The other stock is more risky, but it could return about 9% if the company grows as expected. She wants her total dividend at the end of the year to be at least $300.

   a. Write a system of inequalities that describes the amount invested in the two stocks and the return on that investment.
   $$\begin{cases} x + y \leq 5000 \\ 0.05x + 0.09y \geq 300 \end{cases}$$

   b. Solve the system by graphing. Let the horizontal axis be the "safe" stock and the vertical be the "risky" stock.

   c. Give two different combinations of investment amounts that return at least $300.

   Answers will vary. Some possible (x, y) pairs are $(0, 4000)$, $(1000, 3000)$, and $(2000, 2500)$.

b.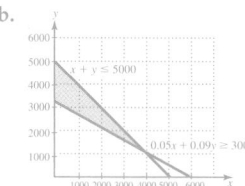

Review Exercises

Exercises 1–4 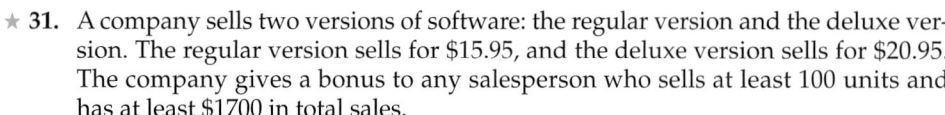 **Expressions**

[1.5] *For Exercises 1–4, evaluate.*

1. 3^2

   9

2. -10^2

   -100

3. $\sqrt{25}$

   5

4. $\sqrt{\dfrac{16}{49}}$

   $\dfrac{4}{7}$

Exercises 5 and 6 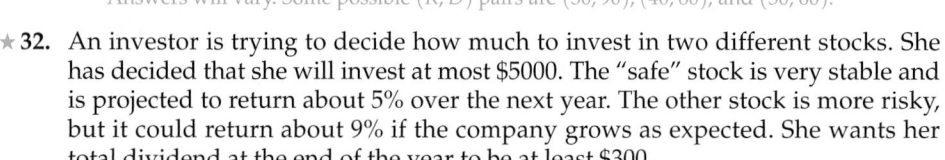 **Equations and Inequalities**

[3.7] **5.** Given $f(x) = 3x - 5$, find $f(3)$.

   4

[3.7] **6.** Given $f(x) = 3x^2 - 2x + 5$, find $f(-2)$.

   21

Chapter 4 Summary and Review Exercises

Complete each incomplete definition, rule, or procedure; study the key examples; and then work the related exercises.

4.1 Solving Systems of Linear Equations Graphically

Definitions/Rules/Procedures	Key Example(s)
A **system of equations** is a(n) ___group of two or more___ equations. A **solution for a system of equations** is an ordered set of numbers that ___makes all of the equations in the system true___. **To verify or check a solution to a system of equations:** 1. Replace each ___variable___ in each equation with its corresponding ___value___. 2. Verify that each equation is ___true___.	Determine whether the ordered pair is a solution to the system. $$\begin{cases} 2x + 3y = 8 \\ y = 5x - 3 \end{cases}$$ **a.** $(-2, 4)$ $\begin{array}{ll} 2x + 3y = 8 & y = 5x - 3 \\ 2(-2) + 3(4) \stackrel{?}{=} 8 & 4 \stackrel{?}{=} 5(-2) - 3 \\ -4 + 12 \stackrel{?}{=} 8 & 4 \stackrel{?}{=} -10 - 3 \\ 8 = 8 & 4 \neq -13 \end{array}$ $(-2, 4)$ is not a solution to the system. **b.** $(1, 2)$ $\begin{array}{ll} 2x + 3y = 8 & y = 5x - 3 \\ 2(1) + 3(2) \stackrel{?}{=} 8 & 2 \stackrel{?}{=} 5(1) - 3 \\ 2 + 6 \stackrel{?}{=} 8 & 2 \stackrel{?}{=} 5 - 3 \\ 8 = 8 & 2 = 2 \end{array}$ $(1, 2)$ is a solution to the system.

Exercises 1 and 2 ▲ Equations and Inequalities

[4.1] *For Exercises 1 and 2, determine whether the given ordered pair is a solution to the given system of equations.*

1. $(4, 3)$; $\begin{aligned} x - y &= 1 \\ -x + y &= -1 \end{aligned}$

yes

2. $(1, 1)$; $\begin{aligned} 2x - y &= 7 \\ x + y &= 8 \end{aligned}$

no

Definitions/Rules/Procedures	Key Example(s)

To solve a system of linear equations graphically:
 	
 1. Graph each equation.
 	
   a. If the lines intersect at a single point, the ___coordinates___ of that point form the solution.
 	
   b. If the lines are parallel, there is ___no solution___.
 	
   c. If the lines are identical, there are a(n) ___infinite number___ of solutions, which are the coordinates of all points on that line.
 	
 2. Check your solution.
	Solve the system of equations graphically. $$\begin{cases} 2x + 3y = 8 \\ y = 5x - 3 \end{cases}$$ Graph the two equations; then find the point of intersection.
	
	 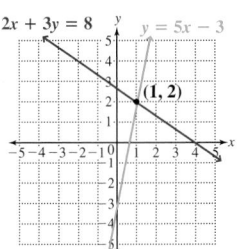
	
	 The two lines intersect at the point $(1, 2)$, so $(1, 2)$ is the solution to the system.

Exercises 3–6 **Equations and Inequalities**

[4.1] *For Exercises 3–6, solve the system graphically.*

3. $\begin{cases} 4x = y \\ 3x + y = -7 \end{cases}$
$(-1, -4)$

4. $\begin{cases} x - y = 4 \\ 2x + 3y = 3 \end{cases}$
$(3, -1)$

5. $\begin{cases} y = -2x - 3 \\ 4x + 2y = 5 \end{cases}$
No solution

6. $\begin{cases} 3x - 2y = 6 \\ y = \dfrac{3}{2}x - 3 \end{cases}$
All ordered pairs that solve $3x - 2y = 6$.

Definitions/Rules/Procedures	Key Example(s)

Definitions/Rules/Procedures

A **consistent system of equations** is a system of equations that has ___at least one solution___ .

An **inconsistent system of equations** is a system of equations that has ___no solution___ .

To classify a system of two linear equations in two unknowns, write the equations in slope–intercept form and compare the slopes and y-intercepts.

1. **Consistent system with independent equations:** The system has a single solution at the point of ___intersection___ . The graphs are ___different___ . They have ___different___ slopes.

2. **Consistent system with dependent equations:** The system has a(n) ___infinite___ number of solutions. The graphs are ___identical___ . They have the ___same___ slope and ___same___ y- intercept.

3. **Inconsistent systems:** The system has ___no___ solution. The graphs are ___parallel___ lines. They have the ___same___ slope but different ___y-intercepts___ .

Key Example(s)

Determine whether the system is consistent with independent equations, consistent with dependent equations, or inconsistent and discuss the number of solutions for the system.

$$\begin{cases} x + 3y = 6 \\ 2x + 6y = -5 \end{cases}$$

Solution: Write the equations in slope–intercept form.

$x + 3y = 6 \qquad\qquad 2x + 6y = -5$
$\quad 3y = -x + 6 \qquad\qquad 6y = -2x - 5$
$\quad\quad y = -\dfrac{1}{3}x + 2 \qquad\qquad y = -\dfrac{1}{3}x - \dfrac{5}{6}$

Because the slopes are the same $\left(\text{both } -\dfrac{1}{3}\right)$ but the y-intercepts are different, this system is inconsistent. This means there is no solution to the system.

Exercises 7–10 **Equations and Inequalities**

[4.1] *For Exercises 7–10:* **a.** *Determine whether the system of equations is consistent with independent equations, consistent with dependent equations, or inconsistent.*
b. *How many solutions does the system have?*

7.

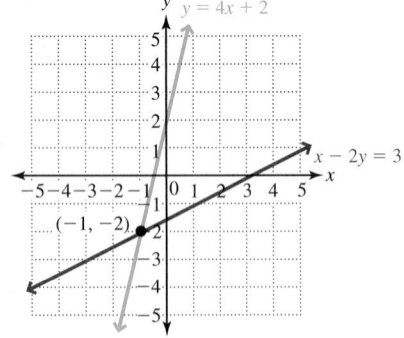

a. consistent with independent equations
b. one solution

8.

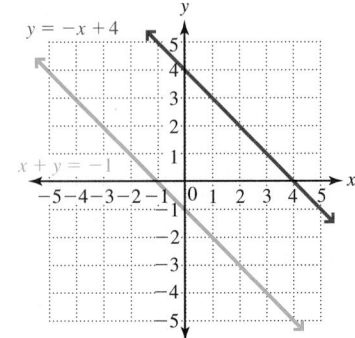

a. inconsistent
b. no solution

9. $\begin{cases} x - 3y = 10 \\ 2x + 3y = 5 \end{cases}$
a. consistent with independent equations
b. one solution

10. $\begin{cases} x - 5y = 10 \\ 2x - 10y = 20 \end{cases}$
a. consistent with dependent equations
b. infinite number of solutions

4.2 Solving Systems of Linear Equations by Substitution; Applications

Definitions/Rules/Procedures	Key Example(s)
To find the solution of a system of two linear equations using the substitution method: 1. Isolate one of the ___variables___ in one of the equations. 2. In the other equation, ___substitute___ the expression you found in step 1 for that variable. 3. ___Solve___ this new equation. (It will now have only one variable.) 4. Using one of the equations containing both variables, ___substitute___ the value you found in step 3 for that variable and solve for the value of the other ___variable___. 5. Check the solution in the original equations.	Solve the system using the substitution method. $$\begin{cases} y = 4x - 1 \\ 5x - y = 2 \end{cases}$$ Because y is already isolated in the first equation, we substitute $4x - 1$ in place of y in the second equation. $\begin{aligned} 5x - y &= 2 \\ 5x - (4x - 1) &= 2 \quad \text{Substitute } 4x - 1 \text{ for } y. \\ 5x - 4x + 1 &= 2 \quad \text{Distribute.} \\ x + 1 &= 2 \quad \text{Combine like terms.} \\ x &= 1 \quad \text{Subtract 1 from both sides.} \end{aligned}$ Now solve for the value of y by substituting 1 in place of x in one of the equations. We will use $y = 4x - 1$. $\begin{aligned} y &= 4x - 1 \\ y &= 4(1) - 1 \quad \text{Substitute 1 for } x. \\ y &= 4 - 1 \\ y &= 3 \end{aligned}$ Solution: $(1, 3)$

Exercises 11–14 ▲ Equations and Inequalities

[4.2] *For Exercises 11–14, solve the system of equations using substitution.*

11. $\begin{cases} 3x + 10y = 2 \\ x - 2y = 6 \end{cases}$
$(4, -1)$

12. $\begin{cases} 2x + 5y = 8 \\ x - 10y = 9 \end{cases}$
$\left(5, -\dfrac{2}{5}\right)$

13. $\begin{cases} 3y - 4x = 6 \\ y = \dfrac{4}{3}x + 2 \end{cases}$
All ordered pairs that solve $3y - 4x = 6$

14. $\begin{cases} 8x - 2y = 12 \\ y - 4x = 3 \end{cases}$
No solution

Definitions/Rules/Procedures	Key Example(s)
To solve a problem with two unknowns using a system of equations: 1. Select a variable to represent each of the ___unknowns___. 2. Translate each relationship to a(n) ___equation___. 3. Solve the system.	The diameter of Neptune's largest moon Triton is 201 miles more than seven times the diameter of Nereid, which is Neptune's second-largest moon. If the difference in their diameters is 1467 miles, find the diameter of each. **Solution:** Let $x =$ the diameter of Triton Let $y =$ the diameter of Nereid *The diameter of Triton is 201 miles more than seven times the diameter of Nereid* translates to $$x = 7y + 201.$$ *The difference of their diameters is 1467 miles* translates to $$x - y = 1467.$$ This gives us the system $$\begin{cases} x = 7y + 201 \\ x - y = 1467 \end{cases}$$

Definitions/Rules/Procedures	Key Example(s)
	Because the first equation is solved for x, we use the substitution method. Substitute $7y + 201$ for x in the second equation.

$$(7y + 201) - y = 1467 \quad \text{Combine like terms.}$$
$$6y + 201 = 1467 \quad \text{Subtract 201.}$$
$$6y = 1266 \quad \text{Divide by 6.}$$
$$y = 211 \quad \text{Diameter of Nereid}$$

Substitute 211 for y in $x = 7y + 201$ to find x.

$$x = 7(211) + 201$$
$$x = 1477 + 201$$
$$x = 1678$$

Therefore, the diameter of Nereid is 211 miles and the diameter of Triton is 1678 miles.

Exercises 15–18 Equations and Inequalities

[4.2] *For Exercises 15–18, solve.*

15. In 2011, a combined total of 34,828 thousand people visited the United States from Canada and Mexico, with 7846 thousand more from Canada than Mexico. Find the number of visitors from each country. (*Source:* U.S. Department of Commerce.)

    Canada: 21,337 thousand; Mexico: 13,491 thousand

16. Western Europe and North America contribute a combined total of 43% of the world's carbon dioxide emissions. If North America contributes 4% less than twice the percentage contributed by Western Europe, what percent does each of these regions contribute?

    Western Europe: $15\frac{2}{3}$%; North America: $27\frac{1}{3}$%

    Of Interest

    The United States alone contributes 23% of the world's carbon dioxide emissions.

17. The smaller of two complementary angles is 10 degrees less than one-third the larger angle. Find the measure of the two angles.

    75°, 15°

18. Yolanda and Dee are traveling west in separate cars on the same highway. Yolanda is traveling 60 miles per hour; Dee, 70 miles per hour. Yolanda passes the I-95 exit at 4:00 P.M. Dee passes the same exit at 4:15 P.M. At what time will Dee catch up with Yolanda?

    5:45 P.M.

4.3 Solving Systems of Linear Equations by Elimination; Applications

Definitions/Rules/Procedures	Key Example(s)
To solve a system of two linear equations using the elimination method:	Solve the system using the elimination method.

1. Write the equations in ___standard___ form $(Ax + By = C)$.

2. Use the multiplication principle of equality to eliminate ___fractions___ or ___decimals___ (optional).

3. If necessary, multiply one or both equations by a number (or numbers) so that they have a pair of terms that are ___additive inverses___.

$$\begin{cases} \dfrac{1}{4}x + \dfrac{1}{2}y = -1 \\ 4x + y = 5 \end{cases}$$

Solution: First, we eliminate the fractions in the first equation.

$$4\left(\frac{1}{4}x + \frac{1}{2}y\right) = 4(-1)$$
$$x + 2y = -4$$

Definitions/Rules/Procedures	Key Example(s)
4. _____Add_____ the equations. The result should be an equation in terms of one variable. **5.** _____Solve_____ the equation from step 4 for the value of that variable. **6.** Using an equation containing both variables, _____substitute_____ the value you found in step 5 for the corresponding variable and solve for the value of the other variable. **7.** Check your solution in the _____original equation_____.	Now we will eliminate y by multiplying the second equation by -2, then add the two equations. $x + 2y = -4$ $\qquad$ $x + 2y = -4$ $4x + y = 5$ $\xrightarrow{\text{Multiply by } -2.}$ $-8x - 2y = -10$ $\qquad$ Add equations. $\quad -7x + 0 = -14$ $\qquad$ Divide both sides by -7. $\quad -7x = -14$ $\qquad\qquad\qquad\qquad x = 2$

Now solve for the value of y by substituting 2 for x in one of the equations. We will use $4x + y = 5$.

$$4x + y = 5$$
$$4(2) + y = 5 \qquad \text{Substitute 2 for } x.$$
$$8 + y = 5 \qquad \text{Multiply.}$$
$$y = -3 \qquad \text{Subtract 8 from both sides.}$$

Solution: $(2, -3)$

A vendor sells small and large drinks for \$2 and \$3, respectively. If the vendor sold 2400 drinks for a total of \$5400, how many of each size were sold?

Let x = the number of small drinks sold.
Let y = the number of large drinks sold.

The total number of drinks sold, 2400, translates to $x + y = 2400$.
The total revenue is \$5400, which translates to $2x + 3y = 5400$.

$$\text{System } \begin{cases} x + y = 2400 \\ 2x + 3y = 5400 \end{cases}$$

To solve the system, we use the elimination method because there are no isolated variables.

$x + y = 2400$ $\quad$ Multiply by -3 $\quad -3x - 3y = -7200$
$2x + 3y = 5400$ $\xrightarrow{\qquad\qquad}$ $\qquad\quad 2x + 3y = 5400$
$\qquad\qquad\qquad$ We eliminated y. $\quad -x + 0 = -1800$
$\qquad\qquad\qquad$ Solve for x. $\qquad\qquad x = 1800$

Substitute 1800 for x in an equation.

$$1800 + y = 2400 \qquad \text{We chose } x + y = 2400.$$
$$y = 600 \qquad \text{Isolate } y.$$

The vendor sold 1800 small and 600 large drinks.

Exercises 19–26 Equations and Inequalities

[4.3] *For Exercises 19–22, solve the system of equations using elimination.*

19. $\begin{cases} x + y = 4 \\ x - y = -2 \end{cases}$
$(1, 3)$

20. $\begin{cases} 3x + 2y = 4 \\ 2x - 3y = 7 \end{cases}$
$(2, -1)$

21. $\begin{cases} 0.25x + 0.75y = 4 \\ x - y = -4 \end{cases}$
$(1, 5)$

22. $\begin{cases} \dfrac{1}{5}x - \dfrac{1}{3}y = 2 \\ x + y = 2 \end{cases}$
$(5, -3)$

[4.3] *For Exercises 23–26, solve.*

23. A movie theater charges $7.00 for adults and $5.50 for children. If 139 tickets were sold for a total of $835, how many of each ticket were sold?

47 adult, 92 children

24. Jose invested part of $5000 in a fund that earned 7% interest and the rest in a fund that earned 9% interest. If he earned $380 interest during the year, how much did he invest in each fund?

$3500 at 7% and $1500 at 9%

25. A plane travels 450 miles in 3 hours with the jet stream. When the plane returns, it takes 5 hours to travel the same distance against the jet stream. What is the speed of the plane in still air?

120 miles per hour

26. How much of a 10% saline solution is needed to combine with a 60% saline solution to obtain 50 milliliters of a 30% saline solution?

30 ml

4.4 Solving Systems of Linear Equations in Three Variables; Applications

Definitions/Rules/Procedures	Key Example(s)
To solve a system of three linear equations with three unknowns using elimination: 1. Write each equation in the form $\underline{Ax + By + Cz = D}$. 2. Eliminate one variable from one pair of equations using the $\underline{\text{elimination}}$ method. 3. If necessary, eliminate the $\underline{\text{same variable}}$ from another pair of equations. 4. Steps 2 and 3 result in two equations with the same $\underline{\text{two variables}}$. Solve these equations using the elimination method. 5. To find the third variable, substitute the values of the variables found in step 4 into any of the three $\underline{\text{original equations}}$ that contain the third variable. 6. Check the ordered triple in $\underline{\text{all three original equations}}$.	Solve the following system of equations. $$\begin{cases} x + 2y - z = -6 & \text{(Equation 1)} \\ 2x - 3y + 4z = 26 & \text{(Equation 2)} \\ -x + 2y - 3z = -18 & \text{(Equation 3)} \end{cases}$$ Eliminate x by adding equation 1 and equation 3. $\begin{aligned} x + 2y - z &= -6 \\ \underline{-x + 2y - 3z} &= \underline{-18} \\ 4y - 4z &= -24 \quad \textbf{Divide both sides by 4.} \\ y - z &= -6 \quad \text{(Eq. 4)} \end{aligned}$ Eliminate x again using equations 2 and 3. $\begin{aligned} 2x - 3y + 4z &= 26 \\ -x + 2y - 3z &= -18 \end{aligned} \longrightarrow \begin{aligned} 2x - 3y + 4z &= 26 \\ \underline{-2x + 4y - 6z} &= \underline{-36} \\ y - 2z &= -10 \quad \text{(Eq. 5)} \end{aligned}$ $\qquad\qquad\qquad$ **Multiply by 2.** Use equations 4 and 5 to solve for z. $\begin{aligned} y - z &= -6 \\ y - 2z &= -10 \end{aligned} \xrightarrow{\textbf{Multiply by } -1.} \begin{aligned} y - z &= -6 \\ \underline{-y + 2z} &= \underline{10} \\ z &= 4 \end{aligned}$ Substitute 4 for z in equation 4 and solve for y. $$\begin{aligned} y - 4 &= -6 \\ y &= -2 \end{aligned}$$ Substitute -2 for y and 4 for z in equation 1 and solve for x. $$\begin{aligned} x + 2(-2) - 4 &= -6 \\ x - 4 - 4 &= -6 \\ x &= 2 \end{aligned}$$ The solution is $(2, -2, 4)$.

Exercises 27–32 ◢◣ **Equations and Inequalities**

[4.4] *For Exercises 27–30, solve the system using the elimination method.*

27. $\begin{cases} x + y + z = -2 \\ 2x + 3y + 4z = -10 \\ 3x - 2y - 3z = 12 \end{cases}$
$(1, 0, -3)$

28. $\begin{cases} x + y + z = 0 \\ 2x - 4y + 3z = -12 \\ 3x - 3y + 4z = 2 \end{cases}$
No solution

29. $\begin{cases} 3x + 4y = -3 \\ -2y + 3z = 12 \\ 4x - 3z = 6 \end{cases}$
$(3, -3, 2)$

30. $\begin{cases} 2x + 2y - 3z = 5 \\ x = -3y + 4z \\ z = -x - 2y + 5 \end{cases}$
$(4, 0, 1)$

[4.4] *For Exercises 31 and 32, solve.*

31. John bought vitamin supplements that cost $8 each, film that cost $4 per roll, and bags of candy that cost $3 per bag. He bought a total of eight items that cost $32. The number of rolls of film was one less than the number of bags of candy. Find the number of each type of item that he bought.

1 vitamin supplement, 3 rolls of film, 4 bags of candy

32. At a swim meet, 5 points are awarded for each first-place finish, 3 points for each second-place finish, and 1 point for each third-place finish. Shawnee Mission South High School scored a total of 38 points. The number of first-place finishes was one more than the number of second-place finishes. The number of third-place finishes was three times the number of second-place finishes. Find the number of first-, second-, and third-place finishes for the school.

4 first place, 3 second place, 9 third place

4.5 Solving Systems of Linear Equations Using Matrices

Definitions/Rules/Procedures	Key Example(s)
A **matrix** is a(n) <u>rectangular array</u> of numbers.	An example of a matrix is $\begin{bmatrix} -4 & 2 \\ 3 & -1 \end{bmatrix}$.
An **augmented matrix** is made up of the <u>coefficients</u> and the <u>constant</u> terms of a system. The <u>constant</u> terms are separated from the <u>coefficients</u> by a dashed vertical line.	The augmented matrix for the system $\begin{cases} 2x - 4y = 5 \\ x + 6y = -4 \end{cases}$ is $\left[\begin{array}{cc:c} 2 & -4 & 5 \\ 1 & 6 & -4 \end{array}\right]$.
An augmented matrix is in **row echelon form** if its coefficient portion has 1's on the diagonal from <u>upper left</u> to <u>lower right</u> and 0s below the 1's.	An augmented matrix in row echelon form is $\left[\begin{array}{cc:c} 1 & 4 & -5 \\ 0 & 1 & 6 \end{array}\right]$.
Row operations: The solution of a system is not affected by the following row operations in its augmented matrix.	Solve $\begin{cases} 2x + 3y = -5 \\ x + 2y = -4 \end{cases}$ using the row echelon method.
1. Any two rows may be <u>interchanged</u>.	The augmented matrix is $\left[\begin{array}{cc:c} 2 & 3 & -5 \\ 1 & 2 & -4 \end{array}\right]$.
2. The elements of any row may be <u>multiplied</u> (or <u>divided</u>) by any nonzero real number.	$\left[\begin{array}{cc:c} 1 & 2 & -4 \\ 2 & 3 & -5 \end{array}\right]$ We need row 1, column 1 to be a 1; so interchange the two rows.
3. Any row may be replaced by a row resulting from <u>adding</u> the elements of that row (or multiples of that row) to a(n) <u>multiple</u> of the elements of any other row.	$-2R_1 + R_2 \rightarrow \left[\begin{array}{cc:c} 1 & 2 & -4 \\ 0 & -1 & 3 \end{array}\right]$ We need row 2, column 1 to be 0; so multiply row 1 by -2 and add it to row 2.
Row operations can be used to solve a system of linear equations by transforming the augmented matrix of the system into an equivalent matrix that is in <u>row echelon</u> form.	$-R_2 \rightarrow \left[\begin{array}{cc:c} 1 & 2 & -4 \\ 0 & 1 & -3 \end{array}\right]$ Multiply row 2 by -1 to get row echelon form.
Note: Systems of three equations with three unknowns are solved in a similar manner.	Row 2 means $y = -3$, and row 1 means $x + 2y = -4$. $x + 2(-3) = -4$ Substitute -3 for y and solve for x. $x = 2$ The solution is $(2, -3)$.

Exercises 33 and 34 ▲▲ **Equations and Inequalities**

[4.5] *For Exercises 33 and 34, solve the system using the row echelon method.*

33. $\begin{cases} x - 4y = 8 \\ x + 2y = 2 \end{cases}$

$(4, -1)$

34. $\begin{cases} x + y + z = 2 \\ 2x + y + 2z = 1 \\ 3x + 2y + z = 1 \end{cases}$

$(-2, 3, 1)$

4.6 Solving Systems of Linear Inequalities

Definitions/Rules/Procedures	Key Example(s)
To solve a system of linear inequalities in two variables, graph all of the inequalities on the same grid. The solution set for the system contains all ordered pairs in the region where the inequalities' solutions sets ___overlap___ along with ordered pairs on the portion of any ___solid___ line that touches the region of overlap.	Solve $\begin{cases} x + y > -2 \\ y \le 3x - 1 \end{cases}$. Solution: 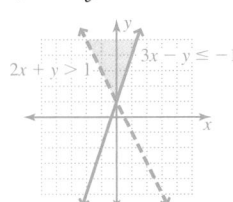 Solution region for the system (including all ordered pairs on the portion of the solid blue line touching this region).

Exercises 35–37 ▲▲ **Equations and Inequalities**

[4.6] *For Exercises 35 and 36, graph the solution set for the system of inequalities.*

35. $\begin{cases} x + y > -5 \\ x - y \ge -1 \end{cases}$

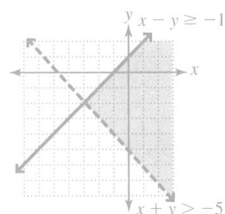

36. $\begin{cases} 2x + y > 1 \\ 3x - y \le -1 \end{cases}$

[4.6] *Solve using a system of linear inequalities.*

37. A real estate course is offered in a continuing education program at a two-year college. The course is limited to 16 students. Also, to run the course, the college must receive at least $2750 in tuition. The college charges $250 per student if he or she lives in the same county as the college and $400 if he or she lives in another county (or out of state).

   a. Write a system of inequalities that describes the number of in-county students, x, enrolled in the course versus the number of out-of-county students, y, and how much they paid.

   b. Solve the system by graphing.

$\begin{cases} x + y \le 16 \\ 250x + 400y \ge 2750 \\ x \ge 0 \\ y \ge 0 \end{cases}$

   c. Give a combination of students that could enroll in the course so that it runs.

Answers may vary. One example is $(8, 4)$.

b.

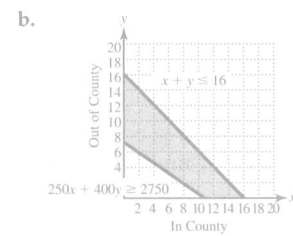

Learning Strategy

The best way to prepare for an exam is study, study, study!

—Ruben C.

Chapter 4 Practice Test

For Extra Help

Step-by-step test solutions are found on the Chapter Test Prep Videos available in MyMathLab® *or on* YouTube.

For Exercises 1 and 2, determine whether the given ordered pair is a solution to the system of equations.

1. $(-1, 3);$ $\begin{cases} 3x + 2y = 3 \\ 4x - y = -7 \end{cases}$

yes [4.1]

2. $(2, 2, 3);$ $\begin{cases} 2x - 3y + z = 1 \\ -2x + y - 3z = 11 \\ 3x + y + 3z = 14 \end{cases}$

no [4.4]

3. Solve by graphing: $\begin{cases} x + 3y = 1 \\ -2x + y = 5 \end{cases}$

$(-2, 1)$ [4.1]

For Exercises 4–9, solve the system of equations using substitution or elimination. Note that some systems may be inconsistent or consistent with dependent equations.

4. $\begin{cases} 2x + y = 15 \\ y = 7 - x \end{cases}$

$(8, -1)$ [4.2]

5. $\begin{cases} x - 2y = 1 \\ 3x - 5y = 4 \end{cases}$

$(3, 1)$ [4.2]

6. $\begin{cases} 3x - 2y = -8 \\ 2x + 3y = -14 \end{cases}$

$(-4, -2)$ [4.3]

7. $\begin{cases} 4x + 6y = 2 \\ 6x + 9y = 3 \end{cases}$

All ordered pairs that solve
$4x + 6y = 2$. [4.3]

8. $\begin{cases} x + 2y + z = 2 \\ x + 4y - z = 12 \\ 3x - 3y - 2z = -11 \end{cases}$

$(-2, 3, -2)$ [4.4]

9. $\begin{cases} x + 2y - 3z = -9 \\ 3x - y + 2z = -8 \\ 4x - 3y + 3z = -13 \end{cases}$

$(-4, 2, 3)$ [4.4]

For Exercises 10 and 11, solve the equations using the row echelon method.

10. $\begin{cases} x + 2y = -6 \\ 3x + 4y = -10 \end{cases}$

$(2, -4)$ [4.5]

11. $\begin{cases} x + y + z = 6 \\ 3x - 2y + 3z = -7 \\ 4x - 2y + z = -12 \end{cases}$

$(-1, 5, 2)$ [4.5]

For Exercises 12 and 13, solve the system of equations using the method of your choice.

12. $\begin{cases} 3x + 4y = 14 \\ 2x - 3y = -19 \end{cases}$

$(-2, 5)$ [4.6]

13. $\begin{cases} x + y + z = -1 \\ 3x - 2y + 4z = 0 \\ 2x + 5y - z = -11 \end{cases}$

$(-4, 0, 3)$ [4.4]

14. Graph the solution set for the system of inequalities.

$\begin{cases} 2x - 3y < 1 \\ x + 2y \le -2 \end{cases}$

[4.6]

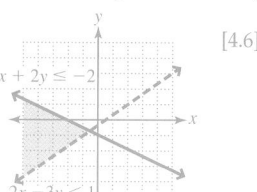

For Exercises 15–20, solve.

15. Excedrin surveyed workers in various professions who get headaches at least once a year on the job. Nine more accountants than waiters/waitresses in the survey got a headache. If the combined number of accountants and waiters/waitresses that got a headache was 163, how many people in each of those professions got a headache? (*Source:* the *State* newspaper.)

86 accountants, 77 waiters/waitresses [4.2]

16. When asked what the Internet most resembled, three times as many people said a library as opposed to a highway. If 240 people were polled, how many people considered the Internet to be a library? (*Source: bLINK magazine.*)

180 [4.2]

17. A boat traveling with the current took 3 hours to go 30 miles. The same boat went 12 miles in 3 hours against the current. What is the rate of the boat in still water?

7 mph [4.3]

18. Janice invested $12,000 in two funds. One of the funds returned 6% interest, and the other returned 8% interest after one year. If the total interest for the year was $880, how much did she invest in each fund?

$4000 at 6%, $8000 at 8% [4.3]

19. Tickets for the senior play at Apopka High School cost $3 for children, $5 for students, and $8 for adults. There were 800 tickets sold for a total of $4750. The number of adult tickets sold was 50 more than two times the number of children tickets. Find the number of each type of ticket.

Children: 150; student: 300; adult: 350. [4.3]

20. A landscaper wants to plan a bordered rectangular garden area in a yard. Because she currently has 200 feet of border materials, the perimeter needs to be, at most, 200 feet. She thinks the garden will look best if the length is at least 10 feet more than the width.

   a. Write a system of inequalities to describe the situation.

$$\begin{cases} 2l + 2w \le 200 \\ l \ge w + 10 \\ l > 0 \\ w > 0 \end{cases}$$

   b. Solve the system by graphing.

   c. Give a combination of length and width that satisfies the requirements for the garden.

   Answers may vary.
   One example is $(60, 20)$. [4.6]

20. b.

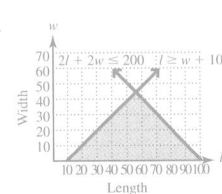

Chapters 1–4 Cumulative Review Exercises

For Exercises 1–3, answer true or false.

[3.4] 1. The graph of $y = -3x + 5$ is a line with slope of $-3x$.

false

[1.5] 2. $-(-4)^3 = -64$

false

[4.1] 3. The ordered pair $(1, 1)$ is a solution for the system of equations
$$\begin{cases} x = y \\ 9y - 13 = -4x. \end{cases}$$

true

For Exercises 4–6, fill in the blank.

[1.7] 4. Complete using the distributive property:
$4(2x + 3) = $ _____.

8x + 12

[2.6] 5. A proportion is an equation in the form
$\dfrac{a}{b} = $ _____.

[3.5] 6. The graph of $y = -2x + 5$ is _____ to the graph of $y = \dfrac{1}{2}x - 3$.

perpendicular

Exercises 7–9 **Expressions**

For Exercises 7–9, simplify.

[1.5] 7. $-(6 - 2^2)^2 + 15$

11

[1.5] 8. $-|2 - 8| + 3(2)$

0

[1.7] 9. $\dfrac{3}{8}x + 4y + 12 - \dfrac{1}{2}x - \dfrac{7}{10}y - 7$

$-\dfrac{1}{8}x + \dfrac{33}{10}y + 5$

Exercises 10–30 **Equations and Inequalities**

For Exercises 10–13, solve.

[2.3] 10. $3.6 - 4(0.7p - 15) = 2.2p - 1.4$

p = 13

[2.3] 11. $2(x - 3) + 4x = 6(x - 2)$

contradiction (no solution)

[2.8] 12. $4z - 32 - 5z < 6z + 13 + 2z$

z > -5

[2.6] 13. $\dfrac{4}{3x - 3} = \dfrac{7}{4x + 1}$

x = 5

[2.4] 14. $A = P + Prt$ for t

$t = \dfrac{A - P}{Pr}$

[1.7] 15. For what value(s) of x is $\dfrac{x}{x + 2}$ undefined?

x = -2

[2.1] 16. A house is situated on a trapezoidal lot as shown below. If the area around the house is to be sodded, find the area to be sodded.

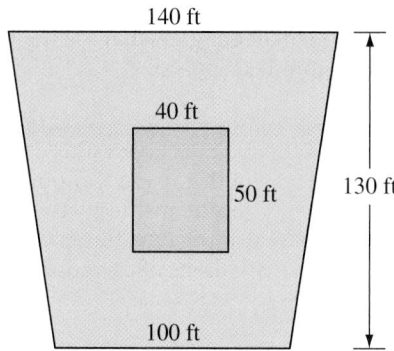

140 ft

40 ft

50 ft   130 ft

100 ft

13,600 ft².

For Exercises 17–20, graph.

[3.5] 17. $y = 3x + 1$

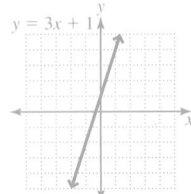

[3.2] 18. $4x + 5y = 10$

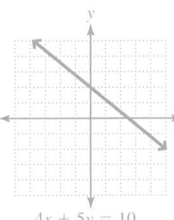

[3.2] 19. $x = -4$

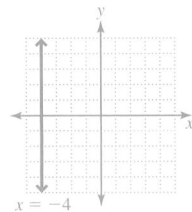

[3.6] 20. $2x - 7y < 14$

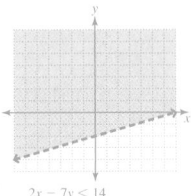

[3.5] *For Exercises 21–23, write the equations of the lines. Write the equation in slope–intercept form.*

21. Passing through $(-2, -1)$, $m = -\dfrac{1}{5}$

$y = -\dfrac{1}{5}x - \dfrac{7}{5}$

22. Passing through $(4, -7)$ and $(-2, -1)$

$y = -x - 3$

23. Passing through $(-4, 2)$ and parallel to the graph of $3x + 4y = 4$. Write the equation in standard form.

$3x + 4y = -4$

For Exercises 24–26, solve the system.

[4.3] 24. $\begin{cases} 4x - 3y = -2 \\ 6x - 7y = 7 \end{cases}$

$(-3.5, -4)$

[4.4] 25. $\begin{cases} x + y + z = 5 \\ 2x + y - 2z = -5 \\ x - 2y + z = 8 \end{cases}$

$(2, -1, 4)$

[4.6] 26. $\begin{cases} 4x - 3y < 12 \\ 3x + y > 6 \end{cases}$

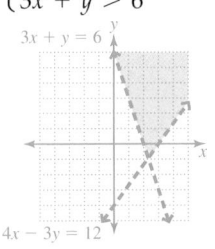

For Exercises 27–30, solve.

[2.3] 27. A concert hall is rectangular in shape and measures 50 feet longer than it is wide. If the perimeter is 420 feet, what are its length and width?

$L = 130$ ft.
$W = 80$ ft.

[4.3] 28. In the construction of a roof frame, a support beam is connected to a horizontal joist, forming two angles. If the greater angle is 15° less than twice the smaller angle, what are the measures of the two angles?

115°; 65°

[4.3] 29. Roxanne received an inheritance of $50,000. She invested part of it at 3% and the remainder at 2.5%. How much did she invest at each rate if the total interest from both investments was $1425 per year?

$35,000 at 3% and $15,000 at 2.5%

[4.2] 30. A paint contractor paid $588 for 24 gallons of paint to paint the inside and outside of a house. If the paint for the inside costs $22 per gallon and the paint for the outside costs $28 per gallon, how many gallons of each did he buy?

10 gallons of outside, 14 gallons of inside

Chapter Overview

The focus of this chapter is on polynomial expressions. Because polynomials involve exponents and are related to numbers written in scientific notation, we also explore rules of exponents and scientific notation. The primary skills in the chapter are as follows:

▶ Use rules of exponents to simplify exponential expression and numbers in scientific notation.

▶ Add, subtract, multiply, or divide polynomial expressions.

Instructor Note

We prefer to teach the rules of exponents as they are needed rather than all at once. If you prefer to teach all of them at the same time, they can be found in the following locations. Section 5.1 presents raising a quotient to a power, zero as an exponent, and negative exponents. Section 5.4 presents the product rule for exponents, raising a power to a power, and raising a product to a power. Section 5.6 presents the quotient rule for exponents and a summary of all the rules of exponents. As an application of the rules of exponents, scientific notation is introduced in Section 5.1, multiplication of numbers in scientific notation in Section 5.4, and division of numbers in scientific notation in Section 5.6.

5.1 Exponents and Scientific Notation

5.2 Introduction to Polynomials

5.3 Adding and Subtracting Polynomials

5.4 Exponent Rules and Multiplying Monomials

5.5 Multiplying Polynomials; Special Products

5.6 Exponent Rules and Dividing Polynomials

5.1 Exponents and Scientific Notation

Objectives

1 Evaluate exponential forms with integer exponents.

2 Write scientific notation in standard form.

3 Write standard form numbers in scientific notation.

Warm-up

[1.5] Evaluate the exponential forms.

1. $\left(\dfrac{2}{3}\right)^4$

2. $(-3)^5$

Objective 1 Evaluate exponential forms with integer exponents.

Positive Integer Exponents

We learned in Section 1.5 that exponents indicate repeated multiplication of a base number.

$$2^4 = \underbrace{2 \cdot 2 \cdot 2 \cdot 2}_{\substack{\text{four factors} \\ \text{of two}}} = 16 \qquad (-5)^3 = \underbrace{(-5)(-5)(-5)}_{\substack{\text{three factors} \\ \text{of negative five}}} = -125$$

We also discovered some rules for evaluating exponential forms with negative bases.

Learning Strategy

When taking notes, develop your own consistent format that is easily readable to you not only during the time you take them, but also afterwards when you need to review. It doesn't matter if your notes conform to any particular style, but what is important is that they speak to you in a consistent way by bearing a structure that you can easily recognize every time you look at them.

—Zheng Alick Z.

> **Rule** **Evaluating Exponential Forms with Negative Bases**
>
> If the base of an exponential form is a negative number and the exponent is even, then the product is positive.
> If the base is a negative number and the exponent is odd, then the product is negative.

What if the base is a fraction, as in $\left(\dfrac{2}{3}\right)^4$?

$$\left(\frac{2}{3}\right)^4 = \frac{2}{3} \cdot \frac{2}{3} \cdot \frac{2}{3} \cdot \frac{2}{3} = \frac{2^4}{3^4} = \frac{16}{81}$$

This example suggests the following conclusion and rule.

Conclusion: If a fraction is raised to a power, we can write its numerator and denominator to that power.

> **Rule** **Raising a Quotient to a Power**
>
> If a and b are real numbers, where $b \neq 0$ and n is a natural number, then
>
> $$\left(\frac{a}{b}\right)^n = \frac{a^n}{b^n}.$$

Example 1 Evaluate each exponential form.

a. $(-3)^5$

Solution: $(-3)^5 = (-3)(-3)(-3)(-3)(-3) = -243$

b. -2^4

Solution: $-2^4 = -[2 \cdot 2 \cdot 2 \cdot 2] = -16$

Warning Notice that -2^4 does not mean that four -2s are to be multiplied. Rather, it means find the additive inverse after multiplying four 2s.

c. $\left(\dfrac{4}{5}\right)^3$

Solution: $\left(\dfrac{4}{5}\right)^3 = \dfrac{4^3}{5^3} = \dfrac{64}{125}$

Answers to Warm-up

1. $\dfrac{16}{81}$

2. -243

Your Turn 1 | Evaluate each exponential form.

a. $(-4)^4$ **b.** -4^4 **c.** $\left(\dfrac{5}{8}\right)^2$

Nonpositive Integer Exponents

So far, we have considered evaluating exponential forms only with positive exponents. However, exponents can be positive, zero, or negative. What does an exponent of 0 indicate? Consider the following pattern.

$2^4 = 2 \cdot 2 \cdot 2 \cdot 2 = 16$

$2^3 = 2 \cdot 2 \cdot 2 \quad = 8$ $\longleftarrow$ Notice that as the exponent decreases, we can divide by 2 to determine the next result; that is, because 2^4 is 16, 2^3 is $16 \div 2$, which is 8.

$2^2 = 2 \cdot 2 \quad\quad = 4$

$2^1 = 2 \quad\quad\quad = 2$

$2^0 = \quad\quad\quad = 1$ $\longleftarrow$ To continue the pattern, when we go from 2^1 to 2^0, we divide the result for 2^1 by 2 to determine the result for 2^0, and $2 \div 2 = 1$.

The pattern suggests the following conclusion and rule.

Conclusion: A nonzero base number raised to the 0 power is equal to 1. It can be shown that any nonzero number raised to the 0 power equals 1.

> **Rule** **Zero as an Exponent**
> If a is a real number and $a \neq 0$, then $a^0 = 1$.

If we continue the pattern, we can see what a negative exponent means.

$2^0 = 1$

$2^{-1} = \dfrac{1}{2}$ $\longleftarrow$ To continue the pattern, when we go from 2^0 to 2^{-1}, we divide the result for 2^0 by 2 to determine the result for 2^{-1}, and $1 \div 2 = \dfrac{1}{2}$.

$2^{-2} = \dfrac{1}{2^2} = \dfrac{1}{4}$ $\longleftarrow$ To continue the pattern, when we go from 2^{-1} to 2^{-2}, we divide the result for 2^{-1} by 2 to determine the result for 2^{-2}, and

$2^{-3} = \dfrac{1}{2^3} = \dfrac{1}{8}$ $\quad\quad\quad \dfrac{1}{2} \div 2 = \dfrac{1}{2} \cdot \dfrac{1}{2} = \dfrac{1}{2^2} = \dfrac{1}{4}.$

Instructor Note You may be accustomed to introducing all of the exponent rules in one section. The exponent rules are developed throughout the chapter as needed.

The pattern suggests the following conclusion.

Conclusion: An exponential form with a negative exponent is equal to its reciprocal with the exponent made positive.

It can be shown that our conclusion is true for all nonzero real-number bases.

> **Rule**
> If a is a real number, where $a \neq 0$ and n is a natural number, then $a^{-n} = \dfrac{1}{a^n}$.

Warning A common error is to confuse negative exponents with negative numbers. The value of $4^{-2} \neq -16$. Also, $4^{-2} \neq (-2)(4)$. The value of $4^{-2} = \dfrac{1}{4^2} = \dfrac{1}{16}$.

Example 2 | Rewrite with a positive exponent; then if the expression is numeric, evaluate it.

a. 10^0

Solution: $10^0 = 1$ $\quad a^0 = 1$ for $a \neq 0$

b. $(-2)^{-4}$

Solution: $(-2)^{-4} = \dfrac{1}{(-2)^4} = \dfrac{1}{(-2)(-2)(-2)(-2)} = \dfrac{1}{16}$

c. $(-5)^{-3}$

Answers to Your Turn 1

a. 256 **b.** -256 **c.** $\dfrac{25}{64}$

Solution: $(-5)^{-3} = \dfrac{1}{(-5)^3} = \dfrac{1}{(-5)(-5)(-5)} = \dfrac{1}{-125} = -\dfrac{1}{125}$

d. x^{-6}

Solution: $x^{-6} = \dfrac{1}{x^6}$ Rewrite using $a^{-n} = \dfrac{1}{a^n}$.

e. $2a^{-3}$

Solution: $2a^{-3} = 2 \cdot \dfrac{1}{a^3} = \dfrac{2}{a^3}$ **Note** Only a is raised to the -3 power.

f. -6^{-2}

Solution: $-6^{-2} = -\dfrac{1}{6^2} = -\dfrac{1}{36}$ **Note** Only 6 is raised to the -2 power, not -6.

Your Turn 2 If necessary, rewrite with a positive exponent; then if the expression is numeric, evaluate it.

a. $(-8)^0$ **b.** $(-3)^{-4}$ **c.** $(-2)^{-3}$ **d.** y^{-5} **e.** $4b^{-5}$ **f.** -2^{-4}

What if a negative exponent is in the denominator of a fraction, as in $\dfrac{1}{2^{-3}}$? If we rewrite 2^{-3} in the denominator using $a^{-n} = \dfrac{1}{a^n}$, we have

$$\frac{1}{2^{-3}} = \frac{1}{\dfrac{1}{2^3}} = \frac{1}{\dfrac{1}{8}} = \frac{8}{1} = 8$$ Rewrite the denominator using $a^{-n} = \dfrac{1}{a^n}$.

Our example suggests the following rule.

> **Connection** Remember that $\dfrac{1}{\dfrac{1}{8}}$ means $1 \div \dfrac{1}{8}$ and that
> $$1 \div \frac{1}{8} = 1 \cdot \frac{8}{1} = \frac{8}{1} = 8.$$

▶

Rule

If a is a real number, where $a \neq 0$ and n is a natural number, then $\dfrac{1}{a^{-n}} = a^n$.

Example 3 Rewrite with a positive exponent; then if the expression is numeric, evaluate it.

a. $\dfrac{1}{3^{-4}}$

Solution: $\dfrac{1}{3^{-4}} = 3^4 = 3 \cdot 3 \cdot 3 \cdot 3 = 81$ Rewrite using $\dfrac{1}{a^{-n}} = a^n$; then simplify.

b. $\dfrac{1}{x^{-5}}$

Solution: $\dfrac{1}{x^{-5}} = x^5$ Rewrite using $\dfrac{1}{a^{-n}} = a^n$.

Your Turn 3 Rewrite with a positive exponent; then if the expression is numeric, evaluate it.

a. $\dfrac{1}{4^{-3}}$ **b.** $\dfrac{1}{x^{-8}}$

Answers to Your Turn 2
a. 1 **b.** $\dfrac{1}{81}$ **c.** $-\dfrac{1}{8}$ **d.** $\dfrac{1}{y^5}$
e. $\dfrac{4}{b^5}$ **f.** $-\dfrac{1}{16}$

Answers to Your Turn 3
a. 64 **b.** x^8

What if a fraction is raised to a negative exponent, as in $\left(\frac{2}{5}\right)^{-2}$? If we rewrite the expression using $a^{-n} = \frac{1}{a^n}$, we have

$$\left(\frac{2}{5}\right)^{-2} = \frac{1}{\left(\frac{2}{5}\right)^2} = \frac{1}{\frac{4}{25}} = \frac{25}{4} = \left(\frac{5}{2}\right)^2$$

Our example suggests that when we evaluate a fraction raised to a negative exponent, we can write the reciprocal of the fraction and change the sign of the exponent.

Rule

If a and b are real numbers, where $a \neq 0$ and $b \neq 0$ and n is a natural number, then $\left(\frac{a}{b}\right)^{-n} = \left(\frac{b}{a}\right)^n$.

Example 4 Rewrite with a positive exponent; then if the expression is numeric, evaluate it.

a. $\left(\frac{3}{4}\right)^{-3}$

Solution: $\left(\frac{3}{4}\right)^{-3} = \left(\frac{4}{3}\right)^3 = \frac{4^3}{3^3} = \frac{64}{27}$ Rewrite using $\left(\frac{a}{b}\right)^{-n} = \left(\frac{b}{a}\right)^n$; then simplify.

b. $\left(\frac{x}{y}\right)^{-4}$

Solution: $\left(\frac{x}{y}\right)^{-4} = \left(\frac{y}{x}\right)^4 = \frac{y^4}{x^4}$ Rewrite using $\left(\frac{a}{b}\right)^{-n} = \left(\frac{b}{a}\right)^n$; then simplify.

Your Turn 4 Rewrite with a positive exponent; then if the expression is numeric, evaluate it.

a. $\left(-\frac{4}{5}\right)^{-3}$ **b.** $\left(\frac{a}{b}\right)^{-5}$

Objective 2 Write scientific notation in standard form.

Sometimes we use very large or very small numbers, such as when we describe the vast distances between stars and galaxies or the tiny size of bacteria and atomic structures. For example, the distance from the Sun to the next nearest star, Proxima Centauri, is about 24,700,000,000,000 miles; a single streptococcus bacterium (photo at left) is about 0.00000075 meter in diameter. The large number of zero digits in these numbers makes them tedious to write. **Scientific notation** gives us a shorthand way to write such numbers.

Definition Scientific notation: A number expressed in the form $a \times 10^n$, where a is a decimal number with $1 \leq |a| < 10$ and n is an integer.

Answers to Your Turn 4
a. $-\frac{125}{64}$ **b.** $\frac{b^5}{a^5}$

The number 3.58×10^4 is in scientific notation because 3.58 is greater than or equal to 1 but less than 10 and the exponent 4 is an integer. The number 35.8×10^3 is not in scientific notation because 35.8 is greater than 10. The number 0.358×10^5 also is not in scientific notation because 0.358 is less than 1.

Instructor Note The definition of scientific notation contains $1 \le |a| < 10$ to account for negative numbers expressed in scientific notation. However, at this level, we consider only positive numbers in scientific notation.

What does 3.58×10^4 mean?

$$
\begin{aligned}
3.58 \times 10^4 &= 3.58 \times 10 \times 10 \times 10 \times 10 \\
&= 35.8 \times 10 \times 10 \times 10 \\
&= 358 \times 10 \times 10 \\
&= 3580 \times 10 \\
&= 35,800
\end{aligned}
$$

Note 3.58×10^4, 35.8×10^3, and 0.358×10^5 all represent the same number, 35,800. Only 3.58×10^4 is in scientific notation.

Notice that each multiplication by 10 moves the decimal point one place to the right. Because we multiplied by four factors of 10, the decimal point moved a total of four places to the right from where it started in 3.58.

$$3.58 \times 10^4 = 3\,5\,8\,0\,0 = 35,800 \quad \text{The decimal point moves four places to the right.}$$

Our example suggests that the exponent of the power of 10 determines the number of places the decimal point moves.

> **Procedure** **Changing Scientific Notation (Positive Exponent) to Standard Form**
>
> To change from scientific notation with a positive integer exponent to standard form, move the decimal point to the right the number of places indicated by the exponent.

Example 5 Write each number in standard form.

a. 4.5×10^6

Solution: Multiplying 4.5 by 10^6 means that the decimal point will move six places to the right.

$$4.5 \times 10^6 = 4,500,000 \quad \blacktriangleleft$$

Connection Because 10^6 is another way to represent 1,000,000, it is popular to read 4.5×10^6 as 4.5 million.

b. 9×10^5

Solution: Multiplying 9 by 10^5 means that the decimal point will move five places to the right.

$$9 \times 10^5 = 900,000$$

Your Turn 5 Write each number in standard form.

a. 7.305×10^6 **b.** 8×10^7

What if the scientific notation contains a negative exponent, as in 8.45×10^{-2}? To evaluate 10^{-2}, we write it with a positive exponent, $\dfrac{1}{10^2}$.

$$
\begin{aligned}
8.45 \times 10^{-2} &= 8.45 \times \frac{1}{10^2} \\
&= \frac{8.45}{10^2} \\
&= \frac{8.45}{100} \\
&= 0.0845
\end{aligned}
$$

Note Multiplying 8.45 by 10^{-2} is equivalent to dividing 8.45 by 10^2. Dividing by 10^2, which is 100, causes the decimal point to move two places to the left.

$\blacktriangleleft 0.0\,8\,4\,5$

Answers to Your Turn 5
a. 7,305,000 **b.** 80,000,000

Our example suggests the following procedure.

Procedure Changing Scientific Notation (Negative Exponent) to Standard Form

To change from scientific notation with a negative exponent to standard form, move the decimal point to the left the same number of places as the absolute value of the exponent.

Example 6 Write each number in standard form.

a. 4.2×10^{-5}

Solution: Multiplying by 10^{-5} is equivalent to dividing by 10^5, which causes the decimal point to move five places to the *left*.

$$4.2 \times 10^{-5} = \frac{4.2}{10^5} = 0.000042$$

b. 7×10^{-9}

Solution: Multiplying by 10^{-9} is equivalent to dividing by 10^9, which causes the decimal point to move nine places to the left.

$$7 \times 10^{-9} = \frac{7}{10^9} = 0.000000007$$

Your Turn 6 Write each number in standard form.

a. 7.24×10^{-6} **b.** 5×10^{-7}

Objective 3 Write standard form numbers in scientific notation.

We mentioned that the distance from the Sun to the next nearest star, Proxima Centauri, is about 24,700,000,000,000 miles. Suppose we want to express 24,700,000,000,000 in scientific notation. Because scientific notation begins with a decimal number greater than or equal to 1 but less than 10, our first step is to determine the position of the decimal point. The only place the decimal point can go to satisfy the criteria for scientific notation is between the 2 and 4 digits.

24,700,000,000,000

The decimal goes here to express a decimal number
greater than or equal to 1 but less than 10.

Notice that there are 13 place values to the right of the decimal point in its new position between the 2 and the 4. To make the scientific notation equal to the original number, we must account for the 13 place values by expressing them as a power of 10.

$$24{,}700{,}000{,}000{,}000 = 2.4700000000000 \times 10^{13}$$

The 13 places to the right of the new decimal
position are expressed as a power of 10.

$$24{,}700{,}000{,}000{,}000 = 2.4700000000000 \times 10^{13} = 2.47 \times 10^{13}$$

These zeros to the right of the 7
are no longer needed.

Procedure Changing Standard Form to Scientific Notation

To write a number greater than 1 in scientific notation:
1. Move the decimal point so that the number is greater than or equal to 1 but less than 10. (*Tip: Place the decimal point to the right of the first nonzero digit*).
2. Write the decimal number multiplied by 10^n, where n is the number of places between the new decimal position and the original decimal position.
3. Delete zeros to the right of the last nonzero digit.

Answers to Your Turn 6
a. 0.00000724 **b.** 0.0000005

Example 7 Write 986,000 in scientific notation.

Solution: Place the decimal to the right of the first nonzero digit, 9. Next, count the places to the right of this decimal position, which is five places. Write the 5 as the exponent of 10. Finally, delete all of the 0s to the right of the last nonzero digit, which is the 6 in this case.

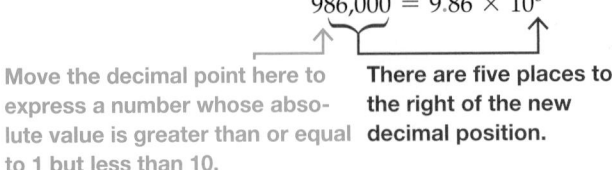

$$986,000 = 9.86 \times 10^5$$

Move the decimal point here to express a number whose absolute value is greater than or equal to 1 but less than 10.

There are five places to the right of the new decimal position.

Your Turn 7 Write the number in scientific notation.

a. 603,000,000
b. 92,500,000,000,000

We learned that a number written in scientific notation that has a negative exponent, such as 5×10^{-2}, has an absolute value less than 1. Therefore, a positive number less than 1, such as 0.00000075, has a negative exponent when written in scientific notation. To write 0.00000075 in scientific notation, we place the decimal between the 7 and 5 digits so that the number is greater than or equal to 1 but less than 10. Moving the decimal point to this position means that we have seven places to account for, which we indicate with a -7 exponent.

$$0.00000075 = 7.5 \times 10^{-7}$$

This suggests the following procedure.

Instructor Note When students are writing positive numbers in scientific notation, some of them find it easier to determine whether the exponent of 10^n is positive or negative by remembering that if the number is greater than 1, n is 0 or positive and if the number is less than 1, n is negative.

Procedure **Changing Standard Form to Scientific Notation**

To write a positive decimal number that is less than 1 in scientific notation:
1. Move the decimal point so that the number is greater than or equal to 1 but less than 10. (*Tip: Place the decimal point to the right of the first nonzero digit.*)
2. Write the decimal number multiplied by 10^n, where n is a negative integer whose absolute value is the number of places between the new decimal position and the original decimal position.
3. Delete zeros to the left of the first nonzero digit.

Example 8 Write 0.0000608 in scientific notation.

Solution: Place the decimal point between the 6 and 0 digits so that you have 6.08, which is a decimal number greater than 1 but less than 10. Because there are five decimal places between the original decimal position and the new position, the exponent is -5.

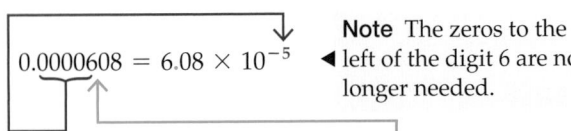

Note The zeros to the left of the digit 6 are no longer needed.

$$0.0000608 = 6.08 \times 10^{-5}$$

There are five decimal places between the original position and the new position, so the exponent is -5.

Move the decimal point here to express a number whose absolute value is greater than or equal to 1 but less than 10.

Answers to Your Turn 7
a. 6.03×10^8 **b.** 9.25×10^{13}

Answers to Your Turn 8
a. 1.72×10^{-4} **b.** 9.034×10^{-7}

Your Turn 8 Write in scientific notation.

a. 0.000172
b. 0.0000009034

5.1 Exercises For Extra Help MyMathLab®

Note: Exercises marked with a ★ represent challenging exercises.

Objective 1

Prep Exercise 1 After simplifying -5^2, is the result positive or negative? Explain.
The result is negative. The order of operations agreement tells us to find the additive inverse *after* multiplying the two fives: $-5^2 = -(5 \cdot 5) = -25$.

For Exercises 1–14, evaluate the exponential form. See Example 1.

1. 4^0
1

2. 3^0
1

3. -2^3
-8

4. -3^2
-9

5. $(-5)^3$
-125

6. $(-2)^5$
-32

7. -4^2
-16

8. -3^4
-81

9. $\left(\dfrac{3}{4}\right)^4$
$\dfrac{81}{256}$

10. $\left(\dfrac{5}{6}\right)^2$
$\dfrac{25}{36}$

11. $(-0.2)^3$
-0.008

12. $(-0.04)^3$
-0.000064

13. $(-2.3)^2$
5.29

14. $(-1.2)^2$
1.44

Prep Exercise 2 Explain how to rewrite an expression with a negative exponent so that its exponent is positive.

If a is a real number, where $a \neq 0$ and n is a natural number, then $a^{-n} = \dfrac{1}{a^n}$.

For Exercises 15–36, rewrite each expression with positive exponents; then if the expression is numeric, evaluate it. See Examples 2–4.

15. a^{-5}
$\dfrac{1}{a^5}$

16. x^{-4}
$\dfrac{1}{x^4}$

17. 2^{-3}
$\dfrac{1}{8}$

18. 5^{-1}
$\dfrac{1}{5}$

19. $(-4)^{-3}$
$-\dfrac{1}{64}$

20. $(-1)^{-5}$
-1

21. -2^{-4}
$-\dfrac{1}{16}$

22. -3^{-2}
$-\dfrac{1}{9}$

★ **23.** $(-0.2)^{-3}$
-125

★ **24.** $(-0.1)^{-4}$
$10{,}000$

25. $5m^{-2}$
$\dfrac{5}{m^2}$

26. $6r^{-4}$
$\dfrac{6}{r^4}$

27. $-6x^{-7}$
$-\dfrac{6}{x^7}$

28. $-8y^{-6}$
$-\dfrac{8}{y^6}$

29. $\left(\dfrac{m}{n}\right)^{-5}$
$\dfrac{n^5}{m^5}$

30. $\left(\dfrac{p}{q}\right)^{-6}$
$\dfrac{q^6}{p^6}$

31. $\left(\dfrac{5}{7}\right)^{-2}$
$\dfrac{49}{25}$

32. $\left(\dfrac{1}{6}\right)^{-3}$
216

33. $\dfrac{1}{b^{-6}}$
b^6

34. $\dfrac{1}{c^{-2}}$
c^2

35. $\dfrac{1}{4^{-3}}$
64

36. $\dfrac{1}{3^{-4}}$
81

📷 *For Exercises 37–44, use a scientific or graphing calculator to evaluate each exponential form.*

37. $(-24)^{-5}$
$-\dfrac{1}{7{,}962{,}624}$

38. $(-45)^{-4}$
$\dfrac{1}{4{,}100{,}625}$

39. $\left(-\dfrac{5}{8}\right)^{-3}$
$-\dfrac{512}{125}$

40. $\left(-\dfrac{4}{9}\right)^{-4}$
$\dfrac{6561}{256}$

41. -0.05^{-4}
$-160{,}000$

42. -0.02^{-2}
-2500

43. $(3.7)^{-6}$
≈ 0.00039

44. $(2.3)^{-4}$
≈ 0.03573

Find ⊗ the Mistake *For Exercises 45–48, find and explain the mistake; then work the problem correctly. See Examples 1 and 2.*

45. $-3^4 = (-3)(-3)(-3)(-3) = 81$
Mistake: The expression was interpreted to mean $(-3)^4$, but it means the additive inverse of 3^4. Correct: -81

46. $(-5)^3 = -75$
Mistake: -75 was obtained by squaring -5, then multiplying by -3 instead of multiplying $(-5)(-5)(-5)$. Correct: -125

47. $5^{-2} = -25$
Mistake: The negative sign in the exponent was passed to the product instead of the original expression being inverted. Correct: $\dfrac{1}{25}$

48. $\left(\dfrac{3}{4}\right)^{-2} = \dfrac{9}{16}$
Mistake: The negative sign was disregarded. The original expression should have been inverted. Correct: $\dfrac{16}{9}$

Objective 2

Prep Exercise 3 To write 7.8×10^5 in standard form, how far and in what direction do you move the decimal point?

5 places to the right

Prep Exercise 4 To write 3.71×10^{-8} in standard form, how far and in what direction do you move the decimal point?

8 places to the left

For Exercises 49–64, write the number in standard form. See Examples 5 and 6.

Of Interest

49. Scientists speculate that the universe is about 1.65×10^{10} years old.

    16,500,000,000

50. The solar system formed approximately 4.6×10^9 years ago.

    4,600,000,000

51. In empty space, light travels at a constant speed of approximately 2.998×10^8 meters per second.

    299,800,000

52. The equatorial radius of the Sun is approximately 6.96×10^8 meters.

    696,000,000

53. The equatorial radius of Earth is approximately 6.378×10^6 meters.

    6,378,000

54. A light-year is the distance that light travels in a year's time and is equal to about 5.76×10^{12} miles.

    5,760,000,000,000

The light from the Sun takes about 8.2 minutes to reach Earth and about 4.26 years to reach Proxima Centauri. We would say that Earth is at a distance of about 8.2 light-minutes from the Sun and Proxima Centauri is at a distance of about 4.26 light-years.

55. Earth is about 9.292×10^7 miles from the Sun.

    92,920,000

56. The next nearest star to ours (the Sun) is Proxima Centauri, which is about 2.47×10^{13} miles from the Sun.

    24,700,000,000,000

57. Mitochondria are organelles that are about 2.5×10^{-6} meter across.

    0.0000025

58. The diameter of human DNA is about 2×10^{-9} meter.

    0.000000002

59. The rest mass of a proton is about 1.67×10^{-27} kilogram.

    0.00000000000000000000000000167

60. Light with a wavelength of 7.6×10^{-11} meter is in the X-ray portion of the spectrum.

    0.000000000076

61. Light with a wavelength of 2.95×10^{-9} meter is in the ultraviolet portion of the spectrum.

    0.00000000295

62. Light with a wavelength of 4.5×10^{-7} meter is in the visible spectrum and is blue in color.

    0.00000045

63. The mass of an alpha particle, which is emitted in the radioactive decay of plutonium 239, is 6.645×10^{-27} kilogram.

    0.000000000000000000000000006645

64. Protons have a positive electrical charge of 1.602×10^{-19} coulomb.
0.0000000000000000001602

Objective 3

Prep Exercise 5 To write 45,900,000 in scientific notation, between what two digits should the decimal point be placed?
Between the 4 and the 5

Prep Exercise 6 To write 0.00000784 in scientific notation, what exponent is used?
−6

For Exercises 65–80, write the number in scientific notation. See Examples 7 and 8.

65. There are approximately 25,000,000,000,000 red blood cells in the average person's bloodstream.
2.5×10^{13}

66. A single red blood cell contains about 250,000,000 molecules of hemoglobin.
2.5×10^{8}

67. Human DNA consists of about 5,300,000,000 nucleotide pairs.
5.3×10^{9}

68. The Andromeda galaxy is about 2,140,000 light-years from the Sun.
2.14×10^{6}

69. Uranium 236 is a radioactive isotope of uranium and has a half-life of 23,420,000 years.
2.342×10^{7}

70. The population of the United States is about 304,000,000.
3.04×10^{8}

71. The most populated country in the world is China, with a population of about 1,300,000,000.
1.3×10^{9}

72. In 2008, the population of Earth was about 6,678,000,000.
6.678×10^{9}

73. Light with a wavelength of 0.000000586 meter is yellow in color.
5.86×10^{-7}

74. Light with a wavelength of 0.000000712 meter is red in color.
7.12×10^{-7}

75. Electrons have a mass of approximately 0.00000000000000000000000000000091094 kilogram.
9.1094×10^{-31}

76. The size of the HIV virus that causes AIDS is about 0.0000001 meter.
1×10^{-7}

77. The most common diameter for the core of a fiber-optic cable is 0.0000625 meter.
6.25×10^{-5}

78. A human hair is about 0.00005 meter wide.
5×10^{-5}

79. Intel introduced an improved process for manufacturing computer chips, which allowed the company to make transistors that are 0.00000005 meter in length.
5×10^{-8}

80. On April 2, 2008, 0.00975 U.S. dollar equaled 1 Japanese yen.
9.75×10^{-3}

Of Interest

Hemoglobin is an iron–protein compound essential for respiration. It is responsible for carrying oxygen and carbon dioxide in the red blood cells. It also regulates blood pressure and acidity in blood.

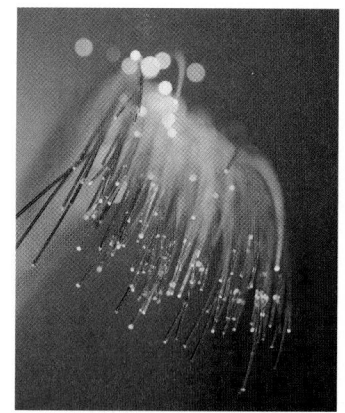

For Exercises 81 and 82, write the numbers in order from smallest to largest.

81. 7.5×10^6, 8.95×10^5, 9×10^7, 1.3×10^8, 7.2×10^6
     8.95×10^5, 7.2×10^6, 7.5×10^6, 9×10^7, 1.3×10^8

82. 2.8×10^{-5}, 3.7×10^{-7}, 6.2×10^{-2}, 9.8×10^{-5}, 6.1×10^{-2}
     3.7×10^{-7}, 2.8×10^{-5}, 9.8×10^{-5}, 6.1×10^{-2}, 6.2×10^{-2}

★ **83.** In Exercises 60 and 61, we saw that light with a wavelength of 7.6×10^{-11} meter is in the X-ray portion of the spectrum and light with a wavelength of 2.95×10^{-9} meter is in the ultraviolet portion of the spectrum. About how many times greater is ultraviolet light's wavelength compared to that of X-rays?
The wavelength of ultraviolet light is about 38.8 times the wavelength of X-rays.

★ **84.** In Exercises 59 and 63, we saw that a proton has a mass of 0.00000000000000000000000000167 kilogram and an alpha particle has a rest mass of about 6.645×10^{-27} kilogram, respectively. Which has the greater mass? About how many times greater is the mass of the larger of the two particles?
Alpha particles are about 4 times bigger.

Review Exercises

Exercises 1 and 2   **Expressions**

[1.3] **1.** Simplify: $-13 + 2$
     -11

[1.7] **2.** Use the distributive property to rewrite the expression $-4(y - 3)$.
     $-4y + 12$

Exercises 3–6   **Equations and Inequalities**

[2.7] **3.** 40 is 15% of what?
     $266\frac{2}{3}$

[3.4] **4.** Find the slope of the line that passes through $(2, -1)$ and $(-3, -2)$.
     $\frac{1}{5}$

[3.4] **5.** Find the slope and y-intercept for $2x - 3y = 6$.
     slope $\frac{2}{3}$; y-intercept: $(0, -2)$

[3.7] **6.** Given the function $f(x) = x^3 - 5x^2 + 7$, find $f(-3)$.
     -65

5.2 Introduction to Polynomials

Objectives

1 Identify monomials.
2 Identify the coefficient and degree of a monomial.
3 Classify polynomials.
4 Identify the degree of a polynomial.
5 Evaluate polynomials.
6 Write polynomials in descending order of degree.
7 Combine like terms.

Warm-up

[1.5] **1.** Evaluate: $40 \div 5(-4) - (-3)^3$
[1.7] **2.** Evaluate: $-16a^2 + b$ when $a = 3$ and $b = 180$

Objective 1 Identify monomials.

We have explored some simple algebraic expressions (Chapter 1) and linear equations and inequalities (Chapters 2–4). We now move back to the expression level of our Algebra Pyramid and explore a particular class of expressions known as *polynomials*.

Answers to Warm-up
  1. -5
  2. 36

The Algebra Pyramid

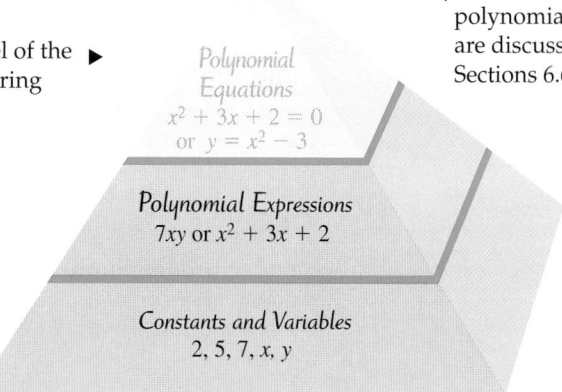

◄ **Note** Two types of polynomial equations are discussed in Sections 6.6 and 6.7.

Note Inequalities are also in the top level of the Algebra Pyramid, but we won't be exploring polynomial inequalities at this point. ►

First, we consider a special type of polynomial expression called a **monomial**.

Definition Monomial: An expression that is a constant, a variable, or a product of a constant and variable(s) that are raised to whole-number powers.

Implied in the definition of a monomial is that there is no addition or subtraction; so there is only one term.

Connection Although no variable factor appears with the constant 7, it can be expressed with a variable factor. Recall from Section 5.1 that any nonzero number raised to the 0 power simplifies to 1. Therefore, 7 can be written as $7x^0$ because $x^0 = 1$ (assuming that $x \neq 0$). We use ► this fact to help us understand the next objective concerning degree.

Example 1   Is the given expression a monomial? Explain.

a. 7

Answer: 7 is a monomial because it is a constant.

b. $3x^2$

Answer: $3x^2$ is a monomial because it is a product of a constant, 3, and a variable, x, which has a whole-number exponent of 2.

c. $9y$

Answer: Remember that a number or variable with no apparent exponent has an understood 1 exponent; so $9y = 9y^1$. Therefore, $9y$ is a monomial because it is a product of a constant, 9, and a variable, y, which has a whole-number exponent of 1.

d. $-0.6xy^2$

Answer: $-0.6xy^2$ is a monomial because it is a product of a constant, -0.6, and variables, x and y^2, which have whole-number exponents. As we saw in Example 1(c), we could write $-0.6xy^2$ as $-0.6x^1y^2$. Both of the variables' exponents, 1 and 2, are whole numbers.

e. y^4

Answer: It might seem that y^4 is not a monomial because no numerical factor is visible. However, y^4 is the same as $1y^4$. We don't need to write the 1 because 1 multiplied by an amount does not affect the amount.

f. $\dfrac{5}{x^2}$

Note A monomial can't have a ► variable as a divisor.

Answer: The expression $\dfrac{5}{x^2}$ can be written as a product by writing the divisor with a negative exponent, $5x^{-2}$. However, when written as a product, the variable's exponent is not a whole number, which means it is not a monomial.

g. $4y^2 + 9y - 3$

Answer: The expression is not a monomial because it is not a product of a constant and variables. Instead, addition and subtraction are involved.

| **Your Turn 1** | Is the given expression a monomial? Explain.

a. x^3 **b.** $-y^2$ **c.** $-\dfrac{2}{3}tu$ **d.** $\dfrac{4m^2}{n}$ **e.** 0.56 **f.** $3y + 7$

Objective 2 Identify the coefficient and degree of a monomial.

In our introduction to expressions in Section 1.7, we defined a **coefficient** as the numerical or constant factor in a term. The same is true for monomials. Another piece of information about a monomial is its **degree**, which relates to the exponents of the variables.

Definitions   Coefficient of a monomial: The numerical factor in a monomial.
Degree of a monomial: The sum of the exponents of all variables in a monomial.

| **Example 2** | Identify the coefficient and degree of each monomial.

a. $-5t^2$

Answer: The coefficient is -5 because it is the numerical factor. The degree is 2 because it is the exponent for the single variable t.

b. $-mn^4$

Answer: We can express $-mn^4$ as $-1m^1n^4$. In this form, we can see that the coefficient is -1 and that the exponents for the variables are 1 and 4. Because the degree is the sum of the variables' exponents, the degree is 5.

c. 7

Answer: Remember that $7 = 7x^0$, where x is any real number except 0. In this alternative form, we can see that 7 is the coefficient and 0 is the degree because it is the variable's exponent.

| **Your Turn 2** | Identify the coefficient and degree of each monomial.

a. $-4x^4$ **b.** $-0.9n^2p^5$ **c.** $\dfrac{5}{8}ab$ **d.** 13

Objective 3 Classify polynomials.

Now we are ready to formally define a **polynomial**.

Definition   Polynomial: A monomial or a sum of monomials.

Examples of polynomials: $4x$, $4x + 8$, $2x^2 + 5xy + 8y$, and $2x^3 - 5x^2 + 3x - 9$

Note Although it seems that the subtractions violate the word *sum* in the definition, the expression $2x^3 - 5x^2 + 3x - 9$ is a polynomial because it can be written as a sum of monomials by rewriting the subtractions as equivalent additions.

$$2x^3 - 5x^2 + 3x - 9 = 2x^3 + (-5x^2) + 3x + (-9)$$

Although those two forms are equivalent, $2x^3 - 5x^2 + 3x - 9$ is preferable because it is the simplest form. That is, it has the fewest symbols.

Notice that the polynomial $2x^3 - 5x^2 + 3x - 9$ has the same variable x in each variable term, whereas $2x^2 + 5xy + 8y$ has a mixture of two variables, x and y. We call a polynomial such as $2x^3 - 5x^2 + 3x - 9$ a **polynomial in one variable**.

Answers to Your Turn 1
a. x^3 is a monomial because it is a product of a constant, 1, and a variable, x, which has a whole-number exponent of 3.
b. $-y^2$ is a monomial because it is a product of a constant, -1, and a variable y, with a whole-number exponent of 2.
c. $-\dfrac{2}{3}tu$ is a monomial because it is a product of a constant, $-\dfrac{2}{3}$, and variables, t and u, both of which have the whole number 1 as an exponent.
d. $\dfrac{4m^2}{n}$ is not a monomial because it has a variable, n, as a divisor.
e. 0.56 is a monomial because 0.56 is a constant.
f. $3y + 7$ is not a monomial because it contains addition.

Answers to Your Turn 2
a. c: -4; d: 4   **b.** c: -0.9; d: 7
c. c: $\dfrac{5}{8}$; d: 2   **d.** c: 13; d: 0

Definition **Polynomial in one variable:** A polynomial in which every variable term has the same variable.

$x^2 - 5x + 2$ is a polynomial in one variable, x.

$x^2 + 8y$ is a polynomial in two variables, x and y.

Special names have been given to some polynomials. We have already discussed monomials such as $4x$, which is a single-term polynomial. Notice that the prefix *mono* in the word *monomial* indicates "one" term and the prefix *poly* in *polynomial* indicates "many" terms. Continuing with the prefixes, a two-term polynomial such as $4x + 8$ is called a **binomial**, and a three-term polynomial such as $2y^2 + 5y + 8$ is called a **trinomial**. No special names are given to polynomials with more than three terms.

Definitions **Binomial:** A polynomial containing two terms.
Trinomial: A polynomial containing three terms.

Example 3 Indicate whether the expression is a monomial, binomial, or trinomial; has no special polynomial name; or is not a polynomial.

a. $2.9xy^3z$

Answer: $2.9xy^3z$ is a monomial because it has a single term.

b. $-7a^2 + b$

Answer: $-7a^2 + b$ is a binomial because it contains two terms.

c. $9t^2 + 3.4t - 8.2$

Answer: $9t^2 + 3.4t - 8.2$ is a trinomial because it contains three terms.

d. $y^3 + 6y^2 - y + 15$

Answer: Although $y^3 + 6y^2 - y + 15$ is a polynomial, it has no special name because it has more than three terms.

e. $6x + \dfrac{3}{y}$

Answer: $6x + \dfrac{3}{y}$ is not a polynomial because $\dfrac{3}{y}$ is not a monomial.

Your Turn 3 Indicate whether the expression is a monomial, binomial, or trinomial; has no special polynomial name; or is not a polynomial.

a. $b + 2$ **b.** $7r^3 - 9r^2 + r - 6$ **c.** $5.7x^4y$ **d.** $x^2 - 7xy - y^2$ **e.** $4m - \dfrac{3}{n^2}$

Objective 4 Identify the degree of a polynomial.

Previously, we defined the degree of a monomial to be the sum of the exponents of the variables. We can also talk about the **degree of a polynomial**.

Definition **Degree of a polynomial:** The greatest degree of any of the terms in the polynomial.

Answers to Your Turn 3
a. binomial
b. no special polynomial name
c. monomial
d. trinomial
e. not a polynomial

Example 4 Identify the degree of each polynomial.

a. $x^3 + 12x^7 - 10x^2 + 9x - 14$

Answer: The degree is 7 because it is the greatest degree of all of the terms.

b. $12x^6 - 3x^5y^3 - xy^3 + 4y^2 - 16$

Answer: To determine the degree of the monomial $-3x^5y^3$, we add the exponents of its variables; so its degree is 8. Likewise, we add the exponents of the variables in $-xy^3$ and find that its degree is 4. Comparing these degrees with the degrees of the other terms, we see that 8 is the greatest degree. Therefore, 8 is the degree of the polynomial.

Your Turn 4 Identify the degree of each polynomial.

a. $y^7 - 3y^2 - 8y^9 + 4y + 5$ **b.** $2x^5 - 9x^6y + x^3y^3 - 15y - 7$

Objective 5 Evaluate polynomials.

In Chapter 1, we learned that we can evaluate an expression. Let's now evaluate a polynomial expression. To evaluate an expression, we replace each variable with its given value and calculate the numerical result.

Example 5 Evaluate $x^3 - 5x^2 + 7$ when $x = -3$.

Solution: $x^3 - 5x^2 + 7$

$(-3)^3 - 5(-3)^2 + 7$ Replace each x with -3.

$= -27 - 5(9) + 7$ Simplify.

$= -27 - 45 + 7$

$= -65$

Example 6 If we neglect air resistance, the polynomial $-16t^2 + h_0$ describes the height of a falling object after falling from an initial height h_0 for t seconds. It is said that Galileo Galilei dropped cannonballs from the Leaning Tower of Pisa to study gravitational effects. If the tower is 180 feet tall, what is the height of a cannonball after it falls 3 seconds?

Solution: Evaluate $-16t^2 + h_0$ when $t = 3$ and $h_0 = 180$.

$-16(3)^2 + 180$ Replace t with 3 and h_0 with 180.

$= -16(9) + 180$

$= -144 + 180$

$= 36$

After 3 seconds, the height of the cannonball is 36 feet.

Your Turn 6 Use the polynomial in Example 6 to find the height of an object after it falls 4 seconds from an initial height of 300 feet.

Of Interest

Galileo revolutionized scientific thought with his contributions to the understanding of gravity and the motion of objects. By constructing his own telescope, he made observations that confirmed Copernicus's theory that Earth and the planets revolve around the Sun. Although these beliefs came under great scrutiny, their accuracy holds true to this day.

Objective 6 Write polynomials in descending order of degree.

Now let's consider some ways to rewrite a polynomial. Recall from Chapter 1 that rewriting an expression means that we write an equivalent expression by applying properties of arithmetic. For example, we can apply the commutative property of addition and rearrange the terms in $-8 + 4x^3 - 7x^2 + 2x$ to get an equivalent expression, $4x^3 - 7x^2 + 2x - 8$. Notice that the signs move with the terms. Also notice that the terms in $4x^3 - 7x^2 + 2x - 8$ are written in order from the greatest degree to the least degree.

Answers to Your Turn 4
a. 9   **b.** 7

Answer to Your Turn 6
44 ft.

$$4x^3 \qquad -7x^2 \qquad +2x \qquad -8$$

Degree 3   Degree 2   Degree 1   Degree 0

A polynomial written this way is said to be in *descending order of degree*.

> **Procedure** **Writing a Polynomial in Descending Order of Degree**
>
> To write a polynomial in descending order of degree, place the highest degree term first, then the next highest degree term, and so on.

Writing a polynomial in more than one variable in descending order of degree can be complicated, so we will write only polynomials in one variable in descending order of degree.

Example 7 Write each polynomial in descending order of degree.

a. $9x^2 + 7x^4 - 2x^3 - 5 + 3x$

Solution: Rearrange the terms so that the highest degree term is first, then the next highest degree, and so on.

$$7x^4 \qquad -2x^3 \qquad +9x^2 \qquad +3x \qquad -5$$

Degree 4   Degree 3   Degree 2   Degree 1   Degree 0

Answer: $7x^4 - 2x^3 + 9x^2 + 3x - 5$

b. $-9y^3 + 2y^6 + y - 3y^4 + 12 - 6y^2$

Answer: $2y^6 - 3y^4 - 9y^3 - 6y^2 + y + 12$

Your Turn 7 Write each polynomial in descending order of degree.

  a. $2n^3 + n^5 - 9n^2 + 8 - 3n^4 + 4n$ **b.** $-4t^2 + 2t + 16 + 5t^3 - 13t^6$

Objective 7 Combine like terms.

Another way we can rewrite a polynomial expression is by combining like terms. We learned in Section 1.7 that combining like terms is one way to rewrite an expression in simplest form. We also learned that like terms have the same variables raised to the same powers.

Example 8 Combine like terms and write the resulting polynomial in descending order of degree.

$$12x^3 + 5x^2 - 2x^3 - 13 + 6x + x^2 + 2 - 6x$$

Solution: $12x^3 + 5x^2 - 2x^3 - 13 + 6x + x^2 + 2 - 6x$

$= 12x^3 - 2x^3 + 5x^2 + x^2 + 6x - 6x - 13 + 2$ Collect like terms as needed.

$= \quad 10x^3 \quad + \quad 6x^2 \quad + \quad 0 \quad -11$ Combine like terms.

$= 10x^3 + 6x^2 - 11$ Drop 0 and bring the terms together.

Warning When combining like terms, we *add only the coefficients*, keeping the exponents the same.

Answers to Your Turn 7
a. $n^5 - 3n^4 + 2n^3 - 9n^2 + 4n + 8$
b. $-13t^6 + 5t^3 - 4t^2 + 2t + 16$

Alternative Solution: Some people prefer to skip the collecting step and strike through like terms as they are combined to keep track of the terms that have been combined. It is helpful to combine the terms in descending order of degree so that the answer will be in descending order of degree.

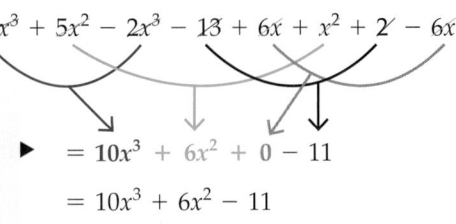

Note We consider the terms in the following order:
Degree 3 terms: $12x^3 - 2x^3 = 10x^3$
Degree 2 terms: $5x^2 + x^2 = 6x^2$
Degree 1 terms: $6x - 6x = 0$
Degree 0 terms: $-13 + 2 = -11$

$$\blacktriangleright \quad = 10x^3 + 6x^2 + 0 - 11$$
$$= 10x^3 + 6x^2 - 11$$

Your Turn 8 Combine like terms and write the resulting polynomial in descending order of degree.

a. $10x^3 - 4x^2 - 3x^3 + 8 - 5x^2 - 9 + 6x$
b. $3y^4 - 12y - y^2 + y - 6 + y^2 + 5y^4$

Remember that polynomials may have multiple variables. Multivariable monomials are like terms if they have the same variables raised to the same exponents.

Example 9 Combine like terms.

$$x^4 + 8x^2y - 7 - 13x^2y + 4y - 2xy^2 + 6y - x^4 + 2$$

Solution: Collecting the like terms and combining them gives us the following:

$$x^4 + 8x^2y - 7 - 13x^2y + 4y - 2xy^2 + 6y - x^4 + 2$$
$$= x^4 - x^4 + 8x^2y - 13x^2y - 2xy^2 + 4y + 6y + 2 - 7$$

$$= \quad 0 \qquad - 5x^2y \qquad - 2xy^2 \qquad + 10y \qquad - 5$$
$$= -5x^2y - 2xy^2 + 10y - 5$$

Alternative Solution: Instead of collecting the like terms, we strike through like terms in the given polynomial as they are combined.

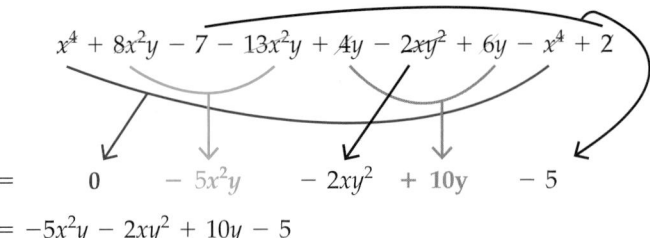

$$= \quad 0 \qquad - 5x^2y \qquad - 2xy^2 \qquad + 10y \qquad - 5$$
$$= -5x^2y - 2xy^2 + 10y - 5$$

Learning Strategy

If you are a tactile learner, you may prefer to collect the like terms first. If you are a visual learner, you may prefer to strike through the terms as you combine them.

Answers to Your Turn 8
a. $7x^3 - 9x^2 + 6x - 1$
b. $8y^4 - 11y - 6$

Answers to Your Turn 9
a. $2x^5 - 3x^3y^2 - 4$
b. $3t^3u - t^2u - 2u$

Your Turn 9 Combine like terms.

a. $-x^5 + 6x^3y^2 + 2y - 9x^3y^2 + 3x^5 - 4 - 2y$
b. $-6t^3u + 12tu - 4t^2u + 9t^3u - u + 3t^2u - 12tu - u$

5.2 Exercises (For Extra Help) MyMathLab®

Note: Exercises marked with a ★ represent challenging exercises.

Objective 1

Prep Exercise 1 What is a monomial?

A constant, variable, or product of a constant and variable(s) that are raised to whole-number powers

For Exercises 1–12, determine whether the expression is a monomial. See Example 1.

1. $-3x^2$
monomial

2. $5y^3$
monomial

3. $\dfrac{1}{5}$
monomial

4. $\dfrac{4}{7}$
monomial

5. $\dfrac{6m}{4y^3}$
not

6. $\dfrac{-8t}{5x}$
not

7. $4x^2 - 3x + 7$
not

8. $p^2 - q^2$
not

9. y
monomial

10. 0.76
monomial

11. $3m^4n^2$
monomial

12. $6u^3v^4w$
monomial

Objective 2

Prep Exercise 2 What is the coefficient of a monomial? The numerical factor in a monomial

Prep Exercise 3 Explain how to determine the degree of a monomial. Find the sum of the exponents on all variables in the monomial.

For Exercises 13–24, identify the coefficient and degree of each monomial. See Example 2.

13. $-5m^3$
c: -5; d: 3

14. $8p^2$
c: 8; d: 2

15. $-xy^4$
c: -1; d: 5

16. $-m^2n^6$
c: -1; d: 8

17. -9
c: -9; d: 0

18. 18
c: 18; d: 0

19. $4.2n^3p$
c: 4.2; d: 4

20. $-6.7uv^7$
c: -6.7; d: 8

21. $16abc$
c: 16; d: 3

22. $-8.1lkm$
c: -8.1; d: 3

23. y
c: 1; d: 1

24. w
c: 1; d: 1

Objectives 3 and 4

Prep Exercise 4 Explain the differences between a monomial, a binomial, and a trinomial.
A monomial contains one term, a binomial contains two terms, and a trinomial contains three terms.

For Exercises 25–36, indicate whether the expression is a monomial, binomial, or trinomial or has no special polynomial name. If the expression is a polynomial, give the degree. See Examples 3 and 4.

25. $6x + 8y + 3z$
trinomial, 1

26. $17.3x^2 - 3x + 2.1$
trinomial, 2

27. $-7m^2n$
monomial, 3

28. $18xy^3$
monomial, 4

29. $5.2x^3 - 3x^2 + 4.1x - 11$
No special polynomial name, 3

30. $x^3 - 4x^2 + 4x + 5$
No special polynomial name, 3

31. $5u^3 - 16u$
binomial, 3

32. $7k + 5k^3$
binomial, 3

33. -21
monomial, 0

34. 36
monomial, 0

35. $\dfrac{25}{x} - x^2$
Not a polynomial

36. $y^2 - \dfrac{16}{y}$
Not a polynomial

Objective 4

Prep Exercise 5 Explain how to determine the degree of a polynomial with two or more terms. Find the greatest degree of any of the terms in the polynomial.

For Exercises 37–46, identify the degree of each polynomial. See Example 4.

37. $19 - 7y^4 + 3y - 2y^3 - 7$
4

38. $6a^5 - 19a^3 + a^9 + 5a - 14$
9

39. $16z^4 + 7z^2 - z^5 - 4z^3 + z$
5

40. $11 + 4t^5 - 5t + 7t^3 - 18t^2 + t^8$
8

41. $2u^2 + 5u^3 - u^7 + 13u^4 - 3u$
7

42. $22j^3 + 5j^2 - 16j^4 + 21 - 14j$
4

43. $3ab^2 - 6a^2b^2 + 4a - 7b^3$
4

44. $6mn - 7m^3n^2 + 4m^2n^5 - 3m^4$
7

45. $-8x^4y + 5x^3y^3 - 5x^2y^6 + 3xy^7 + 2x^6$
8

46. $-2p^3q^3 + 5p^6q - 3p^2q^4 + 7p^4q^4 - 8q^5$
8

Objective 5

Prep Exercise 6 Explain how to evaluate a polynomial.
Replace each variable with its given value and calculate the numerical result.

For Exercises 47–54, evaluate the polynomial using the given values. See Example 5.

47. $-3xy^2; x = -5, y = 2$
60

48. $-2x^2y; x = -1, y = 4$
-8

49. $x^2 - 6x + 1; x = 3$
-8

50. $n^2 - 8n - 3; n = -4$
45

51. $a^3 + 0.5ab + 2.4b;$
$a = -1, b = -2$
-4.8

52. $m^3 + 3.6mn - 0.2n^2;$
$m = -3, n = 4$
-73.4

53. $\frac{1}{2}x^2 + \frac{2}{3}x + 3; x = 6$
25

54. $\frac{1}{4}a^2 + \frac{3}{2}a + 2; a = 4$
12

For Exercises 55–62, evaluate. See Example 6.

55. If we neglect air resistance, the polynomial $-16t^2 + h_0$ describes the height of a falling object after falling from an initial height h_0 for t seconds. A marble is dropped from a tower at a height of 50 feet.
  a. What is its height after 0.5 seconds?
  46 ft.
  b. What is its height after 1.2 seconds?
  26.96 ft.

56. The polynomial $2lw + 2lh + 2wh$ describes the surface area of a box.
  a. Find the surface area of a box with a length of 8 inches, a width of 6.5 inches, and a height of 4 inches.
  220 in.²

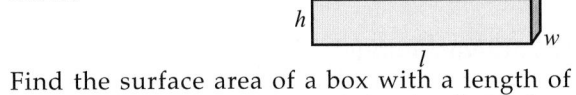

  b. Find the surface area of a box with a length of 4 inches, a width of 3 inches, and a height of 2.5 inches.
  59 in.²

57. The polynomial $5p^2 - 6p + 3$ describes the number of units sold for every 1000 people in a particular region based on the price of the product, which is represented by p.
  a. Find the number of units sold for every 1000 people if the price of each unit is $3. 30 units
  b. Find the number of units sold for every 1000 people if the price of each unit is $3.50. 43 units

58. The polynomial $7r^2 - 2r + 6$ describes the voltage in a circuit, where r represents the resistance in the circuit.
  a. Find the voltage if the resistance is 6 ohms.
  246 V
  b. Find the voltage if the resistance is 8 ohms.
  438 V

59. An engineer is designing a chemical storage tank that is capsule-shaped. The polynomial $\frac{4}{3}\pi r^3 + \pi r^2 h$ describes the volume of the tank.

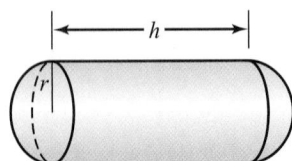

  a. Find the volume of the tank if the radius is 4 feet and the height is 20 feet. Round the result to the nearest tenth. ≈ 1273.4 ft.³
  b. Find the volume of the tank if the radius is 3 feet and the height is 15 feet. Round the result to the nearest tenth. ≈ 537.2 ft.³

60. The polynomial $lwh - \pi r^2 h$ describes the volume of metal remaining in a block after a cylinder has been bored into the block of metal.

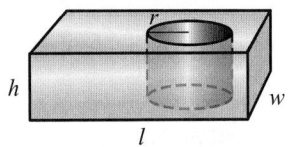

  a. Find the volume of metal remaining if the length is 15 inches, the width is 7 inches, the height is 4 inches, and the radius of the cylinder is 3 inches. Round to the nearest tenth. ≈ 306.9 in.³
  b. Find the volume of metal remaining if the length is 15 inches, the width is 7 inches, the height is 4 inches, and the radius of the cylinder is 2 inches. Round to the nearest tenth. ≈ 369.7 in.³

61. The polynomial $-0.2x^2 + 4.25x + 26$ describes the number of international visitors (in millions) to the United States each year from 1986 on, where x represents the number of years after 1986. ($x = 0$ means 1986, $x = 1$ means 1987, and so on.) (*Source:* Tourism Industries, International Trade Administration, Department of Commerce.)

   a. How many international visitors came to the United States in 1986?

   26 million

   b. How many international visitors came to the United States in 1993?

   45.95 million

   c. How many international visitors came to the United States in 2003?

   40.45 million

   d. If the model holds, predict the number of international visitors in 2012. Do you think the prediction is reasonable? Explain.

   1.3 million; answers will vary, but the prediction is most likely not reasonable because of the increase in world population and unforeseen social factors that might affect travel; it is not likely that the number will decrease that much.

62. The polynomial $28x^3 + 25x^2 + 100x + 340$ describes the number of cell phone subscribers in the United States (in thousands of subscribers) each year since 1985, where x represents the number of years since 1985. ($x = 0$ means 1985, $x = 1$ means 1986, and so on.) (*Source:* The CTIA Semiannual Wireless Industry Survey.)

   a. How many subscribers were in the United States in 1985?

   340,000

   b. How many subscribers were in the United States in 1995?

   31,840,000

   c. How many subscribers were in the United States in 2001?

   123,028,000

   d. If the model holds, predict the number of subscribers in 2015. Do you think the prediction is reasonable? Explain.

   718,840,000; answers will vary, but the prediction is not reasonable because at the current growth rates, the population of the United States will be around 322,000,000 in 2015.

Objective 6

Prep Exercise 7 Given a polynomial in one variable, what does *descending order of degree* mean? Place the highest degree term first, then the next highest degree, and so on.

For Exercises 63–68, write each polynomial in descending order of degree. See Example 7.

63. $7x^4 + 5x^7 - 8x + 14 - 3x^5$
   $5x^7 - 3x^5 + 7x^4 - 8x + 14$

64. $-4y^3 + 7y - 8y^5 - 6y^2 + 9 + 2y^4$
   $-8y^5 + 2y^4 - 4y^3 - 6y^2 + 7y + 9$

65. $4r - 3r^2 + 18r^5 + 7r^3 - 8r^6$
   $-8r^6 + 18r^5 + 7r^3 - 3r^2 + 4r$

66. $u^4 + 7u^2 - u^9 + 15u + 27 - 3u^5 - 8u^3$
   $-u^9 - 3u^5 + u^4 - 8u^3 + 7u^2 + 15u + 27$

67. $20 - w^3 + 5w + 11w^2 - 12w^4$
   $-12w^4 - w^3 + 11w^2 + 5w + 20$

68. $a - 6 + 7a^5 + 3a^2 - 4a^3$
   $7a^5 - 4a^3 + 3a^2 + a - 6$

Objectives 6 and 7

Prep Exercise 8 Explain how to combine like terms. Add the coefficients and leave the variables and exponents the same.

For Exercises 69–88, combine like terms and write the resulting polynomial in descending order of degree. See Example 8.

69. $3x + 4 + 2x - 7$
   $5x - 3$

70. $2y - 9 + 5y + 3$
   $7y - 6$

71. $7 - 6a + 2a - 2$
   $-4a + 5$

72. $9 + 4b - 8b - 3$
   $-4b + 6$

73. $7x^2 + 2x + 4x - 9x^2$
   $-2x^2 + 6x$

74. $4a^3 + 3a^2 - 2a^2 - 7a^3$
   $-3a^3 + a^2$

75. $\frac{1}{2}x^2 + \frac{1}{3}x + \frac{4}{3}x + \frac{3}{2}x^2$
   $2x^2 + \frac{5}{3}x$

76. $\frac{2}{3}y^2 + \frac{1}{4}y + \frac{7}{4}y + \frac{5}{3}y^2$
   $\frac{7}{3}y^2 + 2y$

77. $\frac{5}{3}a^2 + \frac{2}{5}a - \frac{1}{2}a^2 - \frac{5}{3}a$
   $\frac{7}{6}a^2 - \frac{19}{15}a$

78. $\dfrac{9}{4}n^2 - \dfrac{7}{6}n - \dfrac{2}{3}n^2 + \dfrac{3}{4}n$
$\dfrac{19}{12}n^2 - \dfrac{5}{12}n$

79. $2x^2 + 5x - 7x^2 + 8 - 5x + 6$
$-5x^2 + 14$

80. $3m^5 + 7m^2 - 8m + 9m^5 - 7m - 7m^2$
$12m^5 - 15m$

81. $15 - 4y + 8y^2 - 4y - 3y^2 + 7 - y$
$5y^2 - 9y + 22$

82. $6l - 5l^3 - 2l^4 + 7l^4 + 5l - l^3 + 4l^3$
$5l^4 - 2l^3 + 11l$

83. $9k - 5k^2 + 6k^3 + 2k^2 + k^4 - 3k - 3k^4$
$-2k^4 + 6k^3 - 3k^2 + 6k$

84. $11p - p^2 + 12p^3 + 5p^2 - 7p^3 + 20 - 4p$
$5p^3 + 4p^2 + 7p + 20$

85. $7a^2 - 5a^4 + 6a^3 - 5a^4 + 12 - 6a^3 + a + 4$
$-10a^4 + 7a^2 + a + 16$

86. $-6c^9 + 7c^3 - 4c^5 + 8 - 14c^2 + 4c^5 - 7 - c^9$
$-7c^9 + 7c^3 - 14c^2 + 1$

87. $12v^3 + 19v^2 - 20 - v^5 - 5v^3 + 16v^2 - 20v^5 + v^3$
$-21v^5 + 8v^3 + 35v^2 - 20$

88. $b^3 - 13b^2 + b^4 + 6 - 2b^3 - 15b^4 - 17b^3 + 8$
$-14b^4 - 18b^3 - 13b^2 + 14$

Objective 7

For Exercises 89–100, combine like terms. See Example 9.

89. $6a - 3b - 2a + b$
$4a - 2b$

90. $5x - y - 3x - 2y$
$2x - 3y$

91. $3y^2 + 5y - 2y - 7y^2$
$-4y^2 + 3y$

92. $2z^3 - 4z - 6z - 10z^3$
$-8z^3 - 10z$

93. $2x^2 + 3y - 4x^2 + 6y - 5x^2 - 2x^2$
$-9x^2 + 9y$

94. $3a^2 - 5b - 4a^2 + 7b - 5b + 6b$
$-a^2 + 3b$

★95. $\dfrac{1}{3}a^2 + \dfrac{3}{5}a - \dfrac{3}{2}a + \dfrac{5}{4}a^2 + \dfrac{1}{3}a$
$\dfrac{19}{12}a^2 - \dfrac{17}{30}a$

★96. $\dfrac{7}{6}x^2 - \dfrac{3}{4}x - \dfrac{2}{5}x^2 - \dfrac{4}{5}x + \dfrac{4}{3}x$
$\dfrac{23}{30}x^2 - \dfrac{13}{60}x$

97. $y^6 + 2yz^4 - 10yz + 3yz^4 - 3z^5 - 7z^3 + 4yz - 10$
$y^6 + 5yz^4 - 6yz - 3z^5 - 7z^3 - 10$

98. $x^4 + xy^3 - 6xy + xy^3 + 3y^4 - 4y^2 + 3xy - 8$
$x^4 + 3y^4 + 2xy^3 - 4y^2 - 3xy - 8$

99. $-3w^2z - 9 + 6wz^2 + 3w^2 - w^2z + 8 + 9w^2 - 7wz^2$
$-4w^2z - wz^2 + 12w^2 - 1$

100. $-m^3n - 8mn + 4 - 5m^2n^2 + 3m^3n - 7 + mn - 16m^2n^2$
$2m^3n - 21m^2n^2 - 7mn - 3$

Puzzle Problem Rearrange the numbers in the figure shown so that no two consecutive numbers are next to each other horizontally, vertically, or diagonally.

1 and 8 must be in the middle with 7 and 2 on the top and bottom.
3, 4, 5, and 6 are then placed appropriately in the side squares.

Review Exercises

Exercises 1–4 ▰▰▶ **Expressions**

[1.3] *For Exercises 1 and 2, simplify.*

1. $-\dfrac{3}{4} + \dfrac{1}{6}$

$\dfrac{7}{12}$

2. $-14.6 - (-10.3)$

-4.3

[5.1] **3.** Evaluate: -7^0

-1

[5.1] **4.** The Andromeda galaxy (the closest one to our Milky Way galaxy) contains at least 200,000,000,000 stars. Write this number in scientific notation. 2×10^{11}

Exercises 5 and 6 ▰▰▶ **Equations and Inequalities**

[2.3] **5.** Solve and check: $\dfrac{1}{4}x = 3$

12

[4.3] **6.** How many liters of a 40% solution must be added to 200 liters of a 20% solution to obtain a 35% solution?

600 L

5.3 Adding and Subtracting Polynomials

Objectives

1 Add polynomials.

2 Subtract polynomials.

Warm-up

[1.3] **1.** What is the additive inverse of -4.6?

[1.3] **2.** Write $13 - (-6)$ as an equivalent addition and then evaluate.

[1.7] **3.** Combine like terms.

$$12m - 15n + 13 - 17m - 20 + 13n$$

Objective 1 Add polynomials.

Understanding Polynomial Addition

We can add and subtract polynomials in the same way that we add and subtract numbers. In fact, polynomials are like whole numbers that are in an expanded form. Consider the polynomial $4x^2 + 3x + 6$. If we replace the x's with the number 10, we have the expanded form for the number 436.

$$4x^2 \ + \ 3x \ + \ 6$$
$$4 \cdot 10^2 \ + \ 3 \cdot 10 \ + \ 6 \qquad \text{Replacing } x \text{ with 10}$$
$$= 400 \ + \ 30 \ + \ 6$$
$$= 436$$

In our base-ten number system, each place value is a power of 10. We can think of polynomials as a variable-base number system. In other words, the place values are variables, where x^2 is like the hundreds place (10^2) and x is like the tens place (10^1). To add whole numbers, we add the digits in like place values. Polynomials are added in a similar way. However, instead of adding digits in like place values, we add like terms. Consider the following comparison.

Answers to Warm-up
1. 4.6 **2.** $13 + 6 = 19$
3. $-5m - 2n - 7$

Numeric addition:

$436 + 251 = 687$

$$\begin{array}{r} 436 \\ + \ 251 \\ \hline 687 \end{array}$$

Note In numeric addition, like place values are added.

Polynomial addition:

$(4x^2 + 3x + 6) + (2x^2 + 5x + 1) = 6x^2 + 8x + 7$

$$\begin{array}{r} 4x^2 + 3x + 6 \\ + \ 2x^2 + 5x + 1 \\ \hline 6x^2 + 8x + 7 \end{array}$$

Note In polynomial addition, like terms are added.

From this simple case, we see that polynomials are added by combining the like terms.

> **Procedure** Adding Polynomials
>
> To add polynomials, combine like terms.

Adding Polynomials

Although we stacked the preceding polynomials, we do not actually have to stack polynomials to add them.

Note Our thinking flows in this order:

Degree 3 terms:
$2x^3 + 6x^3 = 8x^3$

Degree 2 terms:
$5x^2$ has no like term, so it is copied into the answer.

Degree 1 terms:
$4x + 3x = 7x$

Degree 0 terms:
$1 + 8 = 9$

Example 1 Add and write the resulting polynomial in descending order of degree.

a. $(2x^3 + 5x^2 + 4x + 1) + (6x^3 + 3x + 8)$

Solution: Combine like terms. Notice that combining in order of degree places the resulting polynomial in descending order of degree.

$$(2x^3 + 5x^2 + 4x + 1) + (6x^3 + 3x + 8)$$

$$= 8x^3 + 5x^2 + 7x + 9$$

b. $(y^5 + 9y^3 - 3y - 7) + (2y^5 - 4y^3 + 3y + 5)$

Solution: $(y^5 + 9y^3 - 3y - 7) + (2y^5 - 4y^3 + 3y + 5)$ Combine like terms.

$$= 3y^5 + 5y^3 \qquad + 0 \quad -2$$
$$= 3y^5 + 5y^3 - 2$$

If we had stacked the polynomials in Example 1(a), we would have had to leave a blank space under the $5x^2$ because there was no x^2 term in the second polynomial. It is the same as having a 0 in that place.

$$\begin{array}{r} 2x^3 + 5x^2 + 4x + 1 \\ + \ 6x^3 + \ \ 0 \ + 3x + 8 \\ \hline 8x^3 + 5x^2 + 7x + 9 \end{array}$$

As the polynomials get more complex, these blanks become more prevalent. This is why the stacking method is not a preferred method for adding polynomials.

Your Turn 1 Add and write the resulting polynomial in descending order of degree.

a. $(x^4 + 3x^3 + 4x + 2) + (5x^3 + 2x + 7)$

b. $(14n^5 - 6n^3 - 12n + 15) + (2n^5 - 4n^3 + 8n - 5)$

Answers to Your Turn 1
a. $x^4 + 8x^3 + 6x + 9$
b. $16n^5 - 10n^3 - 4n + 10$

Sometimes the polynomials contain terms that have several variables or coefficients that are fractions or decimals.

Note Our thinking flows like this:

The t^4 terms:
$$7t^4 + (-8t^4) = -1t^4 = -t^4$$

The tu^2 terms:
$$2.9tu^2 + (-7tu^2) = -4.1tu^2$$

The tu terms:
$$-\frac{3}{4}tu + \frac{1}{6}tu =$$
$$-\frac{9}{12}tu + \frac{2}{12}tu = -\frac{7}{12}tu$$

The constant terms:
$$11 + (-5) = 6$$

Example 2 Add and write the resulting polynomial in descending order of degree:

$$\left(7t^4 + 2.9tu^2 - \frac{3}{4}tu + 11\right) + \left(-8t^4 - 7tu^2 + \frac{1}{6}tu - 5\right)$$

Solution: $\left(7t^4 + 2.9tu^2 - \frac{3}{4}tu + 11\right) + \left(-8t^4 - 7tu^2 + \frac{1}{6}tu - 5\right)$ Combine like terms.

$$= -t^4 - 4.1tu^2 - \frac{7}{12}tu + 6$$

Your Turn 2 Add and write the resulting polynomial in descending order of degree.

a. $(12x^3 - 6.2xy^2 - 0.3xy - y^2) + (x^3 + 7xy^2 - 0.9xy + 0.4y^2)$

b. $\left(a^4 - \frac{4}{5}ab^3 - \frac{2}{3}ab + 9\right) + \left(a^4 + \frac{1}{4}ab^3 + ab - 5\right)$

Example 3 Write an expression in simplest form for the perimeter of the rectangle shown.

$3x + 7$ [rectangle]
$5x - 1$

Understand *Perimeter* means the total distance around the shape. Therefore, we need to add the lengths of all of the sides of the shape.

Plan In this case, the lengths of the sides are represented by polynomials. Therefore, we add the polynomials to represent the perimeter.

Execute Perimeter = Length + Width + Length + Width
$$= (5x - 1) + (3x + 7) + (5x - 1) + (3x + 7)$$
$$= 5x + 3x + 5x + 3x - 1 + 7 - 1 + 7$$
$$= 16x + 12$$

Answer The expression for the perimeter is $16x + 12$.

Check To check, we (1) choose a value for x and evaluate the original expressions for length and width, (2) determine the corresponding numeric perimeter, and (3) evaluate the perimeter expression using the same value for x and verify that we get the same numeric perimeter. Let's choose $x = 2$.

Length: $5x - 1$ Width: $3x + 7$ Perimeter:

$5(2) - 1$ $3(2) + 7$ The perimeter of the rectangle with a
$= 10 - 1$ $= 6 + 7$ length of 9 and width of 13 is
$= 9$ $= 13$ $9 + 13 + 9 + 13 = 44$.

Now evaluate the perimeter expression where $x = 2$; you should find that the result is 44.

Perimeter expression: $16x + 12$
$$16(2) + 12$$
$$= 32 + 12$$
$$= 44$$ This agrees with our calculation above.

Answers to Your Turn 2

a. $13x^3 + 0.8xy^2 - 1.2xy - 0.6y^2$

b. $2a^4 - \frac{11}{20}ab^3 + \frac{1}{3}ab + 4$

Your Turn 3 Write an expression in simplest form for the perimeter of the following shape.

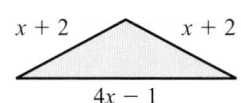

$x + 2$ $x + 2$

$4x - 1$

Objective 2 Subtract polynomials.

Understanding Polynomial Subtraction

Subtracting polynomials is similar to subtracting signed numbers. When we subtract signed numbers, it is often simpler to write the subtraction statement as an equivalent addition statement. Consider the following comparison of numeric subtraction and polynomial subtraction.

Numeric subtraction:	**Polynomial subtraction:**
$975 - 621$	$(9x^2 + 7x + 5) - (6x^2 + 2x + 1)$
We can stack by place:	We can stack like terms:

$$\begin{array}{r} 975 \\ -621 \\ \hline 354 \end{array}$$

$$\begin{array}{r} 9x^2 + 7x + 5 \\ -(6x^2 + 2x + 1) \\ \hline 3x^2 + 5x + 4 \end{array}$$

In numeric subtraction, we subtract digits in the same place value, whereas in polynomial subtraction, we subtract like terms. Note that subtracting 1 from 5 is equivalent to combining 5 and -1, subtracting $2x$ from $7x$ is equivalent to combining $7x$ and $-2x$, and subtracting $6x^2$ from $9x^2$ is equivalent to combining $9x^2$ and $-6x^2$. This suggests that we can write polynomial subtraction as equivalent polynomial addition by changing the sign of each term in the second polynomial (the subtrahend), like this:

$$(9x^2 + 7x + 5) - (6x^2 + 2x + 1)$$
$$= (9x^2 + 7x + 5) + (-6x^2 - 2x - 1)$$
$$= 3x^2 + 5x + 4$$

▲

Note We are allowed to change the signs because of the distributive property. If we disregard the initial polynomial, we have

$$-(6x^2 + 2x + 1) = -1(6x^2 + 2x + 1).$$

The distributive property tells us that we can distribute the -1 (or minus sign) to each term inside the parentheses.

$$= -1 \cdot 6x^2 - 1 \cdot 2x - 1 \cdot 1$$
$$= -6x^2 - 2x - 1$$

Connection The procedure for subtracting polynomials is the same as the procedure for subtracting signed numbers (Section 1.3). In both cases, we write subtraction as equivalent addition by changing the subtrahend to its additive inverse.

Answer to Your Turn 3
$6x + 3$

Subtracting Polynomials

Our exploration suggests the following procedure for subtracting polynomials.

Procedure **Subtracting Polynomials**

To subtract polynomials:
1. Write the subtraction statement as an equivalent addition statement.
   a. Change the operation symbol from a minus sign to a plus sign.
   b. Change the subtrahend (second polynomial) to its additive inverse. To get the additive inverse, change the sign of each term in the polynomial.
2. Combine like terms.

Example 4 Subtract and write the resulting polynomial in descending order of degree: $(7x^3 + 8x^2 + 6x + 9) - (3x^3 + 6x^2 + 5x + 1)$

Solution: Write an equivalent addition statement; then combine like terms.

$$(7x^3 + 8x^2 + 6x + 9) - (\quad 3x^3 + 6x^2 + 5x + 1)$$

Change the minus sign to a plus sign. ↓ ↓ ↓ ↓ ↓ Change all signs in the subtrahend.

$$= (7x^3 + 8x^2 + 6x + 9) + (-3x^3 - 6x^2 - 5x - 1)$$
$$= 4x^3 + 2x^2 + x + 8 \qquad \text{Combine like terms.}$$

Connection The polynomial subtraction in Example 4 is equivalent to the following numeric subtraction:

$$\begin{array}{r} 7869 \\ -3651 \\ \hline 4218 \end{array}$$

Notice that the numeric result 4218 corresponds to the polynomial result $4x^3 + 2x^2 + x + 8$. We could use the stacking method for subtracting polynomials; however, as we saw with adding polynomials, it is not the best method for all cases because we often have missing terms or terms that are not like terms.

You may have noticed that the polynomials in Example 4 contained only plus signs. It is important to recognize that the polynomials may contain a mixture of signs and multiple variables. This is shown in Example 5.

Example 5 Subtract and write the resulting polynomial in descending order of degree.

a. $(15y^3 + 3y^2 + y - 2) - (7y^3 - 6y^2 + 5y - 8)$

Solution: Write an equivalent addition statement; then combine like terms.

$$(15y^3 + 3y^2 + y - 2) - (\quad 7y^3 - 6y^2 + 5y - 8)$$

Change the minus sign to a plus sign. ↓ ↓ ↓ ↓ ↓ Change all signs in the subtrahend.

$$= (15y^3 + 3y^2 + y - 2) + (-7y^3 + 6y^2 - 5y + 8)$$
$$= 8y^3 + 9y^2 - 4y + 6 \qquad \text{Combine like terms.}$$

b. $(9.7x^5 - 2x^2y + xy^2 - 14.6xy - 7y^2) - (x^5 - 5.8xy^2 + 10.3xy - 15y^2)$

Solution: We write an equivalent addition statement, then combine like terms.

$$(9.7x^5 - 2x^2y + xy^2 - 14.6xy - 7y^2) - (\quad x^5 - 5.8xy^2 + 10.3xy - 15y^2)$$

Change the minus sign to a plus sign. ↓ ↓ ↓ ↓ Change all signs in the subtrahend.

$$= (9.7x^5 - 2x^2y + xy^2 - 14.6xy - 7y^2) + (-x^5 + 5.8xy^2 - 10.3xy + 15y^2)$$
$$= 8.7x^5 - 2x^2y + 6.8xy^2 - 24.9xy + 8y^2$$

▲

Note The $-2x^2y$ term had no like term, so we rewrote it in the final expression.

Answers to Your Turn 5

a. $6t^4 + 8t + 3$

b. $-0.6x^4 - 24x^3 + 8x^2 + 5$

c. $5a^5 - 2a^3b^2 + 9a^2b^2 - ab^2 - 5b - 3$

Your Turn 5 Subtract and write the resulting polynomial in descending order of degree.

a. $(8t^4 + 5t^2 + 9t + 7) - (2t^4 + 5t^2 + t + 4)$

b. $(12.5x^4 - 15x^3 + 2x^2 - 9) - (13.1x^4 + 9x^3 - 6x^2 - 14)$

c. $(7a^5 - a^3b^2 + 9a^2b^2 - 2ab^2 + b - 18) - (2a^5 + a^3b^2 - ab^2 + 6b - 15)$

5.3 Exercises For Extra Help MyMathLab®

Note: Exercises marked with a ★ represent challenging exercises.

Objective 1

Prep Exercise 1 Explain how to add two polynomials.
To add polynomials, combine like terms.

Prep Exercise 2 What is the function of parentheses in the addition problem $(3x + 9) + (4x - 2)$?
To identify the different polynomials

For Exercises 1–24, add and write the resulting polynomial in descending order of degree.
See Examples 1 and 2.

1. $(3x + 2) + (5x - 1)$
$8x + 1$

2. $(5y + 4) + (3y + 1)$
$8y + 5$

3. $(8y + 7) + (2y + 5)$
$10y + 12$

4. $(3m - 4) + (5m + 1)$
$8m - 3$

5. $(2x + 3y) + (5x - 3y)$
$7x$

6. $(4a - 3b) + (6a + 3b)$
$10a$

7. $(2x + 5) + (3x^2 - 4x + 7)$
$3x^2 - 2x + 12$

8. $(5p^2 + 3p - 1) + (4p + 8)$
$5p^2 + 7p + 7$

9. $(z^2 - 3z + 7) + (5z^2 - 8z - 9)$
$6z^2 - 11z - 2$

10. $(2w^2 - 5w - 1) + (4w^2 - 5w + 1)$
$6w^2 - 10w$

11. $(3r^2 - 2r + 10) + (2r^2 - 5r - 11)$
$5r^2 - 7r - 1$

12. $(7m^2 + 8m - 1) + (-4m^2 - 3m - 2)$
$3m^2 + 5m - 3$

13. $(4y^2 - 8y + 1) + (5y^2 + 8y + 2)$
$9y^2 + 3$

14. $(5k^2 - k - 1) + (4k^2 + k + 7)$
$9k^2 + 6$

15. $\left(x^2 - \dfrac{2}{3}x + 3\right) + \left(2x^2 + \dfrac{3}{4}x - 2\right)$
$3x^2 + \dfrac{1}{12}x + 1$

16. $\left(3a^2 + 3a - \dfrac{5}{6}\right) + \left(2a^2 - 5a + \dfrac{3}{4}\right)$
$5a^2 - 2a - \dfrac{1}{12}$

17. $(4r^2 + 2.7r - 3.6) + (-2r^2 - 1.3r - 4.2)$
$2r^2 + 1.4r - 7.8$

18. $(5b^2 - 4.6b - 1.7) + (-4b^2 + 3.3b + 5.6)$
$b^2 - 1.3b + 3.9$

19. $(9a^3 + 5a^2 - 3a - 1) + (-4a^3 - 2a^2 + 6a - 2)$
$5a^3 + 3a^2 + 3a - 3$

20. $(4u^3 - 6u^2 + u + 11) + (-5u^3 - 3u^2 + u - 5)$
$-u^3 - 9u^2 + 2u + 6$

21. $(7p^3 - 9p^2 + 5p - 1) + (-4p^3 + 8p^2 + 2p + 10)$
$3p^3 - p^2 + 7p + 9$

22. $(12r^4 - 5r^2 + 8r - 15) + (-7r^4 + 3r^3 + 2r - 9)$
$5r^4 + 3r^3 - 5r^2 + 10r - 24$

23. $(-5w^4 - 3w^3 - 8w^2 + w - 14) + (-3w^4 + 6w^3 + w^2 + 12w + 5)$
$-8w^4 + 3w^3 - 7w^2 + 13w - 9$

24. $(-r^4 + 3r^2 - 12r - 14) + (5r^4 - 2r^3 - 3r^2 + 10r - 5)$
$4r^4 - 2r^3 - 2r - 19$

★ *For Exercises 25–34, add. See Example 2.*

25. $(a^3b^2 - 6ab^2 - 6a^2b + 5ab + 5b^2 - 6) + (-4a^3b^2 + 4ab^2 - 2a^2b - ab - 2b^2 - 2)$
$-3a^3b^2 - 2ab^2 - 8a^2b + 4ab + 3b^2 - 8$

26. $(x^2y^2 + 8xy^2 - 12x^2y - 4xy + 7y^2 - 9) + (-3x^2y^2 + 2xy^2 + 4x^2y - xy + 6y^2 + 4)$
$-2x^2y^2 + 10xy^2 - 8x^2y - 5xy + 13y^2 - 5$

27. $(-2u^4 + 6uv^3 - 8u^2v^3 + 7u^3v - u + v^2 + 9) + (-5u^4 - 6uv^3 + 9u^2v^3 + 8u - 14v^2 - 12)$
$-7u^4 + u^2v^3 + 7u^3v + 7u - 13v^2 - 3$

28. $(-13a^6 + a^3b^2 + 3ab^3 - 8b^3) + (11a^6 - 3a^2b^3 - 4ab^3 + 10ab^2 + 4b^3)$
$-2a^6 + a^3b^2 - 3a^2b^3 - ab^3 + 10ab^2 - 4b^3$

29. $(-8mnp - 6m^2n^2 - 13mn^2p + 14m^2n - 12n + 6) + (-4nmp - 10m^2n^2 + mn^2p - 19n + 3)$
$-12mnp - 16m^2n^2 - 12mn^2p + 14m^2n - 31n + 9$

30. $(14abc - 2a^2b^2 - ab^2c + 4b - 9c - 3) + (-15abc + 5a^2b^2 - 13ab^2c + 7a - 12b + 6)$
$-abc + 3a^2b^2 - 14ab^2c + 7a - 8b - 9c + 3$

31. $\left(4a^4 - \dfrac{2}{3}a^3b + \dfrac{3}{5}ab^3 - 3b^4\right) + \left(-2a^4 + \dfrac{1}{4}a^3b + \dfrac{1}{3}ab^3 + 2b^4\right)$
$2a^4 - \dfrac{5}{12}a^3b + \dfrac{14}{15}ab^3 - b^4$

32. $\left(-5y^4 + \dfrac{3}{4}y^2z^2 - \dfrac{5}{6}yz^3 + 2z^4\right) + \left(3y^4 - \dfrac{1}{6}y^2z^2 - \dfrac{3}{8}yz^3 - 5z^4\right)$
$-2y^4 + \dfrac{7}{12}y^2z^2 - \dfrac{29}{24}yz^3 - 3z^4$

33. $(3.6a^2bc + 2.3ab^2c^2 - 5.7a^2b^2c^2) + (-1.8a^2bc - 4.1ab^2c^2 + 3.3a^2b^2c^2)$
$1.8a^2bc - 1.8ab^2c^2 - 2.4a^2b^2c^2$

34. $(-6.8x^3yz^2 + 4.1x^2y^2z - 8.2xy^2z^2) + (5.2x^3yz^2 - 3.6x^2y^2z - 3.2xy^2z^2)$
$-1.6x^3yz^2 + 0.5x^2y^2z - 11.4xy^2z^2$

For Exercises 35–38, write an expression for the perimeter in simplest form. See Example 3.

35.

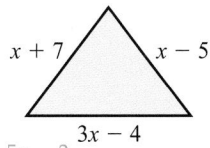

$x + 7$ $x - 5$
$3x - 4$
$5x - 2$

36.

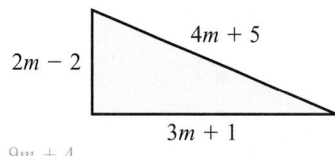

$2m - 2$ $4m + 5$
$3m + 1$
$9m + 4$

37.

$a - 8$
$6a + 2$
$14a - 12$

38.

$5x - 7$
$4x + 1$
$18x - 12$

Objective 2

Prep Exercise 3 Explain how to find the additive inverse of a polynomial. Change the signs of each term in the polynomial.

Prep Exercise 4 Explain how to subtract two polynomials.
To subtract polynomials:
a. Write the subtraction statement as an equivalent addition statement.
  i. Change the operation symbol from a minus sign to a plus sign.
  ii. Change the subtrahend (second polynomial) to its additive inverse. To get the additive inverse, change the sign of each term in the polynomial.
b. Combine like terms.

For Exercises 39–56, subtract and write the resulting polynomial in descending order of degree.
See Examples 4 and 5.

39. $(6x + 3) - (2x + 1)$
$4x + 2$

40. $(5m + 8) - (2m + 3)$
$3m + 5$

41. $(18a^2 + 3) - (5a^2 + 4)$
$13a^2 - 1$

42. $(12b^2 + 4b) - (5b^2 + b)$
$7b^2 + 3b$

43. $(2x^2 - 3x + 4) - (6x + 2)$
$2x^2 - 9x + 2$

44. $(7y^2 - 3y + 2) - (8y + 5)$
$7y^2 - 11y - 3$

45. $(6z^2 - 3z + 1) - (4z^2 + 4z - 8)$
$2z^2 - 7z + 9$

46. $(5y^2 - 3y - 10) - (5y^2 + 2y - 10)$
$-5y$

47. $(8a^2 + 11a - 12) - (-4a^2 + a + 3)$
$12a^2 + 10a - 15$

48. $(y^2 + 3y + 6) - (-5y^2 + 3y - 8)$
$6y^2 + 14$

49. $(6u^3 + 3u^2 - 8u + 5) - (7u^3 + 5u^2 - 2u - 3)$
$-u^3 - 2u^2 - 6u + 8$

50. $(-8t^3 + 3t^2 - 7t - 9) - (-10t^3 - 5t^2 + 3t - 11)$
$2t^3 + 8t^2 - 10t + 2$

51. $(-w^5 + 3w^4 + 8w^2 - w - 1) - (10w^5 + 3w^2 - w - 3)$
$-11w^5 + 3w^4 + 5w^2 + 2$

52. $(-s^4 + 7s^3 + 6s^2 - 13s - 11) - (3s^4 - 2s^2 - s + 5)$
$-4s^4 + 7s^3 + 8s^2 - 12s - 16$

53. $(-5p^4 + 8p^3 - 9p^2 + 3p + 1) - (6p^4 + 5p^3 + 4p^2 - 2p + 7)$
$-11p^4 + 3p^3 - 13p^2 + 5p - 6$

54. $(-6r^5 - 9r^4 + 5r^3 - 2r^2 + 8) - (5r^5 + 2r^4 + 7r^3 - 8r^2 - 9)$
$-11r^5 - 11r^4 - 2r^3 + 6r^2 + 17$

55. $(16v^4 + 21v^2 + 4v - 8) - (5v^4 - 3v^3 - 2v^2 + 9v + 1)$
$11v^4 + 3v^3 + 23v^2 - 5v - 9$

56. $(12q^4 + 5q^3 - 2q + 15) - (8q^4 - 14q^3 - 3q^2 + 15q + 8)$
$4q^4 + 19q^3 + 3q^2 - 17q + 7$

For Exercises 57–64, subtract. See Example 5.

57. $(x^2 + xy + y^2) - (x^2 - xy + y^2)$
$2xy$

58. $(a^2 - 4ab + b^2) - (a^2 + 4ab + b^2)$
$-8ab$

59. $(4xy + 5xz - 6yz) - (10xy - xz - 8yz)$
$-6xy + 6xz + 2yz$

60. $(4ab - 5bc + 6ac) - (10ab - bc - ac)$
$-6ab - 4bc + 7ac$

61. $(18p^3q^2 - p^3q^3 + 6pq^2 - 4pq + 16q^2 - 4) - (-p^3q^2 - p^3q^3 + 4pq^2 - 3pq - 6)$
$19p^3q^2 + 2pq^2 - pq + 16q^2 + 2$

62. $(15a^3y^4 + a^2y^2 + 5ay^3 - 7ay + 12y^2 - 11) - (-a^3y^4 + a^2y^2 + 2ay^3 - ay + 9)$
$16a^3y^4 + 3ay^3 - 6ay + 12y^2 - 20$

63. $(5.6a^2b - 2.3ab^2 - 4.3a^2b^2) - (3.7a^2b - 3.4ab^2 + 2.2a^2b^2)$
$1.9a^2b + 1.1ab^2 - 6.5a^2b^2$

64. $(7.4m^3n - 6.7m^2n^2 + 2.6mn^3) - (3.6m^3n - 8.3m^2n^2 - 4.7mn^3)$
$3.8m^3n + 1.6m^2n^2 + 7.3mn^3$

For Exercises 65–68, solve. (Hint: Profit = Revenue − Cost)

65. The polynomial $54.95x + 92.20y + 25.35z$ describes the revenue that a company generates from the sale of three different portable CD players. The expression $22.85x + 56.75y + 19.34z$ describes the cost of producing each CD player. Write a polynomial in simplest form that describes the company's net profit.

$32.1x + 35.45y + 6.01z$

66. The polynomial $27.50x + 14.70y + 42.38z$ describes the revenue that a company generates from the sale of three different types of printer cartridges. The expression $12.75x + 5.25y + 27.42z$ describes the cost of producing each cartridge. Write a polynomial in simplest form that describes the company's net profit.

$14.75x + 9.45y + 14.96z$

67. The polynomial $6.50a + 3.38b + 25.00c$ describes the revenue that a department store generates from the sale of three different candles. The expression $2.28a + 1.75b + 12.87c$ describes the cost the store pays to sell each candle. Write a polynomial in simplest form that describes the store's net profit.

$4.22a + 1.63b + 12.13c$

68. The polynomial $10.55m + 14.75n + 27.50p$ describes the revenue that a pet store generates from the sale of three different litter boxes. The expression $5.73m + 8.26n + 15.22p$ describes the cost the store pays to sell each product. Write a polynomial in simplest form that describes the store's net profit.

$4.82m + 6.49n + 12.28p$

Find ⊗ the Mistake *For Exercises 69 and 70, find and explain the mistake; then work the problem correctly.*

69.
$$(3x^2 - 5x + 10) - (7x^2 - 3x + 5) = (3x^2 - 5x + 10) + (-7x^2 - 3x + 5)$$
$$= 3x^2 - 5x + 10 - 7x^2 - 3x + 5$$
$$= -4x^2 - 8x + 15$$

Mistake: Only the sign of the first term in the subtrahend was changed instead of all three terms. Correct: $-4x^2 - 2x + 5$

70.
$$(7y^2 - 3y + 1) - (y^2 + 4y - 1) = (-7y^2 + 3y - 1) + (-y^2 - 4y + 1)$$
$$= -7y^2 + 3y - 1 - y^2 - 4y + 1$$
$$= -8y^2 - y$$

Mistake: Signs were changed in both polynomials instead of just the second polynomial (subtrahend). Correct: $6y^2 - 7y + 2$

Review Exercises

Exercises 1–5 Expressions

For Exercises 1–4, evaluate.

[1.4] 1. $(2.3)(-7.5)$

-17.25

[1.4] 2. $(-5)(-2)(1.5)$

15

[5.1] 3. 10^{-4}

0.0001

[1.5] 4. $10^3 \cdot 10^4$

10^7

[5.1] 5. Write 5.89×10^7 in standard form.

$58,900,000$

Exercises 6 Equations and Inequalities

[2.1] 6. The surface of the water in a rectangular swimming pool measures 20 meters by 15 meters. What is the surface area of the water?

300 m^2

5.4 Exponent Rules and Multiplying Monomials

Objectives

1 Multiply monomials.
2 Multiply numbers in scientific notation.
3 Simplify a monomial raised to a power.

Warm-up
For Exercises 1–4, find the product.

[1.4] **1.** $\left(-\dfrac{5}{6}\right)\left(\dfrac{2}{3}\right)$

[1.5] **2.** $(-0.4)^3$

[1.5] **3.** $(2^3)^2$

[1.7] **4.** $(-6)(5x + 3)$

In this section, we learn how to multiply monomials, multiply numbers in scientific notation, and simplify a monomial raised to a power. These simplifications require some new rules for exponents.

Objective 1 Multiply monomials.

Because monomials contain variables in exponential form, to multiply monomials, we must develop a rule for multiplying exponential forms.

Multiplying Exponential Forms

Consider $2^3 \cdot 2^4$, which is a product of exponential forms. To simplify $2^3 \cdot 2^4$, we could follow the order of operations and evaluate the exponential forms first, then multiply.

$$2^3 \cdot 2^4 = 8 \cdot 16 = 128$$

However, there is an alternative. We can write the result in exponential form by first writing 2^3 and 2^4 in their factored forms.

Notice that $2^7 = 128$, which agrees with our earlier calculation. Also notice that in the alternative method, the resulting exponent is the sum of the original exponents.

$$2^3 \cdot 2^4 = 2^{3+4} = 2^7$$

Conclusion: To multiply exponential forms that have the same base, we can add the exponents and keep the same base.

This suggests the following rule.

Rule Product Rule for Exponents

If a is a real number and m and n are integers, then $a^m \cdot a^n = a^{m+n}$.

We apply this rule when multiplying monomials.

Learning Strategy

If we understand why a formula is invented and why it is meaningful as well as what (physical or intuitive) message it's meant to contain, we can remember it more easily and apply it more intuitively than if we only know the formula as a string of letters and numbers.
— Zheng Alick Z.

Example 1 Multiply $(x^5)(x^3)$.

Solution: Because the bases are the same, we can add the exponents and keep the same base.

$$(x^5)(x^3) = x^{5+3} = x^8$$

Answers to Warm-up

1. $-\dfrac{5}{9}$ **2.** -0.064

3. 64   **4.** $-30x - 18$

Learning Strategy

If you are an auditory learner, to solidify a rule in your mind, try saying the rule as you work each problem.

Instructor Note Show how $(4x^3)(2x^6)$ can be expanded to

$$4 \cdot x \cdot x \cdot x \cdot 2 \cdot$$
$$x \cdot x \cdot x \cdot x \cdot x \cdot x$$

so that students see how there really are nine x's.

Instructor Note Instead of regrouping, you can connect the factors to be multiplied as follows:

$$(5x^2)(7xy) = 35x^3y$$

Note Remember that a variable or number with no apparent exponent has an understood exponent of 1.

$$-\frac{5}{6}a^2bc^3 = -\frac{5}{6}a^2b^1c^3$$

Answers to Your Turn 1
a. 3^7 **b.** c^{10} **c.** t^8

Answers to Your Turn 2
a. $45n^{12}$ **b.** $15x^3y^5z^3$

Your Turn 1 Multiply.

a. $3^2 \cdot 3^5$ **b.** $(c^7)(c^3)$ **c.** $t^5(t^3)$

Multiplying Monomials

Now let's consider multiplying monomials that have coefficients, such as $(4x^3)(2x^6)$. Because monomials are products by definition, we can use the commutative property of multiplication and write the following:

$$
\begin{aligned}
(4x^3)(2x^6) &= 4 \cdot x^3 \cdot 2 \cdot x^6 \\
&= 4 \cdot 2 \cdot x^3 \cdot x^6 \\
&= 8 \cdot x^{3+6} \qquad \text{We multiply the coefficients and use the product} \\
&= 8x^9 \qquad \text{rule for exponents to simplify } x^3 \cdot x^6.
\end{aligned}
$$

What if the monomial factors have some variables that are different, as in $(5x^2)(7xy)$? If we expand the monomials in factored form and use the commutative property of multiplication, we have the following:

$$
\begin{aligned}
(5x^2)(7xy) &= 5 \cdot x^2 \cdot 7 \cdot x \cdot y \\
&= 5 \cdot 7 \cdot x^2 \cdot x \cdot y \\
&= 35 \cdot x^{2+1} \cdot y \\
&= 35x^3y
\end{aligned}
$$

Notice that the unlike variable base, y, is simply written unchanged in the result. Our examples suggest the following procedure.

> **Procedure Multiplying Monomials**
>
> To multiply monomials:
> 1. Multiply coefficients.
> 2. Add the exponents of the like bases.
> 3. Write any unlike variable bases unchanged in the product.

Example 2 Multiply.

a. $(9y^5)(3y^8)$

Solution: $(9y^5)(3y^8) = 9 \cdot 3y^{5+8}$ Multiply the coefficients and add the exponents of the like bases.

$$= 27y^{13} \qquad \text{Simplify the exponents.}$$

b. $\left(-\frac{5}{6}a^2bc^3\right)\left(\frac{2}{3}a^5b^4\right)$

Solution: $\left(-\frac{5}{6}a^2bc^3\right)\left(\frac{2}{3}a^5b^4\right) = -\frac{5}{\underset{3}{6}} \cdot \frac{\overset{1}{2}}{3}a^{2+5}b^{1+4}c^3$ Multiply the coefficients, add the exponents of the like bases, and write the unlike variable base c unchanged in the product.

$$= -\frac{5}{9}a^7b^5c^3 \qquad \text{Simplify.}$$

Your Turn 2 Multiply.

a. $(9n^4)(5n^8)$ **b.** $-2.5x^2y^3(-6xy^2z^3)$

Connection Multiplying distance measurements in area and volume is like multiplying monomials.

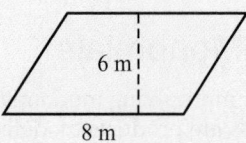

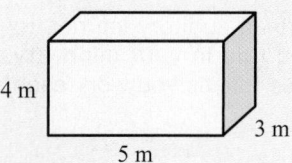

The area of the parallelogram:

$$A = bh$$
$$A = (8\,\text{m})(6\,\text{m})$$
$$A = 48\,\text{m}^2$$

The area calculation is similar to multiplying two monomials with m as the variable.

Multiply: $(8m)(6m) = 48m^2$

The volume of the box:

$$V = lwh$$
$$V = (5\,\text{m})(3\,\text{m})(4\,\text{m})$$
$$V = 60\,\text{m}^3$$

The volume calculation is similar to multiplying three monomials with m as the variable.

Multiply: $(5m)(3m)(4m) = 60m^3$

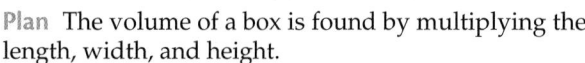

Instructor Note Point out that x represents width; so if we select a value for x, we are selecting a particular width. Evaluating $7x$ and $2x$ gives the corresponding values for length and height.

Example 3 Write an expression in simplest form for the volume of the box shown.

Understand We are given a box with side lengths that are monomial expressions.

Plan The volume of a box is found by multiplying the length, width, and height.

Execute $V = lwh$

$$V = (7x)\,(x)\,(2x)$$
$$V = 14x^3$$

Answer The expression for volume is $14x^3$.

Check Because $(7x)(x)(2x) = 14x^3$ is an identity, we can evaluate $(7x)$, (x), $(2x)$, and $14x^3$ using any chosen value for x and get the same result. Let's choose a value for x, such as 3, and try it.

$$(7x)(x)(2x) = 14x^3$$
$$[7(3)][(3)][2(3)] \stackrel{?}{=} 14(3)^3$$
$$(21)(3)(6) \stackrel{?}{=} 14(27)$$
$$378 = 378 \quad \text{This is true.}$$

Note Choosing only one value does not guarantee that our simplification is correct, but it is a good indicator. Testing several values provides more conclusive assurance.

Objective 2 Multiply numbers in scientific notation.

We multiply numbers in scientific notation using the same procedure we used to multiply monomials. If we replace a with 10 in $(4a^3)(2a^6)$, we have a product of two numbers in scientific notation: $(4 \times 10^3)(2 \times 10^6)$. Following the same procedure for multiplying monomials, we multiply $4 \cdot 2$ to get 8 and then add the exponents of the like bases.

Instructor Note Help students see the advantage of multiplying numbers in scientific notation by multiplying $(4 \times 10^3)(2 \times 10^6)$ in standard form. Show how the exponent accounts for all of the 0s.

$$(4 \times 10^3)(2 \times 10^6)$$
$$= (4000)(2,000,000)$$
$$= 8,000,000,000$$

This product contains an 8 followed by nine zeros.

Monomials:	Scientific notation:
$(4a^3)(2a^6) = 4 \cdot 2a^{3+6}$	$(4 \times 10^3)(2 \times 10^6) = 4 \times 2 \times 10^{3+6}$
$= 8a^9$	$= 8 \times 10^9$

Note The product 17.25×10^{11} is *not* in scientific notation because the absolute value of 17.25 is *not less than* 10. To fix this problem, we must move the decimal point one place to the left and account for this move by increasing the exponent by 1. If you are unconvinced that this is correct, consider the fact that 17.25×10^{11} is equal to 1,725,000,000,000. If we now write that standard form number in scientific notation, the proper position for the decimal point is between the 1 and the 7 digits, which means we must account for 12 places.

▶

Example 4 Multiply and write the answers in scientific notation.

a. $(2.3 \times 10^5)(7.5 \times 10^6)$

Solution: Multiply 2.3 and 7.5; then add the exponents for base 10s.
$$(2.3 \times 10^5)(7.5 \times 10^6) = 2.3 \times 7.5 \times 10^{5+6}$$
$$= 17.25 \times 10^{11}$$
$$= 1.725 \times 10^{12}$$

b. $(4.2 \times 10^4)(6.5 \times 10^{-9})$

Solution: Multiply 4.2 and 6.5 and add the exponents for the base 10s.
$$(4.2 \times 10^4)(6.5 \times 10^{-9}) = 4.2 \times 6.5 \times 10^{4+(-9)}$$
$$= 27.3 \times 10^{-5}$$
$$= 2.73 \times 10^{-4} \quad \text{Rewrite in scientific notation.}$$

Your Turn 4 Multiply and write the answers in scientific notation.

a. $(4.2 \times 10^5)(2.8 \times 10^8)$ **b.** $(2.7 \times 10^{-3})(7.3 \times 10^{10})$

Objective 3 Simplify a monomial raised to a power.

We now discuss how to simplify an expression such as $(3x^2)^4$, which is a monomial raised to a power. To simplify such an expression, we need to discuss two new rules of exponents: (1) raising a power to a power and (2) raising a product to a power.

Raising a Power to a Power

Consider a power raised to a power, such as $(2^3)^2$. To calculate $(2^3)^2$, we could follow the order of operations and evaluate the exponential form 2^3 within the parentheses first, then square the result.

$$(2^3)^2 = (2 \cdot 2 \cdot 2)^2 = 8^2 = 64$$

However, there is an alternative. The outside exponent, 2, indicates to multiply the inside exponential form 2^3 by itself.

$$(2^3)^2 = 2^3 \cdot 2^3 \quad \text{Because this is multiplication of exponential}$$
$$= 2^{3+3} \quad \text{forms that have the same base, we can add}$$
$$= 2^6 \quad \text{the exponents.}$$

Instructor Note Point out that this alternative approach is desirable in situations such as $(7^9)^6$ because calculating a result using the order of operations is impractical.

Notice that $2^6 = 64$. Also, in comparing $(2^3)^2$ with 2^6, notice that the exponent in 2^6 is the product of the original exponents.

$$(2^3)^2 = 2^{3 \cdot 2} = 2^6$$

Conclusion: If an exponential form is raised to a power, we can multiply the exponents and keep the same base.

> **Rule   A Power Raised to a Power**
> If a is a real number and m and n are integers, then $(a^m)^n = a^{mn}$.

Now let's consider a product raised to a power, such as $(3 \cdot 4)^2$. We could write the factored form twice and use the commutative property to rearrange the like bases.

$$(3 \cdot 4)^2 = (3 \cdot 4)(3 \cdot 4) \quad \text{Because there are two factors of 3 and two}$$
$$= 3 \cdot 3 \cdot 4 \cdot 4 \quad \text{factors of 4, we can write each of these in}$$
$$= 3^2 \cdot 4^2 \quad \text{exponential form.}$$

Answers to Your Turn 4
a. 1.176×10^{14} **b.** 1.971×10^8

Conclusion: If a product is raised to a power, we can evaluate the factors raised to that power.

> **Rule Raising a Product to a Power**
> If a and b are real numbers and n is an integer, then $(ab)^n = a^n b^n$.

Raising a Monomial to a Power

We apply these exponent rules to simplify an expression such as $(3x^2)^4$, which is a monomial raised to a power. Because the monomial $3x^2$ expresses a product, we can use the rule for raising a product to a power and raise the factors 3 and x^2 to the 4th power.

$$(3x^2)^4 = (3)^4(x^2)^4$$

To simplify $(x^2)^4$, we use the rule for raising a power to a power.

$$= 81x^{2 \cdot 4}$$
$$= 81x^8$$

Notice that the outside exponent is distributed to the coefficient and the variable(s) within the monomial. This suggests the following procedure.

> **Procedure Simplifying a Monomial Raised to a Power**
> To simplify a monomial raised to a power:
> 1. Evaluate the coefficient raised to that power.
> 2. Multiply each variable's exponent by the power.

Example 5 Simplify.

a. $(5a^6)^3$

Solution: $(5a^6)^3 = (5)^3 a^{6 \cdot 3}$ Write the coefficient, 5, raised to the 3rd power and multiply the variable's exponent by 3.

$$= 125a^{18}$$ Simplify.

b. $\left(-\frac{2}{3}xy^3z^5\right)^4$

> **Note** Recall that $x = x^1$.
> ▼

Solution: $\left(-\frac{2}{3}xy^3z^5\right)^4 = \left(-\frac{2}{3}\right)^4 x^{1 \cdot 4}y^{3 \cdot 4}z^{5 \cdot 4}$ Write the coefficient, $-\frac{2}{3}$, raised to the 4th power and multiply each variable's exponent by 4.

$$= \frac{16}{81}x^4y^{12}z^{20}$$ Simplify.

c. $(-0.4mn^6)^3$

Solution: $(-0.4mn^6)^3 = (-0.4)^3 m^{1 \cdot 3}n^{6 \cdot 3}$
$$= -0.064m^3n^{18}$$

Your Turn 5 Simplify.

a. $(2y^4)^5$ **b.** $\left(-\frac{5}{3}t^2u^6\right)^3$ **c.** $(-0.2a^4b^5c)^6$

Answers to Your Turn 5
a. $32y^{20}$
b. $-\frac{125}{27}t^6u^{18}$
c. $0.000064a^{24}b^{30}c^6$

Now let's consider expressions that require us to use all of the exponent rules.

Example 6   Simplify.

a. $(3x^2)(4x^5)^3$

Solution: Because the order of operations is to simplify exponents before multiplying, we simplify $(4x^5)^3$ first, then multiply the result by $3x^2$.

$$(3x^2)(4x^5)^3 = (3x^2)(4^3x^{5\cdot3})$$
$$= (3x^2)(64x^{15}) \quad \text{Simplify.}$$
$$= 3\cdot64x^{2+15} \quad \text{Multiply coefficients and add}$$
$$\qquad\qquad\qquad\text{exponents of like variables.}$$
$$= 192x^{17}$$

b. $(-0.2a)^3(4ab)(1.5a^3c)^2$

Solution: We follow the order of operations and simplify the monomials raised to a power first, then multiply the monomials.

$$(-0.2a)^3(4ab)(1.5a^3c)^2 = ((-0.2)^3a^{1\cdot3})(4ab)((1.5)^2a^{3\cdot2}c^{1\cdot2})$$
$$= (-0.008a^3)(4ab)(2.25a^6c^2) \quad \text{Simplify.}$$
$$= -0.008\cdot4\cdot2.25a^{3+1+6}bc^2 \quad \text{Multiply coefficients}$$
$$\qquad\qquad\qquad\qquad\qquad\text{and add exponents of}$$
$$\qquad\qquad\qquad\qquad\qquad\text{like variables.}$$
$$= -0.072a^{10}bc^2$$

Instructor Note Remind students that if they ever forget a rule of exponents, they can go back to the definition of an exponent and derive the rule, as we did in the text.

Answers to Your Turn 6
a. $\frac{1}{18}y^{11}$ **b.** $60m^6n^5p^6$

Your Turn 6   Simplify.

a. $\left(\frac{1}{4}y^3\right)^2\left(\frac{8}{9}y^5\right)$

b. $(-5mn^4)(-2mp^2)^3(1.5m^2n)$

5.4 Exercises   For Extra Help   MyMathLab®

Note: Exercises marked with a ★ represent challenging exercises.

Objective 1

Prep Exercise 1 When multiplying exponential forms that have the same base, we __add__ the exponents and keep the same base.

Prep Exercise 2 Explain why $2^3\cdot5^2 \neq 10^5$.
The bases must be the same in order to add the exponents.

Prep Exercise 3
a. In multiplying $(5x^2)(7x^4y)$, what do you do with the coefficients?
   multiply
b. What do you do with the exponents of the two x's?
   add
c. What do you do with the y?
   Write y unchanged in the product.

For Exercises 1–26, multiply. See Examples 1 and 2.

1. $a^3\cdot a^4$
   a^7

2. $b^2\cdot b^3$
   b^5

3. $2^3\cdot2^{10}$
   2^{13}

4. $5^4\cdot5^3$
   5^7

5. $a^3\cdot a^2b$
   a^5b

6. $y^2\cdot y^3z$
   y^5z

7. $2x\cdot3x^2$
   $6x^3$

8. $5b\cdot2b^3$
   $10b^4$

9. $(-2mn)(4m^2n)$
$-8m^3n^2$

10. $(-3uv^2)(2u^3v)$
$-6u^4v^3$

11. $\left(\dfrac{5}{8}st\right)\left(-\dfrac{2}{7}s^2t^7\right)$
$-\dfrac{5}{28}s^3t^8$

12. $\left(\dfrac{3}{8}x^2y\right)\left(\dfrac{4}{9}xy^2\right)$
$\dfrac{1}{6}x^3y^3$

13. $(2.3ab^2c)(1.2a^2b^3c^5)$
$2.76a^3b^5c^6$

14. $(-3.1mn)(2.4m^2n)$
$-7.44m^3n^2$

15. $(r^2)(2r^2)(-3r)$
$-6r^5$

16. $(-2j)(-5j^3)(7j)$
$70j^5$

17. $3xz^2(-4x^3z^2)(z^5)$
$-12x^4z^9$

18. $-5mn^4(-m^3n^9)(4m^3n)$
$20m^7n^{14}$

19. $(5xyz)(9x^2y^4)(2x^3z^2)$
$90x^6y^5z^3$

20. $(6hj^3k)(7h^5k)(hkj)$
$42h^7j^4k^3$

21. $(5q^2r^4s^9)(-q^3rs^2)(-r^5s^{10})$
$5q^5r^{10}s^{21}$

22. $(a^3b^3c^3)(-6a^3b^3c^3)(-a^4c^2)$
$6a^{10}b^6c^8$

★ **23.** $-5a^2\left(-\dfrac{1}{15}abc\right)\left(-\dfrac{3}{4}a^3b^2c\right)$
$-\dfrac{1}{4}a^6b^3c^2$

★ **24.** $\left(-\dfrac{1}{4}l^2m\right)\left(\dfrac{5}{6}lm^3n\right)\left(-\dfrac{2}{3}l^5\right)$
$\dfrac{5}{36}l^8m^4n$

★ **25.** $(0.4wxy)(w^2y^2)(2.5x^2y^2)$
$w^3x^3y^5$

★ **26.** $(0.4u^2v)(-3.2u^2v^3w^3)(0.2vw^9)$
$-0.256u^4v^5w^{12}$

Find 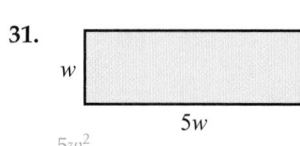 the Mistake   *For Exercises 27–30, find and explain the mistake; then work the problem correctly. See Examples 1 and 2.*

27. $7x^3 \cdot 5x^4y = 35x^{12}y$
Mistake: The exponents of x were multiplied instead of added. Correct: $35x^7y$

28. $-3m^2n \cdot 10m^5n = -30m^7n$
Mistake: The exponents of n were not added.
Correct: $-30m^7n^2$

29. $(9x^3y^2z)(6xy^4) = 54x^3y^6z$
Mistake: The exponents of x were not added.
Correct: $54x^4y^6z$

30. $(a^3b^2c)(ab^5) = a^4b^7$
Mistake: The c in the first factor was not rewritten in the product. Correct: a^4b^7c

For Exercises 31 and 32, write an expression in simplest form for the area of the figure. See Example 3.

31.
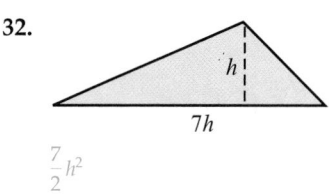
$5w^2$

32.
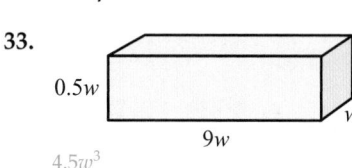
$\dfrac{7}{2}h^2$

For Exercises 33 and 34, write an expression in simplest form for the volume of the figure. See Example 3.

33.
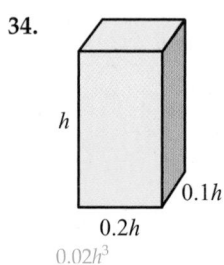
$4.5w^3$

34.

$0.02h^3$

★ **35.** The height of a trapezoid is half the base. The top side of the trapezoid is one-fourth the base.

    a. If b represents the length of the base, write expressions for the height and length of the top side of the trapezoid in terms of b.

      height $= \frac{1}{2}b$, top side $= \frac{1}{4}b$

    b. Using your expressions from part a, write an expression in simplest form for the area of the trapezoid. $\left[\text{Use } A = \frac{1}{2}h(a + b). \right]$

      $A = \frac{5}{16}b^2$

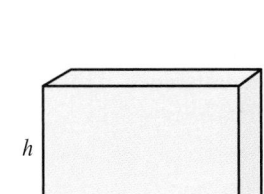

★ **36.** The height of a box is three times the width. The length is five times the width.

    a. If w represents the width, write expressions for the height and length of the box in terms of w.

      height $= 3w$, length $= 5w$

    b. Using your expressions from part a, write an expression in simplest form for the volume of the box.

      $V = 15w^3$

    c. Using your expressions from part a, write an expression in simplest form for the surface area of the box. (Use $SA = 2lw + 2lh + 2wh$.)

      $SA = 46w^2$

Objective 2

Prep Exercise 4 In multiplying the powers of 10 in $(8 \times 10^4)(6 \times 10^5)$, we add the exponents to get 10^9. However, the correct answer in scientific notation is 4.8×10^{10}. Why is the exponent 10 instead of 9?

$8 \cdot 6 = 48$, and 48 must be changed to 4.8 to be in scientific notation. This requires increasing the exponent by 1 to account for moving the decimal point one place to the left.

For Exercises 37–46, multiply and write your answer in scientific notation. See Example 4.

37. $(2 \times 10^5)(6 \times 10^3)$
1.2×10^9

38. $(9 \times 10^6)(5 \times 10^5)$
4.5×10^{12}

39. $(3.2 \times 10^5)(7.5 \times 10^8)$
2.4×10^{14}

40. $(4.5 \times 10^{10})(8.4 \times 10^7)$
3.78×10^{18}

41. $(2.9 \times 10^8)(6.3 \times 10^{-4})$
1.827×10^5

42. $(7.2 \times 10^{-2})(4.6 \times 10^7)$
3.312×10^6

43. $(4.23 \times 10^{-8})(2.7 \times 10^3)$
1.1421×10^{-4}

44. $(3.75 \times 10^6)(4.3 \times 10^{-12})$
1.6125×10^{-5}

45. $(9.2 \times 10^{-3})(6.1 \times 10^{-8})$
5.612×10^{-10}

46. $(8.4 \times 10^{-6})(7.3 \times 10^{-8})$
6.132×10^{-13}

47. Tennessee is roughly shaped like a parallelogram with a base of approximately 350 miles and a height of approximately 120 miles. Calculate the area of Tennessee and write your answer in scientific notation. (Use $A = bh$.)
4.2×10^4 mi.^{2}

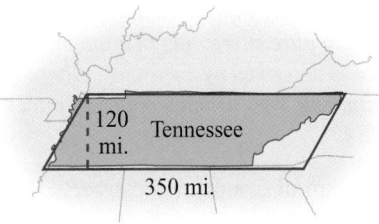

48. Idaho is roughly shaped like a triangle with a base of approximately 320 miles and a height of approximately 520 miles. Calculate the area of Idaho and write your answer in scientific notation. $\left(\text{Use } A = \frac{1}{2}bh. \right)$
8.32×10^4 mi.^{2}

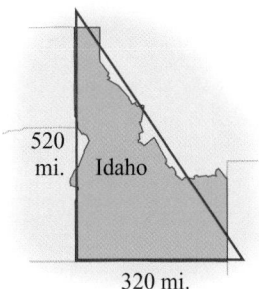

For Exercises 49 and 50, use the formula for the circumference of a circle: $C = 2\pi r$.
(Use 3.14 to approximate π.) Write your answers in scientific notation. See Example 4.

★ **49.** The path of Earth's orbit around the
  Sun is approximately a circle with
  a radius of about 9.3×10^7 miles.
  What is the circumference of Earth's
  orbit?
  $\approx 5.8404 \times 10^8$ mi.

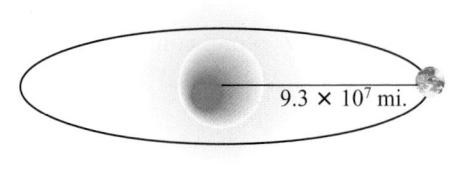

9.3×10^7 mi.

Not to scale.

Of Interest

Earth travels the distance that you
calculated in Exercise 49 in one
year, which translates to about
18.5 miles every second.

★ **50.** The radius of Earth is approximately 6.37×10^6 meters. Calculate
  the circumference of the equator.
  $\approx 4.00036 \times 10^7$ m

For Exercises 51 and 52, use the formula for the volume of a sphere: $V = \dfrac{4}{3}\pi r^3$.

(Use 3.14 to approximate π.) Write your answers in scientific notation. See Example 4.

★ **51.** The radius of the Sun is approximately 6.96×10^8 meters. Calculate the volume of the Sun.
  $\approx 1.412 \times 10^{27}$ m^3

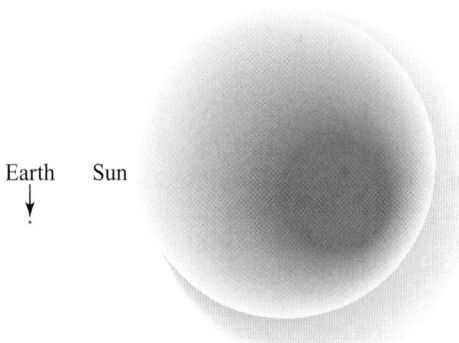

Earth   Sun

Of Interest

The Sun's radius is about 100
times the length of Earth's radius (10^8 versus 10^6). To get a
sense of that scale, if Earth's radius were 1 centimeter, the Sun's
radius would be about 1 meter.
The scale in the picture on this
page is accurate, with Earth being about the width of a period.

★ **52.** The radius of Earth is approximately 6.37×10^6 meters. Calculate the volume of Earth.
  $\approx 1.082 \times 10^{21}$ m^3

For Exercises 53 and 54, use the following information. The energy, in joules, of a single
photon of light can be determined by $E = hv$, where h is the constant 6.626×10^{-34}
joule-seconds and v is the frequency of the light in hertz. Write your answers in scientific
notation. See Example 4.

53. Find the energy of a photon of red light with a frequency of 4.5×10^{14} Hz.
  2.9817×10^{-19} joule

54. Find the energy of a photon of violet light with a frequency of 8.5×10^{14} Hz.
  5.6321×10^{-19} joule

For Exercises 55 and 56, use the following information. Albert Einstein discovered that if a
mass of m kilograms is converted to pure energy (as in a nuclear reaction), the amount of
energy, in joules, that is released is described by $E = mc^2$, where c represents a constant
speed of light, which is approximately 3×10^8 meters per second. Write your answers in
scientific notation. See Example 4.

55. Suppose 2.4×10^{-10} kilogram of uranium is converted
  to energy in a nuclear reaction. How much energy is
  released?
  2.16×10^7 joules

56. Suppose 4.5×10^{-8} kilogram of uranium is converted
  to energy in a nuclear reaction. How much energy is
  released?
  4.05×10^9 joules

Objective 3

Prep Exercise 5 Explain how to simplify $(x^3)^8$.
Use $(a^m)^n = a^{mn}$. So $(x^3)^8 = x^{3 \cdot 8} = x^{24}$.

Prep Exercise 6 Explain how to simplify $(3x^4)^2$.
Use $(a b)^n = a^n \cdot b^n$. So $(3x^4)^2 = 3^2(x^4)^2 = 9x^8$.

For Exercises 57–92, simplify. See Examples 5 and 6.

57. $(x^2)^3$
x^6

58. $(y^3)^5$
y^{15}

59. $(3^2)^4$
3^8

60. $(2^3)^2$
2^6

61. $(-h^2)^4$
h^8

62. $(-j^3)^2$
j^6

63. $(-y^2)^3$
$-y^6$

64. $(-x^4)^5$
$-x^{20}$

65. $(xy)^3$
x^3y^3

66. $(ab)^3$
a^3b^3

67. $(3x^2)^3$
$27x^6$

68. $(4x^4)^2$
$16x^8$

69. $(-2x^3y^2)^3$
$-8x^9y^6$

70. $(-5a^4b^2)^2$
$25a^8b^4$

71. $\left(\dfrac{1}{4}m^2n\right)^3$
$\dfrac{1}{64}m^6n^3$

72. $\left(\dfrac{5}{6}u^3v\right)^2$
$\dfrac{25}{36}u^6v^2$

73. $(-2p^4q^4r^2)^6$
$64p^{24}q^{24}r^{12}$

74. $(-6u^2v^4w^2)^3$
$-216u^6v^{12}w^6$

75. $(-0.2x^2y^5z)^3$
$-0.008x^6y^{15}z^3$

76. $(-0.3r^4st^3)^4$
$0.0081r^{16}s^4t^{12}$

77. $(2x^3)(3x^2)^2$
$18x^7$

78. $(-3y^2)(2y^2)^3$
$-24y^8$

79. $(rs^2)(r^2s)^2$
r^5s^4

80. $(m^2n)^3(mn)$
m^7n^4

81. $(2ab)^2(5a^2b)^2$
$100a^6b^4$

82. $(4xy^3)^2(2x^2y)^2$
$64x^6y^8$

83. $\left(\dfrac{1}{4}a^2b^3c\right)^2\left(\dfrac{1}{4}a^2b^3c\right)^3$
$\dfrac{1}{1024}a^{10}b^{15}c^5$

84. $\left(-\dfrac{1}{2}s^2tu^4\right)^2\left(-\dfrac{1}{2}s^2tu^4\right)^5$
$-\dfrac{1}{128}s^{14}t^7u^{28}$

85. $5(3a)^2(2b)^3$
$360a^2b^3$

86. $2(2x)^4(4y)^3$
$2048x^4y^3$

87. $-3(-2a^4b^2)^3(3a^3b)^2$
$216a^{18}b^8$

88. $-6(-2m^2n^4)^3(-4mn^3)^2$
$768m^8n^{18}$

89. $(6u)^2(u^2v)(-3uv)^2$
$324u^6v^3$

90. $(3x)^2(xy^3)(2x^3yz)^3$
$72x^{12}y^6z^3$

91. $(0.6u)^2(u^2v)(-3uv)^2$
$3.24u^6v^3$

92. $(-0.2w)^4(-w^2z^3)^3(-4w^4z)^2$
$-0.0256w^{18}z^{11}$

Find ✕ the Mistake *For Exercises 93–96, find and explain the mistake; then work the problem correctly. See Example 5.*

93. $(5x^3)^2 = 25x^5$
Mistake: The exponents were added instead of multiplied. Correct: $25x^6$

94. $(-4mn^5)^3 = -12m^3n^{15}$
Mistake: The coefficient was multiplied by 3 instead of cubed. Correct: $-64m^3n^{15}$

95. $(2x^4y)^4 = 8x^8y$
Mistake: The coefficient was multiplied by 4 instead of being raised to the 4th power. Also, the exponents were not multiplied properly. Correct: $16x^{16}y^4$

96. $(-2ab^5)^3 = -8ab^{15}$
Mistake: The a was not raised to the 3rd power. Correct: $-8a^3b^{15}$

Puzzle Problem

a. If there are about 10^{11} stars in a galaxy and about 10^{11} galaxies in the visible universe, about how many stars are in the visible universe?

$10^{11} \cdot 10^{11} = 10^{22}$ stars in the visible universe

b. Suppose one star out of every billion stars has a solar system with one planet that could support life. How many stars in the visible universe would have a solar system with one planet that could support life? (Hint: One out of every billion is one billionth, or 10^{-9}.)

$10^{-9} \cdot 10^{22} = 10^{13}$ stars with a solar system and one planet that could support life

Review Exercises

Exercises 1–6   Expressions

[1.7] 1. Simplify $2(3x + 4)$ using the distributive property.
$6x + 8$

[1.7] 3. Simplify: $-10x - 12x$
$-22x$

[1.3] 5. Add: $(-4) + (-57)$
-61

[5.4] 2. Simplify: $-0.3t^2u \cdot 6t^3u$
$-1.8t^5u^2$

[5.3] 4. Add: $(3x + 4y - z) + (-7x - 4y - 3z)$
$-4x - 4z$

[5.2] 6. What is the degree of the monomial $-2r^5s$?
6

5.5 Multiplying Polynomials; Special Products

Objectives

1 Multiply a polynomial by a monomial.

2 Multiply binomials.

3 Multiply polynomials.

4 Determine the product when given special polynomial factors.

Warm-up

For Exercises 1 and 2, find the products.

[1.7] 1. $-7(5x + 3)$

[5.4] 2. $(-3t^2u)(6t^3u)$

[5.3] 3. Simplify by combining like terms. Write the answer in descending order of degree.

$$x^4 + 2x^3 - 5x^2 - 3x^3 - 6x^2 + 15x + 4x^2 + 8x - 20$$

In this section, we use the rules for multiplying monomials to help us multiply multiple-term polynomials.

Objective 1 Multiply a polynomial by a monomial.

To multiply polynomials, we use the distributive property, which we introduced in Section 1.7 and used in solving linear equations and inequalities in Chapter 2. For example, we have applied the distributive property in situations such as $2(3x + 4)$.

$$2(3x + 4) = 2 \cdot 3x + 2 \cdot 4$$
$$= 6x + 8$$

Notice that $2(3x + 4)$ is a product of a monomial 2 and a binomial $3x + 4$. This suggests the following procedure.

Answers to Warm-up
1. $-35x - 21$
2. $-18t^5u^2$
3. $x^4 - x^3 - 7x^2 + 23x - 20$

> **Procedure** **Multiplying a Polynomial by a Monomial**
>
> To multiply a polynomial by a monomial, use the distributive property to multiply each term in the polynomial by the monomial.

Connection Multiplying a polynomial by a monomial is like multiplying a multidigit number by a single-digit number. Consider $2 \cdot 34$.

$$\begin{array}{r} 34 \\ \times\ 2 \\ \hline 68 \end{array}$$

Notice that in the numerical multiplication, we distribute the 2 to each digit in the 34. In fact, if we expand 34 by expressing it as $30 + 4$, the problem takes the same form as $2(3x + 4)$, where $x = 10$.

$$2(30 + 4) = 2 \cdot 30 + 2 \cdot 4 \qquad 2(3x + 4) = 2 \cdot 3x + 2 \cdot 4$$
$$= 60 + 8 \qquad\qquad\qquad = 6x + 8$$
$$= 68$$

Now let's multiply a polynomial by a monomial that contains variables.

Example 1 Multiply.

a. $2y(3y^2 + 4y + 1)$

Solution: $2y \quad\quad (3y^2 + 4y + 1)$ Multiply each term in the polynomial by $2y$.

$2y \cdot 3y^2$
$2y \cdot 4y$
$2y \cdot 1$

$= 2y \cdot 3y^2 + 2y \cdot 4y + 2y \cdot 1$ ◄
$= 6y^3 + 8y^2 + 2y$

Note When multiplying the individual terms by the monomial, we multiply the coefficients and add the exponents of the like bases.

b. $-7x^2(5x^2 + 3x - 4)$

Solution: $-7x^2 \quad\quad (5x^2 + 3x - 4)$ Multiply each term in the polynomial by $-7x^2$.

$-7x^2 \cdot 5x^2$
$-7x^2 \cdot 3x$
$-7x^2 \cdot (-4)$

$= -7x^2 \cdot 5x^2 - 7x^2 \cdot 3x - 7x^2 \cdot (-4)$
$= -35x^4 - 21x^3 + 28x^2$

Note When we multiply a polynomial by a negative monomial, the signs of the resulting polynomial are opposite the signs in the original polynomial.

$-7x^2(\ 5x^2\ +\ 3x\ -\ 4) \quad\blacktriangleright$
$= -35x^4 - 21x^3 + 28x^2$

c. $-0.3t^2u(6t^3u + tu^2 - 3t^3 + 0.2uv)$

Solution: $-0.3t^2u \quad (\ 6t^3u\ +\ tu^2\ -\ 3t^3\ +\ 0.2uv)$ Multiply each term in the polynomial by $-0.3t^2u$.

$-0.3t^2u \cdot 6t^3u$
$-0.3t^2u \cdot tu^2$
$-0.3t^2u \cdot (-3t^3)$
$-0.3t^2u \cdot 0.2uv$

$= -0.3t^2u \cdot 6t^3u - 0.3t^2u \cdot tu^2 - 0.3t^2u \cdot (-3t^3) - 0.3t^2u \cdot 0.2uv$
$= -1.8t^5u^2 - 0.3t^3u^3 + 0.9t^5u - 0.06t^2u^2v$

▲

Note When multiplying multivariable terms, it is helpful to multiply the coefficients first, then the variables in alphabetical order.

For $-0.3t^2u \cdot 6t^3u$, think $-0.3 \cdot 6 = -1.8; t^2 \cdot t^3 = t^5; u \cdot u = u^2$.
Result: $-1.8t^5u^2$

Learning Strategy

If you are a visual learner, try drawing lines connecting the terms that are to be multiplied.

Answers to Your Turn 1
a. $36x^5 - 28x^4 + 12x^3$
b. $-15x^5yz^2 + 3x^3y^4z^3 - 24x^4yz^2$

Your Turn 1 Multiply.

a. $4x^3(9x^2 - 7x + 3)$ **b.** $-3x^3yz(5x^2z - y^3z^2 + 8xz)$

Objective 2 Multiply binomials.

Understanding Binomial Multiplication

Now that we have seen how to multiply a polynomial by a monomial, we are ready to multiply two binomials, which is like multiplying a pair of two-digit numbers. For example, consider $(12)(13)$ and compare that with $(x + 2)(x + 3)$.

$$
\begin{array}{r}
12 \\
\times\ 13 \\
\hline
36 \\
+\ 12 \\
\hline
156
\end{array}
$$

◀ **Note** We think to ourselves "3 times 2 is 6, then 3 times 1 is 3," which creates the 36. We then move to the 1 in the tens place of the 13 and do the same thing. Because this 1 digit is in the tens place, it means 10; so when we multiply this 10 times 12, it makes 120. We usually omit writing the 0 in the ones place and write 12 in the next two places.

Again, the distributive property is the governing principle. We multiply each digit in one number by each digit in the other number and shift underneath as we move to each new place. The same process applies to the binomials. We can stack them as we just did. Notice, however, that we stack this only once to make the connection to numeric multiplication. As the polynomials get more complex, the stacking method becomes too tedious.

$$
\begin{array}{r}
x + 2 \\
x + 3 \\
\hline
3x + 6 \\
x^2 + 2x \\
\hline
x^2 + 5x + 6
\end{array}
$$

◀ **Note** We think "3 times 2 is 6, then 3 times x is $3x$." Now move to the x and think "x times 2 is $2x$, then x times x is x^2." Notice how we shifted so that the $2x$ and $3x$ line up. This is because they are like terms. It is the same as lining up the tens column when multiplying numbers. Note that the numeric result, 156, is basically the same as the algebraic result, $x^2 + 5x + 6$.

Notice that each term in $x + 2$ is multiplied by each term in $x + 3$. In general, this is how we multiply two polynomials.

> **Procedure** **Multiplying Polynomials**
>
> To multiply two polynomials:
> 1. Multiply each term in the second polynomial by each term in the first polynomial.
> 2. Combine like terms.

Multiply Binomials

Connection In Example 2, every term in both binomials is positive, the like terms are added and both signs in the resulting trinomial are plus signs. Noting these sign patterns will help when we factor in Chapter 6.

 Learning Strategy

The word *FOIL* is a popular way to remember the process of multiplying two binomials. *FOIL* stands for First Outer Inner Last. We will use Example 2 to demonstrate.

First terms: $x \cdot x = x^2$
$(x + 4)(x + 3)$

Outer terms: $x \cdot 3 = 3x$
$(x + 4)(x + 3)$

$(x + 4)(x + 3)$
Inner terms: $4 \cdot x = 4x$

$(x + 4)(x + 3)$
Last terms: $4 \cdot 3 = 12$

| **Example 2** | Multiply $(x + 4)(x + 3)$. |

Solution:

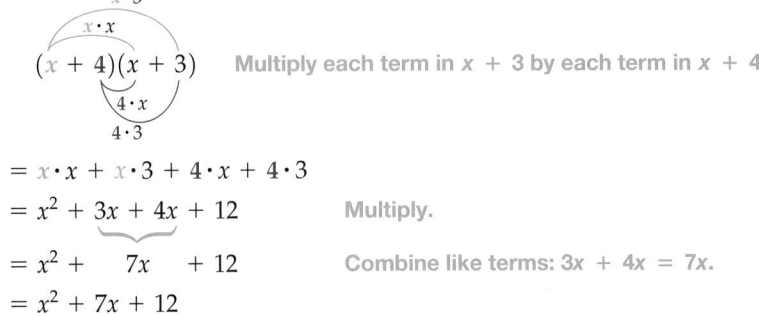

$$
\begin{aligned}
&= x \cdot x + x \cdot 3 + 4 \cdot x + 4 \cdot 3 \\
&= x^2 + 3x + 4x + 12 &&\text{Multiply.}\\
&= x^2 + 7x + 12 &&\text{Combine like terms: } 3x + 4x = 7x.\\
&= x^2 + 7x + 12
\end{aligned}
$$

Warning FOIL makes sense only when you are multiplying two binomials. When multiplying polynomials with more than two terms, just remember to multiply each term in the second polynomial by each term in the first polynomial.

Example 3 Multiply $(3x + 5)(x - 4)$.

Solution:

$(3x + 5) \quad (x - 4)$ Multiply each term in $x - 4$ by each term in $3x + 5$. (Think *FOIL*.)

First   **Outer**   Inner   **Last**

$= 3x \cdot x + 3x \cdot (-4) + 5 \cdot x + 5 \cdot (-4)$

$= 3x^2 \qquad - 12x \quad + 5x \qquad - 20$ Multiply.

$= 3x^2 \qquad\qquad - 7x \qquad\qquad - 20$ Combine like terms: $-12x + 5x = -7x$.

$= 3x^2 - 7x - 20$

Connection Notice that $3x + 5$ has a plus sign, whereas $x - 4$ has a minus sign. When multiplying binomials such as these, the last term of the resulting trinomial is always negative. The middle term can be positive or negative depending on the coefficients of the like terms.

In Example 4, we switch the signs in the binomials from Example 3. Look at how this changes the sign of the middle term in the resulting trinomial.

Example 4 Multiply $(3x - 5)(x + 4)$.

Solution:

$(3x - 5) \quad (x + 4)$ Multiply each term in $x + 4$ by each term in $3x - 5$.

First   Outer   Inner   Last

$= 3x \cdot x + 3x \cdot 4 + (-5) \cdot x + (-5) \cdot 4$

$= 3x^2 \qquad + 12x \quad - 5x \qquad - 20$ Multiply.

$= 3x^2 \qquad\qquad + 7x \qquad - 20$ Combine like terms: $12x - 5x = 7x$.

$= 3x^2 + 7x - 20$

Connection In Example 3, the result was $3x^2 - 7x - 20$, whereas in Example 4, the result was $3x^2 + 7x - 20$. Notice that switching the signs like we did merely changed the sign of the middle term.

Now consider a case when the signs between terms in the binomials are both negative.

Example 5 Multiply $(2x - 3)(4x - 5)$.

Solution:

$(2x - 3) \quad (4x - 5)$ Multiply each term in $4x - 5$ by each term in $2x - 3$. (Think *FOIL*.)

First   Outer   Inner   Last

$= 2x \cdot 4x + 2x \cdot (-5) + (-3) \cdot 4x + (-3) \cdot (-5)$

$= 8x^2 \qquad - 10x \qquad - 12x \qquad + 15$

$= 8x^2 \qquad\qquad - 22x \qquad\qquad + 15$ Combine like terms: $-10x - 12x = -22x$.

$= 8x^2 - 22x + 15$

Connection When multiplying binomials such as $2x - 3$ and $4x - 5$, notice that the last term in the resulting trinomial is positive because the product of two negatives is positive and the middle term is negative because the sum of two negatives is negative.

Keep in mind that Examples 2–5 do not illustrate all sign combinations that you could be given. The point of our discussion of signs is to note some patterns with the signs in the given binomial factors to see how they affect the resulting polynomial. Noticing these sign patterns now will help when we factor in Chapter 6.

Your Turn 5 Multiply.

a. $(n + 6)(n - 2)$ **b.** $(2y - 5)(y - 3)$

Connection The product of two binomials can be shown in terms of geometry. Consider the rectangle shown with a length of $x + 7$ and a width of $x + 5$. We can describe its area in two ways: (1) the product of the length and width, $(x + 7)(x + 5)$, and (2) the sum of the areas of the four internal rectangles, $x^2 + 5x + 7x + 35$. Because these two approaches describe the same area, the two expressions must be equal.

	x	7
x	x^2	$7x$
5	$5x$	35

Length $\cdot$ Width $=$ Sum of the areas of the
four internal rectangles
$(x + 7)(x + 5) = x^2 + 5x + 7x + 35$
$ = x^2 + 12x + 35$ Combine like
terms.

Objective 3 Multiply polynomials.

Now consider an example of multiplication involving a larger polynomial, such as a trinomial. Remember that no matter how many terms are in the polynomials, the process is to multiply every term in the second polynomial by every term in the first polynomial.

Example 6 Multiply.

a. $(x + 2)(3x^2 + 5x - 4)$

Solution:

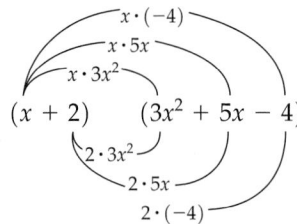

$(x + 2) \quad (3x^2 + 5x - 4)$ Multiply each term in $3x^2 + 5x - 4$ by each term in $x + 2$.

$= x \cdot 3x^2 + x \cdot 5x + x \cdot (-4) + 2 \cdot 3x^2 + 2 \cdot 5x + 2 \cdot (-4)$
$= 3x^3 + 5x^2 -4x + 6x^2 + 10x -8$
$= 3x^3 + 11x^2 + 6x - 8$ Combine like terms.

Warning FOIL does not make sense in Example 6 because the polynomials have too many terms. FOIL handles only the four terms from two binomials.

b. $(2x - 3)(4x^2 + 6x + 9)$

Solution:

$(2x - 3)(4x^2 + 6x + 9)$
$= 2x(4x^2 + 6x + 9) - 3(4x^2 + 6x + 9)$ Multiply each term in $4x^2 + 6x + 9$ by each term in $2x - 3$.
$= 8x^3 + 12x^2 + 18x - 12x^2 - 18x - 27$ Distribute.
$= 8x^3 - 27$ Combine like terms.

Answers to Your Turn 5
a. $n^2 + 4n - 12$
b. $2y^2 - 11y + 15$

c. $(x^2 - 3x + 4)(x^2 + 2x - 5)$

Solution:

$$(x^2 - 3x + 4)(x^2 + 2x - 5)$$

$$= x^2(x^2 + 2x - 5) - 3x(x^2 + 2x - 5) + 4(x^2 + 2x - 5)$$ Multiply each term in $x^2 + 2x - 5$ by each term in $x^2 - 3x + 4$.

$$= x^4 + 2x^3 - 5x^2 - 3x^3 - 6x^2 + 15x + 4x^2 + 8x - 20$$ Distribute.

$$= x^4 - x^3 - 7x^2 + 23x - 20$$ Combine like terms.

Your Turn 6 Multiply.

  a. $(x - 3)(x^2 - 6x + 2)$
  b. $(x + 5)(x^2 - 5x + 25)$
  c. $(y^2 + 4y - 2)(y^2 - 3y - 5)$

Objective 4 Determine the product when given special polynomial factors.

Multiplying Conjugates

In Objective 2, we discussed how the signs in the binomial factors affect the signs in the product in preparing for factoring in Chapter 6. To further prepare for factoring, it is helpful to note some patterns that occur when certain special polynomial factors are multiplied. First, we will multiply **conjugates**.

Definition **Conjugates:** Binomials that differ only in the sign separating the terms.

The following binomial pairs are conjugates.

$$x + 9 \text{ and } x - 9 \qquad 2x + 3 \text{ and } 2x - 3 \qquad -6x + 5 \text{ and } -6x - 5$$

Let's multiply the conjugates $2x + 3$ and $2x - 3$ to see what pattern emerges.

First   Outer   Inner   Last

$$(2x + 3)(2x - 3) = 2x \cdot 2x + 2x \cdot (-3) + 3 \cdot 2x + 3 \cdot (-3)$$

$$= 4x^2 \quad -6x \quad + 6x \quad -9$$

$$= 4x^2 + 0 - 9 \quad \text{Combine like terms: } -6x + 6x = 0.$$

$$= 4x^2 - 9 \blacktriangleleft \textbf{Note} \text{ The sum of the like terms is 0, so the resulting binomial is the square of the first term minus the square of the last term.}$$

When conjugates are multiplied, the like terms are always additive inverses; so their sum is always 0. This suggests the following rule.

Instructor Note Point out that this is a shortcut to FOIL that applies only to conjugates.

Rule **Multiplying Conjugates**

If a and b are real numbers, variables, or expressions, then $(a + b)(a - b) = a^2 - b^2$.

Using this rule allows us to multiply conjugates quickly.

Answers to Your Turn 6
a. $x^3 - 9x^2 + 20x - 6$
b. $x^3 + 125$
c. $y^4 + y^3 - 19y^2 - 14y + 10$

Example 7 Multiply.

  a. $(x + 3)(x - 3)$

Solution: $(x + 3)(x - 3) = x^2 - 3^2$ Use $(a + b)(a - b) = a^2 - b^2$.

$$= x^2 - 9 \quad \text{Simplify.}$$

b. $(3t - 8)(3t + 8)$

Solution: $(3t - 8)(3t + 8) = (3t)^2 - 8^2$ Use $(a + b)(a - b) = a^2 - b^2$.

$\qquad\qquad\qquad\qquad\qquad = 9t^2 - 64$ Simplify.

Your Turn 7   Multiply.

   a. $(m + 7)(m - 7)$ **b.** $(5y - 3)(5y + 3)$

Squaring a Binomial

Now let's look for a pattern when we square a binomial, as in $(3x + 4)^2$. Recall that squaring a number or an expression means to multiply that number or expression by itself; so $(3x + 4)^2 = (3x + 4)(3x + 4)$.

Instructor Note Point out that $(3x + 4)^2 \neq (3x)^2 + 4^2$.

$$
\begin{aligned}
(3x + 4)^2 = (3x + 4)(3x + 4) &= 3x \cdot 3x + 3x \cdot 4 + 4 \cdot 3x + 4 \cdot 4 \\
&= 9x^2 + 12x + 12x + 16 \\
&= 9x^2 + 24x + 16 \quad \text{Combine like terms.} \\
&\qquad\qquad\qquad\qquad 12x + 12x = 24x
\end{aligned}
$$

First   Outer   Inner   Last

Warning Notice that $(3x + 4)^2$ does not equal $9x^2 + 16$.

Notice the pattern.

$$(3x + 4)^2 = 9x^2 + 24x + 16$$

This first term is the square of the first term in the binomial: $(3x)^2 = 9x^2$.

This last term is the square of the second term in the binomial: $(4)^2 = 16$.

This middle term is twice the product of the two terms in the binomial: $2(3x)(4) = 24x$.

Connection The square of a binomial can also be shown in terms of geometry. Consider the square shown, each side being $x + 3$. We can describe the area in two ways: (1) the length of a side squared, $(x + 3)^2$, or (2) the sum of the areas of the four internal rectangles, $x^2 + 3x + 3x + 3^2 = x^2 + 2 \cdot 3x + 9 = x^2 + 6x + 9$.

Side squared = Sum of the areas of the four internal rectangles.

$$
\begin{aligned}
(x + 3)^2 &= x^2 + 2 \cdot 3x + 9 \\
&= x^2 + 6x + 9
\end{aligned}
$$

Squaring a binomial with a minus sign, as in $(3x - 4)^2$, yields a similar pattern.

$$
\begin{aligned}
(3x - 4)^2 = (3x - 4)(3x - 4) &= 3x \cdot 3x + 3x \cdot (-4) + (-4) \cdot 3x + (-4) \cdot (-4) \\
&= 9x^2 - 12x - 12x + 16 \\
&= 9x^2 - 24x + 16 \quad \text{Combine like terms:} \\
&\qquad\qquad\qquad\qquad -12x - 12x = -24x.
\end{aligned}
$$

First   Outer   Inner   Last

Answers to Your Turn 7
a. $m^2 - 49$
b. $25y^2 - 9$

Warning Notice that $(3x - 4)^2$ does *not* equal $9x^2 - 16$ or $9x^2 + 16$.

Notice the pattern.

$$(3x - 4)^2 = 9x^2 - 24x + 16$$

This first term is the square of the first term in the binomial: $(3x)^2 = 9x^2$.

This last term is the square of the second term in the binomial: $(-4)^2 = 16$.

This middle term is twice the product of the two terms in the binomial: $2(3x)(-4) = -24x$.

The patterns from our two examples suggest the following rules.

> **Rules Squaring a Binomial**
>
> If a and b are real numbers, variables, or expressions, then
> $$(a + b)^2 = a^2 + 2ab + b^2$$
> $$(a - b)^2 = a^2 - 2ab + b^2$$

Example 8 Multiply.

a. $(5x + 3)^2$

Solution: $(5x + 3)^2 = (5x)^2 + 2(5x)(3) + (3)^2$ Use $(a + b)^2 = a^2 + 2ab + b^2$.
$$= 25x^2 + 30x + 9$$ Simplify.

b. $(6y - 7)^2$

Solution: $(6y - 7)^2 = (6y)^2 - 2(6y)(7) + (7)^2$ Use $(a - b)^2 = a^2 - 2ab + b^2$.
$$= 36y^2 - 84y + 49$$ Simplify.

Answers to Your Turn 8
a. $9m^2 + 42m + 49$
b. $16t^2 - 40t + 25$

Your Turn 8 Multiply.

a. $(3m + 7)^2$

b. $(4t - 5)^2$

5.5 Exercises For Extra Help MyMathLab®

Note: Exercises marked with a ★ represent challenging exercises.

Objective 1

Prep Exercise 1 What mathematical property is applied when multiplying a polynomial by a monomial?
Distributive property

For Exercises 1–28, multiply the polynomial by the monomial. See Example 1.

1. $5(x + 3)$
$5x + 15$

2. $2(y + 7)$
$2y + 14$

3. $-6(2x - 4)$
$-12x + 24$

4. $-4(3y - 5)$
$-12y + 20$

5. $4n(7n + 2)$
$28n^2 + 8n$

6. $6n(2n + 5)$
$12n^2 + 30n$

7. $9a(a - b)$
$9a^2 - 9ab$

8. $3y(y - z)$
$3y^2 - 3yz$

9. $\frac{1}{8}m(2m - 5n)$
$\frac{1}{4}m^2 - \frac{5}{8}mn$

10. $\frac{1}{4}k(2k - 3l)$
$\frac{1}{2}k^2 - \frac{3}{4}kl$

11. $-0.3p^2(p^3 - 2q^2)$
$-0.3p^5 + 0.6p^2q^2$

12. $-0.4b^2(b^2 - 3c^2)$
$-0.4b^4 + 1.2b^2c^2$

13. $5x(4x^2 + 2x - 5)$
$20x^3 + 10x^2 - 25x$

14. $5a(2a^2 - a + 6)$
$10a^3 - 5a^2 + 30a$

15. $-3x^2(7x^3 - 5x + 1)$
$-21x^5 + 15x^3 - 3x^2$

16. $-4w^3(3w^3 + 7w^2 - 2)$
$-12w^6 - 28w^5 + 8w^3$

17. $-r^2s(r^2s^3 + 3rs^2 - s)$
$-r^4s^4 - 3r^3s^3 + r^2s^2$

18. $-x^2y^3(-2xy^2 + 3x^2y^5 - y)$
$2x^3y^5 - 3x^4y^8 + x^2y^4$

19. $-2a^2b^2(2a^3 - 6b + 3ab - a^2b^2)$
$-4a^5b^2 + 12a^2b^3 - 6a^3b^3 + 2a^4b^4$

20. $-5hk^3(3k^3 - 2h^2k - 3hk^4 + 4h)$
$-15hk^6 + 10h^3k^4 + 15h^2k^7 - 20h^2k^3$

21. $3abc(4a^2b^2c^2 - 2abc + 5)$
$12a^3b^3c^3 - 6a^2b^2c^2 + 15abc$

22. $5xy^2z(2xy^3 - 4x^2z^2 + 3)$
$10x^2y^5z - 20x^3y^2z^3 + 15xy^2z$

23. $4b\left(b^5 - \dfrac{1}{4}b^3 + \dfrac{1}{20}b^2 - \dfrac{1}{8}b + 4\right)$
$4b^6 - b^4 + \dfrac{1}{5}b^3 - \dfrac{1}{2}b^2 + 16b$

24. $5r^2\left(r^4 - \dfrac{1}{20}r^3 - \dfrac{1}{5}r^2 + \dfrac{1}{10}r - 1\right)$
$5r^6 - \dfrac{1}{4}r^5 - r^4 + \dfrac{1}{2}r^3 - 5r^2$

25. $-0.4rt^2(2.5r^3t - 8r^2t^2 + 4.1rt^3 - 2)$
$-r^4t^3 + 3.2r^3t^4 - 1.64r^2t^5 + 0.8rt^2$

26. $-0.25suv^3(10s^2u - 2u^2v - 18sv^4 + 4)$
$-2.5s^3u^2v^3 + 0.5su^3v^4 + 4.5s^2uv^7 - suv^3$

27. $\dfrac{5}{6}x^2(30x^4y - 18x^3y^2 + 60xy^9)$
$25x^6y - 15x^5y^2 + 50x^3y^9$

28. $\dfrac{2}{3}b^4(-30a^2 + 18ab^3 - 12a^2b^2)$
$-20a^2b^4 + 12ab^7 - 8a^2b^6$

Objective 2

Prep Exercise 2 What does each letter in FOIL indicate?
First, Outer, Inner, Last

Prep Exercise 3 Explain, in general, how to multiply two polynomials.
To multiply two polynomials:
a. Multiply each term in the second polynomial by each term in the first polynomial.
b. Combine like terms.

For Exercises 29–46, multiply the binomials. (Use FOIL.) See Examples 2–5.

29. $(x + 3)(x + 4)$
$x^2 + 7x + 12$

30. $(a + 4)(a + 9)$
$a^2 + 13a + 36$

31. $(x - 7)(x + 2)$
$x^2 - 5x - 14$

32. $(n + 7)(n - 5)$
$n^2 + 2n - 35$

33. $(y - 3)(y - 6)$
$y^2 - 9y + 18$

34. $(r - 4)(r - 5)$
$r^2 - 9r + 20$

35. $(2y + 5)(3y + 2)$
$6y^2 + 19y + 10$

36. $(2x + 4)(4x + 3)$
$8x^2 + 22x + 12$

37. $(5m - 3)(3m + 4)$
$15m^2 + 11m - 12$

38. $(2x + 3)(3x - 2)$
$6x^2 + 5x - 6$

39. $(3t - 5)(4t - 2)$
$12t^2 - 26t + 10$

40. $(2t - 7)(3t - 1)$
$6t^2 - 23t + 7$

41. $(5q - 3t)(3q + 4t)$
$15q^2 + 11qt - 12t^2$

42. $(7k - 2j)(3k + 4j)$
$21k^2 + 22jk - 8j^2$

43. $(7y + 3x)(2y + 4x)$
$14y^2 + 34xy + 12x^2$

44. $(2a + 5b)(3a + 4b)$
$6a^2 + 23ab + 20b^2$

45. $(a^2 + b)(a^2 - 2b)$
$a^4 - a^2b - 2b^2$

46. $(x^2 + 4y)(x^2 - 5y)$
$x^4 - x^2y - 20y^2$

For Exercises 47–50, a larger rectangle is formed out of smaller rectangles.

a. Write an expression in simplest form for the length (along the top).
b. Write an expression in simplest form for the width (along the side).
c. Write an expression that is the product of the length and width that you found in parts a and b.
d. Write an expression in simplest form that is the sum of the areas of each of the smaller rectangles.
e. Explain why the expressions in parts c and d are equivalent.

47.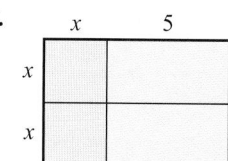
   a. $x + 5$
   b. $2x$
   c. $2x(x + 5)$
   d. $2x^2 + 10x$
   e. They describe the same area.

48.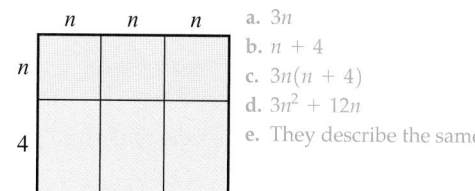
   a. $3n$
   b. $n + 4$
   c. $3n(n + 4)$
   d. $3n^2 + 12n$
   e. They describe the same area.

49.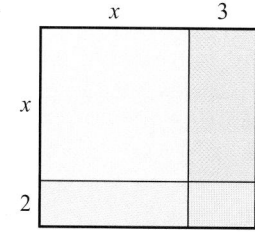
   a. $x + 3$
   b. $x + 2$
   c. $(x + 2)(x + 3)$
   d. $x^2 + 5x + 6$
   e. They describe the same area.

50.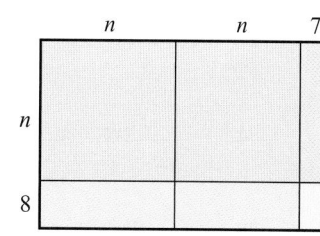
   a. $2n + 7$
   b. $n + 8$
   c. $(2n + 7)(n + 8)$
   d. $2n^2 + 23n + 56$
   e. They describe the same area.

Objective 3

For Exercises 51–66, multiply the polynomials. See Example 6.

51. $(x + 2)(x^2 - 2x + 1)$
   $x^3 - 3x + 2$

52. $(x + 3)(x^2 - 3x + 2)$
   $x^3 - 7x + 6$

53. $(a - 1)(3a^2 - a - 2)$
   $3a^3 - 4a^2 - a + 2$

54. $(x - 1)(3x^2 - 2x + 1)$
   $3x^3 - 5x^2 + 3x - 1$

55. $(3c + 2)(2c^2 - c - 3)$
   $6c^3 + c^2 - 11c - 6$

56. $(2x + 3)(4x^2 + 3x - 5)$
   $8x^3 + 18x^2 - x - 15$

57. $(2f + 3g)(2f^2 - 3fg - 9g^2)$
   $4f^3 - 27fg^2 - 27g^3$

58. $(2m - 3n)(4m^2 - 6mn + 8n^2)$
   $8m^3 - 24m^2n + 34mn^2 - 24n^3$

59. $(3m - 2n)(9m^2 + 6mn + 4n^2)$
   $27m^3 - 8n^3$

60. $(4a - b)(16a^2 + 4ab + b^2)$
   $64a^3 - b^3$

61. $(2a + 5b)(4a^2 - 10ab + 25b^2)$
   $8a^3 + 125b^3$

62. $(5m + 3n)(25m^2 - 15mn + 9n^2)$
   $125m^3 + 27n^3$

63. $(x^2 + xy + y^2)(x^2 + 2xy - y^2)$
   $x^4 + 3x^3y + 2x^2y^2 + xy^3 - y^4$

64. $(u^2 - u - 1)(u^2 + 2u + 1)$
   $u^4 + u^3 - 2u^2 - 3u - 1$

65. $(x^2 + 10x + 25)(x^2 + 4x + 4)$
   $x^4 + 14x^3 + 69x^2 + 140x + 100$

66. $(a^2 + 6a + 9)(a^2 - 8a + 16)$
   $a^4 - 2a^3 - 23a^2 + 24a + 144$

Objective 4

Prep Exercise 4 Explain how to recognize that two binomials are conjugates.
They differ only in the sign separating the terms.

For Exercises 67–74, state the conjugate of the given binomial. See Objective 4.

67. $x + 3$
   $x - 3$

68. $y + 8$
   $y - 8$

69. $4x - 2y$
   $4x + 2y$

70. $2g - 3h$
   $2g + 3h$

71. $4d + 3c$
   $4d - 3c$

72. $5m + 2n$
   $5m - 2n$

73. $-3j - k$
   $-3j + k$

74. $-3x - 5y$
   $-3x + 5y$

Prep Exercise 5 Complete the rule for multiplying conjugates:
$(a + b)(a - b) = \underline{\quad a^2 - b^2 \quad}$

Prep Exercise 6 Complete the rules for squaring binomials:
$(a + b)^2 = \underline{\quad a^2 + 2ab + b^2 \quad}$
$(a - b)^2 = \underline{\quad a^2 - 2ab + b^2 \quad}$

For Exercises 75–96, multiply using the rules for special products. See Examples 7 and 8.

75. $(x - 5)(x + 5)$
$x^2 - 25$

76. $(y + 4)(y - 4)$
$y^2 - 16$

77. $(2m - 5)(2m + 5)$
$4m^2 - 25$

78. $(3p + 2)(3p - 2)$
$9p^2 - 4$

79. $(x + y)(x - y)$
$x^2 - y^2$

80. $(a + b)(a - b)$
$a^2 - b^2$

81. $(8r - 10s)(8r + 10s)$
$64r^2 - 100s^2$

82. $(4b - 5c)(4b + 5c)$
$16b^2 - 25c^2$

83. $(-2x - 3)(-2x + 3)$
$4x^2 - 9$

84. $(-h + 2k)(-h - 2k)$
$h^2 - 4k^2$

85. $(x + 3)^2$
$x^2 + 6x + 9$

86. $(y + 5)^2$
$y^2 + 10y + 25$

87. $(4t - 1)^2$
$16t^2 - 8t + 1$

88. $(9k - 2)^2$
$81k^2 - 36k + 4$

89. $(m + n)^2$
$m^2 + 2mn + n^2$

90. $(a + b)^2$
$a^2 + 2ab + b^2$

91. $(2u + 3v)^2$
$4u^2 + 12uv + 9v^2$

92. $(3r + 7s)^2$
$9r^2 + 42rs + 49s^2$

93. $(9w - 4z)^2$
$81w^2 - 72wz + 16z^2$

94. $(2q - 11c)^2$
$4q^2 - 44cq + 121c^2$

95. $(9 - 5y)^2$
$81 - 90y + 25y^2$

96. $(6 - 7t)^2$
$36 - 84t + 49t^2$

For Exercises 97–100, write an expression in simplest form for the area.

97.
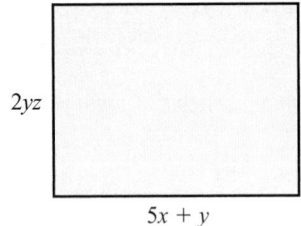
2yz
5x + y
$10xyz + 2y^2z$

98.
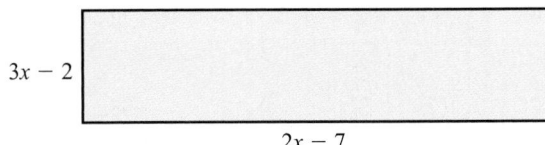
3x − 2
2x − 7
$6x^2 - 25x + 14$

99.
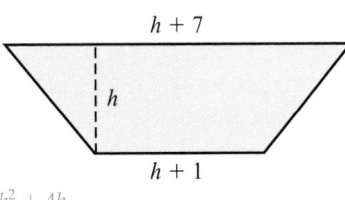
h + 7
h
h + 1
$h^2 + 4h$

100.
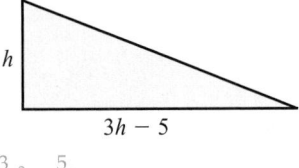
h
3h − 5
$\frac{3}{2}h^2 - \frac{5}{2}h$

101. A rectangular room has a length that is 3 feet more than the width. Using w to represent the width, write an expression in simplest form for the area of the room in terms of w.
$w^2 + 3w$

★ **102.** A circular metal plate for a machine has radius r. A smaller plate with radius $r - 2$ is to be used in another part of the machine. Write an expression in simplest form for the sum of the areas of the two circles in terms of r.
$2\pi r^2 - 4\pi r + 4\pi$

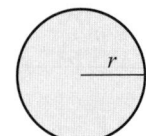
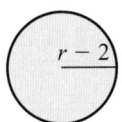

For Exercises 103 and 104, write an expression in simplest form for the volume.

103.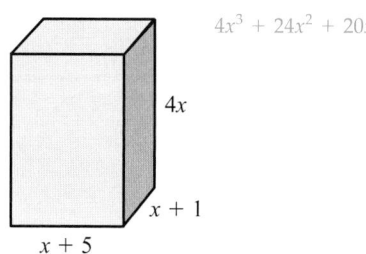

$4x^3 + 24x^2 + 20x$

$4x$

$x + 1$

$x + 5$

104.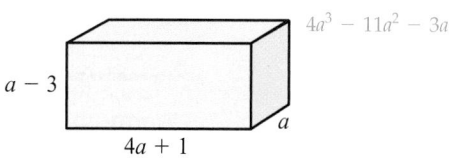

$4a^3 - 11a^2 - 3a$

$a - 3$

a

$4a + 1$

105. A crate is designed so that the length is twice the width and the height is 3 feet more than the length. Let w represent the width. Write an expression for the volume in terms of w.

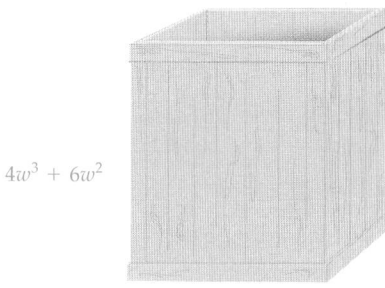

$4w^3 + 6w^2$

106. A fish tank's length is 8 inches more than its width. The height of the tank is 5 inches greater than its width. Let w represent the width. Write an expression for the volume of the tank in terms of w.

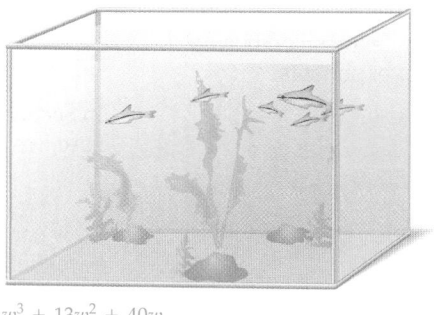

$w^3 + 13w^2 + 40w$

★ **107.** A cylinder has a height that is 2 inches more than the radius. Write an expression for the volume of the cylinder in terms of the radius.

$V = 3.14r^3 + 6.28r^2$, or $\pi r^3 + 2\pi r^2$

★ **108.** A cone has a radius that is 5 centimeters less than the height. Write an expression for the volume of the cone in terms of the height.

$\frac{1}{3}\pi h^3 - \frac{10}{3}\pi h^2 + \frac{25}{3}\pi h$

Review Exercises

Exercises 1–4 Expressions

For Exercises 1 and 2, simplify.

[1.4] 1. $1.08 \div 2.4$

0.45

[1.3] 2. $-6 - (-1)$

-5

[5.1] 3. Write 6.304×10^6 in standard form.

6,304,000

[5.4] 4. Multiply: $x^2 \cdot x^5$

x^7

Exercises 5 and 6 Equations and Inequalities

[2.4] 5. Isolate r in the formula $\dfrac{5 - d}{r} = m$.

$r = \dfrac{5 - d}{m}$

[4.2] **6.** The greater of two positive numbers is four times
[4.3] the smaller number. The difference between the numbers is 27. What are the numbers?

9, 36

5.6 Exponent Rules and Dividing Polynomials

Objectives

1 Divide exponential forms with the same base.

2 Divide numbers in scientific notation.

3 Divide monomials.

4 Divide a polynomial by a monomial.

5 Use long division to divide polynomials.

6 Simplify expressions using rules of exponents.

Warm-up

For Exercises 1–4, simplify. Write answers with positive exponents only.

[1.4] **1.** $\dfrac{1.82}{6.5}$

[5.1] **2.** Write 4.5×10^{-13} in standard form.

[5.4] **3.** $(m^2)^{-3}$

[5.5] **4.** $(x - 6)(3x + 4)$

In Section 5.4, we began by learning the product rule of exponents, which we used in multiplying polynomials. In this section, we follow a similar development and begin by learning the quotient rule of exponents so that we can divide polynomials.

Objective 1 Divide exponential forms with the same base.

In Section 5.4, we found that when we multiply exponential forms that have the same base, we can add the exponents and keep the same base. For example,

$$2^3 \cdot 2^4 = 2^{3+4} = 2^7$$

Now consider how to reverse this process and divide exponential forms with the same base. Remember that multiplication and division are inverse operations, which means that they "undo" each other. A related division statement for the multiplication just shown is $2^7 \div 2^4 = 2^3$, which implies that in the division problem, the exponents can be subtracted. To verify this result, it is helpful to write the division in fraction form and use the definition of exponents.

$$\frac{2^7}{2^4} = \frac{2 \cdot 2 \cdot 2 \cdot 2 \cdot 2 \cdot 2 \cdot 2}{2 \cdot 2 \cdot 2 \cdot 2}$$

◀ **Note** We can divide out four of the common factors. Three 2s are left in the top, which can be expressed in exponential form.

$$= \frac{2 \cdot 2 \cdot 2}{1}$$

$$= 2^3$$

Instructor Note After deducing the quotient rule, pose this question: What happens if the exponents are the same, as in $\dfrac{2^4}{2^4}$? Show that $\dfrac{2 \cdot 2 \cdot 2 \cdot 2}{2 \cdot 2 \cdot 2 \cdot 2} = 1$. Also show it by the quotient rule, $2^{4-4} = 2^0$. This means that $2^0 = 1$, which reinforces the rule from Section 5.1 that $a^0 = 1$. Next, pose this question: What if the numerator's exponent is less than the denominator's, as in $\dfrac{2^4}{2^7}$? Use a similar process to reinforce the rule for negative exponents, $a^{-n} = \dfrac{1}{a^n}$.

This result supports our thinking that we can subtract the exponents.

$$2^7 \div 2^4 = 2^{7-4} = 2^3$$

Notice that the divisor's exponent is subtracted from the dividend's exponent. This order is important because subtraction is not a commutative operation.

Conclusion: To divide exponential forms that have the same base, we can subtract the divisor's exponent from the dividend's exponent and keep the same base.

Note We cannot let a equal 0 because if a were replaced with 0, we would have $0^m \div 0^n$, and in Section 1.4, we concluded that $0 \div 0$ is indeterminate. Recall also that $\dfrac{a^m}{a^n}$ can be written as $a^m \div a^n$.

> **Rule Quotient Rule for Exponents**
>
> If m and n are integers and a is a real number, where $a \neq 0$, then $\dfrac{a^m}{a^n} = a^{m-n}$.

Example 1 Divide. $\dfrac{x^6}{x^2}$

◀ **Note** Throughout the remainder of this section, assume that variables are not replaced by values that cause denominators to be 0.

Solution: Because the exponential forms have the same base, we can subtract the exponents and keep the same base.

$$\frac{x^6}{x^2} = x^6 \div x^2 = x^{6-2} = x^4$$

Answers to Warm-up
1. 0.28
2. 0.00000000000045
3. m^{-6}
4. $3x^2 - 14x - 24$

Your Turn 1 Divide.

a. $\dfrac{t^8}{t^2}$ **b.** $\dfrac{n^5}{n}$ **c.** $5^9 \div 5^2$

We can apply the quotient rule for exponential forms when the exponents are negative. After simplifying, if the exponent in the result is negative, we will rewrite the exponential form so that the exponent is positive. Recall the rule from Section 5.1:

If a is any real number not equal to 0 and n is a natural number, then $a^{-n} = \dfrac{1}{a^n}$.

Example 2 Divide and write the result with a positive exponent.

a. $\dfrac{x^6}{x^{-2}}$

Solution: $\dfrac{x^6}{x^{-2}} = x^{6-(-2)}$ Subtract exponents and keep the same base.

$= x^{6+2}$ ◄ **Note** To evaluate the subtraction, we write an equivalent addition.

$= x^8$

b. $n^{-5} \div n^{-3}$

Solution: $n^{-5} \div n^{-3} = n^{-5-(-3)}$ Subtract the exponents and keep the same base.

$= n^{-5+3}$ Rewrite the subtraction as addition.

$= n^{-2}$ Simplify.

$= \dfrac{1}{n^2}$ Write with a positive exponent.

Your Turn 2 Divide and write the result with a positive exponent.

a. $\dfrac{y^3}{y^7}$ **b.** $m^{-2} \div m^{-6}$ **c.** $\dfrac{7^{-4}}{7^5}$

Answers to Your Turn 1
a. t^6 **b.** n^4 **c.** 5^7

Answers to Your Turn 2

a. $\dfrac{1}{y^4}$ **b.** m^4 **c.** $\dfrac{1}{7^9}$

Answer to Your Turn 3
2.8×10^{-14}

Note 0.45×10^5 is not scientific notation because 0.45 is not a decimal number greater than 1. We must move the decimal point one place to the right. This means that because we have one less factor of 10 to account for with the exponent, we subtract 1 from the exponent, or 0.45×10^5
$= 4.5 \times 10^{-1} \times 10^5$
$= 4.5 \times 10^4$.

Objective 2 Divide numbers in scientific notation.

We can use the quotient rule of exponents to divide numbers in scientific notation.

Example 3 Divide and write the result in scientific notation: $\dfrac{1.08 \times 10^9}{2.4 \times 10^4}$

Solution: The decimal factors and powers of 10 can be separated into a product of two fractions. This allows us to calculate the decimal division and divide the powers of 10 separately.

$\dfrac{1.08 \times 10^9}{2.4 \times 10^4} = \dfrac{1.08}{2.4} \times \dfrac{10^9}{10^4}$ Separate the decimal numbers and powers of 10.

$= 0.45 \times 10^{9-4}$ Divide decimal numbers and subtract exponents.

$= 0.45 \times 10^5$

$= 4.5 \times 10^4$

Your Turn 3 Divide $\dfrac{1.82 \times 10^{-9}}{6.5 \times 10^4}$ and write the result in scientific notation.

Objective 3 Divide monomials.

Dividing monomials is much like dividing numbers in scientific notation. Let's consider $\dfrac{24x^5}{6x^2}$. We can separate the coefficients and variables in the same way that we separated the decimal factors and powers of 10 when dividing numbers in scientific notation.

$$\frac{24x^5}{6x^2} = \frac{24}{6} \cdot \frac{x^5}{x^2}$$

We can now divide the coefficients and subtract the exponents of the like bases.

$$\frac{24x^5}{6x^2} = \frac{24}{6} \cdot \frac{x^5}{x^2} = 4x^{5-2} = 4x^3$$

This suggests the following procedure.

Procedure Dividing Monomials

To divide monomials:
1. Divide the coefficients or simplify them to fractions in lowest terms.
2. Use the quotient rule for the exponents with like bases.
3. Do not change unlike variable bases in the quotient.
4. Write the final expression so that all exponents are positive.

Example 4 | Divide.

a. $\dfrac{16a^4b^7c}{20a^5b^2}$

Solution: $\dfrac{16a^4b^7c}{20a^5b^2} = \dfrac{16}{20} \cdot \dfrac{a^4}{a^5} \cdot \dfrac{b^7}{b^2} \cdot \dfrac{c}{1}$

$= \dfrac{16}{20} \cdot a^{4-5} \cdot b^{7-2} \cdot c$ 　　Divide coefficients and subtract exponents of the like bases. The unlike base, c, will remain unchanged.

$= \dfrac{4}{5} a^{-1}b^5c$

$= \dfrac{4}{5} \cdot \dfrac{1}{a} \cdot \dfrac{b^5}{1} \cdot \dfrac{c}{1}$

$= \dfrac{4b^5c}{5a}$

Note Because $a^{-1} = \dfrac{1}{a}$, the variable a is written in the denominator as a factor with the 5. Because b^5 and c have positive exponents, they are expressed in the numerator as factors with the 4.

▶

b. $\dfrac{10x^3y^2}{24x^3y}$

Solution: $\dfrac{10x^3y^2}{24x^3y} = \dfrac{10}{24} \cdot \dfrac{x^3}{x^3} \cdot \dfrac{y^2}{y}$

$= \dfrac{10}{24} \cdot x^{3-3} \cdot y^{2-1}$ 　　Divide coefficients and subtract exponents of like bases.

$= \dfrac{5}{12} x^0 y^1$

$= \dfrac{5}{12} y$

Your Turn 4 | Divide.

a. $\dfrac{-32m^2n}{8mn}$

b. $\dfrac{56t^2u^6v}{16t^2u^4}$

Answers to Your Turn 4

a. $-4m$ **b.** $\dfrac{7}{2}u^2v$

Objective 4 Divide a polynomial by a monomial.

Let's now consider how to divide a polynomial by a monomial, as in $\dfrac{24x^5 + 18x^3}{6x^2}$.

Recall that when fractions with a common denominator are added (or subtracted), the numerators are added and the denominator stays the same.

$$\frac{1}{7} + \frac{2}{7} = \frac{1 + 2}{7}$$

This process can be reversed so that a sum in the numerator of a fraction can be broken into fractions, with each addend over the same denominator.

$$\frac{1 + 2}{7} = \frac{1}{7} + \frac{2}{7}$$

> **Rule** **Dividing a Polynomial by a Monomial**
> If a, b, and c are real numbers, variables, or expressions with $c \neq 0$, then
> $$\frac{a + b}{c} = \frac{a}{c} + \frac{b}{c}.$$

We can apply this rule to the division problem $\dfrac{24x^5 + 18x^3}{6x^2}$.

$$\frac{24x^5 + 18x^3}{6x^2} = \frac{24x^5}{6x^2} + \frac{18x^3}{6x^2}$$ We now have a sum of monomial divisions, which we can simplify separately.

$$= 4x^{5-2} + 3x^{3-2}$$
$$= 4x^3 + 3x$$

Our illustration suggests the following procedure.

> **Procedure** **Dividing a Polynomial by a Monomial**
> To divide a polynomial by a monomial, divide each term of the polynomial by the monomial.

Example 5 Divide.

a. $\dfrac{32y^6 + 24y^4 - 48y^2}{8y^2}$

Solution: $\dfrac{32y^6 + 24y^4 - 48y^2}{8y^2} = \dfrac{32y^6}{8y^2} + \dfrac{24y^4}{8y^2} - \dfrac{48y^2}{8y^2}$ Divide each term in the polynomial by the monomial.

$$= 4y^4 + 3y^2 - 6y^0$$
$$= 4y^4 + 3y^2 - 6$$

b. $\dfrac{42t^7u^2 - 18t^4u - 10t}{2t^3u}$

Solution: $\dfrac{42t^7u^2 - 18t^4u - 10t}{2t^3u} = \dfrac{42t^7u^2}{2t^3u} - \dfrac{18t^4u}{2t^3u} - \dfrac{10t}{2t^3u}$

Note When t is divided by t^3, because the denominator's exponent is greater than the numerator's exponent, the result is a negative exponent. Because u has no like base in the third term, it is written unchanged.

$$= 21t^4u - 9t^1u^0 - \frac{5t^{-2}}{u}$$

$$= 21t^4u - 9t - \frac{5}{ut^2}$$ Rewrite with positive exponents.

Warning A common error is to write the answer to Example 5(b) as $\dfrac{21t^4u - 9t - 5}{ut^2}$.

Your Turn 5 Divide.

a. $\dfrac{45x^6 - 27x^5 + 9x^3}{9x^3}$

b. $\dfrac{28m^2n^6 + 36m^4n - 40m^2}{4m^2n}$

Objective 5 Use long division to divide polynomials.

To divide a polynomial by a polynomial, we can use long division. Consider the following numeric long division and think about the process.

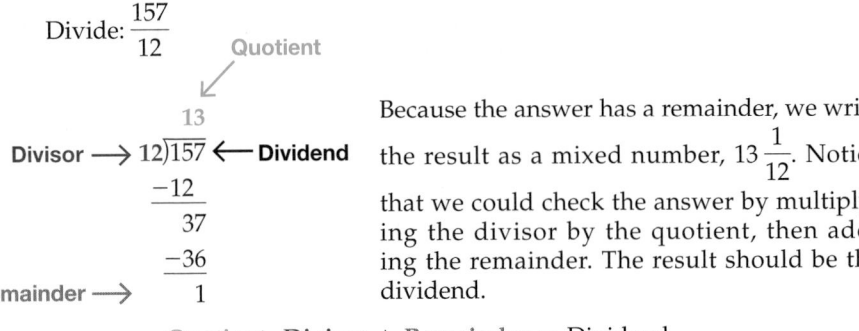

Divide: $\dfrac{157}{12}$

Quotient

$$\begin{array}{r} 13 \\ 12\overline{)157} \\ -12 \\ \hline 37 \\ -36 \\ \hline 1 \end{array}$$

Divisor →   ← Dividend

Remainder →

Because the answer has a remainder, we write the result as a mixed number, $13\frac{1}{12}$. Notice that we could check the answer by multiplying the divisor by the quotient, then adding the remainder. The result should be the dividend.

Quotient · Divisor + Remainder = Dividend

$$13 \quad \cdot \quad 12 \quad + \quad 1 \quad = 157$$

Now let's consider polynomial division, which follows the same long division process.

Example 6 Divide $\dfrac{x^2 + 5x + 7}{x + 2}$.

Solution: Use long division. First, we determine what term will multiply by the first term in the divisor to equal the first term in the dividend. A clever way to determine this first term in the quotient is to divide the first term in the dividend by the first term in the divisor.

$$\begin{array}{r} x \\ x + 2\overline{)x^2 + 5x + 7} \end{array}$$

Divide these first terms to determine the first term in the quotient: $x^2 \div x = x$.

Next, we multiply the divisor $x + 2$ by the x in the quotient.

multiply

$$\begin{array}{r} x \\ x + 2\overline{)x^2 + 5x + 7} \\ x^2 + 2x \end{array}$$

Next, we subtract. Note that to subtract the binomial, we change the signs of the terms and then combine like terms. After combining terms, we bring down the next term in the dividend, which is 7.

$$\begin{array}{r} x \\ x + 2\overline{)x^2 + 5x + 7} \\ -(x^2 + 2x) \end{array}$$

Change signs. →

$$\begin{array}{r} x \\ x + 2\overline{)x^2 + 5x + 7} \\ -x^2 - 2x \\ \hline 3x + 7 \end{array}$$

Combine like terms and bring down the next term.

Answers to Your Turn 5

a. $5x^3 - 3x^2 + 1$

b. $7n^5 + 9m^2 - \dfrac{10}{n}$

Determine the final term of the quotient by dividing $-6x$ by $2x$, which is -3.

$$
\begin{array}{r}
4x^3 + 2x^2 - 6x - 3 \\
2x - 1\overline{)8x^4 + 0x^3 - 14x^2 + 0x - 9} \\
\underline{-8x^4 + 4x^3} \\
4x^3 - 14x^2 \\
\underline{-4x^3 + 2x^2} \\
-12x^2 + 0x \\
\underline{12x^2 - 6x} \\
-6x - 9 \\
\underline{-(-6x + 3)}
\end{array}
\qquad
\begin{array}{r}
4x^3 + 2x^2 - 6x - 3 \\
2x - 1\overline{)8x^4 + 0x^3 - 14x^2 + 0x - 9} \\
\underline{-8x^4 + 4x^3} \\
4x^3 - 14x^2 \\
\underline{-4x^3 + 2x^2} \\
-12x^2 + 0x \\
\underline{12x^2 - 6x} \\
-6x - 9 \\
\underline{+6x - 3} \\
-12
\end{array}
$$

Change signs. $\longrightarrow$ Combine like terms.

Answer: $4x^3 + 2x^2 - 6x - 3 + \dfrac{-12}{2x - 1}$, or $4x^3 + 2x^2 - 6x - 3 - \dfrac{12}{2x - 1}$

Your Turn 8 Divide $\dfrac{7 + 8x^3 - 18x}{4x - 2}$.

Objective 6 Simplify expressions using rules of exponents.

The quotient rule for exponents completes our rules of exponents in this chapter. So far, the problems involving exponent rules have required only that we use one or two rules. Let's now consider problems that require us to use several of the exponent rules. Following is a summary of all of the rules of exponents.

Instructor Note Before doing Example 9, consider working through an example illustrating each of the properties.

Rule Exponents Summary

Assume that no denominators are 0, that a and b are real numbers, and that m and n are integers.

Zero as an exponent: $a^0 = 1$, where $a \neq 0$
$\quad 0^0$ is indeterminate.

Negative exponents: $a^{-n} = \dfrac{1}{a^n}$

$\quad \dfrac{1}{a^{-n}} = a^n$

$\quad \left(\dfrac{a}{b}\right)^{-n} = \left(\dfrac{b}{a}\right)^n$

Product rule for exponents: $a^m \cdot a^n = a^{m+n}$

Quotient rule for exponents: $\dfrac{a^m}{a^n} = a^{m-n}$

Raising a power to a power: $(a^m)^n = a^{mn}$

Raising a product to a power: $(ab)^n = a^n b^n$

Raising a quotient to a power: $\left(\dfrac{a}{b}\right)^n = \dfrac{a^n}{b^n}$

Example 9 Simplify. Write all answers with positive exponents.

a. $\left(\dfrac{n^{-5}}{n^2}\right)^3$

Solution: Following the order of operations, we simplify within the parentheses first, then consider the exponent outside the parentheses.

$$
\begin{aligned}
\left(\frac{n^{-5}}{n^2}\right)^3 &= (n^{-5-2})^3 \quad &&\text{Use the quotient rule for exponents.} \\
&= (n^{-7})^3 \\
&= n^{-7 \cdot 3} \quad &&\text{Use the rule for raising a power to a power.} \\
&= n^{-21} \\
&= \frac{1}{n^{21}} \quad &&\text{Write with a positive exponent.}
\end{aligned}
$$

Answer to Your Turn 8

$2x^2 + x - 4 + \dfrac{-1}{4x - 2}$, or

$2x^2 + x - 4 - \dfrac{1}{4x - 2}$

b. $\dfrac{(m^2)^{-3}}{m^4 \cdot m^{-5}}$

Solution: $\dfrac{(m^2)^{-3}}{m^4 \cdot m^{-5}} = \dfrac{m^{2 \cdot (-3)}}{m^{4+(-5)}}$ In the numerator, use the rule for raising a power to a power. In the denominator, use the product rule for exponents.

$= \dfrac{m^{-6}}{m^{-1}}$

$= m^{-6-(-1)}$ Use the quotient rule for exponents.

$= m^{-6+1}$ Write the subtraction as an equivalent addition.

$= m^{-5}$

$= \dfrac{1}{m^5}$ Write with a positive exponent.

c. $\dfrac{(4y^3)^2}{(2y^4)^5}$

Solution: $\dfrac{(4y^3)^2}{(2y^4)^5} = \dfrac{4^2 y^{3 \cdot 2}}{2^5 y^{4 \cdot 5}}$ Use the rule for raising a product to a power.

$= \dfrac{16y^6}{32y^{20}}$

$= \dfrac{16}{32} \cdot \dfrac{y^6}{y^{20}}$ Separate the coefficients and variables.

$= \dfrac{16}{32} \cdot y^{6-20}$ Use the quotient rule for exponents.

$= \dfrac{1}{2} \cdot y^{-14}$ Simplify the coefficient.

$= \dfrac{1}{2} \cdot \dfrac{1}{y^{14}}$ Write with a positive exponent.

$= \dfrac{1}{2y^{14}}$ Simplify.

Learning Strategy

Even if you don't get a homework grade, make sure you do the homework. Teachers assign homework for a reason; often quizzes are based on the homework. If you get something wrong, go back and see why it is wrong and learn from it.

—Ellyn G.

Answers to Your Turn 9

a. $\dfrac{1}{x^{28}}$ **b.** y^{35} **c.** $\dfrac{y^6}{8x^8}$

Your Turn 9 Simplify. Write all answers with positive exponents.

a. $\left(\dfrac{x^2}{x^{-5}}\right)^{-4}$

b. $(y^{-2} \cdot y^{-3})^{-4} \div y^{-15}$

c. $\dfrac{(2x^4)^{-3}}{(x^2y^3)^{-2}}$

5.6 Exercises For Extra Help MyMathLab®

Note: Exercises marked with a ★ represent challenging exercises.

Objective 1

Prep Exercise 1 When dividing exponential forms that have the same base, _____ and keep the same base.

subtract the divisor's exponent from the dividend's exponent

Prep Exercise 2 What is the rule for rewriting a^{-n}, where a is a real number, $a \neq 0$, and n is a natural number, so that it has a positive exponent?

$a^{-n} = \dfrac{1}{a^n}$

For Exercises 1–26, simplify using the rules of exponents. Write all answers with positive exponents. See Examples 1 and 2.

1. a^{-3}
$\frac{1}{a^3}$

2. n^{-7}
$\frac{1}{n^7}$

3. 2^{-5}
$\frac{1}{2^5}$

4. 3^{-4}
$\frac{1}{3^4}$

5. $\frac{y^8}{y^3}$
y^5

6. $\frac{n^4}{n}$
n^3

7. $\frac{3^4}{3^3}$
3

8. $\frac{6^{11}}{6^5}$
6^6

9. $\frac{4^3}{4^9}$
$\frac{1}{4^6}$

10. $\frac{5}{5^9}$
$\frac{1}{5^8}$

11. $\frac{x^3}{x^5}$
$\frac{1}{x^2}$

12. $\frac{m^4}{m^{10}}$
$\frac{1}{m^6}$

13. $a^8 \div a^6$
a^2

14. $u^5 \div u^3$
u^2

15. $w^7 \div w^{10}$
$\frac{1}{w^3}$

16. $j^6 \div j^{14}$
$\frac{1}{j^8}$

17. $\frac{a^{-2}}{a^5}$
$\frac{1}{a^7}$

18. $\frac{m^{-6}}{m^3}$
$\frac{1}{m^9}$

19. $\frac{r^6}{r^{-4}}$
r^{10}

20. $\frac{u^5}{u^{-12}}$
u^{17}

21. $\frac{y^{-7}}{y^{-15}}$
y^8

22. $\frac{x^{-3}}{x^{-8}}$
x^5

23. $\frac{p^{-5}}{p^{-2}}$
$\frac{1}{p^3}$

24. $\frac{n^{-11}}{n^{-5}}$
$\frac{1}{n^6}$

25. $\frac{t^{-4}}{t^{-4}}$
1

26. $\frac{x^{-6}}{x^{-6}}$
1

Objective 2

Prep Exercise 3 Explain how to divide two numbers in scientific notation.
Separate the decimal factors and powers of 10 into a product of two fractions. Then calculate the decimal division and divide the powers of 10 separately.

For Exercises 27–34, divide and write your answers in scientific notation. See Example 3.

27. $\frac{8.32 \times 10^5}{3.2 \times 10^6}$
2.6×10^{-1}

28. $\frac{3.6 \times 10^6}{2.4 \times 10^9}$
1.5×10^{-3}

29. $\frac{1.26 \times 10^{-4}}{2.1 \times 10^3}$
6×10^{-8}

30. $\frac{1.32 \times 10^{-5}}{6.6 \times 10^{15}}$
2×10^{-21}

31. $\frac{9.088 \times 10^1}{1.28 \times 10^{-9}}$
7.1×10^{10}

32. $\frac{9.964 \times 10^4}{1.88 \times 10^{-2}}$
5.3×10^6

33. $\frac{8.64 \times 10^{-3}}{3.2 \times 10^{-7}}$
2.7×10^4

34. $\frac{7.92 \times 10^{-3}}{2.2 \times 10^{-7}}$
3.6×10^4

For Exercises 35–38, solve. See Example 3.

35. If light travels at 3×10^8 meters per second, how long does it take light to travel the 1.5×10^{11} meters from the Sun to Earth?
500 sec. or $8\frac{1}{3}$ min.

36. In 2012, the population of the United States was approximately 3.14×10^8. Calculate the approximate number of people per square mile in the United States in 2012 if the estimated amount of land in the United States is 3.8×10^6 square miles. (*Source:* U.S. Bureau of the Census.)
83 people per square mile

37. As of August 23, 2012, the national debt was about $\$1.596 \times 10^{13}$. If each of the 3.14×10^8 people in the United States contributed an equal share toward the debt, what would each person have to contribute to pay off the debt? (*Source:* U.S. National Debt Clock.)
$50,828.03

Of Interest

The Great Pyramid is thought to have been built for King Khufu, who ruled Egypt from 2589 to 2566 B.C. Each block of stone was quarried nearby and shaped to fit perfectly in its final place using only copper tools. Although only simple tools and techniques were used in its construction, the precision of the pyramid's shape and position rivals modern construction.

38. There are about 2.3×10^6 stone blocks in the Great Pyramid. If the total weight of the pyramid is about 1.3×10^{10} pounds and we assume that each block is of equal size, how much does each block weigh? (*Source:* Nova Online.)
   ≈ 5652 lb.

★ *For Exercises 39 and 40, use Einstein's formula* $E = mc^2$, *which describes the rest energy contained within a mass m, where c represents the speed of light, which is a constant with a value of* 3×10^8 *meters per second. The units for energy are joules, which are equivalent to* $\frac{kg \cdot m^2}{s^2}$. *See Example 3.*

39. The first atom bomb released an energy equivalent of about 8.4×10^{13} joules. What amount of mass was converted to energy? Write your answer in scientific notation.
   $\approx 9.3 \times 10^{-4}$kg

40. The largest atom bomb detonated was a hydrogen bomb tested by the Soviet Union in 1961. The bomb released 2.4×10^{17} joules of energy. What amount of mass was converted to energy in that explosion? (*Source:* Cesare Emiliani, *The Scientific Companion*; Wiley Popular Science, 1995.)
   ≈ 2.7 kg

Objective 3

Prep Exercise 4

a. When dividing $\frac{(28x^6y)}{(7x^2)}$, what do you do with the coefficients?
   divide

b. What do you do with the exponents of the two x's?
   subtract

c. What do you do with the y?
   Write y unchanged in the numerator.

For Exercises 41–52, divide the monomials. See Example 4.

41. $\frac{15x^5}{3x^2}$
   $5x^3$

42. $\frac{14a^6}{2a^4}$
   $7a^2$

43. $\frac{24x^2}{-6x^5}$
   $\frac{-4}{x^3}$

44. $\frac{-18a^6}{9a^{12}}$
   $\frac{-2}{a^6}$

45. $\frac{9m^3n^5}{-3mn}$
   $-3m^2n^4$

46. $\frac{-42x^5y^3}{6xy^2}$
   $-7x^4y$

47. $\frac{-24p^3q^5}{15p^3q^2}$
   $\frac{8}{5}q^3$

48. $\frac{48t^4u^7}{-18t^4u^2}$
   $\frac{8}{3}u^5$

49. $\frac{56x^2y^6}{42x^7y^2}$
   $\frac{4y^4}{3x^5}$

50. $\frac{60a^8b^4}{66a^2b^9}$
   $\frac{10a^6}{11b^5}$

51. $\frac{12a^4bc^3}{9a^6bc^2}$
   $\frac{4c}{3a^2}$

52. $\frac{28p^5q^4r}{42p^2q^7r}$
   $\frac{2p^3}{3q^3}$

Objective 4

Prep Exercise 5 Explain how to divide a polynomial by a monomial.
Divide each term in the polynomial by the monomial.

For Exercises 53–70, divide the polynomial by the monomial. See Example 5.

53. $\frac{7a + 14b}{7}$
   $a + 2b$

54. $\frac{4m + 16y}{4}$
   $m + 4y$

55. $\frac{12x^3 - 6x^2}{3x}$
   $4x^2 - 2x$

56. $\frac{18m^4 - 27m^2}{9m}$
   $2m^3 - 3m$

57. $\frac{12x^3 + 8x}{4x^2}$
   $3x + \frac{2}{x}$

58. $\frac{24y^3 + 16y}{8y^2}$
   $3y + \frac{2}{y}$

59. $\frac{5x^2y^2 - 15xy^3}{5xy}$
   $xy - 3y^2$

60. $\frac{12k^4l^2 + 15k^2l^3}{3k^2l}$
   $4k^2l + 5l^2$

Of Interest

The first atomic bomb, tested in 1945, used the process of nuclear *fission*, which involves the splitting of atoms of heavy elements such as uranium and plutonium. The split atoms, in turn, split other atoms, resulting in a chain reaction in which some of the atomic mass is converted into energy.

In contrast, the hydrogen bomb uses nuclear *fusion*, which is how the stars release energy. In a fusion reaction, two nuclei are fused together to form a new atom. For example, two hydrogen nuclei can be forced together to form a helium nucleus. During this reaction, some of the mass of the two hydrogen nuclei is converted to pure energy so that the resulting helium nucleus has slightly less mass compared with the two original hydrogen nuclei.

61. $\dfrac{6abc^2 - 24a^2b^2c}{-3abc}$

$-2c + 8ab$

62. $\dfrac{30x^3yz^5 - 15xyz^2}{-5xyz^2}$

$-6x^2z^3 + 3$

63. $\dfrac{24x^3 + 16x^2 - 8x}{8x}$

$3x^2 + 2x - 1$

64. $\dfrac{3x^3 + 6x^2 - 3x}{3x}$

$x^2 + 2x - 1$

65. $\dfrac{6x^3 - 12x^2 + 9x}{3x^2}$

$2x - 4 + \dfrac{3}{x}$

66. $\dfrac{12y^4 - 16y^3 + 8y}{4y^2}$

$3y^2 - 4y + \dfrac{2}{y}$

67. $\dfrac{36u^3v^4 + 12uv^5 - 15u^2v^2}{3u^2v}$

$12uv^3 + \dfrac{4v^4}{u} - 5v$

68. $\dfrac{16hk^4 - 28hk - 4h^2k^2}{4hk^2}$

$4k^2 - \dfrac{7}{k} - h$

69. $\dfrac{y^5 + y^7 - 3y^8 + y}{y^3}$

$y^2 + y^4 - 3y^5 + \dfrac{1}{y^2}$

70. $\dfrac{x^6 + x^8 - 4x^{10} + x^2}{x^4}$

$x^2 + x^4 - 4x^6 + \dfrac{1}{x^2}$

71. The area of the parallelogram shown is described by the monomial $35mn^2$. Find the height.

$\dfrac{5mn}{2}$

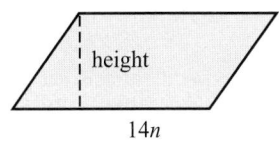

72. The area of the triangle shown is $18x^2y$. Find the base.

$12x$

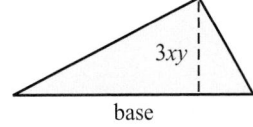

73. An engineer is designing a steel cover plate in the shape of a rectangle. The area of the plate is described by the polynomial $9x^2 - 15x + 12$, and the length must be $3x$.
   a. Find an expression for the width.

   $3x - 5 + \dfrac{4}{x}$

   b. Find the area, length, and width if $x = 2$.

   Area $= 18$, length $= 6$, width $= 3$

74. The voltage in a circuit is described by the polynomial $36s^3 - 24s^2 + 42s$. The resistance is described by $6s$.
   a. Find an expression for the current. (*Hint:* Voltage = Current × Resistance)

   $6s^2 - 4s + 7$

   b. Find the voltage, current, and resistance if $s = 3$.

   Voltage $= 882$, current $= 49$, resistance $= 18$

Objective 5

Prep Exercise 6 Write the result of the following division.

$$
\begin{array}{r}
x + 5 \\
x + 2 \overline{\smash{)}\, x^2 + 7x + 6} \\
\underline{-x^2 - 2x} \\
5x + 6 \\
\underline{-5x - 10} \\
-4
\end{array}
$$

$x + 5 + \dfrac{-4}{x + 2}$ or $x + 5 - \dfrac{4}{x + 2}$

Prep Exercise 7 When do we use a 0 placeholder in the process of dividing two polynomials by long division?

Use a 0 placeholder when the dividend, written in descending order, has a "missing" term.

For Exercises 75–98, use long division to divide the polynomials. See Examples 6–8.

75. $\dfrac{x^2 + 7x + 12}{x + 3}$

$x + 4$

76. $\dfrac{y^2 + 4y + 4}{y + 2}$

$y + 2$

77. $\dfrac{m^2 - 8m + 15}{m - 3}$

$m - 5$

78. $\dfrac{u^2 - 6u + 9}{u - 3}$

$u - 3$

79. $\dfrac{x^2 - 17x + 64}{x - 5}$

$x - 12 + \dfrac{4}{x - 5}$

80. $\dfrac{x^2 + 6x + 10}{x + 3}$

$x + 3 + \dfrac{1}{x + 3}$

81. $\dfrac{3x^3 - 73x - 10}{x - 5}$

$3x^2 + 15x + 2$

82. $\dfrac{2x^3 - 35x - 12}{x + 4}$

$2x^2 - 8x - 3$

83. $\dfrac{2x^3 - x^2 + 5}{x - 2}$

$2x^2 + 3x + 6 + \dfrac{17}{x - 2}$

84. $\dfrac{2x^3 - 2x^2 + 5}{x + 3}$

$2x^2 - 8x + 24 - \dfrac{67}{x + 3}$

85. $\dfrac{p^3 + 125}{p + 5}$

$p^2 - 5p + 25$

86. $\dfrac{x^3 + 8}{x + 2}$

$x^2 - 2x + 4$

87. $\dfrac{12x^2 - 7x - 10}{4x - 5}$

$3x + 2$

88. $\dfrac{6a^2 + 2a - 28}{3a + 7}$

$2a - 4$

89. $\dfrac{14y^2 - 8y + 17}{7y + 3}$

$2y - 2 + \dfrac{23}{7y + 3}$

90. $\dfrac{6x^2 + 7x + 5}{3x - 1}$

$2x + 3 + \dfrac{8}{3x - 1}$

91. $\dfrac{4k^3 + 8k^2 + 6 - k}{k + 2}$

$4k^2 - 1 + \dfrac{8}{k + 2}$

92. $\dfrac{y^3 + 2y^2 + 3 - 11y}{y + 2}$

$y^2 - 11 + \dfrac{25}{y + 2}$

93. $\dfrac{21u^3 - 19u^2 + 14u - 6}{3u + 2}$

$7u^2 - 11u + 12 - \dfrac{30}{3u + 2}$

94. $\dfrac{2v^3 + 5v^2 - 10v + 2}{2v - 3}$

$v^2 + 4v + 1 + \dfrac{5}{2v - 3}$

95. $\dfrac{14b + b^3 - 6b^2 - 12}{b - 3}$

$b^2 - 3b + 5 + \dfrac{3}{b - 3}$

96. $\dfrac{-14x + x^2 + x^3 - 5}{x + 4}$

$x^2 - 3x - 2 + \dfrac{3}{x + 4}$

97. $\dfrac{y^3 - y + 6}{y + 2}$

$y^2 - 2y + 3$

98. $\dfrac{2b^4 + 3b^3 - 4b - 6}{2b + 3}$

$b^3 - 2$

99. The volume of a restaurant's walk-in refrigerator is described by $(6x^2 + 42x + 72)$ cubic feet. The width has to be 6 feet, and the length is described by $(x + 3)$ feet.
a. Find an expression for the height of the refrigerator.
  $x + 4$
b. Find the length, height, and volume of the refrigerator if $x = 4$.
  Length = 7 ft., height = 8 ft., volume = 336 ft.^{3}

100. In an architectural project, the specifications for a rectangular room call for the area to be $(10x^2 + 3x - 18)$ square feet. It is decided that the length should be described by the binomial $(2x + 3)$ feet.
a. Find an expression for the width.
  $5x - 6$
b. Find the length, width, and area of the room if $x = 5$.
  Length = 13 ft., width = 19 ft., area = 247 ft.^{2}

Objective 6

Prep Exercise 8 Complete the rule. $a^m \cdot a^n = $ _____ a^{m+n}

Prep Exercise 9 Complete the rule. $\dfrac{a^m}{a^n} = $ _____ a^{m-n}

Prep Exercise 10 Complete the rule. $(a^m)^n = $ _____ a^{mn}

For Exercises 101–124, simplify. Write all answers with positive exponents. See Example 9.

101. $(6m^2n^{-2})^2$

$\dfrac{36m^4}{n^4}$

102. $(-4a^3b^{-2}c)^2$

$\dfrac{16a^6c^2}{b^4}$

103. $(4x^{-3}y^2)^{-3}$

$\dfrac{x^9}{64y^6}$

104. $(3m^4n^{-3})^{-4}$

$\dfrac{n^{12}}{81m^{16}}$

105. $(4rs^3)^3(2r^3s^{-1})^2$

$256r^9s^7$

106. $(-3abc^{-2})^2(2a^2b^{-3})^3$

$\dfrac{72a^8}{b^7c^4}$

107. $\left(\dfrac{m^6}{m^2}\right)^4$

m^{16}

108. $\left(\dfrac{p^7}{p^4}\right)^5$

p^{15}

109. $\left(\dfrac{x^{-3}}{x^4}\right)^2$

$\dfrac{1}{x^{14}}$

110. $\left(\dfrac{r^{-4}}{r^2}\right)^5$

$\dfrac{1}{r^{30}}$

111. $\left(\dfrac{x^2y^7}{x^5y^{-2}}\right)^3$

$\dfrac{y^{27}}{x^9}$

112. $\left(\dfrac{m^6n^5}{m^{-3}n^8}\right)^2$

$\dfrac{m^{18}}{n^6}$

113. $\left(\dfrac{x^{-3}}{x^4}\right)^{-2}$

x^{14}

114. $\left(\dfrac{p^{-5}}{p^3}\right)^{-3}$

p^{24}

115. $\left(\dfrac{3x^{-1}y^2}{z^2}\right)^3$

$\dfrac{27y^6}{x^3z^6}$

116. $\left(\dfrac{2x^{-2}y}{z^{-1}}\right)^4$

$\dfrac{16y^4z^4}{x^8}$

117. $\dfrac{(x^4)^{-3}}{x^{-5} \cdot x^3}$

$\dfrac{1}{x^{10}}$

118. $\dfrac{(y^3)^{-5}}{y^6 \cdot y^{-8}}$

$\dfrac{1}{y^{13}}$

119. $\dfrac{4xy^{-2}z^2}{x^{-3}y^3z^{-1}}$

$\dfrac{4x^4z^3}{y^5}$

120. $\dfrac{3m^6n^{-1}p^{-2}}{m^{-1}n^{-3}p}$

$\dfrac{3m^7n^2}{p^3}$

121. $\dfrac{(2ab)^3}{12a^4b^5}$

$\dfrac{2}{3ab^2}$

122. $\dfrac{(2xy^2)^2}{16x^2y^3}$

$\dfrac{y}{4}$

123. $\dfrac{(8x^3)^3}{(4x^4)^5}$

$\dfrac{1}{2x^{11}}$

124. $\dfrac{(3y^4)^2}{(6y^7)^3}$

$\dfrac{1}{24y^{13}}$

Puzzle Problem Fill the following blanks with the numbers 1, 2, 3, 4, 5, 6, 7, 8, and 9, using each number once so that the equation is true.

$$\square\square \div \square = \square\square \div \square = \square\square \div \square$$

There is more than one solution. Here are two solutions:
$27 \div 3 = 54 \div 6 = 81 \div 9$ and $49 \div 7 = 56 \div 8 = 21 \div 3$

Review Exercises

Exercises 1–4 Expressions

[1.2] 1. Find the prime factorization of 3024.

$2^4 \cdot 3^3 \cdot 7$

[5.4] 2. Simplify: $-3x^2y \cdot 6x^2y^2$

$-18x^4y^3$

[5.5] 3. Simplify: $y(3x - 4)$

$3xy - 4y$

[1.6] 4. Translate to an algebraic expression: one number plus half of a second number.

$x + \dfrac{1}{2}y$

Exercises 5 and 6 Equations and Inequalities

[2.1] 5. Find the area of a playground that is 17 feet by 20 feet.

340 ft.^2

[2.3] 6. Solve and check: $3 - y = 27$

-24

Chapter 5   Summary and Review Exercises

Complete each incomplete definition, rule, or procedure; study the key examples; and then work the related exercises.

5.1 Exponents and Scientific Notation

Definitions/Rules/Procedures	Key Example(s)
If the base of an exponential form is a negative number and the exponent is even, then the product is ____positive____. If the base is a negative number and the exponent is ____odd____, then the product is negative. If a and b are real numbers, where $b \neq 0$ and n is a natural number, then $\left(\dfrac{a}{b}\right)^n = $ ____$\dfrac{a^n}{b^n}$____. If a is a real number and $a \neq 0$, then $a^0 = $ ____1____. If a is a real number, where $a \neq 0$ and n is a natural number, then $a^{-n} = $ ____$\dfrac{1}{a^n}$____ and $\dfrac{1}{a^{-n}} = $ ____a^n____. If a and b are real numbers, where $a \neq 0$ and $b \neq 0$ and n is a natural number, then $\left(\dfrac{a}{b}\right)^{-n} = $ ____$\left(\dfrac{b}{a}\right)^n$____.	Evaluate. **a.** $(-2)^4 = (-2)(-2)(-2)(-2) = 16$ **b.** $(-2)^5 = (-2)(-2)(-2)(-2)(-2) = -32$ Simplify. **a.** $\left(\dfrac{3}{4}\right)^2 = \dfrac{3^2}{4^2} = \dfrac{9}{16}$ **b.** $7^0 = 1$ **c.** $2^{-4} = \dfrac{1}{2^4} = \dfrac{1}{16}$ **d.** $\dfrac{1}{3^{-2}} = 3^2 = 9$ **e.** $\left(\dfrac{2}{3}\right)^{-4} = \left(\dfrac{3}{2}\right)^4 = \dfrac{81}{16}$

Exercises 1–4 ➤ **Expressions**

[5.1] *For Exercises 1–4, evaluate the exponential form.*

1. $\left(\dfrac{2}{5}\right)^3$

$\dfrac{8}{125}$

2. -4^2

-16

3. 5^{-2}

$\dfrac{1}{25}$

4. $\dfrac{1}{6^{-2}}$

36

Definitions/Rules/Procedures	Key Example(s)
If a number is expressed in scientific notation, it is expressed in the form $a \times 10^n$, where a is a decimal with ____$1 \leq |a| < 10$____ and n is a(n) ____integer____. To change from scientific notation with a positive integer exponent to standard form, move the decimal point to the ____right____ the number of places indicated by the ____exponent____. To change from scientific notation with a negative exponent to standard form, move the decimal point to the ____left____ the same number of places as the ____absolute value____ of the exponent. To write a number greater than 1 in scientific notation: 1. Move the decimal point so that the number is greater than or equal to ____1____ but less than ____10____. (Tip: Place the decimal point to the ____right____ of the first nonzero digit.).	3.26×10^6 and 8.3×10^{-4} are written in scientific notation. Write in standard form. **a.** $3.4 \times 10^5 = 340{,}000$ **b.** $4.2 \times 10^{-5} = 0.000042$ Write in scientific notation. $7{,}230{,}000 = 7.23 \times 10^6$

Definitions/Rules/Procedures	Key Example(s)
2. Write the decimal number multiplied by ____10^n____, where n is the number of places between the new decimal position and the original decimal position.	
3. Delete ____zeros____ to the right of the last nonzero digit.	
To write a positive decimal number that is less than 1 in scientific notation:	
1. Move the decimal point so that the number is greater than or equal to ____1____ but less than ____10____. (Tip: Place the decimal point to the ____right____ of the first nonzero digit.)	Write in scientific notation. $$0.00000056 = 5.6 \times 10^{-7}$$
2. Write the decimal number multiplied by 10^n, where n is a(n) ____negative integer____ whose absolute value is the number places between the new decimal position and the original decimal position.	
3. Delete zeros to the left of the first nonzero digit.	

Exercises 5–8 Expressions

[5.1] *For Exercises 5 and 6, write the number in standard form.*

5. The radioactive half-life of uranium 238 is 4.5×10^9 years.
4,500,000,000

6. The mass of a hydrogen atom is 1.663×10^{-24} gram.
0.000000000000000000000001663

[5.1] *For Exercises 7 and 8, write the number in scientific notation.*

7. Because atoms are so small, their weight is measured in atomic mass units (AMU). One atomic mass unit is 0.0000000000000000000000001661 gram.
1.661×10^{-24}

8. The speed of light is about 300,000,000 meters per second.
3×10^8

5.2 Introduction to Polynomials

Definitions/Rules/Procedures	Key Example(s)
A monomial is an expression that is a(n) ____constant____, a(n) ____variable____, or a product of a(n) ____constant____ and ____variable(s)____ that are raised to whole-number powers.	Examples of monomials are 3, x, $3x$, and $3x^2y$.
The coefficient of a monomial is the ____numerical factor____ in a monomial.	The coefficient of $-5a^3b$ is -5.
The degree of a monomial is the ____sum____ of the exponents of all variables in a monomial.	The degree of $4x^3y^2z$ is $3 + 2 + 1 = 6$.
A polynomial is a monomial or a(n) ____sum of monomials____.	Examples of polynomials are 3, x, $x + 3$, and $2x^2 - 5x + 4$.
A polynomial in one variable is a polynomial in which every variable term has the ____same variable____.	$2x^2 - 5x + 4$ is a polynomial of one variable, x.
A binomial is a polynomial containing ____two____ terms, and a trinomial is a polynomial containing ____three____ terms.	$4a + 3$ is a binomial, and $3a^2 - 5a - 7$ is a trinomial.

Exercises 9–20 ▲ **Expressions**

[5.2] *For Exercises 9–12, determine whether the expression is a monomial.*

9. $3xy^5$
yes

10. $-\dfrac{1}{2}x$
yes

11. $\dfrac{4}{ab^3}$
no

12. $2x^2 - 9$
no

[5.2] *For Exercises 13–16, identify the coefficient and degree of each monomial.*

13. $6x^4$
c: 6; d: 4

14. 27
c: 27; d: 0

15. $-2.6xy^3$
c: −2.6; d: 4

16. $-m$
c: −1; d: 1

[5.2] *For Exercises 17–20, indicate whether the expression is a monomial, binomial, or trinomial; has no special polynomial name; or is not a polynomial.*

17. $4x^2 - 25$
binomial

18. $-st^5$
monomial

19. $3x - \dfrac{4}{y^2}$
Not a polynomial

20. $5x^3 - 6x^2 - 3x + 11$
No special polynomial name

Definitions/Rules/Procedures	Key Example(s)
The degree of a polynomial is the ___greatest degree___ of any of the terms in the polynomial.	The degree of $5y - 6 + 2y^3 - 6y^2$ is 3.
To write a polynomial in descending order of degree, place the ___highest degree term___ first, then the next ___highest degree term___, and so on.	Write in descending order. $$2x^2 + 4x^5 + 9 - 6x^3 = 4x^5 - 6x^3 + 2x^2 + 9$$
A polynomial expression can be simplified by ___combining___ like terms.	Combine like terms and, if possible, write the resulting polynomial in descending order. $$4x^3y - 3xy^2 + 7x^3y - 5 + 6xy^2 - 7$$ $$= 11x^3y + 3xy^2 - 12$$

Exercises 21–28 ▲ **Expressions**

[5.2] *For Exercises 21 and 22, identify the degree of each polynomial and write each polynomial in descending order of degree.*

21. $15 - 2x^9 - 3x + 21x^5 - 19x^3$
9; $-2x^9 + 21x^5 - 19x^3 - 3x + 15$

22. $22j^2 - 19 - j^4 + 5j$
4; $-j^4 + 22j^2 + 5j - 19$

[5.2] *For Exercises 23–28, combine like terms. If possible, write the resulting polynomial in descending order of degree.*

23. $8y^5 - 3y^4 + 2y - 4y^5 - 2y + 8 + 7y^4$
$4y^5 + 4y^4 + 8$

24. $3m - 4 - m^2 - 2m^3 + 7 - 2m + 8m^2$
$-2m^3 + 7m^2 + m + 3$

25. $5xyz^2 - 4x^2yz - 7xyz^2 - 5x^2yz - 2x^2zy$
$-11x^2yz - 2xyz^2$

26. $4a^2bc + 5abc^3 - 8abc - 9a^5 - 2abc^3 - 4a^2bc + 8 + a^5$
$-8a^5 + 3abc^3 - 8abc + 8$

27. $18jk - 2j^4 + 12jk^3 - j^4 - 8jk - 9jk^3 + 12 + k^2$
$-3j^4 + 3jk^3 + k^2 + 10jk + 12$

28. $4m^2 - 3mn + 2mn^2 - 8n^2 - 6mn - 4m^2 + 7 + mn^2$
$3mn^2 - 8n^2 - 9mn + 7$

5.3 Adding and Subtracting Polynomials

Definitions/Rules/Procedures	Key Example(s)
To add polynomials, ___combine like terms___.	Add. $$(5x^3 + 12x^2 - 9x + 1) + (7x^2 - x - 13)$$ $$= 5x^3 + 12x^2 + 7x^2 - 9x - x + 1 - 13$$ $$= 5x^3 + 19x^2 - 10x - 12$$

Definitions/Rules/Procedures	Key Example(s)

To subtract polynomials:
 	
1. Write the subtraction statement as an equivalent <u>addition statement</u>.
 	
  a. Change the operation symbol from a(n) <u>minus sign</u> to a(n) <u>plus sign</u>.
 	
  b. Change the subtrahend (second polynomial) to its <u>additive inverse</u>. To get the <u>additive inverse</u>, change the sign of each term in the polynomial.
 	
2. Combine <u>like terms</u>.
	Subtract.
	
	$$\begin{aligned}&(6y^3 - 5y + 18) - (y^3 - 5y + 7)\\ &= (6y^3 - 5y + 18) + (-y^3 + 5y - 7)\\ &= 6y^3 - y^3 - 5y + 5y + 18 - 7\\ &= 5y^3 + 11\end{aligned}$$

Exercises 29–34   Expressions

[5.3] *For Exercises 29–34, add or subtract and write the resulting polynomial in descending order of degree.*

29. $(8x^2 - 3) + (2x^2 + 3x - 1)$
$10x^2 + 3x - 4$

30. $(5n + 8) - (2n^3 - 3n^2 + n + 5)$
$-2n^3 + 3n^2 + 4n + 3$

31. $(2y - 4) + (4y + 8) - (6y - 5)$
9

32. $(3x^3 - 2x + 8) + (4x^2 - 3x - 1) - (4x^3 - 5x - 10)$
$-x^3 + 4x^2 + 17$

33. $(5x^2 - 2xy + y^2) + (3x^2 + xy + 5y^2)$
$8x^2 - xy + 6y^2$

34. $(m^2 + 5mn + 6) - (4m^2 - 3mn + 8)$
$-3m^2 + 8mn - 2$

5.4 Exponent Rules and Multiplying Monomials

Definitions/Rules/Procedures	Key Example(s)
The product rule for exponents: If a is a real number and m and n are integers, then $a^m \cdot a^n = \underline{a^{m+n}}$. To multiply monomials: 1. Multiply <u>coefficients</u>. 2. <u>Add</u> the exponents of the like bases. 3. Write any unlike variable bases <u>unchanged</u> in the product. If a is a real number and m and n are integers, then $(a^m)^n = \underline{a^{mn}}$. If a and b are real numbers and n is an integer, then $(ab)^n = \underline{a^n b^n}$. To simplify a monomial raised to a power: 1. Evaluate the <u>coefficient</u> raised to that power. 2. Multiply each variable's <u>exponent</u> by the power.	Multiply. **a.** $x^3 \cdot x^4 = x^{3+4} = x^7$ **b.** $-6x^4yz^2 \cdot 5x^5y^3 = -30x^{4+5}y^{1+3}z^2$ $= -30x^9y^4z^2$ Simplify. **a.** $(x^3)^4 = x^{3\cdot4} = x^{12}$ **b.** $(m^2n^3)^5 = m^{2\cdot5}n^{3\cdot5} = m^{10}n^{15}$ **c.** $(3x^2y^5)^4 = 3^4x^{2\cdot4}y^{5\cdot4} = 81x^8y^{20}$

Exercises 35–41   Expressions

[5.4] *For Exercises 35–39, simplify. Write all answers with positive exponents.*

35. $m \cdot m^4$
m^5

36. $2a \cdot 5a^8$
$10a^9$

37. $(-3x^2y)(2x^4y^5)$
$-6x^6y^6$

38. $(5x^2y)^2$
$25x^4y^2$

39. $(-6u^3)^2(2u^4)$
$72u^{10}$

[5.4] 40. A park ranger is studying the number of trees per square meter in a national forest. The area in the study is rectangular and measures 9×10^3 meters by 2.5×10^4 meters. Find the area of this region. Write your answer in scientific notation.
$2.25 \times 10^8 m^2$

[5.4] 41. In a circuit, the current is measured to be 3.5×10^{-3} ampere. If the circuit has a resistance of 8.4×10^{-2} ohm, find the voltage (Hint: Voltage = Current $\times$ Resistance). Write your answer in scientific notation.
$2.94 \times 10^{-4} V$

5.5 Multiplying Polynomials; Special Products

Definitions/Rules/Procedures	Key Example(s)
To multiply a polynomial by a monomial, use the distributive property to __multiply each term in the polynomial by the monomial__.	Multiply. $3x(x^4 - 7x^2y + 9y^2 - 2)$ $= 3x \cdot x^4 - 3x \cdot 7x^2y + 3x \cdot 9y^2 - 3x \cdot 2$ $= 3x^5 - 21x^3y + 27xy^2 - 6x$

Exercises 42–45   Expressions

[5.5] *For Exercises 42–44, multiply.*

42. $4a(a - 3)$
$4a^2 - 12a$

43. $-6b(2b^2 - 4b - 1)$
$-12b^3 + 24b^2 + 6b$

44. $-4abc^2(3a - 4ab^3 + 2abc^2 + 8)$
$-12a^2bc^2 + 16a^2b^4c^2 - 8a^2b^2c^4 - 32abc^2$

45. a. Write an expression in simplest form for the area of the figure shown.
$20x^2 - 28x$

b. Calculate the area if $x = 3$ feet.
96 ft.^2

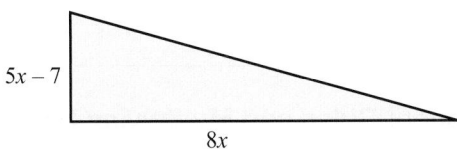

$5x - 7$

$8x$

Definitions/Rules/Procedures	Key Example(s)
To multiply two polynomials: **1.** Multiply each term in the __second polynomial__ by each term in the __first polynomial__. **2.** Combine __like terms__.	$(3x + 5)(2x - 1)$ $= 3x \cdot 2x + 3x \cdot (-1) + 5 \cdot 2x + 5(-1)$ $= 6x^2 - 3x + 10x - 5$ $= 6x^2 + 7x - 5$

Exercises 46–50   Expressions

[5.5] *For Exercises 46–50, multiply.*

46. $(x + 5)(x - 1)$
$x^2 + 4x - 5$

47. $(2m + 5)(6m - 1)$
$12m^2 + 28m - 5$

48. $(5a + 2b)(3a - 2b)$
$15a^2 - 4ab - 4b^2$

49. $(3y - 1)(y^2 - 2y + 4)$
$3y^3 - 7y^2 + 14y - 4$

50. $(a^2 + ab + b^2)(a^2 - 2ab - b^2)$
$a^4 - a^3b - 2a^2b^2 - 3ab^3 - b^4$

Definitions/Rules/Procedures	Key Example(s)
Two binomials are conjugates if they differ only in the __signs__ separating the terms. *Special Products* (*a* and *b* are real numbers, variables, or expressions): Conjugates: $(a + b)(a - b) = $ __$a^2 - b^2$__. Squaring a sum: $(a + b)^2 = $ __$a^2 + 2ab + b^2$__. Squaring a difference: $(a - b)^2 = $ __$a^2 - 2ab + b^2$__.	**a.** $(4x + 3)(4x - 3) = 16x^2 - 9$ **b.** $(3y + 2)^2 = 9y^2 + 12y + 4$ **c.** $(3y - 2)^2 = 9y^2 - 12y + 4$

Exercises 51–54 ➤ Expressions

[5.6] *For Exercises 51–54, multiply.*

51. $(2x + 1)(2x - 1)$
$4x^2 - 1$

52. $(4 - x)(4 + x)$
$16 - x^2$

53. $(x - 6)^2$
$x^2 - 12x + 36$

54. $(3r + 5)^2$
$9r^2 + 30r + 25$

5.6 Exponent Rules and Dividing Polynomials

Definitions/Rules/Procedures	Key Example(s)
Quotient rule for exponents: If m and n are integers and a is a real number, where $a \neq 0$, then $\dfrac{a^m}{a^n} = \underline{\quad a^{m-n} \quad}$.	Divide. $\dfrac{x^7}{x^2} = x^{7-2} = x^5$

Exercises 55–59 ➤ Expressions

[5.6] *For Exercises 58–59, Simplify. Write all answers with positive exponents.*

55. $\dfrac{x^3}{x^{-5}}$
x^8

56. $\dfrac{u^{-1}}{u^{-8}}$
u^7

57. $\dfrac{s^{-5}}{s}$
$\dfrac{1}{s^6}$

58. $\left(\dfrac{1}{x^3}\right)^{-2}$
x^6

59. $\dfrac{x^3}{x^7}$
$\dfrac{1}{x^4}$

Definitions/Rules/Procedures	Key Example(s)
To divide monomials: **1.** Divide the ___coefficients___ or simplify them to fractions in lowest terms. **2.** Use the ___quotient rule___ for the exponents with like bases. **3.** Do not change ___unlike___ variable bases in the quotient. **4.** Write the final expression so that all exponents are ___positive___.	$\dfrac{-39a^6b^5c^2}{13a^4b^5c}$ $= -3a^{6-4}b^{5-5}c^{2-1}$ $= -3a^2b^0c^1$ $= -3a^2(1)c$ $= -3a^2c$

Exercises 60–67 ➤ Expressions

[5.6] *For Exercises 60–67 simplify. Write all answers with positive exponents.*

60. $\dfrac{28z^6}{4z^2}$
$7z^4$

61. $\dfrac{-48x^4y^2}{6x^3y^5}$
$-\dfrac{8x}{y^3}$

62. $\dfrac{24t^3u^6}{-15t^3u^4}$
$-\dfrac{8u^2}{5}$

63. $\dfrac{28a^4b^3z}{42ab^3z^4}$
$\dfrac{2a^3}{3z^3}$

64. $(2a^3b^{-2})^{-3}$
$\dfrac{b^6}{8a^9}$

65. $\dfrac{4hj^4k^{-2}}{(3hj^{-2})^2}$
$\dfrac{4j^8}{9hk^2}$

66. $8x^0 - (2x)^0$
7

67. $(-18x^4y^{-5})^0$
1

Definitions/Rules/Procedures	Key Example(s)
If a, b, and c are real numbers, variables, or expressions with $c \neq 0$, then $\dfrac{a + b}{c} = $ ____$\dfrac{a}{c} + \dfrac{b}{c}$____. To divide a polynomial by a monomial, divide each term of the __polynomial__ by the __monomial__.	$\dfrac{30y^6 + 45y^3}{5y^2} = \dfrac{30y^6}{5y^2} + \dfrac{45y^3}{5y^2}$ $= 6y^{6-2} + 9y^{3-2}$ $= 6y^4 + 9y$

Exercises 68–75 Expressions

[5.6] *For Exercises 68–73, divide.*

68. $\dfrac{4x - 20}{4}$
$x - 5$

69. $\dfrac{5y^2 - 10y + 15}{-5}$
$-y^2 + 2y - 3$

70. $\dfrac{2st + 20s^2t^4 - 4st^5}{2st}$
$1 + 10st^3 - 2t^4$

71. $\dfrac{a^3bc^3 - abc^2 + 2a^2bc^4}{abc^3}$
$a^2 - \dfrac{1}{c} + 2ac$

72. $\dfrac{6a^2b^2 + 3a^2b^3}{3a^2b^2}$
$2 + b$

73. $\dfrac{2xyz^2 - 5x^2y^2z^2 + 10xy^2z}{5xy^3z}$
$\dfrac{2z}{5y^2} - \dfrac{xz}{y} + \dfrac{2}{y}$

74. a. The area of the parallelogram shown is described by the polynomial $54x^2 - 60x$. Find the height.
$9x - 10$

   b. Find the base, height, and area of this parallelogram if $x = 4$ inches.
   24 in., 26 in., and 624 in.²

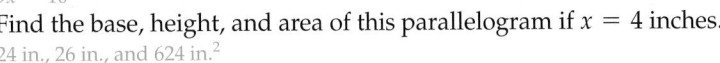

75. Proxima Centauri is approximately 2.47×10^{13} miles from the Sun. If light travels 5.881×10^{12} miles per year, how long does the light from Proxima Centauri take to reach Earth? (*Hint*: Distance = Rate × Time)
≈ 4.2 yr.

Definitions/Rules/Procedures	Key Example(s)
To divide a polynomial by a polynomial, use ____long division____. If there is a remainder, write the result in the following form:____quotient + $\dfrac{\text{remainder}}{\text{divisor}}$____	$\dfrac{6x^2 + 7x - 17}{3x - 4}$ $\begin{array}{r} 2x + 5 \\ 3x - 4\overline{)6x^2 + 7x - 17} \\ \underline{-6x^2 + 8x} \\ 15x - 17 \\ \underline{-15x + 20} \\ 3 \end{array}$ Answer: $2x + 5 + \dfrac{3}{3x - 4}$

Exercises 76–80 Expressions

[5.6] *For Exercises 76–79, use long division to divide the polynomials.*

76. $\dfrac{z^2 + 8z + 15}{z + 3}$
$z + 5$

77. $\dfrac{6m^2 - 10m + 5}{3m - 2}$
$2m - 2 + \dfrac{1}{3m - 2}$

78. $\dfrac{8x^3 - 27}{2x - 3}$
$4x^2 + 6x + 9$

79. $\dfrac{2s^3 - 2s - 3}{s - 1}$
$2s^2 + 2s - \dfrac{3}{s - 1}$

80. A storage room is to have an area described by $4x^2 + 23x - 35$. If the length is described by $x + 7$, find an expression for the width.
$4x - 5$

Exponents Summary

Assume that no denominators are 0, that a and b are real numbers, and that m and n are integers.

Zero as an exponent: $a^0 = \underline{\ 1\ }$ where $a \neq 0$

0^0 is $\underline{\text{indeterminate}}$.

Negative exponents: $a^{-n} = \underline{\dfrac{1}{a^n}}$

$\dfrac{1}{a^{-n}} = \underline{a^n}$

$\left(\dfrac{a}{b}\right)^{-n} = \underline{\left(\dfrac{b}{a}\right)^n}$

Product rule for exponents: $a^m \cdot a^n = \underline{a^{m+n}}$

Quotient rule for exponents: $\dfrac{a^m}{a^n} = \underline{a^{m-n}}$

Raising a power to a power: $(a^m)^n = \underline{a^{mn}}$

Raising a product to a power: $(ab)^n = \underline{a^n b^n}$

Raising a quotient to a power $\left(\dfrac{a}{b}\right)^n = \underline{\dfrac{a^n}{b^n}}$

Evaluate.

a. $3^0 = 1$ **b.** $x^0 = 1, x \neq 0$

c. $3^{-4} = \dfrac{1}{3^4} = \dfrac{1}{81}$

d. $\dfrac{1}{x^{-5}} = x^5$

e. $\left(\dfrac{3}{x}\right)^{-2} = \left(\dfrac{x}{3}\right)^2$

f. $x^3 \cdot x^5 = x^{3+5} = x^8$

g. $\dfrac{r^4}{r^{-2}} = r^{4-(-2)} = r^{4+2} = r^6$

h. $(y^3)^5 = y^{3 \cdot 5} = y^{15}$

i. $(3x)^4 = 3^4 x^4 = 81x^4$

j. $\left(\dfrac{2}{a}\right)^5 = \dfrac{2^5}{a^5} = \dfrac{32}{a^5}$

Chapter 5 Practice Test

For Extra Help

Step-by-step test solutions are found on the Chapter Test Prep Videos available in MyMathLab® or on You Tube.

For Exercises 1 and 2, evaluate the exponential form.

1. 2^{-3}
$\dfrac{1}{8}$ [5.1]

2. $\left(\dfrac{2}{3}\right)^{-2}$
$\dfrac{9}{4}$ [5.1]

3. Write 6.201×10^{-3} in standard form.
0.006201 [5.1]

4. Write 275,000,000 in scientific notation.
2.75×10^8 [5.1]

For Exercises 5 and 6, identify the degree.

5. $-7x^2 y$
3 [5.2]

6. $4x^2 - 9x^4 + 8x - 7$
4 [5.2]

7. Evaluate $-6mn - n^3$, where $m = 4$ and $n = -2$.
56 [5.2]

8. Combine like terms and write the resulting polynomial in descending order of degree.
$-5x^4 + 7x^2 + 6x^2 - 5x^4 + 12 - 6x^3 + x^2$
$-10x^4 - 6x^3 + 14x^2 + 12$ [5.2]

For Exercises 9 and 10, add or subtract and write the resulting polynomial in descending order of degree.

9. $(3x^2 + 4x - 2) + (5x^2 - 3x - 2)$
$8x^2 + x - 4$ [5.3]

10. $(7x^4 - 3x^2 + 4x + 1) - (2x^4 + 5x - 7)$
$5x^4 - 3x^2 - x + 8$ [5.3]

For Exercises 11–18, multiply.

11. $(4x^2)(3x^5)$
$12x^7$ [5.4]

12. $(2ab^3c^7)(-a^5b)$
$-2a^6b^4c^7$ [5.4]

13. $(4xy^3)^2$
$16x^2y^6$ [5.4]

14. $3x(x^2 - 4x + 5)$
$3x^3 - 12x^2 + 15x$ [5.5]

15. $-6t^2u(4t^3 - 8tu^2)$
$-24t^5u + 48t^3u^3$ [5.5]

16. $(n - 1)(n + 4)$
$n^2 + 3n - 4$ [5.5]

17. $(2x - 3)^2$
$4x^2 - 12x + 9$ [5.5]

18. $(x + 2)(x^2 - 4x + 3)$
$x^3 - 2x^2 - 5x + 6$ [5.5]

19. Write an expression in simplest form for the area of the shape shown.
$6n^2 - 7n - 20$ [5.5]

$2n - 5$

$3n + 4$

For Exercises 20–25, simplify.

20. $\dfrac{x^9}{x^4}$
x^5 [5.6]

21. $\dfrac{(x^3)^{-2}}{x^4 \cdot x^{-5}}$
$\dfrac{1}{x^5}$ [5.6]

22. $\dfrac{(3y)^{-2}}{(x^3y^2)^{-3}}$
$\dfrac{x^9y^4}{9}$ [5.6]

23. $\dfrac{24x^5 + 18x^3}{6x^2}$
$4x^3 + 3x$ [5.6]

24. $\dfrac{x^2 - x - 12}{x + 3}$
$x - 4$ [5.6]

25. $\dfrac{15x^2 - 22x + 14}{3x - 2}$
$5x - 4 + \dfrac{6}{3x - 2}$ [5.6]

Chapters 1–5 Cumulative Review Exercises

For Exercises 1–4, answer true or false.

[3.3] 1. The graph of $3x - 4y = -24$ has an x-intercept of $(0, 8)$ and a y-intercept of $(-6, 0)$.

false

[1.6] 2. 6 less than a number can be translated to $6 - x$.

false

[3.5] 3. $y - 2 = -3(x - 5)$ is an example of the point–slope form of a line.

true

[3.2] 4. There is only one solution to any linear equation in two variables.

false

For Exercises 5 and 6, fill in the blank.

[2.3] 5. To use the multiplication principle to eliminate a coefficient:

   a. Determine the coefficient you want to eliminate.

   b. Multiply both sides by the multiplicative inverse of that ___coefficient___ or divide both sides by the ___coefficient___.

[3.3] 6. If an equation can be written in the form $y = mx$, where m is a real number other than 0, the x- and y-intercepts will be at ___(0, 0) or the origin___.

Exercises 7–18 **Expressions**

[1.5] *For Exercises 7 and 8, simplify.*

7. $5 + 2(6 - 3^2)^2$

23

8. $|5 - 10| + 3(2) \div (-1)$

−1

[2.7] 9. 30.2% of what number is 18.12?

60

For Exercises 10–15, simplify.

[5.3] 10. $(3y^2 - 2y + 10) + (2y^2 - 2y - 11)$

$5y^2 - 4y - 1$

[5.3] 11. $(2x + 1) - (-4x + 7)$

$6x - 6$

[5.5] 12. $(2x + 3)(x - 7)$

$2x^2 - 11x - 21$

[5.5] 13. $(2x - 3)^2$

$4x^2 - 12x + 9$

[5.6] 14. $(3m^2 - 6m + 1) \div (3m)$

$m - 2 + \dfrac{1}{3m}$

[5.6] 15. $(x^2 + 5x + 7) \div (x + 2)$

$x + 3 + \dfrac{1}{x + 2}$

[5.6] *For Exercises 16–18, simplify. Leave your answers with positive exponents only.*

16. $(3x^{-2}y^3)^{-3}$

$\dfrac{x^6}{27y^9}$

17. $\left(\dfrac{2x^3}{y}\right)^2$

$\dfrac{4x^6}{y^2}$

18. $(2x^2y^{-3})^2(-3x^4y^2)^3$

$-108x^{16}$

Exercises 19–30 **Equations and Inequalities**

[3.3] 19. Graph: $3x + 2y = -6$

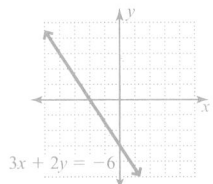

$3x + 2y = -6$

[3.6] 20. Graph: $y > \dfrac{2}{3}x - 2$

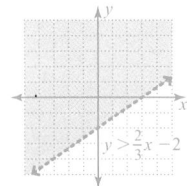

$y > \dfrac{2}{3}x - 2$

For Exercises 21–24, solve.

[2.3] 21. $10x - 3 + 5 = 8x + 26 + 4x$
-12

[2.8] 22. $3(x - 2) > 4(x + 1)$
$x < -10$

[2.6] 23. $\dfrac{16}{9} = \dfrac{x}{20.25}$
36

24. $-\dfrac{3}{7}x = \dfrac{12}{35}$
$-\dfrac{4}{5}$

[2.4] 25. Solve for l in the equation $P = 2l + 2w$.
$l = \dfrac{P - 2w}{2}$

[4.2] 26. Solve the system $\begin{cases} 2x - 3y = 14 \\ y = 3x - 14 \end{cases}$ using the substitution method.
$(4, -2)$

[4.3] 27. Solve the system $\begin{cases} 3x - 4y = 24 \\ 5x + 3y = 11 \end{cases}$ using the elimination method.
$(4, -3)$

For Exercises 28–30, solve.

[4.3] 28. A broker invests $\dfrac{1}{4}$ of a client's money in a government security paying 9% interest annually and the rest in a money market fund paying 13% interest annually. Find the amount invested in each if the total interest earned is $5400.
$33,750 at 13%, $11,250 at 9%

[2.7] 29. Use the accompanying bar graph. If 1600 parents at a local high school were polled, how many could be expected to think that teens should not work?
480

[2.7] 30. In Westchester County, New York, the average temperature is 68° in July and only 31.5° in January. What is the percent of decrease in temperature from July to January?
$\approx 53.7\%$

SHOULD TEENS WORK?

67%   30%   3%

Good Idea   Bad Idea   Don't Know

(*Source:* Yankelovich Partners for Lutheran Brotherhood, U.S. Department of Labor.)

Chapter Overview

In this chapter, we explore polynomials further and learn about factoring. Factoring is another way to rewrite an expression and is one of the most important skills you will acquire in algebra. It is used extensively in simplifying the more complex expressions that we will study in future chapters. The general flow of the chapter is as follows:

▶ Learn various methods for factoring polynomial expressions.

▶ Use factoring to solve equations.

▶ Graph special polynomial equations and functions.

Instructor Note

Factoring can be daunting for many students. Although the sequence of topics is flexible, we have tried assembling the chapter in a way that flows easily. With each method, we reinforce factoring out a monomial GCF as a first step. In Section 6.5, we summarize all of the methods presented in Sections 6.1–6.4 with a procedure to help students identify the appropriate method based on the given polynomial. In Section 6.6, we turn to solving equations using factoring. Call attention to the Algebra Pyramid there to remind students that we are changing gears from expressions to equations. In Section 6.7, we use the approach established in Section 3.7 and show the connection between quadratic equations in two variables and quadratic functions.

6.1 Greatest Common Factor and Factoring by Grouping

6.2 Factoring Trinomials of the Form $x^2 + bx + c$

6.3 Factoring Trinomials of the Form $ax^2 + bx + c$, where $a \neq 1$

6.4 Factoring Special Products

6.5 Strategies for Factoring

6.6 Solving Quadratic Equations by Factoring

6.7 Graphs of Quadratic Equations and Functions

6.1 Greatest Common Factor and Factoring by Grouping

Objectives

1 List all possible factors for a given number.

2 Find the greatest common factor of a set of numbers or monomials.

3 Write a polynomial as a product of a monomial GCF and a polynomial.

4 Factor by grouping.

Warm-up

[1.2] **1.** Find the prime factorization of 90.

[5.5] **2.** Multiply: $12x^2(2y - 5x)$

[5.6] *For Exercises 3 and 4, divide.*

3. $\dfrac{9x^2y^2z}{-3x^2y}$

4. $\dfrac{24x^2y - 60x^3}{12x^2}$

Often, in mathematics, we need to consider the factors of a number or an expression. A number or an expression written as a product of factors is said to be in **factored form**.

Definition Factored form: A number or an expression written as a product of factors.

Following are some examples of factored form.

An integer written in factored form with integer factors: $28 = 2 \cdot 14$

A monomial written in factored form with monomial factors: $8x^5 = 4x^2 \cdot 2x^3$

A polynomial written in factored form with a monomial factor and a polynomial factor: $2x + 8 = 2(x + 4)$

A polynomial written in factored form with two polynomial factors: $x^2 + 5x + 6 = (x + 2)(x + 3)$

Notice that we can check the factored form by multiplying the factors to equal the product. The process of writing an expression in factored form is called *factoring*.

Objective 1 List all possible factors for a given number.

We begin our exploration of factoring by listing natural number factors when given an integer.

Example 1   List all natural number factors of 24.

Solution: To list all natural number factors, we can divide 24 by 1, 2, 3, and so on, writing each divisor and quotient pair as a product until we have all possible combinations.

> **Note** The number 5 does not divide 24 evenly, and 6 is already in the $4 \cdot 6$ pair. When you reach a natural number that you already have listed, you can stop.

$1 \cdot 24$
$2 \cdot 12$
$3 \cdot 8$
▶ $4 \cdot 6$

The natural number factors of 24 are 1, 2, 3, 4, 6, 8, 12, 24.

Your Turn 1   List all natural number factors of each given number.

a. 30   **b.** 42

Answers to Your Turn 1
a. 1, 2, 3, 5, 6, 10, 15, 30
b. 1, 2, 3, 6, 7, 14, 21, 42

Answers to Warm-up
1. $2 \cdot 3^2 \cdot 5$
2. $24x^2y - 60x^3$
3. $-3yz$
4. $2y - 5x$

Objective 2 Find the greatest common factor of a set of numbers or monomials.

When factoring polynomials, the first step is to determine whether there is a monomial factor that is common to all of the terms in the polynomial. Further, we want that monomial factor to be the **greatest common factor** of the terms. Let's first consider how to find the greatest common factor of numbers.

Instructor Note Show students that in checking the GCF by dividing, the quotients have no common factor except the number 1.

These quotients share no common factors except the number 1.

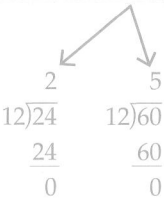

Also show what happens in the check if we find a number that is a common factor but is not the GCF, such as 6.

These quotients are both divisible by 2. This means that 6 is not the greatest common factor.

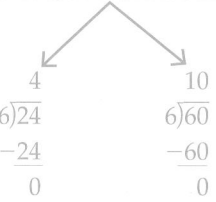

Explain that the 2, a common factor in these quotients, is the very factor by which we were off in mistakenly finding 6 as the GCF. In other words, we merely need to multiply the answer we found, 6, by the common factor of 2 that is appearing in the quotients and we get the correct GCF, 12. The reason this works will become clearer when we look at the prime factorization method for generating GCF.

Learning Strategy

If you are a visual learner, you may prefer to continue circling the common prime factors.

Definition Greatest common factor (GCF): The largest natural number that divides all given numbers with no remainder.

For example, the greatest common factor of 12 and 18 is 6 because 6 is the largest number that divides into both 12 and 18 evenly.

Using Listing to Find the GCF of Numbers

One way to find the GCF of a given set of numbers is by listing factors.

> **Procedure Listing Method for Finding GCF**
>
> To find the GCF of a set of numbers by listing:
> 1. List all possible factors for each given number.
> 2. Search the lists for the largest factor common to all lists.

Example 2   Find the GCF of 24 and 60.

Solution: Factors of 24: 1, 2, 3, 4, 6, 8, 12, 24
Factors of 60: 1, 2, 3, 4, 5, 6, 10, 12, 15, 20, 30, 60
The GCF of 24 and 60 is 12.

Using Prime Factorization to Find the GCF

The listing method is a good method for smaller numbers, but for larger numbers, it is not the most efficient method to use. It turns out that we can use prime factorization, which we explained in Section 1.2, to find the GCF of a given set of numbers. Consider the prime factorizations for the numbers from Example 2.

$$24 = 2 \cdot 2 \cdot 2 \cdot 3 = 2^3 \cdot 3$$
$$60 = 2 \cdot 2 \cdot 3 \cdot 5 = 2^2 \cdot 3 \cdot 5$$

In Example 2, we found the GCF to be 12. Let's look at the prime factorization of 12 to see if we can discover a rule for finding the GCF.

$$12 = 2 \cdot 2 \cdot 3 = 2^2 \cdot 3$$

Notice that the 12 contains two factors of 2 and one factor of 3, which are the factors that are common to 24's and 60's prime factorizations. This suggests that we use only primes that are common to all factorizations involved. The factor 5 is not common to both 24's and 60's prime factorizations, so it is not included in the GCF.

$$24 = 2 \cdot 2 \cdot 2 \cdot 3 = 2^3 \cdot 3$$
$$60 = 2 \cdot 2 \cdot 3 \cdot 5 = 2^2 \cdot 3 \cdot 5$$

$$\text{GCF} = 2 \cdot 2 \cdot 3 = 2^2 \cdot 3 = 12$$

Notice that we could also compare exponents for a given common prime factor to determine how many of that factor should be included in the GCF. For example, 2^2 is included instead of 2^3 because 2^2 has the smaller exponent. This suggests the following procedure.

> **Procedure Prime Factorization Method for Finding GCF**
>
> To find the GCF of a given set of numbers by prime factorization:
> 1. Write the prime factorization of each given number in exponential form.
> 2. Create a factorization for the GCF that includes only those prime factors common to all factorizations, each raised to its smallest exponent in the factorizations.
> 3. Multiply the factors in the factorization created in step 2.
>
> *Note*: If there are no common prime factors, the GCF is 1.

Learning Strategy

If you are an auditory learner, you might find it easier to remember the procedure by thinking about the words in the procedure: To find the **greatest common** factor, you use the **common** primes raised to their **smallest** exponent. Notice that the words *greatest* and *smallest* are opposites.

Example 3   Find the GCF.

a. 54 and 90

Solution: Write the prime factorization of 54 and 90 in factored form.

$$54 = 2 \cdot 3^3$$
$$90 = 2 \cdot 3^2 \cdot 5$$

The common prime factors are 2 and 3. Compare the exponents on each of these common factors. The smallest exponent of 2 is 1, and the smallest exponent of 3 is 2. The GCF is the product of 2^1 and 3^2.

$$GCF = 2^1 \cdot 3^2 = 2 \cdot 9 = 18$$

b. 3024 and 2520

Solution: Write the prime factorization of 3024 and 2520 in factored form.

$$3024 = 2^4 \cdot 3^3 \cdot 7$$
$$2520 = 2^3 \cdot 3^2 \cdot 5 \cdot 7$$

The common prime factors are 2, 3, and 7. Compare the exponents of each of these common factors. The smallest exponent of 2 is 3. The smallest exponent of 3 is 2. The smallest exponent of 7 is 1. The GCF is the product of 2^3, 3^2, and 7.

$$GCF = 2^3 \cdot 3^2 \cdot 7 = 8 \cdot 9 \cdot 7 = 504$$

Your Turn 3   Find the GCF.

   a. 180 and 600   **b.** 63, 84, and 105

GCF of Monomials

We use a similar approach to find the GCF of a set of monomials that have variables. The variables in the monomials are treated like prime factors.

Example 4   Find the GCF.

a. $x^3y^3z^4$ and x^2y^6

Solution: By treating the variables like prime factors, each monomial is already written as a prime factorization. The common prime factors are x and y. The smallest exponent for the factor x is 2, and the smallest exponent for the factor y is 3. The GCF is the product of x^2 and y^3.

$$GCF = x^2y^3$$

b. $24x^2y$ and $60x^3$

Solution: Write the prime factorization of each monomial, treating the variables like prime factors.

$$24x^2y = 2^3 \cdot 3 \cdot x^2 \cdot y$$
$$60x^3 = 2^2 \cdot 3 \cdot 5 \cdot x^3$$

The common prime factors are 2, 3, and x. The smallest exponent of 2 is 2. The smallest exponent of 3 is 1. The smallest exponent of x is 2. The GCF is the product of 2^2, 3, and x^2.

$$GCF = 2^2 \cdot 3 \cdot x^2 = 12x^2$$

Answers to Your Turn 3
  a. 60   **b.** 21

Answers to Your Turn 4
  a. a^2c^3 **b.** 8*ab*   **c.** 1

Your Turn 4   Find the GCF.

   a. $a^3b^4c^3$ and $a^2c^5d^3$ **b.** $32a^2b$ and $40abc$ **c.** $35x^2$ and $18y$

Objective 3 Write a polynomial as a product of a monomial GCF and a polynomial.

Recall that when we multiply a monomial by a polynomial, we apply the distributive property.

$$2(x + 4) = 2 \cdot x + 2 \cdot 4$$
$$= 2x + 8$$

Now let's reverse this process. Suppose we are given the product, $2x + 8$, and we want to write the factored form, $2(x + 4)$. Notice that terms in $2x + 8$ have 2 as a common factor. When writing factored form, we first determine the GCF of the terms. We then create a missing factor statement like this:

$$2x + 8 = 2(?)$$

Notice that the missing factor must be $x + 4$. To determine this missing factor, we divide the product $2x + 8$ by the known factor, 2.

$$2x + 8 = 2 \, (?)$$
$$\frac{2x + 8}{2} = (?)$$
$$\frac{2x}{2} + \frac{8}{2} = (?)$$
$$x + 4 = (?)$$

Connection In Section 5.6, we learned that we divide a polynomial by a monomial by dividing each term in the polynomial by the monomial.

The factored form of $2x + 8$ is $2(x + 4)$, which suggests the following procedure.

Procedure **Factoring a Monomial GCF Out of a Polynomial**

To factor a monomial GCF out of a given polynomial:
1. Find the GCF of the terms in the polynomial.
2. Rewrite the given polynomial as a product of the GCF and the quotient of the polynomial and the GCF.

$$\text{Polynomial} = \text{GCF}\left(\frac{\text{Polynomial}}{\text{GCF}}\right)$$

Instructor Note Point out that answers such as $6x \cdot 2x + 9 \cdot 2x$ are not factorizations of $12x^2 + 18x$.

Note An alternative way of doing Example 5 is by inspection using the distributive property. Think of $12x^2 + 18x = 6x(\_\_ + \_\_)$, where $6x$ times the first blank gives $12x^2$ and $6x$ times the second blank gives $18x$. Because $6x \cdot 2x = 12x^2$ and $6x \cdot 3 = 18x$, the blanks are filled with $2x$ and 3. So $12x^2 + 18x = 6x(2x + 3)$.

Example 5   Factor $12x^2 + 18x$.

Solution:
1. Find the GCF of $12x^2$ and $18x$.
   The GCF of $12x^2$ and $18x$ is $6x$.

   Note 6 is the largest number that divides both 12 and 18 evenly, and x has the smaller exponent of the x^2 and x.

2. Write the given polynomial as the product of the GCF and the quotient of the polynomial and the GCF.

   $$12x^2 + 18x = 6x\left(\frac{12x^2 + 18x}{6x}\right)$$
   $$= 6x\left(\frac{12x^2}{6x} + \frac{18x}{6x}\right) \quad \text{Separate the terms.}$$
   $$= 6x(2x + 3) \quad \text{Divide the terms by the GCF.}$$

Check: We can check by multiplying the factored form using the distributive property.

$$6x(2x + 3) = 6x \cdot 2x + 6x \cdot 3 \quad \text{Distribute } 6x.$$
$$= 12x^2 + 18x \quad \text{The product is the original polynomial.}$$

Connection Keep in mind that when we factor, we are simply writing the original expression in a different form called *factored form*. When written equal to each other, the factored form and product make an identity, which means that every real number is a solution to the equation. Consider $12x^2 + 18x = 6x(2x + 3)$. Let's choose a value such as $x = 2$ and verify that it satisfies the equation. (Remember, we can select any real number.)

$$12x^2 + 18x = 6x(2x + 3)$$
$$12(2)^2 + 18(2) \overset{?}{=} 6(2)(2\cdot 2 + 3)$$
$$12(4) + 36 \overset{?}{=} 12(4 + 3)$$
$$48 + 36 \overset{?}{=} 12(7)$$
$$84 = 84 \qquad \text{It checks.}$$

Example 6 Factor $24x^2y - 60x^3$.

Solution:

1. Find the GCF of $24x^2y$ and $60x^3$.

    We found this GCF in Example 4(b) to be $12x^2$.

2. Write the given polynomial as the product of the GCF and the quotient of the polynomial and the GCF.

Note Again, we could factor using the distributive property by thinking of $24x^2y - 60x^3 = 12x^3(\_\_ + \_\_)$. Because $12x^2 \cdot 2y = 24x^2y$ and $12x^2(-5x) = -60x^3$, the blanks represent $2y$ and $-5x$. So $24x^2y - 60x^3 = 12x^2(2y - 5x)$.

$$24x^2y - 60x^3 = 12x^2\left(\frac{24x^2y - 60x^3}{12x^2}\right)$$
$$= 12x^2\left(\frac{24x^2y}{12x^2} - \frac{60x^3}{12x^2}\right)$$
$$= 12x^2(2y - 5x)$$

Connection Remember that when we divide exponential forms that have the same base, we subtract exponents.

Check: Multiply the factored form using the distributive property.
$$12x^2(2y - 5x) = 12x^2 \cdot 2y - 12x^2 \cdot 5x$$
$$= 24x^2y - 60x^3$$

Your Turn 6 Factor.

   a. $30xy - 45x$ **b.** $16r^2s^4 - 8r^3s^4 + 12rs^3$

Factoring When the First Term Is Negative

Instructor Note Mention that if there is a positive term anywhere in the polynomial to be factored, then a second option is to use the commutative property to rearrange the polynomial so that the positive term is written first.

$$-6x + 10y = 10y - 6x$$
$$= 2(5y - 3x)$$

Indicate that in many cases, this is less desirable because it may cause the polynomial to no longer be in descending order.

Consider factoring the expression $-6x + 10y$. Because the first term of the polynomial is negative, when we factor out the GCF, 2, the first term inside the parentheses is also negative, so that we have $2(-3x + 5y)$. However, it is considered undesirable to have a negative first term inside parentheses. We can avoid a negative first term in the parentheses by factoring the negative of the GCF; so in our example, we factor out -2. This changes the sign of each term inside the parentheses so that we have $-2(3x - 5y)$.

Answers to Your Turn 6
a. $15x(2y - 3)$
b. $4rs^3(4rs - 2r^2s + 3)$

Example 7 Factor by factoring out the negative of the GCF:
$$-18x^4y^3 + 9x^2y^2z - 12x^3y$$

Solution:

1. Find the GCF of $-18x^4y^3$, $9x^2y^2z$, and $-12x^3y$.

    Because the first term in the polynomial is negative, we factor out the negative of the GCF to avoid a negative first term inside the parentheses. We factor out $-3x^2y$.

2. Write the given polynomial as the product of the GCF and the quotient of the polynomial and the GCF.

$$-18x^4y^3 + 9x^2y^2z - 12x^3y = -3x^2y\left(\frac{-18x^4y^3 + 9x^2y^2z - 12x^3y}{-3x^2y}\right)$$

$$= -3x^2y\left(\frac{18x^4y^3}{-3x^2y} + \frac{9x^2y^2z}{-3x^2y} - \frac{12x^3y}{-3x^2y}\right)$$

$$= -3x^2y(6x^2y^2 - 3yz + 4x)$$

Check: Multiply the factored form using the distributive property.

$$-3x^2y(6x^2y^2 - 3yz + 4x) = -3x^2y \cdot 6x^2y^2 - 3x^2y \cdot (-3yz) - 3x^2y \cdot 4x$$

$$= -18x^4y^3 + 9x^2y^2z - 12x^3y$$

Your Turn 7 Factor by factoring out the negative of the GCF.

   a. $-10x^3 - 15x^2 + 25x$ **b.** $-48a^4b^5 - 24a^3b^4 + 16ab^2c$

Before we discuss additional techniques of factoring, it is important to state that no matter what type of polynomial we are asked to factor, we always consider whether a monomial GCF (other than 1) can be factored out of the polynomial.

Factoring When the GCF Is a Polynomial

Sometimes in factoring, the GCF is a polynomial with more than one term.

Instructor Note If students have difficulty factoring out a common binomial factor, have them try substitution. For example, in Example 8, let $z = x + 2$. So $y(x + 2) + 7(x + 2) = yz + 7z = z(y + 7)$. Now substitute $x + 2$ for z: $(x + 2)(y + 7)$.

Example 8 Factor $y(x + 2) + 7(x + 2)$.

Solution: Notice that this expression is a sum of two products, y and $(x + 2)$, and 7 and $(x + 2)$. Further, note that $(x + 2)$ is the GCF of the two products.

$$y(x + 2) + 7(x + 2) = (x + 2)\left(\frac{y(x + 2) + 7(x + 2)}{x + 2}\right)$$

$$= (x + 2)\left(\frac{y(x + 2)}{x + 2} + \frac{7(x + 2)}{x + 2}\right)$$

$$= (x + 2)(y + 7)$$

◄ **Note** The procedure is the same. We rewrite the given expression as the product of the GCF and the quotient of the expression and the GCF.

Your Turn 8 Factor $4a(a - 3) - b(a - 3)$.

Objective 4 Factor by grouping.

The process of factoring out a polynomial GCF, as we did in Example 8, is an intermediate step in a process called *factoring by grouping*, which is a technique that we try when factoring a four-term polynomial such as $xy + 2y + 7x + 14$. The method is called *grouping* because we group pairs of terms and look for a common factor within each group or pair. We begin by pairing the first two terms as one group and the last two terms as a second group.

$$xy + 2y + 7x + 14 = (xy + 2y) + (7x + 14)$$

Note that the first two terms have a common factor of y and the last two terms have 7 as a common factor. If we factor y out of the first two terms and 7 out of the last two terms, we have the same expression that we factored in Example 8.

Answers to Your Turn 7
   a. $-5x(2x^2 + 3x - 5)$
   b. $-8ab^2(6a^3b^3 + 3a^2b^2 - 2c)$

Answer to Your Turn 8
   $(a - 3)(4a - b)$

$$xy + 2y + 7x + 14 = (xy + 2y) + (7x + 14)$$

$$= y(x + 2) + 7(x + 2)$$

$$= (x + 2)(y + 7) \qquad \text{Factor out } (x + 2).$$

Procedure Factoring by Grouping

To factor a four-term polynomial by grouping:
1. Factor out any monomial GCF (other than 1) that is common to all four terms.
2. Group pairs of terms and factor the GCF out of each pair or group.
3. If there is a common binomial factor, factor it out.
4. If there is no common binomial factor, interchange the middle two terms and repeat the process. If there is still no common binomial factor, the polynomial cannot be factored by grouping.

Note Sometimes the GCF is 1 or −1.

Example 9 Factor.

a. $6x^3 - 8x^2 + 3xy - 4y$

Solution: First, we look for a monomial GCF (other than 1). This polynomial does not have one. Because the polynomial has four terms, we now try to factor by grouping.

$$6x^3 - 8x^2 + 3xy - 4y = (6x^3 - 8x^2) + (3xy - 4y) \qquad \text{Factor } 2x^2 \text{ out of } 6x^3 \text{ and } 8x^2;$$
$$\text{then factor } y \text{ out of } 3xy \text{ and } 4y.$$
$$= 2x^2(3x - 4) + y(3x - 4)$$
$$= (3x - 4)(2x^2 + y) \qquad \text{Factor out } 3x - 4.$$

Note When factoring by grouping ▶ and the third term is negative, the GCF of the third and fourth terms is usually negative.

b. $12mn^2 - 20mn - 24n^2 + 40n$

Solution: Note that in this case, there is a monomial GCF, $4n$; so we first factor this GCF out of all four terms.

$$12mn^2 - 20mn - 24n^2 + 40n = 4n(3mn - 5m - 6n + 10)$$

Because the polynomial in the parentheses has four terms, we try to factor by grouping.

Note Now that the expression is in factored form, we no longer need the brackets. ▶

$$= 4n[(3mn - 5m) + (-6n + 10)] \qquad \text{Factor } m \text{ out of } 3\,mn \text{ and } 5m; \text{ then}$$
$$\text{factor } -2 \text{ out of } -6n \text{ and } 10.$$
$$= 4n[m(3n - 5) - 2(3n - 5)]$$
$$= 4n(3n - 5)(m - 2) \qquad \text{Factor out } 3n - 5.$$

Your Turn 9 Factor.

a. $6x^2 + 15xz + 2xy + 5yz$ **b.** $42a^2b - 56a^2 - 12ab + 16a$

Answers to Your Turn 9
a. $(2x + 5z)(3x + y)$
b. $2a(3b - 4)(7a - 2)$

6.1 Exercises (For Extra Help) MyMathLab®

Note: Exercises marked with a ★ represent challenging exercises.

Objective 1

Prep Exercise 1 Explain how to list all possible natural number factors of a number.

Divide by 1, 2, 3, and so on, writing each divisor and quotient pair as a product until we have all possible combinations.

For Exercises 1–16, list all natural number factors of the given number. See Example 1.

1. 9

1, 3, 9

2. 49

1, 7, 49

3. 33

1, 3, 11, 33

4. 21

1, 3, 7, 21

5. 18
1, 2, 3, 6, 9, 18

6. 12
1, 2, 3, 4, 6, 12

7. 16
1, 2, 4, 8, 16

8. 81
1, 3, 9, 27, 81

9. 44
1, 2, 4, 11, 22, 44

10. 45
1, 3, 5, 9, 15, 45

11. 60
1, 2, 3, 4, 5, 6, 10, 12, 15, 20,
30, 60

12. 36
1, 2, 3, 4, 6, 9, 12, 18, 36

13. 56
1, 2, 4, 7, 8, 14, 28, 56

14. 64
1, 2, 4, 8, 16, 32, 64

15. 90
1, 2, 3, 5, 6, 9, 10, 15, 18, 30,
45, 90

16. 84
1, 2, 3, 4, 6, 7, 12, 14, 21, 28,
42, 84

Objective 2

Prep Exercise 2 Define the GCF of a given set of numbers in your own words.
The largest natural number that divides all given numbers with no remainder

Prep Exercise 3 Given a set of terms, after finding the prime factorization of each term, how do you use those prime factors to create the GCF's factorization? The factorization for the GCF includes only those prime factors common to all of the factorizations, each raised to its smallest exponent.

For Exercises 17–28, find the GCF. See Examples 2–4.

17. 21, 30
3

18. 21, 35
7

19. 72, 80
8

20. 35, 60
5

21. 12, 42, 60
6

22. 10, 18, 36
2

23. $4xy, 6xy$
$2xy$

24. $6ab, 9ab$
$3ab$

25. $25h^2, 60h^4$
$5h^2$

26. $25x^4, 10x^3$
$5x^3$

27. $6a^5b, 15a^4b^2$
$3a^4b$

28. $7m^6n, 21m^5n^4$
$7m^5n$

Objective 3

Prep Exercise 4 When factoring a monomial GCF out of the terms of a polynomial, after finding the GCF, explain how to rewrite the polynomial.
Rewrite the given polynomial as a product of the GCF and the quotient of the polynomial and the GCF.
$$\text{Polynomial} = \text{GCF}\left(\frac{\text{Polynomial}}{\text{GCF}}\right)$$

For Exercises 29–56, factor by factoring out the GCF. See Examples 5 and 6.

29. $5c - 20$
$5(c - 4)$

30. $8y - 24$
$8(y - 3)$

31. $8x - 12$
$4(2x - 3)$

32. $15x - 10$
$5(3x - 2)$

33. $x^2 - x$
$x(x - 1)$

34. $y^2 + y$
$y(y + 1)$

35. $18z^6 - 12z^4$
$6z^4(3z^2 - 2)$

36. $15k^3 - 24k^2$
$3k^2(5k - 8)$

37. $18p^3 - 15p^5$
$3p^3(6 - 5p^2)$

38. $18r^4 - 27r^6$
$9r^4(2 - 3r^2)$

39. $6a^2b - 3ab^2$
$3ab(2a - b)$

40. $9m^2n - 3mn^2$
$3mn(3m - n)$

41. $14uv^2 - 7uv$
$7uv(2v - 1)$

42. $12x^2y^3 - 6xy^2$
$6xy^2(2xy - 1)$

43. $25xy - 50xz + 100x$
$25x(y - 2z + 4)$

44. $8ab - 32ac + 40a$
$8a(b - 4c + 5)$

45. $x^2y + xy^2 + x^3y^3$
$xy(x + y + x^2y^2)$

46. $w^3v^2 + w^2v + wv^2$
$wv(w^2v + w + v)$

47. $28ab^3c - 36a^2b^2c$
$4ab^2c(7b - 9a)$

48. $30x^3yz^2 - 24x^2y^3z$
$6x^2yz(5xz - 4y^2)$

49. $20p^2q + 24pq - 16pq^2$
$4pq(5p + 6 - 4q)$

50. $6a^2c + 18abc - 12ac^2$
$6ac(a + 3b - 2c)$

51. $3mn^5p^2 + 18mn^3p - 6mnp$
$3mnp(n^4p + 6n^2 - 2)$

52. $4ax^3y^3 - 6ax^2y^2 + 4axy^2$
$2axy^2(2x^2y - 3x + 2)$

53. $105a^3b^2 - 63a^2b^3 + 84a^6b^4$
$21a^2b^2(5a - 3b + 4a^4b^2)$

54. $21t^4u^3 - 105t^5u^3v + 63t^6u^2v^4$
$21t^4u^2(u - 5tuv + 3t^2v^4)$

55. $18x^4 - 9x^3 + 30x^2 - 12x$
$3x(6x^3 - 3x^2 + 10x - 4)$

56. $10g + 40g^3 - 100g^4 + 120g^5$
$10g(1 + 4g^2 - 10g^3 + 12g^4)$

For Exercises 57–66, factor by factoring out the negative of the GCF. See Example 7.

57. $-3x + 6y$
$-3(x - 2y)$

58. $-8x + 4y$
$-4(2x - y)$

59. $-20a^2 - 15a$
$-5a(4a + 3)$

60. $-30b^2 - 24b$
$-6b(5b + 4)$

61. $-12a^4b + 20a^3b^3$
$-4a^3b(3a - 5b^2)$

62. $-15x^4y + 21x^2y^2$
$-3x^2y(5x^2 - 7y)$

63. $-4x^2 - 8x + 16$
$-4(x^2 + 2x - 4)$

64. $-6x^2 - 18x + 24$
$-6(x^2 + 3x - 4)$

65. $-24x^3y^2z - 30x^2y^3z^4 + 12x^4y^5z^2$
$-6x^2y^2z(4x + 5yz^3 - 2x^2y^3z)$

66. $-21a^4bc^3 - 15a^3b^3c^3 + 27a^2b^2c^4$
$-3a^2bc^3(7a^2 + 5ab^2 - 9bc)$

For Exercises 67–74, factor out the polynomial GCF. See Example 8.

67. $y(a - 3) + 2(a - 3)$
$(a - 3)(y + 2)$

68. $a(x - 4) + 5(x - 4)$
$(x - 4)(a + 5)$

69. $4a(2m + 3n) + b(2m + 3n)$
$(2m + 3n)(4a + b)$

70. $3m(3a + 5b) + n(3a + 5b)$
$(3a + 5b)(3m + n)$

71. $5x(4m - 5) - 2(4m - 5)$
$(4m - 5)(5x - 2)$

72. $3m(3a - 2) - 4(3a - 2)$
$(3a - 2)(3m - 4)$

★**73.** $2r(6p + 5q) - 4s(6p + 5q)$
$2(6p + 5q)(r - 2s)$

★**74.** $6x(3y + 5z) - 3w(3y + 5z)$
$3(3y + 5z)(2x - w)$

Objective 4

Prep Exercise 5 What kinds of polynomials are factored by grouping?
Four-term polynomials

Prep Exercise 6 When factoring by grouping, after factoring out a monomial GCF, what is the next step?
Group together pairs of terms and factor the GCF out of each pair.

For Exercises 75–98, factor by grouping. See Example 9(a).

75. $bx + 2b + cx + 2c$
$(x + 2)(b + c)$

76. $cx + cy + bx + by$
$(x + y)(c + b)$

77. $am - an - bm + bn$
$(m - n)(a - b)$

78. $xy - xw - yz + wz$
$(y - w)(x - z)$

79. $x^3 + 2x^2 - 3x - 6$
$(x + 2)(x^2 - 3)$

80. $y^4 + 4y^3 - by - 4b$
$(y + 4)(y^3 - b)$

81. $1 - m + m^2 - m^3$
$(1 - m)(1 + m^2)$

82. $1 + x + x^2 + x^3$
$(1 + x^2)(1 + x)$

83. $3xy + 5y + 6x + 10$
$(3x + 5)(y + 2)$

84. $2ab + 8a + 3b + 12$
$(b + 4)(2a + 3)$

85. $4b^2 - b + 4b - 1$
$(4b - 1)(b + 1)$

86. $x^3 - 4x^2 + x - 4$
$(x - 4)(x^2 + 1)$

87. $6 + 2b - 3a - ab$
$(3 + b)(2 - a)$

88. $12 + 3y - 4x - xy$
$(3 - x)(4 + y)$

89. $3ax + 6ay + 8by + 4bx$
$(x + 2y)(3a + 4b)$

90. $ac + 2ad + 2bc + 4bd$
$(c + 2d)(a + 2b)$

91. $x^2y - x^2s - ry + rs$
$(y - s)(x^2 - r)$

92. $a^3m - a^3n - bm + bn$
$(m - n)(a^3 - b)$

93. $w^2 + 3wz + 5w + 15z$
$(w + 3z)(w + 5)$

94. $a^2 + 4ab + 3a + 12b$
$(a + 4b)(a + 3)$

95. $3st + 3ty - 2s - 2y$
$(s + y)(3t - 2)$

96. $5ab + 5ac - 3b - 3c$
$(b + c)(5a - 3)$

97. $ax^2 - 5y^2 + ay^2 - 5x^2$
$(x^2 + y^2)(a - 5)$

98. $m^2p - 3n^2 + n^2p - 3m^2$
$(m^2 + n^2)(p - 3)$

For Exercises 99–108, factor completely. See Example 9(b).

99. $2a^2 + 2ab + 6a + 6b$
$2(a + b)(a + 3)$

100. $3m^2 + 3mn + 12m + 12n$
$3(m + n)(m + 4)$

101. $12ab + 18a + 16b + 24$
$2(2b + 3)(3a + 4)$

102. $12pq + 8q + 30p + 20$
$2(3p + 2)(2q + 5)$

103. $24xz - 12xw - 6yz + 3yw$
$3(4x - y)(2z - w)$

104. $12ac + 8ad - 18bc - 12bd$
$2(3c + 2d)(2a - 3b)$

105. $3a^2y - 12a^2 + 9ay - 36a$
$3a(y - 4)(a + 3)$

106. $4x^2y - 8x^2 + 20xy - 40x$
$4x(y - 2)(x + 5)$

107. $2x^3 + 6x^2y + 10x^2 + 30xy$
$2x(x + 3y)(x + 5)$

108. $10uv^2 - 5v^2 + 30uv - 15v$
$5v(2u - 1)(v + 3)$

For Exercises 109 and 110, write an expression for the area of the shaded region; then factor completely.

109. 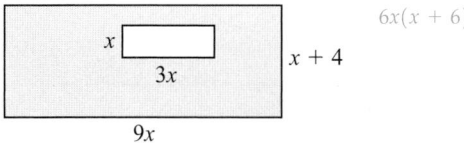 $6x(x + 6)$

110. 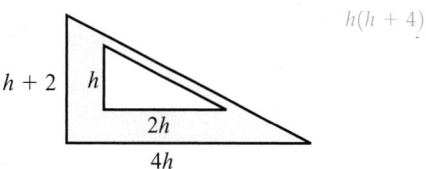 $h(h + 4)$

★ **111.** The diagram shows the floor plan of a room with the hearth of a fireplace. Write an expression in factored form of the area of the room excluding the hearth, which is 4 feet by 1 foot.

$3x(7x + 9)$

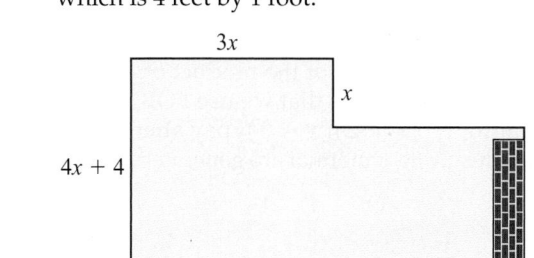

★ **112.** Write an expression in factored form for the area of the side of the house shown excluding the window.

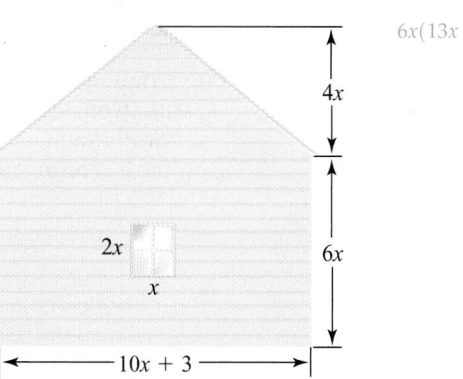 $6x(13x + 4)$

★ *For Exercises 113 and 114, write an expression for the volume of the object shown; then factor completely.*

113.

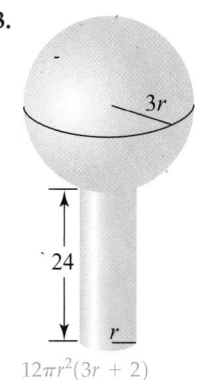

$12\pi r^2(3r + 2)$

114.

$\pi r^2(h + 2)$

Review Exercises

Exercises 1–6   **Expressions**

[5.1] 1. Write 3.74×10^9 in standard form.
3,740,000,000

[5.1] 2. Write 45,600,000 in scientific notation.
4.56×10^7

[5.5] *For Exercises 3–6, multiply.*

3. $(x + 2)(x + 5)$
$x^2 + 7x + 10$

4. $(x - 4)(x + 3)$
$x^2 - x - 12$

5. $(x - 5)(x - 3)$
$x^2 - 8x + 15$

6. $x(x - 7)$
$x^2 - 7x$

6.2 Factoring Trinomials of the Form $x^2 + bx + c$

Objectives

1 Factor trinomials of the form $x^2 + bx + c$.

2 Factor out a monomial GCF; then factor the trinomial of the form $x^2 + bx + c$.

Instructor Note Before beginning this section, you might want to work through several examples using FOIL.

Warm-up

[5.5] *For Exercises 1 and 2, find the products.*
1. $(x + 3)(x - 8)$
2. $3mn(n - 4)(n + 8)$

[6.1] *For Exercises 3 and 4, factor.*
3. $12x^3y^2 - 15xy^5$
4. $x^3 + 8x^2 - 2xy - 16y$

Objective 1 Factor trinomials of the form $x^2 + bx + c$.

In this section, we consider trinomials of the form $x^2 + bx + c$. Notice that the coefficient of the squared term is 1 and the trinomial is in descending order. In Section 6.3, we consider trinomials of this form in which the squared term has a coefficient other than 1. Following are some examples of trinomials of the form $x^2 + bx + c$.

$$x^2 + 5x + 6 \qquad x^2 - 7x + 12 \qquad x^2 + 2x - 15 \qquad x^2 - 5x - 24$$

Trinomials of the form $x^2 + bx + c$ are often the result of the product of two binomials. Consider $(x + 2)(x + 3)$. Recall from Section 5.5 that we use FOIL to multiply two binomials. In the use of FOIL to multiply $(x + 2)(x + 3)$, pay attention to how the last term and the middle term of the resulting trinomial are generated.

$$
\begin{array}{cccc}
\text{F} & \text{O} & \text{I} & \text{L} \\
\end{array}
$$
$$(x + 2)(x + 3) = x^2 + 3x + 2x + 6$$
$$= x^2 \quad + 5x \quad + 6$$

Note These two numbers multiply to equal the last term, 6, and add to equal the coefficient of the middle term, 5.

sum of 2 and 3   product of 2 and 3

When we factor a trinomial of the form $x^2 + bx + c$, we reverse the FOIL process, using the fact that b is the sum of the last terms in the binomials and c is the product of the last terms in the binomials. For example, to factor $x^2 + 5x + 6$, we must find two binomials with last terms whose product is 6 and whose sum is 5.

$$x^2 + 5x + 6 = (x + \underline{\quad})(x + \underline{\quad})$$

The product of these numbers must be 6, and their sum must be 5.

If unsure of what those two numbers are, we can list all factor pairs of 6 and then find the pair whose sum is 5.

Factors Pairs of 6	Corresponding Sum
$1 \cdot 6$	$1 + 6 = 7$
$2 \cdot 3$	$2 + 3 = 5$ ← This is the correct pair.

Notice that the 2 and 3 become the last terms in the binomials so that the factored form for $x^2 + 5x + 6$ looks like this:

$$x^2 + 5x + 6 = (x + 2)(x + 3)$$

This suggests the following procedure.

Answers to Warm-up
1. $x^2 - 5x - 24$
2. $3mn^3 + 12mn^2 - 96mn$
3. $3xy^2(4x^2 - 5y^3)$
4. $(x + 8)(x^2 - 2y)$

> **Procedure** Factoring $x^2 + bx + c$
>
> To factor a trinomial of the form $x^2 + bx + c$:
> 1. Find two numbers whose product is c and whose sum is b.
> 2. The factored trinomial will have the form $(x + \text{first number})$ $(x + \text{second number})$.
>
> *Note:* The signs in the binomial factors can be minus signs depending on the signs of b and c.

In the following examples, we explore cases where minus signs occur. Pay attention to how we determine the signs in the binomial factors.

Example 1	Factor.

a. $x^2 - 7x + 12$

Solution: We must find a pair of numbers whose product is 12 and whose sum is -7. Because the product of the two numbers must be positive and their sum negative, both numbers must be negative. Following is a table listing the products and sums.

Product	Sum	
$(-1)(-12) = 12$	$-1 + (-12) = -13$	This is the correct combination,
$(-2)(-6) = 12$	$-2 + (-6) = -8$	so -3 and -4 are the second
$(-3)(-4) = 12$	$-3 + (-4) = -7$	← terms in each binomial of the factored form.

Answer: $x^2 - 7x + 12 = (x - 3)(x - 4)$

◄ **Note** We could have first written $[x + (-3)][x + (-4)]$ to follow the procedure and then simplify.

Check: We can check by multiplying the binomial factors to see if their product is the original polynomial.

$$(x - 3)(x - 4) = x^2 - 4x - 3x + 12 \quad \text{Multiply the factors using FOIL.}$$
$$= x^2 - 7x + 12 \quad \text{The product is the original polynomial.}$$

b. $x^2 + 2x - 15$

Instructor Note Emphasize that because the coefficient of the middle term is positive, we can tell that the factor with the greater absolute value must be positive. This allows us to quickly eliminate $(1)(-15)$ and $(3)(-5)$ as possibilities.

Solution: We must find a pair of numbers whose product is -15 and whose sum is 2. Because the product is negative, the two numbers must have different signs. Because the sum is positive, the number with the greater absolute value will be positive.

Product	Sum	
$(-1)(15) = -15$	$-1 + 15 = 14$	This is the correct combination, so
$(-3)(5) = -15$	$-3 + 5 = 2$	← -3 and 5 are the second terms in each binomial of the factored form.

Answer: $x^2 + 2x - 15 = (x - 3)(x + 5)$

Check: $(x - 3)(x + 5) = x^2 + 5x - 3x - 15 \quad \text{Multiply the factors using FOIL.}$
$$= x^2 + 2x - 15 \quad \text{The product is the original polynomial.}$$

c. $x^2 - 5x - 24$

Solution: We must find a pair of numbers whose product is -24 and whose sum is -5. Because the product is negative, the two numbers must have different signs. Because the sum also is negative, the number with the greater absolute value will be negative.

Product	Sum
$(1)(-24) = -24$	$1 + (-24) = -23$
$(2)(-12) = -24$	$2 + (-12) = -10$
$(3)(-8) = -24$	$3 + (-8) = -5$

This is the correct combination, so 3 and -8 are the second terms in each binomial of the factored form. ⟵

Answer: $x^2 - 5x - 24 = (x + 3)(x - 8)$

d. $x^2 + 3x + 4$

Solution: We must find a pair of numbers whose product is 4 and whose sum is 3. If two numbers have a positive product and a positive sum, both must be positive.

Product	Sum
$(1)(4) = 4$	$1 + 4 = 5$
$(2)(2) = 4$	$2 + 2 = 4$

Note We listed all positive factors of 4 and found no combination whose sum is 3. This means that $x^2 + 3x + 4$ has no binomial factors with integer coefficients.

A polynomial such as $x^2 + 3x + 4$ that cannot be factored is like a prime number in that its factors are only 1 and the polynomial itself. We say that such a polynomial is *prime*. Notice that the factorization $(x + 4)(x - 1)$ gives the correct middle term, $3x$, but the last term is -4 instead of $+4$.

Note In Examples 1(c)–3, we leave the checks to the reader.

Your Turn 1 Factor.

a. $y^2 - 11y + 18$ **b.** $t^2 + 3t - 28$ **c.** $n^2 - n - 30$ **d.** $x^2 - 2x - 6$

Now let's see how to factor trinomials containing two variables.

Example 2 Factor.

a. $x^2 - 10xy + 25y^2$

Solution: Because y is in the last term, it is helpful to rearrange the order of the variables in the middle term, writing the polynomial as $x^2 - 10yx + 25y^2$, so that we view the "coefficient" of $-10yx$ as $-10y$. We must find a pair of terms whose product is $25y^2$ and whose sum is $-10y$. These terms would have to be $-5y$ and $-5y$.

Answer: $x^2 - 10xy + 25y^2 = (x - 5y)(x - 5y)$ or $(x - 5y)^2$

b. $a^2 + 2ab - 15b^2$

Solution: As in Example 2(a), we can view the coefficient of a in the middle term as $2b$. We must find a pair of terms whose product is $-15b^2$ and whose sum is $2b$. These terms would have to be $-3b$ and $5b$.

Answer: $a^2 + 2ab - 15b^2 = (a - 3b)(a + 5b)$

Your Turn 2 Factor.

a. $m^2 + 7mn - 18n^2$ **b.** $x^2 - xy - 12y^2$

Objective 2 Factor out a monomial GCF; then factor the trinomial of the form $x^2 + bx + c$.

Sometimes there is a monomial GCF (other than 1) in the three terms of a trinomial. Consider $2x^3 + 10x^2 + 12x$. Notice that the monomial $2x$ is the GCF of the terms. Factoring out this monomial, we have

$$2x^3 + 10x^2 + 12x = 2x(x^2 + 5x + 6).$$

Answers to Your Turn 1
a. $(y - 2)(y - 9)$
b. $(t - 4)(t + 7)$
c. $(n + 5)(n - 6)$
d. prime

Answers to Your Turn 2
a. $(m + 9n)(m - 2n)$
b. $(x + 3y)(x - 4y)$

Now we try to factor the trinomial within the parentheses. We look for two numbers whose product is 6 and whose sum is 5. Note that 2 and 3 work, so we can write the following:

$$2x^3 + 10x^2 + 12x = 2x(x^2 + 5x + 6)$$
$$= 2x(x + 2)(x + 3) \quad \leftarrow \boxed{\text{Factored form}}$$

Whenever factoring polynomials, the first step always should be to look for a monomial GCF among the terms.

Instructor Note Remind students to include the common factor as a factor of their answer.

Example 3   Factor.

a. $3mn^3 + 12mn^2 - 96mn$

Solution: First, we look for a monomial GCF (other than 1). Notice that the GCF of the terms is $3mn$. Factoring out this monomial, we have the following:

$$3mn^3 + 12mn^2 - 96mn = 3mn(n^2 + 4n - 32)$$

Now try to factor the trinomial to two binomials. We must find a pair of numbers whose product is -32 and whose sum is 4.

Product	Sum
$(-1)(32) = -32$	$-1 + 32 = 31$
$(-2)(16) = -32$	$-2 + 16 = 14$
$(-4)(8) = -32$	$-4 + 8 = 4$

This is the correct combination, so -4 and 8 are the second $\leftarrow$ terms in each binomial.

Answer: $3mn^3 + 12mn^2 - 96mn = 3mn(n - 4)(n + 8)$

b. $x^4 + 6x^3 + 14x^2$

Solution: First, factor out the monomial GCF, x^2.

$$x^4 + 6x^3 + 14x^2 = x^2(x^2 + 6x + 14)$$

Now, try to factor the trinomial to two binomials by finding a pair of numbers whose product is 14 and whose sum is 6.

Note When $x^2 + 6x + 14$ is factored, the product must be positive 14 and the sum must be positive 6. This means that the second terms of both binomial factors must be positive. So neither -1 and -14 nor -2 and -7 are possible.

Product	Sum
$(1)(14) = 14$	$1 + 14 = 15$
$(2)(7) = 14$	$2 + 7 = 9$
$(-1)(-14) = 14$	$-1 + (-14) = -15$
$(-2)(-7) = 14$	$-2 + (-7) = -9$

Note We listed all factor pairs of 14 and found no combination whose sum is 6. This means that $x^2 + 6x + 14$ has no binomial factors with integer terms; so $x^2(x^2 + 6x + 14)$ is the final factored form.

Answers to Your Turn 3
a. $y^2(y - 9)(y - 1)$
b. $4ab(a - 7)(a + 2)$
c. $4x(y^2 - 2y + 3)$

Your Turn 3   Factor.

a. $y^4 - 10y^3 + 9y^2$ **b.** $4a^3b - 20a^2b - 56ab$ **c.** $4xy^2 - 8xy + 12x$

6.2 Exercises   For Extra Help   MyMathLab®

Note: Exercises marked with a ★ represent challenging exercises.

Objective 1

Prep Exercise 1 To factor $x^2 + 6x + 8$, find two numbers whose product is __8__ and whose sum is __6__.

Prep Exercise 2 To factor $x^2 - 4x - 12$, find two numbers whose product is __−12__ and whose sum is __−4__.

For Exercises 1–8, fill in the missing values in the factors. See Objective 1.

1. $x^2 + 7x + 10 = (x + 2)(x + \boxed{})$
   5

2. $y^2 + 8y + 15 = (y + 5)(y + \boxed{})$
   3

3. $n^2 - 12n + 20 = (n - 10)(n - \boxed{})$
   2

4. $t^2 - 9t + 18 = (t - \boxed{})(t - 3)$
   6

5. $m^2 - 2m - 24 = (m - \boxed{})(m + 4)$
   6

6. $a^2 + a - 12 = (a + 4)(a - \boxed{})$
   3

7. $u^2 + 13u - 30 = (u - \boxed{})(u + 15)$
   2

8. $n^2 - 9n - 36 = (n - \boxed{})(n + 3)$
   12

Prep Exercise 3 Complete the factored form by inserting the appropriate signs.
$$x^2 - 8x + 15 = (x \underline{} 3)(x \underline{} 5)$$

Prep Exercise 4 Complete the factored form by inserting the appropriate signs.
$$x^2 - 5x - 6 = (x \underline{} 6)(x \underline{\pm} 1)$$

Prep Exercise 5 Complete the factored form by inserting the appropriate signs.
$$x^2 + 8x - 20 = (x \underline{\pm} 10)(x \underline{} 2)$$

For Exercises 9–40, factor. If the polynomial is prime, so state. See Example 1.

9. $r^2 + 4r + 3$
   $(r + 3)(r + 1)$

10. $t^2 + 8t + 7$
    $(t + 1)(t + 7)$

11. $x^2 - 8x + 7$
    $(x - 7)(x - 1)$

12. $x^2 - 4x + 3$
    $(x - 3)(x - 1)$

13. $z^2 - 2z - 3$
    $(z - 3)(z + 1)$

14. $y^2 - 4y - 5$
    $(y - 5)(y + 1)$

15. $y^2 + 5y + 6$
    $(y + 2)(y + 3)$

16. $n^2 + 8n + 15$
    $(n + 3)(n + 5)$

17. $u^2 - 6u + 8$
    $(u - 2)(u - 4)$

18. $b^2 - 10b + 21$
    $(b - 3)(b - 7)$

19. $u^2 + u - 6$
    $(u + 3)(u - 2)$

20. $k^2 - 3k - 10$
    $(k - 5)(k + 2)$

21. $a^2 + 7a + 12$
    $(a + 3)(a + 4)$

22. $x^2 + 9x + 18$
    $(x + 3)(x + 6)$

23. $y^2 - 6y + 9$
    $(y - 3)^2$

24. $x^2 - 10x + 25$
    $(x - 5)^2$

25. $w^2 - w - 12$
    $(w - 4)(w + 3)$

26. $b^2 - 3b - 40$
    $(b - 8)(b + 5)$

27. $x^2 - x - 30$
    $(x - 6)(x + 5)$

28. $b^2 - 6b - 16$
    $(b - 8)(b + 2)$

29. $n^2 - 11n + 30$
    $(n - 6)(n - 5)$

30. $y^2 - 7y + 12$
    $(y - 3)(y - 4)$

31. $r^2 - 9r + 18$
    $(r - 3)(r - 6)$

32. $x^2 - 10x + 24$
    $(x - 6)(x - 4)$

33. $x^2 - 5x - 24$
    $(x - 8)(x + 3)$

34. $a^2 - 10a - 24$
    $(a - 12)(a + 2)$

35. $p^2 - 5p - 36$
    $(p - 9)(p + 4)$

36. $u^2 - 8u - 20$
    $(u - 10)(u + 2)$

37. $m^2 - 4m + 6$
    prime

38. $z^2 + 7z + 8$
    prime

39. $x^2 - 6x - 8$
    prime

40. $a^2 + 3a - 20$
    prime

For Exercises 41–48, factor the trinomials containing two variables. If the polynomial is prime, so state. See Example 2.

41. $p^2 - 10pq + 21q^2$
    $(p - 3q)(p - 7q)$

42. $x^2 - 9xy + 14y^2$
    $(x - 2y)(x - 7y)$

43. $a^2 - 6ab - 27b^2$
    $(a - 9b)(a + 3b)$

44. $m^2 + 2mn - 8n^2$
    $(m + 4n)(m - 2n)$

45. $x^2 - 14xy + 24y^2$
    $(x - 12y)(x - 2y)$

46. $t^2 - 11tu + 24u^2$
    $(t - 8u)(t - 3u)$

47. $r^2 - rs - 30s^2$
    $(r - 6s)(r + 5s)$

48. $h^2 - 8hk - 20k^2$
    $(h - 10k)(h + 2k)$

Objective 2

Prep Exercise 6 To factor $6x^2y + 30xy - 36y$, we must first factor out the GCF of all the terms. What is that GCF? $6y$

For Exercises 49–68, factor completely. See Example 3.

49. $4x^2 - 40x + 84$
$4(x - 3)(x - 7)$

50. $3m^2 - 33m + 54$
$3(m - 9)(m - 2)$

51. $2m^3 - 14m^2 + 12m$
$2m(m - 1)(m - 6)$

52. $2k^2y - 18ky + 28y$
$2y(k - 7)(k - 2)$

53. $3a^2b - 15ab - 72b$
$3b(a - 8)(a + 3)$

54. $3a^2y + 6ay - 72y$
$3y(a + 6)(a - 4)$

55. $4x^2 - 24x + 36$
$4(x - 3)^2$

56. $6a^2r + 12ar + 6r$
$6r(a + 1)^2$

57. $n^4 + 5n^3 + 6n^2$
$n^2(n + 2)(n + 3)$

58. $r^4 + 6r^3 + 8r^2$
$r^2(r + 4)(r + 2)$

59. $7u^4 + 42u^3 + 35u^2$
$7u^2(u + 5)(u + 1)$

60. $5x^5 + 20x^4 + 15x^3$
$5x^3(x + 3)(x + 1)$

61. $6a^2b^2c - 36ab^2c + 48b^2c$
$6b^2c(a - 2)(a - 4)$

62. $3a^2b^2c - 24ab^2c + 45b^2c$
$3b^2c(a - 3)(a - 5)$

63. $3x^2 - 21xy + 30y^2$
$3(x - 2y)(x - 5y)$

64. $5m^2 - 30mn + 40n^2$
$5(m - 4n)(m - 2n)$

65. $2a^2b - 6ab^2 - 36b^3$
$2b(a - 6b)(a + 3b)$

66. $7h^2k + 14hk^2 - 168k^3$
$7k(h + 6k)(h - 4k)$

67. $4x^4y - 12x^3y^2 - 60x^2y^3$
$4x^2y(x^2 - 3xy - 15y^2)$

68. $3a^3b^2 + 15a^2b^3 - 12ab^4$
$3ab^2(a^2 + 5ab - 4b^2)$

Find ⊗ the Mistake   *For Exercises 69–72, find and correct the mistake. See Example 1.*

69. $x^2 - 5x - 6 = (x - 3)(x - 2)$
$(x - 6)(x + 1)$

70. $x^2 - 5x + 6 = (x - 6)(x + 1)$
$(x - 2)(x - 3)$

71. $x^2 - 3x - 4 = (x - 1)(x + 4)$
$(x - 4)(x + 1)$

72. $x^2 + x + 2 = (x + 2)(x + 1)$
prime

★ *For Exercises 73–76, find all natural number values of* **b** *that make the trinomial factorable.*

73. $x^2 + bx - 21$
4, 20

74. $x^2 + bx - 10$
3, 9

75. $x^2 + bx + 12$
7, 8, 13

76. $x^2 + bx + 18$
9, 11, 19

★ *For Exercises 77–80, find all natural numbers* **c** *that make the trinomial factorable.*

77. $x^2 - 7x + c$
6, 10, 12

78. $x^2 - 5x + c$
4, 6

79. $x^2 + 10x + c$
9, 16, 21, 24, 25

80. $x^2 + 11x + c$
10, 18, 24, 28, 30

81. The expression $h^2 + 6h + 8$ describes the area of the top (or bottom) of the crate shown, where h represents its height. The unknown expression for the length is the sum of h and an integer, and the expression for the width is the sum of h and a different integer. Find expressions for the length and width.
Length: $h + 4$; width: $h + 2$

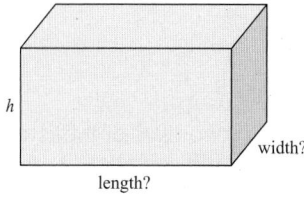

82. The expression $w^2 + w - 2$ describes the area of the top of the bar of gold shown, where w represents the width of the base of the bar. The unknown expression for the length of the top of the bar is the sum of w and an integer. The expression for the width of the top of the bar is the difference of w and an integer. Find the expressions for the length and width of the top of the bar.
Length: $w + 2$; width: $w - 1$

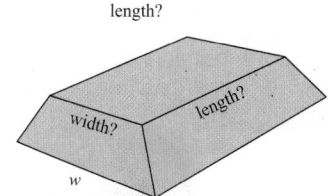

83. The expression $w^2 - 16w + 60$ describes the viewing area in the picture frame shown, where w represents the width of the frame. The unknown expression for the length of the viewing area is the difference of w and an integer. The expression for the width of the viewing area is the difference of w and a different integer. Find expressions for the length and width of the viewing area in the frame.
Length: $w - 6$; width: $w - 10$

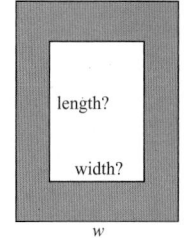

★ **84.** The expression $\pi r^2 - 2\pi r + \pi$ describes the area occupied by the circular base of the papasan chair shown, where r represents the radius of the circle on which the chair rests. The expression that describes the radius of the circle that touches the floor is the difference of r and an integer. Find the expression that describes the radius of the circle that touches the floor.

$r - 1$

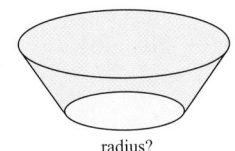

radius?

Review Exercises

Exercises 1–10   **Expressions**

[6.1] 1. What is the GCF of $6x$ and $35y$?

1

[5.5] *For Exercises 2–4, multiply.*

2. $(2x + 3)(5x + 1)$
$10x^2 + 17x + 3$

3. $(3y - 4)(5y + 6)$
$15y^2 - 2y - 24$

4. $4n(6n - 1)(3n - 2)$
$72n^3 - 60n^2 + 8n$

[6.1] *For Exercises 5–10, factor completely.*

5. $10x^2y^3 - 5x^3y - 20x^2y^2$
$5x^2y(2y^2 - x - 4y)$

6. $8a^2b^4 - 16ab^2 - 12a^2b^3$
$4ab^2(2ab^2 - 4 - 3ab)$

7. $x(x + 2) - 3(x + 2)$
$(x + 2)(x - 3)$

8. $z(z + 5) - 6(z + 5)$
$(z + 5)(z - 6)$

9. $15ac - 5ad - 3bc + bd$
$(3c - d)(5a - b)$

10. $6ac + 4ad - 9bc - 6bd$
$(3c + 2d)(2a - 3b)$

6.3 Factoring Trinomials of the Form $ax^2 + bx + c$, where $a \neq 1$

Objectives

1 Factor trinomials of the form $ax^2 + bx + c$, where $a \neq 1$, by trial.

2 Factor trinomials of the form $ax^2 + bx + c$, where $a \neq 1$, by grouping.

Warm-up

[5.5] *For Exercises 1 and 2, find the products.*
1. $(3x + 2)(x + 5)$
2. $4x(5x - 1)(2x - 3)$
[6.1] 3. Factor by grouping: $8x^2 - 2x - 20x + 5$
[6.2] 4. Factor: $x^2 + 4x - 12$

Objective 1 Factor trinomials of the form $ax^2 + bx + c$, where $a \neq 1$, by trial.

In Section 6.2, we factored trinomials in which the coefficient of the squared term was 1. Now we focus on factoring trinomials in which the coefficient of the squared term is other than 1, such as the following:

$$3x^2 + 17x + 10 \qquad 8x^2 + 29x - 12$$

In general, like trinomials of the form $x^2 + bx + c$, trinomials of the form $ax^2 + bx + c$, where $a \neq 1$, may also have two binomial factors. When a is not 1, we must consider

Answers to Warm-up
1. $3x^2 + 17x + 10$
2. $40x^3 - 68x^2 + 12x$
3. $(4x - 1)(2x - 5)$
4. $(x + 6)(x - 2)$

the factors of the ax^2 term and the factors of the c term. First, we develop a trial-and-error method for finding the factored form. Consider $3x^2 + 17x + 10$. Because all of the terms are positive, we know that all of the terms in the binomial factors will be positive.

The *first* terms must multiply to equal $3x^2$. The logical factors are $3x$ and x.

$$3x^2 + 17x + 10 = (\quad + \quad)(\quad + \quad)$$

The *last* terms must multiply to equal 10. These factors could be 1 and 10, or 2 and 5.

We now try various combinations of these *first* and *last* terms in binomial factors and consider the resulting trinomial products. If the sum of the inner and outer terms matches the original trinomial, we know we have the correct combination in the binomial factors.

$(3x + 10)(x + 1) = 3x^2 + 3x + 10x + 10 = 3x^2 + 13x + 10$

$(3x + 1)(x + 10) = 3x^2 + 30x + x + 10 = 3x^2 + 31x + 10$ — Incorrect combinations

$(3x + 5)(x + 2) = 3x^2 + 6x + 5x + 10 = 3x^2 + 11x + 10$

$(3x + 2)(x + 5) = 3x^2 + 15x + 2x + 10 = 3x^2 + 17x + 10$ Correct combination

You may find that drawing lines to connect each product is helpful.

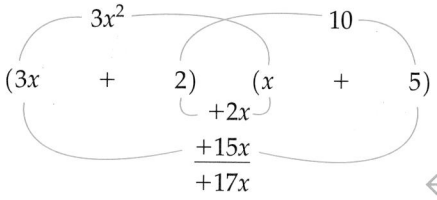

$$(3x \quad + \quad 2) \quad (x \quad + \quad 5)$$
$$+2x$$
$$+15x$$
$$+17x$$

The sum of the *inner* and *outer* products verifies that this is the correct combination.

Our example suggests the following procedure.

Procedure Factoring by Trial and Error

To factor a trinomial of the form $ax^2 + bx + c$, where $a \neq 1$, by trial and error:

1. Factor out any monomial GCF common to all of the terms.

2. Write a pair of *first* terms whose product is ax^2.

$$\overset{ax^2}{(\Box + \quad)(\Box + \quad)}$$

3. Write a pair of *last* terms whose product is c.

$$\overset{c}{(\Box + \Box)(\Box + \Box)}$$

$$(\Box + \Box) \quad (\Box + \Box)$$

4. Verify that the sum of the *inner* and *outer* products is bx (the middle term of the trinomial).

Inner
$$+\text{Outer}$$
$$bx$$

If the sum of the inner and outer products is *not* bx, try the following:

a. Exchange the last terms of the binomials from step 3; then repeat step 4.
b. For each additional pair of last terms, repeat steps 3 and 4.
c. For each additional pair of first terms, repeat steps 2–4.

Example 1   Factor.

a. $3x^2 + 5x + 2$

The *first* terms must multiply to equal $3x^2$.
These must be $3x$ and x.

Solution: $3x^2 + 5x + 2 = (\quad + \quad)(\quad + \quad)$

The last terms must multiply to equal 2, which can only be 1 and 2.
Because the product of the last terms is positive and the sum of the
outside and inside products must be $+$, both signs of the binomial
must be $+$.

$(3x + 1)(x + 2) = 3x^2 + 6x + x + 2 = 3x^2 + 7x + 2$ Incorrect combination
$(3x + 2)(x + 1) = 3x^2 + 3x + 2x + 2 = 3x^2 + 5x + 2$ Correct combination

Answer: $3x^2 + 5x + 2 = (3x + 2)(x + 1)$

b. $2x^2 - 7x + 6$

The *first* terms must multiply to $2x^2$.
These must be $2x$ and x.

Solution: $2x^2 - 7x + 6 = (\quad + \quad)(\quad - \quad)$

The *last* terms must multiply to equal $+6$, which could be 1 and 6, or 2 and 3.
Because the product of the last terms is positive and the sum of the outside and
inside products is $-$, both signs of the binomial must be $-$.

$(2x - 1)(x - 6) = 2x^2 - 12x - x + 6 = 2x^2 - 13x + 6$ Incorrect combination
$(2x - 6)(x - 1) = 2x^2 - 2x - 6x + 6 = 2x^2 - 8x + 6$ Incorrect combination
$(2x - 2)(x - 3) = 2x^2 - 6x - 2x + 6 = 2x^2 - 8x + 6$ Incorrect combination
$(2x - 3)(x - 2) = 2x^2 - 4x - 3x + 6 = 2x^2 - 7x + 6$ Correct combination

Answer: $2x^2 - 7x + 6 = (2x - 3)(x - 2)$

Note Because the original
polynomial does not have a
common factor, no factors can
have a common factor. This means
that $2x - 6$ and $2x - 2$ cannot
be factors because both have a
common factor of 2.

c. $8x^2 + 29x - 12$

Solution:

The *first* terms must multiply to equal $8x^2$.
These could be x and $8x$, or $2x$ and $4x$.

$8x^2 + 29x - 12 = (\quad + \quad)(\quad - \quad)$

The *last* terms must multiply to equal -12. Because -12 is
negative, the last terms in the binomials must have different
signs. We have already written the appropriate signs, so these
factor pairs could be 1 and 12, 2 and 6, or 3 and 4.

Now we multiply binomials with various combinations of these first and last terms
until we find a combination whose inner and outer products combine to equal $29x$.

$(8x + 1)(x - 12) = 8x^2 - 96x + x - 12 = 8x^2 - 95x - 12$
$(8x - 1)(x + 12) = 8x^2 + 96x - x - 12 = 8x^2 + 95x - 12$
$(8x + 12)(x - 1) = 8x^2 - 8x + 12x - 12 = 8x^2 + 4x - 12$ Incorrect
$(8x + 2)(x - 6) = 8x^2 - 48x + 2x - 12 = 8x^2 - 46x - 12$ combinations
$(8x + 6)(x - 2) = 8x^2 - 16x + 6x - 12 = 8x^2 - 10x - 12$
$(8x + 3)(x - 4) = 8x^2 - 32x + 3x - 12 = 8x^2 - 29x - 12$

Note The products
$(8x + 1)(x - 12)$ and
$(8x - 1)(x + 12)$ differ only in
the sign of the middle term. So
unless a factorization is correct
except for the sign of the middle
term of the product, switching
the signs of the binomial factors
will not result in the correct
factorization.

Note In the above factorizations,
the factors $8x + 12$, $8x + 2$, and
$8x + 6$ are not possible because
each has a common factor.

Note This middle term has the correct
absolute value but the incorrect sign. (We
are looking for $+29x$.) Switching the $+$ and
$-$ signs in the binomials will give us $+29x$.

Instructor Note Show that if we multiply the coefficient of the first and last terms in $8x^2 + 29x - 12$, we have $(8)(12) = 96$. With every combination of binomial factors, after multiplying, the coefficients of the like terms are factors of 96 (96 and 1, 8 and 12, 48 and 2, 16 and 6, 32 and 3). Explain that we will use this fact when we consider factoring these trinomials using grouping. Also point out that we did not attempt to factor using first terms of $2x$ and $4x$ because that would always result in factors with a common factor.

$(8x - 3)(x + 4) = 8x^2 + 32x - 3x - 12 = 8x^2 + 29x - 12$ Correct combination

Answer: $8x^2 + 29x - 12 = (8x - 3)(x + 4)$

Your Turn 1   Factor.

a. $5x^2 + 9x + 4$
b. $6c^2 - 19cd + 15d^2$
c. $6y^2 + 13y - 5$

Factoring Out a Monomial GCF First

Remember that we should always look to see if a monomial GCF can be factored out of the terms in the trinomial. This is step 1 for every factoring problem.

Example 2   Factor $40x^3 - 68x^2 + 12x$.

Solution: First, we factor out the monomial GCF, $4x$.

$$40x^3 - 68x^2 + 12x = 4x(10x^2 - 17x + 3)$$

Now we factor the trinomial within the parentheses. The fact that 3 is positive and $-17x$ is negative indicates that both signs of the last terms in the binomial factors will be minus signs.

The *first* terms must multiply to equal $10x^2$.
These could be x and $10x$, or $2x$ and $5x$.

$$4x(10x^2 - 17x + 3) = 4x(\quad - \quad)(\quad - \quad)$$

The *last* terms must multiply to equal 3. Because 3 is a prime number, its factors are 1 and 3.

Now we multiply binomials with various combinations of these first and last terms until we find a combination whose inner and outer products combine to equal $-17x$.

$4x(10x - 3)(x - 1) = 4x(10x^2 - 10x - 3x + 3) = 4x(10x^2 - 13x + 3)$
$4x(10x - 1)(x - 3) = 4x(10x^2 - 30x - x + 3) = 4x(10x^2 - 31x + 3)$ ⎤ Incorrect
$4x(5x - 3)(2x - 1) = 4x(10x^2 - 5x - 6x + 3) = 4x(10x^2 - 11x + 3)$ ⎦ combinations

$4x(5x - 1)(2x - 3) = 4x(10x^2 - 15x - 2x + 3) = 4x(10x^2 - 17x + 3)$ Correct combination

Answer: $40x^3 - 68x^2 + 12x = 4x(5x - 1)(2x - 3)$

Your Turn 2   Factor.

a. $16y^3 - 102y^2 + 36y$
b. $36m^2n - 120mn^2 - 21n^3$

Objective 2 Factor trinomials of the form $ax^2 + bx + c$, where $a \neq 1$, by grouping.

Instructor Note Illustrate that b is equal to the sum of two factors of ac by referring to the list of trial and error for Example 1.

The major drawback in using trial and error to factor is that it can be tedious to find the correct combination. An alternative method is to factor by grouping, which we introduced in Section 6.1. Recall that in the grouping method, we group pairs of terms in a four-term polynomial, then factor out the GCF from each pair of terms. Because a trinomial of the form $ax^2 + bx + c$ has only three terms, we split the bx term into two like terms to create a four-term polynomial that we can factor by grouping. To determine how to split the bx term, we use the fact that if $ax^2 + bx + c$ is factorable, then b will always equal the sum of a pair of factors of the product of a and c.

Answers to Your Turn 1
a. $(5x + 4)(x + 1)$
b. $(3c - 5d)(2c - 3d)$
c. $(2y + 5)(3y - 1)$

Answers to Your Turn 2
a. $2y(y - 6)(8y - 3)$
b. $3n(6m + n)(2m - 7n)$

Consider again the trinomial $3x^2 + 17x + 10$, which we factored by trial and error at the beginning of this section. Notice that $a = 3$, $b = 17$, and $c = 10$; so the product of a and c is $(3)(10) = 30$. To split the bx term, which is $17x$ in this case, we look for a pair of factors of 30 whose sum is 17. It is helpful to list the factor pairs and their corresponding sums in a table.

Factors of ac	Sum of Factors of ac
$(1)(30) = 30$	$1 + 30 = 31$
$(2)(15) = 30$	$2 + 15 = 17$
$(3)(10) = 30$	$3 + 10 = 13$
$(5)(6) = 30$	$5 + 6 = 11$

← Notice that 2 and 15 is the only factor pair of 30 whose sum is 17.

Instructor Note Show that we also can write the four-term polynomial as follows:

$$3x^2 + 2x + 15x + 10$$
$$= x(3x + 2) + 5(3x + 2)$$
$$= (3x + 2)(x + 5)$$

Now we can write $17x$ as $15x + 2x$ and then factor by grouping.

$$3x^2 + 17x + 10 = 3x^2 + 15x + 2x + 10$$
$$= 3x(x + 5) + 2(x + 5)$$
$$= (x + 5)(3x + 2)$$

◄ **Note** We also could have written $17x$ as $2x + 15x$ to get the same result.

Every trinomial factorable by grouping has only one factor pair of ac whose sum is b, which suggests the following procedure.

Procedure Factoring $ax^2 + bx + c$, where $a \neq 1$, by Grouping

To factor a trinomial of the form $ax^2 + bx + c$, where $a \neq 1$, by grouping:
1. Factor out any monomial GCF common to all of the terms.
2. Find two factors of ac whose sum is b.
3. Write a four-term polynomial in which bx is written as the sum of two like terms whose coefficients are the two factors you found in step 2.
4. Factor by grouping.

Example 3 | Factor by grouping.

a. $8x^2 - 22x + 5$

Solution: Notice that for this trinomial, $a = 8$, $b = -22$, and $c = 5$. The product ac is $(8)(5) = 40$. So we need two factors of 40 whose sum is -22. Because the product is positive and the sum is negative, both factors must be negative. It is helpful to list the combinations in a table.

Factors of ac	Sum of Factors of ac
$(-1)(-40) = 40$	$-1 + (-40) = -41$
$(-2)(-20) = 40$	$-2 + (-20) = -22$
$(-4)(-10) = 40$	$-4 + (-10) = -14$
$(-5)(-8) = 40$	$-5 + (-8) = -13$

← Correct

Instructor Note Show that if we write $8x^2 - 20x - 2x + 5$, we can still factor by grouping:

$$8x^2 - 20x - 2x + 5$$
$$= 4x(2x - 5) - 1(2x - 5)$$
$$= (2x - 5)(4x - 1)$$

The difficulty is in recognizing that we can factor out the -1 to get matching binomial factors. Remind students that if they have trouble "seeing" what to factor out of the two groups, they should try exchanging the like terms.

You do not need to list all possible combinations as we did here. We listed them to illustrate that only one combination is correct. We now write $-22x$ as $-2x - 20x$ and then factor by grouping.

$$8x^2 - 22x + 5 = 8x^2 - 2x - 20x + 5 \qquad \text{Write } -22x \text{ as } -2x - 20x.$$
$$= 2x(4x - 1) - 5(4x - 1) \qquad \text{Factor } 2x \text{ out of } 8x^2 - 2x;$$
$$\qquad\qquad\qquad\qquad\qquad\qquad \text{factor } -5 \text{ out of } -20x + 5.$$
$$= (4x - 1)(2x - 5) \qquad \text{Factor out } (4x - 1).$$

Check: $(4x - 1)(2x - 5) = 8x^2 - 20x - 2x + 5 = 8x^2 - 22x + 5$

b. $6a^2 + 13a + 6$

Solution: For this trinomial, $a = 6, b = 13$, and $c = 6$. The product ac is $(6)(6) = 36$. So we need two factors of 36 whose sum is 13. Because the product and sum are positive, both factors are positive.

Factors of ac	Sum of Factors of ac	
$(1)(36) = 36$	$1 + 36 = 37$	
$(2)(18) = 36$	$2 + 18 = 20$	
$(3)(12) = 36$	$3 + 12 = 15$	
$(4)(9) = 36$	$4 + 9 = 13$	← Correct

Now write $13a$ as $4a + 9a$ and then factor by grouping.

$$6a^2 + 13a + 6 = 6a^2 + 4a + 9a + 6 \qquad \text{Write } 13a \text{ as } 4a + 9a.$$
$$= 2a(3a + 2) + 3(3a + 2) \qquad \text{Factor } 2a \text{ out of } 6a^2 + 4a;$$
$$\text{factor 3 out of } 9a + 6.$$
$$= (3a + 2)(2a + 3) \qquad \text{Factor out } (3a + 2).$$

Check: $(3a + 2)(2a + 3) = 6a^2 + 9a + 4a + 6 = 6a^2 + 13a + 6$

c. $18y^4 + 15y^3 - 12y^2$

Solution: Notice that there is a monomial GCF, $3y^2$, that we can factor out.

$$18y^4 + 15y^3 - 12y^2 = 3y^2(6y^2 + 5y - 4)$$

Now we factor the trinomial within the parentheses. For this trinomial, $a = 6$, $b = 5$, and $c = -4$. The product ac is $(6)(-4) = -24$. So we need two factors of -24 whose sum is 5. Because the product is negative, the two factors will have different signs. Because the sum is positive, the factor with the greater absolute value must be positive.

Factors of ac	Sum of Factors of ac	
$(-1)(24) = -24$	$-1 + 24 = 23$	
$(-2)(12) = -24$	$-2 + 12 = 10$	
$(-3)(8) = -24$	$-3 + 8 = 5$	← Correct

Now write $5y$ as $-3y + 8y$ and then factor by grouping.

$$3y^2(6y^2 + 5y - 4) = 3y^2(6y^2 - 3y + 8y - 4) \qquad \text{Write } 5y \text{ as } -3y + 8y.$$
$$= 3y^2[3y(2y - 1) + 4(2y - 1)] \qquad \text{Factor } 3y \text{ out of } 6y^2 - 3y;$$
$$\text{factor 4 out of } 8y - 4.$$
$$= 3y^2(2y - 1)(3y + 4) \qquad \text{Factor out } (2y - 1).$$

Check: The check is left to the reader.

Answers to Your Turn 3
a. $(3x - 2)(2x - 5)$
b. $(3x - 4)(2x + 3)$
c. $4y(2x - 9)(x + 2)$

Your Turn 3 Factor by grouping.

a. $6x^2 - 19x + 10$ **b.** $6x^2 + x - 12$ **c.** $8x^2y - 20xy - 72y$

6.3 Exercises   For Extra Help   MyMathLab®

Note: Exercises marked with a ★ represent challenging exercises.

Objective 1

Prep Exercise 1 In factoring a polynomial, the first step is to factor out any _monomial GCF common to all of the terms_ .

Prep Exercise 2 Given a trinomial of the form $ax^2 + bx + c$, where $a \neq 1$, explain how to determine the first terms in its binomial factors. Write a pair of first terms whose product is ax^2.

Prep Exercise 3 Given a trinomial of the form $ax^2 + bx + c$, where $a \neq 1$, explain how to determine the last terms in its binomial factors. Write a pair of last terms whose product is c.

For Exercises 1–6, fill in the missing values in the factors. See Objective 1.

1. $3y^2 + 14y + 8 = (3y + \quad)(y + \quad)$
   2, 4

2. $6x^2 + 11x + 4 = (2x + \quad)(3x + \quad)$
   1, 4

3. $4t^2 + 19t - 30 = (\quad + 6)(\quad - 5)$
   $t, 4t$

4. $8m^2 - 2m - 15 = (\quad - 3)(\quad + 5)$
   $2m, 4m$

5. $6x^2 - 25x + 14 = (2x - \quad)(\quad - 2)$
   $7, 3x$

6. $9n^2 - 56n + 12 = (9n - \quad)(\quad - 6)$
   $2, n$

For Exercises 7–38, factor completely. If prime, so indicate. See Examples 1 and 2.

7. $2j^2 + 5j + 2$
   $(2j + 1)(j + 2)$

8. $2x^2 + 7x + 3$
   $(2x + 1)(x + 3)$

9. $2y^2 - 3y - 5$
   $(2y - 5)(y + 1)$

10. $3m^2 - 10m + 3$
    $(3m - 1)(m - 3)$

11. $3m^2 - 10m + 8$
    $(3m - 4)(m - 2)$

12. $3y^2 - 17y + 10$
    $(3y - 2)(y - 5)$

13. $6a^2 + 13a + 7$
    $(6a + 7)(a + 1)$

14. $4y^2 + 8y + 3$
    $(2y + 3)(2y + 1)$

15. $6p^2 + 2p + 1$
    prime

16. $6u^2 + 3u - 7$
    prime

17. $4a^2 - 19a + 12$
    $(4a - 3)(a - 4)$

18. $4u^2 - 23u + 15$
    $(4u - 3)(u - 5)$

19. $6x^2 + 19x + 15$
    $(2x + 3)(3x + 5)$

20. $8x^2 + 18x + 9$
    $(4x + 3)(2x + 3)$

21. $16d^2 - 14d - 15$
    $(2d - 3)(8d + 5)$

22. $18a^2 + 9a - 35$
    $(3a + 5)(6a - 7)$

23. $3p^2 + 13pq + 4q^2$
    $(3p + q)(p + 4q)$

24. $2u^2 + 7uv + 6v^2$
    $(2u + 3v)(u + 2v)$

25. $5k^2 - 7kh - 12h^2$
    $(k + h)(5k - 12h)$

26. $3a^2 - 13ab - 30b^2$
    $(3a + 5b)(a - 6b)$

27. $12a^2 - 40ab + 25b^2$
    $(2a - 5b)(6a - 5b)$

28. $6w^2 - 19wv + 10v^2$
    $(2w - 5v)(3w - 2v)$

29. $8m^2 - 27mn + 9n^2$
    $(8m - 3n)(m - 3n)$

30. $9x^2 - 21xy + 10y^2$
    $(3x - 2y)(3x - 5y)$

31. $8x^2y - 4xy - y$
    $y(8x^2 - 4x - 1)$

32. $6a^2b + 3ab + b$
    $b(6a^2 + 3a + 1)$

33. $12y^2 + 24y + 9$
    $3(2y + 1)(2y + 3)$

34. $6a^2 - 20a + 16$
    $2(3a - 4)(a - 2)$

35. $6ab^2 - 20ab + 14a$
    $2a(3b - 7)(b - 1)$

36. $20xy^2 - 50xy + 30x$
    $10x(y - 1)(2y - 3)$

37. $6w^3 + 16w^2v + 8wv^2$
    $2w(3w + 2v)(w + 2v)$

38. $6x^3 + 51x^2y + 90xy^2$
    $3x(x + 6y)(2x + 5y)$

Objective 2

Prep Exercise 4 To factor a polynomial of the form $ax^2 + bx + c$ by grouping, first find two factors of _ac_ whose sum is _b_ .

Prep Exercise 5 What are two factors of 30 whose sum is 11?
6 and 5

Prep Exercise 6 What are two factors of -24 whose sum is -10?
-12 and 2

For Exercises 39–70, factor by grouping. If prime, so state. See Example 3.

39. $3a^2 + 4a + 1$
$(a + 1)(3a + 1)$

40. $5m^2 + 11m + 2$
$(m + 2)(5m + 1)$

41. $2t^2 - 3t + 1$
$(2t - 1)(t - 1)$

42. $3h^2 - 5h + 2$
$(3h - 2)(h - 1)$

43. $3x^2 - 4x - 7$
$(3x - 7)(x + 1)$

44. $3l^2 + 2l - 5$
$(3l + 5)(l - 1)$

45. $2y^2 - y - 6$
$(y - 2)(2y + 3)$

46. $3a^2 + 2a - 8$
$(3a - 4)(a + 2)$

47. $10a^2 - 19a + 7$
$(2a - 1)(5a - 7)$

48. $8m^2 - 10m + 3$
$(2m - 1)(4m - 3)$

49. $8r^2 - 6r - 9$
$(4r + 3)(2r - 3)$

50. $6x^2 + 5x - 6$
$(3x - 2)(2x + 3)$

51. $20x^2 - 23x + 6$
$(5x - 2)(4x - 3)$

52. $12c^2 - 28c + 15$
$(2c - 3)(6c - 5)$

53. $6k^2 + 7jk + 2j^2$
$(2k + j)(3k + 2j)$

54. $9x^2 + 9xy + 2y^2$
$(3x + 2y)(3x + y)$

55. $5x^2 - 26xy + 5y^2$
$(x - 5y)(5x - y)$

56. $2t^2 - 3tu + u^2$
$(2t - u)(t - u)$

57. $10s^2 + st - 2t^2$
$(5s - 2t)(2s + t)$

58. $10x^2 - 33xy - 7y^2$
$(2x - 7y)(5x + y)$

59. $6u^2 + 5uv - 6v^2$
$(3u - 2v)(2u + 3v)$

60. $14u^2 + uv - 4v^2$
$(2u - v)(7u + 4v)$

61. $15y^2 - 13y - 8$
prime

62. $8n^2 - 5n - 9$
prime

63. $4k^2 - 14k + 12$
$2(k - 2)(2k - 3)$

64. $12a^2 + 34a + 20$
$2(6a + 5)(a + 2)$

65. $12m^3 + 10m^2 - 12m$
$2m(2m + 3)(3m - 2)$

66. $36x^2y - 78xy + 36y$
$6y(3x - 2)(2x - 3)$

67. $24x^3 - 64x^2y - 24xy^2$
$8x(x - 3y)(3x + y)$

68. $36x^3 - 93x^2y + 60xy^2$
$3x(4x - 5y)(3x - 4y)$

69. $24x^3 - 20x^2y - 24xy^2$
$4x(2x - 3y)(3x + 2y)$

70. $28m^2n + 2mn - 8n$
$2n(7m + 4)(2m - 1)$

Find ⊗ the Mistake   *For Exercises 71–74, explain the mistake; then factor the trinomial correctly. See Examples 1 and 2.*

71. $6x^2 + 13x + 5 = (6x + 1)(x + 5)$
Mistake: Using *FOIL* to check, we see that the first and last terms check but the inner and outer terms combine to give $31x$ instead of $13x$.
Correct: $(2x + 1)(3x + 5)$

72. $2x^2 + x - 3 = (2x - 3)(x + 1)$
Mistake: Using *FOIL* to check, we see that the first and last terms check but the inner and outer terms combine to give $-x$ instead of x.
Correct: $(2x + 3)(x - 1)$

73. $4n^2 + 12n + 8 = (4n + 8)(n + 1)$
Mistake: Did not factor completely.
Correct: $4(n + 2)(n + 1)$

74. $4h^2 - 3hk - 10k^2 = (h - 2)(4h + 5)$
Mistake: Left out k.
Correct: $(h - 2k)(4h + 5k)$

For Exercises 75 and 76, given the area of the figure, factor to find possible expressions for the length and width. See Example 1.

75.

Area = $15x^2 - 11x + 2$

$(5x - 2)(3x - 1)$

76.

Area = $2x^2 + 13x + 20$

$(2x + 5)(x + 4)$

77. Two adjacent plots of land are for sale. The area of the larger plot is described by $6w^2 - w - 2$ square feet, where w is the width of the smaller plot. Factor to find possible expressions for the dimensions of the larger plot.

$2w + 1$ and $3w - 2$

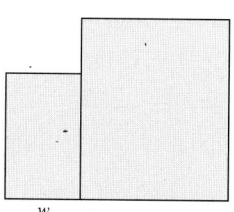

78. A cookie sheet has an area of $5h^2 + 8h - 85$ square inches, where h represents the height of the lip of the sheet. Factor to find possible expressions for the width and length of the cookie sheet.

$5h - 17$ and $h + 5$

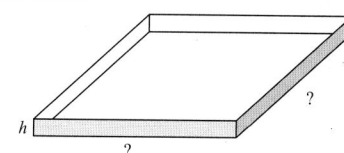

★ *For Exercises 79–84, find all natural numbers that can replace b and make the expression factorable. See Example 1.*

79. $2x^2 + bx + 6$
7, 8, or 13

80. $3x^2 + bx + 10$
11, 13, 17, or 31

81. $5x^2 + bx - 9$
4, 12, or 44

82. $2x^2 + bx - 15$
1, 7, 13, or 29

83. $6x^2 + bx - 7$
1, 11, 19, or 41

84. $4x^2 + bx - 5$
1, 8, or 19

Review Exercises

Exercises 1–7 ◣ **Expressions**

[5.5] *For Exercises 1–3, multiply.*

1. $(3x + 5)(3x - 5)$
$9x^2 - 25$

2. $(x - 2)(x^2 + 2x + 4)$
$x^3 - 8$

3. $(2x - 5)^2$
$4x^2 - 20x + 25$

For Exercises 4–7, factor completely.

[6.1] 4. $16c^3d^2 - 24c^2d^4 + 36cd^2$
$4cd^2(4c^2 - 6cd^2 + 9)$

[6.1] 5. $ab + 5a + 3b + 15$
$(b + 5)(a + 3)$

[6.2] 6. $m^2 + 2m - 24$
$(m - 4)(m + 6)$

[6.2] 7. $3x^3 + 9x^2 - 30x$
$3x(x - 2)(x + 5)$

6.4 Factoring Special Products

Objectives

1 Factor perfect square trinomials.

2 Factor a difference of squares.

3 Factor a difference of cubes.

4 Factor a sum of cubes.

Warm-up

[5.5] *For Exercises 1–3, find the products.*
  1. $(2x - 3)^2$
  2. $(3x + 2)(3x - 2)$
  3. $(4y - 3)(16y^2 + 12y + 9)$
[6.3] 4. Factor: $5x^2 - 12x - 9$

In Section 5.5, we explored some special products found by multiplying conjugates and squaring binomials. In this section, we will learn how to factor those special products.

Objective 1 Factor perfect square trinomials.

First, we consider the product that is a result of squaring a binomial. The trinomial product is a perfect square. Recall the following rules for squaring a binomial that we developed in Section 5.5.

$$(a + b)^2 = a^2 + 2ab + b^2$$
$$(a - b)^2 = a^2 - 2ab + b^2$$

Consider $(2x + 3)^2$ and $(2x - 3)^2$. Using the rules, we have

$$(2x + 3)^2 = (2x)^2 + 2(2x)(3) + (3)^2 \quad (2x - 3)^2 = (2x)^2 - 2(2x)(3) + (3)^2$$
$$= 4x^2 + 12x + 9 \qquad\qquad\qquad = 4x^2 - 12x + 9$$

Answers to Warm-up
1. $4x^2 - 12x + 9$
2. $9x^2 - 4$
3. $64y^3 - 27$
4. $(5x + 3)(x - 3)$

To use these rules when factoring, we look at the terms in the given trinomial to determine whether the trinomial fits the form of a perfect square. If we were asked to factor $4x^2 + 12x + 9$, we would note that the first and last terms are perfect squares and that twice the product of their square roots equals the middle term.

The square root of $4x^2$ is $2x$. The square root of 9 is 3.

$$4x^2 + 12x + 9$$

Twice the product of $2x$ and 3 is the middle term, $12x$.

Because $4x^2 + 12x + 9$ is a perfect square trinomial fitting the form $a^2 + 2ab + b^2$, where $a = 2x$ and $b = 3$, we can write the factored form as $(a + b)^2$, which is $(2x + 3)^2$.

> **Rule** **Factoring Perfect Square Trinomials**
>
> $$a^2 + 2ab + b^2 = (a + b)^2$$
> $$a^2 - 2ab + b^2 = (a - b)^2$$

Instructor Note Some students factor perfect square trinomials using the methods of Sections 6.2 and 6.3. Encourage students not to do so because it takes longer and in Section 7.4, they will have to write a perfect square trinomial as a binomial squared to get the correct LCD.

Example 1 Factor.

a. $9y^2 + 30y + 25$

Solution: This trinomial is a perfect square because the first and last terms are perfect squares and twice the product of their roots is the middle term.

The square root of $9y^2$ is $3y$. The square root of 25 is 5.

$$9y^2 + 30y + 25$$

Twice the product of $3y$ and 5 is $2(3y)(15) = 30y$, which is the middle term.

Answer: $9y^2 + 30y + 25 = (3y + 5)^2$ Use $a^2 + 2ab + b^2 + (a + b)^2$, where $a = 3y$ and $b = 5$.

b. $36x^2 - 84x + 49$

Solution: This trinomial is a perfect square.

The square root of $36x^2$ is $6x$. The square root of 49 is 7.

$$36x^2 - 84x + 49$$

Twice the product of $6x$ and 7 is $2(6x)(7) = 84x$, which is the middle term.

Answer: $36x^2 - 84x + 49 = (6x - 7)^2$ Use $a^2 - 2ab + b^2 = (a - b)^2$, where $a = 6x$ and $b = 7$.

Your Turn 1 Factor.

  a. $25n^2 + 20n + 4$ **b.** $49x^2 - 14x + 1$

Answers to Your Turn 1
  a. $(5n + 2)^2$ **b.** $(7x - 1)^2$

Note that in the perfect square trinomial forms, both a and b can have variables. Also don't forget that as a first step, you should always look for a monomial GCF in the terms.

Example 2 Factor.

a. $16x^2 - 8xy + y^2$

Solution: $16x^2 - 8xy + y^2 = (4x - y)^2$,   Use $a^2 - 2ab + b^2 = (a - b)^2$, where $a = 4x$ and $b = y$.

b. $4x^3y - 36x^2y + 81xy$

Solution:

$$4x^3y - 36x^2y + 81xy = xy(4x^2 - 36x + 81)$$ Factor out the monomial GCF, xy.

$$= xy(2x - 9)^2$$ Factor $4x^2 - 36x + 81$ using $a^2 - 2ab + b^2 = (a - b)^2$, where $a = 2x$ and $b = 9$.

Your Turn 2 Factor.

a. $9m^2 + 48mn + 64n^2$ **b.** $2xy^4 + 24xy^3 + 72xy^2$

Objective 2 Factor a difference of squares.

Another special product that we considered in Section 5.5 was found by multiplying conjugates. Recall that conjugates are binomials that differ only in the sign separating the terms. For example, $3x + 2$ and $3x - 2$ are conjugates. Note that multiplying conjugates produces a product that is a difference of squares.

$$(3x + 2)(3x - 2) = 9x^2 - 4$$

↑   ↑

This term is the square of 3x.   **This term is the square of 2.**

The rule for multiplying conjugates that we developed in Section 5.5 is

$$(a + b)(a - b) = a^2 - b^2.$$

To turn that rule into a rule for factoring a difference of squares, we reverse it.

Warning A *sum* of squares $a^2 + b^2$ is prime and *cannot* be factored.

> **Rule Factoring a Difference of Squares**
> $$a^2 - b^2 = (a + b)(a - b)$$

Example 3 Factor.

a. $25x^2 - 36y^2$

Solution: This binomial is a difference of squares because $25x^2 - 36y^2 = (5x)^2 - (6y)^2$. To factor it, we use the rule $a^2 - b^2 = (a + b)(a - b)$.

$$a^2 - b^2 = (a + b)(a - b)$$
$$25x^2 - 36y^2 = (5x)^2 - (6y)^2 = (5x + 6y)(5x - 6y)$$ Replace a with 5x and b with 6y.

b. $9x^4 - 16$

Solution: This binomial is a difference of squares because $9x^4 - 16 = (3x^2)^2 - (4)^2$.

$$9x^4 - 16 = (3x^2 + 4)(3x^2 - 4)$$ Use $a^2 - b^2 = (a + b)(a - b)$ with $a = 3x^2$ and $b = 4$.

c. $25y^7 - 64y^9$

Solution: The terms in this binomial have a monomial GCF, y^7.

$$25y^7 - 64y^9 = y^7(25 - 64y^2)$$ Factor out the monomial GCF, y^7.

$$= y^7(5 + 8y)(5 - 8y)$$ Factor $25 - 64y^2$ using $a^2 - b^2 = (a + b)(a - b)$ with $a = 5$ and $b = 8y$.

Answers to Your Turn 2
a. $(3m + 8n)^2$
b. $2xy^2(y + 6)^2$

Your Turn 3 Factor.

a. $n^2 - 49$ **b.** $9t^6 - 49u^4$ **c.** $50x^3 - 18x$

Sometimes the factors themselves can be factored.

Example 4 Factor $x^4 - 81$.

Solution: This binomial is a difference of squares, where $a = x^2$ and $b = 9$.

$$
\begin{aligned}
x^4 - 81 &= (x^2 + 9)(x^2 - 9) &&\text{Use } a^2 - b^2 = (a + b)(a - b). \\
&= (x^2 + 9)(x + 3)(x - 3) &&\text{Factor } x^2 - 9 \text{ using } a^2 - b^2 = (a + b)(a - b) \\
&&&\text{with } a = x \text{ and } b = 3.
\end{aligned}
$$

▲

Note The factor $x^2 + 9$ is a sum of squares, which is prime.

Your Turn 4 Factor $y^4 - 16$.

Objective 3 Factor a difference of cubes.

Another common binomial form that we can factor is a difference of cubes. A difference of cubes is a result of a multiplication such as the following:

$$
\begin{aligned}
(x - 2)(x^2 + 2x + 4) &= x^3 + 2x^2 + 4x - 2x^2 - 4x - 8 \\
&= x^3 - 8 \quad \longleftarrow \textbf{Difference of cubes}
\end{aligned}
$$

Writing factored form reverses this process so that we have

$$
x^3 - 8 = (x - 2)(x^2 + 2x + 4)
$$

This suggests the following rule for factoring a difference of cubes.

Rule **Factoring a Difference of Cubes**
$$
a^3 - b^3 = (a - b)(a^2 + ab + b^2)
$$

Example 5 Factor.

a. $64y^3 - 27$

Solution: This binomial is a difference of cubes because $64y^3 - 27 = (4y)^3 - (3)^3$. To factor, we use the rule $a^3 - b^3 = (a - b)(a^2 + ab + b^2)$ with $a = 4y$ and $b = 3$.

$$
\begin{array}{ccccccccc}
a^3 & - & b^3 & = & (a - b) & (a^2 & + & a & b & + & b^2) \\
\end{array}
$$

$$
64y^3 - 27 = (4y)^3 - (3)^3 = (4y - 3)((4y)^2 + (4y)(3) + (3)^2) \quad \begin{array}{l}\text{Replace } a \\ \text{with } 4y \text{ and} \\ b \text{ with } 3.\end{array}
$$

$$
= (4y - 3)(\underbrace{16y^2 + 12y + 9}) \quad \text{Simplify.}
$$

▲

Note This trinomial may seem like a perfect square. However, to be a perfect square, the middle term should be $2ab$. In this trinomial, we have only ab; so it cannot be factored.

b. $27m^4 - 8mn^3$

Solution: The terms in this binomial have a monomial GCF, m.

$$
\begin{aligned}
27m^4 - 8mn^3 &= m(27m^3 - 8n^3) &&\begin{array}{l}\text{Factor out the} \\ \text{monomial GCF, } m.\end{array} \\
&= m(3m - 2n)((3m)^2 + (3m)(2n) + (2n)^2) &&\begin{array}{l}\text{Factor } 27m^3 - 8n^3, \text{ which} \\ \text{is a difference of cubes with}\end{array} \\
&= m(3m - 2n)(9m^2 + 6mn + 4n^2) &&\begin{array}{l}a = 3m \text{ and } b = 2n. \\ \text{Simplify.}\end{array}
\end{aligned}
$$

Answers to Your Turn 3
a. $(n + 7)(n - 7)$
b. $(3t^3 + 7u^2)(3t^3 - 7u^2)$
c. $2x(5x + 3)(5x - 3)$

Answer to Your Turn 4
$(y^2 + 4)(y + 2)(y - 2)$

| Your Turn 5 | Factor. |

a. $27 - u^3$

b. $xy^3 - 64x$

Objective 4 Factor a sum of cubes.

A sum of cubes can be factored using a pattern similar to the difference of cubes. A sum of cubes is a result of a multiplication such as the following:

$$(x + 2)(x^2 - 2x + 4) = x^3 - 2x^2 + 4x + 2x^2 - 4x + 8$$
$$= x^3 + 8$$

Writing factored form reverses this process so that we have

$$x^3 + 8 = (x + 2)(x^2 - 2x + 4)$$

In general, we can write the following rule for factoring a sum of cubes.

> **Rule Factoring a Sum of Cubes**
> $$a^3 + b^3 = (a + b)(a^2 - ab + b^2)$$

 Learning Strategy

A mnemonic for remembering the signs of the factorization of the sum or difference of cubes is SOAP.

$a^3 + b^3 =$
$(a + b)(a^2 - ab + b^2)$
$a^3 - b^3 =$
$(a - b)(a^2 + ab + b^2)$

S	O	A	P
a	p	l	o
m	p	w	s
e	o	a	i
	s	y	t
	i	s	i
	t		v
	e		e

Note Because $1^2 = 1$, $1^3 = 1$, and so on, 1 is a perfect square, a perfect cube, and so on.

The factored form of the sum or difference of cubes always has a binomial and a trinomial factor. The binomial factor consists of the cube roots of the terms of the sum or difference of cubes with the same sign. The first term of the trinomial factor is the square of the first term of the binomial factor, the second term of the trinomial factor is the opposite of the product of the terms of the binomial factor, and the third term of the trinomial factor is the square of the second term of the binomial factor.

| Example 6 | Factor. |

a. $27x^3 + 8$

Solution: This binomial is a sum of cubes because $27x^3 + 8 = (3x)^3 + (2)^3$. To factor, we use the rule $a^3 + b^3 = (a + b)(a^2 - ab + b^2)$ with $a = 3x$ and $b = 2$.

$$a^3 + b^3 = (a + b)(a^2 - a\,b + b^2)$$
$$27x^3 + 8 = (3x)^3 + (2)^3 = (3x + 2)((3x)^2 - (3x)(2) + (2)^2) \quad \text{Replace } a \text{ with } 3x \text{ and } b \text{ with } 2.$$
$$= (3x + 2)(9x^2 - 6x + 4) \quad \text{Simplify.}$$

Note This trinomial cannot be factored.

b. $5h + 40hk^3$

Solution: The terms in this binomial have a monomial GCF, $5h$.

$$5h + 40hk^3 = 5h(1 + 8k^3) \quad \text{Factor out the monomial GCF, } 5h.$$
$$= 5h(1 + 2k)((1)^2 - (1)(2k) + (2k)^2) \quad \text{Factor } 1 + 8k^3, \text{ which is a sum of cubes with } a = 1 \text{ and } b = 2k.$$
$$= 5h(1 + 2k)(1 - 2k + 4k^2)$$

Answers to Your Turn 5
a. $(3 - u)(9 + 3u + u^2)$
b. $x(y - 4)(y^2 + 4y + 16)$

Answers to Your Turn 6
a. $(4 + m)(16 - 4m + m^2)$
b. $x^2(5x + 3)(25x^2 - 15x + 9)$

| Your Turn 6 | Factor. |

a. $64 + m^3$

b. $125x^5 + 27x^2$

6.4 Exercises For Extra Help MyMathLab®

Note: Exercises marked with a ★ represent challenging exercises.

Objective 1

Prep Exercise 1 Explain how to recognize a perfect square trinomial. The first and last terms are perfect squares, and twice the product of their square roots equals the middle term.

Prep Exercise 2 Complete the rules for factoring perfect square trinomials.

$a^2 + 2ab + b^2 = $ $\underline{(a + b)^2}$
$a^2 - 2ab + b^2 = $ $\underline{(a - b)^2}$

For Exercises 1–18, factor the trinomials that are perfect squares. If the trinomial is not a perfect square, write **not a perfect square.** *See Examples 1 and 2.*

1. $x^2 + 14x + 49$
$(x + 7)^2$

2. $y^2 + 6y + 9$
$(y + 3)^2$

3. $b^2 - 8b + 16$
$(b - 4)^2$

4. $m^2 - 6m + 9$
$(m - 3)^2$

5. $n^2 + 12n + 144$
Not a perfect square

6. $y^2 + 6y + 36$
Not a perfect square

7. $25u^2 - 30u + 9$
$(5u - 3)^2$

8. $9m^2 - 24m + 16$
$(3m - 4)^2$

9. $100w^2 + 20w + 1$
$(10w + 1)^2$

10. $25a^2 + 10a + 1$
$(5a + 1)^2$

11. $y^2 + 2yz + z^2$
$(y + z)^2$

12. $w^2 + 2wv + v^2$
$(w + v)^2$

13. $4p^2 - 28pq + 49q^2$
$(2p - 7q)^2$

14. $9r^2 - 12rs + 4s^2$
$(3r - 2s)^2$

15. $16g^2 + 24gh + 9h^2$
$(4g + 3h)^2$

16. $4y^2 - 12by + 9b^2$
$(2y - 3b)^2$

17. $16t^2 + 80t + 100$
$4(2t + 5)^2$

18. $54q^2 - 72q + 24$
$6(3q - 2)^2$

Objective 2

Prep Exercise 3 Complete the rule for factoring a difference of squares.
$a^2 - b^2 = $ $\underline{(a + b)(a - b)}$

Prep Exercise 4 The factors of a difference of squares are binomials known as $\underline{\text{conjugates}}$.

For Exercises 19–38, factor the binomials that are the difference of squares. If prime, so state. See Examples 3 and 4.

19. $x^2 - 4$
$(x + 2)(x - 2)$

20. $a^2 - 121$
$(a + 11)(a - 11)$

21. $16 - y^2$
$(4 + y)(4 - y)$

22. $49 - a^2$
$(7 - a)(7 + a)$

23. $p^2 - q^2$
$(p + q)(p - q)$

24. $m^2 - n^2$
$(m + n)(m - n)$

25. $25u^2 - 16$
$(5u + 4)(5u - 4)$

26. $16x^2 - 49$
$(4x + 7)(4x - 7)$

27. $9x^2 + b^2$
prime

28. $16x^2 + y^2$
prime

29. $64m^2 - 25n^2$
$(8m + 5n)(8m - 5n)$

30. $121p^2 - 9q^2$
$(11p + 3q)(11p - 3q)$

31. $50x^2 - 32y^2$
$2(5x + 4y)(5x - 4y)$

32. $27a^2 - 48b^2$
$3(3a + 4b)(3a - 4b)$

33. $4x^2 + 100y^2$
$4(x^2 + 25y^2)$

34. $16x^2 + 36y^2$
$4(4x^2 + 9y^2)$

35. $x^4 - y^4$
$(x^2 + y^2)(x + y)(x - y)$

36. $a^4 - b^4$
$(a^2 + b^2)(a + b)(a - b)$

37. $x^4 - 16$
$(x^2 + 4)(x + 2)(x - 2)$

38. $b^4 - 1$
$(b^2 + 1)(b + 1)(b - 1)$

Objectives 3 and 4

Prep Exercise 5 Complete the rules for factoring a sum of cubes.

$a^3 + b^3 = \underline{(a + b)(a^2 - ab + b^2)}$

Prep Exercise 6 Complete the rules for factoring a difference of cubes.

$a^3 - b^3 = \underline{(a - b)(a^2 + ab + b^2)}$

For Exercises 39–58, factor the sum or difference of cubes. See Examples 5 and 6.

39. $n^3 - 27$
$(n - 3)(n^2 + 3n + 9)$

40. $y^3 - 64$
$(y - 4)(y^2 + 4y + 16)$

41. $x^3 + 27$
$(x + 3)(x^2 - 3x + 9)$

42. $y^3 + 8$
$(y + 2)(y^2 - 2y + 4)$

43. $x^3 - 1$
$(x - 1)(x^2 + x + 1)$

44. $r^3 - 1$
$(r - 1)(r^2 + r + 1)$

45. $m^3 + n^3$
$(m + n)(m^2 - mn + n^2)$

46. $p^3 + q^3$
$(p + q)(p^2 - pq + q^2)$

47. $27k^3 - 8$
$(3k - 2)(9k^2 + 6k + 4)$

48. $125a^3 - 64$
$(5a - 4)(25a^2 + 20a + 16)$

49. $27k^3 + 8$
$(3k + 2)(9k^2 - 6k + 4)$

50. $125b^3 + 27$
$(5b + 3)(25b^2 - 15b + 9)$

51. $c^3 - 64d^3$
$(c - 4d)(c^2 + 4cd + 16d^2)$

52. $a^3 - 8b^3$
$(a - 2b)(a^2 + 2ab + 4b^2)$

53. $125x^3 + 64y^3$
$(5x + 4y)(25x^2 - 20xy + 16y^2)$

54. $27a^3 + 1000b^3$
$(3a + 10b)(9a^2 - 30ab + 100b^2)$

55. $27x^3 - 64y^3$
$(3x - 4y)(9x^2 + 12xy + 16y^2)$

56. $1000a^3 - 27b^3$
$(10a - 3b)(100a^2 + 30ab + 9b^2)$

57. $8p^3 + q^3z^3$
$(2p + qz)(4p^2 - 2pqz + q^2z^2)$

58. $x^3y^3 + 27z^3$
$(xy + 3z)(x^2y^2 - 3xyz + 9z^2)$

Objectives 1–4

For Exercises 59–74, factor. If prime, so state. See Examples 1–6.

59. $2x^2 - 50$
$2(x + 5)(x - 5)$

60. $5y^2 - 20$
$5(y + 2)(y - 2)$

61. $16x^2 - \dfrac{25}{49}$
$\left(4x + \dfrac{5}{7}\right)\left(4x - \dfrac{5}{7}\right)$

62. $25y^2 - \dfrac{1}{36}$
$\left(5y + \dfrac{1}{6}\right)\left(5y - \dfrac{1}{6}\right)$

63. $2u^3 - 2u$
$2u(u + 1)(u - 1)$

64. $4x - 16x^3$
$4x(1 + 2x)(1 - 2x)$

65. $y^5 - 16y^3b^2$
$y^3(y + 4b)(y - 4b)$

66. $a^4b^2 - a^6$
$a^4(b + a)(b - a)$

67. $50x^3 + 2x$
$2x(25x^2 + 1)$

68. $27m^3 + 363mn^2$
$3m(9m^2 + 121n^2)$

69. $3y^3 - 24z^3$
$3(y - 2z)(y^2 + 2yz + 4z^2)$

70. $2x^3 - 54$
$2(x - 3)(x^2 + 3x + 9)$

71. $c^3 - \dfrac{8}{27}$
$\left(c - \dfrac{2}{3}\right)\left(c^2 + \dfrac{2}{3}c + \dfrac{4}{9}\right)$

72. $y^3 - \dfrac{1}{125}$
$\left(y - \dfrac{1}{5}\right)\left(y^2 + \dfrac{1}{5}y + \dfrac{1}{25}\right)$

73. $16c^4 + 2cd^3$
$2c(2c + d)(4c^2 - 2cd + d^2)$

74. $8d^4 - 64c^3d$
$8d(d - 2c)(d^2 + 2cd + 4c^2)$

Objectives 2–4

★ *For Exercises 75–82, use the rules for a difference of squares, sum of cubes, or difference of cubes to factor completely. See Examples 3–6.*

75. $(2a - b)^2 - c^2$
$(2a - b + c)(2a - b - c)$

76. $(3x - y)^2 - z^2$
$(3x - y + z)(3x - y - z)$

77. $16 - 9(x - y)^2$
$(4 - 3x + 3y)(4 + 3x - 3y)$

78. $25a^2 - 16(b + c)^2$
$(5a - 4b - 4c)(5a + 4b + 4c)$

79. $x^3 + 27(y + z)^3$
$(x + 3y + 3z)(x^2 - 3xy - 3xz + 9y^2 + 18yz + 9z^2)$

80. $m^3 + 8(y + b)^3$
$(m + 2y + 2b)(m^2 - 2my - 2bm + 4y^2 + 8by + 4b^2)$

81. $(x + y)^3 - 64d^3$
$(x + y - 4d)(x^2 + 2xy + y^2 + 4dx + 4dy + 16d^2)$

82. $(t + u)^3 - 64$
$(t + u - 4)(t^2 + 2tu + u^2 + 4t + 4u + 16)$

Objective 1

For Exercises 83–86, find a natural number, b, that makes the expression a perfect square trinomial. See Objective 1.

83. $16x^2 + bx + 9$
24

84. $25x^2 + bx + 4$
20

85. $4x^2 - bx + 81$
36

86. $36x^2 - bx + 49$
84

For Exercises 87–90, find the natural number, c, that completes the perfect square trinomial. See Objective 1.

87. $x^2 + 10x + c$
25

88. $x^2 + 8x + c$
16

89. $9x^2 - 24x + c$
16

90. $4x^2 - 28x + c$
49

For Exercises 91 and 92, write a polynomial for the area of the shaded region; then factor completely.

★ **91.**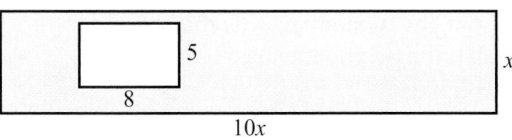

$10(x + 2)(x - 2)$

★ **92.**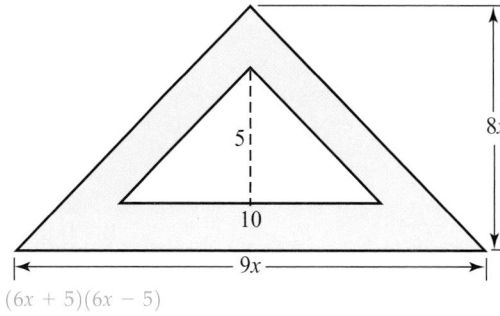

$(6x + 5)(6x - 5)$

For Exercises 93 and 94, write a polynomial for the volume of the shaded region; then factor completely.

★ **93.**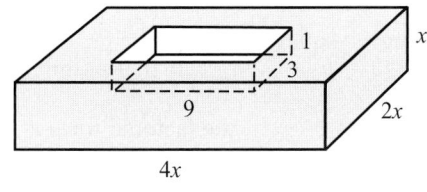

$(2x - 3)(4x^2 + 6x + 9)$

★ **94.**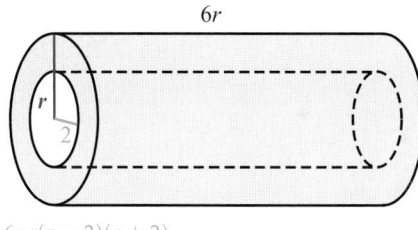

$6\pi r(r - 2)(r + 2)$

Review Exercises

Exercises 1 and 2 ▲ **Constants and Variables**

[6.1] 1. List all natural number factors of 36.
1, 2, 3, 4, 6, 9, 12, 18, 36

[1.2] 2. Find the prime factorization of 100.
$2 \cdot 2 \cdot 5 \cdot 5$

Exercises 3–8 ▲ **Expressions**

For Exercises 3–8, factor completely.

[6.1] 3. $14x^3y^3 - 21x^2y^4 - 7xy^2$
$7xy^2(2x^2y - 3xy^2 - 1)$

[6.1] 4. $x^2 + 4xy + 3x + 12y$
$(x + 4y)(x + 3)$

[6.2] 5. $a^2 - 3ab - 18b^2$
$(a + 3b)(a - 6b)$

[6.2] 6. $3a^2b + 6ab - 45b$
$3b(a + 5)(a - 3)$

[6.3] 7. $6n^2 + 5n - 6$
$(3n - 2)(2n + 3)$

[6.3] 8. $12x^3y - 38x^2y^2 + 20xy^3$
$2xy(2x - 5y)(3x - 2y)$

6.5 Strategies for Factoring

Objective

1 Factor polynomials.

Warm-up

For Exercises 1–4, factor.

[6.1] 1. $6x^3y^2 - 12x^2y^3 + 6x^4y$ **[6.2] 2.** $y^2 + 5yz - 24z^2$

[6.3] 3. $8a^2 + 14a - 15$ **[6.4] 4.** $16x^2 - 9y^2$

Objective 1 Factor polynomials.

One of the difficulties in factoring expressions is determining which technique to use. In this section, we use the following general outline for factoring.

Instructor Note Before working through this section, it would be a good idea to review each of the different types of factoring. You could use the Chapter Summary and Review Exercises for this purpose.

> **Procedure** Factoring a Polynomial
>
> To factor a polynomial, first factor out any monomial GCF; then consider the number of terms in the polynomial. If the polynomial has
>
> **I. Four terms**, try to factor by grouping.
> **II. Three terms**, determine whether the trinomial is a perfect square.
>   **A.** If the trinomial is a perfect square, consider its form.
>   **1.** If in form $a^2 + 2ab + b^2$, the factored form is $(a + b)^2$.
>   **2.** If in form $a^2 - 2ab + b^2$, the factored form is $(a - b)^2$.
>   **B.** If the trinomial is not a perfect square, consider its form.
>   **1.** If in form $x^2 + bx + c$, find two factors of c whose sum is b and write the factored form as $(x + \text{first number})(x + \text{second number})$.
>   **2.** If in form $ax^2 + bx + c$, where $a \neq 1$, use trial and error. Or find two factors of ac whose sum is b; write these factors as coefficients of two like terms that, when combined, equal bx; and factor by grouping.
> **III. Two terms**, determine whether the binomial is a difference of squares, sum of cubes, or difference of cubes.
>   **A.** If given a binomial that is a difference of squares, $a^2 - b^2$, the factors are conjugates and the factored form is $(a + b)(a - b)$. Note that a sum of squares cannot be factored.
>   **B.** If given a binomial that is a sum of cubes, $a^3 + b^3$, the factored form is $(a + b)(a^2 - ab + b^2)$.
>   **C.** If given a binomial that is a difference of cubes, $a^3 - b^3$, the factored form is $(a - b)(a^2 + ab + b^2)$.
>
> **Note** Always check to see if any of the factors can be factored further.

Example 1 Factor.

a. $12x^2 + 28x - 5$

Solution: There is no monomial GCF. Because this expression is a trinomial, we check to see if it is a perfect square. It is not a perfect square because the first and last terms are not perfect squares. It is a trinomial of the form $ax^2 + bx + c$, where $a \neq 1$. Therefore, we use trial and error or the alternative method that involves factoring by grouping. We will use trial and error here.

The *first* terms multiply to equal $12x^2$. The factors could be x and $12x$, $2x$ and $6x$, or $3x$ and $4x$.

$$12x^2 + 28x - 5 = (\quad + \quad)(\quad - \quad)$$

The *last* terms multiply to equal -5. We have already written in the appropriate signs, so these factors could be 1 and 5.

Answers to Warm-up
1. $6x^2y(xy - 2y^2 + x^2)$
2. $(y - 3z)(y + 8z)$
3. $(4a - 3)(2a + 5)$
4. $(4x - 3y)(4x + 3y)$

Note Because $(x + 5)(12x - 1)$ did not give the middle term with the correct absolute value, but the incorrect sign, $(x - 5)(12x + 1)$ will also be incorrect. The same is true of $(12x + 5)(x - 1)$ and $(12x - 5)(x + 1)$.

Now list possible combinations of these first and last factor pairs and check by multiplying to see if the sum of the products of the inner and outer terms equals $28x$.

$$\left. \begin{array}{l} (x + 5)(12x - 1) = 12x^2 - x + 60x - 5 = 12x^2 + 59x - 5 \\ (12x + 5)(x - 1) = 12x^2 - 12x + 5x - 5 = 12x^2 - 7x - 5 \end{array} \right\} \quad \text{Incorrect combinations}$$

$$(2x + 5)(6x - 1) = 12x^2 - 2x + 30x - 5 = 12x^2 + 28x - 5 \quad \begin{array}{l}\text{Correct}\\\text{combination}\end{array}$$

Answer: $12x^2 + 28x - 5 = (2x + 5)(6x - 1)$

b. $7x^3 + 14x^2 - 105x$

Solution:

$$7x^3 + 14x^2 - 105x = 7x(x^2 + 2x - 15) \quad \text{Factor out the monomial GCF, 7x.}$$

The trinomial in the parentheses is not a perfect square and is in the form $x^2 + bx + c$; so we look for two numbers whose product is -15 and whose sum is 2. Because the product is negative, the two factors must have different signs. Because the sum is positive, the factor with the greater absolute value will be the positive factor.

Product	Sum
$(-1)(15) = -15$	$-1 + 15 = 14$
$(-3)(5) = -15$	$-3 + 5 = 2$

$(-3)(5)$ row: **Correct combination**

Answer: $7x^3 + 14x^2 - 105x = 7x(x - 3)(x + 5)$

c. $12m^4 - 48n^2$

Solution:

$$12m^4 - 48n^2 = 12(m^4 - 4n^2) \quad \text{Factor out the monomial GCF, 12.}$$

The binomial in the parentheses is a difference of squares in the form $a^2 - b^2$, where $a = m^2$ and $b = 2n$. Using $a^2 - b^2 = (a + b)(a - b)$ gives us

$$= 12(m^2 + 2n)(m^2 - 2n) \quad \blacktriangleleft \textbf{Note} \text{ Neither of these binomial factors can be factored further.}$$

d. $10y^5 + 40y^3$

Solution:

$$10y^5 + 40y^3 = 10y^3(y^2 + 4) \quad \text{Factor out the monomial GCF, } 10y^3.$$

The binomial in the parentheses is a sum of squares, which cannot be factored.

e. $45x^3y^3 - 120x^2y^4 + 80xy^5$

Solution:

$$45x^3y^3 - 120x^2y^4 + 80xy^5 = 5xy^3(9x^2 - 24xy + 16y^2) \quad \text{Factor out the monomial GCF, } 5xy^3.$$

The trinomial in the parentheses is a perfect square in the form $a^2 - 2ab + b^2$, where $a = 3x$ and $b = 4y$. Using $a^2 - 2ab + b^2 = (a - b)^2$ gives us

$$= 5xy^3(3x - 4y)^2$$

f. $x^5 - x^3 - 8x^2 + 8$

Solution: There is no monomial GCF common to all terms. Because this polynomial has four terms, we try to factor by grouping.

$$x^5 - x^3 - 8x^2 + 8 = x^3(x^2 - 1) - 8(x^2 - 1)$$
$$= (x^2 - 1)(x^3 - 8)$$

The two binomial factors can be factored further. The first binomial, $x^2 - 1$, is a difference of squares with $a = x$ and $b = 1$. The second binomial, $x^3 - 8$, is a difference of cubes with $a = x$ and $b = 2$.

$$= (x + 1)(x - 1)(x - 2)(x^2 + 2x + 4)$$

Answers to Your Turn 1
a. $(3y - 7)(8y + 1)$
b. $2t(t - 6)(t - 3)$
c. $5(3a + b)(3a - b)$
d. $3m^2n(n^2 + 5n + 2)$
e. $4a^2b(2a - 5b)^2$
f. $(2x + 3)(2x - 3)(3x - 2)$

| **Your Turn 1** | Factor. |

a. $24y^2 - 53y - 7$
b. $2t^3 - 18t^2 + 36t$
c. $45a^2 - 5b^2$
d. $3m^2n^3 + 15m^2n^2 + 6m^2n$
e. $16a^4b - 80a^3b^2 + 100a^2b^3$
f. $12x^3 - 8x^2 - 27x + 18$

6.5 Exercises For Extra Help MyMathLab®

Objective 1

Prep Exercise 1 What should you look for first when factoring? A monomial GCF common to all terms

Prep Exercise 2 What are three different types of factorable binomials? Difference of squares, difference of cubes, and sum of cubes

Prep Exercise 3 Describe one type of binomial that cannot be factored. Sum of squares

Prep Exercise 4 What are the two forms that a perfect square trinomial can have? $a^2 + 2ab + b^2$ and $a^2 - 2ab + b^2$

Prep Exercise 5 If a polynomial has four terms, which factoring method should you consider? Factor by grouping

Prep Exercise 6 How might you check a factored form to see if it is correct? Multiply the factors to see if the product is the original polynomial.

For Exercises 1–76, factor completely. If prime, so state. See Example 1.

1. $3xy^2 + 6x^2y$
   $3xy(y + 2x)$

2. $2a + ab$
   $a(2 + b)$

3. $7a(x + y) - b(x + y)$
   $(x + y)(7a - b)$

4. $(u + v)r + (u + v)s$
   $(u + v)(r + s)$

5. $2x^2 - 32$
   $2(x + 4)(x - 4)$

6. $2x^2 - 2y^2$
   $2(x + y)(x - y)$

7. $ax + ay + bx + by$
   $(a + b)(x + y)$

8. $ax + xy + ay + y^2$
   $(a + y)(x + y)$

9. $12a^3b^2c + 3a^2b^2c^2 + 5abc^3$
   $abc(12a^2b + 3abc + 5c^2)$

10. $15m^2n^3p + 2m^2n^2p^2 - 2mn^3p$
    $mnp(15mn^2 + 2mnp - 2n^2)$

11. $x^2 + 8x + 15$
    $(x + 3)(x + 5)$

12. $b^2 + 9b + 14$
    $(b + 2)(b + 7)$

13. $x^4 - 16$
    $(x^2 + 4)(x + 2)(x - 2)$

14. $t^4 - 81$
    $(t^2 + 9)(t + 3)(t - 3)$

15. $x^2 + 25$
    prime

16. $a^2 + 49$
    prime

17. $15x^2 + 7x - 2$
    $(5x - 1)(3x + 2)$

18. $2a^2 - 5a - 12$
    $(2a + 3)(a - 4)$

19. $ax^2 + 4ax + 4a$
    $a(x + 2)^2$

20. $9a^2x + 12ax + 4x$
    $x(3a + 2)^2$

21. $6ab - 36ab^2$
    $6ab(1 - 6b)$

22. $5ax - 25a^2x$
    $5ax(1 - 5a)$

23. $x^2 + x + 2$
    prime

24. $x^2 + 3x + 4$
    prime

25. $x^2 - 49$
    $(x + 7)(x - 7)$

26. $x^2 - 25$
    $(x + 5)(x - 5)$

27. $p^2 + p - 30$
    $(p + 6)(p - 5)$

28. $y^2 - 6y - 40$
    $(y - 10)(y + 4)$

29. $u^3 - u$
    $u(u + 1)(u - 1)$

30. $5m^2 - 45$
    $5(m + 3)(m - 3)$

31. $2b^2 + 14b + 24$
    $2(b + 3)(b + 4)$

32. $4y^2 + 36y + 80$
$4(y + 5)(y + 4)$

33. $6r^2 - 15r^3$
$3r^2(2 - 5r)$

34. $18y^3 - 24y^4$
$6y^3(3 - 4y)$

35. $14u^2 + 7u - 105$
$7(2u - 5)(u + 3)$

36. $6r^2 - 26r - 20$
$2(3r + 2)(r - 5)$

37. $4h^2 + 12h + 4$
$4(h^2 + 3h + 1)$

38. $3m^2 + 9m + 3$
$3(m^2 + 3m + 1)$

39. $5p^2 - 80$
$5(p + 4)(p - 4)$

40. $3r^2 - 108$
$3(r + 6)(r - 6)$

41. $3w^2 + 5w + 2$
$(3w + 2)(w + 1)$

42. $3m^2 - 8m + 5$
$(3m - 5)(m - 1)$

43. $8q^2 - 10q - 3$
$(2q - 3)(4q + 1)$

44. $10r^2 + 9r - 7$
$(5r + 7)(2r - 1)$

45. $12v^2 + 23v + 10$
$(3v + 2)(4v + 5)$

46. $6x^2 + 13x + 6$
$(2x + 3)(3x + 2)$

47. $2 - 50x^2$
$2(1 + 5x)(1 - 5x)$

48. $6 - 24x^2$
$6(1 + 2x)(1 - 2x)$

49. $80k^2 - 20k^2l^2$
$20k^2(2 + l)(2 - l)$

50. $27x^3 - 3x^3y^2$
$3x^3(3 + y)(3 - y)$

51. $2j^2 + j + 3$
prime

52. $3k^2 - 2k + 1$
prime

53. $50 - 20t + 2t^2$
$2(5 - t)^2$

54. $243 - 54x + 3x^2$
$3(9 - x)^2$

55. $3ax - 6ay - 8by + 4bx$
$(x - 2y)(3a + 4b)$

56. $2ax - 6bx - 3by + ay$
$(a - 3b)(2x + y)$

57. $x^3 - x^2 - x + 1$
$(x + 1)(x - 1)^2$

58. $x^3 - 2x^2 - x + 2$
$(x - 2)(x + 1)(x - 1)$

59. $4x^2 - 28x + 49$
$(2x - 7)^2$

60. $4m^2 + 20m + 25$
$(2m + 5)^2$

61. $2x^4 - 162$
$2(x^2 + 9)(x + 3)(x - 3)$

62. $5a^4 - 3125$
$5(a^2 + 25)(a + 5)(a - 5)$

63. $a^2 + 2ab + ab + 2b^2$
$(a + 2b)(a + b)$

64. $x^2 - 3xy + xy - 3y^2$
$(x - 3y)(x + y)$

65. $9 - 4m^2$
$(3 + 2m)(3 - 2m)$

66. $64 - 9r^2$
$(8 - 3r)(8 + 3r)$

67. $x^5 - 4xy^2$
$x(x^2 + 2y)(x^2 - 2y)$

68. $m^5 - 25mn^2$
$m(m^2 + 5n)(m^2 - 5n)$

69. $20x^2 + 3xy + 2y^2$
prime

70. $20a^2 - 9a + 20$
prime

71. $b^3 + 125$
$(b + 5)(b^2 - 5b + 25)$

72. $x^3 + 216$
$(x + 6)(x^2 - 6x + 36)$

73. $3y^3 - 24$
$3(y - 2)(y^2 + 2y + 4)$

74. $4x^3 - 4$
$4(x - 1)(x^2 + x + 1)$

75. $54x - 2xy^3$
$2x(3 - y)(9 + 3y + y^2)$

76. $81n - 3m^3n$
$3n(3 - m)(9 + 3m + m^2)$

For Exercises 77 and 78, given an expression for the area of each rectangle, find the length and width.

77.

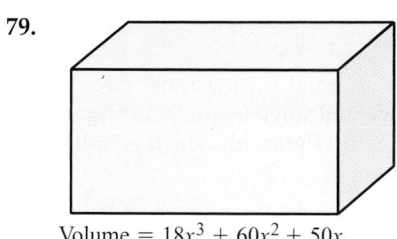

Area = $6x^2 - 11x - 7$

$l = 2x + 1, w = 3x - 7$

78.

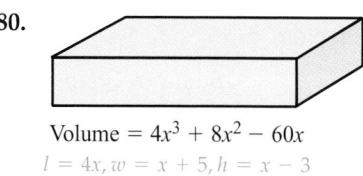

Area = $25n^2 - 49$

$l = 5n + 7, w = 5n - 7$

For Exercises 79 and 80, given an expression for the volume of each box, find the length, width, and height.

79.

Volume = $18x^3 + 60x^2 + 50x$

$l = 2x, w = 3x + 5, h = 3x + 5$

80.

Volume = $4x^3 + 8x^2 - 60x$

$l = 4x, w = x + 5, h = x - 3$

81. A swimmer's goggles fall off a diving board that is 100 meters high. The height of the goggles after t seconds is given by $100 - 16t^2$. Factor $100 - 16t^2$.
    $4(5 + 2t)(5 - 2t)$

82. An object is dropped from a cliff that is 400 feet high. The expression $400 - 16t^2$ gives the height of the falling object after t seconds. Factor $400 - 16t^2$.
    $16(5 + t)(5 - t)$

83. The voltage in a circuit is the product of two factors, the resistance in the circuit and the current. If the voltage in a circuit is described by the expression $6ir + 15i + 8r + 20$, find the expressions for the current and resistance. (The expression for current will contain i, and the expression for resistance will contain r.)
    Current: $3i + 4$; resistance: $2r + 5$

84. If the voltage in the circuit described in Exercise 83 is changed so that it is described by the expression $35ir + 15i - 7r - 3$, what are the expressions for the current and resistance?
    Current: $5i - 1$; resistance: $7r + 3$

Review Exercises

Exercises 1 and 2 ◢ Expressions

[5.6] **1.** Divide: $\dfrac{2x^2 - 7x - 8}{2x + 1}$
    $x - 4 - \dfrac{4}{2x + 1}$

[6.3] **2.** Factor $6y^2 + 23y - 4$.
    $(6y - 1)(y + 4)$

Exercises 3–5 ◢ Equations and Inequalities

[2.1] **3.** Find the area of a rectangular garden that is 14 feet by 16 feet.
    $224\ \text{ft.}^2$

[2.5] **4.** Translate to an equation; then solve. The product of four and the sum of a number and three is equal to twenty. Find the unknown number.
    $4(x + 3) = 20; 2$

[4.2, 4.3] **5.** Find three consecutive integers whose sum is 54.
    $17, 18, 19$

6.6 Solving Quadratic Equations by Factoring

Objectives

1 Use the zero-factor theorem to solve equations containing expressions in factored form.

2 Solve quadratic equations by factoring.

3 Solve problems involving quadratic equations.

4 Use the Pythagorean theorem to solve problems.

Answers to Warm-up

1. $y = -\dfrac{4}{3}$
2. $(2x - 1)(x + 5)$
3. $x(x - 5)(x + 2)$
4. $16(t + 4)(t - 4)$

Warm-up

[2.3] **1.** Solve: $3y + 4 = 0$

[6.5] *For Exercises 2–4, factor.*

2. $2x^2 + 9x - 5$

3. $x^3 - 3x^2 - 10x$

4. $16t^2 - 256$

Objective 1 Use the zero-factor theorem to solve equations containing expressions in factored form.

So far in Chapters 5 and 6, we have rewritten polynomial expressions. We are now ready to work with polynomial equations, which we will solve using factoring. Notice that we are moving to the equation level of our Algebra Pyramid, which is built upon the foundation of expressions, variables, and constants.

The Algebra Pyramid

Note Inequalities are also in ▶ the top level of the Algebra Pyramid, but we will not be exploring polynomial inequalities at this point.

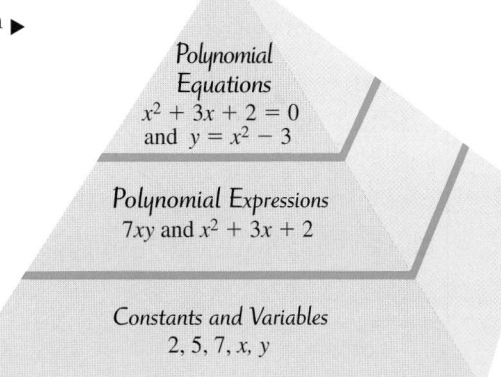

Polynomial Equations
$x^2 + 3x + 2 = 0$
and $y = x^2 - 3$

Polynomial Expressions
$7xy$ and $x^2 + 3x + 2$

Constants and Variables
$2, 5, 7, x, y$

Connection Throughout this book, after developing new types of expressions and ways of rewriting those expressions, we put those skills into the context of solving equations.

When solving polynomial equations by factoring, we make use of the **zero-factor theorem**, which states that if the product of two factors is 0, then one or the other of the two factors or both factors are equal to 0.

> **Rule Zero-Factor Theorem**
>
> If a and b are real numbers and $ab = 0$, then $a = 0$ or $b = 0$.

Example 1 Solve $(x + 2)(x - 3) = 0$.

Solution: In this equation, the product of two factors, $x + 2$ and $x - 3$, is equal to 0. According to the zero-factor theorem, one of the two factors or both factors must equal 0. In solving, we determine what values for x cause each factor to equal 0.

$$x + 2 = 0 \quad \text{or} \quad x - 3 = 0$$

Now solve each of the linear equations to get two solutions.

$$x = -2 \qquad x = 3$$

Check: Verify that -2 and 3 satisfy the original equation, $(x + 2)(x - 3) = 0$.

For $x = -2$: $(-2 + 2)(-2 - 3) \overset{?}{=} 0$ For $x = 3$: $(3 + 2)(3 - 3) \overset{?}{=} 0$
$$(0)(-5) = 0 \qquad\qquad\qquad (5)(0) = 0$$

Both -2 and 3 check; therefore, both are solutions.

Your Turn 1 Solve $(n - 6)(n + 1) = 0$.

Example 1 suggests the following procedure.

> **Procedure Solving Equations with Two or More Factors Equal to 0**
>
> To solve an equation in which two or more factors are equal to 0, use the zero-factor theorem.
> 1. Set each factor equal to zero.
> 2. Solve each of those equations.

Answer to Your Turn 1
$n = 6$ or -1

Example 2 Solve.

a. $y(3y + 4) = 0$

Solution: In this equation, the product of two factors, y and $3y + 4$, is equal to 0. We set each factor equal to 0 and then solve each of those equations.

$$y = 0 \quad \text{or} \quad 3y + 4 = 0$$

Note This equation is already solved. ▲

$3y = -4$ Subtract 4 from both sides to isolate $3y$.

$y = -\dfrac{4}{3}$ Divide both sides by 3 to isolate y.

Check: Verify that 0 and $-\dfrac{4}{3}$ satisfy the original equation, $y(3y + 4) = 0$.

For $y = 0$: $(0)(3(0) + 4) \stackrel{?}{=} 0$ For $y = -\dfrac{4}{3}$: $\left(-\dfrac{4}{3}\right)\left[\dfrac{\overset{1}{3}}{1}\left(-\dfrac{4}{\underset{1}{3}}\right) + 4\right] \stackrel{?}{=} 0$

$(0)(4) = 0$ $\left(-\dfrac{4}{3}\right)[-4 + 4] \stackrel{?}{=} 0$

$\left(-\dfrac{4}{3}\right)(0) = 0$

Both 0 and $-\dfrac{4}{3}$ check; therefore, both are solutions.

b. $(x - 4)^2 = 0$

Solution: Note that $(x - 4)^2 = 0$ can be written as $(x - 4)(x - 4) = 0$. Because the two factors are identical, there is no need to write two separate equations. We set $x - 4$ equal to 0 and then solve.

$$x - 4 = 0$$
$$x = 4 \quad \text{Add 4 to both sides to isolate } x.$$

Check: Verify that 4 satisfies the original equation, $(x - 4)^2 = 0$.

$$(4 - 4)^2 \stackrel{?}{=} 0$$
$$(0)^2 = 0 \quad \text{4 checks; therefore, it is a solution.}$$

c. $n(n + 6)(5n - 3) = 0$

Solution: In this equation, the product of three factors, n, $n + 6$, and $5n - 3$, is 0. We set each factor equal to 0 and then solve each of those equations.

$$n = 0 \quad \text{or} \quad n + 6 = 0 \quad \text{or} \quad 5n - 3 = 0$$
$$n = -6 \qquad\qquad 5n = 3$$
$$n = \dfrac{3}{5}$$

Check: To check, we verify that 0, -6, and $\dfrac{3}{5}$ satisfy the original equation. We leave this check to the reader.

Your Turn 2 Solve.

a. $5x(3x - 2) = 0$ **b.** $(y + 3)^2 = 0$ **c.** $t(4t - 3)(t + 1) = 0$

Answers to Your Turn 2

a. $x = 0 \text{ or } \dfrac{2}{3}$

b. $y = -3$

c. $t = 0, \dfrac{3}{4}, \text{ or } -1$

Objective 2 Solve quadratic equations by factoring.

Now that we have seen how to use the zero-factor theorem to solve equations containing expressions that are in factored form, we are ready to consider equations that have an expression that is not in factored form. More specifically, we will use factoring to solve **quadratic equations in one variable**.

Definition Quadratic equation in one variable: An equation that can be written in the form $ax^2 + bx + c = 0$, where a, b, and c are real numbers and $a \neq 0$.

The fact that a cannot equal zero means that the ax^2 term must be present for the equation to be a quadratic equation. A quadratic equation written in the form $ax^2 + bx + c = 0$ is said to be in *standard form*. Following are some examples of quadratic equations.

$$x^2 + 5x + 6 = 0 \qquad 3x^2 - 48 = 0 \qquad x^2 = x + 6$$

Note This quadratic equation is in standard form.

Note This quadratic equation is in standard form. It has no bx term because $b = 0$.

Note Although not in standard form, this equation is still quadratic. Written in standard form, the equation would be $x^2 - x - 6 = 0$.

In Objective 1, we learned that to use the zero-factor theorem, we need an expression in factored form set equal to 0. So to solve a quadratic equation using factoring, it needs to be in standard form. If given an equation that is not in standard form, such as $x^2 = x + 6$, we first write it in standard form.

$$x^2 - x - 6 = 0 \quad \text{Subtract } x \text{ and 6 from both sides to get standard form.}$$

After factoring $x^2 - x - 6$, we have $(x + 2)(x - 3) = 0$, which is the same equation that we solved in Example 1. This suggests the following procedure.

Instructor Note Point out that if $ax^2 + bx + c$ is prime, the quadratic equation cannot be solved by factoring. We will learn to solve these types of equations in Chapter 10.

Procedure Solving Quadratic Equations Using Factoring

To solve a quadratic equation using factoring:
1. Write the equation in standard form ($ax^2 + bx + c = 0$).
2. Write the variable expression in factored form.
3. Use the zero-factor theorem to solve.

Example 3 Solve $2x^2 + 9x - 5 = 0$.

Solution: This equation is in standard form, so we can simply factor the variable expression.

$$(2x - 1)(x + 5) = 0 \quad \text{Factor } 2x^2 + 9x - 5.$$

Now use the zero-factor theorem to solve.

$$2x - 1 = 0 \quad \text{or} \quad x + 5 = 0$$
$$2x = 1 \qquad\qquad x = -5$$
$$x = \frac{1}{2}$$

Check: To check, we verify that $\frac{1}{2}$ and -5 satisfy the original equation. We leave this check to the reader.

Your Turn 3 Solve $12y^2 + 11y + 2 = 0$.

Answer to Your Turn 3
$y = -\dfrac{2}{3}$ or $-\dfrac{1}{4}$

Note Although $x^3 - 3x^2 - 10x = 0$ is not a quadratic equation, after we factor out the x, the parentheses contain a quadratic form: $x^2 - 3x - 10$.

Sometimes we may have to factor out a monomial GCF.

Example 4 Solve $x^3 - 3x^2 - 10x = 0$.

Solution:

$$x(x^2 - 3x - 10) = 0 \quad \text{Factor out the monomial GCF, } x.$$
$$x(x - 5)(x + 2) = 0 \quad \text{Factor } x^2 - 3x - 10.$$

Now use the zero-factor theorem to solve.

$$x = 0 \quad \text{or} \quad x - 5 = 0 \quad \text{or} \quad x + 2 = 0$$
$$x = 5 \qquad\qquad x = -2$$

Check: Verify that $0, 5$, and -2 each satisfy the original equation. We leave the check to the reader.

Your Turn 4 | Solve $12x^3 + 20x^2 - 8x = 0$.

Rewrite Quadratic Equations in Standard Form

If a quadratic equation is not in standard form ($ax^2 + bx + c = 0$), we put it in standard form, then factor and use the zero-factor theorem.

Example 5 | Solve.

a. $6y^2 - 5y = 4 - 28y$

Solution: We first need to manipulate the equation so that it is in standard form.

$$6y^2 + 23y = 4 \qquad\qquad \text{Add } 28y \text{ to both sides.}$$
$$6y^2 + 23y - 4 = 0 \qquad\qquad \text{Subtract 4 from both sides to get standard form.}$$
$$(6y - 1)(y + 4) = 0 \qquad\qquad \text{Factor using trial and error.}$$
$$6y - 1 = 0 \quad \text{or} \quad y + 4 = 0 \qquad \text{Use the zero-factor theorem to solve.}$$
$$6y = 1 \qquad\qquad y = -4$$
$$y = \frac{1}{6}$$

Check: We leave the check to the reader.

b. $x(x - 5) = -6$

Solution:

Warning We cannot solve $x(x - 5) = -6$ by setting $x = -6$ or $x - 5 = -6$. For the *zero-factor* theorem to be used, one side of the equation must be 0.

> **Note** We could have multiplied $x(x - 5)$ first and then moved the 6.

> **Instructor Note** Compare factoring a polynomial with solving an equation.

$$x(x - 5) + 6 = 0 \qquad\qquad \text{Add 6 to both sides so that the right-hand side is 0.}$$
$$x^2 - 5x + 6 = 0 \qquad\qquad \text{Multiply } x(x - 5) \text{ to get standard form.}$$
$$(x - 2)(x - 3) = 0 \qquad\qquad \text{Factor by looking for two numbers whose product is 6 and sum is } -5.$$
$$x - 2 = 0 \quad \text{or} \quad x - 3 = 0 \qquad \text{Use the zero-factor theorem to solve.}$$
$$x = 2 \qquad\qquad x = 3$$

Check: We leave the check to the reader.

c. $(x + 3)(x - 5) = 20$

Solution:

> **Warning** We cannot solve $(x + 3)(x - 5) = 20$ by setting $x + 3 = 20$ or $x - 5 = 20$.

$$(x + 3)(x - 5) = 20$$
$$x^2 - 2x - 15 = 20 \qquad\qquad \text{Multiply } x + 3 \text{ and } x - 5.$$
$$x^2 - 2x - 35 = 0 \qquad\qquad \text{Subtract 20 from both sides.}$$
$$(x + 5)(x - 7) = 0 \qquad\qquad \text{Factor.}$$
$$x + 5 = 0 \quad \text{or} \quad x - 7 = 0 \qquad \text{Use the zero-factor theorem to solve.}$$
$$x = -5 \qquad\qquad x = 7$$

Check: We leave the check to the reader.

Answer to Your Turn 4
$$x = 0, \frac{1}{3}, \text{ or } -2$$

Answers to Your Turn 5

a. $n = -\dfrac{2}{5}$ or $-\dfrac{4}{3}$ **b.** $x = 6$ or -8

c. $x = -2$ or $x = 5$

Your Turn 5 | Solve.

a. $15n^2 + 20n = -8 - 6n$ **b.** $x(x + 2) = 48$ **c.** $(x + 4)(x - 7) = -18$

Objective 3 Solve problems involving quadratic equations.

Many problems and applications are solved using quadratic equations. See Section 2.5 for a review of key words and their translations.

Example 6 The product of two consecutive odd natural numbers is 195. Find the numbers.

Solution: Odd natural numbers are $1, 3, 5, \ldots$. Note that adding 2 to a given odd natural number gives a consecutive odd natural number, which suggests this pattern:

First odd natural number: x

Consecutive odd natural number: $x + 2$

The word *product* means that the two odd numbers x and $x + 2$ are multiplied to equal 195.

Translation: $x(x + 2) = 195$ — Write an equation that is the product of x and $x + 2$ set equal to 195.

$$x^2 + 2x = 195$$ — Distribute x.

$$x^2 + 2x - 195 = 0$$ — Subtract 195 from both sides to get the form $ax^2 + bx + c = 0$.

$$(x - 13)(x + 15) = 0$$ — Factor by finding a pair of numbers whose product is -195 and sum is 2.

$$x - 13 = 0 \quad \text{or} \quad x + 15 = 0$$ — Use the zero-factor theorem to solve.

$$x = 13 \qquad x = -15$$ ◀ **Note** We can disregard -15 because it is not a natural number.

Answer: The first number is 13, and its consecutive odd natural number is $13 + 2 = 15$.

Check: 13 and 15 are consecutive odd natural numbers, and their product is 195; so the answer is correct.

Your Turn 6 The product of two consecutive natural numbers is 462. Find the numbers.

Example 7 An architect is designing an addition to a house. The addition will be in the shape of a rectangle, and the room's floor will have an area of 270 square feet. The width of the room is to be 3 feet less than the length. Find the dimensions of the floor.

Understand The area of a rectangle is found by multiplying the length and width. We are given a relationship about the width and length, which we can translate as follows:

"The width is to be 3 feet less than the length."
Width $= l - 3$, where l represents length.

Plan Translate to an equation and then solve.

Execute $l(l - 3) = 270$ — The equation has the product of the length, l, and width, $l - 3$, set equal to 270.

$$l(l - 3) - 270 = 0$$ — Subtract 270 from both sides.

$$l^2 - 3l - 270 = 0$$ — Multiply $l(l - 3)$ to get standard form.

$$(l - 18)(l + 15) = 0$$ — Factor.

$$l - 18 = 0 \quad \text{or} \quad l + 15 = 0$$ — Use the zero-factor theorem to solve.

$$l = 18 \qquad l = -15$$

Answer to Your Turn 6
21 and 22

Area $= 113.04 \text{ cm}^2$

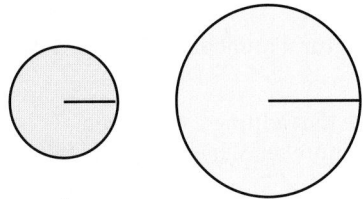

▲
Note This radius is 8 less than the other disc's radius.

Answer Because the problem involves room dimensions, only the positive number makes sense as an answer. If the length is 18 feet and the width is 3 feet less than that, the width is 15 feet.

Check The width, 15 feet, is 3 feet less than the length, 18 feet. Also, the area of a 15-foot by 18-foot rectangle is 270 square feet.

Your Turn 7   The two discs in the margin are designed so that the smaller of the two has an area of 113.04 square centimeters and a radius that is 8 centimeters less than the larger disc's radius. Find the radius of the larger disc. (Use 3.14 to approximate π.)

Example 8   The equation $h = -16t^2 + h_0$ describes the height, h, in feet of an object after falling from an initial height h_0 for a time, t seconds. An object is dropped from a height of 600 feet. When the object reaches 344 feet, how much time has passed?

Understand We are given a formula, and we are to find the time it takes an object to drop from an initial height of 600 feet to a height of 344 feet.

Plan Replace the variables in the formula with the given values to get an equation in terms of t; then solve for t.

Execute		
$344 = -16t^2 + 600$	Replace h with 344 and h_0 with 600.
$-256 = -16t^2 + 0$	Subtract 600 from both sides.
$16t^2 - 256 = 0$	Add $16t^2$ to both sides.
$16(t^2 - 16) = 0$	Factor out the monomial GCF, 16.
$t^2 - 16 = 0$	Divide out the constant factor, 16.
$(t - 4)(t + 4) = 0$	Factor.
$t - 4 = 0 \quad \text{or} \quad t + 4 = 0$	Use the zero-factor theorem to solve.
$t = 4 \qquad\qquad t = -4$	

Answer Because the problem involves the amount of time elapsed after an object was dropped, only the positive value, 4, makes sense. This means that it took the object 4 seconds to fall from 600 feet to 344 feet.

Check Verify that after 4 seconds, an object dropped from an initial height of 600 feet will descend to a height of 344 feet.

$$h = -16(4)^2 + 600$$
$$h = -256 + 600$$
$$h = 344$$

Your Turn 8   An object is dropped from a height of 225 feet above the ground. When the object is 81 feet from the ground, how much time has passed?

Objective 4 Use the Pythagorean theorem to solve problems.

One of the most popular theorems in mathematics is the Pythagorean theorem, named after the Greek mathematician Pythagoras. The theorem relates the side lengths of all right triangles. Recall that in a right triangle, the two sides that form the 90° angle are *legs* and the side directly across from the 90° angle is the *hypotenuse*.

Answer to Your Turn 7
14 cm

Answer to Your Turn 8
3 sec.

Rule Pythagorean Theorem

Given a right triangle, where a and b represent the lengths of the legs and c represents the length of the hypotenuse, $a^2 + b^2 = c^2$. Further, if $a, b,$ and c represent the lengths of the sides of a triangle and $a^2 + b^2 = c^2$, then the triangle is a right triangle with hypotenuse c.

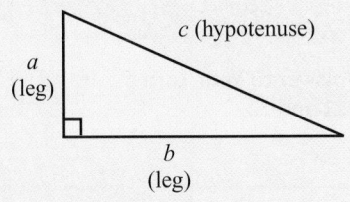

Of Interest

Pythagoras was a Greek mathematician who is believed to have been born around 569 B.C. and died
in 500 B.C. The Pythagorean theorem is named after
Pythagoras not because he discovered the relationship, as commonly thought, but because he and his
followers are the first to have proved that the relationship is true for all right triangles. The relationship was
known and used by other cultures prior to the Greeks;
however, Pythagoras and others in the Greek culture
believed that it was important to prove mathematical
relationships.

We can use the Pythagorean theorem to find a missing length in a right triangle if we
know the other two lengths.

Example 9   Find the length of the missing side.

Solution: Use the Pythagorean theorem, $a^2 + b^2 = c^2$.
The hypotenuse, c, is the missing side length.

$$3^2 + 4^2 = c^2$$ Substitute $a = 3$ and $b = 4$.

$$9 + 16 = c^2$$ Simplify exponential forms.

$$25 = c^2$$ Add.

$$0 = c^2 - 25$$ Subtract 25 from both sides to get standard form.

$$0 = (c - 5)(c + 5)$$ Factor.

$$c - 5 = 0 \quad \text{or} \quad c + 5 = 0$$ Use the zero-factor theorem.

$$c = 5 \qquad\qquad c = -5$$

Because we are dealing with lengths, only the positive solution is sensible; so the
missing length is 5.

Your Turn 9   In the construction of a
roof, three beams are used to form a right
triangle frame. A 12-foot beam and a 5-foot
beam are brought together to form a 90°
angle. How long must the third beam be?

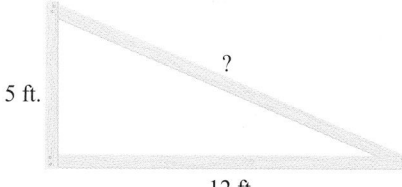

Answer to Your Turn 9
13 ft.

6.6 Exercises   For Extra Help   MyMathLab®

Note: Exercises marked with a ★ represent challenging exercises.

Objective 1

Prep Exercise 1 Explain the zero-factor theorem in your own
words.
If a and b are real numbers and $ab = 0$, then $a = 0$ or $b = 0$.

Prep Exercise 2 Explain how to use the zero-factor theorem
to solve $(x + 3)(x - 4) = 0$.
Solve $x + 3 = 0$ and $x - 4 = 0$.

For Exercises 1–12, solve using the zero-factor theorem. See Examples 1 and 2.

1. $(x + 5)(x + 2) = 0$
$-5, -2$

2. $(x - 7)(x + 3) = 0$
$-3, 7$

3. $(a + 3)(a - 4) = 0$
$-3, 4$

4. $(b - 7)(b + 2) = 0$
$-2, 7$

5. $(3x - 4)(2x + 3) = 0$
$-\dfrac{3}{2}, \dfrac{4}{3}$

6. $(3a + 2)(2a - 5) = 0$
$-\dfrac{2}{3}, \dfrac{5}{2}$

7. $x(x - 7) = 0$
$0, 7$

8. $y(y + 3) = 0$
$0, -3$

9. $m(m - 1)(m + 2) = 0$
$-2, 1, 0$

10. $r(r - 2)(r + 3) = 0$
$-3, 0, 2$

11. $b(b - 2)^2 = 0$
$0, 2$

12. $m(m - 4)^2 = 0$
$0, 4$

Objective 2

Prep Exercise 3 What is the first step in solving a quadratic equation?

Write the equation in standard form ($ax^2 + bx + c = 0$).

Prep Exercise 4 Explain why we cannot write $(x + 2)$ $(x - 2) = 32$ as $x + 2 = 32$ and $x - 2 = 32$.

To use the zero-factor theorem, the product must be 0 and no other number.

For Exercises 13–52, solve the quadratic equations. See Examples 3–5.

13. $y^2 - 4y = 0$
$0, 4$

14. $p^2 - 3p = 0$
$0, 3$

15. $6r^2 + 10r = 0$
$0, -\dfrac{5}{3}$

16. $12c^2 + 9c = 0$
$0, -\dfrac{3}{4}$

17. $x^2 - 4 = 0$
$-2, 2$

18. $y^2 - 25 = 0$
$-5, 5$

19. $x^2 + x - 6 = 0$
$-3, 2$

20. $b^2 - 2b - 8 = 0$
$-2, 4$

21. $n^2 - 6n + 9 = 0$
3

22. $p^2 - 16p + 64 = 0$
8

23. $3a^2 - 11a - 4 = 0$
$4, -\dfrac{1}{3}$

24. $2d^2 + 7d - 4 = 0$
$-4, \dfrac{1}{2}$

25. $6r^2 + 11r - 10 = 0$
$-\dfrac{5}{2}, \dfrac{2}{3}$

26. $8c^2 - 6c - 9 = 0$
$\dfrac{3}{2}, -\dfrac{3}{4}$

27. $x^2 - 4x = 21$
$-3, 7$

28. $u^2 + 6u = 27$
$-9, 3$

29. $p^2 = 3p - 2$
$1, 2$

30. $x^2 = 6x - 8$
$2, 4$

31. $v^2 = 9$
$-3, 3$

32. $m^2 = 16$
$-4, 4$

33. $2a^2 + 18a + 28 = 0$
$-7, -2$

34. $3x^2 - 6x - 45 = 0$
$5, -3$

35. $x^3 - x^2 - 12x = 0$
$0, 4, -3$

36. $x^3 + 7x^2 + 10x = 0$
$0, -5, -2$

37. $12r^3 + 22r^2 - 20r = 0$
$0, -\dfrac{5}{2}, \dfrac{2}{3}$

38. $16c^3 - 12c^2 - 18c = 0$
$0, \dfrac{3}{2}, -\dfrac{3}{4}$

39. $3y^2 + 8 = 14y$
$\dfrac{2}{3}, 4$

40. $6k^2 + 1 = 7k$
$\dfrac{1}{6}, 1$

41. $4x^2 = 60 - x$
$-4, \dfrac{15}{4}$

42. $4n^2 = 5 - 19n$
$-5, \dfrac{1}{4}$

43. $7t = 12 + t^2$
$3, 4$

44. $-9x = -20 - x^2$
$4, 5$

45. $b(b - 5) = 14$
$-2, 7$

46. $m(m + 6) = -9$
-3

47. $4x(x + 7) = -49$
$-\dfrac{7}{2}$

48. $c(2c - 11) = -5$
$\dfrac{1}{2}, 5$

49. $(x + 1)(x + 5) = -3$
$-2, -4$

50. $(a + 3)(a + 6) = -2$
$-4, -5$

51. $(a - 1)(a + 4) = 14$
$3, -6$

52. $(b + 4)(b - 2) = 16$
$4, -6$

Objective 3

For Exercises 53–70, translate to an equation and then solve. See Examples 6 and 7.

53. Find every number such that the square of the number added to 55 is the same as sixteen times the number.
$5, 11$

54. Find every number such that triple the square of the number is equal to four times that number.
$0, \dfrac{4}{3}$

55. The product of two consecutive odd natural numbers is 143. Find the numbers.
$11, 13$

56. The product of two consecutive natural numbers is 306. Find the numbers.
$17, 18$

★ **57.** The sum of the squares of two consecutive natural numbers is 365. Find the numbers.
13, 14

★ **58.** The difference of the squares of two consecutive even natural numbers is 60. Find the numbers.
16, 14

59. The length of a rectangular garden is 9 meters more than the width. If the area is 252 square meters, find the dimensions of the garden.
12 m by 21 m

60. The floor of the central chamber of the Lincoln Memorial has an area of 4440 square feet. If the length of the chamber is 14 feet more than the width, find the dimensions of the chamber floor. (*Source:* National Parks Service.)
60 ft. by 74 ft.

61. The largest billboard in Times Square's history has a length that is 10 feet more than three times its width. The area of the billboard is 9288 square feet. Find the dimensions of the billboard.
172 ft. by 54 ft.

62. The length of a football field is 40 yards less than three times the width. The total area of the football field is 6400 square yards. Find the dimensions of the football field.
120 yd. by $53\frac{1}{3}$ yd.

63. The design of the base of a small building calls for a rectangular shape with dimensions of 22 feet by 28 feet. The architect decides to change the shape of the building to a circle, but wants the base to have the same area. If we use $\frac{22}{7}$ to approximate π, what is the radius of the circular building?
14 ft.

64. A rectangle has a length of 14 centimeters and a width of 11 centimeters. Using $\frac{22}{7}$ as an approximation for π, find the radius of a circle with the same area as the rectangle.
7 cm

65. A design on the front of a marketing brochure calls for a triangle with a base that is 6 centimeters less than the height. If the area of the triangle is to be 216 square centimeters, what are the lengths of the base and height?
Base: 18 cm; height: 24 cm

66. The base of a triangular sign is 4 inches more than the height. If the area of the sign is 160 square inches, find the base and height.
Base = 20 in.; height = 16 in.

67. A steel plate in the shape of a trapezoid has an area of 85.5 square inches. The dimensions are shown. Note that the length of the base is equal to the height. Calculate the height.
9 in.

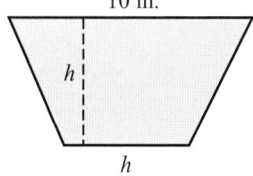

68. The front elevation of one wing of a house is shown. Because of budget constraints, the total area of the front of this wing must be 352 square feet. The height of the triangular portion is 14 feet less than the base. Find the base length.
22 ft.

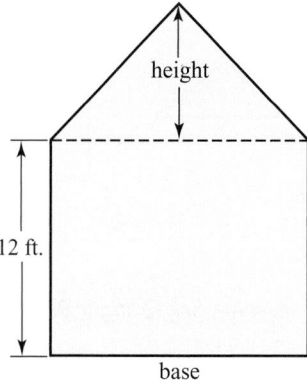

69. Find three consecutive even integers such that the product of the largest and second largest integers is 168.
10, 12, 14 or −16, −14, −12

70. Find three consecutive odd integers such that the product of the largest and second largest integers is 255.
13, 15, 17 or −19, −17, −15

For Exercises 71 and 72, use the formula $h = -16t^2 + v_0t + h_0$, where h is the final height in feet, t is the time of travel in seconds, v_0 is the initial velocity in feet per second, and h_0 is the initial height in feet of an object traveling upward. See Example 8.

★ **71.** A ball is thrown upward at 4 feet per second from a building 29 feet high. When will the ball be 9 feet above the ground?

1.25 sec.

★ **72.** A toy rocket is fired from the ground with an upward velocity of 200 feet per second. How many seconds will the rocket take to return to the ground?

12.5 sec.

For Exercises 73 and 74, use the formula $B = P(1 + \frac{r}{n})^{nt}$, which is used to calculate the final balance of an investment or a loan after being compounded. Following is a list of what each variable represents: See Example 8.

B represents the final balance.

P represents the principal, which is the amount invested.

r represents the annual interest rate as a decimal.

t represents the number of years the principal is compounded.

n represents the number of times per year the principal is compounded.

73. Carlita invests $4000 in an account that is compounded annually. If after two years her balance is $4840, what was the interest rate of the account?

10%

74. Donovan invests $1000 in an account that is compounded semiannually (every six months). If after one year his balance is $1210, what is the interest rate of the account?

20%

Objective 4

For Exercises 75–78, find the length of the hypotenuse. See Example 9.

75.

9
?
12
15

76.

8
?
6
10

77.

?
7
24
25

78.

?
8
15
17

For Exercises 79–84, solve. See Example 9.

79. An artist welds a metal rod measuring 12 centimeters to a second rod measuring 35 centimeters to form a 90° angle. He wants a third rod to be welded to the ends of the first two rods to form a right triangle. What length must the third rod be?

37 cm

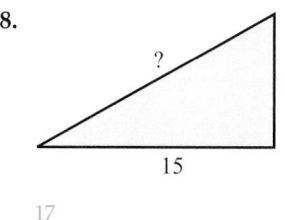

80. A rectangular screen measures 27 inches by 36 inches. Find the diagonal distance between two opposing corners.

45 in.

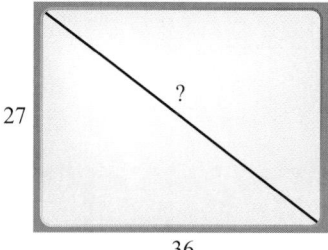

27
?
36

81. A person drives south 18 blocks then turns east and drives 24 blocks. Assuming that all city blocks are equal size and form a grid of streets that intersect at 90°, find the shortest distance in blocks that the person is from her original location.

30 blocks

82. To support a power pole, a wire is to be attached 24 feet from the ground, then staked 10 feet from the base of the pole. How long must the wire be?

26 ft.

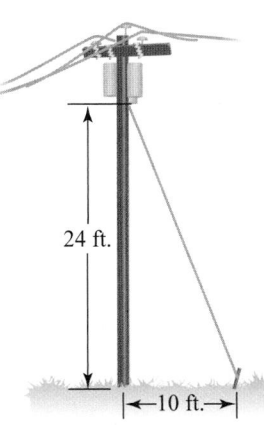

24 ft.

|←10 ft.→|

★ **83.** When hurricanes threaten, a common practice is to tape windows to keep the glass from shattering. If a rectangular window is 24 inches wide and 51 inches diagonally, how long is the window?

45 in.

★ **84.** A sign is in the shape of a rectangle with a stripe painted diagonally across the sign. If the length of the stripe is 26 inches and the sign is 24 inches long, how wide is the sign?

10 in.

Review Exercises

Exercises 1–6   Equations and Inequalities

[3.2] 1. Is the equation $y = x^2 - 1$ a linear equation? Explain.

No, it is not linear because the variable x has an exponent other than 1.

[3.2] 2. Is $(1, -4)$ a solution for $4x + 3y = -8$?

yes

[3.3] 3. Find the x- and y-intercepts for $2x - 5y = -20$.

$(-10, 0), (0, 4)$

[3.5] 4. Write the equation of the line passing through $(-1, 4)$ and $(-3, -2)$ in the form $Ax + By = C$, where $A, B,$ and C are integers and $A > 0$.

$3x - y = -7$

[3.7] 5. For the function $f(x) = x^2 + 2$, find each of the following.
   a. $f(0)$ 2   **b.** $f(-2)$ 6

[3.7] 6. Graph: $f(x) = 2x - 3$

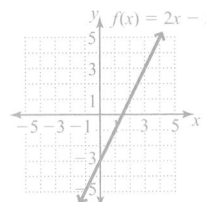

6.7 Graphs of Quadratic Equations and Functions

Objectives

1 Graph quadratic equations in the form $y = ax^2 + bx + c$.

2 Graph quadratic functions.

In Chapter 3, we graphed linear equations in two variables. Now we will graph **quadratic equations in two variables**.

Definition Quadratic equation in two variables: An equation that can be written in the form $y = ax^2 + bx + c$, where a, b, and c are real numbers and $a \neq 0$.

Following are some examples of quadratic equations in two variables.

$$y = x^2 \qquad y = 3x^2 + 2 \qquad y = x^2 - 4x + 1$$

We learned in Section 2.2 that every term in a linear equation has a degree of 1 or 0. Because quadratic equations always have a term with a degree of 2, they are nonlinear, which means that their graphs will not be straight lines. What might their graphs look like?

Objective 1 Graph quadratic equations in the form $y = ax^2 + bx + c$.

Recall that one way to graph an equation is to plot solutions to the equation in the coordinate plane. In Section 3.1, we said that a solution to an equation in two variables is an ordered pair (x, y) that satisfies the equation. For example, $(3, 9)$ is a solution to $y = x^2$ because $9 = (3)^2$ is true. Let's make a table of solutions for $y = x^2$, plot those solutions, and connect the points to see what the graph looks like.

Note Because we know that quadratic equations are nonlinear, we find many solutions to get a good sense of the shape of the graph. ▶

x	y
-3	9
-2	4
-1	1
0	0
1	1
2	4
3	9

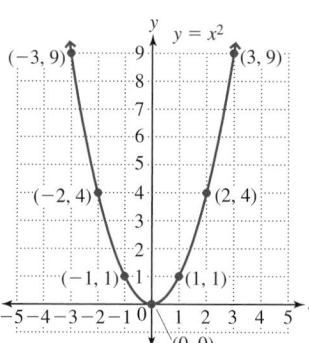

The graph of $y = x^2$ is a *parabola*. In fact, the graph of every quadratic equation in two variables is a parabola.

Note Parabolas are said to open upward or downward, as follows.

upward   downward

In the table of solutions for $y = x^2$, notice that when $x = 1$ or -1, $y = 1$; when $x = 2$ or -2, $y = 4$; and when $x = 3$ or -3, $y = 9$. In general, each x-coordinate and its additive inverse have the same y-coordinate. Looking at the graph, we see that the left side of the y-axis is the mirror image of the right side of the y-axis. Consequently, we say that this graph is symmetrical about the line $x = 0$ (the y-axis). Every parabola will have symmetry about a line called its **axis of symmetry**. Further, a parabola's axis of symmetry always passes through a point called the **vertex**.

Definitions Axis of symmetry: A line that divides a graph into two symmetrical halves. **Vertex of a parabola:** The lowest point on a parabola that opens upward or the highest point on a parabola that opens downward.

For $y = x^2$, the axis of symmetry is the line $x = 0$ (y-axis) and the vertex is $(0, 0)$. In Section 10.4, we will learn more about the vertex and axis of symmetry. For now, we graph by finding enough solutions to see where the parabola turns.

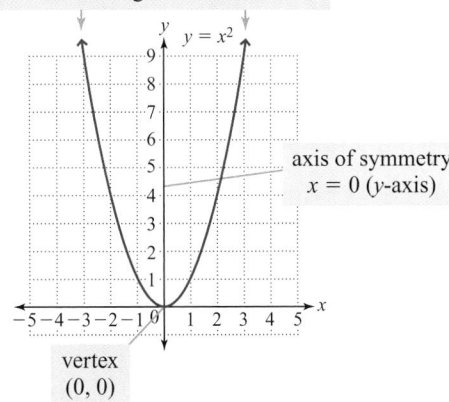

The axis of symmetry divides the graph into two halves that are mirror images of each other.

axis of symmetry
$x = 0$ (y-axis)

vertex
$(0, 0)$

Procedure   Graphing Quadratic Equations

To graph a quadratic equation:
1. Find ordered pair solutions and plot them in the coordinate plane. Continue finding and plotting solutions until the shape of the parabola can be clearly seen.
2. Connect the points to form a parabola.

Example 1   Graph.

a. $y = 3x^2 + 2$

Solution: We complete a table of solutions, plot the solutions in the coordinate plane, and connect the points to form the graph.

Connection In a quadratic equation in the form $y = ax^2 + bx + c$, the constant c indicates the y-intercept just as the constant b does in a linear equation in the form $y = mx + b$.

Note The constant 2 in $y = 3x^2 + 2$ indicates the y-intercept because replacing x with 0 leaves us with $y = 3(0)^2 + 2 = 2$.

x	y
−2	14
−1	5
0	2
1	5
2	14

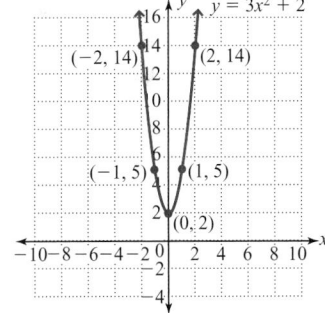

b. $y = x^2 + 2x − 8$

Solution:

Connection To find the x-intercepts of a graph, let $y = 0$. Thus, the solutions of $0 = x^2 + 2x − 8$ are the x-intercepts of the graph of $y = x^2 + 2x − 8$, which are $(−4, 0)$ and $(2, 0)$.

x	y
−4	0
−3	−5
−2	−8
−1	−9
0	−8
1	−5
2	0

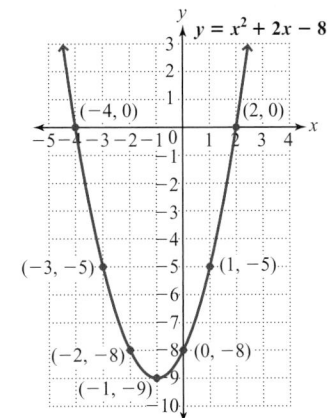

Notice that in each equation we have graphed so far ($y = x^2, y = 3x^2 + 2$, and $y = x^2 + 2x - 8$), the coefficient of x^2 is a positive number. Also notice that the graph of each of those equations was a parabola that opens upward. Now let's consider quadratic equations in which the coefficient of x^2 is a negative number, as in $y = -2x^2 + 6x + 3$. We will see that the graphs of these equations open downward.

Example 2 Graph $y = -2x^2 + 6x + 3$.

Solution: Complete a table of solutions, plot the solutions, and connect the points to form a parabola.

x	y
-1	-5
0	3
1	7
2	7
3	3
4	-5

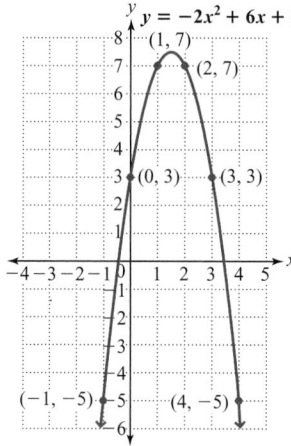

Note The coefficient of x^2 is a negative number, -2, and the graph opens downward. In an equation of the form $y = ax^2 + bx + c$, if a is negative, the parabola opens downward.

Note Because $x = 1$ and $x = 2$ have the same y-values, the x-coordinate of the vertex will be halfway between 1 and 2 at 1.5.

Rule Opening of a Parabola

Given an equation in the form $y = ax^2 + bx + c$, if $a > 0$, then the parabola opens upward; if $a < 0$, then the parabola opens downward.

Answers to Your Turn 2

a.
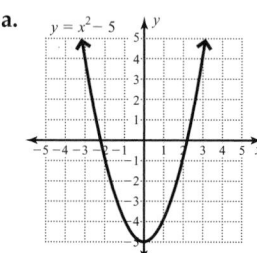

b.
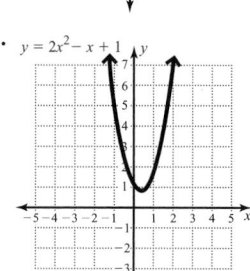

c.
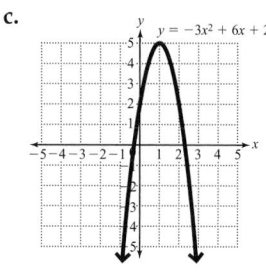

Your Turn 2 Graph.

a. $y = x^2 - 5$ **b.** $y = 2x^2 - x + 1$ **c.** $y = -3x^2 + 6x + 2$

Objective 2 Graph quadratic functions.

In Section 3.7, we learned how to graph linear functions. Now let's graph quadratic functions. A quadratic *equation* in two variables, $y = ax^2 + bx + c$, is a quadratic *function* that can be written in the form $f(x) = ax^2 + bx + c$. Notice that to change the equation in two variables to function notation, we replace y with $f(x)$.

Quadratic equation in two variables: $y = -2x^2 + 6x + 3$

Using function notation: $f(x) = -2x^2 + 6x + 3$

The graph of $f(x) = -2x^2 + 6x + 3$ is the same as $y = -2x^2 + 6x + 3$, which we graphed in Example 2. Recall from Section 3.7 that with function notation, we find ordered pairs by evaluating the function using various values of x. For example, the notation $f(1)$ means to find the value of the function when $x = 1$.

$$f(1) = -2(1)^2 + 6(1) + 3$$
$$= -2 + 6 + 3$$
$$= 7 \qquad \text{The ordered pair is } (1, 7).$$

Connection Finding $f(1)$ for $f(x) = -2x^2 + 6x + 3$ is the same as finding the y-value for $y = -2x^2 + 6x + 3$ when $x = 1$.

We would continue finding ordered pairs in this manner to produce the same table and graph that we produced in Example 2, which suggests the following procedure.

Procedure **Graphing Quadratic Functions**

To graph a quadratic function:
1. Find enough ordered pairs by evaluating the function for various values of x so that when those ordered pairs are plotted, the shape of the parabola can be clearly seen.
2. Connect the points to form the parabola.

Example 3 Graph.

a. $f(x) = 3x^2 - 5$

Solution: Find enough ordered pairs to clearly see the graph; then connect the points to form the parabola.

x	y
−2	7
−1	−2
0	−5
1	−2
2	7

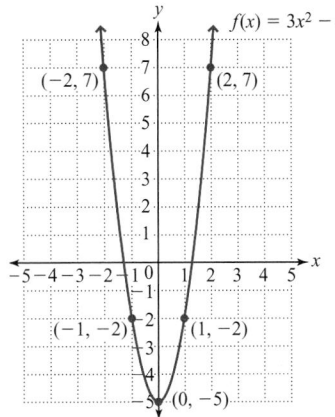

Connection This parabola opens upward. Given a quadratic function in the form $f(x) = ax^2 + bx + c$, if $a > 0$, the parabola opens upward.

b. $f(x) = -x^2 + 2x + 4$

Solution:

x	y
−2	−4
−1	1
0	4
1	5
2	4
3	1
4	−4

Connection This parabola opens downward. Given a quadratic function in the form $f(x) = ax^2 + bx + c$, if $a < 0$, the parabola opens downward.

Answers to Your Turn 3

a.
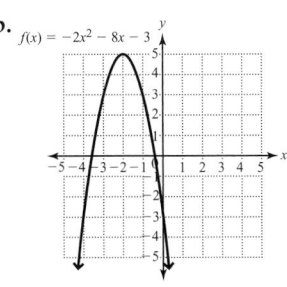

b.

Your Turn 3 Graph.

a. $f(x) = x^2 - 4$

b. $f(x) = -2x^2 - 8x - 3$

In Section 3.7, we learned that a graph represents a function if it passes the vertical line test. (A vertical line through any point in the domain touches the graph at only one point.) Because graphs of quadratic functions are parabolas that open upward or downward, they pass the vertical line test. There are also parabolas that open right or left, which do not pass the vertical line test, and therefore are not functions.

Example 4 Determine whether the graph is the graph of a function. Give the domain and range.

a.

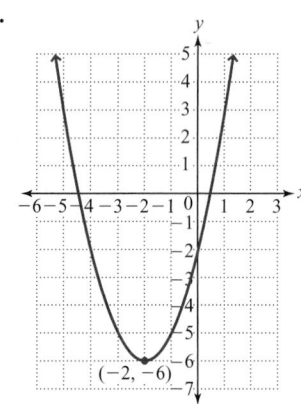

b.

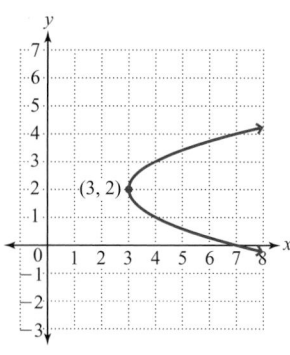

Solution: a. This is a function because any vertical line intersects the graph in at most one point, as illustrated by $x = -4$.

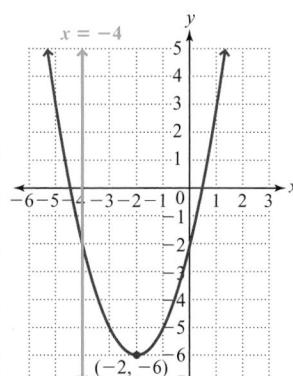

Domain: all real numbers or $(-\infty, \infty)$
Range: $\{y \mid y \geq -6\}$ or $[-6, \infty)$

Solution: b. This is not a function because a vertical line can be drawn that intersects the graph in more than one point, as illustrated by $x = 5$.

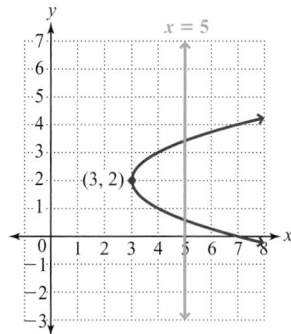

Domain: $\{x \mid x \geq 3\}$ or $[3, \infty)$
Range: all real numbers or $(-\infty, \infty)$

Your Turn 4 Determine whether the graph is the graph of a function. Give the domain and range.

a.

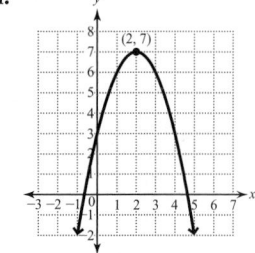

b.

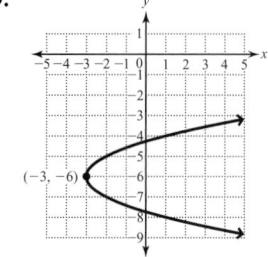

Answers to Your Turn 4
a. is a function; domain: all real numbers or $(-\infty, \infty)$; range: $\{y \mid y \leq 7\}$ or $(-\infty, 7]$
b. not a function; domain: $\{x \mid x \geq -3\}$ or $[-3, \infty)$; range: all real numbers or $(-\infty, \infty)$

6.7 Exercises For Extra Help MyMathLab®

Note: Exercises marked with a ★ represent challenging exercises.

Objective 1

Prep Exercise 1 The axis of symmetry is a line that divides a graph into two __symmetrical__ halves.

Prep Exercise 2 The vertex of a parabola is the __lowest__ point on a parabola that opens up or the __highest__ point on a parabola that opens down.

Prep Exercise 3 Explain what is meant by *symmetry* in a parabola.
If a vertical line is drawn through the vertex, the sides of the parabola are mirror images.

For Exercises 1–4, use the graph of the parabola to determine the coordinates of the vertex. See Objective 1.

1.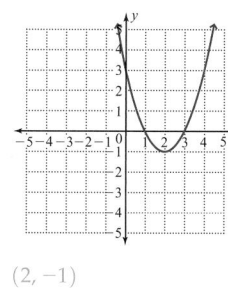

$(2, -1)$

2.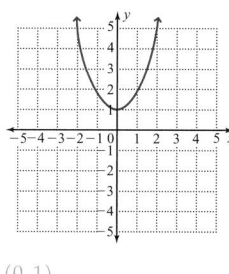

$(0, 1)$

3.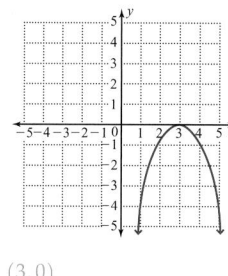

$(3, 0)$

4.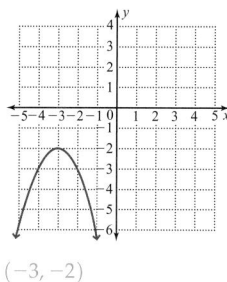

$(-3, -2)$

Prep Exercise 4 Given an equation in the form $y = ax^2 + bx + c$, what indicates whether the graph opens up or down?
The sign of a indicates whether the graph opens upward or downward.

Prep Exercise 5 Given an equation in the form $y = ax^2 + bx + c$, what indicates the y-coordinate of the y-intercept?
The y-coordinate of the y-intercept is c.

For Exercises 5–16, complete the table of solutions and then graph. See Examples 1 and 2.

5. $y = 2x^2$

x	y
-2	8
-1	2
0	0
1	2
2	8

6. $y = 3x^2$

x	y
-2	12
-1	3
0	0
1	3
2	12

7. $y = -x^2$

x	y
-2	-4
-1	-1
0	0
1	-1
2	-4

8. $y = -x^2 + 2$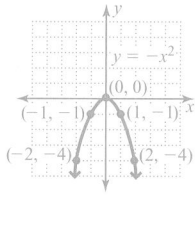

x	y
-2	-2
-1	1
0	2
1	1
2	-2

9. $y = x^2 - 3$

x	y
-2	1
-1	-2
0	-3
1	-2
2	1

10. $y = x^2 + 3$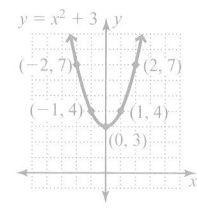

x	y
-2	7
-1	4
0	3
1	4
2	7

11. $y = -x^2 + 2x$

x	y
−1	−3
0	0
1	1
2	0
3	−3

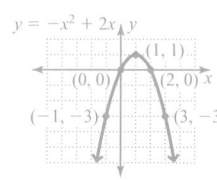

12. $y = -x^2 + 4x$

x	y
0	0
1	3
2	4
3	3
4	0

13. $y = 2x^2 - 4x + 1$

x	y
−1	7
0	1
1	−1
2	1
3	7

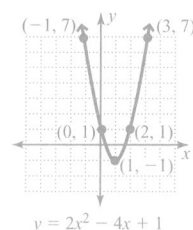

14. $y = 2x^2 + 4x - 2$

x	y
−3	4
−2	−2
−1	−4
0	−2
1	4

15. $y = -3x^2 + 6x + 4$

x	y
−1	−5
0	4
1	7
2	4
3	−5

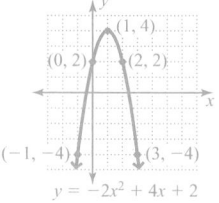

16. $y = -2x^2 + 4x + 2$

x	y
−1	−4
0	2
1	4
2	2
3	−4

For Exercises 17–32, graph. See Examples 1 and 2.

17. $y = x^2 + 1$

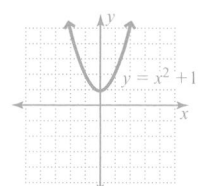

18. $y = x^2 - 4$

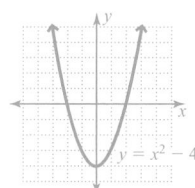

19. $y = -x^2 + 3$

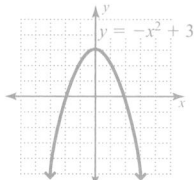

20. $y = -x^2 + 2$

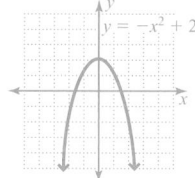

21. $y = 2x^2 - 5$

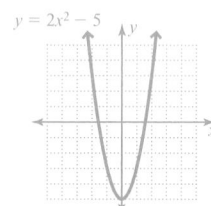

22. $y = 3x^2 - 4$

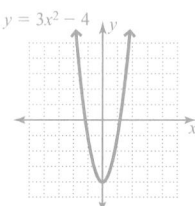

23. $y = -3x^2 + 2$

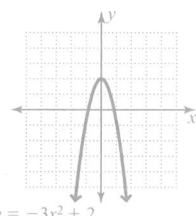

24. $y = -4x^2 + 5$

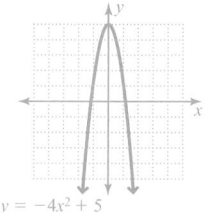

25. $y = x^2 + 4x - 1$

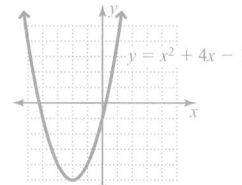

26. $y = x^2 - 6x + 3$

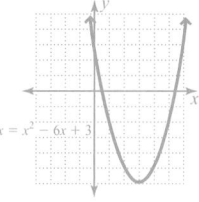

27. $y = -x^2 + 6x - 5$

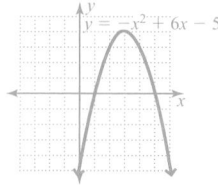

28. $y = -x^2 + 4x - 5$

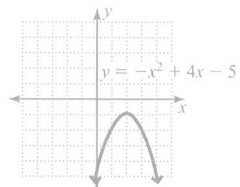

29. $y = 2x^2 - 4x - 3$

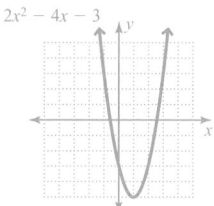

30. $y = 3x^2 + 6x - 1$

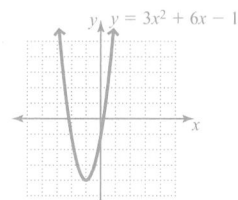

31. $y = -2x^2 + 8x - 5$

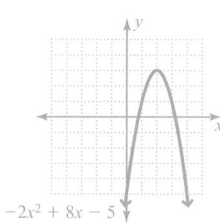

32. $y = -4x^2 - 12x - 3$

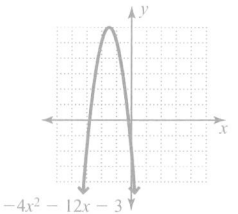

Objective 2

Prep Exercise 6 Given that the graph of $y = ax^2 + bx + c$, where $a \neq 0$, is a parabola that opens upward or downward, explain why $f(x) = ax^2 + bx + c$ is a function.

Parabolas that open up or down pass the vertical line test and are therefore functions.

For Exercises 33–44, graph the function. See Example 3.

33. $f(x) = 3x^2$

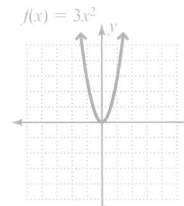

34. $f(x) = 2x^2$

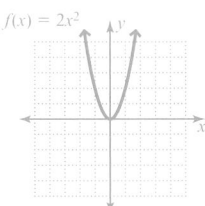

35. $f(x) = -x^2 + 1$

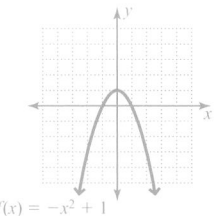

36. $f(x) = -x^2 - 1$

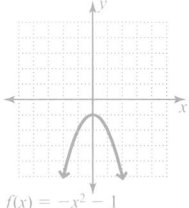

37. $f(x) = x^2 + 4x - 1$

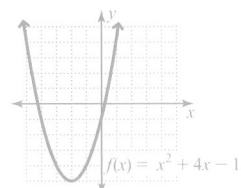

38. $f(x) = x^2 - 6x + 5$

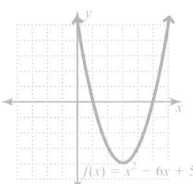

39. $f(x) = -x^2 - 4x - 3$

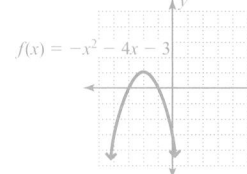

40. $f(x) = -x^2 + 6x - 2$

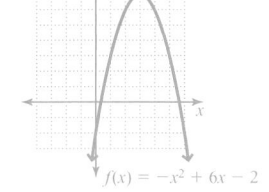

41. $f(x) = 3x^2 - 12x + 5$

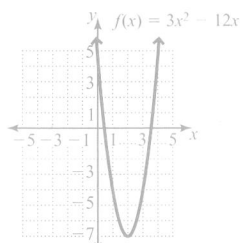

42. $f(x) = 2x^2 + 6x - 1$

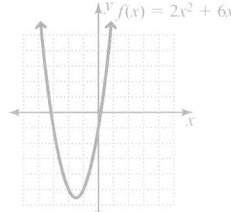

43. $f(x) = -2x^2 - 4x - 1$

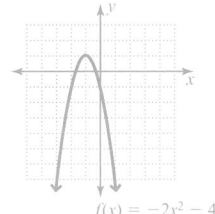

44. $f(x) = -4x^2 + 8x - 3$

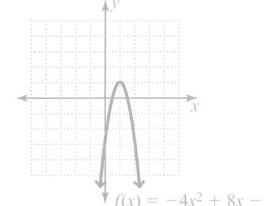

For Exercises 45–52, state whether the graph is the graph of a function. Give the domain and range. See Example 4.

45.

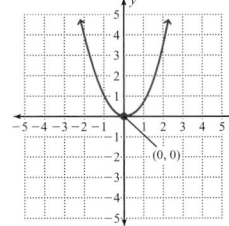

A function
Domain: all real numbers or $(-\infty, \infty)$
Range: $\{y \mid y \geq 0\}$ or $[0, \infty)$

46.

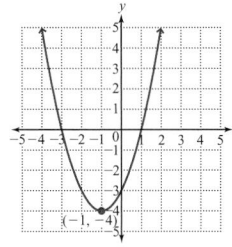

A function
Domain: all real numbers or $(-\infty, \infty)$
Range: $\{y \mid y \geq -4\}$ or $[-4, \infty)$

47.

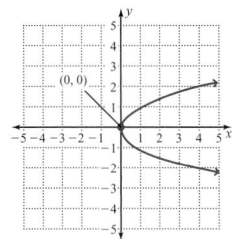

Not a function
Domain: $\{x \mid x \geq 0\}$ or $[0, \infty)$
Range: all real numbers or $(-\infty, \infty)$

48.

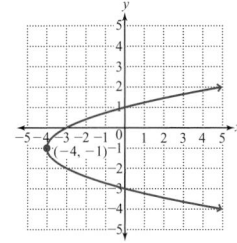

Not a function
Domain: $\{x \mid x \geq -4\}$ or $[-4, \infty)$
Range: all real numbers or $(-\infty, \infty)$

49.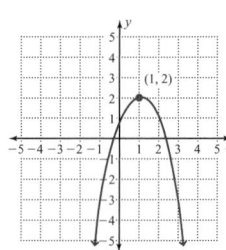

A function
Domain: all real numbers
or $(-\infty, \infty)$
Range: $\{y \mid y \le 2\}$ or
$(-\infty, 2]$

50.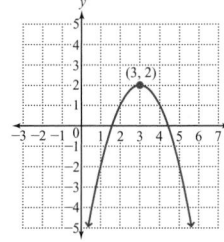

A function
Domain: all real numbers
or $(-\infty, \infty)$
Range: $\{y \mid y \le 2\}$ or
$(-\infty, 2]$

51.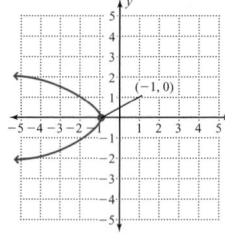

Not a function
Domain: $\{x \mid x \le -1\}$
or $(-\infty, -1]$
Range: all real numbers or
$(-\infty, \infty)$

52.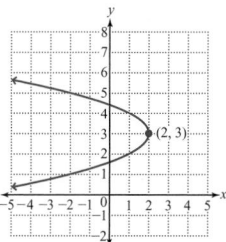

Not a function
Domain: $\{x \mid x \le 2\}$
or $(-\infty, 2]$
Range: all real numbers or
$(-\infty, \infty)$

For Exercises 53–56, use a graphing calculator.

53. a. Enter the equation $y = x^2$; then graph. What are the coordinates of the
y-intercept?
$(0, 0)$

b. Enter a second equation, $y = x^2 + 2$; then graph on the same grid. What are
the coordinates of the y-intercept?
$(0, 2)$

c. On the same grid, graph $y = x^2 + 4$ and compare all three graphs. What do
your explorations suggest about how c affects the graph of an equation in the
form $y = x^2 + c$?
The original graph is moved upward c units.

d. On the same grid, graph $y = x^2 - 3$. Comparing this graph to the first three
graphs, how did changing c to a negative value affect the graph?
The original graph moved downward 3 units.

54. a. Enter the equation $y = x^2$; then graph. What are the coordinates of the vertex?
$(0, 0)$

b. Enter a second equation, $y = 2x^2$. Compare the graph of $y = 2x^2$ with the
graph of $y = x^2$; then discuss their similarities and differences.
The vertexes are the same, but the graph of $y = 2x^2$ is narrower than the graph of $y = x^2$.

c. On the same grid, graph $y = 3x^2$ and compare all three graphs. What does
this exploration suggest about how the size of a in $y = ax^2$ affects the graph?
The greater the value of a, the narrower the graph appears to be.

d. In $y = ax^2$, if $a = \dfrac{1}{2}$, how will the graph be different from the graphs in

parts a–c? Graph $y = \dfrac{1}{2}x^2$.

The graph will appear to be wider than the graphs of parts a–c.

e. What do these explorations suggest about how the size of a affects the graph
of an equation in the form $y = ax^2$?
The value of a indicates the width of the graph.

55. a. Graph $y = x^2$ and $y = -x^2$ on the same grid. What is different about the two
graphs? What does this suggest about the sign of a in an equation of the form
$y = ax^2$?
$y = x^2$ opens upward, and $y = -x^2$ opens downward. The sign of a indicates whether the
parabola opens up or down.

b. Clearing the equations from part a, graph $y = 2x^2$ and $y = -2x^2$ together. Do
these graphs confirm your conclusion from part a?
yes

c. On a new grid, graph $y = \frac{1}{2}x^2$ and $y = -\frac{1}{2}x^2$. What is different about these graphs compared with those in parts a and b?

These graphs are "wider."

d. What do these explorations suggest about how the absolute value of a affects the graph of an equation in the form $y = ax^2$?

The smaller the absolute value of a, the wider the parabola.

56. a. Enter the equation $y = x^2 + x + 1$; then graph. What are the coordinates of the y-intercept?

$(0, 1)$

b. On the same grid, graph $y = x^2 + 2x + 1$. Compare the graph of $y = x^2 + x + 1$ with the graph of $y = x^2 + 2x + 1$; then discuss their similarities and differences.

They have the same y-intercept, but the vertex of $y = x^2 + 2x + 1$ is further left and down.

c. Using the same grid, graph $y = x^2 + 3x + 1$. Compare this graph with the graphs of $y = x^2 + x + 1$ and $y = x^2 + 2x + 1$; then discuss similarities and differences.

The vertex of $x^2 + 3x + 1$ is further left and down. All three graphs open upward, have the same y-intercepts, and have the same shape.

d. On the same grid, graph $y = x^2 - x + 1$, $y = x^2 - 2x + 1$, and $y = x^2 - 3x + 1$. Compare these graphs with your graphs from parts a–c. What do these explorations suggest about how b affects the graph of an equation in the form $y = ax^2 + bx + c$?

It shifts the location of the vertex.

Review Exercises

Exercises 1–4 Expressions

For Exercises 1 and 2, multiply.

[5.4] 1. $3mn(m^2 - 4mn + 5)$

$3m^3n - 12m^2n^2 + 15mn$

[5.5] 2. $(x - 3)(2x + 1)$

$2x^2 - 5x - 3$

For Exercises 3 and 4, factor.

[6.3] 3. $3n^2 - 17n + 20$

$(3n - 5)(n - 4)$

[6.4] 4. $x^2 - 25$

$(x + 5)(x - 5)$

Exercises 5 and 6 Equations and Inequalities

[6.6] *For Exercises 5 and 6, solve.*

5. $h(h - 3) = 0$

$0, 3$

6. $y^2 + 3y - 18 = 0$

$-6, 3$

Chapter 6 Summary and Review Exercises

Complete each incomplete definition, rule, or procedure; study the key examples; and then work the related exercises.

6.1 Greatest Common Factor and Factoring by Grouping

Definitions/Rules/Procedures	Key Example(s)
A number or an expression is written in **factored form** if it is written as a(n) ___product of factors___.	One factorization of 24 is $4 \cdot 6$. All of the natural number factors of 24 are 1, 2, 3, 4, 6, 8, 12, and 24.

Exercises 1–6 **Expressions**

[6.1] *For Exercises 1–6, list all of the natural number factors.*

1. 9
   1, 3, 9

2. 33
   1, 3, 11, 33

3. 81
   1, 3, 9, 27, 81

4. 45
   1, 3, 5, 9, 15, 45

5. 30
   1, 2, 3, 5, 6, 10, 15, 30

6. 60
   1, 2, 3, 4, 5, 6, 10, 12, 15, 20, 30, 60

Definitions/Rules/Procedures	Key Example(s)
The **greatest common factor (GCF)** of a given set of numbers is the ___largest___ natural number that ___divides___ all given numbers with no remainder.	
To find the GCF of a set of numbers by listing:	Find the GCF of 54 and 180 by listing.
1. List all ___possible factors___ for each given number.	Factors of 54: 1, 2, 3, 6, 9, 18, 27, 54
2. Search the lists for the ___largest factor common___ to all lists.	Factors of 180: 1, 2, 3, 4, 5, 6, 9, 10, 12, 15, 18, 20, 30, 36, 45, 60, 90, 180
To find the GCF of a given set of numbers by prime factorization:	GCF = 18
1. Write the prime factorization of each given number in ___exponential form___.	Find the GCF of 54 and 180 by prime factorization.
2. Create a factorization for the GCF that includes only those prime factors ___common___ to all factorizations, each raised to its ___smallest exponent___ in the factorizations.	$54 = 2 \cdot 3^3$ $180 = 2^2 \cdot 3^2 \cdot 5$ $GCF = 2 \cdot 3^2 = 18$
3. ___Multiply___ the factors in the factorization created in step 2.	
Note: If there are no common prime factors, the GCF is 1.	

Exercises 7–10 **Expressions**

[6.1] *For Exercises 7–10, find the GCF.*

7. 21, 30
  3

8. 12, 42, 60
  6

9. $10y^4, 25y^2$
  $5y^2$

10. $15m^2n, 3mn^5$
  $3mn$

Definitions/Rules/Procedures	Key Example(s)
To factor a monomial GCF out of a polynomial:	Factor $54x^3y - 180x^2yz$.
1. Find the ___GCF___ of the terms in the polynomial.	$54x^3y - 180x^2yz$
2. Rewrite the given polynomial as a product of the ___GCF___ and the quotient of the ___polynomial___ and the ___GCF___. $\text{Polynomial} = \text{GCF}\left(\dfrac{\text{Polynomial}}{\text{GCF}}\right).$	$= 18x^2y\left(\dfrac{54x^3y - 180x^2yz}{18x^2y}\right)$ $= 18x^2y\left(\dfrac{54x^3y}{18x^2y} - \dfrac{180x^2yz}{18x^2y}\right)$ $= 18x^2y(3x - 10z)$

Exercises 11–16 ⟹ Expressions

[6.1] *For Exercises 11–16, factor out the GCF.*

11. $4x - 2$
$2(2x - 1)$

12. $5m - 35m^3$
$5m(1 - 7m^2)$

13. $x^2y + xy^2 + x^3y^3$
$xy(x + y + x^2y^2)$

14. $105a^3b^2 - 63a^2b^3 + 84a^6b^4$
$21a^2b^2(5a - 3b + 4a^4b^2)$

15. $18ab^3c - 36a^2b^2c$
$18ab^2c(b - 2a)$

16. $100k^4 + 120k^5 - 10k + 40k^3$
$10k(10k^3 + 12k^4 - 1 + 4k^2)$

Definitions/Rules/Procedures	Key Example(s)
To factor a four-term polynomial by grouping: 1. Factor out any monomial GCF (other than 1) that is ___common___ to all four terms. 2. Group pairs of terms and factor the ___GCF___ out of each pair or group. 3. If there is a common ___binomial___ factor, factor it out. 4. If there is no common ___binomial___ factor, interchange the ___middle two terms___ and repeat the process. If there is still no common ___binomial___ factor, the polynomial cannot be factored by grouping.	Factor by grouping: $$8x^2 - 20x - 6xy + 15y$$ $8x^2 - 20x - 6xy + 15y$ $= (8x^2 - 20x) + (-6xy + 15y)$ $= 4x(2x - 5) - 3y(2x - 5)$ $= (2x - 5)(4x - 3y)$

Exercises 17–24 ⟹ Expressions

[6.1] *For Exercises 17–24, factor by grouping.*

17. $ax + ay + bx + by$
$(x + y)(a + b)$

18. $ax + 2a + bx + 2b$
$(x + 2)(a + b)$

19. $y^3 + 2y^2 + 3y + 6$
$(y + 2)(y^2 + 3)$

20. $y^4 + 4y^3 - by - 4b$
$(y + 4)(y^3 - b)$

21. $xy + y + x + 1$
$(x + 1)(y + 1)$

22. $ax^2 - 5y^2 + ay^2 - 5x^2$
$(x^2 + y^2)(a - 5)$

23. $2b^3 - 2b^2 + b - 1$
$(2b^2 + 1)(b - 1)$

24. $u^2 - 3u + 4uv - 12v$
$(u - 3)(u + 4v)$

6.2 Factoring Trinomials of the Form $x^2 + bx + c$

Definitions/Rules/Procedures	Key Example(s)
To factor a trinomial of the form $x^2 + bx + c$: 1. Find two numbers whose product is ___c___ and whose sum is ___b___. 2. The factored trinomial will have the form $(x + $ ___first number___ $)(x + $ ___second number___ $)$. *Note:* The signs in the binomial factors can be minus signs depending on the signs of b and c.	Factor $x^2 + 9x + 20$. Find two numbers whose product is 20 and whose sum is 9. Product: Sum: $(1)(20) = 20$ $1 + 20 = 21$ $(2)(10) = 20$ $2 + 10 = 12$ $(4)(5) = 20$ $4 + 5 = 9$ Correct combination $x^2 + 9x + 20 = (x + 4)(x + 5)$ Factor $x^2 + 2x - 15$. Product: Sum: $(-1)(15) = -15$ $-1 + 15 = 14$ $(-3)(5) = -15$ $-3 + 5 = 2$ Correct combination $x^2 + 2x - 15 = (x - 3)(x + 5)$

Exercises 25–32 Expressions

[6.2] *For Exercises 25–32, factor.*

25. $x^2 - x - 12$
$(x - 4)(x + 3)$

26. $x^2 + 14x + 45$
$(x + 5)(x + 9)$

27. $n^2 - 6n + 8$
$(n - 2)(n - 4)$

28. $a^2 + 3a - 20$
prime

29. $h^2 + 51h + 144$
$(h + 3)(h + 48)$

30. $y^2 - 10y - 24$
$(y - 12)(y + 2)$

31. $4x^2 - 24x + 36$
$4(x - 3)^2$

32. $3m^2 - 33m + 54$
$3(m - 9)(m - 2)$

6.3 Factoring Trinomials of the Form $ax^2 + bx + c$, where $a \neq 1$

Definitions/Rules/Procedures	Key Example(s)

To factor a trinomial of the form $ax^2 + bx + c$, where $a \neq 1$, by trial and error:
 	
1. Factor out any monomial ____GCF____ common to all of the terms.
 	
2. Write a pair of ____first____ terms whose product is ax^2.
 	
3. Write a pair of last terms whose product is ____c____.
 	
4. Verify that the sum of the ____inner____ and ____outer____ products is bx (the middle term of the trinomial). If the sum of the ____inner____ and ____outer____ products is not bx, try the following:
  a. Exchange the ____last____ terms of the binomials from step 3; then repeat step 4.
  b. For each additional pair of ____last terms____, repeat steps 3 and 4.
  c. For each additional pair of ____first____ terms, repeat steps 2–4.
	Factor $4x^2 - 9x + 5$.
	
	Factors of 4 are 4 and 1, or 2 and 2.
	Factors of 5 are 1 and 5.
	Because the middle term $-9x$ is negative and the last term 5 is positive, we know that the second term in each binomial factor will be negative.
	
	$$(x - 1)(4x - 5) = 4x^2 - 9x + 5$$

To factor a trinomial of the form $ax^2 + bx + c$, where $a \neq 1$, by grouping: 1. Factor out any monomial ____GCF____ common to all of the terms. 2. Find two factors of ____ac____ whose sum is ____b____. 3. Write a four-term polynomial in which ____bx____ is written as the sum of two like terms whose coefficients are the two factors you found in step 2. 4. Factor by ____grouping____.	Factor $18y^3 + 12y^2 - 48y$. Factor out the monomial GCF, $6y$. $$18y^3 + 12y^2 - 48y = 6y(3y^2 + 2y - 8)$$ To factor the trinomial, multiply $3(-8) = -24$; then find two factors of -24 whose sum is 2. Note that -4 and 6 work. Write the middle term, $2y$, as $-4y + 6y$; then factor by grouping. $$= 6y(3y^2 - 4y + 6y - 8)$$ $$= 6y[y(3y - 4) + 2(3y - 4)]$$ $$= 6y(3y - 4)(y + 2)$$

Exercises 33–40 Expressions

[6.3] *For Exercises 33–40, factor completely.*

33. $6x^2 + 3x - 7$
prime

34. $2u^2 + 5u + 2$
$(2u + 1)(u + 2)$

35. $3m^2 - 10m + 3$
$(3m - 1)(m - 3)$

36. $5k^2 - 7kh - 12h^2$
$(5k - 12h)(k + h)$

37. $6a^2 - 20a + 16$
$2(3a - 4)(a - 2)$

38. $8x^2y - 4xy - y$
$y(8x^2 - 4x - 1)$

39. $3p^2 - 13pq + 4q^2$
$(3p - q)(p - 4q)$

40. $24x^2 - 64xy - 24y^2$
$8(3x + y)(x - 3y)$

6.4 Factoring Special Products

Definitions/Rules/Procedures	Key Example(s)
Rules for factoring special products: **Perfect square trinomials:** $a^2 + 2ab + b^2 = \underline{(a+b)^2}$. $ a^2 - 2ab + b^2 = \underline{(a-b)^2}$. **Difference of squares:** $a^2 - b^2 = \underline{(a+b)(a-b)}$. Note: A **sum of squares**, $a^2 + b^2$, cannot be factored. **Difference of cubes:** $a^3 - b^3 = \underline{(a-b)(a^2 + ab + b^2)}$. **Sum of cubes:** $a^3 + b^3 = \underline{(a+b)(a^2 - ab + b^2)}$.	**Example 1:** Factor. **a.** $9x^2 + 30x + 25 = (3x + 5)^2$ **b.** $36m^2 - 60mn + 25n^2 = (6m - 5n)^2$ **c.** $4y^2 - 81 = (2y + 9)(2y - 9)$ **d.** $16p^2 + 1$ is prime. **e.** $8x^3 - 27 = (2x - 3)(4x^2 + 6x + 9)$ **f.** $n^3 + 64 = (n + 4)(n^2 - 4n + 16)$

Exercises 41–48 ➡ Expressions

[6.4] *For Exercises 41–48, factor.*

41. $v^2 - 8v + 16$
$(v - 4)^2$

42. $u^2 + 6u + 9$
$(u + 3)^2$

43. $4x^2 + 20x + 25$
$(2x + 5)^2$

44. $9y^2 - 12y + 4$
$(3y - 2)^2$

45. $x^2 - 4$
$(x + 2)(x - 2)$

46. $25 - y^2$
$(5 + y)(5 - y)$

47. $x^3 - 1$
$(x - 1)(x^2 + x + 1)$

48. $x^3 + 27$
$(x + 3)(x^2 - 3x + 9)$

6.5 Strategies for Factoring

Definitions/Rules/Procedures	Key Example(s)

To factor a polynomial, first factor out any monomial GCF; then consider the number of terms in the polynomial. If the polynomial has
 	
  I. Four terms, try factoring by $\underline{\text{grouping}}$.
 	
  II. Three terms, determine whether the trinomial is a perfect square.
    A. If the trinomial is a perfect square, consider its form.
      1. If in form $a^2 + 2ab + b^2$, the factored form is $\underline{(a+b)^2}$.
      2. If in form $a^2 - 2ab + b^2$, the factored form is $\underline{(a-b)^2}$.
 	
    B. If the trinomial is not a perfect square, consider its form.
      1. If in form $x^2 + bx + c$, find two factors of $\underline{c}$ whose sum is $\underline{b}$ and write the factored form as $(x + \text{first number})(x + \text{second number})$.
      2. If in form $ax^2 + bx + c$, where $a \neq 1$, use $\underline{\text{trial and error}}$. Or find two factors of $\underline{ac}$ whose sum is $\underline{b}$; write these factors as coefficients of two like terms that, when combined, equal bx; and factor by grouping.
	The examples shown do not represent every possible type of problem. Each example contains a monomial GCF other than 1.
	
	Factoring by grouping:
	$8x^3 + 24x^2 - 2x^2y - 6xy$
	$ = 2x(4x^2 + 12x - xy - 3y)$
	$ = 2x[4x(x + 3) - y(x + 3)]$
	$ = 2x(x + 3)(4x - y)$
	
	Factoring a perfect square trinomial:
	$75x^4 + 60x^3y + 12x^2y^2$
	$ = 3x^2(25x^2 + 20xy + 4y^2)$
	$ = 3x^2(5x + 2y)^2$

Definitions/Rules/Procedures	Key Example(s)
III. Two terms, determine whether the binomial is a difference of squares, sum of cubes, or difference of cubes.	Factoring a difference of squares:
A. If given a binomial that is a difference of squares $a^2 - b^2$, the factors are <u> conjugates </u> and the factored form is <u> $(a + b)(a - b)$ </u>. Note that a sum of squares cannot be factored.	$$\begin{aligned} 81y^5 - 16y &= y(81y^4 - 16) \\ &= y(9y^2 + 4)(9y^2 - 4) \\ &= y(9y^2 + 4)(3y + 2)(3y - 2) \end{aligned}$$
B. If given a binomial that is a sum of cubes, $a^3 + b^3$, the factored form is <u> $(a + b)(a^2 - ab + b^2)$ </u>.	Factoring a sum of cubes:
C. If given a binomial that is a difference of cubes, $a^3 - b^3$, the factored form is <u> $(a - b)(a^2 + ab + b^2)$ </u>.	$$\begin{aligned} 54n^4 + 16nm^3 &= 2n(27n^3 + 8m^3) \\ &= 2n(3n + 2m)(9n^2 - 6mn + 4m^2) \end{aligned}$$
Note: Always check to see if any of the factors can be factored further.	

Exercises 49–58   **Expressions**

[6.5] *For Exercises 49–58, factor completely.*

49. $6b^2 + b - 2$
    $(3b + 2)(2b - 1)$

50. $4ab - 24ab^2$
    $4ab(1 - 6b)$

51. $y^2 + 25$
    prime

52. $3x^2 - 3y^2$
    $3(x + y)(x - y)$

53. $x^4 - 81$
    $(x^2 + 9)(x + 3)(x - 3)$

54. $7u^2 - 14u - 105$
    $7(u - 5)(u + 3)$

55. $8x^3 - 27y^3$
    $(2x - 3y)(4x^2 + 6xy + 9y^2)$

56. $2 - 50y^2$
    $2(1 + 5y)(1 - 5y)$

57. $3am - 6an - 8bn + 4bm$
    $(m - 2n)(3a + 4b)$

58. $3m^2 + 9m + 27$
    $3(m^2 + 3m + 9)$

6.6 Solving Quadratic Equations by Factoring

Definitions/Rules/Procedures	Key Example(s)
Zero-Factor Theorem	Solve $(2x - 5)(x + 4) = 0$.
If a and b are real numbers and $ab = 0$, then <u> $a = 0$ </u> or <u> $b = 0$ </u>.	$$2x - 5 = 0 \quad \text{or} \quad x + 4 = 0$$ $$2x = 5 \qquad\qquad x = -4$$ $$x = \frac{5}{2}$$
To solve an equation in which two or more factors are equal to 0, use the zero-factor theorem.	
1. Set each factor equal to <u> zero </u>.	
2. Solve each of those equations.	
A **quadratic equation** in one variable is an equation that can be written in the form <u> $ax^2 + bx + c = 0$ </u>, where a, b, and c are real numbers and $a \neq 0$.	Solve $2x^2 = 3 - x$.
To solve a quadratic equation using factoring:	$$2x^2 = 3 - x$$
1. Write the equation in <u> standard </u> form $(ax^2 + bx + c = 0)$.	$2x^2 + x = 3$ Add x to both sides.
	$2x^2 + x - 3 = 0$ Subtract 3 from both sides.
2. Write the variable expression in <u> factored </u> form.	$(2x + 3)(x - 1) = 0$ Factor.
3. Use the <u> zero-factor </u> theorem to solve.	$2x + 3 = 0 \quad \text{or} \quad x - 1 = 0$ Use the zero-factor theorem.
	$2x = -3 \qquad\qquad x = 1$
	$x = -\dfrac{3}{2}$

Definitions/Rules/Procedures	Key Example(s)
	The length of a small rectangular building is to be 15 feet more than the width. The area of the base is to be 700 square feet. What must the dimensions of the base be?
	Let w represent width. The length will be $w + 15$. Writing an equation for area, we have

$$w(w + 15) = 700$$
$$w^2 + 15w = 700 \qquad \text{Distribute } w.$$
$$w^2 + 15w - 700 = 0 \qquad \text{Subtract 700 from both sides.}$$
$$(w - 20)(w + 35) = 0 \qquad \text{Factor.}$$
$$w - 20 = 0 \quad \text{or} \quad w + 35 = 0 \qquad \text{Use the zero-factor theorem.}$$
$$w = 20 \qquad \qquad w = -35$$

Answer: The width is 20 ft., and the length is 35 ft.

Exercises 59–71 ◣ Equations and Inequalities

[6.6] *For Exercises 59–66, solve.*

59. $(m + 3)(m - 4) = 0$
$-3, 4$

60. $y^2 - 4 = 0$
$-2, 2$

61. $x^2 - 5x + 6 = 0$
$2, 3$

62. $x^2 - 4x = 21$
$-3, 7$

63. $m^2 = 9$
$-3, 3$

64. $x^2 + 3x = 18$
$-6, 3$

65. $y(y + 6) = -9$
-3

66. $3n^2 - 11n = 4$
$-\dfrac{1}{3}, 4$

[6.6] *For Exercises 67–71, translate to an equation; then solve.*

67. Find every number such that five times the square of the number is equal to twice the number.
$5x^2 = 2x; 0, \dfrac{2}{5}$

68. The sum of the squares of two consecutive natural numbers is 61. Find the numbers.
$x^2 + (x + 1)^2 = 61; 5, 6$

69. The product of a number and four times that same number is 100. Find all numbers.
$x \cdot 4x = 100; -5, 5$

70. Find the dimensions of a rectangle whose length is 6 more than its width and whose area is 91 square inches.
$(w + 6)w = 91; 13$ in. by 7 in.

71. The product of two consecutive natural numbers is 110. Find the numbers.
$x(x + 1) = 110; 10, 11$

Definitions/Rules/Procedures	Key Example(s)

Pythagorean Theorem

Given a right triangle, where a and b represent the lengths of the legs and c represents the length of the hypotenuse, $a^2 + b^2 = \underline{c^2}$. Further, if a, b, and c represent lengths of the sides of a triangle and $\underline{\quad a^2 + b^2 = c^2 \quad}$, then the triangle is a right triangle with hypotenuse c.

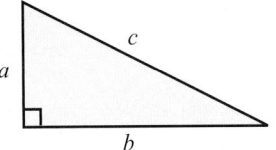

Find the unknown length in the following right triangle.

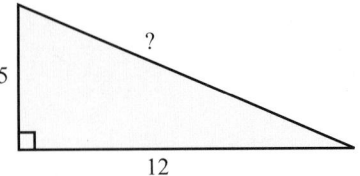

Solution: Use the Pythagorean theorem, $a^2 + b^2 = c^2$.

$5^2 + 12^2 = c^2$	$a = 5, b = 12$
$25 + 144 = c^2$	Simplify exponential forms.
$169 = c^2$	Add.
$0 = c^2 - 169$	Subtract 169 from both sides.
$0 = (c - 13)(c + 13)$	Factor.
$c - 13 = 0, c + 13 = 0$	Use the zero-factor theorem.
$c = 13 \qquad c = -13$	

Answer: The unknown length is 13.

Exercise 72 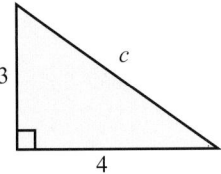 Equations and Inequalities

72. Find the length of the missing side of the right triangle shown.
$3^2 + 4^2 = c^2; 5$

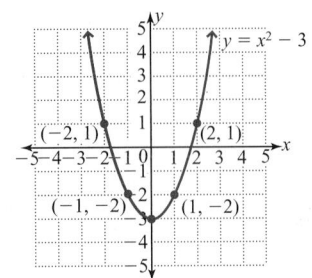

6.7 Graphs of Quadratic Equations and Functions

Definitions/Rules/Procedures	Key Example(s)

A **quadratic equation** in two variables is an equation that can be written in the form $\underline{y = ax^2 + bx + c}$, where a, b, and c are real numbers and $a \neq 0$.

The **axis of symmetry** is a line that divides a graph into two $\underline{\text{symmetrical}}$ halves.

The **vertex of a parabola** is the $\underline{\text{lowest}}$ point on a parabola that opens upward or the $\underline{\text{highest}}$ point on a parabola that opens downward.

To graph a quadratic equation:

1. Find ordered pair $\underline{\text{solutions}}$ and plot them in the coordinate plane. Continue finding and plotting $\underline{\text{solutions}}$ until the shape of the $\underline{\text{parabola}}$ can be clearly seen.

2. Connect the points to form a(n) $\underline{\text{parabola}}$.

Given an equation in the form $y = ax^2 + bx + c$, if $a > 0$, then the parabola opens $\underline{\text{upward}}$; if $a < 0$, then the parabola opens $\underline{\text{downward}}$.

Graph $y = x^2 - 3$.

Definitions/Rules/Procedures	Key Example(s)
To graph a quadratic function: 1. Find enough ordered pairs by evaluating the function for various values of _____*x*_____ so that when those ordered pairs are plotted, the shape of the ___parabola___ can be clearly seen. 2. Connect the points to form the ___parabola___.	Graph $f(x) = -x^2 + 4x - 1$. 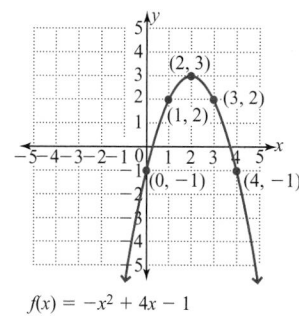 $f(x) = -x^2 + 4x - 1$

Exercises 73–76 Equations and Inequalities

[6.7] *For Exercises 73–76, graph.*

73. $y = -5x^2$

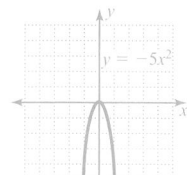

74. $y = x^2 - 3$

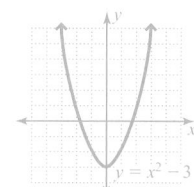

75. $f(x) = x^2 + 2$

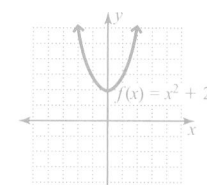

76. $f(x) = -3x^2 + 6x + 2$

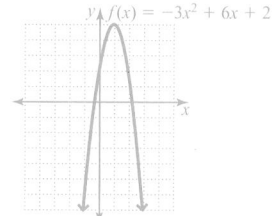

Learning Strategy

When preparing for an exam, give yourself enough time. Study a day or two before the exam and the morning of the exam. Eat breakfast and be calm.

—Biridianna N.

Chapter 6 **Practice Test**

For Extra Help

Step-by-step test solutions are found on the Chapter Test Prep Videos available in MyMathLab® *or on* YouTube.

1. List all possible natural number factors of 80.

   1, 2, 4, 5, 8, 10, 16, 20, 40, 80 [6.1]

2. Find the GCF of $25m^2$ and $40m$.

   5m [6.1]

Factor completely.

3. $5y - 30$

   $5(y - 6)$ [6.1]

4. $6x^2y - 2y^2$

   $2y(3x^2 - y)$ [6.1]

5. $ax - ay - bx + by$

   $(x - y)(a - b)$ [6.1]

6. $m^2 + 7m + 12$

   $(m + 3)(m + 4)$ [6.2]

7. $r^2 - 8r + 16$

   $(r - 4)^2$ [6.4]

8. $4y^2 + 20y + 25$

   $(2y + 5)^2$ [6.3]

9. $3q^2 - 10q + 8$

   $(q - 2)(3q - 4)$ [6.3]

10. $6x^2 - 23x + 20$

   $(3x - 4)(2x - 5)$ [6.3]

11. $ax^2 - 5ax - 24a$

   $a(x + 3)(x - 8)$ [6.2]

12. $10n^3 + 38n^2 - 8n$

   $2n(n + 4)(5n - 1)$ [6.2]

13. $c^2 - 25$

   $(c + 5)(c - 5)$ [6.4]

14. $x^2 + 4$

   prime [6.4]

15. $2 - 50u^2$

   $2(1 + 5u)(1 - 5u)$ [6.4]

16. $-4x^2 + 16$

   $-4(x + 2)(x - 2)$ [6.4]

17. $m^3 + 125$

   $(m + 5)(m^2 - 5m + 25)$ [6.4]

18. $x^3 - 8$

   $(x - 2)(x^2 + 2x + 4)$ [6.4]

Solve.

19. $a(a + 3) = 0$
-3, 0 [6.6]

20. $x^2 - 4x - 12 = 0$
6, -2 [6.6]

21. $2n^2 + 7n = 15$
$\dfrac{3}{2}$, -5 [6.6]

22. The product of two consecutive natural numbers is 72. Find the numbers.
8, 9 [6.6]

23. The width of a rectangle is 5 feet less than the length. If the area is 36 square feet, find the length and width.
9 ft., 4 ft. [6.6]

Graph.

24. $y = -2x^2 + 4$
[6.7]

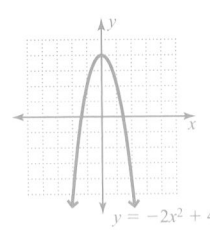

25. $f(x) = x^2 - 6x + 5$
[6.7]

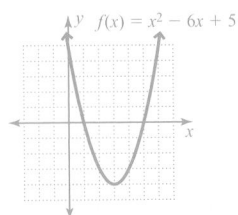

Chapters 1–6 Cumulative Review Exercises

For Exercises 1–4, answer true or false.

[5.4] 1. $x^3 \cdot x^4 = x^{12}$
false

[6.1] 2. The GCF of 12 and 5 is 1.
true

[1.7] 3. The expression $5 + x$ can be simplified to equal $5x$.
false

[6.1] 4. $4x + 12 = 2(2x + 6)$ is factored completely.
false

For Exercises 5 and 6, fill in the blank.

[6.4] 5. To factor a difference of squares, we use the rule $a^2 - b^2 =$ ___$(a + b)(a - b)$___.

[6.6] 6. To solve a quadratic equation using factoring:
  a. Write the equation in ___standard___ form $(ax^2 + bx + c = 0)$.
  b. Write the variable expression in ___factored___ form.
  c. Use the zero-factor theorem to solve.

Exercises 7–19 **Expressions**

For Exercises 7–14, simplify.

[1.5] 7. $14 - 2 \cdot 6 \cdot 3 + 8^2$
42

[1.5] 8. $-6|5 + 3^2| - \sqrt{14 + 11}$
−89

[5.3] 9. $(5x^2 + 6x - 1) - (2x - 3)$
$5x^2 + 4x + 2$

[5.4] 10. $(6x^2)(3x^2y)$
$18x^4y$

[5.4] 11. $(3x)(4x^3)^2$
$48x^7$

[5.5] 12. $(y - 8)(2y + 1)$
$2y^2 - 15y - 8$

[5.6] 13. $\dfrac{16h^3 + 4h^2 - 8h}{4h^2}$
$4h + 1 - \dfrac{2}{h}$

[5.6] 14. $(x^2 + 7x + 12) \div (x + 3)$
$x + 4$

[6.1–6.5] *For Exercises 15–19, factor completely.*

15. $x^2 - 121$
$(x + 11)(x - 11)$

16. $10x^2 + 40$
$10(x^2 + 4)$

17. $x^2 - 8x + 16$
$(x - 4)^2$

18. $2x^2 + 5x - 12$
$(2x - 3)(x + 4)$

19. $x^3 - 5x^2 + 5x - 25$
$(x - 5)(x^2 + 5)$

Exercises 20–30 **Equations and Inequalities**

For Exercises 20–22, solve.

[2.3] 20. $\dfrac{3}{4} = x - \dfrac{2}{3}$
$\dfrac{17}{12}$

[2.3] 21. $2.6 + 7a + 5 = 8a - 5.6$
13.2

[6.6] 22. $x^2 - 3x = 28$
$-4, 7$

For Exercises 23–25, solve.

[2.4] 23. Solve for x in the equation $2x - 3y = 7$.
$x = \dfrac{7 + 3y}{2}$

[3.3] 24. Find the x- and y-intercepts for $y = 2x - 6$.
$(3, 0), (0, -6)$

[3.4] 25. Find the slope of the line containing the points $(6, 2)$ and $(5, 2)$.
0

[3.2] 26. Graph $y = 3x + 2$.

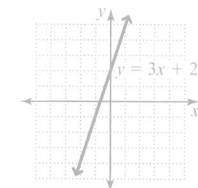

[2.8] *For Exercise 27 solve.*   *a.*   *Write the solution set using set-builder notation.*
                               b.   *Write the solution set using interval notation.*
                                 c.   *Graph the solution set.*

27. $7x - 5 \le 3x + 11$

     a. $\{x \mid x \le 4\}$ b. $(-\infty, 4]$

     c.

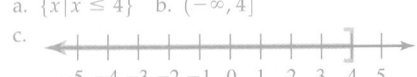

          $-5 \ \ -4 \ \ -3 \ \ -2 \ \ -1 \ \ 0 \ \ 1 \ \ 2 \ \ 3 \ \ 4 \ \ 5$

[4.2] 28. Solve the system $\begin{cases} 3x + y = 2 \\ 6x + 5y = 7 \end{cases}$ using substitution.

     $\left(\dfrac{1}{3}, 1 \right)$

For Exercises 29 and 30, solve.

[4.2] 29. Find three consecutive even integers whose sum is 24.
[4.3]   6, 8, 10

[4.2] 30. Janice invests a total of $8000 in two different accounts. The first account
[4.3]   returns 5%, while the second account returns 8%. If the total interest earned
         after one year is $565, what principal did she invest in each account?
         $2500 in 5%, $5500 in 8%

Chapter Overview

In this chapter, we explore rational expressions, which are like algebraic fractions. The chapter flow is similar to the flow of Chapters 5 and 6. We will:

▶ Define rational expressions.

▶ Perform arithmetic operations on rational expressions.

▶ Solve equations that contain rational expressions.

Instructor Note

Rational expressions may be the most challenging topic students encounter in algebra. To be successful, students must have a clear understanding of polynomial arithmetic and factoring. In this chapter, it is especially helpful to include exercises in each assignment to review essential skills prior to the new material in the next section. Similarly, use the warm-up exercises at the beginning of class to remind students of those skills essential to the material that will be covered in the section.

7.1 Simplifying Rational Expressions

7.2 Multiplying and Dividing Rational Expressions

7.3 Adding and Subtracting Rational Expressions with the Same Denominator

7.4 Adding and Subtracting Rational Expressions with Different Denominators

7.5 Complex Rational Expressions

7.6 Solving Equations Containing Rational Expressions

7.7 Applications with Rational Expressions, Including Variation

7.1 Simplifying Rational Expressions

Objectives

1 Evaluate rational expressions.

2 Find numbers that cause a rational expression to be undefined.

3 Simplify rational expressions containing only monomials.

4 Simplify rational expressions containing multiterm polynomials.

Connection Notice how the definition of a rational expression is like the definition of a rational number. Recall from Chapter 1 that a rational number can be expressed in the form $\frac{a}{b}$, where a and b are integers and $b \neq 0$. Some texts call rational expressions algebraic fractions.

Warm-up

[1.2] **1.** Simplify to lowest terms: $\dfrac{24}{60}$

[5.6] **2.** Divide: $\dfrac{28x^5}{4x^2}$

For Exercises 3 and 4, factor completely.

[6.2] **3.** $y^2 + 4y - 32$ [6.3] **4.** $7x^4 - 14x^3 - 105x^2$

Objective 1 Evaluate rational expressions.

Now that we have explored polynomials, we are ready to develop a new class of expressions called **rational expressions**.

Definition **Rational expression:** An expression that can be written in the form

$$\frac{P}{Q}, \text{ where } P \text{ and } Q \text{ are polynomials and } Q \neq 0.$$

Following are some rational expressions.

$$\frac{3x^5}{18x^2} \qquad \frac{7x}{x^2 - 9} \qquad \frac{x^2 + 2x - 15}{4x + 20}$$

As with any expression, we can evaluate or rewrite rational expressions. First, let's look at evaluating rational expressions. Remember that to evaluate an algebraic expression, we replace the variables with given values and then simplify the resulting numerical expression.

Example 1 Evaluate the expression $\dfrac{3x - 5}{x + 1}$, when

a. $x = 2$

Solution: $\dfrac{3x - 5}{x + 1} = \dfrac{3(2) - 5}{2 + 1}$ Replace *x* with 2.

$$= \frac{6 - 5}{3}$$

$$= \frac{1}{3}$$

b. $x = -0.4$

Solution: $\dfrac{3x - 5}{x + 1} = \dfrac{3(-0.4) - 5}{-0.4 + 1}$ Replace *x* with −0.4.

$$= \frac{-1.2 - 5}{0.6}$$

$$= \frac{-6.2}{0.6}$$

$$= -10.\overline{3} \quad \text{or} \quad -10\frac{1}{3}$$

Answers to Warm-up

1. $\dfrac{2}{5}$ 2. $7x^3$

3. $(y + 8)(y - 4)$

4. $7x^2(x - 5)(x + 3)$

c. $x = -1$

Solution: $\dfrac{3x - 5}{x + 1} = \dfrac{3(-1) - 5}{-1 + 1}$ Replace x with -1.

$$= \dfrac{-3 - 5}{0}$$

$$= \dfrac{-8}{0}, \quad \text{which is undefined}$$

Learning Strategy

The book offers practice problems as well as the solutions. Do the practice problems and check your answers. If they are wrong, keep practicing until you get the correct solutions.

—Ruben C.

Your Turn 1   Evaluate the expression $\dfrac{x^2 - 9}{x - 2}$, when

a. $x = 4$.   **b.** $x = -3$.   **c.** $x = 2$.

Objective 2 Find numbers that cause a rational expression to be undefined.

Note that in Example 1(c), the expression $\dfrac{3x - 5}{x + 1}$ is undefined when $x = -1$ because replacing x with -1 causes the denominator to be 0. We need to avoid values that cause a rational expression to be undefined, which means we need to identify and avoid values that make their denominators 0. This suggests the following procedure.

> **Procedure** **Finding Values That Make a Rational Expression Undefined**
>
> To determine the value(s) that make a rational expression undefined:
> 1. Set the denominator equal to 0.
> 2. Solve the equation.

Example 2   Find every value for the variable that makes the expression undefined.

a. $\dfrac{5y}{3y - 2}$

Note We ignore the numerator ▶ because the rational expression is undefined only when the denominator is 0.

Solution: $3y - 2 = 0$ Set the denominator, $3y - 2$, equal to 0.

$\qquad\qquad 3y = 2$ Add 2 to both sides.

$\qquad\qquad y = \dfrac{2}{3}$ Divide both sides by 3.

Answer: $\dfrac{5y}{3y - 2}$ is undefined if y is replaced with $\dfrac{2}{3}$.

b. $\dfrac{7}{x^2 + x - 12}$

Solution: $x^2 + x - 12 = 0$ Set the denominator equal to 0.

$\qquad\quad (x + 4)(x - 3) = 0$ Factor.

$\qquad\quad x + 4 = 0 \quad \text{or} \quad x - 3 = 0$ Use the zero-factor theorem.

$\qquad\qquad\quad x = -4 \qquad\qquad x = 3$

Answer: $\dfrac{7}{x^2 + x - 12}$ is undefined if x is replaced with -4 or 3.

c. $\dfrac{n - 6}{n^2 + 9}$

Solution: $n^2 + 9 = 0$ Set the denominator equal to 0.

For all real-number values of n, $n^2 \geq 0$. Consequently $n^2 + 9 \neq 0$ for all real numbers. So $\dfrac{n - 6}{n^2 + 9}$ is defined for all real numbers.

Answers to Your Turn 1

a. $\dfrac{7}{2}$ **b.** 0   **c.** undefined

Your Turn 2 Find every value that can replace the variable in the expression and cause the expression to be undefined.

a. $\dfrac{y + 4}{3y - 1}$ b. $\dfrac{2x}{x^2 + 5x}$ c. $\dfrac{m}{m^2 + 1}$

Objective 3 Simplify rational expressions containing only monomials.

Now that we have defined rational expressions, we are ready to begin simplifying them. Recall from Section 1.2 that a fraction is in lowest terms when the only common factor of its numerator and denominator is 1. Also in Section 1.2, we used the rule $\dfrac{an}{bn} = \dfrac{a \cdot 1}{b \cdot 1} = \dfrac{a}{b}$, where $b \neq 0$ and $n \neq 0$, to simplify fractions. We can rewrite this rule so that it applies to rational expressions.

> **Rule**
>
> $\dfrac{PR}{QR} = \dfrac{P \cdot 1}{Q \cdot 1} = \dfrac{P}{Q}$, where P, Q, and R are polynomials and Q and R are not 0.

Instructor Note Elaborate on the Note by explaining that if $x = 0$ in $\dfrac{2x}{3x}$, the expression will be indeterminate. Consequently, we should add that $\dfrac{2x}{3x} = \dfrac{2}{3}$, where $x \neq 0$. However, we streamline our answers by making the assumption that variables will not be replaced with values that cause the rational expressions to be undefined or indeterminate.

The rule indicates that a factor common to both the numerator and denominator can be divided out of a rational expression. Consider the following comparison between simplifying a fraction and simplifying a similar rational expression.

$$\dfrac{10}{15} = \dfrac{2 \cdot 5}{3 \cdot 5} = \dfrac{2 \cdot 1}{3 \cdot 1} = \dfrac{2}{3} \qquad \dfrac{2x}{3x} = \dfrac{2 \cdot x}{3 \cdot x} = \dfrac{2 \cdot 1}{3 \cdot 1} = \dfrac{2}{3}$$

◄ **Note** When working with rational expressions, we assume that the variables will not be replaced with any value that would cause the expression to be undefined or indeterminate.

5 is the common factor here.

x is the common factor here.

This suggests the following procedure for simplifying rational expressions.

> **Procedure** Simplifying Rational Expressions to Lowest Terms
>
> To simplify a rational expression to lowest terms:
> 1. Factor the numerator and denominator completely.
> 2. Divide out all common factors in the numerator and denominator.
> 3. Multiply the remaining factors in the numerator and the remaining factors in the denominator.

We first simplify rational expressions that are a monomial over a monomial.

Connection In Section 5.6 and Chapter 6, we divided monomials using $\dfrac{a^m}{a^n} = a^{m-n}$ and $a^{-n} = \dfrac{1}{a^n}$. In this chapter, we divide monomials (simplify to lowest terms) by factoring the numerator and/or denominator into prime factors, as we did in Section 1.2, for consistency of approach. As we will see later in this section, this approach is the only way rational expressions with multiterm polynomials in the numerator and/or denominator can be simplified to lowest terms.

Example 3 Simplify $\dfrac{3x^5}{15x^2}$.

Solution: Factor the numerator and denominator completely; then divide out all common factors.

$$\dfrac{3x^5}{15x^2} = \dfrac{3 \cdot x \cdot x \cdot x \cdot x \cdot x}{3 \cdot 5 \cdot x \cdot x}$$

Note From here on, we omit this step with the 1's and simply highlight the common factors that are eliminated.

►

$$= \dfrac{1 \cdot x \cdot x \cdot x \cdot 1}{1 \cdot 5 \cdot 1}$$

$$= \dfrac{x^3}{5}$$

◄ **Note** The common factors are a single 3 and two x's. These form the GCF that we divide out, which is $3x^2$.

Answers to Your Turn 2

a. $\dfrac{1}{3}$ b. 0 or -5

c. defined for all real numbers

There are different styles for showing the process of dividing out common factors. For example, we could have used cancellation marks to show that the two x's and one 3 divide out in Example 3.

$$\frac{3x^5}{15x^2} = \frac{\overset{1}{3} \cdot \overset{1}{\cancel{x}} \cdot \overset{1}{\cancel{x}} \cdot x \cdot x \cdot x}{\underset{1}{3} \cdot 5 \cdot \underset{1}{\cancel{x}} \cdot \underset{1}{\cancel{x}}} = \frac{x^3}{5}$$

Develop a style that works best for you, keeping in mind that no matter the style, the bottom line is that you will be dividing out the GCF of the numerator and denominator to get a rational expression in lowest terms.

Connection
Using cancellation marks:

$$-\frac{4x^2}{28x^3} = -\frac{\overset{1}{2} \cdot \overset{1}{2} \cdot \overset{1}{\cancel{x}} \cdot \overset{1}{\cancel{x}}}{\underset{1}{2} \cdot \underset{1}{2} \cdot 7 \cdot \underset{1}{\cancel{x}} \cdot \underset{1}{\cancel{x}} \cdot x}$$

$$= -\frac{1}{7x}$$

Instructor Note Discuss that the GCF we divided out in Example 4(a) is $4x^2$, and in Example 4(b), we divided out $18ab^3$.

| **Example 4** | Simplify. |

a. $-\dfrac{4x^2}{28x^3}$

Solution: $-\dfrac{4x^2}{28x^3} = -\dfrac{2 \cdot 2 \cdot x \cdot x}{2 \cdot 2 \cdot 7 \cdot x \cdot x \cdot x}$

Factor the numerator and denominator completely. Then eliminate the common factors, which are two 2s and two x's. Multiply the remaining factors.

$$= -\frac{1}{7x}$$ ◀ **Note** Because all of the factors in the numerator divided out, the numerator is 1.

b. $\dfrac{36ab^3c}{54a^2b^3}$

Solution: $\dfrac{36ab^3c}{54a^2b^3} = \dfrac{2 \cdot 2 \cdot 3 \cdot 3 \cdot a \cdot b \cdot b \cdot b \cdot c}{2 \cdot 3 \cdot 3 \cdot 3 \cdot a \cdot a \cdot b \cdot b \cdot b}$

Factor the numerator and denominator completely. Then divide out the common factors, which are one 2, two 3s, one a, and three b's. Multiply the remaining factors.

$$= \frac{2c}{3a}$$

| **Your Turn 4** | Simplify. |

a. $-\dfrac{12t^6}{30t^2}$ **b.** $\dfrac{9n}{18n^3}$ **c.** $-\dfrac{72x^2y}{60x^2y^3z}$

Connection Dividing out common factors when simplifying rational expressions with monomial numerators and/or denominators to lowest terms can be viewed as dividing monomials using the rules of exponents, as we did in Section 5.6.

Dividing out common factors:

$$\frac{12x^4y^2}{6x^2y} = \frac{2 \cdot 2 \cdot 3 \cdot x \cdot x \cdot x \cdot x \cdot y \cdot y}{2 \cdot 3 \cdot x \cdot x \cdot y}$$

$$= 2 \cdot x \cdot x \cdot y = 2x^2y$$

Using rules of exponents:

$$\frac{12x^4y^2}{6x^2y} = \frac{12}{6} \cdot \frac{x^4}{x^2} \cdot \frac{y^2}{y}$$

$$= 2x^{4-2}y^{2-1} = 2x^2y$$

The coefficients can be simplified as we do with ordinary fractions. Because we are dividing out two of the four x's and one of the two y's in the numerator, we are left with two x's and one y in the numerator. Subtracting the exponents accounts for eliminating the two common factors of x and the one common factor of y. This connection also applies when the greater exponent is in the denominator.

Dividing out common factors:

$$\frac{15a^2}{10a^5} = \frac{3 \cdot 5 \cdot a \cdot a}{2 \cdot 5 \cdot a \cdot a \cdot a \cdot a \cdot a}$$

$$= \frac{3}{2 \cdot a \cdot a \cdot a} = \frac{3}{2a^3}$$

Using rules of exponents:

$$\frac{15a^2}{10a^5} = \frac{15}{10} \cdot \frac{a^2}{a^5} = \frac{3}{2} \cdot a^{2-5} = \frac{3}{2}a^{-3}$$

$$= \frac{3}{2} \cdot \frac{1}{a^3} = \frac{3}{2a^3}$$

Answers to Your Turn 4
a. $-\dfrac{2t^4}{5}$ **b.** $\dfrac{1}{2n^2}$ **c.** $-\dfrac{6}{5y^2z}$

Notice that when dividing out like bases, the position of the base with the greater exponent determines whether we place the result in the numerator or denominator.

Objective 4 Simplify rational expressions containing multiterm polynomials.

The rational expressions we have considered so far have had monomials in both the numerator and denominator. We now consider rational expressions that contain multiterm polynomials. Remember that we can divide out only common factors, so we must first factor polynomials completely.

Example 5 Simplify $\dfrac{9ab}{3a + 6}$.

Solution: $\dfrac{9ab}{3a + 6} = \dfrac{3 \cdot 3 \cdot a \cdot b}{3 \cdot (a + 2)}$ Factor the numerator and denominator completely. Then divide out the common factor, which is 3. Multiply the remaining factors.

$$= \dfrac{3ab}{a + 2}$$

Your Turn 5 Simplify.

a. $\dfrac{6xy}{3x + 12}$

b. $\dfrac{2a + 8}{10a^2}$

We now consider rational expressions with multiterm polynomials in both the numerator and denominator. Again, because we can divide out only common factors, we must first factor the polynomials.

Example 6 Simplify.

a. $\dfrac{2x^2 + 8x}{3x^2 + 12x}$

Solution: $\dfrac{2x^2 + 8x}{3x^2 + 12x} = \dfrac{2 \cdot x \cdot (x + 4)}{3 \cdot x \cdot (x + 4)}$ Factor the numerator and denominator completely. Then divide out the common factors, x and $x + 4$.

$$= \dfrac{2}{3}$$

b. $\dfrac{y^2 - 16}{y^2 + 4y - 32}$

Solution: $\dfrac{y^2 - 16}{y^2 + 4y - 32} = \dfrac{(y + 4)(y - 4)}{(y + 8)(y - 4)}$ Factor the numerator and denominator completely. Then divide out the common factor, $y - 4$.

$$= \dfrac{y + 4}{y + 8}$$

c. $\dfrac{2x^2 - 3x - 20}{2x^2 + x - 10} = \dfrac{(2x + 5)(x - 4)}{(2x + 5)(x - 2)}$ Factor the numerator and denominator completely. Then divide out the common factor, $2x + 5$.

$$= \dfrac{x - 4}{x - 2}$$

Warning A common error is to divide out terms instead of factors. Recall that the terms of a polynomial are connected by plus and minus signs. You have factors only when multiplying. Below are examples of both the correct and incorrect way of simplifying a rational expression to lowest terms.

Correct: $\dfrac{x^2 + x - 6}{x^2 + 5x + 6} = \dfrac{\cancel{(x + 3)}(x - 2)}{\cancel{(x + 3)}(x + 2)} = \dfrac{x - 2}{x + 2}$ Divide out factors.

Incorrect: $\dfrac{x^2 + x - 6}{x^2 + 5x - 6} = \dfrac{\overset{1}{\cancel{x^2}} + \overset{1}{\cancel{x}} - \overset{1}{\cancel{6}}}{\underset{1}{\cancel{x^2}} + \underset{1}{\cancel{5x}} - \underset{1}{\cancel{6}}} = \dfrac{1}{5}$ Do not divide out terms.

Answers to Your Turn 5

a. $\dfrac{2xy}{x + 4}$ b. $\dfrac{a + 4}{5a^2}$

Your Turn 6 Simplify.

a. $\dfrac{5x^2 + 10x}{7x^2 + 14x}$ b. $\dfrac{y^2 - 36}{y^2 - 2y - 24}$ c. $\dfrac{3n^2 - 23n - 8}{3n^2 + 13n + 4}$

Recall from Chapter 5 that to factor multiterm polynomials completely, we must sometimes factor the first factorization we get, as in Example 7.

Example 7 Simplify.

a. $\dfrac{6y^4 + 2y^3 - 4y^2}{36y^3 - 42y^2 + 12y}$

Solution: $\dfrac{6y^4 + 2y^3 - 4y^2}{36y^3 - 42y^2 + 12y} = \dfrac{2y^2(3y^2 + y - 2)}{6y(6y^2 - 7y + 2)}$ Factor out the monomial GCFs.

$= \dfrac{2 \cdot y \cdot y \cdot (y + 1) \cdot (3y - 2)}{2 \cdot 3 \cdot y \cdot (2y - 1) \cdot (3y - 2)}$ Factor the polynomial factors.

$= \dfrac{y(y + 1)}{3(2y - 1)}$ or $\dfrac{y^2 + y}{6y - 3}$ Divide out the common factors 2, y, and 3y − 2.

b. $\dfrac{3x^2 + 6x - 24}{x^4 - 16}$

Solution: Factor the numerator and denominator completely. Then divide out all common factors.

$\dfrac{3x^2 + 6x - 24}{x^4 - 16} = \dfrac{3(x^2 + 2x - 8)}{(x^2 + 4)(x^2 - 4)}$ Factor 3 out of $3x^2 + 6x - 24$ and factor $x^4 - 16$, which is a difference of squares.

$= \dfrac{3 \cdot (x + 4) \cdot (x - 2)}{(x^2 + 4) \cdot (x + 2) \cdot (x - 2)}$ Factor the polynomial factors. Notice that $x^2 - 4$ is a difference of squares.

$= \dfrac{3(x + 4)}{(x^2 + 4)(x + 2)}$ or $\dfrac{3x + 12}{x^3 + 2x^2 + 4x + 8}$ Divide out the common factor, x − 2.

Your Turn 7 Simplify.

a. $\dfrac{14x^3 + 70x^2 + 84x}{7x^4 - 14x^3 - 105x^2}$ b. $\dfrac{5x^2 - 5x - 60}{x^4 - 81}$

Binomial factors that are additive inverses can be tricky.

Answers to Your Turn 6

a. $\dfrac{5}{7}$ b. $\dfrac{y + 6}{y + 4}$ c. $\dfrac{n - 8}{n + 4}$

Answers to Your Turn 7

a. $\dfrac{2(x + 2)}{x(x - 5)}$ or $\dfrac{2x + 4}{x^2 - 5x}$

b. $\dfrac{5(x - 4)}{(x^2 + 9)(x - 3)}$ or

$\dfrac{5x - 20}{x^3 - 3x^2 + 9x - 27}$

Example 8 Simplify $\dfrac{3x^2 - 7x - 20}{8 - 2x}$.

Solution: Factor the numerator and denominator completely. Then divide out all common factors.

$$\dfrac{3x^2 - 7x - 20}{8 - 2x} = \dfrac{(3x + 5)(x - 4)}{2(4 - x)}$$

It appears that there are no common factors. However, $x - 4$ and $4 - x$ are additive inverses, which is more apparent when $4 - x$ is written in descending order as $-x + 4$. We can get matching binomial factors by factoring −1 out of $x - 4$ or $4 - x$.

$$x - 4 = -1(-x + 4) = -1(4 - x) \quad \text{or} \quad 4 - x = -1(-4 + x) = -1(x - 4)$$

We can use either form. We will use $x - 4 = -1(4 - x)$.

Note We could have factored the -1 out of the denominator:

$$\frac{(3x + 5)(x - 4)}{2(-1)(x - 4)} = \frac{3x + 5}{-2}$$

▶

$$= \frac{(3x + 5)(-1)(4 - x)}{2(4 - x)}$$

$$= \frac{-3x - 5}{2} \quad \text{or} \quad -\frac{3x + 5}{2}$$

Divide out the common factor, $4 - x$; then multiply the remaining factors.

Example 8 also illustrates the rule of sign placement in a fraction or rational expression. In a negative fraction (or rational expression), the minus sign can be placed in the numerator or denominator or aligned with the fraction line. Note, however, that it is generally considered unsightly to write the minus sign in the denominator.

Rule

$$-\frac{P}{Q} = \frac{-P}{Q} = \frac{P}{-Q}, \text{ where } P \text{ and } Q \text{ are polynomials and } Q \neq 0.$$

Answer to Your Turn 8

$$-\frac{x}{x + 6}$$

Your Turn 8 Simplify $\dfrac{2x - x^2}{x^2 + 4x - 12}$.

7.1 Exercises For Extra Help MyMathLab®

Objective 1

Prep Exercise 1 Explain how to evaluate an expression.
Replace the variables with given values and then simplify the numerical expression.

For Exercises 1–8, evaluate the rational expression. See Example 1.

1. $\dfrac{4x^2}{9y}$
    a. when $x = -2, y = 5$
$\dfrac{16}{45}$
    b. when $x = 3, y = -4$
-1
    c. when $x = 0, y = 7$
0

2. $\dfrac{3x}{7y^2}$
    a. when $x = 1, y = 2$
$\dfrac{3}{28}$
    b. when $x = -2, y = 3$
$-\dfrac{2}{21}$
    c. when $x = -4, y = -2$
$\dfrac{3}{7}$

3. $\dfrac{2x}{x + 4}$
    a. when $x = 4$
1
    b. when $x = -4$
undefined
    c. when $x = -2.4$
-3

4. $\dfrac{4x}{x - 3}$
    a. when $x = 3$
undefined
    b. when $x = -3$
2
    c. when $x = 2.5$
-20

5. $\dfrac{2x + 5}{3x - 1}$
    a. when $x = 3$
$\dfrac{11}{8}$
    b. when $x = -2$
$\dfrac{1}{7}$
    c. when $x = -1.3$
$\dfrac{24}{49}$ or ≈ -0.49

6. $\dfrac{3x + 2}{4x - 5}$
    a. when $x = 2$
$\dfrac{8}{3}$
    b. when $x = -1$
$\dfrac{1}{9}$
    c. when $x = 1.6$
$\dfrac{34}{7}$ or ≈ 4.86

7. $\dfrac{x^2 - 4}{x + 3}$

    a. when $x = 0$

      $-\dfrac{4}{3}$

    b. when $x = -1$

      $-\dfrac{3}{2}$

    c. when $x = -4.2$

      $-\dfrac{341}{30}$ or $-11.3\overline{6}$

8. $\dfrac{x^2 + 3}{2x + 1}$

    a. when $x = 1$

      $\dfrac{4}{3}$

    b. when $x = -2$

      $\dfrac{7}{3}$

    c. when $x = -3.5$

      $\dfrac{61}{24}$ or $-2.541\overline{6}$

Objective 2

Prep Exercise 2 Explain how to determine the values that cause a rational expression to be undefined.

Set the denominator equal to 0 and solve the equation.

For Exercises 9–20, find every value for the variable that makes the expression undefined. See Example 2.

9. $\dfrac{2x}{x - 3}$

   3

10. $\dfrac{3y}{y + 5}$

   -5

11. $\dfrac{3}{x^2 - 9}$

   $-3, 3$

12. $\dfrac{4r}{r^2 - 4}$

   $-2, 2$

13. $\dfrac{6}{x^2 - 14x + 45}$

   $5, 9$

14. $\dfrac{2x + 3}{x^2 + 5x + 6}$

   $-2, -3$

15. $\dfrac{2m}{m^2 - 3m}$

   $0, 3$

16. $\dfrac{2p + 3}{p^2 + 6p}$

   $0, -6$

17. $\dfrac{4a}{2a^2 - 3a - 5}$

   $-1, \dfrac{5}{2}$

18. $\dfrac{t}{2t^2 + 7t + 3}$

   $-3, -\dfrac{1}{2}$

19. $\dfrac{x - 5}{x^2 + 16}$

   Defined for all real numbers

20. $\dfrac{3x - 2}{y^2 + 36}$

   Defined for all real numbers

Objectives 3 and 4

Prep Exercise 3 What is the first step in simplifying a rational expression?

Factor the numerator and denominator completely.

Prep Exercise 4 What common factors can be eliminated in $\dfrac{x(x + 3)(x + 5)}{2x(x + 3)(x - 2)}$?

x and $x + 3$

Prep Exercise 5 Explain why you cannot divide out the y's in $\dfrac{x + y}{2y}$.

y is not a factor in the numerator, $x + y$.

For Exercises 21–72, simplify. See Examples 3–8.

21. $\dfrac{9x^2}{12xy}$

   $\dfrac{3x}{4y}$

22. $\dfrac{14h^2k}{21h^3}$

   $\dfrac{2k}{3h}$

23. $-\dfrac{4m^3n}{mn^2}$

   $\dfrac{4m^2}{n}$

24. $-\dfrac{5a^3b}{a^2b^4}$

   $\dfrac{5a}{b^3}$

25. $\dfrac{54x^4y}{36x^6yz^2}$

   $\dfrac{3}{2x^2z^2}$

26. $-\dfrac{48t^5uv^4}{32tuv^2}$

   $\dfrac{3t^4v^2}{2}$

27. $-\dfrac{15m^2n}{5m + 10n}$

   $\dfrac{3m^2n}{m + 2n}$

28. $-\dfrac{28r^3t}{7r + 14t}$

   $\dfrac{4r^3t}{r + 2t}$

29. $\dfrac{7}{21(x + 5)}$

   $\dfrac{1}{3(x + 5)}$

30. $\dfrac{9}{18(a - 1)}$

   $\dfrac{1}{2(a - 1)}$

31. $\dfrac{3(m - 2)}{5(m - 2)}$

   $\dfrac{3}{5}$

32. $\dfrac{7(k + 2)}{15(k + 2)}$

   $\dfrac{7}{15}$

33. $\dfrac{10x - 20}{15x - 30}$

   $\dfrac{2}{3}$

34. $\dfrac{12y + 2}{18y + 3}$

   $\dfrac{2}{3}$

35. $\dfrac{30a + 15b}{18a + 9b}$

   $\dfrac{5}{3}$

36. $\dfrac{8m - 24n}{12m - 36n}$

$\dfrac{2}{3}$

37. $\dfrac{x^2 + 5x}{4x + 20}$

$\dfrac{x}{4}$

38. $\dfrac{ab - b^2}{2a - 2b}$

$\dfrac{b}{2}$

39. $\dfrac{m^2 - 6m}{m - 6}$

m

40. $\dfrac{x^2 + 5x}{x + 5}$

x

41. $\dfrac{5y^2 - 10y - 15}{xy^2 - 2xy - 3x}$

$\dfrac{5}{x}$

42. $\dfrac{3m^3 - 2m^2 + m}{6m^2 - 4m + 2}$

$\dfrac{m}{2}$

43. $\dfrac{x^2 - y^2}{x^2 + 2xy + y^2}$

$\dfrac{x - y}{x + y}$

44. $\dfrac{a^2 - b^2}{a^2 + 2ab + b^2}$

$\dfrac{a - b}{a + b}$

45. $\dfrac{x^2 - 5x}{x^2 - 7x + 10}$

$\dfrac{x}{x - 2}$

46. $\dfrac{x^2 - 2x}{x^2 - 6x + 8}$

$\dfrac{x}{x - 4}$

47. $\dfrac{t^2 - 9}{t^2 + 5t + 6}$

$\dfrac{t - 3}{t + 2}$

48. $\dfrac{k^2 - 4}{k^2 + k - 2}$

$\dfrac{k - 2}{k - 1}$

49. $\dfrac{4a^2 - 4a - 3}{6a^2 - a - 2}$

$\dfrac{2a - 3}{3a - 2}$

50. $\dfrac{3x^2 + 16x - 35}{5x^2 + 33x - 14}$

$\dfrac{3x - 5}{5x - 2}$

51. $\dfrac{6b^2 + b - 12}{10b^2 + 13b - 3}$

$\dfrac{3b - 4}{5b - 1}$

52. $\dfrac{8y^2 - 6y - 9}{12y^2 + y - 6}$

$\dfrac{2y - 3}{3y - 2}$

53. $\dfrac{x^3 - 16x}{x^3 + 6x^2 + 8x}$

$\dfrac{x - 4}{x + 2}$

54. $\dfrac{m^3 - 4m}{m^3 - 4m^2 + 4m}$

$\dfrac{m + 2}{m - 2}$

55. $\dfrac{px + py + qx + qy}{mx - nx + my - ny}$

$\dfrac{p + q}{m - n}$

56. $\dfrac{by - ay + bx - ax}{x^2 + ax + xy + ay}$

$\dfrac{b - a}{x + a}$

57. $\dfrac{6u^3 - 4u^2 - 3u + 2}{3u^2 + u - 2}$

$\dfrac{2u^2 - 1}{u + 1}$

58. $\dfrac{8n^3 + 20n^2 - 2n - 5}{4n^2 + 12n + 5}$

$2n - 1$

59. $\dfrac{x - 3}{3 - x}$

-1

60. $\dfrac{5 - b}{b - 5}$

-1

61. $\dfrac{4x - 4}{1 - x}$

-4

62. $\dfrac{12 - 4m}{m - 3}$

-4

63. $\dfrac{y^2 - 4}{2 - y}$

$-y - 2$

64. $\dfrac{a^2 - b^2}{b - a}$

$-a - b$

65. $\dfrac{3 - w}{w^2 - 2w - 3}$

$\dfrac{1}{w + 1}$

66. $\dfrac{1 - u}{u^2 - 3u + 2}$

$\dfrac{1}{u - 2}$

67. $\dfrac{x^2 - 4}{x^3 + 8}$

$\dfrac{x - 2}{x^2 - 2x + 4}$

68. $\dfrac{r^2 - s^2}{r^3 + s^3}$

$\dfrac{r - s}{r^2 - rs + s^2}$

69. $\dfrac{x^3 + y^3}{x^3 - x^2y + xy^2}$

$\dfrac{x + y}{x}$

70. $\dfrac{x^3 - 8}{x^3 + 2x^2 + 4x}$

$\dfrac{x - 2}{x}$

71. $\dfrac{3 - m}{m^3 - 27}$

$\dfrac{1}{m^2 + 3m + 9}$

72. $\dfrac{4 - x}{x^3 - 64}$

$\dfrac{1}{x^2 + 4x + 16}$

Find ⊗ the Mistake *For Exercises 73–78, find and explain the mistake; then correct the mistake.*

73. $-\dfrac{5x^2y}{10x^3y} = -\dfrac{x}{2}$

Mistake: Placed remaining x in numerator.

Correct: $-\dfrac{1}{2x}$

74. $\dfrac{8x^2y}{4x^3} = 2xy$

Mistake: Placed remaining x in numerator.

Correct: $\dfrac{2y}{x}$

75. $\dfrac{x - \overset{3}{\cancel{6}}}{\underset{1}{\cancel{2}}y} = \dfrac{x - 3}{y}$

Mistake: Divided out a part of a multiterm polynomial.
Correct: Can't simplify

76. $\dfrac{y - \overset{3}{\cancel{12}}}{\underset{1}{\cancel{4}}y - 3} = \dfrac{y - 3}{y - 3} = 1$

Mistake: Divided out a part of a multiterm polynomial.
Correct: Can't simplify

77. $\dfrac{yx - x^2}{x - y} = \dfrac{x(y - x)}{x - y} = x$

Mistake: $y - x$ and $x - y$ are not identical factors until -1 is factored out of one of them.
Correct: $-x$

78. $\dfrac{x^2 - x - 6}{x^3 + 2x^2 - 9x - 18} = \dfrac{(x + 2)(x - 3)}{(x + 2)(x^2 - 9)} = \dfrac{x - 3}{x^2 - 9}$

Mistake: $\dfrac{x - 3}{x^2 - 9}$ is not in simplest form because $x^2 - 9$ can be factored, revealing a common factor, $x - 3$.

Correct: $\dfrac{1}{x + 3}$

For Exercises 79–84, evaluate. See Example 1.

79. A person's body mass index (BMI) is a measure of the amount of body fat based on height, h, in inches, and weight, w, in pounds. The formula for BMI is as follows:

$$\text{BMI} = \frac{704.5w}{h^2}$$

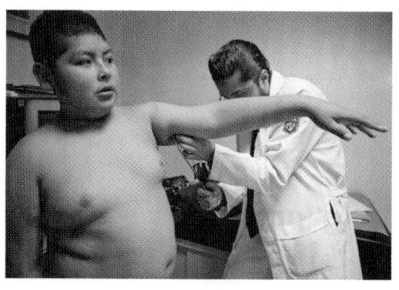

a. Use the formula to find the BMI of each person listed here.

Height	Weight	BMI
5′ 4″	138	23.7
5′ 9″	155	22.9
6′ 2″	220	28.3

b. Use the following table from the National Institutes of Health to classify each person in part a as underweight, normal weight, overweight, or obese.

Underweight:	BMI ≤ 18.5
Normal Weight:	BMI $= 18.5 - 24.9$
Overweight:	BMI $= 25 - 29.9$
Obese:	BMI ≥ 30

The first two are normal weight, and the third is overweight.

Connection In Section 2.5, Exercise 62, you translated the BMI formula from words to symbols.

80. Recall the following formula for slope from Chapter 3:

$$m = \frac{y_2 - y_1}{x_2 - x_1}$$

a. Find the slope of a line passing through the points $(3, 5)$ and $(-1, -4)$.

$\dfrac{9}{4}$

b. Find the slope of a line passing through the points $(-4, 7)$ and $(-2, -1)$.

-4

Connection Although we hadn't formally defined rational expressions in Chapter 3, we were evaluation the rational expression $\dfrac{y_2 - y_1}{x_2 - x_1}$ to find the slope.

81. Given the diameter in inches, d, and the number of revolutions per minute, N, of a revolving tool such as a circular saw or drill, the cutting speed, C, in feet per minute, of the tool can be found by the following formula:

$$C = \frac{\pi d N}{12}$$

(*Source:* Robert D. Smith, *Mathematics for Machine Technology*, 5th ed., Cengage Learning, 2003.)

Use the formula to complete the following table for a table saw. Leave answers in terms of π.

Blade Diameter	rpm	Cutting Speed
8 inches	800 rpm	$533.\overline{3}\pi$
8 inches	1000 rpm	$666.\overline{6}\pi$
10 inches	800 rpm	$666.\overline{6}\pi$
10 inches	1000 rpm	$833.\overline{3}\pi$

82. The cutting time, T, in minutes, that it takes a drill to cut through something depends on its speed of revolution, N, in rpm; the thickness of the material, L, in inches; and the tool feed, F, in inches per revolution. The formula for calculating cutting time is as follows:

$$T = \frac{L}{FN}$$

(*Source:* Robert D. Smith, *Mathematics for Machine Technology*, 5th ed., Cengage Learning, 2003.)

Use the formula to complete the following table for a drill operating at a speed of 480 rpm and a feed of 0.04 inch per revolution.

Material Thickness (inches)	Cut Time (minutes)
0.25	0.013
0.5	0.026
0.75	0.039
1	0.052

83. The formula for the volume of a cone is $V = \dfrac{1}{3}\pi r^2 h$.

    a. Solve this formula for h so that we have a formula for finding the height of a cone given its volume and radius. (The formula for height of a cone will contain a rational expression.)

       $h = \dfrac{3V}{\pi r^2}$

    b. Suppose a cone has a volume of approximately 150.8 cubic centimeters with a radius of 4 centimeters. Find the height of the cone.

       $\approx 9\text{ cm}$

84. The formula for the volume of a pyramid is $V = \dfrac{1}{3}lwh$.

    a. Solve this formula for h so that we have a formula for finding the height of a pyramid given its volume, length, and width.

       $h = \dfrac{3V}{lw}$

    b. Suppose a decorative onyx pyramid has a volume of 126 cubic inches, a width of 3.5 inches, and a length of 4.5 inches. Find the height of the pyramid.

       24 in.

For Exercises 85 and 86, simplify. See Examples 5–7.

85. In analyzing circuits, electrical engineers often must simplify rational expressions. Suppose the following rational expression describes an electrical circuit.

$$\frac{s^2 - 9}{s^2 + 7s + 12}$$

Simplify the expression to lowest terms.

$\dfrac{s - 3}{s + 4}$

86. An engineer derives the following rational expression, which describes a certain circuit.

$$\frac{4s^2 - 6s}{2s^2 + 7s - 15}$$

Simplify the expression to lowest terms.

$\dfrac{2s}{s + 5}$

Review Exercises

Exercises 1–6 ◢ Expressions

[1.4] *For Exercises 1 and 2, multiply.*

1. $\dfrac{2}{3} \cdot \dfrac{6}{4}$

    1

2. $-\dfrac{15}{16} \cdot \dfrac{4}{5}$

    $-\dfrac{3}{4}$

[1.4] *For Exercises 3 and 4, divide.*

3. $\dfrac{3}{8} \div \dfrac{5}{6}$

$\dfrac{9}{20}$

4. $-\dfrac{7}{9} \div \left(-\dfrac{5}{12}\right)$

$\dfrac{28}{15}$

For Exercises 5 and 6, factor completely.

[6.4] 5. $4y^2 - 25$

$(2y + 5)(2y - 5)$

[6.3] 6. $3x^2 + 13x - 10$

$(3x - 2)(x + 5)$

7.2 Multiplying and Dividing Rational Expressions

Objectives

1 Multiply rational expressions.
2 Divide rational expressions.
3 Convert units of measurement using dimensional analysis.

Warm-up

[1.4] 1. Multiply: $\dfrac{9}{10}\left(-\dfrac{5}{12}\right)$

[1.4] 2. Divide: $-\dfrac{7}{5} \div \left(-\dfrac{5}{6}\right)$

For Exercises 3 and 4, factor completely.

[6.1] 3. $12x^2 - 6x$

[6.3] 4. $4x^2 + 9x + 2$

Objective 1 Multiply rational expressions.

In Section 1.4, we used the rule $\dfrac{a}{b} \cdot \dfrac{c}{d} = \dfrac{ac}{bd}$, where $b \neq 0$ and $d \neq 0$, to multiply fractions. We can rewrite this rule so that it applies to multiplying rational expressions.

> **Rule** **Multiplying Rational Expressions**
>
> $\dfrac{P}{Q} \cdot \dfrac{R}{S} = \dfrac{PR}{QS}$, where P, Q, R, and S are polynomials and $Q \neq 0$ and $S \neq 0$.

Also remember that when multiplying fractions, we can simplify after multiplying or we can simplify before multiplying by dividing out factors common to both the numerator and denominator. For example,

Multiply, then eliminate common factors.	or	Eliminate common factors; then multiply.
$\dfrac{8}{15} \cdot \dfrac{3}{4} = \dfrac{24}{60} = \dfrac{2 \cdot 2 \cdot 2 \cdot 3}{2 \cdot 2 \cdot 3 \cdot 5} = \dfrac{2}{5}$		$\dfrac{8}{15} \cdot \dfrac{3}{4} = \dfrac{2 \cdot 2 \cdot 2}{3 \cdot 5} \cdot \dfrac{3}{2 \cdot 2} = \dfrac{2}{5}$

Multiplying rational expressions works the same way.

> **Procedure** **Multiplying Rational Expressions:**
>
> To multiply rational expressions:
> 1. Factor each numerator and denominator completely.
> 2. Divide out any numerator factor with any matching denominator factor.
> 3. Multiply numerator by numerator and denominator by denominator.
> 4. Simplify as needed.

Answers to Warm-up

1. $-\dfrac{3}{8}$

2. $\dfrac{14}{15}$

3. $6x(2x - 1)$

4. $(4x + 1)(x + 2)$

Example 1 Multiply $\dfrac{9x}{5y} \cdot \dfrac{15xy^2}{12}$.

Solution: $\dfrac{9x}{5y} \cdot \dfrac{15xy^2}{12} = \dfrac{3 \cdot 3 \cdot x}{5 \cdot y} \cdot \dfrac{3 \cdot 5 \cdot x \cdot y \cdot y}{2 \cdot 2 \cdot 3}$ Factor the numerators and denominators completely.

$= \dfrac{3 \cdot 3 \cdot x}{1} \cdot \dfrac{x \cdot y}{2 \cdot 2}$ Divide out the common factors, which are 3, 5, and *y*.

$= \dfrac{9x^2 y}{4}$ Multiply the remaining numerator factors and denominator factors.

Your Turn 1 Multiply.

a. $-\dfrac{36m}{8n^4} \cdot \dfrac{14mn^2}{28n}$ b. $-\dfrac{9}{4xy^2} \cdot -\dfrac{20xz}{15y}$

Now consider rational expressions that contain multiterm polynomials.

Example 2 Multiply.

a. $-\dfrac{2x}{3x - 9} \cdot \dfrac{5x - 15}{20x^3}$

Solution: $-\dfrac{2x}{3x - 9} \cdot \dfrac{5x - 15}{20x^3}$

$= -\dfrac{2 \cdot x}{3 \cdot (x - 3)} \cdot \dfrac{5 \cdot (x - 3)}{2 \cdot 2 \cdot 5 \cdot x \cdot x \cdot x}$ Factor the numerators and denominators completely.

$= -\dfrac{1}{3} \cdot \dfrac{1}{2 \cdot x \cdot x}$ Divide out the common factors, which are 2, 5, *x*, and (*x* − 3).

$= -\dfrac{1}{6x^2}$ Multiply the remaining numerator factors and denominator factors.

Instructor Note Ask students what they would get if they substituted 5 (or any other number except 0 or 3) for *x*. You hope they use $\dfrac{-2x}{3}$ and not the original problem.

b. $\dfrac{12 - 4x}{15x} \cdot \dfrac{5x^2}{2x - 6}$

Solution: $\dfrac{12 - 4x}{15x} \cdot \dfrac{5x^2}{2x - 6} = \dfrac{4 \cdot (3 - x)}{3 \cdot 5 \cdot x} \cdot \dfrac{5 \cdot x \cdot x}{2 \cdot (x - 3)}$

$= \dfrac{2 \cdot 2 \cdot (-1)(x - 3)}{3 \cdot 5 \cdot x} \cdot \dfrac{5 \cdot x \cdot x}{2 \cdot (x - 3)}$

$= \dfrac{2 \cdot (-1)}{3} \cdot \dfrac{x}{1}$ Divide out the common factors, which are 2, 5, *x*, and (*x* − 3).

$= \dfrac{-2x}{3}$ Multiply the remaining numerator factors and denominator factors.

Note We can factor 4 further. Also, because 3 − *x* and *x* − 3 are additive inverses, we can factor −1 out of one of them.

c. $\dfrac{x^2 + x - 6}{9x^3} \cdot \dfrac{12x^2 - 6x}{2x^2 - 5x + 2}$

Solution: $\dfrac{x^2 + x - 6}{9x^3} \cdot \dfrac{12x^2 - 6x}{2x^2 - 5x + 2}$

$= \dfrac{(x - 2)(x + 3)}{3 \cdot 3 \cdot x \cdot x \cdot x} \cdot \dfrac{2 \cdot 3 \cdot x \cdot (2x - 1)}{(x - 2)(2x - 1)}$ Factor the numerators and denominators completely.

$= \dfrac{x + 3}{3 \cdot x \cdot x} \cdot \dfrac{2}{1}$ Divide out the common factors, which are (*x* − 2), 3, *x*, and (2*x* − 1).

$= \dfrac{2(x + 3)}{3x^2}$ or $\dfrac{2x + 6}{3x^2}$ Multiply the remaining numerator factors and denominator factors.

Answers to Your Turn 1

a. $-\dfrac{9m^2}{4n^3}$ b. $\dfrac{3z}{y^3}$

Your Turn 2 Multiply.

a. $\dfrac{-8h^2}{2h+10} \cdot \dfrac{6h+30}{18h^4}$ **b.** $-\dfrac{9y-3}{y^2+3y} \cdot \dfrac{2y+6}{2-6y}$ **c.** $\dfrac{m^2-25}{20m} \cdot \dfrac{30m+10}{3m^2+16m+5}$

Objective 2 Divide rational expressions.

In Section 1.4, we learned how to divide fractions using the rule $\dfrac{a}{b} \div \dfrac{c}{d} = \dfrac{a}{b} \cdot \dfrac{d}{c}$, where b, c, and d are not 0, which indicates to multiply by the reciprocal of the divisor. We can rewrite this rule so that it applies to dividing rational expressions.

> **Rule Dividing Rational Expressions**
>
> $\dfrac{P}{Q} \div \dfrac{R}{S} = \dfrac{P}{Q} \cdot \dfrac{S}{R}$, where P, Q, R, and S are polynomials and $Q \neq 0$, $R \neq 0$, and $S \neq 0$.

Following is a procedure for dividing rational expressions.

> **Procedure Dividing Rational Expressions**
>
> To divide rational expressions:
> 1. Write an equivalent multiplication statement with the reciprocal of the divisor.
> 2. Factor each numerator and denominator completely. (Steps 1 and 2 are interchangeable.)
> 3. Divide out any numerator factor with any matching denominator factor.
> 4. Multiply numerator by numerator and denominator by denominator.
> 5. Simplify as needed.

Example 3 Divide $-\dfrac{9x^2y}{28z^3} \div \dfrac{12y}{7z}$.

Solution: $-\dfrac{9x^2y}{28z^3} \div \dfrac{12y}{7z} = -\dfrac{9x^2y}{28z^3} \cdot \dfrac{7z}{12y}$ Write an equivalent multiplication with the reciprocal of the divisor.

$= -\dfrac{3 \cdot 3 \cdot x \cdot x \cdot y}{2 \cdot 2 \cdot 7 \cdot z \cdot z \cdot z} \cdot \dfrac{7 \cdot z}{2 \cdot 2 \cdot 3 \cdot y}$ Factor the numerators and denominators completely.

$= -\dfrac{3 \cdot x \cdot x}{2 \cdot 2 \cdot z \cdot z} \cdot \dfrac{1}{2 \cdot 2}$ Divide out the common factors, which are 3, 7, y, and z.

$= -\dfrac{3x^2}{16z^2}$ Multiply the remaining factors.

Answers to Your Turn 2

a. $\dfrac{4}{3h^2}$ **b.** $\dfrac{3}{y}$ **c.** $\dfrac{m-5}{2m}$

Answers to Your Turn 3

a. $\dfrac{3a^2}{8b^3}$ **b.** $\dfrac{7u}{44t^2v^2}$

Your Turn 3 Divide.

a. $\dfrac{15a^3}{28b^2} \div \dfrac{10ab}{7}$ **b.** $-\dfrac{2t}{33u^3v} \div \left(-\dfrac{8t^3v}{21u^4}\right)$

Now consider dividing rational expressions that contain multiterm polynomials.

Example 4 Divide.

a. $\dfrac{2x - 10}{15x^3} \div \dfrac{x^2 - 5x}{12x}$

Solution: $\dfrac{2x - 10}{15x^3} \div \dfrac{x^2 - 5x}{12x} = \dfrac{2x - 10}{15x^3} \cdot \dfrac{12x}{x^2 - 5x}$

Write an equivalent multiplication with the reciprocal of the divisor.

$= \dfrac{2 \cdot (x - 5)}{3 \cdot 5 \cdot x \cdot x \cdot x} \cdot \dfrac{2 \cdot 2 \cdot 3 \cdot x}{x \cdot (x - 5)}$

Factor the numerators and denominators completely.

$= \dfrac{2}{5 \cdot x \cdot x} \cdot \dfrac{2 \cdot 2}{x}$

Divide out the common factors, which are 3, x, and $(x - 5)$.

$= \dfrac{8}{5x^3}$

Multiply the remaining factors.

b. $\dfrac{12x^2 + 3x}{2x^2 + 3x - 9} \div \dfrac{4x^2 + 9x + 2}{6x^3 + 18x^2}$

Solution: $\dfrac{12x^2 + 3x}{2x^2 + 3x - 9} \div \dfrac{4x^2 + 9x + 2}{6x^3 + 18x^2}$

$= \dfrac{12x^2 + 3x}{2x^2 + 3x - 9} \cdot \dfrac{6x^3 + 18x^2}{4x^2 + 9x + 2}$

Write an equivalent multiplication.

$= \dfrac{3x(4x + 1)}{(2x - 3)(x + 3)} \cdot \dfrac{6x^2(x + 3)}{(4x + 1)(x + 2)}$

Factor the numerators and denominators completely.

$= \dfrac{3 \cdot x \cdot (4x + 1)}{(2x - 3)(x + 3)} \cdot \dfrac{2 \cdot 3 \cdot x \cdot x \cdot (x + 3)}{(4x + 1) \cdot (x + 2)}$

$= \dfrac{3 \cdot x}{2x - 3} \cdot \dfrac{2 \cdot 3 \cdot x \cdot x}{x + 2}$

Divide out the common factors, which are $(4x + 1)$ and $(x + 3)$.

$= \dfrac{18x^3}{(2x - 3)(x + 2)}$

Multiply the remaining factors.

Note It is customary not to multiply out remaining binomial factors. ▶

c. $\dfrac{8x^2 - 6x^3}{6xy} \div \dfrac{9x^2 - 16}{5y^2 - 10y}$

Solution: $\dfrac{8x^2 - 6x^3}{6xy} \div \dfrac{9x^2 - 16}{5y^2 - 10y} = \dfrac{8x^2 - 6x^3}{6xy} \cdot \dfrac{5y^2 - 10y}{9x^2 - 16}$

Write an equivalent multiplication.

$= \dfrac{2x^2(4 - 3x)}{6xy} \cdot \dfrac{5y(y - 2)}{(3x + 4)(3x - 4)}$

Factor the numerators and denominators completely.

$= \dfrac{2 \cdot x \cdot x \cdot (-1) \cdot (3x - 4)}{2 \cdot 3 \cdot x \cdot y} \cdot \dfrac{5 \cdot y \cdot (y - 2)}{(3x + 4) \cdot (3x - 4)}$

Note $4 - 3x$ and $3x - 4$ are additive inverses, so we factored -1 out of $4 - 3x$. ▶

$= \dfrac{x \cdot (-1)}{3} \cdot \dfrac{5 \cdot (y - 2)}{3x + 4}$

Divide out the common factors, which are 2, x, y, and $(3x - 4)$.

$= \dfrac{-5x(y - 2)}{3(3x + 4)}$ or $\dfrac{-5xy + 10x}{9x + 12}$ or $\dfrac{10x - 5xy}{9x + 12}$

Multiply the remaining factors.

Answers to Your Turn 4

a. $\dfrac{9z^2}{4y}$ b. $\dfrac{(n - 3)(n + 1)}{2m^3n}$

c. $\dfrac{y^2(2y - 5)}{3(y + 2)}$ or $\dfrac{2y^3 - 5y^2}{3y + 6}$

d. $-\dfrac{2x^2(3x + 4)}{15}$ or $-\dfrac{6x^3 + 8x^2}{15}$

Your Turn 4 Divide.

a. $\dfrac{6yz^3}{2y - y^2} \div \dfrac{8yz}{6 - 3y}$

b. $\dfrac{mn - 3m}{14m^4} \div \dfrac{n}{7n + 7}$

c. $\dfrac{4y^2 - 25}{9y^2 + 18y} \div \dfrac{2y^2 + 3y - 5}{3y^4 - 3y^3}$

d. $\dfrac{3x^2 + 10x + 8}{18 - 6x} \div \dfrac{10 + 5x}{4x^3 - 12x^2}$

Objective 3 Convert units of measurement using dimensional analysis.

We can convert from one unit of a measurement to another using *dimensional analysis*, which is also called the *factor-label method*. The method involves writing measurement facts as conversion factors. A conversion factor is a ratio of equivalent measures.

Measurement fact:	Conversion factor:
1 foot $=$ 12 inches	$\dfrac{1 \text{ ft.}}{12 \text{ in.}} = \dfrac{12 \text{ in.}}{1 \text{ ft.}} = 1$
1 pound $=$ 16 ounces	$\dfrac{1 \text{ lb.}}{16 \text{ oz.}} = \dfrac{16 \text{ oz.}}{1 \text{ lb.}} = 1$

Note A conversion factor is a ratio of equivalent measures, so it can be inverted without changing its value, which is 1.

Because a conversion factor is equivalent to 1, multiplying by one of them gives an equivalent measurement with different units. For example, multiplying 6 feet by $\dfrac{12 \text{ in.}}{1 \text{ ft.}}$ causes the unit feet to divide out so that the result is equivalent to 6 feet, but now in inches.

$$6 \text{ ft.} = \frac{6 \text{ ft.}}{1} \cdot \frac{12 \text{ in.}}{1 \text{ ft.}} = 72 \text{ in.}$$

Note 72 inches is equivalent to 6 feet because we multiplied 6 feet by 1 in the form of $\dfrac{12 \text{ in.}}{1 \text{ ft.}}$

> **Procedure** **Using Dimensional Analysis to Convert between Units of Measurement**
>
> To convert units using dimensional analysis, multiply the given measurement by conversion factors so that the undesired units divide out, leaving the desired units.

Example 5 Convert.

a. 500 ounces to pounds

Solution: There are 16 ounces to 1 pound. We write this fact as a unit ratio and multiply so that ounces divides out, leaving pounds.

$$500 \text{ oz.} = \frac{500 \text{ oz.}}{1} \cdot \frac{1 \text{ lb.}}{16 \text{ oz.}} = \frac{500}{16} \text{ lb.} = 31.25 \text{ lb.}$$

b. 6 miles to yards

Solution: There are 5280 feet to 1 mile and 3 feet to 1 yard. Thus, we write these facts as unit fractions so that miles and feet divide out, leaving yards.

$$6 \text{ mi.} = \frac{6 \text{ mi.}}{1} \cdot \frac{5280 \text{ ft.}}{1 \text{ mi.}} \cdot \frac{1 \text{ yd.}}{3 \text{ ft.}} = \frac{31{,}680}{3} \text{ yd.} = 10{,}560 \text{ yd.}$$

c. 40 miles per hour to feet per second

Solution: To begin, 40 miles per hour means $\dfrac{40 \text{ miles}}{1 \text{ hour}}$. Notice that we must convert miles to feet and hours to seconds. For the distance conversion, we use the fact that there are 5280 feet to 1 mile. For the time conversion, there are 60 minutes in 1 hour and 60 seconds in 1 minute.

$$\frac{40 \text{ miles}}{1 \text{ hour}} = \frac{40 \text{ mi.}}{1 \text{ hr.}} \cdot \frac{5280 \text{ ft.}}{1 \text{ mi.}} \cdot \frac{1 \text{ hr.}}{60 \text{ min.}} \cdot \frac{1 \text{ min.}}{60 \text{ sec.}} = \frac{211{,}200 \text{ ft.}}{3600 \text{ sec.}} = 58.\overline{6} \text{ ft./sec.}$$

Connection Dividing out common units in dimensional analysis is like dividing out common factors when multiplying rational expressions.

Your Turn 5 Convert.

a. 192 ounces to pounds

b. 540 inches to yards

c. 44 feet per second to miles per hour

Answers to Your Turn 5
a. 12 lb.   **b.** 15 yd.   **c.** 30 mph

7.2 Exercises For Extra Help MyMathLab®

Note: Exercises marked with a ★ represent challenging exercises.

Objective 1

Prep Exercise 1 When multiplying rational expressions, why is it important to factor the numerators and denominators completely? To eliminate any common factors

Prep Exercise 2 What is the unknown factor that makes each of the following statements true?
$$5 - x = \underline{\ ?\ }(x - 5)$$
$$m - n = \underline{\ ?\ }(n - m)$$
$$-2x + 7 = \underline{\ ?\ }(2x - 7)$$
$$-1$$

For Exercises 1–38, multiply. See Examples 1 and 2.

1. $\dfrac{m}{n} \cdot \dfrac{2m}{5n}$
    $\dfrac{2m^2}{5n^2}$

2. $\dfrac{x}{y} \cdot \dfrac{3x}{2y}$
    $\dfrac{3x^2}{2y^2}$

3. $\dfrac{r^3}{s^3} \cdot \dfrac{s^5}{r^6}$
    $\dfrac{s^2}{r^3}$

4. $\dfrac{a^5}{b^3} \cdot \dfrac{b^2}{a^3}$
    $\dfrac{a^2}{b}$

5. $\dfrac{3}{y} \cdot \dfrac{y^3}{3}$
    y^2

6. $\dfrac{y^4}{7} \cdot \dfrac{7}{y^2}$
    y^2

7. $-\dfrac{n}{rs^2} \cdot \dfrac{rs}{mn^3}$
    $-\dfrac{1}{mn^2s}$

8. $\dfrac{ab}{xy} \cdot -\dfrac{x}{a^3b^2}$
    $-\dfrac{1}{a^2by}$

9. $-\dfrac{6r^2s}{5r^2s^2} \cdot -\dfrac{15r^4s^2}{2r^3s^2}$
    $\dfrac{9r}{s}$

10. $-\dfrac{14mn^2}{8m^2n} \cdot -\dfrac{16mz^2}{7n^2z}$
    $\dfrac{4z}{n}$

11. $\dfrac{5n^2}{15m} \cdot \dfrac{n}{2} \cdot \dfrac{3m^2n}{n^2}$
    $\dfrac{mn^2}{2}$

12. $\dfrac{a}{b^2} \cdot \dfrac{3b^2}{4a} \cdot \dfrac{ab^2}{9}$
    $\dfrac{ab^2}{12}$

13. $\dfrac{3b - 12}{14} \cdot \dfrac{21}{5b - 20}$
    $\dfrac{9}{10}$

14. $\dfrac{9}{2a + 4} \cdot \dfrac{3a + 6}{15}$
    $\dfrac{9}{10}$

15. $\dfrac{2x - 6}{3y + 12} \cdot \dfrac{2y + 8}{5x - 15}$
    $\dfrac{4}{15}$

16. $\dfrac{3a + 12}{5b - 30} \cdot \dfrac{3b - 18}{4a + 16}$
    $\dfrac{9}{20}$

17. $\dfrac{4x^2 - 25}{3x - 4} \cdot \dfrac{9x^2 - 16}{2x + 5}$
    $(2x - 5)(3x + 4)$

18. $\dfrac{16x^2 - 9}{4x + 3} \cdot \dfrac{25x^2 - 1}{5x + 1}$
    $(4x - 3)(5x - 1)$

19. $\dfrac{r^2 + 3r}{r^2 - 36} \cdot \dfrac{r^2 - 6r}{r^2 - 9}$
    $\dfrac{r^2}{(r + 6)(r - 3)}$

20. $\dfrac{m^2 + 5m}{m^2 - 16} \cdot \dfrac{m^2 - 4m}{m^2 - 25}$
    $\dfrac{m^2}{(m + 4)(m - 5)}$

21. $\dfrac{12q}{q^2 - 2q + 1} \cdot \dfrac{q^2 - 1}{24q^2}$
    $\dfrac{q + 1}{2q(q - 1)}$

22. $\dfrac{36x^3}{x^2 - 25} \cdot \dfrac{x^2 - 10x + 25}{9x}$
    $\dfrac{4x^2(x - 5)}{x + 5}$

23. $\dfrac{w^2 - 4w - 5}{w^2 - 1} \cdot \dfrac{w^2 - w - 20}{w^2 - 10w + 25}$
    $\dfrac{w + 4}{w - 1}$

24. $\dfrac{y^2 - 2y - 24}{y^2 - 16} \cdot \dfrac{y^2 + 10y + 25}{y^2 - y - 30}$
    $\dfrac{y + 5}{y - 4}$

25. $\dfrac{n^2 - 6n + 9}{n^2 - 9} \cdot \dfrac{n^2 + 4n + 3}{n^2 - 3n}$
    $\dfrac{n + 1}{n}$

26. $\dfrac{4u^2 + 4u + 1}{2u^2 + u} \cdot \dfrac{u^2 + 3u - 4}{2u^2 - u - 1}$
    $\dfrac{u + 4}{u}$

27. $\dfrac{6x^2 + x - 12}{2x^2 - 5x - 12} \cdot \dfrac{3x^2 - 14x + 8}{9x^2 - 18x + 8}$
    1

28. $\dfrac{2x^2 + x - 15}{4x^2 + 11x - 3} \cdot \dfrac{4x^2 - 9x + 2}{2x^2 - 9x + 10}$
    1

29. $\dfrac{ac + 3a + 2c + 6}{ad + a + 2d + 2} \cdot \dfrac{ad - 5a + 2d - 10}{ac + 2c + 3a + 6}$

$\dfrac{d - 5}{d + 1}$

30. $\dfrac{mn + 2m + 4n + 8}{np - 3n + 2p - 6} \cdot \dfrac{pq + 5p - 3q - 15}{mq + 4m + 4q + 16}$

$\dfrac{q + 5}{q + 4}$

31. $\dfrac{x^3 + 27}{x^3 - 3x^2 + 9x} \cdot \dfrac{x^2 - 7x + 10}{x^2 - 2x - 15}$

$\dfrac{x - 2}{x}$

32. $\dfrac{x^2 + 6x + 8}{x^3 + 5x^2 + 25x} \cdot \dfrac{x^3 - 125}{x^2 - x - 20}$

$\dfrac{x + 2}{x}$

33. $\dfrac{x^2 + 7xy + 10y^2}{x^2 + 6xy + 5y^2} \cdot \dfrac{x + 2y}{y} \cdot \dfrac{x + y}{x^2 + 4xy + 4y^2}$

$\dfrac{1}{y}$

34. $\dfrac{x^2 + 6xy + 9y^2}{x^2 + xy - 6y^2} \cdot \dfrac{x + 4y}{x} \cdot \dfrac{x - 2y}{x^2 + 7xy + 12y^2}$

$\dfrac{1}{x}$

35. $\dfrac{4x^2y^3}{2x - 6} \cdot \dfrac{12 - 4x}{6x^3y}$

$-\dfrac{4y^2}{3x}$

36. $\dfrac{6b^3c^3}{70 - 10b} \cdot \dfrac{5b - 35}{8b^2c^5}$

$-\dfrac{3b}{8c^2}$

37. $\dfrac{2x^2 - 11x + 12}{2x^2 + 11x - 21} \cdot \dfrac{3x^2 + 20x - 7}{4 - x}$

$1 - 3x$

38. $\dfrac{3x^2 - 17x + 10}{3x^2 + 7x - 6} \cdot \dfrac{2x^2 + x - 15}{5 - x}$

$5 - 2x$

Objective 2

Prep Exercise 3 Why are $\dfrac{x}{y}$ and $\dfrac{y}{x}$ multiplicative inverses (or reciprocals)?

Their product is 1.

Prep Exercise 4 To divide rational expressions, we write an equivalent multiplication with the ___reciprocal___ ___(or multiplication inverse)___ of the divisor.

For Exercises 39–68, divide. See Examples 3 and 4.

39. $\dfrac{x}{y^4} \div \dfrac{x^3}{y^2}$

$\dfrac{1}{x^2y^2}$

40. $\dfrac{a^3}{b^3} \div \dfrac{a^5}{b^2}$

$\dfrac{1}{a^2b}$

41. $\dfrac{2m}{3n^2} \div \dfrac{8m^3}{15n}$

$\dfrac{5}{4m^2n}$

42. $\dfrac{21c^2}{7d^3} \div \dfrac{8c^4}{12d^2}$

$\dfrac{9}{2c^2d}$

43. $-\dfrac{12w^2y}{5z^2} \div \left(-\dfrac{12w^2}{y}\right)$

$\dfrac{y^2}{5z^2}$

44. $-\dfrac{7a^2b}{2c^2} \div \left(-\dfrac{7a^2}{b}\right)$

$\dfrac{b^2}{2c^2}$

45. $\dfrac{8x^7g^3}{15x^2g^4} \div \left(-\dfrac{3xg^2}{4x^2g^3}\right)$

$-\dfrac{32x^6}{45}$

46. $-\dfrac{10c^5g^5}{12c^2g^4} \div \dfrac{8cg^2}{4c^2g}$

$-\dfrac{5c^4}{12}$

47. $\dfrac{4a - 8}{15a} \div \dfrac{3a - 6}{5a^2}$

$\dfrac{4a}{9}$

48. $\dfrac{6b + 24}{7b^2} \div \dfrac{7b + 28}{14b^3}$

$\dfrac{12b}{7}$

49. $\dfrac{4p + 12q}{3p - 6q} \div \dfrac{5p + 15q}{6p - 12q}$

$\dfrac{8}{5}$

50. $\dfrac{5r + 10s}{3r - 12s} \div \dfrac{6r + 12s}{4r - 16s}$

$\dfrac{10}{9}$

51. $\dfrac{a^2 - b^2}{x^2 - y^2} \div \dfrac{a + b}{x - y}$

$\dfrac{a - b}{x + y}$

52. $\dfrac{x + 4}{y - 3} \div \dfrac{x^2 - 16}{y^2 - 9}$

$\dfrac{y + 3}{x - 4}$

53. $\dfrac{x^2 + 16x + 64}{x^2 - 16x + 64} \div \dfrac{x + 8}{x - 8}$

$\dfrac{x + 8}{x - 8}$

54. $\dfrac{w + 3}{w - 3} \div \dfrac{w^2 + 6w + 9}{w^2 - 6w + 9}$

$\dfrac{w - 3}{w + 3}$

55. $\dfrac{x^2 + 7x + 10}{x^2 + 2x} \div \dfrac{x + 5}{x^2 - 4x}$

$x - 4$

56. $\dfrac{t^2 - 2t}{t + 1} \div \dfrac{t^2 + 3t}{t^2 + 4t + 3}$

$t - 2$

57. $\dfrac{a^2 - b^2}{a^2 + 2ab + b^2} \div \dfrac{a^2 - 3ab + 2b^2}{a^2 + 3ab + 2b^2}$

$\dfrac{a + 2b}{a - 2b}$

58. $\dfrac{3w^2 - 7w - 6}{w^2 - 9} \div \dfrac{3w^2 - 10w - 8}{3w^2 + 7w - 6}$

$\dfrac{3w - 2}{w - 4}$

59. $\dfrac{3k + 2}{3k^4 + 2k^3} \div \dfrac{10k^2 + 3k - 1}{5k^3 - k^2}$

$\dfrac{1}{k(2k + 1)}$

60. $\dfrac{5y - 3}{4y^2 - y - 3} \div \dfrac{5y^3 - 3y^2}{4y^2 + 3y}$

$\dfrac{1}{y(y - 1)}$

61. $\dfrac{8x^3 - 27}{3x^2 + 20x - 7} \div \dfrac{2x^2 - 7x + 6}{x^2 + 5x - 14}$

$\dfrac{4x^2 + 6x + 9}{3x - 1}$

62. $\dfrac{x^2 + 6x - 16}{3x^2 - 10x + 8} \div \dfrac{2x^2 + 13x - 24}{27x^3 - 64}$

$\dfrac{9x^2 + 12x + 16}{2x - 3}$

63. $\dfrac{x^2 - 2x - 8}{x^2 + 6x + 8} \div (x^2 - 3x - 4)$

$\dfrac{1}{(x + 4)(x + 1)}$

64. $\dfrac{u^2 + 7u + 12}{u^2 + u - 6} \div (u^2 + u - 12)$

$\dfrac{1}{(u - 2)(u - 3)}$

65. $\dfrac{ab + 3a + 2b + 6}{bc + 4b + 3c + 12} \div \dfrac{ac - 3a + 2c - 6}{bc + 4b - 4c - 16}$

$\dfrac{b - 4}{c - 3}$

66. $\dfrac{xy - 3x + 4y - 12}{xy + 6x - 3y - 18} \div \dfrac{xy + 5x + 4y + 20}{xy + 5x - 3y - 15}$

$\dfrac{y - 3}{y + 6}$

67. $\dfrac{b^2 - 16}{b^2 + b - 12} \div \dfrac{4 - b}{b^2 + 2b - 15}$

$-b - 5$

68. $\dfrac{y^2 - y - 20}{y^2 - 2y - 24} \div \dfrac{5 - y}{y^2 - 3y - 18}$

$-y - 3$

Objectives 1 and 2

★ *For Exercises 69 –74, perform the indicated operations. (Remember that the order of operations is to multiply or divide from left to right.)*

69. $\dfrac{y^2 + y - 6}{2y^2 - 3y - 2} \cdot \dfrac{2y^2 + 9y + 4}{y^2 + 7y + 12} \div \dfrac{4y + y^2}{7 + y}$

$\dfrac{7 + y}{y(y + 4)}$

70. $\dfrac{2m^2 - 7m - 15}{4m^2 - 9} \cdot \dfrac{2m^2 - 5m - 3}{2m^2 - 7m + 3} \div \dfrac{m^2 - 5m}{2m - 1}$

$\dfrac{2m + 1}{m(2m - 3)}$

71. $\dfrac{6x^2 - 5x - 6}{x^4 - 81} \div \dfrac{3x - 2x^2}{x^3 + 3x^2 + 9x + 27} \div \dfrac{3x + 2}{x^4}$

$\dfrac{x^3}{x - 3}$

72. $\dfrac{12h^2 + 11h - 5}{h^4 - 16} \div \dfrac{h - 3h^2}{h^3 + 4h - 2h^2 - 8} \div \dfrac{4h + 5}{h^3}$

$\dfrac{h^2}{h + 2}$

73. $\dfrac{6x^2}{x^2 - 9} \cdot \dfrac{3x^2 + x - 24}{20y - 6xy} \cdot \dfrac{x - 3}{3x - 8} \div \dfrac{3x^2 + 9x}{3x^2 - x - 30}$

$\dfrac{x}{y}$

74. $\dfrac{8y^3}{3y^2 - 4y - 4} \div \dfrac{8xy - 20x}{2y^2 - 9y + 10} \cdot \dfrac{y + 5}{7 - 4y} \cdot \dfrac{12y^2 - 13y - 14}{2y^2 + 10y}$

$\dfrac{y^2}{x}$

Find 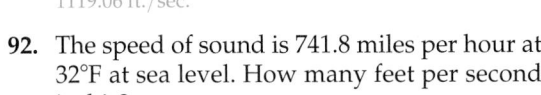 the Mistake   *For Exercises 75–78, explain the mistake; then correct the mistake.*

75. $\dfrac{5y}{9} \div \dfrac{3y}{10} = \dfrac{y}{3} \cdot \dfrac{y}{2} = \dfrac{y^2}{6}$

Mistake: Did not invert the divisor.

Correct: $\dfrac{50}{27}$

76. $\dfrac{12x^3y}{5} \cdot \dfrac{3y}{6x^2} = \dfrac{2xy}{5} \cdot \dfrac{3y}{1} = \dfrac{6y^2}{5}$

Mistake: Left x out of the answer.

Correct: $\dfrac{6xy^2}{5}$

77. $\dfrac{x^2 - 9}{15y} \cdot \dfrac{y - 2}{x + 3} = \dfrac{(x - 3)(x + 3)}{15y} \cdot \dfrac{y - 2}{x + 3}$

$\qquad = \dfrac{x - 3}{15} \cdot (-2)$

$\qquad = \dfrac{-2x + 6}{15}$

Mistake: Divided out y, which is not allowed because y is not a factor in $y - 2$.

Correct: $\dfrac{(x - 3)(y - 2)}{15y}$

78. $\dfrac{x^2 + 3x + 2}{x^2 - 1} \cdot \dfrac{x - 1}{x + 1} = \dfrac{(x + 1)(x + 2)}{1} \cdot \dfrac{1}{x + 1}$

$\qquad = x + 2$

Mistake: Divided out $x - 1$ and $x^2 - 1$, but they are not identical.

Correct: $\dfrac{x + 2}{x + 1}$

Objective 3

Prep Exercise 5 Explain how to convert one unit of measurement to another using dimensional analysis.

Multiply the given measurement by conversion factors so that the undesired units divide out, leaving the desired units.

Prep Exercise 6 What conversion factor would be used to convert 3.8 pounds to ounces?

$\dfrac{16 \text{ oz.}}{1 \text{ lb.}}$

For Exercises 79–84, use dimensional analysis and the following facts to convert units of length. See Example 5.

$$1 \text{ foot} = 12 \text{ inches} \qquad 1 \text{ yard} = 3 \text{ feet} \qquad 1 \text{ mile} = 5280 \text{ feet}$$

79. 18.5 feet to inches

222 in.

80. 24.2 miles to feet

127,776 ft.

81. 6 miles to yards

10,560 yd.

82. 2400 yards to miles

$1.3\overline{6}$ mi.

83. 8 yards to inches

288 in.

84. 216 inches to yards

6 yd.

For Exercises 85–90, use dimensional analysis and the following values to convert units of weight. See Example 5.

$$1 \text{ pound} = 16 \text{ ounces} \qquad 1 \text{ ton} = 2000 \text{ pounds}$$

85. 5.5 pounds to ounces

88 oz.

86. 36 ounces to pounds

2.25 lb.

87. 25,400 pounds to tons

12.7 T

88. 2.5 tons to pounds

5000 lb.

89. 3 tons to ounces

96,000 oz.

90. 48,000 ounces to tons

1.5 T

For Exercises 91–98, use dimensional analysis. See Example 5.

91. The world record for the fastest speed of a wheeled vehicle is 763 miles per hour. How many feet per second is this?

$1119.0\overline{6}$ ft./sec.

92. The speed of sound is 741.8 miles per hour at 32°F at sea level. How many feet per second is this?

1087.973 ft./sec.

93. A sprinter's best time in the 40-yard dash is 4.3 seconds. What is that speed in miles per hour?

≈ 19.03 mi./hr.

Of Interest

The world record for the fastest speed in a wheeled vehicle was set on October 15, 1997, by Andy Green, a former pilot of the British Royal Air Force, in a vehicle called the ThrustSSC. In setting the record, Green became the first person to reach supersonic speed on land.

94. A runner knows that she can run 400 yards in 87.2 seconds. If she can maintain that pace, how many minutes will it take her to run a mile?
≈ 6.39 min.

95. The mean orbital velocity of Earth is 29.78 kilometers per second. How many kilometers does Earth orbit in one year? (Hint: One year is $365\frac{1}{4}$ days.)

    (*Source:* Windows to the Universe.)
    939,785,328 km

96. The mean orbital velocity of Pluto is 4.74 kilometers per second. How many kilometers does Pluto orbit in one year? (*Source:* Windows to the Universe.)
149,583,024 km

★ 97. One light-year is the distance light travels in a year. If light travels at a speed of approximately 186,000 miles per second, how many miles does light travel in one year?
≈ 5.87×10^{12} mi./yr.

★ 98. The light from the Sun takes about 8 minutes to reach Earth. Using the fact that light travels at a speed of about 186,000 miles per second, determine Earth's distance from the Sun in miles.
89,280,000 mi.

Puzzle Problem If $\frac{1}{1000}$ of a light-year is about 9.46×10^{9} kilometers, find the speed of light in meters per second.
≈ 3×10^{8} m/sec.

Of Interest

The nearest star to the Sun is Proxima Centauri, which is about 4.2 light-years away. This means that when we look at Proxima Centauri, we are seeing an image of that star that is 4.2 years old. Some objects we see in the night sky are much farther from Earth. For example, the Andromeda galaxy is about 2,140,000 light-years from Earth, which means that when we look at it, we are seeing it as it was 2,140,000 years ago.

For Exercises 99–104, use dimensional analysis and the following table of exchange rates as of 11/29/12 to convert. See Example 5.

	USD($)	GBP(£)	CAN($)	EUR(€)
USD($)	1	1.6038	1.0084	1.2979
GBP(£)	0.6235	1	0.6285	0.8105
CAN($)	0.9917	1.5911	1	1.2877
EUR(€)	0.7705	1.2338	0.7766	1

Note Each number is the ratio of the row's unit to 1 of the column's ◀ unit. For example, 1.2877 means that there are 1.2877 Can$ to 1€.

(*Source:* Federal Reserve Bank of New York.)

99. Convert $500 into £ (Great Britain pound).
311.75 £

100. Convert $250 into Can$ (Canadian dollars).
252.10 Can$

101. Convert 450 € (euros) into U.S. dollars.
$584.06

102. Convert 700 £ into U.S. dollars.
$1122.66

★ 103. Convert 650 £ to €.
801.97 €

★ 104. Convert 400 Can$ to €.
310.64 €

Review Exercises

Exercises 1–6 Expressions

For Exercises 1–6, simplify.

[1.3] 1. $\frac{1}{3} + \frac{2}{3}$
1

[1.3] 2. $\frac{1}{4} - \frac{3}{4}$
$-\frac{1}{2}$

[1.7] 3. $\frac{1}{2}x + \frac{3}{2}x$
2x

[1.7] 4. $\frac{3}{8}y - 5 - \frac{1}{8}y + 1$
$\frac{1}{4}y - 4$

[5.3] 5. $(6x^2 + x - 1) + (2x^2 - 7x - 3)$
$8x^2 - 6x - 4$

[5.3] 6. $(3n^2 - 5n + 7) - (4n^2 - 5n + 3)$
$-n^2 + 4$

7.3 Adding and Subtracting Rational Expressions with the Same Denominator

Objective

1 Add or subtract rational expressions with the same denominator.

Warm-up

[5.3] *For Exercises 1 and 2, combine the polynomials.*

1. $(x^2 - 5x - 13) + (3x - 11)$ **2.** $(x^3 + 4x - 3) - (4x + 5)$

[1.3] *For Exercises 3 and 4, combine the fractions.*

3. $\dfrac{5}{14} + \dfrac{1}{14}$ **4.** $\dfrac{4}{15} - \left(-\dfrac{8}{15}\right)$

Objective 1 Add or subtract rational expressions with the same denominator.

Recall that when adding or subtracting fractions with the same denominator, we add or subtract the numerators and keep the same denominator.

$$\frac{3}{7} + \frac{1}{7} = \frac{3+1}{7} = \frac{4}{7} \quad \text{and} \quad \frac{7}{9} - \frac{2}{9} = \frac{7-2}{9} = \frac{5}{9}$$

We add and subtract rational expressions the same way.

$$\frac{3x}{7} + \frac{x}{7} = \frac{3x+x}{7} = \frac{4x}{7} \quad \text{and} \quad \frac{7x}{9} - \frac{2x}{9} = \frac{7x-2x}{9} = \frac{5x}{9}$$

> **Procedure** **Adding or Subtracting Rational Expressions (Same Denominator)**
>
> To add or subtract rational expressions that have the same denominator:
> 1. Add or subtract the numerators and keep the same denominator.
> 2. Simplify to lowest terms. (Remember to factor the numerators and denominators completely to simplify.)

Example 1 Add $\dfrac{5x}{14} + \dfrac{x}{14}$.

Solution: $\dfrac{5x}{14} + \dfrac{x}{14} = \dfrac{5x+x}{14}$ Because the rational expressions have the same denominator, we add numerators and keep the same denominator.

$= \dfrac{6x}{14}$ Combine like terms.

$= \dfrac{2 \cdot 3 \cdot x}{2 \cdot 7}$ Factor.

$= \dfrac{3x}{7}$ Divide out the common factor, 2.

Answer to Your Turn 1

$-\dfrac{a^2}{3}$

Answers to Warm-up
1. $x^2 - 2x - 24$ **2.** $x^3 - 8$
3. $\dfrac{3}{7}$ **4.** $\dfrac{4}{5}$

Your Turn 1 Subtract $\dfrac{2a^2}{15} - \dfrac{7a^2}{15}$.

Numerators with No Like Terms

In Example 1, the numerators were like terms. If the numerators have no like terms, we write a multiterm polynomial expression in the numerator of the sum or difference.

Example 2 Subtract.

a. $\dfrac{y}{y+5} - \dfrac{3}{y+5}$

Solution: $\dfrac{y}{y+5} - \dfrac{3}{y+5} = \dfrac{y-3}{y+5}$ Subtract numerators and keep the same denominator. Because y and 3 are not like terms, we express their difference as a polynomial.

Warning Although it may be tempting to do so, we *cannot* divide out the y's because they are terms, not factors.

b. $\dfrac{n^2}{n+3} - \dfrac{9}{n+3}$

Solution: $\dfrac{n^2}{n+3} - \dfrac{9}{n+3} = \dfrac{n^2-9}{n+3}$ ◀ **Note** The numerator can be factored, so we may be able to simplify.

$$= \dfrac{(n-3)(n+3)}{n+3}$$ Factor the numerator.

$$= n-3$$ Divide out the common factor, $n+3$.

c. $\dfrac{4m}{15} - \left(-\dfrac{8}{15}\right)$

Solution: $\dfrac{4m}{15} - \left(-\dfrac{8}{15}\right) = \dfrac{4m-(-8)}{15}$

Note $4m+8$ can be factored.

$$4m+8 = 4(m+2)$$

$$= \dfrac{4m+8}{15}$$ However, because there are no ◀ common factors in the denominator, we cannot simplify further.

Your Turn 2 Add or subtract.

a. $\dfrac{x}{x-4} + \dfrac{6}{x-4}$ b. $\dfrac{16n^2}{4n+5} - \dfrac{25}{4n+5}$ c. $\dfrac{5x^2}{7} - \left(-\dfrac{2x}{7}\right)$

Numerators That Are Multiterm Polynomials

So far, the numerators in the given rational expressions were monomials. Now look at some cases where the numerators in the given rational expressions are multiterm polynomials. Remember from Chapter 5 that to add polynomials, we combine like terms.

Example 3 Add.

a. $\dfrac{7y-13}{5} + \dfrac{y+8}{5}$

Solution: $\dfrac{7y-13}{5} + \dfrac{y+8}{5} = \dfrac{(7y-13)+(y+8)}{5}$ Combine $7y$ with y to get $8y$ and combine -13 with 8 to get -5.

$$= \dfrac{8y-5}{5}$$ ◀ **Warning** It may be tempting to divide out the 5, but the 5 in $8y-5$ is not a factor. Because $8y-5$ cannot be factored, we cannot simplify this rational expression.

Answers to Your Turn 2

a. $\dfrac{x+6}{x-4}$ b. $4n-5$

c. $\dfrac{5x^2+2x}{7}$

b. $\dfrac{x^2 - 5x - 13}{x - 6} + \dfrac{3x - 11}{x - 6}$

Solution: $\dfrac{x^2 - 5x - 13}{x - 6} + \dfrac{3x - 11}{x - 6} = \dfrac{(x^2 - 5x - 13) + (3x - 11)}{x - 6}$

Note Because the numerator can be factored, we may be able to simplify.

▶ $= \dfrac{x^2 - 2x - 24}{x - 6}$ Combine like terms in the numerator.

$= \dfrac{(x - 6)(x + 4)}{x - 6}$ Factor the numerator.

$= x + 4$ Divide out the common factor, $x - 6$.

Your Turn 3   Add.

a. $\dfrac{n + 7}{11n} + \dfrac{2 - 5n}{11n}$ **b.** $\dfrac{3t^2 - 5t}{t - 3} + \dfrac{t^2 - 8t + 3}{t - 3}$

Simplifying When the Denominator Is Factorable

Sometimes you also may be able to factor the denominator. If the denominator can be factored, you will need to write it in factored form to simplify.

Example 4   Add $\dfrac{4x^2 + 9x + 1}{2x^3 - 18x} + \dfrac{6x^2 + 23x + 5}{2x^3 - 18x}$.

Solution: $\dfrac{4x^2 + 9x + 1}{2x^3 - 18x} + \dfrac{6x^2 + 23x + 5}{2x^3 - 18x} = \dfrac{(4x^2 + 9x + 1) + (6x^2 + 23x + 5)}{2x^3 - 18x}$

$= \dfrac{10x^2 + 32x + 6}{2x^3 - 18x}$ Combine like terms in the numerator.

$= \dfrac{2(5x^2 + 16x + 3)}{2x(x^2 - 9)}$ Factor 2 out of the numerator and 2x out of the denominator.

$= \dfrac{2(5x + 1)(x + 3)}{2x(x - 3)(x + 3)}$ Factor the trinomial in the numerator and the difference of squares in the denominator.

$= \dfrac{5x + 1}{x(x - 3)} \text{ or } \dfrac{5x + 1}{x^2 - 3x}$ Divide out the common factors, 2 and $x + 3$.

Your Turn 4   Add $\dfrac{x^3 + 3x^2 - 7x}{3x^4 - 12x^2} + \dfrac{5x - 2x^2}{3x^4 - 12x^2}$.

Answers to Your Turn 3
a. $\dfrac{9 - 4n}{11n}$ **b.** $4t - 1$

Answer to Your Turn 4
$\dfrac{x - 1}{3x(x - 2)} \text{ or } \dfrac{x - 1}{3x^2 - 6x}$

Subtracting with Numerators That Are Multiterm Polynomials

Now let's look at subtraction of rational expressions that have the same denominator and multiterm polynomials in the numerators. Recall from Chapter 5 that when subtracting polynomials, we write an equivalent addition with the additive inverse of the subtrahend.

Instructor Note Use Exercise 34, on page 523, as a subtraction problem and then as a division problem to emphasize how much exercises may look alike but require different steps to simplify.

Example 5 Subtract.

a. $\dfrac{7t^2 - t + 5}{t + 2} - \dfrac{3t^2 - 6t - 2}{t + 2}$

Solution: $\dfrac{7t^2 - t + 5}{t + 2} - \dfrac{3t^2 - 6t - 2}{t + 2} = \dfrac{(7t^2 - t + 5) - (3t^2 - 6t - 2)}{t + 2}$

Note To write an equivalent addition, change the operation symbol from a minus sign to a plus sign and change all of the signs in the subtrahend (second) polynomial.

$$\blacktriangleright = \dfrac{(7t^2 - t + 5) + (-3t^2 + 6t + 2)}{t + 2}$$

$$= \dfrac{4t^2 + 5t + 7}{t + 2} \quad \text{Combine like terms.}$$
$$\qquad\qquad\qquad 4t^2 + 5t + 7 \text{ is prime.}$$

b. $\dfrac{x^3 + 4x - 3}{5x^2 - 10x} - \dfrac{4x + 5}{5x^2 - 10x}$

Solution: $\dfrac{x^3 + 4x - 3}{5x^2 - 10x} - \dfrac{4x + 5}{5x^2 - 10x} = \dfrac{(x^3 + 4x - 3) - (4x + 5)}{5x^2 - 10x}$

$$= \dfrac{(x^3 + 4x - 3) + (-4x - 5)}{5x^2 - 10x} \quad \text{Write the equivalent addition.}$$

Note The numerator is a difference of cubes.

$$\blacktriangleright = \dfrac{x^3 - 8}{5x^2 - 10x} \quad \text{Combine like terms.}$$

$$= \dfrac{(x - 2)(x^2 + 2x + 4)}{5x(x - 2)} \quad \begin{array}{l}\text{Factor the numerator and} \\ \text{denominator. Divide out the} \\ \text{common factor, } x - 2.\end{array}$$

$$= \dfrac{x^2 + 2x + 4}{5x} \quad x^2 + 2x + 4 \text{ is prime.}$$

Your Turn 5 Subtract.

a. $\dfrac{3t^2 - 4t - 7}{t^2 + 10t + 21} - \dfrac{2t^2 - 6t - 4}{t^2 + 10t + 21}$

b. $\dfrac{y^3 + y - 20}{3y^2 - 7y - 6} - \dfrac{y + 7}{3y^2 - 7y - 6}$

Answers to Your Turn 5
a. $\dfrac{t - 1}{t + 7}$ **b.** $\dfrac{y^2 + 3y + 9}{3y + 2}$

7.3 Exercises (For Extra Help) MyMathLab®

Objective 1

Prep Exercise 1 Explain how to add or subtract two rational expressions that have the same denominator.

Add or subtract the numerators and keep the same denominator. Then simplify to lowest terms.

Prep Exercise 2 After adding or subtracting two rational expressions, what is the first step in simplifying the resulting rational expression?

Factor both the numerator and denominator.

Prep Exercise 3 Two rational expressions that have numerators that are multiterm polynomials and the same denominator are to be subtracted. Explain the process.

Write an equivalent addition with the additive inverse of the subtrahend.

Prep Exercise 4 After adding or subtracting two rational expressions, the resulting rational expression contains a multiterm polynomial in the numerator that can be factored. Does this mean that the rational expression can be simplified? Explain.

It can be simplified only if there are factors common to both the numerator and denominator.

For Exercises 1–46, add or subtract. Simplify your answers to lowest terms. See Examples 1–5.

1. $\dfrac{3x}{10} + \dfrac{2x}{10}$

$\dfrac{x}{2}$

2. $\dfrac{2x}{9} + \dfrac{x}{9}$

$\dfrac{x}{3}$

3. $\dfrac{4a^2}{9} + \left(-\dfrac{a^2}{9}\right)$

$\dfrac{a^2}{3}$

4. $\dfrac{6x^3}{11} + \left(-\dfrac{x^3}{11}\right)$

$\dfrac{5x^3}{11}$

5. $\dfrac{6}{a} - \left(-\dfrac{3}{a}\right)$

$\dfrac{9}{a}$

6. $\dfrac{3}{w} - \left(-\dfrac{2}{w}\right)$

$\dfrac{5}{w}$

7. $\dfrac{6x}{7y^2} - \dfrac{2x}{7y^2} - \dfrac{5x}{7y^2}$

$-\dfrac{x}{7y^2}$

8. $\dfrac{9a}{11x^2} - \dfrac{8a}{11x^2} - \dfrac{3a}{11x^2}$

$-\dfrac{2a}{11x^2}$

9. $\dfrac{n}{n+5} + \dfrac{3}{n+5}$

$\dfrac{n+3}{n+5}$

10. $\dfrac{n}{n+2} + \dfrac{7}{n+2}$

$\dfrac{n+7}{n+2}$

11. $\dfrac{2r}{r+2} + \dfrac{r}{r+2}$

$\dfrac{3r}{r+2}$

12. $\dfrac{5z}{b-6} + \dfrac{7z}{b-6}$

$\dfrac{12z}{b-6}$

13. $\dfrac{1-2q}{3q} + \dfrac{5q+2}{3q}$

$\dfrac{q+1}{q}$

14. $\dfrac{6x-3}{4x} + \dfrac{2x+5}{4x}$

$\dfrac{4x+1}{2x}$

15. $\dfrac{19u+6}{5u} - \dfrac{4u+6}{5u}$

3

16. $\dfrac{8y+5}{3y} - \dfrac{2y+5}{3y}$

2

17. $\dfrac{2-2a}{2a-3} + \dfrac{8a-11}{2a-3}$

3

18. $\dfrac{4-2x}{3x-2} + \dfrac{14x-12}{3x-2}$

4

19. $\dfrac{w^2+5}{w+1} + \dfrac{6-w^2}{w+1}$

$\dfrac{11}{w+1}$

20. $\dfrac{x^2+2}{x+1} + \dfrac{6-x^2}{x+1}$

$\dfrac{8}{x+1}$

21. $\dfrac{4m+3}{m-1} - \dfrac{m+4}{m-1}$

$\dfrac{3m-1}{m-1}$

22. $\dfrac{2x+5}{x+1} - \dfrac{x+8}{x+1}$

$\dfrac{x-3}{x+1}$

23. $\dfrac{1-3t}{2t-3} - \dfrac{8t-2}{2t-3}$

$\dfrac{-11t+3}{2t-3}$

24. $\dfrac{3-4a}{5a-2} - \dfrac{2a-6}{5a-2}$

$\dfrac{-6a+9}{5a-2}$

25. $\dfrac{2w+3}{w+5} + \dfrac{5w-2}{w+5} - \dfrac{w+6}{w+5}$

$\dfrac{6w-5}{w+5}$

26. $\dfrac{6q+21}{q+1} - \dfrac{2q+8}{q+1} + \dfrac{6q-5}{q+1}$

$\dfrac{10q+8}{q+1}$

27. $\dfrac{g}{g^2-9} - \dfrac{3}{g^2-9}$

$\dfrac{1}{g+3}$

28. $\dfrac{x}{x^2-49} + \dfrac{7}{x^2-49}$

$\dfrac{1}{x-7}$

29. $\dfrac{z^2+2z}{z-7} + \dfrac{21-12z}{z-7}$

$z-3$

30. $\dfrac{u^2+5u}{u+1} + \dfrac{2-2u}{u+1}$

$u+2$

31. $\dfrac{2s}{s^2+4s+4} - \dfrac{s-2}{s^2+4s+4}$

$\dfrac{1}{s+2}$

32. $\dfrac{v+8}{v^2+8v+16} - \dfrac{4}{v^2+8v+16}$

$\dfrac{1}{v+4}$

33. $\dfrac{y^2}{y^2+3y} - \dfrac{9}{y^2+3y}$

$\dfrac{y-3}{y}$

34. $\dfrac{x^2}{x^2-5x} - \dfrac{25}{x^2-5x}$

$\dfrac{x+5}{x}$

35. $\dfrac{x^2+3x}{x^2-x-12} + \dfrac{2x+6}{x^2-x-12}$

$\dfrac{x+2}{x-4}$

36. $\dfrac{y^2+2y}{y^2+5y+4} + \dfrac{5y+12}{y^2+5y+4}$

$\dfrac{y+3}{y+1}$

37. $\dfrac{t^2-4t}{t^2+10t+21} - \dfrac{-6t+3}{t^2+10t+21}$

$\dfrac{t-1}{t+7}$

38. $\dfrac{v^2+6}{v^2-3v-10} - \dfrac{2v+14}{v^2-3v-10}$

$\dfrac{v-4}{v-5}$

39. $\dfrac{3x^3}{x^2-2x+4} + \dfrac{24}{x^2-2x+4}$

$3(x+2)$

40. $\dfrac{2x^3}{x^2-4x+16} + \dfrac{128}{x^2-4x+16}$

$2(x+4)$

41. $\dfrac{x-5}{x^2-25} + \dfrac{2x+10}{x^2-25} + \dfrac{-2x}{x^2-25}$

$\dfrac{1}{x-5}$

42. $\dfrac{3x-3}{x^2-1} + \dfrac{2x+2}{x^2-1} + \dfrac{-4x}{x^2-1}$

$\dfrac{1}{x+1}$

43. $\dfrac{3x^2 + 2x}{x^2 - 2x - 15} + \dfrac{3x - 8}{x^2 - 2x - 15} + \dfrac{2x + 2}{x^2 - 2x - 15}$

$\dfrac{3x - 2}{x - 5}$

44. $\dfrac{m^2 - 6m}{m^2 - 5m + 6} + \dfrac{2m - 4}{m^2 - 5m + 6} + \dfrac{6m - 4}{m^2 - 5m + 6}$

$\dfrac{m + 4}{m - 3}$

45. $\dfrac{2x^2 + x - 3}{x^2 + 6x + 5} + \dfrac{x^2 - 2x + 4}{x^2 + 6x + 5} - \dfrac{2x^2 + x + 4}{x^2 + 6x + 5}$

$\dfrac{x - 3}{x + 5}$

46. $\dfrac{3x^2 - 2x + 3}{x^2 + x - 12} - \dfrac{x^2 + x - 2}{x^2 + x - 12} - \dfrac{x^2 + 4x - 7}{x^2 + x - 12}$

$\dfrac{x - 4}{x + 4}$

Find ⊗ the Mistake   *For Exercises 47–50, explain the mistake; then find the correct sum or difference.*

47. $\dfrac{4y}{5} + \dfrac{1}{5} = \dfrac{5y}{5} = y$

Mistake: Combined terms that are not like.

Correct: $\dfrac{4y + 1}{5}$

48. $\dfrac{7}{m} + \dfrac{n}{m} = \dfrac{7 + n}{2m}$

Mistake: Added denominators.

Correct: $\dfrac{7 + n}{m}$

49. $\dfrac{x + 1}{3} - \dfrac{x - 5}{3} = \dfrac{(x + 1) - (x - 5)}{3}$

$= \dfrac{(x + 1) + (-x - 5)}{3}$

$= -\dfrac{4}{3}$

Mistake: Did not change sign of -5.

Correct: 2

50. $\dfrac{x}{x - 2} + \dfrac{x - 4}{x - 2} = \dfrac{x^2 - 4}{x - 2}$

$= \dfrac{(x + 2)(x - 2)}{x - 2}$

$= x + 2$

Mistake: Multiplied x's instead of adding.

Correct: 2

For Exercises 51–54, find the perimeter of the figure shown.

51.

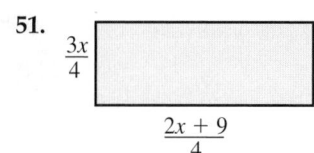

$\dfrac{5x + 9}{2}$

52.

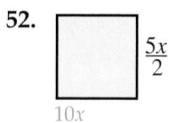

53.

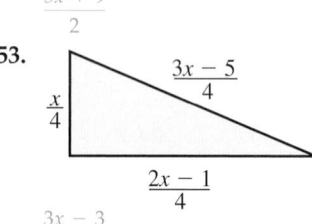

$\dfrac{3x - 3}{2}$

54.

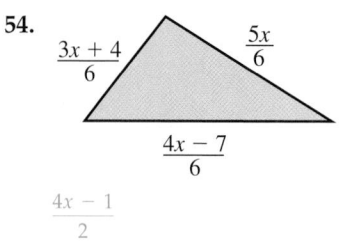

$\dfrac{4x - 1}{2}$

Review Exercises

Exercises 1–6 ⬋ Expressions

For Exercises 1–6, simplify.

[1.3] **1.** $\dfrac{1}{2} + \dfrac{3}{5}$

$\dfrac{11}{10}$

[1.3] **2.** $\dfrac{5}{6} - \dfrac{3}{8}$

$\dfrac{11}{24}$

[1.3] **3.** $\dfrac{4}{7} - \left(-\dfrac{3}{5}\right)$

$1\dfrac{6}{35}$

[1.7] **4.** $\dfrac{1}{3}x - \dfrac{1}{6} + \dfrac{2}{5}x + \dfrac{3}{4}$

$\dfrac{11}{15}x + \dfrac{7}{12}$

[5.3] **5.** $\left(x^2 - 9x - \dfrac{5}{8}\right) + \left(x^2 - 3x + \dfrac{1}{6}\right)$

$2x^2 - 12x - \dfrac{11}{24}$

[5.3] **6.** $\left(3y^2 + \dfrac{3}{4}y - 8\right) - \left(y^2 + \dfrac{4}{5}y + 2\right)$

$2y^2 - \dfrac{1}{20}y - 10$

7.4 Adding and Subtracting Rational Expressions with Different Denominators

Objectives

1 Find the LCD of two or more rational expressions.

2 Given two rational expressions, write equivalent rational expressions with their LCD.

3 Add or subtract rational expressions with different denominators.

Warm-up

[1.3] **1.** Add: $-\dfrac{3}{4} + \dfrac{5}{6}$

[1.7] **2.** Distribute: $x(x-3)$

[5.5] **3.** Find the product: $(x-1)(x-4)$

Adding or subtracting rational expressions with different denominators follows three main stages: (1) finding the LCD, (2) rewriting the expressions with the LCD, and (3) adding or subtracting the numerators. We explore the first two stages in Objectives 1 and 2, then put them back in their proper context in Objective 3.

Objective 1 Find the LCD of two or more rational expressions.

Recall from Chapter 1 that the least common denominator (LCD) is the smallest number that is evenly divisible by all of the denominators. For example, the LCD of $\dfrac{1}{8}$ and $\dfrac{7}{12}$ is 24. The LCD can be found using the prime factors of the denominators. The idea is to use each prime factor the most it appears in the factorizations. Look at the factorizations of 8 and 12.

$$8 = 2^3 \qquad 12 = 2^2 \cdot 3$$

$$\text{LCD} = 2^3 \cdot 3 = 24$$

Note 8 has the most 2s, and 12 has the most 3s. Multiplying those prime factors gives the LCD.

This suggests the following procedure.

> **Procedure** **Finding the LCD**
>
> To find the LCD of two or more rational expressions:
> 1. Find the prime factorization of each denominator.
> 2. Write a product that contains each unique prime factor the greatest number of times it occurs in the factorizations. (If using exponential forms, use each prime factor raised to its greatest exponent in the factorizations.)
> 3. Simplify the product found in step 2.

First, let's consider cases where the denominators are monomials.

Example 1 Find the LCD of $\dfrac{3}{4x^2}$ and $\dfrac{5}{6x^3}$.

Solution: Factor the denominators $4x^2$ and $6x^3$. Then write a product that contains each prime factor raised to its greatest exponent in the factorizations.

$$4x^2 = 2^2 \cdot x^2 \qquad \text{Factor } 4x^2.$$
$$6x^3 = 2 \cdot 3 \cdot x^3 \qquad \text{Factor } 6x^3.$$
$$\text{LCD} = 2^2 \cdot 3 \cdot x^3$$
$$= 12x^3$$

Note 2^2 has the greatest exponent for 2, 3 has the greatest exponent for 3, and x^3 has the greatest exponent for x.

Answers to Warm-up

1. $\dfrac{1}{12}$

2. $x^2 - 3x$

3. $x^2 - 5x + 4$

Your Turn 1 Find the LCD.

a. $\dfrac{y}{12x^5}$ and $\dfrac{5}{9x^2}$

b. $\dfrac{p}{10mn}$ and $\dfrac{7}{15n^2}$

Now consider denominators that are multiterm polynomials.

Example 2 Find the LCD.

a. $\dfrac{7}{5x-15}$ and $\dfrac{5}{2x-6}$

Solution: Factor the denominators $5x-15$ and $2x-6$. Then write a product that contains each prime factor raised to its greatest exponent in the factorizations.

$$5x - 15 = 5(x-3) \quad \text{Factor } 5x - 15.$$
$$2x - 6 = 2(x-3) \quad \text{Factor } 2x - 6.$$
$$\text{LCD} = 2 \cdot 5 \cdot (x-3)$$
$$= 10(x-3) \text{ or } 10x - 30$$

◄ **Note** The most that 2, 5, and $x-3$ appear in any factorization is once. In other words, the greatest exponent for each of them is 1.

b. $\dfrac{7}{x^2+10x+25}$ and $\dfrac{17}{6x^4+30x^3}$

Solution: $x^2 + 10x + 25 = (x+5)^2$ Factor $x^2 + 10x + 25$.

$$6x^4 + 30x^3 = 6x^3(x+5) = 2 \cdot 3 \cdot x^3(x+5) \quad \text{Factor } 6x^4 + 30x^3.$$
$$\text{LCD} = 2 \cdot 3 \cdot x^3 \cdot (x+5)^2$$
$$= 6x^3(x+5)^2$$

◄ **Note** 2 and 3 each have 1 as their greatest exponent, x^3 has the greatest exponent for x, and $(x+5)^2$ has the greatest exponent for $x+5$.

Your Turn 2 Find the LCD.

a. $\dfrac{y}{3y+6}$ and $\dfrac{5}{4y+8}$

b. $\dfrac{9}{x^2-2x+1}$ and $\dfrac{17}{2x^3-2x}$

Objective 2 Given two rational expressions, write equivalent rational expressions with their LCD.

Writing equivalent rational expressions is much like writing equivalent fractions, which we reviewed in Section 1.2. Remember that to write an equivalent fraction, we multiply the numerator and denominator by the same nonzero number.

Example 3 Write $\dfrac{3}{4x^2}$ and $\dfrac{5}{6x^3}$ as equivalent rational expressions with their LCD.

Instructor Note Make it clear that we are multiplying each rational expression by 1, the multiplicative identity. Consequently, we are not changing the value of the rational expression.

Solution: In Example 1, we found the LCD to be $12x^3$. For each rational expression, we multiply both the numerator and denominator by an appropriate factor so that the denominator becomes $12x^3$.

$$\frac{3}{4x^2} = \frac{3 \cdot 3x}{4x^2 \cdot 3x} = \frac{9x}{12x^3}$$

◄ **Note** We can determine the appropriate factor by dividing the LCD by the original denominator. In this case, $12x^3 \div 4x^2 = 3x$.

$$\frac{5}{6x^3} = \frac{5 \cdot 2}{6x^3 \cdot 2} = \frac{10}{12x^3}$$

Answers to Your Turn 1
a. $36x^5$ **b.** $30mn^2$

Answers to Your Turn 2
a. $12(y+2)$
b. $2x(x-1)^2(x+1)$

Now consider rational expressions with denominators that are multiterm polynomials. Rewriting these rational expressions with their LCD is usually easier if we leave their denominators in factored form.

Example 4 Write $\dfrac{7}{x^2 + 10x + 25}$ and $\dfrac{17}{6x^4 + 30x^3}$ as equivalent rational expressions with their LCD.

Solution: In Example 2(b), we found the LCD to be $6x^3(x + 5)^2$.

$$\frac{7}{x^2 + 10x + 25} = \frac{7}{(x + 5)^2}$$ Factor the denominator.

Note Another way to determine the appropriate factor is to think of it as the factor in the LCD that is missing from the original denominator. In this case, $6x^3$ is in the LCD, but not in the original denominator.

$$\blacktriangleright = \frac{7 \cdot 6x^3}{(x + 5)^2 \cdot 6x^3}$$ Multiply the numerator and denominator by the same factor, $6x^3$, to get the LCD, $6x^3(x + 5)^2$.

$$= \frac{42x^3}{6x^3(x + 5)^2}$$ ◄ **Note** Although we could multiply out $6x^3(x + 5)^2$, it is simpler to leave it in this form.

$$\frac{17}{6x^4 + 30x^3} = \frac{17}{6x^3(x + 5)}$$ Factor the denominator.

Note There are two factors of $x + 5$ in the LCD, whereas only one factor of $x + 5$ appears in the original denominator.

$$\blacktriangleright = \frac{17 \cdot (x + 5)}{6x^3(x + 5) \cdot (x + 5)}$$ Multiply the numerator and denominator by the same factor, $x + 5$, to get the LCD, $6x^3(x + 5)^2$.

$$= \frac{17(x + 5)}{6x^3(x + 5)^2}$$

Your Turn 4 Write equivalent rational expressions with their LCD. You found the LCD of these expressions in Your Turns 1b and 2b.

a. $\dfrac{p}{10mn}$ and $\dfrac{7}{15n^2}$ **b.** $\dfrac{9}{x^2 - 2x + 1}$ and $\dfrac{17}{2x^3 - 2x}$

Objective 3 Add or subtract rational expressions with different denominators.

To add or subtract rational expressions with different denominators, we use the same process that we use to add or subtract fractions.

Procedure Adding or Subtracting Rational Expressions with Different Denominators

To add or subtract rational expressions with different denominators:
1. Find the LCD.
2. Write each rational expression as an equivalent expression with the LCD.
3. Add or subtract the numerators and keep the LCD.
4. Simplify.

Example 5 Add or subtract as indicated.

a. $\dfrac{3}{4x^2} + \dfrac{5}{6x^3}$

Solution: We found the LCD for these expressions in Example 1. We then wrote equivalent expressions in Example 3. Now we add the numerators and keep the LCD.

$$\frac{3}{4x^2} + \frac{5}{6x^3} = \frac{9x}{12x^3} + \frac{10}{12x^3}$$ Rewrite with the LCD.

$$= \frac{9x + 10}{12x^3}$$ Add numerators.

Note Because $9x$ and 10 are not like terms, we must indicate the addition as a polynomial expression. Also, ◄ because $9x + 10$ cannot be factored, we cannot simplify this rational expression.

Answers to Your Turn 4

a. $\dfrac{3np}{30mn^2}$ and $\dfrac{14m}{30mn^2}$

b. $\dfrac{18x(x + 1)}{2x(x - 1)^2(x + 1)}$ and $\dfrac{17(x - 1)}{2x(x - 1)^2(x + 1)}$

b. $\dfrac{7}{x^2 + 10x + 25} - \dfrac{17}{6x^4 + 30x^3}$

Solution: We found the LCD for these expressions in Example 2(b) and wrote equivalent expressions in Example 4. Now we combine the numerators and keep the LCD.

$$\dfrac{7}{x^2 + 10x + 25} - \dfrac{17}{6x^4 + 30x^3} = \dfrac{42x^3}{6x^3(x + 5)^2} - \dfrac{17(x + 5)}{6x^3(x + 5)^2} \quad \text{Rewrite with the LCD.}$$

Warning Do not divide out the $(x + 5)$'s because $(x + 5)$ is not a factor of the numerator.

$$= \dfrac{42x^3 - 17(x + 5)}{6x^3(x + 5)^2} \quad \text{Subtract numerators.}$$

$$= \dfrac{42x^3 - 17x - 85}{6x^3(x + 5)^2} \quad \text{Simplify in the numerator.}$$

Note Leaving the LCD in factored form helps us determine whether the final expression can be simplified. Because $42x^3 - 17x - 85$ cannot be factored, we cannot reduce this rational expression. ▶

Now that we have seen all of the pieces of the process spread out over several examples, let's put them together using new rational expressions.

Example 6   Add or subtract as indicated.

a. $\dfrac{2x}{9y^3} + \dfrac{3}{4xy}$

Solution: First find the LCD.

Factored forms: $4xy = 2^2 \cdot x \cdot y; \quad 9y^3 = 3^2 \cdot y^3$

$$\text{LCD} = 2^2 \cdot 3^2 \cdot x \cdot y^3 = 36xy^3$$

$$\dfrac{2x}{9y^3} + \dfrac{3}{4xy} = \dfrac{2x(4x)}{9y^3(4x)} + \dfrac{3(9y^2)}{4xy(9y^2)} \quad \text{Write equivalent rational expressions with the LCD, } 36xy^3.$$

$$= \dfrac{8x^2}{36xy^3} + \dfrac{27y^2}{36xy^3}$$

$$= \dfrac{8x^2 + 27y^2}{36xy^3} \quad \text{Add numerators.}$$

Note The numerator cannot be factored, so we cannot simplify. ▶

b. $\dfrac{4}{m - 5} - \dfrac{3}{m + 2}$

Solution: First find the LCD. Because both $m - 5$ and $m + 2$ are prime, the LCD is their product, $(m - 5)(m + 2)$.

$$\dfrac{4}{m - 5} - \dfrac{3}{m + 2} = \dfrac{4(m + 2)}{(m - 5)(m + 2)} - \dfrac{3(m - 5)}{(m + 2)(m - 5)} \quad \text{Write equivalent rational expressions with the LCD, } (m - 5)(m + 2).$$

$$= \dfrac{4m + 8}{(m - 5)(m + 2)} - \dfrac{3m - 15}{(m - 5)(m + 2)} \quad \text{Multiply numerators.}$$

$$= \dfrac{(4m + 8) - (3m - 15)}{(m - 5)(m + 2)} \quad \text{Subtract numerators.}$$

$$= \dfrac{(4m + 8) + (-3m + 15)}{(m - 5)(m + 2)} \quad \text{Rewrite the subtraction as addition.}$$

$$= \dfrac{m + 23}{(m - 5)(m + 2)} \quad \text{Combine like terms.}$$

c. $\dfrac{x}{x + 2} - \dfrac{10}{x^2 - x - 6}$

Solution: First find the LCD.

Factored forms: $x + 2$ cannot be factored; $x^2 - x - 6 = (x + 2)(x - 3)$
LCD $= (x + 2)(x - 3)$

$$\dfrac{x}{x + 2} - \dfrac{10}{x^2 - x - 6} = \dfrac{x}{x + 2} - \dfrac{10}{(x + 2)(x - 3)}$$ Factor the denominators.

$$= \dfrac{x\,(x - 3)}{(x + 2)(x - 3)} - \dfrac{10}{(x + 2)(x - 3)}$$ Write equivalent rational expressions with the LCD, $(x + 2)(x - 3)$.

$$= \dfrac{x^2 - 3x}{(x + 2)(x - 3)} - \dfrac{10}{(x + 2)(x - 3)}$$ Distribute.

$$= \dfrac{x^2 - 3x - 10}{(x + 2)(x - 3)}$$ Subtract numerators.

$$= \dfrac{(x + 2)(x - 5)}{(x + 2)(x - 3)}$$ Factor the numerator.

$$= \dfrac{x - 5}{x - 3}$$ Divide out the common factor, $x + 2$.

d. $\dfrac{x - 1}{x^2 + 4x} - \dfrac{3x + 1}{x^2 - 16}$

Solution: First find the LCD.

Factored forms: $x^2 + 4x = x(x + 4)$; $x^2 - 16 = (x + 4)(x - 4)$
LCD $= x(x + 4)(x - 4)$

$$\dfrac{x - 1}{x^2 + 4x} - \dfrac{3x + 1}{x^2 - 16} = \dfrac{x - 1}{x(x + 4)} - \dfrac{3x + 1}{(x + 4)(x - 4)}$$ Factor the denominators.

$$= \dfrac{(x - 1)\,(x - 4)}{x(x + 4)\,(x - 4)} - \dfrac{(3x + 1)(x)}{(x + 4)(x - 4)(x)}$$ Write equivalent rational expressions with the LCD, $x(x + 4)(x - 4)$.

$$= \dfrac{x^2 - 5x + 4}{x(x + 4)(x - 4)} - \dfrac{3x^2 + x}{x(x + 4)(x - 4)}$$ Multiply in the numerator.

$$= \dfrac{(x^2 - 5x + 4) - (3x^2 + x)}{x(x + 4)(x - 4)}$$ Subtract numerators.

Note Remember that to subtract polynomials, we write an equivalent addition.

$$= \dfrac{(x^2 - 5x + 4) + (-3x^2 - x)}{x(x + 4)(x - 4)}$$

$$= \dfrac{-2x^2 - 6x + 4}{x(x + 4)(x - 4)}$$

◀ **Note** Although the numerator can be factored to $-2(x^2 + 3x - 2)$, none of those factors match any factor in the denominator; so the result is in lowest terms.

Your Turn 6 Add or subtract as indicated.

a. $\dfrac{5}{6x^2y} + \dfrac{3}{8xy^3}$

b. $\dfrac{7}{x + 3} - \dfrac{3}{x - 4}$

c. $\dfrac{x}{x + 3} - \dfrac{-6x + 24}{x^2 - x - 12}$

d. $\dfrac{2}{x^2 + x} - \dfrac{x - 4}{x^2 - 1}$

Answers to Your Turn 6

a. $\dfrac{20y^2 + 9x}{24x^2y^3}$

b. $\dfrac{4x - 37}{(x + 3)(x - 4)}$

c. $\dfrac{x + 6}{x + 3}$ **d.** $\dfrac{-x^2 + 6x - 2}{x(x + 1)(x - 1)}$

If denominators are additive inverses, we can multiply the numerator and denominator of either rational expression by -1 to obtain the LCD.

Example 7   Add $\dfrac{5x}{x-3} + \dfrac{2}{3-x}$.

Solution: $\dfrac{5x}{x-3} + \dfrac{2}{3-x} = \dfrac{5x}{x-3} + \dfrac{2(-1)}{(3-x)(-1)}$

$= \dfrac{5x}{x-3} + \dfrac{-2}{x-3}$

$= \dfrac{5x-2}{x-3}$

Because $x - 3$ and $3 - x$ are additive inverses, we obtain the LCD by multiplying the numerator and denominator of one of the rational expressions by -1. We chose the second rational expression.

Instructor Note Discuss that we could have used $(x - 3)(3 - x)$ as a common denominator in Example 7, but doing so would have been more complicated than multiplying by -1, as shown in Example 7.

Answers to Your Turn 7
a. $\dfrac{7y+4}{2y-3}$ b. 3

Your Turn 7   Add or subtract as indicated.

a. $\dfrac{7y}{2y-3} - \dfrac{4}{3-2y}$

b. $\dfrac{3x}{x-2} + \dfrac{6}{2-x}$

7.4 Exercises   For Extra Help   MyMathLab®

Note: Exercises marked with a ★ represent challenging exercises.

Objectives 1 and 2

Prep Exercise 1 When finding the LCD of two rational expressions, after finding the prime factorization of each denominator, explain how to use the primes to form the LCD. Write a product that contains each unique prime factor the greatest number of times it occurs in the factorizations.

Prep Exercise 2 Suppose the LCD of two fractions is $48x^3yz$. If one of the fraction's denominators is $8x^2y$, by what factor do you multiply when writing it as an equivalent fraction with the LCD? Explain how you determine this factor. $6xz$; divide $48x^3yz$ by $8x^2y$.

For Exercises 1–26, find the least common denominator for the rational expressions and write equivalent rational expressions with the LCD. See Examples 1–4.

1. $\dfrac{2}{n}, \dfrac{3}{m}$
$mn; \dfrac{2m}{mn}, \dfrac{3n}{mn}$

2. $\dfrac{6}{r}, \dfrac{5}{t}$
$rt; \dfrac{6t}{rt}, \dfrac{5r}{rt}$

3. $\dfrac{2t}{a^2b^5}, \dfrac{3z}{a^3b^3}$
$a^3b^5; \dfrac{2at}{a^3b^5}, \dfrac{3b^2z}{a^3b^5}$

4. $\dfrac{6u}{h^4k^2}, \dfrac{4v}{h^3k^7}$
$h^4k^7; \dfrac{6uk^5}{h^4k^7}, \dfrac{4vh}{h^4k^7}$

5. $\dfrac{2}{3a}, \dfrac{3}{5a^2}$
$15a^2; \dfrac{10a}{15a^2}, \dfrac{9}{15a^2}$

6. $\dfrac{5}{4b^2}, \dfrac{3}{5b}$
$20b^2; \dfrac{25}{20b^2}, \dfrac{12b}{20b^2}$

7. $\dfrac{3}{7x^2y}, \dfrac{5}{14xy^2}$
$14x^2y^2; \dfrac{6y}{14x^2y^2}, \dfrac{5x}{14x^2y^2}$

8. $\dfrac{2}{9mn^3}, \dfrac{7}{27m^3n}$
$27m^3n^3; \dfrac{6m^2}{27m^3n^3}, \dfrac{7n^2}{27m^3n^3}$

9. $\dfrac{8}{15m^3n^3}, \dfrac{7}{18m^2n}$
$90m^3n^3; \dfrac{48}{90m^3n^3}, \dfrac{35mn^2}{90m^3n^3}$

10. $\dfrac{5}{10x^2y^3}, \dfrac{3}{14x^2y^2}$
$70x^2y^3; \dfrac{35}{70x^2y^3}, \dfrac{15y}{70x^2y^3}$

11. $\dfrac{3}{y+2}, \dfrac{4y}{y-2}$
$(y+2)(y-2); \dfrac{3y-6}{(y+2)(y-2)}, \dfrac{4y^2+8y}{(y+2)(y-2)}$

12. $\dfrac{3x}{x-5}, \dfrac{2}{x+1}$
$(x-5)(x+1); \dfrac{3x^2+3x}{(x-5)(x+1)}, \dfrac{2x-10}{(x-5)(x+1)}$

13. $\dfrac{3y}{4y-20}, \dfrac{7y}{6y-30}$
$12(y-5); \dfrac{9y}{12(y-5)}, \dfrac{14y}{12(y-5)}$

14. $\dfrac{2x}{6x-12}, \dfrac{3x}{8x-16}$
$24(x-2); \dfrac{8x}{24(x-2)}, \dfrac{9x}{24(x-2)}$

15. $\dfrac{5}{6x+18}, \dfrac{2}{x^2+3x}$
$6x(x+3); \dfrac{5x}{6x(x+3)}, \dfrac{12}{6x(x+3)}$

16. $\dfrac{3}{4w-8}, \dfrac{5}{w^2-2w}$

$4w(w-2); \dfrac{3w}{4w(w-2)}, \dfrac{20}{4w(w-2)}$

17. $\dfrac{2}{p^2-4}, \dfrac{3+p}{p+2}$

$p^2-4; \dfrac{2}{p^2-4}, \dfrac{p^2+p-6}{p^2-4}$

18. $\dfrac{5}{y^2-36}, \dfrac{y-2}{y-6}$

$y^2-36; \dfrac{5}{y^2-36}, \dfrac{y^2+4y-12}{y^2-36}$

19. $\dfrac{4u}{(u+2)^2}, \dfrac{2}{3(u+2)}$

$3(u+2)^2; \dfrac{12u}{3(u+2)^2}, \dfrac{2u+4}{3(u+2)^2}$

20. $\dfrac{7m}{(x-3)^2}, \dfrac{4}{7(x-3)}$

$7(x-3)^2; \dfrac{49m}{7(x-3)^2}, \dfrac{4x-12}{7(x-3)^2}$

21. $\dfrac{2t-4}{t^2-36}, \dfrac{3t+2}{t^2-7t+6}$

$(t+6)(t-6)(t-1);$

$\dfrac{2t^2-6t+4}{(t+6)(t-6)(t-1)}, \dfrac{3t^2+20t+12}{(t+6)(t-6)(t-1)}$

22. $\dfrac{4a+2}{a^2-9}, \dfrac{2a-1}{a^2-a-6}$

$(a+3)(a-3)(a+2);$

$\dfrac{4a^2+10a+4}{(a+3)(a-3)(a+2)}, \dfrac{2a^2+5a-3}{(a+3)(a-3)(a+2)}$

23. $\dfrac{x+2}{x^2+3x-4}, \dfrac{x-3}{x^2+6x+8}$

$(x+4)(x-1)(x+2);$

$\dfrac{x^2+4x+4}{(x+4)(x-1)(x+2)}, \dfrac{x^2-4x+3}{(x+4)(x-1)(x+2)}$

24. $\dfrac{b-4}{b^2-2b-35}, \dfrac{b+3}{b^2+7b+10}$

$(b+5)(b-7)(b+2);$

$\dfrac{b^2-2b-8}{(b+5)(b-7)(b+2)}, \dfrac{b^2-4b-21}{(b+5)(b-7)(b+2)}$

25. $\dfrac{2}{x-1}, \dfrac{3}{x^2-3x+2}, \dfrac{5x}{x-2}$

$(x-1)(x-2);$

$\dfrac{2x-4}{(x-1)(x-2)}, \dfrac{3}{(x-1)(x-2)}, \dfrac{5x^2-5x}{(x-1)(x-2)}$

26. $\dfrac{3}{y-3}, \dfrac{2}{y^2+2y-15}, \dfrac{4y}{y+5}$

$(y-3)(y+5);$

$\dfrac{3y+15}{(y-3)(y+5)}, \dfrac{2}{(y-3)(y+5)}, \dfrac{4y^2-12y}{(y-3)(y+5)}$

Objective 3

Prep Exercise 3 In general, if the denominators of two rational expressions are multiterm polynomials, what is the first step in finding the LCD? Factor the denominators.

Prep Exercise 4 Instead of using the LCM as a common denominator, a student rewrites $\dfrac{3}{2x-y} + \dfrac{5}{y-2x}$ as $\dfrac{-3}{y-2x} + \dfrac{5}{y-2x}$. Explain what the student did. The student multiplied the numerator and denominator of $\dfrac{3}{2x-y}$ by -1.

Prep Exercise 5 After rewriting two rational expressions with the LCD, suppose the subtraction is at the stage $\dfrac{(7x+21)-(2x-10)}{(x-5)(x+3)}$. Explain the next step. Rewrite the numerator as an equivalent addition sentence.

★ **Prep Exercise 6** After rewriting two rational expressions with the LCD, suppose the subtraction is at the stage $\dfrac{(7x+21)-(2x-10)}{(x-5)(x+3)}$. What might the original problem have been? Explain. $\dfrac{7}{x-5} - \dfrac{2}{x+3}$; we note that $7x+21$ and $2x-10$ factor to $7(x+3)$ and $2(x-5)$. Because the denominator is the product of $(x-5)$ and $(x+3)$, they must be the denominators of two fractions with 7 and 2 as the original numerators.

For Exercises 27–64, add or subtract as indicated. See Examples 5–7.

27. $\dfrac{3a-5}{9} - \dfrac{2a+7}{6}$

$\dfrac{-31}{18}$

28. $\dfrac{3x-4}{6} - \dfrac{2x-1}{4}$

$\dfrac{-5}{12}$

29. $\dfrac{5}{2c} + \dfrac{1}{6c}$

$\dfrac{8}{3c}$

30. $\dfrac{5}{8r} + \dfrac{3}{4r}$

$\dfrac{11}{8r}$

31. $\dfrac{5}{x} - \dfrac{2}{x^3}$

$\dfrac{5x^2-2}{x^3}$

32. $\dfrac{5a}{b^2} - \dfrac{3a}{b}$

$\dfrac{5a-3ab}{b^2}$

33. $\dfrac{3}{y+2} + \dfrac{5}{y-4}$

$\dfrac{8y-2}{(y+2)(y-4)}$

34. $\dfrac{2}{c+4} + \dfrac{3}{c+3}$

$\dfrac{5c+18}{(c+4)(c+3)}$

35. $\dfrac{3}{x-2} - \dfrac{2}{x+2}$

$\dfrac{x+10}{(x-2)(x+2)}$

36. $\dfrac{4}{x+3} - \dfrac{3}{x-3}$

$\dfrac{x-21}{(x+3)(x-3)}$

37. $\dfrac{4}{x^2-25} + \dfrac{2}{x-5}$

$\dfrac{2x+14}{(x+5)(x-5)}$

38. $\dfrac{5}{r^2-9} + \dfrac{4}{r+3}$

$\dfrac{4r-7}{(r+3)(r-3)}$

39. $\dfrac{2p}{p^2 - 2p + 1} + \dfrac{8}{p - 1}$

$\dfrac{10p - 8}{p^2 - 2p + 1}$

40. $\dfrac{u}{u - 1} + \dfrac{2u}{u^2 - 2u + 1}$

$\dfrac{u^2 + u}{(u - 1)^2}$

41. $\dfrac{z + 5}{z^2 + z - 12} - \dfrac{z + 2}{z - 3}$

$\dfrac{-z^2 - 5z - 3}{(z + 4)(z - 3)}$

42. $\dfrac{a + 6}{a^2 + 8a + 15} - \dfrac{a - 3}{a + 3}$

$\dfrac{-a^2 - a + 21}{(a + 5)(a + 3)}$

43. $\dfrac{n + 2}{n - 4} + \dfrac{n - 3}{n + 3}$

$\dfrac{2n^2 - 2n + 18}{(n - 4)(n + 3)}$

44. $\dfrac{y - 8}{y + 5} + \dfrac{y - 9}{y - 6}$

$\dfrac{2y^2 - 18y + 3}{(y + 5)(y - 6)}$

45. $\dfrac{t + 2}{t + 4} - \dfrac{2t - 1}{t + 6}$

$\dfrac{-t^2 + t + 16}{(t + 4)(t + 6)}$

46. $\dfrac{3s + 1}{s - 2} - \dfrac{4s + 1}{s - 3}$

$\dfrac{-s^2 - s - 1}{(s - 2)(s - 3)}$

47. $\dfrac{c + 5}{c^2 + 10c + 25} + \dfrac{3}{2c + 10}$

$\dfrac{5}{2(c + 5)}$

48. $\dfrac{k + 3}{k^2 + 6k + 9} - \dfrac{7}{2k + 6}$

$\dfrac{-5}{2(k + 3)}$

49. $\dfrac{5y + 6}{y^2 + 3y} + \dfrac{y}{y + 3}$

$\dfrac{y + 2}{y}$

50. $\dfrac{2b - 8}{b^2 + 4b} + \dfrac{b}{b + 4}$

$\dfrac{b - 2}{b}$

51. $\dfrac{3r}{r^2 - 1} - \dfrac{2r + 1}{r^2 - 2r + 1}$

$\dfrac{r^2 - 6r - 1}{(r + 1)(r - 1)^2}$

52. $\dfrac{x + 1}{x^2 - 4x + 4} - \dfrac{4x}{x^2 + 3x - 10}$

$\dfrac{-3x^2 + 14x + 5}{(x - 2)^2(x + 5)}$

53. $\dfrac{q + 5}{q^2 - 9} - \dfrac{1}{q^2 + 3q}$

$\dfrac{q + 1}{q(q - 3)}$

54. $\dfrac{x - 2}{x^2 - 16} - \dfrac{1}{x^2 - 4x}$

$\dfrac{x + 1}{x(x + 4)}$

55. $\dfrac{w - 7}{w^2 + 4w - 5} - \dfrac{w - 9}{w^2 + 3w - 10}$

$\dfrac{1}{(w - 1)(w - 2)}$

56. $\dfrac{y - 1}{y^2 - 3y + 2} - \dfrac{y + 2}{y^2 + y - 2}$

$\dfrac{1}{(y - 1)(y - 2)}$

57. $\dfrac{2v + 5}{v^2 - 16} - \dfrac{v - 9}{v^2 - v - 12}$

$\dfrac{v^2 + 16v + 51}{(v + 4)(v - 4)(v + 3)}$

58. $\dfrac{4t}{t^2 - 9} - \dfrac{3t - 2}{t^2 - 8t + 15}$

$\dfrac{t^2 - 27t + 6}{(t - 3)(t - 5)(t + 3)}$

59. $\dfrac{3}{t - 3} + \dfrac{6}{3 - t}$

$\dfrac{-3}{t - 3}$

60. $\dfrac{10}{2r - 1} - \dfrac{6}{1 - 2r}$

$\dfrac{16}{2r - 1}$

61. $\dfrac{4}{2x - 1} + \dfrac{x}{1 - 2x}$

$\dfrac{4 - x}{2x - 1}$

62. $\dfrac{3}{t - 4} + \dfrac{t}{4 - t}$

$\dfrac{3 - t}{t - 4}$

63. $\dfrac{y}{3y - 6} - \dfrac{2}{2 - y}$

$\dfrac{y + 6}{3(y - 2)}$

64. $\dfrac{x}{6x - 2} - \dfrac{3x}{1 - 3x}$

$\dfrac{7x}{2(3x - 1)}$

Find ⊗ the Mistake *For Exercises 65–70, explain the mistake; then find the correct sum or difference.*

65. $\dfrac{3}{x} - \dfrac{x}{2} = \dfrac{3 - x}{x - 2}$

Mistake: Did not find a common denominator.

Correct: $\dfrac{6 - x^2}{2x}$

66. $\dfrac{\cancel{5}}{2x} + \dfrac{3}{\cancel{10}x} = \dfrac{1}{2x} + \dfrac{3}{2x}$

$= \dfrac{4}{2x}$

$= \dfrac{2}{x}$

Mistake: Divided out a common factor of 5 across addition.

Correct: $\dfrac{14}{5x}$

67. $\dfrac{3x}{x + 2} - \dfrac{4x + 7}{x + 2} = \dfrac{3x - (4x + 7)}{x + 2}$

$= \dfrac{3x + (-4x + 7)}{x + 2}$

$= \dfrac{-x + 7}{x + 2}$

Mistake: Did not change the sign of 7.

Correct: $\dfrac{-x - 7}{x + 2}$

68. $\dfrac{3}{x} + \dfrac{2}{x + y} = \dfrac{3(+y)}{x(+y)} + \dfrac{2}{x + y}$

$= \dfrac{3 + y + 2}{x + y}$

$= \dfrac{5 + y}{x + y}$

Mistake: Added y to the first fraction, attempting to make a common denominator.

Correct: $\dfrac{5x + 3y}{x(x + y)}$

69. $\dfrac{2w}{x} + \dfrac{5w}{x} = \dfrac{7w}{2x}$

Mistake: Added denominators.

Correct: $\dfrac{7w}{x}$

70. $\dfrac{2}{x} + \dfrac{3}{y} = \dfrac{6}{xy}$

Mistake: Multiplied the expressions.

Correct: $\dfrac{2y + 3x}{xy}$

For Exercises 71–76, solve.

71. Adam and Candace build and install cabinets in newly constructed homes. When they work together, the expression $\dfrac{t}{2}$ represents the portion of the job completed by Adam and $\dfrac{t}{3}$ represents the portion completed by Candace. Write a rational expression in simplest form that describes their combined work.

$\dfrac{5t}{6}$

72. Two construction teams are combined in an effort to speed construction on a building. The expression $\dfrac{t}{5}$ describes the portion of the job that team 1 can complete and $\dfrac{t}{6}$ describes the portion of the job that team 2 can complete working independently. Write a rational expression in simplest form that describes their combined work.

$\dfrac{11t}{30}$

73. An engineer uses the expression $\dfrac{1}{x+3}$ to describe the width of a rectangular steel plate she is designing for a machine. The expression $\dfrac{1}{x}$ describes the length. Write a rational expression in simplest form for the perimeter of the steel plate.

$\dfrac{4x+6}{x^2+3x}$

74. An electrical engineer is trying to determine the effects of using a thinner layer of a chemical on a circuit board. Currently, the thickness of the chemical layer is described by $\dfrac{1}{n-4}$. If the thinner layer is described by $\dfrac{1}{n-2}$, write a rational expression that describes the difference in the coating thickness.

$\dfrac{2}{n^2-6n+8}$

75. Write a rational expression in simplest form for the area of the shaded region in the figure shown.

$\dfrac{4ah+5bh}{8}$

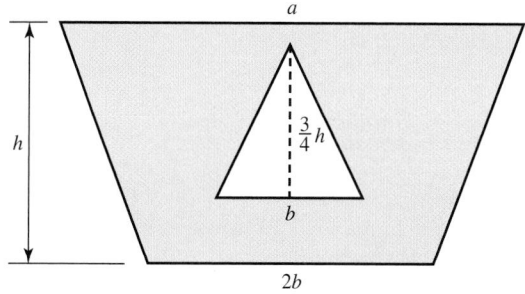

76. Write a rational expression in simplest form for the area of the figure shown.

$\dfrac{3ah+2bh}{6}$

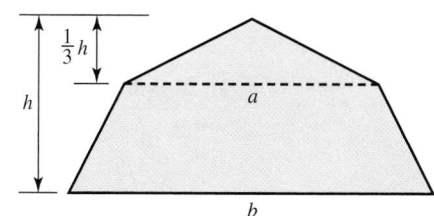

★ *For Exercises 77–86, use the order-of-operations agreement to simplify.*

77. $\dfrac{x}{4x^2-1}+\dfrac{2}{4x-2}-\dfrac{3x+1}{1-2x}$

$\dfrac{6x^2+8x+2}{(2x-1)(2x+1)}$

78. $\dfrac{c}{4c^2-9}-\dfrac{2c+1}{3-2c}+\dfrac{3}{6c+9}$

$\dfrac{4c^2+11c}{(2c+3)(2c-3)}$

79. $(x+y)\div\left(\dfrac{1}{x}+\dfrac{1}{y}\right)$

xy

80. $\left(\dfrac{1}{a}+\dfrac{1}{b}\right)\div(a+b)$

$\dfrac{1}{ab}$

81. $\dfrac{2}{a}+\dfrac{3}{4a+8}\cdot\dfrac{4a}{5}$

$\dfrac{3a^2+10a+20}{5a(a+2)}$

82. $\dfrac{3}{x}+\dfrac{1}{x^2+3x}\cdot\dfrac{2x}{3}$

$\dfrac{11x+27}{3x(x+3)}$

83. $\left(\dfrac{a + b}{a - b} + \dfrac{a - b}{a + b} \right) \div \dfrac{a + b}{a - b}$

$\dfrac{2a^2 + 2b^2}{(a + b)^2}$

84. $\left(\dfrac{2x - y}{2x + y} + \dfrac{2x + y}{2x - y} \right) \div \dfrac{2x - y}{2x + y}$

$\dfrac{8x^2 + 2y^2}{(2x - y)^2}$

85. $\left(\dfrac{1}{x + 2} - \dfrac{3}{x^2 - 4} \right) \div \dfrac{3}{x - 2}$

$\dfrac{x - 5}{3(x + 2)}$

86. $\dfrac{6}{x + 4} \div \left(\dfrac{1}{x - 3} - \dfrac{x + 1}{x^2 + x - 12} \right)$

$2x - 6$

Review Exercises

Exercises 1–6 🔺 **Expressions**

For Exercises 1 and 2, divide.

[1.4] **1.** $\dfrac{1}{4} \div \dfrac{2}{3}$

$\dfrac{3}{8}$

[7.2] **2.** $\dfrac{m}{6} \div \dfrac{18}{m}$

$\dfrac{m^2}{108}$

For Exercises 3 and 4, factor completely.

[6.4] **3.** $4x^2 - 16$

$4(x + 2)(x - 2)$

[6.3] **4.** $6y^3 - 27y^2 - 15y$

$3y(2y + 1)(y - 5)$

[7.1] *For Exercises 5 and 6, simplify.*

5. $\dfrac{x^2 - 7x + 6}{x^2 - 4x - 12}$

$\dfrac{x - 1}{x + 2}$

6. $\dfrac{x + 3}{x^2 - 9}$

$\dfrac{1}{x - 3}$

7.5 Complex Rational Expressions

Objective

1 Simplify complex rational expressions.

Warm-up

[1.4] **1.** Divide: $\dfrac{3}{4} \div \dfrac{5}{6}$

[7.4] **2.** Add: $1 + \dfrac{7}{x} + \dfrac{12}{x^2}$

For Exercises 3 and 4, find the products.

[1.4] **3.** $\dfrac{2}{3} \cdot 24$

[5.5] **4.** $6x\left(5 + \dfrac{x}{3} \right)$

Objective 1 Simplify complex rational expressions.

Sometimes a rational expression may contain rational expressions in its numerator or denominator. Such an expression is called a **complex rational expression** or a **complex fraction**.

Definition **Complex rational expression:** A rational expression that contains rational expressions in the numerator or denominator.

Answers to Warm-up

1. $\dfrac{9}{10}$

2. $\dfrac{x^2 + 7x + 12}{x^2}$

3. 16

4. $30x + 2x^2$

Examples of complex rational expressions:

$$\dfrac{\dfrac{3}{4}}{\dfrac{5}{6}} \qquad \dfrac{x + \dfrac{x}{3}}{\dfrac{y}{2}} \qquad \dfrac{7}{\dfrac{t+2}{t}} \qquad \dfrac{\dfrac{2}{n} + \dfrac{1}{4}}{\dfrac{1}{n} - \dfrac{n}{3}}$$

There are two common methods for simplifying complex rational expressions. One method is to write the complex rational expression as a horizontal division problem. A second method is to multiply the numerator and denominator of the complex fraction by the LCD of the numerator and denominator fractions.

Consider $\dfrac{\dfrac{3}{4}}{\dfrac{5}{6}}$.

Method 1: $\dfrac{\dfrac{3}{4}}{\dfrac{5}{6}} = \dfrac{3}{4} \div \dfrac{5}{6} = \dfrac{3}{\overset{}{\underset{2}{4}}} \cdot \dfrac{\overset{3}{6}}{5} = \dfrac{9}{10}$ Write as a horizontal division. Then simplify.

> **Connection** Notice that the second method is writing an equivalent fraction by multiplying both the numerator and denominator by the same factor.

Method 2: $\dfrac{\dfrac{3}{4} \cdot 12}{\dfrac{5}{6} \cdot 12} = \dfrac{\dfrac{3}{4} \cdot \dfrac{\overset{3}{12}}{1}}{\dfrac{5}{6} \cdot \dfrac{\overset{2}{12}}{1}} = \dfrac{\dfrac{9}{1}}{\dfrac{10}{1}} = \dfrac{9}{10}$ Multiply the numerator and denominator by the LCD of 4 and 6, which is 12. Then simplify.

> **Procedure** **Simplifying Complex Rational Expressions**
>
> To simplify a complex rational expression, use one of the following methods.
>
> **Method 1**
> 1. Simplify the numerator and denominator if needed.
> 2. Rewrite as a horizontal division problem.
>
> **Method 2**
> 1. Multiply the numerator and denominator of the complex rational expression by their LCD.
> 2. Simplify.

Example 1 Simplify.

a. $\dfrac{\dfrac{2}{x}}{\dfrac{y}{3}}$

> **Note** Method 1 is best used when the numerator and denominator are both a single rational expression, as in Example 1(a). Otherwise, method 2 is at least as easy as method 1.

Method 1: $\dfrac{\dfrac{2}{x}}{\dfrac{y}{3}} = \dfrac{2}{x} \div \dfrac{y}{3}$ Write the complex fraction as a division problem.

$= \dfrac{2}{x} \cdot \dfrac{3}{y}$ Write an equivalent multiplication problem.

$= \dfrac{6}{xy}$

Method 2: $\dfrac{\dfrac{2}{x}}{\dfrac{y}{3}} = \dfrac{\dfrac{2}{x} \cdot 3x}{\dfrac{y}{3} \cdot 3x}$ Multiply the numerator and denominator by the LCD, $3x$.

$$= \dfrac{\dfrac{2}{\cancel{x}} \cdot \dfrac{\overset{1}{\cancel{3x}}}{1}}{\dfrac{y}{\cancel{3}} \cdot \dfrac{\overset{1}{\cancel{3x}}}{1}}$$ Divide out the common factors.

$$= \dfrac{6}{xy}$$ Simplify.

b. $\dfrac{\dfrac{1}{4} + \dfrac{2}{3}}{\dfrac{3}{8} - \dfrac{1}{6}}$

Method 1: $\dfrac{\dfrac{1}{4} + \dfrac{2}{3}}{\dfrac{3}{8} - \dfrac{1}{6}} = \dfrac{\dfrac{1(3)}{4(3)} + \dfrac{2(4)}{3(4)}}{\dfrac{3(3)}{8(3)} - \dfrac{1(4)}{6(4)}}$ Write the numerator fractions as equivalent fractions with their LCD, 12, and write the denominator fractions with their LCD, 24.

$$= \dfrac{\dfrac{3}{12} + \dfrac{8}{12}}{\dfrac{9}{24} - \dfrac{4}{24}}$$

$$= \dfrac{\dfrac{11}{12}}{\dfrac{5}{24}}$$ Add in the numerator and subtract in the denominator.

$$= \dfrac{11}{12} \div \dfrac{5}{24}$$ Write the complex fraction as a division problem.

$$= \dfrac{11}{\underset{1}{\cancel{12}}} \cdot \dfrac{\overset{2}{\cancel{24}}}{5}$$ Write an equivalent multiplication problem, divide out the common factors, and then multiply.

$$= \dfrac{22}{5}$$

Method 2: $\dfrac{\dfrac{1}{4} + \dfrac{2}{3}}{\dfrac{3}{8} - \dfrac{1}{6}} = \dfrac{\left(\dfrac{1}{4} + \dfrac{2}{3}\right) \cdot 24}{\left(\dfrac{3}{8} - \dfrac{1}{6}\right) \cdot 24}$ Multiply the numerator and denominator by the LCD, 24.

$$= \dfrac{\dfrac{1}{\cancel{4}} \cdot \dfrac{\overset{6}{\cancel{24}}}{1} + \dfrac{2}{\cancel{3}} \cdot \dfrac{\overset{8}{\cancel{24}}}{1}}{\dfrac{3}{\cancel{8}} \cdot \dfrac{\overset{3}{\cancel{24}}}{1} - \dfrac{1}{\cancel{6}} \cdot \dfrac{\overset{4}{\cancel{24}}}{1}}$$ Distribute the 24 to each fraction and divide out the common factors.

$$= \dfrac{6 + 16}{9 - 4}$$ Simplify the numerator and denominator.

$$= \dfrac{22}{5}$$

c. $\dfrac{5 + \dfrac{x}{3}}{\dfrac{4}{x} - \dfrac{1}{2x}}$

Method 1: $\dfrac{5 + \dfrac{x}{3}}{\dfrac{4}{x} - \dfrac{1}{2x}} = \dfrac{\dfrac{5(3)}{1(3)} + \dfrac{x(1)}{3(1)}}{\dfrac{4(2)}{x(2)} - \dfrac{1(1)}{2x(1)}}$

Write the numerator fractions as equivalent fractions with their LCD, 3, and write the denominator fractions with their LCD, 2x.

$= \dfrac{\dfrac{15}{3} + \dfrac{x}{3}}{\dfrac{8}{2x} - \dfrac{1}{2x}}$

$= \dfrac{\dfrac{15 + x}{3}}{\dfrac{7}{2x}}$

Add in the numerator and subtract in the denominator.

$= \dfrac{15 + x}{3} \div \dfrac{7}{2x}$

Write the complex fraction as a division problem.

$= \dfrac{15 + x}{3} \cdot \dfrac{2x}{7}$

Write an equivalent multiplication problem and then multiply.

$= \dfrac{30x + 2x^2}{21}$

Method 2: $\dfrac{5 + \dfrac{x}{3}}{\dfrac{4}{x} - \dfrac{1}{2x}} = \dfrac{\left(5 + \dfrac{x}{3}\right) \cdot 6x}{\left(\dfrac{4}{x} - \dfrac{1}{2x}\right) \cdot 6x}$

Multiply the numerator and denominator by the LCD, 6x.

$= \dfrac{\dfrac{5}{1} \cdot \dfrac{6x}{1} + \dfrac{x}{3} \cdot \dfrac{\overset{2}{6x}}{1}}{\dfrac{4}{x} \cdot \dfrac{\overset{1}{6x}}{1} - \dfrac{1}{2x} \cdot \dfrac{\overset{3}{6x}}{1}}$

Distribute the 6x to each fraction and divide out the common factors.

$= \dfrac{30x + 2x^2}{24 - 3}$

Simplify the numerator and denominator.

$= \dfrac{30x + 2x^2}{21}$

d. $\dfrac{1 - \dfrac{2}{x} - \dfrac{15}{x^2}}{1 + \dfrac{7}{x} + \dfrac{12}{x^2}}$

Method 1:

$\dfrac{1 - \dfrac{2}{x} - \dfrac{15}{x^2}}{1 + \dfrac{7}{x} + \dfrac{12}{x^2}} = \dfrac{1 \cdot \dfrac{x^2}{x^2} - \dfrac{2}{x} \cdot \dfrac{x}{x} - \dfrac{15}{x^2}}{1 \cdot \dfrac{x^2}{x^2} + \dfrac{7}{x} \cdot \dfrac{x}{x} + \dfrac{12}{x^2}}$

Write the numerator fractions as equivalent fractions with their LCD, x^2, and write the denominator fractions with their LCD, also x^2.

$$= \cfrac{\dfrac{x^2}{x^2} - \dfrac{2x}{x^2} - \dfrac{15}{x^2}}{\dfrac{x^2}{x^2} + \dfrac{7x}{x^2} + \dfrac{12}{x^2}}$$

Simplify.

$$= \cfrac{\dfrac{x^2 - 2x - 15}{x^2}}{\dfrac{x^2 + 7x + 12}{x^2}}$$

Subtract in the numerator and add in the denominator.

$$= \frac{x^2 - 2x - 15}{x^2} \div \frac{x^2 + 7x + 12}{x^2}$$

Write the complex fraction as a division problem.

$$= \frac{x^2 - 2x - 15}{x^2} \cdot \frac{x^2}{x^2 + 7x + 12}$$

Write as an equivalent multiplication problem.

$$= \frac{(x - 5)\,\cancel{(x + 3)}^{1}}{\cancel{x^2}_{1}} \cdot \frac{x^2}{(x + 4)\,\cancel{(x + 3)}_{1}}$$

Factor, divide out the common factors, and then multiply the remaining factors.

$$= \frac{x - 5}{x + 4}$$

Method 2:

$$\frac{1 - \dfrac{2}{x} - \dfrac{15}{x^2}}{1 + \dfrac{7}{x} + \dfrac{12}{x^2}} = \frac{\left(1 - \dfrac{2}{x} - \dfrac{15}{x^2}\right) \cdot x^2}{\left(1 + \dfrac{7}{x} + \dfrac{12}{x^2}\right) \cdot x^2}$$

Multiply the numerator and denominator by the LCD, x^2.

$$= \frac{1 \cdot x^2 - \dfrac{2}{\cancel{x}_{1}} \cdot \dfrac{\cancel{x^2}^{\,x}}{1} - \dfrac{15}{\cancel{x^2}_{1}} \cdot \dfrac{\cancel{x^2}^{\,1}}{1}}{1 \cdot x^2 - \dfrac{7}{\cancel{x}_{1}} \cdot \dfrac{\cancel{x^2}^{\,x}}{1} - \dfrac{12}{\cancel{x^2}_{1}} \cdot \dfrac{\cancel{x^2}^{\,1}}{1}}$$

Distribute the x^2 to each fraction and divide out the common factors.

$$= \frac{x^2 - 2x - 15}{x^2 + 7x + 12}$$

Simplify the numerator and denominator.

$$= \frac{(x - 5)\,\cancel{(x + 3)}}{(x + 4)\,\cancel{(x + 3)}}$$

Factor and divide out the common factor, $x + 3$.

$$= \frac{x - 5}{x + 4}$$

Your Turn 1 Simplify.

a. $\dfrac{\dfrac{a}{5}}{\dfrac{3}{b}}$

b. $\dfrac{2 - \dfrac{3}{4}}{\dfrac{4}{5} + \dfrac{1}{2}}$

c. $\dfrac{2 - \dfrac{1}{y}}{y - \dfrac{1}{3}}$

d. $\dfrac{1 - \dfrac{3}{x} - \dfrac{28}{x^2}}{1 - \dfrac{1}{x} - \dfrac{20}{x^2}}$

Answers to Your Turn 1

a. $\dfrac{ab}{15}$ **b.** $\dfrac{25}{26}$

c. $\dfrac{6y - 3}{3y^2 - y}$ **d.** $\dfrac{x - 7}{x - 5}$

7.5 Exercises

For Extra Help | MyMathLab®

Objective 1

Prep Exercise 1 Identify the numerator and denominator of the complex rational expression $\dfrac{\dfrac{4}{x+3}}{\dfrac{3}{x}}$.

$\dfrac{4}{x+3}$ is the numerator. $\dfrac{3}{x}$ is the denominator.

Prep Exercise 2 Given the complex rational expression $\dfrac{\dfrac{2x}{x-3}}{4}$, is the numerator $2x$ or $\dfrac{2x}{x-3}$? How could the expression be written so that the numerator and denominator are easily identified? Unable to tell; the line between the numerator and denominator could be darker or longer.

Prep Exercise 3 Write $\dfrac{2x}{x-3} \div \dfrac{4-x}{2x-6}$ as a complex rational expression.

$\dfrac{\dfrac{2x}{x-3}}{\dfrac{4-x}{2x-6}}$

Prep Exercise 4 Write $\dfrac{5}{x-1} \div 7x$ as a complex rational expression.

$\dfrac{\dfrac{5}{x-1}}{7x}$

Prep Exercise 5 Which of the two methods discussed in this section would you use to simplify $\dfrac{\dfrac{x}{6}-\dfrac{3}{4}}{\dfrac{2}{3}+x}$? Why?

In this case, multiplying the numerator and denominator by their LCD, 12, (method 2) is preferable because doing so requires fewer steps than simplifying $\dfrac{x}{6}-\dfrac{3}{4}$ and $\dfrac{2}{3}+x$ and then dividing the simplified expressions (method 1). (Answers may vary.)

Prep Exercise 6 Which of the two methods discussed in this section would you use to simplify $\dfrac{\dfrac{x}{7}}{\dfrac{x}{x+1}}$? Why?

Because $\dfrac{x}{7}$ and $\dfrac{x}{x+1}$ are in simplest form, dividing them (method 1) is preferable because doing so requires fewer steps than multiplying them by their LCD and then simplifying (method 2). (Answers may vary.)

For Exercises 1–42, simplify. See Example 1.

1. $\dfrac{\dfrac{3}{4}}{\dfrac{3}{2}}$

$\dfrac{1}{2}$

2. $\dfrac{\dfrac{2}{3}}{\dfrac{3}{2}}$

$\dfrac{4}{9}$

3. $\dfrac{\dfrac{m}{n}}{\dfrac{q}{p}}$

$\dfrac{mp}{nq}$

4. $\dfrac{\dfrac{a}{b}}{\dfrac{c}{d}}$

$\dfrac{ad}{bc}$

5. $\dfrac{\dfrac{1}{2}+\dfrac{3}{5}}{\dfrac{3}{2}-\dfrac{1}{5}}$

$\dfrac{11}{13}$

6. $\dfrac{\dfrac{2}{3}+\dfrac{1}{4}}{\dfrac{4}{3}-\dfrac{3}{4}}$

$\dfrac{11}{7}$

7. $\dfrac{5+\dfrac{2}{a}}{3+\dfrac{4}{a}}$

$\dfrac{5a+2}{3a+4}$

8. $\dfrac{4+\dfrac{1}{x}}{3+\dfrac{5}{x}}$

$\dfrac{4x+1}{3x+5}$

9. $\dfrac{\dfrac{1}{y}-1}{\dfrac{1}{y}+1}$

$\dfrac{1-y}{1+y}$

10. $\dfrac{x-\dfrac{1}{x}}{1+\dfrac{1}{x}}$

$x-1$

11. $\dfrac{r-\dfrac{3}{4}}{\dfrac{1}{2}-r}$

$\dfrac{4r-3}{2-4r}$

12. $\dfrac{\dfrac{5}{6}-x}{x-\dfrac{2}{3}}$

$\dfrac{5-6x}{6x-4}$

13. $\dfrac{\dfrac{3}{x}-1}{\dfrac{9}{x^2}-1}$

$\dfrac{x}{3+x}$

14. $\dfrac{1-\dfrac{2}{p}}{1-\dfrac{4}{p^2}}$

$\dfrac{p}{p+2}$

15. $\dfrac{x+y}{\dfrac{1}{x}+\dfrac{1}{y}}$

xy

16. $\dfrac{\dfrac{2}{a}+\dfrac{3}{b}}{a+b}$

$\dfrac{2b+3a}{a^2b+ab^2}$

17. $\dfrac{x+2}{x^2-4}$

$\dfrac{x}{x-2}$

18. $\dfrac{\dfrac{a}{b}-1}{a^2-b^2}$

$\dfrac{1}{b(a+b)}$

19. $\dfrac{\dfrac{6}{3x-2}}{\dfrac{x}{2x+1}}$

$\dfrac{6(2x+1)}{x(3x-2)}$

20. $\dfrac{\dfrac{5}{2x-1}}{\dfrac{x}{x+1}}$

$\dfrac{5(x+1)}{x(2x-1)}$

21. $\dfrac{\dfrac{a}{a^2-b^2}}{\dfrac{b}{a+b}}$

$\dfrac{a}{b(a-b)}$

22. $\dfrac{\dfrac{x}{4x^2-1}}{\dfrac{5}{2x-1}}$

$\dfrac{x}{5(2x+1)}$

23. $\dfrac{\dfrac{y^2-25}{5y}}{\dfrac{y^2+5y}{2y^3}}$

$\dfrac{2y(y-5)}{5}$

24. $\dfrac{\dfrac{x^2+4x}{y^2}}{\dfrac{x^2-16}{3y}}$

$\dfrac{3x}{y(x-4)}$

25. $\dfrac{2+\dfrac{2}{u-1}}{\dfrac{2}{u-1}}$

u

26. $\dfrac{\dfrac{1}{w+1}-1}{\dfrac{1}{w+1}}$

$-w$

27. $\dfrac{\dfrac{a}{a-2}-1}{1+\dfrac{a}{a-2}}$

$\dfrac{1}{a-1}$

28. $\dfrac{1-\dfrac{x}{x+2}}{\dfrac{x}{x+2}+1}$

$\dfrac{1}{x+1}$

29. $\dfrac{\dfrac{x-3}{x}-1}{\dfrac{x^2-9}{x}-x}$

$\dfrac{1}{3}$

30. $\dfrac{q-\dfrac{q^2-1}{q}}{1-\dfrac{q-1}{q}}$

1

31. $\dfrac{\dfrac{1}{x-1}-\dfrac{1}{x+1}}{\dfrac{1}{x^2-1}}$

2

32. $\dfrac{\dfrac{16}{x^2-16}}{\dfrac{2}{x+4}-\dfrac{2}{x-4}}$

-1

33. $\dfrac{1+\dfrac{1}{y}-\dfrac{2}{y^2}}{\dfrac{1}{y}+\dfrac{2}{y^2}}$

$y-1$

34. $\dfrac{1-\dfrac{3}{x}}{1-\dfrac{2}{x}-\dfrac{3}{x^2}}$

$\dfrac{x}{x+1}$

35. $\dfrac{1-\dfrac{1}{n}}{n-2+\dfrac{1}{n}}$

$\dfrac{1}{n-1}$

36. $\dfrac{y-1-\dfrac{6}{y}}{1+\dfrac{2}{y}}$

$y-3$

37. $\dfrac{1-\dfrac{1}{2y+1}}{\dfrac{1}{2y^2-5y-3}-\dfrac{1}{y-3}}$

$3-y$

38. $\dfrac{1-\dfrac{1}{f-3}}{\dfrac{1}{f^2-f-6}-\dfrac{1}{f+2}}$

$-f-2$

39. $\dfrac{\dfrac{x^2-2x-8}{x^2-16}}{\dfrac{x^2+2x}{x^2+3x-4}}$

$\dfrac{x-1}{x}$

40. $\dfrac{\dfrac{v^2+v-2}{v^2+4v}}{\dfrac{v^2-4}{v^2+2v-8}}$

$\dfrac{v-1}{v}$

41. $\dfrac{\dfrac{2}{x+h}-\dfrac{2}{x}}{h}$

$\dfrac{-2}{x(x+h)}$

42. $\dfrac{h}{\dfrac{3}{y+h}-\dfrac{3}{y}}$

$\dfrac{-y(y+h)}{3}$

Find ✖ the Mistake   *For Exercises 43–50, explain the mistake; then find the correct answer.*

43. $\dfrac{x-\dfrac{1}{2}}{y-\dfrac{1}{2}}=\dfrac{x}{y}$

Mistake: Divided out $\dfrac{1}{2}$, which is not a common factor.

Correct: $\dfrac{2x-1}{2y-1}$

44. $\dfrac{\dfrac{2}{3}}{\dfrac{4}{6}}=\dfrac{\overset{1}{2}}{3}\cdot\dfrac{4}{\underset{3}{6}}$

$=\dfrac{4}{9}$

Mistake: Did not invert $\dfrac{4}{6}$.

Correct: 1

45. $\dfrac{\dfrac{m}{6}}{\dfrac{3}{m}}=\dfrac{\overset{1}{m}}{\underset{2}{6}}\div\dfrac{3}{\underset{1}{m}}$

$=\dfrac{1}{2}$

Mistake: Did not write as an equivalent multiplication.

Correct: $\dfrac{m^2}{18}$

46. $\dfrac{\dfrac{1}{x}+\dfrac{1}{y}}{x^2-y^2}=\dfrac{\dfrac{x}{1}\cdot\dfrac{1}{x}+\dfrac{y}{1}\cdot\dfrac{1}{y}}{x\cdot x^2-y\cdot y^2}$

$=\dfrac{1+1}{x^3-y^3}$

$=\dfrac{2}{x^3-y^3}$

Mistake: Multiplied through by factors other than the LCD.

Correct: $\dfrac{1}{xy(x-y)}$

47. $\dfrac{c+\dfrac{1}{c}}{\dfrac{1}{c}}=\dfrac{c+\dfrac{c}{1}\cdot\dfrac{1}{c}}{\dfrac{c}{1}\cdot\dfrac{1}{c}}$

$=\dfrac{c+1}{1}$

$=c+1$

Mistake: Did not distribute c in the numerator.

Correct: c^2+1

48. $\dfrac{1+\dfrac{2}{r}}{1-\dfrac{4}{r^2}}=\dfrac{r\cdot 1+\dfrac{r}{1}\cdot\dfrac{2}{r}}{r\cdot 1-\dfrac{r}{1}\cdot\dfrac{4}{r^2}}$

$=\dfrac{r+2}{r-4r}$

$=\dfrac{r+2}{-3r}$

Mistake: Multiplied through by an amount other than the LCD and then eliminated the r^2 denominator.

Correct: $\dfrac{r}{r-2}$

49. $\dfrac{\dfrac{3}{x}}{\dfrac{x}{6}} = \dfrac{3}{x} \div \dfrac{x}{6}$

$= \dfrac{3}{x} \cdot \dfrac{6}{x}$

$= \dfrac{18}{2x}$ Mistake: Added the x's instead of multiplying them.

$= \dfrac{9}{x}$ Correct: $\dfrac{18}{x^2}$

50. $\dfrac{\dfrac{1}{x} + \dfrac{1}{y}}{\dfrac{1}{x} - \dfrac{1}{y}} = \dfrac{\dfrac{xy}{1} \cdot \dfrac{1}{x} + \dfrac{xy}{1} \cdot \dfrac{1}{y}}{\dfrac{xy}{1} \cdot \dfrac{1}{x} - \dfrac{xy}{1} \cdot \dfrac{1}{y}}$

$= \dfrac{y + x}{y - x}$

$= -1$

Mistake: Divided out variables that are not factors.

Correct: $\dfrac{y + x}{y - x}$

For Exercises 51 and 52, average rate can be found using the following formula:

$$\text{Average rate} = \frac{\text{Total distance}}{\text{Total time}}$$

51. Suppose a person travels 10 miles in $\dfrac{1}{3}$ of an hour, then returns by traveling that same 10 miles in $\dfrac{1}{4}$ of an hour. What is that person's average rate?

$34\dfrac{2}{7}$ mph

52. Juan travels 50 miles in $\dfrac{3}{4}$ of an hour and then encounters a traffic jam so that the next 15 miles take $\dfrac{1}{3}$ of an hour. The last 40 miles of his trip take him $\dfrac{2}{3}$ of an hour. What was his average rate for the trip?

60 mph

53. Given the area and length of a rectangle, the width can be found using the formula $w = \dfrac{A}{l}$. Suppose the area of a rectangle is described by $\dfrac{5n - 2}{8}$ square feet and the length is described by $\dfrac{n - 1}{6}$ feet. Find an expression for the width.

$\dfrac{3(5n - 2)}{4(n - 1)}$

54. Given the area and base of a triangle, the height can be found using the formula $h = \dfrac{2A}{b}$. Suppose the area of a triangle is described by $\dfrac{n + 1}{6}$ and its base is described by $\dfrac{5n - 2}{9}$. Find an expression for the height.

$\dfrac{3(n + 1)}{5n - 2}$

55. In electrical circuits, if two resistors R_1 and R_2 are wired in parallel, the resistance of the circuit is found using the complex rational expression $\dfrac{1}{\dfrac{1}{R_1} + \dfrac{1}{R_2}}$.

a. Simplify this complex rational expression.

$\dfrac{R_1 R_2}{R_2 + R_1}$

b. If a 100-ohm and 10-ohm resistor are wired in parallel, find the resistance of the circuit. $\approx 9.1\ \Omega$

c. Two 16-ohm speakers are wired in parallel; find the resistance of the circuit. $8\ \Omega$

56. Suppose three resistors R_1, R_2, and R_3 are wired in parallel. The resistance of the circuit is found using the complex rational expression $\dfrac{1}{\dfrac{1}{R_1} + \dfrac{1}{R_2} + \dfrac{1}{R_3}}$.

a. Simplify this complex rational expression.

$$\frac{R_1 R_2 R_3}{R_2 R_3 + R_1 R_3 + R_1 R_2}$$

b. A 20-ohm, a 10-ohm, and a 40-ohm resistor are wired in parallel. Find the resistance of the circuit.

$\approx 5.7\ \Omega$

c. A 300-ohm, a 500-ohm, and a 200-ohm resistor are wired in parallel. Find the resistance of the circuit.

$\approx 96.8\ \Omega$

Review Exercises

Exercises 1–6 Equations and Inequalities

[2.3] *For Exercises 1–4, solve.*

1. $-6x = 42$

-7

2. $9t - 8 = 19$

3

3. $\dfrac{x}{4} = \dfrac{5}{8}$

2.5

4. $\dfrac{3}{4}y - \dfrac{1}{3} = \dfrac{1}{6}y + 2$

4

[2.1] **5.** A runner travels 1.2 miles in 15 minutes. Find the runner's average rate in miles per hour.

4.8 mph

[4.2, 4.3] **6.** An eastbound car and a westbound car pass each other on a highway. The eastbound car is traveling 40 miles per hour, and the westbound car is traveling 60 miles per hour. How much time elapses from the time they pass each other until they are 9 miles apart?

0.09 hr. or 5.4 min.

7.6 Solving Equations Containing Rational Expressions

Objective

1 Solve equations containing rational expressions.

Instructor Note Make clear the difference between solving equations containing rational expressions and adding/subtracting rational expressions. After completing this section, some students will try to add/subtract rational expressions by multiplying by the LCD. See the Warning after Your Turn 5.

Warm-up

[7.2] *For Exercises 1 and 2, find the product and simplify when necessary.*

1. $4x\left(\dfrac{3}{x} - \dfrac{1}{4}\right)$

2. $(x + 2)\left(\dfrac{x^2 + 3}{x + 2} - 3\right)$

For Exercises 3 and 4, solve.

[2.3] **3.** $\dfrac{3}{4}x + \dfrac{1}{5} = \dfrac{1}{2}x - \dfrac{7}{10}$

[6.6] **4.** $3 - 5x = x^2 - 3x$

Objective 1 Solve equations containing rational expressions.

In Section 2.3, we learned that we can eliminate fractions from an equation by multiplying both sides of the equation by their LCD. For example, we can eliminate the fractions in $\dfrac{1}{2}x - \dfrac{3}{4} = \dfrac{2}{3}$ by multiplying both sides by 12.

Answers to Warm-up
1. $12 - x$ **2.** $x^2 - 3x - 3$
3. $-\dfrac{18}{5}$ **4.** -3 or 1

$$\overset{6}{\cancel{12}} \cdot \dfrac{1}{\underset{1}{2}}x - \overset{3}{\cancel{12}} \cdot \dfrac{3}{\underset{1}{4}} = \overset{4}{\cancel{12}} \cdot \dfrac{2}{\underset{1}{3}}$$

$$6x - 9 = 8$$

Because the rewritten equation contains only integers, it is easier to solve. Similarly, we can use the LCD to eliminate rational expressions in an equation.

Example 1   Solve $\dfrac{x}{5} = \dfrac{x}{2} - \dfrac{1}{4}$.

Solution: $20\left(\dfrac{x}{5}\right) = 20\left(\dfrac{x}{2} - \dfrac{1}{4}\right)$ Eliminate the rational expressions by multiplying both sides by the LCD, 20.

$\overset{4}{\cancel{20}} \cdot \dfrac{x}{\underset{1}{5}} = \overset{10}{\cancel{20}} \cdot \dfrac{x}{\underset{1}{2}} - \overset{5}{\cancel{20}} \cdot \dfrac{1}{\underset{1}{4}}$ Use the distributive property to multiply every term by 20 and then divide out the common factors.

$4x = 10x - 5$

$-6x = -5$ Subtract 10x from both sides.

$\dfrac{-6x}{-6} = \dfrac{-5}{-6}$ Divide both sides by −6.

$x = \dfrac{5}{6}$ Simplify both sides.

Check: $\dfrac{1}{5}x = \dfrac{1}{2}x - \dfrac{1}{4}$ Using the original equation, we replace x with $\dfrac{5}{6}$ and see if the equation is true. It is helpful to write $\dfrac{x}{5}$ as $\dfrac{1}{5}x$ and $\dfrac{x}{2}$ as $\dfrac{1}{2}x$.

$\dfrac{1}{5}\left(\dfrac{\overset{1}{\cancel{5}}}{\underset{1}{6}}\right) \overset{?}{=} \dfrac{1}{2}\left(\dfrac{5}{6}\right) - \dfrac{1}{4}$ Replace x with $\dfrac{5}{6}$ and then simplify.

$\dfrac{1}{6} \overset{?}{=} \dfrac{5}{12} - \dfrac{3}{12}$

$\dfrac{1}{6} = \dfrac{2}{12}$ True; therefore, $\dfrac{5}{6}$ is a solution.

Now consider equations that contain rational expressions with variables in the denominator.

Example 2   Solve $\dfrac{1}{2} = \dfrac{3}{x} - \dfrac{1}{4}$.

Solution: $4x\left(\dfrac{1}{2}\right) = 4x\left(\dfrac{3}{x} - \dfrac{1}{4}\right)$ Eliminate the rational expressions by multiplying both sides by the LCD, 4x.

$\overset{2}{\cancel{4}}x \cdot \dfrac{1}{\underset{1}{2}} = 4\overset{1}{\cancel{x}} \cdot \dfrac{3}{\underset{1}{x}} - \overset{1}{\cancel{4}}x \cdot \dfrac{1}{\underset{1}{4}}$ Use the distributive property to multiply every term by 4x and then divide out the common factors.

$2x = 12 - x$

$3x = 12$ Add x to both sides.

$x = 4$ Divide both sides by 3.

Check: We will leave the check to the reader.

Your Turn 2   Solve.

a. $\dfrac{x}{2} = 4 - \dfrac{x}{3}$

b. $\dfrac{1}{x} + \dfrac{3}{8} = \dfrac{5}{6}$

Answers to Your Turn 2
a. $\dfrac{24}{5}$ b. $\dfrac{24}{11}$

Extraneous Solutions

If an equation contains rational expressions with variables in the denominators, we must make sure that no solution causes any of the expressions in the equation to be undefined. In Example 2, if the solution had turned out to be 0, the expression $\frac{3}{x}$ would have been undefined. By inspecting the denominators of each rational expression, you can determine the values that will cause the expressions to be undefined before solving the equation. If by solving an equation you obtain a number that causes an expression in the original equation to be undefined, we say that the number is an **extraneous solution** and we discard it.

Definition **Extraneous solution:** An apparent solution that does not solve its equation.

The following procedure summarizes the process of solving equations containing rational expressions.

Procedure **Solving Equations Containing Rational Expressions**

To solve an equation that contains rational expressions:
1. Eliminate the rational expressions by multiplying both sides of the equation by their LCD.
2. Solve the equation using the methods we learned in Chapters 2 (linear equations) and 6 (quadratic equations).
3. Check your solution(s) in the original equation. Discard any extraneous solutions.

Connection The equation in Example 3 is a proportion. We could have solved the equation by cross multiplying, which leads to the same equation as in the second step of the solution in Example 3.

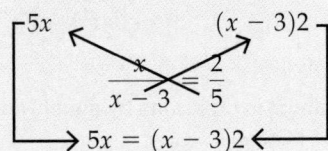

Warning You can use this method only if the equation is a proportion.

Example 3 Solve $\dfrac{x}{x-3} = \dfrac{2}{5}$.

Solution: Notice that if $x = 3$, then $\dfrac{x}{x-3}$ is undefined; so the solution cannot be 3.

$$5(x-3) \cdot \frac{x}{x-3} = \frac{1}{5}(x-3) \cdot \frac{2}{5}$$

Multiply both sides by the LCD, $5(x-3)$; then divide out the common factors.

$$5x = (x-3)2$$

$$5x = 2x - 6$$ Distribute 2 to eliminate the parentheses.

$$3x = -6$$ Subtract 2x from both sides.

$$x = -2$$ Divide both sides by 3.

Check: We will leave the check to the reader.

Your Turn 3 Solve $\dfrac{2}{x+5} = \dfrac{4}{x-1}$.

In Examples 1–3, after eliminating the rational expressions, we were left with linear equations. Now let's look at some equations that transform to quadratic equations after we use the LCD to eliminate the rational expressions. Remember that when solving quadratic equations, we manipulate the equation so that one side is 0; then we factor and use the zero-factor theorem.

Answer to Your Turn 3
-11

Example 4   Solve.

a. $\dfrac{3}{x} - 5 = x - 3$

Solution: Notice that because the denominator is x, the solution cannot be 0.

$$x\left(\dfrac{3}{x} - 5\right) = x(x - 3)$$ Multiply both sides by the LCD, x.

$$\overset{1}{\cancel{x}} \cdot \dfrac{3}{\underset{1}{\cancel{x}}} - x \cdot 5 = x \cdot x - x \cdot 3$$ Distribute x; then divide out common factors.

$$3 - 5x = x^2 - 3x$$ Because this equation is quadratic, we set it equal to 0 by subtracting 3 from and adding 5x to both sides.

$$0 = x^2 - 3x + 5x - 3$$

$$0 = x^2 + 2x - 3$$ Combine like terms.

$$0 = (x + 3)(x - 1)$$ Factor.

$$x + 3 = 0 \quad \text{or} \quad x - 1 = 0$$ Use the zero-factor theorem.

$$x = -3 \qquad\qquad x = 1$$

Check: Verify that both -3 and 1 make the original equation true.

$$\dfrac{3}{-3} - 5 \overset{?}{=} -3 - 3 \qquad\qquad \dfrac{3}{1} - 5 \overset{?}{=} 1 - 3$$

$$-1 - 5 \overset{?}{=} -6 \qquad\qquad 3 - 5 \overset{?}{=} -2$$

$$-6 = -6 \qquad \text{True} \qquad\qquad -2 = -2 \qquad \text{True}$$

Both -3 and 1 are solutions.

b. $\dfrac{x}{x + 1} = \dfrac{6}{3x - 1}$

Solution: We could multiply both sides by the LCD, $(x + 1)(3x - 1)$, but because the equation is a proportion, we choose to cross multiply instead. Note that $\dfrac{1}{3}$ and -1 cannot be solutions.

$$\dfrac{x}{x + 1} = \dfrac{6}{3x - 1}$$

$$\begin{array}{cc} 3x^2 - x & 6x + 6 \end{array}$$

$$\dfrac{x}{x + 1} \diagtimes \dfrac{6}{3x - 1}$$ Cross multiply.

$$3x^2 - x = 6x + 6$$ Because this equation is quadratic, we set it equal to 0 by subtracting 6x and 6 from both sides.

$$3x^2 - 7x - 6 = 0$$

$$(3x + 2)(x - 3) = 0$$ Factor.

$$3x + 2 = 0, x - 3 = 0$$ Use the zero-factor theorem.

$$3x = -2 \quad x = 3$$ Solve.

$$x = -\dfrac{2}{3}$$

Check: We will leave the check to the reader.

Answers to Your Turn 4
a. $-\dfrac{2}{3}$ and -3 **b.** $-\dfrac{3}{4}$ and 2

Your Turn 4   Solve.

a. $\dfrac{1}{3} - \dfrac{2}{x} = x + 4$

b. $\dfrac{x}{x + 1} = \dfrac{6}{4x + 1}$

| **Example 5** | Solve. |

a. $\dfrac{x^2}{x + 2} - 3 = \dfrac{x + 6}{x + 2} - 4$

Solution: By inspecting the denominators, notice that the solution cannot be -2.

$$(x + 2)\left(\dfrac{x^2}{x + 2} - 3\right) = (x + 2)\left(\dfrac{x + 6}{x + 2} - 4\right)$$

    Multiply both sides by the LCD, $x + 2$.

$$(x + 2) \cdot \dfrac{x^2}{x + 2} - (x + 2) \cdot 3 = (x + 2) \cdot \dfrac{x + 6}{x + 2} - (x + 2) \cdot 4$$

    Distribute $x + 2$; then divide out the common factors.

$$x^2 - 3x - 6 = x + 6 - 4x - 8$$

    Combine like terms.

$$x^2 - 3x - 6 = -3x - 2$$

$$x^2 - 4 = 0$$

    Because the equation is quadratic, we set it equal to 0 by adding $3x$ and 2 to both sides.

$$(x + 2)(x - 2) = 0$$ Factor.

$$x + 2 = 0 \quad \text{or} \quad x - 2 = 0$$ Use the zero-factor theorem.

$$x = -2 \qquad\qquad x = 2$$

Check: In the original equation, $\dfrac{x^2}{x + 2}$ and $\dfrac{x + 6}{x + 2}$ are both undefined when $x = -2$; so -2 is extraneous. Therefore, we need to check only 2.

$$\dfrac{2^2}{2 + 2} - 3 \stackrel{?}{=} \dfrac{2 + 6}{2 + 2} - 4$$

    In the original equation, replace x with 2 and then simplify.

$$\dfrac{4}{4} - 3 \stackrel{?}{=} \dfrac{8}{4} - 4$$

$$1 - 3 \stackrel{?}{=} 2 - 4$$

$$-2 = -2$$ True; so 2 is a solution and the only solution.

b. $\dfrac{5}{x - 3} = \dfrac{x}{x - 2} + \dfrac{x}{x^2 - 5x + 6}$

Solution: To determine the LCD of all of the denominators, we first factor quadratic form $x^2 - 5x + 6$.

$$\dfrac{5}{x - 3} = \dfrac{x}{x - 2} + \dfrac{x}{(x - 2)(x - 3)}$$

    Factor the denominator $x^2 - 5x + 6$.

$$(x - 2)(x - 3)\left(\dfrac{5}{x - 3}\right) = (x - 2)(x - 3)\left(\dfrac{x}{x - 2} + \dfrac{x}{(x - 2)(x - 3)}\right)$$

    Multiply both sides by the LCD, $(x - 2)(x - 3)$.

> **Note** Inspecting the denominators after factoring, we see that neither 2 nor 3 can be a solution.

> Distribute $(x - 2)(x - 3)$; then divide out the common factors.

$$(x - 2)(x - 3) \cdot \dfrac{5}{(x - 3)} = (x - 2)(x - 3) \cdot \dfrac{x}{(x - 2)} + (x - 2)(x - 3) \cdot \dfrac{x}{(x - 2)(x - 3)}$$

$$(x - 2)5 = (x - 3)x + x$$

$$5x - 10 = x^2 - 3x + x$$ Distribute to eliminate the parentheses.

$$5x - 10 = x^2 - 2x$$ Combine like terms.

$$0 = x^2 - 7x + 10$$

    Because the equation is quadratic, we set it equal to 0 by subtracting $5x$ from and adding 10 to both sides.

$$0 = (x - 2)(x - 5)$$ Factor.

$$x - 2 = 0 \quad \text{or} \quad x - 5 = 0$$ Use the zero-factor theorem.

$$x = 2 \qquad\qquad x = 5$$

Check: In the original equation, note that $\dfrac{x}{x - 2}$ is undefined when $x = 2$; so 2 is an extraneous solution. Consequently, 5 is the only solution. We will leave the check for 5 to the reader.

Your Turn 5 Solve.

a. $\dfrac{3x + 4}{x - 4} = \dfrac{x^2}{x - 4} + 3$

b. $\dfrac{2}{x + 5} - \dfrac{6}{x^3 + 5x^2} = \dfrac{1}{x}$

Warning Make sure you understand the difference between performing operations with rational expressions and solving equations containing rational expressions. For example, $\dfrac{5}{x} + \dfrac{1}{3}$ is an *expression* that means two rational expressions are to be added. Remember, we cannot solve expressions; we can only evaluate or rewrite them. To rewrite this expression, we begin by writing each rational expression with the LCD, $3x$.

$$\dfrac{5}{x} + \dfrac{1}{3}$$
$$= \dfrac{5(3)}{x(3)} + \dfrac{1(x)}{3(x)}$$
$$= \dfrac{15}{3x} + \dfrac{x}{3x}$$
$$= \dfrac{15 + x}{3x}$$

To solve an *equation* such as $\dfrac{3}{x} + \dfrac{1}{4} = \dfrac{1}{2}$, we first eliminate the rational expressions by multiplying both sides of the equation by the LCD, $4x$.

$$4x\left(\dfrac{3}{x} + \dfrac{1}{4}\right) = 4x\left(\dfrac{1}{2}\right).$$
$$12 + x = 2x$$
$$12 = x \qquad \text{Subtract } x \text{ from both sides.}$$

Answers to Your Turn 5
a. -4, (4 is extraneous)
b. 6 and -1

7.6 **Exercises** (For Extra Help) MyMathLab®

Note: Exercises marked with a ★ represent challenging exercises.

Objective 1

Prep Exercise 1 How are rational expressions eliminated from an equation? Both sides of the equation are multiplied by the LCD of the rational expressions.

Prep Exercise 2 What is the LCD for the rational expressions in $\dfrac{x}{x^2 + 4x - 5} = \dfrac{5}{x} - 2x$? $x(x + 5)(x - 1)$

Prep Exercise 3 What is an extraneous solution for an equation that contains rational expressions? An apparent solution that does not solve its equation

Prep Exercise 4 Why might an equation with rational expressions lead to one or more extraneous solutions? It is possible to have an apparent solution cause the denominator of one or more of the rational expressions in the original equation to be undefined.

Prep Exercise 5 What values might be extraneous solutions in the equation $\dfrac{3}{x - 4} = \dfrac{5x}{x + 2}$? Why? 4 and -2; substituting either of those numbers for x causes an expression in the given equation to be undefined.

Prep Exercise 6 What values might be extraneous solutions in the equation $\dfrac{x}{x^2 + 4x - 5} = \dfrac{5}{x} - 2x$? Why? 0, -5, and 1; substituting one of those numbers for x causes an expression in the given equation to be undefined.

For Exercises 1–6, check the given values to see if they are solutions to the equation.

1. $\dfrac{3}{x} + \dfrac{2}{3} = \dfrac{5}{6}$; $x = 9$
no

2. $\dfrac{3}{4} - \dfrac{1}{x} = \dfrac{2}{3}$; $x = 12$
yes

3. $\dfrac{4}{m} - \dfrac{2}{5} = \dfrac{3}{4m}$; $m = \dfrac{65}{8}$
yes

4. $\dfrac{5}{6n} + \dfrac{1}{4} = \dfrac{2}{3n}$; $n = -\dfrac{2}{3}$
yes

5. $\dfrac{1}{t+2} + \dfrac{1}{4} = \dfrac{t}{16}$; $t = -4$
yes

6. $\dfrac{1}{y-3} - \dfrac{5}{6} = -\dfrac{y}{12}$; $y = -7$
no

For Exercises 7–56, solve and check. Identify any extraneous solutions. See Examples 1–5.

7. $\dfrac{2n}{3} + \dfrac{3n}{2} = \dfrac{13}{3}$
2

8. $\dfrac{3x}{5} - \dfrac{x}{2} = \dfrac{19}{5}$
38

9. $3y - \dfrac{4y}{5} = 22$
10

10. $\dfrac{3t}{4} - 2 = \dfrac{t}{4}$
4

11. $\dfrac{r-7}{2} - 1 = \dfrac{r+9}{9} + \dfrac{1}{3}$
15

12. $\dfrac{d+5}{3} - 4 = \dfrac{d-8}{4} - \dfrac{1}{2}$
−2

13. $\dfrac{4}{x} + \dfrac{3}{2x} = \dfrac{11}{6}$
3

14. $\dfrac{3}{2u} - \dfrac{1}{3} = \dfrac{5}{6u}$
2

15. $\dfrac{4-6t}{2t} + \dfrac{3}{5} = -\dfrac{2}{5t}$
1

16. $\dfrac{2x+3}{2x} = \dfrac{5}{12} + \dfrac{3x+2}{3x}$
2

17. $\dfrac{5}{20} = \dfrac{x-7}{x+2}$
10

18. $\dfrac{y-2}{y+3} = \dfrac{3}{8}$
5

19. $\dfrac{6}{y-2} = \dfrac{5}{y-3}$
8

20. $\dfrac{4}{3x-3} = \dfrac{7}{4x+1}$
5

21. $\dfrac{x+4}{x+2} - 5 = \dfrac{6}{x+2}$
−3

22. $\dfrac{m+5}{m-3} - 5 = \dfrac{4}{m-3}$
4

23. $\dfrac{6f-5}{6} = \dfrac{2f-1}{2} - \dfrac{f+2}{2f+5}$
−1

24. $\dfrac{2x-1}{8} = \dfrac{x}{4} - \dfrac{x-1}{5x-2}$
2

25. $\dfrac{2}{x^2-2x-3} - \dfrac{3}{x-3} = \dfrac{2}{x+1}$
1

26. $\dfrac{5}{t^2-9} = \dfrac{3}{t+3} - \dfrac{2}{t-3}$
20

27. $\dfrac{1}{u-4} + \dfrac{2}{u^2-16} = \dfrac{3}{u+4}$
9

28. $\dfrac{5}{h+5} - \dfrac{2}{h^2+2h-15} = \dfrac{2}{h-3}$
9

29. $\dfrac{x}{2-3x} = \dfrac{1}{3x+2}$
$-2, \dfrac{1}{3}$

30. $\dfrac{5p}{p+4} = \dfrac{p}{p-1}$
$0, \dfrac{9}{4}$

31. $\dfrac{5}{2x} + \dfrac{15}{2x+16} = \dfrac{6}{x+8}$
−5

32. $\dfrac{4}{3x} - \dfrac{2}{x+4} = \dfrac{2}{3x+12}$
4

33. $\dfrac{2}{x^2-1} = \dfrac{-3}{7x+7}$
$-\dfrac{11}{3}$, (−1 is extraneous)

34. $\dfrac{3}{y^2-4} = \dfrac{-2}{5y+10}$
$-\dfrac{11}{2}$, (−2 is extraneous)

35. $\dfrac{3}{x+5} - \dfrac{x-1}{2x} = \dfrac{-3}{2x^2+10x}$
4, −2

36. $\dfrac{-1}{3x^2+9x} = \dfrac{1}{x+3} - \dfrac{x-3}{3x}$
−2, 5

37. $\dfrac{3m+1}{m^2-9} - \dfrac{m+3}{m-3} = \dfrac{1-5m}{m+3}$
$-\dfrac{1}{4}, 5$

38. $\dfrac{2n+10}{n^2-1} - \dfrac{n-1}{n+1} + \dfrac{2n+1}{n-1} = 0$
−2, −5

39. $\dfrac{x+3}{x-1} + \dfrac{2}{x+3} = \dfrac{5}{3}$
−2, 9

40. $\dfrac{m+2}{m-2} - \dfrac{2}{m+2} = -\dfrac{7}{3}$

$-1, \dfrac{2}{5}$

41. $1 - \dfrac{3}{x-2} = -\dfrac{12}{x^2-4}$

1, (2 is extraneous)

42. $\dfrac{1}{x-3} + \dfrac{1}{3} = \dfrac{6}{x^2-9}$

-6, (3 is extraneous)

43. $\dfrac{2r}{r-2} - \dfrac{4r}{r-3} = -\dfrac{7r}{r^2-5r+6}$

$0, \dfrac{9}{2}$

44. $\dfrac{3a}{a^2-2a-15} - \dfrac{a}{a+3} = \dfrac{2a}{a-5}$

$0, \dfrac{2}{3}$

45. $\dfrac{4u^2+3u+4}{u^2+u-2} - \dfrac{3u}{u+2} = \dfrac{-2u-1}{u-1}$

$-3, -\dfrac{2}{3}$

46. $\dfrac{2z}{2z-3} - \dfrac{3z}{2z+3} = \dfrac{15-32z^2}{4z^2-9}$

$-1, \dfrac{1}{2}$

47. $\dfrac{2k}{k+7} - 1 = \dfrac{1}{k^2+10k+21} + \dfrac{k}{k+3}$

-2

48. $\dfrac{p-5}{p+5} - 2 + \dfrac{p+15}{p-5} = \dfrac{50}{p^2-25}$

-10

49. $\dfrac{6}{t^2+t-12} = \dfrac{4}{t^2-t-6} + \dfrac{1}{t^2+6t+8}$

1

50. $\dfrac{7}{v^2-6v+5} - \dfrac{2}{v^2-4v-5} = \dfrac{3}{v^2-1}$

-12

51. $\dfrac{x+1}{x^2-2x-15} - \dfrac{6}{x^2-3x-10} = \dfrac{2}{x^2+5x+6}$

$6, -1$

52. $\dfrac{x+1}{x^2+2x-24} - \dfrac{3}{x^2-x-12} = \dfrac{3}{x^2+9x+18}$

$3, -1$

★ **53.** $\dfrac{1}{d^2-5d+6} + \dfrac{d-2}{d^2-d-6} - \dfrac{d}{d^2-4} = 0$

No solution

★ **54.** $\dfrac{1}{y^2-5y+6} + \dfrac{y}{y^2-2y-3} = \dfrac{y+2}{y^2-y-2}$

No solution

★ **55.** $\dfrac{1}{a-3} - \dfrac{6}{a^2-1} = \dfrac{12}{a^3-3a^2-a+3}$

5, (1 is extraneous)

★ **56.** $\dfrac{1}{b+2} - \dfrac{2}{b^2-1} = \dfrac{-2}{b^3+2b^2-b-2}$

3, (-1 is extraneous)

Find ⊗ the Mistake   *For Exercises 57 and 58, explain the mistake; then find the correct solution(s).*

57.

$$\dfrac{y}{3} + \dfrac{y}{y-3} = \dfrac{3}{y-3}$$

$$3(y-3)\cdot\dfrac{y}{3} + 3(y-3)\cdot\dfrac{y}{y-3} = 3(y-3)\cdot\dfrac{3}{y-3}$$

$$y^2 - 3y + 3y = 9$$

$$y^2 - 9 = 0$$

$$(y+3)(y-3) = 0$$

$$y = -3, 3$$

Mistake: 3 is extraneous.
Correct: -3 is the only answer.

58.

$$\dfrac{x}{x+3} + \dfrac{3}{x+3} = 0$$

$$(x+3)\cdot\dfrac{x}{x+3} + (x+3)\cdot\dfrac{3}{x+3} = (x+3)\cdot 0$$

$$x + 3 = 0$$

$$x = -3$$

Mistake: -3 is extraneous.
Correct: There is no solution.

59. Explain how the LCD is used differently in $\dfrac{3x}{4} - \dfrac{5}{6}$ and in $\dfrac{3x}{4} - \dfrac{5}{6} = \dfrac{x}{2}$.

In an expression, the LCD is used to combine the numerator over a single denominator. In an equation, the LCD is used to eliminate the denominator.

60. Explain how the LCD is used differently in $\dfrac{5x}{8} + \dfrac{1}{3}$ and in $\dfrac{5x}{8} + \dfrac{1}{3} = \dfrac{x}{6}$.

In an expression, the LCD is used to combine the numerator over a single denominator. In an equation, the LCD is used to eliminate the denominator.

For Exercises 61–64, use the formula for the total resistance R in an electrical circuit with resistors $R_1, R_2, R_3, \ldots , R_n$ that are wired in parallel.

$$\frac{1}{R} = \frac{1}{R_1} + \frac{1}{R_2} + \frac{1}{R_3} + \cdots + \frac{1}{R_n}$$

61. Two resistors are wired in parallel, one of which is 100 ohms. If the total resistance is to be 80 ohms, what is the value of the other resistor?

400 Ω

★ **62.** Three resistors are wired in parallel, two of which are equal in value. If the third resistor is 200 ohms and the total resistance is to be 20 ohms, what are the values of the other two resistors?

$44.\overline{4}\Omega$

★ **63.** Two resistors are wired in parallel. One resistor is to have a value that is 10 ohms more than the other resistor, and the total resistance is to be 12 ohms. Find the value of both resistors.

20 and 30 Ω

★ **64.** Two resistors are wired in parallel. One resistor is to have a value that is 20 ohms less than the other resistor, and the total resistance is to be 24 ohms. Find the value of both resistors.

40 and 60 Ω

In optics, a lens can be used to bend light from an object through a focal point to reproduce an inverted image that is larger (or smaller) than the original object. The following formula describes how the image of an object is affected by a lens, where o represents the object's distance from the lens, i represents the image's distance from the lens, and f represents the focal length of the lens.

$$\text{Formula: } \frac{1}{o} + \frac{1}{i} = \frac{1}{f}$$

Of Interest

The lens in our eye focuses images on the retina in the back of the eyeball. The image is inverted on our retina. Our brains then correct the inverted images so that we do not see the world upside down.

65. An object is 40 feet from a lens with a focal length of 2 feet. What is the image's distance from the lens?

≈ 2.1 ft.

66. An object is placed 50 centimeters from your eye. Your lens focuses the image of the object onto your retina, which is about 2.5 centimeters from your lens. What is the focal length of your lens?

≈ 2.38 cm

★ **67.** A lens is being designed so that the focal length will be 5 millimeters less than the image length. Suppose an object is placed 60 millimeters from the lens. What are the focal length and image length?

$f = 15$ mm, $i = 20$ mm

★ **68.** A lens is being designed so that the image length is 9 millimeters more than the focal length. Suppose an object is placed 40 millimeters from the lens. What are the focal length and image length?

$f = 15$ mm, $i = 24$ mm

Of Interest

When you look at an object, muscles in your eye adjust your lens, which changes its focal length so that the image is produced on your retina. The eyes of a nearsighted person cannot adjust the lenses properly for objects far away, and those images are incorrectly focused in front of the retinas. The eyes of a farsighted person incorrectly focus near objects behind the retinas.

Review Exercises

Exercises 1–6   Equations and Inequalities

[2.1] **1.** Andrew notes that he traveled a distance of 120 miles in $1\frac{3}{4}$ hours. What was his average rate in miles per hour?

≈ 68.6 mph

[2.6] **2.** A grocery store has a 10.5-ounce can of soup on sale for $0.88. Write a unit ratio in simplest form of the price to capacity. Interpret the answer.

$\approx \frac{0.084}{1}$; each ounce costs 8.4 cents.

[2.6] 3. Daniel knows that he can paint 800 square feet of wall space in 4 hours. If he were able to maintain that rate, how long would it take him to paint 1200 square feet?

6 hr.

[4.2, 4.3] 5. A rectangular eraser has a perimeter of 36 millimeters. If the length is three less than twice the width, find the length and width.

The length is 11 mm, and the width is 7 mm.

[4.2, 4.3] 4. The sum of two positive consecutive even integers is 146. What are the numbers?

72, 74

[4.2, 4.3] 6. Two joggers pass each other going in opposite directions on a path. If one jogger is traveling at a rate of 6 miles per hour and the other is traveling at a rate of 4 miles per hour, how long will it take them to be 1.5 miles apart?

0.15 hr. (9 min.)

7.7 Applications with Rational Expressions, Including Variation

Objectives

1 Use tables to solve problems with two unknowns involving rational expressions.

2 Solve problems involving direct variation.

3 Solve problems involving inverse variation.

4 Solve problems involving joint variation.

5 Solve problems involving combined variation.

Learning Strategy

If you don't understand the lecture, don't be afraid to ask questions. If you are too intimidated to ask a question in a large lecture, go to all review sessions and see the professor during office hours. Put a star next to problems that you don't understand to get help on later. Sit with a friend in class and ask him or her for help.

—Lauren H.

Answers to Warm-up

1. $t = \dfrac{6}{5}$

2. $r = 4$ or $r = -\dfrac{6}{5}$

3. $k = 40$

4. 3

Warm-up

[7.6] *For Exercises 1–3, solve the equation.*

1. $\dfrac{t}{2} + \dfrac{t}{3} = 1$

2. $\dfrac{3}{r} + \dfrac{3}{r + 2} = \dfrac{5}{4}$

3. $4 = \dfrac{k}{10}$

[1.7] 4. Find the value of $\dfrac{q_1 q_2}{d^2}$ when $q_1 = 16$, $q_2 = 12$, and $d = 8$.

Objective 1 Use tables to solve problems with two unknowns involving rational expressions.

In Chapter 4, we used tables to organize information involving two unknown amounts. Now let's look at some similar problems that lead to equations containing rational expressions.

Problems Involving Work

Tables are helpful in problems involving two or more people (or machines) working together to complete a task. In these problems, we are given each person's rate of work and are asked to find the time for them to complete the task if they work together. For each person involved, the rate of work, time at work, and amount of the task completed are related as follows:

$$\boxed{\begin{array}{c}\text{Person's rate}\\\text{of work}\end{array}} \cdot \boxed{\begin{array}{c}\text{Time}\\\text{at work}\end{array}} = \boxed{\begin{array}{c}\text{Amount of the task}\\\text{completed by that person}\end{array}}$$

Because the people are working together, the sum of each individual's amount of the task completed equals the whole task.

$$\boxed{\begin{array}{c}\text{Amount completed}\\\text{by one person}\end{array}} + \boxed{\begin{array}{c}\text{Amount completed by}\\\text{the other person}\end{array}} = \boxed{\text{Whole task}}$$

Example 1 Karen and Jeff own a cleaning business. Karen can clean an average-size house in 2 hours. Jeff can clean a similar house in 3 hours. How long will it take them to clean an average-size house working together?

Understand Karen cleans at a rate of 1 house in 2 hours, or $\frac{1}{2}$ of a house per hour.

Jeff cleans at a rate of 1 house in 3 hours, or $\frac{1}{3}$ of a house per hour.

Category	Rate of Work	Time at Work	Amount of Task Completed
Karen	$\frac{1}{2}$	t	$\frac{1}{2}t$ or $\frac{t}{2}$
Jeff	$\frac{1}{3}$	t	$\frac{1}{3}t$ or $\frac{t}{3}$

Note Because they are working together on the same job for the same amount of time, we let t represent that amount of time.

Note Multiplying the rate of work and the time at work gives an expression of the amount of work completed. For example, if Karen works 4 hours at a rate of $\frac{1}{2}$ of a house every hour, she can clean $4 \cdot \frac{1}{2} = 2$ houses.

The total job in this case is 1 house, so we can write an equation that combines the individual expressions for the task completed and set this sum equal to 1 house.

Plan and Execute Karen's amount completed + Jeff's amount completed = 1 house

$$\frac{t}{2} + \frac{t}{3} = 1$$

$$6\left(\frac{t}{2} + \frac{t}{3}\right) = 6(1) \qquad \text{Multiply both sides by the LCD, 6.}$$

$$\overset{3}{6} \cdot \frac{t}{\underset{1}{2}} + \overset{2}{6} \cdot \frac{t}{\underset{1}{3}} = 6 \qquad \text{Distribute and then divide out common factors.}$$

$$3t + 2t = 6$$

$$5t = 6 \qquad \text{Combine like terms.}$$

$$t = \frac{6}{5} \qquad \text{Divide both sides by 5.}$$

Answer Working together, it takes Karen and Jeff $\frac{6}{5}$, or $1\frac{1}{5}$, hours to clean an average-size house.

Check Karen cleans $\frac{1}{2}$ of a house per hour; so if she works alone $1\frac{1}{5}$ hours, she cleans $\frac{1}{2} \cdot \frac{6}{5} = \frac{3}{5}$ of a house. Jeff cleans $\frac{1}{3}$ of a house per hour; so in $1\frac{1}{5}$ hours, he cleans $\frac{1}{3} \cdot \frac{6}{5} = \frac{2}{5}$ of a house. Combining their individual amounts, we see that in $1\frac{1}{5}$ hours, they clean $\frac{3}{5} + \frac{2}{5} = \frac{5}{5} = 1$ house.

Your Turn 1 Terrell owns a landscaping company. Every Monday, he services the same group of clients, and it takes him 5 hours to complete the service. On one occasion, he hired a student, Jarod, to service those same clients, and it took Jarod 6 hours. Working together, how much time would it take them to service the clients?

Answer to Your Turn 1

$2\frac{8}{11}$ hr.

Motion Problems

Recall that the formula for calculating distance, given the rate of travel and time of travel, is $d = rt$. If we isolate r, we have $r = \dfrac{d}{t}$. If we isolate t, we have $t = \dfrac{d}{r}$. These equations suggest that we use rational expressions when describing rate or time.

Example 2 Warren runs 3 miles away from the gym, turns around, and runs an average of 2 miles per hour faster on the return trip. If the total time of his run is $1\frac{1}{4}$ hours, what is his speed in the outbound leg and the inbound leg of his run?

Understand We are to find the rates for each leg of Warren's run. Because this situation involves two different rates, we use a table to organize the distance, rate, and time of each leg of the run.

Category	Distance	Rate	Time
Outbound	3 miles	r	$\dfrac{3}{r}$
Inbound	3 miles	$r + 2$	$\dfrac{3}{r + 2}$

Note Warren runs 2 mph faster during the inbound leg of his run; so we let r represent the outbound rate and add 2 mph to r for the inbound rate.

Note Because $d = rt$, to describe time, we divide the distance by the rate: $t = \dfrac{d}{r}$.

Because the total time of the trip is $1\frac{1}{4}$ $\left(\text{or } \dfrac{5}{4}\right)$ hours, we can write an equation that is the sum of the outbound and inbound times.

Plan and Execute Outbound time + Inbound time = $\dfrac{5}{4}$ hours

$$\frac{3}{r} + \frac{3}{r + 2} = \frac{5}{4}$$

$$4r(r + 2)\left(\frac{3}{r} + \frac{3}{r + 2}\right) = 4r(r + 2)\left(\frac{5}{4}\right)$$ Multiply both sides by the LCD, $4r(r + 2)$.

$$4r(r + 2) \cdot \frac{3}{r} + 4r(r+2) \cdot \frac{3}{r+2} = 4r(r + 2) \cdot \frac{5}{4}$$ Distribute and then divide out common factors.

$$12(r + 2) + 12r = 5r(r + 2)$$

$$12r + 24 + 12r = 5r^2 + 10r$$ Distribute.

$$24r + 24 = 5r^2 + 10r$$ Combine like terms.

$$0 = 5r^2 - 14r - 24$$ Subtract $24r$ and 24 from both sides.

$$0 = (5r + 6)(r - 4)$$ Factor.

$$5r + 6 = 0 \quad \text{or} \quad r - 4 = 0$$ Use the zero-factor theorem.

$$5r = -6 \qquad\qquad r = 4$$

$$r = -\frac{6}{5}$$

Answer Although $-\dfrac{6}{5}$ is a solution to the equation, a negative rate does not make sense in this situation; so we do not consider it as an answer. Therefore, the rate of the outbound leg is 4 miles per hour. Because the rate of the inbound leg is 2 miles per hour faster, the rate of speed must be 6 miles per hour.

Check Verify that traveling 3 miles at 4 miles per hour, then 3 miles at 6 miles per hour takes a total of $1\frac{1}{4}$ hours.

$$\text{Outbound time} = \frac{3}{4} \text{ hour} \qquad \text{Inbound time} = \frac{3}{6} = \frac{1}{2} \text{ hour}$$

$$\text{Total time} = \frac{3}{4} + \frac{1}{2} = \frac{3}{4} + \frac{2}{4} = \frac{5}{4} = 1\frac{1}{4} \text{ hour}$$

Your Turn 2 A plane travels 1200 miles against the jet stream, causing its airspeed to be decreased by 20 miles per hour. On the return flight, the plane travels with the jet stream so that its airspeed is increased by 20 miles per hour. If the total flight time of the round trip is $6\frac{1}{3}$ hours, what is the plane's rate in still air?

Objective 2 Solve problems involving direct variation.

In 2013, the Internal Revenue Service allowed a \$0.565 per mile deduction for each mile driven for business purposes. If d is the amount of the deduction and n is the number of miles driven, then $d = 0.565n$. In the following table, we use that formula to determine the deduction for various values of n.

Number of Miles	Deduction ($d = 0.565n$)
1	0.565
2	1.13
3	1.695
4	2.26
5	2.825

From the table, we see that as the number of miles increases, so does the amount of the deduction. Or, more formally, as values of n increase, so do values of d. In $d = 0.565n$, the two variables, d and n, are said to **vary directly**, or are *directly proportional*, and 0.565 is the constant of variation.

Definition Direct variation: Two variables y and x vary directly if $y = kx$. If y varies directly as the nth power of x, then $y = kx^n$, where k is the constant of variation.

In words, direct variation is written as *y varies directly as x* or *y is directly proportional to x*, and these phrases translate to $y = kx$. The expression $y = kx^n$ is translated as *y varies directly as the nth power of x* or *y is directly proportional to the nth power of x*.

In all variation problems in this section, we are given one set of values of the variables, which we use to find the constant of variation, k. We then use this value of k and the equation to find other values of the variable(s).

Procedure Solving Variation Problems

1. Write the equation.
2. Substitute the initial values and find k.
3. Substitute for k in the equation found in step 1.
4. Substitute the additional data into the equation from step 3 and solve for the unknown.

Answer to Your Turn 2
380 mph

Example 3 Solve the direct variations.

a. Suppose y varies directly as x. If $y = 18$ when $x = 5$, find y when $x = 8$.

Solution: $y = kx$ Translate "*y* varies directly as *x*."

$\qquad 18 = k \cdot 5$ Replace *y* with 18 and *x* with 5.

$\qquad 3.6 = k$ Divide both sides by 5.

In $y = kx$, replacing k with 3.6 gives $y = 3.6x$. Now find y when $x = 8$.

$$y = 3.6(8) = 28.8 \quad \text{In } y = 3.6x, \text{ replace } x \text{ with 8.}$$

b. Suppose u varies directly as the square of v. If $u = 54$ when $v = 3$, find u when $v = 5$.

Solution: $u = kv^2$ Translate "*u* varies directly as the square of *v*."

$\qquad 54 = k \cdot 3^2$ Replace *u* with 54 and *v* with 3.

$\qquad 54 = 9k$ Square 3.

$\qquad 6 = k$ Divide both sides by 9.

In $u = kv^2$, replacing k with 6 gives $u = 6v^2$. Now find u when $v = 5$.

$$u = 6(5)^2 \quad \text{In } u = 6v^2, \text{ replace } v \text{ with 5}$$
$$u = 6(25) \quad \text{Square 5.}$$
$$u = 150 \quad \text{Multiply.}$$

Your Turn 3 Suppose m varies directly as the square of n. If $m = 24$ when $n = 2$, find m when $n = -3$.

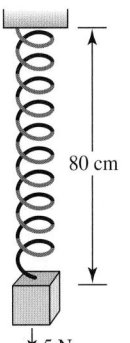

Example 4 In physical science, Hooke's law states that the distance a spring of uniform material and thickness is stretched varies directly with the force applied to the spring. If a force of 5 newtons stretches a spring 80 centimeters, how much force is required to stretch the spring 128 centimeters?

Understand Translating "the distance a spring of uniform material and thickness is stretched varies directly with the force," we write $d = kF$, where d represents distance and F represents the force.

Plan Use $d = kF$, replacing d with 80 centimeters and F with 5 newtons to solve for the value of k. Then use that value in $d = kF$ to solve for the force required to stretch the spring 128 centimeters.

Execute $80 = k \cdot 5$ Replace *d* with 80 and *F* with 5.

$\qquad\quad 16 = k$ Divide both sides by 5.

Replacing k with 16 in $d = kF$, we have $d = 16F$, which we use to solve for F when d is 128 centimeters.

$$128 = 16F \quad \text{Substitute 128 for } d.$$
$$8 = F \quad \text{Divide both sides by 16.}$$

Answer To stretch the spring 128 centimeters, a force of 8 newtons is applied.

Check A force of 8 newtons stretches the spring $d = 16(8) = 128$ centimeters.

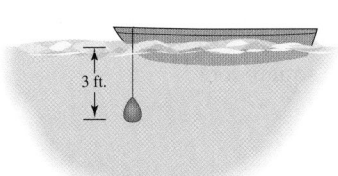

Answer to Your Turn 3
54

Answer to Your Turn 4
512 lb./ft.^{2}

Your Turn 4 The pressure exerted by a liquid varies directly as the depth beneath the surface. If an object is submerged in seawater to a depth of 3 feet, the pressure is 192 pounds per square foot. Find the pressure per square foot if an object is submerged in seawater to a depth of 8 feet.

Objective 3 Solve problems involving inverse variation.

Suppose a campaign worker must stuff 100 envelopes. If r is the rate at which she can stuff the envelopes and t is the time required, then $t = \dfrac{100}{r}$. For example, if she can stuff 2 envelopes per minute, it will take her $\dfrac{100}{2} = 50$ minutes to stuff the envelopes. In the following table, we use $t = \dfrac{100}{r}$ to see the relationship between the rate at which she can stuff envelopes and the amount of time required to complete the job.

Rate r	Time $\left(t = \dfrac{100}{r} \right)$
1 per minute	100 minutes
2 per minute	50 minutes
4 per minute	25 minutes

From the table, we see that as the rate at which she stuffs the envelopes increases, the time required decreases. More formally, as values of r increase, values of t decrease. In $t = \dfrac{100}{r}$, the two variables, t and r, are said to be in **inverse variation**, or are *inversely proportional*, and 100 is the constant of variation.

Definition Inverse variation: Two variables y and x vary inversely if $y = \dfrac{k}{x}$. If $y = \dfrac{k}{x^n}$, then y varies inversely as the nth power of x, where k is the constant of variation.

In words, inverse variation is written as *y varies inversely as x* or *y is inversely proportional to x*, and these phrases translate to $y = \dfrac{k}{x}$. Similarly, *y varies inversely as the nth power of x* or *y is inversely proportional to the nth power of x* translates to $y = \dfrac{k}{x^n}$.

Example 5 Boyle's law states that if the temperature is held constant, the volume of a gas in a closed container is inversely proportional to the pressure applied to it. If a gas has a volume of 4 cubic feet when the pressure is 10 pounds per square foot, find the volume when the pressure is 2.5 pounds per square foot.

Understand Because the volume and pressure vary inversely, we can write $V = \dfrac{k}{P}$, where V represents volume and p represents pressure.

Plan Use the fact that the volume is 4 cubic feet when the pressure is 10 pounds per square foot to find the value of the constant, k. Then use this value of the constant to find the volume when the pressure is 2.5 pounds per square foot.

Execute $\quad 4 = \dfrac{k}{10} \qquad$ Substitute for V and P.

$\qquad 10 \cdot 4 = 10 \cdot \dfrac{k}{10} \qquad$ Multiply both sides by 10.

$\qquad 40 = k$

Replacing k with 40 in $V = \dfrac{k}{P}$, we have $V = \dfrac{40}{P}$, which we use to solve for V when P is 2.5 pounds per square foot.

$$V = \dfrac{40}{2.5} \qquad \text{Substitute 40 for } P.$$

$$V = 16$$

Answer With a pressure of 2.5 pounds per square foot, the volume is 16 cubic feet.

Your Turn 5 The intensity of a light varies inversely as the square of the distance from the light source. If the intensity is 150 foot-candles (fc) when the distance is 2 meters, find the intensity if the distance is 4 meters.

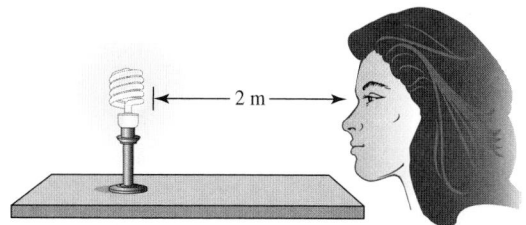

Objective 4 Solve problems involving joint variation.

Often, one quantity will vary as the product of two or more quantities. This is called **joint variation**.

Definition **Joint variation:** If y varies jointly as x and z, then $y = kxz$, where k is the constant of variation.

In words, joint variation is written as *y varies jointly as x and z* or *y is jointly proportional to x and z*, and these phrases translate to $y = kxz$.

Example 6 Suppose y varies jointly with x and z. If $y = 72$ when $x = 4$ and $z = 6$, find y when $x = 2$ and $z = 5$.

Solution:
$$y = kxz \qquad \text{Translate "y varies jointly with x and z."}$$
$$72 = k \cdot 4 \cdot 6 \qquad \text{Replace y with 72, x with 4, and z with 6.}$$
$$72 = 24k \qquad \text{Multiply 4 and 6.}$$
$$3 = k \qquad \text{Divide both sides by 24.}$$

In $y = kxz$, replacing k with 3 gives $y = 3xz$. Now find y when $x = 2$ and $z = 5$.

$$y = 3 \cdot 2 \cdot 5 \qquad \text{In } y = 3xz, \text{ replace x with 2 and z with 5.}$$
$$y = 30 \qquad \text{Multiply.}$$

Your Turn 6 Suppose n varies jointly with p and the square of q. If $n = 90$ when $p = 5$ and $q = 3$, find n when $p = 2$ and $q = 5$.

Objective 5 Solve problems involving combined variation.

Problems involving more than one type of variation are called *combined variation*.

Example 7 Coulomb's law states that the force, F, between two charges q_1 and q_2 in a vacuum varies jointly with the charges and inversely with the square of the distance, d, between them. If the force is 5 dynes when q_1 is 9 electrostatic units and q_2 is 20 electrostatic units and the distance between the charges is 6 centimeters, find the force between the particles if q_1 is 16 electrostatic units, q_2 is 12 electrostatic units, and the distance between the particles is 8 centimeters.

Answer to Your Turn 5
37.5 fc

Answer to Your Turn 6
$n = 100$

Solution: $F = \dfrac{kq_1q_2}{d^2}$ Translate "the force, F, between two charges q_1 and q_2 in a vacuum varies jointly with the charges and inversely with the square of the distance, d, between them."

$5 = \dfrac{k \cdot 9 \cdot 20}{6^2}$ Replace F with 5, q_1 with 9, q_2 with 20, and d with 6.

$5 = \dfrac{180k}{36}$ Multiply.

$5 = 5k$ Simplify.

$1 = k$ Divide both sides by 5.

Replacing k with 1 in $F = \dfrac{kq_1q_2}{d^2}$ gives $F = \dfrac{q_1q_2}{d^2}$. Now find F when $q_1 = 16$, $q_2 = 12$, and $d = 8$.

$F = \dfrac{16 \cdot 12}{8^2}$ In $F = \dfrac{q_1q_2}{d^2}$, replace q_1 with 16, q_2 with 12, and d with 8.

$F = \dfrac{192}{64}$ Multiply.

$F = 3$ Simplify.

Answer: The force is 3 dynes.

$\boxed{\textbf{Your Turn 7}}$ The maximum height, h, obtained by an object that is launched vertically upward varies directly with the square of the velocity, v, and inversely as the acceleration due to gravity, g. If an object that is launched vertically upward with a velocity of 64 feet per second and acceleration due to gravity of 32 feet per second per second obtains a maximum height of 64 feet, find the maximum height obtained by an object launched vertically upward with a velocity of 128 feet per second and acceleration due to gravity of 32 feet per second per second.

Answer to Your Turn 7
256 ft.

7.7 Exercises $\begin{smallmatrix}\text{For}\\\text{Extra}\\\text{Help}\end{smallmatrix}$ MyMathLab®

Note: Exercises marked with a ★ represent challenging exercises.

Objective 1

Prep Exercise 1 If a person can complete a task in x hours, what part of the task can he complete in 1 hour?

$\dfrac{1}{x}$

Prep Exercise 2 If a represents the number of hours for one person to complete a task and b represents the number of hours for a second person to complete the same task, represent the part of the task completed in 1 hour while both people work together.

$\dfrac{1}{a} + \dfrac{1}{b}$

Prep Exercise 3 If a vehicle travels a distance of 100 miles at a rate of r, write an expression for the time, t, it takes the vehicle to travel the 100 miles. Which type of variation is this?

$\dfrac{100}{r}$, inverse

Prep Exercise 4 If Fred can row a canoe x miles per hour in still water and the current in the Wekiva River is 3 miles per hour, write an expression for Fred's rate rowing upstream and an expression for his rate rowing downstream in the river.

Upstream: $x - 3$ mph; downstream: $x + 3$ mph

For Exercises 1–16, use a table to organize the information; then solve. See Examples 1 and 2.

1. Jason can wash and wax his car in 4 hours. His younger sister can wash and wax the same car in 6 hours. Working together, how fast can they wash and wax the car?

$2\frac{2}{5}$ hr.

2. Joe can paint the outside of a house in 6 days working alone. His helper, Frank, can paint the outside of the same house in 9 days working alone. How long will it take them to paint the outside of the house working together?

$3\frac{3}{5}$ days

3. Alicia and Geraldine have volunteered to make quilts for a charity auction. If Alicia can make a quilt in 25 days and Geraldine can make a quilt in 35 days, in how many days can they make a quilt working together?

$14\frac{7}{12}$ days

4. It takes Alice 90 minutes to put a futon frame together, and it takes Maya 60 minutes to put the same type of frame together. If they worked together, how long would it take to put a frame together?

36 min.

5. Working together, it takes two roofers 4 hours to put a new roof on a portable classroom. If the first roofer can do the job by himself in 6 hours, how many hours will it take the second roofer to do the job by himself?

12 hr.

6. Working together, Rita and Tiffany can cut and trim a lawn in 2 hours. If it takes Rita 5 hours to do the lawn by herself, how long will it take Tiffany to do the lawn by herself?

$3\frac{1}{3}$ hr.

7. With both the cold water and hot water faucets open, it takes 9 minutes to fill a bathtub. The cold water faucet alone takes 15 minutes to fill the tub. How long will it take to fill the tub with just the hot water faucet?

22.5 min.

8. The cargo hold of a ship has two loading pipes. Used together, the two pipes can fill the cargo hold in 6 hours. If the larger pipe alone can fill the hold in 8 hours, how many hours will it take the smaller pipe to fill the hold by itself?

24 hr.

9. A bus leaves Valdosta, Georgia, at 10 A.M. traveling north on I-75 at an average rate of 63 miles per hour. At the same time, a car leaves Valdosta traveling south on I-75 at an average rate of 72 miles per hour. At what time will they be 675 miles apart?

3 P.M.

10. Sailing in opposite directions, an aircraft carrier and a destroyer leave their base in Hawaii at 5 A.M. If the destroyer sails at 30 miles per hour and the aircraft carrier sails at 20 miles per hour, at what time will the two ships be 300 miles apart?

11 A.M.

11. Rapid City is 360 miles from Sioux Falls. At 6 A.M., a freight train leaves Rapid City for Sioux Falls, and at the same time, a passenger train leaves Sioux Falls for Rapid City. The two trains meet at 9 A.M. If the freight train travels $\frac{3}{5}$ of the speed of the passenger train, how fast does the passenger train travel?

75 mph

12. Houston and Calgary are about 2100 miles apart. At 2 P.M., an airplane leaves Houston for Calgary, flying 250 miles per hour. At the same time, an airplane leaves Calgary for Houston, flying 450 miles per hour. How long will it be before the two airplanes meet?

3 hr.

13. Jack leaves his home in Atlanta, traveling north on I-75 at an average rate of 45 miles per hour. Two hours later, his wife, Frances, leaves home and takes the same route, traveling at an average rate of 60 miles per hour. How long will it take Frances to catch Jack?

6 hr.

14. A ship leaves port traveling 15 miles per hour. Two hours later, a speedboat leaves the same port traveling 40 miles per hour. How long will it take the speedboat to overtake the ship?

$1\frac{1}{5}$ hr.

15. An airliner flies against the wind from Washington, D.C., to San Francisco in 5.5 hours. It flies back to Washington, D.C., with the wind in 5 hours. If the average speed of the wind is 21 miles per hour, what is the speed of the airliner in still air?

441 mph

16. A river has a current of 3 miles per hour. A boat goes 40 miles upstream in the same time it goes 50 miles downstream. Find the speed of the boat in still water.

27 mph

Objective 2

Prep Exercise 5 Translate "*m* varies directly as *n*."

$m = kn$

Prep Exercise 6 After substituting the initial values into the translated equation, what is the next step?

Find the value of k.

Prep Exercise 7 If *p* varies directly as *q*, then as *q* decreases, what happens to the value of *p*? (Assume that $k > 0$).

p decreases.

For Exercises 17–26, solve the direct variations. See Examples 3 and 4.

17. Suppose *a* varies directly as *b*. If $a = 4$ when $b = 9$, find *a* when $b = 27$.

    12

18. Suppose *r* varies directly as *s*. If $r = 6$ when $s = 9$, find *r* when $s = 15$.

    10

19. Suppose *y* varies directly as the square of *x*. If $y = 100$ when $x = 5$, find *y* when $x = 3$.

    36

20. Suppose *t* varies directly as the square of *u*. If $t = 45$ when $u = 3$, find *t* when $u = -2$.

    20

21. Suppose *m* varies directly as *n*. If $m = 6$ when $n = 8$, what is the value of *n* when $m = 9$?

    12

22. Suppose *x* varies directly as *y*. If $x = 12$ when $y = 15$, what is the value of *y* when $x = 4$?

    5

23. The price of salmon at a fish market is constant, so the cost increases with the quantity purchased. Tamika notices that 2.5 pounds cost $16.25. If she plans to buy 6 pounds of salmon, how much will she pay?

    $39

24. At a produce stand, the cost of zucchini is constant and sells 3 for $0.89. To the nearest cent, find the cost of 5 zucchini.

    $1.48

25. According to Charles's law, if the pressure is held constant, the volume of a gas varies directly with the temperature measured on the Kelvin scale. If the volume of a gas is 288 cubic centimeters when the temperature is 80 K, find the volume when the temperature is 50 K.

    180 cm^3

26. According to Ohm's law, the current, *I*, which is measured in amperes, in an electrical circuit varies directly with the voltage, *V*. If the current is 0.64 ampere when the voltage is 24 volts, find the current when the voltage is 48 volts.

    1.28 amps

In Exercises 27–30, use the fact that, ignoring air resistance, the distance a free-falling body falls is directly proportional to the square of the time it has been falling. See Example 4.

27. On Earth, if an object falls 144 feet in 3 seconds, how many seconds will it take the object to fall 400 feet?

    5 sec.

28. On Earth, if an object falls 39.2 meters in 2 seconds, how far will it fall in 5 seconds?

    245 m

29. On the Moon, an object falls 6.48 meters in 2 seconds. How far will it fall in 5 seconds?

    40.5 m

30. On the Moon, an object falls 14.58 meters in 3 seconds. How long will it take the object to fall 58.32 meters?

    6 sec.

31. The circumference of a circle varies directly with its diameter. If the circumference is 12.56 feet when the diameter is 4 feet, find the diameter when the circumference is 21.98 feet.

    7 ft.

Of Interest

Without air resistance, all objects fall at the same rate. This was demonstrated by astronaut David Scott in 1971 when he dropped a hammer and a feather from the same height on the Moon and they landed on the Moon's surface at the same time.

32. The pressure on an object submerged in a liquid varies directly as the depth beneath the surface. If an object is submerged in gasoline to a depth of 6 feet, the pressure is 253.8 pounds per square foot. Find the pressure if the object is submerged 9 feet.

380.7 lb./ft.²

Objective 3

Prep Exercise 8 Translate "*t* varies inversely as *u*."

$t = \dfrac{k}{u}$

Prep Exercise 9 If *m* is inversely proportional to *n*, then as the *n* quantity increases, what happens to the value of *m*? (Assume that $k > 0$.)

m decreases.

For Exercises 33–40, solve the inverse variations. See Example 5.

33. Suppose *a* varies inversely as *b*. If $a = 3.2$ when $b = 5$, what is *b* when *a* is 8?

2

34. Suppose *p* varies inversely as *q*. If *p* is 8 when *q* is 3.25, what is *q* when *p* is 2?

13

35. Suppose *m* varies inversely as *n*. If *n* is 6 when *m* is 11, what is *n* when *m* is 8?

8.25

36. Suppose *x* varies inversely as *y*. If $y = 12$ when $x = 3$, what is *y* when *x* is 12?

3

37. By Charles's law, if the temperature is held constant, the pressure that a gas exerts against the walls of a container is inversely proportional to the volume of the container. A gas is inside a cylinder with a piston at one end that can vary the volume of the cylinder. When the volume of the cylinder is 20 cubic inches, the pressure inside is 40 psi (pounds per square inch). Find the pressure of the gas if the piston compresses the gas to a volume of 16 cubic inches.

50 psi

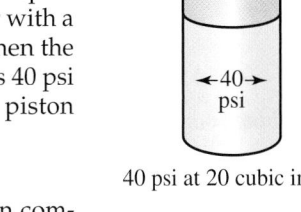
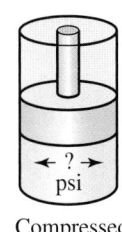

40 psi at 20 cubic inches

Compressed to 16 cubic inches

38. Find the volume in the cylinder from Exercise 37 if the piston compresses the gas to a pressure of 30 psi.

26.7 in.³

39. In an electrical conductor, the current, *I*, which is measured in amperes, varies inversely as the resistance, *R*, which is measured in ohms. If the current is 10 amperes when the resistance is 15 ohms, find the resistance when the current is 25 amperes.

6 Ω

40. In Exercise 39, find the current when the resistance is 15 ohms.

10 A

In Exercises 41 and 42, use the fact that the length of a radio wave varies inversely with its frequency. See Example 5.

41. If the length of a radio wave is 400 meters when the frequency is 900 kilohertz, find the wavelength when the frequency is 600 kilohertz.

600 m

42. If the frequency of a radio wave is 800 kilohertz when the wavelength is 200 meters, find the frequency when the radio wavelength is 500 meters.

320 kHz

43. The f-stop setting for a camera lens is inversely proportional to the aperture, which is the size of the opening in the lens. For a particular lens, the f-stop setting is 5.6 when the aperture is 50 mm. Find the aperture when the f-stop is 2.8.

100 mm

44. The weight of an object varies inversely with the square of the distance from the center of the Earth. At Earth's surface, the weight of the space shuttle is about 4.5 million pounds. At that point, it is 4000 miles from the center of the Earth. How much will the space shuttle weigh when it is in orbit 200 miles above the surface of the Earth?

4.08 million lb.

Objective 4

Prep Exercise 10 Translate "*c* varies jointly with *d* and *e*."

$c = kde$

For Exercises 45–52, solve the joint variations. See Example 6.

45. Suppose *a* varies jointly with *b* and *c*. If *a* = 96 when *b* = 6 and *c* = 4, find *a* when *b* = 2 and *c* = 8.

64

46. Suppose *m* varies jointly with *p* and *q*. If *m* = 70 when *p* = 5 and *q* = 2, find *m* when *p* = 6 and *q* = 8.

336

47. Suppose *a* varies jointly as the square of *b* and *c*. If *a* = 96 when *b* = 2 and *c* = 6, find *a* when *b* = 3 and *c* = 2.

72

48. Suppose *x* varies jointly with *y* and the square of *z*. If *x* = 40 when *y* = 1 and *z* = 2, find *x* when *y* = 2 and *z* = 1.

20

49. If the width of a rectangular solid is held constant, the volume varies jointly with the length and the height. If the volume is 192 cubic inches when the length is 8 inches and the height is 6 inches, find the volume when the length is 7 inches and the height is 12 inches.

336 in.^{3}

50. For a fixed amount of principal, the simple interest varies jointly with the rate and the time. If the simple interest is $1000 when the rate is 4% and the time is 5 years, find the simple interest when the rate is 6% and the time is 10 years.

$3000

51. The volume of a right circular cylinder varies jointly as the square of the radius and the height. If the volume is 301.6 cubic centimeters when the radius is 4 centimeters and the height is 6 centimeters, find the volume when the radius is 3 centimeters and the height is 6 centimeters.

169.65 cm^3

52. The number of units produced varies jointly with the number of workers and the number of hours worked per worker. If 80 units can be produced by 10 workers who work 40 hours each, how many units can be produced by 15 workers who work 30 hours each?

90 units

Objective 5

For Exercises 53–60, solve the combined variations. See Example 7.

53. Suppose *y* varies directly with *x* and inversely as *z*. If *y* = 8 when *x* = 4 and *z* = 6, find *y* when *x* = 5 and *z* = 10.

6

54. Suppose *m* varies directly with *n* and inversely as *p*. If *m* = 27 when *n* = 6 and *p* = 4, find *m* when *n* = 9 and *p* = 6.

27

55. Suppose *y* varies jointly with *x* and *z* and inversely with *n*. If *y* = 81 when *x* = 4, *z* = 9, and *n* = 8, find *n* when *x* = 6, *y* = 8, and *z* = 12.

162

56. Suppose *p* varies jointly with *q* and *r* and inversely with *s*. If *p* = 15 when *q* = 5, *r* = 3, and *s* = 6, find *p* when *q* = 5, *r* = 3, and *s* = 9.

10

57. The resistance of a wire, *R*, varies directly with the length and inversely with the square of the diameter. If the resistance is 7.5 ohms when the wire is 6 meters long and the diameter is 0.02 meter, find the resistance in a wire of the same material if the length is 10 meters and the diameter is 0.04 meter.

3.125 Ω

58. The universal gas law states that the volume of a gas varies directly with the temperature and inversely with the pressure. If the volume of a gas is 1.75 cubic meters when the temperature is 70 K and the pressure is 20 grams per square centimeter, find the volume when the temperature is 80 K and the pressure is 40 grams per square centimeter.

1 m^3

59. Newton's Law of Universal Gravitation states that the force of attraction between two bodies is jointly proportional to their masses and inversely proportional to the square of the distance between them. If the force of attraction between two masses of 4 grams and 6 grams that are 3 centimeters apart is 48 dynes, find the force of attraction between two masses of 2 grams and 12 grams that are 6 centimeters apart.

12 dyn

60. The weight-carrying capacity of a rectangular beam varies jointly with its width and the square of its height and inversely as its length. If a beam is 4 inches wide, 6 inches high, and 10 feet long, it has a carrying capacity of 1400 pounds. Find the carrying capacity of a beam made of the same material if it is 3 inches wide, 5 inches high, and 12 feet long.

607.64 lb.

★ **61.** The table lists the number of miles a car can drive on the given number of gallons of gas.

Number of Miles	Number of Gallons
135	6
180	8
225	10
270	12

   a. Plot the ordered pairs of data as points in the coordinate plane. Connect the points to form a graph.

   b. Are the number of miles driven and the number of gallons of gas directly proportional or inversely proportional? Explain.

      They are directly proportional. As the number of miles increases, so does the number of gallons.

   c. Find the constant of variation. What does it represent?

      $k = 22.5$, which represents the miles per gallon the car gets.

   d. Does the data represent a function? Explain.

      Yes, the data represent a function because for any given number of miles, there will be one quantity of gasoline (assuming that the miles per gallon stays constant).

a.
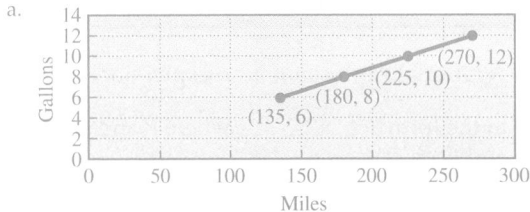

★ **62.** The table lists the time required to make a long trip at various average driving speeds.

Speed (in mph)	Driving Time (in hr.)
60	10
50	12
40	15
30	20
20	30

   a. Plot the ordered pairs of data as points in the coordinate plane. Connect the points to form a graph.

   b. Are speeds and driving times directly proportional or inversely proportional? Explain.

      They are inversely proportional because as speed increases, time decreases.

   c. Find the constant of variation. What does it represent?

      $k = 600$, which represents the total miles driven.

   d. Do the data represent a function? Explain.

      Yes, the data represent a function because for any given speed, there is only one time required to drive the constant 600 miles.

a.
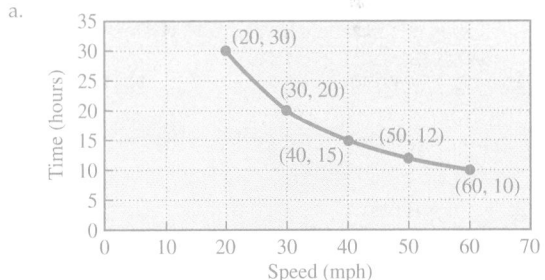

Review Exercises

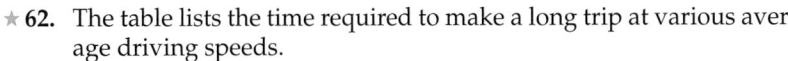

Exercises 1 and 2   Expressions

[1.1] *For Exercises 1 and 2, find the absolute values.*

1. $|-4.3|$

   4.3

2. $\left|\dfrac{5}{8}\right|$

   $\dfrac{5}{8}$

Exercises 3–6   Equations and Inequalities

[2.8] *For Exercises 3 and 4, graph on the number line.*

3. $x \geq 2$

4. $x < -3$

[2.8] *For Exercises 5 and 6, solve.*

5. $2x + 11 > 3$

   $x > -4$

6. $-4x - 3 \leq -11$

   $x \geq 2$

Chapter 7 Summary and Review Exercises

Complete each incomplete formula, definition, rule, or procedure; study the key examples; and then work the related exercises.

7.1 Simplifying Rational Expressions

Definitions/Rules/Procedures	Key Example(s)
A **rational expression** is an expression that can be written in the form $\underline{\frac{P}{Q}}$, where P and Q are $\underline{polynomials}$ and $Q \neq 0$.	$\dfrac{x}{x-4}$ is a rational expression for $x - 4 \neq 0$.
To **evaluate a rational expression,** $\underline{replace}$ the variables with given values and then $\underline{simplify}$ the resulting numerical expression.	**Evaluate** $\dfrac{x}{x-4}$ for $x = 2$. **Solution:** Substitute 2 for x and simplify. $$\frac{x}{x-4} = \frac{2}{2-4} = \frac{2}{-2} = -1$$

Exercises 1 and 2   **Expressions**

[7.1] *For Exercises 1 and 2, evaluate the rational expression.*

1. $\dfrac{4x+5}{2x}$ when

    a. $x = 3$ $\frac{17}{6}$
    b. $x = -1.3$ ≈ 0.077
    c. $x = -2$ $\frac{3}{4}$

2. $\dfrac{5x}{x+2}$ when

    a. $x = 3$ 3
    b. $x = -3$ 15
    c. $x = 1.2$ 1.875

Definitions/Rules/Procedures	Key Example(s)

To determine the value(s) that make a rational expression undefined:
 	
 1. Set the denominator equal to $\underline{0}$.
 	
 2. $\underline{Solve}$ the equation.
	Find every value for the variable that makes the expression undefined.
	
	 a. $\dfrac{5x}{x-6}$
	
	    Solution: $x - 6 = 0$
	                $x = 6$
	
	 b. $\dfrac{n}{n^2 - 4n - 5}$
	
	    Solution: $n^2 - 4n - 5 = 0$
	          $(n-5)(n+1) = 0$
	    $n - 5 = 0$ or $n + 1 = 0$
	       $n = 5$ $n = -1$

Exercises 3–6   **Equations and Inequalities**

[7.1] *For Exercises 3–6, find every value that can replace the variable in the expression and cause the expression to be undefined.*

3. $\dfrac{9y}{5-y}$

     5

4. $\dfrac{2x}{x^2-4}$

     $-2, 2$

5. $\dfrac{6}{x^2+5x-6}$

     $-6, 1$

6. $\dfrac{2x+3}{x^2+6x}$

     $0, -6$

Definitions/Rules/Procedures	Key Example(s)
To simplify a rational expression to lowest terms: 1. ___Factor___ the numerator and denominator completely. 2. Divide out all ___common factors___ in the numerator and denominator. 3. Multiply the ___remaining factors___ in the numerator and the ___remaining factors___ in the denominator.	Simplify. **a.** $\dfrac{18x^3y}{24x^2yz} = \dfrac{2\cdot3\cdot3\cdot x\cdot x\cdot x\cdot y}{2\cdot2\cdot2\cdot3\cdot x\cdot x\cdot y\cdot z}$ $= \dfrac{3\cdot x}{2\cdot2\cdot z}$ $= \dfrac{3x}{4z}$ **b.** $\dfrac{5m^2-15m}{m^2-9} = \dfrac{5\cdot m\cdot(m-3)}{(m+3)(m-3)}$ $= \dfrac{5m}{m+3}$

Exercises 7–14 Expressions

[7.1] *For Exercises 7–14, simplify to lowest terms.*

7. $\dfrac{3(x+5)}{9}$
$\dfrac{x+5}{3}$

8. $\dfrac{2y+12}{3y+18}$
$\dfrac{2}{3}$

9. $\dfrac{x^2+6x}{4x+24}$
$\dfrac{x}{4}$

10. $\dfrac{xy-y^2}{3x-3y}$
$\dfrac{y}{3}$

11. $\dfrac{a^2-b^2}{a^2-2ab+b^2}$
$\dfrac{a+b}{a-b}$

12. $\dfrac{3x+9}{x^4-81}$
$\dfrac{3}{(x^2+9)(x-3)}$

13. $\dfrac{x^2-y^2}{y-x}$
$-(x+y)$

14. $\dfrac{1-w}{w^2+2w-3}$
$\dfrac{1}{w+3}$

7.2 Multiplying and Dividing Rational Expressions

Definitions/Rules/Procedures	Key Example(s)
To multiply rational expressions: 1. Factor each ___numerator___ and ___denominator___ completely. 2. Divide out any numerator ___factor___ with any matching denominator ___factor___. 3. Multiply numerator by ___numerator___ and denominator by ___denominator___. 4. Simplify as needed.	Multiply. **a.** $-\dfrac{8u^3}{3t}\cdot\dfrac{9t^2}{10u^3} = -\dfrac{2\cdot2\cdot2\cdot u\cdot u\cdot u}{3\cdot t}\cdot\dfrac{3\cdot3\cdot t\cdot t}{2\cdot5\cdot u\cdot u\cdot u}$ $= -\dfrac{2\cdot2}{1}\cdot\dfrac{3\cdot t}{5}$ $= -\dfrac{12t}{5}$ **b.** $\dfrac{x^2+4x}{x^2-3x-10}\cdot\dfrac{3x+6}{12x}$ $= \dfrac{x\cdot(x+4)}{(x+2)\cdot(x-5)}\cdot\dfrac{3\cdot(x+2)}{2\cdot2\cdot3\cdot x}$ $= \dfrac{(x+4)}{(x-5)}\cdot\dfrac{1}{2\cdot2}$ $= \dfrac{x+4}{4(x-5)}$ or $\dfrac{x+4}{4x-20}$

Exercises 15–20 ➤ Expressions

[7.2] *For Exercises 15–20, multiply.*

15. $-\dfrac{2n^2}{8m^2n} \cdot \dfrac{24mp}{10n^2p^5}$

$\dfrac{3}{5mnp^4}$

16. $\dfrac{6}{2m + 4} \cdot \dfrac{3m + 6}{15}$

$\dfrac{3}{5}$

17. $\dfrac{n^2 - 6n + 9}{n^2 - 9} \cdot \dfrac{n^2 + 4n + 3}{n^2 - 3n}$

$\dfrac{n + 1}{n}$

18. $\dfrac{4r^2 + 4r + 1}{r + 2r^2} \cdot \dfrac{2r}{2r^2 - r - 1}$

$\dfrac{2}{r - 1}$

19. $\dfrac{4m}{m^2 - 2m + 1} \cdot \dfrac{m^2 - 1}{16m^2}$

$\dfrac{m + 1}{4m(m - 1)}$

20. $\dfrac{w^2 - 2w - 24}{w^2 - 16} \cdot \dfrac{(w + 5)(w + 4)}{w^2 - w - 30}$

$\dfrac{w + 4}{w - 4}$

Definitions/Rules/Procedures	Key Example(s)
To divide rational expressions: 1. Write an equivalent multiplication statement with the ___reciprocal___ of the divisor. 2. Factor each ___numerator___ and ___denominator___ completely. 3. Divide out any numerator ___factor___ with any matching denominator ___factor___. 4. Multiply numerator by ___numerator___ and denominator by ___denominator___. 5. Simplify as needed.	Divide. **a.** $\dfrac{3n^3}{14m} \div \dfrac{12mn^2}{7m} = \dfrac{3n^3}{14m} \cdot \dfrac{7m}{12mn^2}$ $= \dfrac{3 \cdot n \cdot n \cdot n}{2 \cdot 7 \cdot m} \cdot \dfrac{7 \cdot m}{2 \cdot 2 \cdot 3 \cdot m \cdot n \cdot n}$ $= \dfrac{n}{2} \cdot \dfrac{1}{2 \cdot 2 \cdot m}$ $= \dfrac{n}{8m}$ **b.** $\dfrac{r - 4}{3r^2 - 17r + 10} \div \dfrac{r^2 - 3r - 4}{3r^2 + r - 2}$ $= \dfrac{r - 4}{3r^2 - 17r + 10} \cdot \dfrac{3r^2 + r - 2}{r^2 - 3r - 4}$ $= \dfrac{r - 4}{(3r - 2) \cdot (r - 5)} \cdot \dfrac{(3r - 2) \cdot (r + 1)}{(r - 4) \cdot (r + 1)}$ $= \dfrac{1}{r - 5}$

Exercises 21–26 ➤ Expressions

[7.2] *For Exercises 21–26, divide.*

21. $-\dfrac{7y^2}{10b^2} \div -\dfrac{21y^2}{25b^2}$

$\dfrac{5}{6}$

22. $\dfrac{2a + 4}{5} \div \dfrac{4a + 8}{25a}$

$\dfrac{5a}{2}$

23. $\dfrac{x^2 - y^2}{c^2 - d^2} \div \dfrac{x - y}{c + d}$

$\dfrac{x + y}{c - d}$

24. $\dfrac{b^3 - 6b^2 + 8b}{6b} \div \dfrac{2b - 4}{10b + 40}$

$\dfrac{5(b + 4)(b - 4)}{6}$

25. $\dfrac{3j + 2}{5j^2 - j} \div \dfrac{6j^2 + j - 2}{10j^2 + 3j - 1}$

$\dfrac{2j + 1}{j(2j - 1)}$

26. $\dfrac{u^2 - 2u - 8}{u^2 + 3u + 2} \div (u^2 - 3u - 4)$

$\dfrac{1}{(u + 1)^2}$

Definitions/Rules/Procedures	Key Example(s)
To **convert units using dimensional analysis,** multiply the given measurement by the ___conversion factors___ so that the undesired units divide out, leaving the desired units.	Convert. **a.** 20 feet to inches $$20 \text{ ft.} = \frac{20 \text{ ft.}}{1} \cdot \frac{12 \text{ in.}}{1 \text{ ft.}} = 240 \text{ in.}$$ **b.** 6 miles to yards $$6 \text{ mi.} = \frac{6 \text{ mi.}}{1} \cdot \frac{5280 \text{ ft.}}{1 \text{ mi.}} \cdot \frac{1 \text{ yd.}}{3 \text{ ft.}} = 10{,}560 \text{ yd.}$$

Exercises 27–30 Expressions

[7.2] *For Exercises 27–30, use dimensional analysis to convert to the given unit.*

27. 42 inches to feet
   3.5 ft.

28. 2.5 miles to feet
   13,200 ft.

29. 68 ounces to pounds
   4.25 lb.

30. 4 tons to ounces
   128,000 oz.

7.3 Adding and Subtracting Rational Expressions with the Same Denominator

Definitions/Rules/Procedures	Key Example(s)
To add or subtract rational expressions that have the same denominators: **1.** Add or subtract the ___numerators___ and keep the same denominator. **2.** Simplify to lowest terms. (Remember to factor the numerators and denominators completely to simplify.)	Add or subtract. **a.** $\dfrac{7y}{12} + \dfrac{y}{12} = \dfrac{7y + y}{12} = \dfrac{8y}{12} = \dfrac{2y}{3}$ **b.** $\dfrac{u}{u + 2} - \dfrac{3}{u + 2} = \dfrac{u - 3}{u + 2}$ **c.** $\dfrac{7x^2 + 5x}{x + 1} - \dfrac{x + 3}{x + 1} = \dfrac{(7x^2 + 5x) - (x + 3)}{x + 1}$ $\qquad = \dfrac{(7x^2 + 5x) + (-x - 3)}{x + 1}$ $\qquad = \dfrac{7x^2 + 4x - 3}{x + 1}$ $\qquad = \dfrac{(7x - 3)(x + 1)}{x + 1}$ $\qquad = 7x - 3$

Exercises 31–36 Expressions

[7.3] *For Exercises 31–36, add or subtract as indicated.*

31. $\dfrac{5x}{18} + \dfrac{x}{18}$
   $\dfrac{x}{3}$

32. $\dfrac{8m + 5}{6m^2} - \dfrac{10m + 5}{6m^2}$
   $\dfrac{1}{3m}$

33. $\dfrac{2x}{x + y} - \dfrac{x - y}{x + y}$
   1

34. $\dfrac{2r - 5s}{r^2 - s^2} - \dfrac{2r - 6s}{r^2 - s^2}$
   $\dfrac{s}{r^2 - s^2}$

35. $\dfrac{9x^2 - 24x + 16}{2x + 1} + \dfrac{2x^2 + 3x - 9}{2x + 1}$
   $\dfrac{11x^2 - 21x + 7}{2x + 1}$

36. $\dfrac{q^2 + 2q}{q - 7} - \dfrac{12q - 21}{q - 7}$
   $q - 3$

7.4 Adding and Subtracting Rational Expressions with Different Denominators

Definitions/Rules/Procedures	Key Example(s)
To find the LCD of two or more rational expressions:	Find the LCD.
1. Find the <u>prime factorization</u> of each denominator.	**a.** $\dfrac{5}{6t}$ and $\dfrac{1}{8t^2}$
2. Write a product that contains each unique prime factor the <u>greatest number</u> of times it occurs in the factorizations. (If using exponential forms, use each prime raised to its <u>greatest exponent</u> in the factorizations.)	$$6t = 2 \cdot 3 \cdot t$$ $$8t^2 = 2^3 \cdot t^2$$ $$\text{LCD} = 2^3 \cdot 3 \cdot t^2 = 24t^2$$
3. Simplify the product found in step 2.	**b.** $\dfrac{2}{5t - 15}$ and $\dfrac{t}{t^2 - 9}$
	$$5t - 15 = 5(t - 3)$$ $$t^2 - 9 = (t - 3)(t + 3)$$ $$\text{LCD} = 5(t - 3)(t + 3)$$
To write two or more rational expressions with the LCD, multiply both the numerator and denominator of each rational expression by an appropriate factor so that the <u>denominator</u> becomes the LCD.	Write as equivalent rational expressions with the LCD.
	a. $\dfrac{5}{6t}$ and $\dfrac{1}{8t^2}$
	The LCD is $24t^2$.
	$$\frac{5}{6t} = \frac{5 \cdot 4t}{6t \cdot 4t} = \frac{20t}{24t^2}$$
	$$\frac{1}{8t^2} = \frac{1 \cdot 3}{8t^2 \cdot 3} = \frac{3}{24t^2}$$
	b. $\dfrac{2}{5t - 15}$ and $\dfrac{t}{t^2 - 9}$
	The LCD is $5(t - 3)(t + 3)$.
	$$\frac{2}{5t - 15} = \frac{2 \cdot (t + 3)}{5(t - 3) \cdot (t + 3)} = \frac{2(t + 3)}{5(t - 3)(t + 3)}$$
	$$\frac{t}{t^2 - 9} = \frac{t \cdot 5}{(t - 3)(t + 3) \cdot 5} = \frac{5t}{5(t - 3)(t + 3)}$$

Exercises 37–42 ▰▰▱ **Expressions**

[7.4] *For Exercises 37–42, write equivalent rational expressions with their LCD.*

37. $\dfrac{2}{4c}, \dfrac{3}{3c^3}$

$\dfrac{6c^2}{12c^3}, \dfrac{12}{12c^3}$

38. $\dfrac{5}{xy^3}, \dfrac{3}{x^2y^2}$

$\dfrac{5x}{x^2y^3}, \dfrac{3y}{x^2y^3}$

39. $\dfrac{4}{m - 1}, \dfrac{4y}{m + 1}$

$\dfrac{4m + 4}{(m - 1)(m + 1)}, \dfrac{4my - 4y}{(m - 1)(m + 1)}$

40. $\dfrac{3}{4h - 8}, \dfrac{5}{h^2 - 2h}$

$\dfrac{3h}{4h(h - 2)}, \dfrac{20}{4h(h - 2)}$

41. $\dfrac{x}{x^2 - 1}, \dfrac{2 + x}{x + 1}$

$\dfrac{x}{(x + 1)(x - 1)}, \dfrac{x^2 + x - 2}{(x + 1)(x - 1)}$

42. $\dfrac{2}{p^2 - 4}, \dfrac{p}{p^2 + 4p + 4}$

$\dfrac{2p + 4}{(p - 2)(p + 2)^2}, \dfrac{p^2 - 2p}{(p - 2)(p + 2)^2}$

Definitions/Rules/Procedures	Key Example(s)
To add or subtract rational expressions with different denominators: 1. Find the _____LCD_____ . 2. Write each rational expression as a(n) _equivalent expression_ with the LCD. 3. Add or subtract the _numerators_ and keep the LCD. 4. Simplify.	Add or subtract. **a.** $\dfrac{5}{6t} + \dfrac{1}{8t^2} = \dfrac{5(4t)}{6t(4t)} + \dfrac{1(3)}{8t^2(3)}$ $= \dfrac{20t}{24t^2} + \dfrac{3}{24t^2}$ $= \dfrac{20t + 3}{24t^2}$ **b.** $\dfrac{2}{5t - 15} - \dfrac{t}{t^2 - 9}$ $= \dfrac{2}{5(t - 3)} - \dfrac{t}{(t - 3)(t + 3)}$ $= \dfrac{2(t + 3)}{5(t - 3)(t + 3)} - \dfrac{t(5)}{(t - 3)(t + 3)(5)}$ $= \dfrac{2t + 6}{5(t - 3)(t + 3)} - \dfrac{5t}{5(t - 3)(t + 3)}$ $= \dfrac{2t + 6 - 5t}{5(t - 3)(t + 3)}$ $= \dfrac{-3t + 6}{5(t - 3)(t + 3)}$

Exercises 43–48   Expressions

[7.4] *For Exercises 43–48, add or subtract as indicated.*

43. $\dfrac{3a - 4}{18} - \dfrac{2a + 5}{12}$

$-\dfrac{23}{36}$

44. $\dfrac{3}{y - 2} + \dfrac{5}{y - 4}$

$\dfrac{8y - 22}{(y - 2)(y - 4)}$

45. $\dfrac{6t}{(t - 3)^2} - \dfrac{3t}{2t - 6}$

$\dfrac{-3t^2 + 21t}{2(t - 3)^2}$

46. $\dfrac{a + 6}{a^2 + 7a + 12} - \dfrac{a - 3}{a + 3}$

$\dfrac{18 - a^2}{(a + 3)(a + 4)}$

47. $\dfrac{10}{2y - 1} - \dfrac{5}{1 - 2y}$

$\dfrac{15}{2y - 1}$

48. $\dfrac{3}{3x - 12} + \dfrac{15}{x^2 - 16}$

$\dfrac{x + 19}{(x + 4)(x - 4)}$

7.5 Complex Rational Expressions

Definitions/Rules/Procedures	Key Example(s)
A **complex rational expression** is a rational expression that contains _rational expressions_ in the numerator or denominator. **To simplify a complex rational expression,** use one of the following methods.	

Definitions/Rules/Procedures	Key Example(s)

Method 1

1. Simplify the __numerator__ and __denominator__ if needed.

2. Rewrite as a horizontal __division problem__.

Method 2

1. Multiply the numerator and denominator of the complex rational expression by their __LCD__.

2. Simplify.

Simplify $\dfrac{\dfrac{5t}{t+1}}{\dfrac{10}{t^2-1}}$.

$$\dfrac{\dfrac{5t}{t+1}}{\dfrac{10}{t^2-1}} = \dfrac{5t}{t+1} \div \dfrac{10}{t^2-1} \qquad \text{Using method 1}$$

$$= \dfrac{5t}{t+1} \cdot \dfrac{t^2-1}{10}$$

$$= \dfrac{5 \cdot t}{t+1} \cdot \dfrac{(t+1)(t-1)}{2 \cdot 5}$$

$$= \dfrac{t}{1} \cdot \dfrac{t-1}{2}$$

$$= \dfrac{t^2-t}{2}$$

Simplify $\dfrac{\dfrac{5}{6} - \dfrac{t}{2}}{\dfrac{t}{4} + \dfrac{1}{3}}$.

$$\dfrac{\dfrac{5}{6} - \dfrac{t}{2}}{\dfrac{t}{4} + \dfrac{1}{3}} = \dfrac{12\left(\dfrac{5}{6} - \dfrac{t}{2}\right)}{12\left(\dfrac{t}{4} + \dfrac{1}{3}\right)} \qquad \text{Using method 2}$$

$$= \dfrac{\dfrac{\overset{2}{\cancel{12}}}{1}\cdot\dfrac{5}{\cancel{6}} - \dfrac{\overset{6}{\cancel{12}}}{1}\cdot\dfrac{t}{\cancel{2}}}{\dfrac{\overset{3}{\cancel{12}}}{1}\cdot\dfrac{t}{\cancel{4}} + \dfrac{\overset{4}{\cancel{12}}}{1}\cdot\dfrac{1}{\cancel{3}}}$$

$$= \dfrac{10-6t}{3t+4}$$

Exercises 49–56 ◢ **Expressions**

[7.5] *For Exercises 49–56, simplify.*

49. $\dfrac{\dfrac{a}{b}}{\dfrac{x}{y}}$
$\dfrac{ay}{bx}$

50. $\dfrac{\dfrac{1}{3} - \dfrac{1}{2}}{\dfrac{1}{2} - \dfrac{1}{3}}$
-1

51. $\dfrac{\dfrac{5}{3x+5}}{\dfrac{x}{x-2}}$
$\dfrac{5(x-2)}{x(3x+5)}$

52. $\dfrac{\dfrac{x+3}{x^2-9}}{x}$
$\dfrac{x}{x-3}$

53. $\dfrac{\dfrac{x}{4x^2-1}}{\dfrac{5}{2x+1}}$
$\dfrac{x}{5(2x-1)}$

54. $\dfrac{r - \dfrac{r^2-1}{r}}{1 - \dfrac{r-1}{r}}$
1

55. $\dfrac{\dfrac{1}{x^2} + \dfrac{1}{y^2}}{\dfrac{7}{xy}}$
$\dfrac{y^2+x^2}{7xy}$

56. $\dfrac{\dfrac{1}{h+1} - 1}{\dfrac{1}{h+1}}$
$-h$

7.6 Solving Equations Containing Rational Expressions

Definitions/Rules/Procedures	Key Example(s)
An **extraneous solution** is an apparent solution that <u>does not solve the equation</u>. **To solve an equation that contains rational expressions:** 1. Eliminate the rational expressions by multiplying both sides of the equation by their <u> LCD </u>. 2. <u> Solve </u> the equation using the methods we learned in Chapters 2 (linear equations) and 6 (quadratic equations). 3. <u> Check </u> your solution(s) in the original equation. Discard any extraneous solutions.	Solve $$\frac{x^2 - 6}{x - 3} + 2 = \frac{x^2 - 2x}{x - 3} - x.$$ Note that x cannot be 3 because it would cause the denominators to be 0, making those expressions undefined. $$(x - 3)\left(\frac{x^2 - 6}{x - 3} + 2\right) = (x - 3)\left(\frac{x^2 - 2x}{x - 3} - x\right)$$ $$(x - 3) \cdot \frac{x^2 - 6}{x - 3} + (x - 3) \cdot 2$$ $$= (x - 3) \cdot \frac{x^2 - 2x}{x - 3} - (x - 3) \cdot x$$ $$x^2 - 6 + 2x - 6 = x^2 - 2x - x^2 + 3x$$ $$x^2 + 2x - 12 = x$$ $$x^2 + x - 12 = 0$$ $$(x - 3)(x + 4) = 0$$ $$x - 3 = 0 \quad \text{or} \quad x + 4 = 0$$ $$x = 3 \qquad\qquad x = -4$$ 3 is extraneous. -4 is the solution.

Exercises 57–62   Equations and Inequalities

[7.6] *For Exercises 57–62, solve and check. Identify any extraneous solutions.*

57. $\dfrac{3x}{2} - \dfrac{x}{5} = \dfrac{19}{5}$

$\dfrac{38}{13}$

58. $\dfrac{y + 5}{y - 3} - 5 = \dfrac{4}{y - 3}$

4

59. $\dfrac{x}{x - 2} = \dfrac{6}{x - 1}$

3, 4

60. $\dfrac{6g - 5}{6} = \dfrac{2g - 1}{2} - \dfrac{g + 2}{2g + 5}$

-1

61. $\dfrac{1}{p - 4} + \dfrac{1}{4} = \dfrac{8}{p^2 - 16}$

-8, (4 is extraneous)

62. $\dfrac{3}{r - 2} - \dfrac{4}{r + 3} = -\dfrac{6}{r^2 + r - 6}$

23

7.7 Applications with Rational Expressions, Including Variation

Definitions/Rules/Procedures	Key Example(s)
If y varies directly as x, then $y = $ <u> kx </u>. If y varies directly as the nth power of x, then $y = $ <u> kx^n </u>, where k is the constant of variation.	The distance a car can travel varies directly with the number of gallons of gas used. On the highway, a Toyota Corolla traveled 280 miles using 8 gallons of gas. How far can the same car travel on 12 gallons of gas? **Solution:** $d = kn$ — Translate "the distance a car can travel varies directly as the number of gallons of gas used." $280 = k(8)$ — Substitute 280 for d and 8 for n. $\dfrac{280}{8} = \dfrac{8k}{8}$ — Divide both sides by 8. $35 = k$

Definitions/Rules/Procedures	Key Example(s)
If y varies inversely as x, then $y = \underline{\quad \frac{k}{x} \quad}$. If y varies inversely as the nth power of x, then $y = \underline{\quad \frac{k}{x^n} \quad}$, where k is the constant of variation.	In $d = kn$, replacing k with 35 gives $d = 35n$. Now find d when $n = 12$. $$d = (35)(12) = 420 \text{ miles}$$ The intensity of light is inversely proportional to the square of the distance from the source. A light meter 6 feet from a light bulb measures the intensity to be 16 foot-candles. What is the intensity 24 feet from the bulb? **Solution:** $$I = \frac{k}{d^2} \quad \text{Translate "the intensity of light is inversely proportional to the square of the distance from the source."}$$ $$16 = \frac{k}{6^2} \quad \text{Substitute 16 for } I \text{ and 6 for } d.$$ $$36(16) = 36 \cdot \frac{k}{36} \quad \text{Multiply both sides by 36.}$$ $$576 = k$$ In $I = \frac{k}{d^2}$, replacing k with 576 gives $I = \frac{576}{d^2}$. Now Find I when $d = 24$. $$I = \frac{576}{24^2} \quad \text{In } I = \frac{576}{d^2}, \text{ replace } d \text{ with 24.}$$ $$I = \frac{576}{576} = 1 \text{ foot-candle}$$
If y varies jointly as x and z, then $y = \underline{\quad kxz \quad}$, where k is the constant of variation.	Suppose m varies jointly as n and p. If $m = 288$ when $n = 3$ and $p = 6$, find p when $m = 240$ and $n = 5$. **Solution:** $\quad m = knp \quad$ Translate "m varies jointly with m and p." $288 = k(3)(6) \quad$ Substitute 3 for n and 6 for p. $288 = 18k \quad$ Divide both sides by 18. $\quad 16 = k$ In $m = knp$, replacing k with 16 gives $m = 16np$. Now find p when $m = 240$ and $n = 5$. $240 = 16(5)(p) \quad$ In $m = 16np$, replace m with 240 and n with 5. $240 = 80p \quad$ Divide both sides by 80. $\quad 3 = p$

Exercises 63–71 ◣ Equations and Inequalities

[7.7] *For Exercises 63–71, solve.*

63. Suppose p varies directly as q. If $p = 24$ when $q = 4$, find p when $q = 7$.

42

64. Suppose s varies directly as the square of t. If $s = 72$ when $t = 3$, find s when $t = 5$.

200

65. Suppose y varies inversely as x. If $y = 8$ when $x = 3$, find x when $y = 4$.

6

66. Suppose m varies inversely as the square of n. If $m = 16$ when $n = 3$, find m when $n = 4$.

9

67. Suppose y varies jointly with x and z. If $y = 40$ when $x = 2$ and $z = 5$, find y when $x = 4$ and $z = 2$.

32

68. Suppose m varies directly with n and inversely with p. If $m = 2$ when $n = 3$ and $p = 9$, find m when $n = 6$ and $p = 4$.

9

69. The distance a car can travel varies directly with the amount of gas it carries. On a trip, a Chevy Corvette travels 156 miles using 6 gallons of fuel. How many gallons are required to travel 234 miles?

9 gal.

70. If the wavelength of a wave remains constant, the velocity, v, of a wave is inversely proportional to its period, T. In an experiment, waves are created in a fluid so that the period is 7 seconds and the velocity is 4 centimeters per second. If the period is increased to 12 seconds, what is the velocity?

$\dfrac{7}{3}$ cm/sec.

71. The volume of a right circular cylinder varies jointly with the radius squared and the height. If the volume is 62.8 cubic inches when the radius is 2 inches and the height is 5 inches, find the volume when the radius is 4 inches and the height is 2 inches.

100.48 in.^{3}

Chapter 7 Practice Test

For Extra Help

Step-by-step test solutions are found on the Chapter Test Prep Videos available in MyMathLab® *or on* You Tube.

For Exercises 1 and 2, find the value(s) that can replace the variable in the expression and cause the expression to be undefined.

1. $\dfrac{2x}{x - 7}$

7 [7.1]

2. $\dfrac{8 - m}{m^2 - 16}$

$-4, 4$ [7.1]

For Exercises 3 and 4, simplify to lowest terms.

3. $\dfrac{12 - 3x}{x^2 - 8x + 16}$

$-\dfrac{3}{x - 4}$ [7.1]

4. $\dfrac{x - y}{x^2 - y^2}$

$\dfrac{1}{x + y}$ [7.1]

For Exercises 5 and 6, write equivalent rational expressions with their LCD.

5. $\dfrac{6}{fg^3}, \dfrac{2}{f^2}$

$\dfrac{6f}{f^2g^3}, \dfrac{2g^3}{f^2g^3}$ [7.4]

6. $\dfrac{5x}{x^2 - 9}, \dfrac{2}{x^2 + 6x + 9}$

$\dfrac{5x(x + 3)}{(x + 3)^2(x - 3)}, \dfrac{2(x - 3)}{(x + 3)^2(x - 3)}$ [7.4]

For Exercises 7–18, perform the indicated operation.

7. $\dfrac{4a^2b^3}{15x^3y} \cdot \dfrac{25x^5y}{16ab}$

$\dfrac{5ab^2x^2}{12}$ [7.2]

8. $-\dfrac{9x^3y^4}{16ab^2} \div \dfrac{45x^5y^2}{14a^7b^9}$

$-\dfrac{7a^6b^7y^2}{40x^2}$ [7.2]

9. $\dfrac{4ab - 8b}{x^2} \div \dfrac{2b - ab}{x^3}$

$-4x$ [7.2]

10. $\dfrac{12x^2 - 6x}{x^2 + 6x + 5} \cdot \dfrac{2x^2 + 10x}{4x^2 - 1}$

$\dfrac{12x^2}{(x+1)(2x+1)}$ [7.2]

11. $\dfrac{2x}{2x + 3} + \dfrac{5x}{2x + 3}$

$\dfrac{7x}{2x + 3}$ [7.3]

12. $\dfrac{x}{x^2 - 9} - \dfrac{3}{x^2 - 9}$

$\dfrac{1}{x + 3}$ [7.3]

13. $\dfrac{6}{x^2 - 4} + \dfrac{x - 3}{x^2 - 2x}$

$\dfrac{x^2 + 5x - 6}{x(x+2)(x-2)}$ [7.4]

14. $\dfrac{4}{x - 1} + \dfrac{5}{x + 2}$

$\dfrac{9x + 3}{(x-1)(x+2)}$ [7.4]

15. $\dfrac{5}{x - 4} - \dfrac{2}{x + 1}$

$\dfrac{3x + 13}{(x-4)(x+1)}$ [7.4]

16. $\dfrac{x}{2x + 4} - \dfrac{2}{x^2 + 2x}$

$\dfrac{x - 2}{2x}$ [7.4]

17. $\dfrac{2 + \dfrac{1}{y}}{3 - \dfrac{1}{y}}$

$\dfrac{2y + 1}{3y - 1}$ [7.5]

18. $\dfrac{\dfrac{x - y}{2}}{\dfrac{x^2 - y^2}{4}}$

$\dfrac{2}{x + y}$ [7.5]

For Exercises 19–22, solve the equation. Identify any extraneous solutions.

19. $\dfrac{3x - 1}{4} - \dfrac{7}{6} = \dfrac{2}{3}$

$\dfrac{25}{9}$ [7.6]

20. $\dfrac{4}{5m - 1} = \dfrac{2}{2m - 1}$

-1 [7.6]

21. $\dfrac{y}{y + 2} - \dfrac{y + 6}{y^2 - 4} + \dfrac{2}{y - 2} = 0$

-1, (2 is extraneous) [7.6]

22. $\dfrac{8}{g - 1} = \dfrac{8}{g + 2}$

2, 4 [7.6]

For Exercises 23–25, solve.

23. If Erin can paint a room in 3 hours and her husband can paint the same room in 2 hours, how long will it take them to paint the room together?
1.2 hr. [7.7]

24. Jake runs 3 miles away from the gym, turns around, and runs an average of 2 miles per hour slower on the return trip. If the total time of his run is $1\dfrac{1}{4}$ hours, what was his speed in the outbound leg and the inbound leg of his run?
6 mph, 4 mph [7.7]

25. The weight of an object is directly proportional to its mass. Suppose an object with a mass of 40 kilograms weighs 392 newtons. How much would an object with a mass of 54 kilograms weigh?
529.2 N [7.7]

Chapters 1–7 Cumulative Review Exercises

For Exercises 1–3, answer true or false.

[2.2] **1.** The equation $y = x^2 - 3$ is a linear equation.

false

[3.3] **2.** The x-intercept for $y = 2x$ is $(0, 2)$.

false

[7.4] **3.** The LCD of $\dfrac{2}{x^2 - 25}$ and $\dfrac{4x}{x + 5}$ is $(x^2 - 25)(x + 5)$.

false

For Exercises 4–7, fill in the blank.

[7.3] **4.** To add or subtract rational expressions that have the same denominator:

    a. Add or subtract the numerators and keep the same ___denominator___.

    b. Simplify to lowest terms. (Remember to factor the numerators and denominators completely to simplify.)

[7.1] **5.** Explain in your own words what it means to simplify a rational expression to lowest terms.

There are no common factors in the numerator and denominator besides 1.

[5.4] **6.** To multiply monomials:

    a. _____Multiply_____ coefficients.

    b. _____Add_____ the exponents of the like bases.

    c. Write any unlike bases unchanged in the product.

[7.5] **7.** A ___complex rational expression___ is a rational expression that contains rational expressions in the numerator or denominator.

Exercises 8–19 ➡ **Expressions**

[1.5] *For Exercises 8 and 9, simplify.*

8. $5^2 - (-4^2) - 20$

21

9. $16 \div (-8) + 2^3$

6

For Exercises 10 and 11, simplify.

[5.6] 10. $\dfrac{15a^4 b}{-5a^{-9}b}$

$-3a^{13}$

[5.5] 11. $(m + 3)(m^2 - 6m + 1)$

$m^3 - 3m^2 - 17m + 3$

For Exercises 12–14, factor completely.

[6.1] 12. $ax - 2x + a - 2$

$(a - 2)(x + 1)$

[6.3] 13. $3y^2 - 3y - 60$

$3(y - 5)(y + 4)$

[6.5] 14. $-3h^2 + 75$

$-3(h + 5)(h - 5)$

For Exercises 15–19, simplify.

[7.2] 15. $\dfrac{3x-6}{4} \cdot \dfrac{12}{x-2}$

9

[7.2] 16. $\dfrac{8}{m-4} \div \dfrac{4}{m^2-16}$

$2(m+4)$

[7.3] 17. $\dfrac{x}{x^2-9} - \dfrac{3}{x^2-9}$

$\dfrac{1}{x+3}$

[7.4] 18. $\dfrac{4}{x^2-49} + \dfrac{2}{x^2-7x}$

$\dfrac{6x+14}{x(x+7)(x-7)}$

[7.5] 19. $\dfrac{\dfrac{y^2-y-12}{4y}}{\dfrac{y-4}{8}}$

$\dfrac{2(y+3)}{y}$

Exercises 20–30 ▲ **Equations and Inequalities**

[2.3] 20. Solve and check: $20 + 24x = 8 - 3(5 - 9x)$

9

[3.4] 21. For the equation $y = \dfrac{2}{3}x + 1$,

   a. Find the slope.

   $\dfrac{2}{3}$

   b. Graph.

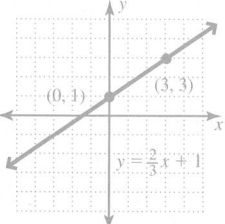

$(0, 1)$ $(3, 3)$ $y = \frac{2}{3}x + 1$

[6.6] 22. Solve $2x^2 - 13x = 7$.

$-\dfrac{1}{2}, 7$

[2.4] 23. Solve $V = \dfrac{1}{3}\pi r^2 h$ for h.

$h = \dfrac{3V}{\pi r^2}$

[7.6] *For Exercises 24 and 25, solve.*

24. $\dfrac{u^2}{u-3} - \dfrac{12}{u-3} = 8$

2, 6

25. $\dfrac{12}{r} = \dfrac{4}{r-4}$

6

[4.3] 26. Solve the system $\begin{cases} 3x + 2y = 2 \\ 4x - 3y = 14 \end{cases}$ using the elimination method.

$(2, -2)$

For Exercises 27–30, solve.

[4.2] 27. Two planes leaving Seattle at the same time are going in opposite directions.
[4.3] If the first plane travels 600 miles per hour and the second plane travels 450 miles per hour, how long will it be until the two planes are 2625 miles apart?
2.5 hr.

[4.2] 28. Jackson invested $7000. He put some of the $7000 in an account earning 9%
[4.3] interest annually and the rest in an account earning 7%. If he earned $600 in interest after one year, how much did he invest at each rate?
$5500 at 9% and $1500 at 7%

[7.7] 9. Jake can clean the garage in 8 hours, and his brother can do it in 6 hours. Working together, how long will it take them to clean the garage?

$3\dfrac{3}{7}$ hr.

[7.7] 30. If y varies directly as x and $y = 7$ when $x = 2$, find x when y is 21.

6

More on Inequalities, Absolute Value, and Functions

Chapter Overview

In this chapter, we build on what we have learned about inequalities, absolute value, and functions. We consider the following topics:

▶ Solve compound inequalities.

▶ Solve absolute value equations and graph absolute value functions.

▶ Solve absolute value inequalities.

We will see that compound inequalities involve more than one inequality. These compound inequalities are used in solving absolute value inequalities.

Instructor Note

The topics in Chapter 8 are important in the study of calculus and other advanced mathematics courses. Sections 8.1 and 8.2 build the foundation for Section 8.3. Graphing absolute value functions is a third objective in Section 8.2, allowing it to be eliminated easily if you do not include functions in your course.

8.1 Compound Inequalities

8.2 Equations Involving Absolute Value; Absolute Value Functions

8.3 Inequalities Involving Absolute Value

8.1 Compound Inequalities

Objectives

1 Solve compound inequalities involving *and*.

2 Solve compound inequalities involving *or*.

Warm-up

[2.8] *For Exercises 1 and 2, solve the inequality.*

1. $2x + 4 \leq 14$ **2.** $-3x + 7 > -2$

In Section 2.8, we learned to solve linear inequalities and represent their solution sets with set-builder notation, interval notation and graphically on number lines. For example, the solution set for $x > 3$ is represented in those three ways here.

Set-builder notation: $\{x \mid x > 3\}$

Interval notation: $(3, \infty)$

Graph:

Note Remember that the parenthesis indicates that 3 is **not** included in the solution set.

In this section, we build on this foundation and explore **compound inequalities**.

Definition Compound inequality: Two inequalities joined by either *and* or *or*.

Examples of compound inequalities are as follows:

$$x > 3 \text{ and } x \leq 8 \qquad -2 \geq x \text{ or } x > 4$$

Objective 1 Solve compound inequalities involving *and*.

Interpreting Compound Inequalities with *and*

First, let's consider inequalities involving *and*, such as $x > 3$ *and* $x \leq 8$. The word *and* indicates that the solution set contains only values that satisfy *both* inequalities. Therefore, the solution set is the **intersection**, or overlap, of the two inequalities' solution sets.

Definition Intersection: For two sets A and B, the intersection of A and B, symbolized by $A \cap B$, is a set containing only elements that are in both A and B.

For example, if $A = \{1, 2, 3, 4, 5\}$ and $B = \{3, 4, 5, 6, 7\}$, then $A \cap B = \{3, 4, 5\}$ because 3, 4, and 5 are the only elements in both A and B. Let's look at our compound inequality $x > 3$ and $x \leq 8$. First, we will graph the two inequalities separately, then consider their intersection.

$x > 3$

Note We use the blue field to indicate the intersection and the dashed lines to indicate the intersection boundaries. ▶

$x \leq 8$

Note The intersection of the two inequalities contains elements that are only in **both** of their solution sets. Notice that 3 is excluded in the intersection because it was excluded in one of the individual graphs.

Intersection of $x > 3$ and $x \leq 8$:

So $x > 3$ and $x \leq 8$ means that x is any number greater than 3 *and* less than or equal to 8. Because this inequality indicates that x is between two values, we can write it without the word *and*: $3 < x \leq 8$. In set-builder notation, we write $\{x \mid 3 < x \leq 8\}$. Using interval notation, we write $(3, 8]$.

Answers to Warm-up

1. $x \leq 5$ **2.** $x < 3$

Example 1 For the compound inequality $x > -2$ and $x < 3$, graph the solution set and write the compound inequality without *and* if possible. Then write in set-builder notation and in interval notation.

Solution: The solution set is the region of intersection.

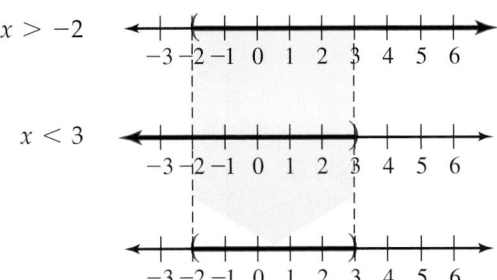

Without *and*: $-2 < x < 3$

Set-builder notation: $\{x \mid -2 < x < 3\}$

Interval notation: $(-2, 3)$

Warning Be careful not to confuse the interval notation $(-2, 3)$ with an ordered pair.

Your Turn 1 For each compound inequality, graph the solution set and write the compound inequality without *and* if possible. Then write in set-builder notation and in interval notation.

a. $x > 4$ and $x < 7$ **b.** $x \geq -8$ and $x \leq -2$

Solving Compound Inequalities with *and*

Now let's consider solution sets for more complex inequalities involving *and*.

Example 2 For the inequality $3x - 1 > 2$ and $2x + 4 \leq 14$, graph the solution set. Then write the solution set in set-builder notation and in interval notation.

Solution:

$3x - 1 > 2$ and $2x + 4 \leq 14$

     $3x > 3$ and $2x \leq 10$ Use the addition principle to isolate each *x* term.

      $x > 1$ and $x \leq 5$ Divide out each coefficient.

The solution set is the intersection of the two individual solution sets.

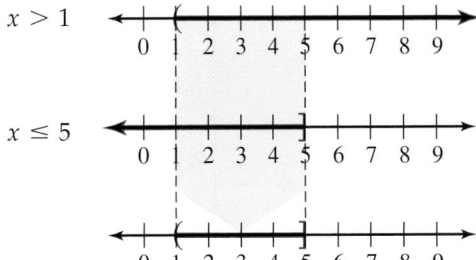

Set-builder notation: $\{x \mid 1 < x \leq 5\}$

Interval notation: $(1, 5]$

Example 2 suggests the following procedure:

Procedure Solving Compound Inequalities Involving *and*

To solve a compound inequality involving *and*:

1. Solve each inequality in the compound inequality.
2. Find the intersection of the individual solution sets.

Answers to Your Turn 1

a.

Without *and*: $4 < x < 7$

Set-builder: $\{x \mid 4 < x < 7\}$

Interval: $(4, 7)$

b.

Without *and*: $-8 \leq x \leq -2$

Set-builder: $\{x \mid -8 \leq x \leq -2\}$

Interval: $[-8, -2]$

Finally, we consider some special situations with compound inequalities involving *and*.

Example 3   For the compound inequality, graph the solution set. Then write the solution set in set-builder notation and in interval notation.

a. $-2 \leq -3x + 7 \leq 1$

Solution:

$-3x + 7 \geq -2$ and $-3x + 7 \leq 1$ Write $-2 \leq -3x + 7 \leq 1$ using *and*.
Remember, $-2 \leq -3x + 7$ is the same as $-3x + 7 \geq -2$.

> **Note** Remember, when we multiply or divide both sides of an inequality by a negative number, we must change the direction of the inequality symbol. ▶

$\quad\quad -3x \geq -9$ and $\quad\quad -3x \leq -6$ Subtract 7 from both sides of each inequality.

$\quad\quad\quad x \leq 3$ and $\quad\quad\quad\quad x \geq 2$ Divide both sides of each inequality by -3, which changes the direction of each inequality.

Remember, $x \leq 3$ and $x \geq 2$ is the same as $2 \leq x \leq 3$.

Solution set graph:

$$\begin{array}{c}\text{number line from 0 to 9 with segment between 2 and 3}\\ 0\ 1\ 2\ 3\ 4\ 5\ 6\ 7\ 8\ 9\end{array}$$

Set-builder notation: $\{x \mid 2 \leq x \leq 3\}$

Interval notation: $[2, 3]$

> **Instructor Note** Emphasize to students that they must do the same thing to *all three parts* of the inequality.

The compound inequality can also be solved in its original form.

$-2 \leq -3x + 7 \leq 1$

$\quad -9 \leq -3x \leq -6$ Subtract 7 from all three parts of the compound inequality.

$\quad\quad 3 \geq x \geq 2$ Divide all three parts of the compound inequality by -3.

Note that $3 \geq x \geq 2$ is the same as $2 \leq x \leq 3$.

b. $2x + 11 \geq 5$ and $2x + 11 > 3$

Solution:

$2x + 11 \geq 5$ and $2x + 11 > 3$

$\quad\quad 2x \geq -6$ and $\quad\quad 2x > -8$ Subtract 11 from both sides of each inequality.

$\quad\quad\quad x \geq -3$ and $\quad\quad\quad x > -4$ Divide both sides of each inequality by 2.

$x \geq -3$ $\begin{array}{c}\text{number line with ray from }-3\text{ rightward}\\ -5\ -4\ -3\ -2\ -1\ 0\ 1\ 2\ 3\ 4\end{array}$

$x > -4$ $\begin{array}{c}\text{number line with open ray from }-4\text{ rightward}\\ -5\ -4\ -3\ -2\ -1\ 0\ 1\ 2\ 3\ 4\end{array}$

Solution set graph: $\begin{array}{c}\text{number line with ray from }-3\text{ rightward}\\ -5\ -4\ -3\ -2\ -1\ 0\ 1\ 2\ 3\ 4\end{array}$

> **Note** The region of intersection for these graphs is from -3 to ∞. ◀

Set-builder notation: $\{x \mid x \geq -3\}$

Interval notation: $[-3, \infty)$

c. $-4x - 3 > 1$ and $-4x - 3 < -11$

Solution:

$-4x - 3 > 1$ and $-4x - 3 < -11$

$\quad\quad -4x > 4$ and $\quad\quad -4x < -8$ Add 3 to both sides of each inequality.

$\quad\quad\quad x < -1$ and $\quad\quad\quad x > 2$ Divide both sides of each inequality by -4, which changes the direction of each inequality.

$x < -1$ $\begin{array}{c}\text{number line with open ray from }-1\text{ leftward}\\ -3\ -2\ -1\ 0\ 1\ 2\ 3\ 4\ 5\ 6\end{array}$

> ◀ **Note** There is no region of intersection for these graphs, so the solution set contains no values and is said to be empty.

$x > 2$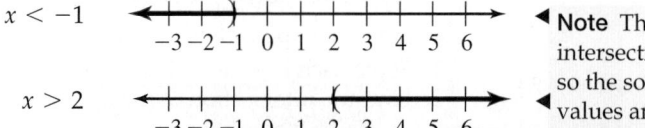

To graph an empty set, we draw a number line with no shading.

Solution set graph: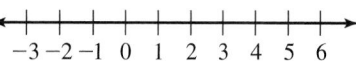

Set-builder notation: { } or ∅

Interval notation: We do not write interval notation because there are no values in the solution set.

> **Your Turn 3** For each compound inequality, graph the solution set. Then write the solution set in set-builder notation and in interval notation.
>
> **a.** $-9 \leq -5x + 11 < 21$ **b.** $3x - 2 \leq 4$ and $5x + 7 \leq 32$
> **c.** $-4x - 3 > 1$ and $-4x - 3 < -5$

Objective 2 Solve compound inequalities involving *or*.

Interpreting Compound Inequalities with *or*

In a compound inequality such as $x > 2$ or $x \leq -1$, the word *or* indicates that the solution set contains values that satisfy *either* inequality. Therefore, the solution set is the **union**, or joining, of the two inequalities' solution sets.

Definition Union: For two sets A and B, the union of A and B, symbolized by $A \cup B$, is a set containing every element in A or in B.

For example, if $A = \{1, 2, 3, 4, 5\}$ and $B = \{3, 4, 5, 6, 7\}$, then $A \cup B = \{1, 2, 3, 4, 5, 6, 7\}$. Consider our compound inequality $x > 2$ or $x \leq -1$. Let's graph the two inequalities separately, then consider their union.

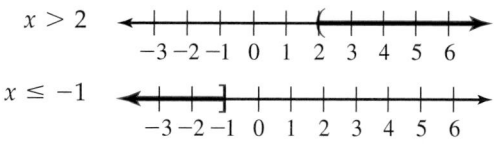

Note This graph is the union of the two individual solution sets because it contains every element in either of the individual solution sets.

Using set-builder notation, we write the solution set as $\{x \mid x > 2 \text{ or } x \leq -1\}$. Interval notation is a little more challenging for compound inequalities such as $x > 2$ or $x \leq -1$ because the solution set is a combination of two intervals. As we scan from left to right, the first interval is $(-\infty, -1]$. The second interval is $(2, \infty)$. To indicate the union of these two intervals, we write $(-\infty, -1] \cup (2, \infty)$.

Solving Compound Inequalities with *or*

Now let's solve more complex inequalities involving *or*. The process is similar to the process we used for solving compound inequalities involving *and*.

> **Procedure Solving Compound Inequalities Involving *or***
>
> To solve a compound inequality involving *or*:
> 1. Solve each inequality in the compound inequality.
> 2. Find the union of the individual solution sets.

Answers to Your Turn 3

a.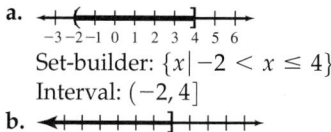
Set-builder: $\{x \mid -2 < x \leq 4\}$
Interval: $(-2, 4]$

b.
Set-builder: $\{x \mid x \leq 2\}$
Interval: $(-\infty, 2]$

c.
Set-builder: { } or ∅
No interval notation

Example 4 For each compound inequality, graph the solution set. Then write the solution set in set-builder notation and in interval notation.

a. $\frac{2}{3}x - 5 \leq -7$ or $\frac{2}{3}x - 5 \geq -1$

Solution:

$$\frac{2}{3}x - 5 \leq -7 \quad \text{or} \quad \frac{2}{3}x - 5 \geq -1$$

$$\frac{2}{3}x \leq -2 \quad \text{or} \quad \frac{2}{3}x \geq 4 \qquad \text{Add 5 to both sides of each inequality.}$$

$$x \leq -3 \quad \text{or} \quad x \geq 6 \qquad \text{Multiply both sides by } \frac{3}{2} \text{ in each inequality.}$$

The solution set is the union of the two individual solution sets.

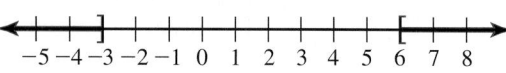

$x \leq -3$

$x \geq 6$

Solution set graph:

Set-builder notation: $\{x \,|\, x \leq -3 \text{ or } x \geq 6\}$

Interval notation: $(-\infty, -3] \cup [6, \infty)$

b. $-3x + 8 < -1$ or $-3x + 8 \leq 20$

Solution: $-3x + 8 < -1 \quad \text{or} \quad -3x + 8 \leq 20$

$$-3x < -9 \quad \text{or} \quad -3x \leq 12 \qquad \text{Subtract 8 from both sides of each inequality.}$$

$$x > 3 \quad \text{or} \quad x \geq -4 \qquad \text{Divide both sides of each inequality by } -3, \text{ which changes the direction of each inequality.}$$

Now we graph the individual solution sets and consider their union.

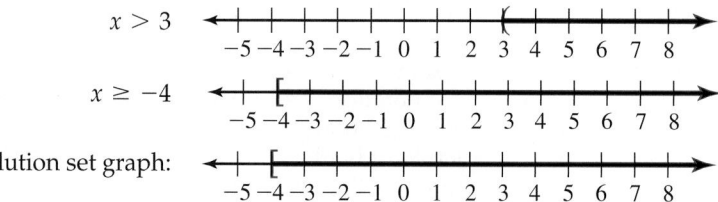

$x > 3$

$x \geq -4$

Solution set graph:

Set-builder notation: $\{x \,|\, x \geq -4\}$

Interval notation: $[-4, \infty)$

c. $9x - 16 > -34$ or $9x - 16 \leq 29$

Solution: $9x - 16 > -34 \quad \text{or} \quad 9x - 16 \leq 29$

$$9x > -18 \quad \text{or} \quad 9x \leq 45 \qquad \text{Add 16 to both sides of each inequality.}$$

$$x > -2 \quad \text{or} \quad x \leq 5 \qquad \text{Divide both sides of each inequality by 9.}$$

Now we graph the individual solution sets and consider their union.

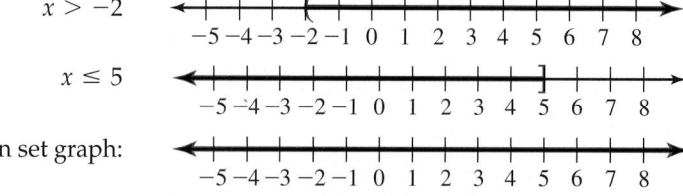

$x > -2$

$x \leq 5$

Solution set graph:

Learning Strategy

If you are a visual learner, imagine placing one graph on top of the other to form their union.

Note When we join the two graphs, the graph of $x \geq -4$ covers all of the graph of $x > 3$; so the union of the two graphs is actually $x \geq -4$. ▶

Note When we join the two graphs, the entire number line is covered; so the union of the two graphs is the set of real numbers. ▶

Answers to Your Turn 4

a.
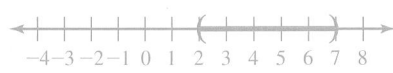
−6−5−4−3−2−1 0 1 2 3

Set-builder:
$\{x \mid x < -4 \text{ or } x \geq -1\}$
Interval: $(-\infty, -4) \cup [-1, \infty)$

b.

−3−2−1 0 1 2 3 4 5 6

Set-builder: $\{x \mid x < 4\}$
Interval: $(-\infty, 4)$

c.

−4−3−2−1 0 1 2 3 4

Set-builder:
$\{x \mid x \text{ is a real number}\}$, or $\mathbb{R}$
Interval: $(-\infty, \infty)$

Set-builder notation: $\{x \mid x \text{ is a real number}\}$, or $\mathbb{R}$.

Interval notation: $(-\infty, \infty)$

Your Turn 4 For each compound inequality, graph the solution set. Then write the solution set in set-builder notation and in interval notation.

a. $-5x - 8 > 12$ or $-5x - 8 \leq -3$

b. $\dfrac{3}{4}x + 5 < -1$ or $\dfrac{3}{4}x + 5 < 8$

c. $5 - 3x \geq 11$ or $5 - 3x < 17$

8.1 Exercises For Extra Help MyMathLab®

Note: Exercises marked with a ★ represent challenging exercises.

Objective 1

Prep Exercise 1 What is a compound inequality? Two inequalities joined by either *and* or *or*

Prep Exercise 2 What two key words are used to denote compound inequalities? *and* and *or*

Prep Exercise 3 Describe the intersection of two sets. For two sets A and B, the intersection of A and B, symbolized by $A \cap B$, is a set containing only elements that are in both A and B.

Prep Exercise 4 Describe how to graph a compound inequality involving *and.* Graph the region of overlap of the two inequalities.

For Exercises 1–4, find the intersection of the given sets. See Objective 1.

1. $A = \{1, 3, 5\}$
$B = \{1, 3, 5, 7, 9\}$
$\{1, 3, 5\}$

2. $A = \{2, 4, 6\}$
$B = \{2, 4, 6, 8, 10\}$
$\{2, 4, 6\}$

3. $A = \{5, 6, 7, 8\}$
$B = \{7, 8, 9\}$
$\{7, 8\}$

4. $A = \{8, 10, 12, 14, 16\}$
$B = \{14, 16, 18\}$
$\{14, 16\}$

For Exercises 5–12, write each inequality without and. See Example 1.

5. $x > -4$ and $x < 5$
$-4 < x < 5$

6. $n \geq -3$ and $n < 7$
$-3 \leq n < 7$

7. $y > 0$ and $y \leq 2$
$0 < y \leq 2$

8. $m \geq 0$ and $m < 3$
$0 \leq m < 3$

9. $w > -7$ and $w < 3$
$-7 < w < 3$

10. $r > -1$ and $r \leq 1$
$-1 < r \leq 1$

11. $u \geq 0$ and $u \leq 2$
$0 \leq u \leq 2$

12. $t > 7$ and $t < 15$
$7 < t < 15$

For Exercises 13–20, graph the compound inequality. See Examples 1–3.

13. $x > 2$ and $x < 7$

−4−3−2−1 0 1 2 3 4 5 6 7 8

14. $x > 3$ and $x < 9$

−2−1 0 1 2 3 4 5 6 7 8 9 10

15. $x > -1$ and $x \leq 5$

−6−5−4−3−2−1 0 1 2 3 4 5 6

16. $x \geq -4$ and $x < 0$
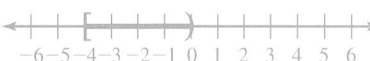
−6−5−4−3−2−1 0 1 2 3 4 5 6

17. $1 \leq x \leq 10$

18. $-2 < x < -1$

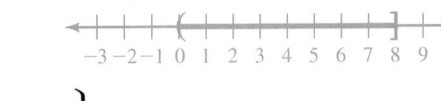

19. $-3 \leq x < 4$

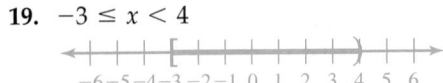

20. $0 < x \leq 8$

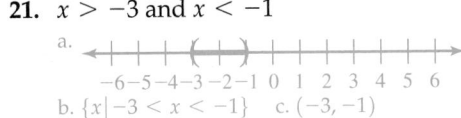

For Exercises 21–40:   **a.**   Graph the solution set.
                   b.   Write the solution set using set-builder notation. $\Big\}$ See Examples 1–3.
                   c.   Write the solution set using interval notation.

21. $x > -3$ and $x < -1$

a.

b. $\{x \mid -3 < x < -1\}$ c. $(-3, -1)$

22. $x > 4$ and $x < 7$

a.

b. $\{x \mid 4 < x < 7\}$ c. $(4, 7)$

23. $x + 2 > 5$ and $x - 4 \leq 2$

a.

b. $\{x \mid 3 < x \leq 6\}$ c. $(3, 6]$

24. $x - 3 \geq 1$ and $x + 2 < 10$

a.

b. $\{x \mid 4 \leq x < 8\}$ c. $[4, 8)$

25. $-x > 1$ and $-2x \leq -10$

a.

b. $\{\ \}$ or $\varnothing$ c. no interval notation

26. $-4x \geq 8$ and $-3x \leq -15$

a.

b. $\{\ \}$ or $\varnothing$ c. no interval notation

27. $3x + 8 \geq -7$ and $4x - 7 < 5$

a.

b. $\{x \mid -5 \leq x < 3\}$ c. $[-5, 3)$

28. $3x + 5 \geq -1$ and $5x - 2 < 13$

a.

b. $\{x \mid -2 \leq x < 3\}$ c. $[-2, 3)$

29. $-3x - 8 > 1$ and $-4x + 5 \leq -3$

a.

b. $\{\ \}$ or $\varnothing$ c. no interval notation

30. $-6x + 4 < -14$ and $-x - 3 \geq -2$

a.

b. $\{\ \}$ or $\varnothing$ c. no interval notation

31. $-3 < x + 4 < 1$

a.

b. $\{x \mid -7 < x < -3\}$ c. $(-7, -3)$

32. $-2 \leq x - 2 \leq 2$

a.

b. $\{x \mid 0 \leq x \leq 4\}$ c. $[0, 4]$

33. $-7 \leq 4x - 3 \leq 5$

a.

b. $\{x \mid -1 \leq x \leq 2\}$ c. $[-1, 2]$

34. $8 < 3x - 4 < 14$

a.

b. $\{x \mid 4 < x < 6\}$ c. $(4, 6)$

35. $0 \leq 2 + 3x < 8$

a.

b. $\left\{x \mid -\dfrac{2}{3} \leq x < 2\right\}$ c. $\left[-\dfrac{2}{3}, 2\right)$

36. $0 \leq -2 + 5x \leq 13$

a.

b. $\left\{x \mid \dfrac{2}{5} \leq x \leq 3\right\}$ c. $\left[\dfrac{2}{5}, 3\right]$

37. $-1 < -2x + 5 \leq 5$

a.

b. $\{x \mid 0 \leq x < 3\}$ c. $[0, 3)$

38. $-3 < 5 - 2x < 1$

a.

b. $\{x \mid 2 < x < 4\}$ c. $(2, 4)$

39. $3 \leq 6 - x \leq 6$

a.

b. $\{x \mid 0 \leq x \leq 3\}$ c. $[0, 3]$

40. $-4 \leq -4 - x \leq 2$

a.

b. $\{x \mid -6 \leq x \leq 0\}$ c. $[-6, 0]$

Objective 2

Prep Exercise 5 Describe the union of two sets. For two sets A and B, the union of A and B, symbolized by $A \cup B$, is a set containing each element in either A or B.

Prep Exercise 6 Describe how to graph a compound inequality involving *or*. Graph the region included in either of the two inequalities.

For Exercises 41–44, find the union of the given sets. See Objective 2.

41. $A = \{c, a, t\}$
$B = \{d, o, g\}$
$\{a, c, d, g, o, t\}$

42. $A = \{l, o, v, e\}$
$B = \{m, a, t, h\}$
$\{a, e, h, l, m, o, t, v\}$

43. $A = \{x, y, z\}$
$B = \{w, x, y, z\}$
$\{w, x, y, z\}$

44. $A = \{a, b, c, d, e\}$
$B = \{c, d, e, f\}$
$\{a, b, c, d, e, f\}$

For Exercises 45–52, graph each inequality. See Objective 2.

45. $x < -2$ or $x > 6$

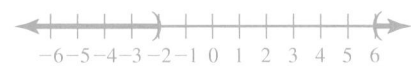

46. $n \leq 3$ or $n > 4$

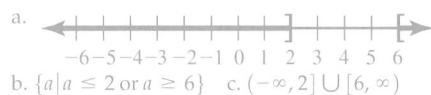

47. $y < -3$ or $y \geq 0$

48. $m \leq 2$ or $m > 8$

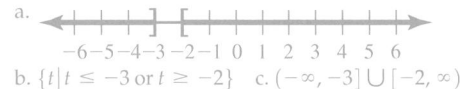

49. $w > -3$ or $w > 2$

50. $r > -3$ or $r \geq -1$

51. $u \geq 0$ or $u \leq 2$

52. $t > 6$ or $t < 13$

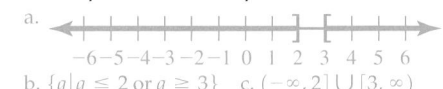

For Exercises 53–68:   **a.** Graph the solution set.
b. Write the solution set using set-builder notation.   } *See Example 4.*
c. Write the solution set using interval notation.

53. $y + 2 < -7$ or $y + 2 > 7$

a.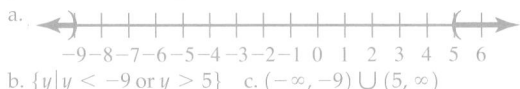

b. $\{y \mid y < -9 \text{ or } y > 5\}$ c. $(-\infty, -9) \cup (5, \infty)$

54. $a - 4 \leq -2$ or $a - 4 \geq 2$

a.

b. $\{a \mid a \leq 2 \text{ or } a \geq 6\}$ c. $(-\infty, 2] \cup [6, \infty)$

55. $4r - 3 < -11$ or $2r - 3 > -1$

a.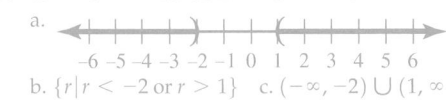

b. $\{r \mid r < -2 \text{ or } r > 1\}$ c. $(-\infty, -2) \cup (1, \infty)$

56. $3t - 5 \leq -14$ or $5t + 5 \geq -5$

a.

b. $\{t \mid t \leq -3 \text{ or } t \geq -2\}$ c. $(-\infty, -3] \cup [-2, \infty)$

57. $-w + 2 \leq -5$ or $-w + 2 \geq 3$

a.

b. $\{w \mid w \leq -1 \text{ or } w \geq 7\}$ c. $(-\infty, -1] \cup [7, \infty)$

58. $-2x - 3 \leq -1$ or $-2x - 3 \geq 5$

a.

b. $\{x \mid x \leq -4 \text{ or } x \geq -1\}$ c. $(-\infty, -4] \cup [-1, \infty)$

59. $7 - 4k \leq -5$ or $6 - 2k \geq 2$

a.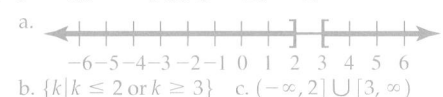

b. $\{k \mid k \leq 2 \text{ or } k \geq 3\}$ c. $(-\infty, 2] \cup [3, \infty)$

60. $8 - 2q \leq 2$ or $8 - 2q \geq 4$

a.

b. $\{q \mid q \leq 2 \text{ or } q \geq 3\}$ c. $(-\infty, 2] \cup [3, \infty)$

61. $\frac{2}{3}x - 4 \geq -2$ or $\frac{2}{5}x - 2 \leq -4$

a.
```
←+[+++++++[+++→
 -6-5-4-3-2-1 0 1 2 3 4 5 6
```
b. $\{x \mid x \leq -5 \text{ or } x \geq 3\}$ c. $(-\infty, -5] \cup [3, \infty)$

62. $\frac{3}{4}x + 3 \geq 6$ or $\frac{3}{5}x - 3 \leq -6$

a.
```
←+[+++++++++[+++→
 -6-5-4-3-2-1 0 1 2 3 4 5 6
```
b. $\{x \mid x \leq -5 \text{ or } x \geq 4\}$ c. $(-\infty, -5] \cup [4, \infty)$

63. $-2(c - 1) < -4$ or $-2(c - 2) < -6$

a.
```
←+++++++++(+++++→
 -6-5-4-3-2-1 0 1 2 3 4 5 6
```
b. $\{c \mid c > 3\}$ c. $(3, \infty)$

64. $-3(y - 1) \leq 6$ or $-4(y - 2) \leq 4$

a.
```
←+++++++[++++++→
 -6-5-4-3-2-1 0 1 2 3 4 5 6
```
b. $\{y \mid y \geq -1\}$ c. $[-1, \infty)$

65. $2(x + 3) + 3 \leq 1$ or $2(x + 1) + 7 \leq -3$

a.
```
←+++]++++++++++→
 -6-5-4-3-2-1 0 1 2 3 4 5 6
```
b. $\{x \mid x \leq -4\}$ c. $(-\infty, -4]$

66. $5(d - 1) + 8 < 8$ or $5(d + 1) - 2 < 18$

a.
```
←++++++++++)++++→
 -6-5-4-3-2-1 0 1 2 3 4 5 6
```
b. $\{d \mid d < 3\}$ c. $(-\infty, 3)$

67. $-3(x - 1) - 1 \leq -1$ or $-3(x + 1) + 5 \geq -2$

a.
```
←++++++++++++++→
 -6-5-4-3-2-1 0 1 2 3 4 5 6
```
b. $\{x \mid x \text{ is a real number}\}$, or $\mathbb{R}$ c. $(-\infty, \infty)$

68. $-4(k - 2) - 1 \geq -1$ or $-4(k - 2) - 1 \leq 5$

a.
```
←++++++++++++++→
 -6-5-4-3-2-1 0 1 2 3 4 5 6
```
b. $\{x \mid x \text{ is a real number}\}$, or $\mathbb{R}$ c. $(-\infty, \infty)$

Objectives 1 and 2

For Exercises 69–80:   *a.* *Graph the solution set.*
                   b. *Write the solution set using set-builder notation.* $\Big\}$ *See Examples 2, 3, and 4.*
                   c. *Write the solution set using interval notation.*

69. $-3 < x + 4$ and $x + 4 < 7$

a.
```
←+(+++++++++)+++→
 -7-6-5-4-3-2-1 0 1 2 3 4 5
```
b. $\{x \mid -7 < x < 3\}$ c. $(-7, 3)$

70. $0 < x - 1$ and $x - 1 \leq 3$

a.
```
←+++++++++(+++]+++→
 -6-5-4-3-2-1 0 1 2 3 4 5 6
```
b. $\{x \mid 1 < x \leq 4\}$ c. $(1, 4]$

71. $4x + 3 < -5$ or $3x + 2 > 8$

a.
```
←+++++)+++(++++→
 -6-5-4-3-2-1 0 1 2 3 4 5 6
```
b. $\{x \mid x < -2 \text{ or } x > 2\}$ c. $(-\infty, -2) \cup (2, \infty)$

72. $6x - 2 < -8$ or $4x - 3 > 9$

a.
```
←++++++)+++(++++→
 -6-5-4-3-2-1 0 1 2 3 4 5 6
```
b. $\{x \mid x < -1 \text{ or } x > 3\}$ c. $(-\infty, -1) \cup (3, \infty)$

73. $4 < -2x < 6$

a.
```
←+++(+)+++++++++→
 -6-5-4-3-2-1 0 1 2 3 4 5 6
```
b. $\{x \mid -3 < x < -2\}$ c. $(-3, -2)$

74. $9 \leq -3x \leq 15$

a.
```
←+[++]++++++++++→
 -6-5-4-3-2-1 0 1 2 3 4 5 6
```
b. $\{x \mid -5 \leq x \leq -3\}$ c. $[-5, -3]$

75. $7 \leq 4x - 5 \leq 19$

a.
```
←++++++++[+++]++→
 -6-5-4-3-2-1 0 1 2 3 4 5 6
```
b. $\{x \mid 3 \leq x \leq 6\}$ c. $[3, 6]$

76. $-9 < 2x - 7 < -3$

a.
```
←++++++(+++)++++→
 -6-5-4-3-2-1 0 1 2 3 4 5 6
```
b. $\{x \mid -1 < x < 2\}$ c. $(-1, 2)$

77. $-5 < 1 - 2x < -1$

a.
```
←+++++++++(+)++++→
 -6-5-4-3-2-1 0 1 2 3 4 5 6
```
b. $\{x \mid 1 < x < 3\}$ c. $(1, 3)$

78. $4 \leq -3x + 1 < 7$

a.
```
←++++++(]++++++→
 -6-5-4-3-2-1 0 1 2 3 4 5 6
```
b. $\{x \mid -2 < x \leq -1\}$ c. $(-2, -1]$

★ **79.** $x + 3 < 2x + 1 < 3x$

a.
```
←+++++++++++(+++→
 -6-5-4-3-2-1 0 1 2 3 4 5 6
```
b. $\{x \mid x > 2\}$ c. $(2, \infty)$

★ **80.** $2x - 2 \leq x + 1 \leq 2x + 5$

a.
```
←++[+++++++]+++→
 -6-5-4-3-2-1 0 1 2 3 4 5 6
```
b. $\{x \mid -4 \leq x \leq 3\}$ c. $[-4, 3]$

For Exercises 81–90, solve. Then: *a.* *Graph the solution set.*
                                       b. *Write the solution set using set-builder notation.*
                                       c. *Write the solution set using interval notation.*

81. If Andrea's current long-distance bill is from $36 to $45 in a month, she can save money by switching to another company. Her current rate is $0.09 per minute. What is the range of minutes that Andrea must use long distance to justify switching to the other company?

a.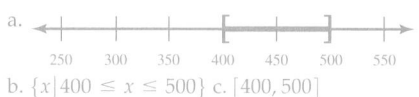
b. $\{x \mid 400 \le x \le 500\}$ c. $[400, 500]$

82. A mail-order music club offers one bonus CD if the total of an order is from $75 to $100. Dayle decides to buy the lowest-priced CDs, which are $12.50 each. In what range would the number of CDs he orders have to be for him to receive a bonus CD?

a.
b. $\{x \mid 6 \le x \le 8\}$ c. $[6, 8]$

83. Juan has taken four of the five tests in his history course. To get a B in the course, his average needs to be at least 80 and less than 90. If his scores on the first four tests are 95, 80, 82, and 88, in what range of values can his score on the fifth test be so that he has a B average?

a.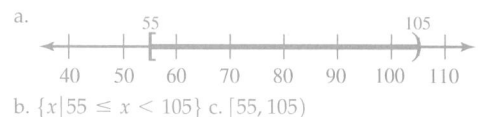
b. $\{x \mid 55 \le x < 105\}$ c. $[55, 105)$

84. Students in a chemistry course receive a C if the average of four tests is at least 70 and less than 80. Suppose a student has the following scores: 76, 72, and 84. What range of scores on the fourth test would result in the student receiving a C?

a.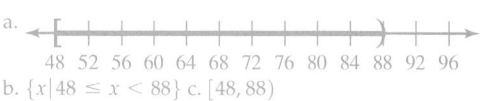
b. $\{x \mid 48 \le x < 88\}$ c. $[48, 88)$

85. To conserve energy, it is recommended that a home's thermostat be set at 5° above 73°F during summer or 5° below 73° during winter. If the heat pump does not run when the temperature is at or between those values, in what range of temperatures is the heat pump off?

a.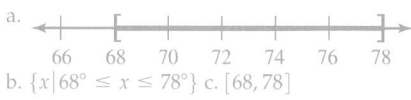
b. $\{x \mid 68° \le x \le 78°\}$ c. $[68, 78]$

86. A house thermostat is set so that the heat pump/air conditioner comes on if the temperature is 2° or more above or below 70°. In what range of temperatures is the heat pump off?

a.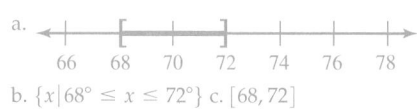
b. $\{x \mid 68° \le x \le 72°\}$ c. $[68, 72]$

87. In the maintenance of a saltwater aquarium, the temperature should be within 4° of 76°. What range of temperatures would be acceptable for saltwater fish? (*Source: The Conscientious Marine Aquarist*, Robert M. Fenner.)

a.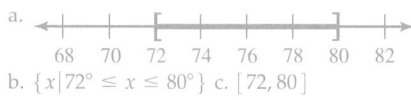
b. $\{x \mid 72° \le x \le 80°\}$ c. $[72, 80]$

88. When driving in Mississippi, motorists risk a ticket if they drive more than 15 miles per hour over or under the 55 miles per hour speed limit on a state highway. In what range of speeds would a motorist not risk receiving a ticket? (*Source:* Mississippi State Code 63-3-509.)

a.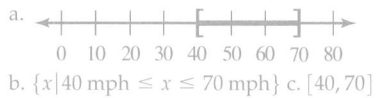
b. $\{x \mid 40 \text{ mph} \le x \le 70 \text{ mph}\}$ c. $[40, 70]$

89. A building is to be designed in the shape of a trapezoid as shown. Find the range of values for the length of the back side of the building so that the square footage is from 6000 to 8000 square feet.

[Use $A = \dfrac{1}{2}h(a + b)$.]

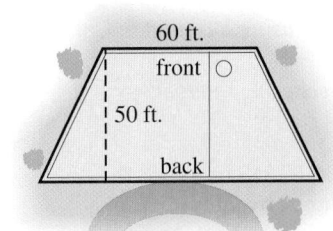

a.
170 180 190 200 210 220 230 240 250 260 270

b. $\{x \mid 180 \text{ ft.} \le x \le 260 \text{ ft.}\}$ c. $[180, 260]$

90. Water is a liquid between 32°F and 212°F. Use a compound inequality to describe the range of values in degrees Celsius for water in its liquid state. [Use $C = \dfrac{5}{9}(F - 32)$.]

a.
−10 0 10 20 30 40 50 60 70 80 90 100 110

b. $\{x \mid 0°C < x < 100°C\}$ c. $(0, 100)$

Puzzle Problem A cyclist is involved in a multiple-day race. She believes that she needs to complete today's 40 kilometers between 1 hour and 45 minutes and 2 hours. Find the range of values her rate can be to complete this leg of the race in her desired time frame.

20 km/hr. $\le r \le$ 22.9 km/hr.

Review Exercises

Exercises 1–3   Expressions

[1.1] 1. Is the absolute value of a number always positive? Explain.

No, the absolute value of zero is zero and zero is neither negative nor positive.

[1.5] *For Exercises 2 and 3, simplify.*

2. $-|2 - 3^2| - |-4|$
−11

3. $-|-|-16||$
−16

Exercises 4–6   Equations and Inequalities

[2.3] *For Exercises 4 and 5, solve.*

4. $3x - 14x = 17 - 12x - 11$
6

5. $\dfrac{3}{5}(25 - 5x) = 15 - \dfrac{3}{5}$
$\dfrac{1}{5}$

[2.5] 6. One-half the sum of a number and 2 is zero. Find the number.
−2

8.2 Equations Involving Absolute Value; Absolute Value Functions

Objectives

1 Solve equations involving absolute value.

2 Graph absolute value functions.

Warm-up

[2.3] *For Exercises 1 and 2, solve the equations.*

    1. $3x - 7 = -(5 - 3x)$ **2.** $8 - 2(4x - 8) = -16$

[1.7] Evaluate $-|x - 2| + 1$ for $x = 4$.

Objective 1 Solve equations involving absolute value.

We learned in Section 1.1 that the absolute value of a number is its distance from zero on a number line. For example, $|4| = 4$ and $|-4| = 4$ because both 4 and -4 are 4 units from 0 on a number line.

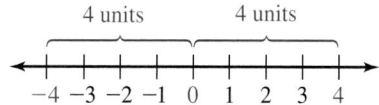

Now we will solve equations in which a variable appears within the absolute value symbols, as in $|x| = 4$. Notice that the solutions are 4 and -4, which suggests the following rule:

> **Rule Absolute Value Property**
>
> If $|x| = a$, where x is a variable or an expression and $a \geq 0$, then $x = a$ or $x = -a$.

Instructor Note An alternative way of deriving the absolute value property is by using the definition of absolute value. Because $|x| = \pm x, |x| = a, a > 0$, means $\pm x = a$. So $x = a$ or $-x = a$, which gives $x = a$ or $x = -a$.

Example 1 Solve.

a. $|2x - 5| = 9$

Solution: $2x - 5 = 9$ or $2x - 5 = -9$ Use the absolute value property to separate into a positive case and a negative case.

                $2x = 14$ or $2x = -4$ In each case, add 5 to both sides.

                 $x = 7$ or $x = -2$ In each case, divide both sides by 2.

The solutions are 7 and -2.

b. $|4x + 1| = -11$

Solution: This equation has the absolute value equal to a negative number, -11. Because the absolute value of every real number is a positive number or zero, this equation has no solution.

From Example 1(b), we can say that an absolute value equation in the form $|x| = a$, where $a < 0$, has no solution.

Note The absolute value property does not apply if $a < 0$. For example, $|x| = -5$ has no solution because there is no real number whose absolute value is -5.

Your Turn 1 Solve.

    a. $|3x + 4| = 8$ **b.** $|2x - 1| = 0$ **c.** $|5x - 2| = -3$

To use the absolute value property, we need the **absolute value isolated**.

Answers to Your Turn 1

a. $\dfrac{4}{3}$ and -4 **b.** $\dfrac{1}{2}$

c. no solution

Answers to Warm-up

1. No solution or $\varnothing$

2. $x = 5$

3. -1

Example 2 Solve $8 - 2|4x - 8| = -16$.

Solution: $-2|4x - 8|$ $= -24$ Subtract 8 from both sides of the equation.

            $|4x - 8|$ $= 12$ Divide both sides by -2 to isolate the absolute value.

$$4x - 8 = 12 \quad \text{or} \quad 4x - 8 = -12 \quad \text{Use the absolute value property.}$$
$$4x = 20 \quad \text{or} \quad 4x = -4$$
$$x = 5 \quad \text{or} \quad x = -1$$

The solutions are 5 and -1.

Examples 1 and 2 suggest the following procedure:

> **Procedure** **Solving Equations Containing a Single Absolute Value**
>
> To solve an equation containing a single absolute value:
> 1. **Isolate the absolute value** so that the equation is in the form $|ax + b| = c$. If $c > 0$, proceed to steps 2 and 3. If $c < 0$, the equation has no solution.
> 2. Separate the absolute value into two equations: $ax + b = c$ and $ax + b = -c$.
> 3. Solve both equations.

Your Turn 2 Solve.

a. $12 - 3|3x - 6| = -6$ b. $|2x + 1| + 6 = 4$

More Than One Absolute Value

Some equations contain more than one absolute value expression, such as $|x + 3| = |2x - 8|$. If two absolute values are equal, they must contain expressions that are *equal* or *opposites*. Some simple examples follow.

Equal		Opposites
$|7| = |7|$	or	$|7| = |-7|$
$|-5| = |-5|$	or	$|-5| = |5|$

For the equation $|x + 3| = |2x - 8|$, therefore, the expressions $x + 3$ and $2x - 8$ must be equal or opposites.

Equal		Opposites
$x + 3 = 2x - 8$	or	$x + 3 = -(2x - 8)$

This suggests the following procedure:

> **Procedure** **Solving Equations in the Form $|ax + b| = |cx + d|$**
>
> To solve an equation in the form $|ax + b| = |cx + d|$:
> 1. Separate the absolute value equation into two equations: $ax + b = cx + d$ and $ax + b = -(cx + d)$.
> 2. Solve both equations.

Example 3 Solve.

a. $|3x + 5| = |x - 9|$

Solution: Separate the absolute value equation into two equations.

Equal		Opposites
$3x + 5 = x - 9$	or	$3x + 5 = -(x - 9)$
$2x + 5 = -9$	or	$3x + 5 = -x + 9$
$2x = -14$	or	$4x + 5 = 9$
$x = -7$	or	$4x = 4$
		$x = 1$

◄ **Note** Think of $-(x - 9)$ as $-1(x - 9)$.

Answers to Your Turn 2
a. 0 and 4 b. no solution

The solutions are -7 and 1.

b. $|3x - 7| = |5 - 3x|$

Solution:

Equal	Opposites
$3x - 7 = 5 - 3x$	or $3x - 7 = -(5 - 3x)$

Separate into two equations.

$$6x - 7 = 5 \qquad \text{or} \quad 3x - 7 = -5 + 3x$$

$$6x = 12 \qquad \text{or} \qquad -7 = -5$$

$$x = 2$$

◄ **Note** Subtracting $3x$ from both sides gives us a false equation; so the "opposites" equation is a contradiction and thus has no solution.

This absolute value equation has only one solution, 2.

Your Turn 3 Solve.

a. $|5x - 3| = |3x - 13|$ **b.** $|9 + 4x| = |1 - 4x|$

Objective 2 Graph absolute value functions.

The basic absolute value function is $f(x) = |x|$. It is the same as $y = |x|$. As we learned with linear and quadratic functions, to graph a function, we plot ordered pairs found by evaluating the function using various values of x. For example, if we let $x = -4$, we have $f(-4) = |-4| = 4$; so the ordered pair is $(-4, 4)$. Below, we complete a table of ordered pairs, plot those ordered pairs, and then connect the points to see that the graph is a V shape.

x	f(x)
−4	4
−2	2
0	0
2	2
4	4

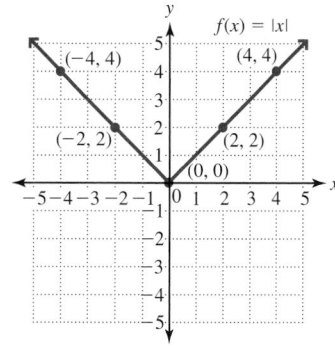

Note $f(x) = |x|$ is equivalent to $f(x) = -x$ when $x < 0$ and to $f(x) = x$ when $x \geq 0$. Notice that $f(x) = -x$ and ◄ $f(x) = x$ are linear functions with opposite slopes, which explains the V shape of the graph.

We can see that $f(x) = |x|$ is a function because it passes the vertical line test. Notice that every real number along the x-axis can be used as an input in the function, so the domain is $(-\infty, \infty)$. The range is $[0, \infty)$ because the minimum y-coordinate on the graph is 0 and the graph continues upward without end.

Example 4 Graph each function. Then give the domain and range.

a. $f(x) = |x| + 1$

Solution:

x	f(x)
−4	5
−2	3
0	1
2	3
4	5

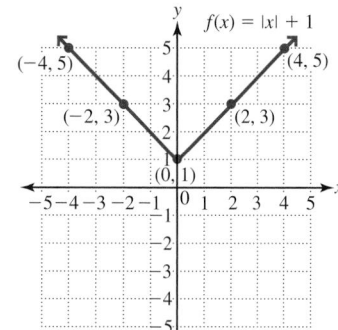

Note Every point on ◄ $f(x) = |x| + 1$ is one higher than every point on $f(x) = |x|$. Adding 1 to $|x|$ shifts the graph of $f(x) = |x|$ up 1 unit.

Domain: $(-\infty, \infty)$
Range: $[1, \infty)$

b. $f(x) = |x - 2| + 1$

Solution:

x	f(x)
−4	7
−2	5
0	3
2	1
4	3

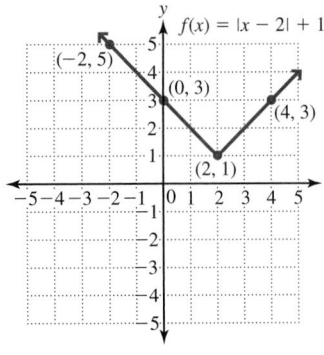

Note This graph is the graph of $f(x) = |x|$ ◀ shifted to the right 2 units and up 1 unit. The $x - 2$ inside the absolute value causes the shift to the right 2 units, and the +1 outside the absolute value causes the upward shift 1 unit.

Domain: $(-\infty, \infty)$
Range: $[1, \infty)$

Answers to Your Turn 4

a.

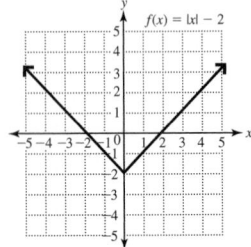

b.

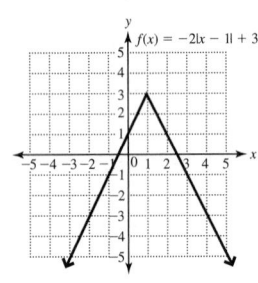

c. $f(x) = -|x - 2| + 1$

Solution:

x	f(x)
−4	−5
−2	−3
0	−1
2	1
4	−1

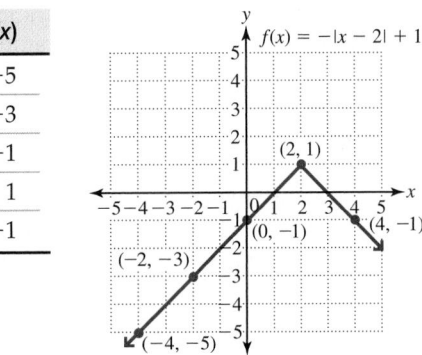

Note The minus sign in front of the absolute value causes the graph to invert so that the **V** is "upside ◀ down." This means that the range goes from a low of $-\infty$ up to 1.

Domain: $(-\infty, \infty)$
Range: $(-\infty, 1]$

Your Turn 4 Graph each function.

a. $f(x) = |x| - 2$

b. $f(x) = -2|x - 1| + 3$

8.2 Exercises

For Extra Help MyMathLab®

Objective 1

Prep Exercise 1 What does absolute value mean in terms of a number line? *The distance a number is from zero.*

Prep Exercise 2 How do we interpret $|x| = 5$? *We are to find numbers that are 5 units from 0.*

Prep Exercise 3 State the absolute value property in your own words. *If $|n| = a$, where n is a variable or an expression and $a \geq 0$, then $n = a$ or $n = -a$.*

Prep Exercise 4 If given an equation in the form $|ax + b| = c$ and $c < 0$, what can you conclude about its solution(s)? *There is no solution.*

For Exercises 1–16, solve using the absolute value property. See Example 1.

1. $|x| = 2$
  $-2, 2$

2. $|y| = 5$
  $-5, 5$

3. $|a| = -4$
  No solution

4. $|r| = -1$
  No solution

5. $|x + 3| = 8$
  $-11, 5$

6. $|w - 1| = 4$
  $-3, 5$

7. $|2m - 5| = 1$
  $2, 3$

8. $|3s + 6| = 12$
  $2, -6$

9. $|6 - 5x| = 1$
$1, \dfrac{7}{5}$

10. $|2 - 3x| = 5$
$-1, \dfrac{7}{3}$

11. $|4 - 3w| = 6$
$-\dfrac{2}{3}, \dfrac{10}{3}$

12. $|1 - 2x| = 2$
$-\dfrac{1}{2}, \dfrac{3}{2}$

13. $|4m - 2| = -5$
No solution

14. $|6p + 5| = -1$
No solution

15. $|4w - 3| = 0$
$\dfrac{3}{4}$

16. $|3m - 5| = 0$
$\dfrac{5}{3}$

For Exercises 17–32, isolate the absolute value; then use the absolute value property.
See Example 2.

17. $|2y| - 3 = 5$
$-4, 4$

18. $|3r| + 7 = 10$
$-1, 1$

19. $|y + 1| + 2 = 4$
$1, -3$

20. $|x + 3| - 1 = 5$
$-9, 3$

21. $|b - 4| - 6 = 2$
$12, -4$

22. $|v - 2| + 2 = 4$
$0, 4$

23. $3 + |5x - 1| = 7$
$-\dfrac{3}{5}, 1$

24. $2 + |5t - 2| = 5$
$-\dfrac{1}{5}, 1$

25. $1 - |2k + 3| = -4$
$-4, 1$

26. $-3 = 1 - |4u - 8|$
$1, 3$

27. $4 - 3|z - 2| = -8$
$-2, 6$

28. $3 - 2|x - 5| = -7$
$0, 10$

29. $6 - 2|3 - 2w| = -18$
$-\dfrac{9}{2}, \dfrac{15}{2}$

30. $15 - 2|5 - 2x| = -5$
$-\dfrac{5}{2}, \dfrac{15}{2}$

31. $|3x - 2(x + 5)| = 10$
$0, 20$

32. $|4y - 2(6 - y)| = 18$
$-1, 5$

Prep Exercise 5 Explain the first step in solving an equation containing two absolute values, such as $|2x + 1| = |3x - 5|$. Separate the absolute value equation into two equations: $ax + b = cx + d$ and $ax + b = -(cx + d)$.

Prep Exercise 6 How can you tell that an absolute value equation containing two absolute values has only one solution? After the absolute value equation is separated into two equations, one with the two expressions equal and the other with the two expressions opposites, one of the equations leads to a contradiction.

For Exercises 33–42, solve by separating the two absolute values into equal and opposite cases. See Example 3.

33. $|2x + 1| = |x + 5|$
$-2, 4$

34. $|2p + 5| = |3p + 10|$
$-3, -5$

35. $|x + 3| = |2x - 4|$
$\dfrac{1}{3}, 7$

36. $|p - 1| = |2p + 8|$
$-9, -\dfrac{7}{3}$

37. $|3v + 4| = |1 - 2v|$
$-5, -\dfrac{3}{5}$

38. $|3 - 6c| = |2c + 3|$
$0, \dfrac{3}{2}$

39. $|2n + 3| = |3 + 2n|$
All real numbers

40. $|2r - 1| = |1 - 2r|$
All real numbers

41. $|2k + 1| = |2k - 5|$
1

42. $|4 + 2q| = |2q + 8|$
-3

For Exercises 43–50, solve. See Examples 1 and 2.

43. $|10 - (5 - h)| = 8$
$3, -13$

44. $|7 - (2 - n)| = 17$
$-22, 12$

45. $\left|\dfrac{b}{2} - 1\right| = 4$
$-6, 10$

46. $\left|\dfrac{u}{3} - 2\right| = 3$
$15, -3$

47. $\left|\dfrac{4 - 3x}{2}\right| = \dfrac{3}{4}$
$\dfrac{5}{6}, \dfrac{11}{6}$

48. $\left|\dfrac{6 - 5w}{6}\right| = \dfrac{2}{3}$
$\dfrac{2}{5}, 2$

49. $\left|2y + \dfrac{3}{2}\right| - 2 = 5$
$-\dfrac{17}{4}, \dfrac{11}{4}$

50. $\left|p + \dfrac{2}{3}\right| - 1 = 7$
$-\dfrac{26}{3}, \dfrac{22}{3}$

Objective 2

Prep Exercise 7 What is the shape of the graph of every absolute value function? How do you graph an absolute value function? The graph of every absolute value function is a V shape. Find many ordered pairs, plot them, and then draw a V-shape through those points.

Prep Exercise 8 Define *domain* and *range*. *Domain*: The set of all input values for a relation *Range*: The set of all output values for a relation

For Exercises 51–62, graph the function. Then give the domain and range.

51. $f(x) = |x| - 4$
Domain: $(-\infty, \infty)$
Range: $[-4, \infty)$

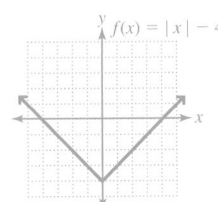

52. $f(x) = |x| + 1$
Domain: $(-\infty, \infty)$
Range: $[1, \infty)$

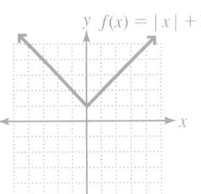

53. $f(x) = |x + 3| - 2$
Domain: $(-\infty, \infty)$
Range: $[-2, \infty)$

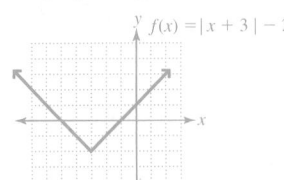

54. $f(x) = |x - 1| + 3$
Domain: $(-\infty, \infty)$
Range: $[3, \infty)$

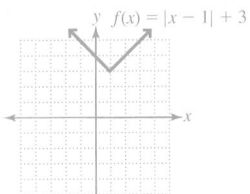

55. $f(x) = 2|x - 1| + 3$
Domain: $(-\infty, \infty)$
Range: $[3, \infty)$

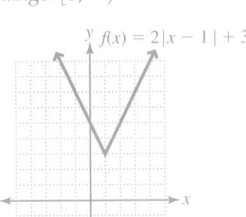

56. $f(x) = 3|x + 1| - 2$
Domain: $(-\infty, \infty)$
Range: $[-2, \infty)$

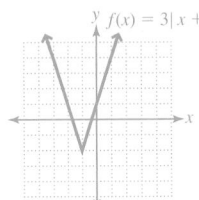

57. $f(x) = -|x| + 3$
Domain: $(-\infty, \infty)$
Range: $[-\infty, 3]$

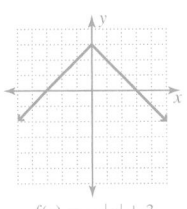

58. $f(x) = -|x| - 1$
Domain: $(-\infty, \infty)$
Range: $[-\infty, -1]$

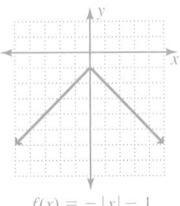

59. $f(x) = -3|x + 2| - 1$
Domain: $(-\infty, \infty)$
Range: $[-\infty, -1]$

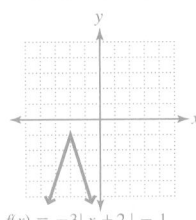

60. $f(x) = -2|x - 1| + 4$
Domain: $(-\infty, \infty)$
Range: $[-\infty, 4]$

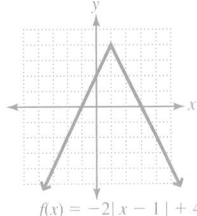

61. $f(x) = -\dfrac{1}{2}|x + 1| + 1$
Domain: $(-\infty, \infty)$
Range: $[-\infty, 1]$

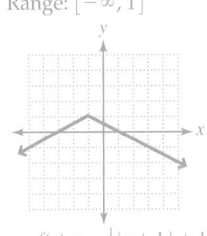

62. $f(x) = -\dfrac{1}{4}|x - 2| + 2$
Domain: $(-\infty, \infty)$
Range: $[-\infty, 2]$

Review Exercises

Exercise 1 **Expressions**

[8.1] 1. Use interval notation to represent the solution set shown on the graph to the right.
$(-2, 0]$

Exercises 2–6 **Equations and Inequalities**

[8.1] 2. Graph the compound inequality: $-1 < x \le 0$

[2.8] *For Exercises 3 and 4, solve.*

3. $6 - 4x > -5x - 1$
$x > -7$

4. $-4x \le 12$
$x \ge -3$

[8.1] 5. Solve the compound inequality: $-1 < \dfrac{3x + 2}{4} < 4$

$-2 < x < \dfrac{14}{3}$

[2.8] 6. Translate to a linear inequality and solve: Two less than a number is less than five.
$n - 2 < 5; n < 7$

8.3 Inequalities Involving Absolute Value

Objectives

1 Solve absolute value inequalities involving less than.

2 Solve absolute value inequalities involving greater than.

Note Solutions for the **equal to** part of $|x| \leq 3$ are 3 and -3. The **less than** part of $|x| \leq 3$ means all numbers between 3 and -3 because their absolute values are less than 3.

Instructor Note This procedure can also be developed using the definition of absolute value. If $|x| \leq a$, then $\pm x \leq a$. So $x \leq a$ and $-x \leq a$, which gives $x \leq a$ and $x \geq -a$. Solving $-x \leq a$ for x explains why $x \geq -a$ has the inequality sign reversed and is the opposite of a.

Warm-up

For Exercises 1 and 2, solve.

[8.1] 1. $-3 < -0.5x + 1 < 3$

[8.2] 2. $|2x - 1| + 3 = 16$

Objective 1 Solve absolute value inequalities involving less than.

Now let's solve inequalities that contain absolute value. Think about $|x| \leq 3$. A solution for this inequality is any number whose absolute value is less than or equal to 3. This means that the solutions are a distance of 3 units or less from 0 on a number line.

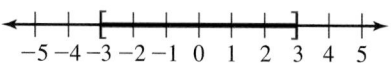

Notice that the solution region corresponds to the compound inequality $x \geq -3$ and $x \leq 3$, which we can write as $-3 \leq x \leq 3$. In set-builder notation, the solution is $\{x | -3 \leq x \leq 3\}$, and in interval notation, we write $[-3, 3]$.

Our examples suggest the following procedure:

Procedure **Solving Inequalities in the Form $|x| < a$, where $a > 0$**

To solve an inequality in the form $|x| < a$, where $a > 0$:

1. Rewrite the inequality as a compound inequality involving *and*: $x > -a$ and $x < a$ (or use $-a < x < a$).
2. Solve the compound inequality.

Similarly, to solve $|x| \leq a$, we write $x \geq -a$ and $x \leq a$ (or $-a \leq x \leq a$).

Example 1 For each inequality, solve, graph the solution set, and write the solution set in both set-builder and interval notation.

a. $|x| < 4$

Solution: $x > -4$ **and** $x < 4$ Rewrite as a compound inequality.

Solution set graph: ◀ **Note** $x > -4$ and $x < 4$ means $-4 < x < 4$.

Set-builder notation: $\{x | -4 < x < 4\}$

Interval notation: $(-4, 4)$

b. $|x - 4| \leq 3$

Solution: $x - 4 \geq -3$ **and** $x - 4 \leq 3$ Rewrite as a compound inequality.

$\qquad\qquad x \geq 1$ **and** $x \leq 7$ Add 4 to both sides of each inequality.

Solution set graph: ◀ **Note** The solution set for $|x - 4| \leq 3$ contains every number whose distance from 4 is 3 units or less.

Set-builder notation: $\{x | 1 \leq x \leq 7\}$

Interval notation: $[1, 7]$

c. $|3x + 1| < 5$

Solution: Instead of $3x + 1 > -5$ and $3x + 1 < 5$, we use the more compact form.

$-5 < 3x + 1 < 5$ Rewrite as a compound inequality.

$-6 < 3x < 4$ Subtract 1 from all three parts of the inequality.

$-2 < x < \dfrac{4}{3}$ Divide all three parts of the inequality by 3 to isolate x.

Solution set graph:

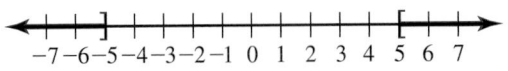

$$-3 \quad -2 \quad -1 \quad 0 \quad 1\frac{4}{3} \quad 2$$

Set-builder notation: $\left\{x \mid -2 < x < \dfrac{4}{3}\right\}$

Interval notation: $\left(-2, \dfrac{4}{3}\right)$

d. $|-0.5x + 1| - 2 < 1$

Solution: Notice that the equation is not in the form $|x| < a$; so our first step is to isolate the absolute value.

$$|-0.5x + 1| < 3 \quad \text{Add 2 to both sides to isolate the absolute value.}$$
$$-3 < -0.5x + 1 < 3 \quad \text{Rewrite as a compound inequality.}$$
$$-4 < -0.5x < 2 \quad \text{Subtract 1 from all three parts of the inequality.}$$
$$8 > x > -4 \quad \text{Divide all three parts of the inequality by } -0.5, \text{ which changes the direction of each inequality.}$$

Solution set graph:

$$-6 \quad -4 \quad -2 \quad 0 \quad 2 \quad 4 \quad 6 \quad 8 \quad 10$$

Set-builder notation: $\{x \mid -4 < x < 8\}$

Interval notation: $(-4, 8)$

e. $\left|\dfrac{1}{3}x - 4\right| + 6 < -1$

Solution: $\left|\dfrac{1}{3}x - 4\right| < -7 \quad \text{Subtract 6 from both sides to isolate the absolute value.}$

Because absolute values cannot be negative, this inequality has no solution; so the solution set is empty.

Solution set graph:

$$-5 \quad -4 \quad -3 \quad -2 \quad -1 \quad 0 \quad 1 \quad 2 \quad 3 \quad 4 \quad 5$$

Set-builder notation: $\{\ \}$ or $\varnothing$

Interval notation: We do not write interval notation because there are no values in the solution set.

Note Any inequality of the form $|x| < a, a \le 0$, or $|x| \le a, a < 0$, has no solution because the absolute value of any real number is nonnegative and, therefore, can only be zero or positive. ▶

| **Your Turn 1** | For each inequality, solve, graph the solution set, and write the solution set in both set-builder notation and interval notation.

a. $|x + 5| < 2$ **b.** $|2x - 3| \le 7$

c. $\left|-\dfrac{1}{3}x + 1\right| - 2 \le 1$ **d.** $\left|\dfrac{1}{2}x - 1\right| + 3 < -2$

Answers to Your Turn 1

a.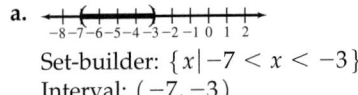
$$-8\,-7\,-6\,-5\,-4\,-3\,-2\,-1\ 0\ 1\ 2$$
Set-builder: $\{x \mid -7 < x < -3\}$
Interval: $(-7, -3)$

b.
$$-4\,-3\,-2\,-1\ 0\ 1\ 2\ 3\ 4\ 5\ 6$$
Set-builder: $\{x \mid -2 \le x \le 5\}$
Interval: $[-2, 5]$

c.
$$-8\,-6\,-4\,-2\ 0\ 2\ 4\ 6\ 8\ 10\,12\,14$$
Set-builder: $\{x \mid -6 \le x \le 12\}$
Interval: $[-6, 12]$

d.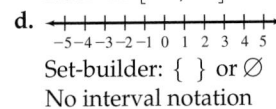
$$-5\,-4\,-3\,-2\,-1\ 0\ 1\ 2\ 3\ 4\ 5$$
Set-builder: $\{\ \}$ or $\varnothing$
No interval notation

Objective 2 Solve absolute value inequalities involving greater than.

Now we consider inequalities with greater than, such as $|x| \ge 5$. A solution for $|x| \ge 5$ is any number whose absolute value is greater than or equal to 5. As the graph shows, the solution set contains all values that are a distance of 5 units or more from 0.

$$-7\,-6\,-5\,-4\,-3\,-2\,-1\ 0\ 1\ 2\ 3\ 4\ 5\ 6\ 7$$

Notice that the solutions are equivalent to $x \le -5$ or $x \ge 5$.

An inequality such as $|x| > 2$ uses parentheses.

Note Solutions for the **equal to** part of $|x| \ge 5$ are 5 and -5. The **greater than** part means all values that are farther from 0 than 5 and -5 because their absolute values are greater than 5.

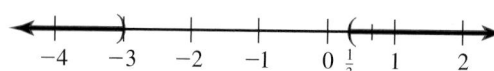

The solutions are equivalent to $x < -2$ or $x > 2$. Our examples suggest that we can split these inequalities into a compound inequality involving *or*.

> **Procedure** **Solving Inequalities in the Form $|x| > a$, where $a > 0$**
>
> To solve an inequality in the form $|x| > a$, where $a > 0$:
> 1. Rewrite the inequality as a compound inequality involving *or*: $x < -a$ or $x > a$.
> 2. Solve the compound inequality.
>
> Similarly, to solve $|x| \geq a$, we write $x \leq -a$ or $x \geq a$.

Warning When solving absolute value inequalities, look carefully at the order symbol. The rules for $<$ and $>$ are very different.

Instructor Note The procedure for solving $|x| > a, a > 0$, can also be developed using the definition of absolute value.

Example 2 For each inequality, solve, graph the solution set, and write the solution set in both set-builder notation and interval notation.

a. $|x - 2| \geq 5$

Solution: $x - 2 \leq -5$ **or** $x - 2 \geq 5$ Rewrite as a compound inequality.

$\qquad\qquad x \leq -3$ **or** $\qquad x \geq 7$ Add 2 to both sides of each inequality.

Solution set graph:

Set-builder notation: $\{x \mid x \leq -3 \text{ or } x \geq 7\}$

Interval notation: $(-\infty, -3] \cup [7, \infty)$

b. $|-3x - 4| > 5$

Solution: $-3x - 4 < -5$ or $-3x - 4 > 5$ Rewrite as a compound inequality.

$\qquad\quad -3x < -1$ or $-3x > 9$ Add 4 to both sides of each inequality.

$\qquad\quad x > \dfrac{1}{3}$ or $x < -3$ Divide both sides of each inequality by -3.

Solution set graph:

Set-builder notation: $\left\{ x \mid x < -3 \text{ or } x > \dfrac{1}{3} \right\}$

Interval notation: $(-\infty, -3) \cup \left(\dfrac{1}{3}, \infty \right)$

c. $\left| \dfrac{3}{4}x - 2 \right| + 1 \geq 6$

Solution: $\left| \dfrac{3}{4}x - 2 \right| \geq 5$ Subtract 1 from both sides to isolate the absolute value.

$\dfrac{3}{4}x - 2 \leq -5$ or $\dfrac{3}{4}x - 2 \geq 5$ Rewrite as a compound inequality.

$\qquad \dfrac{3}{4}x \leq -3$ or $\dfrac{3}{4}x \geq 7$ Add 2 to both sides of each inequality.

$\qquad\quad x \leq -4$ or $x \geq \dfrac{28}{3}$ Multiply both sides of each inequality by $\dfrac{4}{3}$.

Solution set graph:

$$\frac{28}{3}$$

-6 -5 -4 -3 -2 -1 0 1 2 3 4 5 6 7 8 9 10 11

Set-builder notation: $\left\{x \mid x \le -4 \text{ or } x \ge \dfrac{28}{3}\right\}$

Interval notation: $(-\infty, -4] \cup \left[\dfrac{28}{3}, \infty\right)$

d. $|0.2x + 1| - 3 > -7$

Solution: $|0.2x + 1| > -4$ Add 3 to both sides to isolate the absolute value.

This inequality indicates that the absolute value is greater than a negative number. Because the absolute value of every real number is either positive or 0, the solution set is $\mathbb{R}$. The graph is the entire number line.

Note The solutions of any inequality of the form $|x| > a$, $a < 0$, is all real numbers because the absolute value of any real number is nonnegative and, therefore, greater than any negative number.

Solution set graph:

-4 -3 -2 -1 0 1 2 3 4

Set-builder notation: $\{x \mid x \text{ is a real number}\}$, or $\mathbb{R}$

Interval notation: $(-\infty, \infty)$

Answers to Your Turn 2

a.

-2 -1 0 1 2 3 4 5 6 7

Set-builder: $\{x \mid x < 1 \text{ or } x > 5\}$
Interval: $(-\infty, 1) \cup (5, \infty)$

b.

-4 -3 -2 -1 0 1 2 3 4 5

Set-builder:

$\left\{x \mid x \le -\dfrac{5}{2} \text{ or } x \ge 2\right\}$

Interval: $\left(-\infty, -\dfrac{5}{2}\right] \cup [2, \infty)$

c.

-2 -1 0 1 2 3 4 5 6 7

Set-builder: $\{x \mid x \le 0 \text{ or } x \ge 5\}$
Interval: $(-\infty, 0] \cup [5, \infty)$

d.

-4 -3 -2 -1 0 1 2 3 4

Set-builder:
$\{x \mid x \text{ is a real number}\}$
Interval: $(-\infty, \infty)$

Your Turn 2 For each inequality, solve, graph the solution set, and write the solution set in both set-builder notation and interval notation.

a. $|x - 3| > 2$

b. $|4x + 1| \ge 9$

c. $|-0.4x + 1| + 2 \ge 3$

d. $\left|\dfrac{1}{3}x - 4\right| - 2 > -9$

8.3 Exercises

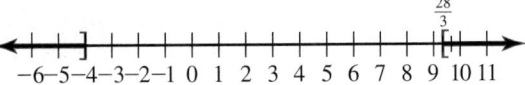

Note: Exercises marked with a ★ represent challenging exercises.

Objective 1

*For Exercises 1–4, assume that **a** is a positive number.*

Prep Exercise 1 What compound inequality is related to $|x| \le a$, where $a > 0$?
$x \ge -a$ and $x \le a$ or $-a \le x \le a$

Prep Exercise 2 How would you characterize the graph of $|x| < a$, where $a > 0$?
Shade between the values of $-a$ and a.

Prep Exercise 3 Under what conditions does $|x| < a$ have no solution?
When $a < 0$

For Exercises 1–14, solve the inequality. Then:
    a. Graph the solution set.
    b. Write the solution set using set-builder notation.
    c. Write the solution set using interval notation.
    See Example 2.

1. $|x| < 5$

a.

-6 -5 -4 -3 -2 -1 0 1 2 3 4 5 6

b. $\{x \mid -5 < x < 5\}$ c. $(-5, 5)$

2. $|y| \le 3$

a.

-6 -5 -4 -3 -2 -1 0 1 2 3 4 5 6

b. $\{y \mid -3 \le y \le 3\}$ c. $[-3, 3]$

3. $|x + 3| \leq 7$

a.
   -12 -10 -8 -6 -4 -2 0 2 4

b. $\{x \,|\, -10 \leq x \leq 4\}$ c. $[-10, 4]$

4. $|m - 5| < 2$

a.
   -4 -3 -2 -1 0 1 2 3 4 5 6 7 8

b. $\{m \,|\, 3 < m < 7\}$ c. $(3, 7)$

5. $|s + 3| + 6 < 9$

a.
   -6 -5 -4 -3 -2 -1 0 1 2 3 4 5 6

b. $\{s \,|\, -6 < s < 0\}$ c. $(-6, 0)$

6. $|p - 3| + 4 \leq 8$

a.
   -4 -3 -2 -1 0 1 2 3 4 5 6 7 8

b. $\{p \,|\, -1 \leq p \leq 7\}$ c. $[-1, 7]$

7. $|2m - 5| - 3 < 6$

a.
   -3 -2 -1 0 1 2 3 4 5 6 7 8

b. $\{m \,|\, -2 < m < 7\}$ c. $(-2, 7)$

8. $|4x + 12| + 2 < 10$

a.
   -6 -5 -4 -3 -2 -1 0 1

b. $\{x \,|\, -5 < x < -1\}$ c. $(-5, -1)$

9. $|-3k + 5| + 7 \leq 8$

a.
   0 1 $\frac{4}{3}$ 2 3

b. $\left\{ k \,\middle|\, \frac{4}{3} \leq k \leq 2 \right\}$ c. $\left[\frac{4}{3}, 2 \right]$

10. $|-5h - 1| - 6 < 8$

a.
   -4 -3 -2 -1 0 1 $\frac{13}{5}$ 2 3

b. $\left\{ h \,\middle|\, -3 < h < \frac{13}{5} \right\}$ c. $\left(-3, \frac{13}{5} \right)$

11. $2|x| + 7 \leq 3$

a.
   -6 -5 -4 -3 -2 -1 0 1 2 3 4 5 6

b. $\{ \, \}$ or $\varnothing$ c. no interval notation

12. $3|u| + 2 < -7$

a.
   -6 -5 -4 -3 -2 -1 0 1 2 3 4 5 6

b. $\{ \, \}$ or $\varnothing$ c. no interval notation

13. $2|w - 3| + 4 < 10$

a.
   -6 -5 -4 -3 -2 -1 0 1 2 3 4 5 6

b. $\{w \,|\, 0 < w < 6\}$ c. $(0, 6)$

14. $4|n + 2| - 3 \leq 9$

a.
   -6 -5 -4 -3 -2 -1 0 1 2 3 4 5 6

b. $\{n \,|\, -5 \leq n \leq 1\}$ c. $[-5, 1]$

Objective 2

Prep Exercise 4 What compound inequality is related to $|x| \geq a$, where $a > 0$?

$x \leq -a$ or $x \geq a$

Prep Exercise 5 How would you characterize the graph of $|x| > a$, where $a > 0$?

Shade to the left of $-a$ and to the right of a.

Prep Exercise 6 Under what conditions does the solution set for $|x| > a$ contain all real numbers?

When $a < 0$

For Exercises 15–28, solve the inequality. Then: **a.** *Graph the solution set.*
  b. *Write the solution set using set-builder notation.* } *See Example 2.*
  c. *Write the solution set using interval notation.*

15. $|c| > 12$

a.
   -12 -10 -8 -6 -4 -2 0 2 4 6 8 10 12

b. $\{c \,|\, c < -12 \text{ or } c > 12\}$ c. $(-\infty, -12) \cup (12, \infty)$

16. $|h| \geq 6$

a.
   -6 -5 -4 -3 -2 -1 0 1 2 3 4 5 6

b. $\{h \,|\, h \leq -6 \text{ or } h \geq 6\}$ c. $(-\infty, -6] \cup [6, \infty)$

17. $|y + 2| \geq 7$

a.
   -9 -8 -7 -6 -5 -4 -3 -2 -1 0 1 2 3 4 5 6

b. $\{y \,|\, y \leq -9 \text{ or } y \geq 5\}$ c. $(-\infty, -9] \cup [5, \infty)$

18. $|a - 4| > 3$

a.
   -6 -5 -4 -3 -2 -1 0 1 2 3 4 5 6 7

b. $\{a \,|\, a < 1 \text{ or } a > 7\}$ c. $(-\infty, 1) \cup (7, \infty)$

19. $|p - 6| - 3 > 5$

a.

-4 -2 0 2 4 6 8 10 12 14 16 18 20

b. $\{p \mid p < -2 \text{ or } p > 14\}$ c. $(-\infty, -2) \cup (14, \infty)$

20. $|x + 5| - 3 \geq 6$

a.

-16 -14 -12 -10 -8 -6 -4 -2 0 2 4 6 8

b. $\{x \mid x \leq -14 \text{ or } x \geq 4\}$ c. $(-\infty, -14] \cup [4, \infty)$

21. $|3x + 6| - 3 \geq 9$

a.

-6 -5 -4 -3 -2 -1 0 1 2 3 4

b. $\{x \mid x \leq -6 \text{ or } x \geq 2\}$ c. $(-\infty, -6] \cup [2, \infty)$

22. $|2x - 5| - 1 \geq 10$

a.

-5 -4 -3 -2 -1 0 1 2 3 4 5 6 7 8 9

b. $\{x \mid x \leq -3 \text{ or } x \geq 8\}$ c. $(-\infty, -3] \cup [8, \infty)$

23. $|-4n - 5| + 3 > 8$

a.

-4 -3 -2 -1 0 1 2

b. $\left\{n \mid n < -\dfrac{5}{2} \text{ or } n > 0\right\}$ c. $\left(-\infty, -\dfrac{5}{2}\right) \cup (0, \infty)$

24. $|-6h + 1| - 7 > 4$

a.

-3 -2 -1 0 1 2 3

b. $\left\{h \mid h < -\dfrac{5}{3} \text{ or } h > 2\right\}$ c. $\left(-\infty, -\dfrac{5}{3}\right) \cup (2, \infty)$

25. $4|v| + 3 \geq 7$

a.

-6 -5 -4 -3 -2 -1 0 1 2 3 4 5 6

b. $\{v \mid v \leq -1 \text{ or } v \geq 1\}$ c. $(-\infty, -1] \cup [1, \infty)$

26. $5|m| + 2 > 12$

a.

-6 -5 -4 -3 -2 -1 0 1 2 3 4 5 6

b. $\{m \mid m < -2 \text{ or } m > 2\}$ c. $(-\infty, -2) \cup (2, \infty)$

27. $4|y + 2| - 1 > 3$

a.

-6 -5 -4 -3 -2 -1 0 1 2 3 4 5 6

b. $\{y \mid y < -3 \text{ or } y > -1\}$ c. $(-\infty, -3) \cup (-1, \infty)$

28. $3|x - 2| - 5 \geq 1$

a.

-6 -5 -4 -3 -2 -1 0 1 2 3 4 5 6

b. $\{x \mid x \leq 0 \text{ or } x \geq 4\}$ c. $(-\infty, 0] \cup [4, \infty)$

Objectives 1 and 2

For Exercises 29–48, solve the inequality. Then: a. *Graph the solution set.*
b. *Write the solution set using set-builder notation.*
c. *Write the solution set using interval notation.*   } *See Examples 1 and 2.*

29. $|4m + 8| - 2 > 10$

a.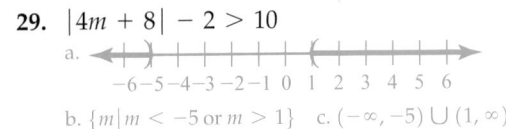

-6 -5 -4 -3 -2 -1 0 1 2 3 4 5 6

b. $\{m \mid m < -5 \text{ or } m > 1\}$ c. $(-\infty, -5) \cup (1, \infty)$

30. $|2x + 4| - 2 \geq 10$

a.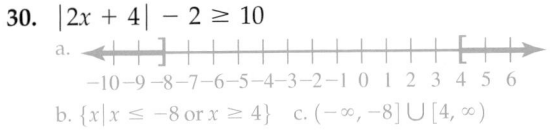

-10 -9 -8 -7 -6 -5 -4 -3 -2 -1 0 1 2 3 4 5 6

b. $\{x \mid x \leq -8 \text{ or } x \geq 4\}$ c. $(-\infty, -8] \cup [4, \infty)$

31. $|-3x + 6| < 6$

a.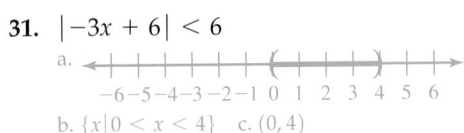

-6 -5 -4 -3 -2 -1 0 1 2 3 4 5 6

b. $\{x \mid 0 < x < 4\}$ c. $(0, 4)$

32. $|-2y + 3| \leq 3$

a.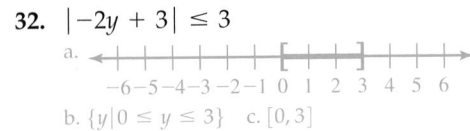

-6 -5 -4 -3 -2 -1 0 1 2 3 4 5 6

b. $\{y \mid 0 \leq y \leq 3\}$ c. $[0, 3]$

33. $|2r - 3| > -3$

a.

-6 -5 -4 -3 -2 -1 0 1 2 3 4 5 6

b. $\{r \mid r \text{ is a real number}\}$ c. $(-\infty, \infty)$

34. $|3b + 7| > -2$

a.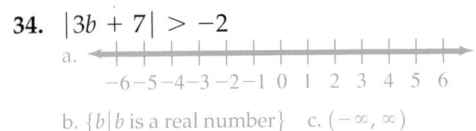

-6 -5 -4 -3 -2 -1 0 1 2 3 4 5 6

b. $\{b \mid b \text{ is a real number}\}$ c. $(-\infty, \infty)$

35. $4 - 2|x + 3| > 2$

a.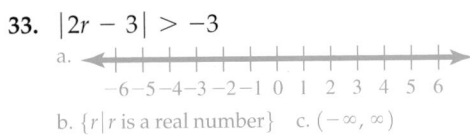

-6 -5 -4 -3 -2 -1 0 1 2 3 4 5 6

b. $\{x \mid -4 < x < -2\}$ c. $(-4, -2)$

36. $5 - 2|u + 4| \geq 1$

a.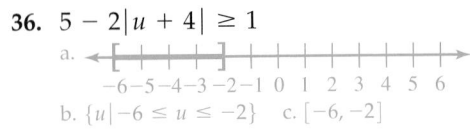

-6 -5 -4 -3 -2 -1 0 1 2 3 4 5 6

b. $\{u \mid -6 \leq u \leq -2\}$ c. $[-6, -2]$

37. $6|2x - 1| - 3 < 3$

a.

$$-6\ -5\ -4\ -3\ -2\ -1\ 0\ 1\ 2\ 3\ 4\ 5\ 6$$

b. $\{x \mid 0 < x < 1\}$ c. $(0, 1)$

38. $4|3p + 6| - 2 < 22$

a.

$$-6\ -5\ -4\ -3\ -2\ -1\ 0\ 1\ 2\ 3\ 4\ 5\ 6$$

b. $\{p \mid -4 < p < 0\}$ c. $(-4, 0)$

39. $5 - |w + 4| > 10$

a.

$$-14\ -12\ -10\ -8\ -6\ -4\ -2\ 0\ 2\ 4\ 6\ 8\ 10$$

b. $\varnothing$ c. no interval notation

40. $4 - |5 + k| > 7$

a.

$$-6\ -5\ -4\ -3\ -2\ -1\ 0\ 1\ 2\ 3\ 4\ 5\ 6$$

b. $\varnothing$ c. no interval notation

41. $\left| 2 - \dfrac{3}{2}k \right| \le 5$

a.

$$-6\ -5\ -4\ -3\ -2\ -1\ 0\ 1\ 2\ 3\ 4\ 5\ 6$$

b. $\left\{ k \mid -2 \le k \le \dfrac{14}{3} \right\}$ c. $\left[-2, \dfrac{14}{3} \right]$

42. $\left| 3 - \dfrac{1}{2}x \right| \le 7$

a.

$$-12\ -8\ -4\ 0\ 4\ 8\ 12\ 16\ 20\ 24$$

b. $\{x \mid -8 \le x \le 20\}$ c. $[-8, 20]$

43. $|0.25x - 3| + 2 > 4$

a.

$$0\ 4\ 8\ 12\ 16\ 20\ 24\ 28$$

b. $\{x \mid x < 4 \text{ or } x > 20\}$ c. $(-\infty, 4) \cup (20, \infty)$

44. $|0.5y - 3| + 4 > 5$

a.

$$1\ 2\ 3\ 4\ 5\ 6\ 7\ 8\ 9$$

b. $\{y \mid y < 4 \text{ or } y > 8\}$ c. $(-\infty, 4) \cup (8, \infty)$

45. $\left| 2.4 - \dfrac{3}{4}y \right| \le 7.2$

a.

$$-8\ -6\ -4\ -2\ 0\ 2\ 4\ 6\ 8\ 10\ 12\ 14$$

b. $\{y \mid -6.4 \le y \le 12.8\}$ c. $[-6.4, 12.8]$

46. $\left| 5.3 - \dfrac{2}{3}w \right| \ge 5.3$

a.

$$-2\ 0\ 2\ 4\ 6\ 8\ 10\ 12\ 14\ 16$$

b. $\{w \mid w \le 0 \text{ or } w \ge 15.9\}$ c. $(-\infty, 0] \cup [15.9, \infty)$

47. $|2p - 8| + 5 > 1$

a.

$$-5\ -4\ -3\ -2\ -1\ 0\ 1\ 2\ 3\ 4\ 5$$

b. $\{p \mid p \text{ is a real number}\}$ c. $(-\infty, \infty)$

48. $|8p + 7| + 4 > 3$

a.

$$-5\ -4\ -3\ -2\ -1\ 0\ 1\ 2\ 3\ 4\ 5$$

b. $\{p \mid p \text{ is a real number}\}$ c. $(-\infty, \infty)$

★ *For Exercises 49–58, write an inequality involving absolute value that describes the graph shown.*

49.

$$-3\ -2\ -1\ 0\ 1\ 2\ 3$$

$|x| < 3$

50.

$$-5\ -4\ -3\ -2\ -1\ 0\ 1\ 2\ 3\ 4\ 5$$

$|x| \le 5$

51.

$$-5\ -4\ -3\ -2\ -1\ 0\ 1\ 2\ 3\ 4\ 5$$

$|x| \ge 4$

52.

$$-3\ -2\ -1\ 0\ 1\ 2\ 3$$

$|x| \ge 2$

53.

$$-2\ -1\ 0\ 1\ 2$$

$|x + 1| > 1$

54.

$$0\ 1\ 2\ 3\ 4\ 5$$

$|x - 3| > 2$

55.

$$0\ 1\ 2\ 3\ 4\ 5$$

$|x - 3| \le 2$

56.

$$-6\ -5\ -4\ -3\ -2\ -1\ 0\ 1\ 2\ 3\ 4$$

$|x + 1| \le 4$

57.

$$-3\ -2\ -1\ 0\ 1$$

$|x| > \text{ any negative number}$

58.

$$-4\ -3\ -2\ -1\ 0\ 1\ 2\ 3\ 4$$

$|x| < \text{ any negative number}$

Review Exercises

Exercises 1–6 🔺 Equations and Inequalities

[2.5] 1. Let x represent the first number and y represent the second number; then translate to an equation. Twice the first number plus three times the second number is 3.

$2x + 3y = 3$

[2.1] 2. Determine whether $x = -4$ is a solution of $2x + 3 = -5$.

yes

For Exercises 3 and 4, solve.

[6.6] 3. $x^2 - 16 = 0$

$-4, 4$

[8.2] 4. $|x + 5| = -8$

no solution

For Exercises 5 and 6, graph.

[3.5] 5. $y = 2x - 3$

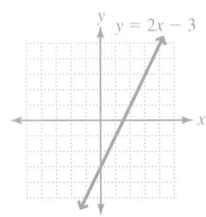

[3.3] 6. $2x - y = 4$

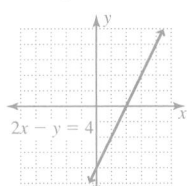

Chapter 8 Summary and Review Exercises

Complete each incomplete definition, rule, or procedure; study the key examples; and then work the related exercises.

8.1 Compound Inequalities

Definitions/Rules/Procedures	Key Example(s)
A **compound inequality** is two inequalities joined by either _____*and*_____ or _____*or*_____.	Examples of compound inequalities are $x > 5$ or $x < 7$, $x \geq -4$ and $x < 8$.
The **intersection of two sets** A and B, symbolized by $A \cap B$, is a set containing only elements that are in _____both A and B_____.	If $A = \{-2, -1, 0, 3, 7\}$ and $B = \{-3, -1, 1, 3\}$, then $A \cap B = \{-1, 3\}$ and $A \cup B = \{-3, -2, -1, 0, 1, 3, 7\}$.
The **union of two sets** A and B, symbolized by $A \cup B$, is a set containing every element in _____A or B_____.	

Exercises 1 and 2 ◢ Expressions

[8.1] *For Exercises 1 and 2, find the intersection and union of the given sets.*

1. $A = \{1, 5, 9\}$ Intersection: $\{1, 5, 9\}$
   $B = \{1, 5, 7, 9\}$ Union: $\{1, 5, 7, 9\}$

2. $A = \{1, 2, 3\}$ Intersection: $\{\ \}$ or $\varnothing$
   $B = \{4, 5, 6, 7\}$ Union: $\{1, 2, 3, 4, 5, 6, 7\}$

Definitions/Rules/Procedures	Key Example(s)
To solve a compound inequality involving *and*: 1. Solve each _____*inequality*_____ in the compound inequality. 2. Find the _____*intersection*_____ of the individual solution sets.	Solve the compound inequality. Then: **a.** Graph the solution set. **b.** Write the solution set using set-builder notation. **c.** Write the solution set using interval notation. $$-3 < -2x - 5 \leq 1$$ Solution: **a.** $-2x - 5 > -3$ and $-2x - 5 \leq 1$ 　　　$-2x > 2$　and　　$-2x \leq 6$ 　　　$x < -1$　and　　　$x \geq -3$ Graph: **b.** Set-builder notation: $\{x \mid -3 \leq x < -1\}$ **c.** Interval notation: $[-3, -1)$

Exercises 3–10 ◢ Equations and Inequalities

[8.1] *For Exercises 3–8, solve the compound inequality. Then:* *a.* *Graph the solution set.*
b. *Write the solution set using set-builder notation.*
c. *Write the solution set using interval notation.*

3. $4 < -2x < 6$
a.
b. $\{x \mid -3 < x < -2\}$ c. $(-3, -2)$

4. $9 \leq -3x \leq 15$
a.
b. $\{x \mid -5 \leq x \leq -3\}$ c. $[-5, -3]$

5. $-3 < x + 4 < 7$
a.
b. $\{x \mid -7 < x < 3\}$ c. $(-7, 3)$

6. $0 < x - 1 \leq 3$
a.
b. $\{x \mid 1 < x \leq 4\}$ c. $(1, 4]$

7. $2x + 3 \geq -1$ and $2x + 3 < 3$

a.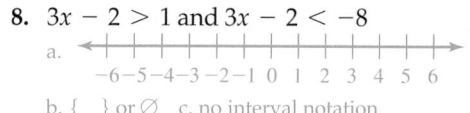
$\begin{array}{c} \leftarrow\!\!\!+\!\!+\!\!+\!\!+\!\![\!\!+\!\!+\!\!)\!\!+\!\!+\!\!+\!\!+\!\!+\!\!+\!\!\rightarrow \\ -6\!-\!5\!-\!4\!-\!3\!-\!2\!-\!1\ 0\ 1\ 2\ 3\ 4\ 5\ 6 \end{array}$

b. $\{x \mid -2 \leq x < 0\}$ c. $[-2, 0)$

8. $3x - 2 > 1$ and $3x - 2 < -8$

a. $\begin{array}{c} \leftarrow\!\!\!+\!\!+\!\!+\!\!+\!\!+\!\!+\!\!+\!\!+\!\!+\!\!+\!\!+\!\!+\!\!+\!\!\rightarrow \\ -6\!-\!5\!-\!4\!-\!3\!-\!2\!-\!1\ 0\ 1\ 2\ 3\ 4\ 5\ 6 \end{array}$

b. $\{ \ \}$ or $\varnothing$ c. no interval notation

[8.1] *For Exercises 9 and 10, solve. Then:*   *a. Graph the solution set.*
   b. Write the solution set using set-builder notation.
   c. Write the solution set using interval notation.

9. Students in a biology course receive a B if the average of four tests is 80 or higher and less than 90. Suppose a student has the following scores: 80, 89, and 83. What range of scores on the fourth test would cause the student to receive a B?

a. $\begin{array}{c} \leftarrow\!\!\!|\!|\!|\![\!\!|\!|\!|\!|\!|\!|\!|\!|\!|\!|\!|\!|\!|\!|\!|\!|\!|\!|\!|\!|\!|\!|\!|\!)\!|\!|\!\rightarrow \\ 65\ 70\ 75\ 80\ 85\ 90\ 95\ 100\ 105\ 110 \end{array}$

b. $\{x \mid 68 \leq x < 108\}$ c. $(68, 108)$

10. The width of a rectangular building is to be 80 feet. What range of values can the length have so that the area of the base of the building is from 12,000 to 16,000 square feet?

a. $\begin{array}{c} \leftarrow\!\![\!\!+\!\!+\!\!+\!\!+\!\!+\!\!+\!\!+\!\!+\!\!]\!\!+\!\!\rightarrow \\ 150\quad 160\quad 170\quad 180\quad 190\quad 200 \end{array}$

b. $\{x \mid 150 \leq x \leq 200\}$ c. $[150, 200]$

Definitions/Rules/Procedures	Key Example(s)
To solve a compound inequality involving *or*: **1.** Solve each ___inequality___ in the compound inequality. **2.** Find the ___union___ of the individual solution sets.	Solve the compound inequality. Then: **a.** Graph the solution set. **b.** Write the solution set using set-builder notation. **c.** Write the solution set using interval notation. $5x - 1 \leq -16$ or $5x - 1 > 9$ Solution: **a.** $\begin{aligned} 5x - 1 &\leq -16 &\text{or}&& 5x - 1 &> 9 \\ 5x &\leq -15 &\text{or}&& 5x &> 10 \\ x &\leq -3 &\text{or}&& x &> 2 \end{aligned}$ Graph: $\begin{array}{c} \leftarrow\!\!\!+\!\!+\!\!]\!\!+\!\!+\!\!+\!\!+\!\!(\!\!+\!\!+\!\!\rightarrow \\ -5\!-\!4\!-\!3\!-\!2\!-\!1\ 0\ 1\ 2\ 3\ 4 \end{array}$ **b.** Set-builder notation: $\{x \mid x \leq -3 \text{ or } x > 2\}$ **c.** Interval notation: $(-\infty, -3] \cup (2, \infty)$

Exercises 11–16   🔺   **Equations and Inequalities**

[8.1] *For Exercises 11–16, solve the compound inequality. Then:*   *a. Graph the solution set.*
   b. Write the solution set using set-builder notation.
   c. Write the solution set using interval notation.

11. $w + 4 \leq -2$ or $w + 4 \geq 2$

a. $\begin{array}{c} \leftarrow\!\!]\!\!+\!\!+\!\!+\!\![\!\!+\!\!+\!\!+\!\!+\!\!+\!\!+\!\!+\!\!\rightarrow \\ -6\!-\!5\!-\!4\!-\!3\!-\!2\!-\!1\ 0\ 1\ 2\ 3\ 4\ 5\ 6 \end{array}$

b. $\{w \mid w \leq -6 \text{ or } w \geq -2\}$ c. $(-\infty, -6] \cup [-2, \infty)$

12. $4w - 3 < 1$ or $4w - 3 > 0$

a. $\begin{array}{c} \leftarrow\!\!+\!\!+\!\!+\!\!+\!\!+\!\!+\!\!+\!\!+\!\!+\!\!+\!\!+\!\!+\!\!+\!\!\rightarrow \\ -6\!-\!5\!-\!4\!-\!3\!-\!2\!-\!1\ 0\ 1\ 2\ 3\ 4\ 5\ 6 \end{array}$

b. $\{x \mid x \text{ is a real number}\}$ c. $(-\infty, \infty)$

13. $2m - 5 < 0$ or $2m - 5 > 5$

a. $\begin{array}{c} \overset{\frac{5}{2}}{} \\ \leftarrow\!\!+\!\!+\!\!+\!\!+\!\!+\!\!+\!\!+\!\!+\!\!)\!\!+\!\!(\!\!+\!\!\rightarrow \\ -6\!-\!5\!-\!4\!-\!3\!-\!2\!-\!1\ 0\ 1\ 2\ 3\ 4\ 5\ 6 \end{array}$

b. $\left\{ m \mid m < \dfrac{5}{2} \text{ or } m > 5 \right\}$ c. $\left(-\infty, \dfrac{5}{2} \right) \cup (5, \infty)$

14. $3x + 2 \leq -2$ or $3x + 2 \geq 8$

a. $\begin{array}{c} \overset{-\frac{4}{3}}{} \\ \leftarrow\!\!+\!\!+\!\!+\!\!+\!\!]\!\!+\!\!+\!\![\!\!+\!\!+\!\!+\!\!+\!\!\rightarrow \\ -6\!-\!5\!-\!4\!-\!3\!-\!2\!-\!1\ 0\ 1\ 2\ 3\ 4\ 5\ 6 \end{array}$

b. $\left\{ x \mid x \leq -\dfrac{4}{3} \text{ or } x \geq 2 \right\}$ c. $\left(-\infty, -\dfrac{4}{3} \right] \cup [2, \infty)$

15. $-x - 6 \le -2$ or $-x - 6 \ge 3$

a.
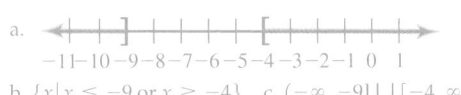
$$-11\text{-}10\text{-}9\text{-}8\text{-}7\text{-}6\text{-}5\text{-}4\text{-}3\text{-}2\text{-}1\;\;0\;\;1$$

b. $\{x \mid x \le -9 \text{ or } x \ge -4\}$ c. $(-\infty, -9] \cup [-4, \infty)$

16. $-4w + 1 \le -3$ or $-4w + 1 \ge 5$

a.
$$-6\text{-}5\text{-}4\text{-}3\text{-}2\text{-}1\;\;0\;\;1\;\;2\;\;3\;\;4\;\;5\;\;6$$

b. $\{w \mid w \le -1 \text{ or } w \ge 1\}$ c. $(-\infty, -1] \cup [1, \infty)$

8.2 Equations Involving Absolute Value; Absolute Value Functions

Definitions/Rules/Procedures	Key Example(s)
Absolute Value Property If $\lvert x \rvert = a$, where x is a variable or an expression and $a \ge 0$, then $x = \underline{\quad a \quad}$ or $x = \underline{\quad -a \quad}$. **To solve an equation containing a single absolute value:** 1. Isolate the absolute value so that the equation is in the form $\underline{\quad \lvert ax + b \rvert = c \quad}$. If $c > 0$, proceed to steps 2 and 3. If $c < 0$, the system has no solution. 2. Separate the absolute value into two equations: $\underline{\quad ax + b = c \quad}$ and $\underline{\quad ax + b = -c \quad}$. 3. Solve both equations.	Solve $\lvert 4x + 1 \rvert - 3 = 2$. **Solution:** $\lvert 4x + 1 \rvert = 5$ Isolate the absolute value. $4x + 1 = 5$ or $4x + 1 = -5$ Use the absolute value property. $\quad 4x = 4$ or $\quad 4x = -6$ $\quad\quad x = 1$ or $\quad\quad x = -\dfrac{3}{2}$

Exercises 17–24 Equations and Inequalities

[8.2] *For Exercises 17–24, solve.*

17. $\lvert x \rvert = 4$
$-4, 4$

18. $\lvert x - 4 \rvert = 7$
$-3, 11$

19. $\lvert 2w - 1 \rvert = 3$
$-1, 2$

20. $\lvert 5r + 8 \rvert = -3$
no solution

21. $\lvert q - 4 \rvert - 3 = 8$
$-7, 15$

22. $\lvert 5w \rvert - 2 = 13$
$-3, 3$

23. $2\lvert 3x - 4 \rvert = 8$
$0, \dfrac{8}{3}$

24. $4 - 2\lvert r - 5 \rvert = -8$
$-1, 11$

Definitions/Rules/Procedures	Key Example(s)
To solve an equation in the form $\lvert ax + b \rvert = \lvert cx + d \rvert$: 1. Separate the absolute value equation into two equations: $ax + b = \underline{\quad cx + d \quad}$ and $ax + b = \underline{\quad -(cx + d) \quad}$. 2. Solve both equations.	Solve $\lvert 5x - 7 \rvert = \lvert 3x + 1 \rvert$. **Solution:** Separate into two equations. **Equal** $\qquad\qquad$ **Opposites** $5x - 7 = 3x + 1$ or $5x - 7 = -(3x + 1)$ $2x - 7 = 1$ $\qquad$ or $5x - 7 = -3x - 1$ $\quad 2x = 8$ $\qquad$ or $\quad 8x = 6$ $\quad\; x = 4$ $\qquad$ or $\quad\;\; x = \dfrac{3}{4}$

Exercises 25–28 Equations and Inequalities

[8.2] *For Exercises 25–28, solve.*

25. $\lvert 3x - 2 \rvert = \lvert x + 2 \rvert$
$0, 2$

26. $\lvert 2x - 1 \rvert = \lvert 3x + 2 \rvert$
$x = -3, -\dfrac{1}{5}$

27. $\lvert -3x + 1 \rvert = \lvert 3 - 2x \rvert$
$-2, \dfrac{4}{5}$

28. $\lvert 9 - 4x \rvert = \lvert 7 - 2x \rvert$
$1, \dfrac{8}{3}$

Definitions/Rules/Procedures	Key Example(s)
An **absolute value function** is of the form $f(x) =$ ___$a\lvert x - h\rvert + k$___. The graph of an absolute value function is a ___V___ shape. To graph, find enough ordered pairs to generate the ___V___ shape.	Graph. $f(x) = \lvert x + 2\rvert$ 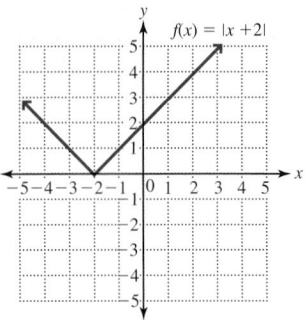

Exercises 29 and 30 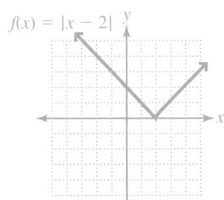 Equations and Inequalities

[8.2] *For Exercises 29 and 30, graph.*

29. $f(x) = \lvert x - 2\rvert$

$f(x) = \lvert x - 2\rvert$

30. $f(x) = -2\lvert x + 1\rvert + 3$

$f(x) = -2\lvert x + 1\rvert + 3$

8.3 Inequalities Involving Absolute Value

Definitions/Rules/Procedures	Key Example(s)

To solve an inequality in the form $\lvert x\rvert < a$, where $a > 0$:
 	
1. Rewrite the inequality as a compound inequality ___$x > -a$___ and ___$x < a$___ (or use $-a < x < a$).
 	
2. Solve the compound inequality.
 	
Similarly, to solve $\lvert x\rvert \le a$, we write ___$x \ge -a$___ and ___$x \le a$___ (or use $-a \le x \le a$).
	For each example, solve the inequality. Then:
	a. Graph the solution set.
	b. Write the solution set using set-builder notation.
	c. Write the solution set using interval notation.
	
	$\lvert 2x - 3\rvert - 1 < 4$
	
	Solution:
	
	a. $\lvert 2x - 3\rvert < 5$ Isolate the absolute value.
	
	  $2x - 3 > -5$ and $2x - 3 < 5$ Separate
	  $2x > -2$ and $2x < 8$ using *and*.
	  $x > -1$ and $x < 4$
	
	  Solution graph:
	
	  $-2\ -1\ \ 0\ \ 1\ \ 2\ \ 3\ \ 4\ \ 5$
	
	b. Set-builder notation. $\{x \mid -1 < x < 4\}$
	
	c. Interval notation: $(-1, 4)$

Exercises 31–34  Equations and Inequalities

[8.3] *For Exercises 31–34, solve the inequality. Then:*   *a. Graph the solution set.*
*b. Write the solution set using
set-builder notation.*
*c. Write the solution set using
interval notation.*

31. $|x| < 5$

   a.

   b. $\{x \mid -5 < x < 5\}$ c. $(-5, 5)$

32. $|2m + 6| < 4$

   a.

   b. $\{m \mid -5 < m < -1\}$ c. $(-5, -1)$

33. $|3s - 1| \leq -2$

   a.

   b. $\{\ \}$ or $\varnothing$ c. no interval notation

34. $7|m + 3| \leq 21$

   a.

   b. $\{m \mid -6 \leq m \leq 0\}$ c. $[-6, 0]$

Definitions/Rules/Procedures	Key Example(s)

To solve an inequality in the form $\|x\| > a$, where $a > 0$:
 	
 1. Rewrite the inequality as a compound inequality
    $\underline{\ \ x < -a\ \ }$ or $\underline{\ \ x > a\ \ }$.
 	
 2. Solve the compound inequality.
 	
 Similarly, to solve $\|x\| \geq a$, we write $\underline{\ \ x \leq -a\ \ }$ or $\underline{\ \ x \geq a\ \ }$.
	For each example, solve the inequality. Then:
	 a. Graph the solution set.
	 b. Write the solution set using set-builder notation.
	 c. Write the solution set using interval notation.
	
	 $\|3x - 5\| - 7 \geq 4$
	
	 Solution:
	
	 a. $\|3x - 5\| \geq 11$ Isolate the absolute value.
	
	    $3x - 5 \leq -11$ or $3x - 5 \geq 11$ Separate using *or*.
	
	      $3x \leq -6$ or $3x \geq 16$
	
	       $x \leq -2$ or $x \geq \dfrac{16}{3}$
	
	 Solution graph:
	
	 b. Set-builder notation: $\left\{ x \mid x \leq -2 \text{ or } x \geq \dfrac{16}{3} \right\}$
	
	 c. Interval notation: $(-\infty, -2] \cup \left[\dfrac{16}{3}, \infty \right)$

Exercises 35–40  Equations and Inequalities

[8.3] *For Exercises 35–40, solve the inequality. Then:*   *a. Graph the solution set.*
*b. Write the solution set using
set-builder notation.*
*c. Write the solution set using
interval notation.*

35. $|p| \geq 4$

   a.

   b. $\{p \mid p \leq -4 \text{ or } p \geq 4\}$ c. $(-\infty, -4] \cup [4, \infty]$

36. $|x - 3| > 7$

   a.

   b. $\{x \mid x < -4 \text{ or } x > 10\}$ c. $(-\infty, -4) \cup (10, \infty)$

37. $5|b| - 2 > 3$

a.

b. $\{b \mid b < -1 \text{ or } b > 1\}$ c. $(-\infty, -1) \cup (1, \infty)$

38. $-2|t - 5| < -10$

a.

b. $\{t \mid t < 0 \text{ or } t > 10\}$ c. $(-\infty, 0) \cup (10, \infty)$

39. $5 - 2|2k - 3| \le -15$

a.

b. $\left\{ k \mid k \le -\dfrac{7}{2} \text{ or } k \ge \dfrac{13}{2} \right\}$ c. $\left(-\infty, -\dfrac{7}{2} \right] \cup \left[\dfrac{13}{2}, \infty \right)$

40. $3 - 7|2p + 4| \le 24$

a.

b. $\{x \mid x \text{ is a real number}\}$ c. $(-\infty, \infty)$

Chapter 8 Practice Test

For Extra Help

Step-by-step test solutions are found on the Chapter Test Prep Videos available in MyMathLab® *or on* You**Tube**.

1. Find the intersection and union of the given sets. $A = \{h, o, m, e\}$
$B = \{h, o, u, s, e\}$

$\{e, h, o\}$
$\{e, h, m, o, s, u\}$ [8.1]

For Exercises 2–6, solve. Then: **a.** *Graph the solution set.*
b. *Write the solution set using set-builder notation.*
c. *Write the solution set using interval notation.*

2. $-3 < x + 4 \le 7$

a.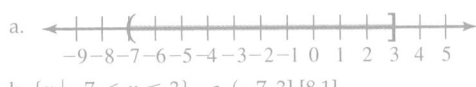

b. $\{x \mid -7 < x \le 3\}$ c. $(-7, 3]$ [8.1]

3. $4 < -2x \le 6$

a.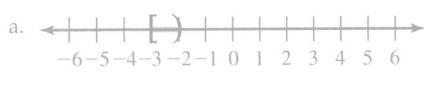

b. $\{x \mid -3 \le x < -2\}$ c. $[-3, -2)$ [8.1]

4. $5x + 2 \le -3$ or $5x + 2 \ge 12$

a.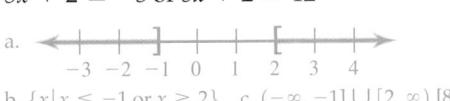

b. $\{x \mid x \le -1 \text{ or } x \ge 2\}$ c. $(-\infty, -1] \cup [2, \infty)$ [8.1]

5. $6 - 2n < 2$ or $6 - 2n < 4$

a.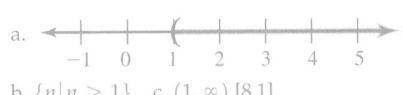

b. $\{n \mid n > 1\}$ c. $(1, \infty)$ [8.1]

6. A mail-order book club offers one bonus book if the total of an order is from \$40 to \$50. If a person were to order only books that cost \$5, what range would the number of books ordered have to be to receive a bonus book?

a.

b. $\{x \mid 8 \le x \le 10\}$ c. $[8, 10]$ [8.1]

For Exercises 7–10, solve.

7. $|x + 3| = 5$
$-8, 2$ [8.2]

8. $3 - |2x - 3| = -6$
$-3, 6$ [8.2]

9. $|2x + 3| = |x - 5|$
$-8, \dfrac{2}{3}$ [8.2]

10. $|5x - 4| = -3$
no solution [8.2]

For Exercises 11 and 12, graph.

11. $f(x) = |x + 1| - 2$

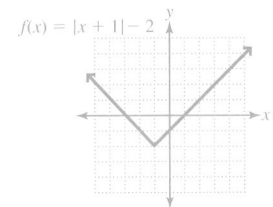

[8.2]

12. $f(x) = -2|x - 3| + 1$

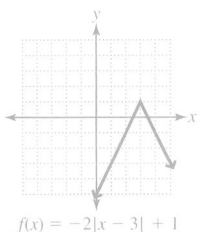

[8.2]

For Exercises 13–18, solve. Then: **a.** *Graph the solution set.*
                                            b. *Write the solutions set using set-builder notation.*
                                            c. *Write the solution set using interval notation.*

13. $|x + 4| < 9$

a.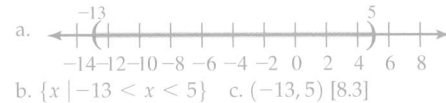

b. $\{x \,|\, -13 < x < 5\}$ c. $(-13, 5)$ [8.3]

14. $2|x - 1| > 4$

a.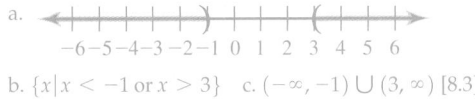

b. $\{x \,|\, x < -1 \text{ or } x > 3\}$ c. $(-\infty, -1) \cup (3, \infty)$ [8.3]

15. $3 - 2|x + 4| > -3$

a.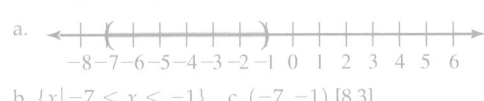

b. $\{x \,|\, -7 < x < -1\}$ c. $(-7, -1)$ [8.3]

16. $|3y - 2| < -2$

a.

b. $\{ \quad \}$ or $\varnothing$ c. no interval notation [8.3]

17. $|8t + 4| \geq -12$

a.

b. $\{x \,|\, x \text{ is a real number}\}$ c. $(-\infty, \infty)$ [8.3]

18. $2|3x - 4| \leq 10$

a.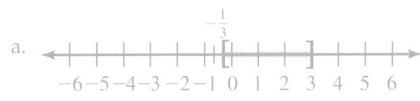

b. $\left\{ x \,\middle|\, -\dfrac{1}{3} \leq x \leq 3 \right\}$ c. $\left[-\dfrac{1}{3}, 3 \right]$ [8.3]

Chapters 1–8 Cumulative Review Exercises

For Exercises 1–4, answer true or false.

[5.2] 1. All terms are monomials.
false

[3.6] 2. To graph $2x - 3y > -6$, you would first graph $2x - 3y = -6$ as a dashed line.
true

[7.6] 3. An extraneous solution of $\dfrac{-2}{3} = \dfrac{5a + 10}{a^2 - 4}$ is -2.
true

[8.2] 4. The solution set of $|4x - 3| = -8$ is $\varnothing$.
true

For Exercises 5–7, fill in the blank.

[1.2] 5. The least common denominator (LCD) is the ___least common multiple (LCM)___ of the denominators of a given set of fractions.

[2.5] 6. Doubling the difference of x and 8 results in -4, which results in the equation ___$2(x - 8) = -4$___.

[3.2] 7. The graph of $x = -3$ is a(n) ___vertical line___.

Exercises 8–16 **Expressions**

For Exercises 8–14, simplify.

[1.5] 8. $10 - 2(6 - 3^2) \div 2 \cdot 3$
19

[5.5] 9. $(3x - y)^2$
$9x^2 - 6xy + y^2$

[5.6] 10. $(4xyz^{-3})^2(-2x^3y^{-4})^3$
$-\dfrac{128x^{11}}{y^{10}z^6}$

[5.6] 11. $\dfrac{x^3 - 5x^2 + 10x}{5x^2}$
$\dfrac{1}{5}x - 1 + \dfrac{2}{x}$

[7.2] 12. $\dfrac{a^2 - b^2}{x^2 - y^2} \div \dfrac{a + b}{x - y}$
$\dfrac{a - b}{x + y}$

[7.5] 13. $\dfrac{\dfrac{10x}{2x^2 + 7x - 4}}{\dfrac{5}{2x - 1}}$
$\dfrac{2x}{x + 4}$

[7.4] 14. $\dfrac{x^2 + 4x}{x^2 + 8x + 16} + \dfrac{3}{x + 4}$
$\dfrac{x + 3}{x + 4}$

For Exercises 15 and 16, factor completely.

[6.1] 15. $12bc - 9b + 20c - 15$
$(4c - 3)(3b + 5)$

[6.5] 16. $2x^3 - 14x^2 + 20x$
$2x(x - 5)(x - 2)$

Exercises 17–30 **Equations and Inequalities**

For Exercises 17–22, solve.

[2.3] 17. $2x - 3(2x - 4) = 4 + 4x$
1

[2.6] 18. $\dfrac{4}{3} = \dfrac{2x + 2}{2x - 1}$
5

[6.6] 19. $3x^2 - 7x - 20 = 0$
$-\dfrac{5}{3}, 4$

[7.6] 20. $\dfrac{x}{x - 2} - \dfrac{4}{x - 1} = \dfrac{2}{x^2 - 3x + 2}$
3, (2 is extraneous)

[8.1] 21. $0 \le -1.5x - 3 \le 4.5$
$-5 \le x \le -2$

[8.3] 22. $|5x - 2| \ge 8$
$x \le -\dfrac{6}{5} \text{ or } x \ge 2$

For Exercises 23 and 24, graph.

[3.7] 23. $f(x) = -\dfrac{2}{3}x + 4$

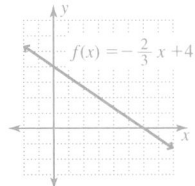

[3.6] 24. $x - 2y \le 4$

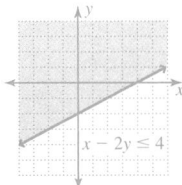

[2.7] 25. What percent of 90 is 32.4?
36%

[3.7] 26. Given $f(x) = \dfrac{x + 4}{x - 2}$, find

    a. $f(0)$ **b.** $f(5)$ **c.** $f(2)$
       -2 3 undefined

For Exercises 27–30, solve.

[4.2] 27. How much of a 40% saline solution must be added
[4.3] to 50 ml of a 15% saline solution so that a 30%
saline solution is created?
75 ml

[4.2] 28. A ship leaves port traveling at 15 mph. Two hours
[4.3] later a speedboat leaves the same port traveling in
the same direction at 40 mph. How long will it take
for the speedboat to overtake the ship?
$1\dfrac{1}{5}$ hr.

[7.7] 29. If y varies jointly as x and the square of z and
$y = 48$ when $x = 4$ and $z = 2$, find y when $x = 3$
and $z = 1$.
9

[6.6] 30. A rock is thrown vertically upward from ground
level with an initial velocity of 96 feet per second.
The equation giving its height above the ground is
$h = 96t - 16t^2$, where h is the height in feet and t
is the number of seconds after it was thrown. Find
the number of seconds when the rock is 80 feet
above the ground.
1 sec., 5 sec.

CHAPTER

9

Rational Exponents, Radicals, and Complex Numbers

Chapter Overview

In this chapter, we explore square roots and radicals in more detail. More specifically, we learn the following skills:

▶ Evaluate and simplify radical expressions.

▶ Connect radicals with rational exponents.

▶ Rewrite expressions containing radicals.

▶ Solve equations containing radicals.

▶ Explore complex numbers.

Instructor Note

The flow of this chapter parallels that of polynomials. After evaluating radicals, we explore arithmetic with radical expressions and then solve equations containing radical expressions. Solving radical equations sets the stage for discussing complex numbers. If pressed for time, you may consider skipping rationalizing numerators.

9.1 Radical Expressions and Functions

9.2 Rational Exponents

9.3 Multiplying, Dividing, and Simplifying Radicals

9.4 Adding, Subtracting, and Multiplying Radical Expressions

9.5 Rationalizing Numerators and Denominators of Radical Expressions

9.6 Radical Equations and Problem Solving

9.7 Complex Numbers

9.1 Radical Expressions and Functions

Objectives

1 Find the *n*th root of a number.
2 Approximate roots using a calculator.
3 Simplify radical expressions.
4 Evaluate radical functions.
5 Find the domain of radical functions.
6 Solve applications involving radical functions.

Warm-up

[1.5] *For Exercises 1 and 2, evaluate.*

1. $\sqrt{49}$

2. $\sqrt{\dfrac{64}{121}}$

[1.7] **3.** Evaluate $3x + 5$ for $x = -3$.

[2.8] **4.** Solve: $-2x + 6 \geq 0$

In this section through Section 9.5, we focus on the expression portion of our Algebra Pyramid and explore square root and radical expressions.

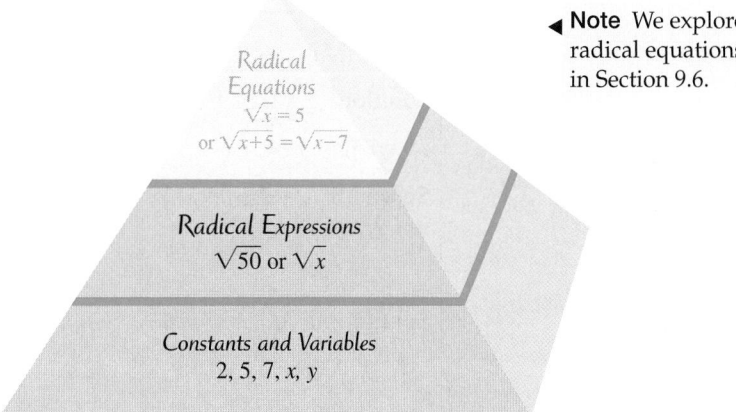

The Algebra Pyramid

◄ **Note** We explore radical equations in Section 9.6.

Objective 1 Find the *n*th root of a number.

In Section 1.5, we learned that a square root of a given number is a number whose square is the given number. We also learned that every positive real number has two square roots, a positive and a negative root. For example, the square roots of 16 are 4 and -4 because $4^2 = 16$ and $(-4)^2 = 16$. We can write the two square roots more compactly as ± 4.

Similarly, a *cube root* of a given number is a number whose cube is the given number. For example, the cube root of 8 is 2 because $2^3 = 8$.

The pattern continues for higher roots. For example, the fourth roots of 81 are ± 3 because $3^4 = 81$ and $(-3)^4 = 81$. Our examples suggest the following definition of **nth root**.

Definition *n*th root: The number b is an *n*th root of a number a if $b^n = a$.

Recall from Section 1.5, that the radical sign, $\sqrt{}$, denotes the principal (nonnegative) square root of a number. For roots other than square roots, we use the symbol $\sqrt[n]{a}$, read "the *n*th root of *a*," where *n*, the *root index*, indicates which root we are to find. The number *a*, called the *radicand*, is the number or expression whose root we are to find. The entire expression is called a *radical,* and any expression containing a radical is called a *radical expression.*

Just as $\sqrt{a}$ denotes the principal (nonnegative) square root of *a*, if *n* is even, then $\sqrt[n]{a}$ denotes the *principal nth root* of *a*. For example, $\sqrt[4]{81} = 3$ because 3 is the nonnegative fourth root of 81. Further, for $\sqrt[n]{a}$ to be a real number when *n* is even, *a* must be nonnegative because no real number can be raised to an even power to equal a negative number. For example, $\sqrt{-4}$ is not a real number because there is no real number whose square is -4.

Answers to Warm-up

1. 7   **2.** $\dfrac{8}{11}$

3. -4 **4.** $x \leq 3$

If n is odd in $\sqrt[n]{a}$, then a can be positive or negative because a positive number raised to an odd power is positive and a negative number raised to an odd power is negative.

> **Rule Evaluating nth Roots**
>
> When evaluating a radical expression $\sqrt[n]{a}$, the sign of a and the index n will determine possible outcomes.
>
> If a is nonnegative, then $\sqrt[n]{a} = b$, where $b \geq 0$ and $b^n = a$.
>
> If a is negative and n is even, then there is no real-number root.
>
> If a is negative and n is odd, then $\sqrt[n]{a} = b$, where b is negative and $b^n = a$.

Note In Section 9.7. we define the square root of a negative number using a set of numbers called the **imaginary numbers**.

Example 1 Evaluate each root if possible.

a. $\sqrt{225}$

Solution: $\sqrt{225} = 15$ Because $15^2 = 225$

b. $-\sqrt{0.81}$

Solution: $-\sqrt{0.81} = -0.9$ Because $(-0.9)^2 = 0.81$

◄ **Note** It can be helpful to think of $-\sqrt{0.81}$ as $-1 \cdot \sqrt{0.81} = -1 \cdot 0.9 = -0.9$.

c. $\sqrt{-9}$

Solution: $\sqrt{-9}$ is not a real number because there is no real number whose square is -9.

d. $\pm\sqrt{121}$

Solution: $\pm\sqrt{121} = \pm 11$ Because $(11)^2 = 121$ and $(-11)^2 = 121$

e. $\sqrt{\dfrac{4}{9}}$

Solution: $\sqrt{\dfrac{4}{9}} = \dfrac{2}{3}$ Because $\left(\dfrac{2}{3}\right)^2 = \dfrac{4}{9}$

f. $\sqrt[3]{8}$

Solution: $\sqrt[3]{8} = 2$ Because $2^3 = 8$

g. $\sqrt[3]{-8}$

Solution: $\sqrt[3]{-8} = -2$ Because $(-2)^3 = -8$

h. $\sqrt[4]{16}$

Solution: $\sqrt[4]{16} = 2$ Because $2^4 = 16$

Your Turn 1 Evaluate each root if possible.

a. $\sqrt{121}$ **b.** $-\sqrt{4}$ **c.** $\sqrt{-64}$ **d.** $\pm\sqrt{0.36}$

e. $\sqrt{\dfrac{25}{36}}$ **f.** $\sqrt[3]{64}$ **g.** $\sqrt[3]{-27}$ **h.** $-\sqrt[4]{81}$

Instructor Note Use a calculator to illustrate how the greater the number of decimal places we include in our approximation, the more accurate the approximation becomes. Checking $\sqrt{2} \approx 1.41$ by squaring 1.41, we get 1.9881. Checking $\sqrt{2} \approx 1.414$, we get 1.999396, which is much closer to 2.

Objective 2 Approximate roots using a calculator.

Each root we have considered so far has been rational, which means a rational number can express its exact value. But some roots, such as $\sqrt{2}$, are irrational, which means no rational number exists that expresses its exact value. In fact, writing $\sqrt{2}$ with the radical sign is how we express its exact value. However, $\sqrt{2}$ can be approximated using a calculator.

Calculator approximation to nine decimal places: $\sqrt{2} \approx 1.414213562$

Approximation to three decimal places: $\sqrt{2} \approx 1.414$

Approximation to two decimal places: $\sqrt{2} \approx 1.41$

◄ **Note** Recall that $\approx$ means "approximately equal to."

Answers to Your Turn 1
a. 11   **b.** -2
c. not a real number   **d.** ± 0.6
e. $\dfrac{5}{6}$ **f.** 4   **g.** -3 **h.** -3

Example 2 Approximate using a calculator. Round to three decimal places.

a. $\sqrt{12}$

Answer: $\sqrt{12} \approx 3.464$

b. $-\sqrt{38}$

Answer: $-\sqrt{38} \approx -6.164$

c. $\sqrt[3]{45}$

Answer: $\sqrt[3]{45} \approx 3.557$

Your Turn 2 Approximate using a calculator. Round to three decimal places.

a. $\sqrt{19}$　　　**b.** $-\sqrt{93}$　　　**c.** $\sqrt[3]{63}$

Objective 3 Simplify radical expressions.

The definition of a root can also be used to find roots with variable radicands. Recall that with an even index, the principal root is nonnegative. So at first, we will assume that all variables represent nonnegative values. We will use $(a^m)^n = a^{mn}$ to verify the roots.

Note Remember that with an even index, the principal root is nonnegative. Because we are assuming that the variables are nonnegative, our result accurately indicates the principal square root.

Example 3 Find the root. Assume that variables represent nonnegative values.

a. $\sqrt{x^2}$

Solution: $\sqrt{x^2} = x$　　Because $(x)^2 = x^2$

b. $\sqrt{a^6}$

Solution: $\sqrt{a^6} = a^3$　　Because $(a^3)^2 = a^6$

c. $\sqrt{16x^8}$

Solution: $\sqrt{16x^8} = 4x^4$　　Because $(4x^4)^2 = 16x^8$

d. $\sqrt{\dfrac{25x^8}{49y^2}}$

Solution: $\sqrt{\dfrac{25x^8}{49y^2}} = \dfrac{5x^4}{7y}$　　Because $\left(\dfrac{5x^4}{7y}\right)^2 = \dfrac{25x^8}{49y^2}$

Connection To raise a power to another power using $(a^m)^n = a^{mn}$, we multiply the powers. To find a root of a power, we divide the index into the exponent.

e. $\sqrt[3]{y^6}$

Solution: $\sqrt[3]{y^6} = y^2$　　Because $(y^2)^3 = y^6$

f. $\sqrt[4]{16x^{12}}$

Solution: $\sqrt[4]{16x^{12}} = 2x^3$　　Because $(2x^3)^4 = 16x^{12}$

Connection In parts c and f, notice the similarity between finding the root of a product and raising a product to a power. To raise a product to a power, we raise each factor to the power. To find a root of a product, we find the root of each factor.

Your Turn 3 Find the root. Assume that variables represent nonnegative values.

a. $\sqrt{x^4}$　　**b.** $\sqrt{9x^{10}}$　　**c.** $\sqrt{36a^{12}}$

d. $\sqrt{\dfrac{100x^4}{81y^6}}$　　**e.** $\sqrt[3]{27y^9}$　　**f.** $\sqrt[4]{b^8}$

Answers to Your Turn 2
a. 4.359　**b.** −9.644　**c.** 3.979

Answers to Your Turn 3
a. x^2　**b.** $3x^5$　**c.** $6a^6$
d. $\dfrac{10x^2}{9y^3}$　**e.** $3y^3$　**f.** b^2

If the variables can represent *any* real number and the index is even, we must be careful to ensure that the principal root is nonnegative by using absolute value symbols. If the root index is odd, however, we do not need absolute value symbols because the root can be positive or negative depending on the sign of the radicand.

Note To illustrate why $\sqrt{x^2} = |x|$, suppose we were to evaluate $\sqrt{x^2}$ when $x = -3$. We would have $\sqrt{(-3)^2} = \sqrt{9} = 3$. Notice that the root, 3, is, in fact, the absolute value of -3. If we had incorrectly stated that $\sqrt{x^2} = x$, then $\sqrt{(-3)^2}$ would have had to equal -3, which is not true.

Note In part d, we do not need absolute value because y^4, with its even exponent, is always nonnegative. In parts e and f, we do not use absolute value because the indices are odd.

| **Example 4** | Find the root. Assume that variables represent any real number. |

a. $\sqrt{x^2}$

Solution: $\sqrt{x^2} = |x|$

b. $\sqrt{25a^6}$

Solution: $\sqrt{25a^6} = 5|a^3|$

c. $\sqrt{(n+1)^2}$

Solution: $\sqrt{(n+1)^2} = |n+1|$

d. $\sqrt{25y^8}$

Solution: $\sqrt{25y^8} = 5y^4$

e. $\sqrt[3]{8n^3}$

Solution: $\sqrt[3]{8n^3} = 2n$

f. $\sqrt[3]{(t-2)^3}$

Solution: $\sqrt[3]{(t-2)^3} = t-2$

| **Your Turn 4** | Find the root. Assume that variables represent any real number. |

a. $\sqrt{x^4}$

b. $\sqrt{9x^{10}}$

c. $\sqrt{36a^{12}}$

d. $\sqrt{1.21u^6t^2}$

e. $\sqrt{\dfrac{100x^4}{81y^6}}$

f. $\sqrt[3]{27y^9}$

g. $\sqrt[4]{b^8}$

Objective 4 Evaluate radical functions.

Now that we have learned about radical expressions, let's examine **radical functions**.

Definition Radical function: A function containing a radical expression whose radicand has a variable.

| **Example 5** |

a. Given $f(x) = \sqrt{3x-2}$, find $f(3)$.

Solution: To find $f(3)$, substitute 3 for x and simplify.

$$f(3) = \sqrt{3(3)-2} = \sqrt{9-2} = \sqrt{7}$$

b. Given $f(x) = \sqrt{2x-6}$, find $f(0)$.

Solution: To find $f(0)$, substitute 0 for x and simplify. $f(0) = \sqrt{2(0)-6} = \sqrt{-6}$, which is not a real number.

| **Your Turn 5** | Given $f(x) = \sqrt{2x+5}$, find each of the following |

a. $f(-1)$

b. $f(-3)$

Objective 5 Find the domain of radical functions.

In Example 5(b), we found that $f(0)$ did not exist because $\sqrt{-6}$ is not a real number. What does this suggest about the domains of functions involving radicals? Consider the graphs of $f(x) = \sqrt{x}$ and $f(x) = \sqrt[3]{x}$, which we can generate by choosing values for x.

x	$f(x) = \sqrt{x}$
-1	Not real
0	0
1	1
4	2

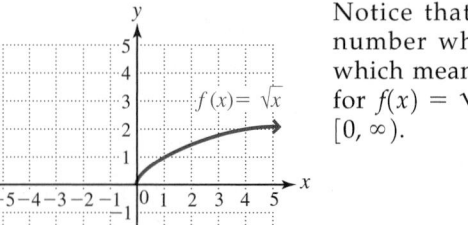

Notice that $\sqrt{x}$ is not a real number when x is negative, which means that the domain for $f(x) = \sqrt{x}$ is $\{x \mid x \geq 0\}$, or $[0, \infty)$.

Answers to Your Turn 4

a. x^2 **b.** $3|x^5|$ **c.** $6a^6$

d. $1.1|u^3t|$ **e.** $\dfrac{10x^2}{9|y^3|}$ **f.** $3y^3$ **g.** b^2

Answers to Your Turn 5

a. $f(-1) = \sqrt{3}$

b. $f(-3)$ is not a real number.

x	$f(x) = \sqrt[3]{x}$
-8	-2
-1	-1
0	0
1	1
8	2

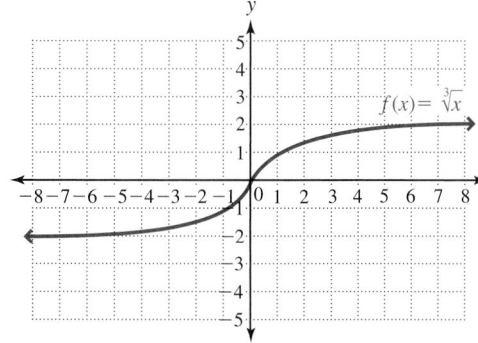

Alternatively, $\sqrt[3]{x}$, with its odd index, is a real number when x is negative; so its domain is all real numbers, or $(-\infty, \infty)$. Our graphs suggest the following conclusion:

Conclusion: The domain of a radical function with an even index must contain values that keep its radicand nonnegative.

Procedure   Finding the Domain of a Radical Function

If the root index is *odd*, the domain is $\{x \mid x \text{ is a real number}\}$, or $(-\infty, \infty)$.
If the root index is *even*, the radicand must be nonnegative. Consequently, the domain is the solution set to the inequality "radicand ≥ 0."

Example 6   Find the domain of each of the following.

a. $f(x) = \sqrt{x - 4}$

Solution: Because the index is even, the radicand must be nonnegative.

$$x - 4 \geq 0 \quad \text{Set the radicand} \geq 0.$$
$$x \geq 4 \quad \text{Add 4 to both sides.}$$

Domain: $\{x \mid x \geq 4\}$, or $[4, \infty)$

b. $f(x) = \sqrt{-2x + 6}$

Solution: $-2x + 6 \geq 0 \quad$ Set the radicand ≥ 0.

$$-2x \geq -6 \quad \text{Subtract 6 from both sides.}$$
$$x \leq 3 \quad \text{Divide both sides by } -2 \text{ and change the direction of the inequality.}$$

Domain: $\{x \mid x \leq 3\}$, or $(-\infty, 3]$

Note Recall that when we divide *both* sides of an inequality by a negative number, we reverse the direction of the inequality.

Your Turn 6   Find the domain of each of the following.

a. $f(x) = \sqrt{2x - 4}$ **b.** $f(x) = \sqrt{-3x - 9}$

Objective 6   Solve applications involving radical functions.

Often, radical functions appear in real-world situations where one variable is a function of another.

Example 7

a. The velocity of a free-falling object is a function of the distance it has fallen. Ignoring air resistance, the velocity of an object, v, in meters per second, can be found after it has fallen h meters by using the formula $v = -\sqrt{19.6h}$. Find the velocity of a stone that has fallen 30 meters after being dropped from a cliff.

Answers to Your Turn 6
a. $\{x \mid x \geq 2\}$, or $[2, \infty)$
b. $\{x \mid x \leq -3\}$, or $(-\infty, -3]$

Solution: $v = -\sqrt{19.6(30)}$ In $v = -\sqrt{19.6h}$, replace h with 30.

$v = -\sqrt{588}$ Multiply within the radical.

$v \approx -24.2$ Evaluate the square root.

Note: A negative velocity indicates that the object is traveling downward.

Answer: ≈ -24.2 m/sec.

b. The period of a pendulum is the amount of time it takes the pendulum to swing from the point of release to the opposite extreme and then back to the point of release. The period is a function of the length. The period, T, measured in seconds, can be found using the formula $T = 2\pi\sqrt{\dfrac{L}{9.8}}$, where L represents the length of the pendulum in meters. Find the period of a pendulum that is 0.5 meter long.

Solution: $T \approx 2(3.14)\sqrt{\dfrac{0.5}{9.8}}$ In $T = 2\pi\sqrt{\dfrac{L}{9.8}}$, substitute 3.14 for π and 0.5 for L.

$T \approx 6.28\sqrt{0.051}$ Simplify.

$T \approx 6.28(0.226)$ Approximate the square root.

$T \approx 1.42$ Multiply.

Answer: ≈ 1.42 sec.

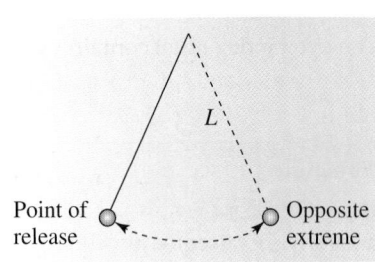

Point of release

Opposite extreme

Your Turn 7

a. A skydiver jumps from a plane and puts her body into a dive position so that air resistance is minimized. Find her velocity after she falls 100 meters. Use the formula in Example 7(a).

b. Find the period of a pendulum that is 0.2 meter long. Round to the nearest thousandth. Use the formula in Example 7(b).

Answers to Your Turn 7
a. ≈ -44.3 m/sec.
b. ≈ 0.898 sec.

9.1 Exercises · For Extra Help · MyMathLab®

Objective 1

Prep Exercise 1 Why are there two square roots for every positive real number?
Squaring a number or its additive inverse results in the same positive number.

For Exercises 1–8, find all square roots of each number. See Objective 1.

1. 36
± 6

2. 64
± 8

3. 121
± 11

4. 81
± 9

5. 196
± 14

6. 400
± 20

7. 225
± 15

8. 289
± 17

Prep Exercise 2 If the index is even, what is the principal root of a number?
The nonnegative root of a number

Prep Exercise 3 Why is an even root of any negative number not a real number?
You cannot raise a number to an even power and get a negative value.

For Exercises 9–50, evaluate each root if possible. See Example 1.

9. $\sqrt{25}$
5

10. $\sqrt{49}$
7

11. $\sqrt{-64}$
Not a real number

12. $\sqrt{-25}$
Not a real number

13. $-\sqrt{25}$
-5

14. $-\sqrt{100}$
-10

15. $\pm\sqrt{25}$
± 5

16. $\pm\sqrt{100}$
± 10

17. $\sqrt{1.44}$
1.2

18. $\sqrt{1.96}$
1.4

19. $\sqrt{-10.64}$
Not a real number

20. $\sqrt{-2.25}$
Not a real number

21. $-\sqrt{0.0121}$
-0.11

22. $-\sqrt{0.0169}$
-0.13

23. $\sqrt{\dfrac{49}{81}}$
$\dfrac{7}{9}$

24. $\sqrt{\dfrac{64}{169}}$
$\dfrac{8}{13}$

25. $-\sqrt{\dfrac{144}{169}}$
$-\dfrac{12}{13}$

26. $-\sqrt{\dfrac{25}{4}}$
$-\dfrac{5}{2}$

27. $\sqrt[3]{27}$
3

28. $\sqrt[3]{125}$
5

29. $\sqrt[3]{-64}$
-4

30. $\sqrt[3]{-216}$
-6

31. $-\sqrt[3]{-216}$
6

32. $-\sqrt[3]{-125}$
5

33. $\sqrt[4]{256}$
4

34. $\sqrt[4]{81}$
3

35. $\sqrt[4]{-625}$
Not a real number

36. $\sqrt[4]{-81}$
Not a real number

37. $-\sqrt[4]{16}$
-2

38. $-\sqrt[4]{625}$
-5

39. $\sqrt[5]{32}$
2

40. $\sqrt[5]{243}$
3

41. $\sqrt[5]{-243}$
-3

42. $\sqrt[5]{-32}$
-2

43. $-\sqrt[5]{-32}$
2

44. $-\sqrt[5]{-3125}$
5

45. $\sqrt[6]{64}$
2

46. $\sqrt[6]{729}$
3

47. $\sqrt[3]{-\dfrac{8}{27}}$
$-\dfrac{2}{3}$

48. $\sqrt[3]{-\dfrac{64}{125}}$
$-\dfrac{4}{5}$

49. $\sqrt[4]{\dfrac{16}{81}}$
$\dfrac{2}{3}$

50. $\sqrt[5]{\dfrac{1}{32}}$
$\dfrac{1}{2}$

Objective 2

Prep Exercise 4 Why is $\sqrt{16}$ rational whereas $\sqrt{17}$ is irrational?
$\sqrt{16}$ is rational because its exact value is a rational number, 4. $\sqrt{17}$ is irrational because its exact value cannot be expressed using a rational number.

Prep Exercise 5 How can we express the exact value of a root that is irrational?
With the radical symbol

For Exercises 51–66, approximate using a calculator. Round to three decimal places. See Example 2.

51. $\sqrt{7}$
2.646

52. $\sqrt{12}$
3.464

53. $-\sqrt{11}$
-3.317

54. $-\sqrt{41}$
-6.403

55. $\sqrt[3]{50}$
3.684

56. $\sqrt[3]{21}$
2.759

57. $\sqrt[3]{-53}$
-3.756

58. $\sqrt[3]{-83}$
-4.362

59. $\sqrt[4]{189}$
3.708

60. $\sqrt[4]{123}$
3.330

61. $-\sqrt[4]{85}$
-3.036

62. $-\sqrt[4]{77}$
-2.962

63. $\sqrt[5]{89}$
2.454

64. $\sqrt[5]{62}$
2.283

65. $\sqrt[6]{146}$
2.295

66. $\sqrt[6]{98}$
2.147

Objective 3

Prep Exercise 6 If x is any real number, explain why $\sqrt{x^2} = |x|$ instead of x (with no absolute value).
$\sqrt{x^2}$ means the nonnegative square root of x^2 even if x is negative.

For Exercises 67–90, find the root. Assume that variables represent nonnegative values. See Example 3.

67. $\sqrt{b^4}$
b^2

68. $\sqrt{r^8}$
r^4

69. $\sqrt{16x^2}$
$4x$

70. $\sqrt{81t^2}$
$9t$

71. $\sqrt{100r^8s^6}$
$10r^4s^3$

72. $\sqrt{121x^8y^{10}}$
$11x^4y^5$

73. $\sqrt{0.25a^6b^{12}}$
$0.5a^3b^6$

74. $\sqrt{0.81r^4s^{14}}$
$0.9r^2s^7$

75. $\sqrt[3]{m^3}$
m

76. $\sqrt[3]{n^6}$
n^2

77. $\sqrt[3]{27a^9b^6}$
$3a^3b^2$

78. $\sqrt[3]{64u^{12}t^9}$
$4u^4t^3$

79. $\sqrt[3]{-64a^3b^{12}}$
$-4ab^4$

80. $\sqrt[3]{-27r^{15}s^3}$
$-3r^5s$

81. $\sqrt[3]{0.008x^{18}}$
$0.2x^6$

82. $\sqrt[3]{0.027r^{12}}$
$0.3r^4$

83. $\sqrt[4]{a^4}$
a

84. $\sqrt[4]{x^{12}}$
x^3

85. $\sqrt[4]{16x^{16}}$
$2x^4$

86. $\sqrt[4]{81t^{20}}$
$3t^5$

87. $\sqrt[5]{32x^{10}}$
$2x^2$

88. $\sqrt[5]{243x^{15}}$
$3x^3$

89. $\sqrt[6]{x^{12}y^6}$
x^2y

90. $\sqrt[7]{s^7t^{21}}$
st^3

For Exercises 91–102, find the root. Assume that variables represent any real number.
See Example 4.

91. $\sqrt{36m^2}$
$6|m|$

92. $\sqrt{9t^6}$
$3|t^3|$

93. $\sqrt{(r-1)^2}$
$|r-1|$

94. $\sqrt{(k+3)^2}$
$|k+3|$

95. $\sqrt[4]{256y^{12}}$
$4|y^3|$

96. $\sqrt[4]{16x^{20}}$
$2|x^5|$

97. $\sqrt[3]{27y^3}$
$3y$

98. $\sqrt[3]{125x^6}$
$5x^2$

99. $\sqrt{(y-3)^4}$
$(y-3)^2$

100. $\sqrt{(x+2)^4}$
$(x+2)^2$

101. $\sqrt[3]{(y-4)^6}$
$(y-4)^2$

102. $\sqrt[3]{(n+5)^9}$
$(n+5)^3$

Objective 4

Prep Exercise 7 A radical function is a function containing a(n) ___radical___ expression whose ___radicand___ has a variable.

For Exercises 103–106, find the indicated value of the function. See Example 5.

103. $f(x) = \sqrt{2x+4}$; find $f(0)$
2

104. $f(x) = \sqrt{3x+2}$; find $f(3)$
$\sqrt{11}$

105. $f(x) = \sqrt{4x+3}$; find $f(3)$
$\sqrt{15}$

106. $f(x) = \sqrt{-2x+3}$; find $f(-2)$
$\sqrt{7}$

Objective 5

Prep Exercise 8 Explain how to determine the domain of a radical function.
Set the radicand ≥ 0. The solution set is the domain.

For Exercises 107–110, find the domain. See Example 6.

107. $f(x) = \sqrt{2x-8}$
$\{x \mid x \geq 4\}$, or $[4, \infty)$

108. $f(x) = \sqrt{3x+12}$
$\{x \mid x \geq -4\}$, or $[-4, \infty)$

109. $f(x) = \sqrt{-4x+16}$
$\{x \mid x \leq 4\}$, or $(-\infty, 4]$

110. $f(x) = \sqrt{-2x+6}$
$\{x \mid x \leq 3\}$, or $(-\infty, 3]$

For Exercises 111–114: *a.* Use a graphing calculator to graph the function.
  See Objective 5.
  b. Find the domain of the function.
  See Objective 5 and Example 6.

111. $f(x) = \sqrt{x-2}$
a.
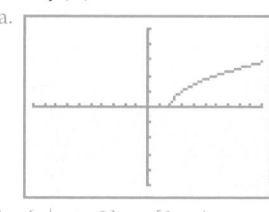
b. $\{x \mid x \geq 2\}$, or $[2, \infty)$

112. $f(x) = \sqrt{x+3}$
a.
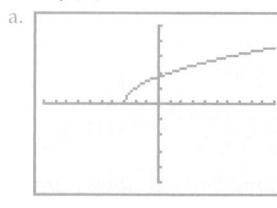
b. $\{x \mid x \geq -3\}$, or $[-3, \infty)$

113. $f(x) = \sqrt[3]{x + 1}$

a.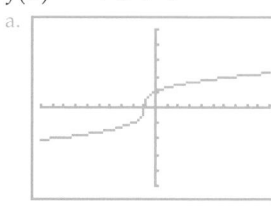

b. $\mathbb{R}$, or $(-\infty, \infty)$

114. $f(x) = \sqrt[4]{x - 3}$

a.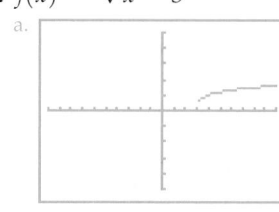

b. $\{x \mid x \geq 3\}$, or $[3, \infty)$

For Exercises 115 and 116, use the following information. Ignoring air resistance, the velocity, v, in meters per second, of an object after falling h meters can be found using the formula $v = -\sqrt{19.6h}$. Round your answers to the nearest thousandth. See Example 7(a).

115. Find the velocity of a rock that has been dropped from a cliff after it falls 16 meters.
≈ -17.709 m/sec.

116. Find the velocity of a ball that has been dropped from a roof after it falls 9 meters.
≈ -13.282 m/sec.

For Exercises 117 and 118, use the following information. The period, T, of a pendulum in seconds can be found using the formula $T = 2\pi\sqrt{\dfrac{L}{9.8}}$, where L represents the length of the pendulum in meters. Round your answers to the nearest thousandth. See Example 7(b).

117. Find the period of a pendulum that is 3 meters long.
≈ 3.476 sec.

118. Find the period of a pendulum that is 6 meters long.
≈ 4.916 sec.

Of Interest

One of the many topics Galileo Galilei studied was pendulums. His interest in them was piqued when he noticed a swinging light fixture in the cathedral of his hometown of Pisa and timed its period using his pulse.

For Exercises 119 and 120, use the following information. The formula $S = \dfrac{7}{2}\sqrt{2D}$ can be used to approximate the speed, S, in miles per hour, that a car was traveling prior to braking and skidding a distance, D, in feet, on asphalt. (Source: Harris Technical Services Traffic Accident Reconstructionists.) Round your answers to the nearest thousandth.

119. Find the speed of a car if the skid distance is 15 feet.
≈ 19.170 mph

120. Find the speed of a car if the skid distance is 40 feet.
≈ 31.305 mph

For Exercises 121 and 122, use the following information. Given the lengths a and b of the sides of a right triangle, we can find the hypotenuse, c, using the formula $c = \sqrt{a^2 + b^2}$.

121. Three pieces of lumber are to be connected to form a right triangle that will be part of the frame for a roof. If the horizontal piece is to be 12 feet and the vertical piece is to be 5 feet, how long must the connecting piece be?
13 ft.

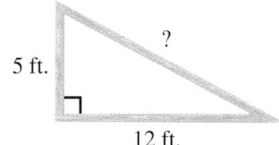

122. A counselor decides to create a ropes course with a zip line. She wants the line to connect from a 40-foot-tall tower to the ground at a point 100 feet from the base of the tower. Assuming that the tower and ground form a right angle, find the length of the zip line. Round your answer to the nearest tenth.
≈ 107.7 ft.

Connection In Section 6.6, we used the Pythagorean theorem, $c^2 = a^2 + b^2$, to find missing side lengths of a right triangle. The formula $c = \sqrt{a^2 + b^2}$ comes from isolating c in that theorem.

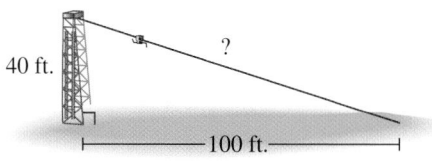

For Exercises 123 and 124, use the following information. Two forces, F_1 and F_2, acting on an object at a 90° angle will pull the object with a resultant force, R, at an angle between F_1 and F_2. (See the figure.) The value of the resultant force can be found using the formula $R = \sqrt{F_1^2 + F_2^2}$.

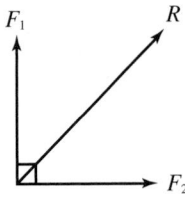

123. Find the resultant force if $F_1 = 9\,\text{N}$ and $F_2 = 12\,\text{N}$.

15 N

◄ **Note** The unit of force in the metric system is the newton (abbreviated N).

124. Find the resultant force if $F_1 = 8\,\text{N}$ and $F_2 = 15\,\text{N}$.

17 N

125. The number of earthquakes worldwide for 2008–2012 with a magnitude of 1.0–1.9 can be approximated using the function $f(x) = 10.5\sqrt{x} + 21$, where x represents the number of years after 2008. (*Hint:* The year 2008 corresponds to $x = 0$.) (*Source:* USGS Earthquake Hazards Program.)

  a. Find the approximate number of earthquakes with a magnitude of 1.0–1.9 that occurred in 2010.

  ≈ 36

  b. Find the approximate number of earthquakes with a magnitude of 1.0–1.9 that occurred in 2011.

  ≈ 39

126. The average mathematics scores for 8th graders in the United States on the International Mathematics and Science Study from 1995 to 2011 can be approximated by the function $f(x) = 4.3\sqrt{x} + 492$, where x represents the number of years after 1995. (*Hint:* The year 1995 corresponds to $x = 0$.) (*Source:* National Center for Education Statistics.)

  a. What was the average mathematics score for 8th graders in 2003? Round to the nearest whole number.

  504

  b. What was the average mathematics score for 8th graders in 2011? Round to the nearest whole number.

  509

Puzzle Problem A telemarketer calls a house and speaks to the mother of three children. The telemarketer asks the mother, "How old are you?" The woman answers, "41." The marketer then asks, "How old are your children?" The woman replies, "The product of their ages is our house number, 1296, and all of their ages are perfect squares." What are the ages of the children?

9, 9, 16 (The ages cannot be 4, 4, 81 because a 41-year-old mother cannot have an 81-year-old child or 4, 9, 36 because a 41-year-old mother cannot have a 36-year-old child.)

Of Interest

In 2011, U.S. 8th graders ranked 9th in mathematics on the International Mathematics and Science Study behind Finland (514), Israel (516), Russian Federation (539), Japan (570), Hong Kong-CHN (586), Chinese Taipei-CHN (609), Singapore (611), and Republic of Korea (613).

Review Exercises

Exercises 1–6 ▲ **Expressions**

[1.5] 1. Follow the order of operations to simplify.

  a. $\sqrt{16 \cdot 9}$ 12

  b. $\sqrt{16} \cdot \sqrt{9}$ 12

For Exercises 2–6, multiply.

[5.4] 2. $x^5 \cdot x^3$

x^8

[5.4] 3. $(-9m^3n)(5mn^2)$

$-45m^4n^3$

[5.5] 4. $4x^2(3x^2 - 5x + 1)$

$12x^4 - 20x^3 + 4x^2$

[5.5] 5. $(7y - 4)(3y + 5)$

$21y^2 + 23y - 20$

[5.5] 6. Write an expression for the area of the figure shown.

$x - 9$

$2x + 1$

$2x^2 - 17x - 9$

9.2 Rational Exponents

Objectives

1 Evaluate rational exponents.

2 Write radicals as expressions raised to rational exponents.

3 Simplify expressions with rational number exponents using the rules of exponents.

4 Use rational exponents to simplify radical expressions.

Objective 1 Evaluate rational exponents.

Rational Exponents with a Numerator of 1

So far, we have seen only integer exponents. However, expressions can have fractional exponents also, as in $3^{1/2}$. The fractional exponent in $3^{1/2}$ is called a **rational exponent**.

Definition Rational exponent: An exponent that is a rational number.

To discover what rational exponents mean, consider the following:

$$(3^{1/2})^2 = 3^{1/2 \cdot 2} = 3^1 = 3 \quad \text{Use } (a^m)^n = a^{mn}, \text{ which we learned in Section 5.1.}$$

By the definition of square root, $\sqrt{3}$ is the number whose square is 3; so

$$(\sqrt{3})^2 = 3$$

Our exploration suggests that $3^{1/2} = \sqrt{3}$. Note that the denominator, 2, of the rational exponent is the index of the root. (Remember that square roots have an unwritten index of 2.) This relationship holds for other roots as well.

> **Rule Rational Exponents with a Numerator of 1**
> $a^{1/n} = \sqrt[n]{a}$, where n is a natural number other than 1.

Note If a is negative and n is odd, the root is negative. If a is negative and n is even, there is no real-number root. Such roots are imaginary and are discussed in Section 9.7.

Example 1   Rewrite using radicals; then simplify if possible. Assume that all variables represent nonnegative values.

a. $36^{1/2}$

Solution: $36^{1/2} = \sqrt{36} = 6$

b. $81^{1/4}$

Solution: $81^{1/4} = \sqrt[4]{81} = 3$

c. $(-125)^{1/3}$

Solution: $(-125)^{1/3} = \sqrt[3]{-125} = -5$

d. $(-64)^{1/4}$

Solution: $(-64)^{1/4} = \sqrt[4]{-64}$ There is no real-number answer.

Note In $(-64)^{1/4}$, the parentheses group the minus sign with the radicand, whereas in $-36^{1/2}$, the minus sign is not part of the radicand.

e. $-36^{1/2}$

Solution: $-36^{1/2} = -\sqrt{36} = -6$

f. $x^{1/5}$

Solution: $x^{1/5} = \sqrt[5]{x}$

g. $(64x^6)^{1/2}$

Solution: $(64x^6)^{1/2} = \sqrt{64x^6} = 8x^3$

h. $12z^{1/4}$

Solution: $12\sqrt[4]{z}$

◀ **Note** Only the z is raised to the one-fourth power.

i. $\left(\dfrac{a^6}{36}\right)^{1/2}$

Solution: $\left(\dfrac{a^6}{36}\right)^{1/2} = \sqrt{\dfrac{a^6}{36}} = \dfrac{a^3}{6}$

Rewrite using radicals; then simplify if possible. Assume that all variables represent positive values.

a. $49^{1/2}$ **b.** $625^{1/4}$ **c.** $(-64)^{1/3}$ **d.** $-81^{1/2}$

e. $y^{1/4}$ **f.** $(25a^8)^{1/2}$ **g.** $18n^{1/5}$ **h.** $\left(\dfrac{49}{x^{10}}\right)^{1/2}$

Rational Exponents with a Numerator Other Than 1

Let's explore how we can rewrite expressions such as $8^{2/3}$ in which the numerator of the rational exponent is a number other than 1. Note that we could write the fraction $\dfrac{2}{3}$ as a product.

$$8^{2/3} = 8^{2(1/3)} \quad \text{or} \quad 8^{2/3} = 8^{(1/3)2}$$

Using the rule of exponents $(n^a)^b = n^{ab}$ in reverse, we can write

$$8^{2/3} = 8^{2(1/3)} = (8^2)^{1/3} \quad \text{or} \quad 8^{2/3} = 8^{(1/3)2} = (8^{1/3})^2$$

Applying the rule $a^{1/n} = \sqrt[n]{a}$, we can write

$$8^{2/3} = 8^{2(1/3)} = (8^2)^{1/3} = \sqrt[3]{8^2} \quad \text{or} \quad 8^{2/3} = 8^{(1/3)2} = (8^{1/3})^2 = (\sqrt[3]{8})^2$$

Notice that the denominator of the rational exponent becomes the index of the radical. The numerator of the rational exponent can be written as the exponent of the radicand or as an exponent for the entire radical.

> **Rule   General Rule for Rational Exponents**
> $a^{m/n} = \sqrt[n]{a^m} = (\sqrt[n]{a})^m$, where $a \geq 0$ and m and n are natural numbers other than 1.

Example 2   Rewrite using radicals; then simplify if possible. Assume that all variables represent nonnegative values.

a. $8^{2/3}$

Solution: $8^{2/3} = \sqrt[3]{8^2}$ Rewrite.   or $8^{2/3} = (\sqrt[3]{8})^2$ Rewrite.

$\qquad\qquad = \sqrt[3]{64}$ Square the radicand. $\qquad\quad = (2)^2$ Evaluate the root.

$\qquad\qquad = 4$ Evaluate the root. $\qquad\qquad\qquad = 4$ Evaluate the exponential form.

Warning In the expression $-9^{5/2}$, the negative sign tells us to find the opposite of the value of the exponential form.

$$-9^{5/2} = -(9^{5/2}) = -(\sqrt{9})^5$$

Or some people find it helpful to think of $-9^{5/2}$ as $-1 \cdot 9^{5/2}$.

$$-9^{5/2} = -1 \cdot 9^{5/2}$$
$$= -1 \cdot (\sqrt{9})^5$$

b. $625^{3/4}$

Solution: We could rewrite $625^{3/4}$ as $\sqrt[4]{625^3}$, but calculating 625^3 is tedious without a calculator; so we use the other form of the rule.

$$625^{3/4} = (\sqrt[4]{625})^3 \quad \text{Rewrite.}$$
$$= (5)^3 \quad \text{Evaluate the root.}$$
$$= 125 \quad \text{Evaluate the exponential form.}$$

c. $-9^{5/2}$

Solution: $-9^{5/2} = -(\sqrt{9})^5 = -(3)^5 = -243$

d. $\left(\dfrac{1}{9}\right)^{5/2}$ **Note** The simplification is usually easier if we write $a^{m/n}$ as $(\sqrt[n]{a})^m$ if a is a constant and as $\sqrt[n]{a^m}$ if a is a variable or a variable expression.

Solution: $\left(\dfrac{1}{9}\right)^{5/2} = \left(\sqrt{\dfrac{1}{9}}\right)^5 = \left(\dfrac{1}{3}\right)^5 = \dfrac{1}{243}$

e. $x^{2/3}$

Solution: $x^{2/3} = \sqrt[3]{x^2}$ ◀ **Note** Because the base is a variable, we put the exponent beneath the radical sign.

f. $(3x - 5)^{3/5}$

Solution: $(3x - 5)^{3/5} = \sqrt[5]{(3x - 5)^3}$

Your Turn 2   Rewrite using radicals; then simplify if possible. Assume that all variables represent nonnegative values.

    a. $32^{3/5}$ **b.** $-49^{3/2}$ **c.** $\left(\dfrac{1}{16}\right)^{3/2}$ **d.** $y^{5/6}$ **e.** $(3a + 4)^{4/5}$

Negative Rational Exponents

Recall from Section 5.1 that $a^{-b} = \dfrac{1}{a^b}$. For example, $2^{-3} = \dfrac{1}{2^3}$. This same rule applies to negative rational exponents.

> **Rule Negative Rational Exponents**
>
> $a^{-m/n} = \dfrac{1}{a^{m/n}}$, where $a \neq 0$ and m and n are natural numbers with $n \neq 1$.

Example 3   Rewrite using radicals; then simplify if possible.

a. $81^{-1/2}$

Solution: $81^{-1/2} = \dfrac{1}{81^{1/2}}$ Rewrite the exponential form with a positive exponent by inverting the base and changing the sign of the exponent.

$= \dfrac{1}{\sqrt{81}}$ Write the rational exponent in radical form.

$= \dfrac{1}{9}$ Evaluate the square root.

b. $16^{-3/4}$

Solution: $16^{-3/4} = \dfrac{1}{16^{3/4}}$ Rewrite the exponential form with a positive exponent by inverting the base and changing the sign of the exponent.

$= \dfrac{1}{(\sqrt[4]{16})^3}$ Write the rational exponent in radical form.

$= \dfrac{1}{(2)^3}$ Evaluate the radical.

$= \dfrac{1}{8}$ Simplify the exponential form.

Connection Although we used $a^{-m/n} = \dfrac{1}{a^{m/n}}$ in part c, we could have used $\left(\dfrac{a}{b}\right)^{-n} = \left(\dfrac{b}{a}\right)^{n}$, which we learned in Chapter 5.

c. $\left(\dfrac{16}{25}\right)^{-1/2}$

Solution: $\left(\dfrac{16}{25}\right)^{-1/2} = \dfrac{1}{\left(\dfrac{16}{25}\right)^{1/2}}$ Rewrite the exponential form with a positive exponent by inverting the base and changing the sign of the exponent.

$= \dfrac{1}{\sqrt{\dfrac{16}{25}}}$ Write the rational exponent in radical form.

Answers to Your Turn 2

a. 8   **b.** -343 **c.** $\dfrac{1}{64}$

d. $\sqrt[6]{y^5}$ **e.** $\sqrt[5]{(3a + 4)^4}$

Of Interest

John Wallis (1616–1703) was one of the first mathematicians to explain rational and negative exponents. Wallis wrote extensively about physics and mathematics; unfortunately, his work was overshadowed by the work of his countryman Isaac Newton. Newton added to what Wallis began with exponents and roots and popularized the notation that we use today. (*Source:* D. E. Smith, *History of Mathematics*, Dover, 1953.)

$$= \frac{1}{\frac{4}{5}} \qquad \text{Evaluate the square root.}$$

$$= \frac{5}{4} \qquad \text{Simplify the complex fraction.}$$

d. $(-64)^{-2/3}$

Solution: $(-64)^{-2/3} = \dfrac{1}{(-64)^{2/3}}$ Rewrite the exponential form with a positive exponent.

$$= \frac{1}{(\sqrt[3]{-64})^2} \qquad \text{Write the rational exponent in radical form.}$$

$$= \frac{1}{(-4)^2} \qquad \text{Evaluate the cube root.}$$

$$= \frac{1}{16} \qquad \text{Square } -4.$$

> **Your Turn 3**   Rewrite using radicals; then simplify if possible.

a. $49^{-1/2}$ **b.** $-27^{-2/3}$ **c.** $\left(\dfrac{16}{81}\right)^{-3/4}$ **d.** $(-8)^{-5/3}$

Objective 2 Write radicals as expressions raised to rational exponents.

In upper-level math courses, it is often necessary to write radical expressions in exponential form. To do so, we use the facts that $\sqrt[n]{a^m} = a^{m/n}$ and $(\sqrt[n]{a})^m = a^{m/n}$.

> **Example 4**   Write each of the following in exponential form. Assume that all variables represent positive values.

a. $\sqrt[4]{x^3}$
Solution: $\sqrt[4]{x^3} = x^{3/4}$

b. $\dfrac{1}{\sqrt[3]{x^2}}$

Solution: $\dfrac{1}{\sqrt[3]{x^2}} = \dfrac{1}{x^{2/3}} = x^{-2/3}$

c. $(\sqrt[5]{z})^3$
Solution: $(\sqrt[5]{z})^3 = z^{3/5}$

d. $\sqrt[6]{(3x - 5)^5}$
Solution: $\sqrt[6]{(3x - 5)^5} = (3x - 5)^{5/6}$

> **Your Turn 4**   Write each of the following in exponential form. Assume that all variables represent positive values.

a. $\sqrt[5]{y^2}$ **b.** $\dfrac{1}{\sqrt[4]{x^3}}$ **c.** $(\sqrt[7]{a})^3$ **d.** $\sqrt[3]{(3y - 2)^5}$

Answers to Your Turn 3
a. $\dfrac{1}{7}$ **b.** $-\dfrac{1}{9}$
c. $\dfrac{27}{8}$ **d.** $-\dfrac{1}{32}$

Answers to Your Turn 4
a. $y^{2/5}$ **b.** $x^{-3/4}$
c. $a^{3/7}$ **d.** $(3y - 2)^{5/3}$

Objective 3 Simplify expressions with rational number exponents using the rules of exponents.

Rational exponents follow the same rules that we established in Chapter 5 for integer exponents, which we review here.

Rule **Rules of Exponents Summary**

(Assume that no denominators are 0, that a and b are real numbers, and that m and n are integers.)

Zero as an exponent: $a^0 = 1$, where $a \neq 0$
0^0 is indeterminate.

Negative exponents: $a^{-n} = \dfrac{1}{a^n}$ and $\dfrac{1}{a^{-n}} = a^n$ $\left(\dfrac{a}{b}\right)^{-n} = \left(\dfrac{b}{a}\right)^{n}$

Product rule of exponents: $a^m \cdot a^n = a^{m+n}$

Quotient rule for exponents: $\dfrac{a^m}{a^n} = a^{m-n}$

Raising a power to a power: $(a^m)^n = a^{mn}$

Raising a product to a power: $(ab)^n = a^n b^n$

Raising a quotient to a power: $\left(\dfrac{a}{b}\right)^n = \dfrac{a^n}{b^n}$

Example 5 Use the rules of exponents to simplify. Write the answers with positive exponents. Assume that all variables represent positive values.

a. $x^{1/5} \cdot x^{3/5}$

Solution: $x^{1/5} \cdot x^{3/5} = x^{1/5 + 3/5}$ Use $a^m \cdot a^n = a^{m+n}$.

$= x^{4/5}$ Add the exponents.

b. $(2a^{1/2})(4a^{1/3})$

Solution: $(2a^{1/2})(4a^{1/3}) = 8a^{1/2 + 1/3}$ Use $a^m \cdot a^n = a^{m+n}$.

$= 8a^{3/6 + 2/6}$ Rewrite the exponents with a common denominator of 6.

$= 8a^{5/6}$ Add the exponents.

c. $\dfrac{5^{2/3}}{5^{3/4}}$

Solution: $\dfrac{5^{2/3}}{5^{3/4}} = 5^{2/3 - 3/4}$ Use $\dfrac{a^m}{a^n} = a^{m-n}$.

$= 5^{8/12 - 9/12}$ Rewrite the exponents with a common denominator of 12.

$= 5^{-1/12}$ Subtract the exponents.

$= \dfrac{1}{5^{1/12}}$ Rewrite with a positive exponent.

d. $(-4y^{-3/5})(5y^{4/5})$

Solution: $(-4y^{-3/5})(5y^{4/5}) = -20y^{-3/5 + 4/5}$ Use $a^m \cdot a^n = a^{m+n}$.

$= -20y^{1/5}$ Add the exponents.

e. $(w^{3/4})^2$

Solution: $(w^{3/4})^2 = w^{(3/4)2}$ Use $(a^m)^n = a^{mn}$.

$= w^{3/2}$ Multiply the exponents.

f. $(2a^{2/3}b^{3/5})^3$

Solution: $(2a^{2/3}b^{3/5})^3 = 2^3(a^{2/3})^3(b^{3/5})^3$ Use $(ab)^n = a^n \cdot b^n$.

$= 8a^{(2/3)3}b^{(3/5)3}$ Use $(a^m)^n = a^{mn}$.

$= 8a^2 b^{9/5}$ Multiply the exponents.

g. $\dfrac{(3x^{4/3})^3}{x^3}$

Solution: $\dfrac{(3x^{4/3})^3}{x^3} = \dfrac{3^3(x^{4/3})^3}{x^3}$ Use $(ab)^n = a^n \cdot b^n$ in the numerator.

$= \dfrac{27x^{(4/3)3}}{x^3}$ Use $(a^m)^n = a^{mn}$.

$= \dfrac{27x^4}{x^3}$ Simplify.

$= 27x^{4-3}$ Use $\dfrac{a^m}{a^n} = a^{m-n}$.

$= 27x$ Simplify.

Your Turn 5 Use the rules of exponents to simplify. Write the answers with positive exponents. Assume that all variables represent positive values.

a. $(x^{2/3})(x^{1/2})$ **b.** $\dfrac{y^{2/7}}{y^{5/7}}$ **c.** $(b^{3/4})^2$ **d.** $(2x^{3/2}y^{2/3})^4$ **e.** $\dfrac{(2z^{5/3})^3}{z^2}$

Objective 4 Use rational exponents to simplify radical expressions.

Often, radical expressions can be simplified by rewriting them with rational exponents, simplifying, and then rewriting as a radical expression.

Example 6 Rewrite as a radical with a smaller root index. Assume that all variables represent nonnegative values.

a. $\sqrt[4]{49}$

Solution: $\sqrt[4]{49} = 49^{1/4}$ Rewrite in exponential form.

$= (7^2)^{1/4}$ Write 49 as 7^2.

$= 7^{2 \cdot 1/4}$ Use $(a^m)^n = a^{mn}$.

$= 7^{1/2}$ Simplify.

$= \sqrt{7}$ Write in radical form.

b. $\sqrt[6]{x^4}$

Solution: $\sqrt[6]{x^4} = x^{4/6}$ Rewrite in exponential form.

$= x^{2/3}$ Simplify to lowest terms.

$= \sqrt[3]{x^2}$ Write in radical form.

c. $\sqrt[6]{a^4b^2}$

Solution: $\sqrt[6]{a^4b^2} = (a^4b^2)^{1/6}$ Rewrite in exponential form.

$= (a^4)^{1/6}(b^2)^{1/6}$ Use $(ab)^n = a^nb^n$.

$= a^{4 \cdot 1/6}b^{2 \cdot 1/6}$ Use $(a^m)^n = a^{mn}$.

$= a^{2/3}b^{1/3}$ Simplify.

$= (a^2b)^{1/3}$ Use $(ab)^n = a^nb^n$.

$= \sqrt[3]{a^2b}$ Write in radical form.

Answers to Your Turn 5

a. $x^{7/6}$ **b.** $\dfrac{1}{y^{3/7}}$ **c.** $b^{3/2}$

d. $16x^6y^{8/3}$ **e.** $8z^3$

Answers to Your Turn 6

a. $\sqrt{6}$ **b.** $\sqrt[4]{x^3}$ **c.** $\sqrt[4]{x^3y}$

Your Turn 6 Rewrite as a radical with a smaller root index. Assume that all variables represent nonnegative values.

a. $\sqrt[4]{36}$ **b.** $\sqrt[8]{x^6}$ **c.** $\sqrt[8]{x^6y^2}$

Writing radicals in exponential form also allows us to multiply and divide radical expressions with different root indices.

> **Procedure** **Multiplying and Dividing Radical Expressions with Different Root Indices**
>
> To multiply or divide radical expressions with different root indices:
> 1. Change from radical to exponential form.
> 2. Multiply or divide using the appropriate rule(s) of exponents.
> 3. Write the result from step 2 in radical form.

Example 7   Perform the indicated operations. Write the result using a radical. Assume that all variables represent positive values.

a. $\sqrt{x} \cdot \sqrt[3]{x^2}$

Solution:

$$\sqrt{x} \cdot \sqrt[3]{x^2} = x^{1/2} \cdot x^{2/3} \qquad \text{Write in exponential form.}$$
$$= x^{1/2 + 2/3} \qquad \text{Use } a^m \cdot a^n = a^{m+n}.$$
$$= x^{3/6 + 4/6} \qquad \text{Rewrite the exponents with their LCD.}$$
$$= x^{7/6} \qquad \text{Simplify.}$$
$$= \sqrt[6]{x^7} \qquad \text{Write in radical form.}$$

Note In Section 9.3, we learn to further simplify expressions such as $\sqrt[6]{x^7}$. ►

b. $\dfrac{\sqrt[4]{x^3}}{\sqrt[3]{x}}$

Solution:

$$\frac{\sqrt[4]{x^3}}{\sqrt[3]{x}} = \frac{x^{3/4}}{x^{1/3}} \qquad \text{Write in exponential form.}$$
$$= x^{3/4 - 1/3} \qquad \text{Use } \frac{a^m}{a^n} = a^{m-n}.$$
$$= x^{9/12 - 4/12} \qquad \text{Rewrite the exponents with their LCD.}$$
$$= x^{5/12} \qquad \text{Simplify.}$$
$$= \sqrt[12]{x^5} \qquad \text{Write in radical form.}$$

c. $\sqrt{3} \cdot \sqrt[3]{2}$

Solution:

$$\sqrt{3} \cdot \sqrt[3]{2} = 3^{1/2} \cdot 2^{1/3} \qquad \text{Write in exponential form.}$$
$$= 3^{3/6} \cdot 2^{2/6} \qquad \text{Write the exponents with their LCD.}$$
$$= (3^3 \cdot 2^2)^{1/6} \qquad \text{Use } a^n b^n = (ab)^n.$$
$$= (27 \cdot 4)^{1/6} \qquad \text{Evaluate } 3^3 \text{ and } 2^2.$$
$$= 108^{1/6} \qquad \text{Multiply.}$$
$$= \sqrt[6]{108} \qquad \text{Rewrite as a radical.}$$

Your Turn 7   Perform the indicated operations. Write the result using a radical. Assume that all variables represent positive values.

a. $\sqrt[3]{x^2} \cdot \sqrt[4]{x}$

b. $\dfrac{\sqrt[4]{a^3}}{\sqrt[3]{a^2}}$

c. $\sqrt[4]{2} \cdot \sqrt{5}$

Answers to Your Turn 7
a. $\sqrt[12]{x^{11}}$ **b.** $\sqrt[12]{a}$ **c.** $\sqrt[4]{50}$

By writing radical expressions using rational exponents, we can also find the root of a root.

Example 8 Write $\sqrt[3]{\sqrt{x}}$ as a single radical. Assume that all variables represent nonnegative values.

Solution: $\sqrt[3]{\sqrt{x}} = (x^{1/2})^{1/3}$ Write in exponential form.

$\qquad\qquad = x^{(1/2)(1/3)}$ Apply $(a^m)^n = a^{mn}$.

$\qquad\qquad = x^{1/6}$ Simplify.

$\qquad\qquad = \sqrt[6]{x}$ Write as a radical.

Answer to Your Turn 8
$\sqrt[12]{y}$

Your Turn 8 Write $\sqrt[4]{\sqrt[3]{y}}$ as a single radical. Assume that all variables represent nonnegative values.

9.2 Exercises For Extra Help MyMathLab®

Objective 1

Prep Exercise 1 If $4^{1/3}$ were written as a radical expression, what would be the radicand? 4

Prep Exercise 2 If $4^{3/5}$ were written as a radical expression, what would be the root index? 5

Prep Exercise 3 If $a^{m/n}$, with n even and m/n in lowest terms, were written as a radical expression, what restrictions would be placed on a? Why? *a* would be nonnegative because with an even index, the radicand must be nonnegative.

Prep Exercise 4 If $a^{m/n}$, with n odd, were written as a radical expression, what restrictions would be placed on a? Why? No restrictions would be placed on *a* because with an odd index, the radicand can be positive or negative.

For Exercises 1–34, rewrite each of the following using radicals; then simplify if possible. Assume that all variables represent nonnegative values. See Examples 1–3.

1. $25^{1/2}$
  $\sqrt{25} = 5$

2. $64^{1/2}$
  $\sqrt{64} = 8$

3. $-100^{1/2}$
  $-\sqrt{100} = -10$

4. $-64^{1/2}$
  $-\sqrt{64} = -8$

5. $27^{1/3}$
  $\sqrt[3]{27} = 3$

6. $216^{1/3}$
  $\sqrt[3]{216} = 6$

7. $(-64)^{1/3}$
  $\sqrt[3]{-64} = -4$

8. $(-125)^{1/3}$
  $\sqrt[3]{-125} = -5$

9. $y^{1/4}$
  $\sqrt[4]{y}$

10. $w^{1/8}$
  $\sqrt[8]{w}$

11. $(144x^8)^{1/2}$
  $\sqrt{144x^8} = 12x^4$

12. $(121z^6)^{1/2}$
  $\sqrt{121z^6} = 11z^3$

13. $18r^{1/2}$
  $18\sqrt{r}$

14. $22a^{1/2}$
  $22\sqrt{a}$

15. $\left(\dfrac{x^4}{81}\right)^{1/2}$
  $\sqrt{\dfrac{x^4}{81}} = \dfrac{x^2}{9}$

16. $\left(\dfrac{n^8}{36}\right)^{1/2}$
  $\sqrt{\dfrac{n^8}{36}} = \dfrac{n^4}{6}$

17. $64^{2/3}$
  $(\sqrt[3]{64})^2 = 16$

18. $16^{3/4}$
  $(\sqrt[4]{16})^3 = 8$

19. $-81^{3/4}$
  $-(\sqrt[4]{81})^3 = -27$

20. $-16^{5/4}$
  $-(\sqrt[4]{16})^5 = -32$

21. $(-8)^{4/3}$
  $(\sqrt[3]{-8})^4 = 16$

22. $(-27)^{5/3}$
  $(\sqrt[3]{-27})^5 = -243$

23. $16^{-3/4}$
  $\dfrac{1}{(\sqrt[4]{16})^3} = \dfrac{1}{8}$

24. $8^{-4/3}$
  $\dfrac{1}{(\sqrt[3]{8})^4} = \dfrac{1}{16}$

25. $x^{4/5}$
  $\sqrt[5]{x^4}$

26. $m^{5/7}$
  $\sqrt[7]{m^5}$

27. $8n^{2/3}$
  $8\sqrt[3]{n^2}$

28. $6a^{5/6}$
  $6\sqrt[6]{a^5}$

29. $(-32)^{-2/5}$
  $\dfrac{1}{(\sqrt[5]{-32})^2} = \dfrac{1}{4}$

30. $(-216)^{-2/3}$
  $\dfrac{1}{(\sqrt[3]{-216})^2} = \dfrac{1}{36}$

31. $\left(\dfrac{1}{25}\right)^{3/2}$
  $\left(\sqrt{\dfrac{1}{25}}\right)^3 = \dfrac{1}{125}$

32. $\left(\dfrac{1}{32}\right)^{3/5}$
  $\left(\sqrt[5]{\dfrac{1}{32}}\right)^3 = \dfrac{1}{8}$

33. $(2a + 4)^{5/6}$
  $\sqrt[6]{(2a + 4)^5}$

34. $(5r - 2)^{5/7}$
  $\sqrt[7]{(5r - 2)^5}$

Objective 2

Prep Exercise 5 When a radical expression is written as an expression with a rational exponent, what number is used as the denominator of the exponent? *The index*

For Exercises 35–50, write each of the following in exponential form. Assume that all variables represent positive values. See Example 4.

35. $\sqrt[4]{25}$
$25^{1/4}$

36. $\sqrt[5]{42}$
$42^{1/5}$

37. $\sqrt[6]{z^5}$
$z^{5/6}$

38. $\sqrt[7]{r^5}$
$r^{5/7}$

39. $\dfrac{1}{\sqrt[6]{5^5}}$
$5^{-5/6}$

40. $\dfrac{1}{\sqrt[7]{6^2}}$
$6^{-2/7}$

41. $\dfrac{5}{\sqrt[5]{x^4}}$
$5x^{-4/5}$

42. $\dfrac{8}{\sqrt[7]{n^9}}$
$8n^{-9/7}$

43. $(\sqrt[3]{5})^7$
$5^{7/3}$

44. $(\sqrt[5]{8})^6$
$8^{6/5}$

45. $(\sqrt[7]{x})^2$
$x^{2/7}$

46. $(\sqrt[3]{m})^8$
$m^{8/3}$

47. $\sqrt[4]{(4a-5)^7}$
$(4a-5)^{7/4}$

48. $\sqrt[7]{(5w+3)^5}$
$(5w+3)^{5/7}$

49. $(\sqrt[5]{2r-5})^8$
$(2r-5)^{8/5}$

50. $(\sqrt[5]{3r-6})^9$
$(3r-6)^{9/5}$

Objective 3

Prep Exercise 6 Complete the rules.

$a^m \cdot a^n = a^?$
$m+n$

$\dfrac{a^m}{a^n} = a^?$
$m-n$

$(a^m)^n = a^?$
mn

For Exercises 51–90, use the rules of exponents to simplify. Write the answers with positive exponents. Assume that all variables represent positive values. See Example 5.

51. $x^{1/5} \cdot x^{3/5}$
$x^{4/5}$

52. $n^{1/7} \cdot n^{5/7}$
$n^{6/7}$

53. $x^{3/2} \cdot x^{-1/3}$
$x^{7/6}$

54. $n^{2/3} \cdot n^{-1/2}$
$n^{1/6}$

55. $a^{2/3} \cdot a^{3/4}$
$a^{17/12}$

56. $r^{5/2} \cdot r^{5/3}$
$r^{25/6}$

57. $(3w^{1/7})(7w^{3/7})$
$21w^{4/7}$

58. $(5p^{1/9})(3p^{4/9})$
$15p^{5/9}$

59. $(-3a^{2/3})(4a^{3/4})$
$-12a^{17/12}$

60. $(8c^{4/5})(-4c^{3/2})$
$-32c^{23/10}$

61. $\dfrac{7^{7/3}}{7^{2/3}}$
$7^{5/3}$

62. $\dfrac{3^{7/9}}{3^{2/9}}$
$3^{5/9}$

63. $\dfrac{x^{1/6}}{x^{5/6}}$
$\dfrac{1}{x^{2/3}}$

64. $\dfrac{y^{2/9}}{y^{5/9}}$
$\dfrac{1}{y^{1/3}}$

65. $\dfrac{x^{3/4}}{x^{1/2}}$
$x^{1/4}$

66. $\dfrac{x^{5/8}}{x^{1/2}}$
$x^{1/8}$

67. $\dfrac{r^{3/4}}{r^{2/3}}$
$r^{1/12}$

68. $\dfrac{m^{3/5}}{m^{1/2}}$
$m^{1/10}$

69. $\dfrac{x^{-3/7}}{x^{2/7}}$
$\dfrac{1}{x^{5/7}}$

70. $\dfrac{v^{-4/5}}{v^{2/5}}$
$\dfrac{1}{v^{6/5}}$

71. $\dfrac{a^{3/4}}{a^{-3/2}}$
$a^{9/4}$

72. $\dfrac{b^{5/6}}{b^{-2/3}}$
$b^{3/2}$

73. $(5s^{-2/7})(4s^{5/7})$
$20s^{3/7}$

74. $(6u^{8/9})(-6u^{-5/9})$
$-36u^{1/3}$

75. $(-6b^{-5/4})(4b^{3/2})$
$-24b^{1/4}$

76. $(-6y^{-5/6})(-7y^{5/3})$
$42y^{5/6}$

77. $(x^{2/3})^3$
x^2

78. $(r^{3/4})^4$
r^3

79. $(a^{5/6})^2$
$a^{5/3}$

80. $(n^{3/8})^4$
$n^{3/2}$

81. $(b^{2/3})^{3/5}$
$b^{2/5}$

82. $(m^{3/2})^{2/5}$
$m^{3/5}$

83. $(2x^{2/3}y^{1/2})^6$
$64x^4y^3$

84. $(2a^{1/4}b^{3/2})^8$
$256a^2b^{12}$

85. $(8q^{3/2}t^{3/4})^{1/3}$
$2q^{1/2}t^{1/4}$

86. $(16x^{2/3}y^{1/3})^{3/4}$
$8x^{1/2}y^{1/4}$

87. $\dfrac{(3a^{5/4})^4}{a^2}$
$81a^3$

88. $\dfrac{(5v^{5/2})^2}{v^3}$
$25v^2$

89. $\dfrac{(9z^{7/3})^{1/2}}{z^{5/6}}$
$3z^{1/3}$

90. $\dfrac{(36x^{3/4})^{1/2}}{x^{1/8}}$
$6x^{1/4}$

Objective 4

Prep Exercise 7 Does $\sqrt[4]{100} = \sqrt{10}$? Why or why not?   Yes, because $100^{1/4} = (10^2)^{1/4} = 10^{1/2}$

Prep Exercise 8 Does $\sqrt[3]{3} \cdot \sqrt[5]{2} = \sqrt[15]{1944}$? Why or why not?   Yes, because $3^{1/3} \cdot 2^{1/5} = 3^{5/15} \cdot 2^{3/15} = (3^5 \cdot 2^3)^{1/15} = 1944^{1/15}$

For Exercises 91–102, represent each of the following as a radical with a smaller root index. Assume that all variables represent nonnegative values. See Example 6.

91. $\sqrt[4]{4}$
$\sqrt{2}$

92. $\sqrt[4]{49}$
$\sqrt{7}$

93. $\sqrt[6]{49}$
$\sqrt[3]{7}$

94. $\sqrt[6]{100}$
$\sqrt[3]{10}$

95. $\sqrt[4]{x^2}$
$\sqrt{x}$

96. $\sqrt[6]{y^3}$
$\sqrt{y}$

97. $\sqrt[8]{r^6}$
$\sqrt[4]{r^3}$

98. $\sqrt[10]{n^6}$
$\sqrt[5]{n^3}$

99. $\sqrt[8]{x^6y^2}$
$\sqrt[4]{x^3y}$

100. $\sqrt[6]{y^2z^4}$
$\sqrt[3]{yz^2}$

101. $\sqrt[10]{m^4n^6}$
$\sqrt[5]{m^2n^3}$

102. $\sqrt[10]{a^2b^8}$
$\sqrt[5]{ab^4}$

For Exercises 103–114, perform the indicated operations. Write the result using a radical. Assume that all variables represent positive values. See Example 7.

103. $\sqrt[3]{x} \cdot \sqrt{x}$
$\sqrt[6]{x^5}$

104. $\sqrt[4]{y} \cdot \sqrt[3]{y^2}$
$\sqrt[12]{y^{11}}$

105. $\sqrt[4]{y^2} \cdot \sqrt[3]{y^2}$
$\sqrt[6]{y^7}$

106. $\sqrt[5]{x^4} \cdot \sqrt[3]{x^2}$
$\sqrt[15]{x^{22}}$

107. $\dfrac{\sqrt[3]{x^4}}{\sqrt[4]{x^2}}$
$\sqrt[6]{x^5}$

108. $\dfrac{\sqrt[3]{y^5}}{\sqrt[4]{y^2}}$
$\sqrt[6]{y^7}$

109. $\dfrac{\sqrt[5]{n^4}}{\sqrt[3]{n^2}}$
$\sqrt[15]{n^2}$

110. $\dfrac{\sqrt[6]{z^4}}{\sqrt{z}}$
$\sqrt[6]{z}$

111. $\sqrt{5} \cdot \sqrt[3]{3}$
$\sqrt[6]{1125}$

112. $\sqrt[3]{4} \cdot \sqrt{5}$
$\sqrt[6]{2000}$

113. $\sqrt[4]{6} \cdot \sqrt[3]{2}$
$\sqrt[12]{3456}$

114. $\sqrt[4]{4} \cdot \sqrt[3]{2}$
$\sqrt[12]{1024}$

For Exercises 115–118, write each as a single radical. Assume that all variables represent nonnegative values. See Example 8.

115. $\sqrt[3]{\sqrt[3]{x}}$
$\sqrt[9]{x}$

116. $\sqrt[3]{\sqrt[5]{m}}$
$\sqrt[15]{m}$

117. $\sqrt{\sqrt[3]{n}}$
$\sqrt[6]{n}$

118. $\sqrt[4]{\sqrt[5]{z}}$
$\sqrt[20]{z}$

Review Exercises

Exercises 1–6   **Expressions**

[1.5]   **1.** Rewrite $2 \cdot 2 \cdot 2 \cdot 2 \cdot x \cdot x \cdot x \cdot y \cdot y$ using exponents.
$2^4x^3y^2$

[1.5] *For Exercises 2 and 3, simplify using the order of operations agreement.*

2. $\sqrt{16} \cdot \sqrt{9}$
12

3. $\sqrt[3]{27} \cdot \sqrt[3]{125}$
15

[5.4] *For Exercises 4 and 5, simplify.*

4. $(2.5 \times 10^6)(3.2 \times 10^5)$ 8×10^{11}

5. $\left(\dfrac{3}{4}x^3y\right)\left(-\dfrac{5}{6}xyz^2\right)$ $-\dfrac{5}{8}x^4y^2z^2$

[5.6]   **6.** Use long division to find the quotient: $\dfrac{2x^3 - 2x^2 - 19x + 18}{x + 3}$

$2x^2 - 8x + 5 + \dfrac{3}{x + 3}$

9.3 Multiplying, Dividing, and Simplifying Radicals

Objectives

1 Multiply radical expressions.

2 Divide radical expressions.

3 Use the product rule to simplify radical expressions.

Warm-up

For Exercises 1 and 2, simplify.

[9.1] 1. $\sqrt{81x^{12}y^4}$

[9.2] 2. $27^{2/3}$

[9.2] *For Exercises 3 and 4, use the rules of exponents to simplify. Write answers with positive exponents. Assume that all variables represent positive values.*

3. $(6x^{3/4})(7x^{4/5})$

4. $\dfrac{(4x^{7/3})^3}{2x^2}$

In this section, we explore some ways to simplify expressions that involve multiplication or division of radicals.

Objective 1 Multiply radical expressions.

Consider the expression $\sqrt{9} \cdot \sqrt{16}$. The usual approach is to find the roots and then multiply those roots. However, we can also multiply the radicands and then find the root of the product.

Find the roots first:

$$\sqrt{9} \cdot \sqrt{16} = 3 \cdot 4 = 12$$

Multiply the radicands first:

$$\sqrt{9} \cdot \sqrt{16} = \sqrt{9 \cdot 16} = \sqrt{144} = 12$$

Both approaches give the same result, suggesting the following rule:

Note To apply this rule, the root indices must be the same. ▶

> **Rule Product Rule for Radicals**
>
> If both $\sqrt[n]{a}$ and $\sqrt[n]{b}$ are real numbers, then $\sqrt[n]{a} \cdot \sqrt[n]{b} = \sqrt[n]{a \cdot b}$.

Example 1 Find the product and simplify. Assume that all variables represent positive values.

a. $\sqrt{3} \cdot \sqrt{27}$

Solution: $\sqrt{3} \cdot \sqrt{27} = \sqrt{3 \cdot 27} = \sqrt{81} = 9$

Connection Example 1(a) illustrates that the product of two irrational numbers can be a rational number.

b. $\sqrt{11} \cdot \sqrt{x}$

Solution: $\sqrt{11} \cdot \sqrt{x} = \sqrt{11x}$

c. $\sqrt[3]{4} \cdot \sqrt[3]{2}$

Solution: $\sqrt[3]{4} \cdot \sqrt[3]{2} = \sqrt[3]{4 \cdot 2} = \sqrt[3]{8} = 2$

d. $\sqrt[3]{3x} \cdot \sqrt[3]{4x}$

Solution: $\sqrt[3]{3x} \cdot \sqrt[3]{4x} = \sqrt[3]{3x \cdot 4x} = \sqrt[3]{12x^2}$

e. $\sqrt[4]{5} \cdot \sqrt[4]{7x^2}$

Solution: $\sqrt[4]{5} \cdot \sqrt[4]{7x^2} = \sqrt[4]{5 \cdot 7x^2} = \sqrt[4]{35x^2}$

f. $\sqrt{\dfrac{5}{x}} \cdot \sqrt{\dfrac{y}{2}}$

Solution: $\sqrt{\dfrac{5}{x}} \cdot \sqrt{\dfrac{y}{2}} = \sqrt{\dfrac{5}{x} \cdot \dfrac{y}{2}} = \sqrt{\dfrac{5y}{2x}}$

Connection Expressions that have a fraction in a radical, such as $\sqrt{\dfrac{5y}{2x}}$, are not considered to be in simplest form. We learn how to simplify them in Section 9.5.

g. $\sqrt{x} \cdot \sqrt{x}$

Solution: $\sqrt{x} \cdot \sqrt{x} = \sqrt{x \cdot x} = \sqrt{x^2} = x$

Answers to Warm-up

1. $9x^6y^2$ **2.** 9

3. $42x^{31/20}$ **4.** $32x^5$

Your Turn 1 Find the product and simplify. Assume that all variables represent positive values.

a. $\sqrt{2} \cdot \sqrt{32}$ **b.** $\sqrt{5} \cdot \sqrt{a}$ **c.** $\sqrt[3]{5x} \cdot \sqrt[3]{2y}$ **d.** $\sqrt{\dfrac{2}{a}} \cdot \sqrt{\dfrac{b}{7}}$ **e.** $\sqrt{a} \cdot \sqrt{a}$

Notice that in Example 1(g), $\sqrt{x} \cdot \sqrt{x} = x$. It is also true that $\sqrt{x} \cdot \sqrt{x} = (\sqrt{x})^2$. Therefore, $(\sqrt{x})^2 = x$. Similarly, $\sqrt[3]{x} \cdot \sqrt[3]{x} \cdot \sqrt[3]{x} = \sqrt[3]{x \cdot x \cdot x} = \sqrt[3]{x^3} = x$. It is also true that $\sqrt[3]{x} \cdot \sqrt[3]{x} \cdot \sqrt[3]{x} = (\sqrt[3]{x})^3$; so $(\sqrt[3]{x})^3 = x$. These examples suggest the following rule:

Rule Raising an nth Root to the nth Power

For any nonnegative real number a, $(\sqrt[n]{a})^n = a$.

Instructor Note You also could develop the rule $(\sqrt[n]{a})^n = a$ as follows:

$$(\sqrt[n]{a})^n = (a^{1/n})^n = a^{1/n \cdot n}$$
$$= a^1 = a$$

This means that $(\sqrt[3]{4})^3 = 4$, $(\sqrt[5]{19})^5 = 19$, and $(\sqrt[4]{3x^2})^4 = 3x^2$.

Objective 2 Divide radical expressions.

Earlier we developed the product rule for radicals. Now we develop a similar rule for quotients such as $\dfrac{\sqrt{100}}{\sqrt{25}}$. We can follow the order of operations and divide the roots, or we can divide the radicands and then find the square root of the quotient.

Find the roots first:

$$\frac{\sqrt{100}}{\sqrt{25}} = \frac{10}{5} = 2$$

Divide the radicands first:

$$\frac{\sqrt{100}}{\sqrt{25}} = \sqrt{\frac{100}{25}} = \sqrt{4} = 2$$

Both approaches give the same result, which suggests the following rule:

Rule Quotient Rule for Radicals

If both $\sqrt[n]{a}$ and $\sqrt[n]{b}$ are real numbers, then $\dfrac{\sqrt[n]{a}}{\sqrt[n]{b}} = \sqrt[n]{\dfrac{a}{b}}$, where $b \neq 0$.

As with all equations, this rule can be used going from left to right or right to left.

Example 2 Simplify. Assume that variables represent positive values.

a. $\sqrt{\dfrac{7}{36}}$

Solution: $\sqrt{\dfrac{7}{36}} = \dfrac{\sqrt{7}}{\sqrt{36}} = \dfrac{\sqrt{7}}{6}$

b. $\dfrac{\sqrt{108}}{\sqrt{3}}$

Solution: $\dfrac{\sqrt{108}}{\sqrt{3}} = \sqrt{\dfrac{108}{3}} = \sqrt{36} = 6$

c. $\sqrt[3]{\dfrac{9}{x^3}}$

Solution: $\sqrt[3]{\dfrac{9}{x^3}} = \dfrac{\sqrt[3]{9}}{\sqrt[3]{x^3}} = \dfrac{\sqrt[3]{9}}{x}$

d. $\dfrac{\sqrt[3]{15}}{\sqrt[3]{5}}$

Solution: $\dfrac{\sqrt[3]{15}}{\sqrt[3]{5}} = \sqrt[3]{\dfrac{15}{5}} = \sqrt[3]{3}$

e. $\sqrt[4]{\dfrac{y}{81}}$

Solution: $\sqrt[4]{\dfrac{y}{81}} = \dfrac{\sqrt[4]{y}}{\sqrt[4]{81}} = \dfrac{\sqrt[4]{y}}{3}$

Answers to Your Turn 1
a. 8 **b.** $\sqrt{5a}$
c. $\sqrt[3]{10xy}$ **d.** $\sqrt{\dfrac{2b}{7a}}$
e. a

Your Turn 2 Simplify. Assume that variables represent positive values.

a. $\sqrt{\dfrac{x}{49}}$ 　　 b. $\dfrac{\sqrt{75}}{\sqrt{5}}$ 　　 c. $\sqrt[4]{\dfrac{6}{x^4}}$ 　　 d. $\dfrac{\sqrt[3]{32}}{\sqrt[3]{4}}$

Objective 3 Use the product rule to simplify radical expressions.

Note In future sections, we ▶ explore other conditions that require simplification.

Several conditions exist in which a radical is not considered to be in simplest form. One such condition is when a radicand has a factor that can be written to a power greater than or equal to the index. For example, $\sqrt[3]{81}$ is not in simplest form because the perfect cube 27 is a factor of 81. Our first step in simplifying $\sqrt[3]{81}$ is to rewrite it as $\sqrt[3]{27 \cdot 3}$ so that we can then use the product rule for radicals.

> **Procedure** **Simplifying *n*th Roots**
>
> To simplify an *n*th root:
> 1. Write the radicand as a product of the greatest possible perfect *n*th power and a number or an expression that has no perfect *n*th power factors.
> 2. Use the product rule $\sqrt[n]{ab} = \sqrt[n]{a} \cdot \sqrt[n]{b}$, where a is the perfect *n*th power.
> 3. Find the *n*th root of the perfect *n*th power radicand.

Connection A list of perfect powers may be helpful.
Perfect squares:

　1, 4, 9, 16, 25, 36, 49, 64,
　81, 100, . . .

Perfect cubes:

　1, 8, 27, 64, 125, 216, . . .

Perfect fourth powers:

　1, 16, 81, 256, 625, . . .

Example 3 Simplify.

a. $\sqrt{18}$

Solution: $\sqrt{18} = \sqrt{9 \cdot 2}$ 　The greatest perfect square factor of 18 is 9, so we write 18 as $9 \cdot 2$.

　　　　$= \sqrt{9} \cdot \sqrt{2}$ 　Use the product rule of roots to separate the factors into two radicals.

　　　　$= 3\sqrt{2}$ 　Simplify the square root of 9.

b. $5\sqrt{72}$

Solution: $5\sqrt{72} = 5 \cdot \sqrt{36 \cdot 2}$ 　The greatest perfect square factor of 72 is 36, so we write 72 as $36 \cdot 2$.

　　　　$= 5 \cdot \sqrt{36} \cdot \sqrt{2}$ 　Use the product rule of roots to separate the factors into two radicals.

　　　　$= 5 \cdot 6 \cdot \sqrt{2}$ 　Simplify the square root of 36.

　　　　$= 30\sqrt{2}$ 　Multiply $5 \cdot 6$.

Note We can use perfect *n*th power factors other than the greatest perfect *n*th power factor. For example, in simplifying $\sqrt{72}$, instead of $\sqrt{72} = \sqrt{36 \cdot 2} = 6\sqrt{2}$, we could write

$$\sqrt{72} = \sqrt{4 \cdot 18} = 2\sqrt{18} = 2\sqrt{9 \cdot 2} = 2 \cdot 3\sqrt{2} = 6\sqrt{2}.$$

Notice that using the greatest perfect *n*th factor saves steps.

c. $\sqrt[3]{40}$

Solution: $\sqrt[3]{40} = \sqrt[3]{8 \cdot 5}$ 　The greatest perfect cube factor of 40 is 8.

　　　　$= \sqrt[3]{8} \cdot \sqrt[3]{5}$ 　Use the product rule of roots.

　　　　$= 2\sqrt[3]{5}$ 　Simplify the cube root of 8.

d. $4\sqrt[4]{162}$

Solution: $4\sqrt[4]{162} = 4\sqrt[4]{81 \cdot 2}$ 　The greatest perfect fourth power factor of 162 is 81.

　　　　$= 4\sqrt[4]{81} \cdot \sqrt[4]{2}$ 　Use the product rule of roots.

　　　　$= 4 \cdot 3 \cdot \sqrt[4]{2}$ 　Simplify the fourth root of 81.

　　　　$= 12\sqrt[4]{2}$

Answers to Your Turn 2

a. $\dfrac{\sqrt{x}}{7}$ 　 b. $\sqrt{15}$

c. $\dfrac{\sqrt[4]{6}}{x}$ 　 d. 2

Your Turn 3 Simplify.

a. $\sqrt{150}$ 　　　b. $6\sqrt{80}$ 　　　c. $\sqrt[3]{108}$ 　　　d. $3\sqrt[4]{80}$

Using Prime Factorization

If the greatest perfect nth power of a particular radicand is not obvious, try using the prime factorization of the radicand. Each prime factor that appears twice will have a square root equal to one of the two factors, each prime factor that appears three times will have a cube root equal to one of the three factors, and so on. The remaining factors stay in the radical sign.

Example 4 Simplify the following radicals using prime factorizations.

a. $\sqrt{375}$

Solution: $\sqrt{375} = \sqrt{5 \cdot 5 \cdot 5 \cdot 3}$ 　　Write 375 as the product of its prime factors.

$= 5\sqrt{3 \cdot 5}$ 　　The square root of the pair of 5s is 5.

$= 5\sqrt{15}$ 　　Multiply the prime factors in the radicand.

b. $\sqrt[3]{324}$

Solution: $\sqrt[3]{324} = \sqrt[3]{3 \cdot 3 \cdot 3 \cdot 3 \cdot 2 \cdot 2}$ 　　Write 324 as the product of its prime factors.

$= 3\sqrt[3]{3 \cdot 2 \cdot 2}$ 　　The cube root of the three 3s is 3.

$= 3\sqrt[3]{12}$ 　　Multiply the prime factors in the radicand.

c. $\sqrt[4]{240}$

Solution: $\sqrt[4]{240} = \sqrt[4]{2 \cdot 2 \cdot 2 \cdot 2 \cdot 3 \cdot 5}$ 　　Write 240 as the product of its prime factors.

$= 2\sqrt[4]{3 \cdot 5}$ 　　The fourth root of the four 2s is 2.

$= 2\sqrt[4]{15}$ 　　Multiply the prime factors in the radicand.

Your Turn 4 Simplify the following radicals using prime factorizations.

a. $\sqrt{294}$ 　　　b. $\sqrt[3]{324}$ 　　　c. $\sqrt[4]{486}$

Instructor Note Because $\sqrt[n]{a^m} = a^{m/n}$, you can simplify $\sqrt[n]{a^m}$ by dividing n into m. The number of times n goes into m is the exponent of a outside the radical of the answer, and the remainder is the exponent of a remaining in the radicand. For example, to simplify $\sqrt[3]{x^{11}}$, $11 \div 3 = 3\,r\,2$; so the answer is $x^3\sqrt[3]{x^2}$.

Simplifying Radicals with Variables

We can use either procedure to simplify radicals whose radicands contain variables. Because $\sqrt[n]{a^m} = a^{m/n}$, we can find an exact root if n divides into m evenly. For example, $\sqrt{x^4} = x^{4/2} = x^2$ and $\sqrt[3]{x^{12}} = x^{12/3} = x^4$. Therefore, to find the nth root, we rewrite the radicand as a product in which one factor has the greatest possible exponent divisible by the index n.

Example 5 Simplify. Assume that variables represent nonnegative values.

a. $\sqrt{x^7}$

Solution: $\sqrt{x^7} = \sqrt{x^6 \cdot x}$ 　　The greatest number smaller than 7 that is divisible by 2 is 6, so write x^7 as $x^6 \cdot x$.

$= \sqrt{x^6} \cdot \sqrt{x}$ 　　Use the product rule of roots.

$= x^3\sqrt{x}$ 　　Simplify $\sqrt{x^6} = x^3$.

b. $3\sqrt{24a^5b^9}$

Solution: $3\sqrt{24a^5b^9} = 3\sqrt{4 \cdot 6 \cdot a^4 \cdot a \cdot b^8 \cdot b}$ 　　Write 24 as $4 \cdot 6$, a^5 as $a^4 \cdot a$, and b^9 as $b^8 \cdot b$.

$= 3\sqrt{4a^4b^8 \cdot 6ab}$ 　　Regroup the factors so that perfect squares are together.

$= 3\sqrt{4a^4b^8} \cdot \sqrt{6ab}$ 　　Use the product rule of roots.

$= 3 \cdot 2a^2b^4 \cdot \sqrt{6ab}$ 　　Simplify $\sqrt{4} = 2$, $\sqrt{a^4} = a^2$, and $\sqrt{b^8} = b^4$.

$= 6a^2b^4\sqrt{6ab}$ 　　Multiply $3 \cdot 2 = 6$.

Connection A list of perfect powers might be helpful.
Perfect squares:

$x^2, x^4, x^6, x^8, \ldots$

Perfect cubes:

$x^3, x^6, x^9, x^{12}, \ldots$

Answers to Your Turn 3
a. $5\sqrt{6}$ 　b. $24\sqrt{5}$
c. $3\sqrt[3]{4}$ 　d. $6\sqrt[4]{5}$

Answers to Your Turn 4
a. $7\sqrt{6}$ 　b. $3\sqrt[3]{12}$ 　c. $3\sqrt[4]{6}$

c. $y^2\sqrt[3]{y^8}$

Solution: $y^2\sqrt[3]{y^8} = y^2\sqrt[3]{y^6 \cdot y^2}$ The greatest number smaller than 8 that is divisible by 3 is 6, so write y^8 as $y^6 \cdot y^2$.

$$= y^2\sqrt[3]{y^6} \cdot \sqrt[3]{y^2}$$ Use the product rule of roots.

$$= y^2 \cdot y^2 \cdot \sqrt[3]{y^2}$$ Simplify $\sqrt[3]{y^6} = y^2$.

$$= y^4\sqrt[3]{y^2}$$ Multiply $y^2 \cdot y^2 = y^4$.

d. $\sqrt[5]{64x^9y^{12}}$

Solution: $\sqrt[5]{64x^9y^{12}} = \sqrt[5]{32 \cdot 2 \cdot x^5 \cdot x^4 \cdot y^{10} \cdot y^2}$ Write 64 as $32 \cdot 2$, x^9 as $x^5 \cdot x^4$, and y^{12} as $y^{10} \cdot y^2$.

$$= \sqrt[5]{32x^5y^{10} \cdot 2x^4y^2}$$ Regroup the factors.

$$= \sqrt[5]{32x^5y^{10}} \cdot \sqrt[5]{2x^4y^2}$$ Use the product rule of roots.

$$= 2xy^2\sqrt[5]{2x^4y^2}$$ Simplify $\sqrt[5]{32} = 2$, $\sqrt[5]{x^5} = x$, and $\sqrt[5]{y^{10}} = y^2$.

Your Turn 5   Simplify. Assume that variables represent nonnegative values.

   a. $\sqrt{n^{11}}$ **b.** $2\sqrt{45r^7s^3}$ **c.** $\sqrt[4]{m^{13}}$ **d.** $\sqrt[3]{a^8b^{10}}$

After using the product or quotient rules, it is often necessary to simplify the results.

Example 6   Find the product or quotient and simplify the results. Assume that variables represent positive values.

a. $\sqrt{3} \cdot \sqrt{6}$

Solution: $\sqrt{3} \cdot \sqrt{6} = \sqrt{18}$ Use the product rule of roots to multiply.

$$= \sqrt{9 \cdot 2}$$ Write 18 as $9 \cdot 2$.

$$= 3\sqrt{2}$$ Simplify $\sqrt{9} = 3$.

b. $5\sqrt{3x^3} \cdot 3\sqrt{15x^2}$

Solution: $5\sqrt{3x^3} \cdot 3\sqrt{15x^2} = 5 \cdot 3\sqrt{3x^3} \cdot \sqrt{15x^2}$ Regroup the factors.

$$= 15\sqrt{45x^5}$$ Multiply.

$$= 15\sqrt{9 \cdot 5 \cdot x^4 \cdot x}$$ Write 45 as $9 \cdot 5$ and x^5 as $x^4 \cdot x$.

$$= 15 \cdot 3x^2\sqrt{5x}$$ Simplify $\sqrt{9} = 3$ and $\sqrt{x^4} = x^2$.

$$= 45x^2\sqrt{5x}$$ Multiply.

c. $\dfrac{\sqrt{288}}{\sqrt{6}}$

Solution: $\dfrac{\sqrt{288}}{\sqrt{6}} = \sqrt{\dfrac{288}{6}}$ Use the quotient rule of roots.

$$= \sqrt{48}$$ Divide the radicand.

$$= \sqrt{16 \cdot 3}$$ Write 48 as $16 \cdot 3$.

$$= 4\sqrt{3}$$ Simplify $\sqrt{16} = 4$.

d. $\dfrac{8\sqrt{756a^8b^5}}{2\sqrt{7a^4b^2}}$

Solution: $\dfrac{8\sqrt{756a^8b^5}}{2\sqrt{7a^4b^2}} = 4\sqrt{\dfrac{756a^8b^5}{7a^4b^2}}$ Divide coefficients and use the quotient rule of radicals.

$$= 4\sqrt{108a^4b^3}$$ Divide the radicand.

$$= 4\sqrt{36a^4b^2 \cdot 3b}$$ Rewrite the radicand with a perfect square factor.

Answers to Your Turn 5
  a. $n^5\sqrt{n}$ **b.** $6r^3s\sqrt{5rs}$
  c. $m^3\sqrt[4]{m}$ **d.** $a^2b^3\sqrt[3]{a^2b}$

$$= 4 \cdot 6a^2b\sqrt{3b} \qquad \text{Find the square roots.}$$
$$= 24a^2b\sqrt{3b} \qquad \text{Multiply.}$$

Your Turn 6 Find the product or quotient and simplify the results. Assume that variables represent positive values.

Answers to Your Turn 6
a. $3\sqrt{10}$ b. $16x^3\sqrt{21x}$
c. $6\sqrt{3}$ d. $21x^2y\sqrt{7xy}$

a. $\sqrt{6} \cdot \sqrt{15}$ b. $4\sqrt{14x^3} \cdot 2\sqrt{6x^4}$ c. $\dfrac{\sqrt{1296}}{\sqrt{12}}$ d. $\dfrac{14\sqrt{315x^{11}y^8}}{2\sqrt{5x^6y^5}}$

9.3 Exercises   For Extra Help   MyMathLab®

Objective 1

Prep Exercise 1 For $\sqrt{8} \cdot \sqrt{18}$, explain the difference between using the product rule for radicals and multiplying the approximate roots of 8 and 18. Multiplying the approximate roots gives $\sqrt{8} \cdot \sqrt{18} \approx 2.828 \cdot 4.243 = 11.999204$, which is tedious and inexact. Using the product rule for radicals gives $\sqrt{8} \cdot \sqrt{18} = \sqrt{8 \cdot 18} = \sqrt{144} = 12$, which is fast and exact.

For Exercises 1–28, find the product and simplify. Assume that variables represent positive values. See Example 1.

1. $\sqrt{2} \cdot \sqrt{32}$
   8

2. $\sqrt{3} \cdot \sqrt{48}$
   12

3. $\sqrt{3x} \cdot \sqrt{27x^5}$
   $9x^3$

4. $\sqrt{8y^3} \cdot \sqrt{2y}$
   $4y^2$

5. $\sqrt{6xy^3} \cdot \sqrt{24xy}$
   $12xy^2$

6. $\sqrt{50u^3v^2} \cdot \sqrt{2uv^4}$
   $10u^2v^3$

7. $\sqrt{5} \cdot \sqrt{13}$
   $\sqrt{65}$

8. $\sqrt{6} \cdot \sqrt{11}$
   $\sqrt{66}$

9. $\sqrt{15} \cdot \sqrt{x}$
   $\sqrt{15x}$

10. $\sqrt{17} \cdot \sqrt{y}$
    $\sqrt{17y}$

11. $\sqrt[3]{3} \cdot \sqrt[3]{9}$
    3

12. $\sqrt[3]{4} \cdot \sqrt[3]{16}$
    4

13. $\sqrt[3]{5y} \cdot \sqrt[3]{2y}$
    $\sqrt[3]{10y^2}$

14. $\sqrt[3]{6m} \cdot \sqrt[3]{2m}$
    $\sqrt[3]{12m^2}$

15. $\sqrt[4]{3} \cdot \sqrt[4]{7}$
    $\sqrt[4]{21}$

16. $\sqrt[4]{7} \cdot \sqrt[4]{5}$
    $\sqrt[4]{35}$

17. $\sqrt[4]{12w^3} \cdot \sqrt[4]{6w}$
    $w\sqrt[4]{72}$

18. $\sqrt[4]{21r^3} \cdot \sqrt[4]{7r}$
    $r\sqrt[4]{147}$

19. $\sqrt[4]{3x^2y} \cdot \sqrt[4]{5xy^2}$
    $\sqrt[4]{15x^3y^3}$

20. $\sqrt[4]{2ab^2} \cdot \sqrt[4]{6ab}$
    $\sqrt[4]{12a^2b^3}$

21. $\sqrt[5]{6x^3} \cdot \sqrt[5]{5x^4}$
    $x\sqrt[5]{30x^2}$

22. $\sqrt[5]{3m^2} \cdot \sqrt[5]{8m^7}$
    $m\sqrt[5]{24m^4}$

23. $\sqrt[6]{4x^2y^3} \cdot \sqrt[6]{2x^3y}$
    $\sqrt[6]{8x^5y^4}$

24. $\sqrt[6]{ab^3} \cdot \sqrt[6]{7a^3b^2}$
    $\sqrt[6]{7a^4b^5}$

25. $\sqrt{\dfrac{7}{2}} \cdot \sqrt{\dfrac{3}{5}}$
    $\sqrt{\dfrac{21}{10}}$

26. $\sqrt{\dfrac{5}{2}} \cdot \sqrt{\dfrac{11}{3}}$
    $\sqrt{\dfrac{55}{6}}$

27. $\sqrt{\dfrac{6}{x}} \cdot \sqrt{\dfrac{y}{5}}$
    $\sqrt{\dfrac{6y}{5x}}$

28. $\sqrt{\dfrac{a}{3}} \cdot \sqrt{\dfrac{7}{b}}$
    $\sqrt{\dfrac{7a}{3b}}$

Objective 2

Prep Exercise 2 How does the quotient rule avoid approximation in simplifying $\dfrac{\sqrt{125}}{\sqrt{5}}$?

The quotient rule transforms the expression to the square root of a perfect square: $\sqrt{\dfrac{125}{5}} = \sqrt{25} = 5$.

For Exercises 29–44, simplify. Assume that variables represent positive values. See Example 2.

29. $\sqrt{\dfrac{25}{36}}$
    $\dfrac{5}{6}$

30. $\sqrt{\dfrac{49}{81}}$
    $\dfrac{7}{9}$

31. $\sqrt{\dfrac{10}{9}}$
    $\dfrac{\sqrt{10}}{3}$

32. $\sqrt{\dfrac{15}{81}}$
    $\dfrac{\sqrt{15}}{9}$

33. $\dfrac{\sqrt{180}}{\sqrt{5}}$
6

34. $\dfrac{\sqrt{243}}{\sqrt{3}}$
9

35. $\dfrac{\sqrt{15}}{\sqrt{5}}$
$\sqrt{3}$

36. $\dfrac{\sqrt{21}}{\sqrt{3}}$
$\sqrt{7}$

37. $\sqrt[3]{\dfrac{4}{w^6}}$
$\dfrac{\sqrt[3]{4}}{w^2}$

38. $\sqrt[3]{\dfrac{7}{v^9}}$
$\dfrac{\sqrt[3]{7}}{v^3}$

39. $\sqrt[3]{\dfrac{5y^2}{27x^9}}$
$\dfrac{\sqrt[3]{5y^2}}{3x^3}$

40. $\sqrt[3]{\dfrac{5a}{8r^6}}$
$\dfrac{\sqrt[3]{5a}}{2r^2}$

41. $\dfrac{\sqrt[3]{320}}{\sqrt[3]{5}}$
4

42. $\dfrac{\sqrt[3]{162}}{\sqrt[3]{6}}$
3

43. $\sqrt[4]{\dfrac{3u^3}{16x^8}}$
$\dfrac{\sqrt[4]{3u^3}}{2x^2}$

44. $\sqrt[4]{\dfrac{3x^2}{81y^4}}$
$\dfrac{\sqrt[4]{3x^2}}{3y}$

Objective 3

Prep Exercise 3 Explain why $\sqrt{28}$ is not in simplest form. The radicand, 28, has the perfect square 4 as a factor. So $\sqrt{28} = 2\sqrt{7}$.

Prep Exercise 4 Explain how to simplify a cube root containing a radicand with a perfect cube factor. Rewrite the expression as a product of two radicals, the first containing the perfect cube and the second containing no perfect cubes. Then simplify the first radical. For example, $\sqrt[3]{54x^8}$ can be rewritten as $\sqrt[3]{27x^6} \cdot \sqrt[3]{2x^2}$, then simplified to $3x^2\sqrt[3]{2x^2}$.

Prep Exercise 5 Explain why the expression $3x^2\sqrt[3]{x^5}$ is not in simplest form. The radicand, x^5, has the perfect cube x^3 as a factor.

For Exercises 45–80, simplify. Assume that variables represent nonnegative values. See Examples 3–5.

45. $\sqrt{98}$
$7\sqrt{2}$

46. $\sqrt{96}$
$4\sqrt{6}$

47. $\sqrt{128}$
$8\sqrt{2}$

48. $\sqrt{180}$
$6\sqrt{5}$

49. $6\sqrt{80}$
$24\sqrt{5}$

50. $4\sqrt{75}$
$20\sqrt{3}$

51. $5\sqrt{112}$
$20\sqrt{7}$

52. $3\sqrt{147}$
$21\sqrt{3}$

53. $\sqrt{a^7}$
$a^3\sqrt{a}$

54. $\sqrt{d^5}$
$d^2\sqrt{d}$

55. $\sqrt{x^2y^4}$
xy^2

56. $\sqrt{a^6b^2}$
a^3b

57. $\sqrt{x^6y^8z^{10}}$
$x^3y^4z^5$

58. $\sqrt{p^4q^8r^8}$
$p^2q^4r^4$

59. $rs^2\sqrt{r^9s^5}$
$r^5s^4\sqrt{rs}$

60. $a^2b^3\sqrt{a^{11}b^3}$
$a^7b^4\sqrt{ab}$

61. $3\sqrt{72x^5}$
$18x^2\sqrt{2x}$

62. $6\sqrt{75d^3}$
$30d\sqrt{3d}$

63. $\sqrt[3]{32}$
$2\sqrt[3]{4}$

64. $\sqrt[3]{128}$
$4\sqrt[3]{2}$

65. $\sqrt[3]{x^7}$
$x^2\sqrt[3]{x}$

66. $\sqrt[3]{b^{11}}$
$b^3\sqrt[3]{b^2}$

67. $\sqrt[3]{x^6y^5}$
$x^2y\sqrt[3]{y^2}$

68. $\sqrt[3]{m^{13}n^9}$
$m^4n^3\sqrt[3]{m}$

69. $\sqrt[3]{128z^8}$
$4z^2\sqrt[3]{2z^2}$

70. $\sqrt[3]{48h^{14}}$
$2h^4\sqrt[3]{6h^2}$

71. $2\sqrt[3]{40}$
$4\sqrt[3]{5}$

72. $4\sqrt[3]{250}$
$20\sqrt[3]{2}$

73. $\sqrt[4]{80}$
$2\sqrt[4]{5}$

74. $\sqrt[4]{162}$
$3\sqrt[4]{2}$

75. $3x^2\sqrt[4]{243x^9}$
$9x^4\sqrt[4]{3x}$

76. $3a^4\sqrt[4]{48a^7}$
$6a^5\sqrt[4]{3a^3}$

77. $\sqrt[5]{486x^{16}}$
$3x^3\sqrt[5]{2x}$

78. $\sqrt[5]{160n^{18}}$
$2n^3\sqrt[5]{5n^3}$

79. $\sqrt[6]{x^8y^{14}z^{11}}$
$xy^2z\sqrt[6]{x^2y^2z^5}$

80. $\sqrt[7]{a^{16}b^9c^{12}}$
$a^2bc\sqrt[7]{a^2b^2c^5}$

For Exercises 81–90, find the product and write the answer in simplest form. Assume that variables represent nonnegative values. See Examples 6(a) and 6(b).

81. $\sqrt{3} \cdot \sqrt{21}$
$3\sqrt{7}$

82. $\sqrt{5} \cdot \sqrt{15}$
$5\sqrt{3}$

83. $5\sqrt{10} \cdot 3\sqrt{6}$
$30\sqrt{15}$

84. $2\sqrt{6} \cdot 5\sqrt{21}$
$30\sqrt{14}$

85. $\sqrt{y^3} \cdot \sqrt{y^2}$
$y^2\sqrt{y}$

86. $\sqrt{m^7} \cdot \sqrt{m^4}$
$m^5\sqrt{m}$

87. $x\sqrt{x^2y^3} \cdot y^2\sqrt{x^4y^4}$
$x^4y^5\sqrt{y}$

88. $x\sqrt{x^5y^2} \cdot y\sqrt{xy^3}$
$x^4y^3\sqrt{y}$

89. $4\sqrt{6c^3} \cdot 3\sqrt{10c^5}$
$24c^4\sqrt{15}$

90. $6\sqrt{15c^2} \cdot 2\sqrt{10c^4}$
$60c^3\sqrt{6}$

For Exercises 91 and 92, write an expression in simplest form for the area of the figure.

91.

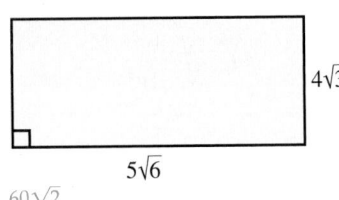

$4\sqrt{3}$

$5\sqrt{6}$

$60\sqrt{2}$

92.

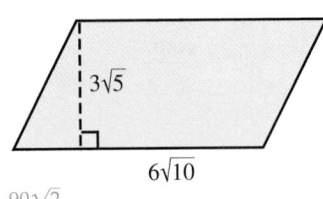

$3\sqrt{5}$

$6\sqrt{10}$

$90\sqrt{2}$

For Exercises 93–104, find the quotient and write the answer in simplest form.
Assume that variables represent positive values. See Examples 6(c) and 6(d).

93. $\dfrac{\sqrt{48}}{\sqrt{6}}$
$2\sqrt{2}$

94. $\dfrac{\sqrt{54}}{\sqrt{3}}$
$3\sqrt{2}$

95. $\dfrac{9\sqrt{160}}{3\sqrt{8}}$
$6\sqrt{5}$

96. $\dfrac{10\sqrt{280}}{2\sqrt{10}}$
$10\sqrt{7}$

97. $\dfrac{\sqrt{c^5d^6}}{\sqrt{cd^3}}$
$c^2d\sqrt{d}$

98. $\dfrac{\sqrt{m^6n^5}}{\sqrt{m^3n^3}}$
$mn\sqrt{m}$

99. $\dfrac{8\sqrt{45a^5}}{2\sqrt{5a}}$
$12a^2$

100. $\dfrac{6\sqrt{48n^7}}{3\sqrt{3n^3}}$
$8n^2$

101. $\dfrac{12\sqrt{72c^5}}{4\sqrt{6c^2}}$
$6c\sqrt{3c}$

102. $\dfrac{15\sqrt{48a^7}}{5\sqrt{2a^2}}$
$6a^2\sqrt{6a}$

103. $\dfrac{36\sqrt{96x^6y^{11}}}{4\sqrt{3x^2y^4}}$
$36x^2y^3\sqrt{2y}$

104. $\dfrac{54\sqrt{240r^{11}s^{10}}}{9\sqrt{5r^6s^4}}$
$24r^2s^3\sqrt{3r}$

For Exercises 105–110, find the product and write the answer in simplest form.
Assume that variables represent nonnegative values.

105. $\sqrt{\dfrac{3}{7}} \cdot \sqrt{\dfrac{8}{7}}$
$\dfrac{2\sqrt{6}}{7}$

106. $\sqrt{\dfrac{8}{5}} \cdot \sqrt{\dfrac{6}{5}}$
$\dfrac{4\sqrt{3}}{5}$

107. $\sqrt{\dfrac{a^3}{2}} \cdot \sqrt{\dfrac{a^5}{2}}$
$\dfrac{a^4}{2}$

108. $\sqrt{\dfrac{c^7}{6}} \cdot \sqrt{\dfrac{c^5}{6}}$
$\dfrac{c^6}{6}$

109. $\sqrt{\dfrac{3x^5}{2}} \cdot \sqrt{\dfrac{15x^5}{8}}$
$\dfrac{3x^5\sqrt{5}}{4}$

110. $\sqrt{\dfrac{5y^3}{3}} \cdot \sqrt{\dfrac{10y^3}{27}}$
$\dfrac{5y^3\sqrt{2}}{9}$

For Exercises 111 and 112, write an expression in simplest form for the area of the figure.

111.

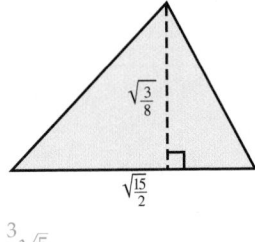

$\sqrt{\dfrac{3}{8}}$

$\sqrt{\dfrac{15}{2}}$

$\dfrac{3}{8}\sqrt{5}$

112.

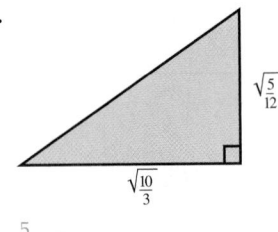

$\sqrt{\dfrac{5}{12}}$

$\sqrt{\dfrac{10}{3}}$

$\dfrac{5}{12}\sqrt{2}$

Review Exercises

Exercises 1–6 ➤ Expressions

[1.7] *For Exercises 1 and 2, use* $a = 6$ *and* $b = 8$.

1. Does $\sqrt{a^2 + b^2} = \sqrt{a^2} + \sqrt{b^2} = a + b$?
no

2. Does $\sqrt{(a + b)^2} = a + b$?
yes

[1.7] **3.** Simplify: $6x^2 - 4x - 3x + 2x^2$
$8x^2 - 7x$

[5.5] *For Exercises 4–6, multiply.*

4. $(2a - 3b)(4a + 3b)$
$8a^2 - 6ab - 9b^2$

5. $(3m + 5n)(3m - 5n)$
$9m^2 - 25n^2$

6. $(2x - 3y)^2$
$4x^2 - 12xy + 9y^2$

9.4 Adding, Subtracting, and Multiplying Radical Expressions

Objectives

1 Add or subtract like radicals.

2 Use the distributive property in expressions containing radicals.

3 Simplify radical expressions that contain mixed operations.

Warm-up

For Exercises 1–4, simplify.

[9.3] **1.** $\sqrt{3} \cdot \sqrt{6}$

[5.5] **3.** $2x(3 + 5x)$

[1.7] **2.** $4x^2 - 5 + 2x^2 + 2$

[5.5] **4.** $(5 - 3x)(5 + 3x)$

Objective 1 Add or subtract like radicals.

Recall that like terms such as $-8a^2$ and $3a^2$ have identical variables with identical exponents. Similarly, **like radicals** have identical radicands and root indices.

Definition **Like radicals:** Radical expressions with identical radicands and identical root indices.

The radicals $3\sqrt{2}$ and $5\sqrt{2}$ are like. The radicals $3\sqrt{2}$ and $5\sqrt{3}$ are unlike because their radicands are different. The radicals $3\sqrt{2}$ and $5\sqrt[3]{2}$ are unlike because their root indices are different.

Adding or subtracting like radicals is essentially the same as combining like terms. Remember that the distributive property is at work when we combine like terms because we factor out the common variable, which leaves a sum or difference of the coefficients.

Like terms:
$$3x + 4x = (3 + 4)x$$
$$= 7x$$

Like radicals:
$$3\sqrt{5} + 4\sqrt{5} = (3 + 4)\sqrt{5}$$
$$= 7\sqrt{5}$$

Our example suggests the following procedure for adding like radical expressions:

Answers to Warm-up
1. $3\sqrt{2}$
2. $6x^2 - 3$
3. $6x + 10x^2$
4. $25 - 9x^2$

Procedure **Adding Like Radicals**

To add or subtract like radicals, add or subtract the coefficients and leave the radical parts the same.

Example 1 Add or subtract. Assume that all variables represent nonnegative values.

a. $7\sqrt{5} + 2\sqrt{5}$

Solution: $7\sqrt{5} + 2\sqrt{5} = (7 + 2)\sqrt{5}$
$$= 9\sqrt{5}$$

b. $6\sqrt[3]{4} + \sqrt[3]{4}$

Solution: $6\sqrt[3]{4} + \sqrt[3]{4} = (6 + 1)\sqrt[3]{4}$
$$= 7\sqrt[3]{4}$$

c. $6x\sqrt[4]{3x} - 2x\sqrt[4]{3x}$

Solution: $6x\sqrt[4]{3x} - 2x\sqrt[4]{3x} = (6x - 2x)\sqrt[4]{3x}$
$$= 4x\sqrt[4]{3x}$$

d. $8\sqrt{3} + 6\sqrt{2} - 5\sqrt{3} + 3\sqrt{2}$

Solution: $8\sqrt{3} + 6\sqrt{2} - 5\sqrt{3} + 3\sqrt{2} = 8\sqrt{3} - 5\sqrt{3} + 6\sqrt{2} + 3\sqrt{2}$ Regroup the terms.
$$= (8 - 5)\sqrt{3} + (6 + 3)\sqrt{2}$$
$$= 3\sqrt{3} + 9\sqrt{2}$$

Warning Never add radicands! Consider $\sqrt{9} + \sqrt{16}$. It might be tempting to add the radicands to get $\sqrt{25}$, but $\sqrt{9} + \sqrt{16} \neq \sqrt{25}$. If we calculate each side separately, we see why they are not equivalent.

$$\sqrt{9} + \sqrt{16} \neq \sqrt{25}$$
$$3 + 4 \neq 5$$
$$7 \neq 5$$

Your Turn 1 Add or subtract. Assume that all variables represent nonnegative values.

a. $5\sqrt{6} + 2\sqrt{6}$ **b.** $5\sqrt[4]{5x^2} + \sqrt[4]{5x^2}$

c. $7y\sqrt[3]{7} - 12y\sqrt[3]{7}$ **d.** $9x\sqrt{5x} - 2y\sqrt{3y} - 6x\sqrt{5x} + 7y\sqrt{3y}$

In a problem involving addition or subtraction of radicals, if the radicals are not like radicals, it may be possible to simplify one or more of the radicals so that they are like.

Example 2 Add or subtract. Assume that variables represent nonnegative values.

a. $7\sqrt{3} + \sqrt{12}$

Solution: $7\sqrt{3} + \sqrt{12} = 7\sqrt{3} + \sqrt{4 \cdot 3}$ Factor out the perfect square factor, 4.
$$= 7\sqrt{3} + \sqrt{4} \cdot \sqrt{3}$$ Use the product rule to separate the radicals.
$$= 7\sqrt{3} + 2\sqrt{3}$$ Simplify.
$$= 9\sqrt{3}$$ Combine like radicals.

b. $3\sqrt[3]{24} - 2\sqrt[3]{3}$

Solution: $3\sqrt[3]{24} - 2\sqrt[3]{3} = 3\sqrt[3]{8 \cdot 3} - 2\sqrt[3]{3}$ Rewrite 24 as $8 \cdot 3$.
$$= 3\sqrt[3]{8} \cdot \sqrt[3]{3} - 2\sqrt[3]{3}$$ Use the product rule.
$$= 3 \cdot 2\sqrt[3]{3} - 2\sqrt[3]{3}$$ Simplify $\sqrt[3]{8}$.
$$= 6\sqrt[3]{3} - 2\sqrt[3]{3}$$ Multiply.
$$= 4\sqrt[3]{3}$$ Combine like radicals.

c. $\sqrt{48x^3} + \sqrt{12x^3}$

Solution: $\sqrt{48x^3} + \sqrt{12x^3} = \sqrt{16x^2 \cdot 3x} + \sqrt{4x^2 \cdot 3x}$ Rewrite $48x^3$ as $16x^2 \cdot 3x$ and $12x^3$ as $4x^2 \cdot 3x$.
$$= \sqrt{16x^2} \cdot \sqrt{3x} + \sqrt{4x^2} \cdot \sqrt{3x}$$ Use the product rule.
$$= 4x\sqrt{3x} + 2x\sqrt{3x}$$ Find $\sqrt{16x^2}$ and $\sqrt{4x^2}$.
$$= 6x\sqrt{3x}$$ Combine like radicals.

Answers to Your Turn 1
a. $7\sqrt{6}$ **b.** $6\sqrt[4]{5x^2}$ **c.** $-5y\sqrt[3]{7}$
d. $3x\sqrt{5x} + 5y\sqrt{3y}$

d. $3\sqrt[4]{32x^5} + \sqrt[4]{162x^9}$

Solution: $3\sqrt[4]{32x^5} + \sqrt[4]{162x^9}$

$$= 3\sqrt[4]{16 \cdot 2 \cdot x^4 \cdot x} + \sqrt[4]{81 \cdot 2 \cdot x^8 \cdot x} \quad \text{Write 32 as } 16 \cdot 2, x^5 \text{ as } x^4 \cdot x, 162 \text{ as}$$
$$\qquad\qquad 81 \cdot 2, \text{ and } x^9 \text{ as } x^8 \cdot x.$$
$$= 3 \cdot 2x\sqrt[4]{2x} + 3x^2\sqrt[4]{2x} \qquad \text{Find the roots.}$$
$$= 6x\sqrt[4]{2x} + 3x^2\sqrt[4]{2x} \qquad \text{Multiply.}$$
$$\blacktriangleright \quad = (6x + 3x^2)\sqrt[4]{2x} \qquad \text{Factor.}$$

Note Although the radicals are like, we cannot combine the coefficients further because they are not like terms.

| **Your Turn 2** | Add or subtract. Assume that variables represent nonnegative values.

a. $4\sqrt{24} - 6\sqrt{54}$ **b.** $4\sqrt{50x^5} - 2\sqrt{18x^5}$
c. $6a\sqrt[3]{54a^4} - 2\sqrt[3]{128a^7}$ **d.** $2\sqrt[3]{81x^7} - 5\sqrt[3]{24x^4}$

Objective 2 Use the distributive property in expressions containing radicals.

Products involving sums and differences of radicals are found in much the same way as products of polynomials. We use the distributive property, multiply binomials using FOIL, and square a binomial (all from Section 5.5).

| **Example 3** | Find the product. Assume that variables represent nonnegative values.

a. $\sqrt{3}(\sqrt{3} + \sqrt{15})$

Solution: $\sqrt{3}(\sqrt{3} + \sqrt{15}) = \sqrt{3} \cdot \sqrt{3} + \sqrt{3} \cdot \sqrt{15}$ Use the distributive property.
$$= \sqrt{3 \cdot 3} + \sqrt{3 \cdot 15} \qquad \text{Use the product rule.}$$
$$= \sqrt{9} + \sqrt{45} \qquad \text{Multiply.}$$
$$= 3 + 3\sqrt{5} \qquad \text{Find } \sqrt{9} \text{ and simplify } \sqrt{45}.$$

b. $2\sqrt{6}(3 + 5\sqrt{5})$

Solution: $2\sqrt{6}(3 + 5\sqrt{5}) = 2\sqrt{6} \cdot 3 + 2\sqrt{6} \cdot 5\sqrt{5}$ Use the distributive property.
$$= 2 \cdot 3\sqrt{6} + 2 \cdot 5\sqrt{6 \cdot 5} \qquad \text{Use the product rule.}$$
$$= 6\sqrt{6} + 10\sqrt{30} \qquad \text{Multiply.}$$

c. $(2 + \sqrt{3})(\sqrt{5} - \sqrt{6})$

Solution: $(2 + \sqrt{3})(\sqrt{5} - \sqrt{6}) = 2\sqrt{5} - 2\sqrt{6} + \sqrt{3} \cdot \sqrt{5} - \sqrt{3} \cdot \sqrt{6}$ Use FOIL.
$$= 2\sqrt{5} - 2\sqrt{6} + \sqrt{15} - \sqrt{18} \qquad \text{Use the product rule}$$
$$= 2\sqrt{5} - 2\sqrt{6} + \sqrt{15} - 3\sqrt{2} \qquad \text{Simplify } \sqrt{18}.$$

d. $(3\sqrt{x} + \sqrt{y})(2\sqrt{x} - 5\sqrt{y})$

Solution: $(3\sqrt{x} + \sqrt{y})(2\sqrt{x} - 5\sqrt{y})$
$$= 3\sqrt{x} \cdot 2\sqrt{x} - 3\sqrt{x} \cdot 5\sqrt{y} + \sqrt{y} \cdot 2\sqrt{x} - \sqrt{y} \cdot 5\sqrt{y} \qquad \text{Use FOIL.}$$
$$= 6x - 15\sqrt{xy} + 2\sqrt{xy} - 5y \qquad \text{Use the product rule.}$$
$$= 6x - 13\sqrt{xy} - 5y \qquad \text{Combine like radicals.}$$

e. $(5 + \sqrt{3})^2$

Solution: $(5 + \sqrt{3})^2 = 5^2 + 2 \cdot 5\sqrt{3} + (\sqrt{3})^2$ Use $(a + b)^2 = a^2 + 2ab + b^2$.
$$= 25 + 10\sqrt{3} + 3 \qquad \text{Simplify.}$$
$$= 28 + 10\sqrt{3} \qquad \text{Add 25 and 3.}$$

Answers to Your Turn 2
a. $-10\sqrt{6}$ **b.** $14x^2\sqrt{2x}$
c. $10a^2\sqrt[3]{2a}$ **d.** $(6x^2 - 10x)\sqrt[3]{3x}$

Answers to Your Turn 3
a. $6\sqrt{11} - 6\sqrt{66}$
b. $2a - 3\sqrt{ab} - 9b$
c. $14 + 6\sqrt{5}$

| **Your Turn 3** | Find the product. Assume that variables represent nonnegative values.

a. $2\sqrt{11}(3 - 3\sqrt{6})$ **b.** $(2\sqrt{a} + 3\sqrt{b})(\sqrt{a} - 3\sqrt{b})$ **c.** $(3 + \sqrt{5})^2$

Radicals in Conjugates

Radical expressions can be conjugates. Like binomial conjugates, conjugates involving radicals differ only in the sign separating the terms. For example, $7 - \sqrt{3}$ and $7 + \sqrt{3}$ are conjugates. Let's explore what happens when we multiply conjugates containing radicals.

Example 4   Find the product.

a. $(2 + \sqrt{3})(2 - \sqrt{3})$

Solution: $(2 + \sqrt{3})(2 - \sqrt{3}) = 2^2 - (\sqrt{3})^2$ Use $(a + b)(a - b) = a^2 - b^2$ (from Section 5.5).

$= 4 - 3$ Simplify.

$= 1$

b. $(\sqrt{5} - 3\sqrt{3})(\sqrt{5} + 3\sqrt{3})$

Solution: $(\sqrt{5} - 3\sqrt{3})(\sqrt{5} + 3\sqrt{3}) = (\sqrt{5})^2 - (3\sqrt{3})^2$ Use $(a - b)(a + b) = a^2 - b^2.$

$= 5 - 9 \cdot 3$ Simplify.

$= 5 - 27$ Multiply.

$= -22$ Subtract.

Note The product of conjugates *always* results in a rational number. This will be useful in Section 9.5 when we rationalize denominators.

Your Turn 4   Find the product.

a. $(4 + \sqrt{10})(4 - \sqrt{10})$ **b.** $(3\sqrt{5} + 2\sqrt{6})(3\sqrt{5} - 2\sqrt{6})$

Objective 3 Simplify radical expressions that contain mixed operations.

Now let's use the order of operations to simplify radical expressions that have more than one operation.

Example 5   Simplify.

a. $\sqrt{2} \cdot \sqrt{10} + \sqrt{3} \cdot \sqrt{15}$

Solution: $\sqrt{2} \cdot \sqrt{10} + \sqrt{3} \cdot \sqrt{15} = \sqrt{2 \cdot 10} + \sqrt{3 \cdot 15}$ Use the product rule.

$= \sqrt{20} + \sqrt{45}$ Multiply.

$= \sqrt{4 \cdot 5} + \sqrt{9 \cdot 5}$ Rewrite 20 as $4 \cdot 5$ and 45 as $9 \cdot 5$.

$= \sqrt{4} \cdot \sqrt{5} + \sqrt{9} \cdot \sqrt{5}$ Use the product rule.

$= 2\sqrt{5} + 3\sqrt{5}$ Find $\sqrt{4}$ and $\sqrt{9}$.

$= 5\sqrt{5}$ Combine like radicals.

b. $\dfrac{\sqrt{54}}{\sqrt{3}} + \sqrt{32}$

Solution: $\dfrac{\sqrt{54}}{\sqrt{3}} + \sqrt{32} = \sqrt{\dfrac{54}{3}} + \sqrt{16 \cdot 2}$ Use the quotient rule and rewrite 32 as $16 \cdot 2$.

$= \sqrt{18} + \sqrt{16} \cdot \sqrt{2}$ Divide and use the product rule.

$= \sqrt{9 \cdot 2} + 4\sqrt{2}$ Rewrite 18 as $9 \cdot 2$ and find $\sqrt{16}$.

$= \sqrt{9} \cdot \sqrt{2} + 4\sqrt{2}$ Use the product rule.

$= 3\sqrt{2} + 4\sqrt{2}$ Find $\sqrt{9}$.

$= 7\sqrt{2}$ Combine like radicals.

Answers to Your Turn 4
a. 6   **b.** 21

Answers to Your Turn 5
a. $34\sqrt{2}$ **b.** $5\sqrt{7}$

Your Turn 5   Simplify.

a. $2\sqrt{3} \cdot \sqrt{6} + 4\sqrt{7} \cdot \sqrt{14}$ **b.** $\sqrt{63} + \dfrac{\sqrt{140}}{\sqrt{5}}$

9.4 Exercises For Extra Help MyMathLab®

Note: Exercises marked with a ★ represent challenging exercises.

Objective 1

Prep Exercise 1 What must be identical in like radicals? What can be different? Like radicals have the same index and the same radicand, but their coefficients may be different.

Prep Exercise 2 Explain how to add $3\sqrt{2} + 2\sqrt{2}$. Add the coefficients, $3 + 2$, and keep the radical, $\sqrt{2}$, the same.

For Exercises 1–14, add or subtract. Assume that variables represent nonnegative values. See Example 1.

1. $9\sqrt{6} - 15\sqrt{6}$
$-6\sqrt{6}$

2. $4\sqrt{5} - 13\sqrt{5}$
$-9\sqrt{5}$

3. $7\sqrt{a} + 2\sqrt{a}$
$9\sqrt{a}$

4. $5\sqrt{y} + 7\sqrt{y}$
$12\sqrt{y}$

5. $4\sqrt{5} - 2\sqrt{6} + 8\sqrt{5} - 6\sqrt{6}$
$12\sqrt{5} - 8\sqrt{6}$

6. $9\sqrt{7} - 4\sqrt{11} - 3\sqrt{7} - 5\sqrt{11}$
$6\sqrt{7} - 9\sqrt{11}$

7. $3a\sqrt{5a} - 4b\sqrt{7b} + 8a\sqrt{5a} + 2b\sqrt{7b}$
$11a\sqrt{5a} - 2b\sqrt{7b}$

8. $12n\sqrt{2n} - 14m\sqrt{5m} - 8n\sqrt{2n} + 18m\sqrt{5m}$
$4n\sqrt{2n} + 4m\sqrt{5m}$

9. $6x\sqrt[3]{9} - 3x\sqrt[3]{9}$
$3x\sqrt[3]{9}$

10. $4y\sqrt[3]{3} - y\sqrt[3]{3}$
$3y\sqrt[3]{3}$

11. $6x^2\sqrt[4]{5x} - 12x^2\sqrt[4]{5x}$
$-6x^2\sqrt[4]{5x}$

12. $3y^3\sqrt[4]{8y} - 9y^3\sqrt[4]{8y}$
$-6y^3\sqrt[4]{8y}$

★ 13. $3x\sqrt{5x} + 4x\sqrt[3]{5x}$
Cannot combine because the radicals are not like

★ 14. $4z\sqrt[4]{2z} - 7z\sqrt{2z}$
Cannot combine because the radicals are not like

For Exercises 15–34, add or subtract. Assume that variables represent nonnegative values. See Example 2.

15. $\sqrt{48} - \sqrt{75}$
$-\sqrt{3}$

16. $\sqrt{24} - \sqrt{96}$
$-2\sqrt{6}$

17. $\sqrt{80y} - \sqrt{125y}$
$-\sqrt{5y}$

18. $\sqrt{27x} + \sqrt{75x}$
$8\sqrt{3x}$

19. $\sqrt{80} - 4\sqrt{45}$
$-8\sqrt{5}$

20. $\sqrt{20} - 2\sqrt{180}$
$-10\sqrt{5}$

21. $3\sqrt{96} - 2\sqrt{54}$
$6\sqrt{6}$

22. $5\sqrt{63} + 2\sqrt{28}$
$19\sqrt{7}$

23. $6\sqrt{48a^3} - 2\sqrt{75a^3}$
$14a\sqrt{3a}$

24. $4\sqrt{98y^5} - 7\sqrt{128y^5}$
$-28y^2\sqrt{2y}$

25. $\sqrt{150} - \sqrt{54} + \sqrt{24}$
$4\sqrt{6}$

26. $\sqrt{20} + \sqrt{125} - \sqrt{80}$
$3\sqrt{5}$

27. $2\sqrt{8} - 3\sqrt{48} + 2\sqrt{98} - \sqrt{75}$
$18\sqrt{2} - 17\sqrt{3}$

28. $3\sqrt{180} - \sqrt{192} - 4\sqrt{80} - \sqrt{48}$
$2\sqrt{5} - 12\sqrt{3}$

29. $\sqrt[3]{128} + \sqrt[3]{54}$
$7\sqrt[3]{2}$

30. $\sqrt[3]{24} + \sqrt[3]{81}$
$5\sqrt[3]{3}$

31. $4\sqrt[3]{135x^5} - 6x\sqrt[3]{320x^2}$
$-12x\sqrt[3]{5x^2}$

32. $3a^2\sqrt[3]{500a^4} + 6a\sqrt[3]{108a^7}$
$33a^3\sqrt[3]{4a}$

33. $-4\sqrt[4]{32x^9} + 2x\sqrt[4]{162x^5}$
$-2x^2\sqrt[4]{2x}$

34. $6y\sqrt[4]{243y^6} - 2y^2\sqrt[4]{48y^2}$
$14y^2\sqrt[4]{3y^2}$

Objective 2

Prep Exercise 3 Explain how to multiply $(\sqrt{5} + 3)(\sqrt{5} + 2)$.

Multiply each term in the second expression by each term in the first (that is, use FOIL)

For Exercises 35–42 use the distributive property. Assume that variables represent nonnegative values. See Examples 3(a) and 3(b).

35. $\sqrt{2}(3 + \sqrt{2})$

$3\sqrt{2} + 2$

36. $\sqrt{5}(4 - \sqrt{5})$

$4\sqrt{5} - 5$

37. $\sqrt{3}(\sqrt{3} - \sqrt{15})$

$3 - 3\sqrt{5}$

38. $\sqrt{6}(\sqrt{6} + \sqrt{3})$

$6 + 3\sqrt{2}$

39. $\sqrt{5}(\sqrt{6} + 2\sqrt{10})$

$\sqrt{30} + 10\sqrt{2}$

40. $\sqrt{7}(\sqrt{5} - 3\sqrt{14})$

$\sqrt{35} - 21\sqrt{2}$

41. $4\sqrt{3x}(2\sqrt{3x} - 4\sqrt{6x})$

$24x - 48x\sqrt{2}$

42. $6\sqrt{2y}(3\sqrt{2y} + 2\sqrt{10y})$

$36y + 24y\sqrt{5}$

For Exercises 43–62, multiply. (Use FOIL.) Assume that variables represent nonnegative values. See Examples 3(c) and 3(d).

43. $(3 + \sqrt{5})(4 - \sqrt{2})$

$12 - 3\sqrt{2} + 4\sqrt{5} - \sqrt{10}$

44. $(2 + \sqrt{5})(6 - \sqrt{3})$

$12 - 2\sqrt{3} + 6\sqrt{5} - \sqrt{15}$

45. $(3 + \sqrt{x})(2 + \sqrt{x})$

$6 + 5\sqrt{x} + x$

46. $(5 - \sqrt{a})(2 + \sqrt{a})$

$10 + 3\sqrt{a} - a$

47. $(2 + 3\sqrt{3})(3 + 5\sqrt{2})$

$6 + 10\sqrt{2} + 9\sqrt{3} + 15\sqrt{6}$

48. $(7 - 3\sqrt{5})(2 - 2\sqrt{10})$

$14 - 14\sqrt{10} - 6\sqrt{5} + 30\sqrt{2}$

49. $(\sqrt{3} + \sqrt{5})(\sqrt{5} + \sqrt{7})$

$\sqrt{15} + \sqrt{21} + 5 + \sqrt{35}$

50. $(\sqrt{5} + \sqrt{2})(\sqrt{2} + \sqrt{7})$

$\sqrt{10} + \sqrt{35} + 2 + \sqrt{14}$

51. $(\sqrt{x} + 3\sqrt{y})(\sqrt{x} - 2\sqrt{y})$

$x + \sqrt{xy} - 6y$

52. $(2\sqrt{a} + \sqrt{b})(\sqrt{a} + 4\sqrt{b})$

$2a + 9\sqrt{ab} + 4b$

53. $(4\sqrt{2} + 2\sqrt{5})(3\sqrt{7} - 3\sqrt{3})$

$12\sqrt{14} - 12\sqrt{6} + 6\sqrt{35} - 6\sqrt{15}$

54. $(8\sqrt{2} - 2\sqrt{3})(2\sqrt{5} + 3\sqrt{10})$

$16\sqrt{10} + 48\sqrt{5} - 4\sqrt{15} - 6\sqrt{30}$

55. $(2\sqrt{a} + 3\sqrt{b})(4\sqrt{a} - \sqrt{b})$

$8a + 10\sqrt{ab} - 3b$

56. $(3\sqrt{m} - \sqrt{n})(2\sqrt{m} + 4\sqrt{n})$

$6m + 10\sqrt{mn} - 4n$

57. $(\sqrt[3]{4} + 5)(\sqrt[3]{4} - 8)$

$2\sqrt[3]{2} - 3\sqrt[3]{4} - 40$

58. $(\sqrt[3]{9} + 5)(\sqrt[3]{9} - 2)$

$3\sqrt[3]{3} + 3\sqrt[3]{9} - 10$

59. $(\sqrt[3]{9} + \sqrt[3]{4})(\sqrt[3]{3} - \sqrt[3]{2})$

$1 - \sqrt[3]{18} + \sqrt[3]{12}$

60. $(\sqrt[3]{5} + \sqrt[3]{9})(\sqrt[3]{25} - \sqrt[3]{3})$

$2 - \sqrt[3]{15} + \sqrt[3]{225}$

★ **61.** $(\sqrt[3]{x} + 2)(\sqrt[3]{x^2} - 2\sqrt[3]{x} + 4)$

$x + 8$

★ **62.** $(\sqrt[3]{r} - 3)(\sqrt[3]{r^2} + 3\sqrt[3]{r} + 9)$

$r - 27$

For Exercises 63–70, find the product. See Example 3(e).

63. $(4 + \sqrt{6})^2$

$22 + 8\sqrt{6}$

64. $(3 + \sqrt{7})^2$

$16 + 6\sqrt{7}$

65. $(4 - \sqrt{2})^2$

$18 - 8\sqrt{2}$

66. $(5 - \sqrt{2})^2$

$27 - 10\sqrt{2}$

67. $(2 + 2\sqrt{3})^2$

$16 + 8\sqrt{3}$

68. $(3 + 2\sqrt{5})^2$

$29 + 12\sqrt{5}$

69. $(2\sqrt{3} + 3\sqrt{2})^2$

$30 + 12\sqrt{6}$

70. $(4\sqrt{2} - 5\sqrt{6})^2$

$182 - 80\sqrt{3}$

For Exercises 71–84, multiply the conjugates. Assume that variables represent nonnegative values. See Example 4.

71. $(4 + \sqrt{3})(4 - \sqrt{3})$
13

72. $(3 + \sqrt{5})(3 - \sqrt{5})$
4

73. $(\sqrt{2} + 4)(\sqrt{2} - 4)$
−14

74. $(\sqrt{7} - 3)(\sqrt{7} + 3)$
−2

75. $(6 + \sqrt{x})(6 - \sqrt{x})$
$36 - x$

76. $(5 + \sqrt{y})(5 - \sqrt{y})$
$25 - y$

77. $(\sqrt{3} + \sqrt{2})(\sqrt{3} - \sqrt{2})$
1

78. $(\sqrt{5} - \sqrt{3})(\sqrt{5} + \sqrt{3})$
2

79. $(\sqrt{x} + \sqrt{y})(\sqrt{x} - \sqrt{y})$
$x - y$

80. $(\sqrt{a} - \sqrt{b})(\sqrt{a} + \sqrt{b})$
$a - b$

81. $(4 + 2\sqrt{3})(4 - 2\sqrt{3})$
4

82. $(7 + 2\sqrt{3})(7 - 2\sqrt{3})$
37

83. $(3\sqrt{7} + \sqrt{13})(3\sqrt{7} - \sqrt{13})$
50

84. $(4\sqrt{5} - 3\sqrt{2})(4\sqrt{5} + 3\sqrt{2})$
62

Objective 3

Prep Exercise 4 Why is $\sqrt{a} + \sqrt{b} \neq \sqrt{a + b}$? Use examples if necessary.

$\sqrt{a} + \sqrt{b} \neq \sqrt{a + b}$ because the order of operations is different. For $\sqrt{a} + \sqrt{b}$, the radicals must be evaluated first and then the addition. For $\sqrt{a + b}$, the addition must be evaluated first and then the radical. Example: $\sqrt{36} + \sqrt{64} \neq \sqrt{36 + 64}$ because $6 + 8 \neq 10$.

For Exercises 85–92, simplify. See Example 5.

85. $\sqrt{3} \cdot \sqrt{15} + \sqrt{8} \cdot \sqrt{10}$
$7\sqrt{5}$

86. $\sqrt{18} \cdot \sqrt{6} + \sqrt{8} \cdot \sqrt{24}$
$14\sqrt{3}$

87. $3\sqrt{3} \cdot \sqrt{18} - 4\sqrt{18} \cdot \sqrt{12}$
$-15\sqrt{6}$

88. $3\sqrt{2} \cdot 2\sqrt{40} - 5\sqrt{12} \cdot \sqrt{15}$
$-6\sqrt{5}$

89. $\dfrac{\sqrt{54}}{\sqrt{3}} + \sqrt{72}$
$9\sqrt{2}$

90. $\dfrac{\sqrt{60}}{\sqrt{5}} + \sqrt{48}$
$6\sqrt{3}$

91. $\dfrac{\sqrt{540}}{\sqrt{3}} - 4\sqrt{125}$
$-14\sqrt{5}$

92. $\dfrac{\sqrt{288}}{\sqrt{6}} - 6\sqrt{108}$
$-32\sqrt{3}$

For Exercises 93 and 94, find the perimeter of the figure in simplest form.

93.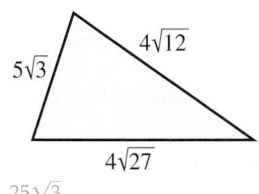
$5\sqrt{3}$ $4\sqrt{12}$ $4\sqrt{27}$
$25\sqrt{3}$

94.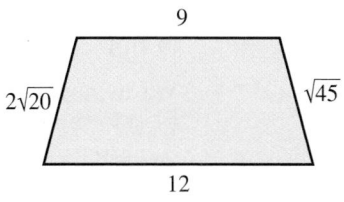
9   $2\sqrt{20}$ $\sqrt{45}$ 12
$7\sqrt{5} + 21$

95. Crown molding, which is placed at the top of a wall, is to be installed around the perimeter of the room shown.
   a. Write an expression in simplest form for the perimeter of the room.
   $42 + 2\sqrt{5}$ ft.
   b. Use a calculator to approximate the perimeter, rounded to the nearest tenth.
   46.5 ft.
   c. If crown molding costs $1.89 per foot length, how much will the crown molding cost for this room?
   $87.89

√5 ft.   9 ft.   √5 ft.

10 ft.

13 ft.

96. A tabletop is to be fitted with veneer strips along the sides.

    a. Write an expression in simplest form for the perimeter of the tabletop.

       $232 + 8\sqrt{2}$ in.

    b. Use a calculator to approximate the perimeter, rounded to the nearest tenth.

       243.3 in.

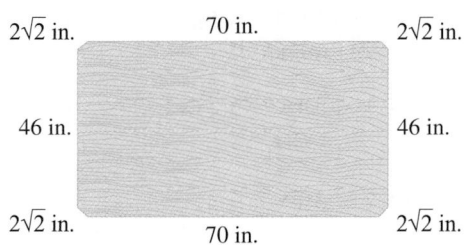

Review Exercises

Exercises 1–4   **Expressions**

[5.5] 1. What is the conjugate of $2x - 5$?

    $2x + 5$

[5.5] 2. Multiply: $(4x + 3)(4x - 3)$

    $16x^2 - 9$

[9.3] 3. What factor could multiply $\sqrt{8}$ to equal $\sqrt{16}$?

    $\sqrt{2}$

[9.3] 4. What factor could multiply $\sqrt[3]{2}$ to equal $\sqrt[3]{8}$?

    $\sqrt[3]{4}$

Exercises 5 and 6 **Equations and Inequalities**

[3.4] 5. For the equation $5y - 2x = 10$, find the slope and the y-intercept.

    Slope is $\dfrac{2}{5}$; y-intercept is $(0, 2)$.

[3.2] 6. Graph: $y = -\dfrac{1}{3}x + 4$

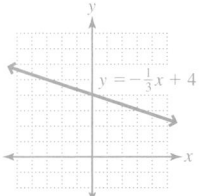

9.5 Rationalizing Numerators and Denominators of Radical Expressions

Objectives

1 Rationalize denominators.

2 Rationalize denominators that have a sum or difference with a square root term.

3 Rationalize numerators.

Warm-up

[9.3] 1. Simplify: $\sqrt{3} \cdot \sqrt{3}$

[9.3] 2. Simplify: $\sqrt[3]{9a^2} \cdot \sqrt[3]{3a}$

[9.3] 3. By what would you multiply $\sqrt[3]{5}$ to get $\sqrt[3]{125}$?

[9.4] 4. Simplify: $(\sqrt{6} + \sqrt{2})(\sqrt{6} - \sqrt{2})$

We are now ready to formalize the conditions for a radical expression that is in simplest form. A radical expression is in simplest form if

1. The radicand has no factor raised to a power greater than or equal to the index.
2. There are neither radicals in the denominator of a fraction nor radicands that are fractions.
3. All possible sums, differences, products, and quotients have been found.

In this section, we explore how to simplify expressions that have a radical in the denominator of a fraction, as in $\dfrac{1}{\sqrt{2}}$.

Objective 1 Rationalize denominators.

Answers to Warm-up

1. 3

2. $3a$

3. $\sqrt[3]{25}$

4. 4

If the denominator of a fraction contains a radical, our goal is to *rationalize* the denominator, which means to rewrite the expression so that it has a rational number in the denominator. In general, we multiply the fraction by a well-chosen 1 so that the radical is eliminated. We determine that 1 by finding a factor that multiplies the nth root in the denominator so that its radicand is a perfect nth power.

Square Root Denominators

In the case of a square root in the denominator, we multiply it by a factor that makes the radicand a perfect square, which allows us to eliminate the square root.

Instructor Note Illustrate why a fraction with a rationalized denominator is preferred over one with an irrational denominator. Consider $\frac{1}{\sqrt{2}}$ and its equivalent $\frac{\sqrt{2}}{2}$. Using a calculator, we can easily find their approximate value:

$$\frac{1}{\sqrt{2}} = \frac{\sqrt{2}}{2} \approx 0.7071067812.$$

However, if we had to find an approximate value of $\frac{1}{\sqrt{2}}$ without a calculator, we would have to divide 1 by a decimal approximation of $\sqrt{2}$. Using long division, dividing by that decimal approximation would not be easy. The equivalent expression $\frac{\sqrt{2}}{2}$ is easier to evaluate because we divide the decimal approximation of $\sqrt{2}$ by 2, which is an easier task using long division.

For that reason, $\frac{\sqrt{2}}{2}$ is considered a simpler form.

For example, to rationalize $\frac{1}{\sqrt{2}}$, we could multiply by $\frac{\sqrt{2}}{\sqrt{2}}$ because the product's denominator is the square root of a perfect square.

$$\frac{1}{\sqrt{2}} = \frac{1}{\sqrt{2}} \cdot \frac{\sqrt{2}}{\sqrt{2}} = \frac{\sqrt{2}}{\sqrt{4}} = \frac{\sqrt{2}}{2}$$

Note We are not changing the value of $\frac{1}{\sqrt{2}}$ because we are multiplying it by 1 in the form of $\frac{\sqrt{2}}{\sqrt{2}}$.

Any factor that produces a perfect square radicand will work. For example, we could have multiplied $\frac{1}{\sqrt{2}}$ by $\frac{\sqrt{8}}{\sqrt{8}}$.

$$\frac{1}{\sqrt{2}} = \frac{1}{\sqrt{2}} \cdot \frac{\sqrt{8}}{\sqrt{8}} = \frac{\sqrt{8}}{\sqrt{16}} = \frac{\sqrt{4 \cdot 2}}{4} = \frac{2\sqrt{2}}{4} = \frac{\sqrt{2}}{2}$$

Notice, however, that multiplying by $\frac{\sqrt{2}}{\sqrt{2}}$ required fewer steps to simplify than does multiplying by $\frac{\sqrt{8}}{\sqrt{8}}$.

| **Example 1** | Rationalize the denominator. Assume that variables represent positive values.

a. $\dfrac{2}{\sqrt{5}}$

Solution: $\dfrac{2}{\sqrt{5}} = \dfrac{2}{\sqrt{5}} \cdot \dfrac{\sqrt{5}}{\sqrt{5}}$ Multiply by $\dfrac{\sqrt{5}}{\sqrt{5}}$.

$\qquad\qquad = \dfrac{2\sqrt{5}}{\sqrt{25}}$ Simplify.

$\qquad\qquad = \dfrac{2\sqrt{5}}{5}$

b. $\sqrt{\dfrac{5}{8}}$

Solution: $\sqrt{\dfrac{5}{8}} = \dfrac{\sqrt{5}}{\sqrt{8}}$ Use the quotient rule of square roots to separate the numerator and denominator into two radicals.

$\qquad\qquad = \dfrac{\sqrt{5}}{\sqrt{8}} \cdot \dfrac{\sqrt{2}}{\sqrt{2}}$ Multiply by $\dfrac{\sqrt{2}}{\sqrt{2}}$.

$\qquad\qquad = \dfrac{\sqrt{10}}{\sqrt{16}}$ Multiply.

$\qquad\qquad = \dfrac{\sqrt{10}}{4}$ Simplify.

◀ **Note** We chose to multiply by $\frac{\sqrt{2}}{\sqrt{2}}$ because it leads to a smaller perfect square than do other choices such as $\frac{\sqrt{8}}{\sqrt{8}}$. Multiplying by $\frac{\sqrt{8}}{\sqrt{8}}$ produces the same final answer but requires more steps.

Warning Although it may be tempting to do so, we cannot divide out the 4 and 10 because 10 is a radicand, whereas 4 is not. We *never* divide out factors common to a radicand and a number that is not a radicand.

c. $\dfrac{3}{\sqrt{2x}}$

Solution: $\dfrac{3}{\sqrt{2x}} = \dfrac{3}{\sqrt{2x}} \cdot \dfrac{\sqrt{2x}}{\sqrt{2x}}$ Multiply by $\dfrac{\sqrt{2x}}{\sqrt{2x}}$.

$= \dfrac{3\sqrt{2x}}{\sqrt{4x^2}}$ Multiply.

$= \dfrac{3\sqrt{2x}}{2x}$ Simplify.

Your Turn 1   Rationalize the denominator. Assume that variables represent positive values.

a. $\dfrac{1}{\sqrt{7}}$ b. $\sqrt{\dfrac{7}{12}}$ c. $\dfrac{3}{\sqrt{10x}}$

*n*th-Root Denominators

If the denominator contains a higher-order root such as a cube root, we multiply appropriately to get a perfect cube radicand in the denominator so that we can eliminate the radical. For example, $\dfrac{2}{\sqrt[3]{5}} = \dfrac{2}{\sqrt[3]{5}} \cdot \dfrac{\sqrt[3]{25}}{\sqrt[3]{25}} = \dfrac{2\sqrt[3]{25}}{\sqrt[3]{125}} = \dfrac{2\sqrt[3]{25}}{5}$. We summarize as follows:

> **Procedure**   **Rationalizing Denominators**
>
> To rationalize a denominator containing a single *n*th root, multiply the fraction by a form of 1 so that the product's denominator has a radicand that is a perfect *n*th power.

Example 2   Rationalize the denominator. Assume that variables represent positive values.

a. $\dfrac{3}{\sqrt[3]{2}}$

Solution: $\dfrac{3}{\sqrt[3]{2}} = \dfrac{3}{\sqrt[3]{2}} \cdot \dfrac{\sqrt[3]{4}}{\sqrt[3]{4}}$ Because $\sqrt[3]{2} \cdot \sqrt[3]{4} = \sqrt[3]{8} = 2$, multiply the fraction by $\dfrac{\sqrt[3]{4}}{\sqrt[3]{4}}$.

$= \dfrac{3\sqrt[3]{4}}{\sqrt[3]{8}}$ Multiply.

$= \dfrac{3\sqrt[3]{4}}{2}$ Simplify.

b. $\dfrac{\sqrt[3]{a}}{\sqrt[3]{b}}$

Solution: $\dfrac{\sqrt[3]{a}}{\sqrt[3]{b}} = \dfrac{\sqrt[3]{a}}{\sqrt[3]{b}} \cdot \dfrac{\sqrt[3]{b^2}}{\sqrt[3]{b^2}}$ Because $\sqrt[3]{b} \cdot \sqrt[3]{b^2} = \sqrt[3]{b^3} = b$, multiply the fraction by $\dfrac{\sqrt[3]{b^2}}{\sqrt[3]{b^2}}$.

$= \dfrac{\sqrt[3]{ab^2}}{\sqrt[3]{b^3}}$ Multiply.

$= \dfrac{\sqrt[3]{ab^2}}{b}$ Simplify.

Answers to Your Turn 1

a. $\dfrac{\sqrt{7}}{7}$ b. $\dfrac{\sqrt{21}}{6}$ c. $\dfrac{3\sqrt{10x}}{10x}$

c. $\sqrt[3]{\dfrac{5}{9a^2}}$

Solution: $\sqrt[3]{\dfrac{5}{9a^2}} = \dfrac{\sqrt[3]{5}}{\sqrt[3]{9a^2}}$ Use the quotient rule to separate the numerator and denominator.

$\qquad\qquad = \dfrac{\sqrt[3]{5}}{\sqrt[3]{9a^2}} \cdot \dfrac{\sqrt[3]{3a}}{\sqrt[3]{3a}}$ Because $\sqrt[3]{9a^2} \cdot \sqrt[3]{3a} = \sqrt[3]{27a^3} = 3a$, multiply the fraction by $\dfrac{\sqrt[3]{3a}}{\sqrt[3]{3a}}$.

$\qquad\qquad = \dfrac{\sqrt[3]{15a}}{\sqrt[3]{27a^3}}$ Multiply.

$\qquad\qquad = \dfrac{\sqrt[3]{15a}}{3a}$ Simplify.

d. $\dfrac{5}{\sqrt[4]{3}}$

Solution: $\dfrac{5}{\sqrt[4]{3}} = \dfrac{5}{\sqrt[4]{3}} \cdot \dfrac{\sqrt[4]{27}}{\sqrt[4]{27}}$ Because $\sqrt[4]{3} \cdot \sqrt[4]{27} = \sqrt[4]{81} = 3$, multiply the fraction by $\dfrac{\sqrt[4]{27}}{\sqrt[4]{27}}$.

$\qquad\qquad = \dfrac{5\sqrt[4]{27}}{\sqrt[4]{81}}$ Multiply.

$\qquad\qquad = \dfrac{5\sqrt[4]{27}}{3}$ Simplify.

Your Turn 2 Rationalize the denominator. Assume that variables represent positive values.

a. $\dfrac{6}{\sqrt[3]{3}}$ **b.** $\sqrt[3]{\dfrac{3}{x^2}}$ **c.** $\dfrac{4}{\sqrt[3]{4y}}$ **d.** $\dfrac{7}{\sqrt[4]{2}}$

Objective 2 Rationalize denominators that have a sum or difference with a square root term.

In Example 4 of Section 9.4, we saw that the product of two conjugates containing square roots does not contain any radicals. Consequently, if the denominator of a fraction contains a sum or difference with a square root term, we can rationalize the denominator by multiplying the fraction by a 1 made up of the conjugate of the denominator. For example, to rationalize $\dfrac{5}{7 - \sqrt{3}}$, we multiply by $\dfrac{7 + \sqrt{3}}{7 + \sqrt{3}}$. Because $7 - \sqrt{3}$ and $7 + \sqrt{3}$ are conjugates, their product will not contain any radicals; so the denominator will be rationalized.

$$\dfrac{5}{7 - \sqrt{3}} = \dfrac{5}{7 - \sqrt{3}} \cdot \dfrac{7 + \sqrt{3}}{7 + \sqrt{3}} = \dfrac{5(7 + \sqrt{3})}{(7)^2 - (\sqrt{3})^2} = \dfrac{35 + 5\sqrt{3}}{49 - 3} = \dfrac{35 + 5\sqrt{3}}{46}$$

Answers to Your Turn 2

a. $2\sqrt[3]{9}$ **b.** $\dfrac{\sqrt[3]{3x}}{x}$

c. $\dfrac{2\sqrt[3]{2y^2}}{y}$ **d.** $\dfrac{7\sqrt[4]{8}}{2}$

Procedure **Rationalizing a Denominator Containing a Sum or Difference**

To rationalize a denominator containing a sum or difference with at least one square root term, multiply the fraction by a form of 1 whose numerator and denominator are the conjugate of the denominator.

Example 3   Rationalize the denominator and simplify. Assume that variables represent positive values.

a. $\dfrac{9}{\sqrt{2} + 7}$

Solution: $\dfrac{9}{\sqrt{2} + 7} = \dfrac{9}{\sqrt{2} + 7} \cdot \dfrac{\sqrt{2} - 7}{\sqrt{2} - 7}$ The conjugate of $\sqrt{2} + 7$ is $\sqrt{2} - 7$, so we multiply by $\dfrac{\sqrt{2} - 7}{\sqrt{2} - 7}$.

$= \dfrac{9(\sqrt{2} - 7)}{(\sqrt{2})^2 - (7)^2}$ Multiply. In the denominator, use the rule $(a + b)(a - b) = a^2 - b^2$.

$= \dfrac{9\sqrt{2} - 63}{2 - 49}$ Simplify.

$= \dfrac{9\sqrt{2} - 63}{-47}$ We can simplify the negative denominator by factoring out -1 in the numerator and denominator.

$= \dfrac{-1(63 - 9\sqrt{2})}{-1(47)}$ After factoring out the -1, the signs of the terms change. Because 63 is now positive, we write it first.

$= \dfrac{63 - 9\sqrt{2}}{47}$ Divide out the common factor -1.

b. $\dfrac{2\sqrt{3}}{\sqrt{6} - \sqrt{2}}$

Solution: $\dfrac{2\sqrt{3}}{\sqrt{6} - \sqrt{2}} = \dfrac{2\sqrt{3}}{\sqrt{6} - \sqrt{2}} \cdot \dfrac{\sqrt{6} + \sqrt{2}}{\sqrt{6} + \sqrt{2}}$ The conjugate of $\sqrt{6} - \sqrt{2}$ is $\sqrt{6} + \sqrt{2}$, so we multiply by $\dfrac{\sqrt{6} + \sqrt{2}}{\sqrt{6} + \sqrt{2}}$.

$= \dfrac{2\sqrt{3}(\sqrt{6} + \sqrt{2})}{(\sqrt{6})^2 - (\sqrt{2})^2}$

$= \dfrac{2\sqrt{18} + 2\sqrt{6}}{6 - 2}$ Multiply in the numerator and evaluate the exponents in the denominator.

$= \dfrac{2\sqrt{9 \cdot 2} + 2\sqrt{6}}{4}$ Simplify $\sqrt{18}$ by factoring out a perfect square factor in 18.

$= \dfrac{2 \cdot 3\sqrt{2} + 2\sqrt{6}}{4}$ Simplify $\sqrt{9 \cdot 2}$ by finding the square root of 9.

$= \dfrac{2(3\sqrt{2} + \sqrt{6})}{4}$ Factor out the common factor 2 in the numerator.

$= \dfrac{3\sqrt{2} + \sqrt{6}}{2}$ Divide out the common factor 2.

Note We cannot combine $3\sqrt{2}$ and $\sqrt{6}$ because they are not like radicals. ▶

c. $\dfrac{6}{\sqrt{x} - 5}$

Solution: $\dfrac{6}{\sqrt{x} - 5} = \dfrac{6}{\sqrt{x} - 5} \cdot \dfrac{\sqrt{x} + 5}{\sqrt{x} + 5}$ The conjugate of $\sqrt{x} - 5$ is $\sqrt{x} + 5$, so we multiply by $\dfrac{\sqrt{x} + 5}{\sqrt{x} + 5}$.

$= \dfrac{6(\sqrt{x} + 5)}{(\sqrt{x})^2 - (5)^2}$ Multiply.

$= \dfrac{6\sqrt{x} + 30}{x - 25}$ Multiply in the numerator and evaluate the exponents in the denominator.

Your Turn 3 — Rationalize the denominator and simplify. Assume that variables represent positive values.

a. $\dfrac{9}{\sqrt{5} + 2}$

b. $\dfrac{\sqrt{2}}{\sqrt{5} - \sqrt{3}}$

c. $\dfrac{3}{\sqrt{x} + 4}$

Objective 3 Rationalize numerators.

In later mathematics courses, you may need to rationalize the numerator. We use the same procedure that we use in rationalizing denominators.

Example 4 — Rationalize the numerator. Assume that variables represent positive values.

a. $\dfrac{\sqrt{5x}}{4}$

Solution: $\dfrac{\sqrt{5x}}{4} = \dfrac{\sqrt{5x}}{4} \cdot \dfrac{\sqrt{5x}}{\sqrt{5x}}$ — To create a perfect square radicand in the numerator, we multiply by $\dfrac{\sqrt{5x}}{\sqrt{5x}}$.

$= \dfrac{\sqrt{25x^2}}{4\sqrt{5x}}$ — Multiply.

$= \dfrac{5x}{4\sqrt{5x}}$ — Simplify.

b. $\dfrac{3 + \sqrt{2x}}{4}$

Solution: $\dfrac{3 + \sqrt{2x}}{4} = \dfrac{3 + \sqrt{2x}}{4} \cdot \dfrac{3 - \sqrt{2x}}{3 - \sqrt{2x}}$ — The conjugate of $3 + \sqrt{2x}$ is $3 - \sqrt{2x}$, so we multiply by $\dfrac{3 - \sqrt{2x}}{3 - \sqrt{2x}}$.

$= \dfrac{3^2 - (\sqrt{2x})^2}{4(3 - \sqrt{2x})}$ — Multiply.

$= \dfrac{9 - 2x}{12 - 4\sqrt{2x}}$ — Simplify.

Answers to Your Turn 3

a. $9\sqrt{5} - 18$ b. $\dfrac{\sqrt{10} + \sqrt{6}}{2}$

c. $\dfrac{3\sqrt{x} - 12}{x - 16}$

Answers to Your Turn 4

a. $\dfrac{3a}{7\sqrt{3a}}$ b. $\dfrac{25 - 3a}{10 + 2\sqrt{3a}}$

Your Turn 4 — Rationalize the numerators. Assume that variables represent positive values.

a. $\dfrac{\sqrt{3a}}{7}$

b. $\dfrac{5 - \sqrt{3a}}{2}$

9.5 Exercises — For Extra Help — MyMathLab®

Note: Exercises marked with a ★ represent challenging exercises.

Objective 1

Prep Exercise 1 Explain why each of the following expressions is not in simplest form.

a. $\sqrt{\dfrac{3}{16}}$

$\sqrt{16}$ is rational.

b. $\dfrac{5}{\sqrt{3}}$

A radical, $\sqrt{3}$, is in the denominator.

Prep Exercise 2 Although $\dfrac{1}{\sqrt{3}}$ and $\dfrac{\sqrt{3}}{3}$ are equal, explain why $\dfrac{\sqrt{3}}{3}$ is considered simplest form.

The denominator in $\dfrac{\sqrt{3}}{3}$ is a rational number, whereas it is irrational in $\dfrac{1}{\sqrt{3}}$.

Prep Exercise 3 Explain how to rationalize a denominator that is the square root of a number $\left(\text{for example, } \dfrac{2}{\sqrt{3}} \text{ or } \dfrac{\sqrt{5}}{\sqrt{7}}\right)$.

Multiply the fraction by a 1 so that the product's denominator has a radicand that is a perfect square.

For Exercises 1–24, rationalize the denominator. Assume that variables represent positive values. See Example 1.

1. $\dfrac{1}{\sqrt{3}}$

$\dfrac{\sqrt{3}}{3}$

2. $\dfrac{1}{\sqrt{7}}$

$\dfrac{\sqrt{7}}{7}$

3. $\dfrac{3}{\sqrt{8}}$

$\dfrac{3\sqrt{2}}{4}$

4. $\dfrac{5}{\sqrt{12}}$

$\dfrac{5\sqrt{3}}{6}$

5. $\sqrt{\dfrac{36}{7}}$

$\dfrac{6\sqrt{7}}{7}$

6. $\sqrt{\dfrac{64}{3}}$

$\dfrac{8\sqrt{3}}{3}$

7. $\sqrt{\dfrac{5}{12}}$

$\dfrac{\sqrt{15}}{6}$

8. $\sqrt{\dfrac{11}{18}}$

$\dfrac{\sqrt{22}}{6}$

9. $\dfrac{\sqrt{7x^2}}{\sqrt{50}}$

$\dfrac{x\sqrt{14}}{10}$

10. $\dfrac{\sqrt{3x^2}}{\sqrt{32}}$

$\dfrac{x\sqrt{6}}{8}$

11. $\dfrac{\sqrt{8}}{\sqrt{56}}$

$\dfrac{\sqrt{7}}{7}$

12. $\dfrac{\sqrt{7}}{\sqrt{42}}$

$\dfrac{\sqrt{6}}{6}$

13. $\dfrac{5}{\sqrt{3a}}$

$\dfrac{5\sqrt{3a}}{3a}$

14. $\dfrac{11}{\sqrt{7b}}$

$\dfrac{11\sqrt{7b}}{7b}$

15. $\sqrt{\dfrac{3m}{11n}}$

$\dfrac{\sqrt{33mn}}{11n}$

16. $\sqrt{\dfrac{5r}{6s}}$

$\dfrac{\sqrt{30rs}}{6s}$

17. $\dfrac{10}{\sqrt{5x}}$

$\dfrac{2\sqrt{5x}}{x}$

18. $\dfrac{28}{\sqrt{7a}}$

$\dfrac{4\sqrt{7a}}{a}$

19. $\dfrac{\sqrt{6x}}{\sqrt{32x}}$

$\dfrac{\sqrt{3}}{4}$

20. $\dfrac{\sqrt{10a}}{\sqrt{18a}}$

$\dfrac{\sqrt{5}}{3}$

21. $\dfrac{3}{\sqrt{x^3}}$

$\dfrac{3\sqrt{x}}{x^2}$

22. $\dfrac{5}{\sqrt{b^5}}$

$\dfrac{5\sqrt{b}}{b^3}$

23. $\dfrac{8x^2}{\sqrt{2x}}$

$4x\sqrt{2x}$

24. $\dfrac{18a^4}{\sqrt{6a}}$

$3a^3\sqrt{6a}$

Find ⊗ the Mistake *For Exercises 25 and 26, explain the mistake; then simplify correctly.*

25. $\dfrac{\sqrt{3}}{\sqrt{2}} = \dfrac{\sqrt{3}}{\sqrt{2}} \cdot \dfrac{2}{2} = \dfrac{2\sqrt{3}}{2}$

Mistake: The product of $\sqrt{2}$ and 2 is not 2. Correct: $\dfrac{\sqrt{6}}{2}$

26. $\sqrt{\dfrac{7}{3}} = \dfrac{\sqrt{7}}{\sqrt{3}} \cdot \dfrac{\sqrt{3}}{\sqrt{3}} = \dfrac{\sqrt{21}}{9}$

Mistake: Multiplied $\sqrt{3} \cdot \sqrt{3}$ incorrectly. Correct: $\dfrac{\sqrt{21}}{3}$

For Exercises 27–46, rationalize the denominators. Assume that variables represent positive values. See Example 2.

27. $\dfrac{5}{\sqrt[3]{3}}$

$\dfrac{5\sqrt[3]{9}}{3}$

28. $\dfrac{7}{\sqrt[3]{5}}$

$\dfrac{7\sqrt[3]{25}}{5}$

29. $\sqrt[3]{\dfrac{5}{2}}$

$\dfrac{\sqrt[3]{20}}{2}$

30. $\sqrt[3]{\dfrac{3}{4}}$

$\dfrac{\sqrt[3]{6}}{2}$

31. $\dfrac{6}{\sqrt[3]{4}}$

$3\sqrt[3]{2}$

32. $\dfrac{9}{\sqrt[3]{9}}$

$3\sqrt[3]{3}$

33. $\dfrac{m}{\sqrt[3]{n}}$

$\dfrac{m\sqrt[3]{n^2}}{n}$

34. $\dfrac{p}{\sqrt[3]{q}}$

$\dfrac{p\sqrt[3]{q^2}}{q}$

35. $\sqrt[3]{\dfrac{a}{b^2}}$

$\dfrac{\sqrt[3]{ab}}{b}$

36. $\sqrt[3]{\dfrac{m}{n^2}}$

$\dfrac{\sqrt[3]{mn}}{n}$

37. $\dfrac{4}{\sqrt[3]{2x}}$

$\dfrac{2\sqrt[3]{4x^2}}{x}$

38. $\dfrac{9}{\sqrt[3]{3a}}$

$\dfrac{3\sqrt[3]{9a^2}}{a}$

39. $\sqrt[3]{\dfrac{6}{25a^2}}$

$\dfrac{\sqrt[3]{30a}}{5a}$

40. $\sqrt[3]{\dfrac{5}{16b^2}}$

$\dfrac{\sqrt[3]{20b}}{4b}$

41. $\dfrac{5}{\sqrt[4]{4}}$

$\dfrac{5\sqrt[4]{4}}{2}$ or $\dfrac{5\sqrt{2}}{2}$

42. $\dfrac{7}{\sqrt[4]{9}}$

$\dfrac{7\sqrt[4]{9}}{3}$ or $\dfrac{7\sqrt{3}}{3}$

43. $\sqrt[4]{\dfrac{3}{x^2}}$

$\dfrac{\sqrt[4]{3x^2}}{x}$

44. $\sqrt[4]{\dfrac{5}{y^3}}$

$\dfrac{\sqrt[4]{5y}}{y}$

45. $\dfrac{9}{\sqrt[4]{3x^3}}$

$\dfrac{3\sqrt[4]{27x}}{x}$

46. $\dfrac{12}{\sqrt[4]{2x^2}}$

$\dfrac{6\sqrt[4]{8x^2}}{x}$

Objective 2

Prep Exercise 4 Explain how to rationalize a denominator that is a sum or difference with a square root term $\left(\text{for example, } \dfrac{3}{5 + \sqrt{2}} \text{ or } \dfrac{2}{\sqrt{x} - \sqrt{y}}\right)$.

Multiply the fraction by a 1 whose numerator and denominator are the conjugate of the denominator.

For Exercises 47–68, rationalize the denominator and simplify. Assume that variables represent positive values. See Example 3.

47. $\dfrac{3}{\sqrt{2} + 1}$

$3\sqrt{2} - 3$

48. $\dfrac{3}{\sqrt{5} + 2}$

$3\sqrt{5} - 6$

49. $\dfrac{4}{2 - \sqrt{3}}$

$8 + 4\sqrt{3}$

50. $\dfrac{2}{4 - \sqrt{15}}$

$8 + 2\sqrt{15}$

51. $\dfrac{5}{\sqrt{2} + \sqrt{3}}$

$5\sqrt{3} - 5\sqrt{2}$

52. $\dfrac{7}{\sqrt{6} + \sqrt{7}}$

$7\sqrt{7} - 7\sqrt{6}$

53. $\dfrac{4}{1 - \sqrt{5}}$

$-1 - \sqrt{5}$

54. $\dfrac{6}{1 - \sqrt{7}}$

$-1 - \sqrt{7}$

55. $\dfrac{\sqrt{3}}{\sqrt{3} - 1}$

$\dfrac{3 + \sqrt{3}}{2}$

56. $\dfrac{\sqrt{5}}{\sqrt{5} - 1}$

$\dfrac{\sqrt{5} + 5}{4}$

57. $\dfrac{2\sqrt{3}}{\sqrt{3} - 4}$

$\dfrac{-6 - 8\sqrt{3}}{13}$

58. $\dfrac{2\sqrt{5}}{\sqrt{5} - 4}$

$\dfrac{-10 - 8\sqrt{5}}{11}$

59. $\dfrac{4\sqrt{3}}{\sqrt{7} + \sqrt{2}}$

$\dfrac{4\sqrt{21} - 4\sqrt{6}}{5}$

60. $\dfrac{2\sqrt{2}}{\sqrt{4} + \sqrt{3}}$

$4\sqrt{2} - 2\sqrt{6}$

61. $\dfrac{8\sqrt{2}}{4\sqrt{2} - \sqrt{6}}$

$\dfrac{32 + 8\sqrt{3}}{13}$

62. $\dfrac{4\sqrt{3}}{4\sqrt{6} + \sqrt{2}}$

$\dfrac{24\sqrt{2} - 2\sqrt{6}}{47}$

63. $\dfrac{6\sqrt{y}}{\sqrt{y}+1}$
$\dfrac{6y-6\sqrt{y}}{y-1}$

64. $\dfrac{4\sqrt{x}}{\sqrt{x}+1}$
$\dfrac{4x-4\sqrt{x}}{x-1}$

65. $\dfrac{3\sqrt{t}}{\sqrt{t}+2\sqrt{u}}$
$\dfrac{3t-6\sqrt{tu}}{t-4u}$

66. $\dfrac{2\sqrt{m}}{\sqrt{n}-3\sqrt{m}}$
$\dfrac{2\sqrt{mn}+6m}{n-9m}$

67. $\dfrac{\sqrt{2y}}{\sqrt{x}-\sqrt{6y}}$
$\dfrac{\sqrt{2xy}+2y\sqrt{3}}{x-6y}$

68. $\dfrac{\sqrt{14h}}{\sqrt{2h}+\sqrt{k}}$
$\dfrac{2h\sqrt{7}-\sqrt{14hk}}{2h-k}$

Objective 3

Prep Exercise 5 Explain how to rationalize the numerator in $\dfrac{2+\sqrt{7x}}{5}$.

Multiply the numerator and denominator by $2-\sqrt{7x}$.

For Exercises 69–80, rationalize the numerator. Assume that variables represent positive values. See Example 4.

69. $\dfrac{\sqrt{3}}{2}$
$\dfrac{3}{2\sqrt{3}}$

70. $\dfrac{\sqrt{7}}{3}$
$\dfrac{7}{3\sqrt{7}}$

71. $\dfrac{\sqrt{2x}}{5}$
$\dfrac{2x}{5\sqrt{2x}}$

72. $\dfrac{\sqrt{7y}}{3}$
$\dfrac{7y}{3\sqrt{7y}}$

73. $\dfrac{\sqrt{8n}}{6}$
$\dfrac{2n}{3\sqrt{2n}}$

74. $\dfrac{\sqrt{20t}}{8}$
$\dfrac{5t}{4\sqrt{5t}}$

75. $\dfrac{2+\sqrt{3}}{5}$
$\dfrac{1}{10-5\sqrt{3}}$

76. $\dfrac{3+\sqrt{2}}{4}$
$\dfrac{7}{12-4\sqrt{2}}$

77. $\dfrac{\sqrt{5x}-6}{9}$
$\dfrac{5x-36}{9\sqrt{5x}+54}$

78. $\dfrac{\sqrt{2x}+7}{3}$
$\dfrac{2x-49}{3\sqrt{2x}-21}$

★79. $\dfrac{5\sqrt{n}+\sqrt{6n}}{2n}$
$\dfrac{19}{10\sqrt{n}-2\sqrt{6n}}$

★80. $\dfrac{4\sqrt{k}-\sqrt{10k}}{5k}$
$\dfrac{6}{20\sqrt{k}+5\sqrt{10k}}$

81. Given $f(x)=\dfrac{5\sqrt{2}}{x}$, find each of the following. Express your answer in simplest form.

   a. $f(\sqrt{6})$
   $\dfrac{5\sqrt{3}}{3}$

   b. $f(\sqrt{10})$
   $\sqrt{5}$

   c. $f(\sqrt{22})$
   $\dfrac{5\sqrt{11}}{11}$

82. Given $g(x)=\dfrac{\sqrt{2}}{x-1}$, find each of the following. Express your answer in simplest form.

   a. $g(\sqrt{5})$
   $\dfrac{\sqrt{10}+\sqrt{2}}{4}$

   b. $g(3\sqrt{2})$
   $\dfrac{6+\sqrt{2}}{17}$

   c. $g(2\sqrt{6})$
   $\dfrac{4\sqrt{3}+\sqrt{2}}{23}$

83. Graph $f(x)=\dfrac{1}{\sqrt{x}}$; then graph $g(x)=\dfrac{\sqrt{x}}{x}$.

   a. What do you notice about the two graphs? What does this indicate about the two functions?
   The graphs are identical. The functions are identical.

   b. Simplify $f(x)$ by rationalizing the denominator. What do you notice?
   $f(x)=g(x)$

84. Graph $f(x)=-\dfrac{1}{\sqrt{x}}$; then graph $g(x)=-\dfrac{\sqrt{x}}{x}$.

   a. What do you notice about the two graphs? What does this indicate about the two functions?
   The graphs are identical. The functions are identical.

   b. Simplify $f(x)$ by rationalizing the denominator. What do you notice?
   $f(x)=g(x)$

85. Previously, we used the formula $T = 2\pi\sqrt{\dfrac{L}{9.8}}$ to determine the period of a pendulum, where T is the period in seconds and L is the length in meters.

    a. Rewrite the formula so that the denominator is rationalized.
    $T = \pi\sqrt{9.8L}/4.9$

    b. Rewrite the formula so that the numerator is rationalized.
    $T = 2\pi L/\sqrt{9.8L}$

86. The formula $t = \sqrt{\dfrac{h}{16}}$ can be used to find the time, t, in seconds for an object to fall a distance of h feet.

    a. Rewrite the formula so that the denominator is rationalized.
    $t = \sqrt{h}/4$

    b. Rewrite the formula so that the numerator is rationalized.
    $t = h/(4\sqrt{h})$

Of Interest

The Great Pyramid was built for King Khufu from about 2589 to 2566 B.C. The pyramid contains approximately 2,300,000 blocks and weighs about 6.5 million tons, with each block averaging about 2.8 tons.

📱 87. The formula $s = \sqrt{\dfrac{3V}{h}}$ can be used to find the length, s, of each side of the base of a pyramid having a square base, volume V in cubic feet, and height h in feet.

    a. Rationalize the denominator in the formula.
    $s = \sqrt{3Vh}/h$

    b. The volume of the Great Pyramid at Giza in Egypt is approximately 83,068,742 cubic feet, and its height is 449 feet. Find the length of each side of its base.
    745 ft.

📱 ★ 88. The formula $s = 2\sqrt{\dfrac{A}{6\sqrt{3}}}$ can be used to find the length, s, of each side of a regular hexagon having an area A.

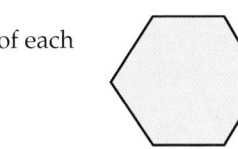

s

    a. Rationalize the denominator in the formula.
    $s = \dfrac{\sqrt{2A\sqrt{3}}}{3}$

    b. If A is 100 square meters, write an expression in simplest form for the side length.
    $\dfrac{10\sqrt{2\sqrt{3}}}{3}$ m

    c. Use a calculator to approximate the side lengths, rounded to three decimal places.
    ≈ 6.204 m

89. In AC circuits, voltage is often expressed as a *root-mean-square*, or *rms*, value. The formula for calculating the rms voltage, V_{rms}, given the maximum voltage, V_m, value is $V_{rms} = \dfrac{V_m}{\sqrt{2}}$.

    a. Rationalize the denominator in the formula.
    $V_{rms} = \dfrac{\sqrt{2}}{2}V_m$ or $\dfrac{\sqrt{2}V_m}{2}$

    b. Given a maximum voltage of 163 V, write an expression for the rms voltage.
    $V_{rms} = \dfrac{163\sqrt{2}}{2}$

    c. Use a calculator to approximate the rms voltage, rounded to the nearest tenth.
    ≈ 115.3

90. The velocity, in meters per second, of a particle can be determined by the formula $v = \sqrt{\dfrac{2E}{m}}$, where E represents the kinetic energy, in joules, of the particle and m represents the mass, in kilograms, of the particle.

    a. Rationalize the denominator in the formula.
    $v = \dfrac{\sqrt{2Em}}{m}$

b. A particle with a mass of 1×10^{-6} kilograms has 2.4×10^{7} joules of kinetic energy. Write an expression for its velocity.

   $4,000,000\sqrt{3}$ m/sec.

c. Use a calculator to approximate the velocity rounded to the nearest tenth.

   $6,928,203.2$ m/sec.

91. The resistance in a circuit is found to be $\dfrac{5\sqrt{2}}{3 + \sqrt{6}}$ ohms. Rationalize the denominator.

   $\dfrac{15\sqrt{2} - 10\sqrt{3}}{3}\Omega$

92. Two charged particles, q_1 and q_2, are separated by a distance of 8 centimeters. The values of the charges are $q_1 = 3 \times 10^{-6}$ coulombs and $q_2 = 1 \times 10^{-6}$ coulombs. Each charged particle exerts an electrical field. At a point between the two particles x centimeters away from q_1, the electric fields cancel each other so that the value of the fields at the point x is 0.

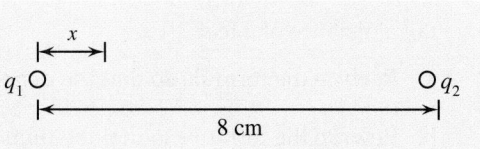

a. Use the formula $x = \dfrac{l}{1 + \sqrt{\dfrac{q_2}{q_1}}}$ to find the distance from q_1 at which the electric field is canceled, where l is the distance separating the particles. Write the distance with a rationalized denominator.

   $12 - 4\sqrt{3}$

b. Use a calculator to approximate the distance, rounded to the nearest tenth.

   5.1 cm

Review Exercises

Exercises 1 and 2   Expressions

[9.3] 1. Simplify: $\pm \sqrt{28}$

   $\pm 2\sqrt{7}$

[6.4] 2. Factor: $x^2 - 6x + 9$

   $(x - 3)^2$

Exercises 3–6   Equations and Inequalities

[2.3] *For Exercises 3 and 4, solve.*

3. $2x - 3 = 5$ 4

4. $2x - 3 = -5$

   -1

[6.6] *For Exercises 5 and 6, solve and check.*

5. $x^2 - 36 = 0$

   $-6, 6$

6. $x^2 - 5x + 6 = 0$

   $2, 3$

9.6 Radical Equations and Problem Solving

Objective

1 Use the power rule to solve radical equations.

Answers to Warm-up

1. $\dfrac{\sqrt{15}}{5}$ **2.** $6x - 1$

3. $x = 4$ **4.** $x = -3, 4$

Warm-up

[9.5] 1. Rationalize the denominator. $\sqrt{\dfrac{3}{5}}$

[9.3] 2. Simplify: $(\sqrt{6x - 1})^2$

For Exercises 2 and 3, solve.

[2.3] 3. $3x + 4 = 16$

[6.6] 4. $x^2 + 4x + 4 = 5x + 16$

We now explore how to solve **radical equations**.

Definition **Radical equation:** An equation containing at least one radical expression whose radicand has a variable.

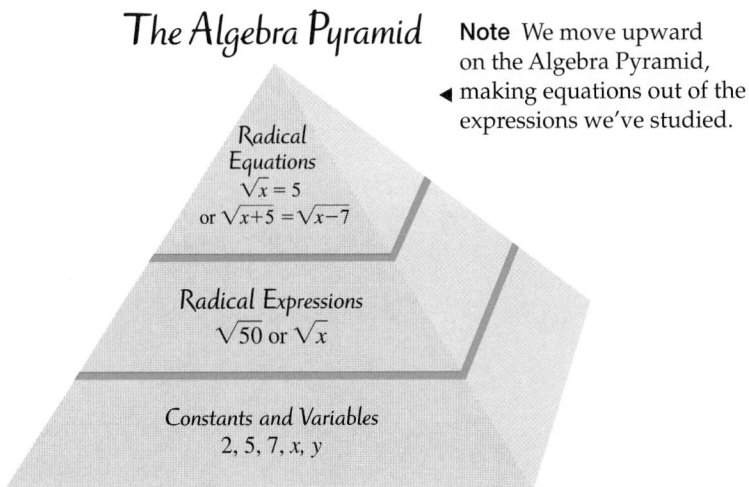

The Algebra Pyramid

Radical
Equations
$\sqrt{x} = 5$
or $\sqrt{x+5} = \sqrt{x-7}$

Radical Expressions
$\sqrt{50}$ or $\sqrt{x}$

Constants and Variables
$2, 5, 7, x, y$

Note We move upward on the Algebra Pyramid, ◄ making equations out of the expressions we've studied.

Objective 1 Use the power rule to solve radical equations.

To solve radical equations, we use a principle of equality called the *power rule.*

Rule **Power Rule for Solving Equations**

If both sides of an equation are raised to the same integer power, the resulting equation contains all solutions of the original equation and perhaps some solutions that do not solve the original equation. That is, the solutions of the equation $a = b$ are contained among the solutions of $a^n = b^n$, where n is an integer.

Isolated Radicals

First, we consider equations such as $\sqrt{x} = 9$ in which the radical is isolated. When we use the power rule, we raise both sides of the equation to the same integer power as the root index, then use the principle $(\sqrt[n]{x})^n = x$, which eliminates the radical, leaving its radicand.

Example 1 Solve.

a. $\sqrt{x} = 9$

Solution: $(\sqrt{x})^2 = (9)^2$ Because the root index is 2, we square both sides.

$x = 81$

Check: $\sqrt{81} \overset{?}{=} 9$

$9 = 9$ True

b. $\sqrt[3]{y} = -2$

Solution: $(\sqrt[3]{y})^3 = (-2)^3$ Because the root index is 3, we cube both sides.

$y = -8$

Check: $\sqrt[3]{-8} \overset{?}{=} -2$

$-2 = -2$ True

Extraneous Solutions

In Section 7.6, we learned that some equations have *extraneous solutions*, which are apparent solutions that do not make the original equation true. Using the power rule can sometimes lead to extraneous solutions, so it is important to check solutions. For example, watch what happens when we use the power rule to solve the equation $\sqrt{x} = -9$.

Warning As we will see in the check, this result is not a solution. $(\sqrt{x})^2 = (-9)^2$ Square both sides.
$\longrightarrow x = 81$

By checking 81 in the original equation, we see that it is extraneous.

$$\sqrt{81} \overset{?}{=} -9$$
$$9 = -9 \quad \text{This equation is false, so 81 is extraneous.}$$

In fact, $\sqrt{x} = -9$ has no real number solution because if x is a real number, then $\sqrt{x}$ must be nonnegative.

Example 2   Solve.

a. $\sqrt{x - 7} = 8$

Solution: $(\sqrt{x - 7})^2 = (8)^2$ Square both sides.
$x - 7 = 64$ Simplify.
$x = 71$ Add 7 to both sides.

Check: $\sqrt{71 - 7} \overset{?}{=} (8)^2$
$\sqrt{64} \overset{?}{=} 8$
$8 = 8$ True. The solution is 71.

b. $\sqrt[3]{x - 3} = -1$

Solution: $(\sqrt[3]{x - 3})^3 = (-1)^3$ Cube both sides.
$x - 3 = -1$ Simplify.
$x = 2$ Add 3 to both sides.

Check: $\sqrt[3]{2 - 3} \overset{?}{=} -1$
$\sqrt[3]{-1} \overset{?}{=} -1$
$-1 = -1$ True. The solution is 2.

Note We can see that this equation has no real-number solution because a principal square root cannot be equal to a negative number. However, we work through the steps to confirm this.

▶   **c.** $\sqrt{2x + 1} = -3$

Solution: $(\sqrt{2x + 1})^2 = (-3)^2$ Square both sides.
$2x + 1 = 9$ Simplify.
$2x = 8$ Subtract 1 from both sides.
$x = 4$ Divide both sides by 2.

Check: $\sqrt{2(4) + 1} \overset{?}{=} -3$
$\sqrt{9} \overset{?}{=} -3$
$3 = -3$ False, so 4 is extraneous. This equation has no real-number solution.

Your Turn 2   Solve.

a. $\sqrt{x - 5} = 6$ **b.** $\sqrt[3]{x + 2} = 3$ **c.** $\sqrt{x + 3} = -7$

Radicals on Both Sides of the Equation

Answers to Your Turn 2
a. 41   **b.** 25
c. no real-number solution

As we will see in Example 3, the power rule can be used to solve equations with radicals on both sides of the equal sign.

Example 3 Solve.

a. $\sqrt{6x - 1} = \sqrt{x + 2}$

Solution: $(\sqrt{6x - 1})^2 = (\sqrt{x + 2})^2$

$$6x - 1 = x + 2 \qquad \text{Square both sides.}$$

$$5x = 3 \qquad \text{Subtract } x \text{ from and add 1 to both sides.}$$

$$x = \frac{3}{5} \qquad \text{Divide both sides by 5.}$$

Check: $\sqrt{6\left(\frac{3}{5}\right) - 1} \stackrel{?}{=} \sqrt{\frac{3}{5} + 2}$

$$\sqrt{\frac{18}{5} - \frac{5}{5}} \stackrel{?}{=} \sqrt{\frac{3}{5} + \frac{10}{5}}$$

$$\sqrt{\frac{13}{5}} = \sqrt{\frac{13}{5}} \qquad \text{True. The solution is } \frac{3}{5}.$$

b. $\sqrt[3]{5x - 2} = \sqrt[3]{3x + 2}$

Solution: $(\sqrt[3]{5x - 2})^3 = (\sqrt[3]{3x + 2})^3$ Cube both sides.

$$5x - 2 = 3x + 2 \qquad \text{Simplify.}$$

$$2x = 4 \qquad \text{Subtract } 3x \text{ and add 2 to both sides.}$$

$$x = 2 \qquad \text{Divide both sides by 2.}$$

Check: $\sqrt[3]{5(2) - 2} \stackrel{?}{=} \sqrt[3]{3(2) + 2}$

$$\sqrt[3]{10 - 2} \stackrel{?}{=} \sqrt[3]{6 + 2}$$

$$\sqrt[3]{8} \stackrel{?}{=} \sqrt[3]{8}$$

$$2 = 2 \qquad \text{True. The solution is 2.}$$

Your Turn 3 Solve.

a. $\sqrt{8x + 5} = \sqrt{2x + 7}$ **b.** $\sqrt[3]{6x + 9} = \sqrt[3]{10x - 3}$

Multiple Solutions

Radical equations may have multiple solutions if, after using the power rule, we are left with a quadratic form.

Example 4 Solve $x + 2 = \sqrt{5x + 16}$.

Solution: $(x + 2)^2 = (\sqrt{5x + 16})^2$ Square both sides.

$$x^2 + 4x + 4 = 5x + 16 \qquad \text{Use FOIL on the left-hand side.}$$

$$x^2 - x - 12 = 0 \qquad \text{Because the equation is quadratic, we set it equal to 0 by subtracting } 5x \text{ and 16 from both sides.}$$

$$(x + 3)(x - 4) = 0 \qquad \text{Factor.}$$

$$x + 3 = 0 \quad \text{or} \quad x - 4 = 0 \qquad \text{Use the zero-factor theorem.}$$

$$x = -3 \qquad\qquad x = 4 \qquad \text{Solve each equation.}$$

Checks:

$$-3 + 2 \stackrel{?}{=} \sqrt{5(-3) + 16} \qquad\qquad 4 + 2 \stackrel{?}{=} \sqrt{5(4) + 16}$$

$$-1 \stackrel{?}{=} \sqrt{-15 + 16} \qquad\qquad 6 \stackrel{?}{=} \sqrt{20 + 16}$$

$$-1 \stackrel{?}{=} \sqrt{1} \qquad\qquad 6 \stackrel{?}{=} \sqrt{36}$$

$$-1 = 1 \quad \text{False} \qquad\qquad 6 = 6 \quad \text{True. The solution is 4.}$$

Because -3 does not check, it is an extraneous solution. The only solution is 4.

Answers to Your Turn 3

a. $\frac{1}{3}$ **b.** 3

Answer to Your Turn 4

2 and 1

Your Turn 4 Solve $\sqrt{9x + 7} = x + 3$.

Radicals Not Isolated

Now we consider radical equations in which the radical term is not isolated. In such equations, we must first isolate the radical term.

Example 5 Solve.

a. $\sqrt{x+1} - 2x = x + 1$

Solution: $\sqrt{x+1} = 3x + 1$ Add $2x$ to both sides to isolate the radical term.

$(\sqrt{x+1})^2 = (3x+1)^2$ Square both sides.

$x + 1 = 9x^2 + 6x + 1$ Use FOIL on the right-hand side.

$0 = 9x^2 + 5x$ Because the equation is quadratic, we set it equal to 0 by subtracting x and 1 from both sides.

$0 = x(9x + 5)$ Factor.

$x = 0$ or $9x + 5 = 0$ Use the zero-factor theorem.

$9x = -5$

$x = -\dfrac{5}{9}$

Checks: $\sqrt{0+1} - 2(0) \overset{?}{=} 0 + 1$

$\sqrt{1} - 0 \overset{?}{=} 1$

$1 = 1$ True. The solution is 0.

$\sqrt{-\dfrac{5}{9} + 1} - 2\left(-\dfrac{5}{9}\right) \overset{?}{=} -\dfrac{5}{9} + 1$

$\sqrt{-\dfrac{5}{9} + \dfrac{9}{9}} - 2\left(-\dfrac{5}{9}\right) \overset{?}{=} -\dfrac{5}{9} + \dfrac{9}{9}$

$\sqrt{\dfrac{4}{9}} + \dfrac{10}{9} \overset{?}{=} \dfrac{4}{9}$

$\dfrac{2}{3} + \dfrac{10}{9} \overset{?}{=} \dfrac{4}{9}$

$\dfrac{6}{9} + \dfrac{10}{9} \overset{?}{=} \dfrac{4}{9}$

$\dfrac{16}{9} = \dfrac{4}{9}$ False

Note This false equation indicates that $-\dfrac{5}{9}$ is an extraneous solution.

The solution is 0.

b. $\sqrt[4]{3x + 4} + 5 = 7$

Solution: $\sqrt[4]{3x + 4} = 2$ Subtract 5 from both sides to isolate the radical term.

$(\sqrt[4]{3x + 4})^4 = 2^4$ Raise both sides to the fourth power.

$3x + 4 = 16$ Simplify both sides.

$3x = 12$ Subtract 4 from both sides.

$x = 4$ Divide both sides by 3.

Check: $\sqrt[4]{3(4) + 4} + 5 \overset{?}{=} 7$

$\sqrt[4]{12 + 4} + 5 \overset{?}{=} 7$

$\sqrt[4]{16} + 5 \overset{?}{=} 7$

$2 + 5 \overset{?}{=} 7$

$7 = 7$ True. The solution is 4.

Your Turn 5 Solve.

a. $\sqrt{5x^2 + 6x - 7} + 3x = 5x + 1$ **b.** $\sqrt[4]{3x + 6} - 7 = -4$

Using the Power Rule Twice

Answers to Your Turn 5
a. 2   **b.** 25

Some equations may require that we use the power rule twice to eliminate all radicals.

Example 6 Solve $\sqrt{x + 21} = \sqrt{x} + 3$.

Solution: $(\sqrt{x + 21})^2 = (\sqrt{x} + 3)^2$ Because one of the radicals is isolated, we square both sides.

$$x + 21 = (\sqrt{x} + 3)(\sqrt{x} + 3)$$ Simplify.

$$x + 21 = x + 3\sqrt{x} + 3\sqrt{x} + 9$$ Use FOIL on the right-hand side.

$$x + 21 = x + 6\sqrt{x} + 9$$ Combine like terms.

$$12 = 6\sqrt{x}$$ Subtract x and 9 from both sides to isolate the remaining radical expression.

$$2 = \sqrt{x}$$ Divide both sides by 6.

$$(2)^2 = (\sqrt{x})^2$$ Square both sides.

$$4 = x$$

Check: $\sqrt{4 + 21} \overset{?}{=} \sqrt{4} + 3$

$$\sqrt{25} \overset{?}{=} 2 + 3$$

$$5 = 5$$ True. The solution is 4.

Our examples suggest the following procedure.

Procedure **Solving Radical Equations**

To solve a radical equation:

1. Isolate the radical if necessary. (If there is more than one radical term, isolate one of the radical terms.)
2. Raise both sides of the equation to the same power as the root index of the isolated radical.
3. If all radicals have been eliminated, solve. If a radical term remains, isolate that radical term and raise both sides to the same power as its root index.
4. Check each solution. Any apparent solution that does not check is an extraneous solution.

Answer to Your Turn 6
0 and 4

Your Turn 6 Solve $\sqrt{2x + 1} = \sqrt{x} + 1$.

9.6 Exercises
For Extra Help | MyMathLab®

Note: Exercises marked with a ★ represent challenging exercises.

Objective 1

Prep Exercise 1 Explain why we must check all potential solutions to radical equations. Some of the solutions may be extraneous.

Prep Exercise 2 What is an extraneous solution? A solution that does not satisfy the original equation

Prep Exercise 3 Explain why there is no real-number solution for the radical equation $\sqrt{x} = -6$. The principal square root of a number cannot equal a negative.

Prep Exercise 4 Show why $(\sqrt{a})^2 = a$, assuming that $a \geq 0$. $(\sqrt{a})^2 = \sqrt{a} \cdot \sqrt{a} = \sqrt{a^2} = a$

For Exercises 1–22, solve. See Examples 1 and 2.

1. $\sqrt{x} = 2$
4

2. $\sqrt{y} = 5$
25

3. $\sqrt{k} = -4$
No real-number solution

4. $\sqrt{x} = -1$
No real-number solution

5. $\sqrt[3]{y} = 3$
27

6. $\sqrt[3]{m} = 4$
64

7. $\sqrt[3]{z} = -2$
-8

8. $\sqrt[3]{p} = -5$
-125

9. $\sqrt{n - 1} = 4$
17

10. $\sqrt{x - 2} = 3$
11

11. $\sqrt{t + 5} = 4$
11

12. $\sqrt{m + 8} = 1$
-7

13. $\sqrt{3x - 2} = 4$
6

14. $\sqrt{2x + 5} = 3$
2

15. $\sqrt{2x + 24} = 4$
-4

16. $\sqrt{3y + 34} = 5$
-3

17. $\sqrt{2n - 8} = -3$
No real-number solution

18. $\sqrt{5x - 1} = -6$
No real-number solution

19. $\sqrt[3]{x - 3} = 2$
11

20. $\sqrt[3]{k + 2} = 4$
62

21. $\sqrt[3]{3y - 2} = -2$
-2

22. $\sqrt[3]{2x + 4} = -4$
-34

Prep Exercise 5 Given the radical equation $\sqrt{x + 2} + 3x = 4x - 1$, what would be the first step in solving the equation? Why?
Subtract $3x$ from both sides to isolate the radical. This allows use of the power rule to eliminate the radical.

For Exercises 23–32 solve. First isolate the radical term. See Example 5.

23. $\sqrt{u - 3} - 10 = 1$
124

24. $\sqrt{y + 1} - 4 = 2$
35

25. $\sqrt{y - 6} + 2 = 9$
55

26. $\sqrt{r - 5} + 6 = 10$
21

27. $\sqrt{6x - 5} - 2 = 3$
5

28. $\sqrt{8x + 4} - 2 = 4$
4

29. $\sqrt[3]{n + 3} - 2 = -4$
-11

30. $\sqrt[3]{x - 4} + 2 = -3$
-121

31. $\sqrt[4]{x - 2} - 2 = -4$
No real-number solution

32. $\sqrt[4]{m + 3} + 2 = 1$
No real-number solution

Prep Exercise 6 Give an example of a radical equation that requires you to use the power rule twice in solving the equation. Explain why the principle must be used twice.
Answers may vary. One example is $\sqrt{x + 4} = \sqrt{2x + 1}$. The radicals must be totally eliminated.

For Exercises 33–58, solve. Identify any extraneous solutions. See Examples 3–6.

33. $\sqrt{3x - 2} = \sqrt{8 - 2x}$
2

34. $\sqrt{m + 2} = \sqrt{2m - 3}$
5

35. $\sqrt{4x - 5} = \sqrt{6x + 5}$
No real-number solution (-5 is an extraneous solution)

36. $\sqrt{3x - 4} = \sqrt{5x + 2}$
No real-number solution (-3 is an extraneous solution)

37. $\sqrt[3]{2r + 2} = \sqrt[3]{3r - 1}$
3

38. $\sqrt[3]{3h - 4} = \sqrt[3]{h + 4}$
4

39. $\sqrt[4]{4x + 4} = \sqrt[4]{5x + 1}$
3

40. $\sqrt[4]{2x + 4} = \sqrt[4]{3x - 2}$
6

41. $\sqrt{2x + 24} = x + 8$
-4 (-10 is an extraneous solution)

42. $\sqrt{2x + 15} = x + 6$
-3 (-7 is an extraneous solution)

43. $y - 1 = \sqrt{2y - 2}$
1, 3

44. $3 + x = \sqrt{7 + 3x}$
$-2, -1$

45. $\sqrt{3x + 10} - 4 = x$
$-3, -2$

46. $\sqrt{6x + 1} - 1 = x$
0, 4

47. $\sqrt{10n + 4} - 3n = n + 1$
$\frac{1}{2} \left(-\frac{3}{8} \text{ is an extraneous solution} \right)$

48. $\sqrt{6x - 1} - 6x = 2 - 9x$
$\frac{1}{3} \left(\frac{5}{3} \text{ is an extraneous solution} \right)$

49. $\sqrt[3]{5x + 2} + 2 = 5$
5

50. $\sqrt[3]{4x - 1} - 4 = -1$
7

51. $\sqrt[3]{n^2 - 2n + 5} = 2$
$-1, 3$

52. $\sqrt[3]{y^2 + y + 7} = 3$
$-5, 4$

53. $1 + \sqrt{x} = \sqrt{2x + 1}$
0, 4

Review Exercises

Exercises 1–6 ➤ **Expressions**

[5.1] *For Exercises 1–3, evaluate.*

1. 3^4
81

2. $(-0.2)^3$
−0.008

3. $\left(\dfrac{2}{5}\right)^{-4}$
39.0625 or $\dfrac{625}{16}$

For Exercises 4–6, use the rules of exponents to simplify.

[5.4] **4.** $(x^3)(x^5)$
x^8

[5.4] **5.** $(n^4)^6$
n^{24}

[5.6] **6.** $\dfrac{y^7}{y^3}$
y^4

9.7 Complex Numbers

Objectives

1 Write imaginary numbers using *i*.

2 Perform arithmetic operations with complex numbers.

3 Raise *i* to powers.

Of Interest

When the idea of finding square roots of negative numbers was first introduced, members of the established mathematics community said that this type of number existed, but only in the imagination of those finding them. From then on, the numbers have been called "imaginary numbers."

Note An imaginary number of ▶ the form *bi* is a pure imaginary number.

Note In a product with the imaginary unit, it is customary to write integer factors first, then *i*, then square root factors. ▶

Warm-up

[9.6] **1.** Solve: $\sqrt{7 - x} = x - 1$
[9.4] *For Exercises 2–4, simplify.*
2. $\left(-1 + 13\sqrt{2}\right) - \left(8 - 3\sqrt{2}\right)$
3. $\left(5\sqrt{3}\right)\left(4 - 7\sqrt{3}\right)$
4. $\left(6 - 5\sqrt{2}\right)\left(6 + 5\sqrt{2}\right)$

We have said that the square root of a negative number is not a real number. In this section, we learn about the *imaginary number system,* in which square roots of negative numbers are expressed using a notation involving the letter *i*.

Objective 1 Write imaginary numbers using *i*.

Using the product rule of square roots, any square root of a negative number can be rewritten as a product of a real number and the **imaginary unit**, which we express as *i*.

Definition   Imaginary unit: The number represented by *i*, where $i = \sqrt{-1}$ and $i^2 = -1$.

A number written with the imaginary unit is called an **imaginary number**.

Definition   Imaginary number: A number that can be expressed in the form $a + bi$, where *a* and *b* are real numbers, *i* is the imaginary unit, and $b \neq 0$.

Example 1   Write each imaginary number as a product of a real number and *i*.

a. $\sqrt{-9}$

Solution: $\sqrt{-9} = \sqrt{-1 \cdot 9}$ Factor out −1 in the radicand.

$= \sqrt{-1} \cdot \sqrt{9}$ Use the product rule of square roots.

$= i \cdot 3$ Replace $\sqrt{-1}$ with *i*.

$= 3i$ Simplify.

b. $\sqrt{-54}$

Solution: $\sqrt{-54} = \sqrt{-1 \cdot 54}$ Factor out −1 in the radicand.

$= \sqrt{-1} \cdot \sqrt{54}$ Use the product rule of square roots.

$= i\sqrt{9 \cdot 6}$ Write 54 as $9 \cdot 6$.

$= 3i\sqrt{6}$ Simplify.

Example 1 suggests the following procedure:

Procedure Rewriting Imaginary Numbers

To write an imaginary number $\sqrt{-n}$ in terms of the imaginary unit i:
1. Separate the radical into two factors, $\sqrt{-1} \cdot \sqrt{n}$.
2. Replace $\sqrt{-1}$ with i.
3. Simplify $\sqrt{n}$.

Your Turn 1   Write each imaginary number as a product of a real number and i.

  a. $\sqrt{-64}$ **b.** $\sqrt{-18}$ **c.** $\sqrt{-48}$

Objective 2 Perform arithmetic operations with complex numbers.

We now have two distinct sets of numbers, the set of real numbers and the set of imaginary numbers. There is yet another set of numbers, called the set of **complex numbers**, that contains both real and imaginary numbers.

Definition   **Complex number:** A number that can be expressed in the form $a + bi$, where a and b are real numbers and i is the imaginary unit.

Note When the coefficient of i is a radical expression, as in $-2.1 - i\sqrt{5}$ and $4i\sqrt{3}$, we write i to the left of the radical sign so that i does not look like it is part of the radicand.

When written in the form $a + bi$, a complex number is said to be in *standard form*. Following are some examples of complex numbers written in standard form.

$$2 + 3i \qquad 4 - 7i \qquad -2.1 - i\sqrt{5}$$

Note that if $a = 0$, then the complex number is purely an imaginary number, such as the following:

$$-6i \qquad 4i\sqrt{3}$$

If $b = 0$, then the complex number is a real number.

The following Venn diagram shows how all complex numbers are either real numbers or imaginary numbers.

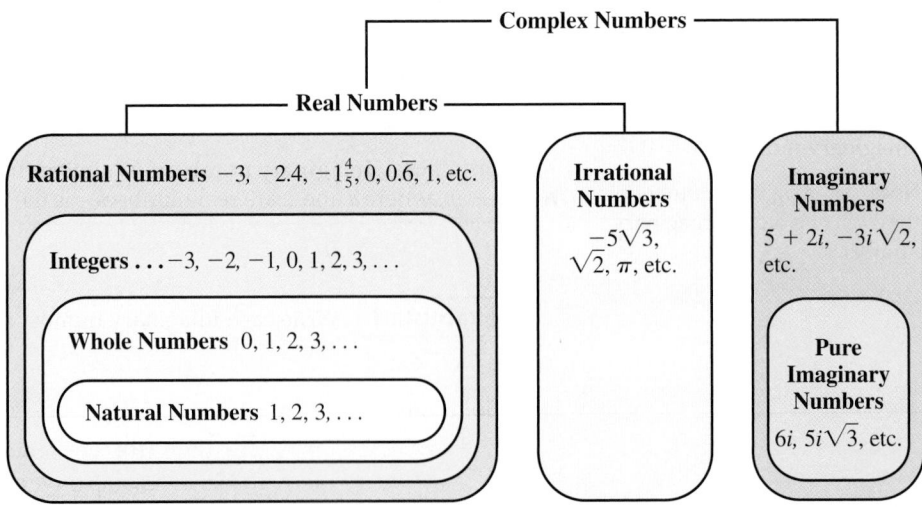

Answers to Your Turn 1
  a. $8i$ **b.** $3i\sqrt{2}$ **c.** $4i\sqrt{3}$

We can perform arithmetic operations with complex numbers. In general, we treat the complex numbers just like polynomials, where i is like a variable.

Adding and Subtracting Complex Numbers

Example 2 Add or subtract.

a. $(-8 + 7i) + (5 - 19i)$

Solution: We add complex numbers just like we add polynomials—by combining like terms.

$$(-8 + 7i) + (5 - 19i) = -3 - 12i$$

b. $(-1 + 13i) - (8 - 3i)$

Solution: We subtract complex numbers just like we subtract polynomials—by writing an equivalent addition and changing the signs in the second complex number.

$$(-1 + 13i) - (8 - 3i) = (-1 + 13i) + (-8 + 3i)$$
$$= -9 + 16i$$

Your Turn 2 Add or subtract.

a. $(-3 - 5i) + (4 - 6i)$ **b.** $(4 - 7i) - (-3 + 2i)$

Multiplying Complex Numbers

We multiply complex numbers the same way that we multiply monomials and binomials. However, we must be careful when simplifying because these products may contain i^2, which is equal to -1.

Example 3 Multiply.

a. $(9i)(-8i)$

Solution: Multiply the same way that we multiply monomials.

$$
\begin{aligned}
(9i)(-8i) &= -72i^2 && \text{Multiply.} \\
&= -72(-1) && \text{Replace } i^2 \text{ with } -1. \\
&= 72
\end{aligned}
$$

b. $(5i)(4 - 7i)$

Solution: Multiply the same way that we multiply a binomial by a monomial.

$$
\begin{aligned}
(5i)(4 - 7i) &= 20i - 35i^2 && \text{Distribute.} \\
&= 20i - 35(-1) && \text{Replace } i^2 \text{ with } -1. \\
&= 20i + 35 \\
&= 35 + 20i && \text{Write in standard form, } a + bi.
\end{aligned}
$$

c. $(8 - 3i)(2 + i)$

Solution: Multiply the same way that we multiply binomials.

$$
\begin{aligned}
(8 - 3i)(2 + i) &= 16 + 8i - 6i - 3i^2 && \text{Use FOIL.} \\
&= 16 + 2i - 3(-1) && \text{Combine like terms and replace } i^2 \text{ with } -1. \\
&= 16 + 2i + 3 && \text{Multiply.} \\
&= 19 + 2i && \text{Write in standard form, } a + bi.
\end{aligned}
$$

d. $(6 - 5i)(6 + 5i)$

Solution: Note that these complex numbers are conjugates.

$$
\begin{aligned}
(6 - 5i)(6 + 5i) &= 36 + 30i - 30i - 25i^2 && \text{Use FOIL.} \\
&= 36 - 25(-1) && \text{Combine like terms and replace } i^2 \text{ with } -1. \\
&= 36 + 25 \\
&= 61
\end{aligned}
$$

Note The product of these two ◄ complex numbers is a real number.

Answers to Your Turn 2
a. $1 - 11i$ **b.** $7 - 9i$

The complex numbers multiplied in Example 3(d) are called **complex conjugates**.

Definition **Complex conjugate:** The complex conjugate of a complex number $a + bi$ is $a - bi$.

Other examples of complex conjugates follow.

$$4 + i \quad \text{and} \quad 4 - i$$
$$9 - 7i \quad \text{and} \quad 9 + 7i$$

The product of complex conjugates is always a real number. In general, $(a + bi)$ $(a - bi) = a^2 - b^2 i^2 = a^2 - b^2(-1) = a^2 + b^2$, which is a real number.

Your Turn 3 Multiply.

a. $(-4i)(-7i)$ b. $(-3i)(6 - i)$ c. $(8 - 3i)(5 + 7i)$ d. $(7 + 3i)(7 - 3i)$

Connection Recall that we rationalize denominators to clear undesired square root expressions from a denominator. The imaginary unit i represents a square root expression, $\sqrt{-1}$, which is why we rationalize denominators that contain i.

Dividing Complex Numbers

When dividing by a pure imaginary number, we use the fact that $i^2 = -1$, and when dividing by a complex number, we use the fact that the product of complex conjugates is a real number. The process is similar to rationalizing denominators. For example, to rationalize the denominator of $\dfrac{3}{2\sqrt{5}}$, we multiply by $\dfrac{\sqrt{5}}{\sqrt{5}}$. Similarly, to divide $\dfrac{3}{2i}$, we multiply by $\dfrac{i}{i}$. To rationalize the denominator of $\dfrac{4}{2 + \sqrt{3}}$, we multiply by $\dfrac{2 - \sqrt{3}}{2 - \sqrt{3}}$, where $2 - \sqrt{3}$ is the conjugate of $2 + \sqrt{3}$. Similarly, to divide $\dfrac{4}{2 + i}$, we multiply by $\dfrac{2 - i}{2 - i}$, where $2 - i$ is the complex conjugate of $2 + i$.

Example 4 Divide. Write in standard form.

a. $\dfrac{9}{2i}$

Solution: Because $i^2 = -1$, multiplying by $\dfrac{i}{i}$ eliminates i from the denominator.

$$\frac{9}{2i} = \frac{9}{2i} \cdot \frac{i}{i} \qquad \text{\textcolor{gray}{Multiply the numerator and denominator by } } i.$$

$$= \frac{9i}{2i^2} \qquad \text{\textcolor{gray}{Multiply.}}$$

$$= \frac{9i}{2(-1)} \qquad \text{\textcolor{gray}{Replace } } i^2 \text{ \textcolor{gray}{with } } -1.$$

$$= \frac{9i}{-2}$$

$$= -\frac{9}{2}i$$

Note We could have thought of $2i$ as $0 + 2i$; so the complex conjugate is $0 - 2i = -2i$ and is multiplied by $\dfrac{-2i}{-2i}$.

$$\frac{9}{2i} = \frac{9}{2i} \cdot \frac{-2i}{-2i} = \frac{-18i}{-4i^2}$$

$$= \frac{-18i}{-4(-1)} = \frac{-18i}{4}$$

$$= -\frac{9}{2}i$$

b. $\dfrac{6 + 5i}{7 - 2i}$

Solution: $\dfrac{6 + 5i}{7 - 2i} = \dfrac{6 + 5i}{7 - 2i} \cdot \dfrac{7 + 2i}{7 + 2i}$ \textcolor{gray}{Multiply the numerator and denominator by the complex conjugate of the denominator, which is $7 + 2i$.}

$$= \frac{42 + 12i + 35i + 10i^2}{49 - 4i^2} \qquad \text{\textcolor{gray}{Multiply.}}$$

Answers to Your Turn 3
a. -28 b. $-3 - 18i$
c. $61 + 41i$ d. 58

$$= \frac{42 + 47i + 10(-1)}{49 - 4(-1)} \qquad \text{Simplify.}$$

$$= \frac{42 + 47i - 10}{49 + 4}$$

$$= \frac{32 + 47i}{53}$$

$$= \frac{32}{53} + \frac{47}{53}i \qquad \text{Write in standard form.}$$

Your Turn 4 Divide. Write in standard form.

a. $\dfrac{8}{3i}$

b. $\dfrac{6 + i}{4 - 5i}$

Objective 3 Raise i to powers.

We have defined i as $\sqrt{-1}$ and have seen that $i^2 = -1$. Raising i to other powers leads to an interesting pattern.

$$i^1 = i \qquad\qquad\qquad i^5 = i^4 \cdot i = 1 \cdot i = i$$
$$i^2 = -1 \qquad\qquad\quad i^6 = i^4 \cdot i^2 = 1 \cdot (-1) = -1$$
$$i^3 = i^2 \cdot i = -1 \cdot i = -i \qquad i^7 = i^4 \cdot i^3 = 1 \cdot (-i) = -i$$
$$i^4 = (i^2)^2 = (-1)^2 = 1 \qquad i^8 = (i^4)^2 = 1^2 = 1$$

Note If the exponent on i is an even number not divisible by 4, the result is -1. If the exponent on i is divisible by 4, the result is 1. If the exponent is an odd number that precedes a number divisible by 4, the result is $-i$.

If we continue the pattern, we get the following:

$$i^1 = i \qquad i^5 = i \qquad i^9 = i$$
$$i^2 = -1 \qquad i^6 = -1 \qquad i^{10} = -1$$
$$i^3 = -i \qquad i^7 = -i \qquad i^{11} = -i$$
$$i^4 = 1 \qquad i^8 = 1 \qquad i^{12} = 1$$

Notice that i to any integer power is i, -1, $-i$, or 1. This pattern allows us to find i to any integer power. We will use the fact that because $i^4 = 1$, $(i^4)^n = 1$ for any integer value of n.

Example 5 Find the powers of i.

a. i^{25}

Solution: $i^{25} = i^{24} \cdot i$ Write i^{25} as $i^{24} \cdot i$ because 24 is the largest multiple of 4 that is smaller than 25.

$$= (i^4)^6 \cdot i \qquad \text{Write } i^{24} \text{ as } (i^4)^6 \text{ because } i^4 = 1.$$

$$= 1^6 \cdot i \qquad \text{Replace } i^4 \text{ with 1.}$$

$$= 1 \cdot i$$

$$= i$$

Note Example 5(b) could have been done as follows:

$$i^{-14} = i^{-12} \cdot i^{-2} = (i^4)^{-3} \cdot i^{-2} =$$

$$(1)^{-3} \cdot \frac{1}{i^2} = 1 \cdot \frac{1}{-1} = -1$$

b. i^{-14}

Solution: $i^{-14} = \dfrac{1}{i^{14}}$ Write i^{-14} with a positive exponent.

$$= \frac{1}{i^{12} \cdot i^2} \qquad \text{Write } i^{14} \text{ as } i^{12} \cdot i^2 \text{ because 12 is the largest multiple of 4 that is smaller than 14.}$$

$$= \frac{1}{(i^4)^3 \cdot (-1)} \qquad \text{Write } i^{12} \text{ as } (i^4)^3 \text{ and replace } i^2 \text{ with } -1.$$

$$= \frac{1}{1(-1)} \qquad (i^4)^3 = 1^3 = 1$$

$$= -1$$

Answers to Your Turn 4

a. $-\dfrac{8}{3}i$ b. $\dfrac{19}{41} + \dfrac{34}{41}i$

Answers to Your Turn 5
  a. $-i$ **b.** -1

 Your Turn 5   Find the powers of i.

  a. i^{43} **b.** i^{-18}

9.7 Exercises   For Extra Help   MyMathLab®

Objective 1

Prep Exercise 1 What does the imaginary unit i represent?
$\sqrt{-1}$

Prep Exercise 2 Is every real number a complex number? Explain.
Yes, every real number can be expressed as $a + 0i$.

Prep Exercise 3 Is every complex number an imaginary number? Explain.
No, every real number is a complex number that is not imaginary. For example, 2 is complex but not imaginary.

Prep Exercise 4 Explain how to write an imaginary number $\sqrt{-n}$ using the imaginary unit i.
Separate into two radicals, $\sqrt{-1} \cdot \sqrt{n}$. Replace $\sqrt{-1}$ with i and simplify $\sqrt{n}$.

For Exercises 1–16, write the imaginary number using i. See Example 1.

1. $\sqrt{-36}$
$6i$

2. $\sqrt{-81}$
$9i$

3. $\sqrt{-5}$
$i\sqrt{5}$

4. $\sqrt{-10}$
$i\sqrt{10}$

5. $\sqrt{-8}$
$2i\sqrt{2}$

6. $\sqrt{-12}$
$2i\sqrt{3}$

7. $\sqrt{-18}$
$3i\sqrt{2}$

8. $\sqrt{-32}$
$4i\sqrt{2}$

9. $\sqrt{-27}$
$3i\sqrt{3}$

10. $\sqrt{-72}$
$6i\sqrt{2}$

11. $\sqrt{-125}$
$5i\sqrt{5}$

12. $\sqrt{-80}$
$4i\sqrt{5}$

13. $\sqrt{-63}$
$3i\sqrt{7}$

14. $\sqrt{-54}$
$3i\sqrt{6}$

15. $\sqrt{-245}$
$7i\sqrt{5}$

16. $\sqrt{-810}$
$9i\sqrt{10}$

Objective 2

Prep Exercise 5 Explain how to add complex numbers.
We add complex numbers just like we add polynomials—by combining like terms.

Prep Exercise 6 Explain how to subtract complex numbers.
We subtract complex numbers just like we subtract polynomials—by writing an equivalent addition and changing the signs in the second complex number.

For Exercises 17–32, add or subtract. See Example 2.

17. $(9 + 3i) + (-3 + 4i)$
$6 + 7i$

18. $(6 + 4i) + (-2 - 2i)$
$4 + 2i$

19. $(6 + 2i) + (5 - 8i)$
$11 - 6i$

20. $(4 + i) + (7 - 6i)$
$11 - 5i$

21. $(-4 + 6i) - (3 + 5i)$
$-7 + i$

22. $(8 - 3i) - (-1 - 2i)$
$9 - i$

23. $(8 - 5i) - (-3i)$
$8 - 2i$

24. $(19 + 6i) - (-2i)$
$19 + 8i$

25. $(12 + 3i) + (-15 - 13i)$
$-3 - 10i$

26. $(14 - 7i) + (-8 - 9i)$
$6 - 16i$

27. $(-5 - 9i) - (-5 - 9i)$
0

28. $(-4 + 2i) - (-4 + 2i)$
0

29. $(10 + i) - (2 - 13i) + (6 - 5i)$
$14 + 9i$

30. $(-14 + 2i) - (6 + i) + (19 + 10i)$
$-1 + 11i$

31. $(5 - 2i) - (9 - 14i) + (16i)$
$-4 + 28i$

32. $(-12i) - (4 - 7i) - (18 + 4i)$
$-22 - 9i$

Prep Exercise 7 What is the value of i^2?
-1

For Exercises 33–48, multiply. See Example 3.

33. $(8i)(3i)$
-24

34. $(9i)(5i)$
-45

35. $(-8i)(5i)$
40

36. $(4i)(-7i)$
28

37. $2i(6 - 7i)$
$14 + 12i$

38. $6i(9 - i)$
$6 + 54i$

39. $-8i(4 - 9i)$
$-72 - 32i$

40. $-7i(5 - 8i)$
$-56 - 35i$

41. $(6 + i)(3 - i)$
$19 - 3i$

42. $(5 - 2i)(4 + i)$
$22 - 3i$

43. $(8 + 5i)(5 - 2i)$
$50 + 9i$

44. $(6 - 3i)(7 + 8i)$
$66 + 27i$

45. $(8 + i)(8 - i)$
65

46. $(4 + 9i)(4 - 9i)$
97

47. $(3 - 4i)^2$
$-7 - 24i$

48. $(5 - 3i)^2$
$16 - 30i$

Prep Exercise 8 Is the expression $\dfrac{5 - 4i}{3}$ in standard form for a complex number? Explain.
No, because it is not in the form $a + bi$

Prep Exercise 9 Explain how to divide $\dfrac{9 + 5i}{3 - 2i}$.
Multiply the numerator and denominator by $3 + 2i$.

For Exercises 49–68, divide and write in standard form. See Example 4.

49. $\dfrac{2}{i}$
$-2i$

50. $\dfrac{6}{-i}$
$6i$

51. $\dfrac{4}{5i}$
$-\dfrac{4i}{5}$

52. $\dfrac{6}{7i}$
$-\dfrac{6i}{7}$

53. $\dfrac{6}{2i}$
$-3i$

54. $\dfrac{12}{3i}$
$-4i$

55. $\dfrac{2 + i}{2i}$
$\dfrac{1}{2} - i$

56. $\dfrac{3 + i}{3i}$
$\dfrac{1}{3} - i$

57. $\dfrac{4 + 2i}{4i}$
$\dfrac{1}{2} - i$

58. $\dfrac{6 - 2i}{6i}$
$-\dfrac{1}{3} - i$

59. $\dfrac{7}{2 + i}$
$\dfrac{14}{5} - \dfrac{7}{5}i$

60. $\dfrac{5}{6 + i}$
$\dfrac{30}{37} - \dfrac{5}{37}i$

61. $\dfrac{2i}{3 - 7i}$
$-\dfrac{7}{29} + \dfrac{3}{29}i$

62. $\dfrac{4i}{5 - 3i}$
$-\dfrac{6}{17} + \dfrac{10}{17}i$

63. $\dfrac{5 - 9i}{1 - i}$
$7 - 2i$

64. $\dfrac{3 + i}{2 - i}$
$1 + i$

65. $\dfrac{3 + i}{2 + 3i}$
$\dfrac{9}{13} - \dfrac{7}{13}i$

66. $\dfrac{1 + 3i}{5 + 2i}$
$\dfrac{11}{29} + \dfrac{13}{29}i$

67. $\dfrac{1 + 6i}{4 + 5i}$
$\dfrac{34}{41} + \dfrac{19}{41}i$

68. $\dfrac{5 - 6i}{2 - 9i}$
$\dfrac{64}{85} + \dfrac{33}{85}i$

Objective 3

Prep Exercise 10 Complete each rule.

$i = \underline{\ ?\ }$ i

$i^2 = \underline{\ ?\ }$ -1

$i^3 = \underline{\ ?\ }$ $-i$

$i^4 = \underline{\ ?\ }$ 1

For Exercises 69–84, find the powers of i. See Example 5.

69. i^{19}
$-i$

70. i^{25}
i

71. i^{42}
-1

72. i^{51}
$-i$

73. i^{38}
-1

74. i^{45}
i

75. i^{60}
1

76. i^{52}
1

77. i^{-20}
1

78. i^{-32}
1

79. i^{-30}
-1

80. i^{-38}
-1

81. i^{-21}
$-i$

82. i^{-45}
$-i$

83. i^{-35}
i

84. i^{-44}
1

Review Exercises

Exercises 1 and 2 **Expressions**

[6.4] 1. Factor: $4x^2 - 12x + 9$
$(2x - 3)^2$

[9.3] 2. Simplify: $\pm \sqrt{48}$
$\pm 4\sqrt{3}$

Exercises 3–6 **Equations and Inequalities**

For Exercises 3 and 4, solve.

[2.3] 3. $3x - 2 = 4$
2

[6.6] 4. $2x^2 - x - 6 = 0$
$-\dfrac{3}{2}, 2$

[3.3] 5. Find the x- and y-intercepts for $2x + 3y = 6$.
$(0, 2), (3, 0)$

[6.7] 6. Graph: $y = x^2 - 3$

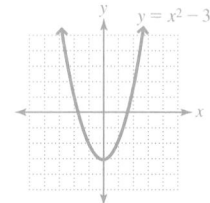

Chapter 9 Summary and Review Exercises

Complete each incomplete definition, rule, or procedure; study the key examples; and then work the related exercises.

9.1 Radical Expressions and Functions

Definitions/Rules/Procedures	Key Example(s)
The number b is an nth root of a number a if $\underline{b^n = a}$. **If a is nonnegative**, then $\sqrt[n]{a} = b$, where $b \geq 0$ and $\underline{b^n = a}$. **If a is negative and n is even**, then there is $\underline{\text{no real-number}}$ root. **If a is negative and n is odd**, then $\sqrt[n]{a} = b$, where b is $\underline{\text{negative}}$ and $b^n = a$.	Find each root if possible. **a.** $\sqrt{36} = 6$ **b.** $\sqrt{-25}$ is not a real number. **c.** $-\sqrt{121} = -11$ **d.** $\sqrt{\dfrac{49}{100}} = \dfrac{7}{10}$ **e.** $\sqrt{16x^6} = 4\lvert x^3 \rvert$ ◀ **Note** If we assume that the variables represent nonnegative numbers, the solution for part e is $\sqrt{16x^6} = 4x^3$. **f.** $\sqrt[3]{-27} = -3$

Exercises 1–22   **Expressions**

[9.1] *For Exercises 1 and 2, find all square roots of the given number.*

1. 121
± 11

2. 49
± 7

[9.1] *For Exercises 3–8, evaluate the square root.*

3. $\sqrt{169}$
13

4. $-\sqrt{49}$
-7

5. $\sqrt{-36}$
Not a real number

6. $\sqrt{\dfrac{1}{25}}$
$\dfrac{1}{5}$

7. Three pieces of lumber are to be connected to form a right triangle that will be part of the roof frame for a small storage building. If the horizontal piece is to be 4 feet and the vertical piece is to be 3 feet, how long must the connecting piece be?
5 ft.

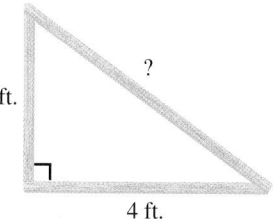
3 ft.   ?   4 ft.

8. In the formula $T = 2\pi\sqrt{\dfrac{L}{9.8}}$, T is the period of a pendulum in seconds and L is the length of the pendulum in meters. Find the period of a pendulum with a length of 2.45 meters.
π sec.

[9.1] *For Exercises 9 and 10, use a calculator to approximate each root to the nearest thousandth.*

9. $\sqrt{7}$
2.646

10. $\sqrt{90}$
9.487

11. The speed of a car can be determined by the length of the skid marks using the formula $S = 2\sqrt{2L}$, where L is the length of the skid mark in feet and S is the speed of the car in miles per hour.

a. Write an expression for the exact speed of the car if the length of the skid is 40 feet.
$8\sqrt{5}$ mph

b. Approximate the speed to the nearest tenth.
17.9 mph

12. In the formula $t = \sqrt{\dfrac{h}{16}}$, t represents the time in seconds it takes an object to fall a distance of h feet.

    a. Write an expression in simplest form of the time an object takes to fall 40 feet.

        $\dfrac{\sqrt{10}}{2}$ sec.

    b. Approximate the time to the nearest hundredth.

        1.58 sec.

[9.1] *For Exercises 13–20, simplify. Assume that variables represent nonnegative values.*

13. $\sqrt{49x^8}$
    $7x^4$

14. $\sqrt{144a^6b^{12}}$
    $12a^3b^6$

15. $\sqrt{0.16m^2n^{10}}$
    $0.4mn^5$

16. $\sqrt[3]{x^{15}}$
    x^5

17. $\sqrt[3]{-64r^9s^3}$
    $-4r^3s$

18. $\sqrt[4]{81x^{12}}$
    $3x^3$

19. $\sqrt[5]{32x^{15}y^{20}}$
    $2x^3y^4$

20. $\sqrt[7]{x^{14}y^7}$
    x^2y

[9.1] *For Exercises 21 and 22, simplify. Assume that variables represent any real number.*

21. $\sqrt{81x^2}$
    $9|x|$

22. $\sqrt[4]{(x-1)^8}$
    $(x-1)^2$

Definitions/Rules/Procedures	Key Example(s)
A **radical function** is a function containing a(n) <u> radical </u> expression whose <u> radicand </u> has a variable.	If $f(x) = \sqrt{4x+3}$, find $f(3)$. Solution: $$f(3) = \sqrt{4(3)+3} = \sqrt{12+3} = \sqrt{15}$$
To evaluate a radical function, replace the variable with the indicated value and <u> simplify </u>.	
Finding the domain of a radical function If the root index is odd, the domain is $\{x \mid$ <u>x is a real number</u>, or (<u>$-\infty$</u> , <u>∞</u>). If the root index is even, the radicand must be <u> nonnegative </u>. Consequently, the domain is the solution set to the inequality "radicand <u> ≥ 0 </u>."	Find the domain of $$f(x) = \sqrt{2x-4}.$$ Solution: Because $2x-4$ must be nonnegative, we solve $2x-4 \geq 0$. $$2x-4 \geq 0$$ $$2x \geq 4$$ $$x \geq 2$$ Domain: $\{x \mid x \geq 2\}$ or $[2, \infty)$

Exercises 23 and 24 ◢◣▲ Equations and Inequalities

[9.1] *For Exercises 23 and 24, find the indicated value of the function.*

23. If $f(x) = \sqrt{4x-4}$, find $f(5)$.
    4

24. If $f(x) = \sqrt[3]{5x+2}$, find $f(-2)$.
    -2

Exercises 25 and 26 ◢◣▲ Expressions

[9.1] *For Exercises 25 and 26, find the domain.*

25. $\sqrt{2x+10}$
    $x \geq -5$

26. $\sqrt{9-3x}$
    $x \leq 3$

9.2 Rational Exponents

Definitions/Rules/Procedures	Key Example(s)
$a^{1/n} = \underline{\quad \sqrt[n]{a} \quad}$, where n is a natural number other than 1. **Note:** If a is negative and n is odd, the root is negative. If a is negative and n is even, there is no real-number root. $a^{m/n} = \underline{\quad \sqrt[n]{a^m} \quad} = \underline{\quad (\sqrt[n]{a})^m \quad}$, where $a \geq 0$ and m and n are natural numbers other than 1. $a^{-m/n} = \underline{\quad \dfrac{1}{a^{m/n}} \quad}$, where $a \neq 0$ and m and n are natural numbers with $n \neq 1$.	Evaluate. **a.** $25^{1/2} = \sqrt{25} = 5$ **b.** $(-27)^{1/3} = \sqrt[3]{-27} = -3$ **c.** $(-16)^{1/4} = \sqrt[4]{-16}$, which is not a real number. Simplify. $$27^{4/3} = (\sqrt[3]{27})^4 = 3^4 = 81$$ Write in exponential form. **a.** $\sqrt[3]{x^2} = x^{2/3}$ **b.** $(\sqrt[6]{x})^5 = x^{5/6}$ Evaluate. $$16^{-3/4} = \frac{1}{16^{3/4}} = \frac{1}{(\sqrt[4]{16})^3} = \frac{1}{2^3} = \frac{1}{8}$$ Write $\sqrt[8]{x^6}$ as a radical expression with a smaller root index. Assume variables represent nonnegative values. **Solution:** $\sqrt[8]{x^6} = x^{6/8} = x^{3/4} = \sqrt[4]{x^3}$

Exercises 27–40   Expressions

[9.2] *For Exercises 27–34, rewrite using radicals; then simplify if possible. Assume that variables represent nonnegative values.*

27. $(-64)^{1/3}$
-4

28. $(24a^4)^{1/2}$
$2a^2\sqrt{6}$

29. $\left(\dfrac{1}{32}\right)^{3/5}$
$\dfrac{1}{8}$

30. $(5r - 2)^{5/7}$
$\sqrt[7]{(5r - 2)^5}$

31. $121^{-1/2}$
$\dfrac{1}{11}$

32. $81^{-3/4}$
$\dfrac{1}{27}$

33. $\left(\dfrac{16}{49}\right)^{3/2}$
$\dfrac{64}{343}$

34. $81x^{3/4}$
$81\sqrt[4]{x^3}$

[9.2] *For Exercises 35–40, write in exponential form.*

35. $\sqrt[8]{33}$
$33^{1/8}$

36. $\dfrac{8}{\sqrt[7]{n^3}}$
$8n^{-3/7}$

37. $(\sqrt[5]{8})^3$
$8^{3/5}$

38. $(\sqrt[3]{m})^8$
$m^{8/3}$

39. $(\sqrt[4]{3xw})^3$
$(3xw)^{3/4}$

40. $\sqrt[3]{(a + b)^4}$
$(a + b)^{4/3}$

Definitions/Rules/Procedures	Key Example(s)
To multiply or divide radical expressions with different root indices: 1. Change from radical to $\underline{\text{ exponential }}$ form. 2. Multiply or divide using the appropriate rule(s) of $\underline{\text{ exponents }}$. 3. Write the result from step 2 in $\underline{\text{ radical }}$ form.	Simplify. Assume variables represent nonnegative values. **a.** $\sqrt[4]{x} \cdot \sqrt{x} = x^{1/4} \cdot x^{1/2} = x^{1/4 + 1/2}$ $\qquad\qquad\qquad\qquad = x^{1/4 + 2/4}$ $\qquad\qquad\qquad\qquad = x^{3/4} = \sqrt[4]{x^3}$ **b.** $\dfrac{\sqrt[3]{x^2}}{\sqrt[4]{x}} = \dfrac{x^{2/3}}{x^{1/4}} = x^{(2/3)-(1/4)}$ $\qquad\qquad = x^{(8/12)-(3/12)} = x^{5/12} = \sqrt[12]{x^5}$ Write $\sqrt[3]{\sqrt[5]{x}}$ as a single radical. Assume variables represent nonnegative values. **Solution:** $\sqrt[3]{\sqrt[5]{x}} = (x^{1/5})^{1/3} = x^{(1/5)\cdot(1/3)}$ $\qquad\qquad\qquad\quad = x^{1/15} = \sqrt[15]{x}$

Exercises 41–47 Expressions

[9.2] *For Exercises 41–47, use the rules of exponents to simplify. Assume that variables represent positive values.*

41. $x^{2/3} \cdot x^{4/3}$
x^2

42. $(4m^{1/4})(8m^{5/4})$
$32m^{3/2}$

43. $\dfrac{y^{3/5}}{y^{4/5}}$
$\dfrac{1}{y^{1/5}}$

44. $\dfrac{b^{2/5}}{b^{-3/5}}$
b

45. $(k^{2/3})^{3/4}$
$k^{1/2}$

46. $(2xy^{1/5})^{3/4}$
$2^{3/4}x^{3/4}y^{3/20}$

47. Write $\sqrt[4]{\sqrt[3]{x^2}}$ as a single radical.
$\sqrt[6]{x}$

9.3 Multiplying, Dividing, and Simplifying Radicals

Definitions/Rules/Procedures	Key Example(s)
Product rule for radicals If both $\sqrt[n]{a}$ and $\sqrt[n]{b}$ are real numbers, then $\sqrt[n]{a} \cdot \sqrt[n]{b} = $ ___$\sqrt[n]{a \cdot b}$___ . **Raising an nth root to the nth power** For any nonnegative real number a, $(\sqrt[n]{a})^n = $ ___a___ .	Simplify. Assume that variables represent nonnegative values. **a.** $\sqrt{2} \cdot \sqrt{50} = \sqrt{2 \cdot 50} = \sqrt{100} = 10$ **b.** $\sqrt{3x} \cdot \sqrt{12x^3} = \sqrt{3x \cdot 12x^3} = \sqrt{36x^4} = 6x^2$ **c.** $\sqrt[3]{5x} \cdot \sqrt[3]{3x} = \sqrt[3]{15x^2}$ Simplify $(\sqrt[4]{x^3})^4$. Assume variables represent nonnegative values. **Solution:** $(\sqrt[4]{x^3})^4 = x^3$

Exercises 48–57 Expressions

[9.3] *For Exercises 48–57, simplify. Assume that variables represent positive values.*

48. $\sqrt{3} \cdot \sqrt{27}$
9

49. $\sqrt{5x^5} \cdot \sqrt{20x^3}$
$10x^4$

50. $\sqrt[3]{2} \cdot \sqrt[3]{4}$
2

51. $\sqrt[4]{7} \cdot \sqrt[4]{6}$
$\sqrt[4]{42}$

52. $\sqrt[5]{3x^2y^3} \cdot \sqrt[5]{5x^2y}$
$\sqrt[5]{15x^4y^4}$

53. $4\sqrt{6} \cdot 7\sqrt{15}$
$84\sqrt{10}$

54. $\sqrt{x^9} \cdot \sqrt{x^6}$
$x^7\sqrt{x}$

55. $4\sqrt{10c} \cdot 2\sqrt{6c^4}$
$16c^2\sqrt{15c}$

56. $a\sqrt{a^3b^2} \cdot b^2\sqrt{a^5b^3}$
$a^5b^4\sqrt{b}$

57. $(\sqrt[4]{6ab^2})^4$
$6ab^2$

Definitions/Rules/Procedures	Key Example(s)
Quotient rule for radicals If both $\sqrt[n]{a}$ and $\sqrt[n]{b}$ are real numbers, then $\dfrac{\sqrt[n]{a}}{\sqrt[n]{b}} = $ ___$\sqrt[n]{\dfrac{a}{b}}$___ where $b \neq 0$.	Simplify. Assume variables represent positive values. **a.** $\dfrac{\sqrt{45}}{\sqrt{5}} = \sqrt{\dfrac{45}{5}} = \sqrt{9} = 3$ **b.** $\dfrac{\sqrt[3]{128x^7}}{\sqrt[3]{2x}} = \sqrt[3]{\dfrac{128x^7}{2x}} = \sqrt[3]{64x^6} = 4x^2$ **c.** $\sqrt[4]{\dfrac{8}{x^4}} = \dfrac{\sqrt[4]{8}}{\sqrt[4]{x^4}} = \dfrac{\sqrt[4]{8}}{x}$

Exercises 58–61 Expressions

[9.3] *For Exercises 58–61, simplify. Assume that variables represent positive values.*

58. $\sqrt{\dfrac{49}{121}}$
$\dfrac{7}{11}$

59. $\sqrt[3]{-\dfrac{27}{8}}$
$-\dfrac{3}{2}$

60. $\dfrac{9\sqrt{160}}{3\sqrt{8}}$
$6\sqrt{5}$

61. $\dfrac{36\sqrt{96x^6y^{11}}}{4\sqrt{3x^2y^4}}$
$36x^2y^3\sqrt{2y}$

Definitions/Rules/Procedures	Key Example(s)
To simplify an *n*th root: 1. Write the radicand as the product of the greatest possible perfect _____*n*th_____ power and a number or an expression that has no perfect _____*n*th_____ power factors. 2. Use the product rule $\sqrt[n]{ab} = \sqrt[n]{a} \cdot \sqrt[n]{b}$, where a is the _____perfect *n*th power_____. 3. Find the *n*th root of the _____perfect *n*th power_____ radicand.	Simplify. Assume that variables represent nonnegative values. **a.** $\sqrt{50} = \sqrt{25 \cdot 2} = \sqrt{25} \cdot \sqrt{2} = 5\sqrt{2}$ **b.** $\sqrt{48x^3} = \sqrt{16x^2 \cdot 3x} = \sqrt{16x^2} \cdot \sqrt{3x} = 4x\sqrt{3x}$ **c.** $\sqrt[4]{48a^9} = \sqrt[4]{16a^8 \cdot 3a} = \sqrt[4]{16a^8} \cdot \sqrt[4]{3a} = 2a^2\sqrt[4]{3a}$ Perform the operation and simplify. $\sqrt{6} \cdot \sqrt{10} = \sqrt{60} = \sqrt{4 \cdot 15} = \sqrt{4} \cdot \sqrt{15} = 2\sqrt{15}$

Exercises 62–65 **Expressions**

[9.3] *For Exercises 62–65, simplify. Assume that variables represent positive values.*

62. $4b\sqrt{27b^7}$
$12b^4\sqrt{3b}$

63. $5\sqrt[3]{108}$
$15\sqrt[3]{4}$

64. $2\sqrt[3]{40x^{10}}$
$4x^3\sqrt[3]{5x}$

65. $2x^4\sqrt[4]{162x^7}$
$6x^5\sqrt[4]{2x^3}$

9.4 Adding, Subtracting, and Multiplying Radical Expressions

Definitions/Rules/Procedures	Key Example(s)
Like radicals are radical expressions with identical _____radicands_____ and identical _____root indices_____. **To add or subtract** like radicals, add or subtract the coefficients and leave the _____radical parts_____ the same.	Add or subtract. Assume that variables represent nonnegative values. **a.** $2\sqrt{x} - 7\sqrt{x} = (2 - 7)\sqrt{x} = -5\sqrt{x}$ **b.** $4\sqrt[3]{5} + 2\sqrt[3]{5} = (4 + 2)\sqrt[3]{5} = 6\sqrt[3]{5}$ Simplify the radicals; then combine like radicals. $9\sqrt{3} - \sqrt{12} + 6\sqrt{75} = 9\sqrt{3} - \sqrt{4 \cdot 3} + 6\sqrt{25 \cdot 3}$ $\qquad = 9\sqrt{3} - 2\sqrt{3} + 6 \cdot 5\sqrt{3}$ $\qquad = 9\sqrt{3} - 2\sqrt{3} + 30\sqrt{3}$ $\qquad = 37\sqrt{3}$

Exercises 66–73 **Expressions**

[9.4] *For Exercises 66–73, add or subtract. Assume that variables represent nonnegative values.*

66. $-5\sqrt{n} + 2\sqrt{n}$
$-3\sqrt{n}$

67. $3y^3\sqrt[4]{8y} - 9y^3\sqrt[4]{8y}$
$-6y^3\sqrt[4]{8y}$

68. $\sqrt{45} + \sqrt{20}$
$5\sqrt{5}$

69. $4\sqrt{24} - 6\sqrt{54}$
$-10\sqrt{6}$

70. $\sqrt{150} - \sqrt{54} + \sqrt{24}$
$4\sqrt{6}$

71. $4\sqrt{72x^2y} - 2x\sqrt{128y} + 5\sqrt{32x^2y}$
$28x\sqrt{2y}$

72. $\sqrt[3]{250x^4y^5} + \sqrt[3]{128x^4y^5}$
$9xy\sqrt[3]{2xy^2}$

73. $\sqrt[4]{48} + \sqrt[4]{243}$
$5\sqrt[4]{3}$

Definitions/Rules/Procedures	Key Example(s)
To simplify radical expressions that contain more than one operation, use the _order of operations_ agreement.	Find the product. $(\sqrt{3} - 4)(\sqrt{3} + 6) = \sqrt{3} \cdot \sqrt{3} + \sqrt{3} \cdot 6 - 4 \cdot \sqrt{3} - 4 \cdot 6$ **Use FOIL.** $= \sqrt{9} + 6\sqrt{3} - 4\sqrt{3} - 24$ **Multiply.** $= 3 + 6\sqrt{3} - 4\sqrt{3} - 24$ **Simplify** $\sqrt{9}$. $= -21 + 2\sqrt{3}$ **Combine like radicals.** Simplify. **a.** $\sqrt{6} \cdot \sqrt{8} + \sqrt{5} \cdot \sqrt{15} = \sqrt{48} + \sqrt{75}$ $= \sqrt{16 \cdot 3} + \sqrt{25 \cdot 3}$ $= 4\sqrt{3} + 5\sqrt{3} = 9\sqrt{3}$ **b.** $\dfrac{\sqrt{60}}{\sqrt{5}} + \sqrt{48} = \sqrt{\dfrac{60}{5}} + \sqrt{48}$ $= \sqrt{12} + \sqrt{48}$ $= \sqrt{4 \cdot 3} + \sqrt{16 \cdot 3}$ $= 2\sqrt{3} + 4\sqrt{3} = 6\sqrt{3}$

Exercises 74–82   **Expressions**

[9.4] *For Exercises 74–82, find the product. Assume that variables represent nonnegative values.*

74. $\sqrt{5}(\sqrt{3} + \sqrt{2})$
$\sqrt{15} + \sqrt{10}$

75. $\sqrt[3]{7}(\sqrt[3]{3} + 2\sqrt[3]{7})$
$\sqrt[3]{21} + 2\sqrt[3]{49}$

76. $3\sqrt{6}(2 - 3\sqrt{6})$
$6\sqrt{6} - 54$

77. $(\sqrt{2} - \sqrt{3})(\sqrt{5} + \sqrt{7})$
$\sqrt{10} + \sqrt{14} - \sqrt{15} - \sqrt{21}$

78. $(\sqrt[3]{2} - 4)(\sqrt[3]{4} + 2)$
$-6 + 2\sqrt[3]{2} - 4\sqrt[3]{4}$

79. $(\sqrt[4]{6x} + 2)(\sqrt[4]{2x} - 1)$
$\sqrt[4]{12x^2} - \sqrt[4]{6x} + 2\sqrt[4]{2x} - 2$

80. $(\sqrt{5a} + \sqrt{3b})(\sqrt{5a} - \sqrt{3b})$
$5a - 3b$

81. $(2\sqrt{3} - \sqrt{5})(2\sqrt{3} + \sqrt{5})$
7

82. $(\sqrt{2} - \sqrt{5})^2$
$7 - 2\sqrt{10}$

9.5 Rationalizing Numerators and Denominators of Radical Expressions

Definitions/Rules/Procedures	Key Example(s)
A radical expression is in simplest form if **1.** The radicand has no factor raised to a power greater than or equal to the _index_. **2.** There are neither radicals in the denominator of a fraction nor radicands that are _fractions_. **3.** All possible _sums_, _differences_, _products_, and _quotients_ have been found. **To rationalize** a denominator containing a single nth root, multiply the fraction by a form of 1 so that the product's denominator has a radicand that is a(n) _perfect nth power_.	Rationalize the denominator. **a.** $\dfrac{7}{\sqrt{3}} = \dfrac{7}{\sqrt{3}} \cdot \dfrac{\sqrt{3}}{\sqrt{3}} = \dfrac{7\sqrt{3}}{3}$ **b.** $\dfrac{4}{\sqrt[3]{3}} = \dfrac{4}{\sqrt[3]{3}} \cdot \dfrac{\sqrt[3]{9}}{\sqrt[3]{9}} = \dfrac{4\sqrt[3]{9}}{\sqrt[3]{27}} = \dfrac{4\sqrt[3]{9}}{3}$

Exercises 83–88 Expressions

[9.5] *For Exercises 83–92, rationalize the denominator and simplify. Assume that variables represent positive values.*

83. $\dfrac{1}{\sqrt{2}}$

$\dfrac{\sqrt{2}}{2}$

84. $\dfrac{3}{\sqrt[3]{3}}$

$\sqrt[3]{9}$

85. $\sqrt{\dfrac{4}{7}}$

$\dfrac{2\sqrt{7}}{7}$

86. $\dfrac{\sqrt[4]{5x^2}}{\sqrt[4]{2}}$

$\dfrac{\sqrt[4]{40x^2}}{2}$

87. $\sqrt[3]{\dfrac{17}{3y^2}}$

$\dfrac{\sqrt[3]{153y}}{3y}$

88. $\dfrac{\sqrt[4]{9}}{\sqrt[4]{27}}$

$\dfrac{\sqrt[4]{27}}{3}$

Definitions/Rules/Procedures	Key Example(s)
To rationalize a denominator containing a sum or difference with at least one square root term, multiply the fraction by a form of 1 whose numerator and denominator are the ___conjugate___ of the denominator.	Rationalize the denominator and simplify. $\dfrac{6}{4-\sqrt{5}} = \dfrac{6}{4-\sqrt{5}} \cdot \dfrac{4+\sqrt{5}}{4+\sqrt{5}}$ $= \dfrac{6\cdot 4 + 6\cdot\sqrt{5}}{16-5}$ $= \dfrac{24+6\sqrt{5}}{11}$

Exercises 89–92 Expressions

[9.5] *For Exercises 89–92, rationalize the denominator and simplify. Assume that variables represent positive values.*

89. $\dfrac{4}{\sqrt{2}-\sqrt{3}}$

$-4\sqrt{2}-4\sqrt{3}$

90. $\dfrac{1}{4+\sqrt{3}}$

$\dfrac{4-\sqrt{3}}{13}$

91. $\dfrac{1}{2-\sqrt{n}}$

$\dfrac{2+\sqrt{n}}{4-n}$

92. $\dfrac{2\sqrt{3}}{3\sqrt{2}-2\sqrt{3}}$

$\sqrt{6}+2$

Definitions/Rules/Procedures	Key Example(s)
The numerator of a rational expression is rationalized the same way a(n) ___denominator___ is rationalized.	Rationalize the numerator. **a.** $\dfrac{\sqrt{6}}{3} = \dfrac{\sqrt{6}}{3}\cdot\dfrac{\sqrt{6}}{\sqrt{6}} = \dfrac{\sqrt{36}}{3\sqrt{6}} = \dfrac{6}{3\sqrt{6}} = \dfrac{2}{\sqrt{6}}$ **b.** $\dfrac{2-\sqrt{3}}{5} = \dfrac{2-\sqrt{3}}{5}\cdot\dfrac{2+\sqrt{3}}{2+\sqrt{3}}$ $= \dfrac{4-3}{5(2+\sqrt{3})} = \dfrac{1}{10+5\sqrt{3}}$

Exercises 93–96 Expressions

[9.5] *For Exercises 93–96, rationalize the numerator. Assume that variables represent positive values.*

93. $\dfrac{\sqrt{10}}{6}$

$\dfrac{5}{3\sqrt{10}}$

94. $\dfrac{\sqrt{3x}}{5}$

$\dfrac{3x}{5\sqrt{3x}}$

95. $\dfrac{2-\sqrt{3}}{8}$

$\dfrac{1}{16+8\sqrt{3}}$

★ **96.** $\dfrac{2\sqrt{t}+\sqrt{3t}}{5t}$

$\dfrac{1}{10\sqrt{t}-5\sqrt{3t}}$

9.6 Radical Equations and Problem Solving

Definitions/Rules/Procedures	Key Example(s)
A **radical equation** is an equation containing at least one ___radical expression___ whose radicand has a(n) ___variable___.	Solve $\sqrt{x-5} = 7$. $$(\sqrt{x-5})^2 = 7^2 \qquad \text{Square both sides.}$$ $$x - 5 = 49 \qquad \text{Multiply.}$$ $$x = 54 \qquad \text{Add 5 to both sides.}$$ Check: $\sqrt{54-5} = 7$ $$\sqrt{49} = 7 \qquad \text{True}$$
Power rule for solving equations If both sides of an equation are raised to the same ___integer power___, the resulting equation contains all solutions of the original equation and perhaps some solutions that do not solve the original equation. That is, the solutions of the equation $a = b$ are contained among the solutions of $a^n = b^n$, where n is an integer.	Solve $\sqrt{n+14} = n + 2$. $$(\sqrt{n+14})^2 = (n+2)^2 \qquad \text{Square both sides.}$$ $$n + 14 = n^2 + 4n + 4$$ $$0 = n^2 + 3n - 10 \qquad \text{Subtract } n \text{ and 14 from both sides.}$$ $$0 = (n+5)(n-2) \qquad \text{Factor.}$$ $$n + 5 = 0 \quad \text{or} \quad n - 2 = 0 \qquad \text{Use the zero-factor theorem.}$$ $$n = -5 \qquad\qquad n = 2$$
To solve a radical equation: 1. Isolate the ___radical___ if necessary. (If there is more than one ___radical term___, isolate one of the ___radical terms___.) 2. Raise both sides of the equation to the same power as the ___root index___ of the isolated radical. 3. If all radicals have been eliminated, solve. If a(n) ___radical term___ remains, isolate it and raise both sides to the same power as its ___root index___. 4. Check each solution. Any apparent solution that does not check is a(n) ___extraneous___ solution.	Check: $\sqrt{-5+14} = -5 + 2 \qquad \sqrt{2+14} = 2 + 2$ $$\sqrt{9} = -3 \quad \text{False} \qquad \sqrt{16} = 4 \quad \text{True}$$ The only solution is 2. (-5 is extraneous.) Solve $3 + \sqrt{t} = \sqrt{t+21}$. $$(3 + \sqrt{t})^2 = (\sqrt{t+21})^2 \qquad \text{Square both sides.}$$ $$9 + 6\sqrt{t} + t = t + 21 \qquad \text{Multiply.}$$ $$6\sqrt{t} = 12 \qquad \text{Subtract 9 and } t \text{ from both sides.}$$ $$\sqrt{t} = 2 \qquad \text{Divide both sides by 6.}$$ $$(\sqrt{t})^2 = 2^2 \qquad \text{Square both sides.}$$ $$t = 4$$ We will leave the check to the reader.

Exercises 97–111 ▲ Equations and Inequalities

[9.6] *For Exercises 97–108, solve. Identify any extraneous solutions.*

97. $\sqrt{x} = 9$
81

98. $\sqrt{y} = -3$
No real-number solution

99. $\sqrt{w-1} = 3$
10

100. $\sqrt[3]{3x-2} = -2$
-2

101. $\sqrt[4]{x-2} - 3 = -1$
18

102. $\sqrt{y+1} = \sqrt{2y-4}$
5

103. $\sqrt{x-6} = x + 2$
No real-number solution

104. $\sqrt[3]{3x+10} - 4 = 5$
$\dfrac{719}{3}$

105. $\sqrt[4]{x+8} = \sqrt[4]{2x+1}$
7

106. $\sqrt{5n-1} = 4 - 2n$
$1 \left(\dfrac{17}{4} \text{ is extraneous.} \right)$

107. $1 + \sqrt{x} = \sqrt{2x+1}$
0, 4

108. $\sqrt{3x+1} = 2 - \sqrt{3x}$
$\dfrac{3}{16}$

109. The speed of a car can be determined by the length of the skid marks using the formula $S = 2\sqrt{2L}$, where L is the length of the skid mark in feet and S is the speed of the car in miles per hour. Find the length of skid marks if the driver brakes hard at a speed of 50 miles per hour.
312.5 ft.

110. In the formula $T = 2\pi\sqrt{\dfrac{L}{9.8}}$, T is the period of a pendulum in seconds and L is the length of the pendulum in meters. If the period of the pendulum is $\dfrac{\pi}{3}$ seconds, find the length to the nearest thousandth.
0.272 sec.

111. In the formula $t = \sqrt{\dfrac{h}{16}}$, t represents the time in seconds it takes an object to fall a distance of h feet. Find the distance an object falls in 0.3 second.

1.44 ft.

9.7 Complex Numbers

Definitions/Rules/Procedures	Key Example(s)
The **imaginary unit** is i, where $i = \underline{\quad \sqrt{-1} \quad}$ and $i^2 = \underline{\quad -1 \quad}$.	
An **imaginary number** is a number that can be expressed in the form $\underline{\quad a + bi \quad}$, where a and b are $\underline{\text{real numbers}}$, i is the $\underline{\text{imaginary unit}}$, and $\underline{\quad b \neq 0 \quad}$.	
A **complex number** is a number that can be expressed in the form $\underline{\quad a + bi \quad}$, where a and b are $\underline{\text{real numbers}}$ and i is the $\underline{\text{imaginary unit}}$.	
The complex conjugate of $a + bi$ is $\underline{\quad a - bi \quad}$.	
To write an imaginary number $\sqrt{-n}$ in terms of the imaginary unit i: **1.** Separate the radical into two factors, $\underline{\sqrt{-1} \cdot \sqrt{n}}$. **2.** Replace $\sqrt{-1}$ with $\underline{\quad i \quad}$. **3.** Simplify $\sqrt{n}$.	Write using the imaginary unit. **a.** $\sqrt{-36} = \sqrt{-1} \cdot \sqrt{36} = 6i$ **b.** $\sqrt{-32} = \sqrt{-1} \cdot \sqrt{32}$ $\qquad = i \cdot 4\sqrt{2} = 4i\sqrt{2}$
To add complex numbers, combine $\underline{\text{like terms}}$.	Add $(5 - 6i) + (9 + 2i)$. $\qquad (5 - 6i) + (9 + 2i) = 14 - 4i$
To subtract complex numbers, write the equivalent addition and change the $\underline{\text{signs}}$ of the second complex number; then combine like terms.	Subtract $(7 - i) - (3 + 5i)$. $\qquad (7 - i) - (3 + 5i) = (7 - i) + (-3 - 5i)$ $\qquad\qquad\qquad\qquad\quad = 4 - 6i$

Exercises 112–115 Expressions

[9.7] *For Exercises 112 and 113, write the imaginary number using i.*

112. $\sqrt{-9}$

$3i$

113. $\sqrt{-20}$

$2i\sqrt{5}$

[9.7] *For Exercises 114 and 115, add or subtract.*

114. $(3 + 2i) + (5 - 8i)$

$8 - 6i$

115. $(7 - 3i) - (-2 + 4i)$

$9 - 7i$

Definitions/Rules/Procedures	Key Example(s)
To multiply complex numbers, follow the same procedures for multiplying $\underline{\text{monomials}}$ and $\underline{\text{binomials}}$. Remember that $i^2 = -1$.	Multiply $(6 + 5i)(2 - 3i)$. $\quad (6 + 5i)(2 - 3i) = 12 - 18i + 10i - 15i^2$ $\qquad\qquad\qquad\qquad = 12 - 8i - 15(-1)$ $\qquad\qquad\qquad\qquad = 12 - 8i + 15$ $\qquad\qquad\qquad\qquad = 27 - 8i$

Exercises 116–119 ◢ **Expressions**

[9.7] *For Exercises 116–119, multiply and simplify.*

116. $(3i)(4i)$
-12

117. $2i(4 - i)$
$2 + 8i$

118. $(6 + 2i)(4 - i)$
$26 + 2i$

119. $(5 - i)^2$
$24 - 10i$

Definitions/Rules/Procedures	Key Example(s)
To divide by a pure imaginary number, multiply by $\dfrac{i}{i}$. **To divide complex numbers,** rationalize the denominator using the __complex conjugate__.	Divide and write $\dfrac{2 - 3i}{4 + 5i}$ in standard form. $\dfrac{2 - 3i}{4 + 5i} = \dfrac{2 - 3i}{4 + 5i} \cdot \dfrac{4 - 5i}{4 - 5i}$ $= \dfrac{8 - 10i - 12i + 15i^2}{16 - 25i^2}$ $= \dfrac{8 - 22i + 15(-1)}{16 - 25(-1)}$ $= \dfrac{-7 - 22i}{41}$ $= -\dfrac{7}{41} - \dfrac{22}{41}i$

Exercises 120–125 ◢ **Expressions**

[9.7] *For Exercises 120–125, divide and write in standard form.*

120. $\dfrac{5}{i}$
$-5i$

121. $\dfrac{3}{-i}$
$3i$

122. $\dfrac{4}{3i}$
$\frac{4i}{3}$

123. $\dfrac{7 + i}{5i}$
$\frac{1}{5} - \frac{7}{5}i$

124. $\dfrac{3}{2 + i}$
$\frac{6}{5} - \frac{3}{5}i$

125. $\dfrac{5 + i}{2 - 3i}$
$\frac{7}{13} + \frac{17}{13}i$

Definitions/Rules/Procedures	Key Example(s)
Powers of i: $i^1 = $ ___i___ $i^2 = $ ___-1___ $i^3 = $ ___$-i$___ $i^4 = $ ___1___	Find the powers of i. **a.** $i^{19} = i^{16} \cdot i^3 = (i^4)^4 \cdot (-i)$ $\qquad = 1^4(-i) = 1(-i) = -i$ **b.** $i^{-21} = \dfrac{1}{i^{21}} = \dfrac{1}{i^{20} \cdot i} = \dfrac{1}{(i^4)^5 \cdot i} = \dfrac{1}{1^5 \cdot i}$ $\qquad = \dfrac{1}{i} = \dfrac{1}{i} \cdot \dfrac{i}{i} = \dfrac{i}{i^2}$ $\qquad = \dfrac{i}{-1} = -i$

Exercises 126 and 127 ◢ **Expressions**

[9.7] *For Exercises 126 and 127, find the powers of i.*

126. i^{20}
1

127. i^{15}
$-i$

Learning Strategy

During exams, I like to work one or two problems that don't count as much and then move on to the bigger problems. That way, if I run out of time, I will have received the most possible credit.

—Jason J.

Chapter 9 Practice Test

For Extra Help

Step-by-step test solutions are found on the Chapter Test Prep Videos available in MyMathLab® *or on* YouTube.

For Exercises 1 and 2, evaluate the square root.

1. $\sqrt{36}$
6 [9.1]

2. $\sqrt{-49}$
7i [9.7]

For Exercises 3–12, simplify. Assume that variables represent positive values.

3. $\sqrt{81x^2y^5}$
$9xy^2\sqrt{y}$ [9.1]

4. $\sqrt[3]{54}$
$3\sqrt[3]{2}$ [9.1]

5. $\sqrt[4]{4x}\cdot\sqrt[4]{4x^5}$
$2x\sqrt[4]{x^2}$, or $2x\sqrt{x}$ [9.3]

6. $-\sqrt[3]{-27r^{15}}$
$3r^5$ [9.1]

7. $\dfrac{\sqrt{5}}{\sqrt{45}}$
$\dfrac{1}{3}$ [9.3]

8. $\dfrac{\sqrt[4]{1}}{\sqrt[4]{81}}$
$\dfrac{1}{3}$ [9.3]

9. $6\sqrt{7}-\sqrt{7}$
$5\sqrt{7}$ [9.4]

10. $(\sqrt{3}-1)^2$
$4-2\sqrt{3}$ [9.4]

11. $x^{2/3}\cdot x^{-4/3}$
$\dfrac{1}{x^{2/3}}$ [9.2]

12. $(\sqrt[3]{2}-4)(\sqrt[3]{4}+2)$
$-6+2\sqrt[3]{2}-4\sqrt[3]{4}$ [9.4]

For Exercises 13 and 14, write in exponential form. Assume that variables represent non-negative values.

13. $\sqrt[5]{8x^3}$
$8^{1/5}x^{3/5}$ [9.2]

14. $\sqrt[3]{(2x+5)^2}$
$(2x+5)^{2/3}$ [9.2]

For Exercises 15 and 16, rationalize the denominator and simplify. Assume that variables represent positive values.

15. $\dfrac{1}{\sqrt[3]{4}}$
$\dfrac{\sqrt[3]{2}}{2}$ [9.5]

16. $\dfrac{\sqrt{x}}{\sqrt{x}+\sqrt{y}}$
$\dfrac{x-\sqrt{xy}}{x-y}$ [9.5]

For Exercises 17–19, solve the equation. Identify any extraneous solutions.

17. $\sqrt{3x-2}=8$
22 [9.6]

18. $\sqrt[4]{x+8}=\sqrt[4]{2x+1}$
7 [9.6]

19. $\sqrt{2x+1}-\sqrt{x+1}=2$
24 (0 is an extraneous solution.) [9.6]

For Exercises 20–22, simplify and write the answer in standard form $(a+bi)$.

20. $(2-i)-(4+3i)$
$-2-4i$ [9.7]

21. $(4-i)(4+i)$
$17+0i$ or 17 [9.7]

22. $\dfrac{2}{4-3i}$
$\dfrac{8}{25}+\dfrac{6}{25}i$ [9.7]

23. Write an expression in simplest form for the area of the figure.
60 m² [9.3]

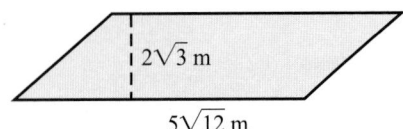

24. The formula $t = \sqrt{\dfrac{h}{16}}$ describes the amount of time t, in seconds, that an object falls a distance of h feet.

   a. Write an expression in simplest form for the exact amount of time an object falls a distance of 12 feet.

       $\dfrac{\sqrt{3}}{2}$ [9.6]

   b. Find the distance an object falls in 2 seconds.

       64 ft. [9.6]

25. The formula $S = \dfrac{7}{4}\sqrt{D}$ can be used to approximate the speed, S, in miles per hour, that a car was traveling prior to braking and skidding a distance D, in feet, on ice. (*Source:* Harris Technical Services, Traffic Accident Reconstructionists.)

   a. Find the exact speed of a car if the skid length measures 40 feet long.

       $3.5\sqrt{10}$ mph [9.6]

   b. Find the length of the skid marks that a car traveling 30 miles per hour makes if it brakes hard and skids to a halt.

       293.88 ft or 14,400/49 ft. [9.6]

Chapters 1–9 Cumulative Review Exercises

For Exercises 1–4, answer true or false

[8.3] **1.** If $|x| > 3$, then $-3 < x < 3$.
false

[4.1] **2.** $(2, -3)$ is a solution of $\begin{cases} 4x + 3y = -1 \\ 2x - 5y = -11. \end{cases}$
false

[9.4] **3.** The conjugate of $4 - 2\sqrt{3}$ is $4 + 2\sqrt{3}$.
true

[9.7] **4.** 3 is a complex number.
true

For Exercises 5–7, fill in the blanks.

[3.3] **5.** Given an equation of the form $ax + by = c$, to find the ___x-intercept___, let $y = 0$.

[4.2] **6.** A system of equations that has no solution is ___inconsistent___.

[7.4] **7.** The LCD for $\dfrac{5}{8x^3y^2}$ and $\dfrac{3}{12x^5y}$ is ___$24x^5y^2$.___.

Exercises 8–18 **Expressions**

[1.5] **8.** Simplify $4 - 6(5 - 3^2) + 12 \div 2 - 6$.
28

[5.6] **9.** Simplify $\dfrac{7.44 \times 10^{-2}}{3.1 \times 10^4}$. Write the answer in standard form.
0.0000024

For Exercises 10–17, simplify. Write the answers with positive exponents only.

[5.6] **10.** $\dfrac{(x^{-2})^3(x^3)^4}{(x^{-3})^3}$
x^{15}

[5.5] **11.** $(3x^2 - 4y)^2$
$9x^4 - 24x^2y + 16y^2$

[5.6] **12.** $\dfrac{8x^3 - 22x + 8}{2x - 3}$
$4x^2 + 6x - 2 + \dfrac{2}{2x - 3}$

[7.2] **13.** $\dfrac{2x^2 + 7x + 3}{x^2 - 9} \div \dfrac{2x^2 + 11x + 5}{x^2 - 3x}$
$\dfrac{x}{x + 5}$

[9.4] **14.** $2\sqrt{5a^2} \cdot 3\sqrt{10a^5}$
$30a^3\sqrt{2a}$

[9.4] **15.** $3x\sqrt[3]{24x^4} - 4x^2\sqrt[3]{81x}$
$-6x^2\sqrt[3]{3x}$

[9.2] **16.** $32^{-3/5}$
$\dfrac{1}{8}$

[9.7] **17.** $(4 + 5i)(2 - 3i)$
$23 - 2i$

[9.5] **18.** Rationalize the denominator of $\dfrac{2 + \sqrt{2}}{4 - \sqrt{2}}$.
$\dfrac{5 + 3\sqrt{2}}{7}$

Exercises 19–30 **Equations and Inequalities**

For Exercises 19–22, solve. Identify any extraneous solutions.

[2.4] **19.** $A = \dfrac{1}{2}h(B + b)$ for b
$b = \dfrac{2A - Bh}{h}$ or $\dfrac{2A}{h} - B$

[6.6] **20.** $(x + 2)(x + 4) = 63$
$5, -11$

[7.6]   **21.** $\dfrac{3y}{y-4} - \dfrac{12}{y-4} = 6$

     No solution (4 is an extraneous solution.)

[9.6]   **22.** $\sqrt{3x+10} = x+2$

     2 (-3 is an extraneous solution.)

[3.4]   **23.** Find the slope and y-intercept of the graph of $5x - 2y = -15$.

     slope $= \dfrac{5}{2}$, y-intercept $= \dfrac{15}{2}$

[3.5]   **24.** Write the equation of the line containing $(-4, 2)$ that is parallel to the graph of $y = 3x - 5$. Leave the answer in slope–intercept form.

     $y = 3x + 14$

For Exercises 25 and 26, graph.

[3.2]   **25.** $y = -\dfrac{5}{3}x + 2$

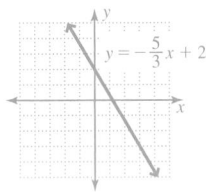

[4.6]   **26.** $\begin{cases} 2x - y > -4 \\ y > -\dfrac{2}{3}x - 2 \end{cases}$

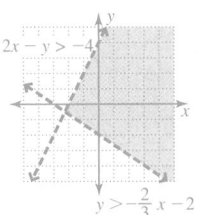

For Exercises 27–30, solve.

[2.3]   **27.** Hernando earns $1\dfrac{1}{4}$ times his normal hourly wage for all hours in excess of 40 hours per week. Last week he worked 46 hours and earned \$782.80. What is his normal hourly wage?

     \$16.48

[7.7]   **28.** A river has a current of 5 miles per hour. A boat can make a trip of 20 miles upstream in the same time it can make a trip of 30 miles downstream. Find the rate of the boat in still water.

     25 mph

[7.7]   **29.** Suppose y varies jointly as x and the square of z. If $y = 72$ when $x = 4$ and $z = 3$, find y when $x = 3$ and $z = 4$.

     96

[9.6]   **30.** The period T, in seconds, of a pendulum whose length is L, in meters, is given by $T = 2\pi\sqrt{\dfrac{L}{9.8}}$. Find the length of a pendulum whose period is 4π seconds.

     39.2 m

Chapter Overview

Previously, we learned to solve quadratic equations by factoring. In this chapter, we learn three additional methods:

▶ The square root principle.

▶ Completing the square.

▶ The quadratic formula.

In addition, we expand our knowledge of graphing quadratic functions, solving inequalities, and working with functions.

Instructor Note

The sections "The Square Root Principle and Completing the Square" and "Graphing Quadratic Functions" are prerequisites for the chapter on conic sections. The function operations of addition, subtraction, multiplication, and division are discussed here in order to transition into the next chapter, which begins with composition of functions.

10.1 The Square Root Principle and Completing the Square

10.2 Solving Quadratic Equations Using the Quadratic Formula

10.3 Solving Equations That Are Quadratic in Form

10.4 Graphing Quadratic Functions

10.5 Solving Nonlinear Inequalities

10.6 Function Operations

10.1 The Square Root Principle and Completing the Square

Objectives

1 Use the square root principle to solve quadratic equations.

2 Solve quadratic equations by completing the square.

Note The expression $\pm\sqrt{a}$ is read "plus or minus the square root of a."

▶

Connection In Section 9.3, we learned how to simplify square roots of numbers that have perfect square factors.

Note The $\pm$ symbol means the two solutions are $2\sqrt{10}$ and $-2\sqrt{10}$. ▶

Answers to Warm-up
1. $\pm7\sqrt{2}$ 2. $\pm3i$
3. $\left(x - \dfrac{7}{2}\right)^2$
4. $x = \dfrac{7 + \sqrt{37}}{2}$

Warm-up *For Exercises 1 and 2, simplify.*
[9.3] **1.** $\pm\sqrt{98}$ [9.7] **2.** $\pm\sqrt{-9}$
[6.4] **3.** Write $x^2 - 7x + \dfrac{49}{4}$ as the square of a binomial.
[2.2] **4.** Solve: $x - \dfrac{7}{2} = \dfrac{\sqrt{37}}{2}$

Objective 1 Use the square root principle to solve quadratic equations.

In Section 6.6, we solved quadratic equations such as $x^2 = 25$ by subtracting 25 from both sides, factoring, and then using the zero-factor theorem. Let's recall that process.

$$x^2 - 25 = 0$$

$(x - 5)(x + 5) = 0$ Factor.

$x - 5 = 0$ or $x + 5 = 0$ Use the zero-factor theorem.

$x = 5$ $x = -5$ Solve for x in each equation.

Another approach to solving $x^2 = 25$ involves square roots. Notice that the solutions to this equation must be numbers that can be squared to equal 25. Those numbers are the square roots of 25, which are 5 and -5. This suggests a new rule called the *square root principle.*

Rule Square Root Principle

If $x^2 = a$, where a is a real number, then $x = \sqrt{a}$ or $x = -\sqrt{a}$.
It is common to indicate the positive and negative solutions by writing $\pm\sqrt{a}$.

For example, if $x^2 = 64$, then $x = \sqrt{64} = 8$ or $x = -\sqrt{64} = -8$. Or we could simply write $x = \pm\sqrt{64} = \pm8$.

Solve Equations in the Form $x^2 = a$

The square root principle is especially useful for solving equations in the form $x^2 = a$ when a is not a perfect square.

Example 1 Solve.

a. $x^2 = 40$

Solution: $x^2 = 40$

$x = \pm\sqrt{40}$ Use the square root principle.

$x = \pm\sqrt{4 \cdot 10}$ Simplify by factoring out a perfect square.

$x = \pm2\sqrt{10}$

We can check the two solutions using the original equation.

Check: $(2\sqrt{10})^2 \overset{?}{=} 40$ Replace x with $2\sqrt{10}$.   **Check:** $(-2\sqrt{10})^2 \overset{?}{=} 40$ Replace x with $-2\sqrt{10}$.

$4 \cdot 10 = 40$ True $4 \cdot 10 = 40$ True

For the remaining examples, the checks will be left to the reader.

Note In this chapter, complex number solutions are allowed. If solutions are to be restricted to real numbers only, we say, "Solve over the real numbers."

b. $x^2 = -9$

Solution: $x^2 = -9$

$$x = \pm\sqrt{-9} \qquad \text{Use the square root principle.}$$
$$x = \pm 3i \qquad \text{Write the imaginary number using } i.$$

Your Turn 1 Solve.

a. $x^2 = 50$ **b.** $x^2 = -100$

Solve Equations in the Form $ax^2 + b = c$

If an equation is in the form $ax^2 + b = c$, we use the addition and multiplication principles of equality to isolate x^2 and then use the square root principle.

Example 2 Solve.

a. $x^2 + 2 = 100$

Solution: $x^2 + 2 = 100$

$$x^2 = 98 \qquad \text{Subtract 2 from both sides to isolate } x^2.$$
$$x = \pm\sqrt{98} \qquad \text{Use the square root principle.}$$
$$x = \pm\sqrt{49 \cdot 2} \qquad \text{Factor out a perfect square.}$$
$$x = \pm 7\sqrt{2} \qquad \text{Simplify using } \sqrt{ab} = \sqrt{a} \cdot \sqrt{b}.$$

b. $5x^2 + 2 = 62$

Solution: $5x^2 + 2 = 62$

$$5x^2 = 60 \qquad \text{Subtract 2 from both sides.}$$
$$x^2 = 12 \qquad \text{Divide both sides by 5.}$$
$$x = \pm\sqrt{12} \qquad \text{Use the square root principle.}$$
$$x = \pm\sqrt{4 \cdot 3} \qquad \text{Factor out a perfect square.}$$
$$x = \pm 2\sqrt{3} \qquad \text{Simplify using } \sqrt{ab} = \sqrt{a} \cdot \sqrt{b}.$$

Your Turn 2 Solve.

a. $x^2 - 6 = 42$ **b.** $2x^2 + 11 = 65$

Solve Equations in the Form $(ax + b)^2 = c$

In an equation in the form $(ax + b)^2 = c$, notice that the expression $ax + b$ is squared. We can use the square root principle to eliminate the square by thinking of this form as follows:

$$\text{If } (ax + b)^2 = c, \text{ then } ax + b = \pm\sqrt{c}.$$

Example 3 Solve.

a. $(x + 6)^2 = 49$

Solution: $(x + 6)^2 = 49$

$$x + 6 = \pm\sqrt{49} \qquad \text{Use the square root principle.}$$
$$x + 6 = \pm 7 \qquad \text{Simplify.}$$
$$x = -6 \pm 7 \qquad \text{Subtract 6 from both sides.}$$
$$x = -6 + 7 \quad \text{or} \quad x = -6 - 7 \qquad \text{Simplify by separating the two solutions.}$$
$$x = 1 \qquad\qquad\qquad x = -13$$

Answers to Your Turn 1
a. $\pm 5\sqrt{2}$ **b.** $\pm 10i$

Answers to Your Turn 2
a. $\pm 4\sqrt{3}$ **b.** $\pm 3\sqrt{3}$

b. $(5x - 1)^2 = 10$

Solution: $(5x - 1)^2 = 10$

$$5x - 1 = \pm \sqrt{10} \qquad \text{Use the square root principle.}$$

$$5x = 1 \pm \sqrt{10} \qquad \text{Add 1 to both sides to isolate } 5x.$$

$$x = \frac{1 \pm \sqrt{10}}{5} \qquad \text{Divide both sides by 5 to solve for } x.$$

c. $(8x + 6)^2 = -32$

Solution: $(8x + 6)^2 = -32$

$$8x + 6 = \pm \sqrt{-32} \qquad \text{Use the square root principle.}$$

$$8x = -6 \pm \sqrt{-16 \cdot 2} \qquad \text{Subtract 6 from both sides and simplify the square root.}$$

$$8x = -6 \pm 4i\sqrt{2} \qquad \text{Rewrite the imaginary number using } i \text{ notation.}$$

$$x = \frac{-6 \pm 4i\sqrt{2}}{8} \qquad \text{Divide both sides by 8 to solve for } x.$$

$$x = -\frac{6}{8} \pm \frac{4\sqrt{2}}{8}i \qquad \text{Write the complex number in standard form } a + bi.$$

$$x = -\frac{3}{4} \pm \frac{\sqrt{2}}{2}i \qquad \text{Simplify.}$$

Your Turn 3 Solve.

a. $(x - 3)^2 = 25$ **b.** $(2x + 1)^2 = 14$ **c.** $(4x - 5)^2 = -12$

Objective 2 Solve quadratic equations by completing the square.

To make use of the square root principle, we need one side of the equation to be a perfect square and the other side to be a constant, as in $(x + 2)^2 = 5$. But suppose we are given an equation such as $x^2 + 6x = 2$, whose left-hand side is an "incomplete" square. We can use the addition principle of equality to add an appropriate number (9 in this case) to both sides so that the left-hand side becomes a perfect square. We call this process *completing the square*.

$$x^2 + 6x + 9 = 2 + 9 \qquad \text{Adding 9 to both sides completes the square on the left side.}$$

We can now factor the left-hand side of the equation, then use the square root principle to finish solving the equation.

$$(x + 3)^2 = 11 \qquad \text{Write the left-hand side in factored form.}$$

$$x + 3 = \pm \sqrt{11} \qquad \text{Use the square root principle to eliminate the square.}$$

$$x = -3 \pm \sqrt{11} \qquad \text{Subtract 3 from both sides to isolate } x.$$

Connection The product of every perfect square in the form $(x + b)^2$ is a trinomial in the form $x^2 + 2bx + b^2$. Notice that half of x's coefficient, $2b$, is b, which is then squared to equal the last term.

But how do we determine the number to add to complete the square? That number is the square of half of the coefficient of x.

$$x^2 + 6x + 9$$

Half of 6 is 3, and 3 squared is 9.

$$\left(\frac{6}{2}\right)^2 = (3)^2$$

Also notice that half of the coefficient of x is the constant in the factored form.

$$x^2 + 6x + 9 = (x + 3)^2$$

Half of 6 is 3, which is the constant in the factored form.

$$\left(\frac{6}{2}\right)^2 = (3)^2$$

Answers to Your Turn 3
a. $8, -2$
b. $\dfrac{-1 \pm \sqrt{14}}{2}$ **c.** $\dfrac{5}{4} \pm \dfrac{\sqrt{3}}{2}i$

Equations in the Form $x^2 + bx = c$

To solve a quadratic equation by completing the square, we need the equation in the form $x^2 + bx = c$ so that the coefficient of x^2 is 1. Once the equation is in the form $x^2 + bx = c$, we complete the square and then use the square root principle.

| **Example 4** | Solve by completing the square. |

a. $x^2 + 12x + 15 = 0$

Solution: We first write the equation in the form $x^2 + bx = c$.

$$x^2 + 12x = -15 \qquad \text{Subtract 15 from both sides to get the form } x^2 + bx = c.$$

$$x^2 + 12x + 36 = -15 + 36 \qquad \text{Complete the square by adding 36 to both sides.}$$

$$(x + 6)^2 = 21 \qquad \text{Factor the left side and simplify the right.}$$

$$x + 6 = \pm\sqrt{21} \qquad \text{Use the square root principle.}$$

$$x = -6 \pm \sqrt{21} \qquad \text{Subtract 6 from both sides to isolate } x.$$

Note We found 36 by squaring half of 12.

$$\left(\frac{12}{2}\right)^2 = 6^2 = 36$$

b. $x^2 - 7x + 8 = 5$

Solution: $x^2 - 7x = -3 \qquad \text{Subtract 8 from both sides to get the form } x^2 + bx = c.$

$$x^2 - 7x + \frac{49}{4} = -3 + \frac{49}{4} \qquad \text{Complete the square by adding } \frac{49}{4} \text{ to both sides.}$$

$$\left(x - \frac{7}{2}\right)^2 = \frac{37}{4} \qquad \text{Factor the left side and simplify the right side.}$$

$$x - \frac{7}{2} = \pm\sqrt{\frac{37}{4}} \qquad \text{Use the square root principle.}$$

$$x = \frac{7}{2} \pm \sqrt{\frac{37}{4}} \qquad \text{Add } \frac{7}{2} \text{ to both sides to isolate } x.$$

$$x = \frac{7}{2} \pm \frac{\sqrt{37}}{2} \qquad \text{Simplify the square root.}$$

$$x = \frac{7 \pm \sqrt{37}}{2} \qquad \text{Combine the fractions.}$$

Note We found $\frac{49}{4}$ by squaring half of -7.

$$\left(\frac{-7}{2}\right)^2 = \frac{49}{4}$$

| **Your Turn 4** | Solve by completing the square. |

a. $x^2 + 8x - 29 = 0$ **b.** $x^2 - 9x - 6 = 5$

Equations in the Form $ax^2 + bx = c$, Where $a \neq 1$

Note Our procedure for completing the square does not work unless the coefficient of the squared term is 1.

So far, our equations have been in the form $x^2 + bx = c$, where the coefficient of the x^2 term is already 1. To solve an equation such as $2x^2 + 12x = 3$, we need to divide both sides of the equation by 2 $\left(\text{or multiply both sides by } \frac{1}{2}\right)$ so that the coefficient of x^2 becomes 1.

$$\frac{2x^2 + 12x}{2} = \frac{3}{2} \qquad \text{Divide both sides by 2.}$$

$$x^2 + 6x = \frac{3}{2} \qquad \text{Simplify.}$$

We can now solve by completing the square.

$$x^2 + 6x + 9 = \frac{3}{2} + 9 \qquad \text{Add 9 to both sides to complete the square.}$$

$$(x + 3)^2 = \frac{21}{2} \qquad \text{Factor the left side and simplify the right side.}$$

Answers to Your Turn 4

a. $-4 \pm 3\sqrt{5}$ **b.** $\dfrac{9 \pm 5\sqrt{5}}{2}$

$$x + 3 = \pm\sqrt{\frac{21}{2}}$$ Use the square root principle.

$$x = -3 \pm \sqrt{\frac{21}{2}}$$ Subtract 3 from both sides to isolate x.

$$x = -3 \pm \frac{\sqrt{21}}{\sqrt{2}} \cdot \frac{\sqrt{2}}{\sqrt{2}}$$ Rationalize the denominator.

$$x = -3 \pm \frac{\sqrt{42}}{2}$$

Note The answer $-3 \pm \dfrac{\sqrt{42}}{2}$ can also be written as $\dfrac{-6 \pm \sqrt{42}}{2}$.

We can now write a procedure for solving any quadratic equation by completing the square.

Connection In Section 10.4, we use the process of completing the square to rewrite quadratic functions in a form that allows us to easily determine features of the graph.

Procedure **Solving Quadratic Equations by Completing the Square**

To solve a quadratic equation by completing the square:
1. Write the equation in the form $x^2 + bx = c$.
2. Complete the square by adding $\left(\dfrac{b}{2}\right)^2$ to both sides.
3. Write the completed square in factored form and simplify the right side, $c + \left(\dfrac{b}{2}\right)^2$.
4. Use the square root principle to eliminate the square.
5. Isolate the variable.
6. Simplify as needed.

Example 5 Solve by completing the square.

a. $2x^2 - 8 = 5x$

Solution:

$$2x^2 - 5x = 8$$ Rewrite in the form $ax^2 + bx = c$.

$$\frac{2x^2 - 5x}{2} = \frac{8}{2}$$ Divide both sides by 2.

$$x^2 - \frac{5}{2}x = 4$$ Simplify.

$$x^2 - \frac{5}{2}x + \frac{25}{16} = 4 + \frac{25}{16}$$ Add $\frac{25}{16}$ to both sides to complete the square.

$$\left(x - \frac{5}{4}\right)^2 = \frac{89}{16}$$ Factor the left side and simplify the right side.

$$x - \frac{5}{4} = \pm\sqrt{\frac{89}{16}}$$ Use the square root principle.

$$x = \frac{5}{4} \pm \frac{\sqrt{89}}{4}$$ Add $\frac{5}{4}$ to both sides and simplify the square root.

$$x = \frac{5 \pm \sqrt{89}}{4}$$ Combine like fractions.

Note To complete the square, square half of $-\dfrac{5}{2}$.

$$\left(\frac{1}{2} \cdot -\frac{5}{2}\right)^2 = \left(-\frac{5}{4}\right)^2 = \frac{25}{16}$$

b. $4x^2 - 8x + 7 = 2$

Solution: $4x^2 - 8x = -5$ Rewrite in the form $ax^2 + bx = c$.

$$\frac{4x^2 - 8x}{4} = -\frac{5}{4}$$ Divide both sides by 4.

$$x^2 - 2x = -\frac{5}{4}$$ Simplify.

$$x^2 - 2x + 1 = -\frac{5}{4} + 1$$ Add 1 to both sides to complete the square.

$$(x - 1)^2 = -\frac{1}{4}$$ Factor the left side and simplify the right side.

$$x - 1 = \pm\sqrt{-\frac{1}{4}}$$ Use the square root principle.

$$x = 1 \pm \frac{1}{2}i$$ Add 1 to both sides and simplify the square root.

Answers to Your Turn 5

a. $\dfrac{-3 \pm 2\sqrt{2}}{3}$ or $-1 \pm \dfrac{2\sqrt{2}}{3}$

b. $4 \pm 3i$

Your Turn 5   Solve by completing the square.

a. $9x^2 + 18x = -1$ **b.** $x^2 - 8x + 11 = -14$

10.1 Exercises   For Extra Help   MyMathLab®

Note: Exercises marked with a ★ represent challenging exercises.

Objective 1

Prep Exercise 1 Explain why there are two solutions to an equation in the form $x^2 = a$.

$x^2 = a$ has two solutions because squaring $\sqrt{a}$ and $-\sqrt{a}$ gives a for every real number a.

For Exercises 1–10, solve and check. See Example 1.

1. $x^2 = 49$
± 7

2. $x^2 = 144$
± 12

3. $y^2 = \dfrac{4}{25}$
$\pm\dfrac{2}{5}$

4. $t^2 = \dfrac{9}{49}$
$\pm\dfrac{3}{7}$

5. $n^2 = 1.44$
± 1.2

6. $p^2 = 1.21$
± 1.1

7. $z^2 = 45$
$\pm 3\sqrt{5}$

8. $m^2 = 72$
$\pm 6\sqrt{2}$

9. $w^2 = -25$
$\pm 5i$

10. $c^2 = -49$
$\pm 7i$

Prep Exercise 2 Write a formula for the solutions of $ax^2 - b = c$ by solving for x.

$x = \pm\sqrt{\dfrac{b + c}{a}}$

For Exercises 11–30, solve and check. Begin by using the addition or multiplication principles of equality to isolate the squared term. See Example 2.

11. $n^2 - 7 = 42$
± 7

12. $y^2 - 5 = 59$
± 8

13. $y^2 - 16 = 65$
± 9

14. $k^2 + 5 = 30$
± 5

15. $4n^2 = 36$
± 3

16. $5y^2 = 125$
± 5

17. $25t^2 = 9$
$\pm\dfrac{3}{5}$

18. $16d^2 = 49$
$\pm\dfrac{7}{4}$

19. $4h^2 = -16$
$\pm 2i$

20. $3k^2 = -108$
$\pm 6i$

21. $\dfrac{5}{6}x^2 = \dfrac{24}{5}$
$\pm\dfrac{12}{5}$

22. $-\dfrac{2}{3}m^2 = -\dfrac{27}{32}$
$\pm\dfrac{9}{8}$

23. $2x^2 + 5 = 21$
$\pm 2\sqrt{2}$

24. $4x^2 - 11 = 97$
$\pm 3\sqrt{3}$

25. $5y^2 - 7 = -97$
$\pm 3i\sqrt{2}$

26. $4n^2 + 20 = -76$
$\pm 2i\sqrt{6}$

27. $\frac{3}{4}y^2 - 5 = 3$
$\pm \frac{4\sqrt{6}}{3}$

28. $\frac{25}{9}m^2 - 2 = 6$
$\pm \frac{6\sqrt{2}}{5}$

29. $0.2t^2 - 0.5 = 0.012$
± 1.6

30. $0.5p^2 + 1.28 = 1.6$
± 0.8

Prep Exercise 3 Write a formula for the solutions of $(ax - b)^2 = c$ by solving for x.

$x = \frac{b \pm \sqrt{c}}{a}$

For Exercises 31–46, solve and check. Use the square root principle to eliminate the square. See Example 3.

31. $(x + 8)^2 = 49$
$-15, -1$

32. $(y + 3)^2 = 36$
$-9, 3$

33. $(5n - 3)^2 = 16$
$-\frac{1}{5}, \frac{7}{5}$

34. $(6h - 5)^2 = 81$
$-\frac{2}{3}, \frac{7}{3}$

35. $(m - 8)^2 = -16$
$8 \pm 4i$

36. $(t - 2)^2 = -4$
$2 \pm 2i$

37. $(4k - 1)^2 = 40$
$\frac{1 \pm 2\sqrt{10}}{4}$

38. $(3x - 4)^2 = 80$
$\frac{4 \pm 4\sqrt{5}}{3}$

39. $(m - 7)^2 = -12$
$7 \pm 2i\sqrt{3}$

40. $(t - 5)^2 = -28$
$5 \pm 2i\sqrt{7}$

41. $\left(y - \frac{3}{4}\right)^2 = \frac{9}{16}$
$0, \frac{3}{2}$

42. $\left(x + \frac{4}{9}\right)^2 = \frac{25}{81}$
$-1, \frac{1}{9}$

43. $\left(\frac{5}{9}d - \frac{1}{2}\right)^2 = \frac{1}{36}$
$\frac{3}{5}, \frac{6}{5}$

44. $\left(\frac{3}{4}h + \frac{4}{5}\right)^2 = \frac{1}{100}$
$-\frac{14}{15}, -\frac{6}{5}$

45. $(0.4x + 3.8)^2 = 2.56$
$-13.5, -5.5$

46. $(0.8n - 6.8)^2 = 1.96$
$6.75, 10.25$

Find ⊗ **the Mistake** *For Exercises 47–50, explain the mistake; then find the correct solutions.*

47. $x^2 - 15 = 34$
$x^2 = 49$
$x = \sqrt{49}$
$x = 7$
Mistake: Gave only the positive solution
Correct: ± 7

48. $x^2 = 20$
$x = \sqrt{20}$
$x = 2\sqrt{5}$
Mistake: Gave only the positive solution
Correct: $\pm 2\sqrt{5}$

49. $(x - 5)^2 = -6$
$x - 5 = \pm\sqrt{6}$
$x = 5 \pm \sqrt{6}$
Mistake: Changed -6 to 6
Correct: $5 \pm i\sqrt{6}$

50. $(x - 1)^2 = -12$
$x - 1 = \pm\sqrt{-12}$
$x = 1 \pm 2\sqrt{3}$
Mistake: Square root of a negative gives an i value.
Correct: $1 \pm 2i\sqrt{3}$

▤ *For Exercises 51–56, solve; then use a calculator to approximate the irrational solutions rounded to three places.*

51. $x^2 = 96$
$\pm 4\sqrt{6} \approx \pm 9.798$

52. $t^2 = 56$
$\pm 2\sqrt{14} \approx \pm 7.483$

53. $y^2 - 15 = 5$
$\pm 2\sqrt{5} \approx \pm 4.472$

54. $x^2 - 13 = 35$
$\pm 4\sqrt{3} \approx 6.928$

55. $(n - 6)^2 = 15$
$6 \pm \sqrt{15} \approx 9.873, 2.127$

56. $(m + 3)^2 = 10$
$-3 \pm \sqrt{10} \approx 0.162, -6.162$

Objective 2

Prep Exercise 4 Given an expression of the form $x^2 + bx$, what must be added to complete the square?

$\left(\frac{b}{2}\right)^2$

Prep Exercise 5 If $x^2 + bx + c$ is a perfect square trinomial, what is its factored form?

$\left(x + \frac{b}{2}\right)^2$

For Exercises 57–64:   **a.** Add a term to the expression to make it a perfect square.
              b. Factor the perfect square. See Objective 2.

57. $x^2 + 14x$
a. $x^2 + 14x + 49$
b. $(x + 7)^2$

58. $c^2 + 8c$
a. $c^2 + 8c + 16$
b. $(c + 4)^2$

59. $n^2 - 10n$
a. $n^2 - 10n + 25$
b. $(n - 5)^2$

60. $a^2 - 12a$
a. $a^2 - 12a + 36$
b. $(a - 6)^2$

61. $y^2 - 9y$

  a. $y^2 - 9y + \dfrac{81}{4}$

  b. $\left(y - \dfrac{9}{2}\right)^2$

62. $m^2 - 11m$

  a. $m^2 - 11m + \dfrac{121}{4}$

  b. $\left(m - \dfrac{11}{2}\right)^2$

63. $s^2 - \dfrac{2}{3}s$

  a. $s^2 - \dfrac{2}{3}s + \dfrac{1}{9}$

  b. $\left(s - \dfrac{1}{3}\right)^2$

64. $y^2 - \dfrac{4}{5}y$

  a. $y^2 - \dfrac{4}{5}y + \dfrac{4}{25}$

  b. $\left(y - \dfrac{2}{5}\right)^2$

Prep Exercise 6 Given an equation in the form $x^2 + bx = c$, explain how to complete the square.

Add $\left(\dfrac{b}{2}\right)^2$ to both sides of the equation.

Prep Exercise 7 Consider the equation $x^2 - 7x + 12 = 0$. Which is a better method for solving the equation, factoring and then using the zero-factor theorem or completing the square? Explain. Because $x^2 - 7x + 12$ is easy to factor, factoring is a better method as it requires fewer steps than completing the square.

Prep Exercise 8 Consider the equation $x^2 + 4x + 5 = 0$. Which is a better method for solving the equation, factoring and then using the zero-factor theorem or completing the square? Explain.

Because $x^2 + 4x + 5$ cannot be factored, completing the square is the only option at this point.

For Exercises 65–76, solve the equation by completing the square. See Example 4.

65. $w^2 + 2w = 15$

  $-5, 3$

66. $p^2 + 8p = 9$

  $-9, 1$

67. $r^2 - 2r + 50 = 0$

  $1 \pm 7i$

68. $c^2 - 6c + 45 = 0$

  $3 \pm 6i$

69. $k^2 = 9k - 18$

  $3, 6$

70. $a^2 = 6a - 8$

  $2, 4$

71. $b^2 - 2b - 11 = 5$

  $1 \pm \sqrt{17}$

72. $x^2 + 2x + 7 = 9$

  $-1 \pm \sqrt{3}$

73. $h^2 - 6h + 3 = -26$

  $3 \pm 2i\sqrt{5}$

74. $j^2 - 4j + 25 = -3$

  $2 \pm 2i\sqrt{6}$

75. $u^2 + \dfrac{1}{2}u = \dfrac{3}{2}$

  $-\dfrac{3}{2}, 1$

76. $y^2 + \dfrac{1}{3}y = \dfrac{2}{3}$

  $-1, \dfrac{2}{3}$

For Exercises 77–88, solve the equation by completing the square. Begin by writing the equation in the form $x^2 + bx = c$. See Example 5.

77. $4x^2 + 16x = 9$

  $-\dfrac{9}{2}, \dfrac{1}{2}$

78. $4m^2 - 8m = 5$

  $-\dfrac{1}{2}, \dfrac{5}{2}$

79. $9x^2 - 18x + 5 = 0$

  $\dfrac{1}{3}, \dfrac{5}{3}$

80. $16x^2 - 32x + 7 = 0$

  $\dfrac{1}{4}, \dfrac{7}{4}$

81. $2n^2 - n - 3 = 0$

  $-1, \dfrac{3}{2}$

82. $2x^2 - 5x - 12 = 0$

  $4, -\dfrac{3}{2}$

83. $4x^2 = -16x + 7$

  $-2 \pm \dfrac{\sqrt{23}}{2}$

84. $4x^2 = 24x + 11$

  $3 \pm \dfrac{\sqrt{47}}{2}$

85. $5k^2 + k - 2 = 0$

  $\dfrac{-1 \pm \sqrt{41}}{10}$

86. $3s^2 - 4s = 2$

  $\dfrac{2 \pm \sqrt{10}}{3}$

87. $3a^2 - 8 = -4a$

  $\dfrac{-2 \pm 2\sqrt{7}}{3}$

88. $3x^2 - 4 = -8x$

  $\dfrac{-4 \pm 2\sqrt{7}}{3}$

Find ⊗ the Mistake *For Exercises 89 and 90, explain the mistake. Find the correct solution.*

89.

$3x^2 + 4x = 2$

$3x^2 + 4x + 4 = 2 + 4$

$(3x + 2)^2 = 6$

$3x + 2 = \pm\sqrt{6}$

$3x = -2 \pm \sqrt{6}$

$x = \dfrac{-2 \pm \sqrt{6}}{3}$

Did not divide by 3 so that x^2 has a coefficient of 1; then wrote an incorrect factored form

Correct: $\dfrac{-2 \pm \sqrt{10}}{3}$

90.

$x^2 + 6x = 7$

$x^2 + 6x + 9 = 7 + 9$

$(x + 3)^2 = 16$

$x + 3 = \sqrt{16}$

$x + 3 = 4$

$x = -3 + 4$

$x = 1$

Need to use $\pm\sqrt{16}$, thereby producing two answers

Correct: $1, -7$

For Exercises 91–102, solve.

91. A square sheet of metal has an area of 196 square inches. What is the length of each side?

  14 in.

92. A severe thunderstorm warning is issued by the National Weather Service for a square area covering 14,400 square miles. What is the length of each side of the square area?
120 mi.

93. A tank is to be made for an aquarium so that the length is 8 feet, the width is twice the height, and the volume is 144 cubic feet. Find the height and width of the tank.
Height: 3 ft.; width: 6 ft.

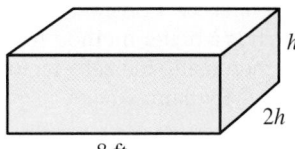

94. The length of a swimming pool is to be three times the width. If the depth is to be a constant 4 feet and the volume is to be 4800 cubic feet, find the length and width.
Length: 60 ft.; width: 20 ft.

95. The Arecibo radio telescope is like a giant satellite dish that covers a circular area of approximately $23{,}256.25\pi$ square meters. Find the diameter of the dish.
305 m

Of Interest

The Arecibo radio telescope, the world's largest radio telescope, analyzes radiation and signals emitted by objects in space. It was built in a natural depression near Arecibo, Puerto Rico.

96. A field is planted in a circular pattern. A watering device is to be constructed with pipe in a line extending from the center of the field to the edge of the field. The pipe is set on wheels so that it can rotate around the field and cover the entire field with water. If the area of the field is 7225π square feet, how long will the watering pipe be?
85 ft.

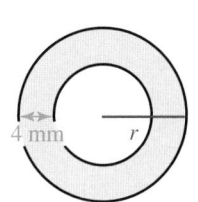

7225π ft.²

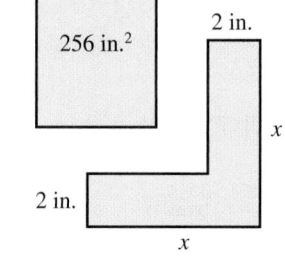

Removed piece

256 in.² 2 in.

2 in.

x

x

97. The corner of a sheet of plywood has been cut out as shown. If the area of the piece that was removed was 256 square inches, find x.
18 in.

98. The area of the hole in the washer shown is 25π square millimeters. Find r, the radius of the washer.
9 mm

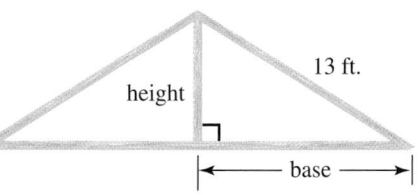

4 mm   r

99. An LCD computer monitor measures 20 inches across its diagonal. If the width is 4 inches less than the length, find the dimensions of the monitor.
16 ft. by 12 ft.

13 ft.

height

base

100. Two identical right triangles are placed together to form the frame of a roof. If the base of each triangle is 7 feet longer than its height and its hypotenuse is 13 feet, what are the dimensions of the base and height?
Base: 12 ft.; height: 5 ft.

★ **101.** A plastic panel is to have a rectangular hole cut as shown.

   a. Find l so that the area remaining after the hole is cut is 1230 square centimeters.
   15 cm

   b. Find the length and width of the plastic panel.
   Length: 45 cm; width: 30 cm

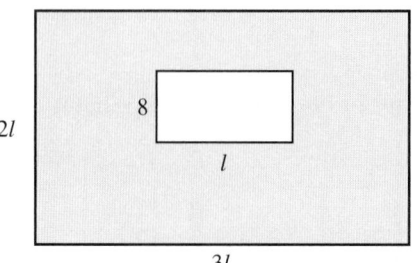

2l

8

l

3l

★ **102.** A 6-inch-wide groove with a height of h is to be cut into a wood block as shown.

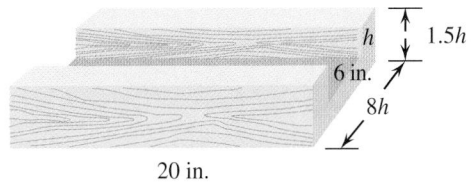

a. Find h so that the volume remaining in the block after the groove is cut is 360 cubic inches.
   1.5 in.

b. Find the height and width of the block.
   Height: 2.25 in.; width: 12 in.

For Exercises 103 and 104, if an object is dropped, the formula $d = 16t^2$ describes the distance d in feet that the object falls in t seconds.

103. Suppose a cover for a ceiling light falls from a 9-foot ceiling. How long does the cover take to hit the floor?
   $\frac{3}{4}$ sec.

104. A construction worker tosses a scrap piece of lumber from the roof of a house. How long does it take the piece of lumber to reach the ground 25 feet below?
   $\frac{5}{4}$ sec.

For Exercises 105 and 106, use the following information. In physics, when an object is in motion, it has kinetic energy. The formula $E = \dfrac{1}{2}mv^2$ is used to calculate the kinetic energy E of an object with a mass m and velocity v. If the mass is measured in kilograms and the velocity is in meters per second, the kinetic energy will be in units called joules (J).

105. Suppose an object with a mass of 50 kilograms has 400 joules of kinetic energy. Find its velocity.
   4 m/sec.

106. In a crash test, a vehicle with a mass of 1200 kilograms is found to have kinetic energy of 117,600 joules just before impact. Find the velocity of the vehicle just before impact.
   14 m/sec.

Puzzle Problem Without using a calculator, which of the following four numbers is a perfect square? (*Hint:* Write a list of smaller perfect squares and look for a pattern.)

9,456,804,219,745,618
2,512,339,789,576,516
7,602,985,471,286,543
4,682,715,204,643,182

The ones digit of every perfect square is 0, 1, 4, 5, 6, or 9; so 2,512,339,789,576,516 is the perfect square.

Review Exercises

Exercises 1–6 Expressions

[6.4] *For Exercises 1 and 2, factor.*

1. $x^2 - 10x + 25$
   $(x - 5)^2$

2. $x^2 + 6x + 9$
   $(x + 3)^2$

[9.1] *For Exercises 3 and 4, simplify. Assume that variables represent nonnegative values.*

3. $\sqrt{36x^2}$
   $6x$

4. $\sqrt{(2x + 3)^2}$
   $2x + 3$

[1.5, 9.1, 9.7] *For Exercises 5 and 6, simplify.*

5. $\dfrac{-4 + \sqrt{8^2 - 4(2)(5)}}{2(2)}$
   $\dfrac{-2 + \sqrt{6}}{2}$ or $-1 + \dfrac{\sqrt{6}}{2}$

6. $\dfrac{-6 - \sqrt{6^2 - 4(5)(2)}}{2(5)}$
   $\dfrac{-3 - i}{5}$ or $-\dfrac{3}{5} - \dfrac{1}{5}i$

10.2 Solving Quadratic Equations Using the Quadratic Formula

Objectives

1 Solve quadratic equations using the quadratic formula.

2 Use the discriminant to determine the number of real solutions that a quadratic equation has.

3 Find the x- and y-intercepts of a quadratic function.

4 Solve applications using the quadratic formula.

Warm-up

[10.1] **1.** Solve by completing the square: $3x^2 + 4x - 2 = 0$

[1.7] **2.** Evaluate $b^2 - 4ac$ for $a = 0.5$, $b = -0.8$, and $c = 2.5$.

[1.5, 9.3] **3.** Simplify: $\dfrac{-8 \pm \sqrt{8^2 - 4(1)(-29)}}{2(1)}$

In this section, we solve quadratic equations using a formula called the *quadratic formula*. Using this formula is much easier than completing the square.

Objective 1 Solve quadratic equations using the quadratic formula.

To derive the quadratic formula, we begin with the general form of the quadratic equation, $ax^2 + bx + c = 0$, and assume that $a > 0$. We follow the procedure for solving a quadratic equation by completing the square.

$$ax^2 + bx + c = 0$$

$$ax^2 + bx = -c \qquad \text{Subtract } c \text{ from both sides.}$$

$$x^2 + \frac{b}{a}x = -\frac{c}{a} \qquad \text{Divide both sides by } a \text{ so that the coefficient of } x^2 \text{ is 1.}$$

$$x^2 + \frac{b}{a}x + \frac{b^2}{4a^2} = -\frac{c}{a} + \frac{b^2}{4a^2} \qquad \text{Complete the square.}$$

$$\left(x + \frac{b}{2a}\right)^2 = -\frac{4ac}{4a^2} + \frac{b^2}{4a^2} \qquad \text{Factor on the left side. On the right side, write } -\frac{c}{a} \text{ with the LCD } 4a^2 \text{ to add the rational expressions.}$$

$$\left(x + \frac{b}{2a}\right)^2 = \frac{b^2 - 4ac}{4a^2} \qquad \text{Add the rational expressions. We rearranged } b^2 \text{ and } -4ac \text{ in the numerator.}$$

$$x + \frac{b}{2a} = \pm\sqrt{\frac{b^2 - 4ac}{4a^2}} \qquad \text{Use the square root principle to eliminate the square.}$$

$$x + \frac{b}{2a} = \pm\frac{\sqrt{b^2 - 4ac}}{2a} \qquad \text{Simplify the square root in the denominator.}$$

$$x = -\frac{b}{2a} \pm \frac{\sqrt{b^2 - 4ac}}{2a} \qquad \text{Subtract } \frac{b}{2a} \text{ from both sides to isolate } x.$$

$$x = \frac{-b \pm \sqrt{b^2 - 4ac}}{2a} \qquad \text{Combine the rational expressions.}$$

This final equation is the *quadratic formula*. It can be used to solve any quadratic equation simply by replacing a, b, and c with the corresponding values from the given equation.

> **Note** To complete the square, square half of $\dfrac{b}{a}$.
>
> $$\left(\frac{1}{2} \cdot \frac{b}{a}\right)^2 = \left(\frac{b}{2a}\right)^2$$
> $$= \frac{b^2}{4a^2}$$

> **Note** The result is the same if $a < 0$.

> **Note** A quadratic equation must be in the form $ax^2 + bx + c = 0$ so that a, b, and c can be identified for use in the quadratic formula.

Answers to Warm-up

1. $\dfrac{-2 \pm \sqrt{10}}{3}$

2. -4.36 3. $-4 \pm 3\sqrt{5}$

Procedure Using the Quadratic Formula

To solve a quadratic equation in the form $ax^2 + bx + c = 0$, where $a \neq 0$, use the quadratic formula:

$$x = \frac{-b \pm \sqrt{b^2 - 4ac}}{2a}$$

Example 1   Solve.

a. $3x^2 + 10x - 8 = 0$

Solution: This equation is in the form $ax^2 + bx + c = 0$, where $a = 3$, $b = 10$, and $c = -8$. So we can use the quadratic formula, $x = \dfrac{-b \pm \sqrt{b^2 - 4ac}}{2a}$.

$$x = \frac{-10 \pm \sqrt{10^2 - 4(3)(-8)}}{2(3)}$$ Replace a with 3, b with 10, and c with -8.

$$x = \frac{-10 \pm \sqrt{196}}{6}$$ Simplify in the radical and the denominator.

$$x = \frac{-10 \pm 14}{6}$$ Evaluate $\sqrt{196}$.

$$x = \frac{-10 + 14}{6} \quad \text{or} \quad x = \frac{-10 - 14}{6}$$ Split up the $\pm$ to calculate the two solutions.

$$x = \frac{4}{6} \qquad\qquad x = \frac{-24}{6}$$

$$x = \frac{2}{3} \qquad\qquad x = -4$$ Simplify to lowest terms.

Connection Because the radicand 196 is a perfect square, the two solutions are rational numbers. As we consider more examples in this section, note how the radicand determines the type of solutions.

b. $x^2 + 8x - 15 = 14$

Solution: $x^2 + 8x - 29 = 0$ Subtract 14 from both sides to get the form $ax^2 + bx + c = 0$.

$$x = \frac{-8 \pm \sqrt{8^2 - 4(1)(-29)}}{2(1)}$$ Use the quadratic formula, replacing a with 1, b with 8, and c with -29.

$$x = \frac{-8 \pm \sqrt{180}}{2}$$ Simplify in the radical and the denominator.

$$x = \frac{-8 \pm \sqrt{36 \cdot 5}}{2}$$ Use the product rule of radicals.

$$x = \frac{-8 \pm 6\sqrt{5}}{2}$$ Simplify the radical.

$$x = \frac{-8}{2} \pm \frac{6\sqrt{5}}{2}$$ Separate into two rational expressions to simplify.

$$x = -4 \pm 3\sqrt{5}$$ Simplify by dividing out the 2.

Connection Because the radicand 180 is not a perfect square, the two solutions are irrational numbers.

Warning Always make sure the radical is simplified *before* trying to simplify the fraction.

c. $9x^2 + 4 = -12x$

Solution: $9x^2 + 12x + 4 = 0$ Add $12x$ to both sides to get the form $ax^2 + bx + c = 0$.

$$x = \frac{-12 \pm \sqrt{12^2 - 4(9)(4)}}{2(9)}$$ Use the quadratic formula, replacing a with 9, b with 12, and c with 4.

$$x = \frac{-12 \pm \sqrt{0}}{18}$$ Simplify in the radical and the denominator.

$$x = -\frac{2}{3}$$ Simplify the radical and simplify the fraction to lowest terms.

Connection Because the radicand is 0, this equation has only one solution.

d. $3x^2 + 9 = -8x + 2$

Solution: $3x^2 + 8x + 7 = 0$ Add $8x$ to and subtract 2 from both sides to get the form $ax^2 + bx + c = 0$.

$$x = \frac{-8 \pm \sqrt{8^2 - 4(3)(7)}}{2(3)}$$ Use the quadratic formula, replacing a with 3, b with 8, and c with 7.

$$x = \frac{-8 \pm \sqrt{-20}}{6}$$ Simplify in the radical and the denominator.

Connection The negative radicand causes the two solutions to be nonreal complex numbers.

Note In standard form, the two complex solutions are $-\dfrac{4}{3} + \dfrac{\sqrt{5}}{3}i$ and $-\dfrac{4}{3} - \dfrac{\sqrt{5}}{3}i$.

▶

$$x = \frac{-8 \pm 2i\sqrt{5}}{6}$$ Simplify the radical.

$$x = \frac{-4 \pm i\sqrt{5}}{3}$$ Simplify to lowest terms.

Your Turn 1 Solve using the quadratic formula.

a. $4x^2 - 5x - 6 = 0$ b. $4x^2 = 15 - 2x$

c. $x^2 - 8x + 16 = 0$ d. $x^2 + 8 = 2 - 4x$

Choosing a Method for Solving Quadratic Equations

We have learned several methods for solving quadratic equations. The following table summarizes the methods and conditions that make each method the best choice.

Methods for Solving Quadratic Equations

Method	When the Method Is Beneficial
1. Factoring (Section 6.6)	Use when the quadratic equation can be easily factored.
2. Square root principle (Section 10.1)	Use when the quadratic equation can be easily written in the form $ax^2 = c$ or $(ax + b)^2 = c$.
3. Completing the square (Section 10.1)	Keep in mind that this is rarely the best method, but it is important for future topics.
4. Quadratic formula (Section 10.2)	Use when factoring is not easy or is not possible with integer coefficients.

Objective 2 Use the discriminant to determine the number of real solutions that a quadratic equation has.

In the Connection boxes for Example 1, we pointed out how the radicand in the quadratic formula affects the solutions to a given quadratic equation. The expression $b^2 - 4ac$, which is the radicand, is called the **discriminant**.

Definition **Discriminant:** The discriminant is the radicand, $b^2 - 4ac$, in the quadratic formula.

We use the discriminant to determine the number and type of solutions to a quadratic equation.

Answers to Your Turn 1

a. $2, -\dfrac{3}{4}$ b. $\dfrac{-1 \pm \sqrt{61}}{4}$

c. 4   d. $-2 \pm i\sqrt{2}$

Note When the discriminant is 0, the solution is $\dfrac{-b \pm \sqrt{0}}{2a} = -\dfrac{b}{2a}$. ▶

Procedure **Using the Discriminant**

Given a quadratic equation in the form $ax^2 + bx + c = 0$, where a, b, and c are rational numbers and $a \neq 0$, to determine the number and type of solutions, evaluate the discriminant $b^2 - 4ac$.

If the **discriminant is positive**, the equation has two real-number solutions. The solutions will be rational if the discriminant is a perfect square and irrational otherwise.

If the **discriminant is 0**, the equation has one rational solution.

If the **discriminant is negative**, the equation has two nonreal complex solutions.

Example 2   Use the discriminant to determine the number and type of solutions.

a. $3x^2 - 7x = 8$

Solution: $3x^2 - 7x - 8 = 0$ Write the equation in the form $ax^2 + bx + c = 0$.

$(-7)^2 - 4(3)(-8)$ In $b^2 - 4ac$, replace a with 3, b with -7, and c with -8.

$= 49 + 96$ Warning 145 is the *value of the discriminant*, not a solution for the equation $3x^2 - 7x = 8$.

$= 145$

> **Note** Follow the order of operations when evaluating $b^2 - 4ac$ and note that $b^2 \geq 0$ if b is a real number.

Because the discriminant is positive, this equation has two real-number solutions. Because 145 is not a perfect square, the solutions are irrational.

b. $x^2 = \dfrac{4}{5}x - \dfrac{4}{25}$

Solution: $x^2 - \dfrac{4}{5}x + \dfrac{4}{25} = 0$ Write the equation in the form $ax^2 + bx + c = 0$.

$25 \cdot x^2 - \dfrac{25}{1} \cdot \dfrac{4}{5}x + \dfrac{25}{1} \cdot \dfrac{4}{25} = 25 \cdot 0$ Eliminate the fractions by multiplying both sides of the equation by the LCD, 25. This will make the discriminant easier to evaluate.

$25x^2 - 20x + 4 = 0$

$(-20)^2 - 4(25)(4)$ In $b^2 - 4ac$, replace a with 25, b with -20, and c with 4.

$= 400 - 400$

$= 0$

> **Note** Multiplying both sides of a quadratic equation by a constant does not change the solutions, but does change the value of the discriminant.

> **Note** Because the discriminant is 0, the solution is
> $$-\dfrac{b}{2a} = -\dfrac{(-20)}{2(25)} = \dfrac{20}{50} = \dfrac{2}{5},$$
> which is a rational number.

Because the discriminant is zero, this equation has only one real solution.

c. $0.5x^2 - 0.8x + 2.5 = 0$

Solution: $(-0.8)^2 - 4(0.5)(2.5)$ In $b^2 - 4ac$, replace a with 0.5, b with -0.8, and c with 2.5.

$= 0.64 - 5$

$= -4.36$

> **Note** We could avoid calculations with decimal numbers by multiplying the original equation through by 10 so that it becomes $5x^2 - 8x + 25 = 0$.

Because the discriminant is negative, there are two nonreal complex solutions.

Your Turn 2   Use the discriminant to determine the number and type of solutions for the equation. If the solutions are real, state whether they are rational or irrational.

a. $5x^2 + 8x - 9 = 0$ **b.** $\dfrac{5}{2}x^2 + \dfrac{5}{6} = \dfrac{4}{3}x$ **c.** $x^2 - 0.12x = -0.0036$

Objective 3 Find the *x*- and *y*-intercepts of a quadratic function.

In Chapter 3, we learned that *x*-intercepts are points where a graph intersects the *x*-axis. Because an *x*-intercept is on the *x*-axis, the *y*-coordinate of the point is 0. In Section 6.7, we graphed quadratic functions, which have the form $f(x) = ax^2 + bx + c$ (or $y = ax^2 + bx + c$), and learned that their graphs are parabolas. Notice that when we replace *y* with 0 to find the *x*-intercepts in $y = ax^2 + bx + c$, we have the quadratic equation $ax^2 + bx + c = 0$, which has two, one, or no real-number solutions. As a result, quadratic functions in the form $y = ax^2 + bx + c$ have two, one, or no *x*-intercepts. The following graphs illustrate the possibilities.

Answers to Your Turn 2
a. discriminant $= 244$; two irrational solutions
b. discriminant $= -236$; two nonreal complex solutions
c. discriminant $= 0$; one rational solution

Two x-intercepts: If $0 = ax^2 + bx + c$ has two real-number solutions, which occurs if $b^2 - 4ac > 0$, then the graph of $y = ax^2 + bx + c$ has two x-intercepts and looks like one of the following graphs.

Note In Section 6.7, we noted that if $a > 0$, the parabola opens up and if $a < 0$, the parabola opens down. ▶

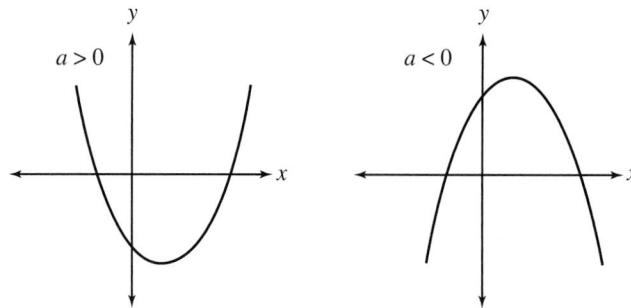

One x-intercept: If $0 = ax^2 + bx + c$ has one real-number solution, which occurs if $b^2 - 4ac = 0$, then the graph of $y = ax^2 + bx + c$ has one x-intercept and looks like one of the following graphs.

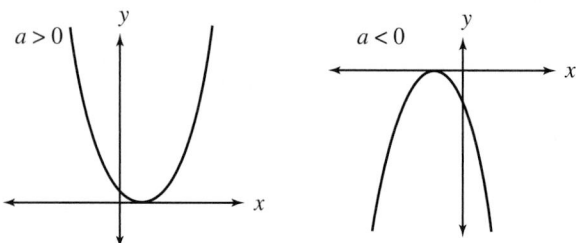

No x-intercepts: If $0 = ax^2 + bx + c$ has no real-number solution, which occurs if $b^2 - 4ac < 0$, then the graph of $y = ax^2 + bx + c$ has no x-intercepts and looks like one of the following graphs.

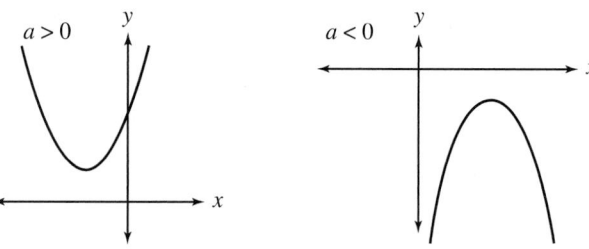

We also have learned that a y-intercept is where a graph intersects the y-axis. To find y-intercepts, we let $x = 0$ and solve for y. Notice that when we replace x with 0 in $y = ax^2 + bx + c$, we have $y = a(0)^2 + b(0) + c = c$; so the y-intercept is $(0, c)$.

Connection We could have solved $0 = x^2 - 2x - 8$ using the quadratic formula.

$$x = \frac{-(-2) \pm \sqrt{(-2)^2 - 4(1)(-8)}}{2(1)}$$

$$= \frac{2 \pm \sqrt{36}}{2} = \frac{2 \pm 6}{2}$$

$$= 1 \pm 3$$

$$= 4 \quad \text{or} \quad -2$$

Note We also could have ▶ calculated the y-intercept:

$$y = (0)^2 + 2(0) - 8 = -8$$

Example 3 Find the x- and y-intercepts of $y = x^2 - 2x - 8$; then graph.

Solution: Find the x-intercepts:

$0 = x^2 - 2x - 8$ Replace y with 0.

$0 = (x - 4)(x + 2)$ Factor.

$x - 4 = 0 \quad \text{or} \quad x + 2 = 0$

$\quad\quad x = 4 \quad\quad\quad\quad\quad x = -2$

x-intercepts: $(4, 0)$ and $(-2, 0)$

y-intercept: $(0, -8)$ The y-intercept is $(0, c)$, and c is -8 in this equation.

Because $a = 1$, which is positive, the graph opens upward. Now we can get a rough sketch of the graph.

Note Although knowing the intercepts and knowing whether the graph opens up or down is helpful, knowing the exact location of the vertex would improve accuracy. We learn how to find the coordinates of the vertex in Section 10.4.

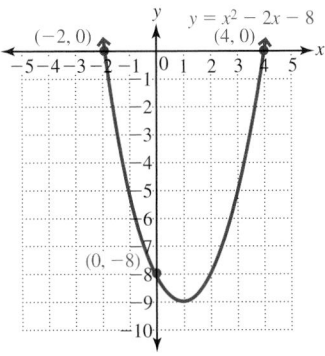

Your Turn 3 Find the x- and y-intercepts. Verify on a graphing calculator.

a. $y = 3x^2 - 5x + 1$ **b.** $y = x^2 - 8x + 16$ **c.** $y = 2x^2 - 3x + 5$

Objective 4 Solve applications using the quadratic formula.

In physics, the general formula for describing the height of an object after it has been thrown upward is $h = \dfrac{1}{2}gt^2 + v_0 t + h_0$, where g represents the acceleration due to gravity, t is the time in flight, v_0 is the initial velocity, and h_0 is the initial height. For Earth, the acceleration due to gravity is -32.2 ft./sec.^{2} or -9.8 m/sec.^{2}

Answers to Your Turn 3
a. x-intercepts:

$$\left(\frac{5 + \sqrt{13}}{6}, 0\right), \left(\frac{5 - \sqrt{13}}{6}, 0\right)$$

y-intercept: $(0, 1)$
b. x-intercept: $(4, 0)$
   y-intercept: $(0, 16)$
c. no x-intercepts
   y-intercept: $(0, 5)$

Answer to Your Turn 4
1.802 sec.

Example 4 In an extreme games competition, a motorcyclist jumps with an initial velocity of 70 feet per second from a ramp height of 25 feet, landing on a ramp with a height of 15 feet. Find the time the motorcyclist is in the air.

Solution:

$15 = \dfrac{1}{2}(-32.2)t^2 + 70t + 25$	In $h = \dfrac{1}{2}gt^2 + v_0 t + h_0$, replace with $h = 15$, $v_0 = 70$, and $h_0 = 25$. Because the units are in feet, we use -32.2 ft./sec.^{2} for g.
$15 = -16.1t^2 + 70t + 25$	Multiply.
$0 = -16.1t^2 + 70t + 10$	Subtract 15 from both sides to get the form $ax^2 + bx + c = 0$.
$x = \dfrac{-70 \pm \sqrt{70^2 - 4(-16.1)(10)}}{2(-16.1)}$	Use the quadratic formula, replacing a with -16.1, b with 70, and c with 10.
$x = \dfrac{-70 \pm \sqrt{5544}}{-32.2}$	Simplify in the radical and the denominator.
$x \approx -0.138$ or 4.486	Approximate the two irrational solutions.

Answer: Because the time cannot be negative, the motorcycle is in the air approximately 4.486 seconds.

Your Turn 4 A ball is thrown from an initial height of 1.5 meters with an initial velocity of 8 meters per second. Find the time for the ball to land on the ground ($h = 0$). Approximate the time to the nearest thousandth.

10.2 Exercises For Extra Help MyMathLab®

Note: Exercises marked with a ★ represent challenging exercises.

Objective 1

Prep Exercise 1 Write the quadratic formula.

$$x = \frac{-b \pm \sqrt{b^2 - 4ac}}{2a}$$

Prep Exercise 2 Discuss the advantages and disadvantages of using the quadratic formula to solve quadratic equations. Answers may vary. One advantage is that it can be used to solve any quadratic equation. One disadvantage is that the steps can be tedious.

Prep Exercise 3 Are there quadratic equations that cannot be solved using factoring? Explain. Yes, if they have irrational or nonreal complex solutions

Prep Exercise 4 Under what conditions will a quadratic equation have only one solution? Modify the quadratic formula to describe this single solution. When the discriminant is zero. The quadratic formula becomes $x = -\dfrac{b}{2a}$.

For Exercises 1–8, rewrite each quadratic equation in the form $ax^2 + bx + c = 0$; then identify a, b, and c. See Objective 1.

1. $x^2 - 3x + 7 = 0$
$a = 1, b = -3, c = 7$

2. $x^2 + 6x - 15 = 0$
$a = 1, b = 6, c = -15$

3. $3x^2 - 9x = 4$
$a = 3, b = -9, c = -4$

4. $2x^2 - 8x = -11$
$a = 2, b = -8, c = 11$

5. $x = 1.5x^2 + 0.2$
$a = 1.5, b = -1, c = 0.2$
or $a = -1.5, b = 1,$
$c = -0.2$

6. $x = 0.8x^2 + 4.5$
$a = 0.8, b = -1, c = 4.5$
or $a = -0.8, b = 1,$
$c = -4.5$

7. $\dfrac{3}{4}x = -\dfrac{1}{2}x^2 + 6$
$a = -\dfrac{1}{2}, b = -\dfrac{3}{4}, c = 6$
or $a = \dfrac{1}{2}, b = \dfrac{3}{4}, c = -6$

8. $\dfrac{1}{4}x = \dfrac{5}{6}x^2 + 6$
$a = \dfrac{5}{6}, b = -\dfrac{1}{4}, c = 6$ or
$a = -\dfrac{5}{6}, b = \dfrac{1}{4}, c = -6$

For Exercises 9–32, solve using the quadratic formula. See Example 1.

9. $x^2 + 9x + 20 = 0$
$-5, -4$

10. $x^2 + 8x + 15 = 0$
$-3, -5$

11. $4x^2 + 5x = 6$
$-2, \dfrac{3}{4}$

12. $2x^2 - x = 3$
$-1, \dfrac{3}{2}$

13. $x^2 - 9x = 0$
$0, 9$

14. $x^2 - 6x = 0$
$0, 6$

15. $x^2 - 8x = -16$
4

16. $x^2 + 14x = -49$
-7

17. $3x^2 + 4x = 4$
$-2, \dfrac{2}{3}$

18. $5x^2 - 13x = 6$
$3, -\dfrac{2}{5}$

19. $x^2 + 2 = 2x$
$1 \pm i$

20. $x^2 + 5 = 4x$
$2 \pm i$

21. $x^2 - x - 1 = 0$
$\dfrac{1 \pm \sqrt{5}}{2}$

22. $x^2 + 3x - 5 = 0$
$\dfrac{-3 \pm \sqrt{29}}{2}$

23. $3x^2 + 10x + 5 = 0$
$\dfrac{-5 \pm \sqrt{10}}{3}$

24. $3x^2 - x - 3 = 0$
$\dfrac{1 \pm \sqrt{37}}{6}$

25. $-4x^2 = 5 - 6x$
$\dfrac{3 \pm i\sqrt{11}}{4}$

26. $-5x^2 = 3 - 4x$
$\dfrac{2 \pm i\sqrt{11}}{5}$

27. $18x^2 + 2 = -15x$
$\dfrac{1}{6}, \dfrac{2}{3}$

28. $10x^2 - 12 = -7x$
$\dfrac{3}{2}, \dfrac{4}{5}$

29. $3x^2 - 4x = -3$
$\dfrac{2 \pm i\sqrt{5}}{3}$

30. $3x^2 - 6x = -8$
$\dfrac{3 \pm i\sqrt{15}}{3}$

31. $6x^2 - 4 = 3x$
$\dfrac{3 \pm \sqrt{105}}{12}$

32. $6x^2 - 6 = 13x$
$\dfrac{13 \pm \sqrt{313}}{12}$

For Exercises 33–44, solve using the quadratic formula. If the solutions are irrational, give an exact answer and an approximate answer rounded to the nearest thousandth. (Hint: You might first clear the fractions or decimals by multiplying both sides by an appropriately chosen number.)

33. $2x^2 + 0.1x = 0.03$
$-0.15, 0.1$

34. $4x^2 + 8.6x - 2.4 = 0$
$-2.4, 0.25$

35. $x^2 + \dfrac{1}{2}x - 3 = 0$
$-2, \dfrac{3}{2}$

36. $x^2 + \dfrac{1}{3}x - \dfrac{2}{3} = 0$

$-1, \dfrac{2}{3}$

37. $x^2 - \dfrac{49}{36} = 0$

$\pm\dfrac{7}{6}$

38. $x^2 - \dfrac{64}{25} = 0$

$\pm\dfrac{8}{5}$

39. $\dfrac{1}{2}x^2 + \dfrac{3}{2} = x$

$1 \pm i\sqrt{2}$

40. $\dfrac{1}{3}x^2 + 2 = \dfrac{2}{3}x$

$1 \pm i\sqrt{5}$

41. $x^2 - 0.5 = -0.06x$

$\dfrac{-3 \pm \sqrt{5009}}{100} \approx -0.738, 0.678$

42. $0.5x^2 - 0.4 = 0.25x$

$\dfrac{5 \pm \sqrt{345}}{20} \approx 1.179, -0.679$

43. $1.2x^2 - 0.6x = -0.5$

$\dfrac{3 \pm i\sqrt{51}}{12}$

44. $2.4x^2 + 4.5 = 6.3x$

$\dfrac{21 \pm i\sqrt{39}}{16}$

Find ⊗ the Mistake   *For Exercises 45–48, explain the mistake; then solve correctly.*

45. Solve $3x^2 - 7x + 1 = 0$ using the quadratic formula.

$$\dfrac{-7 \pm \sqrt{(-7)^2 - (4)(3)(1)}}{2(3)} = \dfrac{-7 \pm \sqrt{49 - 12}}{6}$$

$$= \dfrac{-7 \pm \sqrt{37}}{6}$$

Mistake: Did not evaluate $-b$. Correct: $\dfrac{7 \pm \sqrt{37}}{6}$

46. Solve $2x^2 - 6x - 5 = 0$ using the quadratic formula.

$$\dfrac{6 \pm \sqrt{(-6)^2 - (4)(2)(5)}}{2(2)} = \dfrac{6 \pm \sqrt{36 - 40}}{4}$$

$$= \dfrac{3}{2} \pm \dfrac{1}{2}i$$

Mistake: Substituted 5 for c instead of -5. Correct: $\dfrac{3 \pm \sqrt{19}}{2}$

47. Solve $x^2 - 2x + 3 = 0$ using the quadratic formula.

$$\dfrac{-(-2) \pm \sqrt{(-2)^2 - (4)(1)(3)}}{2(1)} = \dfrac{2 \pm \sqrt{4 - 12}}{2}$$

$$= \dfrac{2 \pm \sqrt{-8}}{2}$$

Mistake: The result was not completely simplified.
Correct: $1 \pm i\sqrt{2}$

48. Solve $x^2 - 8 = 0$ using the quadratic formula.

$$\dfrac{-(-8) \pm \sqrt{(-8)^2 - (4)(1)(0)}}{2(1)} = \dfrac{8 \pm \sqrt{64 - 0}}{2}$$

$$= \dfrac{8 \pm \sqrt{64}}{2}$$

$$= \dfrac{8 \pm 8}{2}$$

$$= 8, 0$$

Mistake: -8 is c, not b. Correct: $\pm 2\sqrt{2}$

Objective 2

Prep Exercise 5 What part of the quadratic formula is the discriminant?
Write the formula for the discriminant.
The discriminant is the radicand in the quadratic formula; $b^2 - 4ac$

For Exercises 49–58, use the discriminant to determine the number and type of solutions for the equation. If the solution(s) are real, state whether they are rational or irrational. See Example 2.

49. $x^2 + 10x = -25$
One rational

50. $2x^2 - 8x + 8 = 0$
One rational

51. $\dfrac{1}{4}x^2 - 4x = -4$
Two irrational

52. $\dfrac{1}{2}x^2 - 5 = -3x$
Two irrational

53. $3x^2 + 10x - 8 = 0$
Two rational

54. $2x^2 - 3x - 2 = 0$
Two rational

55. $x^2 - x + 3 = 0$
Two nonreal complex

56. $x^2 + 4x = -5$
Two nonreal complex

57. $x^2 - 6x + 6 = 0$
Two irrational

58. $3x^2 = 13x - 8$
Two irrational

For Exercises 59–68, indicate which of the following methods is the best choice for solving the given equation: factoring, using the square root principle, or using the quadratic formula. Then solve the equation.

59. $x^2 - 81 = 0$
Square root principle or factoring; ± 9

60. $x^2 - 64 = 0$
Square root principle or factoring; ± 8

61. $2x^2 - 4x - 3 = 0$
Quadratic formula;
$\dfrac{2 \pm \sqrt{10}}{2}$

62. $3x^2 - 6x - 4 = 0$
Quadratic formula;
$\dfrac{3 \pm \sqrt{21}}{3}$

63. $x^2 + 6x = 0$
Factoring; $0, -6$

64. $x^2 + 14x + 45 = 0$
Factoring; $-9, -5$

65. $(x + 7)^2 = 40$
Square root principle;
$-7 \pm 2\sqrt{10}$

66. $(x - 3)^2 = 48$
Square root principle;
$3 \pm 4\sqrt{3}$

67. $x^2 = 8x - 19$
Quadratic formula;
$4 \pm i\sqrt{3}$

68. $x^2 - 4x = -10$
Quadratic formula;
$2 \pm i\sqrt{6}$

Objective 3

Prep Exercise 6 If x-intercepts for a quadratic function are imaginary numbers, what does that indicate about its graph? The graph does not intersect the x-axis.

For Exercises 69–76, find the x- and y-intercepts. See Example 3.

69. $y = x^2 - x - 2$
$(2, 0), (-1, 0), (0, -2)$

70. $y = x^2 - 3x + 2$
$(2, 0), (1, 0), (0, 2)$

71. $y = 4x^2 - 12x + 9$
$\left(\dfrac{3}{2}, 0\right), (0, 9)$

72. $y = 9x^2 + 6x + 1$
$\left(-\dfrac{1}{3}, 0\right), (0, 1)$

73. $y = 2x^2 + 15x - 8$
$\left(\dfrac{1}{2}, 0\right), (-8, 0), (0, -8)$

74. $y = -15x^2 + x + 6$
$\left(-\dfrac{3}{5}, 0\right), \left(\dfrac{2}{3}, 0\right), (0, 6)$

75. $y = -2x^2 + 3x - 6$
No x-intercepts,
$(0, -6)$

76. $y = -3x^2 - 2x - 5$
No x-intercepts,
$(0, -5)$

Objective 4

For Exercises 77–84, translate to a quadratic equation; then solve using the quadratic formula.

77. A positive integer squared plus five times its consecutive integer is equal to 71. Find the integers.
$x^2 + 5(x + 1) = 71; 6, 7$

78. The square of a positive integer minus twice its consecutive integer is equal to 22. Find the integers.
$x^2 - 2(x + 1) = 22; 6, 7$

79. A right triangle has side lengths that are three consecutive integers. Use the Pythagorean theorem to find the lengths of those sides. (Remember that the hypotenuse in a right triangle is always the longest side.)
$x^2 + (x + 1)^2 = (x + 2)^2; 3, 4, 5$

80. A right triangle has side lengths that are consecutive even integers. Use the Pythagorean theorem to find the lengths of those sides. (Remember that the hypotenuse in a right triangle is always the longest side.)
$x^2 + (x + 2)^2 = (x + 4)^2; 6, 8, 10$

81. The length of a rectangular fence gate is 3.5 feet less than three times the width. Find the length and width of the gate if the area is 34 square feet.
$w(3w - 3.5) = 34$; width: 4 ft.; length: 8.5 ft.

82. A small access door for an attic storage area is designed so that the width is 9.5 feet less than two times the length. Find the length and width if the area is 52 square feet.
$l(2l - 9.5) = 52$; length: 8 ft.; width: 6.5 ft.

83. An architect is experimenting with two different shapes of a room as shown.
  a. Find x so that the rooms have the same area.
   $22x = 1.5x^2 + 4x$; 12 ft.
  b. Complete the dimensions for the L-shaped room.
   18 ft., 12 ft., 10 ft., 6 ft., 8 ft., 18 ft. (from bottom clockwise)

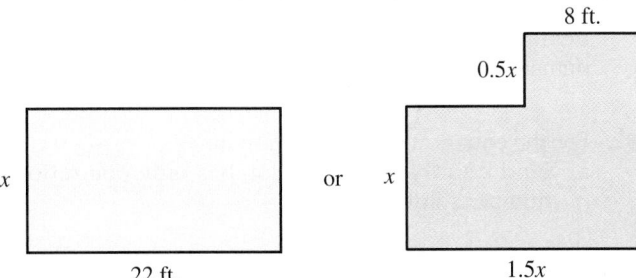

84. A cylinder is to be made so that its volume is equal to that of a sphere with a radius of 9 inches. If a cylinder is to have a height of 4 inches, find its radius. $4\pi r^2 = \frac{4}{3}\pi(9)^3$, $9\sqrt{3}$ in. ≈ 15.6 in.

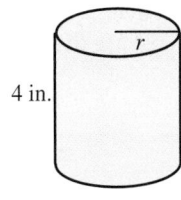
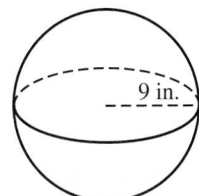

📱 *For Exercises 85–88, use the formula $h = \frac{1}{2}gt^2 + v_0 t + h_0$, where g represents the acceleration due to gravity, t is the time in flight, v_0 is the initial velocity, and h_0 is the initial height. For Earth, the acceleration due to gravity is -32.2 ft./sec.^{2} or -9.8 m/sec.^{2} Approximate irrational answers to the nearest hundredth. See Example 4.*

85. Robbie Knievel jumped the Grand Canyon on a motorcycle. Suppose that on take-off, his motorcycle had a vertical velocity of 48 feet per second and at the end of the launch ramp his altitude was 485 feet above sea level. If he landed on a ramp that was 460 feet above sea level, how long was he in flight?
 ≈ 3.43 sec.

86. In the extreme games competition, Brian Deegan successfully performed the first 360-degree flip ever attempted in a competition. Suppose his motorcycle had a vertical velocity of 12.8 meters per second from the end of a ramp that was 8 meters high and he landed on a ramp at a point 4 meters above the ground. How long was he in the air?
 ≈ 2.89 sec.

87. A platform diver dives from a platform that is 10 meters above the water.
  a. Write the equation that describes her height during the dive. (Assume that her initial velocity is 0.)
   $h = -4.9t^2 + 10$
  b. Find the time it takes the diver to be 5 meters above the water.
   ≈ 1.01 sec.
  c. Find the time it takes the diver to enter the water.
   ≈ 1.43 sec.

88. In a cliff diving championship in Acapulco, a diver dives from a cliff at a height of 70 feet.
  a. Write the equation that describes his height during the dive. (Assume that his initial velocity is 0.)
   $h = -16.1t^2 + 70$
  b. Find the time it takes the diver to be 50 feet above the water.
   ≈ 1.11 sec.
  c. Find the time it takes the diver to enter the water.
   ≈ 2.09 sec.

89. The expression $0.5n^2 + 2.5n$ describes the gross income from the sale of a particular software product, where n is the number of units sold in thousands. The expression $4.5n + 16$ describes the cost of producing the n units. Find the number of units that must be produced and sold for the company to break even. (To *break even* means that the gross income and cost are the same.)

8000 units

90. An economist and a marketing manager discover that the expression $2n^2 + 5n$ models the price of a CD based on its demand, where n is the number of units (in millions) the market demands. The expression $-0.5n^2 + 17.1$ describes the price of the CD based on the number of units (also in millions) supplied to the market. Find the number of units that must be demanded and supplied so that the price based on demand is equal to the price based on supply.

1.8 million units

★ **91.** For the equation $2x^2 - 5x + c = 0$:
   a. Find c so that the equation has only one rational number solution.

$\dfrac{25}{8}$

   b. Find the range of values of c for which the equation has two real-number solutions.

$c < \dfrac{25}{8}$

   c. Find the range of values of c for which the equation has no real-number solution.

$c > \dfrac{25}{8}$

★ **92.** For the equation $4x^2 + 6x + c = 0$:
   a. Find c so that the equation has only one rational number solution.

$\dfrac{9}{4}$

   b. Find the range of values of c for which the equation has two real-number solutions.

$c < \dfrac{9}{4}$

   c. Find the range of values of c for which the equation has no real-number solution.

$c > \dfrac{9}{4}$

★ **93.** For the equation $ax^2 + 12x + 8 = 0$:
   a. Find a so that the equation has only one rational number solution.

$\dfrac{9}{2}$

   b. Find the range of values of a for which the equation has two real-number solutions.

$a < \dfrac{9}{2}$

   c. Find the range of values of a for which the equation has no real-number solution.

$a > \dfrac{9}{2}$

★ **94.** For the equation $3x^2 + bx + 8 = 0$:
   a. Find b so that the equation has only one rational number solution.

$\pm 4\sqrt{6}$

   b. Find all positive values of b for which the equation has two real-number solutions.

$b > 4\sqrt{6}$

   c. Find all positive values of b for which the equation has no real-number solution.

$0 < b < 4\sqrt{6}$

Puzzle Problem In the equation shown, A, B, C, D, and E are five consecutive positive integers where $A < B < C < D < E$. What are the integers?

$$A^2 + B^2 + C^2 = D^2 + E^2$$

$A = 10, B = 11, C = 12, D = 13, E = 14$

Review Exercises

Exercises 1–4 ➤ **Expressions**

For Exercises 1 and 2, factor.

[6.2] 1. $u^2 - 9u + 14$
$(u - 7)(u - 2)$

[6.3] 2. $3u^2 - 2u - 16$
$(3u - 8)(u + 2)$

For Exercises 3 and 4, simplify.

[5.4] 3. $(x^2)^2$
x^4

[9.2] 4. $(x^{1/3})^2$
$x^{2/3}$

Exercises 5 and 6 ➤ **Equations and Inequalities**

For Exercises 5 and 6, solve.

[6.6] 5. $5u^2 + 13u = 6$
$\dfrac{2}{5}, -3$

[7.6] 6. $\dfrac{7}{3u} = \dfrac{5}{u} - \dfrac{1}{u - 5}$
8

10.3   Solving Equations That Are Quadratic in Form

Objectives

1 Solve equations by rewriting them in quadratic form.

2 Solve equations that are quadratic in form by using substitution.

3 Solve application problems using equations that are quadratic in form.

Warm-up

For Exercises 1–4, solve.

[10.2] **1.** $2x^2 + 5 = 6x$

[7.6] **2.** $\dfrac{200}{x} = \dfrac{200}{x+1} + 10$

[9.6] **3.** $\sqrt[3]{x} = -2$

[9.6] **4.** $x - \sqrt{x} - 2 = 0$

Many equations that are not quadratic equations are **quadratic in form** and can be solved using the same methods used to solve quadratic equations.

Definition An equation is **quadratic in form** if it can be rewritten as a quadratic equation $au^2 + bu + c = 0$, where $a \neq 0$ and u is a variable or an expression.

Objective 1 Solve equations by rewriting them in quadratic form.

Equations with Rational Expressions

In Section 7.6, we solved equations containing rational expressions by multiplying both sides of the equation by the LCD. Those rewritten equations are often quadratic. Remember that equations containing rational expressions sometimes have extraneous solutions.

Connection Recall that an apparent solution for an equation with rational expressions is extraneous if it causes one or more of the denominators to equal 0.

Example 1 Solve $\dfrac{3}{x+1} = 1 - \dfrac{3}{x(x+1)}$.

Solution: $x(x+1) \cdot \dfrac{3}{x+1} = x(x+1)\left(1 - \dfrac{3}{x(x+1)}\right)$

 Multiply both sides by the LCD $x(x+1)$.

$x(x+1) \cdot \dfrac{3}{x+1} = x(x+1) \cdot 1 - x(x+1) \cdot \dfrac{3}{x(x+1)}$

 Distribute $x(x+1)$.

$3x = x^2 + x - 3$ Simplify both sides.

$0 = x^2 - 2x - 3$ Subtract $3x$ from both sides to get the quadratic form $ax^2 + bx + c = 0$.

$0 = (x-3)(x+1)$ Factor.

$x - 3 = 0$ or $x + 1 = 0$ Use the zero-factor theorem.

$x = 3$ $x = -1$ Solve each equation.

Checks: $x = 3$ $x = -1$

$\dfrac{3}{3+1} \overset{?}{=} 1 - \dfrac{3}{3(3+1)}$ $\dfrac{3}{-1+1} \overset{?}{=} 1 - \dfrac{3}{-1(-1+1)}$

$\dfrac{3}{4} \overset{?}{=} 1 - \dfrac{3}{12}$ $\dfrac{3}{0} \overset{?}{=} 1 - \dfrac{3}{0}$ ◀ **Note** The expression $\dfrac{3}{0}$ is undefined, so

$\dfrac{3}{4} = \dfrac{3}{4}$ True -1 is extraneous.

The solution is 3. (-1 is extraneous.)

Your Turn 1 Solve $\dfrac{x}{x-2} = \dfrac{6}{x} + \dfrac{4}{x(x-2)}$.

Answer to Your Turn 1
4 (2 is extraneous.)

Answers to Warm-up

1. $\dfrac{3}{2} \pm \dfrac{1}{2}i$

2. $x = -5, x = 4$

3. $x = -8$

4. $x = 4$ (1 is extraneous.)

Equations Containing Radicals

In Section 9.6, we found that after the power rule was used on equations containing radicals, the result was often a quadratic equation. Remember that these radical equations sometimes have extraneous solutions.

Connection Recall that an apparent solution for an equation containing radicals is extraneous if it makes the original equation false.

| **Example 2** | Solve. |

a. $\sqrt{x} + x = 6$

Solution: $\sqrt{x} = 6 - x$ Subtract x from both sides to isolate the radical.

$(\sqrt{x})^2 = (6 - x)^2$ Square both sides.

$x = 36 - 12x + x^2$ Simplify both sides.

$0 = x^2 - 13x + 36$ Subtract x from both sides to write in quadratic form.

$0 = (x - 4)(x - 9)$ Factor.

$x - 4 = 0$ or $x - 9 = 0$ Use the zero-factor theorem.

$x = 4$ $x = 9$ Solve each equation.

Checks: $x = 4$ $x = 9$

$\sqrt{4} + 4 \overset{?}{=} 6$ $\sqrt{9} + 9 \overset{?}{=} 6$

$2 + 4 = 6$ True $3 + 9 = 6$ False; so 9 is extraneous.

The solution is 4. (9 is extraneous.)

b. $\sqrt{x - 1} = 2x - 1$

Solution: $(\sqrt{x - 1})^2 = (2x - 1)^2$ Square both sides of the equation.

$x - 1 = 4x^2 - 4x + 1$ Simplify both sides.

$0 = 4x^2 - 5x + 2$ Subtract x from and add 1 to both sides to write in quadratic form.

We cannot factor $4x^2 - 5x + 2$, so we use the quadratic formula.

Note Written in standard form, these solutions are
$\frac{5}{8} + \frac{\sqrt{7}}{8}i$ and $\frac{5}{8} - \frac{\sqrt{7}}{8}i$.

$$x = \frac{-(-5) \pm \sqrt{(-5)^2 - 4(4)(2)}}{2(4)} = \frac{5 \pm \sqrt{-7}}{8} = \frac{5 \pm i\sqrt{7}}{8}$$

Check: Because the solutions are complex, we will not check them.

| **Your Turn 2** | Solve. |

a. $x - 2\sqrt{x} - 3 = 0$ **b.** $\sqrt{6x - 11} = 3x - 1$

Objective 2 Solve equations that are quadratic in form by using substitution.

Recall that an equation that is quadratic in form can be written as $au^2 + bu + c = 0$, where $a \neq 0$ and u is an *expression*. We now explore a method for solving these equations where we substitute u for an expression. To use substitution, it is important to note a pattern with the exponents of the terms in a quadratic equation. Notice that the degree of the first term, au^2, is 2; the degree of the middle term, bu, is 1; and the third term is a constant. If a trinomial has one term with an expression raised to the second power, a second term with that same expression raised to the first power, and a third term that is a constant, the equation is quadratic in form and we can use substitution to solve it.

Consider the equation $x^4 - 13x^2 + 36 = 0$. Notice that we can rewrite the equation as $(x^2)^2 - 13(x^2) + 36 = 0$, so it is quadratic in form. By substituting u for each x^2, we have a "friendlier" form of quadratic equation, which we can then solve by factoring, completing the square, or using the quadratic formula.

Answers to Your Turn 2
a. 9 (1 is extraneous.)
b. $\dfrac{2 \pm 2i\sqrt{2}}{3}$

Note Given an equation containing a trinomial, ▶
if the degree of the first term is twice the degree
of the middle term and the third term is a
constant, try using substitution.

$$(x^2)^2 - 13(x^2) + 36 = 0$$

$$u^2 - 13u + 36 = 0 \quad \text{Substitute } u \text{ for } x^2.$$

$$(u - 9)(u - 4) = 0 \quad \text{Factor.}$$

$$u - 9 = 0 \quad \text{or} \quad u - 4 = 0 \quad \text{Use the zero-factor theorem.}$$

$$u = 9 \qquad\qquad u = 4 \quad \text{Solve each equation.}$$

Note that these solutions are for u, not x. We must substitute x^2 back in place of u to finish solving for x.

Note The equation
$x^4 - 13x^2 + 36 = 0$ has four ▶
solutions: $3, -3, 2,$ and -2.

$$x^2 = 9 \qquad x^2 = 4 \qquad \text{Substitute } x^2 \text{ for } u.$$

$$x = \pm\sqrt{9} \qquad x = \pm\sqrt{4} \qquad \text{Use the square root principle.}$$

$$x = \pm 3 \qquad x = \pm 2 \qquad \text{Simplify.}$$

Our example suggests the following procedure.

Procedure   **Using Substitution to Solve Equations That Are Quadratic in Form**

To solve equations that are quadratic in form using substitution:
1. Rewrite the equation so that it is in the form $au^2 + bu + c = 0$.
2. Solve the quadratic equation for u.
3. Substitute for u and solve.
4. Check the solutions.

Example 3   Solve.

a. $(x + 2)^2 - 2(x + 2) - 8 = 0$

Solution: If we substitute u for $x + 2$, we have an equation in the form $au^2 + bu + c = 0$.

$$(x + 2)^2 - 2(x + 2) - 8 = 0$$

$$u^2 - 2u - 8 = 0 \quad \text{Substitute } u \text{ for } x + 2.$$

$$(u - 4)(u + 2) = 0 \quad \text{Factor.}$$

$$u - 4 = 0 \quad \text{or} \quad u + 2 = 0 \qquad \text{Use the zero-factor theorem.}$$

$$u = 4 \qquad\qquad u = -2 \qquad \text{Solve each equation for } u.$$

$$x + 2 = 4 \qquad x + 2 = -2 \qquad \text{Substitute } x + 2 \text{ for } u.$$

$$x = 2 \qquad\qquad x = -4 \qquad \text{Solve each equation for } x.$$

Check: Verify that 2 and -4 make $(x + 2)^2 - 2(x + 2) - 8 = 0$ true. We will leave this check to the reader.

b. $x^{2/3} - x^{1/3} - 6 = 0$

Solution: Because $x^{2/3} - x^{1/3} - 6 = 0$ can be written as $(x^{1/3})^2 - x^{1/3} - 6 = 0$, it is quadratic in form. We substitute u for $x^{1/3}$.

$$(x^{1/3})^2 - x^{1/3} - 6 = 0 \qquad \text{Rewrite in quadratic form.}$$

$$u^2 - u - 6 = 0 \qquad \text{Substitute } u \text{ for } x^{1/3}.$$

$$(u - 3)(u + 2) = 0 \qquad \text{Factor.}$$

$$u - 3 = 0 \quad \text{or} \quad u + 2 = 0 \qquad \text{Use the zero-factor theorem.}$$

$$u = 3 \qquad\qquad u = -2 \qquad \text{Solve each equation for } u.$$

$$x^{1/3} = 3 \qquad x^{1/3} = -2 \qquad \text{Substitute } x^{1/3} \text{ for } u.$$

$$(x^{1/3})^3 = 3^3 \qquad (x^{1/3})^3 = (-2)^3 \qquad \text{Cube both sides of the equations.}$$

$$x = 27 \qquad\qquad x = -8 \qquad \text{Simplify.}$$

Checks: $x = 27$ $\qquad\qquad\qquad\qquad\qquad$ $x = -8$

$$27^{2/3} - 27^{1/3} - 6 \overset{?}{=} 0 \qquad\qquad (-8)^{2/3} - (-8)^{1/3} - 6 \overset{?}{=} 0$$

$$(\sqrt[3]{27})^2 - \sqrt[3]{27} - 6 \overset{?}{=} 0 \qquad\qquad (\sqrt[3]{-8})^2 - \sqrt[3]{-8} - 6 \overset{?}{=} 0$$

$$3^2 - 3 - 6 \overset{?}{=} 0 \qquad\qquad\qquad (-2)^2 - (-2) - 6 \overset{?}{=} 0$$

$$9 - 3 - 6 \overset{?}{=} 0 \qquad\qquad\qquad\qquad 4 + 2 - 6 \overset{?}{=} 0$$

$$0 = 0 \quad \text{True} \qquad\qquad\qquad\qquad 0 = 0 \quad \text{True}$$

Your Turn 3 Solve.

a. $x^4 - 10x^2 + 9 = 0$ $\quad$ **b.** $(n - 2)^2 + 4(n - 2) - 12 = 0$ $\quad$ **c.** $x^{2/3} - x^{1/3} - 2 = 0$

Objective 3 Solve application problems using equations that are quadratic in form.

Example 4 Solve.

a. The average speed of a car is 10 miles per hour more than the average speed of a bus. The bus takes 1 hour longer than the car to travel 200 miles. Find how long it takes the car to travel 200 miles.

Understand: Both the car and bus travel 200 miles, but it takes the bus 1 hour longer than it takes the car. The rate of the car is 10 miles per hour more than the rate of the bus.

Plan: We use a table to organize the information, then write an equation, which we can solve.

Execute: We let x represent the time for the car to travel 200 miles. Because the bus travels 1 more hour, $x + 1$ describes the time for the bus to travel 200 miles.

Vehicle	d	t	r
bus	200	$x + 1$	$\dfrac{200}{x + 1}$
car	200	x	$\dfrac{200}{x}$

◄ **Note** We use $r = \dfrac{d}{t}$ to describe each rate.

Because the rate of the car is 10 miles per hour faster than the rate of the bus, we can say that (the rate of the car) $=$ (the rate of the bus) $+ 10$.

$$\frac{200}{x} = \frac{200}{x + 1} + 10$$

$$x(x + 1)\left(\frac{200}{x}\right) = x(x + 1)\left(\frac{200}{x + 1} + 10\right) \qquad \text{Multiply both sides by the LCD } x(x + 1).$$

$$(x + 1)(200) = x(x + 1)\left(\frac{200}{x + 1}\right) + x(x + 1)(10) \qquad \text{Multiply on the left and distribute on the right.}$$

$$200x + 200 = 200x + 10x^2 + 10x \qquad \text{Simplify.}$$

$$0 = 10x^2 + 10x - 200 \qquad \text{Subtract 200x and 200 from both sides to get quadratic form.}$$

$$0 = x^2 + x - 20 \qquad \text{Divide both sides by 10.}$$

$$0 = (x + 5)(x - 4) \qquad \text{Factor.}$$

$$x + 5 = 0 \quad \text{or} \quad x - 4 = 0 \qquad \text{Use the zero-factor theorem.}$$

$$x = -5 \qquad\qquad x = 4 \qquad \text{Solve each equation.}$$

Answers to Your Turn 3
a. $\pm 3, \pm 1$ $\quad$ **b.** ± 4 $\quad$ **c.** $8, -1$

Answer: Because time cannot be negative, it takes the car 4 hours to travel 200 miles.

Check: The rate of the car is $\dfrac{200}{x} = \dfrac{200}{4} = 50$ miles per hour; so in 4 hours, the car travels $50(4) = 200$ miles. The bus travels 200 miles in $4 + 1 = 5$ hours at a rate of $\dfrac{200}{4+1} = \dfrac{200}{5} = 40$ miles per hour; so in 5 hours, the bus travels $5(40) = 200$ miles. Notice also that the car's rate is 10 miles per hour faster than the bus's.

b. Bobby and Pam manufacture and install blinds. Working together, they can install blinds in every window of an average-sized house in $1\frac{1}{3}$ hours. Working alone, Bobby takes 2 hours longer than Pam to do the same installation. Working alone, how long does each person take to install blinds in an average-sized house?

Understand: We are given the time it takes Bobby and Pam working together and are asked to find how long it takes each person individually. We also know that it takes Bobby 2 hours longer than it takes Pam.

Plan: We use a table to organize the information, write an equation, and then solve.

Execute: We let x represent the number of hours it takes Pam to install blinds working alone. Because Bobby takes 2 more hours, his time working alone is represented by $x + 2$.

Worker	Time to Complete the Job Alone	Rate of Work	Time at Work	Part of Job Completed
Pam	x	$\dfrac{1}{x}$	$\dfrac{4}{3}$	$\dfrac{4}{3x}$
Bobby	$x + 2$	$\dfrac{1}{x+2}$	$\dfrac{4}{3}$	$\dfrac{4}{3(x+2)}$

▲ **Note** Multiplying the rate of work and the time at work gives an expression of the part of the job completed.

The total job in this case is 1 average-sized house, so we can write an equation that combines Pam's and Bobby's individual expressions for work completed and set this sum equal to 1.

$$(\text{Part Pam does}) + (\text{Part Bobby does}) = 1 \ (\text{the entire job})$$

$$\frac{4}{3x} + \frac{4}{3(x+2)} = 1$$

$$3x(x+2)\left(\frac{4}{3x} + \frac{4}{3(x+2)}\right) = 3x(x+2)(1) \qquad \text{Multiply both sides by the LCD } 3x(x+2).$$

$$3x(x+2)\left(\frac{4}{3x}\right) + 3x(x+2)\left(\frac{4}{3(x+2)}\right) = 3x^2 + 6x \qquad \text{Distribute on the left and multiply on the right.}$$

$$4(x+2) + 4x = 3x^2 + 6x \qquad \text{Continue simplifying.}$$

$$4x + 8 + 4x = 3x^2 + 6x \qquad \text{Distribute.}$$

$$8x + 8 = 3x^2 + 6x \qquad \text{Combine like terms.}$$

$$0 = 3x^2 - 2x - 8 \qquad \text{Subtract 8x and 8 from both sides.}$$

$$0 = (3x + 4)(x - 2) \qquad \text{Factor.}$$

$$3x + 4 = 0 \quad \text{or} \quad x - 2 = 0 \qquad \text{Use the zero-factor theorem.}$$

$$3x = -4 \qquad\qquad x = 2 \qquad \text{Solve each equation.}$$

$$x = -\frac{4}{3}$$

Answer: Because negative time makes no sense in the context of this problem, it takes Pam 2 hours to install blinds working alone and Bobby $x + 2 = 2 + 2 = 4$ hours working alone.

Check: Because Pam can install blinds in 2 hours, she can install them in $\frac{1}{2}$ of an average-sized house in 1 hour. Because Bobby takes 4 hours to do the same work, he can install blinds in $\frac{1}{4}$ of an average-sized house in 1 hour. Together, they can install blinds in $\frac{1}{2} + \frac{1}{4} = \frac{2}{4} + \frac{1}{4} = \frac{3}{4}$ of an average-sized house in 1 hour; so in $1\frac{1}{3}$ hours, they can install blinds in $1\frac{1}{3} \cdot \frac{3}{4} = \frac{4}{3} \cdot \frac{3}{4} = 1$ average-sized house.

Your Turn 4 Solve.

a. The average speed of a passenger train is 25 miles per hour more than the average speed of a car. The time required for the car to travel 300 miles is 2 hours more than the time required for the train to travel the same number of miles. Find the average speed of the car.

b. Terri and Tommy run a flea market. It takes Terri 2 hours longer to put out the merchandise than it does Tommy. If they can put out the merchandise together in $1\frac{7}{8}$ hours, how long does it take each working alone?

Answers to Your Turn 4
a. 50 mph
b. It takes Tommy 3 hr. and Terri 5 hr.

10.3 Exercises (For Extra Help) MyMathLab®

Note: Exercises marked with a ★ represent challenging exercises.

Objective 1

Prep Exercise 1 What does it mean to say that an equation is quadratic in form? It can be rewritten as a quadratic equation.

Prep Exercise 2 Is $x^{3/4} - x^{1/4} - 6 = 0$ quadratic in form? Why or why not? No, $x^{3/4} = (x^{1/4})^3$ shows a cubic relationship.

For Exercises 1–12, solve the equations with rational expressions. Identify any extraneous solutions. See Example 1.

1. $\dfrac{60}{x + 2} = \dfrac{60}{x} - 5$
$-6, 4$

2. $\dfrac{120}{x + 2} = \dfrac{120}{x} - 5$
$-8, 6$

3. $\dfrac{1}{x} + \dfrac{1}{x + 2} = \dfrac{3}{4}$
$-\dfrac{4}{3}, 2$

4. $\dfrac{1}{x} + \dfrac{3}{x - 4} = \dfrac{7}{8}$
$\dfrac{4}{7}, 8$

5. $\dfrac{1}{p - 4} + \dfrac{1}{4} = \dfrac{8}{p^2 - 16}$
-8 (4 is extraneous.)

6. $\dfrac{1}{y - 5} - \dfrac{10}{y^2 - 25} = -\dfrac{1}{5}$
-10 (5 is extraneous.)

7. $\dfrac{6}{2y + 5} = \dfrac{2}{y + 5} + \dfrac{1}{5}$
$-7.5, 5$

8. $\dfrac{6}{2x + 3} = \dfrac{2}{x - 6} + \dfrac{4}{3}$
$2.25, 3$

9. $1 + 2x^{-1} - 8x^{-2} = 0$
$-4, 2$

10. $1 + 5x^{-1} + 6x^{-2} = 0$
$-3, -2$

11. $3 + 13x^{-1} - 10x^{-2} = 0$
$-5, \dfrac{2}{3}$

12. $3 - 5x^{-1} - 12x^{-2} = 0$
$3, -\dfrac{4}{3}$

For Exercises 13–24, solve the equations with radical expressions. Identify any extraneous solutions. See Example 2.

13. $x - 8\sqrt{x} + 15 = 0$
$9, 25$

14. $x - 3\sqrt{x} + 2 = 0$
$1, 4$

15. $2x - 5\sqrt{x} - 7 = 0$
$\dfrac{49}{4}$ (1 is extraneous.)

16. $3x + 4\sqrt{x} - 4 = 0$
$\dfrac{4}{9}$ (4 is extraneous.)

17. $\sqrt{2a + 5} = 3a - 3$
$2\left(\dfrac{2}{9} \text{ is extraneous.}\right)$

18. $\sqrt{3b + 1} = 5b - 3$
$1\left(\dfrac{8}{25} \text{ is extraneous.}\right)$

19. $\sqrt{2m - 8} - m - 1 = 0$
$\pm 3i$

20. $\sqrt{8x - 9} - x - 4 = 0$
$\pm 5i$

21. $\sqrt{4x + 1} = \sqrt{x + 2} + 1$
$2 \left(-\dfrac{2}{9} \text{ is extraneous.} \right)$

22. $\sqrt{4x + 1} = \sqrt{x + 3} + 2$
$6 \left(-\dfrac{2}{9} \text{ is extraneous.} \right)$

23. $\sqrt{2x + 1} - \sqrt{3x + 4} = -1$
$0, 4$

24. $\sqrt{2x - 1} - \sqrt{4x + 5} = -2$
$1, 5$

Objective 2

Prep Exercise 3 To solve $3\left(\dfrac{x + 2}{3} \right)^2 + 13\left(\dfrac{x + 2}{3} \right) - 10 = 0$ using substitution, what would your substitution be?
$u = \dfrac{x + 2}{3}$

Prep Exercise 4 To solve $x^4 - 7x^2 + 12 = 0$ using substitution, what would your substitution be? $u = x^2$

For Exercises 25–46, solve using substitution. Identify any extraneous solutions. See Example 3.

25. $x^4 - 10x^2 + 9 = 0$
$\pm 3, \pm 1$

26. $x^4 - 13x^2 + 36 = 0$
$\pm 3, \pm 2$

27. $4x^4 - 13x^2 + 9 = 0$
$\pm 1.5, \pm 1$

28. $9x^4 - 37x^2 + 4 = 0$
$\pm \dfrac{1}{3}, \pm 2$

29. $x^4 + 5x^2 - 36 = 0$
$\pm 2, \pm 3i$

30. $x^4 - 12x^2 - 64 = 0$
$\pm 4, \pm 2i$

31. $(x + 2)^2 + 6(x + 2) + 8 = 0$
$-6, -4$

32. $(x - 3)^2 + 2(x - 3) - 15 = 0$
$-2, 6$

33. $2(x + 3)^2 - 9(x + 3) - 5 = 0$
$-3.5, 2$

34. $3(x - 1)^2 + 4(x - 1) - 4 = 0$
$-1, \dfrac{5}{3}$

35. $\left(\dfrac{x - 1}{2} \right)^2 + 8\left(\dfrac{x - 1}{2} \right) + 15 = 0$
$-9, -5$

36. $\left(\dfrac{x + 2}{3} \right)^2 - 5\left(\dfrac{x + 2}{3} \right) - 6 = 0$
$-5, 16$

37. $2\left(\dfrac{x + 2}{2} \right)^2 + \left(\dfrac{x + 2}{2} \right) - 3 = 0$
$-5, 0$

38. $3\left(\dfrac{x - 3}{3} \right)^2 + 5\left(\dfrac{x - 3}{3} \right) + 2 = 0$
$0, 1$

39. $x^{2/3} - 5x^{1/3} + 6 = 0$
$8, 27$

40. $x^{2/3} - 3x^{1/3} + 2 = 0$
$1, 8$

41. $2x^{2/3} - 3x^{1/3} - 2 = 0$
$-\dfrac{1}{8}, 8$

42. $3x^{2/3} - 4x^{1/3} - 4 = 0$
$-\dfrac{8}{27}, 8$

43. $x^{1/2} - 5x^{1/4} + 6 = 0$
$81, 16$

44. $x^{1/2} - 4x^{1/4} + 3 = 0$
$1, 81$

45. $5x^{1/2} + 8x^{1/4} - 4 = 0$
$\dfrac{16}{625}$ (16 is extraneous.)

46. $2x^{1/2} - x^{1/4} - 3 = 0$
$\dfrac{81}{16}$ (1 is extraneous.)

Objective 3

Prep Exercise 5 If Alisha can clean her house in x hours, what part of her house can she clean in 1 hour?
$\dfrac{1}{x}$

Prep Exercise 6 If a car travels 400 miles in x hours, how many miles does it travel in 1 hour?
$\dfrac{400}{x}$

For Exercises 47–58, solve. See Example 4.

47. The average rate of a bus is 15 miles per hour more than the average rate of a truck. The truck takes 1 hour longer than the bus to travel 180 miles. How long does the bus take to travel 180 miles?
3 hr.

48. A charter business has two types of small planes, a jet and a twin propeller plane. The average air speed of the prop plane is 120 miles per hour less than the jet's. The prop plane takes 1 hour longer to travel 720 miles than the jet does. How long does the prop plane take to travel the 720 miles?

3 hr.

★**49.** The average speed of Wesley Korir, who won the Boston Marathon in 2012, was 3.66 miles per hour faster than the person who finished 1 hour behind him. Find the time Wesley took to run the 26-mile course.

≈ 2.21 hr.

★**50.** The average speed of Sharon Cherop, who won first place among women in the 2012 Boston Marathon was 2.91 miles per hour faster than the person who finished 1 hour behind her. Find the time Sharon took to run the 26-mile course.

≈ 2.53 hr.

51. Suppose the time required for a bus to travel 360 miles is 3 hours more than the time required for a motorcycle to travel 300 miles. If the average rate of the motorcycle is 15 miles per hour more than the average rate of the bus, find the average rate of the bus.

45 mph

52. Suppose the time required for a truck to travel 400 miles is 3 hours more than the time required for a car to travel 300 miles. If the average rate of the car is 10 miles per hour more than the average rate of the truck, find the average rate of the truck.

50 mph

53. After training hard for a year, a novice cyclist discovers that she has increased her average rate by 6 miles per hour and can travel 36 miles in 1 hour less than she did a year ago.
  a. What was her old average rate? What is her new average rate?
    Old rate = 12 mph; new rate = 18 mph
  b. What was her old time to travel 36 miles? What is her new time for 36 miles?
    Old time = 3 hr.; new time = 2 hr.

54. A high school track coach determines that his fastest long-distance runner runs 2 miles per hour faster than his slowest runner. Compared to the faster runner, the slower runner takes a half hour longer to run 12 miles.
  a. Find the average rates of both runners.
    Fast rate = 8 mph; slow rate = 6 mph
  b. Find the time for both runners to run 12 miles.
    Fast time = 1.5 hr.; slow time = 2 hr.

55. Billy and Jody are commercial fishermen. Working alone, it takes Billy 2 hours longer to run the hoop nets than it takes Jody working alone. Together they can run the hoop nets in $2\frac{2}{5}$ hours. How long does it take each working alone?

Jody: 4 hr.; Billy: 6 hr.

56. Using a riding mower, Fran can mow the grass at the campground in 4 hours less time than it takes Donnie using a push mower. Together they can mow the grass in $2\frac{2}{3}$ hours. How long does it take each working alone?

Fran: 4 hr.; Donnie: 8 hr.

★**57.** A newspaper has two presses, one of which is older than the other. By itself, the newer press can print all of the copies for a typical day in a half hour less time than the older press takes. When running at the same time, the presses print all of a typical day's copies in 2 hours. How long would it take to print a day's worth if they worked alone?

Older: ≈ 4.27 hr.; newer: ≈ 3.77 hr.

Of Interest

The world record for completing a marathon was set by Geoffrey Mutai of Kenya in the 2011 Boston Marathon with a time of 2:03:02. The world record for women was set by Paula Radcliffe in the 2003 London Marathon with a time of 2:15:25.

(*Source:* marathonguide.com.)

★ **58.** A school's copy center prints the school newsletter in a half hour using two copiers running at the same time. If the center uses only one copier, the faster of the two copiers takes 6 minutes less time than the other copier to print all of the newsletters. How long does each copier take to print the newsletters working alone?

Slower: ≈1.05 hr.; faster: ≈ 0.95 hr.

■ *On a guitar, the frequency of the vibrating string is related to the tension on the string. Suppose two strings of the same diameter and length are placed on an instrument and wound to different tensions. The formula $\dfrac{F_1^2}{F_2^2} = \dfrac{T_1}{T_2}$ describes the relationship between their frequency and tension.*

★ **59.** One string is wound on a guitar to a tension of 50 pounds. A second string of the same diameter is wound to 60 pounds. If the string with the greater tension has a frequency that is 40 vibrations per second greater than the other string, what are the frequencies of the two strings? Round to the nearest ten.

420 vps, 460 vps

★ **60.** Two strings of the same diameter are wound on a banjo to different tensions: one at 80 pounds and the other at 90 pounds. The frequency of the string under less tension is 20 vibrations per second less than the other string. What are the frequencies of the two strings? Round to the nearest ten.

330 vps, 350 vps

Of Interest

In addition to tension, the diameter and length of a string also affect its frequency. Increasing the diameter or length of a string under the same tension decreases its frequency.

Review Exercises

Exercises 1 and 2   Expressions

[1.7] 1. Evaluate $-\dfrac{b}{2a}$ when a is 2 and b is 8.

-2

[1.7] 2. Evaluate $\dfrac{4ac - b^2}{4a}$ when $a = 2$, $b = -8$, and $c = 5$.

-3

Exercises 3–6   Equations and Inequalities

[3.3] 3. Find the x- and y-intercepts $2x + 3y = 6$

(3, 0), (0, 2)

[3.7] 4. If $f(x) = 2x^2 - 3x + 1$, find $f(-1)$.

6

[6.7] *For Exercises 5 and 6, graph.*

5. $f(x) = x^2 - 3$

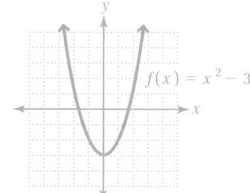

6. $f(x) = 3x^2$

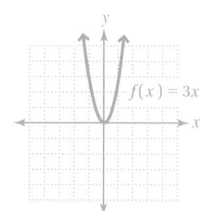

10.4 Graphing Quadratic Functions

Objectives

1 Graph quadratic functions of the form $f(x) = ax^2$.

2 Graph quadratic functions of the form $f(x) = ax^2 + k$.

3 Graph quadratic functions of the form $f(x) = a(x - h)^2$.

4 Graph quadratic functions of the form $f(x) = a(x - h)^2 + k$.

5 Graph quadratic functions of the form $f(x) = ax^2 + bx + c$.

6 Solve applications involving parabolas.

Warm-up

[10.3] **1.** Solve: $x^4 - 10x^2 + 9 = 0$

[1.7] **2.** Evaluate $-\dfrac{b}{2a}$ and $\dfrac{4ac - b^2}{4a}$ for $a = -0.8$, $b = 2.4$, and $c = 6$.

[10.1] **3.** Solve $8 = x^2 - 2x$ by completing the square.

[3.7] **4.** Given $f(x) = 2x^2 - 8x + 5$, find $f(2)$.

In Section 6.7 we learned that quadratic functions have the form $f(x) = ax^2 + bx + c$, where a, b, and c are real numbers and $a \neq 0$. By plotting many ordered pairs, we found that the graphs of these functions are parabolas that open up if $a > 0$ and down if $a < 0$. By replacing x with 0, we found that the y-intercept is $(0, c)$.

By replacing y (or $f(x)$) with 0, we found the x-intercepts. We also learned that every parabola will have symmetry about a line called its *axis of symmetry* and the axis of symmetry always passes through a point called the *vertex*, which is the lowest point on a parabola that opens upward or the highest point on a parabola that opens downward. Look at the following graphs.

x	$f(x) = x^2$
−3	9
−2	4
−1	1
0	0
1	1
2	4
3	9

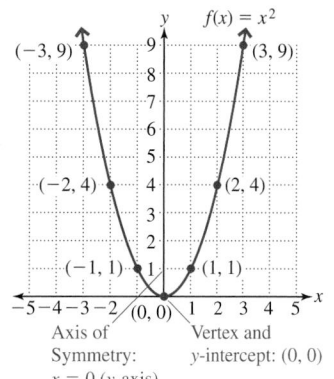

Axis of Symmetry: $x = 0$ (y-axis)

Vertex and y-intercept: $(0, 0)$

x	$f(x) = -x^2 + 4$
−3	−5
−2	0
−1	3
0	4
1	3
2	0
3	−5

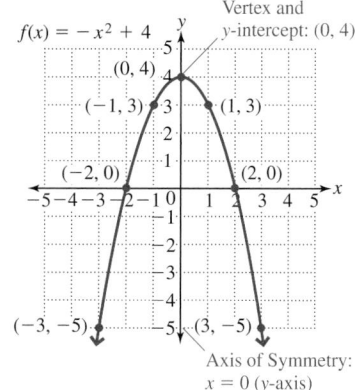

Axis of Symmetry: $x = 0$ (y-axis)

Notice that the graph of $f(x) = x^2$ opens upward because $a = 1$, which is positive, whereas the graph of $f(x) = -x^2 + 4$ opens downward because $a = -1$, which is negative. In this section, we learn more about the graphs of quadratic functions. Later in the section, we revisit the form $f(x) = ax^2 + bx + c$ and learn how to determine a parabola's vertex and axis of symmetry from the values of a and b.

Objective 1 Graph quadratic functions of the form $f(x) = ax^2$.

First, we consider $f(x) = ax^2$ to discover more about how a affects the parabola. We will see that a affects not only whether the parabola opens upward or downward but also how wide the parabola is.

Example 1 Compare the graphs of each function.

a. $f(x) = \dfrac{1}{4}x^2$, $g(x) = \dfrac{1}{3}x^2$, $h(x) = \dfrac{1}{2}x^2$, $k(x) = x^2$, and $m(x) = 2x^2$

Answers to Warm-up
1. $x = \pm 1, \pm 3$
2. $1.5, 7.8$
3. $x = 4, x = -2$
4. -3

Solution: We graph all five functions on the same grid.

Note All of these graphs open upward because a is positive. Also notice that as a increases, the parabolas appear narrower. ▶

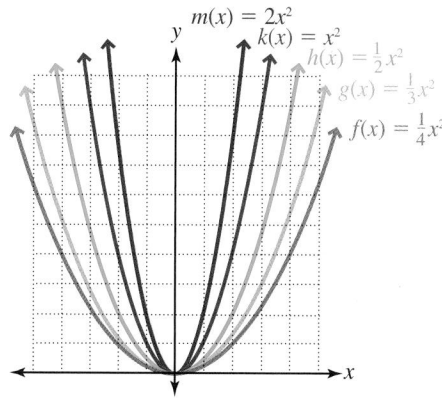

b. $f(x) = -\dfrac{1}{4}x^2$, $g(x) = -\dfrac{1}{3}x^2$, $h(x) = -\dfrac{1}{2}x^2$, $k(x) = -x^2$, and $m(x) = -2x^2$

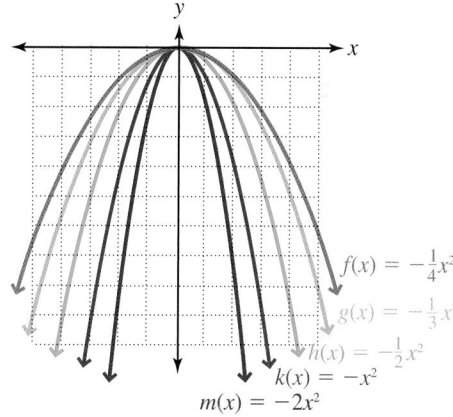

Note All of these graphs open downward because a is negative. Also notice that as $|a|$ increases, the parabolas appear narrower. ▶

Answers to Your Turn 1

a.

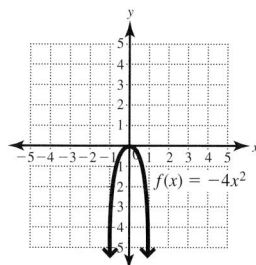

b.

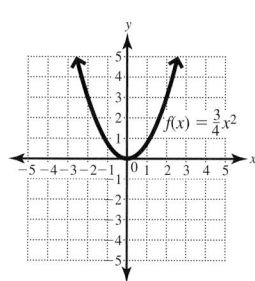

Example 1 suggests the following conclusions about functions of the form $f(x) = ax^2$.

Conclusion: Given a function in the form $f(x) = ax^2$, the axis of symmetry is $x = 0$ (the y-axis) and the vertex is $(0, 0)$. Also, the greater the absolute value of a, the narrower the parabola appears (or the smaller the absolute value of a, the wider the parabola appears).

Your Turn 1 | Graph the following functions.

a. $f(x) = -4x^2$

b. $f(x) = \dfrac{3}{4}x^2$

Objective 2 Graph quadratic functions of the form $f(x) = ax^2 + k$.

We have learned that the vertex of a parabola of the form $f(x) = ax^2$ is at $(0, 0)$ and the axis of symmetry is $x = 0$ (the y-axis). Now let's consider quadratic functions of the form $f(x) = ax^2 + k$, where k is a constant. We will see that for the same value of a, the graphs of $f(x) = ax^2 + k$ and $f(x) = ax^2$ have the same width and shape but the graph of $f(x) = ax^2 + k$ is shifted up or down k units on the y-axis from the origin so that the vertex is at $(0, k)$.

Example 2 Graph the following functions.

a. $g(x) = 2x^2 + 3$ and $h(x) = 2x^2 - 4$

Solution: We graph both functions on the same grid.

Note The vertex of the "basic" function $f(x) = 2x^2$ is the origin $(0, 0)$. The vertex of $g(x) = 2x^2 + 3$ is shifted up 3 and the vertex of $h(x) = 2x^2 - 4$ is shifted down 4 from the origin.

▶

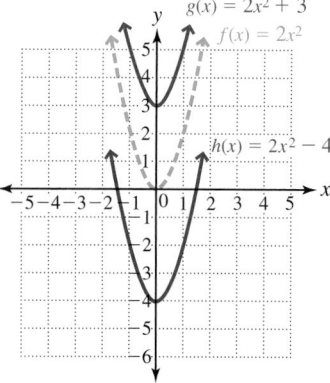

b. $g(x) = -x^2 + 4$ and $h(x) = -x^2 - 2$

Solution:

Note The vertex of the "basic" function $f(x) = -x^2$ is the origin $(0, 0)$. The vertex of $g(x) = -x^2 + 4$ is shifted up 4 and the vertex of $h(x) = -x^2 - 2$ is shifted down 2 from the origin.

▶

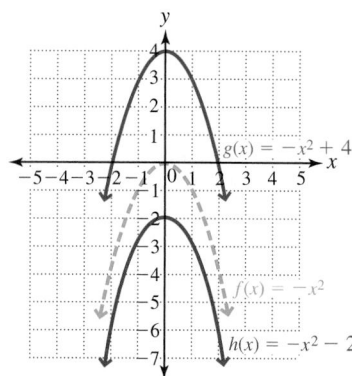

Example 2 suggests the following conclusion about the constant k in functions of the form $f(x) = ax^2 + k$.

Conclusion: Given a function in the form $f(x) = ax^2 + k$, if $k > 0$, then the graph of $f(x) = ax^2$ is shifted k units *up* from the origin. If $k < 0$, then the graph of $f(x) = ax^2$ is shifted $|k|$ units *down* from the origin. The new position of the vertex is $(0, k)$. The axis of symmetry is $x = 0$ (the y-axis).

Answers to Your Turn 2

a.

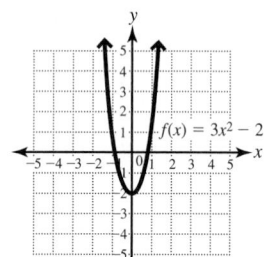

b.

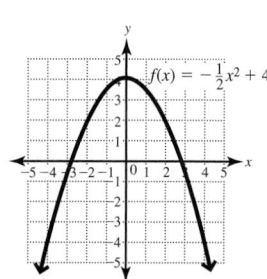

Your Turn 2 Graph the following functions.

a. $f(x) = 3x^2 - 2$

b. $f(x) = -\dfrac{1}{2}x^2 + 4$

Objective 3 Graph quadratic functions of the form $f(x) = a(x - h)^2$.

We now consider the form $f(x) = a(x - h)^2$, and we will see that the constant h causes the parabola to shift right or left.

Example 3 Graph $m(x) = 2(x - 3)^2$ and $n(x) = 2(x + 4)^2$.

Solution:

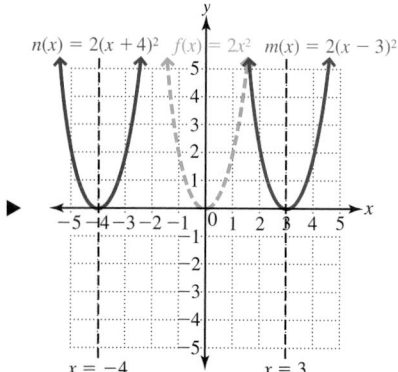

Note The vertex of the "basic" function $f(x) = 2x^2$ is $(0, 0)$, and the axis of symmetry is $x = 0$. The graph of $m(x) = 2(x - 3)^2$ has the same shape as $f(x) = 2x^2$, only the vertex and axis of symmetry are shifted **right** 3 units from the origin along the x-axis so that they are $(3, 0)$ and $x = 3$, respectively. Similarly, the vertex and axis of symmetry of $n(x) = 2(x + 4)^2$ are shifted **left** 4 units from the origin along the x-axis so that they are $(-4, 0)$ and $x = -4$, respectively.

▶

Example 3 suggests the following conclusion about h in $f(x) = a(x - h)^2$.

Conclusion: Given a function in the form $f(x) = a(x - h)^2$, if $h > 0$, then the graph of $f(x) = ax^2$ is shifted h units *right* from the origin. If $h < 0$, then the graph of $f(x) = ax^2$ is shifted $|h|$ units *left* from the origin. The new position of the vertex is $(h, 0)$, and the axis of symmetry is $x = h$.

> **Your Turn 3**   Graph the following functions.
>
> **a.** $f(x) = -2(x + 1)^2$ **b.** $g(x) = \dfrac{1}{3}(x - 4)^2$

Objective 4 Graph quadratic functions of the form
$$f(x) = a(x - h)^2 + k.$$

Examples 2 and 3 suggest that the graph of a function of the form $f(x) = a(x - h)^2 + k$ has the same shape as $f(x) = ax^2$ but the vertex is shifted from the origin to (h, k) and the axis of symmetry is shifted from $x = 0$ to $x = h$. We can summarize all we have learned with the following rule.

Rule Parabola with Vertex (h, k)

The graph of a function in the form $f(x) = a(x - h)^2 + k$ is a parabola with vertex at (h, k). The equation of the axis of symmetry is $x = h$. The parabola opens upward if $a > 0$ and downward if $a < 0$. The larger the $|a|$, the narrower the graph.

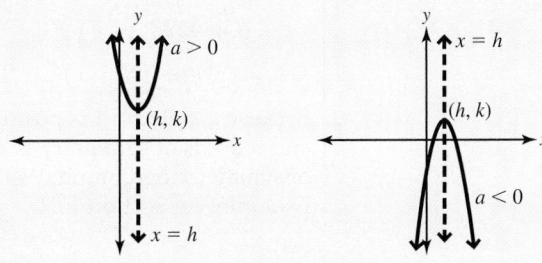

Now let's graph quadratic functions of the form $f(x) = a(x - h)^2 + k$. Recall that the domain of a function is the set of all possible x-values and the range is the set of all possible y-values.

Answers to Your Turn 3

a.

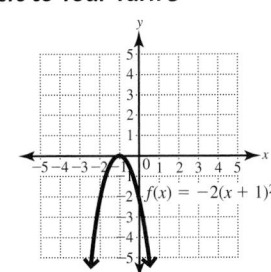

b.

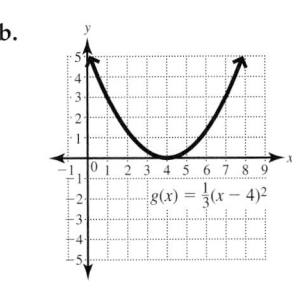

Example 4   Given $f(x) = 2(x - 3)^2 + 1$, determine whether the graph opens upward or downward, find the vertex and axis of symmetry, and draw the graph. Find the domain and range.

Solution: We see that $f(x) = 2(x - 3)^2 + 1$ is in the form $f(x) = a(x - h)^2 + k$, where $a = 2, h = 3$, and $k = 1$. Because a is positive 2, the parabola opens upward. The vertex is at $(3, 1)$, and the axis of symmetry is $x = 3$. To complete the graph, we find a few points on either side of the axis of symmetry.

Note We find these additional points by choosing x-values on either side of the axis of symmetry and using the equation to find the corresponding y-values. ▶

x	y
2	3
1	9
4	3
5	9

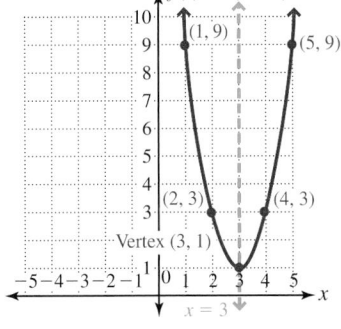

Domain: $\{x \mid x \text{ is a real number}\}$, or $(-\infty, \infty)$
Range: $\{y \mid y \geq 1\}$ or $[1, \infty)$

Your Turn 4 Given $f(x) = -\frac{1}{2}(x-1)^2 + 3$, determine whether the graph opens upward or downward, find the vertex and axis of symmetry, and draw the graph. Find the domain and range.

Objective 5 Graph quadratic functions of the form
$$f(x) = ax^2 + bx + c.$$

The advantage of the form $f(x) = a(x-h)^2 + k$ is that we can "see" the vertex and axis of symmetry and we can "see" whether the parabola opens upward or downward by looking at the equation. If an equation is in the form $f(x) = ax^2 + bx + c$, it can be transformed into $f(x) = a(x-h)^2 + k$ by completing the square.

Example 5 Write the function in the form $f(x) = a(x-h)^2 + k$. Then determine whether the graph opens upward or downward, find the vertex and axis of symmetry, and draw the graph. Find the domain and range.

a. $f(x) = x^2 - 2x - 8$

Solution: $y = x^2 - 2x - 8$ Replace $f(x)$ with y to make manipulations more workable.

$y + 8 = x^2 - 2x$ Add 8 to both sides of the equation to isolate the x terms.

$y + 8 + 1 = x^2 - 2x + 1$ Add 1 to both sides of the equation to complete the square.

$y + 9 = (x - 1)^2$ Factor the right side as a perfect square.

$y = (x - 1)^2 - 9$ Subtract 9 from both sides to solve for y.

Because $a = 1$ and 1 is positive, the graph opens upward. The vertex is at $(1, -9)$, and the axis of symmetry is $x = 1$. Plot a few points on either side of the axis of symmetry and graph. We include the x- and y-intercepts, which we found in Example 3 of Section 10.2.

Note Remember, to find x-intercepts, we replace $f(x)$ (or y) with 0 and solve for x. To find y-intercepts, we replace x with 0 and solve for y.

	x	$f(x)$
x-intercept $\rightarrow$	-2	0
	-1	-5
y-intercept $\rightarrow$	0	-8
vertex $\rightarrow$	1	-9
	2	-8
	3	-5
x-intercept $\rightarrow$	4	0

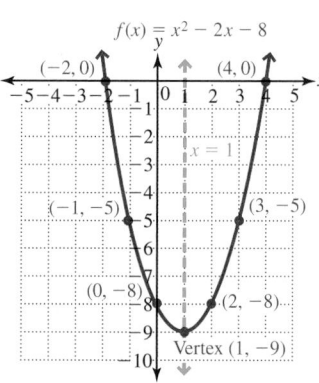

Note The points we found are 1, 2, and 3 units to the left and right of the axis of symmetry.

Domain: $\{x \,|\, x \text{ is a real number}\}$, or $(-\infty, \infty)$
Range: $\{y \,|\, y \geq -9\}$ or $[-9, \infty)$

b. $f(x) = 2x^2 + 12x + 19$

Solution: $y = 2x^2 + 12x + 19$ Replace $f(x)$ with y.

$y - 19 = 2x^2 + 12x$ Subtract 19 from both sides of the equation.

$y - 19 = 2(x^2 + 6x)$ Factor 2 out of the right side of the equation.

$y - 19 + 18 = 2(x^2 + 6x + 9)$ Complete the square inside the parentheses.

Note We added 18 to the left side of the equation because $2 \cdot 9 = 18$ was added to the right side.

$y - 1 = 2(x + 3)^2$ Combine terms on the left and factor on the right.

$y = 2(x + 3)^2 + 1$ Add 1 to both sides.

Because $a = 2$ and 2 is positive, the graph opens upward. The vertex is at $(-3, 1)$, and the axis of symmetry is $x = -3$. Plot a few points on either side of the axis of symmetry and graph.

Answer to Your Turn 4
opens downward; vertex: $(1, 3)$; axis: $x = 1$; domain: $\{x \,|\, x \text{ is a real number}\}$, or $(-\infty, \infty)$; range: $\{y \,|\, y \leq 3\}$ or $(-\infty, 3]$

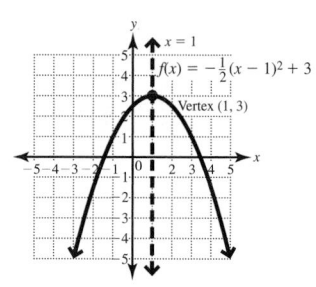

Note The points we found are 1 and 2 units to the right and left of the axis of symmetry.

x	f(x)
−2	3
−1	9
−4	3
−5	9

Note that there are no x-intercepts and the y-intercepts is $(0, 19)$.

Domain: $\{x \mid x$ is a real number$\}$, or $(-\infty, \infty)$

Range: $\{y \mid y \geq 1\}$ or $[1, \infty)$

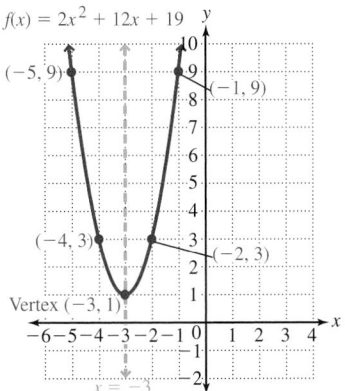

$f(x) = 2x^2 + 12x + 19$

$(-5, 9)$ $(-1, 9)$

$(-4, 3)$ $(-2, 3)$

Vertex $(-3, 1)$

$x = -3$

Your Turn 5   Write the function in the form $f(x) = a(x - h)^2 + k$; determine whether the graph opens upward or downward; find the x- and y-intercepts, vertex, and axis of symmetry; and draw the graph. Find the domain and range.

a. $f(x) = x^2 - 2x - 3$ **b.** $f(x) = -2x^2 + 8x - 6$

We also can find the vertex using a formula that we derive by completing the square on the general form $f(x) = ax^2 + bx + c$.

$y = ax^2 + bx + c$	Replace $f(x)$ with y.
$y - c = ax^2 + bx$	Isolate the x terms.
$y - c = a\left(x^2 + \dfrac{b}{a}x\right)$	Factor out a.
$y - c + \dfrac{b^2}{4a} = a\left(x^2 + \dfrac{b}{a}x + \dfrac{b^2}{4a^2}\right)$	Complete the square.
$y - \dfrac{4ac - b^2}{4a} = a\left(x + \dfrac{b}{2a}\right)^2$	Find the LCD of the two terms on the left. Factor on the right.
$y = a\left(x + \dfrac{b}{2a}\right)^2 + \dfrac{4ac - b^2}{4a}$	Add $\dfrac{4ac - b^2}{4a}$ to both sides.

Note Adding $\left[\dfrac{1}{2}\left(\dfrac{b}{a}\right)\right]^2$ or $\dfrac{b^2}{4a^2}$ completes the square inside the parentheses. Because ◀ those parentheses are multiplied by a, we add $a \cdot \dfrac{b^2}{4a^2} = \dfrac{b^2}{4a}$ to the left side to keep the equation balanced.

Notice that if we rewrite the last line above as $y = a\left(x - \left(-\dfrac{b}{2a}\right)\right)^2 + \dfrac{4ac - b^2}{4a}$, we see that $h = -\dfrac{b}{2a}$ and $k = \dfrac{4ac - b^2}{4a}$ so that the vertex is $\left(-\dfrac{b}{2a}, \dfrac{4ac - b^2}{4a}\right)$. This also means that the axis of symmetry is $x = -\dfrac{b}{2a}$. Also, although the y-coordinate of the vertex is $\dfrac{4ac - b^2}{4a}$, it is usually easier to find the vertex by substituting the x-coordinate into the original function.

Procedure   **Finding the Vertex of a Quadratic Function in the Form**
$$f(x) = ax^2 + bx + c$$

Given an equation in the form $f(x) = ax^2 + bx + c$, to determine the vertex:

1. Find the x-coordinate using the formula $x = -\dfrac{b}{2a}$.

2. Find the y-coordinate by evaluating $f\left(-\dfrac{b}{2a}\right)$ or $\dfrac{4ac - b^2}{4a}$.

Answers to Your Turn 5

a. x-intercepts: $(-1, 0)$, $(3, 0)$
y-intercept: $(0, -3)$
$f(x) = (x - 1)^2 - 4$;
opens upward; vertex: $(1, -4)$;
axis: $x = 1$

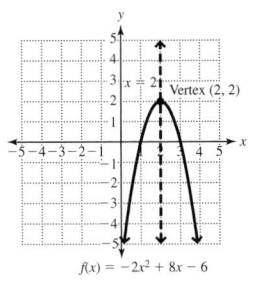

$f(x) = x^2 - 2x - 3$ Vertex $(1, -4)$ $x = 1$

domain: $\{x \mid x$ is a real number$\}$, or $(-\infty, \infty)$; range: $\{y \mid y \geq -4$ or $[-4, \infty)$

b. $f(x) = -2(x - 2)^2 + 2$;
x-intercepts: $(1, 0)$, $(3, 0)$
y-intercepts: $(0, -6)$;
opens downward;
vertex: $(2, 2)$; axis: $x = 2$

$x = 2$ Vertex $(2, 2)$

$f(x) = -2x^2 + 8x - 6$

domain $\{x \mid x$ is a real number$\}$, or $(-\infty, \infty)$;
range: $\{y \mid y \leq 2\}$ or $(-\infty, 2]$

Note Using $\dfrac{4ac - b^2}{4a}$ also gives the y-coordinate:

$$\dfrac{4(2)(5) - (-8)^2}{4(2)} = \dfrac{-24}{8} = -3$$

However, substituting the x-value, 2, into the original function is easier. ▶

Example 6 For the function $f(x) = 2x^2 - 8x + 5$, find the coordinates of the vertex.

Solution: First find the x-coordinate of the vertex using $-\dfrac{b}{2a}$.

$$x\text{-coordinate of the vertex: } -\dfrac{b}{2a} = -\dfrac{(-8)}{2(2)} = \dfrac{8}{4} = 2$$

Now find the y-coordinate by evaluating $f(2)$.

$$y\text{-coordinate of the vertex: } f(2) = 2(2)^2 - 8(2) + 5 = 8 - 16 + 5 = -3$$

The vertex is $(2, -3)$.

Your Turn 6 For the equation $f(x) = 3x^2 + 18x - 19$, find the vertex.

Objective 6 Solve applications involving parabolas.

Because the vertex of a parabola is the highest point in parabolas that open downward or the lowest point in parabolas that open upward, we can find a minimum or maximum value in applications involving quadratic functions.

Example 7 A toy rocket is launched straight up with an initial velocity of 40 feet per second. The equation $h = -16t^2 + 40t$ describes the height, h, of the rocket t seconds after being launched.

a. Find the maximum height the rocket reaches.
b. Find the amount of time the rocket is in the air.
c. Draw the graph of the height of the rocket.

Solution:

Instructor Note Underscore that this graph represents the relationship between height and flight time. It does *not* represent the trajectory, or path, of the flight.

a. Because the graph of $h = -16t^2 + 40t$ is a parabola that opens down ($a = -16$), the maximum height occurs at its vertex.

$$t\text{-coordinate of the vertex: } -\dfrac{b}{2a} = -\dfrac{40}{2(-16)} = -\dfrac{40}{-32} = \dfrac{5}{4} = 1.25 \text{ seconds}$$

To find the h-coordinate of the vertex, we replace t with 1.25 in $h = -16t^2 + 40t$.

$$h = -16(1.25)^2 + 40(1.25) = 25 \text{ feet}$$

Note We could have found the vertex by writing the equation in $y = a(x - h)^2 + k$ form, which would be $h = -16(t - 1.25)^2 + 25$. ▶

The vertex is $(1.25, 25)$, so the maximum height is 25 feet, which occurs 1.25 seconds after the rocket is launched.

b. The time the rocket is in the air is from launch until it returns to the ground. At launch and upon returning to the ground, the rocket's height is 0; so we need to find t when $h = 0$.

$$0 = -16t^2 + 40t \qquad \text{Replace } h \text{ with 0.}$$
$$0 = -8t(2t - 5) \qquad \text{Factor out a common factor of } -8t.$$
$$-8t = 0 \quad \text{or} \quad 2t - 5 = 0 \qquad \text{Use the zero-factor theorem.}$$
$$t = 0 \qquad\qquad 2t = 5$$
$$t = 2.5$$

Answer to Your Turn 6
$(-3, -46)$

This means that height is 0 when $t = 0$ and when $t = 2.5$ seconds, so the rocket is in the air for 2.5 seconds.

Note The graph of
$h = -16t^2 + 40t$ is **not** the flight path of the rocket. Also, because the situation involves a rocket being launched from the ground (0 feet) and returning to the ground, only the portion of the graph in the first quadrant is realistic.

▶

c. The graph of $h = -16t^2 + 40t$ shows that at 0 seconds, the rocket is at 0 feet, which is when it is launched. Its height increases to a maximum of 25 feet 1.25 seconds after launch. Then the height decreases back to 0 feet 2.5 seconds after launch.

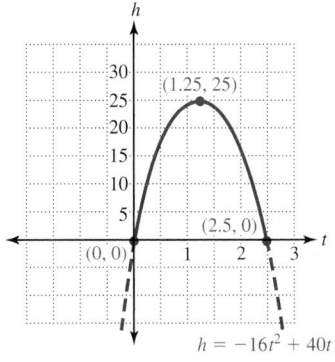

Your Turn 7 A soccer player kicks a ball straight up with an initial velocity of 56 feet per second. The equation $h = -16t^2 + 56t$ describes the height, h, of the ball t seconds after being kicked.

a. After how many seconds is the ball at its maximum height?
b. What is the maximum height the ball reaches?
c. How long is the ball in the air?

If an object is thrown or shot upward and outward, gravity causes its path, or trajectory, to be in the shape of a parabola.

Note x represents the horizontal distance the object travels, and y represents the vertical distance the object travels. ▶

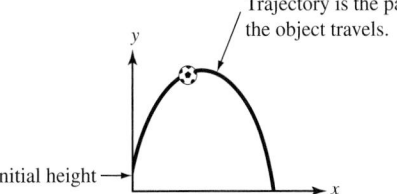

Trajectory is the path the object travels.

Initial height →

Example 8 The equation $y = -0.8x^2 + 2.4x + 6$ when $x \geq 0$ and $y \geq 0$ describes the trajectory of a ball thrown upward and outward from an initial height of 6 feet.

a. What is the maximum height the ball reaches?
b. How far does the ball travel horizontally?
c. Graph the trajectory of the ball.

Solution:

a. The maximum height will be the y-coordinate of the vertex.

$$x\text{-coordinate of the vertex: } -\frac{b}{2a} = -\frac{(2.4)}{2(-0.8)} = 1.5$$

$$y\text{-coordinate of the vertex: } y = -0.8(1.5)^2 + 2.4(1.5) + 6 = 7.8$$

The vertex is $(1.5, 7.8)$, so the maximum height is 7.8 feet.

b. To find how far the ball travels, we need to find the x-value when it hits the ground, which is where $y = 0$.

$$0 = -0.8x^2 + 2.4x + 6 \qquad \text{Substitute 0 for } y \text{ in the equation.}$$

$$x = \frac{-(2.4) \pm \sqrt{(2.4)^2 - 4(-0.8)(6)}}{2(-0.8)} \qquad \begin{array}{l}\text{Substitute } a = -0.8, b = 2.4, \text{ and}\\ c = 6 \text{ into the quadratic formula.}\end{array}$$

$$= \frac{-2.4 \pm \sqrt{24.96}}{-1.6} \approx -1.62 \text{ or } 4.62$$

Answers to Your Turn 7
a. 1.75 sec. **b.** 49 ft. **c.** 3.5 sec.

The negative value does not make sense in the context of this problem, so the distance the ball travels must be approximately 4.62 feet.

c.

Note Because the equation describes the trajectory when $x \geq 0$ and $y \geq 0$, we consider only the part of the graph in the first quadrant. ▶

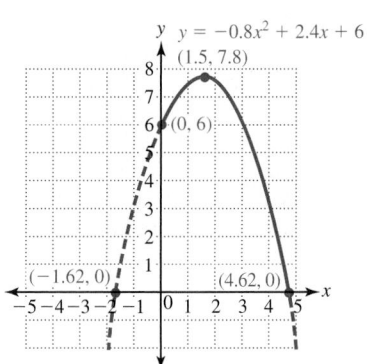

10.4 Exercises For Extra Help MyMathLab®

Note: Exercises marked with a ★ represent challenging exercises.

Objectives 1–4

Prep Exercise 1 In an equation in the form $f(x) = a(x - h)^2 + k$, what determines whether the parabola opens upward or downward?
The sign of a

Prep Exercise 2 What is the axis of symmetry in a parabola that opens up or down?
The axis of symmetry is the vertical line through the vertex.

Prep Exercise 3 In an equation in the form $f(x) = a(x - h)^2 + k$, what are the coordinates of the vertex? What is the equation of the axis of symmetry?
$(h, k); x = h$

For Exercises 1–16, find the coordinates of the vertex and write the equation of the axis of symmetry. See Examples 4–6.

1. $f(x) = 5(x - 2)^2 - 3$
V: $(2, -3)$; axis: $x = 2$

2. $h(x) = 2(x - 1)^2 + 4$
V: $(1, 4)$; axis: $x = 1$

3. $g(x) = -2(x + 1)^2 - 5$
V: $(-1, -5)$; axis: $x = -1$

4. $k(x) = -(x + 4)^2 - 3$
V: $(-4, -3)$; axis: $x = -4$

5. $k(x) = x^2 + 2$
V: $(0, 2)$; axis: $x = 0$

6. $g(x) = x^2 + 5$
V: $(0, 5)$; axis: $x = 0$

7. $h(x) = 3(x + 5)^2$
V: $(-5, 0)$; axis: $x = -5$

8. $f(x) = 5(x + 1)^2$
V: $(-1, 0)$; axis: $x = -1$

9. $f(x) = -0.5x^2$
V: $(0, 0)$; axis: $x = 0$

10. $k(x) = 2.5x^2$
V: $(0, 0)$; axis: $x = 0$

11. $f(x) = x^2 - 4x + 8$
V: $(2, 4)$; axis: $x = 2$

12. $g(x) = x^2 + 6x + 8$
V: $(-3, -1)$; axis: $x = -3$

13. $k(x) = 2x^2 + 16x + 27$
V: $(-4, -5)$; axis: $x = -4$

14. $f(x) = 3x^2 - 6x + 1$
V: $(1, -2)$; axis: $x = 1$

15. $g(x) = -3x^2 + 2x + 1$
V: $\left(\dfrac{1}{3}, \dfrac{4}{3}\right)$; axis: $x = \dfrac{1}{3}$

16. $h(x) = -2x^2 - 3x + 2$
V: $\left(-\dfrac{3}{4}, \dfrac{25}{8}\right)$; axis: $x = -\dfrac{3}{4}$

Objective 5

For Exercises 17–30:   *a.   State whether the parabola opens upward or downward.*
b.   Find the coordinates of the vertex.
c.   Write the equation of the axis of symmetry.
d.   Graph. See Examples 1–4.

17. $h(x) = -3x^2$
a. downward   b. $(0,0)$ c. $x = 0$
d.
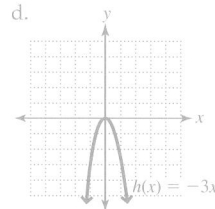

18. $f(x) = 2x^2$
a. upward   b. $(0,0)$ c. $x = 0$
d.
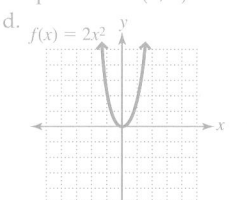

19. $k(x) = \frac{1}{4}x^2$
a. upward   b. $(0,0)$ c. $x = 0$
d.
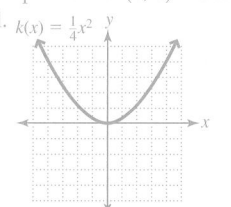

20. $g(x) = -\frac{1}{3}x^2$
a. downward   b. $(0,0)$ c. $x = 0$
d.
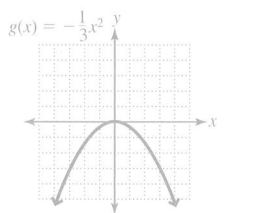

21. $f(x) = 4x^2 - 3$
a. upward   b. $(0,-3)$ c. $x = 0$
d.
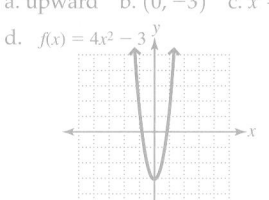

22. $h(x) = -2x^2 + 4$
a. downward   b. $(0,4)$ c. $x = 0$
d.
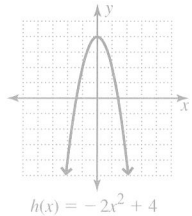

23. $g(x) = -0.5x^2 + 2$
a. downward   b. $(0,2)$ c. $x = 0$
d.
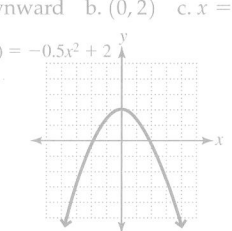

24. $k(x) = 0.5x^2 - 3$
a. upward b. $(0,-3)$ c. $x = 0$
d.
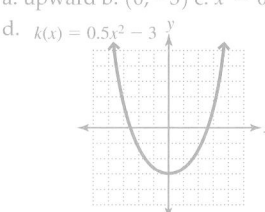

25. $f(x) = (x - 3)^2 + 2$
a. upward   b. $(3,2)$ c. $x = 3$
d.
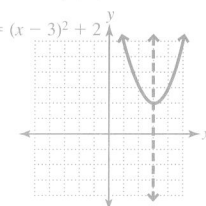

26. $h(x) = (x - 1)^2 - 3$
a. upward   b. $(1,-3)$ c. $x = 1$
d.
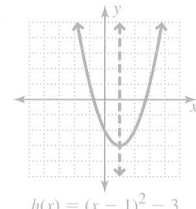

27. $k(x) = -2(x + 1)^2 - 3$
a. downward   b. $(-1,-3)$ c. $x = -1$
d.
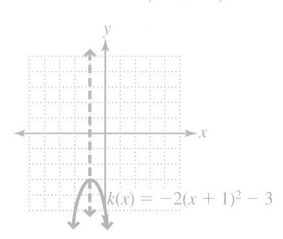

28. $g(x) = -2(x + 3)^2 + 1$
a. downward   b. $(-3,1)$ c. $x = -3$
d.
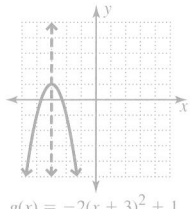

29. $h(x) = \frac{1}{3}(x - 2)^2 - 1$
a. upward   b. $(2,-1)$ c. $x = 2$
d.
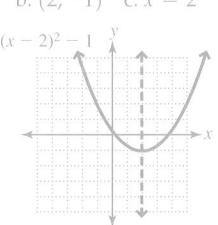

30. $f(x) = -\frac{1}{4}(x - 3)^2 - 2$
a. downward   b. $(3,-2)$ c. $x = 3$
d.
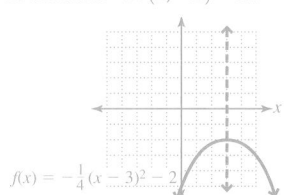

Prep Exercise 4 Suppose the y-intercept for a given quadratic equation is $(0, -5)$. If the vertex is at $(3, -2)$, what can you conclude about the x-intercepts?

There are no x-intercepts.

Prep Exercise 5 If the solutions for a quadratic equation are imaginary numbers, what does that indicate about the graph of the corresponding quadratic function?

There are no x-intercepts.

Prep Exercise 6 Given an equation in the form $y = ax^2 + bx + c$, what is the x-coordinate of the vertex?

$x = -\dfrac{b}{2a}$

For Exercises 31–36:
   a. Find the x- and y-intercepts.
   b. Write the equation in the form $f(x) = a(x - h)^2 + k$.
   c. State whether the parabola opens upward or downward.
   d. Find the coordinates of the vertex.
   e. Write the equation of the axis of symmetry.
   f. Graph. See Examples 5 and 6.
   g. Find the domain and range.

31. $h(x) = x^2 + 6x + 9$
  a. $(-3, 0), (0, 9)$
  b. $h(x) = (x + 3)^2$
  c. upward d. $(-3, 0)$ e. $x = -3$
  f.
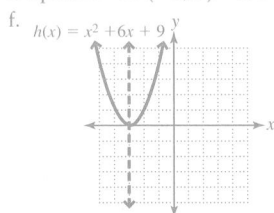
  g. domain: $\{x \mid x \text{ is a real number}\}$,
    or $(-\infty, \infty)$;
    range: $\{y \mid y \geq 0\}$ or $[0, \infty)$

32. $k(x) = x^2 - 4x + 4$
  a. $(2, 0), (0, 4)$ b. $k(x) = (x - 2)^2$
  c. upward d. $(2, 0)$ e. $x = 2$
  f.
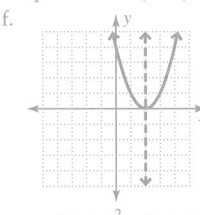
  g. domain: $\{x \mid x \text{ is a real number}\}$,
    or $(-\infty, \infty)$;
    range: $\{y \mid y \geq 0\}$ or $[0, \infty)$

33. $g(x) = -3x^2 + 6x - 5$
  a. no x-intercepts, $(0, -5)$
  b. $g(x) = -3(x - 1)^2 - 2$
  c. downward d. $(1, -2)$ e. $x = 1$
  f.
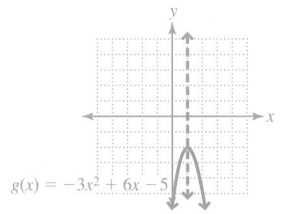
  g. domain: $\{x \mid x \text{ is a real number}\}$,
    or $(-\infty, \infty)$;
    range: $\{y \mid y \leq -2\}$ or $(-\infty, -2]$

34. $k(x) = -x^2 - 2x + 3$
  a. $(-3, 0), (1, 0), (0, 3)$
  b. $k(x) = -(x + 1)^2 + 4$
  c. downward d. $(-1, 4)$ e. $x = -1$
  f.
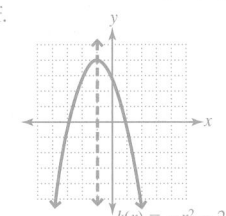
  g. domain: $\{x \mid x \text{ is a real number}\}$,
    or $(-\infty, \infty)$;
    range: $\{y \mid y \leq 4\}$ or $(-\infty, 4]$

35. $f(x) = 2x^2 + 6x + 3$
  a. $\left(\dfrac{-3 + \sqrt{3}}{2}, 0\right), \left(\dfrac{-3 - \sqrt{3}}{2}, 0\right), (0, 3)$
  b. $f(x) = 2\left(x + \dfrac{3}{2}\right)^2 - \dfrac{3}{2}$ c. upward
  d. $\left(-\dfrac{3}{2}, -\dfrac{3}{2}\right)$ e. $x = -\dfrac{3}{2}$
  f.
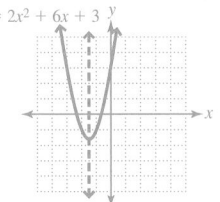
  g. domain: $\{x \mid x \text{ is a real number}\}$,
    or $(-\infty, \infty)$; range:
    $\left\{y \mid y \geq -\dfrac{3}{2}\right\}$ or $\left[-\dfrac{3}{2}, \infty\right)$

36. $h(x) = -3x^2 - 6x + 4$
  a. $\left(\dfrac{-3 + \sqrt{21}}{3}, 0\right), \left(\dfrac{-3 - \sqrt{21}}{3}, 0\right),$
    $(0, 4)$
  b. $h(x) = -3(x + 1)^2 + 7$
  c. downward d. $(-1, 7)$ e. $x = -1$
  f.
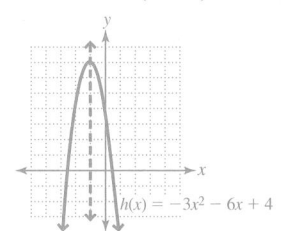
  g. domain: $\{x \mid x \text{ is a real number}\}$,
    or $(-\infty, \infty)$;
    range: $\{y \mid y \leq 7\}$ or $(-\infty, 7]$

Objective 6

For Exercises 37–48, solve. See Examples 7 and 8.

37. A toy rocket is launched with an initial velocity of 45 meters per second. The equation $h = -4.9t^2 + 45t$ describes the height, h, of the rocket in meters t seconds after being launched.
  a. After how many seconds does the rocket reach its maximum height?
    4.59 sec.
  b. What is the maximum height the rocket reaches?
    103.32 m
  c. How long is the rocket in the air?
    9.18 sec.

38. A ball is drop-kicked straight up with an initial velocity of 36 feet per second. The equation $h = -16t^2 + 36t$ describes the height, h, of the ball in feet t seconds after being kicked.
  a. After how many seconds does the ball reach its maximum height?
    1.125 sec.
  b. What is the maximum height the ball reaches?
    20.25 ft.
  c. How long is the ball in the air?
    2.25 sec.

39. The equation $y = -0.8x^2 + 3.2x + 6$ models the trajectory of a ball thrown upward and outward from a height of 6 feet. (Assume that $x \geq 0$ and $y \geq 0$.)
  a. What is the maximum height the ball reaches?
    9.2 ft.
  b. How far does the ball travel horizontally?
    ≈ 5.39 ft.
  c. Graph the trajectory of the ball.

c.
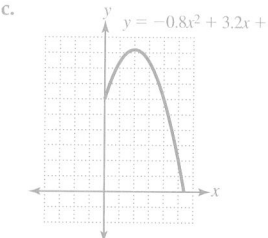

40. The javelin toss is an event in track and field in which participants try to throw a javelin the farthest. Suppose the equation $y = -0.02x^2 + 1.3x + 8$ models the trajectory of one particular throw. (Assume that x and y represent distances in meters and that $x \geq 0$ and $y \geq 0$.)
  a. What is the maximum height the javelin reaches?
    29.125 m
  b. How far does the javelin travel horizontally?
    70.661 m
  c. Graph the trajectory of the javelin.

c.
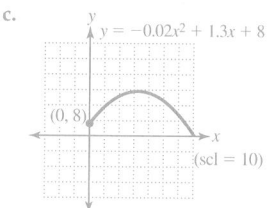

41. A record company discovers that the number of CDs sold each week after release follows a parabolic pattern. The function $n(t) = -200t^2 + 4000t$ describes the number, n, of CDs an artist sold each of t weeks after the release of the album.
  a. Which week had the greatest number of CDs sold?
    The greatest number of CDs were sold in the 10th week.
  b. How many CDs sold that week?
    20,000

42. The function $n(t) = -3t^2 + 42t$ describes the number, n, of tickets sold for a play each of t days after tickets went on sale.
  a. What day had the greatest number of tickets sold?
    The greatest number of tickets were sold on the 7th day.
  b. How many tickets sold that day?
    147

★ **43.** A farmer has enough materials to build a fence with a total length of 400 feet. He wants the enclosed space to be rectangular and wants to maximize the area enclosed. Find the length and width so that the area is maximized.
    The area is maximized if the length and width are 100 ft.

★ **44.** An architect wants the length and width of a rectangular building to be a total of 150 feet long. She also wants to maximize the rectangular base area that the building occupies. Find the length and width so that the base area is maximized.
    75 ft.

45. The function $C(n) = n^2 - 110n + 5000$ describes a company's cost, C, of producing n units of its product. Find the number of units the company should produce to minimize its cost.
    55 units

46. The function $P(t) = 0.001t^2 - 0.24t + 59.90$ roughly models a particular stock's closing price, P, each of t days of trading during one year.
  a. After how many days is the price at its lowest?
    120 days
  b. What was the lowest price during that year?
    $45.50

★ **47.** One integer is 12 more than another. If their product is minimized, find the integers and their product.
    $-6, 6$; the product is -36.

★ **48.** The greater of two integers minus the smaller integer gives a result of 20. Find the two integers so that their product is minimized. Also find their product.
    $-10, 10$; the product is -100.

Puzzle Problem Given the vertex and the x- and y-intercepts of a parabola, reconstruct the equation.

a. Vertex $(-3, -1)$
   Intercepts $(-2, 0)$, $(-4, 0)$, $(0, 8)$
   $y = x^2 + 6x + 8$

b. Vertex $(-1, 0)$
   Intercepts $(-1, 0)$, $(0, 1)$
   $y = x^2 + 2x + 1$

c. Vertex $(-2, 1)$
   Intercepts $(0, 5)$
   $y = x^2 + 4x + 5$

d. Vertex $(1, 0)$
   Intercepts $(1, 0)$, $(0, 1)$
   $y = x^2 - 2x + 1$

Review Exercises

Exercise 1  Expressions

[1.5] **1.** Evaluate: $-|-2 + 3 \cdot 4| - 4^0$
   -11

Exercises 2–6 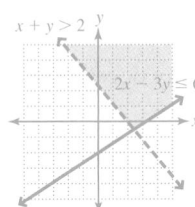 Equations and Inequalities

[6.6] **2.** Solve $x^2 + 2x = 15$ by factoring.
   $-5, 3$

[10.1] **3.** Solve $x^2 = -20$ using the square root principle.
   $\pm 2i\sqrt{5}$

For Exercises 4–6, solve; then graph the solution set.

[2.8] **4.** $4x - 8 \le 2x + 1$
   $x \le 4.5$

[8.3] **5.** $|x - 3| \ge 8$
   $x \le -5 \text{ or } x \ge 11$

[4.6] **6.** $\begin{cases} x + y > 2 \\ 2x - 3y \le 6 \end{cases}$

10.5 Solving Nonlinear Inequalities

Objectives

1 Solve quadratic and other inequalities.
2 Solve rational inequalities.

Warm-up

[10.4] **1.** Graph: $y = x^2 - 4x + 1$

[1.7] **2.** Evaluate $\dfrac{x - 4}{x + 2}$ for $x = 5$.

[2.8] **3.** Write $0 \le x \le 5$ in interval notation and graph on a number line.

[7.6] **4.** Solve: $\dfrac{x + 5}{x - 1} = 4$

Answers to Warm-up

1.

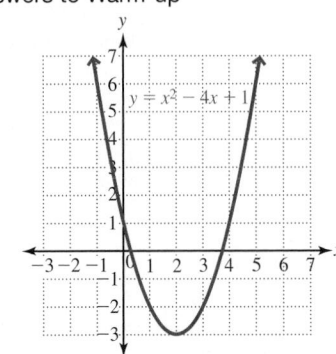

2. $\dfrac{1}{7}$

3. $[0, 5]$;

4. $x = 3$

Objective 1 Solve quadratic and other inequalities.

Now that we have solved quadratic equations, we can learn how to solve **quadratic inequalities**.

Definition Quadratic inequality: An inequality that can be written in the form $ax^2 + bx + c > 0$ or $ax^2 + bx + c < 0$, where $a \neq 0$.
Note: The symbols $<$ and $>$ can be replaced with $\leq$ and $\geq$.

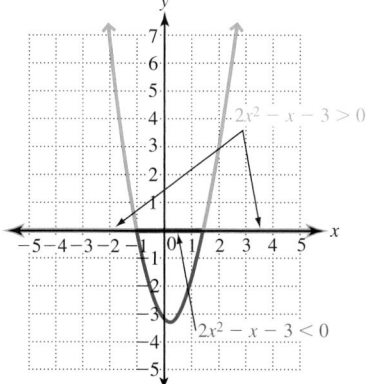

For example, $2x^2 - x - 3 < 0$ and $2x^2 - x - 3 \geq 0$ are quadratic inequalities. Let's see what we can learn about these inequalities by looking at the graph of the corresponding quadratic function: $y = 2x^2 - x - 3$. Note that by letting $y = 0$, we have the corresponding equation $2x^2 - x - 3 = 0$, and solving this equation gives the x-intercepts $x = -1$ and $x = \frac{3}{2}$. Also notice that those x-intercepts divide the x-axis into three intervals: $(-\infty, -1)$, $\left(-1, \frac{3}{2}\right)$, and $\left(\frac{3}{2}, \infty\right)$.

Notice that the parabola is above the x-axis in intervals $(-\infty, -1)$ and $\left(\frac{3}{2}, \infty\right)$, shown in blue. Evaluating $y = 2x^2 - x - 3$ using any x-value in these intervals produces a y-value that is greater than 0; so these intervals are the solution sets for $2x^2 - x - 3 > 0$.

The parabola is below the x-axis in the interval $\left(-1, \frac{3}{2}\right)$, shown in red. Evaluating $y = 2x^2 - x - 3$ using any x-value in this interval produces a y-value that is less than 0; so this interval is the solution set for $2x^2 - x - 3 < 0$.

The following number lines indicate various solution sets based on the inequality.

$2x^2 - x - 3 > 0$

$$\xleftarrow{\hspace{1cm}} -6\ -5\ -4\ -3\ -2\ -1\ 0\ \ 1\ \tfrac{3}{2}2\ \ 3\ \ 4\ \ 5\ \ 6 \xrightarrow{\hspace{1cm}}$$

Solution set: $(-\infty, -1) \cup \left(\frac{3}{2}, \infty\right)$

$2x^2 - x - 3 < 0$

$$\xleftarrow{\hspace{1cm}} -6\ -5\ -4\ -3\ -2\ -1\ 0\ \ 1\ \tfrac{3}{2}2\ \ 3\ \ 4\ \ 5\ \ 6 \xrightarrow{\hspace{1cm}}$$

Solution set: $\left(-1, \frac{3}{2}\right)$

$2x^2 - x - 3 \geq 0$

$$\xleftarrow{\hspace{1cm}} -6\ -5\ -4\ -3\ -2\ -1\ 0\ \ 1\ \tfrac{3}{2}2\ \ 3\ \ 4\ \ 5\ \ 6 \xrightarrow{\hspace{1cm}}$$

Solution set: $(-\infty, -1] \cup \left[\frac{3}{2}, \infty\right)$

$2x^2 - x - 3 \leq 0$

$$\xleftarrow{\hspace{1cm}} -6\ -5\ -4\ -3\ -2\ -1\ 0\ \ 1\ \tfrac{3}{2}2\ \ 3\ \ 4\ \ 5\ \ 6 \xrightarrow{\hspace{1cm}}$$

Solution set: $\left[-1, \frac{3}{2}\right]$

Based on these observations, we use the following procedure to solve quadratic inequalities.

Procedure Solving Quadratic Inequalities

1. Solve the related equation $ax^2 + bx + c = 0$.
2. Plot the solutions of $ax^2 + bx + c = 0$ on a number line. These solutions divide the number line into intervals.
3. Choose a test number from each interval and substitute the number into the inequality. If the test number makes the inequality *true*, then all numbers in that interval will solve the inequality. If the test number makes the inequality *false*, then no numbers in that interval will solve the inequality.
4. State the solution set of the inequality: It is the union of all intervals that solve the inequality. If the inequality symbols are $\leq$ or $\geq$, the values from step 2 are included. If the symbols are $<$ or $>$, they are not solutions.

Instructor Note We can make an interesting connection between the graph of $y = x^2 - x - 6$ by evaluating $x^2 - x - 6$ in factored form. Evaluating $(x - 3)(x + 2)$ for any value in the interval $(-\infty, -2)$ gives two negative factors, which makes all of the y-values positive, which is why the graph is above the x-axis. For example, if we choose $x = -3$, we have

$$y = (-3 - 3)(-3 + 2)$$
$$= (-6)(-1).$$

Evaluating $(x - 3)(x + 2)$ for any value in the interval $(-2, 3)$ gives two factors with opposite signs, which makes all of the y-values negative, which is why the graph is below the x-axis. Evaluating $(x - 3)(x + 2)$ for any value in the interval $(3, \infty)$ gives two positive factors, which makes all of the y-values positive so that the graph is above the x-axis.

Connection Notice that the solution set for $x^2 - x - 6 < 0$ corresponds to the interval where the graph of $f(x) = x^2 - x - 6$ is below the x-axis and the x-intercepts are the endpoints of the interval and are not included in the solution set.

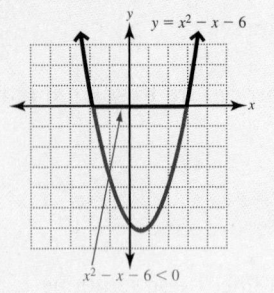

Example 1 Solve. Write the solution set using interval notation; then graph the solution set on a number line.

a. $x^2 - x < 6$

Solution: $x^2 - x = 6$ Write the related equation.

$\quad x^2 - x - 6 = 0$ Subtract 6 from both sides to get quadratic form.

$(x - 3)(x + 2) = 0$ Factor.

$x - 3 = 0 \quad \text{or} \quad x + 2 = 0$ Use the zero-factor theorem.

$\quad\quad x = 3 \quad\quad\quad\quad x = -2$ These are the x-intercepts of the graph of $y = x^2 - x - 6$.

Plot -2 and 3 on a number line (x-axis), which divides the number line into three intervals.

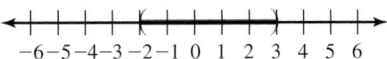

 $-6\ -5\ -4\ -3\ -2\ -1\ 0\ 1\ 2\ 3\ 4\ 5\ 6$

$(-\infty, -2)\quad (-2, 3)\quad\quad (3, \infty)$

◀ **Note** -2 and 3 are not part of the solution set because the inequality symbol is $<$.

Choose a test number from each interval and substitute that value into $x^2 - x < 6$.

For $(-\infty, -2)$, we choose $x = -3$.

$(-3)^2 - (-3) < 6$

$9 + 3 < 6$

$12 < 6$

This is false, so $(-\infty, -2)$ is not in the solution set.

For $(-2, 3)$, we choose $x = 0$.

$0^2 - 0 < 6$

$0 < 6$

This is true, so $(-2, 3)$ is in the solution set.

For $(3, \infty)$, we choose $x = 4$.

$4^2 - 4 < 6$

$16 - 4 < 6$

$12 < 6$

This is false, so $(3, \infty)$ is not in the solution set.

Because $(-2, 3)$ is the only interval that has solutions to the inequality, it is the solution set. Following is the graph of the solution set.

 $-6\ -5\ -4\ -3\ -2\ -1\ 0\ 1\ 2\ 3\ 4\ 5\ 6$

b. $x^2 + 3x \geq 0$

Solution: $x^2 + 3x = 0$ Write the related equation.

$\quad x(x + 3) = 0$ Factor.

$x = 0 \quad \text{or} \quad x + 3 = 0$ Use the zero-factor theorem.

$\quad\quad\quad\quad\quad x = -3$

Plot -3 and 0 on a number line and note the intervals.

 $-6\ -5\ -4\ -3\ -2\ -1\ 0\ 1\ 2\ 3\ 4\ 5\ 6$

$(-\infty, -3)\ (-3, 0)\quad\quad (0, \infty)$

◀ **Note** -3 and 0 are included in the solution set because the inequality symbol is $\geq$.

▲
Note Because we have already determined that -3 and 0 are in the solution set, we do not include them in any of our test intervals, which is why we use parentheses for each interval.

Choose a test number from each interval and test in $x^2 + 3x \geq 0$.

For $(-\infty, -3)$, we choose $x = -4$.

$(-4)^2 + 3(-4) \geq 0$

$16 - 12 \geq 0$

$4 \geq 0$

This is true, so $(-\infty, -3)$ is in the solution set.

For $(-3, 0)$, we choose $x = -1$.

$(-1)^2 + 3(-1) \geq 0$

$1 - 3 \geq 0$

$-2 \geq 0$

This is false, so $(-3, 0)$ is not in the solution set.

For $(0, \infty)$, we choose $x = 1$.

$1^2 + 3(1) \geq 0$

$1 + 3 \geq 0$

$4 \geq 0$

This is true, so $(0, \infty)$ is in the solution set.

Connection Notice that the solution set for $x^2 + 3x \geq 0$ corresponds to the intervals where the graph of $f(x) = x^2 + 3x$ is above the x-axis and the x-intercepts are endpoints in the intervals and are included in the solution set.

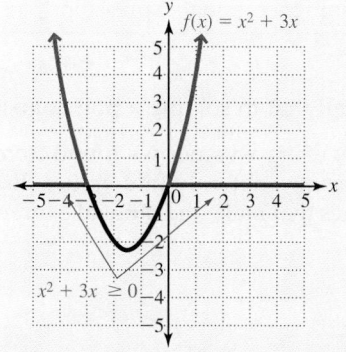

Connection Look at the graph of $f(x) = (x - 1)^2$.

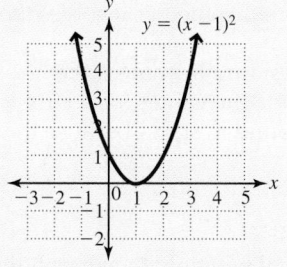

Notice that $f(x)$ is always 0 or positive. So $(x - 1)^2 > -2$ is true for all real x-values and $(x - 1)^2 < -2$ is false for all real x-values.

Instructor Note Have students graph $f(x) = (x + 4)$ $(x + 1)(x - 3)$ on a graphing calculator. Point out that the graph is below the x-axis in the intervals $(-\infty, -4)$ and $(-1, 3)$. It is above the x-axis in $(-4, -1)$ and $(3, \infty)$.

Answers to Your Turn 1
a. $[-2, 4]$

b. $(-\infty, 0) \cup (4, \infty)$

c. $\varnothing$

The solution set is $(-\infty, -3] \cup [0, \infty)$, which is graphed next.

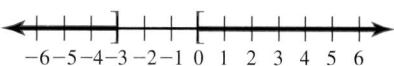

c. $(x - 1)^2 > -2$

Solution: Because $(x - 1)^2$ is always 0 or positive, it is always greater than -2. So every real number is a solution.

$$\text{Solution set: } \mathbb{R}, \text{ or } (-\infty, \infty)$$

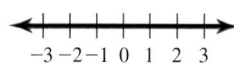

d. $(x - 1)^2 < -2$

Solution: Because $(x - 1)^2$ is never negative, its value can never be less than -2. So there are no real solutions for $(x - 1)^2 < -2$.

$$\text{Solution set: } \varnothing$$

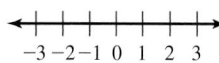

Your Turn 1 Solve. Write the solution set using interval notation; then graph the solution set on a number line.

a. $x^2 - 2x - 8 \leq 0$ **b.** $x^2 - 4x > 0$ **c.** $(3x - 8)^2 < -5$

Other Polynomial Inequalities

We use a similar procedure for expressions with more than two factors. In the following example, we condense the procedure to a table.

Example 2 Solve $(x + 4)(x + 1)(x - 3) \leq 0$. Write the solution set in interval notation; then graph the solution set on a number line.

Solution: $(x + 4)(x + 1)(x - 3) = 0$ Write the related equation.

$x + 4 = 0$ or $x + 1 = 0$ or $x - 3 = 0$ Set each factor equal to 0.

    $x = -4$ $x = -1$ $x = 3$ Solve the equations.

Plot -4, -1, and 3 on a number line and note the intervals.

Note -4, -1, and 3 are included in the solution set because the inequality symbol is $\leq$. Remember that we test intervals around these values, so they are not included in those intervals.

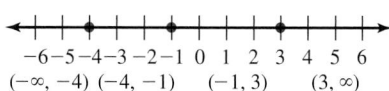

Interval	$(-\infty, -4)$	$(-4, -1)$	$(-1, 3)$	$(3, \infty)$
Test Number	-5	-2	0	4
Test Results	$-32 \leq 0$	$10 \leq 0$	$-12 \leq 0$	$40 \leq 0$
True or False	True	False	True	False

Therefore, the solution set is $(-\infty, -4] \cup [-1, 3]$, which is graphed next.

Note Because the inequality is $\leq$, -4, -1, and 3 are included in the solution set.

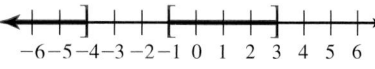

Your Turn 2 Solve. Write the solution set using interval notation; then graph the solution set on a number line.

$$(x + 3)(x - 2)(x + 5) \leq 0$$

Objective 2 Solve rational inequalities.

Now we consider solving **rational inequalities.**

Definition Rational inequality: An inequality containing a rational expression.

For example, $\dfrac{x + 2}{x - 3} > 4$ is a rational inequality. Recall that in solving a polynomial inequality, we divided the number line into intervals using x-values we found from solving its related equation. With rational inequalities, we not only find values that solve the related equation but also must consider values that make the rational expression undefined (denominator $= 0$).

> **Procedure** **Solving Rational Inequalities**
>
> 1. Find all values that make any denominator equal to 0. These values must be excluded from the solution set.
> 2. Solve the related equation.
> 3. Plot the numbers found in steps 1 and 2 on a number line and label the intervals.
> 4. Choose a test number from each interval and determine whether it solves the inequality.
> 5. The solution set is the union of all regions whose test number solves the inequality. If the inequality symbol is $\leq$ or $\geq$, include the values found in step 2. The solution set never includes the values found in step 1 because they make a denominator equal to 0.

Example 3 Solve. Write the solution set using interval notation; then graph the solution set on a number line.

a. $\dfrac{x - 4}{x + 2} \geq 0$

Solution: First, we find the values that make the denominator equal to 0.

$$x + 2 = 0$$
$$x = -2$$

Now solve the related equation.

$$\frac{x - 4}{x + 2} = 0 \qquad \text{Write the related equation.}$$

$$(x + 2)\frac{x - 4}{x + 2} = (x + 2)(0) \qquad \text{Multiply both sides by the LCD, } x + 2.$$

$$x - 4 = 0 \qquad \text{Simplify.}$$

$$x = 4$$

Plot -2 and 4 on a number line and label the intervals.

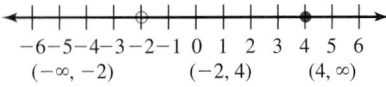

$$\begin{array}{ccc} (-\infty, -2) & (-2, 4) & (4, \infty) \end{array}$$

Answer to Your Turn 2

$(-\infty, -5] \cup [-3, 2]$

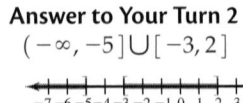

Again, we use a table.

Interval	$(-\infty, -2)$	$(-2, 4)$	$(4, \infty)$
Test Number	-3	0	5
Test Results	$7 \geq 0$	$-2 \geq 0$	$\dfrac{1}{7} \geq 0$
True or False	True	False	True

Note Remember, because -2 makes $\dfrac{x-4}{x+2}$ undefined, it is not included in the solution set. Because the inequality is $\geq$, 4 is included in the solution set. ▶

The solution set is $(-\infty, -2) \cup [4, \infty)$ and is graphed next.

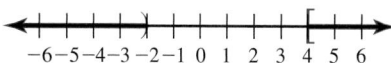

b. $\dfrac{x+5}{x-1} < 4$

Solution: Find the values that make the denominator equal to 0.

$$x - 1 = 0$$
$$x = 1$$

Solve the related equation.

$$\dfrac{x+5}{x-1} = 4 \qquad \text{Write the related equation.}$$

$$(x-1)\dfrac{x+5}{x-1} = (x-1)(4) \qquad \text{Multiply both sides by the LCD, } x - 1.$$

$$x + 5 = 4x - 4 \qquad \text{Simplify.}$$

$$-3x = -9 \qquad \text{Subtract } 4x \text{ and 5 from both sides.}$$

$$x = 3$$

Plot 1 and 3 on a number line and label the intervals.

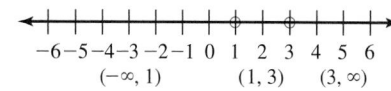

Interval	$(-\infty, 1)$	$(1, 3)$	$(3, \infty)$
Test Number	0	2	4
Test Results	$-5 < 4$	$7 < 4$	$3 < 4$
True or False	True	False	True

Note Because 1 makes $\dfrac{x+5}{x-1}$ undefined, it is not included in the solution set. Also, 3 is not in the solution set because $\dfrac{x+5}{x-1} = 4$ when $x = 3$. ▶

The solution set is $(-\infty, 1) \cup (3, \infty)$ and is graphed next.

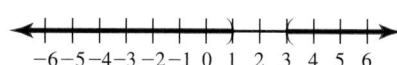

Answers to Your Turn 3

a. $(-3, 1)$

b. $(-\infty, -7] \cup (-4, \infty)$

Your Turn 3 Solve. Write the solution set using interval notation; then graph the solution set on a number line.

a. $\dfrac{x+3}{x-1} < 0$

b. $\dfrac{x-2}{x+4} \leq 3$

10.5 Exercises   For Extra Help   MyMathLab®

Note: Exercises marked with a ★ represent challenging exercises.

Objective 1

Prep Exercise 1 Explain how to solve $x^2 + 2x - 15 \geq 0$. Solve the related equation and then determine which of the three intervals satisfy the inequality by testing a number in each interval. If a test number satisfies the inequality, then all points in that interval satisfy the inequality.

Prep Exercise 2 If the graph of $y = ax^2 + bx + c$ intersects the x-axis at -2 and 2, what intervals need to be tested to solve $ax^2 + bx + c > 0$? $(-\infty, -2), (-2, 2), (2, \infty)$

Prep Exercise 3 Is it possible to have a quadratic inequality whose solution set is the empty set? Explain. Yes, for example, $x^2 + 2 \leq 0$ has an empty solution set because $x^2 + 2$ is always positive.

Prep Exercise 4 Is it possible to have a quadratic inequality whose solution set is one number? Explain. Yes, for example, $(x + 3)^2 \leq 0$ has only $x = -3$ as a solution.

For Exercises 1–8, the graph of a quadratic function is given. Use the graph to solve each equation and inequality. For solution sets that involve intervals, use interval notation. See Objective 1.

1. a. $x^2 + 6x + 5 = 0$
     $x = -5, -1$
  b. $x^2 + 6x + 5 < 0$
     $(-5, -1)$
  c. $x^2 + 6x + 5 > 0$
     $(-\infty, -5) \cup (-1, \infty)$

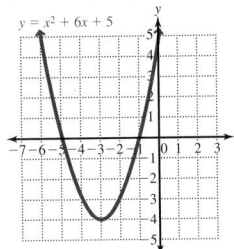

2. a. $x^2 + x - 2 = 0$
     $x = -2, 1$
  b. $x^2 + x - 2 > 0$
     $(-\infty, -2) \cup (1, \infty)$
  c. $x^2 + x - 2 < 0$
     $(-2, 1)$

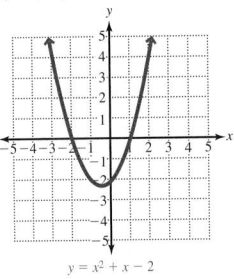

3. a. $-x^2 + 2x + 3 = 0$
     $x = -1, 3$
  b. $-x^2 + 2x + 3 \leq 0$
     $(-\infty, -1] \cup [3, \infty)$
  c. $-x^2 + 2x + 3 \geq 0$
     $[-1, 3]$

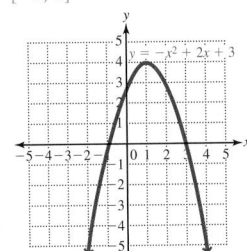

4. a. $-x^2 + 5x = 0$
     $x = 0, 5$
  b. $-x^2 + 5x \leq 0$
     $(-\infty, 0] \cup [5, \infty)$
  c. $-x^2 + 5x \geq 0$
     $[0, 5]$

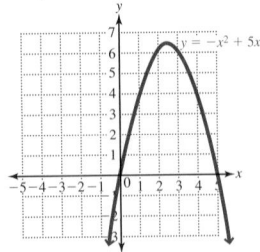

5. a. $x^2 + 4x + 4 = 0$
     $x = -2$
  b. $x^2 + 4x + 4 > 0$
     $(-\infty, -2) \cup (-2, \infty)$
  c. $x^2 + 4x + 4 < 0$
     $\varnothing$

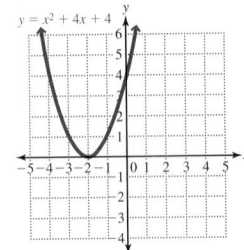

6. a. $x^2 - 6x + 9 = 0$
     $x = 3$
  b. $x^2 - 6x + 9 < 0$
     $\varnothing$
  c. $x^2 - 6x + 9 > 0$
     $(-\infty, 3) \cup (3, \infty)$

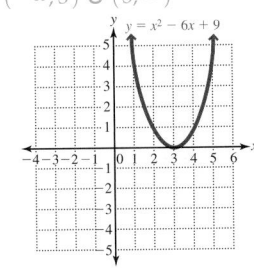

7. a. $x^2 + x + 4 = 0$
   $\varnothing$

   b. $x^2 + x + 4 \leq 0$
   $\varnothing$

   c. $x^2 + x + 4 \geq 0$
   $\mathbb{R}$, or $(-\infty, \infty)$

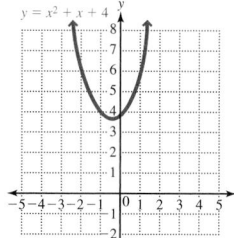

8. a. $-x^2 - x - 2 = 0$
   $\varnothing$

   b. $-x^2 - x - 2 \leq 0$
   $\mathbb{R}$, or $(-\infty, \infty)$

   c. $-x^2 - x - 2 \geq 0$
   $\varnothing$

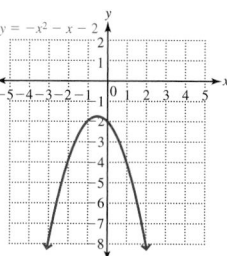

For Exercises 9–38, solve. Write the solution set using interval notation; then graph the solution set on a number line. See Examples 1 and 2.

9. $(x + 4)(x + 2) < 0$ $(-4, -2)$

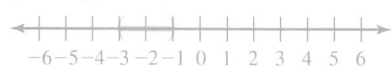

10. $(x + 3)(x + 1) < 0$ $(-3, -1)$

11. $(x - 2)(x - 5) > 0$ $(-\infty, 2) \cup (5, \infty)$

12. $(x + 4)(x - 3) > 0$ $(-\infty, -4) \cup (3, \infty)$

13. $x^2 + 5x + 4 < 0$ $(-4, -1)$

14. $x^2 + 6x + 5 < 0$ $(-5, -1)$

15. $x^2 - 4x + 3 > 0$ $(-\infty, 1) \cup (3, \infty)$

16. $x^2 - 8x + 7 > 0$ $(-\infty, 1) \cup (7, \infty)$

17. $b^2 - 6b + 8 \leq 0$ $[2, 4]$

18. $c^2 - 9c + 14 \leq 0$ $[2, 7]$

19. $y^2 - 5 \geq 4y$ $(-\infty, -1] \cup [5, \infty)$

20. $z^2 - 3 \geq 2z$ $(-\infty, -1] \cup [3, \infty)$

21. $a^2 - 3a < 10$ $(-2, 5)$

22. $b^2 + 4b \leq 21$ $[-7, 3]$

23. $y^2 + 6y + 9 \geq 0$ $(-\infty, \infty)$, or $\mathbb{R}$

24. $x^2 + 10x + 25 \geq 0$ $(-\infty, \infty)$, or $\mathbb{R}$

25. $2c^2 - 4c + 7 \leq 0$ No solution, or $\varnothing$

26. $2a^2 - 3a + 5 \leq 0$ No solution, or $\varnothing$

27. $4r^2 + 21r + 5 > 0$ $(-\infty, -5) \cup (-0.25, \infty)$

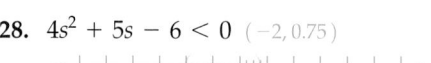

28. $4s^2 + 5s - 6 < 0$ $(-2, 0.75)$

29. $x^2 - 5x > 0$ $(-\infty, 0) \cup (5, \infty)$

30. $x^2 + 3x > 0$ $(-\infty, -3) \cup (0, \infty)$

31. $x^2 \leq 6x$ $[0, 6]$

32. $x^2 \leq 3x$ $[0, 3]$

33. $(x + 1)^2 \geq -9$ $(-\infty, \infty)$, or $\mathbb{R}$

34. $(2x - 1)^2 > -4$ $(-\infty, \infty)$, or $\mathbb{R}$

35. $(x - 4)(x + 2)(x + 4) \geq 0$ $[-4, -2] \cup [4, \infty)$

36. $(x + 5)(x - 3)(x + 1) \geq 0$ $[-5, -1] \cup [3, \infty)$

37. $(x + 2)(x + 6)(x - 1) < 0$ $(-\infty, -6) \cup (-2, 1)$

38. $(x - 3)(x + 3)(x + 1) < 0$ $(-\infty, -3) \cup (-1, 3)$

For Exercises 39–44, solve using the quadratic formula.

39. $3c^2 + 4c - 1 < 0$

$\left(\dfrac{-2 - \sqrt{7}}{3}, \dfrac{-2 + \sqrt{7}}{3} \right)$

40. $5a^2 - 10a + 2 < 0$

$\left(\dfrac{5 - \sqrt{15}}{5}, \dfrac{5 + \sqrt{15}}{5} \right)$

41. $4r^2 + 8r - 3 \geq 0$

$\left(-\infty, \dfrac{-2 - \sqrt{7}}{2} \right] \cup \left[\dfrac{-2 + \sqrt{7}}{2}, \infty \right)$

42. $2y^2 - 6y - 5 \geq 0$

$\left(-\infty, \dfrac{3 - \sqrt{19}}{2} \right] \cup \left[\dfrac{3 + \sqrt{19}}{2}, \infty \right)$

43. $-0.2a^2 - 1.6a - 2 \leq 0$

$(-\infty, -4 - \sqrt{6}] \cup [-4 + \sqrt{6}, \infty)$

44. $-0.1x^2 - 1.2x + 4 > 0$

$(-6 - 2\sqrt{19}, -6 + 2\sqrt{19})$

Objective 2

Prep Exercise 5 The quadratic inequality $(x - 2)(x + 5) < 0$ and the rational inequality $\dfrac{x - 2}{x + 5} < 0$ have the same solution sets. Explain how this is possible.

In both cases, the intervals to check are $(-\infty, -5)$, $(-5, 2)$, and $(2, \infty)$, and in both cases, only x-values in $(-5, 2)$ satisfy the original inequalities.

Prep Exercise 6 To solve $\dfrac{x + 2}{(x + 5)(x - 3)} \geq 0$, what intervals need to be tested?

$(-\infty, -5)$, $(-5, -2)$, $(-2, 3)$, $(3, \infty)$

For Exercises 45–62, solve the rational inequalities. Write the solution set using interval notation; then graph the solution set on a number line. See Example 3.

45. $\dfrac{a + 4}{a - 1} > 0$ $(-\infty, -4) \cup (1, \infty)$

46. $\dfrac{m - 6}{m + 2} \geq 0$ $(-\infty, -2) \cup [6, \infty)$

47. $\dfrac{n + 1}{n + 5} \leq 0$ $(-5, -1]$

48. $\dfrac{b - 2}{b + 3} < 0$ $(-3, 2)$

49. $\dfrac{6}{x + 4} > 0$ $(-4, \infty)$

50. $\dfrac{3}{x - 3} < 0$ $(-\infty, 3)$

51. $\dfrac{c}{c + 3} < 3$ $\left(-\infty, -\dfrac{9}{2}\right) \cup (-3, \infty)$

-6-5-4-3-2-1 0 1 2 3 4 5 6

52. $\dfrac{m}{m - 2} < 2$ $(-\infty, 2) \cup (4, \infty)$

-6-5-4-3-2-1 0 1 2 3 4 5 6

53. $\dfrac{a + 5}{a - 4} > 4$ $(4, 7)$

-1 0 1 2 3 4 5 6 7 8 9 10 11

54. $\dfrac{j + 4}{j - 4} > 5$ $(4, 6)$

-6-5-4-3-2-1 0 1 2 3 4 5 6

55. $\dfrac{p + 3}{p - 3} \geq 4$ $(3, 5]$

-6-5-4-3-2-1 0 1 2 3 4 5 6

56. $\dfrac{c + 1}{c - 4} \geq 6$ $(4, 5]$

-6-5-4-3-2-1 0 1 2 3 4 5 6

57. $\dfrac{(k + 3)(k - 2)}{k - 5} \leq 0$ $(-\infty, -3] \cup [2, 5)$

-6-5-4-3-2-1 0 1 2 3 4 5 6

58. $\dfrac{(m + 1)(m - 3)}{m + 4} \geq 0$ $(-4, -1] \cup [3, \infty)$

-6-5-4-3-2-1 0 1 2 3 4 5 6

59. $\dfrac{(2x - 1)^2}{x} \geq 0$ $(0, \infty)$

-6-5-4-3-2-1 0 1 2 3 4 5 6

60. $\dfrac{(4x + 3)^2}{x} \geq 0$ $(0, \infty)$

-6-5-4-3-2-1 0 1 2 3 4 5 6

61. $\dfrac{x^2 - 7x + 10}{x + 1} < 0$ $(-\infty, -1) \cup (2, 5)$

-6-5-4-3-2-1 0 1 2 3 4 5 6

62. $\dfrac{x^2 - 2x - 8}{x - 1} \geq 0$ $[-2, 1) \cup [4, \infty)$

-6-5-4-3-2-1 0 1 2 3 4 5 6

For Exercises 63–66, solve.

63. If a ball is thrown upward with an initial velocity of 80 feet per second from the top of a building 96 feet high, the height, h, above the ground after t seconds is given by $h = -16t^2 + 80t + 96$, where h is in feet.
  a. After how many seconds will the ball hit the ground? (*Hint*: Think about the value of h when the ball is on the ground.)
    6 sec.
  b. After how many seconds is the ball 192 feet above the ground?
    At 2 sec. and again at 3 sec.
  c. Find the interval of time when the ball is more than 192 feet above the ground.
    $(2, 3)$ sec.
  d. Use the answers to parts a, b, and c to find the intervals of time when the ball is less than 192 feet above the ground.
    $[0, 2) \cup (3, 6]$ sec.

64. If an object is dropped from the top of a cliff that is 256 feet high, the equation giving the height, h, above the ground is $h = 256 - 16t^2$, where h is in feet and t is in seconds.
  a. After how many seconds will the object hit the ground?
    4 sec.
  b. After how many seconds is the object 112 feet above the ground?
    3 sec.
  c. Find the interval of time when the object is more than 112 feet above the ground.
    $[0, 3)$ sec.
  d. Find the interval of time when the ball is less than 112 feet above the ground.
    $(3, 4]$ sec. (0 ft. after 4 sec.)

65. In the parallelogram shown, the height is to be 2 inches less than x. The base is to be 6 inches more than x.

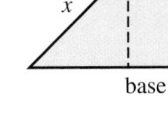

   a. Find the range of values for x so that the area of the parallelogram is at least 20 square inches.

   $x \geq 4$ in.

   b. Find the range of values for the base and the height.

   Base ≥ 10 in.; height ≥ 2 in.

66. An auditorium is to be designed roughly in the shape of a box. The height is set to be 40 feet. It is preferred that the length of the space be 20 feet more than the width.

   a. Find the range of values for the width so that the volume of the space is at least 140,000 cubic feet.

   $w \geq 50$ ft.

   b. Find the range of values for the length.

   $l \geq 70$ ft.

For Exercises 67 and 68, use the formula for the slope of a line: $m = \dfrac{y_2 - y_1}{x_2 - x_1}$.

★ **67.** Suppose a line is to be drawn in the coordinate plane so that it passes through the point at $(2, 5)$. Find the range of values for the second point (x_2, y_2) so that $x_2 = y_2$ and the slope of the line is at most $\dfrac{1}{2}$.

   $2 < x \leq 8$

★ **68.** Suppose a line is to be drawn in the coordinate plane so that it passes through the point at $(-3, 2)$. Find the range of values for the second point (x_2, y_2) so that $x_2 = y_2$ and the slope of the line is at least $\dfrac{1}{4}$.

   $x < -3, x \geq \dfrac{11}{3}$

Review Exercises

Exercises 1 and 2 ◢ **Expressions**

[10.1]   *For Exercises 1 and 2, add a term to the expression to make it a perfect square; then factor the perfect square.*

1. $x^2 - 6x$

   $9, (x - 3)^2$

2. $x^2 + 5x$

   $\dfrac{25}{4}, \left(x + \dfrac{5}{2}\right)^2$

Exercises 3–6 ◢ **Equations and Inequalities**

[9.6]   **3.** Solve: $\sqrt{x - 2} = 4$

   18

[6.7]   **4.** Does the graph of $f(x) = -x^2 + 2x - 3$ open upward or downward? Why?

   Downward, because the coefficient of x^2 is negative

[10.4]   **5.** Find the x-intercepts for $y = x^2 + 4x - 12$.

   $(-6, 0), (2, 0)$

[10.4]   **6.** What are the coordinates of the vertex of the graph of $f(x) = x^2 - 6x + 5$?

   $(3, -4)$

10.6 Function Operations

Objectives

1 Add or subtract functions.
2 Multiply functions.
3 Divide functions.

Note Letters other than f and g can be used to name functions, such as h, p, and q.

Warm-up *For Exercises 1–3, simplify.*

[5.3] 1. $(6x^2 + 9x - 15) - (4x^2 - 10x - 1)$

[5.5] 2. $(x - 8)(x + 3)$

[5.6] 3. $\dfrac{24x^4 - 30x^3}{6x^2}$

So far, we have seen how to evaluate and graph functions. Now we will consider how to add, subtract, multiply, and divide functions. Because we will be considering two different functions in the same problem, we will write one function as f and the second as g.

Objective 1 Add or subtract functions.

Given two functions f and g, their sum or difference is found using the following rules.

> **Rule Adding or Subtracting Functions**
> The sum of two functions, $f + g$, is found by $(f + g)(x) = f(x) + g(x)$.
> The difference of two functions, $f - g$, is found by $(f - g)(x) = f(x) - g(x)$.

Example 1   Given $f(x) = 2x + 1$ and $g(x) = 7x + 3$, find the following.

a. $f + g$
b. $f - g$
c. $(f - g)(-4)$

Solution:

a. $(f + g)(x) = f(x) + g(x)$
$\qquad\qquad = (2x + 1) + (7x + 3)$ Write a sum of the two functions.
$\qquad\qquad = 9x + 4$ Combine like terms.

b. $(f - g)(x) = f(x) - g(x)$
$\qquad\qquad = (2x + 1) - (7x + 3)$ Write a difference of the two functions.
$\qquad\qquad = (2x + 1) + (-7x - 3)$ Write as an equivalent addition.
$\qquad\qquad = -5x - 2$ Combine like terms.

Note We could also find $(f - g)(4)$ by finding $f(4) - g(4)$.

► c. In part b, we found $(f - g)(x) = -5x - 2$; so to find $(f - g)(-4)$, we replace x with -4.

$(f - g)(-4) = -5(-4) - 2$ Replace x with -4.
$\qquad\qquad = 20 - 2$
$\qquad\qquad = 18$

Your Turn 1   Given $h(x) = x^2 - 3x + 5$ and $k(x) = x^2 - 2x - 9$, find the following.

a. $h + k$ b. $h - k$ c. $(h + k)(-2)$

Objective 2 Multiply functions.

Given two functions f and g, their product is found using the following rule.

Answers to Your Turn 1
a. $2x^2 - 5x - 4$
b. $-x + 14$
c. 14

Answers to Warm-up
1. $2x^2 + 19x - 14$
2. $x^2 - 5x - 24$
3. $4x^2 - 5x$

> **Rule Multiplying Functions**
>
> The product of two functions, $f \cdot g$, is found by $(f \cdot g)(x) = f(x)\, g(x)$.

Example 2 Given $f(x) = 3x - 4$ and $g(x) = x + 5$, find $f \cdot g$.

Solution: $(f \cdot g)(x) = f(x)\, g(x)$

$\qquad\qquad = (3x - 4)(x + 5)$ Write a product of the functions.

$\qquad\qquad = 3x^2 + 15x - 4x - 20$ Use FOIL to multiply the binomials.

$\qquad\qquad = 3x^2 + 11x - 20$ Combine like terms.

Your Turn 2 Given $p(x) = 7x - 2$ and $q(x) = 5x - 1$, find the following.

a. $p \cdot q$
b. $(p \cdot q)(2)$

Objective 3 Divide functions.

Given two functions f and g, their quotient is found using the following rule.

> **Rule Dividing Functions**
>
> The quotient of two functions, f/g, is found by $(f/g)(x) = \dfrac{f(x)}{g(x)}$, where $g(x) \neq 0$.

Example 3 Find f/g.

a. $f(x) = 9x^3 - 12x^2 + 6x,\ g(x) = 3x$

Note Because we divided by $3x$, we are assuming that $x \neq 0$ in $(f/g)(x)$.

Solution: $(f/g)(x) = \dfrac{f(x)}{g(x)} = \dfrac{9x^3 - 12x^2 + 6x}{3x}$ Write a quotient of the functions.

$\qquad\qquad = \dfrac{9x^3}{3x} - \dfrac{12x^2}{3x} + \dfrac{6x}{3x}$ Because the divisor is a monomial, we separate the terms.

$\qquad\qquad = 3x^2 - 4x + 2$ Divide.

b. $f(x) = 3x^2 - 13x - 6,\ g(x) = x - 5$

Note The domain of the quotient function contains all real numbers except 5; so we would write the domain as $\{x \mid x \neq 5\}$ or, using interval notation, $(-\infty, 5) \cup (5, \infty)$.

Solution: $(f/g)(x) = \dfrac{f(x)}{g(x)} = \dfrac{3x^2 - 13x - 6}{x - 5}$ Write a quotient of the functions.

To find the quotient, we use long division.

$$
\begin{array}{r}
3x \\
x - 5 \overline{)3x^2 - 13x - 6} \\
-(3x^2 - 15x)
\end{array}
$$
Change signs. $\longrightarrow$
$$
\begin{array}{r}
3x \\
x - 5 \overline{)3x^2 - 13x - 6} \\
-3x^2 + 15x \\
\hline
2x - 6
\end{array}
$$
Combine like terms and bring down the next term.

$$
\begin{array}{r}
3x + 2 \\
x - 5 \overline{)3x^2 - 13x - 6} \\
-3x^2 + 15x \\
\hline
2x - 6 \\
-(2x - 10)
\end{array}
$$
Change signs. $\longrightarrow$
$$
\begin{array}{r}
3x + 2 \\
x - 5 \overline{)3x^2 - 13x - 6} \\
-3x^2 + 15x \\
\hline
2x - 6 \\
-2x + 10 \\
\hline
4
\end{array}
$$
Combine like terms.

Answer: $3x + 2 + \dfrac{4}{x - 5}$

Answers to Your Turn 2
a. $35x^2 - 17x + 2$
b. 108

Answers to Your Turn 3

a. $f/g = 4x^2 + 12x - 7$;
$(f/g)(-2) = -15$

b. $m/n = 6x - 5 + \dfrac{-1}{x+4}$

or $6x - 5 - \dfrac{1}{x+4}$;

$(m/n)(1) = \dfrac{4}{5}$

Your Turn 3

a. Given $f(x) = 12x^4 + 36x^3 - 21x^2$ and $g(x) = 3x^2$, find f/g and $(f/g)(-2)$.
b. Given $m(x) = 6x^2 + 19x - 21$ and $n(x) = x + 4$, find m/n and $(m/n)(1)$.

10.6 Exercises   For Extra Help   MyMathLab®

Objective 1

Prep Exercise 1 If $p(x)$ and $q(x)$ are both polynomial functions, how do you find $p + q$? $(p + q)(x) = p(x) + q(x)$

Prep Exercise 2 If $p(x)$ and $q(x)$ are both polynomial functions, how do you find $p - q$? $(p - q)(x) = p(x) - q(x)$

For Exercises 1–10, find $f + g$ and $f - g$. See Example 1.

1. $f(x) = 2x, g(x) = 3x - 1$
$5x - 1; -x + 1$

2. $f(x) = 3x, g(x) = 4x + 2$
$7x + 2; -x - 2$

3. $f(x) = x - 5, g(x) = 2x - 3$
$3x - 8; -x - 2$

4. $f(x) = 5x + 1, g(x) = x - 8$
$6x - 7; 4x + 9$

5. $f(x) = x^2 - 4x + 7, g(x) = x - 3$
$x^2 - 3x + 4; x^2 - 5x + 10$

6. $f(x) = x^2 - 3x + 1, g(x) = x - 2$
$x^2 - 2x - 1; x^2 - 4x + 3$

7. $f(x) = 2x^2 - 5x - 3, g(x) = x^2 + 5$
$3x^2 - 5x + 2; x^2 - 5x - 8$

8. $f(x) = -x^2 - 8x + 2, g(x) = x^2 + 2$
$-8x + 4; -2x^2 - 8x$

9. $f(x) = -5x^2 + 4x + 8, g(x) = 3x^2 - 2x - 1$
$-2x^2 + 2x + 7; -8x^2 + 6x + 9$

10. $f(x) = -2x^2 + 5x + 3, g(x) = 4x^2 + 2x - 9$
$2x^2 + 7x - 6; -6x^2 + 3x + 12$

Objective 2

Prep Exercise 3 If $p(x)$ and $q(x)$ are both binomial functions, how do you find $p \cdot q$? Use FOIL to multiply the binomials.

For Exercises 11–20, find $f \cdot g$. See Example 2.

11. $f(x) = 2x, g(x) = x - 5$
$2x^2 - 10x$

12. $f(x) = -2x, g(x) = x + 4$
$-2x^2 - 8x$

13. $f(x) = x + 1, g(x) = x - 2$
$x^2 - x - 2$

14. $f(x) = x - 5, g(x) = x + 4$
$x^2 - x - 20$

15. $f(x) = 3x + 2, g(x) = x + 2$
$3x^2 + 8x + 4$

16. $f(x) = 2x - 5, g(x) = 3x - 1$
$6x^2 - 17x + 5$

17. $f(x) = x^2 - 3x + 2, g(x) = x^2 + 2x - 7$
$x^4 - x^3 - 11x^2 + 25x - 14$

18. $f(x) = x^2 - 2x - 1, g(x) = x^2 + 3x + 4$
$x^4 + x^3 - 3x^2 - 11x - 4$

19. $f(x) = 2x^2 - 3x + 1, g(x) = 3x^2 - x - 1$
$6x^4 - 11x^3 + 4x^2 + 2x - 1$

20. $f(x) = 3x^2 - 2x - 1, g(x) = 2x^2 - 2x - 3$
$6x^4 - 10x^3 - 7x^2 + 8x + 3$

Objective 3

Prep Exercise 4 If $p(x)$ is a trinomial function and $q(x)$ is a monomial function, how do you find p/q? Divide each term in the trinomial by the monomial.

For Exercises 21–30, find f/g. See Example 3.

21. $f(x) = 2x^2 - 6x, g(x) = 2x$
$x - 3, x \neq 0$

22. $f(x) = 3x^2 - 9x, g(x) = 3x$
$x - 3, x \neq 0$

23. $f(x) = 6x^2 - 3x + 6, g(x) = 3x$
$2x - 1 + \dfrac{2}{x}, x \neq 0$

24. $f(x) = 4x^2 + 4x - 8, g(x) = 2x$
$2x + 2 - \dfrac{4}{x}, x \neq 0$

25. $f(x) = x^2 - 14x + 45, g(x) = x - 9$
$x - 5, x \neq 9$

26. $f(x) = x^2 + 9x + 20, g(x) = x + 4$
$x + 5, x \neq -4$

27. $f(x) = 2x^2 - 3x - 5, g(x) = 2x - 5$
$x + 1, x \neq \dfrac{5}{2}$

28. $f(x) = 3x^2 - 4x - 7, g(x) = 3x - 7$
$x + 1, x \neq \dfrac{7}{3}$

29. $f(x) = 2x^2 - 5x - 7, g(x) = x + 5$
$2x - 15 + \dfrac{68}{x + 5}, x \neq -5$

30. $f(x) = 6x^2 - 5x + 4, g(x) = 2x - 1$
$3x - 1 + \dfrac{3}{2x - 1}, x \neq \dfrac{1}{2}$

Objectives 1–3

For Exercises 31–34, find: a. $p + q$, b. $p - q$, c. $p \cdot q$, d. p/q. See Examples 1–3.

31. $p(x) = 5x - 1, q(x) = 2x + 3$
a. $7x + 2$ b. $3x - 4$
c. $10x^2 + 13x - 3$ d. $\dfrac{5x - 1}{2x + 3}, x \neq \dfrac{-3}{2}$

32. $p(x) = 3x - 4, q(x) = 5x + 2$
a. $8x - 2$ b. $-2x - 6$
c. $15x^2 - 14x - 8$ d. $\dfrac{3x - 4}{5x + 2}, x \neq -\dfrac{2}{5}$

33. $p(x) = 2x^2 - 3x + 5, q(x) = x - 1$
a. $2x^2 - 2x + 4$ b. $2x^2 - 4x + 6$
c. $2x^3 - 5x^2 + 8x - 5$ d. $2x - 1 + \dfrac{4}{x - 1}, x \neq 1$

34. $p(x) = 4x^2 - 4x - 1, q(x) = 2x + 3$
a. $4x^2 - 2x + 2$ b. $4x^2 - 6x - 4$
c. $8x^3 + 4x^2 - 14x - 3$ d. $2x - 5 + \dfrac{14}{2x + 3}, x \neq -\dfrac{3}{2}$

For Exercises 35–38, find the function values. See Example 1c.

35. Using the functions in Exercise 31, find the following. a. -12 b. 11 c. -6 d. 2
   a. $(p + q)(-2)$ **b.** $(p - q)(5)$ **c.** $(p \cdot q)(-1)$ **d.** $(p/q)(7)$

36. Using the functions in Exercise 32, find the following. a. -18 b. -12 c. 80 d. $\dfrac{1}{6}$
   a. $(p + q)(-2)$ **b.** $(p - q)(3)$ **c.** $(p \cdot q)(-2)$ **d.** $(p/q)(2)$

37. Using the functions in Exercise 33, find the following. a. 16 b. 54 c. 7 d. -8
   a. $(p + q)(3)$ **b.** $(p - q)(-4)$ **c.** $(p \cdot q)(2)$ **d.** $(p/q)(-3)$

38. Using the functions in Exercise 34, find the following. a. 22 b. 14 c. -23 d. $4\dfrac{3}{11}$
   a. $(p + q)(-2)$ **b.** $(p - q)(3)$ **c.** $(p \cdot q)(-2)$ **d.** $(p/q)(4)$

For Exercises 39 and 40: **a.** *Find $f + g$.*
   b. *Complete the table shown.*
   c. *Graph f, g, and $f + g$ on the same graph.*
   d. *What do the table and graphs suggest about the sum of two functions?*

39. $f(x) = 3x + 1, g(x) = x + 5$

   a. $4x + 6$

   b.

x	f(x)	g(x)	(f + g)(x)
−4	−11	1	−10
−2	−5	3	−2
0	1	5	6
2	7	7	14
4	13	9	22

   c.

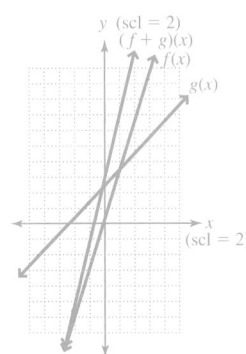

   d. The output values for $f(x)$ added to the output values of $g(x)$ are equal to the output values of $(f + g)(x)$.

40. $f(x) = -x + 3, g(x) = 4x + 3$

   a. $3x + 6$

   b.

x	f(x)	g(x)	(f + g)(x)
−4	7	−13	−6
−2	5	−5	0
0	3	3	6
2	1	11	12
4	−1	19	18

   c.

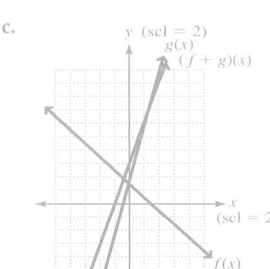

   d. The output values for $f(x)$ added to the output values of $g(x)$ are equal to the output values of $(f + g)(x)$.

For Exercises 41 and 42: **a.** *Find* $h - k$.

                            b. *Complete the table shown.*

                            c. *Graph h, k, and* $h - k$ *on the same graph.*

                            d. *What do the table and graphs suggest about the difference of two functions?*

41. $h(x) = 2x - 9, k(x) = 3x + 5$

   a. $-x - 14$

   b.

x	h(x)	k(x)	(h − k)(x)
−4	−17	−7	−10
−2	−13	−1	−12
0	−9	5	−14
2	−5	11	−16
4	−1	17	−18

   c.

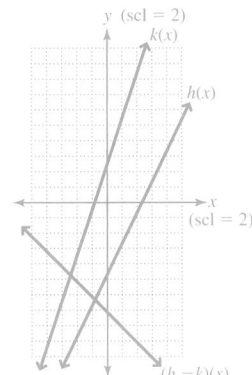

   d. The output values for $k(x)$ subtracted from the output values of $h(x)$ are equal to the output values of $(h - k)(x)$.

42. $h(x) = 4x + 1, k(x) = 2x - 7$

   a. $2x + 8$

   b.

x	h(x)	k(x)	(h − k)(x)
−4	−15	−15	0
−2	−7	−11	4
0	1	−7	8
2	9	−3	12
4	17	1	16

   c.

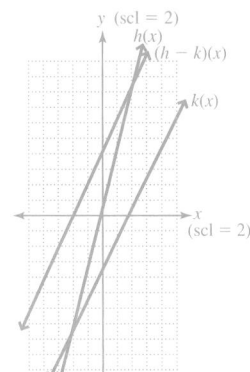

   d. The output values for $k(x)$ subtracted from the output values of $h(x)$ are equal to the output values of $(h - k)(x)$.

For Exercises 43 and 44, the function $w(x)$ describes a company's monthly costs in wages and $p(x)$ describes the cost of production, where x represents the number of units produced.

43. $w(x) = 3x + 2, p(x) = x + 2$
   a. Find the function, $t(x)$, that describes the total cost. $t(x) = 4x + 4$
   b. Calculate the total cost if $x = 200$. $804
   c. Graph all three functions.

43c.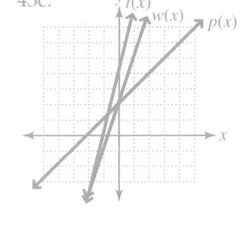

44c.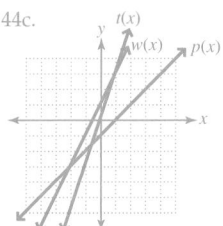

44. $w(x) = 2x + 1, p(x) = x - 1$
   a. Find the function, $t(x)$, that describes the total cost. $t(x) = 3x$
   b. Calculate the total cost if $x = 175$. $525
   c. Graph all three functions.

For Exercises 45 and 46, the function $r(x)$ describes a company's monthly revenue and $c(x)$ describes the monthly costs where x represents the number of units produced and then sold. Use the formula Net profit $=$ Revenue $-$ Cost.

45. $r(x) = x^2 - 3x - 4, c(x) = x + 8$
   a. Find the function, $p(x)$, that describes the net profit or loss. $p(x) = x^2 - 4x - 12$
   b. Calculate the net profit or loss if $x = 100$. Is this value a profit or a loss?
       $9588 profit
   c. Graph all three functions.
   d. How many units must be sold for the company to break even (which means the net $= 0$)?   6 units

45c.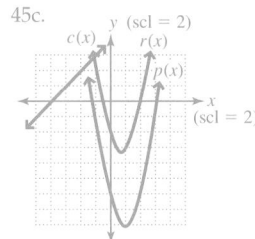

46. $r(x) = x^2 - 4x - 5, c(x) = x + 1$
   a. Find the function, $p(x)$, that describes the net profit or loss. $p(x) = x^2 - 5x - 6$
   b. Calculate the net profit or loss if $x = 300$. Is this value a profit or a loss?
       $88,494 profit
   c. Graph all three functions.
   d. How many units must be sold for the company to break even (which means the net $= 0$)?   6 units

46c.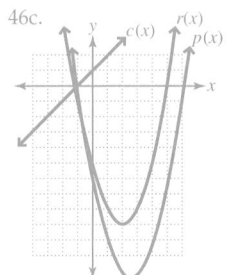

For Exercises 47 and 48, the function $l(x)$ describes the length of a box and $h(x)$ describes the height, where x represents the width.

47. $l(x) = 3x + 2, h(x) = 2x$
   a. Find the function that describes the area of the front face of the box.
   b. Calculate the area of the front face if $x = 3$ feet. a. $A(x) = 6x^2 + 4x$ b. 66 ft.²

48. $l(x) = 5x - 1, h(x) = 3x$
   a. Find the function that describes the area of the front face of the box.
   b. Calculate the area of the front face if $x = 2$ meters. a. $A(x) = 15x^2 - 3x$ b. 54 m²

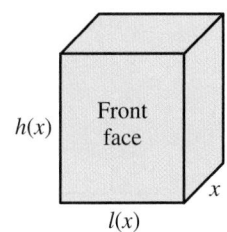

* For Exercises 49 and 50, the function $V(x)$ describes the volume of a box, $l(x)$ describes its length, and $w(x)$ describes its width.

49. $V(x) = 4x^3 - 7x^2 - 14x - 3, l(x) = x - 3, w(x) = x + 1$
   a. Find the function, $h(x)$, that describes the height of the box.
   b. Calculate the height of the box if $x = 4$ centimeters.
   a. $h(x) = 4x + 1$ b. 17 cm

50. $V(x) = 6x^3 + 11x^2 - 47x + 20, l(x) = x + 4, w(x) = 2x - 1$
   a. Find the function, $h(x)$, that describes the height of the box.
   b. Calculate the height of the box if $x = 5$ inches.
   a. $h(x) = 3x - 5$ b. 10 in.

Review Exercises

Exercises 1–5 Equations and Inequalities

[3.7] *Given the function* $f(x) = x^2 + 1$, *find the following:*

1. $f(2)$ 5

2. $f(3a - 2)$ $9a^2 - 12a + 5$

3. $f(\sqrt{a - 1})$ a

For Exercises 4 and 5, solve for x.

[2.4] 4. $y = 2x + 4$ $x = \dfrac{y - 4}{2}$

[9.6] 5. $y = x^3 + 2$ $x = \sqrt[3]{y - 2}$

Chapter 10 Summary and Review Exercises

Complete each incomplete definition, rule, or procedure; study the key examples; and then work the related exercises.

10.1 The Square Root Principle and Completing the Square

Definitions/Rules/Procedures	Key Example(s)
Square root principle If $x^2 = a$, where a is a real number, then $x = \underline{\quad \sqrt{a} \quad}$ or $x = \underline{\quad -\sqrt{a} \quad}$. It is common to indicate the positive and negative solutions by writing $\underline{\quad \pm\sqrt{a} \quad}$.	Solve. **a.** $x^2 = 36$ **Solution:** $x = \pm\sqrt{36}$ $\qquad\quad x = \pm 6$ **b.** $(x - 3)^2 = 12$ **Solution:** $x - 3 = \pm\sqrt{12}$ $\qquad\qquad\quad x = 3 \pm 2\sqrt{3}$

Exercises 1–11 ▲ **Equations and Inequalities**

[10.1] *For Exercises 1–8, solve and check.*

1. $x^2 = 16$
± 4

2. $y^2 = \dfrac{1}{36}$
$\pm\dfrac{1}{6}$

3. $k^2 + 2 = 30$
$\pm 2\sqrt{7}$

4. $3x^2 = 42$
$\pm\sqrt{14}$

5. $5h^2 + 24 = 9$
$\pm i\sqrt{3}$

6. $(x + 7)^2 = 25$
$-12, -2$

7. $(x - 9)^2 = -16$
$9 \pm 4i$

8. $\left(m + \dfrac{3}{5}\right)^2 = \dfrac{16}{25}$
$-\dfrac{7}{5}, \dfrac{1}{5}$

[10.1] *For Exercises 9–11, solve.*

9. A crop circle appeared July 7, 2003, in Windham Hill, England, and covered $22{,}500\pi$ square feet. Find the radius of the circle.
150 ft.

10. Using the formula $E = \dfrac{1}{2}mv^2$, where E represents the kinetic energy in joules of an object with a mass of m kilograms and a velocity of v meters per second, find the velocity of an object with a mass of 50 kilograms and 400 joules of kinetic energy.
4 m/sec.

11. A right circular cylinder is to be constructed so that its volume is equal to that of a sphere with a radius of 9 inches. If the cylinder is to have a height of 4 inches, find the radius of the cylinder.
$9\sqrt{3}$ in.

Definitions/Rules/Procedures	Key Example(s)
To solve a quadratic equation by completing the square: 1. Write the equation in the form $\underline{\quad x^2 + bx = c \quad}$. 2. Complete the square by adding $\underline{\quad \left(\dfrac{b}{2}\right)^2 \quad}$ to both sides. 3. Write the completed square in $\underline{\quad \text{factored} \quad}$ form. 4. Use the $\underline{\quad \text{square root} \quad}$ principle to eliminate the square. 5. Isolate the $\underline{\quad \text{variable} \quad}$. 6. Simplify as needed.	Solve by completing the square. $x^2 + 6x - 7 = 8$ $\quad x^2 + 6x = 15$ $\qquad$ Add 7 to both sides. $x^2 + 6x + 9 = 15 + 9$ $\quad$ Complete the square. $\quad (x + 3)^2 = 24$ $\qquad$ Factor. $\quad x + 3 = \pm\sqrt{24}$ $\qquad$ Use the square root principle. $\qquad\quad x = -3 \pm 2\sqrt{6}$ $\quad$ Subtract 3 from both sides and simplify the square root.

Exercises 12–15 ▲ Equations and Inequalities

[10.1] *For Exercises 12–15, solve by completing the square.*

12. $m^2 + 8m = -7$
$-7, -1$

13. $u^2 - 6u - 12 = 100$
$-8, 14$

14. $2b^2 - 6b + 7 = 0$
$\dfrac{3 \pm i\sqrt{5}}{2}$

15. $u^2 + \dfrac{1}{4}u = \dfrac{3}{4}$
$-1, \dfrac{3}{4}$

10.2 Solving Quadratic Equations Using the Quadratic Formula

Definitions/Rules/Procedures	Key Example(s)
To solve a quadratic equation in the form $ax^2 + bx + c = 0$, where $a \neq 0$, use the quadratic formula: $$x = \frac{-b \pm \sqrt{b^2 - 4ac}}{2a}$$	Solve $3x^2 - 4x + 2 = 0$. **Solution:** In the quadratic formula, replace a with 3, b with -4, and c with 2. $$x = \frac{-(-4) \pm \sqrt{(-4)^2 - 4(3)(2)}}{2(3)}$$ $$= \frac{4 \pm \sqrt{16 - 24}}{6} = \frac{4 \pm \sqrt{-8}}{6}$$ $$= \frac{4 \pm 2\sqrt{-2}}{6} = \frac{2}{3} \pm \frac{\sqrt{2}}{3}i$$

Exercises 16–21 ▲ Equations and Inequalities

[10.2] *For Exercises 16–19, solve using the quadratic formula.*

16. $p^2 - 5 = -2p$
$-1 \pm \sqrt{6}$

17. $3x^2 - 2x + 1 = 0$
$\dfrac{1 \pm i\sqrt{2}}{3}$

18. $2t^2 + t - 5 = 0$
$\dfrac{-1 \pm \sqrt{41}}{4}$

19. $2x^2 + 0.1x - 0.03 = 0$
$-\dfrac{3}{20}, \dfrac{1}{10}$

[10.2] *For Exercises 20 and 21, solve.*

20. The length of a small rectangular shed is 4 feet more than its width. If the area of the base is 285 square feet, what are the dimensions of the base of the shed?
Width: 15 ft.; length: 19 ft.

21. A ramp is constructed so that it is a right triangle with a base that is 9 feet longer than its height. If the hypotenuse is 17 feet, find the dimensions of the base and height. Round to the nearest hundredth.
Height: 6.65 ft.; base: 15.65 ft.

Definitions/Rules/Procedures	Key Example(s)
Given a quadratic equation in the form $ax^2 + bx + c = 0$, where a, b, and c are rational numbers and $a \neq 0$, the discriminant is $\underline{b^2 - 4ac}$. If the **discriminant is positive**, the equation has $\underline{\text{two}}$ real-number solutions. The solutions will be rational if the discriminant is a(n) $\underline{\text{perfect square}}$ and irrational otherwise. If the **discriminant is 0**, the equation has $\underline{\text{one}}$ rational solution. If the **discriminant is negative**, the equation has two $\underline{\text{nonreal complex}}$ solutions.	Use the discriminant to determine the number and type of solutions for $5x^2 - 7x + 8 = 0$. **Solution:** In the discriminant, replace a with 5, b with -7, and c with 8. $$(-7)^2 - 4(5)(8) = 49 - 160$$ $$= -111$$ Because the discriminant is negative, the equation has two nonreal complex solutions.

Exercises 22–25 ▲ Equations and Inequalities

[10.2] *For Exercises 22–25, find the discriminant and determine the number and type of solutions for the equation.*

22. $b^2 - 4b - 12 = 0$
$D = 64$; two rational

23. $6z^2 - 7z + 5 = 0$
$D = -71$; two nonreal complex

24. $k^2 + 6k + 9 = 0$
$D = 0$; one rational

25. $0.8x^2 + 1.2x + 0.3 = 0$
$D = 0.48$; two irrational

10.3 Solving Equations That Are Quadratic in Form

Definitions/Rules/Procedures	Key Example(s)
An equation is **quadratic in form** if it can be rewritten as a quadratic equation $au^2 + bu + c = 0$, where $a \neq 0$ and u is a(n) ___variable___ or a(n) ___expression___. If an equation involves rational expressions, multiply both sides by the ___LCD___. Be sure to check for ___extraneous___ solutions.	Solve $\dfrac{6}{x-2} + \dfrac{6}{x-1} = 5$. **Solution:** First multiply both sides by the LCD, $(x-2)(x-1)$. $6(x-1) + 6(x-2) = 5(x-2)(x-1)$ $\qquad\qquad 12x - 18 = 5x^2 - 15x + 10 \qquad$ Multiply. $\qquad\qquad\qquad 0 = 5x^2 - 27x + 28 \qquad$ Write in $ax^2 + bx + c = 0$ form. $\qquad\qquad\qquad 0 = (5x - 7)(x - 4) \qquad$ Factor. $5x - 7 = 0 \ $ or $\ x - 4 = 0 \qquad$ Use the zero-factor theorem. $\qquad x = \dfrac{7}{5} \qquad\qquad x = 4 \qquad$ Solve each equation. **Check:** Verify that $\dfrac{7}{5}$ and 4 solve the original equation.

Exercises 26–29 ▲ Equations and Inequalities

[10.3] *For Exercises 26–29, solve. Identify any extraneous solutions.*

26. $\dfrac{1}{y} + \dfrac{1}{y+3} = \dfrac{2}{3}$

$\pm\dfrac{3\sqrt{2}}{2}$

27. $\dfrac{1}{u} + \dfrac{1}{u-5} = \dfrac{10}{u^2-25}$

$-\dfrac{5}{2} \, (5 \text{ is extraneous.})$

28. $6 - 5x^{-1} + x^{-2} = 0$

$\dfrac{1}{3}, \dfrac{1}{2}$

29. $2 - 3x^{-1} - x^{-2} = 0$

$\dfrac{3 \pm \sqrt{17}}{4}$

Definitions/Rules/Procedures	Key Example(s)
If an equation has radicals, ___isolate___ a radical and square both sides. Continue the process as needed until all radicals have been eliminated. Solve the resulting equation and check for ___extraneous___ solutions.	Solve $\sqrt{4x+4} = 2x - 2$. $4x + 4 = 4x^2 - 8x + 4 \qquad$ Square both sides. $\qquad 0 = 4x^2 - 12x \qquad$ Write in $ax^2 + bx + c = 0$ form. $\qquad 0 = 4x(x - 3) \qquad$ Factor. $4x = 0 \ $ or $\ x - 3 = 0 \qquad$ Use the zero-factor theorem. $x = 0 \qquad\qquad x = 3 \qquad$ Solve each equation. **Check:** $x = 0 \qquad\qquad\qquad x = 3$ $\sqrt{4(0)+4} = 2(0) - 2 \qquad \sqrt{4(3)+4} = 2(3) - 2$ $\quad \sqrt{0+4} = 0 - 2 \qquad\qquad \sqrt{12+4} = 6 - 2$ $\qquad\qquad 2 = -2 \ $ False $\qquad\qquad\quad 4 = 4 \ $ True $x = 3$ is the only solution. $(x = 0$ is extraneous.$)$

Exercises 30–33 ▲ Equations and Inequalities

[10.3] *For Exercises 30–33, solve. Identify any extraneous solutions.*

30. $14\sqrt{x} + 45 = 0$

No solution

31. $\sqrt{4m} = 3m - 1$

$1 \left(\dfrac{1}{9} \text{ is extraneous.} \right)$

32. $\sqrt{6r+13} = 2r + 1$

$2 \left(-\dfrac{3}{2} \text{ is extraneous.} \right)$

33. $\sqrt{21t+2} + t = 2 + 4t$

$\dfrac{1}{3}, \dfrac{2}{3}$

Definitions/Rules/Procedures	Key Example(s)

To solve equations that are **quadratic in form** using substitution:
 	
1. Rewrite the equation so that it is in the form $\underline{au^2 + bu + c = 0}$.
 	
2. Solve the quadratic equation for $\underline{\quad u \quad}$.
 	
3. $\underline{\text{Substitute}}$ for u and solve.
 	
4. Check the solutions.
	Solve $(a + 2)^2 + 7(a + 2) + 12 = 0$.
	
	$u^2 + 7u + 12 = 0$ Substitute u for $a + 2$.
	$(u + 3)(u + 4) = 0$ Factor.
	$u + 3 = 0$ or $u + 4 = 0$ Use the zero-factor
	     $u = -3$ $u = -4$ theorem.
	$a + 2 = -3$ or $a + 2 = -4$ Substitute $a + 2$ for u;
	     $a = -5$ $a = -6$ solve for a.
	
	Check: Verify that -5 and -6 solve the original equation.

Exercises 34–39 ◣ Equations and Inequalities

[10.3] *For Exercises 34–39, solve the equations using substitution. Identify any extraneous solutions.*

34. $x^4 - 5x^2 + 6 = 0$
$\pm\sqrt{2}, \pm\sqrt{3}$

35. $2m^4 - 3m^2 + 1 = 0$
$\pm 1, \pm\dfrac{\sqrt{2}}{2}$

36. $6(x + 5)^2 - 5(x + 5) + 1 = 0$
$\dfrac{9}{2}, \dfrac{14}{3}$

37. $\left(\dfrac{x - 1}{3}\right)^2 + 10\left(\dfrac{x - 1}{3}\right) + 9 = 0$
$-26, -2$

38. $p^{2/3} - 11p^{1/3} + 24 = 0$
$27, 512$

39. $5a^{1/2} + 13a^{1/4} - 6 = 0$
$\dfrac{16}{625}$ (81 is extraneous.)

10.4 Graphing Quadratic Functions

Definitions/Rules/Procedures	Key Example(s)
The graph of a function in the form $f(x) = a(x - h)^2 + k$ is a parabola with vertex at $\underline{(h, k)}$. The equation of the axis of symmetry is $\underline{x = h}$. The parabola opens upward if $\underline{a > 0}$ and downward if $\underline{a < 0}$. The larger the $\lvert a \rvert$, the narrower the graph.	For $f(x) = -3(x + 1)^2 - 2$: **a.** Determine whether the graph opens upward or downward. **b.** Find the vertex. **c.** Write the equation of the axis of symmetry. **d.** Graph. **Answers:** **a.** downward **b.** $(-1, -2)$ **c.** $x = -1$ **d.**

Exercises 40–43 Equations and Inequalities

[10.4] *For Exercises 40–43:* **a.** *Find the x- and y-intercepts.*
b. *State whether the parabola opens upward or downward.*
c. *Find the coordinates of the vertex.*
d. *Write the equation of the axis of symmetry.*
e. *Graph.*

40. $f(x) = -2x^2$ e.
a. $(0,0)$
b. downward
c. $(0,0)$
d. $x = 0$

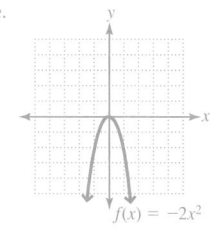

41. $g(x) = \dfrac{1}{2}x^2 + 1$ e.
a. no x-intercepts, $(0,1)$
b. upward
c. $(0,1)$
d. $x = 0$

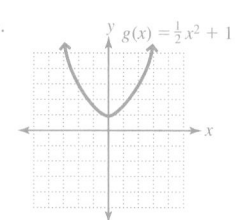

42. $h(x) = -\dfrac{1}{3}(x-2)^2$ e.
a. $(2,0), \left(0, -\dfrac{4}{3}\right)$
b. downward
c. $(2,0)$
d. $x = 2$

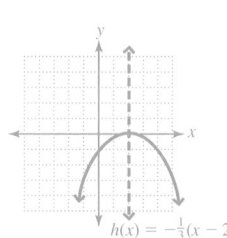

43. $k(x) = 4(x+3)^2 - 2$
a. $\left(-3 + \dfrac{\sqrt{2}}{2}, 0\right), \left(-3 - \dfrac{\sqrt{2}}{2}, 0\right), (0,34)$
b. upward   e.
c. $(-3, -2)$
d. $x = -3$

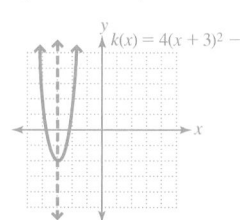

Definitions/Rules/Procedures	Key Example(s)
Given an equation in the form $f(x) = ax^2 + bx + c$, to determine the vertex: **1.** Find the x-coordinate using the formula $x = \dfrac{-\dfrac{b}{2a}}{\rule{2cm}{0.4pt}}$. **2.** Find the y-coordinate by evaluating $f\left(-\dfrac{b}{2a}\right)$ or the formula $\dfrac{\dfrac{4ac - b^2}{4a}}{\rule{2cm}{0.4pt}}$.	For $f(x) = 2x^2 - 12x + 19$: **a.** Determine whether the graph opens upward or downward. **b.** Find the vertex. **c.** Write the equation of the axis of symmetry. **d.** Graph. **Answers:** **a.** upward **b.** For the x-coordinate, use $-\dfrac{b}{2a}$. $$x = -\dfrac{(-12)}{2(2)} = 3 \quad a = 2, b = -12$$ For the y-coordinate, evaluate $f(3)$. $$f(3) = 2(3)^2 - 12(3) + 19$$ $$= 18 - 36 + 19 = 1$$ Vertex: $(3, 1)$ **c.** $x = 3$ **d.** 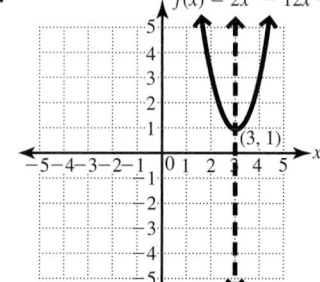 **Note** As an alternative approach, we could have transformed $f(x) = 2x^2 - 12x + 19$ to $f(x) = 2(x-3)^2 + 1$ by completing the square.

Exercises 44–47 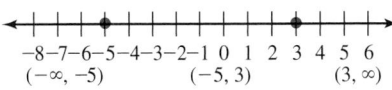 Equations and Inequalities

[10.4] *For Exercises 44 and 45:* **a.** *Write the equation in the form* $f(x) = a(x - h)^2 + k$.
b. *Find the x- and y-intercepts.*
c. *State whether the parabola opens upward or downward.*
d. *Find the coordinates of the vertex.*
e. *Write the equation of the axis of symmetry.*
f. *Graph.*
g. *Find the domain and range.*

44. $m(x) = x^2 + 2x - 1$
a. $m(x) = (x + 1)^2 - 2$
b. $(-1 + \sqrt{2}, 0), (-1 - \sqrt{2}, 0), (0, -1)$ c. upward
d. $(-1, -2)$ e. $x = -1$
f. 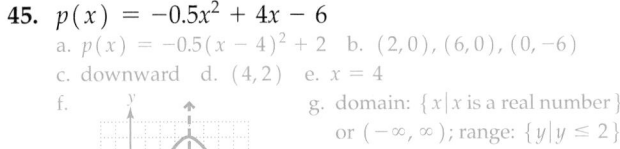   g. domain:
$\{x \mid x \text{ is a real number}\},$
or $(-\infty, \infty)$; range:
$\{y \mid y \geq -2\}$ or $[-2, \infty)$

45. $p(x) = -0.5x^2 + 4x - 6$
a. $p(x) = -0.5(x - 4)^2 + 2$ b. $(2, 0), (6, 0), (0, -6)$
c. downward   d. $(4, 2)$ e. $x = 4$
f. 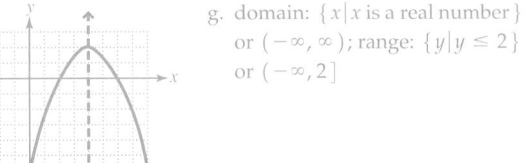   g. domain: $\{x \mid x \text{ is a real number}\}$,
or $(-\infty, \infty)$; range: $\{y \mid y \leq 2\}$
or $(-\infty, 2]$

[10.4] *For Exercises 46 and 47, solve.*

46. With an initial velocity of 24 feet per second, an acrobat is launched upward from one end of a lever. The function $h = -16t^2 + 24t$ describes the height, h, of the acrobat t seconds after being launched.

a. After how many seconds does the acrobat reach maximum height?
0.75 sec.

b. What is the maximum height the acrobat reaches?
9 ft.

c. How long is the acrobat in the air?
1.5 sec.

d. Graph the function.

d.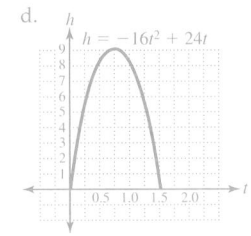

Of Interest

The yardage you found is the horizontal distance the punt carried in the air. It was actually recorded as a 98-yard punt, which includes the amount of roll after the punt landed and was downed.

47. The longest punt on record in the NFL was by Steve O'Neal in a game between the New York Jets and the Denver Broncos on September 21, 1969. The function $y = -0.03x^2 + 2.16x$ models the trajectory of the punt. (Note that x and y are distances in yards.)

a. Find the maximum height the punt reached.
38.88 yd.

b. Find the horizontal distance the punt traveled.
72 yd.

10.5 Solving Nonlinear Inequalities

Definitions/Rules/Procedures	Key Example(s)
A **quadratic inequality** is an inequality that can be written in the form $\underline{ax^2 + bx + c > 0}$ or $\underline{ax^2 + bx + c < 0}$, where $a \neq 0$. Note: The symbols $<$ and $>$ can be replaced with $\leq$ and $\geq$.	Solve $x^2 + 2x - 15 \geq 0$.
	$x^2 + 2x - 15 = 0$ Write the related equation.
	$(x + 5)(x - 3) = 0$ Factor.
Solving quadratic inequalities	$x + 5 = 0$ or $x - 3 = 0$ Use the zero-factor theorem.
1. Solve the related equation $\underline{ax^2 + bx + c = 0}$.	$x = -5$ $x = 3$ Solve each equation.
2. Plot the solutions of $ax^2 + bx + c = 0$ on a(n) $\underline{\text{number line}}$. These solutions divide the $\underline{\text{number line}}$ into intervals.	Plot -5 and 3 on a number line and label the intervals.
	$\begin{array}{c} \longleftarrow\!\!+\!\!+\!\!+\!\!\bullet\!\!+\!\!+\!\!+\!\!+\!\!+\!\!+\!\!+\!\!\bullet\!\!+\!\!+\!\!+\!\!\longrightarrow \\ -8\,{-}7\,{-}6\,{-}5\,{-}4\,{-}3\,{-}2\,{-}1\ 0\ 1\ 2\ 3\ 4\ 5\ 6 \\ (-\infty, -5) \qquad (-5, 3) \qquad (3, \infty) \end{array}$

Definitions/Rules/Procedures	Key Example(s)
3. Choose a test number from each interval and substitute the number into the inequality. If the test number makes the inequality *true*, then all numbers in that interval will <u>solve the inequality</u>. If the test number makes the inequality *false*, then <u>no number</u> in that interval will solve the inequality. 4. State the solution set of the inequality: It is the <u>union</u> of all intervals that solve the inequality. If the inequality symbols are $\leq$ or $\geq$, the values from step 2 are included. If the symbols are $<$ or $>$, they are not solutions.	Choose a number from each interval and test it in the original inequality.

Choose a number from each interval and test it in the original inequality.

Interval	$(-\infty, -5)$	$(-5, 3)$	$(3, \infty)$
Test Value	-6	0	4
Result	$9 \geq 0$	$-15 \geq 0$	$9 \geq 0$
True/False	True	False	True

The solution set is $(-\infty, -5] \cup [3, \infty)$.

$$-8\ -7\ -6\ -5\ -4\ -3\ -2\ -1\ 0\ 1\ 2\ 3\ 4\ 5\ 6$$

Exercises 48–52 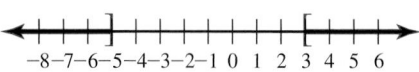 **Equations and Inequalities**

[10.5] *For Exercises 48–51, solve the inequalities.*

48. $(x + 5)(x - 3) > 0$
$(-\infty, -5) \cup (3, \infty)$

49. $n^2 - 6n \leq -8$
$[2, 4]$

50. $x^2 + 9x + 14 < 0$
$(-7, -2)$

51. $(x + 3)(x - 1)(x - 2) \geq 0$
$[-3, 1] \cup [2, \infty)$

[10.5] *For Exercise 52, solve.*

52. In a triangle, the height is to be 2 inches less than x. The base is to be 4 inches more than x.

   a. Find the range of values for x so that the area of the triangle is at least 56 square inches.
   $x \geq 10$ in.

   b. Find the range of values for the base and the height.
   Height ≥ 8 in., base ≥ 14 in.

Definitions/Rules/Procedures	Key Example(s)
A **rational inequality** is an inequality containing a(n) <u>rational expression</u>. **Solving rational inequalities** 1. Find all values that make any denominator equal to <u>0</u>. These values must be excluded from the solution set. 2. Solve the related <u>equation</u>.	Solve $\dfrac{x + 5}{x - 1} > 4$. Find the value(s) that make any denominator equal to 0. $$x - 1 = 0$$ $$x = 1$$ Solve the related equation.

Solve $\dfrac{x + 5}{x - 1} > 4$.

Find the value(s) that make any denominator equal to 0.

$$x - 1 = 0$$
$$x = 1$$

Solve the related equation.

$\dfrac{x + 5}{x - 1} = 4$	Write the related equation.
$(x - 1)\dfrac{x + 5}{x - 1} = (x - 1)(4)$	Multiply both sides by the LCD.
$x + 5 = 4x - 4$	Simplify.
$9 = 3x$	Subtract x from and add 5 to both sides.
$3 = x$	Divide by 3.

Definitions/Rules/Procedures	Key Example(s)
3. Plot the numbers found in steps 1 and 2 on a(n) <u>number line</u> and label the intervals. **4.** Choose a(n) <u>test number</u> from each interval and determine whether it solves the inequality. **5.** The solution set is the <u>union</u> of all regions whose test number solves the inequality. If the inequality symbol is $\leq$ or $\geq$, include the values found in step <u>2</u>. The solution set never includes the values found in step <u>1</u>.	Plot 1 and 3 on the number line and label the intervals. 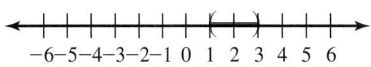 $-6\,-5\,-4\,-3\,-2\,-1\ 0\ 1\ 2\ 3\ 4\ 5\ 6$ $(-\infty,1)\qquad (1,3)\qquad (3,\infty)$ Choose a number from each interval and test it in the original inequality.

	$(\infty, 1)$	$(1, 3)$	$(3, \infty)$
Interval	$(\infty, 1)$	$(1, 3)$	$(3, \infty)$
Test Value	0	2	4
Result	$-5 > 4$	$7 > 4$	$3 > 4$
True/False	False	True	False

The solution set is $(1, 3)$.

$-6\,-5\,-4\,-3\,-2\,-1\ 0\ 1\ 2\ 3\ 4\ 5\ 6$

Exercises 53–56   **Equations and Inequalities**

[10.5] For Exercises 53–56, solve the rational inequalities.

53. $\dfrac{a+3}{a-1} \geq 0$

$(-\infty,-3]\cup(1,\infty)$

54. $\dfrac{r}{r+2} < 2$

$(-\infty,-4)\cup(-2,\infty)$

55. $\dfrac{n-3}{n-4} \leq 5$

$(-\infty,4)\cup\left[\dfrac{17}{4},\infty\right)$

56. $\dfrac{(k+2)(k-3)}{k-5} < 0$

$(-\infty,-2)\cup(3,5)$

10.6 Function Operations

Definitions/Rules/Procedures	Key Example(s)
The **sum** of two functions, $f + g$, is found by $(f + g)(x) = $ <u>$f(x) + g(x)$</u>. The **difference** of two functions, $f - g$, is found by $(f - g)(x) = $ <u>$f(x) - g(x)$</u>. The **product** of two functions, $f \cdot g$, is found by $(f \cdot g)(x) = $ <u>$f(x) \cdot g(x)$</u>. The **quotient** of two functions, f/g, is found by $(f/g)(x) = $ <u>$\dfrac{f(x)}{g(x)}$</u>, where $g(x) \neq 0$.	**Example 1:** Given $f(x) = 2x - 5$ and $g(x) = 3x + 4$, find **a.** $f + g$ **b.** $f - g$ **c.** $f \cdot g$ **a.** $(f + g)(x) = (2x - 5) + (3x + 4)$ $\qquad\qquad\quad = 5x - 1$ **b.** $(f - g)(x) = (2x - 5) - (3x + 4)$ $\qquad\qquad\quad = (2x - 5) + (-3x - 4)$ $\qquad\qquad\quad = -x - 9$ **c.** $(f \cdot g)(x) = (2x - 5)(3x + 4)$ $\qquad\qquad\quad = 6x^2 + 8x - 15x - 20$ $\qquad\qquad\quad = 6x^2 - 7x - 20$ **Example 2:** Given $f(x) = 12x^3 - 20x^2 + 8x$ and $g(x) = 4x$, find f/g. **Solution:** $(f/g)(x) = \dfrac{f(x)}{g(x)} = \dfrac{12x^3 - 20x^2 + 8x}{4x}$ $\qquad\qquad = \dfrac{12x^3}{4x} - \dfrac{20x^2}{4x} + \dfrac{8x}{4x}$ $\qquad\qquad = 3x^2 - 5x + 2, x \neq 0$

Exercises 57–64 ◢ **Equations and Inequalities**

[10.6] *For Exercises 57–60, find a. $f + g$.*
$\quad\quad\quad\quad\quad\quad\quad\quad\quad$ *b. $f - g$.*
$\quad\quad\quad\quad\quad\quad\quad\quad\quad$ *c. $f \cdot g$.*

57. $f(x) = -3x + 9, g(x) = 4x + 1$
$\quad$ a. $x + 10$ b. $-7x + 8$ c. $-12x^2 + 33x + 9$

58. $f(x) = -x - 2, g(x) = 8x + 2$
$\quad$ a. $7x$ b. $-9x - 4$ c. $-8x^2 - 18x - 4$

59. $f(x) = 3x^2 - x + 5, g(x) = x + 2$
$\quad$ a. $3x^2 + 7$ b. $3x^2 - 2x + 3$ c. $3x^3 + 5x^2 + 3x + 10$

60. $f(x) = x^2 - 2x - 1, g(x) = 3x + 1$
$\quad$ a. $x^2 + x$ b. $x^2 - 5x - 2$ c. $3x^3 - 5x^2 - 5x - 1$

For Exercises, 61–62, find f/g.

61. $f(x) = 6x^4 - 15x^3 + 12x^2, g(x) = 3x^2$
$\quad$ $2x^2 - 5x + 4, x \neq 0$

62. $f(x) = 2x^3 - 17x^2 + 36x - 17, g(x) = 2x - 5$
$\quad$ $x^2 - 6x + 3 + \dfrac{-2}{2x - 5}, x \neq \dfrac{5}{2}$

[10.6] *For Exercises 63–64, solve.*

63. The function $w(x) = 3x + 5$ describes a company's monthly expenditure on wages. The function $p(x) = x - 1$ describes the total cost of production. Note that x represents the number of units produced.
$\quad$ **a.** Find the function, $c(x)$, that describes the total cost.
$\quad$ **b.** If $x = 225$, then calculate the total cost.
$\quad$ **c.** Graph all three functions.
$\quad\quad$ a. $c(x) = 4x + 4$ b. 904 c.

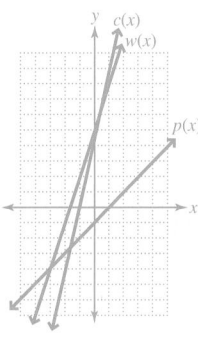

64. The function $V(x) = 6x^3 + 41x^2 + 26x - 24$ describes the volume of a box, $l(x) = 3x + 4$ describes its length, and $w(x) = x + 6$ describes its width.
$\quad$ **a.** Find the function, $h(x)$, that describes the height of the box.
$\quad$ **b.** Calculate the height of the box if $x = 9$ inches.
$\quad\quad$ a. $h(x) = 2x - 1$ b. 17 in.

Learning Strategy

Before taking an exam, make sure you're well rested and eat something beforehand. Make sure you do any homework the teacher assigns. Go back and redo the problems you struggled with in the homework and find similar problems to do.

$\quad\quad\quad\quad\quad\quad\quad\quad\quad\quad\quad\quad\quad\quad\quad\quad\quad\quad\quad$ —Ellyn G.

Chapter 10 **Practice Test**

For Extra Help
Step-by-step test solutions are found on the Chapter Test Prep Videos available in MyMathLab® *or on* You Tube.

For Exercises 1 and 2, use the square root principle to solve and check.

1. $x^2 = 81$
$\quad$ ± 9 [10.1]

2. $(x - 3)^2 = 20$
$\quad$ $3 \pm 2\sqrt{5}$ [10.1]

For Exercises 3 and 4, solve by completing the square.

3. $x^2 - 8x = -4$
$\quad$ $4 \pm 2\sqrt{3}$ [10.1]

4. $3m^2 - 6m = 5$
$\quad$ $\dfrac{3 \pm 2\sqrt{6}}{3}$ [10.1]

For Exercises 5 and 6, solve using the quadratic formula.

5. $2x^2 + x - 6 = 0$
$\quad$ $-2, \dfrac{3}{2}$ [10.2]

6. $x^2 - 8x + 15 = 0$
$\quad$ $3, 5$ [10.2]

For Exercises 7–10, solve using any method.

7. $u^2 - 16 = -6u$

$-8, 2$ [10.1, 10.2]

8. $4w^2 + 6w + 3 = 0$

$\dfrac{-3 \pm i\sqrt{3}}{4}$ [10.1, 10.2]

9. $x^2 + 16 = 0$

$\pm 4i$ [10.1, 10.2]

10. $3k^2 = -5k$

$0, -\dfrac{5}{3}$ [10.1, 10.2]

For Exercises 11–14, solve. Identify any extraneous solutions.

11. $\dfrac{1}{x+2} + \dfrac{1}{x} = \dfrac{5}{12}$

$-\dfrac{6}{5}, 4$ [10.3]

12. $3 - x^{-1} - 2x^{-2} = 0$

$-\dfrac{2}{3}, 1$ [10.3]

13. $9\sqrt{x} + 8 = 0$

No solution [10.3]

14. $\sqrt{x+8} - x = 2$

1 (-4 is extraneous.) [10.3]

For Exercises 15 and 16, solve the equations using substitution. Identify any extraneous solutions.

15. $9a^4 + 26a^2 - 3 = 0$

$\pm\dfrac{1}{3}, \pm i\sqrt{3}$ [10.3]

16. $(x + 1)^2 + 3(x + 1) - 4 = 0$

$-5, 0$ [10.3]

17. For $f(x) = -x^2 + 6x - 4$:

a. Find the x- and y-intercepts.

$(3 + \sqrt{5}, 0)\,(3 - \sqrt{5}, 0), (0, -4)$

b. Write the equation in the form $f(x) = a(x - h)^2 + k$.

$f(x) = -(x - 3)^2 + 5$

c. State whether the parabola opens upward or downward.

downward

d. Find the coordinates of the vertex.

$(3, 5)$

e. Write the equation of the axis of symmetry.

$x = 3$

f. Graph.

g. Find the domain and range.

domain: $\{x \mid x \text{ is a real number}\}$,

or $(-\infty, \infty)$;

range: $\{y \mid y \le 5\}$ or $(-\infty, 5]$

f.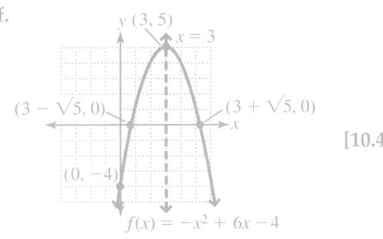

[10.4]

For Exercises 18 and 19: **a.** *Solve the inequality.*
b. *Graph the solution set on a number line.*

18. $(x + 1)(x - 4) \le 0$

$[-1, 4]$

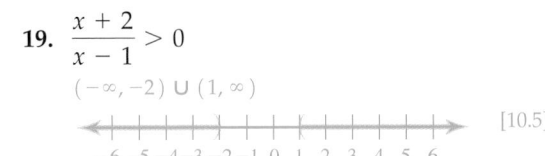

[10.5]

19. $\dfrac{x + 2}{x - 1} > 0$

$(-\infty, -2) \cup (1, \infty)$

[10.5]

20. Given $f(x) = x^2$ and $g(x) = 3x^2 - 2$, find $f + g, f - g,$ and $f \cdot g$.

a. $4x^2 - 2$

b. $-2x^2 + 2$

c. $3x^4 - 2x^2$ [10.6]

21. Given $f(x) = 20x^4 + 36x^3 - 28x^2$ and $g(x) = 4x^2$ find f/g.

$5x^2 + 9x - 7$ [10.6]

For Exercises 22–25, solve.

22. A ball is thrown downward from a window in a tall building. The distance, d, the ball traveled is given by the equation $d = 16t^2 + 32t$, where t is the time traveled in seconds. How long will it take the ball to fall 180 feet?

    2.5 sec. [10.2]

23. An archer shoots an arrow in a field. Suppose the equation $y = -0.02x^2 + 1.3x + 8$ models the trajectory of the arrow. (Assume that x and y represent distances in meters and that $x \geq 0$ and $y \geq 0$.)
    a. What is the maximum height the arrow reaches?

       29.125 m
    b. How far does the arrow travel horizontally?

       70.66 m
    c. Graph the trajectory of the arrow.

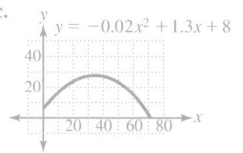

[10.4]

24. A zoo is planning to install a new aquarium tank. The tank is to be in the shape of a box with a height of 12 feet. It is preferred that the length be 15 feet more than the width.
    a. Find the range of values for the width so that the volume of the space is, at most, 12,000 cubic feet.

       $0 < w \leq 25$ ft.
    b. Find the range of values for the length.

       $15 < l \leq 40$ ft. [10.5]

25. The function $A(x) = 36x^2 - x - 2$ describes the area of the front face of a box and $l(x) = 4x - 1$ describes its length, where x represents its width.
    a. Find the function, $h(x)$, that describes the height of the box.
    b. Calculate the height of the box if $x = 15$ cm.

       a. $h(x) = 9x + 2$ b. 137 cm [10.6]

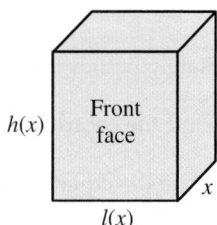

Chapters 1–10 Cumulative Review Exercises

For Exercises 1–3, answer true or false.

[3.1] **1.** The point with coordinates $(45, -12)$ is in quadrant II.   false

[4.1] **2.** A system of equations that has no solution is said to be consistent.   false

[5.2] **3.** The degree of the monomial $5x^3y$ is 4.   true

For Exercises 4–6, fill in the blank.

[6.7] **4.** The graph of a polynomial function of the form $f(x) = ax^2 + bx + c$ is a(n) ____parabola____ that opens up if ____$a > 0$____ .

[9.3] **5.** If $\sqrt[n]{a}$ and $\sqrt[n]{b}$ are both real numbers, then $\sqrt[n]{a} \cdot \sqrt[n]{b} = $ ____$\sqrt[n]{ab}$____ .

[10.1]6. If $x^2 = a$ and a is a real number, then $x = $ ____$\sqrt{a}$____ or $x = -\sqrt{a}$.

Exercises 7–15 ◢ **Expressions**

For Exercises 7–9, simplify.

[5.3] **7.** $(-3x - 4) - (3x + 2)$
$-6x - 6$

[5.4] **8.** $(8n^2)(-7mn^3)$
$-56mn^5$

[5.5] **9.** $(x - 3)(4x^2 - 2x + 1)$
$4x^3 - 14x^2 + 7x - 3$

[6.4] *For Exercises 10 and 11, factor completely.*

10. $m^3 + 8$
$(m + 2)(m^2 - 2m + 4)$

11. $x^2 + 10x + 25$
$(x + 5)^2$

For Exercises 12–14, simplify.

[7.2]12. $\dfrac{4u^2 + 4u + 1}{u + 2u^2} \cdot \dfrac{u}{2u^2 - u - 1}$
$\dfrac{1}{u - 1}$

[7.2]13. $\dfrac{3}{x - 2} - \dfrac{2}{x + 2}$
$\dfrac{x + 10}{(x - 2)(x + 2)}$

[9.2]14. $x^{3/4} \cdot x^{-1/4}$
$x^{1/2}$

[9.5]15. Rationalize the denominator of $\dfrac{\sqrt{n}}{\sqrt{m} - \sqrt{n}}$.
$\dfrac{\sqrt{mn} + n}{m - n}$

Exercises 16–30 ◢ **Equations and Inequalities**

For Exercises 16–20, solve.

[2.3]16. $3n - 5 = 7n + 9$
$-\dfrac{7}{2}$

[8.2]17. $|3x + 4| + 3 = 7$
$-\dfrac{8}{3}, 0$

[6.6]18. $2x^2 + 7x = 15$
$\dfrac{3}{2}, -5$

[7.6]19. $\dfrac{5}{x - 2} - \dfrac{3}{x} = \dfrac{11}{3x}$
8

[9.6]20. $\sqrt{5x - 4} = 9$
17

[10.2]21. Solve $x^2 - 6x + 11 = 0$ using the quadratic formula.
$3 \pm i\sqrt{2}$

For Exercises 22 and 23: **a.** *Graph the solution set on a number line.*
b. Write the solution set in set-builder notation.
c. Write the solution set in interval notation.

[2.8] 22. $-2 < x + 4 < 5$ **a.** 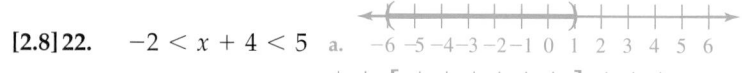   **b.** $\{x | -6 < x < 1\}$ **c.** $(-6, 1)$

[10.5] 23. $x^2 + x \le 12$ **a.**    **b.** $\{x | -4 \le x \le 3\}$ **c.** $[-4, 3]$

[3.7] 24.   Graph $f(x) = -2x$.

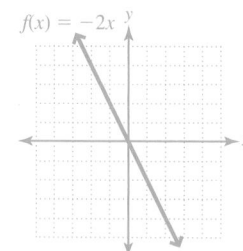

[4.4] 26.   Solve the system: $\begin{cases} x + y + z = 5 \\ 2x + y - 2z = -5 \\ x - 2y + z = 8 \end{cases}$ $(2, -1, 4)$

For Exercises 27–30, solve.

[4.3] 27. A trucking company finds that the average speed of experienced drivers is 10 miles per hour more than the average speed of inexperienced drivers. If inexperienced drivers take 1 hour longer than experienced drivers to drive 300 miles, find how long it takes the experienced driver to drive 300 miles.
5 hr.

[7.7] 29. If the wavelength of a wave remains constant, the velocity, v, of a wave is inversely proportional to its period, T. In an experiment, waves are created in a pool of water so that the period is 8 seconds and the velocity is 6 centimeters per second. If the period is increased to 15 seconds, what is the velocity?
3.2 cm/sec.

[3.5] 25. Write the equation of a line in standard form that passes through the point $(-2, 4)$ and is perpendicular to the line $3x - 2y = 6$.
$2x + 3y = 8$

[4.3] 28. A chemist has a bottle containing 50 ml of 15% saline solution and a bottle of 40% saline solution. She wants a 30% solution. How much of the 40% solution must be added to the 15% solution so that a 30% concentration is created?
75 ml

30. The formula $t = \sqrt{\dfrac{h}{16}}$ describes the amount of time, t, in seconds, that an object falls a distance of h feet.

[9.1] a. Write an expression in simplest form for the exact amount of time an object falls a distance of 24 feet.
$\dfrac{\sqrt{6}}{2}$ sec.

[9.6] b. Find the distance an object falls in 3 seconds.
144 ft.

11

Exponential and Logarithmic Functions

Chapter Overview

Exponential and logarithmic functions are among the most important in mathematics. A few of their applications include the following:

▶ Population growth

▶ Electrical circuits

▶ Sound decibels

▶ pH

▶ Richter scale

▶ Solving of equations

11.1 Composite and Inverse Functions

11.2 Exponential Functions

11.3 Logarithmic Functions

11.4 Properties of Logarithms

11.5 Common and Natural Logarithms

11.6 Exponential and Logarithmic Equations with Applications

Instructor Note

Exponential and logarithmic functions are extremely important in the study of calculus and the sciences. These topics are often part of higher-level courses such as precalculus. However, should you have time to include an introduction to these important topics, your students would benefit from the exposure. We have tried to make the chapter accessible by selecting applications that students find familiar and interesting.

11.1 Composite and Inverse Functions

Objectives

1 Find the composition of two functions.

2 Show that two functions are inverses.

3 Show that a function is one-to-one.

4 Find the inverse of a function.

5 Graph a given function's inverse function.

Objective 1 Find the composition of two functions.

In Section 10.6, we learned to perform the basic operations with functions. We now explore a new operation. Recall that a function pairs elements from an "input" set called the *domain* with elements in an "output" set called the *range*. If the output elements of one function are used as input elements in a second function, the resulting function is the **composition** of the two functions. Composition occurs when one quantity depends on a second quantity, which, in turn, depends on a third quantity.

We can visualize the composition of two functions as a machine with two stages, which are the two functions. For example, some coffee machines have two stages. Roasted coffee beans, which are ground in the first stage (first function) are put in the machine. The grounds are then fed into the second stage (second function), which runs hot water over the grounds to produce coffee.

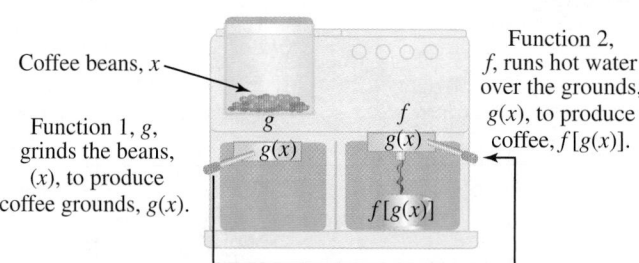

Coffee beans, x —

Function 1, g, grinds the beans, (x), to produce coffee grounds, $g(x)$.

g f

$g(x)$ $g(x)$

$f[g(x)]$

Function 2, f, runs hot water over the grounds, $g(x)$, to produce coffee, $f[g(x)]$.

Grounds, $g(x)$, are placed into second function.

Note The notation $(g \circ f)(x)$ is read "g composed with f of x," or simply "g of f of x."

Warning Do not confuse $(f \circ g)(x)$ with $(f \cdot g)(x)$.

Definition Composition of functions: If f and g are functions, then the composition of f and g is defined as $(f \circ g)(x) = f[g(x)]$ for all x in the domain g for which $g(x)$ is in the domain of f. The composition of g and f is defined as $(g \circ f)(x) = g[f(x)]$ for all x in the domain of f for which $f(x)$ is in the domain of g.

For example, suppose a pebble is dropped in a calm lake, causing ripples to form in concentric circles. If the radius of the outer ripple increases at a rate of 2 feet per second, we can describe the length of the radius as a function of t by using $r(t) = 2t$, where t is in seconds. The area of the circle formed by the outer ripple is also a function described by $A(r) = \pi r^2$. If we substitute the output of $r(t)$, which is $2t$ for r, into $A(r)$, we form the composite function $(A \circ r)(t)$.

$$(A \circ r)(t) = A[r(t)]$$
$$= \pi(2t)^2 \quad \text{Substitute } 2t \text{ for } r \text{ in } A(r) = \pi r^2.$$
$$= 4\pi t^2 \quad \text{Simplify.}$$

The composite function $(A \circ r)(t)$ gives the area of the circle as a function of time.

Example 1 If $f(x) = 2x - 3$ and $g(x) = x^2 + 1$, find the following.

a. $(f \circ g)(2)$

Solution: $(f \circ g)(2) = f[g(2)]$ Definition of composition
$\qquad\qquad\qquad = f(5)$ $g(2) = 2^2 + 1 = 4 + 1 = 5$ ◀
$\qquad\qquad\qquad = 2(5) - 3$ In $f(x)$, replace x with 5.
$\qquad\qquad\qquad = 7$ Simplify.

Note We first find $g(2)$, which is 5, then substitute that result into f.

b. $(g \circ f)(2)$

Solution: $(g \circ f)(2) = g[f(2)]$ Definition of composition

$= g(1)$ $f(2) = 2(2) - 3 = 4 - 3 = 1$

$= 1^2 + 1$ In $g(x)$, replace x with 1.

$= 2$ Simplify.

c. $(f \circ g)(x)$

Solution: $(f \circ g)(x) = f[g(x)]$ Definition of composition

$= f(x^2 + 1)$ Replace $g(x)$ with $x^2 + 1$.

$= 2(x^2 + 1) - 3$ In $f(x)$, replace x with $x^2 + 1$.

$= 2x^2 + 2 - 3$ Multiply.

$= 2x^2 - 1$ Add.

Note We could have found $(f \circ g)(2)$ from part a by first finding $(f \circ g)(x) = 2x^2 - 1$, then substituting 2 for x:
$(f \circ g)(2) = 2(2)^2 - 1$
$= 8 - 1$
$= 7$

▶

d. $(g \circ f)(x)$

Solution: $(g \circ f)(x) = g[f(x)]$ Definition of composition

$= g(2x - 3)$ Replace $f(x)$ with $2x - 3$.

$= (2x - 3)^2 + 1$ In $g(x)$, replace x with $2x - 3$.

$= 4x^2 - 12x + 9 + 1$ Multiply.

$= 4x^2 - 12x + 10$ Add.

Note We could have found $(g \circ f)(2)$ from part b by first finding $(g \circ f)(x) = 4x^2 - 12x + 10$, then substituting 2 for x:
$(g \circ f)(2) = 4(2)^2 - 12(2) + 10$
$= 16 - 24 + 10$
$= 2$

▶

Generally, $(f \circ g)(x) \neq (g \circ f)(x)$, as in Examples 1(c) and (d), but there are special functions for which $(f \circ g)(x) = (g \circ f)(x)$.

| **Your Turn 1** | If $f(x) = x^2 + 2$ and $g(x) = 3x + 5$, find the following.

a. $f[g(-3)]$ **b.** $g[f(2)]$ **c.** $f[g(x)]$

Objective 2 Show that two functions are inverses.

Operations that undo each other, such as addition and subtraction, are called *inverse operations*. For example, if we begin with a number x and add a second number y, we have $x + y$. Now we subtract y (the number we just added) from that result, and we have $x + y - y = x$, which is the original number.

Likewise, two functions that undo each other under composition are called **inverse functions**.

Definition Inverse functions: Two functions f and g are inverses if and only if $(f \circ g)(x) = x$ for all x in the domain of g and $(g \circ f)(x) = x$ for all x in the domain of f.

Loosely speaking, f and g are inverse functions if you evaluate f for a value x in its domain, substitute that result into g and evaluate, and get x again and vice versa.

Answers to Your Turn 1
a. $f[g(-3)] = 18$
b. $g[f(2)] = 23$
c. $f[g(x)] = 9x^2 + 30x + 27$

Procedure Inverse Functions

To determine whether two functions f and g are inverses of each other:
1. Show that $f[g(x)] = x$ for all x in the domain of g.
2. Show that $g[f(x)] = x$ for all x in the domain of f.

Example 2 Verify that f and g are inverses.

a. $f(x) = 3x + 2, g(x) = \dfrac{x - 2}{3}$

Solution: We need to show that $f[g(x)] = x$ and $g[f(x)] = x$.

$$f[g(x)] = f\left(\dfrac{x - 2}{3}\right) \qquad \text{Substitute } \dfrac{x - 2}{3} \text{ for } g(x).$$

$$= 3\left(\dfrac{x - 2}{3}\right) + 2 \qquad \text{Substitute } \dfrac{x - 2}{3} \text{ for } x \text{ in } f(x).$$

$$= x - 2 + 2 \qquad \text{Multiply.}$$

$$= x \qquad \text{So } f[g(x)] = x.$$

$$g[f(x)] = g(3x + 2) \qquad \text{Substitute } 3x + 2 \text{ for } f(x).$$

$$= \dfrac{3x + 2 - 2}{3} \qquad \text{Replace } x \text{ with } 3x + 2 \text{ in } g(x).$$

$$= \dfrac{3x}{3} \qquad \text{Add.}$$

$$= x \qquad \text{So } g[f(x)] = x.$$

Because $f[g(x)] = x$ and $g[f(x)] = x$, f and g are inverses.

b. $f(x) = x^3 + 5$ and $g(x) = \sqrt[3]{x - 5}$

Solution: We need to show that $f[g(x)] = x$ and $g[f(x)] = x$.

$$f[g(x)] = f(\sqrt[3]{x - 5}) \qquad \text{Replace } g(x) \text{ with } \sqrt[3]{x - 5}.$$

$$= (\sqrt[3]{x - 5})^3 + 5 \qquad \text{Replace } x \text{ with } \sqrt[3]{x - 5} \text{ in } f(x).$$

$$= x - 5 + 5 \qquad \text{Apply } (\sqrt[n]{x})^n = x.$$

$$= x \qquad \text{So } f[g(x)] = x.$$

$$g[f(x)] = g(x^3 + 5) \qquad \text{Replace } f(x) \text{ with } x^3 + 5.$$

$$= \sqrt[3]{x^3 + 5 - 5} \qquad \text{Replace } x \text{ with } x^3 + 5 \text{ in } g(x).$$

$$= \sqrt[3]{x^3} \qquad \text{Add.}$$

$$= x \qquad \text{So } g[f(x)] = x.$$

Because $f[g(x)] = x$ and $g[f(x)] = x$, f and g are inverses.

Your Turn 2 Verify that $f(x) = 5x + 6$ and $g(x) = \dfrac{x - 6}{5}$ are inverse functions.

Answer to Your Turn 2

$$f[g(x)] = f\left[\dfrac{x - 6}{5}\right]$$

$$= 5\left(\dfrac{x - 6}{5}\right) + 6$$

$$= x - 6 + 6$$

$$= x$$

$$g[f(x)] = g[5x + 6]$$

$$= \dfrac{5x + 6 - 6}{5}$$

$$= \dfrac{5x}{5}$$

$$= x$$

Because $f[g(x)] = x$ and $g[f(x)] = x$, f and g are inverses.

Before finding inverse functions, we need to determine what types of functions have inverses. Consider the following:

$$f = \{(1, 2), (3, 4), (5, 6), (x, y)\}; \text{domain} = \{1, 3, 5, x\} \text{ and range} = \{2, 4, 6, y\}$$

$$g = \{(2, 1), (4, 3), (6, 5), (y, x)\}; \text{domain} = \{2, 4, 6, y\} \text{ and range} = \{1, 3, 5, x\}$$

Note that the ordered pairs in g are the ordered pairs of f with x and y interchanged. Because $(1, 2)$ is in f, by definition, $f(1) = 2$. Because $(2, 1)$ is in g, $g(2) = 1$. Therefore, $f[g(2)] = f(1) = 2$ and $g[f(1)] = g(2) = 1$. Similarly, $f[g(4)] = f(3) = 4$ and $g[f(3)] = g(4) = 3$. In particular, $f[g(y)] = f(x) = y$ and $g[f(x)] = g(y) = x$ for all real numbers x and y. Consequently, f and g are inverses.

In general, to find the inverse of a function, we reverse the ordered pairs by interchanging x and y. To indicate the inverse of the function f, we use a special notation f^{-1} instead of g. The following figure illustrates inverse functions.

Warning Do not confuse f^{-1} with raising an expression to the negative 1 power. It is usually clear from the context that f^{-1} means an inverse function.

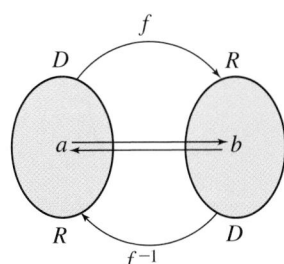

Note Notice that f sends a to b and f^{-1} sends b back to a. Hence, $f^{-1}[f(a)] = a$. Also, f^{-1} sends b to a and f sends a back to b. Hence, $f[f^{-1}(b)] = b$. We also see that the domain of f is the range of f^{-1} and the range of f is the domain of f^{-1}.

Before we can find inverse functions, we need to explore one more concept.

Objective 3 Show that a function is one-to-one.

If we merely reverse the ordered pairs of a function, we do not always get an inverse function. Consider the following sets in which the ordered pairs of B are the ordered pairs of A with x and y interchanged:

$$A = \{(1,2), (2,4), (3,2), (-2,5)\}$$
$$B = \{(2,1), (4,2), (2,3), (5,-2,)\}$$

Note that A represents a function but B does not because 2 in the domain of B is paired with 1 and 3. How did this happen? Because the ordered pairs of B are those of A with x and y interchanged, two ordered pairs of A having the same y-value of 2 $[(1,2)$ and $(3,2)]$ became two ordered pairs with the same x-value in B. Therefore, if the inverse of a function is to be a function, no two ordered pairs of the function may have the same y-value. Such a function is called a **one-to-one function**.

A function is one-to-one if each value in the range corresponds to only one value in the domain. In terms of x and y, two different x-values must result in two different y-values, which suggests the following formal definition.

Definition One-to-one function: A function f is one-to-one if for any two numbers a and b in its domain, when $f(a) = f(b), a = b$ and when $a \neq b, f(a) \neq f(b)$.

It is possible to determine whether a function is one-to-one by looking at its graph. If two ordered pairs have the same y-value, the corresponding points lie on the same horizontal line. Consequently, if a function is one-to-one, the graph cannot be intersected by any horizontal line in more than one point.

Rule Horizontal Line Test for One-to-One Functions

Given a function's graph, the function is one-to-one if every horizontal line that can intersect the graph does so at one and only one point.

Example 3 Determine whether the following graphs are of one-to-one functions.

a. $f(x) = x^3$

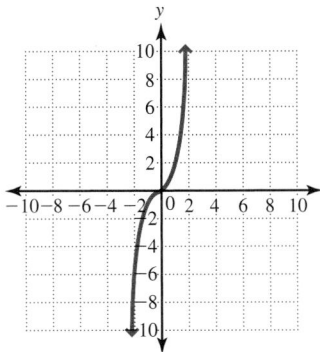

b. $f(x) = x^2 + 2$

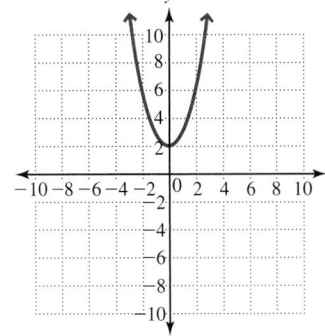

c. $f(x) = \sqrt{36 - x^2}$

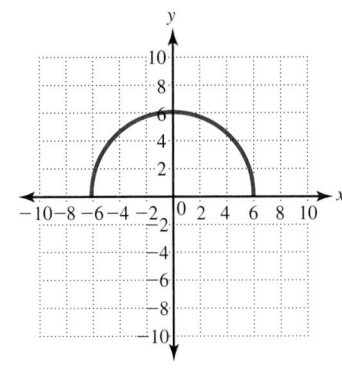

Solution:

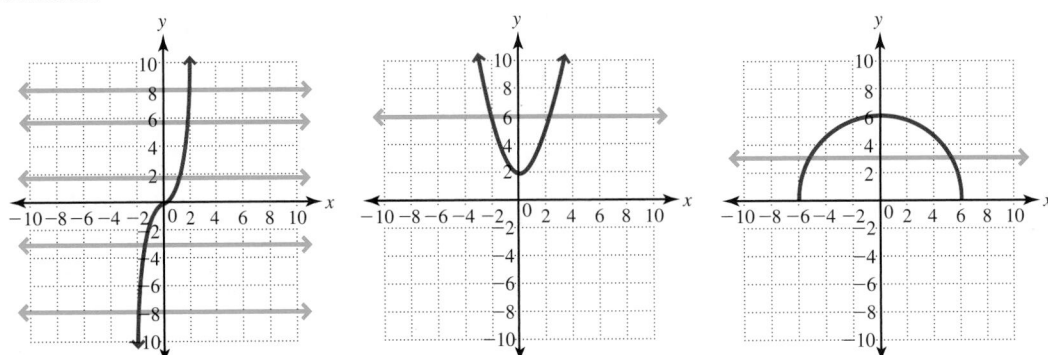

a. Every horizontal line that can intersect this graph does so at one and only one point, so the function is one-to-one.

b. A horizontal line can intersect this graph in more than one point, so the function is not one-to-one.

c. A horizontal line can intersect this graph in more than one point, so the function is not one-to-one.

Your Turn 3 Determine whether the following graphs are of one-to-one functions.

a.

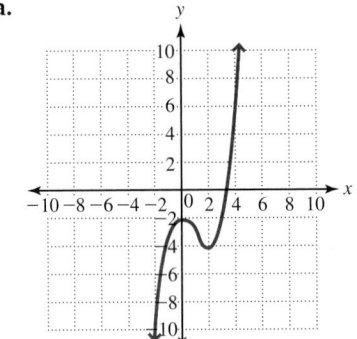

b.

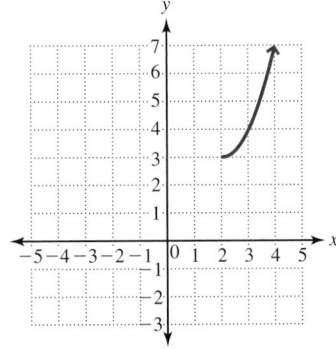

Objective 4 Find the inverse of a function.

Interchanging the ordered pairs of a function that is not one-to-one results in a relation that is not a function. Consequently, we have the following:

> **Rule Existence of Inverse Functions**
>
> A function has an inverse function if and only if the function is one-to-one.

We have already seen that to find the inverse of a function, we need to interchange the x- and y-values of all of the ordered pairs of f. That is, we replace x with y and y with x in the equation defining the function. However, we are not finished. We now have the inverse in the form $f(y) = x$, so we must solve this equation for y to have y as a function of x. The steps are summarized as follows.

Answers to Your Turn 3
a. no   b. yes

> **Procedure Finding the Inverse Function of a One-to-One Function**
>
> 1. If necessary, replace $f(x)$ with y.
> 2. Replace all x's with y's and y's with x's.
> 3. Solve the equation from step 2 for y.
> 4. Replace y with $f^{-1}(x)$.

Example 4 Find $f^{-1}(x)$ for each of the following one-to-one functions.

a. $f(x) = 2x + 4$

The domain and range of f is the set of all real numbers, so the domain and range of f^{-1} is also the set of all real numbers.

Solution: $y = 2x + 4$ Replace $f(x)$ with y.

$\qquad\qquad x = 2y + 4$ Replace x with y and y with x.

$\qquad\qquad \dfrac{x - 4}{2} = y$ Solve for y.

$\qquad\qquad f^{-1}(x) = \dfrac{x - 4}{2}$ Replace y with $f^{-1}(x)$.

To verify that we have found the inverse, we need to show that $f[f^{-1}(x)] = x$ and $f^{-1}[f(x)] = x$.

$$f[f^{-1}(x)] = f\left(\frac{x - 4}{2}\right) = 2\left(\frac{x - 4}{2}\right) + 4 = x - 4 + 4 = x$$

$$f^{-1}[f(x)] = f^{-1}(2x + 4) = \frac{(2x + 4) - 4}{2} = \frac{2x}{2} = x$$

Because $f[f^{-1}(x)] = x$ and $f^{-1}[f(x)] = x$, they are inverses.

b. $f(x) = x^3 + 2$

The domain and range of f is the set of all real numbers, so the domain and range of f^{-1} is also the set of all real numbers.

Solution: $y = x^3 + 2$ Replace $f(x)$ with y.

$\qquad\qquad x = y^3 + 2$ Replace x with y and y with x.

$\qquad\qquad x - 2 = y^3$ Begin solving for y by subtracting 2 from both sides.

$\qquad\qquad \sqrt[3]{x - 2} = y$ Take the cube root of each side to solve for y.

$\qquad\qquad f^{-1}(x) = \sqrt[3]{x - 2}$ Replace y with $f^{-1}(x)$.

We can verify that f and f^{-1} are inverses, as in part a.

c. $f(x) = \sqrt{x - 3}$

The domain of f is $[3, \infty)$, and the range is $[0, \infty)$. Therefore, the domain of f^{-1} is $[0, \infty)$ and the range of f^{-1} is $[3, \infty)$.

Solution: $y = \sqrt{x - 3}$ Replace $f(x)$ with y.

$\qquad\qquad x = \sqrt{y - 3}$ Replace x with y and y with x.

$\qquad\qquad x^2 = y - 3$ Begin solving for y by squaring both sides.

$\qquad\qquad x^2 + 3 = y$ Add 3 to both sides to solve for y.

The domain of $y = x^2 + 3$ is all real numbers, but the domain of f^{-1} is $[0, \infty)$. Therefore, we write $f^{-1}(x) = x^2 + 3, x \geq 0$.

To verify that these are inverses, we need to show that $f[f^{-1}(x)] = x$ and $f^{-1}[f(x)] = x$.

$$f[f^{-1}(x)] = f(x^2 + 3) = \sqrt{(x^2 + 3) - 3} = \sqrt{x^2 + 3 - 3} = \sqrt{x^2}$$

Because the domain of f^{-1} is $[0, \infty)$, $\sqrt{x^2} = x$. So $f[f^{-1}(x)] = x$.

$$f^{-1}[f(x)] = f^{-1}(\sqrt{x - 3}) = (\sqrt{x - 3})^2 + 3 = x - 3 + 3 = x$$

Because $f[f^{-1}(x)] = x$ and $f^{-1}[f(x)] = x$, they are inverses.

Note Recall $\sqrt{x^2} = |x|$ for x a real number.

Answers to Your Turn 4

a. $f^{-1}(x) = \dfrac{x - 2}{5}$

b. $f^{-1}(x) = \dfrac{3}{x - 1}$

Your Turn 4 Find $f^{-1}(x)$ for each of the following one-to-one functions.

a. $f(x) = 5x + 2$ **b.** $f(x) = \dfrac{x + 3}{x}$

Objective 5 Graph a given function's inverse function.

In the following figure, we have plotted pairs of points whose coordinates are interchanged and have graphed the line $y = x$.

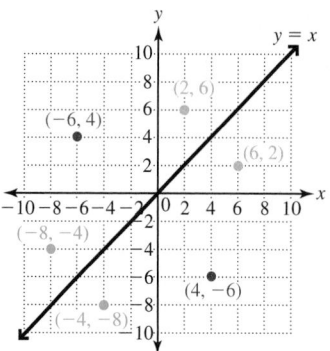

If the graph were folded along the line $y = x$, the points whose coordinates are interchanged would fall on top of each other. These points, therefore, are symmetric with respect to the line $y = x$. It can be shown that the graphs of any two points of the form (a, b) and (b, a) are symmetric with respect to the graph of $y = x$. Similarly, because the ordered pairs of f and f^{-1} have interchanged coordinates, the graphs of f and f^{-1} are symmetric with respect to the line $y = x$.

Learning Strategy

If you are a tactile learner, imagine placing a mirror on the line $y = x$. The graphs of f and f^{-1} are reflections in the line $y = x$.

Rule Graphs of Inverse Functions

The graphs of f and f^{-1} are symmetric with respect to the graph of $y = x$.

Following are the graphs of f and f^{-1} for Example 4 along with the graph of $y = x$.

a. $f(x) = 2x + 4$

$f^{-1}(x) = \dfrac{x - 4}{2}$

b. $f(x) = x^3 + 2$

$f^{-1}(x) = \sqrt[3]{x - 2}$

c. $f(x) = \sqrt{x - 3}$

$f^{-1}(x) = x^2 + 3, \quad x \geq 0$

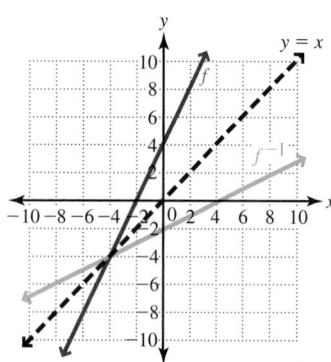
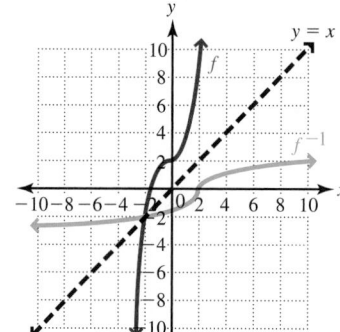
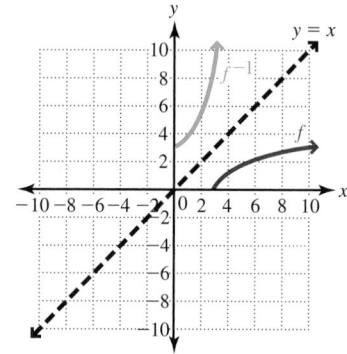

Example 5

Sketch the inverse of the functions whose graphs are shown in parts a and b.

a.

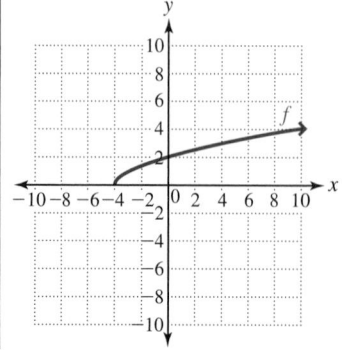

Solution: Draw the line $y = x$ and reflect the graph in the line.

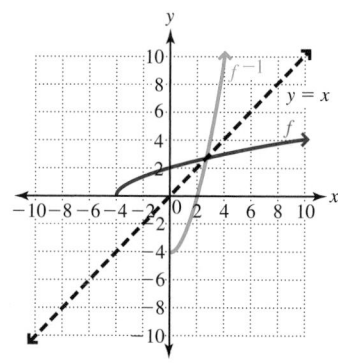

b.

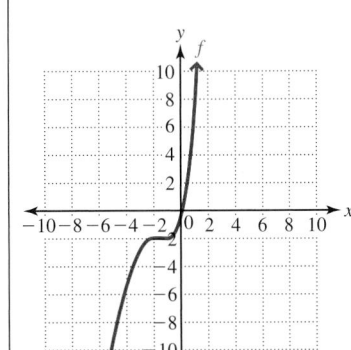

Solution: Draw the line $y = x$ and reflect the graph in the line.

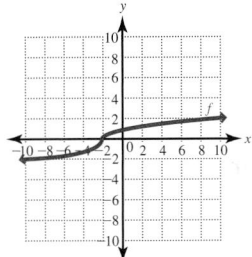

Answers to Your Turn 5

a.

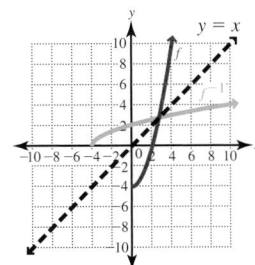

Your Turn 5 Sketch the graph of the inverse of the functions whose graphs are given.

b.

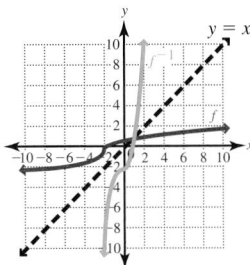

a.

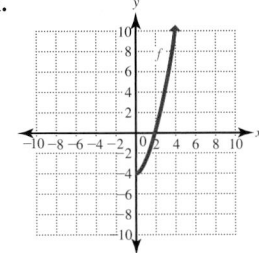

b.

11.1 Exercises For Extra Help MyMathLab®

Objective 1

Prep Exercise 1 If f and g are functions, then the composition of f and g is defined as $(f \circ g)(x) = $ _____$f[g(x)]$_____ for all x in the domain g for which $g(x)$ is in the domain of f.

Prep Exercise 2 If f and g are functions, then the composition of g and f is defined as $(g \circ f)(x) = $ _____$g[f(x)]$_____ for all x in the domain f for which $f(x)$ is in the domain of g.

For Exercises 1–12, if $f(x) = 3x + 5$, $g(x) = x^2 + 3$, and $h(x) = \sqrt{x+1}$, find each composition. See Example 1.

1. $(f \circ g)(0)$
14

2. $(g \circ f)(0)$
28

3. $(h \circ f)(1)$
3

4. $(h \circ g)(3)$
$\sqrt{13}$

5. $(f \circ g)(-2)$
26

6. $(g \circ f)(-1)$
7

7. $(f \circ g)(x)$
$3x^2 + 14$

8. $(g \circ f)(x)$
$9x^2 + 30x + 28$

9. $(f \circ h)(x)$
$3\sqrt{x+1} + 5$

10. $(h \circ g)(x)$
$\sqrt{x^2 + 4}$

11. $(h \circ f)(0)$
$\sqrt{6}$

12. $(g \circ h)(0)$
4

For Exercises 13–22, find $(f \circ g)(x)$ and $(g \circ f)(x)$. See Example 1.

13. $f(x) = 2x - 2, g(x) = 3x + 4$
$(f \circ g)(x) = 6x + 6; (g \circ f)(x) = 6x - 2$

14. $f(x) = 4x + 7, g(x) = 3x + 4$
$(f \circ g)(x) = 12x + 23; (g \circ f)(x) = 12x + 25$

15. $f(x) = x + 2, g(x) = x^2 + 1$
$(f \circ g)(x) = x^2 + 3; (g \circ f)(x) = x^2 + 4x + 5$

16. $f(x) = x^2 - 3, g(x) = x + 5$
$(f \circ g)(x) = x^2 + 10x + 22; (g \circ f)(x) = x^2 + 2$

17. $f(x) = x^2 + 3x - 4; g(x) = 3x$
$(f \circ g)(x) = 9x^2 + 9x - 4; (g \circ f)(x) = 3x^2 + 9x - 12$

18. $f(x) = -3x, g(x) = x^2 + 2x + 4$
$(f \circ g)(x) = -3x^2 - 6x - 12; (g \circ f)(x) = 9x^2 - 6x + 4$

19. $f(x) = \sqrt{x+2}, g(x) = 2x - 5$
$(f \circ g)(x) = \sqrt{2x-3}; (g \circ f)(x) = 2\sqrt{x+2} - 5$

20. $f(x) = 5x + 2, g(x) = \sqrt{x-5}$
$(f \circ g)(x) = 5\sqrt{x-5} + 2; (g \circ f)(x) = \sqrt{5x-3}$

21. $f(x) = \dfrac{x+1}{x}, g(x) = \dfrac{x-3}{x}$
$(f \circ g)(x) = \dfrac{2x-3}{x-3}; (g \circ f)(x) = \dfrac{1-2x}{x+1}$

22. $f(x) = \dfrac{x+4}{x}, g(x) = \dfrac{2-x}{x}$
$(f \circ g)(x) = \dfrac{2+3x}{2-x}; (g \circ f)(x) = \dfrac{x-4}{x+4}$

Objective 2

Prep Exercise 3 How are the domains and ranges of inverse functions related? Why?

The domain of a function is the range of its inverse; the range of a function is the domain of its inverse. This occurs because the ordered pairs of f and f^{-1} have each x and y interchanged.

Prep Exercise 4 If the coordinates of the ordered pairs of a function are interchanged, is the result always the inverse function? Why or why not?

No, the function must be one-to-one. Otherwise, the inverse is not a function.

For Exercises 23 and 24, answer each question. See Objective 2.

23. If the domain of f is $[3, \infty)$ and the range is $[0, \infty)$, what are the domain and range of f^{-1}?

Domain is $[0, \infty)$, and range is $[3, \infty)$.

24. If f and g are inverse functions and $f(2) = 5$, then $g(5) = $ _____2_____.

For Exercises 25–28, determine whether the following functions f and g are inverses. See Objective 2.

25. $f = \{(1,2), (-1,-3), (3,4), (2,-5)\}, g = \{(2,1), (-3,-1), (4,3), (-5,2)\}$
yes

26. $f = \{(4,-3), (1,4), (5,2), (-3,1), (-1,3)\}, g = \{(-3,4), (4,1), (2,5), (1,-3), (3,-1)\}$
yes

27. $f = \{(-2,-2), (3,-3), (-4,4), (-6,-6)\}, g = \{(-2,-2), (-3,3), (-4,4), (-6,-6)\}$
no

28. $f = \{(5,5), (-4,4), (2,-2), (-7,-7)\}, g = \{(-5,-5), (4,-4), (-2,2), (7,7)\}$
no

Prep Exercise 5 What must be shown to prove that the two functions f and g are inverses?

To prove that functions f and g are inverses, it must be shown that $f[g(x)] = x$ and $g[f(x)] = x$.

For Exercises 29–42, determine whether f and g are inverses by determining whether $(f \circ g)(x) = x$ and $(g \circ f)(x) = x$. See Example 2.

29. $f(x) = x + 5, g(x) = x - 5$
yes

30. $f(x) = x - 8, g(x) = x + 8$
yes

31. $f(x) = 6x, g(x) = \dfrac{x}{6}$
yes

32. $f(x) = -\dfrac{x}{3}, g(x) = -3x$
yes

33. $f(x) = 2x - 3, g(x) = \dfrac{x+3}{2}$
yes

34. $f(x) = \dfrac{x-4}{5}, g(x) = 5x + 4$
yes

35. $f(x) = x^3 - 4, g(x) = \sqrt[3]{x + 4}$
yes

36. $f(x) = x^5 + 3, g(x) = \sqrt[5]{x - 3}$
yes

37. $f(x) = x^2, g(x) = \sqrt{x}$
no

38. $f(x) = x^4, g(x) = \sqrt[4]{x}$
no

39. $f(x) = x^2, x \geq 0; g(x) = \sqrt{x}$
yes

40. $f(x) = x^2 + 2, x \geq 0; g(x) = \sqrt{x - 2}$
yes

41. $f(x) = \dfrac{3}{x + 5}, g(x) = \dfrac{3 - 5x}{x}$
yes

42. $f(x) = \dfrac{x}{x + 4}, g(x) = \dfrac{4x}{1 - x}$
yes

For Exercises 43 and 44, answer each question.

43. Is $f(x) = \dfrac{1}{x}$ its own inverse? Explain.
Yes, because $(f \circ g)(x) = (g \circ f)(x) = x$

44. Is $f(x) = x$ its own inverse? Explain.
Yes, because $(f \circ g)(x) = (g \circ f)(x) = x$

Objective 3

Prep Exercise 6 If given the graph of a function, what kind of test can be performed to determine whether it is one-to-one?
Horizontal line test

For Exercises 45–48, determine whether the function is one-to-one. See Example 3.

45.

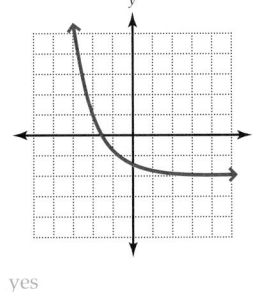

yes

46.

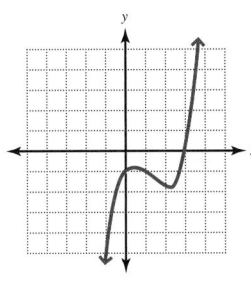

no

47.

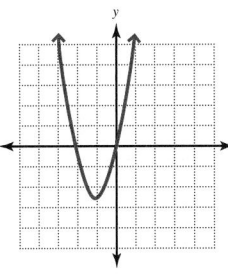

no

48.

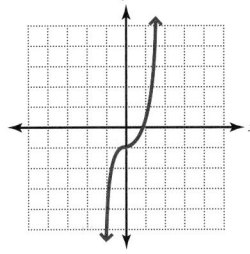

yes

Objective 4

Prep Exercise 7 When finding the inverse of a one-to-one function, after exchanging x and y, what is the next step?
Solve for y.

For Exercises 49–70, find $f^{-1}(x)$ for each of the following one-to-one functions f.
See Example 4.

49. $f = \{(-3, 2), (-1, -3), (0, 4), (4, 6)\}$
$f^{-1} = \{(2, -3), (-3, -1), (4, 0), (6, 4)\}$

50. $f = \{(-4, 2), (-1, -3), (2, 3), (5, 7)\}$
$f^{-1} = \{(2, -4), (-3, -1), (3, 2), (7, 5)\}$

51. $f = \{(7, -2), (9, 2), (-4, 1), (3, 3)\}$
$f^{-1} = \{(-2, 7), (2, 9), (1, -4), (3, 3)\}$

52. $f = \{(1, 2), (4, 5), (-3, 3), (-9, 0)\}$
$f^{-1} = \{(2, 1), (5, 4), (3, -3), (0, -9)\}$

53. $f(x) = x + 6$
$f^{-1}(x) = x - 6$

54. $f(x) = x - 4$
$f^{-1}(x) = x + 4$

55. $f(x) = 2x + 3$
$f^{-1}(x) = \dfrac{x - 3}{2}$

56. $f(x) = 4x - 5$
$f^{-1}(x) = \dfrac{x + 5}{4}$

57. $f(x) = x^3 - 1$
$f^{-1}(x) = \sqrt[3]{x + 1}$

58. $f(x) = x^3 - 3$
$f^{-1}(x) = \sqrt[3]{x + 3}$

59. $f(x) = \dfrac{2}{x + 2}$
$f^{-1}(x) = \dfrac{2 - 2x}{x}$

60. $f(x) = \dfrac{-3}{x - 3}$
$f^{-1}(x) = \dfrac{3x - 3}{x}$

61. $f(x) = \dfrac{x + 2}{x - 3}$

$f^{-1}(x) = \dfrac{3x + 2}{x - 1}$

62. $f(x) = \dfrac{x - 4}{x + 2}$

$f^{-1}(x) = \dfrac{-2x - 4}{x - 1}$

63. $f(x) = \sqrt{x - 2}$

$f^{-1}(x) = x^2 + 2, x \geq 0$

64. $f(x) = \sqrt{2x - 4}$

$f^{-1}(x) = \dfrac{x^2}{2} + 2, x \geq 0$

65. $f(x) = 2x^3 + 4$

$f^{-1}(x) = \sqrt[3]{\dfrac{x - 4}{2}}$

66. $f(x) = 4x^3 - 5$

$f^{-1}(x) = \sqrt[3]{\dfrac{x + 5}{4}}$

67. $f(x) = \sqrt[3]{x + 2}$

$f^{-1}(x) = x^3 - 2$

68. $f(x) = \sqrt[3]{x + 1}$

$f^{-1}(x) = x^3 - 1$

69. $f(x) = 2\sqrt[3]{2x + 4}$

$f^{-1}(x) = \dfrac{x^3}{16} - 2$

70. $f(x) = 4\sqrt[3]{2x - 6}$

$f^{-1}(x) = \dfrac{x^3}{128} + 3$

Objective 5

Prep Exercise 8 If two one-to-one functions are inverses, then their graphs are symmetric about what line? $y = x$

For Exercises 71–82, sketch the graph of the inverse of each of the following functions. See Example 5.

71.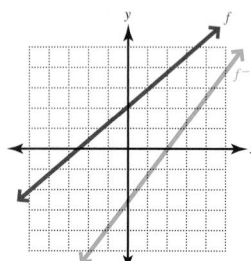

72.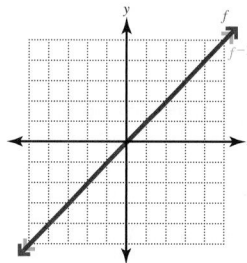

73.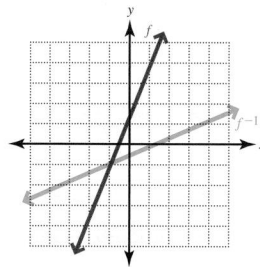

74.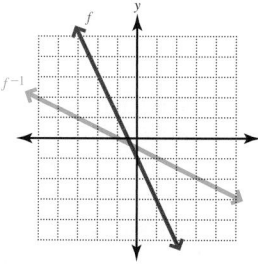

75.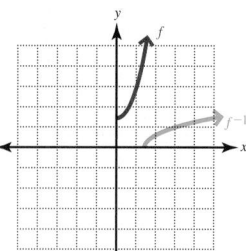

76.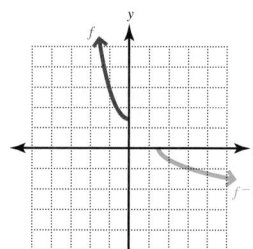

77.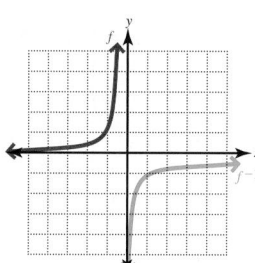

78.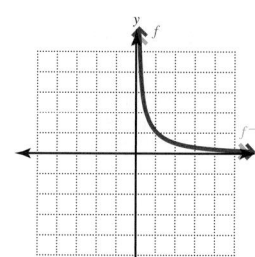

79.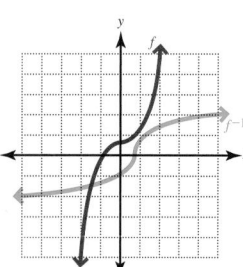

80.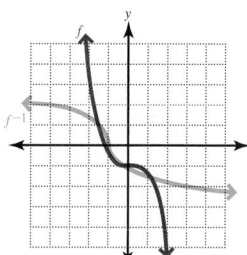

81.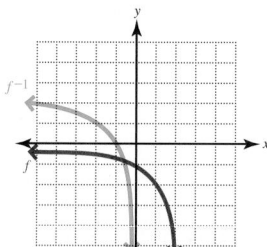

82.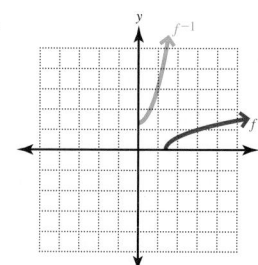

For Exercises 83–86, solve each problem.

83. If a salesperson works for $100 per week plus 5% commission on sales, the weekly salary is $y = 0.05x + 100$, where y represents the salary and x represents the sales.

  a. Find the inverse function.

  $y = \dfrac{x - 100}{0.05}$

  b. What does each variable of the inverse function represent?

  x represents the salary; y represents the sales.

  c. Use the inverse function to find the sales for a week in which the salary was $350.

  $5000

84. If a Toyota Prius averages 50 miles per hour on a trip, then $y = 50x$, where y represents the number of miles traveled and x represents the number of hours.
   a. Find the inverse function.
   $$y = \frac{x}{50}$$
   b. What does each variable of the inverse function represent?
   x represents the number of miles; y represents the number of hours.
   c. Use the inverse function to find the number of hours to travel 210 miles.
   4.2 hr.

85. An office building installed 105 fans, some of which measure 36 inches with the remaining fans measuring 54 inches. If the 36-inch fans cost \$45 each and the 54-inch fans cost \$65 each, the cost of the fans is $y = 45x + 65(105 - x)$, where y represents the cost and x represents the number of 36-inch fans.
   a. Find the inverse function.
   $$y = \frac{6825 - x}{20}$$
   b. What does each variable of the inverse function represent?
   x represents the cost; y represents the number of 36-in. fans.
   c. Use the inverse function to find the number of 36-inch fans if the total cost was \$5225.
   80

86. A painting contractor purchased 24 gallons of paint for the interior and exterior of a house. If he paid \$18 per gallon for the interior paint and \$22 per gallon for the exterior paint, the amount he paid is $y = 18x + 22(24 - x)$, where y represents the amount paid and x represents the number of gallons of interior paint.
   a. Find the inverse function.
   $$y = \frac{528 - x}{4}$$
   b. What does each variable of the inverse function represent?
   x represents the amount paid; y represents the number of gallons of interior paint.
   c. Use the inverse function to find the number of gallons of interior paint if he paid a total of \$488 for the 24 gallons of paint.
   10 gal.

For Exercises 87–90, answer the question.

87. If $(-4, 5)$ is an ordered pair on the graph of g, what are the coordinates of an ordered pair on the graph of g^{-1}?
   $(5, -4)$

88. If $f(2) = 4$ and $f^{-1}(a) = 2$, find a.
   $a = 4$

89. A linear function is of the form $f(x) = ax + b, a \neq 0$. Find $f^{-1}(x)$.
   $$f^{-1}(x) = \frac{x - b}{a}$$

90. The square root function is defined by $f(x) = \sqrt{x}$. Find $f^{-1}(x)$. Be careful!
   $f^{-1}(x) = x^2, x \geq 0$

Review Exercises

Exercises 1–5   Expressions

[1.5] 1. Write 32 as 2 to a power.
   $32 = 2^5$

For Exercises 2–5, evaluate each expression.

[1.5] 2. 2^3
   8

[1.5] 3. $\left(-\dfrac{1}{3}\right)^3$
   $-\dfrac{1}{27}$

[5.1] 4. 4^{-2}
   $\dfrac{1}{16}$

[9.2] 5. $4^{3/2}$
   8

Exercise 6   Equations and Inequalities

[10.4] 6. What are the coordinates of the vertex of $g(x) = (x - 3)^2 + 1$?
   $(3, 1)$

11.2 Exponential Functions

Objectives

1 Define and graph exponential functions.

2 Solve equations of the form $b^x = b^y$ for x.

3 Use exponential functions to solve application problems.

Warm-up

[11.1] **1.** Find the inverse function: $f(x) = x^3 - 2$

[1.5] **2.** Write 81 as 3 to a power.

For Exercises 3 and 4, evaluate.

[5.1] **3.** $\left(\dfrac{1}{4}\right)^{-2}$

[9.2] **4.** $27^{2/3}$

Objective 1 Define and graph exponential functions.

Previously, we defined rational number exponents. For example, we know that
$b^3 = b \cdot b \cdot b$, $b^{2/3} = \sqrt[3]{b^2}$, $b^0 = 1$, and $b^{-3} = \dfrac{1}{b^3}$.

It can be shown that irrational number exponents have meaning as well, and we can approximate expressions such as $2^{\sqrt{3}}$ and 5^{π} by using rational approximations for the exponents. Therefore, the exponential expression b^x has meaning if x is any real number (rational or irrational). So we can define the **exponential function** as follows.

Note The definition of the exponential function has two restrictions on b. If $b = 1$, then $f(x) = b^x = 1^x = 1$, which is a linear function. If $b < 0$, we may get values for which the function is not defined as a real number. For example, if $f(x) = (-4)^x$, then $f\left(\dfrac{1}{2}\right) = (-4)^{1/2} = \sqrt{-4}$, which is not a real number.

▶ **Definition** **Exponential function:** If $b > 0$, $b \neq 1$, and x is any real number, then the exponential function is $f(x) = b^x$.

We graph exponential functions by plotting enough points to determine the graph's shape. We will find that the graph has one typical shape if $b > 1$ and a different typical shape if $0 < b < 1$.

Example 1 Graph.

a. $f(x) = 2^x$ and $g(x) = 4^x$

Solution: Choose some values of x and find the corresponding values of $f(x)$ and $g(x)$ (which are the y-values).

x	−3	−2	−1	0	1	2	3
f(x)	$\dfrac{1}{8}$	$\dfrac{1}{4}$	$\dfrac{1}{2}$	1	2	4	8
g(x)	$\dfrac{1}{64}$	$\dfrac{1}{16}$	$\dfrac{1}{4}$	1	4	16	64

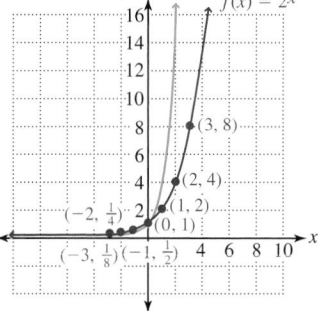

Plotting the ordered pairs for each function gives smooth curves typical of the graphs of $f(x) = b^x$ with $b > 1$. Comparing the graphs of $f(x) = 2^x$ and $g(x) = 4^x$, we can see that the greater the value of b, the steeper the graph.

b. $h(x) = \left(\dfrac{1}{2}\right)^x$ and $k(x) = \left(\dfrac{1}{4}\right)^x$

Answers to Warm-up

1. $f^{-1}(x) = \sqrt[3]{x + 2}$

2. 3^4

3. 16

4. 9

Solution: Choose some values of x and find the corresponding values of $h(x)$ and $k(x)$.

x	−3	−2	−1	0	1	2	3
h(x)	8	4	2	1	$\frac{1}{2}$	$\frac{1}{4}$	$\frac{1}{8}$
k(x)	64	16	4	1	$\frac{1}{4}$	$\frac{1}{16}$	$\frac{1}{64}$

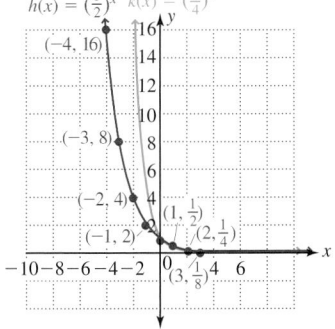

Note We left the points and coordinate labels off the graphs of $g(x)$ and $k(x)$ to avoid additional clutter. From all of the graphs in Example 1, we see that b^x is never negative.

After plotting the ordered pairs for each function, we see graphs that are typical of exponential functions in the form $f(x) = b^x$ with $0 < b < 1$. Comparing the graphs of $h(x) = \left(\frac{1}{2}\right)^x$ and $k(x) = \left(\frac{1}{4}\right)^x$, we see that the smaller the value of b, the steeper the graph.

We summarize the graphs of $f(x) = b^x$ as follows.

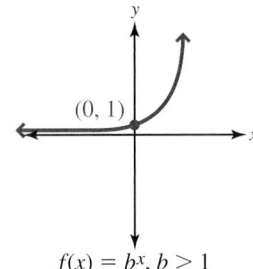

$f(x) = b^x, b > 1$

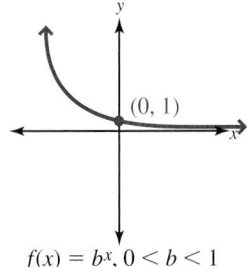

$f(x) = b^x, 0 < b < 1$

Note Any function that increases or decreases along its entire domain is one-to-one because no horizontal line will intersect its graph at more than one point.

The function is increasing from left to right and is one-to-one. The graph always passes through $(0, 1)$. For negative values of x, the graph approaches the x-axis but never touches it. The larger the values of b, the steeper the graph. The domain is $(-\infty, \infty)$, and the range is $(0, \infty)$.

The function is decreasing from left to right and is one-to-one. The graph always passes through $(0, 1)$. For positive values of x, the graph approaches the x-axis but never touches it. The smaller the values of x, the steeper the graph. The domain is $(-\infty, \infty)$, and the range is $(0, \infty)$.

More complicated exponential functions are graphed in the same manner.

Example 2 Graph $f(x) = 3^{2x-1}$.

Note $f(x) = 3^{2x-1}$ is not of the form $f(x) = b^x$. Note that it does not pass through $(0, 1)$.

Solution: Find some ordered pairs, plot them, and draw the graph.

x	y = f(x)
−1	$3^{-3} = \dfrac{1}{27}$
0	$3^{-1} = \dfrac{1}{3}$
1	$3^1 = 3$
2	$3^3 = 27$

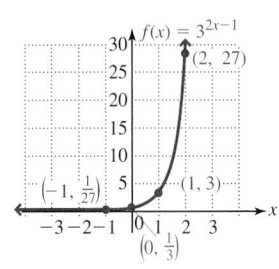

Answer to Your Turn 2

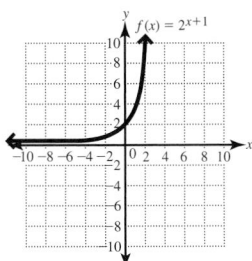

Your Turn 2 Graph $f(x) = 2^{x+1}$.

Objective 2 Solve equations of the form $b^x = b^y$ for x.

Earlier we solved equations containing expressions having a variable base and a constant exponent, such as $x^2 = 4$. Now we solve equations that contain expressions having a constant base and a variable exponent, such as $2^x = 16$. We need a new method to solve these equations. Previously, we noted that the exponential function is one-to-one; so for each x-value, there is a unique y-value and vice versa. Consequently, we have the following rule.

Rule One-to-One Property of Exponentials

Given $b > 0$ and $b \neq 1$, if $b^x = b^y$, then $x = y$.

To solve some types of exponential equations, we use the following procedure.

Procedure Solving Exponential Equations

1. If necessary, write both sides of the equation as a power of the same base.
2. If necessary, simplify the exponents.
3. Set the exponents equal to each other.
4. Solve the resulting equation.

Example 3 Solve.

a. $3^x = 81$

Solution: $3^x = 3^4$ Write 81 as 3^4 so that both sides have the same base.

$\qquad\quad x = 4$ Set the exponents equal to each other.

Check: $3^x = 81$

$\qquad 3^4 = 81$ Replace x with 4.

$\qquad 81 = 81$ True; so $x = 4$ is the solution.

b. $4^x = 32$

Solution: $(2^2)^x = 2^5$ Write 4 as 2^2 and 32 as 2^5 so that both sides have the same base.

$\qquad\quad 2^{2x} = 2^5$ Simplify $(2^2)^x$ by applying $(a^m)^n = a^{mn}$.

$\qquad\quad 2x = 5$ Set the exponents equal to each other.

$\qquad\quad x = \dfrac{5}{2}$ Solve for x.

Note We leave the checks for parts b–d to the reader.

c. $8^{x+4} = 4^{2x+3}$

Solution: $(2^3)^{x+4} = (2^2)^{2x+3}$ Write 8 as 2^3 and 4 as 2^2 so that both sides have the same base.

$\qquad\quad 2^{3x+12} = 2^{4x+6}$ Simplify the exponents by applying $(a^m)^n = a^{mn}$.

$\qquad\quad 3x + 12 = 4x + 6$ Set the exponents equal to each other.

$\qquad\qquad\quad 6 = x$ Solve for x.

d. $\left(\dfrac{1}{5}\right)^x = 25$

Solution: $(5^{-1})^x = 5^2$ Write $\dfrac{1}{5}$ as 5^{-1} and 25 as 5^2 so that both sides have the same base.

$\qquad\quad 5^{-x} = 5^2$ Simplify $(5^{-1})^x$ by applying $(a^m)^n = a^{am}$.

$\qquad\quad -x = 2$ Set the exponents equal to each other.

$\qquad\quad x = -2$ Solve for x.

Your Turn 3 Solve.

   a. $2^{x+1} = 8$ **b.** $27^{x-2} = 9^x$ **c.** $\left(\dfrac{1}{2}\right)^x = 16$

Objective 3 Use exponential functions to solve application problems.

Compound Interest Formula

Exponential functions occur in many areas, especially the sciences and business. If P dollars are invested at an annual interest rate of r (written as a decimal) compounded n times per year for t years, the accumulated amount A in the account is given by the formula $A = P\left(1 + \dfrac{r}{n}\right)^{nt}$.

Example 4 Find the accumulated amount in an account if $5000 is deposited at 6% compounded quarterly for 10 years. Round the answer to the nearest cent.

Solution: $A = 5000\left(1 + \dfrac{0.06}{4}\right)^{4(10)}$ In $A = P\left(1 + \dfrac{r}{n}\right)^{nt}$, substitute 5000 for P, **0.06** for r, **4** for n, and **10** for t.

$A = 5000(1 + 0.015)^{40}$ Simplify.

$A = 5000(1.015)^{40}$

$A = 9070.09$ Evaluate using a calculator and round to the nearest cent (hundredths place).

Answer: After 10 years, the accumulated amount in the account is $9070.09.

Your Turn 4 If $3000 is invested at 4% compounded semiannually (twice per year), how much money is in the account at the end of 8 years? Round to the nearest cent.

Half-Life

The *half-life* of a radioactive substance is the amount of time it takes until only half of the original amount of the substance remains. Suppose we begin with 100 grams of a substance that has a half-life of 10 days. After 10 days (one half-life), 50 grams will remain; after 20 days (two half-lives), 25 grams will remain; after 30 days (three half-lives), 12.5 grams will remain; and so on. The formula $A = A_0\left(\dfrac{1}{2}\right)^{t/h}$ gives the amount remaining, where A_0 is the initial amount, t is the time, and h is the half-life.

Example 5 The isotope ^{45}Ca has a half-life of 165 days. How many grams of a 50-gram sample of ^{45}Ca will remain after 825 days?

Solution: $A = 50\left(\dfrac{1}{2}\right)^{825/165}$ In $A = A_0\left(\dfrac{1}{2}\right)^{t/h}$, substitute 50 for A_0, **825** for t, and **165** for h.

$A = 50\left(\dfrac{1}{2}\right)^{5}$ Simplify.

$A = 1.5625$ Evaluate using a calculator.

Answer: 1.5625 grams of ^{45}Ca will remain after 825 days.

Your Turn 5 The radioactive isotope ^{61}Cr has a half-life of 26 days. How much of a 10-gram sample would remain after 208 days? Give your answer to the nearest thousandth of a gram.

Answers to Your Turn 3
a. 2   **b.** 6   **c.** −4

Answer to Your Turn 4
$4118.36

Answer to Your Turn 5
0.039 g

Aging

Example 6 The number of people in the United States aged 65 and over (in millions) is given in the following table.

Connection Following is a graph of the function $y = 3.29(1.025)^x$ with the point corresponding to the solution of Example 6 indicated.

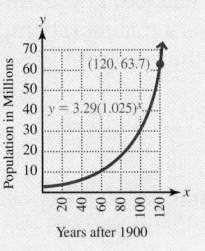

Year	Number 65 and over (in millions)
1900	3.1
1910	4.0
1920	4.9
1930	6.7
1940	9.0
1950	12.4
1960	16.7
1970	20.1
1980	25.5
1990	31.4
2000	35.0
2010	40.3

(*Source:* U.S. Bureau of the Census.)

The data can be approximated by the function $y = 3.29(1.025)^x$, where x is the number of years after 1900 and y is the number of people in millions. Use the model to estimate the number of people in the United States aged 65 and over in the year 2020. Round the answer to the nearest tenth of a million.

Solution:

$x = 2020 - 1900 = 120$ Subtract 1900 from 2020 to find the value of x that corresponds to the year 2020.

$y = 3.29(1.025)^{120}$ Substitute 120 for x in $y = 3.29(1.025)^x$.

$y = 63.7$ million Evaluate using a calculator and round to the nearest tenth.

Answer: According to the function, about 63.7 million people will be aged 65 and over in the year 2020.

Year	Number of Computers (in millions)
1980	3.1
1985	22.2
1990	51.3
1995	90.2
2000	184
2005	244
2010	306

Answer to Your Turn 6
1439.2 million

Your Turn 6 The table in the margin shows the number of computers (in millions) in use in the United States for selected years since 1980.

The data can be approximated by $y = 8.77(1.136)^x$, where x is the number of years after 1980 and y is the number of computers (in millions). Use the model to approximate the number of computers in use in the United States in 2020.

11.2 Exercises For Extra Help MyMathLab®

Objective 1

Prep Exercise 1 Is $f(x) = (-2)^x$ an exponential function? Why or why not? No, because the base is negative

Prep Exercise 2 Find the domain and range of $f(x) = 2^{x-4}$.
The domain is all real numbers; the range is $(0, \infty)$.

Prep Exercise 3 As x gets larger, which graph is steeper: $f(x) = 3^x$ or $g(x) = 1.5^x$? Why? $f(x) = 3^x$ is the steeper graph because it has the greater base.

Prep Exercise 4 Are the graphs of $f(x) = 2^x$ and $g(x) = \left(\dfrac{1}{2}\right)^x$ symmetric with respect to the y-axis? Explain.

Yes, the graphs are symmetric. The point $(1, 2)$ on $f(x)$ corresponds to the point $(-1, 2)$ on $g(x)$, and so on.

For Exercises 1–18 graph. See Examples 1 and 2.

1. $f(x) = 3^x$

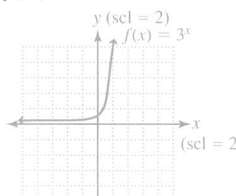

2. $f(x) = 4^x$

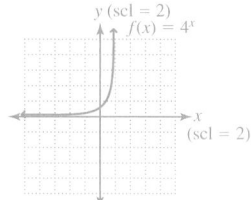

3. $f(x) = 4^x - 3$

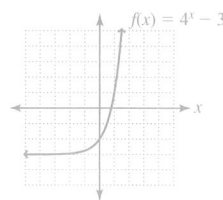

4. $f(x) = 3^x + 1$

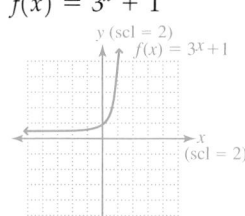

5. $f(x) = \left(\dfrac{1}{3}\right)^x$

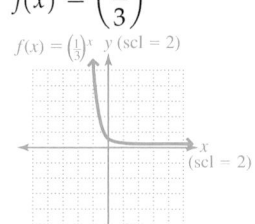

6. $f(x) = \left(\dfrac{1}{4}\right)^x$

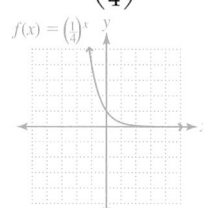

7. $f(x) = \left(\dfrac{2}{3}\right)^x + 2$

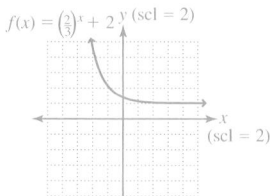

8. $f(x) = \left(\dfrac{3}{2}\right)^x - 1$

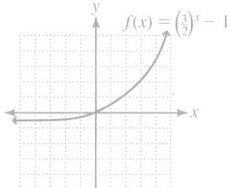

9. $f(x) = -3^x$

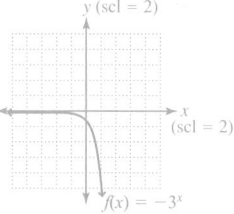

10. $f(x) = -2^x$

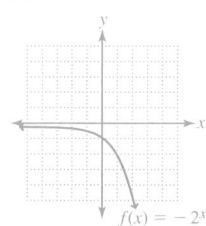

11. $f(x) = 2^{x-2}$

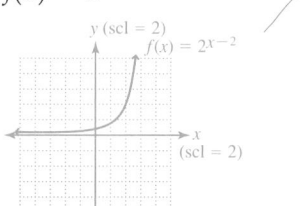

12. $f(x) = 3^{x+1}$

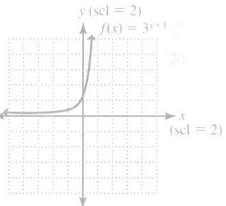

13. $f(x) = 2^{-x}$

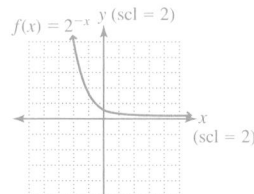

14. $f(x) = 3^{-x}$

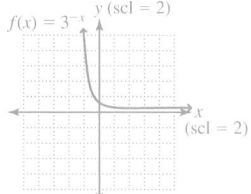

15. $f(x) = 2^{2x-3}$

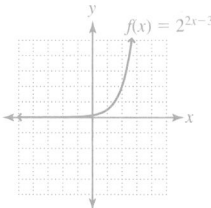

16. $f(x) = 3^{2x+1}$

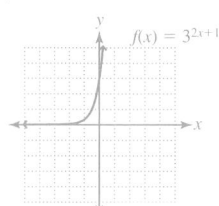

17. $f(x) = 3^{-x+2}$

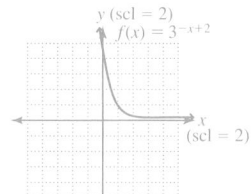

18. $f(x) = 2^{-x-1}$

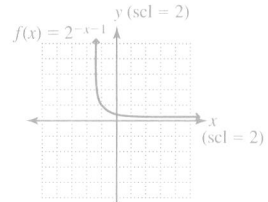

Objective 2

Prep Exercise 5 For $b > 0$ and $b \neq 1$, if $b^x = b^y$, why does $x = y$?
Because the exponential function is one-to-one

For Exercises 19–36, solve each equation. See Example 3.

19. $2^x = 8$
3

20. $3^x = 81$
4

21. $8^x = 32$
$\frac{5}{3}$

22. $27^x = 81$
$\frac{4}{3}$

23. $16^x = 4$
$\frac{1}{2}$

24. $36^x = 216$
$\frac{3}{2}$

25. $6^x = \frac{1}{36}$
-2

26. $3^x = \frac{1}{27}$
-3

27. $\left(\frac{1}{3}\right)^x = 9$
-2

28. $\left(\frac{1}{5}\right)^x = 125$
-3

29. $\left(\frac{2}{3}\right)^x = \frac{8}{27}$
3

30. $\left(\frac{3}{2}\right)^x = \frac{9}{4}$
2

31. $\left(\frac{1}{2}\right)^x = 16$
-4

32. $\left(\frac{1}{3}\right)^x = 27$
-3

33. $25^{x+1} = 125$
$\frac{1}{2}$

34. $9^{x+2} = 243$
$\frac{1}{2}$

35. $8^{2x-1} = 32^{x-3}$
-12

36. $4^{2x-1} = 32^{x+2}$
-12

▤ *For Exercises 37–40, use a graphing utility.*

37. a. Graph $f(x) = 2^x$ and $g(x) = 2^{x+2}$ in the window $[-5, 5]$ for x and $[-1, 10]$ for y.
　　b. How does the graph of g compare with the graph of f?

　a.
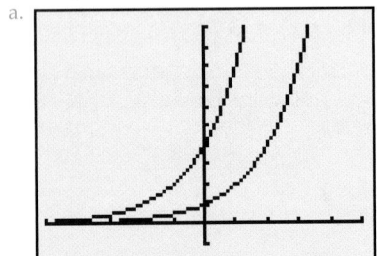

　b. The graph of g is the graph of f shifted 2 units to the left.

38. a. Graph $f(x) = 3^x$ and $g(x) = 3^x - 3$ in the window $[-3, 3]$ for x and $[-3, 9]$ for y.
　　b. How does the graph of g compare with the graph of f?

　a.
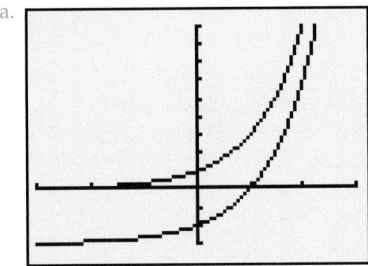

　b. The graph of g is the graph of f shifted 3 units down.

39. a. Graph $f(x) = 2^x$ and $g(x) = 2^{-x}$ in the window $[-5, 5]$ for x and $[-1, 10]$ for y.
　　b. How does the graph of g compare with the graph of f?

　a.
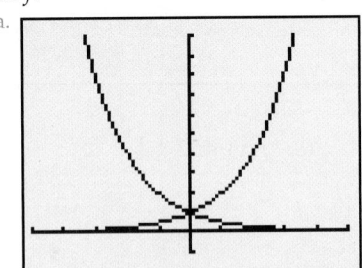

　b. The graph of g is the graph of f reflected about the y-axis.

40. a. Graph $f(x) = 2^x$ and $g(x) = -2^x$ in the window $[-5, 5]$ for x and $[-5, 5]$ for y.
　　b. How does the graph of g compare with the graph of f?

　a.
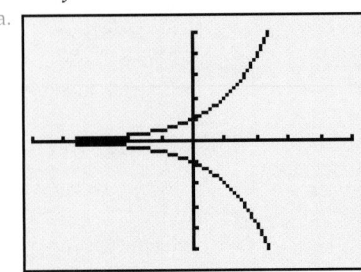

　b. The graph of g is the graph of f reflected about the x-axis.

Objective 3

Prep Exercise 6 If we have 10 grams of a substance that has a half-life of 50 days, how many grams will be present after 200 days? 0.625 g

For Exercises 41 and 42, use the following information.

Under ideal conditions, a culture of *E. coli* bacteria doubles in size every 20 minutes. If A_0 is the initial amount and t is the number of minutes passed, the amount present is A, where $A = A_0(2)^{t/20}$.

41. If a culture of *E. coli* begins with 100 cells, how many cells will be present after 120 minutes?

6400 cells

42. If a culture of *E. coli* currently has 500 cells, how many cells were present 90 minutes earlier? (*Hint:* Let $A_0 = 500$, and the time will be negative.)

Approximately 22 cells

43. Under ideal conditions, human beings could double their population every 50 years. If A_0 is the initial population and t is the number of years passed, the current population is A, where $A = A_0(2)^{t/50}$. If 6 billion humans were on Earth in 2000, how many humans will there be in 2500 if their growth is uncontrolled?

6144 billion people

44. Using the information from Exercise 43, how many human beings were on the Earth in 1900?

1.5 billion people

For Exercises 45 and 46, use the formula $A = P\left(1 + \dfrac{r}{n}\right)^{nt}$. See Example 4.

45. If $10,000 is deposited into an account paying 8% interest compounded quarterly, how much will be in the account after 12 years?

$25,870.70

46. If $15,000 is deposited into an account paying 6% interest compounded semiannually, how much will be in the account after nine years?

$25,536.50

For Exercises 47 and 48, use the formula $A = A_0\left(\dfrac{1}{2}\right)^{t/h}$. See Example 5.

47. Einsteinium (^{254}ES) has a half-life of 270 days. How much of a 5-gram sample would remain after 2160 days? Give the answer to the nearest thousandth of a gram.

0.020 g

48. Nobelium (^{257}No) has a half-life of 23 seconds. How much of a 100-gram sample would remain after 275 seconds? Give the answer to the nearest thousandth of a gram.

0.025 g

For Exercises 49–54, solve. See Example 6.

49. Since 1960, the U.S. gross domestic product can be approximated by $y = 2632.31(1.033)^x$, where y is the gross domestic product and x is the number of years beginning with 1960. ($x = 1$ represents 1960.) Estimate the U.S. gross domestic product in 2018.

≈17,875.41

50. Since 1970, the U.S. consumer price index can be approximated by $y = 46.56(1.045)^x$, where y is the consumer price index and x is the number of years beginning with 1970. ($x = 1$ represents 1970.) Estimate the consumer price index in 2019.

≈420.56

51. Chlorine is frequently used to disinfect swimming pools. The concentration should remain between 1.5 and 2.5 parts per million. On a warm, sunny day, 30% of the chlorine can dissipate into the air or combine with other chemicals. If the initial amount of chlorine is 2.5 parts per million, the function $f(x) = 2.5(0.7)^x$ models the amount of chlorine after x days. How much chlorine is in the pool after 2 days?

1.225 parts per million

52. It is estimated that the value of a car depreciates 20% per year for the first five years. If the original price of a car is P, the value, A, of a car after t years is given by $A = P(0.8)^t$. If a car originally cost \$25,960, find the value of the car after 3 years to the nearest dollar.

\$13,292

53. Between 1993 and 2012, the number of transistors that can be placed on a single chip has grown significantly, as indicated in the table. The data can be approximated by the function $T(x) = 1.97(1.503)^x$, where x is the number of years after 1993 and $T(x)$ is the number of transistors in millions. If the current trend continues, estimate the number of transistors that could be put on a single chip in 2020. (*Source:* http://en.wikipedia.org.)

$\approx 118{,}130$ million transistors

Year	Chip	Transistors (millions)
1993	Pentium	3.3
1995	P6	5.5
1997	Pentium II	7.5
1999	Pentium III	9.5
2000	Pentium IV	42
2004	Pentium IV Prescott	125
2006	Core 2 Duo	586
2008	Quad IV	820
2010	8-core Xeon Nehalem-EX	2300
2012	62-core Xeon Phi	5000

54. Dave makes a cup of coffee with cream and places it on the counter to cool. The temperature of the coffee at various times is given in the table shown.

The data can be approximated by the function $T(t) = 146.9(0.989)^t$, where t is the time in seconds and T is the temperature in °F. Estimate the temperature of the coffee after 1 minute.

≈ 75.6°F

t (seconds)	$T(°F)$
0.2	155.8
8.4	133.2
16.6	117.9
24.8	107.9
33	100.7
41.1	94.9
49.3	90.5

Puzzle Problem What is the greatest number that can be written using three numerals?

9^{9^9}

Review Exercises

Exercises 1–4   **Expressions**

[5.1] *For Exercises 1–3, evaluate.*

1. 5^3

125

2. $\left(\dfrac{1}{3}\right)^{-2}$

9

3. 4^{-3}

$\dfrac{1}{64}$

[9.2] **4.** Write $5^{2/3}$ in radical notation.

$\sqrt[3]{5^2}$ or $(\sqrt[3]{5})^2$

Exercises 5 and 6   **Equations and Inequalities**

[11.1] 5. If $f(x) = 3x + 4$, find $f^{-1}(x)$.

$f^{-1}(x) = \dfrac{x - 4}{3}$

[11.1] 6. Graph the inverse of the function whose graph is shown.

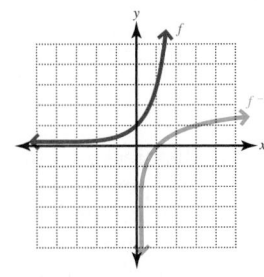

11.3 Logarithmic Functions

Objectives

1 Convert between exponential and logarithmic forms.
2 Solve logarithmic equations by changing to exponential form.
3 Graph logarithmic functions.
4 Solve applications involving logarithms.

Warm-up

[11.2] 1. Solve: $3^{5x+7} = 9^{x-4}$
[5.1] *For Exercises 2 and 3, evaluate.*

2. $\left(\dfrac{1}{2}\right)^{-4}$ 3. $\left(\dfrac{1}{2}\right)^{0}$

[5.1] 3. Write $\dfrac{1}{27}$ as 3 to a power.

Objective 1 Convert between exponential and logarithmic forms.

In Section 11.2, we defined the exponential function as $f(x) = b^x$ with $b > 0$ and $b \neq 1$. Because the exponential function is a one-to-one function, it has an inverse. Following is the graph of $f(x) = 2^x$ and its inverse, which we find by reflecting the graph about the line $y = x$.

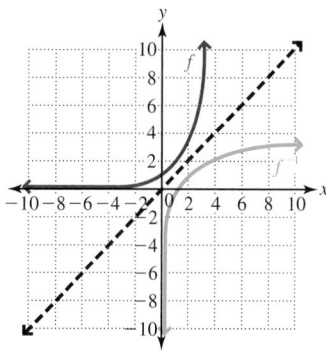

To find the inverse of $f(x) = 2^x$, we use the procedure from Section 11.1.

$$f(x) = 2^x$$
$$y = 2^x \quad \text{Replace } f(x) \text{ with } y.$$
$$x = 2^y \quad \text{Replace } x \text{ with } x \text{ and } x \text{ with } y.$$

The next step is to solve for y, but we haven't learned how to isolate a variable that is an exponent. To rewrite $x = 2^y$, we define a **logarithm**.

Definition **Logarithm:** If $b > 0$ and $b \neq 1$, then $y = \log_b x$ is equivalent to $x = b^y$.

If we apply this definition to $x = 2^y$ in the preceding equation, we get $y = \log_2 x$. Replacing y with $f^{-1}(x)$, we get $f^{-1}(x) = \log_2 x$; so the inverse function for $f(x) = 2^x$ is $f^{-1}(x) = \log_2 x$. To generalize, exponential functions and *logarithmic functions* are inverses. Consequently, if $y = \log_b x$, the domain is $(0, \infty)$ and the range is $(-\infty, \infty)$ with $b > 0$ and $b \neq 1$.

The expression $\log_b x$ is read "the logarithm base b of x" and *is the exponent to which b must be raised to get x.* Compare the two forms.

The exponent is the logarithm.

$$x = b^y \qquad y = \log_b x$$

Note $y = \log_b x$ means that y is the power to which we raise b to get x.

The base of the exponent is the base of the logarithm.

The definition of a logarithm allows us to convert from one form to another. The following table contains pairs of equivalent forms.

Logarithmic Form	Exponential Form
$\log_2 8 = 3$	$2^3 = 8$
$\log_{10} \dfrac{1}{10} = -1$	$10^{-1} = \dfrac{1}{10}$
$\log_5 1 = 0$	$5^0 = 1$
$\log_{16} 4 = \dfrac{1}{2}$	$16^{1/2} = 4$
$\log_{1/2} 32 = -5$	$\left(\dfrac{1}{2}\right)^{-5} = 32$

Note The logarithmic equations are in the form $\log_b x = y$. The values of x are positive numbers only, and the values of y are both positive and negative. ▶

Example 1 Write in logarithmic form.

a. $3^4 = 81$

Solution: $\log_3 81 = 4$ The base of the exponent is the base of the logarithm, and the exponent is the logarithm.

b. $\left(\dfrac{1}{2}\right)^{-3} = 8$

Solution: $\log_{1/2} 8 = -3$ The base of the exponent is the base of the logarithm, and the exponent is the logarithm.

c. $9^{1/2} = 3$

Solution: $\log_9 3 = \dfrac{1}{2}$ The base of the exponent is the base of the logarithm, and the exponent is the logarithm.

Your Turn 1 Write in logarithmic form.

a. $4^3 = 64$ **b.** $2^{-3} = \dfrac{1}{8}$ **c.** $27^{1/3} = 3$

VISUAL

Learning Strategy

If you are a visual learner, try visualizing the following "loop" for rewriting logarithms in exponential form.

equals

$\log_6 36 = 2$

6 raised to 2

Answers to Your Turn 1
a. $\log_4 64 = 3$
b. $\log_2 \dfrac{1}{8} = -3$
c. $\log_{27} 3 = \dfrac{1}{3}$

Answers to Your Turn 2
a. $4^3 = 64$ **b.** $8^{1/3} = 2$
c. $5^{-3} = \dfrac{1}{125}$

Example 2 Write in exponential form.

a. $\log_6 36 = 2$

Solution: $6^2 = 36$ The base of the logarithm is the base of the exponent, and the logarithm is the exponent.

b. $\log_{16} 2 = \dfrac{1}{4}$

Solution: $16^{1/4} = 2$ The base of the logarithm is the base of the exponent, and the logarithm is the exponent.

c. $\log_{1/2} 16 = -4$

Solution: $\left(\dfrac{1}{2}\right)^{-4} = 16$ The base of the logarithm is the base of the exponent, and the logarithm is the exponent.

Your Turn 2 Write in exponential form.

a. $\log_4 64 = 3$ **b.** $\log_8 2 = \dfrac{1}{3}$ **c.** $\log_5 \dfrac{1}{125} = -3$

Objective 2 Solve logarithmic equations by changing to exponential form.

A logarithmic equation in the form $\log_b x = y$ may have b, x, or y as an unknown.

> **Procedure** **Solving Logarithmic Equations**
>
> To solve an equation of the form $\log_b x = y$, where b, x, or y is a variable, write the equation in exponential form, $b^y = x$, and then solve for the variable.

Example 3 Solve.

a. $\log_b 16 = 2$

Solution: $b^2 = 16$ Write in exponential form.

$\quad\quad\quad\quad b = \pm 4$ Find the positive and negative square roots of 16.

$\quad\quad\quad\quad b = 4$ The base must be positive, so $b = 4$.

b. $\log_3 \dfrac{1}{27} = y$

Solution: $3^y = \dfrac{1}{27}$ Write in exponential form.

$\quad\quad\quad 3^y = \dfrac{1}{3^3}$ Write 27 as 3^3.

$\quad\quad\quad 3^y = 3^{-3}$ Write $\dfrac{1}{3^3}$ as 3^{-3}.

$\quad\quad\quad\; y = -3$ Set the exponents equal to each other.

c. $\log_{25} x = \dfrac{1}{2}$

Solution: $25^{1/2} = x$ Write in exponential form.

$\quad\quad\quad\quad 5 = x$ Simplify. $25^{1/2} = \sqrt{25} = 5$

d. $\log_{36} \sqrt[4]{6} = y$

Solution: $\quad 36^y = \sqrt[4]{6}$

$\quad\quad (6^2)^y = 6^{1/4}$ Write 36 as 6^2 and $\sqrt[4]{6}$ as $6^{1/4}$.

$\quad\quad\; 6^{2y} = 6^{1/4}$ $(6^2)^y = 6^{2y}$

$\quad\quad\quad 2y = \dfrac{1}{4}$ Set the exponents equal to each other.

$\quad\quad\quad\; y = \dfrac{1}{8}$ Solve for y.

Your Turn 3 Solve.

a. $\log_b \dfrac{1}{9} = -2$ **b.** $\log_5 \dfrac{1}{25} = y$ **c.** $\log_{27} x = \dfrac{1}{3}$

Answers to Your Turn 3
a. 3   **b.** −2   **c.** 3

If an equation containing a logarithm is not in the form $\log_b a = c$, try using the addition or multiplication principles of equality to rewrite the equation in that form.

Example 4   Solve $5 - 3\log_2 x = -7$.

Solution: Use the addition principle of equality and multiplication principle of equality to write the equation in the form $\log_b a = c$. We can then change the equation to exponential form to solve for x.

$$-3\log_2 x = -12 \quad \text{Subtract 5 from both sides.}$$
$$\log_2 x = 4 \quad \text{Divide both sides by } -3 \text{ to isolate the logarithm.}$$
$$x = 2^4 \quad \text{Write in exponential form.}$$
$$x = 16 \quad \text{Simplify.}$$

Your Turn 4   Solve.

a. $9 + \log_3 x = 13$ **b.** $5\log_n 36 = 10$ **c.** $12 - 7\log_4 t = -9$

The definition of logarithm leads to the following two properties.

> **Rule** **For any real number b, where $b > 0$ and $b \neq 1$:**
> 1. $\log_b b = 1$.   2. $\log_b 1 = 0$.

Based on the definition of a logarithm, $\log_b b = 1$ because $b^1 = b$ and $\log_b 1 = 0$ because $b^0 = 1$.

Example 5   Find the value.

a. $\log_5 5$ **b.** $\log_e e$ **c.** $\log_{10} 1$

Solution: $\log_5 5 = 1$ **Solution:** $\log_e e = 1$ **Solution:** $\log_{10} 1 = 0$

Your Turn 5   Find the value.

a. $\log_{10} 10$ **b.** $\log_{\sqrt{3}} 1$

Objective 3 Graph logarithmic functions.

To graph logarithmic functions, which have the form $f(x) = \log_b x$, where $b > 0$ and $b \neq 1$, we first change to exponential form so that it will be easier to find ordered pairs.

> **Procedure** **Graphing Logarithmic Functions**
> To graph a function in the form $f(x) = \log_b x$:
> 1. Replace $f(x)$ with y and write the logarithm in exponential form $x = b^y$.
> 2. Find ordered pairs that satisfy the equation by assigning values to y and finding x.
> 3. Plot the ordered pairs and draw a smooth curve through the points.

Example 6   Graph.

a. $f(x) = \log_2 x$

Solution: $y = \log_2 x$ Replace $f(x)$ with y.
$$2^y = x \quad \text{Write in exponential form.}$$

Choose values for y and find x.

y	0	1	2	3	-1	-2
x	$2^0 = 1$	$2^1 = 2$	$2^2 = 4$	$2^3 = 8$	$2^{-1} = \dfrac{1}{2}$	$2^{-2} = \dfrac{1}{4}$

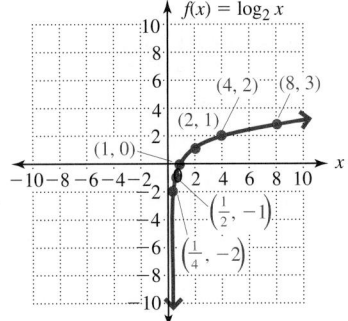

Answers to Your Turn 4
a. 81   **b.** 6   **c.** 64

Answers to Your Turn 5
a. 1   **b.** 0

b. $f(x) = \log_{1/2}x$

Solution: $y = \log_{1/2}x$ Replace $f(x)$ with y.

$$x = \left(\frac{1}{2}\right)^y$$ Write in exponential form.

Choose values for y and find x.

y	0	1	2	3	-1	-2	-3
x	$\left(\frac{1}{2}\right)^0 = 1$	$\left(\frac{1}{2}\right)^1 = \frac{1}{2}$	$\left(\frac{1}{2}\right)^2 = \frac{1}{4}$	$\left(\frac{1}{2}\right)^3 = \frac{1}{8}$	$\left(\frac{1}{2}\right)^{-1} = 2$	$\left(\frac{1}{2}\right)^{-2} = 4$	$\left(\frac{1}{2}\right)^{-3} = 8$

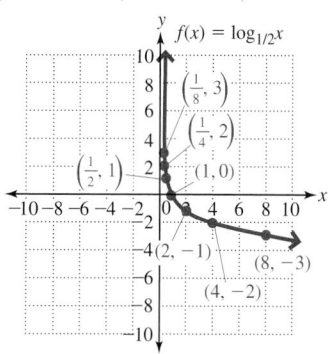

Following is a summary of the key features of the graphs of logarithmic functions.

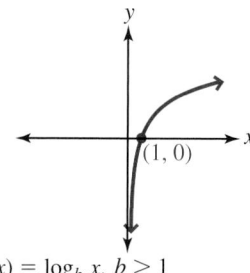

$f(x) = \log_b x,\ b > 1$

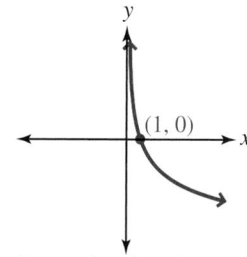

$f(x) = \log_b x,\ 0 < b < 1$

The graph passes through $(1,0)$, approaches the y-axis, and increases. The domain is $(0, \infty)$. The range is $(-\infty, \infty)$.

The graph passes through $(1,0)$, approaches the y-axis, and decreases. The domain is $(0, \infty)$. The range is $(-\infty, \infty)$.

Your Turn 6   Graph $f(x) = \log_3 x$.

Objective 4 Solve applications involving logarithms.

Example 7   The function $P = 95 - 30 \log_2 x$ models the percent, P, of students that recall the important features of a classroom lecture over time, where x is the number of days that have elapsed since the lecture was given. What percent of the students recall the important features of a lecture 8 days after it was given? (*Source: Psychology in the New Millennium*, 8th Edition, Spencer A. Rathus, Thomson Publishing Company.)

Solution: $P = 95 - 30 \log_2 8$ Substitute 8 for x in $P = 95 - 30 \log_2 x$.

$\quad\quad\quad\quad P = 95 - 30(3)$ $\log_2 8 = 3$ because $2^3 = 8$.

$\quad\quad\quad\quad P = 95 - 90$ Multiply.

$\quad\quad\quad\quad P = 5$ Simplify.

Answer: Five percent of the students remember the important features of a lecture 8 days after it is given.

Answer to Your Turn 6

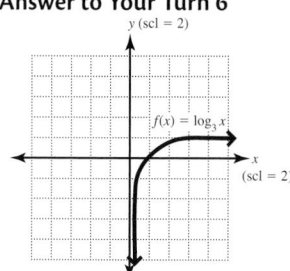

Your Turn 7 Refer to Example 7.

a. Find the percent of students who remember the important features of a lecture 2 days after it was given.

Answers to Your Turn 7
a. 65% **b.** no

b. When the number of days was decreased by a fourth (from 8 to 2), was the amount retained also decreased by one-fourth?

11.3 Exercises For Extra Help MyMathLab®

Objective 1

Prep Exercise 1 If $f(x) = 3^x$, what is $f^{-1}(x)$? Why?
$f^{-1}(x) = \log_3 x$ because logarithmic and exponential functions are inverses.

For Exercises 1–16, write in logarithmic form. See Example 1.

1. $2^5 = 32$
$\log_2 32 = 5$

2. $5^4 = 625$
$\log_5 625 = 4$

3. $10^3 = 1000$
$\log_{10} 1000 = 3$

4. $10^4 = 10{,}000$
$\log_{10} 10{,}000 = 4$

5. $e^4 = x$
$\log_e x = 4$

6. $e^{-2} = z$
$\log_e z = -2$

7. $5^{-3} = \dfrac{1}{125}$
$\log_5 \dfrac{1}{125} = -3$

8. $7^{-4} = \dfrac{1}{2401}$
$\log_7 \dfrac{1}{2401} = -4$

9. $10^{-2} = \dfrac{1}{100}$
$\log_{10} \dfrac{1}{100} = -2$

10. $10^{-4} = \dfrac{1}{10{,}000}$
$\log_{10} \dfrac{1}{10{,}000} = -4$

11. $625^{1/4} = 5$
$\log_{625} 5 = \dfrac{1}{4}$

12. $343^{1/3} = 7$
$\log_{343} 7 = \dfrac{1}{3}$

13. $\left(\dfrac{1}{4}\right)^2 = \dfrac{1}{16}$
$\log_{1/4} \dfrac{1}{16} = 2$

14. $\left(\dfrac{3}{4}\right)^3 = \dfrac{27}{64}$
$\log_{3/4} \dfrac{27}{64} = 3$

15. $7^{1/2} = \sqrt{7}$
$\log_7 \sqrt{7} = \dfrac{1}{2}$

16. $10^{1/2} = \sqrt{10}$
$\log_{10} \sqrt{10} = \dfrac{1}{2}$

Prep Exercise 2 Why are logarithms exponents? Logarithms are exponents because logarithms are inverses of exponential functions.

Prep Exercise 3 If $f(x) = \log_b x$, what are the restrictions on b? $b > 0, b \neq 1$

Prep Exercise 4 If $f(x) = \log_b x$, what are the restrictions on x? $x > 0$

For Exercises 17–32, write in exponential form. See Example 2.

17. $\log_3 81 = 4$
$3^4 = 81$

18. $\log_4 64 = 3$
$4^3 = 64$

19. $\log_4 \dfrac{1}{16} = -2$
$4^{-2} = \dfrac{1}{16}$

20. $\log_3 \dfrac{1}{243} = -5$
$3^{-5} = \dfrac{1}{243}$

21. $\log_{10} 100 = 2$
$10^2 = 100$

22. $\log_{10} 1000 = 3$
$10^3 = 1000$

23. $\log_e a = 5$
$e^5 = a$

24. $\log_e y = -4$
$e^{-4} = y$

25. $\log_e \dfrac{1}{e^4} = -4$

$e^{-4} = \dfrac{1}{e^4}$

26. $\log_e \dfrac{1}{e^2} = -2$

$e^{-2} = \dfrac{1}{e^2}$

27. $\log_{1/8} \dfrac{1}{64} = 2$

$\left(\dfrac{1}{8}\right)^2 = \dfrac{1}{64}$

28. $\log_{1/4} \dfrac{1}{256} = 4$

$\left(\dfrac{1}{4}\right)^4 = \dfrac{1}{256}$

29. $\log_{1/5} 25 = -2$

$\left(\dfrac{1}{5}\right)^{-2} = 25$

30. $\log_{1/3} 81 = -4$

$\left(\dfrac{1}{3}\right)^{-4} = 81$

31. $\log_7 \sqrt{7} = \dfrac{1}{2}$

$7^{1/2} = \sqrt{7}$

32. $\log_6 \sqrt{6} = \dfrac{1}{2}$

$6^{1/2} = \sqrt{6}$

Objective 2

Prep Exercise 5 In $y = \log_b x$, y is the ___power___ to which ___b___ must be raised to get ___x___.

For Exercises 33–52, solve. See Example 3.

33. $\log_2 x = 5$
32

34. $\log_3 x = 4$
81

35. $\log_5 x = -2$
$\dfrac{1}{25}$

36. $\log_4 x = -2$
$\dfrac{1}{16}$

37. $\log_3 81 = y$
4

38. $\log_2 32 = y$
5

39. $\log_5 \dfrac{1}{25} = y$
-2

40. $\log_4 \dfrac{1}{16} = y$
-2

41. $\log_b 1000 = 3$
10

42. $\log_b 10{,}000 = 4$
10

43. $\log_m \dfrac{1}{16} = -4$
2

44. $\log_n \dfrac{1}{36} = -2$
6

45. $\log_{1/2} x = 2$
$\dfrac{1}{4}$

46. $\log_{1/5} x = 3$
$\dfrac{1}{125}$

47. $\log_{1/3} h = -5$
243

48. $\log_{1/6} k = -2$
36

49. $\log_{1/3} \dfrac{1}{9} = y$
2

50. $\log_{1/4} \dfrac{1}{64} = y$
3

51. $\log_{1/2} 64 = t$
-6

52. $\log_{1/5} 125 = u$
-3

For Exercises 53–64, solve. See Example 4.

53. $\log_2 x + 4 = 8$
16

54. $\log_3 x + 2 = 5$
27

55. $\log_{1/4} h - 2 = 1$
$\dfrac{1}{64}$

56. $\log_{1/5} k - 4 = -1$
$\dfrac{1}{125}$

57. $3 \log_b 16 = 12$
2

58. $2 \log_b 81 = 8$
3

59. $\dfrac{1}{3} \log_5 c = 1$
125

60. $\dfrac{1}{4} \log_2 d = 2$
256

61. $3 \log_t 9 + 6 = 12$
3

62. $2 \log_u 125 + 8 = 14$
5

63. $\dfrac{1}{2} \log_4 m - 2 = -3$
$\dfrac{1}{16}$

64. $\dfrac{1}{3} \log_5 n - 4 = -5$
$\dfrac{1}{125}$

Objective 3

Prep Exercise 6 What is the relationship between the graphs of $f(x) = 5^x$ and $g(x) = \log_5 x$? Why? *The graphs are symmetric about the line $y = x$ because they are inverses.*

Exercises 65–68, graph. See Example 6.

65. $f(x) = \log_4 x$

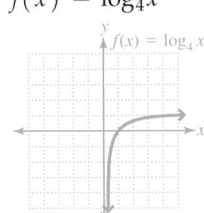

66. $f(x) = \log_5 x$

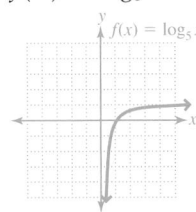

67. $f(x) = \log_{1/3} x$

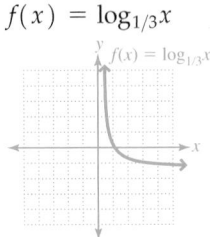

68. $f(x) = \log_{1/4} x$

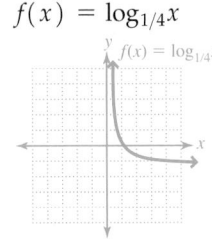

Objective 4

For Exercises 69–72, solve. See Example 7.

69. The percent of adult height attained by a 5- to 15-year-old girl can be approximated by $f(x) = 62 + 35 \log_{10}(x - 4)$, where x is the age in years and $f(x)$ is the percent. At age 14, what percent of her adult height has a girl reached?
97%

70. The percent of adult height attained by a 5- to 15-year-old boy can be approximated by $f(x) = 29 + 48.8 \log_{10}(x + 1)$, where x is the age in years and $f(x)$ is the percent. At age 9, what percent of his adult height has a boy reached?
77.8%

71. In Example 7, we learned that the percent of students who recall the important features of a lecture is given by $P = 95 - 30 \log_2 x$, where P is the percent and x is the number of days that have elapsed since the lecture was given. After how many days will 35% of the students recall the important features?
After 4 days

72. Using the formula from Exercise 71, find the percent of students who recall the important features of a lecture 1 day after it is given.
95%

73. Why does $\log_b b = 1$ for any value of $b > 0$ and $b \neq 1$?
$\log_b b = 1$ because $b^1 = b$.

74. In the definition of a logarithm, $y = \log_b x$, why must $b \neq 1$?
If $b = 1$, then $y = \log_b x = \log_1 x$. Written exponentially, $1^y = x$, which is $x = 1$ for all y.

75. Why does $\log_b 1 = 0$?
$\log_b 1 = 0$ because $b^0 = 1$.

76. If $f(x) = b^x$, the domain is $(-\infty, \infty)$ and the range is $(0, \infty)$. What are the domain and range of $f(x) = \log_b x$? Why?
Domain is $(0, \infty)$ and range is $(-\infty, \infty)$ because $f(x) = b^x$ and $f(x) = \log_b x$ are inverse functions.

Review Exercises

Exercises 1–6 **Expressions**

[5.1] 1. Write $\dfrac{1}{x^6}$ as x to a negative power.
x^{-6}

For Exercises 2–5, simplify using the rules of exponents.

[5.4] 2. $x^4 \cdot x^2$
x^6

[5.4] 3. $(x^3)^5$
x^{15}

[5.6] 4. $\dfrac{x^6}{x^3}$
x^3

[5.6] 5. $\dfrac{(x^3)^2 \cdot x^4}{x^5}$
x^5

[9.2] 6. Write $\sqrt[4]{x^3}$ in exponential form.
$x^{3/4}$

11.4 Properties of Logarithms

Objectives

1 Apply the inverse properties of logarithms.
2 Apply the product, quotient, and power properties of logarithms.

Objective 1 Apply the inverse properties of logarithms.

Earlier we developed two properties of logarithms, $\log_b b = 1$ and $\log_b 1 = 0$, which were based on the definition of a logarithm. The fact that $f[f^{-1}(x)] = x$ and $f^{-1}[f(x)] = x$ gives us the following properties of logarithms.

> **Rule Inverse Properties of Logarithms**
> For any real numbers b and x, where $b > 0$, $b \neq 1$ and $x > 0$:
> 1. $b^{\log_b x} = x.$ 2. $\log_b b^x = x.$

To prove that $b^{\log_b x} = x$, let $f(x) = b^x$ so that $f^{-1}(x) = \log_b x$.

$$f[f^{-1}(x)] = x \quad \text{Composition of a function with its inverse is } x.$$
$$f[\log_b x] = x \quad \text{Replace } f^{-1}(x) \text{ with } \log_b x.$$
$$b^{\log_b x} = x \quad \text{Replace } x \text{ with } \log_b x \text{ in } f(x).$$

To prove that $\log_b b^x = x$, use the definition of a logarithm. $\text{Log}_a b = c$ means that $a^c = b$; so $\log_b b^x = x$ because $b^x = b^x$.

Example 1 Find the value.

a. $3^{\log_3 8}$

Solution: $3^{\log_3 8} = 8$ Use $b^{\log_b x} = x$.

b. $6^{\log_6 x}$

Solution: $6^{\log_6 x} = x$ Use $b^{\log_b x} = x$.

c. $\log_3 3^6$

Solution: $\log_3 3^6 = 6$ Use $\log_b b^x = x$.

d. $\log_5 5^a$

Solution: $\log_5 5^a = a$ Use $\log_b b^x = x$.

Your Turn 1 Find the value.

a. $8^{\log_8 4}$

b. $\log_3 3^{-2}$

Answers to Your Turn 1
a. 4   **b.** -2

Answers to Warm-up
1. 125
2. $\left(\dfrac{a^3}{b}\right)^{1/2}$
3. $x^2 - x - 6$
4. $\dfrac{1}{25}$

Objective 2 Apply the product, quotient, and power properties of logarithms.

Logarithms were invented to perform operations on very large and very small numbers. With the invention of handheld calculators, they are no longer used for this purpose. However, we still use the properties of logarithms, which are based on the fact that logarithms are exponents.

Warning There is no rule for the logarithms of sums or differences. The $\log_b(x + y) \neq \log_b x + \log_b y$.

Note When the product and quotient rules are used, all of the bases of all of the logarithms **must** be the same.

Rule Further Properties of Logarithms

For real numbers x, y, and b, where $x > 0, y > 0, b > 0$, and $b \neq 1$:

Product Rule of Logarithms: $\log_b xy = \log_b x + \log_b y$

(The logarithm of the product of two numbers is equal to the sum of the logarithms of the numbers.)

Quotient Rule of Logarithms: $\log_b \dfrac{x}{y} = \log_b x - \log_b y$

(The logarithm of the quotient of two numbers is equal to the difference of the logarithms of the numbers.)

Power Rule of Logarithms: $\log_b x^r = r \log_b x$

(The logarithm of a number raised to a power is equal to the exponent times the logarithm of the number.)

To prove that $\log_b xy = \log_b x + \log_b y$, let $M = \log_b x$ and $N = \log_b y$.

$x = b^M$ and $y = b^N$ Write each logarithmic equation in exponential form.

$xy = b^M \cdot b^N$ Multiply the left and right sides of the exponential forms.

$xy = b^{M+N}$ Add the exponents.

$\log_b xy = M + N$ Write in logarithmic form.

$\log_b xy = \log_b x + \log_b y$ Substitute $\log_b x$ for M and $\log_b y$ for N, and the proof is complete.

The proofs of $\log_b \dfrac{x}{y} = \log_b x - \log_b y$ and $\log_b x^r = r \log_b x$ are similar.

Example 2   Use the product rule of logarithms to write the expression as a sum of logarithms.

a. $\log_{10} xyz$

Solution: $\log_{10} xyz = \log_{10} x + \log_{10} y + \log_{10} z$

b. $\log_b x(x + 3)$ **Note** $\log_b(x + 3) \neq \log_b x + \log_b 3.$

Solution: $\log_b x(x + 3) = \log_b x + \log_b(x + 3)$

Your Turn 2   Use the product rule of logarithms to write the expression as a sum of logarithms.

a. $\log_a 6x$ **b.** $\log_2 x(3x + 2)$

Example 3   Use the product rule of logarithms in the form $\log_b x + \log_b y = \log_b xy$ to write the expression as a single logarithm.

a. $\log_3 8 + \log_3 2$

Solution: $\log_3 8 + \log_3 2 = \log_3 8 \cdot 2 = \log_3 16$

b. $\log_8 5 + \log_8 x + \log_8(2x - 3)$

Solution: $\log_8 5 + \log_8 x + \log_8(2x - 3) = \log_8 5x(2x - 3)$
$= \log_8(10x^2 - 15x)$

Answers to Your Turn 2
a. $\log_a 6 + \log_a x$
b. $\log_2 x + \log_2(3x + 2)$

Answers to Your Turn 3
a. $\log_6 63$
b. $\log_7(4x^2 + 8x)$

Your Turn 3   Use the product rule of logarithms to write each of the following as a single logarithm.

a. $\log_6 7 + \log_6 9$ **b.** $\log_7 2 + \log_7 x + \log_7(2x + 4)$

Example 4 Use the quotient rule of logarithms to write the expression as a difference of logarithms. Leave the answers in simplest form.

a. $\log_5 \dfrac{5}{11}$

Solution: $\log_5 \dfrac{5}{11} = \log_5 5 - \log_5 11$

$\qquad\qquad = 1 - \log_5 11$ Remember, $\log_5 5 = 1$.

b. $\log_4 \dfrac{x}{x - 5}$

Solution: $\log_4 \dfrac{x}{x - 5} = \log_4 x - \log_4 (x - 5)$ ◄ Warning $\log_4 (x - 5) \neq \log_4 x - \log_4 5$.

Your Turn 4 Use the quotient rule of logarithms to write the expression as a difference of logarithms. Leave your answers in simplest form.

a. $\log_9 \dfrac{9}{10}$

b. $\log_5 \dfrac{x}{x + 2}$

Example 5 Use the quotient rule of logarithms in the form $\log_b x - \log_b y = \log_b \dfrac{x}{y}$ to write the expression as a single logarithm.

a. $\log_9 3 - \log_9 x$

Solution: $\log_9 3 - \log_9 x = \log_9 \dfrac{3}{x}$

b. $\log_{10} x - \log_{10} (x^2 + 4)$

Solution: $\log_{10} x - \log_{10} (x^2 + 4) = \log_{10} \dfrac{x}{x^2 + 4}$

Your Turn 5 Use the quotient rule of logarithms to write each of the following as a single logarithm.

a. $\log_3 15 - \log_3 5$

b. $\log_4 (x^2 + 2) - \log_4 (x + 1)$

Example 6 Use the power rule of logarithms to write the expression as a multiple of a logarithm.

a. $\log_4 a^6$

Solution: $\log_4 a^6 = 6 \log_4 a$

b. $\log_b \sqrt[4]{x^3}$

Solution: $\log_b \sqrt[4]{x^3} = \log_b x^{3/4}$ Write $\sqrt[4]{x^3}$ as $x^{3/4}$.

$\qquad\qquad = \dfrac{3}{4} \log_b x$

c. $\log_b \dfrac{1}{x^3}$

Solution: $\log_b \dfrac{1}{x^3} = \log_b x^{-3}$ Write $\dfrac{1}{x^3}$ as x^{-3}.

$\qquad\qquad = -3 \log_b x$ ◄ Note $\log_b \dfrac{1}{x^3}$ could be rewritten as

$\qquad\qquad\qquad\qquad \log_b 1 - \log_b x^3 = 0 - 3 \log_b x = -3 \log_b x.$

Answers to Your Turn 4
a. $1 - \log_9 10$
b. $\log_5 x - \log_5 (x + 2)$

Answers to Your Turn 5
a. $\log_3 3 = 1$
b. $\log_4 \dfrac{x^2 + 2}{x + 1}$

Your Turn 6 Use the power rule of logarithms to write the expression as a multiple of a logarithm.

a. $\log_a z^4$

b. $\log_a \sqrt[3]{x^2}$

Example 7 Use the power rule of logarithms in the form $r \log_b x = \log_b x^r$ to write the expression as a logarithm of a quantity to a power. Leave the answers in simplest form without negative or fractional exponents.

a. $5 \log_4 x$

Solution: $5 \log_4 x = \log_4 x^5$

b. $-2 \log_b 5$

Solution: $-2 \log_b 5 = \log_b 5^{-2}$

$$= \log_b \frac{1}{5^2} \qquad \text{Write } 5^{-2} \text{ as } \frac{1}{5^2}.$$

$$= \log_b \frac{1}{25} \qquad 5^2 = 25$$

c. $\frac{2}{3} \log_7 y$

Solution: $\frac{2}{3} \log_7 y = \log_7 y^{2/3}$

$$= \log_7 \sqrt[3]{y^2} \qquad \text{Write } y^{2/3} \text{ as } \sqrt[3]{y^2}.$$

Your Turn 7 Use the power rule of logarithms to write the expression as a logarithm of a quantity to a power.

a. $5 \log_7 x$

b. $-3 \log_b z$

Often, it is necessary to use more than one rule to rewrite a logarithmic expression.

Example 8 Write the expression as a sum or difference of multiples of logarithms.

a. $\log_b \frac{z^3}{yz}$

Solution: $\log_b \frac{z^3}{yz} = \log_b z^3 - \log_b yz$ 　　Use the quotient rule.

$$= 3 \log_b z - (\log_b y + \log_b z) \qquad \text{Use the power rule and product rule. Note the use of parentheses.}$$

$$= 3 \log_b z - \log_b y - \log_b z \qquad \text{Remove the parentheses.}$$

b. $\log_3 \sqrt{\frac{a^3}{b}}$

Solution: $\log_3 \sqrt{\frac{a^3}{b}} = \log_3 \left(\frac{a^3}{b}\right)^{1/2}$ 　　Write $\sqrt{\frac{a^3}{b}}$ as $\left(\frac{a^3}{b}\right)^{1/2}$.

$$= \frac{1}{2} \log_3 \left(\frac{a^3}{b}\right) \qquad \text{Use the power rule.}$$

$$= \frac{1}{2} (\log_3 a^3 - \log_3 b) \qquad \text{Use the quotient rule.}$$

$$= \frac{1}{2} (3 \log_3 a - \log_3 b) \qquad \text{Use the power rule.}$$

$$= \frac{3}{2} \log_3 a - \frac{1}{2} \log_3 b \qquad \text{Distribute } \frac{1}{2}.$$

Answers to Your Turn 6

a. $4 \log_a z$　　**b.** $\frac{2}{3} \log_a x$

Answers to Your Turn 7

a. $\log_7 x^5$　　**b.** $\log_b \frac{1}{z^3}$

c. $\log_5 5^2 b^3$

Solution: $\log_5 5^2 b^3 = \log_5 5^2 + \log_5 b^3$ Use the product rule.

$\qquad\qquad\qquad = 2\log_5 5 + 3\log_5 b$ Use the power rule.

$\qquad\qquad\qquad = 2 + 3\log_5 b$ $\log_5 5 = 1$

> **Your Turn 8** Write the expression as a sum or difference of multiples of logarithms.
>
> **a.** $\log_a \dfrac{x^2 y^3}{z}$ **b.** $\log_4 \sqrt[4]{\dfrac{x}{y^2}}$ **c.** $\log_8 8^5 m^6$

> **Example 9** Write the expression as a single logarithm. Leave the answers in simplest form without negative or fractional exponents.

a. $4\log_b 2 - 2\log_b 3$

Solution: $4\log_b 2 - 2\log_b 3 = \log_b 2^4 - \log_b 3^2$ Use the power rule.

$\qquad\qquad\qquad = \log_b \dfrac{2^4}{3^2}$ Use the quotient rule.

$\qquad\qquad\qquad = \log_b \dfrac{16}{9}$ Simplify.

b. $\dfrac{1}{2}(\log_2 5 - \log_2 b)$

Solution: $\dfrac{1}{2}(\log_2 5 - \log_2 b) = \dfrac{1}{2}\log_2 \dfrac{5}{b}$ Use the quotient rule.

$\qquad\qquad\qquad = \log_2 \left(\dfrac{5}{b}\right)^{1/2}$ Use the power rule.

$\qquad\qquad\qquad = \log_2 \sqrt{\dfrac{5}{b}}$ Write $\left(\dfrac{5}{b}\right)^{1/2}$ as $\sqrt{\dfrac{5}{b}}$.

c. $\log_a(x+2) + \log_a(x-3)$

Solution: $\log_a(x+2) + \log_a(x-3) = \log_a(x+2)(x-3)$ Apply $\log_b xy = \log_b x + \log_b y$.

$\qquad\qquad\qquad = \log_a(x^2 - x - 6)$ Multiply.

Answers to Your Turn 8
a. $2\log_a x + 3\log_a y - \log_a z$
b. $\dfrac{1}{4}\log_4 x - \dfrac{1}{2}\log_4 y$
c. $5 + 6\log_8 m$

Answers to Your Turn 9
a. $\log_a \dfrac{16}{x^3}$ **b.** $\log_a \sqrt[3]{xy^2}$
c. $\log_6(x^2 - 2x)$

> **Your Turn 9** Write the expression as a single logarithm. Leave your answers in simplest form without negative or fractional exponents.
>
> **a.** $2\log_a 4 - 3\log_a x$ **b.** $\dfrac{1}{3}(\log_a x + 2\log_a y)$ **c.** $\log_6 x + \log_6(x-2)$

11.4 Exercises
For Extra Help   MyMathLab®

Objective 1

Prep Exercise 1 For any real numbers b and x, where $b > 0$, $b \neq 1$, and $x > 0$, $b^{\log_b x} = $ _____ x _____.

Prep Exercise 2 For any real numbers b and x, where $b > 0$, $b \neq 1$, and $x > 0$, $\log_b b^x = $ _____ x _____.

For Exercises 1–12, find the value. See Example 1.

1. $8^{\log_8 2}$ 2   **2.** $3^{\log_3 7}$ 7   **3.** $a^{\log_a r}$ r   **4.** $b^{\log_b a}$ a   **5.** $a^{\log_a 4x}$ 4x   **6.** $b^{\log_b 5a}$ 5a

7. $\log_3 3^5$ 5   **8.** $\log_7 7^3$ 3   **9.** $\log_e e^y$ y   **10.** $\log_c c^x$ x   **11.** $\log_a a^{7x}$ 7x   **12.** $\log_b b^{6y}$ 6y

Objective 2

Prep Exercise 3 Write the product rule of logarithms:
$\log_b xy = $ ___$\log_b x + \log_b y$___.

Prep Exercise 4 $\log_b(x + y) = $ _____.

$\log_b(x + y)$; there is no rule for the logarithm of a sum.

For Exercises 13–20, use the product rule to write the expression as a sum of logarithms.
See Example 2.

13. $\log_2 5y$
$\log_2 5 + \log_2 y$

14. $\log_3 4z$
$\log_3 4 + \log_3 z$

15. $\log_a pq$
$\log_a p + \log_a q$

16. $\log_b rs$
$\log_b r + \log_b s$

17. $\log_4 mnp$
$\log_4 m + \log_4 n + \log_4 p$

18. $\log_4 pqr$
$\log_4 p + \log_4 q + \log_4 r$

19. $\log_a x(x - 5)$
$\log_a x + \log_a(x - 5)$

20. $\log_b x(x + 6)$
$\log_b x + \log_b(x + 6)$

For Exercises 21–32, use the product rule to write the expression as a single logarithm.
See Example 3.

21. $\log_3 5 + \log_3 8$
$\log_3 40$

22. $\log_6 4 + \log_6 7$
$\log_6 28$

23. $\log_4 3 + \log_4 9$
$\log_4 27$

24. $\log_9 4 + \log_9 21$
$\log_9 84$

25. $\log_a 7 + \log_a m$
$\log_a 7m$

26. $\log_b 2 + \log_b n$
$\log_b 2n$

27. $\log_4 a + \log_4 b$
$\log_4 ab$

28. $\log_6 r + \log_6 m$
$\log_6 rm$

29. $\log_a 2 + \log_a x + \log_a(x + 5)$
$\log_a(2x^2 + 10x)$

30. $\log_a 4 + \log_a y + \log_a(y - 5)$
$\log_a(4y^2 - 20y)$

31. $\log_4(x + 1) + \log_4(x + 3)$
$\log_4(x^2 + 4x + 3)$

32. $\log_6(x - 3) + \log_6(x + 2)$
$\log_6(x^2 - x - 6)$

Prep Exercise 5 Write the quotient rule of logarithms: $\log_b \dfrac{x}{y} = $ ___$\log_b x - \log_b y$___.

For Exercises 33–42, use the quotient rule to write the expression as a difference of logarithms. Leave your answers in simplest form. See Example 4.

33. $\log_2 \dfrac{7}{9}$
$\log_2 7 - \log_2 9$

34. $\log_4 \dfrac{5}{6}$
$\log_4 5 - \log_4 6$

35. $\log_a \dfrac{x}{5}$
$\log_a x - \log_a 5$

36. $\log_b \dfrac{y}{3}$
$\log_b y - \log_b 3$

37. $\log_a \dfrac{a}{b}$
$1 - \log_a b$

38. $\log_b \dfrac{a}{b}$
$\log_b a - 1$

39. $\log_a \dfrac{x}{x - 3}$
$\log_a x - \log_a(x - 3)$

40. $\log_b \dfrac{y}{2y + 5}$
$\log_b y - \log_b(2y + 5)$

41. $\log_4 \dfrac{2x - 3}{4x + 5}$
$\log_4(2x - 3) - \log_4(4x + 5)$

42. $\log_6 \dfrac{4x - 1}{3x + 2}$
$\log_6(4x - 1) - \log_6(3x + 2)$

For Exercises 43–54, use the quotient rule to write the expression as a single logarithm.
Leave your answers in simplest form. See Example 5.

43. $\log_6 24 - \log_6 3$
$\log_6 8$

44. $\log_7 18 - \log_7 3$
$\log_7 6$

45. $\log_2 24 - \log_2 12$
1

46. $\log_3 48 - \log_3 16$
1

47. $\log_a x - \log_a 3$
$\log_a \dfrac{x}{3}$

48. $\log_a r - \log_a 5$
$\log_a \dfrac{r}{5}$

49. $\log_4 p - \log_4 q$
$\log_4 \dfrac{p}{q}$

50. $\log_2 a - \log_2 b$
$\log_2 \dfrac{a}{b}$

51. $\log_b x - \log_b (x - 4)$

$\log_b \dfrac{x}{x-4}$

52. $\log_b y - \log_b (y - 5)$

$\log_b \dfrac{y}{y-5}$

53. $\log_x (x^2 - x) - \log_x (x - 1)$

1

54. $\log_a (a^2 + 2a) - \log_a (a + 2)$

1

Prep Exercise 6 Write the power rule of logarithms: $\log_b x^r =$ _____$r \log_b x$_____.

For Exercises 55–66, use the power rule to write the expression as a multiple of a logarithm.
See Example 6.

55. $\log_4 3^6$

$6 \log_4 3$

56. $\log_3 5^4$

$4 \log_3 5$

57. $\log_a x^7$

$7 \log_a x$

58. $\log_b y^8$

$8 \log_b y$

59. $\log_a \sqrt{3}$

$\dfrac{1}{2} \log_a 3$

60. $\log_a \sqrt{6}$

$\dfrac{1}{2} \log_a 6$

61. $\log_3 \sqrt[3]{x^2}$

$\dfrac{2}{3} \log_3 x$

62. $\log_6 \sqrt[4]{y^3}$

$\dfrac{3}{4} \log_6 y$

63. $\log_a \dfrac{1}{6^2}$

$-2 \log_a 6$

64. $\log_a \dfrac{1}{5^4}$

$-4 \log_a 5$

65. $\log_a \dfrac{1}{y^2}$

$-2 \log_a y$

66. $\log_a \dfrac{1}{x^5}$

$-5 \log_a x$

For Exercises 67–78, use the power rule to write the expression as a logarithm of a quantity
to a power. Leave your answers in simplest form without negative or fractional exponents.
See Example 7.

67. $4 \log_3 5$

$\log_3 5^4$

68. $5 \log_3 4$

$\log_3 4^5$

69. $-3 \log_2 x$

$\log_2 \dfrac{1}{x^3}$

70. $-4 \log_3 y$

$\log_3 \dfrac{1}{y^4}$

71. $\dfrac{1}{2} \log_7 64$

$\log_7 8$

72. $\dfrac{1}{3} \log_3 8$

$\log_3 2$

73. $\dfrac{3}{4} \log_a x$

$\log_a \sqrt[4]{x^3}$

74. $\dfrac{5}{6} \log_b y$

$\log_b \sqrt[6]{y^5}$

75. $\dfrac{2}{3} \log_a 8$

$\log_a 4$

76. $\dfrac{3}{4} \log_a 81$

$\log_a 27$

77. $-\dfrac{1}{2} \log_3 x$

$\log_3 \dfrac{1}{\sqrt{x}}$

78. $-\dfrac{1}{3} \log_2 y$

$\log_2 \dfrac{1}{\sqrt[3]{y}}$

For Exercises 79–90, write the expression as the sum or difference of multiples of
logarithms. See Example 8.

79. $\log_a \dfrac{x^3}{y^4}$

$3 \log_a x - 4 \log_a y$

80. $\log_a \dfrac{x^6}{y^5}$

$6 \log_a x - 5 \log_a y$

81. $\log_3 a^4 b^2$

$4 \log_3 a + 2 \log_3 b$

82. $\log_7 m^3 n^5$

$3 \log_7 m + 5 \log_7 n$

83. $\log_a \dfrac{xy}{z}$

$\log_a x + \log_a y - \log_a z$

84. $\log_a \dfrac{pq}{r}$

$\log_a p + \log_a q - \log_a r$

85. $\log_x \dfrac{a^2}{bc^3}$

$2 \log_x a - \log_x b - 3 \log_x c$

86. $\log_x \dfrac{c^2}{m^2 n}$

$2 \log_x c - 2 \log_x m - \log_x n$

87. $\log_4 \sqrt[4]{\dfrac{x^3}{y}}$

$\dfrac{3}{4} \log_4 x - \dfrac{1}{4} \log_4 y$

88. $\log_5 \sqrt{\dfrac{a^5}{b^3}}$

$\dfrac{5}{2} \log_5 a - \dfrac{3}{2} \log_5 b$

89. $\log_a \sqrt[3]{\dfrac{x^2 y}{z^3}}$

$\dfrac{2}{3} \log_a x + \dfrac{1}{3} \log_a y - \log_a z$

90. $\log_3 \sqrt{\dfrac{c^2 d^3}{m^4}}$

$\log_3 c + \dfrac{3}{2} \log_3 d - 2 \log_3 m$

For Exercises 91–104, write the expression as a single logarithm. Leave your answers in simplest form without negative or fractional exponents. See Example 9.

91. $3\log_3 2 - 2\log_3 4$

$\log_3 \dfrac{1}{2}$

92. $3\log_4 3 - 2\log_4 9$

$\log_4 \dfrac{1}{3}$

93. $4\log_b x + 3\log_b y$

$\log_b x^4 y^3$

94. $5\log_b a + 4\log_b c$

$\log_b a^5 c^4$

95. $\dfrac{1}{2}(\log_a 5 - \log_a 7)$

$\log_a \sqrt{\dfrac{5}{7}}$

96. $\dfrac{1}{2}(\log_4 6 - \log_4 5)$

$\log_4 \sqrt{\dfrac{6}{5}}$

97. $\dfrac{2}{3}(\log_a x^2 + \log_a y^3)$

$\log_a \sqrt[3]{(x^2 y^3)^2}$

98. $\dfrac{3}{4}(\log_a m^3 + \log_a n^2)$

$\log_a \sqrt[4]{(m^3 n^2)^3}$

99. $\log_b x + \log_b(3x - 2)$

$\log_b(3x^2 - 2x)$

100. $\log_a y + \log_a(2y - 4)$

$\log_a(2y^2 - 4y)$

101. $3\log_a(x - 2) - 4\log_a(x + 1)$

$\log_a \dfrac{(x - 2)^3}{(x + 1)^4}$

102. $2\log_b(x + 4) - 3\log_b(2x - 1)$

$\log_b \dfrac{(x + 4)^2}{(2x - 1)^3}$

103. $2\log_a x + 4\log_a z - 3\log_a w - 6\log_a u$

$\log_a \dfrac{x^2 z^4}{w^3 u^6}$

104. $5\log_a b + 2\log_a c - 4\log_a d - 3\log_a e$

$\log_a \dfrac{b^5 c^2}{d^4 e^3}$

Review Exercises

Exercises 1 and 2 ◢ **Expressions**

For Exercises 1 and 2, simplify.

[5.4] 1. $(10^{9.5})(10^{-12})$

$10^{-2.5}$

[5.6] 2. $\dfrac{10^{-3}}{10^{-12}}$

10^9

Exercises 3–6 ◢ **Equations and Inequalities**

[11.2] 3. If $10,000 is deposited at 6% compounded quarterly for five years, how much will be in the account?

$13,468.55

[11.3] 4. Write $10^{1.6990} = 50$ in logarithmic form.

$\log_{10} 50 = 1.6990$

[11.3] *For Exercises 5 and 6, write in exponential form.*

5. $\log_{10} 45 = 1.6532$

$10^{1.6532} = 45$

6. $\log_e 0.25 = -1.3863$

$e^{-1.3863} = 0.25$

11.5 Common and Natural Logarithms

Objectives

1 Define common logarithms and evaluate them using a calculator.

2 Define natural logarithms and evaluate them using a calculator.

3 Solve applications using common logarithms.

4 Solve applications using natural logarithms.

Warm-up

[11.4] 1. Evaluate: $\log_{10} 10^{9.5}$

[11.4] 2. Write as a single logarithm: $2\log_n 5 - 3\log_n 2$

[5.6] 3. Simplify: $\dfrac{10^{-2.5}}{10^{-12}}$

[11.3] 4. Solve $6.723 = 3\log_{10} x$ for x. Round to the nearest tenth.

Objective 1 Define common logarithms and evaluate them using a calculator.

Of all possible bases of logarithms, two are most useful. As previously mentioned, logarithms were invented to do computations on very large and very small numbers. Because our system is a base-10 number system, base-10 logarithms were commonly used for this purpose. Consequently, base-10 logarithms are called **common logarithms** and $\log_{10} x$ is written as $\log x$, where the base 10 is understood. Base-10 logarithms are found in engineering, economics, the social sciences, and the natural sciences.

Definition Common logarithms: Logarithms with a base of 10. $\log_{10} x$ is written as $\log x$. Note that $\log 10 = 1$.

Connection Because logarithmic and exponential functions are inverse functions, if $f(x) = 10^x$, then $f^{-1}(x) = \log x$ and if $g(x) = \log x$, then $g^{-1}(x) = 10^x$.

To evaluate common logarithms, we use a calculator. We will round all results to four places.

Example 1 Use a calculator to approximate each common logarithm to four decimal places.

a. log 23

Solution: $\log 23 \approx 1.3617$

b. log 0.00236

Solution: $\log 0.00236 \approx -2.6271$

Connection Remember that $\log 23 \approx 1.3617$ means that $10^{1.3617} \approx 23$, which provides a way to check. Because we rounded the decimal value, evaluating $10^{1.3617}$ will not give exactly 23.

Your Turn 1 Use a calculator to approximate each common logarithm to four decimal places.

a. log 436
b. log 0.0724

Answers to Warm-up
1. 9.5
2. $\log_n\left(\dfrac{25}{8}\right)$
3. $10^{9.5}$
4. 174.2

Answers to Your Turn 1
a. 2.6395 **b.** -1.1403

Objective 2 Define natural logarithms and evaluate them using a calculator.

The number e is an irrational number whose approximate value is 2.7182818285. It is a universal constant like π. In the natural sciences, compared with base-10 logarithms, base-e logarithms are more prevalent. Because base-e logarithms occur in so many "natural" situations, they are called **natural logarithms**. The notation for $\log_e x$ is ln x, which is read "el en of x."

Definition Natural logarithms: Base-e logarithms are called natural logarithms, and $\log_e x$ is written as ln x. Note that ln $e = 1$.

To find natural logarithms, we use a calculator.

Example 2 Use a calculator to approximate each natural logarithm to four decimal places.

a. ln 83

Solution: ln $83 \approx 4.4188$

b. ln 0.0055

Solution: ln $0.0055 \approx -5.2030$

> **Connection** Remember that ln $83 \approx 4.4188$ means that $e^{4.4188} \approx 83$. Similarly, ln $0.0055 = -5.2030$ means that $e^{-5.2030} \approx 0.0055$.

Your Turn 2 Use a calculator to approximate each natural logarithm to four decimal places.

a. ln 102

b. ln 0.0573

Objective 3 Solve applications using common logarithms.

Common logarithms can be used to calculate sound intensity and runway length.

Example 3

a. Sound intensity can be measured in watts per unit of area or, more commonly, in decibels. The function $d = 10 \log \dfrac{I}{I_0}$ is used to calculate sound intensity, where d represents the intensity in decibels, I represents the intensity in watts per unit of area, and I_0 represents the faintest audible sound to the average human ear (which is 10^{-12} watts per square meter). A motorcycle has a sound intensity of about $10^{-2.5}$ watts per square meter. Find the decibel reading for the motorcycle.

Solution: $d = 10 \log \dfrac{10^{-2.5}}{10^{-12}}$ In $d = 10 \log \dfrac{I}{I_0}$, substitute $10^{-2.5}$ for I and 10^{-12} for I_0.

$d = 10 \log 10^{9.5}$ Subtract exponents $[-2.5 - (-12) = 9.5]$.

$d = 10(9.5)$ Use $\log_b b^x = x$.

$d = 95$ Simplify.

> **Note** The abbreviation for decibels is dB.

Answer: The motorcycle has a decibel reading of 95 dB.

b. The minimum length of an airport runway needed for a plane to take off is related to the weight of the plane. For some planes, the minimum runway length may be modeled by the function $y = 3 \log x$, where x is the plane's weight in thousands of pounds and y is the length of the runway in thousands of feet. Find the minimum length of a runway needed by a Boeing 737 whose maximum take-off weight is 174,200 pounds.

Answers to Your Turn 2
a. 4.6250   **b.** −2.8595

Solution: $x = \dfrac{174,200}{1000} = 174.2$ Divide 174,200 by 1000 to find the value of x corresponding to 174,200 pounds.

$y = 3 \log 174.2$ In $y = 3 \log x$, replace x with 174.2.

$y \approx 6.723$ Evaluate using a calculator.

Answer: Because y is in thousands of feet, the minimum runway length is about $(6.723)(1000) = 6723$ ft.

Your Turn 3

a. The sound intensity of a rock band often exceeds $10^{-0.5}$ watts per square meter. Find the decibel reading of the band.

b. Find the minimum runway length needed for a B-52 Stratofortress whose maximum takeoff weight is 488,000 pounds.

Objective 4 Solve applications using natural logarithms.

If money is deposited into an account and the interest is compounded continuously, then the time t (in years) that it will take an investment of P dollars to grow into A dollars at an interest rate r (written as a decimal) is given by $t = \dfrac{1}{r} \ln \dfrac{A}{P}$.

Example 4
An amount of $5000 is deposited into an account earning 5% annual interest compounded continuously. How many years will it take until the account has reached $10,000?

Solution: $t = \dfrac{1}{0.05} \ln \dfrac{10,000}{5000}$ In $t = \dfrac{1}{r} \ln \dfrac{A}{P}$, replace P with 5000, r with 0.05, and A with 10,000.

$t = 20 \ln 2$ Divide.

$t \approx 13.9$ Multiply using a calculator.

Answer: It will take about 13.9 years for $5000 to grow to $10,000 if it is compounded continuously at 5%.

Answers to Your Turn 3
a. $115\,\text{dB}$ **b.** ≈ 8065 ft.

Answer to Your Turn 4
≈ 11.5 yr.

Your Turn 4
How long will it take $2000 to grow to $5000 at 8% interest if the interest is compounded continuously?

11.5 Exercises   For Extra Help   MyMathLab®

Objectives 1 and 2

Prep Exercise 1 In the common logarithm $\log x$, what is the base? 10

Prep Exercise 2 In the natural logarithm $\ln x$, what is the base? e

Prep Exercise 3 Is $\log x$ positive or negative for $0 < x < 1$? Why? Negative. If $y = \log x$, then $10^y = x$. If $0 < x < 1$, then y must be negative so that $x = \dfrac{1}{10^y}$ where $y > 0$, which results in values of x such that $0 < x < 1$. Also, any positive value of y results in a value of $x > 1$. Also look at the graph.

Prep Exercise 4 Is $\ln x$ positive or negative for $x > 1$? Why? Positive. If $y = \ln x$, then $e^y = x$. If $x > 1$, then $e^y > 1$, which is true if $y > 0$.

For Exercises 1–20, use a calculator to approximate each logarithm to four decimal places. See Examples 1 and 2.

1. log 64
   1.8062

2. log 27
   1.4314

3. log 0.0067
   −2.1739

4. log 0.00087
   −3.0605

5. log 435.6
   2.6391

6. log 785.4
   2.8951

7. $\log(1.5 \times 10^4)$
   4.1761

8. $\log(5.7 \times 10^7)$
   7.7559

9. $\log(1.6 \times 10^{-6})$
   −5.7959

10. $\log(7.5 \times 10^{-5})$
    −4.1249

11. ln 9.34
    2.2343

12. ln 5.33
    1.6734

13. ln 79.2
    4.3720

14. ln 765.4
    6.6404

15. ln 0.034
    −3.3814

16. ln 0.0923
    −2.3827

17. $\ln(5.4 \times e^4)$
    5.6864

18. $\ln(2.4 \times e^3)$
    3.8755

19. log e
    0.4343

20. ln 10
    2.3026

21. Use your calculator to find log 0. What happened? Why?
    Error results because the domain of $\log_a x$ is $(0, \infty)$, so log 0 is undefined.

22. Use your calculator to find ln (−1). What happened? Why?
    Error results because the domain of ln x is $(0, \infty)$, so ln(−1) is undefined.

Prep Exercise 5 If $e^{0.91629} = 2.5$, find ln 2.5.
0.91629

Prep Exercise 6 Without using a calculator, find the exact value of log $10^{\sqrt{5}}$. $\sqrt{5}$

For Exercises 23–34, find the exact value of each logarithm using $\log_b b^x = x$.

23. log 100
    2

24. log 1000
    3

25. $\log \dfrac{1}{100}$
    −2

26. $\log \dfrac{1}{1000}$
    −3

27. $\log \sqrt[3]{10}$
    $\dfrac{1}{3}$

28. $\log \sqrt[4]{10}$
    $\dfrac{1}{4}$

29. log 0.001
    −3

30. log 0.00001
    −5

31. $\ln e^3$
    3

32. $\ln e^5$
    5

33. $\ln \sqrt{e}$
    $\dfrac{1}{2}$

34. $\ln \sqrt[5]{e}$
    $\dfrac{1}{5}$

Objective 3

For Exercises 35–38, use the formula $d = 10 \log \dfrac{I}{I_{00}}$, where $I_0 = 10^{-12}$ watts/m^2. See Example 3.

35. The sound intensity of a firecracker is 10^{-3} watts per square meter. What is the decibel reading for the firecracker?
    90 dB

36. The sound intensity of a race car is 10^{-1} watts per square meter. What is the decibel reading for the race car?
    110 dB

37. If a noisy office has a decibel reading of 60, what is the sound intensity?
    10^{-6} watts/m^2

38. If loud thunder has a decibel reading of 80, what is the sound intensity?
    10^{-4} watts/m^2

For Exercises 39–42, use the following information. In chemistry, the pH of a substance determines whether it is a base (pH $>$ 7) or an acid (pH $<$ 7). To find the pH of a solution, we use the formula pH $= -\log\left[H_3O^+\right]$, where $\left[H_3O^+\right]$, is the hydronium ion concentration in moles per liter. Note that the pH is unitless.

39. Find the pH of vinegar if $[H_3O^+] = 1.6 \times 10^{-3}$ moles per liter.

≈ 2.796

40. Find the pH of maple syrup if $[H_3O^+] = 2.3 \times 10^{-7}$ moles per liter.

≈ 6.638

41. Find the hydronium ion concentration of sauerkraut, which has a pH of 3.5.

$10^{-3.5}$ moles/L

42. Find the hydronium ion concentration of blood, which has a pH of 7.4.

$10^{-7.4}$ moles/L

Objective 4

For Exercises 43 and 44, use the formula $t = \dfrac{1}{r}\ln\dfrac{A}{P}$. See Example 4.

43. Find how long it will take $2000 to grow to $5000 at 4% interest if the interest is compounded continuously.

≈ 22.9

44. Find how long it will take $5000 to grow to $8000 at 5% interest if the interest is compounded continuously.

≈ 9.4 yr.

45. The magnitude of an earthquake is given by the Richter scale, whose formula is $R = \log\dfrac{I}{I_0}$, where I is the intensity of the earthquake and I_0 is the intensity of a minimal earthquake and is used for comparison purposes. The 1906 San Francisco earthquake had a magnitude of 7.8, and the 1964 Alaska earthquake had a magnitude of 8.4. Compare the intensity of the two earthquakes. (*Hint:* Express the intensity of each in terms of I_0.)

The 1964 Alaska earthquake was about four times as severe as the 1906 San Francisco earthquake.

46. Using the formula from Exercise 45, compare the intensity of the 1949 Queen Charlotte Islands earthquake, whose magnitude was 8.1, with the 2004 earth-quake in the Indian Ocean, whose magnitude was estimated at 9.2, that killed more than 225,000 people.

The 2004 earthquake was about 12.6 times as severe as the 1949 earthquake.

Of Interest

This photograph shows the destruction of San Francisco that was caused by the 1906 earthquake.

47. During an earthquake, energy is released in various forms. The amount of energy radiated from the earthquake as seismic waves is given by $\log E_s = 11.8 + 1.5\,M$, where E_s is measured in ergs and M is the magnitude of the earthquake as given by the Richter scale. Vancouver Island had an earthquake whose magnitude was 7.3. How much energy was released in the form of seismic waves?

$10^{22.75}$ ergs

48. Using the formula from Exercise 47, find the energy released in the form of seismic waves from the Double Springs Flat earthquake of 1994, whose magnitude was 6.1 on the Richter scale.

$10^{20.95}$ ergs

49. The purchasing power of $1.00 in 2007 can be approximated by $y = -67.89 + 22.56 \ln x$, where y is the number of years beginning with 1970 ($y = 1$ corresponds with 1970) until 2010 and x is the purchasing power, in cents, of $1.00 in 2010. In what year was the purchasing power of $1.00 in 2010 approximately $0.40?

1986

50. The average number of miles per gallon for light trucks can be approximated by $y = 14.27 + 2.98 \ln x$, where y is the average number of miles per gallon and x is the number of years beginning with 1975. ($y = 1$ corresponds with 1975.) Find the approximate average number of miles per gallon in 2013.

≈ 25.2 mpg

51. Using the formula from Example 3(b), $y = 3 \log x$, find the minimum runway length needed for a Boeing 717 whose maximum weight is 110,000 pounds at takeoff.

≈ 6124 ft.

52. Walking speeds in various cities are a function of the population, because as populations increase, so does the pace of life. Average walking speeds can be modeled by the function $W = 0.35 \ln P + 2.74$, where P is the population, in thousands, and W is the walking speed in feet per second. Find the average walking speed in Chicago, whose population was approximately 2,700,000 in 2010.

≈ 5.51 ft./sec.

Review Exercises

Exercises 1 ◢ **Expressions**

[11.4] 1. Write $\log_3(2x + 1) - \log_3(x - 1)$ as a single logarithm.

$\log_3 \dfrac{2x + 1}{x - 1}$

Exercises 2–6 ◢ **Equations and Inequalities**

[11.3] 2. Write $\log_3(2x + 5) = 2$ in exponential form.

$3^2 = 2x + 5$

For Exercises 3–6, solve.

[2.3] 3. $3x - (7x + 2) = 12 - 2(x - 4)$

-11

[6.6] 4. $x^2 + 2x = 15$

$-5, 3$

[7.6] 5. $\dfrac{5}{x} + \dfrac{3}{x + 1} = \dfrac{23}{3x}$

8

[9.6] 6. $\sqrt{5x - 1} = 7$

10

11.6 Exponential and Logarithmic Equations with Applications

Objectives

1 Solve equations that have variables as exponents.

2 Solve equations containing logarithms.

3 Solve applications involving exponential and logarithmic functions.

4 Use the change-of-base formula.

Warm-up

[11.5] 1. The formula $t = \dfrac{1}{r} \ln \dfrac{A}{P}$ can be used to find the time, in years, for a principal amount, P, invested at interest rate, r, to grow to an amount, A. How many years will it take $2000 invested at 4% interest to grow to $6000?

[11.4] 2. Write $\log_5 x + \log_5(x + 3)$ as a single logarithm.

[6.6] 3. Solve: $x(x + 3) = 4$

[11.3] 4. Write $\log_2 \dfrac{5x + 1}{x - 1} = 3$ in exponential form.

Answers to Warm-up

1. ≈ 27.5 yr.
2. $\log_5 x(x + 3)$
3. $x = -4, 1$
4. $\dfrac{5x + 1}{x - 1} = 2^3$

In Section 11.2, we solved equations that could be put in the form $b^x = b^y$. For example, if $3^x = 81$, then $3^x = 3^4$; so $x = 4$. To use this form, we had to write both sides of the equation as the same base raised to a power. In this section, we solve equations like $3^x = 16$, which is the same as $3^x = 2^4$, where the bases are not the same. To solve this and other exponential and logarithmic equations, we need the following properties.

> **Rule** **Properties for Solving Exponential and Logarithmic Equations**
>
> For any real numbers b, x, and y, where $b > 0$ and $b \neq 1$:
> 1. If $b^x = b^y$, then $x = y$.
> 2. If $x = y$, then $b^x = b^y$.
> 3. For $x > 0$ and $y > 0$, if $\log_b x = \log_b y$, then $x = y$.
> 4. For $x > 0$ and $y > 0$, if $x = y$, then $\log_b x = \log_b y$.
> 5. For $x > 0$, if $\log_b x = y$, then $b^y = x$.

These properties are true because the exponential and logarithmic functions are one-to-one.

Objective 1 Solve equations that have variables as exponents.

To solve equations that have variables as exponents, we use property 4, which says that if two positive numbers are equal, then so are their logarithms.

Example 1 Solve $5^x = 16$.

Solution: $\log 5^x = \log 16$ Use if $x = y$, then $\log_b x = \log_b y$ (property 4).

$\qquad\qquad x \log 5 = \log 16$ Use $\log_b x^r = r \log_b x$.

$$x = \frac{\log 16}{\log 5} \qquad \text{Divide both sides of the equation by log 5.}$$

The exact solution is $x = \dfrac{\log 16}{\log 5}$. Using a calculator, we find $x \approx 1.7227$ correct to four decimal places.

Check: $5^x = 16$

$\qquad 5^{1.7227} = 16$ Substitute 1.7727 for x.

$\quad 15.9998 \approx 16$ The answer is correct.

Your Turn 1 Solve $6^{2x} = 42$ for x. Round the answer to four decimal places.

Answer to Your Turn 1
1.0430

In Example 1, we took the common logarithm of both sides, but we could have used natural logarithms (or logarithms of any other base) instead. If one side of the equation contains a power of e, natural logs are preferred so that we can use the fact that $\log_e e^x = x$ or, more simply, $\ln e^x = x$.

Note If your calculator automatically puts the left parenthesis, this must be entered as $\ln(23) \div 4$. ▶

Example 2 Solve $e^{4x} = 23$.

Solution: $\ln e^{4x} = \ln 23$ Use if $x = y$; then $\log_b x = \log_b y$.

$\qquad\qquad 4x = \ln 23$ Use $\log_b b^x = x$.

$\qquad\qquad x = \dfrac{\ln 23}{4}$ Divide both sides by 4.

$\qquad\qquad x \approx 0.7839$

We will leave the check to the reader.

Your Turn 2 Solve $e^{3x} = 5$ for x. Round the answer to four decimal places.

Objective 2 Solve equations containing logarithms.

Now that we have explored additional properties of logarithms, we can modify the procedure presented in Section 11.3 for solving equations using logarithms.

> **Procedure** **Solving Equations Containing Logarithms**
>
> To solve equations containing logarithms, use the properties of logarithms to simplify each side of the equation and then use one of the following.
>
> If the simplification results in an equation in the form $\log_b x = \log_b y$, use the fact that $x = y$ and then solve for the variable.
>
> If the simplification results in an equation in the form $\log_b x = y$, write the equation in exponential form, $b^y = x$, and then solve for the variable (as we did in Section 11.3).

Instructor Note Some instructors prefer to solve all logarithmic equations using the form $\log_b x = y$. So Example 3(a) could be solved as follows:

$$\log_5 x + \log_5(x + 3) = \log_5 4$$
$$\log_5 x + \log_5(x + 3) - \log_5 4 = 0$$
$$\log_5 \frac{x(x + 3)}{4} = 0$$
$$\frac{x(x + 3)}{4} = 5^0$$
$$\frac{x(x + 3)}{4} = 1$$
$$x^2 + 3x = 4$$
$$x^2 + 3x - 4 = 0$$
$$(x + 4)(x - 1) = 0$$
$$x + 4 = 0, \quad x - 1 = 0$$
$$x = -4, \qquad x = 1$$

Then check the answers.

Answer to Your Turn 2
0.5365

Example 3 Solve.

a. $\log_5 x + \log_5(x + 3) = \log_5 4$

Solution: $\log_5 x(x + 3) = \log_5 4$ Use $\log_b xy = \log_b x + \log_b y$ to simplify the left side.

$\qquad\qquad x(x + 3) = 4$ The equation is in the form $\log_b x = \log_b y$, so $x = y$.

$\qquad\qquad x^2 + 3x = 4$ Simplify.

$\qquad\qquad x^2 + 3x - 4 = 0$ Write in $ax^2 + bx + c = 0$ form.

$\qquad\qquad (x + 4)(x - 1) = 0$ Factor.

$\qquad\qquad x + 4 = 0 \quad \text{or} \quad x - 1 = 0$ Set each factor equal to 0.

$\qquad\qquad x = -4 \quad \text{or} \qquad x = 1$ Solve each equation to find possible solutions.

If -4 is substituted into the original equation, we have $\log_5(-4) + \log_5(-1) = \log_5 4$, but logarithms are defined for positive numbers only. So $x = -4$ is not a solution. A check will show that $x = 1$ is a solution.

b. $\log_3(2x + 5) = 2$

Solution: $3^2 = 2x + 5$ The equation is in the form $\log_b x = y$; so write it in exponential form, $b^y = x$.

$\qquad\qquad 9 = 2x + 5$ Evaluate 3^2.

$\qquad\qquad 4 = 2x$ Subtract 5 from both sides.

$\qquad\qquad 2 = x$ Solve for x.

We will let the reader check that $x = 2$ is a solution.

c. $\log_2(5x + 1) - \log_2(x - 1) = 3$

Solution: $\log_2 \dfrac{5x + 1}{x - 1} = 3$ Use $\log_b \dfrac{x}{y} = \log_b x - \log_b y$ to simplify the left side.

$\dfrac{5x + 1}{x - 1} = 2^3$ The equation is in the form $\log_b x = y$, so $b^y = x$.

$5x + 1 = 8x - 8$ Multiply both sides by $x - 1$.

$9 = 3x$ Isolate the x term.

$x = 3$ Solve for x.

We will let the reader check that $x = 3$ is a solution.

> **Your Turn 3** Solve.
>
> **a.** $\ln x + \ln(x + 2) = \ln 8$ **b.** $\log_3(4x + 2) - \log_3(x - 2) = 2$

Objective 3 Solve applications involving exponential and logarithmic functions.

A wide variety of problems from business and the sciences can be solved using exponential or logarithmic functions. Earlier we used the formula for compound interest, $A = P\left(1 + \dfrac{r}{n}\right)^{nt}$, to find A when given P, r, n, and t. Using logarithms, it is also possible to find t when given A, P, r, and n.

> **Example 4** How long will it take $6000 invested at 6% interest compounded quarterly to grow to $10,000? Round the answer to the nearest tenth of a year.
>
> **Solution:** $10{,}000 = 6000\left(1 + \dfrac{0.06}{4}\right)^{4t}$ In $A = P\left(1 + \dfrac{r}{n}\right)^{nt}$, replace A with 10,000, P with 6000, r with 0.06, and n with 4.
>
> $\dfrac{5}{3} = (1.015)^{4t}$ Divide both sides by 6000 and simplify inside the parentheses.
>
> $\log \dfrac{5}{3} = \log 1.015^{4t}$ Use if $x = y$; then $\log_b x = \log_b y$.
>
> $\log \dfrac{5}{3} = 4t \log 1.015$ Use $\log_b x^r = r \log_b x$.
>
> $\dfrac{\log \dfrac{5}{3}}{4 \log 1.015} = t$ Divide both sides by $4 \log 1.015$.
>
> $8.6 \approx t$ Evaluate using a calculator and round.
>
> **Answer:** It will take about 8.6 years.

> **Connection** Solving $A = Pe^{rt}$ for t gives the formula $t = \dfrac{1}{r} \ln \dfrac{A}{P}$.
>
> $\dfrac{A}{P} = e^{rt}$ Divide both sides by P to isolate e^{rt}.
>
> $\log_e \dfrac{A}{P} = rt$ Write in log form.
>
> $\dfrac{1}{r} \ln \dfrac{A}{P} = t$ Divide both sides by r, and by definition, $\log_e \dfrac{A}{P}$ is $\ln \dfrac{A}{P}$.

> **Your Turn 4** How long will it take $8000 invested at 4% annual interest compounded semiannually to grow to $12,000? Round to the nearest tenth of a year.

Many banks compound interest continuously. The formula for interest compounded continuously is $A = Pe^{rt}$, where A is the amount in the account, P is the amount deposited, r is the interest rate as a decimal, and t is the time in years.

> **Example 5** If $5000 is deposited into an account at 5% interest compounded continuously, how much will be in the account after 9 years?
>
> **Solution:** $A = 5000e^{0.05(9)}$ In $A = Pe^{rt}$, replace P with 5000, r with 0.05, and t with 9.
>
> $A = 5000e^{0.45}$ Apply $(a^m)^n = a^{mn}$.
>
> $A = \$7841.56$ Evaluate using a calculator.
>
> **Answer:** There will be $7841.56 in the account.

Answers to Your Turn 3
a. 2   **b.** 4

Answer to Your Turn 4
10.2 yr.

Example 6 Since the 1980s, the greater Orlando area has been one of the fastest-growing areas in the United States. The following table shows the population of the greater Orlando area for selected years.

Year	Population in Millions
1980	0.805
1985	0.996
1990	1.225
1995	1.428
2000	1.645
2003	1.803
2007	2.033

The data can be approximated by $y = 0.823e^{0.0361t}$, where y is the population in millions and t is the number of years after 1980. (*Source:* U.S. Bureau of the Census.)

a. Assuming that the population continues to grow in the same manner, use the model to estimate the population of Orlando in 2015.

Solution: $t = 2015 - 1980 = 35$
Subtract 1980 from 2015 to determine the value of t that corresponds to 2015.

$$y = 0.823e^{0.0361(35)}$$
In $y = 0.823e^{0.0361t}$, substitute 35 for t.

$$y \approx 2.912$$
Evaluate using a calculator.

Answer: Because y is in millions, the estimated population is 2,912,000 in 2015.

b. Find the year in which the population will be 3,500,000.

Solution: $y = \dfrac{3,500,000}{1,000,000} = 3.5$
Divide 3,500,000 by 1,000,000 to find the value of y that corresponds to 3,500,000.

$$3.5 = 0.823e^{0.0361t}$$
In $y = 0.823e^{0.0361t}$, substitute 3.5 for y.

$$\frac{3.5}{0.823} = e^{0.0361t}$$
Divide both sides by 0.823.

$$\ln \frac{3.5}{0.823} = \ln e^{0.0361t}$$
Use if $x = y$; then $\log_b x = \log_b y$.

$$\ln \frac{3.5}{0.823} = 0.0361t$$
Use $\log_b b^x = x$.

$$\frac{\ln \dfrac{3.5}{0.823}}{0.0361} = t$$
Divide both sides by 0.0361.

$$40.1 \approx t$$
Calculate.

Note Our manipulations suggest the following formula for calculating t given y:

$$t = \frac{1}{0.0361} \ln \frac{y}{0.823}$$

▶

Answer: Because t represents the number of years after 1980, the population will reach 3,500,000 in $1980 + 40.1 = 2020.1$, which means during 2020.

Your Turn 6

a. The bacteria *Bacillus megaterium* increases at a rate of 4% per minute in a sucrose–salts medium. If 4500 bacteria were present initially, the number, A, present after t minutes is given by the equation $A = 4500e^{0.04t}$. (*Source: Todar's Online Textbook of Bacteriology.*) How many bacteria are present after 26 minutes?

b. After how many minutes will 10,000 bacteria be present?

Exponential growth and decay can be represented by the equation $A = A_0 e^{kt}$, where A is the amount present, A_0 is the initial amount, t is the time, and k is a constant that is determined by the substance. Exponential growth is indicated if $k > 0$; exponential decay, if $k < 0$. Recall that the half-life of a substance is the amount of time until only one-half of the original amount is present.

Answers to Your Turn 6
a. ≈12,731 bacteria
b. ≈20 min.

Plutonium-239 is frequently used as fuel in nuclear reactors to generate electricity. One of the problems with using plutonium is disposing of the radioactive waste, which is extremely dangerous for a very long period of time.

Example 7 A nuclear reactor contains 10 kilograms of radioactive plutonium ^{239}P. Plutonium disintegrates according to the formula $A = A_0 e^{-0.0000284t}$.

a. How much will remain after 10,000 years?

Solution: $A = 10e^{-0.0000284(10,000)}$ In $A = A_0 e^{-0.0000284t}$, replace A_0 with 10 and t with 10,000.

$\qquad\qquad A = 10e^{-0.284}$ Simplify.

$\qquad\qquad A \approx 7.53$ Evaluate using a calculator.

Answer: About 7.53 kilograms $\left(\text{or about } \dfrac{3}{4} \text{ of the original amount}\right)$ will remain after 10,000 years.

b. Find the half-life of ^{239}P.

Solution: $5 = 10e^{-0.0000284t}$ In $A = A_0 e^{-0.0000284t}$, replace A with 5 and A_0 with 10.

$\qquad\qquad \dfrac{5}{10} = e^{-0.0000284t}$ Divide both sides by 10.

$\qquad\qquad \ln\dfrac{5}{10} = \ln e^{-0.0000284t}$ Use if $x = y$; then $\log_b x = \log_b y$.

$\qquad\qquad \ln\dfrac{5}{10} = -0.0000284t$ Use $\log_b b^x = x$.

$\qquad\qquad \dfrac{\ln\dfrac{5}{10}}{-0.0000284} = t$ Divide both sides by -0.0000284.

$\qquad\qquad 24,406.59 \approx t$ Calculate.

Note Our manipulations suggest the following formula for calculating t given A and A_0:

$$t = \dfrac{1}{-0.0000284}\ln\dfrac{A}{A_0}$$ ▶

Answer: The half-life of ^{239}P is about 24,400 years.

Your Turn 7 Carbon-14 is a radioactive form of carbon that is present in all living things. Archaeologists and paleontologists frequently use carbon-14 dating in estimating the age of organic fossils. Carbon-14 disintegrates according to the formula $A = A_0 e^{-0.000121t}$.

a. If a sample contains 5 grams of carbon-14, how much will be present after 1500 years?

b. What is the half-life of carbon-14?

Objective 4 Use the change-of-base formula.

Sometimes applications involve logarithms other than common or natural logarithms. For example, earlier we were given the formula $P = 95 - 30\log_2 x$, where P is the percent of students who recall the important features of a lecture after x days. To find the percent after 5 days, we need to calculate $\log_2 5$. Most calculators have only base-10 and base-e logarithms; so to calculate $\log_2 5$ using a calculator, we need to write $\log_2 5$

Answers to Your Turn 7
a. ≈ 4.17 g   **b.** ≈ 5728 yr.

in terms of common or natural logarithms using the change-of-base formula. To derive the change-of-base formula, we let $y = \log_a x$.

$$a^y = x \qquad \text{Write } y = \log_a x \text{ in exponential form.}$$

$$\log_b a^y = \log_b x \qquad \text{Take } \log_b \text{ of both sides.}$$

$$y \log_b a = \log_b x \qquad \text{Use } \log_b x^r = r \log_b x.$$

$$y = \frac{\log_b x}{\log_b a} \qquad \text{Divide both sides by } \log_b a.$$

$$\log_a x = \frac{\log_b x}{\log_b a} \qquad \text{Substitute } \log_a x \text{ for } y.$$

> **Rule Change-of-Base Formula**
>
> In general, if $a > 0$, $a \neq 1$, $b > 0$, $b \neq 1$, and $x > 0$, then $\log_a x = \dfrac{\log_b x}{\log_b a}$.
>
> In terms of common and natural logarithms, $\log_a x = \dfrac{\log x}{\log a} = \dfrac{\ln x}{\ln a}$.

Example 8   Use the change-of-base formula to calculate $\log_5 19$. Round the answer to four decimal places.

Note We could have used ln rather than log.

$$\log_5 19 = \frac{\ln 19}{\ln 5} \approx 1.8295$$

Solution: $\log_5 19 = \dfrac{\log 19}{\log 5} \approx 1.8295$ Use $\log_a x = \dfrac{\log_b x}{\log_b a}$; then evaluate using a calculator.

Check: $5^{1.8295} = 19.0005 \approx 19$, so the answer is correct.

Your Turn 8   Find $\log_6 25$. Round the answer to four decimal places.

Example 9   Use $P = 95 - 30 \log_2 x$ and the change-of-base formula to find the percent of students who retain the main points of a lecture after 5 days.

Solution: $P = 95 - 30 \log_2 5$ Substitute 5 for x in $P = 95 - 30 \log_2 x$.

$$P = 95 - 30 \frac{\ln 5}{\ln 2} \qquad \text{Use } \log_a x = \frac{\log_b x}{\log_b a}.$$

$$P \approx 95 - 30(2.3219) \qquad \text{Evaluate } \frac{\ln 5}{\ln 2} \text{ using a calculator.}$$

$$P \approx 25.34 \qquad \text{Simplify.}$$

About 25% of the students remember the main points of a lecture 5 days later.

Answer to Your Turn 8
1.7965

Answer to Your Turn 9
5%

Your Turn 9   Use the formula from Example 9 to find the percent of students who retain the main points of a lecture 8 days later.

11.6 Exercises For Extra Help MyMathLab®

Objective 1

Prep Exercise 1 If $b^x = b^y$, then ____$x = y$____.

Prep Exercise 2 When solving $10^{x+2} = 45$, would natural or common logarithms be the better choice? Why?

Common logs because the base is 10

Prep Exercise 3 What principle is used to solve $100 = (5)^{2n}$?

If $x = y$, then $\log_b x = \log_b y$.

For Exercises 1–12, solve. Round your answers to four decimal places. See Example 1.

1. $2^x = 9$
3.1699

2. $3^x = 20$
2.7268

3. $5^{2x} = 32$
1.0767

4. $6^{2x} = 48$
1.0803

5. $5^{x+3} = 10$
−1.5693

6. $6^{x+4} = 38$
−1.9698

7. $8^{x-2} = 6$
2.8617

8. $5^{x-3} = 12$
4.5440

9. $4^{x+2} = 5^x$
12.4251

10. $6^{x+4} = 10^x$
14.0303

11. $2^{x+1} = 3^{x-2}$
7.1285

12. $5^{x-3} = 3^{x+1}$
11.6026

For Exercises 13–20, solve. Round your answers to four decimal places. See Example 2.

13. $e^{3x} = 5$
0.5365

14. $e^{2x} = 7$
0.9730

15. $e^{0.03x} = 25$
107.2959

16. $e^{0.07x} = 32$
49.5105

17. $e^{-0.022x} = 5$
−73.1563

18. $e^{-0.032x} = 8$
−64.9825

19. $\ln e^{4x} = 24$
6

20. $\ln e^{5x} = 35$
7

Objective 2

Prep Exercise 4 If $\log_a m = \log_a n$, then ____$m = n$____.

Prep Exercise 5 What principle is used to solve $\log(x - 3) = \log(3x - 13)$?

If $\log_b x = \log_b y$, then $x = y$.

Prep Exercise 6 In solving $\log x + \log(x + 2) = \log 15$, we get possible solutions of $x = 3$ and $x = -5$. Why must $x = -5$ be rejected?

The solution $x = -5$ is rejected because substituting into the equation gives $\log(-5)$ and $\log(-3)$, which are both undefined.

For Exercises 21–54, solve. Give exact answers. See Example 3.

21. $\log_4(x + 5) = 2$
11

22. $\log_3(x + 4) = 2$
5

23. $\log_4(4x - 8) = 2$
6

24. $\log_5(3x + 7) = 2$
6

25. $\log_4 x^2 = 2$
±4

26. $\log_2 x^2 = 6$
±8

27. $\log_6(x^2 + 5x) = 2$
4, −9

28. $\log_4(x^2 + 6x) = 2$
−8, 2

29. $\log(4x - 3) = \log(3x + 4)$
7

30. $\log(4x + 1) = \log(2x + 7)$
3

31. $\ln(3x + 4) = \ln(x - 6)$
No solution

32. $\ln(5x + 6) = \ln(3x - 8)$
No solution

33. $\log_9(x^2 + 4x) = \log_9 12$
−6, 2

34. $\log_8(x^2 + x) = \log_8 30$
−6, 5

35. $\log_4 x + \log_4 8 = 2$
2

36. $\log_6 x + \log_6 4 = 2$
9

37. $\log_2 x - \log_2 5 = 1$
10

38. $\log_5 x - \log_5 3 = 2$
75

39. $\log_3 x + \log_3(x + 6) = 3$
3

40. $\log_2 x + \log_2(x - 3) = 2$
4

41. $\log_3(2x + 15) + \log_3 x = 3$
$\frac{3}{2}$

42. $\log_2(3x - 2) + \log_2 x = 4$
$\frac{8}{3}$

43. $\log_2(7x + 3) - \log_2(2x - 3) = 3$
3

44. $\log_3(3x + 3) - \log_3(x - 3) = 2$
5

45. $\log_2(3x + 8) - \log_2(x + 1) = 2$
4

46. $\log_3(2x + 1) - \log_3(x - 1) = 1$
4

47. $\log_8 2x + \log_8 6 = \log_8 10$
$\frac{5}{6}$

48. $\log_5 3x + \log_5 2 = \log_5 4$
$\frac{2}{3}$

49. $\ln x + \ln(2x - 1) = \ln 10$
$\frac{5}{2}$

50. $\ln x + \ln(3x - 5) = \ln 12$
3

51. $\log x - \log(x - 5) = \log 6$
6

52. $\log x - \log(x - 2) = \log 3$
3

53. $\log_6(3x + 4) - \log_6(x - 2) = \log_6 8$
4

54. $\log_7(5x + 2) - \log_7(x - 2) = \log_7 9$
5

Objective 3

📱 *For Exercises 55–74, solve. See Examples 4–7.*

55. How long will it take $5000 invested at 5% compounded quarterly to grow to $8000? Round your answer to the nearest tenth of a year. 9.5 yr.

56. How long will it take $7000 invested at 6% compounded monthly to grow to $12,000? Round your answer to the nearest tenth of a year. 9.0 yr.

57. Assume that $8000 is deposited into an account at 6% annual interest compounded continuously.
 a. How much money will be in the account after 15 years? $19,676.82
 b. How long will it take the $8000 to grow to $14,000? ≈9.3 yr.

58. Assume that $4000 is deposited into an account at 5% annual interest compounded continuously.
 a. How much money will be in the account after 10 years? $6594.89
 b. How long will it take the $4000 to grow to $10,000? ≈18.3 yr.

59. The probability that a person will have an accident while driving at a given blood alcohol level is approximated by $P(b) = e^{21.5b}$, where b is the blood alcohol level ($0 \leq b \leq 0.4$) and P is the percent probability of having an accident.
 a. What is the probability of an accident if the blood alcohol level is 0.08, which is legally drunk in many states? ≈5.58%
 b. Estimate the blood alcohol level when the probability of an accident is 50%. ≈0.18

60. Atmospheric pressure (in pounds per square inch, psi) is a function of the altitude above sea level and can be modeled by $P(a) = 14.7e^{-0.21a}$, where P is the pressure and a is the altitude above sea level in miles.
 a. Find the atmospheric pressure at the peak of Mount McKinley, Alaska, which is 3.85 miles above sea level. 6.55 psi
 b. If the atmospheric pressure at the peak of Mount Everest in Nepal is 4.68 pounds per square inch, find the height of Mount Everest. 5.45 mi. above sea level

61. The population of a mosquito colony increases at a rate of 4% per day. If the initial number of mosquitoes is 500, the number present, A, at the end of t days is given by $A = 500e^{0.04t}$.
 a. How many mosquitoes are present after 2 weeks? ≈875 mosquitoes
 b. Find the number of days until 10,000 mosquitoes are present. ≈75 days

62. A cake is removed from the oven at a temperature of 210°F and is left to cool on a counter where the room temperature is 70°F. The cake cools according to the function $T = 70 + 140e^{-0.0231t}$, where T is the temperature of the cake and t is in minutes.
 a. What is the temperature of the cake after 40 minutes? ≈126°F
 b. After how many minutes will the cake's temperature be 75°F? ≈144 min.

63. The world population in 2010 was about 6.8 billion and is increasing at a rate of 1.1% per year. The world population after 2010 can be approximated by the equation $A = A_0 e^{0.011t}$, where A_0 is the world population in 2010 and t is the number of years after 2010.
   a. Assuming that the growth rate follows the same trend, find the world population in 2020. ≈7.59 billion
   b. In what year will the world population reach 7 billion? ≈2.6 yr. after 2010, in 2012

64. In 2012, Africa had a population of 1072 million and a natural growth rate of 2.3% (2.1 times the world's growth rate). The population growth can be approximated by $A = A_0 e^{0.023t}$, where A_0 is the population in millions in 2012 and t is the number of years after 2012.

   a. Excluding immigration, what will the population of Africa be in 2025? ≈1445.6 million
   b. Excluding immigration, in what year will Africa's population reach 1.2 billion? ≈4.9 yr. after 2012, in 2016

65. The barometric pressure x miles from the eye of a hurricane is approximated by the function $P = 0.48 \ln(x + 1)$, where P is inches of mercury. (*Source:* A. Miller and R. Anthes, *Meteorology*, 4th Edition, Merrill Publishing.)
   a. Find the barometric pressure 50 miles from the center of the hurricane. ≈1.89 in. of mercury
   b. Find the distance from the center where the pressure is 1.5 inches of mercury. ≈21.8 mi.

66. The first two-year college, Joliet Junior College in Chicago, was founded in 1901. The number of two-year colleges in the United States grew rapidly, especially during the 1960s, but growth has tapered off. The total number of two-year colleges in the United States since 1960 can be approximated by the function $y = 175.6 \ln x + 513$, where x is the number of years after 1960 and y is the number of two-year colleges. Use the model to estimate the number of two-year colleges in the United States in 2015. (*Source:* American Association of Two-Year Colleges.) ≈1217 colleges

67. The percent, $f(x)$, of adult height attained by a boy who is x years old is modeled by $f(x) = 29 + 48.8 \log(x + 1)$, where $5 \le x \le 15$.
   a. At age 10, about what percent of his adult height has a boy reached? ≈79.8%
   b. At what age will a boy attain 75% of his adult height? ≈7.76, so about age 8

68. The annual depreciation rate r of a car purchased for P dollars and worth A dollars after t years can be found by the formula $\log(1 - r) = \frac{1}{t} \log \frac{A}{P}$. Find the depreciation rate of a car that is purchased for $22,500 and is sold 3 years later for $10,000. ≈0.24, or 24%

69. The following table shows the purchasing power of $1.00 in 1970 for subsequent years. For example, in 1980, $2.12 had the same purchasing power as $1.00 in 1970.

Year	Purchasing Power ($)
1970	1.00
1975	1.39
1980	2.12
1985	2.77
1990	3.37
1995	3.92
2000	4.43
2005	5.03
2010	5.61

(*Source:* www.measuringworth.com.)

The data can be approximated by the exponential function $y = 1.244(1.043)^x$, where y is the purchasing power of $1.00 and x is the number of years after 1970. Assuming that the trend continues, find the approximate number of dollars in 2020 that it will take to have the same purchasing power as $1.00 had in 1970. ≈$10.21

70. The following table shows the U.S. public debt, in trillions of dollars, for selected years.

Year	Public Debt (in trillions)
1990	3.233
1992	4.065
1994	4.693
1996	5.224
1998	5.526
2000	5.674
2002	6.228
2004	7.379
2006	8.507
2008	10.025
2010	13.562

(*Source: U.S. Department of the Treasury*)

The data can be approximated by the exponential function $y = 3.385(1.063)^x$, where y is the public debt (in trillions) and x is the number of years after 1990. Assuming that the trend continues, what will be the approximate public debt in 2020? ≈ $21.162 trillion

71. Since 1985, the amount spent on recreation in the United States can be approximated by the exponential equation $y = 78.62(1.092)^x$, where y is the amount spent in billions of dollars and x is the number of years after 1985. Assuming that the trend continues, what is the approximate amount that will be spent on recreation in 2018. (*Source:* U.S. Department of Commerce.) ≈ $1435.1 billion

72. Since 1980, the median income for males in the United States can be approximated by the exponential equation $y = 13192.9(1.0369)^x$, where y is the median income and x is the number of years after 1980. Assuming that the trend continues, find the approximate median income for a male in the United States in 2018. (*Source: U.S. Department of Commerce*) ≈ $52,280.71

Objective 4

For Exercises 73–80, use the change-of-base formula to find the logarithms. Round your answers to four decimal places. See Example 8.

73. $\log_4 12$ 1.7925

74. $\log_5 23$ 1.9482

75. $\log_8 3$ 0.5283

76. $\log_9 5$ 0.7325

77. $\log_{1/2} 5$ −2.3219

78. $\log_{1/3} 4$ −1.2619

79. $\log_{1/4} \frac{3}{5}$ 0.3685

80. $\log_{1/3} \frac{4}{7}$ 0.5094

For Exercises 81 and 82, use $P = 95 - 30\log_2 x$ and the change-of-base formula to find the percentage, P, of the lecture retained after the given number of days. Round your answer to the nearest whole percent. See Example 9.

81. 3 days 47%

82. 7 days 11%

Review Exercises

Exercises 1–6 ◢ **Expressions**

For Exercises 1 and 2, simplify.

[5.3] 1. $(3x + 2) - (2x - 1)$ $x + 3$

[9.4] 2. $(3\sqrt{5} + 2)(2\sqrt{5} - 1)$ $28 + \sqrt{5}$

For Exercises 3–6, use $f(x) = x^2 + 4$ and $g(x) = 2x - 1$.

[3.7] 3. Find $f(-5)$. 29

[10.6] 4. Find $(f + g)(x)$. $x^2 + 2x + 3$

[11.1] 5. Find $(f \circ g)(x)$. $4x^2 - 4x + 5$

[11.1] 6. Find $(f \circ g)(2)$. 13

Chapter 11 Summary and Review Exercises

Complete each incomplete definition, rule, or procedure; study the key examples; and then work the related exercises.

11.1 Composite and Inverse Functions

Definitions/Rules/Procedures	Key Example(s)
Composition of Functions $(f \circ g)(x) = \underline{\ f[g(x)]\ }$ for all x in the domain $\underline{\ g\ }$ for which $g(x)$ is in the domain of $\underline{\ f\ }$. $(g \circ f)(x) = \underline{\ g[f(x)]\ }$ for all x in the domain of $\underline{\ f\ }$ for which $f(x)$ is in the domain of $\underline{\ g\ }$.	If $f(x) = x^2 - 3$ and $g(x) = 3x + 4$, find (a) $(f \circ g)(x)$ and (b) $(g \circ f)(x)$. **a.** $(f \circ g)(x) = f[g(x)]$ $ = f(3x + 4)$ Substitute $3x + 4$ for $g(x)$. $ = (3x + 4)^2 - 3$ Replace x with $3x + 4$ in $f(x)$. $ = 9x^2 + 24x + 13$ Simplify. **b.** $(g \circ f)(x) = g[f(x)]$ $ = g(x^2 - 3)$ Substitute $x^2 - 3$ for $f(x)$. $ = 3(x^2 - 3) + 4$ Replace x with $x^2 - 3$ in $g(x)$. $ = 3x^2 - 5$ Simplify.

Exercises 1–8 ▲▲▲, Equations and Inequalities

[11.1] *For Exercises 1–4, find each composition if* $f(x) = 3x + 4$ *and* $g(x) = x^2 - 2$.

1. $(f \circ g)(3)$
25

2. $(g \circ f)(3)$
167

3. $f[g(0)]$
−2

4. $g[f(0)]$
14

[11.1] *For Exercises 5–8, find* $(f \circ g)(x)$ *and* $(g \circ f)(x)$.

5. $f(x) = 3x - 6, g(x) = 2x + 3$
$(f \circ g)(x) = 6x + 3, (g \circ f)(x) = 6x - 9$

6. $f(x) = x^2 + 4, g(x) = 3x - 7$
$(f \circ g)(x) = 9x^2 - 42x + 53, (g \circ f)(x) = 3x^2 + 5$

7. $f(x) = \sqrt{x - 3}, g(x) = 2x - 1$
$(f \circ g)(x) = \sqrt{2x - 4}, (g \circ f)(x) = 2\sqrt{x - 3} - 1$

8. $f(x) = \dfrac{x + 3}{x}, g(x) = \dfrac{x - 4}{x}$
$(f \circ g)(x) = \dfrac{4x - 4}{x - 4}, (g \circ f)(x) = \dfrac{3 - 3x}{x + 3}$

Definitions/Rules/Procedures	Key Example(s)
To determine whether two functions f and g are **inverses** of each other: **1.** Show that $f[g(x)] = \underline{\ x\ }$ for all x in the domain of $\underline{\ g\ }$. **2.** Show that $g[f(x)] = \underline{\ x\ }$ for all x in the domain of $\underline{\ f\ }$.	Show that $f(x) = 5x - 6$ and $g(x) = \dfrac{x + 6}{5}$ are inverse functions. $f[g(x)] = f\left(\dfrac{x + 6}{5}\right)$ $ = 5\left(\dfrac{x + 6}{5}\right) - 6$ $ = x + 6 - 6 = x$ $g[f(x)] = g(5x - 6)$ $ = \dfrac{5x - 6 + 6}{5}$ $ = \dfrac{5x}{5} = x$ Because $f[g(x)] = g[f(x)] = x, f$ and g are inverse functions.

Exercises 9–12   Equations and Inequalities

[11.1] *For Exercises 9–12, determine whether f and g are inverse functions.*

9. $f(x) = 3x + 2, g(x) = \dfrac{x - 2}{3}$

     yes

10. $f(x) = x^3 + 6, g(x) = \sqrt[3]{x - 6}$

     yes

11. $f(x) = x^2 - 3, x \geq 0; g(x) = \sqrt{x + 2}$

     no

12. $f(x) = \dfrac{x}{x + 4}, g(x) = \dfrac{-4x}{x + 1}$

     no

Definitions/Rules/Procedures	Key Example(s)
A function f is **one-to-one** if for any two numbers a and b in its domain, when $f(a) = f(b)$, $\underline{a = b}$ and when $a \neq b$, $\underline{f(a) \neq f(b)}$. Given a function's graph, the function is one-to-one if every horizontal line that can intersect the graph does so at $\underline{\text{one and only one point}}$. A function has an inverse function if and only if the function is $\underline{\text{one-to-one}}$. The graphs of f and f^{-1} are symmetric with respect to the graph of $\underline{y = x}$.	Determine whether the functions whose graphs follow are one-to-one. a. 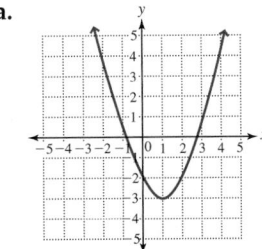 b. 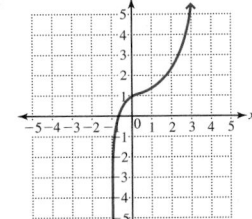 **Solution:** A horizontal line can intersect this graph in more than one point, so the function is not one-to-one. **Solution:** Every horizontal line that can intersect this graph does so at one and only one point, so the function is one-to-one. Sketch the inverse of the function whose graph follows. Solution: Draw the line $y = x$ and reflect the graph in the line. 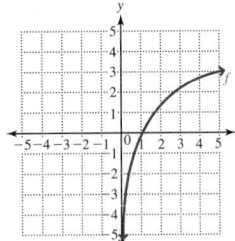 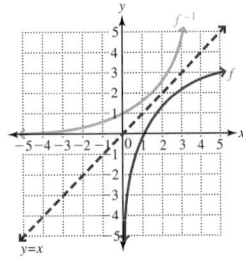

Exercises 13 and 14   Equations and Inequalities

[11.1] *For Exercises 13 and 14, determine whether each is the graph of a one-to-one function. If the function is one-to-one, sketch the graph of the inverse function.*

13.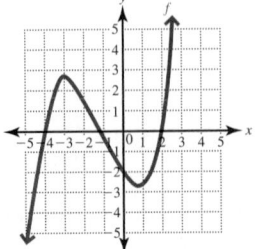

     no

14.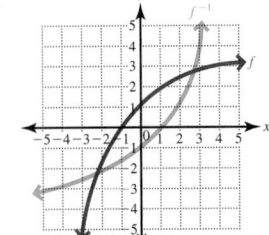

     yes

Definitions/Rules/Procedures	Key Example(s)
Finding the **inverse** of a one-to-one function 1. If necessary, replace $f(x)$ with _____y_____. 2. Replace all x's with _____y's_____ and all y's with _____x's_____. 3. Solve the equation from step 2 for _____y_____. 4. Replace y with _____$f^{-1}(x)$_____.	If $f(x) = 3x - 7$, find $f^{-1}(x)$. **Solution:** $f(x) = 3x - 7$ $\qquad y = 3x - 7$ Replace $f(x)$ with y. $\qquad x = 3y - 7$ Interchange x and y. $\qquad \dfrac{x+7}{3} = y$ Solve for y. $\qquad f^{-1}(x) = \dfrac{x+7}{3}$ Replace y with $f^{-1}(x)$. Verify by showing that $f[f^{-1}(x)] = x$ and $f^{-1}[f(x)] = x$.

Exercises 15–20 ▲ Equations and Inequalities

[11.1] *For Exercises 15–18, find $f^{-1}(x)$ for each of the following one-to-one functions.*

15. $f(x) = 5x + 4$
$f^{-1}(x) = \dfrac{x-4}{5}$

16. $f(x) = x^3 + 6$
$f^{-1}(x) = \sqrt[3]{x - 6}$

17. $f(x) = \dfrac{4}{x + 5}$
$f^{-1}(x) = \dfrac{4 - 5x}{x}$

18. $f(x) = \sqrt[3]{3x + 2}$
$f^{-1}(x) = \dfrac{x^3 - 2}{3}$

19. If $(4, -6)$ is an ordered pair on the graph of f, what ordered pair is on the graph of f^{-1}?
$(-6, 4)$

20. Fill in the blanks. If $f(a) = b$, then $f^{-1}(\rule{1cm}{0.4pt}) = \rule{1cm}{0.4pt}$.
$f^{-1}(b) = a$

11.2 Exponential Functions

Definitions/Rules/Procedures	Key Example(s)
Graphs of **exponential functions** 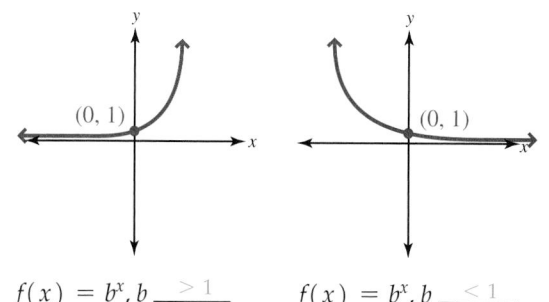 $f(x) = b^x, b \underline{\;>1\;}$ $f(x) = b^x, b \underline{\;<1\;}$	Graph $f(x) = 2^x$. Choose some values for x and find the corresponding values of $f(x)$, which are the y-values. | x	-2	-1	0	1	2 | |---|---|---|---|---|---|
$f(x)$	$\dfrac{1}{4}$	$\dfrac{1}{2}$	1	2	4	 Plot the points and draw the graph. 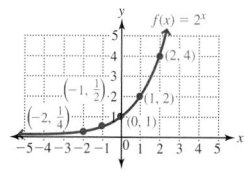

Exercises 21–24 Equations and Inequalities

[11.2] *For Exercises 21–24, graph.*

21. $f(x) = 3^x$

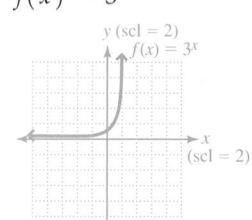

22. $f(x) = \left(\dfrac{1}{2}\right)^x$

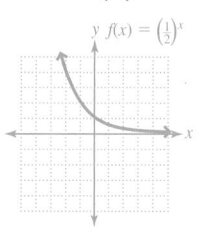

23. $f(x) = 2^{x-3}$

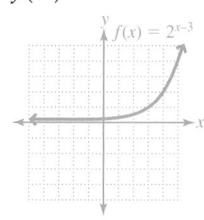

24. $f(x) = 3^{-x+2}$

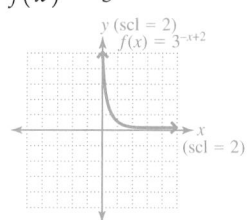

Definitions/Rules/Procedures	Key Example(s)
Given $b > 0$ and $b \neq 1$, if $b^x = b^y$, then ___$x = y$___.	Solve $8^{x-3} = 16^{2x}$.
Solving **exponential equations**	**Solution:** $8^{x-3} = 16^{2x}$
1. If necessary, write both sides of the equation as a power of the ___same base___.	$\left(2^3\right)^{x-3} = \left(2^4\right)^{2x}$ Rewrite 8 and 16 as powers of 2.
	$2^{3x-9} = 2^{8x}$ Multiply exponents.
2. If necessary, simplify the ___exponents___.	$3x - 9 = 8x$ Set exponents equal.
3. Set the ___exponents___ equal to each other.	$-\dfrac{9}{5} = x$ Solve for x.
4. Solve the resulting equation.	

Exercises 25–36 Equations and Inequalities

[11.2] *For Exercises 25–32, solve each equation.*

25. $5^x = 625$

4

26. $16^x = 64$

$\dfrac{3}{2}$

27. $6^x = \dfrac{1}{36}$

-2

28. $\left(\dfrac{3}{4}\right)^x = \dfrac{16}{9}$

-2

29. $4^{x-1} = 64$

4

30. $5^{x+2} = 25^x$

2

31. $\left(\dfrac{1}{3}\right)^{-x} = 27$

3

32. $8^{3x-2} = 16^{4x}$

$-\dfrac{6}{7}$

33. The median doubling time for a malignant tumor is about 100 days. If there are 500 cells initially, then the number of cells, A, after t days is given by $A = A_0 2^{t/100}$. How many cells are present after one year?

≈ 6277 cells

34. If \$25,000 is deposited into an account paying 6% interest compounded monthly, how much will be in the account after eight years?

\$40,353.57

35. The radioactive isotope ^{82}R has a half-life of 107 days. How much of a 50-gram sample remains after 300 days? $\left(\text{Use } A = A_0 \left(\dfrac{1}{2}\right)^{t/h}.\right)$

≈ 7.16 g

36. Since 1990, the hourly minimum wage can be approximated by the exponential equation $y = 4.0144\,(1.028)^x$, where y is the hourly minimum wage and x is the number of years after 1990. Find the approximate minimum wage in 2020.

$\approx \$9.19$

11.3 Logarithmic Functions

Definitions/Rules/Procedures	Key Example(s)
If $b > 0$ and $b \neq 1$, then $y = \log_b x$ is equivalent to $\underline{x = b^y}$.	Write in logarithmic form. **a.** $3^4 = 81$ **b.** $\left(\dfrac{1}{4}\right)^{-3} = 64$ **Solution:** $\log_3 81 = 4$ **Solution:** $\log_{1/4} 64 = -3$ Write in exponential form. **a.** $\log_5 125 = 3$ **b.** $\log_3 \dfrac{1}{27} = -3$ **Solution:** $5^3 = 125$ **Solution:** $3^{-3} = \dfrac{1}{27}$

Exercises 37–44   Expressions

[11.3] *For Exercises 37–40, write in logarithmic form.*

37. $7^3 = 343$
$\log_7 343 = 3$

38. $4^{-3} = \dfrac{1}{64}$
$\log_4 \dfrac{1}{64} = -3$

39. $\left(\dfrac{3}{2}\right)^4 = \dfrac{81}{16}$
$\log_{3/2} \dfrac{81}{16} = 4$

40. $11^{1/3} = \sqrt[3]{11}$
$\log_{11} \sqrt[3]{11} = \dfrac{1}{3}$

[11.3] *For Exercises 41–44, write in exponential form.*

41. $\log_9 81 = 2$
$9^2 = 81$

42. $\log_{1/5} 125 = -3$
$\left(\dfrac{1}{5}\right)^{-3} = 125$

43. $\log_a 16 = 4$
$a^4 = 16$

44. $\log_e c = b$
$e^b = c$

Definitions/Rules/Procedures	Key Example(s)
To solve an equation of the form $\log_b x = y$, where b, x, or y is a variable, write the equation in exponential form, $\underline{b^x = y}$, and then solve for the variable.	Solve. **a.** $\log_2 \dfrac{1}{16} = x$ **Solution:** $2^x = \dfrac{1}{16}$ Change to exponential form. $2^x = \dfrac{1}{2^4}$ $16 = 2^4$ $2^x = 2^{-4}$ $\dfrac{1}{2^4} = 2^{-4}$ $x = -4$ **b.** $\log_x 49 = 2$ **Solution:** $x^2 = 49$ Change to exponential form. $x = \pm 7$ The square roots of 49 are ± 7. $x = 7$ The base must be positive.
For any real number b, where $b > 0$ and $b \neq 1$: **1.** $\log_b b = \underline{\ 1\ }$. **2.** $\log_b 1 = \underline{\ 0\ }$.	Find the following logarithms. **a.** $\log_8 8$ Solution: $\log_8 8 = 1$ **b.** $\log_5 1$ Solution: $\log_5 1 = 0$

Exercises 45–52   Equations and Inequalities

[11.3] *For Exercises 45–52, solve.*

45. $\log_3 x = -4$
$\frac{1}{81}$

46. $\log_{1/2} x = -2$
4

47. $\log_2 32 = x$
5

48. $\log_{1/4} 16 = x$
-2

49. $\log_x 81 = 4$
3

50. $\log_x \frac{1}{1000} = 3$
$\frac{1}{10}$

51. $\log_{3/4} \frac{3}{4} = x$
1

52. $\log_{81} 1 = x$
0

Definitions/Rules/Procedures	Key Example(s)
To **graph** a function in the form $f(x) = \log_b x$: 1. Replace $f(x)$ with ____y____ and write the logarithm in exponential form ___$x = b^y$___. 2. Find ordered pairs that satisfy the equation by assigning values to ____y____ and finding ___x___. 3. Plot the ordered pairs and draw a smooth curve through the points.	Graph $y = \log_3 x$. **Solution:** Write as $3^y = x$ and assign values to y. 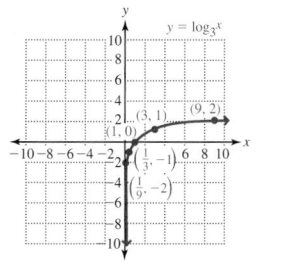

x	y
$\frac{1}{9}$	-2
$\frac{1}{3}$	-1
1	0
3	1
9	2

Exercises 53–56   Equations and Inequalities

[11.3] *For Exercises 53 and 54, graph.*

53. $f(x) = \log_4 x$

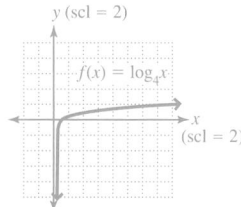

54. $f(x) = \log_{1/3} x$

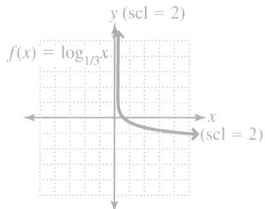

55. If $f(x) = \log_b x$, the domain is $(0, \infty)$, and the range is $(-\infty, \infty)$, what are the domain and range of $g(x) = b^x$? Why?

Domain is $(-\infty, \infty)$ and range is $(0, \infty)$ because $f(x) = \log_b x$ and $g(x) = b^x$ are inverses.

56. The formula for the number of decibels in a sound is $d = 10 \log \frac{I}{I_0}$. Find the decibel reading of a sound whose intensity is $I = 1000 I_0$.

30 dB

11.4 Properties of Logarithms

Definitions/Rules/Procedures	Key Example(s)
For any real numbers b and x, where $b > 0$, $b \neq 1$, and $x > 0$: 1. $b^{\log_b x} = $ ____x____ 2. $\log_b b^x = $ ____x____	Find the value of each. **a.** $6^{\log_6 5}$ Solution: $6^{\log_6 5} = 5$ **b.** $\log_8 8^a$ Solution: $\log_8 8^a = a$

Definitions/Rules/Procedures	Key Example(s)
For real numbers x, y, and b, where $x > 0$, $y > 0$, $b > 0$, and $b \neq 1$: Product Rule: $\log_b xy = \underline{\log_b x + \log_b y}$.	Use the product rule. **a.** Write $\log_4 4x$ as the sum of logarithms. **Solution:** $\log_4 4x = \log_4 4 + \log_4 x$ $\qquad\qquad\quad = 1 + \log_4 x \qquad \log_4 4 = 1$ **b.** Write $\log_7 4 + \log_7 2$ as a single logarithm. **Solution:** $\log_7 4 + \log_7 2 = \log_7 4 \cdot 2$ $\qquad\qquad\qquad\qquad = \log_7 8$
Quotient Rule: $\log_b \dfrac{x}{y} = \underline{\log_b x - \log_b y}$.	Use the quotient rule. **a.** Write $\log_a \dfrac{w}{7}$ as the difference of logarithms. **Solution:** $\log_a \dfrac{w}{7} = \log_a w - \log_a 7$ **b.** Write $\log_7(x + 5) - \log_7(x - 3)$ as a single logarithm. **Solution:** $\log_7(x + 5) - \log_7(x - 3) = \log_7 \dfrac{x + 5}{x - 3}$
Power Rule: $\log_b x^r = \underline{r \log_b x}$.	Use the power rule. **a.** Write $\log_b \sqrt[5]{x^3}$ as a multiple of a logarithm. **Solution:** $\log_b \sqrt[5]{x^3} = \log_b x^{3/5} = \dfrac{3}{5} \log_b x$ **b.** Write $-3 \log_2 y$ as a logarithm of a quantity to a power. **Solution:** $-3 \log_2 y = \log_2 y^{-3}$ $\qquad\qquad\qquad = \log_2 \dfrac{1}{y^3}$ Write $\log_a \dfrac{x^2 y}{z^4}$ as the sum or difference of multiples of logarithms. **Solution:** $\log_a \dfrac{x^2 y}{z^4} = \log_a x^2 y - \log_a z^4 \qquad$ Quotient rule $\qquad\qquad = \log_a x^2 + \log_a y - \log_a z^4 \qquad$ Product rule $\qquad\qquad = 2 \log_a x + \log_a y - 4 \log_a z \qquad$ Power rule

Definitions/Rules/Procedures	Key Example(s)
	Write $\frac{1}{4}(2\log_5 x - 3\log_5 y)$ as a single logarithm. **Solution:** $\frac{1}{4}(2\log_5 x - 3\log_5 y)$ $= \frac{1}{4}(\log_5 x^2 - \log_5 y^3)$ Power rule $= \frac{1}{4}\log_5\frac{x^2}{y^3}$ Quotient rule $= \log_5\left(\frac{x^2}{y^3}\right)^{1/4}$ Power rule $= \log_5\sqrt[4]{\frac{x^2}{y^3}}$

Exercises 57–80 ◣ Expressions

[11.4] *For Exercises 57 and 58, find the value.*

57. $3^{\log_3 8}$
8

58. $\log_4 4^6$
6

[11.4] *For Exercises 59 and 60, use the product rule to write the expression as a sum of logarithms.*

59. $\log_6 6x$
$1 + \log_6 x$

60. $\log_4 x(2x - 5)$
$\log_4 x + \log_4(2x - 5)$

[11.4] *For Exercises 61 and 62, use the product rule to write the expression as a single logarithm.*

61. $\log_3 4 + \log_3 8$
$\log_3(32)$

62. $\log_5 3 + \log_5 x + \log_5(x - 2)$
$\log_5(3x^2 - 6x)$

[11.4] *For Exercises 63 and 64, use the quotient rule to write the expression as a difference of logarithms.*

63. $\log_b \frac{x}{5}$
$\log_b x - \log_b 5$

64. $\log_a \frac{3x - 2}{4x + 3}$
$\log_a(3x - 2) - \log_a(4x + 3)$

[11.4] *For Exercises 65 and 66, use the quotient rule to write the expression as a single logarithm.*

65. $\log_8 32 - \log_8 16$
$\log_8 2$

66. $\log_2(x + 5) - \log_2(2x - 3)$
$\log_2 \frac{x + 5}{2x - 3}$

[11.4] *For Exercises 67–70, use the power rule to write the expression as a multiple of a logarithm.*

67. $\log_3 7^4$
$4\log_3 7$

68. $\log_a \sqrt[3]{x}$
$\frac{1}{3}\log_a x$

69. $\log_4 \frac{1}{a^4}$
$-4\log_4 a$

70. $\log_a \sqrt[5]{a^4}$
$\frac{4}{5}$

[11.4] *For Exercises 71 and 72, use the power rule to write the expression as the logarithm of a quantity to a power. Simplify the answer if possible.*

71. $4\log_a x$
$\log_a x^4$

72. $\frac{3}{5}\log_a y$
$\log_a \sqrt[5]{y^3}$

[11.4] *For Exercises 73–76, write the expression as the sum or differences of multiples of logarithms.*

73. $\log_a x^2 y^3$

$2 \log_a x + 3 \log_a y$

74. $\log_a \dfrac{c^4}{d^3}$

$4 \log_a c - 3 \log_a d$

75. $\log_a \dfrac{x^2 y^3}{z^4}$

$2 \log_a x + 3 \log_a y - 4 \log_a z$

76. $\log_a \sqrt{\dfrac{a^3}{b^4}}$

$\dfrac{3}{2} - 2 \log_a b$

[11.4] *For Exercises 77–80, write the expression as a single logarithm.*

77. $3 \log_x y + 5 \log_x z$

$\log_x y^3 z^5$

78. $3 \log_a 4 - 2 \log_a 3$

$\log_a \dfrac{4^3}{3^2}$

79. $\dfrac{1}{4} (2 \log_a x + 3 \log_a y)$

$\log_a \sqrt[4]{x^2 y^3}$

80. $4 \log_a (x + 5) + 2 \log_a (x - 3)$

$\log_a (x + 5)^4 (x - 3)^2$

11.5 Common and Natural Logarithms

Definitions/Rules/Procedures	Key Example(s)
Base-10 logarithms are called ___common___ logarithms, and $\log_{10} x$ is written as ___$\log x$___.	Evaluate using a calculator and round to four decimal places.
Common logarithms can be evaluated using the ___ LOG ___ key on a calculator.	**a.** $\log 356$ **Solution:** $\log 356 \approx 2.5514$
Base-e logarithms are called ___natural___ logarithms, and $\log_e x$ is written as ___$\ln x$___.	**b.** $\log 0.0059$ **Solution:** $\log 0.0059 \approx -2.2291$
Natural logarithms can be evaluated using the ___ ln ___ key on a calculator.	**c.** $\ln 72$ **Solution:** $\ln 72 \approx 4.2767$
	d. $\ln 0.097$ **Solution:** $\ln 0.097 \approx -2.3330$

Exercises 81–86 ▰▰▸ Expressions

[11.5] *For Exercises 81–84, use a calculator to approximate each logarithm to four decimal places.*

81. $\log 326$

2.5132

82. $\log 0.0035$

-2.4559

83. $\ln 0.043$

-3.1466

84. $\ln 92$

4.5218

[11.5] *For Exercises 85 and 86, find the exact value without using a calculator.*

85. $\log 0.00001$

-5

86. $\ln \sqrt[4]{e}$

$\dfrac{1}{4}$

Exercises 87–90 ▰▰▸ Equations and Inequalities

87. The sound intensity of a clap of thunder was $10^{-3.5}$ watts per square meter. What was the decibel reading? $\left(\text{Use } d = 10 \log \dfrac{I}{I_0}, \text{ where } I_0 = 10^{-12} \text{watts/m}^2. \right)$

85 dB

88. Using $pH = -\log [H_3O^+]$ (the hydronium ion concentration), find the pH of an apple whose $[H_3O^+]$ is 0.001259.

≈ 2.9

89. How long will it take $6000 to grow to $10,000 if it is invested at 3% annual interest compounded continuously? (Use $A = Pe^{rt}$.)

≈ 17 yr.

90. Using $R = \log \dfrac{I}{I_0}$ for Richter scale readings, compare the intensity of an earthquake whose Richter scale reading was 7.8 with one whose reading was 6.8. What do you notice? (*Hint:* Solve for I in terms of I_0.)

The 7.8 earthquake was 10 times as severe.

11.6 Exponential and Logarithmic Equations with Applications

Definitions/Rules/Procedures	Key Example(s)
For any real numbers b, x, and y, where $b > 0$ and $b \neq 1$: **1.** If $b^x = b^y$, then ___$x = y$___. **2.** If $x = y$, then $b^x =$ ___b^y___. **3.** For $x > 0$ and $y > 0$, if $\log_b x = \log_b y$, then ___$x = y$___. **4.** For $x > 0$ and $y > 0$, if $x = y$, then $\log_b x =$ ___$\log_b y$___. **5.** For $x > 0$, if $\log_b x = y$, then ___b^y___ $= x$.	Solve $3^x = 4$. **Solution:** $\begin{aligned} \log 3^x &= \log 4 & &\text{Use property 4.} \\ x \log 3 &= \log 4 & &\text{Power rule} \\ x &= \frac{\log 4}{\log 3} & &\text{Divide by log 3.} \\ x &\approx 1.2619 & &\text{Evaluate.} \end{aligned}$ Solve $e^{3x} = 12$. **Solution:** Because the base is e, take the natural logarithm of both sides. $\begin{aligned} \ln e^{3x} &= \ln 12 & &\text{Use property 4.} \\ 3x &= \ln 12 & &\text{Use } \log_b b^x = x. \\ x &= \frac{\ln 12}{3} \approx 0.8283 & &\text{Divide by 3 and approximate.} \end{aligned}$ How long will it take \$12,000 invested at 4% compounded quarterly to grow to \$15,000? **Solution:** Use $A = P\left(1 + \dfrac{r}{n}\right)^{nt}$ with $A = 15{,}000$, $P = 12{,}000$, $r = 4\% = 0.04$, and $n = 4$. Find t. $\begin{aligned} 15{,}000 &= 12{,}000\left(1 + \frac{0.04}{4}\right)^{4t} & &\text{Substitute.} \\ \frac{5}{4} &= (1.01)^{4t} & &\text{Divide by 12,000.} \\ \log\frac{5}{4} &= \log 1.01^{4t} & &\text{Use property 4.} \\ \log\frac{5}{4} &= 4t \log 1.01 & &\text{Power rule} \\ \frac{\log\frac{5}{4}}{4 \log 1.01} &= t & &\text{Divide by 4 log 1.01.} \\ 5.61 &\approx t & &\text{Evaluate.} \end{aligned}$ It will take about 5.6 years. If \$3000 is invested in an account paying 4.5% compounded continuously, how much will be in the account after 10 years? **Solution:** Use $A = Pe^{rt}$ with $P = 3000$, $r = 0.045$, and $t = 10$. $\begin{aligned} A &= Pe^{rt} \\ A &= 3000e^{(0.045)(10)} & &\text{Substitute.} \\ A &= 3000e^{0.45} & &10(0.045) = 0.45 \\ A &\approx 4704.94 & &\text{Evaluate.} \end{aligned}$ There will be \$4704.94 in the account.

Definitions/Rules/Procedures	Key Example(s)
Exponential growth and decay can be represented by the equation $A = \underline{\quad A_0e^{kt} \quad}$, where A is the amount present, A_0 is the $\underline{\quad \text{initial} \quad}$ amount, t is the time, and k is a constant that is determined by the substance. If $k > 0$, there is exponential $\underline{\quad \text{growth} \quad}$, and if $k < 0$, there is exponential $\underline{\quad \text{decay} \quad}$. The half-life of a substance is the amount of time until only one-half of the original amount is present.	The element bismuth has an isotope, ^{200}Bi, that disintegrates according to the formula $A = A_0e^{-0.0198t}$, where t is in minutes. Find the half-life of ^{200}Bi. **Solution:** After one half-life, $\frac{1}{2}A_0$ of the original A_0 grams remains. $A = A_0e^{-0.0198t}$ $\frac{1}{2}A_0 = A_0e^{-0.0198t}$ Substitute $\frac{1}{2}A_0$ for A. $\frac{1}{2} = e^{-0.0198t}$ Divide by A_0. $\ln\frac{1}{2} = \ln e^{-0.0198t}$ Use if $x = y$; then $\log_b x = \log_b y$. $\ln\frac{1}{2} = -0.0198t$ Use $\log_b b^x = x$. $\dfrac{\ln\frac{1}{2}}{-0.0198} = t$ Divide by -0.0198. $35 \approx t$ Evaluate. The half-life is approximately 35 minutes.

Exercises 91–100   Equations and Inequalities

[11.6] *For Exercises 91–96, solve. Round your answers to four decimal places.*

91. $9^x = 32$
1.5773

92. $3^{5x} = 19$
0.5360

93. $6^{2x-1} = 22$
1.3626

94. $4^{2x-3} = 5^{x+1}$
4.9592

95. $e^{4x} = 11$
0.5995

96. $e^{-0.003x} = 5$
−536.4793

97. How long will it take $15,000 invested at 3% compounded monthly to grow to $18,000? Round your answer to the nearest tenth of a year.
$$\left(\text{Use } A = P\left(1 + \frac{r}{n}\right)^{nt}.\right)$$
6.1 yr.

98. Assume that $7000 is deposited at 7% annual interest compounded continuously
  a. How much will be in the account after eight years? (Use $A = Pe^{rt}$.)
    $12,254.71
  b. How long will it take until $12,000 is in the account?
    ≈ 7.7 yr.

99. The population of an ant colony is 400 and is increasing at the rate of 3% per month. Use $A = A_0e^{0.03t}$ to answer the following questions.
  a. How many ants will be in the colony after one year?
    ≈ 573 ants
  b. After how many months will 800 ants be in the colony?
    ≈ 23 months

100. Bacteria reproduce by cell division, and the amount of time required for the cells to divide is called the generation time, G. The generation time is found using the equation $G = \dfrac{t}{3.3 \log \dfrac{b}{B}}$, where B is the number of bacteria at the beginning of the time interval, b is the number of bacteria at the end of the time interval, and t is the time interval in minutes. Find the generation time of a bacteria population that increases from 1000 to 1,000,000 cells in 4 hours. (*Source: Todar's Online Textbook of Bacteriology.*)
  ≈ 24 min.

Definitions/Rules/Procedures	Key Example(s)
To solve **equations containing logarithms**, use the properties of logarithms to simplify each side of the equation and then use one of the following.	Solve $\log_3 x + \log_3(x - 3) = \log_3 10$.
If the simplification results in an equation in the form $\log_b x = \log_b y$, use the fact that ____ $x = y$ ____ and then solve for the variable.	**Solution:** $\log_3 x(x - 3) = \log_3 10$ Product rule $x(x - 3) = 10$ Use property 3. $x^2 - 3x - 10 = 0$ Multiply; then subtract 10. $(x - 5)(x + 2) = 0$ Factor. $x - 5 = 0$ or $x + 2 = 0$ Use the zero-factor theorem. $x = 5$ or $x = -2$ Solve for x We must reject -2 because it results in $\log_3(-2)$ and $\log_3(-5)$ in the original equation, which do not exist.
If the simplification results in an equation in the form $\log_b x = y$, write the equation in exponential form, ____ $b^y = x$ ____, and then solve for the variable.	Solve $\log_2(2x + 2) - \log_2(x - 4) = 2$. **Solution:** $\log_2 \dfrac{2x + 2}{x - 4} = 2$ Quotient rule $\dfrac{2x + 2}{x - 4} = 2^2$ Use property 5. $\dfrac{2x + 2}{x - 4} = 4$ $2^2 = 4$ $2x + 2 = 4x - 16$ Multiply by $x - 4$. $18 = 2x$ Isolate the x term. $9 = x$ Solve for x.

Exercises 101–108 △ Equations and Inequalities

[11.6] *For Exercises 101–108, solve. Give exact answers.*

101. $\log_4(x + 8) = 2$
8

102. $\log_2(x^2 + 2x) = 3$
−4, 2

103. $\log(3x - 8) = \log(x - 2)$
3

104. $\log 5 + \log x = 2$
20

105. $\log_3 x - \log_3 4 = 2$
36

106. $\log_2 x + \log_2(x - 6) = 4$
8

107. $\log_4 x + \log_4(x + 2) = \log_4 8$
2

108. $\log_3(5x + 2) - \log_3(x - 2) = 2$
5

Definitions/Rules/Procedures	Key Example(s)
Change-of-Base Formula In general, if $a > 0, a \neq 1, b > 0, b \neq 1$, and $x > 0$, then $\log_a x = \dfrac{\log_b x}{\log_b a}$. In terms of common and natural logarithms, $\log_a x = \dfrac{\log x}{\log a} = \dfrac{\ln x}{\ln a}$.	Find $\log_5 16$. **Solution:** Use $\log_a x = \dfrac{\log x}{\log a}$ or $\dfrac{\ln x}{\ln a}$; then evaluate using a calculator. $\log_5 16 = \dfrac{\log 16}{\log 5} \approx 1.7227$ or $\log_5 16 = \dfrac{\ln 16}{\ln 5} \approx 1.7227$

Exercises 109 and 110 △ Expressions

[11.6] *For Exercises 109 and 110, use the change-of-base formula to approximate each logarithm to four decimal places.*

109. $\log_4 15$
1.9534

110. $\log_{1/2} 6$
−2.5850

Chapter 11 Practice Test

For Extra Help

Step-by-step test solutions are found on the Chapter Test Prep Videos available in MyMathLab® *or on* You Tube.

1. If $f(x) = x^2 - 6$ and $g(x) = 3x - 5$, find $f[g(x)]$.

   $f[g(x)] = 9x^2 - 30x + 19$ [11.1]

2. Use $f(x) = 4x - 3$ to answer parts a–c.
   a. Find $f^{-1}(x)$

      $f^{-1}(x) = \dfrac{x+3}{4}$ [11.1]

   b. Verify that $f[f^{-1}(x)] = x$ and $f^{-1}[f(x)] = x$.

      $f[f^{-1}(x)] = f\left[\dfrac{x+3}{4}\right] = 4\left(\dfrac{x+3}{4}\right) - 3 = x + 3 - 3 = x$

      $f^{-1}[f(x)] = f^{-1}(4x - 3) = \dfrac{4x - 3 + 3}{4} = \dfrac{4x}{4} = x$ [11.1]

   c. What is the relationship between the graphs of $f(x)$ and $f^{-1}(x)$?

      The graphs are symmetric about the graph of $y = x$. [11.1]

3. The graph of a function f is shown to the right. Graph f^{-1}.

   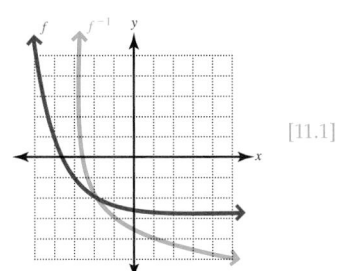

   [11.1]

4. Graph $f(x) = 2^{x-1}$.

   [11.2]

5. Solve $32^{x-2} = 8^{2x}$.

   -10 [11.2]

6. A sum of \$20,000 is invested at 5% compounded quarterly.
   a. Find the amount in the account after 15 years.

      \$42,143.63

   b. Find the number of years until \$30,000 is in the account.

      ≈ 8.2 yr. [11.2]

7. The isotope ^{98}Nb has a half-life of 30 minutes. Use
   $$A = A_0\left(\dfrac{1}{2}\right)^{t/n}.$$

   a. How much of a 100-gram sample remains after 3.6 hours?

      ≈ 0.68 g

   b. How long will it take until only 20 grams are remaining?

      ≈ 1.16 hr. [11.2]

8. The number of people in the United States aged 65 and over has increased rapidly since 1900 and can be approximated by $y = 3.17(1.026)^x$, where y is the number in millions aged 65 and older and x is the number of years after 1900. (*Source:* U.S. Bureau of the Census.)
   a. Find the number aged 65 and older in 1960 and in 2018.

      ≈ 14.8 million in 1960, ≈ 65.5 million in 2018

   b. In what year were 31.4 million people aged 65 and older?

      1989 [11.2]

9. Write $\log_{1/3}81 = -4$ in exponential form.

   $\left(\dfrac{1}{3}\right)^{-4} = 81$ [11.3]

For Exercises 10 and 11, solve. Give exact answers.

10. $\log_6 \dfrac{1}{216} = x$

    -3 [11.3]

11. $\log_x 625 = 4$

    5 [11.3]

12. Graph $f(x) = \log_2 x$.

    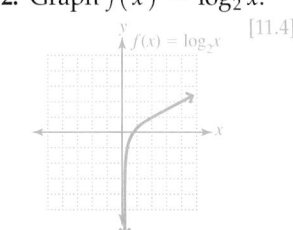

    [11.4]

For Exercises 13 and 14, write as the sum or difference of multiples of logarithms.

13. $\log_b \dfrac{x^4 y^2}{z}$

$4\log_b x + 2\log_b y - \log_b z$ [11.4]

14. $\log_b \sqrt[4]{\dfrac{x^5}{y^7}}$.

$\dfrac{5}{4}\log_b x - \dfrac{7}{4}\log_b y$ [11.4]

15. Write $\dfrac{3}{4}(2\log_b x + 3\log_b y)$ as a single logarithm.

$\log_b \sqrt[4]{x^6 y^9}$ [11.4]

16. The isotope ^{119}Sn disintegrates according to the function $A = A_0 e^{-0.0028t}$, where t is the time in days.
   a. How much of a 300-gram sample remains after 500 days?
   $\approx 74\,\text{g}$
   b. What is the half-life of ^{119}Sn?
   ≈ 247.6 days [11.5]

For Exercises 17–19, solve. If necessary, use a calculator to approximate to four decimal places.

17. $6^{x-3} = 19$
   4.6433 [11.5]

18. $\log_3 x + \log_3(x + 6) = 3$
   3 [11.5]

19. $\log(4x + 2) - \log(3x - 2) = \log 2$
   3 [11.5]

20. The number of generations, n, of a population of bacteria cells is given by
   $n = 3.3\log\dfrac{b}{B}$, where B is the number of bacteria cells at the beginning of a time interval and b is the number at the end of the time interval. (*Source: Todar's Online Textbook of Bacteriology.*)
   a. Find the number of generations in a bacteria population that increases from 100 to 10,000,000 cells.
   16.5 generations
   b. Find the number of bacteria at the end of the time interval if the number at the beginning is 100 and there are 9.9 generations.
   100,000 [11.6]

Chapters 1–11 Cumulative Review Exercises

For Exercises 1–3, answer true or false.

[11.3] **1.** If $f(x) = 3^x$, then $f^{-1}(x) = \log_3 x$.
true

[9.2] **2.** The radical expression $\sqrt[4]{x^3}$ can be written exponentially as $x^{3/4}$.
true

[6.7] **3.** The range of $f(x) = x^2 + 4$ is $(-\infty, 4]$.
false

For Exercises 4–6, fill in the blank.

[5.1] **4.** When 0.0000000135 is written in scientific notation, the exponent of 10 is _____-8_____.

[6.7] **5.** The solutions of $2x^2 - 9x + 10 = 0$ are the ____x-intercept____ of the graph of $f(x) = 2x^2 - 9x + 10$.

[8.2] **6.** If $|x - 3| = 4$, then ____$x - 3 = 4$____ or ____$x - 3 = -4$____.

Exercises 7–16 ⟩ **Expressions**

For Exercises 7–14, simplify. Write your answers with positive exponents only.

[1.5] **7.** $6 - 8 \div 2 \cdot 3^2 - 4(3 - 4 \cdot 2^3)$
86

[5.6] **8.** $\dfrac{(4a^{-4})^3(2a^2)^{-3}}{(2a^{-2})^3}$
$\dfrac{1}{a^{12}}$

[5.6] **9.** $\dfrac{21p^2q^4 - 14p^5q^3 + 6p^3q^7}{7p^3q^5}$
$\dfrac{3}{pq} - \dfrac{2p^2}{q^2} + \dfrac{6q^2}{7}$

[7.4] **10.** $\dfrac{2x + 3}{x^2 + 6x + 9} - \dfrac{x + 4}{x + 3}$
$\dfrac{-x^2 - 5x - 9}{(x + 3)^2}$

[9.3] **11.** $4\sqrt{6c^3} \cdot 3\sqrt{10c^5}$
$24c^4\sqrt{15}$

[9.7] **12.** $(4 + 3i)(2 - 4i)$
$20 - 10i$

[11.4] **13.** $\log_b b^{2x}$
$2x$

[7.2] **14.** $\dfrac{d^2 - d - 12}{2d^2} \div \dfrac{3d^2 + 13d + 12}{d}$
$\dfrac{d - 4}{2d(3d + 4)}$

For Exercises 15 and 16, factor completely.

[6.4] **15.** $x^4 - 9x^2 - 4x^2y^2 + 36y^2$
$(x + 2y)(x - 2y)(x + 3)(x - 3)$

[6.3] **16.** $24x^4y - 30x^3y^2 + 9x^2y^3$
$3x^2y(4x - 3y)(2x - y)$

Exercises 17–30 ⟩ **Equations and Inequalities**

For Exercises 17–25, solve. Identify any extraneous solutions.

[8.2] **17.** $|2x - 3| - 5 = -4$
$1, 2$

[8.3] **18.** $2|6x - 10| > -8$
All real numbers

[4.3] **19.** $\begin{cases} 4x - 5y = -22 \\ 3x + 2y = -5 \end{cases}$
$(-3, 2)$

[10.3] **20.** $(2x - 3)^2 - 2(2x - 3) = 8$
$\dfrac{1}{2}, \dfrac{7}{2}$

[7.6] 21. $\dfrac{x}{x-2} - \dfrac{4}{x-1} = \dfrac{2}{x^2 - 3x + 2}$

3, (2 is extraneous.)

[9.6] 22. $\sqrt{x+4} - \sqrt{x-4} = 4$

No solution (5 is extraneous.)

[11.2] 23. $4^{x+2} = 8$

$-\dfrac{1}{2}$

[11.3] 24. $\dfrac{1}{2}\log_2 x = 2$

16

[11.6] 25. $\log(3x-5) + \log x = \log 12$

3

[3.4] 26. For $2x - 5y = 10$:

    a. find the slope.

        $\dfrac{2}{5}$

    b. find the intercepts.

        $(5,0), (0,-2)$

    c. draw the graph.

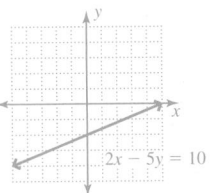

For Exercises 27–30, solve.

[6.6] 27. A frame is in the shape of a right triangle as shown. Find the length of each side.

20, 21, 29

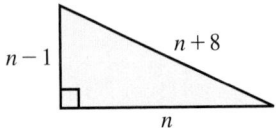

[11.2] 28. If \$15,000 is deposited into an account paying 4% compounded monthly, how much will be in the account after three years?

\$16,909.08

[7.7] 29. Ethan and Pam operate a landscaping company. Ethan can mow, edge, and trim hedges on a quarter-acre lot in 2 hours. Pam takes 1.5 hours to do the same work. How much time would it take them working together?

≈0.857 hr., or ≈51 min.

[4.3] 30. A car travels at an average speed that is 10 miles per hour more than the average speed of a bus. If the bus takes 1 hour longer to travel 300 miles, find how long it takes the car to travel 300 miles.

5 hr.

Chapter Overview

In this chapter, we expand the graphing of second-degree equations to include the conic sections:

- ▶ Parabolas
- ▶ Circles
- ▶ Ellipses
- ▶ Hyperbolas

We also solve systems of equations with lines and conics and graph systems of inequalities that involve conics.

Instructor Note

This chapter introduces the conic sections with the understanding that students will study them in more detail in later courses. Because this is an introduction, some of the forms of ellipses and hyperbolas have been omitted. The application exercises for the conics are particularly interesting because they involve properties of the conics that make them practical in many everyday situations and in the sciences.

12.1 Parabolas and Circles

12.2 Ellipses and Hyperbolas

12.3 Nonlinear Systems of Equations

12.4 Nonlinear Inequalities and Systems of Inequalities

12.1 Parabolas and Circles

Objectives

1. Graph parabolas of the form $x = a(y - k)^2 + h$.
2. Find the distance and midpoint between two points.
3. Graph circles of the form $(x - h)^2 + (y - k)^2 = r^2$.
4. Find the equation of a circle with a given center and radius.
5. Graph circles of the form $x^2 + y^2 + dx + ey + f = 0$.

Instructor Note Use manipulatives to help tactile learners better understand how the inclination of the plane changes the conic section.

Warm-up

[10.4] 1. What are the coordinates of the vertex and the equation of the axis of symmetry of the graph of $y = 2(x - 3)^2 + 1$?

[10.1] *For Exercises 2 and 3, complete the square and write the resulting trinomial as the square of a binomial.*

2. $x^2 - 6x$ **3.** $y^2 + 8y$

[1.5, 1.7] 4. Evaluate the expression $\sqrt{(x_2 - x_1)^2 + (y_2 - y_1)^2}$ for $x_2 = 3, x_1 = -1, y_2 = -1,$ and $y_1 = 2$.

The intersection of a plane with a cone will be a circle, an ellipse, a parabola, or a hyperbola. For that reason, these curves are called **conic sections** or **conics**.

Definition Conic section: A curve in a plane that is the result of intersecting the plane with a cone—more specifically, a circle, an ellipse, a parabola, or a hyperbola.

 Circle    Ellipse    Parabola    Hyperbola

Recall from Section 10.4 that we graphed parabolas in the form $y = a(x - h)^2 + k$. The graph opened upward if $a > 0$, opened downward if $a < 0$, had a vertex at (h, k), and had $x = h$ as the axis of symmetry.

Example 1 For $y = 2(x - 3)^2 + 1$, determine whether the graph opens upward or downward, find the vertex and axis of symmetry, and draw the graph.

Solution: The graph opens upward because $a = 2$ and 2 is positive. We compare the equation with the form $y = a(x - h)^2 + k$ and observe that the vertex is at the point with coordinates $(3, 1)$ and the axis of symmetry is $x = 3$. Plot a few points on either side of the axis of symmetry by letting x have values on either side of 3 and finding y.

x	y
2	3
1	9
4	3
5	9

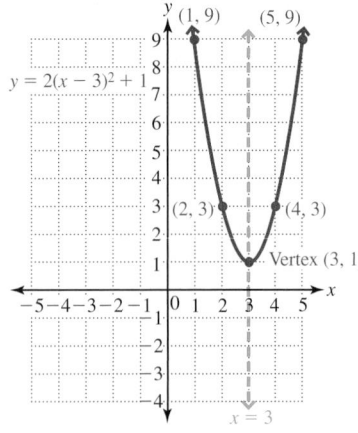

Answer to Your Turn 1
opens downward; vertex: $(1, 3)$; axis of symmetry: $x = 1$

Your Turn 1 For $y = -2(x - 1)^2 + 3$, determine whether the graph opens upward or downward, find the vertex and axis of symmetry, and draw the graph.

Answers to Warm-up
1. $(3, 1), x = 3$
2. $x^2 - 6x + 9, (x - 3)^2$
3. $y^2 + 8y + 16, (y + 4)^2$
4. 5

Objective 1 Graph parabolas of the form $x = a(y - k)^2 + h$.

If we interchange x and y in the equations of parabolas that open upward and downward, we get the equations of parabolas that open to the left or right. To keep the vertex at (h, k), we also interchange h and k.

Rule Equations of Parabolas Opening Left or Right

The graph of an equation in the form $x = a(y - k)^2 + h$ is a parabola with vertex at (h, k). The parabola opens to the right if $a > 0$ and to the left if $a < 0$. The equation of the axis of symmetry is $y = k$.

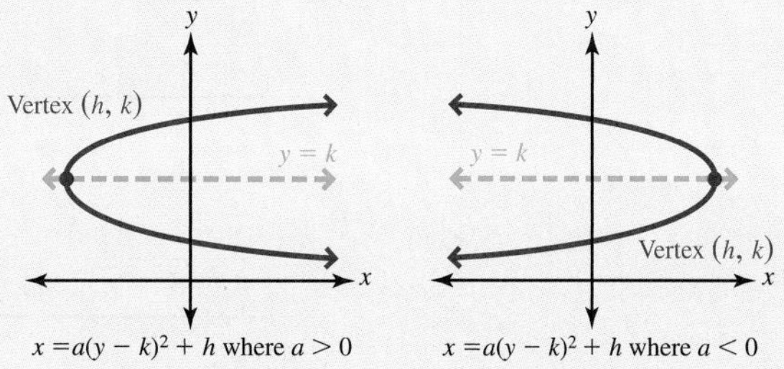

Note Parabolas that open to the left or right are not functions.

$x = a(y - k)^2 + h$ where $a > 0$ $x = a(y - k)^2 + h$ where $a < 0$

Example 2 For each equation, determine whether the graph opens left or right, find the vertex and axis of symmetry, and draw the graph.

a. $x = -2(y + 3)^2 - 2$

Solution: This parabola opens to the left because $a = -2$, which is negative. Rewrite the equation as $x = -2(y - (-3))^2 - 2$. Comparing this equation with $x = a(y - k)^2 + h$, we see that $h = -2$ and $k = -3$. The vertex is at the point with coordinates $(-2, -3)$, and the axis of symmetry is $y = -3$. To graph, plot a few points on either side of the axis of symmetry by letting y equal values on either side of -3 and finding x.

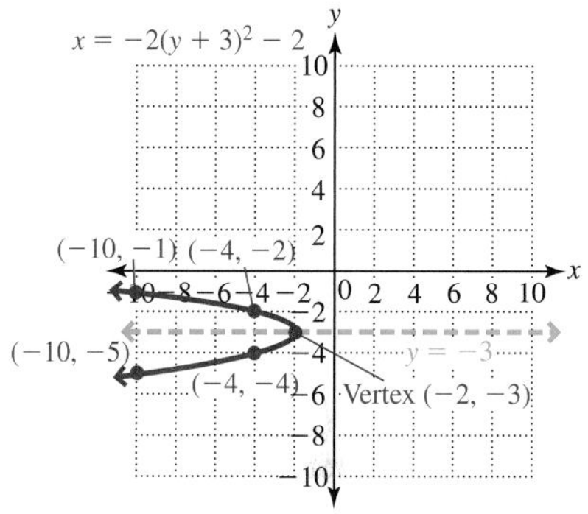

Note In choosing y-values, we went one unit above the axis of symmetry, then two above, then one below, then two below, etc.

x	y
−4	−2
−10	−1
−4	−4
−10	−5

b. $x = 3y^2 + 12y + 8$

Solution: This parabola opens to the right because $a = 3$, which is positive. To find the vertex and axis of symmetry, we need to write the equation in the form $x = a(y - k)^2 + h$.

$x = 3y^2 + 12y + 8$	Original equation
$x - 8 = 3y^2 + 12y$	Subtract 8 from both sides.
$x - 8 = 3(y^2 + 4y)$	Factor out the common factor, 3.
$x - 8 + 12 = 3(y^2 + 4y + 4)$	Complete the square. Note that we added $3 \cdot 4 = 12$ to both sides of the equation.
$x + 4 = 3(y + 2)^2$	Simplify the left side and factor the right side.
$x = 3(y + 2)^2 - 4$	Subtract 4 from both sides of the equation.

The vertex is at the point with coordinates $(-4, -2)$, and the axis of symmetry is $y = -2$.

To complete the graph, let y equal values on either side of -2 and find x.

x	y
-1	-1
8	0
-1	-3
8	-4

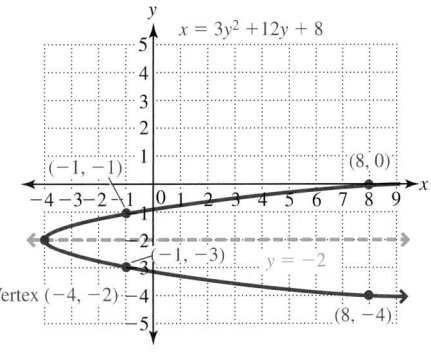

Your Turn 2 For each equation, determine whether the graph opens left or right, find the vertex and axis of symmetry, and draw the graph.

a. $x = -(y - 2)^2 + 1$

b. $x = y^2 + 4y + 3$

Objective 2 Find the distance and midpoint between two points.

To derive the other conic's general equations, we need to be able to find the distance between any two points in the coordinate plane. Consider the two points (x_1, y_1) and (x_2, y_2) shown in the following graph.

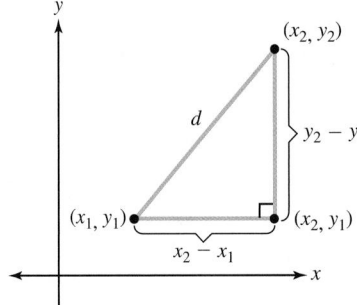

Note If y_2 were 6 and y_1 were 2, the distance between the points would be $6 - 2 = 4$. Therefore, to calculate the length of the vertical leg, we calculate $y_2 - y_1$.

▶

Instructor Note It is helpful to demonstrate this process with actual points first, such as $(1, 2)$ and $(4, 6)$.

Connection Because the distance between the two points is measured along a line segment that is the hypotenuse of a right triangle, we can use the Pythagorean theorem (see below) to find the distance.

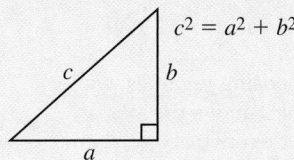

The lengths of the legs of the triangle are $x_2 - x_1$ and $y_2 - y_1$, as illustrated.

▲

Note If x_2 were 4 and x_1 were 1, the distance between the points would be $4 - 1 = 3$. Therefore, to calculate the length of the horizontal leg, we calculate $x_2 - x_1$.

Now we can use the Pythagorean theorem, replacing a with $x_2 - x_1$, b with $y_2 - y_1$, and c with d.

$$d^2 = (x_2 - x_1)^2 + (y_2 - y_1)^2$$
$$d = \pm\sqrt{(x_2 - x_1)^2 + (y_2 - y_1)^2}$$ Use the square root principle to isolate d.

Because d is a distance, it must be positive; so we use only the positive value.

Answers to Your Turn 2
a. opens left; vertex: $(1, 2)$; axis of symmetry: $y = 2$

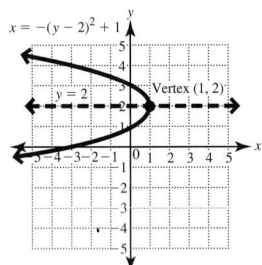

b. opens right; vertex: $(-1, -2)$; axis of symmetry: $y = -2$

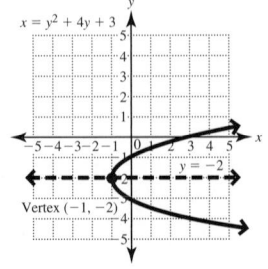

> **Rule Distance Formula**
>
> The distance, d, between two points with coordinates (x_1, y_1) and (x_2, y_2) can be found using the formula
> $$d = \sqrt{(x_2 - x_1)^2 + (y_2 - y_1)^2}.$$

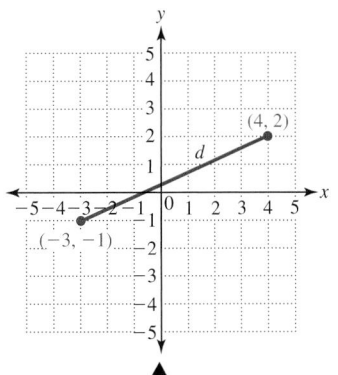

Note The distance formula holds no matter what quadrants the points are in.

Example 3 Find the distance between $(4, 2)$ and $(-3, -1)$. If the distance is an irrational number, also give a decimal approximation rounded to three places.

Solution: $d = \sqrt{(x_2 - x_1)^2 + (y_2 - y_1)^2}$ Use the distance formula.

$d = \sqrt{(-3 - 4)^2 + (-1 - 2)^2}$ Let $(4, 2) = (x_1, y_1)$ and $(-3, -1) = (x_2, y_2)$.

$d = \sqrt{(-7)^2 + (-3)^2}$

$d = \sqrt{49 + 9}$

$d = \sqrt{58}$

$d \approx 7.616$

Note It doesn't matter which ordered pair is (x_1, y_1) and which is (x_2, y_2). Consider Example 3 again with $(-3, -1)$ as (x_1, y_1) and $(4, 2)$ as (x_2, y_2):

$d = \sqrt{(4 - (-3))^2 + (2 - (-1))^2}$

$= \sqrt{(7)^2 + (3)^2} = \sqrt{49 + 9}$

$= \sqrt{58} \approx 7.616$

Your Turn 3 Determine the distance between the given points. If the distance is an irrational number, also give a decimal approximation rounded to three places.

 a. $(8, 2)$ and $(3, -4)$ **b.** $(6, -5)$ and $(0, -1)$

On a line segment, the *midpoint* is located equally distant from each of the endpoints. Equivalently, the distance from either endpoint to the midpoint is half the length of the line segment.

> **Rule Midpoint Formula**
>
> If the coordinates of the endpoints of a line segment are (x_1, y_1) and (x_2, y_2), the coordinates of the midpoint are
>
> $$\left(\frac{x_1 + x_2}{2}, \frac{y_1 + y_2}{2} \right).$$

Note The coordinates of the midpoint of a line segment are the average of the x-coordinates and the average of the y-coordinates of the endpoints.

Example 4 Find the midpoint of the line segment whose endpoints are $(-3, 2)$ and $(5, 6)$.

Solution: Midpoint $= \left(\dfrac{x_1 + x_2}{2}, \dfrac{y_1 + y_2}{2} \right)$ Use the midpoint formula.

$= \left(\dfrac{-3 + 5}{2}, \dfrac{2 + 6}{2} \right)$ Let $(-3, 2) = (x_1, y_1)$ and $(5, 6) = (x_2, y_2)$.

$= \left(\dfrac{2}{2}, \dfrac{8}{2} \right)$

$= (1, 4)$

Your Turn 4 Find the midpoint of the line segments whose endpoints are given.

 a. $(4, -5), (-6, -1)$ **b.** $(3, 1), (-2, -7)$

Answers to Your Turn 3
 a. $\sqrt{61} \approx 7.810$
 b. $2\sqrt{13} \approx 7.211$

Answers to Your Turn 4
 a. $(-1, -3)$ **b.** $\left(\dfrac{1}{2}, -3 \right)$

Objective 3 Graph circles of the form $(x - h)^2 + (y - k)^2 = r^2$.

The second conic section that we consider is the **circle** with **radius** r.

Definitions Circle: A set of points in a plane that are equally distant from a central point. The central point is the center. **Radius:** The distance from the center of a circle to any point on the circle.

If the center of a circle is (h, k) and the radius is r, we can use the distance formula to derive the equation of the circle. If (x, y) is any point on the circle, the distance between (x, y) and (h, k) must be the radius, r.

$$\sqrt{(x_2 - x_1)^2 + (y_2 - y_1)^2} = d$$

$$\sqrt{(x - h)^2 + (y - k)^2} = r \quad \text{Substitute } (x, y)$$
for (x_2, y_2), (h, k)
for (x_1, y_1) and r
for d.

$$(x - h)^2 + (y - k)^2 = r^2 \quad \text{Square both sides.}$$

Rule Standard Form of the Equation of a Circle

The equation of a circle with center at (h, k) and radius r is $(x - h)^2 + (y - k)^2 = r^2$.

Note If the center of a circle is at the ▶ origin, then $(h, k) = (0, 0)$ and the equation of the circle becomes $(x - 0)^2 + (y - 0)^2 = r^2$, which simplifies to $x^2 + y^2 = r^2$.

Example 5   Find the center and radius of each circle and draw the graph.

a. $(x - 3)^2 + (y + 2)^2 = 36$

Solution: $(x - 3)^2 + (y - (-2))^2 = 6^2$ Write in the form $(x - h)^2 + (y - k)^2 = r^2$.

Because $h = 3$ and $k = -2$, the center is $(3, -2)$. Because $36 = 6^2$, the radius is 6.

▲

Note To find the radius, we evaluate the square root of 36. Because the radius is a distance, we give only the principal square root.

$$r = \sqrt{36} = 6$$

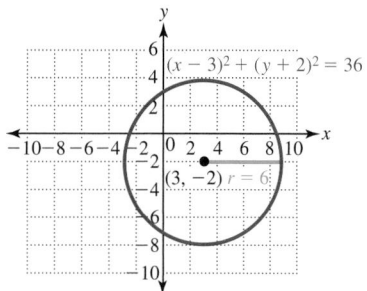

Answers to Your Turn 5

a. center: $(3, -5)$; radius: 2

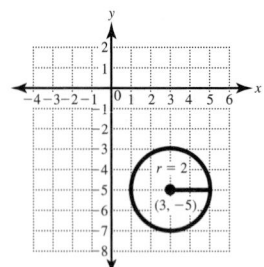

b. center: $(-1, 1)$; radius: $3\sqrt{2}$

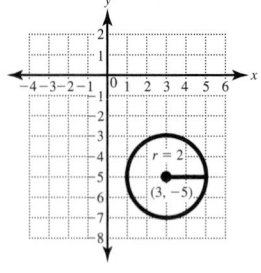

b. $(x + 4)^2 + (y + 1)^2 = 28$

Solution: $(x - (-4))^2 + (y - (-1))^2 = (\sqrt{28})^2$ Write in the form $(x - h)^2 + (y - k)^2 = r^2$.

Because $h = -4$ and $k = -1$, the center is $(-4, -1)$. For this radius, $r = \sqrt{28} = 2\sqrt{7} \approx 5.292$.

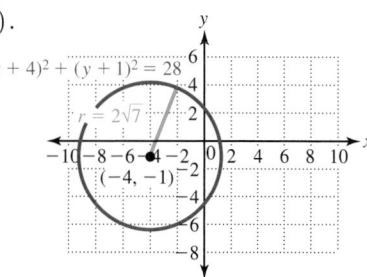

Your Turn 5   Find the center and radius of each circle and draw the graph.

a. $(x - 3)^2 + (y + 5)^2 = 4$ **b.** $(x + 1)^2 + (y - 1)^2 = 18$

Objective 4 Find the equation of a circle with a given center and radius.

We can also use the standard form of a circle to write equations of circles.

Learning Strategy

Note how the signs of the numbers in the parentheses of the general equation are opposite the signs of the center coordinates.

$$(x + 4)^2 + (y - 2)^2 = 64$$

Center: $(-4, 2)$

Example 6 Write the equation of each circle in standard form.

a. Center: $(-4, 2)$; radius: 8

Solution: Because the center is at $(-4, 2)$, $h = -4$ and $k = 2$. Also $r = 8$.

$$(x - h)^2 + (y - k)^2 = r^2 \quad \text{Standard form of a circle.}$$
$$(x - (-4))^2 + (y - 2)^2 = 8^2 \quad \text{Substitute for } h, k, \text{ and } r.$$
$$(x + 4)^2 + (y - 2)^2 = 64 \quad \text{Simplify.}$$

b. Center: $(0, 0)$; radius: 2

Solution: Because the center is at $(0, 0)$, the standard form of the equation is $x^2 + y^2 = r^2$.

$$x^2 + y^2 = 2^2 \quad \text{Substitute for } r.$$
$$x^2 + y^2 = 4 \quad \text{Simplify.}$$

Your Turn 6 Write the equation of each circle in standard form.

a. Center: $(4, -2)$; radius: 5   **b.** Center: $(0, 0)$; radius: 9

Objective 5 Graph circles of the form $x^2 + y^2 + dx + ey + f = 0$.

If the equation of a circle is not given in standard form, we complete the square to write the equation in the form $(x - h)^2 + (y - k)^2 = r^2$.

Example 7 Find the center and radius of the circle whose equation is $x^2 + y^2 - 6x + 8y + 9 = 0$ and draw the graph.

Solution:

$$x^2 + y^2 - 6x + 8y = -9 \quad \text{Subtract 9 from both sides to isolate the variable terms.}$$

$$(x^2 - 6x) + (y^2 + 8y) = -9 \quad \text{Group } x \text{ and } y \text{ terms.}$$

$$(x^2 - 6x + 9) + (y^2 + 8y + 16) = -9 + 9 + 16 \quad \text{Complete the square in } x \text{ and } y \text{ by adding 9 and 16 to both sides of the equation.}$$

$$(x - 3)^2 + (y + 4)^2 = 16 \quad \text{Factor and simplify.}$$

The center is at $(3, -4)$, and the radius is 4.

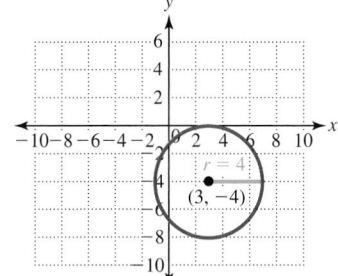

Answers to Your Turn 6
a. $(x - 4)^2 + (y + 2)^2 = 25$
b. $x^2 + y^2 = 81$

Answers to Your Turn 7

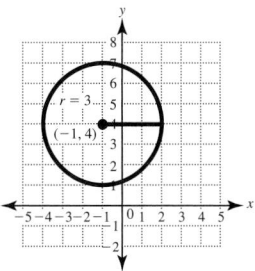

center: $(-1, 4)$; radius: 3

Your Turn 7 Find the center and radius of the circle whose equation is $x^2 + y^2 + 2x - 8y + 8 = 0$ and draw the graph.

Note Some calculators automatically insert the left parenthesis when the root function is used; others do not.

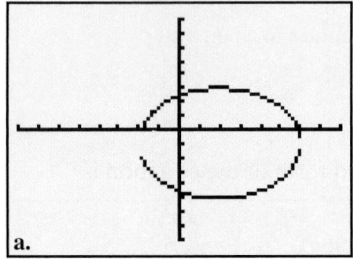

a.

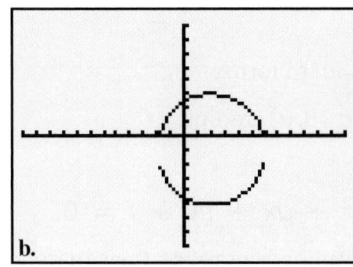

b.

Example 8 Graph $(x - 2)^2 + (y + 1)^2 = 16$ using a graphing calculator.

Solution: The graph of a circle fails the vertical line test, so the equation of a circle is not a function. To graph equations that are not functions, we can solve for y and graph the two resulting functions on the same screen.

$$(x - 2)^2 + (y + 1)^2 = 16$$

$$(y + 1)^2 = 16 - (x - 2)^2 \qquad \text{Subtract } (x - 2)^2 \text{ from both sides.}$$

$$y + 1 = \pm \sqrt{16 - (x - 2)^2} \qquad \text{Apply the square root principle.}$$

$$y = -1 \pm \sqrt{16 - (x - 2)^2} \qquad \text{Subtract 1 from both sides.}$$

This equation defines two functions. On the graphing calculator, define $Y_1 = -1 + \sqrt{(16 - (x - 2)^2)}$ and $Y_2 = -1 - \sqrt{(16 - (x - 2)^2)}$ and graph both in the window $[-8, 8]$ for x and $[-8, 8]$ for y. The resulting graph is labeled "*a*"; note that the graph does not look like the graph of a circle. To make the graph look like a circle, select ZSquare from the $\boxed{\text{ZOOM}}$ menu. This results in the graph labeled "*b*."

12.1 Exercises _{For Extra Help} MyMathLab®

Note: Exercises marked with a ★ represent challenging exercises.

Objective 1

Prep Exercise 1 What are the four conic sections?
Parabola, circle, ellipse, and hyperbola

Prep Exercise 2 In what direction does the parabola defined by $y = 2(x - 1)^2 - 3$ open? What is the vertex?
Up; $(1, -3)$

Prep Exercise 3 In what direction does the parabola defined by $x = -2(y + 3)^2 - 4$ open? What is the vertex?
Left; $(-4, -3)$

Prep Exercise 4 Is $x^2 + 2x - 3 + y = 0$ the equation of a circle or a parabola? Explain.
It is a parabola because only one variable is squared.

For Exercises 1–22, find the direction the parabola opens, the coordinates of the vertex, and the equation of the axis of symmetry and draw the graph. See Examples 1 and 2.

1. $y = (x - 1)^2 + 2$
Opens upward;
vertex: $(1, 2)$;
axis of symmetry: $x = 1$

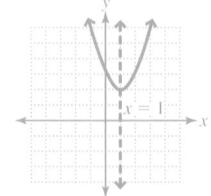

2. $y = (x - 2)^2 + 3$
Opens upward;
vertex: $(2, 3)$;
axis of symmetry: $x = 2$

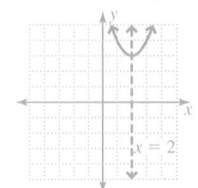

3. $y = -x^2 - 2x + 3$
Opens downward;
vertex: $(-1, 4)$;
axis of symmetry: $x = -1$

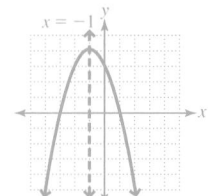

4. $y = -x^2 + 4x - 1$
Opens downward;
vertex: $(2, 3)$;
axis of symmetry: $x = 2$

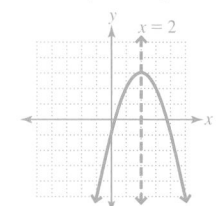

5. $x = (y + 2)^2 - 2$
Opens right;
vertex: $(-2, -2)$;
axis of symmetry: $y = -2$

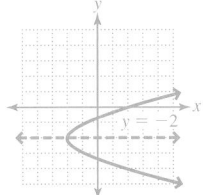

6. $x = (y + 3)^2 + 2$
Opens right;
vertex: $(2, -3)$;
axis of symmetry: $y = -3$

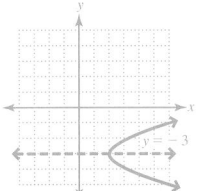

7. $x = -(y - 1)^2 + 3$
Opens left;
vertex: $(3, 1)$;
axis of symmetry: $y = 1$

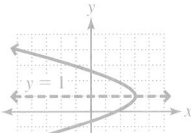

8. $x = -(y - 2)^2 - 1$
Opens left;
vertex: $(-1, 2)$;
axis of symmetry: $y = 2$

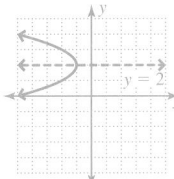

9. $x = 2(y + 2)^2 - 4$
Opens right;
vertex: $(-4, -2)$;
axis of symmetry: $y = -2$

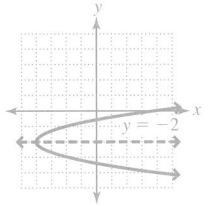

10. $x = 3(y + 3)^2 - 4$
Opens right;
vertex: $(-4, -3)$;
axis of symmetry: $y = -3$

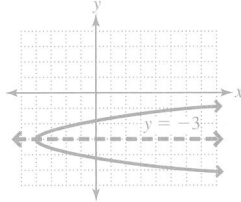

11. $x = -3(y + 2)^2 - 5$
Opens left;
vertex: $(-5, -2)$;
axis of symmetry: $y = -2$

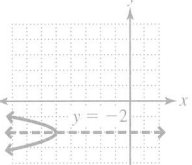

12. $x = -2(y - 4)^2 + 1$
Opens left;
vertex: $(1, 4)$;
axis of symmetry: $y = 4$

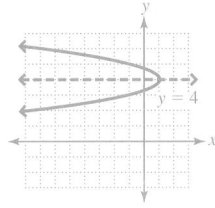

13. $x = y^2 + 4y + 3$
Opens right;
vertex: $(-1, -2)$;
axis of symmetry: $y = -2$

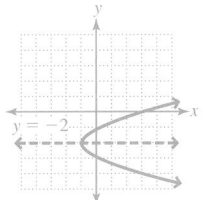

14. $x = y^2 - 2y - 3$
Opens right;
vertex: $(-4, 1)$;
axis of symmetry: $y = 1$

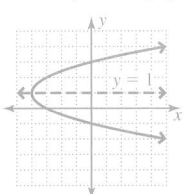

15. $x = -y^2 + 6y - 5$
Opens left;
vertex: $(4, 3)$;
axis of symmetry: $y = 3$

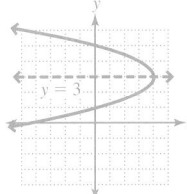

16. $x = -y^2 - 4y - 3$
Opens left;
vertex: $(1, -2)$;
axis of symmetry: $y = -2$

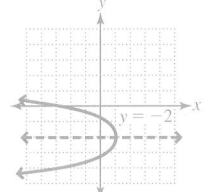

17. $x = 2y^2 + 8y + 3$
Opens right;
vertex: $(-5, -2)$;
axis of symmetry: $y = -2$

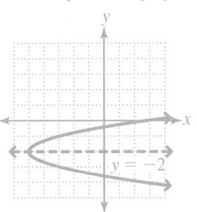

18. $x = 2y^2 - 4y + 1$
Opens right;
vertex: $(-1, 1)$;
axis of symmetry: $y = 1$

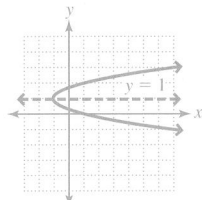

19. $x = 3y^2 - 6y + 1$
Opens right;
vertex: $(-2, 1)$;
axis of symmetry: $y = 1$

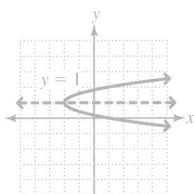

20. $x = 3y^2 - 12y + 9$
Opens right;
vertex: $(-3, 2)$;
axis of symmetry: $y = 2$

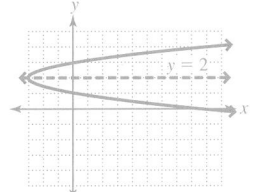

21. $x = -2y^2 + 4y + 5$
Opens left;
vertex: $(7, 1)$;
axis of symmetry: $y = 1$

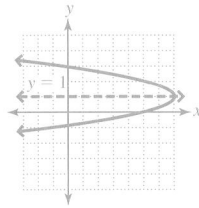

22. $x = -3y^2 - 6y - 2$
Opens left;
vertex: $(1, -1)$;
axis of symmetry: $y = -1$

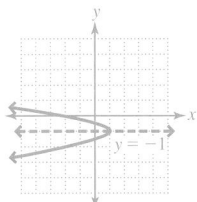

For Exercises 23–26, match the equation with the correct graph.

23. $x = (y + 3)^2 - 2$

c

24. $y = (x + 3)^2 - 2$

b

25. $y = 2x^2 - 8x + 5$

a

26. $x = 2y^2 - 8y + 5$

d

a.

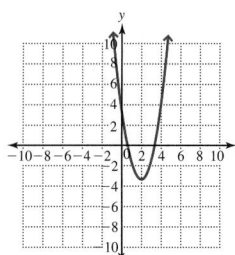

b.

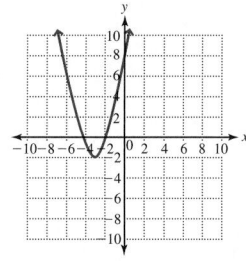

c.

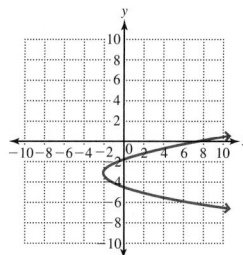

d.

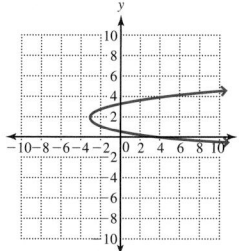

Objective 2

Prep Exercise 5 What formula is used to find the distance between two points (x_1, y_1) and (x_2, y_2)? $d = \sqrt{(x_2 - x_1)^2 + (y_2 - y_1)^2}$

Prep Exercise 6 What formula is used to find the midpoint along a segment with endpoints at (x_1, y_1) and (x_2, y_2)? $\left(\dfrac{x_1 + x_2}{2}, \dfrac{y_1 + y_2}{2} \right)$

For Exercises 27–38, find the distance and midpoint between the two points.
See Examples 3 and 4.

27. $(-4, 2)$ and $(-1, 6)$

$5, \left(-\dfrac{5}{2}, 4 \right)$

28. $(5, -1)$ and $(1, 2)$

$5, \left(3, \dfrac{1}{2} \right)$

29. $(-8, -4)$ and $(-3, 8)$

$13, \left(-\dfrac{11}{2}, 2 \right)$

30. $(-3, 2)$ and $(3, -6)$

$10, (0, -2)$

31. $(-8, -10)$ and $(4, -5)$

$13, \left(-2, -\dfrac{15}{2} \right)$

32. $(4, -6)$ and $(10, 2)$

$10, (7, -2)$

33. $(2, 4)$ and $(4, 8)$

$2\sqrt{5}, (3, 6)$

34. $(-3, 2)$ and $(1, -4)$

$2\sqrt{13}, (-1, -1)$

35. $(-5, 2)$ and $(3, -2)$

$4\sqrt{5}, (-1, 0)$

36. $(3, -4)$ and $(7, 2)$

$2\sqrt{13}, (5, -1)$

37. $(6, -2)$ and $(1, -5)$

$\sqrt{34}, \left(\dfrac{7}{2}, -\dfrac{7}{2} \right)$

38. $(3, -6)$ and $(-2, 1)$

$\sqrt{74}, \left(\dfrac{1}{2}, -\dfrac{5}{2} \right)$

Objectives 3 and 5

Prep Exercise 7 The center of the circle defined by $(x - h)^2 + (y - k)^2 = r^2$ is at _____(h, k)_____, and the radius is _____r_____.

Prep Exercise 8 If the equation of a circle is not given in standard form, we __complete the square__ to write the equation in the form $(x - h)^2 + (y - k)^2 = r^2$.

For Exercises 39–42, the coordinates of the center of a circle and a point on the circle are given. Find the radius of the circle.

39. Center: $(4, 2)$; point on the circle: $(8, -1)$

5

40. Center: $(-4, 6)$; point on the circle: $(2, -2)$

10

41. Center: $(2, -6)$; point on the circle: $(10, -1)$

$\sqrt{89}$

42. Center: $(3, -4)$; point on the circle: $(6, 8)$

$3\sqrt{17}$

For Exercises 43–58, find the center and radius and draw the graph.
See Examples 5 and 7.

43. $(x - 2)^2 + (y - 1)^2 = 4$

Center: $(2, 1)$; radius: 2

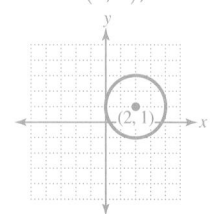

44. $(x - 1)^2 + (y - 3)^2 = 25$

Center: $(1, 3)$; radius: 5

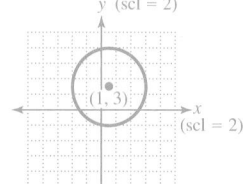

45. $(x + 3)^2 + (y + 2)^2 = 81$

Center: $(-3, -2)$; radius: 9

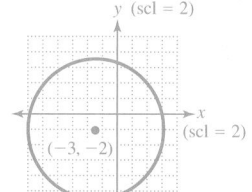

46. $(x + 5)^2 + (y + 4)^2 = 36$
Center: $(-5, -4)$; radius: 6

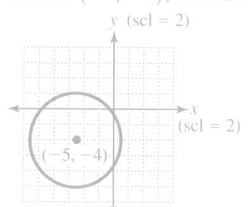

47. $(x - 5)^2 + (y + 3)^2 = 49$
Center: $(5, -3)$; radius: 7

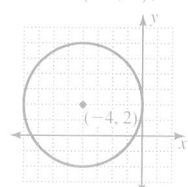

48. $(x + 4)^2 + (y - 2)^2 = 16$
Center: $(-4, 2)$; radius: 4

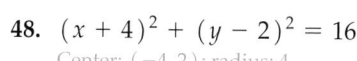

49. $(x - 1)^2 + (y + 1)^2 = 12$
Center: $(1, -1)$; radius: $2\sqrt{3}$

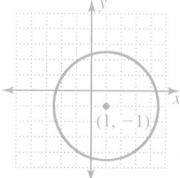

50. $(x + 1)^2 + (y - 3)^2 = 18$
Center: $(-1, 3)$; radius: $3\sqrt{2}$

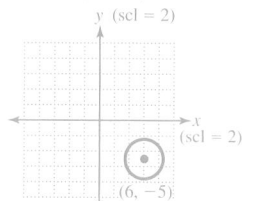

51. $(x + 4)^2 + (y + 2)^2 = 32$
Center: $(-4, -2)$; radius: $4\sqrt{2}$

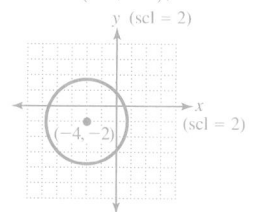

52. $(x - 6)^2 + (y + 5)^2 = 8$
Center: $(6, -5)$; radius: $2\sqrt{2}$

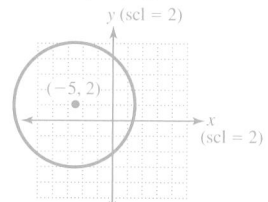

53. $x^2 + y^2 + 8x - 6y + 16 = 0$
Center: $(-4, 3)$; radius: 3

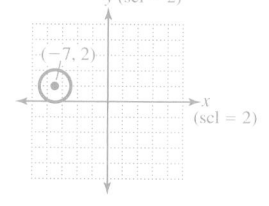

54. $x^2 + y^2 - 2x - 6y - 39 = 0$
Center: $(1, 3)$; radius: 7

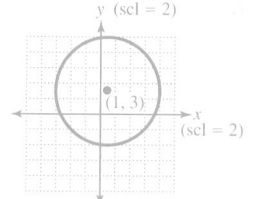

55. $x^2 + y^2 + 10x - 4y - 35 = 0$
Center: $(-5, 2)$; radius: 8

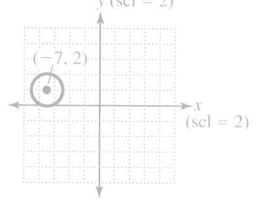

56. $x^2 + y^2 + 12x + 10y + 60 = 0$
Center: $(-6, -5)$; radius: 1

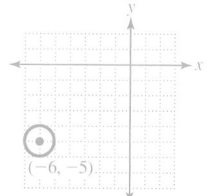

57. $x^2 + y^2 + 14x - 4y + 49 = 0$
Center: $(-7, 2)$; radius: 2

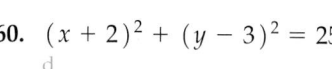

58. $x^2 + y^2 + 8x - 10y + 16 = 0$
Center: $(-4, 5)$; radius: 5

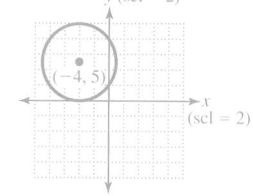

For Exercises 59–62, match the equation with the correct graph.

59. $(x - 2)^2 + (y + 3)^2 = 25$
b

60. $(x + 2)^2 + (y - 3)^2 = 25$
d

61. $x^2 + y^2 + 8x + 2y - 8 = 0$
c

62. $x^2 + y^2 + 2x - 8y - 8 = 0$
a

a. 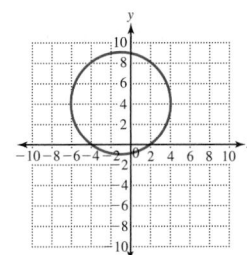 **b.** 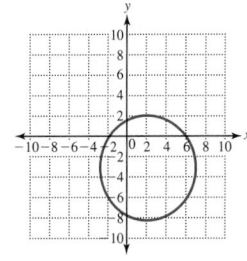 **c.** 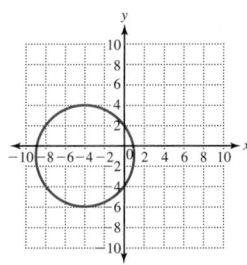 **d.**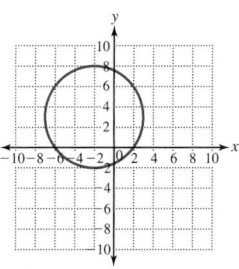

📱 *For Exercises 63–66, graph using a graphing calculator. See Example 8.*

63. $x^2 + y^2 = 49$

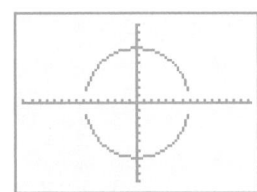

64. $x^2 + y^2 = 36$

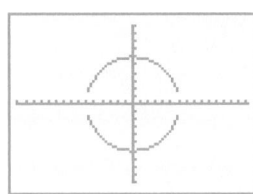

65. $(x - 2)^2 + (y + 3)^2 = 25$

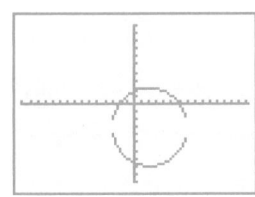

66. $(x + 4)^2 + (y - 1)^2 = 9$

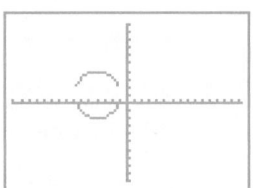

Objective 4

For Exercises 67–74, the center and radius of a circle are given. Write the equation of the circle in standard form. See Example 6.

67. Center: $(4, 2)$; radius: 4
$(x - 4)^2 + (y - 2)^2 = 16$

68. Center: $(3, 2)$; radius: 11
$(x - 3)^2 + (y - 2)^2 = 121$

69. Center: $(-4, -3)$; radius: 5
$(x + 4)^2 + (y + 3)^2 = 25$

70. Center: $(-6, -5)$; radius: 2
$(x + 6)^2 + (y + 5)^2 = 4$

71. Center: $(6, -2)$; radius: $\sqrt{14}$
$(x - 6)^2 + (y + 2)^2 = 14$

72. Center: $(-3, -3)$; radius: $\sqrt{26}$
$(x + 3)^2 + (y + 3)^2 = 26$

73. Center: $(-5, 2)$; radius: $3\sqrt{5}$
$(x + 5)^2 + (y - 2)^2 = 45$

74. Center: $(6, 2)$; radius: $2\sqrt{6}$
$(x - 6)^2 + (y - 2)^2 = 24$

For Exercises 75–78, the center of a circle and a point on the circle are given. Write the equation of the circle in standard form.

75. Center: $(2, 4)$; point on the circle: $(5, 8)$
$(x - 2)^2 + (y - 4)^2 = 25$

76. Center: $(-4, 3)$; point on the circle: $(4, 9)$
$(x + 4)^2 + (y - 3)^2 = 100$

77. Center: $(2, 4)$; point on the circle: $(7, 16)$
$(x - 2)^2 + (y - 4)^2 = 169$

78. Center: $(-6, 8)$; point on the circle: $(-12, 0)$
$(x + 6)^2 + (y - 8)^2 = 100$

79. Write the equation of the set of all points that are a distance of 8 units from $(2, -5)$.
$(x - 2)^2 + (y + 5)^2 = 64$

80. Write the equation of the set of all points that are a distance of 10 units from $(-3, -7)$.
$(x + 3)^2 + (y + 7)^2 = 100$

The diameter of a circle is a line segment whose endpoints lie on the circle and contains the center of the circle. Thus, the center of a circle is the midpoint of a diameter and the diameter is twice the radius.

★ *For Exercises 81–84, the coordinates of the endpoints of a diameter are given. Find the equation of the circle.*

81. $(4, -2), (-2, 6)$
$(x - 1)^2 + (y - 2)^2 = 25$

82. $(6, -6), (-2, 0)$
$(x - 2)^2 + (y + 3)^2 = 25$

83. $(6, -2), (-2, 8)$
$(x - 2)^2 + (y - 3)^2 = 41$

84. $(-8, 6), (-2, 4)$
$(x + 5)^2 + (y - 5)^2 = 10$

85. If a rock is thrown vertically upward from the top of a building 112 feet high with an initial velocity of 96 feet per second, the height, h, above ground level after t seconds is given by $h = -16t^2 + 96t + 112$, where h is in feet and t is in seconds.

   a. What is the maximum height the rock will reach?
   256 ft.

   b. How many seconds will the rock take to reach its maximum height?
   3 sec.

   c. How many seconds will the rock take to hit the ground?
   7 sec.

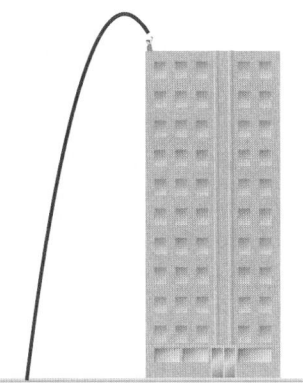

86. The path of a shell fired from ground level is in the shape of the parabola $y = 4x - x^2$, where x and y are given in kilometers.

   a. How high does the shell go?
   4 km

   b. How far has the shell traveled horizontally when it reaches its maximum height?
   2 km

   c. How far from its firing point does the shell land?
   4 km

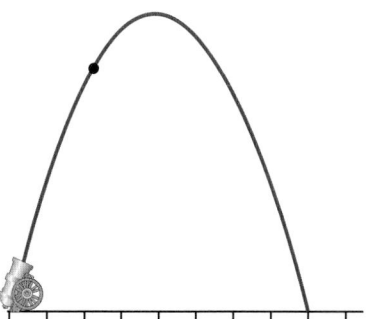

★ **87.** Arches in the shapes of parabolas are often used in construction. Find the equation of a parabolic arc that is 18 feet high at its highest point and 30 feet wide at the base, as illustrated in the following figure. Place the origin at the midpoint of the bridges.

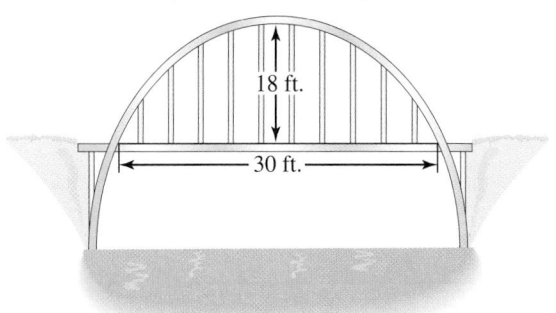

18 ft.

30 ft.

$y = -\dfrac{2}{25}x^2 + 18$ or $y = -0.08x^2 + 18$

★ **88.** The cross sections of satellite dishes are in the shape of parabolas. Find the equation of a dish that is 6 feet across and 1 foot deep if the vertex is at the origin and the parabola is opening upward.

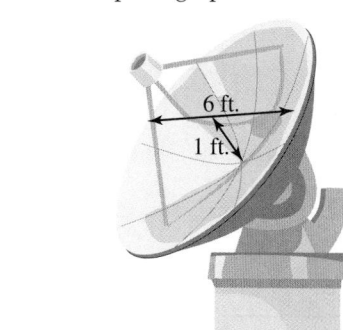

6 ft.

1 ft.

$y = \dfrac{1}{9}x^2$ with origin at base of dish

89. The percent of deaths by age per million miles driven can be approximated by the equation $y = 0.0038x^2 - 0.3475x + 8.316$, where x is the age and y is the percent. Find the percent of deaths per million miles driven for drivers 17 years old. (*Source:* Highway Traffic Safety Administration.)
3.5%

90. The number of drivers involved in fatal accidents for a given blood alcohol content (BAC) can be approximated by $y = -8862.5x^2 + 26622.6x + 332$, where x is the BAC and y is the number of drivers involved in fatal accidents. Find the number of drivers involved in fatal accidents who had a BAC of 0.20. (*Source:* Highway Traffic Safety Administration.)
5302

91. Bill and Don are fishing in the Gulf of Mexico in separate boats that are equipped with radios with a range of 20 miles. If we put Bill's radio at the origin of a coordinate system, what is the equation of all possible locations of Don's boat where the radios would be at their maximum range?
$x^2 + y^2 = 400$

92. A toy plane is attached to a string pinned to the ceiling so that the plane flies in a circle. If the string is 4 feet long, write an equation that describes the path of the plane if the pin is at the origin.
$x^2 + y^2 = 16$

93. A Ferris wheel has a diameter of 200 feet, and the bottom of the Ferris wheel is 10 feet above the ground. Find the equation of the wheel if the origin is placed on the ground directly below the center of the wheel, as illustrated.
$x^2 + (y - 110)^2 = 10,000$

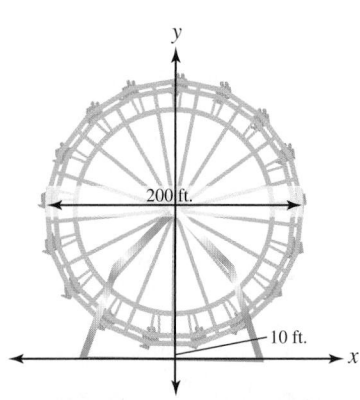

94. The Fermilab tunnel houses the world's largest superconducting synchrotron. A cross section of the tunnel is a circle with radius of 1000 meters. Find the equation of a cross section of the tunnel if the origin and the center of the circle are at the center of the tunnel.
$x^2 + y^2 = 1,000,000$

95. If a satellite is placed in a circular orbit of 230 kilometers above the Earth, what is the equation of the path of the satellite if the origin is placed at the center of the Earth (the radius of the Earth is approximately 6370 kilometers)?
$x^2 + y^2 = 43,560,000$

96. The minute hand of Big Ben is 14 feet long. If the origin is at the center of the clock, what is the equation of the circle swept out by the tip of the hand as it makes one complete revolution?
$x^2 + y^2 = 196$

Puzzle Problem Using only a pencil, you can draw both a rough circle and its center without the point of the pencil losing contact with the paper, resulting in a picture like the one shown. Explain how.

Fold one of the corners to the center as shown to the right. Draw the center and then trace along the back of the folded section to move out to the circle as shown. Once the circle is started, unfold the corner so that the circle can be completed.

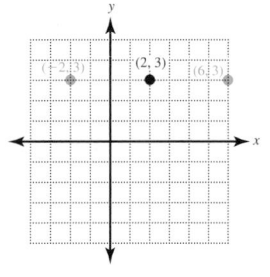

Review Exercises

Exercises 1 and 2  **Expressions**

[3.1] **1.** Following is a coordinate system with the point $(2, 3)$ plotted. Plot the points that are 4 units to the left and right of $(2, 3)$ and give their coordinates.

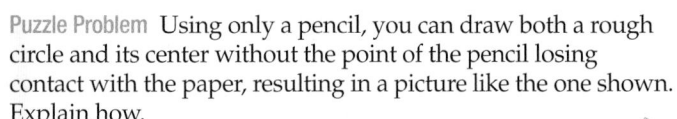

[6.4] **2.** Factor: $25x^2 - 9y^2$
$(5x + 3y)(5x - 3y)$

Exercises 3–6 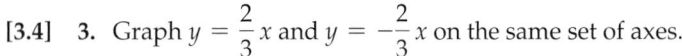 **Equations and Inequalities**

[3.4] **3.** Graph $y = \dfrac{2}{3}x$ and $y = -\dfrac{2}{3}x$ on the same set of axes.

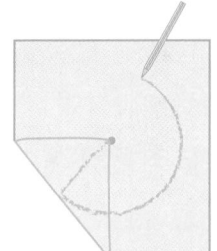

[10.1] **4.** Solve: $\dfrac{x^2}{16} = 1$
± 4

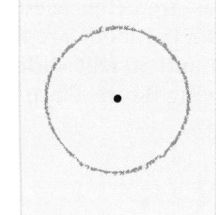

[3.3, 10.1] *For Exercises 5 and 6, find the x- and y- intercepts.*

5. $9x^2 + 16y^2 = 144$
$(4, 0), (-4, 0), (0, 3), (0, -3)$

6. $25x^2 + 9y^2 = 225.$
$(3, 0), (-3, 0), (0, 5), (0, -5)$

12.2 Ellipses and Hyperbolas

Objectives

1 Graph ellipses.
2 Graph hyperbolas.

Instructor Note Use a string and pins on Pegboard to illustrate $c^2 = a^2 - b^2$. The length of the string must be $2c + 2(a - c)$, as shown.

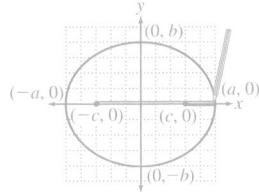

Next, move the point of the pencil to $(0, b)$. Half of the string length is the hypotenuse of a right triangle with legs of length b and c.

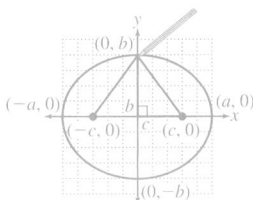

Therefore, the length of the hypotenuse is

$$\frac{2c + 2(a - c)}{2}$$

$$= c + a - c = a.$$

So the length of the string is $2a$. The Pythagorean theorem gives us $b^2 + c^2 = a^2$ or $c^2 = a^2 - b^2$.

Answers to Warm-up

1. $(2, -5), r = 4$
2. $(8, -3), (-4, -3)$
3. $(h, k + b)$
4. $y = \pm 3$

Warm-up

[12.1] 1. Find the center and radius of the circle whose equation is
$x^2 + y^2 - 4x + 10y + 13 = 0$.

[3.1] 2. Find the coordinates of the points that are 6 units to the right and left of the point whose coordinates are $(2, -3)$.

[3.2] 3. Find the coordinates of a point that is b units above the point whose coordinates are (h, k).

[10.1] 4. Solve the equation $\dfrac{y^2}{9} = 1$.

Objective 1 Graph ellipses.

Suppose you drive two nails in a board and tie a string to the two nails. Then you take a pencil, pull the string taut, and draw a figure around the two nails.

The figure you've drawn is called an **ellipse**. The locations of the two nails are the *focal points*.

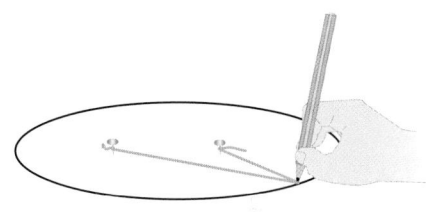

Definition Ellipse: The set of all points the sum of whose distances from two fixed points is constant.

Ellipses occur in many situations. The orbits of the planets about the Sun are elliptical with the Sun at one focal point. The orbits of satellites about the Earth are also elliptical. The cams of compound bows are elliptical, which allows a decrease in the amount of effort required to hold the bow at full draw.

In the definition of an ellipse, the two fixed points are the *foci* (plural of *focus*) and the point halfway between the foci is the *center*. The figure at right shows the graph of an ellipse with foci at $(c, 0)$ and $(-c, 0)$, x-intercepts at $(a, 0)$ and $(-a, 0)$, and y-intercepts at $(0, b)$ and $(0, -b)$. Consequently, the center is at the origin, $(0, 0)$. It can be shown that $c^2 = a^2 - b^2$ for ellipses in which $a > b$.

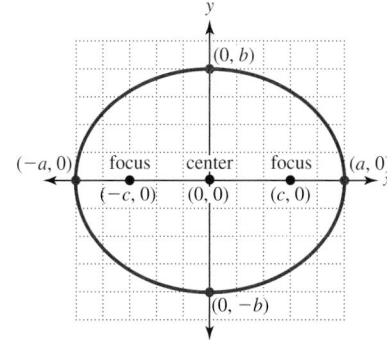

Using the distance formula, it can be shown that an ellipse with these characteristics has the following equation.

> **Rule Equation of an Ellipse Centered at $(0, 0)$**
>
> The equation of an ellipse with center $(0, 0)$, x-intercepts $(a, 0)$ and $(-a, 0)$, and y-intercepts $(0, b)$ and $(0, -b)$ is
>
> $$\frac{x^2}{a^2} + \frac{y^2}{b^2} = 1.$$

Example 1 Graph each ellipse and label the *x*- and *y*-intercepts.

a. $\dfrac{x^2}{16} + \dfrac{y^2}{9} = 1$

Solution: The equation can be rewritten as $\dfrac{x^2}{4^2} + \dfrac{y^2}{3^2} = 1$, so $a = 4$ and $b = 3$. Consequently, the *x*-intercepts are $(4, 0)$ and $(-4, 0)$ and the *y*-intercepts are $(0, 3)$ and $(0, -3)$.

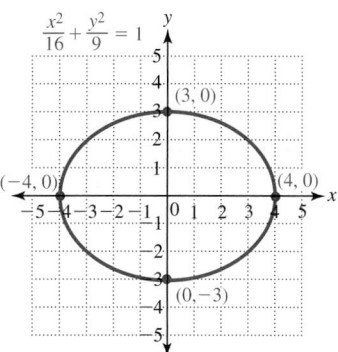

b. $25x^2 + 9y^2 = 225$

Solution: We first need to write the equation in standard form: $\dfrac{x^2}{a^2} + \dfrac{y^2}{b^2} = 1$.

$$\dfrac{25x^2}{225} + \dfrac{9y^2}{225} = \dfrac{225}{225} \qquad \text{Divide both sides by 225.}$$

$$\dfrac{x^2}{9} + \dfrac{y^2}{25} = 1 \qquad \text{Simplify both sides.}$$

We now see that this is an equation of an ellipse with $a = 3$ and $b = 5$, so the *x*-intercepts are $(3, 0)$ and $(-3, 0)$ and the *y*-intercepts are $(0, 5)$ and $(0, -5)$.

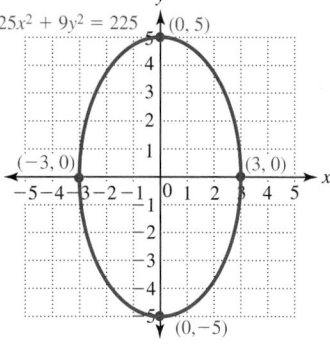

Your Turn 1 Graph the ellipse $\dfrac{x^2}{36} + \dfrac{y^2}{9} = 1$. Label the *x*- and *y*-intercepts.

If the center of an ellipse is not at the origin, it can be shown that the equation has the following form.

Answer to Your Turn 1
$$\dfrac{x^2}{36} + \dfrac{y^2}{9} = 1$$

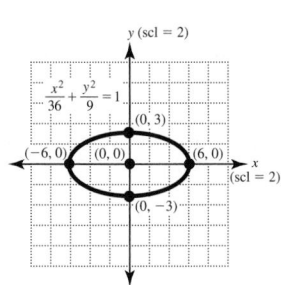

Rule General Equation for an Ellipse

The equation of an ellipse with center (h, k) is $\dfrac{(x - h)^2}{a^2} + \dfrac{(y - k)^2}{b^2} = 1$. The ellipse passes through two points that are *a* units to the right and left of the center and two points that are *b* units above and below the center.

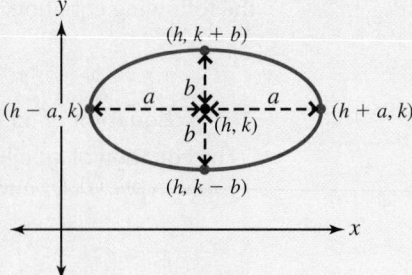

Example 2 Graph the ellipse. Label the center and the points directly above, below, to the left of, and to the right of the center. $\dfrac{(x-2)^2}{36} + \dfrac{(y+3)^2}{16} = 1$

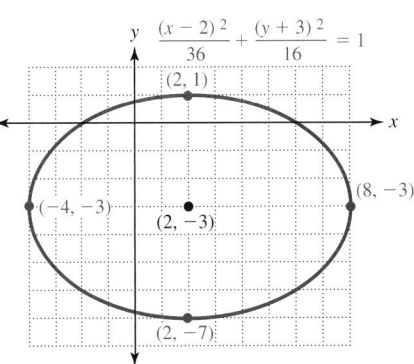

Solution: Because $h = 2$ and $k = -3$, the center of the ellipse is $(2, -3)$. Also, we see that $a = 6$, which means that the ellipse passes through two points 6 units to the right and left of $(2, -3)$. These points are $(8, -3)$ and $(-4, -3)$. Because $b = 4$, the ellipse passes through two points 4 units above and below the center. These points are $(2, 1)$ and $(2, -7)$.

Your Turn 2 Graph the ellipse $\dfrac{(x+1)^2}{25} + \dfrac{(y-2)^2}{36} = 1$. Label the center and the points directly above, below, to the right of, and to the left of the center.

Example 3 Graph the ellipse $\dfrac{x^2}{9} + \dfrac{y^2}{16} = 1$ using a graphing calculator.

Solution: Because an ellipse is not a function, we solve for y to get two functions that we can input.

$\dfrac{y^2}{16} = 1 - \dfrac{x^2}{9}$ Subtract $\dfrac{x^2}{9}$ from both sides.

$y^2 = 16\left(1 - \dfrac{x^2}{9}\right)$ Multiply both sides by 16.

$y = \pm 4 \sqrt{\left(1 - \dfrac{x^2}{9}\right)}$ Apply the square root principle.

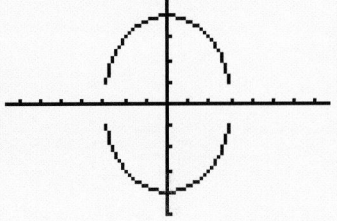

Define $Y_1 = 4\sqrt{\left(1 - \dfrac{x^2}{9}\right)}$ and $Y_2 = -4\sqrt{\left(1 - \dfrac{x^2}{9}\right)}$. Graph in a window $[-5, 5]$ for x and $[-5, 5]$ for y and square the window.

Answer to Your Turn 2
$$\dfrac{(x+1)^2}{25} + \dfrac{(y-2)^2}{36} = 1$$

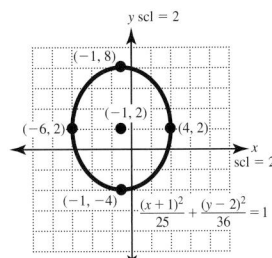

Objective 2 Graph hyperbolas.

The last conic is the **hyperbola**. Applications of hyperbolas include the LORAN navigation system and the orbits of some comets.

Definition Hyperbola: The set of all points the difference of whose distances from two fixed points remains constant.

As with the ellipse, the two fixed points are the *foci* and the point halfway between the foci is the *center*.

Learning Strategy

Noting which variable term is positive in the standard form equation of a hyperbola can help you remember how to orient the graph. If the x^2 term is positive, the hyperbola's intercepts are on the x-axis. If the y^2 term is positive, the hyperbola's intercepts are on the y-axis.

Rule Equations of Hyperbolas in Standard Form

The equation of a hyperbola with center $(0,0)$, x-intercepts $(a,0)$ and $(-a,0)$, and no y-intercepts is

$$\frac{x^2}{a^2} - \frac{y^2}{b^2} = 1.$$

The equation of a hyperbola with center $(0,0)$, y-intercepts $(0,b)$ and $(0,-b)$, and no x-intercepts is

$$\frac{y^2}{b^2} - \frac{x^2}{a^2} = 1.$$

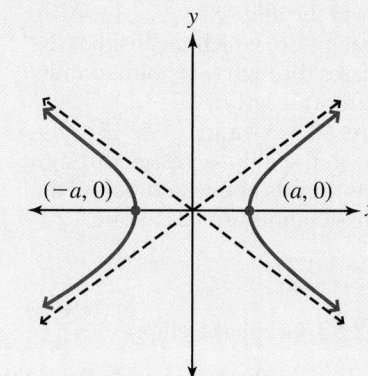 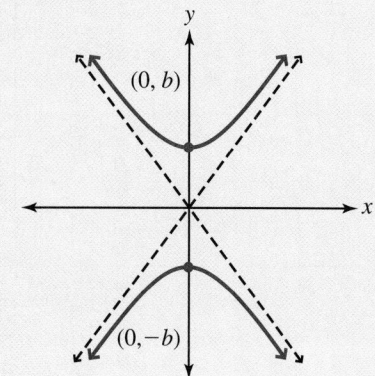

The dashed lines that intersect at the center of a hyperbola are *asymptotes*. They are not a part of the graph, but are used as an aid in graphing. An asymptote is a line that the graph approaches but does not cross as the graph goes away from the origin. The rectangle whose vertices are (a,b), $(-a,b)$, $(a,-b)$, and $(-a,-b)$ is called the *fundamental rectangle,* and the asymptotes are the extended diagonals of the fundamental rectangle, as shown in the following illustration.

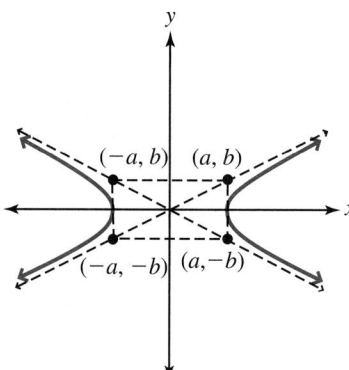

Procedure Graphing a Hyperbola in Standard Form

1. Find the intercepts. If the x^2-term is positive, the x-intercepts are $(a,0)$ and $(-a,0)$ and there are no y-intercepts. If the y^2-term is positive, the y-intercepts are $(0,b)$ and $(0,-b)$ and there are no x-intercepts.
2. Draw the fundamental rectangle. The vertices are (a,b), $(-a,b)$, $(a,-b)$, and $(-a,-b)$.
3. Draw the asymptotes, which are the extended diagonals of the fundamental rectangle.
4. Draw the graph so that each branch passes through an intercept and approaches the asymptotes the farther the branches are from the origin.

Example 4   Graph each hyperbola. Also show the fundamental rectangle with its corner points labeled, the asymptotes, and the intercepts.

a. $\dfrac{x^2}{16} - \dfrac{y^2}{9} = 1$

Solution:

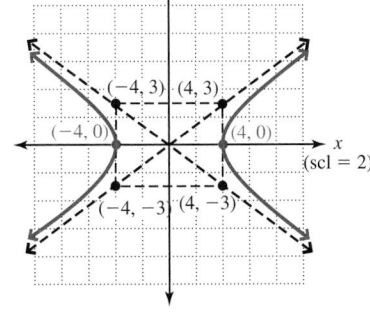

1. This equation can be written as $\dfrac{x^2}{4^2} - \dfrac{y^2}{3^2} = 1$, so $a = 4$ and $b = 3$. Because the x^2-term is positive, the graph has x-intercepts at $(4, 0)$ and $(-4, 0)$.

2. The fundamental rectangle has vertices at $(4, 3)$, $(-4, 3)$, $(4, -3)$, and $(-4, -3)$.
3. Draw the asymptotes.
4. Sketch the graph so that it passes through the x-intercepts and approaches the asymptotes.

b. $\dfrac{y^2}{4} - \dfrac{x^2}{9} = 1$

Solution:

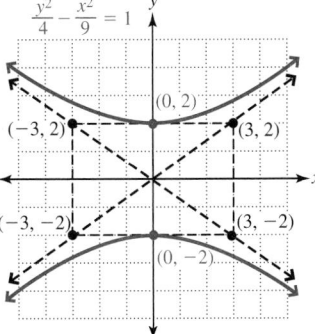

1. This equation can be written as $\dfrac{y^2}{2^2} - \dfrac{x^2}{3^2} = 1$, so $b = 2$ and $a = 3$. Because the y^2-term is positive, the graph has y-intercepts at $(0, 2)$ and $(0, -2)$.

2. The fundamental rectangle has vertices at $(3, 2)$, $(-3, 2)$, $(3, -2)$, and $(-3, -2)$.
3. Draw the asymptotes.
4. Sketch the graph so that it passes through the y-intercepts and approaches the asymptotes.

Your Turn 4   Graph the hyperbola. Also show the fundamental rectangle with its corner points labeled, the asymptotes, and the intercepts.

$$\frac{x^2}{25} - \frac{y^2}{16} = 1$$

Following is a summary of the conic sections.

Answer to Your Turn 4

$\dfrac{x^2}{25} - \dfrac{y^2}{16} = 1$

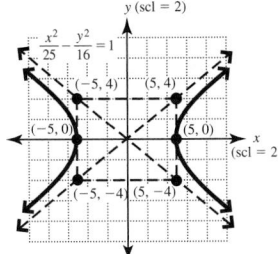

Standard Forms		
Parabola	$y = a(x - h)^2 + k, a > 0$	$y = a(x - h)^2 + k, a < 0$

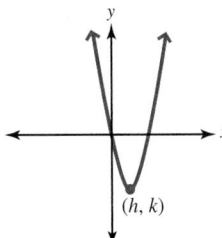

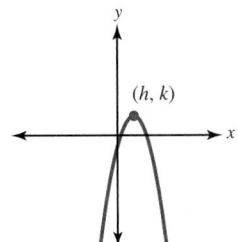

(Continued)

Standard Forms

$$x = a(y - k)^2 + h, a > 0$$

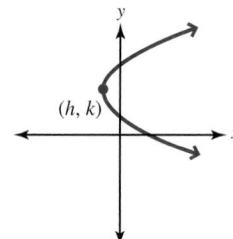

$$x = a(y - k)^2 + h, a < 0$$

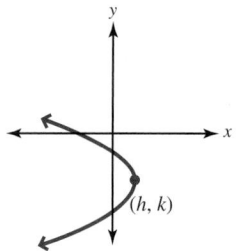

Circle

$$x^2 + y^2 = r^2$$

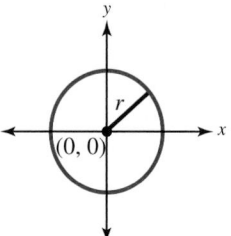

$$(x - h)^2 + (y - k)^2 = r^2$$

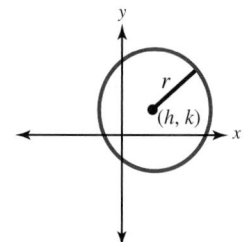

Ellipse

$$\frac{x^2}{a^2} + \frac{y^2}{b^2} = 1$$

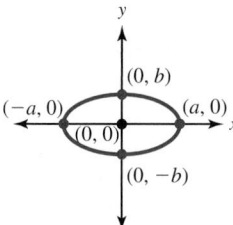

$$\frac{(x - h)^2}{a^2} + \frac{(y - k)^2}{b^2} = 1$$

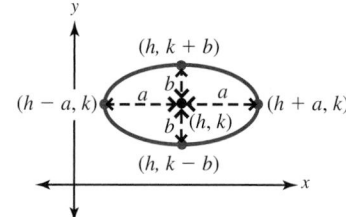

Hyperbola

$$\frac{x^2}{a^2} - \frac{y^2}{b^2} = 1$$

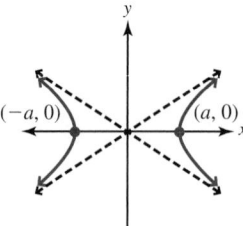

$$\frac{y^2}{b^2} - \frac{x^2}{a^2} = 1$$

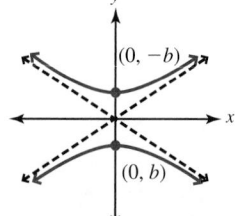

12.2 Exercises For Extra Help MyMathLab®

Objective 1

Prep Exercise 1 What is the definition of an ellipse?

An ellipse is the set of all points the sum of whose distances from two fixed points is constant.

Prep Exercise 2 In the definitions of the ellipse and hyperbola, two fixed points are mentioned. What are these two points called?

The fixed points are the foci.

Prep Exercise 3 Answer the following questions about the ellipse defined by
$$\frac{x^2}{9} + \frac{y^2}{25} = 1.$$
a. What is the center?

$(0,0)$

b. How many units to the left and right of center do you place a point?

3 units left and right

c. How many units above and below center do you place a point?

5 units up and down

For Exercises 1–8, graph each ellipse. Label the x- and y-intercepts. See Example 1.

1. $\dfrac{x^2}{81} + \dfrac{y^2}{64} = 1$

2. $\dfrac{x^2}{25} + \dfrac{y^2}{4} = 1$

3. $\dfrac{x^2}{36} + y^2 = 1$

4. $\dfrac{x^2}{9} + \dfrac{y^2}{16} = 1$

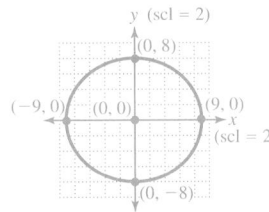
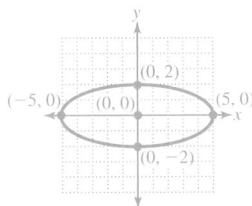
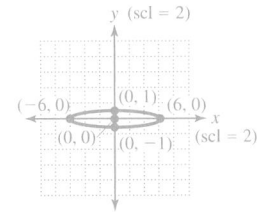
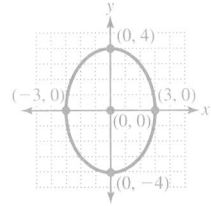

5. $4x^2 + 9y^2 = 36$

6. $25x^2 + 4y^2 = 100$

7. $36x^2 + 4y^2 = 144$

8. $x^2 + 9y^2 = 36$

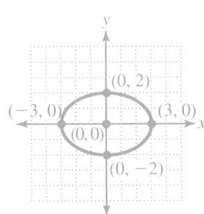
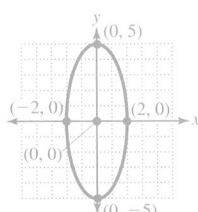
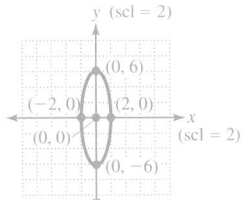
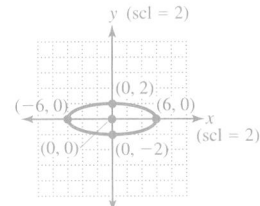

For Exercises 9–12, graph each ellipse. Label the center and the points directly above, below, to the left of, and to the right of the center. See Example 2.

9. $\dfrac{(x-1)^2}{49} + \dfrac{(y+3)^2}{25} = 1$

10. $\dfrac{(x+3)^2}{25} + \dfrac{(y-2)^2}{64} = 1$

11. $\dfrac{(x-4)^2}{4} + \dfrac{(y+3)^2}{36} = 1$

12. $\dfrac{(x+5)^2}{9} + \dfrac{(y+3)^2}{25} = 1$

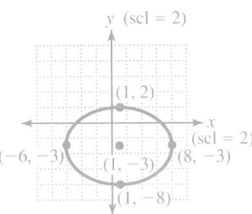
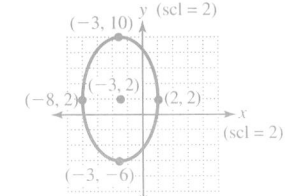
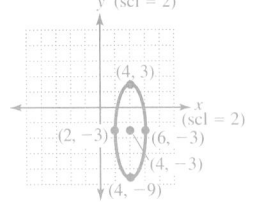
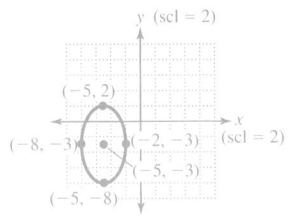

For Exercises 13–16, graph using a graphing calculator. See Example 3.

13. $\dfrac{x^2}{25} + \dfrac{y^2}{4} = 1$

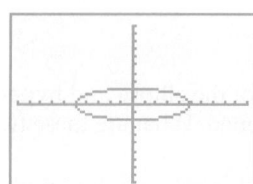

14. $\dfrac{y^2}{12} + \dfrac{x^2}{8} = 1$

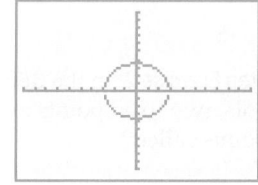

15. $\dfrac{(y-2)^2}{36} + \dfrac{(x+4)^2}{25} = 1$

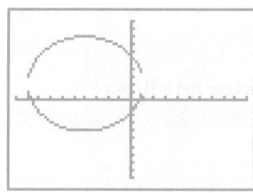

16. $\dfrac{(x+2)^2}{4} + \dfrac{(y+1)^2}{16} = 1$

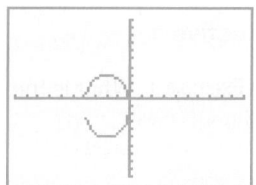

17. If the x-intercepts of an ellipse are $(-3,0)$ and $(3,0)$ and the y-intercepts are $(0,4)$ and $(0,-4)$, what is the equation of the ellipse?

$\dfrac{x^2}{9} + \dfrac{y^2}{16} = 1$

18. If the x-intercepts of an ellipse are $(-6,0)$ and $(6,0)$ and the y-intercepts are $(0,3)$ and $(0,-3)$, what is the equation of the ellipse?

$\dfrac{x^2}{36} + \dfrac{y^2}{9} = 1$

Objective 2

Prep Exercise 4 How do you determine whether a hyperbola opens up and down or left and right?

If the x^2-term is positive, the graph has x-intercepts and opens left and right. If the y^2-term is positive, the graph has y-intercepts and opens up and down.

Prep Exercise 5 Answer the following questions about the hyperbola defined by $\dfrac{x^2}{9} - \dfrac{y^2}{25} = 1$.

a. What is the center?
$(0,0)$

b. What are the x-intercepts?
$(-3,0)$ and $(3,0)$

c. What are the y-intercepts?
It has no y-intercepts.

d. What are the vertices of the fundamental rectangle?
$(3,5), (-3,5), (3,-5), (-3,-5)$

Prep Exercise 6 Answer the following questions about the hyperbola defined by $\dfrac{y^2}{16} - \dfrac{x^2}{81} = 1$.

a. What is the center?
$(0,0)$

b. What are x-intercepts?
It has no x-intercepts.

c. What are the y-intercepts?
$(0,-4)$ and $(0,4)$

d. What are the vertices of the fundamental rectangle?
$(9,4), (-9,4), (9,-4), (-9,-4)$

For Exercises 19–26, graph each hyperbola. Also show the fundamental rectangle with its corner points labeled, the asymptotes, and the intercepts. See Example 4.

19. $\dfrac{x^2}{9} - \dfrac{y^2}{4} = 1$

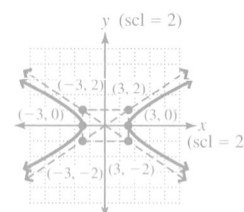

20. $\dfrac{x^2}{16} - \dfrac{y^2}{25} = 1$

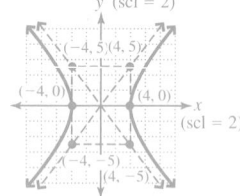

21. $\dfrac{y^2}{36} - \dfrac{x^2}{9} = 1$

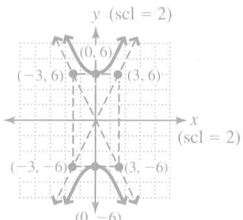

22. $\dfrac{y^2}{4} - \dfrac{x^2}{25} = 1$

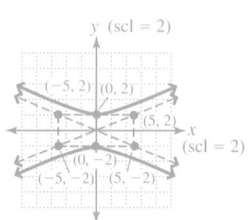

23. $9x^2 - y^2 = 36$

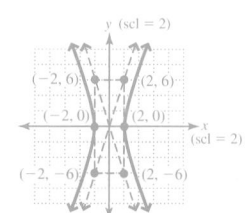

24. $x^2 - 4y^2 = 16$

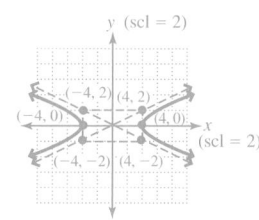

25. $9y^2 - 25x^2 = 225$

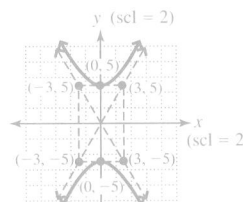

26. $16y^2 - 4x^2 = 64$

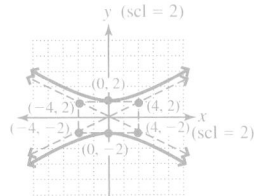

📱 *For Exercises 27 and 28, graph using a graphing calculator. (Hint: Solve the equation for y to find two functions.)*

27. $\dfrac{x^2}{36} - \dfrac{y^2}{4} = 1$

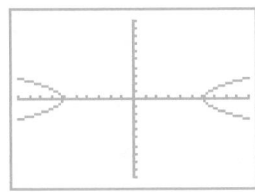

28. $\dfrac{y^2}{9} - \dfrac{x^2}{25} = 1$

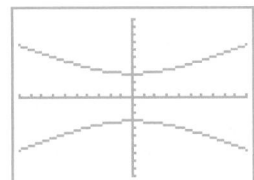

29. If a hyperbola opens left and right and the vertices of the fundamental rectangle are $(3, 2)$, $(3, -2)$, $(-3, 2)$, and $(-3, -2)$, what is the equation of the hyperbola?

$\dfrac{x^2}{9} - \dfrac{y^2}{4} = 1$

30. If a hyperbola opens up and down and the vertices of the fundamental rectangle are $(5, 3)$, $(5, -3)$, $(-5, 3)$, and $(-5, -3)$, what is the equation of the hyperbola?

$\dfrac{y^2}{9} - \dfrac{x^2}{25} = 1$

For Exercises 31–34, match the equation with the graph.

31. $\dfrac{x^2}{9} + \dfrac{y^2}{25} = 1$ c

32. $\dfrac{x^2}{9} - \dfrac{y^2}{25} = 1$ a

33. $\dfrac{x^2}{25} + \dfrac{y^2}{9} = 1$ b

34. $\dfrac{x^2}{25} - \dfrac{y^2}{9} = 1$ d

a.

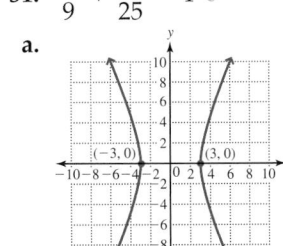

b.

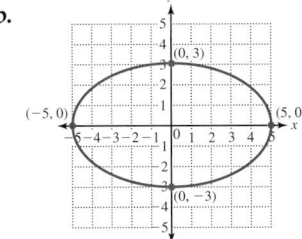

c.

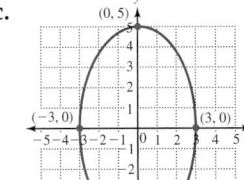

d.

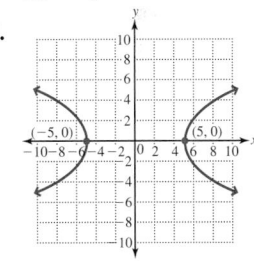

Prep Exercise 7 Which conic has an equation that has x^2 or y^2 but not both?

The parabola has only one squared term.

Prep Exercise 8 How can you tell the equation of an ellipse from the equation of a hyperbola?

The equation of an ellipse is a sum; the equation of a hyperbola is a difference.

For Exercises 35–38, determine whether the graph of the equation is a circle, a parabola, an ellipse, or a hyperbola. Do not draw the graph.

35. $9x^2 + 16y^2 = 144$

ellipse

36. $16y^2 - 9x^2 = 144$

hyperbola

37. $x^2 + y^2 - 6x + 8y - 75 = 0$

circle

38. $2x^2 - 12x + 23 - y = 0$

parabola

For Exercises 39–52, indicate whether the graph of the given equation is a circle, a parabola, an ellipse, or a hyperbola; then draw the graph.

39. $(x - 2)^2 + (y + 2)^2 = 49$ **40.** $x^2 + y^2 = 81$

circle

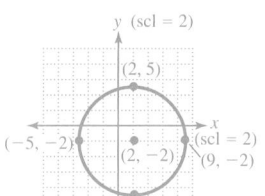

circle

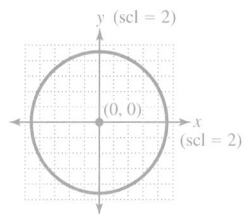

41. $y = 2(x + 1)^2 + 3$

parabola

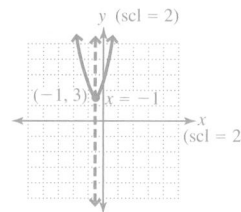

42. $x = (y + 3)^2 - 4$

parabola

43. $\dfrac{x^2}{36} + \dfrac{y^2}{81} = 1$

ellipse

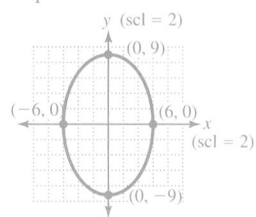

44. $\dfrac{x^2}{36} + \dfrac{y^2}{16} = 1$

ellipse

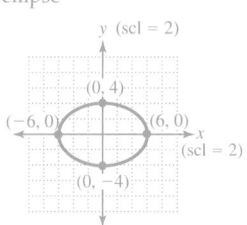

45. $\dfrac{x^2}{4} - \dfrac{y^2}{25} = 1$

hyperbola

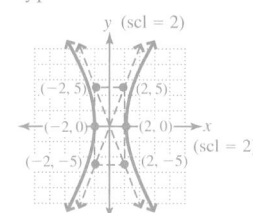

46. $\dfrac{y^2}{49} - \dfrac{x^2}{4} = 1$

hyperbola

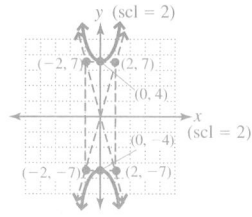

47. $y = 2x^2 + 8x + 6$

parabola

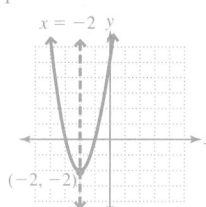

48. $x = y^2 + 6$

parabola

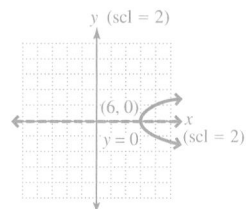

49. $\dfrac{x^2}{16} - \dfrac{y^2}{16} = 1$

hyperbola

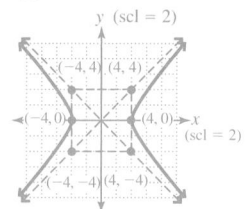

50. $\dfrac{y^2}{25} - \dfrac{x^2}{25} = 1$

hyperbola

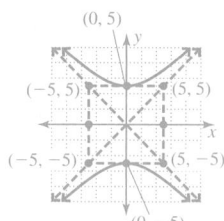

51. $\dfrac{(x + 3)^2}{9} + \dfrac{(y + 2)^2}{36} = 1$

ellipse

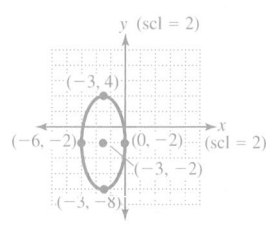

52. $\dfrac{(x - 1)^2}{4} + (y + 3)^2 = 1$

ellipse

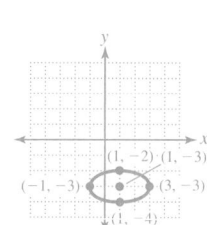

For Exercises 53–60, solve.

53. A bridge over a waterway has an arch in the form of half an ellipse. The equation of the ellipse is $400x^2 + 256y^2 = 102{,}400$.

   a. A sailboat, the top of whose mast is 18 feet above the water, is approaching the arch. Will the mast clear the bridge? Why or why not?

   Yes, the sailboat will clear the bridge. The height of the bridge at the center is 20 feet, and the boat's mast is only 18 feet above the water.

   b. How wide is the base of the arch?

   The bridge is 32 feet wide at the base of the arch.

54. A highway passes beneath an overpass that is in the shape of half an ellipse. The overpass is 15 feet high at the center and 40 feet wide at the base.

   a. What is the equation of the ellipse?

   $\dfrac{x^2}{400} + \dfrac{y^2}{225} = 1$ or $225x^2 + 400y^2 = 90{,}000$

   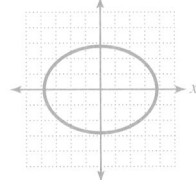

   b. A truck that is 10 feet wide carrying a load that is 14 feet above the level of the road is approaching the bridge. If the truck goes down the middle of the road, will the load clear the bridge? Why or why not?

   Yes, the truck will clear the overpass. If the truck is 10 feet wide and goes down the center of the road, it extends 5 feet on either side. Substituting $x = 5$ into the equation yields $y = 14.523$ feet, so the truck's 14-foot load will just barely clear the overpass.

55. The comet Epoch has an orbit in the shape of an ellipse with the Sun at one of the foci. The equation is approximately $\dfrac{x^2}{3.6^2} + \dfrac{y^2}{2.88^2} = 1$, where x and y are in astronomical units. (An astronomical unit is 93,000,000 miles.) Sketch the graph of the comet Epoch. (*Source: Orbital Motion*, A. E. Roy, Institute of Physics Publishing, London.)

56. The dwarf planet Pluto has an orbit in the shape of an ellipse with the Sun at one of the foci. The equation is approximately $\dfrac{x^2}{39.4^2} + \dfrac{y^2}{38.2^2} = 1$, where x and y are measured in astronomical units. Sketch the graph of the orbit of Pluto. (*Hint:* Make each unit on the x- and y-axes equal to 10.) (*Source: Orbital Motion*, A. E. Roy, Institute of Physics Publishing, London.)

57. Compound bows have elliptical cams that decrease the amount of effort required to hold the bow at full draw. If a cam on a bow is 4 inches from top to bottom and 3 inches across, what is the equation of the ellipse if the origin is at the center of the cam?

   $\dfrac{x^2}{2.25} + \dfrac{y^2}{4} = 1$

   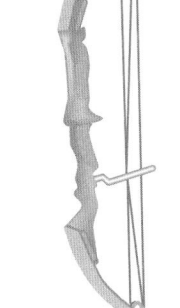

58. One of the most popular exercise machines is an elliptical trainer in which your foot moves in an elliptical path. On a typical machine, the length of the stride is about 19 inches and the height varies from 3 to 5 inches depending on the settings. Write the equation for the path that your foot takes if the total length of the stride is 19 inches and the total height is 4 inches. (*Source:* Precor National Headquarters.)

   $\dfrac{x^2}{90.25} + \dfrac{y^2}{4} = 1$

   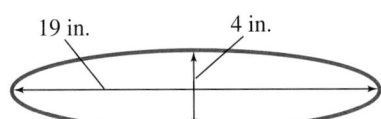

   19 in.   4 in.

59. If a source of light or sound is placed at one focal point of an elliptic reflector, the light or sound is reflected through the other focal point. This principle is used in a lithotripter that uses sound waves to crush kidney stones by placing a source of sound at one focal point and the kidney stone at the other. If the elliptic reflector is based on the ellipse $\dfrac{x^2}{25} + \dfrac{y^2}{16} = 1$, how many units from the center should the kidney stone be placed? (*Hint:* $c^2 = a^2 - b^2$.)

   3 units

60. The same principle used in Exercise 59 is also used in whispering rooms. The room is in the shape of an elliptic reflector where one person speaks at one focal point and the other person places his or her ear at the other focal point and can hear sounds as faint as a whisper. If a whispering room is based on the ellipse $\dfrac{x^2}{225} + \dfrac{y^2}{144} = 1$, how many units from the center of the ellipse would the two people stand? (*Hint:* $c^2 = a^2 - b^2$.)

   9 units

Puzzle Problem Suppose an ellipse is drawn using the method described on p. 847. Assuming that the ellipse is centered at the origin, if the nails are 8 inches apart and the string is 10 inches in length, what are the x- and y-intercepts?

$(5, 0), (-5, 0), (0, 3), (0, -3)$

Review Exercises

Exercises 1–6 Equations and Inequalities

[2.4] 1. Solve $3x^2 + y = 6$ for y.
$y = -3x^2 + 6$

[4.1] 2. Solve the following system using the graphical method. $\begin{cases} x + y = 3 \\ 2x + y = 4 \end{cases}$
$(1, 2)$

[4.2] 3. Solve the following system using the substitution method. $\begin{cases} 2x + y = 1 \\ 3x + 4y = -6 \end{cases}$
$(2, -3)$

[4.3] 4. Solve the following system using the elimination method. $\begin{cases} 2x + 3y = 6 \\ 3x - 4y = -25 \end{cases}$
$(-3, 4)$

For Exercises 5 and 6, solve.

[6.6] 5. $3x^2 + 10x - 8 = 0$
$-4, \dfrac{2}{3}$

[10.1] 6. $4x^2 = 36$
± 3

12.3 Nonlinear Systems of Equations

Objectives

1 Solve nonlinear systems of equations using substitution.
2 Solve nonlinear systems of equations using elimination.

Warm-up

[12.2] 1. Graph: $9x^2 + 16y^2 = 144$
[2.4] 2. Solve $4x - y = 1$ for y.

[4.2] 3. Solve using the substitution method: $\begin{cases} x - 3y = 10 \\ 4x + 5y = 6 \end{cases}$

[4.3] 4. Solve using the elimination method: $\begin{cases} 5x + 2y = 13 \\ 2x - 5y = 11 \end{cases}$

We solved systems of linear equations in Chapter 4. Now we solve **nonlinear systems of equations.**

Definition Nonlinear system of equations: A system of equations that contains at least one nonlinear equation.

The types of equations in a nonlinear system determine the number of solutions that are possible for the system. For example, a system containing a quadratic equation and a linear equation can have 0, 1, or 2 points of intersection and therefore 0, 1, or 2 solutions, respectively, as shown by the following figures.

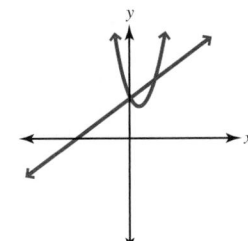

No points of intersection: no solutions

One point of intersection: one solution

Two points of intersection: two solutions

Answers to Warm-up

1.

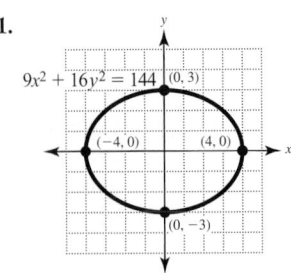

2. $y = 4x - 1$ **3.** $(4, -2)$
4. $(3, -1)$

Connection In Chapter 4, we learned that the number of solutions for a system of linear equations depends on the relative positions of the graphs. We see a similar relationship between the graphs and the number of solutions here.

Objective 1 Solve nonlinear systems of equations using substitution.

When solving nonlinear systems, if one equation is linear (or one of the variables in an equation is isolated), the substitution method is usually preferred.

Example 1   Solve using the substitution method.

a. $\begin{cases} y = 2(x + 2)^2 - 2 \\ 2x + y = -2 \end{cases}$ ◄ **Note** The graphs are a parabola and a line; so there will be 0, 1, or 2 solutions.

Solution: It is easier to solve the linear equation for one of its variables and substitute into the nonlinear equation. Because the coefficient of y is 1, we solve $2x + y = -2$ for y and get $y = -2x - 2$.

$-2x - 2 = 2(x + 2)^2 - 2$	Substitute $-2x - 2$ for y in $y = 2(x + 2)^2 - 2$.
$-2x - 2 = 2(x^2 + 4x + 4) - 2$	To solve this quadratic equation, we need to write it in the form $ax^2 + bx + c = 0$. First, we square $x + 2$.
$-2x - 2 = 2x^2 + 8x + 6$	Distribute 2; then combine like terms.
$0 = 2x^2 + 10x + 8$	Add 2x and 2 to both sides.
$0 = x^2 + 5x + 4$	Divide both sides by 2.
$0 = (x + 1)(x + 4)$	Factor.
$0 = x + 1$ or $0 = x + 4$	Use the zero-factor theorem.
$-1 = x$ or $-4 = x$	Solve each equation.

To find y, substitute -1 and -4 for x in $y = -2x - 2$.

$$y = -2(-1) - 2 \qquad\qquad y = -2(-4) - 2$$
$$y = 2 - 2 \qquad\qquad\qquad y = 8 - 2$$
$$y = 0 \qquad\qquad\qquad\qquad y = 6$$

The solutions are $(-1, 0)$ and $(-4, 6)$, as verified by the graph to the left.

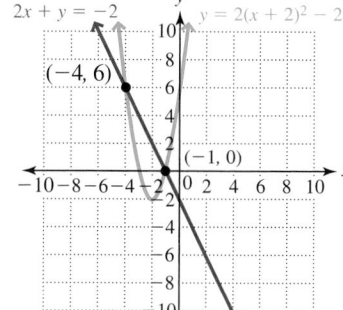

b. $\begin{cases} y = \sqrt{x + 2} \\ x^2 + y^2 = 8 \end{cases}$

Solution: Because $y = \sqrt{x + 2}$ is already solved for y, substitute $\sqrt{x + 2}$ for y in $x^2 + y^2 = 8$.

$x^2 + (\sqrt{x + 2})^2 = 8$	Substitute $\sqrt{x + 2}$ for y in $x^2 + y^2 = 8$.
$x^2 + x + 2 = 8$	Square $\sqrt{x + 2}$.
$x^2 + x - 6 = 0$	Because the equation is now quadratic, we subtract 8 from both sides to get the form $ax^2 + bx + c = 0$.
$(x + 3)(x - 2) = 0$	Factor.
$x + 3 = 0$ or $x - 2 = 0$	Use the zero-factor theorem.
$x = -3$ or $x = 2$	Solve each equation.

To find y, substitute -3 and 2 for x in $y = \sqrt{x + 2}$.

$$y = \sqrt{-3 + 2} \qquad\qquad y = \sqrt{2 + 2}$$
$$y = \sqrt{-1} \qquad\qquad\qquad y = \sqrt{4}$$
$$\qquad\qquad\qquad\qquad\qquad y = 2$$

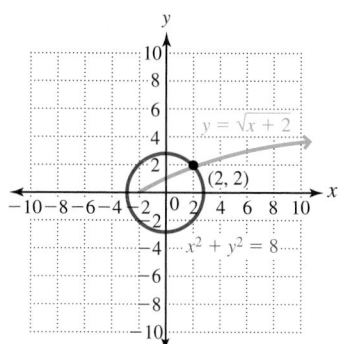

Because $\sqrt{-1}$ is imaginary, the only real solution is $(2, 2)$, as verified by the graph to the left.

Note A system containing a root function and a circle, as in Example 1 (*b*), can have 0, 1, or 2 solutions, as shown.

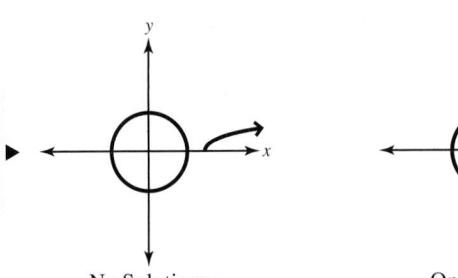

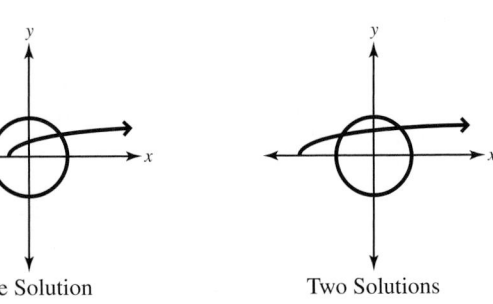

No Solutions   One Solution   Two Solutions

Your Turn 1 Solve using the substitution method.

a. $\begin{cases} 2x - y = 1 \\ x^2 + y^2 = 1 \end{cases}$ b. $\begin{cases} xy = 3 \\ 4x - y = 1 \end{cases}$

Objective 2 Solve nonlinear systems of equations using elimination.

If neither equation contains a radical expression or both equations contain the same powers of the variables, the elimination method can be used.

Example 2 Solve using the elimination method: $\begin{cases} 9x^2 - 4y^2 = 20 \\ x^2 + y^2 = 8 \end{cases}$

Solution: $\begin{cases} 9x^2 - 4y^2 = 20 & (\text{Equation 1}) \\ x^2 + y^2 = 8 & (\text{Equation 2}) \end{cases}$

To eliminate y^2, multiply Equation 2 by 4 and add the equations.

$$9x^2 - 4y^2 = 20 \qquad\qquad 9x^2 - 4y^2 = 20$$
$$x^2 + y^2 = 8 \xrightarrow[\text{Multiply by 4.}]{} \quad \underline{4x^2 + 4y^2 = 32} \quad \text{Add the equations.}$$
$$13x^2 \qquad = 52$$

$$x^2 = 4 \qquad \text{Divide both sides by 13.}$$
$$x = \pm 2 \qquad \text{Find the square roots of 4.}$$

To find y, substitute 2 and -2 for x in one of the original equations. We will use $x^2 + y^2 = 8$.

Substitute $x = 2$.   Substitute $x = -2$.

$2^2 + y^2 = 8$ $(-2)^2 + y^2 = 8$

$4 + y^2 = 8$ $4 + y^2 = 8$

$y^2 = 4$ $y^2 = 4$

$y = \pm 2$ $y = \pm 2$

Therefore, $(2, 2)$ and $(2, -2)$ Therefore, $(-2, 2)$ and $(-2, -2)$
are solutions.   are solutions.

This system has four solutions: $(2, 2)$, $(2, -2)$, $(-2, 2)$, and $(-2, -2)$, as verified by the graph to the left.

Answers to Your Turn 1

a. $\left(\dfrac{4}{5}, \dfrac{3}{5}\right)$, $(0, -1)$

b. $\left(-\dfrac{3}{4}, -4\right)$, $(1, 3)$

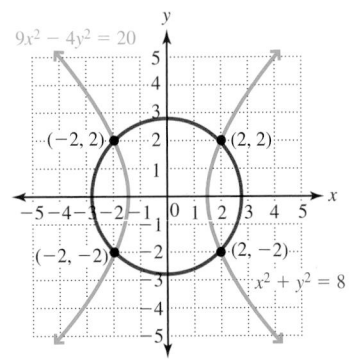

Note If the graphs are a hyperbola and a circle both centered at the origin as in Example 2, there will be 0, 2, or 4 solutions, as shown. ▶

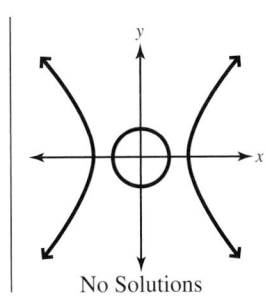

No Solutions

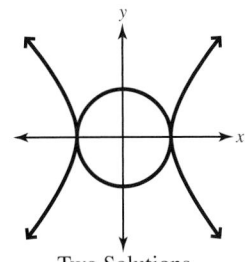

Two Solutions

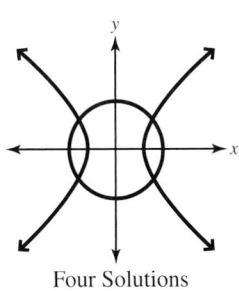

Four Solutions

Answers to Your Turn 2
$(3, 2), (3, -2), (-3, 2), (-3, -2)$

| **Your Turn 2** | Solve using the elimination method: $\begin{cases} 4x^2 + 9y^2 = 72 \\ x^2 + y^2 = 13 \end{cases}$

12.3 Exercises   For Extra Help   MyMathLab®

Objectives 1 and 2

Prep Exercise 1 a. How many real solutions are possible for a system of two equations whose graphs are a circle and a parabola? **b.** Draw a figure illustrating such a system with two solutions.

a. 0, 1, 2, 3, or 4 solutions are possible.

b. Answers may vary.

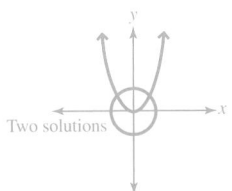

Two solutions

Prep Exercise 2 a. How many real solutions are possible for a system of two equations whose graphs are a line and a hyperbola? **b.** Draw a figure illustrating such a system with one solution.

a. 0, 1, or 2 solutions are possible.

b. Answers may vary.

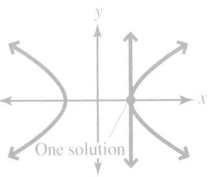

One solution

Prep Exercise 3 Which method (substitution or elimination) would you use to solve the system $\begin{cases} 2x + y = 8 \\ 4x^2 + 3y^2 = 24 \end{cases}$?
Why? (Do not attempt to solve the system.)

Substitution because the first equation has a linear term and is easy to solve for y.

Prep Exercise 5 Without solving the system, what is the number of possible solutions of the system $\begin{cases} 3x - y = 6 \\ 4x^2 + 9y^2 = 36 \end{cases}$?

The graph of the first equation is a line, and the second is an ellipse; so the system could have 0, 1, or 2 solutions.

Prep Exercise 4 Which method (substitution or elimination) would you use to solve the system $\begin{cases} 3x^2 - 4y^2 = 12 \\ 2x^2 + 3y^2 = 24 \end{cases}$?
Why? (Do not attempt to solve the system.)

Elimination because all variables are squared.

Prep Exercise 6 Is it possible for a system of two equations whose graphs are an ellipse and a hyperbola (both centered at the origin) to have three solutions? Why or why not?

No. If both graphs are centered at the origin, then because of the symmetry, there could be only 0, 2, or 4 solutions.

For Exercises 1–36, solve. See Examples 1 and 2.

1. $\begin{cases} x^2 + 2y = 1 \\ 2x + y = 2 \end{cases}$
$(3, -4), (1, 0)$

2. $\begin{cases} y = 2x^2 \\ 2x + y = 4 \end{cases}$
$(-2, 8), (1, 2)$

3. $\begin{cases} y = x^2 + 4x + 4 \\ 3x - y = -6 \end{cases}$
$(-2, 0), (1, 9)$

4. $\begin{cases} y = 6x - x^2 \\ 2x - y = -3 \end{cases}$
$(3, 9), (1, 5)$

5. $\begin{cases} x^2 + y^2 = 25 \\ x - y = -1 \end{cases}$
$(-4, -3), (3, 4)$

6. $\begin{cases} x^2 + y^2 = 25 \\ x - 7y = -25 \end{cases}$
$(3, 4), (-4, 3)$

7. $\begin{cases} x^2 + y^2 = 10 \\ 3x + y = 6 \end{cases}$
$(2.6, -1.8), (1, 3)$

8. $\begin{cases} x^2 + y^2 = 13 \\ 2x + y = 7 \end{cases}$
$(3.6, -0.2), (2, 3)$

9. $\begin{cases} x^2 + 2y^2 = 4 \\ x + y = 5 \end{cases}$
No solution

10. $\begin{cases} 3x^2 + y^2 = 9 \\ 2x + y = 11 \end{cases}$
No solution

11. $\begin{cases} y = 2x^2 + 3 \\ 2x - y = -3 \end{cases}$
$(0, 3), (1, 5)$

12. $\begin{cases} y = -3x^2 - 2 \\ 3x + y = -2 \end{cases}$
$(0, -2), (1, -5)$

13. $\begin{cases} y = (x - 3)^2 + 2 \\ y = -(x - 2)^2 + 3 \end{cases}$
$(3, 2), (2, 3)$

14. $\begin{cases} y = 2(x + 4)^2 + 2 \\ y = -2(x + 3)^2 + 4 \end{cases}$
$(-4, 2), (-3, 4)$

15. $\begin{cases} x^2 + y^2 = 20 \\ x^2 - y^2 = 12 \end{cases}$
$(4, 2), (4, -2), (-4, -2), (-4, 2)$

16. $\begin{cases} x^2 + y^2 = 48 \\ x^2 - y^2 = 24 \end{cases}$
$(6, 2\sqrt{3}), (-6, 2\sqrt{3}),$
$(-6, -2\sqrt{3}), (6, -2\sqrt{3})$

17. $\begin{cases} 4x^2 + 9y^2 = 72 \\ x^2 + y^2 = 13 \end{cases}$
$(3, 2), (3, -2), (-3, -2), (-3, 2)$

18. $\begin{cases} 9x^2 + 4y^2 = 145 \\ x^2 + y^2 = 25 \end{cases}$
$(3, 4), (-3, 4), (-3, -4), (3, -4)$

19. $\begin{cases} 4x^2 - y^2 = 15 \\ x^2 + y^2 = 5 \end{cases}$
$(2, 1), (2, -1), (-2, -1), (-2, 1)$

20. $\begin{cases} 9x^2 - 4y^2 = 32 \\ x^2 + y^2 = 5 \end{cases}$
$(2, 1), (2, -1), (-2, -1), (-2, 1)$

21. $\begin{cases} x^2 + 3y^2 = 36 \\ x = y^2 - 6 \end{cases}$
$(-6, 0), (3, 3), (3, -3)$

22. $\begin{cases} 4x^2 + y^2 = 16 \\ y = x^2 - 4 \end{cases}$
$(0, -4), (2, 0), (-2, 0)$

23. $\begin{cases} 9x^2 + 4y^2 = 36 \\ 4x^2 - 9y^2 = 36 \end{cases}$
No solution

24. $\begin{cases} 9x^2 - 16y^2 = 144 \\ 4x^2 + 9y^2 = 36 \end{cases}$
No solution

25. $\begin{cases} y = x^2 \\ x^2 + y^2 = 20 \end{cases}$
$(2, 4), (-2, 4)$

26. $\begin{cases} 16x^2 + y^2 = 128 \\ y = 2x^2 \end{cases}$
$(2, 8), (-2, 8)$

27. $\begin{cases} 25x^2 - 16y^2 = 400 \\ x^2 + 4y^2 = 16 \end{cases}$
$(4, 0), (-4, 0)$

28. $\begin{cases} 9y^2 - 25x^2 = 225 \\ 4y^2 + 25x^2 = 100 \end{cases}$
$(0, 5), (0, -5)$

29. $\begin{cases} xy = 4 \\ 2x^2 - y^2 = 4 \end{cases}$
$(2, 2), (-2, -2)$

30. $\begin{cases} xy = 2 \\ 4x^2 + y^2 = 8 \end{cases}$
$(1, 2), (-1, -2)$

31. $\begin{cases} y = x^2 - 2x - 3 \\ y = -x^2 + 6x + 7 \end{cases}$
$(5, 12), (-1, 0)$

32. $\begin{cases} y = \dfrac{1}{3}x^2 - 2 \\ 3x^2 + 9y^2 = 36 \end{cases}$
$(0, -2), (3, 1), (-3, 1)$

33. $\begin{cases} 4x^2 + 5y^2 = 36 \\ 4x^2 - 3y^2 = 4 \end{cases}$
$(2, 2), (2, -2), (-2, -2), (-2, 2)$

34. $\begin{cases} 4x^2 + 7y^2 = 64 \\ 4x^2 - 3y^2 = 24 \end{cases}$
$(3, 2), (3, -2), (-3, -2), (-3, 2)$

35. $\begin{cases} x = -y^2 + 2 \\ x^2 - 5y^2 = 4 \end{cases}$
$(-7, 3), (-7, -3), (2, 0)$

36. $\begin{cases} x = y^2 + 2 \\ 9x^2 - 45y^2 = 36 \end{cases}$
$(2, 0), (3, 1), (3, -1)$

37. Create a system of two equations whose graphs are a circle and a line for which there is no solution. Include the graphs.
Answers may vary, but one
possible system is $\begin{cases} x + y = 4 \\ x^2 + y^2 = 1 \end{cases}$.
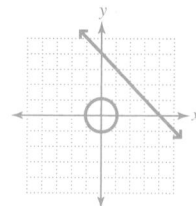

38. Create a system of two equations whose graphs are a circle and a hyperbola for which there are exactly two solutions. Include the graphs.
Answers may vary, but one
possible system is $\begin{cases} x^2 - y^2 = 1 \\ x^2 + y^2 = 1 \end{cases}$.
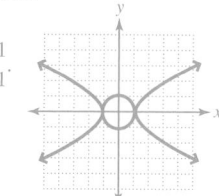

39. The sum of the squares of two integers is 34, and the difference of their squares is 16. Find the integers. There is more than one solution.

$(5,3), (5,-3), (-5,3), (-5,-3)$

40. The difference of the squares of two integers is 32, and their product is 12. Find the integers. There is more than one solution.

$(6,2), (-6,-2)$

41. A computer keyboard has an area of 144 square inches and a perimeter of 52 inches. Find the length and width.

Length: 18 in.; width: 8 in.

42. A rectangular living room has an area of 48 square meters and a perimeter of 28 meters. Find the dimensions.

8 m by 6 m

43. If p is in dollars and x is in hundreds of units, the demand function for a certain style of chair is given by $p = -3x^2 + 120$ and the supply function is given by $p = 11x + 28$. The *market equilibrium* occurs when the number produced is equal to the number demanded. Find the number of chairs and the price per chair when market equilibrium is reached.

At equilibrium, there should be 400 chairs at $72 per chair.

44. If y is in dollars and x is the number of cell phones manufactured (in thousands), a cell phone manufacturer has determined that the cost y to manufacture x cell phones is given by $y = 5x^2 + 30x + 50$ and the revenue from the sales is given by $y = 13x^2$. The *break-even point* is the point (x, y) for which the cost equals the revenue. Solve the system to find the number of units necessary to break even.

To break even requires manufacturing 5000 cell phones.

For Exercises 45–48, use a graphing calculator to verify the results of the exercise given.

45. Exercise 11

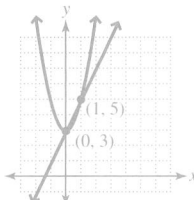

46. Exercise 12

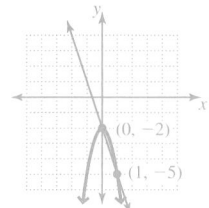

47. Exercise 13

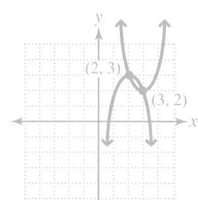

48. Exercise 14

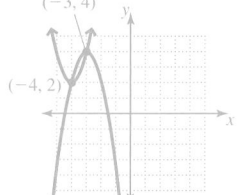

Review Exercises

Exercises 1–6   Equations and Inequalities

[3.6] *For Exercises 1 and 2, determine whether the ordered pair is a solution for $x + 3y \leq 6$.*

1. $(0,0)$

yes

2. $(3,4)$

no

[3.6] *For Exercises 3–6, graph.*

3. $y \geq 3$

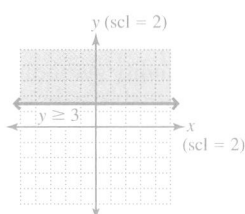

4. $y < 2x + 3$

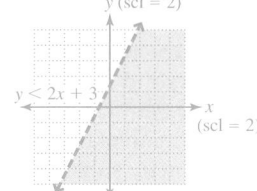

5. $2x + 3y < -6$

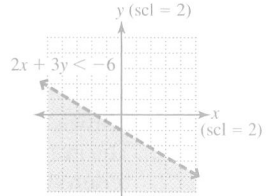

6. $x \geq -2$

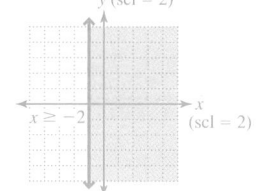

12.4 Nonlinear Inequalities and Systems of Inequalities

Objectives

1 Graph nonlinear inequalities.

2 Graph the solution set of a system of nonlinear inequalities.

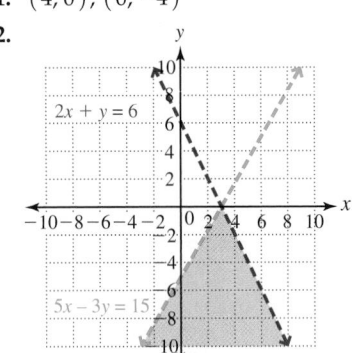
Warm-up

[12.3] **1.** Solve: $\begin{cases} x^2 + y^2 = 16 \\ x - y = 4 \end{cases}$

[4.6] **2.** Graph the solution set of the system of inequalities:
$$\begin{cases} 5x - 3y > 15 \\ 2x + y < 6 \end{cases}$$

Objective 1 Graph nonlinear inequalities.

In Section 3.6, we graphed linear inequalities like $x + 2y > 6$ by graphing the line corresponding to $x + 2y = 6$ and then shading the appropriate region on one side of that boundary line. Recall that we used a dashed line for $<$ and $>$ and a solid line for $\leq$ and $\geq$. We determined which side to shade by using a test point on one side of the boundary.

We use a similar procedure to graph nonlinear inequalities such as $x^2 + y^2 \leq 25$. The boundary is the graph of $x^2 + y^2 = 25$, which is a circle with center at $(0, 0)$ and radius of 5. The solution set of $x^2 + y^2 \leq 25$ contains all ordered pairs on the circle (boundary) along with all ordered pairs inside it or all ordered pairs outside it. To determine which of those two regions is correct, we choose an ordered pair from one of the regions to test in the inequality. Let's test $(0, 0)$.

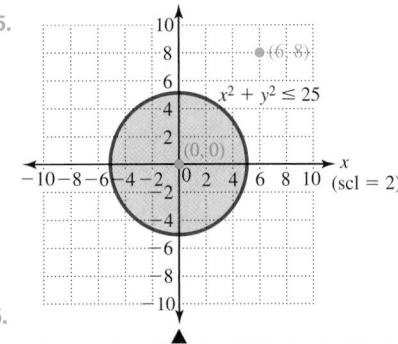

$$0^2 + 0^2 \leq 25 \quad \text{Substitute (0, 0) into } x^2 + y^2 \leq 25.$$

$$0 \leq 25 \quad \text{Simplify. The inequality is true.}$$

Because $0 \leq 25$ is true, all ordered pairs in the region containing $(0, 0)$ are in the solution set; so we shade that region.

Ordered pairs in the region outside the circle are not in the solution set because they do not solve the inequality. To illustrate, let's test $(6, 8)$.

$$6^2 + 8^2 \leq 25 \quad \text{Substitute (6, 8) into } x^2 + y^2 \leq 25.$$

$$100 \leq 25 \quad \text{Simplify. The inequality is false.}$$

Because $100 \leq 25$ is false, the region containing $(6, 8)$ is not in the solution set; so we do not shade that region.

Note If the inequality had been $x^2 + y^2 < 25$, we would have drawn a dashed circle.

Our example suggests the following procedure for graphing nonlinear inequalities.

Procedure **Graphing Nonlinear Inequalities**

1. Graph the related equation (the boundary curve). If the inequality symbol is $\leq$ or $\geq$, draw the graph as a solid curve. If the inequality symbol is $>$ or $<$, draw a dashed curve.

2. The graph divides the coordinate plane into at least two regions. Test an ordered pair from each region by substituting it into the inequality. If the ordered pair satisfies the inequality, shade the region containing that ordered pair.

Example 1 Graph the inequality.

a. $y > 2(x - 1)^2 + 3$

Solution: We first graph the related equation,
$y = 2(x - 1)^2 + 3$.

> **Note** The parabola is dashed because the inequality is $>$. Also notice that the graph divides the coordinate plane into two regions: region 1 *above* the boundary and region 2 *below*.

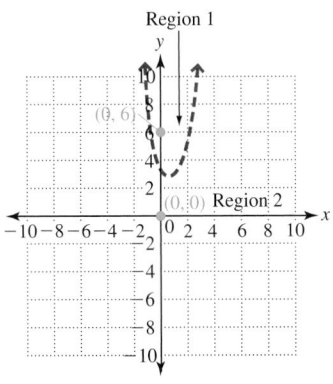

We now test an ordered pair from each region.

> **Note** To make computations easier, choose ordered pairs with 0 for at least one of the coordinates.

Region 1: We choose $(0, 6)$.

$6 > 2(0 - 1)^2 + 3$ Substitute.

$6 > 2(-1)^2 + 3$ Simplify.

$6 > 5$

True; so region 1 is in the solution set.

Region 2: We choose $(0, 0)$.

$0 > 2(0 - 1)^2 + 3$ Substitute.

$0 > 2(-1)^2 + 3$ Simplify.

$0 > 5$

False; so region 2 is not in the solution set.

Because ordered pairs only in region 1 solve the inequality, we shade only that region. The solution set contains all ordered pairs in region 1. Ordered pairs on the parabola are not in the solution set.

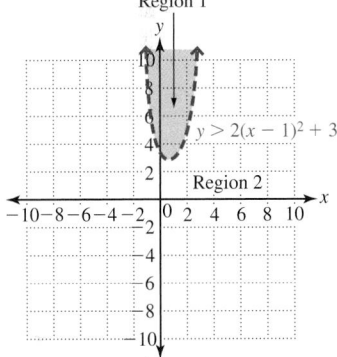

b. $\dfrac{x^2}{9} - \dfrac{y^2}{25} \le 1$

Solution: Graph the related equation,

$\dfrac{x^2}{9} - \dfrac{y^2}{25} = 1$.

> **Note** The hyperbola is solid because the inequality is $\le$. Also notice that the hyperbola divides the coordinate plane into three regions.

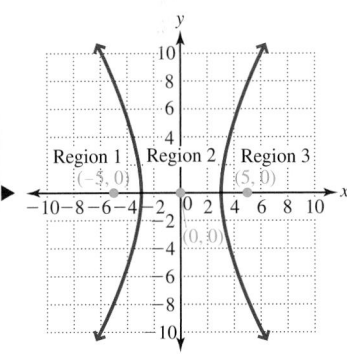

We test an ordered pair from each region.

Region 1: We choose $(-5, 0)$.

$\dfrac{(-5)^2}{9} - \dfrac{0^2}{25} \le 1$ Substitute.

$\dfrac{25}{9} \le 1$ Simplify.

False; so region 1 is not in the solution set.

Region 2: We choose $(0, 0)$.

$\dfrac{0^2}{9} - \dfrac{0^2}{25} \le 1$ Substitute.

$0 \le 1$ Simplify.

True; so region 2 is in the solution set.

Region 3: We choose $(5, 0)$.

$$\frac{(5)^2}{9} - \frac{0^2}{25} \leq 1 \quad \text{Substitute.}$$

$$\frac{25}{9} \leq 1 \quad \text{Simplify.}$$

False; so region 3 is not in the solution set.

Because region 2 is the only region containing ordered pairs that solve the inequality, we shade only that region. The solution set contains all ordered pairs on the hyperbola along with all ordered pairs in region 2.

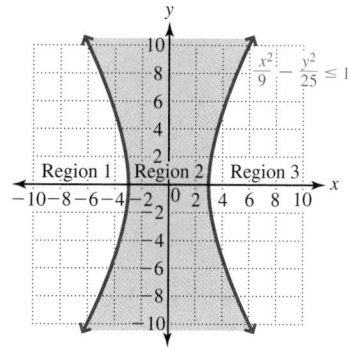

Your Turn 1   Graph the inequality.

a. $y \geq (x + 1)^2 - 3$ **b.** $\dfrac{x^2}{25} + \dfrac{y^2}{9} < 1$

Objective 2 Graph the solution set of a system of nonlinear inequalities.

In Section 4.6, we solved systems of linear inequalities by graphing each inequality on the same grid. The solution set of the system is the intersection of the solution sets of the individual inequalities. We use a similar procedure for systems of nonlinear inequalities.

Example 2   Graph the solution set of the system of inequalities.

a. $\begin{cases} y \geq x^2 - 4 \\ 2x - y < 2 \end{cases}$

Solution: We begin by graphing $y \geq x^2 - 4$ and $2x - y < 2$.

Note The boundary graph, $y = x^2 - 4$, is a parabola opening up with vertex at $(0, -4)$. The test point $(0, 0)$ gives the true statement $0 \geq -4$, so we shade the region containing $(0, 0)$.

Note The boundary graph, $2x - y = 2$, is a dashed line. The test point $(0, 0)$ gives a true statement $0 < 2$, so we shade the region containing $(0, 0)$.

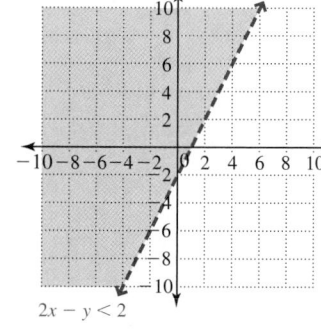

If we place both graphs on the same grid, their intersection (purple shading) is the solution region for the system. In addition to ordered pairs in the purple shaded region, the solution set also contains all ordered pairs on the portions of the parabola that touch the purple shaded region.

Note Remember, ordered pairs on dashed lines or curves are not in the solution set for a system of inequalities.

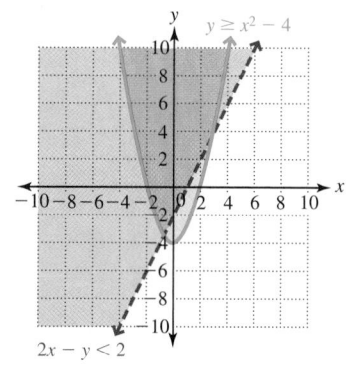

Answers to Your Turn 1

a.

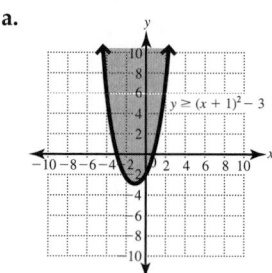

b.

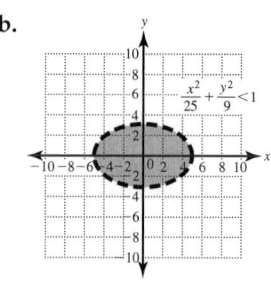

b. $\begin{cases} x^2 + y^2 \le 49 \\ \dfrac{x^2}{16} - \dfrac{y^2}{9} < 1 \\ y \ge 2x + 2 \end{cases}$

Answers to Your Turn 2

a.

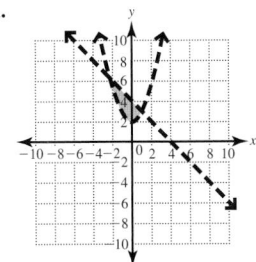

b.

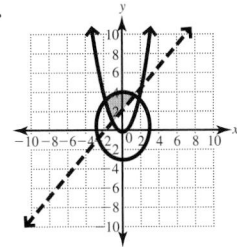

Solution: Graph each inequality on the same coordinate system.

The graph of $x^2 + y^2 \le 49$ is a circle and its interior.

The graph of $\dfrac{x^2}{16} - \dfrac{y^2}{9} < 1$ is the region between the branches of the hyperbola with the curve dashed. The graph of $y \ge 2x + 2$ is a solid line and the region above the line. The solution set for the system contains all ordered pairs in the region where the three graphs overlap (purple shaded region) together with all ordered pairs on the portion of the circle and the line that touches the purple shaded region.

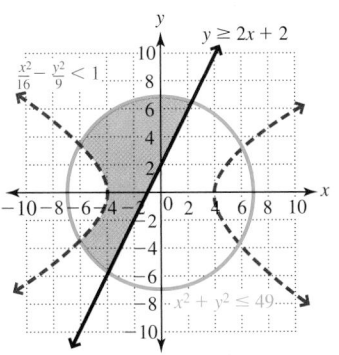

Your Turn 2   Graph the solution set of the system of inequalities.

a. $\begin{cases} y > x^2 + 2 \\ x + y < 4 \end{cases}$

b. $\begin{cases} \dfrac{x^2}{9} + \dfrac{y^2}{16} \le 1 \\ y \ge x^2 \\ -x + y > 2 \end{cases}$

12.4 Exercises For Extra Help MyMathLab®

Objective 1

Prep Exercise 1 What is the boundary curve for the inequality $9x^2 + 4y^2 > 36$? Is the graph drawn solid or dashed?

$9x^2 + 4y^2 = 36$; the ellipse is drawn dashed.

Prep Exercise 2 What is the boundary curve for the inequality $y \ge x^2 + 3$? Is the curve drawn solid or dashed?

The boundary curve is the parabola whose equation is $y = x^2 + 3$; the parabola is drawn solid.

Prep Exercise 3 Describe the procedure used to graph $\dfrac{x^2}{9} - \dfrac{y^2}{16} > 1$.

Draw a dashed hyperbola defined by $\dfrac{x^2}{9} - \dfrac{y^2}{16} = 1$. Then choose test points not on the curve to determine the shaded solution region.

For Exercises 1–20, graph the inequality. See Example 1.

1. $x^2 + y^2 \le 9$

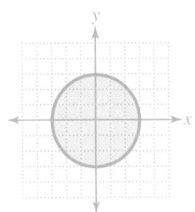

2. $x^2 + y^2 \ge 4$

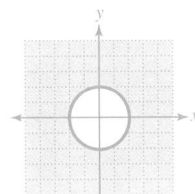

3. $y > x^2$

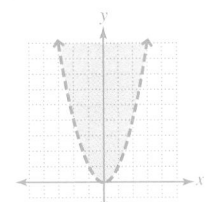

4. $y < -x^2$

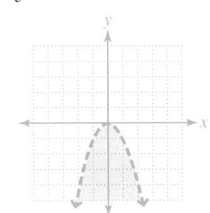

5. $y > 2(x + 2)^2 - 3$

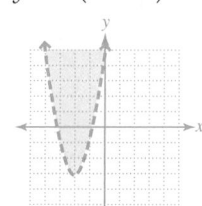

6. $y < (x - 1)^2 + 2$

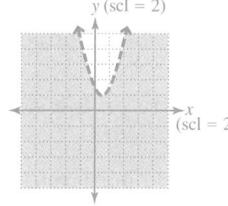

7. $\dfrac{x^2}{16} + \dfrac{y^2}{9} \le 1$

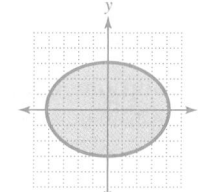

8. $\dfrac{x^2}{4} + \dfrac{y^2}{9} \ge 1$

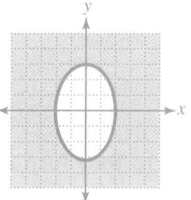

9. $\dfrac{x^2}{25} - \dfrac{y^2}{4} > 1$

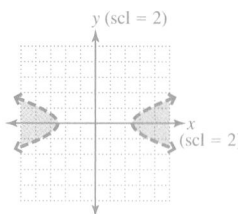

10. $\dfrac{x^2}{36} - \dfrac{y^2}{16} < 1$

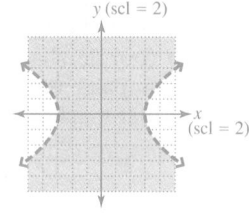

11. $x^2 + y^2 < 16$

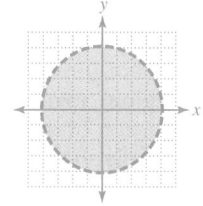

12. $x^2 + y^2 > 25$

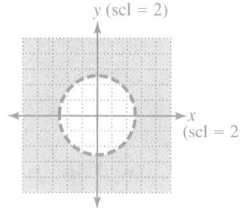

13. $y \ge -2(x - 3)^2 + 3$

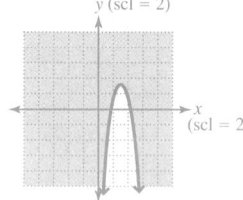

14. $y < -(x + 3)^2 - 2$

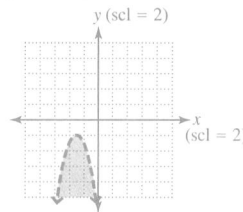

15. $\dfrac{x^2}{16} + \dfrac{y^2}{4} \le 1$

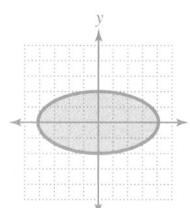

16. $\dfrac{x^2}{49} + \dfrac{y^2}{25} > 1$

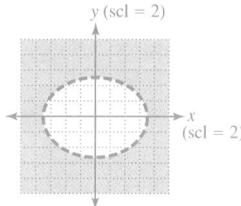

17. $\dfrac{y^2}{25} - \dfrac{x^2}{4} \le 1$

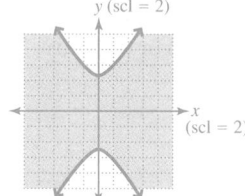

18. $\dfrac{y^2}{36} - \dfrac{x^2}{25} > 1$

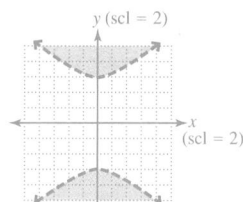

19. $y < x^2 + 4x - 5$

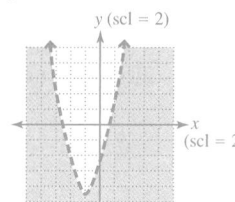

20. $y > x^2 - 6x - 6$

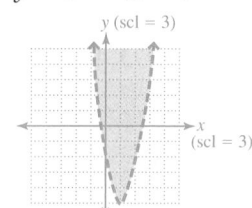

Objective 2

Prep Exercise 4 For the graph of the two inequalities in a system of inequalities, explain how to determine the solution region.

The solution region for the system is the intersection of the solution regions of the inequalities.

Prep Exercise 5 Why does the system $\begin{cases} x^2 + y^2 < 1 \\ x^2 + y^2 > 4 \end{cases}$ have no solution?

The graph of $x^2 + y^2 < 1$ is the region inside the circle with radius 1. The graph of $x^2 + y^2 > 4$ is the region outside the circle with radius 2. Thus, there is no common region of solution.

For Exercises 21–46, graph the solution set of each system of inequalities. See Example 2.

21. $\begin{cases} y < -x^2 \\ 2x - y < 4 \end{cases}$

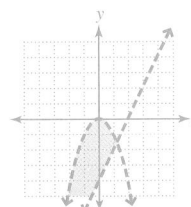

22. $\begin{cases} y \geq x^2 \\ x + y \leq 3 \end{cases}$

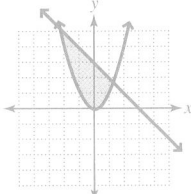

23. $\begin{cases} 3x + 2y \geq -6 \\ x^2 + y^2 \leq 25 \end{cases}$

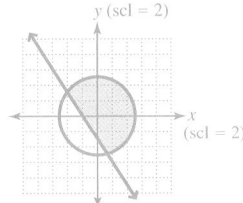

24. $\begin{cases} x - 2y < -4 \\ x^2 + y^2 < 16 \end{cases}$

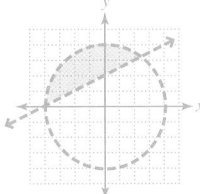

25. $\begin{cases} x^2 + y^2 > 4 \\ x^2 + y^2 > 9 \end{cases}$

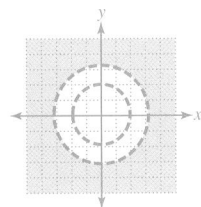

26. $\begin{cases} x^2 + y^2 \geq 9 \\ x^2 + y^2 \leq 25 \end{cases}$

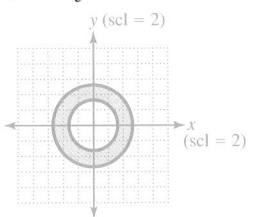

27. $\begin{cases} y > x^2 + 1 \\ 2x + y < 3 \end{cases}$

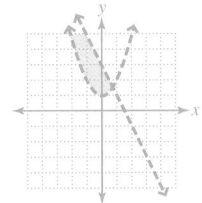

28. $\begin{cases} 3x - y \leq 2 \\ y \leq -x^2 + 2 \end{cases}$

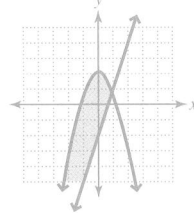

29. $\begin{cases} y < -x^2 + 3 \\ y > x^2 - 2 \end{cases}$

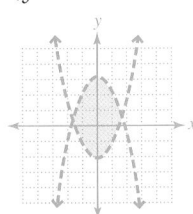

30. $\begin{cases} y > -x^2 + 2 \\ y < x^2 + 5 \end{cases}$

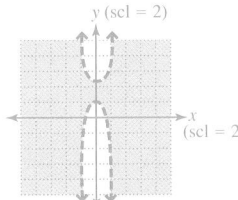

31. $\begin{cases} \dfrac{x^2}{25} + \dfrac{y^2}{9} \leq 1 \\ x^2 + y^2 \geq 4 \end{cases}$

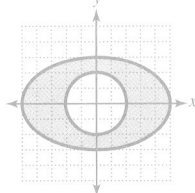

32. $\begin{cases} \dfrac{x^2}{9} + \dfrac{y^2}{4} \leq 1 \\ x^2 + y^2 \leq 4 \end{cases}$

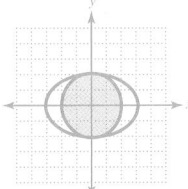

33. $\begin{cases} \dfrac{x^2}{25} + \dfrac{y^2}{9} < 1 \\ y > x^2 + 1 \end{cases}$

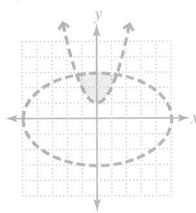

34. $\begin{cases} \dfrac{x^2}{9} + \dfrac{y^2}{4} < 1 \\ y < -x^2 + 2 \end{cases}$

35. $\begin{cases} \dfrac{x^2}{9} - \dfrac{y^2}{4} \leq 1 \\ \dfrac{x^2}{25} + \dfrac{y^2}{9} \leq 1 \end{cases}$

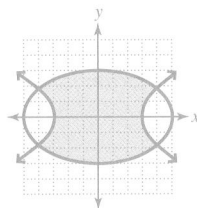

36. $\begin{cases} \dfrac{x^2}{16} - \dfrac{y^2}{9} \geq 1 \\ \dfrac{x^2}{36} + \dfrac{y^2}{16} \leq 1 \end{cases}$

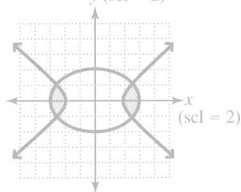

37. $\begin{cases} \dfrac{x^2}{4} - \dfrac{y^2}{4} > 1 \\ y > 2 \end{cases}$

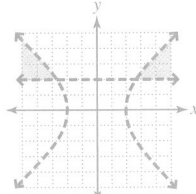

38. $\begin{cases} \dfrac{x^2}{9} - \dfrac{y^2}{9} > 1 \\ x > 3 \end{cases}$

39. $\begin{cases} 3x + 2y \leq 6 \\ x - y > -3 \\ x + 6y \geq 2 \end{cases}$

40. $\begin{cases} 2x + 3y < 6 \\ x - 2y > -4 \\ x + 5y \geq -4 \end{cases}$

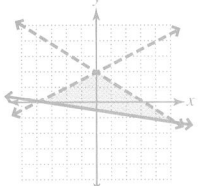

41. $\begin{cases} \dfrac{x^2}{16} + \dfrac{y^2}{4} \le 1 \\ x^2 + y^2 \le 9 \\ y \le x \end{cases}$

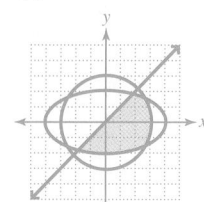

42. $\begin{cases} \dfrac{x^2}{9} + \dfrac{y^2}{16} < 1 \\ x^2 + y^2 < 9 \\ y > x + 1 \end{cases}$

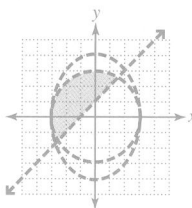

43. $\begin{cases} \dfrac{x^2}{49} - \dfrac{y^2}{16} \le 1 \\ \dfrac{x^2}{64} + \dfrac{y^2}{36} \le 1 \\ 2x - y \le -3 \end{cases}$

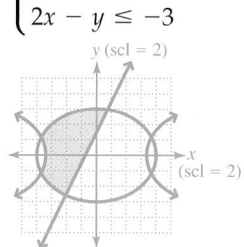

44. $\begin{cases} \dfrac{x^2}{9} + \dfrac{y^2}{49} \le 1 \\ y \ge x^2 + 3 \\ x + y \le 6 \end{cases}$

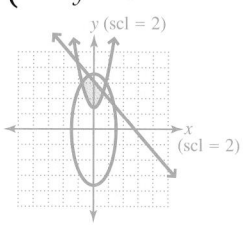

45. $\begin{cases} y < 2x^2 + 8 \\ 2x + y > 3 \\ x - y < 4 \end{cases}$

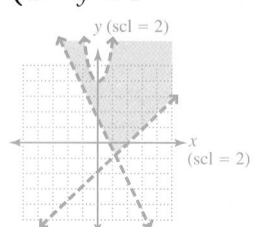

46. $\begin{cases} y < x^2 + 2 \\ 2x + y < 4 \\ 2x - y > -5 \end{cases}$

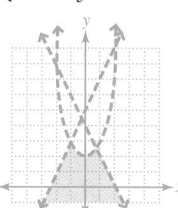

Review Exercises

Exercises 1–5   **Expressions**

[1.5] 1. Simplify: $(3 \cdot 1 + 2) + (3 \cdot 2 + 2) + (3 \cdot 3 + 2) + (3 \cdot 4 + 2)$

  38

[1.7] 2. Evaluate $\dfrac{n}{2}(a_1 + a_n)$, when $n = 12$, $a_1 = -8$, and $a_n = 60$.

  312

[1.7] 3. Evaluate $a_1 + (n - 1)d$, when $a_1 = -12$, $n = 25$, and $d = -3$.

  −84

[1.7] *For Exercises 4 and 5, evaluate the expression for $n = 1, 2, 3,$ and 4.*

4. $2n^2 - 3$
  $-1, 5, 15, 29$

5. $\dfrac{(-1)^n}{3n - 2}$
  $-1, \dfrac{1}{4}, -\dfrac{1}{7}, \dfrac{1}{10}$

Exercise 6   **Equations and Inequalities**

[4.2] 6. The sum of three consecutive odd integers is 207. Find the integers.
  67, 69, 71

Chapter 12 Summary and Review Exercises

Complete each incomplete definition, rule, or procedure; study the key examples; and then work the related exercises.

12.1 Parabolas and Circles

Definitions/Rules/Procedures	Key Example(s)
A **conic section** is a curve in a plane that is the result of intersecting the plane with a(n) ___cone___—more specifically, a circle, an ellipse, a parabola, or a(n) ___hyperbola___. The graph of an equation in the form $x = a(y - k)^2 + h$ is a parabola with vertex at ___(h, k)___. The parabola opens to the right if ___$a > 0$___ and to the left if ___$a < 0$___. The equation of the axis of symmetry is ___$y = k$___. 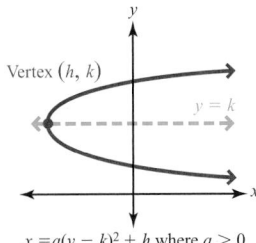 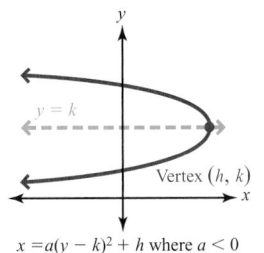 $x = a(y - k)^2 + h$ where $a > 0$ $x = a(y - k)^2 + h$ where $a < 0$ If necessary, complete the square to put the equation in the proper form.	For $x = y^2 - 4y - 5$, determine whether the graph opens left or right, find the vertex and axis of symmetry, and draw the graph. **Solution:** Because $a = 1$ and $1 > 0$, the graph opens to the right. To find the vertex and axis of symmetry, complete the square. $$x = y^2 - 4y - 5$$ $$x + 5 = y^2 - 4y \qquad \text{Add 5 to both sides.}$$ $$x + 5 + 4 = y^2 - 4y + 4 \qquad \text{Add 4 to both sides.}$$ $$x + 9 = (y - 2)^2 \qquad \text{Simplify.}$$ $$x = (y - 2)^2 - 9 \qquad \text{Subtract 9 from both sides.}$$ The vertex is $(-9, 2)$, and the equation of the axis of symmetry is $y = 2$. To graph, choose values of y near the axis of symmetry and plot the graph. <table><tr><th>x</th><th>y</th></tr><tr><td>-5</td><td>0</td></tr><tr><td>-8</td><td>1</td></tr><tr><td>-8</td><td>3</td></tr><tr><td>-5</td><td>4</td></tr></table> 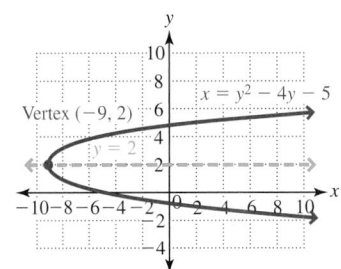

Exercises 1–9   **Equations and Inequalities**

[12.1] *For Exercises 1–4, find the direction that the parabola opens, the coordinates of the vertex, and the equation of the axis of symmetry. Draw the graph.*

1. $y = 2(x - 3)^2 - 5$
Opens upward;
vertex: $(3, -5)$;
axis of symmetry: $x = 3$

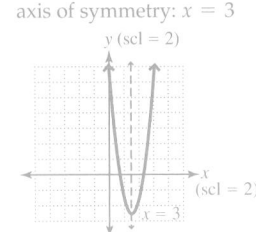

2. $y = -2(x + 2)^2 + 3$
Opens downward;
vertex: $(-2, 3)$;
axis of symmetry: $x = -2$

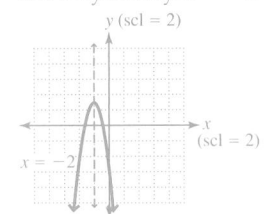

3. $x = -(y - 2)^2 + 4$
Opens left;
vertex: $(4, 2)$;
axis of symmetry: $y = 2$

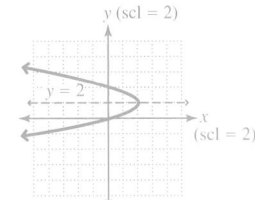

4. $x = 2(y + 3)^2 - 2$
Opens right;
vertex: $(-2, -3)$;
axis of symmetry: $y = -3$

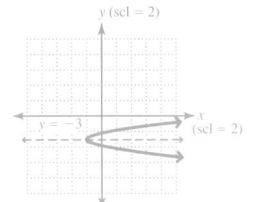

[12.1] *For Exercises 5–8, find the direction that the parabola opens, the coordinates of the vertex, and the equation of the axis of symmetry. Draw the graph.*

5. $x = y^2 + 6y + 8$
Opens right;
vertex: $(-1, -3)$;
axis of symmetry: $y = -3$

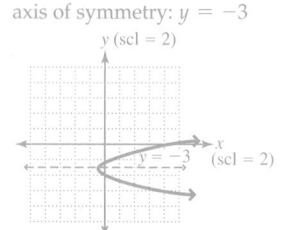

6. $x = 2y^2 - 8y - 6$
Opens right;
vertex: $(-14, 2)$;
axis of symmetry: $y = 2$

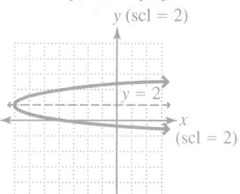

7. $x = -y^2 - 2y + 3$
Opens left;
vertex: $(4, -1)$;
axis of symmetry: $y = -1$

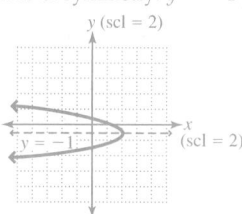

8. $x = -3y^2 - 12y - 9$
Opens left;
vertex: $(3, -2)$;
axis of symmetry: $y = -2$

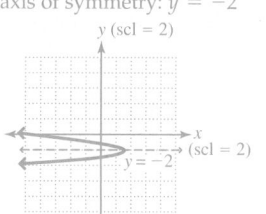

9. If a heavy object is thrown vertically upward with an initial velocity of 32 feet per second from the top of a building 128 feet high, its height above the ground after t seconds is given by $h = -16t^2 + 32t + 128$, where h is the height in feet and t is the time in seconds.

    a. What is the maximum height the object will reach?
      144 ft.

    b. How many seconds will the object take to reach its maximum height?
      1 sec.

    c. How many seconds will the object take to strike the ground?
      4 sec.

Definitions/Rules/Procedures	Key Example(s)
The **distance**, d, between two points with coordinates (x_1, y_1) and (x_2, y_2) can be found using the formula $$d = \underline{\sqrt{(x_2 - x_1)^2 + (y_2 - y_1)^2}}.$$	Find the distance and midpoint between $(-4, 3)$ and $(2, -1)$. **Solution:** Let $(x_1, y_1) = (-4, 3)$ and $(x_2, y_2) = (2, -1)$. $$d = \sqrt{(x_2 - x_1)^2 + (y_2 - y_1)^2}$$ $$d = \sqrt{(2 - (-4))^2 + (-1 - 3)^2} \quad \text{Substitute.}$$ $$d = \sqrt{52} \quad \text{Simplify.}$$ $$d = \sqrt{4 \cdot 13}$$ $$d = 2\sqrt{13}$$
If the coordinates of the endpoints of a line segment are (x_1, y_1) and (x_2, y_2), the coordinates of the midpoint are $$\underline{\left(\frac{x_1 + x_2}{2}, \frac{y_1 + y_2}{2}\right)}.$$	$$\text{Midpoint} = \left(\frac{x_1 + x_2}{2}, \frac{y_1 + y_2}{2}\right)$$ $$\text{Midpoint} = \left(\frac{-4 + 2}{2}, \frac{3 + (-1)}{2}\right) \quad \text{Substitute.}$$ $$\text{Midpoint} = \left(\frac{-2}{2}, \frac{2}{2}\right) \quad \text{Simplify.}$$ $$\text{Midpoint} = (-1, 1)$$

Exercises 10 and 11 ⬛ Expressions

[12.1] *For Exercises 10 and 11, find the distance and midpoint between the two points.*

10. $(-1, -2)$ and $(-5, 1)$
$5, \left(-3, -\frac{1}{2}\right)$

11. $(2, -5)$ and $(-2, 3)$
$4\sqrt{5}, (0, -1)$

Definitions/Rules/Procedures	Key Example(s)
A **circle** is a set of points in a plane that are <u>equally distant</u> from a central point. The central point is the <u>center</u>. The radius of a circle is the distance from the <u>center of a circle</u> to <u>any point on the circle</u>. The **equation of a circle** with center (h, k) and radius r is <u>$(x - h)^2 + (y - k)^2 = r^2$</u>.	Find the center and radius of the circle whose equation is $(x - 2)^2 + (y + 4)^2 = 25$ and draw the graph. **Solution:** Rewrite the equation as $(x - 2)^2 + (y - (-4))^2 = 5^2$. Because $h = 2$ and $k = -4$, the center is $(2, -4)$ and $r = 5$. 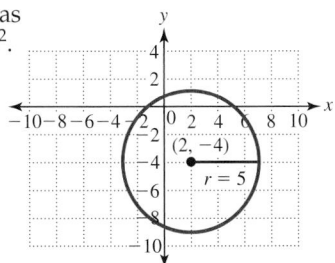
	Write the equation of the circle whose center is $(5, -6)$ and whose radius is 8. **Solution:** Substitute for h, k, and r in $$(x - h)^2 + (y - k)^2 = r^2$$ $$(x - 5)^2 + (y - (-6))^2 = 8^2$$ $$(x - 5)^2 + (y + 6)^2 = 64 \quad \text{Simplify.}$$
To graph circles of the form $x^2 + y^2 + dx + ey + f = 0$, <u>complete the square</u> in x and y to write the equation in the form $(x - h)^2 + (y - k)^2 = r^2$.	Find the center and radius of the circle whose equation is $x^2 + y^2 + 8x - 2y + 8 = 0$ and draw the graph. **Solution:** Complete the square in x and y. $$x^2 + 8x + y^2 - 2y = -8 \quad \text{Group } x \text{ and } y \text{ terms; subtract 8.}$$ $$x^2 + 8x + 16 + y^2 - 2y + 1 = -8 + 16 + 1 \quad \text{Complete the square.}$$ $$(x + 4)^2 + (y - 1)^2 = 9 \quad \text{Factor and simplify.}$$ The center is $(-4, 1)$, and the radius is 3. 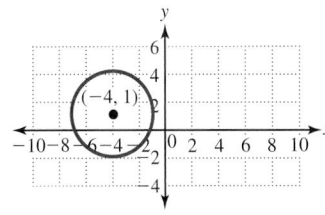

Exercises 12–20 **Equations and Inequalities**

[12.1] 12. The center of a circle is at $(6, -4)$, and the circle passes through $(-2, 2)$. What is the radius of the circle?

10

[12.1] *For Exercises 13–16, find the center and radius and draw the graph.*

13. $(x - 3)^2 + (y + 2)^2 = 25$

Center: $(3, -2)$; radius: 5

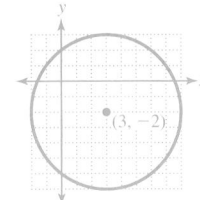

14. $(x + 5)^2 + (y - 1)^2 = 4$

Center: $(-5, 1)$; radius: 2

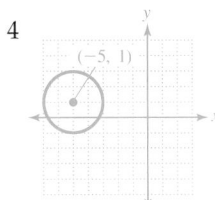

15. $x^2 + y^2 - 4x + 8y + 11 = 0$

Center: $(2, -4)$; radius: 3

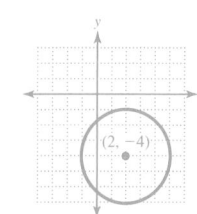

16. $x^2 + y^2 + 10x + 2y + 22 = 0$

Center: $(-5, -1)$; radius: 2

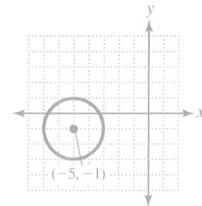

[12.1] *For Exercises 17 and 18, the center and radius of a circle are given. Write the equation of each circle in standard form.*

17. Center: $(6, -8)$, $r = 9$

$(x - 6)^2 + (y + 8)^2 = 81$

18. Center: $(-3, -5)$, $r = 10$

$(x + 3)^2 + (y + 5)^2 = 100$

19. To lay out the border of a circular flower bed, sticks are tied to each end of a rope that is 20 feet long. One person holds one of the sticks stationary while another person uses the other stick to trace out a circular path while keeping the rope taut. If the center of the circle is at the stationary stick, what is the equation of the circle?

$x^2 + y^2 = 400$

20. The center of a circle is at $(-6, 8)$, and the circle passes through $(3, -4)$. What is the equation of the circle?

$(x + 6)^2 + (y - 8)^2 = 225$

12.2 Ellipses and Hyperbolas

Definitions/Rules/Procedures	Key Example(s)
An **ellipse** is the set of all points the ___sum___ of whose distances from __two fixed points__ is constant.	Graph $\dfrac{x^2}{49} + \dfrac{y^2}{36} = 1$.

The equation of an ellipse with center $(0, 0)$, x-intercepts $(a, 0)$ and $(-a, 0)$, and y-intercepts $(0, b)$ and $(0, -b)$ is

$$\dfrac{x^2}{a^2} + \dfrac{y^2}{b^2} = 1.$$

Solution: The equation can be rewritten as $\dfrac{x^2}{7^2} + \dfrac{y^2}{6^2} = 1$, so $a = 7$ and $b = 6$. Consequently, the x-intercepts are $(7, 0)$ and $(-7, 0)$ and the y-intercepts are $(0, 6)$ and $(0, -6)$.

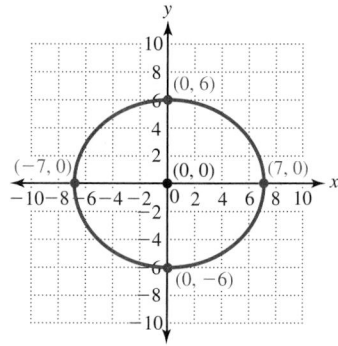

The equation of an ellipse with center (h, k) is

$$\dfrac{(x - h)^2}{a^2} + \dfrac{(y - k)^2}{b^2} = 1.$$

Graph $\dfrac{(x + 2)^2}{9} + \dfrac{(y - 1)^2}{25} = 1$.

Solution: The equation can be rewritten as

$$\dfrac{(x - (-2))^2}{3^2} + \dfrac{(y - 1)^2}{5^2} = 1;$$ so $h = -2$, $k = 1$, $a = 3$, and $b = 5$. The center is $(-2, 1)$. To find other points on the ellipse, go 3 units right and left of the center and 5 units up and down from the center.

The ellipse passes through two points that are ___a___ units to the left and right of the center and two points that are ___b___ units above and below the center.

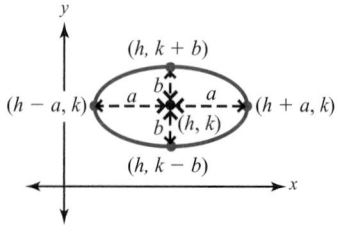

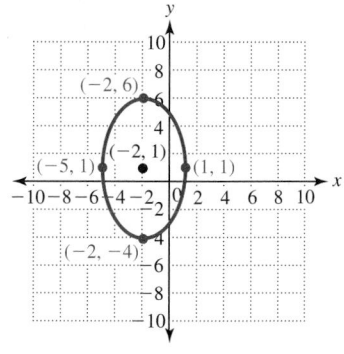

Exercises 21–28 ◣⬛▲ Equations and Inequalities

[12.2] *For Exercises 21–26, graph each ellipse. Label the center and the points directly above, below, to the left of and to the right of the center.*

21. $\dfrac{x^2}{49} + \dfrac{y^2}{25} = 1$

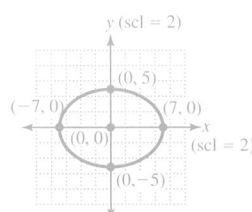

22. $\dfrac{x^2}{9} + \dfrac{y^2}{25} = 1$

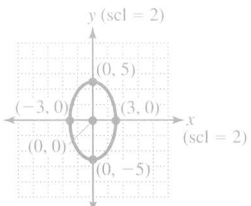

23. $4x^2 + 9y^2 = 36$

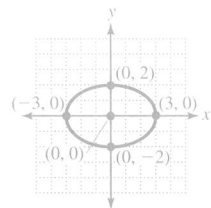

24. $25x^2 + 9y^2 = 225$

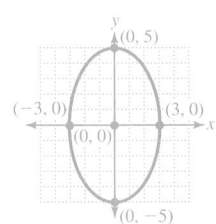

25. $\dfrac{(x + 2)^2}{16} + \dfrac{(y - 3)^2}{4} = 1$

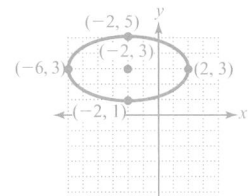

26. $\dfrac{(x - 1)^2}{9} + \dfrac{(y + 4)^2}{25} = 1$

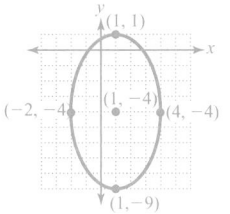

27. A bridge over a canal is in the shape of half an ellipse. If the highest point of the bridge is 15 feet above the water and the base of the bridge is 50 feet across, what is the equation of the ellipse, half of which forms the bridge?

$\dfrac{x^2}{625} + \dfrac{y^2}{225} = 1$

28. The cam of a compound bow is elliptical and is 4 inches long and 3.25 inches wide, as shown in the figure. What is the equation of the ellipse forming the shape of the cam?

$\dfrac{x^2}{1.625^2} + \dfrac{y^2}{4} = 1$

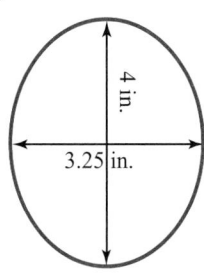

Definitions/Rules/Procedures	Key Example(s)
A **hyperbola** is the set of all points the _____difference_____ of whose distances from __two fixed points__ remains constant. The equation of a hyperbola with center $(0,0)$, x-intercepts $(a,0)$ and $(-a,0)$, and no y-intercepts is $\dfrac{x^2}{a^2} - \dfrac{y^2}{b^2} = 1$. 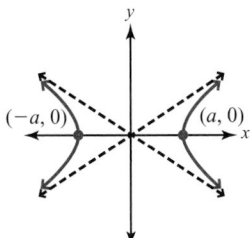	

Definitions/Rules/Procedures	Key Example(s)

The equation of a hyperbola with center $(0,0)$, y-intercepts $(0,b)$ and $(0,-b)$, and no x-intercepts is

$$\underline{\frac{y^2}{b^2} - \frac{x^2}{a^2} = 1}.$$

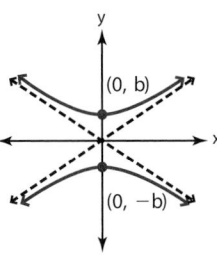

To graph a hyperbola

1. Find the intercepts. If the x^2-term is positive, the x-intercepts are ___(a,0)___ and ___(-a,0)___ and there are no ___y___-intercepts. If the y^2-term is positive, the y-intercepts are ___(0,b)___ and ___(0,-b)___ and there are no ___x___-intercepts.

2. Draw the fundamental rectangle. The vertices are ___$(a,b), (-a,b), (a,-b)$, and $(-a,-b)$___.

3. Draw the asymptotes, which are the extended diagonals of the ___fundamental rectangle___.

4. Draw the graph so that each branch passes through a(n) ___intercept___ and approaches the ___asymptotes___ the farther the branches are from the origin.

Graph $\dfrac{x^2}{36} - \dfrac{y^2}{16} = 1$.

Solution: The equation can be rewritten as $\dfrac{x^2}{6^2} - \dfrac{y^2}{4^2} = 1$, so $a = 6$ and $b = 4$. Because the x^2-term is positive, the x-intercepts are $(6,0)$ and $(-6,0)$ and there are no y-intercepts. The fundamental rectangle has vertices $(6,4)$, $(-6,4)$, $(6,-4)$, and $(-6,-4)$.

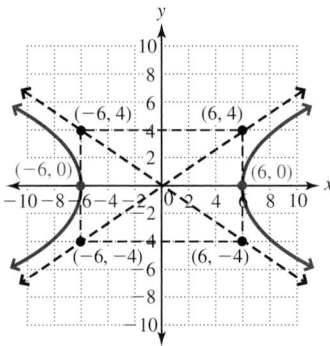

Exercises 29–32 ◢◣ Equations and Inequalities

[12.2] *For Exercises 29–32, graph each hyperbola. Also show the fundamental rectangle with its corner points labeled, the asymptotes, and the intercepts.*

29. $\dfrac{x^2}{25} - \dfrac{y^2}{16} = 1$

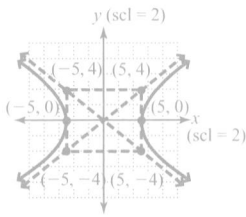

30. $\dfrac{x^2}{36} - \dfrac{y^2}{9} = 1$

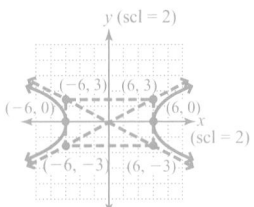

31. $y^2 - 9x^2 = 36$

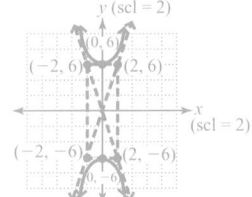

32. $25y^2 - 9x^2 = 225$

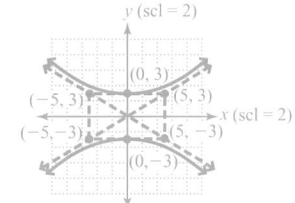

12.3 Nonlinear Systems of Equations

Definitions/Rules/Procedures	Key Example(s)
A **nonlinear system of equations** is a system of equations that contains at least one ___nonlinear___ equation. If a nonlinear system has a linear equation, the ___substitution___ method is usually preferred. Solve the linear equation for one of its variables and ___substitute___ for that variable into the nonlinear equation.	Solve the system $\begin{cases} y = (x-3)^2 + 2 \\ 2x - y = -4 \end{cases}$ by substitution. (**Note:** This is a parabola and a line.) **Solution:** $\begin{aligned} y &= 2x + 4 & &\text{Solve } 2x - y = -4 \text{ for } y. \\ 2x + 4 &= (x-3)^2 + 2 & &\text{Substitute } 2x + 4 \text{ for } y. \\ 2x + 4 &= x^2 - 6x + 11 & &\text{Simplify.} \\ 0 &= x^2 - 8x + 7 & &\text{Write in form} \\ & & &ax^2 + bx + c = 0. \\ 0 &= (x-7)(x-1) & &\text{Factor.} \end{aligned}$ $x - 7 = 0$ or $x - 1 = 0$ Use the zero-factor theorem. $x = 7$ or $x = 1$ Solve each equation. To find y, substitute 7 and 1 into $y = 2x + 4$. $\begin{aligned} y &= 2(7) + 4 & y &= 2(1) + 4 \\ y &= 18 & y &= 6 \end{aligned}$ Solutions: $(7, 18)$, $(1, 6)$
If neither equation is linear, the ___elimination___ method is often preferred.	Solve the system $\begin{cases} 4x^2 + 5y^2 = 36 & (\text{Equation 1}) \\ x^2 + y^2 = 8 & (\text{Equation 2}) \end{cases}$ by elimination. (**Note:** This is an ellipse and a circle.) **Solution:** To eliminate x^2, multiply equation 2 by -4 and add the equations. $\begin{array}{ll} 4x^2 + 5y^2 = 36 & 4x^2 + 5y^2 = 36 \\ x^2 + y^2 = 8 \longrightarrow & \underline{-4x^2 - 4y^2 = -32} \\ \quad\text{Multiply by } -4. & y^2 = 4 \\ & y = \pm 2 \end{array}$ Substitute 2 and -2 for y into either equation to find x. Let's use $x^2 + y^2 = 8$. $\begin{aligned} x^2 + 2^2 &= 8 & x^2 + (-2)^2 &= 8 \\ x^2 + 4 &= 8 & x^2 + 4 &= 8 \\ x^2 &= 4 & x^2 &= 4 \\ x &= \pm 2 & x &= \pm 2 \end{aligned}$ Solutions: $(2, 2)$, $(2, -2)$, $(-2, 2)$, and $(-2, -2)$

Exercises 33–44 Equations and Inequalities

[12.3] *For Exercises 33–40, solve.*

33. $\begin{cases} y = 2x^2 - 3 \\ 2x - y = -1 \end{cases}$
$(2, 5), (-1, -1)$

34. $\begin{cases} x^2 + y^2 = 17 \\ x - y = 3 \end{cases}$
$(-1, -4), (4, 1)$

35. $\begin{cases} 25x^2 + 3y^2 = 100 \\ 2x - y = -3 \end{cases}$
$(-73/37, -35/37), (1, 5)$

36. $\begin{cases} x^2 + y^2 = 64 \\ x^2 - y^2 = 64 \end{cases}$
$(8, 0), (-8, 0)$

37. $\begin{cases} x^2 + y^2 = 25 \\ 25y^2 - 16x^2 = 256 \end{cases}$
$(3,4), (3,-4), (-3,-4), (-3,4)$

38. $\begin{cases} x^2 + 5y^2 = 36 \\ 4x^2 - 7y^2 = 36 \end{cases}$
$(4,2), (4,-2), (-4,-2), (-4,2)$

39. $\begin{cases} y = x^2 - 2 \\ 4x^2 + 5y^2 = 36 \end{cases}$
$(2,2), (-2,2)$

40. $\begin{cases} y = x^2 - 1 \\ 4y^2 - 5x^2 = 16 \end{cases}$
$(2,3), (-2,3)$

41. The sum of the squares of two integers is 89, and the difference of their squares is 39. Find the integers.
$(8,5), (8,-5), (-8,-5), (-8,5)$

42. A rectangular rug has a perimeter of 36 feet and an area of 80 square feet. Find the dimensions of the rug.
8 ft. by 10 ft.

[12.3] *For Exercises 43 and 44, use a graphing calculator to verify the results of the exercise given.*

43. Exercise 33

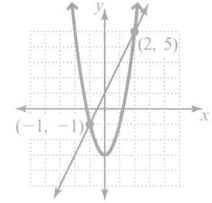

44. Exercise 39

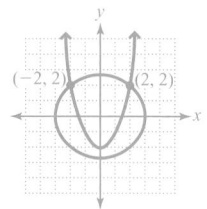

12.4 Nonlinear Inequalities and Systems of Inequalities

Definitions/Rules/Procedures	Key Example(s)
Graphing **nonlinear inequalities** 1. Graph the related ____equation____ (the boundary curve). If the inequality symbol is ≤ or ≥, draw the graph as a(n) ____solid____ curve. If the inequality symbol is < or >, draw a(n) ____dashed____ curve. 2. The graph divides the coordinate plane into at least two regions. Test a(n) ____ordered pair____ from each region by substituting it into the inequality. If the ordered pair satisfies the inequality, shade ____the region containing that ordered pair____.	Graph $y \geq (x - 2)^2 - 1$. **Solution:** The related equation is a parabola with vertex $(2, -1)$ that opens upward. Because the inequality is $\geq$, all ordered pairs on the parabola are in the solution set; so we draw it with a solid curve. We choose $(0, 0)$ as a test point. $0 \geq (0 - 2)^2 - 1$ $0 \geq 3$, which is false; so $(0,0)$ is not in the solution set. Choose $(2, 0)$ as a test point. $0 \geq (2 - 2)^2 - 1$ $0 \geq -1$, which is true; so $(2, 0)$ is in the solution set. Shade the region that contains $(2, 0)$. 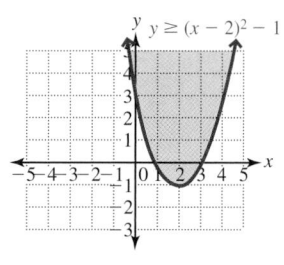

Exercises 45–48 ▲, Equations and Inequalities

[12.4] *For Exercises 45–48, graph the inequality.*

45. $x^2 + y^2 \leq 64$

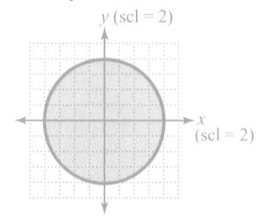

46. $y < 2(x - 3)^2 + 4$

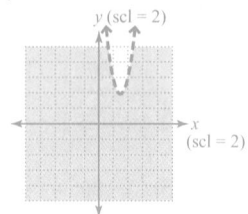

47. $\dfrac{y^2}{25} + \dfrac{x^2}{49} > 1$

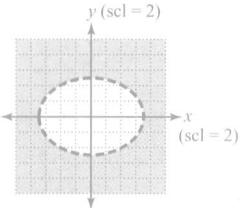

48. $\dfrac{x^2}{25} - \dfrac{y^2}{36} < 1$

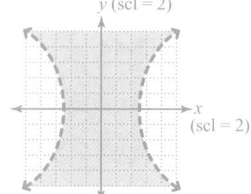

Definitions/Rules/Procedures	Key Example(s)
To graph a **system of nonlinear inequalities**, graph the solution set of each inequality on the same grid. The solution set is the intersection of <u>the solution sets of the individual inequalities</u>.	Graph the solution set of $$\begin{cases} \dfrac{x^2}{4} + \dfrac{y^2}{16} \le 1 \\ x^2 + y^2 \ge 9 \end{cases}.$$ **Solution:** The graph of $\dfrac{x^2}{4} + \dfrac{y^2}{16} \le 1$ is an ellipse and all points inside the ellipse. The graph of $x^2 + y^2 \ge 9$ is a circle and all points outside the circle. So the solution set of the system is the set of all points inside the ellipse and outside the circle. 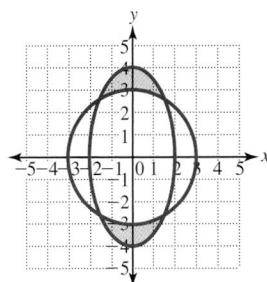

Exercises 49–54 ▲ Equations and Inequalities

[12.4] *For Exercises 49–54, graph the solution set of the system of inequalities.*

49. $\begin{cases} y \ge x^2 - 3 \\ 2x + y < 2 \end{cases}$

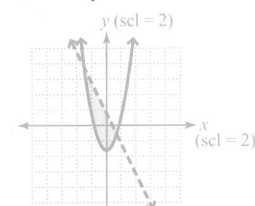

50. $\begin{cases} x + 2y < 4 \\ x^2 + y^2 \le 25 \end{cases}$

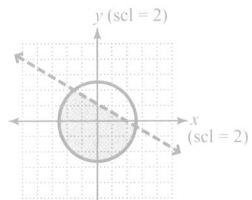

51. $\begin{cases} y > x^2 - 2 \\ y < -x^2 + 1 \end{cases}$

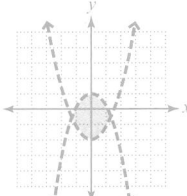

52. $\begin{cases} \dfrac{x^2}{9} + \dfrac{y^2}{25} \le 1 \\ x^2 + y^2 \ge 9 \end{cases}$

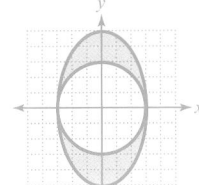

53. $\begin{cases} \dfrac{y^2}{4} - \dfrac{x^2}{9} \le 1 \\ \dfrac{x^2}{9} + \dfrac{y^2}{25} \le 1 \end{cases}$

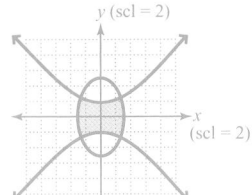

54. $\begin{cases} \dfrac{y^2}{4} + \dfrac{x^2}{16} \le 1 \\ x^2 + y^2 \le 9 \\ y \le x + 1 \end{cases}$

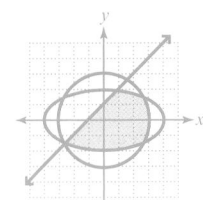

Chapter 12 Practice Test

For Extra Help

Step-by-step test solutions are found on the Chapter Test Prep Videos available in MyMathLab® *or on* You Tube.

For Exercises 1–3, find the direction the parabola opens, the coordinates of the vertex, and the equation of the axis of symmetry. Draw the graph.

1. $y = 2(x + 1)^2 - 4$

Opens upward;
vertex: $(-1, -4)$;
axis of symmetry: $x = -1$

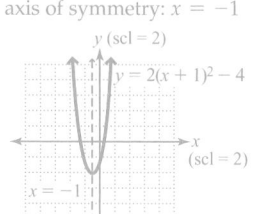

[12.1]

2. $x = -2(y - 3)^2 + 1$

Opens left;
vertex: $(1, 3)$;
axis of symmetry: $y = 3$

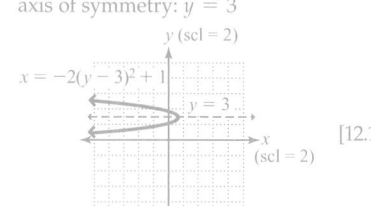

[12.1]

3. $x = y^2 + 4y - 3$

Opens right;
vertex: $(-7, -2)$;
axis of symmetry: $y = -2$

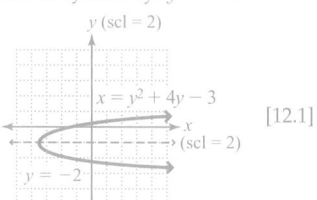

[12.1]

4. Find the distance and midpoint between the points $(2, -3)$ and $(6, -5)$.

$2\sqrt{5}, (4, -4)$ [12.1]

For Exercises 5 and 6, find the center and radius and draw the graph.

5. $(x + 4)^2 + (y - 3)^2 = 36$

Center: $(-4, 3)$; radius: 6

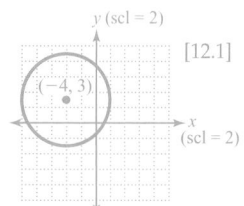

[12.1]

6. $x^2 + y^2 + 4x - 10y + 20 = 0$

Center: $(-2, 5)$; radius: 3

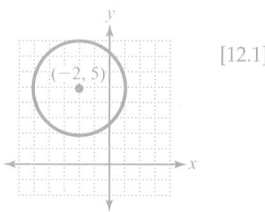

[12.1]

7. Write the equation of the circle with center $(2, -4)$ that passes through the point $(-4, 4)$.

$(x - 2)^2 + (y + 4)^2 = 100$ [12.1]

For Exercises 8–11, graph the equation and label relevant points. If the graph is a hyperbola, show the fundamental rectangle with its corner points labeled, the asymptotes, and the intercepts.

8. $16x^2 + 36y^2 = 576$

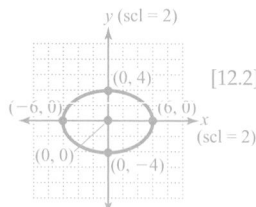

[12.2]

9. $\dfrac{(x + 2)^2}{4} + \dfrac{(y - 1)^2}{25} = 1$

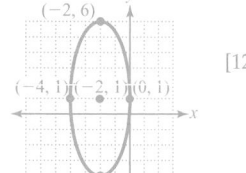

[12.2]

10. $\dfrac{x^2}{49} - \dfrac{y^2}{25} = 1$

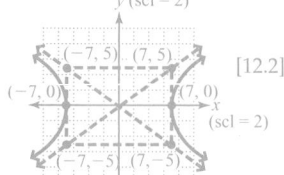

[12.2]

11. $\dfrac{y^2}{9} - \dfrac{x^2}{16} = 1$

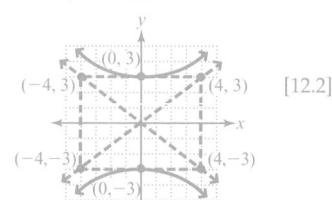

[12.2]

For Exercises 12–15, solve.

12. $\begin{cases} y = (x + 1)^2 + 2 \\ 2x + y = 8 \end{cases}$

$(-5, 18), (1, 6)$ [12.3]

13. $\begin{cases} 3x - y = 4 \\ x^2 + y^2 = 34 \end{cases}$

$(3, 5), \left(-\dfrac{3}{5}, -\dfrac{29}{5}\right)$ [12.3]

14. $\begin{cases} x^2 + y^2 = 13 \\ 3x^2 + 4y^2 = 48 \end{cases}$

$(2, 3), (2, -3), (-2, -3), (-2, 3)$ [12.3]

15. $\begin{cases} x^2 - 2y^2 = 1 \\ 4x^2 + 7y^2 = 64 \end{cases}$

$(3, 2), (3, -2), (-3, -2), (-3, 2)$ [12.3]

For Exercises 16 and 17, graph.

16. $y \leq -2(x+3)^2 + 2$

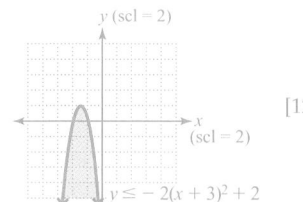

[12.4]

17. $\dfrac{x^2}{4} + \dfrac{y^2}{9} > 1$

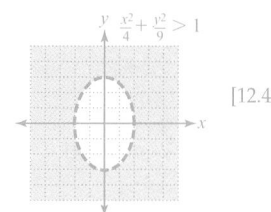

[12.4]

18. Graph the solution set: $\begin{cases} y \geq x^2 - 4 \\ \dfrac{x^2}{9} + \dfrac{y^2}{16} \leq 1 \end{cases}$

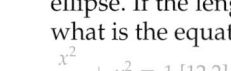

[12.4]

19. An arch is in the shape of a parabola as shown in the figure. If we place the origin, O, as indicated, find the height of the arch and the distance across the base if the equation is $y = -\dfrac{1}{2}x^2 + 18$.

Height is 18; distance across the base is 12. [12.1]

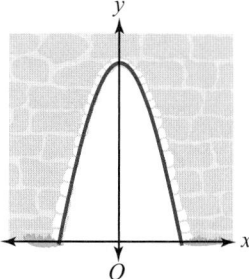

20. The path of a person's foot using an elliptical trainer on a particular setting is an ellipse. If the length of the stride is 19 inches and the height of the stride is 2 inches, what is the equation of the path of the foot?

$\dfrac{x^2}{9.5^2} + y^2 = 1$ [12.2]

2 in.

19 in.

Chapters 1–12 Cumulative Review Exercises

For Exercises 1–3, answer true or false.

[5.1]   **1.** A positive base raised to a negative exponent simplifies to a negative number.
    false

[2.6]   **2.** If $\dfrac{a}{b} = \dfrac{c}{d}$, then $ad = bc$ where $b \neq 0$ and $d \neq 0$.

    true

[11.3]   **3.** Logarithms are exponents.
    true

For Exercises 4–6, fill in the blanks.

[4.5]   **4.** If the row echelon form of a matrix for a system of equations is $\begin{bmatrix} 1 & -3 & \vdots & 7 \\ 0 & 1 & \vdots & -5 \end{bmatrix}$,

    then $x = $ _____ -8 _____ and $y = $ _____ -5 _____.

[9.7]   **5.** $\sqrt{-1} = $ ___ i ___.

[12.1]   **6.** The formula for finding the distance between two points (x_1, y_1) and (x_2, y_2)
    is ___ $d = \sqrt{(x_2 - x_1)^2 + (y_2 - y_1)^2}$ ___.

Exercises 7–11 ▲, **Expressions**

For Exercises 7–9, simplify.

[1.5]   **7.** $6 - (-3) + 2^{-4}$

    $9\dfrac{1}{16}$

[5.4]   **8.** $(5m^3 n^{-2})^{-3}$

    $\dfrac{n^6}{125 m^9}$

[9.7]   **9.** $\sqrt{-20}$

    $2i\sqrt{5}$

For Exercises 10 and 11, factor completely.

[6.1]   **10.** $ax + bx + ay + by$

    $(x + y)(a + b)$

[6.4]   **11.** $k^4 - 81$

    $(k^2 + 9)(k + 3)(k - 3)$

Exercises 12–30 ▲, **Equations and Inequalities**

For Exercises 12–17, solve.

[8.2]   **12.** $|2x + 1| = |x - 3|$

    $-4, \dfrac{2}{3}$

[9.6]   **13.** $\sqrt{x - 3} = 7$

    52

[10.2]   **14.** $4x^2 - 2x + 1 = 0$

    $\dfrac{1 \pm i\sqrt{3}}{4}$

[10.3]   **15.** $9 + 24x^{-1} + 16x^{-2} = 0$

    $-\dfrac{4}{3}$

[11.2]   **16.** $9^x = 27$

    1.5

[11.3]   **17.** $\log_3(x + 1) - \log_3 x = 2$

    $\dfrac{1}{8}$

[10.5]   **18.** Solve the inequality $5m^2 - 3m < 0$; then:
    a. Graph the solution set.
    b. Write the solution set in set-builder notation.
    c. Write the solution set in interval notation.

    a. [number line graph from 0 to 3 with shading between 0 and 3/5]
       $0 \quad \frac{3}{5} \quad 1 \quad\quad 2 \quad\quad 3$
    b. $\left\{ m \,\middle|\, 0 < m < \dfrac{3}{5} \right\}$
    c. $\left(0, \dfrac{3}{5} \right)$

[10.6, 19. Given $f(x) = 2x + 1$ and $g(x) = x^2 - 1$, find the following.
11.1]

    a. $(f + g)(x)$ **b.** $(f - g)(x)$ **c.** $(f \cdot g)(x)$

        $x^2 + 2x$ $-x^2 + 2x + 2$ $2x^3 + x^2 - 2x - 1$

    d. $(f/g)(x)$ **e.** $(f \circ g)(x)$ **f.** $(g \circ f)(x)$

        $\dfrac{2x + 1}{x^2 - 1}; x \neq \pm 1$ $2x^2 - 1$ $4x^2 + 4x$

[11.3] 20. Write $5^3 = 125$ in logarithmic form.

    $\log_5 125 = 3$

[11.4] 21. Write as a sum or difference of multiples of logarithms: $\log_5 x^5 y$

    $5 \log_5 x + \log_5 y$

[12.1] 22. Write the equation of a circle in standard form with a center at $(2, 4)$ passing
    through the point $(7, 16)$.

    $(x - 2)^2 + (y - 4)^2 = 169$

For Exercises 23–25, graph.

[4.6] 23. $\begin{cases} y < -x + 2 \\ y \geq x - 4 \end{cases}$ **[11.3] 24.** $f(x) = \log_3 x$ **[12.2] 25.** $\dfrac{x^2}{4} + \dfrac{y^2}{9} = 1$

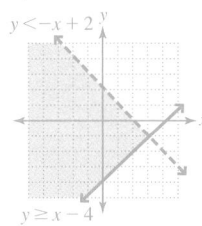   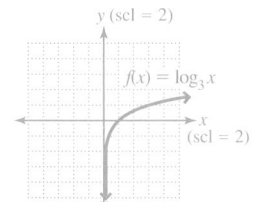   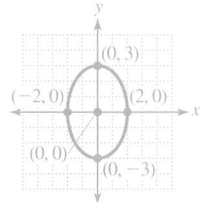

For Exercises 26–30, solve.

[2.4] 26. The volume of a cylinder can be found using the formula $V = \pi r^2 h$.
    a. Solve the formula for h.

        $h = \dfrac{V}{\pi r^2}$

    b. Suppose the volume of a cylinder is 15π cubic inches. Find its height.

        $\dfrac{15}{r^2}$

[4.2] 27. During 2008, shrimp overtook tuna as the top-selling seafood in the Unit-
    ed States. It is estimated that each person consumed a combined total of 6.3
    pounds of these two seafoods and that each person ate $^1/_2$ pound more shrimp
    than tuna. How many pounds of each were consumed?

    2.9 lb. of tuna and 3.4 lb. of shrimp

[4.4] 28. A total of $5000 is invested in three funds. The money market fund pays 5%,
    the income fund pays 6%, and the growth fund pays 3%. The total annual in-
    terest from the three accounts is $255, and the amount in the growth fund is
    $500 less than the amount invested in the money market account. Find the
    amount invested in each of the funds.

    $1500 at 5%, $2500 at 6%, $1000 at 3%

[11.5] **29.** The magnitude of an earthquake was defined in 1935 by Charles Richter as $M = \log\frac{I}{I_0}$. We let I be the intensity of the earthquake measured by the amplitude of a seismograph reading taken 100 km from the epicenter of the earthquake, and we let I_0 be the intensity of a "standard earthquake" whose amplitude is 10^{-4} cm. On January 26, 2001, an earthquake in Rann of Kutch, Gujarat, was measured with an intensity of $10^{2.9}$. What was the magnitude of this earthquake?

6.9

[12.2] **30.** A bridge over a canal is in the shape of half an ellipse. If the highest point of the bridge is 20 feet above the water and the base of the bridge is 150 feet across, what is the equation of the ellipse, half of which forms the bridge?

$\frac{x^2}{5625} + \frac{y^2}{400} = 1$

Elementary Algebra Review

R.1 Review of Expressions

Objective 1 Evaluate an expression.

For a review of this objective, see Section 1.7, page 66.

For Exercises 1–6, evaluate the expressions using $x = -2$, $y = 4$, and $z = -3$.

1. $3y - 2(3x + z)$　　　**2.** $-y^2 - (3z^2 - 4x)$　　　**3.** $\dfrac{2x^3 - 3y^2}{3x^3 - 2z^2}$

4. $|4x^2 - 3y(z^3 + 2y)|$　　**5.** $\sqrt{y} - |-3x + 2z|$　　**6.** $\sqrt{y^2 + z^2} - 2|5xy^2z|$

For additional exercises of this type, see Exercises 1–26 on pages 71 and 72.

Objectives

1 Evaluate an expression.

2 Apply the distributive property.

3 Combine like terms.

Objective 2 Apply the distributive property.

For a review of this objective, see Section 1.7, pages 67 and 68.

For Exercises 7–14, use the distributive property to write an equivalent expression.

7. $3(2a + 4)$　　　　　**8.** $4(3x + 6)$　　　　　**9.** $-4(r - 5)$

10. $-7(-3a + 5b)$　　**11.** $2.3x(4.2x - 1.7)$　　**12.** $3.5x(2.4x - 6.2)$

13. $\dfrac{3}{4}\left(2b + \dfrac{5}{6}\right)$　　**14.** $\dfrac{5}{8}\left(6x - \dfrac{4}{3}\right)$

For additional exercises of this type, see Exercises 35–42 on page 73.

Objective 3 Combine like terms.

For a review of this objective, see Section 1.7, pages 69–71.

For Exercises 15–22, combine like terms.

15. $8x + 3x$　　　　　　　　　　**16.** $-13a + 4a$

17. $9xy + 5x - 3xy + x$　　　　　**18.** $8y - 4k + 2 - 4y + 3 - 9k$

19. $9.3a - 7.6b - 5.7a + 2.4b$　　　**20.** $6.3t - 2.8u - 4.7t - 5.7u$

21. $\dfrac{5}{8}x^2 + 4.6y - \dfrac{3}{4}x^2 - 3.8y$　　　**22.** $\dfrac{2}{3}x^2y - \dfrac{3}{5}xy^2 + \dfrac{1}{4}x^2y - \dfrac{7}{10}xy^2$

For additional exercises of this type, see Exercises 53–78 on pages 73 and 74.

R.2 Review of Solving Linear Equations

Objectives

1 Verify solutions to equations.
2 Solve linear equations using the addition principle.
3 Solve linear equations using the multiplication principle.
4 Solve equations using both the addition and multiplication principles.

Objective 1 Verify solutions to equations.

For a review of this objective, see Section 2.1, pages 88 and 89.

For Exercises 1–8, check to see if the given number is a solution for the given equation.

1. $9y - 5 - 8y = 5y + 10 - 3y; y = -15$
2. $2x + 6x - 8 = -23 + 5x; x = -5$
3. $5(u - 2) + 7(3 - u) = 9u; u = 1$
4. $6(p - 4) - 3p = 18 - 3p; p = 6$
5. $x(x + 3) = 40; x = 5$
6. $a(a + 7) = -12; a = -3$
7. $8x^2 - 15 = 14x; x = -\dfrac{3}{4}$
8. $6x^2 - 7x = 3; x = \dfrac{1}{3}$

For additional exercises of this type, see Exercises 1–16 on pages 95 and 96.

Objective 2 Solve linear equations using the addition principle.

For a review of this objective, see Section 2.2, pages 101–107.

For Exercises 9–16, solve and check.

9. $7 + 5 = 6r + 8 - 5r$
10. $3 + 6 + 6a = 7a + 4$
11. $16x - 5(3x - 2) = 12$
12. $17x - 8 = 6 + 4(4x - 3)$
13. $-4(5x - 4) + 7(3x - 2) - 6 = 8 - 2$
14. $3 = 2(4x + 5) - 7(x + 2)$
15. $-2(4.7y - 5.4) + 10.4y = 3.1$
16. $-3(1.6b - 2.4) + 5.8b = -6.2$

For additional exercises of this type, see Exercises 17–68, on pages 109 and 110.

Objective 3 Solve linear equations using the multiplication principle.

For a review of this objective, see Section 2.3, pages 112–114.

For Exercises 17–24, solve and check.

17. $5x = -20$
18. $-6x = 24$
19. $-1.2z = 4.8$
20. $3.4a = -10.2$
21. $\dfrac{3}{4}x = 9$
22. $\dfrac{8}{5}x = -32$
23. $\dfrac{w}{-4} = -6$
24. $\dfrac{4}{5}r = -\dfrac{12}{25}$

For additional exercises of this type, see Exercises 1–12 on page 119.

Objective 4 Solve equations using both the addition and multiplication principles.

For a review of this objective, see Section 2.3, pages 114–117.

For Exercises 25–34, solve and check.

25. $-7x - 6 = 8$
26. $3x + 9 = -6$
27. $3x + 4 = -5x + 28$
28. $18z - 3.8 = 5z + 4$
29. $2.3y - 1.6 - 0.8y = 4.4$
30. $3a - 4a + 9 = -12a + 18 - 6a$
31. $14(w - 2) + 13 = 4w + 5$
32. $5(a - 1) - 9(a - 2) = -3(2a + 1) - 2$
33. $4(b - 4) + 2(b + 2) = 2(b - 2)$
34. $0.3n + 0.2(4n - 1) = 0.4 + 0.5n$

For additional exercises of this type, see Exercises 13–76 on pages 119–121.

R.3 Review of Graphing Linear Equations

Objectives

1 Plot points in the coordinate plane.
2 Find solutions for equations in two unknowns.
3 Graph linear equations by plotting solutions.
4 Graph linear equations using intercepts.
5 Graph vertical and horizontal lines.

Objective 1 Plot points in the coordinate plane.

For a review of this objective, see Section 3.1, pages 191 and 192.

For Exercises 1 and 2, plot the point described by the coordinates.

1. **a.** $(4, 5)$ **b.** $(0, 3)$ **c.** $(-2, -3)$

2. **a.** $(-3, 2)$ **b.** $(2, 0)$ **c.** $(0, 4)$ **d.** $(1, -5)$

For additional exercises of this type, see Exercises 5–8 on pages 194 and 195.

Objective 2 Find solutions for equations in two unknowns.

For a review of this objective, see Section 3.2, pages 198 and 199.

For Exercises 3–6, find three solutions for each equation. (Answers may vary.)

3. $x + 4y = 8$

4. $3x + 2y = 6$

5. $y = -2x - 5$

6. $y = -\dfrac{2}{3}x + 4$

For additional exercises of this type, see the first part of Exercises 17–52 on pages 203–205.

Objective 3 Graph linear equations by plotting solutions.

For a review of this objective, see Section 3.2, pages 200–202.

For Exercises 7–10, find three solutions for the given equations and graph.

7. $x + 2y = 4$

8. $3x + 2y = 6$

9. $y = -2x - 2$

10. $y = \dfrac{1}{3}x + 1$

For additional exercises of this type, see Exercises 17–52 on pages 203–205.

Objective 4 Graph linear equations using intercepts.

For a review of this objective, see Section 3.3, pages 210–212.

For Exercises 11–16, graph using the x- and y-intercepts.

11. $x + 2y = 4$ 12. $3x - y = 3$ 13. $y = -2x$

14. $y = \dfrac{2}{3}x$ 15. $y = \dfrac{1}{2}x + 2$ 16. $y = \dfrac{2}{3}x - 2$

For additional exercises of this type, see Exercises 27–58 on pages 214 and 215.

Objective 5 Graph vertical and horizontal lines.

For a review of this objective, see Section 3.2, pages 201 and 202 and Section 3.3, page 210.

For Exercises 17–20, graph.

17. $y = 2$ 18. $y = -1$

19. $x = -3$ 20. $x = 4$

For additional exercises of this type, see Exercises 47–50 on pages 204 and 205 and Exercises 51–58 on pages 214 and 215.

R.4 Review of Polynomials

Objectives

1 Add and subtract polynomials.
2 Multiply polynomials.
3 Divide polynomials.

Objective 1 Add and subtract polynomials.

For a review of this objective, see Section 5.3, pages 372–375.

For Exercises 1–4, add or subtract.

1. $(4a^2 - 7a - 4) + (2a^2 + 9a - 6)$

2. $(-3a^3 + 8a^2 - 7a - 3) + (4a^3 + 6a^2 - 9a + 7)$

3. $(6r^2 - 2r + 7) - (4r^2 - 6r + 6)$

4. $(7m^3 - m^2 + 4m - 7) - (-5m^3 + 4m^2 - 9m + 3)$

For additional exercises of this type, see Exercises 1–64 on pages 376–378.

Objective 2 Multiply polynomials.

For a review of this objective, see Section 5.4, page 381, and Section 5.5, pages 390–397.

For Exercises 5–8, multiply the monomials.

5. $(3a^3b^2)(-6a^5b^6c^2)$

6. $(-2p^2q^4)(-6p^7q^3r^5)(3q^4r^5)$

7. $(-2.3m^8n^4)(0.4m^2n^3)$

8. $\left(\frac{2}{3}x^3y^5\right)\left(\frac{9}{5}x^4y^2\right)$

For additional exercises of this type, see Exercises 1–26 on pages 385–386.

For Exercises 9–12, multiply the polynomial by the monomial.

9. $4a^3(6a^2 - 4a + 3)$

10. $-4x^2y(3x^3y^2 + 6xy^3 - 4x^5 + 5)$

11. $4x^2\left(-3x^5 - \frac{1}{8}x^3 + \frac{1}{12}x^2 - \frac{1}{20}x\right)$

12. $4.5a^3b^2c(2.1a^2b^2 - 3.5a^3b^5c^2 + 4)$

For additional exercises of this type, see Exercises 1–28 on pages 397 and 398.

For Exercises 13–18, multiply the polynomials.

13. $(3x - 4)(2x + 3)$

14. $(3y + 5)(y - 4)$

15. $(3p - 4q)(2p - 3q)$

16. $(4a - 5b)(3a + 2b)$

17. $(a + 2)(4a^2 + 3a - 5)$

18. $(3y - 2)(2y^2 - 4y - 5)$

For additional exercises of this type, see Exercises 29–46 on page 398 and 51–66 on page 399.

For Exercises 19 and 20, multiply the conjugates.

19. $(2y + 5)(2y - 5)$

20. $(3m - 7n)(3m + 7n)$

For additional exercises of this type, see Exercises 75–84 on page 400.

For Exercises 21 and 22, square the binomials.

21. $(7x + 2y)^2$

22. $(5a - 3b)^2$

For additional exercises of this type, see Exercises 85–96 on page 400.

Objective 3 Divide polynomials.

For a review of this objective, see Section 5.6, pages 405–409.

For Exercises 23 and 24, divide the polynomial by the monomial.

23. $\dfrac{18x^4 - 12x^3 + 24x^2 - 30x}{6x^2}$

24. $\dfrac{45u^3v^2 - 20u^4v + 15u^2v^2}{5u^2v}$

For additional exercises of this type, see Exercises 53–70 on pages 412 and 413.

For Exercises 25 and 26, divide the polynomials.

25. $\dfrac{16x - 12x^2 + 3x^3 - 12}{x - 2}$

26. $\dfrac{2x^3 - 14x + 5}{2x - 4}$

For additional exercises of this type, see Exercises 75–98 on pages 413 and 414.

R.5 Review of Factoring

Objectives

1. Write a polynomial as a product of a GCF and a polynomial.
2. Factoring by grouping.
3. Factor trinomials of the form $x^2 + bx + c$.
4. Factoring trinomials of the form $ax^2 + bx + c$ where $a \neq 1$.
5. Factor special products

Objective 1 Write a polynomial as a product of a GCF and a polynomial.

For a review of this objective, see Section 6.1, pages 431–433.

For Exercises 1–4, factor by factoring out the GCF.

1. $18c^3d^4 - 12c^2d^6$
2. $16c^3d^2 - 24c^2d^4 + 36cd^2$
3. $-10x^3 - 15x^2 + 25x$
4. $-24m^2n - 28mn + 16mn^2$

For additional exercises of this type, see Exercises 29–74 on pages 435 and 436.

Objective 2 Factoring by grouping.

For a review of this objective, see Section 6.1, pages 433 and 434.

For Exercises 5–8, factor by grouping.

5. $a^3 + 4a^2 + 3a + 12$
6. $8x^2 - 2xy + 12x - 3y$
7. $6ac + 4ad - 9bc - 6bd$
8. $15ac - 5ad - 3bc + bd$

For additional exercises of this type, see Exercises 75–98 on page 436.

Objective 3 Factor trinomials of the form $x^2 + bx + c$.

For a review of this objective, see Section 6.2, pages 438–441.

For Exercises 9–12, factor the trinomial of the form $x^2 + bx + c$.

9. $z^2 + 11z + 18$
10. $y^2 + y - 42$
11. $x^2 - 12xy + 35y^2$
12. $2a^2 + 20ab - 48b^2$

For additional exercises of this type, see Exercises 9–68 on pages 442 and 443.

Objective 4 Factoring trinomials of the form $ax^2 + bx + c$ where $a \neq 1$.

For a review of this objective, see Section 6.3, pages 444–447.

For Exercises 13–16, factor the trinomial of the form $ax^2 + bx + c$.

13. $9x^2 - 18x + 8$
14. $6c^2 - 13cd + 6d^2$
15. $18c^2 - 57cd + 45d^2$
16. $12a^2 + 7ab - 12b^2$

For additional exercises of this type, see Exercises 7–38 on page 450.

Objective 5 Factor special products.

For a review of this objective, see Section 6.4, pages 452–456.

For Exercises 17–20, factor completely the perfect square trinomials.

17. $a^2 - 16a + 64$
18. $25x^2y - 10xy + y$
19. $4x^2 + 20xy + 25y^2$
20. $36m^2 - 84mn + 49n^2$

For additional exercises of this type, see Exercises 1–18 on page 457.

For Exercises 21–23, factor the difference of squares.

21. $4x^2 - 25y^2$
22. $y^4 - 81$
23. $36x^2 + 49y^2$

For additional exercises of this type, see Exercises 19–38 on page 457.

For Exercises 24 and 25, factor the sum or difference of cubes.

24. $r^3 - 125$
25. $27x^3 + 64$

For additional exercises of this type, see Exercises 39–58 on page 458.

R.6 Review of Functions

Objectives

1 Identify the domain and range of a relation and determine whether the relation is a function.

2 Find the value of a function.

3 Graph functions.

Objective 1 Identify the domain and range of a relation and determine whether the relation is a function.

For a review of this objective, see Section 3.7 on pages 248–253.

For Exercises 1–8, identify the domain and range of each relation and determine whether it is a function.

1. $\{(2,1), (-3,2), (0,5), (1,8)\}$

2. $\{(-3,2), (4,1), (-3,5), (2,8)\}$

3. Most Popular USA Travel Destinations for Romantics

Rank	Destination
1	Hawaii
2	Las Vegas
3	Florida
4	California
5	Texas

(*Source:* http://honeymoons.about.com.)

4. Percent of the workforce leaving for work commute

Percent (%)	Time of Departure
2	Work at home
3	Midnight to 4:59 A.M. *and* 10 A.M. to 11:59 A.M.
5	9 A.M. to 9:59 A.M.
14	Noon to 11:59 P.M.
15	8 A.M. to 8:59 A.M.
26	5 A.M. to 6:59 A.M.
33	7 A.M. to 7:59 A.M.

(*Source:* U.S. Census 2000.)

5.

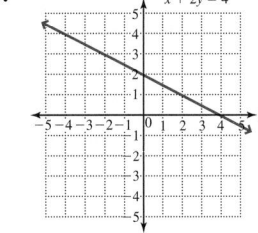

6.

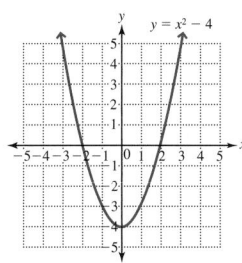

7.

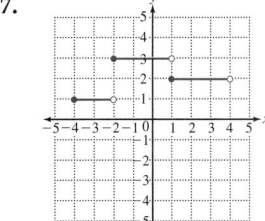

8.

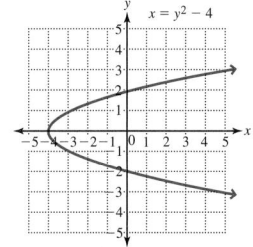

For additional exercises of this type, see exercises 1–40 on pages 256–259.

Objective 2 Find the value of a function.

For a review of this objective, see Section 3.7, pages 254 and 255.

For Exercises 9–12, given $f(x) = -4x + 1$, find each of the following.

9. $f(2)$ 10. $f(-1)$ 11. $f(0)$ 12. $f\left(\dfrac{1}{2}\right)$

For Exercises 13–16, given $f(x) = 2x^2 - 3x + 1$, find each of the following.

13. $f(-2)$ 14. $f(0)$ 15. $f\left(\dfrac{1}{3}\right)$ 16. $f(1.5)$

For Exercises 17–20, given $f(x) = \dfrac{2x}{x-1}$, find each of the following.

17. $f(1)$ 18. $f(-2)$ 19. $f(1.5)$ 20. $f\left(\dfrac{2}{3}\right)$

For Exercises 21 and 22, use the graph to determine the value of the function. (*Hint:* The function value is the *y*-value for the given *x*-value.)

21.
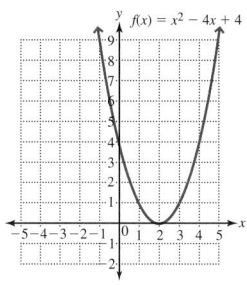

22.
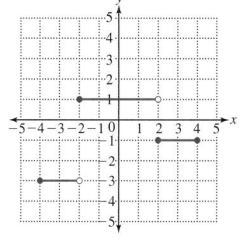

a. $f(3)$ b. $f(-1)$ c. $f(4)$

a. $f(-3)$ b. $f(-2)$ c. $f(2)$

For additional exercises of this type, see exercises 41–68 on pages 259–261.

Objective 3 Graph functions.

For a review of this objective, see Section 3.7, pages 255 and 256 and Section 6.7, pages 478 and 479.

For Exercises 23–28, graph.

23. $f(x) = 2x + 3$ 24. $f(x) = -x + 4$ 25. $f(x) = \dfrac{3}{4}x - 2$

26. $f(x) = x^2 + 4$ 27. $f(x) = -x^2 + 1$ 28. $f(x) = x^2 - 4x + 3$

For additional exercises of this type, see exercises 69–78 on page 261 and exercises 33–44 on page 483.

Mean, Median, and Mode

Objectives

1 Find the mean.
2 Find the median.
3 Find the mode.

Objective 1 Find the mean.

Numbers can be used to describe characteristics of data. A number used in this way is called a **statistic**.

Definition   Statistic: A number used to describe some characteristic of a set of data.

Oftentimes, we use statistics, such as averages, which describe the *middle* or *central tendency* in a set of data. One such statistic is the **arithmetic mean**. Many people use the words *average* and *mean* interchangeably. Actually, the mean is only one type of average, called the **arithmetic average**. Other averages are the median and mode, which we will discuss later in this appendix.

Definition   Mean or **arithmetic average:** The sum of all given numbers divided by the number of numbers.

> **Procedure** **To find the mean, or arithmetic average, of a given set of numbers:**
> 1. Calculate the sum of all the given numbers.
> 2. Divide the sum by the number of numbers.

The variable used to indicate the mean is $\bar{x}$.

Example 1 Clerical specialist salaries within a company are listed below. What is the mean salary of a clerical specialist within the company?

$24,000	$21,900
$19,500	$19,500
$20,400	$25,200

Solution: Divide the sum of the salaries by the number of salaries, which is 6.

$$\bar{x} = \frac{\$24{,}000 + \$19{,}500 + \$20{,}400 + \$21{,}900 + \$19{,}500 + \$25{,}200}{6}$$

$$\bar{x} = \frac{\$130{,}500}{6} \quad \text{Add the salaries.}$$

$$\bar{x} = \$21{,}750 \quad \text{Divide.}$$

Objective 2 Find the median.

Another average or statistic that describes the *middle* or *central tendency* of a set of numbers is the **median**.

Definition   Median: The middle number in an ordered set of numbers.

Consider the set of numbers 4, 5, 9, 3, 2, 10, 9.

The definition of *median* indicates that the set *must be ordered*. This means that we must put the numbers in order from smallest to largest or largest to smallest. We must also write *each* repetition of a number.

$$2, 3, 4, 5, 9, 9, 10$$

median

Note Because 5 is the middle number in this ordered set, it is the ◀ median.

Notice that the median is easy to find in a set that contains an odd number of numbers. Suppose the set contains an even number of numbers, as in 2, 3, 4, 5, 9, 10.

$$2, 3, 4, 5, 9, 10$$

median

Note The middle of this set of ordered numbers is *between* 4 and 5. In fact, the number exactly halfway between 4 and 5 is the mean of 4 and 5.

$$\text{Median} = \text{Mean of 4 and 5} = \frac{4 + 5}{2} = 4.5$$

> **Procedure** **To find the median of a set of numbers:**
>
> 1. Arrange the numbers in order from least to greatest or greatest to least.
> 2. If the list has an odd number of numbers, the median is the middle number. If the list has an even number of numbers, the median is the mean of the middle two numbers.

Example 2   Find the median salary for clerical specialists at the company described in Example 1. The salaries were

$24,000	$21,900
$19,500	$19,500
$20,400	$25,200

Solution: Arrange the salaries in order from least to greatest; then locate the middle salary in the ordered list.

$$\$19{,}500 \quad \$19{,}500 \quad \$20{,}400 \quad \$21{,}900 \quad \$24{,}000 \quad \$25{,}200$$

Note There are an even number of numbers in the list; so the median is the mean of the middle two numbers, 20,400 and 21,900.

$$\text{Median} = \frac{\$20{,}400 + \$21{,}900}{2} = \frac{\$42{,}300}{2} = \$21{,}150$$

In Example 2, notice that half of the people have salaries greater than the median salary and half have salaries less than the median salary. Median does not indicate how much greater or less the other salaries might be. Also, it is possible that no one earns exactly the mean or median salary.

Objective 3 Find the mode.

Another statistic we may consider with data sets is called the **mode**.

Definition   **Mode:** The number that occurs most often in a set of numbers.

Consider the set of numbers 2, 5, 9, 12, 9, 15, 9, 12.

Because 9 is the number that occurs most often, it is the mode of this set. If no number is repeated, then there is no mode. If there is a tie between numbers that occur most often, then we list each as a mode.

Procedure   **To find the mode**

To find the mode of a set of numbers, count the number of repetitions of each number.

The number with the most repetitions is the mode.

Note If no numbers are repeated, then there is no mode. If there is a tie between numbers that occur most often, then list each number as a mode.

| **Example 3** | Find the mode. |

a. Twelve players in a golf tournament post the following numbers:

$$69, 66, 74, 72, 69, 70, 72, 71, 75, 65, 72, 71$$

Note Ordering the data helps spot repeated numbers. ▶

Solution: Order the data and count the number of repetitions of each number. The number with the most repetitions is the mode.

$$65, 66, \underbrace{69, 69}_{\text{twice}}, 70, \underbrace{71, 71}_{\text{twice}}, \underbrace{72, 72, 72}_{\text{three times}}, 74, 75$$

▲
72 is the number with the most repetitions, so it is the mode.

b. Final grades for a class of 15 students:

$$85, 86, 72, 65, 80, 91, 62, 76, 80, 85, 76, 78, 80, 96, 85$$

Solution: Order the data and count the number of repetitions of each number. The number with the most repetitions is the mode.

$$62, 65, 72, \underbrace{76, 76}_{\text{twice}}, 78, \underbrace{80, 80, 80}_{\text{three times}}, \underbrace{85, 85, 85}_{\text{three times}}, 86, 91, 96$$

▲   ▲
85 and 80 both occur three times. This means they are *both* modes.

c. Birth weights for a group of 6 siblings:

7.4 pounds 7.5 pounds 8.2 pounds 7.6 pounds 7.8 pounds 8.1 pounds

Solution: Because no weight is repeated, there is no mode for this set of weights.

Appendix A Exercises
For Extra Help   MyMathLab®

Objectives 1–3

Prep Exercise 1 What is the mean of a set of data?
The sum of all given numbers divided by the number of numbers

Prep Exercise 2 What is the median of a set of data?
The middle number in an ordered set of numbers

Prep Exercise 3 Explain how to find the median of a set containing an even number of numbers.
1. Arrange the numbers in order from least to greatest or greatest to least.
2. Find the mean of the middle two numbers.

Prep Exercise 4 What is the mode of a set of data?
The number that occurs most often in the set of data

For Exercises 1–10, find the mean, median, and mode(s).

1. Following is a list of nurse salaries on the surgical floor of a hospital. Find the mean, median, and mode of the salaries.

$25,500	$28,700
$28,700	$27,500
$26,450	$24,200

mean = $26,841.6̄; median = $26,975; mode = $28,700

2. Following is a list of the hourly wages for employees at a packaging plant. Find the mean, median, and mode of the wages.

$11.00	$12.00	$11.00
$15.00	$18.25	$12.50
$13.50	$16.25	$18.00

mean = $14.1̄6̄; median = $13.50; mode = $11.00

3. Find the mean, median, and mode for the test scores of students in a history class.

80	92	64	78	88
80	82	74	72	60
55	96	100	71	82
75	82	90	86	58

mean = 78.25; median = 80; mode = 82

4. A basketball team has the following final scores. Find the mean, median, and mode of the scores.

| 76 | 82 | 80 | 78 | 85 | 75 |
| 78 | 80 | 72 | 70 | 84 | 88 |

mean = 79; median = 79; modes = 78 and 80

5. A marine biologist studying leatherback sea turtles measures and records their lengths. Following is a list of lengths for the last 20 turtles. Find the mean, median, and mode of the lengths.

2.2 m	1.8 m	1.9 m	2.3 m	2.1 m
1.6 m	1.5 m	1.8 m	1.2 m	2.0 m
2.1 m	1.4 m	1.2 m	2.1 m	1.7 m
2.2 m	2.0 m	1.6 m	1.8 m	1.9 m

mean = 1.82 m; median = 1.85 m; modes = 1.8 m and 2.1 m

6. Following is a list of heights of players on a basketball team. Find the mean, median, and mode of the heights.

| 6.5 ft. | 6.75 ft. | 6.25 ft. | 6.5 ft. | 6.8 ft. | 6.75 ft. |
| 6.25 ft. | 7 ft. | 6.75 ft. | 6.25 ft. | 6 ft. | 6.2 ft. |

mean = 6.5 ft.; median = 6.5 ft.; modes = 6.25 ft. and 6.75 ft.

7. Following is a list of rainfall amounts for one month. Each amount is the total rainfall in one 24-hour period. Calculate the mean, median, and mode.

0.4 in.	0.8 in.	1.2 in.	3.4 in.
0.6 in.	1.5 in.	0.6 in.	1.0 in.
1.4 in.	2.3 in.	1.3 in.	0.5 in.

mean = 1.25 in.; median = 1.1 in.; mode = 0.6 in.

8. The table lists daily high and low temperatures (°F) for the last two weeks. Find the mean, median, and mode of the high temperatures. Then find the mean, median, and mode of the low temperatures.

	Sun.	Mon.	Tue.	Wed.	Thu.	Fri.	Sat.
High	91	92	95	93	90	89	88
Low	72	74	78	75	72	70	71
High	90	94	96	95	92	90	91
Low	72	74	78	76	73	70	72

Highs: mean ≈ 91.9; median = 91.5; mode = 90
Lows: mean ≈ 73.4; median = 72.5; mode = 72

9. The table lists each month's electric and gas charges for a family over a two-year period. Find the mean, median, and mode of the electric and gas charges each year.

Month	Year 1	Year 2
January	$158.92	$165.98
February	$147.88	$162.85
March	$125.90	$130.45
April	$108.40	$112.55
May	$87.65	$90.45
June	$114.58	$125.91
July	$145.84	$137.70
August	$142.78	$140.19
September	$90.25	$96.15
October	$104.12	$115.75
November	$136.62	$145.21
December	$158.18	$160.25

year 1: mean = $126.76; median = $131.26; no mode
year 2: mean = $131.95; median = $134.08; no mode

10. The table lists water consumption, in cubic feet, by a family over a two-year period. Find the mean, median, and mode for each year.

Month	Year 1	Year 2
January	500	550
February	525	540
March	600	600
April	650	650
May	840	900
June	1100	1430
July	1000	1200
August	1400	1250
September	840	750
October	620	700
November	600	550
December	550	500

year 1: mean = 768.75 ft.^{3}; median = 635 ft.^{3}; mode = 600 ft.^{3} and 840 ft.^{3}
year 2: mean = 801.$\overline{6}$ ft.^{3}; median = 675 ft.^{3}; mode = 550 ft.^{3}

Appendix B

Arithmetic Sequences and Series

Objectives

1 Find the terms of a sequence when given the general term.

2 Define and write arithmetic sequences, find their common difference, and find a particular term.

3 Define and write series, find partial sums, and use summation notation.

4 Write arithmetic series and find their sums.

The word **sequence** is used in mathematics much as it is in everyday life. For example, the classes you attend on a given day occur in a particular order, or sequence. As such, a sequence is an ordered list. Suppose we have a bacteria colony that has an initial population of 10,000 and increases at a rate of 10% each day.

On the second day, we have
$10,000 + 0.10(10,000) = 10,000 + 1000 = 11,000$.
On the third day, we have $11,000 + 0.10(11,000) = 11,000 + 1100 = 12,100$.
On the fourth day, we have $12,100 + 0.10(12,100) = 12,100 + 1210 = 13,310$.
On the fifth day, we have $13,310 + 0.10(13,310) = 13,310 + 1331 = 14,641$.

The number of bacteria present after each day forms a sequence that we can summarize as follows.

Days	1	2	3	4	5
Number of Bacteria	10,000	11,000	12,100	13,310	14,641

Based on this, we have the following definition.

Definition Sequence: A function list whose domain is $1, 2, 3, \ldots, n$.

The *domain* of our bacteria example is the numbers of the days $\{1, 2, 3, 4, 5\}$. Each number in the *range* of a sequence is called a *term*. So the terms of our bacteria sequence are the numbers of bacteria: 10,000, 11,000, 12,100, 13,310, 14,641. Sequences can be finite or infinite depending on the number of terms. Our bacteria example is a **finite sequence** because it has a finite number of terms. The sequence 2, 4, 6, 8, 10, . . . is an **infinite sequence** because it has an infinite number of terms.

Definitions Finite sequence: A function with a domain that is the set of natural numbers from 1 to n.
Infinite sequence: A function with a domain that is the set of natural numbers.

Objective 1 Find the terms of a sequence when given the general term.

Because a sequence is a function, we could describe sequences using functional notation. Instead, we use a different notation that emphasizes the fact that the domain is a subset of the natural numbers. We think of the terms of a sequence as $a_1, a_2, a_3, \ldots, a_n$, where the subscript gives the number of the term. Thus, a_n is the nth or *general term* of the sequence and we represent a sequence by giving a formula for a_n. Consider the following sequence.

Term	a_1	a_2	a_3	a_4	a_5	a_n
Term of Sequence	1	4	9	16	25	n^2

We represent this sequence by writing $a_n = n^2$, which means $a_1 = 1^2, a_2 = 2^2$, and so on.

Example 1   Find the first three terms of the following sequences and the 25th term.

a. $a_n = 3n - 1$

Solution: We let $n = 1, 2, 3,$ and 25 and evaluate.

$$a_1 = 3(1) - 1 = 3 - 1 = 2 \qquad \text{Let } n = 1 \text{ and evaluate.}$$
$$a_2 = 3(2) - 1 = 6 - 1 = 5 \qquad \text{Let } n = 2 \text{ and evaluate.}$$
$$a_3 = 3(3) - 1 = 9 - 1 = 8 \qquad \text{Let } n = 3 \text{ and evaluate.}$$
$$a_{25} = 3(25) - 1 = 75 - 1 = 74 \qquad \text{Let } n = 25 \text{ and evaluate.}$$

The first three terms of the sequence are 2, 5, and 8. The 25th term is 74.

b. $a_n = \dfrac{(-1)^n}{n^2 + 1}$

Solution: We let $n = 1, 2, 3,$ and 25 and evaluate.

$$a_1 = \frac{(-1)^1}{1^2 + 1} = \frac{-1}{1 + 1} = -\frac{1}{2} \qquad \text{Let } n = 1 \text{ and evaluate.}$$

$$a_2 = \frac{(-1)^2}{2^2 + 1} = \frac{1}{4 + 1} = \frac{1}{5} \qquad \text{Let } n = 2 \text{ and evaluate.}$$

$$a_3 = \frac{(-1)^3}{3^2 + 1} = \frac{-1}{9 + 1} = -\frac{1}{10} \qquad \text{Let } n = 3 \text{ and evaluate.}$$

$$a_{25} = \frac{(-1)^{25}}{25^2 + 1} = \frac{-1}{625 + 1} = -\frac{1}{626} \qquad \text{Let } n = 25 \text{ and evaluate.}$$

The first three terms of the sequence are $-\dfrac{1}{2}, \dfrac{1}{5},$ and $-\dfrac{1}{10}.$ The 25th term is $-\dfrac{1}{626}.$

Objective 2 Define and write arithmetic sequences, find their common difference, and find a particular term.

In the sequence $-3, 1, 5, 9, 13, \ldots,$ notice that each term after the first is found by adding 4 to the previous term. This is an example of an **arithmetic sequence** or *arithmetic progression*. Any two successive terms of an arithmetic sequence differ by the same amount, which is called the **common difference** and is denoted by $d.$ To find $d,$ choose any term (except the first) and subtract the previous term.

Definition Arithmetic sequence: A sequence in which each term after the first is found by adding the same number to the previous term.

Common difference of an arithmetic sequence: The value of d is found by $d = a_n - a_{n-1},$ where a_n is any term in the sequence (except the first) and a_{n-1} is the previous term.

Example 2   Write the first four terms of the following arithmetic sequences.

a. The first term is 2, and the common difference is 5.

Solution: Begin with 2 and find each successive term by adding 5 to the previous term.

$$a_1 = 2, a_2 = 2 + 5 = 7, a_3 = 7 + 5 = 12, a_4 = 12 + 5 = 17$$

The first four terms of the sequence are 2, 7, 12, 17.

b. $a_1 = -1, d = -2$

Solution: Begin with -1 and find each successive term by adding -2 to the previous term.

$$a_1 = -1, a_2 = -1 - 2 = -3, a_3 = -3 - 2 = -5, a_4 = -5 - 2 = -7$$

The first four terms of the sequence are $-1, -3, -5, -7.$

Example 3   Find the common difference, d, for the following arithmetic sequences.

a. $-4, -1, 2, 5, 8, \ldots$

Solution: Pick any term (except the first) and subtract the term before it.

$$d = -1 - (-4) = -1 + 4 = 3$$

Or we could use $d = 5 - 2 = 3$ and so on.

b. $8, 6, 4, 2, 0, \ldots$

Solution: Pick any term (except the first) and subtract the term before it.

$$d = 6 - 8 = -2$$

Or we could use $d = 2 - 4 = -2$ and so on.

If the first term of an arithmetic sequence is a_1 and the common difference is d, then the arithmetic sequence can be written as $a_1, a_1 + d, a_1 + 2d, a_1 + 3d, a_1 + 4d, \ldots$. Note that the coefficient of d is one less than the number of the term; so we have the following rule.

> **Rule nth Term of an Arithmetic Sequence**
> The formula for finding the nth term of an arithmetic sequence is $a_n = a_1 + (n - 1)d$, where a_1 is the first term and d is the common difference.

Example 4   Find the 23rd term and an expression for the nth term of an arithmetic sequence in which $a_1 = -8$ and $d = 3$.

Solution:

$$a_n = a_1 + (n - 1)d$$
$$a_{23} = -8 + (23 - 1)(3) \quad \text{Substitute 23 for } \textbf{n}, -8 \text{ for } a_1, \text{ and 3 for } \textbf{d}.$$
$$a_{23} = 58 \quad \text{Evaluate.}$$
$$n\text{th term: } a_n = -8 + (n - 1)3 \quad \text{Substitute for } a_1 \text{ and } d.$$
$$a_n = -11 + 3n \quad \text{Simplify.}$$

If we know the first term and one other term, we can use the formula for the nth term of an arithmetic sequence to find the common difference. Consequently, we can find the sequence.

Example 5   The first term of an arithmetic sequence is 1, and the 20th term (a_{20}) is 58. Find the common difference and the first four terms of the sequence.

Solution: We use the fact that we know a_{20} and the formula for the nth term to find d.

$$a_n = a_1 + (n - 1)d$$
$$a_{20} = 1 + (20 - 1)d \quad \text{Substitute 1 for } a_1 \text{ and 20 for } \textbf{n}.$$
$$58 = 1 + 19d \quad \text{Substitute 58 for } a_{20} \text{ and solve.}$$
$$3 = d$$

The first four terms are $1, 1 + 3 = 4, 4 + 3 = 7, 7 + 3 = 10$.

Objective 3 Define and write series, find partial sums, and use summation notation.

When the terms of a sequence are added, the sum is called a **series**.

Definition Series: The sum of the terms of a sequence.

Given a sequence $a_1, a_2, a_3, \ldots, a_n$, then $a_1 + a_2 + a_3 + \cdots + a_n$ is the corresponding series. Finite series correspond to finite sequences, and infinite series correspond to infinite sequences. If an expression for the nth term is known, we can write the series in *summation notation* using the capital Greek letter *sigma* (Σ) as follows.

$$a_1 + a_2 + a_3 + \cdots + a_n = \sum_{i=1}^{n} a_i$$

The i is called the *index of summation*; 1 is the *lower limit* of i, and n is the *upper limit* of i. If the upper limit is a natural number, the series is finite, and if the upper limit is ∞, the series is infinite. To find a finite series from summation notation, replace the index of summation (usually i) with its lower limit and evaluate that term, replace i with the lower limit plus 1 and evaluate that term, and so on, until you reach the upper limit. This gives the series. To find the sum, add the terms of the series.

Example 6

Warning Do not confuse the use of i in sigma notation, as in $\sum_{i=1}^{5} (3i + 2)$, with its use as the imaginary unit, as in $7 \pm 5i$. In sigma notation, i is a variable representing natural numbers, whereas in the imaginary unit, i represents $\sqrt{-1}$.

a. Write the terms of $\sum_{i=1}^{5} (3i + 2)$ and find the sum of the series.

Solution: Replace i with 1, 2, 3, 4, and 5. Then find the sum.

$$\sum_{i=1}^{5} (3i + 2) = (3 \cdot 1 + 2) + (3 \cdot 2 + 2) + (3 \cdot 3 + 2) + (3 \cdot 4 + 2) + (3 \cdot 5 + 2)$$

$$= 5 + 8 + 11 + 14 + 17$$

$$= 55$$

b. Write the first five terms of the infinite series $\sum_{i=1}^{\infty} 2i^2$.

Solution: Replace i with 1, 2, 3, 4, and 5. Then evaluate.

$$\sum_{i=1}^{\infty} 2i^2 = 2 \cdot 1^2 + 2 \cdot 2^2 + 2 \cdot 3^2 + 2 \cdot 4^2 + 2 \cdot 5^2 + \cdots$$

$$= 2 + 8 + 18 + 32 + 50 + \cdots$$

Objective 4 Write arithmetic series and find their sums.

When the terms of an arithmetic sequence are added, it is called an **arithmetic series**.

Definition Arithmetic series: The sum of the terms of an arithmetic sequence.

Consequently, an arithmetic series has the form $a_1 + (a_1 + d) + (a_1 + 2d) + (a_1 + 3d) + \cdots + [a_1 + (n - 1)d]$.

Adding a finite number of terms of an infinite series gives a *partial sum*. The symbol S_n is used to indicate the sum of the first n terms. For example, S_5 means add the first five terms. Let's derive a formula for S_n for an arithmetic series as follows.

$$S_n = a_1 + (a_1 + d) + (a_1 + 2d) + (a_1 + 3d) + \cdots + a_n$$

We also need to include the terms between $(a_1 + 3d)$ and a_n. So we write another version of S_n beginning with the last term, a_n, and subtracting the common difference, d, from the previous term.

$$S_n = a_n + (a_n - d) + (a_n - 2d) + (a_n - 3d) + \cdots + a_1$$

To describe the entire sum, we add our two versions of S_n.

$$
\begin{array}{rl}
S_n = a_1 & + (a_1 + d) + (a_1 + 2d) + (a_1 + 3d) + \cdots + a_n \\
+ S_n = a_n & + (a_n - d) + (a_n - 2d) + (a_n - 3d) + \cdots + a_1 \\
\hline
2S_n = (a_1 + a_n) & + (a_1 + a_n) + (a_1 + a_n) + (a_1 + a_n) + \cdots + (a_1 + a_n)
\end{array}
$$

Because S_n has n terms, there are n terms of $(a_1 + a_n)$; so

$$
2S_n = n(a_1 + a_n), \quad \text{or} \quad S_n = \frac{n}{2}(a_1 + a_n)
$$

> **Rule Partial Sum, S_n, of an Arithmetic Series**
>
> The sum of the first n terms of an arithmetic series, S_n, called the nth partial sum, is given by
>
> $$S_n = \frac{n}{2}(a_1 + a_n),$$
>
> where n is the number of terms, a_1 is the first term, and a_n is the nth term.

Example 7   Find the sum of the first 20 terms (S_{20}) of the arithmetic series $-6 - 2 + 2 + 6 + \cdots$.

Solution: To find the S_{20}, we first need to find the 20th term. Because $d = 4$,

$$a_{20} = -6 + (20 - 1)(4) = 70$$

$$S_n = \frac{n}{2}(a_1 + a_n)$$

$$S_{20} = \frac{20}{2}(-6 + 70) \quad \text{Substitute \textbf{20} for } \textbf{\textit{n}}, -6 \text{ for } a_1, \text{ and } \textbf{70} \text{ for } a_n.$$

$$S_{20} = 10(64) = 640 \quad \text{Evaluate.}$$

Appendix B Exercises
For Extra Help MyMathLab®

Objective 1

Prep Exercise 1 What is a sequence?
A sequence is a function list whose domain is $1, 2, 3, \ldots, n$.

Prep Exercise 2 A sequence with an unlimited number of terms is called a(n) ___infinite___ sequence.

Prep Exercise 3 What is an arithmetic sequence?
An arithmetic sequence is a sequence in which any two successive terms differ by the same amount.

For Exercises 1–8, write the first four terms of the sequence and the indicated term.
See Example 1.

1. $a_n = 2n + 1$, 20th term
$3, 5, 7, 9, 41$

2. $a_n = 3n - 4$, 18th term
$-1, 2, 5, 8, 50$

3. $a_n = n^2 + 2$, 15th term
$3, 6, 11, 18, 227$

4. $a_n = n^2 - 3$, 12th term
$-2, 1, 6, 13, 141$

5. $a_n = \dfrac{n}{n + 2}$, 22nd term
$\dfrac{1}{3}, \dfrac{1}{2}, \dfrac{3}{5}, \dfrac{2}{3}, \dfrac{11}{12}$

6. $a_n = \dfrac{2n}{n + 3}$, 10th term
$\dfrac{1}{2}, \dfrac{4}{5}, 1, \dfrac{8}{7}, \dfrac{20}{13}$

7. $a_n = \dfrac{(-1)^n}{n^2 + 1}$, 15th term
$-\dfrac{1}{2}, \dfrac{1}{5}, -\dfrac{1}{10}, \dfrac{1}{17}, -\dfrac{1}{226}$

8. $a_n = \dfrac{(-1)^n}{n^2 - 3}$, 26th term
$\dfrac{1}{2}, -1, -\dfrac{1}{6}, \dfrac{1}{13}, \dfrac{1}{673}$

Objective 2

Prep Exercise 4 Explain how to find the common difference of an arithmetic sequence.

To find the common difference of an arithmetic sequence, subtract $a_n - a_{n-1}$.

For Exercises 9–14, find the common difference, d, for each arithmetic sequence. See Example 3.

9. $2, 7, 12, 17, \ldots$
5

10. $3, 11, 19, 27, \ldots$
8

11. $25, 22, 19, 16, \ldots$
-3

12. $42, 36, 30, 24, \ldots$
-6

13. $-12, -5, 2, 9, \ldots$
7

14. $-24, -27, -30, -33, \ldots$
-3

Prep Exercise 5 What is the formula for finding the nth term of an arithmetic sequence?

$a_n = a_1 + (n - 1)d$, where a_1 is the first term and d is the common difference

For Exercises 15–20, find the indicated term and an expression for the nth term of the given arithmetic sequence. See Example 4.

15. a_{14} if $a_1 = 14$ and $d = 4$
$a_n = 4n + 10, a_{14} = 66$

16. a_{24} if $a_1 = 16$ and $d = 6$
$a_n = 6n + 10, a_{24} = 154$

17. a_{28} if $a_1 = -8$ and $d = -3$
$a_n = -3n - 5, a_{28} = -89$

18. a_{30} if $a_1 = -7$ and $d = -6$
$a_n = -6n - 1, a_{30} = -181$

19. a_{34} of $-5, -1, 3, 7, \ldots$
$a_n = 4n - 9, a_{34} = 127$

20. a_{21} of $8, 15, 22, 29, \ldots$
$a_n = 7n + 1, a_{21} = 148$

For Exercises 21–26, write the first four terms of the arithmetic sequence with the given characteristics. See Examples 2 and 5.

21. $a_1 = -6, d = 7$
$-6, 1, 8, 15$

22. $a_1 = -2, d = 6$
$-2, 4, 10, 16$

23. $a_1 = -5, d = -3$
$-5, -8, -11, -14$

24. $a_1 = -13, d = -5$
$-13, -18, -23, -28$

25. $a_1 = 7, a_{18} = 75$
$7, 11, 15, 19$

26. $a_1 = 6, a_{22} = 153$
$6, 13, 20, 27$

27. Find the first term of an arithmetic sequence if $a_{45} = 143$ and $d = 3$.
$a_1 = 11$

28. Find the first term of an arithmetic sequence if $a_{39} = 181$ and $d = 6$.
$a_1 = -47$

29. Find the common difference of an arithmetic sequence if the first term is -110 and the 29th term is 2.
$d = 4$

30. Find the common difference of an arithmetic sequence if the first term is 78 and the 15th term is -76.
$d = -11$

For Exercises 31–34, write the first four terms of the arithmetic sequence with the given d and a_n.

31. $d = 7, a_8 = 41$
$-8, -1, 6, 13$

32. $d = 4, a_{11} = 27$
$-13, -9, -5, -1$

33. $d = -3, a_7 = 9$
$27, 24, 21, 18$

34. $d = -6, a_{10} = -36$
$18, 12, 6, 0$

Objective 3

Prep Exercise 6 What is a series?

A series is the sum of the terms of a sequence.

For Exercises 35–42, write the series and find the sum. See Example 6.

35. $\displaystyle\sum_{i=1}^{6} i^2$
$1 + 4 + 9 + 16 + 25 + 36 = 91$

36. $\displaystyle\sum_{i=1}^{3} 3i^2$
$3 + 12 + 27 = 42$

37. $\displaystyle\sum_{i=1}^{4} (2i - 5)$
$-3 - 1 + 1 + 3 = 0$

38. $\sum\limits_{i=1}^{5}(4i+1)$

$5 + 9 + 13 + 17 + 21 = 65$

39. $\sum\limits_{i=1}^{3}(3i^2-4)$

$-1 + 8 + 23 = 30$

40. $\sum\limits_{i=1}^{4}(-2i^2+5)$

$3 - 3 - 13 - 27 = -40$

41. $\sum\limits_{i=3}^{6}(4i-3)$

$9 + 13 + 17 + 21 = 60$

42. $\sum\limits_{i=2}^{5}(4i-2)$

$6 + 10 + 14 + 18 = 48$

Objective 4

Prep Exercise 7 What does S_n represent?

Sum of the first n terms of series

Prep Exercise 8 What is the formula for finding S_n of an arithmetic series?

$S_n = \dfrac{n}{2}(a_1 + a_n)$, where n is the number of terms, a_1 is the first term, and a_n is the nth term.

For Exercises 43–50, find the given S_n for the arithmetic series. See Example 7.

43. If $a_1 = 10$ and $d = 4$, find S_{25}.

1450

44. If $a_1 = -8$ and $d = 2$, find S_{30}.

630

45. If $a_1 = 12$ and $d = -3$, find S_{22}.

-429

46. If $a_1 = 18$ and $d = -2$, find S_{30}.

-330

47. $3 + 9 + 15 + 21 + \cdots$. Find S_{15}.

675

48. $5 + 9 + 13 + 17 + \cdots$. Find S_{25}.

1325

49. $54 + 46 + 38 + 30 + \cdots$. Find S_{12}.

120

50. $44 + 39 + 34 + 29 + \cdots$. Find S_{18}.

27

For Exercises 51–58, answer each question.

51. Find the sum of the first 100 natural numbers.

5050

52. Find the sum of the even integers 2 through 200.

10,100

53. A concert hall has 60 seats in the first row, 64 in the second, 68 in the third, and so on.
  a. How many seats are in the 22nd row?

  144 seats

  b. How many seats are in the concert hall if there are 35 rows?

  4480 seats

54. Johanna bought a car for $24,000 with a special no-interest loan and made monthly payments of $400. Using 24,000 as the first term, write the first five terms of an arithmetic sequence that gives the amount she still owes at the end of each month. How much will she owe at the end of the 25th month?

$24,000, $23,600, $23,200, $22,800, $22,400; $14,400

55. Fence posts are arranged in a triangular stack with 25 on the bottom row, 24 on the next row, 23 on the next row, and so on, until a single post is on the top.
  a. How many posts are on the 10th row from the bottom?

  16 posts

  b. How many posts are in the stack?

  325 posts

56. Carlos is doing sit-ups to strengthen his abdominal muscles. He did 25 the first night and plans to add 1 each night.
  a. Write the first five terms of an arithmetic sequence that gives the number of sit-ups he does each night.

  25, 26, 27, 28, 29

  b. Find the number he does on the 30th night.

  54 sit-ups

  c. Find the total number of sit-ups he has done after 30 nights.

  1185 sit-ups

57. Tanisha takes a job that pays $28,000 the first year and gives a raise of $1500 per year.
  a. Write the first five terms of an arithmetic sequence that gives her salary at the end of each year.

  $28,000, $29,500, $31,000, $32,500, $34,000

  b. What will her salary be for the 10th year?

  $41,500

  c. What are her total earnings for her first 10 years?

  $347,500

58. You have a choice of two jobs. Job A has a starting salary of $25,000 with raises of $900 per year, and job B has a starting salary of $28,000 with raises of $600 per year.
  a. Which job will pay the most during the 10th year?

  B at $33,400

  b. Which job will pay the biggest total amount during the first 15 years?

  B at $483,000

Geometric Sequences and Series

Objectives

1 Write a geometric sequence and find its common ratio and a specified term.
2 Find partial sums of geometric series.
3 Find the sums of infinite geometric series.
4 Solve applications using geometric series.

Note In geometric sequences, the first term is usually denoted as a rather than a_1. ▶

Objective 1 Write a geometric sequence and find its common ratio and a specified term.

In Appendix B Arithmetic Sequences and Series, we generated an arithmetic sequence by adding the same number to each term to get the next term. In this section, we multiply each term by the same number to generate a **geometric sequence**. The number we multiply by is called the **common ratio** and is denoted as r. We can find r by dividing any term (except the first) by the term before it.

Definition **Geometric sequence:** A sequence in which every term after the first is found by multiplying the previous term by the same number, called the **common ratio**, r, where $r = \dfrac{a_n}{a_{n-1}}$.

If the first term is a and the common ratio is r, a geometric sequence has the form $a, ar, ar^2, ar^3, ar^4, \ldots, ar^{n-1}$. Thus, the general term of a geometric series is $a_n = ar^{n-1}$.

We notice that the exponent of r is one less than the number of the term, so we have the following rule.

> **Rule** nth **Term of a Geometric Sequence**
>
> The formula for finding the nth term of a geometric sequence is $a_n = ar^{n-1}$, where a is the first term and r is the common ratio.

Example 1 A geometric sequence has a first term of 3 and a common ratio of 4.

a. Write the first five terms.

Solution: To find each term, multiply the term before it by 4. So

$$a_1 = 3, a_2 = 3(4) = 12, a_3 = 12(4) = 48, a_4 = 48(4) = 192, a_5 = 192(4) = 768.$$

The first five terms of the sequence are 3, 12, 48, 192, 768.

b. Find the 10th term.

Solution: $a_n = ar^{n-1}$ Formula for a_n

$a_{10} = 3(4)^{10-1}$ Substitute **10** for **n**, 3 for a, and **4** for **r**.

$a_{10} = 786{,}432$ Evaluate.

c. Find the general term.

Solution: $a_n = ar^{n-1}$ Formula for a_n

$a_n = 3(4)^{n-1}$ Substitute for a and r.

Example 2 Given the geometric sequence $32, -16, 8, -4, \ldots$, find the common ratio r.

Solution: To find r, divide any term (except the first) by the term before it.

$$r = \frac{-16}{32} = -\frac{1}{2} \quad \text{or} \quad r = \frac{8}{-16} = -\frac{1}{2} \text{ and so on.}$$

Geometric sequences often occur in populations and other applications.

Example 3   The population of rabbits in a large pen increases at a rate of 12% per month. If there are currently 50 rabbits, find the population after 15 months.

Solution: Let P_0 be the initial number of rabbits, P_1 the number after 1 month, P_2 the number after 2 months, and so on. The number of rabbits at the end of each month is the number at the beginning of the month plus an increase of 12% of that number. So the number at the end of each month can be found as follows.

$$P_1 = P_0 + 0.12P_0 = 1.12P_0$$
$$P_2 = P_1 + 0.12P_1 = 1.12P_1 = 1.12(1.12P_0) = (1.12)^2P_0$$
$$P_3 = P_2 + 0.12P_2 = 1.12P_2 = 1.12[(1.12)^2P_0] = (1.12)^3P_0$$

The number of rabbits at the end of each month can be represented by the sequence $P_0, P_1, P_2, P_3, \ldots P_n = P_0, 1.12P_0, (1.12)^2P_0, (1.12)^3P_0, \ldots, (1.12)^{n-1}P_0$, which is a geometric sequence whose first term is P_0 and $r = 1.12$. Consequently, the nth term is $P_0(1.12)^{n-1}$. The number of rabbits at the end of 15 months is the 16th term of the sequence; so

$$P_{16} = P_0(1.12)^{16-1}$$
$$P_{16} = 50(1.12)^{15}$$
$$P_{16} = 273.68$$

Answer: There are about 274 rabbits at the end of 15 months.

Note It can be shown using this procedure that if a population is growing at $p\%$ per unit time and the initial population is P_0, then the population at the end of each unit of time forms a geometric sequence whose first term is P_0 and whose common ratio is $(1 + p)$, where p is $p\%$ written as a decimal. So the geometric sequence is $P_0, P_0(1 + p),$ $P_0(1 + p)^2, P_0(1 + p)^3, \ldots,$ $P_0(1 + p)^{n-1}$. The population after n time periods is the $(n + 1)$st term of the sequence, which is $P_0(1 + p)^n$.

Objective 2   Find partial sums of geometric series.

If we add the terms of an arithmetic sequence, we get an arithmetic series. Likewise, if we add the terms of a geometric sequence, we get a **geometric series**.

Definition Geometric series: The sum of the terms of a geometric sequence.

A geometric series is of the form $a + ar + ar^2 + \cdots + ar^{n-1}$ if the series is finite and $a + ar + ar^2 + \cdots$ if the series is infinite.

In the geometric series $3 + 6 + 12 + 24 + \cdots$, we have $a = 3$ and $r = 2$. In $243 - 81 + 27 - 9 + \cdots$, we have $a = 243$ and $r = -\dfrac{1}{3}$. As with arithmetic series, we can find a formula for partial sum, S_n. Begin with

$$\begin{aligned} S_n &= a + ar + ar^2 + \cdots + ar^{n-1} \\ -rS_n &= \quad\ - ar - ar^2 - \cdots - ar^{n-1} - ar^n \end{aligned}$$ Multiply both sides of the equation by $-r$.

$$S_n - rS_n = a - ar^n$$ Add the equations.
$$S_n(1 - r) = a(1 - r^n)$$ Factor both sides.
$$S_n = \frac{a(1 - r^n)}{1 - r}$$ Divide both sides by $1 - r$.

Rule Partial Sum, S_n, of a Geometric Series

The sum of the first n terms of a geometric series, S_n, called the nth partial sum, is given by

$$S_n = \frac{a(1 - r^n)}{1 - r},$$

where n is the number of terms, a is the first term, and r is the common ratio $(r \neq 1)$.

Example 4   Find the sum of the first 10 terms of the geometric series:
$1 - 3 + 9 - 27 + \cdots$

Solution: We first find r: $r = \dfrac{-3}{1} = -3$.

$$S_n = \frac{a(1 - r^n)}{1 - r} \qquad \text{Formula for } S_n$$

$$S_{10} = \frac{1(1 - (-3)^{10})}{1 - (-3)} \qquad \text{Substitute 10 for } n, 1 \text{ for } a, \text{ and } -3 \text{ for } r.$$

$$S_{10} = -14{,}762 \qquad \text{Evaluate.}$$

Objective 3 Find the sums of infinite geometric series.

If the common ratio satisfies $|r| > 1$, the partial sums become infinitely large as n becomes infinitely large. However, if $|r| < 1$, the partial sums approach a value as n becomes infinitely large. This value is called the *limit* of the partial sums and is the sum of the infinite series.

For example, for the series $2 + 1 + \dfrac{1}{2} + \dfrac{1}{4} + \cdots$, we have

$$S_5 = \frac{2\left(1 - \left(\dfrac{1}{2}\right)^5\right)}{1 - \dfrac{1}{2}} = 3.875$$

$$S_{10} = \frac{2\left(1 - \left(\dfrac{1}{2}\right)^{10}\right)}{1 - \dfrac{1}{2}} \approx 3.9960938$$

$$S_{15} = \frac{2\left(1 - \left(\dfrac{1}{2}\right)^{15}\right)}{1 - \dfrac{1}{2}} \approx 3.9998779$$

Notice that the greater the value of n, the closer the sum is to 4. The partial sums approach 4 because as n gets larger, $\left(\dfrac{1}{2}\right)^n$ approaches 0. In the preceding example, $\left(\dfrac{1}{2}\right)^5 = 0.03125$, $\left(\dfrac{1}{2}\right)^{10} \approx 0.000977$, and $\left(\dfrac{1}{2}\right)^{15} \approx 0.0000305$. In general, if $|r| < 1$, then r^n approaches 0 as n becomes large. Consequently, if $|r| < 1$, the formula $S_n = \dfrac{a(1 - r^n)}{1 - r}$ becomes $S_\infty = \dfrac{a(1 - 0)}{1 - r} = \dfrac{a}{1 - r}$ as n becomes infinitely large. We denote the sum as S_∞ rather than S_n.

Rule Sum of an Infinite Geometric Series

If $|r| < 1$, the sum of an infinite geometric series, S_∞, is given by the formula

$$S_\infty = \frac{a}{1 - r},$$

where a is the first term and r is the common ratio. If $|r| \geq 1$, S_∞ does not exist.

Example 5   Find the sum of the infinite geometric series:

$$2 + 1 + \frac{1}{2} + \frac{1}{4} + \cdots$$

Solution: We know that $a = 2$ and $r = \frac{1}{2}$. Because $|r| < 1$, the sum exists.

$$S_\infty = \frac{a}{1 - r}$$

$$S_\infty = \frac{2}{1 - \frac{1}{2}} \qquad \text{Substitute 2 for } a \text{ and } \frac{1}{2} \text{ for } \textbf{\textit{r}}.$$

$$S_\infty = 4 \qquad \text{Evaluate.}$$

Infinite geometric series also provides us with a method of changing a repeating decimal into a fraction.

Example 6   Write as a fraction: $0.\overline{37}$

Solution: First write $0.\overline{37}$ as an infinite geometric series as follows:

$$0.\overline{37} = 0.37 + 0.0037 + 0.000037 + \cdots$$

$$0.\overline{37} = \frac{37}{100} + \frac{37}{10{,}000} + \frac{37}{1{,}000{,}000} + \cdots$$

The last line is an infinite geometric series with $a = \frac{37}{100}$ and $r = \frac{1}{100}$. Because $|r| < 1$, the sum of this series exists.

$$S_\infty = \frac{a}{1 - r}$$

$$S_\infty = \frac{\frac{37}{100}}{1 - \frac{1}{100}} \qquad \text{Substitute } \frac{37}{100} \text{ for } a \text{ and } \frac{1}{100} \text{ for } r.$$

$$S_\infty = \frac{\frac{37}{100}}{\frac{99}{100}} \qquad \text{Simplify.}$$

$$S_\infty = \frac{37}{99}$$

Answer: $0.\overline{37} = \frac{37}{99}$

Objective 4 Solve applications using geometric series.

Example 7   To save for their child's college education, the McBride family put $1000 into a savings account the first year. Each year thereafter they deposited 10% more than the previous year.

a. Write the first five terms of the geometric sequence that gives the amount of money deposited in the account each year.

Solution: From the note next to Example 3, this is a geometric series in which $a = 1000$ and $r = (1 + 0.10) = 1.1$.

$$a_1 = 1000, a_2 = 1000(1.1) = 1100, a_3 = 1000(1.1)^2 = 1210,$$
$$a_4 = 1000(1.10)^3 = 1331, a_5 = 1000(1.10)^4 = 1464.10$$

The first five terms of the sequence are 1000, 1100, 1210, 1331, 1464.10.

b. How much money will the McBrides have deposited in the account at the end of 15 years?

Solution: The series $1000 + 1100 + 1210 + 1331 + 1464.10 + \cdots$ is geometric with $a = 1000$ and $r = 1.1$.

$$S_n = \frac{a(1 - r^n)}{1 - r}$$

$$S_{15} = \frac{1000(1 - 1.1^{15})}{1 - 1.1} \qquad \text{Substitute } \mathbf{15} \text{ for } \mathbf{n}, \mathbf{1000} \text{ for } \mathbf{a}, \text{ and } \mathbf{1.1} \text{ for } \mathbf{r}.$$

$$S_{15} = 31{,}772.48 \qquad \text{Evaluate.}$$

Answer: They will have deposited \$31,772.48 in the account after 15 years.

Appendix C Exercises

For Extra Help   MyMathLab®

Objective 1

Prep Exercise 1 What is a geometric sequence?

A sequence in which every term after the first is found by multiplying the previous term by the same number, called the common ratio

Prep Exercise 2 Explain how to find the common ratio of a geometric sequence.

Divide any term (except the first) by the previous term.

Prep Exercise 3 What is the formula for finding the nth term of a geometry sequence?

$a_n = ar^{n-1}$

For Exercises 1–8:   *a.* Find the common ratio, r, for the given geometric sequence. See Example 2.
b. Find the indicated term. See Example 1.
c. Find an expression for the general term, a_n. See Example 1.

1. 1, 3, 9, 27, . . . ;
9th term
a. 3   b. $a_9 = 6561$
c. $a_n = 1(3)^{n-1}$

2. 7, 14, 28, 56, . . . ;
11th term
a. 2   b. $a_{11} = 7168$
c. $a_n = 7(2)^{n-1}$

3. $-2, 4, -8, 16, \ldots$;
15th term
a. -2 b. $a_{15} = -32{,}768$
c. $a_n = -2(-2)^{n-1}$

4. $-8, 24, -72, 216, \ldots$;
9th term
a. -3 b. $a_9 = -52{,}488$
c. $a_n = -8(-3)^{n-1}$

5. 243, 81, 27, 9, . . . ;
10th term
a. $\frac{1}{3}$ b. $a_{10} = \frac{1}{81}$
c. $a_n = 243\left(\frac{1}{3}\right)^{n-1}$

6. 8, 4, 2, 1, . . . ;
8th term
a. $\frac{1}{2}$ b. $a_8 = \frac{1}{16}$
c. $a_n = 8\left(\frac{1}{2}\right)^{n-1}$

7. $128, -32, 8, -2, \ldots$;
7th term
a. $-\frac{1}{4}$ b. $a_7 = \frac{1}{32}$
c. $a_n = 128\left(-\frac{1}{4}\right)^{n-1}$

8. $125, -25, 5, -1, \ldots$;
9th term
a. $-\frac{1}{5}$ b. $a_9 = \frac{1}{3125}$
c. $a_n = 125\left(-\frac{1}{5}\right)^{n-1}$

For Exercises 9–12:   *a.* Write the first five terms of the geometric sequence, satisfying the given conditions. See Example 1.
b. Find the indicated term. See Example 1.

9. $a = -4, r = -3$;
8th term
a. $-4, 12, -36, 108, -324$
b. 8748

10. $a = -6, r = -2$;
10th term
a. $-6, 12, -24, 48, -96$
b. 3072

11. $a = 243, r = \frac{1}{3}$;
7th term
a. 243, 81, 27, 9, 3
b. $\frac{1}{3}$

12. $a = 256, r = -\frac{1}{2}$;
10th term
a. $256, -128, 64, -32, 16$
b. $-\frac{1}{2}$

For Exercises 13–18:   **a.**   *Use the formula for the nth term to find r.*
   b.   *Write the first four terms of the geometric sequence.*

13. $a = 1, r > 0, a_5 = 16$
   a. 2   b. 1, 2, 4, 8

14. $a = 3, a_4 = -24$
   a. −2   b. 3, −6, 12, −24

15. $a = -6, r > 0, a_5 = -96$
   a. 2   b. −6, −12, −24, −48

16. $a = -4, a_4 = -32$
   a. 2   b. −4, −8, −16, −32

17. $a = 128, a_4 = 2$
   a. $\dfrac{1}{4}$ b. 128, 32, 8, 2

18. $a = -243, a_6 = 1$
   a. $-\dfrac{1}{3}$ b. −243, 81, −27, 9

19. Find the first term of a geometric sequence in which $r = 3$ and the 5th term is −486.
   −6

20. Find the first term of a geometric sequence in which $r = -\dfrac{1}{2}$ and the 5th term is $-\dfrac{1}{8}$.
   −2

Objective 2

Prep Exercise 4 What is a geometric series?
The sum of the terms of a geometric sequence

Prep Exercise 5 What is the formula for finding the sum of the first *n* terms of a geometric series?
$S_n = \dfrac{a(1 - r^n)}{1 - r}$, where *n* is the number of terms, *a* is the first term, and *r* is the common ratio ($r \neq 1$).

For Exercises 21–26, find the sum of the first n terms of each geometric series for the given value of n. See Example 4.

21. $3 + 9 + 27 + 81 + \cdots, n = 11$
   265,719

22. $-1 - 2 - 4 - 8 - \cdots, n = 9$
   −511

23. $32 + 16 + 8 + 4 + \cdots, n = 9$
   63.875

24. $81 + 27 + 9 + 3 + \cdots, n = 7$
   $121\dfrac{4}{9}$ or $121.\overline{4}$

25. $128 - 32 + 8 - 2 + \cdots, n = 9$
   102.4

26. $625 - 125 + 25 - 5 + \cdots, n = 8$
   ≈ 520.832

Objective 3

Prep Exercise 6 What must be true about the common ratio, *r*, in order to use the formula $S_\infty = \dfrac{a}{1 - r}$ to find the sum of an infinite geometric series?
$|r| < 1$

For Exercises 27–32, find the sum of the infinite geometric series if possible. If it is not possible, explain why. See Example 5.

27. $27 + 9 + 3 + \cdots$
   40.5

28. $8 + 4 + 2 + \cdots$
   16

29. $15 - 9 + \dfrac{27}{5} - \cdots$
   9.375

30. $15 - 10 + \dfrac{20}{3} - \cdots$
   9

31. $9 + 12 + 16 + \cdots$
   The sum does not exist because $|r| \geq 1$.

32. $16 + 20 + 25 + \cdots$
   The sum does not exist because $|r| \geq 1$.

For Exercises 33–36, write each repeating decimal as a fraction. See Example 6.

33. $0.\overline{4}$
   $\dfrac{4}{9}$

34. $0.\overline{7}$
   $\dfrac{7}{9}$

35. $0.\overline{17}$
   $\dfrac{17}{99}$

36. $0.\overline{25}$
   $\dfrac{25}{99}$

For Exercises 37 and 38, answer each question.

37. If $a_n = 500(1.04)^n$,
  a. Find the first five terms of the sequence.
  520, 540.8, 562.432, 584.92928, 608.3264512
  b. Find the 10th term of the sequence.
  740.122142459

38. If $a_n = 350(1.06)^n$,
  a. Find the first five terms of the sequence.
  371, 393.26, 416.8556, 441.866936, 468.37895216
  b. Find the 8th term of the sequence.
  557.846826086

Objective 4

For Exercises 39–44, solve. See Examples 3 and 7.

39. A population of mink is increasing at a rate of 8% per month. The current mink population is 100.
  a. Using 100 as the first term, find the first four terms of the geometric sequence that gives the number of mink at the beginning of each month.
  100, 108, 117, 126
  b. Find the number of mink present at the beginning of the 8th month.
  171 mink
  c. Find the expression for the general term, a_n.
  $a_n = 100(1.08)^{n-1}$

40. The generation time (the time required for the number present to double) for a particular bacteria is 1 hour. Suppose initially one bacteria was present.
  a. Using 1 as the first term, write the first five terms of the geometric sequence giving the number of bacteria present after each hour.
  1, 2, 4, 8, 16
  b. Find an expression for the number present after the nth hour.
  $a_n = 2^n$
  c. How many bacteria are present after 1 day?
  16,777,216 bacteria.

41. Suppose you took a job for a month (20 working days) that paid $0.01 the first day and your salary doubled each day.
  a. Write the first five terms of the geometric sequence that gives your salary each day.
  $0.01, $0.02, $0.04, $0.08, $0.16
  b. Find an expression for the amount earned on the nth day.
  $a_n = 0.01(2)^{n-1}$
  c. How much would you earn on the 20th day?
  $5242.88
  d. What are your total earnings for the month?
  $10,485.75

42. Damarys deposits $200 in the bank. Each month thereafter she deposits 5% more than the month before. She does this for 1 year.
  a. Find an expression for the amount she deposits in the nth month.
  $a_n = 200(1.05)^{n-1}$
  b. Write the first four terms of the geometric sequence that gives the amount of her deposit each month.
  $200, $210, $220.50, $231.53
  c. How much did she deposit in the 10th month?
  $310.27
  d. How much does she deposit for the year?
  $3183.43

43. A new boat costs $20,000 and depreciates by 7% each year. What will the boat be worth in 8 years?
  $11,191.64

44. The isotope $_{15}P^{33}$ has a half-life of 25 days. A sample has 400 grams.
  a. Find the first five terms of the geometric sequence that gives the amount present at the end of each half-life.
  200 g, 100 g, 50 g, 25 g, 12.5 g
  b. Find an expression for the amount present after the nth half-life.
  $a_n = 400\left(\frac{1}{2}\right)^n$
  c. Find the amount present after the 10th half-life.
  ≈ 0.39 g

Appendix D

The Binomial Theorem

Objectives

1. Expand a binomial using Pascal's triangle.
2. Evaluate factorial notation and binomial coefficients.
3. Expand a binomial using the binomial theorem.
4. Find a particular term of a binomial expansion.

In this section, we will learn to raise binomials to natural-number powers.

Objective 1 Expand a binomial using Pascal's triangle.

We begin by writing out $(a + b)^n$, where n is a natural number, and look for patterns. These products are called *binomial expansions*.

$$(a + b)^0 = 1$$
$$(a + b)^1 = a + b$$
$$(a + b)^2 = a^2 + 2ab + b^2$$
$$(a + b)^3 = a^3 + 3a^2b + 3ab^2 + b^3$$
$$(a + b)^4 = a^4 + 4a^3b + 6a^2b^2 + 4ab^3 + b^4$$
$$(a + b)^5 = a^5 + 5a^4b + 10a^3b^2 + 10a^2b^3 + 5ab^4 + b^5$$

Conclusions Several patterns can be observed from the preceding expansions.

1. The 1st term in the expansion, a, is raised to the same power as the binomial, and the power of a decreases by 1 in each successive term. Note that the last term contains a^0.
2. The exponent of b is 0 in the 1st term and increases by 1 on each successive term.
3. The sum of the exponents of the variables of each term equals the exponent of the binomial.
4. The number of terms in the expansion is one more than the exponent of the binomial.

Now consider the coefficients of the terms in these expansions.

Coefficients of Expansions

$(a + b)^0$ 1

$(a + b)^1$ 1   1

$(a + b)^2$ 1   2   1

$(a + b)^3$ 1   3   3   1

$(a + b)^4$ 1   4   6   4   1

$(a + b)^5$ 1   5   10   10   5   1

If we arrange the coefficients of each expansion in a triangular array, we see an interesting pattern. Each row begins and ends with 1. Each number inside a row is the sum of the two numbers in the row above it. For example, each 10 in the bottom row comes from adding the 4 and 6 directly above.

This triangular array of numbers is called *Pascal's triangle* in honor of the French mathematician Blaise Pascal. When these observations are used, the expansion of $(a + b)^6$ has seven terms and the variable portions of the terms are $a^6, a^5b, a^4b^2, a^3b^3, a^2b^4, ab^5, b^6$. By continuing the pattern in Pascal's triangle, we can find the coefficients for each of those terms.

$(a + b)^5$ 1   5   10   10   5   1

$(a + b)^6$ 1   6   15   20   15   6   1

Using the coefficients from the last line of Pascal's triangle and the variables previously listed gives us

$$(a + b)^6 = a^6 + 6a^5b + 15a^4b^2 + 20a^3b^3 + 15a^2b^4 + 6ab^5 + b^6$$

Objective 2 Evaluate factorial notation and binomial coefficients.

Although Pascal's triangle is easy to use, it isn't practical, especially for binomials raised to large powers. Consequently, another method called the *binomial theorem* is often used. Before introducing the binomial theorem, we need **factorial notation**.

Definition **Factorial notation:** For any natural number n, the symbol $n!$ (read "n factorial") means $n(n-1)(n-2)\ldots 3\cdot 2\cdot 1$.

$0!$ is defined to be 1, so $0! = 1$.

| **Example 1** | Evaluate the following factorials. |

a. 5!

Solution: $5! = 5\cdot 4\cdot 3\cdot 2\cdot 1 = 120$

b. 7!

Solution: $7! = 7\cdot 6\cdot 5\cdot 4\cdot 3\cdot 2\cdot 1 = 5040$

Sometimes we may not write all of the factors of a factorial. In such cases, the last desired factor is written as a factorial. Below are some alternative ways to write 7!

$$7! = 7\cdot 6\cdot 5\cdot 4\cdot 3\cdot 2\cdot 1 = 7\cdot 6! \quad \text{or} \quad 7! = 7\cdot 6\cdot 5! \quad \text{or} \quad 7! = 7\cdot 6\cdot 5\cdot 4!$$

The coefficients of a binomial expansion can be expressed in terms of factorials using a special notation called the **binomial coefficient**. We will see how the binomial coefficient is used later.

Definition **Binomial coefficient:** A number written as $\binom{n}{r}$ and defined as $\dfrac{n!}{r!(n-r)!}$.

| **Example 2** | Evaluate the following binomial coefficients. |

a. $\binom{6}{2}$

Note The factorials could have been evaluated as follows:

$$\frac{6!}{2!\cdot 4!} = \frac{6\cdot 5\cdot 4!}{2\cdot 1\cdot 4!} = \frac{6\cdot 5}{2}$$

$$= \frac{30}{2} = 15$$

Solution: $\binom{6}{2} = \dfrac{6!}{2!(6-2)!}$ Substitute 6 for n and 2 for r in $\dfrac{n!}{r!(n-r)!}$.

$\qquad = \dfrac{6\cdot 5\cdot 4\cdot 3\cdot 2\cdot 1}{(2\cdot 1)(4\cdot 3\cdot 2\cdot 1)}$ Expand the factorials.

$\qquad = 15$ Simplify.

b. $\binom{8}{5}$

Solution: $\binom{8}{5} = \dfrac{8!}{5!(8-5)!}$ Substitute 8 for n and 5 for r in $\dfrac{n!}{r!(n-r)!}$.

$\qquad = \dfrac{8!}{5!\cdot 3!}$ Simplify.

$\qquad = \dfrac{8\cdot 7\cdot 6\cdot 5!}{5!\cdot 3\cdot 2\cdot 1}$ Rewrite 8! as $8\cdot 7\cdot 6\cdot 5!$ and simplify.

$\qquad = 56$ Evaluate.

Using the formula for evaluating binomial coefficients, we can prove two special cases: $\binom{n}{0} = 1$ and $\binom{n}{n} = 1$.

Objective 3 Expand a binomial using the binomial theorem.

Let's look at the expansion of $(a + b)^6$ and make another observation.

$$(a + b)^6 = a^6 + 6a^5b + 15a^4b^2 + 20a^3b^3 + 15a^2b^4 + 6ab^5 + b^6$$

Look at the 3rd term of the expansion, $15a^4b^2$. The coefficient is 15, and from Example 2(a), we see that $\binom{6}{2} = 15$. The coefficient of the 4th term is 20, and $\binom{6}{3} = 20$. In both cases, the n value of the binomial coefficient, $\binom{n}{r}$, is the exponent of the binomial and the r value is the exponent of b. This observation leads to the **binomial theorem**.

Definition Binomial theorem: For any positive integer n,

$$(a + b)^n = \binom{n}{0}a^n + \binom{n}{1}a^{n-1}b + \binom{n}{2}a^{n-2}b^2 + \binom{n}{3}a^{n-3}b^3 + \cdots + \binom{n}{n}b^n.$$

Note This is the same result we got earlier using Pascal's triangle. Also note that

$$\binom{6}{3} = \frac{6!}{3!3!} = \frac{6 \cdot 5 \cdot 4 \cdot 3!}{3!3!}$$

$$= \frac{6 \cdot 5 \cdot 4 \cdot 3!}{3 \cdot 2 \cdot 1 \cdot 3!}$$

$$= \frac{6 \cdot 5 \cdot 4}{3 \cdot 2 \cdot 1} = 20, \text{ etc.}$$

Note $\binom{5}{2} = \frac{5!}{2!3!} = \frac{5 \cdot 4 \cdot 3!}{2 \cdot 1 \cdot 3!}$

$$= \frac{5 \cdot 4 \cdot 3!}{2 \cdot 1 \cdot 3!} = \frac{5 \cdot 4}{2}$$

$$= \frac{20}{2} = 10, \text{ etc.}$$

Example 3 Expand each of the following binomials using the binomial theorem.

a. $(a + b)^6$

Solution:

$$(a + b)^6 = \binom{6}{0}a^6 + \binom{6}{1}a^5b + \binom{6}{2}a^4b^2 + \binom{6}{3}a^3b^3 + \binom{6}{4}a^2b^4 + \binom{6}{5}ab^5 + \binom{6}{6}b^6$$

$$= \frac{6!}{0!6!}a^6 + \frac{6!}{1!5!}a^5b + \frac{6!}{2!4!}a^4b^2 + \frac{6!}{3!3!}a^3b^3 + \frac{6!}{4!2!}a^2b^4 + \frac{6!}{5!1!}ab^5 + \frac{6!}{6!0!}b^6$$

$$= a^6 + 6a^5b + 15a^4b^2 + 20a^3b^3 + 15a^2b^4 + 6ab^5 + b^6$$

b. $(a - 3b)^5$

Solution: Write $(a - 3b)^5$ as $[a + (-3b)]^5$

$$[a + (-3b)]^5 = \binom{5}{0}a^5 + \binom{5}{1}a^4(-3b) + \binom{5}{2}a^3(-3b)^2 +$$

$$\binom{5}{3}a^2(-3b)^3 + \binom{5}{4}a(-3b)^4 + \binom{5}{5}(-3b)^5$$

$$= a^5 + 5a^4(-3b) + 10a^3(9b^2) + 10a^2(-27b^3) + 5a(81b^4) + (-243b^5)$$

$$= a^5 - 15a^4b + 90a^3b^2 - 270a^2b^3 + 405ab^4 - 243b^5$$

Objective 4 Find a particular term of a binomial expansion.

Sometimes it is necessary to find only a specific term of a binomial expansion without writing out the entire expansion. Look again at the binomial expansion. Note that the 3rd term (which we will call the $(2 + 1)$st term) is $\binom{n}{2}a^{n-2}b^2$ and the 4th term (which we will call the $(3 + 1)$st term) is $\binom{n}{3}a^{n-3}b^3$. Similarly, the $(m + 1)$st term is $\binom{n}{m}a^{n-m}b^m$. These observations lead to the following.

Rule Finding the $(m + 1)$st Term of a Binomial Expansion

The $(m + 1)$st term of the expansion $(a + b)^n$ is $\binom{n}{m}a^{n-m}b^m$.

Example 4   Find the indicated term of each of the following binomial expansions.

a. $(a + b)^{11}$, 7th term

Solution: Use the formula for the $(m + 1)$st term with $n = 11$ and $m = 6$ (to find the 7th term).

$$\binom{n}{m}a^{n-m}b^m = \binom{11}{6}a^{11-6}b^6 = 462a^5b^6$$

b. $(2x - 5y)^8$, 4th term

Solution: Write $(2x - 5y)^8$ as $[2x + (-5y)]^8$. Use the formula for the $(m + 1)$st term with $n = 8$ and $m = 3$ (to find the 4th term), $a = 2x$, and $b = -5y$.

$$\binom{n}{m}a^{n-m}b^m = \binom{8}{3}(2x)^{8-3}(-5y)^3 = 56(32x^5)(-125y^3) = -224{,}000x^5y^3$$

Appendix D Exercises
For Extra Help MyMathLab®

Objectives 1 and 2

Prep Exercise 1 How many terms are in the expansion of $(5a + b)^{12}$?
13

Prep Exercise 2 What is the sum of the exponents on x and y for any term in the expansion of $(x + y)^9$?
9

Prep Exercise 3 How is the symbol 8! read?
8 factorial

Prep Exercise 4 What does 8! mean? What is its value?
$8 \cdot 7 \cdot 6 \cdot 5 \cdot 4 \cdot 3 \cdot 2 \cdot 1 = 40{,}320$

For Exercises 1–12, evaluate each expression. See Example 1.

1. 4!
24

2. 7!
5040

3. $(4!)(3!)$
144

4. $(3!)(2!)$
12

5. $(6!)(5!)$
86,400

6. $(4!)(7!)$
120,960

7. $\dfrac{8!}{10!}$
$\dfrac{1}{90}$, or $0.0\overline{1}$

8. $\dfrac{7!}{9!}$
$\dfrac{1}{72}$, or $0.013\overline{8}$

9. $\dfrac{10!}{9!}$
10

10. $\dfrac{12!}{11!}$
12

11. $\dfrac{8!}{6!(8 - 6)!}$
28

12. $\dfrac{10!}{6!(10 - 6)!}$
210

Prep Exercise 5 What is the formula for evaluating the binomial coefficient $\binom{n}{r}$?
$\dfrac{n!}{r!(n - r)!}$

For Exercises 13–20, evaluate each binomial coefficient. See Example 2.

13. $\binom{7}{3}$
35

14. $\binom{5}{2}$
10

15. $\binom{10}{4}$
210

16. $\binom{6}{5}$
6

17. $\binom{7}{7}$
1

18. $\binom{4}{4}$
1

19. $\binom{8}{0}$
1

20. $\binom{9}{0}$
1

Objective 3

Prep Exercise 6 Using the binomial theorem to expand $(x + 3)^6$, what is the coefficient of the 4th term?

$\binom{6}{3} = 20$

Prep Exercise 7 What is the exponent of b in the 6th term of the expansion of $(a + b)^{13}$? What is the exponent of a in that term?

5, 8

For Exercises 21–32, use the binomial theorem to expand each of the following.
See Example 3.

21. $(a + b)^5$
$a^5 + 5a^4b + 10a^3b^2 + 10a^2b^3 + 5ab^4 + b^5$

22. $(a + b)^7$
$a^7 + 7a^6b + 21a^5b^2 + 35a^4b^3 + 35a^3b^4 + 21a^2b^5 + 7ab^6 + b^7$

23. $(x - y)^4$
$x^4 - 4x^3y + 6x^2y^2 - 4xy^3 + y^4$

24. $(x - y)^3$
$x^3 - 3x^2y + 3xy^2 - y^3$

25. $(2a + b)^3$
$8a^3 + 12a^2b + 6ab^2 + b^3$

26. $(a + 2b)^3$
$a^3 + 6a^2b + 12ab^2 + 8b^3$

27. $(x - 2y)^5$
$x^5 - 10x^4y + 40x^3y^2 - 80x^2y^3 + 80xy^4 - 32y^5$

28. $(x - 3y)^4$
$x^4 - 12x^3y + 54x^2y^2 - 108xy^3 + 81y^4$

29. $(2m + 3n)^6$
$64m^6 + 576m^5n + 2160m^4n^2 + 4320m^3n^3 + 4860m^2n^4 + 2916mn^5 + 729n^6$

30. $(3c + 2d)^5$
$243c^5 + 810c^4d + 1080c^3d^2 + 720c^2d^3 + 240cd^4 + 32d^5$

31. $(3x - 4y)^4$
$81x^4 - 432x^3y + 864x^2y^2 - 768xy^3 + 256y^4$

32. $(4a - b)^7$
$16,384a^7 - 28,672a^6b + 21,504a^5b^2 - 8960a^4b^3 + 2240a^3b^4 - 336a^2b^5 + 28ab^6 - b^7$

Objective 4

Prep Exercise 8 What is the formula for finding the $(m + 1)$st term of the expansion $(a + b)^n$?

$\binom{n}{m}a^{n-m}b^m$

For Exercises 33–40, find the indicated term of each binomial expansion.
See Example 4.

33. $(x + y)^8$, 5th term
$70x^4y^4$

34. $(a + b)^9$, 4th term
$84a^6b^3$

35. $(a - b)^{10}$, 3rd term
$45a^8b^2$

36. $(m - n)^7$, 2nd term
$-7m^6n$

37. $(4x + y)^9$, 6th term
$32,256x^4y^5$

38. $(3a + b)^{11}$, 7th term
$112,266a^5b^6$

39. $(3m - 2n)^7$, 4th term
$-22,680m^4n^3$

40. $(5x - 3y)^{12}$, 5th term
$15,662,109,375x^8y^4$

Appendix E

Synthetic Division

Objective

1 Use synthetic division to divide a polynomial by a binomial in the form $x - c$.

Objective 1 Use synthetic division to divide a polynomial by a binomial in the form $x - c$.

We saw in Section 5.6 that long division can be a tedious process. When a polynomial is divided by a binomial in the form $x - c$, where c is a constant, we can streamline the process by focusing on just the coefficients in the polynomials using a method called *synthetic division*.

To get a sense of how synthetic division looks, let's look again at the long division for $\dfrac{x^2 + 5x + 7}{x + 2}$, which was Example 6 in Section 5.6. To the right of that long division, we gradually remove bits so that we are left with the bare essentials for finding the quotient.

Long Division	**Variables Removed**	**Bare Essentials**
$x + 3$	$1 + 3$	$1 + 3$
$x + 2\overline{)x^2 + 5x + 7}$	$1 + 2\overline{)1 + 5 + 7}$	$1 - (-2)\overline{)1 + 5 + 7}$
$\underline{-x^2 - 2x}$	$\underline{-1 - 2}$	$\underline{-2}$
$3x + 7$	$3 + 7$	$+7$
$\underline{-3x - 6}$	$\underline{-3 - 6}$	$\underline{-6}$
1	1	1

Note In polynomial long division, the terms in the positions that we've highlighted in blue are always additive inverses. Because they always cancel, we can remove them to get the "bare-essentials" version, as seen to the right.

Note We have rewritten $x + 2$ as $x - (-2)$ so that we have the form $x - c$.

We can place these "bare-essential" values in an even more compact form, which is the form for synthetic division.

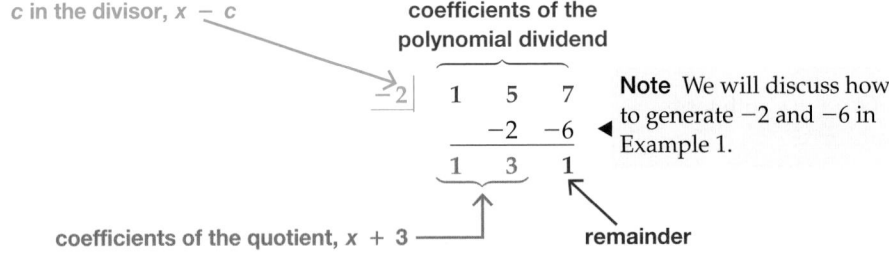

c in the divisor, $x - c$

coefficients of the polynomial dividend

Note We will discuss how to generate -2 and -6 in Example 1.

coefficients of the quotient, $x + 3$

remainder

Now let's examine the steps of synthetic division in Example 1.

Example 1 Use synthetic division to divide.

a. $\dfrac{x^2 + 5x + 7}{x + 2}$

Solution: First we need the divisor to be in the form $x - c$. So we rewrite $x + 2$ as $x - (-2)$, and we see that $c = -2$. We then set up the synthetic division with c and the coefficients of the dividend, as shown.

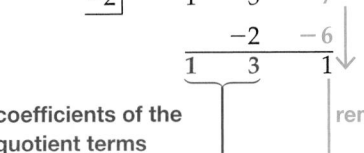

$$
\begin{array}{c|ccc}
-2 & 1 & 5 & 7 \\
  & & & \\
\hline
  & 1 & & \\
\end{array}
$$

Our first step is to bring down the leading coefficient.

$$
\begin{array}{c|ccc}
-2 & 1 & 5 & 7 \\
  & & -2 & \\
\hline
  & 1 & & \\
\end{array}
$$

Multiply 1 by -2 and place the result underneath the next coefficient, 5.

$$
\begin{array}{c|ccc}
-2 & 1 & 5 & 7 \\
  & & -2 & \\
\hline
  & 1 & 3 & \\
\end{array}
$$

Add: $5 - 2 = 3$.

$$
\begin{array}{c|ccc}
-2 & 1 & 5 & 7 \\
  & & -2 & -6 \\
\hline
  & 1 & 3 & \\
\end{array}
$$

Multiply 3 by -2 and place the result underneath the next coefficient, 7.

$$
\begin{array}{c|ccc}
-2 & 1 & 5 & 7 \\
  & & -2 & -6 \\
\hline
  & 1 & 3 & 1 \\
\end{array}
$$

Add: $7 - 6 = 1$, which is the remainder.

coefficients of the quotient terms   remainder

Note In synthetic division, the degree of the quotient polynomial is always one less than the degree of the dividend polynomial. Because the degree of the dividend $x^2 + 5x + 7$ is 2, the degree of the quotient, $x + 3$, is 1. ▶

Answer: $\qquad 1x + 3 + \dfrac{1}{x + 2} \quad$ or $\quad x + 3 + \dfrac{1}{x + 2}$

b. $(x^4 - 6x^3 + 35x - 17) \div (x - 4)$

Solution: Looking at the divisor, $x - 4$, we see that $c = 4$.

$$
\begin{array}{c|ccccc}
4 & 1 & -6 & 0 & 35 & -17 \\
  & & & & & \\
\hline
  & 1 & & & & \\
\end{array}
$$

List the coefficients in the dividend and bring down the leading coefficient.

$$
\begin{array}{c|ccccc}
4 & 1 & -6 & 0 & 35 & -17 \\
  & & 4 & & & \\
\hline
  & 1 & -2 & & & \\
\end{array}
$$

Multiply 1 by 4 and place the result underneath the next coefficient, -6. Then add.

$$
\begin{array}{c|ccccc}
4 & 1 & -6 & 0 & 35 & -17 \\
  & & 4 & -8 & & \\
\hline
  & 1 & -2 & -8 & & \\
\end{array}
$$

Multiply -2 by 4 and place the result underneath the next coefficient, 0. Then add.

$$
\begin{array}{c|ccccc}
4 & 1 & -6 & 0 & 35 & -17 \\
  & & 4 & -8 & -32 & \\
\hline
  & 1 & -2 & -8 & 3 & \\
\end{array}
$$

Multiply -8 by 4 and place the result underneath the next coefficient, 35. Then add.

Note Because the polynomial dividend has no x^2 term, we place a 0 in the degree-2 position as a placeholder. ▶

$$
\begin{array}{c|ccccc}
4 & 1 & -6 & 0 & 35 & -17 \\
  & & 4 & -8 & -32 & 12 \\
\hline
  & 1 & -2 & -8 & 3 & -5 \\
\end{array}
$$

Multiply 3 by 4 and place the result underneath the next coefficient, -17. Then add.

Answer: $x^3 - 2x^2 - 8x + 3 - \dfrac{5}{x - 4}$

Appendix E Exercises

For Extra Help   MyMathLab®

Note: Exercises marked with a ★ represent challenging exercises.

Objective 1

Prep Exercise 1 What form does the divisor need to have so that synthetic division can be used?

$x - c$

Prep Exercise 2 Suppose you are to use synthetic division to divide $\dfrac{x^3 - 8x^2 - 5x + 9}{x + 6}$. What is the value of c?

-6

Prep Exercise 3 If you were to use synthetic division for $\dfrac{x^4 - 5x^2 - 3x + 11}{x - 2}$, what list of coefficients would you write?

$1, 0, -5, -3, 11$

Prep Exercise 4 Explain how to finish the synthetic division shown.

$$\begin{array}{r|rrrr} 2 & 1 & 4 & -8 & -9 \\ & & 2 & 12 & \\ \hline & 1 & 6 & 4 & \end{array}$$

Multiply 4 by 2. Put the resulting 8 underneath the -9. Add $-9 + 8 = -1$, which is the remainder.

Prep Exercise 5 Where does the remainder appear in synthetic division?

It is the last number in the last row.

Prep Exercise 6 When using synthetic division, how do you determine the degree of the first term in the quotient expression?

The degree of the quotient polynomial is 1 less than the degree of the dividend polynomial.

For Exercises 1–6: *a. Write the binomial divisor.*
b. Write the polynomial dividend.
c. Write the polynomial quotient with its remainder (if there is a remainder). See Objective 1.

1.
$$\begin{array}{r|rrr} -3 & 1 & 7 & 12 \\ & & -3 & -12 \\ \hline & 1 & 4 & 0 \end{array}$$
a. $x + 3$ b. $x^2 + 7x + 12$
c. $x + 4$

2.
$$\begin{array}{r|rrr} 9 & 1 & -14 & 45 \\ & & 9 & -45 \\ \hline & 1 & -5 & 0 \end{array}$$
a. $x - 9$ b. $x^2 - 14x + 45$
c. $x - 5$

3.
$$\begin{array}{r|rrrr} 3 & 1 & 4 & -25 & 7 \\ & & 3 & 21 & -12 \\ \hline & 1 & 7 & -4 & -5 \end{array}$$
a. $x - 3$ b. $x^3 + 4x^2 - 25x + 7$
c. $x^2 + 7x - 4 - \dfrac{5}{x - 3}$

4.
$$\begin{array}{r|rrrr} -2 & 1 & -2 & -6 & 7 \\ & & -2 & 8 & -4 \\ \hline & 1 & -4 & 2 & 3 \end{array}$$
a. $x + 2$ b. $x^3 - 2x^2 - 6x + 7$
c. $x^2 - 4x + 2 + \dfrac{3}{x + 2}$

5.
$$\begin{array}{r|rrrrr} 4 & 2 & -8 & -5 & 17 & 10 \\ & & 8 & 0 & -20 & -12 \\ \hline & 2 & 0 & -5 & -3 & -2 \end{array}$$
a. $x - 4$ b. $2x^4 - 8x^3 - 5x^2 + 17x + 10$
c. $2x^3 - 5x - 3 - \dfrac{2}{x - 4}$

6.
$$\begin{array}{r|rrrrr} -5 & 3 & 16 & 0 & -27 & -9 \\ & & -15 & -5 & 25 & 10 \\ \hline & 3 & 1 & -5 & -2 & 1 \end{array}$$
a. $x + 5$ b. $3x^4 + 16x^3 - 27x - 9$
c. $3x^3 + x^2 - 5x - 2 + \dfrac{1}{x + 5}$

For Exercises 7–32, divide using synthetic division. See Example 1.

7. $\dfrac{x^2 - 5x + 6}{x - 2}$

$x - 3$

8. $\dfrac{x^2 + 8x - 9}{x - 1}$

$x + 9$

9. $\dfrac{2x^2 - x - 5}{x - 4}$

$2x + 7 + \dfrac{23}{x - 4}$

10. $\dfrac{3x^2 - 11x - 7}{x - 5}$

$3x + 4 + \dfrac{13}{x - 5}$

11. $\dfrac{x^3 - x^2 - 5x + 6}{x - 2}$

$x^2 + x - 3$

12. $\dfrac{2x^3 - 6x^2 - 5x + 15}{x - 3}$

$2x^2 - 5$

13. $\dfrac{3x^3 + 8x^2 + 3x - 2}{x + 2}$

$3x^2 + 2x - 1$

14. $\dfrac{5x^3 + 32x^2 + 52x + 16}{x + 2}$

$5x^2 + 22x + 8$

15. $(3x^3 - x^2 - 22x + 24) \div (x + 3)$

$3x^2 - 10x + 8$

16. $(8x^3 - 10x^2 - x + 3) \div (x - 1)$
$8x^2 - 2x - 3$

17. $\dfrac{2x^3 - 5x^2 - x + 6}{x + 2}$
$2x^2 - 9x + 17 - \dfrac{28}{x + 2}$

18. $\dfrac{3x^3 - 5x^2 - 16x + 12}{x - 2}$
$3x^2 + x - 14 - \dfrac{16}{x - 2}$

19. $(x^3 - 2x^2 + x - 3) \div (x - 2)$
$x^2 + 1 - \dfrac{1}{x - 2}$

20. $(x^3 + 6x^2 + 2x - 3) \div (x + 6)$
$x^2 + 2 - \dfrac{15}{x + 6}$

21. $\dfrac{2x^3 + x^2 - 4}{x + 3}$
$2x^2 - 5x + 15 - \dfrac{49}{x + 3}$

22. $\dfrac{4x^3 - 2x + 5}{x - 2}$
$4x^2 + 8x + 14 + \dfrac{33}{x - 2}$

23. $\dfrac{3x^3 - 4x^2 + 2}{x - 3}$
$3x^2 + 5x + 15 + \dfrac{47}{x - 3}$

24. $\dfrac{x^3 + 4x^2 - 7}{x + 3}$
$x^2 + x - 3 + \dfrac{2}{x + 3}$

25. $(x^3 - 7x + 6) \div (x - 2)$
$x^2 + 2x - 3$

26. $(x^3 + 2x + 135) \div (x + 5)$
$x^2 - 5x + 27$

27. $\dfrac{x^3 + 27}{x - 3}$
$x^2 + 3x + 9 + \dfrac{54}{x - 3}$

28. $\dfrac{x^3 - 8}{x + 2}$
$x^2 - 2x + 4 - \dfrac{16}{x + 2}$

29. $\dfrac{x^4 + 16}{x + 2}$
$x^3 - 2x^2 + 4x - 8 + \dfrac{32}{x + 2}$

30. $\dfrac{x^4 + 81}{x - 3}$
$x^3 + 3x^2 + 9x + 27 + \dfrac{162}{x - 3}$

★ **31.** $\dfrac{6x^3 + x^2 - 11x - 6}{x + \dfrac{2}{3}}$
$6x^2 - 3x - 9$

★ **32.** $\dfrac{2x^3 - 13x^2 + 27x - 18}{x - \dfrac{3}{2}}$
$2x^2 - 10x + 12$

33. The area of the parallelogram shown is described by $9x^2 - 10x - 16$.
   a. Find an expression for the length of the base.
   $9x + 8$
   b. Find the base, height, and area of the parallelogram if $x = 12$ inches.
   Base: 116 in.; height: 10 in.; area: 1160 in.^{2}

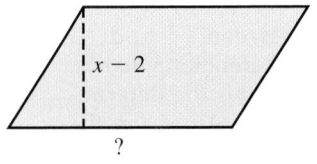

34. The area of the rectangle shown is $12n^2 + 65n - 42$.
   a. Find the length.
   $12n - 7$
   b. Find the length, width, and area of the rectangle if $n = 9$ centimeters.
   Length: 101 cm; width: 15 cm; area: 1515 cm^2

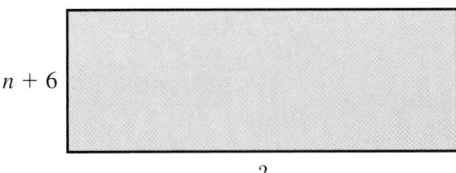

35. The volume of the walk-in freezer shown is described by $2y^3 - 7y^2 - 38y + 88$.
   a. Find an expression for the height of the room.
   $2y - 11$
   b. Find the length, width, height, and volume of the room if $y = 10$ feet.
   Length: 14 ft.; width: 8 ft.; height: 9 ft.; volume: 1008 ft.^{3}

36. The volume of the storage box shown is described by $2x^3 - 168x - 320$.
   a. Find an expression for the length of the box.
   $2x + 4$
   b. Find the length, width, height, and volume of the box if $x = 16$ inches.
   Length: 36 in.; width: 24 in.; height: 6 in.; volume: 5184 in.^{3}

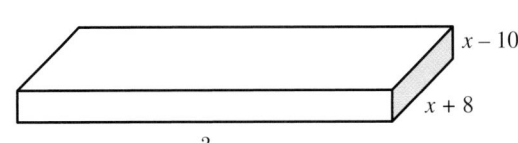

Appendix F

Solving Systems of Linear Equations Using Cramer's Rule

Objectives

1 Evaluate determinants of 2×2 matrices.

2 Evaluate determinants of 3×3 matrices.

3 Solve systems of equations using Cramer's Rule.

Objective 1 Evaluate determinants of 2×2 matrices.

Systems of linear equations can be solved using a special type of matrix called a **square matrix**.

Definition **Square matrix:** A matrix with an equal number of rows and columns.

For example, $\begin{bmatrix} 2 & 3 \\ 1 & -4 \end{bmatrix}$ and $\begin{bmatrix} 1 & -3 & 5 \\ -7 & 2 & 9 \\ 0 & 4 & -6 \end{bmatrix}$ are square matrices.

Every square matrix has a *determinant*. We write the determinant of a matrix, A, as $\det(A)$ or $|A|$. The method used to find the determinant of a matrix depends on its size.

Rule **Determinant of a 2×2 Matrix**

If $A = \begin{bmatrix} a_1 & b_1 \\ a_2 & b_2 \end{bmatrix}$, then $\det(A) = \begin{vmatrix} a_1 & b_1 \\ a_2 & b_2 \end{vmatrix} = a_1b_2 - a_2b_1$.

Warning Be careful to note the difference between [ ] and | |. The notation [ ] denotes a matrix, whereas | | denotes the determinant of a matrix. Also, from the context, we know that | | does not mean absolute value.

Notice that the determinant contains diagonal products, as illustrated by the following diagram.

$$\begin{vmatrix} a_1 & b_1 \\ a_2 & b_2 \end{vmatrix} = a_1b_2 - a_2b_1$$

Note Because subtraction is not commutative, make sure you note the order of the two products.

  Learning Strategies

The products in the determinant are arranged with the downward cross product subtracting the upward cross product. An easy way to remember this order is that you have to "fall down before you can get up."

Note An alternative notation for $\det(A)$ is $|A|$.

Example 1 Find the determinants of the following matrices.

a. $A = \begin{bmatrix} 3 & -2 \\ 2 & 4 \end{bmatrix}$

Solution: $\det(A) = $ 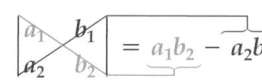 $= (3)(4) - (2)(-2) = 12 + 4 = 16$

b. $B = \begin{bmatrix} -3 & 5 \\ 4 & -2 \end{bmatrix}$

Solution: $\det(B) = (-3)(-2) - (4)(5) = 6 - 20 = -14$

c. $M = \begin{bmatrix} 1 & 3 \\ 3 & 9 \end{bmatrix}$

Solution: $\det(M) = (1)(9) - (3)(3) = 9 - 9 = 0$

Your Turn 1 Find the determinant of $\begin{bmatrix} 1 & -3 \\ -4 & 2 \end{bmatrix}$.

Answer to Your Turn 1
-10

Objective 2   Evaluate determinants of 3 × 3 matrices.

There are various methods of evaluating the determinant of a 3 × 3 matrix. One of the most common methods is *expanding by minors*. Each element of a square matrix has a number called the **minor** for that element.

Definition   Minor of an element of a matrix: The determinant of the remaining matrix when the row and column in which the element is located are ignored.

Example 2   Find the minor of 2 in $\begin{bmatrix} 2 & -3 & -6 \\ -1 & 5 & -2 \\ 3 & -4 & 1 \end{bmatrix}$.

Solution: To find the minor of 2, we ignore its row and column (shown in blue) and evaluate the determinant of the remaining matrix (shown in red).

$$\begin{bmatrix} 2 & -3 & -6 \\ -1 & 5 & -2 \\ 3 & -4 & 1 \end{bmatrix}$$

$$\begin{vmatrix} 5 & -2 \\ -4 & 1 \end{vmatrix} = (5)(1) - (-4)(-2) = 5 - 8 = -3$$

Your Turn 2   Find the minor of 6 in $\begin{bmatrix} 1 & -3 & 6 \\ -2 & 2 & 0 \\ 4 & -1 & 5 \end{bmatrix}$.

To evaluate the determinant of a 3 × 3 matrix, we will expand by minors along the first column.

Note We can expand by minors along *any* row or column to find the determinant of a 3 × 3 matrix. For simplicity, we have chosen to show expanding by minors only along the first column.

Rule   Evaluating the Determinant of a 3 × 3 Matrix

$$\begin{vmatrix} a_1 & b_1 & c_1 \\ a_2 & b_2 & c_2 \\ a_3 & b_3 & c_3 \end{vmatrix} = a_1 \begin{pmatrix} \text{minor} \\ \text{of } a_1 \end{pmatrix} - a_2 \begin{pmatrix} \text{minor} \\ \text{of } a_2 \end{pmatrix} + a_3 \begin{pmatrix} \text{minor} \\ \text{of } a_3 \end{pmatrix}$$

$$= a_1 \begin{vmatrix} b_2 & c_2 \\ b_3 & c_3 \end{vmatrix} - a_2 \begin{vmatrix} b_1 & c_1 \\ b_3 & c_3 \end{vmatrix} + a_3 \begin{vmatrix} b_1 & c_1 \\ b_2 & c_2 \end{vmatrix}$$

Warning Notice that the second term of the expansion has a negative sign.

Example 3   Find the determinant of $\begin{bmatrix} 2 & -3 & -4 \\ -1 & 2 & -2 \\ 3 & -4 & 1 \end{bmatrix}$.

Solution: Using the rule for expanding by minors along the first column, we have the following:

$$\begin{vmatrix} 2 & -3 & -4 \\ -1 & 2 & -2 \\ 3 & -4 & 1 \end{vmatrix} = 2\begin{pmatrix} \text{minor} \\ \text{of } 2 \end{pmatrix} - (-1)\begin{pmatrix} \text{minor} \\ \text{of } -1 \end{pmatrix} + 3\begin{pmatrix} \text{minor} \\ \text{of } 3 \end{pmatrix}$$

$$= 2\begin{vmatrix} 2 & -2 \\ -4 & 1 \end{vmatrix} - (-1)\begin{vmatrix} -3 & -4 \\ -4 & 1 \end{vmatrix} + 3\begin{vmatrix} -3 & -4 \\ 2 & -2 \end{vmatrix}$$

$$= 2(2 - 8) + 1(-3 - 16) + 3(6 + 8)$$

$$= -12 + (-19) + 42$$

$$= 11$$

Answer to Your Turn 2
$\begin{vmatrix} -2 & 2 \\ 4 & -1 \end{vmatrix} = -6$

Your Turn 3 Find the determinant of $\begin{bmatrix} 3 & -2 & 4 \\ -2 & 3 & 1 \\ 2 & -4 & 2 \end{bmatrix}$.

Objective 3 Solve systems of equations using Cramer's Rule.

Now we can use **Cramer's Rule**, which uses determinants to solve systems of equations. To derive Cramer's Rule, we solve a general system of equations using the elimination method. We show the derivation for a system of two equations in two unknowns.

$$\begin{cases} a_1 x + b_1 y = c_1 & (\text{Equation 1}) \\ a_2 x + b_2 y = c_2 & (\text{Equation 2}) \end{cases}$$

We eliminate y by multiplying equation 1 by b_2 and equation 2 by $-b_1$.

$a_1 b_2 x + b_1 b_2 y = b_2 c_1$	Multiply Equation 1 by b_2.
$-a_2 b_1 x - b_1 b_2 y = -b_1 c_2$	Multiply Equation 2 by $-b_1$.
$a_1 b_2 x - a_2 b_1 x \qquad\quad = b_2 c_1 - b_1 c_2$	Add the equations.
$(a_1 b_2 - a_2 b_1)x = b_2 c_1 - b_1 c_2$	Factor out x from the left side.
$x = \dfrac{b_2 c_1 - b_1 c_2}{a_1 b_2 - a_2 b_1}$	Divide by $a_1 b_2 - a_2 b_1$.

Notice that the numerator is $\begin{vmatrix} c_1 & b_1 \\ c_2 & b_2 \end{vmatrix}$ and the denominator is $\begin{vmatrix} a_1 & b_1 \\ a_2 & b_2 \end{vmatrix}$; so $x = \dfrac{\begin{vmatrix} c_1 & b_1 \\ c_2 & b_2 \end{vmatrix}}{\begin{vmatrix} a_1 & b_1 \\ a_2 & b_2 \end{vmatrix}}$.

If we repeat the same process and solve for y, we get $y = \dfrac{\begin{vmatrix} a_1 & c_1 \\ a_2 & c_2 \end{vmatrix}}{\begin{vmatrix} a_1 & b_1 \\ a_2 & b_2 \end{vmatrix}}$.

A similar approach is used to derive the rule for a system of three equations in three unknowns.

Of Interest

Cramer's Rule is named after Gabriel Cramer (1704–1752), who was chair of the mathematics department at Geneva, Switzerland. In one of his books, he gave an example that required finding an equation of degree two whose graph passed through five given points. The solution led to a system of five linear equations in five unknowns. So readers could solve the system, Cramer referred them to an appendix, which explained what we now call Cramer's Rule.

Rule Cramer's Rule

The solution to the system of linear equations $\begin{cases} a_1 x + b_1 y = c_1 \\ a_2 x + b_2 y = c_2 \end{cases}$ is

$$x = \frac{\begin{vmatrix} c_1 & b_1 \\ c_2 & b_2 \end{vmatrix}}{\begin{vmatrix} a_1 & b_1 \\ a_2 & b_2 \end{vmatrix}} = \frac{D_x}{D} \quad \text{and} \quad y = \frac{\begin{vmatrix} a_1 & c_1 \\ a_2 & c_2 \end{vmatrix}}{\begin{vmatrix} a_1 & b_1 \\ a_2 & b_2 \end{vmatrix}} = \frac{D_y}{D}.$$

The solution to the system of linear equations $\begin{cases} a_1 x + b_1 y + c_1 z = d_1 \\ a_2 x + b_2 y + c_2 z = d_2 \text{ is} \\ a_3 x + b_3 y + c_3 z = d_3 \end{cases}$

$$x = \frac{\begin{vmatrix} d_1 & b_1 & c_1 \\ d_2 & b_2 & c_2 \\ d_3 & b_3 & c_3 \end{vmatrix}}{\begin{vmatrix} a_1 & b_1 & c_1 \\ a_2 & b_2 & c_2 \\ a_3 & b_3 & c_3 \end{vmatrix}} = \frac{D_x}{D}, \quad y = \frac{\begin{vmatrix} a_1 & d_1 & c_1 \\ a_2 & d_2 & c_2 \\ a_3 & d_3 & c_3 \end{vmatrix}}{\begin{vmatrix} a_1 & b_1 & c_1 \\ a_2 & b_2 & c_2 \\ a_3 & b_3 & c_3 \end{vmatrix}} = \frac{D_y}{D}, \quad \text{and} \quad z = \frac{\begin{vmatrix} a_1 & b_1 & d_1 \\ a_2 & b_2 & d_2 \\ a_3 & b_3 & d_3 \end{vmatrix}}{\begin{vmatrix} a_1 & b_1 & c_1 \\ a_2 & b_2 & c_2 \\ a_3 & b_3 & c_3 \end{vmatrix}} = \frac{D_z}{D}.$$

Note Each denominator, D, is the determinant of a matrix containing only the coefficients in the system. To find D_x, we replace the column of x-coefficients in the coefficient matrix with the constants from the system. To find D_y, we replace the column of y-coefficients in the coefficient matrix with the constant terms. We do likewise to find D_z.

Answer to Your Turn 3
26

Example 4   Use Cramer's Rule to solve $\begin{cases} 2x + 3y = -5 \\ 3x - y = 9 \end{cases}$.

Solution: First, we find D, D_x, and D_y.

Note These are the coefficients of the system.

$$\blacktriangleright \ D = \begin{vmatrix} 2 & 3 \\ 3 & -1 \end{vmatrix} = (2)(-1) - (3)(3) = -2 - 9 = -11$$

Note Replace the x-coefficients with the constants.

$$\blacktriangleright \ D_x = \begin{vmatrix} -5 & 3 \\ 9 & -1 \end{vmatrix} = (-5)(-1) - (9)(3) = 5 - 27 = -22$$

Note Replace the y-coefficients with the constants.

$$\blacktriangleright \ D_y = \begin{vmatrix} 2 & -5 \\ 3 & 9 \end{vmatrix} = (2)(9) - (3)(-5) = 18 + 15 = 33$$

Note If $D = 0$ and D_x and $D_y \neq 0$, the system is inconsistent (no solution). If D, D_x, and $D_y = 0$, the system is dependent (infinite number of solutions).

Now we can find x and y.

$$x = \frac{D_x}{D} = \frac{-22}{-11} = 2 \qquad y = \frac{D_y}{D} = \frac{33}{-11} = -3$$

The solution is $(2, -3)$, which we can check by verifying that it satisfies both equations in the system. We will leave the check to the reader.

Your Turn 4   Use Cramer's Rule to solve $\begin{cases} 3x - 2y = -16 \\ x + 3y = 2 \end{cases}$.

Example 5   Use Cramer's Rule to solve $\begin{cases} x + 2y - 2x = -7 \\ 3x - 2y + z = 15 \\ 2x + 3y - 3z = -9 \end{cases}$.

Note We find the determinant of a 3 × 3 matrix by expanding by minors along the first column. $\blacktriangleright$

Solution: We need to find D, D_x, D_y, and D_z.

$$D = \begin{vmatrix} 1 & 2 & -2 \\ 3 & -2 & 1 \\ 2 & 3 & -3 \end{vmatrix} = (1)\begin{vmatrix} -2 & 1 \\ 3 & -3 \end{vmatrix} - (3)\begin{vmatrix} 2 & -2 \\ 3 & -3 \end{vmatrix} + (2)\begin{vmatrix} 2 & -2 \\ -2 & 1 \end{vmatrix}$$

$$= 1(6 - 3) - 3(-6 + 6) + 2(2 - 4)$$
$$= 3 - 0 + (-4)$$
$$= -1$$

$$D_x = \begin{vmatrix} -7 & 2 & -2 \\ 15 & -2 & 1 \\ -9 & 3 & -3 \end{vmatrix} = (-7)\begin{vmatrix} -2 & 1 \\ 3 & -3 \end{vmatrix} - (15)\begin{vmatrix} 2 & -2 \\ 3 & -3 \end{vmatrix} + (-9)\begin{vmatrix} 2 & -2 \\ -2 & 1 \end{vmatrix}$$

$$= -7(6 - 3) - 15(-6 + 6) + (-9)(2 - 4)$$
$$= -21 - 0 + 18$$
$$= -3$$

$$D_y = \begin{vmatrix} 1 & -7 & -2 \\ 3 & 15 & 1 \\ 2 & -9 & -3 \end{vmatrix} = (1)\begin{vmatrix} 15 & 1 \\ -9 & -3 \end{vmatrix} - (3)\begin{vmatrix} -7 & -2 \\ -9 & -3 \end{vmatrix} + (2)\begin{vmatrix} -7 & -2 \\ 15 & 1 \end{vmatrix}$$

$$= 1(-45 + 9) - 3(21 - 18) + 2(-7 + 30)$$
$$= -36 - 9 + 46$$
$$= 1$$

$$D_z = \begin{vmatrix} 1 & 2 & -7 \\ 3 & -2 & 15 \\ 2 & 3 & -9 \end{vmatrix} = (1)\begin{vmatrix} -2 & 15 \\ 3 & -9 \end{vmatrix} - (3)\begin{vmatrix} 2 & -7 \\ 3 & -9 \end{vmatrix} + (2)\begin{vmatrix} 2 & -7 \\ -2 & 15 \end{vmatrix}$$

$$= 1(18 - 45) - 3(-18 + 21) + 2(30 - 14)$$
$$= -27 - 9 + 32$$
$$= -4$$

Answer to Your Turn 4
$(-4, 2)$

Note After finding the values of two of the variables, you could find the value of the third variable by substituting these values back into any one of the original equations.

▶

$$x = \frac{D_x}{D} = \frac{-3}{-1} = 3, \qquad y = \frac{D_y}{D} = \frac{1}{-1} = -1, \qquad z = \frac{D_z}{D} = \frac{-4}{-1} = 4$$

The solution is $(3, -1, 4)$. We will leave the check to the reader.

Answer to Your Turn 5
$(1, 1, -3)$

Your Turn 5   Use Cramer's Rule to solve $\begin{cases} 2x + 3y - 2z = 11 \\ 3x + y + 4z = -8 \\ x - 3y - 2z = 4 \end{cases}$.

Appendix F
Exercises For Extra Help **MyMathLab®**

Note: Exercises marked with a ★ represent challenging exercises.

Objectives 1 and 2

Prep Exercise 1 What is a square matrix?
A matrix with an equal number of rows and columns

Prep Exercise 2 What is the formula for evaluating
$$\det(A) = \begin{vmatrix} a_1 & b_1 \\ a_2 & b_2 \end{vmatrix}?$$
$a_1 b_2 - a_2 b_1$

Prep Exercise 3 Is it possible to find the determinant of $\begin{bmatrix} 1 & 2 & 5 \\ 3 & 6 & 2 \end{bmatrix}$? Why or why not?
No, because it is not a square matrix

Prep Exercise 4 Explain the difference between a matrix and a determinant.
A matrix is a rectangular arrangement of numbers. A determinant is a single-value evaluation of a square matrix.

Prep Exercise 5 How do you find the minor of an element of a 3×3 matrix?
Eliminate the elements in the same row and column as the element and evaluate the determinant of the remaining matrix.

For Exercises 1–30, find the determinant. See Examples 1–3.

1. $\begin{bmatrix} 3 & 2 \\ 1 & 5 \end{bmatrix}$
13

2. $\begin{bmatrix} 4 & 1 \\ 3 & 7 \end{bmatrix}$
25

3. $\begin{bmatrix} -3 & 5 \\ 2 & 4 \end{bmatrix}$
-22

4. $\begin{bmatrix} 5 & 4 \\ 2 & -6 \end{bmatrix}$
-38

5. $\begin{bmatrix} 3 & -6 \\ 2 & 4 \end{bmatrix}$
24

6. $\begin{bmatrix} 2 & 8 \\ -3 & 2 \end{bmatrix}$
28

7. $\begin{bmatrix} -3 & 4 \\ -2 & 5 \end{bmatrix}$
-7

8. $\begin{bmatrix} -3 & -5 \\ 4 & 6 \end{bmatrix}$
2

9. $\begin{bmatrix} -2 & -3 \\ -4 & 5 \end{bmatrix}$
-22

10. $\begin{bmatrix} -6 & -2 \\ 3 & -4 \end{bmatrix}$
30

11. $\begin{bmatrix} 0 & 3 \\ -5 & 7 \end{bmatrix}$
15

12. $\begin{bmatrix} -5 & 0 \\ -4 & 2 \end{bmatrix}$
-10

13. $\begin{bmatrix} 1 & 2 & 1 \\ 3 & 1 & 4 \\ 2 & 3 & 2 \end{bmatrix}$
1

14. $\begin{bmatrix} 3 & 1 & 4 \\ 2 & 2 & 3 \\ 1 & 4 & 3 \end{bmatrix}$
3

15. $\begin{bmatrix} -1 & 2 & 0 \\ -3 & 2 & 4 \\ -4 & 2 & 3 \end{bmatrix}$
-12

16. $\begin{bmatrix} -2 & 0 & -1 \\ 3 & -2 & 4 \\ -3 & 2 & 1 \end{bmatrix}$
20

17. $\begin{bmatrix} 2 & 1 & -3 \\ 0 & -3 & 2 \\ 4 & 1 & -3 \end{bmatrix}$
-14

18. $\begin{bmatrix} 3 & -2 & 4 \\ 3 & 0 & 2 \\ -4 & -2 & 2 \end{bmatrix}$
16

19. $\begin{bmatrix} 0 & 4 & -2 \\ 3 & 2 & 0 \\ -1 & 4 & 3 \end{bmatrix}$
-64

20. $\begin{bmatrix} 3 & -5 & 0 \\ 2 & -4 & 1 \\ -2 & 0 & -3 \end{bmatrix}$
16

21. $\begin{bmatrix} 0.3 & -0.5 \\ 1.3 & -0.6 \end{bmatrix}$

0.47

22. $\begin{bmatrix} -0.4 & 1.6 \\ -4.7 & 3.1 \end{bmatrix}$

6.28

■ 23. $\begin{bmatrix} -0.4 & 0.7 & -1.2 \\ 3.1 & 1.5 & -3.2 \\ 1.6 & -2.2 & -1.5 \end{bmatrix}$

14.451

■ 24. $\begin{bmatrix} 1.7 & -3.2 & 4.1 \\ 5.3 & -6.2 & -1.1 \\ -1.3 & 2.3 & -4.5 \end{bmatrix}$

−12.232

25. $\begin{bmatrix} \dfrac{1}{2} & -\dfrac{1}{3} \\ \dfrac{2}{5} & \dfrac{3}{5} \end{bmatrix}$

$\dfrac{13}{30}$

26. $\begin{bmatrix} -\dfrac{3}{4} & -\dfrac{3}{5} \\ \dfrac{3}{2} & \dfrac{2}{5} \end{bmatrix}$

$\dfrac{3}{5}$

★ 27. $\begin{bmatrix} \dfrac{1}{2} & \dfrac{3}{4} & \dfrac{2}{5} \\ \dfrac{1}{3} & \dfrac{1}{5} & -\dfrac{3}{2} \\ -\dfrac{3}{4} & \dfrac{1}{2} & \dfrac{3}{5} \end{bmatrix}$

$-\dfrac{317}{2400}$

★ 28. $\begin{bmatrix} -\dfrac{1}{4} & -\dfrac{3}{2} & \dfrac{4}{3} \\ \dfrac{1}{5} & -\dfrac{5}{4} & \dfrac{1}{2} \\ -\dfrac{5}{3} & \dfrac{1}{4} & -\dfrac{4}{5} \end{bmatrix}$

$-\dfrac{13,823}{7200}$

★ 29. $\begin{bmatrix} x & y & 1 \\ 2 & -1 & 3 \\ -2 & 0 & 1 \end{bmatrix}$

$-x - 8y - 2$

★ 30. $\begin{bmatrix} x & y & 1 \\ -3 & -2 & 4 \\ 3 & -2 & 2 \end{bmatrix}$

$4x + 18y + 12$

Objective 3

Prep Exercise 6 How do you find D_y when solving a system of equations using Cramer's Rule?

In the coefficient matrix, replace the y-column values with the constant-column values and evaluate the determinant of the resulting matrix.

For Exercises 31–44, solve using Cramer's Rule. See Example 4.

31. $\begin{cases} x + y = -5 \\ x - 2y = -2 \end{cases}$

$(-4, -1)$

32. $\begin{cases} x + 3y = 1 \\ x + y = -3 \end{cases}$

$(-5, 2)$

33. $\begin{cases} 2x - 3y = -6 \\ x - y = -1 \end{cases}$

$(3, 4)$

34. $\begin{cases} 2x + 5y = -7 \\ x - y = -7 \end{cases}$

$(-6, 1)$

35. $\begin{cases} -x + 2y = -12 \\ 2x - 3y = 20 \end{cases}$

$(4, -4)$

36. $\begin{cases} -x + 2y = -9 \\ 4x + 5y = -3 \end{cases}$

$(3, -3)$

37. $\begin{cases} 4x - 6y = 7 \\ 6x - 9y = 8 \end{cases}$

No solution

38. $\begin{cases} 6x - 9y = 17 \\ 8x - 12y = 7 \end{cases}$

No solution

39. $\begin{cases} 8x - 3y = 10 \\ 4x + 3y = 14 \end{cases}$

$(2, 2)$

40. $\begin{cases} 2x - y = 10 \\ 4x - 5y = 32 \end{cases}$

$(3, -4)$

41. $\begin{cases} \dfrac{1}{2}x - \dfrac{1}{4}y = 0 \\ \dfrac{3}{4}x + \dfrac{5}{2}y = \dfrac{23}{2} \end{cases}$

$(2, 4)$

42. $\begin{cases} \dfrac{2}{3}x + \dfrac{1}{4}y = \dfrac{1}{2} \\ \dfrac{3}{4}x + \dfrac{4}{3}y = -\dfrac{23}{4} \end{cases}$

$(3, -6)$

43. $\begin{cases} 0.2x + 0.5y = 3.4 \\ 0.7x - 0.3y = -0.4 \end{cases}$

$(2, 6)$

44. $\begin{cases} 1.2x - 0.6y = -2.4 \\ 3.1x + 1.3y = -11.9 \end{cases}$

$(-3, -2)$

For Exercises 45–56, solve using Cramer's Rule. See Example 5.

45. $\begin{cases} x + y + z = 6 \\ 2x - 4y + 2z = 6 \\ 3x + 2y + z = 11 \end{cases}$

$(2, 1, 3)$

46. $\begin{cases} 2x + y - 3z = -1 \\ x + 2y - 2z = -3 \\ -3x - 4y + z = -3 \end{cases}$

$(3, -1, 2)$

47. $\begin{cases} 3x + y - z = -4 \\ 2x - y + 2z = -7 \\ x - 3y + z = -6 \end{cases}$

$(-2, 1, -1)$

48. $\begin{cases} x - y + 3z = -10 \\ 5x + 4y - z = -7 \\ 2x + y - z = -4 \end{cases}$

$(-4, 3, -1)$

49. $\begin{cases} 4x + 2y + 3z = 9 \\ 2x - 4y - z = 7 \\ 3x - 2z = 4 \end{cases}$
$(2, -1, 1)$

50. $\begin{cases} 2x - y = -1 \\ 5x - 3y + 2z = 0 \\ 3x + 2y - 3z = -8 \end{cases}$
$(-1, -1, 1)$

51. $\begin{cases} 3x + 2y = -12 \\ 3y + 10z = -16 \\ 6x - 2z = 3 \end{cases}$
$\left(\dfrac{2}{3}, -7, \dfrac{1}{2}\right)$

52. $\begin{cases} 4x + 2z = 7 \\ 8x - 2y = -7 \\ 10y - 2z = -5 \end{cases}$
$\left(-\dfrac{3}{4}, \dfrac{1}{2}, 5\right)$

53. $\begin{cases} \dfrac{1}{2}x + \dfrac{1}{3}y + \dfrac{3}{4}z = \dfrac{25}{12} \\ \dfrac{1}{3}x + \dfrac{2}{9}y + \dfrac{1}{2}z = \dfrac{25}{18} \\ \dfrac{3}{4}x + \dfrac{1}{4}y - \dfrac{2}{3}z = -\dfrac{7}{4} \end{cases}$
Infinite solutions

54. $\begin{cases} \dfrac{3}{4}x - \dfrac{1}{2}y + \dfrac{1}{2}z = \dfrac{1}{4} \\ \dfrac{7}{6}x + \dfrac{10}{9}y - \dfrac{2}{5}z = -\dfrac{623}{90} \\ \dfrac{7}{4}x + \dfrac{5}{3}y - \dfrac{3}{5}z = -\dfrac{623}{60} \end{cases}$
Infinite solutions

55. $\begin{cases} 0.3x + 0.4y - 0.6z = 2.6 \\ 0.5x - 0.2y + 0.7z = -0.8 \\ 1.4x + 1.3y - 2.2z = 9.8 \end{cases}$
$(2, 2, -2)$

56. $\begin{cases} 1.2x + 2.1y - 0.5z = 7.3 \\ 3.2x - 2.4y + 1.3z = 6.1 \\ 2.5x + 1.3y - 1.7z = 8.4 \end{cases}$
$(3, 2, 1)$

★ *For Exercises 57–60, find x.*

57. $\begin{vmatrix} 9 & x \\ -6 & 5 \end{vmatrix} = 21$
$x = -4$

58. $\begin{vmatrix} 12 & 2 \\ x & -3 \end{vmatrix} = -22$
$x = -7$

59. $\begin{vmatrix} 2 & -1 & 0 \\ 0 & x & 1 \\ -2 & 0 & 4 \end{vmatrix} = -38$
$x = -5$

60. $\begin{vmatrix} 1 & -2 & 4 \\ -1 & 0 & 2 \\ 0 & x & 5 \end{vmatrix} = -28$
$x = 3$

★ *For Exercises 61–66 , use the following information. Suppose a triangle has vertices of (x_1, y_1), (x_2, y_2), and (x_3, y_3), as shown in the graph to the right. The area of the triangle is given by $A = \dfrac{1}{2}\left|\det\begin{bmatrix} x_1 & y_1 & 1 \\ x_2 & y_2 & 1 \\ x_3 & y_3 & 1 \end{bmatrix}\right|$. For example, the area of the triangle whose vertices are $(1, 2)$, $(3, -2)$, and $(-2, 5)$ is $A = \dfrac{1}{2}\left|\det\begin{bmatrix} 1 & 2 & 1 \\ 3 & -2 & 1 \\ -2 & 5 & 1 \end{bmatrix}\right|$.*

Note This notation indicates ◀ half of the absolute value of the determinant.

Expanding about the first column, we have

$$A = \dfrac{1}{2}\left|1 \det\begin{bmatrix} -2 & 1 \\ 5 & 1 \end{bmatrix} - 3 \det\begin{bmatrix} 2 & 1 \\ 5 & 1 \end{bmatrix} - 2 \det\begin{bmatrix} 2 & 1 \\ -2 & 1 \end{bmatrix}\right|$$

$$= \dfrac{1}{2}|(-2 - 5) - 3(2 - 5) - 2(2 + 2)|$$

$$= \dfrac{1}{2}|-7 + 9 - 8| = \dfrac{1}{2}|-6| = 3 \ square \ units.$$

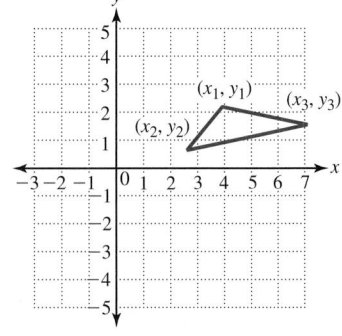

Find the area of the triangles with vertices at the given points.

61. $(2, 4) \ (4, 0) \ (6, 5)$
9 sq. units

62. $(0, 2) \ (3, -2) \ (5, 5)$
14.5 sq. units

63. $(-3, 1) \ (2, -3) \ (4, 4)$
21.5 sq. units

64. $(-3, -2) \ (-1, 4) \ (3, -4)$
20 sq. units

65. $(-4, -1) \ (1, 3) \ (3, -3)$
19 sq. units

66. $(-3, -3) \ (2, 2) \ (4, -1)$
12.5 sq. units

For Exercises 67–72, translate the problem to a system of equations; then solve using Cramer's Rule.

67. The two heaviest known meteorites to be found on Earth's surface are the Hoba West, which was found in Namibia, and the Ahnighito, which was found in Greenland. The total weight of the two meteorites is 90 tons. The Hoba West is twice as heavy as the Ahnighito. Find the weight of each. (*Source: Webster's New World Book of Facts*)

Hoba is 60 tons; Ahnighito is 30 tons.

68. The two largest expenses for the average American family are federal taxes and housing, including household expenses. Together these two items total 43.3% of the average family's income. The amount spent on taxes is 12.1% more than the amount spent on housing. Find the percent spent on each. (*Source: Numbers: How Many, How Long, How Far, How Much*)

27.7% on taxes, 15.6% on housing

69. A restaurant makes a soup that includes garbanzo and black turtle beans. The manager purchased 10 pounds of beans at a cost of $8.80. If the garbanzo beans cost $1.00 per pound and the black turtle beans cost $0.70 per pound, how many pounds of each did he purchase?

6 lb. of garbanzo, 4 lb. of black turtle

70. The perimeter of a triangle is 21 inches, and two sides are of equal length. The length of the third side is 3 inches less than the length of the two equal sides. Find the length of each side of the triangle.

8 in., 8 in., 5 in.

71. Coinage bronze is made up of zinc, tin, and copper. The percent of tin is 4 times the percent of zinc. The percent of copper is 19 times the sum of the percents of zinc and tin. Find the percent of zinc, tin, and copper in coinage bronze. (*Source: Webster's New World Book of Facts*)

1% zinc, 4% tin, 95% copper

72. In winning the 2009 men's NCAA basketball championship, North Carolina scored a total of 56 times in their 89 to 72 victory over Michigan State. The sum of the number of 3-point field goals and 2-point field goals was equal to the number of free throws (1 point each). How many of each did North Carolina score? (*Source:* espn.go.com)

3-point field goals: 5; 2-point field goals: 23, free throws: 28

Photo Credits

Cover
Quintanilla/Shutterstock

Chapter 1
p. 1: Igorsky/Shutterstock; p. 19: Lawrence Lawry/Getty Images; p. 21: Comstock Images; p. 28: Getty Images; p. 31: Jupiter Images; p. 33 (bl): Library of Congress Prints and Photographs Division [LC-USZ62-117117DLC]; p. 33 (br): Library of Congress Prints and Photographs Division [LC-USZ62-13004DLC]; p. 33 (tl): Library of Congress Prints and Photographs Division [LC-USZ62-117116DLC]; p. 33 (tr): Library of Congress Prints and Photographs Division [LC-USZ62-13002DLC]; p. 46: Francois Gohier/VWPics/Alamy; pp. 61: Getty Images; p. 64: Karl Weatherly/Getty Images

Chapter 2
p. 87: Fotolia; p. 93: Auremar/Shutterstock; p. 94: Dennis MacDonald/PhotoEdit, Inc.; p. 95: Doug Menuez/Getty Images; p. 99: Mark Karrass/Corbis; p. 100: Corbis; p. 110: Nick Rowe/Getty Images; p. 124: Jack Hollingsworth/Getty Images; p. 137: Getty Images; p. 144: Jeff Maloney/Getty Images; p. 146: Steve Mason/Getty Images; p. 147: Hishman F. Ibrahim/Getty Images; p. 148: Getty Images; p. 155: Chapple/ThinkStock/Jupiter Images; p. 156: Marje Cannon/iStockphoto; p. 159: Library of Congress Prints and Photographs Division [LC-B2-2015-5]; p. 178: Tim Barnett/Getty Images

Chapter 3
p. 189: Fotolia; p. 191: Library of Congress Prints and Photographs Division [LC-USZ62-61365]; p. 205 (b): Getty Images; p. 205 (t): Dave Thompson/Getty Images; p. 228: Stephan Savoia/AP Images; p. 247: Getty Images

Chapter 4
p. 279: Stuart Jenner/Shutterstock; p. 306: Getty Images; p. 311: Keith Brofsky/Getty Images; p. 317: Chuck Savage/Corbis; p. 328: Getty Images; p. 329: Bill Shettle/CSM/Landov

Chapter 5
p. 349: Simon Grosset/Alamy; p. 353: Dr. Gary Gaugler/Science Source/Photo Researchers, Inc.; p. 358 (b): Getty Images; p. 358 (t): Digital Vision/Jupiter Images; p. 359: Getty Images; p. 360: Kim Jae-Hwan/Getty Images; p. 364: David Buffington/Getty Images; p. 371: Getty Images; p. 411: Hishman F. Ibrahim/Getty Images

Chapter 6
p. 427: Jeff J Daly/Alamy; p. 469: Getty Images; p. 471: North Wind Picture Archives/Alamy; p. 474 (b): Doug Menuez/Getty Images; p. 474 (t): Bill Aron/PhotoEdit, Inc.

Chapter 7
p. 497: Ryan McVay/Photodisc/Getty Images; p. 507 (b): Getty Images; p. 507 (t): Benedicte Desrus/Alamy; p. 517: Dusan Vranic/AP Images; p. 518: Stephen Mcsweeny/Shutterstock; p. 533: Design Pics/LJM Photo/Getty Images; p. 554: Nick Rowe/Getty Images

Chapter 8
p. 577: Dmac/Alamy; p. 587: Fotolia; p. 588: Dorling Kindersley, Ltd.

Chapter 9
p. 612: Carlos Caetano/Shutterstock; p. 626: Bettmann/Corbis

Chapter 10
p. 689: Rudi/Fotolia; p. 698: Don Farrall/Getty Images; p. 705: ZUMA Press, Inc/Alamy; p. 709 (b): Richard Eyre/iStockphoto; p. 709 (t): Frank Micelotta/Getty Images; p. 718 (b): Justin Lane/EPA/Newscom; p. 718 (t): Charles Krupa/AP Images

Chapter 11
p. 763: David J. Green/Alamy; p. 775 (b): Kim Steele/Getty Images; p. 775 (t): Paul Sancya/AP Images; p. 783: Reed Kaestner/Corbis; p. 802: Antony Nettle/Alamy; p. 805: Library of Congress Prints and Photographs Division [LC-D4-19236]; p. 814: ZUMA Press/Alamy; p. 815: NASA

Chapter 12
p. 833: Andre Jenny/Alamy; p. 846: Andrew Ward/Getty Images; p. 856: Claudio Zaccherini/Shutterstock

Appendices
p. APP-35: Jonathan Blair/Corbis

Collaborative Exercises

Section 1.3 Dollars and Sense

Your group is to prepare a simple monthly household budget. The budget should be for a "typical" student or for a new employee in your field of study. (However, you should not use any of your personal information.) Remember to be realistic about both income and expenditures.

1. Discuss and agree upon an amount of income you think an average working student makes each month. (Or use a starting salary for a person in your field of study.)

2. Estimate average expense amounts for each of the following. If your group believes that an expense does not apply, explain why.

   Taxes
   Housing
   Electricity
   Water
   Telephone (home and cell)
   Cable TV
   Internet service
   Food
   Household supplies

   Medicine

   Transportation/Gasoline
   Child care
   Loan repayments
   Hair care
   Clothing
   Magazine/Newspaper subscriptions
   Allowance
   Movies
   Other entertainment (renting DVDs, attending concerts, eating out)

3. Using your estimates in Exercises 1 and 2, calculate the monthly net.

4. Discuss and list ways that a person could cut expenses to improve his or her monthly net.
   All answers will vary.

Section 2.4 Where Does Speeding Get You?

1. The formula that relates distance, rate, and time is $d = rt$. Solve this formula for t.
   $$t = \frac{d}{r}$$

2. Use the formula you found in Problem 1 to determine how long it will take you to travel 20 miles (an average daily commute) at a rate of 55 miles per hour. Note that your result will be in hours. Convert the time to minutes by multiplying by 60 and round to the nearest tenth of a minute.
   ≈ 21.8 min.

3. Now use the formula from Problem 1 to determine how long it will take you to travel 20 miles at 65 miles per hour. Again, convert the time to minutes and round to the nearest tenth of a minute.
   ≈ 18.5 min.

4. How much time does speeding save you? To find out, subtract the time you found in Problem 3 from the time you found in Problem 2.
   ≈ 3.3 min.

5. Discuss in your group whether the additional time is worth the additional risk that comes with increasing the speed by 10 miles per hour.
   Answers will vary. (Most likely it is not worth it.)

6. Complete the table, rounding all values to the nearest tenth. (You've already calculated the values for the 20-mile trip.)

Trip Distance	Time at 55 mph	Time at 65 mph	Time Saved
20 miles	21.8 min.	18.5 min.	3.3 min.
30 miles	32.7 min.	27.7 min.	5 min.
40 miles	43.6 min.	36.9 min.	6.7 min.

7. Is the time saved on longer trips worth the additional risk of increasing your speed by 10 miles per hour? Why or why not? Does the longer distance also increase risk? Why or why not?
   Answers will vary.

Section 2.7 Occupation Growth

Complete the following table by calculating the amount of change and the percent of the increase for each occupation; then answer the questions. Round to the nearest tenth of a percent.

The Ten Fastest-Growing Occupations, 2002–2012

Occupation	Employment		Change	
	2002	**2012**	**Amount**	**Percent**
Medical assistants	364,600	579,400	214,800	58.9%
Network systems and data communications analysts	186,000	292,000	106,000	57.0%
Physician assistants	63,000	93,800	30,800	48.9%
Social and human service assistants	305,200	453,900	148,700	48.7%
Home health aides	579,700	858,700	279,000	48.1%
Medical records and health information technicians	146,900	215,600	68,700	46.8%
Physical therapist aides	37,000	54,100	17,100	46.2%
Computer software engineers, applications	394,100	573,400	179,300	45.5%
Computer software engineers, systems software	281,100	408,900	127,800	45.5%
Physical therapist assistants	50,200	72,600	22,400	44.6%

(Source: Bureau of Labor Statistics, Office of Occupational Statistics and Employment Projections.)

1. In 2002, in which occupation were the greatest number of people employed?
   Home health aides

2. By 2012, which occupation is projected to have the greatest number of people employed?
   Home health aides

3. Explain why the two occupations you listed in Exercises 1 and 2 are not at the top of the list.
   Although home health aides have the greatest employment, it is not the fastest-growing occupation

4. Which is a better indicator of the demand for people in a particular occupation: the number of people employed in a particular year, the amount of change projected from 2002 to 2012, or the percent of increase in employment from 2002 to 2012?
   Percent of increase

5. Based on your conclusions in Exercise 4, what occupation will have the greatest demand? What college majors might have the greatest potential for employment in that occupation? Write your conclusions and present them to the class.
   Medical assistants

Section 3.4 Population Growth

The following data show the total U.S. population by decade.

Decade	1900	1910	1920	1930	1940	1950	1960	1970	1980	1990	2000
Population (in millions)	76.2	92.2	106.0	123.2	132.2	151.3	179.3	203.3	226.5	248.7	281.4

1. Plot each ordered pair as a point on a graph with the decades along the horizontal axis and the population along the vertical axis.

2. Do the points have a linear relationship? no

3. Draw a straight line that connects the point at 1900 to the point at 1950. Find the slope of this line. 1.502

4. Draw a straight line that connects the point at 1950 to the point at 2000. Find the slope of this line. 2.602

5. Compare the slopes of the two lines you have drawn. Which has the greater slope? What does this indicate about population growth in the United States during the 20th century?
   The slope is greater between 1950 and 2000. The population is growing at a faster rate during the last half of the 20th century.

6. Use the line you drew from 1950 to 2000 to predict the population at 2110. 567.62 million

1.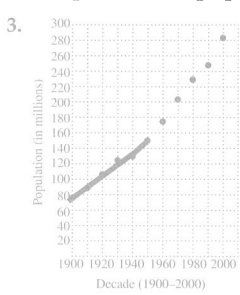

3.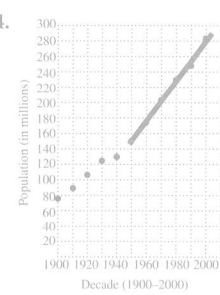

4.

Section 4.6 Maximizing the Profit

Linear programming is an area of mathematics that solves problems like the one described below. One of the fundamental elements of a linear programming model is the constraint. A constraint is simply an inequality that describes some limited resource required in the problem. For example, suppose you make two types of bicycles, style A and style B. Also suppose that style A requires 40 minutes of welding for each bicycle and style B requires 25 minutes of welding. If you have only 600 minutes of welding time available, the welding-time constraint would be $40A + 25B \leq 600$, where A represents the number of style A produced and B represents the number of style B produced.

An aspiring artists' group is preparing for a fund-raising sale. The artists design and construct two types of antique reproduction tables: a semicircular foyer table and a side table. The materials cost \$100 for each foyer table and \$200 for each side table. Both types of table require 2 hours of cutting and carving. The foyer table requires 3.5 hours to assemble, sand, and finish, and the side table requires 1.75 hours to assemble, sand, and finish. The group has \$3300 to purchase all materials; 40 hours available for cutting and carving; and 63 hours available for assembling, sanding, and finishing. For each foyer table they sell, their profit will be \$350. For each side table, their profit will be \$215. How many of each table should they make to maximize their profit?

1. For the artists' fund-raising problem, there are three basic constraints: (1) cost of materials, (2) cutting and carving, and (3) assembling/sanding/finishing. For each of these constraints, write an inequality similar to the one in the bicycle example. Let x represent the number of foyer tables and y the number of side tables.

$100x + 200y \leq 3300, 2x + 2y \leq 40, 3.5x + 1.75y \leq 63$

2. Because x and y represent numbers of tables, they cannot represent negative values. Write inequalities for these two additional constraints.

$x \geq 0, y \geq 0$

3. Now graph the system of inequalities described by these five constraints.

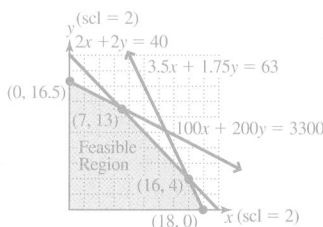

4. The region that is the solution of the system of inequalities is called the *feasible region*, and it includes all points that satisfy the system. The goal is to find the optimal solution, which, according to linear programming, is one of the corner points of the feasible region. There should be five corner points: one at the origin, one on each axis, and two more in quadrant I. Determine the coordinates of each corner point. Note that a corner point lying on an axis is the x- or y-intercept for the line passing through that point. Because two lines intersect to form a corner point that is not on an axis, make a system out of their two equations; then solve the system to find their point of intersection.

$(0,0), (16,4), (7,13), (0,16.5), (18,0)$

5. Next, develop an algebraic expression describing the profit gained for selling x of the foyer tables and y of the side tables. This is called the *objective function*.

$350x + 215y$

6. Using the objective function, test each of the points found in step 4 to see which one yields the maximum profit.

The point $(16, 4)$ yields a maximum profit of \$6460.

7. Using your solution, answer the following questions.

   a. How many of each type of table should the group produce?

     16 foyer tables, 4 side tables

   b. How many hours will be spent in each phase of production?

     40 hours cutting and carving, 63 hours sanding and finishing

   c. How much money will the group need to purchase the materials?

     $2400 in materials

   d. What amount of profit will the group receive?

     $6460 profit

Section 5.3 Building Furniture and Profits

Suppose your group operates a furniture manufacturing company that produces two different sizes of coffee table. Each larger coffee table sells for $350, and each smaller table sells for $280.

1. Write a monomial expression for the revenue if x is the number of large tables sold per month.

   $350x$

2. Write a monomial expression for the revenue if y is the number of small tables sold per month.

   $280y$

3. Write a polynomial expression that describes the total revenue generated from the sale of the two tables per month.

   $350x + 280y$

4. Write a monomial expression for the cost of producing x number of large tables per month.

   $225x$

Suppose it costs $225 to produce each large table and $135 to produce each small table.

5. Write a monomial expression for the cost of producing y number of small tables per month.

   $135y$

6. Your company also spends $18,500 per month to pay for the lease, utilities, and salaries. Write a polynomial expression that describes the total cost involved in production.

   $225x + 135y + 18,500$

7. Using your polynomials for revenue and cost, write a polynomial in simplest form for the profit if x number of large tables are produced and sold and y number of small tables are produced and sold.

   Profit $= (350x + 280y) - (225x + 135y + 18,500)$

           $= 125x + 145y - 18,500$

8. What does each coefficient in the polynomial for profit indicate?

   Each coefficient is the profit made from selling one of the corresponding tables.

9. Suppose in one month, 200 large tables are produced and sold and 400 small tables are produced and sold. Find the revenue, cost, and profit for the month.

   Revenue = $182,000; cost = $117,500; profit = $64,500

10. Suppose in one month, 56 large tables are produced and sold and 65 small tables are produced and sold. Find the revenue, cost, and profit for the month. Explain the meaning of your answer.

   Revenue = $37,800; cost = $39,875; profit = −$2075; the negative profit indicates a loss of $2075 for the month.

11. Suppose in one month, only large tables are produced and sold. How many would have to be produced and sold to break even (profit is 0)?

   $125x + 145(0) - 18,500 = 0$

             $125x = 18,500$

                $x = 148$ large tables need to be sold to break even

12. Suppose only small tables are produced and sold in one month. How many would have to be produced and sold to break even?

   $125(0) + 145y - 18,500 = 0$

             $145y = 18,500$

                $y \approx 128$

   So 128 small tables must be sold to break even. Note that there will actually be a slight profit. (127 tables would yield a slight loss.)

Section 6.7 What Goes Up Must Come Down

Many sets of variables have a relationship that is roughly modeled by a parabolic shape. The following data give the high temperature (in degrees Fahrenheit, °F) and peak power load (in megawatts, MW) for 31 days throughout the calendar year.

Temp (°F)	Peak Load (MW)	Temp (°F)	Peak Load (MW)
69	94	95	141
71	95	46	142
59	100	43	147
65	100	98	150
67	102	96	151
60	104	97	155
74	105	100	158
79	105	43	165
56	112	38	173
84	113	102	176
52	114	36	193
88	115	103	199
50	120	105	204
90	123	106	226
92	132	34	228
92	132		

1. On a sheet of graph paper, draw and label axes with the temperature along the horizontal axis and the peak load on the vertical axis. Then plot the points.
   See the graph below.
2. Draw a smooth parabola through the points that might best fit the data.
   See the graph below.
3. At what temperature does the minimum peak load seem to occur?
   69°
4. In your own words, describe what happens to the peak load as the temperature increases and decreases and explain why.
   Extreme temperatures (high or low) have the greater peak loads because extreme temperatures cause the use of either heat or cooling.
5. Use the model that you drew in part 2 to estimate the peak load on a day when the maximum temperature is 85.
   ≈110 MW
6. Give another example of a situation that might be best modeled by a parabola.
   Answers will vary.

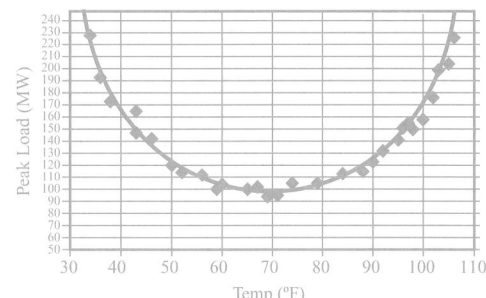

Section 7.1 Graphs of Rational Expressions

Use a graphing utility.

1. Graph $f(x) = \dfrac{x}{x - 3}$. The vertical line you see is called an *asymptote*.
   a. What is the equation of the asymptote?
      $x = 3$
   b. What do you think the asymptote indicates?
      It is the value that causes the function to be undefined.

2. Graph $f(x) = \dfrac{5}{(x + 2)(x - 4)}$.
   a. Why are there two asymptotes?
      The two values cause the function to be undefined.
   b. What are the equations of the two asymptotes?
      $x = -2$ and $x = 4$

   Note The suggested window settings in Exercise 3 are for a TI-83 or TI-84 including plus models. If a different graphing utility is used, the settings may need to be changed to see the hole.

3. Graph $f(x) = \dfrac{x^2 - x - 6}{x^2 + 3x + 2}$ with these window settings: $x_{min} = -4.7$, $x_{max} = 4.7$, $y_{min} = -10$, and $y_{max} = 10$.
   a. Are there any asymptotes? If so, write the equation for each.
      Yes, $x = -1$
   b. Rewrite $f(x) = \dfrac{x^2 - x - 6}{x^2 + 3x + 2}$ so that the numerator and denominator are in factored form.
      $f(x) = \dfrac{(x + 2)(x - 3)}{(x + 2)(x + 1)}$
   c. Find $f(-2)$.
      indeterminate

d. Look at the graph of $f(x) = \dfrac{x^2 - x - 6}{x^2 + 3x + 2}$ in the second quadrant. The blank space you see in the graph where $x = -2$ is a *removable discontinuity* and is often called a *hole*. How is a hole different from an asymptote?

   Answers will vary. One possible response is that a hole is a missing point on a graph that, if filled in, would complete the graph, whereas an asymptote is a vertical or horizontal line that a graph approaches but never intersects.

4. Writing $\dfrac{x^2 - x - 6}{x^2 + 3x + 2}$ in simplest form, we have $\dfrac{x - 3}{x + 1}$. Graph $f(x) = \dfrac{x - 3}{x + 1}$. Compare the graph of $f(x) = \dfrac{x - 3}{x + 1}$ with the graph of $f(x) = \dfrac{x^2 - x - 6}{x^2 + 3x + 2}$. What do you notice about the graphs?

   The graphs are the same except that $f(x) = \dfrac{x^2 - x - 6}{x^2 + 3x + 2}$ has a hole where $x = -2$, whereas $f(x) = \dfrac{x - 3}{x + 1}$ has no hole where $x = -2$.

5. Given a rational function, how can you identify its asymptotes and holes?

   Asymptotes occur where the function is undefined. Holes occur where the function is indeterminate.

Section 8.2 Dollars and Sense

*In the following exercises, you will use a graphing utility to explore and discuss how numbers can **shift, reflect,** or **stretch** a basic function. We will use the graph of $f(x) = |x|$ as our basic function.*

1. Graph $f(x) = |x + 2|$ and compare it with the graph of $f(x) = |x|$. What effect does adding 2 to x within the absolute value have on the graph?

   Adding 2 to x moves the graph of $f(x) = |x|$ to the left 2 units.

2. Graph $f(x) = |x - 3|$ and compare it with the graph of $f(x) = |x|$. What effect does subtracting 3 from x within the absolute value have on the graph?

   Subtracting 3 from x moves the graph of $f(x) = |x|$ to the right 3 units.

3. Graph $f(x) = |x| + 4$ and compare it with the graph of $f(x) = |x|$. What effect does adding 4 to $|x|$ have on the graph?

   Adding 4 to $|x|$ moves the graph of $f(x) = |x|$ up 4 units.

4. Graph $f(x) = |x| - 1$ and compare it with the graph of $f(x) = |x|$. What effect does subtracting 1 from $|x|$ have on the graph?

   Subtracting 1 from $|x|$ moves the graph of $f(x) = |x|$ down 1 unit.

5. Notice that adding or subtracting in a function *shifts* the position of the basic function. Without graphing, how would you expect the graph of $f(x) = |x + 1| - 3$ to be shifted as compared with that of the basic function $f(x) = |x|$?

   The graph of $f(x) = |x + 1| - 3$ is 1 unit to the left and 3 units down as compared with that of $f(x) = |x|$.

6. Graph $f(x) = -|x|$. How is the graph of $f(x) = -|x|$ different from that of $f(x) = |x|$?

   The graph of $f(x) = -|x|$ opens down and is the mirror image of that of $f(x) = |x|$ across the x-axis.

7. $f(x) = -|x|$ is the same as $f(x) = -1 \cdot |x|$. Multiplying the basic function by a negative number causes a *reflection* of the function about the x-axis. Without graphing, how would you describe the differences between the graph of $f(x) = -|x - 2| + 1$ and that of the basic function $f(x) = |x|$?

   The graph of $f(x) = -|x - 2| + 1$ opens down and is 2 units to the right and 1 unit up compared with that of $f(x) = |x|$.

8. Graph $f(x) = 2|x|$. How is the graph of $f(x) = 2|x|$ different from that of $f(x) = |x|$?

   The graph of $f(x) = 2|x|$ opens narrower than that of $f(x) = |x|$.

   Instructor Note You might also point out that in a function in the form $f(x) = m|x - h| + k$, m and $-m$ are the slopes of each line that form the graph.

9. Graph $f(x) = \dfrac{1}{2}|x|$. How is the graph of $f(x) = \dfrac{1}{2}|x|$ different from that of $f(x) = |x|$?

   The graph of $f(x) = \dfrac{1}{2}|x|$ opens wider than that of $f(x) = |x|$.

10. Notice that multiplying the basic function by a number other than 1 causes the graph to be *stretched*, so that it is wider or narrower than the basic function's graph. Without graphing, describe the differences in each of the following compared with the graph of $f(x) = |x|$.

   a. $f(x) = 3|x| + 1$

   The graph of $f(x) = 3|x| + 1$ is narrower and 1 unit up as compared with that of $f(x) = |x|$.

   b. $f(x) = -2|x - 4|$

   The graph of $f(x) = -2|x - 4|$ opens down, is narrower, and is 4 units to the right as compared with that of $f(x) = |x|$.

   c. $f(x) = \dfrac{1}{3}|x + 4| - 2$

   The graph of $f(x) = \dfrac{1}{3}|x + 4| - 2$ is wider, is 4 units to the left, and is 2 units down as compared with that of $f(x) = |x|$.

For further investigation, change the basic function to $f(x) = x^2$ and work through Exercises 1–10 again. For example, Exercise 1 would read as follows: Graph $f(x) = (x + 2)^2$ and compare it with the graph of $f(x) = x^2$. What effect does adding 2 to x have on the graph? Exercise 3 would read as follows: Graph $f(x) = (x^2 + 4)$ and compare it with the graph of $f(x) = x^2$. What effect does adding 4 to x^2 have on the graph?

Section 9.6 Building Time

You are constructing a grandfather clock that operates using a pendulum. The period of the pendulum is the time required to complete one full swing. The formula giving the relationship between the length, L (in meters), and the period, T (in seconds), is $T = 2\pi\sqrt{\dfrac{L}{9.8}}$.

1. Find the length of the pendulum if the period is 1 second.

≈ 0.248 m

2. Suppose you want the pendulum to complete one period in 2 seconds. Would you need to increase or decrease the length of the pendulum found in Exercise 1? Use the same formula to determine the length of the pendulum so that the period is 2 seconds.

Increase the length to approximately 0.993 m.

3. Based on the results of Exercises 1 and 2, what can you conclude about the required length of the pendulum as the period increases?

The length increases.

Section 10.4 Arch Span

The Gateway Arch, located in St. Louis, Missouri, was built from 1963 to 1965 and is the nation's tallest memorial. Although the arch is not a parabola, the equation $h(x) = -0.0063492063x^2 + 630$ can be used to approximate the height of the arch, where x represents the distance from the axis of symmetry to the arch and $h(x)$ represents its height above the ground.

1. Using the equation, find the maximum height of the structure.

630 ft.

2. The span of the arch at a given height is the horizontal distance between the two opposing points on the arch at that height. Find the span of the arch at ground level. (*Hint:* Think of ground level as the x-axis. The height is 0 along the x-axis.)

630 ft.

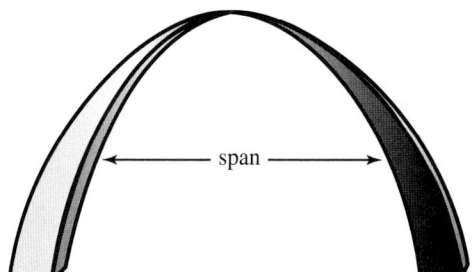

3. How does the span at ground level relate to the maximum height of the arch?

The span and height are equal.

4. The arch has foundations 60 feet below ground level. What is the span of the arch between its foundations?

≈ 659.3 ft.

Section 11.5 Exploring Graphs of Logarithms

Using a graphing calculator, draw the graph of $f(x) = \log x$ and $g(x) = \log(10x)$ in the window $[0, 100]$ for x and $[-2, 4]$ for y.

1. Press the ⬚TRACE⬚ key. Using up or down arrow keys, move from one graph to the other and observe the y-values of several points with the same x-values. What do you observe?

The y-values of the graph of $g(x) = \log(10x)$ are 1 larger than the y-values of $f(x) = \log x$.

2. How is the graph of $g(x) = \log(10x)$ related to the graph of $f(x) = \log x$? Why?

The graph of $g(x) = \log(10x)$ is the graph of $\log x$ shifted up 1 unit. $g(x) = \log(10x) = \log 10 + \log x = 1 + \log x = 1 + f(x)$

3. In general, how is the graph of $g(x) = \log(kx)$ for $k > 0$ related to the graph of $f(x) = \log x$?

The graph of $g(x) = \log(kx)$ is the graph of $f(x) = \log x$ shifted up $\log k$ units.

4. Repeat step 1 using $f(x) = \log x$ and $g(x) = \log\dfrac{x}{10}$.

The y-values of $g(x) = \log\dfrac{x}{10}$ are 1 less than the y-values of $f(x) = \log x$.

5. How is the graph of $g(x) = \log \dfrac{x}{10}$ related to the graph of $f(x) = \log x$? Why?

The graph of $g(x) = \log \dfrac{x}{10}$ is the graph of $f(x) = \log x$ shifted down 1 unit.

$g(x) = \log \dfrac{x}{10} = \log x - \log 10 = \log x - 1 = f(x) - 1$

6. In general, how is the graph of $g(x) = \log \dfrac{x}{k}$ for $k > 0$ related to the graph of $f(x) = \log x$?

The graph of $g(x) = \log \dfrac{x}{k}$ is the graph of $f(x) = \log x$ shifted down $\log k$ units.

Section 12.2 The Elliptical Tablecloth

Recall that an ellipse can be drawn by fixing the ends of a string to the foci and then tracing out the ellipse. By considering the following figures, we can determine an expression for the length of the string and a relationship between a, b, and c.

1. Use the figure to write a formula for the length of the string.

$2c + 2(a - c)$

2. In the figure at right, notice that the string forms an isosceles triangle and the y-axis splits that triangle into two identical right triangles.

   a. Find the length of the hypotenuse of each of those right triangles.

     a

   b. What expression describes the length of the string?

     $2a$

   c. Use the Pythagorean theorem to write a formula relating a, b, and c.

     $a^2 = b^2 + c^2$

   d. Solve the formula for c.

     $c = \pm\sqrt{a^2 - b^2}$

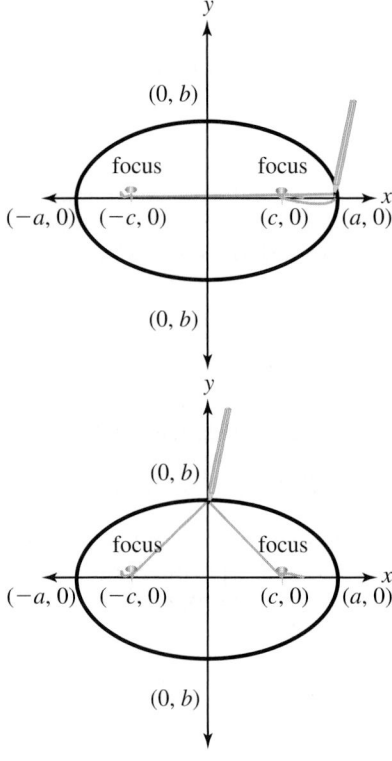

Suppose we are to make an elliptical tablecloth for an elliptical table that is 76 inches long and 58 inches wide. The tablecloth is to drape 6 inches over the edge of the table all the way around, and we need an additional inch for the hem. We have a large rectangular piece of cloth to make the tablecloth. To trace the ellipse on the cloth, we need to know the string length and the location of the foci.

3. Find the dimensions of the elliptical tablecloth, taking into account the amount it needs to drape and the hem.

$a = 45$ in., $b = 36$ in.

4. How long must the string be so that the ellipse can be traced?

90 in.

5. How far from the center are the foci located?

± 27 in.

Answers

Chapter 1
Exercise Set 1.1
Prep Exercises **1.** A constant is a symbol that does not vary in value, whereas a variable is a symbol that varies in value.
2. calculation **3.** equal sign **4.** $<, >, \leq, \geq, \neq$ **5.** Equations and Inequalities; Expressions; Constants and Variables **6.** braces,
$\{\ \}$ **7.** A rational number can be expressed as a ratio of integers, but an irrational number cannot. **8.** rational, irrational
9. positive, 0 **10.** right

Exercises **1.** $\{$ Sunday, Monday, Tuesday, Wednesday, Thursday, Friday, Saturday $\}$ **3.** $\{$ a, e, i, o, u $]$ **5.** $\{5, 10, 15, 20, \ldots\}$
7. $\{9, 11, 13, 15, \ldots\}$ **9.** $\{-6, -5, -4, -3\}$ **11.** rational **13.** rational **15.** irrational **17.** rational **19.** rational **21.** true

23. true **25.** False. All real numbers are either rational or irrational. **27.**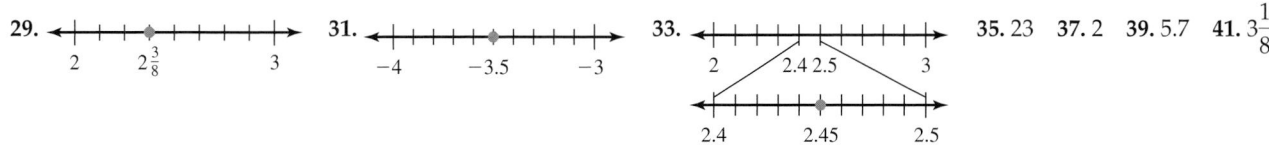

29. **31.** **33.** **35.** 23 **37.** 2 **39.** 5.7 **41.** $3\dfrac{1}{8}$

43. 0 **45.** > **47.** > **49.** > **51.** < **53.** > **55.** > **57.** < **59.** = **61.** < **63.** = **65.** > **67.** > **69.** < **71.** <
73. $-4.7, -2.56, 5.4, \left|-7\dfrac{1}{2}\right|, |8.3|$ **75.** $-0.6, -0.44, 0, |-0.02|, 0.4, \left|1\dfrac{2}{3}\right|, 3\dfrac{1}{4}$

Exercise Set 1.2

Prep Exercises **1.** $\dfrac{a}{b}$ **2.** Multiply or divide the numerator and denominator by the same nonzero number. **3.** No, a multiple of 12
is found by multiplying 12 by an integer, and we cannot multiply 12 by an integer to get 4. **4.** least common multiple
5. 1; itself **6.** No, 4 is not a prime number. The correct prime factorization is $2 \cdot 2 \cdot 2 \cdot 2 \cdot 7$. **7.** The numerator and the denominator
have no common factors other than 1. **8.** all prime factors common to the numerator and denominator

Exercises **1.** $\dfrac{3}{10}$ **3.** $\dfrac{1}{3}$ **5.** $\dfrac{3}{4}$ in. **7.** $\dfrac{3}{8}$ in. **9.** $\dfrac{15}{16}$ in. **11.** 12 **13.** -35 **15.** 16 **17.** 15 **19.** $\dfrac{12}{15}$ and $\dfrac{10}{15}$ **21.** $\dfrac{8}{18}$ and $\dfrac{21}{18}$ **23.** $-\dfrac{44}{72}$ and

$-\dfrac{51}{72}$ **25.** $-\dfrac{9}{144}$ and $-\dfrac{60}{144}$ **27.** $2 \cdot 2 \cdot 11$ **29.** $2 \cdot 2 \cdot 3 \cdot 3$ **31.** $2 \cdot 2 \cdot 2 \cdot 2 \cdot 2 \cdot 2$ **33.** $2 \cdot 5 \cdot 5 \cdot 5$ **35.** $\dfrac{4}{5}$ **37.** $\dfrac{7}{11}$ **39.** $-\dfrac{3}{4}$ **41.** $-\dfrac{8}{15}$
43. Incorrect. You may divide out only factors, not addends. **45.** Incorrect. The prime factorization of 240 should be
$2 \cdot 2 \cdot 2 \cdot 2 \cdot 3 \cdot 5$. **47.** $\dfrac{7}{8}$ **49.** $\dfrac{12}{23}$ **51.** $\dfrac{5}{21}$ **53.** $\dfrac{2}{3}$ **55.** $\dfrac{13}{40}$ **57.** $\dfrac{27}{40}$ **59. a.** 1991 **b.** $\dfrac{177}{500}$ **61.** $\dfrac{7}{9}$ **63.** $\dfrac{6}{73}$ **65.** $\dfrac{2}{3}$ **67.** $\dfrac{193}{435}$ **69.** $\dfrac{4}{15}$

Review Exercises: **1.** $\{$ Mercury, Venus, Earth, Mars $\}$ **2.** It is a rational number because it can be written as a ratio of the integers 8
and 10. **3.** **4.** It is an expression because it has no $=$ sign. **5.** 27 **6.** $=$

Exercise Set 1.3
Prep Exercises **1.** The commutative property of addition changes the order, while the associative property of addition changes
the grouping. **2.** Adding 0 to a number or an expression does not change the identity of the number or expression. **3.** When adding two numbers that have the same sign, add their absolute values and keep the same sign. **4.** When adding two numbers that
have different signs, subtract the smaller absolute value from the greater absolute value and keep the sign of the number with the
greater absolute value. **5.** To add (or subtract) fractions, (1) write each fraction as an equivalent fraction with the LCD, (2) add (or
subtract) the numerators and keep the LCD, and (3) simplify. **6.** The sum of 4 and -4 is 0. **7.** To write a subtraction statement
as an equivalent addition statement, change the operation symbol from a minus sign to a plus sign and change the subtrahend to
its additive inverse. **8.** Subtraction is not commutative, so $6 - 5$ and $5 - 6$ are not equivalent. However, $-6 - 5$ and $-5 - 6$ are
equivalent because $-6 - 5 = -6 + (-5)$ and by the commutative property of addition, $-6 + (-5) = -5 + (-6)$, which is
equivalent to $-5 - 6$.

Exercises **1.** Commutative property of addition **3.** Additive identity **5.** Additive inverse **7.** Associative property of

addition **9.** Commutative property of addition **11.** Additive inverse **13.** 21 **15.** -18 **17.** 9 **19.** -9 **21.** 14 **23.** -16 **25.** $\dfrac{3}{4}$

27. $-\dfrac{2}{3}$ **29.** $-\dfrac{2}{3}$ **31.** $\dfrac{11}{12}$ **33.** $-\dfrac{3}{4}$ **35.** $-\dfrac{9}{14}$ **37.** 0.26 **39.** 6.52 **41.** -4.38 **43.** -28 **45.** 3.18 **47.** $\dfrac{29}{24}$ **49.** -5 **51.** 12

53. 0 **55.** $-\dfrac{5}{6}$ **57.** 0.29 **59.** x **61.** $-\dfrac{m}{n}$ **63.** 2 **65.** -4 **67.** -4 **69.** -12 **71.** -9 **73.** -13 **75.** 7 **77.** -6 **79.** 0 **81.** $\dfrac{13}{10}$

83. $-\dfrac{18}{35}$ **85.** 0.36 **87.** -5.75 **89.** -1.6 **91.** -10 **93.** -0.9 **95.** \$2,036,000,000 **97.** -268.2 N; the negative indicates that the beam is moving downward. **99.** $25.95 **101.** 12,653.12 **103.** $-196 - (-208)$; 12 **105. a.** $18.6 - 18.8$ **b.** -0.2 **c.** The negative difference indicates that the mean composite score in 1989 was less than the score in 1986. **107.** $17,229 **109.** Doctorate; $33,571

Review Exercises: **1.** {Washington, Adams, Jefferson, Madison} **2.** 7 **3.** $2 \cdot 2 \cdot 5 \cdot 5$ **4.** $\frac{6}{7}$ **5.** 5 **6.** <

Exercise Set 1.4

Prep Exercises **1.** $6(-7) = -42$ shows multiplication, whereas $6 - 7 = -1$ shows subtraction. **2.** factors **3.** positive **4.** negative **5.** positive **6.** positive **7.** Their product is 1. **8.** There is no quotient that we can multiply by 0 to get 5. **9.** There is no unique quotient. **10.** Write the division problem as an equivalent multiplication problem that has the dividend multiplied by the multiplicative inverse of the divisor.

Exercises **1.** Distributive property **3.** Multiplicative identity **5.** Multiplicative property of 0 **7.** Commutative property of multiplication **9.** Associative property of multiplication **11.** Commutative property of multiplication **13.** -18 **15.** -50
17. -54 **19.** 20 **21.** 42 **23.** $-\frac{1}{4}$ **25.** $\frac{1}{2}$ **27.** $-\frac{4}{9}$ **29.** -38 **31.** 6.72 **33.** -6.536 **35.** 54 **37.** -48 **39.** -168 **41.** -144 **43.** 720
45. -48 **47.** $\frac{3}{2}$ **49.** $-\frac{2}{5}$ **51.** $-\frac{1}{5}$ **53.** There is no multiplicative inverse of 0. **55.** -3 **57.** 14 **59.** -2 **61.** 9 **63.** 0
65. undefined **67.** indeterminate **69.** 9 **71.** $-\frac{2}{3}$ **73.** $\frac{3}{4}$ **75.** $-\frac{14}{15}$ **77.** 2.1 **79.** 91.8 **81.** $-46.\overline{6}$ **83.** $\frac{1}{2}$ **85.** $-$8680 **87.** ≈ 612.8
89. -402.5 lb.; the force is downward. **91.** 169,202.2245 kg **93.** -51.2 V **95.** 3600 W

Review Exercises: **1.** irrational **2.**

$-8 \qquad -7.2 \quad -7$

3. $-\frac{2}{3}$ **4.** 6.8 **5.** -22 **6.** $-\frac{11}{24}$

Exercise Set 1.5

Prep Exercises **1.** Two cubed **2.** Three squared **3.** 4; 5 **4.** positive **5.** Squaring a number means to multiply the number by itself; finding its square root means to find a number whose square is the given number. **6.** positive; negative **7.** principle (nonnegative) **8.** Divide 40 by 2. **9.** We must add first, then find the square root of the sum. **10.** sum; n

Exercises **1.** Base: 7; exponent: 2; "seven squared" **3.** Base: -5; exponent: 3; "negative five cubed" **5.** Base: 2; exponent: 7; "additive inverse of two to the seventh power" **7.** 81 **9.** 64 **11.** -64 **13.** -125 **15.** -125 **17.** 8 **19.** -1 **21.** $\frac{1}{25}$ **23.** $-\frac{27}{64}$
25. 0.008 **27.** 16.81 **29.** ± 11 **31.** No real-number square roots exist. **33.** ± 14 **35.** ± 16 **37.** 4 **39.** 12 **41.** 0.7 **43.** Not a real number **45.** $\frac{8}{9}$ **47.** 5 **49.** 19 **51.** 0 **53.** 10 **55.** -11 **57.** -4 **59.** -40 **61.** -20 **63.** 17 **65.** -7 **67.** -28 **69.** 26.8 **71.** -97.4
73. -53 **75.** $-10\frac{4}{5}$ **77.** 78 **79.** 19 **81.** -29 **83.** $-\frac{253}{300}$ **85.** 22 **87.** 12 **89.** 0 **91.** 4 **93.** 7 **95.** undefined **97.** Associative property of multiplication; The multiplication was not performed from left to right. **99.** Distributive property; The parentheses were not simplified first. **101.** Mistake: Multiplied before dividing. Correct: 1 **103.** Mistake: Found the square roots of the subtrahend and minuend in square root of a difference. Correct: 24 **105.** 77.2 **107.** 96 **109.** $588.41 **111.** $74.34

Review Exercises: **1.** {0, 1, 2, 3, 4, 5, 6, 7, 8, 9, 10} **2.**

$-1 \ -\frac{4}{5} \qquad 0$

3. $-\frac{1}{9}$ **4.** It is an expression because it has no = sign. **5.** 24 **6.** 3

Exercise Set 1.6

Prep Exercises **1.** Sum, plus, added **2.** Difference, minus, less **3.** Product, times, twice **4.** Quotient, divided, ratio **5.** Less than **6.** Divided into **7.** Some number plus seven, seven added to some number, the sum of some number and seven, seven more than some number, some number increased by seven **8.** Some number minus five, five subtracted from some number, the difference of some number and five, five less than some number, some number decreased by five, some number less five **9.** Seven divided by some number, the quotient of seven and some number, the ratio of seven and some number **10.** Three times some number, the product of three and some number, some number multiplied by three

Exercises **1.** $4x$ **3.** $4x + 16$ **5.** $7x - 8$ **7.** $-4 \div y^3$ or $\frac{-4}{y^3}$ **9.** $8p - 4$ **11.** $\frac{m}{14}$ **13.** $x^4 + 5$ **15.** $7w - \frac{1}{5}$ **17.** $-3(n - 2)$
19. $(4 + n)^5$ **21.** $mn - 5$ **23.** $4 \div n - 2$ or $\frac{4}{n} - 2$ **25.** $-27 - (a + b)$ **27.** $0.6 - 4(y - 2)$ **29.** $(p - q) - (m + n)$
31. $\sqrt{y} - mn$ **33.** $n - 3(n - 6)$ **35.** Mistake: Order is incorrect. Correct: $3t - 17$ **37.** Mistake: Multiplied x by 9 instead of the sum of x and y. Correct: $9(x + y)$ **39.** $w + 5$ **41.** $3w$ **43.** $\frac{1}{2}d$ **45.** $42 - n$ **47.** $t + \frac{1}{4}$ **49.** $2w + 2l$ **51.** $\frac{1}{3}\pi r^2 h$ **53.** mc^2
55. $\sqrt{(x_2 - x_1)^2 + (y_2 - y_1)^2}$ **57.** Mistake: Could be translated as $3(x + 4)$. Correct: Four more than three times a number

59. Mistake: Could be translated as $5x - 1$. Correct: Five times the difference of x and one   **61.** Mistake: Could be translated as $(n + 5)(n - 6)$. Correct: n plus the product of five and the difference of n and six   **63.** One-half of the product of the base and height   **65.** The product of the length, width, and height   **67.** The product of the length and width added to the product of the length and height added to the product of the width and height, all doubled   **69.** The ratio of the difference of y_2 and y_1 to the difference of x_2 and x_1

Review Exercises: **1.** $2 \cdot 5 + 2 \cdot 2 = 14$ **2.** 2   **3.** $\dfrac{5}{12}$ **4.** -28 **5.** 12   **6.** -9

Exercise Set 1.7

Prep Exercises   **1.** To evaluate an expression, (1) replace the variables with their corresponding given values and (2) calculate the numerical expression using the order of operations.   **2.** An expression is undefined when the denominator is equal to 0.
3. Evaluating an expression involves replacing variables with numbers, then calculating. Rewriting an expression involves using mathematical properties to write the expression in an altered but equivalent form.   **4.** The coefficient is the numerical factor in a term.   **5.** Like terms are constant terms or variable terms that have the same variable(s) raised to the same exponents.
6. To combine like terms, add or subtract the coefficients and keep the variables and their exponents the same.

Exercises   **1.** 9   **3.** 3.2   **5.** 23   **7.** $\dfrac{11}{2}$ **9.** 4.6   **11.** -11 **13.** 3   **15.** -12 **17.** -21 **19.** -48 **21.** $\dfrac{3}{13}$ **23. a.** 25 **b.** -12 **25. a.** 2
b. $\dfrac{2}{3}$ **27.** -5 **29.** $-1, 3$ **31.** 0   **33.** $\dfrac{2}{3}$ **35.** $6a + 12$ **37.** $-32 + 24y$ **39.** $\dfrac{7}{16}c - 14$ **41.** $0.6n - 1.6$ **43.** -6 **45.** 1   **47.** -1
49. $-\dfrac{2}{3}$ **51.** $\dfrac{1}{5}$ **53.** $8y$ **55.** $-5a$ **57.** $11x$ **59.** $-11r$ **61.** $-1.9n$ **63.** $-\dfrac{1}{3}b^2$ **65.** $-26c$ **67.** $5x - 1$ **69.** $-10x + 4y - 1$
71. $-9m + 5n + y - 11$ **73.** $-1.3x - 0.4$ **75.** $\dfrac{5}{6}c + \dfrac{5}{2}d + \dfrac{1}{7}$ **77.** $\dfrac{25}{4}a + \dfrac{18}{35}b^2 + 4$ **79. a.** $14 + (6n - 8n)$ **b.** $14 - 2n$ **c.** 20

Review Exercises: **1.** $-3, -2.5, 4.2, 4\dfrac{5}{8}, |-6|$ **2.** $\dfrac{6}{35}$ **3.** -22 **4.** 50   **5.** 1   **6.** 3

Chapter 1 Summary and Review Exercises

1.1 Definitions/Rules/Procedures   can vary; does not vary; describes a calculation; equal sign; inequality symbol
Exercises   **1.** Equations and Inequalities; Expressions; Constants and Variables   Definitions/Rules/Procedures   collection; braces; { }
Exercises   **2.** { March, May }   **3.** $\{ 2, 4, 6, \ldots \}$ Definitions/Rules/Procedures $\dfrac{a}{b}$; integers; rational; rational; irrational
Exercises   **4.** Real Numbers; Rational Numbers; Integers; Whole Numbers; Natural Numbers; Irrational Numbers
5. 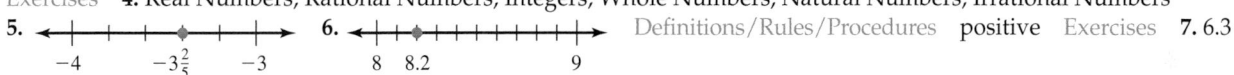   **6.**   Definitions/Rules/Procedures   positive   Exercises   **7.** 6.3

$-4 \quad -3\frac{2}{5} \quad -3 \qquad 8 \quad 8.2 \qquad\qquad 9$

8. $2\dfrac{1}{6}$ Definitions/Rules/Procedures   right   Exercises   **9.** $=$ **10.** $<$

1.2 Definitions/Rules/Procedures $\dfrac{a}{b}$ where $b \neq 0$; multiplying; dividing; product; an integer; smallest number; multiple; each number in a given set of numbers; least common multiple of denominators of a given set of fractions   Exercises   **11.** 12   **12.** 5
13. $\dfrac{20}{24}$ and $\dfrac{9}{24}$ **14.** $-\dfrac{14}{30}$ and $\dfrac{27}{30}$ Definitions/Rules/Procedures $a \cdot b = c$; 1; the number itself; prime factors
Exercises   **15.** $2 \cdot 2 \cdot 5 \cdot 5$ **16.** $2 \cdot 2 \cdot 3 \cdot 7$ Definitions/Rules/Procedures   lowest terms; their prime factorizations; common;
Multiply   Exercises   **17.** $\dfrac{19}{25}$ **18.** $\dfrac{2}{3}$

1.3 Definitions/Rules/Procedures   0; add; the same sign; subtract; greater absolute value; add; keep the same denominator; LCD; numerators; minus sign; plus sign; the subtrahend to its additive inverse   Exercises   **19.** Associative property of addition
20. Commutative property of addition   **21.** -27 **22.** $\dfrac{5}{12}$ **23.** -3.8 **24.** 15.1   **25.** $-\$503.59$ **26. a.** -0.44 **b.** loss of \$0.48

1.4 Definitions/Rules/Procedures   positive; negative; positive; negative; $\dfrac{ac}{bd}$; whole numbers; the total number of decimal places in the factors   Exercises   **27.** Commutative property of multiplication   **28.** Distributive property   **29.** Commutative property of multiplication   **29.** Associative property of multiplication   **30.** Associative property of multiplication   **31.** 91   **32.** -48
33. $-\dfrac{5}{24}$ **34.** 2.6   Definitions/Rules/Procedures   positive; negative; 0; undefined; indeterminate; 1; $\dfrac{a}{b} \cdot \dfrac{d}{c}$; whole number; directly above its position in the dividend; right; integer; the same; divisor; dividend; whole numbers; directly above its new position in the dividend
Exercises   **35.** -5 **36.** 10   **37.** $\dfrac{14}{9}$ or $1\dfrac{5}{9}$ **38.** -4.2 **39. a.** \$11872 **b.** \$88 loss, or $-\$88$ **40.** ≈ 6.2

1.5 Definitions/Rules/Procedures   how many times to use the base as a factor; repeatedly multiplied; base; exponent; positive; negative   Exercises   **41.** 9   **42.** −36   Definitions/Rules/Procedures   positive, negative; no; principle (nonnegative)   Exercises   **43.** ±7

44. No real-number square roots exist   **45.** 14   **46.** $\frac{5}{9}$ Definitions/Rules/Procedures   grouping; Exponents/Roots; Multiplication/

Division; Addition/Subtraction   Exercises   **47.** 160   **48.** 1   **49.** $-\frac{28}{3}$ **50.** −1   **51.** 27   **52.** 7   **53.** −103   **54.** 3   **55.** 24

Definitions/Rules/Procedures   the sum of the numbers by n.   Exercises   **56.** $54,158$\frac{1}{3}$, or $54,158.33

1.6 Definitions/Rules/Procedures   unknowns; constants; key words   Exercises   **57.** $14 - 2n$ **58.** $\frac{y}{7}$ **59.** $y - 2(y + 4)$
60. $7(m - n)$

1.7 Definitions/Rules/Procedures   variables; order of operations agreement   Exercises   **61.** 61   **62.** 52   **63.** −8   **64.** 7   **65.** 5
66. −4   Definitions/Rules/Procedures $b = 0$ Exercises   **67.** −6   **68.** −3, 4   Definitions/Rules/Procedures $ab + ac$

Exercises   **69.** $5x + 30$ **70.** $-15n + 24$ **71.** $2y + \frac{3}{4}$ **72.** $2.7m - 1.26$ Definitions/Rules/Procedures   addends in an expression

that is a sum; numerical factor in a product; are the same variables raised to the same exponents; coefficients; variables and their exponents   Exercises   **73.** $-3x - 3y - 15$ **74.** $4y^2 + 7y$ **75.** $2xy$ **76.** $-2x^3 - 6x^2$ **77.** $-8m$ **78.** $-14x - 6y + 6$

Chapter 1 Practice Test

1.   [1.1]   **2.** 3.67 [1.1]   **3.** $2 \cdot 2 \cdot 5 \cdot 5$ [1.2]   **4.** $\frac{1}{3}$ [1.2]   **5.** 12 [1.2]   **6.** Distributive property [1.4]

7. Commutative property of addition [1.3]   **8.** 4 [1.3]   **9.** $1\frac{17}{24}$ or $\frac{41}{24}$ [1.3]   **10.** 0.6 [1.4]   **11.** $-\frac{5}{4}$ [1.4]   **12.** −64 [1.5]   **13.** −81 [1.5]
14. −8 [1.5]   **15.** 2 [1.5]   **16.** 5 [1.5]   **17.** 10 [1.5]   **18.** $2(m + n)$ [1.6]   **19.** $3w - 5$ [1.6]   **20.** $250 [1.3]   **21.** 5.583 [1.5]
22. −11 [1.7]   **23.** 10 [1.7]   **24.** $-20y - 45$ [1.7]   **25.** $5.6x + 1.3$ [1.7]

Chapter 2
Exercise Set 2.1
Prep Exercises   **1.** An equation has an equal sign.   **2.** A solution to an equation is a number that makes the equation true when it replaces the variable in the equation.   **3.** (1) Replace the variable in the equation with the value, and (2) if the resulting equation is true, the value is a solution.

Exercises   **1.** no   **3.** yes   **5.** no   **7.** yes   **9.** no   **11.** yes   **13.** yes   **15.** yes   **17. a.** 66 ft. **b.** 6 **c.** $53.94   **19.** ≈ 12.57 km

21. 7000 ft.^{2} **23.** $975   **25. a.** 78 ft.^{2} **b.** 8 bales **c.** $36   **27. a.** 260 in.^{2} **b.** $3900   **29. a.** 3208 ft.^{2} **b.** $2138\frac{2}{3}$ft.^{2} **c.** 5

d. $530   **31.** ≈ 6.9 ft.^{2} **33.** 82.25 ft.^{2} **35.** 47.5 m^3 **37.** ≈ 1005.3 cm^3 **39.** 83,068,741.$\overline{6}$ ft.^{3}or $83,068,741\frac{2}{3}$ ft.^{3} **41.** 64.5 mph

43. ≈ 54.3 mph   **45.** 1747.5 mi.   **47.** −52 V   **49.** 93.2°F   **51.** −297.4°F   **53.** −60°C

Review Exercises: **1.** −51   **2.** −34   **3.** $\frac{1}{3}x + 2$ **4.** $-4w + 6$ **5.** $8x - 5$ **6.** $5.4y + 5.5$

Exercise Set 2.2
Prep Exercises   **1.** 1   **2.** The addition principle of equality says that we can add (or subtract) the same amount on both sides of an equation without affecting its solution(s).   **3.** Add $-4x$ to (or subtract $4x$ from) both sides of the equation.   **4.** Add 9 to both sides of the equation.   **5.** Use the addition principle of equality to get the variable terms together on the same side of the equal sign.
6. Distribute to eliminate the parentheses.

Exercises   **1.** yes   **3.** no   **5.** no   **7.** yes   **9.** yes   **11.** no   **13.** yes   **15.** yes   **17.** 9   **19.** 5   **21.** −7   **23.** 5   **25.** −10   **27.** −25
29. $-\frac{3}{40}$ **31.** $-\frac{11}{24}$ **33.** −8.2   **35.** 4.6   **37.** −9   **39.** 7   **41.** 13   **43.** 6   **45.** −1   **47.** −2   **49.** 13.2   **51.** 36   **53.** −10   **55.** 16
57. 4   **59.** 9.7   **61.** All real numbers   **63.** No solution   **65.** No solution   **67.** All real numbers   **69.** $1947 + x = 2373$; $426
71. $16 + x = 42$; 26 mi.   **73.** $1741.62 - 1286.65 = 16.82 + 150.88 + 192.71 + x$; $94.56   **75.** $x + 12.4 + 16.3 + 27.2 = 67.2$;
11.3 cm   **77.** $x + 19 + 10 = 54$; 25 ft.   **79.** $x + 900 + 3 \cdot 1500 + 2 \cdot 1245 = 8500$: $610; yes, because $610 is less than all of his prior

weeks.   **81.** $x + \frac{1}{3} + \frac{2}{5} = 1$; $\frac{4}{15}$

Review Exercises: **1.** $-\dfrac{5}{4}$ **2.** -33.2 **3.** -9 **4.** $-6x + 1$ **5.** $8x^2 - 5x - 9$ **6.** $4x - 9$

Exercise Set 2.3

Prep Exercises **1.** We can multiply (or divide) both sides of an equation by the same amount without affecting its solution(s). **2.** Divide both sides of the equation by the coefficient. **3.** Multiply both sides of the equation by the multiplicative inverse of the coefficient. **4.** Negative because the goal is to get a positive 1 coefficient for the variable. **5.** Multiply both sides of the equation by a common multiple of the denominators. Using the LCD results in an equation with the smallest integers possible. **6.** Multiply both sides by the appropriate power of 10 to eliminate decimals.

Exercises **1.** 4 **3.** 3 **5.** 16 **7.** -20 **9.** $\dfrac{7}{6}$ **11.** $-\dfrac{4}{3}$ **13.** 7 **15.** 5 **17.** 5 **19.** -3 **21.** $-\dfrac{3}{2}$ **23.** 16 **25.** 0 **27.** -9 **29.** -5 **31.** 2

33. 16 **35.** 2 **37.** $\dfrac{20}{3}$ **39.** -2 **41.** 1 **43.** 2 **45.** No solution **47.** 2 **49.** $-\dfrac{15}{2}$ **51.** All real numbers **53.** -9 **55.** No solution

57. $\dfrac{15}{8}$ **59.** $\dfrac{48}{13}$ **61.** 1 **63.** $-\dfrac{2}{3}$ **65.** -8 **67.** 9 **69.** All real numbers **71.** -9 **73.** 12 **75.** 6.5 **77.** Mistake: The minus sign was dropped in front of the 11. Correct: 2 **79.** Mistake: In the second line, the minus sign was not distributed to the 2. Correct: $\dfrac{13}{3}$ **81.** 14 ft. **83.** 19 cm **85.** $l = 54$ ft., $w = 52$ ft. **87.** 110 ft. **89.** 2 in. **91.** ≈ 6374.8 km **93.** ≈ 4.8 m **95.** $w = 11$ ft., $x = 2$ ft., $w + 2 = 13$ ft., $y = 23$ ft. **97.** -9.5 A **99.** 4 m/sec.^{2} **101.** 945.5 kg **103.** 413.6 m/sec. **105.** $51.10 **107.** 15 ft./sec.

Review Exercises: **1.** 70 **2.** $3.8x^2 - 7.63x + 1.5$ **3.** $-\dfrac{9}{2}x - 6$ **4.** 20 **5.** -25 **6.** -10

Exercise Set 2.4

Prep Exercises **1.** Add 2 to both sides. Because 2 is subtracted from w, adding 2 eliminates the 2, leaving w isolated. **2.** Divide both sides by n. Because p is multiplied by n, dividing by n divides out the n, thereby isolating p. **3.** Subtract $2x$ from both sides. Subtracting $2x$ isolates the $3y$ term, which contains the variable we want to isolate. **4.** Substitute the same values for the variables in the original formula and in your answer; then evaluate to see if you got the same ending value.

Exercises **1.** $t = 4u + v$ **3.** $y = -\dfrac{x}{5}$ **5.** $x = \dfrac{b + 3}{2}$ **7.** $m = \dfrac{y - b}{x}$ **9.** $y = \dfrac{8 - 3x}{4}$ **11.** $Y = \dfrac{mn}{4} - f$ **13.** $c = \dfrac{m - np - 12d}{6}$

15. $y = \dfrac{15 - 5x}{3}$ **17.** $n = c\left(\dfrac{3}{4} + 5a\right)$ **19.** $p = \dfrac{A - P}{tr}$ **21.** $a = P - b - c$ **23.** $l = \dfrac{A}{w}$ **25.** $d = \dfrac{C}{\pi}$ **27.** $r^2 = \dfrac{3V}{\pi h}$ **29.** $b = \dfrac{2A}{h}$

31. $w = \dfrac{P - 2l}{2}$ **33.** $l = \dfrac{2S - na}{n}$ **35.** $w = \dfrac{\pi r^2 h - V}{lh}$ **37.** $C = R - P$ **39.** $r = \dfrac{I}{Pt}$ **41.** $C = nP$ **43.** $r = \dfrac{A - P}{Pt}$ **45.** $t = \dfrac{d}{r}$

47. $d = \dfrac{W}{F}$ **49.** $t = \dfrac{W}{P}$ **51.** $t = \dfrac{v - v_0}{-32}$ or $t = \dfrac{v_0 - v}{32}$ **53.** $C = \dfrac{5}{9}(F - 32)$ **55.** $m = \dfrac{FR^2}{GM}$ **57.** Mistake: Subtracted the coefficient 7 instead of dividing. Correct: $t = \dfrac{54 - 3n}{7}$ **59.** Mistake: Multiplied 5 by -2. Correct: $m = \dfrac{5nk + 2}{3}$

Review Exercises: **1.** $7n + 4$ **2.** $3(x + 2) - 9$ **3.** 0.5 **4.** $-\dfrac{5}{6}$ **5.** $-\dfrac{3}{4}$ **6.** -80

Exercise Set 2.5

Prep Exercises **1.** Sum, plus, added **2.** Difference, minus, less **3.** Of, times, twice **4.** Quotient, divided, ratio **5.** The order of the subtraction is different. **6.** The word *sum* indicates the answer to addition. The parentheses ensure that the sum is found before multiplication is done.

Exercises **1.** $y + 3 = -8$; -11 **3.** $6 - x = -3$; 9 **5.** $11m = -99$; -9 **7.** $m \div 4 = 1.6$; 6.4 **9.** $\dfrac{4}{5}x = \dfrac{5}{8}$; $\dfrac{25}{32}$ **11.** $7 + 3w = 34$; 9

13. $5a - 9 = 76$; 17 **15.** $8(8 + t) = 160$; 12 **17.** $-3(x - 2) = 12$; -2 **19.** $\dfrac{1}{3}(g + 2) = 1$; 1 **21.** $3m - 11 = 5 + m$; 8

23. $10 = \dfrac{r}{4} - 1$; 44 **25.** $2a - 3(a + 4) = -3$; -9 **27.** $(2d - 8) + (d - 12) = 13$; 11 **29.** $5\left(x + \dfrac{2}{3}\right) = 3x - \dfrac{2}{3}$; -2

31. $2x - 4 = 16 + 3x$; -20 **33.** $\dfrac{2}{5}x = \dfrac{1}{2}x - 2$; 20 **35.** $\dfrac{n - 3}{3} = \dfrac{n - 1}{4}$; 9 **37.** $-4(2 - x) = 2(4 - 3x) - 6$; 1 **39.** Three added to four times a number is seven. **41.** Six times the sum of a number and four is equal to the product of negative ten and the number. **43.** One-half of the difference of a number and three will result in two-thirds of the difference of the number and eight. **45.** Five-hundredths of a number added to six-hundredths of the difference of the number and eleven is twenty-two. **47.** The sum of two-thirds, three-fourths, and one-half of the same number will equal ten. **49.** Mistake: Subtraction translated in reverse order. Correct: $n - 10 = 40$ **51.** Mistake: Multiplied 5 times the unknown number instead of the difference, which requires parentheses. Correct: $5(x - 6) = -2$ **53.** Mistake: Subtracted the unknown number instead of the sum, which requires parentheses. Correct: $2t - (t + 3) = -6$ **55.** Translation: $P = 2(l + w)$ **a.** 84 ft.   **b.** 61 cm   **c.** $37\frac{1}{4}$ in.   **57.** Translation: $P = b + 2s$ **a.** 27 in.   **b.** 2.7 m   **c.** $34\frac{1}{4}$ in.   **59.** Translation: $I = Prt$ **a.** \$400   **b.** \$7.50   **c.** \$30   **61.** Translation: $t = \dfrac{e - d}{153.8}$ **a.** ≈ 49 sec.   **b.** ≈ 52 sec.   **c.** ≈ 62 sec.

Review Exercises: **1.** True; 19 is equal to 19.   **2.** $=$ **3.** $>$ **4.** -5 **5.** $-\dfrac{10}{9}$ **6.** -25

Exercise Set 2.6

Prep Exercises   **1.** The quantity preceding *to* (longest side) is written in the numerator, and the quantity following the word *to* (shortest side) is written in the denominator.   **2.** A proportion is an equation in the form $\dfrac{a}{b} = \dfrac{c}{d}$, where $b \neq 0$ and $d \neq 0$.

3. $\dfrac{5}{8} \cdot \dfrac{x}{10}$ is a product, whereas $\dfrac{5}{8} = \dfrac{x}{10}$ is a proportion.   **4.** $ad = bc$ **5.** To solve a proportion using cross products, (a) calculate the cross products, (b) set the cross products equal to each other, and (c) use the multiplication principle of equality to isolate the variable.   **6.** To solve proportion problems, (a) set up a proportion in which the numerators and denominators of the ratios correspond in a logical manner and (b) solve using cross products.

Exercises   **1. a.** $\dfrac{4}{15}$ **b.** $\dfrac{6}{11}$ **c.** $\dfrac{2}{1}$ **3.** $\dfrac{1}{2}$ **5.** $\dfrac{2}{3}$ **7.** $\dfrac{63.5}{1}$; he drives an average of 63.5 miles per hour.   **9.** $\dfrac{0.125}{1}$; each banana cost 12.5 cents.   **11.** $\approx \dfrac{15.64}{1}$; the price of the stock is \$15.64 for every \$1 earned in 2012.   **13.** The 24-oz. container is better because it costs less per ounce. (The unit ratio of price to quantity is less.)   **15. a.** $\approx \dfrac{0.18}{1}$ **b.** $\approx \dfrac{0.46}{1}$ **c.** The ratio of aggravated assault hate crimes to total hate crimes was greater in 2010 than in 2008.   **17. a.** $\approx \dfrac{1.26}{1}$; people in the 25–34 age group with an associate degree earn \$1.26 for every \$1.00 earned by high school graduates.   **b.** $\approx \dfrac{1.36}{1}$ **c.** In general, people 25–34 years of age who have bachelor's degree make more money than do high school graduates or people with associate degrees.   **19.** yes   **21.** no   **23.** no   **25.** yes   **27.** no   **29.** 10.5   **31.** -14 **33.** 10.5   **35.** -8 **37.** 115   **39.** $3\frac{3}{5}$ **41.** 4   **43.** 7   **45.** $183\frac{1}{3}$ mi.   **47.** \$1350.65   **49.** \$9450   **51.** ≈ 422.25 pounds   **53.** 4.875 in.   **55.** $2\frac{1}{2}$ tsp.   **57.** 432 highway miles   **59.** ≈ 64.3 sec.   **61.** 225 lb.   **63.** \$85.91   **65.** ≈ 66 deer   **67.** 617,647   **69.** 11.25 cm   **71.** $a = 4\frac{8}{19}$ ft.; $b = 5\frac{1}{4}$ ft.; $c = 5\frac{17}{19}$ ft.   **73.** 300 m   **75.** 26.4 m

Review Exercises: **1.** Yes, because it can be written as a ratio of integers, $\dfrac{58}{100}$ **2.** $\dfrac{3}{25}$ **3.** 3.06   **4.** 0.452   **5.** 320   **6.** 0.4 or $\dfrac{2}{5}$

Exercise Set 2.7

Prep Exercises   **1.** To write a percent as a decimal or fraction, (a) write the percent as a ratio with 100 in the denominator and (b) simplify to the desired form.   **2.** To write a decimal or fraction as a percent, (a) multiply by 100% and (b) simplify.   **3.** A percent of a whole is a part of the whole.   **4.** multiplication   **5.** Because the percent is a ratio with 100 as the denominator, we can write the percent as one of the ratios in a proportion. The equivalent ratio is also some part to a whole amount.   **6.** In a word-for-word translation, the division yields a decimal number that must be written as a percent. When using the proportion method, the decimal number is multiplied by 100, which gives the percent.   **7.** The percent, whole, and part   **8.** First find the amount of the increase by subtracting the initial amount from the final amount. Then write a proportion in the form $\dfrac{P}{100} = \dfrac{\text{amount of increase}}{\text{initial amount}}$ and solve for P.

Exercises   **1.** $0.2, \dfrac{1}{5}$ **3.** $0.15, \dfrac{3}{20}$ **5.** $0.148, \dfrac{37}{250}$ **7.** $0.0375, \dfrac{3}{80}$ **9.** $0.455, \dfrac{91}{200}$ **11.** $0.\overline{3}, \dfrac{1}{3}$ **13.** 60%   **15.** 37.5%   **17.** 83.3%   **19.** 66.7%   **21.** 96%   **23.** 80%   **25.** 9%   **27.** 120%   **29.** 2.8%   **31.** 405.1%   **33.** 28   **35.** 3.1   **37.** 5.92   **39.** 103.2   **41.** 50   **43.** 80   **45.** 62.4   **47.** 20%   **49.** 105%   **51.** $66.\overline{6}$%   **53.** 80%   **55.** 43   **57.** \$4800; \$400   **59.** \$898.80   **61.** ≈ 5783.5 tg   **63.** \$42.50   **65.** \$300   **67.** ≈ 45.2%   **69.** ≈ 71.6%   **71.** ≈ 8.0%   **73.** United States ≈ 24.4%, China ≈ 2.7%, Russia ≈ 53.6%; Russia   **75.** ≈ 23.7%   **77.** 20%   **79.** \$6.22; \$130.67   **81.** \$94.90; \$854.10   **83.** \$2500   **85.** \$1950; 75%   **87.** ≈ 11.6%   **89.** $34.1\overline{6}$%   **91.** ≈ 41.7%   **93.** 2800%   **95.** 400,000   **97.** ≈ 2246.77

Review Exercises: **1.** -1 **2.** $-3x - 21$ **3.** $-xy + 4x + 3y$ **4.** $3(n + 9) = n - 11$; -19 **5.** 2.1   **6.** 10.5 cm

Exercise Set 2.8
Prep Exercises **1.** Any number that can replace the variable(s) in the inequality and make it true. **2.** $x < 8$ means any number less than 8; $x \leq 8$ means any number less than 8 *or* equal to 8. **3.** Any number greater than -5 and -5 itself. **4.** Multiplying or dividing by a negative number.

Exercises **1. a.** $\{x|x \geq -3\}$ **b.** $[-3, \infty)$ **c.** **3. a.** $\{h|h < 6\}$ **b.** $(-\infty, 6)$

$-5\,-4\,-3\,-2\,-1\ \ 0\ \ 1$

c. **5. a.** $\left\{n|n < -\dfrac{2}{3}\right\}$ **b.** $\left(-\infty, -\dfrac{2}{3}\right)$ **c.** **7. a.** $\{t|t \geq 2.4\}$ **b.** $[2.4, \infty)$

$0\ 1\ 2\ 3\ 4\ 5\ 6\ 7\ 8$

$-\dfrac{4}{3}\quad -1\quad -\dfrac{2}{3}\quad -\dfrac{1}{3}$

c. **9. a.** $\{x|-3 < x < 6\}$ **b.** $(-3, 6)$ **c.** **11. a.** $\{n|0 \leq n \leq 5\}$

$2\ \ 2.1\ 2.2\ 2.3\ 2.4\ 2.5\ 2.6$

$-3\,-2\,-1\,0\ 1\ 2\ 3\ 4\ 5\ 6\ 7$

b. $[0, 5]$ **c.** **13.** $\{x|x \geq -4\}, [-4, \infty)$ **15.** $\{x|x < 5\}, (-\infty, 5)$ **17.** $\{x|-4 \leq x < 3\}, [-4, 3)$

$-1\ 0\ 1\ 2\ 3\ 4\ 5\ 6$

19. a. $n > 5$ **b.** $\{n|n > 5\}$ **c.** $(5, \infty)$ **d.** **21. a.** $z \leq -6$ **b.** $\{z|z \leq -6\}$ **c.** $(-\infty, -6]$

$2\ \ 3\ \ 4\ \ 5\ \ 6\ \ 7\ \ 8$

d. **23. a.** $y \geq 2$ **b.** $\{y|y \geq 2\}$ **c.** $[2, \infty)$ **d.** **25. a.** $x \leq 4$

$-8\ -7\ -6\ -5\ -4\ -3$

$0\quad 1\quad 2\quad 3\quad 4$

b. $\{x|x \leq 4\}$ **c.** $(-\infty, 4]$ **d.** **27. a.** $x \geq 6$ **b.** $\{x|x \geq 6\}$ **c.** $[6, \infty)$

$-1\ 0\ 1\ 2\ 3\ 4\ 5$

d. **29. a.** $m > 8$ **b.** $\{m|m > 8\}$ **c.** $(8, \infty)$ **d.**

$3\ \ 4\ \ 5\ \ 6\ \ 7\ \ 8\ \ 9$

$4\ \ 5\ \ 6\ \ 7\ \ 8\ \ 9\ 10\ 11$

31. a. $y > 3$ **b.** $\{y|y > 3\}$ **c.** $(3, \infty)$ **d.** **33. a.** $x > -4$ **b.** $\{x|x > -4\}$ **c.** $(-4, \infty)$

$0\ 1\ 2\ 3\ 4\ 5\ 6$

d. **35. a.** $a < 1$ **b.** $\{a|a < 1\}$ **c.** $(-\infty, 1)$ **d.** **37. a.** $f \leq -2$

$-5\,-4\,-3\,-2\,-1\ 0\ \ 1$

$0\quad 1\quad 2\quad 3\quad 4$

b. $\{f|f \leq -2\}$ **c.** $(-\infty, -2]$ **d.** **39. a.** $u \leq 2$ **b.** $\{u|u \leq 2\}$ **c.** $(-\infty, 2]$ **d.**

$-4\,-3\,-2\,-1\ 0$

$-2\ -1\ \ 0\ \ 1\ \ 2\ \ 3$

41. a. $c > -12$ **b.** $\{c|c > -12\}$ **c.** $(-12, \infty)$ **d.** **43. a.** $w \geq 3$ **b.** $\{w|w \geq 3\}$ **c.** $[3, \infty)$

$-14\,-13\,-12\,-11\,-10\,-9$

d. **45. a.** $x \leq -13$ **b.** $\{x|x \leq -13\}$ **c.** $(-\infty, -13]$ **d.**

$0\ 1\ 2\ 3\ 4\ 5$

$-15\ -14\ -13\ -12\ -11\ -10$

47. a. $n < 7$ **b.** $\{n|n < 7\}$ **c.** $(-\infty, 7)$ **d.** **49. a.** $l < 10$ **b.** $\{l|l < 10\}$ **c.** $(-\infty, 10)$

$4\ \ 5\ \ 6\ \ 7\ \ 8\ \ 9$

d. **51. a.** $t \leq 3.\overline{3}$ **b.** $\{t|t \leq 3.\overline{3}\}$ **c.** $(-\infty, 3.\overline{3}]$ **d.**

$5\ \ 6\ \ 7\ \ 8\ \ 9\ 10\ 11$

$2\qquad 3\ 3.\overline{3}\ 3.\overline{6}\ 4\qquad 5$

53. $n - 4 > 24; n > 28$ **55.** $\dfrac{4}{9}x \leq -8; x \leq -18$ **57.** $8y - 36 \geq 60; y \geq 12$ **59.** $\dfrac{1}{2}a + 5 \leq 2; a \leq -6$ **61.** $4x - 8 < 2x; x < 4$

63. $25 \geq 6x + 7; x \leq 3$ **65.** $l \leq 10$ ft. **67.** $h \geq 15$ in. **69.** $r \leq 23.89$ in. **71.** $t \geq 6\dfrac{4}{13}$ hr. **73.** $R \geq \$1,475,000$ **75.** $x \geq 98$

77. $i \leq 1.5$ A

Review Exercises: **1.** $\{1, 2, 3, 4, 5, 6, 7, 8, 9, 10\}$ **2.** $\dfrac{35}{2}$ **3.** $2(x - 5) + 6$ **4.** 16 **5.** $-\dfrac{135}{4}$ or $-33\dfrac{3}{4}$ **6.** 1426.208

Chapter 2 Summary and Review Exercises

Formulas $2l + 2w; \pi d; 2\pi r; bh; lw; \dfrac{1}{2}bh; \dfrac{1}{2}h(a + b); \pi r^2; 2lw + 2lh + 2wh; lwh; \dfrac{1}{3}lwh; \pi r^2 h; \dfrac{1}{3}\pi r^2 h; \dfrac{4}{3}\pi r^3; rt; \dfrac{d}{t}; iR; \dfrac{5}{9}(F - 32);$

$\dfrac{9}{5}C + 32; R - C$ Procedure Understand; Plan; Execute; Answer; Check 2.1 Definitions/Rules/Procedures two expressions set
equal; a number that makes an equation true when it replaces the variable in the equation; variable; value; true

Exercises   **1.** yes   **2.** no   **3.** no   **4.** yes   Definitions/Rules/Procedures   mathematical relationship; the distance around the figure; the total number of square units that fill the figure; the total number of cubic units that fill a space; the distance around the circle; the distance from the center of a circle to any point on the circle; the distance across a circle through its center; variables; unknown value   Exercises   **5.** 54 m^2 **6.** $\approx$ 200.96 in.^{2} **7.** 150 cm^3 **8.** $\approx$ 100.48 in.^{2} Definitions/Rules/Procedures   add the areas; subtract the area of the smaller figure from the area of the larger figure   Exercises   **9.** 19.75 ft.^{2} **10.** 102.6 in.^{2}

2.2 and 2.3 Definitions/Rules/Procedures   1; $ax + b = c$; $b + c$; additive inverse; bc; multiplicative inverse; the coefficient; parentheses; LCD; LCD; most decimal places; addition principle; multiplication principle; identity; every real number; contradiction; no solution   Exercises   **11.** -12 **12.** $-\dfrac{1}{12}$ **13.** -5 **14.** 18   **15.** 6   **16.** 5   **17.** -2 **18.** $\dfrac{9}{5}$ **19.** No solution   **20.** All real numbers   **21.** 2   **22.** 18   **23.** $\dfrac{9}{4}$ **24.** $-\dfrac{1}{2}$ **25.** $\dfrac{3}{4}$ **26.** 5.25   **27.** All real numbers   **28.** No solution   **29.** 80 in.   **30.** 7.1 in.   **31.** 1.2 mi.   **32.** 15 in.

2.4 Definitions/Rules/Procedures   constants   Exercises   **33.** $x = 1 - y$ **34.** $d = \dfrac{C}{\pi}$ **35.** $h = \dfrac{2A}{b}$ **36.** $m = \dfrac{Fr}{v^2}$ **37.** $h = \dfrac{3V}{\pi r^2}$ **38.** $m = \dfrac{y - b}{x}$ **39.** $w = \dfrac{P - 2l}{2}$ **40.** $h = \dfrac{2A}{a + b}$

2.5 Definitions/Rules/Procedures   unknown(s); constants; key words   Exercises   **41.** $6x = -18$ $x = -3$ **42.** $\dfrac{1}{2}m - 3 = \dfrac{1}{4}m - 9$ $m = -24$

43. $2(v + 4) - 1 = 1$ $v = -3$ **44.** $2(y - 1) + 3y = 6y - 20$ $y = 18$

2.6 Definitions/Rules/Procedures   quotient; 1; $\dfrac{a}{b} = \dfrac{c}{d}$, where $b \neq 0$ and $d \neq 0$; $ad = bc$ Exercises   **45.** $\dfrac{3}{8}$ **46.** $\approx \dfrac{\$0.085}{1 \text{ oz}}$; each ounce costs 8.5 cents.   **47.** no   **48.** yes   Definitions/Rules/Procedures   Find; cross products; multiplication   Exercises   **49.** 25   **50.** 22   **51.** $78\dfrac{3}{4}$ **52.** $-21.1\overline{6}$ Definitions/Rules/Procedures   numerators; denominators; cross products   Exercises   **53.** \$9.10   **54.** $\approx$ 220.5 mi.   **55.** 12 in.   **56.** $a = 4\dfrac{5}{7}$in.; $b = 4\dfrac{2}{3}$in.; $c = 13\dfrac{1}{2}$in.

2.7 Definitions/Rules/Procedures $\dfrac{n}{100}$ Exercises   **57.** 0.15, $\dfrac{3}{20}$ **58.** 0.825, $\dfrac{33}{40}$ **59.** 0.125, $\dfrac{1}{8}$ **60.** $0.\overline{3}$, $\dfrac{1}{3}$ Definitions/Rules/Procedures   100%   Exercises   **61.** 40%   **62.** $36\dfrac{4}{11}$% or $36.\overline{36}$%   **63.** 35%   **64.** 201.6%   Definitions/Rules/Procedures   variable; equal sign; multiplication; division; Part; Whole   Exercises   **65.** 14.56   **66.** $\approx$ 289.7   **67.** 32.5%   **68.** 56.25%   **69.** 478   **70.** \$805.59   **71.** $\approx$ 59.3%   **72.** $\approx$ 36.1%

2.8 Definitions/Rules/Procedures   1; bracket (or solid circle); left; right; parenthesis (or open circle); left; right; right of the indicated number; left of the indicated number; parentheses; LCD; Combine; one side of the inequality; the other side; reverse the direction   Exercises   **73. a.** $x < -3$ **b.** $\{x | x < -3\}$ **c.** $(-\infty, -3)$ **d.**   **74. a.** $x \geq -1$ **b.** $\{x | x \geq -1\}$ **c.** $[-1, \infty)$ **d.**   **75. a.** $z \leq -1$ **b.** $\{z | z \leq -1\}$ **c.** $(-\infty, -1]$ **d.**   **76. a.** $v > -4$ **b.** $\{v | v > -4\}$ **c.** $(-4, \infty)$ **d.**   **77. a.** $m > 4$ **b.** $\{m | m > 4\}$ **c.** $(4, \infty)$ **d.**   **78. a.** $c \geq -9$ **b.** $\{c | c \geq -9\}$ **c.** $[-9, \infty)$ **d.**   **79.** $13 > 3 - 10p$ $p > -1$ **80.** $3 + 2z < 3(z - 5)$ $z > 18$ **81.** $-\dfrac{1}{2}x \geq 4$ $x \leq -8$ **82.** $-2 \leq 1 - \dfrac{1}{4}k$ $k \leq 12$ **83.** $d \leq 3.18$ ft.   **84.** $R \geq \$825,000$

Chapter 2 Practice Test

1. Equation because there is an equal sign. [2.1]   **2.** Nonlinear because there is a variable raised to an exponent other than 1. [2.2]   **3.** yes [2.1]   **4.** 13.2 [2.2]   **5.** 3 [2.3]   **6.** No solution [2.3]   **7.** 4 [2.3]   **8.** $x = \dfrac{7 - y}{2}$ [2.4]   **9.** $h = \dfrac{2A}{b}$ [2.4]   **10.** $3\dfrac{1}{3}$ [2.6]   **11.** 0.22, $\dfrac{11}{50}$ [2.7]   **12.** $0.0\overline{3}$, $\dfrac{1}{30}$ [2.7]   **13.** 320% [2.7]   **14.** 40% [2.7]

15. Set-builder notation: $\{x \mid x \geq 3\}$
Interval notation: $[3, \infty)$
Graph: [2.8]
$\quad 0 \; 1 \; 2 \; 3 \; 4 \; 5 \; 6 \; 7$

16. Set-builder notation: $\{x \mid -1 \leq x < 4\}$
Interval notation: $[-1, 4)$
Graph: [2.8]
$\quad -2 \; -1 \; 0 \; 1 \; 2 \; 3 \; 4 \; 5$

17. a. $m < 3$ **b.** $\{m \mid m < 3\}$ **c.** $(-\infty, 3)$
d. [2.8]
$\quad 0 \; 1 \; 2 \; 3 \; 4$

18. a. $x \geq -8$ **b.** $\{x \mid x \geq -8\}$ **c.** $[-8, \infty)$
d. [2.8]
$\quad -10 \; -9 \; -8 \; -7 \; -6 \; -5$

19. a. $p > -\dfrac{2}{5}$ **b.** $\left\{p \mid p > -\dfrac{2}{5}\right\}$ **c.** $\left(-\dfrac{2}{5}, \infty\right)$
d. [2.8]
$\quad -1 \;\; -\frac{4}{5} \; -\frac{3}{5} \; -\frac{2}{5} \; -\frac{1}{5} \;\; 0$

20. a. $l \leq -\dfrac{5}{3}$ **b.** $\left\{l \mid l \leq -\dfrac{5}{3}\right\}$ **c.** $\left(-\infty, -\dfrac{5}{3}\right]$
d. [2.8]
$\quad -2 \;\; -\frac{5}{3} \;\; -\frac{4}{3} \;\; -1$

21. $\dfrac{2}{3}n - \dfrac{1}{6} = 2n$
$n = -\dfrac{1}{8}$ [2.5]

22. $5(n - 2) - 3 = 10 - 4(n - 1)$
$n = 3$ [2.5]

23. 20% [2.7] **24.** 175 [2.7] **25.** $1 - n > 2n$
$n < \dfrac{1}{3}$ [2.8]
26. 20 ft. [2.1] **27.** 68.25 in.^{2} [2.1] **28.** \$530 [2.1] **29.** $\dfrac{2}{5}$ [2.6]

30. $\dfrac{0.26}{1}$; 1 ounce of cereal costs 26 cents. [2.6] **31.** 12.96 cm [2.6] **32.** 186 [2.7] **33.** 12.5% [2.7] **34.** $x \geq 101$ [2.8]

Chapters 1–2 Cumulative Review Exercises

1. true **2.** false **3.** false **4.** true **5.** itself **6.** LCD **7.** $2^3 \cdot 3 \cdot 5$ **8.** 3 **9.** Distributive property **10.** Set the cross products equal to each other and solve for the variable. **11.** $\dfrac{4}{9}$ **12.** 7 **13.** $\dfrac{1}{10}$ **14.** -14 **15.** $\dfrac{11}{10}$ **16.** -26.48 **17.** $-\dfrac{1}{2}x + 27$ **18.** 9 **19.** $\dfrac{2}{5}$

20. All real numbers **21.** 5 **22.** $b = \dfrac{2A}{h}$ **23.** $c = P - a - b$ **24. a.** $h \geq -9$ **b.** $\{h \mid h \geq -9\}$ **c.** $[-9, \infty)$ **d.**
$\quad -10 \; -9 \; -8 \; -7 \; -6 \; -5$

25. a. $m < -5$ **b.** $\{m \mid m < -5\}$ **c.** $(-\infty, -5)$ **d.**
$\quad -8 \; -7 \; -6 \; -5 \; -4 \; -3$
26. 99 cm^2 **27.** 1376 ft. **28.** 6.345 **29.** $t \geq 88$ **30.** 300 mi.

Chapter 3

Exercise Set 3.1

Prep Exercises **1.** Horizontal-axis coordinate **2.** Beginning at the origin, move to the left or right along the x-axis by the amount indicated by the first coordinate. From that position on the x-axis, move up or down by the amount indicated by the second coordinate. **3.**

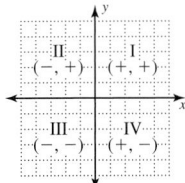

4. Points that can be connected to form a straight line are said to have a *linear* relationship.

Exercises **1.** $A(4, 3), B(-3, 1), C(0, -2), D(2, -5)$ **3.** $A(-4, 0), B(2, 4), C(-2, -3), D(4, -4)$

5.
$(5, 4)$
$(2, 0)$
$(-1, -3)$ $(2, -3)$

7.
$(-2, 0)$ $(2, 2)$
$(-1, -5)$ $(3, -3)$

9. II **11.** I **13.** IV **15.** III **17.** y-axis **19.** x-axis **21.** linear **23.** nonlinear

25. linear **27.** nonlinear **29.** linear **31.** Answers may vary. Some possible answers are $(-1, -5), (0, -3),$ and $(2, 1)$.
33. $A: (-6, -4); B: (-3, 2); C: (5, 2); D: (8, -4)$ **35. a.** $(-3, 1), (2, 1), (1, -3), (-4, -3)$ **b.** $(0, 2), (5, 2), (4, -2), (-1, -2)$
c. $(x + 3, y + 1)$ **37. a.** **b.** 20 units **c.** 21 square units

Review Exercises: **1.**
$\quad -4 \; -3 \; -2 \; -1 \; 0 \;\; 1 \;\; 2$
2. 2 **3.** 12 **4.** 0 **5.** $\dfrac{15}{2}$ **6.** $-\dfrac{9}{10}$

Exercise Set 3.2

Prep Exercises   **1. a.** Replace the variables in the equation with the corresponding coordinates. **b.** Verify that the equation is true.
2. No, because linear equations with two variables have an infinite number of solutions.   **3. a.** Choose a value for one of the variables. **b.** Replace the corresponding variable with your chosen value. **c.** Solve the equation for the value of the other variable.
4. The graph of an equation represents all of the solutions to the equation.   **5.** A minimum of two ordered pairs are needed because two points determine a line.   **6. a.** Find at least two solutions to the equation. **b.** Plot the solutions as points in a rectangular coordinate system. c. Connect the points to form a straight line.

Exercises   **1.** yes   **3.** yes   **5.** no   **7.** yes   **9.** no   **11.** yes   **13.** no   **15.** no

17. $(-1, -1)$
$(0, 0)$
$(1, 1)$
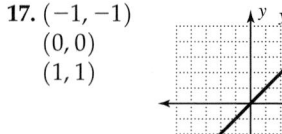

19. $(-1, -2)$
$(0, 0)$
$(1, 2)$
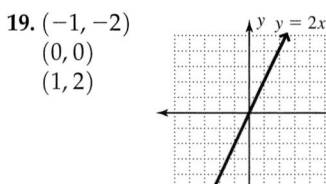

21. $(-1, 5)$
$(0, 0)$
$(1, -5)$
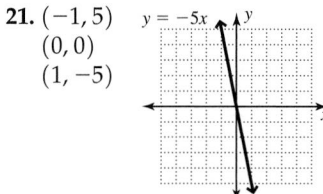

23. $(0, -3)$
$(2, -1)$
$(4, 1)$
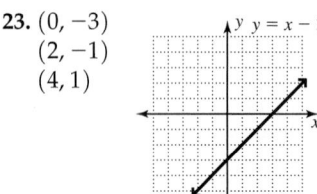

25. $(0, -2)$
$(4, -6)$
$(-4, 2)$
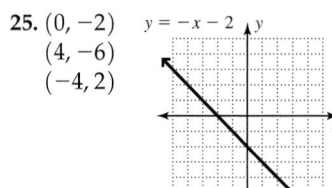

27. $(0, -5)$
$(1, -3)$
$(2, -1)$
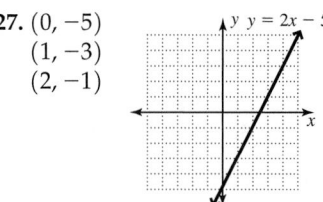

29. $(-1, 6)$
$(0, 4)$
$(1, 2)$
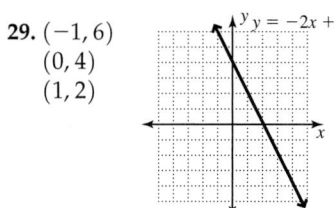

31. $(-2, -1)$
$(0, 0)$
$(2, 1)$
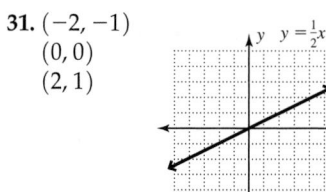

33. $(0, 0)$
$(-3, 2)$
$(3, -2)$
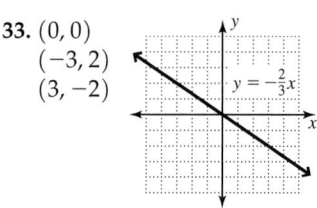

35. $(-3, 6)$
$(0, 4)$
$(3, 2)$
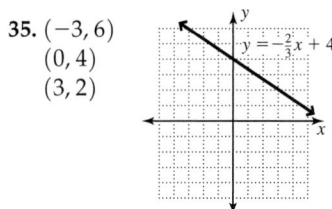

37. $(3, -5)$
$(5, -3)$
$(6, -2)$
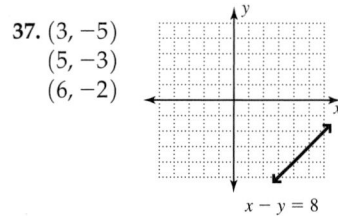

39. $(1, 4)$
$(3, 0)$
$(5, -4)$
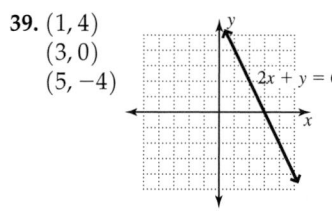

41. $(0, 4)$
$(6, 0)$
$(3, 2)$
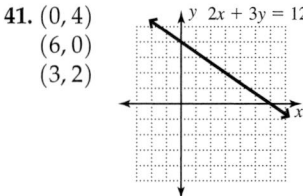

43. $(0, 4)$
$(-3, 0)$
$(3, 8)$
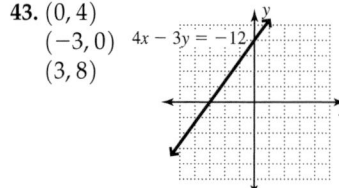

45. $(1, -4)$
$(2, 0)$
$(3, 4)$
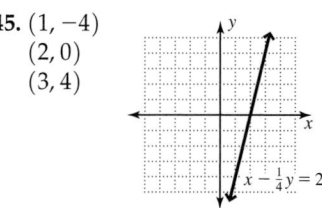

47. $(-1, -5)$,
$(0, -5)$,
$(1, -5)$
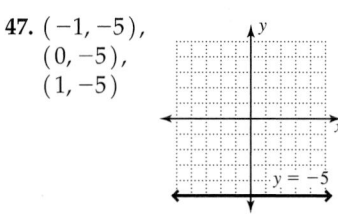

49. $(7, -1)$,
$(7, 0)$,
$(7, 1)$
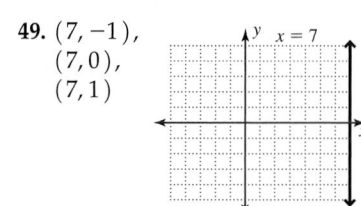

51. $(-1, -2.9)$,
$(0, -2.5)$,
$(1, -2.1)$
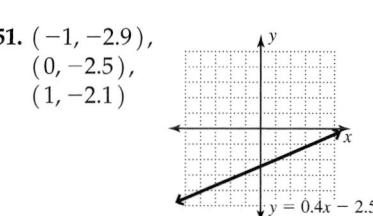

53. As a gets larger, the graph gets steeper.   **55.** The graph will shift up b units on the y-axis.   **57. a.** $160 **b.** 4 hr.

c.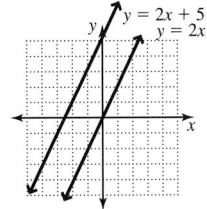

d. It represents the initial charge, which is $80.   **59. a.** $18   **b.** 500 copies **c.**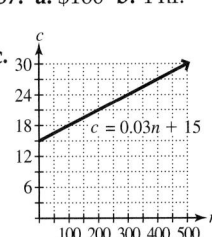

61. a. 412.5 gal. **b.** $\approx$ 71.4 min. **c.** $\approx$ 143 min. **d.**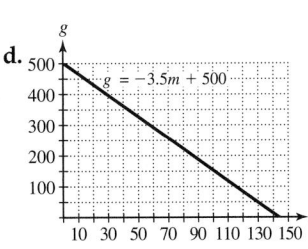

e. It represents the original amount of water in the hot tub, which is 500 gallons. **f.** It represents the amount of time it takes the hot tub to empty (0 gallons of water).

63. a. $1200 **b.** 4 yr. **c.** 8 yr. **d.**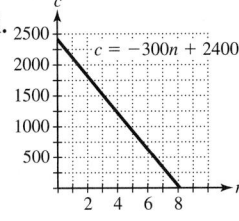

Review Exercises: **1.** 4   **2.** 31.5 in.   **3.** 5.67 sq. in.   **4.** $\dfrac{5}{3}$ **5.** $x = \dfrac{v - k}{m}$ **6.**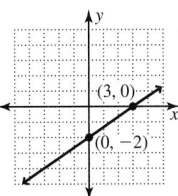

Exercise Set 3.3

Prep Exercises   **1.** A point where a graph intersects the x-axis   **2.** A point where a graph intersects the y-axis.   **3.** (a) Replace y with 0 in the given equation. (b) Solve for x.   **4.** (a) Replace x with 0 in the given equation. (b) Solve for y.   **5.** $(0, b)$
6. no x-intercept; $(0, c)$ **7.** $(c, 0)$; no y-intercept   **8.** The graph of $y = 2x + 5$ is a line parallel to $y = 2x$ shifted 5 units upward.

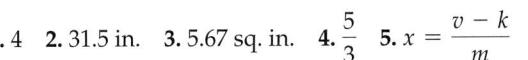

Exercises   **1.** $(1, 0), (0, 3)$ **3.** $(-3, 0), (0, 4)$ **5.** $(3, 0)$; no y-intercept   **7.** no x-intercept; $(0, -2)$ **9.** $(4, 0), (0, -4)$

11. $(3, 0), (0, 2)$ **13.** $\left(-\dfrac{10}{3}, 0\right), \left(0, \dfrac{5}{2}\right)$ **15.** $(2, 0), (0, -6)$ **17.** $\left(-\dfrac{5}{2}, 0\right), (0, 5)$ **19.** $(0, 0)$ for both   **21.** $(0, 0)$ for both

23. $(5, 0)$; no y-intercept   **25.** no x-intercept; $(0, -2)$

27. 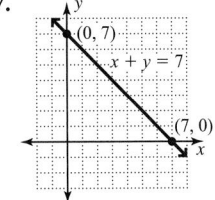   **29.** 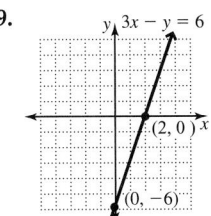   **31.** 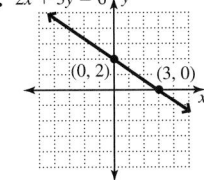   **33.** 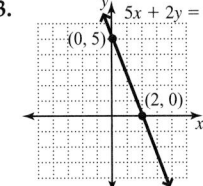   **35.**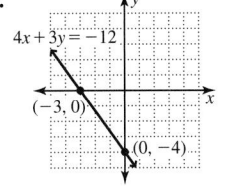

37.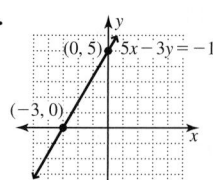

39.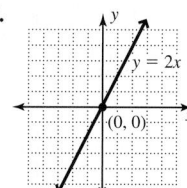

41.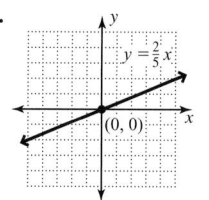

43.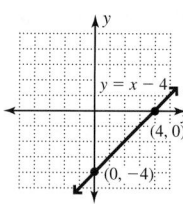

45.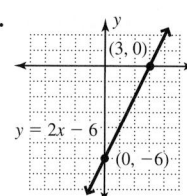

47.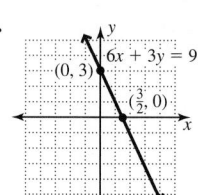

49.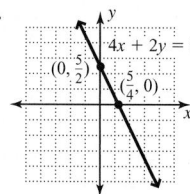

51.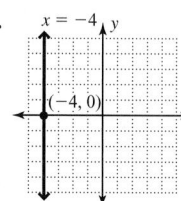

53.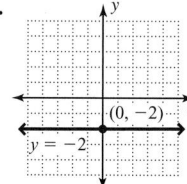

55.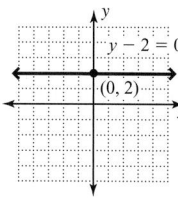

57.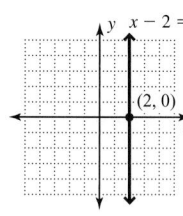

59. d because the x-coordinate of the x-intercept is positive and the y-coordinate of the y-intercept is negative   **61.** c because the x-coordinate of the x-intercept is positive and the y-coordinate of the y-intercept is negative   **63.** $(-2.7, 0)$, $(2.7, 0)$, $(0, -2.7)$, $(0, 2.7)$

65. a. $15x + 20y$ **b.** $15x + 20y = 2000$ **c.** $\left(133\frac{1}{3}, 0\right)$, $(0, 100)$ **d.** The number of units required to meet the goal if only one size is sold.

e. Answers may vary. Some possibilities are $(20, 85)$, $(40, 70)$, and $(60, 55)$. **f.**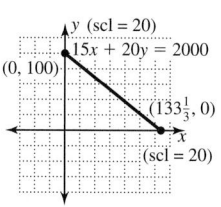

Review Exercises: **1.** -2 **2.** undefined   **3.** $w = \dfrac{P - 2l}{2}$ **4.** $4 + 3x = 5(x + 6)$; -13 **5.** On the y-axis   **6.** no

Exercise Set 3.4

Prep Exercises   **1.** How steep a line is   **2.** Uphill because the slope, 3, is positive.   **3.** Downhill because the slope, -2, is negative.

4. m **5.** $(0, b)$ **6.** "Rise" up 3 units from $(0, 1)$ and then "run" to the right 4 units.   **7.** $m = \dfrac{y_2 - y_1}{x_2 - x_1}$ **8.** Horizontal because for

any two points on the line, the y-coordinates are the same. This causes the numerator of the slope formula to be 0, which simplifies to 0.   **9.** Vertical because for any two points on the line, the x-coordinates are the same. This causes the denominator of the slope formula to equal 0, which is undefined.   **10.** $y = mx + b$

Exercises   **1.**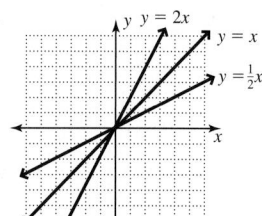

3.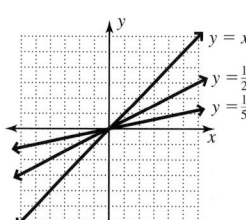

5.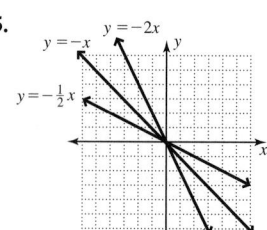

7.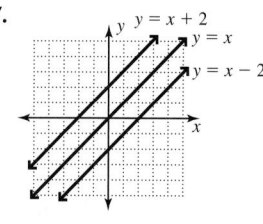

9. $m = 2$ $(0, 3)$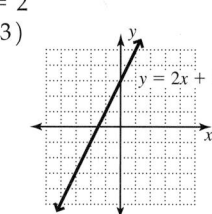

11. $m = -2$ $(0, -1)$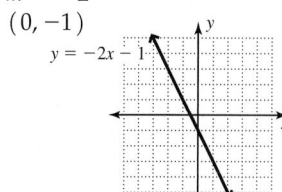

13. $m = \dfrac{1}{3}$ $(0, 2)$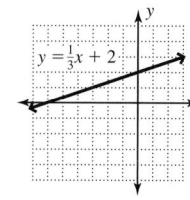

15. $m = \dfrac{3}{4}$ $(0, -2)$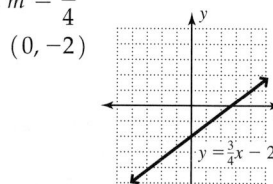

17. $m = -\dfrac{2}{3}$
$(0, 8)$

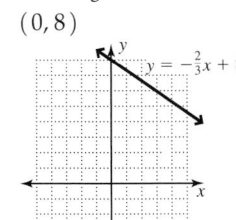

19. $m = -2$
$(0, 4)$

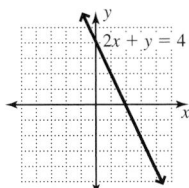

21. $m = 3$
$(0, 1)$

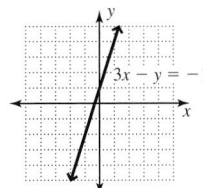

23. $m = -\dfrac{2}{3}$
$(0, 2)$

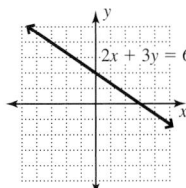

25. $m = -\dfrac{1}{2}$
$(0, -2)$

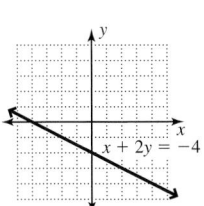

27. $m = \dfrac{3}{2}$
$\left(0, \dfrac{7}{2}\right)$

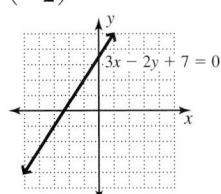

29. $m = -\dfrac{3}{2}$
$\left(0, \dfrac{5}{2}\right)$

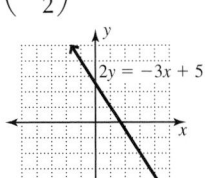

31. $m = 3$
$(0, -5)$

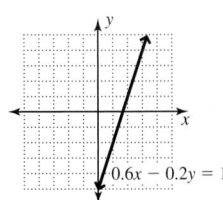

33. d   **35.** a   **37.** f   **39.** g   **41.** h   **43.** $\dfrac{5}{2}$ **45.** $-\dfrac{2}{3}$ **47.** $\dfrac{2}{7}$ **49.** $-\dfrac{4}{3}$ **51.** $\dfrac{2}{3}$ **53.** $-\dfrac{5}{4}$ **55.** 0   **57.** undefined   **59.** $y = 2x + 3$

61. $y = -3x - 9$ **63.** $y = \dfrac{3}{4}x + 5$ **65.** $y = -\dfrac{2}{5}x + \dfrac{7}{8}$ **67.** $y = 0.8x - 5.1$ **69.** $y = -3x$ **71.** $m = 2; (0, 3); y = 2x + 3$

73. $m = -1; (0, 2); y = -x + 2$ **75.** $m = \dfrac{2}{3}; (0, -1); y = \dfrac{2}{3}x - 1$ **77.** $m = -\dfrac{1}{4}; (0, 3); y = -\dfrac{1}{4}x + 3$

79. m is undefined; no y-intercept; $x = 2$ **81.** $m = 0; (0, 3); y = 3$ **83.** $y = -2, x = 3$ **85.** $y = 0$

87. a.
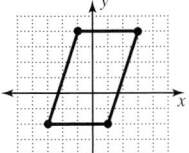
b. 3 and 0 **c.** They are the same.   **89.** $0.58\overline{3}, \dfrac{7}{12}$ **91.** $\dfrac{293}{230} \approx 1.27$

Review Exercises: **1.** $-\dfrac{3}{2}$ **2.** $11x + 7$ **3.** $-2x - 14$ **4.** $x = \dfrac{C - By}{A}$ **5.**
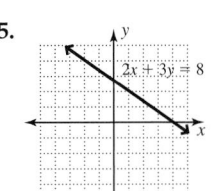
6. $\left(\dfrac{7}{3}, 0\right), \left(0, -\dfrac{7}{5}\right)$

Exercise Set 3.5
Prep Exercises   **1.** $y - y_1 = m(x - x_1)$ **2.** Manipulate the point–slope form so that the variables are on one side of the equation and a constant is on the other side.   **3.** Find the slope.   **4.** $m = \dfrac{y_2 - y_1}{x_2 - x_1}$ **5.** No, both points produce the same equation.
6. equal   **7.** They are reciprocals with opposite signs.

Exercises   **1.** $y = 3x - 13$ **3.** $y = -x - 2$ **5.** $y = \dfrac{2}{3}x - 1$ **7.** $y = -\dfrac{3}{4}x - \dfrac{23}{4}$ **9.** $y = -\dfrac{4}{3}x + \dfrac{4}{3}$ **11.** $y = 2x$ **13.** $2x - 3y = -2$

15. $x + 2y = 3$ **17.** $y = -2x + 5$ **19.** $y = \dfrac{5}{2}x - \dfrac{13}{2}$ **21.** $y = 7x$ **23.** $x = 8$ **25.** $y = -1$ **27.** $y = 3x + 0.8$ **29.** $y = 2x - 2$

31. $y = -\dfrac{6}{5}x + 3$ **33.** $5x - 2y = 4$ **35.** $3x - 2y = 6$ **37.** $4x + 11y = -27$ **39.** $2x + 3y = 0$ **41.** $x = -1$ **43.** $y = -1$

45. a. $y = 4x - 14$ **b.** $4x - y = 14$ **47. a.** $y = -3x - 5$ **b.** $3x + y = -5$ **49. a.** $y = \dfrac{1}{3}x + \dfrac{1}{3}$ **b.** $x - 3y = -1$

51. a. $y = -\dfrac{3}{4}x - \dfrac{15}{4}$ **b.** $3x + 4y = -15$ **53. a.** $y = -2x + 1$ **b.** $2x + y = 1$ **55. a.** $y = \dfrac{3}{4}x + 5$ **b.** $3x - 4y = -20$

57. a. $y = -\dfrac{1}{2}x - 3$ **b.** $x + 2y = -6$ **59. a.** $y = -4x + 8$ **b.** $4x + y = 8$ **61. a.** $y = \dfrac{1}{2}x - 5$ **b.** $x - 2y = 10$

63. a. $y = -\dfrac{2}{3}x + \dfrac{5}{3}$ **b.** $2x + 3y = 5$ **65. a.** $y = -\dfrac{5}{2}x - \dfrac{17}{2}$ **b.** $5x + 2y = -17$ **67. a.** $y = \dfrac{3}{2}x - \dfrac{17}{2}$ **b.** $3x - 2y = 17$

69. parallel   **71.** perpendicular   **73.** neither   **75.** parallel   **77.** perpendicular   **79.** neither   **81.** perpendicular   **83.** parallel

85. a. $c = 0.3n + 7.75$ **b.** \$19.75   **c.**   **87. a.**   **b.** 0.039 **c.** $p = 0.039n + 2.78$

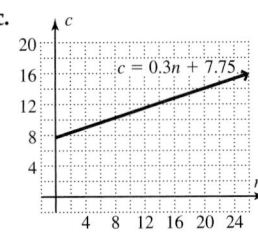

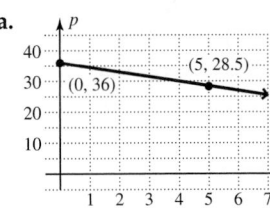

d. 2.975 million **e.** 3.482 million   **89. a.**   **b.** -1.5 **c.** $p = -1.5n + 36$

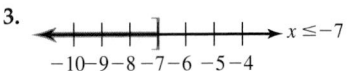

d. $(0, 36)$; the p-intercept indicates the initial price. **e.** $(24, 0)$; the n-intercept indicates that on the 24th day, the price will be \$0.
f. \$24

Review Exercises: **1.** $<$ **2.** $\dfrac{36}{5}$ **3.**    $x \le -7$ **4.** $x \le \dfrac{4}{3}$

5. $m = -\dfrac{3}{4}; (0, 2)$ **6.** $(7, 0) (0, -2)$

Exercise Set 3.6

Prep Exercises **1.** Replace the variables with the corresponding coordinates. If the resulting inequality is true, the ordered pair is a solution. **2.** Replace the inequality symbol with an equal sign. **3.** Draw a dashed line when the inequality is $>$ or $<$. Draw a solid line when the inequality is $\ge$ or $\le$. **4.** Choose an ordered pair on one side of the boundary line and test this ordered pair in the inequality. If the ordered pair satisfies the inequality, shade the region that contains it. If the ordered pair does not satisfy the inequality, shade the region on the other side of the boundary line.

Exercises **1.** yes   **3.** no   **5.** no   **7.** yes

9. $y \le -x + 1$ **11.** $y > -2x + 5$ **13.** $y > 2x$ **15.** $y > \frac{1}{3}x$ **17.** $y \ge \frac{3}{4}x + 2$

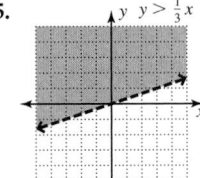

19. $3x + y \le 9$ **21.** $5x - 2y < 10$ **23.** $3x + y \ge 8$ **25.** $2x + 3y < 9$ **27.** $3x + y \ge 0$

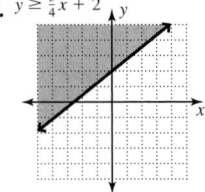

29. $x > -4$ **31.** $y \le 2$ **33.** $x + 2 \ge 0$ **35.** $y - 3 < 0$

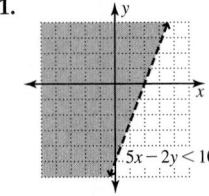

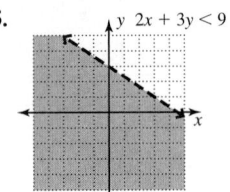

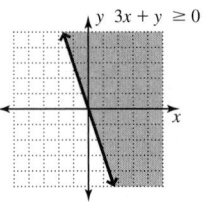

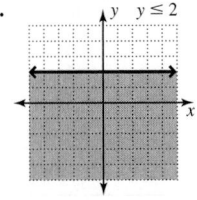

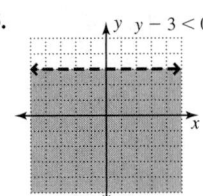

37. a. x represents the number of large bottles sold; y represents the number of small bottles sold. **b.**

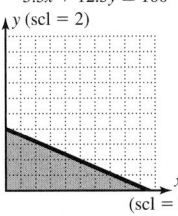

c. The boundary line represents combinations of bottle sizes sold to produce a revenue of exactly $280,000 (breakeven). The shaded region represents combinations that produce a revenue greater than $280,000 (profit). **d.** Answers will vary; three possible combinations are $(80{,}000, 80{,}000)$ and $(20{,}000, 160{,}000)$ and $(50{,}000, 120{,}000)$. **e.** Answers will vary; three possible combinations are $(150{,}000, 0)$ and $(90{,}000, 100{,}000)$ and $(80{,}000, 120{,}000)$ **f.** No, only combinations of whole numbers are possible because the company cannot sell fractions of a bottle of hand lotion.

39. a. $5.5x + 12.5y \leq 100$ **b.** $5.5x + 12.5y \leq 100$ **c.** The combinations of plant purchases that would cost exactly $100

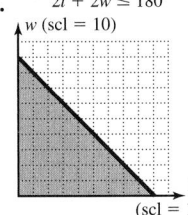

d. The combinations of plants purchased that would cost less than $100 **e.** Answers will vary; some combinations are $(2, 4), (6, 2)$, and $(8, 4)$. **f.** Yes, $(0, 8)$. Because she cannot purchase fractional numbers of plants, the combination must be whole numbers and $(0, 8)$ is the only combination of whole numbers that costs exactly $100. **41. a.** $2l + 2w \leq 180$

b. $2l + 2w \leq 180$ **c.** The combinations of lengths and widths that yield a perimeter of exactly 180 ft.

d. The combinations of length and width that yield a perimeter that is less than 180 ft. **e.** Answers will vary; some combinations are $(10, 80), (20, 70)$, and $(40, 50)$. **f.** Answers will vary; some combinations are $(10, 10), (20, 20)$, and $(30, 30)$.

Review Exercises: **1.** $\{a, e, i, o, u\}$ **2.** 1 **3.** -13 **4.** 20 ft. by 24 ft. **5.** $\left(\frac{8}{3}, 0\right), (0, -4)$ **6.**

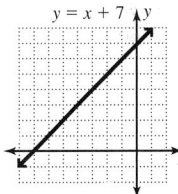

Exercise Set 3.7

Prep Exercises **1.** The domain is a set containing all x-values. **2.** The range is a set containing all y-values. **3.** False; a function is a special type of relation. **4.** Draw or imagine a vertical line through every point in the domain. If each vertical line intersects the graph in at most one point, the relation is a function. **5.** Trail along the x-axis and determine what values of x have a corresponding y-value. **6.** Trail along the y-axis and determine what values of y have a corresponding x-value.

Exercises **1.** Domain: $\{2, 3, 1, -1\}$; range: $\{1, -2, 4, -1\}$ **3.** Domain: $\{4, 2, 3, -4\}$; range: $\{1, -5, 0\}$ **5.** Domain: $\{0, -3, 2, 8\}$; range: $\{0, -3, 9, -16\}$ **7.** yes **9.** no **11.** yes **13.** no **15.** Yes, every element in the domain is assigned to one element in the range. **17.** No, an element in the domain is assigned to more than one element in the range. (8 is assigned to both 2 and -1.) **19.** Yes, every element in the domain is assigned to one element in the range. **21.** No, an element in the domain is assigned to more than one element in the range. **23.** No, it fails the vertical line test. **25.** Yes, it passes the vertical line test. **27.** No, it fails the vertical line test **29.** Yes, it passes the vertical line test. **31.** Domain: $\{x \mid 1980 \leq x \leq 2010\}$ or $[1980, 2010]$; range: $\{y \mid 17.7 \text{ million} \leq y \leq 104.1 \text{ million}\}$ or [17.7 million, 104.1 million]; function **33.** Domain $\{1900, 1910, 1920, 1930, 1940, 1950, 1960, 1970, 1980, 1990, 2000, 2010\}$; range: $\{4.7, 5.4, 6.2, 6.9, 8.0, 8.8, 10.4, 11.6, 12.9, 13.2, 13.6, 14.7\}$; function **35.** Domain: $\{x \mid -4 \leq x \leq 4\}$ or $[-4, 4]$; range: $\{y \mid -3 \leq y \leq 2\}$ or $[-3, 2]$; function **37.** Domain: $\{x \mid x \leq 0\}$ or $(-\infty, 0]$; range: all real numbers or $(-\infty, \infty)$; not a function **39.** Domain: all real numbers or $(-\infty, \infty)$; range: $\{y \mid y \geq -1\}$ or $[-1, \infty)$; function **41. a.** -5 **b.** -3 **c.** -9 **d.** -4 **43. a.** 7 **b.** 5 **c.** 11 **d.** 17 **45. a.** $\sqrt{3}$ **b.** Not a real number **c.** $\sqrt{2}$ **d.** $\sqrt{a+3}$ **47. a.** 0 **b.** Not a real number **c.** 0 **d.** $\sqrt{5}$ **49. a.** -1 **b.** undefined **c.** $-\frac{1}{2}$ **d.** $\frac{1}{a-1}$ **51. a.** -3 **b.** -3.19 **c.** $a^2 - 2a - 3$

d. $a^2 + 2a - 3$ **53. a.** -2 **b.** $-\dfrac{5}{3}$ **c.** $-\dfrac{7}{3}$ **d.** $\dfrac{1}{3}a - 2$ **55. a.** $-\dfrac{3}{4}$ **b.** undefined **c.** $-\dfrac{1}{6}$ **d.** -6 **57. a.** undefined **b.** $\dfrac{1}{\sqrt{3}}$

c. Not a real number **d.** $\dfrac{1}{\sqrt{a + 2}}$ **59. a.** 2 **b.** 4 **c.** 6 **d.** 10 **61. a.** 0 **b.** $-\dfrac{1}{2}$ **c.** 20 **d.** 7 **63. a.** $8a^2 + 6a + 1$ **b.** $18a^2 - 9a + 1$

c. $\dfrac{1}{2}a^2 + \dfrac{3}{2}a + 1$ **d.** $0.02a^2 + 0.3a + 1$ **65. a.** 3 **b.** 4 **c.** 2 **67. a.** 0 **b.** -2 **c.** undefined **69.**

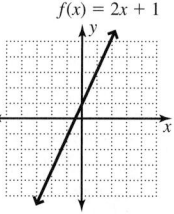

$f(x) = 2x + 1$

71.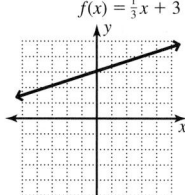

$f(x) = \frac{1}{3}x + 3$

73.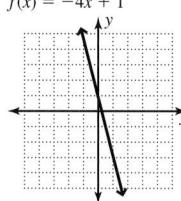

$f(x) = -4x + 1$

75.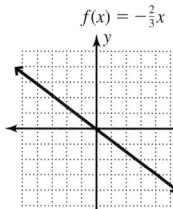

$f(x) = -\frac{2}{3}x$

77.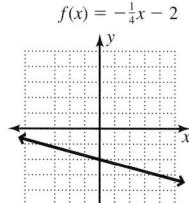

$f(x) = -\frac{1}{4}x - 2$

79. a. \$350 **b.** \$450

c. The cost of producing 30 items is \$750. **81. a.** \$13,500 **b.** \$9000 **c.** The value of the car after ten years is \$3000.

Review Exercises: 1. $-\dfrac{13}{30}$ **2.** $-4x - 2$ **3.** Commutative property of addition **4.** 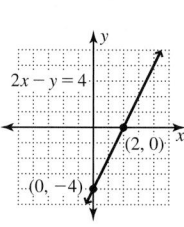 **5.** $y = \dfrac{3}{2}x - 3$ **6.** parallel

$2x - y = 4$ (2, 0) (0, -4)

Chapter 3 Summary and Review Exercises

Forms of Linear Equations $mx + b$; $m(x_1 - x_2)$; By; $\dfrac{y_2 - y_1}{x_2 - x_1}$ 3.1 Definitions/Rules/Procedures *x*-axis; *x*-axis; horizontal;

second coordinate; right; left; up; down; dot Exercises **1.** A $(4, 3)$, B $(-2, 4)$, C $(-4, -2)$, D $(2, -1)$ **2.** A $(3, 1)$, B $(0, 4)$,

C $(-3, 0)$, D $(0, -2)$ **3.** 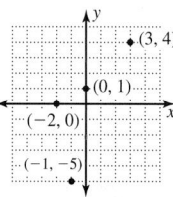 **4.** Definitions/Rules/Procedures $(+, +)$; $(-, +)$; $(-, -)$;

(3, 4) (0, 1) (-2, 0) (-1, -5)

(-2, 2) (4, 1) (5, 0) (0, -3)

$(+, -)$ Exercises **5.** II **6.** III **7.** I **8.** IV

3.2 Definitions/Rules/Procedures coordinates; true Exercises **9.** yes **10.** yes **11.** no **12.** no

Definitions/Rules/Procedures value; variable; variable; two; points; straight line Exercises **13.** $(0, 0)$ $(1, -2)$ $(-1, 2)$

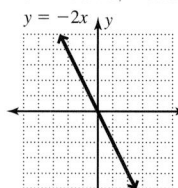

$y = -2x$

14. $(0, 7)$ $(1, 6)$ $(3, 4)$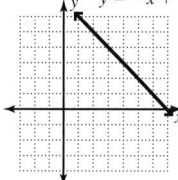

$y = -x + 7$

15. $(0, 2)$ $(3, 0)$ $(-3, 4)$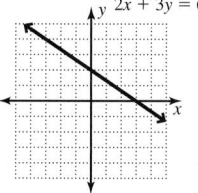

$2x + 3y = 6$

16. $(0, 3)$ $(3, 2)$ $(-3, 4)$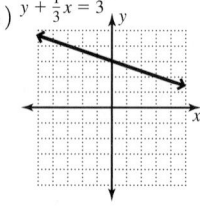

$y + \frac{1}{3}x = 3$

17. a. \$2200 **b.** $(0, 1000)$; this indicates the person's gross pay if he has \$0 in sales during a month.

c.

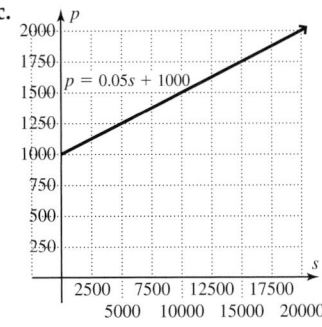

Definitions/Rules/Procedures   horizontal; $(0, c)$; vertical; $(c, 0)$.

Exercises   **18.**   **19.**   **20.**   **21.**

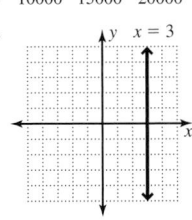

3.3 Definitions/Rules/Procedures   intersects; intersects; y; x; x; y; the origin, $(0, 0)$; $(0, b)$; $(0, c)$; x-axis; y-intercept; y-axis   Exercises

22. $(2, 0), (0, 3)$ **23.** $(0, 0)$ for both   **24.** $(15, 0) \; (0, 3)$ **25.** $(0, -2), (5, 0)$

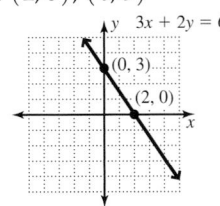 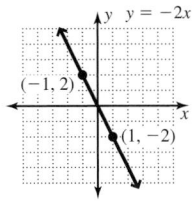 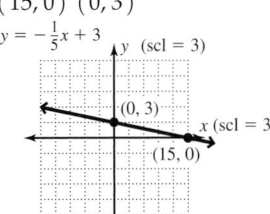 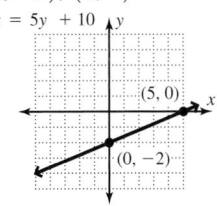

26. no x-intercept, $(0, 7)$ **27.** $(-1, 0)$, no y-intercept

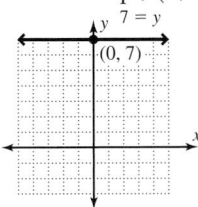 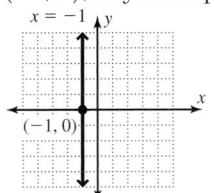

3.4 Definitions/Rules/Procedures   vertical; horizontal; uphill; downhill; y-intercept; rising; running; straight line   Exercises

28. $m = -2; (0, 3)$ **29.** $m = -\dfrac{1}{4}; (0, 2)$ **30.** $m = 3; (0, 0)$ **31.** $m = -1; (0, 5)$

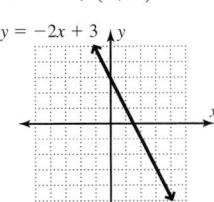 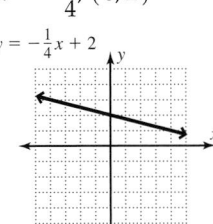 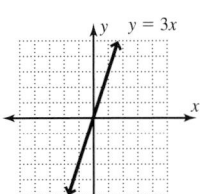 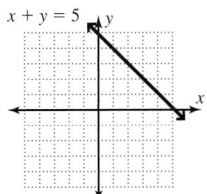

32. $m = \dfrac{1}{3}; \left(0, -\dfrac{7}{3}\right)$ **33.** $m = \dfrac{2}{3}; \left(0, -\dfrac{8}{3}\right)$ Definitions/Rules/Procedures $\dfrac{y_2 - y_1}{x_2 - x_1}$; 0; undefined   Exercises   **34.** 3

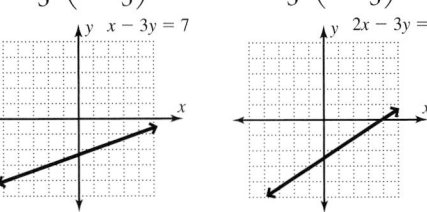

35. $-\dfrac{3}{5}$ **36.** 0   **37.** undefined   Definitions/Rules/Procedures   slope-intercept   Exercises   **38.** $y = -x + 7$ **39.** $y = -\dfrac{1}{5}x - 8$

40. $y = 0.2x + 6$ **41.** $y = x$

3.5 Definitions/Rules/Procedures $y - y_1 = m(x - x_1)$; $y - y_1 = m(x - x_1)$ Exercises **42.** $y = -2x + 9$ **43.** $y = -\dfrac{2}{5}x + \dfrac{21}{5}$

44. $y = -x + 4$ $x + y = 4$ **45.** $y = \dfrac{5}{4}x - \dfrac{37}{4}$ $5x - 4y = 37$ **46. a.**

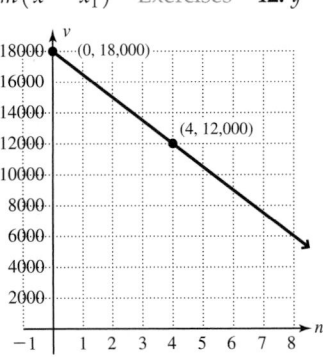

b. -1500

c. $v = -1500n + 18{,}000$ **d.** 2015 **e.** 2021 Definitions/Rules/Procedures slopes; y-intercepts; $-\dfrac{b}{a}$ Exercises

47. $y = 2x - 2$ **48.** $y = -\dfrac{2}{3}x + \dfrac{17}{3}$ **49.** $y = -\dfrac{5}{3}x - \dfrac{11}{3}$ **50.** $y = \dfrac{5}{2}x + 9$

3.6 Definitions/Rules/Procedures true Exercises **51.** yes **52.** yes **53.** yes **54.** no Definitions/Rules/Procedures solid; dashed; contains it; other side Exercises **55.**

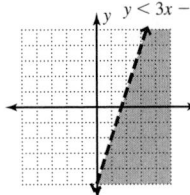

56.

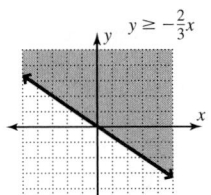

57.

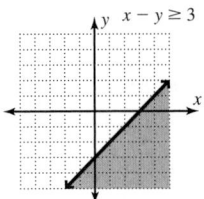

58.

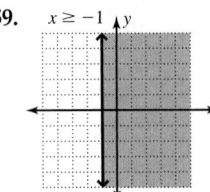

59.

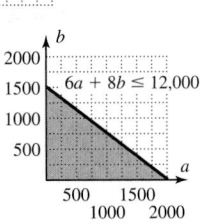

60. a. $6a + 8b \le 12{,}000$ **b.**

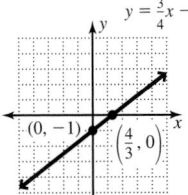

c. Answers may vary; some combinations are (400, 1200), (800, 900), and (1200, 600).

d. Answers may vary; some combinations are (500, 100), (1000, 200), and (200, 1100).

3.7 Definitions/Rules/Procedures ordered pairs; input; output; y-value; x-value Exercises **61.** Domain: $\{2, -1, 3, -3\}$; range: $\{3, 5, 6, -4\}$ **62.** Domain: { Volkswagen, Honda, Toyota, Subaru, Nissan }; range: $\{52.2, 49.7, 49.0, 47.8, 45.8\}$ **63.** no **64.** yes Definitions/Rules/Procedures domain; one; two; more Exercises **65.** Domain: all real numbers or $(-\infty, \infty)$; range: all real numbers or $(-\infty, \infty)$; it is a function. **66.** Domain: $\{x \mid x \le 0\}$ or $(-\infty, 0]$; range: all real numbers or $(-\infty, \infty)$; it is not a function. **67.** Domain: $\{x \mid -4 \le x \le 5\}$ or $[-4, 5]$; range: $\{1, 2, 3\}$; it is a function. Definitions/Rules/Procedures a Exercises **68. a.** 1 **b.** 3 **c.** 3 **d.** 2 **69. a.** 0 **b.** 2 **c.** -10 **70. a.** 2 **b.** undefined **c.** $\dfrac{5}{8}$

Chapter 3 Practice Test

1. A (2, 4) B (-3, 2) C (0, -5) D (4, -2) [3.1] **2.**

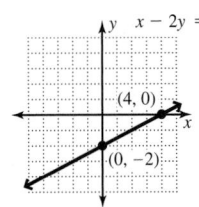

[3.1] **3.** I [3.1] **4.** no [3.2]

5. (4, 0), (0, -2) [3.3]

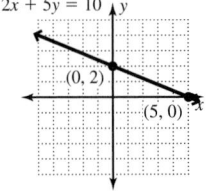

6. (5, 0), (0, 2) [3.3]

7. (0, 0) [3.3]

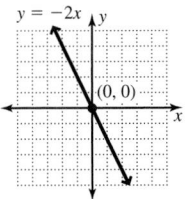

8. $\left(\dfrac{4}{3}, 0\right)$, (0, -1) [3.3]

9. $m = \dfrac{3}{4}$; (0, 11) [3.4] **10.** $m = \dfrac{5}{3}$; $\left(0, -\dfrac{8}{3}\right)$ [3.4] **11.** $-\dfrac{10}{3}$ [3.4] **12.** 0 [3.4]

13. $y = \frac{3}{5}x + 4$ [3.4]   **14.** $y = -\frac{2}{5}x + \frac{11}{5}$ [3.5]   **15.** $2x - 3y = 1$ [3.5]   **16.** $x - 3y = 10$ [3.5]   **17. a.**

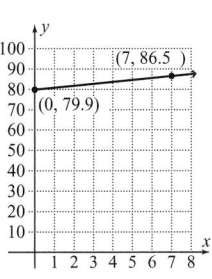

b. 0.943 [3.5]   **c.** $y = 0.943x + 79.9$ [3.5]   **d.** 90.3 million [3.6]   **18.** yes [3.6]   **19.**

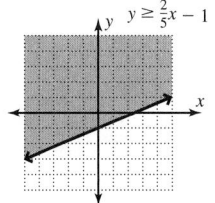

20.

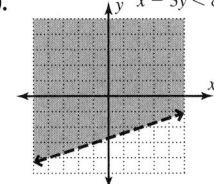

$x - 3y < 8$ [3.6]   **21. a.** $10x + 8y \geq 360$ [3.6]   **b.**

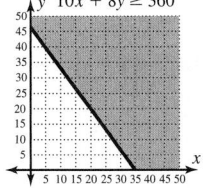

[3.6]   **c.** Answers may vary. One possible answer is $(0, 45)$. [3.6]   **d.** Answers may vary. One possible answer is $(10, 35)$. [3.6]

22. yes [3.7]   **23.** no [3.7]   **24. a.** -2 [3.7] **b.** undefined [3.7] **c.** $-\frac{9}{7}$ [3.7]   **25. a.** Domain: $\{x \mid -3 \leq x \leq 3\}$ or $[-3, 3]$; range: $\{-2, 1, 2\}$ [3.7] **b.** 2 [3.7]

Chapter 3 Cumulative Review Exercises

1. false   **2.** true   **3.** false   **4.** false   **5.** identity   **6.** 100   **7.** $\frac{33}{200}$ **8.** Multiply all of the terms by the power of 10 that will clear the decimal from the number with the most decimal places.   **9.** -23 **10.** 15   **11.** -18 **12.** -81

13. -41 **14.** 12   **15.** $(3, 0), (0, -2)$ **16.** Answers may vary. Three solutions are $(1, 1), (0, 4),$ and $(-1, 7)$

17.

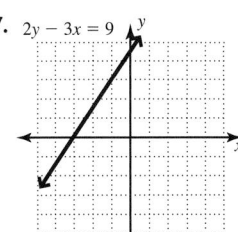

18. $m = -\frac{1}{3}$ **19.** $y = \frac{1}{2}x + \frac{1}{2}$ **20.**

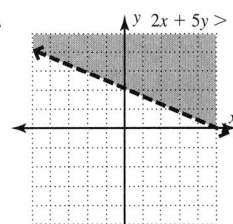

21. $\frac{85}{6}$ **22.** -12 **23.** -12.5 **24.** $r = \frac{C}{2\pi}$

25. 230   **26.** $\approx 1.9\%$ **27.** 420 miles   **28.** $\approx 216\%$ **29.** 75 phones   **30.** 800 miles

Chapter 4
Exercise Set 4.1

Prep Exercises   **1. a.** Replace each variable in each equation with its corresponding value.   **b.** Verify that each equation is true.
2. The solution is the ordered pair of the point of intersection of the graphs.   **3.** The graphs are parallel.   **4.** Their graphs are identical.
5. Graphing by hand can be imprecise, and not all solutions are integers.   **6.** The graphs intersect at a single point.
7. The graphs are identical.   **8.** The graphs are parallel lines.

Exercises   **1.** yes   **3.** no   **5.** yes   **7.** no   **9.** yes   **11.** $(4, 1)$ **13.** $(-3, 4)$ **15.** $(3, -2)$ **17.** $(-1, 3)$ **19.** $(3, -4)$ **21.** $(4, -3)$
23. $(2, -3)$ **25.** No solution   **27.** All ordered pairs along $2x + 3y = 2$ **29.** $(2, 6)$ **31.** $(-2, 2)$ **33.** $(-2, 3)$ **35.** $(5, -2)$
37. a. inconsistent **b.** no solution   **39. a.** consistent with independent equations **b.** one solution   **41. a.** consistent with dependent equations **b.** infinite number of solutions   **43. a.** consistent with independent equations **b.** one solution   **45. a.** inconsistent
b. no solution   **47. a.** consistent with dependent equations **b.** infinite number of solutions   **49. a.** consistent with independent equations **b.** one solution   **51. a.** 5000 units **b.** \$4000 **c.** $u > 5000$ **53. a.** The One-Rate plan **b.** 375 min.   **c.** More than 375 min.

Review Exercises: **1.** $-3y - 35$ **2.** $y = \frac{10 - x}{3}$ **3.** $x = 30 + 6y$ **4.** $\frac{40\pi}{3}$ in.^{3} or ≈ 41.89 in.^{3} **5.** -4 **6.** $2x + 17$

Exercise Set 4.2

Prep Exercises   1. When using the graphing method, if either coordinate in the solution is a fraction, we may have to guess the value. Substitution requires no guessing.   **2. a.** y **b.** $x - 6$ **3.** y in the first equation because it can be isolated without using division   **4.** x in the second equation because it can be isolated without using division.   **5.** The resulting equation will no longer contain variables and becomes a false equation.   **6.** The resulting equation will no longer contain variables and becomes a true equation.   **7.** Two

Exercises   1. $(4, 2)$ **3.** $(12, -4)$ **5.** $(-10, -24)$ **7.** All ordered pairs along $y = -2x$ (dependent)   **9.** $(-3, 2)$ **11.** $(2, 9)$

13. $(-1, -3)$ **15.** $(3, 4)$ **17.** $(1, 1)$ **19.** $\left(3, -\dfrac{1}{2}\right)$ **21.** No solution (inconsistent)   **23.** $(1, -2)$ **25.** $(4, -2)$ **27.** $\left(-\dfrac{2}{5}, \dfrac{2}{3}\right)$

29. $(2, 5)$ **31.** $(1, 2)$ **33.** Mistake: 3 was not distributed to $3y$. Correct: $\left(\dfrac{27}{11}, -\dfrac{13}{11}\right)$ **35.** $-6, -2$ **37.** Jon is 20; Tony is 7.

39. Length: 94 ft.; width: 50 ft.   **41.** Length: 2300 ft.; width: 150 ft.   **43.** $135°, 45°$ **45.** Checking: \$840; savings: \$3360   **47.** U2; \$292 million; Bon Jovi; \$191 million   **49.** 27 3-hour courses, 15 5-hour courses   **51.** 4:45 P.M.   **53.** 1:33 P.M.

Review Exercises: 1. $5x - 4y$ **2.** $-3x + 21y$ **3.** $9x$ **4.** $13y$ **5.** 80 **6.** $-\dfrac{3}{2}$

Exercise Set 4.3

Prep Exercises   1. The elimination method is advantageous over graphing when the solution involves fractions. The elimination method is advantageous over substitution when no coefficients are 1.   **2.** y because $-y$ and y are additive inverses.   **3. a.** x because multiplying only the second equation by -2 allows the x to be eliminated; whereas to eliminate y, we need to multiply each equation by a number. **b.** Multiply $3x + 5y = 7$ by -2 and then add the resulting equation to $6x - 2y = 1$.   **4. a.** y because $4y$ and $-6y$ have opposite signs, which means we can multiply each equation by a positive number; whereas to eliminate x, we need to multiply one equation by a positive number and the other by a negative number. **b.** Multiply the first equation by 3 and the second by 2 and then add the resulting equations.   **5.** Both variables have been eliminated, and the resulting equation is false. **6.** Both variables have been eliminated, and the resulting equation is true.

Exercises   1. $(6, -8)$ **3.** $(2, 3)$ **5.** $(-1, -2)$ **7.** $(2, 3)$ **9.** $(3, -1)$ **11.** $(-4, 1)$ **13.** $(-4, 3)$ **15.** All ordered pairs along $2x + y = -2$ (dependent).   **17.** No solution (inconsistent)   **19.** $(1, 1)$ **21.** $(4, -4)$ **23.** $\left(-\dfrac{2}{5}, \dfrac{2}{3}\right)$ **25.** $(0.2, 0.3)$ **27.** $(6, 4)$

29. $(7, -9)$ **31.** $(-4, -13)$ **33.** $\left(2, \dfrac{5}{2}\right)$ **35.** $(4, 4)$ **37.** Mistake: Did not multiply the right side of $x - y = 1$ by -9. Correct: $(5, 4)$

39. 22 in., 8 in.   **41.** 10 mph in still water; current is 2 mph   **43.** 470 mph in still air, 30 mph winds   **45.** \$12,000 at 4%, \$3500 at 3%   **47.** \$11,100 at 5%, \$2150 at 8%   **49.** \$10,250 at 2.9%, \$6000 at 6%   **51.** 120 ml   **53.** 200 ml of 15%, 100 ml of 45%   **55.** 11 lb. of Italian Roast, 9 lb. of GCB

Review Exercises: 1. $20x + 28y - 4z$ **2.** $x - 3y$ **3.** $y = -\dfrac{1}{3}x + 2$ **4.** $x = 6y + 30$ **5.** $x = 2$ **6.** $y = 0$

Exercise Set 4.4

Prep Exercises   1. It doesn't matter which equations are chosen as long as the second pair is different from the first pair.   **2.** We eliminate the same variable from two pairs of equations to get two equations with the same two variables.   **3.** Answers may vary. A good choice is to eliminate z using Equations 1 and 2 and 1 and 3. Multiplying Equation 1 by -3 and then adding this result to Equations 2 and 3 takes very few steps.   **4.** After adding two equations, if the resulting equation has no variables and is false, the system has no solution.   **5.** Two parallel planes intersecting a third plane   **6.** It will be on the line of intersection of two planes but not on the third plane.

Exercises   1. yes   **3.** no   **5.** yes   **7.** $(2, -1, 4)$ **9.** $(-1, 0, 3)$ **11.** $(-1, 1, 3)$ **13.** No solution (inconsistent)

15. Infinite number of solutions (dependent equations)   **17.** $(2, -3, 1)$ **19.** $\left(-2, -\dfrac{1}{2}, 1\right)$ **21.** $(2, -1, 1)$ **23.** $(-3, 7, -4)$

25. $(2, -4, 1)$ **27.** $(4, -4, 2)$ **29.** $105°, 35°, 40°$ **31.** Burger: \$2.50; fries: \$1.50; drink: \$1.00   **33.** Three 3-point field goals, one 2-point field goal, three free throws   **35.** Ham: 3 lb.; turkey: 5 lb.; beef: 2 lb.   **37.** \$2000 at 4%, \$2500 at 6%, \$3500 at 7%   **39.** Bicycling: 420; walking: 300; climbing stairs: 600

Review Exercises: **1.** $-6x + 12y + 3z - 27$ **2.** $y = -4$ **3.** -1 **4.** $(0, 1)$ **5.** $m = -2$ **6.**

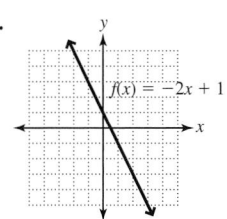

Exercise Set 4.5

Prep Exercises **1.** 4 rows, 2 columns **2.** Arrange each equation so that all variables are in the same order on the left side of the equal sign and the constant term is on the right side of the equal sign. Omit all of the variables and the equal signs. Place a vertical dashed line where the equal signs were so that it is between the coefficients of the variables and the constants. **3.** The dashed line corresponds to the equal signs in the equations. **4.** The rules are the same for both processes. With matrices, we use only the coefficients and the constants. With the elimination method, we also use the variables and the equal sign. **5.** A matrix is in row echelon form when the coefficient portion of the augmented matrix has 1's on the diagonal from upper left to lower right and 0's below the 1's. **6.** The bottom equation represents the value of the last variable. Substitute the known value into the equation above to determine the value of the next variable. Continue until all values for the variables are found.

Exercises **1.** $\begin{bmatrix} 14 & 7 & | & 6 \\ 7 & 6 & | & 8 \end{bmatrix}$ **3.** $\begin{bmatrix} 7 & -6 & | & 1 \\ 0 & -2 & | & 5 \end{bmatrix}$ **5.** $\begin{bmatrix} 1 & -3 & 1 & | & 4 \\ 2 & -4 & 2 & | & -4 \\ 6 & -2 & 5 & | & -4 \end{bmatrix}$ **7.** $\begin{bmatrix} 4 & 6 & -2 & | & -1 \\ 8 & 3 & 0 & | & -12 \\ 0 & -1 & 2 & | & 4 \end{bmatrix}$ **9.** $(4, 2)$ **11.** $(-6, -5, 1)$

13. $\begin{bmatrix} 1 & 3 & | & -1 \\ 0 & 11 & | & 4 \end{bmatrix}$ **15.** $\begin{bmatrix} 1 & -2 & 4 & | & 6 \\ 0 & 2 & -1 & | & -5 \\ 0 & 0 & -2 & | & 17 \end{bmatrix}$ **17.** $\begin{bmatrix} 1 & 2 & | & -2.5 \\ -1 & 3 & | & 2 \end{bmatrix}$ **19.** Replace R_2 with $3R_1 + R_2$. **21.** Replace R_3 with $-2R_2 + R_3$.

23. $(2, -3)$ **25.** $(1, 2)$ **27.** $(-3, -2)$ **29.** $(-2, 0)$ **31.** $(4, 6)$ **33.** $(4, -4)$ **35.** $(-1, 1, 3)$ **37.** $(2, 1, -2)$ **39.** $(2, -3, 1)$
41. $(-1, -1, -2)$ **43.** $(-3, 1, 5)$ **45.** $(8, -2, -4)$ **47.** $(3, -4)$ **49.** $(-6, 8)$ **51.** $(4, -3, 5)$ **53.** $(7, -6, 5)$ **55.** Mistake: In the

second step, $4R_1 + R_2$ is calculated incorrectly. Correct: The correct calculation is $\begin{bmatrix} 1 & 3 & | & 13 \\ 0 & 11 & | & 26 \end{bmatrix}$. The solution is $(65/11, 26/11)$.
57. Chicken sandwich: $3.50; drink: $1.20 **59.** Nile: 4150 mi.; Amazon: 3900 mi. **61.** $6400 in CD, $3600 in money market
63. CD: $16; book: $12, DVD: $28 **65.** 4 touchdowns, 4 extra points, 1 field goal **67.** $v_1 = 9, v_2 = 6, v_3 = 2$

Review Exercises: **1.** 8 **2.** 11 **3.** 2 **4.** 2 **5.** 3 **6.** -1

Exercise Set 4.6

Prep Exercises **1.** Select a point in the solution region and verify that it makes both inequalities true. **2.** After graphing each
inequality, determine the region of overlap. **3.** Parallel lines with shaded regions that do not overlap **4.** $\begin{cases} x > 0 \\ y > 0 \end{cases}$ **5.** $\begin{cases} x < 0 \\ y < 0 \end{cases}$
6. $\begin{cases} x > 0 \\ y < 0 \end{cases}$ **7.** $\geq$ **8.** $\leq$

Exercises

1.
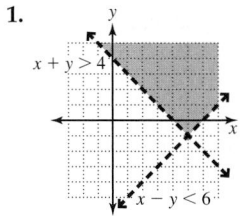

3.
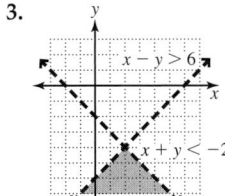

5.
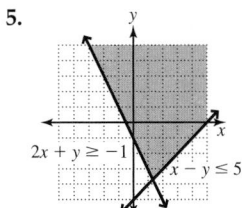

7.
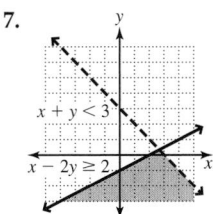

9.
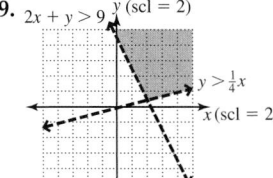

11.
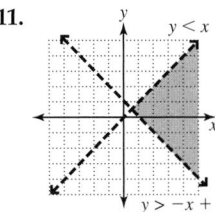

13.
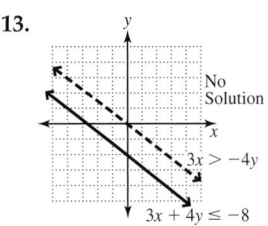

15.
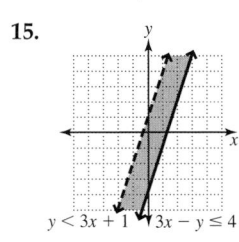

17. 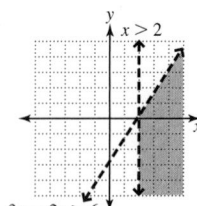   **19.** 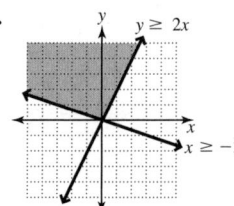   **21.** 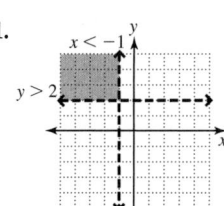   **23.**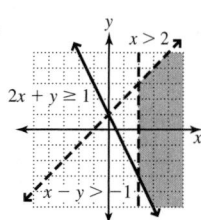

25. $x + y = 3$ should be a dashed line.   **27.** The wrong area is shaded. It should be the region containing the point $(1, 5)$.
29. a. $V + M \geq 1100$ $M \geq 500$ **b.** $V \leq 800$ $M \leq 800$ **c.**   **d.** Answers will vary. Some possible

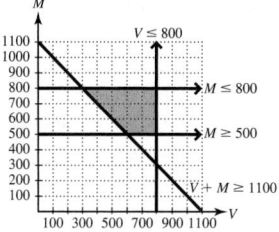

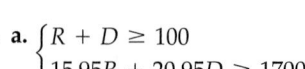

(V, M) pairs are $(500, 700)$, $(600, 500)$, and $(700, 600)$.   **31. a.** $\begin{cases} R + D \geq 100 \\ 15.95R + 20.95D \geq 1700 \end{cases}$ **b.**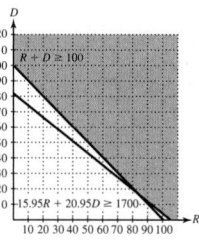

c. Answers will vary. Some possible (R, D) pairs are $(30, 90)$, $(40, 80)$, and $(50, 60)$.

Review Exercises: **1.** 9   **2.** -100 **3.** 5   **4.** $\dfrac{4}{7}$ **5.** 4   **6.** 21

Chapter 4 Summary and Review Exercises
4.1 Definitions/Rules/Procedures   group of two or more; makes all of the equations in the system true; variable; value; true
Exercises   **1.** yes   **2.** no   Definitions/Rules/Procedures   coordinates; no solution; infinite number   Exercises   **3.** $(-1, -4)$
4. $(3, -1)$ **5.** No solution   **6.** All ordered pairs that solve $3x - 2y = 6$.   Definitions/Rules/Procedures   at least one solution; no
solution; intersection; different; different; infinite; identical; same; same; no; parallel; same; y-intercepts   Exercises   **7. a.** consistent
with independent equations **b.** one solution   **8. a.** inconsistent **b.** no solution   **9. a.** consistent with independent equations
b. one solution   **10. a.** consistent with dependent equations **b.** infinite number of solutions

4.2 Definitions/Rules/Procedures   variables; substitute; Solve; substitute; variable   Exercises   **11.** $(4, -1)$ **12.** $\left(5, -\dfrac{2}{5}\right)$ **13.** All

ordered pairs that solve $3y - 4x = 6$ **14.** No solution   Definitions/Rules/Procedures   unknowns; equation   Exercises
15. Canada: 21,337 thousand; Mexico: 13,491 thousand   **16.** Western Europe: $15\dfrac{2}{3}$%; North America: $27\dfrac{1}{3}$%   **17.** 75°, 15°   **18.** 5:45 P.M.

4.3 Definitions/Rules/Procedures   standard; fractions; decimals; additive inverses; Add; Solve; substitute; original equation
Exercises   **19.** $(1, 3)$ **20.** $(2, -1)$ **21.** $(1, 5)$ **22.** $(5, -3)$ **23.** 47 adult, 92 children   **24.** $3500 at 7% and $1500 at 9%
25. 120 miles per hour   **26.** 30 ml

4.4 Definitions/Rules/Procedures $Ax + By + Cz = D$; elimination; same variable; two variables; original equations; all three
original equations   Exercises   **27.** $(1, 0, -3)$ **28.** No solution.   **29.** $(3, -3, 2)$ **30.** $(4, 0, 1)$ **31.** 1 vitamin supplement, 3 rolls of
film, 4 bags of candy   **32.** 4 first place, 3 second place, 9 third place

4.5 Definitions/Rules/Procedures   rectangular array; coefficients; constant; constant; coefficients; upper left; lower right;
interchanged; multiplied; divided; adding; multiple; row echelon   Exercises   **33.** $(4, -1)$ **34.** $(-2, 3, 1)$

4.6 Definitions/Rules/Procedures   overlap; solid   Exercises   **35.** 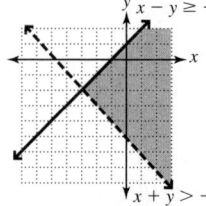   **36.**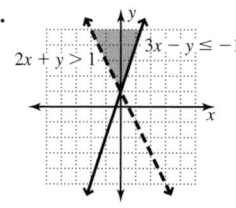

37. a. $\begin{cases} x + y \le 16 \\ 250x + 400y \ge 2750 \\ x \ge 0 \\ y \ge 0 \end{cases}$ **b.** 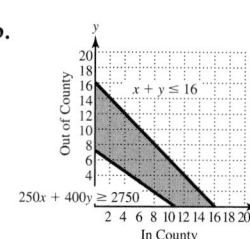 **c.** Answers may vary. One example is $(8, 4)$.

1. yes [4.1] **2.** no [4.4] **3.** $(-2, 1)$ [4.1] **4.** $(8, -1)$ [4.2] **5.** $(3, 1)$ [4.2] **6.** $(-4, -2)$ [4.3] **7.** All ordered pairs that solve $4x + 6y = 2$. [4.3] **8.** $(-2, 3, -2)$ [4.4] **9.** $(-4, 2, 3)$ [4.4] **10.** $(2, -4)$ [4.5] **11.** $(-1, 5, 2)$ [4.5] **12.** $(-2, 5)$ [4.6] **13.** $(-4, 0, 3)$ [4.4] **14.** 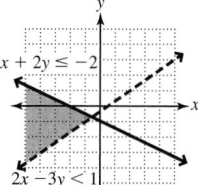 [4.6] **15.** 86 accountants, 77 waiters/waitresses [4.2] **16.** 180 [4.2] **17.** 7 mph [4.3]

18. $4000 at 6%, $8000 at 8% [4.3] **19.** Children: 150; student: 300; adult: 350. [4.3] **20. a.** $\begin{cases} 2l + 2w \le 200 \\ l \ge w + 10 \\ l > 0 \\ w > 0 \end{cases}$

b. 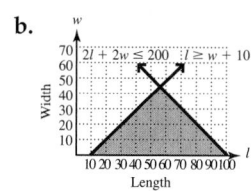 **c.** Answers may vary. One example is $(60, 20)$. [4.6]

1. false **2.** false **3.** true **4.** $8x + 12$ **5.** $\dfrac{c}{d}$ **6.** perpendicular **7.** 11 **8.** 0 **9.** $-\dfrac{1}{8}x + \dfrac{33}{10}y + 5$ **10.** $p = 13$
11. contradiction (no solution) **12.** $z > -5$ **13.** $x = 5$ **14.** $t = \dfrac{A - P}{Pr}$ **15.** $x = -2$ **16.** 13,600 ft^2.

17. 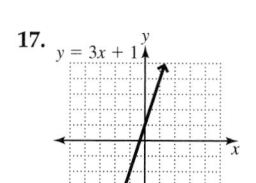 **18.** 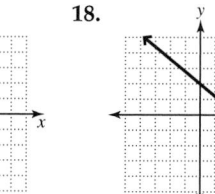 **19.**   **20.** 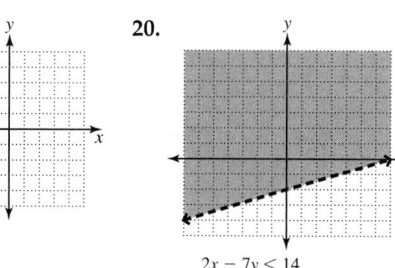 **21.** $y = -\dfrac{1}{5}x - \dfrac{7}{5}$

22. $y = -x - 3$ **23.** $3x + 4y = -4$ **24.** $(-3.5, -4)$ **25.** $(2, -1, 4)$ **26.** 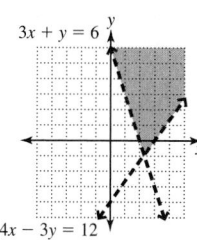 **27.** $L = 130$ ft. $W = 80$ ft.

28. $115°; 65°$ **29.** $35,000 at 3% and $15,000 at 2.5% **30.** 10 gallons of outside, 14 gallons of inside

Prep Exercises **1.** The result is negative. The order of operations agreement tells us to find the additive inverse *after* multiplying the two fives: $-5^2 = -(5 \cdot 5) = -25$. **2.** If a is a real number, where $a \ne 0$ and n is a natural number, then $a^{-n} = \dfrac{1}{a^n}$. **3.** 5 places to the right **4.** 8 places to the left **5.** Between the 4 and the 5 **6.** -6

Exercises **1.** 1 **3.** -8 **5.** -125 **7.** -16 **9.** $\dfrac{81}{256}$ **11.** -0.008 **13.** 5.29 **15.** $\dfrac{1}{a^5}$ **17.** $\dfrac{1}{8}$ **19.** $-\dfrac{1}{64}$ **21.** $-\dfrac{1}{16}$ **23.** -125 **25.** $\dfrac{5}{m^2}$

27. $-\dfrac{6}{x^7}$ **29.** $\dfrac{n^5}{m^5}$ **31.** $\dfrac{49}{25}$ **33.** b^6 **35.** 64 **37.** $-\dfrac{1}{7{,}962{,}624}$ **39.** $-\dfrac{512}{125}$ **41.** $-160{,}000$ **43.** ≈ 0.00039 **45.** Mistake: The expression
was interpreted to mean $(-3)^4$, but it means the additive inverse of 3^4. Correct: -81 **47.** Mistake: The negative sign in the
exponent was passed to the product instead of the original expression being inverted. Correct: $\dfrac{1}{25}$ **49.** $16{,}500{,}000{,}000$

51. $299{,}800{,}000$ **53.** $6{,}378{,}000$ **55.** $92{,}920{,}000$ **57.** 0.0000025 **59.** $0.0000000000000000000000000167$ **61.** 0.00000000295
63. $0.0000000000000000000000000006645$ **65.** 2.5×10^{13} **67.** 5.3×10^9 **69.** 2.342×10^7 **71.** 1.3×10^9 **73.** 5.86×10^{-7}
75. 9.1094×10^{-31} **77.** 6.25×10^{-5} **79.** 5×10^{-8} **81.** $8.95 \times 10^5, 7.2 \times 10^6, 7.5 \times 10^6, 9 \times 10^7, 1.3 \times 10^8$ **83.** The wavelength of
ultraviolet light is about 38.8 times the wavelength of X-rays.

Review Exercises: **1.** -11 **2.** $-4y + 12$ **3.** $266\dfrac{2}{3}$ **4.** $\dfrac{1}{5}$ **5.** slope $\dfrac{2}{3}$; y-intercept: $(0, -2)$ **6.** -65

Exercise Set 5.2
Prep Exercises **1.** A constant, variable, or product of a constant and variable(s) that are raised to whole-number powers
2. The numerical factor in a monomial **3.** Find the sum of the exponents on all variables in the monomial. **4.** A monomial
contains one term, a binomial contains two terms, and a trinomial contains three terms. **5.** Find the greatest degree of any of the
terms in the polynomial. **6.** Replace each variable with its given value and calculate the numerical result. **7.** Place the highest
degree term first, then the next highest degree, and so on. **8.** Add the coefficients and leave the variables and exponents the same.

Exercises **1.** monomial **3.** monomial **5.** not **7.** not **9.** monomial **11.** monomial **13.** c: -5; d: 3 **15.** c: -1; d: 5
17. c: -9; d: 0 **19.** c: 4.2; d: 4 **21.** c: 16; d: 3 **23.** c: 1; d: 1 **25.** trinomial, 1 **27.** monomial, 3 **29.** No special polynomial name, 3
31. binomial, 3 **33.** monomial, 0 **35.** Not a polynomial **37.** 4 **39.** 5 **41.** 7 **43.** 4 **45.** 8 **47.** 60 **49.** -8 **51.** -4.8
53. 25 **55. a.** 46 ft. **b.** 26.96 ft. **57. a.** 30 units **b.** 43 units **59. a.** ≈ 1273.4 ft.^{3} **b.** ≈ 537.2 ft.^{3} **61. a.** 26 million **b.** 45.95 million
c. 40.45 million **d.** 1.3 million; answers will vary, but the prediction is most likely not reasonable because of the increase in world
population and unforeseen social factors that might affect travel; it is not likely that the number will decrease that much.
63. $5x^7 - 3x^5 + 7x^4 - 8x + 14$ **65.** $-8r^6 + 18r^5 + 7r^3 - 3r^2 + 4r$ **67.** $-12w^4 - w^3 + 11w^2 + 5w + 20$ **69.** $5x - 3$

71. $-4a + 5$ **73.** $-2x^2 + 6x$ **75.** $2x^2 + \dfrac{5}{3}x$ **77.** $\dfrac{7}{6}a^2 - \dfrac{19}{15}a$ **79.** $-5x^2 + 14$ **81.** $5y^2 - 9y + 22$ **83.** $-2k^4 + 6k^3 - 3k^2 + 6k$
85. $-10a^4 + 7a^2 + a + 16$ **87.** $-21v^5 + 8v^3 + 35v^2 - 20$ **89.** $4a - 2b$ **91.** $-4y^2 + 3y$ **93.** $-9x^2 + 9y$ **95.** $\dfrac{19}{12}a^2 - \dfrac{17}{30}a$
97. $y^6 + 5yz^4 - 6yz - 3z^5 - 7z^3 - 10$ **99.** $-4w^2z - wz^2 + 12w^2 - 1$

Review Exercises: **1.** $-\dfrac{7}{12}$ **2.** -4.3 **3.** -1 **4.** 2×10^{11} **5.** 12 **6.** 600 L

Exercise Set 5.3
Prep Exercises **1.** To add polynomials, combine like terms. **2.** To identify the different polynomials **3.** Change the signs of each
term in the polynomial. **4.** To subtract polynomials: **a.** Write the subtraction statement as an equivalent addition statement.
i. Change the operation symbol from a minus sign to a plus sign. ii. Change the subtrahend (second polynomial) to its additive
inverse. To get the additive inverse, change the sign of each term in the polynomial. **b.** Combine like terms.

Exercises **1.** $8x + 1$ **3.** $10y + 12$ **5.** $7x$ **7.** $3x^2 - 2x + 12$ **9.** $6z^2 - 11z - 2$ **11.** $5r^2 - 7r - 1$ **13.** $9y^2 + 3$

15. $3x^2 + \dfrac{1}{12}x + 1$ **17.** $2r^2 + 1.4r - 7.8$ **19.** $5a^3 + 3a^2 + 3a - 3$ **21.** $3p^3 - p^2 + 7p + 9$ **23.** $-8w^4 + 3w^3 - 7w^2 + 13w - 9$

25. $-3a^3b^2 - 2ab^2 - 8a^2b + 4ab + 3b^2 - 8$ **27.** $-7u^4 + u^2v^3 + 7u^3v + 7u - 13v^2 - 3$

29. $-12mnp - 16m^2n^2 - 12mn^2p + 14m^2n - 31n + 9$ **31.** $2a^4 - \dfrac{5}{12}a^3b + \dfrac{14}{15}ab^3 - b^4$ **33.** $1.8a^2bc - 1.8ab^2c^2 - 2.4a^2b^2c^2$

35. $5x - 2$ **37.** $14a - 12$ **39.** $4x + 2$ **41.** $13a^2 - 1$ **43.** $2x^2 - 9x + 2$ **45.** $2z^2 - 7z + 9$ **47.** $12a^2 + 10a - 15$
49. $-u^3 - 2u^2 - 6u + 8$ **51.** $-11w^5 + 3w^4 + 5w^2 + 2$ **53.** $-11p^4 + 3p^3 - 13p^2 + 5p - 6$ **55.** $11v^4 + 3v^3 + 23v^2 - 5v - 9$
57. $2xy$ **59.** $-6xy + 6xz + 2yz$ **61.** $19p^3q^2 + 2pq^2 - pq + 16q^2 + 2$ **63.** $1.9a^2b + 1.1ab^2 - 6.5a^2b^2$ **65.** $32.1x + 35.45y + 6.01z$
67. $4.22a + 1.63b + 12.13c$ **69.** Mistake: Only the sign of the first term in the subtrahend was changed instead of all three terms.
Correct: $-4x^2 - 2x + 5$

Review Exercises: **1.** -17.25 **2.** 15 **3.** 0.0001 **4.** 10^7 **5.** $58{,}900{,}000$ **6.** 300 m^2

Exercise Set 5.4
Prep Exercises **1.** add **2.** The bases must be the same in order to add the exponents. **3. a.** multiply **b.** add **c.** Write y
unchanged in the product. **4.** $8 \cdot 6 = 48$, and 48 must be changed to 4.8 to be in scientific notation. This requires increasing the
exponent by 1 to account for moving the decimal point one place to the left. **5.** Use $(a^m)^n = a^{mn}$. So $(x^3)^8 = x^{3 \cdot 8} = x^{24}$.
6. Use $(ab)^n = a^n \cdot b^n$. So $(3x^4)^2 = 3^2(x^4)^2 = 9x^8$.

Exercises **1.** a^7 **3.** 2^{13} **5.** a^5b **7.** $6x^3$ **9.** $-8m^3n^2$ **11.** $-\dfrac{5}{28}s^3t^8$ **13.** $2.76a^3b^5c^6$ **15.** $-6r^5$ **17.** $-12x^4z^9$ **19.** $90x^6y^5z^3$

21. $5q^5r^{10}s^{21}$ **23.** $-\dfrac{1}{4}a^6b^3c^2$ **25.** $w^3x^3y^5$ **27.** Mistake: The exponents of x were multiplied instead of added. Correct: $35x^7y$

29. Mistake: The exponents of x were not added. Correct: $54x^4y^6z$ **31.** $5w^2$ **33.** $4.5w^3$ **35. a.** height $=\dfrac{1}{2}b$, top side $=\dfrac{1}{4}b$

b. $A=\dfrac{5}{16}b^2$ **37.** 1.2×10^9 **39.** 2.4×10^{14} **41.** 1.827×10^5 **43.** 1.1421×10^{-4} **45.** 5.612×10^{-10} **47.** 4.2×10^4 mi.^{2}

49. $\approx 5.8404\times 10^8$ mi. **51.** $\approx 1.412\times 10^{27}$ m^3 **53.** 2.9817×10^{-19} joule **55.** 2.16×10^7 joules **57.** x^6 **59.** 3^8 **61.** h^8 **63.** $-y^6$

65. x^3y^3 **67.** $27x^6$ **69.** $-8x^9y^6$ **71.** $\dfrac{1}{64}m^6n^3$ **73.** $64p^{24}q^{24}r^{12}$ **75.** $-0.008x^6y^{15}z^3$ **77.** $18x^7$ **79.** r^5s^4 **81.** $100a^6b^4$ **83.** $\dfrac{1}{1024}a^{10}b^{15}c^5$

85. $360a^2b^3$ **87.** $216a^{18}b^8$ **89.** $324u^6v^3$ **91.** $3.24u^6v^3$ **93.** Mistake: The exponents were added instead of multiplied.
Correct: $25x^6$ **95.** Mistake: The coefficient was multiplied by 4 instead of being raised to the 4th power. Also, the exponents were
not multiplied properly. Correct: $16x^{16}y^4$

Review Exercises: **1.** $6x+8$ **2.** $-1.8t^5u^2$ **3.** $-22x$ **4.** $-4x-4z$ **5.** -61 **6.** 6

Exercise Set 5.5
Prep Exercises **1.** Distributive property **2.** First, Outer, Inner, Last **3.** To multiply two polynomials: **a.** Multiply each term in
the second polynomial by each term in the first polynomial. **b.** Combine like terms. **4.** They differ only in the sign separating the
terms. **5.** a^2-b^2 **6.** $a^2+2ab+b^2$; $a^2-2ab+b^2$

Exercises **1.** $5x+15$ **3.** $-12x+24$ **5.** $28n^2+8n$ **7.** $9a^2-9ab$ **9.** $\dfrac{1}{4}m^2-\dfrac{5}{8}mn$ **11.** $-0.3p^5+0.6p^2q^2$

13. $20x^3+10x^2-25x$ **15.** $-21x^5+15x^3-3x^2$ **17.** $-r^4s^4-3r^3s^3+r^2s^2$ **19.** $-4a^5b^2+12a^2b^3-6a^3b^3+2a^4b^4$

21. $12a^3b^3c^3-6a^2b^2c^2+15abc$ **23.** $4b^6-b^4+\dfrac{1}{5}b^3-\dfrac{1}{2}b^2+16b$ **25.** $-r^4t^3+3.2r^3t^4-1.64r^2t^5+0.8rt^2$

27. $25x^6y-15x^5y^2+50x^3y^9$ **29.** $x^2+7x+12$ **31.** $x^2-5x-14$ **33.** $y^2-9y+18$ **35.** $6y^2+19y+10$ **37.** $15m^2+11m-12$

39. $12t^2-26t+10$ **41.** $15q^2+11qt-12t^2$ **43.** $14y^2+34xy+12x^2$ **45.** $a^4-a^2b-2b^2$ **47. a.** $x+5$ **b.** $2x$ **c.** $2x(x+5)$

d. $2x^2+10x$ **e.** They describe the same area. **49. a.** $x+3$ **b.** $x+2$ **c.** $(x+2)(x+3)$ **d.** x^2+5x+6 **e.** They describe the

same area. **51.** x^3-3x+2 **53.** $3a^3-4a^2-a+2$ **55.** $6c^3+c^2-11c-6$ **57.** $4f^3-27fg^2-27g^3$ **59.** $27m^3-8n^3$

61. $8a^3+125b^3$ **63.** $x^4+3x^3y+2x^2y^2+xy^3-y^4$ **65.** $x^4+14x^3+69x^2+140x+100$ **67.** $x-3$ **69.** $4x+2y$ **71.** $4d-3c$

73. $-3j+k$ **75.** x^2-25 **77.** $4m^2-25$ **79.** x^2-y^2 **81.** $64r^2-100s^2$ **83.** $4x^2-9$ **85.** x^2+6x+9 **87.** $16t^2-8t+1$

89. $m^2+2mn+n^2$ **91.** $4u^2+12uv+9v^2$ **93.** $81w^2-72wz+16z^2$ **95.** $81-90y+25y^2$ **97.** $10xyz+2y^2z$ **99.** h^2+4h

101. w^2+3w **103.** $4x^3+24x^2+20x$ **105.** $4w^3+6w^2$ **107.** $V=3.14r^3+6.28r^2$, or $\pi r^3+2\pi r^2$

Review Exercises: **1.** 0.45 **2.** -5 **3.** 6,304,000 **4.** x^7 **5.** $r=\dfrac{5-d}{m}$ **6.** 9, 36

Exercise Set 5.6
Prep Exercises **1.** subtract the divisor's exponent from the dividend's exponent **2.** $a^{-n}=\dfrac{1}{a^n}$ **3.** Separate the decimal factors and

powers of 10 into a product of two fractions. Then calculate the decimal division and divide the powers of 10 separately.
4. a. divide **b.** subtract **c.** Write y unchanged in the numerator. **5.** Divide each term in the polynomial by the monomial.

6. $x+5+\dfrac{-4}{x+2}$ or $x+5-\dfrac{4}{x+2}$ **7.** Use a 0 placeholder when the dividend, written in descending order, has a "missing"

term. **8.** a^{m+n} **9.** a^{m-n} **10.** a^{mn}

Exercises **1.** $\dfrac{1}{a^3}$ **3.** $\dfrac{1}{2^5}$ **5.** y^5 **7.** 3 **9.** $\dfrac{1}{4^6}$ **11.** $\dfrac{1}{x^2}$ **13.** a^2 **15.** $\dfrac{1}{w^3}$ **17.** $\dfrac{1}{a^7}$ **19.** r^{10} **21.** y^8 **23.** $\dfrac{1}{p^3}$ **25.** 1 **27.** 2.6×10^{-1}

29. 6×10^{-8} **31.** 7.1×10^{10} **33.** 2.7×10^4 **35.** 500 sec. or $8\dfrac{1}{3}$ min. **37.** \$50,828.03 **39.** $\approx 9.3\times 10^{-4}$kg **41.** $5x^3$ **43.** $\dfrac{-4}{x^3}$

45. $-3m^2n^4$ **47.** $-\dfrac{8}{5}q^3$ **49.** $\dfrac{4y^4}{3x^5}$ **51.** $\dfrac{4c}{3a^2}$ **53.** $a+2b$ **55.** $4x^2-2x$ **57.** $3x+\dfrac{2}{x}$ **59.** $xy-3y^2$ **61.** $-2c+8ab$ **63.** $3x^2+2x-1$

65. $2x-4+\dfrac{3}{x}$ **67.** $12uv^3+\dfrac{4v^4}{u}-5v$ **69.** $y^2+y^4-3y^5+\dfrac{1}{y^2}$ **71.** $\dfrac{5mn}{2}$ **73. a.** $3x-5+\dfrac{4}{x}$ **b.** Area $=18$, length $=6$, width $=3$.

75. $x+4$ **77.** $m-5$ **79.** $x-12+\dfrac{4}{x-5}$ **81.** $3x^2+15x+2$ **83.** $2x^2+3x+6+\dfrac{17}{x-2}$ **85.** $p^2-5p+25$ **87.** $3x+2$

89. $2y - 2 + \dfrac{23}{7y + 3}$ **91.** $4k^2 - 1 + \dfrac{8}{k + 2}$ **93.** $7u^2 - 11u + 12 - \dfrac{30}{3u + 2}$ **95.** $b^2 - 3b + 5 + \dfrac{3}{b - 3}$ **97.** $y^2 - 2y + 3$

99. a. $x + 4$ **b.** Length $= 7$ ft., height $= 8$ ft., volume $= 336$ ft.^{3} **101.** $\dfrac{36m^4}{n^4}$ **103.** $\dfrac{x^9}{64y^6}$ **105.** $256r^9s^7$ **107.** m^{16} **109.** $\dfrac{1}{x^{14}}$

111. $\dfrac{y^{27}}{x^9}$ **113.** x^{14} **115.** $\dfrac{27y^6}{x^3z^6}$ **117.** $\dfrac{1}{x^{10}}$ **119.** $\dfrac{4x^4z^3}{y^5}$ **121.** $\dfrac{2}{3ab^2}$ **123.** $\dfrac{1}{2x^{11}}$

Review Exercises: **1.** $2^4 \cdot 3^3 \cdot 7$ **2.** $-18x^4y^3$ **3.** $3xy - 4y$ **4.** $x + \dfrac{1}{2}y$ **5.** 340 ft.^{2} **6.** -24

Chapter 5 Summary and Review Exercises

5.1 Definitions/Rules/Procedures   positive; odd; $\dfrac{a^n}{b^n}$; 1; $\dfrac{1}{a^n}$; a^n; $\left(\dfrac{b}{a}\right)^n$ Exercises   **1.** $\dfrac{8}{125}$ **2.** -16 **3.** $\dfrac{1}{25}$ **4.** 36

Definitions/Rules/Procedures $1 \le |a| < 10$; integer; right; exponent; left; absolute value; 1; 10; right; 10^n; zeros; 1; 10; right; negative integer   Exercises   **5.** $4{,}500{,}000{,}000$ **6.** $0.0000000000000000000000001663$ **7.** 1.661×10^{-24} **8.** 3×10^8

5.2 Definitions/Rules/Procedures   constant; variable; constant; variable(s); numerical factor; sum; sum of monomials; same variable; two; three   Exercises   **9.** yes   **10.** yes   **11.** no   **12.** no   **13.** c: 6; d: 4   **14.** c: 27; d: 0   **15.** c: -2.6; d: 4   **16.** c: -1; d: 1   **17.** binomial   **18.** monomial   **19.** Not a polynomial   **20.** No special polynomial name   Definitions/Rules/Procedures   greatest degree; highest degree term; highest degree term; combining   Exercises   **21.** 9; $-2x^9 + 21x^5 - 19x^3 - 3x + 15$ **22.** 4; $-j^4 + 22j^2 + 5j - 19$ **23.** $4y^5 + 4y^4 + 8$ **24.** $-2m^3 + 7m^2 + m + 3$ **25.** $-11x^2yz - 2xyz^2$ **26.** $-8a^5 + 3abc^3 - 8abc + 8$ **27.** $-3j^4 + 3jk^3 + k^2 + 10jk + 12$ **28.** $3mn^2 - 8n^2 - 9mn + 7$

5.3 Definitions/Rules/Procedures   combine like terms; addition statement; minus sign; plus sign; additive inverse; additive inverse; like terms   Exercises   **29.** $10x^2 + 3x - 4$ **30.** $-2n^3 + 3n^2 + 4n + 3$ **31.** 9   **32.** $-x^3 + 4x^2 + 17$ **33.** $8x^2 - xy + 6y^2$ **34.** $-3m^2 + 8mn - 2$

5.4 Definitions/Rules/Procedures a^{m+n}; coefficients; Add; unchanged; a^{mn}; a^nb^n; coefficient; exponent   Exercises   **35.** m^5 **36.** $10a^9$ **37.** $-6x^6y^6$ **38.** $25x^4y^2$ **39.** $72u^{10}$ **40.** $2.25 \times 10^8 \text{m}^2$ **41.** $2.94 \times 10^{-4}\text{V}$

5.5 Definitions/Rules/Procedures   multiply each term in the polynomial by the monomial   Exercises   **42.** $4a^2 - 12a$ **43.** $-12b^3 + 24b^2 + 6b$ **44.** $-12a^2bc + 16a^2b^4c^2 - 8a^2b^2c^4 - 32abc^2$ **45. a.** $20x^2 - 28x$ **b.** 96 ft.^{2} Definitions/Rules/Procedures   second polynomial; first polynomial; like terms   Exercises   **46.** $x^2 + 4x - 5$ **47.** $12m^2 + 28m - 5$ **48.** $15a^2 - 4ab - 4b^2$ **49.** $3y^3 - 7y^2 + 14y - 4$ **50.** $a^4 - a^3b - 2a^2b^2 - 3ab^3 - b^4$ Definitions/Rules/Procedures   signs; $a^2 - b^2$; $a^2 + 2ab + b^2$; $a^2 - 2ab + b^2$ Exercises   **51.** $4x^2 - 1$ **52.** $16 - x^2$ **53.** $x^2 - 12x + 36$ **54.** $9r^2 + 30r + 25$

5.6 Definitions/Rules/Procedures a^{m-n} Exercises   **55.** x^8 **56.** u^7 **57.** $\dfrac{1}{s^6}$ **58.** x^6 **59.** $\dfrac{1}{x^4}$ Definitions/Rules/Procedures   coefficients; quotient rule; unlike; positive   Exercises   **60.** $7z^4$ **61.** $-\dfrac{8x}{y^3}$ **62.** $-\dfrac{8u^2}{5}$ **63.** $\dfrac{2a^3}{3z^3}$ **64.** $\dfrac{b^6}{8a^9}$ **65.** $\dfrac{4j^8}{9hk^2}$ **66.** 7   **67.** 1   Definitions/Rules/Procedures $\dfrac{a}{c} + \dfrac{b}{c}$; polynomial; monomial   Exercises   **68.** $x - 5$ **69.** $-y^2 + 2y - 3$ **70.** $1 + 10st^3 - 2t^4$

71. $a^2 - \dfrac{1}{c} + 2ac$ **72.** $2 + b$ **73.** $\dfrac{2z}{5y^2} - \dfrac{xz}{y} + \dfrac{2}{y}$ **74. a.** $9x - 10$ **b.** 24 in., 26 in., and 624 in.^{2} **75.** ≈ 4.2 yr.

Definitions/Rules/Procedures   long division; quotient $+ \dfrac{\text{remainder}}{\text{divisor}}$ Exercises   **76.** $z + 5$ **77.** $2m - 2 + \dfrac{1}{3m - 2}$

78. $4x^2 + 6x + 9$ **79.** $2s^2 + 2s - \dfrac{3}{s - 1}$ **80.** $4x - 5$ Exponents Summary   1; indeterminate; $\dfrac{1}{a^n}$; a^n; $\left(\dfrac{b}{a}\right)^n$; a^{m+n}; a^{m-n}; a^{mn}; a^nb^n; $\dfrac{a^n}{b^n}$

Chapter 5 Practice Test

1. $\dfrac{1}{8}$ [5.1]   **2.** $\dfrac{9}{4}$ [5.1]   **3.** 0.006201 [5.1]   **4.** 2.75×10^8 [5.1]   **5.** 3 [5.2]   **6.** 4 [5.2]   **7.** 56 [5.2]   **8.** $-10x^4 - 6x^3 + 14x^2 + 12$ [5.2]
9. $8x^2 + x - 4$ [5.3]   **10.** $5x^4 - 3x^2 - x + 8$ [5.3]   **11.** $12x^7$ [5.4]   **12.** $-2a^6b^4c^7$ [5.4]   **13.** $16x^2y^6$ [5.4]   **14.** $3x^3 - 12x^2 + 15x$ [5.5]
15. $-24t^5u + 48t^3u^3$ [5.5]   **16.** $n^2 + 3n - 4$ [5.5]   **17.** $4x^2 - 12x + 9$ [5.5]   **18.** $x^3 - 2x^2 - 5x + 6$ [5.5]   **19.** $6n^2 - 7n - 20$ [5.5]
20. x^5 [5.6]   **21.** $\dfrac{1}{x^5}$ [5.6]   **22.** $\dfrac{x^9y^4}{9}$ [5.6]   **23.** $4x^3 + 3x$ [5.6]   **24.** $x - 4$ [5.6]   **25.** $5x - 4 + \dfrac{6}{3x - 2}$ [5.6]

Chapters 1–5 Cumulative Review Exercises

1. false   **2.** false   **3.** true   **4.** false   **5.** coefficient, coefficient   **6.** $(0, 0)$ or the origin   **7.** 23   **8.** -1 **9.** 60   **10.** $5y^2 - 4y - 1$

11. $6x - 6$ **12.** $2x^2 - 11x - 21$ **13.** $4x^2 - 12x + 9$ **14.** $m - 2 + \dfrac{1}{3m}$ **15.** $x + 3 + \dfrac{1}{x + 2}$ **16.** $\dfrac{x^6}{27y^9}$ **17.** $\dfrac{4x^6}{y^2}$ **18.** $-108x^{16}$

19. 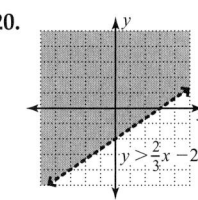   **20.**   **21.** -12 **22.** $x < -10$ **23.** 36   **24.** $-\dfrac{4}{5}$ **25.** $l = \dfrac{P - 2w}{2}$ **26.** $(4, -2)$ **27.** $(4, -3)$

28. \$33,750 at 13%, \$11,250 at 9%   **29.** 480   **30.** $\approx 53.7\%$

Chapter 6
Exercise Set 6.1
Prep Exercises   **1.** Divide by 1, 2, 3, and so on, writing each divisor and quotient pair as a product until we have all possible combinations.   **2.** The largest natural number that divides all given numbers with no remainder.   **3.** The factorization for the GCF includes only those prime factors common to all of the factorizations, each raised to its smallest exponent.   **4.** Rewrite the given polynomial as a product of the GCF and the quotient of the polynomial and the GCF. Polynomial $= \text{GCF}\left(\dfrac{\text{Polynomial}}{\text{GCF}}\right)$

5. Four-term polynomials.   **6.** Group together pairs of terms and factor the GCF out of each pair.

Exercises   **1.** 1, 3, 9   **3.** 1, 3, 11, 33   **5.** 1, 2, 3, 6, 9, 18   **7.** 1, 2, 4, 8, 16   **9.** 1, 2, 4, 11, 22, 44   **11.** 1, 2, 3, 4, 5, 6, 10, 12, 15, 20, 30, 60
13. 1, 2, 4, 7, 8, 14, 28, 56   **15.** 1, 2, 3, 5, 6, 9, 10, 15, 18, 30, 45, 90   **17.** 3   **19.** 8   **21.** 6   **23.** $2xy$ **25.** $5h^2$ **27.** $3a^4b$ **29.** $5(c - 4)$
31. $4(2x - 3)$ **33.** $x(x - 1)$ **35.** $6z^4(3z^2 - 2)$ **37.** $3p^3(6 - 5p^2)$ **39.** $3ab(2a - b)$ **41.** $7uv(2v - 1)$ **43.** $25x(y - 2z + 4)$
45. $xy(x + y + x^2y^2)$ **47.** $4ab^2c(7b - 9a)$ **49.** $4pq(5p + 6 - 4q)$ **51.** $3mnp(n^4p + 6n^2 - 2)$ **53.** $21a^2b^2(5a - 3b + 4a^4b^2)$
55. $3x(6x^3 - 3x^2 + 10x - 4)$ **57.** $-3(x - 2y)$ **59.** $-5a(4a + 3)$ **61.** $-4a^3b(3a - 5b^2)$ **63.** $-4(x^2 + 2x - 4)$
65. $-6x^2y^2z(4x + 5yz^3 - 2x^2y^3z)$ **67.** $(a - 3)(y + 2)$ **69.** $(2m + 3n)(4a + b)$ **71.** $(4m - 5)(5x - 2)$
73. $2(6p + 5q)(r - 2s)$ **75.** $(x + 2)(b + c)$ **77.** $(m - n)(a - b)$ **79.** $(x + 2)(x^2 - 3)$ **81.** $(1 - m)(1 + m^2)$
83. $(3x + 5)(y + 2)$ **85.** $(4b - 1)(b + 1)$ **87.** $(3 + b)(2 - a)$ **89.** $(x + 2y)(3a + 4b)$ **91.** $(y - s)(x^2 - r)$
93. $(w + 3z)(w + 5)$ **95.** $(s + y)(3t - 2)$ **97.** $(x^2 + y^2)(a - 5)$ **99.** $2(a + b)(a + 3)$ **101.** $2(2b + 3)(3a + 4)$
103. $3(4x - y)(2z - w)$ **105.** $3a(y - 4)(a + 3)$ **107.** $2x(x + 3y)(x + 5)$ **109.** $6x(x + 6)$ **111.** $3x(7x + 9)$
113. $12\pi r^2(3r + 2)$

Review Exercises: **1.** 3,740,000,000   **2.** 4.56×10^7 **3.** $x^2 + 7x + 10$ **4.** $x^2 - x - 12$ **5.** $x^2 - 8x + 15$ **6.** $x^2 - 7x$

Exercise Set 6.2
Prep Exercises   **1.** $8; 6$ **2.** $-12; -4$ **3.** $-; -$ **4.** $-; +$ **5.** $+; -$ **6.** $6y$

Exercises   **1.** 5   **3.** 2   **5.** 6   **7.** 2   **9.** $(r + 3)(r + 1)$ **11.** $(x - 7)(x - 1)$ **13.** $(z - 3)(z + 1)$ **15.** $(y + 2)(y + 3)$
17. $(u - 2)(u - 4)$ **19.** $(u + 3)(u - 2)$ **21.** $(a + 3)(a + 4)$ **23.** $(y - 3)^2$ **25.** $(w - 4)(w + 3)$ **27.** $(x - 6)(x + 5)$
29. $(n - 6)(n - 5)$ **31.** $(r - 3)(r - 6)$ **33.** $(x - 8)(x + 3)$ **35.** $(p - 9)(p + 4)$ **37.** prime   **39.** prime
41. $(p - 3q)(p - 7q)$ **43.** $(a - 9b)(a + 3b)$ **45.** $(x - 12y)(x - 2y)$ **47.** $(r - 6s)(r + 5s)$ **49.** $4(x - 3)(x - 7)$
51. $2m(m - 1)(m - 6)$ **53.** $3b(a - 8)(a + 3)$ **55.** $4(x - 3)^2$ **57.** $n^2(n + 2)(n + 3)$ **59.** $7u^2(u + 5)(u + 1)$
61. $6b^2c(a - 2)(a - 4)$ **63.** $3(x - 2y)(x - 5y)$ **65.** $2b(a - 6b)(a + 3b)$ **67.** $4x^2y(x^2 - 3xy - 15y^2)$ **69.** $(x - 6)(x + 1)$
71. $(x - 4)(x + 1)$ **73.** 4, 20   **75.** 7, 8, 13   **77.** 6, 10, 12   **79.** 9, 16, 21, 24, 25   **81.** Length: $h + 4$; width: $h + 2$ **83.** Length: $w - 6$; width: $w - 10$

Review Exercises: **1.** 1   **2.** $10x^2 + 17x + 3$ **3.** $15y^2 - 2y - 24$ **4.** $72n^3 - 60n^2 + 8n$ **5.** $5x^2y(2y^2 - x - 4y)$
6. $4ab^2(2ab^2 - 4 - 3ab)$ **7.** $(x + 2)(x - 3)$ **8.** $(z + 5)(z - 6)$ **9.** $(3c - d)(5a - b)$ **10.** $(3c + 2d)(2a - 3b)$

Exercise Set 6.3
Prep Exercises   **1.** monomial GCF common to all of the terms   **2.** Write a pair of first terms whose product is ax^2.   **3.** Write a pair of last terms whose product is c.   **4.** $ac; b$ **5.** 6 and 5   **6.** -12 and 2

Exercises   **1.** 2, 4   **3.** $t, 4t$ **5.** $7, 3x$ **7.** $(2j + 1)(j + 2)$ **9.** $(2y - 5)(y + 1)$ **11.** $(3m - 4)(m - 2)$ **13.** $(6a + 7)(a + 1)$
15. prime   **17.** $(4a - 3)(a - 4)$ **19.** $(2x + 3)(3x + 5)$ **21.** $(2d - 3)(8d + 5)$ **23.** $(3p + q)(p + 4q)$
25. $(k + h)(5k - 12h)$ **27.** $(2a - 5b)(6a - 5b)$ **29.** $(8m - 3n)(m - 3n)$ **31.** $y(8x^2 - 4x - 1)$ **33.** $3(2y + 1)(2y + 3)$
35. $2a(3b - 7)(b - 1)$ **37.** $2w(3w + 2v)(w + 2v)$ **39.** $(a + 1)(3a + 1)$ **41.** $(2t - 1)(t - 1)$ **43.** $(3x - 7)(x + 1)$
45. $(y - 2)(2y + 3)$ **47.** $(2a - 1)(5a - 7)$ **49.** $(4r + 3)(2r - 3)$ **51.** $(5x - 2)(4x - 3)$ **53.** $(2k + j)(3k + 2j)$
55. $(x - 5y)(5x - y)$ **57.** $(5s - 2t)(2s + t)$ **59.** $(3u - 2v)(2u + 3v)$ **61.** prime   **63.** $2(k - 2)(2k - 3)$
65. $2m(2m + 3)(3m - 2)$ **67.** $8x(x - 3y)(3x + y)$ **69.** $4x(2x - 3y)(3x + 2y)$ **71.** Mistake: Using *FOIL* to check, we see that the first and last terms check but the inner and outer terms combine to give $31x$ instead of $13x$. Correct: $(2x + 1)(3x + 5)$
73. Mistake: Did not factor completely. Correct: $4(n + 2)(n + 1)$ **75.** $(5x - 2)(3x - 1)$ **77.** $2w + 1$ and $3w - 2$
79. 7, 8, or 13   **81.** 4, 12, or 44   **83.** 1, 11, 19, or 41

Review Exercises: **1.** $9x^2 - 25$ **2.** $x^3 - 8$ **3.** $4x^2 - 20x + 25$ **4.** $4cd^2(4c^2 - 6cd^2 + 9)$ **5.** $(b + 5)(a + 3)$ **6.** $(m - 4)(m + 6)$
7. $3x(x - 2)(x + 5)$

Exercise Set 6.4

Prep Exercises **1.** The first and last terms are perfect squares, and twice the product of their roots equals the middle term.
2. $(a + b)^2(a - b)^2$ **3.** $(a + b)(a - b)$ **4.** conjugates **5.** $(a + b)(a^2 - ab + b^2)$ **6.** $(a - b)(a^2 + ab + b^2)$

Exercises **1.** $(x + 7)^2$ **3.** $(b - 4)^2$ **5.** Not a perfect square. **7.** $(5u - 3)^2$ **9.** $(10w + 1)^2$ **11.** $(y + z)^2$ **13.** $(2p - 7q)^2$
15. $(4g + 3h)^2$ **17.** $4(2t + 5)^2$ **19.** $(x + 2)(x - 2)$ **21.** $(4 + y)(4 - y)$ **23.** $(p + q)(p - q)$ **25.** $(5u + 4)(5u - 4)$
27. prime **29.** $(8m + 5n)(8m - 5n)$ **31.** $2(5x + 4y)(5x - 4y)$ **33.** $4(x^2 + 25y^2)$ **35.** $(x^2 + y^2)(x + y)(x - y)$
37. $(x^2 + 4)(x + 2)(x - 2)$ **39.** $(n - 3)(n^2 + 3n + 9)$ **41.** $(x + 3)(x^2 - 3x + 9)$ **43.** $(x - 1)(x^2 + x + 1)$
45. $(m + n)(m^2 - mn + n^2)$ **47.** $(3k - 2)(9k^2 + 6k + 4)$ **49.** $(3k + 2)(9k^2 - 6k + 4)$ **51.** $(c - 4d)(c^2 + 4cd + 16d^2)$
53. $(5x + 4y)(25x^2 - 20xy + 16y^2)$ **55.** $(3x - 4y)(9x^2 + 12xy + 16y^2)$ **57.** $(2p + qz)(4p^2 - 2pqz + q^2z^2)$

59. $2(x + 5)(x - 5)$ **61.** $\left(4x + \dfrac{5}{7}\right)\left(4x - \dfrac{5}{7}\right)$ **63.** $2u(u + 1)(u - 1)$ **65.** $y^3(y + 4b)(y - 4b)$ **67.** $2x(25x^2 + 1)$

69. $3(y - 2z)(y^2 + 2yz + 4z^2)$ **71.** $\left(c - \dfrac{2}{3}\right)\left(c^2 + \dfrac{2}{3}c + \dfrac{4}{9}\right)$ **73.** $2c(2c + d)(4c^2 - 2cd + d^2)$ **75.** $(2a - b + c)(2a - b - c)$

77. $(4 - 3x + 3y)(4 + 3x - 3y)$ **79.** $(x + 3y + 3z)(x^2 - 3xy - 3xz + 9y^2 + 18yz + 9z^2)$
81. $(x + y - 4d)(x^2 + 2xy + y^2 + 4dx + 4dy + 16d^2)$ **83.** 24 **85.** 36 **87.** 25 **89.** 16 **91.** $10(x + 2)(x - 2)$
93. $(2x - 3)(4x^2 + 6x + 9)$

Review Exercises: **1.** 1, 2, 3, 4, 6, 9, 12, 18, 36 **2.** $2 \cdot 2 \cdot 5 \cdot 5$ **3.** $7xy^2(2x^2y - 3xy^2 - 1)$ **4.** $(x + 4y)(x + 3)$ **5.** $(a + 3b)(a - 6b)$
6. $3b(a + 5)(a - 3)$ **7.** $(3n - 2)(2n + 3)$ **8.** $2xy(2x - 5y)(3x - 2y)$

Exercise Set 6.5

Prep Exercises **1.** A monomial GCF common to all terms **2.** Difference of squares, difference of cubes, and sum of cubes
3. Sum of squares **4.** $a^2 + 2ab + b^2$ and $a^2 - 2ab + b^2$ **5.** Factor by grouping **6.** Multiply the factors to see if the product is the
original polynomial.

Exercises **1.** $3xy(y + 2x)$ **3.** $(x + y)(7a - b)$ **5.** $2(x + 4)(x - 4)$ **7.** $(a + b)(x + y)$ **9.** $abc(12a^2b + 3abc + 5c^2)$
11. $(x + 3)(x + 5)$ **13.** $(x^2 + 4)(x + 2)(x - 2)$ **15.** prime **17.** $(5x - 1)(3x + 2)$ **19.** $a(x + 2)^2$ **21.** $6ab(1 - 6b)$
23. prime **25.** $(x + 7)(x - 7)$ **27.** $(p + 6)(p - 5)$ **29.** $u(u + 1)(u - 1)$ **31.** $2(b + 3)(b + 4)$ **33.** $3r^2(2 - 5r)$
35. $7(2u - 5)(u + 3)$ **37.** $4(h^2 + 3h + 1)$ **39.** $5(p + 4)(p - 4)$ **41.** $(3w + 2)(w + 1)$ **43.** $(2q - 3)(4q + 1)$
45. $(3v + 2)(4v + 5)$ **47.** $2(1 + 5x)(1 - 5x)$ **49.** $20k^2(2 + l)(2 - l)$ **51.** prime **53.** $2(5 - t)^2$ **55.** $(x - 2y)(3a + 4b)$
57. $(x + 1)(x - 1)^2$ **59.** $(2x - 7)^2$ **61.** $2(x^2 + 9)(x + 3)(x - 3)$ **63.** $(a + 2b)(a + b)$ **65.** $(3 + 2m)(3 - 2m)$
67. $x(x^2 + 2y)(x^2 - 2y)$ **69.** prime **71.** $(b + 5)(b^2 - 5b + 25)$ **73.** $3(y - 2)(y^2 + 2y + 4)$ **75.** $2x(3 - y)(9 + 3y + y^2)$
77. $l = 2x + 1, w = 3x - 7$ **79.** $l = 2x, w = 3x + 5, h = 3x + 5$ **81.** $4(5 + 2t)(5 - 2t)$ **83.** Current: $3i + 4$; resistance: $2r + 5$

Review Exercises: **1.** $x - 4 - \dfrac{4}{2x + 1}$ **2.** $(6y - 1)(y + 4)$ **3.** 224 ft.^{2} **4.** $4(x + 3) = 20; 2$ **5.** 17, 18, 19

Exercise Set 6.6

Prep Exercises **1.** If a and b are real numbers and $ab = 0$, then $a = 0$ or $b = 0$. **2.** Solve $x + 3 = 0$ and $x - 4 = 0$. **3.** Write the
equation in standard form $(ax^2 + bx + c = 0)$. **4.** To use the zero-factor theorem, the product must be 0 and no other number.

Exercises **1.** $-5, -2$ **3.** $-3, 4$ **5.** $-\dfrac{3}{2}, \dfrac{4}{3}$ **7.** $0, 7$ **9.** $-2, 1, 0$ **11.** $0, 2$ **13.** $0, 4$ **15.** $0, -\dfrac{5}{3}$ **17.** $-2, 2$ **19.** $-3, 2$ **21.** 3 **23.** $4, -\dfrac{1}{3}$

25. $-\dfrac{5}{2}, \dfrac{2}{3}$ **27.** $-3, 7$ **29.** $1, 2$ **31.** $-3, 3$ **33.** $-7, -2$ **35.** $0, 4, -3$ **37.** $0, -\dfrac{5}{2}, \dfrac{2}{3}$ **39.** $\dfrac{2}{3}, 4$ **41.** $-4, \dfrac{15}{4}$ **43.** $3, 4$ **45.** $-2, 7$

47. $-\dfrac{7}{2}$ **49.** $-2, -4$ **51.** $3, -6$ **53.** $5, 11$ **55.** 11, 13 **57.** 13, 14 **59.** 12 m by 21 m **61.** 172 ft. by 54 ft. **63.** 14 ft. **65.** Base: 18 cm;
height: 24 cm **67.** 9 in. **69.** 10, 12, 14 or $-16, -14, -12$ **71.** 1.25 sec. **73.** 10% **75.** 15 **77.** 25 **79.** 37 cm **81.** 30 blocks
83. 45 in.

Review Exercises: **1.** No, it is not linear because the variable x has an exponent other than 1. **2.** yes **3.** $(-10, 0), (0, 4)$
4. $3x - y = -7$ **5. a.** 2 **b.** 6 **6.**

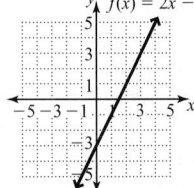

Exercise Set 6.7

Prep Exercises **1.** symmetrical **2.** lowest; highest **3.** If a vertical line is drawn through the vertex, the sides of the parabola are mirror images. **4.** The sign of a indicates whether the graph opens upward or downward. **5.** The y-coordinate of the y-intercept is c. **6.** Parabolas that open up or down pass the vertical line test and are therefore functions.

Exercises

1. $(2, -1)$ **3.** $(3, 0)$ **5.** 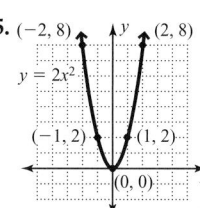 **7.** 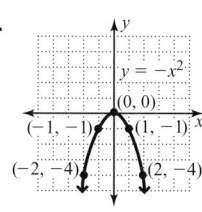 **9.** $y = x^2 - 3$ 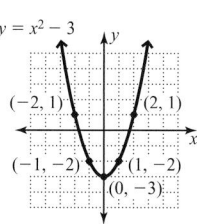 **11.** $y = -x^2 + 2x$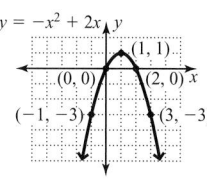

13. 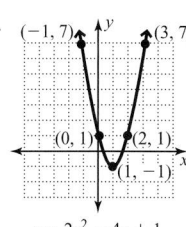 **15.** 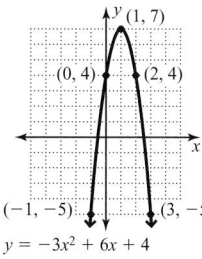 **17.** 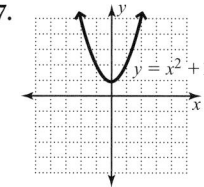 **19.** 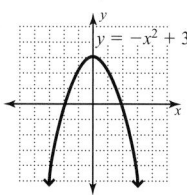 **21.** $y = 2x^2 - 5$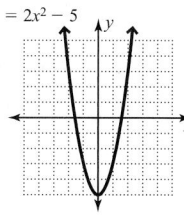

23. 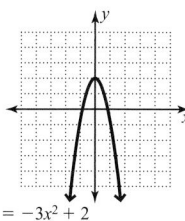 **25.** 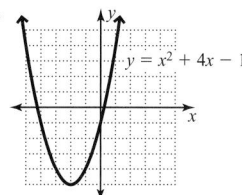 **27.** 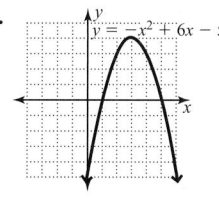 **29.** $y = 2x^2 - 4x - 3$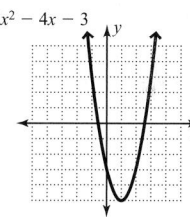

31. 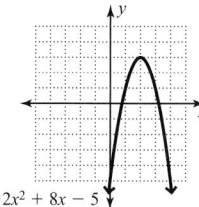 **33.** $f(x) = 3x^2$ 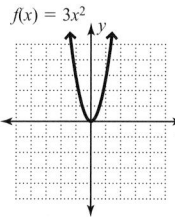 **35.** 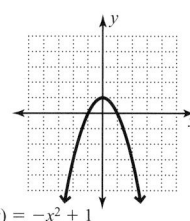 **37.**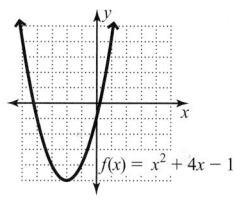

39. 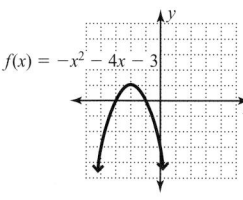 **41.** 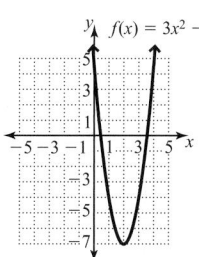 **43.** 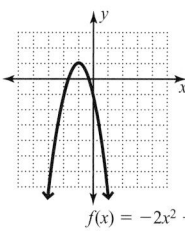 **45.** A function Domain: all real numbers or $(-\infty, \infty)$ Range: $\{y \mid y \geq 0\}$ or $[0, \infty)$

47. Not a function; Domain: $\{x \mid x \geq 0\}$ or $[0, \infty)$; Range: all real numbers or $(-\infty, \infty)$ **49.** A function; Domain: all real numbers or $(-\infty, \infty)$; Range: $\{y \mid y \leq 2\}$ or $(-\infty, 2]$ **51.** Not a function; Domain: $\{x \mid x \leq -1\}$ or $(-\infty, -1]$; Range: all real numbers or $(-\infty, \infty)$ **53. a.** $(0, 0)$ **b.** $(0, 2)$ **c.** The original graph is moved upward c units. **d.** The original graph moved downward 3 units. **55. a.** $y = x^2$ opens upward, and $y = -x^2$ opens downward. The sign of a indicates whether the parabola opens up or down. **b.** yes **c.** These graphs are "wider." **d.** The smaller the absolute value of a, the wider the parabola.

Review Exercises **1.** $3m^3n - 12m^2n^2 + 15mn$ **2.** $2x^2 - 5x - 3$ **3.** $(3n - 5)(n - 4)$ **4.** $(x + 5)(x - 5)$ **5.** $0, 3$ **6.** $-6, 3$

Chapter 6 Summary and Review Exercises

6.1 Definitions/Rules/Procedures product of factors Exercises **1.** $1, 3, 9$ **2.** $1, 3, 11, 33$ **3.** $1, 3, 9, 27, 81$ **4.** $1, 3, 5, 9, 15, 45$ **5.** $1, 2, 3, 5, 6, 10, 15, 30$ **6.** $1, 2, 3, 4, 5, 6, 10, 12, 15, 20, 30, 60$ Definitions/Rules/Procedures largest; divides; possible factors; largest factor common; exponential form; common; smallest exponent; Multiply Exercises **7.** 3 **8.** 6 **9.** $5y^2$ **10.** $3mn$

Definitions/Rules/Procedures GCF; GCF; polynomial; GCF; $GCF\left(\dfrac{\text{Polynomial}}{\text{GCF}}\right)$ Exercises **11.** $2(2x - 1)$

12. $5m(1 - 7m^2)$ **13.** $xy(x + y + x^2y^2)$ **14.** $21a^2b^2(5a - 3b + 4a^4b^2)$ **15.** $18ab^2c(b - 2a)$ **16.** $10k(10k^3 + 12k^4 - 1 + 4k^2)$

Definitions/Rules/Procedures   common; GCF; binomial; binomial; middle two terms; binomial   Exercises   **17.** $(x + y)(a + b)$
18. $(x + 2)(a + b)$ **19.** $(y + 2)(y^2 + 3)$ **20.** $(y + 4)(y^3 - b)$ **21.** $(x + 1)(y + 1)$ **22.** $(x^2 + y^2)(a - 5)$
23. $(2b^2 + 1)(b - 1)$ **24.** $(u - 3)(u + 4v)$

6.2 Definitions/Rules/Procedures c; b; first number; second number   Exercises   **25.** $(x - 4)(x + 3)$ **26.** $(x + 5)(x + 9)$
27. $(n - 2)(n - 4)$ **28.** prime   **29.** $(h + 3)(h + 48)$ **30.** $(y - 12)(y + 2)$ **31.** $4(x - 3)^2$ **32.** $3(m - 9)(m - 2)$

6.3 Definitions/Rules/Procedures   GCF; first; c; inner; outer; inner; outer; last; last terms; first; GCF; ac; b; bx; grouping
Exercises   **33.** prime   **34.** $(2u + 1)(u + 2)$ **35.** $(3m - 1)(m - 3)$ **36.** $(5k - 12h)(k + h)$ **37.** $2(3a - 4)(a - 2)$
38. $y(8x^2 - 4x - 1)$ **39.** $(3p - q)(p - 4q)$ **40.** $8(3x + y)(x - 3y)$

6.4 Definitions/Rules/Procedures $(a + b)^2$; $(a - b)^2$; $(a + b)(a - b)$; $(a - b)(a^2 + ab + b^2)$; $(a + b)(a^2 - ab + b^2)$
Exercises   **41.** $(v - 4)^2$ **42.** $(u + 3)^2$ **43.** $(2x + 5)^2$ **44.** $(3y - 2)^2$ **45.** $(x + 2)(x - 2)$ **46.** $(5 + y)(5 - y)$
47. $(x - 1)(x^2 + x + 1)$ **48.** $(x + 3)(x^2 - 3x + 9)$

6.5 Definitions/Rules/Procedures   grouping; $(a + b)^2$; $(a - b)^2$; c; b; trial and error; ac; b; conjugates; $(a + b)(a - b)$;
$(a + b)(a^2 - ab + b^2)$; $(a - b)(a^2 + ab + b^2)$ Exercises   **49.** $(3b + 2)(2b - 1)$ **50.** $4ab(1 - 6b)$ **51.** prime
52. $3(x + y)(x - y)$ **53.** $(x^2 + 9)(x + 3)(x - 3)$ **54.** $7(u - 5)(u + 3)$ **55.** $(2x - 3y)(4x^2 + 6xy + 9y^2)$
56. $2(1 + 5y)(1 - 5y)$ **57.** $(m - 2n)(3a + 4b)$ **58.** $3(m^2 + 3m + 9)$

6.6 Definitions/Rules/Procedures $a = 0$; $b = 0$; zero; $ax^2 + bx + c = 0$; standard; factored; zero-factor   Exercises   **59.** $-3, 4$

60. $-2, 2$ **61.** $2, 3$ **62.** $-3, 7$ **63.** $-3, 3$ **64.** $-6, 3$ **65.** -3 **66.** $-\dfrac{1}{3}, 4$ **67.** $5x^2 = 2x; 0, \dfrac{2}{5}$ **68.** $x^2 + (x + 1)^2 = 61; 5, 6$

69. $x \cdot 4x = 100; -5, 5$ **70.** $(w + 6)w = 91$; 13 in. by 7 in.   **71.** $x(x + 1) = 110; 10, 11$ Definitions/Rules/Procedures
c^2; $a^2 + b^2 = c^2$ Exercise   **72.** $3^2 + 4^2 = c^2; 5$

6.7 Definitions/Rules/Procedures $y = ax^2 + bx + c$; symmetrical; lowest; highest; solutions; solutions; parabola; parabola; upward;
downward; x;   Exercises   **73.**   **74.**   **75.**   **76.**
parabola; parabola

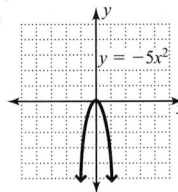   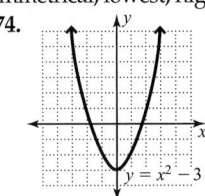   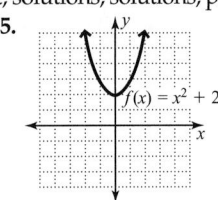   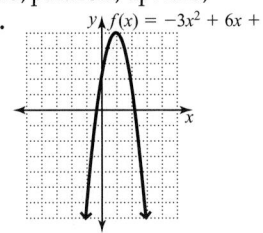

Chapter 6 Practice Test

1. 1, 2, 4, 5, 8, 10, 16, 20, 40, 80 [6.1]   **2.** $5m$ [6.1]   **3.** $5(y - 6)$ [6.1]   **4.** $2y(3x^2 - y)$ [6.1]   **5.** $(x - y)(a - b)$ [6.1]
6. $(m + 3)(m + 4)$ [6.2]   **7.** $(r - 4)^2$ [6.4]   **8.** $(2y + 5)^2$ [6.3]   **9.** $(q - 2)(3q - 4)$ [6.3]   **10.** $(3x - 4)(2x - 5)$ [6.3]
11. $a(x + 3)(x - 8)$ [6.2]   **12.** $2n(n + 4)(5n - 1)$ [6.2]   **13.** $(c + 5)(c - 5)$ [6.4]   **14.** prime [6.4]   **15.** $2(1 + 5u)(1 - 5u)$
[6.4]   **16.** $-4(x + 2)(x - 2)$ [6.4]   **17.** $(m + 5)(m^2 - 5m + 25)$ [6.4]   **18.** $(x - 2)(x^2 + 2x + 4)$ [6.4]   **19.** $-3, 0$ [6.6]

20. $6, -2$ [6.6]   **21.** $\dfrac{3}{2}, -5$ [6.6]   **22.** $8, 9$ [6.6]   **23.** 9 ft., 4 ft. [6.6]   **24.**   [6.7]   **25.**   [6.7]

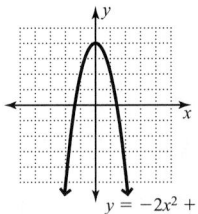   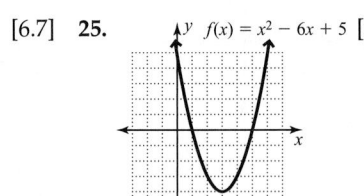

Chapters 1–6 Cumulative Review Exercises

1. false   **2.** true   **3.** false   **4.** false   **5.** $(a + b)(a - b)$ **6. a.** standard **b.** factored   **7.** 42   **8.** -89 **9.** $5x^2 + 4x + 2$

10. $18x^4y$ **11.** $48x^7$ **12.** $2y^2 - 15y - 8$ **13.** $4h + 1 - \dfrac{2}{h}$ **14.** $x + 4$ **15.** $(x + 11)(x - 11)$ **16.** $10(x^2 + 4)$ **17.** $(x - 4)^2$

18. $(2x - 3)(x + 4)$ **19.** $(x - 5)(x^2 + 5)$ **20.** $\dfrac{17}{12}$ **21.** 13.2   **22.** $-4, 7$ **23.** $x = \dfrac{7 + 3y}{2}$ **24.** $(3, 0), (0, -6)$ **25.** 0

26.   **27. a.** $\{x \mid x \le 4\}$ **b.** $(-\infty, 4]$ **c.**   **28.** $\left(\dfrac{1}{3}, 1\right)$ **29.** 6, 8, 10

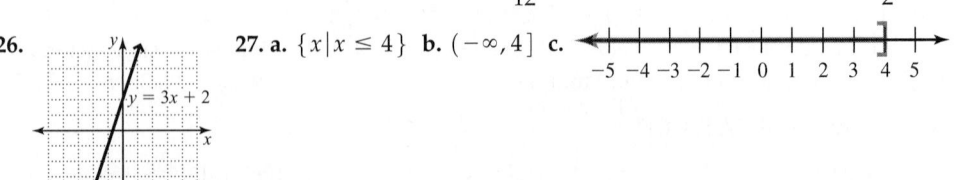

30. $2500 in 5%, $5500 in 8%

Chapter 7
Exercise Set 7.1

Prep Exercises   **1.** Replace the variables with given values and then simplify the numerical expression.   **2.** Set the denominator equal to 0 and solve the equation.   **3.** Factor the numerator and denominator completely.   **4.** x and $x + 3$ **5.** y is not a factor in the numerator, $x + y$.

Exercises   **1. a.** $\dfrac{16}{45}$ **b.** -1 **c.** 0 **3. a.** 1 **b.** undefined **c.** -3 **5. a.** $\dfrac{11}{8}$ **b.** $-\dfrac{1}{7}$ **c.** $-\dfrac{24}{49}$ or ≈ -0.49 **7. a.** $-\dfrac{4}{3}$ **b.** $-\dfrac{3}{2}$

c. $-\dfrac{341}{30}$ or $-11.3\overline{6}$ **9.** 3 **11.** $-3, 3$ **13.** $5, 9$ **15.** $0, 3$ **17.** $-1, \dfrac{5}{2}$ **19.** Defined for all real numbers.   **21.** $\dfrac{3x}{4y}$ **23.** $-\dfrac{4m^2}{n}$

25. $\dfrac{3}{2x^2z^2}$ **27.** $-\dfrac{3m^2n}{m + 2n}$ **29.** $\dfrac{1}{3(x + 5)}$ **31.** $\dfrac{3}{5}$ **33.** $\dfrac{2}{3}$ **35.** $\dfrac{5}{3}$ **37.** $\dfrac{x}{4}$ **39.** m **41.** $\dfrac{5}{x}$ **43.** $\dfrac{x - y}{x + y}$ **45.** $\dfrac{x}{x - 2}$ **47.** $\dfrac{t - 3}{t + 2}$

49. $\dfrac{2a - 3}{3a - 2}$ **51.** $\dfrac{3b - 4}{5b - 1}$ **53.** $\dfrac{x - 4}{x + 2}$ **55.** $\dfrac{p + q}{m - n}$ **57.** $\dfrac{2u^2 - 1}{u + 1}$ **59.** -1 **61.** -4 **63.** $-y - 2$ **65.** $-\dfrac{1}{w + 1}$ **67.** $\dfrac{x - 2}{x^2 - 2x + 4}$

69. $\dfrac{x + y}{x}$ **71.** $-\dfrac{1}{m^2 + 3m + 9}$ **73.** Mistake: Placed remaining x in numerator. Correct: $-\dfrac{1}{2x}$ **75.** Mistake: Divided out a part of a multiterm polynomial. Correct: Can't simplify   **77.** Mistake: $y - x$ and $x - y$ are not identical factors until -1 is factored out of one of them. Correct: $-x$ **79. a.** 23.7; 22.9; 28.3   **b.** The first two are normal weight, and the third is overweight.

81. $533.\overline{3}\pi$; $666.\overline{6}\pi$; $666.\overline{6}\pi$; $833.\overline{3}\pi$ **83. a.** $h = \dfrac{3V}{\pi r^2}$ **b.** ≈ 9 cm   **85.** $\dfrac{s - 3}{s + 4}$

Review Exercises: **1.** 1 **2.** $-\dfrac{3}{4}$ **3.** $\dfrac{9}{20}$ **4.** $\dfrac{28}{15}$ **5.** $(2y + 5)(2y - 5)$ **6.** $(3x - 2)(x + 5)$

Exercise Set 7.2

Prep Exercises   **1.** To eliminate any common factors   **2.** -1 **3.** Their product is 1.   **4.** reciprocal (or multiplication inverse)
5. Multiply the given measurement by conversion factors so that the undesired units divide out, leaving the desired units.
6. $\dfrac{16 \text{ oz.}}{1 \text{ lb.}}$

Exercises   **1.** $\dfrac{2m^2}{5n^2}$ **3.** $\dfrac{s^2}{r^3}$ **5.** y^2 **7.** $-\dfrac{1}{mn^2s}$ **9.** $\dfrac{9r}{s}$ **11.** $\dfrac{mn^2}{2}$ **13.** $\dfrac{9}{10}$ **15.** $\dfrac{4}{15}$ **17.** $(2x - 5)(3x + 4)$ **19.** $\dfrac{r^2}{(r + 6)(r - 3)}$

21. $\dfrac{q + 1}{2q(q - 1)}$ **23.** $\dfrac{w + 4}{w - 1}$ **25.** $\dfrac{n + 1}{n}$ **27.** 1 **29.** $\dfrac{d - 5}{d + 1}$ **31.** $\dfrac{x - 2}{x}$ **33.** $\dfrac{1}{y}$ **35.** $-\dfrac{4y^2}{3x}$ **37.** $1 - 3x$ **39.** $\dfrac{1}{x^2y^2}$ **41.** $\dfrac{5}{4m^2n}$

43. $\dfrac{y^2}{5z^2}$ **45.** $-\dfrac{32x^6}{45}$ **47.** $\dfrac{4a}{9}$ **49.** $\dfrac{8}{5}$ **51.** $\dfrac{a - b}{x + y}$ **53.** $\dfrac{x + 8}{x - 8}$ **55.** $x - 4$ **57.** $\dfrac{a + 2b}{a - 2b}$ **59.** $\dfrac{1}{k(2k + 1)}$ **61.** $\dfrac{4x^2 + 6x + 9}{3x - 1}$

63. $\dfrac{1}{(x + 4)(x + 1)}$ **65.** $\dfrac{b - 4}{c - 3}$ **67.** $-b - 5$ **69.** $\dfrac{7 + y}{y(y + 4)}$ **71.** $-\dfrac{x^3}{x - 3}$ **73.** $-\dfrac{x}{y}$ **75.** Mistake: Did not invert the divisor.

Correct: $\dfrac{50}{27}$ **77.** Mistake: Divided out y, which is not allowed because y is not a factor in $y - 2$. Correct: $\dfrac{(x - 3)(y - 2)}{15y}$

79. 222 in.   **81.** 10,560 yd.   **83.** 288 in.   **85.** 88 oz.   **87.** 12.7 T   **89.** 96,000 oz.   **91.** 1119.0$\overline{6}$ ft./sec.   **93.** $\approx$ 19.03 mi./hr.
95. 939,785,328 km   **97.** $\approx 5.87 \times 10^{12}$ mi./yr.   **99.** 311.75 £   **101.** $584.06   **103.** 801.97 €

Review Exercises: **1.** 1 **2.** $-\dfrac{1}{2}$ **3.** $2x$ **4.** $\dfrac{1}{4}y - 4$ **5.** $8x^2 - 6x - 4$ **6.** $-n^2 + 4$

Exercise Set 7.3

Prep Exercises   **1.** Add or subtract the numerators and keep the same denominator. Then simplify to lowest terms.   **2.** Factor both the numerator and denominator.   **3.** Write an equivalent addition with the additive inverse of the subtrahend.   **4.** It can be simplified only if there are factors common to both the numerator and denominator.

Exercises   **1.** $\dfrac{x}{2}$ **3.** $\dfrac{a^2}{3}$ **5.** $\dfrac{9}{a}$ **7.** $-\dfrac{x}{7y^2}$ **9.** $\dfrac{n + 3}{n + 5}$ **11.** $\dfrac{3r}{r + 2}$ **13.** $\dfrac{q + 1}{q}$ **15.** 3 **17.** 3 **19.** $\dfrac{11}{w + 1}$ **21.** $\dfrac{3m - 1}{m - 1}$ **23.** $\dfrac{-11t + 3}{2t - 3}$

25. $\dfrac{6w - 5}{w + 5}$ **27.** $\dfrac{1}{g + 3}$ **29.** $z - 3$ **31.** $\dfrac{1}{s + 2}$ **33.** $\dfrac{y - 3}{y}$ **35.** $\dfrac{x + 2}{x - 4}$ **37.** $\dfrac{t - 1}{t + 7}$ **39.** $3(x + 2)$ **41.** $\dfrac{1}{x - 5}$ **43.** $\dfrac{3x - 2}{x - 5}$

45. $\dfrac{x - 3}{x + 5}$ **47.** Mistake: Combined terms that are not like. Correct: $\dfrac{4y + 1}{5}$ **49.** Mistake: Did not change sign of -5. Correct: 2

51. $\dfrac{5x + 9}{2}$ **53.** $\dfrac{3x - 3}{2}$

Review Exercises: **1.** $\dfrac{11}{10}$ **2.** $\dfrac{11}{24}$ **3.** $1\dfrac{6}{35}$ **4.** $\dfrac{11}{15}x + \dfrac{7}{12}$ **5.** $2x^2 - 12x - \dfrac{11}{24}$ **6.** $2y^2 - \dfrac{1}{20}y - 10$

Exercise Set 7.4

Prep Exercises   1. Write a product that contains each unique prime factor the greatest number of times it occurs in the factorizations.   2. $6xz$; divide $48x^3yz$ by $8x^2y$.   3. Factor the denominators.   4. The student multiplied the numerator and denominator of $\dfrac{3}{2x-y}$ by -1.   5. Rewrite the numerator as an equivalent addition sentence.   6. $\dfrac{7}{x-5}-\dfrac{2}{x+3}$; we note that $7x+21$ and $2x-10$ factor to $7(x+3)$ and $2(x-5)$. Because the denominator is the product of $(x-5)$ and $(x+3)$, they must be the denominators of two fractions with 7 and 2 as the original numerators.

Exercises   1. mn; $\dfrac{2m}{mn}$, $\dfrac{3n}{mn}$ 3. a^3b^5; $\dfrac{2at}{a^3b^5}$, $\dfrac{3b^2z}{a^3b^5}$ 5. $15a^2$; $\dfrac{10a}{15a^2}$, $\dfrac{9}{15a^2}$ 7. $14x^2y^2$; $\dfrac{6y}{14x^2y^2}$, $\dfrac{5x}{14x^2y^2}$ 9. $90m^3n^3$; $\dfrac{48}{90m^3n^3}$, $\dfrac{35mn^2}{90m^3n^3}$

11. $(y+2)(y-2)$; $\dfrac{3y-6}{(y+2)(y-2)}$, $\dfrac{4y^2+8y}{(y+2)(y-2)}$ 13. $12(y-5)$; $\dfrac{9y}{12(y-5)}$, $\dfrac{14y}{12(y-5)}$

15. $6x(x+3)$; $\dfrac{5x}{6x(x+3)}$, $\dfrac{12}{6x(x+3)}$ 17. p^2-4; $\dfrac{2}{p^2-4}$, $\dfrac{p^2+p-6}{p^2-4}$ 19. $3(u+2)^2$; $\dfrac{12u}{3(u+2)^2}$, $\dfrac{2u+4}{3(u+2)^2}$

21. $(t+6)(t-6)(t-1)$; $\dfrac{2t^2-6t+4}{(t+6)(t-6)(t-1)}$, $\dfrac{3t^2+20t+12}{(t+6)(t-6)(t-1)}$ 23. $(x+4)(x-1)(x+2)$;

$\dfrac{x^2+4x+4}{(x+4)(x-1)(x+2)}$, $\dfrac{x^2-4x+3}{(x+4)(x-1)(x+2)}$ 25. $(x-1)(x-2)$; $\dfrac{2x-4}{(x-1)(x-2)}$, $\dfrac{3}{(x-1)(x-2)}$, $\dfrac{5x^2-5x}{(x-1)(x-2)}$

27. $\dfrac{-31}{18}$ 29. $\dfrac{8}{3c}$ 31. $\dfrac{5x^2-2}{x^3}$ 33. $\dfrac{8y-2}{(y+2)(y-4)}$ 35. $\dfrac{x+10}{(x-2)(x+2)}$ 37. $\dfrac{2x+14}{(x+5)(x-5)}$ 39. $\dfrac{10p-8}{p^2-2p+1}$

41. $\dfrac{-z^2-5z-3}{(z+4)(z-3)}$ 43. $\dfrac{2n^2-2n+18}{(n-4)(n+3)}$ 45. $\dfrac{-t^2+t+16}{(t+4)(t+6)}$ 47. $\dfrac{5}{2(c+5)}$ 49. $\dfrac{y+2}{y}$ 51. $\dfrac{r^2-6r-1}{(r+1)(r-1)^2}$

53. $\dfrac{q+1}{q(q-3)}$ 55. $\dfrac{1}{(w-1)(w-2)}$ 57. $\dfrac{v^2+16v+51}{(v+4)(v-4)(v+3)}$ 59. $\dfrac{-3}{t-3}$ 61. $\dfrac{4-x}{2x-1}$ 63. $\dfrac{y+6}{3(y-2)}$ 65. Mistake: Did not

find a common denominator. Correct: $\dfrac{6-x^2}{2x}$ 67. Mistake: Did not change the sign of 7. Correct: $\dfrac{-x-7}{x+2}$ 69. Mistake: Added

denominators. Correct: $\dfrac{7w}{x}$ 71. $\dfrac{5t}{6}$ 73. $\dfrac{4x+6}{x^2+3x}$ 75. $\dfrac{4ah+5bh}{8}$ 77. $\dfrac{6x^2+8x+2}{(2x-1)(2x+1)}$ 79. xy 81. $\dfrac{3a^2+10a+20}{5a(a+2)}$

83. $\dfrac{2a^2+2b^2}{(a+b)^2}$ 85. $\dfrac{x-5}{3(x+2)}$

Review Exercises: 1. $\dfrac{3}{8}$ 2. $\dfrac{m^2}{108}$ 3. $4(x+2)(x-2)$ 4. $3y(2y+1)(y-5)$ 5. $\dfrac{x-1}{x+2}$ 6. $\dfrac{1}{x-3}$

Exercise Set 7.5

Prep Exercises   1. $\dfrac{4}{x+3}$ is the numerator. $\dfrac{3}{x}$ is the denominator.   2. Unable to tell; the line between the numerator and denomina-

tor could be darker or longer.   3. $\dfrac{\dfrac{2x}{x-3}}{\dfrac{4-x}{2x-6}}$ 4. $\dfrac{5}{x-1}{7x}$ 5. In this case, multiplying the numerator and denominator by their LCD, 12,

(method 2) is preferable because doing so requires fewer steps than simplifying $\dfrac{x}{6}-\dfrac{3}{4}$ and $\dfrac{2}{3}+x$ and then dividing the simplified

expressions (method 1). (Answers may vary.)   6. Because $\dfrac{x}{7}$ and $\dfrac{x}{x+1}$ are in simplest form, dividing them (method 1) is preferable

because doing so requires fewer steps than multiplying them by their LCD and then simplifying (method 2). (Answers may vary.)

Exercises   1. $\dfrac{1}{2}$ 3. $\dfrac{mp}{nq}$ 5. $\dfrac{11}{13}$ 7. $\dfrac{5a+2}{3a+4}$ 9. $\dfrac{1-y}{1+y}$ 11. $\dfrac{4r-3}{2-4r}$ 13. $\dfrac{x}{3+x}$ 15. xy 17. $\dfrac{x}{x-2}$ 19. $\dfrac{6(2x+1)}{x(3x-2)}$ 21. $\dfrac{a}{b(a-b)}$

23. $\dfrac{2y(y-5)}{5}$ 25. u 27. $\dfrac{1}{a-1}$ 29. $\dfrac{1}{3}$ 31. 2   33. $y-1$ 35. $\dfrac{1}{n-1}$ 37. $3-y$ 39. $\dfrac{x-1}{x}$ 41. $\dfrac{-2}{x(x+h)}$ 43. Mistake: Divided

out $\dfrac{1}{2}$, which is not a common factor. Correct: $\dfrac{2x-1}{2y-1}$ 45. Mistake: Did not write as an equivalent multiplication. Correct: $\dfrac{m^2}{18}$

47. Mistake: Did not distribute c in the numerator. Correct: c^2+1 49. Mistake: Added the x's instead of multiplying them.

Correct: $\dfrac{18}{x^2}$ 51. $34\dfrac{2}{7}$ mph   53. $\dfrac{3(5n-2)}{4(n-1)}$ 55. a. $\dfrac{R_1R_2}{R_2+R_1}$ b. $\approx 9.1\ \Omega$ c. $8\ \Omega$

Review Exercises: 1. -7 2. 3   3. 2.5   4. 4   5. 4.8 mph   6. 0.09 hr. or 5.4 min.

Exercise Set 7.6

Prep Exercises **1.** Both sides of the equation are multiplied by the LCD of the rational expressions. **2.** $x(x+5)(x-1)$ **3.** An apparent solution that does not solve its equation **4.** It is possible to have an apparent solution cause the denominator of one or more of the rational expressions in the original equation to be undefined. **5.** 4 and -2; substituting either of those numbers for x causes an expression in the given equation to be undefined. **6.** 0, -5, and 1; substituting one of those numbers for x causes an expression in the given equation to be undefined.

Exercises **1.** no **3.** yes **5.** yes **7.** 2 **9.** 10 **11.** 15 **13.** 3 **15.** 1 **17.** 10 **19.** 8 **21.** -3 **23.** -1 **25.** 1 **27.** 9 **29.** $-2, \dfrac{1}{3}$ **31.** -5

33. $-\dfrac{11}{3}$, $(-1$ is extraneous$)$ **35.** 4, -2 **37.** $-\dfrac{1}{4}, 5$ **39.** $-2, 9$ **41.** 1, (2 is extraneous) **43.** 0, $\dfrac{9}{2}$ **45.** $-3, -\dfrac{2}{3}$ **47.** -2 **49.** 1

51. 6, -1 **53.** No solution. **55.** 5, (1 is extraneous) **57.** Mistake: 3 is extraneous. Correct: -3 is the only answer. **59.** In an expression, the LCD is used to combine the numerator over a single denominator. In an equation, the LCD is used to eliminate the denominator. **61.** 400 Ω **63.** 20 and 30 Ω **65.** ≈ 2.1 ft. **67.** $f = 15$ mm, $i = 20$ mm

Review Exercises: **1.** ≈ 68.6 mph **2.** $\approx \dfrac{0.084}{1}$; each ounce costs 8.4 cents. **3.** 6 hr. **4.** 72, 74 **5.** The length is 11 mm, and the width is 7 mm. **6.** 0.15 hr. (9 min.)

Exercise Set 7.7

Prep Exercises **1.** $\dfrac{1}{x}$ **2.** $\dfrac{1}{a} + \dfrac{1}{b}$ **3.** $\dfrac{100}{r}$, inverse **4.** Upstream: $x - 3$ mph; downstream: $x + 3$ mph. **5.** $m = kn$ **6.** Find the value of k. **7.** p decreases. **8.** $t = \dfrac{k}{u}$ **9.** m decreases. **10.** $c = kde$

Exercises **1.** $2\dfrac{2}{5}$ hr. **3.** $14\dfrac{7}{12}$ days **5.** 12 hr. **7.** 22.5 min. **9.** 3 P.M. **11.** 75 mph **13.** 6 hr. **15.** 441 mph **17.** 12 **19.** 36

21. 12 **23.** $39 **25.** 180 cm³ **27.** 5 sec. **29.** 40.5 m **31.** 7 ft. **33.** 2 **35.** 8.25 **37.** 50 psi **39.** 6 Ω **41.** 600 m **43.** 100 mm

45. 64 **47.** 72 **49.** 336 in.³ **51.** 169.65 cm³ **53.** 6 **55.** 162 **57.** 3.125 Ω **59.** 12 dyn

61. a.

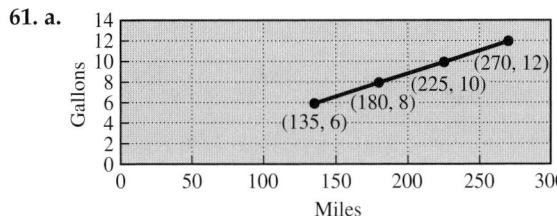

b. They are directly proportional. As the number of miles increases, so does the number of gallons. **c.** $k = 22.5$, which represents the miles per gallon the car gets. **d.** Yes, the data represent a function because for any given number of miles, there will be one quantity of gasoline (assuming that the miles per gallon stays constant).

Review Exercises: **1.** 4.3 **2.** $-\dfrac{5}{8}$ **3.** 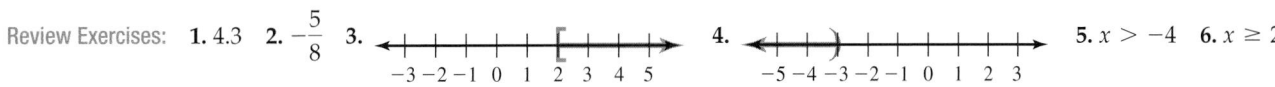 **4.** **5.** $x > -4$ **6.** $x \ge 2$

Chapter 7 Summary and Review Exercises

7.1 Definitions/Rules/Procedures $\dfrac{P}{Q}$; polynomials; replace; simplify Exercises **1. a.** $\dfrac{17}{6}$ **b.** ≈ 0.077 **c.** $\dfrac{3}{4}$ **2. a.** 3 **b.** 15

c. 1.875 Definitions/Rules/Procedures 0; Solve Exercises **3.** 5 **4.** $-2, 2$ **5.** $-6, 1$ **6.** 0, -6 Definitions/Rules/

Procedures Factor; common factors; remaining factors; remaining factors Exercises **7.** $\dfrac{x+5}{3}$ **8.** $\dfrac{2}{3}$ **9.** $\dfrac{x}{4}$ **10.** $\dfrac{y}{3}$ **11.** $\dfrac{a+b}{a-b}$

12. $\dfrac{3}{(x^2+9)(x-3)}$ **13.** $-(x+y)$ **14.** $-\dfrac{1}{w+3}$

7.2 Definitions/Rules/Procedures numerator; denominator; factor; factor; numerator; denominator Exercises **15.** $-\dfrac{3}{5mnp^4}$

16. $\dfrac{3}{5}$ **17.** $\dfrac{n+1}{n}$ **18.** $\dfrac{2}{r-1}$ **19.** $\dfrac{m+1}{4m(m-1)}$ **20.** $\dfrac{w+4}{w-4}$ Definitions/Rules/Procedures reciprocal; numerator; denominator;

factor; factor; numerator; denominator Exercises **21.** $\dfrac{5}{6}$ **22.** $\dfrac{5a}{2}$ **23.** $\dfrac{x+y}{c-d}$ **24.** $\dfrac{5(b+4)(b-4)}{6}$ **25.** $\dfrac{2j+1}{j(2j-1)}$

26. $\dfrac{1}{(u+1)^2}$ Definitions/Rules/Procedures conversion factors Exercises **27.** 3.5 ft. **28.** 13,200 ft. **29.** 4.25 lb.

30. 128,000 oz.

7.3 Definitions/Rules/Procedures numerators Exercises **31.** $\dfrac{x}{3}$ **32.** $-\dfrac{1}{3m}$ **33.** 1 **34.** $\dfrac{s}{r^2 - s^2}$ **35.** $\dfrac{11x^2 - 21x + 7}{2x + 1}$ **36.** $q - 3$

7.4 Definitions/Rules/Procedures prime factorization; greatest number; greatest exponent; denominator

Exercises **37.** $\dfrac{6c^2}{12c^3}, \dfrac{12}{12c^3}$ **38.** $\dfrac{5x}{x^2y^3}, \dfrac{3y}{x^2y^3}$ **39.** $\dfrac{4m + 4}{(m-1)(m+1)}, \dfrac{4my - 4y}{(m-1)(m+1)}$ **40.** $\dfrac{3h}{4h(h-2)}, \dfrac{20}{4h(h-2)}$

41. $\dfrac{x}{(x+1)(x-1)}, \dfrac{x^2 + x - 2}{(x+1)(x-1)}$ **42.** $\dfrac{2p + 4}{(p-2)(p+2)^2}, \dfrac{p^2 - 2p}{(p-2)(p+2)^2}$ Definitions/Rules/Procedures LCD;

equivalent expression; numerators Exercises **43.** $-\dfrac{23}{36}$ **44.** $\dfrac{8y - 22}{(y-2)(y-4)}$ **45.** $\dfrac{-3t^2 + 21t}{2(t-3)^2}$ **46.** $\dfrac{18 - a^2}{(a+3)(a+4)}$ **47.** $\dfrac{15}{2y - 1}$

48. $\dfrac{x + 19}{(x+4)(x-4)}$

7.5 Definitions/Rules/Procedures rational expressions; numerator; denominator; division problem; LCD Exercises **49.** $\dfrac{ay}{bx}$

50. -1 **51.** $\dfrac{5(x-2)}{x(3x+5)}$ **52.** $\dfrac{x}{x-3}$ **53.** $\dfrac{x}{5(2x-1)}$ **54.** 1 **55.** $\dfrac{y^2 + x^2}{7xy}$ **56.** $-h$

7.6 Definitions/Rules/Procedures does not solve the equation; LCD; Solve; Check Exercises **57.** $\dfrac{38}{13}$ **58.** 4

59. 3, 4 **60.** -1 **61.** -8, (4 is extraneous) **62.** 23

7.7 Definitions/Rules/Procedures kx; kx^n; $\dfrac{k}{x}$; $\dfrac{k}{x^n}$; kxz Exercises **63.** 42 **64.** 200 **65.** 6 **66.** 9 **67.** 32 **68.** 9 **69.** 9 gal.

70. $\dfrac{7}{3}$ cm/sec. **71.** 100.48 in.^{3}

Chapter 7 Practice Test

1. 7 [7.1] **2.** $-4, 4$ [7.1] **3.** $-\dfrac{3}{x-4}$ [7.1] **4.** $\dfrac{1}{x+y}$ [7.1] **5.** $\dfrac{6f}{f^2g^3}, \dfrac{2g^3}{f^2g^3}$ [7.4] **6.** $\dfrac{5x(x+3)}{(x+3)^2(x-3)}, \dfrac{2(x-3)}{(x+3)^2(x-3)}$ [7.4]

7. $\dfrac{5ab^2x^2}{12}$ [7.2] **8.** $-\dfrac{7a^6b^7y^2}{40x^2}$ [7.2] **9.** $-4x$ [7.2] **10.** $\dfrac{12x^2}{(x+1)(2x+1)}$ [7.2] **11.** $\dfrac{7x}{2x+3}$ [7.3] **12.** $\dfrac{1}{x+3}$ [7.3]

13. $\dfrac{x^2 + 5x - 6}{x(x+2)(x-2)}$ [7.4] **14.** $\dfrac{9x + 3}{(x-1)(x+2)}$ [7.4] **15.** $\dfrac{3x + 13}{(x-4)(x+1)}$ [7.4] **16.** $\dfrac{x-2}{2x}$ [7.4] **17.** $\dfrac{2y+1}{3y-1}$ [7.5] **18.** $\dfrac{2}{x+y}$ [7.5]

19. $\dfrac{25}{9}$ [7.6] **20.** -1 [7.6] **21.** -1, (2 is extraneous) [7.6] **22.** 2, 4 [7.6] **23.** 1.2 hr. [7.7] **24.** 6 mph, 4 mph [7.7] **25.** 529.2 N [7.7]

Chapters 1–7 Cumulative Review Exercises

1. false **2.** false **3.** false **4. a.** denominator **5.** There are no common factors in the numerator and denominator besides 1. **6. a.** Multiply **b.** Add **7.** complex rational expression **8.** 21 **9.** 6 **10.** $-3a^{13}$ **11.** $m^3 - 3m^2 - 17m + 3$

12. $(a-2)(x+1)$ **13.** $3(y-5)(y+4)$ **14.** $-3(h+5)(h-5)$ **15.** 9 **16.** $2(m+4)$ **17.** $\dfrac{1}{x+3}$ **18.** $\dfrac{6x + 14}{x(x+7)(x-7)}$

19. $\dfrac{2(y+3)}{y}$ **20.** 9 **21. a.** $\dfrac{2}{3}$ **b.**

22. $-\dfrac{1}{2}, 7$ **23.** $h = \dfrac{3V}{\pi r^2}$ **24.** 2, 6 **25.** 6 **26.** $(2, -2)$ **27.** 2.5 hr.

28. \$5500 at 9% and \$1500 at 7% **29.** $3\dfrac{3}{7}$ hr. **30.** 6

Chapter 8
Exercise Set 8.1
Prep Exercises **1.** Two inequalities joined by either *and* or *or* **2.** *and* and *or* **3.** For two sets A and B, the intersection of A and B, symbolized by $A \cap B$, is a set containing only elements that are in both A and B. **4.** Graph the region of overlap of the two inequalities. **5.** For two sets A and B, the union of A and B, symbolized by $A \cup B$, is a set containing each element in either A or B. **6.** Graph the region included in either of the two inequalities.

Exercises   **1.** $\{1,3,5\}$ **3.** $\{7,8\}$ **5.** $-4 < x < 5$ **7.** $0 < y \le 2$ **9.** $-7 < w < 3$ **11.** $0 \le u \le 2$

13.

$-4\;-3\;-2\;-1\;\;0\;\;1\;\;2\;\;3\;\;4\;\;5\;\;6\;\;7\;\;8$

15.

$-6\;-5\;-4\;-3\;-2\;-1\;\;0\;\;1\;\;2\;\;3\;\;4\;\;5\;\;6$

17.

$-1\;\;0\;\;1\;\;2\;\;3\;\;4\;\;5\;\;6\;\;7\;\;8\;\;9\;\;10\;11$

19.

$-6\;-5\;-4\;-3\;-2\;-1\;\;0\;\;1\;\;2\;\;3\;\;4\;\;5\;\;6$

21. a.

$-6\;-5\;-4\;-3\;-2\;-1\;\;0\;\;1\;\;2\;\;3\;\;4\;\;5\;\;6$

b. $\{x \mid -3 < x < -1\}$

c. $(-3,-1)$ **23. a.**

$-5\;-4\;-3\;-2\;-1\;\;0\;\;1\;\;2\;\;3\;\;4\;\;5\;\;6\;\;7$

b. $\{x \mid 3 < x \le 6\}$ **c.** $(3,6]$

25. a.

$-6\;-5\;-4\;-3\;-2\;-1\;\;0\;\;1\;\;2\;\;3\;\;4\;\;5\;\;6$

b. $\{\ \}$ or $\varnothing$ **c.** no interval notation   **27. a.**

$-6\;-5\;-4\;-3\;-2\;-1\;\;0\;\;1\;\;2\;\;3\;\;4\;\;5\;\;6$

b. $\{x \mid -5 \le x < 3\}$ **c.** $[-5,3)$ **29. a.**

$-6\;-5\;-4\;-3\;-2\;-1\;\;0\;\;1\;\;2\;\;3\;\;4\;\;5\;\;6$

b. $\{\ \}$ or $\varnothing$ **c.** no interval notation

31. a.

$-7\;-6\;-5\;-4\;-3\;-2\;-1\;\;0\;\;1\;\;2\;\;3\;\;4\;\;5$

b. $\{x \mid -7 < x < -3\}$ **c.** $(-7,-3)$ **33. a.**

$-6\;-5\;-4\;-3\;-2\;-1\;\;0\;\;1\;\;2\;\;3\;\;4\;\;5\;\;6$

b. $\{x \mid -1 \le x \le 2\}$ **c.** $[-1,2]$ **35. a.**

$-\frac{2}{3}$

$-6\;-5\;-4\;-3\;-2\;-1\;\;0\;\;1\;\;2\;\;3\;\;4\;\;5\;\;6$

b. $\left\{x \mid -\dfrac{2}{3} \le x < 2\right\}$ **c.** $\left[-\dfrac{2}{3}, 2\right)$

37. a.

$-6\;-5\;-4\;-3\;-2\;-1\;\;0\;\;1\;\;2\;\;3\;\;4\;\;5\;\;6$

b. $\{x \mid 0 \le x < 3\}$ **c.** $[0,3)$ **39. a.**

$-6\;-5\;-4\;-3\;-2\;-1\;\;0\;\;1\;\;2\;\;3\;\;4\;\;5\;\;6$

b. $\{x \mid 0 \le x \le 3\}$ **c.** $[0,3]$ **41.** $\{a,c,d,g,o,t\}$ **43.** $\{w,x,y,z\}$ **45.**

$-6\;-5\;-4\;-3\;-2\;-1\;\;0\;\;1\;\;2\;\;3\;\;4\;\;5\;\;6$

47.

$-6\;-5\;-4\;-3\;-2\;-1\;\;0\;\;1\;\;2\;\;3\;\;4\;\;5\;\;6$

49.

$-6\;-5\;-4\;-3\;-2\;-1\;\;0\;\;1\;\;2\;\;3\;\;4\;\;5\;\;6$

51.

$-5\;-4\;-3\;-2\;-1\;\;0\;\;1\;\;2\;\;3\;\;4\;\;5$

53. a.

$-9\;-8\;-7\;-6\;-5\;-4\;-3\;-2\;-1\;\;0\;\;1\;\;2\;\;3\;\;4\;\;5\;\;6$

b. $\{y \mid y < -9 \text{ or } y > 5\}$ **c.** $(-\infty, -9) \cup (5, \infty)$

55. a.

$-6\;-5\;-4\;-3\;-2\;-1\;\;0\;\;1\;\;2\;\;3\;\;4\;\;5\;\;6$

b. $\{r \mid r < -2 \text{ or } r > 1\}$ **c.** $(-\infty, -2) \cup (1, \infty)$

57. a.

$-6\;-5\;-4\;-3\;-2\;-1\;\;0\;\;1\;\;2\;\;3\;\;4\;\;5\;\;6\;\;7\;\;8$

b. $\{w \mid w \le -1 \text{ or } w \ge 7\}$ **c.** $(-\infty, -1] \cup [7, \infty)$

59. a.

$-6\;-5\;-4\;-3\;-2\;-1\;\;0\;\;1\;\;2\;\;3\;\;4\;\;5\;\;6$

b. $\{k \mid k \le 2 \text{ or } k \ge 3\}$ **c.** $(-\infty, 2] \cup [3, \infty)$

61. a.

$-6\;-5\;-4\;-3\;-2\;-1\;\;0\;\;1\;\;2\;\;3\;\;4\;\;5\;\;6$

b. $\{x \mid x \le -5 \text{ or } x \ge 3\}$ **c.** $(-\infty, -5] \cup [3, \infty)$

63. a.

$-6\;-5\;-4\;-3\;-2\;-1\;\;0\;\;1\;\;2\;\;3\;\;4\;\;5\;\;6$

b. $\{c \mid c > 3\}$ **c.** $(3, \infty)$ **65. a.**

$-6\;-5\;-4\;-3\;-2\;-1\;\;0\;\;1\;\;2\;\;3\;\;4\;\;5\;\;6$

b. $\{x \mid x \le -4\}$ **c.** $(-\infty, -4]$ **67. a.**

$-6\;-5\;-4\;-3\;-2\;-1\;\;0\;\;1\;\;2\;\;3\;\;4\;\;5\;\;6$

b. $\{x \mid x \text{ is a real number}\}$, or $\mathbb{R}$

c. $(-\infty, \infty)$ **69. a.**

$-7\;-6\;-5\;-4\;-3\;-2\;-1\;\;0\;\;1\;\;2\;\;3\;\;4\;\;5$

b. $\{x \mid -7 < x < 3\}$ **c.** $(-7,3)$

71. a.

$-6\;-5\;-4\;-3\;-2\;-1\;\;0\;\;1\;\;2\;\;3\;\;4\;\;5\;\;6$

b. $\{x \mid x < -2 \text{ or } x > 2\}$ **c.** $(-\infty, -2) \cup (2, \infty)$

73. a.

$-6\;-5\;-4\;-3\;-2\;-1\;\;0\;\;1\;\;2\;\;3\;\;4\;\;5\;\;6$

b. $\{x \mid -3 < x < -2\}$ **c.** $(-3,-2)$ **75. a.**

$-6\;-5\;-4\;-3\;-2\;-1\;\;0\;\;1\;\;2\;\;3\;\;4\;\;5\;\;6$

b. $\{x \mid 3 \le x \le 6\}$ **c.** $[3,6]$ **77. a.**

$-6\;-5\;-4\;-3\;-2\;-1\;\;0\;\;1\;\;2\;\;3\;\;4\;\;5\;\;6$

b. $\{x \mid 1 < x < 3\}$ **c.** $(1,3)$

79. a.

$-6\;-5\;-4\;-3\;-2\;-1\;\;0\;\;1\;\;2\;\;3\;\;4\;\;5\;\;6$

b. $\{x \mid x > 2\}$ **c.** $(2, \infty)$ **81. a.**

$250\quad 300\quad 350\quad 400\quad 450\quad 500\quad 550$

b. $\{x \mid 400 \le x \le 500\}$ **c.** $[400, 500]$ **83. a.**

55 105

$40\quad 50\quad 60\quad 70\quad 80\quad 90\quad 100\;110$

b. $\{x \mid 55 \le x < 105\}$ **c.** $[55, 105)$

85. a.

$66\quad 68\quad 70\quad 72\quad 74\quad 76\quad 78$

b. $\{x \mid 68° \le x \le 78°\}$ **c.** $[68, 78]$ **87. a.**

$68\quad 70\quad 72\quad 74\quad 76\quad 78\quad 80\quad 82$

b. $\{x \mid 72° \le x \le 80°\}$ **c.** $[72, 80]$ **89. a.**

$170\;180\;190\;200\;210\;220\;230\;240\;250\;260\;270$

b. $\{x \mid 180 \text{ ft.} \le x \le 260 \text{ ft.}\}$ **c.** $[180, 260]$

Review Exercises: **1.** No, the absolute value of zero is zero and zero is neither negative nor positive.   **2.** -11 **3.** -16

4. 6 **5.** $\dfrac{1}{5}$ **6.** -2

Exercise Set 8.2

Prep Exercises   **1.** The distance a number is from zero.   **2.** We are to find numbers that are 5 units from 0.   **3.** If $|n| = a$, where n is a variable or an expression and $a \geq 0$, then $n = a$ or $n = -a$.   **4.** There is no solution.   **5.** Separate the absolute value equation into two equations: $ax + b = cx + d$ and $ax + b = -(cx + d)$.   **6.** After the absolute value equation is separated into two equations, one with the two expressions equal and the other with the two expressions opposites, one of the equations leads to a contradiction.   **7.** The graph of every absolute value function is a V shape. Find many ordered pairs, plot them, and then draw a V-shape through those points.   **8.** *Domain*: The set of all input values for a relation *Range*: The set of all output values for a relation

Exercises   **1.** $-2, 2$ **3.** No solution   **5.** $-11, 5$ **7.** $2, 3$ **9.** $1, \dfrac{7}{5}$ **11.** $-\dfrac{2}{3}, \dfrac{10}{3}$ **13.** No solution   **15.** $\dfrac{3}{4}$ **17.** $-4, 4$ **19.** $1, -3$

21. $12, -4$ **23.** $-\dfrac{3}{5}, 1$ **25.** $-4, 1$ **27.** $-2, 6$ **29.** $-\dfrac{9}{2}, \dfrac{15}{2}$ **31.** $0, 20$ **33.** $-2, 4$ **35.** $\dfrac{1}{3}, 7$ **37.** $-5, -\dfrac{3}{5}$ **39.** All real numbers   **41.** 1

43. $3, -13$ **45.** $-6, 10$ **47.** $\dfrac{5}{6}, \dfrac{11}{6}$ **49.** $-\dfrac{17}{4}, \dfrac{11}{4}$ **51.** Domain: $(-\infty, \infty)$; Range: $[-4, \infty)$ **53.** Domain: $(-\infty, \infty)$;

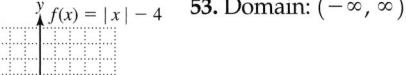

Range: $[-2, \infty)$ 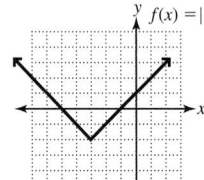   **55.** Domain: $(-\infty, \infty)$; Range: $[3, \infty)$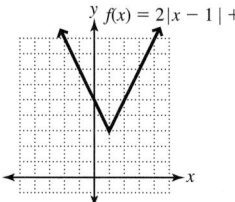

57. Domain: $(-\infty, \infty)$; Range: $[-\infty, 3]$ 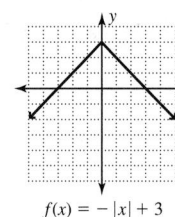   **59.** Domain: $(-\infty, \infty)$; Range: $[-\infty, -1]$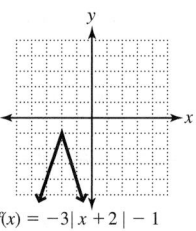

61. Domain: $(-\infty, \infty)$; Range: $[-\infty, 1]$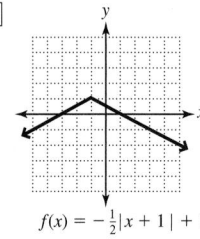

Review Exercises: **1.** $(-2, 0]$ **2.** 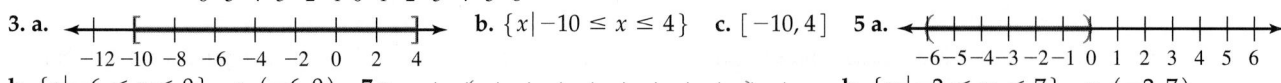   **3.** $x > -7$ **4.** $x \geq -3$ **5.** $-2 < x < \dfrac{14}{3}$
6. $n - 2 < 5; n < 7$

Exercise Set 8.3
Prep Exercises   **1.** $x \geq -a$ and $x \leq a$ or $-a \leq x \leq a$ **2.** Shade between the values of $-a$ and a.   **3.** When $a < 0$
4. $x \leq -a$ or $x \geq a$ **5.** Shade to the left of $-a$ and to the right of a.   **6.** When $a < 0$

Exercises   **1. a.**   **b.** $\{x | -5 < x < 5\}$ **c.** $(-5, 5)$
3. a.   **b.** $\{x | -10 \leq x \leq 4\}$ **c.** $[-10, 4]$ **5 a.**
b. $\{s | -6 < s < 0\}$ **c.** $(-6, 0)$ **7 a.**   **b.** $\{m | -2 < m < 7\}$ **c.** $(-2, 7)$

9 a. 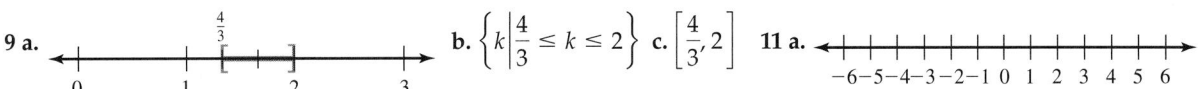 **b.** $\left\{ k \,\middle|\, \dfrac{4}{3} \le k \le 2 \right\}$ **c.** $\left[\dfrac{4}{3}, 2 \right]$ **11 a.**

b. { } or $\varnothing$ **c.** no interval notation **13 a.** **b.** $\{ w \,|\, 0 < w < 6 \}$ **c.** $(0, 6)$

15. a. **b.** $\{ c \,|\, c < -12 \text{ or } c > 12 \}$ **c.** $(-\infty, -12) \cup (12, \infty)$

17. a. **b.** $\{ y \,|\, y \le -9 \text{ or } y \ge 5 \}$ **c.** $(-\infty, -9] \cup [5, \infty)$

19. a. **b.** $\{ p \,|\, p < -2 \text{ or } p > 14 \}$ **c.** $(-\infty, -2) \cup (14, \infty)$

21. a. **b.** $\{ x \,|\, x \le -6 \text{ or } x \ge 2 \}$ **c.** $(-\infty, -6] \cup [2, \infty)$

23. a. **b.** $\left\{ n \,\middle|\, n < -\dfrac{5}{2} \text{ or } n > 0 \right\}$ **c.** $\left(-\infty, -\dfrac{5}{2} \right) \cup (0, \infty)$

25. a. **b.** $\{ v \,|\, v \le -1 \text{ or } v \ge 1 \}$ **c.** $(-\infty, -1] \cup [1, \infty)$

27. a. **b.** $\{ y \,|\, y < -3 \text{ or } y > -1 \}$ **c.** $(-\infty, -3) \cup (-1, \infty)$

29. a. **b.** $\{ m \,|\, m < -5 \text{ or } m > 1 \}$ **c.** $(-\infty, -5) \cup (1, \infty)$

31. a. **b.** $\{ x \,|\, 0 < x < 4 \}$ **c.** $(0, 4)$ **33. a.**

b. $\{ r \,|\, r \text{ is a real number} \}$ **c.** $(-\infty, \infty)$ **35 a.** **b.** $\{ x \,|\, -4 < x < -2 \}$ **c.** $(-4, -2)$

37. a. **b.** $\{ x \,|\, 0 < x < 1 \}$ **c.** $(0, 1)$ **39 a.**

b. $\varnothing$ **c.** no interval notation **41 a.** **b.** $\left\{ k \,\middle|\, -2 \le k \le \dfrac{14}{3} \right\}$ **c.** $\left[-2, \dfrac{14}{3} \right]$

43. a. **b.** $\{ x \,|\, x < 4 \text{ or } x > 20 \}$ **c.** $(-\infty, 4) \cup (20, \infty)$

45. a. **b.** $\{ y \,|\, -6.4 \le y \le 12.8 \}$ **c.** $[-6.4, 12.8]$

47. a. **b.** $\{ p \,|\, p \text{ is a real number} \}$ **c.** $(-\infty, \infty)$ **49.** $|x| < 3$ **51.** $|x| \ge 4$

53. $|x + 1| > 1$ **55.** $|x - 3| \le 2$ **57.** $|x| >$ any negative number

Review Exercises: **1.** $2x + 3y = 3$ **2.** yes **3.** $-4, 4$ **4.** no solution **5.** **6.**

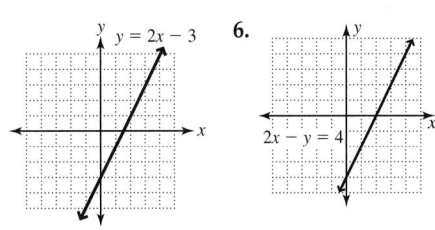

Chapter 8 Summary and Review Exercises

8.1 Definitions/Rules/Procedures *and; or*; both A and B; A or B Exercises **1.** Intersection: $\{ 1, 5, 9 \}$ **2.** Intersection: { } or $\varnothing$
Union: $\{ 1, 5, 7, 9 \}$ Union: $\{ 1, 2, 3, 4, 5, 6, 7 \}$

Definitions/Rules/Procedures inequality; intersection Exercises **3. a.**

b. $\{x \mid -3 < x < -2\}$ **c.** $(-3, -2)$ **4. a.** (number line from -6 to 6, bracket at -5 to -3)   **b.** $\{x \mid -5 \le x \le -3\}$

c. $[-5, -3]$ **5. a.** (number line from -8 to 4, parentheses from -7 to 3)   **b.** $\{x \mid -7 < x < 3\}$ **c.** $(-7, 3)$

6. a. (number line from -6 to 6, $(1$ to $4]$)   **b.** $\{x \mid 1 < x \le 4\}$ **c.** $(1, 4]$ **7. a.** (number line from -6 to 6, bracket $[2$ to $4)$)

b. $\{x \mid -2 \le x < 0\}$ **c.** $[-2, 0)$ **8. a.** (number line from -6 to 6, no shading)   **b.** $\{\ \}$ or $\varnothing$ **c.** no interval notation

9. a. (number line from 65 to 110)   **b.** $\{x \mid 68 \le x < 108\}$ **c.** $[68, 108)$ **10. a.** (number line from 150 to 200)

b. $\{x \mid 150 \le x \le 200\}$ **c.** $[150, 200]$ Definitions/Rules/Procedures   inequality; union

Exercises   **11. a.** (number line from -6 to 6)   **b.** $\{w \mid w \le -6 \text{ or } w \ge -2\}$ **c.** $(-\infty, -6] \cup [-2, \infty)$

12. a. (number line from -6 to 6, all shaded)   **b.** $\{x \mid x \text{ is a real number}\}$ **c.** $(-\infty, \infty)$

13. a. (number line from -6 to 6, mark at $\tfrac{5}{2}$)   **b.** $\left\{m \mid m < \dfrac{5}{2} \text{ or } m > 5\right\}$ **c.** $\left(-\infty, \dfrac{5}{2}\right) \cup (5, \infty)$

14. a. (number line from -6 to 6, mark at $-\tfrac{4}{3}$)   **b.** $\left\{x \mid x \le -\dfrac{4}{3} \text{ or } x \ge 2\right\}$ **c.** $\left(-\infty, -\dfrac{4}{3}\right] \cup [2, \infty)$

15. a. (number line from -11 to 1)   **b.** $\{x \mid x \le -9 \text{ or } x \ge -4\}$ **c.** $(-\infty, -9] \cup [-4, \infty)$

16. a. (number line from -6 to 6)   **b.** $\{w \mid w \le -1 \text{ or } w \ge 1\}$ **c.** $(-\infty, -1] \cup [1, \infty)$

8.2 Definitions/Rules/Procedures $a; -a; |ax + b| = c; ax + b = c; ax + b = -c$ Exercises   **17.** $-4, 4$ **18.** $-3, 11$ **19.** $-1, 2$

20. no solution   **21.** $-7, 15$ **22.** $-3, 3$ **23.** $0, \dfrac{8}{3}$ **24.** $-1, 11$ Definitions/Rules/Procedures $cx + d; -(cx + d)$

Exercises   **25.** $0, 2$ **26.** $x = -3, -\dfrac{1}{5}$ **27.** $-2, \dfrac{4}{5}$ **28.** $1, \dfrac{8}{3}$ Definitions/Rules/Procedures $a|x - h| + k; \text{V}; \text{V}$

Exercises   **29.** $f(x) = |x - 2|$ 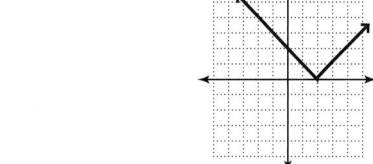   **30.** $f(x) = -2|x + 1| + 3$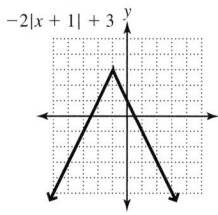

8.3 Definitions/Rules/Procedures $x > -a; x < a; x \ge -a; x \le a$ Exercises   **31. a.** (number line from -6 to 6)

b. $\{x \mid -5 < x < 5\}$ **c.** $(-5, 5)$ **32. a.** (number line from -6 to 6)   **b.** $\{m \mid -5 < m < -1\}$ **c.** $(-5, -1)$

33. a. (number line from -6 to 6)   **b.** $\{\ \}$ or $\varnothing$ **c.** no interval notation   **34. a.** (number line from -6 to 6)

b. $\{m \mid -6 \le m \le 0\}$ **c.** $[-6, 0]$ Definitions/Rules/Procedures $x < -a; x > a; x \le -a; x \ge a$

Exercises   **35. a.** (number line from -6 to 6)   **b.** $\{p \mid p \le -4 \text{ or } p \ge 4\}$ **c.** $(-\infty, -4] \cup [4, \infty]$

36. a. (number line from -6 to 11)   **b.** $\{x \mid x < -4 \text{ or } x > 10\}$ **c.** $(-\infty, -4) \cup (10, \infty)$

37. a. (number line from -6 to 6)   **b.** $\{b \mid b < -1 \text{ or } b > 1\}$ **c.** $(-\infty, -1) \cup (1, \infty)$

38. a. (number line from -2 to 11)   **b.** $\{t \mid t < 0 \text{ or } t > 10\}$ **c.** $(-\infty, 0) \cup (10, \infty)$

39. a. (number line from -6 to 8, marks at $-\tfrac{7}{2}$ and $\tfrac{13}{2}$)   **b.** $\left\{k \mid k \le -\dfrac{7}{2} \text{ or } k \ge \dfrac{13}{2}\right\}$ **c.** $\left(-\infty, -\dfrac{7}{2}\right] \cup \left[\dfrac{13}{2}, \infty\right)$

40. a. (number line from -6 to 6)   **b.** $\{x \mid x \text{ is a real number}\}$ **c.** $(-\infty, \infty)$

Chapter 8 Practice Test

1. $\{e, h, o\}$ $\{e, h, m, o, s, u\}$ [8.1] **2 a.** 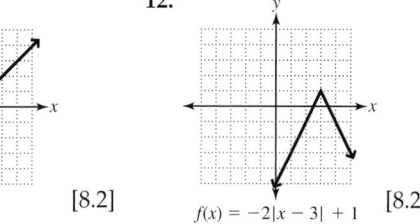 **b.** $\{x \mid -7 < x \le 3\}$ **c.** $(-7, 3]$ [8.1]

3 a. (number line from −6 to 6) **b.** $\{x \mid -3 \le x < -2\}$ **c.** $[-3, -2)$ [8.1] **4 a.** (number line from −3 to 4)

b. $\{x \mid x \le -1 \text{ or } x \ge 2\}$ **c.** $(-\infty, -1] \cup [2, \infty)$ [8.1] **5 a.** (number line from −1 to 5) **b.** $\{n \mid n > 1\}$ **c.** $(1, \infty)$ [8.1]

6 a. (number line from 0 to 12) **b.** $\{x \mid 8 \le x \le 10\}$ **c.** $[8, 10]$ [8.1] **7.** $-8, 2$ [8.2] **8.** $-3, 6$ [8.2] **9.** $-8, \dfrac{2}{3}$ [8.2]

10. no solution [8.2] **11.**

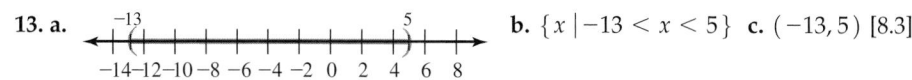

$f(x) = |x + 1| - 2$ [8.2] **12.** $f(x) = -2|x - 3| + 1$ [8.2]

13. a. (number line −14 to 8) **b.** $\{x \mid -13 < x < 5\}$ **c.** $(-13, 5)$ [8.3]

14 a. (number line −6 to 6) **b.** $\{x \mid x < -1 \text{ or } x > 3\}$ **c.** $(-\infty, -1) \cup (3, \infty)$ [8.3]

15 a. (number line −8 to 6) **b.** $\{x \mid -7 < x < -1\}$ **c.** $(-7, -1)$ [8.3]

16 a. (number line −6 to 6) **b.** $\{\quad\}$ or $\varnothing$ **c.** no interval notation [8.3]

17 a. (number line −6 to 6) **b.** $\{x \mid x \text{ is a real number}\}$ **c.** $(-\infty, \infty)$ [8.3]

18. a. (number line −6 to 6) **b.** $\left\{x \mid -\dfrac{1}{3} \le x \le 3\right\}$ **c.** $\left[-\dfrac{1}{3}, 3\right]$ [8.3]

Chapters 1–8 Cumulative Review Exercises

1. false **2.** true **3.** true **4.** true **5.** least common multiple (LCM) **6.** $2(x - 8) = -4$ **7.** vertical line **8.** 19

9. $9x^2 - 6xy + y^2$ **10.** $-\dfrac{128x^{11}}{y^{10}z^6}$ **11.** $\dfrac{1}{5}x - 1 + \dfrac{2}{x}$ **12.** $\dfrac{a - b}{x + y}$ **13.** $\dfrac{2x}{x + 4}$ **14.** $\dfrac{x + 3}{x + 4}$ **15.** $(4c - 3)(3b + 5)$

16. $2x(x - 5)(x - 2)$ **17.** 1 **18.** 5 **19.** $-\dfrac{5}{3}, 4$ **20.** 3, (2 is extraneous) **21.** $-5 \le x \le -2$ **22.** $x \le -\dfrac{6}{5}$ or $x \ge 2$

23. $f(x) = -\dfrac{2}{3}x + 4$ **24.** $x - 2y \le 4$ **25.** 36% **26. a.** -2 **b.** 3 **c.** undefined **27.** 75 ml **28.** $1\dfrac{1}{5}$ hr. **29.** 9 **30.** 1 sec., 5 sec.

Chapter 9

Exercise Set 9.1

Prep Exercises **1.** Squaring a number or its additive inverse results in the same positive number. **2.** The nonnegative root of a number **3.** You cannot raise a number to an even power and get a negative value. **4.** $\sqrt{16}$ is rational because its exact value is a rational number, 4. $\sqrt{17}$ is irrational because its exact value cannot be expressed using a rational number. **5.** With the radical symbol **6.** $\sqrt{x^2}$ means the nonnegative square root of x^2 even if x is negative. **7.** radical; radicand **8.** Set the radicand ≥ 0. The solution set is the domain.

Exercises **1.** ± 6 **3.** ± 11 **5.** ± 14 **7.** ± 15 **9.** 5 **11.** Not a real number **13.** -5 **15.** ± 5 **17.** 1.2 **19.** Not a real number

21. -0.11 **23.** $\dfrac{7}{9}$ **25.** $-\dfrac{12}{13}$ **27.** 3 **29.** -4 **31.** 6 **33.** 4 **35.** Not a real number **37.** -2 **39.** 2 **41.** -3 **43.** 2 **45.** 2

47. $-\dfrac{2}{3}$ **49.** $\dfrac{2}{3}$ **51.** 2.646 **53.** -3.317 **55.** 3.684 **57.** -3.756 **59.** 3.708 **61.** -3.036 **63.** 2.454 **65.** 2.295 **67.** b^2 **69.** $4x$

71. $10r^4s^3$ **73.** $0.5a^3b^6$ **75.** m **77.** $3a^3b^2$ **79.** $-4ab^4$ **81.** $0.2x^6$ **83.** a **85.** $2x^4$ **87.** $2x^2$ **89.** x^2y **91.** $6|m|$ **93.** $|r-1|$

95. $4|y^3|$ **97.** $3y$ **99.** $(y-3)^2$ **101.** $(y-4)^2$ **103.** 2 **105.** $\sqrt{15}$ **107.** $\{x \mid x \geq 4\}$, or $[4, \infty)$ **109.** $\{x \mid x \leq 4\}$, or $(-\infty, 4]$

111. a. 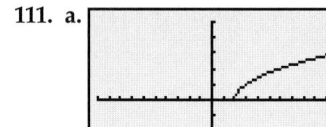 **b.** $\{x \mid x \geq 2\}$ or $[2, \infty)$ **113. a.** 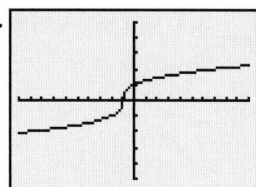 **b.** $\mathbb{R}$, or $(-\infty, \infty)$

115. ≈ -17.709 m/sec. **117.** ≈ 3.476 sec. **119.** ≈ 19.170 mph **121.** 13 ft. **123.** 15 N **125. a.** ≈ 36 **b.** ≈ 39

Review Exercises: **1. a.** 12 **b.** 12 **2.** x^8 **3.** $-45m^4n^3$ **4.** $12x^4 - 20x^3 + 4x^2$ **5.** $21y^2 + 23y - 20$ **6.** $2x^2 - 17x - 9$

Exercise Set 9.2

Prep Exercises **1.** 4 **2.** 5 **3.** a would be nonnegative because with an even index, the radicand must be nonnegative.
4. No restrictions would be placed on a because with an odd index, the radicand can be positive or negative. **5.** The index
6. $m + n$; $m - n$; mn **7.** Yes, because $100^{1/4} = (10^2)^{1/4} = 10^{1/2}$ **8.** Yes, because $3^{1/3} \cdot 2^{1/5} = 3^{5/15} \cdot 2^{3/15} = (3^5 \cdot 2^3)^{1/15} = 1944^{1/15}$

Exercises **1.** $\sqrt{25} = 5$ **3.** $-\sqrt{100} = -10$ **5.** $\sqrt[3]{27} = 3$ **7.** $\sqrt[3]{-64} = -4$ **9.** $\sqrt[4]{y}$ **11.** $\sqrt{144x^8} = 12x^4$ **13.** $18\sqrt{r}$

15. $\sqrt{\dfrac{x^4}{81}} = \dfrac{x^2}{9}$ **17.** $(\sqrt[3]{64})^2 = 16$ **19.** $-(\sqrt[4]{81})^3 = -27$ **21.** $(\sqrt[3]{-8})^4 = 16$ **23.** $\dfrac{1}{(\sqrt[4]{16})^3} = \dfrac{1}{8}$ **25.** $\sqrt[5]{x^4}$ **27.** $8\sqrt[3]{n^2}$

29. $\dfrac{1}{(\sqrt[5]{-32})^2} = \dfrac{1}{4}$ **31.** $\left(\sqrt{\dfrac{1}{25}}\right)^3 = \dfrac{1}{125}$ **33.** $\sqrt[6]{(2a+4)^5}$ **35.** $25^{1/4}$ **37.** $z^{5/6}$ **39.** $5^{-5/6}$ **41.** $5x^{-4/5}$ **43.** $5^{7/3}$ **45.** $x^{2/7}$

47. $(4a-5)^{7/4}$ **49.** $(2r-5)^{8/5}$ **51.** $x^{4/5}$ **53.** $x^{7/6}$ **55.** $a^{17/12}$ **57.** $21w^{4/7}$ **59.** $-12a^{17/12}$ **61.** $7^{5/3}$ **63.** $\dfrac{1}{x^{2/3}}$ **65.** $x^{1/4}$ **67.** $r^{1/12}$

69. $\dfrac{1}{x^{5/7}}$ **71.** $a^{9/4}$ **73.** $20s^{3/7}$ **75.** $-24b^{1/4}$ **77.** x^2 **79.** $a^{5/3}$ **81.** $b^{2/5}$ **83.** $64x^4y^3$ **85.** $2q^{1/2}t^{1/4}$ **87.** $81a^3$ **89.** $3z^{1/3}$ **91.** $\sqrt{2}$

93. $\sqrt[3]{7}$ **95.** $\sqrt{x}$ **97.** $\sqrt[4]{r^3}$ **99.** $\sqrt[4]{x^3y}$ **101.** $\sqrt[5]{m^2n^3}$ **103.** $\sqrt[6]{x^5}$ **105.** $\sqrt[6]{y^7}$ **107.** $\sqrt[6]{x^5}$ **109.** $\sqrt[15]{n^2}$ **111.** $\sqrt[6]{1125}$

113. $\sqrt[12]{3456}$ **115.** $\sqrt[9]{x}$ **117.** $\sqrt[6]{n}$

Review Exercises: **1.** $2^4x^3y^2$ **2.** 12 **3.** 15 **4.** 8×10^{11} **5.** $-\dfrac{5}{8}x^4y^2z^2$ **6.** $2x^2 - 8x + 5 + \dfrac{3}{x+3}$

Exercise Set 9.3

Prep Exercises **1.** Multiplying the approximate roots gives $\sqrt{8} \cdot \sqrt{18} \approx 2.828 \cdot 4.243 = 11.999204$, which is tedious and inexact.
Using the product rule for radicals gives $\sqrt{8} \cdot \sqrt{18} = \sqrt{8 \cdot 18} = \sqrt{144} = 12$, which is fast and exact. **2.** The quotient rule

transforms the expression to the square root of a perfect square: $\sqrt{\dfrac{125}{5}} = \sqrt{25} = 5$. **3.** The radicand, 28, has the perfect square

4 as a factor. So $\sqrt{28} = 2\sqrt{7}$. **4.** Rewrite the expression as a product of two radicals, the first containing the perfect cube and the
second containing no perfect cubes. Then simplify the first radical. For example, $\sqrt[3]{54x^8}$ can be rewritten as $\sqrt[3]{27x^6} \cdot \sqrt[3]{2x^2}$, then
simplified to $3x^2\sqrt[3]{2x^2}$. **5.** The radicand, x^5, has the perfect cube x^3 as a factor.

Exercises **1.** 8 **3.** $9x^3$ **5.** $12xy^2$ **7.** $\sqrt{65}$ **9.** $\sqrt{15x}$ **11.** 3 **13.** $\sqrt[3]{10y^2}$ **15.** $\sqrt[4]{21}$ **17.** $w\sqrt[4]{72}$ **19.** $\sqrt[4]{15x^3y^3}$ **21.** $x\sqrt[5]{30x^2}$

23. $\sqrt[6]{8x^5y^4}$ **25.** $\sqrt{\dfrac{21}{10}}$ **27.** $\sqrt{\dfrac{6y}{5x}}$ **29.** $\dfrac{5}{6}$ **31.** $\dfrac{\sqrt{10}}{3}$ **33.** 6 **35.** $\sqrt[3]{3}$ **37.** $\dfrac{\sqrt[3]{4}}{w^2}$ **39.** $\dfrac{\sqrt[3]{5y^2}}{3x^3}$ **41.** 4 **43.** $\dfrac{\sqrt[4]{3u^3}}{2x^2}$ **45.** $7\sqrt{2}$

47. $8\sqrt{2}$ **49.** $24\sqrt{5}$ **51.** $20\sqrt{7}$ **53.** $a^3\sqrt{a}$ **55.** xy^2 **57.** $x^3y^4z^5$ **59.** $r^5s^4\sqrt{rs}$ **61.** $18x^2\sqrt{2x}$ **63.** $2\sqrt[3]{4}$ **65.** $x^2\sqrt[3]{x}$ **67.** $x^2y\sqrt[3]{y^2}$

69. $4z^2\sqrt[3]{2z^2}$ **71.** $4\sqrt[3]{5}$ **73.** $2\sqrt[4]{5}$ **75.** $9x^4\sqrt[4]{3x}$ **77.** $3x^3\sqrt[5]{2x}$ **79.** $xy^2z\sqrt[6]{x^2y^2z^5}$ **81.** $3\sqrt{7}$ **83.** $30\sqrt{15}$ **85.** $y^2\sqrt{y}$ **87.** $x^4y^5\sqrt{y}$

89. $24c^4\sqrt{15}$ **91.** $60\sqrt{2}$ **93.** $2\sqrt{2}$ **95.** $6\sqrt{5}$ **97.** $c^2d\sqrt{d}$ **99.** $12a^2$ **101.** $6c\sqrt{3c}$ **103.** $36x^2y^3\sqrt{2y}$ **105.** $\dfrac{2\sqrt{6}}{7}$ **107.** $\dfrac{a^4}{2}$

109. $\dfrac{3x^5\sqrt{5}}{4}$ **111.** $\dfrac{3}{8}\sqrt{5}$

Review Exercises: **1.** no **2.** yes **3.** $8x^2 - 7x$ **4.** $8a^2 - 6ab - 9b^2$ **5.** $9m^2 - 25n^2$ **6.** $4x^2 - 12xy + 9y^2$

Exercise Set 9.4

Prep Exercises **1.** Like radicals have the same index and the same radicand, but their coefficients may be different. **2.** Add the coefficients, $3 + 2$, and keep the radical, $\sqrt{2}$, the same. **3.** Multiply each term in the second expression by each term in the first (that is, use FOIL) **4.** $\sqrt{a} + \sqrt{b} \neq \sqrt{a + b}$ because the order of operations is different. For $\sqrt{a} + \sqrt{b}$, the radicals must be evaluated <u>first and</u> then the addition. For $\sqrt{a + b}$, the addition must be evaluated first and then the radical. Example: $\sqrt{36} + \sqrt{64} \neq \sqrt{36 + 64}$ because $6 + 8 \neq 10$.

Exercises **1.** $-6\sqrt{6}$ **3.** $9\sqrt{a}$ **5.** $12\sqrt{5} - 8\sqrt{6}$ **7.** $11a\sqrt{5a} - 2b\sqrt{7b}$ **9.** $3x\sqrt[3]{9}$ **11.** $-6x^2\sqrt[4]{5x}$ **13.** Cannot combine because the radicals are <u>not</u> like **15.** $-\sqrt{3}$ **17.** $-\sqrt{5y}$ **19.** $-8\sqrt{5}$ **21.** $6\sqrt{6}$ **23.** $14a\sqrt{3a}$ **25.** $4\sqrt{6}$ **27.** $18\sqrt{2} - 17\sqrt{3}$ **29.** $7\sqrt[3]{2}$ **31.** $-12x\sqrt[3]{5x^2}$ **33.** $-2x^2\sqrt[4]{2x}$ **35.** $3\sqrt{2} + 2$ **37.** $3 - 3\sqrt{5}$ **39.** $\sqrt{30} + 10\sqrt{2}$ **41.** $24x - 48x\sqrt{2}$ **43.** $12 - 3\sqrt{2} + 4\sqrt{5} - \sqrt{10}$ **45.** $6 + 5\sqrt{x} + x$ **47.** $6 + 10\sqrt{2} + 9\sqrt{3} + 15\sqrt{6}$ **49.** $\sqrt{15} + \sqrt{21} + 5 + \sqrt{35}$ **51.** $x + \sqrt{xy} - 6y$ **53.** $12\sqrt{14} - 12\sqrt{6} + 6\sqrt{35} - 6\sqrt{15}$ **55.** $8a + 10\sqrt{ab} - 3b$ **57.** $2\sqrt[3]{2} - 3\sqrt[3]{4} - 40$ **59.** $1 - \sqrt[3]{18} + \sqrt[3]{12}$ **61.** $x + 8$ **63.** $22 + 8\sqrt{6}$ **65.** $18 - 8\sqrt{2}$ **67.** $16 + 8\sqrt{3}$ **69.** $30 + 12\sqrt{6}$ **71.** 13 **73.** -14 **75.** $36 - x$ **77.** 1 **79.** $x - y$ **81.** 4 **83.** 50 **85.** $7\sqrt{5}$ **87.** $-15\sqrt{6}$ **89.** $9\sqrt{2}$ **91.** $-14\sqrt{5}$ **93.** $25\sqrt{3}$ **95. a.** $42 + 2\sqrt{5}$ ft. **b.** 46.5 ft. **c.** \$87.89

Review Exercises: **1.** $2x + 5$ **2.** $16x^2 - 9$ **3.** $\sqrt{2}$ **4.** $\sqrt[3]{4}$ **5.** Slope is $\dfrac{2}{5}$; y-intercept is $(0, 2)$. **6.**

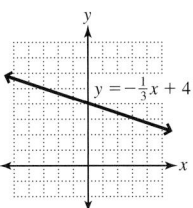

Exercise Set 9.5

Prep Exercises **1. a.** $\sqrt{16}$ is rational. **b.** A radical, $\sqrt{3}$, is in the denominator. **2.** The denominator in $\dfrac{\sqrt{3}}{3}$ is a rational number, whereas it is irrational in $\dfrac{1}{\sqrt{3}}$. **3.** Multiply the fraction by a 1 so that the product's denominator has a radicand that is a perfect square. **4.** Multiply the fraction by a 1 whose numerator and denominator are the conjugate of the denominator. **5.** Multiply the numerator and denominator by $2 - \sqrt{7x}$.

Exercises **1.** $\dfrac{\sqrt{3}}{3}$ **3.** $\dfrac{3\sqrt{2}}{4}$ **5.** $\dfrac{6\sqrt{7}}{7}$ **7.** $\dfrac{\sqrt{15}}{6}$ **9.** $\dfrac{x\sqrt{14}}{10}$ **11.** $\dfrac{\sqrt{7}}{7}$ **13.** $\dfrac{5\sqrt{3a}}{3a}$ **15.** $\dfrac{\sqrt{33mn}}{11n}$ **17.** $\dfrac{2\sqrt{5x}}{x}$ **19.** $\dfrac{\sqrt{3}}{4}$ **21.** $\dfrac{3\sqrt{x}}{x^2}$

23. $4x\sqrt{2x}$ **25.** Mistake: The product of $\sqrt{2}$ and 2 is not 2. Correct: $\dfrac{\sqrt{6}}{2}$ **27.** $\dfrac{5\sqrt[3]{9}}{3}$ **29.** $\dfrac{\sqrt[3]{20}}{2}$ **31.** $3\sqrt[3]{2}$ **33.** $\dfrac{m\sqrt[3]{n^2}}{n}$ **35.** $\dfrac{\sqrt[3]{ab}}{b}$

37. $\dfrac{2\sqrt[3]{4x^2}}{x}$ **39.** $\dfrac{\sqrt[3]{30a}}{5a}$ **41.** $\dfrac{5\sqrt[4]{4}}{2}$ or $\dfrac{5\sqrt{2}}{2}$ **43.** $\dfrac{\sqrt[4]{3x^2}}{x}$ **45.** $\dfrac{3\sqrt[4]{27x}}{x}$ **47.** $3\sqrt{2} - 3$ **49.** $8 + 4\sqrt{3}$ **51.** $5\sqrt{3} - 5\sqrt{2}$ **53.** $-1 - \sqrt{5}$

55. $\dfrac{3 + \sqrt{3}}{2}$ **57.** $\dfrac{-6 - 8\sqrt{3}}{13}$ **59.** $\dfrac{4\sqrt{21} - 4\sqrt{6}}{5}$ **61.** $\dfrac{32 + 8\sqrt{3}}{13}$ **63.** $\dfrac{6y - 6\sqrt{y}}{y - 1}$ **65.** $\dfrac{3t - 6\sqrt{tu}}{t - 4u}$ **67.** $\dfrac{\sqrt{2xy} + 2y\sqrt{3}}{x - 6y}$

69. $\dfrac{3}{2\sqrt{3}}$ **71.** $\dfrac{2x}{5\sqrt{2x}}$ **73.** $\dfrac{2n}{3\sqrt{2n}}$ **75.** $\dfrac{1}{10 - 5\sqrt{3}}$ **77.** $\dfrac{5x - 36}{9\sqrt{5x} + 54}$ **79.** $\dfrac{19}{10\sqrt{n} - 2\sqrt{6n}}$ **81. a.** $\dfrac{5\sqrt{3}}{3}$ **b.** $\sqrt{5}$ **c.** $\dfrac{5\sqrt{11}}{11}$

83. a. The graphs are identical. The functions are identical. **b.** $f(x) = g(x)$ **85. a.** $T = \pi\sqrt{9.8L}/4.9$ **b.** $T = 2\pi L/\sqrt{9.8L}$

87. a. $s = \sqrt{3Vh}/h$ **b.** 745 ft. **89. a.** $V_{\text{rms}} = \dfrac{\sqrt{2}}{2}V_{\text{m}}$ or $\dfrac{\sqrt{2}V_{\text{m}}}{2}$ **b.** $V_{\text{rms}} = \dfrac{163\sqrt{2}}{2}$ **c.** ≈ 115.3 **91.** $\dfrac{15\sqrt{2} - 10\sqrt{3}}{3}\Omega$

Review Exercises: **1.** $\pm 2\sqrt{7}$ **2.** $(x - 3)^2$ **3.** 4 **4.** -1 **5.** $-6, 6$ **6.** $2, 3$

Exercise Set 9.6

Prep Exercises **1.** Some of the solutions may be extraneous. **2.** A solution that does not satisfy the original equation **3.** The principal square root of a number cannot equal a negative. **4.** $(\sqrt{a})^2 = \sqrt{a} \cdot \sqrt{a} = \sqrt{a^2} = a$ **5.** Subtract $3x$ from both sides to isolate the radical. This allows use of the power rule to eliminate the radical. **6.** Answers may vary. One example is $\sqrt{x} + 4 = \sqrt{2x + 1}$. The radicals must be totally eliminated.

Exercises **1.** 4 **3.** No real-number solution **5.** 27 **7.** -8 **9.** 17 **11.** 11 **13.** 6 **15.** -4 **17.** No real-number solution **19.** 11

21. -2 **23.** 124 **25.** 55 **27.** 5 **29.** -11 **31.** No real-number solution **33.** 2 **35.** No real-number solution $(-5$ is an extraneous solution$)$ **37.** 3 **39.** 3 **41.** -4 $(-10$ is an extraneous solution$)$ **43.** 1, 3 **45.** $-3, -2$ **47.** $\dfrac{1}{2}\left(-\dfrac{3}{8}$ is an extraneous solution$\right)$

49. 5 **51.** $-1, 3$ **53.** 0, 4 **55.** $\dfrac{3}{16}$ **57.** 3 $(-1$ is an extraneous solution$)$ **59.** Mistake: You cannot take the principal square root of a number and get a negative. Correct: No real-number solution **61.** Mistake: The binomial $x - 3$ was not squared correctly. Correct: $x = 6$ with $x = 1$ an extraneous solution **63.** 9.8 m **65.** 0.6125 m **67.** 1.44 ft. **69.** 1 ft. **71.** 36.73 ft. **73.** 82.65 ft. **75.** 4 N

77. 6 N   **79. a.** 25; 3; 4   **b.**

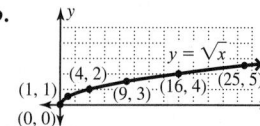

c. No, the x-values must be 0 or positive because real square roots exist only when $x \geq 0$. The y-values must be 0 or positive because, by definition, the principal square root is 0 or positive.   **d.** Yes, because it passes the vertical line test.

81. The graph becomes steeper from left to right.   **83.** The graph rises or lowers according to the value of the constant.

Review Exercises: **1.** 81   **2.** -0.008 **3.** 39.0625 or $\dfrac{625}{16}$ **4.** x^8 **5.** n^{24} **6.** y^4

Exercise Set 9.7

Prep Exercises   **1.** $\sqrt{-1}$ **2.** Yes, every real number can be expressed as $a + 0i$.   **3.** No, every real number is a complex number that is not imaginary. For example, 2 is complex but not imaginary.   **4.** Separate into two radicals, $\sqrt{-1} \cdot \sqrt{n}$. Replace $\sqrt{-1}$ with i and simplify $\sqrt{n}$.   **5.** We add complex numbers just like we add polynomials—by combining like terms.   **6.** We subtract complex numbers just like we subtract polynomials—by writing an equivalent addition and changing the signs in the second complex number.   **7.** -1 **8.** No, because it is not in the form $a + bi$.   **9.** Multiply the numerator and denominator by $3 + 2i$.   **10.** $i, -1, -i, 1$

Exercises   **1.** $6i$ **3.** $i\sqrt{5}$ **5.** $2i\sqrt{2}$ **7.** $3i\sqrt{2}$ **9.** $3i\sqrt{3}$ **11.** $5i\sqrt{5}$ **13.** $3i\sqrt{7}$ **15.** $7i\sqrt{5}$ **17.** $6 + 7i$ **19.** $11 - 6i$ **21.** $-7 + i$ **23.** $8 - 2i$ **25.** $-3 - 10i$ **27.** 0 **29.** $14 + 9i$ **31.** $-4 + 28i$ **33.** -24 **35.** 40 **37.** $14 + 12i$ **39.** $-72 - 32i$ **41.** $19 - 3i$ **43.** $50 + 9i$ **45.** 65 **47.** $-7 - 24i$ **49.** $-2i$ **51.** $-\dfrac{4i}{5}$ **53.** $-3i$ **55.** $\dfrac{1}{2} - i$ **57.** $\dfrac{1}{2} - i$ **59.** $\dfrac{14}{5} - \dfrac{7}{5}i$ **61.** $-\dfrac{7}{29} + \dfrac{3}{29}i$ **63.** $7 - 2i$ **65.** $\dfrac{9}{13} - \dfrac{7}{13}i$ **67.** $\dfrac{34}{41} + \dfrac{19}{41}i$ **69.** $-i$ **71.** -1 **73.** -1 **75.** 1 **77.** 1 **79.** -1 **81.** $-i$ **83.** i

Review Exercises: **1.** $(2x - 3)^2$ **2.** $\pm 4\sqrt{3}$ **3.** 2   **4.** $-\dfrac{3}{2}, 2$ **5.** $(0, 2), (3, 0)$ **6.**

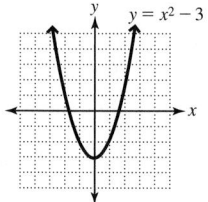

Chapter 9 Summary and Review Exercises

9.1 Definitions/Rules/Procedures $b^n = a$; $b^n = a$; no real-number; negative   Exercises   **1.** ± 11 **2.** ± 7 **3.** 13   **4.** -7 **5.** Not a real number   **6.** $\dfrac{1}{5}$ **7.** 5 ft.   **8.** π sec.   **9.** 2.646   **10.** 9.487   **11. a.** $8\sqrt{5}$ mph   **b.** 17.9 mph   **12. a.** $\dfrac{\sqrt{10}}{2}$ sec.   **b.** 1.58 sec.   **13.** $7x^4$ **14.** $12a^3b^6$ **15.** $0.4mn^5$ **16.** x^5 **17.** $-4r^3s$ **18.** $3x^3$ **19.** $2x^3y^4$ **20.** x^2y **21.** $9|x|$ **22.** $(x - 1)^2$ Definitions/Rules/Procedures   radical; radicand; simplify; x is a real number; $-\infty$; ∞; nonnegative; ≥ 0 Exercises   **23.** 4   **24.** -2 **25.** $x \geq -5$ **26.** $x \leq 3$

9.2 Definitions/Rules/Procedures $\sqrt[n]{a}$; $\sqrt[n]{a^m}$; $(\sqrt[n]{a})^m$; $\dfrac{1}{a^{m/n}}$ Exercises   **27.** -4 **28.** $2a^2\sqrt{6}$ **29.** $\dfrac{1}{8}$ **30.** $\sqrt[7]{(5r - 2)^5}$ **31.** $\dfrac{1}{11}$ **32.** $\dfrac{1}{27}$ **33.** $\dfrac{64}{343}$ **34.** $81\sqrt[4]{x^3}$ **35.** $33^{1/8}$ **36.** $8n^{-3/7}$ **37.** $8^{3/5}$ **38.** $m^{8/3}$ **39.** $(3xw)^{3/4}$ **40.** $(a + b)^{4/3}$ Definitions/Rules/Procedures   exponential; exponents; radical   Exercises   **41.** x^2 **42.** $32m^{3/2}$ **43.** $\dfrac{1}{y^{1/5}}$ **44.** b **45.** $k^{1/2}$ **46.** $2^{3/4}x^{3/4}y^{3/20}$ **47.** $\sqrt[6]{x}$

9.3 Definitions/Rules/Procedures $\sqrt[n]{a \cdot b}$; a Exercises   **48.** 9   **49.** $10x^4$ **50.** 2   **51.** $\sqrt[4]{42}$ **52.** $\sqrt[5]{15x^4y^4}$ **53.** $84\sqrt{10}$ **54.** $x^7\sqrt{x}$ **55.** $16c^2\sqrt{15c}$ **56.** $a^5b^4\sqrt{b}$ **57.** $6ab^2$ Definitions/Rules/Procedures $\sqrt[n]{\dfrac{a}{b}}$ Exercises   **58.** $\dfrac{7}{11}$ **59.** $-\dfrac{3}{2}$ **60.** $6\sqrt{5}$ **61.** $36x^2y^3\sqrt{2y}$

Definitions/Rules/Procedures nth; nth; perfect nth power; perfect nth power   Exercises   **62.** $12b^4\sqrt{3b}$ **63.** $15\sqrt[3]{4}$ **64.** $4x^3\sqrt[3]{5x}$ **65.** $6x^5\sqrt[4]{2x^3}$

9.4 Definitions/Rules/Procedures   radicands; root indices; radical parts   Exercises   **66.** $-3\sqrt{n}$ **67.** $-6y^3\sqrt[4]{8y}$ **68.** $5\sqrt{5}$ **69.** $-10\sqrt{6}$ **70.** $4\sqrt{6}$ **71.** $28x\sqrt{2y}$ **72.** $9xy\sqrt[3]{2xy^2}$ **73.** $5\sqrt[4]{3}$ Definitions/Rules/Procedures   order of operations   Exercises   **74.** $\sqrt{15} + \sqrt{10}$ **75.** $\sqrt[3]{21} + 2\sqrt[3]{49}$ **76.** $6\sqrt{6} - 54$ **77.** $\sqrt{10} + \sqrt{14} - \sqrt{15} - \sqrt{21}$ **78.** $-6 + 2\sqrt[3]{2} - 4\sqrt[3]{4}$ **79.** $\sqrt[4]{12x^2} - \sqrt[4]{6x} + 2\sqrt[4]{2x} - 2$ **80.** $5a - 3b$ **81.** 7   **82.** $7 - 2\sqrt{10}$

9.5 Definitions/Rules/Procedures   index; fractions; sums; differences; products; quotients; perfect nth power   Exercises   **83.** $\dfrac{\sqrt{2}}{2}$ **84.** $\sqrt[3]{9}$ **85.** $\dfrac{2\sqrt{7}}{7}$ **86.** $\dfrac{\sqrt[4]{40x^2}}{2}$ **87.** $\dfrac{\sqrt[3]{153y}}{3y}$ **88.** $\dfrac{\sqrt[4]{27}}{3}$ Definitions/Rules/Procedures   conjugate   Exercises   **89.** $-4\sqrt{2} - 4\sqrt{3}$ **90.** $\dfrac{4 - \sqrt{3}}{13}$ **91.** $\dfrac{2 + \sqrt{n}}{4 - n}$ **92.** $\sqrt{6} + 2$ Definitions/Rules/Procedures   denominator   Exercises   **93.** $\dfrac{5}{3\sqrt{10}}$ **94.** $\dfrac{3x}{5\sqrt{3x}}$ **95.** $\dfrac{1}{16 + 8\sqrt{3}}$ **96.** $\dfrac{1}{10\sqrt{t} - 5\sqrt{3t}}$

9.6 Definitions/Rules/Procedures radical expression; variable; integer power; radical; radical term; radical terms; root index; radical term; root index; extraneous **Exercises** **97.** 81 **98.** No real-number solution **99.** 10 **100.** -2 **101.** 18 **102.** 5 **103.** No real-number solution **104.** $\dfrac{719}{3}$ **105.** 7 **106.** 1 $\left(\dfrac{17}{4}\text{ is extraneous.}\right)$ **107.** 0, 4 **108.** $\dfrac{3}{16}$ **109.** 312.5 ft. **110.** 0.272 sec. **111.** 1.44 ft.

9.7 Definitions/Rules/Procedures $\sqrt{-1}$; -1; $a + bi$; real numbers; imaginary unit; $b \neq 0$; $a + bi$; real numbers; imaginary unit; $a - bi$; $\sqrt{-1} \cdot \sqrt{n}$; i; like terms; signs **Exercises** **112.** $3i$ **113.** $2i\sqrt{5}$ **114.** $8 - 6i$ **115.** $9 - 7i$ **Definitions/Rules/Procedures** monomials; binomials **Exercises** **116.** -12 **117.** $2 + 8i$ **118.** $26 + 2i$ **119.** $24 - 10i$ **Definitions/Rules/Procedures** $\dfrac{i}{i}$; complex conjugate **Exercises** **120.** $-5i$ **121.** $3i$ **122.** $-\dfrac{4i}{3}$ **123.** $\dfrac{1}{5} - \dfrac{7}{5}i$ **124.** $\dfrac{6}{5} - \dfrac{3}{5}i$ **125.** $\dfrac{7}{13} + \dfrac{17}{13}i$ **Definitions/Rules/Procedures** i; -1; $-i$; 1 **Exercises** **126.** 1 **127.** $-i$

Chapter 9 Practice Test

1. 6 [9.1] **2.** $7i$ [9.1] **3.** $9xy^2\sqrt{y}$ [9.1] **4.** $3\sqrt[3]{2}$ [9.1] **5.** $2x\sqrt[4]{x^2}$, or $2x\sqrt{x}$ [9.3] **6.** $3r^5$ [9.1] **7.** $\dfrac{1}{3}$ [9.3] **8.** $\dfrac{1}{3}$ [9.3] **9.** $5\sqrt{7}$ [9.4]

10. $4 - 2\sqrt{3}$ [9.4] **11.** $\dfrac{1}{x^{2/3}}$ [9.2] **12.** $-6 + 2\sqrt[3]{2} - 4\sqrt[3]{4}$ [9.4] **13.** $8^{1/5}x^{3/5}$ [9.2] **14.** $(2x + 5)^{2/3}$ [9.2] **15.** $\dfrac{\sqrt[3]{2}}{2}$ [9.5]

16. $\dfrac{x - \sqrt{xy}}{x - y}$ [9.5] **17.** 22 [9.6] **18.** 7 [9.6] **19.** 24 (0 is an extraneous solution.) [9.6] **20.** $-2 - 4i$ [9.7] **21.** $17 + 0i$ or 17 [9.7]

22. $\dfrac{8}{25} + \dfrac{6}{25}i$ [9.7] **23.** 60 m^2 [9.3] **24. a.** $\dfrac{\sqrt{3}}{2}$ [9.6] **b.** 64 ft. [9.6] **25. a.** $3.5\sqrt{10}$ mph [9.6] **b.** 293.88 ft or 14,400/49 ft. [9.6]

Chapters 1–9 Cumulative Review Exercises

1. false **2.** false **3.** true **4.** true **5.** x-intercept **6.** inconsistent **7.** $24x^5y^2$ **8.** 28 **9.** 0.0000024 **10.** x^{15} **11.** $9x^4 - 24x^2y + 16y^2$

12. $4x^2 + 6x - 2 + \dfrac{2}{2x - 3}$ **13.** $\dfrac{x}{x + 5}$ **14.** $30a^3\sqrt{2a}$ **15.** $-6x^2\sqrt[3]{3x}$ **16.** $\dfrac{1}{8}$ **17.** $23 - 2i$ **18.** $\dfrac{5 + 3\sqrt{2}}{7}$ **19.** $b = \dfrac{2A - Bh}{h}$ or

$\dfrac{2A}{h} - B$ **20.** 5, -11 **21.** No solution (4 is an extraneous solution.) **22.** 2 (-3 is an extraneous solution.) **23.** slope $= \dfrac{5}{2}$,

y-intercept $= \dfrac{15}{2}$ **24.** $y = 3x + 14$ **25.**

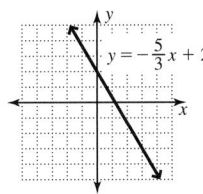

$y = -\dfrac{5}{3}x + 2$

26.

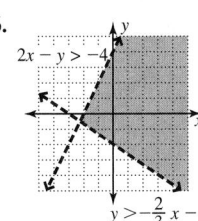

$2x - y > -4$

$y > -\dfrac{2}{3}x - 2$

27. $16.48 **28.** 25 mph **29.** 96 **30.** 39.2 m

Chapter 10
Exercise Set 10.1

Prep Exercises **1.** $x^2 = a$ has two solutions because squaring $\sqrt{a}$ and $-\sqrt{a}$ gives a for every real number a. **2.** $x = \pm\sqrt{\dfrac{b + c}{a}}$

3. $x = \dfrac{b \pm \sqrt{c}}{a}$ **4.** $\left(\dfrac{b}{2}\right)^2$ **5.** $\left(x + \dfrac{b}{2}\right)^2$ **6.** Add $\left(\dfrac{b}{2}\right)^2$ to both sides of the equation. **7.** Because $x^2 - 7x + 12$ is easy to factor, factoring is a better method as it requires fewer steps than completing the square. **8.** Because $x^2 + 4x + 5$ cannot be factored, completing the square is the only option at this point.

Exercises **1.** ± 7 **3.** $\pm\dfrac{2}{5}$ **5.** ± 1.2 **7.** $\pm 3\sqrt{5}$ **9.** $\pm 5i$ **11.** ± 7 **13.** ± 9 **15.** ± 3 **17.** $\pm\dfrac{3}{5}$ **19.** $\pm 2i$ **21.** $\pm\dfrac{12}{5}$ **23.** $\pm 2\sqrt{2}$

25. $\pm 3i\sqrt{2}$ **27.** $\pm\dfrac{4\sqrt{6}}{3}$ **29.** ± 1.6 **31.** $-15, -1$ **33.** $-\dfrac{1}{5}, \dfrac{7}{5}$ **35.** $8 \pm 4i$ **37.** $\dfrac{1 \pm 2\sqrt{10}}{4}$ **39.** $7 \pm 2i\sqrt{3}$ **41.** $0, \dfrac{3}{2}$ **43.** $\dfrac{3}{5}, \dfrac{6}{5}$

45. $-13.5, -5.5$ **47.** Mistake: Gave only the positive solution; Correct: ± 7 **49.** Mistake: Changed -6 to 6; Correct: $5 \pm i\sqrt{6}$ **51.** $\pm 4\sqrt{6} \approx \pm 9.798$ **53.** $\pm 2\sqrt{5} \approx \pm 4.472$ **55.** $6 \pm \sqrt{15} \approx 9.873, 2.127$ **57. a.** $x^2 + 14x + 49$ **b.** $(x + 7)^2$

59. a. $n^2 - 10n + 25$ **b.** $(n - 5)^2$ **61. a.** $y^2 - 9y + \dfrac{81}{4}$ **b.** $\left(y - \dfrac{9}{2}\right)^2$ **63. a.** $s^2 - \dfrac{2}{3}s + \dfrac{1}{9}$ **b.** $\left(s - \dfrac{1}{3}\right)^2$ **65.** $-5, 3$

67. $1 \pm 7i$ **69.** $3, 6$ **71.** $1 \pm \sqrt{17}$ **73.** $3 \pm 2i\sqrt{5}$ **75.** $-\dfrac{3}{2}, 1$ **77.** $-\dfrac{9}{2}, \dfrac{1}{2}$ **79.** $\dfrac{1}{3}, \dfrac{5}{3}$ **81.** $-1, \dfrac{3}{2}$ **83.** $-2 \pm \dfrac{\sqrt{23}}{2}$ **85.** $\dfrac{-1 \pm \sqrt{41}}{10}$

87. $\dfrac{-2 \pm 2\sqrt{7}}{3}$ **89.** Did not divide by 3 so that x^2 has a coefficient of 1; then wrote an incorrect factored form; Correct: $\dfrac{-2 \pm \sqrt{10}}{3}$

91. 14 in.   **93.** Height: 3 ft.; width: 6 ft.   **95.** 305 m   **97.** 18 in.   **99.** 16 ft. by 12 ft.   **101. a.** 15 cm   **b.** Length: 45 cm; width: 30 cm

103. $\dfrac{3}{4}$ sec.   **105.** 4 m/sec.

Review Exercises: **1.** $(x - 5)^2$ **2.** $(x + 3)^2$ **3.** $6x$ **4.** $2x + 3$ **5.** $\dfrac{-2 + \sqrt{6}}{2}$ or $-1 + \dfrac{\sqrt{6}}{2}$ **6.** $\dfrac{-3 - i}{5}$ or $-\dfrac{3}{5} - \dfrac{1}{5}i$

Exercise Set 10.2

Prep Exercises   **1.** $x = \dfrac{-b \pm \sqrt{b^2 - 4ac}}{2a}$ **2.** Answers may vary. One advantage is that it can be used to solve any quadratic equation. One disadvantage is that the steps can be tedious.   **3.** Yes, if they have irrational or nonreal complex solutions   **4.** When the discriminant is zero. The quadratic formula becomes $x = -\dfrac{b}{2a}$.   **5.** The discriminant is the radicand in the quadratic formula; $b^2 - 4ac$ **6.** The graph does not intersect the x-axis.

Exercises   **1.** $a = 1, b = -3, c = 7$ **3.** $a = 3, b = -9, c = -4$ **5.** $a = 1.5, b = -1, c = 0.2$ or $a = -1.5, b = 1, c = -0.2$
7. $a = -\dfrac{1}{2}, b = -\dfrac{3}{4}, c = 6$ or $a = \dfrac{1}{2}, b = \dfrac{3}{4}, c = -6$ **9.** $-5, -4$ **11.** $-2, \dfrac{3}{4}$ **13.** $0, 9$ **15.** 4 **17.** $-2, \dfrac{2}{3}$ **19.** $1 \pm i$ **21.** $\dfrac{1 \pm \sqrt{5}}{2}$
23. $\dfrac{-5 \pm \sqrt{10}}{3}$ **25.** $\dfrac{3 \pm i\sqrt{11}}{4}$ **27.** $-\dfrac{1}{6}, -\dfrac{2}{3}$ **29.** $\dfrac{2 \pm i\sqrt{5}}{3}$ **31.** $\dfrac{3 \pm \sqrt{105}}{12}$ **33.** $-0.15, 0.1$ **35.** $-2, \dfrac{3}{2}$ **37.** $\pm\dfrac{7}{6}$ **39.** $1 \pm i\sqrt{2}$
41. $\dfrac{-3 \pm \sqrt{5009}}{100} \approx -0.738, 0.678$ **43.** $\dfrac{3 \pm i\sqrt{51}}{12}$ **45.** Mistake: Did not evaluate $-b$. Correct: $\dfrac{7 \pm \sqrt{37}}{6}$ **47.** Mistake: The result was not completely simplified. Correct: $1 \pm i\sqrt{2}$ **49.** One rational   **51.** Two irrational   **53.** Two rational   **55.** Two nonreal complex
57. Two irrational   **59.** Square root principle or factoring; ± 9 **61.** Quadratic formula; $\dfrac{2 \pm \sqrt{10}}{2}$ **63.** Factoring; $0, -6$

65. Square root principle; $-7 \pm 2\sqrt{10}$ **67.** Quadratic formula; $4 \pm i\sqrt{3}$ **69.** $(2, 0), (-1, 0), (0, -2)$ **71.** $\left(\dfrac{3}{2}, 0\right), (0, 9)$

73. $\left(\dfrac{1}{2}, 0\right), (-8, 0), (0, -8)$ **75.** No x-intercepts, $(0, -6)$ **77.** $x^2 + 5(x + 1) = 71; 6, 7$
79. $x^2 + (x + 1)^2 = (x + 2)^2; 3, 4, 5$ **81.** $w(3w - 3.5) = 34$; width: 4 ft.; length: 8.5 ft.   **83. a.** $22x = 1.5x^2 + 4x$; 12 ft.   **b.** 18 ft., 12 ft., 10 ft., 6 ft., 8 ft., 18 ft. (from bottom clockwise)   **85.** ≈ 3.43 sec.   **87. a.** $h = -4.9t^2 + 10$ **b.** ≈ 1.01 sec.   **c.** ≈ 1.43 sec.
89. 8000 units   **91. a.** $\dfrac{25}{8}$ **b.** $c < \dfrac{25}{8}$ **c.** $c > \dfrac{25}{8}$ **93. a.** $\dfrac{9}{2}$ **b.** $a < \dfrac{9}{2}$ **c.** $a > \dfrac{9}{2}$

Review Exercises: **1.** $(u - 7)(u - 2)$ **2.** $(3u - 8)(u + 2)$ **3.** x^4 **4.** $x^{2/3}$ **5.** $\dfrac{2}{5}, -3$ **6.** 8

Exercise Set 10.3

Prep Exercises   **1.** It can be rewritten as a quadratic equation.   **2.** No, $x^{3/4} = (x^{1/4})^3$ shows a cubic relationship.   **3.** $u = \dfrac{x + 2}{3}$
4. $u = x^2$ **5.** $\dfrac{1}{x}$ **6.** $\dfrac{400}{x}$

Exercises   **1.** $-6, 4$ **3.** $-\dfrac{4}{3}, 2$ **5.** -8 (4 is extraneous.)   **7.** $-7.5, 5$ **9.** $-4, 2$ **11.** $-5, \dfrac{2}{3}$ **13.** $9, 25$ **15.** $\dfrac{49}{4}$ (1 is extraneous.)
17. $2 \left(\dfrac{2}{9} \text{ is extraneous.}\right)$ **19.** $\pm 3i$ **21.** $2 \left(-\dfrac{2}{9} \text{ is extraneous.}\right)$ **23.** $0, 4$ **25.** $\pm 3, \pm 1$ **27.** $\pm 1.5, \pm 1$ **29.** $\pm 2, \pm 3i$ **31.** $-6, -4$
33. $-3.5, 2$ **35.** $-9, -5$ **37.** $-5, 0$ **39.** $8, 27$ **41.** $-\dfrac{1}{8}, 8$ **43.** $81, 16$ **45.** $\dfrac{16}{625}$ (16 is extraneous.)   **47.** 3 hr.   **49.** ≈ 2.21 hr.
51. 45 mph   **53. a.** Old rate $= 12$ mph; new rate $= 18$ mph   **b.** Old time $= 3$ hr.; new time $= 2$ hr.   **55.** Jody: 4 hr.; Billy: 6 hr.
57. Older: ≈ 4.27 hr.; newer: ≈ 3.77 hr.   **59.** 420 vps, 460 vps

Review Exercises: **1.** -2 **2.** -3 **3.** $(3, 0), (0, 2)$ **4.** 6   **5.**   **6.**

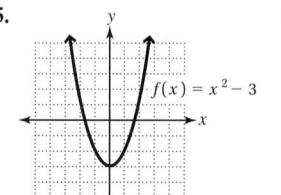

$f(x) = x^2 - 3$

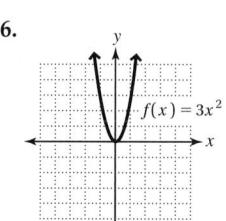

$f(x) = 3x^2$

Exercise Set 10.4

Prep Exercises **1.** The sign of a **2.** The axis of symmetry is the vertical line through the vertex. **3.** (h, k); $x = h$ **4.** There are no
x-intercepts. **5.** There are no x-intercepts. **6.** $x = -\dfrac{b}{2a}$

Exercises **1.** V: $(2, -3)$; axis: $x = 2$ **3.** V: $(-1, -5)$; axis: $x = -1$ **5.** V: $(0, 2)$; axis: $x = 0$ **7.** V: $(-5, 0)$; axis: $x = -5$

9. V: $(0, 0)$; axis: $x = 0$ **11.** V: $(2, 4)$; axis: $x = 2$ **13.** V: $(-4, -5)$; axis: $x = -4$ **15.** V: $\left(\dfrac{1}{3}, \dfrac{4}{3}\right)$; axis: $x = \dfrac{1}{3}$ **17. a.** downward
b. $(0, 0)$ **c.** $x = 0$ **d.**
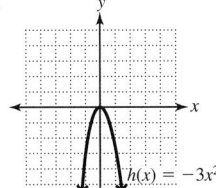
19. a. upward **b.** $(0, 0)$ **c.** $x = 0$ **d.**
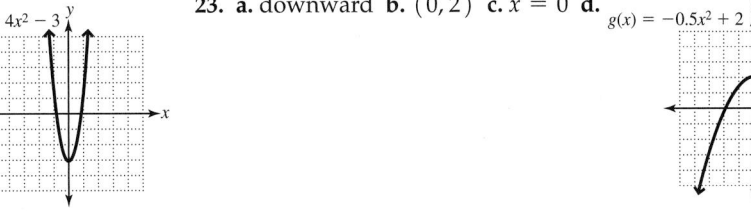
21. a. upward **b.** $(0, -3)$

c. $x = 0$ **d.**
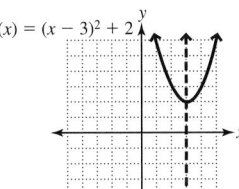
23. a. downward **b.** $(0, 2)$ **c.** $x = 0$ **d.**

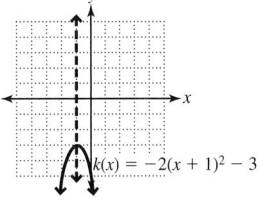
25. a. upward **b.** $(3, 2)$

c. $x = 3$ **d.**
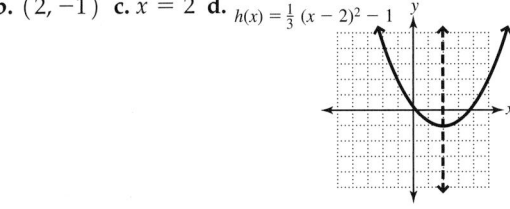
27. a. downward **b.** $(-1, -3)$ **c.** $x = -1$ **d.**

29. a. upward

b. $(2, -1)$ **c.** $x = 2$ **d.**
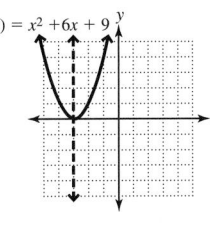
31. a. $(-3, 0)$, $(0, 9)$ **b.** $h(x) = (x + 3)^2$ **c.** upward **d.** $(-3, 0)$ **e.** $x = -3$

f. **g.** domain: $\{x \,|\, x \text{ is a real number}\}$, or $(-\infty, \infty)$; range: $\{y \,|\, y \geq 0\}$ or $[0, \infty)$

33. a. no x-intercepts, $(0, -5)$ **b.** $g(x) = -3(x - 1)^2 - 2$ **c.** downward **d.** $(1, -2)$ **e.** $x = 1$ **f.**
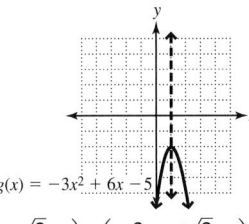

g. domain: $\{x \,|\, x \text{ is a real number}\}$, or $(-\infty, \infty)$; range: $\{y \,|\, y \leq -2\}$ or $(-\infty, -2]$ **35. a.** $\left(\dfrac{-3 + \sqrt{3}}{2}, 0\right)$, $\left(\dfrac{-3 - \sqrt{3}}{2}, 0\right)$, $(0, 3)$

b. $f(x) = 2\left(x + \frac{3}{2}\right)^2 - \frac{3}{2}$ **c.** upward **d.** $\left(-\frac{3}{2}, -\frac{3}{2}\right)$ **e.** $x = -\frac{3}{2}$ **f.** 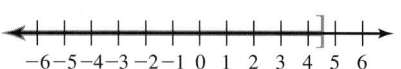 **g.** domain: $\{x \mid x \text{ is a real number}\}$,

or $(-\infty, \infty)$; range: $\left\{y \mid y \geq -\frac{3}{2}\right\}$ or $\left[-\frac{3}{2}, \infty\right)$ **37. a.** 4.59 sec. **b.** 103.32 m **c.** 9.18 sec. **39. a.** 9.2 ft. **b.** ≈ 5.39 ft.

c. 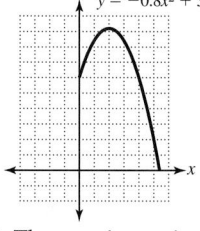 **41. a.** The greatest number of CDs were sold in the 10th week. **b.** 20,000

43. The area is maximized if the length and width are 100 ft. **45.** 55 units **47.** $-6, 6$; the product is -36.

Review Exercises: **1.** -11 **2.** $-5, 3$ **3.** $\pm 2i\sqrt{5}$ **4.** $x \leq 4.5$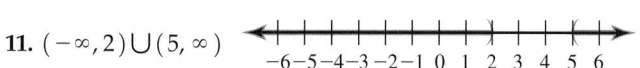

5. $x \leq -5$ or $x \geq 11$ 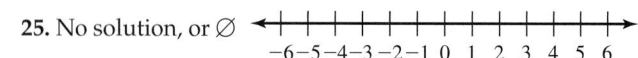 **6.**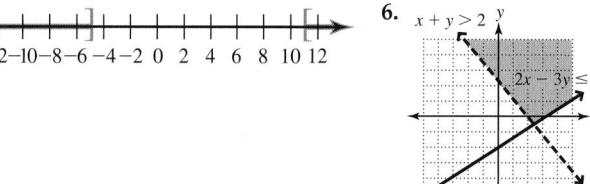

Exercise Set 10.5

Prep Exercises **1.** Solve the related equation and then determine which of the three intervals satisfy the inequality by testing a number in each interval. If a test number satisfies the inequality, then all points in that interval satisfy the inequality.
2. $(-\infty, -2), (-2, 2), (2, \infty)$ **3.** Yes, for example, $x^2 + 2 \leq 0$ has an empty solution set because $x^2 + 2$ is always positive. **4.** Yes, for example, $(x + 3)^2 \leq 0$ has only $x = -3$ as a solution. **5.** In both cases, the intervals to check are $(-\infty, -5), (-5, 2),$ and $(2, \infty)$, and in both cases, only x-values in $(-5, 2)$ satisfy the original inequalities.
6. $(-\infty, -5), (-5, -2), (-2, 3), (3, \infty)$

Exercises **1. a.** $x = -5, -1$ **b.** $(-5, -1)$ **c.** $(-\infty, -5) \cup (-1, \infty)$ **3. a.** $x = -1, 3$ **b.** $(-\infty, -1] \cup [3, \infty)$ **c.** $[-1, 3]$ **5. a.** $x = -2$
b. $(-\infty, -2) \cup (-2, \infty)$ **c.** $\emptyset$ **7. a.** $\emptyset$ **b.** $\emptyset$ **c.** $\mathbb{R}$, or $(-\infty, \infty)$ **9.** $(-4, -2)$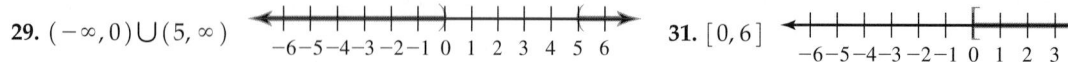

11. $(-\infty, 2) \cup (5, \infty)$ **13.** $(-4, -1)$

15. $(-\infty, 1) \cup (3, \infty)$ **17.** $[2, 4]$

19. $(-\infty, -1] \cup [5, \infty)$ **21.** $(-2, 5)$

23. $(-\infty, \infty)$, or $\mathbb{R}$ **25.** No solution, or $\emptyset$

27. $(-\infty, -5) \cup (-0.25, \infty)$

29. $(-\infty, 0) \cup (5, \infty)$ **31.** $[0, 6]$

33. $(-\infty, \infty)$, or $\mathbb{R}$ **35.** $[-4, -2] \cup [4, \infty)$

37. $(-\infty, -6) \cup (-2, 1)$ 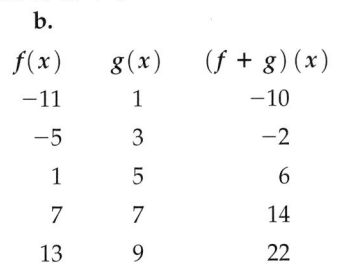 **39.** $\left(\dfrac{-2 - \sqrt{7}}{3}, \dfrac{-2 + \sqrt{7}}{3}\right)$

41. $\left(-\infty, \dfrac{-2 - \sqrt{7}}{2}\right] \cup \left[\dfrac{-2 + \sqrt{7}}{2}, \infty\right)$ **43.** $(-\infty, -4 - \sqrt{6}] \cup [-4 + \sqrt{6}, \infty)$

45. $(-\infty, -4) \cup (1, \infty)$ **47.** $(-5, -1]$

49. $(-4, \infty)$ **51.** $\left(-\infty, -\dfrac{9}{2}\right) \cup (-3, \infty)$

53. $(4, 7)$ **55.** $(3, 5]$

57. $(-\infty, -3] \cup [2, 5)$ **59.** $(0, \infty)$

61. $(-\infty, -1) \cup (2, 5)$ **63. a.** 6 sec. **b.** At 2 sec. and again at 3 sec. **c.** $(2, 3)$ sec.
d. $[0, 2) \cup (3, 6]$ sec. **65. a.** $x \geq 4$ in. **b.** Base ≥ 10 in.; height ≥ 2 in. **67.** $2 < x \leq 8$

Review Exercises: **1.** $9, (x - 3)^2$ **2.** $\dfrac{25}{4}, \left(x + \dfrac{5}{2}\right)^2$ **3.** 18 **4.** Downward, because the coefficient of x^2 is negative
5. $(-6, 0), (2, 0)$ **6.** $(3, -4)$

Exercise Set 10.6
Prep Exercises **1.** $(p + q)(x) = p(x) + q(x)$ **2.** $(p - q)(x) = p(x) - q(x)$ **3.** Use FOIL to multiply the binomials.
4. Divide each term in the trinomial by the monomial.

Exercises **1.** $5x - 1; -x + 1$ **3.** $3x - 8; -x - 2$ **5.** $x^2 - 3x + 4; x^2 - 5x + 10$ **7.** $3x^2 - 5x + 2; x^2 - 5x - 8$
9. $-2x^2 + 2x + 7; -8x^2 + 6x + 9$ **11.** $2x^2 - 10x$ **13.** $x^2 - x - 2$ **15.** $3x^2 + 8x + 4$ **17.** $x^4 - x^3 - 11x^2 + 25x - 14$
19. $6x^4 - 11x^3 + 4x^2 + 2x - 1$ **21.** $x - 3, x \neq 0$ **23.** $2x - 1 + \dfrac{2}{x}, x \neq 0$ **25.** $x - 5, x \neq 9$ **27.** $x + 1, x \neq \dfrac{5}{2}$
29. $2x - 15 + \dfrac{68}{x + 5}, x \neq -5$ **31. a.** $7x + 2$ **b.** $3x - 4$ **c.** $10x^2 + 13x - 3$ **d.** $\dfrac{5x - 1}{2x + 3}, x \neq \dfrac{-3}{2}$ **33. a.** $2x^2 - 2x + 4$
b. $2x^2 - 4x + 6$ **c.** $2x^3 - 5x^2 + 8x - 5$ **d.** $2x - 1 + \dfrac{4}{x - 1}, x \neq 1$ **35. a.** -12 **b.** 11 **c.** -6 **d.** 2 **37. a.** 16 **b.** 54 **c.** 7 **d.** -8

39. a. $4x + 6$
b.

$f(x)$	$g(x)$	$(f + g)(x)$
-11	1	-10
-5	3	-2
1	5	6
7	7	14
13	9	22

c.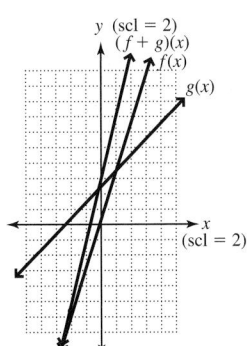

d. The output values for $f(x)$ added to the output values of $g(x)$ are equal to the output values of $(f + g)(x)$.

41. a. $-x - 14$
b.

$h(x)$	$k(x)$	$(h - k)(x)$
-17	-7	-10
-13	-1	-12
-9	5	-14
-5	11	-16
-1	17	-18

c.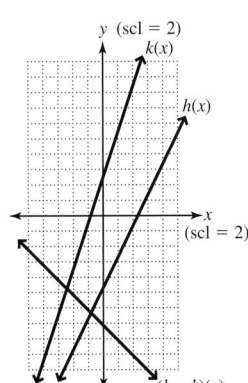

d. The output values for $k(x)$ subtracted from the output values of $h(x)$ are equal to the output values of $(h - k)(x)$.

43. a. $t(x) = 4x + 4$ **b.** \$804 **c.**

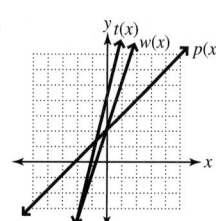

45. a. $p(x) = x^2 - 4x - 12$ **b.** \$9588 profit **c.**

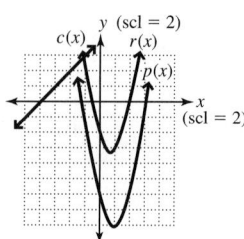

d. 6 units **47. a.** $A(x) = 6x^2 + 4x$ **b.** 66 ft.^{2} **49. a.** $h(x) = 4x + 1$ **b.** 17 cm

Review Exercises: **1.** 5 **2.** $9a^2 - 12a + 5$ **3.** a **4.** $x = \dfrac{y - 4}{2}$ **5.** $x = \sqrt[3]{y - 2}$

Chapter 10 Summary and Review Exercises

10.1 Definitions/Rules/Procedures $\sqrt{a}; -\sqrt{a}; \pm\sqrt{a}$ Exercises **1.** ± 4 **2.** $\pm\dfrac{1}{6}$ **3.** $\pm 2\sqrt{7}$ **4.** $\pm\sqrt{14}$ **5.** $\pm i\sqrt{3}$ **6.** $-12, -2$
7. $9 \pm 4i$ **8.** $-\dfrac{7}{5}, \dfrac{1}{5}$ **9.** 150 ft. **10.** 4 m/sec. **11.** $9\sqrt{3}$ in. Definitions/Rules/Procedures $x^2 + bx = c; \left(\dfrac{b}{2}\right)^2$; factored;
square root; variable Exercises **12.** $-7, -1$ **13.** $-8, 14$ **14.** $\dfrac{3 \pm i\sqrt{5}}{2}$ **15.** $-1, \dfrac{3}{4}$

10.2 Definitions/Rules/Procedures $\dfrac{-b \pm \sqrt{b^2 - 4ac}}{2a}$ Exercises **16.** $-1 \pm \sqrt{6}$ **17.** $\dfrac{1 \pm i\sqrt{2}}{3}$ **18.** $\dfrac{-1 \pm \sqrt{41}}{4}$
19. $-\dfrac{3}{20}, \dfrac{1}{10}$ **20.** Width: 15 ft.; length: 19 ft. **21.** Height: 6.65 ft.; base: 15.65 ft. Definitions/Rules/Procedures $b^2 - 4ac$; two;
perfect square; one; nonreal complex Exercises **22.** $D = 64$; two rational **23.** $D = -71$; two nonreal complex **24.** $D = 0$; one
rational **25.** $D = 0.48$; two irrational

10.3 Definitions/Rules/Procedures variable; expression; LCD; extraneous Exercises **26.** $\pm\dfrac{3\sqrt{2}}{2}$ **27.** $-\dfrac{5}{2}$ (5 is extraneous.)
28. $\dfrac{1}{3}, \dfrac{1}{2}$ **29.** $\dfrac{3 \pm \sqrt{17}}{4}$ Definitions/Rules/Procedures isolate; extraneous Exercises **30.** No solution
31. 1 $\left(\dfrac{1}{9}\text{ is extraneous.}\right)$ **32.** 2 $\left(-\dfrac{3}{2}\text{ is extraneous.}\right)$ **33.** $\dfrac{1}{3}, \dfrac{2}{3}$ Definitions/Rules/Procedures $au^2 + bu + c = 0; u$; Substitute
Exercises **34.** $\pm\sqrt{2}, \pm\sqrt{3}$ **35.** $\pm 1, \pm\dfrac{\sqrt{2}}{2}$ **36.** $-\dfrac{9}{2}, -\dfrac{14}{3}$ **37.** $-26, -2$ **38.** 27, 512 **39.** $\dfrac{16}{625}$ (81 is extraneous.)

10.4 Definitions/Rules/Procedures $(h, k); x = h; a > 0; a < 0$ Exercises **40. a.** $(0, 0)$ **b.** downward **c.** $(0, 0)$ **d.** $x = 0$
e. **41. a.** no x-intercepts, $(0, 1)$ **b.** upward **c.** $(0, 1)$ **d.** $x = 0$ **e.** **42. a.** $(2, 0), \left(0, -\dfrac{4}{3}\right)$

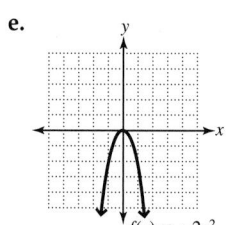

b. downward **c.** $(2, 0)$ **d.** $x = 2$ **e.** **43. a.** $\left(-3 + \dfrac{\sqrt{2}}{2}, 0\right), \left(-3 - \dfrac{\sqrt{2}}{2}, 0\right), (0, 34)$ **b.** upward

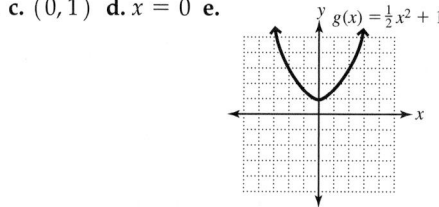

c. $(-3, -2)$ **d.** $x = -3$ **e.** Definitions/Rules/Procedures $-\dfrac{b}{2a}; \dfrac{4ac - b^2}{4a}$

Exercises **44. a.** $m(x) = (x + 1)^2 - 2$ **b.** $(-1 + \sqrt{2}, 0)$, $(-1 - \sqrt{2}, 0)$, $(0, -1)$ **c.** upward **d.** $(-1, -2)$ **e.** $x = -1$
f. 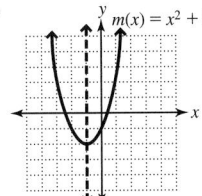 **g.** domain: $\{x \mid x \text{ is a real number}\}$ or $(-\infty, \infty)$; range: $\{y \mid y \geq -2\}$ or $[-2, \infty)$

45. a. $p(x) = -0.5(x - 4)^2 + 2$ **b.** $(2, 0)$, $(6, 0)$, $(0, -6)$ **c.** downward **d.** $(4, 2)$ **e.** $x = 4$ **f.**
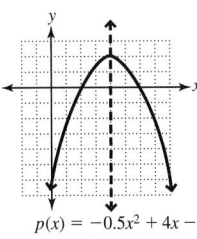

$p(x) = -0.5x^2 + 4x - 6$

g. domain: $\{x \mid x \text{ is a real number}\}$ or $(-\infty, \infty)$; range: $\{y \mid y \leq 2\}$ or $(-\infty, 2]$ **46. a.** 0.75 sec. **b.** 9 ft. **c.** 1.5 sec.
d. 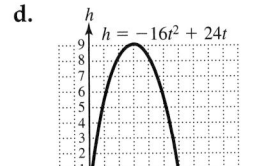   **47. a.** 38.88 yd. **b.** 72 yd.

$h = -16t^2 + 24t$

10.5 Definitions/Rules/Procedures $ax^2 + bx + c > 0$; $ax^2 + bx + c < 0$; $ax^2 + bx + c = 0$; number line; number line; solve the
inequality; no number; union   Exercises **48.** $(-\infty, -5) \cup (3, \infty)$ **49.** $[2, 4]$ **50.** $(-7, -2)$ **51.** $[-3, 1] \cup [2, \infty)$ **52. a.** $x \geq 10$ in.
b. Height ≥ 8 in., base ≥ 14 in.   Definitions/Rules/Procedures   rational expression; 0; equation; number line; test number; union; 2; 1
Exercises **53.** $(-\infty, -3] \cup (1, \infty)$ **54.** $(-\infty, -4) \cup (-2, \infty)$ **55.** $(-\infty, 4) \cup \left[\dfrac{17}{4}, \infty\right)$ **56.** $(-\infty, -2) \cup (3, 5)$

10.6 Definitions/Rules/Procedures $f(x) + g(x)$; $f(x) - g(x)$; $f(x) \cdot g(x)$; $\dfrac{f(x)}{g(x)}$ Exercises **57. a.** $x + 10$ **b.** $-7x + 8$
c. $-12x^2 + 33x + 9$ **58. a.** $7x$ **b.** $-9x - 4$ **c.** $-8x^2 - 18x - 4$ **59. a.** $3x^2 + 7$ **b.** $3x^2 - 2x + 3$ **c.** $3x^3 + 5x^2 + 3x + 10$
60. a. $x^2 + x$ **b.** $x^2 - 5x - 2$ **c.** $3x^3 - 5x^2 - 5x - 1$ **61.** $2x^2 - 5x + 4, x \neq 0$ **62.** $x^2 - 6x + 3 + \dfrac{-2}{2x - 5}, x \neq \dfrac{5}{2}$
63. a. $c(x) = 4x + 4$ **b.** \$904 **c.**   **64. a.** $h(x) = 2x - 1$ **b.** 17 in.

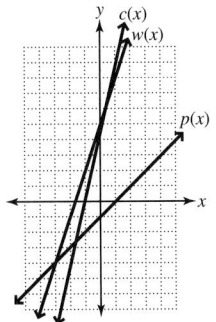

Chapter 10 Practice Test

1. ± 9 [10.1] **2.** $3 \pm 2\sqrt{5}$ [10.1] **3.** $4 \pm 2\sqrt{3}$ [10.1] **4.** $\dfrac{3 \pm 2\sqrt{6}}{3}$ [10.1] **5.** $-2, \dfrac{3}{2}$ [10.2] **6.** $3, 5$ [10.2] **7.** $-8, 2$ [10.1, 10.2]

8. $\dfrac{-3 \pm i\sqrt{3}}{4}$ [10.1, 10.2] **9.** $\pm 4i$ [10.1, 10.2] **10.** $0, -\dfrac{5}{3}$ [10.1, 10.2] **11.** $-\dfrac{6}{5}, 4$ [10.3] **12.** $-\dfrac{2}{3}, 1$ [10.3] **13.** No solution [10.3]

14. 1 (-4 is extraneous.) [10.3] **15.** $\pm \dfrac{1}{3}, \pm i\sqrt{3}$ [10.3] **16.** $-5, 0$ [10.3] **17. a.** $(3 + \sqrt{5}, 0)$ $(3 - \sqrt{5}, 0)$, $(0, -4)$

b. $f(x) = -(x - 3)^2 + 5$ **c.** downward **d.** $(3, 5)$ **e.** $x = 3$ **f.** [10.4]

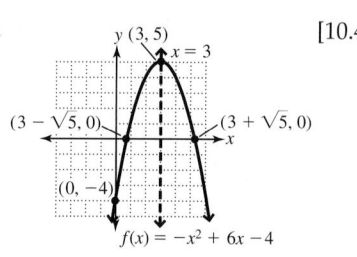

g. domain: $\{x \mid x \text{ is a real number}\}$, or $(-\infty, \infty)$; range: $\{y \mid y \leq 5\}$ or $(-\infty, 5]$

18. $[-1, 4]$ 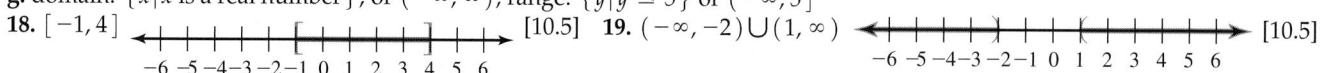 [10.5] **19.** $(-\infty, -2) \cup (1, \infty)$ [10.5]

20. a. $4x^2 - 2$ **b.** $-2x^2 + 2$ **c.** $3x^4 - 2x^2$ [10.6] **21.** $5x^2 + 9x - 7$ [10.6] **22.** 2.5 sec. [10.2] **23. a.** 29.125 m **b.** 70.66 m

c. 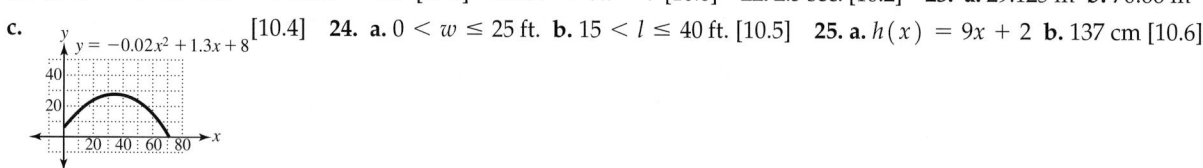 [10.4] **24. a.** $0 < w \leq 25$ ft. **b.** $15 < l \leq 40$ ft. [10.5] **25. a.** $h(x) = 9x + 2$ **b.** 137 cm [10.6]

Chapters 1–10 Cumulative Review Exercises

1. false **2.** false **3.** true **4.** parabola; $a > 0$ **5.** $\sqrt[n]{ab}$ **6.** $\sqrt{a}$; $-\sqrt{a}$ **7.** $-6x - 6$ **8.** $-56mn^5$ **9.** $4x^3 - 14x^2 + 7x - 3$

10. $(m + 2)(m^2 - 2m + 4)$ **11.** $(x + 5)^2$ **12.** $\dfrac{1}{u - 1}$ **13.** $\dfrac{x + 10}{(x - 2)(x + 2)}$ **14.** $x^{1/2}$ **15.** $\dfrac{\sqrt{mn} + n}{m - n}$ **16.** $-\dfrac{7}{2}$

17. $-\dfrac{8}{3}, 0$ **18.** $\dfrac{3}{2}, -5$ **19.** 8 **20.** 17 **21.** $3 \pm i\sqrt{2}$ **22. a.** 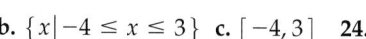 **b.** $\{x \mid -6 < x < 1\}$ **c.** $(-6, 1)$

23. a. 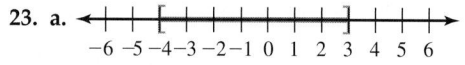 **b.** $\{x \mid -4 \leq x \leq 3\}$ **c.** $[-4, 3]$ **24.** $f(x) = -2x$ 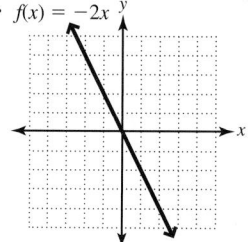 **25.** $2x + 3y = 8$

26. $(2, -1, 4)$ **27.** 5 hr. **28.** 75 ml **29.** 3.2 cm/sec. **30. a.** $\dfrac{\sqrt{6}}{2}$ sec. **b.** 144 ft.

Chapter 11
Exercise Set 11.1

Prep Exercises **1.** $f[g(x)]$ **2.** $g[f(x)]$ **3.** The domain of a function is the range of its inverse; the range of a function is the domain of its inverse. This occurs because the ordered pairs of f and f^{-1} have each x and y interchanged. **4.** No, the function must be one-to-one. Otherwise, the inverse is not a function. **5.** To prove that functions f and g are inverses, it must be shown that $f[g(x)] = x$ and $g[f(x)] = x$. **6.** Horizontal line test **7.** Solve for y. **8.** $y = x$

Exercises **1.** 14 **3.** 3 **5.** 26 **7.** $3x^2 + 14$ **9.** $3\sqrt{x + 1} + 5$ **11.** $\sqrt{6}$ **13.** $(f \circ g)(x) = 6x + 6$; $(g \circ f)(x) = 6x - 2$
15. $(f \circ g)(x) = x^2 + 3$; $(g \circ f)(x) = x^2 + 4x + 5$ **17.** $(f \circ g)(x) = 9x^2 + 9x - 4$; $(g \circ f)(x) = 3x^2 + 9x - 12$
19. $(f \circ g)(x) = \sqrt{2x - 3}$; $(g \circ f)(x) = 2\sqrt{x + 2} - 5$ **21.** $(f \circ g)(x) = \dfrac{2x - 3}{x - 3}$; $(g \circ f)(x) = \dfrac{1 - 2x}{x + 1}$ **23.** Domain is
$[0, \infty)$, and range is $[3, \infty)$. **25.** yes **27.** no **29.** yes **31.** yes **33.** yes **35.** yes **37.** no **39.** yes **41.** yes **43.** Yes, because
$(f \circ g)(x) = (g \circ f)(x) = x$ **45.** yes **47.** no **49.** $f^{-1} = \{(2, -3), (-3, -1), (4, 0), (6, 4)\}$ **51.** $f^{-1} = \{(-2, 7), (2, 9),$
$(1, -4), (3, 3)\}$ **53.** $f^{-1}(x) = x - 6$ **55.** $f^{-1}(x) = \dfrac{x - 3}{2}$ **57.** $f^{-1}(x) = \sqrt[3]{x + 1}$ **59.** $f^{-1}(x) = \dfrac{2 - 2x}{x}$
61. $f^{-1}(x) = \dfrac{3x + 2}{x - 1}$ **63.** $f^{-1}(x) = x^2 + 2, x \geq 0$ **65.** $f^{-1}(x) = \sqrt[3]{\dfrac{x - 4}{2}}$ **67.** $f^{-1}(x) = x^3 - 2$ **69.** $f^{-1}(x) = \dfrac{x^3}{16} - 2$

71. 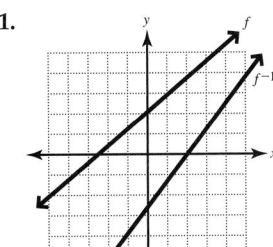   **73.** 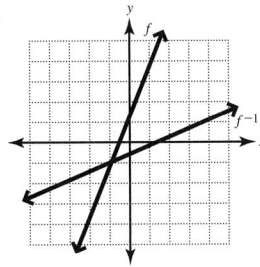   **75.** 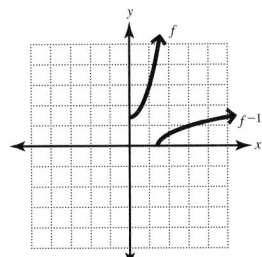   **77.**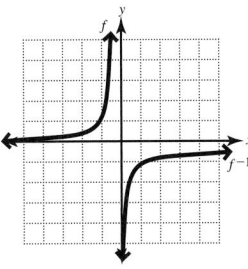

79. 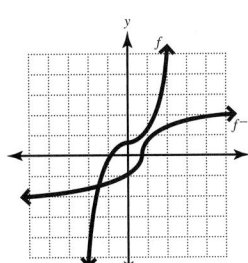   **81.**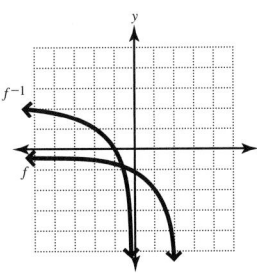

83. a. $y = \dfrac{x - 100}{0.05}$ **b.** x represents the salary; y represents the sales.
c. \$5000   **85. a.** $y = \dfrac{6825 - x}{20}$ **b.** x represents the cost; y represents
the number of 36-in. fans. **c.** 80   **87.** $(5, -4)$ **89.** $f^{-1}(x) = \dfrac{x - b}{a}$

Review Exercises: **1.** $32 = 2^5$ **2.** 8   **3.** $-\dfrac{1}{27}$ **4.** $\dfrac{1}{16}$ **5.** 8   **6.** $(3, 1)$

Exercise Set 11.2

Prep Exercises   **1.** No, because the base is negative   **2.** The domain is all real numbers; the range is $(0, \infty)$.   **3.** $f(x) = 3^x$ is the
steeper graph because it has the greater base.   **4.** Yes, the graphs are symmetric. The point $(1, 2)$ on $f(x)$ corresponds to the point
$(-1, 2)$ on $g(x)$, and so on.   **5.** Because the exponential function is one-to-one   **6.** 0.625 g

Exercises   1. 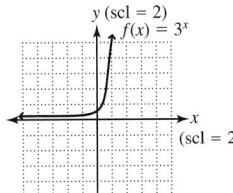   **3.** 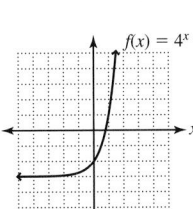   **5.** 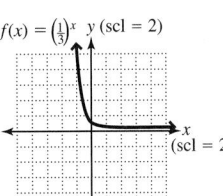   **7.**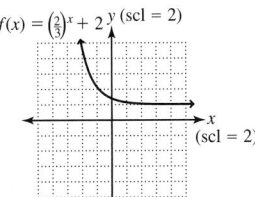

9. 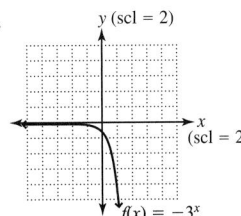   **11.** 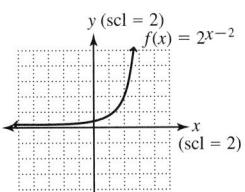   **13.** 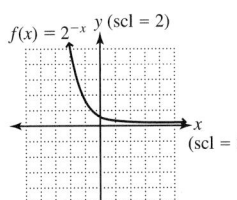   **15.**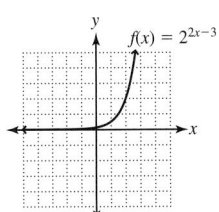

17. 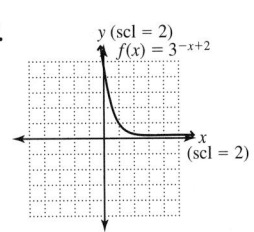   **19.** 3   **21.** $\dfrac{5}{3}$ **23.** $\dfrac{1}{2}$ **25.** -2 **27.** -2 **29.** 3   **31.** -4 **33.** $\dfrac{1}{2}$ **35.** -12 **37. a.**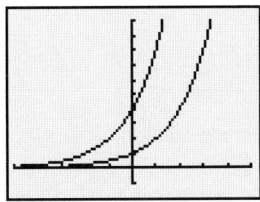

b. The graph of g is the graph of f shifted 2 units to the left.   **39. a.** 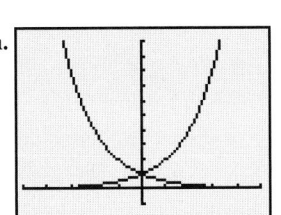   **b.** The graph of g is the graph of f
reflected about the y-axis.   **41.** 6400 cells   **43.** 6144 billion people   **45.** \$25,870.70   **47.** 0.020 g   **49.** $\approx 17,875.41$ **51.** 1.225 parts per
million   **53.** $\approx 118,130$ million transistors

Review Exercises: **1.** 125   **2.** 9   **3.** $\dfrac{1}{64}$ **4.** $\sqrt[3]{5^2}$ or $(\sqrt[3]{5})^2$ **5.** $f^{-1}(x) = \dfrac{x-4}{3}$ **6.**

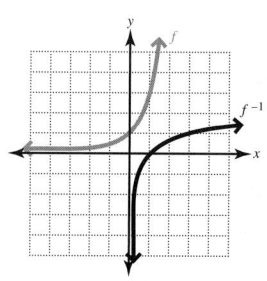

Exercise Set 11.3

Prep Exercises   **1.** $f^{-1}(x) = \log_3 x$ because logarithmic and exponential functions are inverses.   **2.** Logarithms are exponents because logarithms are inverses of exponential functions.   **3.** $b > 0, b \neq 1$ **4.** $x > 0$ **5.** power, b, x **6.** The graphs are symmetric about the line $y = x$ because they are inverses.

Exercises   **1.** $\log_2 32 = 5$ **3.** $\log_{10} 1000 = 3$ **5.** $\log_e x = 4$ **7.** $\log_5 \dfrac{1}{125} = -3$ **9.** $\log_{10} \dfrac{1}{100} = -2$ **11.** $\log_{625} 5 = \dfrac{1}{4}$ **13.** $\log_{1/4} \dfrac{1}{16} = 2$

15. $\log_7 \sqrt{7} = \dfrac{1}{2}$ **17.** $3^4 = 81$ **19.** $4^{-2} = \dfrac{1}{16}$ **21.** $10^2 = 100$ **23.** $e^5 = a$ **25.** $e^{-4} = \dfrac{1}{e^4}$ **27.** $\left(\dfrac{1}{8}\right)^2 = \dfrac{1}{64}$ **29.** $\left(\dfrac{1}{5}\right)^{-2} = 25$

31. $7^{1/2} = \sqrt{7}$ **33.** 32   **35.** $\dfrac{1}{25}$ **37.** 4   **39.** -2 **41.** 10   **43.** 2   **45.** $\dfrac{1}{4}$ **47.** 243   **49.** 2   **51.** -6 **53.** 16   **55.** $\dfrac{1}{64}$ **57.** 2   **59.** 125

61. 3   **63.** $\dfrac{1}{16}$ **65.**

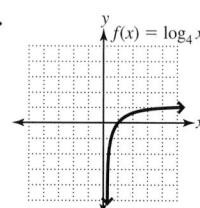

$f(x) = \log_4 x$

67.

$f(x) = \log_{1/3} x$

69. 97%   **71.** After 4 days.   **73.** $\log_b b = 1$ because $b^1 = b$.

75. $\log_b 1 = 0$ because $b^0 = 1$.

Review Exercises: **1.** x^{-6} **2.** x^6 **3.** x^{15} **4.** x^3 **5.** x^5 **6.** $x^{3/4}$

Exercise Set 11.4

Prep Exercises   **1.** x **2.** x **3.** $\log_b x + \log_b y$ **4.** $\log_b(x+y)$; there is no rule for the logarithm of a sum.   **5.** $\log_b x - \log_b y$ **6.** $r \log_b x$

Exercises   **1.** 2   **3.** r **5.** $4x$ **7.** 5   **9.** y **11.** $7x$ **13.** $\log_2 5 + \log_2 y$ **15.** $\log_a p + \log_a q$ **17.** $\log_4 m + \log_4 n + \log_4 p$
19. $\log_a x + \log_a(x-5)$ **21.** $\log_3 40$ **23.** $\log_4 27$ **25.** $\log_a 7m$ **27.** $\log_4 ab$ **29.** $\log_a(2x^2 + 10x)$ **31.** $\log_4(x^2 + 4x + 3)$
33. $\log_2 7 - \log_2 9$ **35.** $\log_a x - \log_a 5$ **37.** $1 - \log_a b$ **39.** $\log_a x - \log_a(x-3)$ **41.** $\log_4(2x - 3) - \log_4(4x + 5)$ **43.** $\log_6 8$
45. 1   **47.** $\log_a \dfrac{x}{3}$ **49.** $\log_4 \dfrac{p}{q}$ **51.** $\log_b \dfrac{x}{x-4}$ **53.** 1   **55.** $6 \log_4 3$ **57.** $7 \log_a x$ **59.** $\dfrac{1}{2}\log_a 3$ **61.** $\dfrac{2}{3}\log_3 x$ **63.** $-2 \log_a 6$ **65.** $-2 \log_a y$
67. $\log_3 5^4$ **69.** $\log_2 \dfrac{1}{x^3}$ **71.** $\log_7 8$ **73.** $\log_a \sqrt[4]{x^3}$ **75.** $\log_a 4$ **77.** $\log_3 \dfrac{1}{\sqrt{x}}$ **79.** $3 \log_a x - 4 \log_a y$ **81.** $4 \log_3 a + 2 \log_3 b$
83. $\log_a x + \log_a y - \log_a z$ **85.** $2 \log_x a - \log_x b - 3 \log_x c$ **87.** $\dfrac{3}{4}\log_4 x - \dfrac{1}{4}\log_4 y$ **89.** $\dfrac{2}{3}\log_a x + \dfrac{1}{3}\log_a y - \log_a z$ **91.** $\log_3 \dfrac{1}{2}$
93. $\log_b x^4 y^3$ **95.** $\log_a \sqrt{\dfrac{5}{7}}$ **97.** $\log_a \sqrt[3]{(x^2 y^3)^2}$ **99.** $\log_b(3x^2 - 2x)$ **101.** $\log_a \dfrac{(x-2)^3}{(x+1)^4}$ **103.** $\log_a \dfrac{x^2 z^4}{w^3 u^6}$

Review Exercises: **1.** $10^{-2.5}$ **2.** 10^9 **3.** \$13,468.55   **4.** $\log_{10} 50 = 1.6990$ **5.** $10^{1.6532} = 45$ **6.** $e^{-1.3863} = 0.25$

Exercise Set 11.5

Prep Exercises   **1.** 10   **2.** e **3.** Negative. If $y = \log x$, then $10^y = x$. If $0 < x < 1$, then y must be negative so that $x = \dfrac{1}{10^y}$ where $y > 0$, which results in values of x such that $0 < x < 1$. Also, any positive value of y results in a value of $x > 1$. Also look at the graph.   **4.** Positive. If $y = \ln x$, then $e^y = x$. If $x > 1$, then $e^y > 1$, which is true if $y > 0$.   **5.** 0.91629   **6.** $\sqrt{5}$

Exercises   **1.** 1.8062   **3.** -2.1739 **5.** 2.6391   **7.** 4.1761   **9.** -5.7959 **11.** 2.2343   **13.** 4.3720   **15.** -3.3814 **17.** 5.6864   **19.** 0.4343
21. Error results because the domain of $\log_a x$ is $(0, \infty)$, so log 0 is undefined.   **23.** 2   **25.** -2 **27.** $\dfrac{1}{3}$ **29.** -3 **31.** 3   **33.** $\dfrac{1}{2}$
35. 90 dB   **37.** 10^{-6} watts/m^2 **39.** ≈ 2.796 **41.** $10^{-3.5}$ moles/L   **43.** ≈ 22.9 **45.** The 1964 Alaska earthquake was about four times as severe as the 1906 San Francisco earthquake.   **47.** $10^{22.75}$ ergs   **49.** 1986   **51.** ≈ 6124 ft.

Review Exercises: **1.** $\log_3 \dfrac{2x+1}{x-1}$ **2.** $3^2 = 2x + 5$ **3.** -11 **4.** $-5, 3$ **5.** 8   **6.** 10

Exercise Set 11.6

Prep Exercises **1.** $x = y$ **2.** Common logs because the base is 10 **3.** If $x = y$, then $\log_b x = \log_b y$. **4.** $m = n$
5. If $\log_b x = \log_b y$, then $x = y$. **6.** The solution $x = -5$ is rejected because substituting into the equation gives $\log(-5)$
and $\log(-3)$, which are both undefined.

Exercises **1.** 3.1699 **3.** 1.0767 **5.** -1.5693 **7.** 2.8617 **9.** 12.4251 **11.** 7.1285 **13.** 0.5365 **15.** 107.2959 **17.** -73.1563 **19.** 6
21. 11 **23.** 6 **25.** ± 4 **27.** 4, -9 **29.** 7 **31.** No solution **33.** $-6, 2$ **35.** 2 **37.** 10 **39.** 3 **41.** $\frac{3}{2}$ **43.** 3 **45.** 4 **47.** $\frac{5}{6}$ **49.** $\frac{5}{2}$
51. 6 **53.** 4 **55.** 9.5 yr. **57. a.** \$19,676.82 **b.** $\approx$9.3 yr. **59. a.** $\approx$5.58% **b.** $\approx$0.18 **61. a.** $\approx$875 mosquitoes **b.** $\approx$75 days
63. a. $\approx$7.59 billion **b.** $\approx$2.6 yr. after 2010, in 2012 **65. a.** $\approx$1.89 in. of mercury **b.** $\approx$21.8 mi. **67. a.** $\approx$79.8% **b.** $\approx$7.76, so about
age 8 **69.** $\approx$\$10.21 **71.** $\approx$\$1435.1 billion **73.** 1.7925 **75.** 0.5283 **77.** -2.3219 **79.** 0.3685 **81.** 47%

Review Exercises: **1.** $x + 3$ **2.** $28 + \sqrt{5}$ **3.** 29 **4.** $x^2 + 2x + 3$ **5.** $4x^2 - 4x + 5$ **6.** 13

Chapter 11 Summary and Review Exercises

11.1 Definitions/Rules/Procedures $f[g(x)]; g; f; g[f(x)]; f; g$ Exercises **1.** 25 **2.** 167 **3.** -2 **4.** 14
5. $(f \circ g)(x) = 6x + 3, (g \circ f)(x) = 6x - 9$ **6.** $(f \circ g)(x) = 9x^2 - 42x + 53, (g \circ f)(x) = 3x^2 + 5$
7. $(f \circ g)(x) = \sqrt{2x - 4}, (g \circ f)(x) = 2\sqrt{x - 3} - 1$ **8.** $(f \circ g)(x) = \dfrac{4x - 4}{x - 4}, (g \circ f)(x) = \dfrac{3 - 3x}{x + 3}$ Definitions/Rules/
Procedures $x; g; x; f$ Exercises **9.** yes **10.** yes **11.** no **12.** no Definitions/Rules/Procedures $a = b; f(a) \neq f(b)$; one and
only one point; one-to-one; $y = x$ Exercises **13.** no **14.** yes Definitions/Rules/Procedures $y; y\text{'s}; x\text{'s}; y; f^{-1}(x)$

Exercises **15.** $f^{-1}(x) = \dfrac{x - 4}{5}$ **16.** $f^{-1}(x) = \sqrt[3]{x - 6}$ **17.** $f^{-1}(x) = \dfrac{4 - 5x}{x}$ **18.** $f^{-1}(x) = \dfrac{x^3 - 2}{3}$ **19.** $(-6, 4)$
20. $f^{-1}(b) = a$

11.2 Definitions/Rules/Procedures $> 1; < 1$ Exercises **21.** 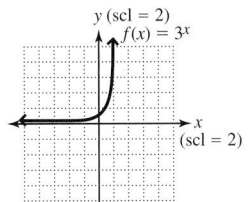 **22.**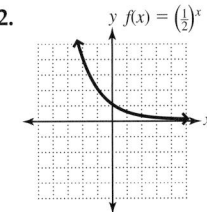

23. 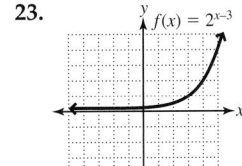 **24.** 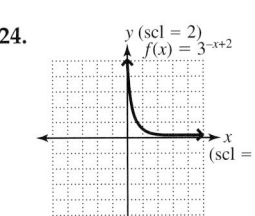 Definitions/Rules/Procedures $x = y$; same base; exponents; exponents

Exercises **25.** 4 **26.** $\dfrac{3}{2}$ **27.** -2 **28.** -2 **29.** 4 **30.** 2 **31.** 3 **32.** $-\dfrac{6}{7}$ **33.** $\approx$6277 cells **34.** \$40,353.57 **35.** $\approx$7.16 g **36.** $\approx$\$9.19

11.3 Definitions/Rules/Procedures $x = b^y$ Exercises **37.** $\log_7 343 = 3$ **38.** $\log_4 \dfrac{1}{64} = -3$ **39.** $\log_{3/2} \dfrac{81}{16} = 4$ **40.** $\log_{11} \sqrt[3]{11} = \dfrac{1}{3}$
41. $9^2 = 81$ **42.** $\left(\dfrac{1}{5}\right)^{-3} = 125$ **43.** $a^4 = 16$ **44.** $e^b = c$ Definitions/Rules/Procedures $b^x = y; 1; 0$ Exercises **45.** $\dfrac{1}{81}$
46. 4 **47.** 5 **48.** -2 **49.** 3 **50.** $\dfrac{1}{10}$ **51.** 1 **52.** 0 Definitions/Rules/Procedures $y; x = b^y; y; x$ Exercises **53.**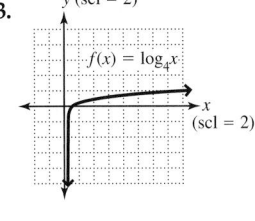

54. 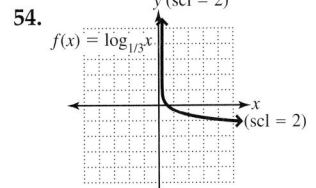 **55.** Domain is $(-\infty, \infty)$ and range is $(0, \infty)$ because $f(x) = \log_b x$ and $g(x) = b^x$ are inverses.
56. 30 dB

11.4 Definitions/Rules/Procedures x; x; $\log_b x + \log_b y$; $\log_b x - \log_b y$; $r\log_b x$ Exercises   **57.** 8   **58.** 6   **59.** $1 + \log_6 x$
60. $\log_4 x + \log_4 (2x - 5)$ **61.** $\log_3 (32)$ **62.** $\log_5 (3x^2 - 6x)$ **63.** $\log_b x - \log_b 5$ **64.** $\log_a (3x - 2) - \log_a (4x + 3)$
65. $\log_8 2$ **66.** $\log_2 \dfrac{x+5}{2x-3}$ **67.** $4\log_3 7$ **68.** $\dfrac{1}{3}\log_a x$ **69.** $-4\log_4 a$ **70.** $\dfrac{4}{5}$ **71.** $\log_a x^4$ **72.** $\log_a \sqrt[5]{y^3}$ **73.** $2\log_a x + 3\log_a y$
74. $4\log_a c - 3\log_a d$ **75.** $2\log_a x + 3\log_a y - 4\log_a z$ **76.** $\dfrac{3}{2} - 2\log_a b$ **77.** $\log_x y^3 z^5$ **78.** $\log_a \dfrac{4^3}{3^2}$ **79.** $\log_a \sqrt[4]{x^2 y^3}$
80. $\log_a (x + 5)^4 (x - 3)^2$

11.5 Definitions/Rules/Procedures   common; $\log x$; LOG ; natural; $\ln x$; ln   Exercises   **81.** 2.5132   **82.** -2.4559
83. -3.1466 **84.** 4.5218   **85.** -5 **86.** $\dfrac{1}{4}$ **87.** 85 dB   **88.** ≈ 2.9 **89.** ≈ 17 yr.   **90.** The 7.8 earthquake was 10 times as severe.

11.6 Definitions/Rules/Procedures $x = y$; b^y; $x = y$; $\log_b y$; b^y; $A_0 e^{kt}$; initial; growth; decay   Exercises   **91.** 1.5773   **92.** 0.5360
93. 1.3626   **94.** 4.9592   **95.** 0.5995   **96.** -536.4793 **97.** 6.1 yr.   **98. a.** \$12,254.71 **b.** ≈ 7.7 yr.   **99. a.** ≈ 573 ants **b.** ≈ 23 months
100. ≈ 24 min.   Definitions/Rules/Procedures $x = y$; $b^y = x$ Exercises   **101.** 8   **102.** $-4, 2$ **103.** 3   **104.** 20   **105.** 36   **106.** 8
107. 2   **108.** 5   Definitions/Rules/Procedures x; a; x; a; x; a Exercises   **109.** 1.9534   **110.** -2.5850

Chapter 11 Practice Test

1. $f[g(x)] = 9x^2 - 30x + 19$ [11.1]   **2. a.** $f^{-1}(x) = \dfrac{x+3}{4}$ [11.1] **b.** $f[f^{-1}(x)] = f\left[\dfrac{x+3}{4}\right] = 4\left(\dfrac{x+3}{4}\right) - 3 = x + 3 - 3 = x$;
$f^{-1}[f(x)] = f^{-1}(4x - 3) = \dfrac{4x - 3 + 3}{4} = \dfrac{4x}{4} = x$ [11.1]   **c.** The graphs are symmetric about the graph of $y = x$. [11.1]
3. 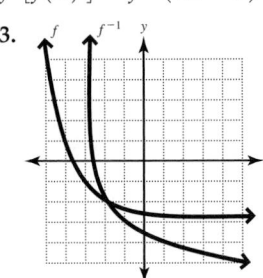 [11.1]   **4.** $f(x) = 2^{x-1}$ [11.2]   **5.** -10 [11.2]   **6. a.** \$42,143.63 **b.** ≈ 8.2 yr. [11.2]   **7. a.** ≈ 0.68 g

b. ≈ 1.16 hr. [11.2]   **8. a.** ≈ 14.8 million in 1960, ≈ 65.5 million in 2018 **b.** 1989 [11.2]   **9.** $\left(\dfrac{1}{3}\right)^{-4} = 81$ [11.3]   **10.** -3 [11.3]
11. 5 [11.3]   **12.** 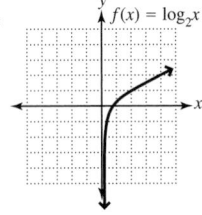 $f(x) = \log_2 x$ [11.4]   **13.** $4\log_b x + 2\log_b y - \log_b z$ [11.4]   **14.** $\dfrac{5}{4}\log_b x - \dfrac{7}{4}\log_b y$ [11.4]   **15.** $\log_b \sqrt[4]{x^6 y^9}$ [11.4]

16. a. ≈ 74 g **b.** ≈ 247.6 days [11.5]   **17.** 4.6433 [11.5]   **18.** 3 [11.5]   **19.** 3 [11.5]   **20. a.** 16.5 generations **b.** 100,000 [11.6]

Chapters 1–11 Cumulative Review Exercises

1. true   **2.** true   **3.** false   **4.** -8.   **5.** x-intercept   **6.** $x - 3 = 4$; $x - 3 = -4$.   **7.** 86   **8.** $\dfrac{1}{a^{12}}$ **9.** $\dfrac{3}{pq} - \dfrac{2p^2}{q^2} + \dfrac{6q^2}{7}$ **10.** $\dfrac{-x^2 - 5x - 9}{(x+3)^2}$
11. $24c^4\sqrt{15}$ **12.** $20 - 10i$ **13.** $2x$ **14.** $\dfrac{d - 4}{2d(3d + 4)}$ **15.** $(x + 2y)(x - 2y)(x + 3)(x - 3)$ **16.** $3x^2 y(4x - 3y)(2x - y)$
17. 1, 2   **18.** All real numbers.   **19.** $(-3, 2)$ **20.** $\dfrac{1}{2}, \dfrac{7}{2}$ **21.** 3, (2 is extraneous.)   **22.** No solution (5 is extraneous.)   **23.** $-\dfrac{1}{2}$
24. 16   **25.** 3   **26. a.** $\dfrac{2}{5}$ **b.** $(5, 0), (0, -2)$ **c.**   **27.** 20, 21, 29   **28.** \$16,909.08   **29.** ≈ 0.857 hr., or ≈ 51 min.
30. 5 hr.

Chapter 12
Exercise Set 12.1

Prep Exercises **1.** Parabola, circle, ellipse, and hyperbola **2.** Up; $(1, -3)$ **3.** Left; $(-4, -3)$ **4.** It is a parabola because only one variable is squared. **5.** $d = \sqrt{(x_2 - x_1)^2 + (y_2 - y_1)^2}$ **6.** $\left(\dfrac{x_1 + x_2}{2}, \dfrac{y_1 + y_2}{2}\right)$ **7.** $(h, k); r$ **8.** complete the square

Exercises **1.** Opens upward; vertex: $(1, 2)$; axis of symmetry: $x = 1$

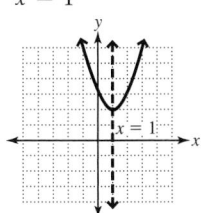

3. Opens downward; vertex: $(-1, 4)$; axis of symmetry: $x = -1$

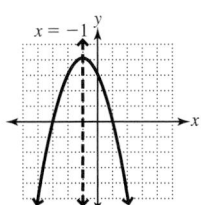

5. Opens right; vertex: $(-2, -2)$; axis of symmetry: $y = -2$

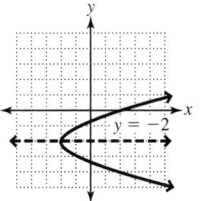

7. Opens left; vertex: $(3, 1)$; axis of symmetry: $y = 1$

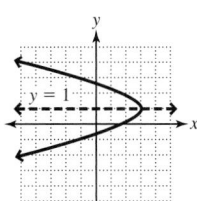

9. Opens right; vertex: $(-4, -2)$; axis of symmetry: $y = -2$

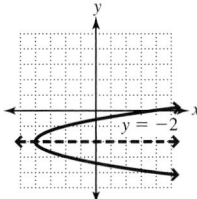

11. Opens left; vertex: $(-5, -2)$; axis of symmetry: $y = -2$

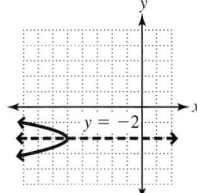

13. Opens right; vertex: $(-1, -2)$; axis of symmetry: $y = -2$

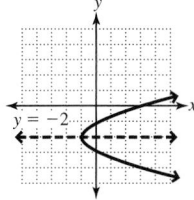

15. Opens left; vertex: $(4, 3)$; axis of symmetry: $y = 3$

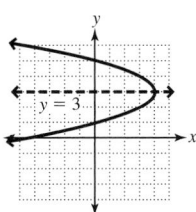

17. Opens right; vertex: $(-5, -2)$; axis of symmetry: $y = -2$

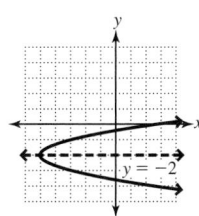

19. Opens right; vertex: $(-2, 1)$; axis of symmetry: $y = 1$

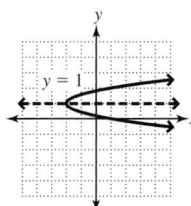

21. Opens left; vertex: $(7, 1)$; axis of symmetry: $y = 1$

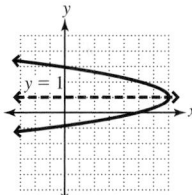

23. c **25.** a **27.** $5, \left(-\dfrac{5}{2}, 4\right)$ **29.** $13, \left(-\dfrac{11}{2}, 2\right)$ **31.** $13, \left(-2, -\dfrac{15}{2}\right)$ **33.** $2\sqrt{5}, (3, 6)$ **35.** $4\sqrt{5}, (-1, 0)$ **37.** $\sqrt{34}, \left(\dfrac{7}{2}, -\dfrac{7}{2}\right)$ **39.** 5

41. $\sqrt{89}$ **43.** Center: $(2, 1)$; radius: 2

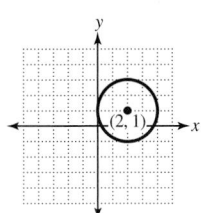

45. Center: $(-3, -2)$; radius: 9

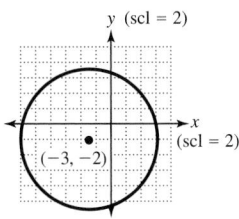

47. Center: $(5, -3)$; radius: 7

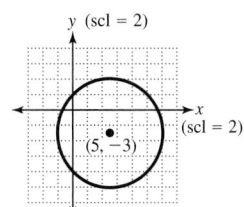

49. Center: $(1, -1)$; radius: $2\sqrt{3}$

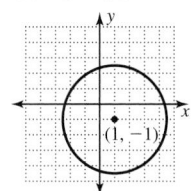

51. Center: $(-4, -2)$; radius: $4\sqrt{2}$

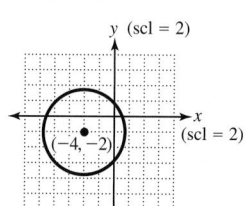

53. Center: $(-4, 3)$; radius: 3

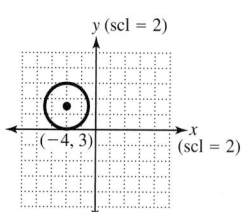

55. Center: $(-5, 2)$; radius: 8

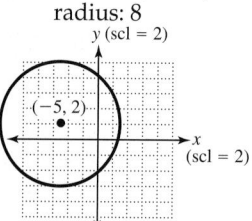

57. Center: $(-7, 2)$; radius: 2

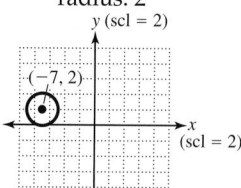

59. b   **61.** c   **63.**

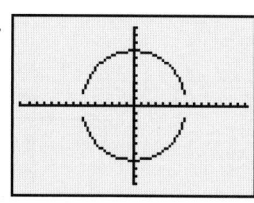

65.

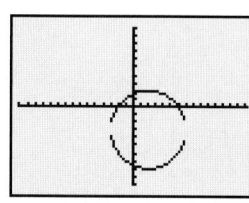

67. $(x - 4)^2 + (y - 2)^2 = 16$

69. $(x + 4)^2 + (y + 3)^2 = 25$ **71.** $(x - 6)^2 + (y + 2)^2 = 14$ **73.** $(x + 5)^2 + (y - 2)^2 = 45$ **75.** $(x - 2)^2 + (y - 4)^2 = 25$
77. $(x - 2)^2 + (y - 4)^2 = 169$ **79.** $(x - 2)^2 + (y + 5)^2 = 64$ **81.** $(x - 1)^2 + (y - 2)^2 = 25$ **83.** $(x - 2)^2 + (y - 3)^2 = 41$

85. a. 256 ft.   **b.** 3 sec.   **c.** 7 sec.   **87.** $y = -\dfrac{2}{25}x^2 + 18$ or $y = -0.08x^2 + 18$ **89.** 3.5%   **91.** $x^2 + y^2 = 400$
93. $x^2 + (y - 110)^2 = 10,000$ **95.** $x^2 + y^2 = 43,560,000$

Review Exercises: **1.**

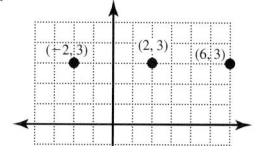

2. $(5x + 3y)(5x - 3y)$ **3.**

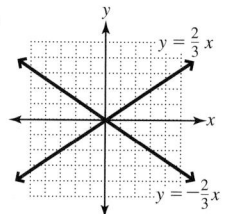

4. ± 4

5. $(4, 0), (-4, 0), (0, 3), (0, -3)$ **6.** $(3, 0), (-3, 0), (0, 5), (0, -5)$

Exercise Set 12.2

Prep Exercises   **1.** An ellipse is the set of all points the sum of whose distances from two fixed points is constant.   **2.** The fixed points are the foci.   **3. a.** $(0, 0)$ **b.** 3 units left and right   **c.** 5 units up and down   **4.** If the x^2-term is positive, the graph has x-intercepts and opens left and right. If the y^2-term is positive, the graph has y-intercepts and opens up and down.   **5. a.** $(0, 0)$ **b.** $(-3, 0)$ and $(3, 0)$ **c.** It has no y-intercepts.   **d.** $(3, 5), (-3, 5), (3, -5), (-3, -5)$ **6. a.** $(0, 0)$ **b.** It has no x-intercepts. **c.** $(0, -4)$ and $(0, 4)$ **d.** $(9, 4), (-9, 4), (9, -4), (-9, -4)$ **7.** The parabola has only one squared term.   **8.** The equation of an ellipse is a sum; the equation of a hyperbola is a difference.

Exercises   **1.**

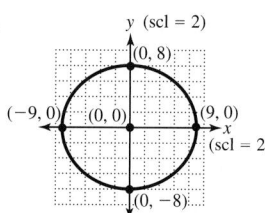

3.

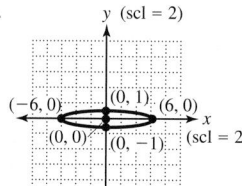

5.

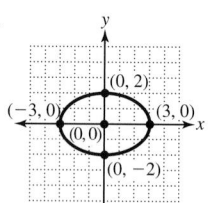

7.

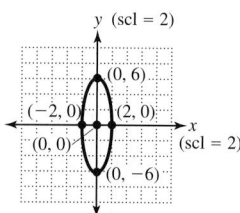

9.

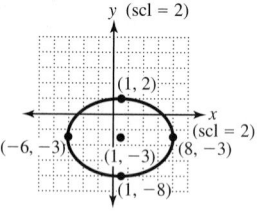

11.

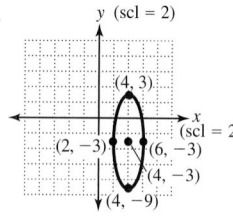

13.

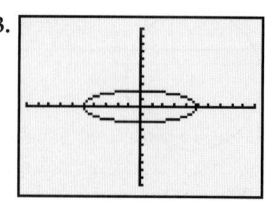

15.

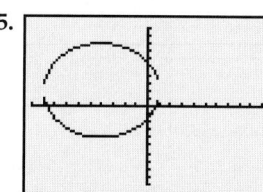

17. $\dfrac{x^2}{9} + \dfrac{y^2}{16} = 1$ **19.**
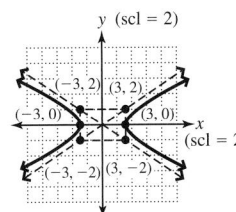

21.
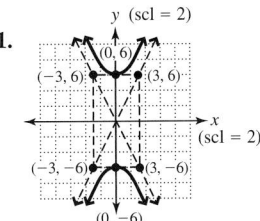

23.
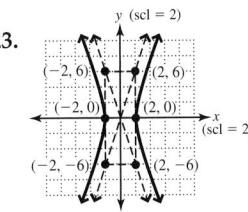

25.
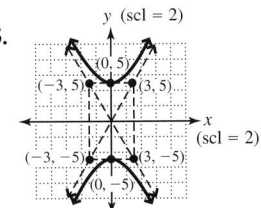

27.

29. $\dfrac{x^2}{9} - \dfrac{y^2}{4} = 1$ **31.** c   **33.** b   **35.** ellipse   **37.** circle

39. circle
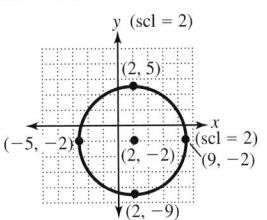

41. parabola
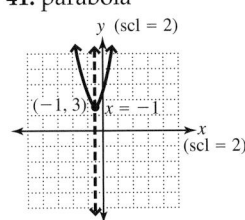

43. ellipse
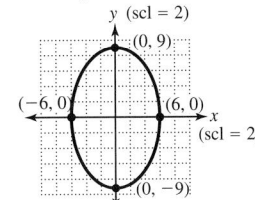

45. hyperbola
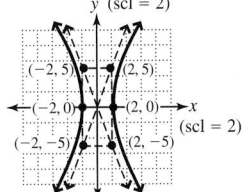

47. parabola
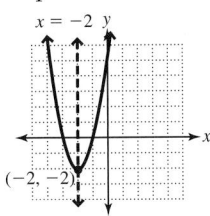

49. hyperbola
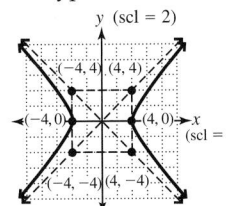

51. ellipse
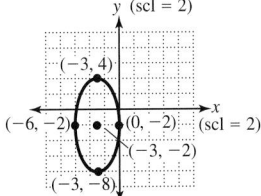

53. a. Yes, the sailboat will clear the bridge. The height of the bridge at the center is 20 feet, and the boat's mast is only 18 feet above the water. **b.** The bridge is 32 feet wide at the base of the arch.   **55.**
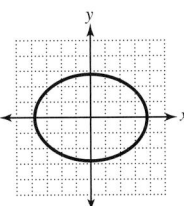
57. $\dfrac{x^2}{2.25} + \dfrac{y^2}{4} = 1$ **59.** 3 units

Review Exercises: **1.** $y = -3x^2 + 6$ **2.** $(1, 2)$ **3.** $(2, -3)$ **4.** $(-3, 4)$ **5.** $-4, \dfrac{2}{3}$ **6.** ± 3

Exercise Set 12.3
Prep Exercises   **1. a.** 0, 1, 2, 3, or 4 solutions are possible.
b. Answers may vary.

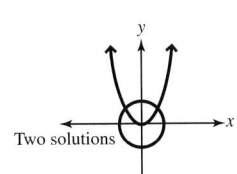

2. a. 0, 1, or 2 solutions are possible.
b. Answers may vary.

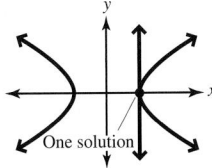

3. Substitution because the first equation has a linear term and is easy to solve for y.   **4.** Elimination because all variables are squared.   **5.** The graph of the first equation is a line, and the second is an ellipse; so the system could have 0, 1, or 2 solutions.   **6.** No. If both graphs are centered at the origin, then because of the symmetry, there could be only 0, 2, or 4 solutions.

Exercises   **1.** $(3, -4), (1, 0)$ **3.** $(-2, 0), (1, 9)$ **5.** $(-4, -3), (3, 4)$ **7.** $(2.6, -1.8), (1, 3)$ **9.** No solution   **11.** $(0, 3), (1, 5)$
13. $(3, 2), (2, 3)$ **15.** $(4, 2), (4, -2), (-4, -2), (-4, 2)$ **17.** $(3, 2), (3, -2), (-3, -2), (-3, 2)$
19. $(2, 1), (2, -1), (-2, -1), (-2, 1)$ **21.** $(-6, 0), (3, 3), (3, -3)$ **23.** No solution   **25.** $(2, 4), (-2, 4)$ **27.** $(4, 0), (-4, 0)$
29. $(2, 2), (-2, -2)$ **31.** $(5, 12), (-1, 0)$ **33.** $(2, 2), (2, -2), (-2, -2), (-2, 2)$ **35.** $(-7, 3), (-7, -3), (2, 0)$

37. Answers may vary, but one possible system is $\begin{cases} x + y = 4 \\ x^2 + y^2 = 1 \end{cases}$.   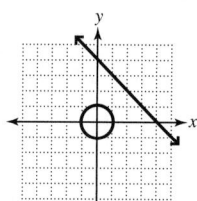   **39.** $(5, 3), (5, -3), (-5, 3), (-5, -3)$

41. Length: 18 in.; width: 8 in.   **43.** At equilibrium, there should be 400 chairs at \$72 per chair.

45.   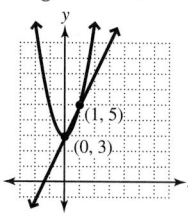   **47.**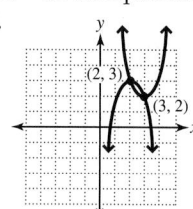

Review Exercises:

1. yes   **2.** no   **3.** 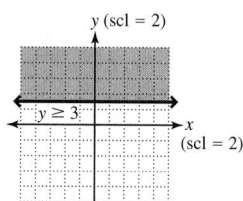   **4.** 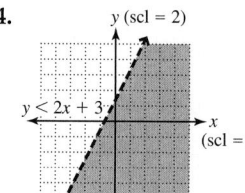   **5.** 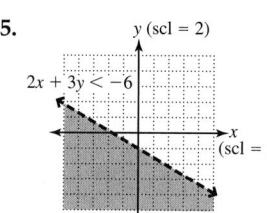   **6.**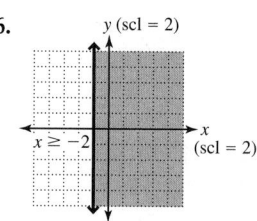

Exercise Set 12.4

Prep Exercises   **1.** $9x^2 + 4y^2 = 36$; the ellipse is drawn dashed.   **2.** The boundary curve is the parabola whose equation is
$y = x^2 + 3$; the parabola is drawn solid.   **3.** Draw a dashed hyperbola defined by $\dfrac{x^2}{9} - \dfrac{y^2}{16} = 1$. Then choose test points not on the
curve to determine the shaded solution region.   **4.** The solution region for the system is the intersection of the solution regions of
the inequalities.   **5.** The graph of $x^2 + y^2 < 1$ is the region inside the circle with radius 1. The graph of $x^2 + y^2 > 4$ is the region
outside the circle with radius 2. Thus, there is no common region of solution.

Exercises

1. 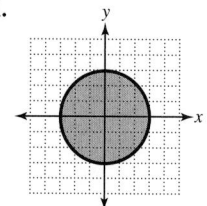   **3.** 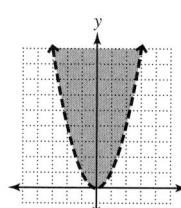   **5.** 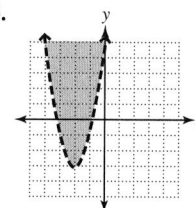   **7.** 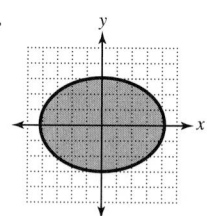   **9.**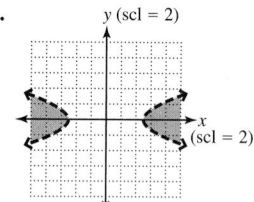

11. 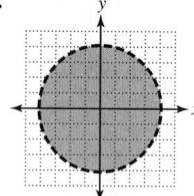   **13.** 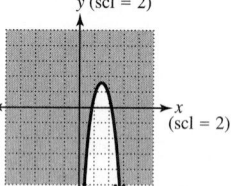   **15.** 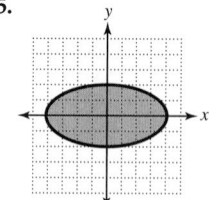   **17.**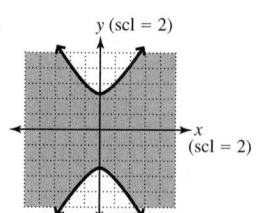

19. 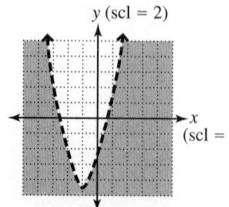   **21.** 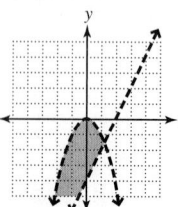   **23.** 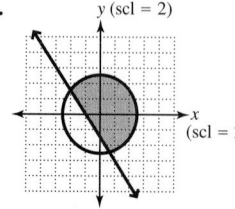   **25.** 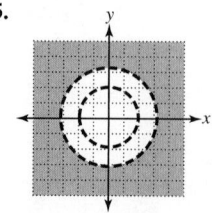   **27.**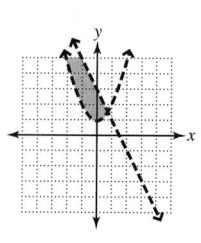

29. 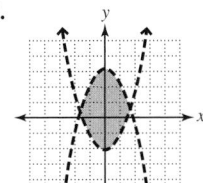 **31.** 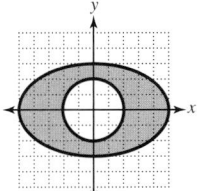 **33.** 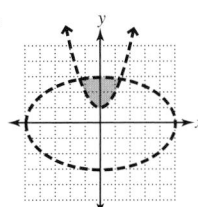 **35.** 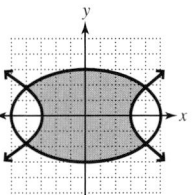 **37.**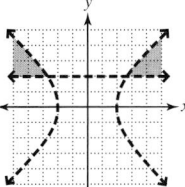

39. 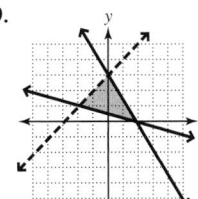 **41.** 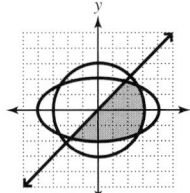 **43.** 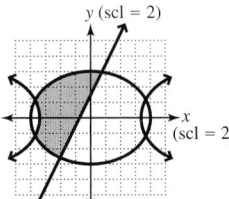 **45.**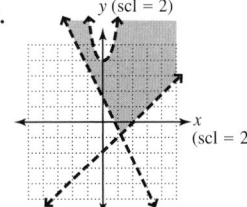

Review Exercises: **1.** 38 **2.** 312 **3.** −84 **4.** −1, 5, 15, 29 **5.** −1, $\frac{1}{4}$, −$\frac{1}{7}$, $\frac{1}{10}$ **6.** 67, 69, 71

Chapter 12 Summary and Review Exercises

12.1 Definitions/Rules/Procedures   cone; hyperbola; (h, k); $a > 0$; $a < 0$; $y = k$ Exercises

1. Opens upward;
vertex: $(3, -5)$;
axis of symmetry: $x = 3$

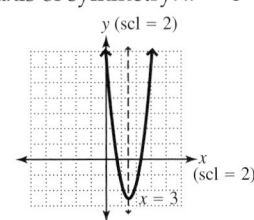

2. Opens downward;
vertex: $(-2, 3)$;
axis of symmetry: $x = -2$

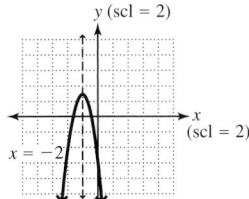

3. Opens left;
vertex: $(4, 2)$;
axis of symmetry: $y = 2$

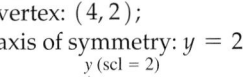
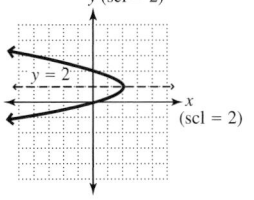

4. Opens right;
vertex: $(-2, -3)$;
axis of symmetry: $y = -3$

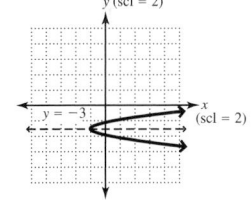

5. Opens right;
vertex: $(-1, -3)$;
axis of symmetry: $y = -3$

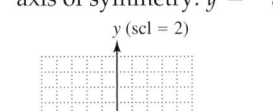
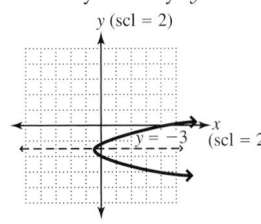

6. Opens right;
vertex: $(-14, 2)$;
axis of symmetry: $y = 2$

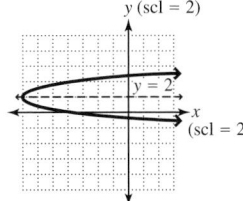

7. Opens left;
vertex: $(4, -1)$;
axis of symmetry: $y = -1$

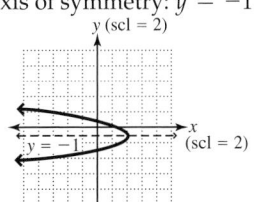

8. Opens left;
vertex: $(3, -2)$;
axis of symmetry: $y = -2$

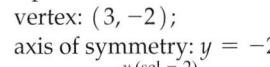
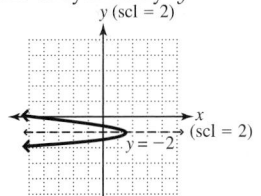

9. a. 144 ft. **b.** 1 sec. **c.** 4 sec.   Definitions/Rules/Procedures $\sqrt{(x_2 - x_1)^2 + (y_2 - y_1)^2}$; $\left(\dfrac{x_1 + x_2}{2}, \dfrac{y_1 + y_2}{2}\right)$

Exercises **10.** 5, $\left(-3, -\dfrac{1}{2}\right)$ **11.** $4\sqrt{5}$, $(0, -1)$ Definitions/Rules/Procedures   equally distant; center; center of a circle; any point on the circle; $(x - h)^2 + (y - k)^2 = r^2$; complete the square   Exercises   **12.** 10

13. Center: $(3, -2)$; radius: 5 **14.** Center: $(-5, 1)$; radius: 2 **15.** Center: $(2, -4)$; radius: 3 **16.** Center: $(-5, -1)$; radius: 2

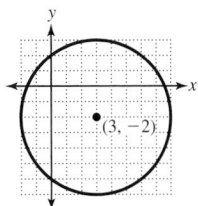 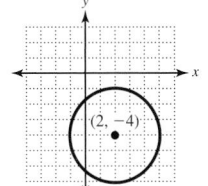 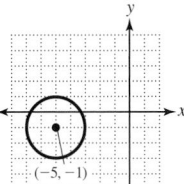

17. $(x - 6)^2 + (y + 8)^2 = 81$ **18.** $(x + 3)^2 + (y + 5)^2 = 100$ **19.** $x^2 + y^2 = 400$ **20.** $(x + 6)^2 + (y - 8)^2 = 225$

12.2 Definitions/Rules/Procedures   sum; two fixed points; $\dfrac{x^2}{a^2} + \dfrac{y^2}{b^2} = 1$; $\dfrac{(x-h)^2}{a^2} + \dfrac{(y-k)^2}{b^2} = 1$; a; b

Exercises   **21.** 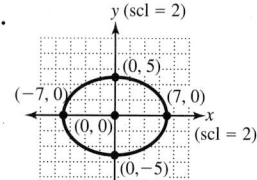   **22.** 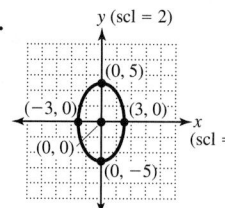   **23.** 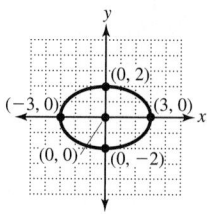   **24.**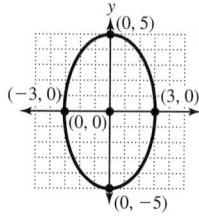

25. 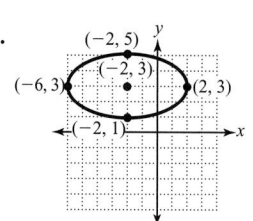   **26.** 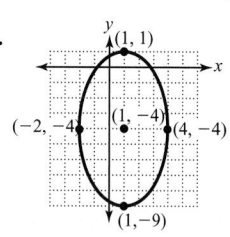   **27.** $\dfrac{x^2}{625} + \dfrac{y^2}{225} = 1$ **28.** $\dfrac{x^2}{1.625^2} + \dfrac{y^2}{4} = 1$

Definitions/Rules/Procedures   difference; two fixed points; $\dfrac{x^2}{a^2} - \dfrac{y^2}{b^2} = 1$; $\dfrac{y^2}{b^2} - \dfrac{x^2}{a^2} = 1$; $(a, 0)$; $(-a, 0)$; y; $(0, b)$; $(0, -b)$; x; (a, b), $(-a, b)$, $(a, -b)$, and $(-a, -b)$; fundamental rectangle; intercept; asymptotes

Exercises   **29.** 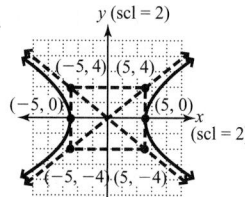   **30.** 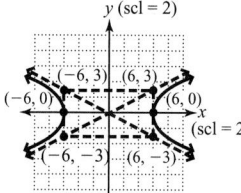   **31.** 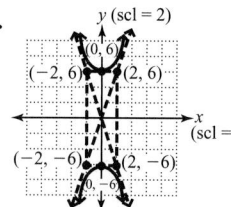   **32.**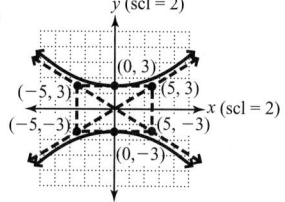

12.3 Definitions/Rules/Procedures   nonlinear; substitution; substitute; elimination   Exercises   **33.** $(2, 5)$, $(-1, -1)$
34. $(-1, -4)$, $(4, 1)$ **35.** $(-73/37, -35/37)$, $(1, 5)$ **36.** $(8, 0)$, $(-8, 0)$ **37.** $(3, 4)$, $(3, -4)$, $(-3, -4)$, $(-3, 4)$
38. $(4, 2)$, $(4, -2)$, $(-4, -2)$, $(-4, 2)$ **39.** $(2, 2)$, $(-2, 2)$ **40.** $(2, 3)$, $(-2, 3)$ **41.** $(8, 5)$, $(8, -5)$, $(-8, -5)$, $(-8, 5)$
42. 8 ft. by 10 ft.   **43.** 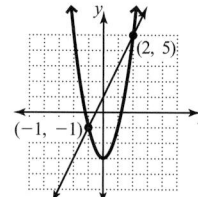   **44.**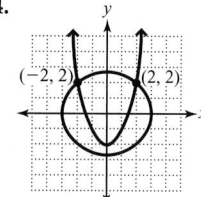

12.4 Definitions/Rules/Procedures   equation; solid; dashed; ordered pair; the region containing that ordered pair
Exercises   **45.** 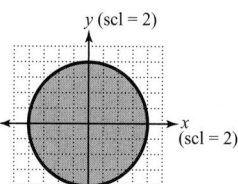   **46.** 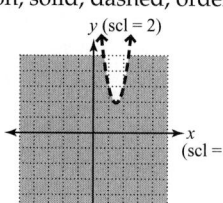   **47.** 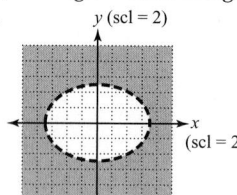   **48.**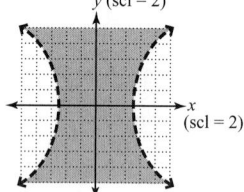

Definitions/Rules/Procedures   the solution sets of the individual inequalities   Exercises   **49.**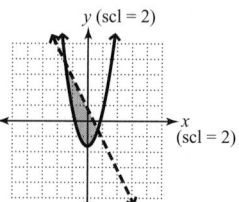

50.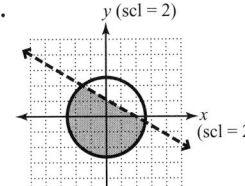

51.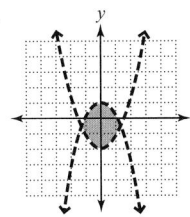

52.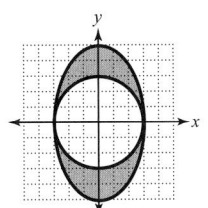

53.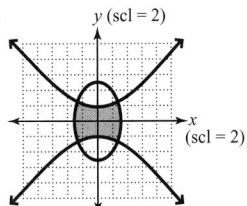

54.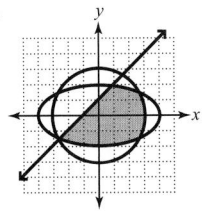

Chapter 12 Practice Test

1. Opens upward; vertex: $(-1, -4)$;
axis of symmetry: $x = -1$

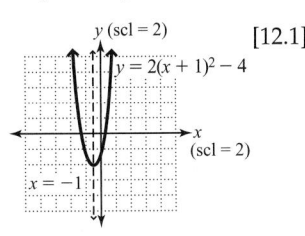 [12.1]

2. Opens left; vertex: $(1, 3)$;
axis of symmetry: $y = 3$ [12.1]

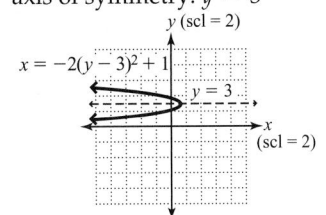

3. Opens right; vertex: $(-7, -2)$;
axis of symmetry: $y = -2$ [12.1]

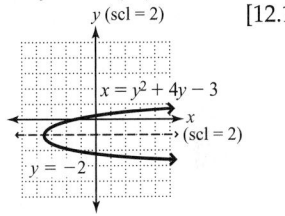

4. $2\sqrt{5}$, $(4, -4)$ [12.1]

5. Center: $(-4, 3)$; radius: 6 [12.1]

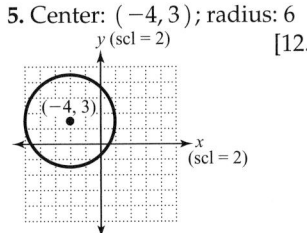

6. Center: $(-2, 5)$; radius: 3 [12.1]

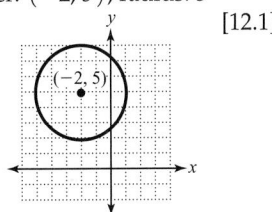

7. $(x - 2)^2 + (y + 4)^2 = 100$ [12.1]

8. 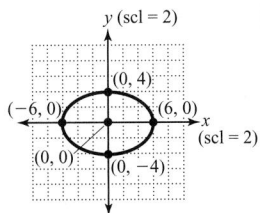 [12.2]

9. [12.2]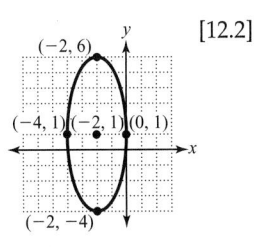

10. 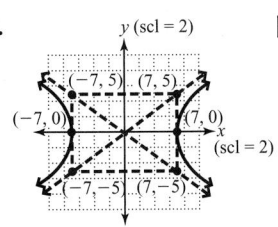 [12.2]

11. 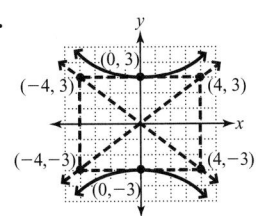 [12.2]

12. $(-5, 18), (1, 6)$ [12.3]

13. $(3, 5), \left(-\dfrac{3}{5}, -\dfrac{29}{5}\right)$ [12.3]

14. $(2, 3), (2, -3), (-2, -3), (-2, 3)$ [12.3]

15. $(3, 2), (3, -2), (-3, -2), (-3, 2)$ [12.3]

16. [12.4]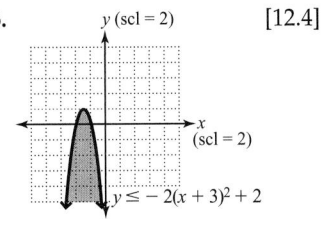

17. 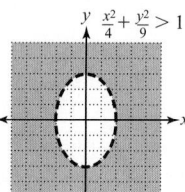 $\frac{x^2}{4} + \frac{y^2}{9} > 1$ [12.4] **18.** 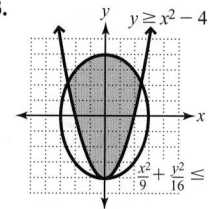 $y \geq x^2 - 4$ [12.4] **19.** Height is 18; distance across the base is 12. [12.1]

20. $\dfrac{x^2}{9.5^2} + y^2 = 1$ [12.2]

Chapters 1–12 Cumulative Review Exercises

1. false **2.** true **3.** true **4.** $-8; -5$ **5.** i. **6.** $d = \sqrt{(x_2 - x_1)^2 + (y_2 - y_1)^2}$. **7.** $9\dfrac{1}{16}$ **8.** $\dfrac{n^6}{125m^9}$ **9.** $2i\sqrt{5}$ **10.** $(x + y)(a + b)$

11. $(k^2 + 9)(k + 3)(k - 3)$ **12.** $-4, \dfrac{2}{3}$ **13.** 52 **14.** $\dfrac{1 \pm i\sqrt{3}}{4}$ **15.** $-\dfrac{4}{3}$ **16.** 1.5 **17.** $\dfrac{1}{8}$ **18. a.**

b. $\left\{ m \mid 0 < m < \dfrac{3}{5} \right\}$ **c.** $\left(0, \dfrac{3}{5} \right)$ **19. a.** $x^2 + 2x$ **b.** $-x^2 + 2x + 2$ **c.** $2x^3 + x^2 - 2x - 1$ **d.** $\dfrac{2x + 1}{x^2 - 1}; x \neq \pm 1$

e. $2x^2 - 1$ **f.** $4x^2 + 4x$ **20.** $\log_5 125 = 3$ **21.** $5\log_5 x + \log_5 y$ **22.** $(x - 2)^2 + (y - 4)^2 = 169$

23. $y < -x + 2$ 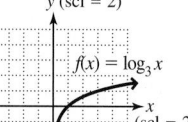 $y \geq x - 4$ **24.** 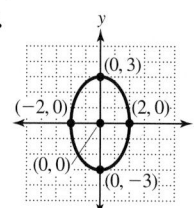 $f(x) = \log_3 x$ **25.** **26. a.** $h = \dfrac{V}{\pi r^2}$ **b.** $\dfrac{15}{r^2}$

27. 2.9 lb. of tuna and 3.4 lb. of shrimp **28.** \$1500 at 5%, \$2500 at 6%, \$1000 at 3% **29.** 6.9 **30.** $\dfrac{x^2}{5625} + \dfrac{y^2}{400} = 1$

Glossary

Absolute value: A given number's distance from 0 on a number line.

Additive inverses: Two numbers whose sum is 0.

Area: The total number of square units that fill a figure.

Arithmetic average or mean: The sum of all given numbers divided by the number of numbers.

Arithmetic sequence: A sequence in which each term after the first is found by adding the same number to the previous term.

Arithmetic series: The sum of the terms of an arithmetic sequence.

Augmented matrix: A matrix made up of the coefficients and the constant terms of a system. The constant terms are separated from the coefficients by a dashed vertical line.

Axis of symmetry: A line that divides a graph into two symmetrical halves.

Base: The number that is repeatedly multiplied.

Binomial: A polynomial containing two terms.

Binomial coefficient: A number written as $\binom{n}{r}$ and defined as $\dfrac{n!}{r!(n-r)!}$.

Binomial theorem: For any positive integer n,

$$(a+b)^n = \binom{n}{0}a^n + \binom{n}{1}a^{n-1}b + \binom{n}{2}a^{n-2}b^2 + \binom{n}{3}a^{n-3}b^3 + \ldots + \binom{n}{n}b^n.$$

Circle: A set of points in a plane that are equally distant from a central point. The central point is the center.

Circumference: The distance around a circle.

Coefficient: The numerical factor in a term.

Coefficient of a monomial: The numerical factor in a monomial.

Common difference of an arithmetic sequence: The value of d is found by $d = a_n - a_{n-1}$, where a_n is any term in the sequence (except the first) and a_{n-1} is the previous term.

Common logarithms: Logarithms with a base of 10. $\log_{10} x$ is written as $\log x$. Note that $\log 10 = 1$.

Complex conjugate: The complex conjugate of a complex number $a + bi$ is $a - bi$.

Complex number: A number that can be expressed in the form $a + bi$, where a and b are real numbers and i is the imaginary unit.

Complex rational expression: A rational expression that contains rational expressions in the numerator or denominator.

Composition of functions: If f and g are functions, then the composition of f and g is defined as $(f \circ g)(x) = f[g(x)]$ for all x in the domain g for which $g(x)$ is in the domain of f. The composition of g and f is defined as $(g \circ f)(x) = g[f(x)]$ for all x in the domain of f for which $f(x)$ is in the domain of g.

Compound inequality: Two inequalities joined by either *and* or *or*.

Congruent angles: Angles that have the same measure. The symbol for congruent is $\cong$.

Conic section: A curve in a plane that is the result of intersecting the plane with a cone—more specifically, a circle, an ellipse, a parabola, or a hyperbola.

Conjugates: Binomials that differ only in the sign separating the terms.

Consistent system of equations: A system of equations that has at least one solution.

Constant: A symbol that does not vary in value.

Contradiction: An equation that has no real-number solution.

Degree of a monomial: The sum of the exponents of all variables in a monomial.

Degree of a polynomial: The greatest degree of any of the terms in the polynomial.

Diameter: The distance across a circle through its center.

Direct variation: Two variables y and x vary directly if $y = kx$. If y varies directly as the nth power of x, then $y = kx^n$, where k is the constant of variation.

Discriminant: The discriminant is the radicand, $b^2 - 4ac$, in the quadratic formula.

Domain: The set of all input values (x-values) for a relation.

Ellipse: The set of all points the sum of whose distances from two fixed points is constant.

Equation: A mathematical relationship that contains an equal sign.

Exponent: A symbol written to the upper right of a base number that indicates how many times to use the base as a factor.

Exponential function: If $b > 0$, $b \neq 1$, and x is any real number, then the exponential function is $f(x) = b^x$.

Expression: A constant, a variable, or any combination of constants, variables, and arithmetic operations that describes a calculation.

Extraneous solution: An apparent solution that does not solve its equation.

Factored form: A number or an expression written as a product of factors.

Factorial notation: For any natural number n, the symbol $n!$ (read "n factorial") means $n(n-1)(n-2)\ldots 3 \cdot 2 \cdot 1$. $0!$ is defined to be 1, so $0! = 1$.

Factors: If $a \cdot b = c$, then a and b are factors of c.

Finite sequence: A function with a domain that is the set of natural numbers from 1 to n.

Formula: An equation that describes a mathematical relationship.

Fraction: A quotient of two numbers or expressions a and b having the form $\dfrac{a}{b}$, where $b \neq 0$.

Function: A relation in which every value in the domain is paired with (or assigned to) exactly one value in the range.

Geometric sequence: A sequence in which every term after the first is found by multiplying the previous term by the same number, called the **common ratio**, r, where $r = \dfrac{a_n}{a_{n-1}}$.

Geometric series: The sum of the terms of a geometric sequence.

Greatest common factor (GCF): The largest natural number that divides all given numbers with no remainder.

Hyperbola: The set of all points the difference of whose distances from two fixed points remains constant.

Identity: An equation that has every real number as a solution (excluding any numbers that cause an expression in the equation to be undefined).

Imaginary number: A number that can be expressed in the form $a + bi$, where a and b are real numbers, i is the imaginary unit, and $b \neq 0$.

Imaginary unit: The number represented by i, where $i = \sqrt{-1}$ and $i^2 = -1$.

Inconsistent system of equations: A system of equations that has no solution.

Inequality: A mathematical relationship that contains an inequality symbol ($\neq, <, >, \leq,$ or $\geq$).

Infinite sequence: A function with a domain that is the set of natural numbers.

Intersection: For two sets A and B, the intersection of A and B, symbolized by $A \cap B$, is a set containing only elements that are in both A and B.

Inverse functions: Two functions f and g are inverses if and only if $(f \circ g)(x) = x$ for all x in the domain of g and $(g \circ f)(x) = x$ for all x in the domain of f.

Inverse variation: Two variables y and x vary inversely if $y = \dfrac{k}{x}$. If $y = \dfrac{k}{x^n}$, then y varies inversely as the nth power of x, where k is the constant of variation.

Irrational number: Any real number that is not rational.

Joint variation: If y varies jointly as x and z, then $y = kxz$, where k is the constant of variation.

Least common denominator (LCD): The least common multiple of the denominators of a given set of fractions.

Least common multiple (LCM): The smallest number that is a multiple of each number in a given set of numbers.

Like radicals: Radical expressions with identical radicands and identical root indices.

Like terms: Constant terms or variable terms that have the same variable(s) raised to the same exponents.

Linear equation: An equation in which each variable term contains a single variable raised to an exponent of 1.

Linear equation in one variable: An equation that can be written in the form $ax + b = c$, where a, b, and c are real numbers and $a \neq 0$.

Linear inequality: An inequality containing expressions in which each variable term contains a single variable with an exponent of 1.

Logarithm: If $b > 0$ and $b \neq 1$, then $y = \log_b x$ is equivalent to $x = b^y$.

Lowest terms: Given a fraction $\dfrac{a}{b}$ and $b \neq 0$, if the only factor common to both a and b is 1, then the fraction is in lowest terms.

Matrix: A rectangular array of numbers.

Mean or arithmetic average: The sum of all given numbers divided by the number of numbers.

Median: The middle number in an ordered set of numbers.

Minor of an element of a matrix: The determinant of the remaining matrix when the row and column in which the element is located are ignored.

Mode: The number that occurs most often in a set of numbers.

Monomial: An expression that is a constant, a variable, or a product of a constant and variable(s) that are raised to whole-number powers.

Multiple: A multiple of a given integer n is the product of n and an integer.

Multiplicative inverses: Two numbers whose product is 1.

Natural logarithms: Base-e logarithms are called natural logarithms, and $\log_e x$ is written as $\ln x$. Note that $\ln e = 1$.

Nonlinear system of equations: A system of equations that contains at least one nonlinear equation.

nth root: The number b is an nth root of a number a if $b^n = a$.

One-to-one function: A function f is one-to-one if for any two numbers a and b in its domain, when $f(a) = f(b), a = b$ and when $a \neq b, f(a) \neq f(b)$.

Percent: A ratio representing some part out of 100.

Perimeter: The distance around a figure.

Polynomial: A monomial or a sum of monomials.

Polynomial in one variable: A polynomial in which every variable term has the same variable.

Prime factorization: A factorization that contains only prime factors.

Prime number: A natural number that has exactly two different factors, 1 and the number itself.

Proportion: An equation in the form $\dfrac{a}{b} = \dfrac{c}{d}$, where $b \neq 0$ and $d \neq 0$.

Quadratic equation in one variable: An equation that can be written in the form $ax^2 + bx + c = 0$, where a, b, and c are real numbers and $a \neq 0$.

Quadratic equation in two variables: An equation that can be written in the form $y = ax^2 + bx + c$, where a, b, and c are real numbers and $a \neq 0$.

Quadratic inequality: An inequality that can be written in the form $ax^2 + bx + c > 0$ or $ax^2 + bx + c < 0$, where $a \neq 0$. *Note:* The symbols $<$ and $>$ can be replaced with $\leq$ and $\geq$.

Quadratic in form: An equation that can be rewritten as a quadratic equation $au^2 + bu + c = 0$, where $a \neq 0$ and u is a variable or an expression.

Radical equation: An equation containing at least one radical expression whose radicand has a variable.

Radical function: A function containing a radical expression whose radicand has a variable.

Radius: The distance from the center of a circle to any point on the circle.

Range: The set of all output values (y-values) for a relation.

Ratio: A comparison of two quantities using a quotient.

Rational exponent: An exponent that is a rational number.

Rational expression: An expression that can be written in the form $\dfrac{P}{Q}$, where P and Q are polynomials and $Q \neq 0$.

Rational inequality: An inequality containing a rational expression.

Rational number: Any real number that can be expressed in the form $\frac{a}{b}$, where a and b are integers and $b \neq 0$.

Real number: Any rational or irrational number.

Relation: A set of ordered pairs.

Row echelon form: An augmented matrix whose coefficient portion has 1's on the diagonal from upper left to lower right and 0s below the 1's.

Scientific notation: A number expressed in the form $a \times 10^n$, where a is a decimal number with $1 \leq |a| < 10$ and n is an integer.

Sequence: A function list whose domain is 1, 2, 3, …, n.

Series: The sum of the terms of a sequence.

Set: A collection of objects.

Similar figures: Figures with congruent angles and proportional side lengths.

Slope: The ratio of the vertical change to the horizontal change between any two points on a line.

Solution: A number that makes an equation true when it replaces the variable in the equation.

Solution for a system of equations: An ordered set of numbers that makes all equations in the system true.

Square matrix: A matrix with an equal number of rows and columns.

Statistic: A number used to describe some characteristic of a set of data.

System of equations: A group of two or more equations.

Terms: Expressions that are the addends in an expression that is a sum.

Trinomial: A polynomial containing three terms.

Unit ratio: A ratio with a denominator of 1.

Union: For two sets A and B, the union of A and B, symbolized by $A \cup B$, is a set containing every element in A or in B.

Variable: A symbol that can vary in value.

Vertex of a parabola: The lowest point on a parabola that opens upward or the highest point on a parabola that opens downward.

Volume: The total number of cubic units that fills a space.

x-intercept: A point where a graph intersects the x-axis.

y-intercept: A point where a graph intersects the y-axis.

Index of Applications

A

Astronomy/Aerospace
Andromeda galaxy, 359, 371, 518
Asteroids on a collision course, 61
Circumference of Earth's equator, 388
Diameter of Venus, 119
Distance from the Earth to the Sun, 358, 518
Distance from the Sun to a star, 353, 359
Equatorial radius of the Sun, 358
Light-year, 358, 518
Orbital velocity of a planet, 518
Orbit of Pluto, 857
Orbit of the comet Epoch, 857
Path of a satellite, 846
Path of Earth's orbit around the Sun, 388
Planet that supports life, 390
Radius of Earth, 123, 358
Solar system, 358
Stars in the visible universe, 390
Temperatures on a planet, 28, 100
Time it takes for light from Proxima Centauri to reach Earth, 422
Time it takes for light to travel from the Sun to Earth, 411
Volume of Earth, 99, 388
Volume of the Sun, 388

Automotive
Cars holding their value for resale, 272
Crash test, 699
Drivers involved in fatal accidents, 845
Flat tire, 110
Gallons of gasoline needed, 142
Gas mileage, 563, 571, 573
Skid length, 621, 665, 675, 682, 686
Sources of distraction among inattentive drivers, 257
Value of a car, 262
Washing and waxing a car, 559

B

Biology
Age of the universe, 358
Atomic mass unit, 417
Bacteria, 810, 827, 830
Culture of E. coli, 783
Diameter of a streptococcus bacterium, 353
Human DNA, 359
Mass of a blue whale, 46
Mass of a hydrogen atom, 417
Mass of an alpha particle, 358
Mass of an electron, 359
Mass of a proton, 359
Mitochondria, 358
Number of deer in a forest, 148
Number of trees in a National forest, 420

Population of a mosquito colony, 814
Population of an ant colony, 827
Population of trout, 148
Positive electrical charge of protons, 359

Business
Audit on a small business, 161
Break-even point, 245, 287, 710, 863
CD sales, 731
Cell phone manufacturer, 863
Coffee shop, 320
Company spending on equipment failure, 141
Company's stock losses, 45
Cookware consultant, 296
Copy center, 206
Cost of producing items, 261
Defective products, 147
Delicatessen, 320
Gross income from software products, 710
Home improvement store, 295
Home interiors store, 333
Home office space, 18
Manufacturing computer chips, 359
Market equilibrium, 863
Maximizing profit, 336
Minimizing cost, 731
Net profit or loss, 31, 379
Number of units produced, 562
Production, 246, 272
Profit, 171, 184, 237, 245, 379
Revenue, 171, 184, 245, 379
Sales, 160, 270, 368
Sales quota, 111
Selling facial cleansing lotion, 216
Small business owner buying a computer, 207
Small company costs, 160
Software sales, 171, 246, 336
Wal-Mart stock, 58
Walt Disney Company's financial report, 31
Wireless company, 206

C

Chemistry
Acid solution, 42–43, 311
Alcohol solution, 311
Atoms in a molecule, 19, 144
Carbon-14 dating, 811
Concentration of chlorine in a swimming pool, 783
Converting uranium to energy, 388
The element bismuth, 827
Half-life, 359, 417, 779, 783, 811, 820, 827, 829, 830
HCl solution, 160, 307–8, 311
Hydronium ion concentration, 805, 825

Mixing a solution, 371
pH of a substance, 805
Saline solution, 342, 611, 762
Temperatures in an experiment, 28, 32, 100
Temperatures of silver, 171
Water in a liquid state, 588

Construction
Arches, 845, 881
Architect designs, 709
Blueprint, 111
Bridge over a waterway or canal, 856, 875, 884
Cardboard needed to make a box, 188
Carpeting a room, 93
Covering a tabletop with Formica, 92
Crown molding, 647
Cutting a groove in a wooden block, 699
Cutting plywood, 698
Cutting time of a drill, 507–8
Designing a building, 333, 588
Designing an addition to a house, 469–70, 473
Designing a platform for an auditorium, 336
Distance around a wall, 91–92
Front elevation of a house, 473
Great Pyramid at Giza, 228
Height of a structure, 149
Highway passing beneath an overpass, 857
Installation, 96, 97, 533, 775
Kitchen floor, 98
Longest vehicular tunnel, 329
New roof on a portable classroom, 559
Painting, 97, 98, 348, 551, 559, 574, 775
Putting a futon frame together, 559
Revolutions of a saw or drill, 507
Roofing, 144, 228, 348, 471, 621, 675
Siding on a house, 94, 98
Steps of the Pyramid of the Sun, 228
Stone blocks in the Great Pyramid, 412
Support beam, 299
Supporting a power pole, 475
Tabletop fitted with veneer strips, 648
Two construction teams, 533
Wallpapering, 92, 96
Welding together steel beams, 247
Wood required to trim a window, 186

Consumer
Amount spent on recreation, 816
Average home prices, 257
Better buy, 145
Buying a car stereo, 107–8
Buying paint, 148
Buying plants, 246
Carpet-cleaning charges, 148
Car rental costs, 124, 261

Cell phone subscribers, 369
Conserving energy, 587
Consumer Price Index (CPI), 783
Cost of a plumber, 205
Cost of carpet, 97
Cost of CDs, books, and DVDs, 329
Cost of covering a flower bed, 97
Cost of crown molding, 96
Cost of school supplies, 319
Cost of sod, 97
Cost of tickets, 159
Cost of wallpaper, 96
Discount price, 156, 160
Food costs, 159, 319, 328, 560
Holiday gift spending, 278
House thermostat, 587
Internet purchase, 160
Long-distance phone plans, 288, 587
Mail-order clubs, 587, 608
Monthly electric and gas charges, 160
Painting charges, 148
Phone costs, 124
Postage costs, 258
Prices, 139, 145, 159, 160, 273, 316–17
Price to capacity, 550
Purchasing vs. renting a home, 288
Seafood consumption, 883
Service charge, 158
Taxi charges, 237
Toll plans, 288
Total cost of a television, 186
Water bill, 147

E

Economics
Converting U.S. dollars to Japanese yen, 359
Deducting a computer on taxes, 238
Depreciation, 265, 784, 815
Dow Jones Industrial Average, 32, 58, 160
Energy use, 160
Exchange rates, 147, 518
Family's taxable income, 31
Median house price, 156
NASDAQ, 253
Property taxes, 158
Purchasing power of one dollar, 805, 815
Sales tax, 155–56, 160
Stock prices, 31, 63, 85, 161, 193, 238, 731
Stock's price-to-earnings ratio, 137, 145
Supply and demand, 710, 863
Taxes on a house, 147, 158
U.S. gross domestic product, 783
U.S. per capita income, 276
Education
ACT scores, 32
Associate degrees, 622
Bachelor's degrees awarded to women, 160

Cashier's office at a college, 299
College bookstore, 299
College tuition, 256
Courses taught by an instructor, 273
Credit hour costs, 299
English students receiving an A, 155
Enrollment in a program, 259
Final grades, 171
Grades on a calculus test, 257
Multiple-choice test, 158
Points scored in a college course, 18
Quiz average, 57
Real estate course, 344
SAT score requirements, 335
Small college has 2450 students, 138
Student loans, 310
Students not receiving a passing grade, 108
Students recalling a classroom lecture, 790, 792
Test scores, 53, 57, 171, 186, 587
Tutoring, 205
Two-year colleges in the United States, 815
Engineering
AC circuit, 657
Analyzing circuits, 508
Aperture and f-stop of a camera lens, 561
Arecibo radio telescope, 698
Atom bomb releasing energy, 412
Bookcase design, 108
Building a model F-16 jet, 147
Chemical on a circuit board, 533
Cross section of a satellite dish, 845
Cross section of a tunnel, 846
Current, 124, 172
Current and resistance, 464
Current in a circuit, 560
Designing a lens, 550
Designing a steel plate, 413
Electrical circuit, 46, 329
Estimating the distance across a river, 150
Fiber-optic cable, 359
Forces on a steel structure, 329
Grade of a road, 144
Hoover Dam supplying electricity, 149
Laser beam, 299
Nuclear reaction, 388
Nuclear reactor, 811
Resistance of an electrical circuit, 46, 124, 541, 542, 550, 561, 562, 657
Resultant force, 31, 622, 666
Transistors on a chip, 784
Value of resistors, 550
Voltage of a current, 99
Weight-carrying capacity of a rectangular beam, 562
Width of a rectangular steel plate, 533
Environment
Area of forest remaining, 207
Atmospheric pressure, 814

Barometric pressure from the eye of a hurricane, 815
Carbon dioxide emissions, 340
Carbon monoxide concentration, 161
Erosion rate for Niagara Falls, 148
Longest rivers, 328
Magnitude of an earthquake, 622, 805, 825, 884
Rainfall, 54, 257
Severe thunderstorm warning, 698
Snowfall, 86
Temperatures, 100, 257, 426
U.S. greenhouse gas emissions, 159

F

Finance
Balance of an account, 21, 26
Checking account balance, 110, 159, 299
Coins, 278
Compound interest, 779, 783, 800, 809, 814, 820, 826, 827, 829, 832
Continuous compounding, 803, 805, 809, 814, 825, 827
Credit cards, 21, 26, 45, 84, 86, 108, 110, 299, 310
Debt, 22–23, 27–28, 45, 85, 158, 783
Down payment for a car, 110
Exchanging money, 147
Growth of an account, 809, 814, 825, 826, 827
Household budget, 33
Interest earned, 137
Interest rate, 473
Investment, 160, 306–7, 310, 320, 329, 336, 340, 342, 346, 348, 426, 473, 496, 576, 809, 883
Lender's guidelines, 158
Loans, 310
Mortgage payments, 158, 288
Savings account, 299
Simple interest, 562

G

Geometry
Angle measurement, 278, 319, 340, 348
Area of a circular region, 98, 400
Area of an entertainment center, 175
Area of a rectangular region, 96, 97, 98, 415, 463
Area of a room, 93, 96, 400, 437, 473
Area of a shaded region, 175, 188, 347, 437, 459, 533
Area of a sign, 93
Area of a state, 387
Area of a triangular region, 97
Area of the side of a house, 106, 437
Base and height of a triangle, 473, 698, 751
Circumference of a circular region or object, 96
Diagonal of a rectangular shaped object, 474
Diameter of a circular region or object, 123, 171, 184, 560

Dimensions of a box-shaped object, 118, 123, 171, 317, 413, 698, 742
Dimensions of a cylindrical object, 123
Dimensions of a parallelogram, 422, 742
Dimensions of a rectangular region, 118, 122, 123, 171, 175, 186, 188, 247, 294, 299, 310, 348, 443, 451, 469, 473, 474, 491, 494, 541, 551, 698, 699, 708, 731, 751, 863, 878
Dimensions of a room, 124, 414, 469, 709, 863
Dimensions of a trapezoid, 111, 123, 143, 178, 588
Height of a box, 118, 123, 171
Height of a cylinder, 123, 884
Height of a pyramid, 122
Height of a trapezoid, 473
Height of a triangle, 541
Length of a side of a triangle, 111, 122, 319, 492, 708, 832
Length of the side of a regular hexagon, 657
Length of the side of a square, 698
Length of the sides of the base of a pyramid, 657
Maximum area, 731
Maximum length of a garden, 168
Maximum perimeter of a building, 247
Minimum surface area, 168
Perimeter of an isosceles triangle, 136
Perimeter of a rectangular region, 92, 96, 136
Perimeter of a semicircle, 136
Radius of a circular region, 171, 444, 470, 473, 698, 750
Radius of a cylinder, 709, 750
Surface area of a box, 136, 368
Surface area of the water in a swimming pool, 379
Volume of a box-shaped object, 98, 175, 382, 401, 413, 463
Volume of a cone, 99, 289, 401, 508
Volume of a cylinder, 99, 401, 561, 562, 573
Volume of an object, 437
Volume of a pyramid, 99, 508
Volume of a rectangular solid, 562, 751
Volume of a shaded region, 459
Volume of a storage tank, 368
Volume of metal remaining in a block, 368

Government
Local referendum, 138–39
National debt, 411
U.S. House of Representatives, 19

H
Health
Body mass index (BMI), 137, 507
Burning calories, 320
Elliptical trainer, 857, 881
National health expenditures, 250–51

Nutrition label, 18
Vitamin supplements, 343

L
Labor
Average weekly earnings, 57
Benefits package, 18
Commission, 265
Cutting jobs, 155
Employees with medical flexible spending accounts, 33
Hourly wage, 688
Hours worked, 276
Median annual income of year-round full-time workers, 33, 145
Minimum wage, 161, 820
Number of people unemployed in the U.S., 57
Occupation growth, 161
Salary, 262
Sales commission, 774
Working alone to complete a task, 559, 716, 718
Working together to complete a task, 552, 559, 574, 576, 832

M
Medical
Experimental medication, 111
Hemoglobin in a red blood cell, 359
Lithotripter, 857
Malignant tumor, 820
Mass of a person, 124
New medication, 16
Patient receiving medications, 110
Red blood cells, 359
Side effects of a new drug, 160
Size of the HIV virus, 359
Whispering rooms, 857
Miscellaneous
Age, 298
Appraisal on a marble-top table, 160
Artist welding a metal rod, 474
Battery power in a laptop, 206
Birth dates, 251
Cake temperature, 814
Castillo at Chichen Itza, 299
Checking e-mail, 278
Circular flower bed, 874
Cleaning, 206, 576, 716
Coffee temperature, 784
Comic book collector, 216
Company picnic attendance, 43
Compound bows, 857, 875
Cutting and trimming a lawn, 559
Fastest recital of Hamlet's soliloquy, 148
Filling a tub, 559
Grade of a hill, 228
Guessing the number of marbles in a jar, 100
Height of girls and boys, 792, 815
Hours spent at an activity, 18
International visitors to the United States, 369

Landscaping, 94, 97, 346
Length of guitar strings, 45
Making quilts for a charity auction, 559
Mass of the Liberty Bell, 46, 124
Maximum load of an elevator, 172
Minute hand of Big Ben, 846
Mixing food or drinks, 311, 576
Number of U.S. radio stations, 257
Organizing a fund-raiser, 246
Recipes, 145, 147
Reflecting Pool at the National Mall, 299
Rye grass seed, 148
Temperature of a salt-water aquarium, 587
Top brands of ice cream, 257
Visitors to the United States from Canada and Mexico, 340
Wavelength of a musical note, 124
Weight of an object, 147, 412, 561, 574
What the Internet most resembles, 345
Width of a human hair, 359
Workers who get headaches, 345
World record for fastest typing, 148

P
Physics
Acceleration, 124
Ball thrown downward, 760
Boyle's Law, 556
Centripetal acceleration, 64
Charged particles, 658
Charles's Law, 560, 561
Coulomb's Law, 557–58
Distance an object travels, 665, 676, 686, 760
Energy of a particle at rest, 64
Energy of a photon of light, 388
Focal length of a lens, 550
Force due to gravity, 43, 46
Free-falling object, 124, 560
Frequency of a radio wave, 561
Frequency of a vibrating string, 719
Gravitational effect between two objects, 61
Height of a falling object, 188, 364, 368, 470
Hooke's law, 555
Intensity of light, 557, 572
Kinetic energy, 657–58, 699, 750
Length of a radio wave, 561
Mass of a load, 43
Mass of an object, 124
Maximum height of an object launched upward, 558, 731
Newton's Law of Universal Gravitation, 562
Object dropped, 464, 699, 741
Object's initial velocity, 125, 473, 705, 726, 741
Object thrown upward, 611, 727–28, 741, 845, 872

Ohm's Law, 560
Path of a shell, 845
Path of a toy plane, 846
Period of a pendulum, 618, 621, 657, 665, 675, 682, 688
Pressure exerted by a liquid, 555, 561
Pressure of a gas, 561
Sound intensity, 320, 802–3, 804, 825
Speed of light, 358, 411, 417
Speed of sound, 124, 517
Theory of relativity, 64
Time for an object to fall, 657, 676, 686
Toy rocket launched, 726–27, 730
Trajectory of an object, 193, 249–50, 731
Universal gas law, 562
Velocity of a free-falling object, 618, 621
Velocity of an object, 194, 230–31
Velocity of a particle, 657–58
Voltage in a circuit, 368, 413, 420, 464, 657
Wavelength of a wave, 573, 762
Wavelength of light, 359

S

Sports/Entertainment
    Acrobat, 755
    Archer, 760
    Batting average, 159
    Bicycling, 297, 299, 588
    Boston Marathon, 718
    Cliff diving championship, 709
    College basketball court, 299
    Cyclist training, 718
    Daytona 500 race, 251
    Extreme games competition, 705, 709
    Fastest speed of a wheeled vehicle, 517
    Ferris wheel, 846
    Field goal attempts, 159
    Fishing, 845
    Golf tournament, 257
    High school athletics, 237
    Hiking, 95
    Home runs, 258
    Javelin toss, 731

Jogging, 64, 297, 299, 551
Jumping the Grand Canyon on a motorcycle, 709
Longest punt, 755
Movie ticket sales, 342
NASCAR championships, 258
NCAA basketball championship, 329
Platform diver, 709
Points scored in a basketball game, 320
Ropes course, 621
Running, 299, 517, 518, 542, 553–54, 574, 718
Skydiving, 137, 618
Snowskiing, 178
Soccer player kicking a ball, 727
Super Bowl, 278, 329
Swim meet, 343
Swimmer's rate, 306
Tennis court, 295
Ticket sales to a play, 328, 346, 731
Top-five cable TV networks, 257
Top-grossing American movie, 299
Top two money makers in the music industry, 299
Tour de France, 99
Track meet, 320
Volleyball court, 299
Yahtzee, 110
Statistics/Demographics
    Computers in use in the United States, 780
    Foreign-born population, 259
    hate crimes in U.S., 145
    Increase in number of divorces, 111
    Median income for males in the United States, 816
    Number of Catholics in the United States, 258
    Number of people aged 65 or older in the United States, 780, 829
    Number of single fathers, 161
    People per square mile in the United States, 411
    Percent of deaths by age per million miles driven, 845

Poll or survey results, 45, 111, 426
Population, 161, 228, 237–38, 359, 783, 815
Population of Africa, 815
Population of the greater Orlando area, 810
Probability of an accident while driving at a given blood alcohol level, 814
U.S. households headed by married couples, 159
U.S. population 21 years of age or older, 159
U.S. public debt, 816
Victims of property crime, 15–16, 18

T

Transportation
    Average rate, 94–95, 99, 171, 541, 551, 716, 717, 762, 832
    Canoeing, 559
    Cargo hold of a ship, 559
    Distance apart, 542, 560, 576
    Distance from original location, 475
    Distance traveled, 99, 110, 147, 148, 716, 775
    Driving time, 563
    Jet fuel prices, 156
    Map scale, 147
    Minimum runway length, 802–3, 806
    Number of miles per gallon, 805
    Plane's speed in still air, 305–6, 310, 342, 554, 559
    Speed, 560
    Speeding ticket, 588
    Speed of a boat in still water, 310, 346, 559
    Time for a person or vehicle to catch up to another, 296, 299, 340, 559
    Time for a person or vehicle to overtake another, 559, 611
    Time for vehicles or people to meet, 559
    Traveling for leisure, 276
    Travel time, 64, 171, 178, 559, 714–16, 717, 775
    Walking speeds, 806

Index

A

Absolute value
  defined, 6
  graphing functions, 591–92
  inequalities involving, 595–98
  more than one, 590–91
  solving equations involving, 589–90
Absolute value property, 589
Absolute zero, 32
Addends, 20
Addition
  associative property of, 21
  commutative property of, 20, 21
  of complex numbers, 669
  of decimal numbers, 26–28
  distributive property of multiplication
    over, 34
  of fractions
    with different denominators,
      24–25
    with same denominator, 23–24
  of functions, 743
  of integers, 20–25
  of like radicals, 641–43
  of numbers
    with different signs, 22–23
    with same sign, 21–22
  polynomial, 372–74
  problem solving involving, 131–32
  properties of, 21
  of rational expressions
    with different denominators,
      525–30
    with same denominator, 519
  translating word phrases to
    expressions in, 59
Addition principles
  applying to formulas, 125–27
  of equality, 101–8
  of inequality, 164–65
  isolating variable using, 126
  solving linear equations using, 101–2,
    114
Additive identity, 20, 21
Additive inverses, 26–27, 301–3
  multiplying each equation by number
    to create, 302–3
  multiplying one equation by number
    to create, 301–2
Aging, 780
Algebra, structure of, 2
Algebraic expression, evaluating, 66
Algebra pyramid, 3, 88, 361, 465, 613, 659
Amazon River, 150
And
  interpreting compound inequalities
    with, 578–79
  solving compound inequalities with,
    579–81

Applications, solving
  addition principle in, 107
  common logarithms in, 802–3
  elimination in, 305–8
  exponential and logarithmic functions
    in, 809–11
  formulas in, 118
  involving logarithms, 789–90
  natural logarithms in, 803
  parabolas in, 726–28
  problems that translate to system of
    three linear equations, 316–17
  problems using equations that are
    quadratic in form, 714–16
  problems using matrices, 324–25
  radical functions in, 617–18
  systems of equations in, 294–97
  systems of linear inequalities in, 333
Area
  calculating, for composite figures,
    92–94
  defined, 90
Arecibo radio telescope, 698
Arithmetic average
  defined, App–1
  finding, of given set of numbers,
    App–1
Arithmetic mean, 53
Arithmetic operations, performing, with
  complex numbers, 668
Arithmetic sequence
  common differences of, App–6–7
  nth term of, App–7
Arithmetic series, App–8–9
  partial Sum, S_n, of, App–9
Associative property
  of addition, 21
  of multiplication, 34
Atomic bomb, 412
Augmented matrix
  defined, 321
  solving system of linear equations by
    transforming to row echelon form,
    322–24
  writing system of equation as, 321
Axis
  of symmetry, 476
  x-, 190
  y-, 190

B

Back-end ratio, 158
Bases
  defined, 47
  division of exponential forms with
    same, 402
  evaluating exponential forms with
    negative, 48
Base-ten number system, 371

Batting averages, 159
Binomial coefficients, App–20
  defined, App–20
  evaluating, App–20
Binomials
  defined, 363
  expanding
    using binomial theorem, App–21
    using Pascal's triangle, App–19
  finding particular term of, App–21–22
  multiplication of, 392
  squaring, 396–97
Binomial theorem, App–19–22
  defined, App–21
  expanding binomial using, App–21
Blue whales, 46
BMI, analyzing, 137
Boston Marathon (2011), 718
Box, geometric formulas for, 91
Braces in indicating a set, 3
British Royal Air Force, 517

C

Calculators
  approximating roots using, 614–15
Cartesian, coordinate system, 190
Celsius scale, 32
Change-of-base formula, 811–12
Chicxulub crater, 96
Circles
  defined, 838
  examples of, 834
  finding equation of, with given center
    and radius, 838–39
  geometric formulas for, 91
  graphing of the form $(x - h)^2 + (y - k)^2$
    $= r^2$, 837–38
  graphing of the form $x^2 + y^2 + dx + ey +$
    $f = 0$, 839–40
  standard form of equation of, 839, 852
Circumference, 90
Coefficient
  binomial, App–20
  defined, 69
  of monomial, 362
Common factor, eliminating, in fraction,
  14–15
Common logarithms
  defined, 801
  solving applications using, 802–3
Common ratio, App–12
Commutative property
  of addition, 20, 21
  of multiplication, 34
Completing the square, 689
Complex fraction, 534
Complex numbers, 667–72
  addition of, 669
  defined, 668

Complex numbers (*Continued*)
   division of, 670–71
   multiplication of, 669–70
   performing arithmetic operations
     with, 668
   subtraction of, 669
Complex rational expressions, 534–38
   defined, 534
   simplifying, 534–38
Composite figures, 92
   calculating area of, 92–94
Composite numbers, 13
Composition of functions, 764–71
Compound inequalities, 164, 578–83
   defined, 578
   interpreting
     with *and*, 578–79
     with *or*, 581
   solving
     with *and*, 579–81
     with *or*, 581–83
Compound interest formula, 779
Cone, geometric formulas for, 91
Conic section, 834
Conjugates
   defined, 395
   multiplying, 395–96
   radicals in, 644
Consecutive integers, 65
Consistent systems
   with dependent equations, 284, 293–97
   of equations, 284
   with independent equations, 284
Constant, 2
Contradiction
   defined, 107
   recognizing, 107
Coordinate plane, plotting points in, 191–92
Coordinates
   determining, of given point, 190–91
   identifying, of a point, 190
Copernicus's theory, 364
Cramer, Gabriel, App–30
Cramer's rule, App–30–32
   solving systems of linear equations
     using, App–28–32
Cross products, 139
Cube root, 613
Cubes
   factoring a difference of, 455
   factoring a sum of, 456
Cylinder, geometric formulas for, 91

D
Decimals
   addition of, 26–28
   division of, 41–42
   eliminating, in equation, 117–19
   multiplication of, 38–39
   in system, 303–4
   writing, as percent, 151–52
   writing percent as, 150–51
Declaration of Independence, 46

Decrease, percent problems involving,
   155–56
Deforestation, rate of, 207
Degree
   of monomial, 362
   of polynomial, 363
Denominators, 10
   addition of fractions
     with different, 24–25
     with same, 23–24
   addition of rational expressions with
     different, 525–30
   least common, 12
   *n*th-root, 650–51
   rationalization of, 648–51
     containing difference, 651–52
     containing sum, 651–52
   simplifying fraction with same
     nonzero, 14
   simplifying when factorable, 521
   subtraction of rational expressions
     with different, 525–30
Dependent equations
   consistent systems with, 284, 293–97
   inconsistent systems and, 304–5
Descartes, René, 190, 191
Determinants
   of 2×2 matrices, App–28
   of 3×3 matrices, App–29–30
Diameter, 90
Difference, 27, 28
   rationalizing denominator containing,
     651–52
   square root of, 52
Dimensional analysis, converting units of
   measurement using, 513
Direct variation
   defined, 554
   solving problems involving, 554–55
Discriminant
   defined, 702
   in determining number of real solu-
     tions that quadratic equation has,
     702–3
Distance
   finding between two point, 836–37
   formula for, 836
Distributive property, 392
   of multiplication over addition, 34,
     67–68
   using, in expressions containing
     radicals, 643
Dividing out, defined, 15
Division
   of complex numbers, 670–71
   of decimal numbers, 41–42
   of fractions, 41
   of functions, 744–45
   involving 0, 40–41
   of monomials, 404
   of numbers in scientific notation, 403
   of polynomials by monomials, 405–6
   of radical expressions, 634

   of rational expressions, 511–12
   of rational numbers, 39
   sign rules for, 40
   synthetic, App–24–25
   translating word phrases to
     expressions in, 59
Domain, 764
   defined, 248
   determining, with graph, 248–49
   finding, of radical function, 616–17

E
Earth
   gravity of, 124
   revolution of, 19
   tilt of, 100
   travel of, 388
Eiffel Tower, 149
Element
   of matrix, App–29
   of set, 3
Elimination
   solving nonlinear systems of equa-
     tions using, 860–61
   solving systems of linear equations
     using, 300–308, 314–16
Ellipses, 3
   defined, 847
   equation of, centered at (0, 0), 847
   examples of, 834
   general equation for, 848
   graphing, 847–49
   standard form of, 852
Embedded parentheses, 50–51
Equality
   addition principle of, 101–8
   multiplication principle of, 112–19
Equations
   clearing fractions in, 116
   containing radicals, 712
   defined, 2, 88
   eliminating decimals in, 117–19
   equivalent, 102
   examples of, 2
   graphing, in slope–intercept form,
     219–21
   with infinite number of solutions,
     106–7
   linear, 101
   nonlinear, 101, 858
   with no solution, 107–8
   of parabolas, 835–36
   radical, 659
   radicals on both sides of, 660–61
   with rational expressions, 711
   rewriting
     in form $Ax + By = C$, 304
   simplifying first, 115–16
   solving
     containing logarithms, 808–9
     containing rational expressions,
       542–47
     involving absolute value, 589–90

that are quadratic in form, 711–16
with variables as exponents, 807–8
with variable terms on both sides, 114–15
translating word sentences to, 130–34
verifying solutions to, 88–89
Equivalent equations, 102
Equivalent fractions, writing, 11
Eratosthenes, 13
Exponential equations
properties for solving, 807
solving, 778–79
solving applications involving, 809–11
Exponential forms
defined, 47
division of, with same bases, 402
evaluating
with negative bases, 350
numbers in, 47–48
multiplying, 380–81
zero as, 351
Exponential functions, 776–80
defined, 776
graphing, 776–77
Exponents
nonpositive integer, 351
one-to-one property of, 778
positive integer, 350
product rule for, 380
quotient rule for, 402–3
rational, 623
solving equations that have variables as, 807–8
Expressions
defined, 2
examples of, 2
translating word phrases to, 58–61
Extraneous solutions, 659–60
defined, 544
Eyes
lens in our, 550
muscles in your, 550

F

Factored form, 428
Factorial notation
defined, App–20
evaluating, App–20
Factoring
difference
of cubes, 455
of squares, 454–55
greatest common factor and factoring by grouping, 428–34
by grouping, 433–34
perfect square trinomials, 452–54
solving quadratic equations by, 464–71
strategies for, 460–62
sum of cubes, 456
by trial and error, 445
trinomials to form $ax^2 + bx + c$, where a^1, 444–49
trinomials to form $x^2 + bx + c$, 438–41

when first term is negative, 432–33
when greatest common factor is polynomial, 433–34
Factors
defined, 13
eliminating common, in fraction, 14–15
listing, for given number, 428
Factor trees in finding prime factorizations, 13–14
Felicitations, 425–37
function on 3 November 2002, 426
Finite sequence, App–5
First Outer Inner Last (FOIL), 394
Formulas
applying addition and multiplication principles to, 125–27
change-of-base, 811–12
compound interest, 779
defined, 90
distance, 836
geometric, 91
midpoint, 837
problems involving combinations of, 92–94
problems involving nongeometric, 94–95
problems involving single, 91–92
in problem solving, 89–95
slope, 221
using, 90
Fourth root, 613
Fraction lines, 52–54
Fractions
additions of
with different denominators, 24–25
with same denominator, 23–24
clearing, in equation, 116
complex, 534
defined, 10
denominator in, 10
division of, 41
eliminating common factor in, 14–15
multiplication of, 37
numerator in, 10
simplifying, to lowest terms, 14–16
in system, 303–4
writing, as percent, 151–52
writing equivalent, 11
writing percent as, 150–51
Front-end ratio, 158
Function notation, 253–54
Functions
addition of, 743
composition of, 764–71
defined, 250
division of, 744–45
exponential, 776–80
finding value of, 254–55
inverse, 765–67
logarithmic, 785–90

multiplication of, 743–44
one-to-one, 767–68
subtraction of, 743

G

Galileo Galilei, 364, 621
Geometric formulas, 91
Geometric sequence
finding common ratio and specified term, App–12–13
nth term of, App–12–13
Geometric series, App–13
finding partial sums of, App–13–14
solving applications using, App–15–16
Giant sequoia tree, 123
Graphing
absolute value functions, 591–92
finding value of function given its, 255–56
intercepts in, 208–12
linear equations, 197–202
linear inequalities, 239–43
logarithmic functions, 788–89
nonlinear inequalities, 864–66
quadratic functions, 477–78, 720–28
form $f(x) = ax^2$, 720–21
of the form $f(x) = ax^2 + bx + c$, 724–26
of the form $f(x) = ax^2 + k$, 721–22
of the form $f(x) = a(x - h)^2$, 722–23
of the form $f(x) = a(x - h)^2 + k$, 723–24
solution set of system of linear inequalities, 330–33
in solving systems of linear equations, 280–85
Graphing calculators
graph of circle on, 840
Graphs
determining set of data points, 193–94
of exponential functions, 776–77
of inverse function, 770–71
linear, 193
nonlinear, 193
of quadratic equations and functions, 476–80
representing solutions to inequalities using, 162–63
Greater than, 167
solving absolute value inequalities involving, 596–98
Greater than or equal to, 167
Greatest common factor (GCF)
defined, 429
finding set of numbers or monomials, 428–29
of monomials, 430–32
using listing to find, of numbers, 429
using prime factorization to find, 429–30
Great Pyramid, 99, 411, 657
Green, Andy, 517
Grouping, factoring by, 433–34

Grouping symbols
    fraction lines as, 52–54
    parentheses as, 50
    radical symbols as, 51

H

Half-life, 779
Hemoglobin, 359
Horizontal lines, 202
Horizontal line test for one-to-one
    functions, 767
Hyperbolas
    defined, 849
    equations of, in standard form, 850
    examples of, 834
    graphing, 849–51
        in standard form, 851–52
    standard form of, 852

I

i, raising to powers, 671–72
Identity
    additive, 20, 21
    defined, 106
    multiplicative, 34
    recognizing, 106
Imaginary numbers
    defined, 667
    rewriting, 668
    writing, using i, 667–68
Imaginary unit, 667
Inconsistent systems, 284, 332
    dependent equations and, 304–5
    of equations, 284, 293
Increase, percent problems involving,
    155–56
Independent equations, consistent
    system with, 284
Index of summation, App–8
Inequalities
    addition principle of, 164–65
    compound, 164, 578–83
    defined, 2
    graphing, 163
    involving absolute value, 595–98
    multiplication principle of,
        165–66
Inequality symbols, 3
Infinite geometric series
    finding sums of, App–14–15
    sum of, App–14
Infinite sequence, App–5
    defined, App–5
Initial velocity, 127
Integer exponents
    nonpositive, 351
    positive, 350
Integers
    addition of, 20–25
    consecutive, 65
    multiplication of, 34
    set of, 4
Intel, 360

Intercepts
    graphing linear equations using, 210–12
    graphing using, 208–12
Interest
    formula for calculating simple, 126
Interpretation
    compound inequalities with *and*,
        578–79
Intersection, defined, 578
Interval notation, 162
Inverse functions, 765–67
    existence of, 768
    finding, of one-to-one function, 768–69
    graphs of, 770–71
Inverse operations, 765
Inverse properties of logarithms, 793
Inverses
    additive, 26–27, 301–3
    multiplicative, 39
Inverse variation
    defined, 556
    solving problems involving, 556
Irrational number, 5

J

Joint variation, solving problems
    involving, 557

K

Kelvin scale, 32
Khufu, King, 99, 411, 657

L

Lactose, 144
    chemical formula for, 19
Learning strategies, 3, 6, 89
    auditory learners, 105, 381, 430, 850,
        App–28
    tactile learners, 192, 252, 366, 375, 770,
        850, App–28
    visual learners, 103, 131, 140, 212, 222,
        366, 375, 391, 429, 582, 839, 850,
        App–28
Least common denominator (LCD)
    defined, 12
    finding, 525–26
    writing equivalent fractions with, 12
Least common multiple (LCM), 12
Lens in our eyes, 550
Less than, 167
    solving absolute value inequalities
        involving, 595–96
Less than or equal to, 167
Liberty Bell, 46, 124
Like radicals
    addition of, 641–43
    defined, 641
    subtraction of, 641–43
Like terms
    collecting, 70–71
    combining, 69–70
    defined, 69
    rewriting an expression by combining,
        68–69

Linear equations
    defined, 101
    graphing, 197–202
    in one variable, 101
    solving, 105, 114, 117
        using multiplication principle,
        112–13
    writing, in standard form, 230–32
Linear inequalities
    defined, 162
    graphing, 239–43
    solving, 162–68
Lines
    comparing, with different slopes,
        217–19
    horizontal, 202
    parallel, 232
    perpendicular, 233
    using point–slope form to write
        equations of, 229–30
    using slope–intercept form to write
        equation of, 219
    vertical, 202
Listing, using, to find greatest common
    factor of numbers, 429
Logarithmic equations
    properties for solving, 807
Logarithmic functions, 785–90
    graphing, 788–89
    solving, by changing to exponential
        form, 787–88
    solving applications involving, 809–11
Logarithms
    applications solving, involving, 789–90
    common, 801
    defined, 785
    inverse properties of, 793
    natural, 802
    power rule of, 794
    product rule of, 793–97
    properties of, 793–97
    quotient rule of, 794
    solving equations containing, 808–9
London Marathon (2003), 718
Long division
    in dividing polynomials, 406–8
    using place marker in, 408–9
Lowest terms, 14
    simplifying fraction to, 14–16
    simplifying rational expressions to, 500

M

Matrices
    augmented, 321
    defined, 321
    determinants of 2×2, App–28
    determinants of 3×3, App–29–30
    minor of element of, App–29
    solving
        application problems using, 324–25
        systems of linear equations using,
        321–25
    square, App–28

Mean
 arithmetic, 53
 defined, App–1
 finding, of given set of numbers, App–1
Measurement, converting units of, using
 dimensional analysis, 513
Median
 defined, App–1–2
 finding, of set of numbers, App–2
Method (Descartes), 191
Midpoint
 finding between two points, 836–37
 formula for, 837
Minor of matrix, App–29
Minuend, 27
Mode
 defined, App–2
 finding, App–3
Monomial greatest common factor (GCF)
 factoring, out of polynomials, 431–32
 factoring out first, 447–49
Monomials
 coefficient of, 362
 defined, 361–62
 degree of, 362
 division of, 404
  polynomials by, 405–6
 finding greatest factor of set of
  numbers or, 428–29
 greatest common factor of, 430–32
 multiplication of polynomial by, 390–91
 multiplying, 381–83
 raising, to a power, 384–85
 simplifying, when raised to a power, 383
 simplifying rational expressions
  containing only, 500
Motion problems, 553–54
Multiple
 defined, 12
 least common, 12
 solutions, 661
Multiplication
 associative property of, 34
 binomials, 392
 commutative property of, 34
 of complex numbers, 669–70
 of conjugates, 395–96
 of decimal numbers, 38–39
 distributive property of, over addition, 34, 67–68
 of each equation by number to create
  additive inverses, 302–3
 of exponential forms, 380–81
 of fractions, 37
 of functions, 743–44
 of integers, 34
 of monomials, 381–83
 more than two numbers, 36–37
 with negative factors, 36
 of numbers
  with different signs, 35
  with same sign, 35–36
 of numbers in scientific notation, 382–83

of one equation by number to create
 additive inverses, 301–2
of polynomial, by a monomial, 390–91
of polynomials, 394–95
problem solving involving, 132
properties of, 34
of radical expressions, 633–34
 with different root indices, 629–30
of rational expressions, 509–10
and simplification, 37–38
translating word phrases to expressions in, 59
Multiplication principles
 applying to formulas, 125–27
 of equality, 112–19
 of inequality, 165–66
 isolating variables using, 126
 solving linear equation using, 112–13, 114
Multiplication statement, 34
Multiplicative identity, 34
Multiplicative inverses, 39
Multiplicative property of 0, 34
Multiterm polynomials
 numerators that are, 520–21
 simplifying rational expressions
  containing, 502–4
 subtracting with numerators that are,
  521–22
Mutai, Geoffrey, 718

N

National Institutes of Health, 137
Natural logarithms, 802
 defined, 802
 solving applications using, 803
Natural numbers, set of, 3
Negative bases, evaluating exponential
 forms with, 48, 350
Negative factors, multiplication with, 36
Negative rational exponents, 625–26
Newton, Isaac, 31
Nongeometric formulas, problems
 involving, 94–95
Nonlinear equations, 101
Nonlinear inequalities
 graphing, 864–66
  solution set of system of, 866–67
 solving, 732–37
Nonlinear system of equations
 defined, 858
 solving
  using elimination, 860–61
  using substitution, 859–60
Nonpositive integer exponents, 351
Notation
 function, 253–54
 interval, 162
nth of arithmetic sequence, App–7
nth power, raising nth root to, 634
nth-root denominators, 650–51
nth roots, 613
 evaluating, 614
 finding, of number, 613–14

raising, to nth power, 634
 simplifying, 635
nth term, App–12
 of geometric sequence, App–12–13
Nuclear fission, 412
Nuclear fusion, 412
Number line, 4
 graphing rational numbers on, 5–6
Numbers
 absolute value of, 6
 addition of
  with different signs, 22–23
  with same sign, 21–22
 comparing, 7
 complex, 667–72
 composite, 13
 evaluating, in exponential form,
  47–48
 finding nth root of, 613–14
 irrational, 5
 multiplication of, 35
  more than two, 36–37
  with same sign, 35–36
 natural, 3
 prime, 13
 prime factorization of, 13–14
 rational, 4–5
 real, 4, 5
 of a set, 3
Number sets
 classification of, 3–4
 symbols of, 5
Numerators, 10
 with no like terms, 519–20
 rational exponents with
  of 1, 623–24
  other than 1, 624–25
 rationalizing, 653
 simplifying fraction with same
  nonzero, 15
 that are multiterm polynomials, 520–21

O

Of Interest, 5, 13, 19, 31, 32, 45, 46, 96, 99,
 100, 123, 124, 126, 127, 137, 144, 149,
 150, 158, 191, 207, 328, 340, 358, 359,
 360, 364, 388, 411, 412, 471, 517, 518,
 550, 560, 621, 622, 626, 657, 698, 718,
 719, 755, 805, App–30
One-to-one functions, 767–68
 finding inverse function of, 768–69
 horizontal line test for, 767
One-to-one property of exponential, 778
Opening of a parabola, 476–77
Operations
 inverse, 765
 problem-solving involving more than
  one, 132
Or
 interpreting compound inequalities
  with, 581
 solving compound inequalities with,
  581–83

Ordered pair, 239
  determining quadrant for a given, 192–93
Ordered triples, determination of, as solution to system of equation, 312
Order-of-operations agreement in simplifying numerical expressions, 50

P

Parabolas
  examples of, 834
  graphing
    of the form $x = a(y - k)^2 + h$, 834–36
  opening of, 476–77
  solving applications involving, 726–28
  standard form of, 851
  with vertex (h, k), 723
Parallel lines, 232
Parallelogram, geometric formulas for, 91
Parentheses
  embedded, 50–51
  translating phrases involving, 61
  translations requiring, 133
Partial sums, finding, of geometric series, App–13–14
Pascal's triangle, expanding binomial using, App–19
Pendulums, 621
Percents, 150–56
  defined, 150
  rewriting, 151
  simple percent sentences with unknown, 154
  writing as fraction or decimal number, 150–51
  writing fraction or decimal as, 151–52
Percent sentences
  simple
    translating and solving, 152
    with unknown part, 153
    with unknown percent, 154
    with unknown whole, 153–54
Perfect cubes, 635
Perfect squares, 635
Perfect square trinomials, factoring, 452–54
Perimeter, 373
  defined, 90
Perpendicular lines, 233
Place marker, using, in long division, 408–9
Plotting a point, 191–92
Point(s)
  determining coordinates of given, 190–91
  finding distance and midpoint between two, 836–37
  plotting a, 191–92
Point–slope form, 229–34
Polya, George, 89
Polynomial addition, 372–74

Polynomial inequalities, 735–36
Polynomials
  defined, 362
  degree of, 363–64
  division of, by monomials, 405–6
  evaluating, 364
  factoring monomials greatest common factor out of, 431–32
  factoring when greatest common factor is, 433–34
  long division in dividing, 406–8
  multiplication of, 394–95
    by monomial, 390–91
  in one variable, 362–63
  subtraction of, 374–76
  synthetic division of, by binomial in the form $x - c$, App–24–25
  writing, in descending order of degree, 364–65
Positive integer exponents, 350
Power
  i raising to, 671–72
  raising, to a power, 383
  raising monomial to, 384–85
  raising quotient to, 350
Power rule
  of logarithms, 794
  for solving radical equations, 659
  in solving radical equations, 659
  using twice, 662–63
Prime factorization, 636
  defined, 13
  using, to find greatest common factor, 429–30
  writing, 13–14
Prime number, 13
Principal square root, 49–50
Problem solving
  applications, using equations that are quadratic in form, 714–16
  combined variation in, 557–58
  direct variation in, 554–55
  formulas in, 89–95
  indirect variation in, 556
  involving motion, 553–54
  involving work, 551–52
  joint variation in, 557
  percents, 150–56
  Pythagorean theorem in, 471
  ratios and proportion in, 138–43
  translating word phrases to expressions, 58–61
  translating word sentences to equations in, 130–34
Product rule
  for exponents, 380
  of logarithms, 793–97
  to radical expressions, 635
  for radicals, 633–34
Products, 34
  cross, 139
  raising, to power, 384
  square root of, 51

Profit, formula for calculating, 126
Proportions, 139
  solving, 139–41
    for unknown number in, 139–40
  in solving for missing lengths in figures that are similar, 142
Proxima Centauri, 358, 518
Puzzle problems, 65, 74, 100, 112, 137, 172, 207, 247, 311, 370, 390, 415, 518, 588, 622, 699, 710, 732, 784, 846, 857
Pyramids, geometric formulas for, 91
Pythagoras, 471
Pythagorean theorem in problem solving, 471

Q

Quadrants
  determining, for a given ordered pair, 192–93
  identifying, 192
Quadratic equations
  choosing method for solving, 702
  graphing, 477–78
  graphs of, 476–80
  in one variable, 466–67
  rewriting, in standard form, 468–70
  solving
    by factoring, 464–71
    using quadratic formula, 700–705
  square root principle in solving, 690–95
    in the form $(ax + b)^2 = c$, 691–92
    in the form $ax^2 + b = c$, 691
    in the form $ax^2 + bx = c$, where a^1, 693–95
    in the form $x^2 = a$, 690–91
    in the form $x^2 + bx = c$, 693
  in two variables, 476
Quadratic formulas
  defined, 700
  solving applications using, 705
  solving quadratic equations using, 700–705
Quadratic functions
  finding vertex of, in form $f(x) = ax^2 + bx + c$, 725
  finding x- and y-intercepts of, 703–5
  graphing, 477–78, 720–28
    form $f(x) = ax^2$, 720–21
    of the form $f(x) = ax^2 + bx + c$, 724–26
    of the form $f(x) = ax^2 + k$, 721–22
    of the form $f(x) = a(x - h)^2$, 722–23
    of the form $f(x) = a(x - h)^2 + k$, 723–24
Quadratic inequality, 733
Quotient
  raising, to power, 350
  square root of, 51
Quotient rules
  for exponents, 402–3
  of logarithms, 794
  for radicals, 634

R

Radcliffe, Paula, 718
Radical equations
    defined, 659
    power rule for solving, 659
    solving, 663
    using power rule to solve, 659
Radical expressions, 613
    division of, 634
    multiplication of, 633–34
    simplifying, 615–16
        that contain mixed operations, 644
    using product rule to simplify, 635
Radical functions
    defined, 616
    evaluating, 616
    finding domain of, 616–17
    solving applications involving, 617–18
Radicals, 49, 613
    on both sides of the equation, 660–61
    in conjugates, 644
    equations containing, 712
    isolated, 659
    not isolated, 662
    product rule for, 633–34
    quotient rule for, 634
    simplifying, with variables, 636–38
    using distributive property in expressions containing, 643
    writing, as expressions raised to rational exponents, 626
Radical sign, 613
    square roots involving, 49
Radical symbols, 51–52
Radicand, 49, 613
Radius, 90, 838
Range
    defined, 248
    determining, with graph, 248–49
Ratio
    back-end, 158
    common, App–12
    defined, 138
    front-end, 158
    unit, 139
Rational exponents, 623–30
    general rule for, 624–25
    negative, 625–26
    with numerator of 1, 623–24
    with numerator other than 1, 624–25
    using, to simplify radical expressions, 628–29
    writing radicals as expressions raised to, 626
Rational expressions
    addition of
        with different denominators, 525–30
        with same denominator, 519
    applications with, 551–58
    complex, 534–38
    defined, 498

dividing, 511–12
    equations with, 711
    evaluating, 498–99
    finding values that make undefined, 499
    multiplying, 509–10
    simplifying, 498–504
        containing multiterm polynomials, 502–4
        containing only monomials, 500
        to lowest terms, 500
    solving equations containing, 542–47
    subtraction of, with different denominators, 525–30
Rational inequalities
    defined, 736
    solving, 736–37
Rational numbers, 4–5
    division of, 39
    graphing, on a number line, 5–6
    simplifying expressions with, using rules of exponents, 626–28
    subtraction of, 27–28
Ratios, solving problems involving, 139
Real numbers, 4
    defined, 5
Rectangles, geometric formulas for, 91
Rectangular coordinate system, 190–94
Relation, defined, 248
Root indices, 613
    multiplying and dividing radical exponents in simplifying, 629–30
Roots, approximating, using calculator, 614–15
Row echelon form
    defined, 322
    solving system of linear equations by transforming its augmented matrix to, 322–24
Row operations, 322
Rules of exponents, 627
    simplifying expressions using, 409–10, 626–28

S

San Francisco 1906 earthquake, 805
Scientific notation
    changing, to standard form, 354–55
    changing standard form to, 355–56
    defined, 353
    division of numbers in, 403
    multiplying numbers in, 382–83
    writing, in standard form, 353–55
Scott, David, 560
Sequence
    defined, App–5
    finding terms of, when given the general term, App–5–6
    finite, App–5
    infinite, App–5
Series
    arithmetic, App–8–9
    defined, App–8

Set notation, representing solutions to inequalities using, 162–63
Sets
    defined, 3
    numbers or elements of, 3
    writing, 3
Sierra Nevada range, 123
Signed numbers
    applications of, 42–43
    division of, 40
Sign rules for division, 40
Similar figures, 143
Simplification
    multiplication and, 37–38
    of rational expressions, 498–504
    when the denominator is factorable, 521
Slope
    comparing lines with different, 217–19
    defined, 219
    formula for, 221
    undefined, 222
    zero, 222
Slope–intercept form, 217–23
Solution(s)
    checking possible, 89
    defined, 88
    equations with infinite number of, 106–7
    equations with no, 107–8
    extraneous, 544, 659–60
    multiple, 661
    for system of equations, 280
    verifying, to equations, 88–89
Solution set
    graphing of system of linear inequalities, 330–33
    types of, for systems of three equations, 313
Sphere, geometric formulas for, 91
Square matrix
    defined, App–28
Square root(s), 5, 613
    of difference, 52
    evaluating, 48–49
    geometric formulas for, 91
    involving radical sign, 49
    principal, 49–50
    of product, 51
    of quotient, 51
    of sum, 52
Square root denominators, 649–50
Square root principle, 690
    in solving quadratic equations, 690–95
        in the form $(ax + b)^2 = c$, 691–92
        in the form $ax^2 + b = c$, 691
        in the form $ax^2 + bx = c$, where a^1, 693–95
        in the form $x^2 = a$, 690–91
        in the form $x^2 + bx = c$, 693
Squaring, binomial, 396–97
SRAM (static random access memory) chips, 360

The Algebra Pyramid:
A Guide to the Development of Algebra Topics

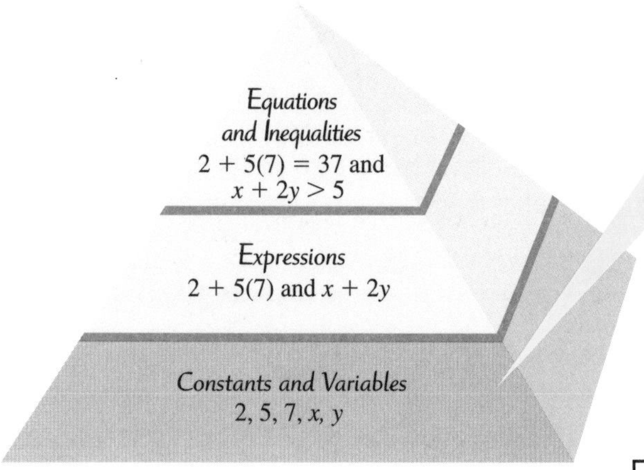

*Equations
and Inequalities*
$2 + 5(7) = 37$ and
$x + 2y > 5$

Expressions
$2 + 5(7)$ and $x + 2y$

Constants and Variables
$2, 5, 7, x, y$

Constants and variables are the foundation of algebra. The following diagram summarizes the entire complex number system (all constants). These constants, along with variables such as x or y, are the basic building blocks that give meaning to the expressions, equations, and inequalities.

— Complex Numbers —

— Real Numbers —

Rational Numbers $-3, -2.4, -1\frac{4}{5}, 0, 0.\overline{6}, 1$, etc.

Integers $\ldots -3, -2, -1, 0, 1, 2, 3, \ldots$

Whole Numbers $0, 1, 2, 3, \ldots$

Natural Numbers $1, 2, 3, \ldots$

Irrational Numbers
$-5\sqrt{3}$,
$\sqrt{2}, \pi$, etc.

Imaginary Numbers
$5 + 2i, -3i\sqrt{2}$,
etc.

Pure Imaginary Numbers
$6i, 5i\sqrt{3}$, etc.

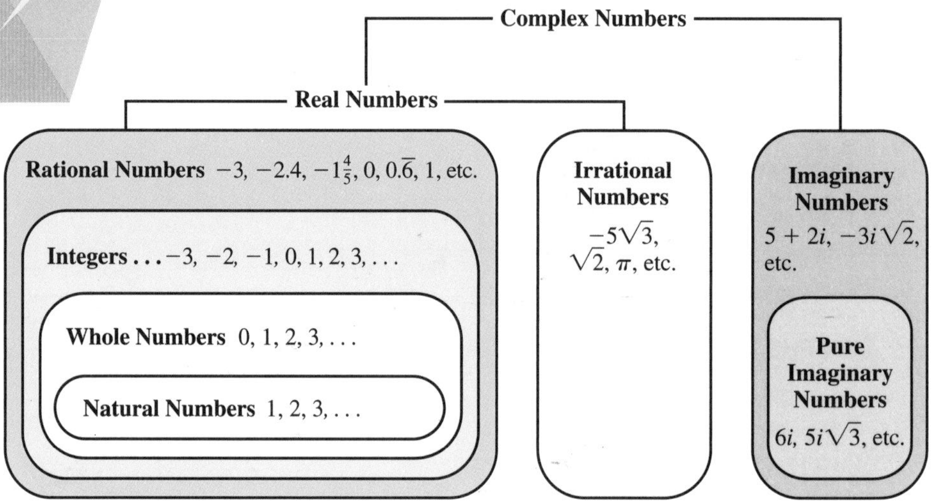

Expressions relate constants and variables using symbols such as operations of arithmetic to describe a calculation. We **evaluate** or **rewrite** expressions. To evaluate, we replace variables with numbers and perform the calculation. To rewrite, we use rules to write an alternative (and usually simpler) form.

The following list outlines the development of expressions in this text:

Chapter 1	Order of Operations
	Translating Word Phrases to Expressions
	Evaluating and Rewriting Expressions
Chapter 5	Exponent Rules and Scientific Notation
	Adding, Subtracting, Multiplying, and Dividing Polynomials
Chapter 6	Factoring Polynomials
Chapter 7	Simplifying, Adding, Subtracting, Multiplying, and Dividing Rational Expressions
	Simplifying Complex Rational Expressions
Chapter 9	Simplifying, Adding, Subtracting, Multiplying, and Dividing Radical Expressions
	Rational Exponents
	Rationalizing Numerators and Denominators of Radical Expressions
	Adding, Subtracting, Multiplying, and Dividing Complex Numbers

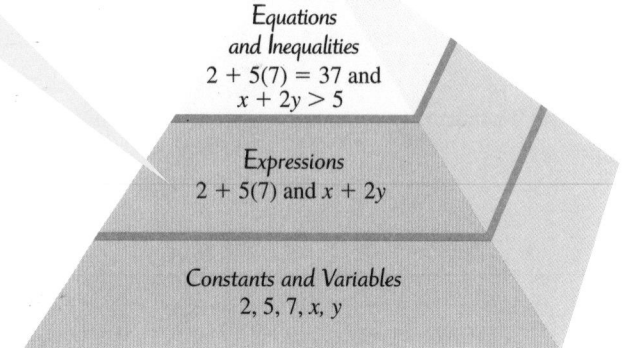

*Equations
and Inequalities*
$2 + 5(7) = 37$ and
$x + 2y > 5$

Expressions
$2 + 5(7)$ and $x + 2y$

Constants and Variables
$2, 5, 7, x, y$